From
Engineering
To
Sustainability

tall buildings

Sixth International Conference on Tall Buildings
Mini Symposium on Sustainable Cities
Mini Symposium on Planning, Design and Socio-Economic Aspects of Tall Residential Living Environment

Hong Kong, China
6 – 8 December 2005

Editors

Y K Cheung • K W Chau
The University of Hong Kong

Organisers

The University of Hong Kong
Tongji University
China Academy of Building Research

World Scientific

NEW JERSEY · LONDON · SINGAPORE · BEIJING · SHANGHAI · HONG KONG · TAIPEI · CHENNAI

Published by

World Scientific Publishing Co. Pte. Ltd.
5 Toh Tuck Link, Singapore 596224
USA office: 27 Warren Street, Suite 401-402, Hackensack, NJ 07601
UK office: 57 Shelton Street, Covent Garden, London WC2H 9HE

British Library Cataloguing-in-Publication Data
A catalogue record for this book is available from the British Library.

TALL BUILDINGS
From Engineering to Sustainability

ISBN 981-256-620-1

Printed in Singapore.

PREFACE

The Sixth International Conference on Tall Buildings (ICTB-VI) is a continuation from a series of international conferences (five in total) which has proven to be a very hot topic. Since the last conference in 1998, numerous landmark skyscrapers have been built in the Asia Pacific rim and all over the world. The timing of the Conference is impeccable as it allows experts and researchers to share information pertinent to the latest practices, lessons and research outcomes of tall buildings. The Conference provides a forum for all stakeholders in the construction industry to exchange ideas on how to further advance the development and management of tall buildings so as to fulfil the needs of the society, clients and end-users.

This Conference is different from the previous conferences as there are two mini symposia: Mini Symposium on Sustainable Cities; and Mini Symposium on Planning, Design and Socio-Economic Aspects of Tall Residential Living Environment are held concurrently with ICTB-VI. It thus provides a forum for exchange of information and ideas between engineers, architects, planners and estate managers.

The initial call for papers received submissions of over 180 abstracts. After rigorous review and selection process, 166 full papers are accepted for oral presentation and inclusion in the conference proceedings. The proceedings consists of papers for the keynote and theme lectures and contributing papers.

On behalf of the members of the Organising Committee, I would like to express my sincere gratitude to the keynote and theme speakers and all contributors. Without their contributions, we could not have produced this valuable and state-of-the-art proceedings. Thanks are also due to the members of the International Advisory Committee, the Technical Committee, and the Task Groups of the Mini Symposia for their assistance and valuable support. The assistance of the Organising Committee, Fund-raising Sub-committee and Local Conference Secretariat — Centre for Asian Tall Buildings and Urban Habitat, are gratefully acknowledged. We are grateful to the financial sponsors for their generous support, and all those who have devoted their time and effort to the organisation of the Conference.

Y. K. Cheung
Chairman
6th International Conference on Tall Buildings
December 2005

Organisers

The University of Hong Kong, Hong Kong, China
China Academy of Building Research, Beijing, China
Tongji University, Shanghai, China

International Advisory Committee

Organising Committee

Chairmen
Y K Cheung, The University of Hong Kong
Y S Li, Tongji University
Y W Wang, China Academy of Building Research

Vice-Chairman
P K K Lee, The University of Hong Kong

Honorary Secretaries
H C Chan, The University of Hong Kong
T S T Ng, The University of Hong Kong

Honorary Treasurer
K S Law, The Hong Kong Institution of Engineers

Members
M Arnold, Hongkong Land Ltd
P Ayres, Ove Arup & Partners Hong Kong Ltd
K W Chau, The University of Hong Kong
J T He, South China University of Technology
T H K Ho, Mass Transit Railway Corporation
T O S Ho, Gammon Skanska Ltd
P S S Lau, Design Consultants Ltd
S S Y Lau, The University of Hong Kong
D Levett, Concord Construction Consultancy Group
X L Lu, Tongji University
F S H Ng, DLS Management Ltd
O Poon, ATAL Engineering Ltd
V K C Tse, Parsons Brinkeroff (Asia) Ltd
B W H Wong, Paul Y.-ITC Construction Holdings Ltd
H S H Wu, COINS Asia-Pacific Ltd
A G O Yeh, The University of Hong Kong
Y Yuan, Tongji University
C H Yue, Architectural Services Department
Q L Zhang, Tongji University

Fund Raising Subcommittee
F S H Ng, DLS Management Ltd
K S Law, The Hong Kong Institution of Engineers
T S T Ng, The University of Hong Kong
O Poon, ATAL Engineering Ltd
V K C Tse, Parsons Brinkeroff (Asia) Ltd
H S H Wu, COINS Asia-Pacific Ltd

Technical Committee

Chairman
K W Chau, The University of Hong Kong

Members
H C Chan, The University of Hong Kong
Y K Cheung, The University of Hong Kong
P K K Lee, The University of Hong Kong
X L Lu, Tongji University
T S T Ng, The University of Hong Kong
E O W Wong, HKU SPACE
A G O Yeh, The University of Hong Kong
J D Zhao, China Academy of Building Research

Mini Symposium on Sustainable Cities Task Group

Chairman
C H Yue, Architectural Services Department

Members
R S Chin, Architectural Services Department
A Y S Fung, Housing Department
S S Y Lau, The University of Hong Kong
K K Li, Architectural Services Department
T S T Ng, The University of Hong Kong
H W Pang, Housing Department
S H Sat, Architectural Services Department
E O W Wong, HKU SPACE
R C P Yip, HKU SPACE

Supporting Organisations

American Society of Heating, Refridgerating and Air-Conditioning Engineers, Hong Kong Chapter
Architectural Services Department
Asian Institute of Intelligent Buildings
The Association of Consulting Engineers of Hong Kong
Building Services Operation and Maintenance Executives Society
Buildings Department
Business Environment Council
The Chartered Institute of Building (Hong Kong)
The Chartered Institution of Building Services Engineers HK Branch
Construction Industry Institute – Hong Kong
Council on Tall Buildings and Urban Habitat
Electrical and Mechanical Services Department
Energy Institute
HK-BEAM Society
The Hong Kong Construction Association Ltd
The Hong Kong Institute of Architects
Hong Kong Institute of Construction Managers
The Hong Kong Institute of Planners
The Hong Kong Institute of Surveyors
The Hong Kong Institution of Engineers - Structural Division
Hong Kong Polytechnic University
Housing Department
The Institution of Civil Engineers
The Institution of Electrical Engineers Hong Kong
The Institution of Structural Engineers
Institution of Fire Engineers (Hong Kong Branch)
Institution of Mechanical Engineers HK Branch
International Commission of Illumination (Hong Kong)
International Council for Research and Innovation in Building and Construction
Planning Department
Research Group on Sustainable Cities, HKU
The Real Estate Developers Association of Hong Kong

CONTENTS

Framed Structure

Seismic Engineering

Case Studies

Wind Engineering

Transfer Structure

Computer Modelling and Analysis

Innovative Technology

RC Structure

Steel Structure

Management Issues

Composite Structure

Other Structural Aspects

Urban Planning and Development

Urban Living Space

Energy Efficiency, New and Renewable Energy

Sustainable Development — Cases

Sustainability Assessment

Sustainable Design

Post-Occupancy Operation, Maintenance and Management

Living Environment

Opening Address & Keynote Papers

THE REFORM AND DEVELOPMENT OF THE ENGINEERING EDUCATION IN CHINA

WU QIDI
Deputy Minister, Ministry of Education of China, PRC

With the Chinese government always attaching great attention to the development of higher education, higher education in China has made great progresses, especially after the adoption of opening-up policy and reform at the end of 1970s. At present, facing the increasingly severe challenges of resources, energy and environment and the reality of economic development in China, the Chinese government has put forward the strategy of accelerating the industrialization process and adopting new approaches with Chinese characteristics for this purpose. At the same time, with the strategic objective of improving the overall competitiveness and building up an innovative country, higher education bears the task of producing a large number of talents with high quality. Therefore, higher education in China bears important and historic responsibilities.

1. The Recent Reform and Development of Chinese Higher Education

Since the adoption of reform and opening-up policy, the reform of Chinese higher education has achieved remarkable successes. The scale of higher education is growing very rapidly and disciplinary structure has been optimized. The reform of higher education has been deepened continuously with a steady improvement of its quality and efficiency.

In 2004, the regular Higher Education Institutions (HEIs) recruited 4,473,400 students, 651,700 more than that of last year, representing a 17.05% increase. The total number of students enrolled in all types of HEIs amounted up to 20,000,000, with the gross enrollment rate of 19%, 9.2% higher than that of 1998. We have generally transferred from an elite higher education to a massive higher education.

In order to promote the teaching reform comprehensively, the Ministry of Education of China (MoE) has made and implemented some national programmes or plans such as 'Plan of Reform on Teaching Contents and Curriculum of Higher Education Facing the 21st Century', 'The Teaching Reform Project for Higher Education in the New Century', and 'The Programme of Teaching Reform and Teaching Quality in Higher Education Institutions'. Almost 20,000 experts and scholars have been involved in the research of these reform strategies and projects. As a result, a large number of exemplary outcomes have been witnessed, providing us with a group of training models and experimental strategies for teaching which are approved by real practices. In addition to that, 1,000 books of 21st century excellent teaching materials have been published, laying a solid foundation to improve the teaching quality of higher education in China.

In order to increase the international influence and competitiveness of Chinese higher education, MoE launched a program to establish world-class universities and top universities in China (which is also called '985 Project') in 1999. Over 30 best Chinese universities have been selected for this purpose. It is expected that after years' efforts, several of them will become world-class or almost meet the equivalent standards.

With the fast development of higher education, their function of promoting economic and social development has been significantly increased. HEIs have become the main force in the field of basic scientific research, applied scientific research, high-tech industrialization, and social science development. According to incomplete statistics, two thirds of our national key laboratories are located in HEIs and one third of academicians from Chinese Academy of Science (CAS) and Chinese Academy of Engineering (CAE) are from universities.

At the same time, Chinese HEIs have been actively involved in international cooperation and exchanges. Since the implementation of reform and opening-up policy, we have developed educational cooperative relationships with 175 other countries and regions, and have signed more than 100 agreements for bilateral or multilateral educational cooperation. For instance, we have signed the 'Convention of mutual recognition of studies, diplomas and degrees in higher education in Asia-Pacific' with other concerned countries in Asia-Pacific region. We also signed Agreements on Mutual Recognition of Academic Degrees and Credentials with Germany, France, the U.K., Australia and New Zealand. More than 800,000 Chinese students have been sent to 108 countries and regions for further studies, and there are more than 400 universities in China receiving international students. In the academic year of 2003/04, 78,000 international students from 175 countries came to China for study.

Although higher education has achieved remarkable outcomes, it also faces new challenges. First of all, China is trying to provide the biggest scale of education on a weak and poor base. Secondary, there is still a severe conflict between the increasing educational needs of the public and the insufficient education provision, particularly education with high quality. In order to further promote the reform and development of higher education, MoE included the plan of higher education reform and development into the 'Action Plan of Rejuvenating Education in China 2003-2007'. This plan will promote the sustainable development of higher education with a good quality.

2. Engineering Education in China

Engineering education has enjoyed a history of over 100 years in China and since the founding of the People's Republic of China, it witnesses a rapid development. During different stages of the national economic development, engineering education has made remarkable contribution in terms of social and economical development. Depending on the talents produced by engineering education, we have master the systematic technology in many important key projects such as Daqing Oil Plant development, GezhouBa Project, Three Gorges Project and the big bridge building projects. In addition, we also have mastered the innovative technology in main industries such as coal, mining, petrol chemistry, manufacturing, power supplying and transportation.

Moreover, in some high-tech R&D and engineering technologies, China has developed its own features and advantages. For instance, we have been in the international leadership in household electrical appliance, big communication equipment, high-speed optic fiber transaction, and super computer development. The laser editing of Chinese words, the R&D of genetic medicine, and mapping of genes of hybrid rice all enjoy the highlight as well. The development of satellites and man-shipped Shenzhou spacecrafts has demonstrated the significant breakthroughs in aerospace technology. In addition, many projects which are being built have been included into the world grand programmes in terms of technology and scale. For example, Qingzang Railway lasts from Xining to Lasa with 1956 km, the longest railway with the highest altitude. The West-East Electricity Transmission Project has established three channels in the north, central and south. The scale, distance and volume involved in energy production and transmission is very unusual even in the world. Furthermore, the West-East Gas Transmission Project involves a total input of nearly 150 billion RMB. In 2005 when the pipelines are open, the gas production in China will be increased by over 40%. In all the above projects, the main players are the engineers trained by our own engineering education.

It is reasonable to assume that millions of engineering experts have made great contribution to the national development in China. We are proud of ourselves for the internationally famous engineers and engineering educators such as Mao Yisheng, Li Siguang, Qian Xuesen and others. The millions of qualified engineers trained by engineering education, are contributing their intelligence and knowledge to the modernization in China. Some of them are also contributing to the development of engineering science of the world.

Currently, there are 1,528 HEIs providing engineering education, among which, 604 offering undergraduate programmes (4-5 years) and 924 diploma courses (2-3 years). The total number of engineering students in HEIs is 4.7 million, which composes 33% of the total enrollment of higher education of China. In terms of structure, the engineering students can be awarded with four kinds of degrees and diploma, i.e. doctoral, master, bachelor degrees and diplomas. The numbers of students at each level are: 69,000; 248,000; 2,422,000; and 1,950,000 respectively, representing one of the largest engineering education in the whole world.

The new century brings a series of changes. Many new disciplines and cross-disciplines emerged along with the development of info-technology, biotechnology and nano-technology. The world industry has gong through a revolutionary change as well. China is entering a new phase of development that aiming at building a well-off society in an all-around way under the market economy. Challenges are coming from the internationalization of personnel competition, the market-directed economy, the growing scale of mass education and the variety of entities in running schools, in addition to the domestic needs for technological innovation required by China's industrial development. Correspondingly, reforms on engineering education must be carried out in order to take the above challenges successfully. During recent years, a series of measures has been taken by MoE to promote the reform and development of engineering education.

2.1 *Consistently reinforce the reform and construction of engineering curriculum*

In the year 2002, the MoE launched a program that is to develop 1500 excellent courses by the following years, with the aim of pushing forward the reform and construction of the curriculum in higher education institutions, and optimizing the resources of higher education curriculum. Moreover, it is also expected in this programme that with the wide application of ICT, the sharing of education resources of high quality will further expanded so as to promote a balanced development of higher education. By now, 451 excellent causes have been developed, which cover 167 courses in engineering programs.

Since 1997, in order to expedite comprehensively the reform and development of curriculum, the MoE has been making efforts to build 56 national teaching bases of 7 categories for fundamental curriculum in engineering, including basics of mechanical engineering, electrical engineering, mechanics, mathematics, physics, chemistry and engineering diagram. After 8-year's efforts, great progress has been achieved in the establishment of the teaching bases. The excellent curriculum newly developed has optimized the teaching content and curriculum system. The success of the teaching bases has become a model, which help to enhance the overall teaching quality of the engineering causes.

2.2 *Prioritizing the improvement of students' application capacity, and emphasizing the improvement of the condition for internship and the reform on practical teaching*

Engineering education, with technical sciences as its basic elements, is a professional education targeting at training engineers with the ability to transform science and technology into productive force. As an integration of education system and engineering system, engineering education shares a common ground with general education and at the same time has its own special features of engineering. The enhancement of students' application ability is a core part of the training of engineering talents. Therefore, MoE pays much attention to carrying out internships in HEIs, to the promotion of laboratories, training center for engineering and internship and training bases, and to encouraging the training of the application capacity of college students. Meanwhile, measures have been taken during the construction of basic laboratories and the internship and training bases, such as using World Bank loans, establishing partnership with other governmental departments, and encouraging industries to cooperate with universities. All these interventions have provided both software and hardware for the training of personnel in terms of their practical activity improvement.

2.3 *Accelerating the cooperation with enterprises; improving the integration among education, industry and research through different approaches*

Integration among education, research and industry is an effective way to establish connections between engineering education and enterprises, and to fully make use of the school's advantage in human capital and the enterprises' advantage in resources. In order to develop the cooperation between schools and enterprises in a comprehensive way, MoE has establish some national internship bases with the partnership of industry, and 44 science parks within universities, as well as 48 national engineering research centers with concentrated resources.

2.4 *Enhancing the students' innovativeness and encourage innovative activities among university students*

Entrusted by the MOE, some professional associations have held various national competitions for university students in electronic design, mechanical innovation design, and structural design, etc. The purpose of these activities is not only to enhance the students' innovative consciousness and team spirit, but also to improve their ability in creative designing, comprehensive scientific skills and engineering internships. In addition, the activities are also beneficial for the improvement of the teaching reform, and for the combination of theoretical education and internships and the combination of in-class teaching and extracurricular study.

2.5 *Speeding up the training of the greatly needed personnel in accordance with the development of industrial development*

In accordance with the national economic development and the needs for personnel in various areas at all levels, the MoE is making efforts to push forward the adjustment of disciplinary structure of higher education. In 2001, with the approval of MoE, 35 national software colleges and other 35 software vocational schools have been established, which will be committed to cultivating high quality, internationalized personnel in the field of software engineering. Also, 36 national training bases for life science and technology have been established in order to meet the needs of the rapid development of the biotechnology. 14 national training bases for integrated circuit have been established to meet the needs of the development of the semiconductor and IC industry. Besides, in line with the requirement set by rapid industrial development, we have also intensified our efforts in the training in some fields which is in great shortage of experts such as information security, e-business, digital control technology, automobile services and innovative industry.

3. The Quality Assurance System in China's Post-secondary Engineering Education

The comprehensive implementation of reform in post-secondary engineering education serves as the intrinsic stimulus and effective mechanism for quality assurance in talent training. In the meantime, strengthened macro management and evaluation are the driving force and external guarantee for a high-quality engineering education in China. After nearly a decade's pilots and practices, a system that assures the quality of our engineering education has gradually taken shape.

3.1 *Macro quality monitoring by expert panels*

In April 2001, 29 engineering Steering Committees for Teaching were reestablished by MoE. In the meantime, 11 Steering Committees for Disciplinary Teaching consisting of about 1,500 experts in the disciplines of transportation, civil engineering, hydro-engineering, oil-drilling and surveying and mapping, were also set up by relevant offices of the State Council. These committees have attested to their capability in guiding the research and practice in relevant fields. They have provided advisories and suggestions to various educational administrative authorities and HEIs, formed and drafted academic discipline development strategies and standards, and assisted the relevant educational administrative authorities in performing monitoring and evaluation roles to assure education quality, in addition to engaging in macro management and guidance in various subjects and disciplines.

3.2 *Implementation of educational evaluations to monitor the level and quality of education in HEIs*

Since 1994, based on the evaluation of post-secondary engineering education, the evaluations of teaching and learning in higher education have been carried out in full scale. Instrumental to the government's macro monitoring of educational quality and the establishment of the quality monitoring system, these evaluations focus both on the educational processes and the educational effectiveness, with an emphasis on educational processes. Evaluations on past performance and future developments are both carried out, with an emphasis on future developments. The overall objectives of these evaluations are to help HEIs find out definitive paths for development, further strengthen investment in educational infrastructures, better manage themselves and ultimately, further improve the quality of education. The evaluations contain criteria in different levels, with 7 criteria in level 1, 19 in level 2, and 44 assessment points further below. Emphases are laid on such fields as educational conceptions, faculty number and structure, teaching infrastructure, funding, subjects, curricula, teaching practice, quality control, basic theories and skills, and graduation theses/designs.

The evaluations have promoted the development of fundamental infrastructures for undergraduate education, especially in such fields as basic experimental education, bases for practices, and quality monitoring and assurance systems in teaching and learning. The quality awareness in undergraduate education was greatly increased and the quality of higher education has been effectively guaranteed.

4. Accreditation of Engineers' Qualifications

Engineers play pivotal roles in the development of industrialization, mass production and construction.

For Chinese economy to embrace globalization and contribute talents for international industrial development, it's vitally essential for China to accelerate the international engineering certification process. This is particularly true in China's post-WTO entry context.

Engineer certification is a powerful enabler in that it bridges between project and talent, ensuring project quality by enforcing economic responsibilities upon practitioners through regulations in the rights, obligation and responsibilities of certified engineers.

The certification process is essential to education in engineering in three important ways. First, the certification system points out the direction for future reforms in engineering education and enhances its practical relevancy. Second, the certification system will promote the application of engineering theories to real-life practice; hence fostering engineers' pragmatic approach to solving real-work problems. Last but not the least, the certification system promotes the internationalization of engineering education since the establishment of a nationally uniform and internationally recognized certification system naturally entails the education of engineering talents following international standards.

Under the leadership of the State Council, the 'National Coordination Group for Engineer System Reform' is founded through the joint efforts of relevant offices. MoE, as one of the group leaders, has been an active participant by founding the 'Committee for Accreditation Of Engineering Education in HEIs' and carrying out extensive research work in a variety of accreditation systems already instituted, such as ABET of the United States, ASIIN of Germany, ASSIIN based EUR-ACE of the EU, and HKIE etc. And the professionalized engineering accreditation is at the top of the agenda after the institution of the engineers' certification system.

Professional accreditation in engineering education is the foundation and prerequisite for the implementation of certification of engineers. Following the implementation of the system for engineer certification, the institution of a system of professional accreditation in engineering education to meet its need has become a matter of great urgency. Due to the features of different countries, the professional evaluation and accreditation for engineering education vary accordingly. It's therefore natural and necessary for China to take an approach of its own. Drawing from our successful experience accumulated in the fields of civil engineering, urban planning evaluations, and evaluations of higher education, which have been carried out in a broad scope, we are now in a very good position to face the urgent and demanding challenges of instituting evaluation systems of professional engineering education and promoting the international recognition of the qualification of Chinese engineers.

To accelerate the modernization of our industries, it's important that we adhere to a rational philosophy for development, seek out new approaches for industrialization, and encourage excellence and variety in training and education. It's also vital that we tailor the engineering educational structure to the needs of the shift in industrial structures, consolidate the linkage between education and practice, promote more cooperative education and interactions between enterprises and academia, and assist engineering education in playing a more vital role in the promotion of science and economy. We should further promote engineering education evaluations and engineering accreditation, to meet the increasing demands for talents in the international engineering marketplace.

5. Conclusion

To conclude, the engineering educational sector in China is keen to join hands with colleagues from all over the world in promoting the development of global industries and human civilization. We hope that many of our universities will remain as the cradle for excellent engineers.

SUSTAINABLE DEVELOPMENT OF TALL BUILDINGS IN HONG KONG

LO YIU-CHING*

Permanent Secretary for the Environment, Transport and Works (Works), Government of Hong Kong SAR

The importance of sustainability is highlighted and the environmental, economic and social aspects of tall buildings are discussed. Sustainable construction of tall buildings in different aspects are delineated, including the building environmental assessment methods, the structures and services of tall buildings as well as energy efficiency and renewable energy to achieve sustainability.

1. Introduction

Hong Kong, as one of the world's renowned international commercial centres, needs to ensure that demand for office space can be met within the central business area. Even land for residential use is in high demand. In this perspective, tall buildings are necessary in order to make the best use of the limited land available. The reasons why Hong Kong has so many tall buildings can be attributed to the following factors:

- Demand for more office accommodation and residential flats;
- Taller buildings are more efficient in making use of the inadequate land supply;
- Infrastructural development facilitates the emergence of tall buildings;
- User demand for better view, more access to daylight and fresh air which can be achieved by increasing the building height; and
- Tall buildings confer prestige and ensure economic competitiveness.

2. Opportunities for Change

The emergence of high-rise residential buildings in Hong Kong is two-pronged. Historically, the current ultra high urban density of over 6300 persons per km^2 was sparked off by an influx of emigrants from Mainland in the mid 20th Century. This was followed by an rapid increase in population between 1950 and 2000 whereby the Hong Kong Government had launched massive housing schemes to accommodate the sharp increase in housing demand. In 1973, the New Town Development Programme was first started and later extended to develop nine new towns to cope with the increase in population and to improve the living environment by de-centralizing the population from the over-crowded urban districts. Meanwhile, the high-density population in urban areas has facilitated the building of many high-rise residential buildings in Kowloon and on Hong Kong Island. Mei Foo Sun Chuen built from mid 1960s has been widely regarded as a watershed both from property development and construction perspectives. This successful model is a multiple intensive land use development in the form of a self-contained township which has been further developed into the nowadays residential super-high-rise towers sitting on top of a multi-storey podium with all kinds of facilities providing comfort, convenience and connectivity for thousands of residents.

3. Sustainable Development Strategy

Sustainable development is a key parameter for action. New tall buildings have to fit within this context. The design, construction and operation of tall buildings has incorporated new local and international standards in relation to sustainable development. The first Sustainable Development Strategy for Hong Kong promulgated by the Council for Sustainable Development in May this year has included solid waste management, renewable energy and urban living space as the three pilot areas whereas the need to balance and integrate social, economic and environmental factors is recognized.

---

* BSc(Hon), MSc(Eng), CEng; FICE; FIStructE; FHKIE

4. Tall Buildings and Transport Provision

Transportation system becomes very effective and economical in the context of Hong Kong where we have a clustering of high rise buildings in densely populated areas. In Hong Kong, railways form the backbone of the transportation system. The fast and efficient railway system of Mass Transit Railway (MTR) and Kowloon-Canton Railway Corporation (KCRC) connecting various districts of Hong Kong Island, Kowloon and New Territories in addition to the extensive highway network provide Hong Kong with the best transport infrastructure in the world in terms of efficiency. This greatly facilitates the movement of people between the high rise towers.

5. Urban Design of Tall Buildings

The Government has issued "Urban Design Guidelines" to maintain views to ridgelines, peaks or water body and to provide breezeways by lowering building height when required. High-rise nodes are allowed at selected strategic locations instead, say, further inland rather than on the waterfront.

While occupants of tall buildings may enjoy more daylighting and better views, their neighbours can also have equal opportunities to enjoy daylight in these same aspects. The Building Regulations of Hong Kong prescribe, among others, a minimum distance between buildings for lighting and ventilation based on the concept of sustained (obstruction) angle requirements.

6. Building Environmental Assessment Methods

Lifecycle assessment of buildings and their construction materials is now gaining wider attention. The Environment, Transport and Works Bureau has commissioned Electrical and Mechanical Services Department (EMSD), with the support of other relevant departments, to carry out a study on the Life Cycle Energy Analysis of building construction in order to promote the concept of sustainable construction. Developers and designers can then use the new software tool to assess the life cycle impact of their buildings for minimization of any consequent negative environmental impact.

Since early 1990s, a number of building environmental assessment methods (BEAMs) have emerged, including the Building Research Establishment Environmental Assessment Method (BREEAM) in the United Kingdom (UK) and the Leadership in Energy & Environmental Design (LEED) in the United States (US). These methods normally allocate points or credits to various assessment aspects and, depending on the score, different award levels or grades are given to the building under assessment. The assessment criteria may vary for different categories of buildings. These assessment methods now extend to cover the concept of embodied energy which is an inventory of materials used in the construction of a building. In Hong Kong, the pioneering scheme is the HK-BEAM which has extended its coverage in 2004 to include good practices in the design and operation of buildings, like initiatives of the Government in indoor air quality, energy efficiency, water quality and waste recycling, etc.; whereas another scheme called the Comprehensive Environmental Performance Assessment Scheme (CEPAS) has now gained a much attention of the industry. The BEAMs just discussed mainly concentrate on the environmental aspects of sustainability. Nevertheless, it is observed that some of these BEAMs have begun to embrace some social and economic criteria in their assessment.

7. Towards Sustainable Construction of Tall Buildings

Buildings can be more sustainable if there is integration between structure and building services and they can make the best use of the site conditions.

The Government has been advocating construction of green buildings and has formally issued the "Joint Practice Notes" to encourage adoption of green features in new private building development since 2001, including balconies; wider common corridors and lift lobbies; communal sky gardens; communal podium gardens; acoustic fins; sunshades and reflectors; wing wall, wind catchers and funnels; non-structural prefabricated external walls; utility platforms and noise barriers.

The thermal and visual performance of a building is affected by the thermal insulation (U-values) of external walls, thickness of floor slabs and window areas, including the fabrics of the facade and the roof. The Architectural Services Department have commissioned separate studies to come up with a Facade Rating Scheme in Hong Kong, in particular for glazed facades, and Sustainability Rating for metal roof systems. For tall buildings, the wind pressure can be very high on the upper floors rendering air tightness of the facade a major issue. Double skin facades could have advantages over single skin. When the space between the two layers of glazing is ventilated through natural uprising thermal air currents, the amount of heat transferred into the buildings could be reduced. The double skin facades could also offer the additional advantages of noise reduction.

The span of floor beams in new buildings, which is usually in the range of 10-15m as compared with 5-8m in the past, provides more flexibility and is less obstructive to the use of space. Floor systems with concrete floor slab supported by steel beams are increasingly popular, resulting in reduced depth of beams and less materials used. Depending on the building height, different structural forms are used. Steel is commonly used in the US for tall structures and different structural systems are used for buildings of different heights. In other parts of the world where steel is not easily available, concrete frame systems are more common. As in the case of steel buildings, different systems of concrete structural frames are used in buildings of different height. It has been shown that a hybrid system combining the advantages of both concrete and steel is most efficient. However, the choice of a structural system is very much influenced, if not governed, by the intended use of the building, in terms of plan layout and use of space. Naturally heavy foundations have to be used in high-rise buildings. This often involves driving piles deep into the ground unless the bedrock or suitable bearing stratum is found at shallow depth. Noise and vibration generated by piling are environmental issues which must be addressed.

The concept of embodied energy has gained increasing attention in recent years. It is an inventory of materials used in the construction of a building. With known energy for producing a unit volume of material, we can then calculate the total energy used in the whole building, including the energy consumed in the construction process. This embodied energy may account for 10-20% of the energy used in the lifetime of the building.

In the context of choice of materials, there is nothing better than timber which has the lowest embodied energy. However, because of strength and serviceability requirements, we seldom have that luxury. Steel manufacturing or recycling consume a lot of energy, and hence steel has a high embodied energy per unit volume. However because of its superior strength to weight ratio, it remains as one of the most common construction materials. The same can be said about reinforced concrete.

8. Solid Waste and Waste Water

The construction and demolition (C&D) of buildings generate a lot of waste known as C&D waste. The management of C&D waste is focused on the 3 R's, i.e. reduce, reuse and recycle. While reducing the use of material in the design is a matter of common sense, reuse and recycling of materials achieve economic benefits while at the same time release the pressure on our landfills, the outlets for the disposal of C&D materials.

For solid wastes such as domestic refuse, mechanical biological treatment is sometimes used overseas whereby the waste goes through four major phases of dry sorting, anaerobic digestion, wet separation and dewatering. For the long term sustainability of Hong Kong, the plausibility of high temperature incineration is now being evaluated.

The same 3 R's are also applicable to the water side. Development of biotechnology enables maximizing the recycle of grey water by micro-organisms. Recycled grey water can then be reused for toilet flushing, irrigation of plants and washing hands. The grey water generated from Waste Water Treatment Plant in EMSD Headquarters for flushing and the waste water treatment plants in Science Parks and Ma On Shan Sport Ground used for irrigation of plants are good examples.

9. Services in Tall Buildings

9.1 *Indoor environment*

People now spend most of their time indoors and the quality of indoor environment becomes increasingly important. The term "sick building syndrome" (SBS) is used to describe situations in which building occupants experience acute health and comfort effects that appear to be linked to time spent in a building, but no specific illness or cause can be identified. The cause of the symptoms is not known and most of the complainants report relief soon after leaving the building. To improve the indoor air quality (IAQ) and promote public awareness of the importance of IAQ, the Government launched a voluntary IAQ Certification Scheme for Offices and Public Places in 2003. Many government premises, particularly tall buildings, and a significant number of private premises have actively participated in this scheme. When participating premises achieve good to excellent class, it means a better indoor environment is provided for the occupants.

9.2 *Air-conditioning*

In 2000, the Government launched a pilot scheme on the wider use of fresh water for water-cooled air conditioning in non-domestic buildings in Hong Kong. The scheme allows non-domestic premises to use fresh water for evaporative cooling. As cooling towers significantly reduce electricity consumption in air-conditioning systems when compared to air-cooled condenser systems, the scheme offers an excellent opportunity for building owners to switch to use cooling towers to reap the saving in electricity.

Although IAQ is affected by the choice of building materials and furniture, the design of air-conditioning system is also critical. Adequate fresh air and sufficient ventilation rate and suitable choice of filters are essential. Energy efficient devices have been adopted in tall building design to reclaim waste energy and recycle energy, such as thermal wheels, heat recovery chillers and photovoltaic (PV) panels with a view to improving energy efficiency, enhancing the quality of air conditioning, and trimming down pollution to the environment.

9.3 *Electric lighting and daylight*

With taller buildings, the footprint can be reduced for the same aggregate floor area and allow for maximizing the use of daylight. For perimeter zones of the building with daylight penetration, daylighting can be integrated with electric lighting to reduce energy consumption. Photocell sensors, infrared sensors and dimmable controls can be used to switch off unnecessary lightings for energy savings. Energy efficient devices, e.g. electronic ballasts and T5 fluorescent lamps with effective lighting control, are more commonly adopted to save energy costs.

For tall buildings, other issues may need to be considered. External lighting or internal lighting with glazed facade provides building identification on flight paths but light pollution must be minimized, even when the portrait of the building visual appearance is in question.

9.4 *Fire safety*

The event of 911 changed the perceptions of building designers, owners and occupiers with respect to safety issues in tall buildings. The suitability of current regulations for very tall buildings is being reviewed by leading construction industry and academic institutions. Due to the building height and number of occupants, early warning and means of escape is also very important. Some issues would be of major concern and critically important to the design of sustainable tall buildings, in particular during emergency or disaster situations. Some new developments in the international arena adopt a threat and risk assessment to quantify real risk for developing suitable mitigation measures, phased and total evacuation of building occupants, and structural integrity in case of intended explosion or fire. In Hong Kong, fire safety design for tall buildings can be based on either the prescriptive fire codes or the fire engineering approach.

9.5 *Vertical transportation in tall buildings*

Vertical transportation is important to the efficiency of occupants' movement inside tall buildings. Lifts of higher speeds tend to be used in the taller buildings but the issue of lift ride comfort will arise at very high speeds. Smoothing the acceleration and deceleration of the lift car can improve passenger comfort and reduce energy use.

A recent development for tall buildings is the adoption of double-deck lifts which occupy less building core space than traditional single-deck lifts do to cater for the same level of lift traffic requirements. In tall buildings, this allows for much greater efficient use of space as the floor area required by lifts would tend to be quite significant.

10. Buildings and Energy Use

Electricity generation currently accounts for the emission of 89%, 45% and 37% of sulphur dioxide, nitrogen oxides and repirable suspended particulates respectively in Hong Kong. It is also responsible for 62% of the greenhouse gases produced locally. Between 1999 and 2004, the annual combined electricity and gas consumption in the commercial sector of Hong Kong rose by an average of about 3.3% although the annual office stock growth rate was about 1.8%. Reducing the consumption of electricity can directly improve the local air quality and help combat the problem of global warming. In order to reduce global warming and to tackle the insecurity in long-term energy supplies, it is imperative to minimize energy consumption. Therefore, we have to study all kinds of energy use in buildings to look for better ways of conserving energy.

Air-conditioning accounts for roughly one-third of the total power consumption of Hong Kong each year and costs some $10 billion. To echo the World Environment Day this year, ETWB have urged the public to adjust their air-conditioning system so that the room temperature is maintained at 25.5°C with a view to reducing emission of carbon dioxide, lessening global warming effect and maximizing energy savings. Meanwhile, the Government has set an example to the community by setting the air conditioning temperature of government offices at or above 25.5 °C in summer months and appointing energy wardens to ensure strict implementation of energy-saving measures.

The attitude of building occupants should also be changed so as to achieve reduction in energy consumption. Using energy efficient electric lights and air-conditioning systems can considerably lessen the environmental impact while maintaining the same or even higher productivity levels. By changing the habits of use and improving the design, construction, operation and maintenance of buildings, energy use per unit building floor area can be reduced.

The effective use of renewable energy resources will help to reduce reliance on fossil fuels and also to reduce greenhouse gas emissions arising from the use of fossil fuels. The Government has been advocating ongoing initiatives on environmentally responsible development by enhancing the use of energy efficient devices and promoting wider adoption of renewable energy in public works projects.

The government projects have adopted the Overall Energy Approach and there are guidelines on the maximum overall thermal transfer value (OTTV) for new government buildings. The Government has also published four prescriptive and one performance-based building energy codes to promote energy efficiency in buildings in Hong Kong.

In Hong Kong, solar energy is used to provide hot water for a number of swimming pools and for the slaughterhouse in Sheung Shui. Some small-scale PV and wind systems have been installed in remote areas to generate nominal electrical power for lighting and on-site data recording equipment. A common kind of renewable energy source is PV installations which converts solar energy to electricity. While PV panels are usually installed on the roof, Building Integrated Photovoltaic (BIPV) panels can replace the building facade and reduce the overall costs. A number of government and institutional buildings have already installed the PV or BIPV panels, including offices and parks. Waste heat from the air-conditioning system has also been utilized (e.g. by installing water-to-water heat pump in the condensing water circuit) in conjunction with solar water heating to generate hot water for the central hot water systems in office buildings, hospitals and game halls.

In addition to the above, we are mapping out strategies by studying the use of wind energy in the territory and also taking a closer look at the possibility of using alternative renewable energy resources such as hydrogen fuel cell technology.

11. Conclusions

The adoption of energy efficient designs and inclusion of renewable energy sources into new buildings and infrastructure projects have become a global trend in striving for a sustainable future. Tall buildings, because of their high compaction, dense population of occupants and complexity of services, should place more emphasis during their design on green features, sustainability, and economic benefits in terms of cost savings in material and energy use. The cost effectiveness of good design can go far beyond these tangible factors. All in all, any tall building proposal should be examined for its sustainability in the widest sense, by considering its economic, environmental and social impacts, based upon life cycle costs and benefits.

References

1. Arup, *Developing the Role of Fire Engineering.* (2005)
2. Arup Facade Engineering, *Metal Roof Systems: Sustainability Rating (Draft 7).* (2003)
3. Building Services Branch, Architectural Services Department, *Guidelines on Wider Use of Renewable Energy Technologies and Energy Efficient Features in Building Projects.* (2005).
4. Buildings Department, *Environmental Report 2003.*
5. R. Chin, *Overall Energy Approach in Achieving an Acceptable Indoor Air Quality in Tomorrow's Buildings.* (2001)
6. Council for Sustainable Development, *First Sustainable Development Strategy for Hong Kong.* (2005)
7. A. Crompton and A. Wilson, *Sustainable Tall Buildings – Fact or Fiction?* (2002)

8. English Heritage and the Commission for Architecture and the Built Environment, *Guidance on Tall Buildings*. (2003)
9. K. K. Li, M. T. Suen and S. Y. Mui, *Hong Kong Science Park – Phase 1*. (2004).
10. A. Lui, *Urban Design Sustainable Development*. (2004)
11. A. Ng, *Forum on Urban Living Space: Grow Out or Grow Smart?* (2004)
12. E. Ng, *A Study on the Accuracy of Daylighting Simulation of Heavily Obstructed Buildings in Hong Kong*. (2001)
13. W. Pank, H. Girardet and G. Cox, *Tall Buildings and Sustainability*. (2002).
14. X. Wang and S. Y. Lau, *Pursuing New Urban Living Environment in the New Millennium: Projecting the Future of High-Rise and High Density Living in Hong Kong*. (2002)
15. R. Wong, *Construction of Residential Buildings: Developments and Trends in Methods and Technology*. (2000)

FUTURE DIRECTIONS FOR THE DEVELOPMENT OF TALL BUILDINGS

PETER AYRES
Ove Arup & Partners HK Ltd.

1. Introduction

My objective in this address is to try to encourage all parts of our industry to work together to drive change through the whole of the design and construction process. Given that tall buildings provide some of the greatest challenges so to can they also provide the best opportunities.

I will look at the way greater advantage can be taken of existing technologies and skills, and areas where more development effort is needed. I will conclude by looking at the role that each party in the process will have to play if we are to be successful. And by way of an introduction I will use my own personal experience over 27 years in Hong Kong to highlight typical developments that have taken place in tall building design over that period.

2. Personal Experience

Before coming to Hong Kong in 1978 my experience had been limited to low-rise institutional, retail and office structures in UK, Africa and Iran. The tallest of these was the Shahyad Ariamehr Monument in Tehran at 45 metres. We were also in the early days of practical computing and the use of slide-rules was still commonplace. Approximate design methods were often employed and personal judgement played a very important part in the final solution. In reviewing my experience in Hong Kong I will concentrate on structural developments because that is what I know best.

Building design in the seventies could be typified by Hopewell Centre and Luk Yeung Sun Chuen in Tsuen Wan. These were constructed of traditional reinforced concrete, slipformed in the case of Hopewell Centre. Vertical elements were designed as fully coupled creating effective tubes rather than beam/column arrangements and transfer plates were introduced where necessary.

In the eighties, typified by Exchange Square, HSBC Headquarters and Queen's Garden/Dynasty Court we see the introduction of structural steel alongside reinforced concrete, outrigger as well as transfer structures, some modularization and the introduction of curtain walling.

In the nineties, typified by Central Plaza, International Finance Centre and Les Saisons, Aldrich Bay, we see the use of much higher strength concretes, composite construction, more sophisticated curtain walling and a fuller understanding of structural dynamics, movement and control.

At the beginning of this millennium we are continuing these developments in creating ever taller buildings of which International Commerce Centre and the Waterfront, both at Union Square, Kowloon, are good examples.

3. Way Ahead

Whilst I might have been less sure ten years ago it is now clear that we will continue to build tall buildings and they will get ever taller. The reasons include city densities, operational efficiencies and "ego" and the demand will be particularly strong in the many rapidly developing cities in Asia. As designers and builders we need to understand the challenges and therefore the opportunities that this scenario presents us.

If we look at the developments that have taken place over the past century our record is not that impressive. In 1913 the world's tallest was the Woolworth Building already achieving a height of 240metres and in 1931 the Empire State Building was opened, 381metres high and completed in 14 months. The materials and construction methods employed were not that different to those in use today. This is in stark contrast to the achievements in the automotive and aeronautical industries over the same period.

We therefore need to develop an appropriate response to the challenges which taller and taller buildings will offer us and which is appropriate to the 21st century. This will inevitably involve significant changes on the part of all participants and also a significant increase in the investment we make in R&D. It is interesting to note that US design and construction invests 0.5% of total revenues in R&D which is 1/7 of average investment amongst mature US industries and probably typical of industry worldwide.

4. Areas for Development

We all need to make changes in the way that we approach these projects in the future. There follow some examples of opportunities we have and things that we should do better.

- Understanding – we now have sophisticated readily available modeling and analysis software to help us understand static, dynamic, social and environmental issues. This better understanding has to be translated into more refined approaches including regulatory ones.
- Form – we no longer have to be constrained by simple forms, although I do not advocate complexity for it's own sake. The form of our buildings can now respond to their function.
- Habitat – super-tall buildings will need to house communities, places to live, work and play. We must recognise that a single building could represent a "township" of 50000 people.
- Materials – we must be seriously exploring the use of new materials and construction systems for our industry as well as continuing to develop our existing ones. We must have a more open mind to new ideas.
- Energy - we must identify the real needs of users as opposed to their desires and use the most energy efficient methods to deliver. Green and sustainable have to underpin our future projects.
- People Movement – we will have to integrate better the various uses within our tall townships using current technology and developing new. We must consider the social needs of the users.
- Security – we must adopt a performance based approach to robustness and means of escape. Proper management plans, training and communications systems must be in place. We must properly understand risks and offer practical solutions.
- Adaptability – we need to take current multi-use concepts forward to allow for future change in both use and technology throughout the building's life. We must recognize that these projects will have a huge impact on society over a very long period.
- Integration – these very complex projects will require that we all work together constructively, understanding and sharing the challenges as one team. Our approach must be interdisciplinary at all levels.

If we can truly face up to these challenges then we can believe in a sustainable future for our developing cities.

5. Actions

In order to maximize the benefits that can come from the huge challenge of the construction of taller and taller city towers and to ensure that the legacy we leave succeeding generations is responsible and sustainable we all need to take action:

- Governments – need to formulate planning approaches which recognize that whole new communities are being created. These should encourage multi-use and social integration.
- Owners – must understand that these buildings are very long term investments for society and they need to be able to respond to the inevitable changes in use, technology and expectation.
- Regulators – must develop methods which encourage innovation and new techniques. In many cases this will involve a significant change in mindset which will need to be explained to society.
- Educators – must understand the big picture for which they are preparing their students, the role these students will aspire to play and the need for much better interdisciplinary and social understanding.
- Construction industry, builders and designers – must respond to the opportunity by becoming braver and making a very significant increase in its commitment to R&D, the costs of which must be accepted by society at large.

A new millennium is a grand opportunity to take stock. We have two choices:

- Carry on as we are making small incremental improvements and developments – we can see where we will be in 50 years and it won't be very impressive I'm sure.
- Rise to this challenge now and together, and it has to be together, start to make a real difference to this vital industry of ours.

I hope we can take the second of these. It will of course benefit the whole industry but such change is always best led by the greatest challenges in my view. If we succeed even in small part society will be the richer and we will be much better placed to regain the enthusiasm of the young for our very exciting business.

CARBON FUELS AND INDUSTRIALISATION: REDUCTION OF ENERGY DEMAND WITH SPECIAL REFERENCE TO TALL BUILDINGS

MAX FORDHAM*

Max Fordham LLP, 42/43 Gloucester Crescent, London NW1 7PE, England

A summary of the development of the world and the thermo-dynamics are of industrialisation is given.to justify the need to reduce the use of carbon fuel in our society. The passive design of buildings to provide light, thermal capacity, and ventilation , so as to provide good control of the internal environment without mechanical and electrical systems is described. The influence of the height of the building on the passive principles is outlined.

1. Introduction

I am committed to the idea that our use of carbon fuels has to be sharply reduced not only to mitigate global warming, but also because we will have to learn how to do without them as they are a finite resource.

The benefits of our industrial society are directly caused by the use of the carbon fuels that enabled an economy to be based on the use of energy instead of human labour. I want to start by explaining something about thermodynamics, and how industrial activity can be seen in that context.

Let us think about the earth for a moment. It has been in existence for about 5 billion years. Kelvin calculated that it would only have taken a few thousand years to cool from the temperature of melting rock to the temperature we observe. He sparked off a vicious debate about the age of the world, which he lost.

The current temperature is reached as an equilibrium, whereby energy leaves the sun's surface at 5000°C, and arrives at the earth, carrying 1000 W per square meter. The projection of the earth onto the flux of light is a circle with the radius of the earth about 7000 kilometres, and it is easy to work out how much energy arrives at the earth. This energy is then reradiated to the universe, which is at few degrees above absolute zero. In the late 19th century Arhenius worked out that the earth should be good deal colder than it is. He arrived at the correct explanation, that the atmosphere is transparent to the incoming short wave radiation from the sun, but partially opaque to long wave radiation radiated from the cold surface of the earth. The Earth has to heat up to enable the equilibrium to be maintained. Now we are increasing the carbon dioxide content of the atmosphere and making it more difficult for the solar radiation to be retransmitted to the universe. We can observe the rise in temperature of the world and we can see why it must be happening.

As the earth cooled there came a stage when water condensed and the chemistry and biology that we can see did evolve.

The equator is hotter than the Poles and convection currents in the sea and the wind distribute heat to the poles. The winds are produced by the heat engine of the heat on the earth going from the equator to the Poles. Water evaporates from the sea into the atmosphere. Any air at ground level which is hot and wet tends to rise reducing in pressure and so it expands and cools. The water condenses and falls as rain. Although the sea has become very polluted by salts washed off the land. It is the source of a continuous supply of distilled water.

The Earth could easily be a lifeless sphere receiving solar radiation at one temperature and reradiating it to the universe at another. There would be strong currents in the air and sea. These are observed on Mars.

In fact a process has evolved, Photosynthesis, which traps the energy from the sun and uses it to make carbon dioxide and water taken from the air combine to produce plant material. This process starts with sugar, which polymerises to starch, and on to cellulose. The organising principle of photosynthesis is genetic. The genetic process produces the cellulose, and it has also produced a process for making cellulose revert to carbon dioxide and water again. The reversion process is metabolism by animals. Animals which range from

* OBE MA (Cantab) FREng FCIBSE MConsE Hon FRIBA
Tel +44(0)20 7267 5161 max@maxfordham.com

large mammals to small bacteria, use the energy in the plant material to enable each animal to reproduce and continue its existence.

The carbon cycle taking carbon dioxide from the atmosphere, producing plants, and decomposing them back to the atmosphere again circulates about 25000 tonnes per second of carbon dioxide.

Not all of the plant material decomposes. Look at a stagnant pond. There are black carbonising leaves at the bottom, there are a few bubbles of methane, coming up from the bottom, and sometimes there is an oily sheen. The pond only has to be involved in an earthquake and taken down under the ground to be heated up and compressed to become oil, coal and methane as we know them. The process does not have to be very fast. If you take the total carbon reserves of the world and divide by the age of the world, then the rate of production is about one hundred grams the second. We are burning them at 700 tonnes per second.

That is we have traced the activity of the solar energy as it travels from the sun to the earth, and out to the universe. It produces currents in the sea and the atmosphere. It produces the water circulation from the sea falling as rain back down the rivers to the sea, and it produces a carbon circulation from the atmosphere and back again. Through our use of carbon fuels it produces our industrial activity. There are some other energy flows on the earth, but they do not dominate the natural world as the sun does.

By the 17th century the use coal as a source of energy had been noted. Refining iron and making steel needs an intense source of heat and charcoal was needed to refine iron and steel. It is not easy to make stoneware with a wood fired kiln.

The industrialisation, which was triggered by the use of coal, could not have been based on renewable energy. The point is that the high intensity of the carbon fuels produced high temperature flames, and enabled industrial processes. The high intensity of the energy in the carbon fuels also makes them useful and easy to transport so that they displaced the early use of Hydro Power. Now we have oil and gas, which will simply flow along pipes, and this helps to make the industrial processes straightforward and easy. The intense energy, that has been laid down over many billions of years, is now being used up by us in a few hundred years. The industrial processes can be seen as means for extracting carbon fuel from the earth, and using them to extract more carbon fuel. There is then the by-product that is the output of our industry and is a benefit for humankind. We simply supervise the operation of this process, and complain that we do not have enough work to do.

Since 1974 I have been committed to reducing the use of carbon fuels by buildings. My understanding of the energy cycle on the earth has come about because I have always found that the economic system favours the use of more energy derived from fresh carbon fuels rather than using energy from renewable and low temperature sources.

For example, on one project where a new building was to be built there were electricity generators, which were running regularly to provide electricity for pumping water. The cost of the generators was attributed to the need for having a reliable stand-by electricity supply. Clearly the waste heat from the cooling jackets of the engines was more than adequate for the building. This heat for the building was free. However, the cost of recovering the heat from the cooling jackets, circulating it to the new building, and slightly increasing the surface area of the heat transfer surfaces in the building did not provide an adequate yield on capital. We wrote a report, which asked for an increased capital budget for the new building in order to reduce the heating costs. The increase was 10 times the fuel saving and the project went ahead. There were some additional costs that were simply absorbed in the budget for the building as a whole.

Energy sources are graded. As the Industrial Revolution was starting and heat was being used to make things move. The theory of heat and work had to be developed. The development was that it could be seen in the universe that heat went from hot places like the stars to cold places like the universe. The result is that the cold places get warmer and the hot places get colder. Eventually everywhere will be at the same temperature and nothing will happen. A measure for this process was devised and called entropy. As heat flows from a hot place to a cold place entropy is changed. The change depends on the quantity of heat and the temperature difference. We have defined entropy so that it increases as heat flows from a hot place to a cold one. When the heat flow processes all stop entropy change will also stop. Entropy will have reached its maximum value. "The entropy of a closed system tends to a maximum" is a statement of the second law of thermodynamics, which C.P. Snow thought was as important as a knowledge of the works of Shakespeare for an educated person. Energy that is available as work can make things move, and it does not need to involve the flow of heat from a hot place to a cold place. It is possible to use heat and convert some of it into work, but some of the heat flows to a cold place. The total of energy is conserved. We start with heat at a high temperature say 5000 °C at the sun's surface. Some of that energy is trapped to convert CO_2 and water into sugar. The energy of the photons, which are involved in the chemical reactions in the leaves, is all available in the sugar. We then burn the plant. The flame cannot be as hot as the original source of the energy. The best we can do is to make sure that the plant material is dry and then make sure that we don't heat too much material when combustion takes place. Usually fires use air which is 20% oxygen (needed for the combustion process) and 80% nitrogen which, has to be heated by the process and, prevents the flame reaching its maximum potential temperature. For a really hot flame burn a carbon fuel in pure oxygen as in Oxy- Acetylene welding. The flame burning a fuel in a boiler with an excess of air only melts parts of a boiler. The grate of a coal boiler or even a wood boiler can be made to get a bit soft. When Whittle was trying to develop the jet engine one of the main problems was that the turbine blades rotating in the flame were exposed to high forces and heated up to a very high temperature. The high temperature was a part of the improvement expected from a jet compared to more old fashioned engines making work from heat. A flame in a boiler can be used to boil water at a high pressure and make steam at about 1000 °C. The steam then expands and makes a turbine rotate. The rotation is harnessed to make electricity, and finally the steam is condensed so that it can be returned to the boiler for boiling to steam again. When the steam is condensed it gives up heat in those large cooling towers. The heat is then at a cool place the surface of the earth and it will be reradiated from earth back to the universe.

I hope this thinking about energy use in a universal and global context explains why we should consider how to reduce our reliance on carbon fuels. Renewable energy is difficult to get in large quantities, and we could easily need as much as the carbon cycle represents if we are too extravagant.

Of course photosynthesis captures light and stores the energy as sugar. The idea of making organic chemistry capture light directly and releasing electrons as electricity is feasible and could be a solution to producing electricity easily from sunlight. That is the reason why a good strategy for reducing our reliance on carbon fuels is to start by examining our demand for energy and seeing what we can do to reduce the demand so that renewable sources can be made to meet the new reduced demand.

2. Buildings

Buildings generate about half the carbon dioxide for an industrialised society. If we remember that travel between buildings covers another quarter then we see that the designing and planning of buildings is very important. The industrial processes them selves that are a large part of the benefit that we want to retain are not susceptible to influence by building designers, but I hope that industrialists will see what they can do to save the planet.

Modern buildings provide comfort and well being for the occupants. One of the benefits of industrialisation is the comfort and clean health provided by the built environment. We do not want to give this up; indeed we want it to be spread around the world so that 10 billion people have access rather than the 1.3 billion as now.

The energy demand for buildings can be reduced significantly at a very much lower cost than using what is needed by tapping renewable resources. Hence attention should be paid to low energy architecture.

Buildings have to provide: tolerable thermal comfort, adequate light for seeing, and ventilation for health.

The most important aspects of buildings which govern their energy demand are: thermal capacity, metabolism, lighting, thermal resistance and air exchange.

The thermal capacity means that the buildings do not follow the extremes of outside temperature exactly, but smooth them out.

The metabolism refers to heat release. Completely dark buildings where nothing happens have no metabolism, but in buildings needing light and where animals live there is a certain amount of heat being liberated. Activities like cooking, and working with a computer also produce heat that is the metabolism of the building. This heat release has to be able to heat the building during cold weather. The heat release is intermittent and so the thermal capacity of the building has to be able to integrate the heat release over time while limiting the variation in temperature. In summer the heat has to be rejected whenever possible to prevent the temperature rising too high. Mechanical systems have been developed to maintain the thermal balance continuously at a constant internal temperature. We have to aim to do without these systems.

LIGHT IS needed for activities inside the building. Before the invention of electric light we had to rely on natural light because the sources of artificial light were so inconvenient. Windows were difficult to make and caused thermal discomfort. Internal light levels had to be low around 100 lux. By 1973 design light levels for offices had reached 1000 lux. Even now that they have halved Carbon emissions for lighting can be half the emission of an office building. Light is one of the main benefits and it leads to the generation of carbon. Natural light is available when the sun is above the horizon and we should be designing buildings so that natural light meets our needs whenever it is day-light. If we need a high level of light we need large windows that must be close to the places where light is needed. The large windows lead to large heat losses and heating loads in winter, and they lead to large heat gains and air conditioning loads in summer. A low demand for light reduces the demand for artificial light and for the thermal load of windows. Thus the level of light needed for a building is one of the most important design parameters.

Heat flow through the envelope of a building can be inhibited using thermal insulating materials. For the UK modern buildings can easily be made so that the basic metabolism is more or less equal to the heat loss on the coldest design day. The heat balance of glazing is important because we need glass for natural light, and it is easy to show that glass can save carbon fuel.

	Vertical window		Horizontal roof-light	
	Day	Night	Day	Night
hours	8	16	8	16
Light Lux	1800	0	4500	0
Light heat input w/sqm	18	0	45	0
Glass resistance	0.7	0.7	0.7	0.7
Good blind resistance	0	2	0	2
Resulting U value	1.4	0.4	1.4	0.4
Temperature difference	18	22	18	22
Heat loss whrs	206	130	206	130
heat gain from light whrs	144	0	360	0
24 hour balance loss whrs	192		-24	
notional average U	0.44		-0.06	
Carbon fuel lost w hrs	-96		-744	

Note negative loss is a positive benefit

3. Light

Short wave radiation from the sun is about half visible light, partly ultraviolet with infra red amounting to almost half.

The light content of direct sunlight is 100,000 lux, and the energy content is 1000 w/sqm. For each watt of energy there are 100 lumens of light. If we could filter out the infrared then 100 lumens of light, would have an energy content of half a watt. There are special glasses to filter out infrared. They tend to colour the remaining light. They reduce the summer cooling load, but they also reduce the in-flow of useful winter heat.

Modern fluorescent light produces about 100 lumens per watt of electricity. Incandescent light sources produce about 10 lumens per watt. In order to produce electricity from carbon fuels about one third of the carbon energy content is converted into electricity. So that when we are thinking about carbon consumption for lighting the conversion rate is about 3 to 30 lumens per watt of carbon fuel. During winter energy input from lighting contributes to heating inside the building and is probably useful. During the summer the energy input contributes to the cooling load and has to be minimised.

During daytime there is a good supply of natural light even on cloudy days. Current design lighting levels inside buildings range between 50 to 500 lux. In the early seventies 1000 lux was quite common, but the rise in oil price has brought design light levels down. In a climate like the U K, which is predominantly cloudy 500 lux requires a daylight factor (DF) of about 10%.

This means horizontal transparent rooflights comprising 10% of the roof in a single storey building or windows of about 30% of the floor area for a side lit building. We did design Lords indoor cricket school with roof-lights providing a 20% daylight factor, and we are currently building two offices with a design of 5% DF using rooflights. This is not a realistic attitude for tall buildings. There is a consensus in Europe that naturally lit office buildings should not be more than six metres deep from the window wall. The daylight factor at the back of such an office is 1--2%. The contrast between the front of the office and the back is about 20 to one. On an overcast day the light level at the back is 50 lux; well below any modern office specification. The result is that the electric lights are on for almost the whole working day. A tall office building is designed to maximise the ratio of useful office to central core. At the World Trade Centre Ref (17) the distance from the window wall to the back of the office was about 20 metres. At Canary Wharf in London Ref (14) the distance is 14 metres. Offices this deep are not naturally lit. The idea of using light collectors and light pipes might be tried in climates where direct bright sun light is reliable.

A very important issue with lighting design is to ensure that the eye is in a reasonably uniformly lit space. The eye can adapt to a halving of the light level every three seconds. Looking up from a working surface to a sky with a bright area of about 5000 lux, and then back to the working surface is very uncomfortable. It is very common for windows to be fitted with blinds or curtains, which are kept closed to avoid glare and then the electric lights are used to provide uniform lighting. There are very few examples of good natural lighting in offices, which do actually work. The National Farmers Union building in Britain is one of the best, but it is not completely successful. I hope our new roof-lit buildings will do better.

To help light penetrate into a side lit building the head of the window has to be about as high as the depth of the room. For Canary Wharf that would produce a 14 m story height. This is not a realistic proposal.

A building with fluorescent light 500 lux 10 hrs/day 220 days/yr 15w/sqm will produce about 25kg/ sqm/yr of CO_2 . If the lights are not controlled and left on for cleaning all night the emission is 90kg /sqm/yr.

The energy has to be rejected by a cooling plant with a coefficient of performance of 2 then an extra 12 to 45 kg is produced.

I am working on an office competition where the Carbon energy target is 90 kwh/sqmyr. We are aiming for 30kwh/sqmyr. . The office is 6m deep and we need to rely on electric light. The background lighting is proposed as 150 lux and each occupant has a 13w task light. The background light will produce 5kg sqm/yr of CO_2 and the task light 1 kgsqm/yr of CO_2 . It is a Scandinavian competition so there will not be air conditioning. We hope our low energy figures will help to win. This standard of lighting is unusual, but I think it is good for computing. The lighting demand will be provided by the annual output of 1 sqm of PV cell.

I am sorry about the muddle of the units. In the UK we are concentrating on Carbon dioxide emission. While in Scandinavia where energy is sold as electricity and heat they concentrate on kWh of delivered energy.

1 kWh of heat from methane combustion produces about 0.2 kg of CO_2

1 kWh of electricity from a thermal power station produces about 0.7 kg of CO_2

With lighting there is another problem. Windows should be rationally sized for the need for light. As bright sunlight brings in 100,000 lux, and on overcast day a vertical window admits 2000 lux, air conditioning engineers try to avoid 100,000 lux, (1000 W per square meter), entering the room and specify glass which restricts the energy input. In my view, the problem is the range of light levels, which fluctuates from time to time, and the solution is movable blinds or shades. This solution has led to the double facade.

Reference to "less light more vision" WREC Florence

The double facade allows a blind or shutter to be protected from the severities of the weather; it allows heat to be intercepted by a blind and the heat can be put to use in cold weather or rejected in hot weather.

Housing is a slightly different problem. The rooms are small, and the demand for light is low. 50 to 100 lux can easily be provided by quite small windows. Even windows which give adequate light on an overcast day will allow much too much direct sunlight to overheat the flat and good automatic blinds are needed.

We have recently completed and monitored the energy use of a housing development called Casper II where there is virtually zero demand for heating in the UK and certainly no cooling. Ref. (3)

The conclusion of this section on light is that tall offices are likely to need electric lighting. The lighting design standards and solutions should be very carefully considered. Electricity for lighting can be kept to about 100 kg per person per year and should be part of a sustainable budget. Renewable energy should meet this demand.

Refer to Heelis for the National Trust Swindon Ref (4)

4. Ventilation

In a building occupied by people there needs to be fresh air to provide oxygen for metabolism, to remove the carbon dioxide waste product, and to remove odours from the building. The odour requirement is the major component and Ole Fanger has done research which leads to a method of relating the rate of ventilation to the odour production in a building. ref (5) The method involves calibrating materials used in a building. Any materials not calibrated should be deemed to be very smelly. It is pity that his proposals are not being followed up. However , the research has led to a consensus that something like 8 to 10 litres per second per person of fresh air , leading to a carbon dioxide content of about 1000 parts per million as a marker should be adopted. A person producing 100 W of sensible heat can raise the temperature of 10 litres per second air by about 10°C. If the external temperature is very low or very high then the ventilation air can be preheated or precooled by a heat exchanger. At Casper, Ref (3) ventilation air was preheated by a heat exchanger in a mechanical ventilation system.

If a mechanical system of ventilation with heat reclaim is proposed. it is very important that the building the itself should be air tight. A heat reclaim system will still need some heat to be provided to the ventilation air. The infiltration leakage of the building is then an additional load , and that if it is adequate for the occupants then the load of the mechanical system is simply an extra not.

We have designed and built an example of these ideas at Casper II See Ref (3).

Where there is the risk of overheating, and no cooling is to be provided then during hot weather heat has to be rejected during cool nights even if people are not in occupation. This means the windows should be automatically operated to provide ventilation. Our culture usually expects opening windows for ventilation and I have got into trouble with clients when opaque panels were expected to be operated.

For the Hong Kong climate it should be possible to design buildings so that they can be comfortable during the winter by adjusting the ventilation. As in the United Kingdom it may be possible to make calculation which show some optimum day time ventilation rate at which the internal temperature may be marginally below the peak external. However a very high ventilation rates around 100 A/C hr provide comfortable air movement over the skin and should figure in the design. Ref (6)

During the hot summer with the high enthalpy and moisture content mechanical ventilation with heat reclaim, using an enthalpy wheel, should be combined with a small cooling plant to meet the 24 hour mean heat gain. I trust this energy use will be able to be satisfied with renewable energy from wind turbines or something equally benign.

The high rates of ventilation, which are appropriate when small temperature differences are the design aim, need to be achieved by natural ventilation using wind power or stack effects, so that high grade energy is not needed to drive fans.

When it seems advisable to boost the natural ventilation flow in a duct rather than inhibit the natural flow with a high impedance fan. it is easy to mount a small propeller fan at one side of the duct to impart some momentum as needed to augment the flow. Road tunnels are often ventilated in this way.

In Hong Kong it is usual for tall buildings as housing to be naturally ventilated. The problems of exposure at height are solvable it seems. These lessons should be applied to commercial buildings.

VENTILATION represents one of the most complicated issues to be considered. Ventilation provides people with fresh oxygen for living and it removes carbon dioxide. It flushes out polluting chemicals that are perceived as odour.

Odour removal defines the minimum amount of ventilation needed for people. The amount of ventilation is currently undergoing much critical evaluation, but low rates of ventilation in the mid 70s lead to criticism of buildings and the term "sick building syndrome" was coined. There is currently a consensus that about 10l/s per person is needed for odour control, and that this can be controlled by setting the Carbon dioxide content of the air to be 1000 ppm. This concentration of Carbon dioxide must not be seen as a critical parameter in itself. It is simply a marker for the amount of metabolism taking place and so a measure of the number of people present. It does have an advantage in that it gives more weight to active people than inactive ones. Having set a minimum ventilation rate with respect to odour the next problem is using ventilation to control heat exchange processes between the building and the environment. Traditional buildings used very high rates of ventilation in some circumstances. Ref (6)

Air exchange is necessary during cold weather to provide fresh air for people, but it is not needed when the building is empty. If a building is warm enough on a cold overcast day then it will tend to be too hot on any other day. Heat will have to be rejected and variable control of the ventilation is needed to reject heat even in cool weather. During the hottest weather people make adjustments. In some climates the ventilation is best restricted at mid-day keeping the air temperature down, but reducing air movement over the skin, and in others the air movement is encouraged up to about 1m/s with the air temperature allowed to be equal to outside. Very high rates of ventilation with open walls are traditional in hot humid climates, and then the climate seldom requires heat to be conserved at any time of year.

Buildings have to be designed so that they can be sealed very thoroughly for extreme weather conditions and openings for ventilation need to be designed so that large amounts of heat can be exchanged with very small temperature differences between inside and outside.

5. Comfort

I raise this issue because sometimes the word "comfort" is introduced. As an engineer it is possible to provide industrial means whereby the occupants of a building can choose any degree of light, temperature control, cleanliness and moisture content of the air. Industrialisation has enabled us to side-step the question of the limits of environmental temperature by providing people the choice of any desired condition. So it is common to see people wearing heavy clothes in cooled buildings in hot climates, and light shirt sleeves in heated buildings in cold climates. Research has concentrated on the optimum temperature in an environment. What is needed is an understanding of the adaptability of human beings. Ref (7). We have tried to define optimum conditions of all these parameters to use as a design basis. Human beings have evolved in an environment with a range of temperatures at the surface from about -50°C to about plus 50°C. Light from about 200,000 lux for a bright day on a snow field to close to dark, no light, on a dark overcast night with no moon. Moisture content of the air from saturated to very dry say dewpoint -50C. These extremes are not comfortable, but a temperature of 22°C at 50% relative humidity, and 500 lux is a very tight definition of what we should expect to provide. We do achieve it and the result is boring, extravagant, and damaging.

The principals and the important parameters are easily understood see the heat stress Index (Belding and Hatch Ref 8), but the relationship of extremes of thermal stress to tolerable or even pleasant comfort are more difficult to define.

During the hottest weather people reject heat to their surroundings inside the building. Heat flow to the skin and secretion of sweat are controlled involuntarily. Clothes are controlled by conscious acts. The environment in the building can be controlled by adjusting shading, and by adjusting ventilation.

5.1 *Warm environment light clothes for comfort- 30°C?*

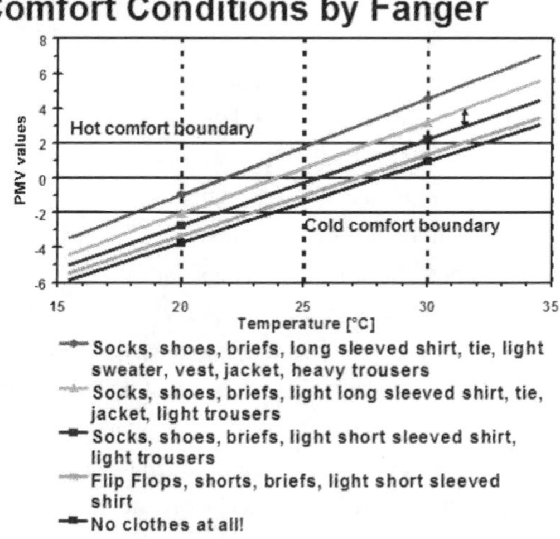

This chart, adapted from Ole Fanger, (Ref 9) shows how the predicted mean vote PMV is related to the clothes we wear and a parameter of temperature. You can see that making quite a small change of clothing produces a change in the PMV of one unit. That change in comfort is quite small. I have suggested tolerable limits of +/- 2 PMV the work of Fanger is very on the high gear of maintaining thermal equilibrium. We should stick to that. A fever of 1°C is not a comfortable situation, and it results from an out of balance in the metabolism of 1 W for 12 hours. However Fanger also assumes a skin temperature of 35°C, and a degree of dryness of the skin.

Years ago I tried to apply the Belding and Hatch index to situations in Doha. I had to adjust the heat source temperature to be 37°C, and then adjust the thermal resistance from blood to air. It was then possible to show that people in Doha could remain alive. When we come to office comfort, I sometimes think that an objective criterion, would be that if we wrote over the place on the paper, where our palm had rested and the ink ran then we could say it is too hot.

The research concentrates on the word comfort as though it represented some optimum condition. Humans are very adaptable and enjoy variability. Research is needed to explore the variability.

For buildings we need to be able to define a winter and a summer temperature. This is a current topic in the UK where we are in danger of advocating air conditioning for climates here while comparable conditions are apparently tolerable elsewhere. I find it difficult to persuade clients that temperatures occasionally reaching 27°C are tolerable although I spent 1 week at this university perfectly comfortably at 27°C 55% RH.

I suggest that a tolerable range of temperature for people is ± 2 or 3°C over a day.

For winter a sensible central temperature and a range with reasonably warm adjustable clothes has to be agreed so that the thermal capacity of the building can be designed to absorb swings in the available heat allowing the temperature to vary within comfortable limits. With a well insulated and sealed building it is usually easy to maintain a temperature constant to +/- 2°C

Next we come to the summer that must not get too hot. I believe that we can rely on the research carried out by the Building Research Establishment in 1966 that investigated overheating of buildings. Ref (10). They calculated how hot a selection of buildings would get on a hot summer's day and then asked people for their perception of the comfort of the building. If the calculations showed the building did not get hotter inside than outside the population would probably be satisfied with the building. If the inside was more than 2°C hotter inside than outside then the perception was that the building was uncomfortably hot. This sets up a very straightforward design process. For summer in the UK we at MFLLP design without cooling to maintain a 24 hr mean internal temperature of 22°C with 24hr mean 19°C outside while the inside temperature fluctuates by +/- 3°C from 19°C to 25°C so that the outside peak is equal to the inside. The 24 hr mean heat gain is rejected by ventilation or by some cooling plant. The thermal capacity of the building is tailored so that the fluctuations in the thermal load can be absorbed by allowing the temperature to swing. A building with internal walls, furniture, light carpet on a screeded floor and a coffered exposed concrete soffit can absorb a swing in heat gain of 54 w with a swing in temperature of ± 3°C. This represents a mean heat gain of 54 w/sqm with a dawn minimum zero and a peak of 108w/sqm. We are able to design for these criterea as at:

BRE building at Garston Ref Ref (11)
RMC Egham Ref (12)
British Embassy Dublin Ref (13)
and several other buildings.

We should design buildings so that no heating is needed on the coldest day. Then on any other day which is warmer there is the need to reject heat to prevent the building from getting hotter than is needed.

As the outside weather temperature rises increased ventilation should be able to maintain the thermal equilibrium at a comfortable temperature.

However when the peak outside temperature is higher than comfortable heat cannot be rejected using ventilation air. The strategy is then to wait for lower night time temperatures and reject the excess heat at night. During the day heat is stored in the structure allowing the building temperature to rise. For this strategy to work the building has to have enough thermal capacity to allow the day- time heat flow to be stored without allowing the temperature to rise too much during the day. The occupants' tolerance of temperature change must be greater than the amount of change caused by a day's heat release.

In hot dry climates with a wide diurnal range of temperature the temperature inside can be kept well below the peak outside and ventilation rates need not be very high, but in less extreme dry bulb temperature regimes the best strategy is to aim for equal inside and outside temperatures with high rates of air flow over the skin to maintain comfort. Even in the UK ventilation rates of 100 A/C per hour can give wind speeds over the skin of below 1 m/s with comfort at 30°C. Ref 9

The relationship between ventilation and thermal mass is particularly interesting when we consider really heavy buildings like a pyramid, Gothic Cathedral, an earth bermed modern building, all with time constants much longer than 24 hours, with times like several weeks, months, or even a year in the case of a pyramid.

In the UK we have a temperate climate with rather less daylight than any other environment but with the winter and summer temperatures less extreme than almost anywhere else in the world. Thus the challenge for lighting is most extreme, but we can aspire to design buildings without mechanical heating or cooling with more chance of success than other people. We are learning lessons from experience of implementing these ideas.

Thermal capacity at high level is particularly useful because a small vertical temperature gradient say ½°C per meter enables the high level structure to absorb extra heat. There is an issue about the acoustic environment in an office. An open space should not be too hushed. If it is very absorbent a raised voice can be heard throughout the space. We have designed several buildings where a carpet and a concrete ceiling have combined to meet the thermal and acoustic requirements.

This discussion on comfort has strayed off in to design and represents the core of this argument.

6. The Envelope

Finally the envelope needs to be insulated. The amount of insulation should be put into context with all the other thermal loads of a building. For Casper II in Britain we used 100 millimetres of foamed polystyrene and that left the triple glazed windows as the biggest thermal load. The windows are the most important part of the envelope. They must be designed to provide plenty of light as required by the use of the building. For housing at Casper II we spent some time trying to meet our target and persuading the architect to reduce the area of glass. Optimising the windows for thermal load and natural light is the most important design activity. Controlling the windows, so that they provide an enough light on overcast day's, but do not provide an excessive heat gain when the skies are clear is the chief problem. The analysis of the benefits of insulated blinds or shutters above shows that glass should be provided to give adequate light and then covered at night to prevent heat loss. Heat gain from direct sun when the energy flux is 50 times greater than an overcast sky can be controlled with reflecting insulating opaque blinds to give a shade factor better than 0.1. This is a hot development topic.

Finally the envelope has to be airtight. Masonry is not airtight. Cracks have to be limited and can be calculated. If a test of 2 air changes per hour at 50 Pa is specified that represents an air flow for Casper II of 150 cubm/hr at about 9m/s. The area of crack is easily calculated and can then be a design parameter. No letter box, sealed doors, lobby, no overflow pipe, and so on.

7. Tall Buildings in Conclusion

The principles I have been discussing do apply widely to all climates. In the UK we are just being able to generate designs to give good comfort with out cooling or heating with the temperature range outside between -5°C to 30°C winter to summer. An air tight envelope, mechanical ventilation with heat reclaim, good transparent insulating glass, automatic insulating blinds, and adequate thermal mass can maintain winter comfort with normal random heat gains.

In summer the windows must be openable, fitted with shades, and controlled automatically to ensure that operation is always optimum.

For more extreme climates the construction must be designed to maximise the period when the building can select its response to the ambient environment. Then during extreme climatic conditions the building excludes the outside and relies on very efficient mechanical systems with heat reclaim. The mechanical systems are designed to meet the 24 hour mean heat loads. For HK air conditioning should be able to be switched off for the winter eight months or so. In the UK we have a heating season, and in HK you will have a cooling system season.

7.1 *The issue of height*

Exposure to unshaded sun and high level wind speeds are more severe with high buildings. All natural ventilation schemes have to close when the wind is strong. Mechanical systems provide minimum ventilation associated with extremes of climate. When high ventilation rates are needed, it is common to design for "windless" anticyclonic conditions, say one metre per second, to give large windows. Then the windows are designed to provide ventilation for strong winds. The openings should be chosen so that large casements can be opened fully in low wind conditions and a small slot left exposed at ceiling level designed for 5 to 50 m/s wind speeds.

It would be possible to design a ventilator with: silencing, and a baffle to break down the velocity if it were necessary for frequent, violent, windy, and hot conditions.

There is the issue of maintaining an efficient floor plate , which I have mentioned, citing the World Trade Centre Ref (17) and Canary Wharf Ref(14) as examples. It seems to me that one of the disadvantages of these building is that electric light is inevitable. It should at least be as efficient as possible, and consideration be given to using individual task lights for working light. The idea of spaces formed inside the enclosure as practised at the Frankfort bank Ref (15) building and the Swiss Re Ref (16) building in London is an attempt to bring light and fresh air into the building. The problem with these buildings is that high rates of natural ventilation are not possible through the contorted air paths that are provided. The next issue is that 1% daylight factor is not adequate. Electric lights are used all the time anyway. The aim should be a daylight factor which provides more natural light than the electric lights.

The highl density development achieved by tall buildings enables the public transport system to be pleasant and efficient. Perhaps the advantages outweigh the increase in energy to run the escalators, and the lights in the deep tall buildings.

References

1. J Pringle, 'Praise the Lord's, RIBA Journal, October 1995, pp 40 – 46
2. H. Coch, R. Serra, A. Isalgue, "The Mediterranean Blind: Less Light, Better vision" WREC 5 Renewable Energy 15 (1998) 431-436 PERGAMON 1998
3. K. Powell, "Core Values", The Architects' Journal, 3/10, pp 30-35 August 2000,
4. M. Spring, "National Treasure" Building 1 July 2005
5. P. O. Fanger, "Introduction of the Olf and Decipol to quantify air pollution...", Energy & Building,12 pp 7-19, 1988
6. Max Fordham "Natural Ventilation" WREC 5 Renewable Energy 19 (2000) 17-37, Elsivier Science Ltd 1999
7. Prof. Fordham S.M. "Do you want to control the temperature?" RSA Lectures, John Adam St., London 1999
8. Belding, H.S & Hatch, T.F. "Index for evaluating heat stress in terms of resulting physiological strains". Heating, Piping and Air Conditioning 207:239 1955
9. P.O. Fanger, "Comfort" Building Services Engineer, 41 pp 313-324, 1973
10. BRE Digest 68 2nd. Series "Window design and solar heat gain" HMSO 1966.
11. R Bunn, 'Green Demo', Building Services Journal, pp. 18-23, March 1997
12. R Bunn, "A Building for All Seasons", Building Services Journal, pp 23 - 26 ,
13. November 1990
14. D Pennington et al, "A Diplomatic Solution", The Architects' Journal, 23 , pp 43-52 November 1995
15. P. Bill, "The New Tower of London", Builder, p14-20, October 1991
16. R. Lacayo, "Buildings that breath." Time Magazine: August 26th 2002,
17. M.Spring, "Gherkin Interior" Building, pp 49-54, 30 April 2004, World Trade center.

Theme Papers

POLICIES AND PLANNING OF TALL BUILDINGS IN HONG KONG

MARCO WU
Director, Buildings Department, HKSAR

Hong Kong's realm of tall buildings makes the place not only magnificent to look at but also exciting to be in. As the authority for enforcing building laws, the Buildings Department has provided the appropriate framework and facilities to secure and promote the construction of tall buildings that are not only safe and healthy but also environmentally friendly and sustainable.

1. The Hong Kong Story

Less than two centuries ago, on this piece of "barren rock" along the South China coast, lies a quiet fishing village. Its topography was so treacherous that early trading settlement in 1842 was compelled to align itself along a narrow strip of foreshore fronting the harbour.

Within 9 years of the city's foundation, the first of many reclamations from the sea was undertaken. Throughout the years, the amount of usable land has multiplied with subsequent extensions of the shoreline. On this reclaimed harbour frontage springs many splendid buildings that have since been replaced, reshaped and rebuilt by bigger, taller and more imposing towers. These tall buildings together with the dramatic harbour in the forefront and the rocky mountains at the backdrop have created one of the world's most instantly recognizable panoramas.

2. The Need for Tall Buildings

Erecting tall buildings in Hong Kong does not only answer the developer's zealous pursuit for the spectacular or satisfy the designers' expression of architectural and engineering excellence, it also makes, to a certain extent, practical sense, economically, socially and environmentally.

2.1 *Economic benefits*

For any places on this planet, land is a precious resources. This is especially so in Hong Kong with its rugged terrain and scarcity of usable building land. Buildings have to be tall in order to maximize the development potential and to obtain the best returns. Reduced travelling time in a highrise compact city also contributes to make a city more efficient and economically more viable.

2.2 *Social Benefits*

When World War II ends, Hong Kong has a population of 1.8 million only. Since then, this figure has grown at a rate of almost 1 million every decade. In order to house a fast growing population with limited land resources, building high-rise has become the inevitable solution. By 1960s, many public housing estates rose up to 16 storeys high. Taller and bigger buildings also help to satisfy the rising aspirations for larger living space. Furthermore, high density built environment meets the general public demand for proximity from home to place of employment. It makes life more convenient. Homes, offices, shops, utilities, transportation nodes are all packaged into one convenient neighbourhood, or located in districts not far from each other.

2.3 *Environmental benefits*

A high density development can obviate the problems otherwise associated with urban sprawls such as those commonly found in some countries, such as those in Northern America. It enables more efficient utilities services and infrastructural support. As a result, it reduces overall energy use and traffic fumes. It also helps the city with limited land resources to save land for parks and recreational use. Indeed, Hong Kong is blessed with large expanses of country park with lush woodlands, and some with sparkling waterfalls that lie beyond our city margins but yet easily accessible to its 6 million residents.

3. The Perils of Tall Buildings

Despite all the benefits offered by tall buildings, they also contribute to the environmental problems and, at times, pose threats to public safety and health.

Tons of construction and demolition waste are generated each day which make up a fair proportion of the total municipal waste in Hong Kong. Buildings consume over 50% of all the electricity used in Hong Kong for lighting and space-conditioning alone, and energy used locally be it gas, petroleum or other forms are all imported.

Densely packed tall buildings trap air and dust in the city. They reflect sunlight and heat to its surroundings, causing glare and compounding the heat island effect. Wind travels through street canyon and tunnel effect to scour our plazas and gardens. Busy traffic on narrow roads makes pedestrian walking an unpleasant experience. Housing units facing each other and stacked one above the next deny our psychological need for privacy. Monotonous tower blocks dull our minds.

The recent spate of incidences of falling concrete, rendering, windows etc. highlighted the danger related to dilapidated structures, and in particular, dilapidated TALL structures. The outbreak of SARS in 2003 also reminded us, inhabitants of tall buildings, the need to properly use and maintain our buildings. Indeed, we have come to know that it does not require total collapse for buildings to cause catastrophic effects that can be detrimental to the environment as well as our health and safety.

4. Facilitating the Design and Construction of Tall Buildings

4.1 *Building (planning) regulations*

Since the major overhaul of the Buildings (Planning) Regulations in 1962, the size of development is no longer determined by the width of the street using volume calculation. Instead, it is linked to the height of a building. So, the building law in effect has been promoting the construction of tall buildings.

4.2 *Performance based requirements*

In step with the development of new technologies, the Buildings Department has undertaken major reviews of the Buildings Ordinance, its subsidiary regulations, Codes of Practice, Practice Notes and Guidelines.

The review has to address the two major aspects of tall buildings, namely safety and serviceability, making use of a risk analysis and reliability check approach. Moreover, the review must be able to facilitate innovation, minimize wastage of energy and material resources and promote sustainability. We are therefore modernising our standards from an essentially prescriptive regime to one based on performance and results.

In the last two years, we have completed the review and drafting of various new Codes of Practice including those relating to demolition of buildings, loadings, structural use of concrete and steel, pre-cast concrete construction, foundations and wind effects on buildings. These Codes have incorporated the latest local data and have introduced new concepts such as dynamic response under the new Wind Code. They have plugged deficiencies in previous standards, such as building acceleration and vibration in concrete. The Codes have been well received by the building industry, and have become new milestones in advancing the application of structural engineering in building control.

We are currently undertaking other consultancy studies including the review of three fire codes, and the plumbing and drainage regulations. Once completed, these will no doubt help to guide designers in creating tall buildings that are safer, healthier and easier to service and maintain.

5. Promoting Sustainable Development

5.1 *Strengthen our buildings laws and requirements*

Throughout the years, we have strengthened our buildings laws to promote quality living and sustainable high rise development. Examples include new site supervision requirements to ensure the quality control and safe operation of construction sites; the restriction on overall thermal transfer value for hotels and offices to conserve energy; the provision of material recovery chamber to assist the recovery of waste; the provision of access facilities for telecommunications and broadcasting services to enable occupiers of a high rise building to enjoy the full range of broadcasting, telecommunications and Internet services at their choice.

We are currently preparing a Code of Practice on the design, construction and maintenance of drainage system that will address the issues concerning durability of the system. We are also close to completing the consultancy study on the Design Manual for Barrier Free Access that aims to improve the accessibility of tall buildings to persons with a disability as well as for the elderly. Thus, our buildings will be better designed to cater for the needs of our aging population.

5.2 SARS

Since the outbreak of SARS in 2003, we have required widening the width of re-entrants in buildings, the provision of a protective coating in all cost iron pipes and improving the quality of cast iron pipes and siting U-traps on external walls of buildings. We have also encouraged the provision of better ventilation in toilets and internal corridors, provisions of automatic refilling device for U-traps and other measures to facilitate the access to and maintenance of drainage systems.

5.3 *Promote green designs and construction*

Removing barriers is an important strategy that we pursue in the advancement and implementation of environmentally friendly and innovative tall structures. We have also set up a dedicated Building Innovation Unit to spearhead the Green Building Movement within the industry and the Department.

Through the issue of practice notes and guidelines, we encourage the increased use of recycled aggregates and pulverised fuel ash in concrete to reduce consumption of material resources. We relax the minimum requirement on the volume of flushing water to pave way for the dual flushing system for water conservation. We build in greater flexibility in the control of site coverage and open space provisions to facilitate innovative designs.

5.4 *Provide incentives*

Since year 2001, the Buildings Department has put in place an incentive scheme to encourage the incorporation of green features in buildings. The incentives we offer in the two Joint Practice Notes exempt a number of green features from gross floor area and site coverage calculations. Since implementation, the scheme has been well received and we have already given approvals to over 200 building proposals bearing one or more of such green features.

We also promote building set back by granting bonus plot ratio and site coverage. With greater set back, the resultant widened pavement can ease the pedestrian flow, enhance greening at street level and improve the micro-climate, air quality and ventilation at street level.

Furthermore, to meet the aspirations for modern living, facilitate easy management and maintenance and to increase durability of buildings and facilities, covered walkways, recreational facilities, plant rooms and floor height over and above the norm may also be accepted or exempted from gross floor area calculations.

5.5 *CEPAS – a green labelling scheme*

After wide consultation with the industry, we have completed the study on the development of a Comprehensive Environmental Performance Assessment Scheme (CEPAS) for buildings in Hong Kong. This assessment scheme for rating and benchmarking the environmental performance of individual buildings will enable us to identify building developments with special merits in environmental performance. It will give an added impetus to the momentum for the construction of green buildings. The Provisional Construction Industry Coordination Board is considering setting up a special task force to hammer out the implementation details for the scheme.

5.6 *Arrest urban decay*

Environmentally friendly structures do not stop short at the design and construction stage. Neither can our tall buildings be sustained in the absence of proper maintenance.

Since year 2001, we have intensified our enforcement actions and embarked on a number of large-scale operations to clear unauthorized building works and repair defective drains and buildings. We provide a wide range of social, technical and financial support to assist building owners and owners' corporations to maintain their buildings in a safe and sanitary condition. We have since removed over 170,000 unauthorized building works, issued over 5,700 repair orders and assisted owners in 6,000 buildings to repair their defective drains.

In addition, to obtain a community consensus on a long-term solution to the problem, the government has conducted a first stage public consultation on building management and maintenance in 2004. Results of the first stage consultation which were released in January this year, point to a community consensus on owners' responsibility to upkeep their buildings, including bearing the necessary financial commitment. The community also generally favour some form of mandatory requirement on owners to ensure the discharge of their maintenance responsibility. We will soon conduct the second public consultation on a proposed mandatory building inspection scheme. This scheme, if taken forward, will help make Hong Kong a clean, safe and healthy living place.

6. The Challenges Ahead

To conclude, tall building development has definite economical, social and environmental advantages. It may well be a necessity for cities like Hong Kong, which has scarce buildable land resources, and which has limited opportunity for expanding into its hinder land. It, therefore, behoves upon us as regulators, town planners, architects, city engineers and builders alike to take up the challenge of overcoming the problems in tall buildings so that we can sustain the advantages offered by living high rise. We believe good designs, durable building materials, quality construction methods, well supervised site construction work, good maintenance and management, all help in providing tall buildings that can withstand the test of time.

As a facilitating regulator, we provide a conducive flexible environment and the kick starting mechanism. Building professionals and developers are invited to let the idea flourish by taking up the challenge to invest, design and produce tall buildings that benefit not just the occupants but the human race at large.

THE SHIFTING PARADIGMS OF HIGH-RISE LIVING

BELINDA YUEN

National University of Singapore

This paper reviews the changing conceptualizations and prospects of high-rise living in contemporary cities. The argument is not for or against high-rise. Rather effort is to situate the modern high-rise in perspective at a moment where many cities around the world are re-discovering and building ever increasingly taller housing for their population in the wake of globalization and technological advancement. Using data largely from Singapore where 90% of the population resides in high-rise, this paper explores the issue of high-rise and everyday life. Intention is to raise residents' perception and acceptability issues not just from the Asian perspective, but also to explore the wider urban representations and performance pathways in the endeavour towards better quality of life with this housing style.

1. Introduction

High-rise living is increasingly experienced in cities across the world, from America to England, Asia and Australia. In spite of negative discourses on high-rise and the strong push to stop the growth of high-rise public housing by the 1970s, there appears to be a revisit of high-rise housing in the last 10-20 years. High-rise housing is fast becoming an alternative option in many inner city redevelopments under the current narratives of urban sustainability and globalization [1]. Melbourne, for example, has in the 1990s witnessed a construction boom in high-rise housing in its revitalized central business district [2]. Elsewhere, even as local authorities in the UK are demolishing many of the 1960s high-rise public housing, London is building 50-storey 'green technology' tower blocks (Skyhouses designed by D Marks and J Barfield) as the latest answer to the country's housing shortage. In intensively urban Asian cities such as Hong Kong and Singapore, high-rise is the common lifestyle of much of the population. Notwithstanding that several urban analysts including Marcuse [3] may have predicted the end of high-rise right after the Sep 11 2001 World Trade Centre destruction, all indications seem to converge towards a continued revival of high-rise housing and its celebration as a symbol of affluent modern living. Against the new and rising images of high-rise, how much do we really understand about this contemporary style of living? What are the attractions and concerns of high-rise living? How do residents in high-rise living perceive and relate to this form of housing?

Available evidence seems to indicate that research into high-rise living has in the most part been based on studies of western cities and population. The unit of study has primarily been the high-rise public housing that emerged in the 1960s. At the time, high-rise proliferated as state provided housing in many countries as a viable option to housing large numbers of people. This housing solution was celebrated as 'a panacea for social problems' [4]. However, by the early 1970s the discourse surrounding high-rise was increasingly filled with negative comments, especially that it was an unhealthy living environment for families and children [5]. The pressure and resultant response was to abandon high-rise public housing. With the construction of such housing being halted for some time, housing research too shifted to emphasize more on other issues (such as affordability and finance).

Notwithstanding the recent renaissance of the high-rise, it would appear that much of our knowledge of high-rise living remains decades-old basing on situations that are not necessarily relevant today. For some, the latest high-rise still resonates with anti-child and family sentiment and the target population is in the main singles, childless 'professional' couples and empty-nesters lacking childcare responsibilities [6]. But the high-rise of today is not that of the past. To quote Greg Lynn of FORM [7], 'Tall buildings are turning into urban fabrics...Architects are thinking about how to pull the qualities of the street into the building.' Many of the new high-rise is what Ali and Armstrong [8] call as a supertall building. Many of the world's tallest buildings are found increasingly not just in Europe or USA but in Asia [9]. By and large, it would seem that research has not kept pace with current realities. In particular, McNeill [10] astutely observes, 'In debates over the future of urban form, existing Western-biased theories and models are of questionable relevance.' Against this unfolding trend, there is urgent need to re-examine and re-define high-rise living in order to give a more accurate interpretation of the way high-rise is experienced every day, especially in Asia where high-rise is mushrooming [11]. As Lefebve [12] argues, to know more about what is will help us to understand more about what is likely to be,

It will help us to grasp how societies generate their (social) space and time – their representational space and their representations of space. This would allow us, not to foresee the future, but to bring relevant factors to bear on the future in prospect—on the project, in other words, of another space and another time in another (possible or impossible) society.

The following section will review the changing conceptualizations and discourses surrounding high-rise with the aim to establish present understanding of high-rise. The paper will also explore issues of resident's living experience drawing on empirical data from Singapore, and highlight the areas for further research.

2. What We Know - The Trends and Forces

The phenomenon of high-rise living is not new. As Ford [13] traces, high-rise living has been expanding in large American cities since early 1900s. According to Douglas [14], a key factor for the growth of tall buildings is rapidly increasing urban land values. As Huxtable [15], Abel [16] and others have contended, the tall building is the direct product of more people wanting to occupy the same choice city sites than could be accommodated by smaller buildings. As a societal expression, tall buildings have come to project the image of a rapidly industrializing and progressive nation or as Huxtable [17] observes, 'It romanticizes power and the urban condition'. The staying power of this is perhaps best encapsulated by Lacayo [18] post-Sep 11 2001, 'The only clients still interested in building them were in nations that wanted a symbol of their arrival as a contender in the global market, mostly in Asia's Pacific Rim.'

From this perspective, high-rise is often seen as an icon of a developed (western) society and a part of urban boosterism especially American who leads the world in the construction of the first modern tall building and its subsequent development as a model for the shape of cities in the 20th century. As Crilley [19] observes, high-rise buildings 'are charged with representational responsibilities to act, by virtue of their towering height, as markers of place, sculptors of the city silhouette and as conveyors of public image.' To this, Mies van der Rohe and Le Corbusier have inspired with their skyscraper visions. Le Corbusier [20] in particular is widely cited for the way he has enshrined the tall building as the landmark in the design of the contemporary city. His monumental towers have been celebrated as a new model of multiple uses that has dominated much of post-war urban renewal in United States and infused urban design thinking in many planning and architectural schools [21].

According to Helleman and Wassenberg [22], the peak of high-rise living was reached during the period of 1960s and the mid-1970s, and there has never since been a period in house building in which similarities between countries have been as great. Largely provided as public housing, high-rise is seen as the solution to housing the masses and to arrest the growth of sub-standard housing. Held as the ultimate, ideal and modern dwelling form, high-rise public housing construction is however quickly contested by documentation (much of the 1970s) relating to the downsides of living in high-rise. Very much the same across many western cities, the failings of the high-rise often concern dystopia of functionality in particular high-rise housing is unsuitable for families with children. Research details the negative effects on children's safety and mental health, impersonalization and loneliness among the adult population [23]. Some such as Stevenson et al [24] have described high-rise housing as 'gaols in the sky' and 'prisons for families'. Others including Doxiades [25] have on reflection concluded that high-rise housing is a mistake,

My greatest crime was the construction of high-rise buildings. The most successful cities of the past were those where people and buildings were in a certain balance with nature. But high-rise buildings work against nature, or, in modern terms, against the environment. High-rise buildings work against man himself, because they isolate him from others, and this isolation is an important factor in the rising crime rate. Children suffer even more because they lose their direct contacts with nature, and with other children. High-rise buildings work against society because they prevent the units of social importance—the family...the neighbourhood, etc – from functioning as naturally and as normally as before. High-rise buildings work against networks of transportation, communication, and of utilities, since they lead to higher densities, to overload roads, to more extensive water supply systems---and, more importantly, because they form vertical networks which create many additional problems—crime being just one of them.

However, just as high-rise housing is being criticized, stopped and demolished in many cities, there is new enthusiasm in recent years in several European cities to revive high-rise in inner city as a residential model for the new urban affluent. Some such as London sees high-rise as part of its global city growth strategy, 'London must continue to grow and maintain its global pre-eminence in Europe. London must continue to reach for the skies.' [26] In contrast to much of the earlier high-rise construction, many of the recent high-rise is by private sector and in prime city locations.

Integral to this--and notwithstanding negative perceptions, is emerging evidence to suggest that high-rise living is valued for the spectacular views and sensation of height [27], privacy and quietness [28] and prestige and status [29]. In particular, Johnson [30] has noted that urban high-rise lofts are attracting 'echo boomers', grounding that attraction to the glamour of high-rise living on recent television shows like 'Frasier' and the hip grunge fashion of cosmopolitanism. Others in the real estate industry have predicted rising demand for high-rise from baby boomers that are fast becoming empty nesters [31]. High-rise supporters have argued that 'it is possible to make tower blocks cater successfully for these groups' [32]. As a symbol of modernization, high-rise seems to gain more power and persuasion now than before. It fits perfectly with the notions of smart growth.

High-rise intensifies urban living, takes up less land and supports high-density building form. As Harada and Yeang [33] would contend, 'super-tall skyscraper or 'hypertower' (being a building of over 500 metres high) offers even greater opportunities for ecologically-responsive solutions and enables the freeing of land at the ground plane for ecological succession (i.e. return to natural flora and fauna states through natural processes of recolonisation)'. Against the narratives of sustainable development and new urbanism, high-rise has re-entered housing policy agenda to offer the best alternative to suburban sprawl [34]. It is also arguably the better setting to facilitate global economic development than the familiar low-density decentralised suburban structure. As Sassen [35] explains, 'strategic, creative activities – whether economic, cultural or political – thrive on density...There is an economic logic, then, to thick, dense places.' Supported by advancement in technology, the deepening trend is to build taller - the supertall buildings, which Ali and Armstrong [36] define as the growing interest and expanding need. What is remarkable in the high-rise trajectory is not just the vertical skywards growth in height but also how its diffusion has globalised, perhaps nowhere more so than in Asia, notwithstanding the limitations of the paradigm.

In rapidly urbanizing Asia, cities such as Hong Kong and Singapore have embraced high-rise as the inevitable option for housing a growing population on limited land. High-rise has become the lifestyle for the majority of the population. Over the past 40 years, not only has high-rise emerged as the dominant building form but its height has been rising. In Hong Kong many of the new housing is 60-storey. Equally in Singapore, high-rise has displaced the traditional low-rise 2-3-storey shop-house as the norm. At present, some 92 per cent of its population lives in apartment housing - 86 per cent in public housing and 6 per cent in private housing. While the tallest public housing in Singapore is at present 40-storey, new housing, especially private, is set to rise to 50- and 70-storey in the central business district [37]. However, in contrast to much of western experience, public acceptability of high-rise living in these cities appears to be high [38]. Take the case of Singapore, recent statistics on its public housing residential mobility reveal a significant 82.5 per cent of all households living in public housing indicating that they would be content to always live in those flats [39]. Such residential satisfaction with high-rise can be viewed as a challenge to cities adopting high-rise living as an urban solution. At the same time, Singapore and Hong Kong with its decades-old tradition of building ever-taller housing for a broadening segment of population offer source-ground for closer analysis to augment our understanding of life in high-rise especially in the light of growing international interest in tall housing amid lingering criticisms that 'Tower blocks are fundamentally flawed as a general form of housing' [40].

With rapid urbanization (and more than half of the world's population is to live in urban areas), tall(er) housing seems an unexceptional component of the future city, especially when building outwards and downwards are not the best options. Against the heritage of discourse fragmentation and the growing trend of more people living in tall(er) buildings [41], there is scholarly motivation to consider liveability from the everyday subjectivity of the occupant, the 'user' of the high-rise. The rationale is perhaps best summed up in the words of Mumford [42], 'It is ultimately the person whose foot has to be fitted with the shoe who knows whether it pinches or not'.

3. Some Key Questions – Going Beyond the Physical to Social

Clearly, with more people living in taller buildings, there is a need in research to go beyond the physical and engineering concerns to consider liveability and impact of that environment on its users. There is no denying the importance of engineering considerations in the future of high-rise especially post-Sep 11. As Easterling [43] observes, it is the elevator technology that will continue to drive the future high-rise. High-rise residents are dependent on the lift everyday to get into and out of their homes. If something goes wrong, they are highly vulnerable. People are anxious of being trapped by a power failure, unable to escape and fearful of crimes in the lifts such as rape, murder and robbery [44]. The catastrophe of the World Trade Centre towers further heightens general concern about the lift and safety of tall buildings. However, as we all know, a house that is adequate from the engineering or design dimensions may not necessarily be considered adequate or satisfactory by the occupants. From an early time, housing observers have argued that the housing unit is but one in a chain of factors of housing habitability that determine people's relative satisfaction with their accommodation [45]. The adequacy of the housing unit, as determined by the internal space, the structural quality, the household facilities and other such housing amenities and qualities within the wider housing environment, will influence the extent to which the inhabitant is satisfied with the unit [46].

The debate on the connection between residential space and satisfaction is not new. It has been pursued in several studies on high-density living [47]. Williamson [48], for example, has studied 530 high-rise apartments in Germany and found that overall residential satisfaction was strongly related to the physical attributes of the building, especially spaciousness, room arrangement and quality of construction. These points are picked up by Odeleye and Jogun [49] in a study of 14 and 12-storey high-rise housing in Africa. They found that most of their respondents were not averse to high-rise living even though they would prefer to move to low-rise if there is a choice. In view of the centrality of residents in the satisfaction outcome, we obviously need to obtain and analyze information about the perceivers themselves. How people feel about living in the tall buildings is central to the liveability index.

Yet, despite the growing literature on the positive and negative views of high-rise living, there has been relatively little documentation on the attitudes and lived experiences of people who have personally chosen high-rise living [50]. Even less scholarship is on the perceived quality, willingness and acceptability of super high-rise living notwithstanding its growing presence internationally. Several key questions therefore remain: do we really understand the nature of the impact of super tall buildings on our city? What do such developments actually mean in terms of the lives of those people who live in them? More pointedly, perhaps, how do residents perceive these dwellings and their neighbourhood? What are the affective and cognitive reactions evoked by the condition of such residential environment? What is the extent of residential annoyance and satisfaction? At what height do they think/feel an apartment building begins to feel tall? Is there any relationship between how tall a person thinks/feels a tall building is and his/her past living arrangement, as well as his/her demographic characteristics? In short, are certain people more suited to high(er)-rise living? As things stand, there is a perennial danger that we may yet follow a similar cycle and resurrect the earlier period of negative visions of high-rise if we do not question expectations and reality. While perceptions may be fraught terrains in the world of scientific, technical engineering inquiry, the case remains that careful analysis of residents' lived experiences offers insights into the social impacts of high(er)-rise living that may yield pathways to strengthen the potential benefits while minimizing the dissatisfactions. It is also a way of shoring up opportunities for residents to stake claims to their living environment and city [51].

Where scholarship must surely turn its attention soon is to occupant research. This is the departing point for enhancing the efficacy and liveability of tall(er) housing. Internationally, following the Sep 11 2001 collapse of the World Trade Centre Towers, the International Council for Research and Innovation in Building and Construction (CIB) has re-examined the use and construction of tall buildings and re-defined its priorities to include a focus on occupant research. The CIB task group on tall buildings (TG50) was then formed to spearhead this research and included on its research agenda are questions relating to societal issues/public expectation, in particular, will people continue to occupy existing tall buildings and will new ones be built? What are the tolerable level(s) of individual, business and societal risk in relation to very tall buildings [52]? Locally, a group of us in universities in Singapore and Hong Kong have begun to explore and extend our understanding of residents' lived experience in high-rise since 1999 [53]. This is a rich field of research where the answers are pertinent to a better understanding of the factors to be addressed in the development of liveable and sustainable tall buildings. Much awaits us still to research and push forward our understanding of high(er)-rise living.

References

1. E. Burton, The compact city, Urban Studies 37(2):1969-2007 (2000).
2. V. Ridge, Asian buyer interest wanes, The Age, 1 Oct, pp2 (1997).
3. P. Marcuse, Urban life will change: Reflections on the consequences of Sep 11, Trialog 70(3):46-7, (2001).
4. G. Tibbits, The enemy within our gates: Slum clearance and high-rise flats in R. Howe (ed) New Houses for Old: Fifty Years of Public Housing in Victoria, 1938-1988, Melbourne, Ministry of Housing and Construction, pp150, (1988).
5. See, for example, P. Jephcott and H. Robinson, Homes in High-rise Flats: Some of the Human Problems involved in Multi-story Housing, Edinburgh, Oliver and Boyd (1971).
6. L. Costello, From prisons to penthouses: The changing image of high-rise living in Melbourne, Housing Studies 20(1):49-62, (2005).
7. Interview reported in Architecture: Going up...and up: When height is all that matters, Newsweek Dec 27, pp106, (2004).
8. M. M. Ali and P. J. Armstrong (ed) Architecture of Tall Buildings, New York, McGraw-Hill, Chp 1, (1995).
9. See www.emporis.com.
10. D. McNeill, Skyscraper geography, Progress in Human Geography 29(1), pp42, (2005).
11. G. Binder, Tall Buildings of Asia and Australia, Mulgrave, Victoria, The Images Publishing Group, (2001).
12. H. Lefebvre, The Production of Space, translated by D. Nicholson-Smith, Oxford: Basil Blackwell, pp92, (1991).
13. L. R. Ford, Cities and Buildings: Skyscrapers, Skid Rows and Suburbs, Baltimore: John Hopkins University Press, (1994).
14. G. H. Douglas, Skyscrapers: A Social History in America, North Carolina, McFarland and Company Inc, (1996).
15. L. Huxtable, The Tall Building Artistically Reconsidered, New York, Pantheon, (1984).
16. Abel, Sky High: Vertical Architecture, London, Royal Academy of Arts, (2003).
17. Huxtable, op cit, pp11.
18. R. Lacayo, Going up...and up: When height is all that matters, Newsweek Dec 27, pp104, (2004).
19. D. Crilley, Megastructures and urban change: Aesthetics, ideology and design in P. L. Knox (ed) The Restless Urban Landscape, Englewood Cliffs, NJ, Prentice Hall, pp145, (1993).
20. Le Corbusier, Towards a New Architecture, London, Architectural Press, (1946); Le Corbusier The Radiant City, London, Faber and Faber, (1967); Le Corbusier The City of Tomorrow, London, Architectural Press, (1971).
21. See D. Gosling and B. Maitland, Concepts of Urban Design, New York, The St Martin's Press, (1984); J. Lang, Urban Design, New York, John Wiley and Sons, (1994).
22. G. Helleman and F. Wassenberg, The renewal of what was tomorrow's idealistic city: Amsterdam's Bijlmermeer high-rise, Cities 21(1), pp4, (2004).
23. See, for example, Jephcott and Robinson, op cit; S. Young, Social and Psychological Effects of Living in High-Rise Buildings, Ian Buchan Fell Research Project on Housing, Sydney, University of Sydney, (1976); The National Tower Blocks Directory, Community Links, London, (1992).
24. A. Stevenson, E. Martin and J. O'Neill, High Living: A Study of Family Life in Flats, Melbourne, Melbourne University Press, pp1, (1967).
25. Doxiades cited by P. Blake, Form Follows Fiasco, Boston, Little Brown, pp82, (1977).
26. Foreword by Ken Livingstone, Mayor of London, Interim strategic planning guidance on tall buildings, strategic views and the skyline in London, Greater London Authority, (2001).
27. G. M. Haber, The impact of tall buildings on users and neighbours in D. J. Conway (ed) Human

Response to Tall Buildings, Stroudsburg, Dowden, Hutchinson and Ross Inc, (1977); E. D. Benson, J. L. Hansen, J. A. L. Schwartz and G. T. Smersh, Pricing residential amenities: The value of a view, Journal of Real Estate Finance and Economics 16(1):55-74, (1998).

28. See, for example, J. Conway and B. Adams, The social effects of living off the ground, Habitat International 2(5/6): 595-614, (1977).
29. B. Johnson, Living the High Life, National Real Estate Investor, Jan, Atlanta, (2002).
30. Ibid.
31. L. Copeland, Nice view: Cities of all sizes embracing high-rise living, The USA Today, Feb 15, (2005).
32. C. Church and T. Gale, Streets in the Sky: Towards improving the quality of life in tower blocks in the UK, the first report of the National Sustainable Tower Blocks Initiative, pp1, (2000).
33. K. Yeang, The Green Skyscraper, New York, Prestel, pp27, (1999).
34. Skyscrapers are 'model of sustainable' building, Planning 26 April, Royal Town Planning Institute, London, (2002).
35. S. Sassen, How downtown can stand tall and step lively again, New York Times article: The case for Skyscrapers, Jan 26, (2003).
36. Ali and Armstrong, op cit.
37. The Straits Times Oct 1 2004.
38. See Y. M. Yeung, Cities that work: Hong Kong and Singapore in R. J. Fuchs, G. W. Jones and E. M. Pernia (ed) Urbanisation and Urban Policies in Pacific Asia, London, Westview Press (1987); Y. M. Yeung and T. K. Y. Wong (ed) Fifty Years of Public Housing in Hong Kong, Hong Kong, Hong Kong Housing Authority, (2003); Housing and Development Board, Singapore, Annual Reports, various years.
39. Housing and Development Board, Singapore, Residential Mobility and Housing Aspirations, Singapore, (2000).
40. BBC News 10 Feb 2003.
41. Abel, op cit.
42. Quoted in Conway and Adams, op cit, pp595.
43. K. Easterling, Conveyance 'germs': Elevators, automated vehicles and the shape of global cities in A. Goetz (ed) Up Down Across: Elevators, Escalators and Moving Sidewalks, London, Merrell, (2003).
44. Conway and Adams, op cit.
45. See, for example, C. Bauer, Social questions in housing and community planning, Journal of Social Issues, 7:1-34, (1951); W. Michelson, Man and His Urban Environment, Reading, Massachusetts, Addison-Wesley, (1970).
46. See, for example, Kennedy Smith Associates, Housing Study, Isolated Communities and Indian Reserves, Prairie Province, Winnipeg, Kennedy Smith Associates, (1967); Y. K. Chan, Density, crowding and factors intervening in their relationship: Evidence from a hyper-dense metropolis, Social Indicators Research 48: 103-124, (1999).
47. R. C. Williamson, Adjustment to the high-rise variables in a German sample, Environment and Behavior ,13(3): 289-310, (1981); W. Odeleye and D. Jogun, Some social significance of high-rise living on Lagos, Nigeria in High-rise, High Density Living SPC Convention: Selected Papers, Singapore Professional Center, pp203-207, (1983); A.G. O. Yeh, The planning and management of a better high density environment in A. G. O. Yeh and M. K. Ng (ed) Planning For A Better Urban Living Environment In Asia. Ashgate, Aldershot, (2000).
48. Williamson, op cit.
49. Odeleye and Jogun, op cit.
50. Young, op cit.
51. A. Rappoport, House Form and Culture, New Jersey, Prentice Hall, (1969).
52. CIB task group on tall buildings TG50, The 2nd CIB Global Leaders Summit on Tall Buildings, Kuala Lumpur, Oct 20-21, (2003).
53. B. Yuen, S. J. Appold, A. Yeh, G. Earl, J. Ting and K. L. Kurnianingrum, Living Experience in Super Tall Residential Buildings, Final Report, The National University of Singapore, Singapore, (2003).

THE CHALLENGE OF MANAGING THE 88-STOREY TWO INTERNATIONAL FINANCE CENTRE IN HONG KONG

THOMAS H.K. HO

Property Director, MTR Corporation Limited, Hong Kong

Building management has become a highly sophisticated profession in the last decade, and the management of mega towers is particularly challenging for building managers. Completed in July 2003, the 88-storey Two International Finance Centre (Two IFC) is currently Hong Kong's tallest building with the most up-to-date technological provisions, making it an unprecedented challenge faced by the building manager, the MTR Corporation. The Corporation also faces stringent customer expectation as tenants of Two IFC are all prominent players in the global financial market. By blending five-star hotel services and security standard into property management, the MTR Corporation has been able to meet all the challenges, and is also dedicated to exceeding customers' expectation at all times.

1. Introduction

International Finance Centre (IFC) is the premier business address in Hong Kong and the new focus of our central business district. It is a 'city within a city', made up of two office towers, One IFC and Two IFC, an 800,000 square foot shopping centre, IFC Mall, in addition to the 1,100 room Four Seasons Hotel and Suite Hotel. This 4.5 million square foot complex is located directly above the Airport Railway Hong Kong Station and its in-town check in facilities, to which Hong Kong International Airport is only a 23-minute train ride away.

IFC is also surrounded by an extensive road network which will be extended to take advantage of the Central-Wan Chai bypass. Three basement levels provide convenient on-site parking for 1,800 cars. The MTR, which connects to all major districts in Hong Kong and the rail network to China, is a short air-conditioned walk from IFC. Direct covered access is provided to all modes of transport, including the Star Ferry and Hong Kong Island's main ferry piers for the Outlying Islands.

The skyline of Hong Kong is instantly recognisable. Two IFC is the 88-storey office tower of this commercial complex, a dramatic addition to Hong Kong's skyline and an icon reaffirming the status of Hong Kong as the financial capital of Asia. Designed by the world-renowned architect Cesar Pelli, this 2 million square feet Grade A office building has attracted some of the finest and most reputable investment banks, financial organisations and professional services firms, as well as the Hong Kong Monetary Authority, which is the territory's de-facto central bank. These top-tier companies have been drawn to the building's physical design, technical specifications, environmental performance, security and building management, as well as its location and surrounding amenities. Collectively,

Two IFC, which represents a new generation of office buildings, confidently caters to the most demanding multi-national tenants.

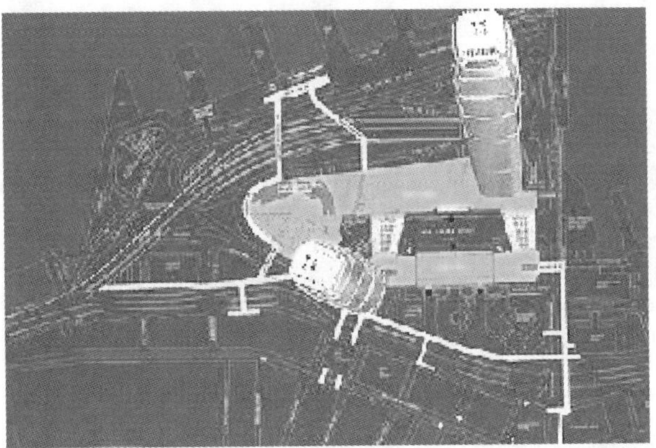

Completed in July 2003, Two IFC was developed by a consortium including Sun Hung Kai Properties, Henderson Land, Hongkong and China Gas and Bank of China Group. The day-to-day management has been taken over by the MTR Corporation's "Premier Management Services", which provides dedicated specialists in security, technical, operational and functional support. One of the greatest challenges for Premier has been to find a balance between providing the unparalleled technical standard of management and security with highest quality of service normally found in the hospitality industry.

This paper aims to share the hands on experience Premier has gained in Two IFC and describes the challenges faced in managing this 2 million square feet, 88-storey office tower, which is firmly established as Hong Kong's tallest and most prestigious landmark building. At Two IFC, the MTR Corporation made the decisive stop of moving away from a conventional building management style by blending five-star hotel service and security standards into property management, and has created a unique and distinguished product that is at the forefront of world-class standards.

2. Key Design Features and Technical Specifications of Two IFC

Two IFC was designed to cater to the demands of investment banks and financial institutions, which are renowned for having the most stringent occupational requirements.

The building contains virtually column free floor plates of approximately 23,000 square feet featuring a 15 metre span from core wall to window, ensuring an efficient space usage. All the tenantable floors provide raised floor access for flexible cabling management. Two IFC offers advanced passenger and service lift systems, flexible air-conditioning and a dual power supply, complemented by a comprehensive communications and information technology infrastructure. There are 22 specifically design trading floors containing an extra high floor to ceiling height of 3.3 metres and providing all the necessary technology and infrastructure to facilitate a bank's trading functions.

The Asian Institute of Intelligent Buildings chose Two IFC as the Building of Year 2004 and awarded it the highest ever score (95 points out of 100). Some of the key design features and technical specifications are summarised below:

2.1 *Building structure – solid as a rock*

The foundations of this 420 metre high building are embedded in solid rock. In contrast to conventional tube-in-tube structuring, Two IFC stands on eight mega-columns, reinforced by three outriggers connected to the central core. The outrigger system minimises the amount of structure required at the outside face of the tower and therefore maximises the views of Victoria Harbour and the city.

Adequate provisions are made in the structural system to ensure that the building will remain intact even with two mega-columns removed. Substantial redundancy is built into Two IFC's structure to cater for the world's one-in-fifty year mega typhoon, ie. mean wind speed of 230 km/h.

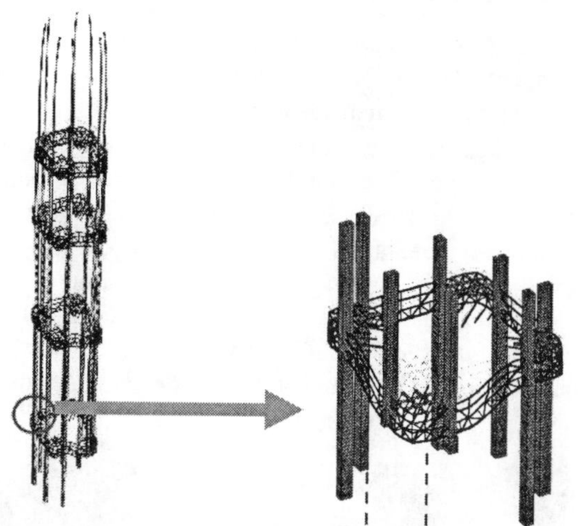

Figure 1. The building is supported by eight mega-columns, and will remain intact even with two mega mega-columns removed.

2.2 *Vertical transportation*

Two IFC contains a total of 62 lifts serving the seven zones of the building and the basement car parking levels, of which 14 are double-deck ultra-high speed shuttle lifts. There are 42 passenger lifts serving the office floors and three service/fireman's lifts. The average lift waiting time of office passenger and shuttle lifts is less than 30 seconds while the handling capacity is higher than 12%.

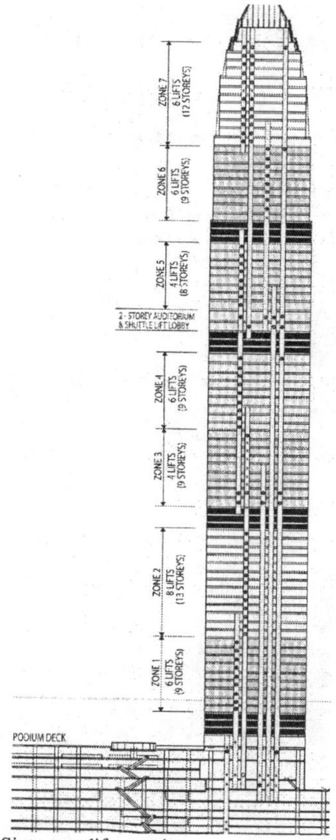

Figure 2. Sixty two lifts serving seven zones of the building and the basement car parking levels.

2.3 *Office floors*

Two IFC contains 49 typical office floors and 22 trading floors. Each of these 71 floors have floor plates averaging about 23,000 square feet and have a quoted efficiency of approximately 94% (lettable to net area) meaning that they are very space efficient for flexible planning.

The distance from the lift core to glass curtain wall is nominally 15 metres; there is generous raised floor access and extra high ceiling heights. Each floor includes male and female bathrooms, a toilet for handicapped access and an executive toilet with shower.

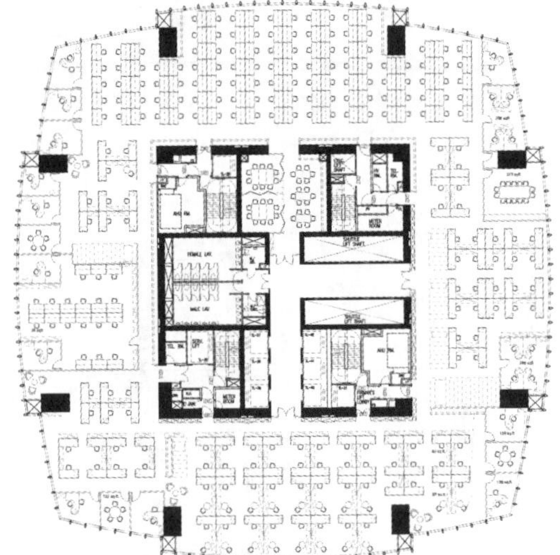

Figure 3. Efficient floor plate for flexible space planning.

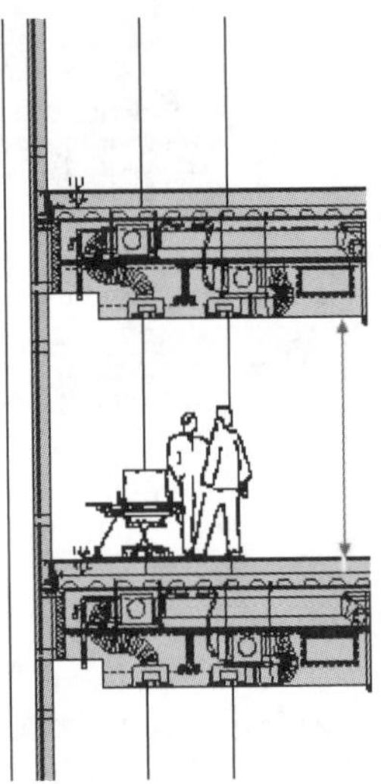

	Typical Floor	Trading Floor
Ceiling Height	2.7m	3.24m
Raised Floor	190mm	340mm
Floor Loading	4 kN/m^2	5 kN/m^2
	10 kN/m^2 for 3m zone around core	

Figure 4. Generous raised floor access, extra high ceiling heights and sufficient floor loading.

2.4 *Heating, ventilation and air-conditioning*

Each floor of Two IFC is equipped with a fully-fitted Variable Air Volume (VAV) air-conditioning system serviced by two air-handling units per floor, providing ventilation, cooling, as well as perimeter zone heating during winter to provide a comfortable ambience. The flexibility to adjust service hours, temperature and airflow levels is achieved through the Direct Digital Control at the fully automated Building Management System. Four pairs of chilled water valves are provided at each floor for tenants' supplementary and essential equipment cooling on a 24-hour basis, which is backed up by a standby air-cooled chiller and generator at 17W/m^2 per floor for a typical office floor and 51W/m^2 per floor for a trading floor.

2.5 *Electrical specifications*

Incoming power supply is provided by multiple independent Hong Kong Electric Co Ltd Zone Substations through dual feeders, each of which is able to support 100% of the entire building electrical demand load. To maintain electricity provision in case of major power failure, Two IFC is equipped with seventeen standby diesel generators with a combined capacity of 25,100 kVA providing more than 50% back-up to the normal power supply. Tenants may also install their own Uninterruptible Power Supply (UPS) system either at the Building's mechanical floors or on their own floor(s).

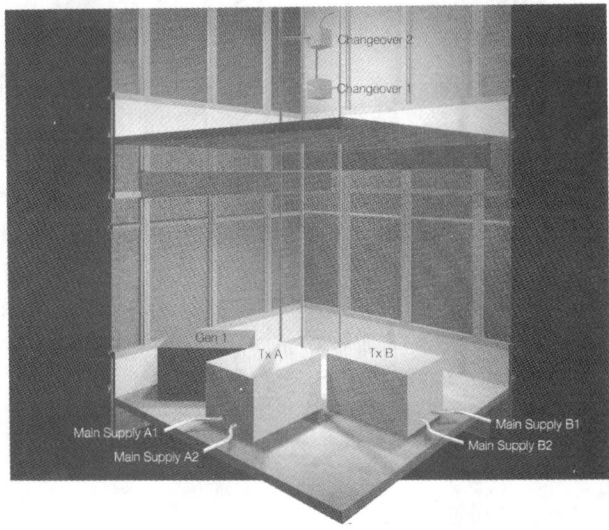

Figure 5. Multiple incoming power supply, coupled with standby diesel generators.

2.6 *Telecommunications*

Two IFC is equipped with fibre-optic cables providing tenants with the flexibility to efficiently accommodate specialised telecommunications and data installations. Satellite communications, satellite TV and cable TV reception is readily available through the provision of a roof satellite and aerial provision. Four spaces for 6 m diameter satellite dish are also available for tenants to install their own satellite dish.

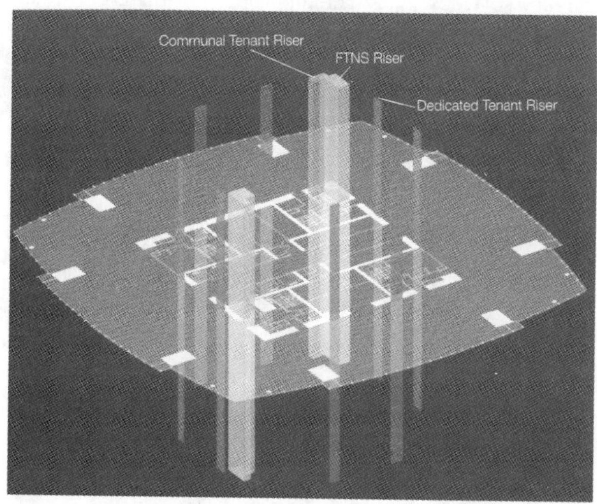

Figure 6. Telecommunications systems

3. Challenges of Managing Two IFC

In recent years, building management has become an increasingly sophisticated profession due to rising customer expectation together with growing technical complexity of new buildings. Occupiers' emphases on security and hygiene posted further pressure on the building manager. Being the tallest building in Hong Kong and perhaps one of the most technically advanced office towers in the world, managing Two IFC has been a major challenge faced by the MTR Corporation during its 25 years' of property management experience in Hong Kong. These challenges are articulated in the following.

3.1 *Critical mass of Two IFC*

Located at the forefront of Central, Two IFC is designed as the new focus of Hong Kong's core business district housing the city's major financial and other professional institutions. The 88-storey mega tower, with a gross floor area of 2 million square feet, accommodates a total of 15,000 tenants and business visitors on a daily average. Building maintenance to cater frequent people movement within the mega tower is thus important to upkeep the quality image of the building. This includes maintenance of technical plants and equipment, building materials, finishes, and last but not the least, the need to ensure cleanliness of the external curtain wall system which is substantially larger than any other building in Hong Kong. The sheer size of the building is already a major challenge faced by the building manager.

3.2 *Prestigious tenant profile*

As its name suggests, Two IFC is the heart of international financial activities and many of its tenants are prominent players in the global financial market, including Hong Kong Monetary Authority, UBS AG, BNP Paribas, Ernst & Young, Lehman Brothers, States Street, Bank of America, and more. The office of these multinational tenants reflects their status and understandably, these triple-A tenants would look for top-quality services from their building manager, ranging from reliability in technical provisions down to the slightest details of cleanliness in common areas.

3.3 *Rising customer expectation*

Added to the building manager's challenges is tenants' ever-increasing expectation, who nowadays look for more diversified services offering. Mere routine maintenance to upkeep the normal functioning of building facilities is far from sufficient. Service quality, proactiveness, communication and efficiency are the key attributes that differentiate leaders from followers. Building managers must be able to predict customers' changing needs and respond to them before such needs develop into deficiencies. A clear understanding on tenants' operation together with an attentiveness to details are the key factors that the building manager must demonstrate in order to meet and exceed customers' rising expectation.

3.4 *Stringent security requirement*

The global political climate has made landlords and tenants more alerted to security management. In particular, the locational and functional prominence of Two IFC have made security the building manager's prime concern, as any slightest loophole in the security system, be it in access control or in IT security, may lead to serious loss and damage to tenants' property. The building manager must be proactive in identifying any potential risks and attempts to resolve them before these potential risks turn into genuine crisis.

3.5 *Technical complexity*

To support the multinational financial activities and to enhance efficiency in serving the needs of the group of 15,000 tenants and visitors, Two IFC is equipped with a number of state-of-the-art building management systems which oversee different aspects of building operation. Further increasing the technical complexity is the interface between the main building systems with tenants' own systems controlling access and other building services provisions in their premises. The building manager must be able to efficiently master all systems and interfaces in order to ensure an uninterruptible operation of the building.

4. Managing the Challenges – Premier Management Services

As being a pioneer of incorporating the world-class hotel service standards into property management, Premier Management Services focuses on delivering unconventional management services to Two IFC by its young and energetic management team. The ultimate goal of Premier is not only to provide diversified services with extraordinary quality comparable to worldwide luxury hotels, but also to furnish various assistances that commensurate with or exceed other prestigious commercial property management companies in the world.

4.1 *Concierge service*

Premier is the pioneer in introducing a complete concierge concept into the property management industry that one may not regularly find in any other landmark buildings around the globe. Premier is keen on providing tenants with an unprecedented level of one-stop hotel-like management services rather than following the conventional way of processing all inquiries through the management office. The concierge desk is located at all prominent entrances with personnel standing out in front offices allowing convenient personal contacts in a more professional manner.

The role of concierge is very important as it provides a very first impression to guests and visitors through processing the building access registration. Apart from being an information centre to convey daily information, concierge also links up various functional departments in providing maintenance support, cleaning, security services and other services after receiving end-users' requests. Good communication and interpersonal skills are crucial to the delivery of superior personalized service. Regular liaisons, through meetings and customer surveys, enable Premier to build a long term partnership and mutual trust with tenants so as to achieve continuous service improvement.

The commitment and nature of service pledges have became very dynamic owing to keen competition amongst service providers in the global hospitality industry. Premier sets no boundaries in its services in order to meet tenants' needs to the fullest extent. At Two IFC, the concierge provides a wide range of business services, for example transportation / accommodation / airlines reservation, visa application, delivery services, etc. Tenants' needs at different times of the day are well taken care of through round the clock service hotlines and duty manager shifts.

Figure 7. Concierge desk allowing convenient personal contacts in a more professional way.

Figure 8. The role of concierge is very important as it provides a very first impression to guests and visitors.

4.2 *Security management*

Premier has devoted extraordinary effort towards ensuring a secure and uninterrupted environment for the prestigious tenants in Two IFC. A professional and dedicated in-house security unit comprising of ex-Army Gurkhas and experienced local security personnel has been developed to serve the mega tower. The security management blueprint has been benchmarked internationally, including New York City, Chicago,

Canary Wharf and many other buildings around the world to deliver a total security solution.

Figure 9. A professional and dedicated in-house security unit.

Premier provides a fully integrated security regime by proactive and preventive methods and swift response to any event of emergency or crisis. A detailed Crisis Management Manual has been developed, containing "Action Cards" for a wide range of possible incidents ranging from fires, serious criminal incidents through to catastrophic scenarios. The security team members are well trained in first aid and fire-fighting techniques in case of emergencies. To further tighten security control, a five-level Security Alert Code system is enforced in Two IFC, ranging from Alert Code 5 for the most acute situation, to Alert Code 1 when no significant security threat is apparent. The enforcement of different alert codes is tied with the global security situation together with an assessment of risk factors in the territory. Security measures taken in the building are strengthened as the alert code level escalates.

As a primary security barrier to outsiders, Premier has implemented an inhouse-developed building access registration system at Two IFC, which works in conjunction with the optical turnstiles installed at all lift lobbies 24 hours a day. Tenants and authorised visitors are identified by sweeping their registered smart card ("Octopus") to activate the optical turnstiles. Other visitors are required to register and be verified by concierge staff before entering the lift lobbies. In addition, over 1,000 digital CCTV cameras are installed at different parts of the building. These cameras are connected to the central control room and are monitored by security staff 24 hours daily. Proper record of all images from the CCTVs are maintained and can be retrieved easily when required, thus significantly tightening security surveillance.

Figure 10. Optical turnstiles

Figure 11. CCTVs monitored by security staff 24 hours daily.

Premier's well-know garrisons, three bomb search dogs, which are the first of its kind in Hong Kong, have aroused many interests from tenants, local media, together with other property management companies.

Figure 12. Bomb search dog.

In addition to the above complimentary services, the following security provisions are also adopted in Two IFC to further strengthen security control:

- Provision of security escort services for VIPs
- Provision of security crowd control services in special circumstances
- Security scenario drills, including evacuation exercises

4.3 *Hygiene management*

To address general public's growing awareness on environmental hygiene, Premier has established a stringent cleaning standard that is equivalent to that of world-class hotels. The housekeeping team strives to provide a clean, healthy, efficient and comfortable environment by maintaining the top service standard. All cleaning and disinfecting agents are carefully selected to ensure effectiveness and safety to tenants and visitors. The team focuses not only on common areas but also tenants' area as well, and daily internal cleaning to tenants' area is included as part of Premier's management services package. Apart from routine daily cleaning, the housekeeping team also offers special cleaning to tenants' area upon request. A contingency plan for infectious disease outbreak is in place to optimize business continuity for tenants.

Figure 13. Curtain wall cleaning.

4.4 *Auxiliary technology support*

A number of state-of-the-art IT systems are critical to the efficient and effective management of Two IFC. First and foremost, a fully automated Building Management System (BMS) is used to centrally monitor all and control some of the building services, ranging from lifts, escalators, lightings, air-conditioning, fire services installations, plumbing and draining etc. The building management team can simply control and monitor all building services in the BMS Room. The system is able to detect any abnormalities and generate alarm signals, not only on the computer screen but also routed to pagers so that the alarm signals can be addressed and the abnormalities rectified promptly.

By the evolution of the Octopus Access Control System in 1999, over 12 million octopus cards have been issued and are under active circulation among Hong Kong's 7 million population, and usage of the card has extended to all walks of life, from shopping, public transportation to other unprecedented possibilities. On building management of Two IFC, the Octopus Card is used in building access, elevator access, carpark control as well as attendance management, bringing much convenience to both the building manager and the tenants. This revolutionary system effectively monitors visitors' admissions and efficiently streamlines the physical registration processes. It further reduces operating costs and enhances productivity, service quality as well as tenants' satisfaction.

The Electronic Monitoring System is used by building management staff in the areas of security and housekeeping. This system ensures designated points are attended by the housekeeping supervisors or patrolled by the security officers. Staff are required to record each visit to a designated point on the Electronic Monitoring System through a Pocket PC. The system further assists the housekeeping manager in monitoring the cleanliness of the building and thus controlling staff performance.

Figure 14. Electronic Monitoring System

Another IT system developed by the MTR Corporation, PM21, is used by concierge staff in visitor registration and permit issuance. It records visitor's name, destination, together with both the check-in and check-out time of each visit for internal security reason.

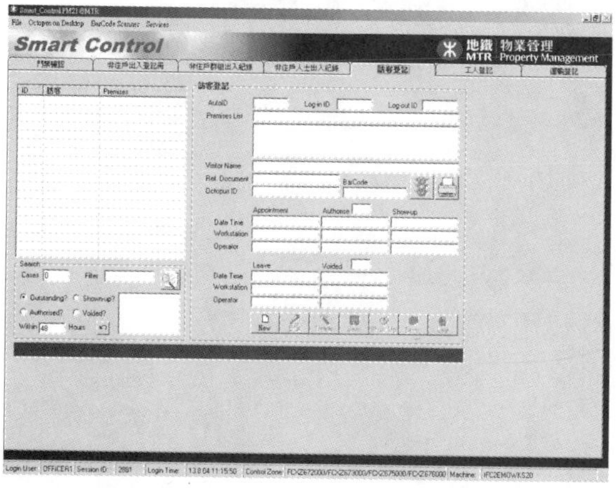

Figure 15. PM21 used on visitor registration and permit issuance.

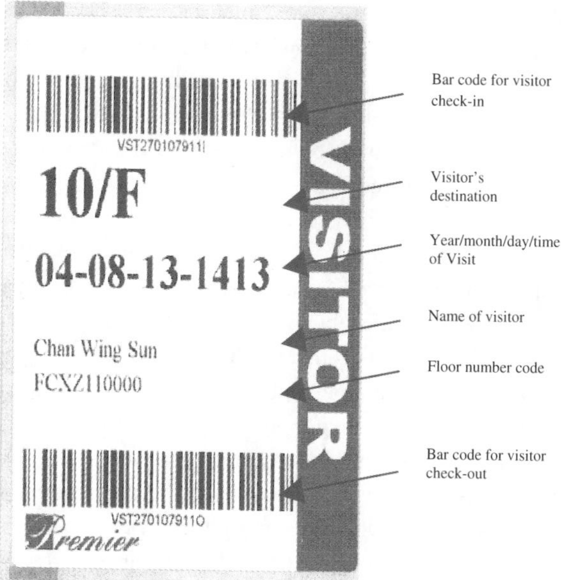

Figure 16. Sample visitor's pass

5. Conclusion

The management of multi-tenant buildings has become an increasingly sophisticated profession in recent years. Tenants' rising expectations, global consciousness in security and hygiene management, and increasing technical complexity of new buildings have all put the building manager in a position of unprecedented challenges.

The management of the 88-storey mega tower, Two IFC, is a classical example demonstrating the challenges faced by the building manager in today's property management context. The massive scale of the building, accommodating 15,000 tenants and visitors on a daily average, has already made routine cleaning and facility management challenging tasks for the building manager. In addition, tenants of Two IFC are world-renowned bankers and other financial institutions who demand management services of triple-A standard. Adding to the building manager's worklist is the global concern on security management, which is also the top concern for the key players of the world financial market housed in Two IFC.

The MTR Corporation has taken up all the challenges by the establishment of Premier Management Services, which differentiates itself from other commercial building managers through providing hospitality-like property management services. The team continuously listens to tenants' needs with the ultimate aim of meeting and exceeding customers' expectation. It also attends to every detail of building management, striving to provide a hassle-free environment for tenants' business operation. To further enhance efficiency and service quality, a number of state-of-the-art auxiliary IT systems are deployed in the daily building operations, making MTR Corporation one of the leaders in the commercial property management field.

Last but not the least, it is worth mentioning that building management should not be a process that commences only after building completion. For a mega project like Two IFC, the building manager should be involved in the planning and design process right from the beginning, providing input to the project team from building management and maintenance point of view. This approach yields considerable benefits, from budget minimization to avoiding unnecessary wastage due to unsuitable choice of building materials in the design process. Building managers nowadays must be prepared to develop multi-skills throughout the real estate development process, in order to meet the increasing challenges in their daily work.

A CITY BEYOND TALL BUILDINGS - REDISCOVER CITY VALUE AND PEOPLE VALUE

MEI NG

UN Global 500 Laureate
Director, Friends of the Earth (HK)

Space value, habitat value and people value are the essences that make up city value. Tall buildings and high density living, to certain degree, compromise lifestyle, life quality and social cohesion. Yet tall buildings, in contrast to spread out habitation, release land for public space and greening. From a compensatory aspect, public space and public facilities should be zoned and designed to remedy the constraints in urban high-density living. In another words, high density could be humanized by the humanness of public common space and facilities that welcome the newborn, that celebrate the newly wed, that energize the newly retired, that honor the newly departed. Romanticism and humanism could add value to a city of tall buildings. Climate responsive, environmentally informed, hazard alert and resource efficient, building designers, builders, managers and occupiers have a role to play in enhancing city sustainability.

City for the people and people for the city, such is the essence of city making. Whether a city is sustainable depends how it bonds inhabitants and place. Whether a city is livable, lovable, laudable or affordable is a measure of humanity and sustainability. Such is intimately linked to birth, living, aging, dying and departing.

My ideal city is a city that cares, shares, informs and integrates. A city that says hello (")... a city that unites (:)...a city that continues (,) ... a city that links (;)... a city that remembers ('').

1. A City and a Beginning (Birth)

The newborn is a new arrival to a city. It is interesting to note how cities say their first hello ("). Look around for the prenatal, maternity and postnatal facilities in a typical city. In Hong Kong, maternity wards have to share different floors of a public hospital, or different wings of a private hospital. Prenatal and postnatal facilities have to get out of the way of emergency services and increasing demand for long stay complications. Giving birth in a populated city may not necessarily be a dignified affair or a harrowing experience. What matters beyond the moment of birth, is the quality of prenatal and postnatal services as well as pediatric facilities to care for the newborn and mother before and after birth. Whether a city is civilized, depends how we handle birth, death and the needy.

Giving birth is a Nature process. Imagine purpose built maternity facilities nested in tranquil countryside, where mothers-to-be check in to prepare themselves mentally, psychologically and physically for the arrival of their first child or an additional family member. Integrated facilities where mothers-to-be learn about prenatal care, postnatal care, parenting and homemaking. Taking walks in the landscaped gardens, practicing breathing exercise infused with the scents of flowers, waking up to birds singing, inspired by every sunrise and sunset, exchanging motherhood experience on the lawn, breast feeding under the canopy of a tree, taking family portraits outdoor, learning about healthy baby diets and nutrition, bond a lifelong relationship between newborn and Nature, between mother and Mother Nature.

Children hospital is more than a place for health check-ups and remedies. It should be a place for learning and discovery about growing up, bodybuilding, illness prevention, health maintenance, habits forming, lifestyle choices and self-care. Look around for children's medical facilities in a typical city, it says a lot about how much importance a city attaches to nurturing a healthy future generation of inhabitants.

Children hospitals should be located near Nature. An average child will have so much to learn about the self-healing magic of Nature and all the remedies that have been derived from plants, insects and species. Children hospital is a classroom to learn about life and living. Children hospital is a happy place for regular visits. Children hospital is a place for piloting green parenting. Programs could include children organic gardening, healthy meal preparation, fitness training and habit tuning.

A city is where life journeys start. A city and a beginning are a lifelong bonding and a reason for celebration.

2. A City and a Bonding (Marriage)

Getting married in a city is also a lifelong bonding. Look around for wedding facilities in a typical city. How a city celebrates wedlock says a lot about the romanticism or lack of it. In Hong Kong, pre-wed couples have to queue in line at the often packed waiting rooms of official registries or fighting to book a church wedding climaxed by fancy banquets in expensive hotel ballrooms or banquet halls. Where is the dignity and romance?

A city of business could also be a city of romance. Where and how could city people celebrate the beginning of a new life partnership? I have previously proposed to revive the "Love is a Many Spendoured Thing" nostalgia of Hong Kong's beautiful Repulse Bay Beach by dedicated it as a wedding paradise. Imagine the newly weds take their vows facing the ocean and the blue sky in a semi-outdoor wedding registry or small chapel? Imagine after hosting an outdoor cocktail reception on the sand, and posing for wedding portraits against the romantic waves and palm trees, the newly weds stroll along the beach to watch the sunset and board a sailing junk that takes them out to sea.

Imagine a wedding facility nestled inside a nature reserve or a country park? Imagine turning a disused water reservoir into a wedding place? Imagine dedicate an old ferry pier and a stretch of the harbour front to become a wedding registry and celebration piazza?

Historical and heritage buildings could be recycled to become romantic places and wedding facilities. Antique decors and nostalgic setting could bring out fantasy of a different era and life stages. Celebration of happiness could be coupled with celebration of eternal love and bonding.

A city that unites will be loved forever. A city that bonds will be remembered for each wedding anniversary. Romance sustainability is a many splendid thing.

3. A City and a Continuity (Retirement)

Growing up in a city and growing old in a city is about life journeys and milestones. Home leaving and home coming have different reasons and justifications. Look around for senior facilities and pensioner services in a typical city. How a city cares about senior citizens says a lot about its sense of belonging.

Retirement communities, pensioner neighborhoods, senior condominiums, elder villages could be located outside the hectic city districts. From passive inaction mode of lifestyle, the retirees should be engaged into more active pastimes such as organic gardening, tea making, baking, weaving, skill learning and fitness training which includes brisk walking and gentle hiking. Regular dose of fresh air and exercise is beneficial to the aging pensioners. Locating the retirement communities to countryside has a certain advantage. Developing new hobbies such as nature photography, landscape illustration, bird watching, herbal growing, flower potting, nature art crafting, poetry and more could be a beneficial kind of therapy.

Hobby gardens and green roofs serve as meeting place and skill learning facilities as well as nature photography subjects.

Linking city greening and urban tree planting to senior citizens could enhance the sense of belonging. Naming rights of big tree planters or landscape gardens to honor senior citizens is one way to recognize the existence and contribution of residents in the city. Name a tree planter to honor the oldest in the city, or a roof garden to honor the longest serving community volunteer or dedicate a landscape patch to honor the model citizen of her era. Thus the senior citizens honored become a part of the city and also a part of the city's legacy. The senior citizens' children and grandchildren will be motivated to "adopt" the tree planter or the landscape garden or the green roof. They serve as volunteer gardeners and caretakers. It will save municipal manpower.

A city that remembers will be able to harness the sense of belonging and community involvement.

4. A City and an Ending (Death)

The living and the dead are fighting for land space in a congested city. Burial space is at a premium in big cities. Cemeteries are running out of space and the waiting time is becoming longer. It is time to rethink about cemeteries and burial planning.

Take a good look at the existing funeral facilities and cemeteries in typical cities. How a city bids farewell says a lot about how much a city honors its departing or departed citizens. It is quite unbelievable to witness the situation in Hong Kong. The privately or public run funeral parlours are relatively small and constrained since most of them are situated close to busy districts. Rental and burial services are at a premium. Bookings by the hour and by the half days are making life difficult for the mourning families. They have to rush through services and rituals. The acoustics are so bad that one could easily overheard the weeping in the parlour next door. Wreaths are thrown out and dumped into the dustbins on the roadside when the time is up. Burning of incense and paraphernalia generate smoke and exhaust which aggravate the air quality inside a small parlour.

Old generation Chinese prefer land burial. In a land shortage city like Hong Kong, procuring burial space is a big challenge. The situation is so acute that old graves have to be dug up to make room for newer burials. Where is the sense of dignity and humanity?

Many developed cities have advocated cremation and green burials. Beijing, Shanghai, Zhejiang, Shanxi and land restrictive cities in China have already launched tree burial schemes. Stockholm's Woodland Cemetery (1915-61) has led the way in burial reforms in the early twentieth century. By enhancing the natural attributes of the site—ridge and valley, earth and sky, forest and clearing, meadow and marsh—to evoke associations of death and rebirth in a landscape of spiritual dimension. From dust cometh, to dust returneth. Cremated ash was added to the soil to plant trees. Woodland, small chapel, natural landscape and bird songs have created a place of remembrance for mourning families to honor their departed loved ones in peace, tranquility and dignity.

Integrating funeral service, burial and nature conservation is a win-win socio-eco-cultural city planning option. Green belts, conservation areas as well as rehabilitated quarries, deserted farmlands and disused mills are possible locations. Linking spiritual, cultural and ecological values to cemetery planning and design highlight the principle of sustainable development. A city that knows how to bid farewell is a city that remembers.

5. A City that Cares (Self-help)

Banking of community service, bartering of community currency and integration with issue of citizen's smart cards have been tested in Japan. The function of public smart cards serves not only as a tool for providing administrative services, but also as a tool for citizen participation and community development. In the City of Yamato, smart cards are issued as identification as well as entitlement to self-help and mutual-help activities. The LOVES system allows the reservation of sports and educational facilities, courses, events, conferences, electronic information. The LOVES system also serves to register unwanted goods ("value things"), volunteer activities ("value people"), shopping information and the management of community currency accounts.

According to a study by Takashi Kobayashi and Yasuo Hibata of Keio University, Kanagawa, Japan, the LOVES transactions is conducted as in a bank transfer, by transferring LOVE from the LOVE account of the person receiving the service to the account of the person providing the service. The price of a service is negotiated between the provider and the recipient. LOVE coupons issued by a Support Center was used for volunteer activities. The LOVES system has won the support and participation of women and senior citizens who have been active in promoting community service and urban space events. Terminals installed throughout the city, including public venues such as government offices, community activity centers, shops and convenient stores.

Quality people make a Quality City. Enhancing the sense of neighborhood and mutual help within a congested city goes beyond communal gardens, lobby spaces, clubhouses, playgrounds and shopping malls. It is equally important to develop the software for mobilizing mutual care and mutual help. Taiwan and some parts of China have been experimenting the community currency program with certain degree of success too. Hunan has launched a Good Deed Banking Scheme in the early nineties.

Mobilizing citizens to clean-up, recycling, greening, composting and stewardship of public open space such as adopt-a-tree, adopt-a-pavement, adopt-a-bin, adopt-a-slope, adopt-a-playground, adopt-a-campsite, adopt-a-woodland, adopt-a-beach, adopt-a-coastline, adopt-a-stretch-of-harbor-front, adopt-a-bay and adopt-a-neighborhood is the essence of putting sustainable development into practice. Sharing and caring enhance the sense of belonging and neighborhood.

5.1 *A city that is climate responsive*

Hong Kong buildings are super chillers? A survey released by Friends of the Earth (HK) in July 2005 listed the commercial and residential buildings in Hong Kong as second on the list of thermal offenders reported by the general public. Average office temperatures are set between 18 to 20 degrees Celsius. Average home temperatures are set around 22 degrees Celsius. Reporters of the thermal offenders have been quoted: "I feel like working in a mortuary" "I need to wear a wind breaker to work." "The tourist guidebook warns us to bring a woollen sweater when visiting Hong Kong".

Building managers have excused themselves by blaming it on the design of the air-conditioning system. They are half right. Most buildings' ventilation systems have been designed with minimal sensitivity for thermal comfort and climate responsiveness. With regard to central air conditioning, building managers often claim that "none or all" is the only option when asked to switch off air-conditioning units at empty work areas or at over-chilled sections. Bad installation and inappropriate choice of locations for fresh air intake and air outlets have also been mentioned as the other explanation for the artic-chilling conditions pervasive in most commercial buildings, restaurants and shopping malls in Hong Kong. The mismatch of the capacity of the air-conditioning units to the size of the space to be ventilated is often due to ignorance of the sales agent, the procurer or the building designer. Excessive lighting that generates excessive heat is another anomaly that is pervasive in most Hong Kong workspace and leisure facilities.

On top of this is the misconception of the general public about air-conditioning. Instead of thermal comfort, the general Hong Kong public equates air-conditioning to coldness. Being cold is being cool. Feeling cool is the answer to air quality improvement and the answer to over-seating capacity.

The biggest hurdle to change this misconception is linked to the building management fee. Most commercial and residential buildings are managed by private building management companies who charge their fees based on services that they provide which may include water or air-cooling for central air-conditionings. The incentive to reduce electricity consumption does not reflect on the management fees. It is desirable to consider a rebate for actual energy and water consumption saved and the cost of savings to be shared between property owners /renters and the building management service providers.

Buildings that offer buyers and renters free electricity appliances should be issued guidelines to procure energy efficient equipments that have been officially labeled by certification bodies. Wherever possible considerations should be given to procure equipments that could be powered by renewable energies such as solar water heaters and photovoltaic panels. In this case, incentive for developers and flat buyers is a determining factor. A Green Credit scheme could be introduced as part of the demand side management measures that utility companies need to adopt as integral to licensing requirements as suppliers or service providers. Green Credits could be redeemed as rebates or discounts for procuring energy efficient appliances as well as entitlement to free maintenance services.

5.2 *A city that informs & empowers*

Tall buildings should provide "birth certificates" to document the full life cycle design and construction process. There have been initiatives undertaken by both government and the private sector to conduct overall environmental performance of buildings with regard to its impact on air, water, energy, waste and sustainability. Performance assessed before and after occupation will provide property buyers, owners and renters a comprehensive picture of a building's environmental footprint and related implications. Such information will help to enhance the environmental awareness and conservation commitment of the occupiers and building managers.

Regular "health check" report should go beyond water, energy and material consumption to reflect the status of lighting, ventilation, thermal comfort, indoor air quality, drainage, wall and floor conditions as well as greening and pest control measures. This could serve as a timely reminder to property owners to step up maintenance and efficiency. Property sellers, buyers and renters have the right to know about the status of the properties they plan to own or plan to live in. Well-being of the buildings will enhance the well-being of the inhabitants.

Equally important is to raise awareness about the health hazards and environmental costs due to breakdowns and malfunctions of home fittings and appliances such as leaking taps, clogged drains, high flush toilets, over chilled air-conditioning, smelly carpets, inflammable curtains, bad lighting, energy intensive appliances, pesticide abuse, adhesive and chemical toxicity and improper garbage disposal. Informed response is the best prevention of hazards and wastage.

Do-It-Yourself maintenance and repair skill transfer help residents of buildings to understand the life cycle of the building that they live in. Basic building management workshops and basic repair training courses would enhance DIY skills of simple check-up, maintenance and repair. Set up building management hotlines, resource centers, exhibitions and professional think tank to advise residents on the ABC of building management which should include the basic understanding of building design, building material, building maintenance and owners/renters' responsibility and liabilities.

5.3 *A city that detoxifies*

Pest control misapplication is a hidden health risk often overlooked. Indiscriminate spraying of pesticides inside buildings and public open space could be harmful to people, species and the environment. Fumigation of pests in tall buildings requires extra care because of the possible migration of toxic fumes through central air conditioning and the air ducts or corridors. Residues of pest control chemicals in gutters, drains, cracks, floor and wall surface have never been monitored for their potential impact on health and ambient air quality. The frequency, dosage, scope, methodology and precaution related to pest control application need to be regulated and monitored. Full disclosure of chemical contents and health risks of the pesticides used should also be enforced.

Added concern comes with the over-reaction to the upsurge of infectious diseases due to global warming and outbreak of epidemics. The increase in the occurrence of dengue fever, malaria, West Nile virus, encephalitis, SARS, bird flu, typhoid and cholera have triggered panic reaction. It leads to excessive use of pesticides, fumigation, chlorination and disinfection. A panel of medical, ecology and toxicology experts should be enlisted on the building management team to advise on the application of pest control and alternative eco-safe options.

An annual assessment of the pest problem inside the building and the justification for routine pest control measures adopted together with proposed contingency measures during high risk periods such as flu and malaria epidemic should be included in the regular "health report" to be prepared and submitted by the building management service providers. There is also the added responsibility to raise building residents' awareness, to encourage reporting of pesticide poisoning symptoms and to develop routine monitoring and response protocol.

Indoor air pollution due to renovation and retrofitting work inside tall buildings is another hidden health hazards being overlooked. The emission of chemical volatile compounds produced during the application of paints, sealants, adhesives, cleaning and varnishing agents at the renovation sites could be highly toxic and health threatening. The migration of toxic fumes, particulates, odor and chemicals inside a tall building aggravate the public health risk. There should be clear guidelines, safety standards, penalties, monitoring program to regulate renovation and retrofitting works within the building. Renovators should be asked to provide information on the profile of the types of chemicals to be used, the methodology of application, the precautions to be taken, the disposal of chemical containers and leftovers. Blacklist and de-list irresponsible renovators and repeated offenders should be considered as an effective measure to safeguard indoor air quality and residents' health.

According to NASA tests, some plants can effectively remove indoor air pollutants such as benzene and formaldehyde. Simply grow a Rubber Plant, Black-eyed Susan, Spider Plant, Aloe Aristata or Peace Lily and you can improve the indoor air quality. Indoor greening and outdoor landscaping of buildings should go beyond beautification. Tall building sky gardens, roof gardens, podium and ground level landscaping should also consider the beneficial aspects of detoxification. It is worthwhile to employ ecologists, botanists and toxicologists to advise on the type of plants, trees, shrubs and turfs to be considered with regard to their air purifying qualities and water consumption demand due to irrigation needs. The experts could also advise on the innovative options of rainwater harvesting for irrigation and food leftover composting for planting.

City greening goes beyond beautification and shading. Certain species of trees clear the air and kill viruses. Camphor, tea trees, pine, cinnamon, lemon trees, silver birch and *willow* have detoxification functions. Greening goes more than planting the fastest growing species and the exotic specimens, there is the need to research into the scientific beneficial qualities of different tree types to improve the general health of a city and immunized it against the increasing occurrence of infectious disease and outbreak of epidemics.

6. A City that Envisions

Community planning and citizen participatory envisioning is a dynamic process to create and identify city value and people value. City making is about space value, life value, community value, green value, creative value and sustainability value.

'If you design a neigbourhood playground, don't forget the washing machines! ' The reckoning came from the design of a children's playground for a low-income area in American Midwest. Mothers are reluctant to let their children go there unsupervised. Finally, the problem was solved by building a small greenhouse equipped with second-hand washing machines and a dryer funded by a neighbourhood party. Mothers gathered there to do their washing, socialize and watch over their children at play.

A City with a human touch is a city that welcomes (birth), that bonds (marriage), that continues (retirement), that remembers (death), that promotes self-help (volunteerism), that empowers through information, skills, incentives and preventive thinking to promote sustainability thinking and responsibility.

www.foe.org.hk
meing@foe.org.hk

SUSTAINABLE TALL BUILDINGS – FROM AN ENERGY PERSPECTIVE

OTTO POON
Chairman, Energy Advisory Committee
Government of Hong Kong SAR

In modern cities, tall buildings are now looked upon in terms of their sustainability. Efficient use of energy in buildings, either in daily operation or as embodied energy in materials, is an important consideration of sustainability. While energy efficient devices are increasingly used in building services systems, an integrated approach should be adopted to achieve an overall reduction in energy consumption. With Hong Kong importing all its materials and fuels from abroad, energy efficiency could bring about benefits to the local economy and help to ease the environmental burden on the producer countries. There should also be greater efforts in exploring alternative renewable energy sources and in promoting energy efficiency.

1. Introduction

In modern cities, tall buildings make positive contribution to city life and signify the image of a city. Tall building makes efficient use and enhances the value of the land, and creates interesting landmarks. To ensure the demand for space from their inhabitants can be met, tall buildings are become increasingly adoptable as a result of limited available land. Many tall buildings are built for commercial purpose. Usually, the developers are not the end-user of the buildings and have lesser interest in the idea of energy-efficiency. Even occupiers themselves concern not much on energy costs since its contribute only as small percentage to the total running costs of a building.

2. Buildings and Energy Use

Tall buildings in the cities consume energy mainly for ventilation system, lighting and electrical equipment for business or living comfort. Today, we are of storage of suffering from global warming and facing uncertainty long-term energy supplies. It is essential to seek ways to improve the situations. For this purpose, we need to analyse all aspects of a building's energy use to workout practical solution. However, energy cost is usually only a tiny proportion of the total business costs of an office building.

3. Constructions and Use of Materials

Good design and construction of tall buildings can make significant improvement on the aspect of energy efficiency, economically and environmentally. Building professionals are striving explore savings in materials through new design. The selection of façade materials, although governed largely by architectural concerns, can greatly influence the thermal performance of buildings. As methodologies for the life cycle assessment of cladding materials develop, environmental awareness is growing among building professionals. The materials for architectural finishes can now be chosen with improved understanding of their relative merits in sustainable terms.

4. Building Materials and its Environmental Impact

Buildings are the main destination for the electricity power produced. It can be seen that the buildings in use consume much more energy annually than in the production and assembly of new buildings. For a typical office building, the embodied energy accounts for 10-20 per cent of the total energy consumption over its average 50-year life. Energy consumed and emissions are not the only recognized sustainability indicators by which to assess the environmental impact of buildings. The construction and use of buildings deplete natural resources, and have physical effects in the form of mineral extraction, the use and pollution of water and of air, disposal to landfill sites, and incineration. The manufacture of materials has an impact on ecology and bio-diversity, and causes deforestation, ozone depletion and climate change. These environmental indicators reflect some of the effects of the spread of industrialisation and the built environment across the planet. Whilst people attach varying values to the critical issues of sustainable development, not all are quantifiable or directly comparable. The embodied energy of materials is, though, a useful indicator for determining the total energy used to construct a building. Using analytical methods, tall buildings can be compared with low-rise buildings to evaluate their relative impacts.

5. Building Services in Tall Buildings

Sufficient capacity in the public utilities services of power, water and sewerage is a pre-requisite for sustainable development of tall buildings, where resource input and waste output are extremely high and concentrated. Sophisticated building services systems are equipped in tall buildings in contrast to low-rise ones to provide a safe and comfortable working environment for up to several thousands people. Certain aspects of the design of services in tall buildings impact upon sustainability more than low-rise buildings. There is additional burden on consumption from the need for lifts in towers; delivery of water at height, and handling the removal of large concentrations of waste.

As a result, it is natural that more cost-effective energy saving methods has been widely adopted in particular for tall buildings. Usually, an integrated approach to the provision of efficient building services systems includes features such as variable volume chilled water distribution using direct drive variable pumping and variable volume air distribution for air-conditioning, energy-efficient lamp technologies and lighting control systems integrating with daylight etc., all intended to provide reductions in overall consumption.

Vertical Transportation is also one of the most important aspects of tall buildings affecting the efficiency of people's movements and contributing to overall energy consumption. Manufacturers have recently developed lifting solutions that maximize efficiency and minimize energy usage for high-rise buildings. As an example, the double-decker lifts in the Petronas Towers in Kuala Lumpur and the IFC Two in Hong Kong demonstrated the achievements attained by the lift industry in recent years.

For high-rise developments, the need to reduce energy drives the use of high performance building envelopes and hence reduce perimeter cooling/heating required. Capital expenditure on insulated cladding is often cost-effective over the life of a building, with added benefits of lower energy consumption.

There also exist a fine balance between maximizing glazing in the façade of a building to harness daylight and minimizing the heat transmission which influence thermal comfort conditions and heating/cooling loads. Along with the need to maximize the penetration by diffuse daylight, at certain times it is also necessary to minimize penetration of direct sunlight disturbing the visual and thermal environments. In tall buildings, unobstructed views of the skyline are desirable while the low altitude of the sun in winter can cause glare problems. The heating effect of lower-intensity winter sun can, however, be maximized to reduce energy costs if glare is managed appropriately. New window systems have been developed that include both light maximizing and minimizing elements. External solar shading can be added to lower sections of a window, while higher sections may reflect sunlight and daylight on to the ceiling plane.

6. Energy Efficiency in Tall Buildings

Energy is a double edged sword. It supports economic growth and a higher standard of living. At the same time, it creates a host of environmental problems. Since opportunity costs of reserves are not considered and environmental costs are not internalized, prices of fossil fuels are relatively inexpensive resulting in lots of waste and inefficiency. In the city, many developers strive for short term financial gains, sacrificing long term benefits. Decisions are often made, based on initial capital costs and neglecting the associated environmental impacts and their life cycle costs. If developers only look for low cost and high selling price, it would be left to the tenants and operators to pay for the cost of inefficiency. The same situation can equally apply to fixed installations and appliances. Equipment and plant are often designed and manufactured to lower selling prices without taking the operating cost or life cycle costs into consideration.

In the recent years, the public is widely aware of energy-efficiency in buildings through promotion. The definition of an energy efficient building is a building which provides the required internal environment and services with minimum energy use in a cost effective and environmentally sensitive manner. Here, we are talking about institutional frameworks and behavior of people. We are talking about civil responsibility of individuals and corporations. I believe to encourage energy efficiency, we need:

a) to set energy efficiency benchmarks for different types of appliances, equipment and plants. Life cycle costs should be taken into consideration,
b) to set energy intensity indices for different types of buildings. There may be incentives in the tariff of electricity to encourage and reward efficiency, and discourage wastage,
c) to monitor the primary energy intensity (primary energy per Gross Domestic Product (GDP) and per Purchasing Power Parity (PPP)) and set progressively lower medium and long term targets,
d) to educate the public and to promote civil responsibility.

As Hong Kong imports all its materials and fuels, the environmental burdens in producing these resources are borne by the source countries. When we talk about energy efficiency and to be correct, we also need to consider the embedded energy of the materials for the construction and demolition of our equipment, buildings and infrastructures on top of operational efficiency. From energy efficiency perspective, designers, contractors and developers would need to optimize the embedded energy of the projects to achieve life cycle energy efficiency. In addition to the financial and environmental benefits, energy efficiency can provide a healthier environmental for the building occupants. Again, design guidelines of embedded energy intensities would be most useful.

For energy efficiency to be implemented effectively, the building operators need to get commitment to make their responsible buildings energy efficient, understand the operation and maintenance programme, communicate and educate the tenants on how to run an energy-efficient building.

In this era of globalization, movement of materials, goods and services across national boundaries is very much part of the world economy. In Hong Kong, we virtually have to import everything we consume and the environmental burdens are borne by producer countries, we naturally cannot take a local view on the issue of sustainable development.

7. Tall Buildings in Sustainable Cities

Sustainable development means a balance of social, economic and environmental costs and benefits. Energy efficiency would at the most provide only a small contribution towards sustainable development of any community. From energy perspective, it means a balance of supply and demand of energy in the wealth creation process, against the impacts on the quality of life and on the environment.

When we consider the supply and demand, most of the world's energy needs are satisfied by fossil fuels. With increase in demand on a finite reserve, the situation is unsustainable on the supply side of the equation. We would one day use up all the fossil fuels. From another angle, the discharge of pollutants such as CO2 is exceeding the assimilative capacity of the earth. It is again unsustainable.

Energy efficiency will help towards conservation which will extend the life of fossil fuels. This is demand side management. Renewable energy will reduce the reliance on fossil fuels. This is supply side management.

The truth is that with growing population most of us aspire to improve our standards of living driven by energy consumption, this two prong approach of conservation and limited availability of renewable energy will only make a small dent to the problem.

As a first step and an intermediate measure to achieve sustainability in energy supply and demand, we need to develop:

- significant advancements to improve the efficiency of and cost lowering of solar energy generated by photovoltaic devices
- lower threshold of wind speeds to improve the availability and efficiency of the wind turbines and generators
- massive energy storage devices for solar and wind farms to improve their availability
- more efficient and environmentally friendly incineration plants to convert municipal wastes into energy
- extensive use of other forms of renewable energies such as hydro, geothermal, biomass, bio-fuels, etc.
- institutional framework to encourage conservation and penalize excesses
- new economic model to recognize opportunity and environmental costs of fossil fuels

Naturally, much greater efforts and more resources have to be devoted to continuous education and promotion on the need for energy conservation and development of renewable energies; as well as on the concept and practice of sustainable development.

8. Conclusions

The adoption of good energy efficient design for tall buildings in particular could introduce economic benefits in terms of cost savings in material, energy and water use. The cost effectiveness of good design can go far beyond these tangible factors.

Building professionals have responsibility for the safety, comfort and health of tall building occupants in the cities. In seeking to achieve this, energy-efficiency is one of the primary factors while designing buildings. Tall building must be sustainable in the broadest sense, taking into account its physical, social, economic and environmental impacts based on the life cycle costs and benefits. As the concern for global issues relating to the environment increases, the role of the building professionals becomes ever more important in the design and operation of tall buildings and their use of energy. Therefore, it is essential to understand the opportunities which exist for improving the effectiveness with which energy is used in buildings.

Integrating environmental considerations into building design is not just a worldwide trend but also a vital step in achieving sustainable development in our cities. Therefore, environmentally friendly design of buildings should be widely adopted while standards and regulations appropriate to the local context should be developed for Hong Kong.

References

1. Ir Otto Poon, *Energy Efficiency, Renewable Energy and Sustainable Development* (2005).
2. Energy Advisory Committee, *Working Group Report on Improvement of Urban Environment* (2005).
3. Will Pank, Herbert Girardet and Grey Cox, Corporation of London, *Tall Buildings and Sustainability* (2002)

BRIEF INTRODUCTION TO CHINESE REVISED TECHNICAL SPECIFICATION FOR CONCRETE STRUCTURES OF TALL BUILDING

XIAO-KUN HUANG

China Academy of Building Research, No 30, Beisanhuandonglu, Beijing 100013, China

The revised provisions of Technical Specification for Concrete Structures of Tall Building (JGJ3-2002) are briefly introduced in the paper. The main contents include the actions, the structure regularities and the principal requirements for structural analysis and design of high-rise buildings. Some requirements for complicated structures are stressed.

1. Introduction

1.1 *High-rise buildings in Mainland China*

The development of high-rise buildings in mainland China can be divided into three stages. The first stage is from 1910 to 1949.The limited amount of tall buildings were constructed mainly in Shanghai, and the stories of the buildings are almost below 20 in that period.

The second one is from 1950 to 1979, the first thirty years after the foundation of new China. The tall buildings were mainly located in some biggest cities such as Beijing, Shanghai, Guangzhou, Tianjin, Shenyang, etc, including the famous ten biggest buildings along the Chang'an avenue in Beijing. Most of the building's height is lower than 100m in that period.

The third one is from 1980 to present, i.e. the reform and open policy stage. The theory and practice of analysis, design and construction of tall buildings at this stage have achieved great progress. According to the newest bulletin by the Ministry of Construction of China, there have 14.9 billion squire meters buildings in all towns and cities at the end of 2004, among which about 9.6 billion squire meters or 64.5 percent are for residential. Based on incomplete statistic data, about 25% of total amount are high-rise buildings, and among which over 90% are made of reinforced concrete.

1.2 *Standard system in China*

The centralized administrative system combined with respective responsibility of relative official departments and civil associations has been adopted in China. The technical standards have been classified into four levels: national, professional, local and enterprise respectively.

All national, professional and local standards are divided into mandatory standards and voluntary standards, and most of them are voluntary. Some provisions or clauses selected from the existing national and professional standards were put together and have become compulsory provisions from year 2000.

1.3 *Specifications for design of tall buildings*

There are several kinds of technical standards related to the design, construction, usage and maintenance of tall buildings, such as the codes for actions, architectural, structural and fire safety design. The Technical Specification for Concrete Structures of Tall Building [1] (TSCSTB in short) and the Technical Specification for Steel Structure of Tall Buildings [2] are two main professional standards widely used in high-rise building structural design and construction in China.

The TSCSTB has had three versions from its founding up to now and has a serial number of JZ102-79, JGJ3-91 and JGJ3-2002 respectively. The newly revised version, JGJ3-2002, was approved and put into effect by the Ministry of Construction of China on June 3, 2002.The discussion will be stressed on TSCSTB (JGJ3-2002) in this paper.

2. Main Contents of JGJ3-2002

In TSCSTB (JGJ3-2002) there are totally 459 clauses, relating to the design and construction of concrete tall building structures, which compose 13 chapters and 5 appendices. From the total clauses, 32 were selected as compulsory provisions by the central government. The main contents of JGJ3-2002 consist of: General principles, Terms and symbols, Actions, Fundamental stipulations of structural design, Structural computation and analysis, Design of frame structures, Design of shear wall structures, Design of frame-shear wall structures, Design of tube structures, Design of complicated structures, Design of mixed structures, Foundation design, Stipulations for tall building construction and 5 relevant appendices.

3. Important Provisions and Revisions

3.1 *Scope*

Besides traditional reinforced concrete (RC) structures, mixed or hybrid structures made of steel and RC or steel reinforced concrete (SRC) members are included, such as steel frame or SRC frame plus RC core or SRC core systems. The applicable scope of building height or stories is changed to not less than 28m or 10 stories. The previous scope was not less than 8 stories in the old version.

3.2 *Actions*

The design reference period for variable actions and time dependent properties of structural materials is unified as 50 years. Ordinarily, the weight load, the wind load and the seismic load shall be taken into account in the design of high-rise buildings.

The wind load is evaluated according to the national code for the load of building structures [3]. For the very important or wind sensitive tall buildings, the recurrence interval or return period of wind load shall be taken as 100 years. When the building height is higher than 200m or the building height exceeds 150m and the building's configuration or wind environment is complicated, the wind tunnel test should be done to evaluate the wind load.

The earthquake action is considered based on the national code for seismic design of buildings [4]. Two earthquake-prone regions with design basic acceleration of ground motion of 0.15g and 0.30g were newly considered as listed in Table 1.

Table 1. Relationship between seismic fortification intensity and design basic acceleration of ground motion

Seismic fortification intensity	6	7	8	9
Design basic acceleration of ground motion	0.05g	0.10g (0.15g)	0.20g (0.30g)	0.40g

Note: g is the acceleration of gravity.

The site is classified into four categories for seismic design, namely Ⅰ, Ⅱ, Ⅲ and Ⅳ, according to the site covering thickness of the profile and the soil effective shear wave velocity of soil layers considered. Usually, the soil covering thickness is measured from the ground level to the top surface of a soil layer that has shear wave velocity larger than 500m/s.

All architectures in seismic resistant zone are classified into four categories, namely A, B, C, and D, based on the *importance* and the nature of *occupancy*. Most of tall buildings belong to class C. For different class of buildings in the same seismic fortification region, the earthquake action and detailing measures must conform to the different requirements by the code. According to the building height and the extent of irregularities, both plan and vertical, the earthquake action of tall buildings can be evaluated by three means: the simplified base shear method, the response spectral analysis method based on mode superposition, and the time-history method. The horizontal seismic influence coefficient, α, based on the elastic acceleration response spectrum, is introduced to determine the magnitude of earthquake action for structures in elastic and elasto-plastic design phases. The curve of seismic coefficient, α, is shown in Fig. 1. The maximum seismic coefficient, $\alpha_{\max}$, is determined by Table 2. The design characteristic period of the ground motion, T_g, is defined by Table 3. From Fig.1 it can be seen that the long period structures and ones with different damping ratios are considered.

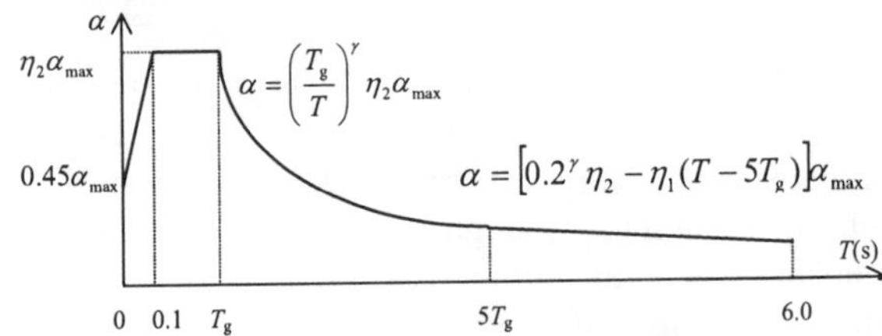

Fig. 1 Curve of seismic influence coefficient

Table 2. The maximum seismic coefficient $\alpha_{\max}$

Earthquake considered	6	7	8	9
Frequent	0.04	0.08 (0.12)	0.16 (0.24)	0.32
Rare	/	0.50 (0.72)	0.90 (1.20)	1.40

Table 3. The design characteristic period T_g (s)

Design earthquake group	Site class			
	Ⅰ	Ⅱ	Ⅲ	Ⅳ
Group 1	0.25	0.35	0.45	0.65
Group 2	0.30	0.40	0.55	0.75
Group 3	0.35	0.45	0.65	0.90

The accidental torsion effects of earthquake action shall be computed by considering the eccentricity of 5% of the story width perpendicular to the earthquake direction. For structures that have distinct irregularities of mass and/or stiffness distribution, the earthquake effects of the two orthogonal directions shall be combined on a square root of the sum as Eq. (1) or Eq. (2), and the larger one must be used for structural design.

$$S=\sqrt{S_x^2+(0.85S_y)^2} \quad (1)$$

or $$S = \sqrt{S_y^2 + (0.85S_x)^2} \quad (2)$$

where, S is the earthquake effect of internal force or displacement; The subscript x or y represents the direction of earthquake acting onto the structure.

The structures shall be provided with a complete lateral-force-resisting system designed to resist the minimum story seismic shear force given by Eq. (3):

$$V_{Eki} \geq \lambda \sum_{j=i}^{n} G_j \quad (3)$$

where, G_j is the representative gravity load assigned to level j; λ is the coefficient of minimum story seismic shear force conforming to Table 4.

Table 4. The coefficient λ of minimum story seismic shear force

Seismic fortification intensity	7	8	9
Structure with distinct torsion or with fundamental period less than 3.5s	0.016 (0.024)	0.032 (0.048)	0.064
Structure with fundamental period larger than 5.0s	0.012 (0.018)	0.024 (0.032)	0.040

Note: Use straight-line interpolation for the values when structure's fundamental period is between 3.5s and 5.0s.

3.3 *Basic stipulations for structural design*

3.3.1 *Design method*

The limit state design method based on the probability theory is employed. The ultimate capacity limit state and the serviceability limit state must be verified for all tall building structures and portions thereof.

3.3.2 *Structural systems*

The frame structures, frame-shear wall structures, shear wall structures and tubular structures should be used.

The column-slab with shear wall system and its design requirements are included in the new version. For seismic design of this kind of system, the shear walls along the direction considered shall be designed to totally resist the story shear force, story-by-story, induced by elastic horizontal earthquake in the same direction, and all columns shall be designed to bear the extra 20 percent of the story shear force.

The single-spanned frame structure should not be applied in seismic regions because of the severe damages in earthquakes.

3.3.3 *Building height limits*

All tall buildings are classified into Height Grade A and Height Grade B buildings based on the building height measured from the grade to the top level of main roof. According to the structural system, material and seismic intensity, the applicable maximum building height is limited for Height Grade A and Height Grade B building respectively. For Height Grade B buildings, the structural regularity and the computational and detailing requirements are stricter.

3.3.4 *Structural analysis*

Action effects on whole structure and parts thereof shall be determined by methods of structural analysis that take into account equilibrium, general stability, geometric compatibility and, in certain circumstances, both short- and long-term material properties.

The total lateral action shall be distributed to the vertical elements of the lateral-force-resisting system in proportion to their rigidities considering the rigidity of the horizontal diaphragm. For some buildings with irregular floor diaphragms, such as floor with narrow-necked configuration or large openings, the in-plane deformation of the diaphragm should be considered.

For Height Grade B buildings and complicated buildings stated in chapter 10, both elastic response spectral method and time-history method shall be employed for overall structural analysis using at least two kinds of structural software with different mechanical models.

3.3.5 *Load combinations*

The tall building structures and portions thereof shall be designed to resist the most critical effects from all combinations of factored loads by Eq. (4) for non-seismic resistant buildings, or by both Eq. (4) and Eq. (5) for seismic resistant buildings.

$$S = \gamma_G S_{Gk} + \psi_Q \gamma_Q S_{Qk} + \psi_w \gamma_w S_{wk} \quad (4)$$

$$S = \gamma_G S_{GE} + \gamma_{Eh} S_{Ehk} + \gamma_{Ev} S_{Evk} + \psi_w \gamma_w S_{wk} \quad (5)$$

where, S is the combined effect of actions considered; γ is partial factor for the action that is larger than or equal to 1.0; ψ is combination coefficient that is less than or equal to 1.0. The partial factors and combination coefficients shall take the specified values defined in chapter 5 for different design conditions and different limit states.

3.3.6 *Structure regularities*

The requirements of structural regularity are revised for both plan layout and vertical arrangement of structural members. There are limitations for plan configurations that contain reentrant corners, convex corners, and/or diaphragms with abrupt discontinuity or variations in stiffness such cutouts or openings. Some of the plan limitations are shown in Fig. 2 and Table 5.

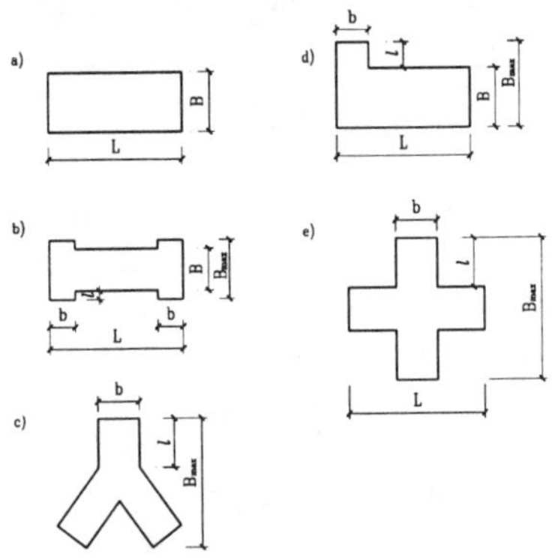

Fig. 2 Building plans and dimensions

Table 5. Dimensional Limits for plan configuration

Seismic intensity	*L/B*	L/B_{max}	*l/b*
6、7	≤6.0	≤0. 35	≤2. 0
8、9	≤5. 0	≤0. 30	≤1. 5

The torsional irregularity shall be considered to exist when the maximum story drift, computed including accidental torsion, at one end of the structure transverse to an axis is more than 1.2 times the average of the story drifts at the two ends of the structure. The maximum story drift shall not be more than 1.5 times the average of the story drifts for Height Grade A buildings or 1.4 times for Height Grade B buildings.

In seismic region, the ratio of the torsional fundamental period to the translational fundamental period shall not more than 0.9 for Height Grade A buildings or 0.85 for Height Grade B buildings.

To avoid soft story within tall building structure, the story stiffness should not be less than 70 percent of that in the story above or 80 percent of the average stiffness of the three stories above.

Besides the soft story, the weak story in the building is considered one kind of vertical structural irregularities. A weak story is one in which the story lateral strength is less than 80% of that in the story above. To control the discontinuity in capacity, the story lateral strength is not allowed to less than 65% and 75% of that in the story above for Height Grade A and B buildings respectively.

3.3.7 *Story drift limits*

The lateral displacement caused by wind load and/or earthquake is limited by controlling the story drift angle. For elastic analysis, when the building height is equal to or larger than 250m, the allowable elastic story drift ratio shall be taken as 1/500. The allowable elastic story drift ratio for buildings with height lower than or equal to 150m are listed in Table 6. For building with height between 150m and 250m, the allowable elastic story drift ratio may be straight-line interpolated.

Table 6. The allowable elastic story drift ratio [θ_e]

Structure system	[θ_e]
Frame	1/550
Frame-shearwall, Frame-corewall, Column-slab with shearwall	1/800
Shearwall, Tube in tube	1/1000
Story with discontinuity in vertical lateral-force-resisting members	1/1000

For Importance Class A buildings, some of the Importance Class B buildings and seismically isolated structures, the elasto-plastic analysis shall be required to evaluate the plastic story drift and therefore to search for the weak floors of the structure considered. The allowable elasto-plastic story drift ratios for different structural systems are listed in Table7.

The static push-over method and dynamic time-history method can be employed to do this for all kinds of structures, while the simplified method shall be used only for structures not higher than 12-story and with uniformly distributed story mass and lateral stiffness.

Table 7. The allowable elasto-plastic story drift ratio [θ_p]

Structure system	[θ_p]
Frame	1/50
Frame-shearwall, Frame-corewall, Column-slab with shearwall	1/100
Shearwall, Tube in tube	1/120
Story with discontinuity in vertical lateral-force-resisting members	1/120

3.3.8 *Comfortableness*

The requirements for comfortableness under the action of wind load are included. For building with height larger than 150m, the wind induced vibrating acceleration at top level is limited to $0.15m/s^2$ for residential building or $0.25m/s^2$ for public one.

3.3.9 *Seismic design grade of RC members*

According to the seismic intensity, structural system and building height, all reinforced or steel reinforced concrete members is divided into five seismic grades, namely SA, A, B, C and D respectively, for structural seismic design. The seismic design grade SA is specially defined in JGJ-2002 for members in Height Grade B buildings and some critical members in complicated tall buildings, that need stricter computational and detailing requirements.

3.3.10 *Overall stability*

Tall building structures shall be designed properly so that the level-by-level stiffness to weight ratios are reasonable. If the weight of the structure is high in proportion to the lateral stiffness, the contributions from the second-order effects, i.e. P-Delta effects herein, are highly amplified and, under certain circumstances, will eventually introduce structure instability. According to the lateral deformation characteristics of tall building structures, the approximate simplified approach to assure and control the overall stability of the structure is put forward, as shown by Eq. (6) for bending-type-deformation structure or Eq. (7) for shear-type-deformation structure. If the lateral stiffness of a structure does not conform to the requirement by Eq. (8) for bending-type-deformation structure or Eq. (9) for shear-type-deformation structure, the P-Delta effects shall be taken into account both in elastic and elasto-plastic design phases. Where the P-Delta effects need to be considered, the approximate formulas of amplifying coefficients for the displacements and internal forces of the structure are provided respectively as alternative.

$$EJ_{\mathrm{d}} \geq 1.4H^2 \sum_{i=1}^{n} G_i \tag{6}$$

$$D_i \geq 10 \sum_{j=i}^{n} G_j / h_i \quad (i = 1, 2, \cdots, n) \tag{7}$$

$$EJ_{\mathrm{d}} \geq 2.7H^2 \sum_{i=1}^{n} G_i \tag{8}$$

$$D_i \geq 20 \sum_{j=i}^{n} G_j / h_i \quad (i = 1, 2, \cdots, n) \tag{9}$$

where, EJ_{d} is the effective elastic lateral stiffness along the main axis direction considered; H is height of the structure; G_i and G_j is the design value of the representative gravity load of level i and level j respectively; h_i is the story height of level i; D_i is the effective elastic lateral stiffness of level i, which can be evaluated by the story shear and story drift; and n is the total levels of the structure.

3.3.11 *Overturning*

Every structure shall be designed to resist the overturning effects caused by lateral actions such as wind load and/or earthquake load specified in chapter 3. For building with the height to width ratio not more than 4.0, the "zero stress" area of subgrade right below the foundation underside shall not be more than 15 percent of the total area of the undersurface of the foundation. For building with the height to width ratio larger than 4.0, the "zero stress" of subgrade right below the foundation underside should not be permitted.

3.3.12 *Frame shear force adjustment of dual systems*

For dual structural systems, such as frame-shear wall structure and frame-corewall structure, the possibility of redistribution of earthquake forces between frames and shear walls shall be considered. Under the action of horizontal frequent earthquake, if the seismic shear force, $V_{\mathrm{f,i}}$, resisted by the framing columns in story *i*, is less than 20 percent of the total seismic base shear force, V_0, then the value of $V_{\mathrm{f,i}}$ shall be adjusted to the smaller one of $0.2V_0$ and $1.5V_{\mathrm{f,max}}$, where $V_{\mathrm{f,max}}$ is the maximum shear force resisted by framing columns among all stories before adjustment.

3.4 *The concepts for seismic design of reinforced concrete members*

The basic requirements for strength checking and detailing of structural reinforced concrete members of all kinds of tall building structures shall conform to the stipulations of the national code for design of concrete structures [5].

The seismic strength design of RC members is based on the combined effects as defined by Eq. (5) and the frequent earthquake action is assumed. To obtain the specified behaviors under the action of moderate earthquake and rare earthquake, the computational and detailing measures must be taken according to the seismic design grade of the member considered. The basic seismic design concepts of "strongest joints, stronger columns and weaker beams" and "stronger shear resistant strength and weaker bending resistant strength" are employed. To meet the design concepts, some of the combined forces of members computed by Eq. (5) shall be adjusted to a higher level according to the specified seismic grades of members. For example, to comply with the stipulations of the specification, the bending moment and shear force of a framing column shall be amplified by multiplying the proper coefficient listed in Table 8 and Table 9 respectively.

Table 8. The amplifying coefficient for combined moment of frame columns

Seismic design grade		SA	A	B	C
Bottom section of first fl.	Corner column	1.98	1.65	1.38	1.27
	Other column	1.80	1.50	1.25	1.15
Top and bottom sections of other fl.	Corner column	1.85	1.54	1.32	1.21
	Other column	1.68	1.40	1.20	1.10

Table 9. The amplifying coefficient for combined shear force of frame columns

Seismic design grade		SA	A	B	C
Bottom section of first fl.	Corner column	3.33	2.31	1.65	1.39
	Other column	3.02	2.10	1.50	1.27
Top and bottom sections of other fl.	Corner column	3.10	2.15	1.58	1.33
	Other column	2.82	1.96	1.44	1.21

3.5 *The special consideration for complicated tall building structures*

The complicated tall building structures herein include the following five types:

(a) The structure with structural transfer story in lower portion of the elevation;

(b) The structure with outriggers and/or stiffening peripheral belt at some story or stories;

(c) The structure with considerable staggered floor diaphragms;

(d) The structure with connecting structural part or portions between two or more separate towers in the upper part of the elevation; and

(e) The structure with two or more separate towers upon a relatively large plinth of one story or several stories.

All complicated structures above are vertically irregular with discontinuity of lateral stiffness. Some more effective measures, both computational and detailing, shall be taken into account for the critical members of the buildings. The following special requirements are some of the measures.

(1) Except the type (e), the complicated structures shall not be applied in seismic regions with design seismic intensity 9. The type (d) should not be applied for Height Type B buildings in seismic intensity 7 and 8 regions.

(2) The building height for type (c) shall not be higher than 80m and 60m in regions of seismic intensity 7 and 8 respectively.

(3) To avoid causing multiple soft and/or weak portions, more than two types of complicated structures shall not be applied within a tall building simultaneously in seismic intensity 7 and 8 regions.

(4) For partial column-supported shear wall structures, the level of structural transfer story should not be over the fifth story and third story in regions of seismic intensity 7 and 8 respectively.

(5) In seismic intensity 8 regions, the vertical earthquake effects shall be considered in design of the transfer members in type (a) structure and the connecting portion in type (d) structure.

(6) The seismic design grade of the critical members shall be upgraded by one level with the exception of grade SA.

(7) The combined internal forces of critical members shall be amplified by multiplying the specified coefficients according to the member's seismic design grade.

4. Conclusions

Some main provisions and revised contents of Chinese Specification JGJ3-2002 are introduced in this paper. It is incomplete, and misunderstandings of the original clauses may exist. So the paper may be a reference to promote the understanding of the Specification and the situation of Chinese codes.

References

1. Technical specification for concrete structures of tall building (JGJ3-2002). China Architecture & Building Press. Beijing, 2002. (In Chinese)
2. Technical specification for steel structure of tall buildings (JGJ99-98). China Architecture & Building Press. Beijing, 1998. (In Chinese)
3. Load code for the design of building structures (GB50009-2001). China Architecture & Building Press. Beijing, 2002. (In Chinese and English)
4. Code for seismic design of buildings (GB50011-2001). China Architecture & Building Press. Beijing .2001. (In Chinese)
5. Code for design of concrete structures (GB50011-2002). China Architecture & Building Press. Beijing .2002. (In Chinese and English)

Framed Structure

BEHAVIOR OF CONCRETE FRAME-COREWALL STRUCTURE WITH NON-CONTINUOUS EXTERIOR FRAME BEAMS

X.G. HE

Institute of Architectural Design & Research
Tsinghua University, Beijing 100084, China

A design project of a concrete frame-corewall tall building is presented in this paper. Due to the required inner balconies with 3-story-height at the two opposite corners on some floors, the corresponded edge beams cannot be connected to the two corner columns which are in a diagonal line. The layout affects the lateral stiffness of the structure. The behavior of the structure is compared with the ordinary frame-corewall structure which has continuous exterior frame beams; and in another way, comparison of behaviors of similar structures, in which inner balconies are with different-story-heights, is carried out. It is concluded that the lacking of some corner beams of the introduced structure has little influence on its overall behavior. The influence is only about 3~4%, and the seismic behavior of the structure is quite well. The related section of the Technical Specification for Concrete Structures of Tall Building of China is also discussed in this paper.

1. Introduction

For good mechanical behavior and being economical, the concrete structural system, composed of central shear wall core and the surrounding outside columns, is an ideal kind of structural system for the tall buildings such as offices, hotels, apartments, etc. [1]. Based on the amount of the surrounding outside columns and the relevant behavior, this structural system is divided into two types by the Technical Specification for Concrete Structures of Tall Building of China (JGJ3-2002) [2], i.e., frame-corewall system and tube-in-tube system, while the latter is suitable for more stories.

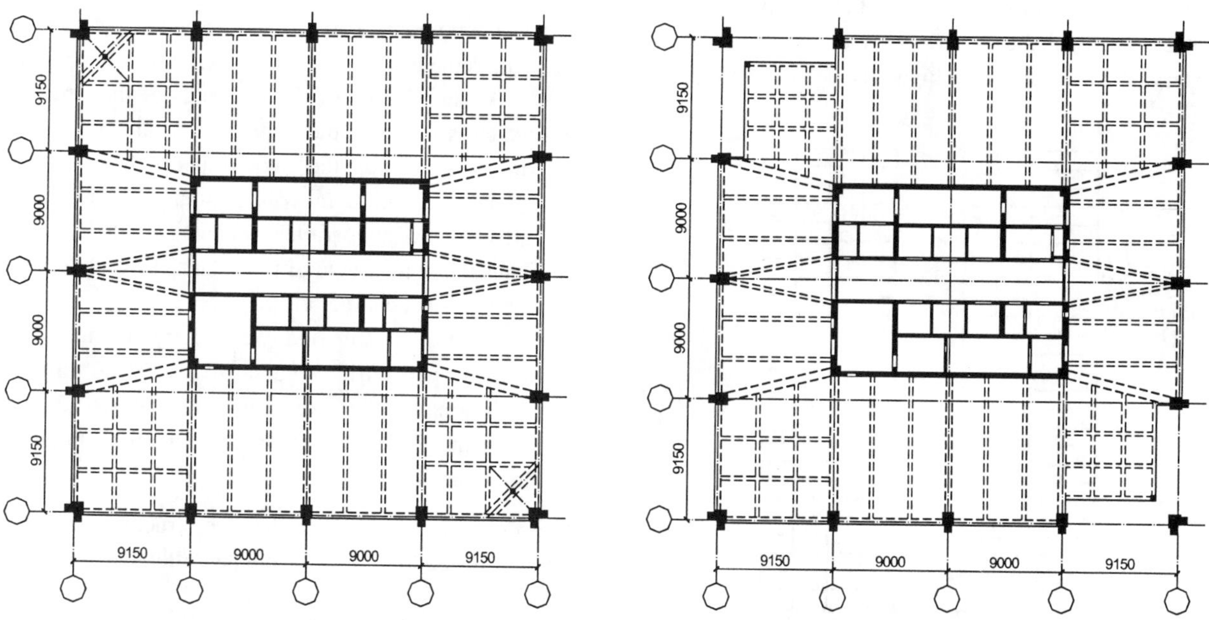

(a) typical floors with balconies

(b) typical floor without balconies

Figure 1. Plan layout of the typical floors (above 5th floor).

The Tsinghua Science & Technology Plaza is located in the Tsinghua Science Park in Beijing, with a total floor area of 190 thousand square meters. Its ground area is 128 by 124 m with 3 levels and 17 m depth below ground. Above the ground, it consists of four office buildings with 25 stories and 99.9 m height, and a 4 stories affiliated building. The tall buildings are separated from the adjacent 4-story affiliated building to avoid collision which might be caused by earthquake vibration. The concrete frame-corewall system is adopted for the four tall buildings, and the one on the northwest of the ground is introduced in this paper. The introduced building has a plan of 36.3 by 36.3 m, where the plan of the core is 18.7 by 14.5 m and the column-to-column span is 9 (9.15) m. Its typical plan layouts, with and without inner balconies, are shown in Figure 1. This building is different from those ones adopted the traditional frame-corewall system. Its upper 21 stories are divided into 7 vertical elements. Due to the requirement of architects, at each vertical element, which is composed of 3 stories, the frame beams connected to the two opposite corner columns of the 2 upper floors are cancelled, and the related slabs are lost, as shown in Figure 1b. This arrangement can form two inner balconies with 3-story-height at the lower floor for each vertical element. The schematic elevation of its exterior frame is shown in Figure 2a.

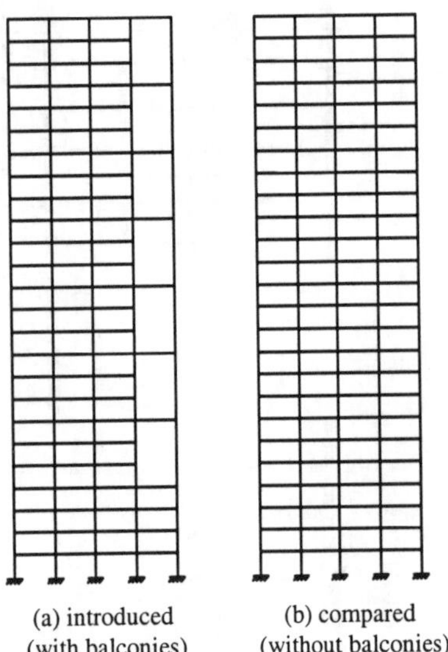

Figure 2. Schematic elevation of the exterior frames.

The L-type corner columns with the overall size of 1500 by 750 mm are keeping constant cross section from bottom to top. The section size of the other columns is 1500~1100 by 750 mm, the thickness of the corewall is 500~200 mm, and the concrete grade is from C60 to C30, all of them are changing from the ground to the top to keep a well-distributed decrement of the lateral stiffness along the vertical direction. From the 5th floor, two 400 by 400 mm concrete columns are set at corners of the L-type balconies to support the upper slabs, as shown in Figure 1. The layout of the lower 4 floors is just like that shown in Figure 1a except that no 400 by 400 mm columns are set.

2. Structural Regularities

The earthquake intensity degree is 8 (0.20gal) in Beijing. According to the requirements of the relevant Sections of the Code for Seismic Design of Buildings of China (GB50011-2001) [3] and the Technical Specification for Concrete Structures of Tall Building of China (JGJ3-2002), the structural regularities of the introduced building are analyzed in the first, which is also the requirement of the Uniform Building Code of USA (1997) [4]. The following conclusions are obtained from the computation and calculation in the preliminary design stage:

1. The maximum story drift at one end of the structure transverse to an axis is about 1.1 times the average of the story drifts of the two ends of the structure, which is less than the requirement value of 1.2. In the same time, the requirements of re-entrant corners, diaphragm discontinuity, out-of-plane offsets, and nonparallel systems are all satisfied. Thus the plan structural irregularities do not exist in the introduced structure.
2. There are no soft and weak story, no mass and vertical geometric irregularities, and no in-plane discontinuity in vertical lateral-force-resisting element. These indices show no vertical structural irregularities.

The above results indicate that the structural layout of the introduced building is regular, although the frame beams at floors without balconies (6~7th, 9~10th, 12~13th, 15~16th, 18~19th, 21~22nd, 24~25th) are not closed around the edges of these floors. Its dynamic parameters also illustrate the same trend. The periods and story drift ratios of the introduced building are listed in Table 1.

Table 1. The periods and story drift ratios of the introduced building and the compared one.

	Period (sec)			Story drift ratio	
	1st(NS)	2nd(tor.)	3rd(EW)	(EW)	(NS)
Introduced	2.66	1.68	1.66	1/1276	1/849
Compared	2.65	1.72	1.67	1/1278	1/911

The conceptual earthquake-resistant design of the introduced building satisfies the requirements of plan and vertical structural regularities. However, it seems that its plan layout might not satisfy the requirement of Section 9.2.4 of the Technical Specification for Concrete Structures of Tall Building of China (JGJ3-2002), i.e., "frame beams must be set between the exterior columns of frame-corewall structure". From the viewpoint of the author, this Section should be understood as follows,

1. This section is stipulated just for the structural system of corewall plus simple slab-column system, which aims to strength the overall stiffness and the anti-earthquake capability of the simple slab-column part. This point can be obviously drawn from the commentary of Section 9.2.4, "when simple slab-column system is adopted for the frame corewall structure, frame beams must be set between the exterior columns at each floor". The frame part of the introduced building does not adopt the simple slab-column system, and its overall stiffness is much larger than that of the simple slab-column system.
2. There are frame beams in each floor of the introduced building, in which the exterior frame beams at floors of 1~5th, 8th, 11th, 14th, 17th, 20th, 23rd, and the top, are closed around the floor edges. The only difference is that frame beams do not form closed ones at edges of the other floors. This layout cannot be regarded as not coinciding with the requirement of Section 9.2.4.
3. Cancellation of the frame beams connected to the two opposite corner columns at floors without balconies might be possible to affect the lateral stiffness of the structure to some extent. Although the influence would not be significant, it should be studied carefully. Appropriate evaluation of the influence should be conducted to assure the seismic behavior of the building.

3. Comparative Analyses

Two kinds of comparative analytical models are established to study the influence on lateral stiffness of the structure caused by the lacking of closed edge frame beams at floors without balconies of the introduced building.

3.1 *Comparison of the introduced building with the one without cancellation of the corresponded beams*

A compared structural model similar to the introduced one is established, in which the differences are that the cancelled frame beams are connected to the two opposite corner columns and the corresponded slabs lost at the typical floor without balconies of the introduced building (Figure 1b) are restored. The schematic elevation of exterior frame of the compared model is shown in Figure 2b, while Figure 2a is that of the introduced one.

The comparative results illustrate that there is little difference of the overall lateral stiffness between the two structural models, as shown in Figure 3, where only the lateral stiffness in NS direction is given for its less lateral stiffness of the corewall in this direction. The reason is obviously that the percentage of the lateral stiffness of the corewall is over 70% of the overall lateral stiffness, which leads the influence of the lateral stiffness change of the frame part being very little. The average lateral stiffness of the frame part of each story of the introduced building (7.2×10^5kN/m) decreases only 12% than that of the compared one (8.2×10^5kN/m), and the decrement is averagely only 3~4% of the overall lateral stiffness at each story. The conclusion could be drawn from the results that the canceling of the frame beams connected to the two opposite corner columns and the losing of the corresponded slabs have little influence on the overall stiffness of the introduced building. The periods and story drift ratios of the compared model are also listed in Table 1.

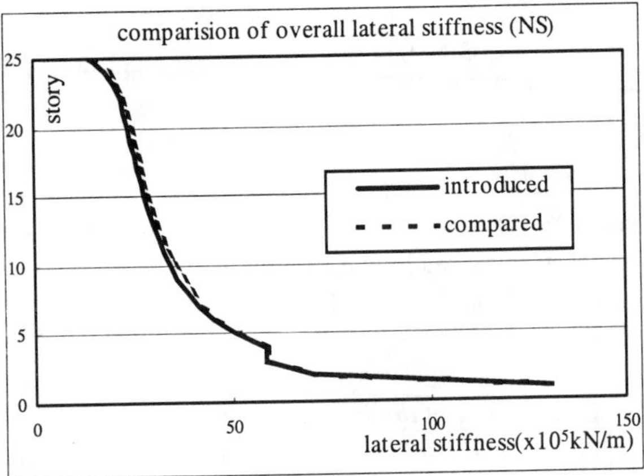

Figure 3. Comparison of the lateral stiffness.

3.2 *Comparison of buildings with inner balconies of different-story-height*

A series of structural models similar to the introduced building, including typical floors with and without balconies as shown in Figure 1, are established. The difference is that the 24 stories above the 2nd floor are divided into different vertical elements. The first model is like a traditional frame-corewall structure with closed exterior frame beams at each floor, just like that in Figure 2b. In other 7 cases, the vertical elements are with inner balconies of 2, 3, 4, 6, 8, 12, 24-story-height, as shown in Figure 4a~g. In the last case, the L-type corner columns above 2nd floor at the location of balconies as appeared in the other cases are cancelled, as shown in Figure 4h.

The comparative analyses of the 9 cases show the results similar to that of Section 3.1. Like in Section 3.1, only the lateral stiffness in NS direction is given. The difference of the average lateral stiffness of each story (above 2nd floor) of the 9 cases is very little, as listed in Table 2, and their distributions of lateral stiffness in the vertical direction are just like that shown in Figure 3. The reason is the same, the percentage of the lateral stiffness of the corewall is about 80% of the overall lateral stiffness, which leads the influence of the lateral stiffness change of the frame part being very little. Two extreme cases, the traditional frame corewall system, 2b* without inner balconies (its difference from Figure 2b is that there is no 400 by 400 mm columns at the location of the balconies), and 4h without L-type columns, can be compared to obtain an obvious result. The difference of their average lateral stiffness of each story of the frame part (5.96×10^5kN/m and 4.86×10^5kN/m) is about 18 percentage, which is only 3~4% of the overall lateral stiffness of each story. The same conclusion could be also drawn from the results that the canceling of the frame beams connected to the two opposite corner columns and the losing of the corresponded slabs has little influence on the overall stiffness.

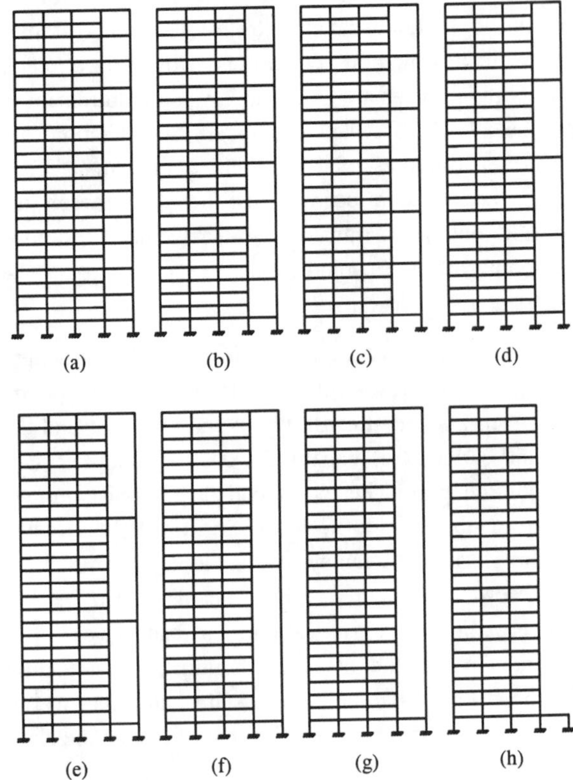

Figure 4. Schematic elevation of the exterior frames of the buildings with balconies of different-story-height.

Table 2. Average lateral stiffness per story (above 2nd floor) of the different cases (10^5kN/m).

	Frame part	Overall	Percentage of frame
2b*	5.96	32.59	22.1%
4a	5.61	32.20	21.2%
4b	5.39	31.86	20.6%
4c	5.28	31.70	20.3%
4d	5.15	31.40	20.0%
4e	5.09	31.27	19.9%
4f	4.99	30.75	19.9%
4g	4.93	30.47	19.5%
4h	4.86	30.23	19.0%

Table 3. The periods and story drift ratios of the 9 comparative models.

	Period (sec)			Story drift ratio	
	1st(NS)	2nd(tor.)	3rd(EW)	(EW)	(NS)
2b*	2.68	1.69	1.64	1/1363	1/848
4a	2.66	1.67	1.62	1/1371	1/857
4b	2.69	1.67	1.63	1/1304	1/802
4c	2.70	1.67	1.63	1/1216	1/792
4d	2.70	1.66	1.63	1/1081	1/746
4e	2.70	1.66	1.63	1/918	1/677
4f	2.71	1.66	1.63	1/522	1/528
4g	3.41(tor.)	3.38(tor.)	2.70(NS)	1/263	1/194
4h	2.68	1.61	1.61	1/1378	1/845

The periods and story drift ratios of the 9 comparative models are listed in Table 3. It could be found that all the models, except case 4g, demonstrate good dynamic behaviors. For case 4g, due to the too slenderness of the L-type corner columns, the torsion index and the story drift ratio are not well enough to satisfy the limitation of Code. When the corner columns above the 2nd floor of case 4g are abandoned, good behavior could be obtained, as that of case 4h listed in Table 3.

4. Conclusion

It could be concluded from the analysis and comparative studies in this paper, for tall building adopted the concrete frame corewall structural system, some exterior frame beams not being continuous, even not connecting to the corner columns, has some influence on the lateral stiffness of the structure. However, this influence is not significant to the dynamic behavior of the structure. For the project introduced in this paper, the decreasing of lateral stiffness is only about 3~4% of the overall lateral stiffness of the structure, about 10% of that of the frame part.

The tall building introduced in this paper has been finished and has been in service in June 2005.

References

1. D. H. Liu and C. R. Yang, *Optimistic Structural Design of Tall Buildings.* China Architectural & Building Press, Beijing (1996).
2. JGJ3-2002, *Technical Specification for Concrete Structures of Tall Building,* China Architectural & Building Press, Beijing (2002).
3. GB50011-2001, *Code for Seismic Design of Buildings.* China Architectural & Building Press, Beijing (2001).
4. *Uniform Building Code.* International Conference of Building Officials, California (1997).

TEST RESEARCH ON BEAM-WALL SEMI-RIGID JOINT IN SUPER HIGH-RISE HYBRID STRUCTURE

DEYUAN ZHOU
State Key Laboratory for Disaster Reduction in Civil Engineering
Tongji University, 1239 Siping Road, Shanghai 200092 , China

AIMIN LIU XUAN ZOU HAO ZHU
State Key Laboratory for Disaster Reduction in Civil Engineering
Tongji University, 1239 Siping Road, Shanghai 200092, China

This paper puts forward the design model of beam-wall semi-rigid joint in super high-rise hybrid structure. Five groups of joint models are tested to analyze the mechanics performance of beam-wall joint under low cyclic reverse loading. Test results manifest that this semi-rigid joint has greater bearing capacity and better seismic behavior, thus it can make up the deficiency of hinge joint's seismic behavior. Based on the tests, this paper adopts ANSYS program to do nonlinear finite element analysis of test models, in which quite a lot of nonlinear traits are put into considerations, such as the crack and crush of concrete, the shear force transferring function when the concrete crevice open and close, bonding and slippage between the joint's anchoring connector and concrete, etc. The results of calculation and analysis fit the test quite well.

1. Introduction

Compared with traditional concrete structure and steel structure, hybrid structure has fine lateral resistance capacity and the advantage of relatively low cost. However, present knowledge on the seismic behavior of this kind of structure still cannot meet the needs of design and construction. In U.S.A and Japan, there are certain limits on the use of hybrid structure in high-rise buildings and super high-rise buildings in seismic areas.

Steel beam-concrete wall joint is one of the joints most widely used, and also is very important in super high-rise hybrid structure. It is crucial for fine cooperative work of interior tube and exterior frame. However, the design of this joint has various limitations. Firstly, for super high-rise buildings, the constructions of interior tube and exterior frame are often not coincident, as concrete tube employs sliding form construction method, while exterior frame begins being worked on after the construction of tube to certain height, which raises some limitation to joint construction. Secondly, one end of the joint is steel beam, while the other end is concrete shear wall. In order to design rigid joint, stronger construction measures on the joint should be adopted to ensure the joint rigidity. However, too big rigidity of the joint will have unfavorable effect on the bearing capacity of shear wall. What's more, as the great dead weight difference of concrete tube and exterior frame in super high-rise will lead to difference in settlement between interior tube and exterior frame, the beam-wall joint ought to have certain deformation capacity to minimize the joint's initial stress caused by settlement. For this reason, the joint is mostly designed into hinge point.

In America's Northridge earthquake of 1994 and Japan's Kobe earthquake of 1995, the steel beam-concrete wall joints in hybrid structure basically suffered from brittle failures, and most of the joints were destructed when the steel beam was even in elastic state. From then on, many scholars of various countries have started the research on steel beam-concrete wall joint. However, at present the research still mainly focuses on the situation of hinge joint.

The author holds that the research on beam-wall joint in hybrid structure cannot be limited to hinge joint, but should consider how to build up joint connection method with better seismic behavior if the construction condition and construction method are permitted.

2. Semi-rigid Joint Design Model

There are always certain deficiencies in joint connection method of steel beam-concrete shear wall in high-rise hybrid structure, no matter whether it adopts hinge connection or rigid connection. Based on the analysis of the advantages and disadvantages of these two connection methods, this paper puts forward semi-rigid joint design model, which is described in figure 1. This design model has the following characteristics:

Firstly, this kind of connection is somewhat between hinge connection and rigid connection. It can bear moment, but needn't be made into rigid connection. If necessary, by adjusting the size and arrangement of connector, it also can approach or even meet the needs of rigid joint.

Secondly, on the basis of shear connector in traditional hinge joint, anchoring connector whose length is designed according to the standard anchoring requirements of tension reinforcement is added. It functions as bearing moment and axial force transferred from the beam end. The reason why it is called anchoring connector is that its tensile capacity is mainly supported by the bonding force between connector and concrete. If there are embedded steel structure parts (embedded steel framed concrete beam or column) in the concrete wall where the joint lies, welding with it can be considered; if there is not, it only needs to be designed according to the standard anchoring length.

What's more, add built-up steel connecting plate (i.e. T-shape connecting plate frequently used in steel structure beam-column joint) to the above and below flanges of steel beam. Connect the connecting plate with steel beam by bolting, and with embedding steel plate by welding. Thanks to the connecting plate, the joint can bear moment to a certain extent. However, different from standard rigid joint or inserting the steel beam directly into concrete, it permits the joint rotate to some extent in steel beam plane through adjusting the constructing time and some constructional arrangements. It differs from the traditional rigid joint that it has rotation angle and can bear moment. It is similar to adding a rotating spring with certain rigidity in beam end of the joint, thus the steel beam's initial stress caused by constructing error and difference in settlement can be reduced in the process of construction. At the same time, the section height of steel beam is increased because of the addition of built-up steel plate connector. In this way, it can avoid the situation that beam end section of the joint would easily be the most fragile section when the floor concrete cracks under the seismic load.

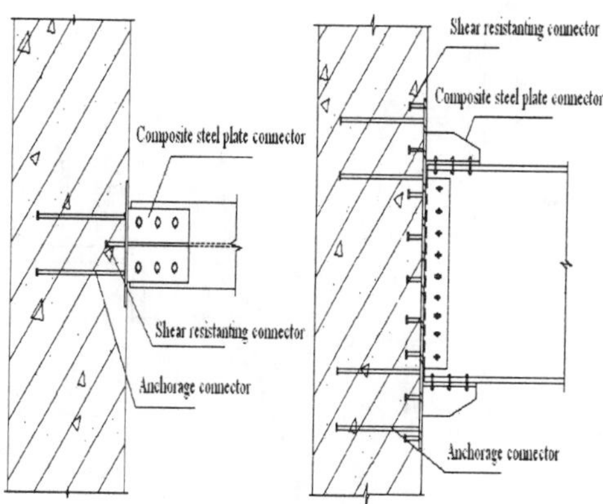

Figure 1. Beam-wall semi-rigid joint design model

3. Beam-wall Joint Seismic Behavior Test

In order to learn the seismic behavior of semi-rigid joint, five test models SRJ-01~SRJ-05 are designed, among which SRJ-01~SRJ-03 are semi-rigid joint models while SRJ-04~SRJ -05 are rigid joint models.

3.1 *Model design*

Taking into consideration of three beam-wall joints in different positions, three semi-rigid connection models are made: SRJ-01 stands for ordinary beam-wall joint; in SRJ-02, behind the shear wall in the place of joint, there is another shear wall which is vertical with and intersects that one; in SRJ-03 joint model, there is elevator or stairs opening under the joint. Meanwhile, because of the differences in position, these three models are different from each other in the way of making joint connector. The joint SRJ-04 and SRJ-05 are respectively corresponding with SRJ-01 and SRJ-02, these joints are constructed in the way that is same as the rigid joint in actual project.

According to the requirements of Specification of Testing Methods for Earthquake Resistant Building and actual condition of the laboratory, the scale of joint model is 1:2. The strength grade of model concrete is C60, the steels are Q235 carbon steel, the shear wall reinforcement adopts gradeII . A detail drawing of SRJ-01 can be seen in figure 2. The detail drawing and size of connector are indicated in figure 3 and table 1.

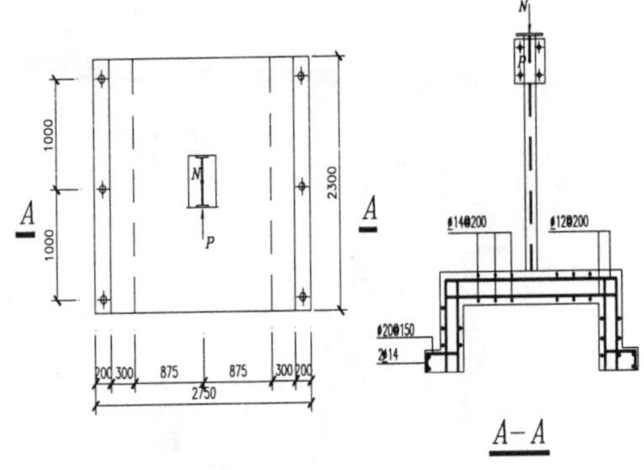

Figure 2. Detail drawing of model SRJ-01

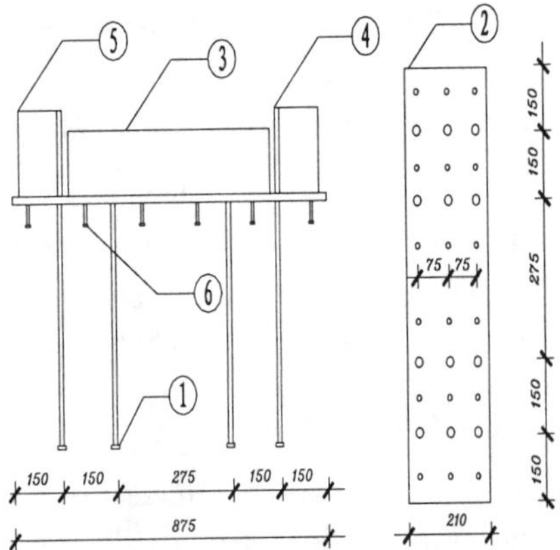

Figure 3. Detail drawing of SRJ-01 connector

Table 1. Size of joint connector and elastic modulus

NO.	size
①	Tensile connector, grade II reinforcement ,length 700mm
②	-210×875×6
③	-133×430×6
④	-150×300×10,4 row bolt-fixed
⑤	-90×300×10
⑥	Shear connector, rivet Φ9, length 120mm

3.2 *Loading installing and loading pattern*

The joint's load can be divided into two parts: vertical shear force at the end of beam and axial force N. In order to simulate the high axial press ratio in concrete wall of high-rise building, prestress is exerted on two ends of the concrete wall, thus the condition of concrete in joint area would be closer to real working condition. In this test, the total prestress exerted amounts to 12,000kN and the axis press ratio of concrete wall achieves 0.6.

In the formerly conducted tests, as the joint can't bear moment, the cyclic reverse loading are all in horizontal shear direction. In this test, the cyclic reverse loading is exerted in vertical shear direction. The horizontal shear loading increases gradually in step form. The cyclic reverse loading P is exerted by Schenck structure test system which transfers the counterforce to counterforce wall. The vertical loading is exerted by 4,000KN hydraulic jack, controlled by oil pump manually. The counter force of hydraulic jack is transferred to test pedestal by steel counterforce frame. The loading pattern is indicated in figure 5 and figure 6.

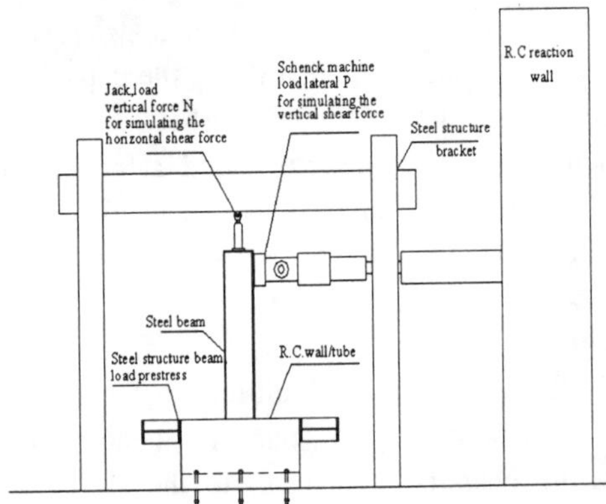

Figure 4. Mode loading installation drawing

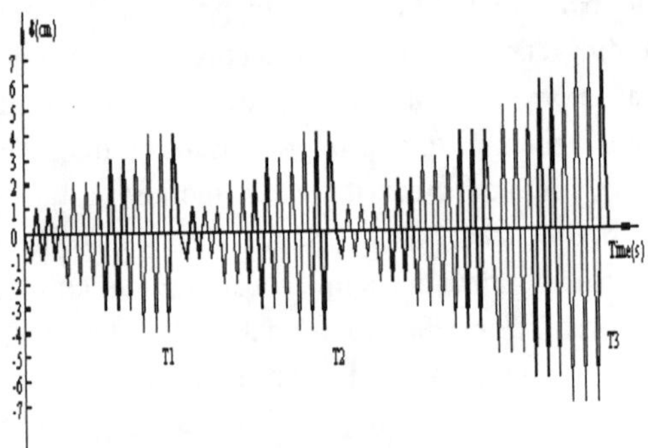

Figure 5. Vertical shear force P loading pattern

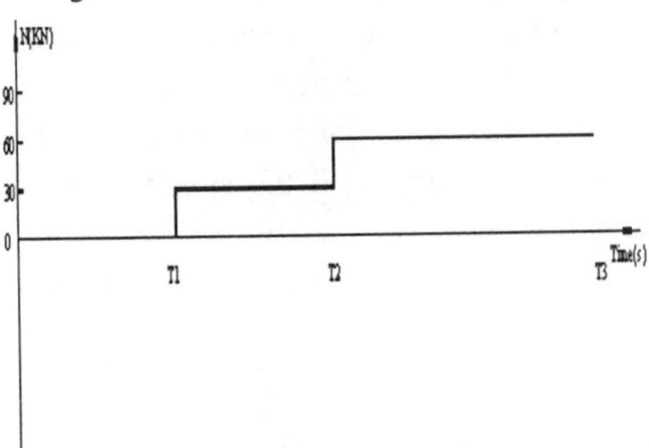

Figure 6. Horizontal shear force N loading pattern

3.3 *Test phenomenon description*

Semi-rigid joint: according to the need of actual constructing, there is certain residual (about 10 mm) in one direction of bolt hole in the connecting position between steel beam and embedding parts. When the horizontal force is firstly exerted, the trace of bolt slippage can be seen in force-displacement curve. When the displacement increases to 20mm or 30mm, we can see clearly the obvious increase of the curvature of horizontal force-displacement curve after the slippage has stopped. When the horizontal displacement achieves 60mm, steel embedded parts used to connect with steel beam begins to separate from concrete in the two ends, and yielding begins to emerge in the flange position of the root. With the enlarging of controlling displacement, the yielding extends to the place of web and also extends upwards. Meanwhile, it increases horizontally, yet the increasing range reduces gradually. But when the horizontal displacement achieves to 100mm, the steel embedded parts are pulled up. The details can be seen in figure 7.

Figure 7. Photograph of semi-rigid joint failure

Rigid joint: when the horizontal displacement is relatively small, the hysteresis loop region is very narrow. As the horizontal displacement reaches 30mm, yielding begins to emerge in the flange position of the root. As the horizontal displacement reaches 40mm, the yielding range of flange in beam root extends to the place of web. The connecting area between beam and concrete wall suffers severe deformation. The bearing capacity decreases, with failures of the joint. The details can be seen in figure 8.

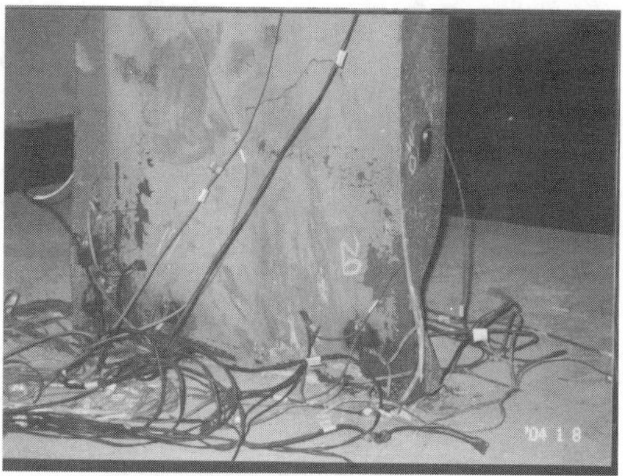

Figure 8. Photograph of rigid joint failure

3.4 *Analysis of the test result*

3.4.1 *Hysteresis loop and envelope curve*

Figure 9 and Figure 10 are the hysteresis loops of semi-rigid joint and rigid joint. From the figures we can see that because of the differences in construction and failure form, these two hysteresis loops have obvious distinctions. In the semi-rigid joint's hysteresis loop, it is obvious to see the effect of bolt's slippage. It also can be seen that there is a pretty long extension of semi-rigid joint after yielding, which proves that the joint has quite good ductility. On the other hand, the rigid joint's hysteresis loop is rather full. The joint has pretty good energy dissipation capacity.

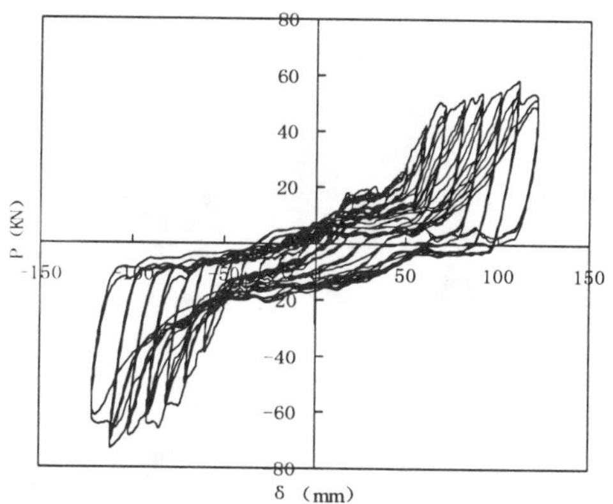

Figure 9. Hysteresis loop of semi-rigid joint

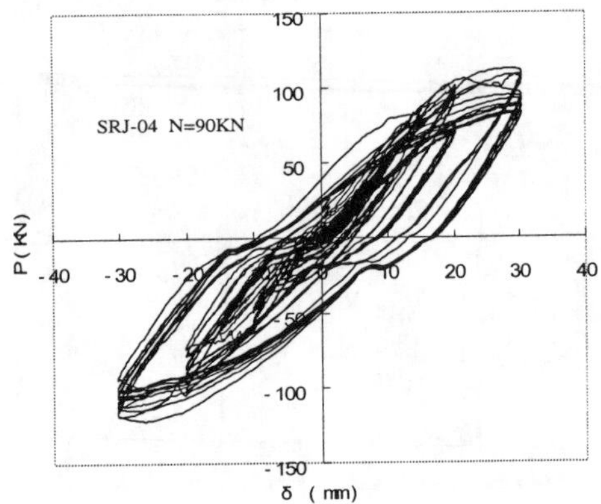

Figure 10. Hysteresis loop of rigid joint

Figure 11 and Figure 12 are the envelope curves of semi-rigid joint and rigid joint. In the semi-rigid joint envelope curve, the affecting elements caused by bolt's slippage are taken into consideration. It is evident that both yielding displacement and failure displacement of semi-rigid joint are larger than those of rigid joint.

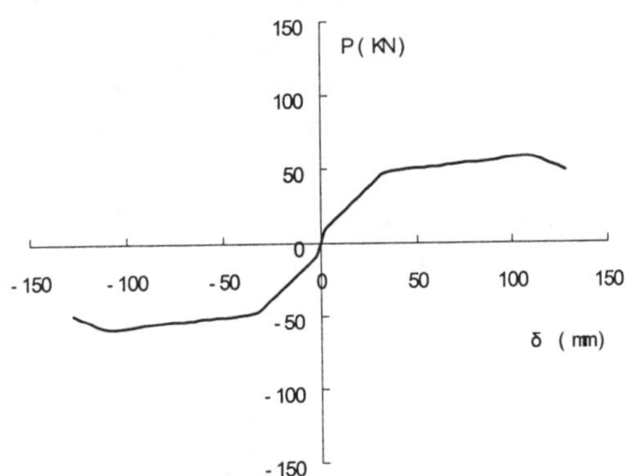

Figure 11. Envelope curve of semi-rigid joint

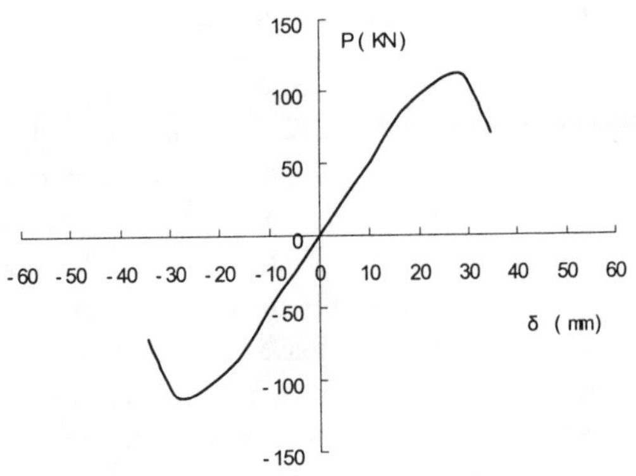

Figure 12. Envelope curve of rigid joint

3.4.2 *Bearing capacity of the joint*

Through analyzing the test result, it is found that in this test , the yielding moment of semi-rigid joint is about 47 % of rigid joint, while the ultimate moment 56%, which illustrates that semi-rigid joint has better bending resistance capacity.

3.4.3 *Ductility and energy dissipation of the joint*

Ductility is an important index to reflect the plastic deformation capacity and is also one of the indexes to judge the seismic behavior. Structure ductility is generally indicated by ductility factor which refers to the ratio of structure and member's maximum deformation value to the corresponding deformation value when the yielding emerges. The sample's energy dissipation capacity is another important index of its seismic behavior. In this paper, the sample's energy dissipation capacity is expressed by its equivalent viscosity damping coefficient he. From table 2, we can see that compared with rigid joint, semi-rigid joint has better ductility, but its energy dissipation capacity is not as good as that of rigid joint.

Table 2. Sample's ductility factor and equivalent viscosity damping coefficient

NO.	Name of sample	Pu(KN)	0.85Pu	δ y (mm)	δ u (mm)	μ	he
1	SRJ-01	57.8	49.13	18.4	125	6.79	0.108
2	SRJ-02	63.47	53.95	17.2	106.3	6.18	0.107
3	SRJ-03	63.84	54.26	16.1	92	5.71	0.103
4	SRJ-04	112.19	95.36	7.79	38.3	4.91	0.232
5	SRJ-05	110.11	93.59	7.87	37.9	4.82	0.216

4. Finite Element Nonlinear Analysis

4.1 *Element selecting*

In the computation and analysis of this paper, the element selecting is as the following: concrete-SOLID65, shear wall reinforcement-PIPE20, steel beam and connector-SOLID45, prestress on bolt clamping-PRETS179, bonding of steel plate with shear wall-CONTA174, bonding and slippage between reinforcement (rivet) and concrete-COMBIN39.

4.2 *Material parameter and finite element model*

In order to compare with the result of model test, the selecting of material parameter in this paper agrees with that in test model. Concrete: grade NO. C60, f_{cu} =61N/mm2 , E=3.6×104 N/mm2 , v = 0.18 , f_c' = 48.9 N/mm2 , f_r = 4.35 N/mm2 , β_t , 0.25(crevice open), 0.65(crevice close) ; steels: plate and section, Q235 , f_y = 235 N/mm2 , f_u = 375 N/mm2 , E = 2.06×105 N/mm2 , bolt, σ = 940 N/mm2 , σ_b = 1140 N/mm2 , E = 2.06 × 105 N/mm2 , grade II reinforcement, f_y = 310 N/mm2 , f_u = 335 N/mm2 , E = 2.0×105 N/mm2 , rivet: σ = 240 N/mm2 , σ_b = 450 N/mm2 , E = 2.06×105 N/mm2.

The built-up of finite element analytic model is also based on the five groups of models SRJ-01~STJ05 which are adopted in joint model test. The detailed drawing of test model SRJ-01 is showed in figure 13. Figure 14 is the corresponding finite element computation model.

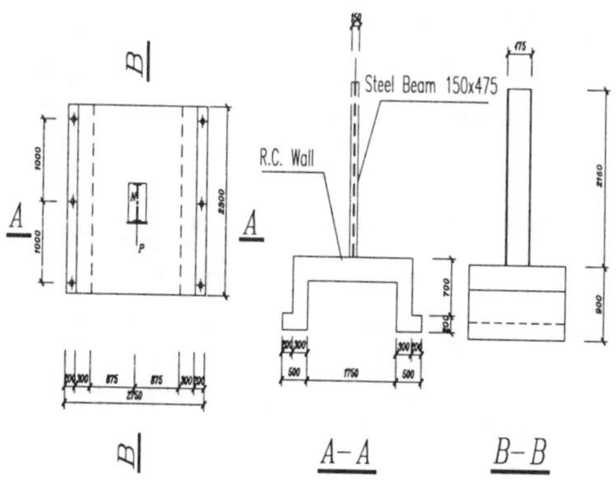

Figure 13. Detail drawing of model SRJ-01

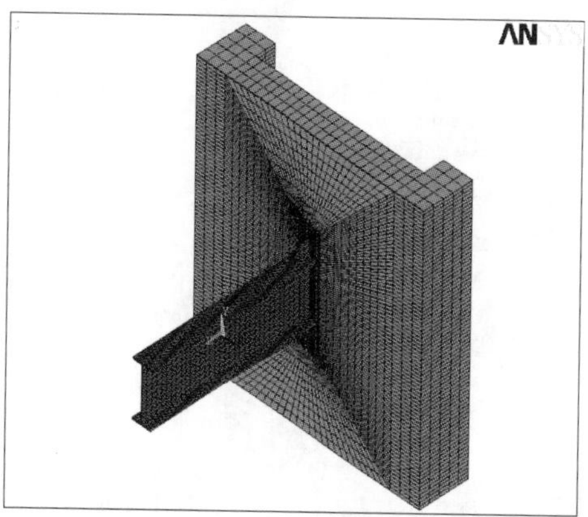

Figure 14. Finite element computation model corresponding to model SRJ-01

4.3 *Comparison with test result*

4.3.1 *Load displacement curve P-* δ

Figure 15 and figure 16 are comparisons between the testing load-displacement curve of model SRJ-01 and SRJ-04 with the computational load-displacement curve of computational analysis model JS-SRJ-01 and JS-SRJ-04.

From figure 15 to figure 16, we can see that the computational result is quite similar to the test result in general shapes, and the computational value before yielding and in the beginning stage of yielding is even very close to the testing value. But in the strengthening stage, especially when near the ultimate strength, there is a 15% ~ 20% difference between computational value and testing value.

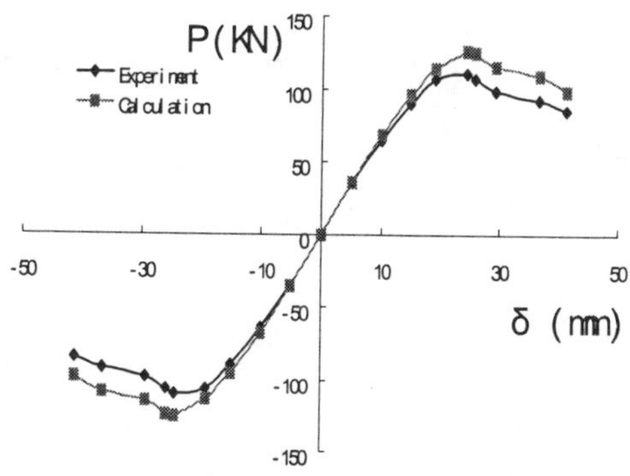

Figure 15. Comparison of SRJ-01 computational result and testing value

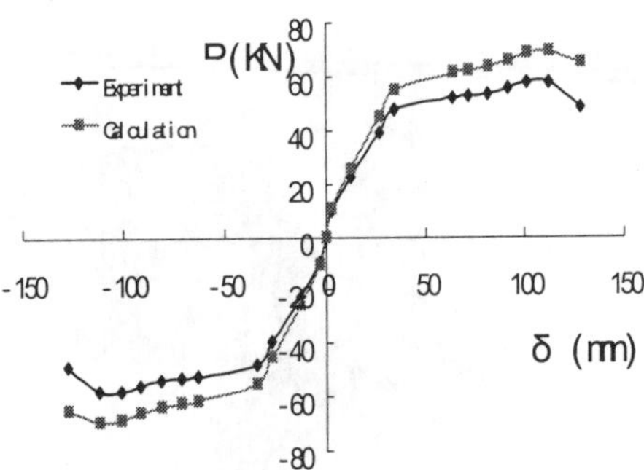

Figure 16. Comparison of SRJ-04 computational result and testing value

4.3.2 *Comparison of steel plate stress value*

In the test, stress test is carried out on the added steel plate connector 4 to find out its stress-strain development situation. Figure 17 is a comparison of stress-strain measuring point's testing result and computational result under ultimate load. This measuring point is in the connecting steel plate on one flank of SRJ-01. From this comparison, It is found that the computational result of ANSYS is rather close to the measured value. The error between them enlarges gradually with the closing of ultimate loading, which is quite similar to the situation in envelope curve.

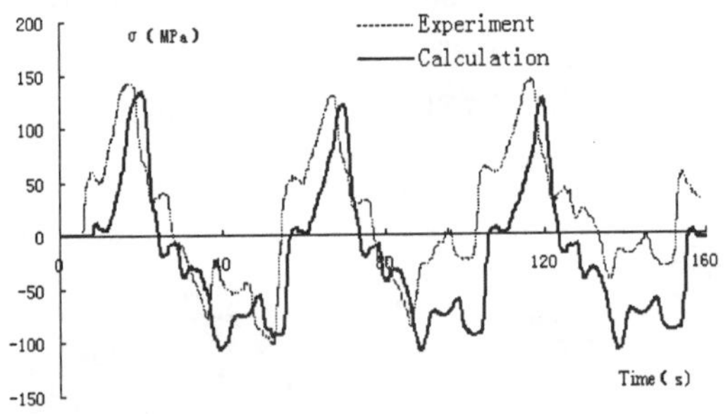

Figure 17. Comparison of testing value and computational value in SRJ-01 measuring point

4.3.3 *Joint ultimate load and ductility factor*

In table 3, a comparison is made between the testing value and computational value of joint ultimate load and ductility factor. From the compared result, it is obvious that both computational values are lager than testing value. Among the errors of ultimate load, error of rigid joints SRJ-04 and SRJ-05 are both quite small. The errors of ductility factor are all within 8%. In yielding displacement, there is no great difference between computational value and testing value, but computational value is greater than testing value in ultimate displacement.

From the comparison in envelope curve, joint's steel plate stress-strain curve, joint's ultimate load and ductility factor, it can be seen that the error between ANSYS computational value and testing value is very small in the initial loading stage. With the closing of ultimate load, the error enlarges gradually. But their general results are quite close to each other, which prove that the adopted elements and material model are suitable, and the computational results are quite accurate.

5. Conclusions

Through model test and finite element nonlinear analysis, it can be found that beam-wall semi-rigid joint model in super high-rise hybrid structure, which is put forward in this paper, has the following advantages:

1. It is feasible in constructing. By its construction, it can solve the problems caused by rigid connection.
2. Semi-rigid joint not only has quite good bearing capacity, but also has fine ductility and energy dissipation capacity.
3. Through the comparison with testing result, it can be found that to make analysis of beam-wall joint in super high-rise hybrid structure by ANSYS program is feasible. The computational result fit the testing value quite well.

Table 3. Comparison of computational value and testing value of joint's ultimate load and ductility factor

NO.	Name of sample	Pu (KN)			μ		
		Testing value	Computational value	Error	Testing value	Computational value	Error
1	SRJ-01	57.8	69.36	16.7%	6.79	7.33	7.4%
2	SRJ-02	63.47	75.35	15.8%	6.18	6.31	2.1%
3	SRJ-03	63.84	72.35	11.8%	5.71	5.89	3.1%
4	SRJ-04	112.19	126.76	11.5%	4.91	5.33	7.9%
5	SRJ-05	110.11	125.52	12.3%	4.82	5.11	5.7%

Reference

1. Bahram M. Shahrooz (2004a), Jeremy Deason T, Gokhan Tunc. Outrigger beam-wall connections I : Component testing and development of design model. Journal of Structural Engineering/ASCE, Vol. 130, No. 2, February 2004, pp. 253-261
2. Bahram M.Shahrooz (2004b), Gokhan Tunc , Jeremy Deason T. Outrigger beam-wall connections II: Subassembly testing and further modeling enhancements. Journal of Structural Engineering/ASCE, Vol. 130, No. 2, February 2004, pp. 262-270
3. Chao-Kuang Ku (1999), Theodor Krauthammer. Numerical assessment of reinforced forced concrete knee-joints under explosively applied loads. ACI Structural Journal / Vol. 96, No. 2, March-April 1999, pp. 239-247
4. Isao Nishiyama (2004), Toshiaki Fujimoto, Toshiyuki Fukumoto, Kenzo Yoshioka. Force-deformation response of joint shear panels in beam-column moment connections to concrete-filled tubes. Journal of Structural Engineering, Vol. 130, No. 2, February 2004, pp. 244-252
5. G.A. Rassati (2004), R.T. Leon, S. Noe. Component modeling of partially restrained composite joints under cyclic and dynamic loading. Journal of Structural Engineering
6. /ASCE, Vol. 130, No. 2, February 2004, pp.343-351
7. Liu Aiming, (2004). Study of beam-wall joint in super high-rise structure, Doctor Dissertation paper of Tongji University, 2004.9 (in Chinese)
8. M. Kattuer (2000) and M.Crisinel. Finite element modeling of semi-rigid composite joints. Computers and structures, Vol. 78, No. 1-3, 2000, pp. 341-35

APPLICATION OF STRUT-AND-TIE METHOD ON OUTRIGGER BRACED CORE WALL BUILDINGS *

R.K.L. SU

Department of Civil Engineering, University of Hong Kong, Pokfulam Road, Hong Kong, China

P.C.W. WONG

Ove Arup & Partners Hong Kong Ltd, 5/F Festival Walk, 80 Tat Chee Avenue, Kowloon Hong Kong, China

A.M. CHANDLER

Department of Civil Engineering, University of Hong Kong, Pokfulam Road, Hong Kong, China

Outrigger braced core wall structure often forms the major lateral load-resisting system to provide the required moment resistance and lateral stiffness for high-rise buildings and skyscrapers. Due to practical considerations, outrigger braces are usually made of structural steel and cast into the reinforced concrete core walls for the buildings in Hong Kong. The correct structural analysis and design of the connections are vitally important to enable it to carry huge amount of internal forces. This paper aims to enhance practicing engineers to understand the general structural behaviour of outrigger braced core wall system. Strut-and-tie method is applied to analyze the whole lateral structural system. The complete load transfer mechanism between the outrigger brace and the core wall is displayed. Many practical concerns in design including the structural behaviours of different configurations of outriggers, the effect of openings through the core wall adjacent to the outrigger brace, the arrangement of shear studs on outrigger brace and the shear link arrangement in core wall are briefly discussed. Some common deficiencies in numerical modelling of this system are pointed out and a set of recommendations related to finite element modelling is given for reference.

1. Introduction

Outrigger braced system has been widely employed to reduce the core wall moment and to increase the lateral stiffness of the skyscrapers. This system usually comprises of a core wall connected at one or more levels to the peripheral columns by horizontal cantilevers. When compared with its competitive alternative of tube-in-tube design concept, outrigger system is usually preferable for skyscrapers as larger window openings on the facades can be provided which could substantially boost up the pricing of rental property. Since the completion of the first building using outrigger system (namely Place Victoria Building) in Montreal in 1962, outrigger braced tall building structures, with building height normally over 200m, have been constructed all over the world (see Table 1). It is observed that general construction practices of outrigger system are varied between countries and regions. In North American regions, core walls, columns and outriggers are often made of steel. In Mainland China, reinforced concrete is commonly used for those structural components. In Hong Kong, composite construction using composite exterior columns, steel outrigger trusses and reinforced concrete core wall is often employed. To enable proper activation of the composite action, steel outriggers are often connected to the concrete wall by welding shear studs on the surface of the steel members.

Early research [1-7] on outrigger system mainly focused on the investigation of the optimum location of the outriggers for reducing the lateral drift as well as bending moments that the core would have been subjected to if it had been free standing. Approximate analysis of a multioutrigger-braced structure using continuum approach was also proposed [8]. The aims of recent studies [9-12] have been changed to study the seismic effects of reinforced concrete core walls with outrigger deep beams. The effect of abrupt change in shear force and bending moment in the core wall at the outrigger levels was highlighted. Shaking table analysis

* This work is supported by Research Grants Council of Hong Kong SAR (Project Nos. HKU7129/03 & 7110/05E).

of seismic attack of a 38-storey reinforced concrete building with two levels of outriggers indicated substantial shear-flexural damage at core wall just above and below the outrigger deep beams [12]. Detailing requirements and design criteria for reinforced concrete core structures with outriggers subject to seismic loads were proposed [9, 12].

In most of the above studies, outriggers were assumed to rigidly attach to the core wall. This assumption is easy to justify for pure steel or reinforced concrete construction. However, for composite construction, the steel outrigger truss needs to rigidly anchor into the reinforced concrete core to enable it to carry huge amount of internal forces, which creates unique challenge to the building designers. The interaction between the outrigger and the concrete core wall as well as the internal load path has rarely been studied and discussed in the past. In this paper, the whole lateral structural system will be idealized as a strut-and-tie system. The complete load transfer mechanism between the outrigger brace and the core wall will be displayed. The common deficiencies in numerical modelling of this type of structures will also be pointed out.

Table 1. Some tall buildings with outrigger braced system.

Name of the building	Building height (m)	Year of Completion
China		
Union Square*	480	2007
Nina Tower *	324	2006
International Finance Centre II	415	2003
Cheung Kong Centre	290	1999
Jin Mao Tower	421	1998
Diwang Mansion	390	1996
Tianjin Hua Xin Building	206	1996
Citybank Plaza	206	1992
Guangdong International Hotel	200	1990
Exchange Square	182	1985
Australia		
Governor Phillip Tower	227	1993
Chifley Tower	215	1992
Malaysia		
Plaza Rakyat Office Tower*	382	2006
USA		
One Houston Center	208	1978
Firstar Center	183	1973
Canada		
Place Victoria Building	190	1962

* Under construction

2. Structural Behaviour of Outrigger Braced Building

When an outrigger braced building deflects under wind or seismic loads, the outrigger trusses, which are rigidly connected to the core wall, force the windward and leeward peripheral columns into axial tension and compression respectively, thus mobilizing them to take part in the structural action. The force couple creates a reverse moment in the peripheral columns to which the outriggers connect (see Figure 1). The overturning moments in the core at the outrigger level, the uplifting and net tension forces throughout the core wall and foundation system are greatly reduced. Moreover, the lateral deflection of the building can also be decreased significantly.

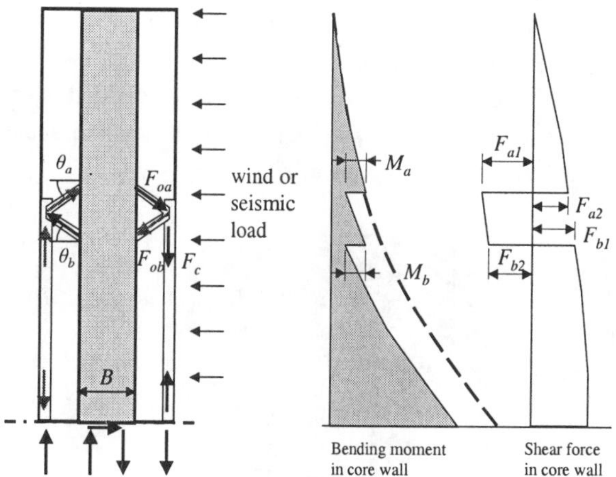

Figure 1. Outrigger braced tall building under lateral load and the corresponding bending moment and shear force in core wall.

Figure 1 shows the bending and shear force distributions at the core wall. It can be seen that abrupt change in moment (M_a+M_b) and shear forces (F_{a1}+F_{a2} and F_{b1}+F_{b2}) occurs at the outrigger level. Reversal of shear forces at the outrigger level implies that the horizontal force exerted from the outrigger truss to the core wall is much greater than the cumulative external shear force (F_{a2}) above the outrigger. Assuming that the inclined struts at the outrigger have similar structural properties and all the connections are pinned, the sudden changes in moment and shear force can be determined by the following equations,

$$M_a = F_{oa} \sin\theta_a B \qquad M_b = F_{ob} \sin\theta_b B \tag{1}$$

$$F_{a1} + F_{a2} = 2F_{oa} \cos\theta_a \qquad F_{b1} + F_{b2} = 2F_{ob} \cos\theta_b \tag{2}$$

Where F_{oa} and F_{ob} are the strut forces at the outrigger truss and can be expressed as,

$$F_{oa} = F_c \frac{\cos\theta_b}{(\cos\theta_b \sin\theta_a + \sin\theta_b \cos\theta_a)}$$
$$F_{ob} = F_c \frac{\cos\theta_a}{(\cos\theta_b \sin\theta_a + \sin\theta_b \cos\theta_a)} \quad (3)$$

The axial forces at the peripheral columns can be determined by finite element analysis.

3. Conventional Analysis of Outrigger Truss

Conventional simplified analysis of outrigger trusses often assumes the truss is simply supported with cantilevered trusses projected from each side of the support subject to a pair of vertical forces from the peripheral columns (as shown in Figure 2). As a result, the shear forces (F_{a2} and F_{b2}) transferred from the core wall are neglected in the design. This assumption would probably lead to over design of the embedded steel members in the core wall but under design the shear reinforcement in the core wall.

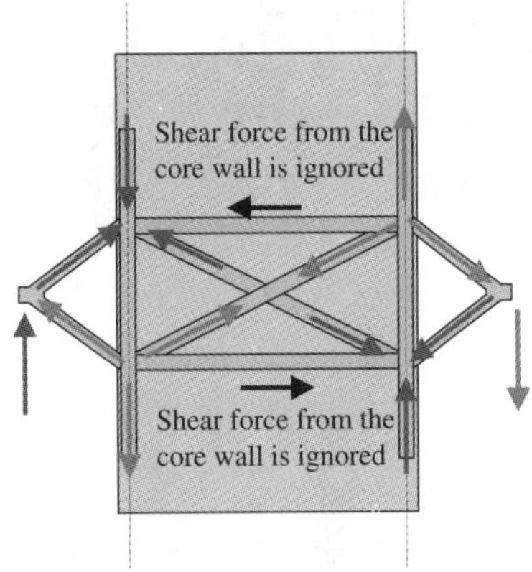

Figure 2. Shear forces from the core wall are ignored in the conventional design for outrigger.

4. Strut-and-tie Model for Outrigger Braced Structure

Based on the load path approach [13-15], the idealized strut-and-tie model for outrigger truss is developed and expressed in Figure 3. Reversal of shear forces at the outrigger level and the force transferred from the core wall to the outrigger are clearly presented.

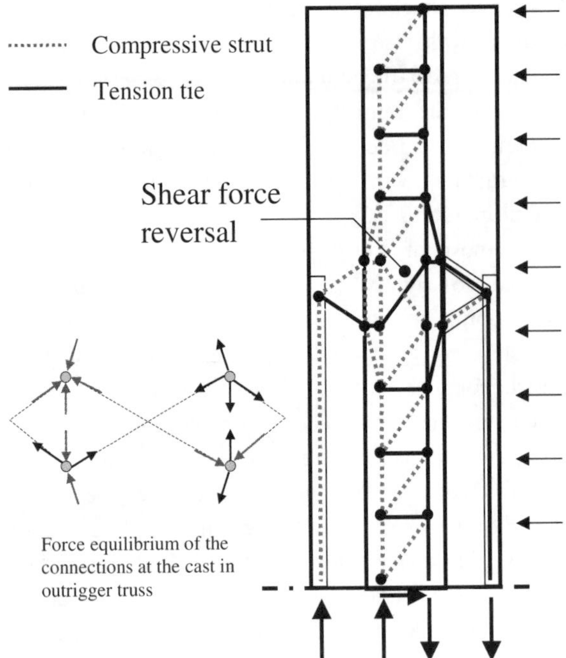

Figure 3. Idealized strut-and-tie model showing the interaction between the core wall and outrigger truss.

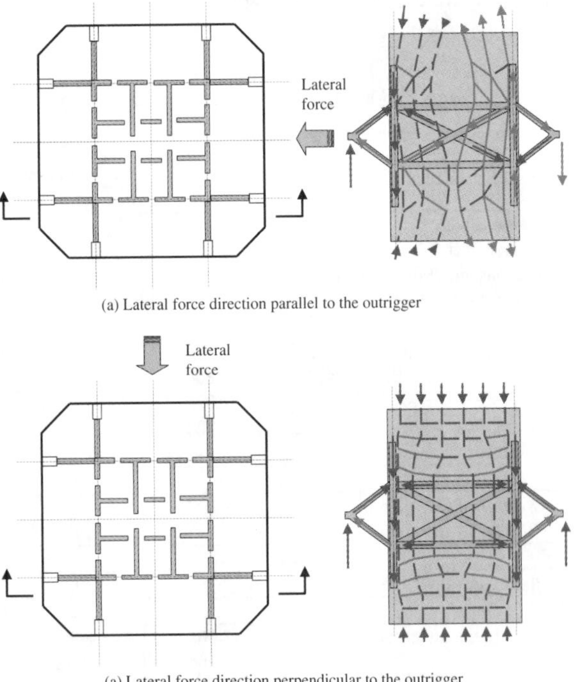

Figure 4. Interaction between outrigger and core wall and the internal force flows inside the core wall.

The internal force flows inside the core wall at outrigger level due solely to lateral loads is shown in Figure 4. Under wind loads from the two principal directions (North and East), vertical and horizontal tensile forces (denoted as solid lines) would be generated inside the core wall around the outrigger. It may argue that the vertical compressive forces in core wall from gravity loads can counter-balance the vertical tensile forces induced from lateral loading, however, the induced horizontal tensile forces still require sufficient tension reinforcement (links) to take up the forces and to control concrete cracking.

The problem may become more acute when major wall openings are provided near the outrigger, as they would significantly affect the internal load transfer between the concrete core wall and steel outrigger. St. Venant's Principle [13] stated that the localized effects caused by any load acting on the body would dissipate or smooth out within regions that are sufficiently away from the location of the stress concentration. Hence, the size of disturbance region (D-region) is estimated and shown in Figure 5. Major openings through the core wall within the D-region should be avoided if possible. Otherwise, such openings should be carefully designed to provide alternative routes for the internal force transfer. Furthermore, to reduce the adverse effect caused by stress concentration around the openings, local thickening of the walls at the D-region may help to alleviate the problem.

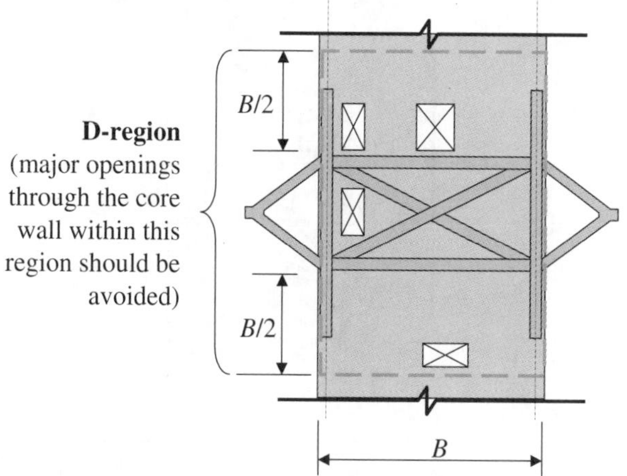

Figure 5. Definition of the D-region.

Shear stud arrangement on the outrigger steel members may also be considerably affected by the wall openings. Those openings would break the normal internal force transfer route inside the concrete. Thus the shear studs provided around the openings may be ineffective to receive the forces from concrete. Hence, the number of shear studs provided may be increased. Alternatively non-uniform arrangement of the shear studs may be considered, i.e. providing less and more studs on the parts of the embedded outrigger, which are close to and away from the openings, respectively.

5. Discussion on Various Configurations of Outrigger Trusses

Various configurations of outrigger trusses that have been proposed and/or adopted in the local construction industry are shown in Figure 6. Moments from core wall are transferred to the outrigger in form of two couples above and below the outrigger. The transfer of compressive forces to the outrigger, by means of both bearing and shear studs, is much more effective than the transfer of tensile forces that rely on studs alone. As concrete would crack under tension, it might further weaken the force transfer process. Hence Cases *A* and *B*, which have steel posts extended to the core wall to enable the necessary force transfer between the steel truss and the concrete core are preferred.

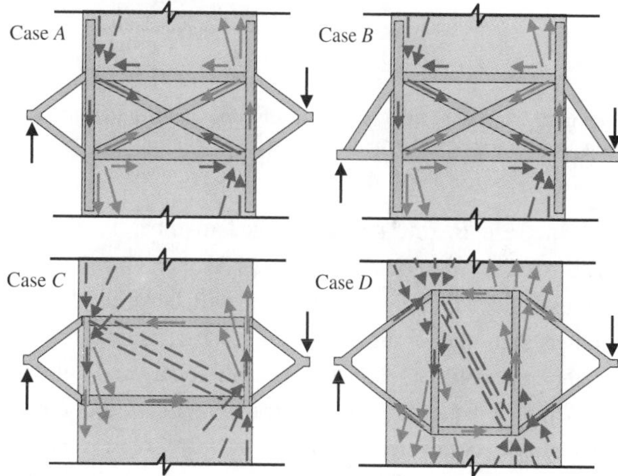

Figure 6. Various forms of outrigger trusses and their interaction with the external and internal loads.

In Cases *C* and *D*, the steel posts located within the outrigger truss region are less effective in transferring tensile forces. The internal strut-and-tie model shown in Figure 7 further reveals that, to bridge the essential load path, same amount of reinforcement has to be extended into the outrigger region and additional horizontal steel has to be provided, despite the significant reduction of core wall moment within the outrigger region. Failure to provide sufficient tensile reinforcement within that zone would weaken the load transfer process and might, in the worst situation, double the compressive load in the leeward columns. Furthermore, it would cause overloading of the outrigger and foundation, flexural cracking of the core wall at the tension zones just above and below the outrigger truss, and excessive lateral movement of the building.

In Cases *C* and *D*, no inclined steel members are provided within the cast in place outrigger. As a result, a diagonal compressive strut has to be formed inside the concrete core. Major openings through the core wall that may intersect with the diagonal compressive strut should be avoided.

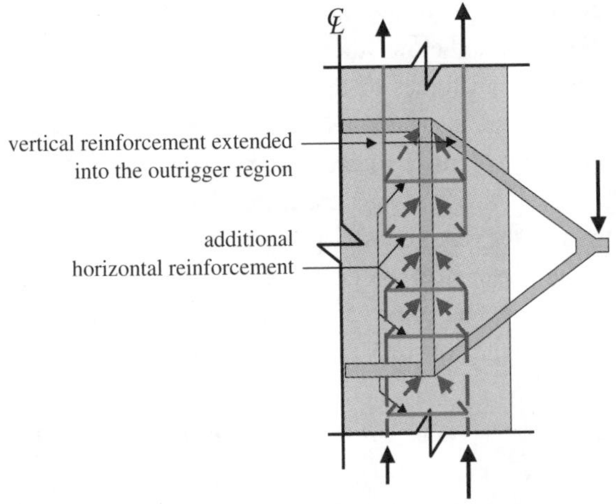

Figure 7. Internal strut-and-tie model for transfer tensile and compressive forces to the outrigger truss.

In Case *B*, one single inclined member is used at each side of the outrigger truss. As the compressive forces induced in those members will be around 40% higher than the counterparts in Case *A*, thicker steel members would need to prevent local and global buckling. Furthermore, this arrangement would cause two single concentrated loads (in form of a couple) exerted at the top part of the truss to counter-balance the core wall moment. It would no doubt enhance the undesirable stress concentration effect in the core wall. Hence configuration Case *A* is more often used in practice than that of Case *B*.

6. Discussion on Numerical modeling of Outrigger Braced Structures

Commercial finite element packages such as ETABS, SAP and TBSA are often employed to analyze outrigger-braced buildings under lateral loads. When modelling the floor slabs, rigid floor diaphragm assumption is not recommended. It is because, for super-high-rise buildings, the in-plane stiffness from the actual floors is usually much weaker than the flexural stiffness of the peripheral columns and the core wall. Rigid floor diaphragm assuming that there is no relative in-plane movement between any two points on the slab is hard to be justified, especially at the outrigger levels. Any small movements between the columns and core wall would easily crack or deform the slab and cause lesser force induced in the slab when compared to the rigid slab. The load paths (as shown in Figure 8) assuming rigid links between different vertical elements are therefore invalid. At outrigger level, there may have larger horizontal relative movements between the interior and exterior walls. Concrete or composite floor slab may act as a deep horizontal ring beam to restrain the wall movements and to carry the shear forces between them.

To correctly simulate the true interactions among all the vertical structural members, shell elements with in-plane membrane stiffness and out-of-plane bending stiffness should be used to model the floor slabs. Major floor openings such as lift-shafts and stairwells have to be correctly represented in the model. The slabs should be designed for out-of-plane bending and normal shear as well as in-plane stresses (in particular at the outrigger levels).

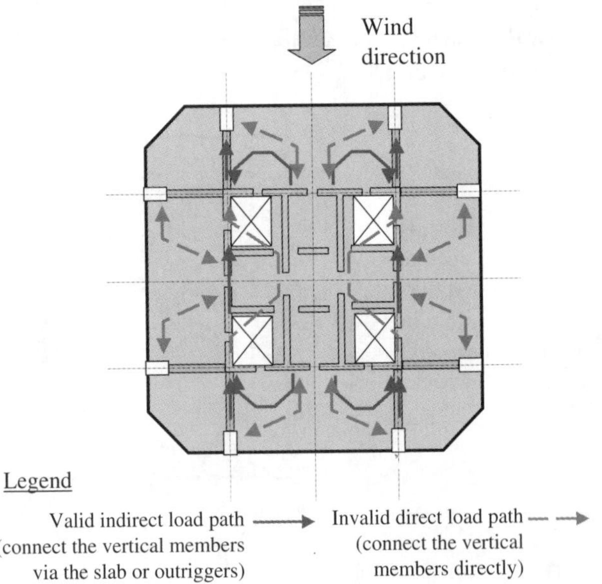

Figure 8. Valid and invalid load paths for transfer shear forces from interior walls to outrigger truss.

The assumption of plane section remains plane after deformation is often adopted in the design of reinforced concrete wall panels. This assumption may not be valid at which large concentrated forces are applied to and from the outrigger and wall openings. Fine finite element mesh (as shown in Figure 9) has to be used to simulate and to display the local stress distribution at the core wall in the vicinity of the outrigger and openings. Local abrupt changes of stresses in the wall panels have to be checked and designed properly. As mentioned in Section 4, moment, axial forces together with shear force have to be designed properly in the wall panels. Depending on various configurations of the outrigger truss, additional horizontal and vertical steel may have to add to the wall panel within the outrigger region to take up the induced tensile forces.

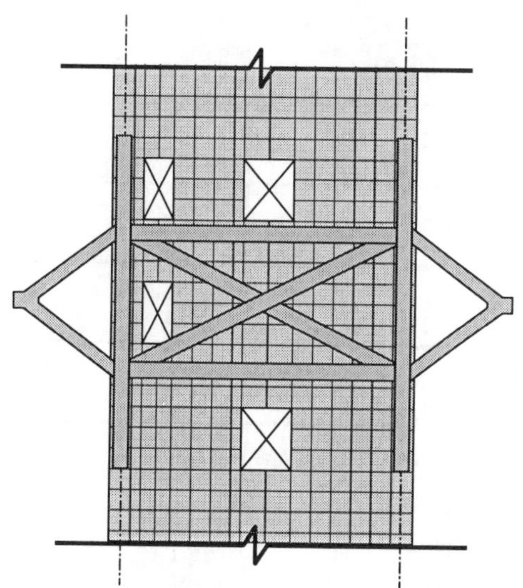

Figure 9. Fine mesh is used to model the local effect due to concentrated loads and wall openings at the structural walls.

It is noted that when modelling the peripheral columns, if the windward columns are under tension, the original uncracked section have to change to cracked section. This process may have to repeat until a converged result has been found. It is noted that this iteration would affect the lateral deflection as well as the internal load distribution of the building.

7. Conclusions

Idealized strut-and-tie models have been applied to analyze outrigger-braced buildings subject to lateral load. Complete load transfer mechanism between the outrigger brace and the core wall has been presented. Many practical concerns in design including the structural behaviour of different outrigger configurations, the effect of openings through the core wall adjacent to the outrigger brace, the shear stud arrangement on outrigger and the shear link arrangement in core wall have been discussed. Some common deficiencies in numerical modelling of this type of structures have been pointed out and a set of numerical modelling recommendations has been given for design reference.

References

1. B.S. Taranath, *Engineering Journal AISC*, **11**, 18-21 (1974)
2. J.W. McNabb and B.B. Muvdi, *Engineering Journal 3nd Quarter*, 88-91 (1975)
3. B. Stafford Smith and I.O. Nwaka, *Behavior of Multi-Outrigger Braced Tall Building Structures subjected to Wind and Earthquake Forces, Publ. SP 63-21,ACI, Detroit*, (1980).
4. B. Stafford Smith and I. Salim, *Journal of the Structural Division-Proceedings of the ASCE*, **107**, 2001-2014 (1981)
5. B. Stafford Smith and I. Salim, *Computers & Structures*, **17**, 45-50 (1983)
6. A. Rutenberg and D. Tal, *Engineering Structures*, **9**, 53-67 (1987)
7. J.R. Wu and Q.S. Li, *Structural Design of Tall and Special Buildings*, **12**, 155-176 (2003)
8. A. Coull and W.H.O. Lau, *Journal of Structural Engineering-ASCE*, **115**, 1811-1815 (1989)
9. X.Y. Fu, *Building Structures*, **10**, 44-47 (1999) [in Chinese]
10. Z.G. Zhang, X.Y. Fu, J.J. Wang and Y.N. Wei, *Journal of Building Structures*, **17**, 2-9 (1996) [in Chinese].
11. P. Gao, *Building Structure*, **10**, 20-25 (1998) [in Chinese].
12. P.F. Xu, J.F. Huang, C.Z. Xiao, G.Y. Li and S.M. Huang, *Journal of Building Structures*, **20**, 2-10 (1999) [in Chinese].
13. J. Schlaich, K. Schäfer and M. Jennewein, *Journal of Prestressed Concrete Institute*, **32**(3), 74-150 (1987).
14. R.K.L. Su and A.M. Chandler, *Progress in Structural Engineering and Materials* **3**(3), 288-298 (2001).
15. R.K.L. Su, P.C.W. Wong and A.M. Chandler, *Transactions of The Hong Kong Institution of Engineers*, **10**(1), 31-37 (2003).

ANALYSIS OF SHEARING STIFFNESS OF COLUMN RELATED TO EXTERNAL FORCES IN FRAMES

WU BANGDA

Wuzhou Engineering Design & Research Institute, China
P.O. Box 55, Beijing Post Code 100053

It is very important to calculate the shearing stiffness of column in frames. But at present, it is seldom concerned with the effects induced by external forces. Therefore, the results of calculation may contain some errors to a certain extent, sometimes being larger. In this paper, taking an example into consideration, the current D-value method will be applied to explain it.

1. Introduction

At present, the current calculating method for solving shearing stiffness of column in frames is D-value methad[1], we may take it for example into consideration. One of its assumptions is that the sidesway angles among each story are equal to each other. D-value is a shape constant, which is decided only by the sectional shape and the restricted condition at ends of column, thus it is not related to external forces. The result of calculation may be conduced wrongly to a certain extent, sometimes the errors may be larger. If the sidesway angles are unequal, D-value is not a shape constant, related to external forces, the errors may be smaller.

2. Shearing Stiffness being a Shape Constant

In a uniform frame (Fig.1), from the assumptions[1] of D-value method, the joint rotational angles and sidesway angles among each story are equal to each other respectively, the shearing stiffness of column ik on r-th story may be obtained as follows.

Under the horizontal forces acting on the joints of the frame, the end moment of column ik is[2]

$$M_{ik} = 2K_{ik}\left(2\theta_i + \theta_k + 3\beta_r\right) \tag{1}$$

The joint rotational angle is

$$\theta_i = \sum_{(i)} \mu_{ik}\left(\theta_k + 3\alpha_{ik}\beta_r\right) \tag{2}$$

The r-th story sidesway angle is

$$\beta_r = \sum_{(r)} \eta_{ik}\left(\theta_i + \theta_k\right) + \overline{\beta_r} \tag{3}$$

where

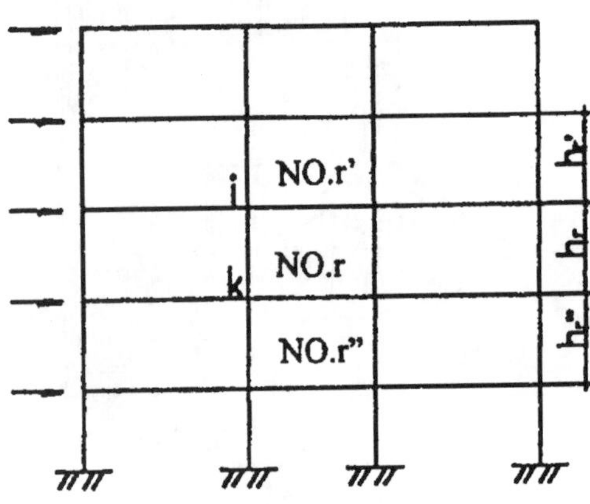

Fig.1

$$\mu_{ik} = -\frac{K_{ik}}{2\sum_{(i)} K_{ik}}, \quad \eta_{ik} = -\frac{\alpha_{ik} K_{ik}}{2\sum_{(r)} \alpha_{ik}^2 K_{ik}}$$

$$\beta_r = -\frac{\Delta_r}{h_r}, \quad \overline{\beta}_r = -\frac{\overline{M}r}{12\sum_{(r)} \alpha_{ik}^2 K_{ik}}$$

$$\overline{M}r = Q_r h_r, \quad K_{ik} = \frac{EI_{ik}}{h_{ik}}, \quad \alpha_{ik} = \frac{h_r}{h_{ik}}$$

Δ_r is the r-th story sidesway, Q_r is the sum of horizontal external forces acting on stories above it, M, θ and β (orΔ) being positive are taken as clockwise, E is the Young's modulus, I_{ik} is the moment of inertia of that member, h_r is the calculative height of a story whose height is unequal.

From the assumption the joint rotational angles being equal, i.e. $\theta_i = \theta_k$, we have

$$\theta_i = \frac{3\sum_{(i)} \mu_{ik}^c \alpha_{ik} \beta_r}{1 - \sum_{(i)} \mu_{ik}} \tag{4}$$

$$\theta_k = \frac{3\sum_{(k)} \mu_{ki}^c \alpha_{ki} \beta_r}{1 - \sum_{(k)} \mu_{ki}} \tag{4a}$$

where μ_{ik}^c is the μ_{ik} of column.

The shearing force on the top of column ik is

$$Q_{ik} = -\frac{M_{ik} + M_{ki}}{h_{ik}}$$

$$= -\frac{6k_{ik}^c (\theta_i + \theta_k + 2\beta_{ik})}{h_{ik}} \tag{5}$$

where K_{ik}^c is the K_{ik} of column.

Substituting eq. (4) and eq. (4a) into eq. (5), from the assumption, the sidesway angles among each story being equal to each other, i.e. $\beta_{r'} = \frac{\alpha_{ik}\beta_r}{\alpha_{ikr'}}$, $\beta_{r''} = \frac{\alpha_{ik}\beta_r}{\alpha_{ikr''}}$, and taking out the common factor β_r, we have

$$\overline{D}_r = \frac{Q_{ik}}{\Delta_r}$$

$$= \frac{12K_{ik}^c \alpha_{ik}^2}{h_r^2}[1 + \frac{1.5(\mu_{ik}^c + \frac{\mu_{ikr'}}{\alpha_{ikr'}}}{1 - \sum_{(i)} \mu_{ik}} + \frac{1.5(\mu_{ki}^c + \frac{\mu_{kir''}^c}{\alpha_{kir''}}}{1 - \sum_{(k)} \mu_{ki}} \tag{6}$$

While the Column ik is placed on the middle story, eq. (6) may be

$$\overline{D_r} = \frac{12K_{ik}^c \alpha_{ik}^2}{h_r^2} \overline{\alpha} \tag{7}$$

$$\overline{\alpha} = [1 + (\mu_{ik}^c + \mu_{ikr'}^c / \alpha_{ikr'}) + (\mu_{ki}^c + \mu_{kir''}^c / \alpha_{ikr''})] \tag{7a}$$

If the heights of each story being equal respectively, all of them are $\alpha_{ik} = \alpha_{ikr'} = \alpha_{ikr''} = 1$.

In a uniform frame, if the sections, heights and spans are equal to each other respectively, we have

$$\overline{D_r} = \frac{12K_{ik}^c}{h_r^2}\left(1 + 4\mu_{ik}^c\right) \tag{8}$$

or

$$D_r = \overline{D_r} = \frac{12K_{ik}^c}{h_r^2}\overline{\alpha} \tag{8a}$$

$$\overline{\alpha} = \frac{\overline{K}}{2 + \overline{K}} \tag{8b}$$

$$\overline{K} = \Sigma \frac{K_{ik}^b}{K_{ik}^c}$$

where K_{ik}^b is the K_{ik} of beam.

Eq. (8a) is a fundamental formula of D-value method for calculating the shearing stiffness of columns, and D_r is a shape constant, that is not related to external forces.

3. Shearing Stiffness not being Shape Constant

If the sidesway angles among each story not being assumed to be equal, D_r cannot be obtained from eq. (6), and the new shearing stiffness can be obtained as follows.

From eq. (5), each member divided by β_r, substituting eq. (3) into β_r in the right-hand member and assuming θ_i and θ_k to be equal in each floor respectively, we have

$$R_r = \frac{Q_{ik}}{\Delta_r} = \frac{12\alpha_{ik}^2 K_{ik}^c}{h_r^2}\overline{\lambda} \tag{9}$$

$$\overline{\lambda} = 1 - \frac{1}{2s} \tag{10}$$

$$s = \alpha_{ik}(0.5 - \frac{\overline{\beta_r}}{\theta_i + \theta_k}) \tag{11}$$

θ_i 、 θ_k may be given alone by A-value approximate method[2], that will simplify the stiffness matrix from higher order [A] to single A, θ_i 、 θ_k and $\overline{\beta_r}$ are produced by external forces, hence R_r is related to them. Examples show that R_r-value is more accurate than D_r–value.

4. An Example

Taking shearing stiffness of column to calculate the sidesway of each story on the frame, as shown in Fig.2. The linear stiffness factor of column K:1st ~ 6th story being 5.4 and 2.8,6 ~ 12th story being 2.6 and 1.06 ,and that of beams of each story being the same as 2.57 and 2.85.

Solution: The partial calculating results in table 1. In this table, C_{ik} is the numbering column, C_r is the accurate value, and e(%) is the percentage of the error. The maximum errors of those of D_r and R_r are +17.06% and +7.90%(the rest are less than ±4%), and Δ_r are -9.19% and −2.74% respectively. Since D_r is larger than R_r, and it would be less safe than R_r.

Table 1

$\frac{r}{C_{ik}}$	D_r R_r C_r ($\times10^2$kN/cm)	e_D e_R (%)	Δ_r (×Pcm)	e_Δ (%)
$\frac{1}{C_{iA}(C_{2B})}$	1.006(2.175) 1.109(2.227) 1.107(2.221)	-9.12(-2.07) +0.18(+0.27) 0	0 . 1 8 8 7 0.1799 0.1803	+4.66 -0.22 0
$\frac{2}{C_{31}(C_{42})}$	0.690(1.551) 0.792(1.424) 0.734(1.422)	-5.99(+9.07) +7.90(+0.14) 0	0.2454 0.2484 0.2552	-3.84 -2.74 0
$\frac{6}{C_{119}(C_{1210})}$	0.690(1.551) 0.702(1.343) 0.710(1.325)	- 2.82(+17.06) -1.13(+1.36) 0	0.1562 0.1711 0.1720	-9.19 -0.52 0
$\frac{7}{C_{1311}(C_{1412})}$	0.456(1.098) 0.449(1.002) 0.458(0.991)	- 0.44(+10.80) -1.97(+1.11) 0	0.1930 0.2067 0.2071	-6.81 -0.19 0
$\frac{11}{C_{2119}(C_{2220})}$	0.456(1.098) 0.456(0.995) 0.470(0.980)	- 2.98(+12.04) -2.98(+1.53) 0	0.0643 0.0689 0.0690	-6.81 -0.14 0
$\frac{12}{C_{2321}(C_{2422})}$	0.456(1.095) 0.456(0.995) 0.467(0.957)	- 2.36(+14.73) -2.36(+3.97) 0	0.0322 0.0344 0.0351	-8.26 -1.99 0

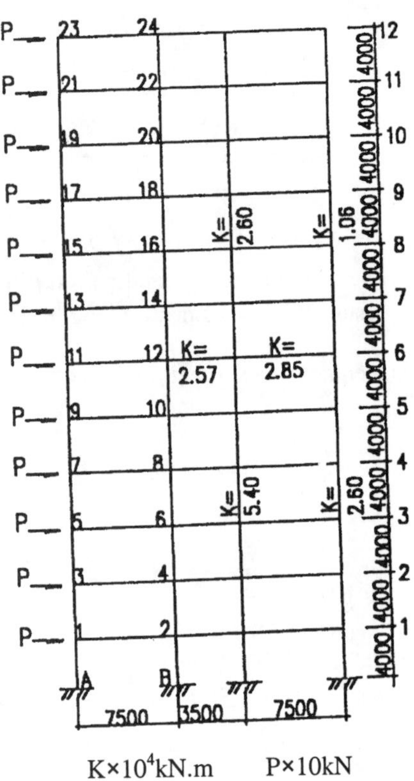

Fig.2

5. Bending Inflexional Point of Column

Let h_o be the height of bending inflexional point of column, and $i_O = h_o / h_r$. Assuming θ_i 、 θ_k on i or k floor (Fig.1) to be equal to each other respectively, we have

$$i_O = \frac{M_{ki}}{M_{ik} + M_{ki}} = \frac{1}{2} \frac{(1+n)}{\bar{\lambda}} \qquad (12)$$

$$n = \frac{\theta_i + 2\theta_k}{3\alpha_{ik}[0.5(\theta_i + \theta_k) + \overline{\beta_r}]} \qquad (13)$$

where M_{ik} is the end bending moment of column ik, $\bar{\lambda}$ can be obtained in eq.(10).

When $\theta_i = \theta_k$, or $\theta_i = \theta_k = 0$, then $i_O = 0.5$.

From eq. (12), i_O is not a shape constant too, and the example (Fig.2) shows that is a more accurate value. The maximum of the error is only 1.6%.

6. Conclusion

The shearing stiffness of column in frames, related to the external forces has been derived above, and that will be helpful to consider the seismic fortification. Since the earthquake actions are variable according to seismic intensity, the shearing stiffness of column in frames would be influenced by seismic intensity too. R_r will be decreased while the frame being placed in the high intensity area, but D_r is not variable, and would be less safe. It may be larger and less safe too that the sheering stiffness of column is a shape constant solved by other methods, not related to exerternal forces or deformations.

References

1. Bao Shihua and Fang Erhua , Design of tall building structrues , Tsinghua Press , 1985.
2. Wu Bangda , New calculating method of frame lateral deformation, Building Structure, 1996.

TIME DEPENDENT DIFFERENTIAL SHORTENING OF REINFORCED CONCRETE COLUMNS IN TALL BUILDINGS

K.W. LEE S.H. LO

Department of Civil Engineering, The University of Hong Kong, Pokfulam Road, Hong Kong

Columns under various magnitudes of load are not uncommon in tall buildings. Resulting from differential loads, uneven shortening of two adjacent columns will induce moment, shear and crack in the connecting slab or beam. The magnitude of this induced moment, shear and crack will increase with time due to the effects of creep and shrinkage. As the creep strain depends on the magnitude of the load, the higher is the load applied to the column, the more is the shortening. In this paper, a systematic time dependent analysis method will be introduced to investigate this problem. This method, which is developed on the concept of age adjusted effective modulus approach, accounts for not only the material nonlinearities but also the geometric nonlinearities of the structural members. Time step analysis is implemented in the analysis in order to simulate the loading situation of typical constructions. Numerical results are analyzed and compared to previously published results by other researchers.

1. Introduction

Vertical elements in tall buildings like column will be subjected to load with various magnitudes. Due to concrete creep and shrinkage, strains in concrete and reinforcement may vary with time. As a result, the length of structural concrete columns also changes considerably with time because of the creep and shrinkage of concrete. Thus uneven loads together with creep and shrinkage further increase the differential shortening problem in tall building.

Time dependent analysis of a concrete structure involves the determination of strains, stresses, curvatures, member's length and deflections at critical points and at critical times during the life of the structure. Engineers are often interested in the final deformation and internal actions at the proposed lifetime of the structure due to creep and shrinkage. The analytical methods developed for engineering practice always contain rough approximation, e.g., linear stress strain materials relationship. Methods using linear concrete stress strain assumption include Effective Modulus Method, Age-adjusted Effective Modulus Method, Rate of Creep Method and Improved Dischinger Method. However, this linear assumption may cause considerable error when stresses in concrete approach 40% of the compressive strength. Under such a high compressive stress, concrete will not only behave nonlinearly but also reach the stress range of nonlinear creep according to EC2 and CEB-FIP MC90. Therefore, a general nonlinear analytical method is necessary.

In this paper, a brief description of the nonlinear approach is presented next. A numerical comparison between the linear and nonlinear approach is followed. Finally, the nonlinear approach is applied to analyze the shortened column of an existing tall building, in which field data is previously published.

2. Concept of concrete relaxation approach

A section with non-zero external force at time t_0 is considered in the following analysis.

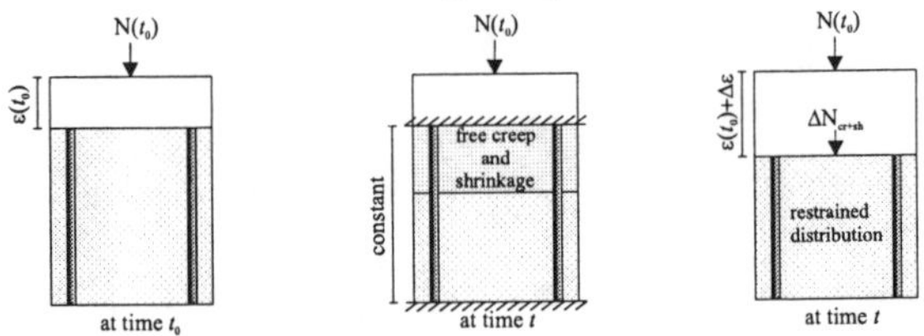

Figure 1. Diagram for relaxation consideration on creep and shrinkage of a loaded member

Initially, instantaneous strain $\varepsilon(t_0)$ is resulted from an external applied force N(t_0). Applying the principle of concrete relaxation, the strain state is kept constant at any time t throughout the analysis, hence creep and shrinkage strains are allowed to vary without the restraint of the reinforcement steel bar from time t_0 to t. As a result, instantaneous strain component should reduce with the same magnitude of unrestrained strains due to creep and shrinkage. This reduction of instantaneous strain component caused by ΔN_{cr+sh} will affect the state of equilibrium. The equilibrium will maintain when an equal and opposite force apply to the restrained reinforced concrete section, the resultant strain computed by this force returns the final restrained strain redistribution due to creep and shrinkage.

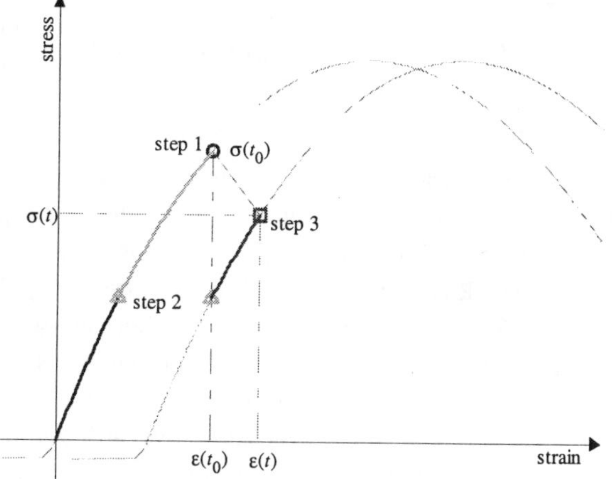

Figure 2. Load path for nonlinear relaxation approach

Details of this approach are elaborated in figure 2. When a load N(t_0) apply to the reinforced concrete section at time t_0, the correspondent stress path at the concrete portion moves from the origin to the circular point (step 1). Provided creep and shrinkage are free to occur without restraint from the reinforcement, it will cause a reduction of instantaneous strain and stress. This is shown by the line starting from the circular point (step 1) to the triangular point (step 2). The reduction due to the combined influence from creep and shrinkage can be imagined as a result of the application a tensile force at time t_0. This tensile force is equivalent to a gradual applied force from zero at time t_0 to its full magnitude at time t. As explained, this tensile force will alter the initial state of equilibrium when the constant strain constraint is removed, a compressive force (same magnitude as the tensile force) should apply to the reinforced concrete section, this will return the final redistribution of stress and strain that follows the line which begins with the triangular mark and ends with square mark in figure 2. This square marked point represents the final redistributed concrete stress and strain of a reinforced concrete section due to the effect of creep and shrinkage from time t_0 to time t.

2.1 *Step by step formation*

The strain and curvatures at time t_0 due to the external forces is given. When creep and shrinkage are free to occur from t_0 to t, it implies that the hypothetical change of strain at any point (y,z) within this period can be expressed as:

$$\Delta\varepsilon_{hypo}(y,z,t) = \varepsilon_{sh}(t,t_s) + \phi(t,t_0)\cdot\left[\varepsilon_o(t_0) + \kappa_y(t_0)z + \kappa_z(t_0)y\right] \quad (1)$$

This change of strain at any point can be imagined as a gradually applied artificial stress from 0 at t_0 to a magnitude $\Delta\sigma_a(y,z,t)$ at t. Since the load is gradually applied to the section, aging should be considered. The change of strain can be expressed as:

$$\Delta\varepsilon_{hypo}(y,z,t) = \left[1+\chi\phi(t,t_0)\right] \times \left[\begin{array}{l} f_\varepsilon(\sigma_c(y,z,t_0) - \Delta\sigma_a(y,z,t_0)) \\ -f_\varepsilon(\sigma_c(y,z,t_0)) \end{array}\right] \quad (2)$$

where $f_\varepsilon(\sigma)$ is a function to return strain value of a given stress. The expressions with $f_\varepsilon(\sigma)$ inside the square brackets in equation (2) is the instantaneous strain results from $\Delta\sigma_a(y,z)$. The entire equation can be read as instantaneous strain due to $\Delta\sigma_a(y,z)$ plus aging creep strain due to gradual application of load $\Delta\sigma_a(y,z)$. Rearrangement of equation (2) returns the instantaneous strain component $\Delta\varepsilon_{ai}(y,z)$ produced by the artificial stress $\Delta\sigma_a(y,z,t)$ applying at time t_0:

$$\Delta\varepsilon_{ai}(y,z) = \frac{\Delta\varepsilon_{hypo}(y,z,t)}{1+\chi\phi(t,t_0)} \quad (3)$$

A function expression $f_\sigma(\varepsilon)$ is now introduced, this function returns the stress $\sigma(y,z)$ at any point when a strain value $\varepsilon(y,z)$ of that point is provided, hence the required artificial forces on the net concrete to prevent change of strain in net concrete section as formulated in equation (1) can be obtained by stress integration over the concrete area as follows:

$$\Delta N_{creep} = \oint\!\!\oint_{A_c} f_\sigma(\varepsilon(y,z,t_0) - \Delta\varepsilon_{ai}(y,z,t))dydz - \oint\!\!\oint_{A_c} f_\sigma(\varepsilon(y,z,t_0))dydz \quad (4)$$

$$\Delta M_{ycreep} = \oint\!\!\oint_{A_c} f_\sigma(\varepsilon(y,z,t_0) - \Delta\varepsilon_{ai}(y,z,t))zdydz - \oint\!\!\oint_{A_c} f_\sigma(\varepsilon(y,z,t_0))zdydz \quad (5)$$

$$\Delta M_{zcreep} = \oint\!\!\oint_{A_c} f_\sigma(\varepsilon(y,z,t_0) - \Delta\varepsilon_{ai}(y,z,t))ydydz - \oint\!\!\oint_{A_c} f_\sigma(\varepsilon(y,z,t_0))ydydz \quad (6)$$

These forces cause the load path to move from the circular point (step 1) to the triangular point (step 2) in figure 2. The system of equilibrium equations to solve the final change of strain and curvatures can be set up by equating respectively the force and moments to the summation on the change of the internal concrete and steel stress integration. The system of equation can be written as:

$$\begin{aligned}
&\Delta N_{creep} = \\
&\iint_{A_c} f_\sigma\left(\varepsilon(y,z,t_0) - \frac{\Delta_{hypo}(y,z,t)+\Delta(y,z,t)}{1+\chi\phi(t,t_0)} \right) dydz \\
&- \iint_{A_c} f_\sigma\left(\varepsilon(y,z,t_0) - \frac{\Delta_{hypo}(y,z,t)}{1+\chi\phi(t,t_0)} \right) dydz \\
&+ \sum_{A_s} f_\sigma\big(\varepsilon(y_{Ri},z_{Ri},t_0)+\Delta\varepsilon(y_{Ri},z_{Ri},t)\big)A_{sRi} \\
&- \sum_{A_s} f_\sigma\big(\varepsilon(y_{Ri},z_{Ri},t_0)\big)A_{sRi}
\end{aligned} \tag{7}$$

$$\begin{aligned}
&\Delta M_{ycreep} = \\
&\iint_{A_c} f_\sigma\left(\varepsilon(y,z,t_0) - \frac{\Delta_{hypo}(y,z,t)+\Delta(y,z,t)}{1+\chi\phi(t,t_0)} \right) zdydz \\
&- \iint_{A_c} f_\sigma\left(\varepsilon(y,z,t_0) - \frac{\Delta_{hypo}(y,z,t)}{1+\chi\phi(t,t_0)} \right) zdydz \\
&+ \sum_{A_s} f_\sigma\big(\varepsilon(y_{Ri},z_{Ri},t_0)+\Delta\varepsilon(y_{Ri},z_{Ri},t)\big)z_{Ri}A_{sRi} \\
&- \sum_{A_s} f_\sigma\big(\varepsilon(y_{Ri},z_{Ri},t_0)\big)z_{Ri}A_{sRi}
\end{aligned} \tag{8}$$

$$\begin{aligned}
&\Delta M_{zcreep} = \\
&\iint_{A_c} f_\sigma\left(\varepsilon(y,z,t_0) - \frac{\Delta_{hypo}(y,z,t)+\Delta(y,z,t)}{1+\chi\phi(t,t_0)} \right) ydydz \\
&- \iint_{A_c} f_\sigma\left(\varepsilon(y,z,t_0) - \frac{\Delta_{hypo}(y,z,t)}{1+\chi\phi(t,t_0)} \right) ydydz \\
&+ \sum_{A_s} f_\sigma\big(\varepsilon(y_{Ri},z_{Ri},t_0)+\Delta\varepsilon(y_{Ri},z_{Ri},t)\big)y_{Ri}A_{sRi} \\
&- \sum_{A_s} f_\sigma\big(\varepsilon(y_{Ri},z_{Ri},t_0)\big)y_{Ri}A_{sRi}
\end{aligned} \tag{9}$$

The only unknown from equations (7) to (9) is $\Delta\varepsilon(y,z,t)$, that is equal to $\Delta\varepsilon_0(t,t_0)+z\Delta\kappa_y(t,t_0)+y\Delta\kappa_z(t,t_0)$. Hence three unknowns can be obtained by solving the system of equations using iterative method. The final redistributed stress and strain at a particular point are:

$$\varepsilon(y,z,t) = \varepsilon(y,z,t_0) + \Delta\varepsilon(y,z,t) \tag{10}$$

$$\begin{aligned}
&\sigma(y,z,t) = \\
&f_\sigma\big(\varepsilon(y,z,t_0) - \varepsilon_{hypo}(y,z,t_0) + \Delta\varepsilon(y,z,t)\big)
\end{aligned} \tag{11}$$

2.2 *Nonlinear creep*

In the previous section, the concrete creep coefficient is assumed to distribute constantly over the cross section during the period of loading. However, this is true only for stresses less than about 40% of the strength (EC2 specify 45%), which happens to fall within the service stress range of normal structures. For higher stress ranges, the behavior of concrete creep is strongly nonlinear. When stress reaches above 40% of the compressive strength, pronounced differences between two models are found, the differences are extending when the stress is increasing. Within this stress range, a linear analysis together with a linear creep assumption will generate considerable inaccuracy. And it is not a good solution to include nonlinear creep to a linear analytical approach, for example AEMM. Therefore it is indeed necessary to develop a general nonlinear creep analysis approach here, which means both materials nonlinear stress strain relationship and nonlinear creep properties are considered simultaneously. Consider a arbitrary cross-section with an initial strain of $\varepsilon_0(t_0)$, $\kappa_y(t_0)$ and $\kappa_z(t_0)$ as shown in Figure 3.

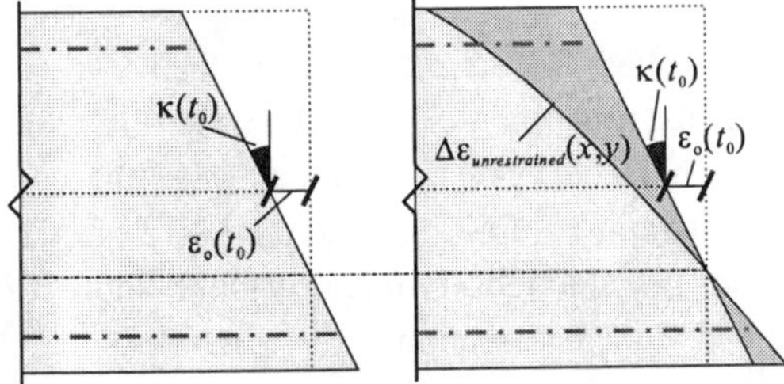

Figure 3. Nonlinear distribution of free creep strain

Part of the induced stress is over 45% of the compressive strength, which is beyond the constant creep range as per EC2. When concrete at any point of the section is free to deform, the creep-induced strain (ignore shrinkage) at that point is:

$$\begin{aligned}
&\Delta\varepsilon_{unrestrained}(y,z) \\
&= \phi(t,t_0,y,z)\cdot\big(\varepsilon_o(t_0)+\kappa_y(t_0)z+\kappa_z(t_0)y\big)
\end{aligned} \tag{12}$$

Unlike the previous derived equations, the creep coefficient is now position dependent. The unrestrained strain due to creep is composed of linear and nonlinear part. The change of forces due to nonlinear creep can be obtained by the integration of the stress distribution over the net concrete section. Although this involves the integration of the nonlinear part, the analysis is exactly the same as the previous section. Thus instantaneous strain component results from $\Delta\sigma_{unrestrained}$ can be expressed as:

$$\Delta\varepsilon_i(y,z)=\frac{\phi(t,t_0,y,z)\cdot\lfloor\varepsilon_o(t_0)+\kappa_y(t_0)z+\kappa_z(t_0)y\rfloor}{1+\chi\phi(t,t_0,y,z)} \quad (13)$$

Integration of stress $\Delta\sigma_{unrestrained}(y,z)$ over the net concrete section gives:

$$\Delta N_{creep} = \oiint_{A_c} f_\sigma\left(\varepsilon(y,z,t_0)-\frac{\Delta\varepsilon_{unrestrained}(y,z)}{1+\chi\phi(t,t_0,y,z)}\right)dydz - \oiint_{A_c} f_\sigma(\varepsilon(y,z,t_0))dydz \quad (14)$$

Referring to the load path diagram in figure 2, this force is from the circular point (step 1) to the triangular point (step 2). Similarly moment about y-axis and z-axis can be obtained in this manner. And the equilibrium equation can be written in a similar way as presented in section 2.1. The final redistribution can be determined by applying the forces obtained above to the reinforced concrete section. It can be graphically represented at figure 4.

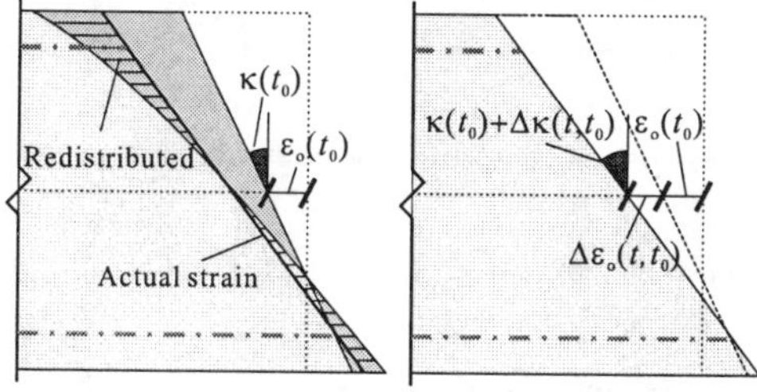

Figure 4. Final redistribution due to restrain of rebar

2.3 *Column axial shortening*

Time dependent axial shortening of column can be determined by simply integrating the strain value $\varepsilon_0(t,t_0)$ over the entire length as:

$$\Delta l(t,t_0) = \int_0^l \varepsilon_o(t,t_0)dx \quad (15)$$

3. Numerical Study

This section presents a parametric study of a reinforced concrete square cross section under various magnitudes of loads such that a comparison between different nonlinear approaches as well as the creep and shrinkage behavior of uncracked section can be made. The first part studies and compares the behavior of linear creep while the second part studies the behavior of nonlinear creep. Details of the cross section can be referred to figure 5.

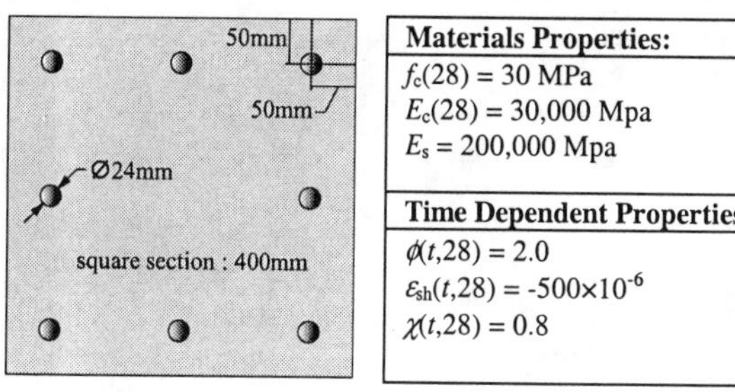

Figure 5. Section details of concrete cross section

3.1 *Axial load + bending moment [full compression]*

Table 1. Instantaneous Behaviour at t_0

$N(t_0)$ = 1200 kN (compression)			
$M_y(t_0)$ = +60 kNm			
$M_z(t_0)$ = 0 kNm			
Unit for curvature [$\times10^{-6}$ mm^{-1}]			
	Bilinear	Parabolic	EC2
$\varepsilon_o(t_0)$	-0.2216 ‰	-0.2360 ‰	-0.2283 ‰
$\kappa_y(t_0)$	+0.8066	+0.8949	+0.8707
$\kappa_z(t_0)$	0	0	0
Steel Stress:			
Top	-68.52 Mpa	-74.04 Mpa	-71.77 Mpa
Middle	-44.32 Mpa	-47.19 Mpa	-45.65 Mpa
Bottom	-20.12 Mpa	-20.34 Mpa	-19.53 Mpa

The graphical results below provide a clear picture on how the stress strain distribution across the cross section varies from time t_0 to t. In the figures, the dotted line represents the unrestrained distribution of creep and shrinkage strain when they are free to deform. The stress strain distribution filled with solid color represents its state at time t_0. The solid line represents the stress strain distribution at time t.

To show the significance of the nonlinear analysis and the application of aging coefficient to account for the variation of stress history, the cross section is re-analyzed without the consideration of aging. Figure 7 shows the results of different time dependent approaches with aging coefficient equal to 1.0. Referring to figure 6 and 7, the results between the nonlinear approach and the stress-strain modification method shows that the CEB-FIP MC1990 method underestimates the final redistribution. This is due to the increasing stress history in the net concrete section. It can be noticed that for an increasing stress history (reduction of absolute value to compressive stress), ignored the changing stress history will underestimate creep strain. While for a decreasing stress history (addition of absolute value to compressive stress), the final redistributed strain will be overestimated.

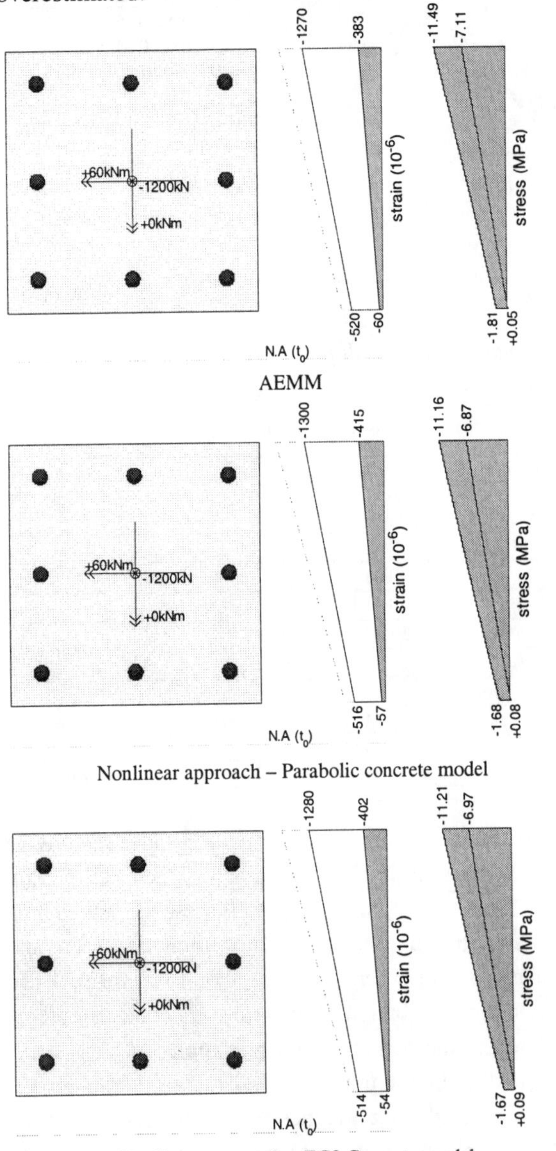

AEMM

Nonlinear approach – Parabolic concrete model

Nonlinear approach – EC2 Concrete model

Figure 6. Results comparison between various analytical approaches with different concrete model.

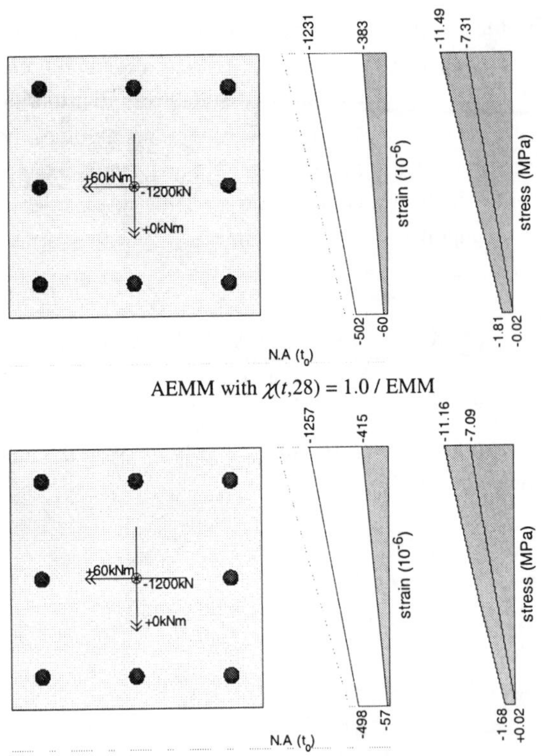

AEMM with $\chi(t,28) = 1.0$ / EMM

Nonlinear Approach/CEB-FIP MC1990 – Parabolic Concrete Model with $\chi(t,28) = 1.0$

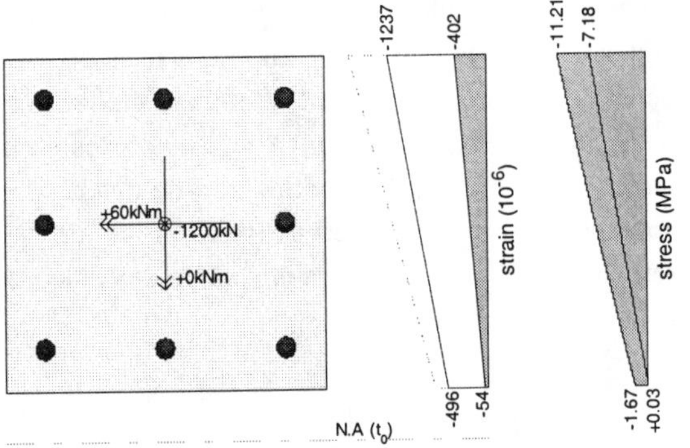

Nonlinear Approach - Eurocode2 concrete model with $\chi(t,28) = 1.0$

Figure 7. Results comparison between various analytical approaches without consideration of aging.

3.2 *Stress in nonlinear creep range*

Table 2. Instantaneous Behavior at t_0

$N_o(t_0)$ = 1800 kN (compression)			
$M_y(t_0)$ = +120 kNm			
$M_z(t_0)$ = 0 kNm			
Unit for curvature [$\times 10^{-6}$ mm^{-1}]			
R.C. Section Strain		Steel Stress	
Parabolic Concrete Model		Bi-linear Steel Model	
$\varepsilon_o(t_0)$	-0.3730 ‰	Top	-131.94 MPa
$\kappa_y(t_0)$	+1.9112	Middle	-74.60 MPa
$\kappa_z(t_0)$	0	Bottom	-17.27 MPa

Table 3. Summary of numerical results for nonlinear creep

Unit for curvature [$\times 10^{-6}$ mm^{-1}]				
		Parabolic Linear Creep	Parabolic Nonlinear Creep	AEMM Linear Creep
Change	$\Delta\varepsilon_o(t,t_0)$	-0.8331 ‰	-0.8787 ‰	-0.8313 ‰
	$\Delta\kappa_y(t,t_0)$	+2.1564	+2.6421	+2.1356
	$\Delta\kappa_z(t,t_0)$	0	0	0
Final	$\varepsilon_o(t,t_0)$	-1.2061 ‰	-1.2517 ‰	-1.1637 ‰
	$\kappa_y(t,t_0)$	+4.0676	+4.5533	+3.7489
	$\kappa_z(t,t_0)$	0	0	0
Reinforcement Steel Stress:				
Change	Top	-231.31 MPa	-255.01 MPa	-230.33 MPa
	Middle	-166.61 MPa	-175.74 MPa	-166.26 MPa
	Bottom	-101.92 MPa	-96.48 MPa	-102.19 MPa
Final	Top	-363.25 MPa	-386.95 MPa	-345.21 MPa
	Middle	-241.22 MPa	-250.35 MPa	-232.74 MPa
	Bottom	-119.19 MPa	-113.75 MPa	-120.27 MPa

This example demonstrates the significance on the consideration of nonlinear creep. The dotted line represents unrestrained creep and shrinkage. The final redistributed strain of the section strain is drawn with solid line. It can be found that the nonlinear part of the unrestrained creep increases in an exponential rate. The final redistribution of strain is restrained by the presence of reinforcement steel bar and plane strain assumption. The results show that a significant reduction of concrete stress within the part of nonlinear creep is observed. For the nonlinear approach without the consideration of nonlinear creep, it can be noticed that the curvature value has been underestimated by 12% if it compares to the approach that has taken nonlinear creep into consideration. The difference even extends to 22% when AEMM is used. This will generate considerable error when deflection or shortened length is to be calculated by these approaches.

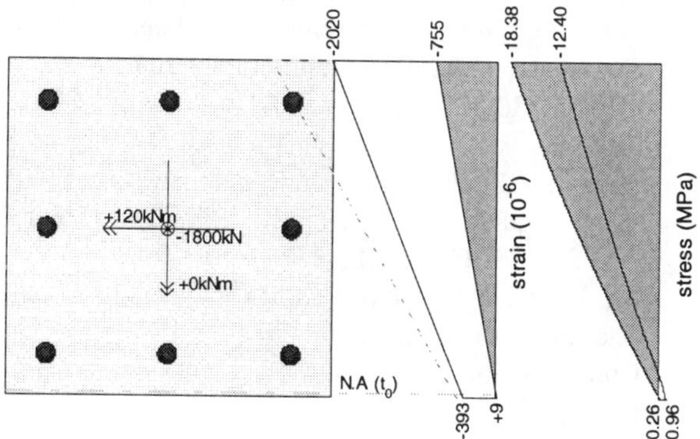

Parabolic Concrete Model – Linear Creep

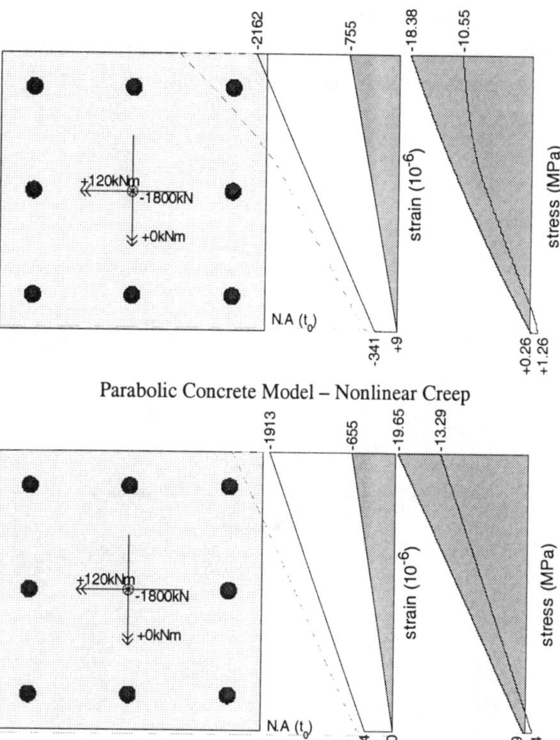

AEMM

Figure 8. Results comparison between various analytical approaches without consideration of aging.

4. Field Data Comparison

Field data with detailed concrete properties and loading schedule from a tall building in America was previously published. Two columns with pronounced load difference are selected for analysis. Details of each measured column are summary in table 4, where loading schedule is presented graphically in figure 9.

Table 4. Properties of measured columns

Location and column ID:	Time dependent properties:
Column mark: C'3 & B'C3¾ Location: 32M-32	Shrinkage: 695$\times 10^{-6}$ Creep: $\phi(28)$=1.466 [sealed] Creep: $\phi(28)$=2.757 [unsealed] Age at initial loading: 43 days
Column dimension/Reinforcement ratio:	
C'3: 460mm×1370mm [0.82%] B'C3¾: 460mm×915mm [2.16%]	
Concrete properties:	
Strength: 53 MPa [Measured field data] Modulus of elasticity: 31580 MPa [Measured field data]	

The analysis strictly follows the parameter given in table 4 and the loading schedule in figure 9. The loading after 586 days is assumed constant with the last recorded cumulative value. The value of aging coefficient is assumed as 0.8 in the nonlinear analysis. The results are summarized graphically in figure 10 and figure 11.

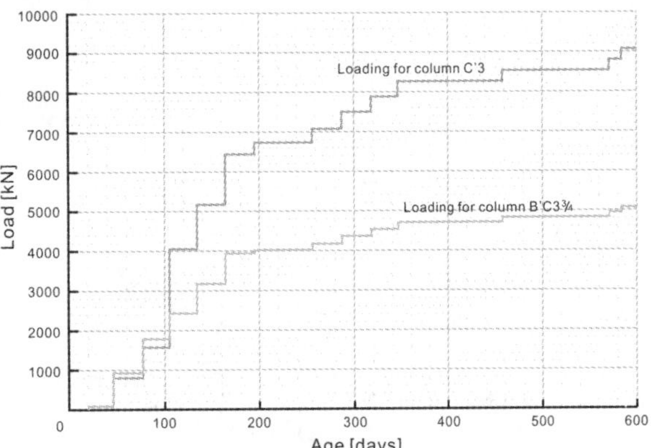

Figure 9. Loading schedule for column C'3 and B'C3¾

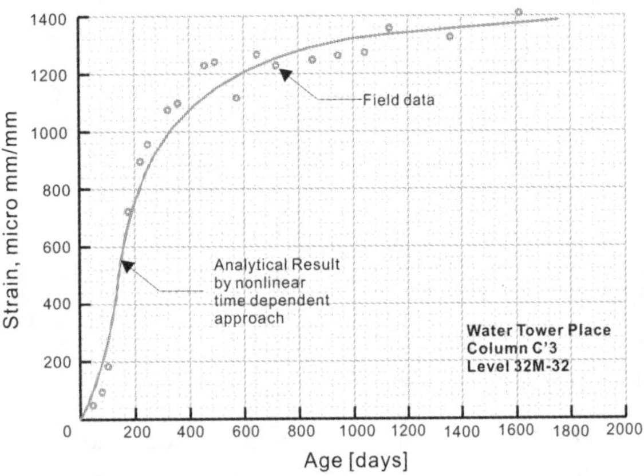

Figure 10. Analytical and measured strains for column C'9

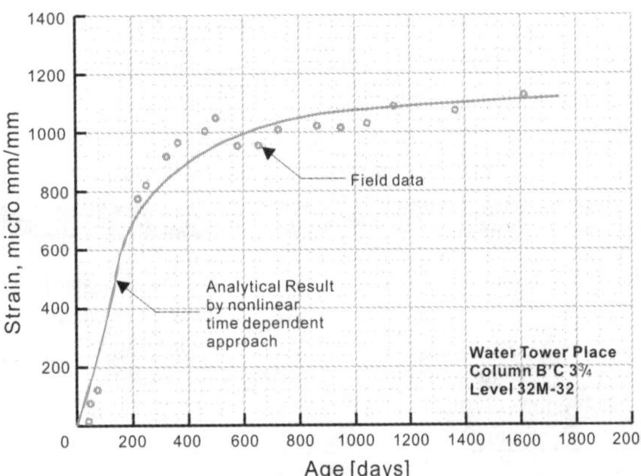

Figure 11. Analytical and measured strains for column B'C3¾

The analytical results on both columns show good agreement with the field data. It can attribute the satisfactory result to several reasons. First of all, the nonlinear approach makes use of a more realistic materials behavior as input parameter. In addition, aging is incorporated into the nonlinear approach. Without the consideration of aging, the entire analysis will underestimate the deformation around 10% to 15% depending on the age of initial loading. Finally, the implementation of discretized time step analysis in the nonlinear approach is also favorable to the final numerical result.

5. Conclusion

A nonlinear analytical approach has been present in this paper to determine the time dependent shortening of column under axial load and biaxial moment. The numerical result shows good agreement with field data. Though the approach is not simplified for hand calculation, with the assistance form microcomputer, the nonlinear approach is a good reference to structural engineers who involved in tall building design.

Acknowledgement

The financial support from the CERG research grant, HKU7117/04E "Analysis of transfer plate structures using high-performance solid 3D hybrid stress hexahedral elements" is greatly appreciated.

References

1. K.W. Lee, Nonlinear Time Dependent Design and Analysis of Slender Reinforced Concrete Columns, Cuvillier Verlag Göttingen (2004).
2. Eurocode 2: Design of concrete structures – Part 1: General rules and rules for buildings, European Committee for Standardization, prEN 1992-1, (October 2001).
3. Comité Euro-International du Béton (CEB) – Fédération Internationale de la Précontrainte (FIP) : Model code for concrete structures (MC90), London, Thomas Telford (1993).
4. H. Trost, Auswirkungen des Superpositionsprinzip auf Kriech und Relaxations – Probleme bei Beton und Spannbeton. Beton und Stahlbetonbau, 62, No. 10, pp 230-238, (1967) and No. 11, pp 261-269, (1967).
5. Bazant, Z.P. Prediction of Concrete Creep Effects using Age-Adjusted Effective Modulus Method. ACI Journal, 69, pp 212-217 (April 1972).
6. M. Fintel, S.K. Ghosh and H. Iyengar, Column Shortening in Tall Structures – Prediction and Compensation, Portland Cement Association, (1987).

Seismic Engineering

DESIGN OF ENERGY DISSIPATION-SEISMIC REDUCTION OF ONE TALL BUILDING AND IT'S SHAKING TABLE TEST VERIFICATION

SI CHAO

Architectural Design & Research Institute of Tong Ji University, 1239 SiPing Road , Shanghai 200092 , China

FULEI YANG WEI QING

Architectural Design & Research Institute of Tong Ji University, 1239 SiPing Road , Shanghai 200092 , China

Energy dissipation-seismic reduction design is a new way in structure earthquake resistant; it has been placed in China's code for seismic design of buildings (GB 50011-2001). But it is introduced briefly in this code and cannot be used facilely, as well lacked practical project referenced to. So this paper cited one tall building with irregular floors (its outside braces worked with velocity-depended nonlinear dampers) to discuss its energy dissipation-seismic reduction design, and compared it to the result of its model shaking table test. This design process and experience could be referred to by the similar projects.

1. General situation of the project

This building is the general teaching and scientific research building of Tongji University, which takes up 48.6m×48.6m with one story underground and twenty-one stories aboveground. The outline of this structure is a regular prism but its floors looks like "L" in plane view for its irregular shapes, and the floor rotates clockwise every three stories (figure 1). So the coupled lateral-torsional vibration characteristic of the structure is apparent under the seismic effect. To improve the whole stiffness and aseismic behavior of the structure, the outer braces with dampers were set on the original structure to form an energy dissipation system.

2. Design Process of the Energy Dissipation Structure

Based on the aseismic code of China, the energy dissipation structure is aimed to control the anticipative displacement under the rare occurrence earthquake. So the controlled target-maximal elasto-plastic story drift [ψ] could be set as 1/80. Figure 2 shows the design process.

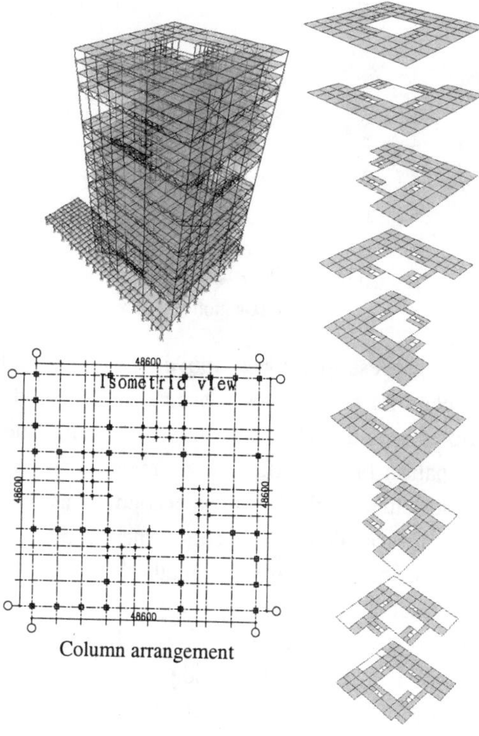

Figure 1 view of the structure

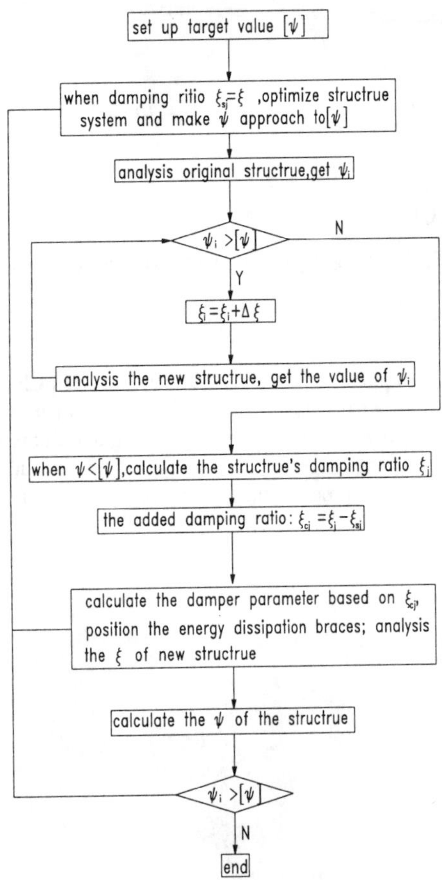

Figure2 Design process of the energy dissipation structure

In the process mentioned above, we need to adjust and calculate for many times to ascertain the added damping ratio to guide choosing and disposing the energy dissipation braces. After the dampers position fixed on the structure, Calculating is needed to assure that the damper parameters and location meet the design needs. Besides, the mode analysis method is used to analysis the structure presently. So forced decoupling of the structure dynamic equation is needed to calculate the natural frequency, the mode of vibration and the damping ratio of the structure. Some domestic and overseas researches show that the error of the result of decoupling compared to the accurate ideal solution is within 5% when the energy dissipation dampers are disposed equably and the added damping ratio is less than 0.20. So the added damping ratio of the energy dissipation braces should be controlled within 20%. The design parameters of the damper, like the restoring force, should be ascertained by experiments.

3. Restoring Force of the Damper and Experimental Verification

Based on the experiences of the projects by past and the primary analysis of the original structure, the maximal damping force and the maximal displacement of the damper in this project is set as 500kN and 50mm. The physical equation of the damper is shown in equation (1):

$$F = C_0 |\dot{x}|^{\alpha} sign(\dot{x}) \qquad (1)$$

Where F is the damping force of the damper, C_0 is the damping constant ($KN \cdot (s/mm)^{\alpha}$); α is the velocity index, constant meet the design needs which varies from 0.1 to 1.0; $\dot{x}$ is the relative velocity between the stopcock of the thrust member and the crust of the damper(mm/s).

The damper in equation (1) is similar to the one in ETABS program. The design parameters drawn up are $\alpha = 0.15$ and $C_v = 250KN/(mm/s)^{\alpha}$.

Considering the shaking table test later and to be combined with the structure model similarity ratio - 1/15, the experimental damping control parameters are converted to $\alpha = 0.15$ and $C_v = 0.3KN/(mm/s)^{\alpha}$. We held the pseudo dynamic test of the experimental damper under the sine wave excitation. Figure 3 is the real hysteretic curve and theoretical curve of the relative displacement and damping force of the stopcock of the damper.

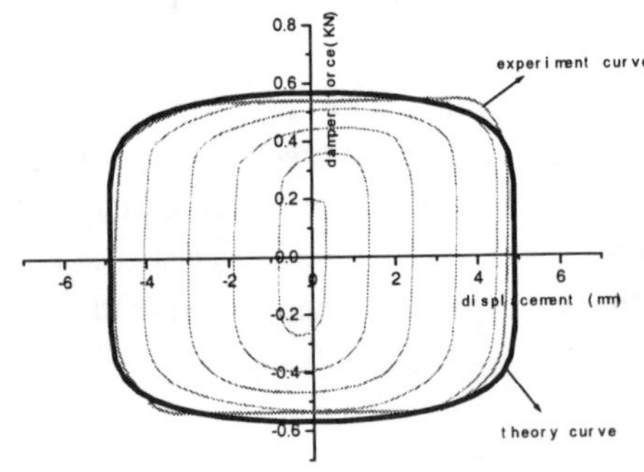

Figure3 real and theoretical curve of the experimental damper

Figure 3 shows the well consistency of the two curves and convince the design parameters are credible.

4. Energy Dissipation Calculating of the Energy Dissipation Braces

The biggest influence of the damper to the structure is the added damping. To simplify the calculating, the aseismic code of China use the ratio of the energy absorbed by the energy dissipation members and the total seismic deformation energy of the structure with the dampers as the added damping. As to the displacement-depended dampers and the velocity-depended nonlinear dampers, the energy dissipated under the horizontal seismic effect is calculated with the equation (2):

$$W_C = \sum A_j \qquad (2)$$

where W_C is the energy dissipated in a circle by the whole energy dissipation members under the anticipated displacement; A_j is the area of the hysteretic curve of the No.j damper under the horizontal displacement Δu_j; Δu_j is the relative displacement of the two end of the No.j damper.

The code prescribes the equation (3) to calculate effective damping ξ_a, which is added to the structure by the dampers.

$$\xi_a = W_c/(4\pi W_s) \qquad (3)$$

where W_s is the total strain energy of the energy dissipation structure at the anticipated displacement. Leave out of account of the torsion, the total strain energy of the dampers under the horizontal seismic effect is calculated with the equation (4):

$$W_s = (1/2)\sum F_i u_i \qquad (4)$$

where F_i is the horizontal seismic force on the $No.i$ point, u_i is the displacement of the $No.i$ point under the characteristic value of the horizontal seismic effect.

So the final damping ratio of the energy dissipation structure ξ is calculated with the equation (5):

$$\xi = \xi_0 + \xi_a \qquad (5)$$

Where ξ_0 is the damping ratio of the original structure.

Based on the analysis process in the chapter 2, 56 dampers are disposed on periphery of the original structure. Calculating with the equation (2) to (5), the added damping ratio of the braces is 8% to 11% and less than 20%.

5. Calculating and Analysis of the Whole Structure and the Shaking Table Test Verification

The structure is analyzed in two work cases to gain the seismic performance at strong and weak earthquake with shanghai got-up wave time history SHW1 in the X-axis and Y-axis. The maximal acceleration of the weak and strong earthquake is 35cm/s^2 and 200 cm/s^2.

The comparison of the story drift between the original structure and the energy dissipation structure under the weak earthquake is shown in figure 4 and figure 5.

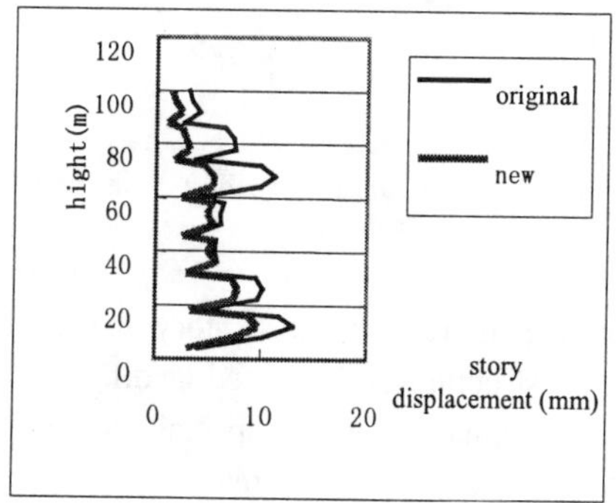

Figure 4 the maximal story drift with weak earthquake in X-axis

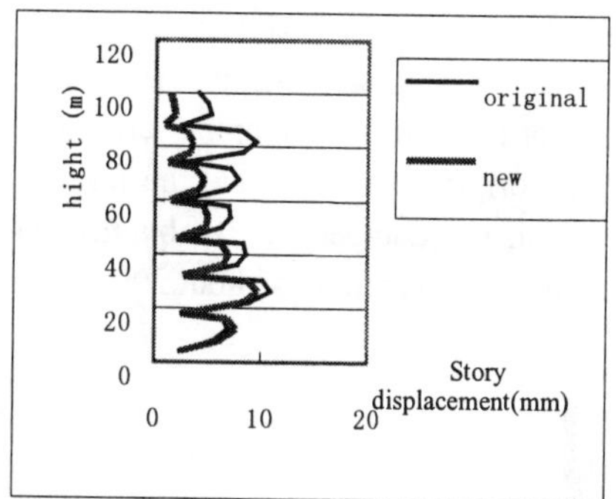

Figure 5 the maximal story drift with weak earthquake in Y-axis

The comparison of the story drift between the original structure and the energy dissipation structure under the strong earthquake is shown in figure 6 and figure 7.

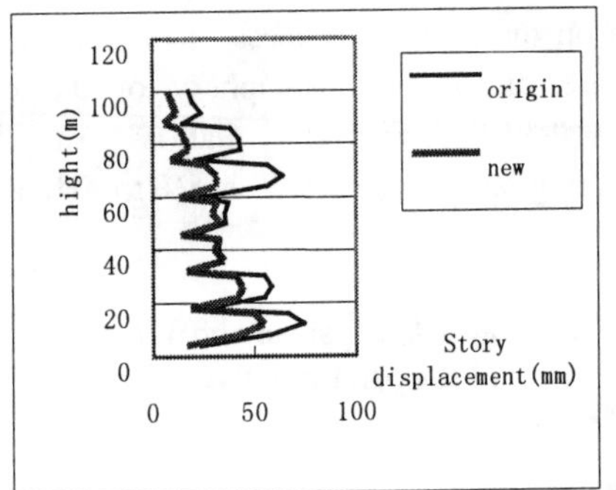

Figure 6 the maximal story drift with strong earthquake in X-axis

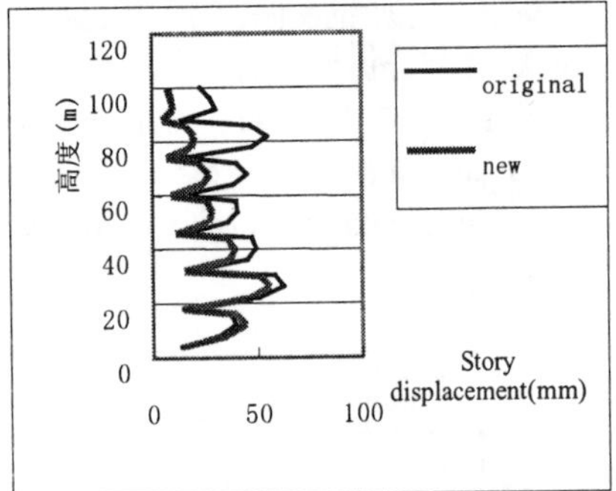

Figure 7 the maximal story drift with strong earthquake in Y-axis

From the figures above, the story drift of the energy dissipation structure is decreased at different degree to that of the original structure, especially at the peak value. With the energy dissipation brace, the story drift of the structure under strong earthquake is controlled within the bound 1/80 in the code.

The shaking table test was held in Tongji University to compare the seismic performance of the structure before and after disposing the braces. Figure 8 shows the variety of the structure damping ratio, where the tint presents the original structure and the fuscous color the structure with the dampers. it is obvious that the first mode damping ratio increases about 20%.

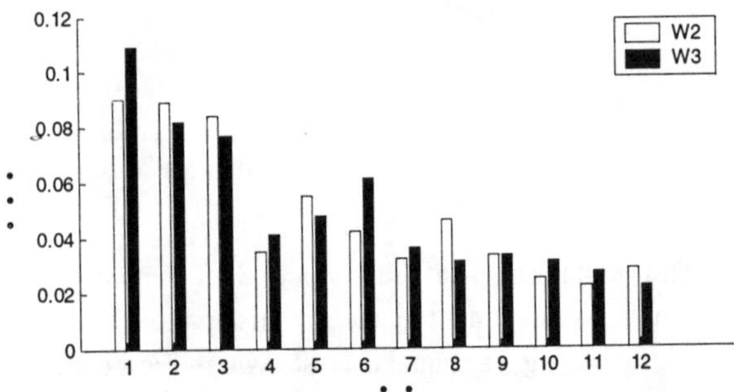

Figure 8 comparison of the structure damping ratio

Figure 9 and figure 10 are the story drift of the structure under the seven-degree frequent occurrence earthquake before and after disposing the braces. The data show that under the seven-degree occurrence earthquake , the displacement to the ground of the model with dampers decreased and the maximal displacement of the top story decreased 12.0%.it is obviously that the seismic reduction effect is apparent.

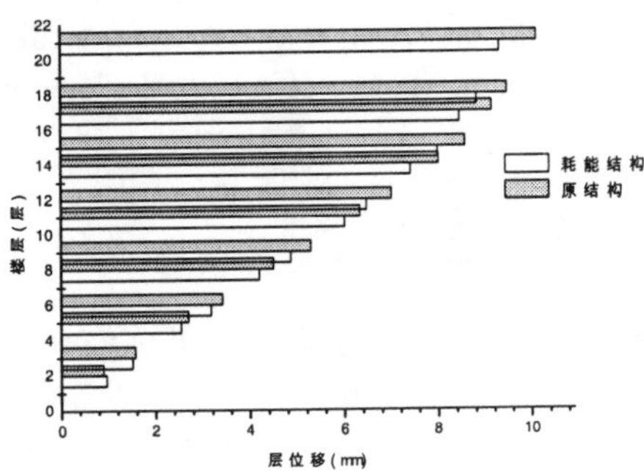

Figure 9 Displacements in X-axis under X-axis weak earthquake

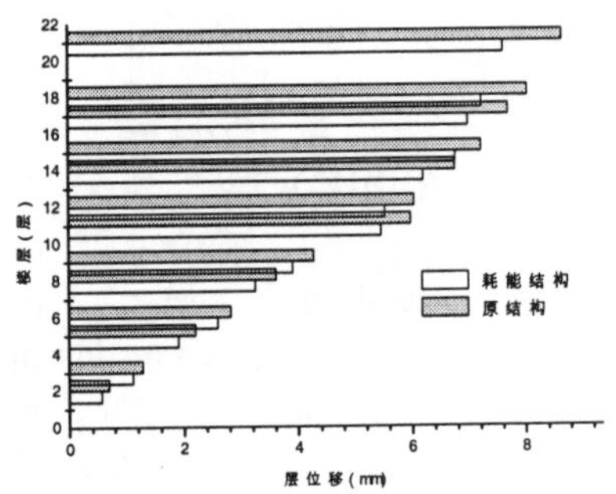

Figure 10 Displacement in Y-axis under Y-axis weak earthquake

6. Conclusion

Due to the insufficiency of the research of the design method of the energy dissipation structure and practical project experience, this paper tends to provide a reference to similar structures through the design introduction of a practical energy dissipation structure and its shaking table test verification.

Reference

1. GB50011-2001. 《Code for seismic design of buildings》,2002.
2. Lin Xinyang, Zhou Fulin , Theory and application of energy dissipation-seismic reduction technology . World Earthquake Engineering , 2002 , 18 (3) .
3. Chang K C , Lai M L , Soong T T , etc . . Seismic behaviours and design guidelines for steel frame structures with added viscoelastic dampers. Technical Report NCEER-93-0009.SUNY at Buffalo .
4. Edward L.Wilson, 《Three-Dimensional Static and Dynamic Analysis of Structrues》,CSI, January 2002;
5. Weng Dagen , Lu Xilin , Study on design parameters of energy dissipation structures with experiment verification , earthquake engineering and engineering vibration , 2004.Vol 24,No.2.

DYNAMIC CHARACTERISTICS AND SEISMIC RESPONSE ANALYSIS OF MULTI-TOWER STRUCTURE CONNECTED WITH CIRCULAR CORRIDOR

YIMIN ZHENG

Architectural Design and Research Institute, Tongji University,1239 Siping Road Shanghai 200092, China

YUTANG CHEN

Department of Building Engineering, Tongji University, 1239 Siping Road Shanghai 200092, China

Based on a practical project, a three-dimension FEM model is established to analyze the dynamic characteristics and seismic response of the Six-tower Connected Structure. The effects on the structure behavior due to the long span connections stiffness and two kinds of connecting method between the corridors and the towers are studied in this paper. The paper establishes a foundation for further researches.

1. Introduction

The corridor among towers not only facilitates the association of different towers, but also can be used as view corridor and cafe. Dynamic characteristics and seismic response analysis of unsymmetrical connected double-towers structure have been carried out in the paper [1-2]; seismic response of symmetrical double-tower is analyzed in the paper [3]; seismic analysis of double towers with big base and double towers with connections is proceeded in the paper [4]. Only double-towers with connections is studied in the upper thesis, dynamic characteristics and seismic response of the six-towers structure with circular corridors are studied in this paper.

2. Computation Model

2.1 *Project general conditions*

The main structure of Hangzhou Citizen Center is composed of six arc-shaped towers distributed around the site center, the towers are connected to each other through a long span circular closed corridor at the 23th~25th story (84.6~92.0m) of the each tower. The planar graph of towers and corridors is shown in Figure 1.

The seismic fortification intensity is 7 degree and the max influence factor is 0.12.The design characteristic period of ground motion is 0.45s.

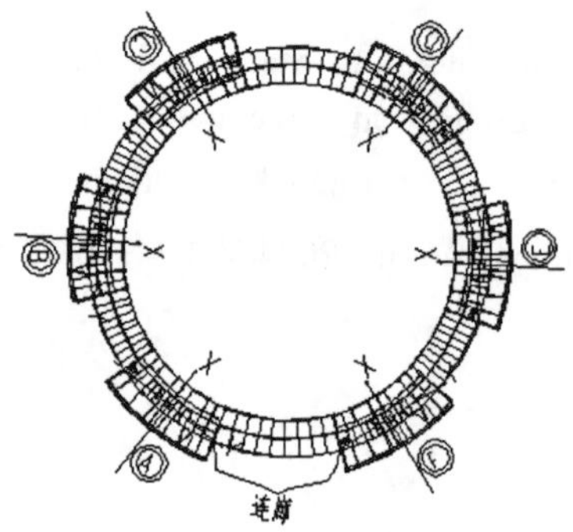

Figure 1 planar graph of towers and corridors

2.2 *Building model*

Space beam element and shell element are used in the three-dimensional FEM model. Two horizontal freedom and a torsion freedom are considered in every story with the hypothesis that the floor stiffness in the plane is rigid and using the static elimination method. Elastic floor is considered in the corridor for analysis. Every story mass is assembled at the level of the floor. Because of the long span, the truss structure form is suitable for the corridor.

3. The Structure Modal Analysis

The mode of A,B,C,D,E,F tower is the same because of their similar structure. The first nine modes of one single tower are listed in the following Figure 2.

From the above figure, it can be found that if mode sorted by the mode curve inflection point, the first mode is the mode without infection point, and the number of mode is equal to the amount of inflection point plus one.so each mode has three vibration patterns—the first pattern is translation in X direction, the second is translation in Y direction and the third is torsion.

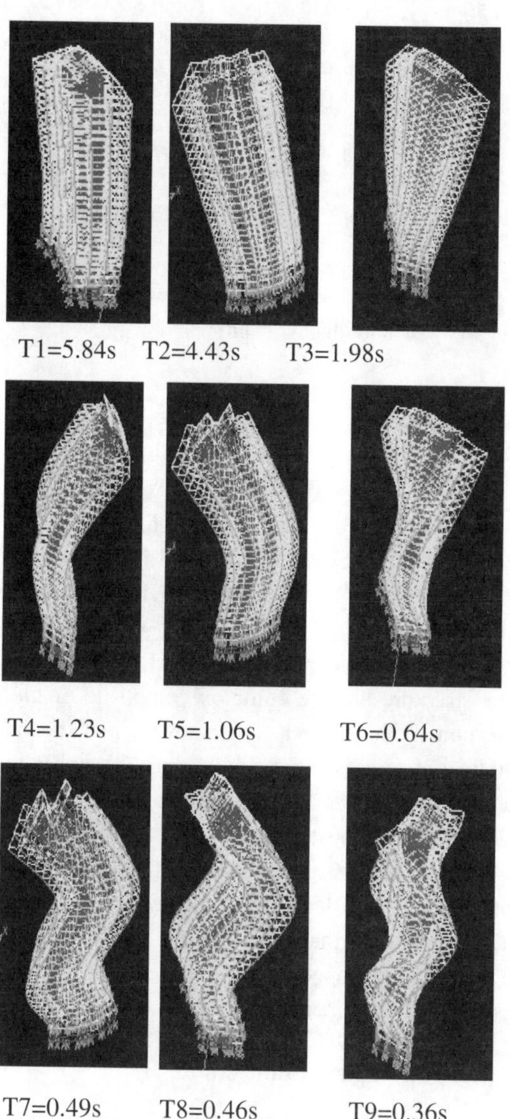

Figure 2. Mode and Periods of A Tower

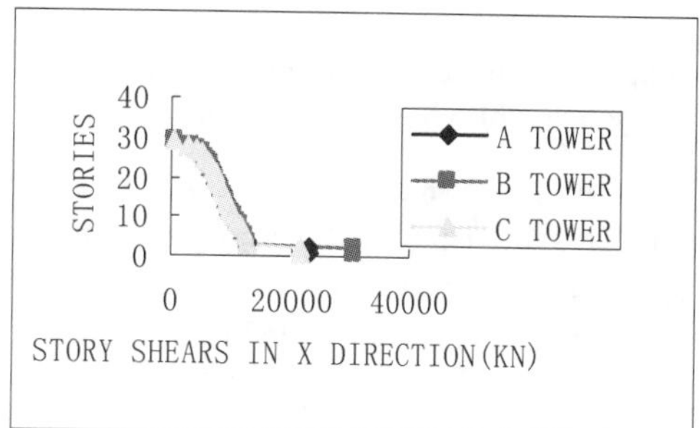

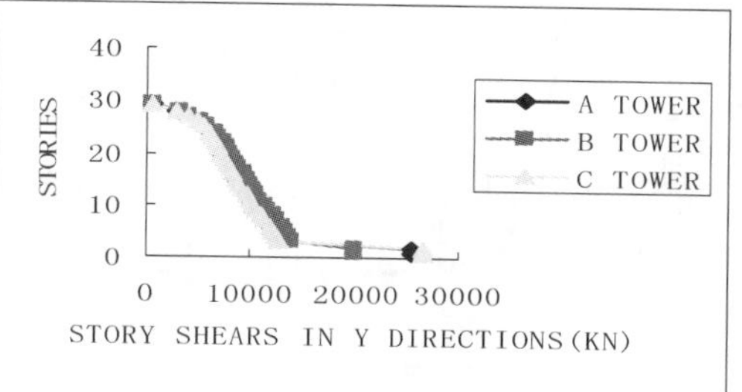

Figure 3 story shears of the single tower

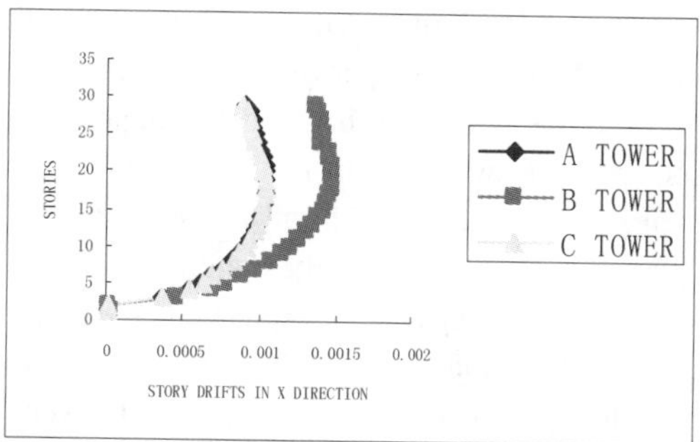

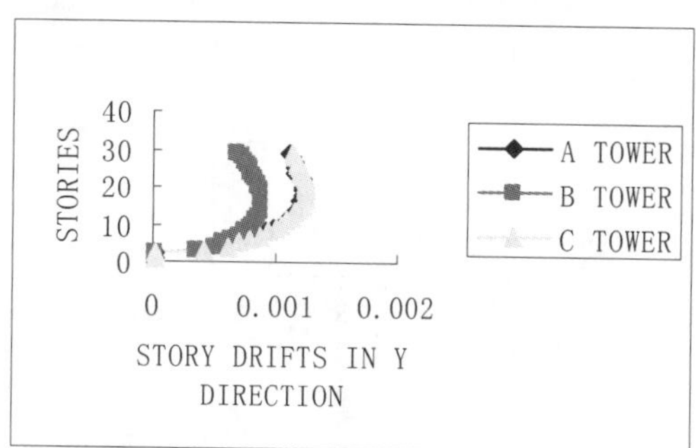

Figure 4 story drifts of the single tower

4. The Seismic Shear of Single Tower

From Figure 3, it can be seen that story shear forces distribute uniformly in vertical direction because the single tower mass and stiffness distribute uniformly in vertical direction. Shear forces change abruptly at the second story just for large restraints at the basement.

The local x axis of the B tower is in accordance with global X. The story drift of B tower is maximal in X direction because the local x of the tower is the weak axis. The story drifts of both A and C towers are smaller than B tower in X direction in Figure 4 because the weak axis of them is in an angle of 55 degree with the global X direction. According to the same reason, the story drift of B tower is minimum in Y direction than A and C tower.

5. Dynamic Characteristics and Seismic Response while Setting Corridors

After setting a connection, there is a great change in dynamic characteristics and seismic response of the structure. The following analysis is on the condition that towers and corridors are connected with hinge.

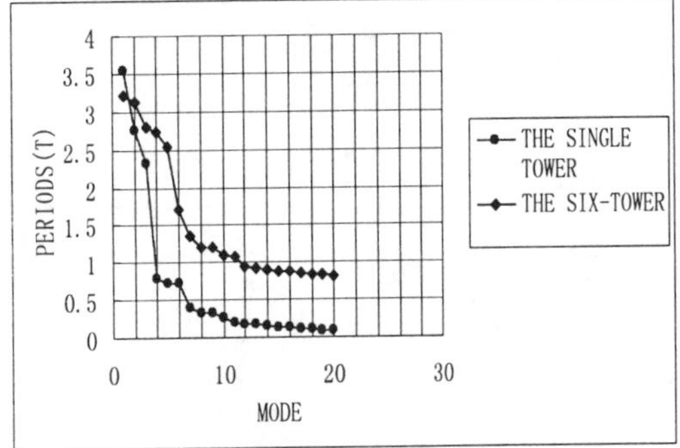

Figure 5. The periods of the single tower and the six-tower

We conclude from Figure 5 that the first period of the single tower is longer than that of the six-tower with connection. Then other periods of the six-tower with connection is longer than those of the single tower. The periods change evidently from the first to forth mode and vary gently from the fifth to twentieth mode.

Figure 6 shows that the maximum drift of the story in x and y direction of all towers is at the 20th story. Because of symmetry, A, B and C tower is correspondingly in coincidence with F,E,D tower. There is a tendency that the story drifts in x and y directions of all towers gradually reduce from the 21st to the 29th story.

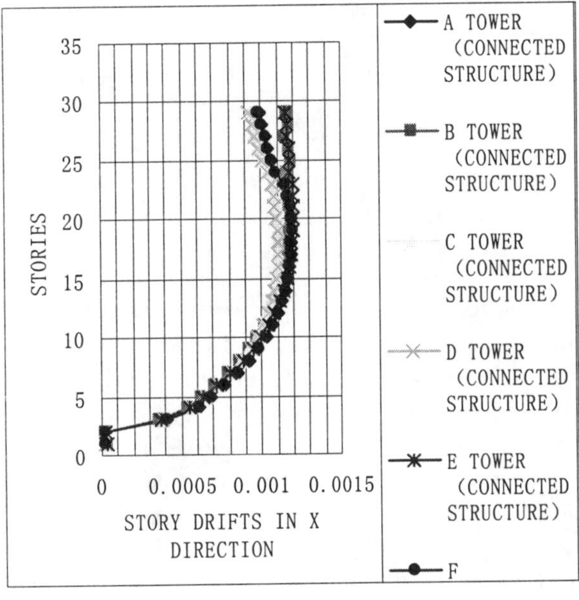

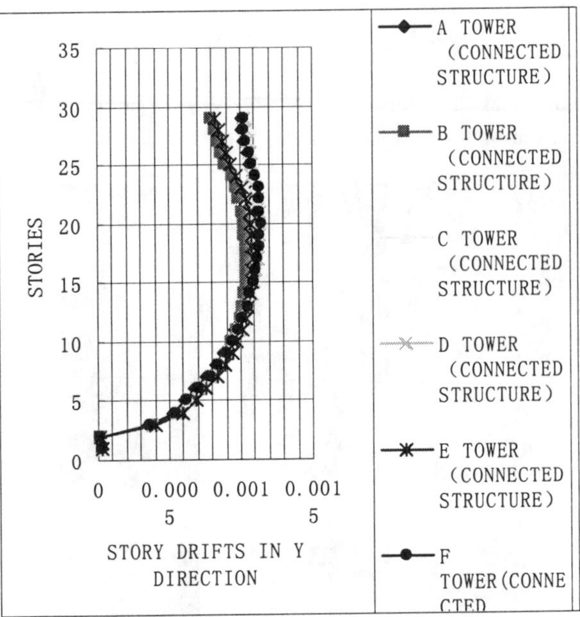

Figure 6. Story drifts of the structure with corridors

The story drifts are almost the same in two directions for all towers below the 15th story and then the difference is more and more obvious from 15th to 29th story.

Because of the corridor, the first fifty mode patterns of the structure are the complex composite of the first translation mode of each single tower, and the rate of periodical change is more slowly than the single tower. The torsion mode is not obvious and displacement of torsion mode is almost zero.

The maximal displacement of structure with hinge connection is 0.1015m in the x direction which is 0.0175m less than that of the single tower x direction. So it is advantageous to decrease the maximal translation displacement by setting a connection.

6. Rigid and Hinge Connection

The story drifts in x direction of the structure with the different connecting method are shown in the following table 1. For the story drifts, there is a little difference between the rigid connection and hinge connection. Though a little difference in story drifts of the whole structure, there is local moment at the joint of the tower and corridor.

There is the first seven mode of the structures in Table 2.The period of the structure with hinge is 0.001s longer than that of it with rigid connection.

From the table 3, it is concluded that the base reaction of structure with hinge connection is smaller than that of structure with rigid connection. The smaller

Table 1 The 29th story drifts in x direction

A TOWER	B TOWER	C TOWER	DTOWER	E TOWER	F TOWER	Connecting method
0.00098	0.001159	0.000934	0.000927	0.001155	0.000986	rigid
0.000992	0.001159	0.000936	0.000929	0.001155	0.000989	hinge

Table 2 The first seven mode of structure(s)

mode	1	2	3	4	5	6	7	Connecting method
Period	3.212	3.1236	2.802	2.513	2.4306	1.4712	1.228	rigid
Period	3.213	3.127	2.805	2.74	2.5315	1.7004	1.3477	hinge

Table 3 Base Reaction

Fl(KN)	F2(KN)	F3	M1(KN.M)	M2(KN.M)	M3(KN.M)	The stiffness of connection	connecting method
76049	77306	0	5535848	5475107	7484907	weak	hinge
71348	62513	0	4176475	5065186	6026594	strong	hinge
76632	77675	0	5541868	5480870	7527227	weak	rigid
72891	65339	0	4469890	5213927	6285575	strong	rigid

the base reaction is, the weaker the connection. So the base reaction decreases as the connection weaken.

The mode period of the structure with hinge connection is longer than that of the structure with rigid connection, because the weaker effect of hinge connection.

The maximal joint displacement of the corridor is 0.0154m in the z direction when the tower and corridor connected with hinge, while the maximum joint displacement of corridor is 0.0127m when the tower and corridor are rigid connection. It is effective to decrease the deflection of the long span corridor by rigid connection, but adversely produce a large local moment at the joint of towers and corridors.

7. The Different Stiffness of Corridor

In order to study the affluence caused by different stiffness of corridor to seismic response of the structure, the stiffness of the slab is not considered. The figure 6 is that story drifts of the structure under this condition.

From the figure 6, it is obvious that the maximum drift of the story in x and y direction of all towers is at the 18th story, just like the story drifts of the single tower under the seismic action. The difference between the A tower and B tower is more and more great from the 4th story. The maximal story drift is 0.0014 in x direction, while it is 0.0012 in figure 4.

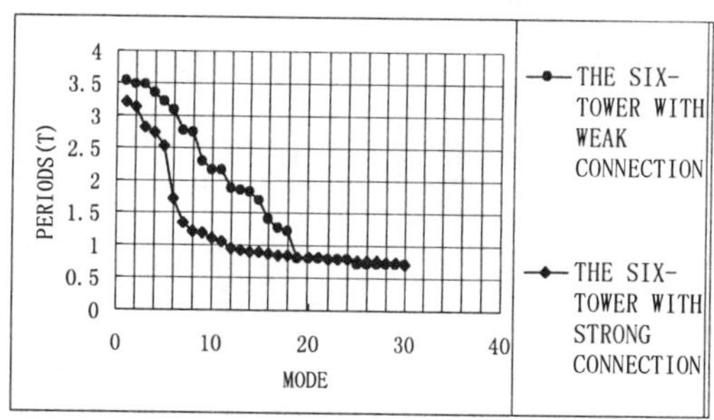

Figure 7 The periods of the six-tower

We conclude from the figure 7 that the period of the six-tower is longer with the weakening of the connection stiffness. The difference is the largest in the 19th mode. There is little difference after the 19th mode.

8. Conclusions

Through the analysis of the single tower and the six-tower connected structure, it is concluded that the vibration mode of the connected structure is more complex than the single tower because of the affluence of the corridor. There is a little difference in the story shears and story drifts under seismic action between the structure with hinge connection and rigid connection. Rigid connection is effective to decrease the deflection for the long span corridor, but adversely produce the large local moment at the joint of towers and corridors. With the stiffness of corridor weakened, the maximal story drift of B and E tower is increased.

References

1. Huang Binkung, SunBin-nan, dynamic Characteristic and Seismic Response Analysis of Unsymmetrical Connected Double-tower Structure Industrial Construction, 2001, vol31, no 8, P27~29
2. Sun Bin-nan, Lou Wenjuan, Dynamic Analysis for Asymmetry Double-tower Building, National Conference on Tall Building,17th session, 2002, P344~350
3. Huang Kunyao, Sun Bin-nan etc., The Influence of Corridor Stiffness on Seismic Response of Connected Double-tower Structure, Journal of Building Structure Vol.22, No3, June, 2001, P21~42
4. Lou Yu, Wang Hongqing, Vibration Analysis of Two Towers and Connected Double-tower Structure With Enlarged Base, building structure 1999, 4(4), P9~12

FINITE ELEMENT ANALYSIS ON SEISMIC PROPERTIES OF MID-HIGHRISE STEEL BRACED FRAME

Y. HUANG
Department of civil engineering, Tsinghua University
Beijing, 100084, China

Y.Q. WANG
Department of civil engineering, Tsinghua University
Beijing, 100084, China

H. CHEN
Architectural Design and Research Institute of Tsinghua University, Tsinghua University
Beijing, 100084, China

Y.J .SHI
Department of civil engineering, Tsinghua University
Beijing, 100084, China

Recently, a large number of mid-highrise buildings appeared in large cities in China, and they are very popular among the constructions for residents and commerce. Steel braced frame is one of the widely used structural forms in mid-highrise building. In the paper, one nine-storey braced frame in Langfang is investigated as an example by the finite element software ANSYS. Three-dimensional model is proposed in which the seismic response is analyzed including modal analysis, response spectrum analysis, elastic/elastoplastic time-history analysis. Futhermore, the lateral deformation of the structure and the distribution of the braces are computed, and the study can be served as the references of similar structure design.

1. Introduction

Mid-highrise residential buildings (7-storey to 12-storey usually) can meet the requirements of modern residence, satisfy the willing of inhabitants, and raise the utilization ratio of the land to economize the land resource. So many superiorities compared with multi-storey and highrise buildings make mid-highrise buildings the trend of the residence developing in large and medium cities.

On the basis of pure frame system, braced frame system increases the building stiffness, reduces horizontal deformation in the transverse force, which economizes the consume of steel. The bending moment of beam-column joints is small, thus the constructional detail can be simplified. And now braced frame system is widely adopted in mid-highrise buildings.

Nowadays, much work has been done on seismic behavior of mid-highrise structures. Many softwares have been used in structural design. But to some buildings whose elevation view is irregular, the non-linear time-history analysis method is still immature. If we just use the specific software for simplified calculation, some error may occur, and the analysis is inadequate in some local constitution. Finite Element software is developed earlier with fewer assumptions, except that its development is restrained by the limitation of the computer hardware. With the renovation of the computer hardware, it becomes possible for finite element analysis to be served to analyze the seismic behaviors of the structures for civil engineers. Here, we choose ANSYS as the calculate software, which transient analysis type can be used in time-history analysis.

Now, taking a mid-highrise braced frame structure as an example, the dynamic properties and seismic behaviors will be analyzed by ANSYS.

2. Background of the Project

A comprehensive service center, which is a steel braced frame structure in Langfang, is computed as a design case. The building is 9 stories, 38 meters high, and its west-east dimension is 38.4 meters while north-south dimension is 16.2 meters. All columns are in box shape and all beams are in I shape. All connections are rigid. The material of steel is Q345(f_y=345N/mm^2, E_X=206000 N/mm^2). On the floor the dead load 4.5kN/m^2, the live load is 4.5kN/m^2, while on the roof both are 0.5kN/m^2. The building is built at the region where the earthquake protected intensity is 8 and the construction site is class III.

The plan view of the first floor and the elevation view are shown in Figure.1. The dimension of primary components is shown as table 1.

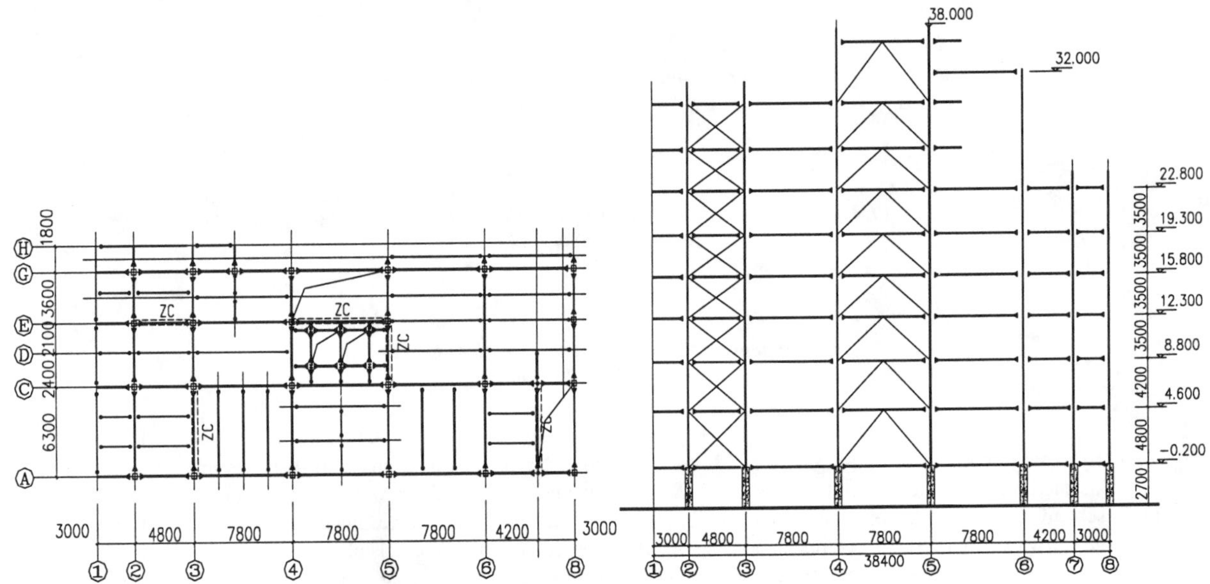

Figure 1. the plan view of the first floor and the elevation view

Table 1. the dimension of primary components

floor	column (box shape)	main beam (I shape)	junior beam (I shape)	brace
1	350×350×18			
2				
3				
4	350×350×16			
5		400×200×8×12	350×200×8×10	2[20a
6				
7	350×350×14			
8				
9				

In the design process, the static analysis and spectrum analysis have been done by the software PKPM, and the static calculation has been checked by ANSYS. Now, we will use ANSYS to get dynamic characteristics and seismic properties of the building further.

3. The Properties of the Model

3.1 *Element choice*

According to the characteristic and requirement of the structure, four element type is chosen after consulting the manual book of element library. They are beam188 (column, beam), shell181 (diaphragm), mass21 (additional mass), link8 (brace).

The choice of material model: the steel used in construction satisfies von-Mises criteria, while flow code is Prantl-Ruses code. The perfect elastoplastic material model is taken for simplicity.

3.2 *Model building*

The elevation of the structure is irregular. The center of mass and the center of the stiffness are not coincident, so the torsional mode can't be ignored. The finite element model is built in three dimensions according to the actual size. The model is built storey by storey for effective check. The dimension of the model is totally 1099 nodes, 2274 elements, 4 kinds of elements. The finite element model is shown in Figure.2.

Figure 2. 3D model

3.3 *Model analysis*

The low order of the vibration mode plays the controlled part in seismic analysis. So in the example, the analysis only considers the low order modes and ignores the contributions of high order modes. The lower tenth modes are taken into account, and lanczos method is chosen in this analysis. Analyze the mode of the structure, the periodic time of each order is shown as table 2.

Table 2. the periodic time of each order

mode	ANSYS	PKPM	experiential formula	Eurocode	Illustration of each mode
1	0.852	1.147	0.72-1.08	0.77	Translation in Y axis and torsion
2	0.795	1.087			Translation in X axis and Y axis
3	0.680	0.784			torsion
4	0.357	0.481			Torsion of the top floor
5	0.321	0.390			Torsion of the top three floors

According to the experiential formula in China, $T_1 = (0.08\sim0.12)N$, where N is the number of total floors.

According to Eurocode 8, the period of the structure whose height is below 40m can be calculated by simplified formula: $T_1 = C_t H^{3/4}$, C_t is 0.05 for braced frame structure.

The period calculated by ANSYS is smaller than by PKPM, which has relation to the parameter value of the whole building. The result by ANSYS is in the range of experiential formula in China and the simplified formula in Eurocode, and in the first five vibration modes two softwares agree quite well.

4. Analysis of Seismic Behavior

4.1 *Response spectrum analysis*

According to seismic code, the seismic influence coefficient is taken as 0.16, and the viscous damping ratio ζ is 0.035. The characteristic period of the site is 0.45s.

In the response spectrum analysis, the bidirectional earthquake response is considered (because in the small span analysis, the effect of vertical earthquake is ignored). The square-root-of-the-sum-of-the-squares (SRSS) method is adopted. The first to tenth mode is considered. The result is shown in tables below.

Table 3. The results by response analysis in unidirectional earthquake

Item	Base shear force (KN)	horizontal force / vertical force	The displacement of the top floor		The maximum value of interstorey drift		
			Δ (mm)	Δ/H	δ (mm)	δ/h	location
X direction	1870.6	7.4%	20.409	1/1803	3.866	1/1242	Base floor
Y direction	1852.3	7.3%	22.191	1/1658	5.27	1/1328	Top floor

Table 4. The results by response analysis in bidirectional earthquake

Item	Base shear force (KN)	horizontal force / vertical force	The displacement of the top floor		The maximum value of interstorey drift		
			Δ (mm)	Δ/H	δ (mm)	δ/h	location
X direction	2082.4	8.2%	22.693	1/1622	4.29	1/1119	Base floor
Y direction	2149.1	8.5%	25.611	1/1437	5.92	1/1182	Top floor

4.2 *Elastic time-history analysis*

Response spectrum analysis is determined by the most violent phase in the seismic acceleration record, which can't externalize the duration of the earthquake well. So we take time-history method and make the direct numerical integration to get results. In the paper, the representative record El-centro record and Kobe record are calculated and declared as example. The maximal value of the acceleration in the record curve is 70cm/s^2. The transient dynamics block is used in ANSYS, and New-mark method is taken as time integration method. Control amplitude method is taken, set $\gamma = 0.05$ as integral parameter. The damp of the structure is set as Rayleigh damp in approximation.

Comparing the interstorey drift calculated by response analysis and time-history analysis (Figure.3), we can conclude that the response analysis method is the statistical result in certain security. But in some specific situation, the time-history analysis result may be larger. In the fig.4, the displacement of the top floor is compared by the spectrum method, greatly exceed occurs in some certain points.

Observing the horizontal displacement along height from fig.3, the structure is main shear deformation. The whisper effect is obviously, which is effected by the fourth vibration mode (the local displacement of the top floor). So we should pay enough attention to the constitution of the top floor.

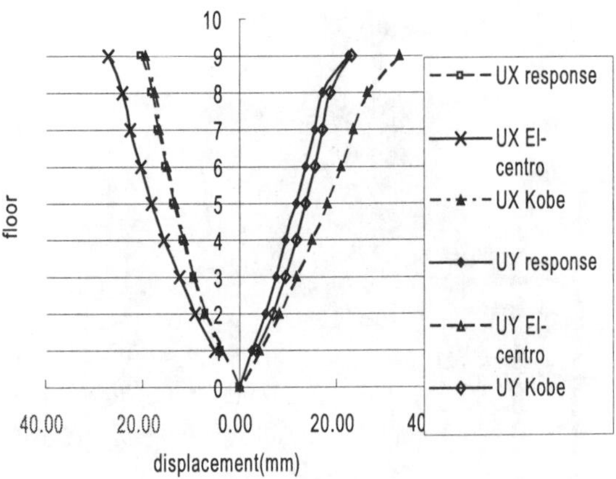

Figure 3. The displacement of the mass center in each floor by different method

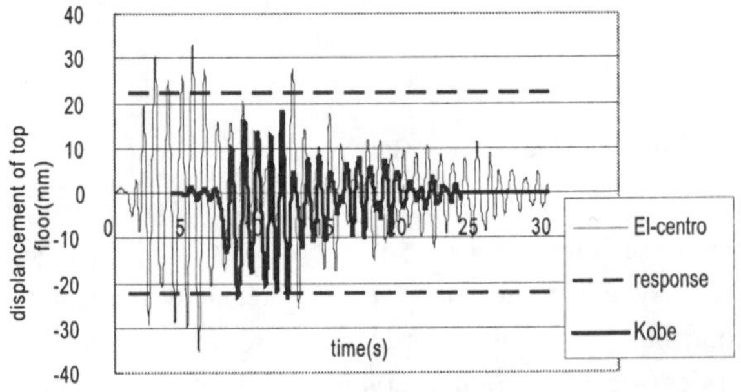

Figure 4. the comparison of time-history analysis and response analysis

4.3 *Elastoplatic time-history analysis*

In china, the building should satisfy the non-collapse requirement in strong earthquake, which is one of the protecting purposes. We will analyze the response of the structure in the earthquake whose reference probability of exceedance in 50 years is 3% according to the code. The maximal value of the acceleration in the record curve is 400 cm/s^2.

In the strong earthquake, there is some local yield in the structure. The horizontal displacement of the structure is shown in table 5, which satisfies the allowable interstorey drift index (1/50) of design code.

Table 5. The results by response analysis in bidirectional earthquake

item	The displacement of the top floor		The maximum value of interstorey drift		
	Δ(mm)	Δ/H	δ(mm)	δ/h	location
X direction	138.2	1/266	31.94	1/144	Base floor
Y direction	161.6	1/227	42	1/166	Top floor

5. The Contribution of the Brace

5.1 *The contribution to seismic effect coefficient*

The braces added in frame structures raise the stiffness of the structure and shorten the vibration period. In a general way, the periodic time of the mid-high rise structure is from 0.8s to 2.0s, in the range of T_g-$5T_g$ of the seismic effect coefficient curve. In this range, the seismic effect coefficient changes in large slope, and greatly impacted by the foundational period.

Table 6. Comparison period and seismic effect coefficient

	Pure frame	Braced frame
Periodic time	1.234s	0.852s
Seismic effect coefficient	0.696	0.980

When the braces are added, the base shear force of the structure will be 1.41 times of pure frame, increasing nearly 34%. Thus, the seismic behavior of braced frame needs special notice.

5.2 *The contribution to horizontal displacement*

The brace can stiffen the structure effectively, and reduce the displacement of the building in horizontal forces (wind or earthquake), which make the structure satisfy the performance-based design.

In the case, the consume of steel of beams and columns is almost 94 percent of the total and brace is only about 6 percent. The consume of steel added is very little, but it can reduce the horizontal displacement effectively (figure 5), which reflects the superiority of the braced frame structures.

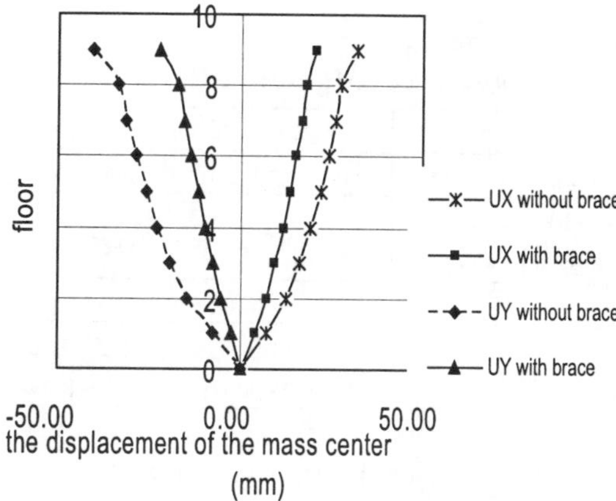

Figure 5. the comparison of the displacement of each floor

5.3 *The contribution to the inner force of columns*

In braced frame structures, the brace can bear a large part of interstory shear force. The proportion of the shear force borne by braces and by columns can be got by response spectrum analysis, the result is shown in table 7.

From table 7, we can see that the existence of braces can reduces the shear forces in columns to a large extent. It is advantageous to the column, but on the other hand, it will enhance the axial force of columns which is adverse. In table 8, the inner force calculation before bracket is in small earthquake, while in bracket is in strong earthquake. From it we can conclude that the magnitude of the additional axial force is about hundreds KN in small earthquake, even several thousand KN in strong earthquake, which may cause the column yield in earlier time, this situation should be pretended.

Table 7. The base shear force partaken by components

	The horizontal force borne by braces (N)	The horizontal force borne by columns (N)	Total horizontal force (N)
X direction	968276	902324	1870600
	52%	48%	
Y direction	1259804	592496	1852300
	68%	32%	

Table 8. the additional axial force to columns by brace

	X direction	*Y direction*
The max axial force N	379kN(1729kN)	423kN(1729kN)
The angle *α* between brace and horizontal plane	30 degree～50 degree	
sin *α*	0.5～0.766	
N×sin *α*	190～290kN (865～1324kN)	212～324kN (865～1324kN)

The dimension of the structure in X direction is larger than Y direction, so the stiffness in X direction is bigger, and the deformation is smaller. Then in X direction the braces function is relatively smaller, which agrees to the data in table 3.

5.4 *The contribution to seismic behavior*

The energy-dissipating capability of concentrically braced frames is weaker than ecconcentrically braced frame. And the ductility is worse correspondingly. The comparison of hysteric property is given between braced frame and pure frame. In El-centro earthquake, the relative curve of displacement of the top floor and base shear force is shown as Figure. 6.

From the comparison between figure6a and figure 6b we can get the result that in the same earthquake the stiffness of the pure frame is small and the displacement of the top floor is pretty big. Because of the flexibility of the structure, the base shear force applied to pure frame is small, which meet the analysis above.

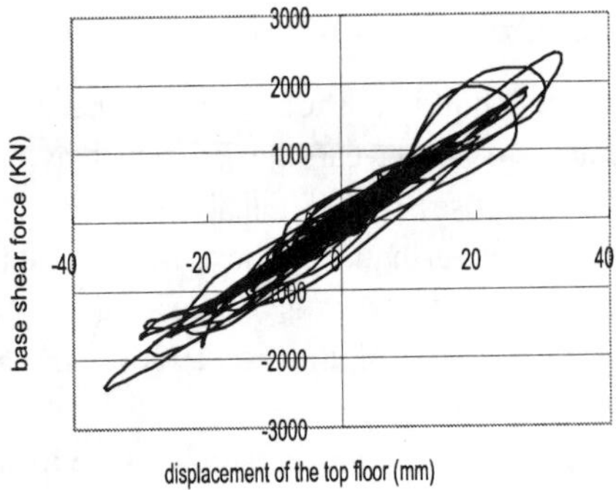

Figure 6a. with brace

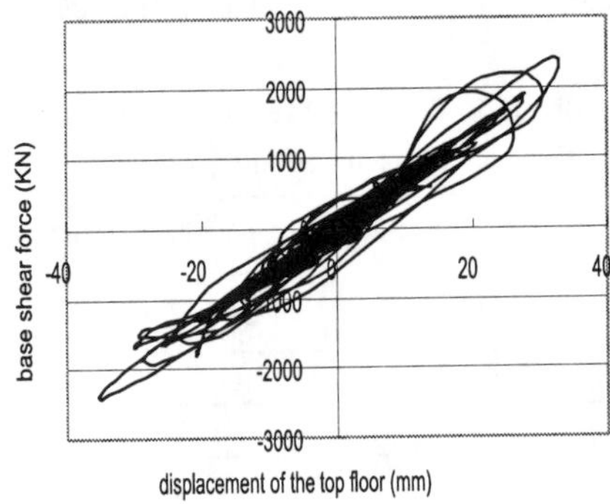

Figure 6b. without brace

Figure 6. base shear force-displacement of top floor curve

The energy-dissipating capacity of braced frame and pure frame is in the same order of magnitude. For the good ductility of steel, the hysteric curve of both is plump. The stiffness of braced frame is large. While the horizontal force is enlarged to a certain distance, the brace will yield first and then its mechanic property is same to pure frame. This will embody the design concept of dual defense of braced frame.

6. Conclusion

The structure satisfies the seismic requirement. The results of response spectrum analysis, time-history analysis in small/large earthquake are in the safety range according to the code, which verifies the results by PKPM software.

The simplification of finite element software is less, which afford an effective and reliable calculative platform for the comprehensive analysis of complex and large-scale structures.

The influence of braces to the seismic effect coefficient of the whole building, the inner forces of the component and the horizontal displacement of each floor in the dynamic loads is analyzed.

To complex structure, the results by response spectrum analysis may not be partial to the security. In this case time-history analysis should be done as verification.

References

1. Shen Jumin, Gao Xiaowang, Liu Jingbo, *Aseismic engineering.* 108 (2000).
2. Wang Xucheng, *the basic principle and digital method of finite element.* 443 (1997).
3. *APDL programming guide.* ANSYS Inc, 40 (2000).
4. *Chinese seismic code: Code for seismic design of building(GB50011-2001),* China construction Industry Press (2002).
5. *Eurocode8: Design of structures for earthquake resistance,* European Committee for Standardization (2003).

COMPARATIVE RESEARCH ON DISPLACEMENT CONTROL SCHEME OF HIGH-RISE BUILDING IN HIGH SEISMIC INTENSITY AREA

GU WEIJIAN

(Beijing CanBao Institute of Architectural Design, Beijing 100850, China)

Research work into the structural system schemes of displacement control are summarized and the corresponding research aimed to establish some available methods of displacement control of high rise buildings in high seismic intensity area based on analytical results from a practical project has been done. It is revealed that the structural displacement can be controlled by improving both vertical and horizontal structural element stiffness. The conclusion from this research can be used as a reference for relative engineering design.

1. Summary

In some high seismic cities which need seismic fortification, with the increase of building height and under side force action of earthquake load, side displacement of structure (or distortion) becomes bigger and bigger. With new regulation implementation of high-rise building design, there are some new regulations for limited value of high-rise building displacement and comprehensive requirements are higher and higher. How to seek for optimal technical approach to control side distortion of high-rise building in high seismic intensity area under the action of earthquake load during structural design of high-rise buildings? More and more civil engineering scientific research and design personnel attach great importance on the issue.

2. Project Summaries

A project in east suburb of Beijing with construction area of 70,200 sq.m. (including 47, 150 sq.m. of superstructure area and 23,050 sq.m. of substructure area). The project has two floors of underground construction, four floors of skirt building, and thirty floors of tower building, total construction height is 98.2 meters, east-west length of skirt building is 132 meters and south-north length is 46 meters.

Tower building is located in the middle part of skirt building basically. Seismic fortification intensity is Level Eight（0.2g）, designed earthquake group is Group No.1 and the site is Class □.

Adopt piled raft foundation as foundation of tower building and use separate column bases with water-proof bottom panel as other foundations. Adopt

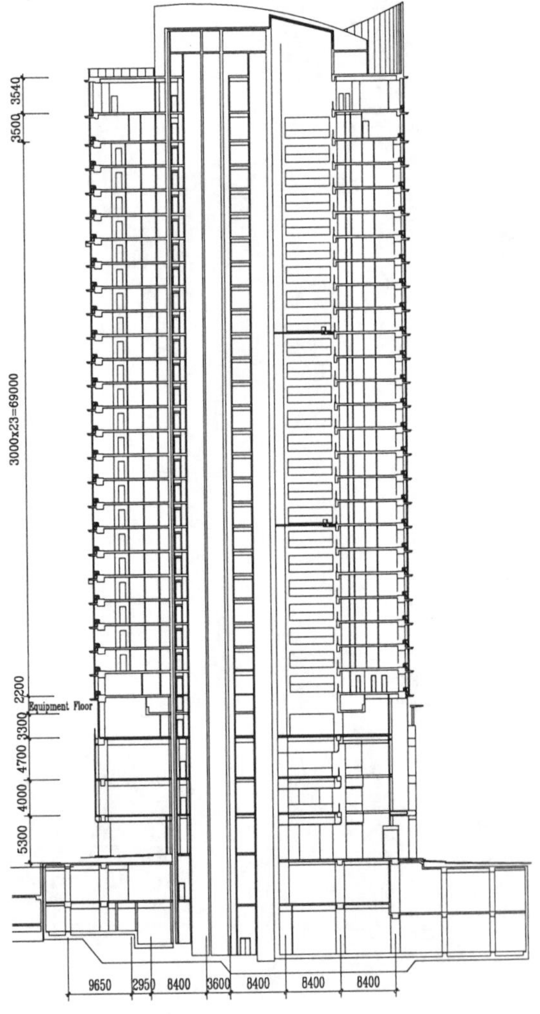

Profile figure

Figure 1

frame-shear wall structure as main structure. Standard floor outline of tower building looks like a bullet head; shear wall is centralized in core tube and sparse frame columns are separated in the periphery of core tube.

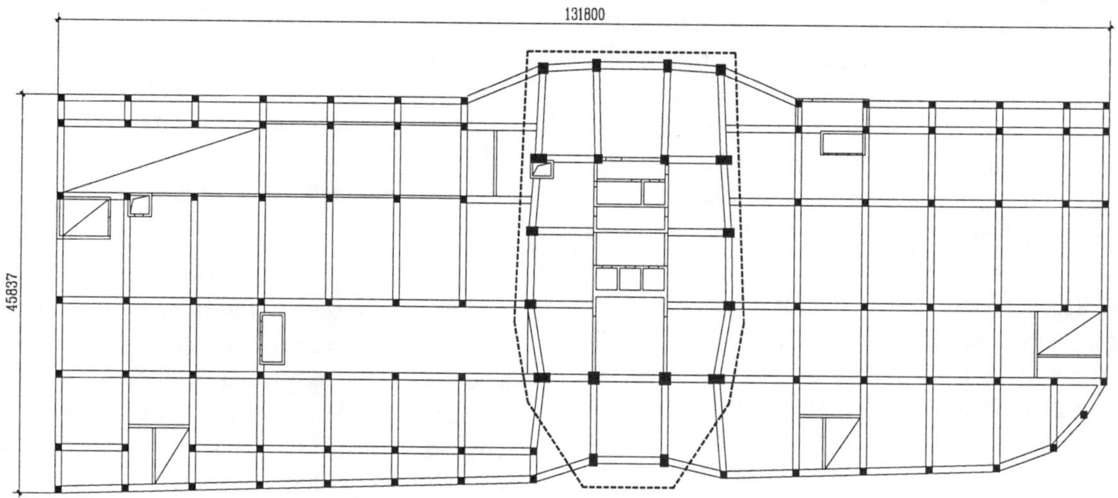

Relationship figure of skirt building and tower building

Figure 2

3. Comparison of Displacement Control of High-rise Building

3.1 *Comparison of displacement control scheme*

Different structural systems have different displacement control methods and effects.

Rod bending distortion is the principal part of frame structure (whole shear distortion is the principle part of side shift; bending distortion of column and beam is main reason of frame side shift.) Layer side shift increases layer by layer from top to bottom. Although column and beam section of frame increases layer by layer from top to bottom, increasing rate of column and beam bending stiffness is lower than increasing rate of floor shearing force. The trend of layers side shift of frame grows from top to bottom still exists. The whole side shift curve of frame is similar to sinusoid with zero structure bottoms. Because of the frame attribute of "shearing distortion", the minimum layers side shift is at the top of structure and the maximum is at the bottom or several layers of structure.

Pure shear wall structure, with less building floors, height and width rate of wall body is less than 1 under horizontal load, shear distortion should give priority to wall body with smaller proportion of bending distortion and side shift curve of wall body should be shear-type. With more building floors, height and width rate of wall body is more than 4, bending distortion should give priority to side shift of wall body and shearing force distortion should be at a secondary position, side shift curve of wall body should close to curve type. When height and width rate of wall is between above two numbers, shearing distortion and bending distortion should occupy a certain proportion separately and side shift curve should be both shearing and bending type.

When earthquake occurs, with regard to frame supported frame-shear wall structure, there is small damage in all levels of upper part when using vertical and horizontal wall body to support the weight; when adopting bottom which using frame to bear the weight, seriously damage happened because floor push resistance rigidity and yield intensity is decreased suddenly and then results in concentrated distortion and centralized energy. Floor yield strength distribution is an important factor which affects distortion of structural layers.

Frame—shear wall core tube structure, under the action of horizontal load, bending distortion becomes the principal part of side distortion and shearing force distortion should be at a secondary position. The side shift curve is close to side shift curve of bending roller. Side shift of frame and core tube in frame - core tube system tends to be consistent because it is restricted by all floors of building cover level rigidity. Originally, upper and lower displacement angles of structural parts are uneven badly, but it has been improved currently. Equal strength design of whole structure and distortion requirements are more superior to pure frame and pure shear wall of single structural system.

Structural system of frame-core tube with rigid arm, vertical and horizontal rigid arms which have been added in the system can reduce side shift of whole structure remarkably. Comparing with frame-core tube system, reduced range of structural side shift of frame-core tube system with rigid arm grows with the increase of rigid arm. However, as for the action of one rigid arm, with increase of rigid arm quantity, effect of structural side shift should be reduced gradually. Rigidity ratio of core tube and exterior beam should affect numerical value of structural side shift directly. If only one rigid arm is placed, the effect is most remarkable and value of structural peak side shift is least when rigid arm is placed at 0.6H. With the increase of ratio of core tube and exterior beam, rigid arm will reduce effect of structural side shift gradually. The aim of putting up rigid arm is to make exterior beam and core tube shape up as a whole and resist overturning moment arising from horizontal load together. Therefore, when the vertical bending rigidity of rigid arm is unlimited, exterior beam will affect whole bending effect of structure most fully; with the decrease of vertical rigidity of rigid arm, vertical shearing and bending of rigid arm increases gradually, so that exterior beam will weaken whole bending effect of structure gradually.

Distortion of tube-in-tube structural system is similar to frame-shear wall structural system. The kind of system provides main contribution of distortion control: reduces maximum layers side shift of exterior tube which is located in lower part of structure and maximum layers displacement of interior tube in upper part of structure to make layers side shift of exterior tube and interior tube symmetrical from top to bottom.

Take actual requirements of project construction function and location of high seismic intensity area of Degree Eight and Class III site into account, structural system has adopted core tube with wide frame of sparse frame column - shear wall structural system outside.

After first calculation, we find that floor maximum displacement angle under the effect of earthquake load can't meet the requirements of regulation; therefore, we should adjust structural layout and displacement control properly.

3.2 *Calculated result analysis of different adjustment schemes*

Please find Figure 3.1 for standard plane of the project. For structure specialty, following critical limitations exist in construction plane layout for displacement control: (1) Construction plane layout is not symmetrical; basically, vertical axis is symmetrical and horizontal axis is not symmetrical; (2) Rigidity and mass centre of structure are not coincided; there are much more numerical value differences; structural eccentricity is very big; (3) It's difficult for core tube of inner reinforced concrete and scheme for installing sparse columns in the outer frame to control displacement in high seismic intensity area effectively; (4) Floors with larger displacement is located in the upper floor of structure, there are obvious comparison between skirt building with big basement and embedded effect of basement; structural rigidity is softer obviously.

To meet limited value requirements of ratio of maximum displacement and storey height of Technical Specification for Concrete Structures of Tall Building (JGJ3 - 2002) , that is $\Delta h/h \leq 1/800$, we put forward displacement control schemes which enhances vertical rigidity of structural parts such as Figure 3.2, Figure 3.3, Figure 3.4 and Figure 3.5 based on the structural system. Scheme II is adding four pieces of reinforced concrete coupling walls each in upper and lower of core tube; Scheme III is adding one piece of reinforced concrete coupling wall in lower part of core tube; Scheme IV is increasing rigidity of horizontal structural parts and establishing reinforced concrete crossbeam in the position of Scheme III instead of adding coupling wall. Scheme V is combining with Scheme II and Scheme III and adding five pieces of coupling walls. Table 3.1 is structural system comparison of all kinds of schemes, Table 3.2—Table 3.4 is statistical conditions of single tower structure of all kinds of calculated results after calculated by SATWE program.

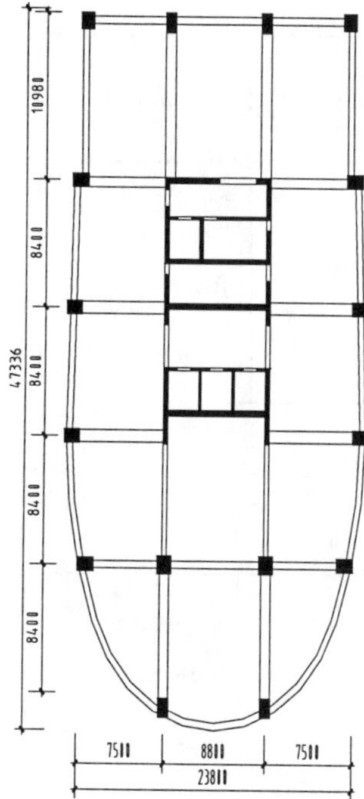

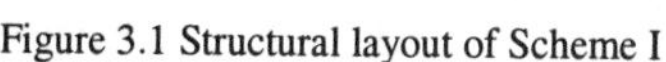

Figure 3.1 Structural layout of Scheme I

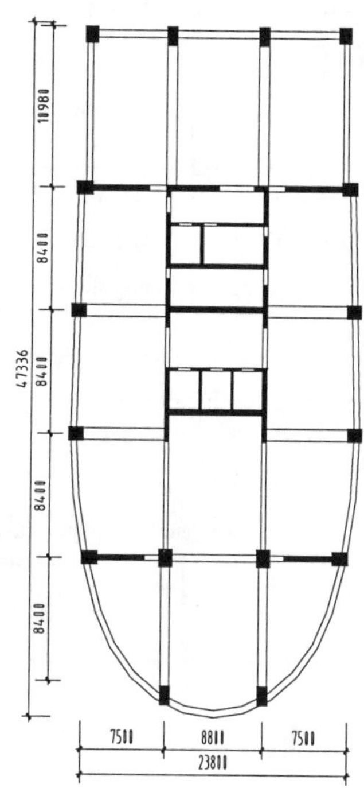

Figure 3.2 Structural layout of Scheme II

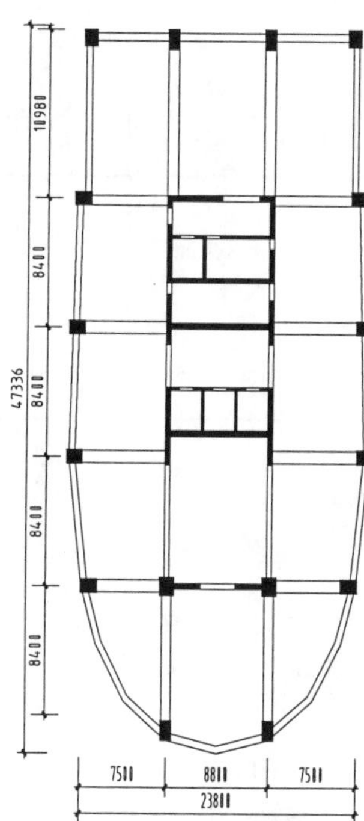

Figure 3.3 Structural layout of Scheme III

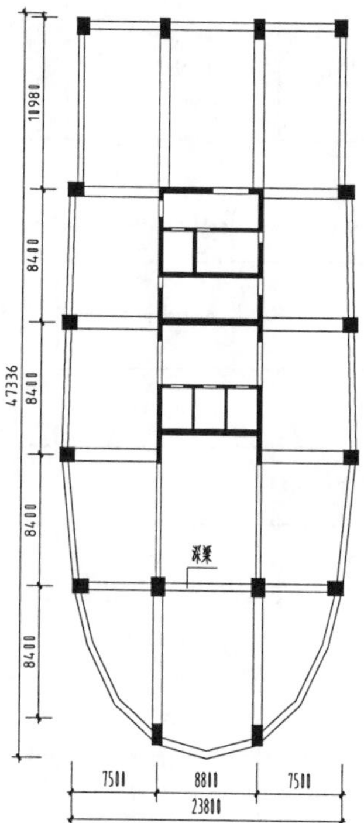

Figure 3.4 Structural layout of Scheme IV

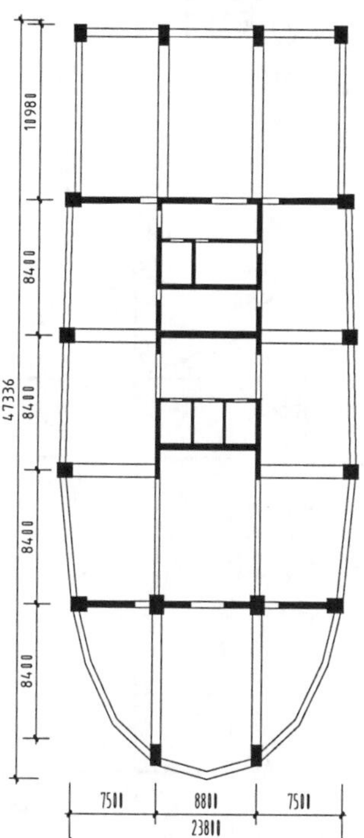

Figure 3.5 Structural layout of Scheme V

Table 3.1 Structural system of all kinds of schemes

Scheme / Content	Scheme I.	Scheme II.	Scheme III.	Scheme IV.	Scheme V.
Structure system	Frame - shear wall	Partial frame-supported frame-shear wall	Partial frame-supported frame-shear wall	Frame - shear wall	Partial frame-supported frame-shear wall
Overlimit level	Level A	Level B overlimit	Level B overlimit	Level A	Level B overlimit

Table 3.2 Vibration period when taking turning couple into account (sec)

Scheme / Period	Scheme I.	Scheme II.	Scheme III.	Scheme IV.	Scheme V.
First period (Torsion coefficient)	1.914(0.11)	1.929(0.00)	1.946(0.01)	1.914(0.00)	1.900(0.00)
Second period (Torsion coefficient)	1.878(0.43)	1.731(0.94)	1.884(0.84)	1.764(0.47)	1.650(1.00)
Third period (Torsion coefficient)	1.687(0.46)	1.526(0.06)	1.768(0.15)	1.680(0.52)	1.489(0.00)

Table 3.3 Foundation shear force, bending moment and shear force weight ratio

Scheme / Content		Scheme I.	Scheme II.	Scheme III.	Scheme IV.	Scheme V.
Foundation shear force (KN)	X	17315.34	22651	19697	19089.93	23117
	Y	17782.48	19363	18380	17887.32	19168
Bending moment (KN-m)	X	836412.81	1174587	983718	961856	1222043
	Y	939557.44	1023031	971731	945436	1014318
Shear weight ratio	X	4.08%	5.02%	4.43%	4.47%	5.10%
	Y	4.19%	4.29%	4.19%	4.19%	4.23%

Table 3.4 Maximum displacement angle of floor under the action of earthquake

Content \ Scheme		Scheme I.	Scheme II.	Scheme III.	Scheme IV.	Scheme V.
X direction (Number of layers)	X	1/1040 (13)	1/1237(16.17)	1/1106 (15.16)	1/1168 (15)	1/1421 (20)
	X-5%	1/1006 (11)	1/1049 (15)	1/1058 (14.15)	1/1081 (14)	1/1198 (17)
	X+5%	1/915 (12)	1/1244 (17)	1/939 (13.14)	1/996 (14)	1/1153 (17)
Y direction (Number of layers)	Y	1/1102 (12)	1/1101 (11)	1/1080 (11)	1/1084 (11)	1/1120 (11)
	Y-5%	1/1061 (11)	1/1050 (11)	1/1019 (11)	1/1073 (11)	1/1068 (11)
	Y+5%	1/1039 (11)	1/1046 (11)	1/1031 (11)	1/1028 (11)	1/1070 (11)

From Table 3.2, we know that with the increase of vertical rigidity of the whole structure (the quantity of shear wall increases); self-vibration period of structure becomes shorter. Torsion effect of Scheme II, Scheme III and Scheme V is much more remarkable. From variation conditions of periods, we can conclude that vibration period is located in the same value basically with smaller variation range; while variation of vibration period in Y direction changes bigger, period of torsion accounts for bigger proportion.

From Table 3.3, we can see clearly that earthquake shear force of X direction increases in direct proportion with increase of structural vertical rigidity. Scheme V with maximum rigidity increase is 25% more than Scheme I in X direction and reach to peak of increase. Earthquake force variation increases in Y direction (within 8%). For variation range of foundation bending moment, the maximum difference is 31.6% in X direction and 8.2% in Y direction.

Maximum displacement angle in X direction is taken into account when it is calculated as combined model of main tower building and skirt building, Scheme I can't meet the regulation requirements of less than 1/800. Currently, calculated results in Table 3.4 are calculated result of single tower structure of main tower building. Numerical value of displacement angle is 25% smaller than combined models. Therefore, maximum displacement angle of actual structure should be improved limitation of structural distortion by means of adjusting and strengthening structural layout of Scheme I. From the calculated result, we can analyze that Scheme V has optimal control result of structural distortion because of maximum rigidity increase. But its weakness is usage space layout of structure is restricted, for example, lots of wall bodies should fall on the ground and construction layout isn't flexible. From effect degree of original scheme, we can analyze that Scheme IV is the best; it only need add one higher frame beam. Through adjusting rigidity of part structural elements and then affect whole distortion characteristics to reach the aim of controlling maximum displacement angle of structure effectively.

In conclusion, we select reinforced scheme of Scheme IV finally. Under the premise of structural system doesn't break through current Level A frame--shear wall, we can reach control of structural displacement angle through mutual coordination with architectural specialty and enhance rigidity of horizontal structural elements.

4 . Conclusions

Through comparative research on displacement control of high-rise building in high seismic intensity area for different structural adjustment schemes, both vertical rigidity which can enhance structure and horizontal structural elements rigidity which can increase structure elements are having an effect on floor displacement control of high-rise building. There is remarkable effect when comparing effect of enhancing vertical rigidity with effect of increasing horizontal structural elements rigidity. For sparse outer frame column—frame of inner core tube-shear wall system, if construction using functions are permitted, rigidity of vertical structural elements should be increased to adjust structural scheme to meet the regulation requirements of building floors displacement.

References

1. Liu Dahai, Optimal Selection of High-rise Building Structure Scheme, Shanxi Science Technology Press, December 1992
2. Xu Peifu, Effect of placement Height of Transform layer on Earthquake Prevention Performance of Frame-supported Shear Wall Structure, Collection of 16th China High-rise Construction Structure Science Convention, Shanghai, October 2000

DESIGN PRINCIPLES OF ASYMMETRICAL BUILDING SUBJECT TO HORIZONTAL EARTHQUAKE AND APPLICATION IN SHEN ZHEN INTERNATIONAL MAYOR BUILDING

WEI LIAN WANG SEN
Shen Zhen division, Shang Hai Wei Lian Structural Design Ltd.
Shang Hai, China

YUAN CHENG MING
Shen Zhen Hua Zhu Structural Design Ltd.
Shen Zhen, China

This paper points out that the main design principles of an asymmetrical building subject to horizontal earthquake are to take into consideration of (1) the torsional angle of the floor (rigid diaphragm) not to induce torsion exceeding the allowable, (2) the inter-story displacement at utmost end frame or shear wall should be less than that required by design code, (3) stresses in plane of the slab shall be controlled within acceptable extent under different intensity earthquakes. Current seismic design code only utilizes the torsional displacement ratio to control the floor torsion. This seems not reasonable enough since its definition is the multiple of the floor torsional angle and the distance of floor mass center till the utmost end frame or shear wall. The International Mayor Building of Shenzhen City is 26 storied high, the tower is L-Shaped in plan and with a transfer beam-slab system at 9th floor. The earthquake action will induce evident torsion to the structure. This paper introduces the seismic design considerations of the building.

1. Introduction

Previous earthquakes have indicated that asymmetrical buildings were widely damaged and the destructive feature was obvious due to torsional effect. Therefore, it is important to make a structure to possess good torsional capacity in seismic design in order that the seismic safety of the structure will be raised.

2. Regulations on Torsional Effects

Item 4.3.5 of the code 《The Technical regulations for concrete tall building》 of China (JGJ3-2002) stipulates that " The torsional affects should be abated as much as possible in the structural layout. The maximum horizontal displacement and inter-story displacement of columns and shear walls under earthquake action considering the accidental torsion, should not exceed the average value multiplied by 1.2, and must not exceed the average value multiplied by 1.5 (for tall buildings of A Class) or 1.4 (for tall buildings of B Class)." The torsional displacement ratio is stipulated in this item of JGJ3-2002. Some other codes, such as American code UBC, Eurocode-8 have the same torsional displacement ratio control values as JGJ3-2002.

3. Disadvantageous Torsional Effects on Structure under Horizontal Earthquake

The basic feature of asymmetrical building different from symmetrical ones is the torsional irregularity of the structure. The more significant the torsional irregularity of the structure, the bigger the floor torsional deformation take place under horizontal earthquake. The disadvantageous influence of floor torsional displacement to the structure are as follows:

1. As shown in fig. 1, the floor is regarded as rigid in the slab plan, the torsional angle of ith floor is θ_{ti} , so the columns C_{ji} and shear walls W_{ji} in ith floor will rotate an angle of

$$\theta_{tji}^{C} = \theta_{tji}^{W} = \theta_{ti} \qquad (1)$$

if the torsional angle of (i-1)th floor is $\theta_{t,i-1}$, the inter-story torsional angle of column C_{ji} and shear wall W_{ji} in ith floor is

$$\Delta\theta_{tji}^{C} = \Delta\theta_{tji}^{W} = \Delta\theta_{ti} = \theta_{ti} - \theta_{ti-1} \qquad (2)$$

The additional torque of columns and shear walls caused by floor torsion is

$$\Delta M_{tji} = K_{tji}\Delta\theta_{ti} \qquad (3)$$

How to provide design measure against the additional torque is not mentioned in the former codes. If the inter-story torsional angle is controlled in a suitable value, it is beneficial to decrease the torque of the vertical members (columns and shear walls) and to increase the structure safety.

2. As shown in fig. 1, the torsional angle of ith floor causes the additional horizontal displacement in the vertical members. The additional displacement in X and Y direction of vertical member S_{ji} in i^{th} floor is:

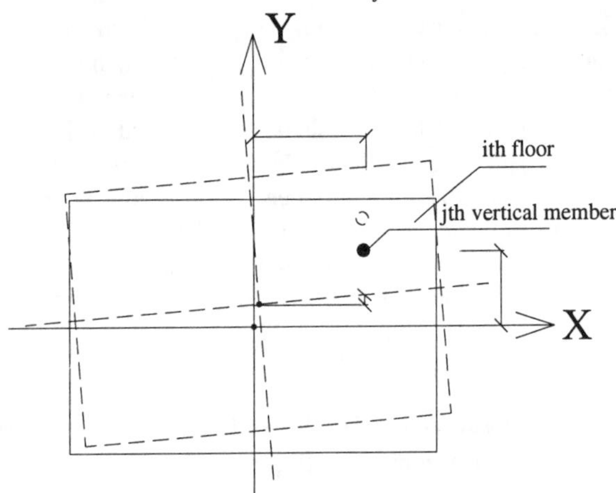

Figure 1.

$$\Delta_{xji} = \theta_{ti} y_{ji} \qquad (4)$$

$$\Delta_{yji} = -\theta_{ti} x_{ji} \qquad (5)$$

Correspondingly the additional displacement in X and Y direction of vertical member S_{ji-1} in (i-1)th floor is:

$$\Delta_{xji-1} = \theta_{ti-1} y_{ji-1} \qquad (6)$$

$$\Delta_{yji-1} = -\theta_{ti-1} x_{ji-1} \qquad (7)$$

Therefore, if the plan dimensions of two asymmetrical buildings are different, even the torsional angle of the two structures is identical, the additional displacements and inter-story displacements of the utmost end vertical member of the two structures may have obvious difference. But the additional displacement which increases the total displacement and inter-story displacement of the vertical member is disadvantage to seismic safety of the structure, so it must be controlled effectively.

3. The floor of an asymmetrical building plays the role to coordinate the deformation of the vertical members under horizontal earthquake, thus significant stress might occur within the slab plan. During the 1972 Manakua earthquake in South America, one of the serious damage features of the Central Bank Building is the occurence of significant crack, and the maximum width of crack reaches 10mm.

The inter-story torsional angle will cause torque and additional displacement of the vertical member directly, so when the inter floor torsion rigidity changes abruptly, it may arise large inter-story torsional angle at the soft torsion rigidity floor. More attention should be paid to control the ratio of inter floor torsion rigidity of the structure in seismic design.

4. Control of the Torsional Displacement Ratio

The torsional displacement ratio can be defined in the following,

$$\mu_t^* = \frac{\Delta_{\max}}{\Delta_a} = \frac{(\Delta_a + \Delta_{t\max})}{\Delta_a} = 1 + \frac{\Delta_{t\max}}{\Delta_a} = 1 + \frac{\theta_t x_m}{\Delta_a} \qquad (8)$$

In formula (8), $\Delta_{\max}$ is maximum displacement of the utmost end member; Δ_a is the average displacement of the vertical member; $\Delta_{t\max}$ is the additional displacement of the utmost end member by structure torsion, $\Delta_{t\max} = \theta_t x_m$; θ_t is the torsional angle; x_m is the distance form the centre of the floor mass to the utmost end member. Let

$$\mu_t^* = 1 + \mu_t \qquad (9)$$

be the new torsional displacement ratio

$$\mu_t = \frac{\Delta_{t\max}}{\Delta_a} = \frac{\theta_t x_m}{\Delta_a} \qquad (10)$$

The following can be obtained from formula (10):

(1) To control μ_t in design is same as to control the product of torsion angle θ_t and x_m / Δ_a, so when x_m / Δ_a is a certain value, to control μ_t is same as to control the torsion angle and inter-story torsion angle θ_t of the floor under a limit value, but the allowable value of the inter-story torsion angle in the current codes is not solved, in fact the designer can calculate torque and check the size and reinforcement of the vertical member to control the torsion angle within a suitable and allowable value.

(2) μ_t is the ratio of displacement of the utmost end member and average displacement, so when θ_t is a certain value, to control μ_t is same as to control the displacement and inter-story displacement of the utmost end member under an allowable value, it is obvious that if θ_t of two structure is identical the wider or longer structure which x_m or y_m is bigger will be more disadvantageous than the other.

(3) If μ_t is equal, the value of average displacement Δ_a effects the conclusion greatly. The average displacement Δ_a is

$$\Delta_a = \frac{\Delta_{\min} + \Delta_{\max}}{2} \qquad (11)$$

In formula (11), $\Delta_{\min}$ is the minimum displacement of utmost end member on one side, and $\Delta_{\max}$ is the maximum displacement of utmost end member on other side,

$$\Delta_{\max} = \Delta_a + \Delta_{t\max} \qquad (12)$$

let Δ_u be limit value stipulated in the code and express Δ_a as follows,

$$\Delta_a = \zeta \Delta_u \qquad (13)$$

thus the new torsional displacement ratio is

$$\mu_t = \frac{\Delta_{t\max}}{\zeta \Delta_u} \qquad (14)$$

Because the maximum displacement of the utmost end member must be less than the inter-story displacement limit value Δ_u given by in codes,

$$\Delta_a + \Delta_{t\max} \leq \Delta_u \qquad (15)$$

when summation of Δ_a and $\Delta_{t\max}$ reaches the maximum value Δ_u ,thus

$$\Delta_{t\max} = \Delta_u - \Delta_a = \Delta_u(1-\zeta) \qquad (16)$$

$$\zeta = \frac{1}{1+\mu_t} \qquad (17)$$

$$\mu_t = \frac{1}{\zeta} - 1 \qquad (18)$$

$$\mu_t^* = \frac{1}{\zeta} \qquad (19)$$

Therefore when μ_t is 0.4, ζ is 0.714, and when μ_t is 0.5 , ζ is 0.666. When ζ is 0.5, which means the average displacement Δ_a is half of Δ_u , the maximum additional displacement of the utmost end member will be $\Delta_{t\max} = 0.5\Delta_u$, accordingly μ_t =1.0(μ_t^* =2.0). Then the value μ_t and μ_t^* is bigger than the limit given by codes, 0.4 ~ 0.5(μ_t^* =1.4 ~ 1.5).

According to the above analysis, the main function of controlling torsional displacement ratio is to control inter-story torsional angle of the floor and the inter-story displacement of the utmost end member, but obviously only controlling torsional displacement ratio simply will lead to problems as follows:

(1) If the average displacement Δ_a is very small, the inter-story displacement of the utmost end vertical member caused by torsion is limited to $\Delta_{t\max} = 0.4 \sim 0.5\Delta_a$ because the ratio μ_t code defined is limited to 0.4 ~ 0.5, it shows that the bigger the value of Δ_a , the more loosely control and the smaller the value of Δ_a , the more strictly control in design, this results disobey the conventional as concept of structure safety.

(2) The limit value of average displacement is not mentioned in the codes. This may evoke misunderstanding that the ratio satisfied to the code control value simply means all very well, in fact it is not true. If the denominator of average displacement Δ_a is big, such as $\mu_t = 0.4$ or 0.5, it may derive unsafety result which can be seen from the following relationship

$$\Delta_{\max} = \Delta_{t\max} + \Delta_a = (0.4或0.5)\Delta_u + \zeta\Delta_u = [\zeta + (0.4或0.5)]\Delta_u$$

According to formula (18) and (19), the torsioal displacement ratio is different as ζ is varied, so we suggest the allowable torsioal displacement ratio, as shown in table 1.

Table 1. torsioal displacement ratio suggested

ζ	≤ 0.3	0.4	0.45	0.5	0.55	0.6	0.65	0.7	0.8
μ_t^*	≥ 2.0	1.9	1.8	1.7	1.60	1.55	1.45	1.40	1.20
μ_t	≥ 1.0	0.9	0.8	0.7	0.60	0.55	0.45	0.40	0.20

5. Application in Shenzhen International Mayor Building

5.1 *Structure*

The International Mayor Building of Shen Zhen City is 26 storeyed high, L-Shaped in plan and with a transfer beam-slab system at nineth floor. It is a multi-use building with office 、 shopping area and housing area and the total height is 109 meters. Below and above the transfer floor the structural system and plan shape are different. Below is frame-shear wall structure with fan-shaped plan and above is shear wall structure with L-shaped plan. It is a typical asymmetrical tall building. The tower plan is shown in figure 2.

5.2 *Calculation results*

5.2.1 *Natural periods*

The preceding three natural periods of the structure is T_1 =1.90S, T_2 =1.79S, T_3 =1.48S. The torsion period ratio is 0.78 (1.48/1.90) which is smaller than 0.85 the code stipulated.

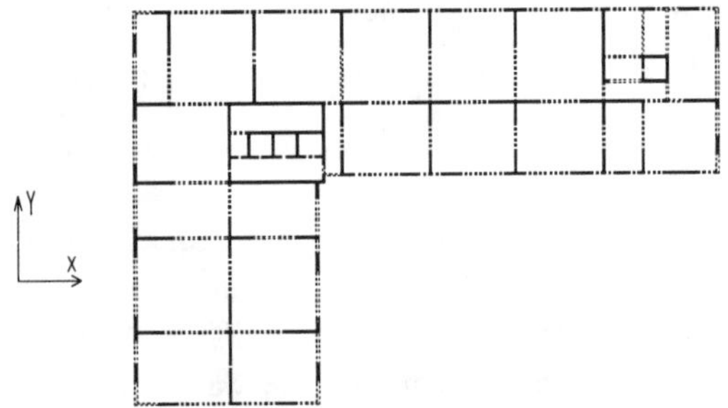

Figure 2. the tower plan

5.2.2 *Torsional displacement ratio*

The torsional displacement ratio is 1.52 at 9th floor under Y direction horizontal earthquake owing to maximum inter-story displacement is 0.85mm and the average inter-story displacement is 0.56mm, which exceeds code limit 1.4.

Based on the opinion in this paper, the rational torsional displacement ratio should be related to the average inter-story displacement Δ_a. The ratio of the inter-story displacement 1/3747 and the inter-story displacement limit value Δ_u limit in code 1/1000 is 0.267, according the ratio value shown in table 1, the allowable ratio is more than 2.0, so the torsional displacement ratio 1.52 is acceptable.

5.2.3 *Others*

Calculation results by ETABS show that stresses in floor plan are less than 1.3 MPa in most of floor areas and locally near the core a little bit higher.

Additional torque on vertical members of the structure caused by floor torsion is calculated and the strength is checked.

5.3 *Shaking table test*

Shaking table test of a structural model scaled 1/25 has been conducted. Test results obviously tell that the transfer beam system and the columns as well as shear walls below it are quite safe as compared with walls above the transfer beams and slab when the earthquake input for the test equivalent to and bigger than middle earthquake action, damage occur first at shear walls in the story just above the transfer beam floor with story torsional rigidity 0.5 due to torsional effect. This damage phenomenon means the importance of a reasonable seismic design for an asymmetrical building subject to horizontal earthquake and the structure design of the building was improved according to the test results obtained.

6. Conclusion

The following points should be paid more attention in designing an asymmetrical building subject to horizontal earthquake:

1. The additional torque in any vertical member which arise from floor torsion is not mentioned in the codes. If the inter-story torsional angle is controlled in a suitable value, it is beneficial to decrease the torque of the vertical members (columns and shear walls) and to increase the structure safety.

2. If the plan dimensions of two asymmetrical buildings are different, even the torsional angle are identical, the additional displacements and inter-story displacements of the utmost end vertical member of the two structures, may have more difference. But the additional displacement increases the total displacement and inter-story displacement of the vertical member is disadvantageous to seismic safety of the structure, so it must be controlled effectively.

3. The floor of an asymmetrical building coordinates the deformation of all the vertical members under horizontal earthquake, thus significant stress can occur within the slab in case of evident asymmetricity of the structure. It is important to calculate stresses in floor plan and take seismic measures to assure safety of the structure.

4. Controlling torsional displacement ratio cannot simply control the torsional angle of both the floor and the vertical members. The same inter-story displacement ratio stipulated for different average inter-story displacement can not reflect the same structure safety, and controlling torsional displacement ratio only can not expose the stress derived within slab plan. The above are problems appearing in design of a stricture under earthquake action when the torsional displacement ratio is only controlled.

References

1. The Technical regulations for concrete tall building (JGJ3-2002), 2002.
2. Wei Lian, Wang Sen, design principles of asymmetrical building subject to horizontal earthquake, structural building, 2005.6.

ELASTIC SEISMIC ANALYSIS OF RC CORE-TUBE STEEL FRAME HYBRID STRUCTURES WITH DIFFERENT DAMPING MODELS

WANG YIQUN

College of Civil Engineering, Tianjin University
Tianjin, China

WEN HONGXING

Division of Building Science and Engineering , City University of Hong Kong
Hong Kong, China

WANG FUZHI

College of Civil Engineering, Tianjin University
Tianjin, China

The reinforced concrete core wall-steel frame hybrid structure is a non-proportional damping system. In practical structural designs, equivalent damping ratio is used and specified in the *Technical Specification for Concrete Structures of Tall Building JGJ3-2002 of China*. In this paper, a reinforced concrete core wall-steel frame hybrid structure under frequently-occur earthquake is calculated by using time history method in elastic range, with the non-proportional damping model and the equivalent damping model, respectively. It is found that there are great differences between the displacement and inner force responses by the two damping models. Some suggestions for using equivalent damping ratio of the structures is put forward.

1. Introduction

Generally, the damping ratios of steel structures and reinforced concrete structures are assumed 0.02 and 0.05 respectively, so the steel frames and concrete core-tube hybrid structures are the non-proportional damping systems. As the analysis of a non-proportional damping system is difficult, an equivalent damping ratio is used in design. For the steel and concrete hybrid structures the equivalent damping ratio is set 0.04 by the *Technical Specification for Concrete Structures of Tall Building JGJ3-2002 of China*.

What is the different between the calculated results by equivalent damping ratio (model 1) and non-proportional damping (model 2)? To answer this question, one steel frames and concrete core-tube hybrid structures under frequently-occur earthquake are analyzed with the two damping models, respectively.

It is found that there are great differences between the displacements and inner forces response results by the two damping models. The comparison of the two results shows that, if the displacement response almost the same with some value of equivalent damping ratio by model 1, the storey shear forces by model 1 are much less than that by model 2 on the low part of the structures and just the contrary on top of the structures. If the inner forces are close to each other with another value of equivalent damping ratio by model 1, the maximum displacements at top of the building by the two models are also different.

As the value of the equivalent damping ratio affected the responses of the two models. A suggested value of equivalent damping ratio is put forward at the end of the paper.

2. Damping Matrix of Non-Proportional Damping System

Supporting the system is made up with n substructures. The damping property is in accordance with Rayleigh's supposition for each of the substructures. That is

$$[C_i]=\alpha_i[M_i]+\beta_i[K_i] \quad (1\le i\le n) \tag{1}$$

Where: $[C_i]$, $[M_i]$, $[K_i]$ are the damping, mass, and stiffness matrices of the i th substructure, respectively;

α_i , β_i are the Rayleigh's coefficients of the i th substructure.

The mass, stiffness and damping matrixes of the global structure are:

$$[M]=\sum_{i=1}^{n}[M_i] \tag{2}$$

$$[K(t)]=\sum_{i=1}^{n}[K_i(t)] \tag{3}$$

$$[C]=\sum_{i=1}^{n}[C_i]=\sum_{i=1}^{n}\alpha_i[M_i]+\sum_{i=1}^{n}\beta_i[K_i] \tag{4}$$

For the structures with materials of two different damping properties, appointing the first kind of material be the one occupying the majority part of the structure, the other one be the second. So the damping matrix of the whole structure can be seen as the one made up with the first kind of material, but some change must be done for the part of structure made up with the second kind of material.

The step by step direct integral algorithm is listed in [1], and adopted in the computer software NDAS2D (Non-linear Dynamic Analysis of Structures in 2 Dimensions), which is developed by the authors.

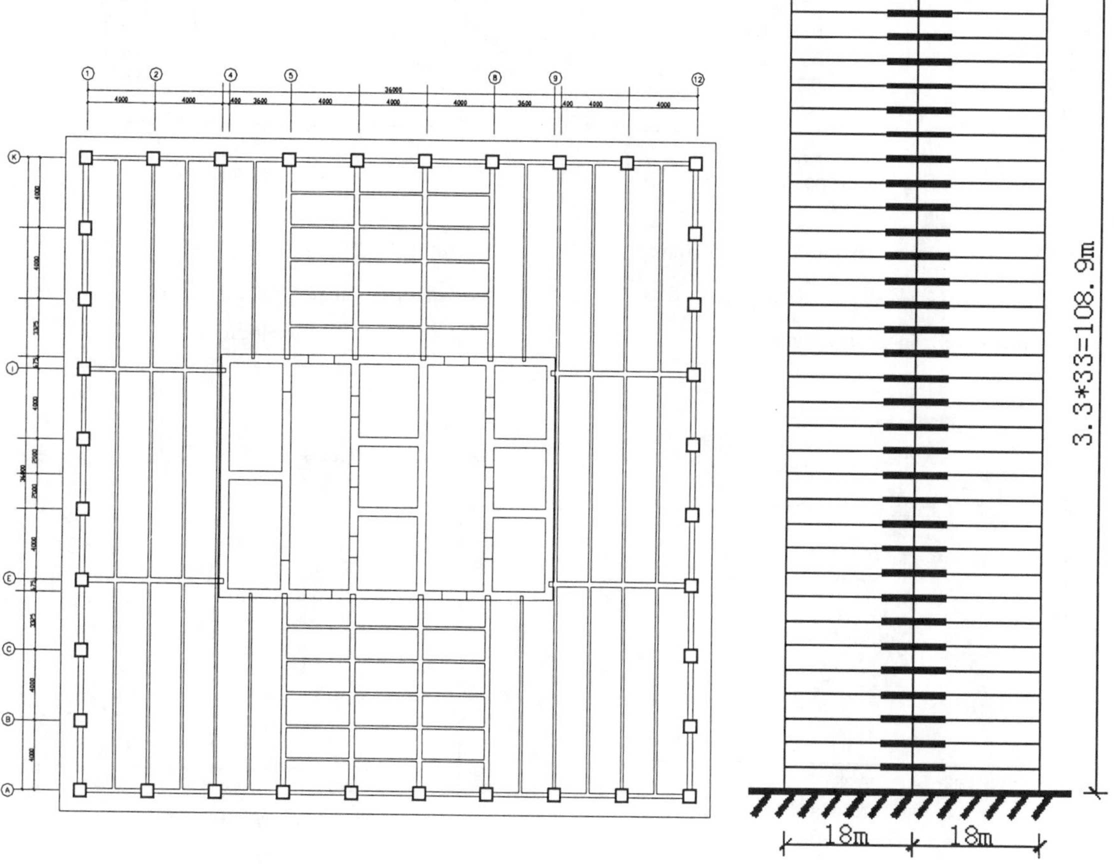

Figure 1 standard plane of structures

Figure 2 calculation model

3. Numerical Example

A real project of 33 storey steel frames and concrete core-tube hybrid structure with total height 108.9m is calculated. The structure built on class II soil and in region of earthquake fortification intensity 7. The building plane is in square shaped (Fig. 1) with plan dimensions 38.4m×38.4m. The storey height is 3.3m. The dimension of the concrete core-tube in plan is 13.35m×19.2m. The dimension of the steel frame in plan is 36m×36m. The building is belonging to earthquake resistant design grade I, according to JGJ3-2002.

This is a symmetric and regular structure. A plane calculation model is taken, as in Fig. 2. The horizontal members are assumed as beam elements; and the vertical members are assumed as beam-column elements. The thick line in the beam elements is the rigid zone, for simulating the concrete wall width effects. The connections between the steel beams and concrete wall are assumed as rigid.

Two damping models, the equivalent damping model (model 1) and the non-proportional damping model (model 2), are used for the example. The later can be called with 'the real damping model', that is from Eq. (1). The damping ratio value 0.02 is set to the steel frames and steel beams which connecting the frames and RC core-tube, 0.05 is set to the RC core-tube. The value of equivalent damping is taken 0.04, 0.03, 0.02, etc. respectively, to examine which value is appropriate.

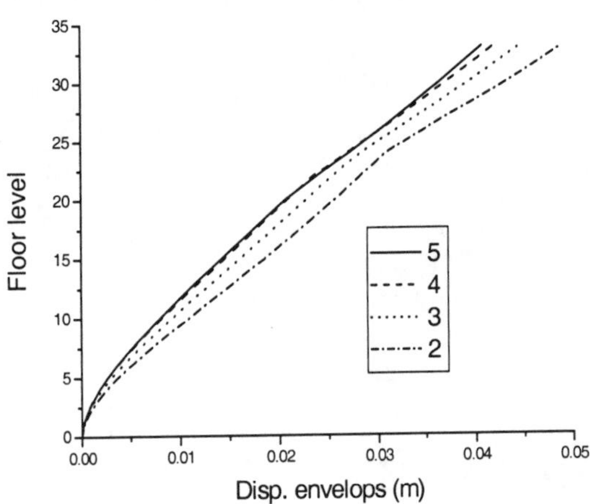

Figure 3 Displacement envelops under El Centro wave

2、3、4 for equivalent damping value 0.02、0.03、0.04 respectively; 5 for damping model 2

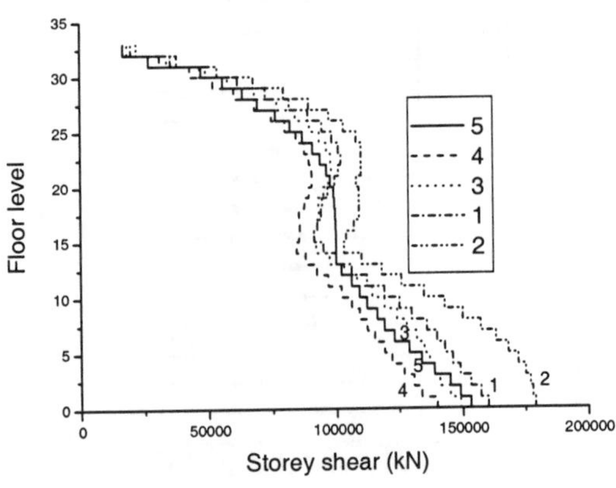

Figure 4 Storey shear envelops under El Centro wave

1、2、3、4 for equivalent damping value 0.025、0.02、0.03、0.04; 5 for damping model 2

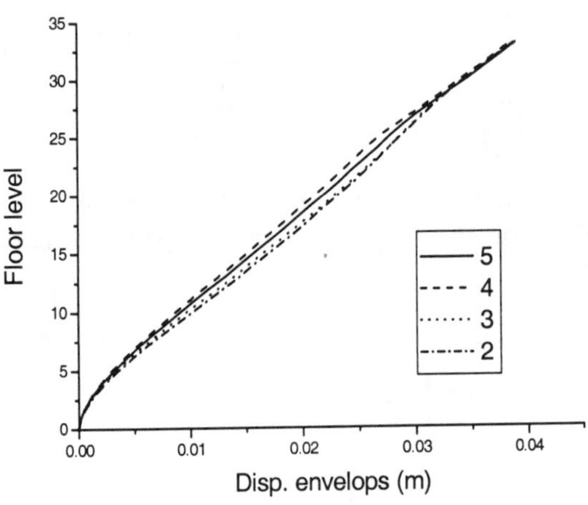

Figure 5 Displacement envelops under Tianjin wave

2、3、4 for equivalent damping value 0.02、0.03、0.04 respectively; 5 for damping model 2

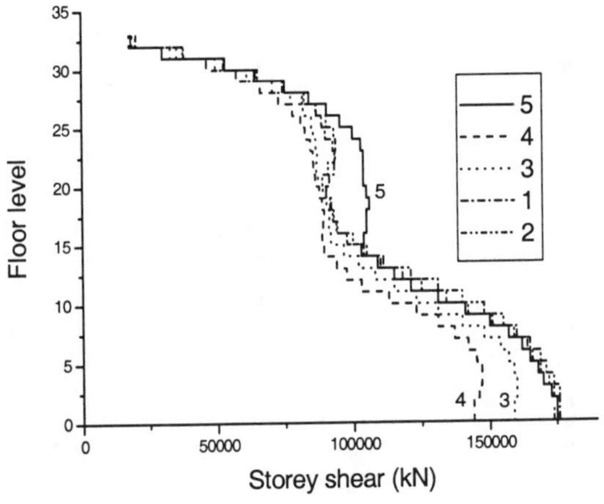

Figure 6 Storey shear envelops under Tianjin wave

1、2、3、4 for equivalent damping value 0.025、0.02、0.03、0.04; 5 for damping model 2

The first and second vibration periods of this structure are 2.32 sec. and 0.580 sec. respectively.

The example is subjected to 1940 El-Centro and 1976 Tianjin earthquake ground motions scaled to 35 cm/s^2, respectively. The example with two damping models is calculated using the software NDAS2D, respectively. The main calculating results are shown in Fig. 3 to Fig. 6.

From Fig. 3 to Fig. 6 we can see that great difference exist between the results by damping model 1 and damping model 2. For the displacement envelops, the results of damping model1 with 0.04 equivalent damping ratio value are equivalent to the results of damping model 2. But for the structural inner forces, such as the first storey shear forces (Fig. 4 and Fig. 6), the results of model 1(0.04) are smaller than the damping model 2 by 8.5%, 16.0%, respectively. Another point to be noted is that the distribution of the storey shear differences between the RC core wall and steel frames is not the same. That is to say that the inner forces of the RC core wall are with great difference between the two damping models, while the inner forces of the steel frame are almost the same for the two damping model. The inner forces difference on RC core wall varies with the height of the building. At upper part of the building, the inner forces of model 1 are great than that of model 2, but the difference is not large on basement of the building.

Fig. 4 and Fig. 6 show the storey shear forces calculated by damping model 1 with several different damping ratio values comparing with that by model 2. From this figure, we can see that when the equivalent damping ratio value is 0.025 the inner forces are approached that of model 2. So we suggest the suitable damping ratio value should be 0.025. This value is equivalent to the value of a similar structure obtained from a complex model method [2], though the displacements of it are larger than that of model 2, form Fig. 3 and Fig. 5. As model 1 is a simplified method, the difference of it with the accurate method (model 2) are that when the equivalent damping ratio is large, the displacements are similar, but the inner forces at bottom of structures are smaller than the accurate results, this will cause unsafe consequence; when the equivalent damping ratio is taken a small value, the inner forces are closed to the accurate results, but the displacements at the upper part of the structure are much greater than the accurate ones (Fig. 3). As safety is more important, a small value of equivalent damping ratio should be used.

4. Conclusions

A steel frames and concrete core-tube hybrid structures under frequently-occur earthquake are analyzed by step and step direct integrate method, with the two damping models, respectively. As it is just one example, we cannot obtain a general conclusion from it. Some suggestions are listed below for reference.

1. For the steel frames and RC core wall hybrid structures, the non-proportional damping model should be used in the time history analysis under frequently-occur earthquake.
2. When using the response spectra analysis in design with equivalent damping model, the value of damping ratio should be determined by inner forces equivalent principle not by building top's displacement equivalent principle.
3. Under seldom-occur earthquake, as the great differences between the elastic limitations of the steel and concrete materials, the concrete members are in elasto-plastic states at first, and the damping ratio of it will be increased more. While the steel members are still in elastic states, the damping ratio of it maintains at 0.02. The difference between the damping ratios of the two material members is greater than it in elastic state. The difference of structural inner forces due to difference of damping ratios will be more obvious than it under frequently-occur earthquake. From this, it can be estimated that the RC core wall will bear the most basement shear force under seldom-occur earthquake action. This is consistent with some of design specification in US [3]. More studies on this field should be conducted, to guarantee the safety of steel and concrete hybrid structures in high seismic intensity zones.

References

1. Y. Wang and Z. Li, *Earthquake Engineering and Engineering Vibration. V19*, *No*.2, 76 (1999). In Chinese.
2. B. Huang, *Shanghai Journal of Mechanics.* ***V19***, *No*.2, 142 (1998). In Chinese.
3. B. M. Shahrooz, B. Gong, G. Tunc and J. T. Deason, *Prog. Struct. Engng. Mater.* ASCE ***No. 3***, 149 (2001)

VARIABLE FRICTION DAMPERS FOR EARTHQUAKE PROTECTION OF COUPLED BUILDING STRUCTURES

C.L. NG

Department of Civil and Structural Engineering, The Hong Kong Polytechnic University, Hung Hom, Kowloon Hong Kong, China

Y.L. XU

Department of Civil and Structural Engineering, The Hong Kong Polytechnic University, Hung Hom, Kowloon Hong Kong, China

This study mainly focuses on the development and application of variable friction damper as a joint control device to enhance the seismic resistance of a main building with podium structure. Shaking table tests using small scale building structures and damper were performed to provide a realistic demonstration and evaluation of effectiveness of the proposed control scenario. The real-time adjustment of the small scale variable friction damper was mechanically realized by using a piezoelectric actuator. A local feedback controller was proposed and implemented in the coupled building system. The improved performance over the uncontrolled cases, including uncoupled and rigid coupled configuration, was verified by considering four simulated historical earthquakes. The test results showed that prompt regulation of friction force by the piezoelectric actuator can be achieved, and the desired control performance and robustness can be guaranteed. The proposed local feedback controller also demonstrated an increased seismic resistance to the coupled system.

1. Introduction

Associated with economic development of modern cities, more building structures appear with stylistic shape or extraordinary height as a landmark of the cities. Main building at medium- to high-rise level constructed with an auxiliary podium structure, based upon both functional and architectural considerations, is also perceived as a growing construction scheme. However, it is intuitive that this building complex is in a form of setback configuration because both structures are normally integrated with rigid members. For such types of structures with vertical irregular layout, concentration of inelastic action at the level of setback was observed during some past earthquakes and resulted in a poor performance of the structures (Arnold [1]; Arnold and Elsesser [2]).

Undoubtedly, installation of energy damping devices as bracing assembly at every stories of the buildings is a very sensible choice and a performance guaranteed practice (Soong and Dargush [3]; Constantinou et al. [4]). It is, however, generally recognized that ineffective energy dissipation under such configuration might be resulted in a stiff structural system, which is probably the case of podium structure, because of relative small drifts and interstory velocities.

There have been, in fact, a number of research studies demonstrating that the developing method of coupling adjacent buildings with passive/active control devices is an effective alternative in reducing seismic responses of structures (Seto et al. [5]; Christenson et al. [6]). Most importantly, practical application of coupled control in Japan, for instances, implementation of passive yielding elements and passive visco-elastic dampers in a coupled low-rise office complex and a headquarter building of Konoike in Osaka respectively, were also noted. Additionally, Authors [7] have experimentally verified control performance of passive control in a building system of main building with podium structure recently. Results obtained also positively supported the feasibility of coupled building control.

Inspired by this successful experiment, this paper proposed an application of smart (or semiactive) dampers to realize an adaptive coupled building system to withstand earthquake. Shaking table tests on 12- and 3-story steel frames which replicate the coupled buildings were conducted. A piezo-driven variable friction damper, being a joint device, was exclusively designed for this test. A local feedback control strategy was also proposed for the application of friction type control device. Smart damping control is shown to provide increased performance and structural safety to both structures compared to the rigid coupled configuration.

2. Description of Test Configuration

The test configuration, as shown in Fig. 1, consisted of a uniaxial seismic simulator and a coupled building system of 12- and 3-story steel frames which were interconnected by a piezo-driven variable friction damper (PVFD) at their third floor to provide semiactive control force applied on both of the structures. The 12- and 3-story building frames were respectively 2.4 and 0.6 m tall. The fundamental frequencies of corresponding building were 3.7 and 9.9 Hz. These two structures were uncoupled in a basic configuration, and they became a rigid and damper coupled system when they were connected by rigid members and control device, respectively. The details of building models were discussed in Ref. [7]. The level of friction force could be maintained at a fixed value or regulated in real time by varying the clamping force of piezo actuator integrated in the damper (Fig. 2).

The acceleration response signals of all floor levels and the seismic simulator platform were recorded during experimental testing. Additional signals of relative displacement between buildings at the floor where damper was mounted and control force of PVFD were also measured by linear variable displacement transducer (LVDT) and load cell, respectively. The management of signals was totally performed by a dSPACE real-time data acquisition and feedback control system. This system was operated with a processor board of DS1005 which features a PowerPC FX processor at 800 MHz to provide computing power for real-time and also to function as an interface to the I/O hardware. The analog input signals were quantized to digital one by passing through I/O board of DS2002. The command control signal, which was determined from the local feedback controller programmed by Simulink and complied in the dSPACE, to PVFD was conveyed by I/O board of DS2102 (Fig. 3).

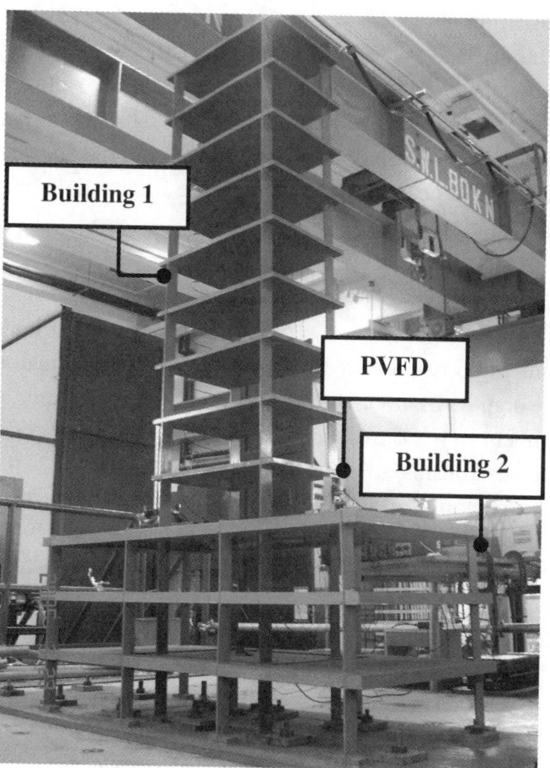

Figure 1. Experimental configuration of the PVFD damper-coupled building system.

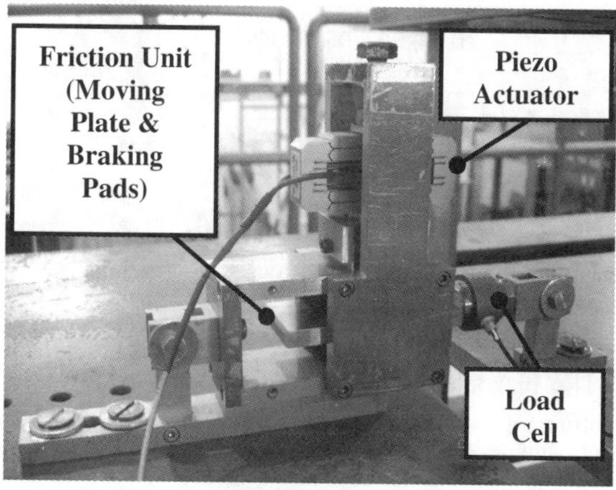

Figure 2. The piezo-driven variable friction damper (PVFD).

3. Identification of Piezo-Driven Variable Friction Damper Properties

The PVFD is simply a mechanical integration of a piezoelectric actuator and a friction generation unit in our design (Fig. 4). The piezoelectric actuator is the key component for being a normal clamping force regular. Perpendicular friction force level can be changed proportionally to the variation of clamping force, which is adjusted by varying the supply voltage, when the sandwiched steel plate slides between the two closely contacted braking pad friction materials. Thus a constant friction coefficient with a time-dependent normal clamping force is used to represent the frictional damper force:

$$F_d(t) = \mu N(t)\text{sgn}[\dot{x}(t)] \qquad (1)$$

where μ is friction coefficient and $N(t)$ is variable clamping force actuated from the piezoelectric actuator. The sign of friction force is consistent to the velocity of damper $\dot{x}(t)$. Consequently, the level of damping force can be adjusted to particular level in response to the voltage supply.

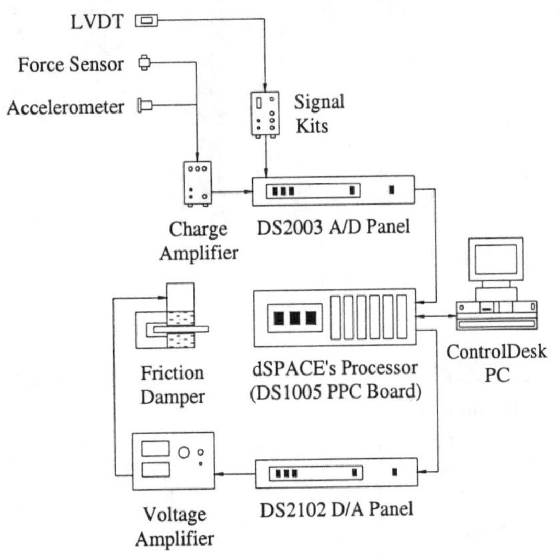

Figure 3. Experimental instrumentation diagram of real-time control system.

The piezoelectric actuator developed by Dynamic Structures and Materials, LLC (DSM) is a leverage-based piezoactuator. A model of FPA-850 of the employed flextensional piezoelectric actuator features an ultimate blocking force of 370 N at maximum voltage supply of 150 V. Its unload resonant frequency and stiffness is 275 Hz and 0.39 N/μm, respectively. The actuator is driven by a linear piezo amplifier, which mainly serves three purposes: (1) amplify an input (or command) voltage signal (0-7.5 V) to an operative level (0-150 V) for driving the actuator; (2) provide a high control resolution of a piezoelectric actuator by minimizing signal-to-noise ratio of the driving system; and (3) enable high control rate and prompt driving of the actuator at a broad frequency bandwidth to be satisfactorily met.

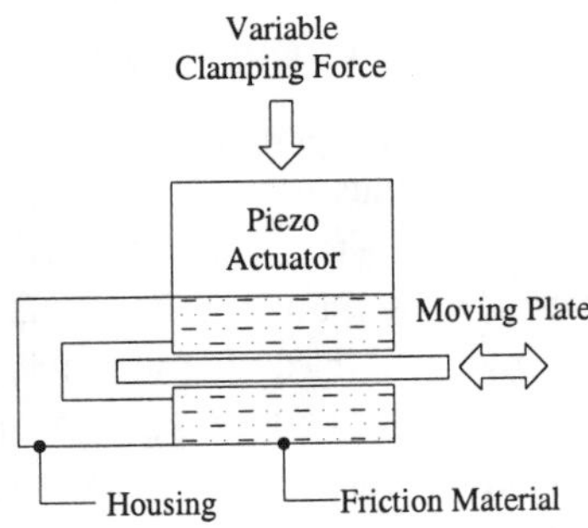

Figure 4. Schematic diagram of PVFD.

The friction damping force F_d of PVFD was directly characterized in a MTS 810 series load frame over a series of sinusoidal stroke displacements and frequencies within the voltage range of 0-150 V supplied to the actuator. At every voltage step of 20 V, the variable friction damper was tested with all combinations of displacement range (2, 3, 4 mm) and frequency range (3, 4, 5 Hz) considered. The variation of damper force was more closely examined by adopting a finer voltage step (5 V) in the case of 3 mm stroke displacement, given the displacement is independent to friction force (Fig. 5), for all three frequency levels. These three sets of test results were particularly used to establish the force-voltage (FV) lookup chart for the later application in closed-loop control of coupled building system.

Typical force-displacement hysteresis loop of the variable friction damper is shown in Fig. 5. Results shown are hysteresis loops recorded at voltage supply of 90, 110, 130 and 150 V at the testing frequency of 3 Hz and displacement of 2, 3, 4 mm. The results display that the loops are effectively maintained in a rectangular shape, which highly follows the Coulomb friction model as written in Eq. (1), at various voltage levels. The weak correlation between friction force and slip displacement, which is also one of the assumption in the Coulomb friction model, is additionally observed. Feature of motion independent between friction force and slipping frequency can also be appreciated by comparing hysteresis loops across different frequencies.

Peak friction forces at all tested voltage levels were summarized into a quasi-static FV chart (Fig. 6). A "zero-voltage" friction force is approximately between 5-10 N, and it is about 350 N at maximum voltage. A curve was fitted among all test data for real-time control operation. To enhance the force tracking capability of PVFD, a proportional controller was implemented with the PVFD to compensate actuated control force discrepancy due to inaccuracy of FV chart. As depicted in Fig. 7, total output voltage V is a summation of original command voltage v_c determined directly from the FV chart and feedback voltage v_b. The proportional gain k_p applied in this study was one.

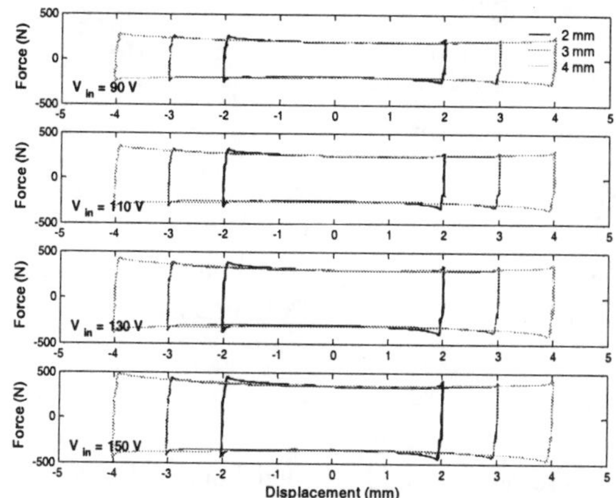

Figure 5. Force-displacement hysteresis loop in different excitation amplitudes.

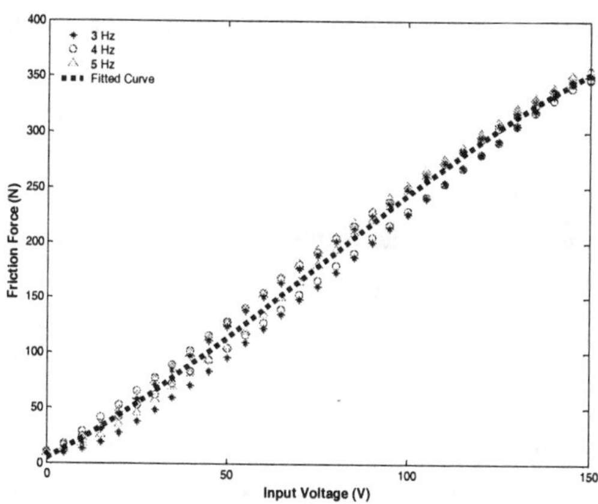

Figure 6. Force-voltage chart of PVFD.

4. Design of Control Algorithm

No matter which type of feedback controller is employed in a closed-loop control with semiactive control device, it is, in fact, a semi-active control strategy because the control force generated by the control device is only a dissipative force (i.e. passive in nature) whereas its magnitude could be adaptive to the vibration level of structural responses and ground motions. Linear quadratic (LQ) controller is one of the mostly adopted active control strategies for semiacitve control devices. This is a kind of global-feedback controller which encompasses the entire building states formed with an objective function to be minimized. Given that the designed optimal control force is very close to a dissipative force, the control effectiveness with semiacitve device will be as high as determined by the active control strategy. Other prerequisite to develop an efficient controller is an accurate mathematical model of the structural system which is not easy to be obtained if the degree of freedom of as-built structure is becoming large.

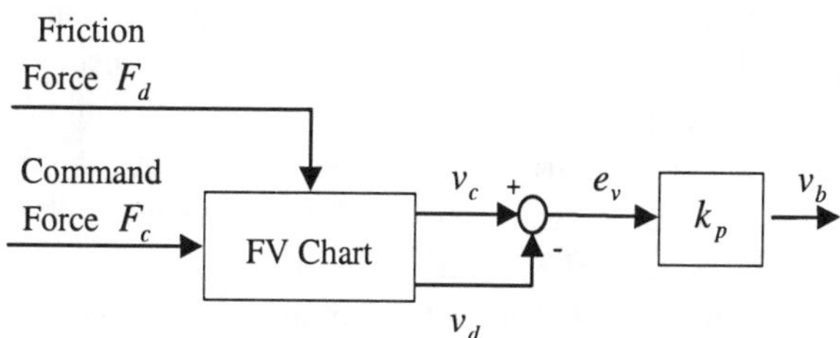

Figure 7. Proportional controller of PVFD.

Hence, a local-feedback controller is proposed for the variable friction damper in the coupled building system to avoid too many design constraint or complexity. In view of friction damper, there is no energy being dissipated in a sticking status and no extra damping can be provided consequently. Based upon this rationale, friction force should be continuously adjusted to various levels that always allow slipping of contact surface to maximize energy dissipation.

A non-sticking friction control force can be achieved by incorporating slip velocity of the damper as feedback signal which is locally obtainable. The friction force around zero crossing of velocity of the damper should be accordingly zero to maintain continuously slipping and thus a smooth transition of force reversal. The non-sticking friction (NSF) controller is proposed and clamping force is expressed as

$$N(t) = N_{\max}\left|\tanh[\alpha\dot{x}_{rel}(t)]\right| \qquad (2)$$

where $N_{\max}$ is the maximum design clamping force of variable friction damper; α is a velocity coefficient and it defines the effective range of smooth boundary. The actual clamping force $N(t)$ is, in fact, a deduction of the $N_{\max}$ by a factor of $\left|\tanh[\alpha\dot{x}_{rel}(t)]\right|$. The semi-active damper force is then given by

$$F_d(t) = \mu N_{\max} \tanh[\alpha\dot{x}_{rel}(t)] \tag{3}$$

$N_{\max}$ and α are two parameters to be designed. The magnitude of $N_{\max}$ determines the ultimate capacity of device required for control system. The value of α will determine the boundary of friction near zero velocity. The smaller the value of α, the less rapid decrease of friction force will be.

5. Experimental Results and Discussion

The performance of the adaptive coupled building system was evaluated experimentally by conducting a number of shaking table tests. The coupled structures were subjected to the following four earthquake records: (1) the N-S component of El Centro recorded at the Imperial Valley Irrigation District substation in El Centro, California, earthquake of May 18, 1940; (2) the N-S component of Hachinohe recorded at Hachinohe City during the Tokachi-oki earthquake of May 16, 1968; (3) The N-S component of Northridge recorded at Sylmar Country Hospital parking lot in Sylmar, California, during the Northridge, California, earthquake of January 17, 1994; and (4) the N-S component of Kobe recorded at the Kobe Japanese Meteorological Agency (JMA) station during Hyogo-ken Nanbu earthquake of January 17, 1995. All the time histories were time-scaled by a factor of 1/3 and the tested intensities included 0.1, 0.15 and 0.2g.

The tests were conducted in four different system configurations.

- Case 1. Building structures were uncoupled.
- Case 2. Building structures were coupled at all three floors by rigid steel members.
- Case 3. Building structures were coupled with single passive friction damper at the 3rd floor.
- Case 4. Building structures were coupled with single variable friction damper at the 3rd floor with semiactive control strategy.

Case 1 and 2 represent two uncontrolled cases. Case 3 denotes passive control of coupled building system. The voltage supply to the piezoelectric actuator was maintained at a constant level under excitation. The proposed semiacitve control strategy was implemented in Case 4 to realize an adaptive coupled building system.

Response evaluation of the coupled building system is based on the four performance indices. The indices include the maximum peak and rms responses ratio of floor displacement (J_1 and J_2) and acceleration (J_3 and J_4):

$$J_1 = \frac{\max_i \left[x_i^{peak}\right]}{x_{uncoupled}^{peak}}, \quad J_2 = \frac{\max_i \left[x_i^{rms}\right]}{x_{uncoupled}^{rms}} \tag{4a}$$

$$J_3 = \frac{\max_i \left[\ddot{x}_i^{peak}\right]}{\ddot{x}_{uncoupled}^{peak}}, \quad J_4 = \frac{\max_i \left[\ddot{x}_i^{rms}\right]}{\ddot{x}_{uncoupled}^{rms}} \tag{4b}$$

where x_i^{peak} and x_i^{rms} ($\ddot{x}_i^{peak}$ and $\ddot{x}_i^{rms}$) are the peak and rms displacement (acceleration) of ith story of either buildings ($i = 3$ or 12); the maximum peak and rms displacement (acceleration) of uncoupled case are denoted by $x_{uncoupled}^{peak}$ and $x_{uncoupled}^{rms}$ ($\ddot{x}_{uncoupled}^{peak}$ and $\ddot{x}_{uncoupled}^{rms}$), respectively. These performance indices are presented in Tables 1 and 2.

Regarding the rigid-coupled building system, it is clearly indicated that the maximum peak and rms acceleration of the 3-story building could be moderately reduced in almost four earthquakes records. The acceleration responses of 12-story building in this configuration are, however, another story. The response in terms of peak value is amplified by a quite large percentage for all earthquakes and intensities. Its rms value is reduced in some earthquakes, but a larger jump of response in other earthquakes could be seen. Overall the mean value of peak acceleration (J_3) for four earthquakes and intensities will be approximately reduced by 25% for 3-story building, a 47% amplification is, however, observed for 12-story building. The displacement is neither better controlled in this configuration. A roughly 20% increase of peak displacement response for the 12-story building in the Kobe earthquake can be observed. An even large provoking of displacement response of 3-story building is again noted, the largest percentage increase of peak and rms value can be up to 100% and 300% in the Kobe earthquake. These results suggest that there is a higher possibility to increase the seismic response of the both buildings in a rigid-coupled building system.

Table 1. Response ratios of rigid-coupled building system in four earthquake records.

Response	Earthquake	Earthquake Intensity 0.10g	0.15g	0.20g
J_1	El Centro	0.941 (1.494)	0.947 (1.456)	0.954 (1.438)
	Hachinohe	0.700 (0.859)	0.692 (0.854)	0.693 (0.853)
	Northridge	0.944 (0.591)	0.931 (0.581)	0.925 (0.576)
	Kobe	1.216 (2.051)	1.259 (2.052)	1.261 (1.818)
J_2	El Centro	1.100 (1.901)	1.064 (1.842)	1.051 (1.818)
	Hachinohe	0.524 (0.974)	0.529 (0.959)	0.544 (0.954)
	Northridge	0.545 (0.611)	0.527 (0.613)	0.524 (0.610)
	Kobe	0.903 (3.129)	0.936 (2.995)	0.932 (2.999)
J_3	El Centro	1.279 (0.728)	1.250 (0.693)	1.259 (0.712)
	Hachinohe	1.127 (0.659)	1.131 (0.653)	1.093 (0.651)
	Northridge	1.854 (0.620)	1.829 (0.588)	1.854 (0.591)
	Kobe	1.578 (1.014)	1.665 (0.804)	1.667 (0.806)
J_4	El Centro	1.511 (0.831)	1.444 (0.804)	1.409 (0.806)
	Hachinohe	0.864 (0.710)	0.869 (0.698)	0.868 (0.683)
	Northridge	0.897 (0.447)	0.881 (0.449)	0.868 (0.448)
	Kobe	1.289 (0.943)	1.332 (0.943)	1.327 (0.942)

1. Value in parenthesis represents the response ratio of 3-story building.

Having the controller parameters of N_{max} and α selected to 150 and 10, the responses of either buildings are generally suppressed under all earthquakes in any intensities (Table 2). The average peak displacement and acceleration of 12-story building are roughly reduced by 19%. More remarkable reduction is reached by average rms response, they are about 48% for both responses. Not only 12-story building gains a reduction in responses, there are 15 and 16% average reduction on the peak displacement and acceleration of 3-story building. Similarly, corresponding reductions in terms of rms value are 25 and 33%, respectively. Fairly steady response reductions of the buildings are maintained as earthquake intensity varies (Table 2). This observation likely suggests that the proposed local-feedback control strategy could allow the coupled building system to be adaptive to different magnitudes of earthquakes. Under the Northridge earthquake, which is a near-field earthquake, it is observed that there is a slight increase of peak acceleration (7%). This increase is perhaps attributed to its characteristic of large-amplitude but short-duration (pulse-like) motion. The efficiency to suppress peak acceleration of the structure due to far-field earthquake, such as El Centro earthquake, is acceptable. Nevertheless, there control performance in terms of rms is equally satisfactory.

Table 2. Response ratios of damper-coupled building system in four earthquake records.

Response	Earthquake	Earthquake Intensity 0.10g	0.15g	0.20g
J_1	El Centro	0.735 (0.942)	0.763 (0.912)	0.769 (0.950)
	Hachinohe	0.717 (0.774)	0.729 (0.763)	0.750 (0.818)
	Northridge	0.924 (0.836)	0.946 (0.829)	0.945 (0.860)
	Kobe	0.786 (0.773)	0.829 (0.821)	0.844 (0.804)
J_2	El Centro	0.566 (0.636)	0.568 (0.671)	0.571 (0.649)
	Hachinohe	0.575 (0.687)	0.590 (0.679)	0.618 (0.690)
	Northridge	0.414 (0.524)	0.415 (0.525)	0.427 (0.559)
	Kobe	0.462 (0.828)	0.502 (0.789)	0.531 (0.841)
J_3	El Centro	0.600 (0.831)	0.604 (0.711)	0.633 (0.799)
	Hachinohe	0.802 (0.819)	0.821 (0.831)	0.766 (0.851)
	Northridge	1.060 (0.861)	1.083 (0.880)	1.063 (0.910)
	Kobe	0.712 (0.843)	0.759 (0.950)	0.780 (0.968)
J_4	El Centro	0.541 (0.721)	0.541 (0.764)	0.532 (0.700))
	Hachinohe	0.571 (0.774)	0.582 (0.769)	0.589 (0.726)
	Northridge	0.480 (0.580)	0.461 (0.569)	0.461 (0.592)
	Kobe	0.472 (0.979)	0.498 (0.882)	0.534 (0.919)

1. Value in parenthesis represents the response ratio of 3-story building.

Control performance of 12-story building is highlighted by two peak response profiles (Figs. 8 and 9). Results from four tested cases being subjected to four earthquakes at 0.2g are illustrated. In case 3, the passive friction control force was selected to 100 N, and control parameters are again $N_{max} = 150$ and $\alpha = 10$ for case 4. Both passive and semiactive control strategies are effective to limit the peak vibration mostly below uncoupled configuration. Semiactive strategy clearly shows a better control towards acceleration compared to passive counterpart. Although peak displacement in semiactive controlled case is slightly larger than that by passive scheme, it achieves a better control balance of both responses. Reduction of displacement will be reduced below a level which semiactive can provide, while limited improvement in acceleration reduction is seen from other tested results if a lower force level of passive damper was used.

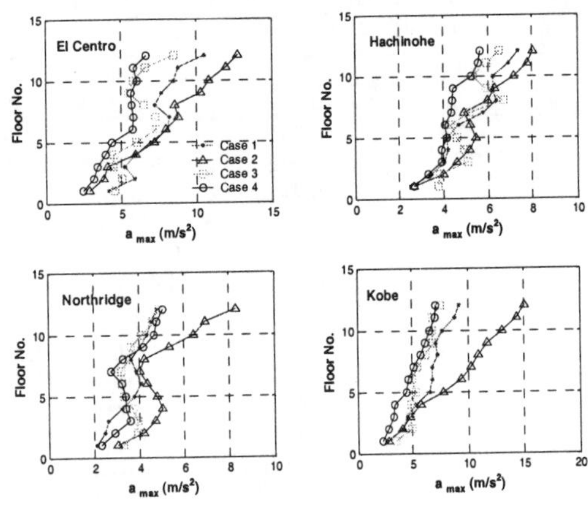

Figure 8. Peak acceleration profile of 12-story building subjected to 4 earthquakes at 0.2g.

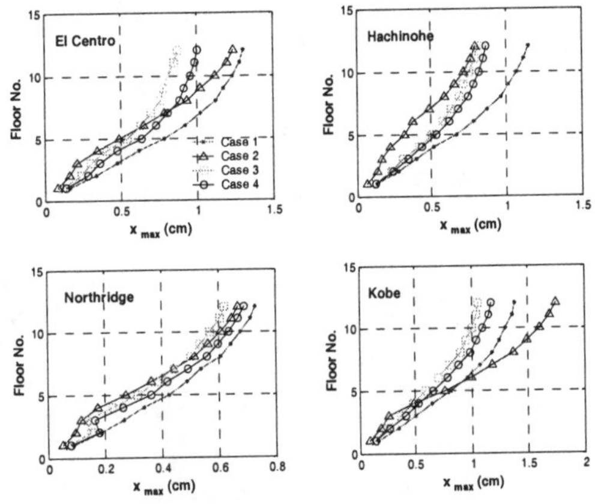

Figure 9. Peak displacement profile of 12-story building subjected to 4 earthquakes at 0.2g.

6. Conclusions

A feasibility study on realizing an adaptive coupled building system with variable friction damper was experimentally evaluated. Result suggested that seismically provoked vibrations in rigid-coupled case were suppressed in a damper-coupled configuration. The local-feedback controller satisfactorily controlled the peak displacement and acceleration in four earthquakes, and demonstrated a steady control performance in different seismic intensities. Limited acceleration control effectiveness in passive control case was improved by the semiactive strategy.

Acknowledgments

The first author gratefully acknowledges the support of this research by The Hong Kong Polytechnic University through a postgraduate scholarship and the financial support through its Area Strategic Development Program in Structural Control and Intelligent Buildings is also appreciated.

References

1. C. Arnold, Building configuration: Characteristics for seismic design, *Proc. Seventh World Conf. on Earthquake Engrg.*, Istanbul, Turkey, **4**, 589-582 (1980).
2. C. Arnold and E. Elsesser, Building configuration: Problems and solutions, *Proc. Seventh World Conf. on Earthquake Engrg.*, Istanbul, Turkey, **4**, 153-160 (1980).
3. T. T. Soong and G. F. Dargush, *Passive energy dissipation systems in structural engineering*, Wiley, Chichester, U.K. (1997).
4. M. C. Constantinou, T. T. Soong and G. F. Dargush, *Passive energy dissipation systems for structural design and retrofit*, Monograph series, MCEER, Buffalo, N.Y. (1998).
5. K. Seto, A structural control method of the vibration of flexible buildings in response to learge earthquakes and strong winds, *Proc. of the 35th Conf. on Decision Control*, Kobe, Japan, 658-669 (1996).
6. R. E. Christenson, B. F. Spencer Jr., N. Hori and K. Seto, Coupled Building Control using Acceleration Feedback, *Computer-Aided Civil and Infrastructure Engineering*, **18**, 4-18 (2003).
7. C. L. Ng and Y. L. Xu, Experimental study on seismic response mitigation of complex structure using passive friction damper, *2004 ANCER Annual Meeting, Networking of Young Earthquake Engineering Researchers and Professionals*, Honolulu, Hawaii, CD-ROM, 14 pages (2004).

THE APPLICATION OF PUSH-OVER ANALYSIS IN SEISMIC DESIGN OF BUILDING STRUCTURES PROCEDURES

C.S. FU

*Shanghai CHINAFU structural design Inc., Rm.1501 No.2 Lane 525 FangXie Rd.
Shanghai, 200011, China*

J. YING

*Shanghai CHINAFU structural design Inc., Rm.1501 No.2 Lane 525 FangXie Rd.
Shanghai, 200011, China*

The development of the seismic analysis procedures has proven that the displacement-based seismic design method will be gradually incorporated in the seismic design codes in near future. Recently, push-over analysis, as the kernel of the displacement-based design method, has been required in the Chinese code for seismic design of buildings (GB 50011-2001) for analysis of the structures with plan or vertical irregularities. Following a brief review on the history of seismic analysis procedures in China, the seismic design method of ~~a~~ tall buildings with out-of-plane offset, diaphragm discontinuity and in-plane discontinuity in vertical lateral-force-resisting elements is given in this paper to illustrate the application of push-over analysis. The results of push-over analysis show that the structure yielded earlier at the edge of long narrow cut and large deformations would consequently occur as the pushing force increased. The long slit is a weak place to resist the seismic ground motion and has great effects to structural capability of damage restriction and collapse prevention after the corner elements around it yielded. It could be concluded that the push-over analysis should be used for seismic design of the structures with irregularities. It will help structural engineers in identifying hidden weak points in a structure when conducting the seismic design.

1. Chinese Codes for Seismic Design

The first and second generations of codes for seismic design in China were issued in 1964 and 1974. The modal analysis and response spectrum method were adopted in both editions. The design earthquake was based on the earthquake with 10% probability of exceedance in 50 years. The structural reduction coefficient, $C<1$, was introduced for different structural systems to reduce the seismic effects obtained by the elastic response spectrum method. The structures were considered to be in elastic condition. However, only strength was considered in the codes when to verify whether the structures were safe to resist the design earthquake ground motion. No provisions were prescribed about deformation limitations or any requirement of non-linear analysis. The procedure presented in the codes belonged to the strength-based seismic design method.

In 1989, China issued the third generation of the seismic design code (GBJ 11-89). The theory of seismic design improved greatly. The structural reduction coefficient was removed and three seismic use levels were introduced. The earthquake with 63% probability of exceedance in 50 years was defined as the design earthquake (small earthquake) for level 1. The earthquake with 10% probability of exceedance in 50 years was defined as the demanded earthquake (median earthquake) for level 2. And for level 3, the earthquake with 2% probability of exceedance in 50 years was defined as the maximum considered earthquake (severe earthquake). Structures are in elastic condition under the design earthquake. Both strength and elastic story drift were required to be examined. There were also provisions stipulated for seismic detailing requirements to guarantee that the designed structures possessed adequate ductility to avoid severe damage under the demanded earthquake and to prevent collapse under the severe earthquake. Structural engineers should carry out seismic design of building structure in two steps as follows: step 1, conducting elastic analysis of a structure by computers; and step 2, design of the seismic detailing in accordance with the requirements prescribed in the code. The design procedure of three seismic use levels and two steps presented in the code (GBJ 11-89) can be categorized as the strength and ductile detailing requirement-based one, which were similar to those adopted in the countries of seismic active region.

The new generation of the seismic design code (GB 50011-2001) has been recently issued in China. Besides maintaining the concept of the three seismic use levels and two design steps, two important points are introduced in the new code. The provisions are prescribed for the first time that the geometrical and torsional irregularities of building structures are defined and limited. The push-over analysis, as the kernel of the displacement-based design method, is required for analysis of the structures with plan or vertical irregularities. The way has been paved in the new Chinese seismic design code to completely introduce the displacement-based seismic design methodology. The theory of seismic design has been developed towards maturity in China [1,2].

2. Capacity Spectrum Method and Push-over Analysis

Although the concept of ductile design is well carried out by means of the seismic detailing measurements, the seismic responses of structures could not effectively be estimated within the elasto-plastic range. However, it is particularly important in the seismic design of buildings to determine the non-linear dynamic behavior and seismic responses of structures during the maximum considered earthquake. The consensus has been reached that the present codes should be improved to incorporate the performance/ displacement-based design procedures. In fact, it has been incorporated in U.S. codes such as SEAOC 2000, ATC-40 and FEMA-273. In Japan, the Building Standard Law will be revised into a performance/ displacement-based design format.

The kernels of performance/displacement-based design procedure are the demand-capacity spectrum method and the push-over analysis. The capacity method graphically compares the capacity of a structure with the demands of maximum considered earthquake on the structure. The capacity of the structure is represented by a force-displacement curve obtained from push-over analysis. The non-linear curve is converted into the spectral accelerations called capacity spectrum. The demands of earthquake on the structure are defined by response spectrum called demand spectrum. They are plotted in the same graph and compared. If the capacity spectrum curve goes across the curve representing demand spectrum, the structure will be considered to have adequate ductility and capability to resist the maximum considered earthquake. The intersection of the spectral curves gives an estimation of non-linear displacement demand, [3-5] as indicated in Fig.1.

The push-over analysis is a static non-linear analysis

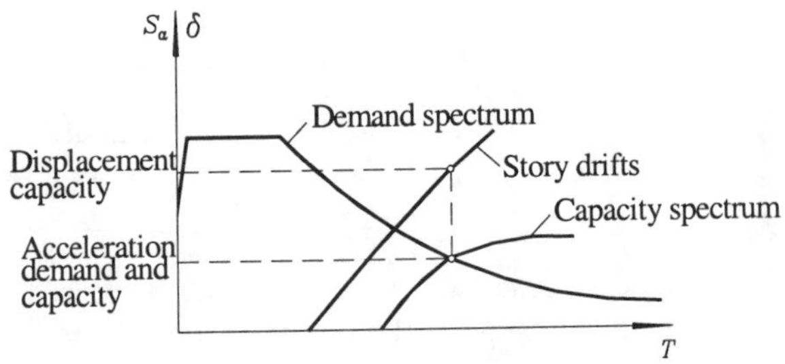

Figure 1. Capacity spectrum method

method, where the lateral forces in some pre-defined pattern are statically applied onto a structure, step by step, until a structural damage or collapse occurs. The non-linear seismic responses is then evaluated quantitatively and effectively. With the increase in the magnitude of the loading, weak links and failure modes of the structure could be found. The aims of the push-over analysis in seismic design are as follows.

1. The non-linear inter-story drifts of a structure will be determined. The structural capability of damage restriction and collapse prevention will be estimated.
2. The positions, at which the structure yielded early, will be located to judge if there is soft story or weak links in building structure in terms of seismic design.
3. The magnitude of applied lateral forces, corresponding to the first structural yield, will be determined to assess the elastic strain energy dissipation capacity to withstand the design earthquake ground motion and damage severity under the demanded earthquake ground motion.
4. The beam/column-plastic hinges will be observed to predict the damage or collapse pattern of the structure. The relationship between the deformations of elements and the inter-story drifts will be established to judge if the seismic design of the structure is reasonable and reliable.

3. Application of Push-over Analysis

An engineering project is given to illustrate the application of push-over analysis. Figure 2 shows the three-dimensional overview of the designed buildings in Shanghai. The building standing on the right hand is a residential mansion of 26 stories and 90.20 meters high. The structural plans are shown in Fig.3. The plan configuration contains a re-entrant corner where the projection of the structure beyond the re-entrant corner is 47% of the overall plan dimension of the structure in the given direction. At the height of 4.7 meters, a mezzanine floor is designed with a diaphragm of open area equal to 43% of the gross enclosed floor diaphragm area. There is also in-plane discontinuity in shear walls at the height of 8.0 meters. Hence, the mansion is categorized as a tall building with the irregularities of out-of-plane offset, diaphragm discontinuity and in-plane discontinuity in vertical lateral-force-resisting elements. The structure was designed in accordance with the seismic use intensity 7 of the national standard of PRC, equivalent to the assigned seismic use group II and the spectral response acceleration equal to 0.23g of IBC. The local site class is IV.

Figure 2. Shanghai JINLIN TIANDI project

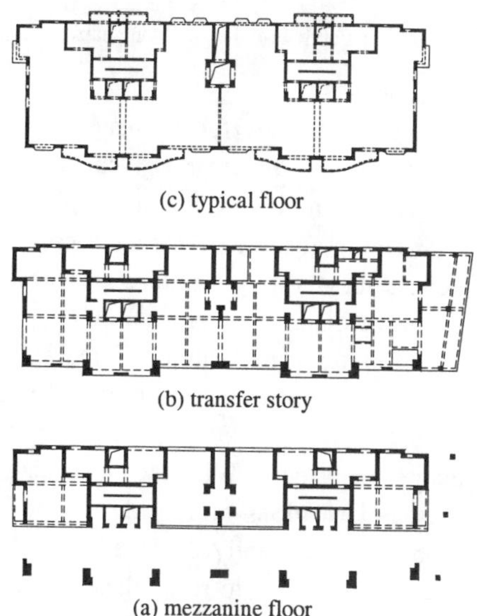

Figure 3. Structural plans of JINLIN TIANDI Residential mansion

The eigen values and torsional contributions in corresponding modes obtained by modal analysis are listed in table 1. The translational and torsional vibration modes of the structure are minimum coupled. The ratio of the fundamental period to the period of torsional mode is 0.74-0.77. The structure was elastically analyzed by using the programs SATWE, PMSAP and ETABS. The curves for ratio of maximum inter-story drift at one end of the structure and the average of the story drifts, $\delta_{max}/\delta_{avg}$, are plotted in Fig.4. The ETABS curve in the figure is the envelop of the results including ±5% accidental torsion. The maximum value is not greater than 1.2, which means the structure falls into torsional regularity. The calculated results show that the dynamic behavior of the structure is regular in the elastic range. However, a number of effects of the long narrow slit to dynamic behaviors of the structure could be found in the non-linear analysis. They are described as follows:

Table 1 Results of modal analysis

			SATWE	PMSAP	ETABS
Modal analysis	Mode 1	Period T_1 (sec.)	2.094	2.122	2.018
		Torsion (%)	0.00	0.00	0.00
	Mode 2	Period T_2 (sec.)	1.727	1.787	1.668
		Torsion (%)	0.00	0.01	0.00
	Mode 3	Period T_t (sec.)	1.563	1.642	1.493
		Torsion (%)	1.00	0.99	1.00
	Ratio of periods(T_t/T_1)		0.75	0.77	0.74

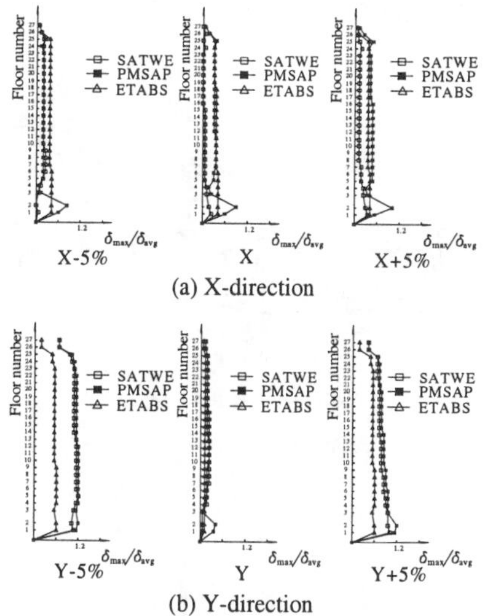

Figure 4. Torsional effects

On the basis of elastic analysis, the structure is pushed-over along the X-direction by using PUSH/EPDA, which is a software developed by China Academy of Building Research, based on the finite element principle and capacity spectrum method for elasto-plastic seismic response analysis of structures[6]. Three cases are considered to investigate the effects of the long narrow slit. There are no beam and slab in the slit for case 1. There are two beams and no slab for case 2 and for case 3, both beams and slab are present. For all three cases, the inverted triangle load pattern is adopted for the lateral pushing forces acting onto the structure along the height. The curves of capacity and demand spectra for the case 2 are plotted in Fig.5. The intersection point is found, which means that the structure would not collapse after the shock of the maximum considered earthquake. Figure 6 shows the displacement of top floor and the inter-story drifts curves corresponding to the three seismic use levels. The maximum plastic story drift is 1/173, less than the limitation, 1/100 that is required in Chinese code. Figure 7 indicates the positions of first plastic hinges and the corresponding deformations. For case 1, the structure first yielded in the elements at the edge of the long narrow slit. With the increase in magnitude of the lateral forces, deformations rapidly developed and the members consequently damaged as shown in Fig.8. For case 2, the deformation of the cut is greatly improved. Only a little plastic hinges took place around it. As expected, the case 3 shows the best behavior. Figure 9 shows the final construction drawing, which has been strengthened using the results obtained by the push-over analysis.

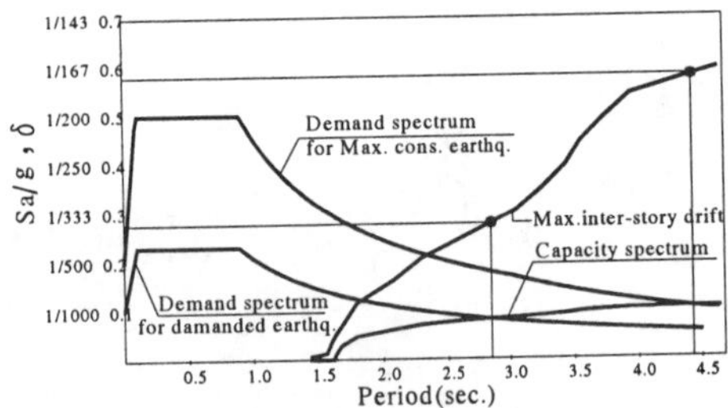

Figure 5. Capacity and demand spectra for case 2

(Push over in X-direction)

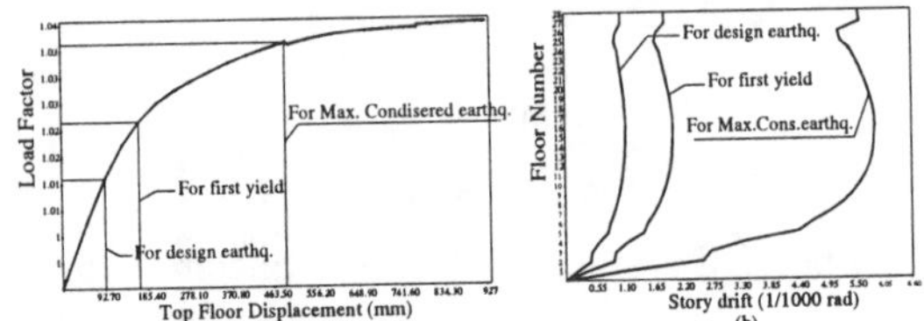

Figure 6. Disp. responses in X-direction for case 2

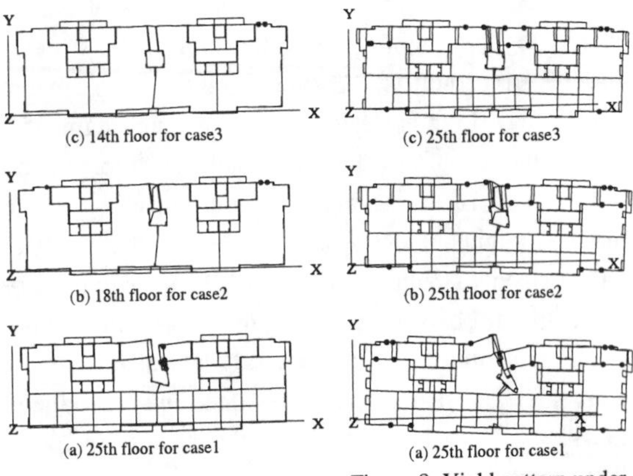

Figure 7. First yield pattern

Figure 8. Yield pattern under max. cons. earthq.

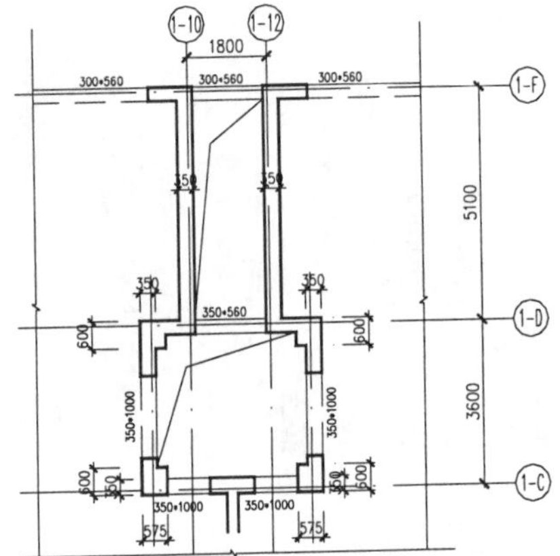

Figure 9. Construction drawings of long narrow slit

4. Conclusions

Brief reviewing of the developing history of seismic analysis procedure in China was given. The seismic design of a tall building with out-of-plane offset, diaphragm discontinuity and in-plane discontinuity in vertical lateral-force-resisting elements was presented in this paper to illustrate the application of push-over analysis. Under the design earthquake, the structure is dynamically regular. However, the results of push-over analysis showed that the structure yielded earlier at the edge of long narrow cut. With the increasing of the lateral forces, deformations rapidly developed and the corner members consequently damaged. The long slit was found as a weak place to resist the seismic ground motion and had great effects to structural capability of damage restriction and collapse prevention after the corner elements around it yielded. It could be concluded that the push-over analysis should be used as required for seismic design of the structures with irregularities. It will help structural engineers in identifying hidden weak points in a structure when conducting the seismic design.

References

1. National Standard of PRC, Code for seismic design of buildings, GB 50011-2001, (2001), (in Chinese).
2. C.S.Fu *et al*, The application of static non-linear analysis in seismic design, *J. of Structural Engineers*, **Vol.20**, **No.3**, 32 (2004), (in Chinese).
3. Freeman S A, Nicoletti J P, Tyrell J V, Evaluations of existing buildings for seismic risk—a case study of Puget Sound Naval Shipyard, Bremerton, Washington, *Proc.1st US National Conf. Earthquake Engng.*, EERI, Berkeley, 113 (1975).
4. Fajfar P, Capacity spectrum method based on inelastic demand spectra, *Earthquake Engineering and Structural Dynamics*, **28**, 979 (1999).
5. Ye L.Y. *et al*, The principle of nonlinear static analysis (push-over) and numerical examples, *J. of Building structures*, **Vol.21**, **No.1**, 37 (2000), (in Chinese).
6. J.F.Huang. *et al*, An introduction to program
7. PUSH/EPDA for push-over analysis of structures, *Symp. of 18th National Conf. on Tall Buildings of China,* 832 (2004).

A NEW METHOD OF EVALUATING DAMAGE OF RC FRAME STRUCTURE UNDER EARTHQUAKE FORCE

WEITAO LI

Institute of Civil Engineering, Tongji University, 1239 Siping Road
Shanghai 200092, China

HUIQING YING

Institute of Civil Engineering, Tongji University, 1239 Siping Road
Shanghai 200092, China

Considering the phenomenon of stiffness degradation of components in RC frame structure under load, the structure is simplified as model of rods system, Damage model of component of structure is build and damage index of component is proposed. At last the damage index is applied to estimate the damage behavior of two spans and three stories RC structure

1. Preface

The damage of structure can be divided into two kinds. One is the prophase damage occurred before the structure is complete, the other is the evening damage occurred after the structure is finished .The former is usually caused by such reasons as damage of materials or quality defection of construction. For example concrete is lack of enough time of conserving. However fire, earthquake and explosive are usually responsible for the latter damage. At present, there are few research on former damage because it is interfered with many random factors and the index of damage is difficult to be defined. As for latter, many damage index of frame structure are supposed base on different aspects such as stiffness、displacement and energy, However, most of them are difficult to be applied in practice because the parameters are difficult to be defined. In this paper, considering stiffness degradation on section of component in structure, new damage index of components is developed to evaluate the damage extent of structure under earthquake force.

2. Damage Model of Component

2.1 *Simplification of component damaged*

Under earthquake force, the damage of structure directly behaved with degradation of stiffness, which is reflected by restoring force model of section of component. Under earthquake force, once moment acted on section of component exceeds its cracking moment, the stiffness of section will degrade, in turn; the stiffness of component will degrade. From the form of load acted on frame structure, the moment acted on end of component is bigger than that acted on the middle of component. So stiffness of section in middle of component degrade severer than that at end of component .Then the damaged component of structure can be simplified as three- stiffness bar (see Figure.1).

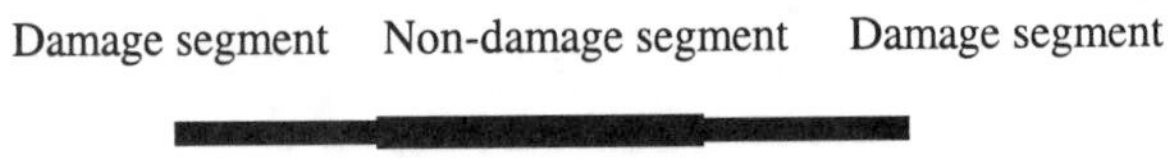

Damage segment Non-damage segment Damage segment

Figure 1 Three- stiffness bar

2.2 *Damage index*

The conception of rotational stiffness should be introduced firstly before the index of damage of component is defined, rotational stiffness denotes that magnitude of moment acted on section in end of component when section have rotation angel 1. Base on rotation stiffness, index of damage of component is proposed.

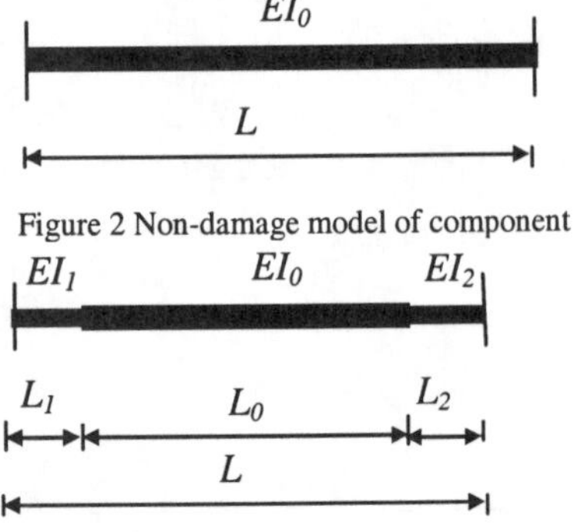

Figure 2 Non-damage model of component

Figure 3 Damage model of component

Non-damage model of component is depicted by Figure2 and damage model of component is depicted by Figure3, in which, EI_0, L_0 denote the moment of inertia and length in non-damage component, EI_1, EI_2 denote moment of inertia in damage segment of component. L_1, L_2 denotes length of damage segment of component

Rotational stiffness of left end of non-damage component is:

$$K_0 = 4\frac{EI_0}{L} \tag{1}$$

Rotational stiffness of left end of damage component is

$$K_S = \frac{1}{c(a_1b_1 + a_2b_2 + 1) - \frac{3}{4}c\frac{(a_1^2b_1 + a_2^2b_2 + 1 + 2a_1 + 2a_2b_2(1+a_1))^2}{a_1^3b_1 + a_2^3b_2 + a_1^3 + 3(a_1+1)(a_1 + a_2b_2(1+a_1+b_1))}} \tag{2}$$

In which:

$$a_1 = \frac{L_1}{L_0} \quad b_1 = \frac{EI_0}{EI_1} \quad a_2 = \frac{L_2}{L_0} \quad b_2 = \frac{EI_0}{EI_2} \quad c = \frac{L_0}{EI_0}$$

Damage index of of component is defined by

$$d = 1 - \frac{K_S}{K_0} \tag{3}$$

Where

K_S —rotational stiffness of end section of damage component

K_0 —rotational stiffness of end section of non-damage component

Both sections at end of component are optional when d is defined. In formula 3 above, the section located in left end of component is selected. K_S , K_0 must be calculated based on the same section.

The formula (2) shows damage index of component is a function of Length of damage segment (L_1,L_2) and stiffness of damage segment (EI_1,EI_2) .The method of defining parameters L_1、L_2、EI_1、EI_2 is introduced as following

2.3 *Definition of parameters in damage index*

2.3.1 *Length of damage section in component*

The Figure 4 depicts the moments act on single frame structure under earthquake force.

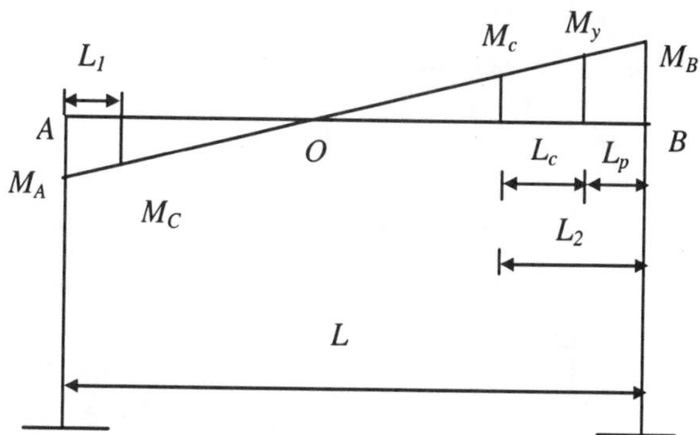

Figure 4 Moment of component under earthquake force

See Figure4, M_A,M_B are moment act on end of beam element, MC is cracking moment of beam element, O is reverse bending point, L is length of beam, L_1,L_2 are length of damage segment of beam .it is suppose that no damage occurs in beam component before beam cracks, in another words, the crack of components is sign of damage. The Figure4 show when moment acted on end of beam lower than yielding moment, just like end A, the damage segment is composed of crack segment. However, it includes both crack segment (Lc) and yielding segment (Lp) when moment acted on end of beam exceeds yield moment, just like end B.

By knowledge of geometry, the length of damage segment of component is obtained by below formula:

$$L_1 = (1 - \frac{M_c}{M_A}) \cdot OA$$
$$L_2 = (1 - \frac{M_C}{M_B}) \cdot OB \tag{4}$$

in which:

$$OA = \frac{M_A \cdot L}{M_A + M_B}$$

$$OB = \frac{M_B \cdot L}{M_A + M_B}$$

Length of undamaged segment of component is

$$L_0 = L - L_1 - L_2 \tag{5}$$

Length of yielding segment of component is

$$L_c = \frac{M_y - M_c}{M_B - M_c} \cdot L_2 \tag{6}$$

Length of crack segment of component is

$$L_p = \frac{M_B - M_y}{M_B - M_c} \cdot L_2 \tag{7}$$

2.3.2 *Stiffness of damage section*

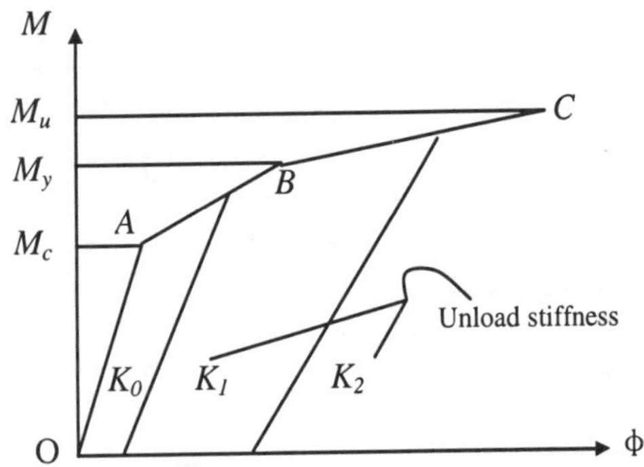

Figure 5 Restoring force model of section of component (part)

Two kinds of stiffness are reflected by restoring force model of section of component shown as Figure5: one is load stiffness and the other is unloading stiffness. Both of them can embody the damage of component for their reduction under earthquake force. In past studies, the degradation of load stiffness is often used to describe damage of section.. However, the degradation of load stiffness is instantaneous. After the load acted on component is removed, the reload stiffness of section should be equal to previous unload stiffness of section, not previous load stiffness of section. Under unload stiffness of section, the new damage will not occur unless the moment acted on section surpass the maximum moment in history. Hence, the unload stiffness indicate resistant capable of section. It can describe the damage of component well. All following stiffness means unload stiffness in this paper. It can be obtained by restoring rule.

According to form of moment acted on component, the simplification of stiffness of damage section is divided into two conditions.

1) Stiffness of damage segment before end of component yield

Before end of component yield, the moment acted on damage segment is between Mc and My .So the unload stiffness of damage sections of component is K1, which is shown in Figure5. In order to simplify calculation, the unload stiffness of all damage sections is denoted with that of damage section located in end of component.

2) Stiffness of damage segment after end of component yield

After yield, the section still has a certainly capability to bear moment .So it is possible that moment acted on the section located in end of component surpass *My*. Under this condition, damage segment of component include yielding segment (*Lp*) and crack segment (L_C), which is shown in Figure 4. Because unload stiffness of section located in yielding segment has difference with that located in cracking segment, the unload stiffness of damage section is denoted with equivalent stiffness which is calculated base on "the principle of equivalent stiffness of rotation".

The damage model of component with yielding segment is depicted in Figure6a. *EIp* and EI_C denote stiffness of section which located in yielding segment and crack segment, *Lp* and *Lc* denote length of yielding segment and crack segment; L_S denote length of whole damage segment. The equivalent damage model of component is depicted in Figure6b. *EIe* is equivalent stiffness of damage segment.

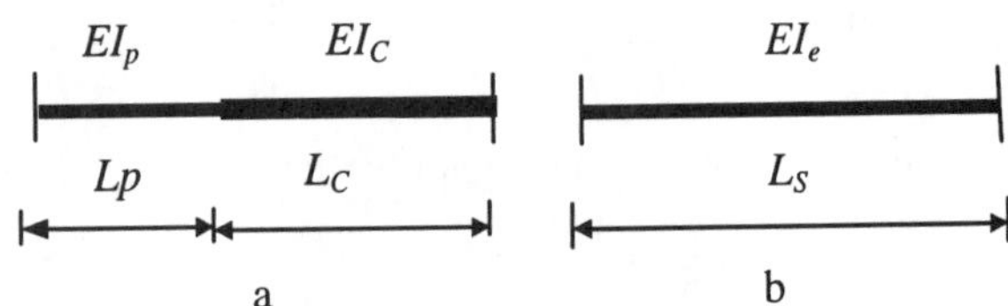

Figure 6 Calculation of equivalent stiffness

According to the principle of equivalent stiffness of rotation, the rotational stiffness of right end of component is (see in Figure6a)

$$K_p = \frac{1}{c(ab+1) - \frac{3}{4}c\frac{(a^2b+1+2a)^2}{a^3b+1+3(a+1)a}} \tag{8}$$

Where: $a = \frac{L_P}{L_c}$ $\quad b = \frac{EI_c}{EI_p}$ $\quad c = \frac{L_c}{EI_c}$

As for equivalent damage model of component shown the rotational stiffness of right end of component is

$$K_e = 4\frac{EI_e}{L_s} \tag{9}$$

Where $L_s = L_p + L_c$

Let $K_p = K_e$, then equivalent stiffness is

$$EI_e = \frac{1}{c} \cdot \frac{L_s}{4(ab+1) - 3\frac{(a^2b+1+2a)^2}{a^3b+1+3(a+1)a}} \quad (10)$$

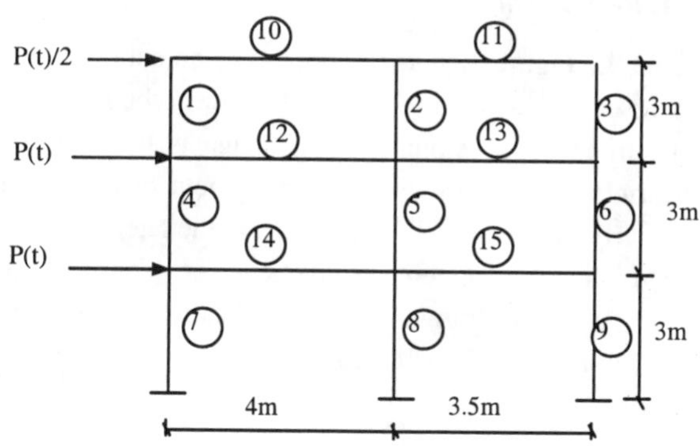

Figure 7 The reinforced concrete frame structure

3. Example Analysis

Base on the damage index of component, the procedure is components under earthquake. Take RC frame structure of 2 spans and 3 stories for example, the damage of structure is analyzed under cycle load with procedure

3.1 *Introduction of RC frame structure*

Table 1. Detailing of structure

Number of floor	Height of floor (m)	Length of beam (m)		Size of section (cm)		Steel contained in component		Remark
		First span	Second span	Pillar	Beam	Pillar	Beam	
F-1	3	4	3.5	25×25	15×20	4Φ16	4Φ12	thickness of conserve layer of steel in section is 25mm
F-2	3	4	3.5	25×20	15×20	4Φ16	4Φ12	
F-3	3	4	3.5	25×15	15×20	4Φ16	4Φ12	

3.2 *Design cycle load*

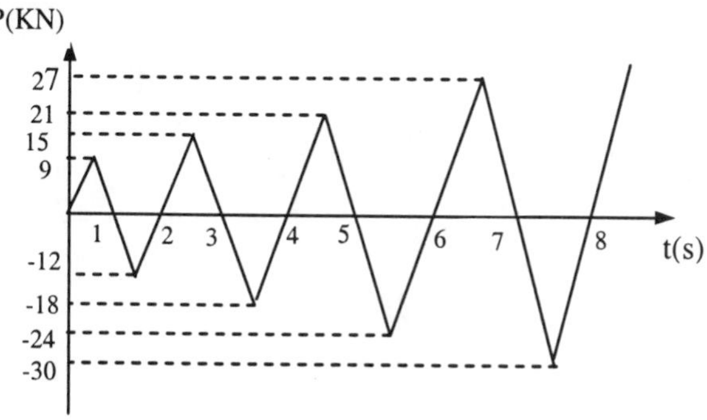

Figure 8 Design cycle load

3.3 *Damage of components in structure*

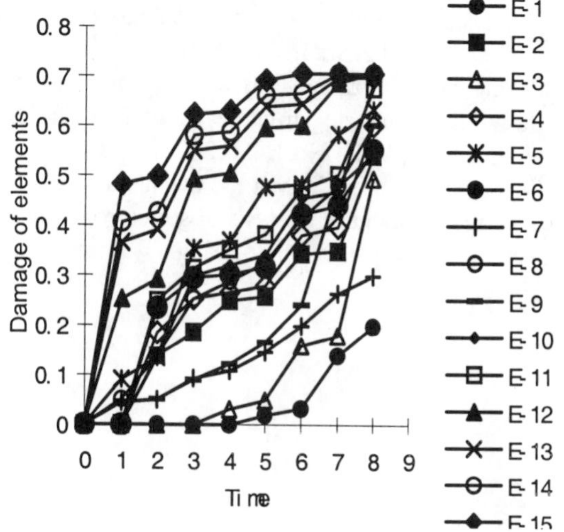

Figure 9 Curve of damage of components

3.4 *Result analysis*

From the Figure 7, as for beam, development of damage in the early time is faster, however slow in the later time. The trend of development of damage is like parabola protrudes up. As for pillar, development of damage in the early time is slow, however fast in the later time. The trend of development of damage is like parabola concave up. Additional damage of beams is series than that of pillar for the strength of pillars is better than that of beams

4. Prospect

In fact, damage of structure is complicate non-linear mechanics phenomena influenced by many factors. At present, most research focus on seeking a reasonable index to evaluate the extent of damage of structure. The mechanism of damage of structure is not clear. The relation between damage of component and that of structure should be researched further. In addition, different type of structure has different damage behavior, even the same type of structure; it is different for the different work conditions. Hence, the research of damage should focus on not only index of damage but also the interior mechanism of damage of structure. Thus the behavior of damage of structure can be explained very well and it is possible to build all-purpose model of damage of structure.

References

1. Shen Zuyan,Dong Bao Development and Evaluation of Researches on Damage Accumulation Analysis for Building Structures, Journal of Tongji university,Vol 2.No.4,137(1997)
2. Wang Mengpu Zhou Xiyuan study on criterion of failure of concrete structure .Earthquake resistant engineering, Vol 4.No.1,456(2002)
3. Wu Bo Li Hui, Li Yuhua the mechanic method for damage for damage analysis of structure earthquake Engineering and Engineering Vibration, Vol 5.No.4,14(1997)

STUDY ON OPTIMAL SHEAR WALL QUANTITY OF FRAME-SHEAR WALL STRUCTURES IN EARTHQUAKE ZONE

SHIMIN HUANG KE SONG

China Academy of Building Research, Beijing, 100013, China

The method for figuring the optimal shear wall quantity for frame-shear wall tall buildings with variable lateral rigidity along the height in earthquake zone is presented in this paper by means of dispersal-serial interaction method. It is the first time to put forward the method for figuring the optimal amount of shear wall in frame-shear wall structure by classifying the vertical members. The definition of Inter-Story Drift is made clear and definite by the author. A 20-storey framed-shear wall tall building is numerically examined and the results show that the method is accurate enough for use in the preliminary stage of design.

1. Introduction

How to determine the shear wall quantity of framed-shear wall tall building structures economically, reasonably and accurately is a major consideration in the preliminary stage of design. Sufficient stiffness must be provided to ensure that the seismic excited inter-story drift does not exceed the drift index limit given by Code. In terms of serviceability limit states, drift must be maintained at a sufficient low level to allow proper functioning of non-structure components, and to avoid distress in the structure. For tall buildings, lateral loads are quite prominent and govern the design of various components. A structure engineer has to choose a moderate amount of shear wall quantity, which would effectively resist the loads and still achieve the economy. Obviously, any design produced must result in a stable structure with all elements being capable of resisting the applied loads. In this particular instance it was intended that a cost effective solution was found which also satisfied a set of criteria based on story drift.

A number of researchers, e.g. L.G. Jaeger & YouShuang He, have investigated and demonstrated methods for the shear wall quantity of framed-shear wall tall building structures. Unfortunately, all these reported studies adopted the assumption that the earthquake action is obtained from the Equivalent Base Shear Method, the stiffness and the mass of the structure are uniform throughout the height. These assumption can only be used in tall buildings which height less than 40m according to China Code for Seismic Design of Buildings. However, with the development of high-rise buildings, the frame-shear wall structure buildings can be built over 100m high when the earthquake fortification intensity is 7 or 8 degree, the stiffness and the mass of the structure are usually vary along the height of the building. So it will lead to a great error by using former methods to determine the shear wall quantity of framed-shear wall tall building structures.

The method for figuring the optimal shear wall quantity of frame-shear wall tall buildings with variable lateral rigidity along the height in earthquake zone is presented in this paper by means of structural optimization. The dispersal-serial interaction method is adopted. Based on the continuum hypothesis of the framed-shear wall elements, the structure is dispersed between regions with different lateral rigidity, and then the Matrix-Displacement method is used. In analyzing, the effects of lateral rigidity, mass and shear deformation of the structure are considered respectively. Compared with the results based on the hypothesis of constant lateral rigidity, the result obtained by using the method presented in this paper in more economical, reasonable and accurate. The definition of inter-story drift is made clear and definite by author. In this paper, it is for the first time to put forward the method for figuring the optimal amount of shear wall in framed-shear wall structure by classifying the vertical members. The method presented in this paper can be used in buildings more than 40 meters high due to the normal response spectrum method for Model Analysis.

2. Establishment of the Mechanical Models

First, we study on lateral deflection of lateral loads excited frame-shear wall tall building structures with constant flexural rigidity by means of compatibility analysis. The analysis is based on the following simplifying assumptions:

(1) The structure behaves linear elastically.

(2) The floor slabs are rigid in-plane and without stiffness out-of-plane. The rotation deflection in-plane is neglected.

The calculation model is shown in Fig.1.

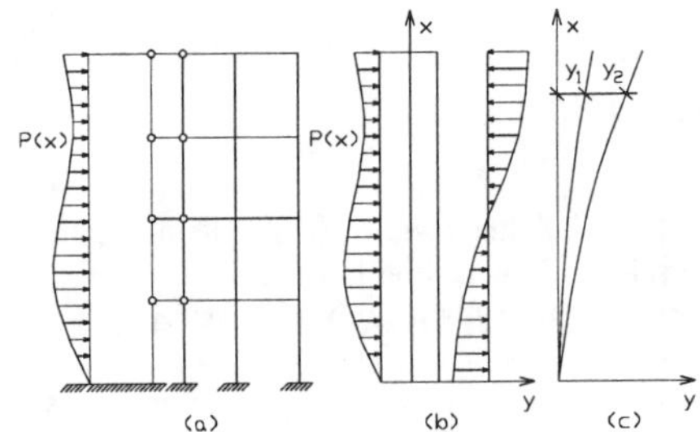

Fig. 1 simplified model for framed-shear wall structures

The lateral deflection of framed-shear wall tall building structure can be written as follows: $y=y_1+y_2$, In which, y_1 is the lateral deflection due only to bending deformation. y_2 is the lateral deflection due only to shear deformation. Differentiate upper equation twice, we get formulas as follows:

$$\frac{\partial^2 y}{\partial x^2}=\frac{\partial^2 y_1}{\partial x^2}+\frac{\partial^2 y_2}{\partial x^2} \tag{1}$$

Think over $\frac{\partial^2 y_1}{\partial x^2}=-\frac{M_w(x)}{E_w I_w}$,the angle due to shear deformation is: $\gamma=\frac{\partial y_2}{\partial x}=\frac{\mu V_w(x)}{GA_w}$.

Denote $\zeta=\frac{x}{H}$; $S=1+\frac{\mu C_F}{GA_w}$; $\lambda=H\sqrt{\frac{C_F}{SE_w I_w}}$.

Substituting all this parameters into Eq.(1), the differential equation can be express as follow[1][2]

$$\frac{d^4 y}{d\zeta^4}-\lambda^2\frac{d^2 y}{d\zeta^2}=-\frac{H^2\mu}{SGA_w}\frac{\partial^2 P(x)}{\partial\zeta^2}+\frac{H^4}{SE_w I_w}P(x) \tag{2}$$

Where,

$E_w I_w$ is the combining of shear wall flexural rigidity.

C_F is the combining of frame rigidity.

H, A_w is the building altitude and section area of shear walls, respectively.

$P(x)$ is the distributed lateral load.

Letting $P(x)=0$, Eq.(2) can be written as:

$$\frac{d^4 y}{d\zeta^4}-\lambda^2\frac{d^2 y}{d\zeta^2}=0 \tag{3}$$

This is the deflection differential equation of frame-shear wall tall building structures with no lateral load, and the solution is:

$$y=A_1+A_2\zeta+A_3 sh(\lambda\zeta)+A_4 ch(\lambda\zeta) \tag{4}$$

Fig.2 shows a segment of framed-shear wall structures with constant rigidity along the height, the positive direction of bending moments, shear forces and displacements are also shown in Fig.2.

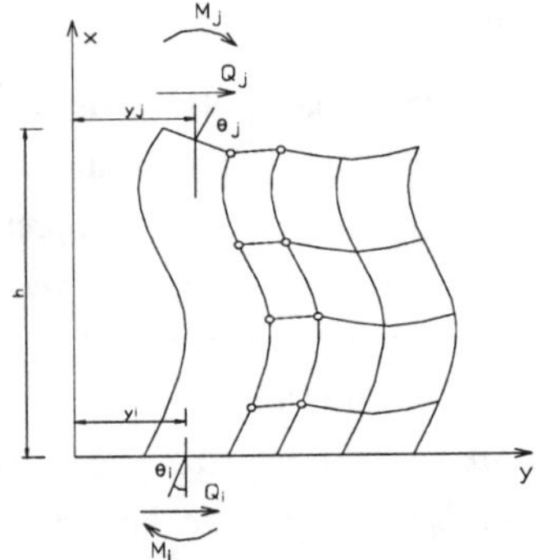

Fig. 2 The calculating model for structures

Substituting boundary conditions into Eq.(4), the equation can be rewrite in matrix form as follows:

$$\{P\}=[K]^e\{\delta\} \tag{5}$$

Where,

$$\{P\}=\{Q_i \quad M_i \quad Q_j \quad M_j\}^T,$$

$$\{\delta\}=\{y_i \quad \theta_i \quad y_j \quad \theta_j\}^T,$$

$[K]^e$ is the stiffness matrix of lateral-resistant structural elements.

Now we divide the framed-shear wall tall building structure into several segments according to the variation of mass and lateral stiffness, each segment has a constant rigidity along the height, as shown in Fig.3. Assembling the element stiffness matrix achieved by Eq.(5) together, the complete element stiffness matrix is obtained:

$$\{F\}=[K]\{\delta\} \tag{6}$$

Where, $[K]$ is (n+1)×(n+1) positive definite(or half positive definite)symmetrical stiffness matrix of the total structural system(n is the total story number of the structure);

$$\{F\} = \{Q_1 \quad M_1 \quad Q_2 \quad M_2 \ldots Q_{n+1} \quad M_{n+1}\}^T$$

$$\{\delta\} = \{y_1 \quad \theta_1 \quad y_2 \quad \theta_2 \ldots y_{n+1} \quad \theta_{n+1}\}^T$$

If the external lateral load is concentrated load acts at the floor level of the structure, Eq.(6) can be solved to obtain the displacements and rotation of the structure at each dispersal nodes.

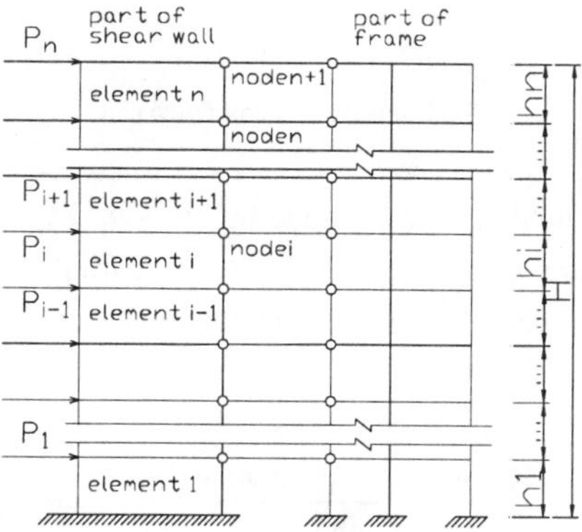

Fig. 3 Scheme of the division model for structure

For the tall buildings with rigid diaphragms under the action of the lateral forces, the sidesway of every lateral-resistant structural elements are equal. The System of Lumped Mass in Series may be taken as its mechanical model for the seismic analysis. The equations of motion of the System of Lumped Mass in Series during an earthquake playing in the undamped free vibration may be written in the familiar forms as follows:

$$[\delta][M]\left[\ddot{y}\right] + [y] = 0 \tag{7}$$

Where,

$[\delta]_{n\times n}$ is the flexibility matrix of the total structural system, obtained by inverse the stiffness matrix $[K]$;

We can use the QR-method (Rong Qin, 1996) to solve for eigenvalues and eigenvectors, and the resultant maximum member forces of the frames and shear walls may be obtained by combining the modal member forces by means of the SRSS (Square Root of the Sum of Squares) method.

3. Vertical Members Classification Method for Frame-Shear Wall Structures

For tall buildings, lateral loads are quite prominent and govern the design of various components. In terms of serviceability limit states, drift must be maintained at a sufficient low level to allow proper functioning of non-structure components (such as curtain walls, decoration components), and to avoid distress in the structure. Researches indicated[3] that the deflection limitation of a rigid frame structure is governed mainly by the damage effect of non-structure components, and for a shear wall structure, it is governed mainly by the crack of the shear walls. When a framed-shear wall tall building structure is loaded laterally, the free deflected forms of the walls and the frames are different: the shear wall part have a distinct rigid rotation about horizontal axis in addition to horizontal deflection, for frame part only the shear deformed deflection be considered and the rigid rotation about horizontal axis is so small that can be neglected. The rigid rotation about horizontal axis, however, which is not caused by internal forces has no effect on structures elements and non-structure components, it can only affect the margin of lateral displacement between two adjacent floor slabs.

Inter-story drift δ_{isd}, defined as the maximum deflection caused by internal forces between two adjacent floor slabs, and the story drift margin δ_{sdm}, defined as the maximum margin of lateral displacement between two adjacent floor slabs, are two simple parameter to estimate lateral stiffness of a building. The relationship between δ_{isd}, δ_{sdm} is shown in Fig.4

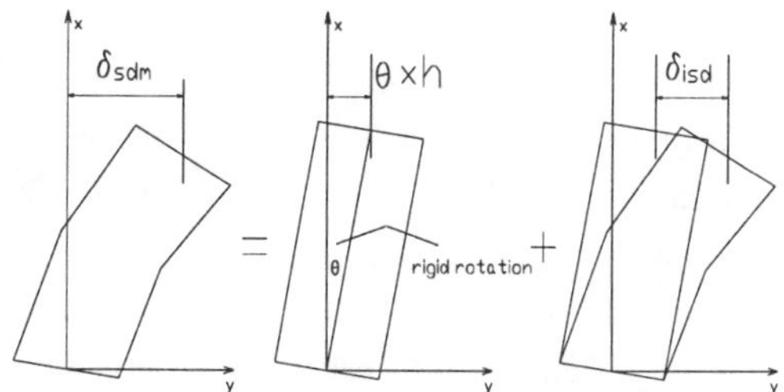

Fig. 4 Relationship between δ_{isd} & δ_{sdm}

When a frame-shear wall tall building structure is loaded laterally, the different free deflected forms of the walls and the frames cause them to interact horizontally through the floor slabs, consequently, Story Drift Margin δ_{sdm} is the same for both frame part and shear wall part of the structure, but Inter-Story Drift δ_{isd} is distinctly different, may be a margin of several times. So, we classify the vertical members into frame part and shear wall part, take δ_{isd} as drift criteria for the part of shear walls and take δ_{sdm} as drift criteria for the part of frames.

To facilitate a numerical solution of the shear wall quantity design optimization question, the two-stage analysis method is proposed. Its analysis procedure is described as follows:

(1) For the part of shear walls, take δ_{isd} as drift criteria and compare with the drift index of shear wall structures (1/1000) given by "China Code for Seismic Design of Buildings", we derive the optimal quantity of shear wall $E_{w1}I_{w1}$;

(2) For the part of frames, take δ_{sdm} as drift criteria and compare with the drift index of frame structures (1/550) given by "China Code for Seismic Design of Buildings", we derive the optimal quantity of shear wall $E_{w2}I_{w2}$;

The final optimal amount of shear wall in frame-shear wall structures E_wI_w is the bigger value from $E_{w1}I_{w1}$, $E_{w2}I_{w2}$. E_wI_w can fulfil simultaneously the proper functioning of non-structure components and uncrack in the structure elements.

4. Optimal Design Problem

Unlike the conventional trial-and-error design method, the structural optimization technique proposed is a systematic goal-oriented design synthesis approach that automatically seeks the most optimal design whilst satisfying specified design criteria. One needs to define explicitly the optimal design problem in terms of a set of design variables. Once he optimal design problem has been formulated, a rigorously derived Optimality Criteria method can be employed to provide the solution. Computer program named FWOP is compiled and the results show that the method is accurate enough for the conceptual design stage where an efficient and economic configuration for structural system is to be selected out of various alternatives, which should satisfy the lateral drift limits.

5. Numerical Examples

In this example a twenty-storey framed-shear wall tall building structure under the action of seismic loading is analyzed. The typical floor layout is shown in Fig.5. The total height is 62m, the vertical load of each floor(including dead load and live load)is 18Kn/m^2; the seismic intensity is 8 degree, site earth class is 2; The sectional dimension of beams is 30×50cm, columns: 65×65cm(1st-10th), 60 × 60cm(11th-20th). Find the final optimal amount of shear wall.

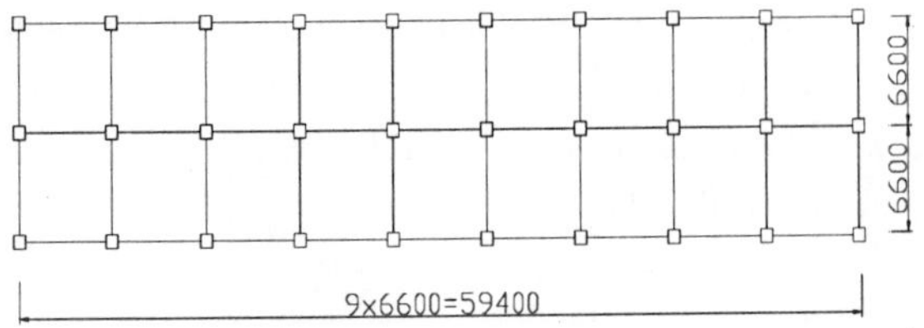

Fig. 5 Typical floor layout of frame part

The optimal analysis is done by program FWOP, and the results are listed in Table 1 by comparing with the results which derived from the international general finite

Table 1 Comparison between structural optimization results

	FWOP method		China Seismic Design Code Method	SAP2000 Method E_wI_w=8.03 ×10^8
	For the part of frames	For the part of shear walls		
Code drift index δ/h	1/550	1/1000	1/800	Frame 1/550 Shear wall 1/1000
Maximum Inter-Story Drift	1/554	1/1001	1/803	Frame 1/556 Shear wall 1/2214
The first three periods (second)	T_1=1.85 T_2=0.49 T_3=0.21	T_1=2.27 T_2=0.71 T_3=0.37	T_1=1.26 T_2=0.28 T_3=0.11	T_1=1.88 T_2=0.48 T_3=0.22
Total earthquake action (kN)	9335	8491	12288	9862
Final optimal amount of shear wall E_wI_w (kNm^2)	$E_{w2}I_{w2}$ 8.03× 10^8	$E_{w1}I_{w1}$ 1.23× 10^8	5.18× 10^9	8.03×10^8
	bigger value of $E_{w1}I_{w1}$, $E_{w2}I_{w2}$ is E_wI_w 8.03× 10^8			

element program SAP2000. From the FWOP results, the lateral Drift Index given by "China Code for Seismic Design of Buildings" can be fulfilled by laying four 6.6m long shear walls with the thickness of 30cm in transverse direction. The results of SAP2000 are given based on this arrangement.

6. Conclusions

Based on the above discussion and example analysis, it can be found out that for the shear wall part in frame-shear wall tall building structures, the Inter-Story Drift differs greatly from the Story Drift Margin and this difference increase by a wide margin with the increasing altitude of the building. As for the numerical example in this paper, if we determine the optimal amount of shear wall in frame-shear wall structures according to Drift Index(1/800) given by "China Code for Seismic Design of Buildings", a margin of several times will occur compare with this paper method. At the present time, almost all of the design softwares take only Story Drift Margin into consideration in their calculation and this sometimes leads to lateral displacement exceed the allowable drift limit specified by Chinese seismic design code. In fact, the Inter-Story Drift due to forces is always small enough to fulfil the allowable drift limit.

The method for figuring the optimal amount of shear wall in frame-shear wall structures by classifying the vertical members can fulfil simultaneously the proper functioning of non-structure components and uncrack in the structure elements. Compared with the results based on the hypothesis of constant lateral rigidity, the result obtained by using the method presented in this paper in more economical, reasonable and accurate. This method can be used in buildings more than 40 meters high due to the normal response spectrum method for Model Analysis.

Lateral deflection limit is a major design consideration for the serviceability requirements in the design of frame-shear wall tall buildings. The present work proposes a model that raptly evaluates the inter-story drift, a measure of lateral deflections of tall structural system. The model is useful at the conceptual design stage where an efficient and economic configuration for structural system is to be selected out of various alternatives, which should satisfy the lateral drift limits.

References

1. ShiMin Huang and Lian Wei. 1997. The Computation Method for Earthquake-Excited Frame-Shear Wall Structure with Variable Rigidity along Height, *Building Science*, No.4, 16-20.
2. ShiMin Huang, Lian Wei, ShaoGe Cheng, HongJian Yi&Bin Zhao. 2003. A study on the optimal quantity of shear walls for frame-shear wall structures in earthquake zone. *Earthquake Resistant Engineering*, No.1, 1-5.
3. Ministry of Construction, P.R.China(2002). *Code for seismic design of buildings*, Beijing, China.
4. ChengJi Wei, JingWei Liu, ZhenHua Qi and YongMing Fang. 1994. Some proposals concerning control of inter-story drift of tall buildings. *Earthquake Resistant Engineering*, No.2, 1-3.
5. Jaefer L.G., Mufti A.A. and Mamet T.C.1973. The structure analysis of tall buildings having irregularly positioned shear walls. *Building-Science*, No.1, 15-19.

MULTI-MODE CONTROL OF COUPLED LATERAL-TORSIONAL VIBRATION OF ASYMMETRICAL TALL BUILDINGS

SHIMIN HUANG KE SONG LEI JIANG HONGJIAN YI

China Academy of Building Research, Beijing, 100013, China

In this paper, coupled vibration characteristics and dynamic response of asymmetrical tall buildings under seismic loads are analysed. Based on the basic theory of structure control, the principle and method of computation are given for asymmetrical tall buildings with LTTMD control system, and computer program is compiled.

1. Introduction

Researches on structural control have been developed rapidly and a lot of achievements have been obtained since the concept of structural control was proposed by Yao[1]. Among these researches, passive control is an important branch. Some effective passive control devices have been established, such as Tuned Mass Damper(TMD), Tuned Liquid Damper(TLD) and so on. The Tuned Mass Damper(TMD) is a classical engineering device consisting of a mass, a spring and a viscous damper attached to a vibrating main system in order to attenuate any undesirable vibrations. Because the natural frequency of the damper is tuned to a frequency near the natural frequency of the main system, the vibration of the main system causes the damper to vibrate in resonance, dissipating the vibration energy through the damping in the tuned-mass damper.

In this paper, based on the existing theoretical studies[2] [3] [4],a new passive control system—LTTMD (Lateral-Torsional Tuned Mass Damper) is utilized to investigate on damper control for coupled lateral and torsionral seismic responses of tall buildings. Objective of this study are: (1)To investigate the principle and method of computation for asymmetrical tall buildings with LTTMD control system, study the effect of parameters of LTTMD control system and optimally select its positions on building; (2)A single LTTMD can only control the mode for which it is designed. Therefore, the concept of multi-tuned mass dampers, i.e., having a separate LTTMD for every structural mode appears to be worth investigating. Much of the research in the area of multi-tuned mass dampers [5][6]has been done with the aim of controlling a single mode only. In this paper, analyses are carried out with the purpose of controlling multiple modes. This will prove the effectiveness, reliability and practicable applicability of LTTMD and provide an ideal construction scheme for the actual engineering application of LTTMD.

2. Scheme, Model and Seismic Analysis

LTTMD is a kind of passive control device with the characteristics of simple construction, convenient installation and easy maintenance, and its scheme of construction is illustrated in Fig.1. LTTMD is mainly constructed with moving mass, inertial pendulum, viscous damper and coil springs. Because the natural frequency of the Lateral-Torsional Tuned Mass Damper is tuned to a frequency near the natural frequency of the main structure, the vibration of the main structure causes the damper to vibrate in resonance, dissipating the vibration energy through the damping in the LTTMD.

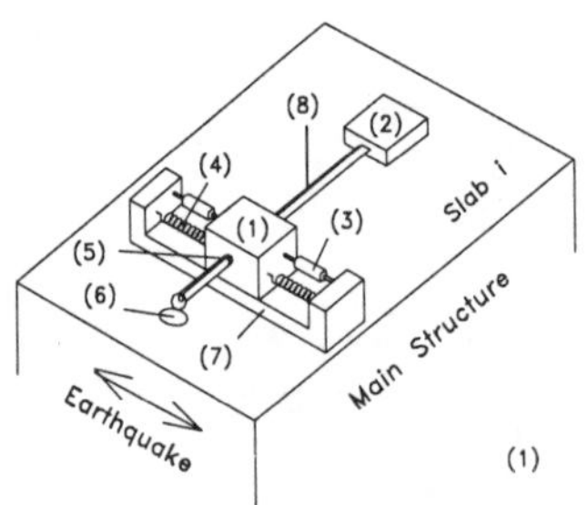

(1)Moving Mass (2)Inertial Pendulum (3)Viscous Damper
(4)Coil Spring (5)Sliding Sleeve (6)Rotating shaft
(7)Mass Guide Rail (8)Pivoted Lever

(a)

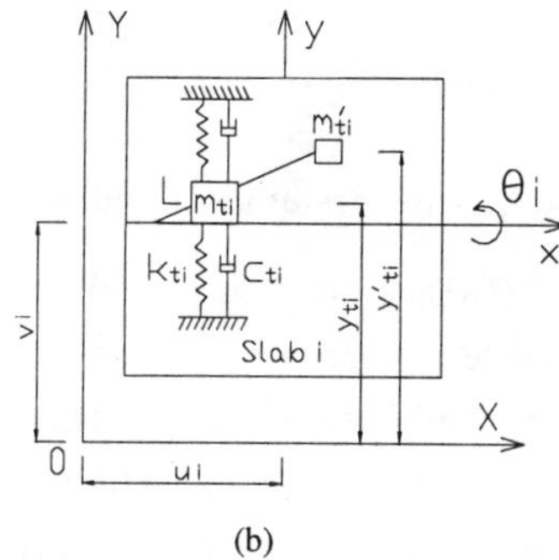

(b)

Fig. 1 Construction Scheme & Mathematical Model

The energy of the system shown in Fig.1(b) can be expressed as:

$$U_t = U_s + \frac{1}{2}k_{ti}(y_{ti} - v_i)^2 \quad (2.1a)$$

$$T_t = T_s + \frac{1}{2}m_{ti}(\dot{y}_{ti} - \dot{v}_g)^2 + \frac{1}{2}m_{ti}\dot{u}_i^2$$

$$+\frac{1}{2}m'_{ti}\left[(1+\lambda)\dot{y}_{ti} - \lambda\dot{v}_i + \lambda l\dot{\theta}_i + \dot{v}_g\right]^2$$

$$+\frac{1}{2}\lambda^2 l^2 m'_{ti}\left(\dot{\theta}_i + \frac{\dot{y}_{ti}}{l} - \frac{\dot{v}_i}{l}\right)^2$$

$$+\frac{1}{2}m'_{ti}\dot{u}_i^2 \quad (2.1b)$$

$$F_t = F_s + \frac{1}{2}C_{ti}(y_{ti} - v_i)^2 \quad (2.1c)$$

in which, U_t, T_t and F_t are the kinetic, potential, and dissipative energy respectively, U_s, T_s and F_s are the kinetic, potential, and dissipative energy of the main structure only. $\lambda = \frac{l_1}{l_2}$, l_1 is the length from the inertial pendulum to the moving mass, l_2 is the length from the rotating shaft to the moving mass.

Using the Lagrange equations to formulate the equations of motion:

$$M_T\ddot{U}_T + C_T\dot{U}_T + K_T U_T = -M_{gT}\ddot{U}_{gT} \quad (2.2)$$

in which , M_T、C_T and K_T are the total mass, damping, and lateral stiffness matrices respectively (including the LTTMD system), and U_T=the column vector of horizontal floor displacements relative to the ground.

$$M_T =$$

$$\begin{bmatrix} [M_1] & \cdots & 0 & \cdots & 0 & 0 \\ \vdots & \ddots & & & \vdots & \vdots \\ 0 & & [M_i]+[M_i]_A & & 0 & [M_{iy}]_T \\ \vdots & & & \ddots & \vdots & \vdots \\ 0 & \cdots & 0 & \cdots & [M_n] & 0 \\ 0 & \cdots & [M_{yi}]_T & \cdots & 0 & Tm \end{bmatrix}$$

in which,

$$[M_i]_A = \begin{bmatrix} m_{ti}+m'_{ti} & 0 & 0 \\ 0 & 2\lambda^2 m'_{ti} & -2\lambda^2 l m'_{ti} \\ 0 & -2\lambda^2 l m'_{ti} & 2\lambda^2 l^2 m'_{ti} \end{bmatrix}$$

$$Tm = m_{ti} + (2\lambda^2 + 2\lambda + 1)m'_{ti};$$

$$[M_{yi}]_T = [M_{iy}]_T^T$$

$$= [0 \quad -\lambda(2\lambda+1)m'_{ti} \quad \lambda l(2\lambda+1)m'_{ti}]$$

$$K_T =$$

$$\begin{bmatrix} [K_{11}] & \cdots & [K_{1i}] & \cdots & [K_{1n}] & 0 \\ \vdots & \ddots & & & \vdots & \vdots \\ [K_{i1}] & & [K_{ii}]+k_{ti} & & [K_{in}] & [K_{iy}]_T \\ \vdots & & & \ddots & \vdots & \vdots \\ [K_{n1}] & \cdots & [K_{ni}] & \cdots & [K_{nn}] & 0 \\ 0 & \cdots & [K_{yi}]_T & \cdots & 0 & K_{ti} \end{bmatrix}$$

$$[K_{yi}]_T = [K_{iy}]_T^T = [0 \quad -k_{ti} \quad 0]$$

$$C_T =$$

$$\begin{bmatrix} [C_{11}] & \cdots & [C_{1i}] & \cdots & [C_{1n}] & 0 \\ \vdots & \ddots & & & \vdots & \vdots \\ [C_{i1}] & & [C_{ii}]+c_{ti} & & [C_{in}] & [C_{iy}]_T \\ \vdots & & & \ddots & \vdots & \vdots \\ [C_{n1}] & \cdots & [C_{ni}] & \cdots & [C_{nn}] & 0 \\ 0 & \cdots & [C_{yi}]_T & \cdots & 0 & C_{ti} \end{bmatrix}$$

$$[C_{yi}]_T = [C_{iy}]_T^{\,T} = [0 \quad -c_{ti} \quad 0]$$

$$U_T = [\{x_1\} \quad \{x_2\} \quad \cdots \quad \{x_n\} \quad y_{ti}]^T$$

$$\{x_i\} = [u_i \quad v_i \quad \theta_i]^T \text{ ; (i=1~n)}$$

$$M_{gT} =$$

$$\begin{bmatrix} [M_1] & \cdots & 0 & \cdots & 0 & 0 \\ \vdots & \ddots & & & \vdots & \vdots \\ 0 & & [M_i]_{gT} & & 0 & 0 \\ \vdots & & & \ddots & \vdots & \vdots \\ 0 & \cdots & 0 & \cdots & [M_n] & 0 \\ 0 & \cdots & 0 & \cdots & 0 & Tm_{gT} \end{bmatrix};$$

$$[M_i]_{gT} = [M_i] + [M_i]_{gA}$$

$$[M_i]_{gA} = \begin{bmatrix} 0 & 0 & 0 \\ 0 & -\lambda m'_{ti} & 0 \\ 0 & 0 & \lambda l m'_{ti} \end{bmatrix};$$

$$Tm_{gT} = m_{ti} + (1+\lambda) m'_{ti}$$

By now, the equations of motion of a general damped structure-LTTMD system with a LTTMD placed at the *j*th floor of the structure are derived. Therefore, the equations of motion of multi-tuned mass dampers place any floor of the structure can also be derived using above method.

For the numerical integration of Eq.(2.2), the constant average acceleration procedure known as "Wilson θ-method" has been used. This method is a convenient and relatively efficient procedure for nonlinear structural analysis.

3. Multi-Lateral-Torsional-Tuned Mass Dampers

When a LTTMD system is installed to control a particular mode, properties of the finally obtained system differ from those of the original structure. Now, if an additional LTTMD system tuned to another mode is also to be installed, it may not perform as expected because of this effective change in structural parameters. Also, the addition of a LTTMD system may affect the performance of LTTMD system(s) already present. This problem of model interaction is discussed with the help of harmonic base excitation analysis. The following rules are taken:

1. To effectively control any particular mode, a separate LTTMD system, specifically tuned to that mode, must be provided.
2. The appropriate mass (for all the dampers) distribution among the various LTTMD systems must first be determined. According to harmonic base excitation analysis, the response values indicate the relative importance of various modes in determining the overall structural response and LTTMD masses are distributed in the peak response floors (floors which undergo largest steady-state deflection according to modal matrix) .

4. Example and Parametric Study

As an example, a 20-story asymmetrical tall building structure is simulated, the plane drawing of structure is shown in Fig. 2.

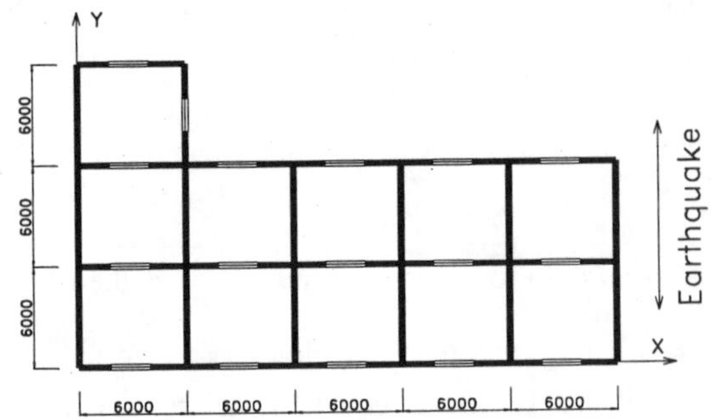

Fig. 2 Typical floor plane of asymmetrical building

To facilitate further discussion, the following additional notations are introduced:

ζ_{si}、ω_{si}: ith-mode modal damping ratio and frequency of main mass, respectively.

ζ_{ti}、ω_{ti}: ith-mode modal damping ratio and frequency of LTTMD systems, respectively.

η_i: Ratio of ith-mode frequency of LTTMD system to ith-mode modal frequency of main mass,

$$\eta_i = \frac{\omega_{ti}}{\omega_{si}}$$

$$\gamma_y = \frac{Y_{o\max} - Y_{t\max}}{Y_{0\max}}$$: Peak response reduction ratio. In which, $Y_{t\max}$、$Y_{0\max}$ are the peak response of floor with and without LTTMD systems under seismic loads.

The structure (with multi-LTTMD system) is designed and investigated with the help of time-history analysis under SAN FERNANDO earthquake using one, two and three LTTMD systems, respectively. Results of these analyses are summarized in Fig.3. to Fig.5.

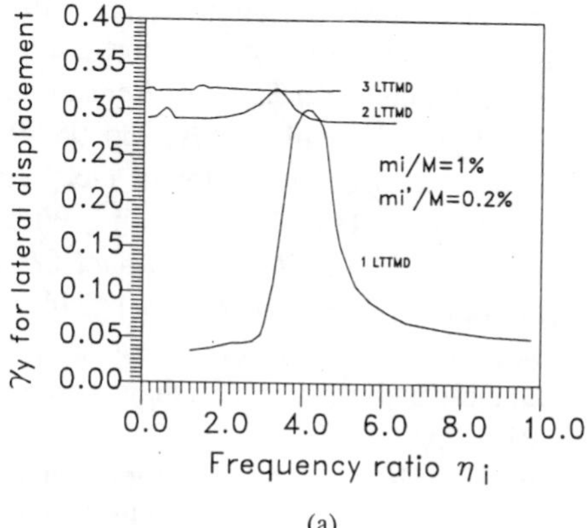

(a)

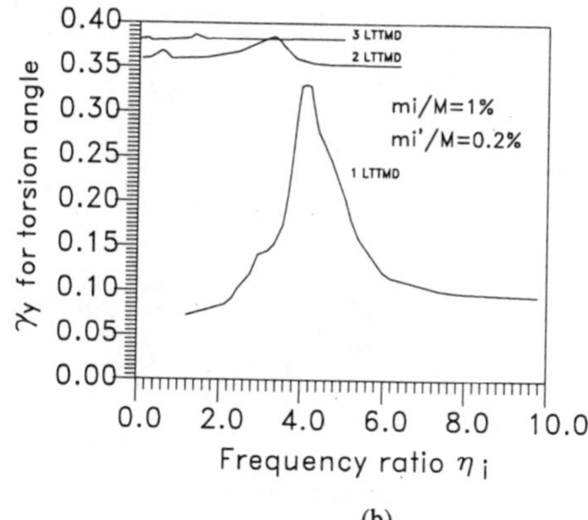

(b)

Fig.3 Peak response reduction ratio for various frequency ratio

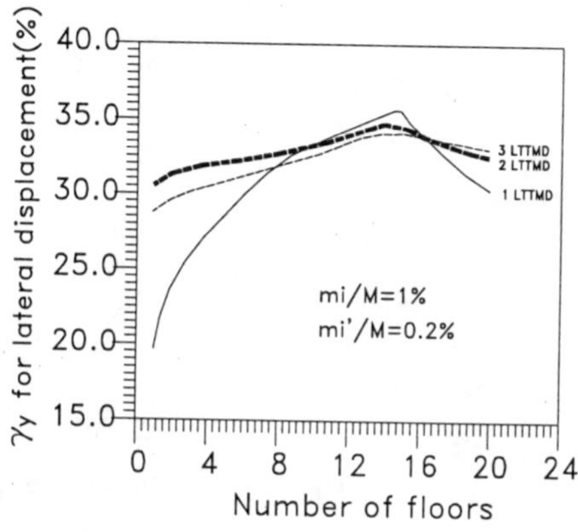

(a)

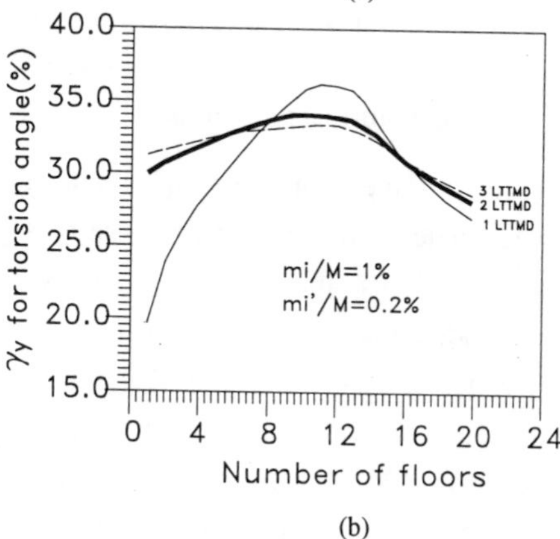

(b)

Fig.4 Peak response reduction ratio for various story

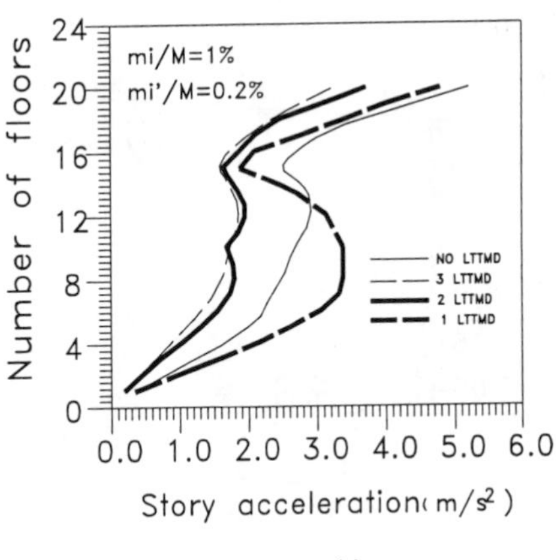

(a)

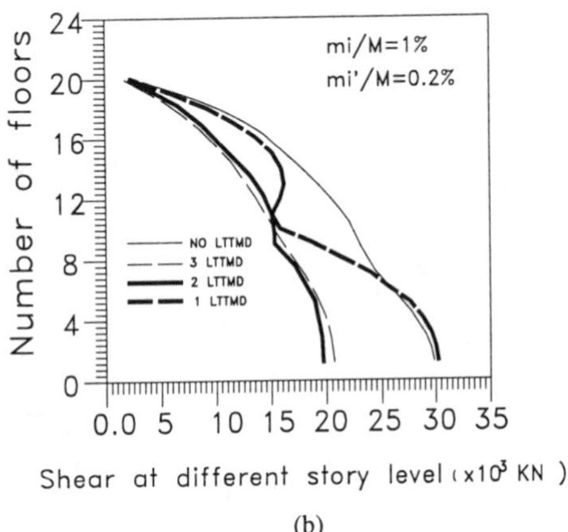

(b)

Fig.5 Peak response at different story level

5. Summary and Conclusions

1. Through the analysis of coupled vibration frequencies and dynamic response of asymmetrical tall buildings under seismic loads, it can be concluded that LTTMD control system is able to effectively control not only the lateral but also the torsion-coupled seismic response of tall buildings.
2. For usual TMD system, it is clear that the maximum response reduction occurs when the TMD is tuned close to the building frequency, in this case the maximum relative displacement occurs with the tuned condition. But for LTTMD system, the optimum frequency ratio η_i is several fold increased, this causes a distinctly decrease in relative displacement of the damper to building and therefore the spring stress of control system is strikingly cut down,
3. In this example, the presents of first-mode LTTMD causes the reduction of top floor acceleration, but increases middle floors acceleration and the base shear force of the building. It is clear that the presence of first-mode LTTMD maybe causes some increase in the higher-mode response. At the same time, to control higher modes, a separate LTTMD must be provided. So, it is reasonable place 2-3 LTTMD control system to control the first two or three mode of the structure from the view of base excitation response.
4. It is also observed in Fig.3- Fig.5 that, the controlling effect of structure response is getting marginally increased due to the increasing of LTTMD control systems from one to three. As for the peak response reduction ratio, the effect of first-mode LTTMD control system is much better than that of second- and third-mode LTTMD control system.

References

1. Yao, J.T.P., Concept of structural control" *Journal of structural division,* ASCE, 98(12), 1972, pp1567-1574.
2. A.J. Clork. Multiple Passive Tuned Mass Dampers for Reducing Earthquake Induced Building Motion, .Proc. of 9th WCEE,1988,Tokyo,Vol.5.
3. HSIANG-CHUAN TSAI & GUAN-CHENG LIN, Optimum Tuned-Mass Dampers for Minimizing Steady-State Response of Support-Excited and Damped Systems, *Earthquake Engineering and Structural Dynamics,* Vol.22, 1993, pp. 957-973.
4. Sinji Ayabe & Sadamitsu Honda etc., Vibration Control System of High-Rise Building With Tuned Liquid Damper, *The Fifth International Conference on Tall Buildings, Hong Kong* 1998, pp776-781.
5. Kareem, A. and Kline, S., Performance of multiple mass dampers under random loading, *J. Struct. Eng.,* 1995 121(2), pp348-361.
6. Yamaguchi, H. and Harnpornchai, N., Foundamental characteristics of multiple tuned mass dampers for suppressing harmonically forced oscillations., *Earthquake Eng. Struct. Dynamics.,* 1993 22, pp51-62.

DISPLACEMENT-BASED SEISMIC DESIGN OF SHEAR WALL STRUCTURE IN TALL BUILDINGS*

LIANG XINGWEN† DENG MINGKE LI BO YANG KEJIA

School of Civil Engineering, Xi'an University of Architecture and Technology, Shanxi Xi'an 710055, China

According to the characteristic of shear wall structure , its performance level is divided into three levels, which are serviceability, life-safety, and collapse prevention. The three levels are qualified with storey drift ratio. Applying the inverted triangular distribution of lateral force to the cantilever wall with identical section, the displaced shape is regarded as the initial lateral displacement mode of shear wall structure. For the serviceability and collapse prevention performance levels, it is assumed that the curvature at the bottom cross section of the first storey of the shear wall structure reaches yield curvature and ultimate curvature, respectively. Then, the target displacement mode of the two performance levels can be obtained. Based on the target displacement mode of shear wall structure in the serviceability performance level, the effective parameter, the base shear V_b and horizontal earthquake force F_i at each floor of the building can be determined, and then the structural members can be designed to satisfy strength demand and details of seismic design. Finally, the designed structure is analyzed with nonlinear static analysis procedure and is adjusted by the pushover analysis result until it is satisfied.

1. Introduction

The Chinese Code for Seismic Design of Buildings (CCSDB), GB50011-2001 [1] and other design codes use two stage method for seismic design. First, it is necessary to calculate the earthquake action and seismic action effect of building structure under minor but frequently occurring earthquakes, combine it with gravity load effect, and check the bearing capacity of the section based on the combined internal force of critical section of members. Then, the designer proceeds to the elastoplastic deformability checking computations for structures under rare but major earthquakes. This procedure is a force-based seismic design method, in which the force is regarded as the design parameter at the beginning.

Site investigation on actual buildings before and after an earthquake shows that the force-based seismic design method mentioned above can ensure that buildings resist minor earthquakes without damage and undergo major earthquakes without collapse. However, in the 1990s, the earthquakes occurring in metropolis of developed countries brought few earthquake casualties, but resulted in tremendous economic losses for the buildings invested heavily in equipment and decoration because of excessive structural damage, although they didn't collapse. Thus, the researchers are aware that simply emphasizing that the structure is not severely damaged and collapsed under the earthquake action is not a perfect seismic design concept and can not satisfy the demand of seismic design of modern engineering structures. Under this background, the scholars of America and Japan put forward the idea of performance-based seismic design. In recent years, much progress has been made on the approach of performance-based seismic design [5][6], and the method has been partially adopted by the modern seismic design codes in many counties.

Experimental investigations show that the performance of engineering structures in each stage is closely related to the deformation capacity, that is, the quantificational relationship between structural performance and deformation index can be established. Hence, the displacement is regarded as the design parameter in performance-based seismic design, and then, the displacement-based seismic design approach [2-8] is presented. In this paper, a direct displacement-based design method for shear wall structure is presented.

* This work is supported by grant 2001C26 of the Shanxi Province Science Foundation.

† E-mail: Lxw000Lxw@yahoo.com.cn

There are two basic problems in displacement-based design for shear wall structure. Firstly, it is necessary to determine the corresponding target drift curve based on the performance objectives of structure, calculate the base shear and the horizontal earthquake force at each floor, and check the bearing capacity of the shear wall section. Secondly, according to the deformation demand of structure under a certain intensity of earthquake action, it should proceed to the deformation capacity design of shear wall section, and ensure that the structure is able to reach the prospective performance objective. Only the first problem is analyzed and studied in this paper.

2. The Performance Level and Quantification of Shear Wall Structure

In displacement-based seismic design, the relationship between performance level and target displacement should be determined. In order to correspond to the earthquake fortification objective in CCSDB and control the deformation behavior for a structure, the performance level of building structures is divided into three: serviceability, life-safety, and collapse prevention in this paper. In a certain levels of ground motions, these three performance levels are consistent with the specified level of without damage, repairable and without collapse in CCSDB, respectively. But in CCSDB, seismic design has been carried out based on the philosophy that buildings should be able to resist minor earthquakes without damage, should continue to function with repairable damage when subject to moderate earthquakes, and should not collapse when subject to major earthquakes. Thus, this is a performance-based earthquake fortification objective, which ensures the safety of the occupants of structure.

However, in performance-based seismic design, various seismic fortification criterion, such as without damage to structure when subject to moderate or major earthquakes, can be adopted according to the importance of the building and the requirement of the owner. It is different from the traditional seismic design methods.

It has been recognized that the storey drift ratio of a multi-storey building, defined as the ratio of the storey drift to the storey height, reflects the total deformation of members in the storey for reinforced concrete structure and can be a good measure of structural and nonstructural damage of the building under various levels of ground motions. Thus, the performance levels can be qualified by storey drift ratio in performance-based seismic design procedure. Based on extensive experimental investigations, the storey drift ratio limits of shear wall structure corresponding to the three performance levels can be given as follows: in the serviceability performance level, taking [θ]=1/1000; in the life safety performance level, taking [θ]=1/250; in the collapse prevention performance level, taking [θ]=1/120.

3. The Initial Lateral Displacement Mode of Shear Wall Structure

In direct displacement-based seismic design, the lateral displacement mode of the structure should be determined at first. Next, the corresponding target drift curve for the structure can be obtained according to the performance objective. Then, the equivalent parameter and base shear can be calculated. For shear wall structure, applying the inverted triangular distribution of lateral force to the cantilever wall with identical section, as shown in figure 1, the displaced shape can be regarded as the initial lateral displacement mode. That is,

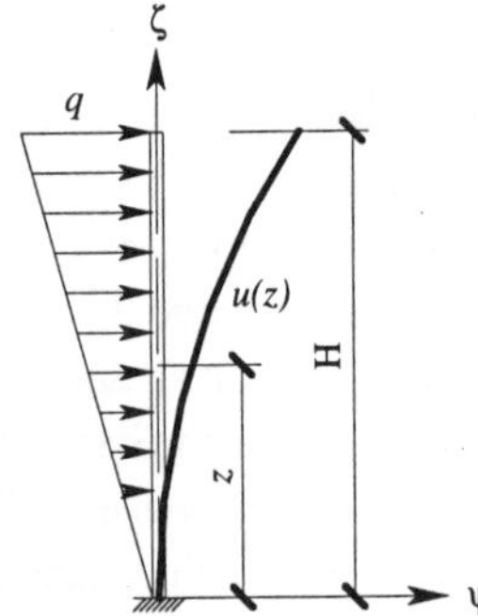

Figure 1. The initial lateral displacement mode

$$u(\xi) = \frac{\phi H^2}{40}(20\xi^2 - 10\xi^3 + \xi^5) \quad (1)$$

where ϕ denotes the curvature at the bottom critical section of the first storey of shear wall structure, H denotes total height of the building, and ξ is the height ratio of storey, which is expressed as $\xi = z / H$.

In serviceability performance level, assuming that the curvature at the bottom critical section of the first storey of the shear wall structure reaches yield curvature, which is corresponding to the elastic limit state, the corresponding target drift curve is expressed as

$$u(\xi) = \frac{\phi_y H^2}{40}(20\xi^2 - 10\xi^3 + \xi^5) \quad (2)$$

where ϕ_y denotes the yield curvature of the section, which can be determined as [7]

$$\phi_y = 2\varepsilon_y / h_w \tag{3}$$

in which ε_y is the yield strain of the longitudinal reinforcement at the end of shear wall section, and h_w is the depth of section.

4. Displacement Response Spectra

In a traditional force-based design, the period of natural vibration of a structure is estimated based on previous trial design, and design pseudo-acceleration response spectra are entered to determine the elastic design force level on the structure. However, in displacement-based seismic design, the required design displacement demand for the structure can be specified, and the displacement response spectra can be used to determine the required period of vibration, provided that the structure has been modeled assuming a linear behavior and a viscous damping equivalent to the actual non-linear response. Hence, the displacement response spectra, for a given level of damping, are the basis of direct displacement-based seismic design. The smoothed elastic response spectra can be determined by two methods:

(1) Based on large numbers of strong ground motion record (or earthquake wave), the relationship between the period of vibration of a structure T and the spectral displacement S_d can be obtained by numerical integration.

(2) Based on the design pseudo-acceleration response spectra $S_a(T)$ in CCSDB, the displacement response spectra can be expressed as

$$S_d = (\frac{T}{2\pi})^2 S_a \tag{4}$$

It has to be pointed out, however, that the spectral pseudo-accelerations is magnified for moderate or long period structures in order to avoid that the earthquake action of the structure is too small. Thus, the displacement response spectra obtained from equation (4) will be different from the real displacement response for moderate or long period structures. Whereas, the difference of period for shear wall structure can be decreased because the corresponding spectral acceleration fall into the range of short or moderate period structures. Consequently, on the condition of without displacement response spectra in accordance with the fact, the design spectra for displacement can be determined by equation (4) to calculate the earthquake displacement response of shear wall structure.

Based on the smoothed acceleration response spectra in CCSDB [1], from equation (4), we obtain the period of vibration expressed as

$$T^2[0.45 + 10(\eta_2 - 0.45)T] = \frac{4\pi^2}{\alpha_{\max} g} S_d \qquad (T \le 0.1s) \tag{5a}$$

$$T = 2\pi\sqrt{\frac{S_d}{\eta_2 \alpha_{\max} g}} \qquad (0.1s \le T \le T_g) \tag{5b}$$

$$T = \left(\frac{4\pi^2}{T_g^\gamma} \cdot \frac{S_d}{\eta_2 \alpha_{\max} g}\right)^{\frac{1}{2-\gamma}} \qquad (T_g \le T \le 5T_g) \tag{5c}$$

$$T^2\left[0.2^\gamma \eta_2 - \eta_1 (T - 5T_g)\right] = \frac{4\pi^2}{\alpha_{\max} g} S_d \qquad (5T_g \le T \le 6.0s) \tag{5d}$$

$$\gamma = 0.9 + \frac{0.05 - \zeta}{0.5 + 5\zeta}$$

$$\eta_1 = 0.02 + \frac{0.05 - \zeta}{8}$$

$$\eta_2 = 1 + \frac{0.05 - \zeta}{0.06 + 1.7\zeta}$$

where $\alpha_{\max}$ denotes the maximum value of horizontal seismic influence coefficient. For frequent, low-intensity and infrequent severe ground motions, it can be obtained from current seismic design code. For moderate ground motion, taking $\alpha_{\max}$ =0.23, 0.45 and 0.90 when the earthquake fortification intensity is VII,VIII and IX,respectively.

When the equivalent displacement u_{eff}, which is equivalent to S_d in equation (5), is obtained, the parameters, such as seismic fortification level, site classification, damping ratio and so on, are determined. Then, the corresponding effective period can be calculated by equation (5).

5. Direct Displacement-based Design Approach for Shear Wall Structure

In this paper, according to the characteristic of shear wall structure, the direct displacement-based design procedure is outlined as follows.

5.1 *Design procedure according to serviceability performance level*

(1) Preliminary design of the structure, including selection of concrete and reinforcement grades and determination of member size of shear wall.

(2) Determine the performance level based on the significance of structure or the requirement of owner. Then, provide the corresponding allowable value of storey drift ratio.

(3) Determine the target displacement pattern of shear wall structure based on equation (2). Then, calculate the storey drift ratio in each storey and satisfy $\theta_i \le [\theta]$.

(4) Calculate the floor displacement at each floor from previous steps, and determine the effective displacement u_{eff} and effective mass M_{eff} of the equivalent single-degree-of-freedom system from equations in the following.

$$u_{eff} = \frac{\sum_{i=1}^{n} m_i u_i^2}{\sum_{i=1}^{n} m_i u_i} \tag{6}$$

$$M_{eff} = \frac{\sum_{i=1}^{n} m_i u_i}{u_{eff}} \tag{7}$$

where m_i denotes the mass at each floor.

(5) According to the seismic fortification level, effective damping ratio ζ_{eff} and effective displacement u_{eff}, determine the effective period T_{eff} from equation (5), in which the effective damping ratio ζ_{eff} is given by [9]

$$\zeta_{eff} = \zeta_0 + 0.2(1 - 1/\sqrt{u}) \tag{8}$$

where ζ_0 represents the viscous damping ratio in elastic system, which can be given as 0.05 for reinforced concrete structure, μ denotes the displacement ductility demand, which can be determined by the seismic fortification level. For example, under a certain level of ground motions, give $\mu = 1.0$ for serviceability performance level and give $\mu = 3.0 \sim 4.0$ for collapse prevention performance level, and so on.

(6) Determine the effective stiffness K_{eff} of the equivalent single-degree-of-freedom system, the base shear V_b of the structure and the horizontal earthquake action F_i at each storey level, namely,

$$K_{eff} = \left(\frac{2\pi}{T_{eff}}\right)^2 M_{eff} \tag{9}$$

$$V_b = K_{eff} \cdot u_{eff} \tag{10}$$

$$F_i = \frac{m_i u_i}{\sum_{j=1}^{n} m_j u_j} V_b \tag{11}$$

(7) Calculate the horizontal earthquake action effect and corresponding gravity load effect of the structure. Then, check the bearing capacity of shear wall section and adopt the details of seismic design.

(8) The designed structure is analyzed with pushover method. If the pushover drift curve, corresponding to the curvature at the bottom cross section of the first storey of the shear wall structure reaches yield curvature, is identical with the initial assumed shape, the design is valid. Otherwise, regard the corresponding pushover drift curve as the modified target lateral displacement curve, and then, recalculate the base shear and horizontal earthquake action, and check the bearing capacity of shear wall section until it is satisfied, based on the method mentioned above.

Based on the method presented above, if the structure is required to satisfy the serviceability performance level when subject to minor but frequently occurring earthquakes, from equation (5), the effective period can be determined based on the corresponding ground motion parameter. Then, other design parameters and horizontal earthquake action can be obtained from equations (6) to (11). Similarly, if it is required to satisfy the serviceability performance level when subject to moderate or rare but major earthquakes, the effective period can be determined according to the corresponding ground motion parameter of moderate or major. Then, the other design parameters can be determined from equations (6) to (11).

5.2 *Life safety and collapse prevention performance level*

Under the frequently occurred earthquake action, the structure is required to satisfy the serviceability performance level. Furthermore, the displacement of the structure should be controlled in the life-safety and collapse prevention performance level.

These two performance levels correspond to the performance of building structure under moderate intensity earthquake and seldomly occurred earthquake. On the action of seldomly occurred earthquake, assume that the curvature at the bottom cross section of shear wall reaches the ultimate curvature ϕ_u. The corresponding target lateral displacement mode can be determined from equation (1) and the storey drift ratio θ_i can be obtained. Ensure that θ_i is satisfy the allowable value, namely, $\theta_i \le [\theta]$. Then, the corresponding displacement u and base shear V_b can be obtained from equation (6) and (10). Under the action of moderate earthquake, assume that the curvature at the bottom cross section of shear wall reaches $\phi_u/2$. Similarly, the corresponding displacement u and base shear V_b can be calculated.

Finally, the base shear and displacement corresponding to the three performance levels can be drawed in a same $V-u$ coordinate, such as point A, B and C in figure 2. By linking these points, the base shear versus displacement curve represents the demand curve under serviceability, life-safety and collapse prevention performance level.

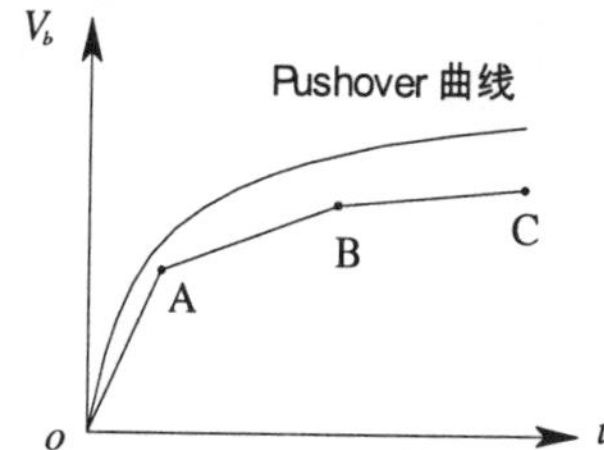

Figure 2. The demand curve and pushover curve

Afterward, the designed structure is analyzed with nonlinear static procedure. Place the pushover curve and demand curve on a same coordinate, as shown in figure 2. According to the relationship between demand curve and pushover curve, the design result can be modified by the following method.

(1) If the pushover curve is consistent with or above the demand curve, as shown in figure 3a, the designed structure satisfies the performance levels.

(2) If the pushover curve is below the demand curve, as shown in figure 3b, the designed structure does not satisfy the performance levels and should be redesigned.

(3) If point B is lower than point A in the demand curve, as shown in figure 3c, it is indicated that the seismic demands in moderate earthquake is too low. Then, the earthquake resistant demand should be improved.

(4) If point B is much higher than point A in the demand curve, as shown in figure 3d, two problems may be exist in the designed structure as follows.

1) The initial stiffness of the structure is too small. The size of the structure member should be adjusted and the initial stiffness should be increase to appropriate range.

2) The performance demand for moderate earthquake is too high, and it should be decreased.

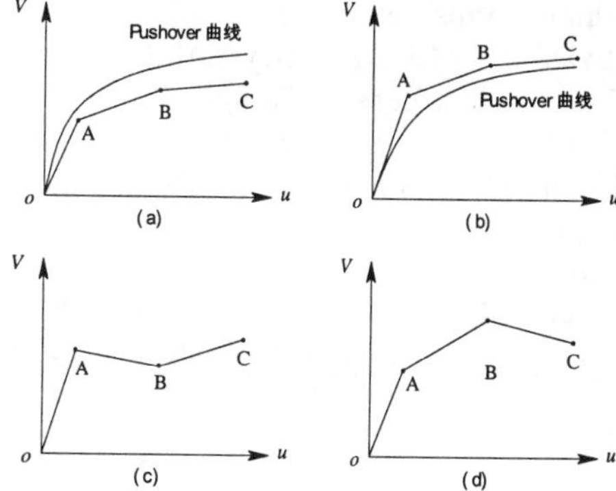

Figure 3. Several demand curve and pushover curve

Based on the displacement demand of the moderate and major earthquakes, the corresponding pushover drift curve can be obtained. Consequently, the damage information and plastic hinge distribution under these two conditions can be obtained. Then, the local reinforcement of the structure can be adjusted according to the information.

6. Conclusion

Above all, the characteristic of the method in this paper is listed as follows:

(1) The plastic hinge of shear wall structure formed at the bottom. Thus, the target lateral displacement curve is determined by assuming that the curvature at the bottom cross section of the first storey of shear wall structure reaches yield curvature, which reflects the fact that the critical section of the structure reaches a certain limit state firstly.

(2) When determine the initial lateral displacement curve, the shear wall structure is regarded as a cantilever wall with identical section. It is not in accordance with the fact. The modified lateral displacement curve, which is obtained from the pushover drift curve corresponding to the state that the bottom cross section of shear wall structure reaches yield curvature, reflects the change of strength and stiffness over the height of the structure.

(3) Based on the procedure in this paper, not only the performance of the structure is controlled under minor and major earthquakes, but also it is controlled under moderate earthquakes. This is the objective that should be achieved in performance-based seismic design.

References

1. The Chinese Code for Seismic Design of Buildings, GB50011-2001, Beijing: China Architecture and Building Press, 2001
2. Kowalsky M.J., Priestley M.J.N. Macrae GA. Displacement-Based Design of RC Bridge Columns in Seismic Regions. *Earthquake Engineering and Structural Dynamics*, 1995;24: 1623~1643
3. Luo Wenbing, Qian Jiaru. Displacement-based Seismic Design for RC Frames. China Civil Engineering Journal, 2003, vol 36(5), 22-29
4. Priestley M.J.N. Performance-Based Seismic Design. *12WCEE*, 2000,Paper 2831,
5. Medhekar M.S., Kennedy D.T.L. Displacement-Based Seismic Design of Building-Theory. *Engineering Structures*, 2000; 22: 201~209
6. Medhekar M.S., Kennedy D.T.L. Displacement-Based Seismic Design of Building- Application. *Engineering Structures*, 2000; 22: 210~221
7. Kowalsky MJ. RC Stuctual Walls Designed According to UBC and Displacement-Based Method. *Journal of Structural Engineering*, 2001;127: 506~516
8. Liang Xingwen, Huang Yajie. Research on Displacement-based Seismic Design Method of RC Frames. China Civil Engineering Journal (In press)
9. Miranda E, Garcia JG. Evaluation of Approximate Methods to Estimate Maximum Inelastic Displacement Demands. *Earthquake Engineering and Structural Dynamics*, 2002;31:539~560

THE UPDATE PHILOSOPHY OF ASEISMIC DESIGN FOR CAST-IN-SITU RC BUILDING STRUCTURES

FU XUEYI

Architectural Design and Research Institute of Shenzhen University
Shen Zhen 518060,China

Based on the investigation of earthquake damages on building structures, through the studies of several projects elastic-plastic static and dynamic analyses and table vibration tests, referring to IBC2003&ATC40, for the most popular cast-in-situ RC building structures, this paper presents the update aseismic design philosophy to achieve this target-no damage while mild earthquake, little damage while moderate earthquake, no collapse while severe earthquake. The update philosophy is consisted of these new concepts: moderate or severe earthquake action for vertical members (columns and walls), mild earthquake action for horizontal members (secondary beams and spandrel beams) and the control standards for minimum ductility. Several projects practices revealed that, these new concepts and methods are coordinated to GB 50011-2001&JGJ3-2002 well and easy to operate. Especially for complicated tall building structures, they are able to provide the realizable platform to do further elastic-plastic static and dynamic analyses and evaluations.

1. Introduction

Based on investigation on the earthquake damages of building structures subjected to Chi-chi earthquake, Hualian earthquake, Kobe earthquake and Alaska earthquake, the most earthquake damages of cast-in-situ RC multiple stories and tall building structures subjected to severe earthquake were the columns and walls cracked and damaged due to shear and buckling, even collapsed. Whereas, it was not found that the reinforcement of the cast-in-situ RC floors and secondary beams yield and form plastic hinge. Only a few spandrel beams without stirrups or with little stirrups and beam-column joints appeared diagonal shear cracks. The common characteristics of those earthquake damages were that the collapses of structures were due to the brittle failure of vertical members such as columns and walls, the secondary beams and the spandrel beams did not reach to elastic-plastic states. These damage phenomena are more realistic and reliable than the results of numerical analyses and model tests. It is a good reference for the aseismic design of cast-in-situ RC multiple stories and tall building structures.

2. Causes and Enlightenment of Earthquake Damages

The cause of the earthquake damages of building structures can be summarized as:

a. 'Unload' and 'reinforcement' effects of cast-in-situ RC floors

'Unload': because of the presence of stiffness in plane and out plane of Cast-in-situ RC slabs, the actual moment of supports of horizontal members (secondary beams and girders connected to slabs) subjected to gravity load are smaller than that calculated from frame element model (around 50% for girders, 30% for secondary beams respectively) so that the flexural strength of the supports of secondary beams using frame element model is on the safe side.

'Reinforcement': because of the presence of load carrying capacity in plane and out plane of cast-in-situ RC slabs, the actual flexural carrying capacity provided by secondary beams, spandrel beams and slabs connected to horizontal members subjected to earthquake are larger than that of supports of secondary beams and spandrel beams using frame element model.

Therefore, when severe earthquake occurs, the secondary beams, spandrel beams and Cast-in-situ RC slabs are not easy to yield firstly and form plastic hinge. Whereas, the vertical members (shear walls and columns) are more weak than horizontal members, forming the 'strong floors-weak columns and walls' system.

b. Shortage of load carrying capacity and ductility of vertical members

At present, the aseismic design of building structures is based on moderate earthquake combined with ductility requirement in most countries, and mild earthquake combined with ductility requirement in China, all don't consider the "unload" and "reinforcement" effects. So during severe earthquake, it is difficult to satisfy the ductility and energy dissipation demand. The horizontal members (secondary beams and spandrel beams) don't yield, but a few vertical members are vulnerable to seismic damage due to the shortage of ultimate load carrying capacity that resulting in brittle shear and buckling failure, even collapse due to ductility absence.

To meet the goal of 'no collapse while severe earthquake', from causes of earthquake damages, we can obtain enlightenment for structures aseismic design as:

(1) The ultimate strength of vertical elements (shear walls and columns) must meet the requirement of load combination effect with severe earthquake;

(2) The vertical members (shear walls and columns) must meet the ductility and detail requirement to avoid brittle shear failure and buckling failure during severe earthquake;

(3) The strength of horizontal members (secondary beams and spandrel beams) meets the elastic requirement of load combination effect with mild earthquake to form plastic hinge during severe earthquake, reducing the earthquake effect. The ductility requirement of 'strong shear-weak bending' for horizontal members must also be satisfied to avoid brittle shear failure.

3. Check on Ultimate Strength of Vertical Members under Severe Earthquake

3.1 *Objective*

The check on ultimate strength of vertical members subjected to the load combination with severe earthquake is propitious to reveal clearly the actual behavior of vertical members of structures during severe earthquake. For a real structures subjected to earthquake, the stress in each vertical member is non-uniform distributed, especially for irregular structures. The load combination effect with severe earthquake is benefit to the reveal the behavior of the vertical members being in severe stress states. So to achieve the objective of 'no collapse while severe earthquake', we can rearrange the structures to improve the non-uniformity of stress, also, we can increase the section dimension or the reinforcement through the check on ultimate strength of vertical members to satisfy the requirement of the load combination effect with severe earthquake.

3.2 *Earthquake effect analysis*

From code for seismic design of buildings (GB 50011-2001), the maximum horizontal earthquake effect coefficients $\alpha_{\max}$ and the peak accelerations $a_{\max}$ in the acceleration time histories corresponding to severe, moderate and mild earthquakes are list in Table 1.

Table 1

Design intensity		7	7.5	8	8.5
Mild earthquake	$\alpha_{\max}^{mild}$	0.08	0.12	0.16	0.24
	$a_{\max}^{\text{mild}}(cm/s^2)$	35	55	70	110
Moderate earthquake	$\alpha_{\max}^{\text{mod}\,erate}$	0.23	0.33	0.46	0.66
	$a_{\max}^{\text{moderate}}(\text{cm}/\text{s}^2)$	100	150	200	300
Severe earthquake	$\alpha_{\max}^{\text{severe}}$	0.5	0.72	0.9	1.2
	$a_{\max}^{\text{severe}}(cm/s^2)$	220	310	400	620

If the elastic analysis is performed for a given structure on a given type of site, the load effect magnifying coefficients β_{severe} and β_{moderate} corresponding to severe and moderate earthquakes can be written as

$$\beta_{\text{severe}} = \alpha_{\max}^{\text{severe}} / \alpha_{\max}^{\text{mild}} \approx a_{\max}^{\text{severe}} / a_{\max}^{\text{mild}} \tag{1}$$

$$\beta_{\text{moderate}} = \alpha_{\max}^{\text{moderate}} / \alpha_{\max}^{\text{mild}} \approx a_{\max}^{\text{moderate}} / a_{\max}^{\text{mild}} \tag{2}$$

β_{severe} and β_{moderate} obtained from Table 1, Equations (1) and (2) are listed in Table 2.

Table 2

Design intensity	7	7.5	8	8.5
β_{severe}	6.25	6	5.625	5
β_{moderate}	2.86	2.73	2.86	2.73

Considering a low probability of occurrence of severe earthquake and economy of structures seismic design demand, the plastic hinges forming in secondary beams and spandrel beams must be designed reasonably so that the structures subjected to severe earthquake can dissipate significant energy as the structures behave inelastically. Then, the structures stiffness degrades and the damping ratio increases, mitigating the severe earthquake effect. Thus, the severe earthquake effect magnifying coefficient β_{severe} used to check the load carrying capacity of the members and listed in Table 2 should be decreased properly. Based on the results of static and dynamic inelastic analysis for a few real multiple stories and tall buildings, β_{severe} can be replaced by β^*_{severe} as

$$\beta^*_{\text{severe}} = (0.7\text{~}0.8)\,\beta_{\text{severe}} \tag{3}$$

3.3 *Block diagram*

Finding in references [2-5], the block diagram for checking the ultimate strength of vertical members subjected to load combination with severe earthquake is shown in Figure 1.

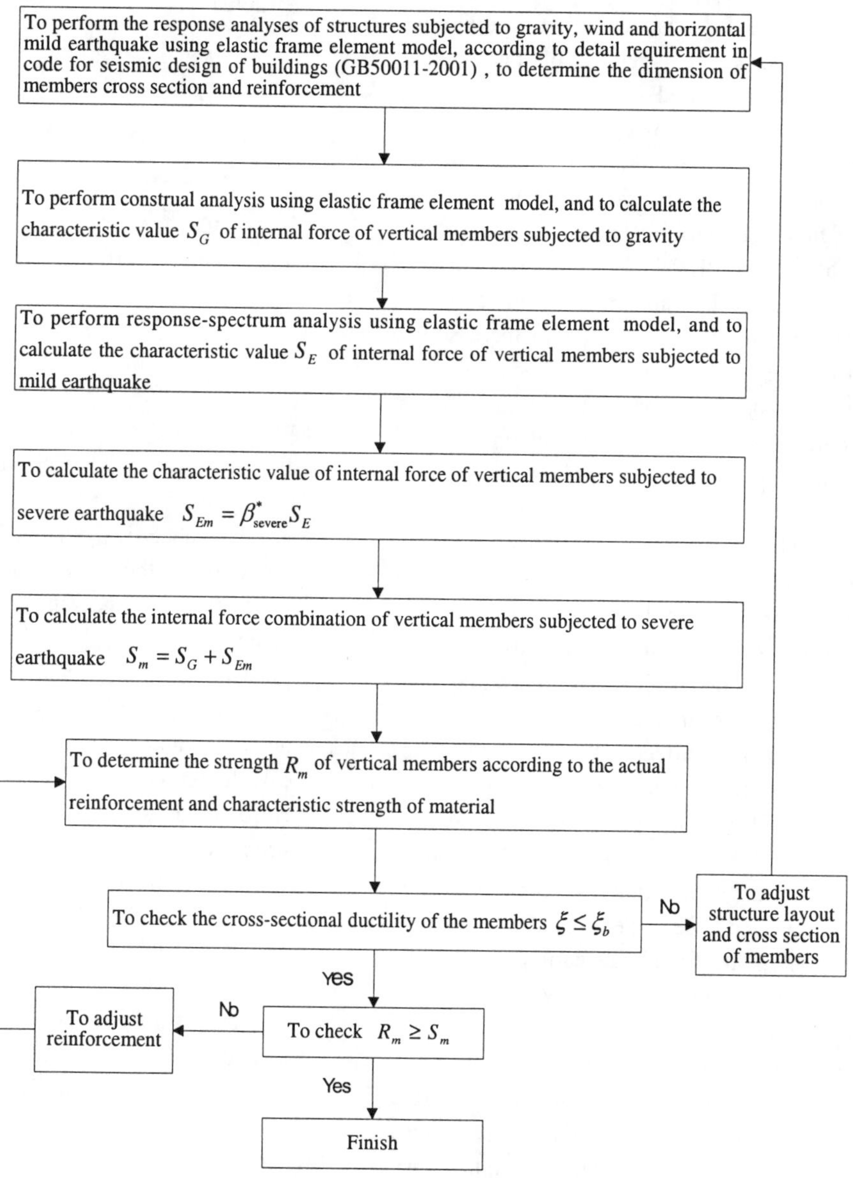

Figure 1. The block diagram for checking the ultimate strength of vertical members subjected to load combination with severe earthquake

3.4 *Feasibility*

(1) Explaining about the check of examples

Several complicated tall building structures designed by authors are checked on ultimate strength of vertical members subjected to severe earthquake by the above proposed method. The design conditions of these tall buildings are the earthquake design intensity of 7 and 8, the site type of II, III and IV, and different seismic zones. The check results show,

For design intensity of 7, site type II, and seismic zone group I, the detail requirement satisfies the code for seismic design of buildings (GB 50011-2001), the strength of the regular structures are also satisfied.

For design intensity of 8, site type II and III, and design intensity of 7, site type III and IV, the detail requirement satisfy the code for seismic design of buildings (GB 50011-2001). The flexural strength of shear walls of the regular structures is satisfied, some shear walls need to increase the horizontal-distributed reinforcement or increase the strength of concretes to satisfy the shear strength requirement. The shear strength of the frame columns are satisfied, some frame columns need to increase the reinforcement, or increase the section dimension or increase the strength of concretes to satisfy the flexural strength requirement.

The results of the examples show that the above proposed method links up well with the method specified in the code for seismic design of buildings (GB 50011-2001) in checking the ultimate strength of structural members based on the load combination effect with mild earthquake. Also, the proposed method has a good operability and continuity. The proposed method is a base for the further check on the static and dynamic elastic-plastic analytical results of complicated tall buildings subjected to severe earthquake.

(2) Aseismic design under moderate earthquake

Most of the countries in the world, currently, aseismic design for building structures are performed by elastic response spectrum method and the peak acceleration corresponding to moderate earthquake (earthquake design intensity), combination with ductility detail requirement. Based on consideration of safety and reliability, the ultimate strength of structural members is calculated by load combination effects with the gravity and seismic action.

Based on the code for seismic design of buildings (GB 50011-2001), considering the protected objective of 'little damage while moderate earthquake', it is suggested that the vertical members maintain elastic states under moderate earthquake. It also can be explained as the design strength of vertical members satisfied the requirement of the load combination effect with moderate earthquake, as shown in Equation (4).

For load combination effect with moderate earthquake

$$S_{\text{moderate}} = S_G + \beta_{\text{moderate}} S_{\text{E}} \leq R_S \tag{4}$$

where, R_S is the design strength of vertical members determined by the design strength of material based on the code for seismic design of buildings (GB 50011-2001); β_{moderate} is the magnifying coefficient of moderate earthquake effect; S_G and S_{E} are the characteristic values of gravity load effect and mild earthquake effect, respectively.

(3) Elastic response spectrum analysis under mild earthquake

The analytical results by the elastic response spectrum method under mild earthquake are the basic data for checking on moderate and severe earthquake effects of building structures. The precision of the analytical results for mild earthquake will affect the reliability of the check on moderate and severe earthquake effects. So, the calculating model for mild earthquake is required as close as possible to the actual behavior of the structures.

According to the code for seismic design of buildings (GB 50011-2001), considering the protected object of 'no damage while mild earthquake', the design strength of vertical members satisfied the requirement of the elastic load combination effect with mild earthquake, as shown in the Equation (5).

For load combination effects with mild earthquake

$$1.25 S_G + 1.3 S_{\text{E}} \leq \frac{R_S}{\gamma_{RE}} \tag{5}$$

where, γ_{RE} is the earthquake-resistance capacity adjustment factor of structural members (for eccentric compression, $\gamma_{RE}=0.8$)

(4) A comparison of check on load combinations effects with severe, moderate and mild earthquake

From Figure 1, Equations (4) and (5), considering vertical members subject to eccentric compression, the relationship between their characteristic strength R_m calculated from characteristic value of material strength and their design strength R_s calculated from design value of material strength can be shown that

$$R_m = \Psi(f_{ck}/f_c, f_{yk}/f_y)R_s \quad (6)$$

$$\Psi(f_{ck}/f_c, f_{yk}/f_y) \approx$$

$$(f_{ck}/f + f_{yk}/f_y)/2 \approx (1.4+1.1)/2 \approx 1.25 \quad (7)$$

$$R_m \approx 1.25 R_s \quad (8)$$

Then, the ratio of load combination effects with severe and moderate earthquake to load combination effect with mild earthquake of eccentrically compressed members can be expressed respectively as

$$\alpha_{severe} = \frac{S_G + \beta^*_{severe} S_E}{1.25(S_G + S_E)} \quad (9)$$

$$\alpha_{moderate} = \frac{S_G + \beta_{moderate} S_E}{S_G + S_E} \quad (10)$$

where, α_{severe} and $\alpha_{moderate}$ are the ratio of load combination effects with severe and moderate earthquake to load combination effect with mild earthquake, respectively.

We define the ratio of the characteristic value S_E of mild earthquake effect to the characteristic value $S_G + S_E$ of load combination effects with gravity and mild earthquake as γ

$$\gamma = S_E/(S_G + S_E) \quad (11)$$

Then, Equations (9) and (10) can be expressed as

$$\alpha_{severe} = 0.8[1+(\beta^*_{severe}-1)\gamma] \quad (12)$$

$$\alpha_{moderate} = 1+(\beta_{moderate}-1)\gamma \quad (13)$$

Form Equations (12), (13) and Table 2, letting $\beta^*_{severe} = 0.75\,\beta_{severe}$, $\alpha_{moderate}$ and α_{severe} can be obtained, and are listed in Table 3.

Table 3

		γ	0	0.05	0.1	0.15	0.2	0.3	0.4	0.5	0.6	0.7	0.8	0.9	1
Design intensity	7	$\alpha_{moderate}$	1	1.09	1.19	1.28	1.37	1.56	1.74	1.93	2.12	2.3	2.49	2.67	2.86
		α_{severe}	0.8	0.95	1.1	1.24	1.39	1.68	1.98	2.28	2.57	2.87	3.16	3.46	3.75
	8	$\alpha_{moderate}$	1	1.09	1.19	1.28	1.37	1.56	1.74	1.93	2.12	2.3	2.49	2.67	2.86
		α_{severe}	0.8	0.93	1.06	1.19	1.32	1.57	1.83	2.09	2.35	2.6	2.86	3.12	3.38

From Table 3, it may be concluded that

(1) The check on moderate earthquake effect is stricter than that on the mild earthquake effect. As earthquake effect ratio γ increases, $\alpha_{moderate}$ increases, and check requirement is more strict;

(2) As earthquake effect ratio γ increases, the ratio of severe earthquake effect to mild earthquake effect α_{severe} increases more quickly, the strict grade of check also increases more quickly. As $\gamma < 0.067$(design intensity7) or 0.078(design intensity8), $\alpha_{severe} < 1$, the check on members will not be dominated by the severe earthquake effect;

(3) As $\gamma < 0.2$(design intensity 7) or 0.3(design intensity 8), $\alpha_{mederate} > \alpha_{severe}$, the check on members will be

dominated by the moderate earthquake effect; as $\gamma > 0.2$(design intensity 7), or 0.3(design intensity 8) $\alpha_{\text{severe}} > \alpha_{mederate}$, the check on members will be dominated by the severe earthquake effect;

(4) α_{severe} is generally close to α_{moderate}. Based on the above design method, as the check on members are dominated by the severe or moderate earthquake effect, the strength of the members subjected to severe earthquake can be ensured, and the behavior of structures can have a good transition from 'little damage while moderate earthquake' (few secondary beams and spandrel beams behave plastically) to 'no collapse while severe earthquake' (the strength of the vertical members meet requirement).

From the above analysis the following conclusion can be stated: to control mild earthquake effect have a weakness that the exact behavior of vertical members subjected to severe earthquake is concealed, especially the vertical members being in severe stress states due to diminution of earthquake effect and magnification of gravity effect. Correspondingly, to control moderate and severe earthquake effects can reveal clearly the exact behavior of vertical members being in severe stress states, and also control explicitly the building with an irregular configuration.

4. Ductility of Vertical Members

4.1 *Objective and significance*

Presently limited by the development of computer programs and their amounts of computation, the elastic response spectrum analysis of structures subjected to mild earthquakes generally adopt the space-frame element model. However, the model does not entirely agree with actual structures, particularly super columns and beams. In addition, the stiffness in plane and out of plane, as well as load-carrying capacity of cast-in-situ RC slabs is not considered in the model analysis. Consequently, the elastically analytical results for mild earthquake have inevitable errors. Actual earthquakes are complex and the present horizontal earthquake response spectrum analyses ignore complex factors like: the torsion component, phase component, long period component and vertical component of seismic actions. With these components not accounted for, it is greatly possible that the effect caused by a severe earthquake could exceed the load-carrying capacity of the vertical members. Adequate ductility of vertical members to avoid brittle shear and buckling failure is another fundamental measure to ensure 'no collapse while severe earthquake'.

4.2 *First condition to control the ductility of vertical members: for large eccentric compression*

Limiting the frame columns and shear walls in load combination effect with severe earthquake in large eccentric compression to brittle failure of small eccentric compression as

$$\xi = (N_G + N_E \cdot \beta^*_{\text{severe}})/1.2 f_{ck} A \le \xi_b \tag{11}$$

where, N_G is the characteristic value of axial force under gravity load; N_E is the characteristic value of axial force under mild earthquake; β^*_{severe} is the magnifying coefficient of the severe earthquake effect; ξ_b is the dividing point between large eccentricities and small eccentricities ($\xi_b \approx 0.5$) ; f_{ck} is the characteristic compressive strength of concrete; A is cross-section area of a vertical member.

The frame columns and shear walls which meet the requirement of the ratio of axial compressive force to axial compressive strength of cross section in the code for seismic design of buildings (GB 50011-2001) are always in large eccentric compression under the load combination effect with severe earthquakes. The ratio of axial compressive force to axial compressive strength of cross section for side columns and corner columns dissipating large amounts of earthquake energy should be properly controlled and the ratio of axial compressive force to axial compressive strength of cross section for the shear walls of non-strengthened parts at the bottom should meet the requirements in GB50011-2001. The concrete compressive strength and cross-section dimension of shear walls of upper columns should change evenly.

4.3 *The second condition to control the ductility of vertical members: strong Shear-weak Bending*

The safety of shear strength is to be higher than that of axial strength and moment capacity for the frame columns and shear walls subjected to severe earthquakes to avoid brittle shear failure.

The ultimate strength of vertical members under severe earthquake should satisfy the following equation:

$$M_u \geq N_{\text{sereve}} e_{\text{sereve}} \tag{12}$$

$$V_u \geq V_{\text{sereve}} \tag{13}$$

while simultaneously the cross section must meet the following requirement:

$$V_{\text{severe}} \leq 0.25 f_c b h_0 \tag{14}$$

where M_u is the moment capacity of vertical members in large eccentric compression, calculated by characteristic strength of material; $N_{\text{severe}} = N_G + N_E \cdot \beta^*_{\text{severe}}$ is the axial force in load combination effect with severe earthquake; $e_{\text{severe}} = \dfrac{M_G + M_E \cdot \beta^*_{\text{severe}}}{N_G + N_E \cdot \beta^*_{\text{severe}}} + \dfrac{h}{2} - a$ is the eccentricity of load combination with severe earthquake; V_u is the shear strength of vertical members in large eccentric compression, calculated by characteristic strength of material; $V_{\text{severe}} = V_G + V_E \cdot \beta^*_{\text{severe}}$ is the shear force in load combination effect with severe earthquake; V_G and M_G are characteristic values of shear force and moment under gravity load, respectively; V_E and M_E are characteristic values of shear force and moment under mild earthquake respectively; f_c is the design compressive strength of concrete; bh_0 is the effective shearing area of vertical member; h is the cross-sectional height of vertical member; a is the nearest distance from the center of reinforcement bar to margin of the cross-section.

The safety of the shear strength of frame columns and shear walls should meet the following requirement:

$$V_U / V_{\text{severe}} \geq 1.2 M_U / N_{\text{severe}} e_{\text{severe}} \tag{15}$$

4.4 *The third condition to control the ductility of vertical members : strong anchorage*

The requirement of reinforcement ratio per unit volume, stirrup ratio and anchorage for frame columns in the code for seismic design of buildings (GB 50011-2001) is to ensure the ductility of the columns and avoid the buckling of reinforcement bars as well as the brittle failure of concrete. It should be carried out strictly.

The requirement of stirrup ratio per unit volume, reinforcement ratio, the horizontal and vertical distributed reinforcement ratio and anchorage for shear walls and the edge restricted members of tube system in the code for seismic design of buildings (GB 50011-2001) is to ensure the ductility of the shear walls and avoid the buckling of reinforcement bars as well as the brittle failure of concrete. It should be carried out strictly.

4.5 *Feasibility*

According to the above ductility requirements, the checking results of examples indicate that on earthquake design intensity 7, site type III、IV and earthquake design intensity 8, site type II and III in the code for seismic design of buildings (GB 50011-2001), only part of the frame columns need to increase the cross-section dimension and the reinforcement or increase the concrete strength to meet the requirement of the ductility for large eccentric compression. And part of shear walls need to increase the thickness and horizontal distributed reinforcement or properly increase concrete strength to meet the ductility requirement about large eccentric compression and strong shear-weak bending.

5. Ductility of Horizontal Members

5.1 *Objective and significance*

Results from earthquake damage studies showed that due to the presence of cast-in-situ RC slabs, horizontal members with slabs have good earthquake-resisting capacity. In general, flexural strength of horizontal members should meet requirement for load combination effect with mild earthquake so that they can yield during severe earthquake, reaching structural ductility goal. Meanwhile, brittle shear failure should be avoided under load combination with earthquake to assure the formation of plastic hinge at beam ends.

5.2 *Spandrel beams and secondary beams*

Check on design strength of spandrel beams and secondary beams should meet the requirement of elastic response spectrum analysis under load combination with mild earthquake:

Strong shear-weak bending: design shear strength should meet the requirement for load combination effect with moderate earthquake, as shown in Equation (4). Furthermore, the following condition should be satisfied,

$$V_{\text{moderate}} \leq 0.25 f_c b h_0 \qquad (16)$$

where, V_{moderate} is shear force at beam ends under load combination with moderate earthquake; f_c is the design compression strength of concrete; bh_0 is the effective shear cross-section area of beam.

The details requirement for shear design described in the code for seismic design of buildings (GB 50011-2001) also should be satisfied.

5.3 *Over-reinforcement of spandrel beams*

To avoid the over-reinforcement of spandrel beams during structural analysis using frame element model, the following measures should be adopted respectively.

For load combination effect with gravity, the measure is effective that 'release' combined with 'resistant'.

'Release': adjusting the layout of members, to reduce the vertical deformation difference between adjacent vertical members under gravity load; setting cast-after region to release the vertical construction load effects.

'Resistant': reducing cross-section height, increasing cross-section width and reinforcement and stirrup.

For load combination effect with earthquake,

'Release': mitigating the earthquake effect of spandrel beams by decreasing stiffness or by limit analysis.

'Resistant': reducing cross-section height, increasing cross-section width and reinforcement and stirrup.

5.4 *Transfer girder*

Check on ultimate strength:

Transfer girders are critical members in structures with transfer storey to ensure 'no collapse while severe earthquake', and their strength should satisfied the specification for load combination effect with severe earthquake. The Block diagram of calculation is shown in Figure 1.

Strong shear-weak bending:

The strength of transfer girders should meet the following requirement:

$$V_U / V_{\text{severe}} \geq 1.2 M_U / M_{\text{severe}} \qquad (17)$$

where, M_U and V_U are respectively flexural strength and shear strength of transfer girder, calculated by characteristic strength of material; M_{severe} and V_{severe} are moment and shear force in load combination effect with severe earthquake.

Furthermore, the cross-section dimension of transfer girders should satisfy the following condition:

$$V_{\text{severe}} \leq 0.25 f_c b h_0 \qquad (18)$$

Details:

For transfer girders, reinforcement ratio, stirrup ratio, stirrup details and horizontal distributed reinforcement should meet the requirements in the code for seismic design of buildings (GB 50011-2001).

6. Conclusions

The investigation studies of earthquake damage show that the seismic design of cast-in-situ RC tall building structures has their own features and rules. In this paper, aiming at their characteristic, the aseismic design philosophy and method suggested by the author only is an attempt. The author expects to be more effective, economical and reliable to meet the earthquake design goal of 'no damage while moderate earthquake and no collapse while severe earthquake' by the method.

References

1. Fu Xueyi and Wu Bing. The behavior studies on beamless floor of Cast-in-situ RC building structures subjected to gravity load. The eighteenth symposium on tall building structures of China, Chongqing, 2004.10

2. Fu Xueyu and Jia Jianying. The update philosophy of building structures aseismic design: elastic response spectrum analysis under severe earthquake. The seventeenth symposium on tall building structures of China, Hangzhou, 2002. 11
3. Seismic design code and commentary for building structures (Taiwan). 2001.7
4. International Building Code , IBC2003. International Code Council.
5. Seismic Evaluation and Retrofit of Concrete Building, Report No. ATC40, Applied Technology Council, Redwood City, California, 1996

RELIABILITY-BASED SEISMIC PERFORMANCE DESIGN OPTIMIZATION OF REINFORCED CONCRETE BUILDINGS *

XIAO-KANG ZOU

Department of Building and Construction, The City University of Hong Kong, Kowloon, Hong Kong, China

Q. S. LI

Department of Building and Construction, The City University of Hong Kong, Kowloon, Hong Kong, China

Future performance-based seismic design, directly addressing the inelastic deformation induced in buildings by earthquake ground motions, should be based on probabilistic approaches. In recent years, reliability-based structural optimization has gained much research attention due to its advantages of accounting for the various sources of the uncertainties and approximation in design process. However, there is yet no effective method for reliability-based design optimization of concrete buildings subject to seismic performance criteria. This paper therefore presents an effective reliability-based design optimization approach for reinforced concrete buildings subjected to earthquake loading in which the structural performance is assessed at the system level by push-over analysis. Such developed approach, which integrates structural reliability theories with optimization technique, can provide a good balance between the structural reliability and the steel reinforcement cost under specified performance requirements.

1. Introductions

It has been well recognized that earthquakes are random excitations with relatively large uncertainties. Structures resting on ground or connected to ground-based structures are subjected to random loadings during earthquakes. The response of such structures to earthquakes is, therefore, a random process. Performance-based seismic design represents the future of seismic retrofit design. Such design procedure should be based on probabilistic approaches, which account for the various sources of uncertainties and approximation. Specifically, the peak interstory drift response over the lifetime of a structure due to earthquakes is uncertain and so a performance parameter can be chosen to be directly related to the interstory drift reliability.

Although structural optimization has been widely applied to deterministic structural design, the deterministic optimised structure may usually have higher failure probabilities. In order to optimise structures to copy with the uncertainties, structural reliability theory should be incorporated into the structural optimization process. In recent years, reliability-based structural optimization has gained much research attention and many developments due to its advantages over a deterministic design optimization. One of the main advantages is that it permits the smoothing out of the inconsistencies in reliability inherent in the deterministic optimization. Reliability-based structural design optimization which integrates structural reliability theories with structural design optimization can provide a good balance between the structural reliability and initial cost or other objectives, where specified performance requirements are satisfied.

Li et al. [1-4] addressed wind-induced dynamic responses of structures with parametric uncertainties and proposed the formulas for structural dynamic reliability analysis considering the randomness of structural resistance and loading. Zou [5] and Chan and Zou [6] developed a computer-based deterministic optimization technique for drift performance design of reinforced concrete (RC) buildings subject to seismic performance criteria. However, there is still a lack of research about reliability-based seismic performance design optimization of concrete buildings. It is necessary to integrate design optimization with reliability analysis for nonlinear response of concrete buildings.

The paper presents an effective indeterministic reliability-based optimization approach. The procedure of the optimal design is decomposed into three parts: nonlinear structural analysis, reliability analysis and design optimization. To consider the uncertainty of

* This work is supported by a grant from Research Grant Council of Hong Kong Special Administrative Region, China (Project No. CityU 093/02).

earthquake loading involved in designing a structure, the structural response interstory drift is modelled as a random variable by defining its probability distribution type, mean value and standard deviation. In this probabilistic approach, the probability of failure is defined as the occurrence that an interstory drift response is larger than its allowable limit. The reliability-based optimization is formulated by minimizing the initial cost subject to the constraints of a prescribed structural drift response reliability. The constraint specifies that the reliability of the structure has to fulfil certain minimum value. Based on the results of the reliability analysis, an explicit formulation of interstory drift reliability index constraints is expressed with respect to deterministic design variables. The Optimality Criteria (OC) method can then be employed to solve this reliability-based design optimization problem. Finally, a practical frame example is presented to illustrate the effectiveness of the reliability-based optimization approach.

2. Reliability-Based Optimal Design Problem Statements

2.1 *Design variables*

For the design optimization of inelastic concrete buildings, deterministic design variables are the steel reinforcement ratios of all members, as presented in Zou [5] and Zou and Chan [7,8]. Indeterministic variables deal with uncertainties. In this research, the earthquake action, structural drift responses and allowable inelastic story drift limit are treated as random variables. In the nonlinear pushover analysis, the dynamic earthquake action which is a stochastic process is transformed into the equivalent static action which is also considered as a random process. Compared with the uncertainties of the structure properties such as the structural stiffness, the uncertainties of the external earthquake loading actions are relatively larger [9]. Therefore, the member stiffness in the formulation of the finite element method can be approximately treated to be deterministic, resulting in that the probability distributions of the structural displacement responses are determined by those of the external actions. That is to say, the statistical properties and the probability distributions of the structural responses (e.g. interstory drift) caused by earthquake actions are the same as those of the external actions, which is usually described by the extreme value distribution Type II [10, 11]. Thus, the interstory drift responses $\Delta\bar{u}$, taken as indeterministic design variables, follow the extreme value Type II distribution, whose cumulative distribution function $F_{\mathrm{II}}(\Delta\bar{u})$ and probability density function $f_{\mathrm{II}}(\Delta\bar{u})$ are defined by

$$F_{\mathrm{II}}(\Delta\bar{u}) = \exp\left[-\left(\frac{\alpha}{\Delta\bar{u}}\right)^{k}\right] \tag{1}$$

$$f_{\mathrm{II}}(\Delta\bar{u}) = F_{\mathrm{II}}^{'}(\Delta\bar{u}) = \left(\frac{k}{\alpha}\right)\cdot\left(\frac{\alpha}{\Delta\bar{u}}\right)^{k+1}\exp\left[-\left(\frac{\alpha}{\Delta\bar{u}}\right)^{k}\right] \tag{2}$$

where α is the largest characteristic value of the initial variable $\Delta\bar{u}$; k is the shape parameter; $\mu_{\Delta\bar{u}}$ and $\sigma_{\Delta\bar{u}}$ are the mean value and standard deviation of the random variable $\Delta\bar{u}$, which are given respectively by

$$\mu_{\Delta\bar{u}} = \alpha\Gamma\left(1-\frac{1}{k}\right) \qquad (k>1) \tag{3}$$

$$\sigma_{\Delta\bar{u}}^{2} = \alpha^{2}\left[\Gamma\left(1-\frac{2}{k}\right)-\Gamma^{2}\left(1-\frac{1}{k}\right)\right] \qquad (k>2) \tag{4}$$

On the basis of Eqs. (3) and (4), the solution of k is determined by

$$\frac{\Gamma\left(1-\frac{2}{k}\right)}{\Gamma^{2}\left(1-\frac{1}{k}\right)} - \left(1+\frac{\sigma_{\Delta\bar{u}}^{2}}{\mu_{\Delta\bar{u}}^{2}}\right) = 0 \tag{5}$$

Substituting the value of k into Eq. (3) to solve the parameter α as

$$\alpha = \frac{\mu_{\Delta\bar{u}}}{\Gamma\left(1-\frac{1}{k}\right)} \tag{6}$$

Upon the known values of α and k as well as the standard value Δu obtained from the pushover analysis, the probabilistic distribution of the random variable $\Delta\bar{u}$ can be determined on the basis of Eqs. (1) and (2). Since the probability of the random variable $\Delta\bar{u}$ follows the extreme value Type II distribution rather than normal distribution, it is necessary to first transform this extreme value Type II distribution into an equivalent normal distribution. Namely, the equivalent normal mean value, $\sigma_{\Delta\bar{u}}^{N}$, and the equivalent normal standard deviation, $\mu_{\Delta\bar{u}}^{N}$, are defined by

$$\sigma_{\Delta\bar{u}}^{N} = \frac{\phi\{\Phi^{-1}[F_{\mathrm{II}}(\Delta\bar{u})]\}}{f_{\mathrm{II}}(\Delta\bar{u})} \tag{7}$$

$$\mu_{\Delta\bar{u}}^{N} = \Delta\bar{u} - \sigma_{\Delta\bar{u}}^{N}\cdot\Phi^{-1}[F_{\mathrm{II}}(\Delta\bar{u})] \tag{8}$$

where $\Phi(.)$ represents the standard normal probability distribution function; $\Phi^{-1}(.)$ is the inverse of standard normal distribution function; and $\phi(.)$ is the standard normal density function.

The allowable story drift limit is also taken as a random variable, denoted by $\bar{d}$, follows the extreme value distribution Type I [11]. The cumulative distribution function $F_I(\bar{d})$ and the corresponding probability density function $f_I(\bar{d})$ of Type I for $\bar{d}$ can be defined by

$$F_I(\bar{d}) = \exp\left[-\exp(-\frac{\bar{d}-k}{\alpha})\right] \tag{9}$$

$$f_I(\bar{d}) = F_I'(\bar{d}) = \frac{1}{\alpha}\exp\left[-\exp(-\frac{\bar{d}-k}{\alpha}) - \frac{\bar{d}-k}{\alpha}\right] \tag{10}$$

In Eqs. (9) and (10), the parameters, α and k, can be determined by the mean values and standard deviation of the random variables as follows.

$$\alpha = \frac{\sigma_{\bar{d}}}{1.2825} \tag{11}$$

$$k = \mu_{\bar{d}} - 0.5772\alpha \tag{12}$$

In Eqs. (11) and (12), $\mu_{\bar{d}}$ and $\sigma_{\bar{d}}$ are the mean value and standard deviation of the random variable $\bar{d}$. d is defined as a standard value of the random variable $\bar{d}$. Similarly, once values of α and k are obtained from statistical analysis, the probabilistic distribution of the random variable $\bar{d}$ can be determined on the basis of Eqs. (9) and (10). The equivalent normal mean value $\sigma_{\bar{d}}^N$ and the equivalent normal standard deviation $\mu_{\bar{d}}^N$ of the random variable $\bar{d}$ are determined like Eqs.(7) and (8).

2.2 *Objective function*

In seismic design, it is commonly assumed that a building behaves linear-elastically under minor earthquakes and may respond nonlinear-inelastically when subjected to moderate and severe earthquakes. Under such an assumption, the entire design optimization process can therefore be decomposed into two phases [5-8]: elastic design optimization where concrete member sizes are considered as the only design variables; inelastic design optimization where steel reinforcements are the only effective material that provides ductility to an RC building structure beyond first yielding. Therefore, for an RC building having i=1, 2, 3, ..., N_i members and 2N_i plastic hinges (assuming one hinge at each end of a member), the tension steel reinforcement ratio, ρ_i, and the compression steel reinforcement ratio, ρ_i', for a rectangular cross section are taken as design variables in this inelastic design optimization study, whereas the member sizes, B_i (width) and D_i (depth), are fixed. If the topology of a building's structural system is predefined, the steel reinforcement cost of the RC framework is minimized as

$$\text{Minimize:} \quad \text{steel cost} = \sum_{i=1}^{N_i} w_{si}(L_{si}\rho_i + L_{si}'\rho_i') \tag{13}$$

where w_{si} is the cost coefficient for steel reinforcements; and L_{si} and L_{si}' are respectively the lengths of the tension and compression steel reinforcements for member i, which can be predefined based on code requirements.

In theory, the uncertainties of the objective function should be properly considered in reliability-based optimization. Due to the lack of reliable cost data for assessing properly the reliability of the design objective function, it is commonly acceptable to consider the cost function as a deterministic function. In this research, the uncertainties of the design objective function are not considered.

2.3 *Design constraints*

In the reliability-based structural optimization under earthquake loadings, uncertainties are accounted for through probabilistic constraints. As described in the references [5-8], interstory drift is an important indicator that measures the level of damages to a building during earthquakes and the peak interstory drift over the lifetime of the structure due to earthquakes is uncertain. Therefore, the major design constraint specifies that the interstory drift reliability must satisfy a minimum level of reliability as represented by specifying the reliability index β_j of the jth interstory drift to be larger than its corresponding target value $\bar{\beta}_j$ as follows.

$$\beta_j \geq \bar{\beta}_j \qquad (j = 1, 2, 3, ..., N_j) \tag{14}$$

where N_j is the number of interstory drift reliability constraints. It should be noted that however in the present design codes there is no specific value of the target reliability index corresponding to interstory drift response. In this research, the First-Order Second-Moment (FOSM) method is used to approximate the reliability index corresponding to a specified interstory drift limit. The reliability index β of the equivalent normal probability distribution is given as follows.

$$\beta = \frac{\mu_{\bar{d}}^{N} - \mu_{\Delta\bar{u}}^{N}}{\sqrt{\left(\sigma_{\bar{d}}^{N}\right)^{2} + \left(\sigma_{\Delta\bar{u}}^{N}\right)^{2}}} \tag{15}$$

where $\sigma_{\Delta\bar{u}}^{N}$ and $\mu_{\Delta\bar{u}}^{N}$ can be expressed respectively by Eqs. (7) and (8) in terms of the random variable $\Delta\bar{u}$; $\sigma_{\bar{d}}^{N}$ and $\mu_{\bar{d}}^{N}$ can be determined similarly. As a result, the reliability index β can be formulated directly in terms of $\Delta\bar{u}$. Further, $\Delta\bar{u}$ should be explicitly expressed in terms of the deterministic design variable, the steel reinforcement ratio ρ in order to facilitate numerical solution of the design optimization problem.

2.4 *Explicit inelastic drift formulations*

Details of explicit inelastic displacement formulations are presented in the references [5-8]. Using the Principle of Virtual Work and the Taylor series approximation, the interstory drift under a specified earthquake ground motion at an initial design point ρ^0 is explicit expressed by

$$\Delta u_j(\rho_i) = \Delta u_j\Big|_{\rho_i=\rho_i^0} + \sum_{i=1}^{N_i} \frac{\partial \Delta u_j}{\partial \rho_i}\Big|_{\rho_i=\rho_i^0}(\rho_i - \rho_i^0) + \frac{1}{2}\sum_{i=1}^{N_i} \frac{\partial^2 \Delta u_j}{\partial \rho_i^2}\Big|_{\rho_i=\rho_i^0}(\rho_i - \rho_i^0)^2$$

$$(j = 1, 2, 3, \ldots, N_j) \tag{16}$$

The reliability index constraints on given in Eq. (14) can be explicitly formulated on the basis of Eqs. (7) and (8), (15) and (16). Finally, the reliability-based inelastic structural design optimization problem can be explicitly expressed in terms of the deterministic design variable, ρ_i. The OC method is employed to solve the reliability-based optimization problem.

3. Illustrative Example

A 10-story, 2-bay planar frame is used to illustrate the practicality and effectiveness of the proposed reliability-based optimal design method. The structural geometry of the frame is given in Fig. 1. The pushover analysis of the structure is conducted for estimating the structural performance. The loads considered in the pushover analysis are lateral seismic loads and vertical gravity loads as shown in Fig. 1. The initial lateral loads applied are proportional to the product of the story mass and the first mode shape of the structure. Flexural moment hinges and axial-moment hinges are assigned to the end locations of the beams and columns, respectively.

On the basis of the research results of Gao and Bao [11], during a design period of 50 years, it is found from statistical study that the mean value $\mu_{\Delta\bar{u}}$ is related to the standard value Δu such as $\mu_{\Delta\bar{u}} = 0.597\Delta u$; the coefficient of variation $\delta_{\Delta\bar{u}}$ (which is the ratio of the standard deviation $\sigma_{\Delta\bar{u}}$ over the mean value $\mu_{\Delta\bar{u}}$) is 1.267. Thereby, $\sigma_{\Delta\bar{u}} = 0.756\Delta u$; $k = 2.35$; $\alpha = 0.385\Delta u$. $F_{\mathrm{II}}(\Delta\bar{u})$ and $f_{\mathrm{II}}(\Delta\bar{u})$ given in Eqs. (1) and (2) can be rewritten in terms of the random variable $\Delta\bar{u}$. In addition, the mean value $\mu_{\bar{d}}$ is related to the standard value d such as $\mu_{\bar{d}} = 1.02d$; the coefficient of variation is 0.399 [11], i.e. $\sigma_{\bar{d}} = 0.399\mu_{\bar{d}}$. Similarly, $F_{\mathrm{I}}(\bar{d})$ and $f_{\mathrm{I}}(\bar{d})$ in Eqs.(9) and (10) can be rewritten in terms of $\bar{d}$. Reliability analysis is conducted using the FOSM method. The target index is assumed to be 1.2.

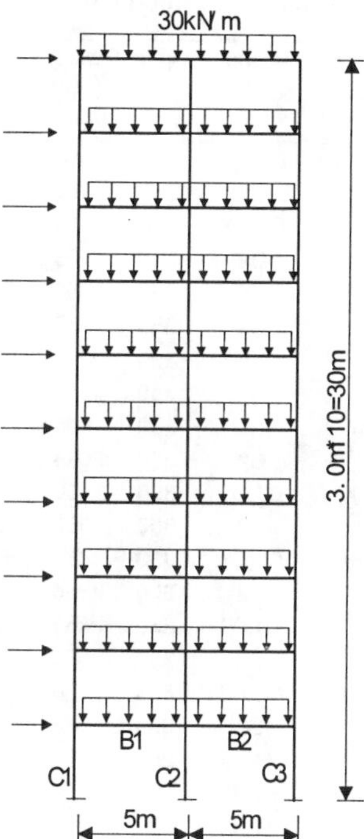

Fig. 1 A planar concrete frame

In this example, the design objective is to minimize the steel reinforcement cost subject to the reliability index constraints corresponding to inelastic interstory drift responses. An allowable inelastic interstory drift ratio limit is assumed to be 1%. Two cases are considered to investigate the effect of reliability-based optimization on an inelastic building. Case 1: the strength-based steel reinforcement ratios for all members are taken as initial values and the deterministic design optimization subject to multiple inelastic interstory drift constraints is carried out. Case 2: the steel reinforcement ratios from the final results of Case 1 are taken as initial values and the reliability-based design optimization subject to the constraint of interstory drift reliability index corresponding to a target value of 1.20 on each story is conducted.

Table 1 shows optimal steel reinforcement ratios in two cases. It is found that there is an increase in steel reinforcement ratios from Cases 1 to 2 particularly for beams. The main reason lies on that the target reliability index specified for the reliability-based design optimization (i.e. Case 2) is indeed higher than the reliability index found for deterministic drift design (i.e. Case 1), resulting in a higher value of steel reinforcement ratio.

Table 1 Optimal steel reinforcement ratios of two cases

Element type	Story level	Mem. group	Optimal sizes width (mm)	Optimal sizes depth (mm)	Case 1 ρ_{opt} (%)	Case 2 ρ_{opt} (%)
	9th~10th	C1,C3	350	350	1.096	1.096
		C2	350	350	1.278	1.386
	8th	C1,C3	350	350	0.860	0.849
		C2	350	475	0.958	0.958
Columns	7th	C1,C3	350	350	1.166	1.456
		C2	350	475	1.092	1.092
	6th	C1,C3	350	400	0.854	0.874
		C2	350	575	0.831	0.831
	5th	C1,C3	350	400	1.004	1.100
		C2	350	575	1.002	1.002
	4th	C1,C3	350	450	0.833	0.868
		C2	350	650	0.739	0.739
	3rd	C1,C3	350	450	0.876	1.077
		C2	350	650	0.857	0.857
	2nd	C1,C3	350	450	1.044	1.127
		C2	350	575	1.224	1.225
	1st	C1,C3	350	450	1.514	1.514
		C2	350	575	1.844	1.844
	9th~10th	B1,B2	200	400	0.800	0.800
	8th	B1,B2	200	450	0.941	0.999
Beams	7th	B1,B2	200	450	1.320	1.549
	6th	B1,B2	251	450	1.310	1.473
	5th	B1,B2	251	450	1.493	1.773
	4th	B1,B2	300	450	1.457	1.705
	3rd	B1,B2	300	450	1.484	1.729
	2nd	B1,B2	300	450	1.512	1.775
	1st	B1,B2	300	450	1.006	1.294

Fig. 2 presents the comparisons of the final reliability indices in two cases. It is found that most of the final indices are about 1.0 in the deterministic design optimization Case 1; while the final values in Case 2 are close to the target index of 1.20. Such a result indicates that in the reliability-based inelastic structural optimal design, the lateral load resisting system can still be improved by the OC procedure such as to distribute the steel reinforcements of all members so that lateral inelastic drifts are reduced and reliability index constraints are satisfied. Furthermore, it is demonstrated that deterministic design optimization cannot guarantee to ensure the design with a satisfactory reliability level.

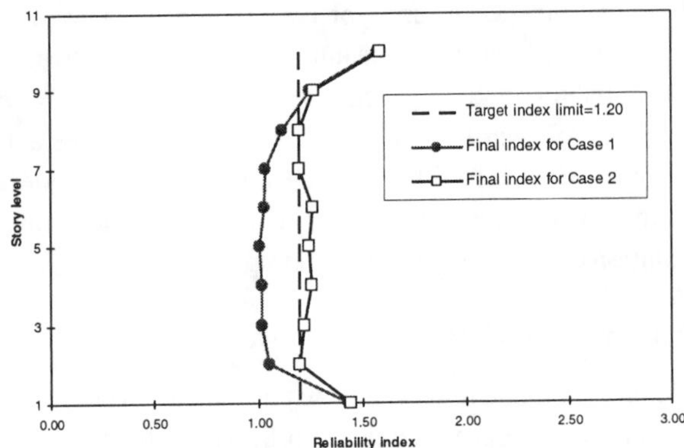

Fig. 2 Reliability indices

Fig. 3 shows the final interstory drift responses for the two cases. It is found that in Case 1, the final interstory drift ratios from the second to seventh story floors are close to the ratio limit of 1%; however, they are smaller the drift limit of 1% in Case 2. It is explained that the reliability-based designs impose higher levels of reliability, thereby resulting in a stiffer structure with less values of interstory drifts. Furthermore, it is indicated that steel reinforcements of members are very effective to improve the inelastic interstory drift values and further drift reliability indices of the building.

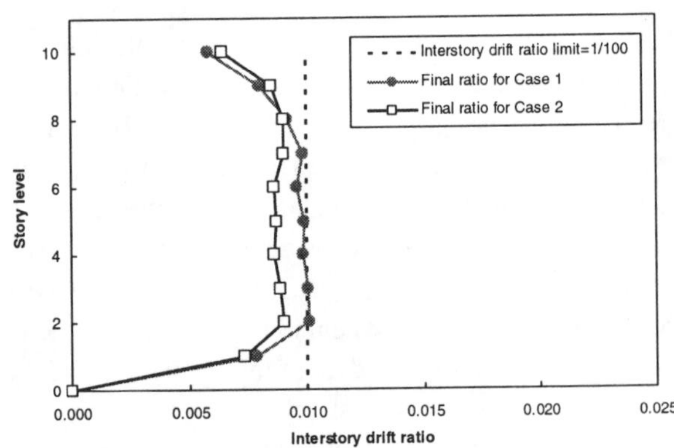

Fig. 3 Inelastic interstory drift ratios

4. Conclusions

The reliability-based optimization problem of an RC building has been formulated such that not only can it minimize the steel cost, but also it can satisfy the required reliability constraints due to the uncertainties of earthquake loadings. The proposed reliability-based optimization technique integrates inelastic structural analysis using the pushover analysis method, reliability analysis using the FOSM method and optimization techniques using the OC method. Although structural analysis, reliability analysis and optimization are three independent procedures, reliability-based optimization requires repeated applications of these three procedures until the solution convergence is achieved.

The illustrative example shows that the OC technique developed is able to optimise a reliability-based concrete structure so as to distribute its lateral stiffness to satisfy interstory drift reliability index constraints with minimum steel cost. Conventional deterministic design optimization cannot ensure designs with satisfactory reliability level. In contrast, the reliability-based design optimization can ensure optimal reliability designs of building structures through the consideration of design constraints that fulfil a target value of the required reliability index for each story. It is believed that the proposed optimal design methodology will provide a rational basis for incorporating other uncertainties such as structural modelling uncertainties, aside from the uncertainties in seismic loading.

References

1. G. Q. Li, H. Cao, Q. S. Li, *Theory and Its Application of Structural Dynamic Reliability*. Earthquake Press, Beijing (1993).
2. G. Q. Li and Q. S. Li, *Theory And Its Application of Time-Dependent Reliability of Engineering Structures*. Science Press, Beijing (2001).
3. Q. S. Li, J. Q. Fang and D. K. Liu, "Evaluation of structural dynamic responses by stochastic finite element method". *Structural Engineering And Mechanics*. **8** (5): 477-490 (1999).
4. Q. S. Li, G. Q. Li, "Time-dependent reliability analysis of glass cladding under wind action". *Engineering Structures*, in press, (2005).
5. X. K. Zou, *Optimal seismic performance-based design of reinforced concrete buildings*. Ph.D. Dissertation, Hong Kong University of Science and Technology, Hong Kang, P.R. China (2002).
6. C. M. Chan and X. K. Zou, "Elastic and inelastic drift performance optimization for reinforced concrete building under earthquake loads". *Earthquake Engrg. and Struct. Dynamics*. **33**(8), 929-950 (2004).
7. X. K. Zou and C. M. Chan, "An Optimal Resizing Technique for Seismic Drift Design of Concrete Buildings Subjected to Response Spectrum and Time History Loadings". *Computers & Structures*. **83**, 1689-1704 (2005a).
8. X. K. Zou and C. M. Chan, "Optimal Seismic Performance-Based Design of Reinforced Concrete Buildings Using Nonlinear Pushover Analysis". *Engineering Structures*. **27**, 1289-1302 (2005b).
9. G. D. Cheng, G. Li and Y. Cai, "Reliability-based structural optimization under hazard loads". *Structural Optimization*. **16**, 128-135 (1998).
10. National Standard of the People's Republic of China, *Chinese Code for Seismic Design of Buildings* (GBJ11-89). Beijing, China (1989).
11. X. W. Gao and A. B. Bao, "Probabilistic Model and Its Statistical Parameters for Seismic Load". *Earthquake Eng. & Eng. Vibration*. **5**(1), 13-22 (1985).

OPTIMAL SEISMIC PERFORMANCE DESIGN OF REINFORCED CONCRETE FRAMES BASED ON LIFE-CYCLE COST*

XIAO-KANG ZOU

Department of Building and Construction, The City University of Hong Kong, Kowloon, Hong Kong, China

CHUN-MAN CHAN[1] GANG LI[2]

[1]*Department of Civil Engineering, Hong Kong University of Science and Technology, Kowloon, Hong Kong, China.*
[2]State Key Laboratory of Structural Analysis of Industrial Equipment, Dalian Univ. *of Technology, Dalia, 116024,* China.

An optimization technique by incorporating the performance-based seismic design of concrete building structures in the form of multiobjective optimization has been developed to meet the emerging trend of performance-based design approach. Specifically, the life-cycle cost of a reinforced concrete (RC) building frame is minimized subject to multiple levels of seismic performance design criteria. Explicit formulation of design constraints involving inelastic drift response performance caused by pushover loading is expressed with the consideration of the occurrence of RC plasticity and the formation of plastic hinges. Due to the fact that the initial construction cost and the expected damage loss are conflicting by nature, the life cycle cost of a building structure can be posed as a multiobjective optimization problem and solved by the ε–constraint method to produce a Pareto optimal set, from which the best compromise solution can be selected. The methodology for each Pareto optimal solution is fundamentally based on the Optimality Criteria (OC) approach.

1. Introductions

The traditional prescriptive design approach based on linear elastic techniques has been developed for many years. One major drawback of this design approach is that it does not directly address structural inelastic seismic responses and thus cannot effectively deal with damage loss due to structural and nonstructural failure during earthquakes. As a result, the long-term risk and benefit implications cannot be assessed using the traditional design approach. Recent earthquakes have shown that structural or non-structural damage or failure loss can be enormous and plays an important role in the total life-cycle cost analysis of structures. A new performance-based methodology for seismic design and construction of buildings has appeared to be the future direction of seismic design codes [1-5].

In the new approach, the design objective function considers the total life cycle structural cost, which consist of the initial construction cost and the expected future failure loss caused by earthquakes. The final design should be a good balance between the initial structural cost and the loss expectation. Unlike the traditional cost of a structure, the detailed quantification of the damage loss is rather difficult since it involves not only the repair cost, but also the costs associated with social, economic, political, cultural and ethical aspects. Numerous research efforts had established a number of different damage loss models which involve realistic modeling of the loading and resistance uncertainty, initial construction cost, damage cost, failure consequence cost, maintenance cost and discount cost of distant future failure. Many studies indicated the importance of consideration of a life-cycle cost in making rational design decisions.

The paper presents a numerical multiobjective optimization technique for performance-based design of seismically excited building frames. Attempts have been made to develop the optimization technique by incorporating the seismic performance-based design of concrete building structures in the form of multiobjective optimization. Specifically, the life-cycle cost of a reinforced concrete (RC) building frame is to be minimized subject to multiple levels of seismic performance design criteria. In formulating the total life-cycle cost, the initial construction cost can be expressed in terms of the element sizing design variables [2,4] and the expected damage loss can be stated as a function of seismic performance levels and their associated failure probability by the means of a statistical model. Explicit formulation of design constraints involving inelastic drift response performance caused by pushover loading

* This work was partially supported by a grant from the Research Grants Council of Hong Kong (Project No. DAG03/04.EG04).

is to be expressed with the consideration of the occurrence of RC plasticity and the formation of plastic hinges. Due to the fact that the initial construction cost and the expected damage loss are conflicting by nature, the life cycle cost of a building structure can be posed as a multiobjective optimization problem which can be solved by the ε–constraint method to produce a Pareto optimal set, from which the best compromise solution can be selected. The methodology for each Pareto optimal solution is fundamentally based on the Optimality Criteria (OC) approach. A 10-story framework example is presented to illustrate the effectiveness of the proposed optimal design method

2. Multiobjective Optimization Problem

2.1 *Formulation of objective function*

Using F to denote the life-cycle cost function including the initial construction cost f_1 and the expected future damage loss f_2, the multi-objective function can be expressed as

Minimize: $$F = \{f_1, f_2\} \qquad (1)$$

In order to facilitate the numerical solution of the optimization problem, the implicit objective function in Eq. (1) needs to be first expressed explicitly in terms of the design variables.

2.1.1 *Construction cost* f_1

The total construction cost f_1 of an RC framework is simply the summation of the concrete material cost f_{1c} and the steel reinforcement cost f_{1s}, as

$$f_1 = f_{1c} + f_{1s} \qquad (2)$$

For an RC building having i=1, 2, …, N_i members with rectangular cross sections, the concrete material cost f_{1c} can be expressed in terms of the width B_i and the depth D_i in the elastic design optimization and will be kept constant in the inelastic design optimization [2-4].

The tension steel reinforcement ratio ρ_i and compression steel reinforcement ratio ρ_i' are taken as design variables in the inelastic design optimization. For a building structure with predefined topology, the steel cost f_{1s} of the structural framework can be expressed as

$$f_{1s}(\rho_i, \rho_i') = \sum_{i=1}^{N_i} w_{si}(L_{si}\rho_i + L_{si}'\rho_i') \qquad (3)$$

where w_{si} is the cost coefficient for steel reinforcements; and L_{si} and L_{si}' are the lengths of the tension and compression steel reinforcements for member *i*.

2.1.2 *Expected future structural loss* f_2

The estimation of structural loss is usually very complex involving not only engineering analysis but also many other considerations. The failure of each structural performance may lead to a different failure loss. According to the work by Li [6], the total loss expectation can be defined by the summation of the product of the occurrence probability of earthquake and the system failure loss. That is, for N_r performance levels, the loss expectation function f_2 is stated as

$$f_2 = \sum_{r=1}^{N_r} P_r * L_r \qquad (4)$$

where r denotes the seismic design level and $r = 1, 2, ..., N_r$; P_r is the occurrence probability of an earthquake at the *r*th seismic design level, which can be determined from specified code requirements; L_r is the structural failure loss including direct and indirect losses under the *r*th seismic design level

2.1.3 *Structural damage loss* L_r

Seismic damage of a structure can be commonly classified into five levels, namely negligible, slight, moderate, severe and complete damage [7]. Building structures can be categorized into four classes based on their importance according to the Chinese Standard for Classification of Seismic Protection of Buildings [8]. The interstory drift index τ defined by $\Delta u \times R / h$ (where Δu is the interstory drift; h is the story height; R is equal to 500 for concrete frames) is proposed in this research for estimating the expected loss value L_r related to the *r*th seismic design level. As a result, the damage loss consisting of the direct and indirect losses can be expressed in terms of τ as a function of Δu, which can be written later as a function of the steel reinforcement design variables [1-2].

Upon formulating the explicit expression of the damage loss $L_r(\tau_j)$ with respect to the *j*th interstory drift at the *r*th performance level, the total structural damage loss f_2 presented in Eq. (1) can be expressed in terms of the variable $\tau(\Delta u)$ as

$$f_2 = f_1 \sum_{r=1}^{N_r} \left\{ P_r \times \left[\sum_{j=1}^{N_j} \left(\alpha_1 \cdot \tau_j(\Delta u) + \alpha_2 \right) \right]_r \right\} \quad (5)$$

2.2 *Formulation of Design Constraints*

2.2.1 *Design Constraints*

The inelastic drift responses generated by moderate and large earthquakes must not exceed the appropriate limits corresponding to a given performance objective. Namely, inelastic story drift ratio should comply with the following requirement as

$$\frac{\Delta u_j}{h_j} = \frac{u_j - u_{j-1}}{h_j} \le \psi_j \quad (6)$$

where Δu_j is the interstory drift of the *j*th story; u_j and u_{j-1} are the respective story displacements of two adjacent *j* and *j*-1 floor levels; h_j is the *j*th story height and ψ_j is the specified limit on the interstory drift ratio.

Besides checking the system responses of the interstory drift ratios for a given performance level, the local restriction on the plastic rotational at the critical sections of each element must also be considered. The plastic rotation, θ_{ph}, at the *h*th end of member *i* (where the subscript *h* represents one end of a member and *h*=1,2) should be constrained in the optimization by

$$\theta_{ph} \le \theta_p^U \quad (7)$$

where θ_p^U is the rotation limit of member *i* for a specific performance level. In addition to the performance-based constraints mentioned above, design limits are imposed on the steel reinforcement ratios as

$$\rho_i^L \le \rho_i \le \rho_i^U \; ; \quad \rho_i'^L \le \rho_i' \le \rho_i'^U \quad (8a, b)$$

where the superscripts *L* and *U* denote the minimum and maximum limits of the design variables, ρ_i and ρ_i', respectively. In order to facilitate a numerical solution of the optimal design problem, it is necessary that the implicit drift constraint in Eq. (6) and plastic rotation constraint in Eq. (7) be expressed explicitly in terms of the design variables, ρ_i and ρ_i'. Details of the explicit drift formulation can be found in the references [2, 3, 5]. Only a brief description is given in the following section.

2.2.2 *Explicit Drift Formulation*

Based on the internal element forces and moments of the structure obtained from the pushover analysis, the principle of virtual work and a second-order Taylor series approximation can be used to explicitly express the pushover displacement u_j in terms of design variables ρ_i as

$$u_j(\rho_i) = u_j\Big|_{\rho_i=\rho_i^0} + \sum_{i=1}^{N_i} \frac{\partial u_j}{\partial \rho_i}\Big|_{\rho_i=\rho_i^0} (\rho_i - \rho_i^0) + \frac{1}{2} \sum_{i=1}^{N_i} \frac{\partial^2 u_j}{\partial \rho_i^2}\Big|_{\rho_i=\rho_i^0} (\rho_i - \rho_i^0)^2 \quad (9)$$

where the tension steel ratio, ρ_i, is herein considered as the major design variable while the compression steel ratio, ρ_i', is assumed to be linearly related to ρ_i for simplicity. The gradient, $\partial\theta_p / \partial\rho_i$, and the second-order term, $\partial^2\theta_p / \partial\rho^2$, can be analytically calculated as given in the references [2, 3, 5].

2.3 *Explicit multi-objective problem formulation*

Upon obtaining the objective function (i.e., the construction cost f_1 in Eq. (2) and the expected damage loss f_2 in Eq. (5)), and the inelastic drift design constraints in Eq. (9), the multi-objective design optimization problem can be written explicitly in terms of the design variables ρ_i as

Minimize:

$$F(\rho_i) = \{f_1, f_2\} \quad (10)$$

Subject to:

$$g_j(\rho_i) = \frac{1}{h_j \psi_j} \left[\begin{array}{l} \Delta u_j\Big|_{\rho_i=\rho_i^0} + \sum_{i=1}^{N_i} \alpha_{1i} (\rho_i - \rho_i^0) \\ + \frac{1}{2} \sum_{i=1}^{N_i} \alpha_{2i} (\rho_i - \rho_i^0)^2 \end{array} \right] \le 1$$

$$(j = 1, 2, ..., N_j) \quad (11)$$

$$\rho_i^L \le \rho_i \le \rho_i^U \quad (i = 1, 2, ..., N_i) \quad (12)$$

where

$$\alpha_{1i} = \frac{\partial \Delta u_j}{\partial \rho_i}\Big|_{\rho_i=\rho_i^0} \; ; \quad \alpha_{2i} = \frac{\partial \Delta u_j^2}{\partial \rho_i^2}\Big|_{\rho_i=\rho_i^0} \; i \quad (13a, b)$$

Eq. (10) defines the life-cycle cost function F, which consists of the construction cost f_1 and the damage loss f_2; Eq. (11) defines the inelastic interstory drift constraint for a specified ground motion; Eq. (12)

defines the lower and upper size bounds specified for the design variables, ρ_i.

3. Multiobjective Optimization Algorithm

The multi-objective optimization problem given in Eqs. (10)~(12) can be written in a general form as

Minimize: $$F(\mathrm{x}) = \{f_1(\mathrm{x}), f_2(\mathrm{x})\} \quad (14)$$

Subject to: $$G(\mathrm{x}) \leq 0 \quad (15)$$

$$\mathrm{x} \in S \quad (16)$$

where the design variable vector x is an element of the design variable set S; $f_1(\mathrm{x})$ and $f_2(\mathrm{x})$ stand for the dual scalar objectives of construction cost and damage loss; $G(\mathrm{x})$ is the set of normalized constraint vector.

The ε–constraint method is a technique that transforms a multi-criteria objective function into a single criterion by retaining one selected objective function as the primary criterion to be optimized and treating the remaining criteria as constraints. The method provides direct control of generating the set Pareto optima and generally provides an efficient method for searching for the solution of the Pareto set. Herein, the construction cost, $f_1(\mathrm{x})$, is taken as a primary objective, while the damage loss, $f_2(\mathrm{x})$, is transformed into a design constraint. The reason for this consideration is that the damage loss $f_2(\mathrm{x})$ in Eq. (5) and the interstory drift constraints in Eq. (6) are directly written in terms of the interstory drifts. The similarity may bring advantages in numerical calculations. Thus, the problem in Eqs. (14)~(16) can be transformed as

Minimize: $$f_1(\mathrm{x}) \quad (17)$$

Subject to: $$f_2(\mathrm{x}) \leq \varepsilon_2 \quad (18)$$

$$G(\mathrm{x}) \leq 1 \quad (19)$$

$$\mathrm{x} \in S \quad (20)$$

By varying the upper level, ε_2, of the objective $f_2(\mathrm{x})$, all Pareto optimal points with minimum value of the objective $f_1(\mathrm{x})$, are, in principle, attainable. In order to find the reasonable range of the upper bound, ε_2, and to seek a good initial point for Eqs. (17)~(20), a single objective optimization problem is defined as

Minimize: $$f_1(\mathrm{x}) \quad (21)$$

Subject to: $$G(\mathrm{x}) \leq 1 \quad (22)$$

$$\mathrm{x} \in S \quad (23)$$

$\Rightarrow$ Optimum solution $\mathrm{x}_0 \in S$

Furthermore the maximum value, $\varepsilon_{2,\max}$, of the upper bound, ε_2, can be found from x_0. The maximum value, $\varepsilon_{2,\max}$ can be defined as

$$\varepsilon_{2,\max} = f_2(\mathrm{x}_0) \quad (24)$$

Thus, the upper bound value, ε_2, is

$$\varepsilon_2 = \varepsilon_{2,\max} - e_2 \quad (25)$$

where e_2 is a relatively small positive real number that can be altered. Upon establishing ε_2 from Eq. (25) and the initial values x_0 of design variables from Eqs. (21)~(23), Eqs. (17)~(20) can be solved using the OC approach to obtain a Pareto solution [2]. Subsequently, a Pareto optimal set is found by altering e_2. In fact, the solution of the single-objective optimization given in Eqs. (21)~(23) is employed as a pre-processor for solving the multiobjective optimization problem Eqs. (17)~(20) using the ε–constraint method.

4. An Illustrative Example

A 10-story, 2-bay planar frame is used to illustrate the practicality and effectiveness of the proposed multi-objective optimal design method. The structural geometry of the frame is given in Fig. 1.

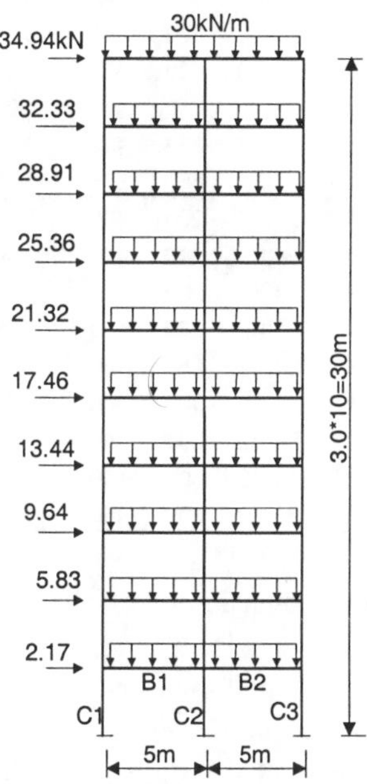

Fig. 1 A ten-story, two-bay concrete frame

Two levels of earthquake loads are considered in this example. One represents a minor earthquake load with a peak acceleration of 0.32*g* according to the acceleration response spectrum of the Chinese seismic design code [7]. Another load level represents a severe earthquake with an initial peak acceleration of 1.4*g*. In the elastic phase of the optimization, the concrete cost of the structure is minimized subject to the elastic spectral drift constraints under the minor earthquake loading condition. The unit cost of concrete is assumed to be US$90/m^3. Elastic interstory drift constraints are taken into account with an allowable interstory drift ratio limit of 1/450. The final optimal member sizes are given in Table 1.

Table 1. Initial and optimal steel ratios at points C^0 and C^*

Element type	Story level	Member group	Member sizes width (mm)	Member sizes depth (mm)	At Point C^0 ρ_{opt} (%)	At Point C^* ρ_{opt} (%)
	9th~10th	C1,C3	350	350	1.096	1.100
		C2	350	350	1.278	1.386
	8th	C1,C3	350	350	0.860	1.034
		C2	350	475	0.958	0.958
Columns	7th	C1,C3	350	350	1.166	1.354
		C2	350	475	1.092	1.092
	6th	C1,C3	350	400	0.854	0.854
		C2	350	575	0.831	0.831
	5th	C1,C3	350	400	1.004	1.012
		C2	350	575	1.002	1.002
	4th	C1,C3	350	450	0.833	0.833
		C2	350	650	0.739	0.739
	3rd	C1,C3	350	450	0.876	0.928
		C2	350	650	0.857	0.857
	2nd	C1,C3	350	450	1.044	1.066
		C2	350	575	1.224	1.225
	1th	C1,C3	350	450	1.514	1.514
		C2	350	575	1.844	1.844
	9th~10th	B1,B2	200	400	0.800	0.800
	8th	B1,B2	200	450	0.941	1.098
Beams	7th	B1,B2	200	450	1.320	1.646
	6th	B1,B2	251	450	1.310	1.631
	5th	B1,B2	251	450	1.493	1.967
	4th	B1,B2	300	450	1.457	1.895
	3rd	B1,B2	300	450	1.484	1.894
	2nd	B1,B2	300	450	1.512	1.915
	1st	B1,B2	300	450	1.006	1.482

In the inelastic phase of the optimization, the design objective is to minimize the steel reinforcement cost subject to the performance-based inelastic drift constraints under the severe earthquake loading condition. The unit construction cost of steel reinforcement is assumed to be US$962/tonne including the costs of the steel material and labor. Inelastic interstory drift constraints are considered with an allowable interstory drift ratio limit of 1%. Initial reinforcement ratios are calculated based on the strength requirements of members after the elastic phase design process. Such strength-based reinforcement ratios are taken as the lower bounds for the inelastic design process. Their upper bounds are assumed to be 4.0% for columns and 2.0% for beams. Flexural moment hinges are assigned to the end locations of the beams and the ultimate plastic rotation, θ_p^U, is assumed to be 0.02 radian; whereas axial-moment hinges are assigned to the end locations of the columns and the ultimate plastic rotation, θ_p^U, is assumed to be 0.015 radian. The P-Delta effect is considered and initial lateral loads applied in pushover analysis are shown in Fig. 1. The occurrence probability P_r for the severe earthquake is assumed to be 0.039.

Fig. 2 presents a Pareto optimal set, which provides a balance between the construction cost, f_1, and the damage loss, f_2. The minimum construction cost, f_1, for the single objective optimal design problem given in Eqs. (21)~(23), i.e., corresponding to point C^0 in Fig. 2, is first obtained such that $US\$4184$ and the corresponding damage loss, f_2, is then obtained such that $US\$7522$. The optimal steel reinforcement ratios corresponding to point C^0 are taken as initial steel ratios of members in the subsequent multi-objective inelastic design optimization (see Eqs. (14)~(16) or (17)~(20)) and the optimal Pareto set curve can be generated.

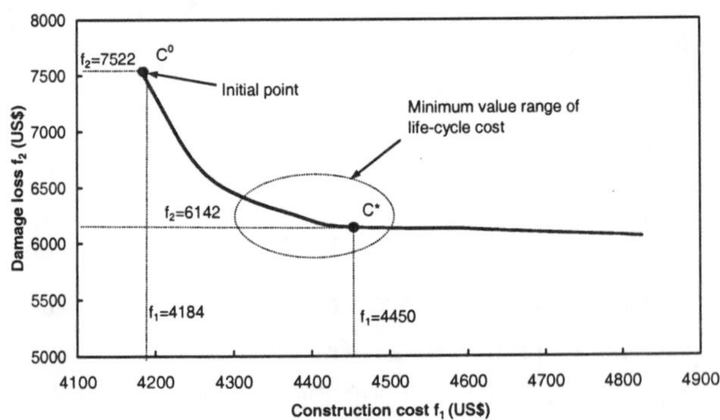

Fig. 2. A Pareto optimal set

Firstly, it is observed from Fig. 2 that an improvement in damage loss leads to an increase in construction cost of the structure. From the Pareto optimal set, a designer can have an overview of the tradeoffs. Secondly, any point in the Pareto optimal set can be selected as an optimum design, which represents a designer's preference. A designer has many choices rather than only one and he/she may finally choose a particular solution, which he/she thinks, should be realized. Thirdly, the life-cycle cost can be further written as $F = \phi_1 f_1 + \phi_2 f_2$, where ϕ_1 and ϕ_2 are the weighting factors and their magnitudes are dependent on a designer's decision. A preferred balance of the initial and damage losses can therefore be obtained. It is worth noting that different values of ϕ_1 and ϕ_2 will lead to different minimum values of life-cycle cost.

For instance, when the importance of f_1 is assumed to be the same as that of f_2 (i.e., $\phi_1 = \phi_2 = 1$), as shown in Fig. 3, the Pareto solution C* (corresponding to $f_1 = US\$4450$ and $f_2 = US\$6142$) is a minimum point of the total life-cycle cost with $F = US\$10593$. Compared with the initial optimal solution (i.e., point C^0), an increase of 6% in the construction cost f_1 (i.e., from US$4184 at point C^0 to US$4450 at point C*) leads to a decrease of 18% in the damage loss f_2 (i.e., from US$7522 at point C^0 to US$6142 at point C*). As a result, the total life-cycle cost is reduced over 9% (i.e., from US$11706 at point C^0 to US$10593 at point C*).

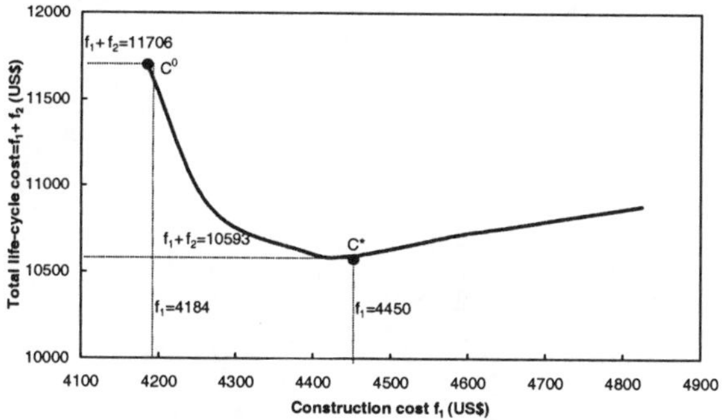

Fig. 3. Life-cycle cost function $F = \phi_1 f_1 + \phi_2 f_2$ where $\phi_1 = \phi_2 = 1$

Table 1 lists the optimal steel reinforcement ratios at points C^0 and C*, respectively. The steel reinforcement ratios of beams are largely increased from the first to the eighth floors from points C^0 to C* (i.e., an increase of 17%~47%), while there is no change of steel reinforcement ratios in most columns due to no occurrence of plastic hinges. Such an increase indicates that steel reinforcement has a significant effect on improving inelastic drift performance and further reducing damage loss.

Fig. 4 presents the results of the interstory drift ratios at points C^0 and C*, respectively. It is found that all the inelastic interstory drift constraints are satisfied corresponding to either point C^0 or C*. In the optimal solution of the single objective design optimization (i.e., corresponding to point C^0), the interstory inelastic drift constraints from the second to the seventh floors are close to the limiting ratio of 1%. However, in the multi-objective design optimization (i.e., point C*), all final drift constraints are not necessarily found to be close to the limiting ratio 1%. Such results indicate that the steel ratios of the structure members are resized to minimize the life-cycle cost by reaching the best balance between the construction cost and damage loss such that the interstorey drift responses may not necessarily have to be fully constrained. The optimization technique developed is capable of seeking for the best balance so that lateral drift constraints are satisfied simultaneously with the least life-cycle cost.

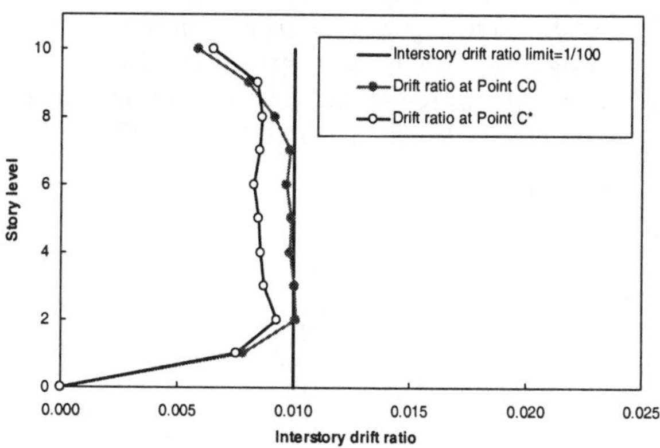

Fig.4. Comparison of inelastic interstory drift ratios

5. Conclusions

The optimization technique developed for the nonlinear multi-objective design of inelastic drift performance of RC frameworks under pushover loadings provides a very effective way to simultaneously optimize the structural life-cycle cost while satisfying all drift performance design criteria. Specifically, an appropriate damage loss model has been developed in this study. A multi-objective optimization algorithm, the ε–constraint method, has been effectively applied to handle the conflicts between the construction cost and the damage loss by producing a Pareto optimal set, from which a decision maker can directly select the best compromise solution. Furthermore, the method developed for specifying appropriate upper bounds for objective functions is also found to be very effective.

The final optimal results from the illustrative example indicate that, for the multi-objective design optimization, the lateral load resisting system can be automatically improved by the optimization procedure to seek for the best balance between construction cost and damage loss. In this balance, not only are the specified lateral drift constraints satisfied, but also the structural life-cycle cost can be minimized by effectively distributing the steel reinforcements of all structural members using the optimization procedure developed.

References

1. X. K. Zou and C. M. Chan, "Optimal Drift Performance Design for Nonlinear Pushover Response of Concrete Structures". *WCSMO-4: Proceedings of the Fourth Would Congress of Structural and Multidisciplinary Optimization*, June 4-8, Dalian, China (2001).
2. X. K. Zou, *Optimal seismic performance-based design of reinforced concrete buildings.* Ph.D. Dissertation, Hong Kong University of Science and Technology, Hong Kang, P.R. China (2002).
3. C. M. Chan and X. K. Zou, "Elastic and inelastic drift performance optimization for reinforced concrete building under earthquake loads." *Earthquake Engrg. and Struct. Dynamics.* **33**(8), 929-950 (2004).
4. X. K. Zou and C. M. Chan, "An Optimal Resizing Technique for Seismic Drift Design of Concrete Buildings Subjected to Response Spectrum and Time History Loadings." *Computers & Structures.* **83**, 1689-1704 (2005).
5. X. K. Zou and C. M. Chan, "Optimal Seismic Performance-Based Design of Reinforced Concrete Buildings Using Nonlinear Pushover Analysis." *Engineering Structures.* **27**, 1289-1302 (2005b).
6. G. Li, *Reliability and performance based optimal design for seismic high-rising structures.* Ph.D. Dissertation. Dalian University of Technology, China, (1998).
7. National Standard of the People's Republic of China, *Chinese Code for Seismic Design of Buildings* (GBJ11-89). Beijing, China (1989).
8. National Standard of the People's Republic of China, *Chinese Standard for Classification of Seismic Protection of Buildings* (GB50223-95). Beijing, China (1995).

MODELING AND SEISMIC BEHAVIOR OF CONCRETE-FILLED RECTANGULAR STEEL TUBULAR COLUMNS IN TALL BUILDINGS*

XUEPING LI

Architectural Design and Research Institute, Tongji University, 1239 Siping Road. Shanghai 200092, China

XILIN LU

State Key Laboratory for Disaster Reduction in Civil Engineering, Tongji University, 1239 Siping Road. Shanghai 200092, China

Based on experimental investigation, the interface bond-slip model of CFRT was put forward, and the hysteretic model of in-filled concrete and steel considering Bauschinger Effect was also built. With fiber section model, the analytical model for CFRT columns was constructed. Analysis program for CFRT columns considering interface slip was accomplished. Full range analysis of the CFRT columns was conducted. The simulation results were compared with test records at the levels of load-displacement and load-strain. To investigate the mechanism of bond-slip interface, simulated stress-strain of fibers was also presented. The effect of the interface bond strength on the hysteretic behavior of CFRT columns was evaluated. At the end, design suggestions were presented for structural application of CFRT columns in high-rise buildings.

1. Introduction

Concrete-filled rectangular steel tubular (CFRT) structure has become increasingly popular in structural application of high-rise buildings. This is due to its good structural behavior, regular shape and easiness for construction. On its mechanics, the steel tube laterally offers confinement to in-filled concrete and improves its ductility, while the in-filled concrete provides support to the steel tube laterally and has its buckling capacity increased. Steel tube plates also act as shear resistant element, which may protect in-filled concrete from shear brittle failure. On construction, the steel tube serves both as erection frame for the floor construction and as formwork for the in-filled concrete; the enhanced behavior of tube plates allows the use of thinner steel plates than those in pure steel columns, which can make the fabrication and site construction easy.

Many experimental studies have been carried out to investigate the static behavior of CFRT columns [1~4]. On its seismic behavior, most of past experimental studies were focused on concrete filled square tubes (CFST). Ge et al. [5] performed cyclic tests on CFST columns. Lu et al. [6~8] carried experimental and analytical study on the seismic behavior of CFST columns and structures. Varma et al. [9] conducted experimental study on high strength concrete filled square tubular beam-columns. But, the work on CFRT columns is fairly scarce.

Although columns with square section are frequently used in building structures, rectangular columns with non-equal sides also have their roles in practical engineering. Considering this, seismic behavior of CFRT columns was experimentally studied by Lu and Li[10] at Tongji University. The test parameters include the axial compression ratio (0.4, 0.6), steel ratio, section-depth to width ratio (1.0, 1.5, 2.0), and loading direction (along strong or weak axis).

Based on the experimental investigations on CFRT columns, analytical study on their seismic behavior was presented in this paper. Here, the following research route was followed, that is: construction of material model, building analytical model, programming the analytical tools, seismic behavior simulation, and design suggestions presentation.

Section fiber model was adopted in this paper. After discretization of section fibers, the tangent bond-slip element at the interface was introduced herein, and the effect of the interface slip was also considered during analysis.

* This work is sponsored by National Natural Science Foundation of China, Grant No. 50025821, 50321803.

Besides that, the following hypothesis was assumed. Firstly, sections of the steel tube and the in-filled concrete columns respectively remain plane all the time. Secondly, shear stiffness remains elastic. Thirdly, the shrink and creep effect was neglected during analysis, for it has little effect on CFT members [11].

2. Description of the Interface Model and Material Models

2.1 *The hysteretic bond-slip model at interface of CFRT columns*

Among current investigations, only monotonic loading tests were conducted on CFT specimens [12~15]. Aval et al. [16] has constructed analytical element for CFT, in which the bond-slip model is elastic-plastic. Based on the achievements of former researchers, simplified model for the tangent bond-slip at interface of CFRT was constructed as Figure 1, in which the effect of the normal stress on the interface was neglected.

The values for the model are set as following. The peak stress of point "B" was calculated from the suggested equation [15]. Other values referred to the characteristic of tested bond-slip curves [14] as following. The stress of point "A" is 0.6 times of peak stress, while the stiffness of elasticity was $1.6 \times 10^9 N/mm^3$. The residential stress of point "C" is 0.7 times of the peak stress, while according slippage is 2.0 times of the peak slippage at point "B".

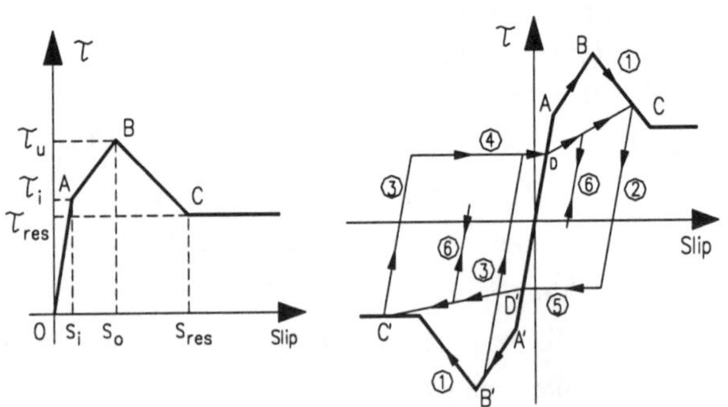

Figure 1. Shear stress-slippage model at the interface

Six states for bond-slip are assumed for the analytical procedure: ①loading along the skeleton; ② unloading in tension zone; ③unloading in compression zone; ④ reloading in tension zone; ⑤ reloading in compression zone; ⑥sub-hysteretic loop. Here it should be pointed out that "tension" means counter-clockwise direction and "compression" means clockwise direction.

2.2 *Stress-strain hysteretic model of steel*

Steel is a kind of idealistic elastic-plastic material. Researchers have built precise hysteretic model with curves. But, for the point of convenience, simplified models with multiple straight lines were also frequently used. When performing analysis of RC structures, idealistic model with two lines was often adopted. Lu and Chen [17] have used this model when analyzing the hysteretic behavior of moment-curvature of CFRT beam-columns. Compared to common RC structures, the steel ratio is high in CFRT columns. So, contribution from the steel is more prominent, and the simulation model for steel should be constructed more precisely.

The hysteretic model for steel was showed as Figure 2. The skeleton was two-lines with kinetic hardening. When loaded at opposite direction, the Bauschinger Effect was also addressed.

Figure 2. Stress-strain hysteretic model of steel

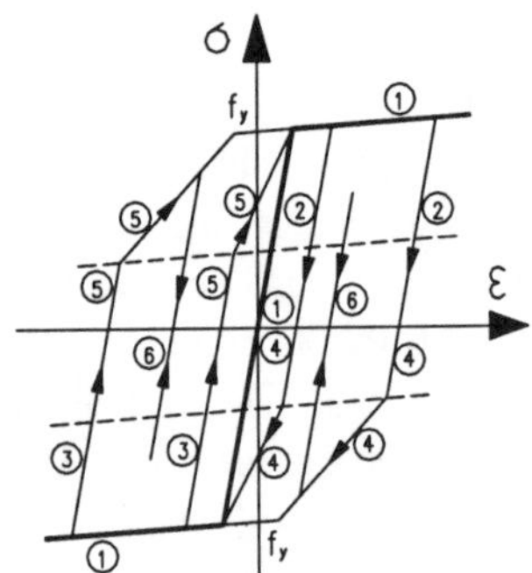

In this paper, the kinetic hardening modulus of steel was set as E_t=0.001E_{s0}, which was lower than that of reinforcement in RC structures [18]. This is because there was no concrete pasted outside the steel tube plate, so the tension-stiffening effect for steel is weak in CFRT structures. Besides that, during test process, the maximum strain of steel was within the yielding phase of steel. When loaded in opposite direction, the yielding stress is set as $0.15f_y$ (f_y is the yielding stress for steel).

The stress-strain process was set to 6 states, which were: ①loaded on skeleton; ②unloaded in tension zone; ③unloaded in compression zone; ④reloaded in tension zone; ⑤reloaded in compression zone; ⑥sub-hysteretic loop.

2.3 *Hysteretic model for in-filled concrete of CFRT*

At ultimate state under axial compression, the in-filled concrete of CFRT stub could be divided to strongly confined zone and weakly confined zone [19]. But, for the steel ratios frequently used in engineering, the strength increase from the confinement effect is limit compared with the member capacity. The confinement effect was more embodied by the improved ductility of in-filled concrete. Uniform strength for the in-filled concrete of the section was assumed here to simplify the material model. Based on past research work [20] and by introducing section depth-to-width ratio, the simplified skeleton for the in-filled concrete was put forward as Figure 3a. In this model, the main variable is the equivalent confinement index ξ_{eq} [19], which describes such characteristics as the section shape and the ratio for steel and in-filled concrete.

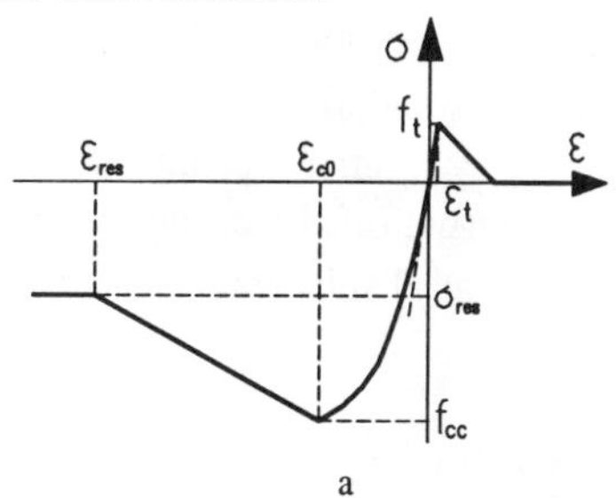

a

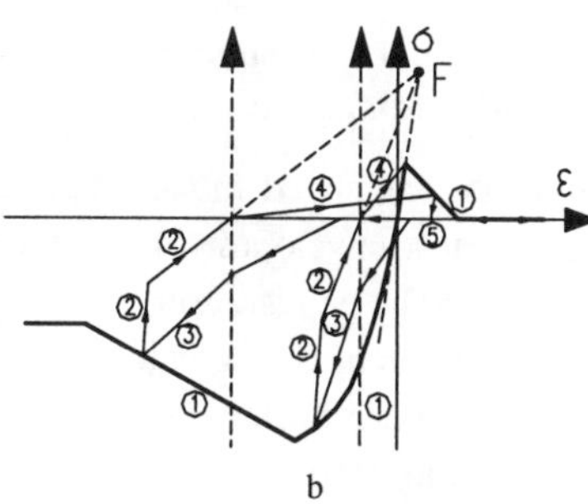

b

Figure 3. hysteretic model of stress-strain for in-filled concrete

The skeleton of the model was shown in Figure3a. The cubic equation proposed by Saenz was introduced herein to describe the ascending phase. Two straight lines were adopted to describe the softening process. The skeleton was expressed as Eq. (1), in which, E_s is the secant modulus, i.e. $E_s=f_{cc}/\varepsilon_{c0}$.

f_{cc} is the peak strength of in-filled concrete, which is computed from Eq. (2), in which f_c is 0.95 times of the cylinder strength (f_c'), and f_{ck} is the characteristic strength of in-filled concrete ($f_{ck}=0.88f_c$). ε_{c0} is the strain at peak strength of in-filled concrete, which is computed from equation Eq. (3). σ_{res} in Eq.(1) is the residential strength of in-filled concrete, which is computed from Eq. (4), and the according strain is $\varepsilon_{res}=3\varepsilon_{c0}$.

$$\sigma_c=\begin{cases} f_{cc}\dfrac{\dfrac{\varepsilon_c}{\varepsilon_{c0}}\cdot\dfrac{E_{c0}}{E_s}}{1+\left(\dfrac{E_{c0}}{E_s}-2\right)\dfrac{\varepsilon_c}{\varepsilon_{c0}}+\left(\dfrac{\varepsilon_c}{\varepsilon_{c0}}\right)^2} & (\varepsilon_c\le\varepsilon_{c0}) \\ f_{cc}-(f_{cc}-\sigma_{res})\dfrac{\varepsilon_c-\varepsilon_{c0}}{2.0\varepsilon_{c0}} & (\varepsilon_{c0}<\varepsilon_c\le 3.0\varepsilon_{c0}) \\ \sigma_{res} & (\varepsilon_c>3.0\varepsilon_{c0}) \end{cases} \quad (1)$$

$$f_{cc}=f_c\left[1.0+0.1\xi_{eq}\left(\frac{30}{f_{ck}}\right)^{0.5}\right] \quad (2)$$

$$\varepsilon_{c0}=\left(0.0019+0.0022\xi_{eq}^{0.75}\right) \quad (3)$$

$$\sigma_{res}=f_{cc}\left(0.34+0.3\xi_{eq}^{0.75}\right)\left(\frac{30}{f_{ck}}\right)^{0.25} \quad (4)$$

The equivalent confinement index of CFRT column with the section of B×D×t is computed with Eq. (5).

$$\xi_{eq}=\left[\xi_D\left(\eta^2-1\right)+\xi_B\right]/\eta^2 \qquad (\eta=D/B) \quad (5)$$

$$\xi_D=\frac{4f_y(D-t)t}{f_{ck}(D-2t)^2}$$

$$\xi_B=\frac{4f_y(B-t)t}{f_{ck}(B-2t)^2}$$

In the above equations, ξ_B is the confinement index of CFST column with the section of B×B×t, ξ_D is the confinement index of CFST column with the section of D×D×t. η is the section depth-to-width ratio.

The above skeleton has been calibrated with the experimental data of axially loaded CFRT columns [19]. The results showed that the above equation could describe the nonlinear characteristic of CFRT columns, and have the simple expression.

The hysteretic model for in-filled concrete was shown as Figure 3b, in which the procedure obeys following rules. When unloaded in tension zone, it moves with the initial tangent rigidity. When unloaded in compression zone, it moves to the half of current stress with the initial rigidity, and then aimed at the focal point F(-0.4f_{cc}, -0.4f_{cc}/E_{c0}). Here the two-lines were used to describe the unloading process. After unloaded to zero stress point, it aimed at the historically farthest point in tension zone. When reloaded in compression zone, the crack interface effect was addressed [21].

During programming, the stress-strain process was set to 6 states, which were: ①loaded along skeleton; ② unloaded in tension zone; ③unloaded in compression zone; ④ reloaded in tension zone; ⑤ reloaded in compression zone.

3. Description of the Analytical Model

3.1 *Discretization of the columns and section*

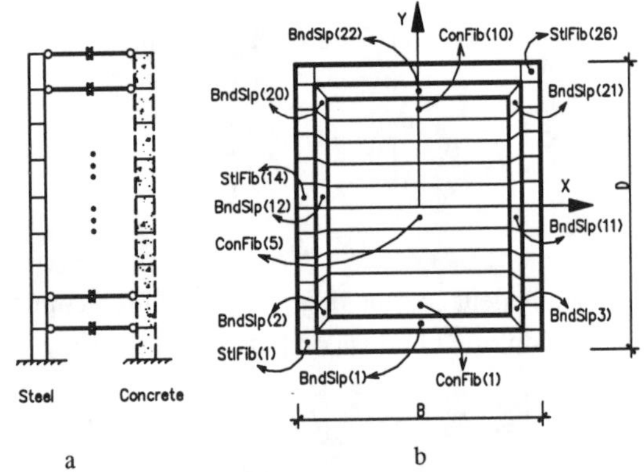

Figure 4. Discretization of the columns and section

Model of cantilever column was adopted during analysis procedure. The member was discretized to nine elements (Figure 4a). When performing nonlinear analysis on R.C. columns, the bottom element length is critical for output results. Similar to that, deformation mostly comes from the bottom element during the nonlinear phase of CFRT columns. According to Varma's tests [9], the length of the bottom element was set as 1.0Depth (when loaded along the strong axis) or 1.0B (when loaded along the minor axis) .

Section was discretized to concrete and steel fibers, and the interface was discretized to bond-slip element, which were shown as Figure 4b.

3.2 *Load-displacement full range analysis procedure*

When analyzing the force-displacement relation, section moment-curvature should be computed firstly. In this paper, uniform curvature hypothesis was adopted within an element. So the section stiffness was also uniform along the element length. When enough number of elements were discretized along the member, enough precision can also be acquired.

According to test procedure, steel tube and the in-filled concrete resist initial axial force together. During cyclic loading, they resist the axial force respectively. The interface slip comes from longitudinal displacement gap and the curvature gap between the steel tube and in-filled concrete, in which the former is the main factor [16]. The prescribed curvatures were exerted at cantilever bottom step by step during the hysteretic full range analysis. The bottom curvature for steel tube and in-filled concrete kept consist all the time. When integrated to the displacement at column top, inversion of section stiffness was avoided with this loading method, and the softening phase of load-displacement could be attained conveniently.

4. Comparison of the Experimental and Analytical Results

4.1 *Comparison of the experimental skeletons and analytical load-displacement*

In Figure5, the calculated monotonic curves were compared with the test envelopes. The capacity agreed well with each other. The gap is lower than 0.15 times of test capacity. On the ductility, the calculated ones were better than tested ones, especially for those with high axial compression ratio (0.6). This is maybe because the stress-strain model for in-filled concrete comes from the monotonic test data. For the specimens subjected to cyclic loading, the capacity of in-filled concrete degraded more seriously. Besides that, the local buckling of the tube plates and the tearing of corner welding had not been simulated in the steel model. From the test process, it was known that these have evident effect during final phase of test.

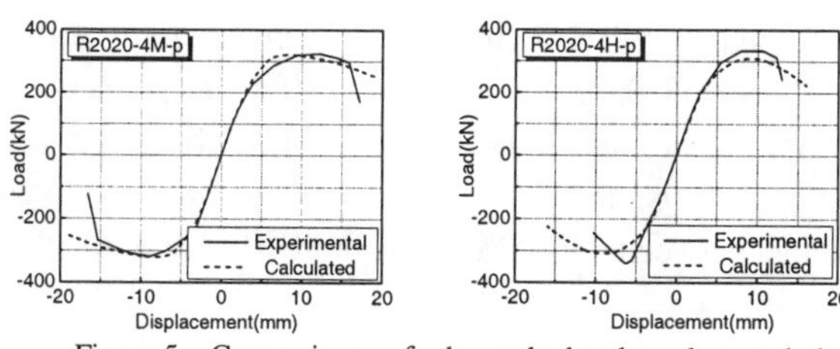

Figure 5. Comparison of the calculated and tested load-displacement envelope curves

4.2 *Hysteretic full range analysis and comparisons*

The full range analytical results were listed in following figures (Figure 6~Figure 8), where the test records were also given for comparison. Comparison of the load-displacement results was showed in Figure 6. It seemed that the models in this paper can be applied to simulate the failure process under cyclic loading, and the nonlinear development of CFRT columns was also manifested. The comparison of load-strain was listed in Figure 7. It seemed that the calculated load-strain development agreed well with the recorded ones. The comparisons of the load-axis shortening process were showed in Figure 8, from which it can be seem that the central fibers were under compression all the time. But, the compression strain development on the specimen with axial compression ratio (0.6) was faster than that of the specimen with the axial compression ratio (0.4). The strain development was concerned not only with axial compression ratio, but also with the cyclic loading process. From the comparison, it can be seen that the analytical results can simulate the hysteretic development process on macro or micro levels, and can describe most of the nonlinear characteristics of the CFRT columns under cyclic loading, i.e. the stiffness degradation, strength degradation, and hysteretic energy dissipation, etc.

From the above, it was shown that the analysis was reasonable and acceptable in this paper. So, the stress-strain simulation of steel and concrete fibers was also presented in Figure 9 & Figure 10. From the simulation results, it can be seen that the flange plates of specimen R1625-5M-p experienced tension yielding, while those of specimen R1625-5H-p have not. That maybe was the inherent reason for the two specimens with different ductility.

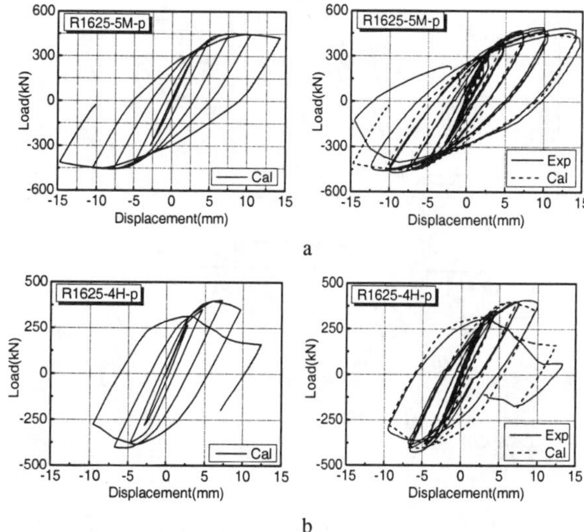

Figure 6. Comparison of the calculated and experimental load-displacement curves

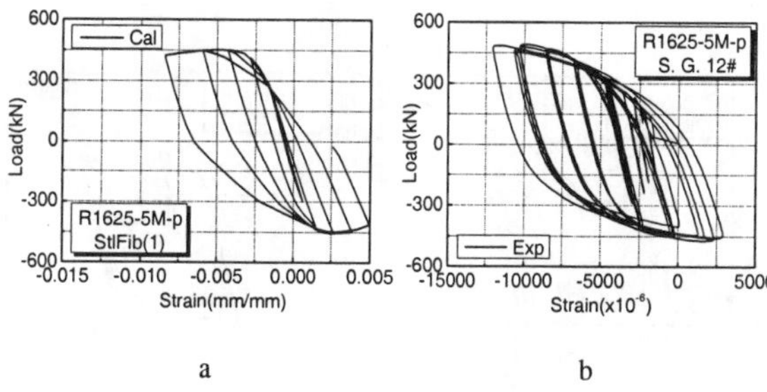

Figure 7. Comparison of the calculated and experimental load-strain curves

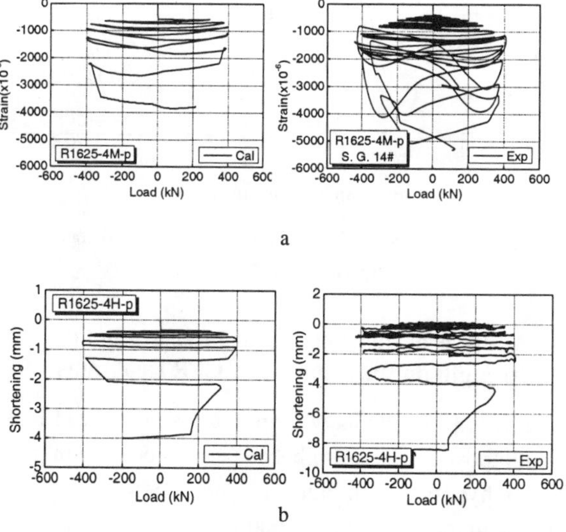

Figure8 Shortening process of calculated and experimental data

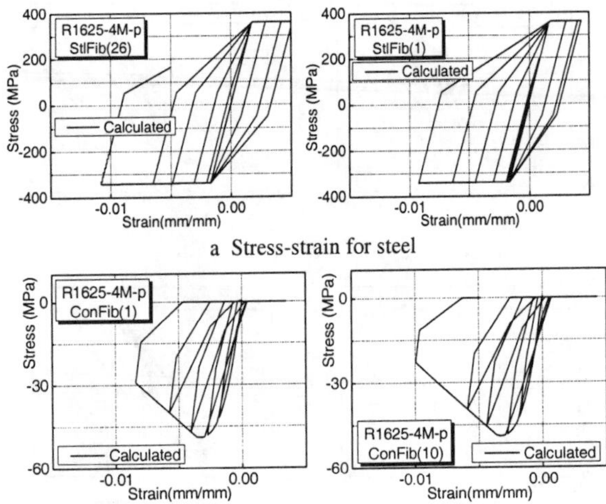

b Stress-strain for in-filled concrete

Figure 9. Calculated stress-strain for steel and in-filled concrete of specimen R1625-4M-p

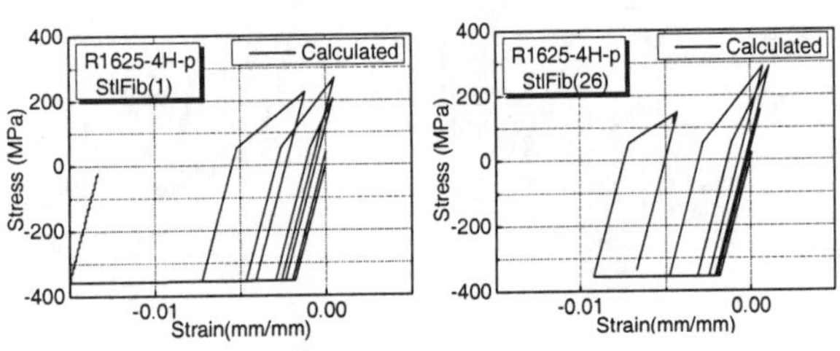

a Stress-strain for steel

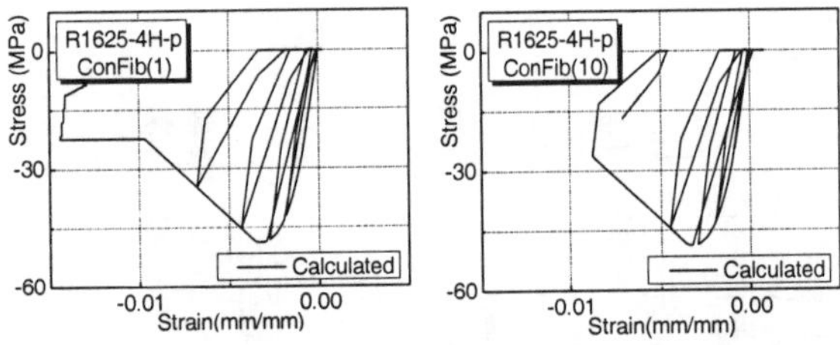

b Stress-strain for in-filled concrete

Figure 10. Calculated stress-strain for steel and in-filled concrete of specimen R1625-4H-p

5. Analysis of the effect of the interface bond strength on the behavior of CFRT columns

Here, it is to known how large bond strength can make the composite columns to behave consistently with plane section hypothesis. In Figure11, the hysteresis curves of the specimen with various bond strength were presented. From the figure, it can be seen that the peak loads (532kN & 534Kn) are nearly the same when the bond strength τ_u is 0.005MPa or 0.5MPa. When the bond strength τ_u descended to 0.002MPa, the peak load also descended to 515kN. But, for the specimens with the bond strength 0.002MPa or 0.001MPa the peak loads are nearly the same (492kN & 487kN). So, from the above analysis, it can be seen that the bond strength various between 0.005MPa or 0.001MPa has distinct effect on the peak load.

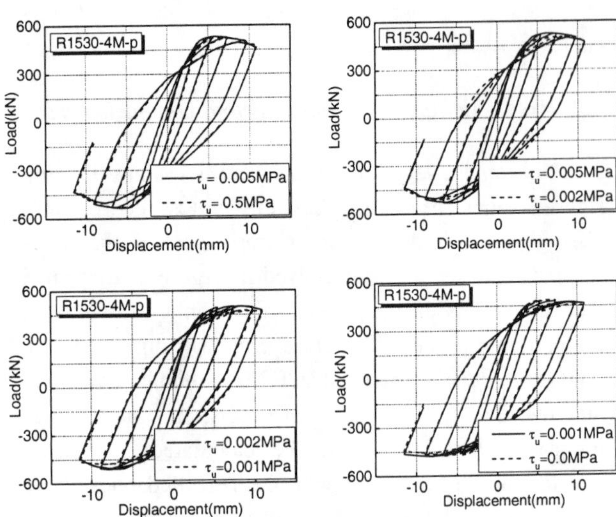

Figure 11. Effect of bond strength on the seismic behavior of R1530-4M-p

6. Design suggestions

When used in tall buildings in seismic zone, the suggested limit axial compression ratio for CFRT columns is 0.5, which can provide definite ductility. In practical engineering, CFRT columns with square sections should be preferentially selected. The concrete grade should be consistent with the steel tube grade, e.g. C30~C50 for Q235 and C40~C60 for Q345, etc.

If rigid inner diagram was welded at panel zone, the composite section capacity can be acquired, i.e. the interface bond-slip effect on the strength of CFRT columns could be neglected. But, with CFRT section depth greater than 800mm, studs or longitudinal ribs should be set at the inner side of tube plates to avoid transverse shrinkage effect of the in-filled concrete, so that the section composite stiffness and strength can be attained.

7. Conclusions

Based on the above analytical study on CFRT columns, the following conclusions can be drawn.

The models proposed in this paper were reasonable, and could be adopted in nonlinear analysis of tall buildings with CFRT members.

The analytical model of this paper could simulate nonlinear development process on member and material levels, and could be used to study the behavior of composite members.

For the CFRT members with 0.4~0.6 axial compression ratio, the interface bond strength 0.5MPa could lead to acceptable composite section strength.

References

1. H. Shakir-Khalil and J. Zeghiche, *The Structural Engineer.* **67(19)**, 346 (1989).
2. H. Shakir-Khalil and M. Mouli, *The Structural Engineer.* **68(20)**, 405 (1990).
3. H. Shakir-Khalil, *Structural Engineering Review.* **6(2)**, 85 (1994).
4. K. Sakino, H. Nakahara, and S. Morino, et al, *J. Struct. Engrg., ASCE,* **130(2)**, 180 (2004).
5. Ge, H., and Usami, T., *J. Struct. Engrg., ASCE,* **122(10)**, 1169 (1996).
6. X. Lu, Y. Yu, and K. Tanaka, et al, *Structural Engineering and Mechanics,* **10(6)**, 577 (2000).
7. X. Lu, and Y. Yu, *Structural Engineering and Mechanics.* **9(4)**, 365 (2000).
8. X. Lu, and W. Lu, 12th WCEE, paper no. 1416 (2000).
9. A. Varma, H. Ricles, and J. M. Sause, et al, *Journal of Constructional Steel Research,* **(58)**, 725 (2002).
10. X. Lu, X. P. Li, et al, 'Experimental study on the seismic behavior of CFRT columns with middle to high axial compression ratio'. *The First International Conference on Advances in Experimental Structural Engineering, AESE* 2005, Nagoya University, Japan.
11. L. H. Ichinose, E. Watanabe, H. Nakai, *Journal of Constructional Steel Research,* **(57)**, 453 (2001).
12. H. Shakir-Khalil, *The Structural Engineer.* **71(13)**, 230 (1993).
13. H. Shakir-Khalil, *The Structural Engineer.* **71(13)**, 233 (1993).
14. L. H. Xue, S. H. Cai, *Building Science,* **(3)**, 22 (1996) (in Chinese).
15. L. H. Xue, S. H. Cai, *Building Science,* **(4)**, 19 (1996) (in Chinese).
16. S. B. B. Aval, M. A. Saadeghvaziri, A. A. Golafshani, *Journal of Engineering Mechanics,* **128(4)**, 428 (2002).
17. Z. Lu, Y. Chen, *Steel structures,* **(2)**, 41 (2002).
18. X. L. Gu, F. F. Sun. 'Computer simulation of RC Structures'. The Publishing House of Tongji University, (2002) (in Chinese).
19. X. P. Li, 'Study on the Mechanical Behavior of Concrete-Filled Rectangular Steel Tubular Columns and the Application Technique of their Connections to Beams'. Ph. D Dissertation, Tongji University, (2004) (in Chinese).
20. L. H. Han, 'Concrete filled tubular structures'. Science Publishing House, (2000) (in Chinese).
21. B. L. Zhu, Z. X. Dong, 'Nonlinear analysis of R.C. structures'. The Publishing House of Tongji University, (1984) (in Chinese)

SHAKING TABLE TEST AND ANALYSIS OF A COMPLICATED CFRT FRAME STRUCTURE

CHUNGUANG MENG

State Key Laboratory for Disaster Reduction in Civil Engineering, Tongji University, 1239 Siping Road Shanghai 200092, China

XILIN LU

State Key Laboratory for Disaster Reduction in Civil Engineering, Tongji University, 1239 Siping Road Shanghai 200092, China

WENSHENG LU

State Key Laboratory for Disaster Reduction in Civil Engineering, Tongji University, 1239 Siping Road Shanghai 200092, China

BIN ZHAO

State Key Laboratory for Disaster Reduction in Civil Engineering, Tongji University, 1239 Siping Road Shanghai 200092, China

In this paper, a 1/15 scale model of concrete filled square steel tubular frame structure with dampers is tested on shaking table to study the dynamic characteristics and seismic response, especially the effect of dampers on seismic behavior of the structure. Test results show that the structure has no obvious weak story, the torsion of the whole structure is small, and the dampers play important role in energy dissipation, but the lateral stiffness of structure is not enough, because the Elasto-Plastic inter-story drift is not meet the code requirement(1/80) for energy dissipation structures. So according to the test results, the structure design is improved to meet the code requirement.

1. Introduction

This paper presents the results of a shaking table experimental study of a 1/15 scale model of a high-rise building in Shanghai that is a 21-storey complicated CFRT frame structure with 56 viscous dampers. The building is 49m by 49m in plan, with an overall height of 100m. Two floors with L-shaped installation of slabs and one with square form a unit that turns 90 degrees every three floors from bottom to top of the building, which resulting in many slender columns without lateral support of slabs in three-floor height. The dampers are installed rotating along with the unit. The typical plan views and the typical elevation view are respectively shown in Figure 1 and Figure 2.

Because of the complexity of the structure that exceeds the applicable scope of Chinese Design codes [1], the State Key Laboratory for Disaster Reduction in Civil Engineering at Tongji University is commissioned to conduct the shaking table test to study the dynamic characteristics , seismic response of the structure , especially the effect of the dampers on seismic behaviors of the building and provided some suggestions to structure design.

2. Design and Construction of Model

2.1 *Similitude relation*

Considering limitations of the shaking table and conditions of model construction, the similitude relation presented in Table 1 is chose. But, for small scale model of tall buildings, it is impossible to satisfy the similitude relation absolutely due to some reasons of technique and economy. Because the main intention of the test is to acquire the seismic response and seismic behavior of the whole structure, the similitude relation can be satisfy mainly from the perspective of members and stories rather than materials [2-3]. So additional mass-blocks are placed on every floor to meet the similitude relation of storey mass horizontally and the similitude rules of main members of the structure are as follows.

1. CFRT columns are simulated according to the axial and bending bearing capacity.
2. Steel beams are simulated according to the bending bearing capacity.

3. Braces are simulated according to the axial bearing capacity.
4. Dampers are simulated according to the energy dissipation capacity and the equation calculating the damping force is $F_d = C_d |V_c|^{\alpha} \times sign(V_c)$, where F_d is the damping force, V_c is the relative velocity of two ends of damper, C_d is damping coefficient, α is exponent of velocity. For model, $\alpha = 0.15$ and $C_d = 0.3kN/(mm/s)^{\alpha}$.

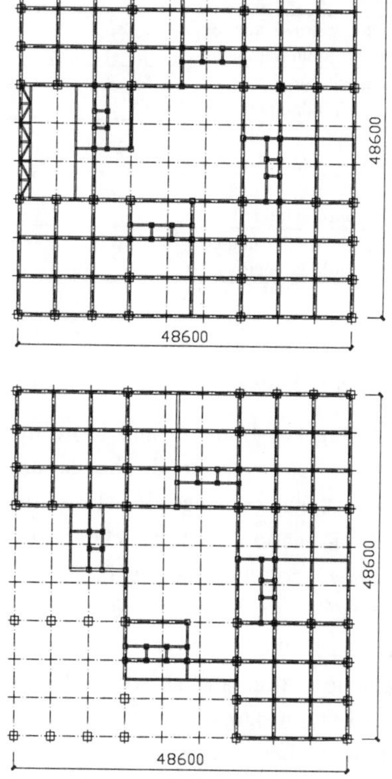

Figure 1. Typical plan views

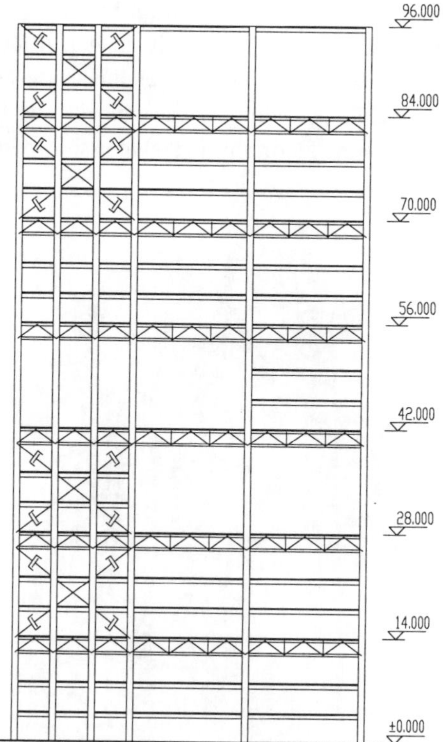

Figure 2. Typical elevation view

Table 1. Similitude relation.

Parameters	Geometry dimension	Elastic modulus	Time	Acceleration
Similitude relation	1/15	1/5	1/6.708	3

2.2 *Construction of model*

According to the similitude relation, the material of the model should have as low elastic modulus as possible and as high gravity as possible, moreover, the stress-strain relation of model material should be similar to prototype material as possible. So, micro concrete and copper are adopted respectively to simulate concrete and steel of the prototype structure. The principle of the model damper is the same to the prototype damper basically and the size of the model damper is small enough to be installed on the model structure. The construction sequence of the model is as follows. First, the copper tubes were located and fixed on the base, and then the copper beams were welded on the copper tube columns floor by floor. After all beams of one floor were welded, wire nettings were welded on the beams and micro concrete was poured to form floor slab. When all floors constructed, the micro concrete in copper tube columns was poured continuously. Final, the braces and the dampers were installed on the model. After finished, the model is 6.7m in height including base and 19t in weight including additional mass blocks and base. The photo of the model on shaking table is shown in Figure 3.

Figure 3. View of the model on shaking table

3. Test Plan

The test is conducted on MTS shaking table of the State Key Laboratory for Disaster Reduction in Civil Engineering at Tongji University. The inputted earthquake records are El Centro, Pasadena and the second Shanghai artificial earthquake wave with frequent, basic and seldom intensity for 7 degree seismic action inputted along two directions of the structure and seldom intensity for 8 degree seismic action inputted only along one direction of the structure due to the capacity limitations of the shaking table. For the stage of frequent intensity for 7 degree seismic action, tests are conducted respectively before and after the dampers are installed in order to study comparatively the effect of the dampers on the seismic response of the structure. The test stage number and the acceleration peaks of inputted seismic excitation at every test stage are shown in Table 2. The white noise scanning test is conducted before the shaking table test and after every test stage to measure the frequency, mode shapes and damping ratio of the model.

Three types of sensor are arranged on the model according to the characteristic of the structure. Eight displacement sensors are used to record the displacement response of key floors. Thirty five acceleration sensors are adopted to record the acceleration response of different floors which can be converted into displacement response though integral. Among twenty four strain sensors, sixteen are installed on important lateral force resisting members and eight are located at the connecting bars of the dampers to calculate the damping force and the energy dissipation of the dampers.

Table 2. Acceleration peaks of inputted seismic excitation

Test stage	Stage number	Acceleration peak (g)
Frequent intensity for 7 degree seismic action (without dampers)	F7-0	0.11
Frequent intensity for 7 degree seismic action (with dampers)	F7-1	0.11
Basic intensity for 7 degree seismic action (with dampers)	B7	0.30
Seldom intensity for 7 degree seismic action (with dampers)	S7	0.66
Seldom intensity for 8 degree seismic action (with dampers)	S8	1.2

4. Test Results

4.1 *Test phenomenon*

At the test stage of frequent and basic intensity for 7 degree seismic action, the model has no any damage in appearance. At the test stage of seldom intensity for 7 degree seismic action, bottom chords of many belt trusses between columns having large spacing have a little deformation out-of-plane and some welded seams at joints crack. At the test stage of seldom intensity for 8 degree seismic action, the structure damages badly in middle and bottom part of the structure, where bottom chords of many belt trusses between columns having large spacing buckle out-of-plane, oblique web members of many belt trusses between columns having small spacing buckle or break and three braces on the 8th and the 11th floor break. Results of white noise scanning tests also indicate that the frequencies begin to decrease at the stage of seismic seldom intensity for 7 degree seismic action. The typical damages are shown in Figure 4.

4.2 *Dynamic properties of model structure*

Figure 5 and Figure 6 presents the results of white noise scanning tests before and after dampers are installed, which show that after dampers installed, the damping

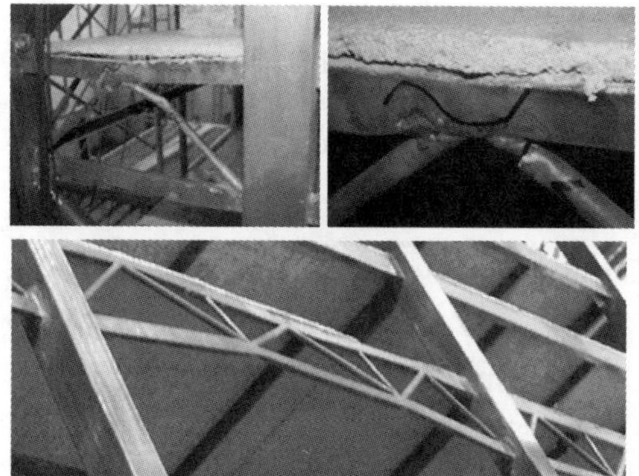

Figure 4. Typical damaged members

ratio of the model increase, but the frequencies have little change. The results of comparison indicate that the installation of viscous dampers have little effect on the stiffness of the structure, but produce additional damping for the structure.

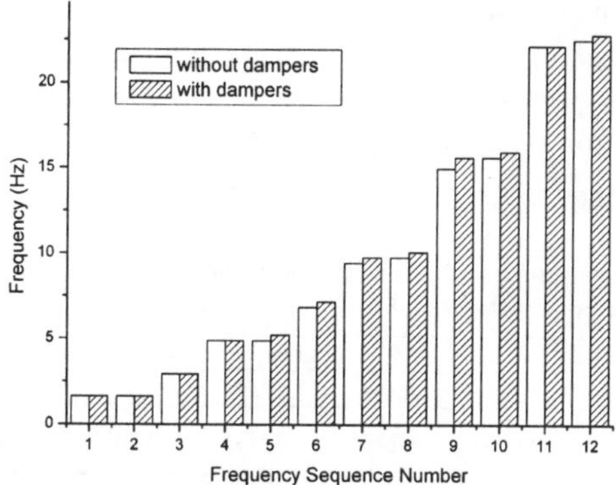

Figure 5. Comparison of frequencies of the model structure with and without dampers

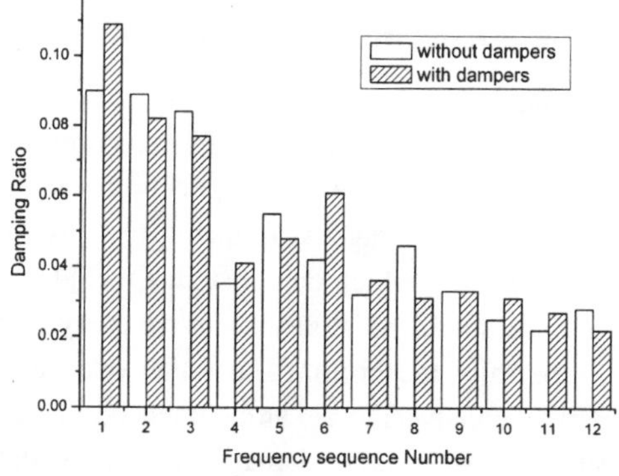

Figure 6. Comparison of damping ratios of the model structure with and without dampers

The frequencies, damping ratio and mode types of the model structure after every stage of test are listed in Table 3, which are get from the analysis of the results of white noise scanning tests. The first frequency decreases by 20% after the stage of seldom intensity for 7 degree seismic action and decreases by 40% after the stage of seldom intensity for 8 degree seismic action, which indicates that the lateral stiffness of the structure drops significantly and the structure damages seriously.

Table 3. First three frequencies, damping ratios and mode types of the model structure with dampers after every test stage.

Test stage		Frequency (Hz)	Damping ratio	Mode type
Initial state		1.628	0.109	Translation
		1.628	0.082	Translation
		2.930	0.077	Torsion
7 degree	F7-1	1.628	0.109	Translation
		1.628	0.085	Translation
		2.930	0.094	Torsion
	B7	1.628	0.142	Translation
		1.628	0.222	Translation
		2.930	-	Torsion
	S7	1.320	0.203	Translation
		1.320	0.125	Translation
		2.604	-	Torsion
8 degree	S8	0.977	0.180	Translation
		1.302	0.204	Translation
		-	-	Torsion

4.3 *Acceleration response of model structure*

Under seismic action of every level, the acceleration responses of the model structure in the X direction and the Y direction are the same basically. Figure 7 presents the envelope curves of acceleration amplification factors in the X direction of the structure, which shows that the acceleration response of the roof is much stronger than the floors at every test stage. With the increase of the inputting intensity, the acceleration amplification factors decrease due to the damage of the structure.

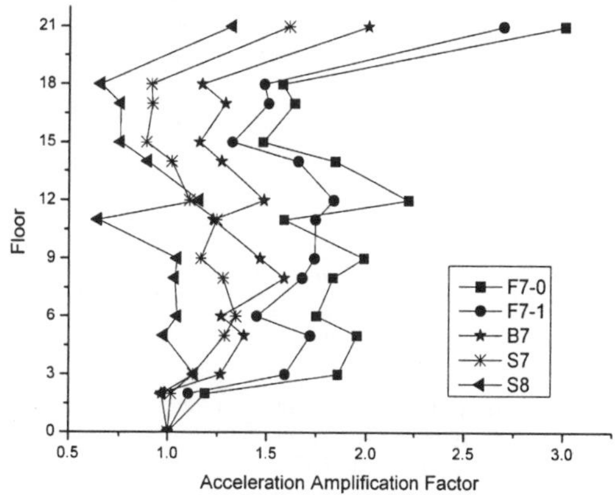

Figure 7. Acceleration amplification factor envelope curve of the model structure in the X direction at every test stage

At the test stage of frequent intensity for 7 degree seismic action, the acceleration response of the structure with dampers is smaller than the response of the structure without dampers and the maximum acceleration amplification factor of the roof decreases by 28%.

4.4 *Displacement response of model structure*

At every test stage, the displacement responses of the model structure in the X direction and the Y direction are the same basically and the lateral displacement curves, a type of flexure-shear, are smooth, which indicate that the structure has no sudden change of lateral stiffness vertically. The displacement envelope curve of the model structure at every test stage is shown in Figure 8.

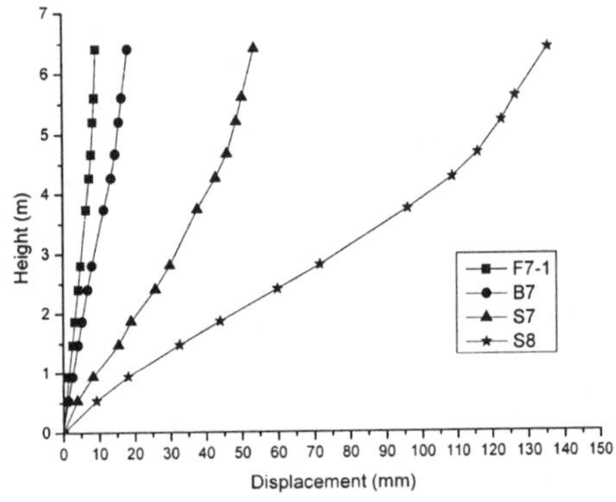

Figure 8. Displacement envelope curve of the model structure with dampers in the X direction at every test stage

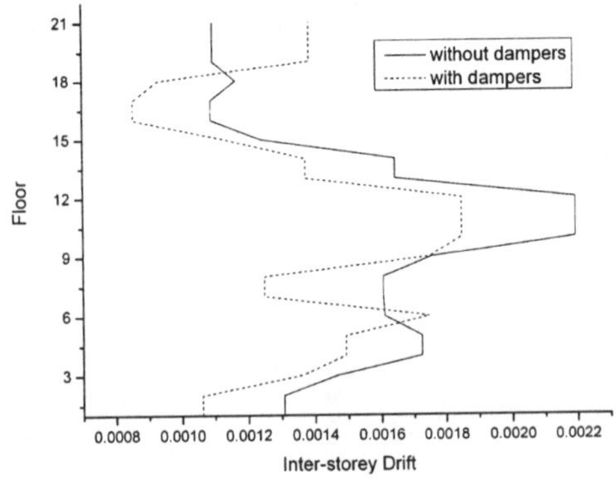

Figure 9. Inter-storey drift envelope curves of the model structure with and without dampers in the X direction at the test stage of frequent intensity for 7 degree seismic action

At the test stage of frequent intensity for 7 degree seismic action, the maximum drift of the roof and the maximum inter-storey drift respectively decrease by 12% and 19% after the dampers are installed. The inter-storey drift envelope curves of the structure with and without dampers at the stage of frequent intensity for 7 degree seismic action are shown in Figure 9.

4.5 *Torsion response of model structure*

The torsion angle of one floor can be calculated from the displacement of two tested points of the floor. Table 4 presents the maximum torsion angle and the torsion effect coefficient (the ratio between maximum torsion angle and maximum displacement angle) of the roof at every test stage [4], which shows that with the increase of seismic action, the torsion response of the structure increase, but the ratio between maximum torsion angle and maximum displacement angle of the roof decrease, which indicates that the torsion response of the structure is relative small at every test stage.

At the test stage of frequent intensity for 7 degree seismic action, after the dampers are installed, the torsion angle of roof decrease by 27%, which indicates that installing dampers is an effective approach to decrease the torsion response of the structure.

Table 4. The maximum torsion angle and torsion effect coefficient of the roof at every test stage.

Test stage	7 degree				8 degree
	F7-0	F7-1	B7	S7	S8
Torsion angle	1/1790	1/2467	1/935	1/404	1/204
torsion effect coefficient	0.34	0.30	0.35	0.36	0.23

4.6 *Storey shear response of model structure*

The storey shear distributions in two direction of the structure are similar. Figure 10 presents the storey shear distribution of the structure in X direction with and without dampers at the stage of frequent intensity for 7 degree seismic action, which demonstrates that after the dampers are installed, the storey shear of every storey decrease, especially the bottom shear that decrease by 7%. From above analysis, it is can be deduced that after the dampers are installed, the damage of the structure is reduced due to the decrease of the storey shear which results in the delay of the plastic development of the structure. Figure 11 presents the storey shear distribution of the structure with dampers in the X direction under seismic action of different levels.

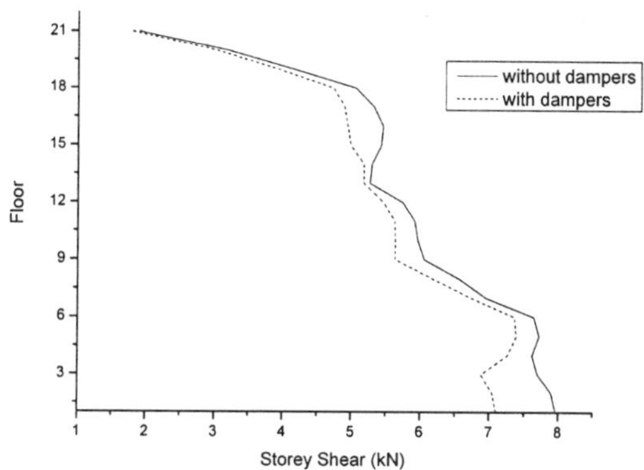

Figure 10. storey shear envelope curves of the model structure with and without dampers in the X direction at the test stage of frequent intensity for 7 degree seismic action

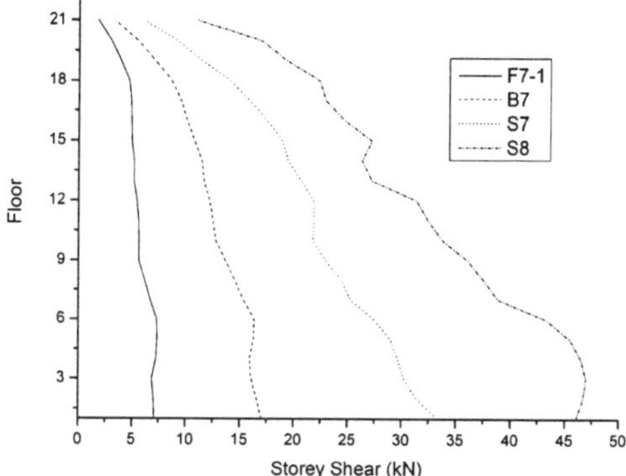

Figure 11. Storey shear envelope curve of the model structure with dampers in the X direction at every test stage

5. Seismic Behavior of Prototype Structure

The seismic properties and seismic response of the prototype structure under seismic action of every level can be calculated from the shaking table test results according to the similitude relation, and then the seismic behavior of the prototype structure can be evaluated.

5.1 *Effect of dampers on structure*

Comparing results between tests with dampers and without at the stage of frequent intensity for 7 degree seismic action, the effect of dampers on the seismic behavior of the structure is as follows.

1. After the dampers are installed, the stiffness and the frequencies of the structure have little change, but the damping ratio increase.
2. After the dampers are installed, the acceleration response of the structure decrease, which result in the decrease of the storey shear.
3. After the dampers are installed, the storey drifts, the inter-storey drifts and the torsion angle of the structure decrease.

The results of test shows that the ratio of the energy dissipation of the dampers to the energy input of the structure are 20%~25%, 28%~33%, 15%~21% and 10%~15% respectively at the test stage of frequent, basic and seldom intensity for 7 degree seismic action and seldom intensity for 8 degree seismic action, which indicate that when the structure comes into elasto-plastic state, the energy dissipation effect of the dampers is weakened.

Generally speaking, installation of the dampers improves seismic resistance capacity of the structure effectively.

5.2 *Dynamic characteristics and response of prototype structure*

According to the similitude relation, both of the first two natural frequencies of the prototype structure are 0.243Hz, corresponding to translational motion. With increase of input intensity after the stage of seldom intensity for 7 degree seismic action, the stiffness, frequencies and dynamic amplification factor decrease, but the damping ratio increase due to the damage of the structure.

Under frequent intensity for 7 degree seismic action, the maximum elastic inter-storey drift is 1/541 in the Y direction and 1/460 in the X direction that meet the code requirement (1/300) for elastic response of frame structures [1]. But under seldom intensity for 7 degree seismic action, the maximum elasto-plastic inter-story drift is 1/82 in the Y direction and 1/57 in the X direction that do not meet the code requirement (1/80) for energy dissipation frame structures [1].

6. Conclusion and Suggestion

From above analysis, it can be concluded that the structure has no obvious weak storey, the torsion response of the whole structure is small and the dampers play important role in energy dissipation , but the lateral stiffness of the structure is not enough and parts of local elements and connections are weak. So the suggestions to design are proposed as follows.

1. Increasing the lateral stiffness of the bottom parts of the structure.
2. Increasing the out-of-plane stiffness of bottom chords of belt trusses between columns having

large spacing

3. Improving connections of braces and dampers
4. Increasing oblique web members of belt trusses between columns having small spacing and improving its connections

Up to now, the structure design is improved to meet the code requirement according to the test results and above suggestions.

References

1. Code for seismic design of buildings (GB 50011-2001), China, 2001.
2. Y. Zhou, W. S. Lu and X. L. Lu, Structural Engineers. China, 14 (2003)
3. L. Xu, W. X. Shi and J. Zhang, Journal of Building Structures. China, 22 (2001).
4. B. Zhao, W. S. Lu, X. L. Lu, H. Zhang and S.C. Gu, Journal of Building Structures. China, 25 (2004).

PRELIMINARY EXPERIMENTAL RESEARCH ON THE FEASIBILITY OF POURING R.C. FRAME JOINTS WITH THE SAME STRENGTH CONCRETE AS BEAMS

LI YINGMIN
College of Civil Engineering, Chongqing University
Chongqing 400045, P. R. China

ZHENG NINA, JI SHUYAN, ZHAO XUAN
College of Civil Engineering, Chongqing University
Chongqing 400045, P. R. China

SHAN LIANG
Ove Arup & Partners Shenzhen Office, Shenzhen 518001, P. R. China

It has been a practical construction scheme to make column beam joints poured with the same strength concrete as beams. Three groups spacial column beam joints, which included 2 interior joints, 2 exterior joints, and 2 corner joints, which only joint concrete strength different in each group, are tested under cyclic loading. The failure pattern, bearing capacity, dissipative capacity and the displacement ductility parameter were investigated. The test results shown that, on the one hand, it is fesible to pour the joint with the same strength concrete as beams, on the other hand, it is necessary to check the axial compression and shear capacity of such joints. Finally, the further investigations on this field are prospected.

1. Introduction

At present, with the rapid development of tall building, the concrete of higher strength in columns is considered to be used in order to meet the needs of the cross-section of columns and the axial compression ratio under the limited value, but for beams and slabs mainly bearing bending moment, higher strength grade of concrete is not needed and suitable, [1,2] so it generally occurs that the strength of concrete in columns is higher than that in beams. The concrete in joint areas is generally as the same strength as columns and poured separately, on the one hand, which makes some troubles in construction, such as supporting forms and pouring joints,[3] on the other hand, which causes the actual strength of concrete in the end of beam higher than its designed strength, so that the mechanism of "strong column and weak beam" can not be formed indeed. Contrarily, if the strength of concrete in joint areas is equal to that in beams, we can pour the joint, beam, and slab concrete simultaneously, the difficulties of construction will be reduced and the time will be economized, and such construction scheme has become an urgent need in the engineering field, and its present difficulties and solving schemes are actively discussed in the academic field.[4,5]

According to the regulations of current codes of China for the frame joints design, for the joint concrete as the same strength as column, only bearing capacity checking in seismic shear of the core is needed. But for the joint concrete strength far lower than column, the bearing capacity of shear, axial compression, and eccentric compression need to be checked carefully. [6] It indicates that when the checking calculation satisfied the rules of codes, the joints concrete strength can be lower than column and the construction scheme is practicable. But the checking method has not been regulated in codes for bearing capacity eccentric compression, and experimental and systematical theoretical analysis is still lack about confined effects on joints of deformed beams and slabs and so on.

According location of joints, RC frame joints can be classified to four kinds which are knee joints, top story joints, exterior joints and internal joints. Generally speaking, the stress of joint in intermediate story is more complex than in top story. As a part of series researches, seismic performance of RC frame joints poured with beams' has been investigated through cyclic loading tests of six joints, the feasibility of such construction scheme of joints has been discussed.

2. Experiments

At first, axial load is applied to the column step by step till its axial compression ratio up to the requirement.

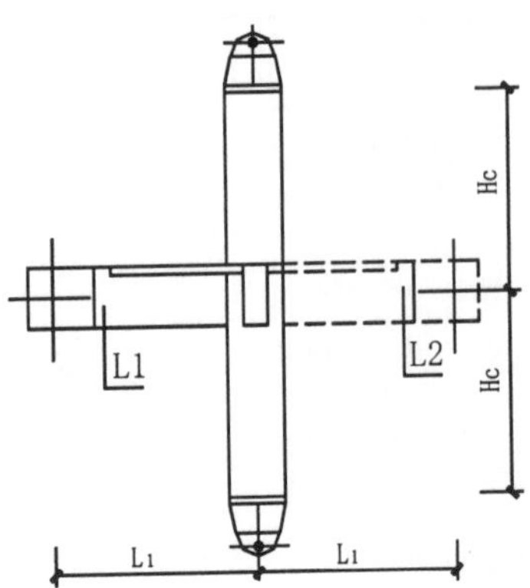

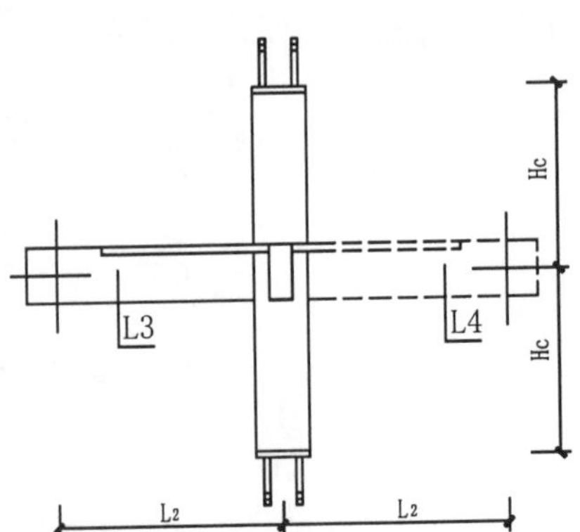

Figure 1 Figure of the specimens

Design of Specimen

Three contrast groups that six space frame joints specimens were designed, which are J-1, J-2 (the interior joints), J-3, J-4 (exterior joints), and J-5, J-6 (corner joints). Each group contain two specimens which different only in the strength grade of concrete in joint. Joint concrete of specimen J-1, J-3 and J-5 are as same strength as columns, while joint concrete of specimen J-2, J-4 and J-6 are as same strength as beams and slabs. Main parameters of specimens are shown in Table 1 and figure of the specimens are shown in Figure 1. Slabs of 50 millimeter were considered in all specimens, and concrete strength of slabs is same as beams.

2.1 *Loading and testing*

This test was done by the MTS Electro-hydraulic Servo-System test device. The loading device is shown in Figure 2, the skeleton of test in Figure 3. An oil actuated jack of 1500kN was taken to apply compress on the upper end of the column, tensile-compression jacks with 600 millimeter range and 250kN maximum force were taken to apply cyclic loading on the end of beam. Tensile or-and compression sensor were taken to measure displacement.

Shear deformations of joints were measured through cross dial indicator in joint core. Strain gauges are uniformly pasted on each limb of stirrups in joint core and also on the longitudinal reinforcement in the beams and columns in joint zone, and plastic hinge zone.

Secondly, static load is applied step by step until the longitudinal bars in beams yield, and then axial force of the column should be filled. Thirdly, low circle load is applied. Load application is controlled compound by loading force before specimen yielded and by displacement after specimen yielded.

When the peak load declined to the value less than 85% of the maximum value of the envelop curve of P-△, the specimen is considered failure.

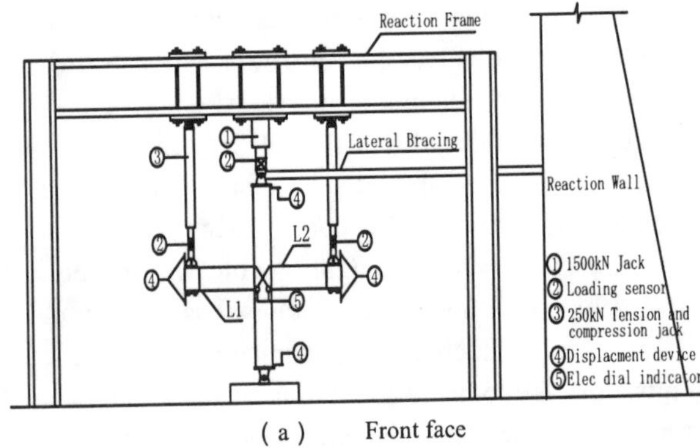

(a) Front face

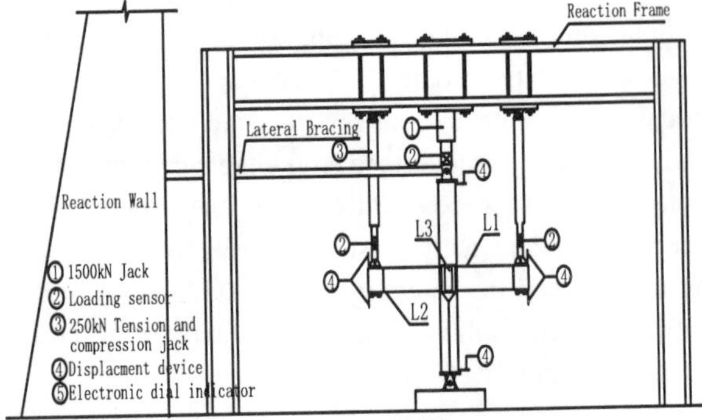

(b) Posterior face

Figure 2 Testing installation

Table 1 Parameters and reinforcement of specimens

Component	Specimen	J-1	J-2	Component	Specimen	J-3	J-4	Component	Specimen	J-5	J-6
Column	$b_c \times h_c$	350×350	350×350	Column	$b_c \times h_c$	350×350	350×350	Column	$b_c \times h_c$	350×350	350×350
	A_s	3Φ14	3Φ14		A_s	3Φ14+2Φ12	3Φ14+2Φ12		A_s	3Φ14	3Φ14
	Stirrup	φ8@100	φ8@100		Stirrup	φ6.5@75	φ6.5@75		Stirrup	φ8@100	φ8@100
	f_{cu} (MPa)	52.2	49.2		f_{cu} (Mpa)	28.7	30.2		f_{cu} (MPa)	37.6	33.3
	μ_N	0.28	0.28		μ_N	0.37	0.37		μ_N	0.31	0.32
	H_c(mm)	1175	1175		H_c(mm)	1580	1580		H_c(mm)	1200	1200
Beam L1 Beam L2	$b_b \times h_b$	170×350	170×350	Beam L1 Beam L2	$b_b \times h_b$	150×300	150×300	Beam L1	$b_b \times h_b$	250×400	250×400
	Upper A	3Φ18+2Φ16	3Φ18+2Φ16		Upper A	3Φ14+2Φ12	3Φ14+2Φ12		Upper A	6Φ18	6Φ18
	Lower A	3Φ18	3Φ18		Lower A	3Φ14	3Φ14		Lower A	6Φ18	6Φ18
	Stirrup	φ8@100	φ8@100		Stirrup	φ6.5@75	φ6.5@75		Stirrup	φ10@80	φ10@80
	f_{cu} (MPa)	33.6	35.7		f_{cu} (MPa)	19.8	21.3		f_{cu} (MPa)	21.2	19.2
	L_1(mm)	1225	1225		L_1(mm)	1225	1225		L_1(mm)	1225	1225
Beam L3 Beam L4	b×h	170×350	170×350	Beam L3	b×h	150×300	150×300	Beam L2	b×h	170×350	170×350
	Upper A	2Φ14	2Φ14		Upper A	3Φ14	3Φ14		Upper A	2Φ14	2Φ14
	Lower A	2Φ14	2Φ14		Lower A	2Φ12	2Φ12		Lower A	2Φ14	2Φ14
	Stirrup	φ8@100	φ8@100		Stirrup	φ6.5@75	φ6.5@75		Stirrup	φ8@100	φ8@100
	f_{cu} (MPa)	33.6	35.7		f_{cu} (MPa)	19.8	21.3		f_{cu} (MPa)	21.2	19.2
	L_2(mm)	1475	1475		L_2(mm)	1475	1475		L_2(mm)	1475	1475
Column-beam Joint	Stirrup	5Φ12	7Φ12	Column-beam Joint	Stirrup	4φ6.5	4φ6.5	Column-beam Joint	Stirrup	4φ8	4Φ12
	λ_V	0.083	0.169		λ_V	0.055	0.073		λ_V	0.045	0.171
	f_{cu} (MPa)	52.2	35.7		f_{cu} (MPa)	28.7	21.3		f_{cu} (MPa)	37.6	19.2
	μ_V	0.167	0.244		μ_V	0.148	0.199		μ_V	0.141	0.276
η		1.0	1.38	η		1.0	1.42	η		1.774	1.734

NOTES: μ_V —Shear Compression Ratio, $\mu_V = V_{jh}/f_c b_j h_j$, where, V_{jh} is the sheer force of joint; b_j is effective width of column-beam joint, is not less than width of column or width of beam plus half width of column; h_j is the effective height of column-beam joint, and is equal to the height of column; f_c is the design value of axial compressive strength of concrete of column-beam joint. μ_N —Axial Force Ratio; λ_V —Reinforcement bars Volume Ratio of Stirrups; η —ratio of f_{cu} between column and beam. Φ—HRB 400, Φ—HRB 335, φ—HPB 235. f_{cu} —the mean measured value of concrete.

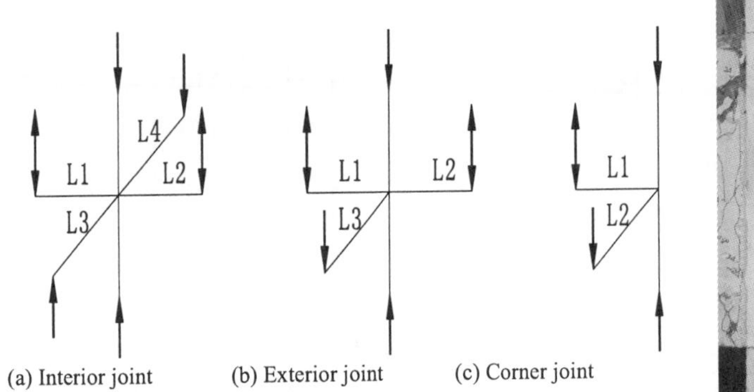

Figure 3 Skeleton of tests

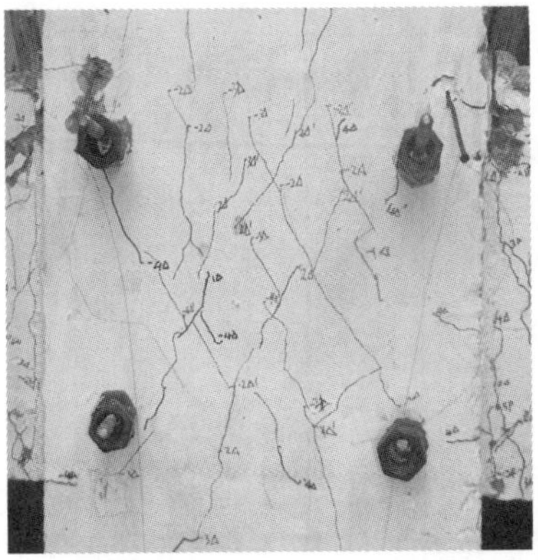

Figure 5 Final failure pattern of specimen J-4

Test Phenomenon

In the test, when longitudinal bars in beams yielded, there were some bending cracks and incline bending-shear cracks at the ends of beams. While applying positive load in the first cycle, incline shear cracks appeared in the front of the joint region and extended to the upper column end. While applying reversal loading in the first cycle, reversal shear cracks appeared in the front of the joint and extended to the top column. Cracks appeared and developed constantly with the cyclic times increasing.

When the specimen failed, for specimen J-1 to 4, the joint core concrete were not crushed, but for specimen J-5 and J-6, the bars in column had been yielded, while the whole joint region looked like a lantern. In the course of test, the split cracks of columns caused concrete on the cover of the joints separating from the core concrete.

The final failure patterns of some specimens are shown in Figure 4~Figure 7.

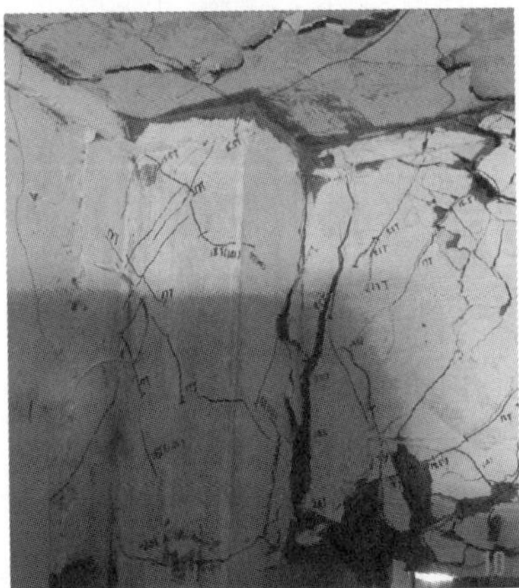

Figure 4 Final failure pattern of specimen J-1

Figure 6 Final failure pattern of specimen J-5

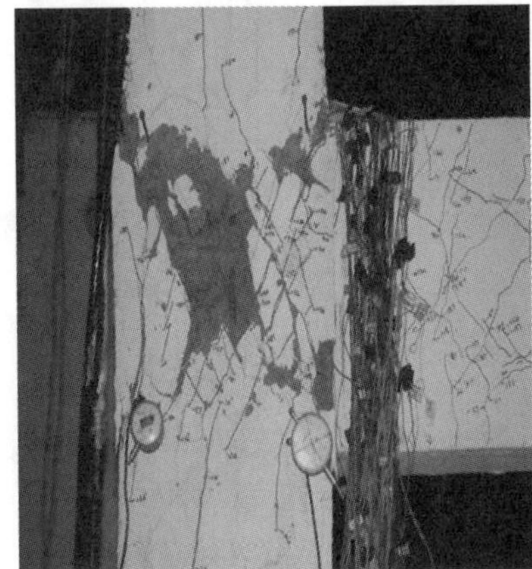

Figure 7 Final failure pattern of specimen J-6

Results and Analysis

In order to analyze more clearly of the test, the failure pattern as well as the ductility parameter of the specimen are given in Table 2. The P-△ curves of contrast specimen are shown in Figure 8 and 9. The hysteretic curves of each specimen are shown in Figure 10 to 15.

Table 2 Main test results

Specimen	Load at first crack in joint	Failure pattern	ductility factor
J-1	Beam L2→10 Ton	Shearing fail of beam	>2
J-2	Beam L2→7 Ton	Shearing fail of beam	3
J-3	First μ_Δ=-2	Flexural fail of beam	8
J-4	First μ_Δ=1	Flexural fail of beam	>6
J-5	Beam L1→15 Ton	Shearing fail of joint after beam yielding	4.3
J-6	Beam L1→15 Ton	Shearing fail of joint after beam yielding	3.6

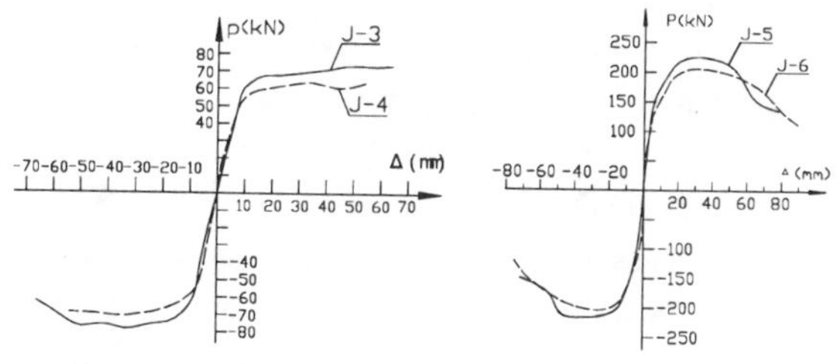

Figure 8 P-Δ curves of specimen J-3 and J-4

Figure 9 P-Δ curves of specimen J-5 and J-6

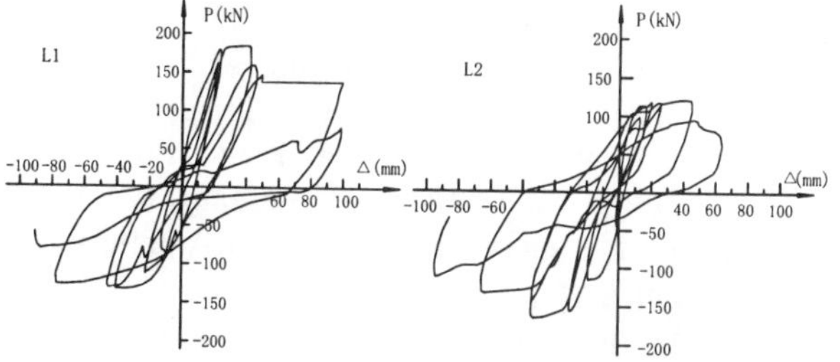

Figure 10 Hysteretic loops of specimen J-1

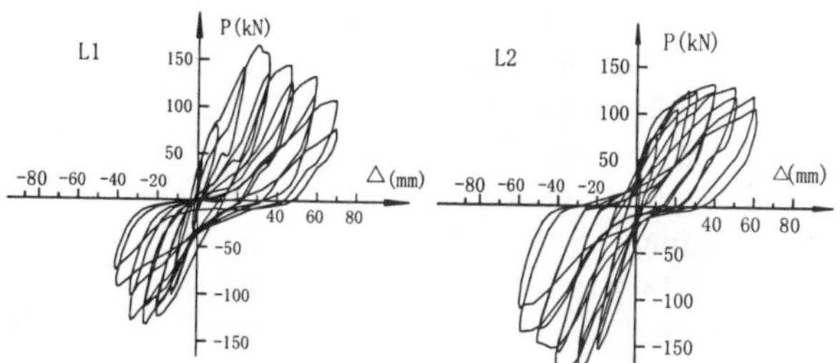

Figure 11 Hysteretic loops of specimen J-2

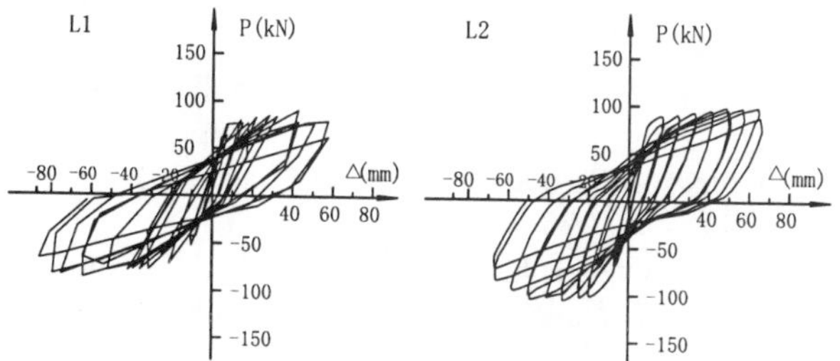

Figure 12 Hysteretic loops of specimen J-3

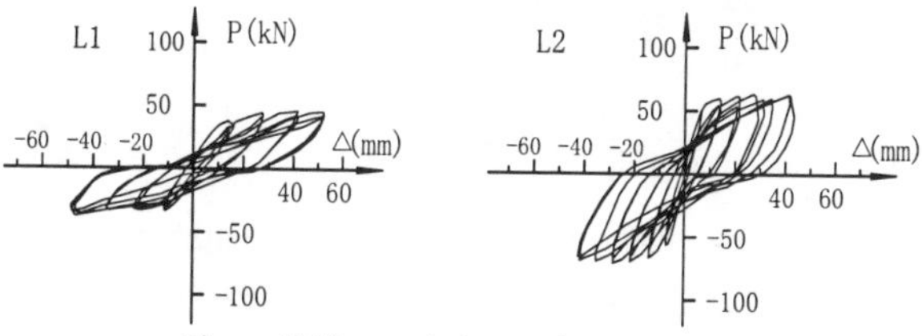

Figure 13 Hysteretic loops of specimen J-4

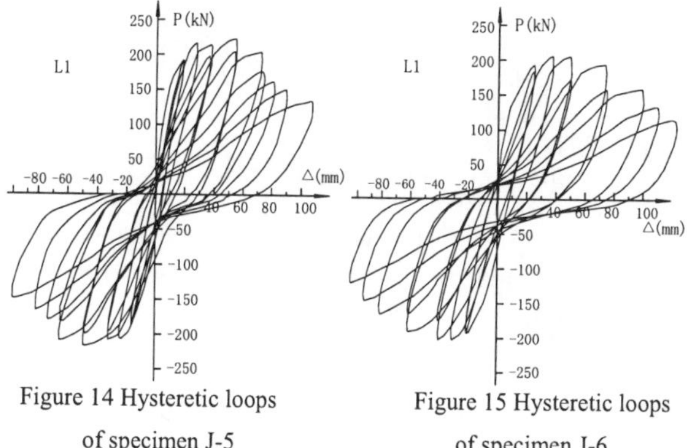

Figure 14 Hysteretic loops of specimen J-5

Figure 15 Hysteretic loops of specimen J-6

The following results can be drawn through the tests. It is necessary to note that: joint concrete of specimen J-1, J-3 and J-5 are as same strength as columns, while J-2, J-4 and J-6 are as same strength as beams and slabs.

1. Failure pattern

The failure patterns of the pair of comparison specimens are basically same. Although specimen J-1and J-2 both failed by shear failure of beam, J-3 and J-4 both failed because of bending failure of beam, it is tribally worthy to pour the joint concrete with the same strength as beams and slabs, with compress and shear capacity check design of joint.

2. Bearing capacity of joint

Comparing with specimen J-1, 3, and 5, bend and shear capacities of beams in specimen J-2, 4, and 6 declined a little, 10~15% approximately, without any additional measures.

3. Bond deterioratio of longitudinal reinforcement

In specimen J-2, 4, and 6, the bond deterioratio of longitudinal reinforcement in columns and beams aggravated slightly, but not notability, so it should not be the main influencing factor to shear capacity of joint.

4. Stirrup stress

there are not obvious difference of the stirrup stress in two contrast specimen, for the stirrup were layout by shear design, so that the area ratio of stirrup to joint of specimen J-2, 4, and 6 are higher than specimen J-1, 3, and 5. On the other hand, the stirrup stress in joint core was mainly affected by the constraint of beams and slabs to joint, stronger constraint, lower stress.

5. Dissipative capacity of joint

According to the results of the contrast tests, the dissipative capacity of J-2, 4, and 6 did not obviously felled down. More tests are needed to testify if it is a general rule.

6. Displacement ductility

Comparing with specimen J-1, 3, and 5, the dehiscence of joint core were earlier in specimen J-2, 4, and 6, and the bonding degeneracy of longitudinal bars in columns and beams were much severely. So that the yielded displacement of the end of beams in specimen J-2, 4, and 6 are higher and the corresponding displacement ductility coefficients were smaller (table 2). It indicated that the checking design of joint is necessary to ensure require seismic performance when joint concrete is as the same strength as beams and slabs.

Conclusions

Some conclusions can be addressed as following from the preliminary 3 groups of tests:

1. It's a feasible construction method to pour the joint concrete with the same strength as beams and slabs, and it is worthy to make further studies on it.
2. When the joints are designed according to the present Chinese Codes, for joints poured concrete with the same strength as beams and slabs, the failure process and the final failure pattern has little difference with the conventional joints, but the bearing capacity, seismic performance, and the displacement ductility has declined a little.
3. It is necessary to check the axis-compression capacity and shear capacity of joint core, when the concrete strength in joints is same as beams and slabs.
4. When pouring the joint concrete with the same strength as beams and slabs, the key problem is to ensure the seismic performance of the joint with high axis-compression ratio and high shear-compression ratio. It is difficult to make axis-compression ratio and shear capacity of the joint to meet the requirement of codes. While the shear capacity of joints core can be easily sufficient by increasing stirrup, the high axis-compression ratio is hard to satisfy for the ratio of column is up to the limited value in practice. And there need some strengthen measures, such as adding short bars and horizontal haunch for beams to improve axial compress capacity of joint core.
5. It is necessary to make further tests and studies to answer series problems, such as how to check the axis-compression ratio of joint, how to definite the strength of restrained concrete in joint core, and so on.

References

1. Technical specification for concrete structures of tall building (JGJ3-2002) (in Chinese), Beijing, China Architecture & Building Press, 2002
2. Code for Seismic Design of Buildings (GB50011-2001) (in Chinese), Beijing, China Architecture & Building Press, 2001
3. Qiu Yushen, Disposal of the Joint Core of the RC Frame during Construction(in Chinese), Construction technology, 32(2),96-97,2001
4. Zhang Yuankun, Strength Check and Construction Disposal of Column Joint of Frame of Tall Building (in Chinese). Gungdong Architecture Civil engineering, (12),3-6,2002
5. Cheng Maokun, Disposal method of beam-column joint of high strength concrete column (in Chinese). Building Structure, 31(5),3-5,2001
6. Code for design of concrete structures(in Chinese), Beijing, China Architecture & Building Press, 2002
7. Fu Jianping, You Yun, Bai Shaoliang, Load transfer analysis of RC seismic frame joints (in Chinese). Journal of Chongqing Jianzhu University, (2),43-51,1996
8. Fu Jianping, You Yuan, Bai Shaoliang, Sort of stress characteristic of RC seismic frame joints (in Chinese), Journal of Chongqing Jianzhu University, (2) 86-91,1996
9. You Yuan, Fu Jianping, Bai Shaoliang, Tang Hua, Ductility design criteria of beam-column joint of RC seismic frame(in Chinese), Journal of Chongqing Jianzhu University, (4) 12-20,1996

10. Fu Jianping, Research of earthquake resistance and design methodology of RC seismic frame joint(in Chinese), Thesis submitted to Chongqing Jianzhu University for Doctor Degree, 2002
11. T.Paulay, M.J.N.Priestley, Earthquake-resistance Design of Reinforced Concrete and Bricking-up Structure(in Chinese), China Architecture & Building Press, 1996,6
12. Shan Liang, Experimental Research and Analysis of Seismic Performance of Frame Joints Poured with Beams' Concrete, Thesis submitted to Chongqing Jianzhu University for Master Degree, 2004

EXPERIMENTAL STUDY ON SUPER HIGH RISING BUILDING

XILIN LU

State Key Laboratory for Disaster Reduction in Civil Engineering Tongji University, Shanghai 20092, China

YUN ZOU

Civil Engineering department Southern Yangtze University, Wuxi Jiangsu 214122, China
State Key Laboratory for Disaster Reduction in Civil Engineering Tongji University, Shanghai 20092, China

WENSHENG LU

State Key Laboratory for Disaster Reduction in Civil Engineering Tongji University, Shanghai 20092, China

BIN ZHAO

State Key Laboratory for Disaster Reduction in Civil Engineering Tongji University, Shanghai 20092, China

The height of 101-story Shanghai World Financial Center Tower is 492m above ground making it possible the tallest building in the world when completed. Three parallel structural systems including mega-frame structure, the reinforced concrete and braced steel services core and the outrigger trusses, are combined to resist lateral loads. The building could be classified as a vertically irregular structure due to a number of stiffened and transfer stories in the building. According to Chinese Code —Technical specifications for concrete structures of tall building (JGJ3-2002), the height of the building clearly exceeds the stipulated maximum height of 190m for a composite frame/reinforced concrete core building. The aspect ratio of height to width also exceeds the stipulated limit of 7 for seismic fortification intensity 7. A 1/50 scaled model is made and tested on shaking table under a series of two-dimensional base excitations with gradually increasing acceleration amplitudes. This paper presents the dynamic characteristics, the seismic responses and the failure mechanism of the building. The test results demonstrate that the structural system is the optimum solution to withstand earthquakes. The inter-story drift and the overall behavior meet the requirements of Chinese Code. Furthermore, weak positions under seldom-occurred earthquakes of seismic fortification intensity 8 are pointed out based on the visible damages, and corresponding suggestions are proposed for the engineering design purposes.

1. Introduction

Since the structural design for the slender building is dominated primarily by the effect of lateral load derived from wind and earthquake, Shanghai World Financial Center Tower (SHWFC Tower) employs a new structural system with a mega-frame structure frame and a reinforced concrete core connected by outrigger trusses. While all of the components of the structural system for the building have been used in other tall buildings, the precise combination of systems now planned for the Shanghai World Financial Center Tower is quite different from others.

SHWFC Tower could be classified as a vertically irregular structure due to a number of stiffened and transfer stories along the height. The peculiar design of its services core, mega-diagonals and outrigger trusses, which will be described in detail in the latter part, make SHWFC Tower become an exceptional example. According to Chinese Code —Technical specifications for concrete structures of tall building (CCSDB, JGJ3-2002)[1], the height of Shanghai World Financial Center clearly exceeds the stipulated maximum height of 190m for a composite frame/reinforced concrete core building. The aspect ratio of height to width also exceeds the stipulated limit of 7 for seismic fortification intensity value of 7 at the building site (The Chinese earthquake intensity is roughly equivalent to the Modified Mecalli Intensity MMI). Thus, it is significant to completely understand the overall structural behavior under moderate and strong earthquakes when designing the building. In this regard, shaking table model test plays an important role in obtaining the general dynamic characteristics, seismic responses and failure mechanism of targeted structure. It should be noted that relatively small number of shaking table tests on tall buildings are executed and published due to the difficulties in modeling scaled material properties, the financial restrictions and limitations of specimen size and capacity of available shaking tables. This paper presents a detailed shaking table test of scaled model on SHWFC Tower. The study attempts to provide some insights into the general dynamic behavior of the new structural system and accumulates the experimental evidence for establishing related design guidelines for such complicated structures in the future.

2. Description of Structure

101-story SHWFC Tower is a mixed-use skyscraper with height of 492m. It is located in Lujiazui Financial and Trade district, Shanghai, China. The structure is diagonally symmetrical with a square base plan of 57.95m by 57.95m. The aspect ratio of height to width is 8.49. Several important characteristics of the structural layout are illustrated as follows.

1. Three parallel structural systems, the mega-frame structure consisting of the mega-columns, the mega-diagonals, and the belt trusses; the reinforced concrete and braced steel services core; and the outrigger trusses, which create interactions between the services core and the mega-structure columns, are combined to resist lateral loads (Figure 1).

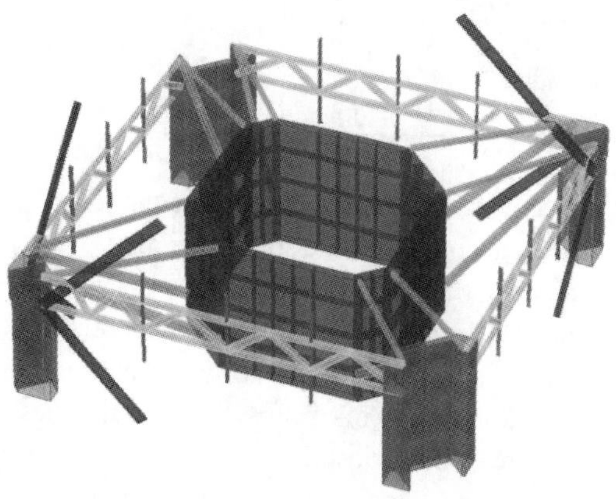

Figure 1. Three parallel structural systems

2. Perimeter concrete wall locates at lower levels from floor 1 to floor 5, and mega-columns are designed at the corners of the building from floor 6.
3. A number of stiffened and transfer stories in the building are regularly spaced throughout the height of the building. One-story high belt-trusses and core transfer trusses are placed at each twelve-story interval, whereas three 3-story outrigger trusses spanning between the mega-structure columns and the corners of the concrete services core are distributed along the height.
4. The continuity of services core is broken by three basic configurations of the walls. The lower (floor 1 to floor 59) and the middle (floor 60 to floor 79) services core are of reinforced concrete services core. Embedded structural steel columns from the middle services core into the lower services core and the boundary elements extend at least 12 floors beyond the overlap where the core geometry changes. Strengthened floor diaphragms are provided from floor 55 through floor 59 to enable proper shear transfer between the middle and the lower services cores. The upper services core is of structural steel with concrete encasement at the two ends of the core (above floor 79). Steel trusses adopted along the long walls will reduce the structural self-weight compared with concrete wall. Structural steel columns and truss-work at the two ends of the services core are encased in concrete. A strengthened floor diaphragm is provided at floor 79 in order to transfer proper shear between the upper and middle services core.
5. The outrigger truss system consisting of three three-story height space frames are again regularly distributed to engage the mega-frame structure with the services core. Due to constraints imposed by the architecture, it is impractical for outrigger trusses to pass the services core directly. They are connected to the embedded core perimeter trusses and the mega-columns at two ends. At the outrigger floors, a core perimeter truss is embedded in the core walls to provide the necessary back-spans for the outrigger trusses. The corner columns of the embedded core perimeter truss extend throughout the height of the services core. The outrigger truss and the core perimeter truss consist of welded structural steel sections.
6. The mega-diagonals, extending to the top of the tower, are an important feature of the three-dimensional braced frame. These mega-diagonal members consist of concrete-filled built-up steel box sections. The single-diagonal system is selected for a more desirable interior space and a more aesthetically pleasing exterior facade.

7. Another important aspect of the mega-frame structure layout is the connection of the braced frames on the four vertical faces of the building. The addition of bracing on the curved façade is also undesirable to the exterior appearance of the building, and would interfere with the view from the interior. Therefore, only single-diagonal system is adopted on the vertical faces in the mega- frame structure instead of providing braced frames at the exterior curved surfaces of the building.

3. Model Experiment

3.1 *Description of the shaking table*

Shaking table test model is carried out using MTS shaking table facility at State Key Laboratory for Disaster Reduction in Civil Engineering, Tongji University, Shanghai, China. The table can input three-dimensional and six degree-of-freedom motions. The dimension of the table is 4m × 4m, and the maximum payload is 25,000kg. The shaking table can vibrate with two maximum horizontal direction accelerations of 1.2g and 0.8g, with a maximum acceleration of 0.7g vertically. Its frequency ranges from 0.1 Hz to 50Hz and there are 96 channels available for data acquisition during testing progress.

3.2 *Model similitude and materials*

Model is designed by scaling down the geometric and material properties from prototype structure. The model structure is constructed using micro-concrete, fine wires, copper to simulate concrete, reinforcing bars and steel, respectively. The basic model similitude rules are established from the scaling theory, which is generally known[3] and therefore there is no need for further explanation of the theory. Since the dynamic behavior of a structure is fully described by means of three basic quantities, only three independent parameters can be selected when designing a model. Subsequently, other parameters are expressed in terms of the basic scale factors chosen. Considering the capacity and the size of the shaking table to be used at Tongji University, the dimension scaling parameter (S_l) is chosen as 1/50. Since the general scaling stress parameter for concrete ranges feasibly froml/5 to 1/3 under laboratory environment by gradually adjusting fine-aggregate grade and the general scaling stress ratio of copper to steel is roughly equal to 1/2, the preliminary scaling stress parameter is set to be 1/3.6 for model deign. Subsequently, the third parameter (S_a) concerning the horizontal acceleration should be determined. Foregoing experiment evidences indicate that properly amplified acceleration are demanded to ensure the accurate measurement. Thus, S_a is selected to be 2.5 in this test. In order to obtain the realistic material properties, an extensive model material investigation is conducted after model fabrication. The final scaling stress parameter ($S_{\sigma con}$) for concrete of 1/3.846 is obtained, whereas the stress scaling parameter ($S_{\sigma cop}$) of copper is 1/1.873. Since the prototype structure is a concrete/steel composite structure, the final stress scaling parameter (S_σ) is determined as 1/3.125 by consideration two main model materials properties. The final general scaling stress parameter is used to modified the related other scaling parameters and the test program is designed on the basis of the final similitude scaling relationship. The main scaling parameters are presented in Table1.

Table I Similitude scale factors for the test model

Parameter	Relationship	Model/prototype
Length	S_l	1/50
Young's Modulus	S_E	0.32
Stress	$S_\sigma = S_E$	0.32
Strain	S_σ / S_E	1
Density	$S_\sigma / (S_a \cdot S_l)$	6.4
Force	$S_\sigma \cdot S_l^2$	1.28E-04
Frequency	$S_l^{-0.5} \cdot S_a^{0.5}$	1/11.18
Acceleration	S_a	2.5

It is noteworthy that the following equation should be established according to the similitude rule:

$$a_m / a_p = g_m / g_p \quad (1)$$

Where, "a" and "g" represent the horizontal acceleration and the gravitational acceleration, respectively. Subscript "m" refers to the quantities of model structure, while subscript "p" corresponds to the prototype structure. However, the Equation (1) is no longer feasible unless the model is excited in a centrifuge. In this case, adoption of the artificial mass becomes the only choice to fulfill the similitude requirements. In practical testing, this is achieved by addition of suitably distributed weights, which are attached to the model in a manner that does not change the strength and stiffness characteristics. In Chinese Code of Technical specifications for concrete structures of tall building (JGJ3-2002)[1], occasional eccentricity between the mass centroid and stiffness center of each floor should be considered in structural design even for symmetric structure in order to take into account the torsional effects. The eccentricity (e_i) is calculated as the following equation:

$e_i=0.05L_i$ (2)

Where L_i refers to the maximum length of each plan. Therefore, The amount of additional floor weight is calculated according to the similitude laws and placed on the floor levels in a manner to meet the requirement of Equation (2) in the test model. Apart from the artificial mass on floors, additional mass is also added to mega-diagonals by attaching leaden sticks to preserve a consistency of mass similitude relationship for important structural components. The total additional mass being attached to the model is 14,700 kg, while the self-weight of the model accounts for 3,500kg.

3.3 *Test set-up and procedure*

To ensure an effective transmission of the table motion to the base of the test structure, the model base plate was firmly mounted on the shaking table through bolt connections. Figure 2 demonstrates an overview of the model structure after the test set-up.

Figure 2. Shaking table test model

The instrumentation is organized so that both overall and local responses of interest could be measured, including accelerations measured by accelerometers, displacements measured by LVDTs and strains measured by strain gauges. Total of 40 accelerometers are placed at floor levels along the height. Apart from those accelerometers on base, stiffened, transfer stories and roof level, other accelerometers distribute as regularly as possible throughout the height. Two accelerometers are set at one corner of the floor along the directions of X and Y of shaking table at each measured level, respectively. Additional two accelerometers along the direction of X are arranged at the opposite corer on roof and floor 60, respectively, for measurement of torsional effects. 9 LVDTs are adopted in the similar manner as that for accelerometers. Amount of 25 strain gauges are placed on mega-columns, belt trusses both at top and bottom chords and mega-diagonals near the joint sections. Therefore a total of 74 channels of data are recorded. All the test data were collected by a computer-controlled data acquisition system and can be transferred to other PC computers for further analysis.

In order to evaluate the dynamic performance under different seismic motions, three earthquake time histories of acceleration are selected as the input data: El Centro wave (1940, NS), San Fernando (1971) and Shanghai design code specified artificial earthquake accelerogram (SHW2, 2003), which is specified in Shanghai Code For Seismic Design of Buildings (SHCSDB, DGJ08-9-2003)[4]. Three different requirements related to the three levels, are set to evaluate the overall capacity of structure under corresponding intensity. As mentioned in above section, the fortification intensity in Shanghai is specified as 7. The test is carried out in four phases representing frequent, basic and seldom occurrences of fortification intensity 7, and seldom occurrence of fortification intensity 8, respectively. The last phase is programmed peculiarly for the further investigation of dynamic responses of the targeted tall building under especially strong earthquakes. The gradually increasing amplitudes of base excitation, which are determined according to SHCSDB, are input successively in a manner of two-dimensional time-scaled earthquake waves. After different series of ground acceleration are input, white-noise is scanned to determine the natural frequencies and the damping ratios of the model structure. The peak value of the white-noise input is selected to 0.05g in order to keep the model in linear elastic deformation. It can be observed that the programmed input acceleration amplitudes differ slightly from measured accelerations of shaking table due to limited abilities of shaking table control system.

3.4 *Test results*

3.4.1 *Cracking and failure*

The test model survived from frequent occurred 7 to seldom-occurred intensity 7 without visible damage. Nevertheless, the gradual decreases of the natural frequencies of the model structure measured suggest that the intrinsic damage within the structure progressively developed during phases of basic intensity 7 and seldom-occurred 7. Major damages eventually occurred and propagated under seldom-occurred earthquake intensity 8, which are demonstrated in Figure 3.

Figure 3. Concrete on Mega-Column crush

3.4.2 *Dynamic characteristics*

The natural frequencies and damping ratios are obtained from white-noise scan tests that preceded main test programs. The fundamental frequencies of model structural are presented in Table 2. The first two modes are translations of the direction X and Y, respectively, and the third mode is overall torsion. The frequency remains constant during the first series of tests, which reveals that no damage occurred in the structure. After the model withstands the second series of tests referring to basic intensity 7, the first natural frequency decreases slightly, which suggests that the structure still remain in elastic state. In the third stage, the structure sustains a strong earthquake resulting in 12% decrease of the first natural frequencies, which demonstrates that the intrinsic damage occurs even though no visible crack is observed in the model surface. The first natural frequency drops faster after the input of seldom-occurred earthquake intensity 8, which is in good correlation with the observed damage to the testing model.

During the testing runs referring to three occurred earthquakes of seismic fortification intensity 7, the first two natural frequencies are equal to each other, which indicates that the overall rigidities of direction X and Y are equivalent.

Taking into account Chinese Technical Specification for Concrete Structures of Tall Building (JGJ3-2002)[1], the ratios of the first two periods in two horizontal directions to the first-order torsional period are required within 0.9 to prevent the excessive torsional effect in the whole structural vibration. In this test, there are,

$$T_3/T_1 = T_3/T_2 = 2.306/5.290=0.436 <0.90 \quad (3)$$

Where T_1, T_2 , T_3 corresponds to the first three periods, respectively. Thus, the studied structure meets the related provision specified in Chinese code of CCSDB.

Table 2. Natural frequency, damping ratio and vibration mode

Initial	$Frequency_m$	2.306	2.306	5.290
	mode	X translation	Y translation	Torsion
	$Frequency_p$	0.185	0.185	0.424
Frequent 7	$Frequency_m$	2.306	2.306	5.290
	mode	X translation	Y translation	Torsion
	$Frequency_p$	0.185	0.185	0.424
Basic 7	$Frequency_m$	2.170	2.170	7.687
	mode	X translation	Y translation	Translation of X
	$Frequency_p$	0.174	0.174	0.688
Seldom 7	$Frequency_m$	2.035	2.035	7.189
	mode	X translation	Y translation	Translation of X
	$Frequency_p$	0.163	0.163	0.576
Seldom 8	$Frequency_m$	1.728	1.623	3.847
	mode	X translation	Y translation	Torsion
	$Frequency_p$	0.130	0.141	0.308

Note :subscript m and p refer to frequency of model structure and prototype structure, respectively.

4. Dynamic response of prototype building

Modeling scale factors, given in Table1, can be used for referring the model test results to prototype building. Natural frequencies of prototype structure are calculated and listed in Table2. The values of story shear are calculated on the basis of the measured accelerations. The maximum story displacement distributions along the height under three earthquake waves are demonstrated in Figure 4.

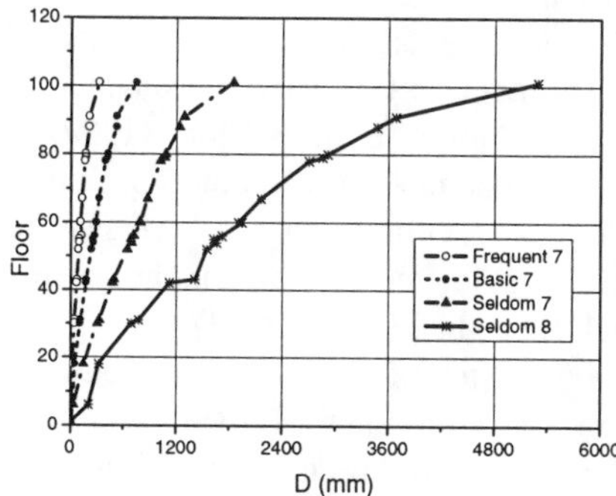

Figure 4. Maximum story displacement of direction X

Generally, the story displacement responses in direction X are approximate to those of direction Y, which preserve a consistency of measured natural frequencies in direction X and Y. The maximum story displacement curves are relatively smooth without obvious inflexions, suggesting that the distribution of equivalent rigidities along height is well proportioned. No extensive deformations at transfer stories of services core are detected, indicating that adopted measures of strengthened floor diaphragm are desirable designs in resisting lateral load.

5. Conclusions

Seismic behavior of Shanghai World Financial Center Tower has been experimentally investigated. A 1/50 scaled model of a 101-storey structure, which consists of mega-frame structure, the reinforced concrete and braced steel services core and the outrigger trusses, is tested on a shaking table by subjecting it to a series of simulated seismic ground motions with increased intensity of shaking in each successive test run. The following conclusions can be drawn from the test results:

1. The model test results indicate that the prototype structure is able to withstand frequent occurred, basic intensity and seldom-occurred earthquakes of intensity 7 without sever damage. The employed structural system demonstrated good quality in resisting earthquakes.
2. The prototype structure remains in elastic stage after subjected to earthquake waves of frequent phase. The total displacement /height in direction X is 1/1522, and that in direction Y is 1/1578. The maximum inter-story drift in direction X and Y are 1/539 and 1/707, respectively, which are smaller than the allowable value of 1/500 according to CCSDB. The structure conforms to the requirement of CCSDB for no damage under frequent earthquakes.
3. The natural frequencies and equivalent rigidities decay very slightly after the basic intensity earthquake, which indicates that the prototype structure still remains in elastic stage.
4. After the seldom-occurred earthquake of intensity 7, macro-cracks occur and the natural frequencies and equivalent rigidities decrease moderately. The total displacement /height in direction X is 1/266, and that in direction Y is 1/261. The maximum inter-story drift in direction X and Y are 1/127 and 1/151, respectively, which are smaller than the allowable value of 1/100 according to CCSDB. The structure meets the requirements of CCSDB for no collapse under seldom-occurred earthquakes.
5. The inter-story drift at strengthened levels decrease by comparison with their neighbor levels, demonstrating that the equivalent rigidities at these special levels are strengthened obviously.
6. Under seldom occurrence of intensity 8, amount of concrete spall in mega-columns at floor 6, as well as perimeter steel columns buckle at floor 6. These damages suggest that the transfer floor 6 is weak level. Measures to increase the ductility and strength of floor 6 are needed to avoid extensive deformations.
7. Strain values measures by strain gauges indicate that the stress on mega-diagonals near the joint of belt trusses and mega-columns are relatively bigger compared with others. It is necessary to attach important to joints connecting details.

Acknowledgement

The financial support of National Natural Science Foundation of China through Grant 50025821 and Grant 50321803 is gratefully acknowledged.

References

1. China Ministry of Construction. Technical specifications for concrete structures of tall building (JGJ3-2002). China Architecture & Building Press, Beijing, China.
2. State Key Laboratory for Disaster Reduction in Civil Engineering, Tongji University. Introduction

of Shaking Table Testing Division; Shanghai, China (2003).

3. Sabnis,G.M., Harris, H.G., White, R.N. and Mirza, M.S. Structural modeling and experimental techniques. Prentis-hall, Inc., Englewood Cliffs, N.T. (1983)
4. Shanghai Government Construction and Management Commission. Code for seismic design of buildings (DGJ08-9-2003). Shanghai Standardization Office, Shanghai, China.
5. Lu, X.L., Zhang, H.Y., Hu, Z.L., Lu, W.S. Shaking table testing of a U-shaped plan building model. Canadian Journal of Civil Engineering; 26: 746-759(1999).
6. Lu, X.L. Application of identification methodology to shaking table tests on reinforced concrete columns. Engineering Structures; 17(7): 505-511(1995).
7. Research Institute of Resistance to Vibration and Explosion. Report No. 5 of structural model shaking table test research; Tsinghua University Publishing House, Beijing, China (1990).
8. Research Institute of Structural Engineering and Disaster Reduction, Tongii University. Report on shaking table testing of Jinghuayuan building model; Shanghai, China (2003).
9. China Ministry of Construction. Code for seismic design of buildings (GBS0011-2001). China Architecture & Building Press, Beijing, China.
10. Roko Zarnic, Samo Gostic, Adam J. Crewe, Colin A. Taylor. Shaking table tests of 1:4 reduced-scale models of masonry infilled reinforced concrete frame buildings. Earthquake Engineering & Structural Dynamics 2001; 30(6): 819-834(2001).

EXPERIMENTAL STUDY ON SHEAR STRENGTH OF EXTERIOR BEAM-COLUMN JOINTS WITH DIFFERENT TYPES OF BEAM BAR ANCHORAGES*

H.F. WONG† J.S. KUANG

Department of Civil Engineering, Hong Kong University of Science and Technology, Clear Water Bay, Kowloon, Hong Kong

Reversed cyclic-load tests on full-scale RC exterior beam-column joints, which are fabricated to simulate those in as-built reinforced concrete framed buildings designed to BS 8110, are carried out. Emphasis of the study is placed on the effects of the types of beam bar anchorage on the seismic behaviour and shear strength of reinforced concrete exterior joints subjected to simulated earthquake load. The behaviour is investigated with different types of anchorage in beam reinforcement. It has been shown from the experiment that all the specimens have low ductility capacity and poor energy dissipation ability. The test results are compared with two preseismic design codes (BS 8110 and EC 2). It is shown that the types of reinforcement anchorage in beams have significant effects on the shear capacity and hysteretic behaviour of exterior beam-column joints. Present codes of practice do not accurately predict the shear strength of the non-seismically detailed joints. It is indicated that ignoring the design of beam-column joints may lead to potential damage of reinforced concrete framed buildings in an unexpected low to moderate earthquake.

1. Introduction

Building design codes of practice in regions of nil or low to moderate seismicity, such as Hong Kong [1], do not generally include any provision for consideration of seismic resistance. A building for which seismic performance is not considered but designed and detailed only based on gravity load and other incidental horizontal loads would have to rely on its inherent ductility to respond acceptably to an unexpected seismic excitation [2]. Indeed, inherent shortcomings of the gravity load design philosophy imply high susceptibility of the building structures to anticipated seismic risk [3, 4].

An improved understanding of seismic behaviour of structures without seismic considerations in design and detailing is essential to evaluate the existing non-seismically designed buildings in regions of moderate seismicity, which could need strengthening or retrofitting if necessary. This may eventually lead to modification of the current non-seismic design codes for moderately earthquake-resistant design of buildings without resorting to a full seismic design.

When a reinforced concrete moment-resisting frame is subjected to seismic load, the possible inelasticity is concentrated in either beams or the beam-column joint regions. This is dependent upon not only the flexural capacity ratio of beam to column, but also the detailing of both longitudinal and transverse reinforcement in the joints, which affects significantly the shear strength of the beam-column connection. To avoid sudden degradation of the strength and stiffness of the frame, it is principally necessary to maintain the integrity of beam-column joints. Damage of the joints in a frame can result in instability of the structure and poor overall structural response. It has been shown that severe damage and/or collapse of many reinforced concrete framed buildings in recent earthquakes are due to poor reinforcement detailing of beam-column joints, including absence of transverse reinforcement and inappropriate anchorage of longitudinal beam bars in columns.

This paper presents reversed cyclic-load tests on full-scale RC exterior beam-column joints, simulating those in as-built reinforced concrete framed buildings designed to BS 8110 [5]. The primary objective of this experimental study is to investigate the effects of types of beam bar anchorage on the shear strength and hysteretic behaviour of reinforced concrete exterior beam-column joints subjected to simulated seismic loading caused by a moderate earthquake. The experimental results are compared with two preseismic design codes of practice, BS 8110 [5] and EC 2 [6], in order to evaluate the validity of code-prescribed methods for predicting the shear strength of these joints under moderate seismic load. It is shown that the types

* This work partially supported by grant HKUST6005/02E of the Hong Kong Research Grant Council.

† Corresponding author: cewhf@ust.hk

of beam bar anchorage have significant effects on shear capacity and hysteretic behaviour of the exterior joints. It is indicated that the design of reinforced concrete beam-column joints should not be ignored even for the cases of low to medium earthquakes, though such design is generally not required in non-seismic design codes of practice.

2. Experimental Programme

2.1 *Test specimens*

Four reinforced concrete exterior beam-column joints with section dimensions 260 × 450 mm for beams and 300 × 300 mm for columns are fabricated. The typical shape and dimensions are shown in Figure 1. The test reference codes start with BS, indicating that reinforcement of the specimens is detailed in accordance with the BS 8110 detailing practice for buildings.

The beam is reinforced with an equal amount of top and bottom longitudinal bars 3T20, considering moment reversals. The column is reinforced with 4T25. There is no transverse reinforcement in beam-column connections typifying that of a conventional preseismic/non-seismic detailing technique. Material strengths and reinforcement anchorage details of the specimens are summarised in Table 1.

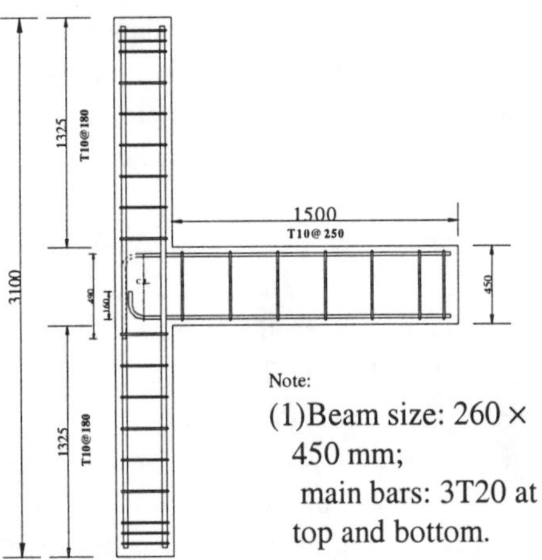

Figure 1. Typical shape and dimensions of four specimens in the laboratory testing.

2.2 *Test setup & loading arrangement*

The experimental set-up and loading system are shown in Figure 2. For convenience of applying loading and testing, the T-shape specimen is rotated 90 degrees, so that the column member is in the horizontal position and the beam member is in the vertical position. Proper boundary conditions are provided in the set-up to simulate the actual working situation of the beam-column joint as if it is a part of the frame structure.

Table 1. Material strengths and reinforcement anchorage details of the specimens tested.

Specimen	f_{cu} MPa	f_y MPa	Anchorage Type
BS-L	38.6	520	Both tension and compression Beam bars bent into joint
BS-OL	38.6	520	Both tension and compression Beam bars bent away from joint
BS-LL	52.6	520	Tension beam bars bent away from joint and compression beam bars bent into joint
BS-U	38.8	520	U-anchorage; laps in beam bars at end zone of beam

In the test, columns of all the specimens are subjected to axial loading of about 15% of the column squash capacity, which is considered as a practical range in laboratory testing as well as in real framed buildings [7]. As shown in Figure 2, rollers are provided near the ends of columns to simulate inflection points in the structure, and axial load is applied to the column by a 1000-kN hydraulic jack located at the steel bearing. A 460-kN hydraulic actuator is employed to apply reversible cyclic loading at the beam end.

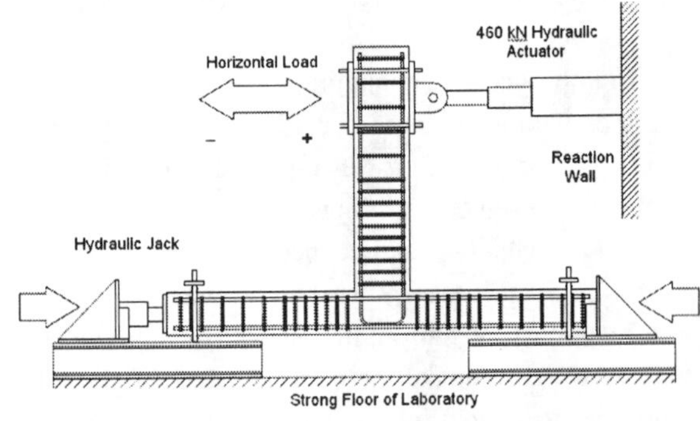

Figure 2. Experimental setup for cyclic-load test on exterior joints.

The reversed cyclic load applied to specimens is shown in Figure 3. In the test, both load and displacement controls are adopted at different loading stages. The load-control method is used at the early loading stages; one cycle of horizontal loading of $\pm 0.5P_i$ and then $\pm 0.75P_i$ are applied, where the load P_i is the cyclic applied load at the top of the specimen when the beam reaches its ultimate flexural strength M_u. The value of M_u is determined using BS 8110 rectangular stress block for concrete at ultimate limit state without partial safety factors. It is well recognised [8] that the yield displacement Δ_y can be defined and calculated based on the stiffness when the lateral load is $\pm 0.75 P_i$, which is then extrapolated linearly to $\pm P_i$, as shown in Figure 3. Thus,

$$\Delta_y = \frac{|\Delta_1| + |\Delta_2|}{2}, \tag{1}$$

where Δ_1 and Δ_2 are horizontal displacements corresponding to P_i and $-P_i$, respectively. The reversed cyclic loading arrangement is then switched to the displacement control, in which the test specimens are subjected to two cycles of reversed loading gradually to achieve $\pm\Delta_y$, $\pm 2\Delta_y$, $\pm 3\Delta_y$, etc.

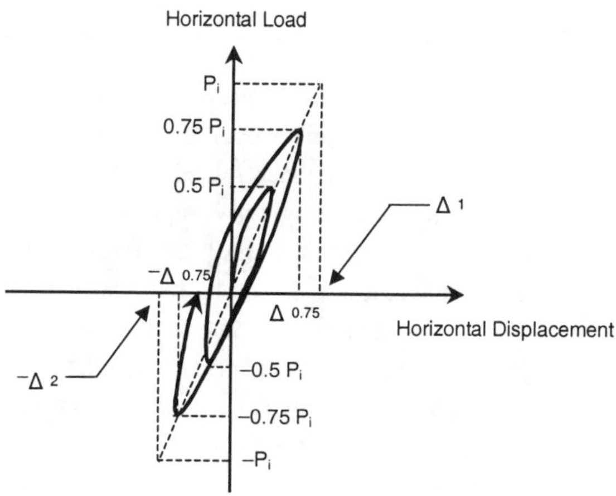

Figure 3. The reversed cyclic load applied on the specimens.

3. Experimental Results

3.1 *Hysteretic behaviour*

Maximum applied loads for the specimens are presented in Table 2. Experimental hysteresis loops and crack patterns at failure of specimens are shown in Figure 4. In general, except specimen BS-U the specimens perform poorly in terms of shear strength and hysteretic behaviour under reversed cyclic loading. The load-deflection hysteretic response for specimen BS-LL is seriously asymmetrical, as shown in Figure 4(c), due to the asymmetric steel detailing of anchorage in the joint. In the tests, all specimens fail with the joint shear failure mode shown in Figure 4 and when the beams reach only 50% to 75%, on the average, of their flexural strength, as summarised in Table 2. No buckling of the longitudinal reinforcement in columns is observed, though there is no transverse reinforcement in joint cores.

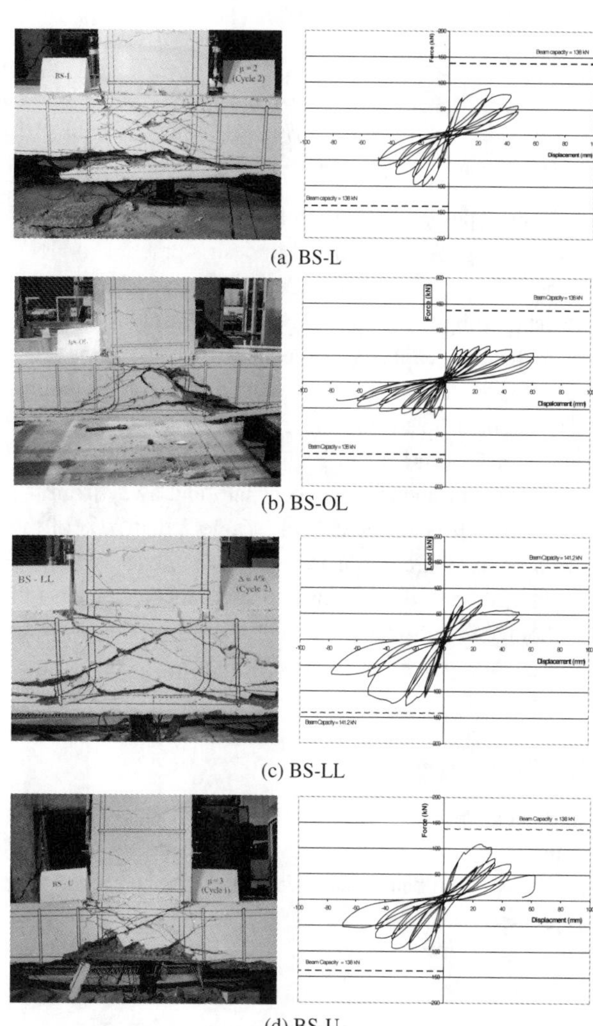

Figure 4. Hysteresis loops and crack pattern of specimens.

It can be seen from Table 2 that for specimen BS-L, which is detailed with the traditional type of anchorage, about 70% of beam flexural strength is reached on both loading sides when the joint fails. For specimen BS-U, nearly 80% of the beam flexural capacity is observed while the other side can still remain as high as 70%, when the joint fails. It indicates that the joint with U-type anchorage performs better than that with L-shape anchorage when subjected to reversed cyclic-loading.

Table 2. Experimental results.

Specimen	Maximum Load: kN		% of M_u	
	-ve direction	+ve direction	-ve direction	+ve direction
BS-L	100.9	90.2	73	65
BS-OL	68.9	70.1	50	50
BS-LL	127.5	83.5	90	59
BS-U	96.2	109.1	70	79

The types of anchorage for specimens BS-OL and BS-LL are generally adopted in design practice in Hong Kong [1], and also in New Zealand [9] and Japan [10]. It is surprising that for specimen BS-OL only 50% of the beam flexural capacity is reached when the joint fails, but this load can be maintained until the corresponding displacement up to 60 mm (i.e. the corresponding drift ratio ≈ 4%). Specimen BS-LL involves both types of reinforcement anchorage – bent into and away from the joint core. It is also seen from Table 2 that 90% of the beam flexural strength is reached in the negative side of loading, though the reinforcing steel does not have full anchorage length. However, the loading on the opposite side is greatly reduced to only 59% of the beam flexural strength.

3.2 *Shear strength*

The shear force in the joint core of an exterior beam-column can be calculated by considering the force in tension steel in beam and shear force in column and it is expressed by

$$V_j = T - V_{col}\,. \tag{2}$$

The shear strengths of joints are presented in Table 3. Expect that of specimen BS-OL, the maximum joint shear stresses, which are associated with the effective joint area A_j are in a range of $0.63\sqrt{f_c'}$ to $0.68\sqrt{f_c'}$.

Possible shear transfer mechanisms in exterior beam-column joints for the different types of beam bar anchorage are shown in Figure 5. For specimen BS-L, reinforcing bars in the beam are bent into the joint core and sufficient reinforcement anchorages are provided. The concrete strut shown in Figure 5(a) may be developed in diagonal sense effectively.

On the other hand, by observing the crack pattern of specimen BS-OL shown in Figure 5(b), it can be seen that the main diagonal crack propagates from the beam compression zone down to and along the column main reinforcement, stopping at the location between the second and third stirrups in the column counted from the joint core. This may explain that the diagonal compression within the joint core cannot be stabilised by the hooks, which are bent away from the joint, but the compression is directly applied on the column reinforcement, as shown in Figure 5(b).

Table 3. Experimental results in terms of joint shear.

Specimen	Maximum shear force in joints: kN		Ratio of joint shear strength to $A_j\sqrt{f_c'}$	
	-ve direction	+ve direction	-ve direction	+ve direction
BS-L	315.5	282.1	0.63	0.56
BS-OL	215.5	219.2	0.43	0.44
BS-LL	398.7	261.1	0.68	0.45
BS-U	300.8	341.2	0.60	0.68

In fact, this compression presses down the column longitudinal steel and causes eventually the destruction of the bond between the concrete and the steel shown in Figure 5(b). Similar observation is reported in an earlier study [9]. This may be one of the reasons for the specimen to have lower strength than those of other specimens. Specimen BS-LL shows a combination of both types of beam bar anchorage and the superior behaviour is obtained on the side with bars bent into the joint core, as shown in Figure 4(c).

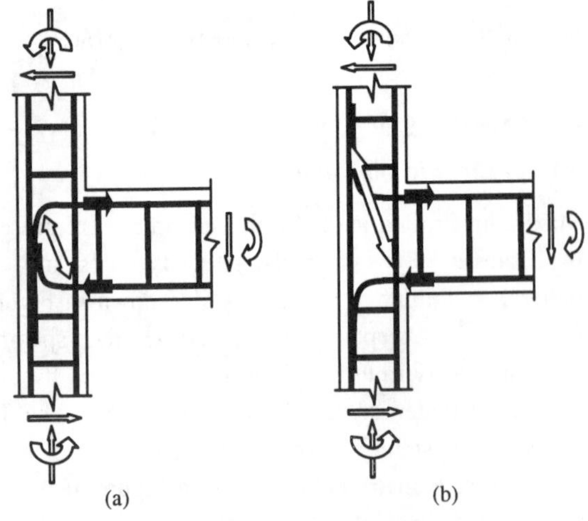

Figure 5. Possible shear transfer mechanisms for exterior joints. (a) BS-L and BS-U; (b) BS-LL and BS-OL.

4. Comparison with Design Codes

To evaluate the validity of existing codes of practice in predicting the shear strength of exterior beam-column joints with non-seismic details under reversed cyclic loading, the experimental results are compared with the prescribed limiting values predicted by BS 8110 [5] and EC 2 [6].

4.1 *Analysis using BS8110*

A beam-column joint can be defined as a portion of the column within the beam depth [11]. The shear strength of the joint may therefore be determined by considering the column that is subjected to axial loads and bending moment transferred from the beam. In BS 8110, the shear strength of a member subjected to axial loading is determined by

$$V = v_c'bd + A_{sv} f_{yv}\left(\frac{d}{s_v}\right), \qquad (3)$$

where v_c' is the shear strength of concrete with axial loading (N/mm^2),

$$v_c' = v_c + 1.33\times 0.6\frac{N}{A_c}\left(\frac{Vh}{M}\right), \qquad (4)$$

in which N is the axial load (N), A_c is the overall cross-sectional area of the section (mm^2), V is the design shear force (N), h is the overall depth of the section (mm), M is the design bending moment (Nmm), and Vh/M is taken to be not greater than 1; and b and d are the width and effective depth (mm) of the section, respectively; A_{sv} is the cross-sectional area of column stirrups (mm^2), f_{yv} is the strength of the stirrup reinforcement, and s_v is the stirrup spacing. In this study, the partial safety factors of 1.25 for material and 1.33 [11] for the coefficient of 0.6 are removed.

4.2 *Analysis using EC 2*

In EC 2, the shear strength of a column is calculated by

$$V_{Rd1} = \left[\tau_{Rd}k(1.2+40\rho)+(0.15\sigma_{cp})\right]bd, \qquad (5)$$

where τ_R is the shear strength of concrete (N/mm^2); $k = (1.6 - d)$, in which d is in metres; ρ_l is the tension steel ratio, which is ≤ 0.02; σ_{cp} is the longitudinal stress of the column (N/mm^2), A_{sw} and f_{yw} are the cross-sectional area (mm^2) and strength of the web reinforcement, respectively; and s is the stirrup spacing. In the analysis of this study, the partial safety factors of 1.5 for material and 1.5 [13] for the coefficient of 0.15 are removed. The upper limit of shear strength is V_{R2}.

5. Initial Diagonal Tension Cracking Criterion

Hakuto et al [8] and Priestley and Calvi [14] suggested that the ultimate shear strength of beam-column joints without shear reinforcement can be reached when the initial diagonal tension crack of the joint core occurs. From Mohr's circle, the principle tensile strength is expressed as

$$\sigma_t = -\frac{\sigma}{2}+\sqrt{\left(\frac{\sigma}{2}\right)^2+v_{jh}}\,. \qquad (6)$$

It is suggested that the cracking strength be $k\sqrt{f_c'}$, where k is an empirically determined constant, taken approximately as 0.3 [14]. Eq. (6) is then become

$$v_{jh} = 0.3\sqrt{f_c'}\sqrt{1+\frac{\sigma}{0.3\sqrt{f_c'}}}\,. \qquad (7)$$

6. Conclusion

6.1 *Discussions*

Experimental shear strengths of the specimens and comparison with those predicted by different codes and the criterion of initial diagonal crack are presented in Table 4, where partial safety factors have been removed in the calculation.

Table 4. Experimental shear strength of specimens & comparison to design codes and IDCC (Initial Diagonal Cracking Criteria)

Specimen	Experimental joint strength V_{exp} : kN	V_{exp} / V_{BS}	V_{exp} / V_{EC2}	V_{exp} / V_{IDCC}
BS-L	315.5	0.77	1.70	1.01
BS-OL	219.2	0.53	1.18	0.70
BS-LL	398.7	0.97	1.94	1.16
BS-U	341.2	0.83	1.84	1.09

In both BS 8110 and EC 2, the same shear mechanism for an ordinary flexural beam with consideration of axial loading is adopted to predict the shear capacity of beam-column joints. The basic shear resistance provided by concrete is obtained from the empirical values given in the codes. This shear resistance is considered as a combined action of shear forces in the uncracked concrete compression zone, from the dowel action of longitudinal reinforcement and due to aggregate interlock. It is also be seen from Table 4 that BS 8110 gives much better predictions of the shear capacity to all the specimens than EC 2 does. Predictions of EC 2 underestimates significantly the actual shear strengths of the joints, and are only approximately 50%-60% of the experimental results.

The criterion of initial diagonal tension cracking gives very good correlations with the test data in this study, as shown in Table 4. However, it can be seen from Eq. (7) that the shear strength determined by this criterion is highly dependent on the level of the axial loads applied. It has been shown [15] that in most of cases the shear strength of beam-column joints does not change significantly with the axial loading. This criterion of failure needs to be investigated further.

6.2 *Conclusions*

Full-scale RC exterior beam-column joints with non-seismic reinforcement details and different types of beam bar anchorage are tested under reversed cyclic loading. Based on the findings from this experimental investigation and the comparison of test results with those predicted by BS 8110, EC 2, and by the initial diagonal tension-cracking criterion, the following conclusions are drawn.

1. The type of beam reinforcement anchorage has significant effect on the load-displacement hysteretic behaviour and shear resistance of the exterior beam-column joints. It has been shown that comparing to specimen BS-L (conventional type of anchorage), the seismic performance of both specimens BS-OL (both tension and compression bars bent away from the joint) and BS-LL (tension bars bent away from the joint) are poor under reversed loading. Moreover, the shear strengths of these two specimens are much lower than and only about two-third of that of specimen BS-L. Therefore, the types of beam bar anchorage adopted in specimens BS-OL and BS-LL are not recommended to use in practice for both the non-seismic and seismic designs.
2. Shear failure of a beam-column joint may occur before the beam section reaches its ultimate flexural strength. The worst scenario in this investigation is that the shear failure of a joint occurs when the beam strength reaches only as low as 50% of the design flexural capacity, showing that the failure of a joint occurs when the beam is only under service loads. It therefore reveals that the design of beam-column joints should not be ignored even for low to medium earthquakes. In fact, such design is not required in most of the non-seismic design codes. Ignoring the design of beam-column joints may lead to potential damage of reinforced concrete framed buildings in an unexpected low to moderate earthquake.
3. The values predicted by BS 8110 are about 20%-40% higher than the experimental results. On the other hand, shear strengths of the joints are significantly underestimated by EC 2, and the experimental results can be nearly double the predictions of EC 2.
4. Although the shear strength of a beam-column joint predicted by the criterion of initial diagonal cracking is highly dependent on the level of axial loads applied on the column, this model gives very good correlations with all the test data in this study.

References

1. H.C. Chan et al., *Transactions HKIE* **5**, 6 (1998).
2. J.S. Kuang, *J. Earth. Engg Soc. of Korea* **2**, 111 (1998).
3. J.S. Kuang and H.F. Wong, *ICE Struct. Bldgs* **158**, 13 (2005).
4. J.S. Kuang and A.I. Atanda, *ICE Struct. Bldgs* (2005).
5. BSI, BS 8110 (1997).
6. European Com. Stand. T250, Eurocode 2 (1995).
7. Y. Hui and P. Irawan, *Civ. Engg Res.* NTU, 44 (2001).
8. R. Park, *Bull. New Zealand Nat. Soc. Earth. Engg* **22**, 155 (1989).
9. S. Hakuto, R. Park and H. Tanaka, *ACI Struct. J.* **97**, 11 (2000).
10. K. Kanada et al., *Transactions JCI* **6**, 433 (1984).
11. Joint ACI-ASCE Com. 352, ACI 352R-02, (2002).
12. R.H. Scott et al. *Struct. Engr* **72**, 55 (1994).
13. D.E. Parker and P.J.M. Bullman, *Struct. Engr* **75**, 53 (1997).
14. M.J.N. Priestley and G.M. Calvi, *Earth. Spectra* **7**, 413 (1991).
15. R.L. Vollum, PhD thesis, University of London, (1998).

Foundation

A STUDY ON LARGEST SCALE SUPER-HIGH RIVERSCAPE DELUXE RESIDENTIAL BUILDING -FOUNDATION INTERACTION IN CHINA

JIAN GONG
Shanghai No.1 Construction Limited Corporation, Shanghai 200120, China

XIHONG ZHAO
College of Civil Engineering, Tongji University, 1239 Siping road, Shanghai 200092, China

BAOLIANG ZHANG
Shanghai COSCO-Laoximen Property Development Co, Shanghai 200011, China

Shimao Riviera Garden is the largest scale super-high riverscape deluxe residential building in China, located in Pudong, Shanghai and comprised seven 49-to-55-storey high deluxe apartments with 169m high, 2.0m thick bottom slab of box foundation and bored piles of φ 850mm driven into silty sand to a depth of 58m. In one apartment the settlement, loads on pile group and stress in bottom slab of box foundation are analyzed using the theory of superstructure-foundation interaction. The analytical results have been shown that the theory presented in this paper is comparatively accurate and feasible in practice.

1. Introduction

Up to the present Shimao Riviera Garden is the largest scale super-high riverscape deluxe residential building in China and comprised seven 49-to-55-storey high deluxe apartments with 169m high, 2.0m thick bottom slab of box foundation and bored piles of φ 850mm driven into silty sand to a depth of 58m. The total number of piles is 464 and the allowable bearing capacity of pile is taken as 500kN according to the loading test. The superstructure is of reinforced concrete shear wall. The foundation is a piled-box foundation with piles distributed fairly evenly. The shape of foundation with area about 2700m^2 is irregular. The embedment of foundation is 8.27m, in which the behavior of piled box foundation for an apartment will be analyzed using the theory of superstructure-foundation interaction.

The prospective view and the plan of basement of Shimao Riviera Garden are shown in Figure 1 and Figure 2, respectively.

Figure 1. Prospective view of Shimao Riviera Garden

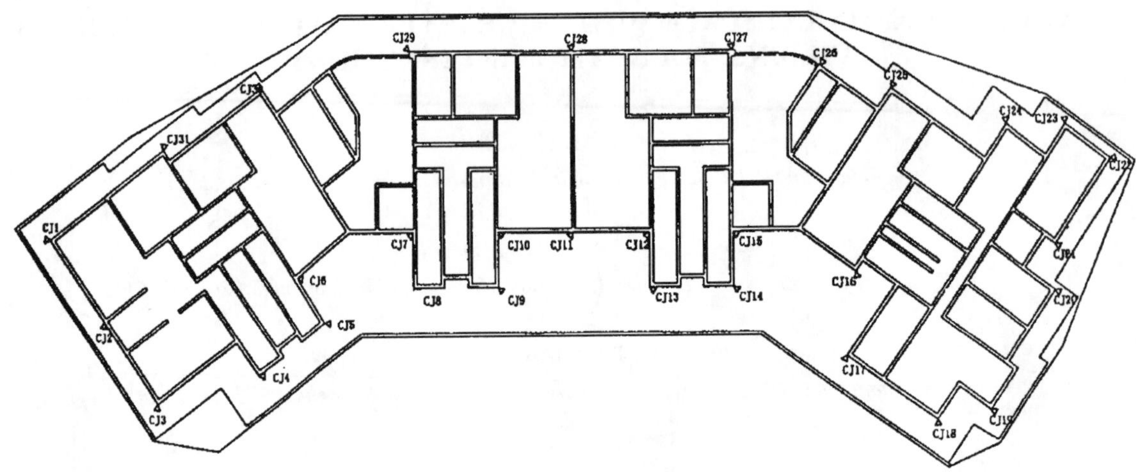

Figure 2. Plan of basement and layout of referent points for settlement.

2. Soil Condition

The soil condition of this building is shown in Table 1.

Table 1. Physical and mechanical index properties of soil layers

No. Soil layer	Soil	Thick-ness (m)	Water content w(%)	Unit weighty (kN/m3)	Void ratio e	Compression coefficient(cm/s)		Consolidated Undrained shear	
						α_{1-2} Mpa^{-1}	E_{1-2} Mpa	C (kPa)	φ (°)
2	Brown yellow silty clay	1.5	33.0	18.9	.0.921	0.45	4.27	24	15°15'
3	Grey very soft silty clay	2.25	42.5	17.7	1.198	0.82	2.74	11	18°45'
3夹	Grey clayey silt	0.85	33.1	18.8	0.919	0.20	9.59	8	30°45'
4	Grey very soft clay	7.48	50.6	17.0	1.427	1.11	2.17	10	9°45'
5-1a	Grey clay	2.5	43.7	17.7	1.225	0.72	3.11	14	10°15'
5-1b	Grey silty clay	7.53	34.4	18.5	0.983	0.42	4.72	19	15°30'
6	Dark green-yellow silty clay	3.6	23.3	20.3	0.658	0.19	8.74	50	20 ° 20'
7-1	Grey green-yellow sandy silt	6.98	30.9	19.0	0.860	0.16	11.66	3	34°15'
7-2	Grey-yellow silt	32.95	27.2	19.4	0.764	0.11	16.02	3	35°15'
9-1	Grey silty fine sand	10.88	25.9	19.5	0.737	0.13	13.34	3	36°15'
9-2	Grey silty fine sand with gravel	29.75	27.5	19.3	0.770	0.14	12.88	3	35°30'

Note: Thickness of fill is 1.3m and under ground water level is 0.5m.

Table 2. Summary of settlement data

Storey No.	G-1	1	2	3	4	5	7	9	11	13	15	17	19	21	23	25	27	29
Date	3/30	4/12	4/24	4/28	5/4	5/14	6/6	6/26	7/12	7/20	7/30	8/8	8/16	9/2	9/13	9/26	10/4	10/12
1	0	0	0	0	0	1	3	5	6	8	9	11	12	13	15	16	17	17
2	1	1		1	1	1	1	3	5	8	9	10	12	12	15	17	17	17
3	0	1		0	1	1	4	6	7	9	11	12	13	14	16	17	19	19
4	1	1		0	0	1	2	5	6	8	10	12	13	14	15	16	18	19
5	1	2		1	1	2	2	destroyed										
6	0	0		0	0	1	2	4	5	8	10	11	12	13	15	17	19	20
7	2	2		0	0	0	1	3	5	7	8	10	12	13	14	16	18	19
8	0	1		1	1	1	2	5	7	10	11	12	13	13	15	17	19	20
9	1	2		0	0	0	2	4	6	8	9	11	11	13	16	18	19	20
10	1	1		0	1	2	3	destroyed										
11	1	1		0	0	1	3	5	7	9	10	11	12	14	16	18	19	19
12	0	1		0	0	0	1	3	5	7	9	11	13	13	16	18	19	20
13	2	2		1	1	1	3	4	6	8	9	11	12	13	15	17	19	19
14	0	1		0	0	0	1	3	5	7	9	11	13	14	16	18	19	20
15	0	1		1	1	1	2	destroyed										
16				0	1	2	4	5	7	9	10	11	12	13	16	18	19	19
17	1	1		1	1	1	1	3	5	7	10	12	12	14	16	19	20	21
18	1	2		0	0	1	2	2	4	7	9	10	12	14	16	17	17	18
19	0	1		1	1	1	3	5	6	8	10	11	12	13	15	17	17	17
20	1	1		0	0	1	2	5	7	8	9	11	12	14	14	16	17	18
21	0	1		1	1	1	2	4	6	8	10	12	13	14	16	19	19	19
22	1	1		0	0	0	2	4	6	8	10	11	12	14	14	16	16	17
23				0	0	1	2	5	7	9	10	12	13	13	15	18	18	18
24				0	1	1	2	4	6	8	10	12	12	13	15	17	17	19
25	1	2		1	1	2	3	destroyed										
26	0	1		0	1	1	3	5	6	9	11	12	13	14	17	19	21	22
27	1	1		1	1	1	1	3	5	7	9	11	12	13	15	19	20	21
28				0	1	1	2	4	6	8	10	11	12	14	17	19	21	22
29	1	1		0	0	1	2	5	7	9	11	12	13	15	17	19	21	22
30	0	1		1	1	1	1	4	6	8	10	11	12	14	15	17	18	20
31	0	1		0	1	1	2	5	7	9	10	12	13	13	15	17	19	20

Table 2. Summery of settlement data (continued)

Storey No.	31	33	35	37	39	41	43	45	47	49	51	53	Decoration---------------------------					
Date	10/29	11/6	11/14	11/23	12/2	12/6	12/21	12/31	1/11	1/16	1/27	2/1	3/11	4/16	5/18	6/13	8/17	10/28
1	17	17	17	18	18	18	20	23	25	27	27	28	28	29	30	31	31	32
2	17	17	17	18	18	18	20	23	25	27	27	29	30	31	31	32	33	34
3	20	20	20	21	22	23	24	27	29	30	30	31	31	32	32	33	33	34
4	19	20	21	22	22	23	25	28	30	31	31	33	33	33	34	34	35	35
5							destroyed											
6	21	22	24	27`	27	27	29	32	33	33	33	36	36	36	37	37	38	38
7	20	20	21	21	22	22	24	26	28	29	31	34	35	36	36	36	37	38
8	22	23	24	25	26	27	29	31	32	34	34	36	37	38	38	39	40	41
9	21	22	23	24	26	28	31	34	34	35	35	38	38	38	38	39	40	40
10							destroyed											
11	20	21	21	22	23	25	26	28	29	30	30	32	32	33	34	35	36	37
12	21	22	23	24	26	27	29	30	31	32	34	37	38	38	39	40	41	42
13	20	20	21	21	22	24	27	29	30	32	32	36	36	37	38	38	39	40
14	20	21	21	23	25	27	29	31	32	34	35	37	37	38	38	39	39	40
15							destroyed											
16	20	22	22	23	24	26	28	31	33	34	34	37	38	39	39	40	41	42
17	22	22	23	24	25	26	28	30	33	34	34	36	36	36	37	37	38	39
18	19	19	19	19	20	21	23	25	27	27	27	29	30	31	31	32	32	33
19	17	19	19	19	20	22	23	24	26	26	26	27	28	29	30	31	32	33
20	18	19	19	19	19	19	21	24	27	29	29	29	29	30	30	31	32	33
21	19	19	19	19	20	21	23	26	28	30	31	31	32	32	33	34	35	35
22	17	18	18	19	19	20	22	24	25	25	27	30	31	31	32	32	33	34
23	18	18	18	19	19	19	22	24	27	30	30	30	31	32	34	34	35	35
24	20	20	21	21	21	22	25	28	31	33	33	34	35	35	35	36	37	37
25							destroyed											
26	23	25	25	27	27	28	31	34	36	37	37	38	38	39	40	40	42	42
27	22	25	27	29	30	31	33	36	37	38	38	39	39	40	41	42	43	44
28	23	25	27	29	30	31	32	34	36	37	37	40	40	40	41	42	42	42
29	23	25	25	27	29	30	32	33	35	36	36	40	40	40	41	41	42	43
30	20	21	21	23	24	25	27	29	31	32	32	36	37	38	40	41	42	42
31	21	23	23	24	24	25	28	29	31	32	32	36	37	37	38	39	39	40

3. Analysis using Theory of Building-foundation Interaction

For the sake of analysis simplification, the equivalent rectangle is considered as the irregular shape of foundation, as shown in Figure 2, and interaction between pile and raft is also considered only.

3.1 *Settlement*

3.1.1 *Measured settlement*

This building was completed on 1st February 2002. The measurement of settlement for building started from 21st March 2001 to 28th October 2002 for 36 times. The detailed data settlements of building are shown in Table 2 and Figure 1. The maximum settlements on the north side No. CJ26, CJ27, CJ28, CJ29 and CJ30 are 42mm, 44mm, 42mm, 43mm and 42mm, respectively; and on the opposite (south) side, No. CJ3 to CJ19; most of them are less than 40mm. This settlement condition is reasonable. It can be said that the settlement is approaching to stable. For example, the maximum settlement, No.CJ27 was 44mm at the last measured time (28th October 2002), and at the completion of structure (1st February 2002) the settlement was 39mm, the settlement rate is 0.02 mm/day.

3.1.2 *Computed settlement*

The computed maximum settlement is 46.5mm, as shown in Figure 3. Thus, the computed maximum settlement is agreeable to the measured one.

3.2 Moment

The moment distribution along longitudinal profile is shown in Figure 4, in which the maximum moment is 3432kN-m, the corresponding stress in foundation is 2288 kN/m^2. It is very safe.

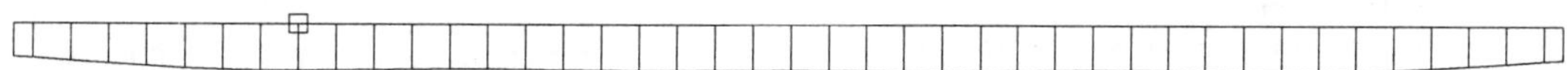

Figure 3. Computed settlement along longitudinal profile

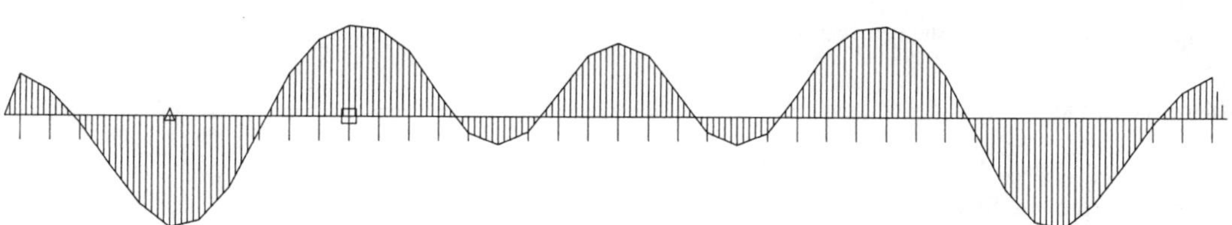

Figure 4. Moment distribution along longitudinal profile

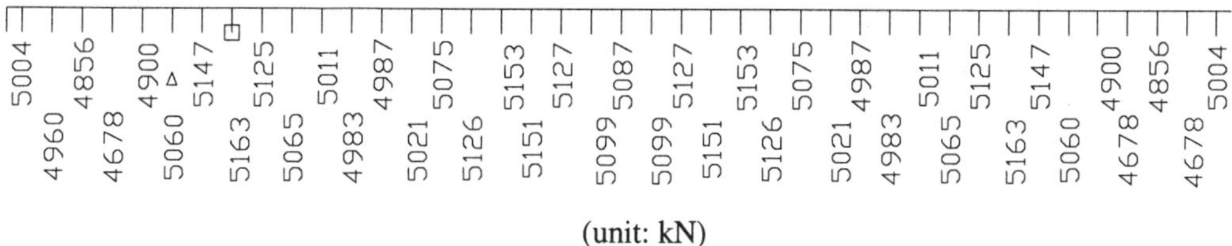

Figure 5. Load distribution on pile group along longitudinal profile

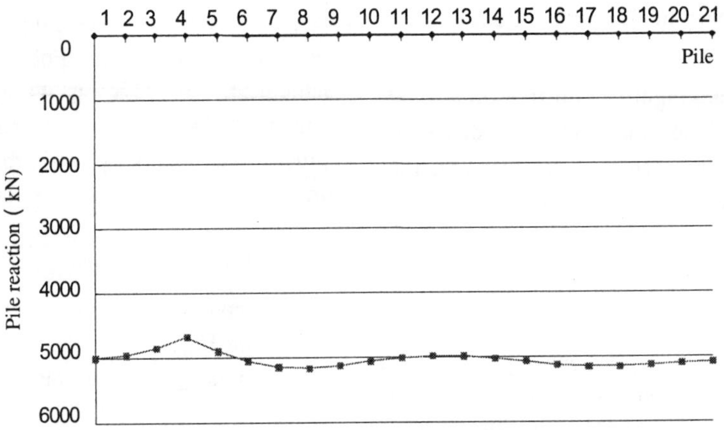

Figure 6. Load distribution on pile group along the symmetric axis on left side

3.3 *Load on pile group*

The load distribution on pile group along longitudinal profile is shown in Figure 5, in which the maximum load on pile is 5163kN near core part. This value is just over the allowable bearing capacity of pile. The load distribution on pile group along the symmetric axis on left side is shown in Figure 6. It should be pointed out that in the normal design of piled raft or box foundation, engineer does not consider the load-sharing between pile and raft or box and even does not consider the whole water buoyancy. Evidently, such design is conservative.

4. Concluding remarks

Based on the analysis of piled box foundation for Shimao Riviera Garden in Shanghai mentioned above, the theory of superstructure-foundation interaction is feasible and practical.

References

1. Zhao Xi Hong et al. 《Theory of Design of Piled Raft and Piled Box Foundations for Tall Buildings in Shanghai》 (in English), Tongji University Press, 1998.

REAL-TIME CONTROL OF INTERNAL FORCE OF RETAINING STRUCTURE FOR THE DEEP PIT

SHAOHUI LI

College of Civil Engineering, Tongji University, 1239 Siping Road
Shanghai 200092, China

MIN SUN

College of Civil Engineering, Tongji University, 1239 Siping Road
Shanghai 200092, China

WEI XU

College of Civil Engineering, Tongji University, 1239 Siping Road
Shanghai 200092, China

The diameter of the south anchor pit of Yangluo Yangtze river bridge is about 70m and the depth of it is 45m. During the process of construction, the internal force of the support for this pit must be controlled strictly. Informational construction technology such as numerical method and observation system is used to control excavation process. Based on the appropriate material constitutive equation, failure criterion and model of cracking, a nonlinear FEM model of retaining structure is presented. The stress of diaphragm wall is presented by back analysis based on in-situ displacement of retaining structure. The result fit the in-situ data well and provide theoretical basis for the safety evaluation of the retaining structure.

1. Guidelines

As the development of bridge engineering in China in recent years, many deep pits appeared during the construction of suspension bridge[1,2]. The representative project is the south anchor pit of Yangluo Yangtze river bridge which is constructing now. The retaining structure is round underground diaphragm wall socketed in bedrock with round liner preventing the displacement of the diaphragm wall. During the process of excavation, the internal force of retaining structure must be controlled strictly. Traditional method to control the stress is measure the stress of the reinforced bar of diaphragm wall, but the number of reinforcement meter is finite and difficult to reflect the stress of the whole retaining structure[2]. Because the in-situ data of the displacement of diaphragm wall is more accurate and abundant than the in-situ stress of reinforcement bar, based on the in-situ displacement, the finite element method is used in back analysis to commutate the stress of the retaining structure in this paper. The retaining structure is shown in Figure.1. The mechanical property of the soil is listed in Table 1. The thickness of diaphragm wall is 1.5m. The dimension of the liner varied in diversity depth, listed in Table 2.The thickness of the first and second liner is 1.5m; 3~6th liner is 2m; 7~14th liner is 2.5m.

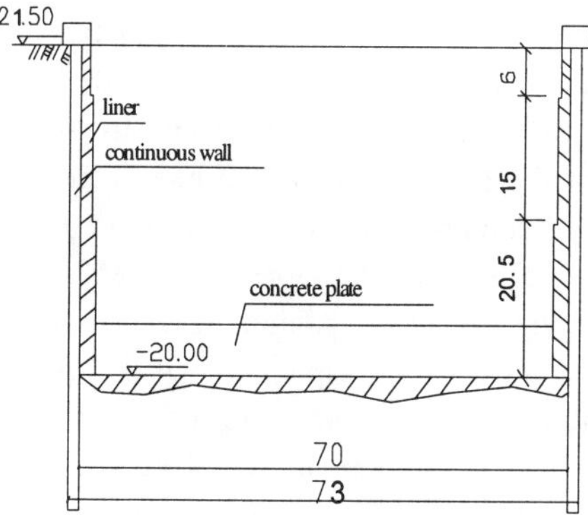

Figure 1 Retaining structure

Table 1 Mechanical property of the soil

Soil	E/MPa	υ	γ (KN/m³)	C(kPa)	ϕ	Thickness(m)
Clayey	3	0.45	18.5	12	10	1.5
Clay	2.6	0.45	18.2	10	8	6.0
Sand clay	3.2	0.42	18.3	6	18	7.5
Sand clay	4.5	0.40	19.2	5	32	26.5
Conglomerate	10	0.38	20.0	6	40	9.0
Rock	600	0.35	20.0	20	55	5.0
Rock	6000	0.32	20.0	40	60	15

Table2 Excavation steps

Excavation step	Lcase1	Lcase2	Lcase3	Lcase4	Lcase5	Lcase6	Lcase7	Lcase8
Depth(m)	3	6	9	12	15	18	21	24
Liner		1	1~2	1~3	1~4	1~5	1~6	1~7
Excavation step	Lcase9	Lcase10	Lcase11	Lcase12	Lcase13	Lcase14	Lcase15	
Depth(m)	27	30	33	36	39	41.5	bedrock	
Liner	1~8	1~9	1~10	1~11	1~12	1~13	1~14	

2. Observation system of diaphragm wall

The radial displacement and stress of reinforcement bar of diaphragm wall are observed during the excavation. We located 8 plastic pipes(P1~P8) in the diaphragm wall to measure the displacement. Because the retaining structure is round, the separation between the plastic pipes is equal. Figure 2 showed the location of the plastic pipes, and the length is equal to the depth of the diaphragm wall. We used reinforcement meter to measure the stress of diaphragm wall, the observation location is showed in Figure3.

In order to control the length, in this paper, only the data of P1 and G-1A/B are presented. Figure4 and Figure5 show the stress got from construction site. The stress of the bar is very low, in most zone, the tensile stress and compression stress is not more than 10MPa, the maximal stress is no more than 35MPa. Obviously, the arching of the retaining structure restrains the vertical stress of diaphragm wall. Because the diaphragm wall is socketed into the bedrock, the stress concentrated on the top surface of bedrock. The radial displacement of the diaphragm wall during excavation showed in Figure6. The bedrock restrains the displacement effectively and the maximal displacement is no more than 30mm.

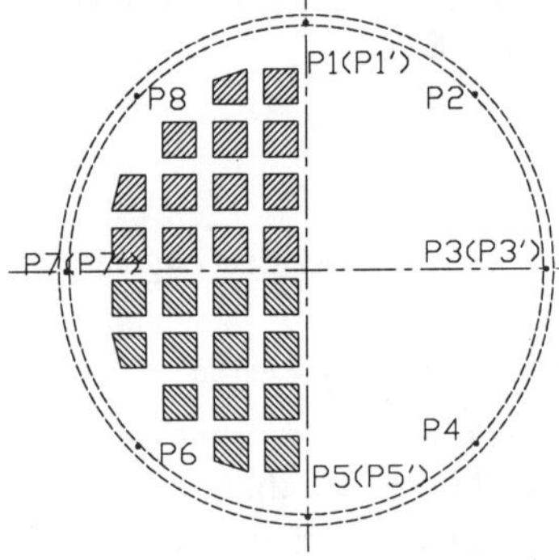

Figure 2 Displacement observation arrangement

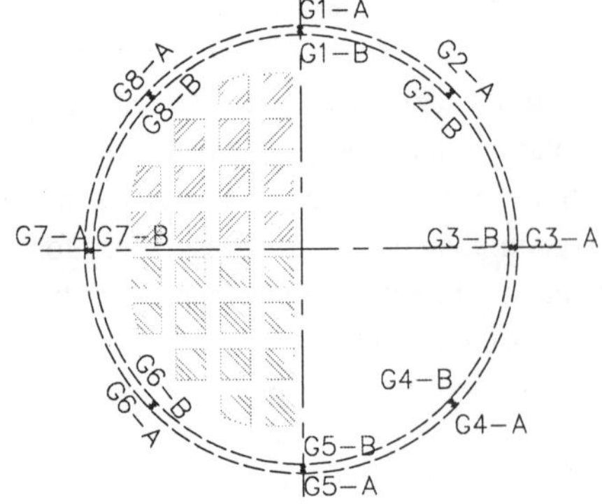

Figure 3 Stress observation arrangement

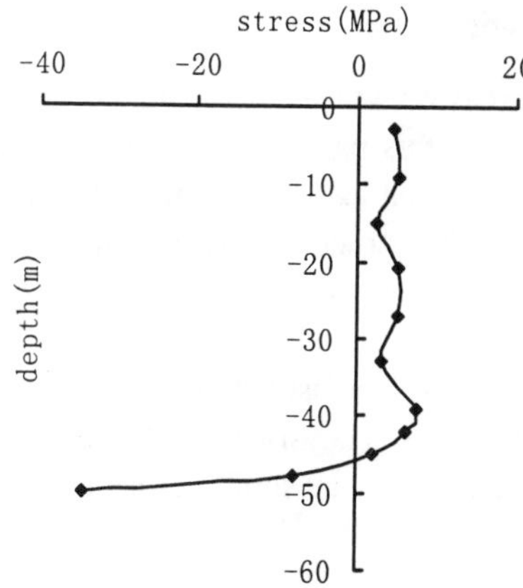

Figure 4 Stress of inner side bar

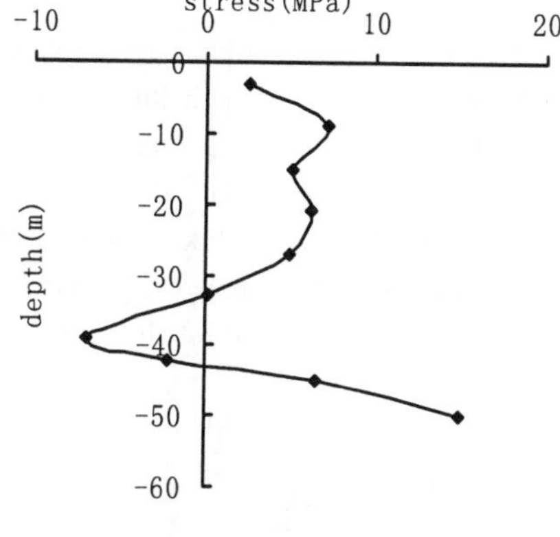

Figure 5 Stress of outer side bar

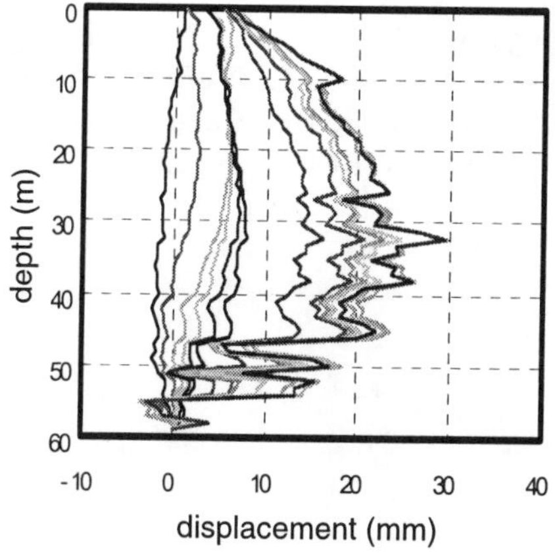

Figure 6 Displacement of diaphragm wall during excavation

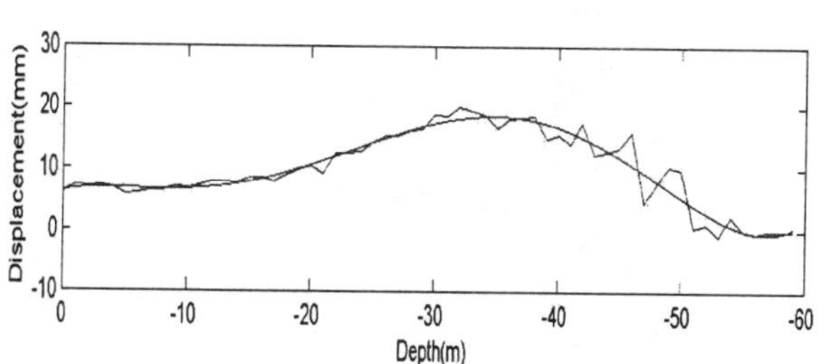

Figure 7 Method of fitting data

3. Back analysis based on in-situ displacement

3.1 *Fitting method of in-situ data*

The in-situ data are in disorder because of the complex working environment of plastic pipe such as the temperature variation and disturbance caused by construction. Considering the continuity of displacement, the in-situ data must be fitted before used in analysis. In this paper, we used the method of minimum square to fit the data. The course showed in the Figure7.

3.2 *FEM model*

We used ANSYS to build the FEM model. The concrete is simulated by solid65[3],which can simulate the crack of concrete. The constitutive relation is defined by Equation (1); the failure criterion is Willam & Warnke criterion. The reinforcement bar is simulated by link8[3],which ignored the lateral shear effect of the bar. The constitutive relation is linear elasticity because the displacement is very low. The soil is simulated by solid45[3]and the constitutive relation is Drucker-Prager model.

$$x \le 1 \quad y = ax + (3-2a)x^2 + (a-2)x^3 \tag{1}$$

$$x \ge 1 \quad y = \frac{x}{\alpha(x-1)^2 + x} \tag{2}$$

$$x = \frac{\varepsilon}{\varepsilon_0} \quad y = \frac{\sigma}{\sigma_0} \tag{3}$$

3.3 *Result of FEM analysis*

Figure8 showed the stress of outer side bar of diaphragm wall, compared it with Figure5, we can know that the result of back analysis fit the in-situ data well. The Figure9 showed the distribution of the crack of diaphragm wall, we know that the crack only appeared only on the top surface of bedrock. So the retaining structure has the enough ability to resist the water and soil pressure.

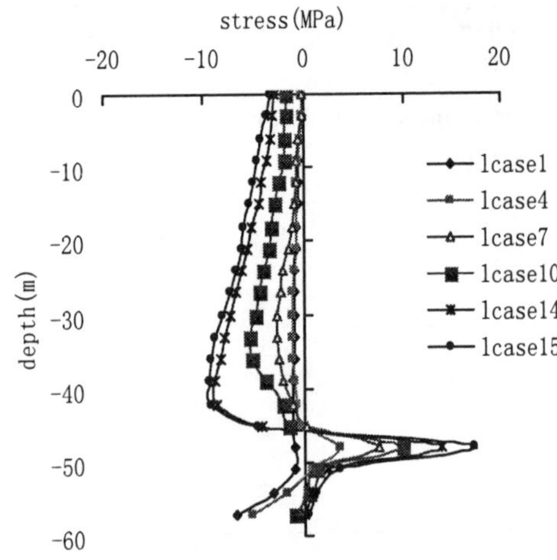

Figure 8 Stress of outer side bar

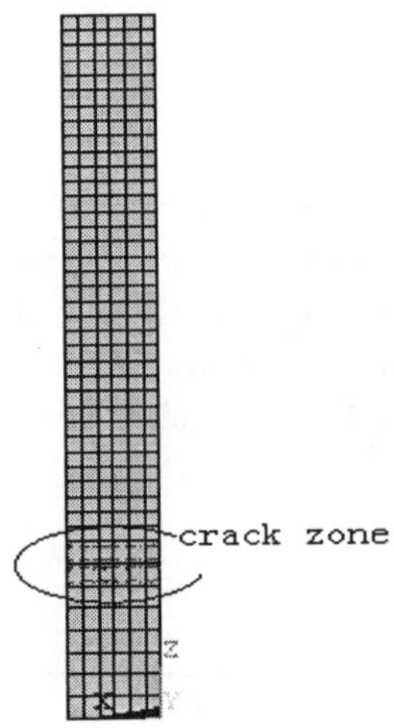

Figure 9 Crack zone of diaphragm wall

4. Conclusions

During the excavation of the deep pit, we used the back analysis given in this paper and in-situ data to monitor the stress of the diaphragm wall. It was the first time that the round retaining structure was used in such deep pit as the south anchor pit of Yangluo bridge, so the construction enterprise had little experience to treat the emergency during construction. The observation experience and back analysis method introduced in this paper can be applied in the similar project.

References

1. Lin ming, Zhang hong and Xu wei. Construction Technology of North Anchor of Runyang Bridge, (2003).
2. Guo huiguang, Sun min and Xu wei. Gigantic Slurry Wall -Analysis of Load -Bearing Properties in Construction Section of Circular Slurry Wall, BUILDING CONSTRUCTION (2004)
3. Ansys company, Ansys Element Reference.*(1999)*

PREDICTION OF BEHAVIOR OF PILED RAFT FOUNDATION FOR SHANGHAI WORLD FINANCIAL CENTER OF 101-STOREY USING COMPARISON CONCEPT

JIAN GONG

Shanghai No.1 Construction Limited Corporation, Shanghai 200120, China

XIHONG ZHAO

College of civil engineering, Tongji University, 1239 Siping Road, Shanghai 200092, China

BAOLIANG ZHANG

Shanghai COSCO-Laoximen Property Development Co. Shanghai 200011, China

Based on the experience of Jimao Building of 88-storey in Shanghai, this paper attempts to predict the behavior of piled raft foundation for Shanghai World Financial Center of 101-storey, such as the bearing capacity of pile and its increase with time, load-sharing between piles and raft, thickness of raft, settlement and theory of design of piled raft foundation. Finally, some comments on these two buildings have been made for examining the theory of super-long pile foundation design for super-tall building in Shanghai.

1. Introduction

Shanghai World Financial Center (SWFC, see Photo 1)[1] will be the tallest building in Shanghai, which located in Pudong, Shanghai. This building is composed of a main tower---a 101-storey reinforced concrete tube-frame structure, 492m high with 4-storey basement, and a five-story podium with 3-storey basement. Now it is under construction. The plan of building and layout of piles for tower are shown in Figure 1.

This building is beside the exiting Jinmao Building (JB) of 88-storey, 420.6m high (see Photo 2)[1]. The plan of building and layout of piles for tower are shown in Figure 2.

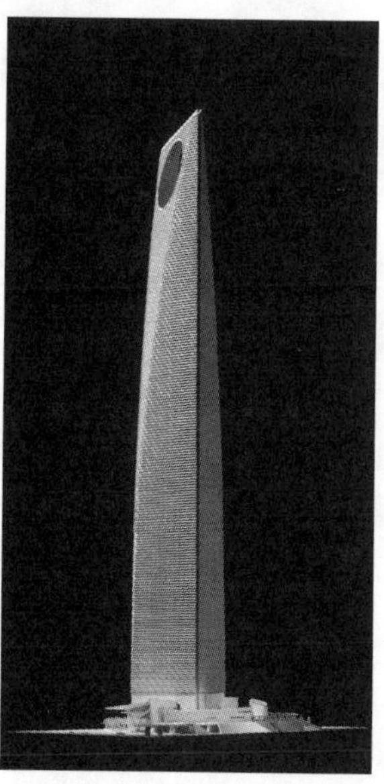

Photo 1. Shanghai World Financial Center

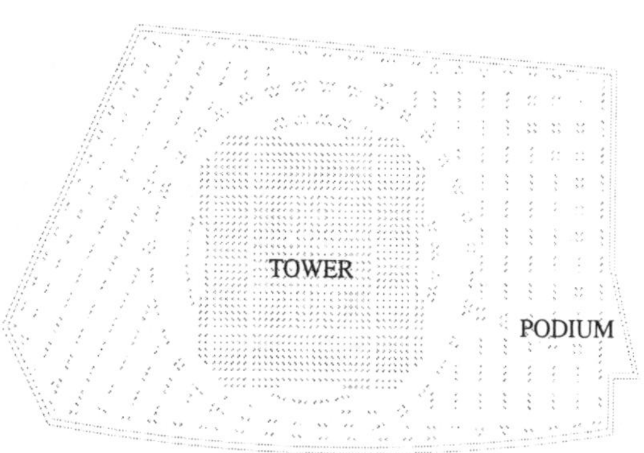

Figure 1. Plan of Shanghai World Financial Center

Photo 2. Jinmao Building

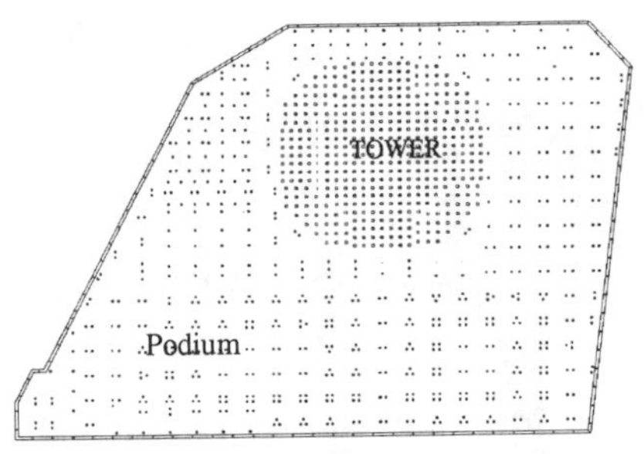

Figure 2. Plan of Jinmao Building

Table 1. Basic data of piled raft foundations for two buildings

Building	Height (m) Number of storeys	Thickness of raft (m) Embedment depth (m)	Number of piles Length of pile (m)	Average load on pile (kN) Allowable bearing capacity of pile (kN)	Total load (kN) Water Buoyancy (kN)	Area of foundation (m^2) Foundation pressure (kN/m^2)
Jinmao Building	420.5 88	4.0 19.65	429 83	7000 7500	3,000,000 0	3519 852.5
SWFC	492.0 101	4.5 18.45	1177 79	3221 4300	4, 400,000 608,220	6200 709.7

The basic data of piled raft foundations for two buildings is listed in Table 1.

This paper, based on the experience of Jinmao Building [1], attempts to predict the behaviour of piled raft foundation for Shanghai World Financial Center using comparison concept.

2. Soil Conditions for Two Buildings

At first let us compare these soil conditions for two buildings as follows.

2.1 *Soil condition for Jinmao Building*

The physical and mechanical properties of soil are shown in Table 2. The average ground water level is 0.5m below the ground surface.

2.2 *Soil condition for Shanghai World Financial Center*

The physical and mechanical properties of soil are shown in Table 3. The average ground water level is the same as that for Jinmao Building.

From Table 1 and Table 2 it can be seen that there are many similarities between these two buildings except that there no soil layer 8 in Shanghai World Financial Center. Such case provides a good condition for predicting the behavior of piled raft foundation.

Table 2. Physical and mechanical properties of soil for Jinmao Building

No.	Soil layers	Thickness (m)	w (%)	γ (kN/m^3)	e	Permeability coefficient		CU test	
						k_h (cm/s)	k_v(cm/s)	C (kPa)	φ (°)
1	Fill	0.900						11.0	17.00
2	Silty clay	2.175	35.3	18.5	1.00	5.11E-5		17.0	20.80
3	Very soft silty clay	4.200	39.6	18.1	1.11	1.77 E-4	2.81E-6	11.5	22.00
4	Very soft clay	9.720	49.0	17.3	1.37	1.64 E-6	2.47 E-7	14.0	13.50
5	Silty clay	8.590	34.4	18.5	0.98	1.33 E-7	2.49 E-5	13.0	20.00
6	Silty clay	3.215	23.0	20.1	0.67			51.0	21.00
7-1	Sandy silt	6.940	31.2	18.6	0.91			4.29	32.70
7-2	Fine sand	28.32	26.9	18.9	0.80			0.00	33.57
8	Sandy silt	3.0~11	32.1	18.5	0.93				
9-1	Sandy silt	3.0~6.0	28.9	18.9	0.84				

Table 3. Physical and mechanical properties of soil for Shanghai World Financial Center

No	Soil layers	Thickness (m)	w (%)	γ (kN/m3)	e	Permeability coefficient		CU test	
						k_h (cm/s)	k_v(cm/s)	C (kPa)	φ (°)
1	Fill	2.9							
2	Clay with silty clay	1.4	36.1	18.5	1.02	7.37E-07	1.30E-07	16.0	12.0
3	Very soft silty clay	4.2	39.7	18.1	1.10	1.10E-04	1.75E-05	6.0	20.3
4	Very soft clay	10.4	48.6	17.3	1.36	8.24E-07	1.32E-07	10.0	10.0
5	Silty clay	7.6	32.7	18.8	0.92	2.34E-06	2.50E-07	10.0	14.0
6	Silty clay	3.9	23.2	20.0	0.68	7.70E-05	2.80E-06	36.0	12.7
7-1	Sandy silt with fine sand	12.6	30.0	19.1	0.83	5.36E-04	3.90E-04	2.0	25.8
7-2	Fine sand	20.8	28.2	19.7	0.72	8.21E-04	1.10E-03	0.0	26.8
7-3	Sandy silt	9.3	30.1	19.3	0.82			2.0	25.0
9-1	Silty sand with silty clay	5.7	27.7	19.5	0.76			2.0	25.0
9-2	Medium coarse sand with gravel	14.5	19.1	20.9	0.53		4.20E-04	1.0	25.0
9-3	Fine sand	54	24.3	20.0	0.68				
10	Silty clay	18.7	26.3	19.8	0.75				

3. Bearing Capacity of Pile for Two Buildings

3.1 *Loading tests of pile for Jinmao Building*

In order to obtain the reliable data of bearing capacity for the design of the piled raft foundation before the commencement of construction the loading tests were carried out on site by 6 steel pipe piles with Φ914. From the comprehensive analysis of data, the allowable bearing capacity of single steel pipe pile is taken as 7500kN.

3.2 *Loading test of pile for Shanghai World Financial Center*

For the design of the piled raft foundation the loading tests were carried out on site by 15 steel pipe piles with Φ700. From data of 15 loading tests of piles (depth of 80m: H3~H9; depth of 79m: H1; depth of 60m: H2, L2~L4; depth of 48m: L1, L5, L6), the ultimate bearing capacities of typical piles for H1(79m), H2(60m), H5(80m) and L2(60m) are 11,000 kN, 9,600 kN, 10,960kN and 7,520kN, respectively.

3.3 *Increase of bearing capacity of piles with time for SWFC*

Loading tests for piles started from July 15th1996 to September 17th 1996, up to now the piles have experienced nearly 8 years. How to estimate the increase of bearing capacity for the piles?

3.3.1 *Summary of study on the increase of bearing capacity of piles with time in Shanghai*

Since 1950's researchers and engineers have paid attention to investigate the increase of bearing capacity of pile in Shanghai soil with time, many valuable data has been accumulated. These data has shown that the increased rate of bearing capacity for driven piles varies from 20% to 60%. Harder the soil in the tip of pile, bigger the increased rate. On the contrary, Softer the soil in the tip of pile, lower the increased rate. Since the driven pile has this inherent characteristic in Shanghai, some old buildings supported by piled raft foundation had added several storeys, all these buildings are in good condition.

For safety sake the increased rate of bearing capacity of pile is taken as 30%. Thus, the increased ultimate bearing capacity of pile minus the ultimate friction force corresponding the embedment depth, then, this result is divided by safe factor of 2. Thus, the allowable bearing capacities of steel pipe pile of Φ700, No. H1(79m), H2(60m), H5(80m) and L2(60m) are 6917 kN, 5232 kN, 5677 kN and 3912 kN, respectively. It can be seen from this comparison, shown in Table 4 that there is a high safety factor in SWFC.

4. Load-sharing between pile and raft for two buildings

The typical data of load-sharing between pile and raft at home and abroad has been summarized in reference 12, in which four methods for calculating load-sharing between piles and raft or box are presented by our research group. Here two practical methods only are used to analyze the load-sharing between piles and raft for Jinmao Building and SWFC.

4.1 *Analysis of load-sharing between pile and raft for Jinmao Building*

4.1.1 *Simple practical method*[3]

$$Pp = P - [pw + (5\%\sim10\%)p]A$$

Where Pp = load carried by piles

Table.4 Comparison between allowable bearing capacity of piles proposed by this paper and LERA

No. of loading test of piles	(1) Allowable bearing capacity of piles proposed by this paper (kN)	(2) Allowable bearing capacity of piles determined by *Leslie E. Robertson Associate* (kN)	(1)/(2)
H1 (79m)	6917	5800	1.193
H2 (60m)	5232	4300	1.217
H5 (80m)	5677	5700	0.996
L2 (60m)	3912	3200	1.223

p = average pressure at bottom of raft = P/A

A = total area of bottom of raft = 3,519m^2.

P = total load = 3,000,000kN = 300,000t

pw = water buoyancy

1. If water buoyancy is considered, when the embedment is 19.65m, for the sake of simplification the water pressure is taken as 19t/m2 , then Pp/P = 0.697 ≈ 70%. When the load carried by raft is taken as 8% of total load.
2. The filtered layer was set under podium, the buoyancy is not considered, then Pp/P = 92%.

Thus, water buoyancy is an important problem.

4.1.2 *Semi-theoretical and semi-empirical method*[3]

$$P_P = \frac{SndE_0}{C}$$

where S = settlement of pile foundation=0.048m

n = number of piles = 429

d = diameter of steel pipe pile =0.914m

E0 = modulus of pile-soil system = 3300t/m2

C = based on *L/d*, *Sp/d* , *n* and Table, we have C=2.93.

Then, *Pp/P* = 0.707 = 70.7%.

So, *Pp/P* calculated using these two methods are almost equal.

4.2 *Analysis of load-sharing between pile and raft for SWFC*

4.2.1 *Simple practical method*[1]

Similarly, for the sake of simplification the water pressure is taken as 18t/m2 , then Pp = 0.67 = 67%.

4.2.2 *Semi-theoretical and sem-empirical method* [3]

Similarly, Pp/P=0.77=77%.

So, the magnitude of Pp/P depends on consideration degree of water buoyancy.

5. Thickness of rafts for two buildings

In 1980s the thickness of raft is taken as 10cm per storey according to a traditional habit in a foreign country which influenced on the design for thickness of raft and bottom slab in Shanghai. For example, the designed thickness of raft was 4m for a building with 43-storey. However, the thickness of raft for Jimao Building with 88-storeydesigned by SOM only was 4m, practice has shown this design is correct and advanced.

According to China foundation code to check the thickness of raft for SWEC, the designed thickness of 4.5m is satisfied.

Nevertheless, how to determine thickness of raft will be discussed in other paper.

6. Settlement for Two Bbuildings

6.1 *Settlement analysis for Jinmao Building*

6.1.1 *Measured settlement for Jinmao building*

This building was completed on 28th Augest1997. The measurement of settlement for tower started from 5th October 1995 to 1st April 2003 for 149 times. The settlements of tower are shown in Figure 3 and Figure 4.

The maximum settlement at center of core-tube area, M7 is 82mm while the minimum settlement at corner, M1, 44mm.Based on the settlements, M1~M13, the settlements are comparatively symmetric, the shape is like a pan.

Average settlement of M1~M13 is 59.4mm while the average settlement of M7, M4, M6, M10 and M8 at the core part, 77.4mm.

According to the regulation of stable settlement for Shanghai code, now the settlement is considered as stable.

6.1.2 *Comprehensive analysis of settlement for Jinmao Building*

Here at first the simple theoretical method, semi-theoretical and semi-empirical method, and statistical method are used to estimate the settlement, and then, the comprehensive method is used to judge the settlement.

The rang of calculated settlement = 91mm ~ 122mm.

Now, the measured maximum settlement is more than 82mm

6.2 *Predicted settlement for Shanghai World Financial Center*

Similarly, the rang of calculated settlement = 81mm ~ 132mm.

The predicted maximum settlement will be more than 82mm.

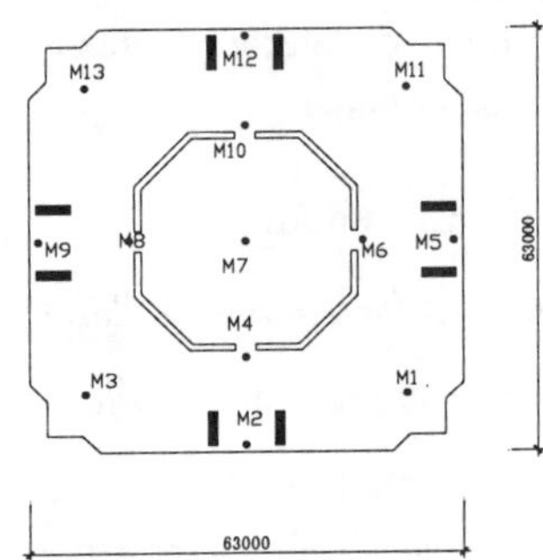

Figure 3. Plan of settlement markers

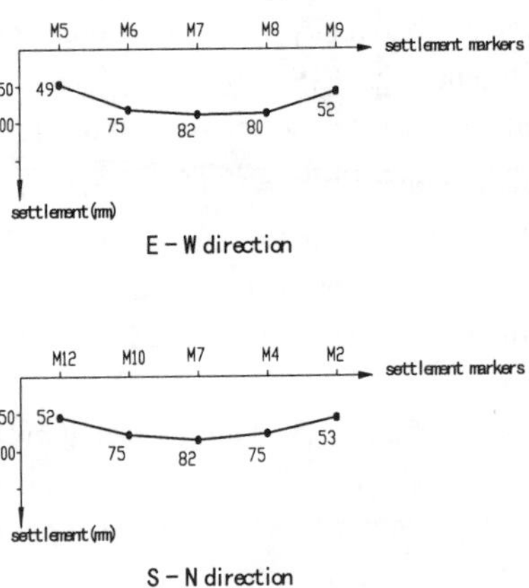

Figure 4. Section of settlement

6.3 *Comment on settlement for two buildings*

Bearing capacity of pile could not be mobilized for SWFC because of L/D >100 and high moment (28,000,000kN-m), especially, the area of foundation is greater than that in Jinmao Building 1.7 times, thus, the effect depth on settlement is more than that in Jnmao Building, the analytical data of piled raft foundations for two buildings in Table 8 has also indicated that the estimated settlement will be greater than that in Jinmao Building.

7. Analysis using theory of design of piled raft foundation interaction

A theory on interaction between pile-foundation and superstructure [3] had been presented and was used to analyze the super-long pile-foundation interaction for super-tall building in Shanghai[11]. Now, for simplification, interaction between pile and raft is considered only and load-sharing between raft and piles is not considered to make a comparison between Jinmao Building and Shanghai World Financial Center briefly.

The computed results on load on pile group and moment in raft are shown in Figure 5 and Figure 6.

From the comparison of loads on pile group and stresses in raft between Jinmao Building and SWFC, it can be seen that the design of piled raft foundation for SWFC is conservative.

8. Concluding remarks

8.1 *Bearing capacity of piles*

Engineering practice has shown that the allowable bearing capacity of pile for Jinmao Building is reasonable.

For Shanghai World Financial Center the increase of bearing capacity of pile with time taken account is also reasonable. Although for pile of Φ700 with L/d>100 , such a pile maybe effect the capacity of resistance to moment. Based on the preliminary design[10] , the moment caused by wind and so on is28,000,000kN-m, the additional vertical load increases up to 1/3 of static load, under this condition the design should consider the high safety factor. For example, the total buoyancy of water is only taken as 60%.

8.2 *Prediction of load-sharing between piles and raft*

1. For Jinmao Building the tower and podium is connected with hinge joints and the filtered layer is set under podium, it is still necessary to consider the water buoyancy. However, the load-sharing between piles and raft has been affected.
2. For Shanghai World Financial Center for sake of safety it is necessary to consider 60% of water buoyancy. The methods for calculating the load-sharing between piles and raft presented in this paper not only provide a basis for reducing the number of piles, but also a reference to judge safety factor of engineering.

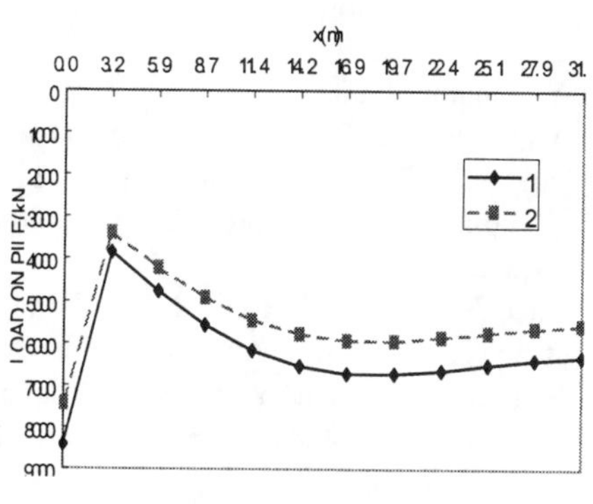

(a) Jinao Building

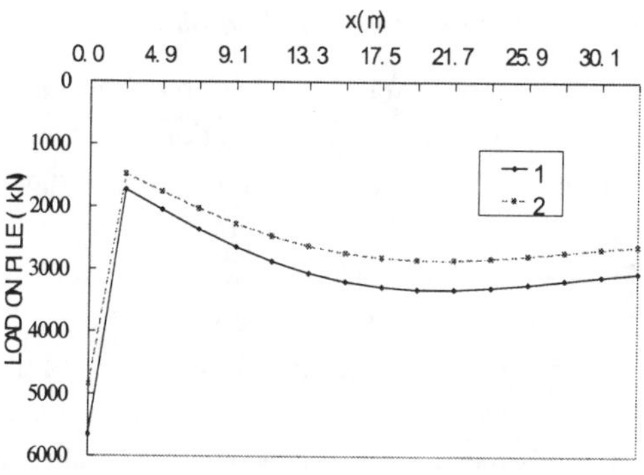

(b) SWFC

(1.without considering water buoyancy ; 2.considering water buoyancy)

Figure 5. Load on pile group along center line

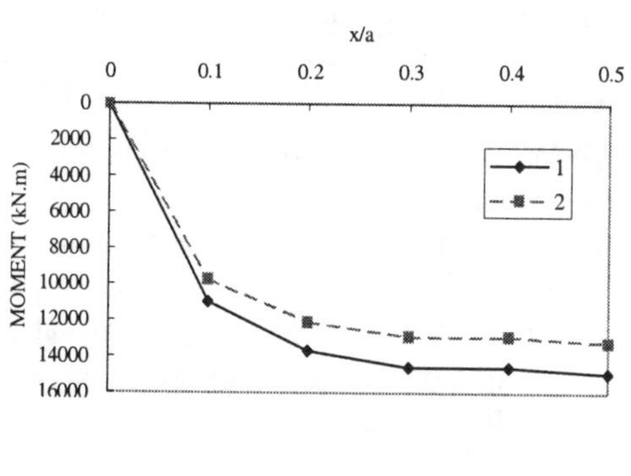

(a) Jinao Building

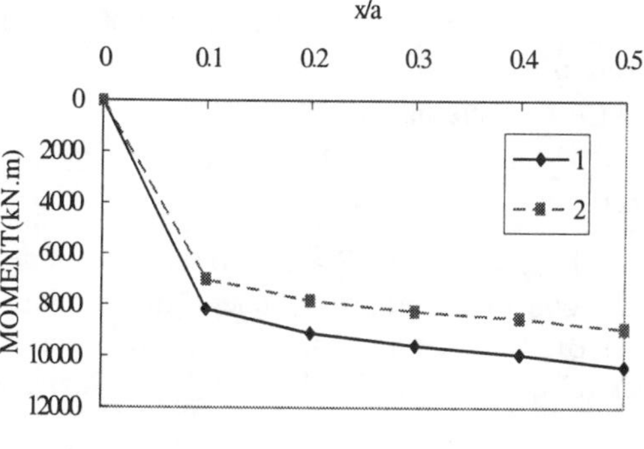

(b) SWFC

(1.without considering water buoyancy ; 2.considering water buoyancy)

Figure 6 Moment distribution along center line

8.3 *Thickness of raft foundation*

1. For Jinmao Building based the design of SOM, the thickness of 4 m is safe as well as advanced design. It has played an important role to expedite the improvement of Chinese code.
2. For Shanghai World Financial Center there is no any data of field experiment for thickness of raft more than 4m, in that case the 4.5m thick is necessary and adequate.

Now in China the thickness of design is conservative. For this purpose of reasonable design the field experiments for super-tall building and super-long piled raft foundation with 4m~6m thick in Shanghai are being conducted.

8.4 *Prediction of settlements*

1. The comprehensive analysis of settlement presented in this paper to predict the settlement for Jinmao Building is feasible and practical.
2. Based on the measured settlement and comprehensive analysis of settlement for Jinmao Building and the comprehensive analysis of settlement for Shanghai World Financial Center, the settlement for SWFC will be about 10cm.

8.5 *Theory of design for piled raft foundation* [11]

Based on the analysis for these two buildings in the design of super-long piled raft foundation for super-tall building it is necessary to control the stress state in elastic static. Under this condition the theory of interaction between superstructure and foundation can be used to the design of piled raft foundation. The comparison between Jinmao Building and SWFC in section 7 has shown that the theory of interaction presented in this paper is reasonable and practical.

If the load-sharing between piles and raft can be considered, the design of piled raft foundation will get more economical benefit.

Based the comprehensive analysis mentioned above, in one word, Jinmao Building is in very good condition, there are more potential of safety factor in the engineering of piled raft foundation in Shanghai World Financial Center, certainly it will be safe.

References

1. Gong J. Research & Practice on Foundation & Excavation Engineering for Super-Tall Building & Super-Big Structure in Shanghai, Tongji University Doctoral Dissertation (in Chinese), 2003, Shanghai, China
2. Hooper JA. Observations on the behavior of a piled raft foundation in London Clay. Proc. ICE, 1973, 55(2), 855-877
3. Zhao XH. et al. Theory of Design of Piled Raft and Box Foundation for Tall Buildings in Shanghai (in Chinese, 1989; Enlarged edition in English ,1998), Tongji University Press
4. Zhuang GM., Lee IK. & Zhao XH. Interaction Analysis of Behavior of Raft-Pile Foundation. Proc. Inter. Conf. on Geotechnical Engineering for Coastal Development-theory & Practice on Soft Ground-(GEO COAST"91), Sept.3~6,1991, Yakohama, Japan, 759~764.
5. Yang JB. and Zhao XH. Optimum design of piled raft (box) foundations for tall building. Computation Mechanics [J] (in Chinese), 1997. 14(2)241~244
6. Zhao XH, Cao MB, Lee IK and Valliappan S. Soil-structure Interaction Analysis and its Application to Foundation Design in Shanghai, Proc. 5th Int. Conf. on Numerical Methods in Geomechanics. Nagoya, Japan, 1985. 805 ~ 811
7. Katzenbach R et al. Soil-structure-interaction of the 300m high Commerzbank tower in Frankfurt am Main measurements and numerical studies, Proc. 14th ICSMFE, Hamburg, 1997,V.2.1081~1084
8. Hu CL et al. Field monitoring and comprehensive analysis of piled raft foundation for tall building under complex conditions, Rock and Soil Mechanics, J (in Chinese). 2003, 24(4):673-676
9. Jinmao Building — Strategic Decision, Design & Construction Ed. by Zhang GL. and L WS, Construction Industry Press (in Chinese), 2000, China, 58-73
10. MORI, KPF, LERA & ECADI . Shanghai World Financial Center--Structural Preliminary Design Report , Dec. 2000
11. Zhao XH, Gong J and Zhang BL A study on super-long pile-foundation interaction for super-tall building in Shanghai, Computers and Geotechnics J, 2004.in press
12. Zhao XH, Gong J Load-sharing between pile and raft or box in field experiment, calculation and mechanism analysis Rock and Soil Mechanics [J], (in Chinese), 2005. 26(3)337~3

PILED RAFT DESIGN PROCESS FOR A HIGH-RISE BUILDING ON THE GOLD COAST, AUSTRALIA

PARAN MOYES
Senior Geotechnical Engineer, Coffey Geosciences Pty Ltd
8/12 Mars Road, Lane Cove West NSW 2066

HARRY G POULOS
Senior Principal, Coffey Geosciences Pty Ltd
Emeritus Professor. Department of Civil Engineering, University of Sydney, Australia

JOHN C. SMALL
Professor. Department of Civil Engineering, University of Sydney, Australia

FRANCES BADELOW
Principal, Coffey Geosciences Pty Ltd

This paper describes the process of design of a piled raft foundation for a high rise residential building on the Gold Coast in Queensland, Australia. The design process comprised an initial stage of geotechnical site characterization using the results of a series of investigation boreholes to prepare a subsurface model and derive geotechnical parameters for raft and pile design from empirical correlations. Following this a preliminary analysis was undertaken using a combination of elastic theory and allowances for non linear behaviour of the piled raft system to assess the viability of such a foundation system and any potential advantages of a piled raft over conventional fully piled foundation systems. Finally, a detailed analysis was undertaken using the GARP8 computer program. These detailed analyses were used to design a more efficient piled raft system and to provide design actions for structural design of the foundation system for a variety of load combinations.

1. Introduction

The development of tall buildings on land previously regarded as unsuitable for large structures presents several geotechnical problems with regard to the design and assessment of foundations. Design of the buildings must take into account both the short and long term deformations of the foundations (serviceability limit state) and the strength of the foundations at ultimate loading (ultimate limit state). Piled raft foundations utilise piled support for control of settlements with piles providing most of the stiffness at serviceability loads, and the raft element providing additional capacity at ultimate loading. A geotechnical assessment for design of such a foundation system therefore needs to consider not only the capacity of the pile elements and the raft elements, but their combined capacity and interaction under serviceability loading. This paper presents details of the geotechnical assessment and design carried out for a piled raft foundation system for a residential tower development located in Surfers Paradise, Queensland Australia. This assessment required the evaluation of a piled raft system subjected to large ultimate loadings with consideration also being given to the satisfaction of stringent differential settlement criteria.

2. Residential Tower Project

The project comprises a 30 storey, 176 unit residential tower located in Surfers Paradise, Queensland. The construction of the development commenced in late 2004 with anticipated completion in late 2005.

2.1 *Geological and geotechnical conditions*

The Surfers Paradise area is underlain by alluvial sediments comprising sands and clays. Beneath the alluvial deposits is a residual soil strata of Silty Clay overlying the meta siltstone bedrock.

2.1.1 *Site geotechnical conditions*

The geotechnical investigations at the site comprised nine boreholes drilled in two phases of fieldwork. Figure 1 illustrates the locations of the geotechnical investigation boreholes. The first phase of investigation comprised the drilling of six boreholes (BH1 to BH6) that were terminated within about 1m penetration into a sandy gravel layer, some 5m above the weathered rock layer. A second phase of investigation boreholes followed ("X" series) which were extended through the sandy gravel layer and encountered extremely weathered rock, typically described as sandy clay/clayey sand and sandy gravel.

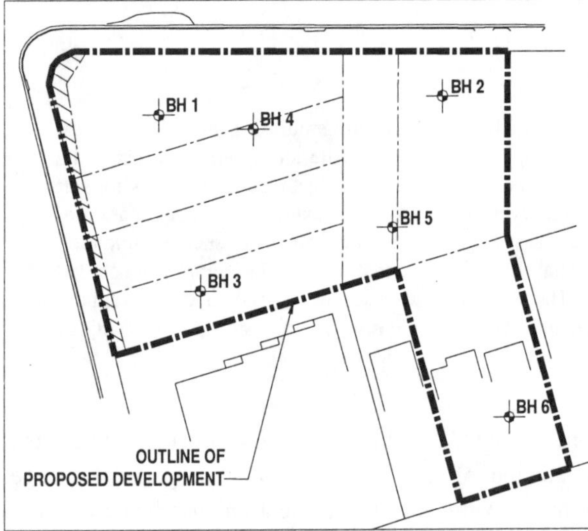

Figure 1. Borehole locations

The ground conditions typically consist of a relatively thick layer of dense to very dense sand, underlain by a layer of peat with some sand present.

Below the peat there is another dense sand layer and then a layer of stiff to very stiff, silty clay with some sand. Underlying this layer is extremely weathered rock, which is described as gravel in some of the boreholes. The extremely weathered rock is underlain by slightly weathered metasiltstone rock at between 40.5m and 42.5m depth below existing ground level.

Based on the subsurface information obtained from boreholes X1, X3, BH1, BH4 and BH5, a geotechnical summary of the borehole details was formulated (refer Figures 2 and 3). The quantitative information is limited and consists of SPT data. The existing ground surface level is generally about RL+5.5m, the ground water table is at about RL+0.7m, and the upper 6m or so of soil was to be excavated beneath the tower to allow for construction of the basement.

3. Geotechnical Model

On the basis of the information in Figures 2 and 3, a geotechnical model was developed in order to analyse the piled raft foundation option. As the conditions at BH5 were the least favourable for foundation design, this borehole was used as the main basis for the geotechnical model. Figure 3 shows the stratigraphy adopted for the model and the relevant geotechnical parameters selected for the various strata appropriate for either the raft or the pile design, where:

E_s(raft) = soil modulus for assessment of raft behaviour

p_u = ultimate bearing capacity of the raft

E_s (pile) = soil modulus for assessment of pile behaviour

f_s = ultimate pile shaft friction

f_b = ultimate pile end bearing capacity.

In assessing these parameters, use was made of the correlations between SPT data and foundation stiffness and capacity parameters suggested by Decourt (1995), together with the authors' previous experience in parameter selection for use in piled raft design.

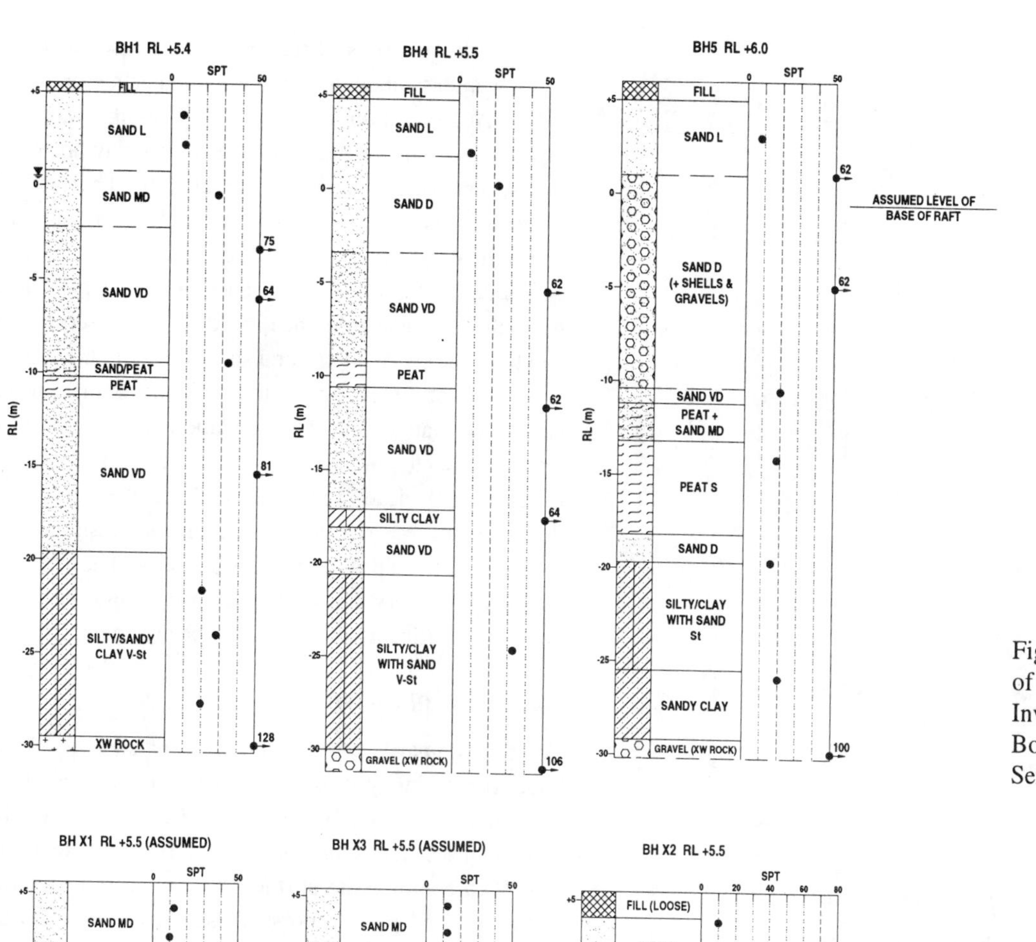

Figure 2: Summary of Geotechnical Investigation Boreholes – BH Series

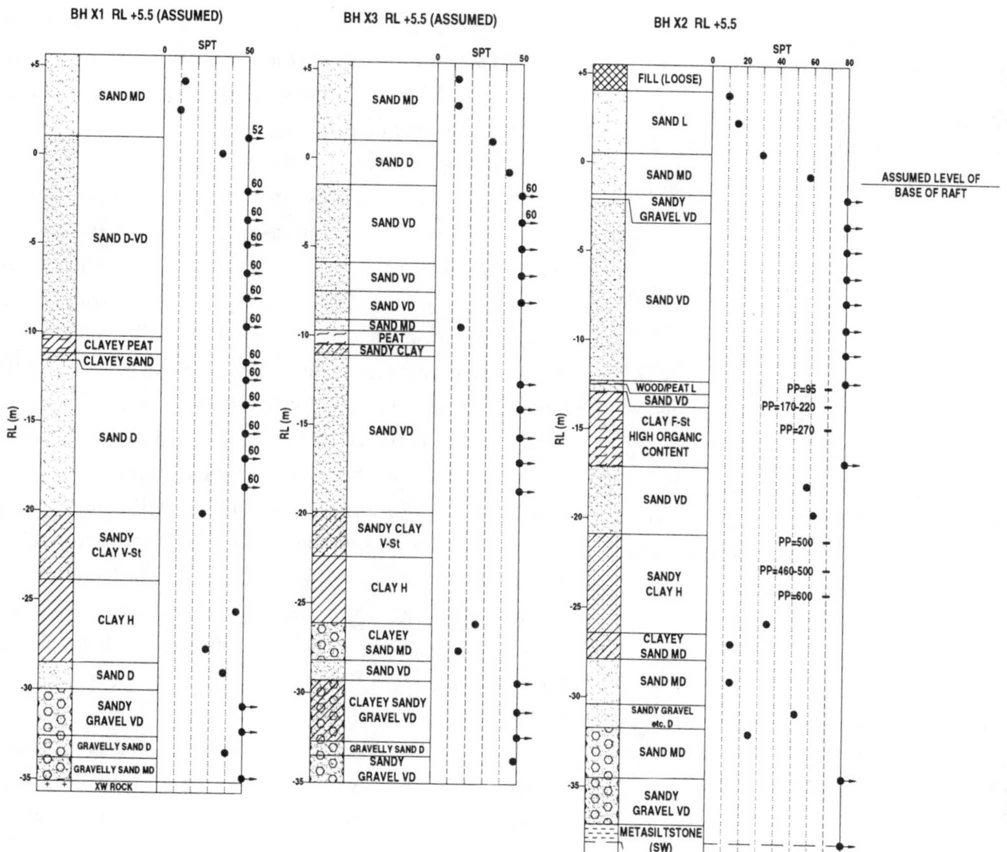

Figure 3: Summary of Geotechnical Investigation Boreholes – X Series

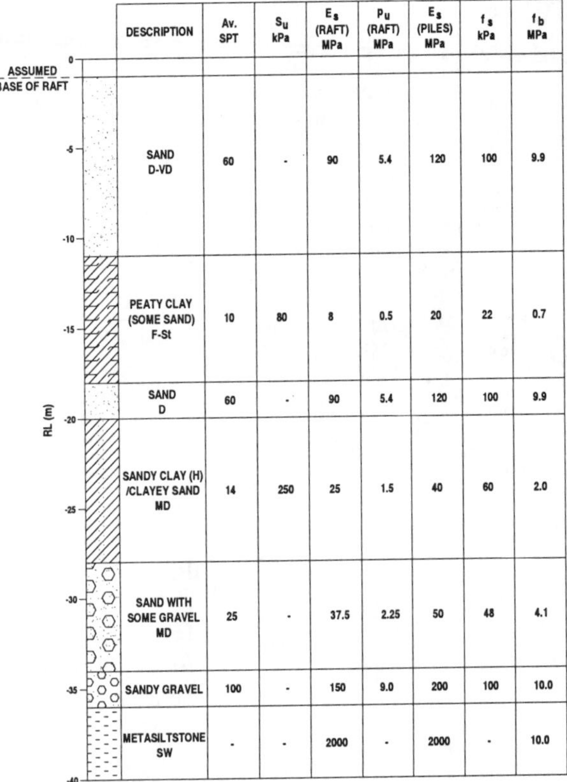

DESCRIPTION	Av. SPT	S_u kPa	E_s (RAFT) MPa	p_u (RAFT) MPa	E_s (PILES) MPa	f_s kPa	f_b MPa
SAND D-VD	60	-	90	5.4	120	100	9.9
PEATY CLAY (SOME SAND) F-St	10	80	8	0.5	20	22	0.7
SAND D	60	-	90	5.4	120	100	9.9
SANDY CLAY (H) /CLAYEY SAND MD	14	250	25	1.5	40	60	2.0
SAND WITH SOME GRAVEL MD	25	-	37.5	2.25	50	48	4.1
SANDY GRAVEL	100	-	150	9.0	200	100	10.0
METASILTSTONE SW	-	-	2000	-	2000	-	10.0

Figure 4. Stratigraphic and Geotechnical Parameters Adopted

In addition to the parameters shown in Figure 4, it was assumed that, below the raft, the soil modulus (E_s) for reloading is 3 times the value for initial loading (shown in Figure 3). This assumption was made to evaluate the benefits of excavation of the upper 6m of soil, and the consequent partial compensation that this excavation provides.

4. Preliminary Geotechnical Assessment

4.1 *Introduction*

Prior to the detailed geotechnical assessment, a feasibility assessment was conducted of various foundation schemes. A geotechnical assessment was carried out for the following foundation schemes:

- A raft alone, without piles;
- A raft with 50 piles;
- A raft with 70 piles;
- A raft with 140 piles (approximating of the proposed design as per the concept foundation drawing prepared by the piling contractor).

For the purposes of the preliminary assessment, the piles were assumed to be 0.7m diameter Continuous Flight Auger (CFA) piles extending to the thin sand layer directly above the stiff to very stiff clay layer, and having an average length of about 18m. The raft was taken to be a square area 50m by 24m in plan, and approximately 0.8m thick. The thickness is of little consequence for the overall load-settlement behaviour, but will of course influence the differential settlements and the bending moments and shear forces in the raft.

Based on the results of the initial assessment it was concluded that while a raft foundation alone would have an overall factor of safety of more than 10 with respect to combined dead and live loading, the foundation design would be governed by settlement considerations, rather than by ultimate bearing capacity. The presence of a dense to very dense sand deposit allows the raft to develop significant vertical load capacity and stiffness.

4.2 *Method of analysis*

For preliminary assessment, an extension of an approximate analysis of the piled raft load-settlement behaviour described by Poulos (2002) was used. This method uses the equations developed by Randolph (1994) to compute the stiffness of a piled raft, in terms of the raft and pile group stiffness values, and also the load sharing between the raft and the piles. A tri-linear load-settlement curve is derived from this process.

An extension to this procedure was developed by Poulos (2005) for compensated piled rafts. In this extended procedure, account is also taken of the increase in soil stiffness during the re-loading phase after excavation has been carried out. This increase in soil stiffness increases the stiffness of the raft during the reloading process, and consequently, the raft carries a greater proportion of the load until the soil pressure below the raft reaches the previous pressure at which virgin loading conditions prevail; this will usually occur when the raft pressure balances the pressure reduction due to the excavation. Thereafter, the raft stiffness is controlled by the virgin loading soil stiffness, which will be smaller than the reloading stiffness. Thus, at that stage, the piles will then tend to take a larger proportion of the further applied load.

For the preliminary assessment, it was assumed (somewhat conservatively) that the soil modulus for reloading was 3 times the value for first or virgin loading. The analysis was carried out using a MATHCAD worksheet developed by the authors.

4.3 *Analysis results*

The results of the preliminary geotechnical assessment are illustrated in Figures 5 and 6. The computed load-settlement relationships are shown in Figure 5, for loads up to 3 times the serviceability load. This figure shows that the settlement tends to decrease as the number of piles increases, but at a decreasing rate. This figure also indicates the transition from reloading to virgin loading conditions at relatively low load levels, and the consequent reduction in stiffness once virgin loading conditions are re-established.

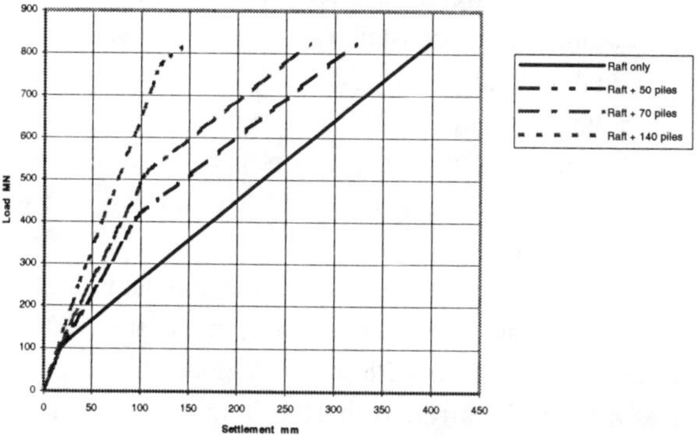

Figure 5: Computed Load Settlement Curves from Preliminary Analysis

Table 1 summarizes the computed settlement of the foundation system at 1, 2 and 3 times the serviceability loads, while Figure 6 plots these settlements versus the number of piles.

Figure 6: Effect of Number of Piles on Settlement from Preliminary

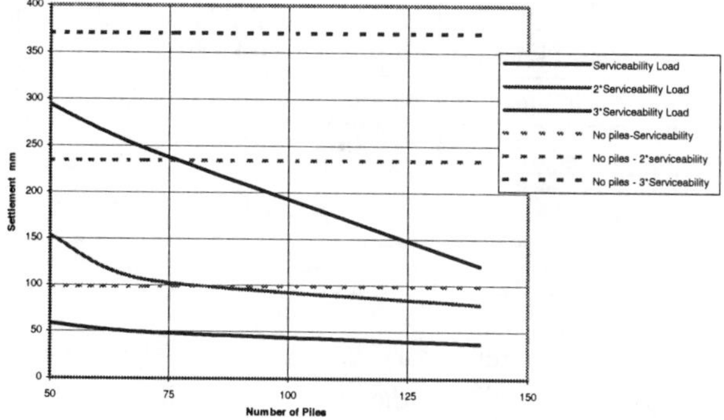

Analysis

Table 1. Summary of Computed Settlements

Number of Piles Below Raft	Average Settlement at Serviceability Load (257.8 MN)	Average Settlement at 2*Serviceability Load (515.6 MN)	Average Settlement at 3*Serviceability Load (773.4 MN)
0	98 mm	234 mm	370 mm
50	58 mm	153 mm	294 mm
70	49 mm	106 mm	247 mm
140	37 mm	79 mm	121 mm

The following observations can be made:

1. The average settlement of the raft alone is likely to be excessive;
2. The average settlement at the serviceability load decreases significantly when a relatively small number of piles are included in the foundation;
3. There is relatively little benefit, at the serviceability load, in doubling the number of piles from 70 to 140. However, the settlement at higher load levels is decreased significantly.

Only the overall load-settlement behaviour was considered in this preliminary assessment. A number of other issues including differential settlements, the thickness of the raft, and the size, length and required locations of the piles were not considered in detail at this stage, but were addressed at the detailed design stage.

4.4 *Recommendations from preliminary geotechnical analysis*

Based on the results of the preliminary geotechnical assessment, it was recommended that a piled raft foundation system should provide a cost-effective solution for the tower. The average settlement at the serviceability load was estimated to be of the order of 35-60mm, depending on the number of piles.

5. Detailed Foundation Design

Following the preliminary geotechnical assessment, a piled raft foundation was adopted for the proposed building. The foundation design comprises a 0.8m thick reinforced concrete raft founded at RL -1m AHD. Beneath the raft, 136 piles were initially located to support loading from columns, the building core and a number of walls. The number of piles was modified on the basis of the analysis undertaken.

The pile design that was assessed to be appropriate for this foundation system comprised 18m long, 0.7m diameter Continuous Flight Auger (CFA) piles. These piles were designed to be founded on the relatively thin, dense sand layer, but allowance was made in the analyses for the possibility of the pile tip extending into the underlying sandy clay/clayey sand layer. Accordingly, the analysis carried out assumed the pile tip capacity to be that for the sandy clay/clayey sand. The ultimate geotechnical capacity of each pile was assessed to be about 4.2MN.

5.1 *Loading*

The load data supplied by the project structural engineer was used to carry out a number of geotechnical assessments based on different combinations of loading. The loading data provided comprised a series of vertical loadings at column locations. Table 2 shows the loading combinations that were assessed based on Section 2.1 of the Australian Loading Code AS1170.0-2002.

Table 2. Loading Cases for Piled Raft Assessment

Limit State	Load Factors			
	Dead Load (G)	Live Load (Q)	Positive Wind Load (permissible) (W_p+)*	Positive Wind Load (permissible) (W_p-)
Serviceability	1	0.4	-	-
Ultimate	1.2	1.5	-	-
	1.2	0.4	1.35	-
	1.2	0.4	-	1.35
	0.9	-	1.35	-
	0.9	-	-	1.35

* *A load factor of 1.35 was adopted to factor the permissible wind load to ultimate wind load*

5.2 *Serviceability limit state (SLS) assessment of piled raft*

The load data supplied by the structural engineer was used to carry out a geotechnical assessment based on serviceability loading for assessment of deflections of the proposed piled raft foundation. The serviceability loading combination was assessed, based on the information provided in Section 2.1 of AS1170.0-2002. The geotechnical assessment of the piled raft was undertaken using the GARP8 program initially developed by the Centre for Geotechnical Research at Sydney University for analysis of piled raft foundations.

GARP (General Analysis of Rafts with Piles) is based on a finite element analysis of the raft, and a boundary element analysis of the piles. The contact stress that acts between the raft and the soil is assumed to be made up of a series of uniform blocks of pressure that act over each element in the raft. Each of the piles is assumed to apply a reaction to the raft at a point (corresponding to a node in the raft).

The boundary element analysis is used to calculate the interaction of pairs of piles, or of a pile with the raft. In doing this, it is assumed that the soil is an elastic material. If the soil is layered, a weighted average of the properties of the soil layers is used in determining the equivalent elastic properties of the overall soil mass.

If the behaviour of the piles is non-linear, this is modeled by allowing the stiffness of the piles to reduce with load level according to a hyperbolic law. However, the interaction between piles and between the piles and raft is assumed to be constant (i.e. to have the values for the original pile stiffness).

Loading on the raft can include point loads, uniformly distributed loads, and moment loadings. As well, the raft can have different thicknesses assigned to the elements that make up the mesh.

The deflections, shear forces and moments in the raft and the vertical loads on the piles due to the loading were assessed. The GARP8 analysis modelled the piled raft as a 0.8m thick raft with 18m long, 0.7m diameter piles located as per the initial foundation concept drawing.

An initial assessment was carried out using a piled raft foundation with 136 piles. The results of the initial assessment are provided in Table 3.

Table 3. Results of Preliminary GARP Analysis for Settlement Under Serviceability Loads

Criteria	Value
Maximum settlement	36 mm
Minimum settlement	0.4 mm
Rotation $(\theta)_x$ max	0.0014 radians (1 in 714)
θ_x min	-0.0020 radians (1 in 500)
θ_y max	0.0025 radians (1 in 400)
θ_Y min	-0.0024 radians (1 in 416)

Following the preliminary assessment, analyses were undertaken to refine the piled raft design. This comprised the removal of some of the piles and subsequent assessment of the performance of the revised design under the serviceability loading. The revised foundation design reduced the number of piles from 136 piles to 123 piles.

Table 4. Results of GARP Analysis for Settlement Under Serviceability Loads

Criteria	Value
Maximum settlement	44 mm
Minimum settlement	1 mm
Maximum differential settlement between adjacent columns	10mm (1/444)

It should be noted from Table 4 that the maximum computed settlement of 44mm was less than the maximum specified allowable value of 50mm, while the maximum differential settlement between adjacent columns was in the order of 1/400, which occurred between the a relatively lightly loaded exterior column and a highly loaded interior column. Following discussion with the piling contractor, further assessment was carried out with an additional pile located at the centre of the pile group beneath one of the most heavily loaded columns to reduce the differential settlement. The results of the further assessment indicated a differential settlement of 1/444 or 0.00225. The calculated settlement contours for this case are shown in Figure 7.

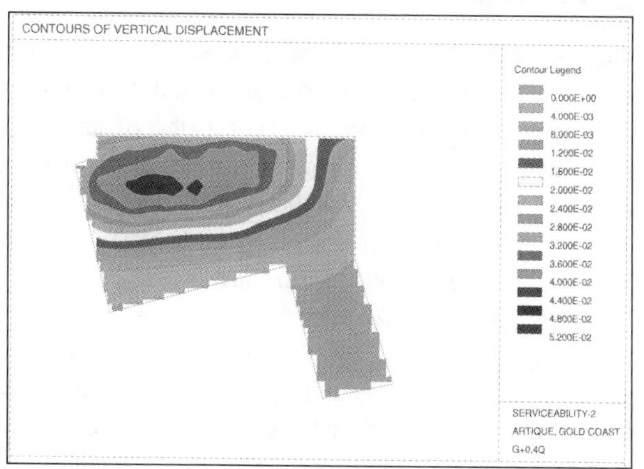

Figure 7. Calculated Settlement Contours (in m) – Serviceability Loading

This revised pile layout was then used to in the assessment of the ultimate limit state design.

5.3 *Ultimate limit state (ULS) assessment of piled raft*

Using the Ultimate Limit State (ULS) loading combinations summarized in Table 2, assessment of the piled raft performance was made. The resultant computed maximum and minimum values for the various structural actions are summarized in Table 5.

Table 5. Results of GARP Analysis for ULS Load Cases

Item	Value	Loading
M_x max	2.66 MNm/m	$1.2G + 1.35W_p + \Psi Q$
M_x min	-0.89 MNm/m	$1.2G + 1.35W_p + \Psi Q$
M_y max	3.27 MNm/m	$1.2G + 1.35W_p + \Psi Q$
M_y min	-1.11 MNm/m	$1.2G + 1.35W_p + \Psi Q$
M_{xy} max	0.82 MNm/m	$1.2G + 1.35W_p + \Psi Q$
M_{xy} min	-0.78 MNm/m	$1.2G + 1.35W_p + \Psi Q$
V_x max	7.59 MN/m*	$1.2G + 1.35W_p + \Psi Q$
V_x min	-7.11MN/m*	$1.2G + 1.35W_p + \Psi Q$
V_y max	7.02 MN/m*	$1.2G + 1.35W_p + \Psi Q$
V_y min	-6.75 MN/m*	$1.2G + 1.35W_p + \Psi Q$
Pile Load (compression)	6.37 MN	$1.2G + 1.5Q$
Pile Load (tension)	-0.24 MN	$0.9G + 1.35W_p$

The values of moment and shear force are for the raft slab. The GARP8 program also provided contour plots of the moments and shears calculated for the various ultimate load cases, and distributions of bending moment along selected sections of the piled raft for use in the structural design of the raft slab.

Some of the pile loads calculated by GARP8 exceeded the maximum design capacity of 4.2MN. These results were based on the GARP8 analysis being elastic only, whereas the actual behaviour of the piles in the piled raft system would restrict the pile from carrying more load than its ultimate capacity. Subsequent upgrades have been made to the GARP program, so that when a pile reaches its design capacity, the program redistributes any excess reactions in the piled raft system.

5.4 *Sensitivity Assessment*

In addition to the serviceability and ultimate limit state assessments detailed previously, additional assessment was undertaken to evaluate the sensitivity of the proposed piled raft design to variability in the foundation system. The sensitivity assessment was carried out by considering the effect of a pile of reduced or increased stiffness located below the most heavily loaded column. The pile was modelled with a ±40% change in pile stiffness and was applied for both the serviceability limit case and also the ultimate case where the pile had previously been assessed as having the greatest load (Load Case ART 3b -1.2G + 1.35W_p + ΨQ). Table 6 provides a summary of the effect of the ±40% change in pile stiffness. The deflections were evaluated using the serviceability load while the moments were assessed using the ultimate load case.

Table 6. Results of GARP Analysis for Sensitivity Assessment – Serviceability Case

Item	Standard Pile Stiffness	Increase Pile Stiffness (+40%)	Decreased Pile Stiffness (-40%)
Deflection beneath load	39mm	38mm	38mm
Differential between column and edge of raft	9mm	8mm	8mm

The results of the assessment illustrate that the redundancy inherent in a piled raft foundation system redistributes the loading and settlement across the piled raft with relatively little influence on the overall behaviour.

Table 7 provides a summary of the resultant maximum and minimum values for the various structural actions with the changed pile stiffnesses.

The redistribution of the structural actions is reflected in the minor differences in the assessed values of moment and shear forces in the ultimate loading cases. The minor differences between the standard pile stiffness case and the reduced and increased pile stiffness cases can again be attributed to the effect of the 0.8m thick raft redistributing stresses and forces.

Table 7. Results of GARP Analysis for Sensitivity Assessment – Ultimate Cases

	Standard Pile Stiffness	Increased Pile Stiffness (+40%)	Decreased Pile Stiffness (-40%)
Item	Value	Value	Value
M_x max	2.66 MNm/m	2.67 MNm/m	2.67 MNm/m
M_x min	-0.86 MNm/m	-0.86 MNm/m	-0.86 MNm/m
M_y max	3.27 MNm/m	3.26 MNm/m	3.26 MNm/m
M_y min	-1.11 MNm/m	-1.11 MNm/m	-1.11 MNm/m
M_{xy} max	0.82 MNm/m	0.84 MNm/m	0.84 MNm/m
M_{xy} min	-0.79 MNm/m	-0.75 MNm/m	-0.75 MNm/m
V_x max*	7.59 MN/m	7.59 MN/m	7.59 MN/m
V_x min*	-7.10 MN/m	-7.10 MN/m	-7.10 MN/m
V_y max*	7.02 MN/m	7.02 MN/m	7.02 MN/m
V_y min*	-6.73 MN/m	-6.73 MN/m	-6.73 MN/m
Pile Load* (compression)	6.23 MN	6.22 MN	6.22 MN
Pile Load (tension)	-0.10 MN	-0.10 MN	-0.10 MN

** Pile load calculated based on elastic theory only. Maximum pile loads equal to maximum design capacity of 4.2MN.*

6. Conclusions

This paper has illustrated the process of design of a piled raft foundation for a large residential development using a three stage procedure, consisting of an initial assessment of the feasibility of the design, a middle stage of refining pile locations and depths, and a detailed design stage of assessing the behaviour of the foundation under various loading cases.

The GARP program provided an efficient computational method for the analysis of a complex geotechnical and structural problem and delivered design actions from the analysis which could be readily used by structural engineers for structural design of the piled raft foundation.

The utilization of a piled raft foundation versus a conventional piled only foundation delivered the required serviceability performance with regard to total and differential settlements while providing cost savings estimated to be of the order of 30% versus the original pile-only solution.

References

1. Decourt, L. (1995). "Prediction of load settlement relationships for foundations on the basis of the SPT-T". Ciclo de Conferencias Inter. "Leonardo Zeevaert", UNAM, Mexico, 85-104.
2. Poulos, H.G. (2005). "Piled raft and compensated piled raft foundations for soft soil sites" Geotechnical Spec Publications No 129, ASCE, 214 – 234.
3. Poulos, H.G. (2002). "Simplified design procedure for piled raft foundations". Deep Foundations 2002, Ed. M.W. O'Neill & F.C. Townsend, ASCE Spec. Geot. Pub. 116, 1:441-458.
4. Randolph, M.F. (1994). "Design methods for pile groups and piled rafts". State of the Art Report, 13 ICSMFE, New Delhi, 5: 61-82.

A FIELD STUDY ON SUPER-TALL BUILDING/SUPER-LONG PILE/THICK RAFT INTERACTION IN SHANGHAI

QINGGUO FAN   BIAOBING DAI   WENLONG DENG

Shanghai Construction Group, Shanghai 200120, China

ZHIYONG AI   XIHONG ZHAO

College of civil engineering, Tongji University, 1239 Siping Road , Shanghai 200092, China

It is the first time in China to carry out such a field experimental study to provide the information on the frame-shear wall structure/super-long pile/thick raft interaction in soft soils. Chang-feng Market is of 60-storey, 238m high, with 4.5m~6.25m thick raft and bored piles of Φ850mm drilled into silty sand layer with medium and coarse sands (i.e. 9-2 soil layer in Shanghai) to a depth of 72.5m. This building is composed of a 60-storey building and a 10-storey podium. Both have 4-storey basement. The depth of embedment is 18.95m. In order to obtain the more information on interaction, after completion of excavation engineering, on July 11, 2004 the following transducers for different purposes were installed in foundation with construction stage, respectively. 19 rebar stress gauges and 19 strain gauges on the top of piles and 36 earth pressure cells beneath the foundation base were installed to measure the pile load and foundation contact pressure. 28 rebar stress gauges were attached to the top and bottom reinforcing steel at 7 locations in the raft to measure the steel stresses. 5 piezometers were installed about 0.5~1.0m below the base of excavation to observe the lift and dissipation of excess pore water pressure and the effect of buoyancy on the bottom. To get the special information for checking the real load on the mega columns and corresponding contribution of every storey stiffness to foundation, 5 rebar stress gauges and 5 strain gauges were attached 5 typical columns. Meanwhile, for the same purpose, to record the settlements of the building with respect to time, 21 permanent reference points were established on columns and walls. Based on the change of settlement type it will provide an important evidence to analyze the contribution of every storey stiffness to foundation.Now the building is still under construction. However, some valuable data have obtained. For example, load-sharing between pile and raft is an objective fact even for super-long pile foundation. The main structure will be completed in the coming September. It is expected that the field experimental and analytical results will be satisfied.

1. Introduction

Chang-feng market is a multi-function and comprehensive modern building, including shopping, dining, office, hotel, et al; which is located at Chang-ning District in Shanghai, China, near Zhong-shan Park station of the second subway line, and cross-point of Chang-ning Road, Hui-chuang Road and Kai-xuan Road.

The structure of the market has a main building of 60-storey frame-shear structure and a 10-storey podium. The main building is 238m high, with piled raft foundation. The prospective view and cross section of the main building and its podium are shown in Figure 1 and Figure 2, respectively. The total area of construction for Chang-feng market is 310000m^2, in which the main building, 97000m^2 and the podium, 125000m^2. The main building and podium both have 4-storey basements. The levels of ground floor of the basement, the first floor of the basement, the second floor, the third floor and the forth floor are located at -0.10m, -5.30m, -10.70m, -13.70m and -16.75m, respectively. The area of each storey basement is 22000m^2. The bottom level of raft in the main building is designed at -20.75m for the thickness of 4m, -21.75m. for the thickness of 5m and -23.00m for the thickness of 6.25m, respectively. The bored piles of Φ850mm are driven into 9-2 soil layer to a depth of 72.5m. The valid length of the bored pile is48m and the total number of piles in main building is 416. The plan of the piles of the main building and podium is shown in Figure 3. The excavation depth of the foundation pit is 18.95m. The external wall of basement is adopted as the diaphragm wall of the 800mm thickness and 1000mm thickness (near the second subway line), respectively. The total area of the foundation pit is about 25890m^2. The environment around the market is complex, and the foundation pit is close to the roads. South side of foundation pit is Chang-ning Road, and near the Zhong-shan Park station of the second subway line, the hall of the station is 10m deep and has 2 storey basements. The west side of foundation pit is next to Kai-xuan Road, which is new road and is full of lots of ground pipe lines. The tramway station of Pearl line is of 40m distance from the western external wall. The north and the east sides of the foundation pit is Hui-chuan Road, which is also full of ground pipe lines. A 6-storey residential building is of 40m distance from the northern external wall, and Guo-Mei market of 12-storey frame structure is of 30m distance from the eastern external wall, and so on.

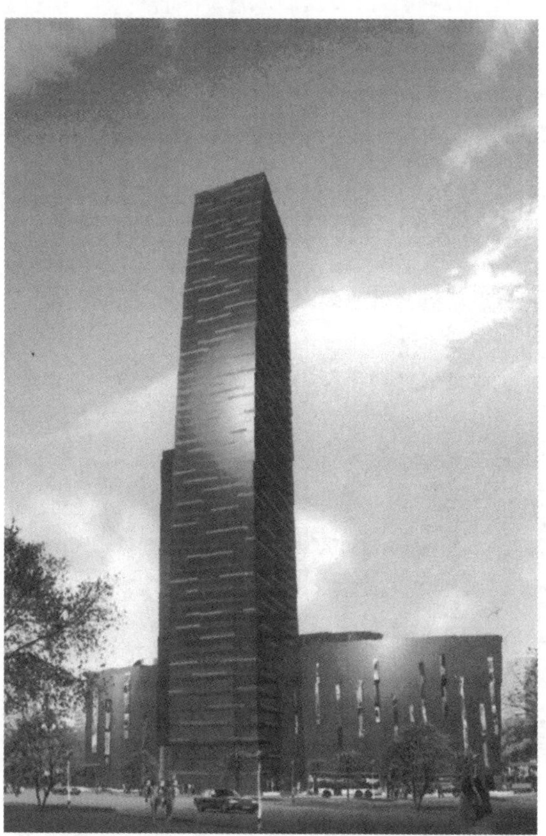

Figure 1. Prospective view of Chang-feng market.

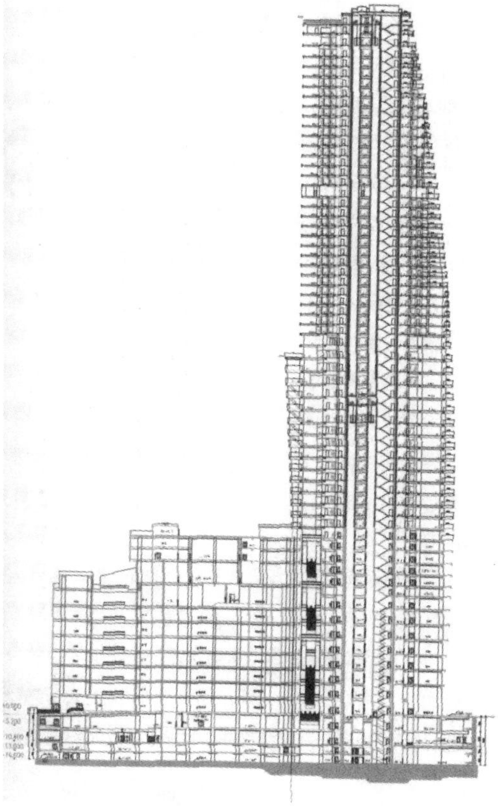

Figure 2. Cross section of Chang-feng market.

2. Purpose and Content of Field Test

The purpose of field test is to investigate the interaction between the super structure and the piled raft foundation in the main building of Chang-feng market, the emphases are to study the interaction between pile and raft foundation. The contents of the field test are listed as follows:

1. The relationship of settlement and loads on pile group;
2. The load distribution of pile group (i.e. the coefficient of distribution among the corner pile, edge pile, internal pile and central pile.);
3. Load-sharing between pile and raft;
4. Relations between stiffness of raft and floors of the building;
5. Affects of the variation for floors and pore water pressures on settlement, the pile loads, load-sharing between pile and raft, stiffness of raft, stress in raft, load distribution of columns;
6. Inner force variation in the raft;
7. Relation between real load and designed load in typical edge columns and the columns next to the centre;
8. Elevation and dissipation of pore water pressures;
9. Relation between water level and soil pressures according to ground water level fluctuation;

In addition, the problem is how to make use of the settlement of the main building and its podium to the study on interaction.

3. Layout of Instruments

The layout of instruments is shown in Figure 4.

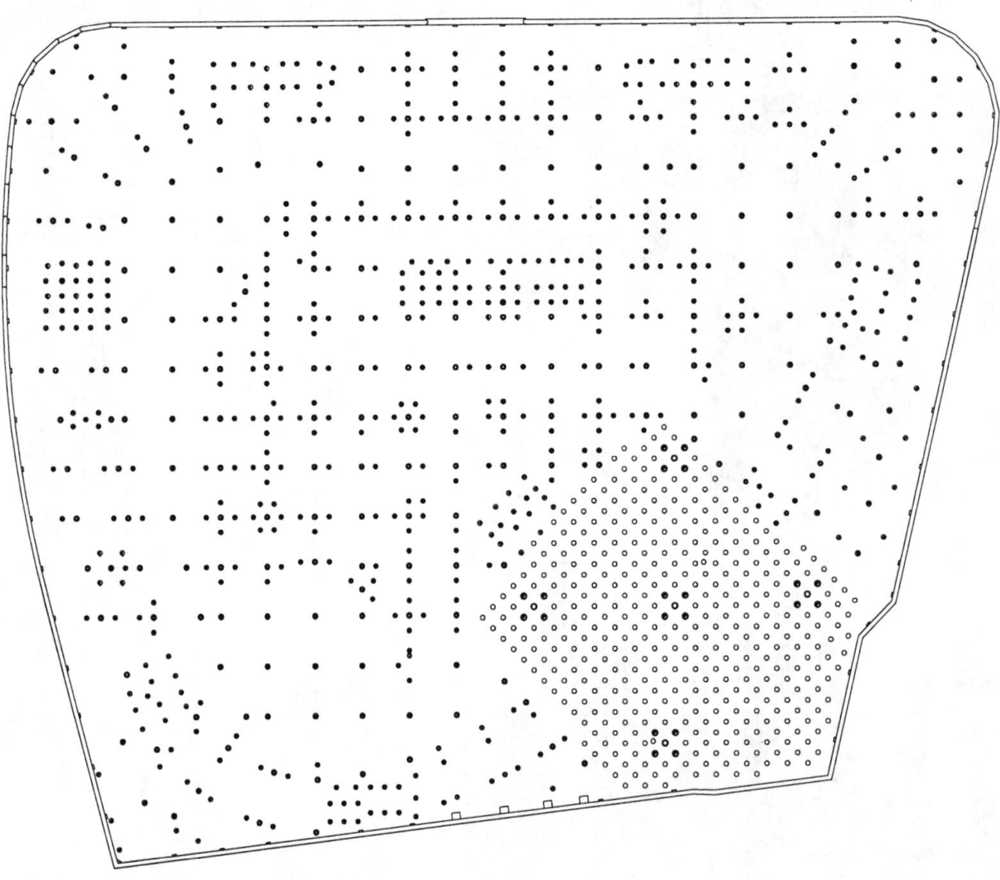

Figure 3. Plan of piles of main building and podium.

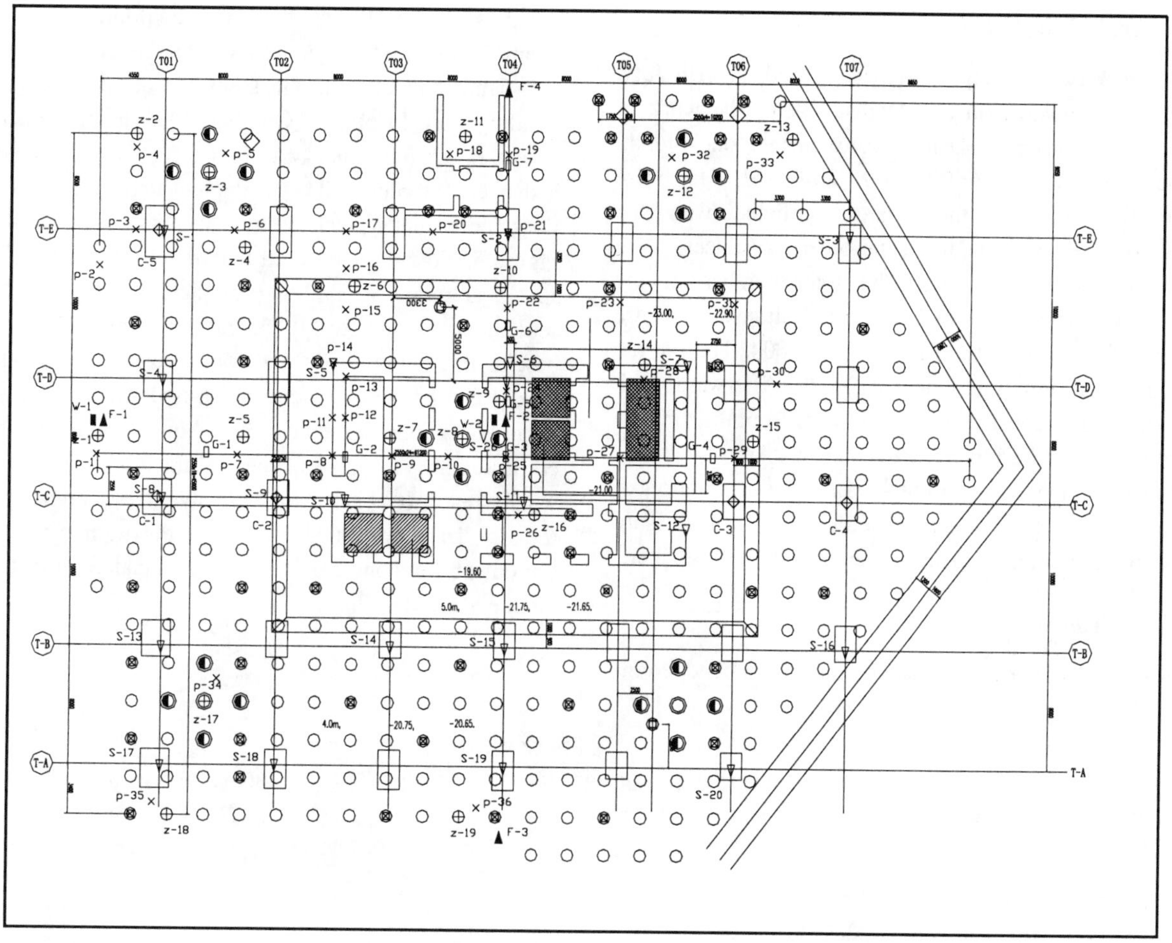

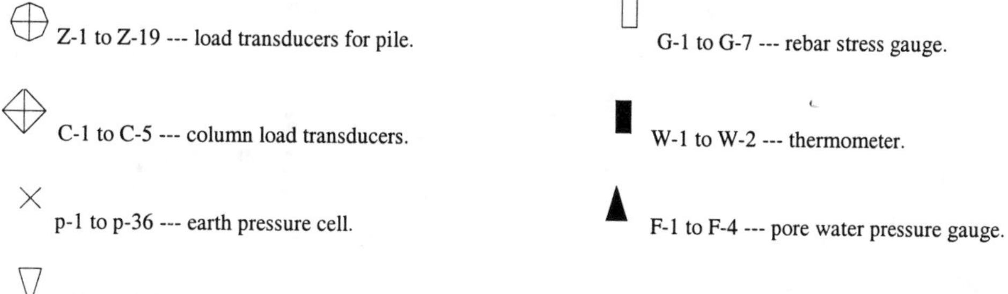

Figure 4. Layout of instrumentation

4. Analysis of test results

The work of installing instrument started on 11th July, 2004 and finished on 5th August, 2004. The work of installing instrument lasted approximately one month. The transducers for P28, Z9, Z14 and Z15 were cancelled, because they were conflicted with the constructions of foundation, the other transducers were all successfully installed. The total number of the instruments is 113, in which 102 transducers are successful by the date of August 6th 2004. That is, the ratio of surviving instrument is 90.3%. Afterward, with developing of construction many wires from instruments had been destroyed by several accidents. Anyway, a lot of test data have been obtained. Due to limitation of time the analytical results of settlement, earth pressure and loads on pile group are presented only in this paper.

4.1 *Settlement*

S-1~S-20 settlement reference points were embedded in columns and walls at the ground floor with their locations, as shown in Figure 4 by triangle symbols.

Unfortunately, the measurement of settlement for building started from 30th October, 2004, in that time the basements of four floors and fourth floor on the ground had been completed. The settlement versus number of floors curve for typical reference point S-5 is shown in Figure 5. Now the measured maximum settlement of S-12 is 33mm (36th floor completed). In fact, the settlement caused by thick raft of 4m~6.25m, the basements of four floors and four floors on the ground should be taken into account. Based on engineering experience, the settlement is about 10mm in that construction period. Thus, the real maximum settlement would reach to 43mm, namely average settlement is roughly 1mm per floor. However, the normal rate of settlement for concrete structure is equal or less than 1mm per floor in Shanghai.

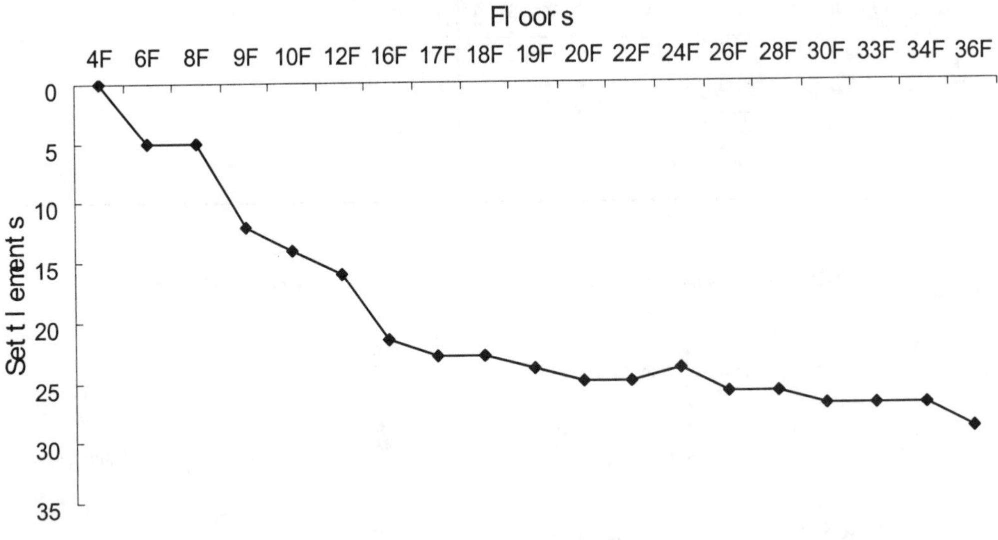

Figure 5. Settlement versus floor for S-5

4.2 *Earth pressure*

In order to measure contact pressure beneath the bottom of raft, P1~P36 earth pressure cells were installed at different locations, as shown in Figure4 by × symbols. The earth pressure started to measure since the time the earth pressure cells were installed on August 2nd 2004. The typical earth pressure of P29 versus number of floors is shown in Figure 6, in which the average data is also listed.

From the data shown in Figure 6 it can be seen that there are three characteristics at three construction stages as follows:

The first construction stage (B4~B0): In this construction period of B4 concrete of raft with 4m~6.25m thick had been poured. The earth pressure versus number of floors curve is just a straight line, the average earth pressure of 35kPa increases to 188.5kPa rapidly.

The second construction stage (B0~F10): This construction period of F10 can be considered as overburden pressure stage (self weight of whole excavation of 18.95m). The earth pressure increases with number of floors slowly, i.e. the average earth pressure of 188.5kPa increases to 213kPa.

The third construction stage (F10~F39): This construction period between F10~F39 can be considered as net foundation pressure stage. The earth pressure basically keeps constant. The average earth pressure can be also taken as 213kPa.

These characteristics can provide useful basis for evaluating the load carried by pile groups versus numbers of floors (see 4.3).

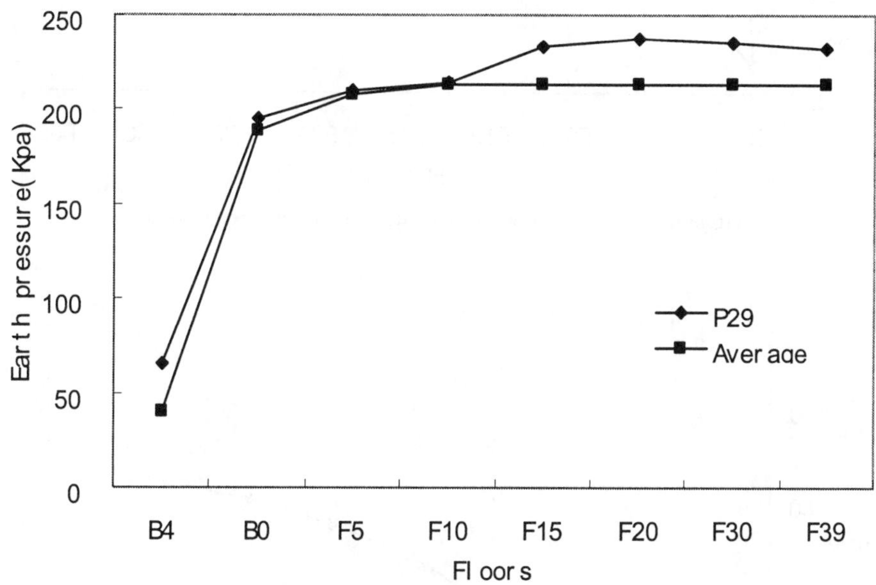

Figure 6. Earth pressure versus number of floors

B4---Fourth floor of basement (08.22.04); B0---First floor on ground (10.03.04)

F5---Fifth floor on ground (11.02.04); F39---Thirty ninth floor on ground (06.24.05).

4.3 ***Loads on pile group***

To study the load distribution on top of pile group Z-1~Z-19 load transducers for pile were installed on the top of piles at different location, as shown in Figure 4 by ⊕ symbols.

It is well known that the load distribution on the top of pile groups is very complex. Now a lot of data is still being analyzed. In this case the load carried by pile groups can be expressed in the simple form as

Ppile = Ptotal - Psoil

Where

Ppile = load carried by pile groups;

Ptotal = total load;

Psoil = load carried by soil (raft).

Then, load carried by pile groups versus number of floors is easily shown in Figure 7.

Figure 7 can be expressed in other form as the ratio of load of pile group to total load, i.e. η versus number of floors, as shown in Figure 8.

Evidently, from Figure 8 it can be seen that the ratio of pile load to total load, η is about 67% during completion of 39th floor. It can be predicted that η will be 75% roughly at the completion of this building.

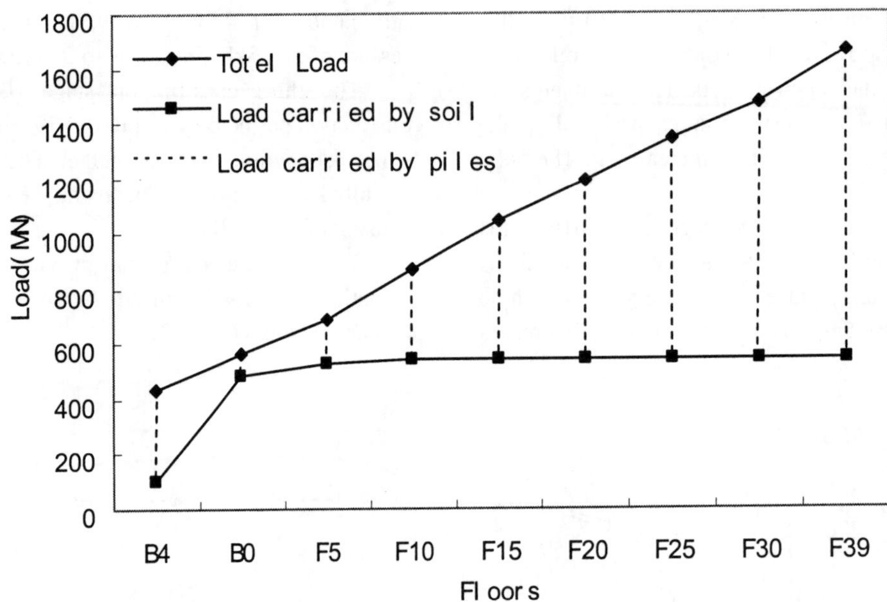

Figure 7. Load-sharing between piles and soil (raft) vs number of floors

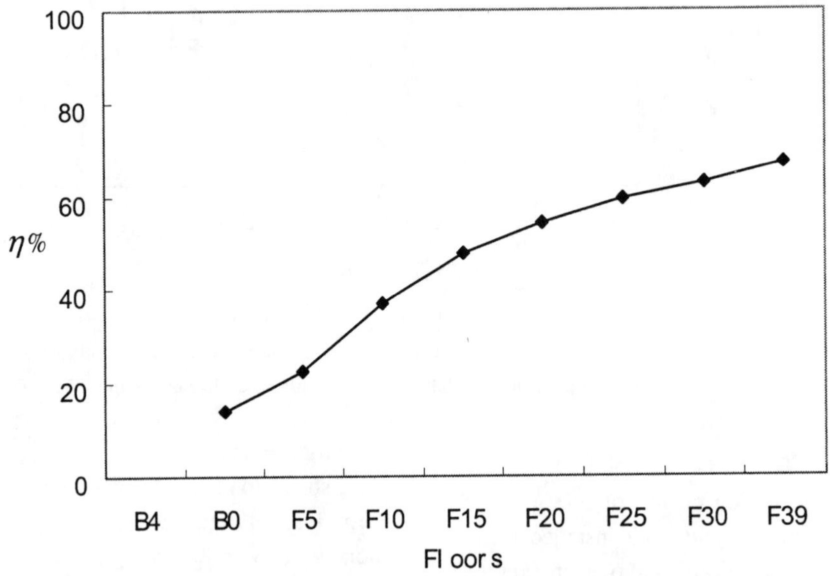

Figure 8. Ratio of load of pile group to total load vs number of floors

5. Concluding Remarks

It is the first time in China to carry out such a field experimental study on the frame-shear wall structure/super-long pile/thick raft interaction in soft soils. Some valuable data have been obtained although this building is still under construction and the field experiment is continuing. However, two preliminary results can be drawn as follows:

1. Based on the engineering experience for concrete structure of tall building in Shanghai, the settlement

rate is one of key points for checking the engineering quality. The settlement rate should be controlled within 1mm per floor for safety, it can be said this building is in a good condition.

2. Based on the measured load carried by soil (raft), at the completion of this building the load carried by pile group will be about 75% of total load.

In the design of this building the allowable bearing capacity of pile is taken as 5000kN, at the completion of this building the total load will be about 2000000kN, then, the average load for single pile will be 4800kN. So, some valuable data obtained in this field experiment will help the designer improve the design level. For example, load-sharing between pile and raft should be taken into account in the design of piled raft foundation for tall building.

References

1. Zhao, X. H. et al. Theory of Design of Piled Raft and Box Foundation for Tall Buildings in Shanghai (in Chinese, 1989; Enlarged edition in English, 1998). Tongji University Press.
2. Katzenbach, R et al. Soil-structure-interaction of the 300m high Commerzbank tower in Frankfurt am Main measurements and numerical studies. Proc. 14th ICSMFE, Hamburg, 1997, Vol.2:1081~1084.
3. Hu, C.L. et al. Field monitoring and comprehensive analysis of piled raft foundation for tall building under complex conditions. Rock and Soil Mechanics. [J] (in Chinese), 2003, 24(4):673-676.
4. Zhao, X. H., Gong, J. Load-sharing between pile and raft or box in field experiment, calculation and mechanism analysis. Rock and Soil Mechanics. [J] (in Chinese), 2005. 26(3):337~341.

SIMPLIFIED DYNAMIC FINITE-ELEMENT ANALYSIS FOR THREE-DIMENSIONAL PILE-GROUPED-RAFT-HIGH-RISE BUILDINGS*

HUI XIONG[a] XILIN LU[a] LIANG HUANG[b]

[a]Research Institute of Structural Engineering and Disaster Reduction, Tongji University, Shanghai 200092,China

[b]Department of Civil Engineering, Hunan University, Changsha 410082,China

Combined with the respective advantages in S-R(Sway-Rocking) impedance concept and finite-element method, a simplified 3D structural dynamic FEM considering composite pile-group-soil effects is presented. The structural members including piles are modeled by space beam or shell elements, and raft-base is divided into thick-shell elements with its spring-dashpot boundary coefficient obtained by impedance backcalculated. The mass-spring elements for soil between piles are set to simulate vertical, horizontal pile-group effects by strata-equivalent approach. The soil beside composite body is separated into near-field and far-field parts. The former is modeled by nonlinear spring-dashpot elements based on Winkler's hypothesis, while the latter is modeled by a series of linear mass-spring-dashpots. With the effects of boundary track forces and energy radiation, the presented model enables researchers to conduct the time-domain nonlinear analysis in a relatively simple manner which avoids sophisticated boundary method and solid-element mesh bringing with tremendous computational cost. The seismic effect on dynamic interaction of pile-soil-complicated structures would be efficiently annotated from two structural engineering and geotechnical engineering aspects and the numerical calculation effort would be drastically decreased too. The complete procedure is mainly performed using the parametric design language assembled in the Finite Element Code Ansys. With the dynamic analysis of foundation and superstructure for a pile-supported 15-storey building, the influence of the participant effect on structural dynamic response will be depicted by various dynamic parameters of pile-soil-raft foundation in detail. Not only do the results have an agreement with some conclusions drawn by the general interaction theory, but also certain of phenomena which would be disagree with that by general analysis is involved. Even with the finite-element meshes for 68 piles, the time-history analysis procedure for PGSS (Pile-Group-Soil-Superstructure) system and the qualitative evaluation with various SSI parameters can be also fulfilled efficiently and rapidly by presented means. These results may be of help to the designers to quickly assess the significance of interaction effect for the high-rise buildings resting on any type or layout of pile-group foundation.

1. Introduction

Currently, many work about dynamic behavior of piles have been developed on the assumption of viscous-elastic linearity. The concept of interaction factors describing approximately the static response of pile groups were initiated by Poulos based on linear superposition technique [1]. Sheta and Novak introduced equivalent coefficients in static interaction factors to deal with vibration problems of pile-groups. Subsequently, the dynamic interaction factors accounting for the pile-soil-pile three-step effect pattern were conducted based on physical wave field theory by Gazetas and Makris [2][3], which established the simplified theory foundation for dynamic pile-group effect. Making use of hybrid boundary element and rigorous axisymmetric element method, Mamoon and Banerjee [4] implemented the dynamic time-domain and frequency-domain analysis of single-pile and pile-groups characterized by different section types embedded in homogeneous elastic half-space and obtained the consistent results with exact solutions by Laplace transient transform-equations [4]. In fact, the soil-foundation often experiences a high level of strain that exceeds the linear limit as suffering from drastic excitation or seismic action and its nonlinear characteristic features should be rationally considered for complicated pile-soil interaction problems. In general, the pile-soil interaction research is carried out mainly focusing on the simulation of soil around piles, and the response of pile foundation is frequently analyzed by two traditional approaches including simple beam-element method (static *p-y* curve method and dynamic Winkler beam foundation theory) and finite-element method. The former firstly presented by Matlock and Reese [5] is deemed that the vertical and horizontal resisting forces of soil around piles could be regarded as a series of irrelevant distributed springs, and this approach was later extended to dynamic case to deal with energy dissipation problem of pile-soil reactions and to obtain the frequency-dependent soil stiffness by Novak [5]. Ashour and Norris [6] considered the resisting zone of soil near pile shaft as a passive fan-cone body and described the load-deformation nonlinear relationship of piles under lateral loads by SSCW(Stress-Strain Cone Wedge) model [6]. The inertial dynamic effect and radiation damping effect were simulated through adding lumped-mass of soil field to spring-dashpots of pile-soil by Penzien [7]. Nogami divide the region around pile shaft into three Winkler soil model zones with respective properties and presents a nonlinear pile-soil interaction model applicable to both frequency- and time-domain dynamic lateral response analysis [8]. Because of its simplicity and convenience, the simple method is favored in practiced design analyses over the analytical method (known as Pile Foundation Engineering Handbook, press in China ; ATC-3 ;ASCE 20-96). On the other hand, the latter-finite element method may be capable of accommodating both the dynamic and nonlinear conditions such as elastoplastic property of soil-field, the geometric nonlinearity at pile-soil interface and apt to performing a integration analysis combined with superstructure in a logical manner. However, this approach for the three-dimensional nonlinear analysis for pure solid-element mesh model requires a tremendous computation and the nodes/elements degrees of freedom increase greatly which may cause to be failed in executing programs under the current computing capability. So the detailed degree of actual finite element model for pile-soil-structure systems could make its choice on the ground of different precision and efficiency demands [9][10]. In this context, a 3D FE model of Pile-Group-Soil-Superstructure (PGSS) system for coupled with simplified linear/nonlinear dynamic characteristics of pile-soil foundation is developed. The effect of pile-soil-pile interaction on pile-groups is taken into account by introducing corresponding S-R impedance spring-mass-dashpot elements as the feedback boundary for pile-to-pile, pile-soil, soil-raft and near field soil-far field soil. The flexibility participation of pile -soil foundation can be effectively reflected in time-history schemes on SSI evaluation and the solution size of FE model can be largely decreased too by present model.

2. Idealization of FE Model Considering SSI Effect

2.1 *Simulation of lateral vibration of pile-soil-pile*

Due to the radiating and scattering effect of soil, a large truncation area which will contain numerous finite-element meshes or intricate artificial boundaries will be often adopted in describing the dynamic behavior of soil-pile-structure interaction systems. As for 3d solid elements applied to meshing pile and soil bodies, the number of nodes and degrees of freedom will be unavoidably enlarged greatly. On the condition including the considerable embedded piles, the work on dividing elements in regular meshes will become tedious and on the other hand it is also difficult to accommodating the diversification of the pile-group distribution in plane position and the superstructure type and satisfying the demand for quick evaluation on structural design with SSI characters. In the paper, a series of pile-to-pile, pile-soil, soil-pile and soil-raft impedance spring-dashpot elements are introduced for describing the participation effect on system response with the varied pile-groups and strata-soil status. The element sketch is shown in Fig.1.(a)~(e).

As for two piles located in the pile-group(see Fig.1(b)), the spring stiffness (complex impedance $\wp$) depicting the pile-soil-pile interaction is expressed by

$$\frac{1}{\wp^{i}_{ks,s-p}} = \frac{u^{i}_{kk}(R_0)}{Q^{i}_{k\to k}} + \frac{u^{i}_{lk}(R_{kl})}{Q^{i}_{l\to k}} \qquad (k,l=1\sim N) \qquad (1)$$

where N is the number of piles; i the soil-stratum number; R_0, R_{kl} the pile radius and the radial distance between pile k and pile l, respectively; $u^{i}_{kk}(R_0)$, $Q^{i}_{k\to k}$ and $u^{i}_{lk}(R_{kl})$, $Q^{i}_{l\to k}$ are the internal displacement, force as self-single pile k and the interaction displacement, force from pile l to pile k. The dynamic differential equilibrium equation of pile elements under horizontal vibration can be resulted from the procedure on the dynamic inertial interaction of pile-groups in layered-soil[11]. The deformation and force solution about $k\to k$ self-pile-shafts can be derived from the following Eqs.(2)-(3).

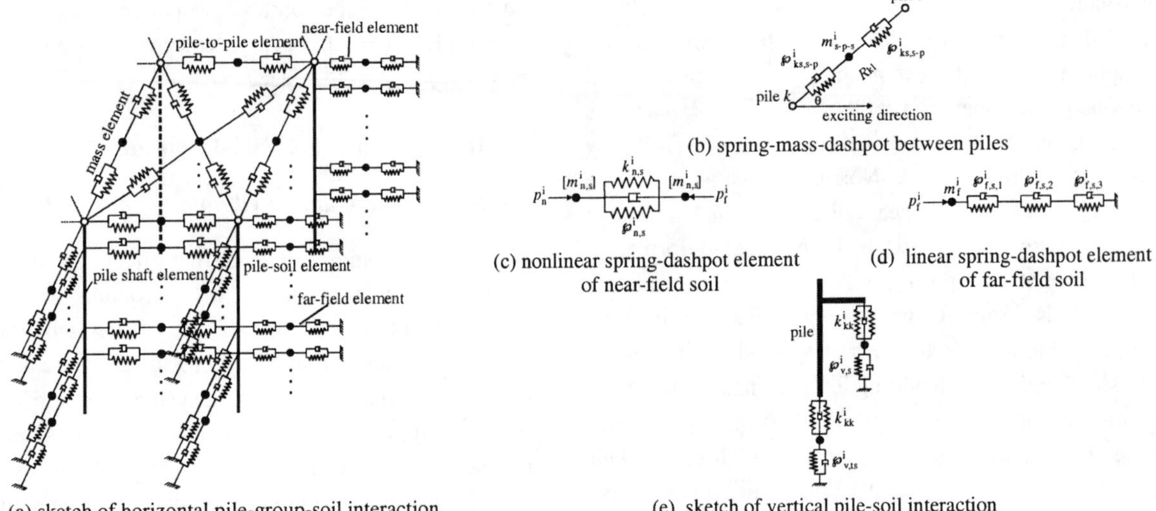

(a) sketch of horizontal pile-group-soil interaction

(e) sketch of vertical pile-soil interaction

Fig.1 Model of pile-group-soil foundation

$$\left.\begin{aligned} u^i_{kk}(z) &= C^i_{u,1}e^{\lambda^i_1 z} + C^i_{u,2}e^{-\lambda^i_1 z} + C^i_{u,3}e^{\lambda^i_2 z} + C^i_{u,4}e^{-\lambda^i_2 z} \\ \varphi^i_{kk}(z) &= C^i_{u,1}\lambda^i_1 e^{\lambda^i_1 z} - C^i_{u,2}\lambda^i_1 e^{-\lambda^i_1 z} + C^i_{u,3}\lambda^i_2 e^{\lambda^i_2 z} - C^i_{u,4}\lambda^i_2 e^{-\lambda^i_2 z} \\ Q^i_{k\to k}(z) &= E_p I_p (C^i_{u,1}\lambda^{i3}_1 e^{\lambda^i_1 z} - C^i_{u,2}\lambda^{i3}_1 e^{-\lambda^i_1 z} + C^i_{u,3}\lambda^{i3}_2 e^{\lambda^i_2 z} - C^i_{u,4}\lambda^{i3}_2 e^{-\lambda^i_2 z} \\ M^i_{k\to k}(z) &= E_p I_p (C^i_{u,1}\lambda^{i2}_1 e^{\lambda^i_1 z} + C^i_{u,2}\lambda^{i2}_1 e^{-\lambda^i_1 z} + C^i_{u,3}\lambda^{i2}_2 e^{\lambda^i_2 z} + C^i_{u,4}\lambda^{i2}_2 e^{-\lambda^i_2 z} \end{aligned}\right\} \tag{2}$$

$$\begin{Bmatrix} u^i_{kk}(h_i) \\ \varphi^i_{kk}(h_i) \\ Q^i_{k\to k}(h_i) \\ M^i_{k\to k}(h_i) \end{Bmatrix} = S_{u,i}\begin{Bmatrix} C^i_{u,1} \\ C^i_{u,2} \\ C^i_{u,3} \\ C^i_{u,4} \end{Bmatrix} = S_{u,i}S_{u,i-1}\cdots S_{u,1}S_{u,0}\begin{Bmatrix} u^1_{kk}(0) \\ \varphi^1_{kk}(0) \\ Q^1_{k\to k}(0) \\ M^1_{k\to k}(0) \end{Bmatrix} = S^*_{u,i}\begin{Bmatrix} u^1_{kk}(0) \\ \varphi^1_{kk}(0) \\ Q^1_{k\to k}(0) \\ M^1_{k\to k}(0) \end{Bmatrix},\ i = 1,\cdots,n \tag{3}$$

While the deformation and force solution about $l\to k$ pile-shafts under the pile-to-pile condition can be also obtained by Eq.(4)

$$\left.\begin{aligned} u^i_{lk}(z) &= E^i_1 e^{\lambda^i_1 z} + E^i_2 e^{-\lambda^i_1 z} + E^i_3 e^{\lambda^i_2 z} + E^i_4 e^{-\lambda^i_2 z} + \Gamma_i\psi(R,\theta)_i u^i_{kk}(z) \\ \varphi^i_{lk}(z) &= E^i_1\lambda^i_1 e^{\lambda^i_1 z} - E^i_2\lambda^i_1 e^{-\lambda^i_1 z} + E^i_3\lambda^i_2 e^{\lambda^i_2 z} - E^i_4\lambda^i_2 e^{-\lambda^i_2 z} + \Gamma_i\psi(R,\theta)_i \varphi^i_{kk}(z) \\ Q^i_{l\to k}(z) &= E_p I_p[E^i_1\lambda^{i3}_1 e^{\lambda^i_1 z} - E^i_2\lambda^{i3}_1 e^{-\lambda^i_1 z} + E^i_3\lambda^{i3}_2 e^{\lambda^i_2 z} - E^i_4\lambda^{i3}_2 e^{-\lambda^i_2 z} + \Gamma_i\psi(R,\theta)_i Q^i_{k\to k}(z)/E_p I_p] \\ M^i_{l\to k}(z) &= E_p I_p[E^i_1\lambda^{i2}_1 e^{\lambda^i_1 z} + E^i_2\lambda^{i2}_1 e^{-\lambda^i_1 z} + E^i_3\lambda^{i2}_2 e^{\lambda^i_2 z} + E^i_4\lambda^{i2}_2 e^{-\lambda^i_2 z} + \Gamma_i\psi(R,\theta)_i M^i_{k\to k}(z)/E_p I_p] \end{aligned}\right\} \tag{4}$$

$$\begin{Bmatrix} u^i_{lk}(h_i) \\ \varphi^i_{lk}(h_i) \\ Q^i_{l\to k}(h_i) \\ M^i_{l\to k}(h_i) \end{Bmatrix} = [ST]_i\begin{Bmatrix} E^i_1 \\ E^i_2 \\ E^i_3 \\ E^i_4 \end{Bmatrix} + \begin{Bmatrix} \Gamma_i\psi(R,\theta)_i u^i_{ll}(h_i) \\ \Gamma_i\psi(R,\theta)_i \varphi^i_{ll}(h_i) \\ \Gamma_i\psi(R,\theta)_i Q^i_{l\to l}(h_i) \\ \Gamma_i\psi(R,\theta)_i M^i_{l\to l}(h_i) \end{Bmatrix} = [ST]_i\begin{Bmatrix} E^i_1 \\ E^i_2 \\ E^i_3 \\ E^i_4 \end{Bmatrix} + \begin{Bmatrix} \vartheta^i_1 \\ \vartheta^i_2 \\ \vartheta^i_3 \\ \vartheta^i_4 \end{Bmatrix} \tag{5}$$

where

$$S_{u,i}=[ST]_i=\begin{bmatrix} e^{\lambda_1^i h_i} & e^{-\lambda_1^i h_i} & e^{\lambda_2^i h_i} & e^{-\lambda_2^i h_i} \\ \lambda_1^i e^{\lambda_1^i h_i} & -\lambda_1^i e^{-\lambda_1^i h_i} & \lambda_2^i e^{\lambda_2^i h_i} & -\lambda_2^i e^{-\lambda_2^i h_i} \\ E_p I_p \lambda_1^{i3} e^{\lambda_1^i h_i} & -E_p I_p \lambda_1^{i3} e^{-\lambda_1^i h_i} & E_p I_p \lambda_2^{i3} e^{\lambda_2^i h_i} & -E_p I_p \lambda_2^{i3} e^{-\lambda_2^i h_i} \\ E_p I_p \lambda_1^{i2} e^{\lambda_1^i h_i} & E_p I_p \lambda_1^{i2} e^{-\lambda_1^i h_i} & E_p I_p \lambda_2^{i2} e^{\lambda_2^i h_i} & E_p I_p \lambda_2^{i2} e^{-\lambda_2^i h_i} \end{bmatrix} \quad (6)$$

and in the above formulas, h_i represents the thickness of the ith soil profile and its bottom coordinate; $C^i_{1,2,3,4}$, $E^i_{1,2,3,4}$ and $\vartheta^i_{1,2,3,4}$ are integral parameters charactered by recursive equations; the lateral vibration scalar $\lambda^i_{1,2}=\sqrt{-2\delta_i \mp 2\sqrt{\delta_i^2-4\Lambda_i}}\,/2$, $\delta_i=N_i/(E_pI_p)$, N_i is the axial force pile elements, E_p the modulus of elasticity of pile, I_p the moment of inertial of pile, A_p the pile area, ρ_p the pile density, m_p is the mass per unit length of pile, $m_p=\rho_pA_p$; ω the frequency of exciting, $\Lambda_i=(f^R_{ui}+i\omega f^I_{ui}-m_\rho\omega^2)/E_pI_p$, $f^R_{ui}+if^I_{ui}$ is the resistance complex stiffness per unit length of pile, i=sqrt(-1); $\Gamma_i=(f^R_{ui}+i\omega f^I_{ui})/(f^R_{ui}+i\omega f^I_{ui}-m_\rho\omega^2)$. It is noted that $Q^1_{k\to k}(0)$, $M^1_{k\to k}(0)$ or $Q^1_{l\to l}(0)$, $M^1_{l\to l}(0)$ of self-single pile should be regarded as known quantity and the variable with subscripts ll, $l\to k$ should be of the same derivable procedure with those with subscripts kk, $k\to k$. Details can be found in [11]. So Eq.(1) becomes

$$\frac{1}{\wp^i_{ks,s-p}}=\frac{\sum_{m=1}^{4}s^{i*}_{u,1m}\chi^1_{kk}(0)}{\sum_{m=1}^{4}s^{i*}_{u,3m}\chi^1_{kk}(0)}+\frac{\sum_{m=1}^{4}st^{i*}_{u,1m}E^i_m+\Gamma_i\psi(R,\theta)_iu^i_{ll}(h_i)}{\sum_{m=1}^{4}st^{i*}_{u,3m}E^i_m+\Gamma_i\psi(R,\theta)_iQ^i_{l\to l}(h_i)} \quad (7)$$

where $s^{i*}_{u,pm}$, $st^{i*}_{u,pm}$ (p,q=1~4)are the corresponding elements of recursive matrix $S^*_{u,i}$, $[ST]_i$, respectively ; For m=1~4, $\chi^1_{kk}(0)$ is on behalf of $u^1_{kk}(0)$, $\varphi^1_{kk}(0)$, $Q^1_{kk}(0)$ and $M^1_{kk}(0)$. $\psi(R,\theta)_i$ is the frequency dependent attenuation function computed using Eqs.(8)-(9)

$$\psi(R,\theta)_i=\psi(R,0)_i\cos^2\theta+\psi(R,\pi/2)_i\sin^2\theta \quad (8)$$

$$\left.\begin{aligned}\psi(R_{kl},0)_i&=(2R_{lk}/R_0)^{-\frac{1}{2}}e^{-(\xi_i+i)(\frac{R_{lk}}{R_0}-\frac{1}{2})\frac{V_{si}}{V_{Lai}}a_{0i}}\\ \psi(R_{kl},\pi/2)_i&=(2R_{lk}/R_0)^{-\frac{1}{2}}e^{-(\xi_i+i)(\frac{R_{lk}}{R_0}-\frac{1}{2})a_{0i}}\end{aligned}\right\} \quad (9)$$

where ξ_{si}, ρ_{si} and V_{si} =material damping , mass of unit volume and shear wave velocity of stratum-soil; a non-dimensional parameter $a_0=\omega R_0/V_{si}$; the Lysmer's analog velocity $V_{Lai}=3.4V_{si}/[\pi(1-v_i)]$, v_i the Poisson's ratio of soil.

The concerned vibration mass of soil between two piles may be defined by Eq. (10). Finally, the spring-dashpot-element coefficients connecting piles can be written as Eq. (11),

$$m^i_{s-p-s}=\frac{\pi\rho_{si}R_0^2}{4}(\frac{R_{kl}}{2R_0}-1)h_i \quad (10)$$

$$\wp^i_{ks,s-p}=\frac{\sum_{m=1}^{4}s^{i*}_{u,3m}\chi^1_{kk}(0)\sum_{m=1}^{4}st^{i*}_{u,3m}E^i_m+\Gamma_i\psi(R,\theta)_iQ^i_{l\to l}(h_i)\sum_{m=1}^{4}s^{i*}_{u,3m}\chi^1_{kk}(0)}{\sum_{m=1}^{4}s^{i*}_{u,1m}\chi^1_{kk}(0)[\sum_{m=1}^{4}st^{i*}_{u,3m}E^i_m+\Gamma_i\psi(R,\theta)_iQ^i_{l\to l}(h_i)]+\sum_{m=1}^{4}s^{i*}_{u,3m}\chi^1_{kk}(0)[\sum_{m=1}^{4}st^{i*}_{u,1m}E^i_m+\Gamma_i\psi(R,\theta)_iu^i_{ll}(h_i)]} \quad (11)$$

2.2 *Near-field soil and far-field soil model around pile-group*

A nonlinear Winker soil-pile model described by a coupled spring-dashpot-mass element style is applied to simulating the nonlinear behavior and hysteretic damping effect of near field soil around piles(see Fig.1.(c)). The consistent stiffness and mass matrix can be defined by

$$[K^i_{n,s}]=\begin{bmatrix} k^i_{n,s} & -k^i_{n,s} \\ -k^i_{n,s} & k^i_{n,s}\end{bmatrix},[M^i_n]=$$
$$\frac{\pi\rho_{si}R_0^2}{6}(\kappa-1)\begin{bmatrix}\kappa+3 & 3\kappa+1\\ 3\kappa+1 & \kappa+1\end{bmatrix}=\begin{bmatrix} m^i_{n,11} & m^i_{n,12}\\ m^i_{n,21} & m^i_{n,22}\end{bmatrix} \quad (12)$$

where $\kappa=R_1/R_0$. The general force-displacement(p-u) relationship are governed by

$$\begin{Bmatrix} p^i_n(\omega) \\ p^i_f(\omega)' \end{Bmatrix}=([K^i_{n,s}]-\omega^2[M^i_n])\begin{Bmatrix} u^i_n(\omega)\\ u^i_f(\omega)\end{Bmatrix} \quad (13)$$

i.e.

$$\left.\begin{aligned} p^i_n(\omega)&=(k^i_{n,s}-\omega^2 m^i_{n,11})u^i_n(\omega)-(k^i_{n,s}+\omega^2 m^i_{n,12})u^i_f(\omega)\\ p^i_f(\omega)'&=-(k^i_{n,s}+\omega^2 m^i_{n,12})u^i_n(\omega)+(k^i_{n,s}-\omega^2 m^i_{n,22})u^i_f(\omega)\end{aligned}\right\}$$

The interaction force between near-field(at R_0) element and far-field(at R_1) element can be also expressed as

$$p^i_f(\omega)=(\wp^i_{f,s}(\omega)-\omega^2 m^i_f)u^i_f(\omega) \quad (14)$$

where subscripts n, f denote near field and far field, respectively; the participant vibration mass of far-field soil $m^i_f=\pi\rho_{si}R_1^2\zeta_m(v_{si})$. Substituting Eq.(14) into the second equation of Eq.(13), so the integrated impedance value between nodes of piles and outside soil can be obtained(it is noted that $p^i_f(\omega)'=-p^i_f(\omega)$)

$$p^i_n(\omega)=\wp^i_{n,s}(\omega)u^i_n(\omega)$$
$$=\frac{k^i_{n,s}-\omega^2 m^i_{n,11}-(k^i_{n,s}+\omega^2 m^i_{n,12})^2}{k^i_{n,s}+\wp^i_{f,s}(\omega)-\omega^2(m^i_f+m^i_{n,22})} \quad (15)$$

where the near-field static spring stiffness

$$k^i_{n,s}=[\frac{1}{k_{ie,static}(0)}-\frac{1}{k^i_{f,static}(0)}]^{-1}$$
$$=\frac{k^i_{f,static}(0)k_{ie,static}(0)}{k^i_{f,static}(0)-k_{ie,static}(0)} \quad (16)$$

where $k_{ie,static}(0)$ is defined by variable stiffness in inelastic range from a p-y curve. Its imaginary and real parts are relevant to the slope of backbone curves and the area of hysteretic circles respectively.

$$k_{ie,static}(0)=p^i_u/u_{i,\max}(R_0)+iA_d/[\pi u^2_{i,\max}(R_0)] \quad (17)$$

Assumed to be an infinitely long vertical rigid bar in elastic medium subjected to a horizontal force Q^i_0, the stiffness for the pile body with distance R_1 can be defined by[8]

$$k^i_{f,static}(0)=\frac{u_i(R_0)}{Q^i_0}-\frac{u_i(R_0)-\overline{u}_i(R_1)}{Q^i_0}$$
$$=[\frac{1}{k_{e,static}(0)}-\frac{3-4v_{si}}{8\pi G_{si}(1-v_{si})}\ln\kappa]^{-1} \quad (18)$$

where p^i_u is the ultimate resistance determined by p-y method; $u_i(R_0)$ is the static displacement of piles, approximately $u_{i,\max}(R_0)=3R_0/80$; A_d can be get from [12] or by model tests; G_{si} is the shear modulus of soil; $k_{e,static}$ is the linear elastic value in p-y stiffness range; the average displacement at far-field point $\overline{u}_i(R_1)=[u_i(R_1,0^\circ)+u_i(R_1,90^\circ)]/2$, where the static displacements can be gained by traditional finite-element routine.

In order to reflect the radiation damping effect and the boundary effect, the far-field impedance is assembled through three seriate numerical spring-dashpots(see Fig.1(d)). So the equivalent far-field impedance is written as

$$\wp^i_{f,s}(\omega)=$$
$$(\frac{1}{k^i_{s,1}+i\omega c^i_{s,1}}+\frac{1}{k^i_{s,2}+i\omega c^i_{s,2}}+\frac{1}{k^i_{s,3}+i\omega c^i_{s,3}})^{-1} \quad (19)$$

where

$$\left.\begin{aligned} k^i_{s,1}+i\omega c^i_{s,1}&=3.518G_{si}\zeta(v_{si})+i113.0973\omega G_{si}R_1\zeta(v_{si})/V_{si}\\ k^i_{s,2}+i\omega c^i_{s,2}&=3.518G_{si}\zeta(v_{si})+i25.133\omega G_{si}R_1\zeta(v_{si})/V_{si}\\ k^i_{s,3}+i\omega c^i_{s,3}&=5.529G_{si}\zeta(v_{si})+i9.362\omega G_{si}R_1\zeta(v_{si})/V_{si}\end{aligned}\right\} \quad (20)$$

where $\zeta(v_{si})$ and $\zeta_m(v_{si})$ are the parameters dependent on v_{si}. For familiar v_{si}=0.35, 0.40, 0.45 in this paper, there are $\zeta(0.35)=1.476$, $\zeta(0.40)=1.580$, $\zeta(0.45)=1.741$ and $\zeta_m(0.35)=0.0352$, $\zeta_m(0.40)=0.1428$, $\zeta_m(0.45)=0.3740$ [13][14].

2.3 *Idealization of pile-soil vertical vibration*

As the same assumption with horizontal pile-soil interaction, the soil around the pile is also divided into two vertical regions: a nonlinear inner region adjacent to the pile and a linear region for the far field (see Fig.1(e)). The following relationship may be used for relating the shear stress τ_0^i of soil to the shear deformation $\gamma_{v,s}^i$ in vertical direction (z-direction) [15]

$$\gamma_{v,s}^i / \gamma_{r,v,s}^i = \frac{\tau_0^i / \tau_u^i}{1 - \tau_0^i / \tau_u^i} \tag{21}$$

where τ_u^i is the ultimate shear strength; the reference shear strain $\gamma_{r,v,s}^i = \tau_u^i / G_{si0}$; G_{si0} the initial tangent shear modulus; τ_0^i / τ_u^i presents the stress level. So the vertical deformation at pile-soil interface point (R_0) can be given as

$$w_{kk}^i = \int_{R_0}^{R_1} \gamma_{v,s}^i dR = \tau_0^i R_0 \ln[(\kappa - \tau_0^i / \tau_u^i)/(1 - \tau_0^i / \tau_u^i)] / G_{si} \tag{22}$$

and then, the vertical force per unit length at the pile shaft is

$$P_{kk}^i = 2\pi R_0 \tau_0^i = \frac{2\pi G_{si}}{\ln[(\kappa - \tau_0^i / \tau_u^i)/(1 - \tau_0^i / \tau_u^i)]} w_{kk}^i \tag{23}$$

Therefore, the inner-field stiffness $k_{kk}^i = 2\pi G_{si} / \ln[(\kappa - \tau_0^i / \tau_u^i)/(1 - \tau_0^i / \tau_u^i)]$; the elastic value without regard to stress level $k_{e,kk}^i = 2\pi G_{si} / \ln \kappa$.

Additionally, the vertical outer-field soil impedance are defined as

$$\wp_{v,s}^i = K_{v,s}^i + i\omega C_{v,s}^i = G_{si} S_{v1} + i G_{si} R_0 S_{v2} / V_{si} \tag{24}$$

where the impedance coefficient S_{v1}, S_{v2} can be determined by common Novak plane-strain condition[5].

Soil resistance at the pile bottom is usually represented by a linear spring-dashpot to model the reaction of a rigid massless disk undergoing harmanic vibration in elastic half-space. The element constants are given by Lysmer and Richart[7]

$$\begin{aligned} \wp_{v,ts} &= K_{v,ts} + i\omega C_{v,ts} \\ &= 4 G_{ts} R_0 / (1 - \nu_{ts}) + i\omega 3.4 R_0 \sqrt{G_{ts} \rho_{ts}} / (1 - \nu_{ts}) \end{aligned} \tag{25}$$

where subscripts t denote the soil parameters at the pile tip.

From the above derivation, the near-field element and the far-field element in the horizontal or vertical vibration mode may be substituted by nonlinear spring-dashpot elements with slipping, gap properties and linear spring-dashpot elements, and there is only one direction DOF in every kind of element.

2.4 *Three-dimensional pile-group-soil-superstructure system model*

The finite element technique has the advantage of the other method in dealing with dynamic interaction problems in time-domain. The precision demand for superstructure members could be satisfied and also the coupled inertial effect of pile foundation as the medium of seismic wave propagation and the supported body could be rationally described in aseismatic design of PGSS systems. McCallen developed a dynamic nonlinear program for two-dimensional soil-building analysis based on continuum structure model, and described the uplift effect of foundations suffering strong loads by geometrical large-displacement equations[9].The task team led by Lü Xilin (2001) coupled the shaking table model test and general the finite element program to offer a new idea for research on SSI under seismic loads and reproduced the liquefaction of sand soil and the destructive procedure of pile foundation and superstructure [11]. A 3D dynamic integrated Pile-Group -Soil-Superstructure system shown in Fig.2(a) is presented in this paper, combined with the aforementioned pile-soil interaction pattern. The structure body selected by any type may be composed of beam/column elements and shell/plate elements and raft foundation may be modeled by thick shell elements. The above so-called soil-pile, pile-to-pile, soil-soil and soil-raft interaction spring-dashpot coefficientsare frequency dependent, nevertheless, it can be assumed that the complex dynamic impedance matrix is calculated with adequate precision based on the predominant frequency obtained by eigenmode analysis in order to avoid the disadvantage that the FE code does not support frequency dependent elements. The complete procedure is mainly fulfilled with the APDL in the finite-element software-Ansys.

3. Dynamic Finite-element Computation undergoing Seismic Loads

A 15-storied pile-supported frame building is illustrated in Fig.2. The height of bottom storey is 4.5m and the others are all 3.5m. The relevant sectional properties of the structure are given in Table.1. The outline scheme of the PGSS system and the plane layout of pile-raft-soil elements are presented in Fig.2(b), respectively. The seismic input (SFD-Wave and PAK-Wave) and soil profile properties is shown in Fig.2(c). The concrete material of beams, columns and raft foundation are selected with C_{30}, and the material of pile shafts is C_{25}(E_p=26.5Gpa). The number of piles N=68, the length of piles is L=15m, R_0=0.6m, the far-field distance ratio R_1/R_0=20. The raft thickness is 1.2m and the number of soil-raft spring-dashpot elements distributed along the edge of raft is 62. The pile is divided into 15 beam elements and the number of spring-dashpot elements of the pile-to-pile, near-field soil, far -field soil is 3420(x direction-310 × 3; y direction-280 × 3; xy inclining direction-550×3), 1440(x direction-60×3; y direction-80× 3; z direction-340 × 3); 2280(x direction-180 × 3; y direction-240×3; z direction-340×3), respectively. The front three-order frequency of the fix-based structure ω_1=11.328rad/s (T_1=0.554s), ω_2=13.778rad/s(T_2=0.456s), ω_3=17.441rad/s(T_3=0.360s) by modal anlysis. The frequency independent impedance coefficients determined by the referenced frequency ϖ =14.182rad/s. The structural damping ratio ξ =0.05, so the corresponding Rayleigh damping factors $\alpha=2\xi_1\omega_1\omega_2/(\omega_1+\omega_2)$= 0.622 and $\beta=2\xi/(\omega_1+\omega_2)$=0.00398, respectively.

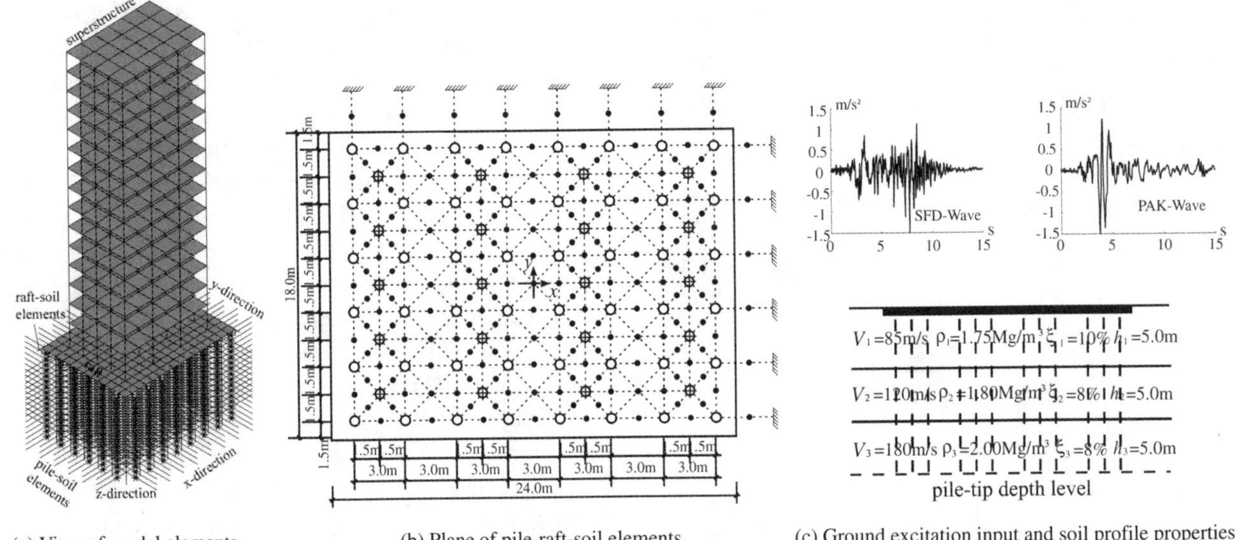

(a) View of model elements (b) Plane of pile-raft-soil elements (c) Ground excitation input and soil profile properties

Fig.2. Scheme of 3D PGSS system model

(⊕ Column-pile elements; ○ Non-column-pile elements; ● Soil-field mass elements; ---- Spring-dashpot elements)

Table 1. Basic parameters of structural members

Storey	Beams			Columns			Piles		
	A (m²)	I (m⁴)	m' (kg/m)	A (m²)	I (m⁴)	m' (kg/m)	L (m)	R_0 (m)	m' (kg/m)
1~5	0.24	0.0144	1735	0.36	0.0108	900			
6~10	0.21	0.0126	1670	0.25	0.0052	625	15	0.6	1256
11~15	0.18	0.0108	1610	0.16	0.0021	400			

(Note: the distribution mass m' of beams includes the mass of slabs, the thickness of slabs is100mm)

Comparing with the corresponding results with fixed-base condition, the basic response quantities of a certain member of structure are shown in Table 2 and the dynamic response curves and the corresponding envelop values are plotted in Fig.3-Fig.4, respectively. The response of structural internal forces will be of the different reduction considering SSI with various input seismic waves and the reduction (20%~40%) with PAK-wave is more than the reduction (10% ~ 30%) with SFD-wave under the combination of pile-soil condition (R_0=0.3m, G_{s1}=12000kPa), but the latter reduction in bottom storey is very small. Because of the participation of pile-soil flexibility, as a whole the storey displacement will increase and the storey acceleration will decrease to some extent comparing with fixed-base condition. The maximum increment in roof displacement is about up to

59%(SFD), 18%(PAK), while the maximum reduction in roof acceleration is about up to 30%(SFD), 20%(PAK). It seems that the results of this example are very consistent with the recommendation in ATC-3 and GB50011-2001 (in China). Additionally, the time-history response results under a stiffer pile-soil condition (R_0=0.45, G_{s1}=24000kPa) are presented in above figures to reflect the SSI effect on various influence parameters.

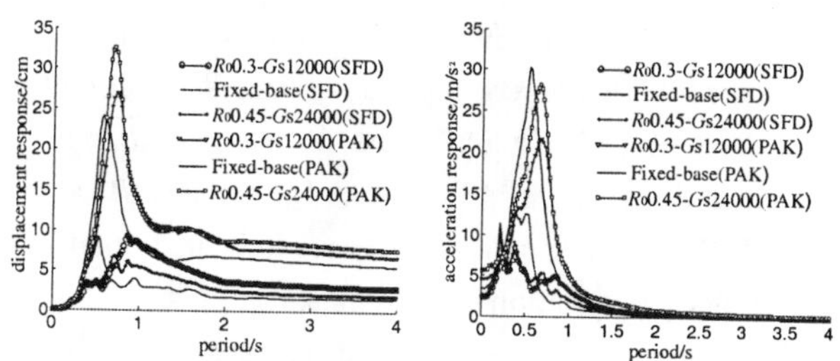

Fig.3. Response spectrum curves of roof

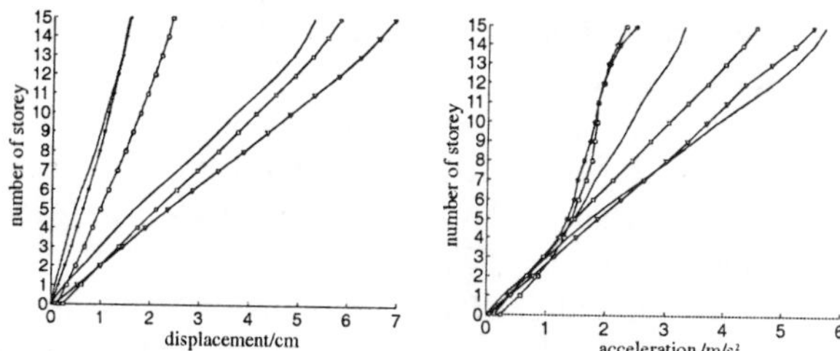

Fig.4. Storey response envelope

Table 2. Peak response quantities for a column member of superstructure with SSI and with fixed-base condition

Storey	Axial force (N_i/kN)				Shear force (Q_i/kN)				Bending moment (M_i/kN·m)			
	SFD		PAK		SFD		PAK		SFD		PAK	
	Fixed	SSI	Fixed	SSI	Fixed	SSI	Fixed	SSI	Fixed	SSI	Fixed	SSI
15	55.65	49.96	74.53	61.97	105.0	85.04	105.0	85.11	74.52	72.85	87.29	80.58
14	83.17	58.75	149.1	101.7	241.9	200.1	248.6	221.8	126.9	94.24	215.2	151.6
13	109.5	73.01	213.7	140.9	371.1	308.4	490.6	406.8	177.8	122.4	332.7	223.3
12	124.4	78.69	272.1	172.1	531.2	438.4	803.4	632.4	209.3	132.5	438.0	281.0
11	140.8	92.6	344.5	211.7	722.5	572	1177	891.2	256.7	164.3	606.5	375.8
10	140.6	100.6	365.9	223.5	934.3	727.6	1631	1202	241.6	169.8	600.3	373.4
9	162.6	111.3	449.1	262.6	1150	906.1	2144	1548	270.2	185.6	734.9	437.8
8	175.6	116.5	491.0	278.6	1363	1089	2718	1923	298.5	197.3	820.5	473.7
7	192.8	124.8	539.1	297.3	1580	1276	3345	2324	327.1	211.8	915.1	512.3
6	212.2	140.3	598.9	325.4	1815	1473	4006	2746	392.7	254.6	1110	602.3
5	207.4	151.4	583.9	314.2	2061	1705	4741	3232	366.0	255	1016	555.3
4	221.1	166.1	660.4	354.7	2328	1955	5500	3756	386.8	275.8	1121	595.7
3	220.7	188.9	705.7	395.8	2604	2241	6296	4333	397.4	322.8	1228	694.6
2	201.7	174.2	689.6	341.5	2872	2554	7119	4951	358.0	275.3	1160	549.1
1	276.1	280.4	910.9	566.6	3138	3031	7996	5834	757.0	743.0	2523	1560

4. Qualitative Parameters Study of Interaction

The dynamic behaviour of SSI system may be partly determined by changing the interaction parameters of soil-pile foundation. The numerical results from spectrum analysis or time-history analysis would be influenced by the various fundamental frequency which is the key design factor to calculating structural reaction forces and deformations. In this paper, the pile-group effect with the alterative pile/soil properties on dynamic interaction is investigated and the influenced degree is reflected by the ratio of SSI period $\overline{T}_i / T_i$ (i=1~3, $\overline{T}$ is the SSI period, T_i the fixed-base period). From the computational data as shown in Table.3, the front three period ratio decrease from 1.294s↓ 1.072s, 1.217s↓ 1.041s, 0.735↓ 0.681s respectively with G_s=4500kPa and R_0 from 0.2m↑ 0.45m. When R_0=0.3m and G_s from 4500kPa↑ 45000kPa, the fundamental period decreases from 1.177s↓ 0.715s and the decrease in corresponding

Table 3. Vibration Features of SSI system considering change in pile-soil parameters

G_{s1} \ R_0	0.2	0.25	0.3	0.35	0.4	0.45
	2.336^1	2.216^1	2.124^1	2.050^1	1.987^1	1.935^1
4500	2.668^2	2.563^2	2.478^2	2.405^2	2.340^2	2.282^2
	2.042^3	2.011^3	1.980^3	1.950^3	1.920^3	1.891^3
	1.796^1	1.710^1	1.648^1	1.602^1	1.567^1	1.538^1
12000	2.009^2	1.929^2	1.871^2	1.828^2	1.794^2	1.768^2
	1.447^3	1.422^3	1.401^3	1.382^3	1.366^3	1.353^3
	1.574^1	1.513^1	1.469^1	1.437^1	1.412^1	1.391^1
24000	1.778^2	1.724^2	1.687^2	1.659^2	1.638^2	1.621^2
	1.283^3	1.270^3	1.259^3	1.249^3	1.241^3	1.234^3
	1.379^1	1.327^1	1.290^1	1.263^1	1.241^1	1.224^1
45000	1.605^2	1.564^2	1.535^2	1.514^2	1.498^2	1.485^2
	1.204^3	1.194^3	1.185^3	1.178^3	1.172^3	1.166^3

period ratio is about 39%, 38%, 40%, respectively. As compared with the corresponding values on condition that R_0=0.6m, G_s=4500kPa, the decrease of $\overline{T}_{1,2,3}/T_{1,2,3}$ is about 48%, 44%, 43% on condition that R_0=0.9m, G_s=45000kPa respectively. It is indicated that the flexibility participation of the pile-soil foundation should be enhanced with the combination of softer soil and smaller piles. The more flexible the pile-soil foundation is, the more strong the dynamic interaction effect on PGSS system is.

5. Conclusion

The present study makes an effort to evaluate the effect of SSI on primary dynamic character of three-dimensional high-rise buildings with pile-raft foundation and attempts to provide rational schemes regarding this issue. Whereas the number of elements is still large in presented pile-group-soil model, there is only one direction DOF in all the corresponding spring-dashpot elements. As a whole, the computational degrees of freedom in PGSS model will be much less than those in solid finite-element model. The numerical results are presented in an efficient and quick manner to show the trend in the effect of variation of dynamic characteristics owing to the incorporation of the interaction effect of pile-soil-structure system. The system period augments to some extent with the participation of pile-soil flexibility. The storied-displacement increases and the storied-acceleration decreases under SSI comparing with the corresponding value under fixed-base condition. In addition to the features of input wave, the internal-force response of structural members are determined by the structural characteristics and pile-soil interaction etc. From the qualitative study of SSI parameters, the more hard the soil is or the more rigid the stiffness of piles are, the less obvious the interaction effect are. The presented analysis is well in agreement with the above trends. Furthermore, the dynamic FEM analysis including any type superstructure in SSI system may be developed with material and geometric nonlinear properties by this model with current computing capability owing to the rational decreasing work for the relatively simplified soil-foundation interaction model.

References

1. H. G. Poulos and E. H. Davis, Pile foundation analysis and design, New York, John Wiley, 1980.
2. G. Gazetas and N. Makris, Dynamic pile-soil-pile interaction.Part I: Analysis of axial vibration. 20(2): 115~132 (1991).
3. N. Makris and G. Gazetas, Dynamic pile-soil-pile interaction. Part II: lateral and seismic response. Earthquake Engineering and Structural Dynamics, 21(2):145~161 (1992).
4. S. M. Mamoon, A. M. Kaynia and P. K. Banerjee, Frequency domain dynamic analysis of piles and pile groups. Journal of the Engineering Mechanics, ASCE, 116(10): 2237~2257 (1990).
5. M. Novak, T. Nogami and Aboul-Ella, Dynamic soil reactions for plane strain case. Journal of the Engin eering Mechanics Division, ASCE, 104(4): 953~969 (1978).
6. M. Ashour and G. Norris, Modeling lateral soil-pile response based on soil-pile interaction. Journal of Geotechnical and Geoenvironmental Engineering, ASCE, 126(5): 420~427 (2000).
7. E. Richart, J. R. Hall and R. D. Woods, Vibration of soils and foundation. Englewood, Cliffs. NJ. 1970.
8. T. Nogami, J. Otani and K.Konagai, Nonlinear soil-pile interaction model for dynamic lateral motion. Journal of Geotechnical Engineering, ASCE, 118(1):89~106 (1992).
9. D. B. McCallen, K. M. Romstad, Nonlinear model for building-soil systems. Journal of the Engineering Mechanics, ASCE, 120(5): 1129~1152 (1994).
10. Xilin lu, Bo Chen, Peizhen Li, et al, Numerical analysis of tall buildings considering dynamic soil-structure interaction. Journal of Asian Architecture and Building Engineering, 2(1):1~8 (2003).
11. Hui Xiong, Dynamic analysis and optimizing design of interactive effect on upper-lower structural parts in multilayer-soil field. Ph.D.Thesis, Hunan University of China, 2003.
12. I. M. Idriss, R. Dobry and R. D. Singh, Nonlinear behavior of soft clays during cyclic loading. Journal of Geotechnical Engineering ASCE, 104 (12):1427~1447 (1978).
13. T. Nogami, Dynamic group effect in axial responses of grouped piles. Journal of Geotechnical Engineering, ASCE, 109(2):228~243 (1983).
14. T. Nogami, Flexural responses of grouped piles under dynamic loading. Earthquake Engineering and Structural Dynamics, 13(3):321~336 (1985).
15. H. Mohamed, Ei. Naggara and M.Novak. Non-linear model for dynamic axial pile response. Journal of Geotechnical Engineering, ASCE, 120(2): 308~331 (1994).

SOIL RESPONSE AND GROUP INTERACTIONS IN TORSIONALLY LOADED LARGE-DIAMETER BORED PILE GROUPS*

LIMIN ZHANG

Department of Civil Engineering,
Hong Kong University of Science and Technology, Clear Water Bay, Hong Kong.

CARMEN Y.M. TSANG

Department of Civil Engineering,
Hong Kong University of Science and Technology, Clear Water Bay, Hong Kong

In a pile group subjected to lateral or moment loading, the soil movements are primarily along the loading direction. In a torsionally load pile group, however, the primary soil movements are rotational. The movements of soils in the pile group can be in arbitrary directions. As a result, the soil response in a torsionally loaded pile group is different from that in a laterally loaded pile group. In this paper, three-dimensional finite-difference analyses of a torsionally loaded single pile, a torsionally loaded pile group, a laterally loaded single pile and a laterally loaded pile group are conducted to study soil responses and group interactions in torsionally loaded pile groups. The large-diameter bored piles studied are 2 m in diameter and 13.2 m in length. The pile groups are 2x2 groups with a three-diameter spacing. The soil response is represented by horizontal stresses in front of the piles and *p-y* curves derived from calculated bending moments along the piles. The *p-y* curves of the piles in the torsionally loaded pile group are lower than those of the laterally loaded single pile and the leading pile in the laterally loaded pile group. In addition, the ultimate subgrade reaction is mobilized at a much smaller local displacement (say, 20 mm) compared with the laterally loaded single pile (> 60 mm). The horizontal normal stresses caused by lateral loading and the horizontal shear stresses caused by torsion interact and reduce the lateral pile resistance in the torsionally loaded pile group.

1. Introduction

Large-diameter bored piles supporting tall buildings are subjected to lateral, moment and torsional loads. In a pile group subjected to lateral or moment loading, the load effect is resisted by lateral and axial resistances of the individual piles. The soil movements are primarily along the loading direction. The lateral resistance of an individual pile is influenced by pile group interactions, particularly, the induced stresses exerted by other group piles that move in the same direction as the pile concerned (shadowing effect). The shadowing effect can be characterized by interaction factors (Poulos 1971) or multiplier factors (Brown et al. 1988; McVay et al. 1998; Zhang et al. 1999). In a torsionally load pile group, however, the primary soil movements are rotational. As a result, the soil response in a torsionally loaded pile group may be different from that in a laterally loaded pile group. The related soil responses and group interactions in a torsionally loaded pile group has not been reported.

In this paper, three-dimensional finite-difference analyses of a torsionally loaded pile group, a laterally loaded single pile, and a laterally loaded pile group are conducted to study soil responses and group interactions in a torsionally loaded pile group. The soil response is represented by horizontal stresses in front of the piles and *p-y* curves derived from calculated bending moments along the piles.

2. Numerical Model

The soil responses in a 3-diameter spacing, 2x2 large-diameter bored pile group under torsion and the same pile group under lateral loading are studied in this paper. The piles are 2 m in diameter and 13.2 m in length, and are fixedly attached to the pile cap. Fig. 1 shows the 3D view and the cross section of the pile group model.

* This work is supported by a grant from the Research Grants Council of the Hong Kong Special Administrative Region (Project No. HKUST6037/01E).

The concrete piles and the pile cap are simulated by an elastic model assuming linear elastic properties. The soil is simulated by an elastic perfectly plastic model following the Mohr-Coulomb yielding criterion with a tension cutoff. The contact between the pile and the soil is modeled by interface elements to represent the possible separation and slippage between the pile and the soil. The interface elements are characterized by Coulomb sliding and/or tensile and shear bonding. The 3D finite-difference program Flac3D (Itasca, 1997) is employed to conduct these analyses.

The soil is assumed to be homogenous. The Poisson ratio v and Young's modulus E of the soil assumed to be 0.3 and 30 MPa, respectively, which represent the properties of a dense completely decomposed granite commonly found in Hong Kong. An internal friction angle $\phi'=38^\circ$ and a unit weight γ = 1700 N/m^3 are adopted following guidelines by GCO (1982). The modulus and Poisson's ratio of the pile and pile cap concrete are taken to be E =26 GPa and v =0.3, respectively. The cohesion, tensile strength, and angle of dilation of the interface model are all specified as zero. The interface friction angle ϕ_i is assumed to be 0.8 times the soil's internal friction angle (GCO 1982).

The numerical simulation is divided into three major steps; namely, pile excavation, formation of piles and the pile cap, as well as torsional or lateral loading. Details of the geometry, the boundary conditions, the constitutive models, and the simulation process have been described in the companion paper (Zhang and Tsang 2005).

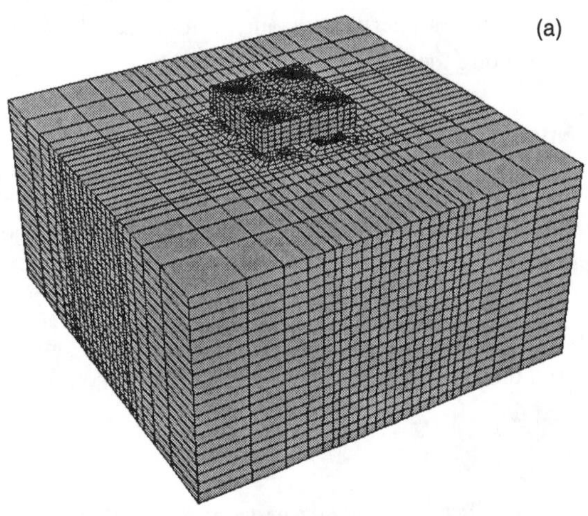

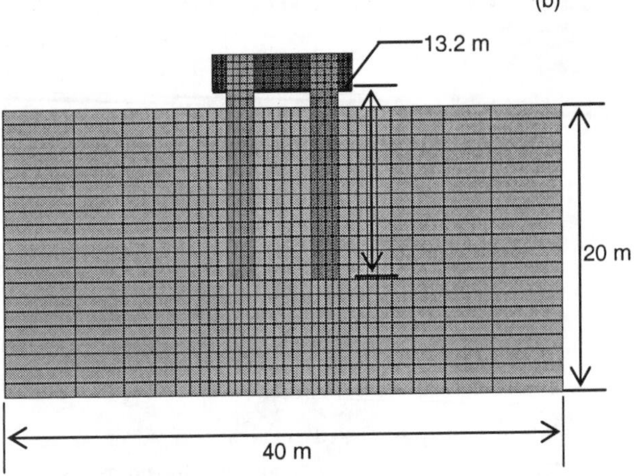

Figure 1. (a) 3D view and (b) cross section of the torsionally or laterally loaded pile group.

3. Soil Movements in Torsionally Loaded and Laterally Loaded Pile Groups

In a pile group subjected to lateral or moment loading, the load effect is resisted by lateral and axial resistances of the group piles. The soil movements are primarily along the loading direction in Fig. 2(a). The lateral resistance of an individual pile is influenced by pile group interactions (shadowing effect); that is, the lateral resistance of one pile is affected by the induced stresses exerted by other group piles that move in the same direction as the pile concerned. The shadowing effect can be characterised by interaction factors (Poulos 1971) or multiplier factors (Brown et al. 1988). In a torsionally load pile group shown in Fig. 2(b), however, the primary soil movements are rotational. The movements of the soils in the pile group can be in arbitrary directions with respect to the movement direction of the pile concerned. As a result, the soil response in a torsionally loaded pile group is different from that in a laterally loaded pile group.

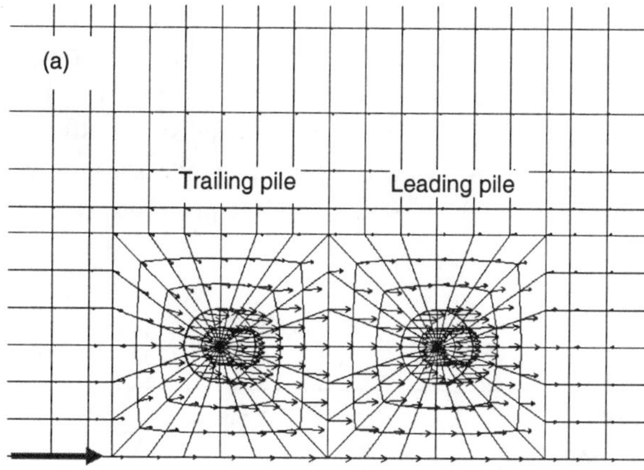

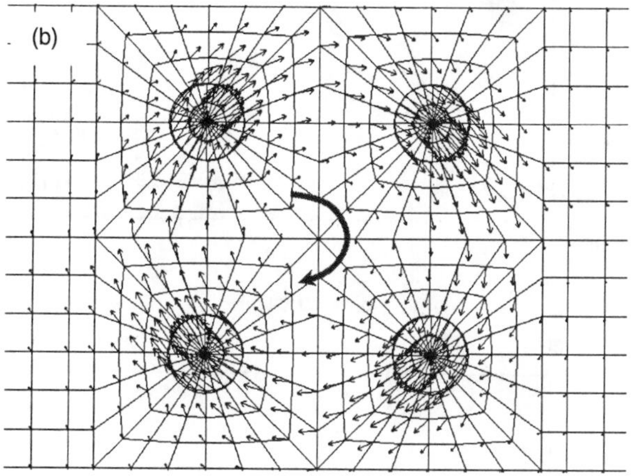

Figure 2. Soil movements in (a) a laterally loaded pile group and (b) a torsionally loaded pile group.

4. Comparison between Torsionally Loaded and Laterally Loaded Pile Groups

4.1 *Shear-deflection curve*

As shown in the companion paper, the applied torque to a pile group is resisted by both the torsional and lateral resistance of the group piles. Accordingly, shear forces and lateral soil resistance develop in the group piles although the pile group is only subjected to a torque. The shear force at the pile head is plotted against the pile head deflection along the tangential direction in Fig. 3. The shear forces in an equivalent laterally loaded single pile and in a laterally loaded pile group with the same geometry are also plotted in Fig. 3 for comparison purposes. Comparing with the laterally loaded group piles, the torsionally loaded group pile bears the largest shear force at the same deflection. The reasons for this will be explained later in discussing *p-y* curves. As expected, the shear forces in the leading pile in the laterally loaded pile group and in the single pile are similar to those in a single laterally loaded pile. The trailing pile in the pile group bears the smallest shear force because of the shadowing effect (Brown et al., 1987).

4.2 *Mean stress-deflection curve*

The mean stress in the soil, $\sigma_m = (\sigma_1 + \sigma_2 + \sigma_3)/3$, in front of the pile along the pile movement direction is computed for both the torsionally and laterally loaded piles in the groups. Fig. 4(a) shows the relationship between the mean stress at 1.5 m below the ground and the local lateral deflection. The mean stresses in front of the trailing pile in the laterally loaded pile group and a pile in the torsionally loaded pile group are comparable,

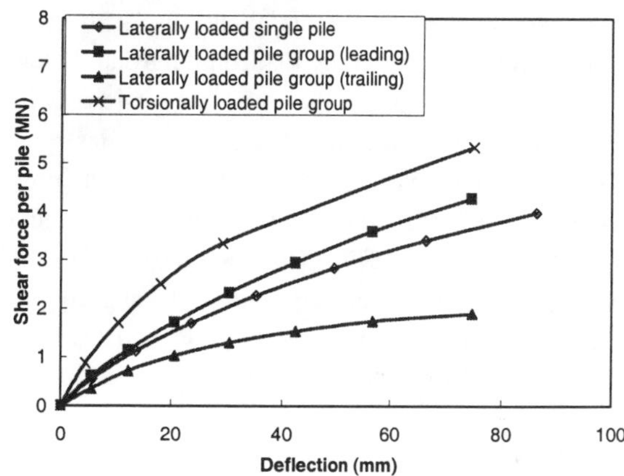

Figure 3. Development of shear forces in piles with pile deflection in laterally and torsionally loaded piles.

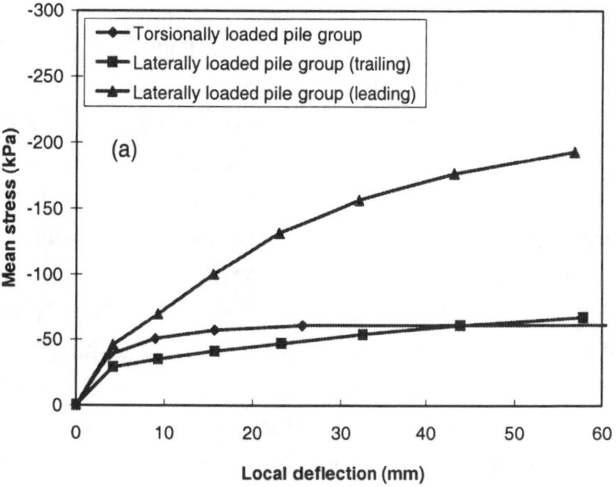

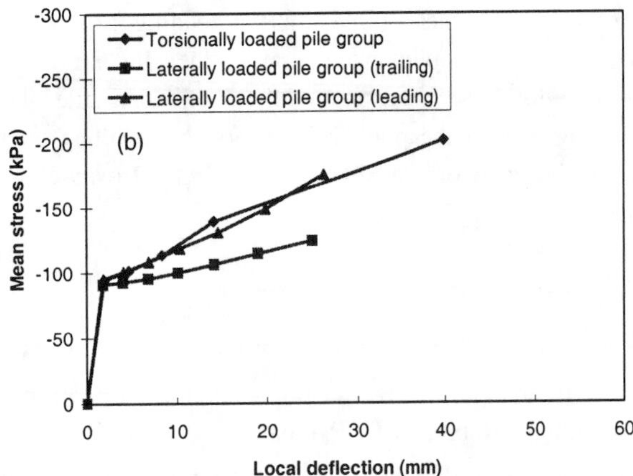

Figure 4. Mean stresses vs. local deflection for torsionally and laterally loaded pile group at (a) 1.5m and (b) 6.5 m below the ground surface.

and are considerably smaller than those in front of the leading pile in the laterally loaded pile group. It suggests that "shadowing effect" also exist in a torsionally loaded pile group, and the torsionally load piles generate reaction forces that are smaller than those in front of a single laterally loaded pile at the same local deflection.

At larger depth [Fig. 4(b)], the mean stress in front of the torsionally loaded pile is only slightly smaller than those in front of the laterally loaded leading pile. Therefore the shadowing effects are less pronounced.

4.3 *p-y Curves*

The lateral soil resistance of a pile can better be represented by subgrade reaction than the normal stress because the normal stress varies around the perimeter of the pile. Subgrade reaction p refers to the lateral soil resistance per unit length and is related to the shear force V along the pile by $p=dV/dx$. Hence the p value along the pile can be obtained from the slope of the shear force distribution along the pile. The distributions of p along a leading pile in the laterally loaded pile group and a pile in the torsionally loaded pile group are shown in Figs. 5(a) and (b), respectively. It can be seen that, for both the laterally and torsionally loaded piles, the subgrade reaction increases from zero at the ground surface to the maximum value at a certain depth and then decreases to small values at the pile toe. In a laterally loaded pile group, the shear strengths of the soils at shallow depths are mobilized. As the applied load increases, the location of the maximum subgrade reaction shifts downwards slightly. In the torsionally loaded pile group, the maximum p value occurs near the ground surface at small torques but at much deeper locations as the applied torque increases. Thus the shear strengths of the soils at large depths can also be mobilized. This explains why larger pile shear forces can develop in the torsionally loaded pile group (see Fig. 3).

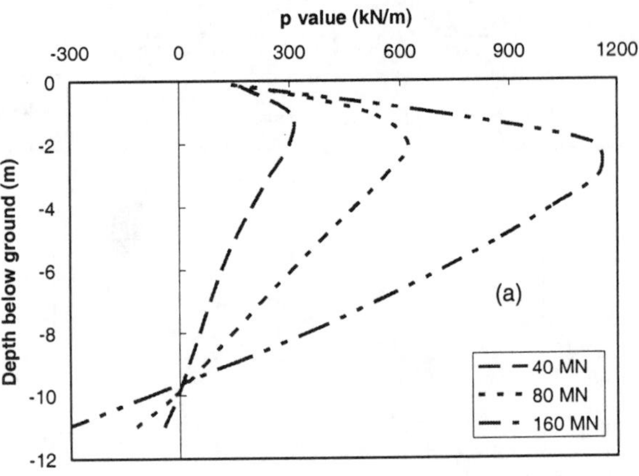

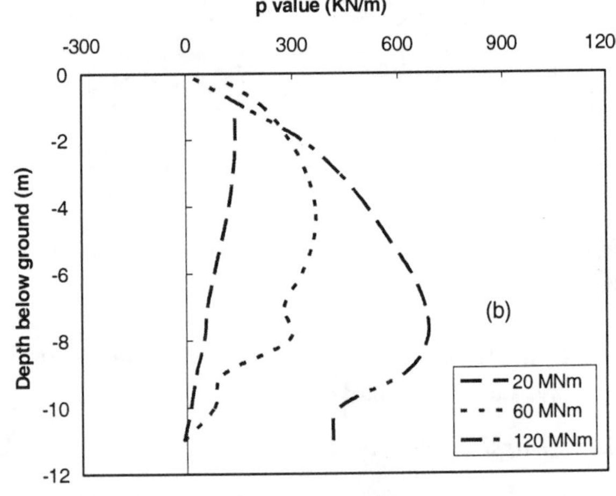

Figure 5. Subgrade reaction p along (a) a laterally loaded group pile and (b) a torsionally loaded group pile.

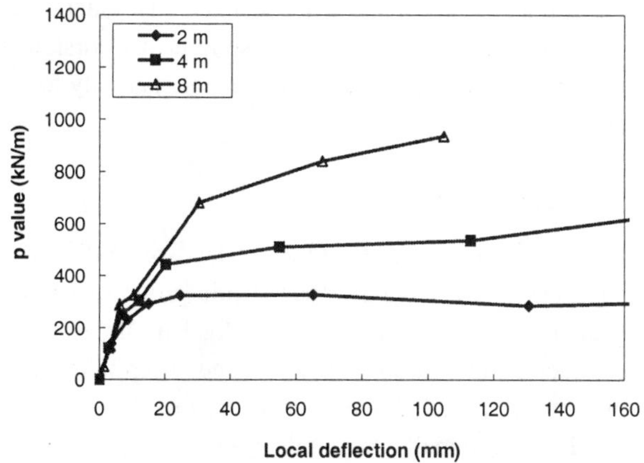

Figure 6. *p-y* curve of a torsionally loaded group pile.

The *p-y* curves of a torsionally loaded group pile, at three depths, 2, 4 and 8 m below the ground surface, are shown in Fig. 6. As expected, a curve at a larger depth is stiffer than that at shallower depth (Reese et al. 1974). These curves are all nonlinear, and the mobilization degree of the lateral resistance is high at shallow depths and low at large depths. The ultimate lateral resistance is mobilized at approximately 20 mm deflection.

Fig. 7 compares the *p-y* curves of a torsionally loaded group pile, a laterally loaded group pile, and a laterally loaded single pile. The *p-y* curves of the laterally loaded single pile are the stiffest, followed by the curves of the leading pile in the laterally loaded pile group. The *p-y* curves of the torsionally loaded pile are the weakest except at very large depths. In addition, the lateral resistance of the torsionally loaded pile is mobilized at a relative small local deflection of about 20 mm, far smaller than the displacements that are needed to mobilize the lateral resistance of a pile without subject to torsion. In the Reese et al. (1974) model for laterally loaded piles, it is assumed that the ultimate lateral resistance, derived from the passive equilibrium of the soil wedge in front of the pile, occurs at a deflection of 3B/80, i.e., 75 mm for the 2 m-diameter piles studied in this paper.

4.4 *Interaction between torsional and lateral loading*

Two factors may affect the *p-y* response of the piles in a torsionally loaded pile group; namely, the shadowing effect and the interaction between stresses caused by torsional and lateral loading.

The shadowing effects in a torsionally loaded pile group and a laterally loaded pile group are shown in Fig. 8. In a laterally loaded pile group [Fig. 8(a)], the direction of soil movements is primarily along the loading direction. If the soil movement in front of a pile is assumed to be limited within a wedge, then the

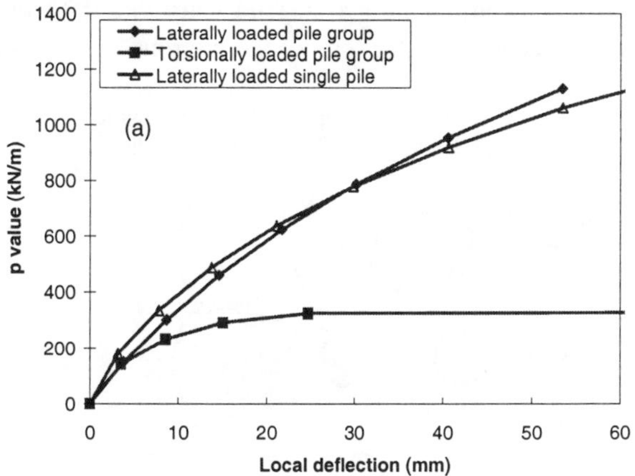

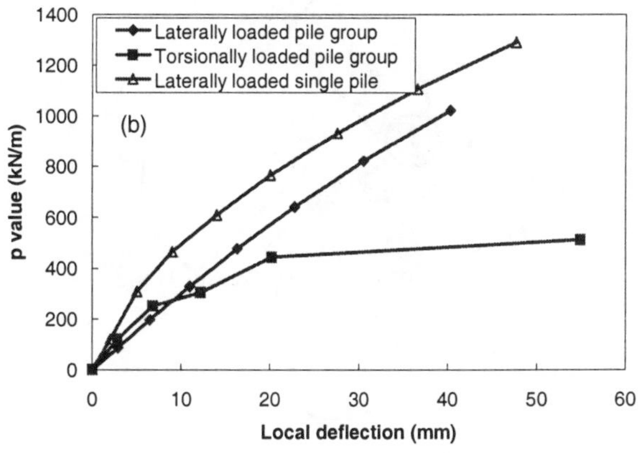

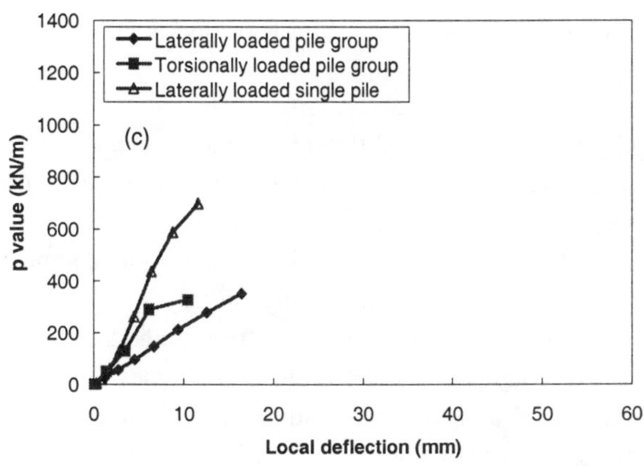

Figure 7. *p-y* curves of torsionally and laterally loaded piles at (a) 2 m, (b) 4 m and (c) 8 m below the ground surface.

wedges in front of the group piles will overlap. Due to the stress overlapping effect, the lateral resistance of the group piles will be smaller than that of a single pile. In a pile group subjected to torsion along, the soil movements are along the tangential direction [see Fig. 2(b)]. Accordingly, the four wedges in front of the group piles move along four different directions. The degree of stress overlap, if any, is smaller than the case of the laterally loaded pile group in Fig. 8(a). This would suggest that the *p-y* curves of the torisonally loaded piles be less affected by the shadowing effect.

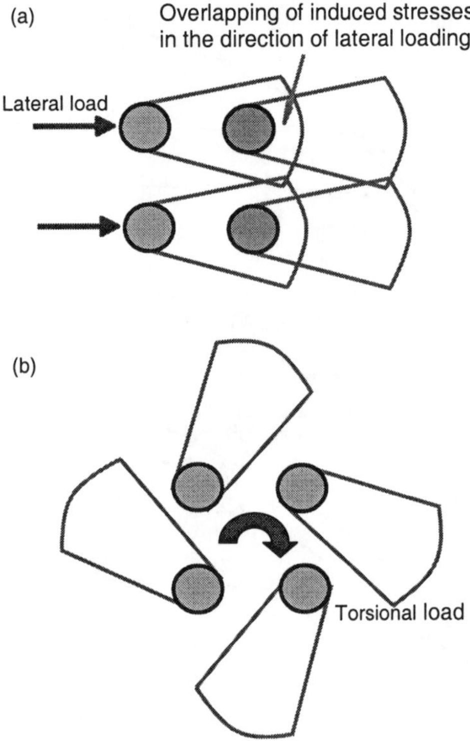

Figure 8. Shadowing effects in laterally and torsionally loaded pile groups.

On the other hand, the soil in front of the torsionally loaded group pile is affected by shear stresses caused by both lateral and torsional loading. The shear failure of the soil in front of a laterally loaded pile is caused by the increase in the horizontal stress while the vertical stress remains constant. If the pile is subjected to torsion and lateral loading concurrently, then an additional shear stress in the horizontal plane is also imposed on the soil around the pile in addition to the increase in the horizontal stress. This will decrease the lateral soil resistance. Therefore, the *p-y* response of a torsionally loaded pile will be weaker than a single laterally loaded pile.

5. Conclusions

In this paper, three-dimensional finite-difference analyses of a torsionally loaded pile group, a laterally loaded single pile, and a laterally loaded pile group are conducted to study soil responses and group interactions in torsionally loaded pile groups.

In a torsionally load pile group, the primary soil movements are rotational. The movements of soils in the pile group can be in arbitrary directions with respect to the movement direction of the pile concerned. As a result, the soil response in a torsionally loaded pile group is different from that in a laterally loaded pile group.

The mean soil stress in front of a torsionally loaded group pile is comparable to that in front of the trailing laterally loaded group pile, and is lower than that in a front of the leading laterally loaded group pile. This suggests that shadowing effect exists in a torsionally loaded pile group, and the group piles will generate reaction forces that are smaller than those in front of a single laterally loaded pile at the same local deflection level.

Similar interaction effects are observed in the lateral subgrade reaction as reflected by *p-y* curves. The *p-y* curves of a pile subjected to torsional loading are weaker than that of a single pile subjected to lateral loading only. The decreased shear strength of soil due to the shear stresses caused by torsion might contribute to a significant loss of lateral resistance, as the shadowing effects in a torsionally loaded pile group are not as severe as those in a laterally loaded pile group.

Finally, due to the torsional shear effect, the lateral resistance of soils at large depths can be mobilized. Therefore large shear forces can still develop in a torsionally loaded pile group.

Acknowledgments

This research was substantially supported by a grant from the Research Grants Council of the Hong Kong SAR (Project No. HKUST6037/01E).

References

1. Brown, D.A., Morrison, C., and Reese L.C. (1988). Lateral load behavior of pile group in sand, *J. Geotech. Eng.*, ASCE **114**(11), 1261 – 1276.
2. GCO (1982). *Guide to Retaining Wall Design, Geoguide 1.* 1993 edition, Geotechnical Control Office, Hong Kong.
3. Itasca (1997). FLAC3D *Fast Largragian Analysis of Continua in Three Dimensions: Version 2.0: User's Manual*, Itasca Consulting Group Inc., Minnesota, U.S.A.
4. McVay, M.C., Zhang, L.M., Molnit, T., and Lai, P.W. (1998). Centrifuge testing of large laterally loaded pile groups in sands. *J. Geotech. and Geoenviron. Eng.*, ASCE **124**(10), 1016-1026.
5. Poulos, H.G. (1971). Behavior of laterally loaded piles: I -single piles. *J. Soil Mech. Founda. Engrg.*, ASCE **97**(5), 711-731.
6. Reese, L.C., Cox, W.R., and Koop, F.D. (1974). Analyses of laterally loaded piles in sand. *Proceeding, 6th Annual Offshore Technology Conference,* Dallas, Tex., 473-483.
7. Zhang, L.M., McVay, M.C., and Lai, P. (1999). Numerical analysis of laterally loaded 3x3 to 7x3 pile groups in sands. *J. Geotech. Geoenviron. Eng.*, ASCE **125**(11), 936-946.

GROUND MOVEMENTS DUE TO EXCAVATION WITH LATERAL SUPPORTING SYSTEMS

TONY W. LAU

Black &Veatch Hong Kong Limited;
Department of Civil Engineering, The University of Hong Kong, Hong Kong SAR, China

J. YANG

Department of Civil Engineering, The University of Hong Kong, Hong Kong SAR, China

The control of ground movements is one of the most critical restraints in modern shoring design, especially when the excavation is carried out in vicinity of buildings, utilities and tunnels. In this study, several lateral supporting systems are investigated, with an effort to identify the key factors contributing to ground movements. It is shown that the ground movement is sensitive to the excavation depth and ground conditions, yet seems to be relatively less sensitive to the wall and strut stiffness. It is also noted that the location of the first layer of strut can be critical to the wall deflection, and an optimum strut arrangement can be achieved to minimize the wall deflection and subsequently the ground movement.

1. Introduction

Construction of tall buildings usually involves deep excavation that places a great demand for a rational and safe shoring support design. Conventionally, calculations used in design and required by codes of practice are mainly based on limit equilibrium methods in which attention is placed on balancing stresses and no explicit account is taken of ground deformation (Simpson, 1992). In the philosophy of modern geotechnical design, however, calculations should be based on the requirements of limit states at which the behavior of the earth structures concerned ceases to be acceptable.

In the case of retaining walls, the behavior can be rationally defined in terms of wall deflection and the associated ground displacement (Bolton et al., 1990; Simpson, 1992). Experience indicates that wall displacements may be larger than normally expected in concrete or steel structures. As a result, ground movements could be large enough to cause severe distress that may bring the supporting system to an unacceptable state. It is apparent that ground displacement should be one of the most critical restraints in shoring design, especially when the excavation is carried out in vicinity of buildings, utilities and tunnels.

In this paper, three lateral supporting systems, i.e. cantilever system, single strut system, and multi-strut system, are analyzed, with an effort made to identify the key factors contributing to the ground movement for each system and subsequently to achieve a better design.

2. Ground Movements due to Excavation

The magnitude and distribution of ground movements caused by excavation depend on many factors, such as the ground condition, excavation depth, and wall stiffness. A number of case histories have indicated that there are two general types of ground movement profiles (Clough and O'Rourke, 1990; Hsieh and Ou, 1998): spandrel type and concave type, as illustrated in Fig. 1.

The spandrel type of displacement has a maximum surface settlement occurring very close to the wall. This displacement profile usually occurs if a large amount of wall deflection occurs at the first stage of excavation and the wall deflection is relatively small at the subsequent excavations or the wall has a cantilever type deflection.

The concave type of displacement has a maximum surface settlement occurring at a distance away from the retaining wall. Such a displacement profile occurs if a relatively small amount of wall deflection occurs at the initial stages of excavation or the deflection in the upper part of the wall is retrained by installation of support.

Based on a large number of case histories, some empirical methods have been proposed for evaluating both types of ground movements (Hsieh and Ou, 1998; Long, 2001). Generally, these empirical methods predict the maximum ground settlement δ_{vm} with reference to the maximum wall deflection δ_{hm}. In most cases, the maximum ground settlement δ_{vm} is found to be in a range of (0.5-0.75) δ_{hm}. Hsieh and Ou (1998) suggest that the following normalized relationship can be used:

$$0.5\ \delta_{hm} / H_e < \delta_{vm}/H_e < \delta_{hm} / H_e \qquad (1)$$

where H_e is the excavation depth.

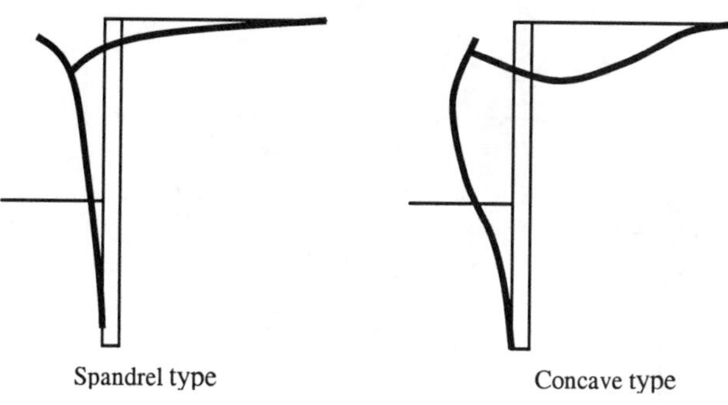

Figure 1. Two types of ground movement profiles

3. Method of Analysis

To effectively control ground movements due to excavation, various factors that may influence the wall deflection need to be identified for different conditions. In this study, the behavior of three lateral supporting systems as shown in Fig. 2 is analyzed using FREW, a program specifically for lateral deformation analysis. The program is capable of computing wall deflections, earth pressures, bending moments and strut forces occurring during each stage of excavation.

In the analysis, factors that may influence the wall deflection are divided into two groups: (a) non-design factors, and (b) design factors.

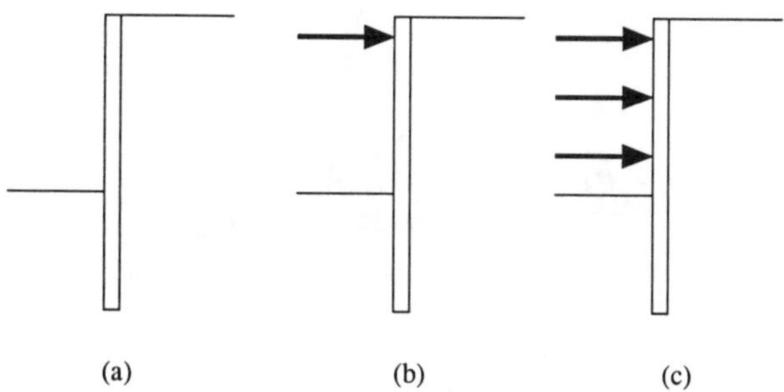

Figure 2. Supporting systems analyzed: (a) cantilever system; (b) single strut system; (c) multi-strut system

The non-design factors refer to those that are not altered in normal shoring design, such as soil properties, ground water table, and excavation depth. The design factors refer to those factors that can be altered by modifying the design, such as wall stiffness, strut stiffness, and strut arrangement. The study of the influence of the design factors on the performance of supporting systems would be helpful for controlling and minimizing the ground movements due to excavation.

Table 1 summarizes the ranges of parameters assumed in the analysis, which cover the common values of the factors in shoring design.

Table 1. Values of different factors used in analysis

Parameter	Cantilever	Single Strut	Multi-strut (Triple-strut)
Excavation depth (Ex)	3m ± 2m	5m ± 2m	10m
Soil internal friction angle	35° ± 5°		
Ground Water Table	No GWT – 0.5 Ex		
Wall Stiffness	68,000 kNm^2/m ± 10%		
Strut Stiffness	36,700 kNm^2/m ± 10%		
Strut arrangement	NA	0.1 – 0.5Ex	0.05 – 0.3Ex

4. Results and Discussion

4.1 *Effects of non-design factors*

The effect of excavation depth on wall deflection for the cantilever and single strut systems is shown in Fig. 3. In the calculation, soil friction angle is assumed as 35°, the ground water table is assumed to be equal to the excavation level, and the wall stiffness is taken as 68,000 kNm^2/m. For the single strut system, the level of the strut is assumed to be 0.1 times the excavation depth, and the strut stiffness is taken as 36,700 kNm^2/m.

The effect of soil friction angle on the wall deflection for cantilever, single strut and triple strut systems is shown in Fig. 4. The effect of ground water table on the wall deflection for the three types of supporting systems is shown in Fig.5. In all the figures the wall deflection is normalized with the excavation depth.

A graphical summarization of the sensitivity of wall deflection to the non-design factors is presented in Figs. 6-8 for the cantilever, single strut, and triple strut systems, respectively.

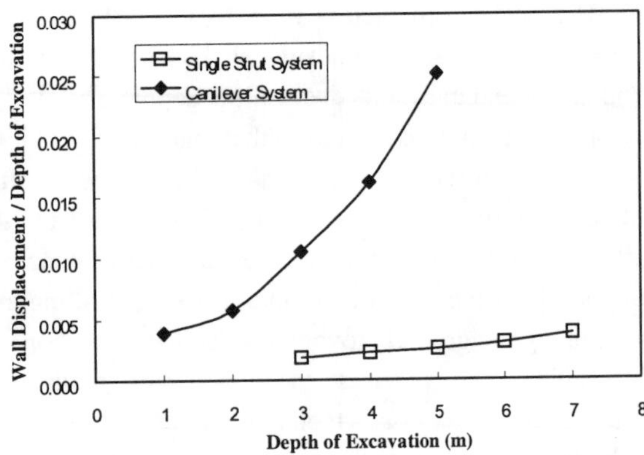

Figure 3. Effect of excavation depth on wall deflection

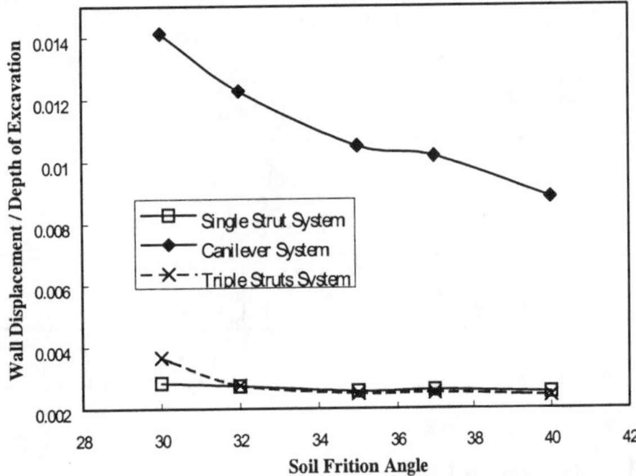

Figure 4. Effect of soil friction angle on wall deflection (excavation depth=3m for cantilever system, 5m for single strut system and 10m for triple strut system; ground water table=excavation level, wall stiffness= 68,000 kNm^2/m, strut stiffness=36,700 kNm^2/m)

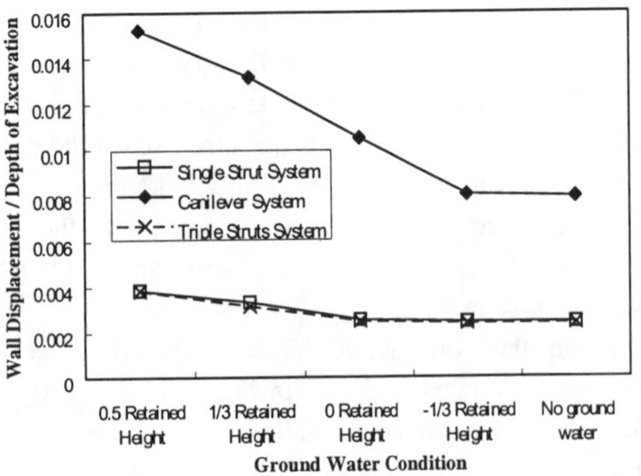

Figure 5. Effect of ground water table on wall deflection (excavation depth=3m for cantilever system, 5m for single strut system and 10m for triple strut system; soil friction angle=35°, wall stiffness=68,000 kNm^2/m, strut stiffness= 36,700kNm^2/m)

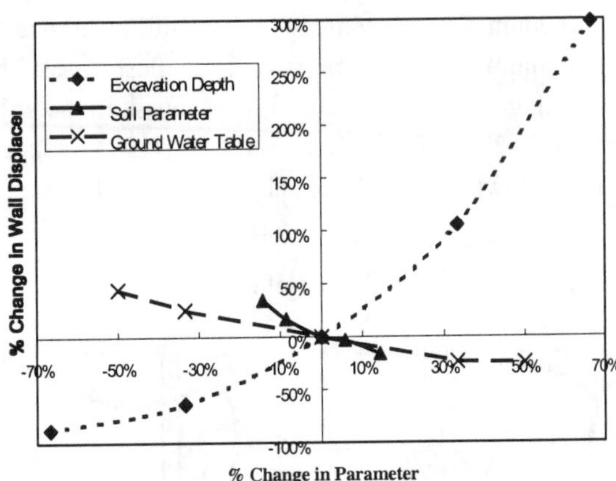

Figure 6. Sensitivity of wall deflection to non-design factors for cantilever system (Reference case: excavation depth=3m; soil friction angle=35°; ground water table=excavation depth)

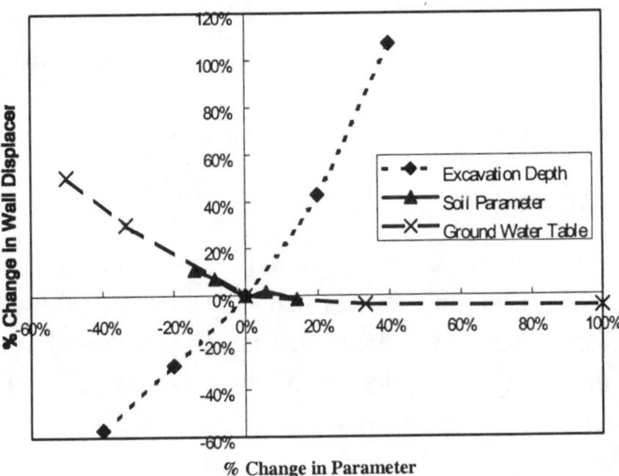

Figure 7. Sensitivity of wall deflection to non-design factors for single strut system (Reference case: excavation depth=5m; soil friction angle=35°; ground water table =excavation depth)

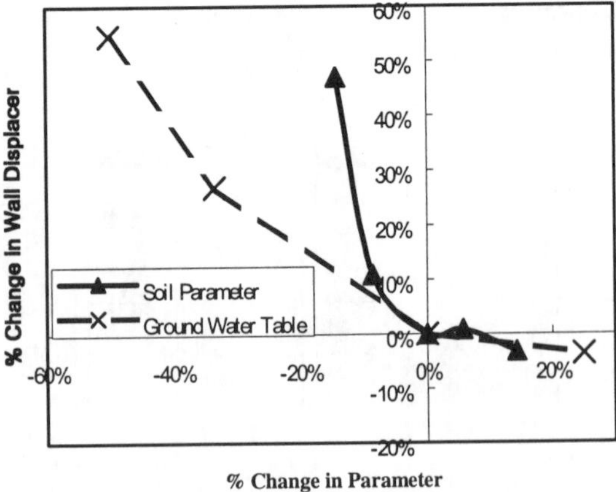

Figure 8. Sensitivity of wall deflection to non-design factors for triple strut system (Reference case: soil friction angle=35°; ground water table= excavation depth)

4.2 *Effects of design factors*

The sensitivity of wall deflection to the non-design factors, including the wall stiffness, strut stiffness, and the location of the first layer of strut, is shown in Figs. 9-11 for the three types of supporting systems.

It can be seen that wall deflection is generally not sensitive to the stiffness of the structural elements. This is in agreement with the field observations of Clough and O'Rourke (1990), who note that in the presence of a stable base, the increase of wall stiffness does not significantly reduce the movements of wall systems.

However, compared with the strut stiffness, increasing the wall stiffness would be relatively more effective in reducing the wall deflection and the ground settlement. A further analysis is carried out to compare the deflection of a diaphragm wall and those of flexible walls. The results are shown in Fig 12.

For strutted systems, installing a diaphragm wall may reduce the wall deflection by approximately 20%. For cantilever systems, the wall deflection can be reduced by approximately 25%. Note that the stiffness of a diaphragm wall can usually be twice or even three times the stiffness of a flexible wall. It would be still effective to construct diaphragm walls to control wall deflections and subsequently the ground movements.

The location (level) of the first layer of strut may be critical to the wall deflection of strutted systems, as shown in Fig. 13. This finding is qualitatively similar with that of Wong et al. (1997), who conclude that the placement of the first prop at or close to the ground surface level is an effective method to control the movement of the wall.

However, it is noted from Fig 13 that when the level of the first layer of strut is lowered down, the wall deflection will decrease until an optimum location is arrived. If we keep lowering the first layer of strut beyond the optimum point, the wall deflection will increase dramatically, implying that rather than putting the strut at the uppermost level, we should figure out the optimum position of the strut so as to minimize the wall deflection.

The rationale for the observed behavior is illustrated in Fig. 14: the wall portion above the top strut behaves as a cantilever while the lower portion behaves as a cantilever propped at the far end. If the prop location were very high, the propped cantilever portion (Portion B) would be much longer than the cantilever portion. Then the deflection of the propped cantilever portion would be greater and dominate.

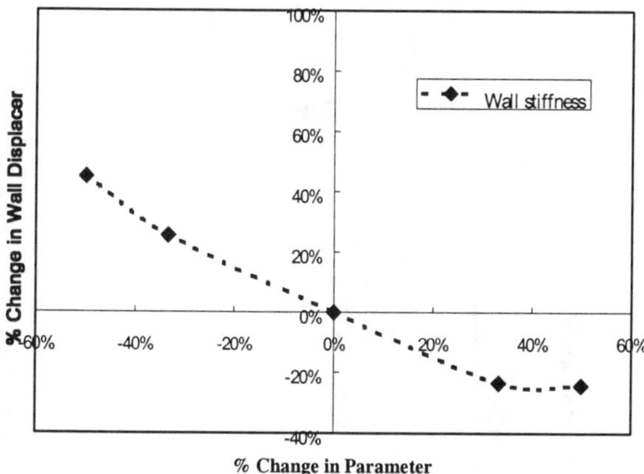

Figure 9. Sensitivity of wall deflection to design factors for cantilever system (Reference case: wall stiffness=68,000 kNm^2/m)

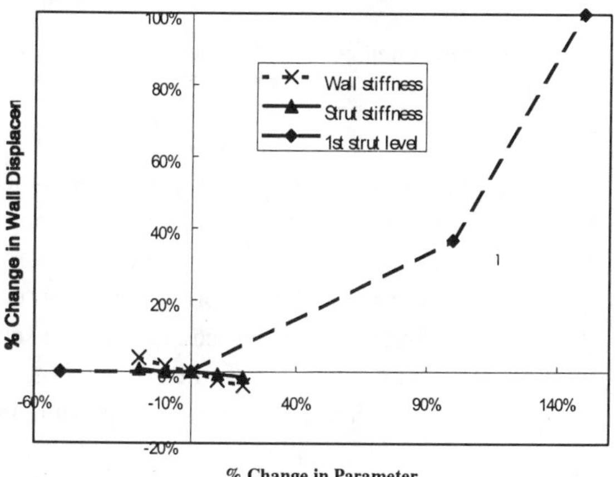

Figure 10. Sensitivity of wall deflection to design factors for single strut system (Reference case: wall stiffness=68,000 kNm^2/m; strut stiffness=36,700 kNm^2/m; 1st strut level = 0.2 x depth of excavation)

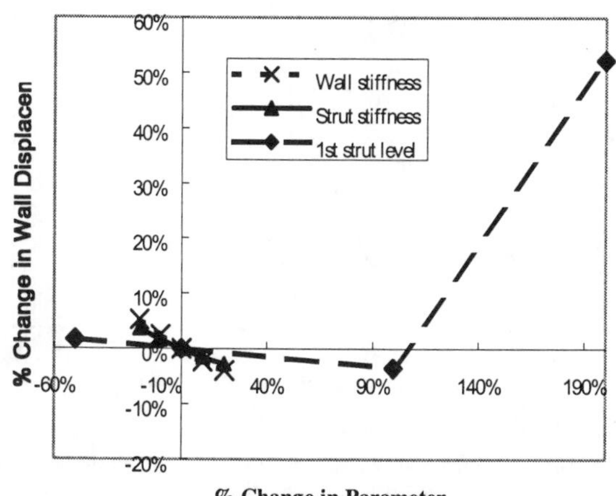

Figure 11. Sensitivity of wall deflection to design factors for triple strut system (Reference case: wall stiffness=68,000kNm^2/m; strut stiffness=36,700 kNm^2/m; 1st strut level =0.1 x depth of excavation)

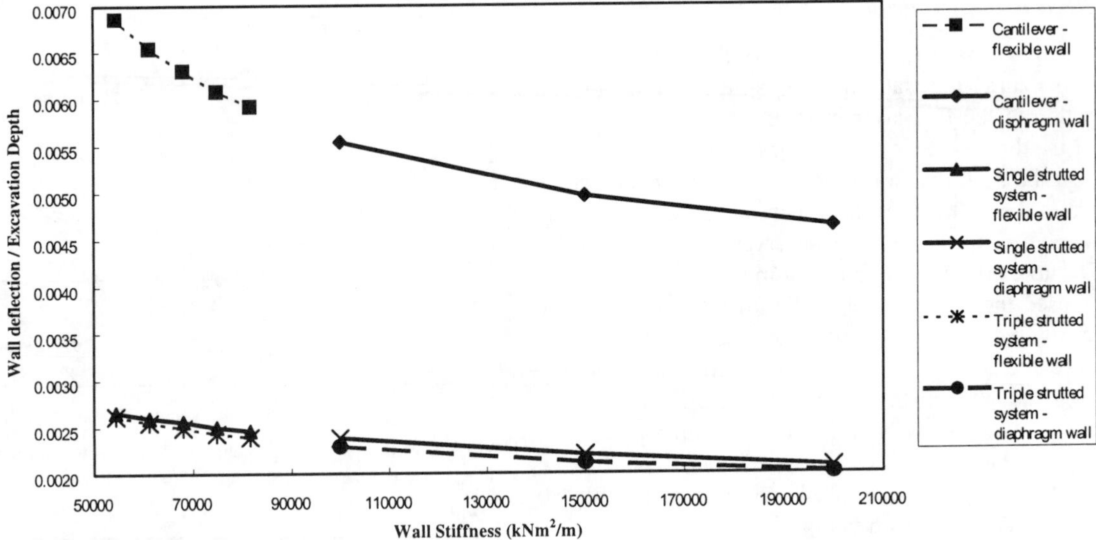

Figure 12. Effect of wall stiffness on wall deflection (excavation depth=3m for cantilever system, 5m for single strut system, and 10m for triple strut system; soil parameter=35°; ground water table = excavation depth; strut stiffness=36,700 kNm^2/m)

On the other hand, when the prop location is too low, the cantilever portion (Portion A) would be longer and the deflection of the cantilever portion would be greater and dominate the deflection of the whole system. In this regard, if the top strut is placed at a location such that the deflection of the cantilever portion is comparable to that of the propped cantilever portion, the wall deflection would be minimized (Fig 15).

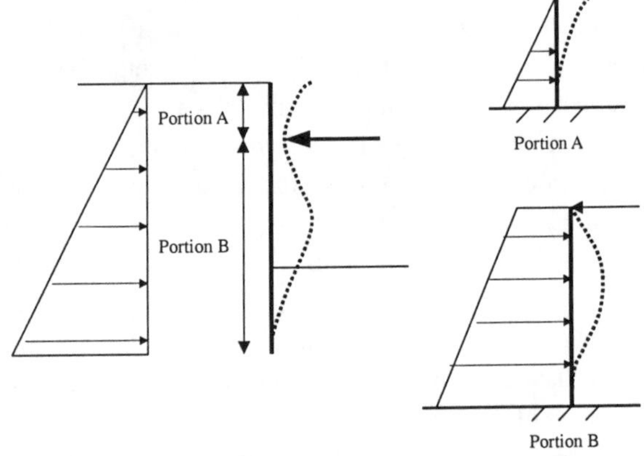

Figure 14. Separation of a strutted system: Portion A and Portion B

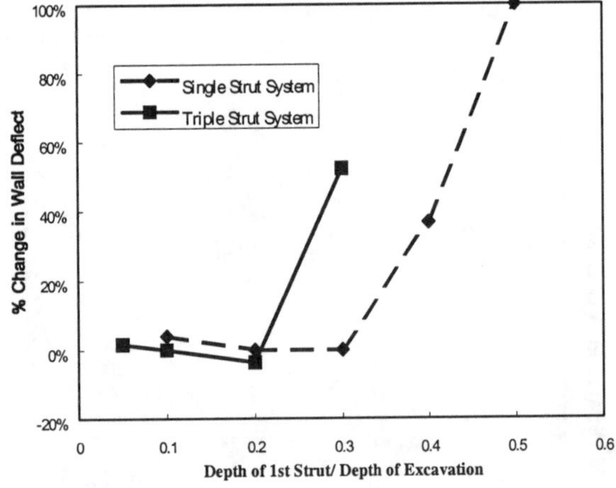

Figure 13. Wall deflection against level of 1st strut for single and triple strut systems (excavation depth=5m for single strut system, and 10m for triple strut system; soil parameter=35°; ground water table=excavation depth; wall stiffness=68,000 kNm^2/m, strut stiffness=36,700 kNm^2/m)

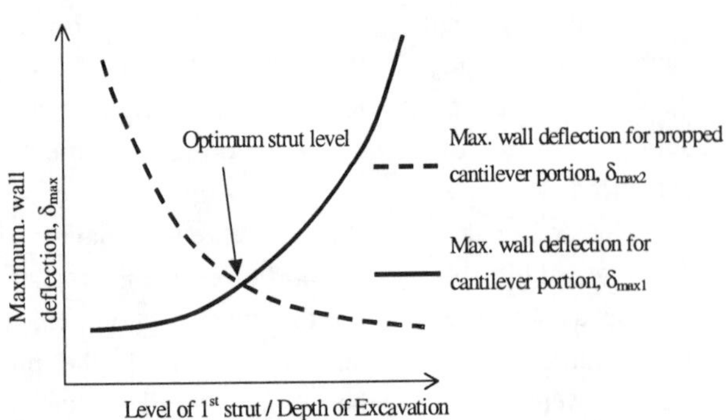

Figure 15. Locating of optimum strut level

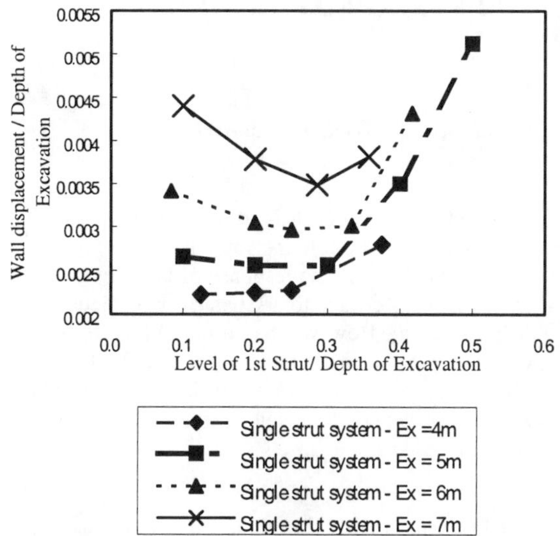

Figure 16. Effect of the location of 1st strut for single strut system (soil friction angle=35°; ground water table=excavation depth, wall stiffness=68,000 kNm²/m; strut stiffness=36,700 kNm²/m)

Further analysis is also performed to investigate four single strut systems of different excavation depths, varying from 4m to 7m. The results are shown in Fig. 16. It is clear that the wall deflection for all the four systems attains its minimum when the strut level is at 0.2–0.4 times the depth of excavation. This observation is consistent with the result of structural analysis, where the optimum strut level is found to be at approximately 0.37 times the depth of excavation.

5. Conclusions

Ground movement is one of the most critical restraints in the shoring design for excavation. The magnitude and distribution of ground movement is closely related to the deflection of retaining walls. For three lateral supporting systems, a series of numerical analyses has been carried out to identify the major factors that may influence the wall deflection.

The results indicate that the wall deflection is sensitive to the excavation depth, soil friction angle, and ground water table, and the excavation depth is the most significant factor. While the wall deflection is less sensitive to the stiffness of structural members (walls and struts), an increase of the wall stiffness is relatively more effective than an increase of the strut stiffness in reducing the wall deflection. Under otherwise identical conditions, using a rigid diaphragm wall can reduce the wall deflection by approximately 20% compared to using a conventional flexible wall.

The level of the first layer of strut is critical to the wall deflection. Rather than placing the first strut at or close to the ground level, an optimum strut level giving a minimum wall deflection can be figured out. The optimum strut level is found to be at 0.2-0.4 times the depth of excavation.

References

1. M. D. Bolton, W. Powrie and I. F. Symons, The design of stiff walls retaining overconsolidated clay: Part I, short term behaviour. *Ground Engrg* 23(1), 34-40 (1990).
2. G. W. Clough and T. D. O'Rourke, Construction-induced movements of in situ walls. *Proceedings, Design and Performance of Earth Retaining Structures*, ASCE Special Conference, Ithaca, New York, pp. 439-470 (1990).
3. P. G. Hsieh and C. Y. Ou, Shape of ground surface settlement profiles caused by excavation. *Can. Geotech. J.* 35, 1004-1017 (1998).
4. M. Long, Database for retaining wall and ground movements due to deep excavation. *J. Geotech. Geoenvir. Engrg.*, ACSE 127(3), 203-224 (2001).
5. B. Simpson, Retaining structures: displacement and design. *Gëotechnique* 42 (4), 51 – 576 (1992).
6. I. H. Wong, T. Y. Poh and H. L. Chuah, Performance of excavations for depressed expressway in Singapore. *J. Geotech. Geoenvir. Engrg.*, ACSE 123(7), 617-625 (1997).

LIMITING FORCE PROFILE AND LATERALLY LOADED PILE GROUPS

WEI DONG GUO[†]

School of Engineering, Griffith University, Parklands Avenue, Southport, QLD 4125, Australia

To evaluate the response of pile groups under lateral loading, a widely accepted approach is to use the curve of soil resistance, p versus local pile deflection, y *(p-y curve)* together with p-multiplier. The limiting force per unit length p_u (limit of p) is often attained within a depth normally up to $8d$ (d = pile diameter), and the response of laterally loaded piles is generally controlled by the mobilized p_u. Thus, determination of this p_u is critical to a satisfactory prediction of the pile response. Various expressions have been developed previously for the p_u. However, based on 70 field tests on single piles and 20 tests on pile groups, an extensive study conducted recently by the author and his team indicates that only three parameters are sufficient to describe the limiting force profile (LFP), while the parameters themselves may be modified to account for other factors. This paper addresses the effect of pile-head conditions, and group interaction on the LFP by comparing measured with predicted response of a few typical pile tests.

1. Introduction

Piles are frequently used in groups to support high-rise buildings, onshore and offshore structures. Besides axial load transferred from superstructures, the pile groups are often subjected to lateral load exerted by soil, wave, wind and/or traffic forces etc.

To estimate the group response under lateral loading, a widely accepted approach is to use the curve of soil resistance, p and local pile deflection, y *(p-y curve)*. The limiting force per unit length p_u (limit of p) is often attained within a depth normally up to $8d$ (d = pile diameter), and the response of laterally loaded piles is generally controlled by the mobilized p_u. Thus, determination of this p_u is critical to a satisfactory prediction of the pile response. Various expressions have been developed previously for the p_u, together with p-multiplier [1-4]. However, based on 70 field tests on single piles and 20 tests on pile groups, an extensive research carried out recently by the author and his team indicates that only three parameters are generally required to produce a sufficiently accurate limiting force profile (LFP). The investigation was conducted for single piles tested in clay, sand and calcareous sand, and model pile groups tested in sand. In other words, the existing methods are probably unnecessarily complicated, and hence uneconomical for design.

Using a generic LFP, elastic-plastic solutions have been developed for response of a laterally loaded, free-head [5] or fixed- head pile by the author [6]. These solutions allow response of single piles to be readily predicted using a spreadsheet program operating in EXCEL™ called GASLFP. They have also been utilized in conjunction with p-multiplier to allow response of a laterally loaded pile group to be predicted, which is undertaken through a spreadsheet called GASLGROUP [6]. These solutions compare well with finite element analysis (FEA) and measured data [5,6].

In this paper, against measured response reported previously, and using the two programs, the effect of head constraints on LFPs for single piles and pile groups is assessed. The shadowing effect due to pile-pile interaction is examined, and presented as p-multiplier. Incorporating these effects in the LFP for a single pile, prediction of nonlinear response of a few typical field and model pile tests is provided.

2. Closed-form Solutions for Fixed-head Piles

The solutions [6] for laterally loaded infinitely long, fixed-head piles, were developed using a coupled model as schematically described in Figure 1a. The pile-head is restrained from rotating but free to move laterally. The pile-soil interaction at each point is depicted by an elastic-perfectly plastic spring, with the p-y curve shown in Figure 1b. Within elastic state, each spring has a subgrade modulus, k, which is the slope of the p-y curve. The springs are linked together by a fictitious membrane with a tension, N_p to cater for the coupled effect. Constant with depth, the k and N_p are functions of modified Bessel functions of a non-dimensional factor [7] that is dependent of pile slenderness ratio (length, L over diameter, d), pile-soil relative stiffness, E_p/G_s (E_p = Young's modulus of an equivalent solid pile; G_s = Shear

[†] This project is supported by the Australian Research Council Discovery Grant, grant number DP0209027.

modulus of soil); and pile-head conditions. Where the soil resistance reaches the limiting force, relative slip takes place along the pile-soil interface, which extends to a depth called slip depth, x_p. Within the x_p, interactions among the springs may be ignored by utilizing $N_p = 0$. A generalized form of LFP may be given by [5]

$$p_u = A_L(\alpha_0 + x)^n \quad (1)$$

where p_u = limiting force per unit length [FL^{-1}]; A_L = gradient of the LFP [FL^{-1-n}]; α_0 = a constant to include the resistance at ground surface [L]; x = depth below ground surface [L]; n = power to the sum of α_0 and x. For piles in sand, it follows: $A_L = N_g \gamma_s d^{2-n}$, in which γ_s = effective unit weight of the sand; d = diameter of the pile; N_g = a limiting force factor, $N_g = 3K_p \sim K_p^2$, with $K_p = \tan^2(45 + \phi/2)$, ϕ = friction angle of the sand. For piles in clay, $A_L = N_g C_u d^{1-n}$ in which C_u = undrained strength. The α_0 is approximately equal to zero, except for driven piles. Real values of the set of n, α_0, and N_g may be back-figured through matching the predicted with the measured responses of displacement, w_g, maximum bending moment M_{max} to the pile-head load, P, and M_{max} versus its depth of occurrence x_{max} relationship [5].

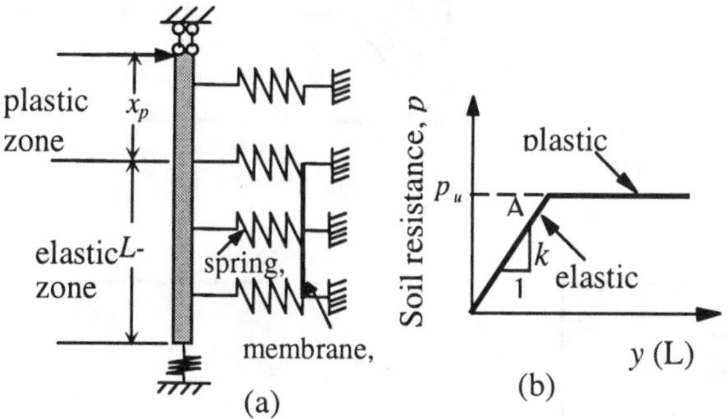

Figure 1. Modeling of fixed-head piles [6]

3. Prediction of Response of Pile Groups

Lateral response of a pile group can be estimated using the closed-form solutions for a single pile. The shadowing effect on each row of piles is catered for by using the p-multiplier suggested by Brown et al [1]. For any a displacement, the p for a pile in a group may be evaluated as the product of the *p-multiplier* and the p obtained from a p-y curve of a single pile [1-4]. Using closed-form solutions, assuming a pile displacement, the slip depth can be obtained, which in turn allows the pile-head load for each pile in a row to be estimated. Multiplying the load by number of piles in the row gives the total load on the row. This calculation is repeated for each row in a group, keeping in mind that each row generally uses a different *p-multiplier*, thus the total load on the group is calculated. Given a series of displacements, the total load is calculated accordingly, thus load-displacement curve is obtained. This procedure has been implemented into the program GASLGROUP [6].

4. Comparison with Field Test Results

4.1 *Piles in clay*

Matlock et al [8] performed lateral loading tests on a single pile, five-pile and ten-pile circular groups embedded in uniform soft clay having a C_u of 20 kPa. The tubular steel piles were 13.4 m in length, 84 mm in external radius, 7.1 mm in wall thickness, and had a flexural stiffness E_pI_p of 2326.17 kN-m^2. They were driven to a depth of 11.6m.The center-to-center spacings for the five-pile and ten-pile groups were $6.8r_o$ and $3.6r_o$ (r_o = outside radius of the pile), respectively. During the tests, deflections of the groups were enforced at two elevations to simulate fixed-head restraints that are typical in offshore foundations.

The response of pile groups are predicted using GASLGROUP assuming free-head and fixed-head, respectively, although the tested pile was regarded having fixed-head. This is to investigate the effect of head condition on the limiting force profile. The shear modulus G_s are taken as (30 ~ 100)C_u and N_g taken as 4.0 for the analysis assuming free-head, while G_s = (15 ~ 50)C_u and N_g =1.0 using fixed-head solutions. Parameter n and α_o are assumed as independent of head condition, but they are affected by group effect. These parameters are provided in Table 1. The LFPs are accordingly plotted in Figure 2 (a), (d) and (g). The prediction using the two programs compares well with the measured load-displacement, and displacement - maximum bending moment curves; so does the prediction using simplified closed-form solutions [6]. The predicted M_{max} for the mudline level has been reduced by the magnitude due to the eccentricity of 0.35 m in order to compare with the reported bending moment for the lower support level.

However, pile deflection at the lower support level is assumed to be identical to that at mudline level. This leads to a slight overestimation of the M_{max}, as the deflection may be slightly higher due to the rotation of the free-length.

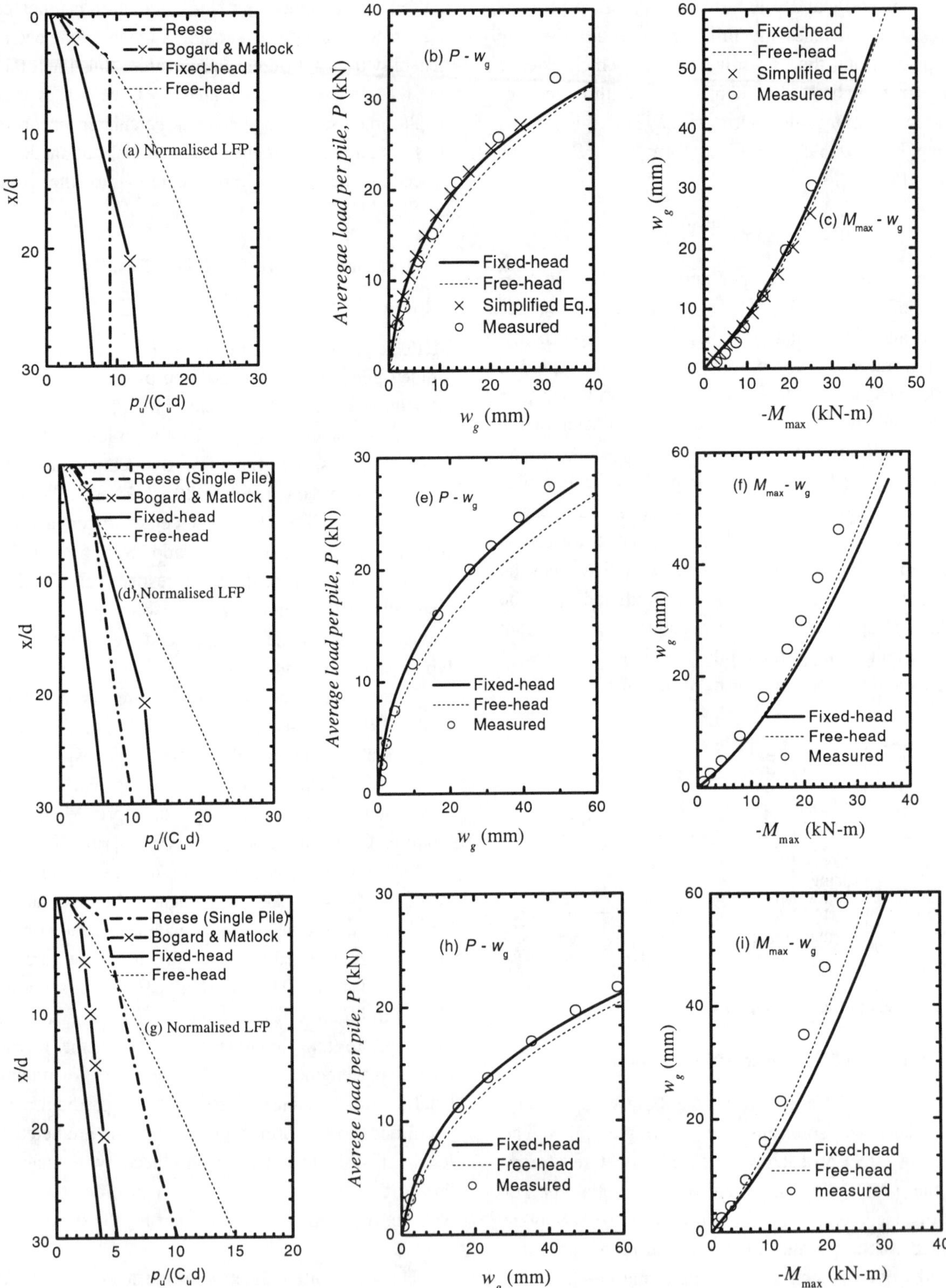

Figure 2 Predicted vs. measured [8] response of: Single pile (a)~(c), 5-pile group (d)~(f), and 10-pile group (g) ~ (i).

Table 1. Case study on piles reported in [8]

Items	Single pile	5-pile	10-pile
	Input parameter		
G_s/C_u	50(100)*	50(100)	17(33)
n	0.55	0.85	0.85
α_o	0.05	0.05	0.25
N_g	1.0(4.)		
p-multiplier	1.0	0.333	0.20
	Calculated values		
N_{co}	.51 (2.05)	.36 (1.43)	1.40(1.43)
x_p/d at P_{max}	15.1(8.0)	19.6(11.3)	23.2(12.1)

**Note values are for both head-conditions. The values without and with parenthesis are for fixed-head, and free-head, respectively, e.g. 50(100) implies G_s/C_u = 50 and 100 for fixed-head, and free-head, respectively.*

The analysis shows the following key points:

(a) To yield good comparison with the measured data, limiting force is much higher for free-head than fixed-head piles. The LFP used by 'Bogard & Matlock' lies in between the two LFPs used herein.

(b) Under fixed-head condition, limiting force was mobilised to a maximum slip depth x_p of 15.1d, 19.6d and 23.2d respectively for a single-pile, a pile in the five-pile, and ten-pile groups under the maximum reported load.

(c) The maximum x_p under fixed head is twice as much as under free-head, given identical pile response.

(d) The power n increases about 55% for those piles acted in groups.

(e) α_o increases from 0.005 for a single pile to 0.25 for piles in the 10-pile groups (N_{co} = 0.5 ~ 1.4).

(f) The p-multiplier of 0.333 and 0.2 [8] works well here for the 5-pile, and 10-pile groups.

(g) G_s/C_u for fixed-head piles is 50% that for free-head piles. It reduces by 66.7% from single pile to the ten-pile group.

4.2 *Piles and pile groups in sand*

The two programs were employed to predict response of model piles reported previously [9]. Two aluminium pipe piles were driven into a depth of 0.5 m below the sand surface. One was tested under fixed-head condition, with the head being constrained by a wooden plank reinforced with a mild steel strip, and the other was tested under free-head condition. Each pile of 0.75m in length, 18.2 mm in outside diameter, and 0.75 mm in wall thickness had a flexural stiffness E_pI_p of 0.086 kN-m^2 (The difference between embedded length and pile length can only have influence on prediction of response of a short pile). The fine to medium sand was clean and dry, with a frictional angle of 36.3 degree, Poisson's ratio of 0.25, and relative density of 60%. The unit weight was 14.6 ~ 17.3 kN/m^3. A shear modulus of 300 kPa was assumed due to low confining pressure, which allows the modulus of subgrade reaction to be obtained as 711.7 and 831.1 kPa for fixed-head and free-head respectively [7].

For the LFP, N_g = 15.23, N_{co} = 0, and n = 1.35 [6] was used for the fixed-head pile. The n value is lower than 1.6 ~1.7 used in COM624P[10]. For the free-head, driven pile in sand, the factor α_o was taken as 0.05 (N_{co} = 59.6) due to apparent cohesion. These values allowed the A_L to be computed as 240.2 ~ 253.2kPa/m, and the LFPs to be plotted in Figure 3(a). The pile cap was kept 10 mm above the sand surface. The predicted against measured responses are plotted in Figure 3(b), and 3(c) in form of the load ~ mudline displacement, and load ~ maximum bending moment curves. The '(FeH)' and '(FxH)' refer to free-head, and fixed-head piles. M_{max} is overestimated for the fixed-head pile against the measured point denoted as 'Mea (FxH)'. Within the x_p of 10d, the LFP has a lower resistance for the fixed-head pile than the free-head one, which is consistent with observation from model pile tests [11]. The slip was initiated at P_e = 0, and 0.02 kN for fixed-head and free-head piles, respectively. At P = 0.332 kN and for the fixed-head case, a normalised slip depth x_p/d was found as 13.5, with the corresponding $\bar{x}_p$ = 1.49. For the free-head case at P = 0.255 kN, x_p/d is 11.71, and $\bar{x}_p$ = 1.35. These LFPs used here can be fully justified only when one more measured response becomes available [5]. The profiles of deflection, shear force, rotation and bending moment along the pile were predicted for a head-displacement of 10 mm, and are shown in Figure 4. The predicted moment profile compares well with the measured one for the free-head pile. However, it is about 40% higher than the measured data for the fixed-head one, which may be attributed to the fact that the pile-head was not ideally fixed.

Figure 4 (a)~(c) shows that the current predicted profiles for the fixed-head pile are quite consistent with those obtained from COM624P, although slight difference in the shear force profiles is noted, which in turn may be attributed to the difference in the LFP adopted in the two solutions noted in Figure 3(a). The response of the fixed-head pile was predicted using the

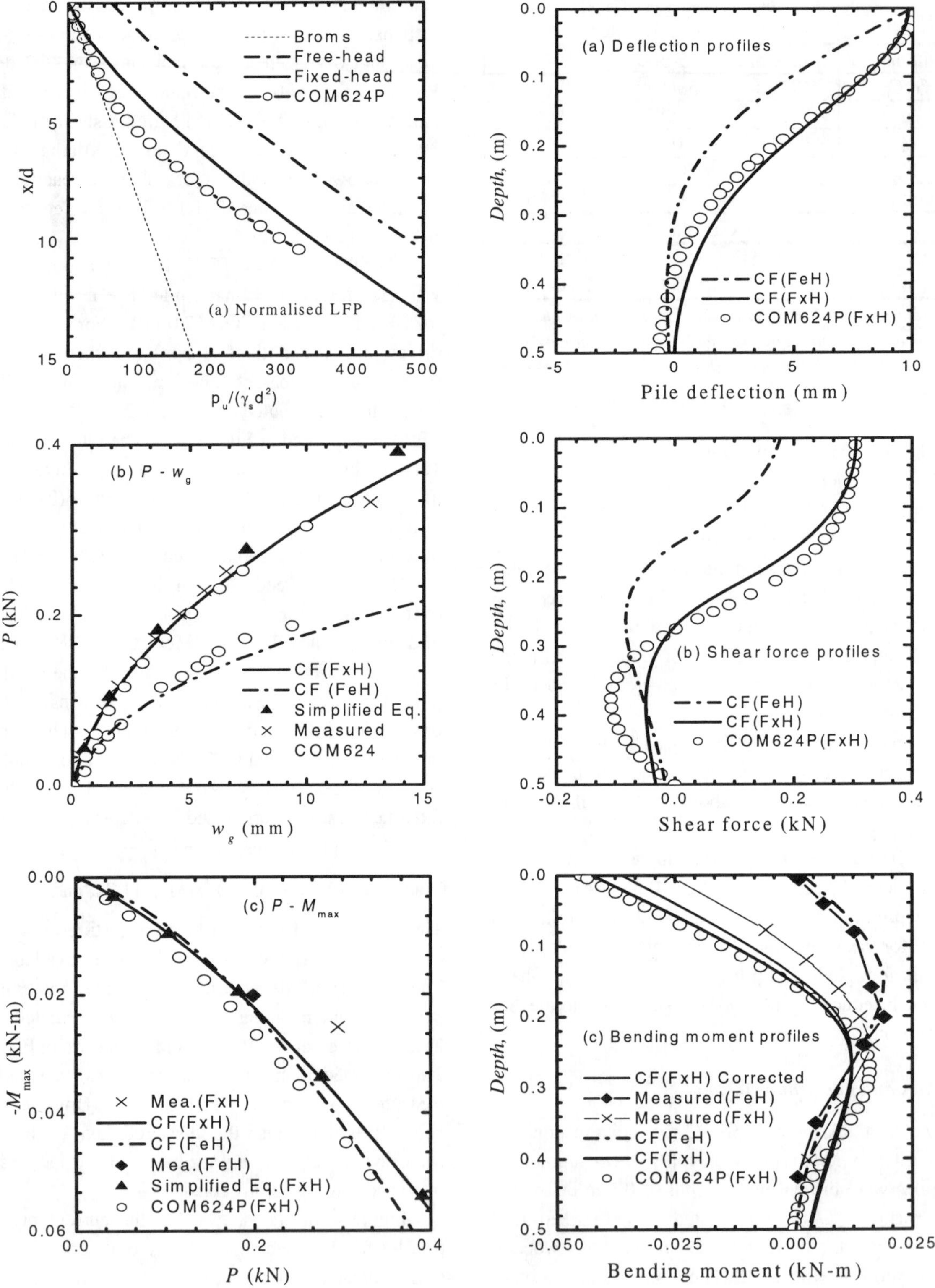

Figure 3 Predicted vs. measured [9] response of the driven piles

Figure 4 Predicted vs. measured [9] response of the driven pile at 10 mm head displacement

simplified closed-form solutions [6], and it is shown in the figure as well.

Good predictions against measured load - displacement relationships were made [12] for all the 20 pile groups reported [9] at s/d = 4, 6 and 8 using GASLGROUP. It was noted that (1) shear modulus was generally identical to that for single pile, except for the groups 1×2, 3×1, 2×2, and 2×3 where up to 10 times higher modulus was found mainly due to densification. (2) Values of $\alpha_o = 0$, $N_g = K^2_p = 15.23$ are identical to those used here for single piles. Values of p-multiplier were reported previously. In particular, it is just noted that the predicted distribution of load on each individual pile in the 3×3 group closely resemble the measured one of pile groups embedded in clay [4], which however cannot be well simulated using FEA [13], indicating the capacity of the current solutions and programs.

5. Conclusions

In this paper, the spreadsheet program GASLGROUP based on elastic-plastic closed-form solutions for free-, and fixed- head piles was employed to investigate the static responses of laterally loaded pile groups. Together with previous study on 22 pile groups, it is noted (1) GASLGROUP is at least equally powerful as other approaches. (2) p- multiplier concept is generally valid; and (3) LFP for a capped pile may well lie in between that for free and fixed head conditions.

Acknowledgments

The work reported herein is currently supported by Australian Research Council Discovery (No. DP0209027). This financial support is gratefully acknowledged.

References

1. D. A. Brown, C. Morrison, and L.C. Reese, *Lateral load behaviour of pile group in sand.* J. of Geotech. and Geoenviron. Engrg. Div., ASCE, 114(11): 1261-1276(1998).
2. K. M. Rollins, K.T. Peterson, and T.J. Weaver, *Lateral load behaviour of full-scale pile group in clay.* J. of Geotech. and Geoenviron. Engrg. Div., ASCE, 124(6): 468-478 (1998).
3. McVay, M., et al., *Centrifuge testing of large laterally loaded pile.* J. of Geotech. and Geoenviron. Engrg. Div., ASCE, 124(10): 1016-1026(1998).
4. Ilyas, T., Leung, C. F., Chow, Y. K., and Budi, S. S., *Centrifuge model study of laterally loaded pile groups in clay.* J. of Geotech. and Geoenviron. Engrg. Div., ASCE, 130(3): 274-283 (2004).
5. W. D. Guo, A simplified approach for piles due to soil movement. In Culligan, Einstein and Whittle (eds), Proc. 12th Pan-American Conf. on Soil Mechanics and Geotech. Engrg., 2: 2215-2220 (2003). Verlag Glückauf GmbH.
6. W. D. Guo, "Fixed-head laterally loaded pile," *International Journal of Numerical and Analytical Methods in Geomechanics.* Under review. (2004)
7. W. D. Guo, and F. H. Lee, *Theoretical load transfer approach for laterally loaded piles.* Int. J. for Numerical and Analytical Methods in Geomechanics, 25(11): 1101-1129 (2001).
8. H. Matlock, B. Wayne, A. E. Kelly, and D. Board, Field tests of the lateral load behaviour of pile groups in soft clay. in Paper No. OTC. 3871, Proc. 12th Ann. Offshore Techno. Conf. (1980). Houston, Texas.
9. S. R. Gandhi, and S. Selvam, Group effect on driven piles under lateral load. J. of Geotech. and Geoenviron. Engrg. Div., ASCE, 123(8): 702-709(1997).
10. FHWA, COM624P-laterally loaded pile analysis program for the microcomputer, in Report No. FHWA-SA-91-048(1993): Washington, D.C. USA.
11. L. Yan, and P.M. Byrne, *Lateral pile response to monotonic pile head loading.* Can. Geotech. J., **29**: 955-97(1992).
12. W. D. Guo, and B. T. Zhu, Non-linear response of individual piles in a group in sand. Proc. 18th Australian Conf. on Mechanics of Struct. & Materials, Perth, Australia, 1-3 Dec. 1005-1012 (2004).
13. Z.Yang, and B. Jeremic, *Numerical study of group effects for pile groups in sand.* Int. J. Numer. and Anal. Meth. in Geomech., 2003. **27**: 1255-1276 (2003).

PRACTICE OF '2 IN 1' TECHNOLOGY IN DESIGN AND CONSTRUCTION OF DEEPLY EXCAVATED FOUNDATION FOR HIGH-RISE BUILDINGS

SHIMING CHEN †

School of Civil Engineering, Tongji University
1239 Siping Road, Shanghai 200092, China

XIAOMING XU

Shanghai Modern Architectural Design (Group) Co. Ltd.
Shanghai 200041, China

In the soft soil ground area, the underground continual walls (or underground concrete diaphragm slurry walls) together with the temporarily horizontal-shoring systems are found effective in deep excavation for high-rise buildings. The temporary shoring will be demolished at the consequent stage when the structures of the underground basement are formed, and in most cases, the underground continual walls remain underground, enclosed with the walls of basement inside. The creative idea of "2 in 1" technology refers to the integration of underground continual walls and the external walls of basement in deep foundation design. The technology has been proved to be effective in control of deformation and settlement of soils during the construction of foundation, and having very light influences on the environmental surroundings. The shoring system could be used as permanent structural skeletons for the basement structures if it is well designed, and the underground continual walls also act as the external walls of basement for buildings, so that construction cost can be reduced. In the paper, the art and key procedures of the technology are assessed and the application of the technology in Phase 2 Project of Shanghai City Hotel is summarized.

1. Introduction

Deep excavation in soft clay normally causes a large wall deflection and large ground surface settlement. Excessive ground surface settlement frequently damages the adjacent properties in urban areas. Concrete diaphragm slurry wall technique, which was early introduced in Europe in the 1950s for the construction of underground reinforced concrete structural wall system, had been demonstrated an sufficient way to minimize influences of wall deflection and ground surface settlement on the surroundings. Permanent underground concrete diaphragm walls are also an ideal solution for structures requiring deep basements, particularly where a high groundwater table is present. Since the 1990s, the technique has been wildly used in the deep excavation in soft soil for tall buildings in the urban areas of Shanghai. Underground concrete diaphragm walls provide the following advantages for urban construction:

1. Temporary / permanent groundwater cutoff;
2. Zero lot line construction;
3. Stiff structural capacity and superior resistance to movements;
4. Easily adapted to both anchors and internal structural bracing systems;
5. Expedited construction, because only interior columns and slabs need to be built.

The idea of "2 in 1" technology developed refers to the integration of underground concrete diaphragm slurry wall (or continual walls) and the external walls of deep basement in the foundation design for a tall building. The technology has been proved to be effective in control of deformation and ground surface settlement during construction of the deep foundation, and have very light affection on the environmental surroundings. The internal horizontal shoring components, which supports the walls laterally could be used as permanent structural skeletons for the basement structures if they are well designed, and the underground continual walls also act as the external walls of basement for buildings, so that construction cost can be reduced. In the paper, the art and key procedures of the technology are assessed and the application of the technology in Phase 2 Project of Shanghai City Hotel is summarized.

† corresponding author, E-mail: chensm@mail.tongji.edu.cn.

2. The Art and Methodologies

Deep underground basements that are integrated into urban development projects early in the overall project design offer many inherent improvements to the overall quality and value of the project and its surrounding community. Diaphragm walls being combined into a single foundation unit, with the functions of temporary shoring, permanent basement walls, hydraulic cutoff, and vertical support elements and shear walls have proven to be an economical alternative in many circumstances [1].

In construction practice in urban areas of Shanghai, concrete diaphragm slurry walls are normally used as essential measures functioning as temporary and permanent groundwater cutoff, and temporary shoring structures to minimize deformation and ground surface settlement during the construction of the deep foundation, when the excavation depth is greater than 10 m or the underground stories are 3 stories beyond. Two approaches could be followed in design practice of concrete diaphragm slurry walls for deep underground basements. (1) temporary shoring structures and groundwater cutoff; (2) an integration of temporary shoring, groundwater cutoff and permanent basement walls etc..

In design and construction of concrete diaphragm slurry walls as the integration of temporary shoring, groundwater cutoff and permanent basement walls, Water tightness is particularly important for permanent diaphragm walls and the degree of water tightness that can be achieved depends primarily on the type of joint between individual panels. The joints generally are classified as flexible joints and rigid joints as shown in Figure 1.

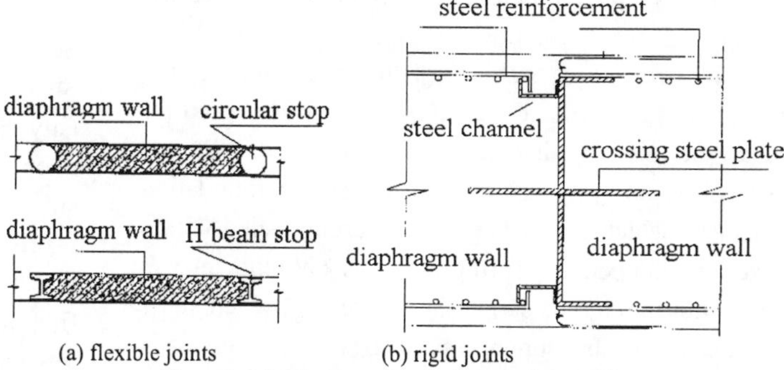

Figure 1 Joints between the diaphragm slurry walls

Another high importance for the integrated structural system is to ensure the reliability of connections between the concrete diaphragm walls and the structural components of the permanent basement. Stiff connectors and flexible connectors conventionally used in practice are showed in Figure 2.

To assess the contribution of diaphragm slurry walls to the bearing capacity to support gravity load of the superstructures, the general piling model proposed in the foundation specification [2] could be adopted as shown in Figure 3. The gravity bearing capacity shared by the diaphragm slurry wall is expressed as:

$$R_d = \frac{b\sum f_{si} l_i}{\gamma_s} + \frac{f_p ab}{\gamma_p} \qquad (1)$$

where f_{si} is the design friction inherent at the interface of between the diaphragm slurry wall and soil at the ith layer of soil; f_p is the bearing strength of soil at the bottom of the wall; a and b are the thickness and width

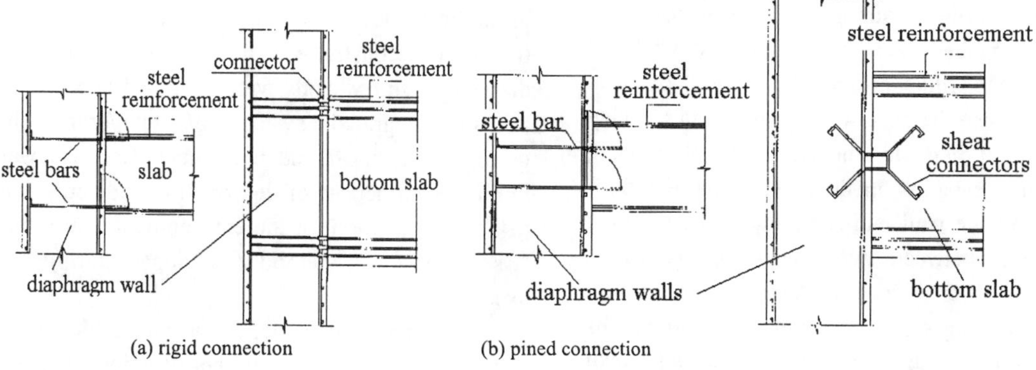

Figure 2 Types of connection between diaphragm wall and slab

of the diaphragm slurry wall respectively, γ_s and γ_p are the factors corresponding to friction and bearing strength.

The overturning stability of the temporary shoring system during open excavation is also to be checked for each individual diaphragm slurry wall. Staged excavation analyses use a beam on elastic foundation model, in which the earth pressures are modeled with a series of independent spring supports (Winkler elastic foundation model) as shown in Figure 4. At each stage of excavation or support system, the spring loads change as soil, water and support system loads are applied or removed and lateral wall displacement occurs. The soil springs load-displacement relationship is determined by the input soil stiffness and governs the spring displacement until the limiting value of active or passive pressure is reached.

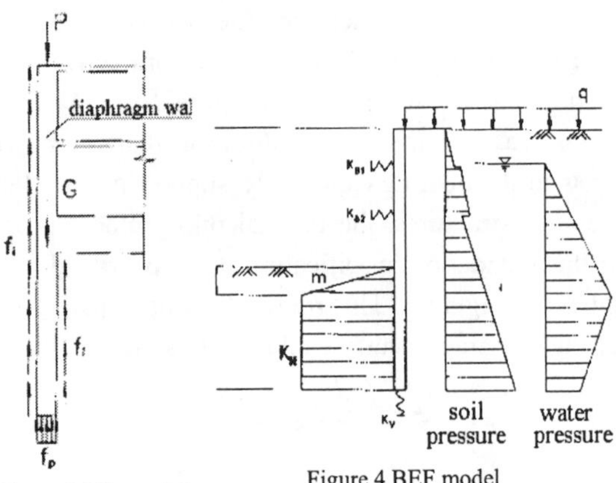

Figure 3 Pile model

Figure 4 BEF model

One criterion in design of deep excavation in soft soil is the sequence of excavation. Most of the cases reported in the literature were constructed using the bottom-up excavation method. This method uses the temporary steel struts to support the excavation wall. The displacement behavior of the supported wall and soil may change little during the period of strut installation because the pore-water pressure in the clay typically does not dissipate quickly, On the other hand, the top-down excavation method uses concrete floor slabs to support the wall and sometimes requires long periods of time between two successive excavation stages to construct the floor slab. Construction of only underground framing system other than full floor slabs in the excavation stage enables a semi-top-down excavation, which makes the excavation easy and a stiff excavation wall.

3. Construction practice

The sample project is located at the west to the Shuanxi Southern Road and Yanan Road, downtown of Shanghai. It is the second phase project of Shanghai City Hotel, the excavation area is 2083 m^2, and the maximum excavation depth is 13.1 m. The construction site is adjacent to the south and the west of several residential apartment buildings, to the north of a school, and only 0.8 m to the external basement walls of the first phase of the hotel in the east, which was completed few yeas ago. The plane of the hotel is shown in Figure 5.

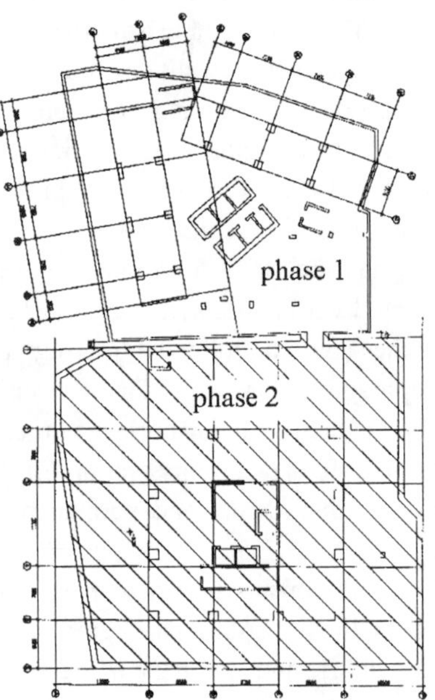

Figure 5 Plane of Shanghai City Hotel

The geological profile of the area claims a typical multi-layer silt clay soft soil with very high groundwater level, as shown in Figure 6. Restricted to the narrow construction space, underground concrete diaphragm slurry walls were designed and constructed as an integrated system of temporary shoring, groundwater cutoff and permanent basement walls, to minimize influences of deformation and settlement due to deep excavation on the environmental surroundings. A semi-top-down method was adopted during the deep excavation.

The prediction of deflection and settlement of the excavation walls due to the construction of diaphragm wall supported deep excavation was considered in the four separate stages as the follows:

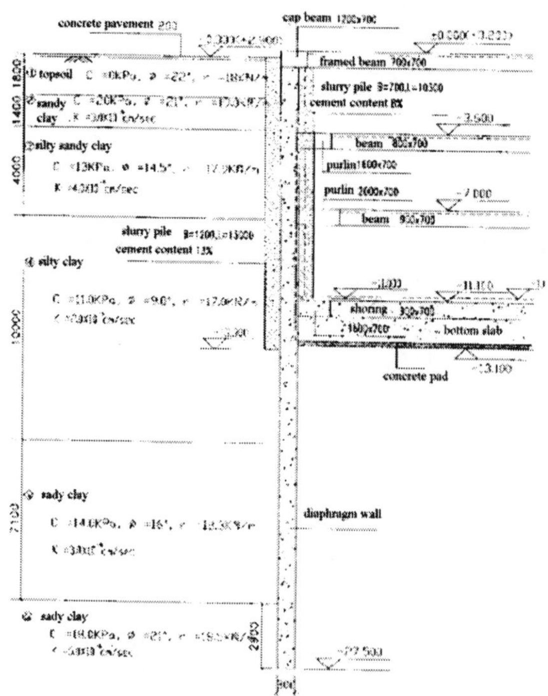

Figure 6 Section of multi-layer soft soil and slurry wall shoring

Stage 1: excavation depth up to –4.3 m when design strength reached for the concrete bracing at the first shoring level;

Stage 2: excavation depth up to –7.7 m when design strength reached for the concrete bracing at the second shoring level;

Stage 3: excavation depth up to –10.3 m when design strength reached for the concrete bracing at the third shoring level;

Stage 4: excavation depth up to –13.1 m when design strength reached for the concrete bracing at the fourth shoring level.

The predicted maximum moment envelop and deflection of the diaphragm wall are shown in Figure 7. Field observation results [3] indicate that a relative large settlement occurs during the formation of the diaphragm walls, however the settlement during the excavation is small and the maximum settlement in the adjacent apartment buildings developed during the whole construction period is 4.6 mm. and 9.1 mm for the phase 1 building. Compared with the predicted deflection of the slurry wall, the observed maximum deflection developed at about 2 m above the bottom of the pit is about half of the predicted deflection This is owing to the consolidation effects on the silt contributed by the additional deep slurry piles used each side of the walls, which is not taken into account in the calculation, however it is on the safe aspect.

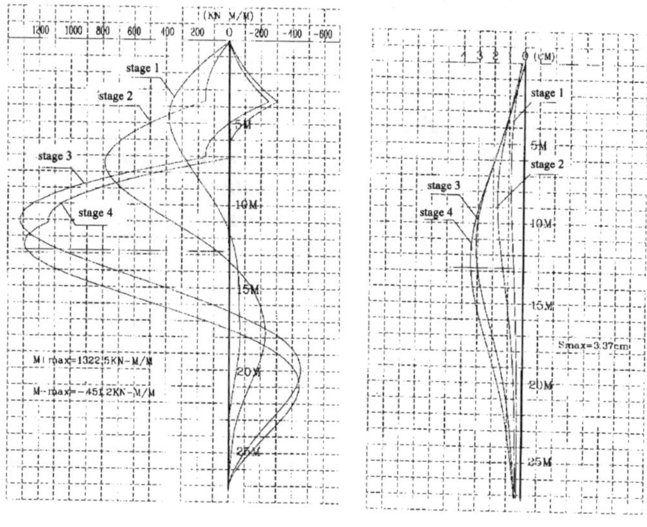

(a) maximum moment envelops (b) deflection curves

Figure 7 Maximum moment envelops and deflection curves of the diaphragm wall for the four stages

4. Conclusions

Practice of '2 in 1' technology or integrating construction of temporary shoring, groundwater cutoff and permanent basement walls was applied in construction of the deep foundation of Phase 2 Project of Shanghai City Hotel. Significant benefits are summarized as the follows:

1. The shoring walls and bracing components are designed as a part of permanent structure, so that the time required to demolish the temporary shoring/bracings is reduced, and the construction cost is also cut down.
2. The construction site area is effectively reduced because the temporary shoring wall and basement wall are combined into one wall, so that the architectural area is fully occupied. For instance, it makes additional areas of 540 m^2 in the basement
3. Environmental aspects: the settlement and deformation of the surrounding buildings are well controlled and minimized, besides, there is no more noise which may occur in the demolishing of concrete shoring/bracings

References

1. Seth L. Pearlman, Michael P. Walker, and Marco D. Boscardin, *Deep underground basements for major building construction,* ASCE Conf. Proc. 135, 61 (2004).
2. Shanghai construction specification, *Specification of soil and foundation design.* (DGJ08-11-1999), 1999, Shanghai.
3. Xu Xiaoming, *Research on '2 in 1'' Technology in deeply excavated foundation design and construction measure,* Master thesis of Tongji University, 2004.

NUMERICAL ANALYSIS OF THE SOIL-FOUNDATION-SUPPORTING PILE-HIGH-RISE BUILDING DYNAMIC INTERACTION SYSTEM*

PEIZHEN LI XILIN LU HONGMEI ZHANG

State Key Laboratory for Disaster Reduction in Civil Engineering, Tongji University, 1239 Siping Road, Shanghai 200092, China

YUEQING CHEN

Wuhan Urban Construction Investment & Development Group Co. Ltd, Wuhan 430050, China

Based on the finite element analysis program ANSYS, the soil-foundation-supporting pile-superstructure dynamic interaction system was analyzed in this report. It was found out that the existing of supporting piles affects the aseismic behavior of high-rise building. Commonly used equivalent linearity model is chosen as constitutive relation of soil. Viscous boundary of soil is implemented in ANSYS program. The parameter study includes the length and rigidity of the supporting piles, the various rigidity of superstructure and the foundation type.

1. Introduction

During the period of high-rise building construction, the supporting pile was usually used as the braced structure for foundation pit, and usually could not be demolished after construction. The effect of the supporting pile on the seismic performance of the high-rise building was usually ignored during the high-rise building design. The supporting piles may increase the rigidity of ground base by intuitive analysis.

In order to analyze the effect of the supporting pile on seismic performance of the high-rise building, combining general-purpose finite element program ANSYS, three-dimensional finite element analysis on soil-foundation-supporting pile-superstructure dynamic interaction system has been carried out. The modeling method had been verified to be coorect in previous study by comparing the results of the finite element analysis with the data from shaking table tests[1, 2].

2. Brief Description of a Practical Engineering

A cast-in-place frame structure supported on pile-raft foundation is studied in this paper. The layout of column grid is shown in Figure 1. The frame structure has 14 stories above ground and two stories underground. The height of underground floor is 4.5m, while the height of aboveground floor is 3.6m. The thickness of cast-in-place floorslab is 120mm, the dimensions of column, boundary beam, and walkway beam are 600×600mm, 250×600mm, and 250×400mm, respectively. The box foundation and pile-box foundation are used in this structure, the upper plate is 400mm in thickness, bottom plate is 600mm, outer wall of the box foundation is 500mm, inner wall is 300mm. The dimension of pile is 450×450mm, and the length of pile is 33m with 0.7m entering the bearing stratum. The layout of pile-raft foundation is shown in Figure 2. The length of the supporting pile is 20m, and the diameter is 1.0m. The

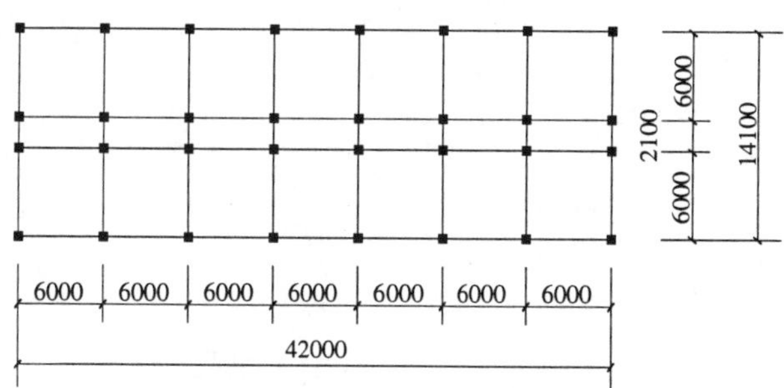

Figure 1 Layout of column grid

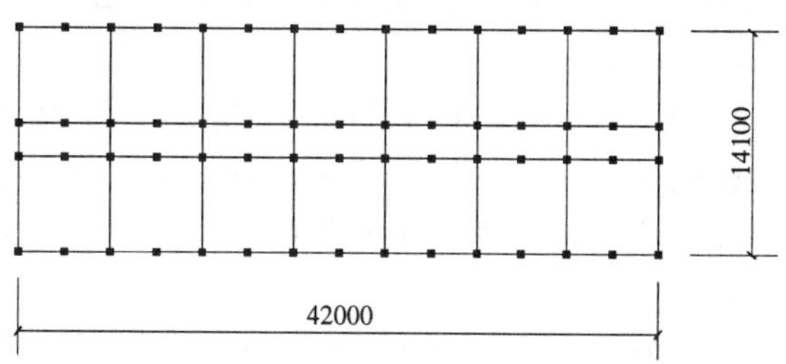

Figure 2 Layout of pile-raft foundation

* This work is sponsored by National Science Foundation of China, Grant No. 50321803, 50308018.

distribution of soil near the First Shimen Road of Shanghai is used[3]. The layers of soil from top to bottom are ①fill, ③very soft gray silty clay, ④very soft gray clay, ⑤-1 gray clay, ⑤-2 gray silty clay, ⑤-3 terreverte clay, ⑦strawyellow-gray silty sand. The shear wave velocity and mass density of soil are shown in Table 1.

Table 1 Soil property

No.	Bottom depth (m)	Mass density (kg/m^3)	Shear wave velocity (m/s)
①	3.5	1800	90
③	8.0	1740	140
④	17.6	1700	140
⑤-1	26.5	1770	160
⑤-2	35.2	1810	180
⑤-3	41.3	1990	210
⑦	>41.3	1960	>210

3. Modeling Method

3.1 *Dynamic constitutive model of soil*

The soil's skeleton curve of Davidenkov model is adopted in this paper[4], and the relationship of $G_d/G_{max} - \gamma_d$ is shown as Equation 1.

$$\frac{G_d}{G_{max}} = 1 - H(\gamma_d) \tag{1}$$

Where

$$H(\gamma_d) = \left[\frac{(|\gamma_d|/\gamma_r)^{2B}}{1+(|\gamma_d|/\gamma_r)^{2B}}\right]^A \tag{2}$$

$$G_{max} = \rho V_s^2 \tag{3}$$

$$\gamma_r = \gamma_r'(0.01\sigma_0')^{1/3} \tag{4}$$

G_{max} is the maximum dynamic shear modulus of soil, ρ is the mass density of soil, and V_s is the shear wave velocity of soil.

γ_r is a shear strain for reference. σ_0' is the average effective confining pressure of soil, and its unit is *kPa*.

The values of parameter *A*, *B* and γ_r' are shown in Table 2.

The hysteresis loop of soil $D/D_{max} - \gamma_d$ is expressed as following empircal formula.

$$\frac{D}{D_{max}} = (1 - \frac{G_d}{G_{max}})^{\beta} \tag{5}$$

D_{max} is the maximum damping ratio of soil. β is the shape factor of curve $D/D_{max} - \gamma_d$, and 1.0 is chosen as β for soft soil of Shanghai area. The value of D_{max} is reference to Table 2.

Table 2 Parameter of Davidenkov model for Shanghai soil

Soil type	*A*	*B*	D_{max}	γ_r' (10^{-3})
clay	1.62	0.42	0.30	0.6
silt	1.12	0.44	0.25	0.8
sand soil	1.10	0.48	0.25	1.0
medium coarse sand	1.10	0.48	0.25	1.2

In ANSYS program, there is a kind of parametric design language named APDL, which is a scripting language. Users can use it to automate common tasks or even build models in terms of parameters. The equivalent linearity model is realized in ANSYS program by using the APDL, and the calculation of material nonlinearity is realized automatically.

3.2 *Damping*

In SSI systems, such as the one considered here, the soil damping ratio is usually larger than that of the concrete superstructure. In order to account for this difference, a material-damping input method available in ANSYS is used. As it is well known, Rayleigh damping ratio can be calculated by equation 6:

$$\xi_i = \alpha/2\omega_i + \beta\omega_i/2 \tag{6}$$

Where: ξ_i is the ratio of actual damping to critical damping for a particular mode of vibration, i. ω_i is the natural circular frequency of mode i.

In many practical structural problems, alpha damping (or mass damping) may be ignored ($\alpha = 0$). In such cases, β can be evaluated from known values of ξ_i and ω_i, as

$$\beta = 2\xi_i/\omega_i \tag{7}$$

Material-dependent damping allows to specify beta damping (β) as a material property. Different damping ratios can be input for different materials by this method[5] .

3.3 *Artificial boundary*

The artificial boundary conditions can be classified as viscous boundary, superposition boundary, paraxial boundary, extrapolation boundary and so on[6]. The viscous boundary is the most commonly used boundary conditions in practice as it has a simple form suitable for finite element formulation and nonlinear analysis. The viscous boundary is adopted in this paper.

Viscous boundary is equivalent to setting a series of dampers on artificial boundary to absorb wave energy. The damping coefficients of dampers have no relation to frequency. In this paper, viscous boundary is implemented by spring-damper element in ANSYS program.

3.4 *Input excitation*

El Centro record is adopted as input excitation. The wave traces and associated Fourier spectrum for El Centro records are shown in Figure 3.

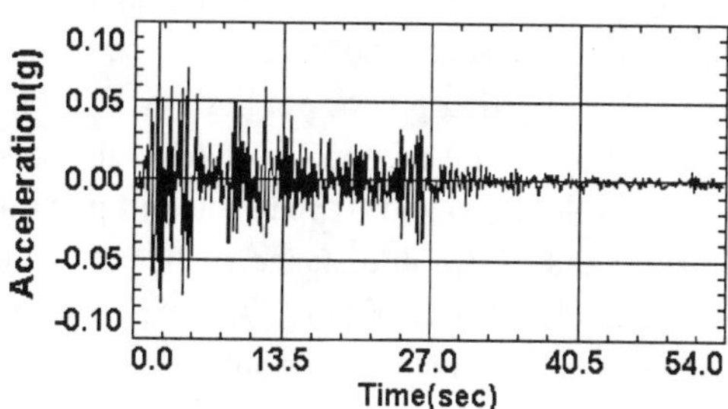

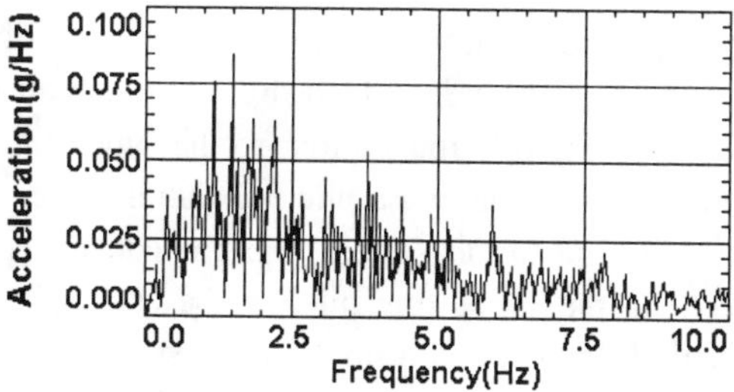

Figure 3 The wave traces and associated Fourier Spectrum for El Centro records

4. Numerical Analysis of the Soil-pile Foundation-supporting Pile-high-rise Building Dynamic Interaction System

Numerical analysis of the soil-pile foundation-supporting pile-high-rise building dynamic interaction

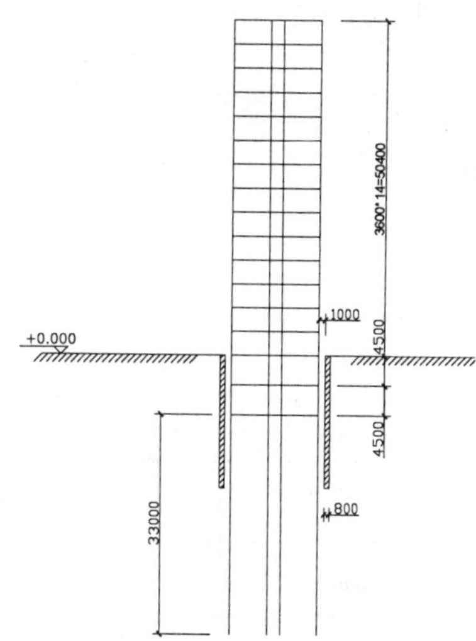

Figure 4 The dynamic interaction

system is introduced in this section. The contribution of the supporting pile to the seismic response of the dynamic interaction system is studied. The interaction system is showed in Figure 4.

4.1 *Calculating region*

According to the numerical analysis already done by the author, it is better to set the soil boundary 10 times to the structure boundary, and set a viscous artificial boundary at the lateral boundary, which can better simulate the infinite soil[7]. The boundary and the model grid is shown in Figure 5. For the model is symmetry, half of the structure along the earthquake input direction is chosen. The frame column, beam and pile are simulated by beam element, the slab by shell element, and the soil by solid element.

4.2 *Natural frequency of SSI system*

The comparision of the natural frequency of the interaction system with and without supporting pile is showed in Table 3. From the table we can see, as to the soil-pile foundation-supporting pile-high-rise building dynamic interaction system, the supporting pile has little effect on the natural frequency of the system.

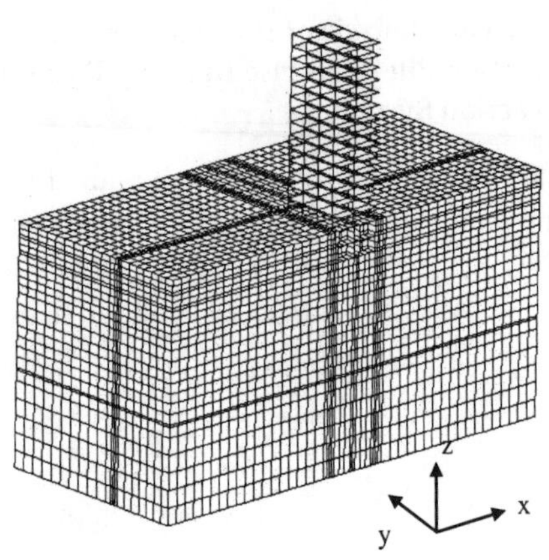

Figure 5 Meshing of SSI system

Table 3 Natural frequency of SSI system with and without supporting pile

No.	A: without supporting pile	B: with supporting pile	
	Frequency (Hz)	Frequency (Hz)	Error to A (%)
1	0.3904	0.3946	1.09
2	0.4631	0.4671	0.86
3	0.6588	0.6623	0.54
4	0.7209	0.7245	0.49
5	0.8901	0.8940	0.45

4.3 *Seismic response of structure*

El Centro earthquake record, whose peak values of acceleration is adjusted to 0.1g, is inputted from the bottom of soil along X axis. The comparison showed in Figure 6. A15 is the midpoint on the top floor, A1 is the mid point of the bottom floor, S18 is the mid point of soil which is 41.3m far from to the surface of the soil. From Figure 6, we can see, the seismic response of the interaction system has small difference with or without supporting pile.

5. Numerical Analysis of the Soil-box Foundation-supporting Pile-high-rise Building Dynamic Interaction System

In this section, the numerical analysis of the soil-box foundation-supporting pile-high-rise building dynamic interaction system is discussed, and contribution of the supporting pile to the dynamic property of the interaction system is studied.

5.1 *Natural frequency of SSI system*

The natural frequency is compared in Table 4, which is given under two condition of with or without supporting pile. As to the system, the supporting pile has some contribution to the natural frequency of the interaction system. The reason may be that the supporting pile make the soil could not deform easy, and increase the soil rigidity, so to enlarge the natural frequency of the system.

Table 4 Natural frequency of SSI system with and without supporting pile

No.	A: without supporting pile	B: with supporting pile	
	Frequency (Hz)	Frequency (Hz)	Error to A (%)
1	0.3536	0.3790	7.20
2	0.4603	0.4660	1.24
3	0.6586	0.6622	0.55
4	0.7207	0.7244	0.51
5	0.8897	0.8942	0.51

5.2 *Seismic response of structure*

El Centro earthquake record, whose peak values of acceleration is adjusted to 0.1g, is inputted from the bottom of soil. The acceleration response comparison of consider the supporting pile or not is showed in Figure 7. A15 is the midpoint on the top floor, A1 is the midpoint of the bottom floor, S6 and S18 are the mid point of soil which is 9m and 41.3m in distance to the surface of the soil respectively. From Figure 7, we can see, the supporting pile has some contribution to the seismic response of the superstructure, but no to that of the soil. It is different to the pile foundation, the reason may be that, the pile foundation system has more great rigidity than the supporting pile, which dips into the soil more shallow than the pile foundation, so the supporting pile has little contribution to the pile foundation system. But to the box foundation system, the embedded depth is lighter than the supporting pile, so the supporting pile may have more great contribution to the whole seismic response of the interaction system.

The acceleration and the displacement peak value of the superstructure under two condition of that considering the supporting pile or not are showed in Figure 8 and Figure 9. From the Figure 8, we can find out, when the supporting pile is in consideration, the acceleration peak value increased, to the top floor, it increased by 15.3%. In Figure 9, the displacement increased by 12.7%. The increased value both exceed the engineering allowable variation. So it is necessary to consider the contribution of the supporting pile in the soil-box foundation-supporting pile-high-rise building dynamic interaction system.

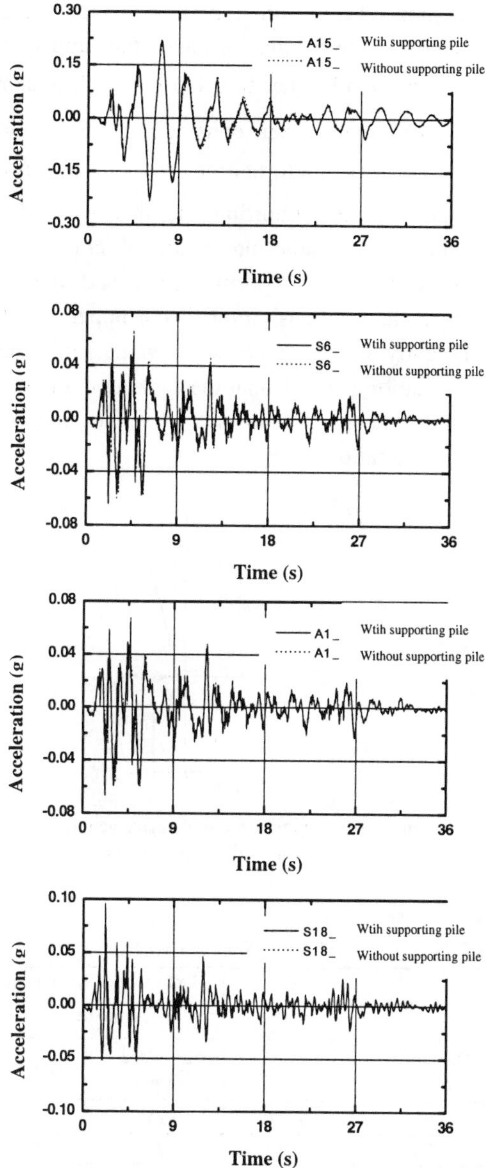

Figure 6 Comparison of acceleration time-history with and without supporting pile

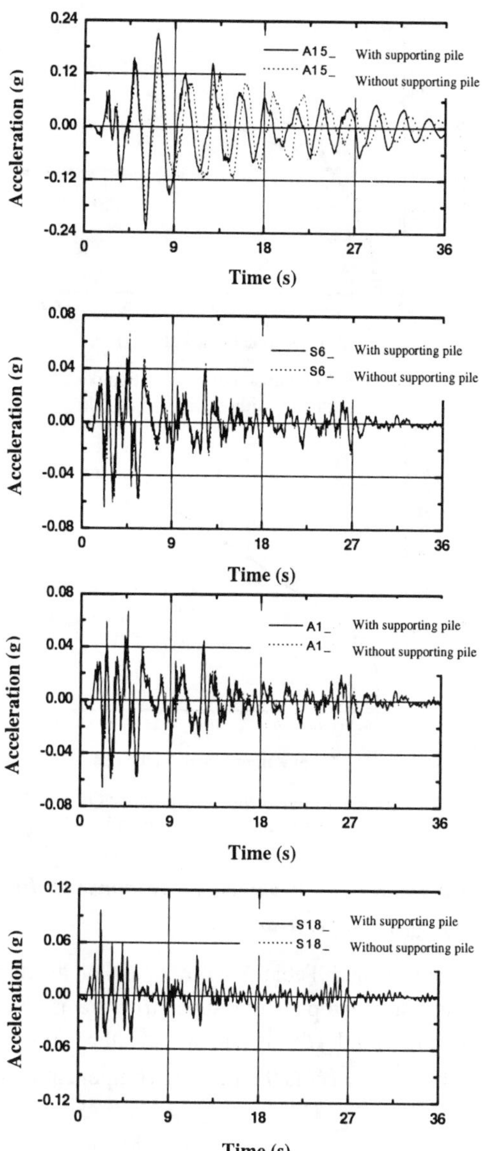

Figure 7 Comparison of acceleration time-history with and without supporting pile

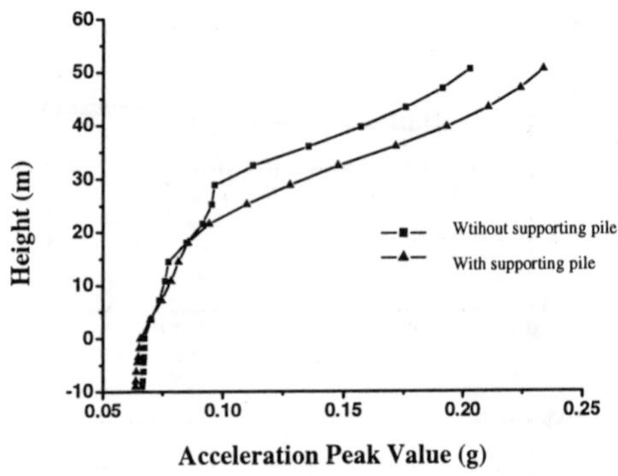

Figure 8 Comparison of acceleration peak value with and without supporting pile

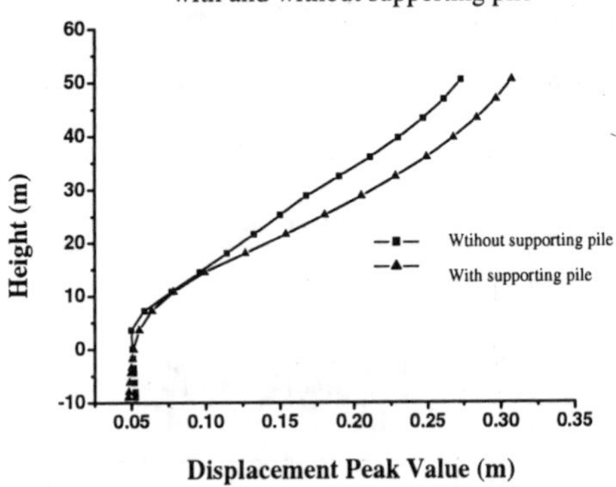

Figure 9 Comparison of displacement peak value with and without supporting pile

5.3 *Component of displacement response at the top of the superstructure*

In Figure 10 and Table 5, component of displacement response at the top of the superstructure is discussed. u is the total displacement at the top of the superstructure, $H\theta$ is the rocking component caused by foundation rotation, u_g is the translation displacement caused by foundation displacement, u_e is the elastic deformation of the upper structure.

It is showed in Table 5 that the rocking component $H\theta$ decreased when considering the supporting pile than not. The reason may lies in the supporting pile formed a hoop, that limited the upper structure's swaying. While the translation displacement u_g has small difference when considering the supporting pile. As to elastic deformation u_e , it increased when considering the supporting pile, that mainly because the rigidity of the whole interaction system increased when considering the supporting pile, and the earthquake energy can be transmitted to the upper structure more efficiently. The increased value of the elastic deformation is great than the rocking component, so the total displacement decreased when the supporting pile is in consideration.

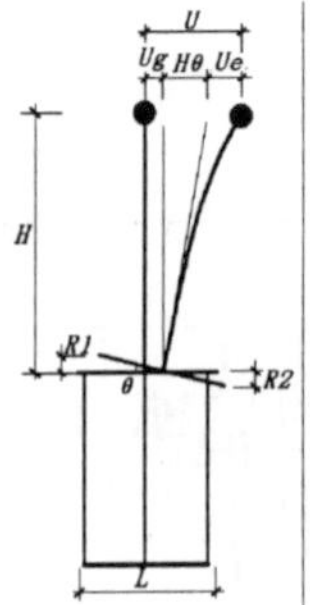

Figure 10 The component of displacement response at the top of the superstructure

Table 5 The component of displacement response at the top of the superstructure (m)

With or without supporting pile	Total displace-ment u	Rocking component $H\theta$		Swing component u_g		Deformation component of the superstructure u_e	
		Displace ment	Ratio to u	Displace-ment	Ratio to u	Displace-ment	Ratio to u
With	0.3073	0.0535	17.4%	0.0514	16.7%	0.2560	83.3%
Without	0.2725	0.0811	29.8%	0.0506	18.6%	0.2068	75.9%

5.4 *Interstory shear amplitude and overturning moment amplitude of structure*

The interstory shear amplitude and the overturning moment amplitude of the upper structure increased when considering the supporting pile. Figure 11 shows that the interstory shear amplitude force increased. The interstory shear amplitude increased by 18.9% in the bottom story. The reason is the same as that of the increase of the acceleration and the displacement in section 5.2. The increased value of the interstory shear amplitude and the overturning moment amplitude both exceed the engineering allowable variation, so it is not safety if the supporting pile is not in consideration. Furthermore, it is necessary to consider the supporting pile in the soil-box foundation-supporting pile-high-rise building dynamic interaction system.

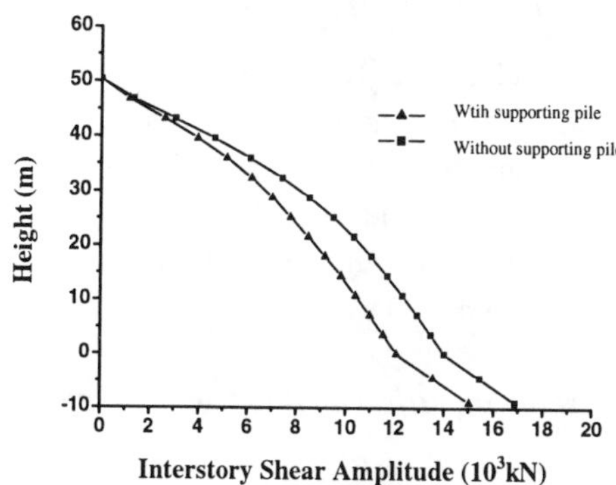

Figure 11 Comparison of interstory shear amplitude with and without supporting pile

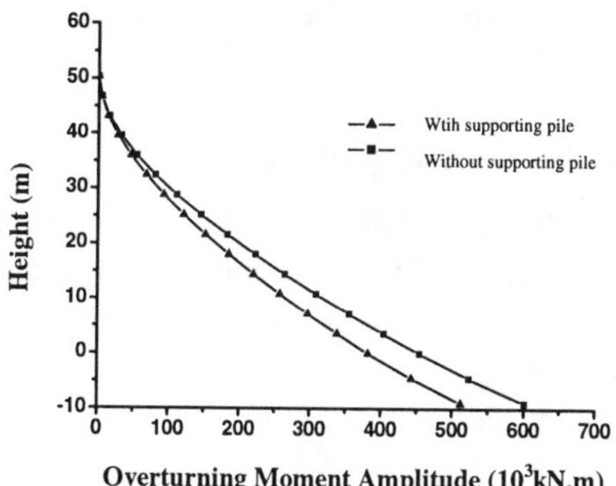

Figure 12 Comparison of overturning moment amplitude with and without supporting pile

6. Parameter Study on SSI System Considering Supporting Pile

There are many factors may contribute to the seismic response of the interaction system. In this section, the parameter considering the supporting pile is analyzed. The length, diameter, the concrete grade of the supporting pile, and the rigidity of the upper structure are respectively changed while the others keep the same. The contribution of above factors to the seismic response of the interaction system is discussed.

6.1 *SSI study on different length of supporting pile*

Keep the other parameter not changed, vary the length of the supporting pile, that is 17.6m, 20m and 23.25m. The influence of the length of the supporting pile on the dynamic property and the seismic response of the interaction system is studied.

6.1.1 *Natural frequency of SSI system*

The natural frequency of the interaction system when the length of the supporting pile vary are compared in Table 6. The influence of the length of the supporting pile on the natural frequency is very small.

6.1.2 *Seismic response of structure*

Figure 13 and Figure 14 shows the acceleration peak value and the displacement peak value along the earthquake direction respectively when the supporting pile length is different. When the length of the supporting pile vary from 17.6m to 20m, the acceleration peak and the displacement peak increased 1%~3%. When the length vary from 20m to 23.5m, the acceleration peak and the displacement peak increase 1%~3% too. That is to say, the length of the supporting pile has not much influence on the response of the upper structure.

Table 6 Natural frequency of SSI system with different length of supporting pile

No. \ Length	A: 17.6m		B: 20m	C: 23.25m	
	Frequency (Hz)	Error to B (%)	Frequency (Hz)	Frequency (Hz)	Error to B (%)
1	0.3765	-0.65	0.3790	0.3818	0.74
2	0.4650	-0.23	0.4660	0.4674	0.28
3	0.6615	-0.10	0.6622	0.6629	0.10

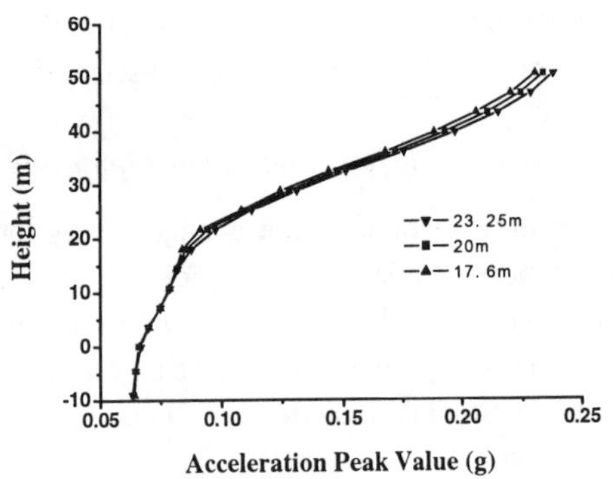

Figure 13 Comparison of acceleration peak value with different length of supporting pile

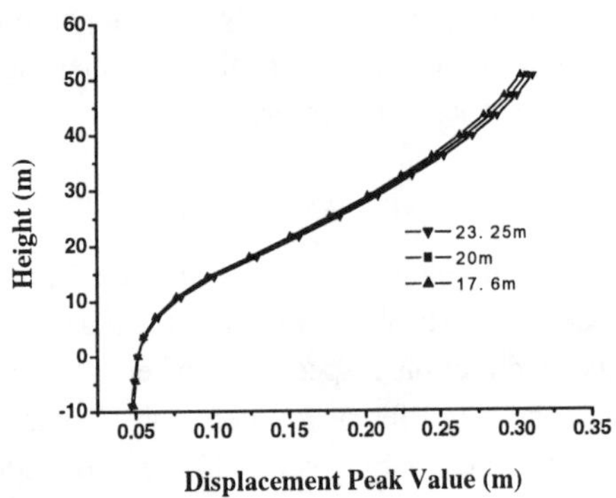

Figure.14 Comparison of displacement peak value with different length of supporting pile

6.2 *SSI study on different concrete grade of supporting pile*

Keep other parameters not changed, vary the concrete grade of the supporting pile, that is C20, C30 and C40. The influence of the concrete grade of the supporting pile on the dynamic property and the seismic response of the interaction system is studied.

6.2.1 *Natural frequency of SSI system*

When the concrete grade of the supporting pile is different, the natural frequency of the interaction system is compared in Table 7. The concrete grade of the supporting pile has not much influence on the natural frequency of the interaction system.

6.2.2 *Seismic response of structure*

Figure 15 and Figure 16 shows the acceleration peak value and the displacement peak value along the earthquake direction respectively. These figures show that the concrete grade of the supporting pile has not much influence on the natural frequency of the interaction system.

6.3 *SSI study on different diameter of supporting pile*

Keep other parameters not changed, vary the diameter of the supporting pile, that is 0.8m, 1.0m and 1.2m. The influence of the diameter of the supporting pile on the dynamic property and the seismic response of the interaction system is studied.

6.3.1 *Natural frequency of SSI system*

When the diameter of the supporting pile is different, the natural frequency of the interaction system is compared in Table 8. The diameter of the supporting pile has not much influence on the natural vibration frequency of the interaction system.

Table 7 Natural frequency of SSI system with different concrete grade of supporting pile

No. \ Concrete grade	A: C20		B: C30	C: C40	
	Frequency (Hz)	Error to B (%)	Frequency (Hz)	Frequency (Hz)	Error to B (%)
1	0.3792	0.05	0.3790	0.3791	0.03
2	0.4662	0.04	0.4660	0.4661	0.02
3	0.6622	0.00	0.6622	0.6622	0.01

Table 8 Natural frequency of SSI system with different diameter of supporting pile

No. \ Diameter	A: 0.8m		B: 1.0m	C: 1.2m	
	Frequency (Hz)	Error to B (%)	Frequency (Hz)	Frequency (Hz)	Error to B (%)
1	0.3776	-0.36	0.3790	0.3804	0.37
2	0.4656	-0.08	0.4660	0.4665	0.10
3	0.6619	-0.04	0.6622	0.6624	0.03

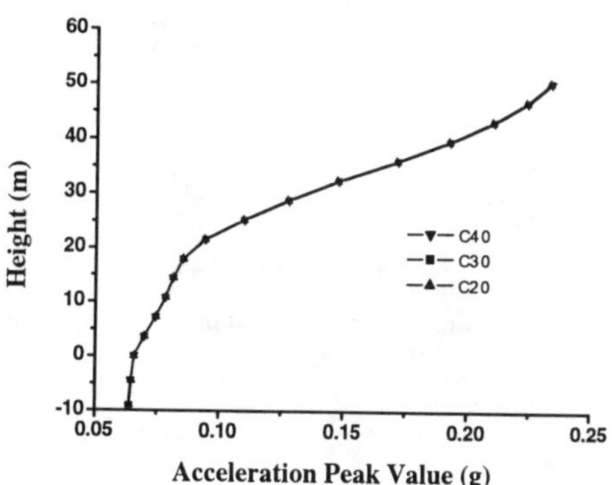

Figure 15 Comparison of acceleration peak value with different concrete grade of supporting pile

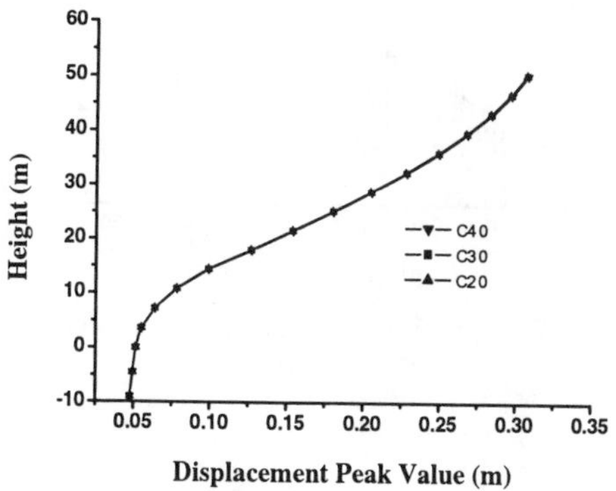

Figure 16 Comparison of displacement peak value with different concrete grade of supporting pile

6.3.2 Seismic response of structure

Figure 17 and Figure 18 shows the acceleration peak value and the displacement peak value along the earthquake direction respectively. When the diameter of the supporting pile vary from 0.8m to 1.0m, the acceleration peak value and the displacement peak value increased 1%~2%. When the length vary from 1.0m to 1.2m, the acceleration peak value and the displacement peak value increased 1%~2% too. So to say, the diameter of the supporting pile has not much influence on the response of the upper structure.

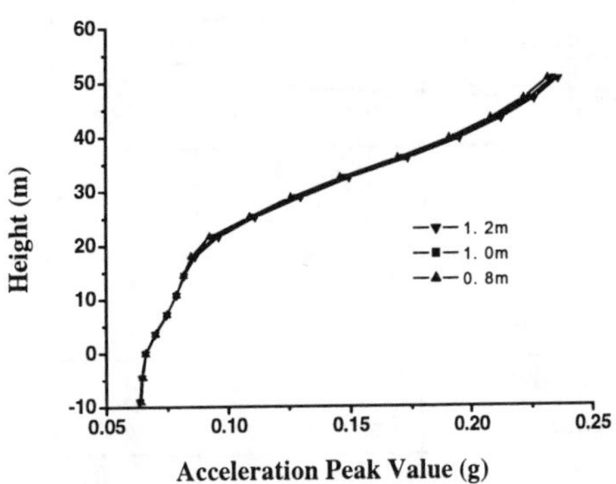

Figure 17 Comparison of acceleration peak value with different diameter of supporting pile

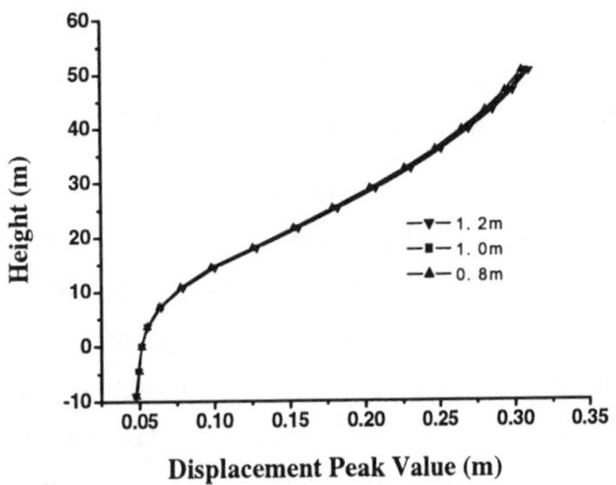

Figure 18 Comparison of displacement peak value with different diameter of supporting pile

6.4 *SSI study on different superstructure*

Keep other parameters not changed, vary the concrete grade of the upper structure, that is C20, C30 and C40. The influence of the concrete grade of the upper structure on the dynamic property and the seismic response of the interaction system is studied.

6.4.1 *Natural frequency of SSI system*

When the concrete grade of the upper structure is different, the natural frequency of the interaction system is compared in Table 9. The concrete grade of the upper structure has not much influence on the natural frequency of the interaction system.

6.4.2 *Seismic response of structure*

Figure 19 and Figure 20 shows the acceleration peak value and the displacement peak value along the earthquake direction respectively. From these figures, we can see, when the concrete grade of the upper structure increase, the acceleration peak value and the displacement peak value also increase, most to 15%. The reference[8] shows that the greater the rigidity of the upper structure is the larger the displacement of considering interaction is. In this section the result is the same.

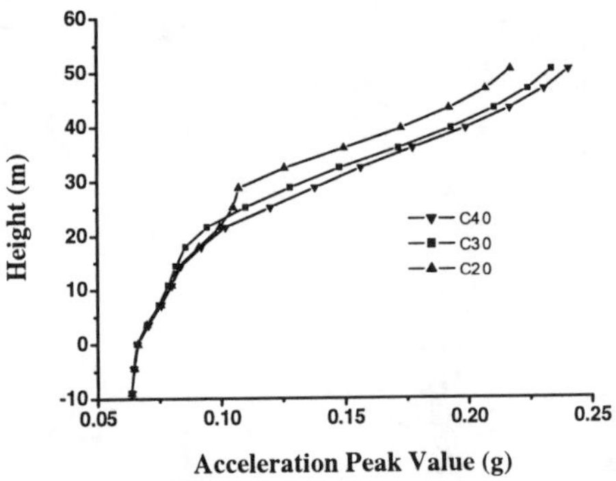

Figure 19 Comparison of acceleration peak value with different superstructure

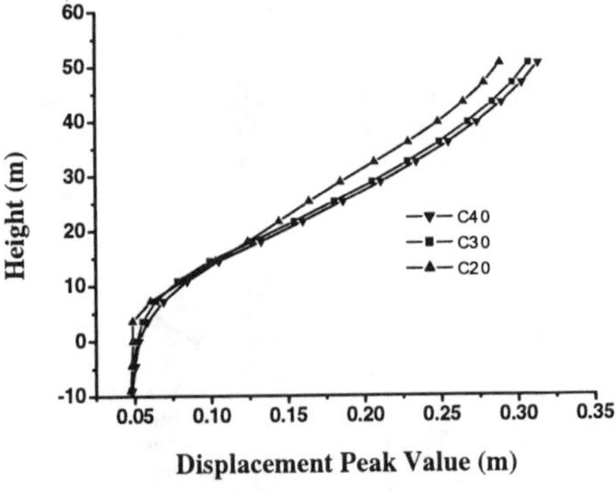

Figure 20 Comparison of displacement peak value with different superstructure

Table 9 Natural frequency of SSI system with different superstructure

No.	A: C20		B: C30	C: C40	
	Frequency (Hz)	Error to B (%)	Frequency (Hz)	Frequency (Hz)	Error to B (%)
1	0.3573	-5.73	0.3790	0.3894	2.75
2	0.4637	-0.50	0.4660	0.4677	0.36
3	0.6617	-0.07	0.6622	0.6625	0.04

7. Conclusion

Based on the finite element analysis program ANSYS, the soil-foundation-supporting pile-high-rise building dynamic interaction system was analyzed in this paper. It was found that the existing of supporting piles affects the aseismic behavior of high-rise building. The parameter study includes the length and rigidity of the supporting piles, the various rigidity of upper structure and the foundation type.

1. As to the soil-pile foundation-supporting pile-high-rise building dynamic interaction system, the seismic response of the SSI system has small difference with or without the supporting pile
2. As to the soil-box foundation-supporting pile-high-rise building dynamic interaction system, the natural frequency has some increase when the supporting pile is in consideration. The acceleration peak value and the displacement peak value on the top of the structure can increase by 15%. So it is necessary to consider the contribution of the supporting pile to the seismic response in the soil-box foundation-supporting pile-high-rise building dynamic interaction system.
3. The length, diameter, the concrete grade of the supporting pile have small influence on the natural frequency of the SSI system. When the length or the diameter increased, the acceleration and the displacement response increased a little. While the concrete grade of the supporting pile has almost no influence on the dynamic response of the upper structure.
4. When the concrete grade of the upper structure increased, the natural frequency had some increase, and the acceleration peak value and the displacement peak value of upper structure also increased.

Reference

1. X.L. Lu, P.Z. Li, B. Chen, and Y.Q. Chen. Numerical analysis of dynamic soil-box foundation-structure interaction system, *Journal of Asian Architecture and Building Engineering*, **1(2):** 9-16 (2002)
2. X.L. Lu, P.Z. Li, B. Chen, and Y.Q. Chen. Computer simulation of the dynamic layered soil-pile-structure interaction system, Cananical. Geotechnical Journal, **42(3)**: 742-751 (2005)
3. Foundation Design Code (DGJ08-11-1999). Shanghai, China, (1999)
4. Y. Huang, Z.C. Chen, and H.B. Zhou, Dynamic calculation model of shanghai soft soil, *Journal of Tongji University*, **(3)**, 359-363 (2000)
5. ANSYS Inc., User's Manual for ANSYS 5.7 (2000)
6. Lysmer J., Kulemeyer R.L.. Finite Dynamic Model for Infinite Media. *J. Eng. Mech. Div., ASCE*; **Vol.95**: 759-877 (1969)
7. X.L. Lu, P.Z. Li, Y.Q. Chen and B. Chen, Shaking Table Model Testing on Dynamic Soil-Structure Interaction System, *13th World Conference on Earthquake Engineering*, Vancouver, Canada, August 1-6, No. **3231** (2004)
8. L.J. Dou, B.P. Yang, G.H. Liu, Some practical problems on dynamic soil-structure interaction, *World Information on earthquake Engineering*, **15(4)**, 62-68 (1999)

AN ANALYTICAL MODEL TO ESTIMATE LOAD CAPACITY POSSESSED BY SUPPORTING SOIL FOR PILED RAFT FOUNDATIONS*

CIMIAN ZHU

School of Civil Engineering, Tongji University, 1239 Siping Road Shanghai 200092, China

XIANGHONG YAN

Architectural Design and Research Institute of Jiangsu Province, 8-1 Gongyuan Road Suzhou 215006, China

SHENGGUO BI

School of Civil Engineering, Tongji University, 1239 Siping Road Shanghai 200092, China

While piled raft foundations are widely adopted in the construction of tall or multistory buildings, an accurate estimation and control of load capacity possessed by supporting soil become essential for the safety of buildings and construction cost reduction, especially in the case on soft soil subgrades. This paper presents an effective analytical model to estimate load capacity possessed by supporting soil for piled raft foundations. In this approach the bending plate is approximately assumed to have linear elastic properties and is modeled by the generalized conforming finite element method while the supporting soil subgrade is modeled by the finite layer method(FLM). Each pile is represented by a single element, and its nonlinear stiffness is evaluated through a load vs. pile head displacement curve obtained from static loading tests, resulting in the governing system of equations for plate-pile-soil interaction problems. The FLM is not only capable of representing the layered subgrade behavior, but it is also of low computation cost. Several real piled raft foundations on the soft soil subgrade have been analysed. The numerical results indicate that, in general, for tall buildings, the load capacities possessed by supporting soil are about 8~15% of the whole one; and that for multistory buildings, where the piles are often used to prevent excessive settlements, are often more than 25%. It can be seen from the numerical results that the load capacity possessed by supporting soil is more affected by stiffness of piles, and increases with the settling of the building, its full utilization in design would evidently reduce the cost of a piled raft foundation construction.

1. Introduction

Piled raft foundations are widely adopted in the construction of tall or multistory buildings. It is common in early piled raft foundation design to consider the entire design loads are resisted only by the piles. During the past three decades, the load capacity possessed by supporting soil has been paid an increasing attention, and in the past few years there has been a recognition that the use of piles to reduce raft settlements and differential settlements can lead to considerable economy without compromising the safety and performance of the foundation. Such a foundation makes use of both the raft and the piles, so gives rises to plate-pile-soil interaction problems.

Plate-raft-soil interaction has been investigated by various methods. Hemsley [1,2] used finite element method to analyse plate-soil interactions. Messafer and Coates[3] developed a combination of finite element method (FEM) and boundary element method (BEM), where a quadrilateral finite element called ACM was used to model the bending plate and a BEM was applied to soil.

To analyse pile-soil interactions, Banerjee developed a linear boundary element formulation using Mindlin's solution. After that, linear solutions for piles and pile groups in non-homogenous soils were presented in Refs.[4]. Faruque and Desai[5] attempted the use of finite element method in 3D nonlinear analysis of piles. Another way of numerical analysis is a load transfer method that models the soil through a set of independent springs attached to the piles. Shen, Chow and Yong[6,7] presented variational approaches in which the soil was modeled with the use of load-transfer curves and as a half-space, respectively. And then, the tractions of soil-

* This work is supported by the Chinese Civil Air Defence Office.

pile interaction were approximated by finite series and the soil contributions were obtained with the help of Mindlin's solution. Kücükarslan, Banerjee and Bildik[8] developed an inelastic analysis of pile-soil structure interaction.

For the analysis of piled rafts, in which all the interactions between the plate, the pile and the soil are simultaneously taken into account, Mendonca and Paiva[9] developed a BEM, in which both the soil and the plate were represented by integral equations. Three years later, they[10] presented an elastostatic analysis of vertically loaded piled raft foundations, in which the FEM and the BEM were coupled. In Ref.[11], an optimal pile arrangement scheme for minimizing the differential settlements of piled raft foundations is proposed. However, these powerful numerical methods require a considerable amount of engineering judgment and computation cost to completely solve the physical problem. Because of these reasons, research in this area has focused on simplified and economical models to characterize the problem[8]. Besides, many experimental researches were also conducted to determine the sources of nonlinearities in the entire system.

This paper presents an effective analytical model to estimate load capacity possessed by supporting soil for piled raft foundations. In this approach the bending plate is approximately assumed to have linear elastic properties and is modeled by the generalized conforming finite element method while the supporting soil subgrade is modeled by the finite layer method(FLM). Each pile is represented by a single element, and its nonlinear stiffness is evaluated through a load vs. pile head displacement curve obtained from static loading tests, resulting in the governing system of equations for plate-pile-soil interaction problems. The FLM is not only capable of representing the layered subgrade behavior, but also of low computation cost. Several real piled raft foundations on the soft soil subgrade have been analysed. The numerical results indicate that, in general, the load capacity possessed by supporting soil is considerably affected by stiffness of piles, and increases with the settling of buildings.

2. Analytical Formulation

2.1 *Soil model*

The soil domain is modeled as a finite continuum with the dimensions which make the boundary displacements be negligible. The layered subgrades are assumed to be elastic and modeled by FLM[12], in which the elastic properties are isotropic only in the horizontal direction but different and variable in the vertical direction. Accordingly, the elasticity matrix can be written as Eq. (1) [13].

$$\boldsymbol{D} = \begin{bmatrix} d_1 & d_2 & d_3 & 0 & 0 & 0 \\ d_2 & d_1 & d_3 & 0 & 0 & 0 \\ d_3 & d_3 & d_4 & 0 & 0 & 0 \\ 0 & 0 & 0 & d_5 & 0 & 0 \\ 0 & 0 & 0 & 0 & d_6 & 0 \\ 0 & 0 & 0 & 0 & 0 & d_6 \end{bmatrix} \tag{1}$$

where

$$\left.\begin{aligned} & d_1 = \lambda n(1-n\mu_2^2), \qquad d_2 = \lambda n(\mu_1 + n\mu_2^2) \\ & d_3 = \lambda n\mu_2(1+\mu_1), \qquad d_4 = \lambda(1-\mu_1^2) \\ & d_5 = \frac{E_1}{2(1+\mu_1)}, \qquad d_6 = G_2 \\ & \lambda = \frac{E_2}{(1+\mu_1)(1-\mu_1-2n\mu_2^2)}, \qquad n = \frac{E_1}{E_2} \end{aligned}\right\} \tag{2}$$

in which E_1 and μ_1 are the deformation modulus and Poisson's ratio in the plane *xy*, respectively; E_2 is the deformation modulus in the *z* direction; μ_2 is the Poisson's ratio between plane *xy* and *z* axis; G_2 is the shear modulus in the plane perpendicular to the plane *xy*.

For vertically loaded foundations, boundary conditions of the soil domain can be represented as Figure 1.

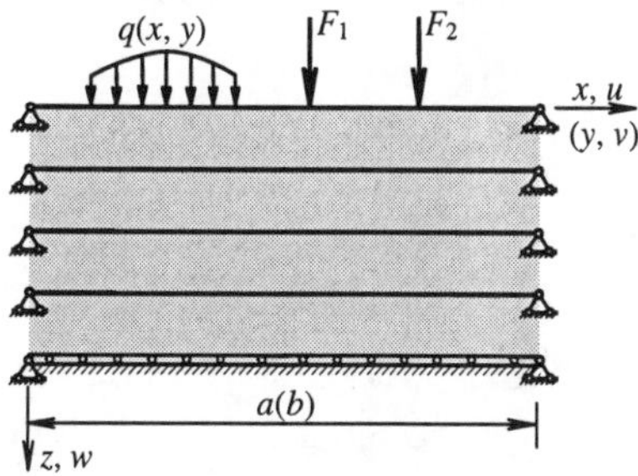

Figure 1. Boundary conditions

A general layer element *ij* is shown in Figure 2, where *a* and *b* are the length and width of the modeled soil domain, respectively; *c* is the thickness of the layer element.

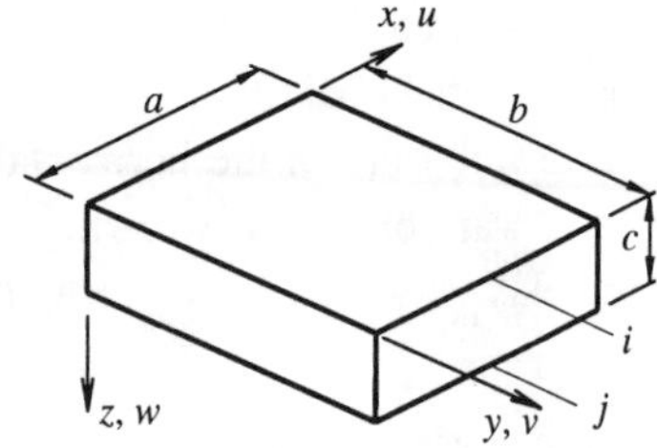

Figure 2. Layer element

The displacement functions of the layer element are assumed as Eq. (3).

$$\left.\begin{aligned} u(x,y,z) &= \sum_{m=1}^{\infty}\sum_{n=1}^{\infty} f_{mn}^{u}(z)\cos k_m x \sin k_n y \\ v(x,y,z) &= \sum_{m=1}^{\infty}\sum_{n=1}^{\infty} f_{mn}^{v}(z)\sin k_m x \cos k_n y \\ w(x,y,z) &= \sum_{m=1}^{\infty}\sum_{n=1}^{\infty} f_{mn}^{w}(z)\sin k_m x \sin k_n y \end{aligned}\right\} \quad (3)$$

where $k_m = \dfrac{m\pi}{a}$; $k_n = \dfrac{n\pi}{b}$; $f_{mn}^{u}(z)$, $f_{mn}^{v}(z)$ and $f_{mn}^{w}(z)$ are linear functions, and can be represented as Eq. (4).

$$\left.\begin{aligned} f_{mn}^{u}(z) &= a_{1,mn} + a_{2,mn} z \\ f_{mn}^{v}(z) &= a_{3,mn} + a_{4,mn} z \\ f_{mn}^{w}(z) &= a_{5,mn} + a_{6,mn} z \end{aligned}\right\} \quad (4)$$

The nodal surface displacements are defined as

$$\boldsymbol{\delta}_{mn} = \begin{bmatrix} u_{mn}^{i} & v_{mn}^{i} & w_{mn}^{i} & u_{mn}^{j} & v_{mn}^{j} & w_{mn}^{j} \end{bmatrix}^{\mathrm{T}}$$

Denoting the displacements at any point within the element by

$$\boldsymbol{\delta} = \begin{bmatrix} u & v & w \end{bmatrix}^{\mathrm{T}}$$

the relationship between the element displacements and nodal surface displacements of the element can be obtained as Eq. (5).

$$\boldsymbol{\delta} = \sum_{m=1}^{r}\sum_{n=1}^{s} \boldsymbol{N}_{mn}\boldsymbol{\delta}_{mn} \quad (5)$$

in which r and s are the number of adopted terms in the series shown in Eq. (3).

Denoting by

$$\boldsymbol{\varepsilon} = \begin{bmatrix} \varepsilon_x & \varepsilon_y & \varepsilon_z & \gamma_{xy} & \gamma_{yz} & \gamma_{zx} \end{bmatrix}^{\mathrm{T}}$$

the element strain components, we have

$$\boldsymbol{\varepsilon} = \sum_{m=1}^{r}\sum_{n=1}^{s} \boldsymbol{B}_{mn}\boldsymbol{\delta}_{mn} \quad (6)$$

where

$$\boldsymbol{B}_{mn} = \boldsymbol{L}(\boldsymbol{N}_{mn}) \quad (7)$$

with the operator

$$\boldsymbol{L} = \begin{bmatrix} \frac{\partial}{\partial x} & 0 & 0 \\ 0 & \frac{\partial}{\partial y} & 0 \\ 0 & 0 & \frac{\partial}{\partial z} \\ \frac{\partial}{\partial y} & \frac{\partial}{\partial x} & 0 \\ 0 & \frac{\partial}{\partial z} & \frac{\partial}{\partial y} \\ \frac{\partial}{\partial z} & 0 & \frac{\partial}{\partial x} \end{bmatrix} \quad (8)$$

Using the vector

$$\boldsymbol{\sigma} = \begin{bmatrix} \sigma_x & \sigma_y & \sigma_z & \tau_{xy} & \tau_{yz} & \tau_{zx} \end{bmatrix}^{\mathrm{T}}$$

for representing the stress components, we have

$$\boldsymbol{\sigma} = \boldsymbol{D}\sum_{m=1}^{r}\sum_{n=1}^{s} \boldsymbol{B}_{mn}\boldsymbol{\delta}_{mn} \quad (9)$$

The mn-th stiffness submatrix of the layer element can be derived using the principle of virtual work. It can be expressed as

$$\boldsymbol{k}_{mn} = \int_V \boldsymbol{B}_{mn}^{\mathrm{T}}\boldsymbol{D}\boldsymbol{B}_{mn}\mathrm{d}V = \begin{bmatrix} s_1 & & & & & \\ s_2 & s_3 & & \text{symtetric} & & \\ s_4 & s_5 & s_6 & & & \\ s_7 & s_8 & s_9 & s_{10} & & \\ s_{11} & s_{12} & s_{13} & s_{14} & s_{15} & \\ s_{16} & s_{17} & s_{18} & s_{19} & s_{20} & s_{21} \end{bmatrix} \quad (10)$$

where

$$\left.\begin{aligned}
&s_1 = s_{10} = \Delta(\frac{d_1}{3}k_m^2 + \frac{d_5}{3}k_n^2 + \frac{d_6}{c^2}), s_2 = s_{14} = 2s_8 \\
&s_3 = s_{15} = \Delta(\frac{d_5}{3}k_m^2 + \frac{d_1}{3}k_n^2 + \frac{d_6}{c^2}), s_4 = -s_{19} = \Delta(d_3 - d_6)\frac{k_m}{2c} \\
&s_5 = -s_{20} = \Delta(d_3 - d_6)\frac{k_n}{2c}, s_6 = s_{21} = \Delta(\frac{d_6}{3}k_m^2 + \frac{d_1}{3}k_n^2 + \frac{d_6}{c^2}) \\
&s_7 = \Delta(\frac{d_1}{6}k_m^2 + \frac{d_5}{6}k_n^2 - \frac{d_6}{c^2}), s_8 = s_{11} = \Delta(d_2 + d_5)\frac{k_m k_n}{6} \\
&s_{12} = \Delta(\frac{d_5}{6}k_m^2 + \frac{d_1}{6}k_n^2 - \frac{d_6}{c^2}), s_{16} = -s_9 = \Delta(-d_3 - d_6)\frac{k_m}{2c} \\
&s_{18} = \Delta(\frac{d_6}{6}k_m^2 + \frac{d_6}{6}k_n^2 - \frac{d_4}{c^2}), s_{17} = -s_{13} = \Delta(-d_3 - d_6)\frac{k_n}{2c} \\
&\Delta = abc/4
\end{aligned}\right\} \quad (11)$$

2.2 *Raft model*

A raft is modeled as a plate based on the Mindlin theory. The bending plate is assumed to have linear elastic properties and is modeled by the generalized conforming FEM[14].

A general quadrilateral element is shown in Figure 3. The element nodal displacement vector $\boldsymbol{a}$ is

$$\boldsymbol{a} = \left[w_1\ \phi_{x1}\ \phi_{y1}\ w_2\ \phi_{x2}\ \phi_{y2}\ w_3\ \phi_{x3}\ \phi_{y3}\ w_4\ \phi_{x4}\ \phi_{y4}\right]^{\mathrm{T}}$$

The final nodal variables used in the formulation are represented by two rotations and one translation in each corner of the quadrileteral.

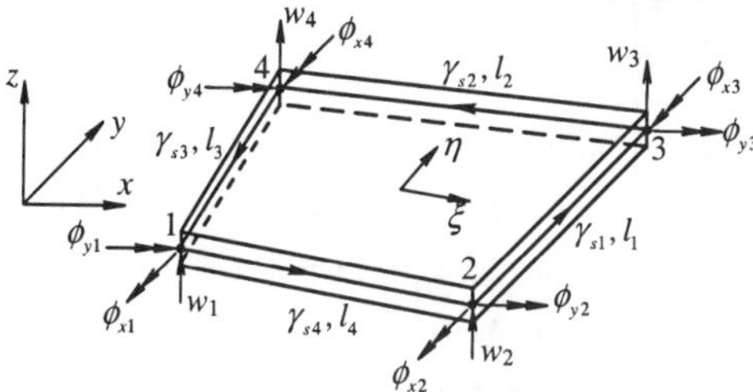

Figure 3. Quadrileteral element.

Based on Timoshenko's theory, the element stiffness matrix can be represented as Eq. (12).

$$\boldsymbol{k}^e = \boldsymbol{k}_b^e + \boldsymbol{k}_s^e \quad (12)$$

Here, $\boldsymbol{k}_b^e$ and $\boldsymbol{k}_s^e$ are the element flexural and shear stiffness matrix, respectively, and are defined as follows.

$$\boldsymbol{k}_b^e = \iint_{\Omega_e} \boldsymbol{B}_b^{\mathrm{T}} \boldsymbol{D}_b \boldsymbol{B}_b \mathrm{d}A = \int_{-1}^{1}\int_{-1}^{1} \boldsymbol{B}_b^{\mathrm{T}} \boldsymbol{D}_b \boldsymbol{B}_b |\boldsymbol{J}| \mathrm{d}\xi \mathrm{d}\eta \quad (13)$$

$$\boldsymbol{k}_s^e = \iint_{\Omega_e} \boldsymbol{B}_s^{\mathrm{T}} \boldsymbol{D}_s \boldsymbol{B}_s \mathrm{d}A = \int_{-1}^{1}\int_{-1}^{1} \boldsymbol{B}_s^{\mathrm{T}} \boldsymbol{D}_s \boldsymbol{B}_s |\boldsymbol{J}| \mathrm{d}\xi \mathrm{d}\eta \quad (14)$$

in which $\boldsymbol{B}_b$ and $\boldsymbol{B}_s$ are the element flexural and shear strain matrix, respectively, and $\boldsymbol{J}$ is the Jacobian matrix.

2.3 *Pile model*

Piles are assumed to be only virtically loaded. Each pile is modeled by a single element, and its nonlinear performance can be evaluated through a sectionally linearized load vs. pile head displacement curve obtained from static loading tests. It is illustrated in Figure 4 in which P_1, P_2 and P_3 are corresponding to 0.5, 0.9, 1.0 time the maximum load capacity of the test pile, respectively.

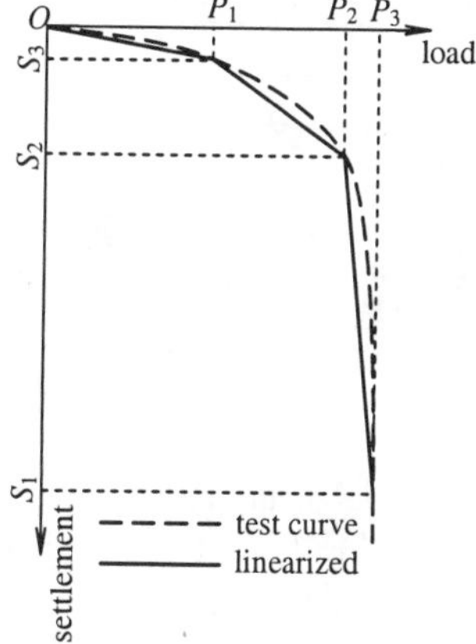

Figure 4. Linearization of the load-settlement curve.

2.4 *Final system of equations*

The systems of equations for FEM and FLM are coupled by eliminating the unknown plate-soil subgrade traction. By introducing piles stiffness, the final system of equations can be written in incremental form as Eq. (15).

$$(\boldsymbol{K}_r + \boldsymbol{K}_s + \boldsymbol{K}_p)\dot{\boldsymbol{U}} = \dot{\boldsymbol{F}} \quad (15)$$

where $\dot{\boldsymbol{U}}$ represents the increment of the nodal displacement vector; $\dot{\boldsymbol{F}}$ is the increment of the equivalent nodal force vector; $\boldsymbol{K}_r$ and $\boldsymbol{K}_s$ represent, respectively, raft and soil stiffness and $\boldsymbol{K}_p$ is nonlinear pile system stiffness. The increment of the nodal displacement vector consists of the vertical settlement of the raft, w, and two rotational angles with respect to x-axias and y-axis, θ_x and θ_y, at each node.

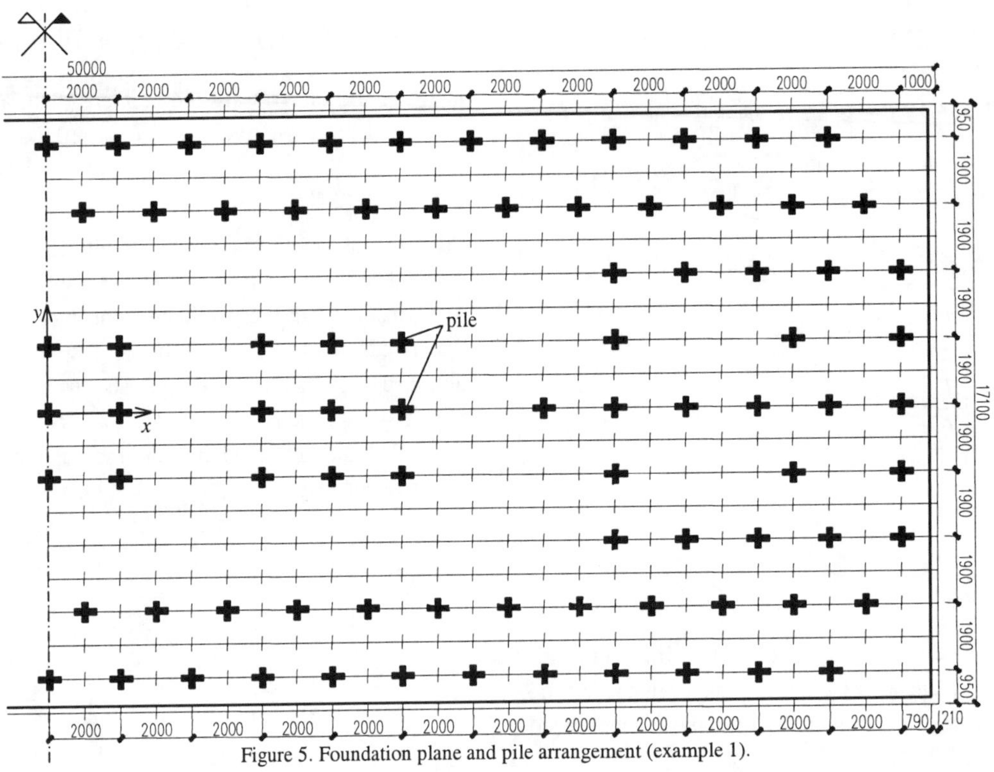

Figure 5. Foundation plane and pile arrangement (example 1).

3. Numerical examples

The proposed analytical model is applied to the following practical piled raft foundations on soft clay with deformation modulus $Es=5Es_{0.1\sim0.2}$ and Poisson's ratio $\mu=0.4$.

3.1 *Commercial tall building in Shanghai, China*

It is a piled raft foundation with a 50m × 17.1m rectangular raft and 166 piles for a 21-story reinforced concrete building on soft clay in Shanghai. Figure 5 shows the foundation plane and pile arrangement. The thickness of the raft is 2m, and bored concrete piles are 43m long and 0.65m in diameter.

Table 1. The geotechnical conditions and the soil compressive moduluses(example 1).

Soil No.	Layer thickness (m)	Elevation of layer bottom (m)	Compressive modulus $Es_{0.1\sim0.2}$ (MPa)
②	0.80 ~ 1.20	0.91 ~ 0.68	
③	4.00 ~ 4.30	-3.29 ~ -3.46	4.73
④	10.00 ~ 10.30	-13.32 ~ -13.69	2.43
⑤-1	4.40 ~ 5.40	-18.06 ~ -19.09	3.60
⑤-2	23.70 ~ 24.70	-42.09 ~ -43.69	4.64
⑤-3	2.60 ~ 4.30	-46.09 ~ -46.36	7.84
⑦	27.60 ~ 28.00	-73.89 ~ -74.12	13.68
⑨			14.28

Table 1 gives the geotechnical conditions and the soil compressive moduluses obtained from laboratory testing and Figure 6 shows the load vs. head displacement curve of the test pile.

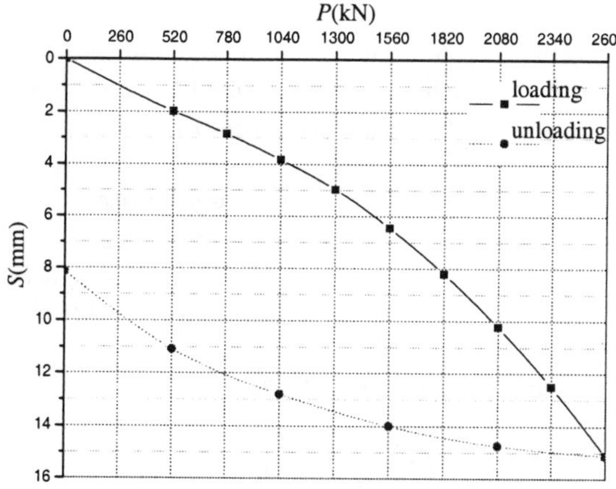

Figure 6. P-S curve of the test pile(example 1).

The numerical results obtained by the FEM/FLM approach with a discretization of the raft into 50×18 finite elements(see Figure 5) indicate that under the design load of 283900kN, the load capacity possessed by supporting soil for the piled raft foundation is about 5.9%.

3.2 *12-story resident building in Shanghai, China*

The building is a R.C. shear wall structure with a piled raft foundation. The building footprint and pile arrangement are shown in Figure 7. The piled raft with dimensions 35.16m×11.76m×0.4m is supported on soil and on 99 fabricated tube piles which are 35m long and 0.4m in diameter.

The geotechnical conditions and corresponding soil compressive moduluses are given in Table 2. The test pile performances are plotted in Figure 8.

Table 2. The geotechnical conditions and the soil compressive moduluses(example 2).

Soil No.	Layer thickness (m)	Elevation of layer bottom (m)	Compressive modulus $Es_{0.1\sim0.2}$ (MPa)
①-1	0.50 ~ 2.30	3.48 ~ 1.51	
①-2	0.55 ~ 2.60	2.44 ~ 0.60	
②	0.30 ~ 1.70	2.54 ~ 1.14	4.14
③	2.60 ~ 6.80	-1.58 ~ -5.27	6.02
④	8.00 ~ 13.90	-11.60 ~ -16.87	2.15
⑤	5.80 ~ 16.50	-20.66 ~ -29.12	4.24
⑥	1.20 ~ 5.30	-23.56 ~ -32.23	7.58
⑦	1.00 ~ 8.00	-29.41 ~ -35.07	10.09

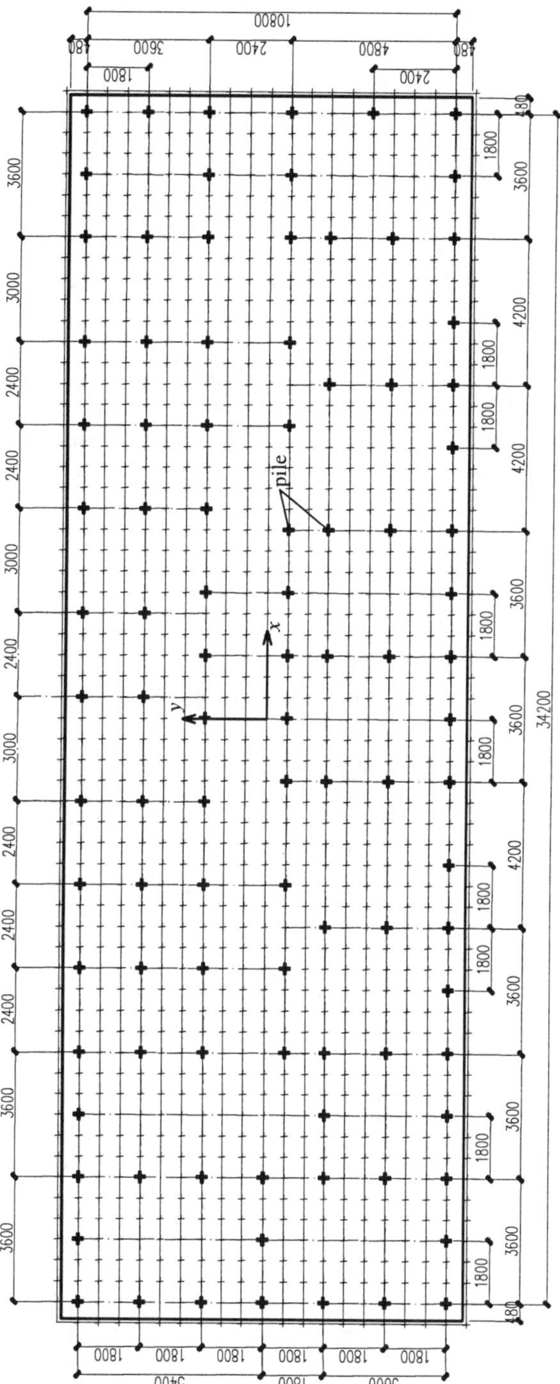

Figure 7. Foundation plane and pile arrangement(example 2).

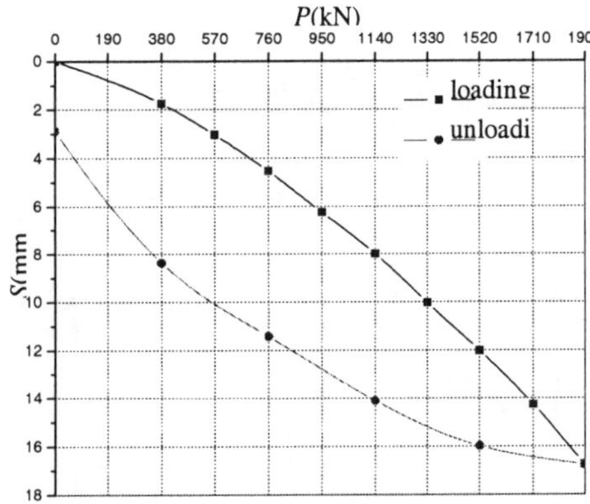

Figure 8. *P-S* curve of the test pile(example 2).

The numerical results of the present approach indicate that the load capacity possessed by supporting soil increases with the settling of the building, and is about 15% under the design load of 92070kN.

3.3 *Multistory public building in Shanghai,China*

It is a biaxial symmetric piled raft foundation for a 5-story reinforced concrete frame on soft clay in Shanghai. The raft, with dimensions 64.1m × 43.4m × 0.5m, is supported on soil and on 316 fabricated piles which are 25m long and with sides 0.3m×0.3m. The foundation plane and pile arrangement are shown in Figure 9.

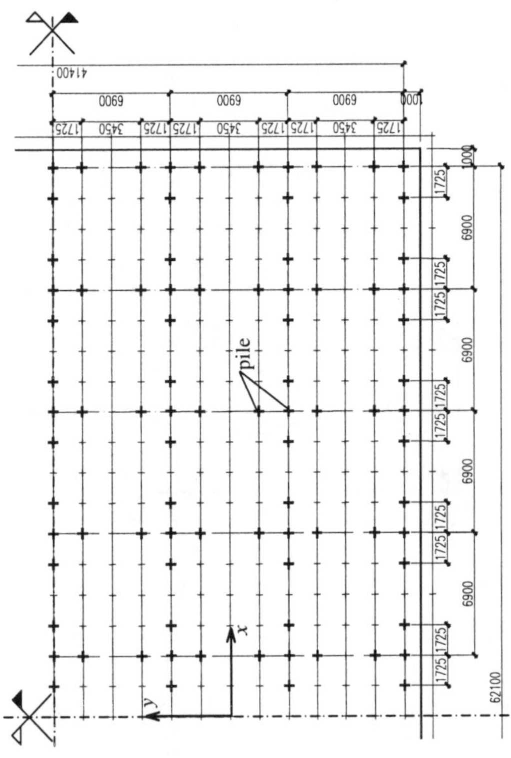

Figure 9. Foundation plane and pile arrangement(example 3a).

The geotechnical conditions, soil compressive moduluses and the pile test results of this example are given in Table 3. and Figure 10, respectively.

Table 3. The geotechnical conditions and the soil compressive moduluses(example 3).

Soil No.	Layer thickness (m)	Elevation of layer bottom (m)	Compressive modulus $Es_{0.1\sim0.2}$ (MPa)
②$_{-1}$	0.60 ~ 1.50	0.70 ~ 1.80	5.79
②$_{-2}$	0.50 ~ 0.90	0.20 ~ 1.10	3.52
③$_{-2}$	2.00 ~ 3.20	-1.20 ~ -2.50	6.58
④	5.00 ~ 7.00	-6.40 ~ -8.40	2.22
⑤$_{-1}^{1}$	4.50 ~ 8.00	-11.70 ~ -16.40	2.55
⑤$_{-1}^{2}$	4.40 ~ 16.50	-19.50 ~ -32.90	4.44
⑤$_{-2}$	2.60 ~ 10.70	-26.00 ~ -32.30	5.91

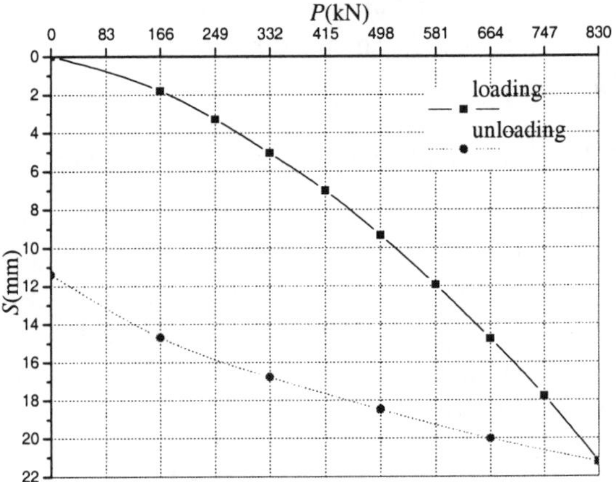

Figure 10. *P-S* curve of the test pile(example 3).

Two other kinds of pile arrangement which contain only 246 and 190 piles, respectively, are shown in Figure 11 and Figure 12. Table 4 shows the numerical results for the three pile arrangement situations under the design load of 173800kN.

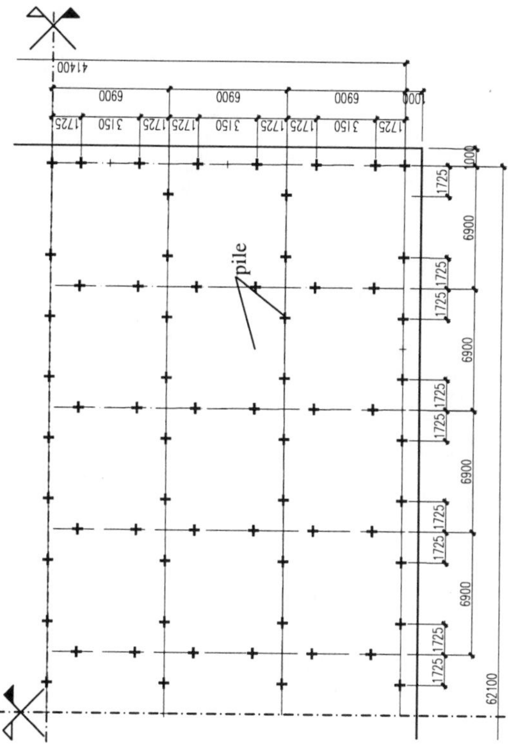

Figure 11. Foundation plane and pile arrangement (example 3b).

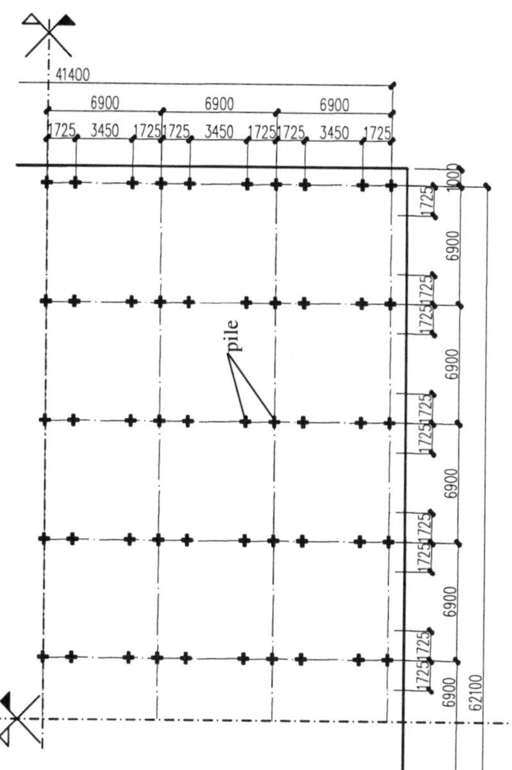

Figure 12. Foundation plane and pile arrangement (example 3c).

Table 4. Numerical results for three kinds of pile arrangement.

No. of piles	Load capacity possessed by soil	Average pile reaction (kN)	Average settlement (mm)
316	18%	453.98	14.20
246	23%	536.36	16.91
190	31%	650.00	20.00

It can be seen from Table 4, when the number of piles reduced from 316 to 190, the raft average settlement increases and the load capacity possessed by supporting soil increases from 18% to 31%. Here, both the average pile reaction and raft settlement are still in the allowable range.

4. Conclusions

A coupled FEM/FLM formulation for the analysis of flexible piled rafts has been presented in which plate-soil and plate-pile interactions are considered and each pile is simplified as a nonlinear spring whose stiffness is evaluated through static loading test. Using the presented analytical model, the entire governing system of equations contains relatively fewer unknowns, and makes the model economical and practical as well as effective.

Numerical results indicate that, for a tall building of more than 20 storys with piled raft foundation on soft clay, the load capacity possessed by supporting soil is generally less than 10%, and that for a tall building of 10~20 storys reaches 10~20%. For a piled raft foundation of multistory building, the load capacity possessed by supporting soil is often more than 30%. It can be seen from the numerical results, that the load capacity possessed by supporting soil is considerably affected by stiffness of piles, and increases with the settling of the building. Its full utilization in the design would evidently reduce the construction cost of a piled raft foundation, especially for a multistory building with piled raft foundation.

References

1. Hemsley JA. Elastic solutions for large matrix problems in foundation interaction analysis. Proc Inst Civil Engnr, Part2 1993;89:471-94.
2. Hemsley JA. Application of large matrix interaction analysis to raft foundations. Proc Inst Civil Engnr, Part2 1993;89:495-526.
3. Messafer T, Coates LE. An application of FEM/BEM coupling to foundation analysis. In:

Brebbia, Connor, editors. Advances in boundary methods. Southampton: CMP; 1993. p .211-21

4. Banerjee PK, Davies TC. The behavior of axially and laterally loaded piles embedded in nonhomogeneous soils. Geotechnique 1978; 28(3):309-26
5. Faruque MO, Desai CS. 3D material and geometric nonlinear analysis of piles. In: Second International Conference on Numerical Methods in Offshore Piling, Texas. 1982. p. 553-7
6. Shen WY, Chow YK, Yong KY. A variational approach for vertical deformation analysis pile groups. Int J Numer Anal Meth Geomech 1997; 21(11):741-52
7. Shen WY, Chow YK, Yong KY. A variational approach for vertically loaded pile groups in an elastic half-space. Geotechnique 1999;49(2):199-213
8. Kücükarslan S, Banerjee PK, Bildik N. Inelastic analysis of pile soil structure interaction. Engineering Structures 2003;25:1231-39
9. Mendonca AV, Paiva JB. A boundary element method for the static analysis of raft foundations on piles. Engng Anal Bound Elem 2000;24:237-47
10. Mendonca AV, Paiva JB. An elastostatic FEM/BEM analysis of vertically loaded raft and piled raft foundation. Engng Anal Bound Elem 2003;27:919-33
11. Kim KN, Lee SH, Kim KS, Chung CK, Kim MM, Lee HS. Optimal pile arrangement for minimizing differential settlements in piled raft foundations. Computers and Geotechnics 2001;28:235-53
12. Zai JM, Zai JZ. Analysis and design for foundations of tall buildings. Beijing: China Architecture & Building Press, 1993
13. Zienkewicz OC, Cheung YK. Finite element method in structural and continuum mechanics. New York: Mc Graw-Hill, 1967
14. Long ZF, Cen S. New theory of the finite element method. Beijing: China Waterpub Press, 2001

THREE-DIMENSIONAL ANALYSIS OF TORSIONALLY LOADED LARGE-DIAMETER BORED PILE GROUPS*

LIMIN ZHANG

Department of Civil Engineering,
Hong Kong University of Science and Technology, Clear Water Bay, Hong Kong.

CARMEN Y.M. TSANG

Department of Civil Engineering,
Hong Kong University of Science and Technology, Clear Water Bay, Hong Kong

Large-diameter bored piles are commonly used in groups to support high-rise buildings or bridges. In addition to vertical, lateral, and moment loads, these piles have to resist torsional loadings from the superstructure by eccentricity of lateral loads. Compared with studies on responses of pile groups subjected to vertical, lateral and moment loadings, only limited study on response of torsionally loaded pile groups has been reported. In this paper, three-dimensional analyses are conducted to investigate the torsional response of a single pile and a 2x2 large-diameter bored pile group. The piles are 2 m in diameter, 13.2 m in length, and 6 m in spacing. For the relatively rigid piles studied, the relative twisting angle between the pile head and the pile toe is small for the torsionally loaded single pile and the individual piles in the pile group. However, the movements of the single pile and the pile group under torsion are quite different. The single pile only shows a pure torsional response while the group piles not only twist but also deflect in the direction of torsional movements. The applied torque is resisted by the torsional resistance and the lateral resistance of the group piles. The torsional resistance provided by the individual piles is mobilized at relatively small movements but the lateral resistance is mobilized continuously with the pile group rotation. At large deformations, the applied torque increment is mostly taken by the lateral resistance. The torsional resistance of individual piles is approximately 30% of the total resistance at a twist angle of 1.0 degree.

1. Introduction

Large-diameter bored piles are commonly used in groups to support high-rise buildings or bridges. These piles have to resist vertical, lateral, moment and torsional loadings transmitted from the superstructure. Vertical loads include dead weights of the superstructure and various vertical live or transient loadings. Lateral loads include horizontal loads from various sources such as earthquakes, wind effects, high-speed vehicles, ocean waves, and ship impacts. Moment loads are usually caused by eccentricity of vertical loadings and torsional loads are usually caused by eccentricity of lateral loadings. The response of single piles and pile groups subjected to vertical, lateral and moment loadings have been studied by many researchers. However, only limited study on the response of torsionally loaded pile groups has been reported in the literature.

Poulos (1975) presented an elastic approach to analyze the response of a single pile subjected to torsion. Randolph (1981) presented closed-form solutions for calculating the torsional stiffness of a pile. Chow (1985) presented a discrete element method for the analysis of torsionally loaded single piles in non-homogenous soils. The pile-soil system is discretized into a series of elements connected at the nodes. Yet, little systematical study on torsionally loaded pile groups has been reported in the literature. Such study is required as most piles are used in groups.

It is preferable to study the response of torsionally loaded large-diameter bored pile groups by full-scale tests. However such tests are likely more costly than conventional axial load tests or lateral load tests. In this paper, three-dimensional finite-difference analyses are conducted to investigate the torsional response of a single pile and a fixed-head, large-diameter bored pile group. Using the numerical model, the response of the torsionally loaded pile group is investigated in detail, including the relationship between the pile group twist angle and the applied torque, transfer of the applied torque between the lateral resistance and the torsional

* This work is supported by a grant from the Research Grants Council of the Hong Kong Special Administrative Region (Project No. HKUST6037/01E).

resistance of the group piles, and variations of internal forces in the individual piles along depth.

2. Numerical Model

2.1 *Boundary conditions*

A single large-diameter bored pile and a 2x2 large-diameter bored pile group are studied in this paper. The piles are 2 m in diameter and 13.2 m in length. For the pile group, the pile spacing is 6 m, i.e., three times the pile diameter. The pile cap is 2.5 m thick and fixedly attached to the piles. The three-dimensional (3D) finite-difference program Flac3D (Itasca 1997) is used in the study. Figs. 1 and 2 show the 3D views and the cross sections of the single pile and the pile group models. The single pile model consists of 8856 zones and the pile group model consists of 21024 zones. The dimensions of the models are 20 m in depth, 40 m in length, and 40 m in width. The origin of the coordinate axes is located at the center of the models at the ground surface, with the z-axis pointing upward. The z-displacement boundary is fixed at the base of the model (z = -20 m). Similarly, the x-displacement is fixed at boundaries x = -20 m and x =20 m and the y-displacement is fixed at boundaries y = -20 m and y = 20 m. For simplicity, the ground water table is assumed to be far below the soil layer.

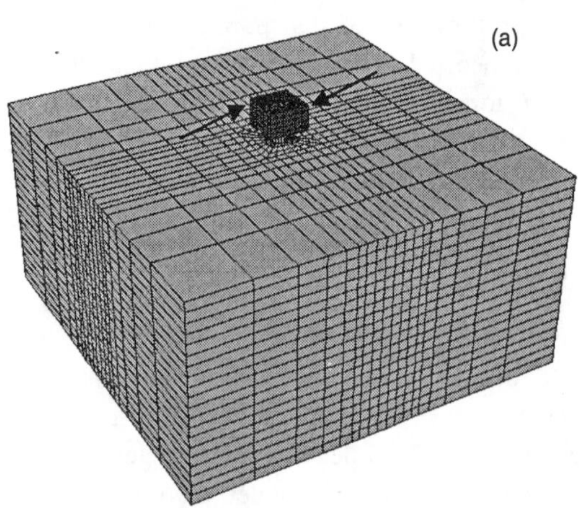

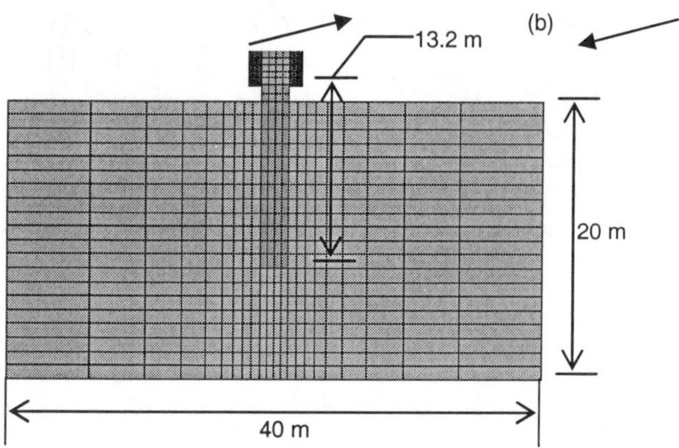

Figure 1. (a) 3D view and (b) cross section of the torsionally loaded single pile.

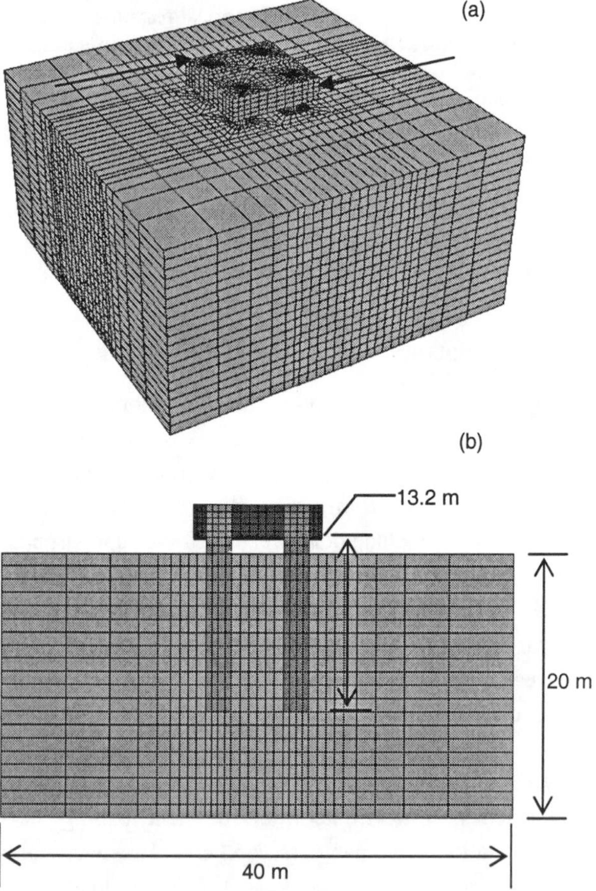

Figure 2. (a) 3D view and (b) cross section of the torsionally loaded pile group.

2.2 *Constitutive models and model parameters*

The concrete piles and the pile cap are assumed to be linear elastic. For the foundation soil, an elastic perfectly plastic model following the Mohr-Coulomb yielding criterion with a tension cutoff has been employed. In addition, the contact between the pile and the soil is modeled by interface elements to represent the possible separation and slippage between the pile and the soil. The interface elements are characterized by sliding and/or tensile and shear bonding (Itasca 1997). The parameters for the interface model include interface friction angle, cohesion, dilation, normal and shear stiffnesses, and tensile and shear bond strengths. During each time step, the absolute normal penetration and the relative shear velocity are calculated for each interface node and its contacting target face. Both of these values are then used by the interface constitutive model to calculate a normal force and a shear-force vector. The Coulomb shear-strength criterion limits the shear force F_{smax} by the following relation,

$$F_{sman} = cA + F_n \tan\phi_i \tag{1}$$

where c is the cohesion along the interface, A is the representative area associated with the interface node, ϕ_i is the friction angle of the interface surface and F_n is the normal force. When the shear force exceeds the maximum shear force, sliding will occur and the shear displacement may cause an increase in the effective normal stress on the joint according to

$$\Delta\sigma_n = \frac{|F_s| - F_{sman}}{Ak_s} k_n \tan\psi \tag{2}$$

where ψ is the dilation angle of the interface surface; $|F_s|$ is the magnitude of shear force before the above correction is made; k_s and k_n are the shear stiffness and normal stiffness, respectively.

The soil is assumed to be homogenous. The Poisson ratio v and Young's modulus E of the soil are assumed to be 0.3 and 30 MPa, respectively, which represent the properties of a dense completely decomposed granite commonly found in Hong Kong. An internal friction angle $\phi' = 38^\circ$ and a unit weight of γ = 1700 N/m^3 are adopted following guidelines by GCO (1982).

Grade 45 concrete is commonly used for construction of bored piles and pile caps. The modulus and Poisson's ratio are therefore E =26 GPa and v =0.3, respectively. The unit weight of concrete is assumed to be 2500 N/m^3.

The behaviour of the interface elements is not sensitive to the normal stiffness and the shear stiffness (Itasca 1997). The stiffness values are set to be very high, i.e., 30 GN/m (Itasca 1997). In order to be consistent with the soil model, the cohesion, tensile strength and angle of dilation of the interface model are all specified as zero. The interface friction angle ϕ_i is assumed to be 0.8 times the soil's internal friction angle (GCO 1982)

2.3 *Construction and loading processes*

The numerical simulation is divided into three major steps; namely, pile bore excavation, formation of piles and the pile cap, and torsional loading. After reaching the initial equilibrium state, excavation and pile installation are simulated. The soil in the pile zones is removed and replaced with the pile material. Then the piles are generated, with dimensions 2 m in diameter, 13.2 m in length, and 1.2 m in the free length above the ground surface. This 1.2 m free length prevents the interaction between the pile cap and the ground. After that, the pile cap is put on top of the piles. The pile cap is designed to be 2.5 m thick and have 1 m over-hang on each side. Therefore, the dimensions of the pile cap are 4 m in length, 4 m in width and 2.5 m in depth for the single pile and 10 m in length, 10 m in width and 2.5 m in depth for the pile group. The center-to-center spacing of the piles in the group is three times the pile diameter, which is 6 m. The detailed geometry of the torsionally loaded pile group model can be referred to Fig. 2. Finally, clockwise torque is imposed on the pile cap by applying two point loads at the corners of the pile cap diagonally (see Figs. 1 and 2).

3. Response of Single Pile under Torsion

The rotation of the pile under torsion may result in shear failure of the adjacent soil and torsional failure of the pile itself. It will also increase the torsional movements of the superstructure the pile supports. Fig. 3 shows the pile-head twisting angle of the single pile versus the applied torque. The curve is highly nonlinear with a distinct turning point at a twisting angle of approximately 0.01 radian. At this twist angle, the tangential pile movement is 10 mm. When the pile torsion exceeds this value, the torsional stiffness will decrease substantially.

Fig. 4 shows the variations of twisting angle along the depth of the single pile under three different applied torques 2, 6 and 8 MNm. The pile-head twisting angle is only slightly smaller than the pile toe twisting angle, with only a small relative rotation of the pile shaft. This is valid for the relatively rigid pile studied in this paper, but may not be valid for very long and flexible piles. Because of the nonlinear response of the pile, the increment in pile rotation when the applied torque increases from 6 MNm to 8 MNm is significantly greater than that from 2 to 6 MNm.

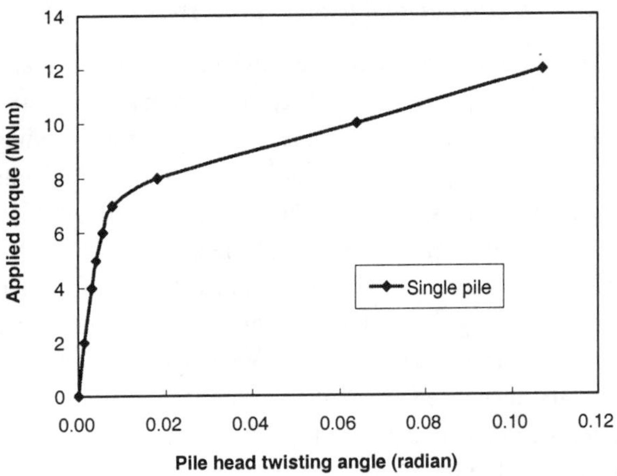

Figure 3. Twisting angle at pile head versus applied torque, torsionally loaded single pile.

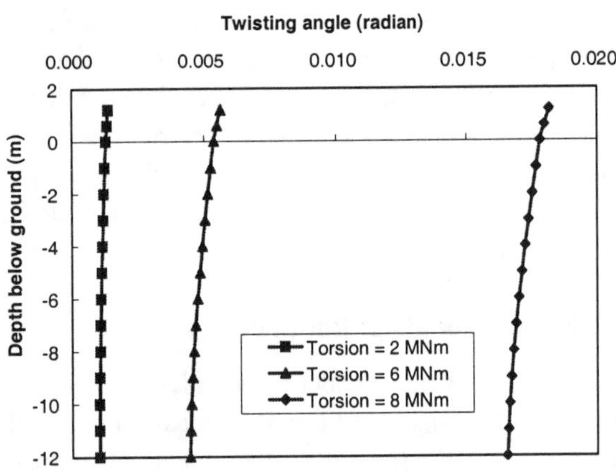

Figure 4. Variation of twisting angle along the shaft of a torsionally loaded single pile under torques 2 MNm, 6 MNm and 8 MNm.

The torsional resistances provided by the soil around the pile are also found and plotted against the pile head twisting angle in Fig. 5. The total torsional resistance comprises a resistance component from the pile shaft and a resistance component from the toe. The shaft resistance is mobilized at a twist angle of about 0.01 radian. At this angle, the mobilized toe resistance is minimal. At larger twist angles, the additional applied torque is essentially resisted by continued mobilization of the toe resistance. Yet the toe resistances at large twist angles are still small. The torque provided by the base is only about 1% at a twist angle of 1.0 degree.

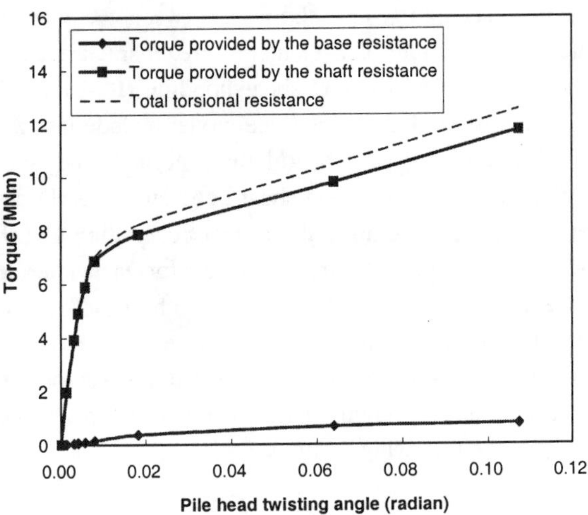

Figure 5. Torsional resistances provided by the shaft and by the base.

4. Response of Pile Group under Torsion

Using the numerical model, the response of the torsionally loaded pile group is investigated in detail, including the relationship between the pile-group twist angle and the applied torque, movements of the soil ground, transfer of the applied torque between the lateral resistance and the torsional resistance of the group piles, variations of internal forces along depth, as well as stresses in soils around the group piles.

4.1 *Pile group movements*

The movements of a single pile and a pile in a pile group under torsion are quite different. When a torque is applied to a single pile, the pile is only subjected to pure torsion. On the other hand, when a pile group is subjected to torsion, the piles in the group will twist and deflect in the torsional direction. Fig. 6(a) shows the relationship between the applied torque and the twisting angle of the pile group. If the pile-cap connection is rigid, then it can be proved that the twist angle of the individual piles is the same as the twist angle of the pile cap. It can be seen in Fig. 6(a) that the relationship is nonlinear, but no distinct turning point as that shown

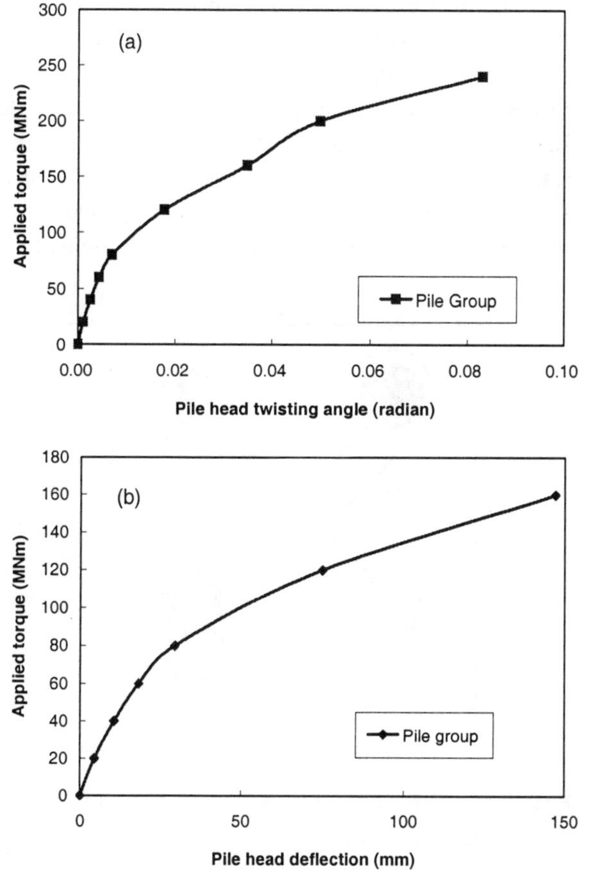

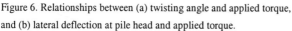

Figure 6. Relationships between (a) twisting angle and applied torque, and (b) lateral deflection at pile head and applied torque.

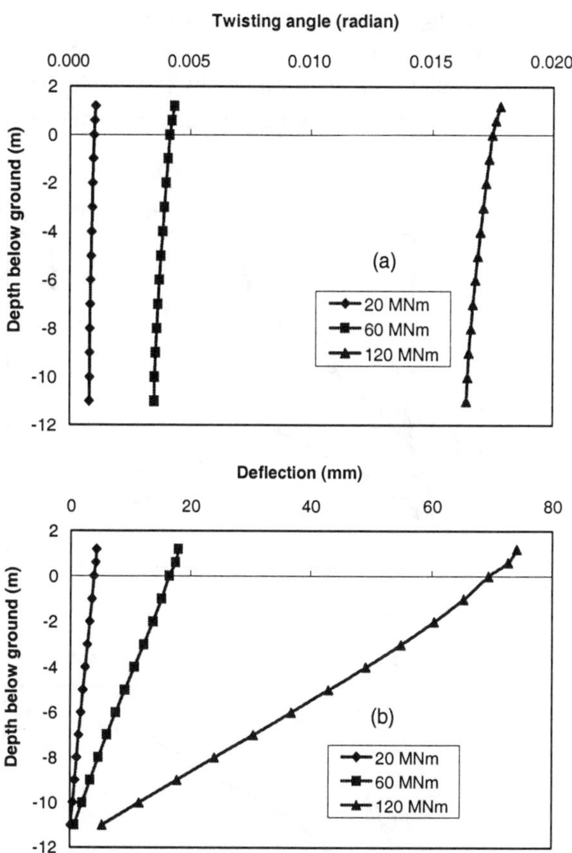

Figure 7. Changes of (a) twisting angle and (b) lateral deflection along depth in a torsionally loaded pile group under applied toques of 20, 60 and 120 MNm.

in Fig. 5 can be found. The piles in the group not only twist but also deflect laterally along the tangential direction of the pile cap movement. Fig. 6(b) shows the relationship between the pile head deflection and the applied torque. In Figs. 6(a) and (b), the pile head deflection is about 150 mm when the twisting angle is 0.035 radians (2 degrees).

The variations of the twisting angle and the lateral deflection along the pile under applied torques of 20, 60 and 120 MNm are shown in Figs. 7(a) and (b). Similar to the torsionally loaded single pile, the relative twisting angle of the piles in the group with respect to the pile toe is very small comparing with the pile toe twisting angle. As a result, the pile toe twist angle is rather large at high torques. On the other hand, Fig. 7(b) shows that the lateral deflection of the piles decreases with the depth and the deflection at the toe is minimal. The curvature near the pile head is large. Accordingly, large bending moments and shear forces at the pile head are expected.

4.2 *Transfer of applied torque into the ground*

Following the discussions on the pile-head movements of torsionally loaded piles in the group, the internal force of piles are also studied. Fig. 8 shows the distributions of the internal shear force and bending moment with the depth. Since the four group piles deflect in the clockwise direction, shear forces and bending moments develop in the piles. The distributions of shear forces and bending moments along the depth at applied torques of 20, 60 and 120 MNm are shown in Figs. 8(a) and (b), respectively. The shear force at a pile section is calculated from the stresses acting on the cross section and the bending moment at the section is then calculated from the shear force distribution. It can be seen that both the shear force and the bending moment decrease with depth. The maximum shear force and bending moment all occur at the pile head and, as expected, the bending moment at the pile toe is zero.

The transfer of the applied torque on the pile group into the ground can be illustrated by the simple diagram shown in Fig. 9. When two point loads are applied to the pile cap, each of the group piles is subjected to a torque T_S and a lateral shear P. If the applied load is F, the total applied torque will be $T_G = FL$ where L is the moment arm of the applied point loads. The group torsional resistance T_G consists of the torsional resistance provided by the four individual piles $4T_S$, and the torsional resistance provided by the lateral resistance of the four piles $4Pl$:

$$T_G = 4T_S + 4Pl \tag{3}$$

where l is the distance between a pile and the center of the pile group.

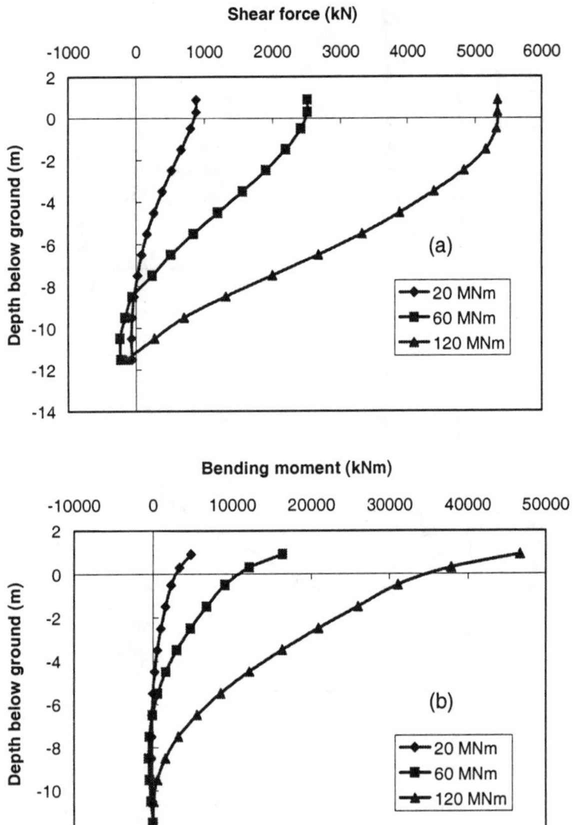

Figure 8. Changes of (a) shear and (b) moment with depth in the torsionally loaded pile group.

The two resistance components are plotted against the pile-head twisting angle in Fig. 10. It can be seen that both the lateral resistance and the torsional resistance of the pile group increase at twisting angles smaller than 0.01 radian. At this angle the torsional resistance provided by the four piles has been substantially mobilized. After that, the applied torque increments are resisted by continued mobilization of the lateral pile resistance. At a twist angle of 1.0 degree (0.01 radian), the torsional resistance of the individual piles contributes to approximately 30% of the total resistance.

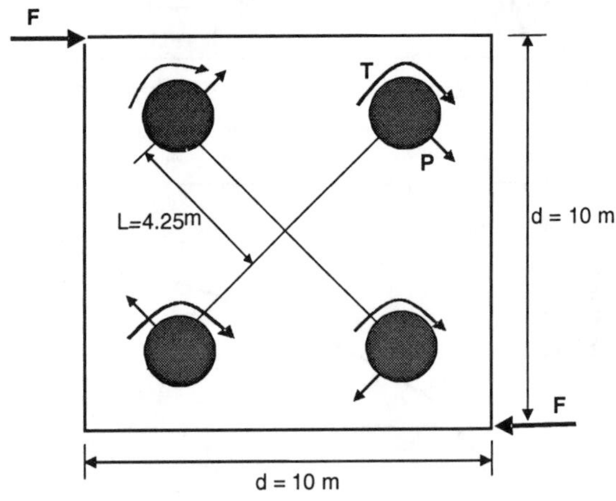

Figure 9. Simple diagram to show the load transfer mechanism of the torsionally loaded pile group.

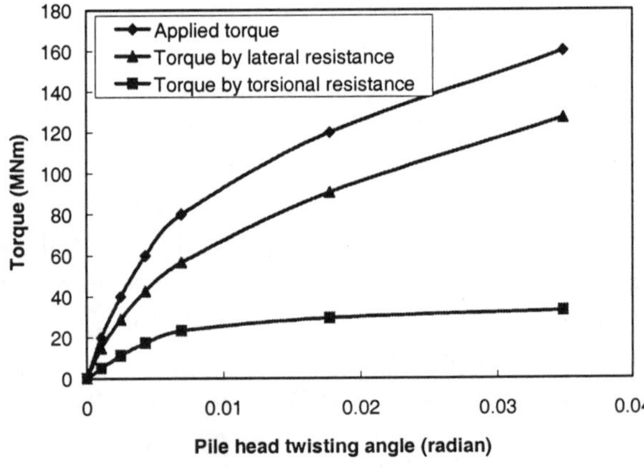

Figure 10. Torsional load transfer in the torsionally loaded pile group.

5. Conclusions

Numerical analyses are conducted to simulate the response of a single pile and a pile group subjected to torsional loading. The torsional movements, sharing of the applied torque between the lateral resistance and the torsional resistance of the piles, and the variations of internal forces along depth are investigated.

For the relatively rigid piles studied, the relative rotations of the torsionally loaded single pile and the groups are small. Yet, the movements of the single pile and the pile group under torsion are quite different. The single pile only shows torsional response while the group piles not only twist but also deflect in the horizontal direction.

The torsional resistance of a single pile is mobilized at a small twisting angle of about 0.01 radian. The torsional resistance is mostly from the shaft resistance. The base resistance is only 1% of the total resistance at a twisting angle of 1 degree.

In the torsionally loaded pile group, the applied torque is resisted by a torsional component and a component contributed by the lateral resistance of the group piles. The torque-twist curve is less nonlinear compared with single pile curve. The torsional resistance provided by the four individual piles is mobilized first. After that, the applied torque increments are resisted by continued mobilization of lateral pile resistance. At a twist angle of 1.0 degree, the torsional resistance of the individual piles contributes to approximately 30% of the total resistance.

References

1. Chow, Y.K. (1985). Torsional response of piles in nonhomogeneous soil, *J. Geotech. Eng.,* ASCE **111**(7), 942-947.
2. GCO (1982). *Guide to Retaining Wall Design, Geoguide 1.* 1993 edition, Geotechnical Control Office, Hong Kong.
3. Itasca (1997). FLAC3D *Fast Largragian Analysis of Continua in Three Dimensions: Version 2.0: User's Manual,* Itasca Consulting Group Inc., Minnesota, U.S.A.
4. Poulos, H. G. (1975). Torsional response of piles. *J. Geotech. Eng.,* ASCE **101**(10), 1019-1035.
5. Randolph M. F. (1981). Piles subjected to torsion. *J. Geotech. Eng.,* ASCE **107**(8), 1095-1111.

DESIGN AND CONSTRUCTION OF THE JOINT NODE USED IN THE DEEP FOUNDATION OF TOP-DOWN CONSTRUCTION METHOD

QI ZHAO
School of Civil Engineering, Tongji Universiy, 1239 Siping Road, Shanghai 200092, China

JIANBIN XIE
School of Civil Engineering, Tongji Universiy, 1239 Siping Road, Shanghai 200092, China

Top-down construction method is a rapidly developing construction technology which used in the deep foundation of high-rise Buildings. As the diaphragm wall or the row of piles can act as both the basement wall and the embosom, and the strutbracings are the floor slabs underground. So the joint nodes of pile and column, column and slab, slab and diaphragm are difficult to deal with. In this article, some practical and effective joint nodes are introduced through some project practice.

1. Introduction

Top-down construction method is a rapidly developing technology which used in the deep foundation of high-rise buildings. This technology has been great utilized in recent years both domestic and overseas. In this kind of construction method, the slurry wall or the row of piles can act as both the basement wall and the embosom, and the strutbracings are the floor slabs underground.

The top-down construction method is an opposite direction construction method from up to down when construct the underground structure. And it has the advantages of shortening the work period, reducing the costs, decreasing the deformation of the foundation pit, decreasing the influence on the environment during the construction of underground structure. As the top-down construction method is a special method different from the usual construction method. So the joint nodes of column and slab, slab and slurry wall are difficult to deal with. In this article, some practical and effective joint nodes are introduced through some project practices.

2. The Design and Construction Demands of the Joint Nodes used in Top-down Construction Method

The joint nodes used in top-down construction method should meet the requirements as below:

- The design of the joint nodes should meet the demands of the state in which the structure withstand the permanent load, also they should meet the demands of the state in which the structure bear the construction load. And the design of the joint nodes should be on the basis of both structure design code and construction condition.
- The design of the joint nodes should be simple and practicable. These joint nodes should meet the demands of construction capability on precondition of bearing loads.
- The joint nodes should have good inpermeability and can be waterproofing.
- The joint nodes should not influence the function of use of the building, for instance they can not take up too big space.

3. The Joint Nodes of Support Column and Slab and Beam

In top-down construction, the design and construction of the joint nodes of support column and beam and slab are mainly overcome the difficulties as below. How the reinforcement bars pass the middle support column and how they connect with support columns to guarantee that after the complex node is finished, the node quality and distributed internal force are unanimous with the designed and calculated ones. This nodal design and construction depend on the structure form of the middle support column mainly.

When the support column adopt the form of steel lattice column, the middle support column is connected with reinforcement bars of beam by borehole reinforcing bar connect law or transfer steel plate connect law. The process of borehole reinforcing bar connect law is to hole on the middle support column steel section where the reinforcing bar pass through the middle support column, then the reinforcing bar pass through. This kind of connect method has the advantages of simple form and easy to pour the concrete of the joint node. But the disadvantage is that the holes have weaken the steel section of the column and it will reduce the bearing capacity of the support column. The process of transfer steel plate connect law is to weld transfer steel plates on the flange of the lattice column where the reinforcing bars connect the middle support column, then weld the reinforcing bars to the transfer plates to form the joint node which can transter the force(1). This method can solve the problem that the reinforcing bars are too many to pass through and too many hole to localization. But it also has its own shortcoming. Large material consumption, construction craft is complicated, welding seam below the transfer plate is difficult to weld, the concrete of the joint node are difficult to pour and its quality is difficult to guarantee, and when pour this part of concrete, it is need to hole air vents on the steel plate.

When the middle support column adopts the form of the concrete post of steel tube, the above two methods can also be used. Mainly the latter. Forms adopted while adopting the latter are a lot of, the commonly used are two kinds of forms of steel corbel and adaptor. And also the joint node should transfer bending moment and shear force. The steel corbel form is to weld steel corbel on the steel tube surface, and then connect the reinforcing bars and the tube by this steel corbel. Another form is to weld a ring type stencil plate on the tube surface to transfer the shear force. And the reinforcing bar adaptors directly weld on the tube surface to transfer the bending moment.

These two method also have heavy defect in nowadays construction. First, welding can affect the bearing capacity of the steel tube, because welding is carry out on the spot while the steel tube is under load. Second, it can influence time limit for a project, the quality and the cost. Because each reinforcing bar needs a steel corbel or a adaptor, the amount of reinforcements used in underground structure is too large, it becomes the key circuit which influences time limit for a project that the node is dealt with, the cost invested is high too. It is relatively has bad condition to weld underground, the quality is difficult to control also. Third, a lot of welding work would influence the environment of the construction site. Because a lot of waste gas will discharge and the underground ventilation condition is bad. So a new type of joint node using single column and double beams combining structure node begins the application in the project too. This nodal design principle is to make node simple and can factory-made.

For example, YouYou Plaze is covered an area of more than 30,000 square meters, two floor underground, the depth of the basement is up to about 17m, the foundation pit plane is an irregular polygon. Project adopt half top-down construction method. The embosom adopts the compound form mixed bored piles act as retaining wall and stirring piles act as waterstop coat. The middle support columns use 450*450mm steel lattice columns. The joint nodes of support column and beam used in this project adopt borehole reinforcing bar connect law. The detail is as below, see Figure 1. Some principles we have to obey in the project construction:

- Reinforcing bars of the beams must be continuous-vision arranged and should be anchored into support according to designing requirement, forbidden cutting off directly in the lattice column place.
- Let the reinforcing bars cross the support column by the side of the lattice column, in crossing nodal place proper haunching treatment can be adopted.
- When the beam reinforcing bars are too intensive to cross the column by its side, proper holes and notches can be done on the angle steel of the lattice column under some strengthening measures according to the design demand.

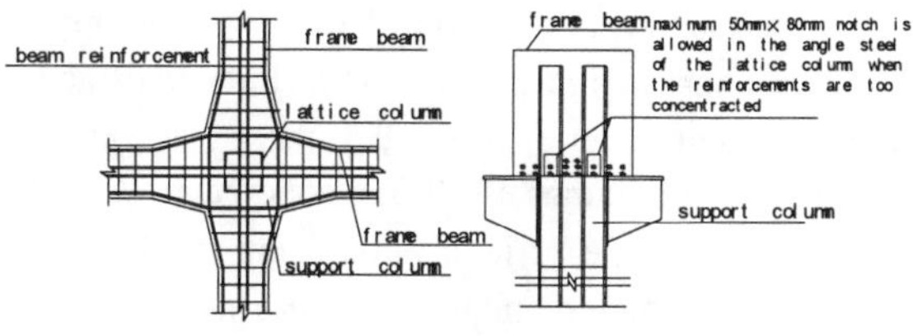

Figure 1. The joint node of lattice column and beam used in YouYou Plaza project

For example, XingYe Mansion lies in No.183 block in Huangpu district, covers an area of 7856. 8 square meters, 19 storeys of the main building, height is 82. 5 meters, three floor underground, the depth of the basement is up to about 14m. The main building has the whole construction area of 55783 square meters on the ground, the underground part is with the whole construction area of 18889 square meters. This project adopt top-down construction method. The slurry wall acts as both basement exterior wall and the embosom. The middle support columns use Φ609 concrete post of steel tube. The joint nodes of support column and beam used in this project adopt "T" shape steel corbel to connect reinforcing bars and the steel tube. The detail is as below, see Figure 2.

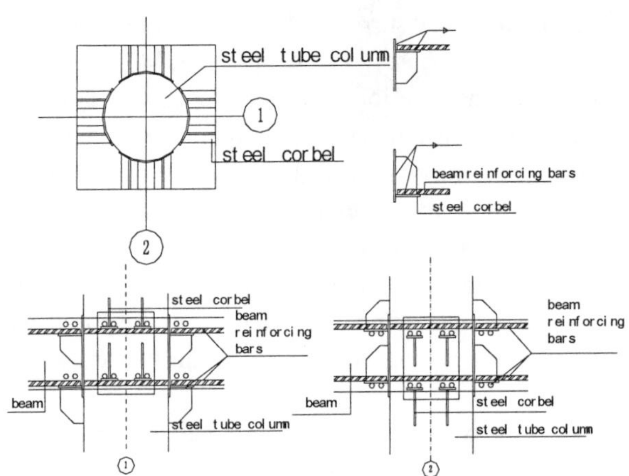

Figure 2. The joint node of steel tube column and beam used in XingYe Mansion project

Now comes the new type of the joint nodes used in ChangFeng Plaze. This project lies quite near to the Zhongshan Park station of Subway Line No. 2. It covers an area of 22000 square meters, 60 storeys of the main building, four floor underground, the depth of the basement is up to about 24m. This project adopt half top-down construction method. The slurry wall acts as both basement exterior wall and the embosom. The middle support columns use Φ550 concrete post of steel tube. So a new type of joint node using single column and double beams combining structure node begins to application in this project. The process is as below:

- When the steel tubes are manufactured in the factory, the Φ22 ring arranged reinforcing bars and their anchor reinforcements are welded on the surface of the steel tube according to its dimension. And these tubes are then used as the support column. When the connect work should be done, the Φ22 ring arranged reinforcing bars will be bended to 90 degree to connect the reinforcing bars in the double beams. This kind of design can make the node transfer the shear force.
- The design to make the node transfer the bending moment is to change the original single beam to double beams pass across by the side of the steel tube to ensure the reinforcing bars keep continuous across the joint node. This kind of node is simple and easy to control and can reduce cost and can guarantee the quality of the node. The detail is as below, see Figure 3.

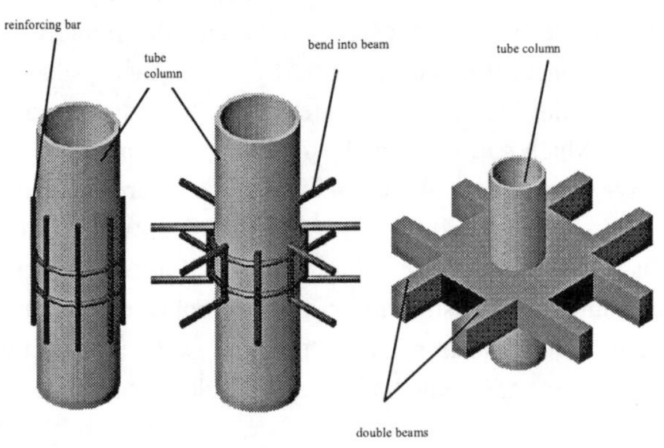

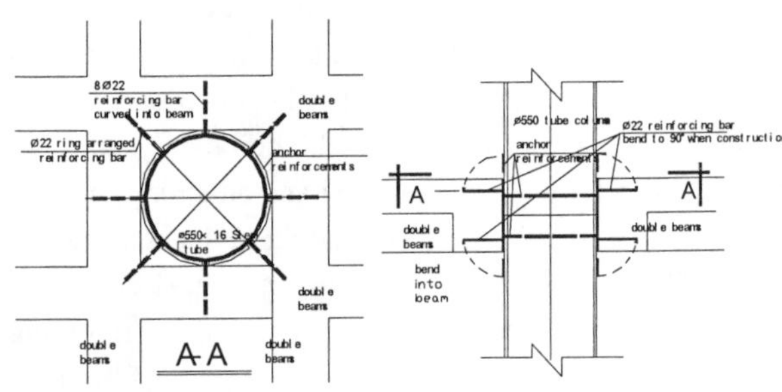

Figure 3. The joint node of steel tube column and beam used in ChangFeng Plaze project

4. The Joint Nodes of Slab (Beam) and Embosom Wall

In the top-down construction method, the strutbracings are the floor beam and slabs underground. In the structure design, this node is considered as consolidated node. So the reliability of this node is quite important. The rigid joint, the hinge joint and half rigid joint are usally used (2). If the slab (beam) and the embosom wall have proper rigidity, and the joint node bear great bending moment, and the slab (beam) can have enough reinforcing bars to connect, the rigid joint can be used. The common form is to embed reinforcing bars or adaptor. If the slab (beam) is much thinner than the embosom wall, and the joint node bear small bending moment, the hinge joint can be used. The common form is to embed reinforcing bars or shear force adaptor, the surrounding side beam can also be used. If the slab (beam) and the embosom wall have proper rigidity, and the joint node bear great bending moment, but the slab (beam) can not have enough reinforcing bars to connect, the half rigid joint can be used. The common forms are as follow: embedding reinforcing bars to connect; embedding steel plates to connect; embedding shear force adaptor to connect; profile adaptor to connect; tapered screw reinforcing bars to connect.

For example, In the construction of ChangFeng Plaze, the surrounding side beams are used to connect slurry wall and the beam(slab), Φ 12@200 two lines of reinforcing bars are embedded in the slurry wall, when construct the slab, the embedded reinforcements are bended and welded into the side beam. The detail is as below, see Figure 4.

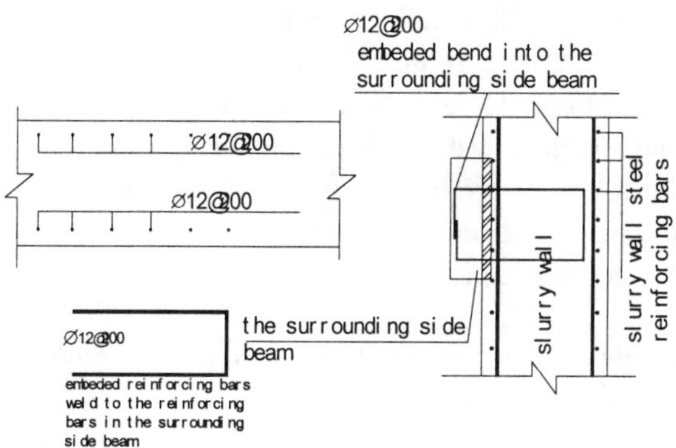

Figure 4. The joint node of slab(beam) and slurry wall used in ChangFeng Plaze project

Same joint nodes are used in in XingYe Mansion project. The detail is as below, see Figure 5.

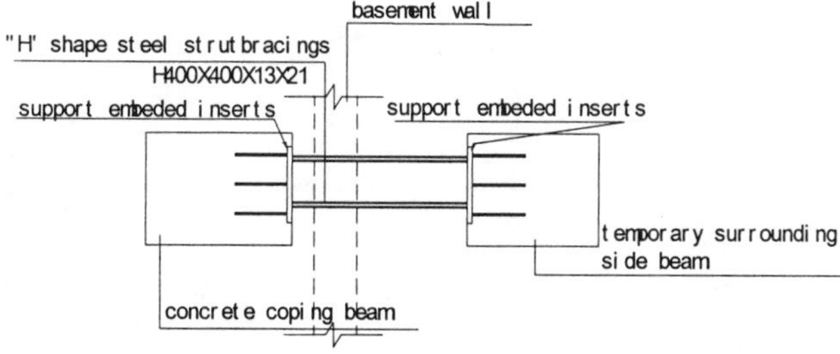

Figure 5. The joint node of steel tube column and beam used in XingYe Mansion project

For another example, in YouYou Plaze project, temporary boundary beams are designed in the side span of the frame. These boundary beams have some distance from the basement wall. During top-down construction course, these beams connect the wale frame by "H" shape steel. These horizontal shape steel can transfer horizontal strutbracing force. And some reinforcements are embedded in these boundary beams to connect with other frame beams. When the top-down construction course end, and the basement wall begin to build from down to top, some holes are allowed in walls and the shape steels to let them pass through. The detail is as below, see Figure 6.

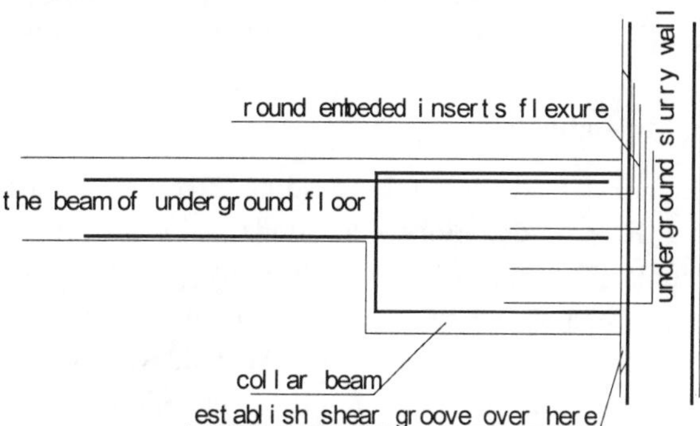

Figure 6. The joint node of slab(beam) and slurry wall used in YouYou Plaze project

5. Conclusion

These joint nodes applied in the deep foundation of top-down construction method have developed constantly through the practice in projects. The only aim is try to set up various kinds of reasonable and simple connection between different structures in top-down construction method. At present, these nodal forms have guaranteed the development and application of the top-down construction method.

References

1. Zhao Xi Hong, Lu Rui Ming, Chen Zhi Ming & Gong Jian, Non-linear space theory &method for excavation engineering &application. 2001 Proc. 15th Int. Conf. on SMGE, Istanbul, Turkey, 27-31 Aug. Vol. 2.138~141.
2. Zhao Xi Hong, Chen Zhi Ming, Gong Jian, Zhou Guo Ran, Zhang Qi Hui & Li Bei Deep Excavation for Tall Building in Shanghai, 2000, Proc. Int. Symposium of Civil Engineering in 21st Century, Beijing, China, 11~13 Oct: 517~525.

DESIGN PRINCIPLE FOR LARGE-AREA THICK RAFT FOUNDATION UNDER TALL BUILDINGS

J.F. GONG

Institute of Foundation Engineering, China Academy of Building Research
Beijing,100013, China

X.L. HUAN Y.J. TENG

Institute of Foundation Engineering, China Academy of Building Research
Beijing,100013, China

Model test and in-situ monitoring results show that in common structure form, large-area thick raft foundation can diffuse the load of tall building but in the limited way. By the interaction of supper structure, thick raft and soil, the effect of the tall building on its podium is limited to the area around the main tower, where an effective range of interaction is produced with the tall building focused. Design for the thick raft foundation under multiple tall or low buildings shall take the characteristic into full account.

1. Introduction

Technical Code for Tall Building Box Foundations and Raft Foundations, the trade standard of the People's Republic of China, JGJ6-99, stipulates that local bending moment can be calculated with the coefficient of reaction given by the code when design of box foundations and raft foundations, while overall bending moment can be ignored. This approach takes the effect of upper structural rigidity into consideration; in calculation, it attaches great importance to the evenness of settlement and settlement amount. For tall buildings with total bending of 0.1 to 0.3‰ and a basement bottom surface almost same with the projected area of superstructure, the practice in the past few years proves it to be correct.

With the development and utilization of the underground space, large-area frame structure with thick raft foundation under tall buildings with a bottom area of over 10,000 square meters is widely used. On such foundation, various stories of tall buildings are built and there is great difference of load and structural rigidity between tall buildings and podium.

With the interaction of the superstructure, the whole basement is unable to form the integral rigidity, so the previous calculation methods, which lay particular stress on settlement uniformity and settlement magnitude, are unable to precisely solve the problem of contact pressure and settlement within the overall scope of raft.

2. Model Test Research

To solve the above-mentioned problem, based on existing data and structural characteristics, serious of indoor large-scale model tests had been done in Institute of Foundation Engineering, China Academy of Building Research. The purpose of the tests is to study the internal force of raft, load dispersion range through frame structure with thick raft foundation under local vertical load, interference between tall buildings and the deformation characteristics of large-area frame structure with thick raft foundation.

2.1 *The pressure diffusion of thick raft foundation under local vertical load*

To study the dispersion range and characteristics of the tall building, the tests[1] were carried out for the circumstances when the two-story frame structure with thick raft foundation at two ends of tall building were of one, two and three spans respectively.

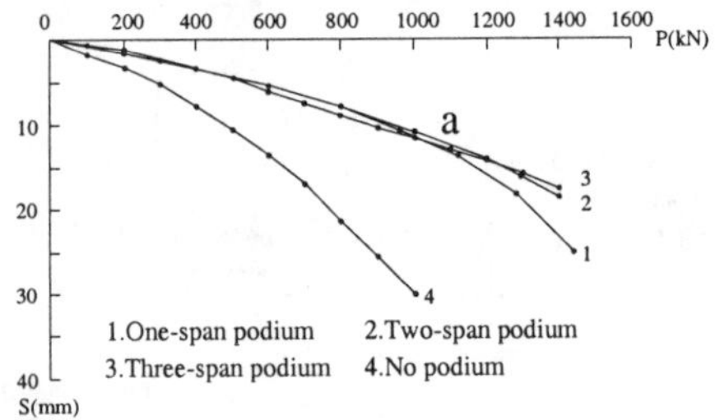

Figure 1. Curves of various model loads-maximum settlement measured values

From the tests, it can be seen:

When the podium at two ends of tall building was one span, the contact pressure of the foundation was evenly distributed, which is indicated in Fig. 2. It shows that the two-story frame thick raft had good load-transfer capability. The pressure could be calculated on average with the area of further adding a span when load value was not over the proportional limit value got from loading test (point a in Fig. 1).

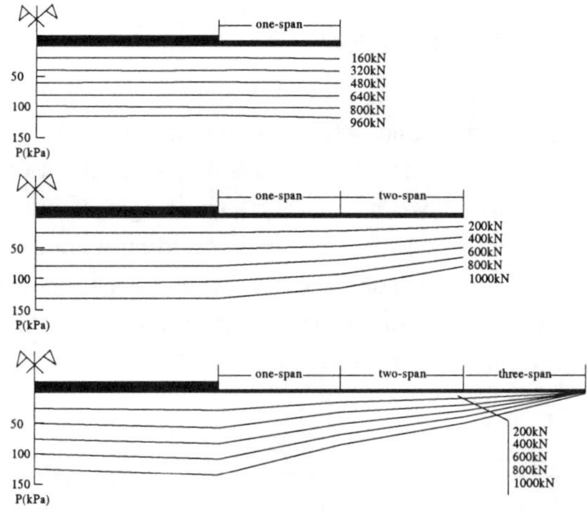

Figure 2.Curve of model test foundation pressure

When the podium at two ends of tall building was three spans, the contact pressure was transferred to the third span and decreased to zero proportionally as shown in Fig. 2; the value of contact pressure diffused was similar to that when the podium was one span.

This test shows that the dispersion through the thick raft foundation for the tall building load is limited to a certain extent under serviceability limit state.

2.2 *Interference between tall buildings*

Table 1.

Test Number	Loading Path	Load of Tall building A (kN)	Load of Tall building B (kN)	Remark
1	Synchronic Loading	0→800	0→800	Curve 1
2	Loading A	800→1600		Curve 2
3	Loading B		800→1600	Curve 3

The effects of settlement and pressure of a tall building on that of the neighboring one are extremely important factors in foundation design. The effects may be reduced due to the pressure diffusion in thick raft foundation. The tests[2] of two-abreast-towers and two-diagonal-towers set out on the two-story frame structure with thick raft foundation respectively were carried out for further investigation of the interaction between tall buildings.

Loading path of two-abreast-towers test is shown in Table 1, and the deformation and contact pressure curve is seen in Figure 3.

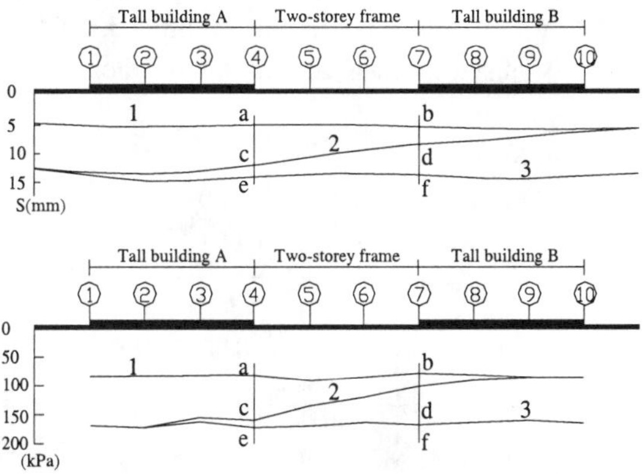

Figure 3.Deformation & contact pressure curve of bi-tall building test with different loading path

The results of two-abreast-towers test show that the contact pressure and settlement of foundation were related to loading path. Settlement and pressure of tall building A had light effect on that of tall building B when the space between two tall buildings was three spans.

Figure 4 is the iso-settlement line diagram for the model test of two-diagonal-towers when the load of two towers A and B is 800kN each. The diagram displays that the subgrade deformation is respectively with the zone in the certain range under the two towers A and B as the centers, radially attenuates outwards and is formed by mutual superposition. Figure 5 is the iso-settlement line diagram of the foundation when the tower A is singularly loaded to 1800kN after unloading. It can be seen that the raft slab deformation is with the zone in the certain range under the tower A as the center and attenuates radially outwards until the deformation value attenuates to zero at a certain distance from the center. This explains that the deformation range of subgrade and foundation under the action of a single tower is only limited to within the neighboring zone with towers as centers.

Results of the above-mentioned tests show that for a large-area thick raft foundation, the deformation of the raft slab is related to the arrangement of the tall building and the magnitude of its loading. Generally, the deformation of the thick raft foundation has the irregular deformation characteristic under various tall buildings as centers of deformation, and the mutual influence between tall buildings is relevant with the load progress. The load of tall building could be dispersed through the frame thick raft structure. The raft slab under the tall building together with its dispersion part is an integral bent plane, its influential range is limited, and this provides requisite for the reaction superposition.

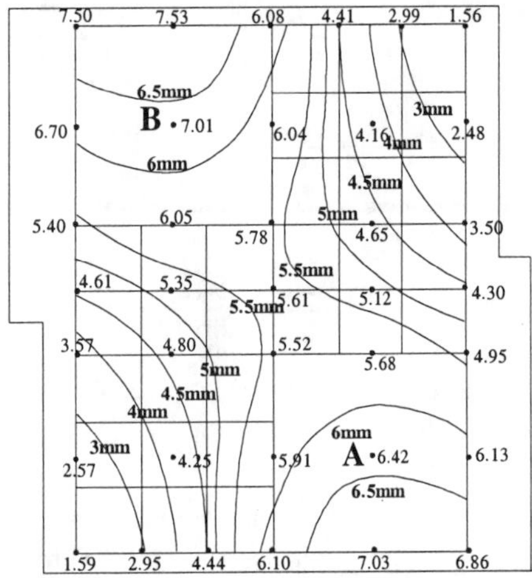

Figure 4 Iso-settlement line diagram of foundation for test.

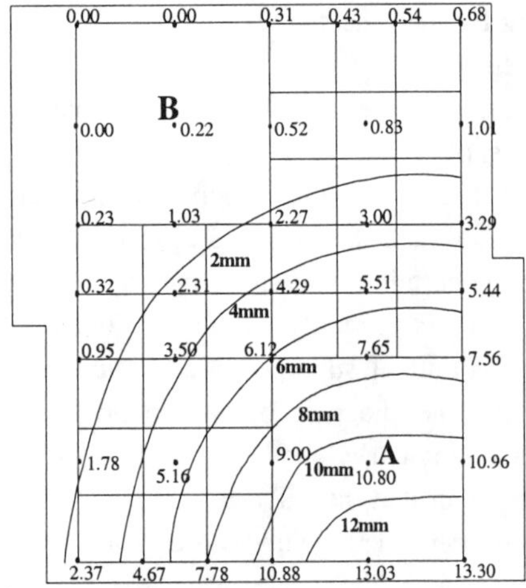

Figure 5 Iso-settlement line diagram of recompressed foundation after rebounce for test.

The adjacent one-span wing podium linking to the tall building can diffuse the load of the tall building, so it can play an important role in adjusting the foundation rigidity of the tall building. Tests show that after the tall building is integrally linked to the podium, the foundation rigidity within the scope of tall building varies from rigid to semi-rigid, and the contact pressure tends to distribute uniformly, so the tall building and its adjacent one-span podium are defined as "the effective range of interaction".

3. Analysis on Engineering In-situ Tests

3.1 *Analysis on settlement measurement of Beijing LG Plaza*

3.1.1 *General information about the project*

Beijing LG Plaza has 31 stories above the ground and 4 stories underground while its podium have 6 stories above the ground, which is a side-by-side twin-tower structure as shown in Figure 6. The height of building above ground is 140m. The main building is reinforced concrete core cylinder enclosed with steel framework. The foundation is of 24-meter deep natural soil raft type. The raft thickness varies as shown in Figure 7.

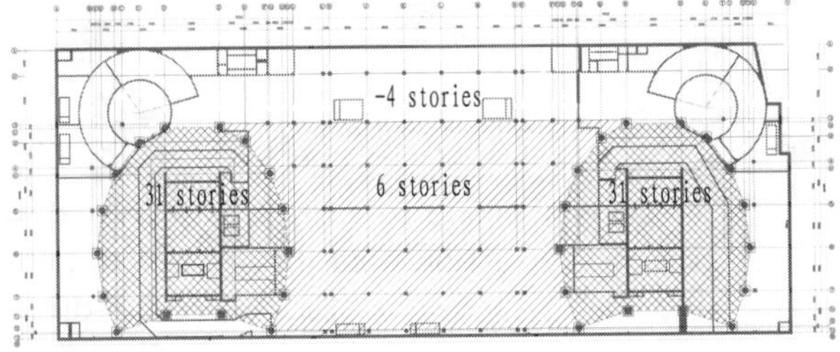

Figure 6 LG Plaza Plan

Figure 7 delay poured strips and the raft thickness varies

3.1.2 *Geological conditions*

The foundation is composed of the alternating of sand, gravel, silty clay and mid-dense to dense common soil of the Quaternary Period.

3.1.3 *Raft's capability to adjust differential settlement*

For solving the differential settlement between tall buildings and podia, two delay poured strips have been set in the large-area thick raft foundation during construction. The delay poured strips are located one span to the edge of each tall building, which can make sure there is a expanded basement under the main building, as shown in Figure 7. To explore the raft's capability to adjust differential settlement, let us make an analysis on the west tower (Figure 8).

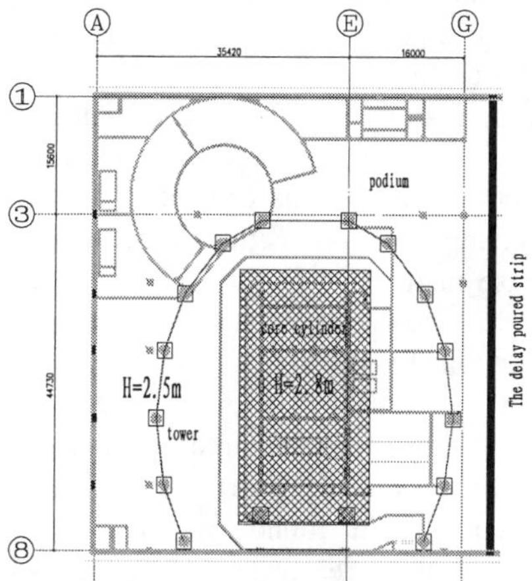

Figure 8 Plan of west tower

The delay poured strip divides the west tower into a single-tower system on a large-area thick raft foundation, and the basement east side extends 6m (one-span)from the edge of the tower, the west side 8.5m (one-span) , the north side 15m (two-span). The distances from the edge of core cylinder to the far ends of east, west and north basements are respectively 19m (two-span), 23m (three-span) and 24m (three-span).

Table 2.

	Loading way 1	Loading way 2	Loading way 3
Basement	–4- story→ 0-story	0-story	0-story
Steel framework outside the core cylinder	–4-story→ 0-story	0-story	0-story→ 31-story
Core cylinder	–4-story→ 19-story	19-story→ 23-story	23-story→ 31-story

According to the construction situation of the site, the loading way for west tower during construction is shown in Table 2. Figure 9 indicates E-axis settlement curves in case of different stage of loading.

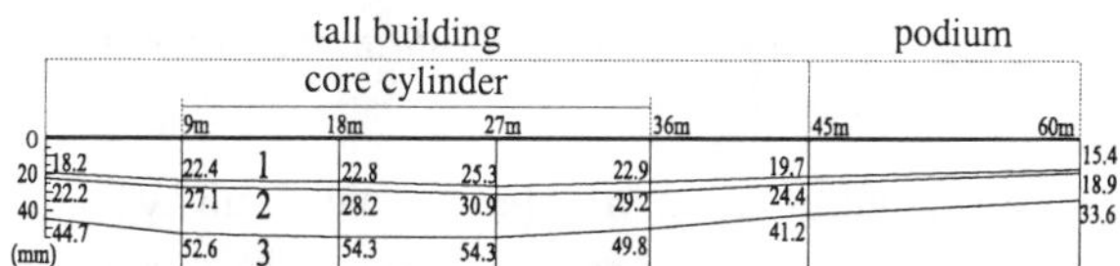

Figure 9 E-axis settlement curves in case of loading way 1, 2, and 3

It can be seen from the settlement curve:

When the basement and the steel framework outside the core cylinder of tall building are constructed to 0-storey, and the core cylinder is constructed to 19-storey(Curve1), from the E-axis settlement curve, it can be seen that the whole deflection of raft is smooth and continuous, the differential settlement of the raft between the edge of core cylinder and the points at 24m far away from the edge of core cylinder is only 7.5mm , the deflection of whole raft is 0.14‰. When the basement and the steel framework outside the core cylinder of tall building are constructed to 0-storey, and the core cylinder is constructed to 23-storey(Curve2), the differential settlement of the raft between the edge of core cylinder and the points at 24m far away from the edge of core cylinder is only 10.3mm and the deflection of whole raft is 0.17‰. In case the loads of the basement and the steel framework outside the core cylinder of tall building remain the same, with the increase of core cylinder load from 19-storey to 23-storey, the settlement of raft far away from the edge of core cylinder is synchronously increasing, its settlement increment is 50% of that at the core cylinder edge, which shows the tall building load is transferred through the raft to the far end (24m) of basement, but settlement attenuation is obvious.

When the tall building roof is sealed (Curve 3), the deflection of raft at E-axis is 0.25‰, and the whole deflection of raft is smooth and continuous. The differential settlement of the raft between the edge of core cylinder and the points at 15m far away from the edge of core cylinder is not more than 1cm. In case the basement load remains the same, with the increase of tall building load, the raft's settlement at the edge of the basement is synchronously increasing, which shows the tall building load is transferred through the raft to the far end of basement (15m).

It can been seen from above descriptions that the distance of tall building load diffused by thick raft can reach 24m (three-span), the in-situ measurement and the model test results show the same load diffusion law.

3.2 *Scope of influence by tall building settlement on the podium*

There is a tall building in Beijing, which with large-area thick raft foundation, it has a three-story basement. The space between the pillar network in the podium is 8.4m× 8.4m. The settlement upon completion is observed as indicated in Figure 10.

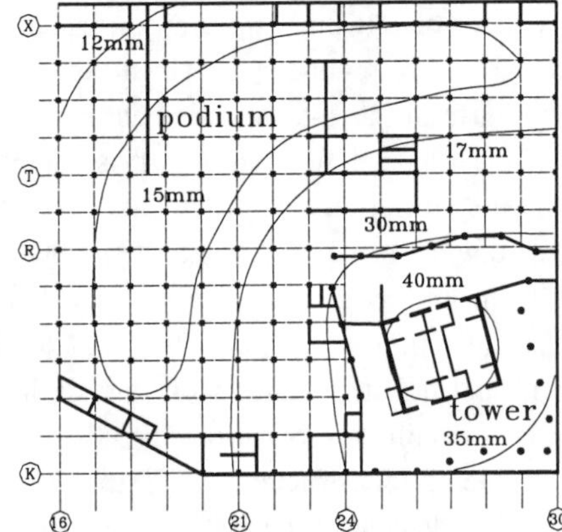

Figure 10 Settlement measurement curves

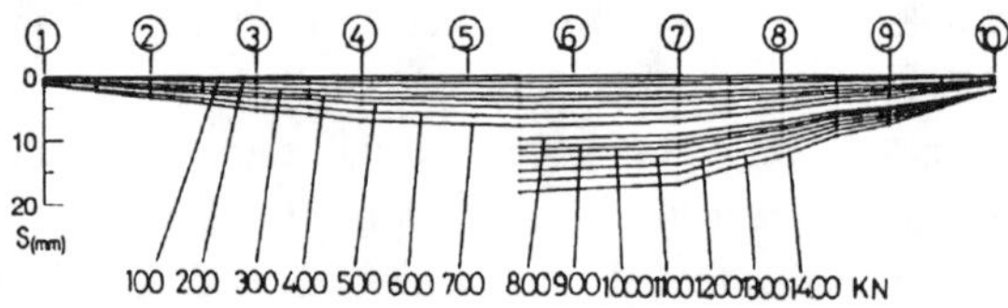

Figure 11 Settlement curves (scope of tall building: from axis 4 to 7)

From the observed settlement measurement, it can be seen that the impact of tall building settlement is concentrated within three spans out side of the tall building, and there is almost no impact at points beyond three spans, this result is consistent with that got from tests [1] [2] (Figure 11, 5). The limitedness of the load diffused by the raft of the podium provides conditions for application of the superposition principle under serviceability limit state.

4. Design Principle for Large-Area Thick Raft Foundation

On the basis of model test results and in-situ settlement monitoring analyses of engineering, the following conclusions can be drawn:

1. The raft of podium can diffuse the load of tall building, loading spreading through raft is limited, the slab that under tall building and the range of raft loading spreading exhibit a deformation characteristic of generally uniform flexible slab, the raft foundation under tall building is still limited rigidity.
2. The effect of raft loading spreading is primarily in the range of 3-span (about 24m), and becomes negligible beyond 3-span.
3. It is necessary to extend the thick raft of tall building outside one-span podium to decrease the additional stress under the foundation of tall building, and to increase the foundation's stability of tall building.
4. In case many tall and low-rise buildings are built on the same large-area integral thick raft foundation, the settlement shall be calculated according to the interaction of superstructure, foundation and soil.
5. In case many high-rise and low-rise buildings are built on the same large-area integral thick raft foundation, by taking each building as the center, and the raft thickness within the one and half span from the side of the tall building remains unchanged, the thickness shall meet the punch and shear requirements; the foundation raft thickness of the podium linking each tall building and low-rise building is determined by the structure pattern and load of the podium, meanwhile, the foundation integrity and foundation raft crack resistance requirements shall be met. When the deflection of the raft meets the requirements, and the variable-thickness raft design is needed, the raft thickness change points shall be within the second span from the side of the tall building and the gradating way is adopted for transition.

References

1. X.Yuan, Study for deformation contact pressure of frame structure with large thick raft plinth under partial vertical load of tall building[D], Beijing: *China Academy of Building Research, 1996*
2. J.F.Gong, Calculation on deformation and contact pressure of frame structure with thick raft plinth under tall buildings[D], Beijing: *China Academy of Building Research, 1999*
3. J.F.Gong, X.L.Huang & Y.J.Teng, Rigidity characteristic and deformation calculation of large-area thick raft foundation, 16th ICSMGE, Osaka, 2005

VERTICALLY LOADED SINGLE PILES IN SAND SUBJECTED TO LATERAL SOIL MOVEMENT

WEI DONG GUO
School of Engineering, Griffith University, Parklands Avenue, Southport, QLD 4125, Australia

HONG YU QIN
School of Engineering, Griffith University, Parklands Avenue, Southport, QLD 4125, Australia

The behavior of a free-head pile under lateral soil movement and vertical load was investigated using a newly developed experimental apparatus. Typical test results are presented to demonstrate the similarity and difference in the response of the pile subjected to triangular and rectangular soil movement profiles. It was found that the response such as distribution of bending moment, and pile displacement, etc. is significantly affected by the soil movement profiles.

1. Introduction

Structures built on piles are vulnerable to lateral forces caused by soil movements, which may be seen when they are used in slope stabilization, to support bridge abutments, and as foundations of tall buildings. Lateral loads generated by soil movements induce additional deflections and bending moments in piles or pile groups, which may compromise structural integrity. Although the characterization has been extensively researched through laboratory and field tests (Poulos and Chen 1995, Bransby and Springman, 1997), numerical and theoretical analysis (Viggiani, 1981, Poulos 1995), the effect of the pattern of soil movements on the response of piles has not been investigated, particularly, when it is coupled with pile-head fixity condition and axial load.

An apparatus was developed to investigate the response of a pile or pile groups (Guo and Ghee, 2004, 2005). The apparatus includes a model scale (1 m x 1 m x 1m) shear box, with the upper part allowing various (triangular, rectangular, and parabolic) soil movement profiles to be simulated. Axial load can be applied on the model pile simultaneously when the soil movement is enforced. Using the apparatus, a number of tests have been performed so far on single piles and small pile groups in sand (Guo and Ghee, 2004, 2005).

In this paper, typical test results are presented for instrumented piles subjected to triangular and rectangular profiles of lateral soil movement, respectively, which enable the effect of soil movement pattern to be approximately assessed. Multivariate analysis indicates that the response such as distribution of bending moment, and pile displacement, etc is significantly affected by the shape of the profiles, and the magnitude of soil movement.

2. Model Pile Tests

A vertical jack was used to drive the pile into the shear box. In order to simulate a free-headed condition, an axial load of either 0 or 294 N was applied on the pile-head using different weights of 10 kg each. The selected load of 294 N was less than 60% the ultimate capacity of the pile determined using the recorded jack-in pressure of 3.45 MPa. The weights were secured on the pile-head by using a connector, and they were about 500 mm above the sand surface. A schematic diagram of the test system can be found in the companion paper by Guo and Ghee (2005).

The response profile along the pile is measured via strain gauges, and at pile-head via linear variable differential transducer (LVDT) displacement transducers. The data are collected and processed via a data acquisition system connected with a computer. A general description of the experimental apparatus and procedure may be found in the previous publications by Guo and Ghee (2004, 2005).

The soil used in this study was oven-dried quartz sand. The particle size distribution was determined using the dry sieving method, which is shown in Figure 1. Sand density was achieved using a desired falling height of 600 mm. Key characteristics of the sand are given in Table 1. The pipe pile was 50mm in outside diameter, 2mm in wall thickness and 1,200mm in length. The pile was made of aluminum pipe, instrumented with 10 levels of strain gauges (Figure 2) which was calibrated for bending moment along the pile shaft. Strain gauges were coated with about 1mm epoxy and wrapped by tapes for protection. A summary of the model tests performed is presented in Table 2 and Figure 3. The soil movement was applied at an increment of 10mm (measured at the soil surface), until it reached a total of 120mm. This allowed the ultimate maximum bending moment for all the tested piles to have been attained.

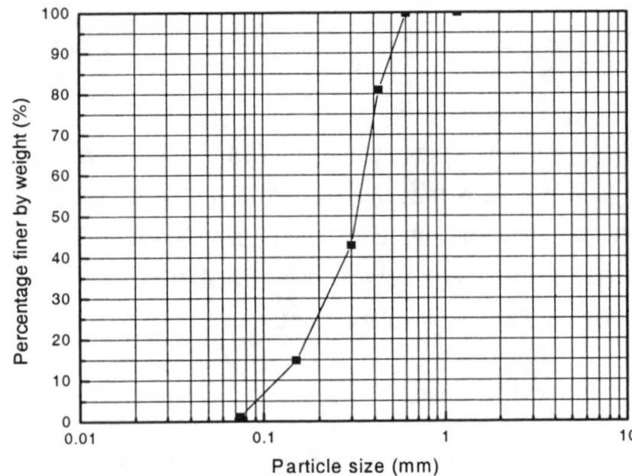

Figure 1. Particle size distribution of sand

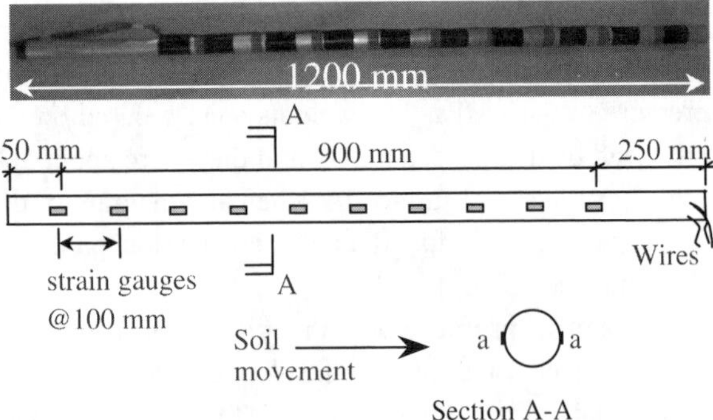

Figure 2. An instrumented model pile

Table 1 Characteristics of the quartz sand

Parameters	Value
Effective particle size, D_{10} (mm)	0.12
Coefficient of uniformity, C_u	2.92
Coefficient of curvature , C_c	1.15
Unit weight density, (kN/m^3)	16.27
Relative density index	0.89
Internal friction angle, (deg)	38

Table 2 Summary of test details

Tests	L_m (mm)	L_s (mm)	Axial load (N)	Soil movement profile
1	200	600	0	Triangular
2	200	600	294	
3	200	600	0	Rectangular
4	200	600	294	

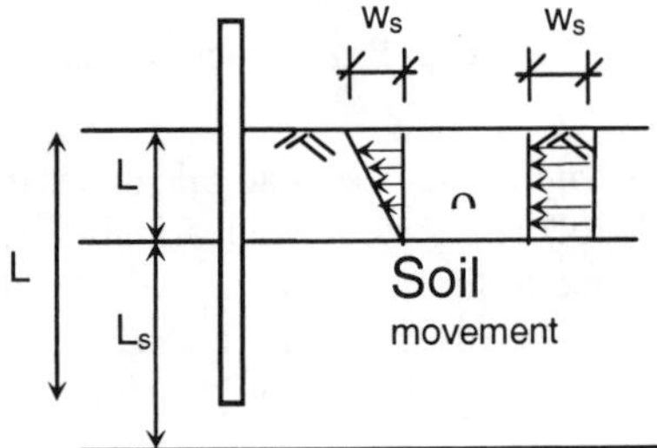

Figure 3. Schematic diagram of a pile under test

3. Test Results

A spreadsheet program was written using Microsoft Excel VBA to process and analyze the data obtained from strain gauges and LVDTs. A bending moment was determined from each pair of strain gauges (Figure 2) for each depth. The inclination, and deflection profiles along the pile were derived, respectively from 1st and 2nd order numerical integration (trapezoidal rule) of the bending moment profile; while the shear force and soil reaction profiles were derived, respectively, by using single and double numerical differentiation (finite difference method) of the moment. The five profiles of bending moment, shear force, soil reaction, rotation and deflection are presented subsequently for each test, which are derived from positive bending moment

3.1 *Test 1 (Axial load = 0 N, Triangular)*

Test 1 was undertaken using a triangular loading block and without axial load. Using the block, the test is referred to having 'triangular soil movement profile'. Figure 4 presented the 'five' profiles during the test for a soil movement up to 120mm. The 'SD' indicates that the thickness of the sliding soil was 200mm. The bending moment, shear force, soil reaction, rotation and deflection increase sharply from soil movement, w_s of 40 to 80mm, afterwards it remained constant (w_s/d = 1.6, where d = pile diameter). Note that the w_s was measured at the loading frame, which initially was about 500 mm away from the pile. The maximum bending moment, M_{max}, located at a depth of 380mm measured from the sand surface.

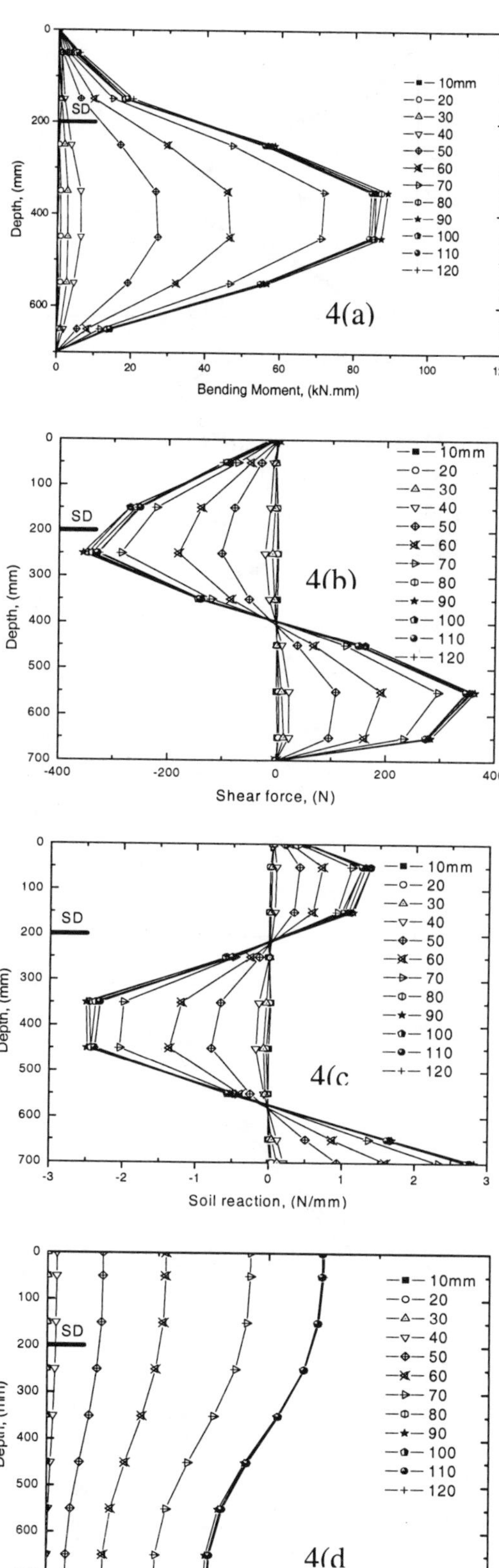

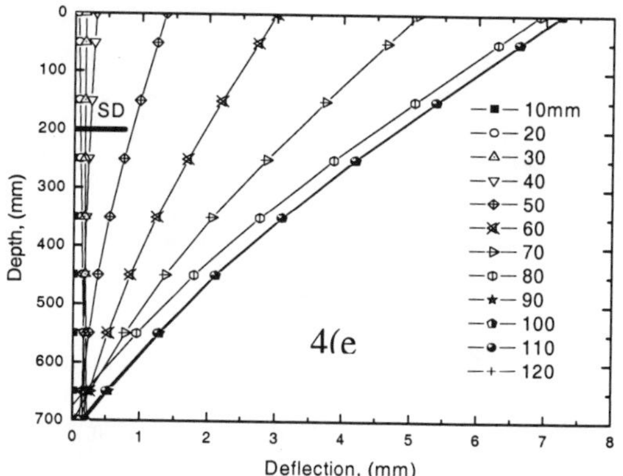

Figure 4 Response of pile during Test 1

Figure 4(e) indicates that the pile rotates about the depth of 650mm. Pile deflection at the surface reaches 7.3mm when soil movement is 120mm. The actual movement of the soil at the pile location is less than 120 mm. However, during the test, it is noted that the sand in the moving layer has flowed around the pile.

3.2 *Test 2 (Axial load = 294 N, Triangular)*

Test 2 was performed under identical conditions to Test 1, but with an axial load of 294N applied at the pile head. Presented in Figure 5 are the five profiles of the pile during the test. A very similar trend to those observed in Test 1 is noted. The magnitudes of each pile response increase rapidly from a soil movement, w_s of 30 to 80mm. They then become invariable. The maximum bending moment M_{max}, is 115.7 kNmm that took place at a depth of 400mm with w_s = 80mm. At the ground level, the pile deflection is about 7.2 mm.

Tests 1 and 2 were for a triangular soil movement profile. The effect of the shape of the profile was examined in Tests 3 & 4 by changing the triangular loading block to a rectangular one, so as to produce 'a rectangular soil movement profile'.

3.3 *Test 3 (Axial load = 0 N, Rectangular)*

Test 3 was performed without axial load. The results are presented in Figure 6. As shown in the figure, the majority increase in bending moment occurs at the first 20mm soil movement. The maximum bending moment is 61.7kNmm at w_s = 60mm, which located at a depth of 400mm. The figure indicates that the pile rotates around pile tip and behaves a bit flexible. It moves about 2.5mm at soil surface when w_s is 20mm.

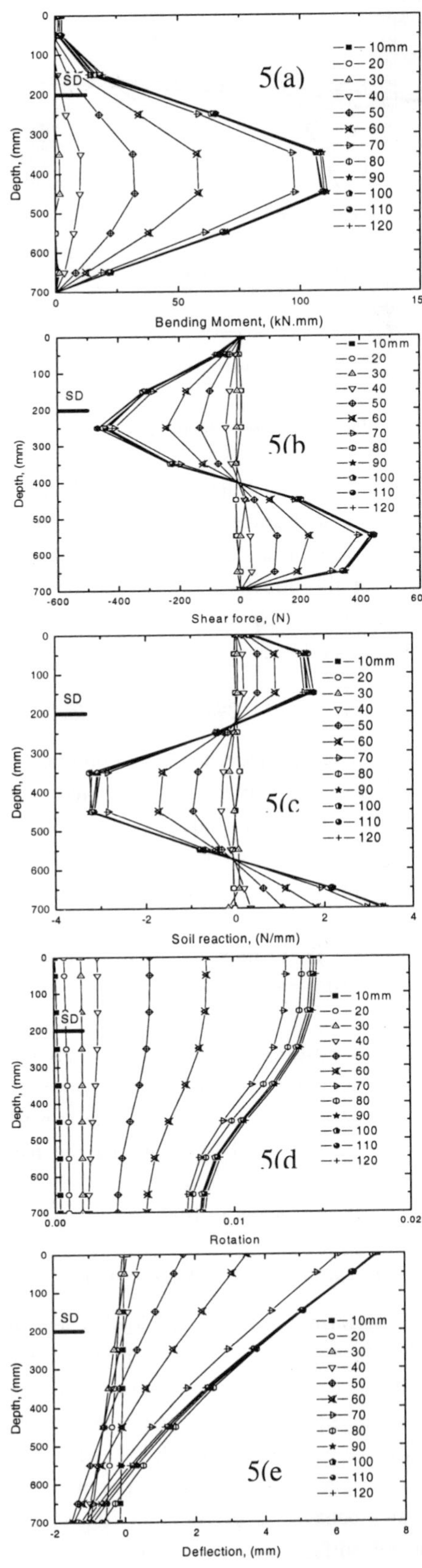

Figure 5 Response of pile during Test 2

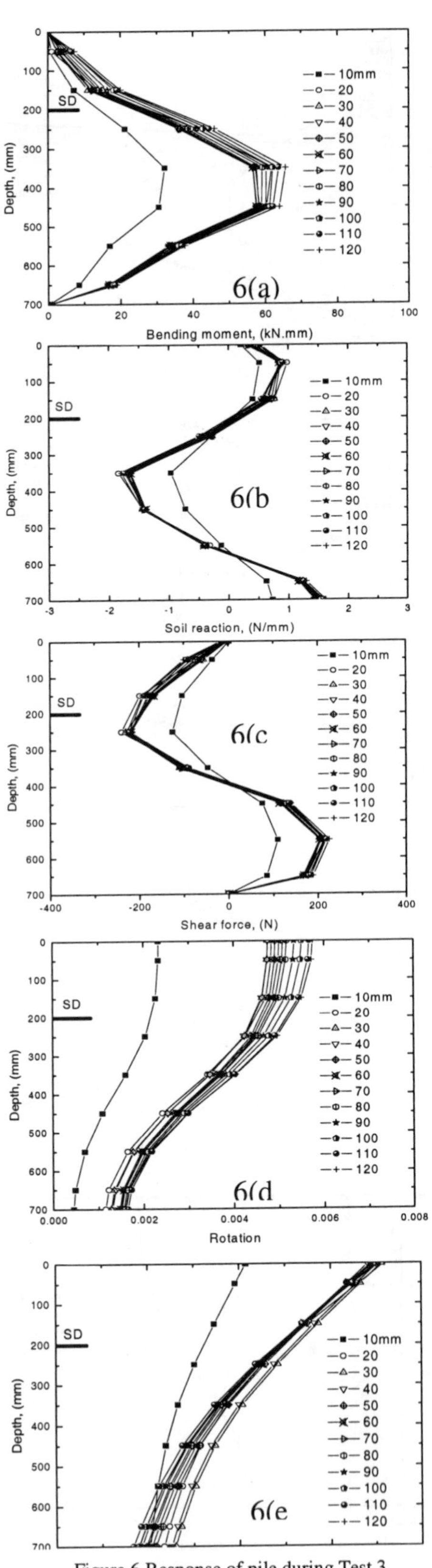

Figure 6 Response of pile during Test 3

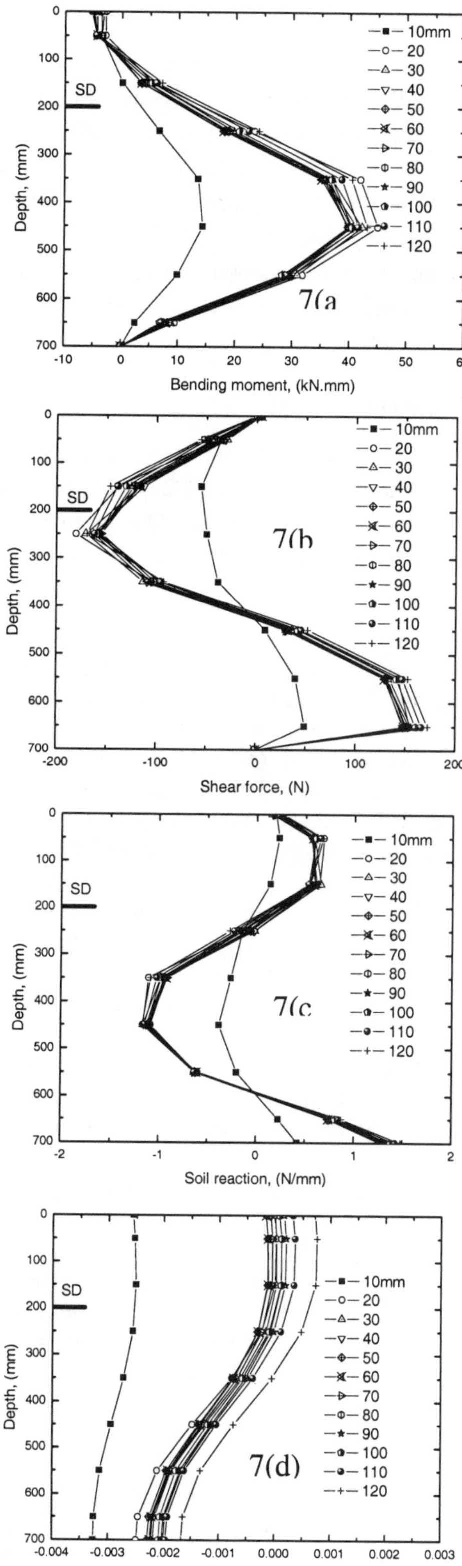

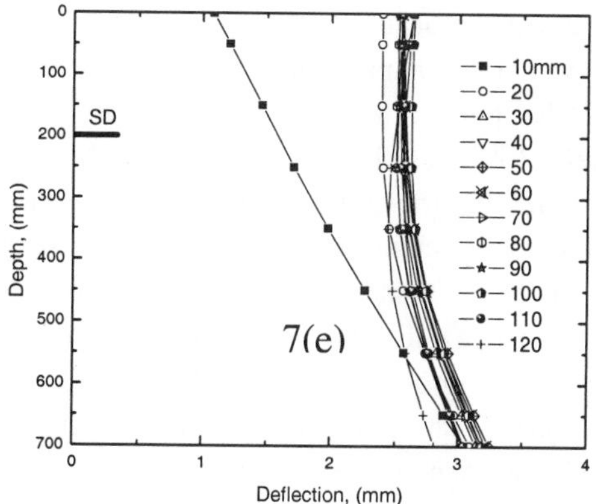

Figure 7 Response of pile during Test 4

3.4 *Test 4 (Axial load = 294 N, Rectangular)*

Under the same condition as Test 3, Test 4 was undertaken albeit with an axial load of 294N. The responses of the pile are shown in Figure 7. Maximum positive bending moment was 42.64 kNmm, which occurred at a depth of 430mm at w_s = 20mm. Negative bending moment was noted at the pile at the sand surface, which is about 12% the maximum positive bending moment. The magnitudes of both the maximum bending moment and shear force are always less than the counterparts without axial load. The opposite orientation of the deflection of the pile at w_s = 10mm indicates that the initial position of the pile was not completely vertical. Under the axial load, the pile deflects mainly by rotation.

Table 3 Summary of test results

Test no.	Depth of M_{max} (mm)	M_{max} (kN.mm)	Max. shear force in L_s (N)	Max. pile def. (mm)	Max. soil reaction (N/mm)		M_{max} at w_s (mm)
					In layer L_m	In layer L_s	
1	380	89.02	363.97	7.3	2.82	-2.49	90
2	400	115.57	445.42	7.2	3.36	-3.24	90
3	400	67.23	225.53	2.6	1.60	-1.75	20
4	430	42.64	172.32	2.5	1.42	-1.12	20

4. Discussion

Table 3 shows the critical responses of maximum bending moment, shear force, soil reaction and pile deflection. Figure 8 shows the development of the responses with soil movement. Given a triangular soil movement profile, the bending moment increases with soil movement w_s from 30 mm onward, and reaches maximum at 90 mm. In contrast, given a rectangular profile, the moment reaches maximum at a much lower soil movement of 20mm. The maximum bending moment of the pile without axial load (Test 1) is approximately 30% less than that with a 294N axial load (Test 2) for the triangular profile. However, it is 20% higher for the rectangular profile as indicated by the curves of Tests 3 and 4. The deflection profile indicates that piles mainly rotate, and behave more stiff for the triangular profile in comparison with the rectangular one. The ultimate maximum bending moment under a triangular soil movement profile is about twice as large as that under a rectangular one.

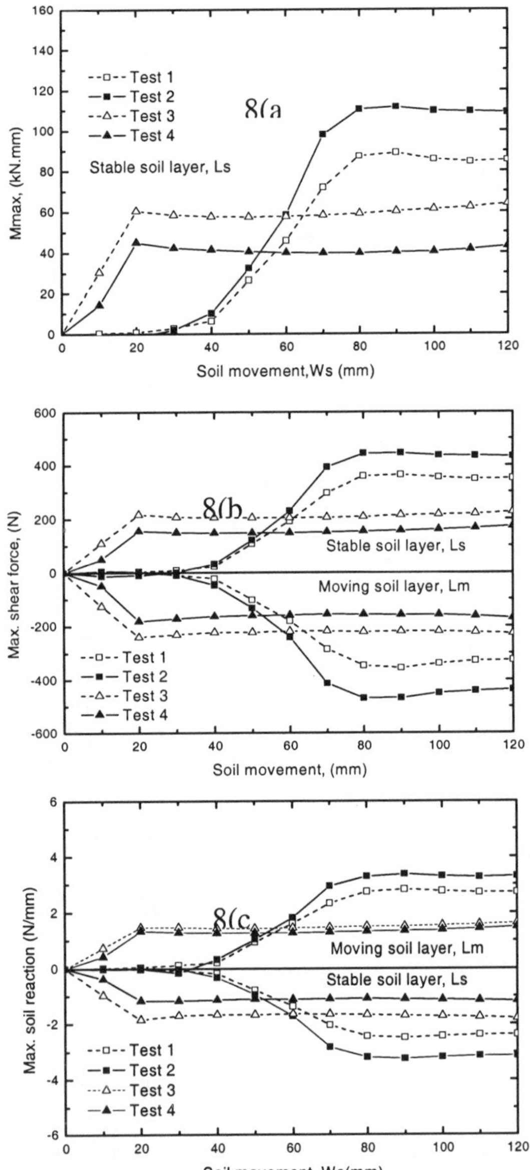

Figure 8 Comparison of response of the tested piles

5. Conclusion

Model tests on piles subjected to lateral soil movement and axial load were undertaken in sand. Preliminary analysis indicates that a single pile may behave more stiff when subjected to a triangular soil movement profile than to a rectangular one. The ultimate maximum bending moment, and pile displacement under a triangular profile is about 2.0 and 2.8 times that under a rectangular one, respectively. More in-depth exploration is underway.

Acknowledgments

The financial support from Australian Research Council (Grant no. DP 0209027) is gratefully acknowledged. The authors would like to thank Mr. Enghow Ghee for his contribution to the experiment.

References

1. M. F. Bransby, and S. M. Springman, Centrifuge modelling of pile groups adjacent to surcharge loads. Soils and Foundations. **37**(2): 39-49 (1997).
2. W. D. Guo, and E. H. Ghee, Model tests on single piles in sand subjected to lateral soil movement. Proceedings of the 18th Australasian Conference on the Mechanics of Structures and Materials, **2**: 997-1003 (2004).
3. W. D. Guo, and E. H. Ghee. Response of axially loaded pile groups subjected to lateral soil movement- an experimental investigation. 6th International Conference on Tall Buildings (ICVB-VI) (2005).
4. H. G. Poulos, L. T. Chen, and T. S. Hull. Model tests on single piles subjected to lateral soil movement. Soils and Foundations. **35**(4): 85-92 (1995).
5. H. G. Poulos. Design of reinforcing piles to increase slope stability. Canadian Geotechnical Journal. **32**(5):808-818 (1995).
6. C. Viggiani. Ultimate lateral load on piles used to stabilise landslide. Proc. 10th Int. Conference on Soil Mechanics and Foundation Engineering, Stockholm. **3**. 555-560 (1981).

RESPONSE OF AXIALLY LOADED PILE GROUPS SUBJECTED TO LATERAL SOIL MOVEMENT - AN EXPREMENTAL INVESTIGATION

WEI DONG GUO[†]

School of Engineering, Griffith University, Parklands Avenue, Southport, QLD 4125, Australia

ENG HOW GHEE

School of Engineering, Griffith University, Parklands Avenue, Southport, QLD 4125, Australia

A new experimental apparatus has been developed, which allows lateral soil movements and vertical load to be applied simultaneously on a pile. A number of tests have been undertaken, which offer consistent results. In this paper, a brief description of the apparatus was presented. Results from two model tests were reported, which were conducted on two instrumented pile groups embedded in sand subjected to uniform lateral soil movement. Preliminary analysis indicated that the increase in bending moment, shear force, soil reaction and the change of pile deflection mode due to axial load on the pile groups.

1. Introduction

Pile foundations designed to support tall buildings structures and services may be subject to lateral soil movements and axial loads simultaneously. There have been many studies on piles subjected to vertical loads, and some research on piles subjected to lateral soil movements. However, little information is available for evaluating the response of vertically loaded piles due to soil movement. These piles often found nearby deep excavation activities, tunneling operations, pile driving operation, unstable slopes, settling embankments and liquefiable soils. Studies on piles due to liquefaction induced lateral soil movement have found that axial load on the piles can cause (1) additional bending moment; (2) additional compression stress, and (3) additional lateral displacement (e.g. Bhattacharya, 2003). The first two phenomena are consistent with those observed from model tests due to soil movement (Guo and Ghee, 2004). However, with axial load, lateral displacement of a free-head pile was found to have reduced rather than increased. On the other hand, lateral soil movement may increase the axial capacity of piles (Chen at el, 2002), which is similar to a laterally loaded pile subjected to axial load noted previously in experiments by Anagnostopoulos & Georgiadis (1993) and in numerical simulations by Trochanis et al (1991). This increase must be capped by an allowable maximum bending moment of the pile that needs to be clarified.

With the support of the Australian Research Council, a new apparatus was developed to investigate the response of piles/pile groups subjected to axial load and lateral soil movement. It mainly consists of a shear box, and a loading system that allow different soil movement profiles and vertical loading to be applied simultaneously. The responses are measured using strain gauges, and LVDTs, which are then recorded via a data acquisition system.

In this paper, typical test results are presented for free-standing pile groups. The instrumented piles were embedded in sand. They were tested under two different axial loads and a uniform lateral soil movement.

2. Test Description

Information regarding the tests has been reported previously by Guo & Ghee (2004). Thus relevant parts are only briefly introduced herein.

2.1 *Shear box*

The shear box has internal dimensions of 1 m by 1 m, and 0.8 m in height as shown in Figure 1. The upper movable part of the box consists of a desired number of 25 mm thick square laminar aluminium frames to achieve a thickness of L_m (< 400 mm). They are moved together by a rectangular loading block (for this research) in order to generate a uniform lateral soil movement. The lower fixed section of the box is made of a timber box 400 mm in height, and a desired number

[†] This project is supported by the Australian Research Council Discovery Grant, grant number DP0209027.

of laminar aluminium frames to achieve a stable sand layer of thickness L_s (L_m+L_s = 800mm). The rate of the movement of the upper shear box (thus the soil) is controlled by a hydraulic pump, and a flow control valve.

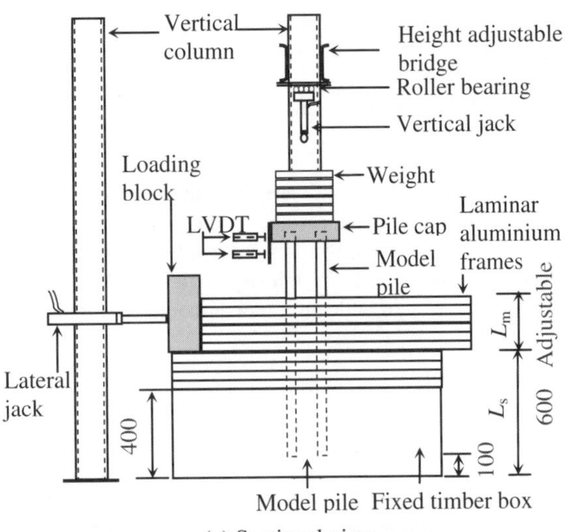

(a) Sectional view

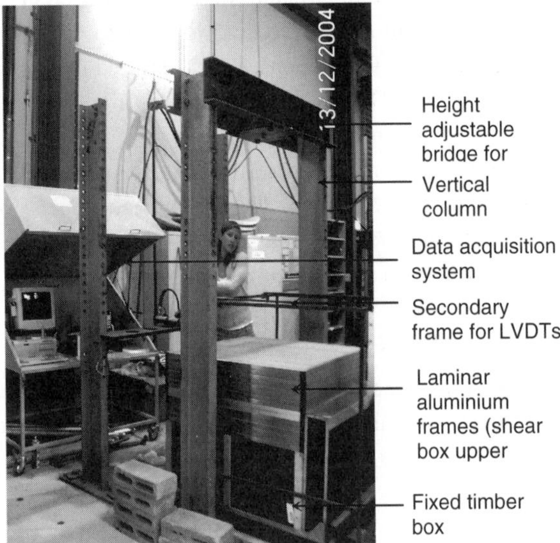

(b) Photograph of test apparatus

Figure 1. Apparatus for testing pile groups

As shown in Figure 1(a), pile cap was first connected with the piles. The pile group is then driven into sand using a vertical jack. The cap was fabricated from solid aluminium block of 50 mm thick. There are two 32±0.5 mm diameter holes (center to center spacing of 3d, d = diameter of the pile) that allow the piles to be socketed into the cap. Axial load was applied using six weights of 10 *kg* each. The load selected was less than 60% the ultimate capacity of the pile determined using the recorded jack-in pressure of 6.0 MPa. The weights were secured on the pile-head by using a connector, and located 500 mm above the sand surface.

2.2 *Sand, model pile & experimental procedure*

The sand used in this study was an oven dried medium grained quartz, Queensland sand. It was well graded with little to no fines. In order to maintain a uniform density of the sand within the shear box and from one test to another, a sand rainer was used. The rainer was fabricated from plywood, having internal dimensions of 1 m by 1 m, and a height of 150 *mm*. The base of the sand rainer was fabricated from a piece of square, timber plate of 18 mm in thickness that is underlaid by a moveable plastic plate of 6 mm thickness. Both pieces were perforated with 5 mm diameter holes in a 35 mm × 50 mm grid pattern. Sand density was controlled by selecting different falling heights. For the current test, a falling height of 600 *mm* was used, which corresponds to a dry density of 16.27kN/m^3, a relative density index of 0.89, and an internal frictional angle of 38° (Guo and Ghee, 2004). The shear modulus at the middle depth of the shear box was about 220 kPa as determined from oedometer test. The model pile used in the tests was made of aluminium tube, 1200 mm in length, 32 mm in outer diameter, and 1.5 mm in wall thickness. The calculated bending stiffness of the pile was 1.28 × 106 $kNmm^2$. Each pile in the group was instrumented with ten pairs of strain gauges mounted along the shaft as shown in Figure 2.

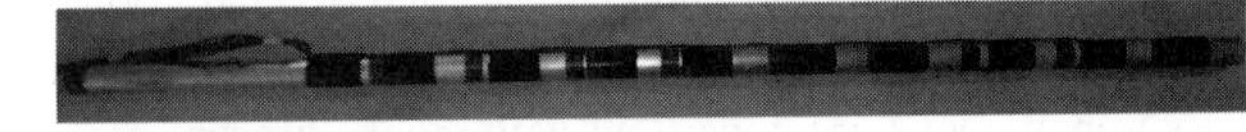

Figure 2. An instrumented model pile

2.3 *Loading conditions in the tests*

A number of tests have been conducted on the model pile groups subjected to a uniform lateral soil movement and axial load. Presented here are tests on two pile groups with an embedment length, *L* of 700 *mm*, as summarized in Table 1 and Figure 3. The groups were tested without any constraint but soil resistance. Pile center-centre spacing was 3d. The uniform soil movement was applied using the rectangular loading block at an increment of 10 mm until a total of 60 *mm* was reached. Strain gauge readings have been used to obtain a bending moment profile for each increment of soil movement, and for both piles A and B.

Table 1. Tests on pile groups subjected to a uniform lateral soil movement together with different axial load.

Test	L_m (mm)	L_s (mm)	Axial load (*N*)	Pile
1	200	600	0	A & B
2	200	600	588	A & B

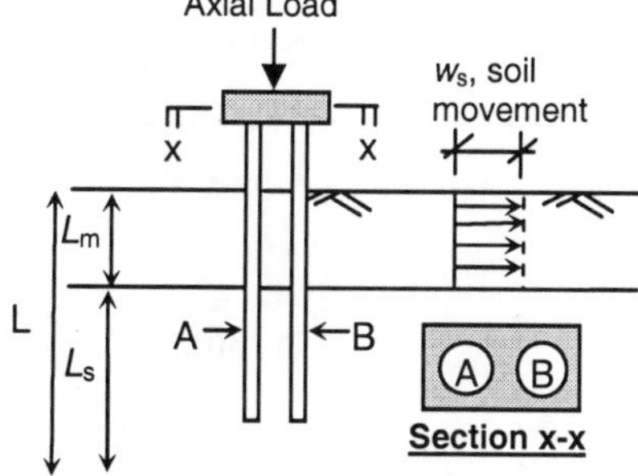

Figure 3. Information about the pile tests

3. Test Results

A spreadsheet program was written to analyse and process the data obtained from strain gauges. The profiles of inclination and deflection along the pile were derived, respectively from 1st and 2nd order numerical integration (trapezoidal rule) of the bending moment profiles, while those of shear force, and the soil reaction were derived by single and double numerical differentiation (finite difference method) of the moment, respectively. The integration and differentiation methods were found to offer consistent results presented in this paper. The five profiles of bending moment, shear force, soil reaction, rotation and deflection were presented subsequently for each pile of the two tests.

3.1 *Test 1 (without axial load), Pile A*

For the back Pile A in Test 1, Figure 4 presents the 'five' profiles under the six magnitudes of soil movement, w_s of 10, 20, 30, 40, 50 and 60 mm. The SD indicates sliding depth of 200 mm. Figure 4a shows that the maximum bending moment in the stable layer, M_{max} occurs at a depth of 460 mm [at this depth the shear force is indeed zero (Fig. 4b)]. The bending moment, shear force and soil reaction reach maxima and changes slightly beyond w_s of 40 mm (w_s/d = 1.25) (Fig. 4a, b, c). Figure 4e indicates that the pile translates at w_s = 10 ~ 60 *mm*. The maximum pile head deflection of 9.7 mm is well below the corresponding w_s of 60 *mm*. Thus the soil in the sliding layer flowed around the Pile A.

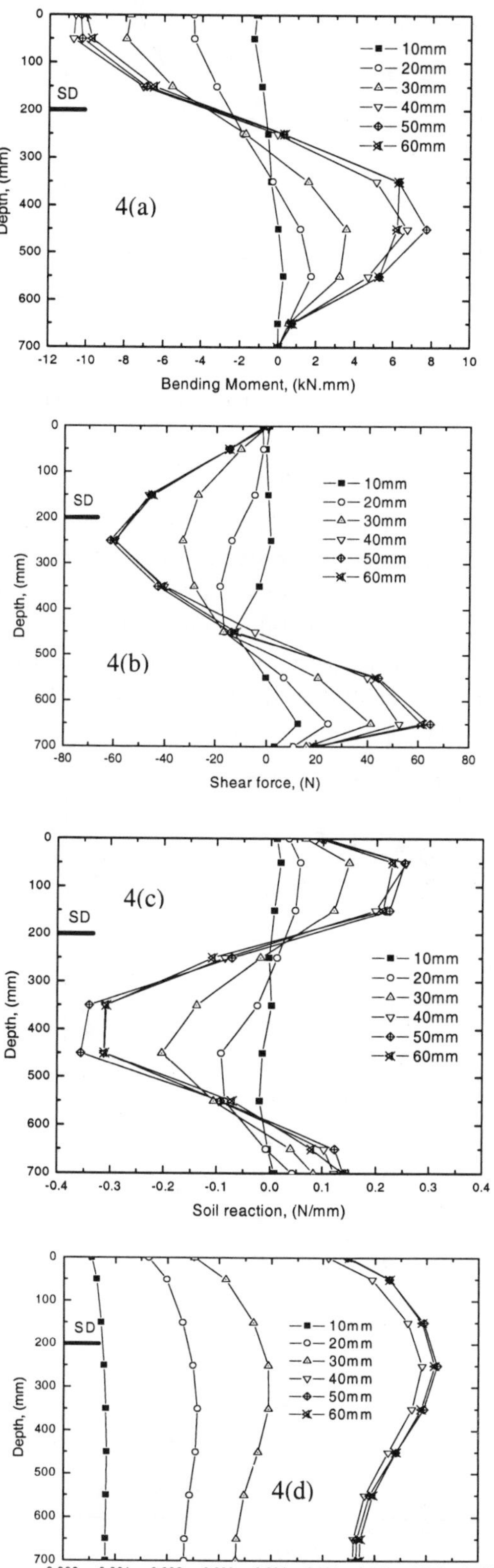

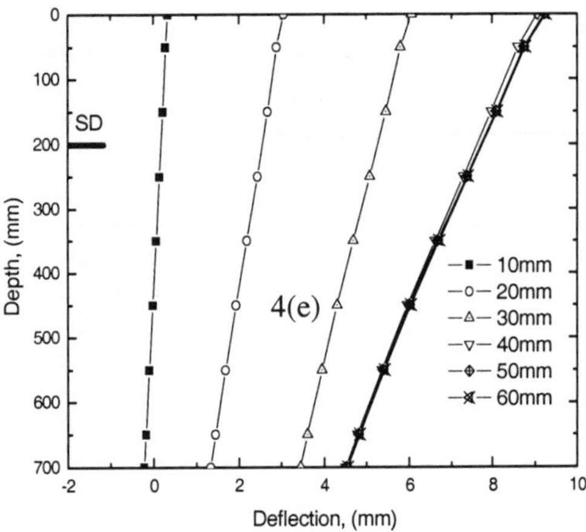

Figure 4. Response of the Pile A during Test 1 (no axial load)

3.2 *Test 1 (without axial load), Pile B*

For the front Pile B in Test 1, Figure 5 provides the same five profiles for the six values of soil movement. Figure 5(a) indicates that the depth of M_{max} is 540 mm, against 460 mm observed in Pile A (Table 2). The bending moment, shear force and soil reaction reach maxima at w_s of 40 mm ($w_s/d = 1.25$) similar to Pile A.

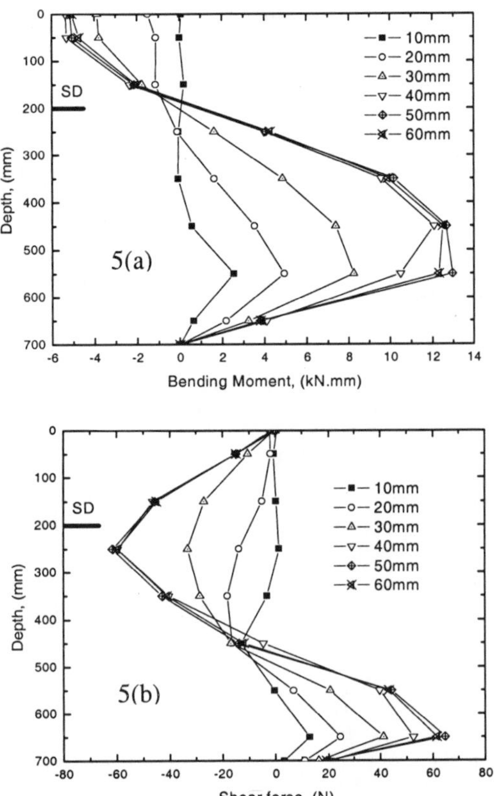

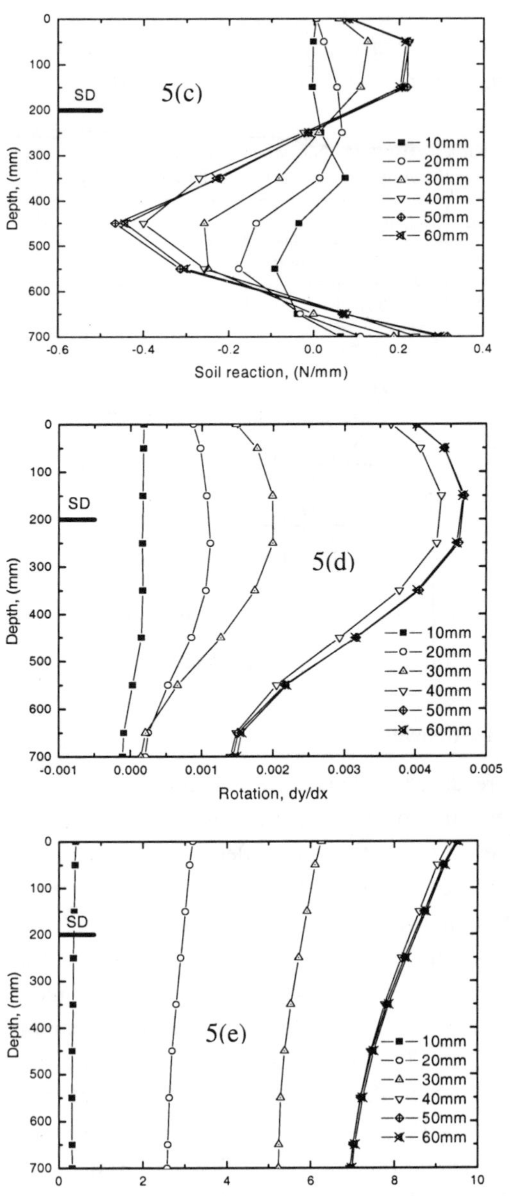

Figure 5. Response of the Pile B during Test 1 (no axial load)

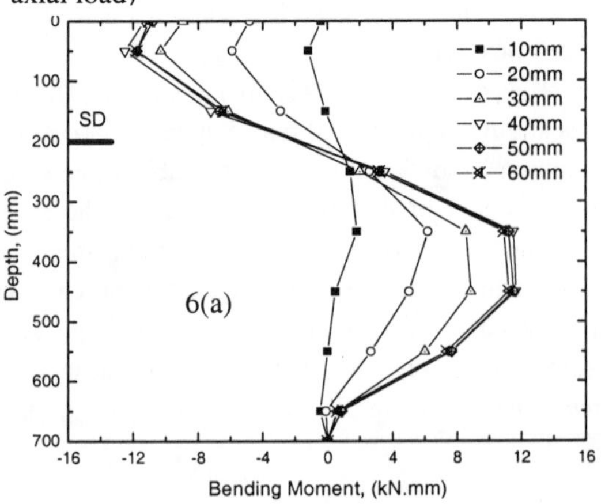

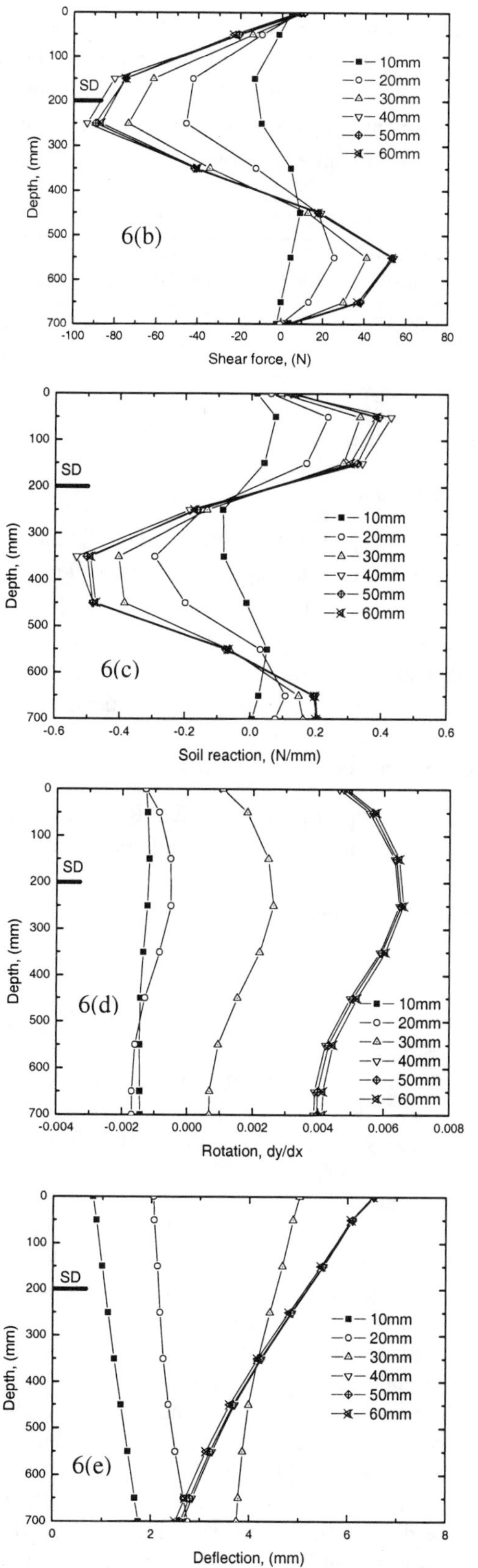

Figure 6. Response of the Pile A during Test 2 (Axial load = 588 N)

3.2 *Test 2 (axial load = 588 N), Pile A*

Test 2 was performed under identical conditions to Test 1, but with an axial load of 588 N applied on top of the pile cap. Figure 6 provides the five profiles for Pile A under the six values of soil movement. Figure 6(a) indicates that the depth of M_{max} in stable layer is 390 mm, against 460 mm observed in Test 1 (Table 2). The bending moment, shear force and soil reaction reach maximum values at w_s of 40 mm (w_s/d = 1.25) similar to Test 1.

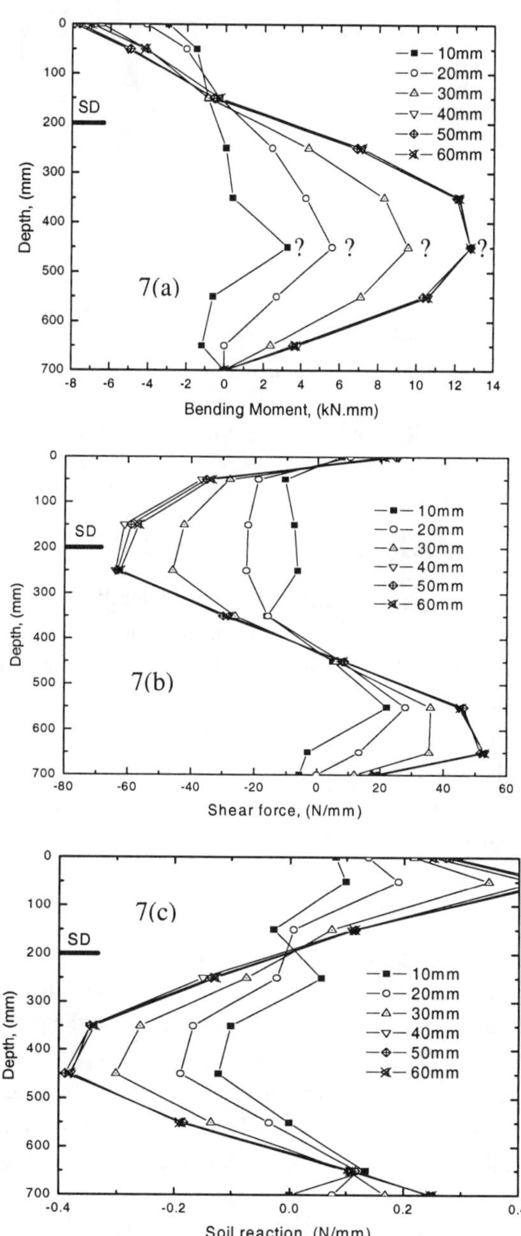

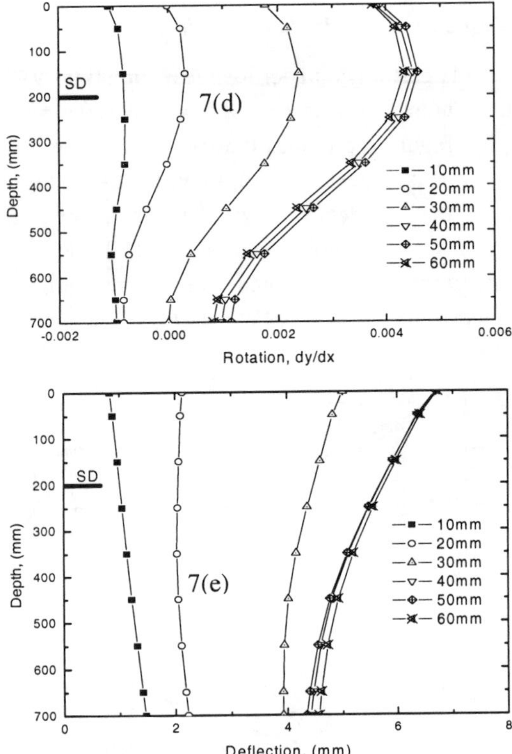

Figure 7. Response of the Pile B during Test 2 (Axial load = 588 N)

3.3 *Test 2 (axial load = 588 N), Pile B*

For Pile B in Test 2, Figure 7 provides the same five profiles for the six values of soil movement. The depth of M_{max} is 460 mm, against 540 mm observed in Pile A (Table 2). The bending moment, shear force and soil reaction reach maxima at w_s of 40 mm (w_s/d = 1.25) similar to Pile A. The question mark '?' in Figure 7(a) indicates the particular strain gauge may not work properly.

4. Discussions

Table 2 compares the critical, maximum responses of bending moment, shear force, and pile deflection. An increase in the axial load from 0 to 588 *N* on the pile cap leads to: (1) a 24% and 14% increase in M_{max} (in L_s) for Piles A and B, respectively; (2) a 18 % decrease in depth of M_{max} (in L_s); (3) a 34% decrease in maximum pile deflection at groundline. Also the M_{max} for Pile A in Test 1 is 10.62 *Nm* that occurs at groundline. The deflection profile for Test 1 indicates that piles mainly translate with slight rotation; while in Test 2 with axial load, the piles translate and rotate. In comparison with those reported earlier for single free-head piles (Guo and Ghee, 2004), a reduction of 30~60% in maximum bending moment, and 40~60% in shear force are noted, respectively for the piles in groups.

Table 2. Summary of test results (w_s = 60 mm)

Test no.	Pile	Depth of M_{max} (mm)	M_{max} in L_s (Nm)	Max. shear force in L_s (*N*)	Max. pile def. (mm)	Max. soil reaction (N/mm)	
						Within L_m	Within L_s
1	A	460	7.72	52.71	9.71	0.25	-0.36
	B	540	12.96	34.80		0.23	-0.47
2	A	390	11.76	93.51	7.22	0.43	-0.53
	B	460	14.79*	60.54		0.27	-0.39
	* Corrected value						

5. Conclusions

Tests on model pile groups subjected to simultaneous uniform lateral soil movement and axial load were undertaken. With an axial load on the pile cap, preliminary analysis shows an increase in bending moment, shear force and soil reaction, but a decrease in the deflection. More research is currently under way.

Acknowledgments

The financial support from Australian Research Council is gratefully acknowledged. The authors would like to thank Tejasvaraj Toor and Hongyu Qin for their contribution to the experiment.

References

1. Anagnostopoulos, C. & Georgiadis, M., 1993, Interaction of axial and lateral pile responses. Journal of Geotechnical Engineering Vol. 119, No.4: 793-798.
2. Bhattacharya, S., 2003. Pile instability during earthquake liquefaction. PhD Thesis, University of Cambridge.
3. Chen C.C., Takahashi, A. & Kusakabe O., 2002, Change in vertical bearing capacity of pile due to horizontal ground movement. Physical Modelling in Geotechnics: 459-464.
4. Guo, W. D. & Ghee, E. H., 2004. Model tests on single piles in sand subjected to lateral soil movement. Proceedings of the 18th Australasian Conference on the Mechanics of Structures and Materials, Vol. 2: 997-1003.
5. Trochanis, A. M., Bielak, J. & Christiano, P., 1991 Three dimensional nonlinear study of piles, Journal of Geotechnical Engineering, Vol. 117: No 3, 448-466.

DEVELOPMENT OF THE APPARATUS FOR QUALITY CONTROL OF RUSHED STONE PILE

LONGAN CHEN ZHAO LIU

School of Mechanical Engineering, Tongji University, 1239 Siping Road Shanghai 200092, China

BINGLON WANGG JIANGUO LIU

School of Mechanical Engineering, Tongji University, 1239 Siping Road Shanghai 200092, China

According to the latest studies of quality control of rushed stone pile, this paper first fixed on the functional aim and measuring parameter of the apparatus. And then on the base of the research of sensors distributed widely, analogue and digital signal processing and data storage, the circuits are designed and a kind of embedded computer with very much storage and a sort of operating system used as the heart of the apparatus to increase the capacity of storing data and to be seasoned with atrocious environment. The apparatus also posed a kind of UATR serial communication port and a kind of device driver program was designed in order to fit to the different kinds of distributed monitoring system. With characteristic of the smallest code size and real time operating, the system software of the apparatus introduced many of the modern technologies usually used in the current operating system development and software architecture and other software develop methods. Except for this, the mechanical apartment used to install the sensors was also designed through mechanical module technology and might be installed on all kinds of cranes without modification of the structure. The result of testing the apparatus in the construction of Da Yang mountain harbor proved it to be easy operating and poses high reliability and satisfy the designing objective.

1. The Development of the Techniques at Home and Abroad

In recent years, vibrator is widely used in engineering construction as an advanced groundwork managing method, but the quality of the construction is largely influenced because of the ragged technical equipment and knowledge of constructional company. In our country has random sampled the construction quality of 1000 rushed stone piles which have been used, 21 piles of which have been convicted to limitation in the result of the investigation. Statistic result is shown in table1. We can find from the table, that the inadequate or uneven density of the piles is the major quality problems in recent.

It can be easily found, that the controlling equipment problem of the density current is the major factor to influence the density problem. Because the construction quality of vibration lies on the density of the pile, and the density of pile rests with the designed density current and the actual density current. In the past construction, the density current is controlled manually through ammeter in the manipulating table. So there are too many manual factors.

Technical innovation has carried through against the density current controlling problem during the vibrating process. Then a kind of vibration device can automatically give an alarm to clue to satisfying the design demand when it reach the vibrating density current and guarantee some time to remain vibration, shown in Figure 1

This kind of vibrator is widely used in our country. But it still exists shortage for the reason that it can't wholly trace the current change during the working of rushed stone pile in the entire length of pile and the measuring data is recorded manually. The system can't directly and graphically reflect the constructing process of the rushed stone piles and the man-made factor can not be eliminated during the constructing control..

Dae Yang Geotechnic Pte Ltd in Korea has exploited the vibrating control device against the vibrator of our company, so that the machining hole current, density current, density time to remain vibration can be tested and controlled during the construction. But this company doesn't promulgate related technique of device to outside in order to remain the technical advantage in bid process.

With the technical improve of the groundwork reinforcing which uses vibrator, the research of monitoring apparatus adapting to our engineering character and cooperated mechanical apparatus is very urgent.

Table 1. The result of random sample of construction quality of rushed stone pile.

Code	item	number	frequent (%)
A	The inadequate or uneven density of the piles	12	57
B	The fall short of the length of piles	3	14
C	The diameter of piles is a little smaller	3	10
D	The clay containing of piles is large	2	5
E	The inadequate of stone input	1	5
	Total	21	100

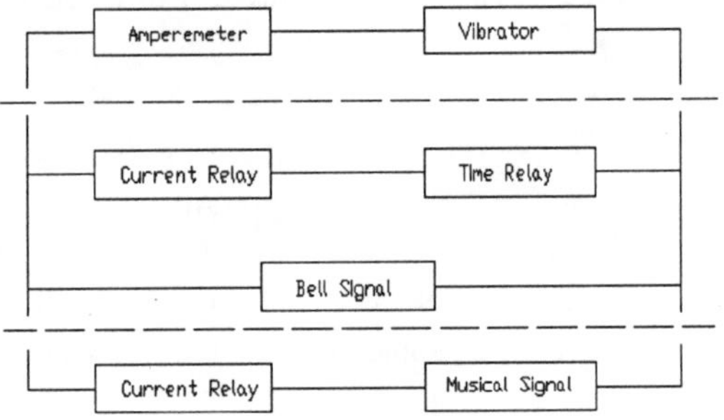

Figure 1. Control systemic structural diagram for density current of vibrator used currently.

2. The Technological Demand and Basic Parameter of the Monitoring Apparatus

2.1 *The technological demand and basic scheme*

Vibratile broken-stone piles are used to stabilize foundation by the high vibration and high-pressure water from vibrator that dropped to the schedule depth through vibration. The hole is first formed, which is flapped by pressuring water and then filled with broken stone about 30cm~50cm and vibrator dropped into the hole vibrating to make the broken stone compacting each times, that process is repeated until the hole is all filled with broken stone.

The quality of vibroflotation is influenced by the following two factors: the first is how many piles there are and how much length of pile is each, and the second is the quality of the pile itself which include sustaining time, compacting current and length and pressure water. The apparatus is used to check the length of it and control its quality concerning hole water pressure and compacting electronic current and sustaining time.

According to the above, the device must have such function that current and water pressure and displacement measurement are worked on and some data can be told to the apparatus. The designing apparatus must have serial port to communicate to the other device to exchange data each other and sensors which can lie apart away from it and a special kind of storage memory to store the acquired data which can't be modified by builder. The main body of the intelligent device is formed of a kind of embedded system and its panel is supplied to enter important data such as sustaining time and compacting current and different water pressure and compacting length by technician not operator to gain quality of the construction. The device may can imply the builder that there are some problems to make builders do work to solve some potent problems. The graphs of the acquired data against time and other gained data against displacement are also displayed on it.

2.2 *The premier parameter*

On the base of the investigation on the native vibraflotation, the premier parameter is as following::

1. Forming hole current and water pressure can be seen from the table 2.

Table 2. Forming hole parameter

Power of vibrator (Kw)	Forming hole current (A)	Forming hole water pressure (Kpa)
30	50~60	400~600
75	100~140	400~800

2. The compacting electrical current and water pressure are shown in the table 3.

Table 3. Compacting parameter

Power of vibrator (Kw)	Compacting current (A)	Compacting water pressure (Kpa)
30	45~60	400~600
75	80~120	400~600

3. Sustaining time is the interval of from 5s to 15s;

4. Compacting length is about from 300mm to 500mm according to JGJ790-91 China.
5. Clearing and forming hole water pressure is about from 400 to 800 KPa.
6. The length of the pile is about

3. The Design of the Monitoring Apparatus

The hardware system mainly include sensors, power supply, embedded computer, keyboard and LCD. All the parts have complicated relation to each other as shown in figure 2. The power supply gives energy to different compositions of the system and keyboard is used to input different parameter and the display LCD is to give implication to builder.

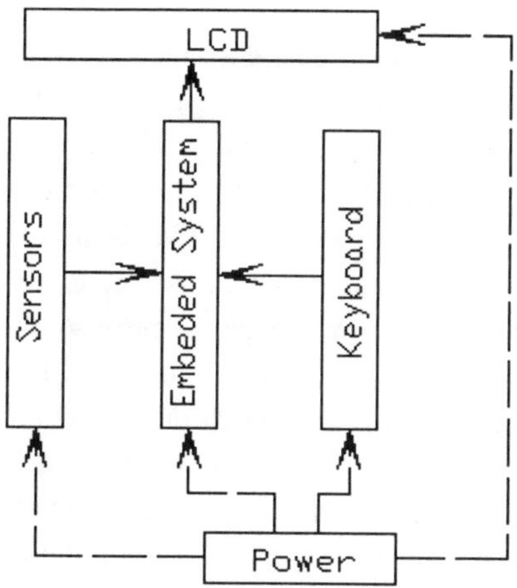

Figure 2. Relation to each other for the apparatus.

3.1 *Hardware designing*

3.1.1 *Sensors*

On the base of the studies of the measured object, all sensors lie very far apart from embedded system and each other, which include electrical current sensors, pressure sensors and displacement sensors. The first two kinds of sensors are to change the measured physics quantity to the current of 4 to 20 milliampere in order to avoid different kinds of interference signal and the last sensor has a kind of long driving outlet to avoid the measured signal to deform.

The forming hole current and compacting current are all the three-phase alternate current and are very different as vibrator is changed. Different types of hall-effect current sensors are used to measure the alternative current for the reason that the sensors have high immunity of noising signals and isolate the embedded system from the alternative current in order to increase the reliability of the whole system and satisfy the different forming hole current and compacting current with the value of from 40 to 200 ampere. The current sensor of BLYK200-S6A/SP22 is selected.

Water pressure of the outlet pump is measured and about 100 meter apart from the embedded system. The pressure sensor of CY-DB11 is selected because it transmits the physical quality of pressure to the current of 4 to 20 milliampere and makes the electrical signal not to be distorted.

The displacement of the vibrator is very tempestuously changed from 20 meters to 300 millimeter. In order to solve the problem, the angle sensor (1) of LBM05-406a200BZ/05L is selected which changes the angle displacement to the pulse signal and each pulse signal presents the angle of about 1.8 degree. The mechanical system must be used to make line displacement of the vibrator (5) to angle displacement as shown in figure 3. To reach the precision of 10 millimeter, the diameter D3 of friction wheel (3) must be less than 630 millimeter. The spring (6) is used to control the force acting on the friction wheel (3) for the reason that the weight of vibrator is about more than 100 kilogram while the diameter of axis of angle sensor (1) is only about 6 millimeter and the diameter of the shaft of friction wheel (3) only about 20 millimeter. For the same reason as the other sensors, the angle sensor has the outlet of long line driving which can transmit the pulse signal to the embedded system being about 100 meters apart from the angle sensor.

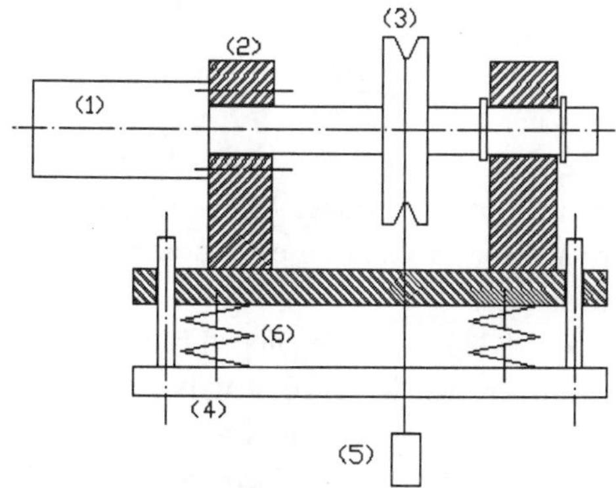

Figure 3. Mechanical erecting device for displacement sensor

3.1.2 *Power*

The interference sources of the monitoring apparatus are input and output channel, space magnetic field and power supply system. The interfering signals caused through input channels and output channels, can be reduced by special electrical circuit, such as filter and photoelectrical isolating circuit while the interference signals from magnetic field can also be avoided through good shield and grounding. So the power supply system must be carefully be designed in order to eliminate noise signals from it.

The power supply system is shown in figure 4, the switching power supply of which converts 220VAC to 24VDC that is input voltage of the pressure sensors and all DC-DC's. The DC-DC's transform 24VDC to other voltage power supplies, 5VDC of which is the input of the embedded system and LCD, while +/-15VDC of which is also the input of the embedded system and the current sensors. The grounds of all power supplies are connected at the same point to remove noise signals coming from them and advance the reliability of the monitoring apparatus.

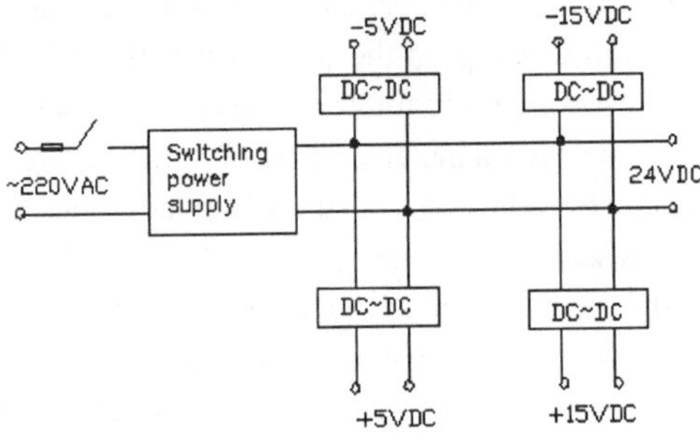

Figure 4. Power supply system

3.1.3 *Embedded system*

The Embedded system is the heart of the monitoring apparatus that mainly includes PCM-3336, LM32C043, keyboard and special circuit used to process all kinds of electrical signals.

The PCM-3336 comes equipped with an embedded micro-controller ALI M6117C that is Intel 386SX-40 compatible. In addition, it comes with two serial ports (RS-232). The PCM-3336 incorporates a PC/104 connector into its design and features an SVGA interface that supports CRT and Flat Panel (TFT, STN, Mono and EL displays), with 1 MB onboard display memory. Except for these, it includes the DiskOnChip 2000 family of products that is a solid-state disk with no moving parts, resulting in a significant reduction in power consumption and an increase in reliability. The DiskOnChip is a small plug and play flash disk. It is easy to use. The DiskOnChip 2000 package is pin-to-pin compatible with a standard 32-pin EPROM device. The DiskOnChip of 16MB is utilized to store acquired data and application program.

LM32C043 is a kind of LCD that can be directly connected to the PCM-3336 and used to display implication of the action to the builder. Keyboard consists of different keys, such as number, letter and direction that are used to input pile number and length, forming hole and compacting water pressure, forming hole and compacting current, compacting length and so on.

The special circuit contains analog circuit, that processes current and pressure signals, and digital circuit that converts analog signal to digital signal and counts pulse signal and determines the direction of the vibrator movement and processes the keyboard event.

The circuit shown in the figure 5 is used to processes water pressure and current signals because they are about from 4 to 20 milliampere current. Electrical transformer converts the current to the voltage of from 0V to 1V and low-pass filter removes unwanted signal and amplifier enlarges the voltage signal to the extent so that A/D converts the analog quantity to digital quantity for the signals to be processed by PCM-3336. According to Shannon theorem, the upper cut-off frequency of the low filter must be equal to the highest frequency in the desired signal and less than moiety of the sampling frequency.

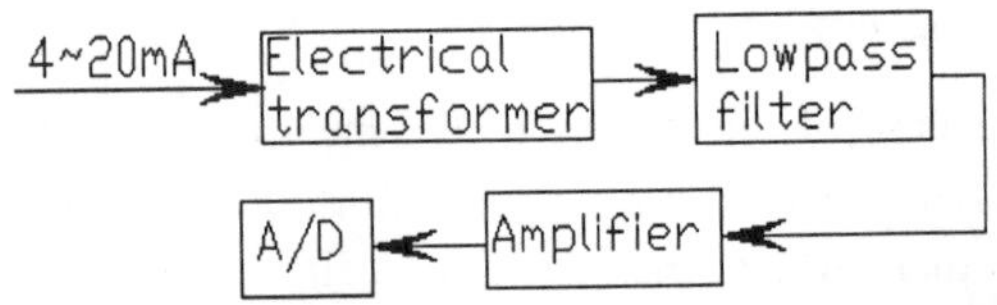

Figure 5. Analogue signal processing circuit block diagram

The circuit shown in the figure 6 is used to convert the pulse signal to another pulse signal and directing signal. Receiver converts differential signals A1 (B1) and A2 (B2) to TTL signal A (B). Sense finder and double frequency multiplier transforms the displacement pulse signals (A, B) to directing state signal (D) and another kind of displacement pulse signal (P) whose frequency is four times as great as original frequency and so increases the displacement resolution. The counter calculates the angle pulse and transforms it to the PCM-3336.

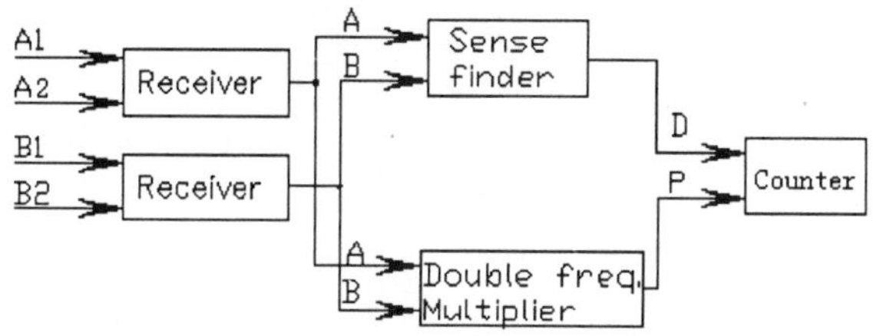

Figure 6. Pulse signal processing circuit block diagram

Keyboard drive circuit consists of a parallel input and a parallel output every one of that have 8 bits. Each bit of them is connected to grounding through a resistance of about 4.7 kilo-ohms in order to avoid troublesome noise interference when no switches are being pressed. Each pair of input and output bits are connected together through switches whose joggling is removed using software delay when keys are impressed. The parallel input and output may form sixty-four keys, but only thirty-four keys of them are used in the keyboard circuit.

3.2 *System software design*

The system software of the monitor and control equipment is developed on a special platform using Borland C++, which features friendly user interface and real timing character and whose size is very small. It contains three layers and two kinds of database.

One kind of the database is used to store calibration factor and other parameters, such as current factor, displacement factor, pressure factor, compacting current, compacting length and so on, the content of which is relatively stable. Another database is utilized to store measured data record, which comprised of electric current, pressure and displacement at the same time and their times. Its records are different for all different piles and can never be changed. These records may be transformed to a host computer by serial communication in order to analyze the quality of the piles offline. And then, construction of the gravel piles may be direct and errors of the process of forming piles removed at their germination. The three layers communicate each other to finish their tasks and the last layers must write and read data to or from the database and acquire data out-world through drive program.

The drive program contains A/D drive software module, counter drive software module, keyboard drive software module. A timing interrupted sever program manages all these drive module. The A/D drive module is executed every sample period and other drive modules are executed every 10 milliseconds. They store their data in themselves' storage and any one of other modules can get data from the buffer by a semaphore flag. The flag may be 1 or 2. When the flag is 1, the store can be operated on. Or it can neither be read nor written. The semaphore protects the shared store to be disrupted. Every module operating on the shared store must first check the flag and if the flag is 1, it modifies the flag to 2 and reads or writes the store and then returns it to 1 and returns or it must returns immediately. The semaphore can only fit to the modules with the same priority level. So another semaphore must be found to satisfy modules with different priority level.

To make all modules not to be dependant on each other, a kind of module is designed between some two modules. For example, A/D drive module gets data from the hardware but the data is explained by another module in order that A/D drive module may not be while measured object changes. All these software technologies make the software system to be reliability and manageable and have very high software code proficient.

4. Conclusions

The monitoring apparatus can measure the forming hole's water pressure and electrical current, compacting water pressure and electrical current, compacting length, pile length and so on. Except for these, the displacement sensor's erecting device can be installed on any crane without modifying the structure of the crane. It can be utilized on any other place where electrical current, water pressure and displacement must be measured. The result of testing the apparatus in the construction of Da Yang mountain harbor proved it to be easy operating and poses high reliability and satisfy the designing objective.

But there are some things to be done in the future. For example, how water pump and three-phase alternating current motor are controlled by the apparatus, in order to provide active controllability of the construction of the gravel piles and how the apparatus is connected to other equipment to form a complete manage system.

References

1. Editorial Committee of 'The Foundation Processing Manual', The Foundation Processing Manual (The Second Edition). Bei Jing: China Architecture and Building Press , 2000, 22-25
2. He Guang Ne, Compound Foundation of Rushed Stone Pile. Bei Jing: China Communication Press , 2001, 4-5
3. Baumann V, Bauer G E A, The Performance of Foundations on Various Soils Stabilized by the Vibro-compaction Method. Canadian Geotechnical Journal, 1974, 3(2); 509-530
4. Mary Shaw, David Garlan. Software Architecture: Perspectives on an Emerging Discipline. Englewood Cliffs, NJ: Prentice Hall, 1996.
5. Irvine, R. G., Operational Amplifiers Characteristics and Applications. Englewood Cliffs, NJ: Prentice Hall, 1994.

EXPERIMENTAL STUDY ON REINFORCED-AT-HIGH-STRESS-REGION SOIL-CEMENT RETAINING STRUCTURE FOR DEEP FOUNDATION PIT

DAYU ZHU

Department of Structural Engineering, Tongji University, 1239 Siping Road Shanghai 200092, China

Rapid development of high-rise and underground buildings makes exact demands on retaining structures for deep foundation pit. As a retaining structure, reinforced soil-cement mixture has more advantages compared with Soil Mixing Wall. But both experimental study and designing theory for it are far behind practice. By model testing, distribution and variety of stress in soil-cement mixture and in shaped steel during loading process are studied in this paper. Strong restraint of shaped steel on cracks is found by analysis of crack development, which can increase the cracking load and the rigidity of specimen obviously. By the data analyzing, effects of factors to structural behaviors are discussed, and some suggestions are made towards designing. By analyzing the experimental data, carrying capacity expression is derived theoretically.

1. Introduction

Rapid development of high-rise and underground buildings makes exact demands on retaining structures for deep foundation pit. Reinforced-At-High-Stress-Region Soil-Cement Retaining Structure is a kind of retaining wall composed of 2 to 4 rows of Soil-Cement piles with I-shape steel (commonly 140 mm to 200mm in height) embedded in before the prehardening of cement. Dual role of retaining and water sealing can be completed at the same time. The typical form of this retaining structure see Figure 1.

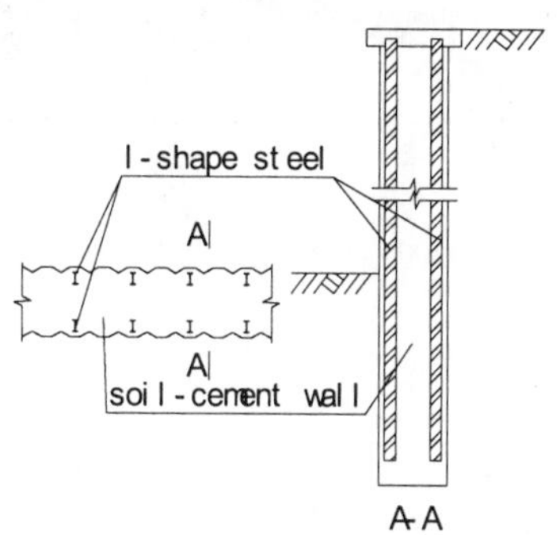

Figure 1. The common form of Reinforced-At-High-Stress-Region Soil-Cement Retaining Structure.

As a retaining structure, reinforced soil-cement mixture has more advantages compared with Soil Mixing Wall. However studies focus mostly on material characteristics of soil-cement mixture now, and few studies has ever been done on reinforced soil-cement mixture as a retaining structure.

In this paper, the behavior of reinforced soil-cement as retaining structure is explored through model test, distribution and variety of stress in soil-cement mixture and in shaped steel are studied during loading process; strong restraint of shaped steel on cracks is found by analysis of crack development. Accordingly the effects of factors to structural behaviors are discussed, and some suggestions are made towards designing. By analyzing the experimental data, carrying capacity equation is derived theoretically.

2. Model Test

2.1 *Design of specimens*

The test adopted similitude coefficient S=1/2 to simulate the biaxial soil-cement pile 700mm in diameter. Table 1 gives the Parameters of the specimens.

Table 1. Parameters of the specimens.

No.	A-1	A-2	A-3	A-4	A-5a~c	A-6	A-7	A-8a~c
Cement content	8%	13%	17%	8%	13%	17%	8%	13%
Number of specimen	1	1	1	1	3	1	1	3
Section of specimen (b×h) (mm^2)	250×600			500×600			500×600	
Arrangement of shaped steel	Unilateral						Bilateral	
Section of shaped steel	6.3							
Span (m)	1.8							
Kind of cement	Ordinary Portland cement (42.5)							
Water-cement ratio	0.5							

2.2 *Preparation of specimens*

Clay in Shanghai region was adopted as the soil of the specimens, which was natural desiccated and sifted out. Water content of the natural-desiccated soil was measured and the cement content and water addition were decided by calculation. The raw material was mixed in mortar mixer and cast in wooden formwork.

2.3 *Design of loading*

Uniformly distributed load applying on the specimens was substituted equivalently by eight-points concentrated load imposed though jack and steel beam (Figure 2). Axial force and lateral restraint was imposed to simulate structure weight and interaction between adjacent elements.

Figure 2. Loading equipment.

2.4 *Load-displacement process*

Figure 3 and Figure 4 show the measured relationships of load and deflection in the middle of span. The curve can be divided into three phases. The first phase is before cracking, and the displacement keeps The plane cross-section assumption. Once cracking, the specimen enters the second phase. Test data show that the initial crack appears in the middle of span for all specimens. During the loading process, expanding of the cracks rest at shaped steel for some time and this phenomenon indicates that shaped steel make a strong restriction on cracking of soil-cement mixture. Along with load increasing, cleavage cracks occur at the bottom surface of specimens, and obliqueangled cracks at the quarter of span expand rapidly on the side surface. The specimen enters the third phase. When cleavage cracks extends to support, load - deflection curve reaches the peak. After the peak, obliqueangled cracks burst in close to supports, expand suddenly and intersect the obliqueangled cracks at the quarter of span at upper part of the side surface. The following exfoliation of the soil-cement between the two obliqueangled cracks indicates the failure of the specimen.

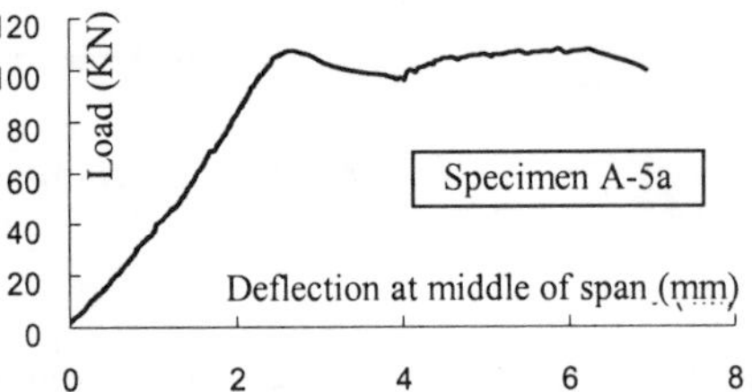

Figure 3. Load-deflection curves of specimen A-5a

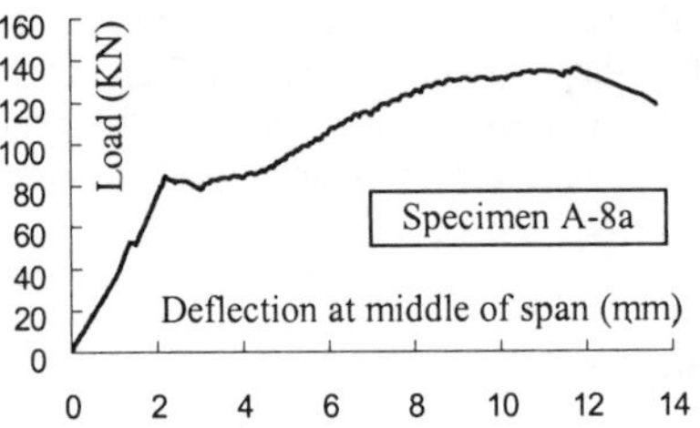

Figure 4. Load-deflection curves of specimen A-8a

2.5 *Failure type*

The disfeaturement of specimens (see Figure 5) indicates a kind of shear failure.

(a) Exfoliation of the soil-cement.

(b) Cleavage cracks at the bottom surface.

Figure 5. The disfeaturement of specimens.

3. Analysis of the Test Results

3.1 *Strain analysis of soil-cement*

Measured strain of soil-cement section in the middle of span (Figure 6) shows that strain of the whole section develops towards the direction of compression. After the obliqueangled cracks at the quarter of span occurs, the whole section is in a complete compressive state. During the loading process, the strain above neutral axis increases most rapidly and keeps a maximum value. Based on the stress trajectory (Figure 7), we can suppose that the section above neutral axis fills the role of an arch through which load was mainly transferred.

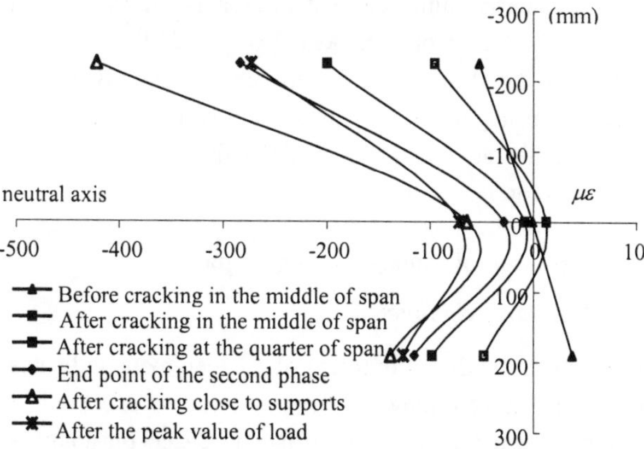

(a) Specimen with unilateral arrangement of shaped steel.

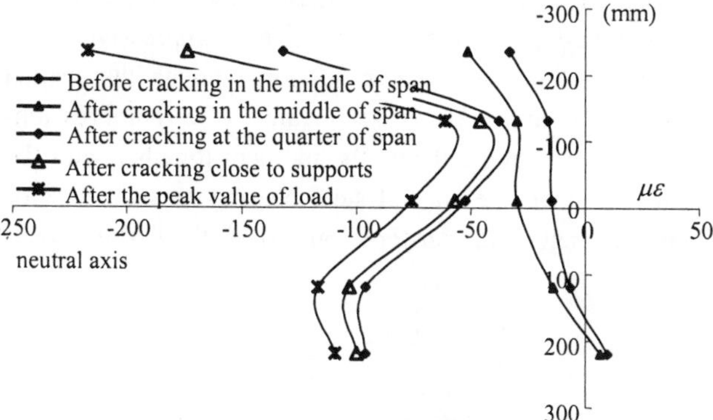

(b) Specimen with bilateral arrangement of shaped steel.

Figure 6. Measured strain of soil-cement section at middle of span.

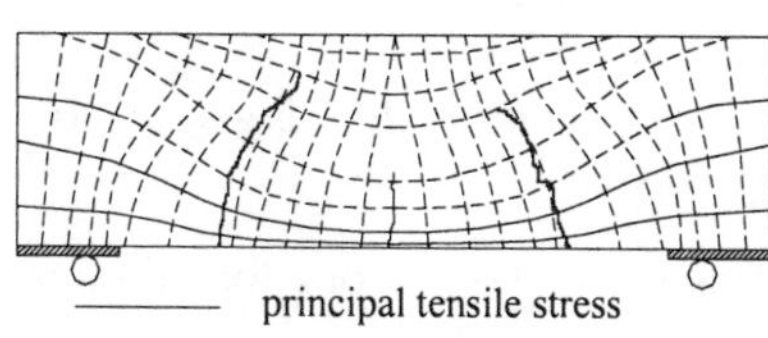

Figure 7. Stress trajectory after cracking at the quarter of span.

3.2 *Analysis of shaped steel*

Figure 8 shows measured strain of shaped steel. The three curves denote the strain variety of the section at the quarter of span, and close to supports. It is obvious that the strain of the section in the middle of span has an abrupt change after cracking and the strain of the section at the quarter of span increase rapidly to a value larger than in the middle of span after cracking. This also proves that the stressed state of specimen turns into an arch with a tension bar after cracking at the quarter of span.

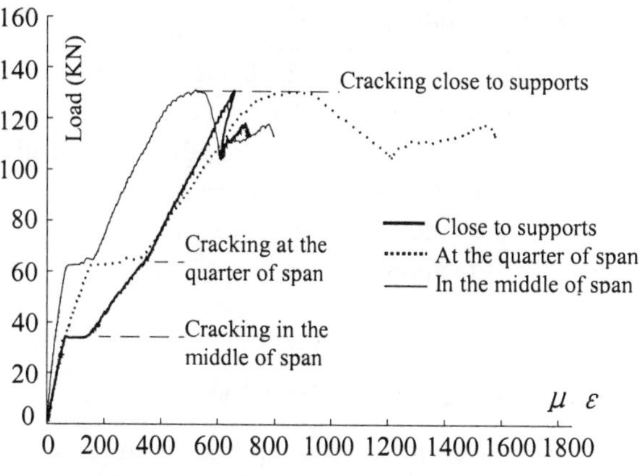

(a) Specimen with unilateral arrangement of shaped steel.

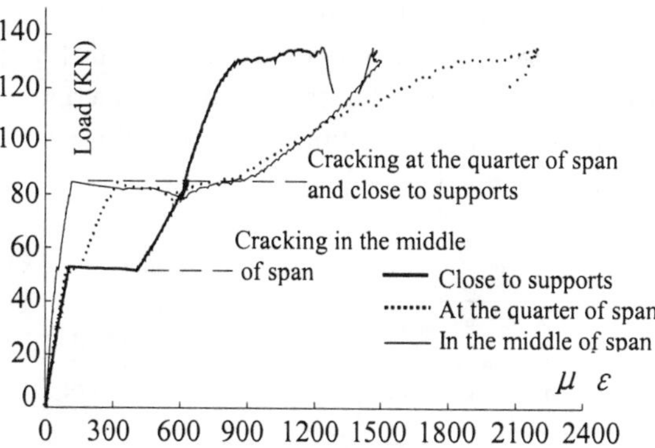

(b) Specimen with bilateral arrangement of shaped steel.

Figure 8. Strain of shaped steel.

3.3 *Analysis of test parameter*

3.3.1 *Cement content*

Table 2. Effect of cement content on structure performance.

Section of specimen (b×h) (mm^2)	Arrangement of shaped steel	Cement content (α_w)	Cracking load (P_{cr}) (KN)	Ultimate load (P_u) (KN)
250×600	Unilateral	8%	9.80	72.27
		13%	12.44	91.89
		17%	21.63	148.73
500×600		8%	18.55	59.49
		13%	35.85	139.37
		17%	38.67	107.21
500×600	Bilateral	8%	29.42	80.25
		13%	50.98	135.32

As shown in Table 2, higher cement content increases both cracking and ultimate load when other parameters are fixed. So a piece of advice can be given that higher cement content should be used in practice to improve the impermeability and ultimate bearing capacity of structures.

3.3.2 *Shaped steel spacing*

When other parameter are fixed, Table 3 shows:

1. Larger shaped steel spacing leads to lower ultimate bearing resistance.
2. Larger shaped steel spacing leads to lower steel ratio and better ductility.

Another piece of advice can be given that shaped steel spacing should be chosen by considering both carrying capacity and ductility of structure. Increasing or decreasing it blindly is mistaken.

Table 3. Effect of shaped steel spacing on structure performance.

Shaped steel spacing (mm)		250	500
Ultimate load per meter in width of specimen (P_u) (KN/m)	α_w=8%	289.08	118.98
	α_w=13%	367.56	278.74
	α_w=17%	594.92	214.42
Ductility factor(μ)	α_w=8%	1.27	2.38
	α_w=13%	1.52	2.09
	α_w=17%	1.6	2.68

3.3.3 *Arrangement of shaped steel*

Test result indicates that the specimens with unilateral and bilateral arrangement of shaped steel have almost the same carrying capacity, but the latter has a higher cracking load in both middle and quarter of span, moreover, it has better ductility (see Table 4).

Table 4. Effect of Arrangement of shaped steel on structure performance.

Arrangement of shaped steel		Unilateral	Bilateral
Cracking load at middle of span (KN)	α_w =8%	18.55	29.42
	α_w =13%	35.85	50.98
Cracking load at quarter of span (KN)	α_w =8%	37.05	64.8
	α_w =13%	66.47	76.61
Ductility factor(μ)	α_w =8%	2.38	4.17
	α_w =13%	2.09	6.17

In addition, load–deflection curve of specimens with bilateral arrangement of shaped steel descends minutely after reaching the first peak load (P1), then appears an ascending segment and reaches to the second peak load (P2). Common P2=(1.075~1.212) P1 in test data. So the third piece of advice can be given that bilateral arrangement of shaped steel should be adopted to restrain cracks and increase the ductility factor and safety factor.

4. Equation of Shear-bearing Capacity

4.1 *Mechanical model*

In our test, the disfeaturement of all specimens indicates a kind of shear failure, so the key issue to this kind of retaining structure is shear resistant. Based on analysis of test data, the load- transferring pattern can be assumed as an arch with a tension bar (Figure 9): the under part (II) bears obliqueangled compressive stress and transfers the stress to supports though the shaped steel; the arch (I) works mainly as a compression-bearing part and balance the moment together with the shaped steel; the shaped steel bearing tensile stress of arch springer and transfer shear though the foxtail effect.

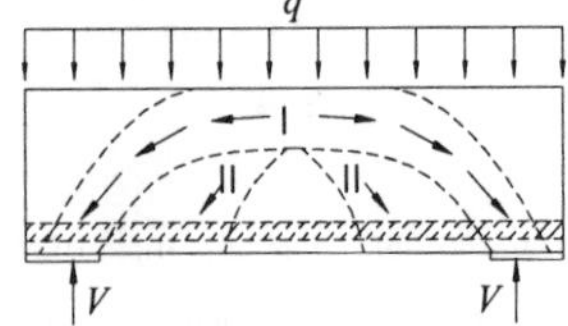

Figure 9. Mechanical model

4.2 *Analysis and equation deriving of shear-bearing capacity*

Because shaped steel has a strong foxtail effect, the total shear-bearing capacity (V_u) is composed of two parts: the shear-bearing capacity of the arch (V_a) and of the foxtail effect of shaped steel (V_s). Superposition method is adopted in calculation, which can be expressed as follows:

$$V_u = V_a + V_s \tag{1}$$

4.2.1 *The shear-bearing capacity of the arch (V_a)*

The diagram for analysis see Figure 10(a).[1] according to the mechanical characteristic, the specimen can be divided into four parts: part I bears vertical uniform compressive stress (q/b), where b is the width of the specimen; part II bears bliqueangled compressive stress (σ_c), and imposes vertical compressive stress and horizontal shearing stress on shaped steel, where α is the included angle between the direction of σ_c and bottom of specimen; part III is a curvilinear zone, and soil-cement bears single-dimension compressive stress in this zone; part IV is a peripheral zone near support in which soil-cement bears compressive stress, too.

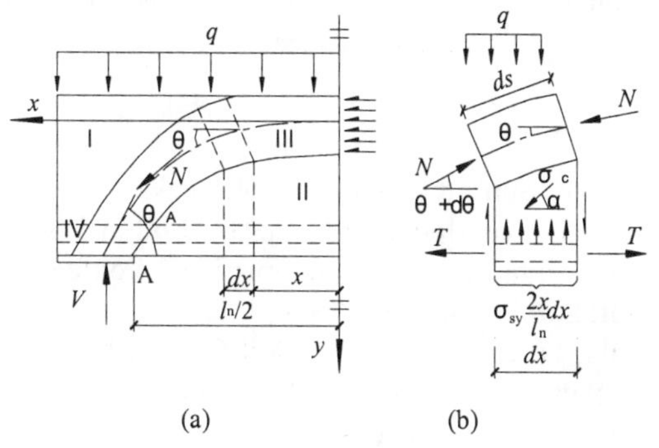

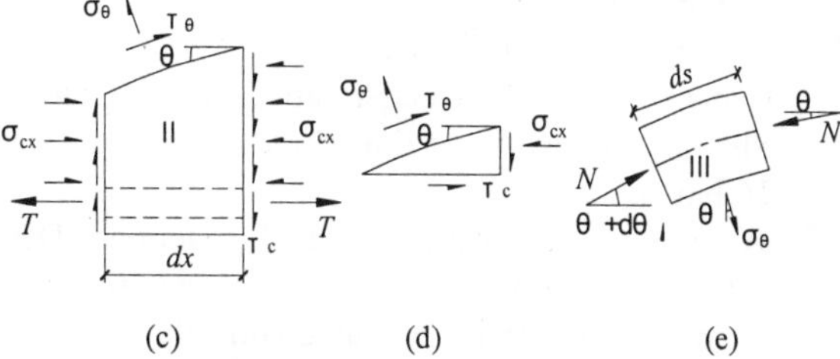

Figure 10. Diagram for analysis of shear-bearing capacity

(a) Force state of segmental specimen; (b) Force state of element; (c) Equilibrium of element in II; (d) Equilibrium of element on interface between II and III; (e) Equilibrium of element in III.

According to the method in reference 1, we get the following two equations at point A:

$$y_A = \frac{\sigma_{sy} b}{N} \cdot \frac{l_n^{\,2}}{24} - \frac{q}{N} \cdot \frac{l_n^{\,2}}{8} \tag{2}$$

$$y_A' = \mathrm{tg}\,\theta_A = \frac{\sigma_{sy} b}{N} \cdot \frac{l_n}{4} - \frac{q}{N} \cdot \frac{l_n}{2} \tag{3}$$

When compressive stress in part III equals to the compressive strength of soil-cement (f_c), the shear-bearing capacity is reached. At this time, the shear at point A (V) can be expressed as the following Eq.

$$V = f_c b (h - y_A) \sin\theta_A \cos\theta_A \tag{4}$$

Considering the decreasing of the effective contact area because of construction deviation, add a factor of supporting condition (α_s) in the right part of Eq. (4) to discount the contact area, then substitute Eq. (2) and Eq. (3) into Eq. (4), yields:

$$V = \alpha_s f_c b h \left[1 + \frac{\sigma_{sy} b}{N} \cdot \frac{l_n^{\,2}}{48h} - \mathrm{tg}\,\theta_A \left(\frac{l_n}{4h} \right) \right] \sin\theta_A \cos\theta_A \tag{5}$$

Determine the maximum value of V using $dV/d\theta_A=0$, then substitute l/h for l_n/h and let $h=1.16h_0$, where l is the theoretical span, h is the height of the section, h_0 is the effective height of the section, we get the following equation:

$$V = 0.58\alpha_s \left[\sqrt{\left(\frac{l}{4h} \right)^2 + m^2} - \left(\frac{l}{4h} \right) \right] f_c b h_0 \tag{6}$$

where

$$m = 1 + \frac{\sigma_{sy} b}{N} \cdot \frac{l^2}{48h} \tag{7}$$

To take off the effect of shaped steel, let $\sigma_{sy}=0$, thus $m=1$, and Eq. (6) turns to the following form:

$$V_a = 0.58\alpha_s \left[\sqrt{\left(\frac{l}{4h} \right)^2 + 1} - \left(\frac{l}{4h} \right) \right] f_c b h_0 \tag{8}$$

Eq. (8) is the shear-bearing capacity of the arch.

4.2.2 *The shear-bearing capacity of the foxtail effect of shaped steel (V_s)*

Before the obliqueangled cracks cross the whole section of specimen, the shape steel at top of specimen bears compressive stress together with the soil-cement around it and have minute foxtail effect, so we only reckon in the foxtail effect of shaped steel at bottom. Its shear-bearing capacity can be defined as:

$$V_s = \alpha_d t_w h_w f_v \tag{9}$$

in which h_w and t_w is the height and thickness of web, respectively, f_v is shearing strength of steel, α_d is reduction factor of shear-bearing capacity to reckon in unyielding of shaped steel.

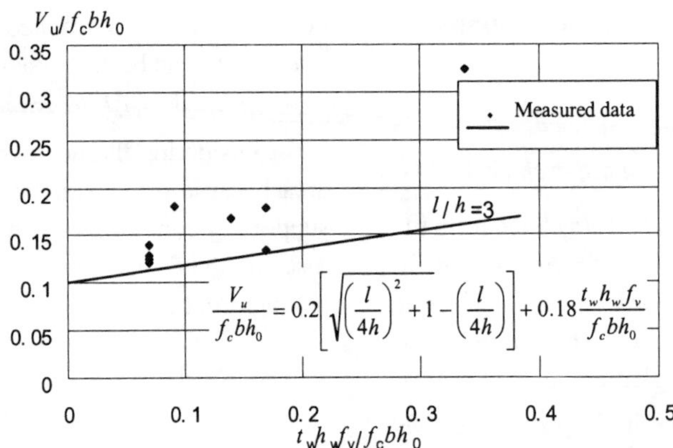

Figure 11. Comparison between measured data and equation of shear-bearing capacity

4.2.3 *Equation of shear-bearing capacity*

Based on above analysis, equation of shear-bearing capacity for Reinforced-At-High-Stress-Region Soil-cement Retaining Structure is provided as following:

$$V_u = 0.58\alpha_s\left[\sqrt{\left(\frac{l}{4h}\right)^2+1}-\left(\frac{l}{4h}\right)\right]f_cbh_0+\alpha_dt_wh_wf_v \quad (10)$$

The factor α_d should be determined through test. The measured data and equation of shear-bearing capacity are shown in figure 11. We can see that when equation is the lower envelope line of the measured data, the value of α_d is 0.18.

5. Conclusion

The main conclusions may be drawn as follows:

1. The strain of the whole section of soil-cement develops towards the direction of compression. the stressed state of specimen turns into an arch with a tension bar after cracking at quarter of span. The section above neutral axis fills the role of an arch through which load is transferred mainly. The shaped steel bearing tensile stress of arch springer and transfer shear though the foxtail effect.
2. Higher cement content can increase both cracking and ultimate load; Larger shaped steel spacing can lead to lower ultimate bearing resistance but better ductility. Retaining structures with unilateral and bilateral arrangement of shaped steel have almost same carrying capacity, but the latter has a higher cracking load and better ductility.
3. Based on analysis of test data and supposed analytical model, the expression of shear-bearing capacity is provided. Value of factor α_d in the expression is determined under experimental condition. This expression can be used as a guideline in design practice.

References

1. De'an Sun & Haijime Matsuaka, A Simple Constitutive Model for Cemented Soil in 3-D Stress, Proceedings of Second International Conference on Soft Soil Engineering, 1996, Vol. 1.
2. Taki, O. & Yang, D. S., Soil-Cement Mixed Wall Technique, Geotechnical engineering congress 1991: geotechnical special publication No. 27, ASCE, New York, N. Y., Vol. 1.
3. Liu Lixin, An Unified Calculation Method for Shear Capacity of R. C. Deep Beams, Short Beams and Shallow Beams, Journal of Building Structures, Vol. 16 No. 4.
4. Wang Jian, Xia Mingyao and Fu Deming, Design and Calculation of Composite Structure with H Shaped Steel and Cemented-soil-pile, Journal of tongji university, Vol. 26 No. 6.

CONCEPT OF PLASTICALLY BEARING PILE AND ITS ENGINEERING APPLICATION*

ZAI JINMIN[1)†] MEI GUOXIONG[4)] WANG XUDONG[5)] ZHOU FENG[6)]
College of Civil Engineering, Nanjing University of Technology
Nanjing, 210009, P.R.C.

PEI JIE[2)]
Wei Lian Design Group
Shanghai, 200092, P.R.C.

LIAO HESHAN[3)]
Construction Office of Xiamen
Xiamen, 361000, P.R.C.

In this paper, the ultimate bearing capacity of single pile is used for the pile, and the loads acted on the composite pile foundations are distinctly shared by the piles and the soil under the cap. Based on the theory of composite pile foundations, the concept of plastically bearing pile is established. The methods of settlement calculation, the design principals of application and some problems of nonlinear design of composite pile foundations interaction are presented. These methods and concept have been used for the design of a 9-story library and a 30-story high-rise. The success of the field experimental studies has fully verified the feasibility and advance of the research and it has a broad prospect in practical application.

1. Concept of Plastically Bearing Pile and its Settlement Calculation

The key concept of composite pile foundations is that let working load of single pile be equal to or close to the ultimate bearing capacity of single pile P_u. For composite pile foundations with extraordinary spacing, such as bigger than or equal to 9d (d-pile diameter), working condition of piles turns from elastically bearing to plastically bearing: 1. bear the load of P_u all the time even though total load still increases; 2.do not provide new bearing rigidity, that's, the settlement condition is controlled by the soil; 3.can stop at any point of a、b、c、d etc in Fig. 1, when appear a little bearing of external force. Composite pile foundations with extraordinary spacing is called plastically bearing pile. It follows the characters of:

1. autonomicity: bearing load of plastically bearing pile is always equal to P_u;
2. adaptability and concordability: pile top settlement under raft S_{0P} is equal to the soil settlement under raft S_{0S}:

$$S_{0P} = S_{0S} \qquad (1)$$

Formula (1) means that when $Q_P = nP_u < Q$, n pieces of plastically bearing piles can unload Q_p=the pressure calculates nPu, and the settlement of foundations $p_s = (Q - nP_u)/A$.

In short, the concept of plastically bearing pile is similar to that of plastic hinge in statically indeterminate structures.

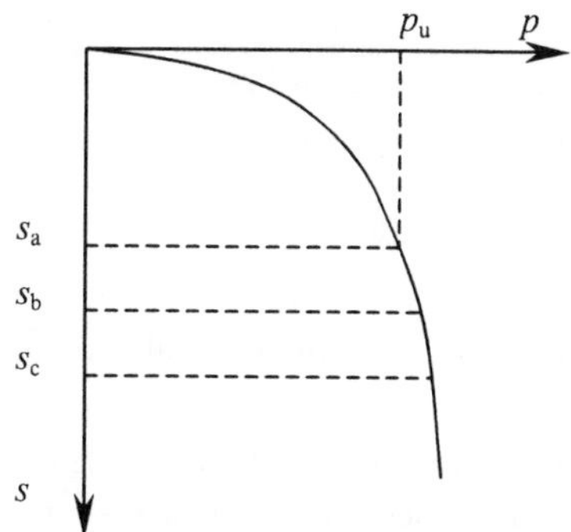

Fig.1 Concept of plastically bearing pile

* This work is supported by National Natural Science Foundation 50478005.
† Work partially supported by grant 04KJB560048.

2. Settlement Calculation of Plastically Bearing Pile

Formula (1) shows that the settlement of plastically bearing pile is controlled by the soil settlement under raft S_{0S}. At the same time, the measured result of a model test shows that for small pile groups with extraordinary spacing, for example, with $8d$ in the test, compression of soil between piles covers more than 95% of total settlement and appears within 1/3 of pile length beneath pile cap. Therefore, it can be calculated by :

1. layer-wise summation method according to additional base pressure of pile cap $p_0 = \xi f - \gamma_0 D$.
2. settlement formula of rigid plate.

$$S_{os} = m_r \cdot \omega_r \cdot b \cdot p_0 \cdot \frac{1-\mu_0^2}{E_0} \tag{2}$$

3. pile and soil nonlinear rigidity method

3. Fundamental Design Principles of Plastically Bearing Pile

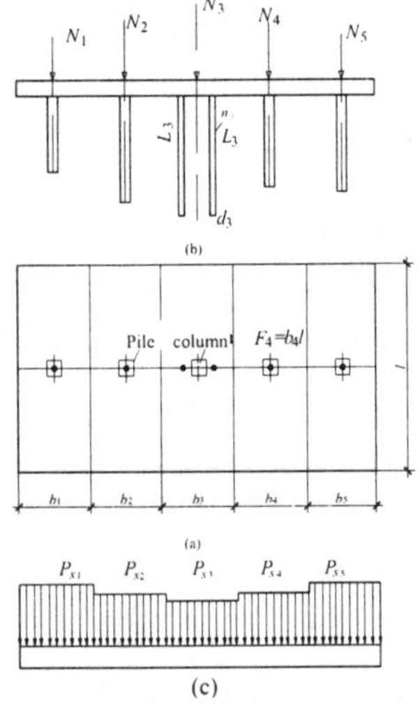

Fig. 2 Case designed by plastically bearing pile

Fig. 2 shows the local plan and section of a raft foundation with piles. Both the loads of column 1~ column 5 $N_1 \sim N_5$ and bottom areas distributed by column $F_1 \sim F_5$ are not equal each other. Plastically bearing piles, different from diameter, length and number, are assigned under columns. The fundamental principles of pile layout are:

$$\left.\begin{aligned} &Q = Q_s + Q_p \text{ , } K = Q_u / Q \geq 2.0 \\ 1.\ &Q_p = \sum \zeta n_i P_{ui} \text{ , } Q_s = Q - Q_p \leq 0.5 fA \\ &\zeta = 0.9 \text{ , } Q_u = 2.5 fA + \sum n_i P_{ui} \end{aligned}\right\} \tag{3}$$

$$2.\ p_{si} = \frac{N_i - \zeta n_i P_{ui}}{F_i} \text{ , } F_i = b_i l \tag{4}$$

p_{si} is close to each other in order to eliminate local differential settlement.

3. The distribution of p_{si} is a little bigger at the corner and margin and a little smaller in the center in order to minimize or eliminate the whole differential settlement.

Moreover, considering the barrier effect of piles and the group efficiency, total coefficient of safety K will be increased by about $5 \sim 12\%$.

4. Some Problems of Nonlinear Design of Composite Pile Foundations Interaction

4.1 *Layout of pile and control of settlement distribution*

The principle of layout of pile is to reduce the differential settlement and interior force of rafts. Position of pile accords with the position of bearing walls and columns is as near as possible. And when load is big (small), piles arranged are many (bit). When structure loads distribute uniform and footing beam is not arranged owing to smaller spacing of piles, piles distribute uniform fundamentally (densely in the interior; sparsely in the exterior); or footing beams are arranged, piles are arranged under the columns and beams (not arranged under the slab). Otherwise, rafts under the tube become thick and piles under tube become denser too. The center of plan of pile group accords with that of gravity of structure as near as possible.

4.2 *Adjusting of local differential settlement*

Although loads and foundations bottom areas shared by every columns are different, the back pressure value ($p_i = (N_i - n_i p_u)/F_i$) of bottom of slab should be made fundamentally the same in order to reduce local differential settlement. When plan of raft-foundation is regulation and rigidity of structure is big, we can make p_i at the corner and margin a bit bigger than that at the central area in order to eliminate completely the integral differential settlement.

4.3 *Design of footing beam*

The section of footing beam adopts trapezoidal section to substitute pile caps and we tamp soils under the beam in order to concentrate the backpressure to the bottom of beams.

4.4 *Design of raft-plate*

The backpressure of bottom of raft can be estimated by summation of soil backpressure and water pressure p_w:

$$p = \frac{\Sigma N_i - \Sigma n_i p_u}{\Sigma F_i} + p_w \qquad (5)$$

5. Engineering Example Analysis

Plastically bearing pile can be used in friction pile (penetration deformation on pile tip) and end-bearing pile engineering (penetration deformation on pile top).

5.1 *Friction pile engineering--9-story library*

5.1.1 *Engineering introduction*

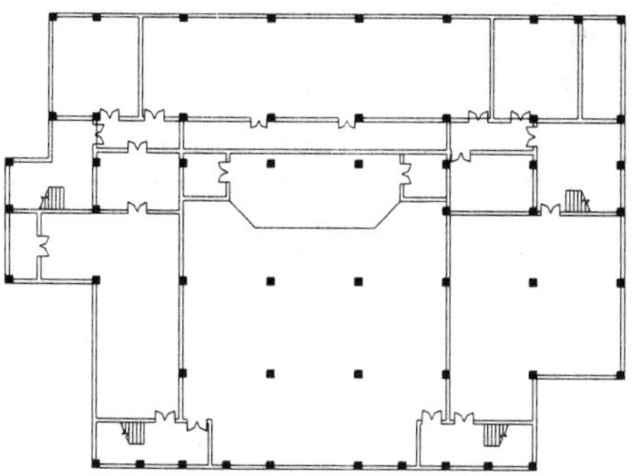

Fig. 3 Plan of library basement

The building in question is a 9storeys library with a 1storey basement. The superstructure is a framework. Bottom area is 1655m^2. Total load is 190780kN. Fig. 3 shows library plan of basement and Fig. 4 shows section of the library. Because the storey number of the library in the north decreases owing to usage, the building is eccentric to the south.

5.1.2. *Project Design*

Average foundations pressure p is 115.3kPa. Embedded depth of foundations D is 2.8m. The stratums are stratified soil which properties are shown in Table 1. It appears that the stratum 2-1 should be used fully, and the stratum 2-3 is a good supporting layer for the friction piles. So the stratums for this building are very convenient to adopt the composite foundation and use the concept of plastically bearing piles. In fact, if the stratums are selected for a natural foundation, the settlement will be about 30cm and it is too large to acceptable.

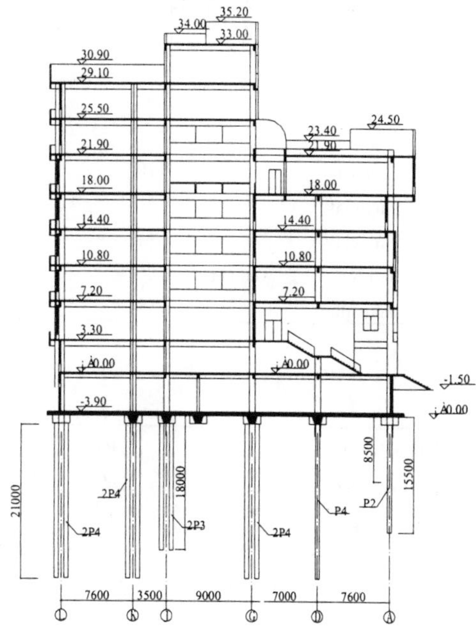

Fig. 4 Section of library

In order to reduce the settlement and minimize the differential settlement, 7 kinds of plastically bearing piles and their combination are adopted to reduce and uniform the back pressure (40~50kPa), and a little bigger (20kPa) at the corner and margin. Table 2 shows the unloading of piles. The piles are static press piles which section sides and lengths are shown in Table 2 also.In order to minimize the settlement, giving p_s=50kPa, so ξ=50/140=0.357. Giving ζ=1.0, ψ=140/115=1.21 and K=1+(2.5-0.357)×1.21= 3.59 > 2, Q_s=Ap_s=1655×50=82750, Q_p/Q=56.6%. In actual piles unload 60% of total load. Fig. 5 shows the distribution of plastically bearing pile. Fig. 6 shows the Calculated p_s distribution of foundations. The final unloading is 119000kN, unloading rate is 62%, and the average soil back pressure $\overline{p}_s$ = (190780-119000) /1655=43.3kPa。Calculated settlement is S=2.89cm according to the layer-wise summation method.

5.1.3 *Measured results and analysis*

Fig. 7 shows soil backpressures and Fig.8 shows settlements. The mean settlement is about 14mm. Because of heavy rain during construction, the foundation soils along the southern margin have been disturbed heavily, so the max differential settlement is different, 3mm in the longitude and 8mm in the latitude. Comparing the Fig. 7 with Fig. 6, it is clear that the back pressures of soils and their distribution shapes are very close to each other.

Fig. 9 shows the measured s~t curve of library. The agreement of the measured curve with that of theoretical curve of foundations soil under box foundations proves the reasonability and security of plastically bearing piles.

Table 1. Engineering properties of stratums

Stratums	Name of soil	Thickness(m)	W(%)	γ (kN/m^3)	e	I_P	I_L	f_k(kPa)	E_s(MPa)
1	Fill	2							
2-1	Silty sand	6	29.0	18.8	0.84			140	10.6
2-2	Silt	13	35.1	18.3	0.98			110	7.7
2-3	Silty sand	10	29.1	18.6	0.87			160	8.5
2-4	Silty clay	5	37.1	18.7	0.91	11.2	0.88	120	4.8
2-5	Silty clay	11	32.1	18.2	0.97	11.1	0.99	100	4.1
3-1	Heavy weathering sandstone	2						350	

Table 2. Unloading of all kinds of combination.

Combination	2P4	2P3	2P2	P4	P3	P2	P1
Unloading (MN)	3.4	2.8	2	1.7	1.4	1	0.5
Pile length(m)	21	18	15.5	21	18	15.5	8.5
Section sides(cm)	45 × 45	45 × 45	35 × 35	45 × 45	45 × 45	35 × 35	30 × 30
Segment combination (m)	13+8	10+8	7+8.5	13+8	10+8	7+8.5	8.5

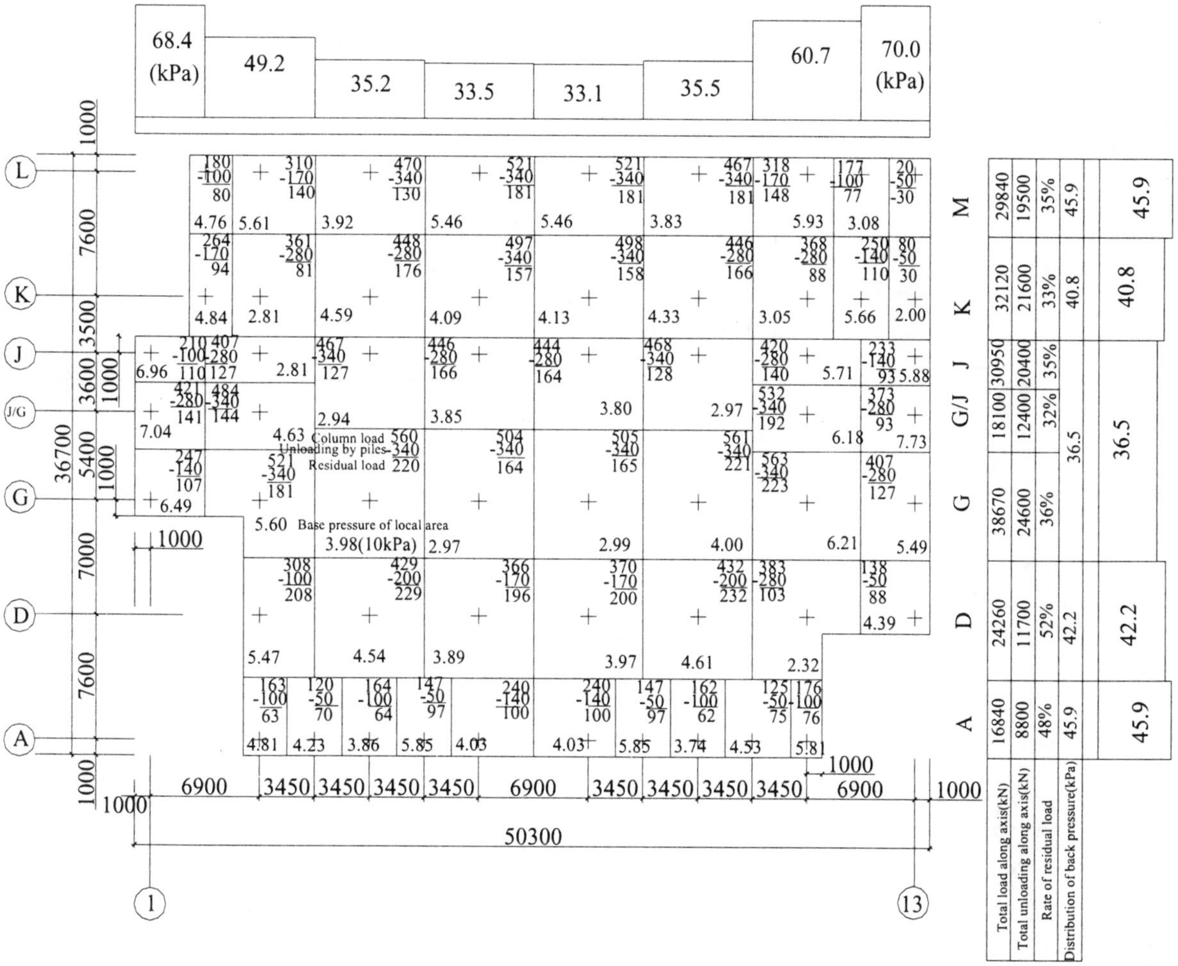

Axis	Total load along axis(kN)	Total unloading along axis(kN)	Rate of residual load	Distribution of back pressure(kPa)	
M	29840	19500	35%	45.9	45.9
K	32120	21600	33%	40.8	40.8
J	30950	20400	35%		
G/J	18100	12400	32%	36.5	36.5
G	38670	24600	36%		
D	24260	11700	52%	42.2	42.2
A	16840	8800	48%	45.9	45.9

Fig. 6 Calculated p_s distribution of foundations

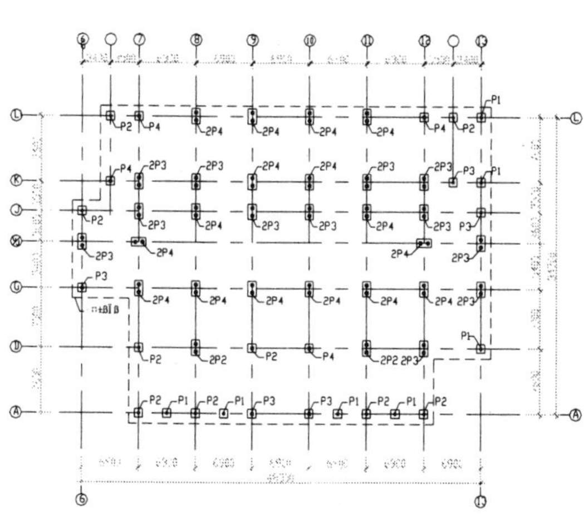

Fig. 5 Layout of plastically bearing pile

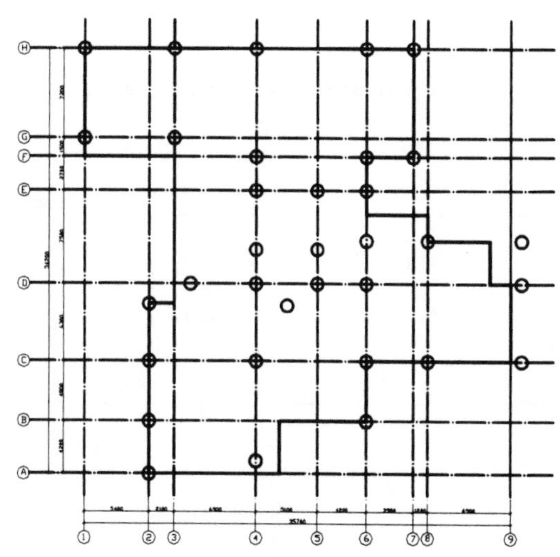

Fig. 10 Layout of plastically bearing pile

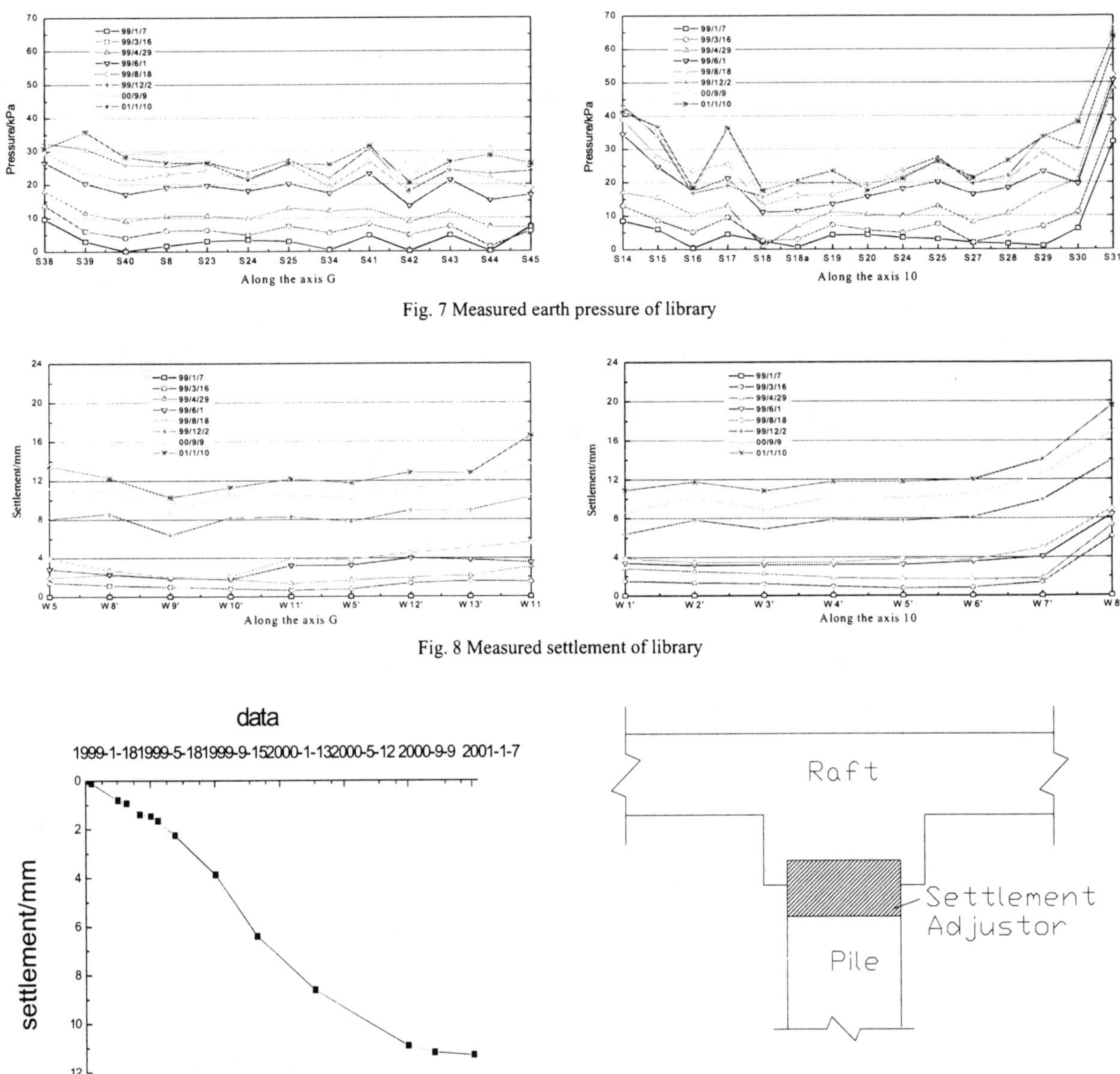

Fig. 7 Measured earth pressure of library

Fig. 8 Measured settlement of library

Fig. 9 Measured $s\sim t$ curve of the library

Fig. 11 Settlement Adjustor

Table 3 Engineering properties of stratums

Code	γ(kN/m^3)	W(%)	I_L	N	e	E_m(MPa)	E_{1-3}(MPa)	E_{3-5}(MPa)	E_{5-7}(MPa)	E_0(MPa)	f_k (kPa)
A	18.5	30	0.12	12	0.985	14.1	5.4	8.5	11.8	12	250
B	19.4	25	0.10	21	0.743	34.8	6.2	10	14	25	300
C	19.9	20	0.01	32	0.631	76.1	7.0	11	16	38	400

5.1.4. Economic Analysis

It is valuable to point that using the methods in the paper, the concrete volume of piles is by 63% less than that according to the conventional method (any contribution of the raft or pile cap being ignored).

5.2 *End-bearing pile engineering—30-story high-rise*

5.2.1 *Engineering introduction*

The construction in Xiamen contains two symmetrical 30-storey-buildings. The bottom is connected by 2–storey-basement and 3-storey- annex basement. Total load is 1,200,000kN. The raft area of the main building is $2200m^2$. Fig. 10 shows one of the two buildings. The field of the construction is very special, beneath which is granite residual soil. Granite residual soil has low residual strength when disturbed. The field also contains lots of atactic granite rocks. The usual design method cannot satisfy the construction.

5.2.2 *Project design*

Average foundations pressure is 545kPa. Embedded depth of foundations is 10.5m. The stratum 1-5 is in the bound of the excavation, so the main affecting factor is the geological condition of granite residual soil. The soil properties of the stratums are shown in Table 3. The modified bearing capacity value of the natural base is 408kPa. Table 3 shows that the soil condition of the field is good, the satisfying rate of the natural base : ψ $=408.05/487.9=0.84>0.5$. So it is very convenient to adopt the composite foundation and use the concept of plastically bearing piles. Considering different settlements of different piles, we add settlement adjustor in the construction. Fig.11 shows the details.

5.2.3 *Measured results and analysis*

The maximal settlement is about 38.1mm,and the mean settlement is about 27mm. The results are very close to our estimate value 3~5cm. Fig.12 shows the details of the settlements.

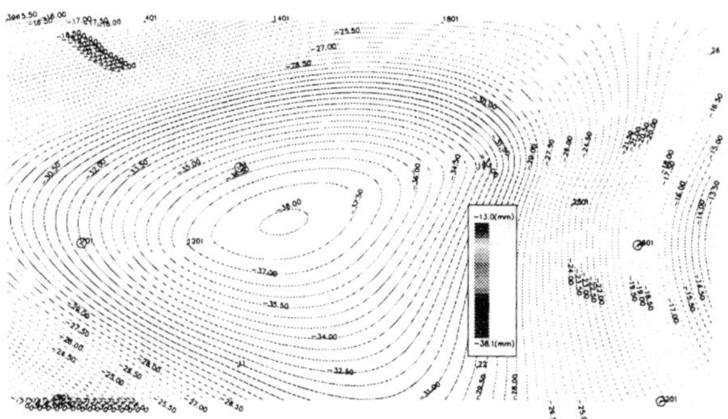

Fig.12 Contour of the settlement

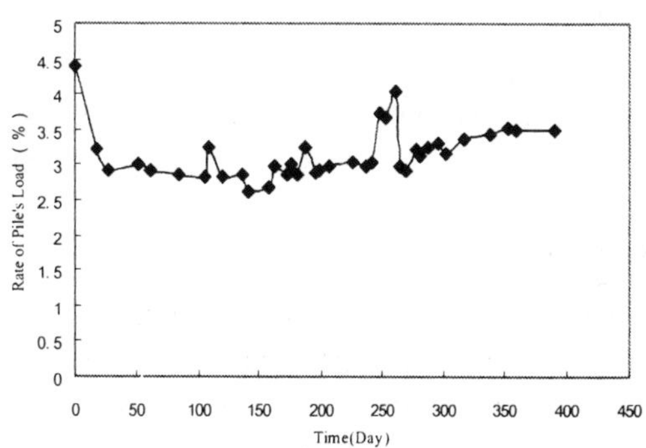

Fig.13 Rate of pile's load ~ time curve

Fig.13 shows the change of the rate of loading, which piles bears and the time. The backpressure of the pile's top and back pressure of the base bottom increase linearly with time. The maximal backpressure of the base bottom is 300kPa,which is close to the bearing capacity value of natural base. So the bearing capacity value of natural base is used sufficiently.

Fig 14 and 15 shows details of the deformation of the adjustor. The mean deformation is 12mm now, which achieves the goal of controlling the differential settlement.

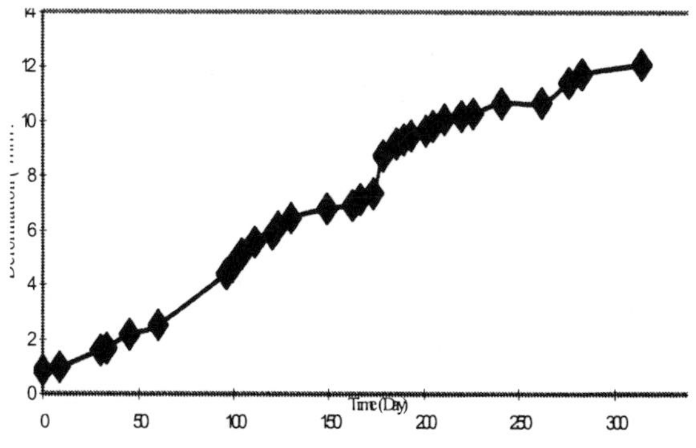

Fig 14 Settlement of adjustor ~ time curve

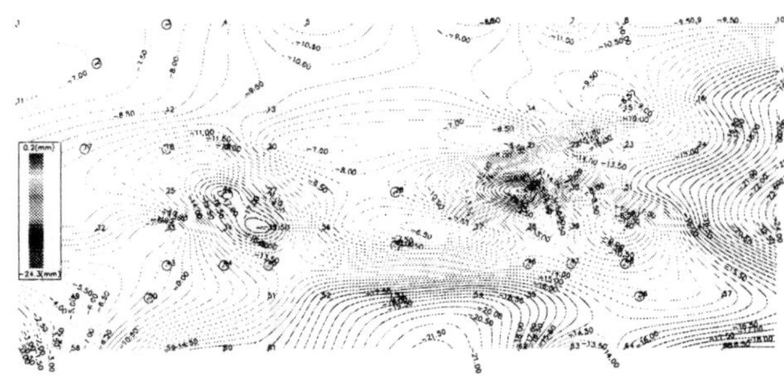

Fig 15 Contour of the settlement adjustor

5.2.4. *Economic Analysis*

It is valuable to point the methods in the paper. It is not only more convenient than the conventional method, but also saves 10,000,000 yuan (RMB) in the construction.

6. Conclusions

The ultimate bearing capacity of single pile is used for the pile, and the loads acted on the composite pile foundations are distinctly shared by the piles and the soil under the cap. Based on the theory of composite pile foundations, the concept of plastically bearing pile is established. The methods of settlement calculation, the design principals of application and some problems of nonlinear design of composite pile foundations interaction are presented. These methods and concept have been used for the design of a 9-story library and a 30-story high-rise. The success of the field experimental studies has fully verified the feasibility and advance of the research and it has a broad prospect in practical application.

References

1. Zai J. M., Zai J. Z. (1993) Foundations Analysis and Design for High Building (in Chinese), China Building Industry Press
2. Zai J. M. (1996) Semi-analytical and semi-numerical method for analyzing nonlinear interaction of pile group-soil-pile cap system. Journal of building structures, 1996(2)
3. Zai J. M. (2001) Concept of Plastically Bearing Pile and Its Practical Application, Chinese Journal of Geotechnical Engineering, 2001(3)
4. Zai J. M. (2001) Engineering verification and field experiment research of plastically bearing pile, Journal of building structures, 2001(5)
5. Zai J. M., Yang R. C. (1997) Analytic solutions to increment of ultimate bearing capacity of soil under caps of composite pile foundations, Chinese Journal of Geotechnical Engineering, 1995(5)
6. Zai J. M.(2004), Theory and application of composite pile foundation, Intellectual property publishing house/China water power press

Case Studies

TUBULAR SYSTEMS FOR TALL OFFICE BUILDINGS WITH SPECIAL CASES FROM TURKEY

AYSIN SEV

Department of Architecture, Mimar Sinan Fine Arts University, Address
Findikli, Istanbul, 34427, Turkey

The advent of computers in conjunction with a boom in the construction industry following the Great Depression and World War II facilitated the development of new structural systems and forms. It is now possible to analyse and investigate different structural systems and concepts with the aid of the computer, which has never been possible before. This is in fact, a primary reason why conventional rigid-frame systems have been the prevalent structural systems for tall buildings until then. Fazlur Khan developed and refined the revolutionary tubular building concept, and the buildings reached to an unimaginable height with unimaginable amount of material. Here the building skeleton comprises closely T aced perimeter columns that provide much greater lateral resistance than is obtained with conventional systems because of the three-dimensional response of the building to lateral loads. This paper gives a brief explanation of tubular systems with a number of case studies from Turkey and abroad. In this context, the historical development of tubular systems is given firstly. Then, the tube concept is identified from the structural point of view as well as architectural point of view, and types of tubular systems – such as framed tubes, trussed tubes, and bundled tubes– are considered. While giving information about various kinds of tubular systems, a number of tall buildings are also presented.

1. Introduction

From the structural point of view, a building can be defined tall, when its height creates different conditions in the design, construction, and use than the conditions that exist for the lower buildings [1, 2]. These conditions are manifest when the effects of lateral loads begin to influence its design. For example, in the design of tall buildings, in addition to the requirements of strength, stiffness and stability, the lateral deflections due to wind or seismic loads should be controlled to prevent both structural and nonstructural damage. Also the wind response of top floors in terms of their accelerations during frequent windstorms should be kept within acceptable limits to minimize motion perception and discomfort to building occupants.

The recent trend in high-rise structures includes the tubular system, which has been developed by Fazlur R. Khan in 1960s [3, 4]. The tube concept has been employed on a number of high-rise office buildings as an efficient framing system. It describes a structural system in which the perimeter of the building, consisting of vertical supports interconnected by beams or bracing members, acts as a giant vertical, internally stiffened tube, resisting the horizontal forces from wind or earthquake, and providing lateral support to all vertical supporting members against buckling [5]. This paper gives brief information about several types of tubular systems, as well as providing information about a number of examples from Turkey.

2. Development of Tubular Concept

The development of the initial generation of tubular systems for high-rise buildings can be traced to the concurrent evolution of reinforced concrete construction following World War II. Prior to the early 1960s, reinforced concrete was utilized primarily for low-rise construction of only a few stories in height. These buildings were characterized by planar Vierendeel beam and column arrangements with wide spacing between members: The basic inefficiency of the frame system for reinforced concrete buildings of more than about 15 stories resulted in member proportions of prohibitive size and structural material cost premiums, and thus such systems were economically inviable. Concrete shear wall systems arranged within the building interior could be utilized, but they were often of insufficient size for stiffness and resistance against overturning. This led to the development of structural systems with a higher degree of efficiency toward lateral load resistance for taller buildings. The notion of a fully three-dimensional structural system utilizing the entire building inertia to resist lateral loads began to emerge at this time. Khan being the main proponent of this design trend systematically pursued a logical evolution of tall building structural systems [2, 6]:

The tubular concept operates as an inherently stiffened three-dimensional framework, where the entire building works to resist overturning moments. Tubes can encompass shear walls, columns and beams attempting to make them act as one unit. The main future of a tube is closely spaced exterior columns connected by deep spandrels that form a spatial skeleton and are advantageous for resisting lateral loads in a three dimensional structural space. Window openings usually cover about 50 % of exterior wall surface. Larger openings such as retail store and garage entries are accommodated by large transfer girders. The tubular concept is both structurally and architecturally applicable to concrete buildings as is evident from DeWitt-Chestnut Apartment Building in Chicago completed in 1965, the first known building engineered as a tube by Khan [3, 4, 7, 8, 9, 10].

The adoption of the framed tube in steel required an examination of fabrication methods. While concrete is field-molded, steel needs to be welded, which is not cost effective in the field. The development of a framed tube module involving one column and half spandrel beams, which can be field-bolted, made possible the application of the framed tube principle to steel. The tree units are shop-fabricated in jigs where welding can be done under controlled conditions. The erection of this unit is both highly efficient and faster than normal construction. The 60 State Street Building of Skidmore, Owings and Merill, designed in 1977, in Boston is one of the many existing examples of framed tube construction in steel (Figure 1) [11, 12]. With this concept, one can examine the freedom for forming exterior surfaces. The shape was derived from consideration of massing with respect to neighboring tall buildings and visual impact. The building was conceived as a concrete frame tube, but was later changed to steel, thus attesting to the interchangeability of materials in this concept.

Figure 1. 60 State Street Building, Boston.

3. Types of Tubular Structures

The shape of the tubes may be designed in a number of ways depending on the layout of the building. Several types may be distinguished from the point of view of structural design and the layout of walls. They will be lined up according to their effectiveness and the implied suitability for large or small heights or slenderness ratios as follows.

3.1 *Framed tube system*

The framed tube system consists of closely spaced exterior columns and spandrel beams, which are rigidly connected together [2, 4, 13]. The monolithic nature of reinforced concrete is ideally suited for such a system. Depending on the height and dimensions of the building, exterior column spaces should be on the order of 1.5 to 4.5 m on center maximum. Spandrel beam depths for office or residential buildings are typically 600 to 1200 mm. The resulting system approximates a tube cantilevered from the ground with openings punched through the tube walls. The closely spaced columns and deep spandrels have a secondary benefit related to the exterior cladding for reinforced concrete constructions. Exterior columns eliminate the need for intermediate vertical mullion elements of the curtain walls, partially or totally. An early example of the framed tube system is the DeWitt Chestnut Apartments in Chicago, as mentioned earlier.

The tube is suitable for both steel and reinforced concrete construction and has been used for buildings ranging from 40 to 100 stories [14]. The highly repetitive pattern of the frames lends itself to prefabrication in steel, and to the use of rapidly movable gang forms in concrete, which allows rapid construction. The framed tube in structural steel requires welding of the beam-column joint to develop rigidity and continuity. The formation of fabricated tree elements, where all welding is performed in the shop in a horizontal position, has made the steel frame tube system more practical and efficient (Figure 2) [2].

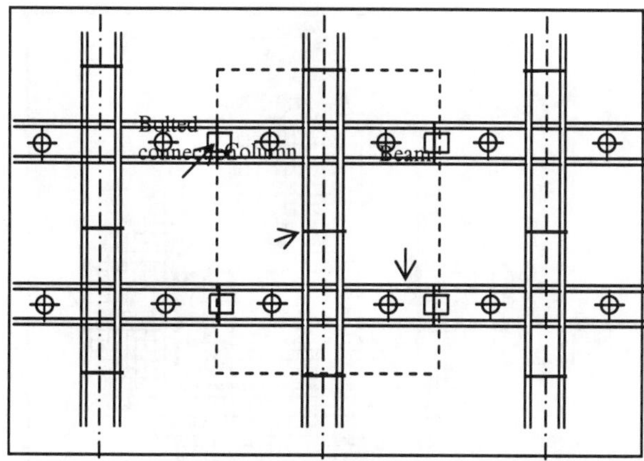

Figure 2. Typical tree erection unit..

The closely spacing of columns throughout the height of a framed tube is usually unacceptable at the entrance levels for architectural reasons. Therefore a limited number of columns can be transferred with little, if any, structural premium because the vierendeel action of the façade frame is generally sufficient to transfer the load. However, if the transfer is too severe requiring removal of a large number of columns, a 1- or 2-story deep transfer girder or truss may be necessary. Temporary shoring is required to support the dead and construction loads until a sufficient number of vierendeel frames are constructed, or in concrete construction, until the girder has achieved the design strength (Figure 3) [1].

Maya Akar Business Center in Istanbul, Turkey is a typical example of framed tube systems. The Business Center consists of two towers, one of which is 19– and the other 34 stories (Figure 4). The plan shape of the 19-story A Block is a rectangle, whereas the plan shape of the 34-story B Block is a square (Figure 5). The exterior columns are spaced at 3.5 m centers and the spandrels connecting the columns are 400 mm deep. The towers also have inner cores, designed to resist the gravity loads only. The width of load-bearing walls are 350 mm and 600 mm respectively in the A and B Blocks [10].

Another recent example of the framed tube systems is the Is Bank General Headquarters, which was completed in late 2000 (Figure 6). The complex consists of three towers, two of which are 36-story and the third is 52 story. Designed by Dogan Tekeli and Sami Sisa Architect, the towers are the tallest buildings of Turkey. The structural engineer, Irfan Balioglu, designed the towers to resist an earthquake magnitude of 9.0 on the Richter Scale. The towers resist the lateral loads by a framed tube at the perimeter (Figure 7a-b). The sizes of the columns are 600 x 600 mm on the lower towers and 600 x 900 m on the taller tower, and spaced 3.5 m on the centers. The spandrel beams of the higher tower is designed as flat beams with a height of 350 mm, whereas the sizes of the spandrel beams of the lower towers are 600 x 750 mm. The towers also contain inner cores, which consists of shear walls. The width of these shear walls in the 36-story tower is 400mm, whereas the width of the shear walls in the 52-story tower is 600 mm [9, 10, 15, 16, 17].

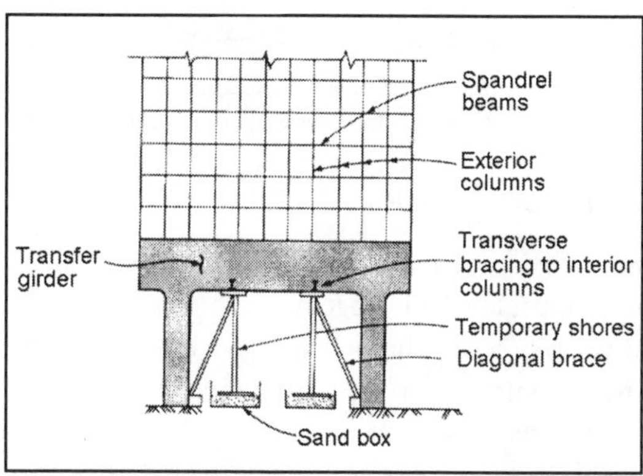

Figure 3. Shoring system for a tube structure.

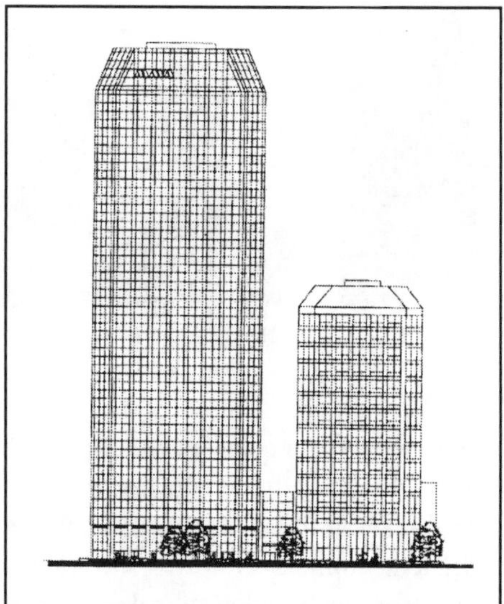

Figure 4. Maya Akar Business Center.

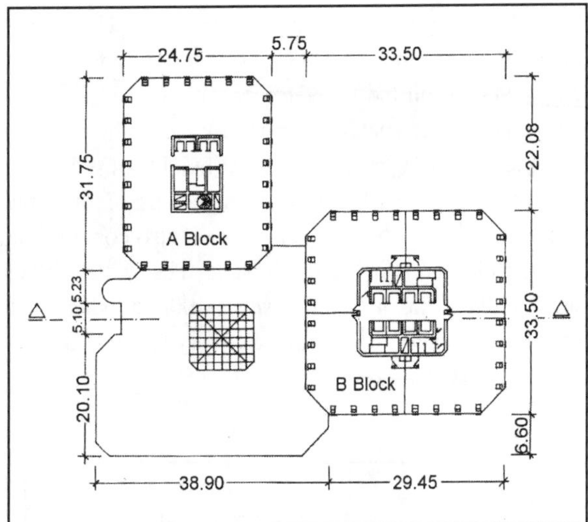

Figure 5. Maya Akar Business Center, typical floor plans

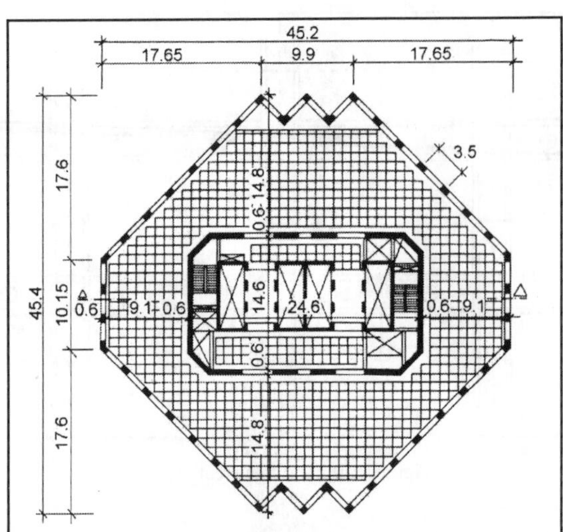

Figure 7. Is Bank General Headquarters, structural framing: (a) The 52-story tower.

3.2 *Tube in tube system*

This variation of the framed tube consists of an outer-framed tube, the "hull" together with an internal elevator and service core. The hull and the core act jointly in resisting both gravity and lateral loading. In a steel structure the core may consists of braced frames, whereas in a concrete structure it would consist of an assembly of shear walls.

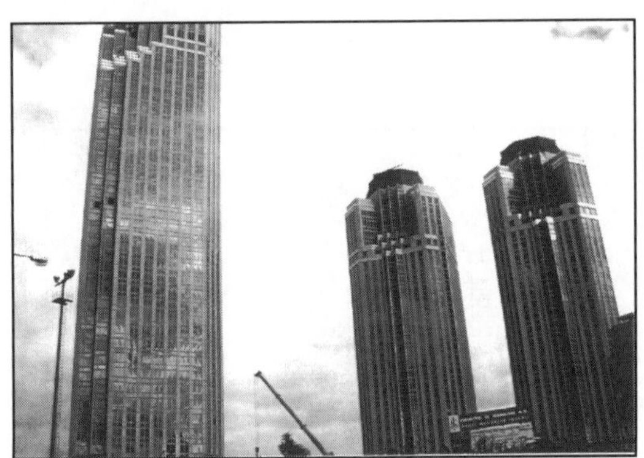

Figure 6. Is Bank General Headquarters, Istanbul.

To some extent, the outer frame tube and the inner core interact horizontally as the shear and flexural components of a wall-frame structure, with the benefit of increased lateral stiffness. However, the structural tube usually adopts a highly dominant role due to its much greater structural depth [4, 10, 14].

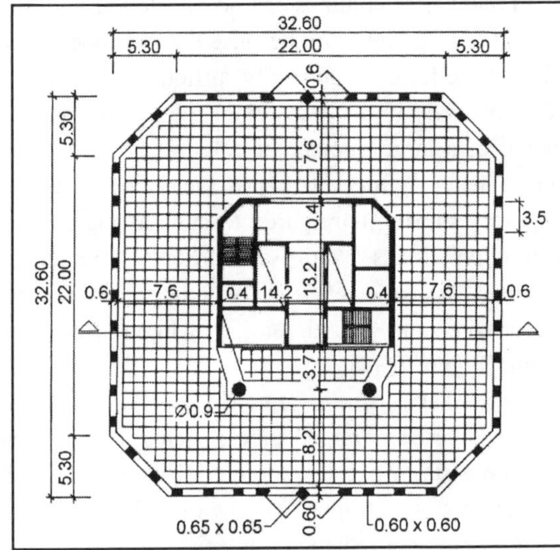

Figure 7. Is Bank General Headquarters, structural framing: (b) The 36 story tower.

3.3 *Trussed tube system*

A trussed tube system represents a classic solution for improving the efficiency of framed tube by increasing its potential for use to even greater heights as well as allowing greater spacing between the columns. This is achieved by adding diagonal bracing to the face of the tube to virtually eliminate the shear lag effects in both flange and web frames [1, 14] This arrangement was first used in a steel structure in 1969, in Chicago's John Hancock Center.

Although the trussed tube was initially developed for structural steel construction, Fazlur Khan applied similar principles to reinforced concrete construction. He visualized a concrete version of the diagonal trussed tube consisting of exterior columns spaced at about 3.04 m centers with block out windows at each floor to create a diagonal pattern on the façade. The diagonals could then be designed to carry the shear forces, thus eliminating bending in the tube columns and girders. Currently there exist two high-rises, which are constructed using this approach. The first is a 50-story office structure located on Third Avenue in N.Y., and the second is a mixed-use building located on Michigan Avenue in Chicago. The first example is a combination of a framed and trussed tube interacting with a system of interior core walls (Figure 8). All the three subsystems, consisting of a framed tube, trussed tube, and shear walls, are designed to carry both lateral as the 780 Third Avenue Building and vertical loads. The building is 173.73 m high with an unusually high height-to-width ratio of 8:1. The diagonals created by filling in the windows serve a dual function. First, they increase the efficiency of the tube by diminishing the shear lag, and second they reduce the differential shortening of the exterior columns by redistributing the gravity loads. A stiffer and much more efficient structure is realized with the addition of diagonals. The idea of diagonally bracing this structure was suggested by Fazlur Khan to the firm of Robert Rossenwasser Associates, who executed the structural design for the building [1, 10]. The Chicago version of the system is a 60-storey multi-use project, named as Onterie Center (Figure 9). The building rises in two tubular segments above a flared base. According to the designer, diagonal bracing was used primarily to allow maximum flexibility in the interior layout needed for mixed use. In contrast to the building in New York, which is clad with polished granite, Onterie Center has an exposed concrete framing and bracing [2, 10, 18]. Citicorp Center is a remarkable example of trussed tube system, which is constructed in steel. The 60-story office building has a 47.8 m square plan, with all of its corners jutting out 23 m unsupported from only four exterior columns, one centered on each side. The central core also supports the tower. The most direct and economical way to achieve the 23-m corner cantilevers on each face of the tower was to provide a steel-framed braced tube with a system of columns and diagonals in compression, channeling the buildings gravity loads into a 1.5-m wide mast columns in the center of each face (Figure 10). The main diagonals repeat in eight-story modules [2, 9, 19].

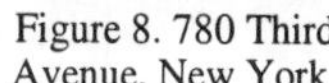

Figure 8. 780 Third Avenue, New York.

Figure 9. Onterie Center, Chicago.

3.4 *Modular (bundled) tube system*

The most efficient plan shape for a framed tube is a square or a circle, whereas a triangular shape has the least inherent efficiency. The high torsional stiffened characteristic of the exterior tubular system is advantageous in structurally unsymmetrical shapes. However, for buildings with significant vertical offsets, the discontinuity in the tubular frame introduces some serious inefficiency. A modular or bundled tube configured with many cells, on the other hand has the ability to offer vertical offsets in buildings without loss in efficiency.

The modular tube system allows for wider spacing of columns in the tubular walls than would be possible with only an exterior framed tube. It is this spacing, which makes it possible to place interior space planning [1]. In principle, any closed-formed shape may be used to create a modular tube. The ability to modulate the cells vertically can create a powerful vocabulary for a variety of dynamic shapes. The bundled tube principle therefore offers great latitude in the architectural planning of a very tall building [2]. The most remarkable example of the modular tube system in steel is the 110-story Sears Tower in Chicago. With a height of 443 m, the tower consists of 9 tubes. Each tube is 22.9 m square and make up a typical lower floor for overall floor dimensions of 68.6 m. This square plan shape extends to the fiftieth floor, where the first tube terminations occur. Other terminations occur at floors

66 and 90 (Figure 11). The structure acts as a vertical cantilever fixed at the base to resist lateral loads. Nine square tubes of varying heights are bundled together to create the larger overall tube. Each tube comprises columns at 4.58-m centers connected by stiff beams. Two adjacent tubes share one set of columns and beams. All column-to-beam connections are fully welded. At three levels, the tubes incorporate trusses, provided to make the axial column loads more uniform where tube drop-offs occur. These trusses occur below floors 66 and 90 and between floors 29 and 31 [1, 2, 4, 9, 10].

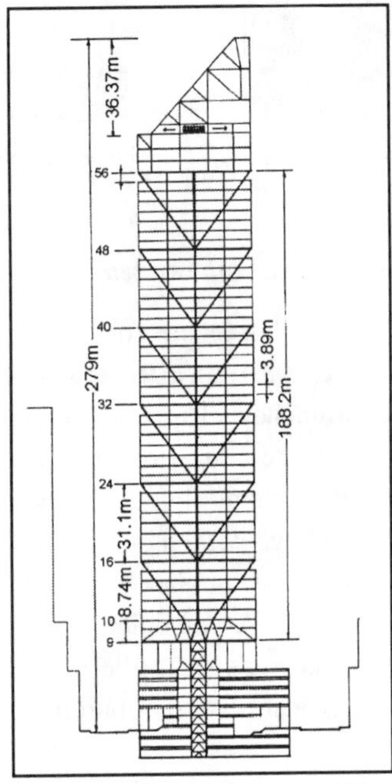

Figure 10. Citicorp Center, elevation.

Another example of the bundled tube system is the 63-story Rialto Building in Melbourne, Australia. A number of structural systems for the Rialto Building were initially investigated and a reinforced concrete structural system was finally adopted, with speed of construction being a prime consideration in the development of formwork and reinforcement details. The external frame of columns and beams, while being designed for the direct and live loads applicable, acts as an external tube in resisting lateral loads. Although plan shape is unsymmetrical and the columns are 5 m apart, analysis of the load transfer around the corners indicated reasonable three-dimensional action. The tube effect also provides for some lateral distribution of load from the more heavily loaded columns (Figure 12) [2, 20].

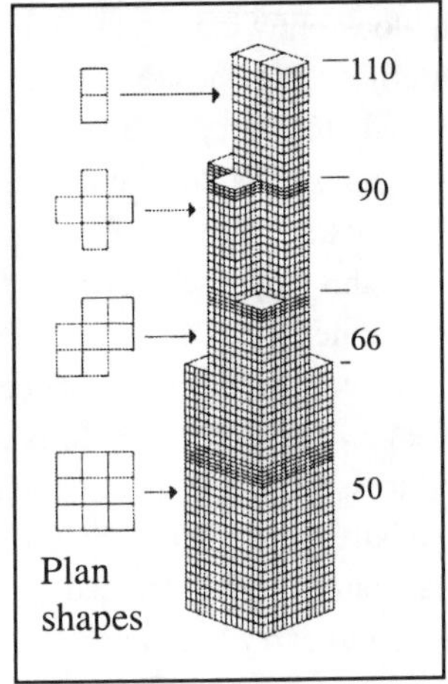

Figure 11. Sears Tower.

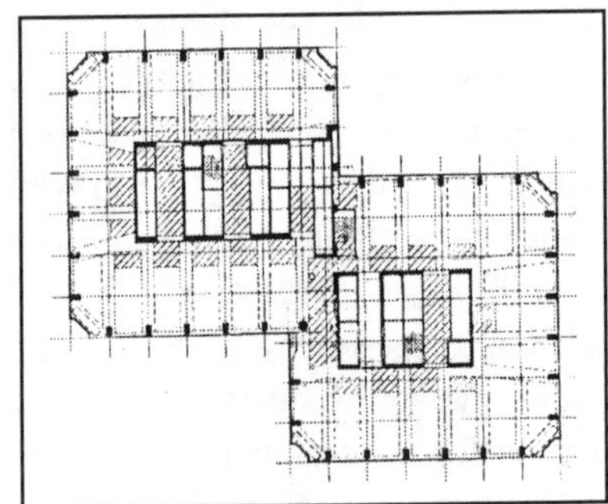

Figure 12. Rialto Towers, Melbourne, structural framing.

4. Conclusions

In the design of tall buildings, in addition to the requirements of strength, stiffness and stability the lateral deflections due to wind or seismic loads should be controlled to prevent structural and nonstructural damage and occupants' discomfort. The recent trends in tall building design include tubular systems, which have been developed by Fazlur Khan in 1960 and have been efficiently employed on a number of buildings since then. The term, in the usual building terminology suggests a system of closely spaced columns tied together with relatively deep spandrels. The system is a fully three dimensional system that utilizes the entire building perimeter to resist lateral loads. Since the entire lateral load is resisted by the perimeter frame, the interior floor plan is kept relatively free of core bracing and large columns, thus increasing the net leasable area of the building. The interior framing can be designed only for resistance to gravity loads. As a trade-off, views from the interior of the building may be hindered by closely spaced exterior columns. This issue is considered to be the best advantage of tubular systems from the architectural point of view.

The tube system can be constructed of reinforced concrete, structural steel, or a combination of the two materials, which is named to be composite construction. Also the type of the tubular system, such as framed, trussed, modular tube or tube-in-tube mostly depends on the layout of the building, as well as the height of the building and the loads effecting on the structure. Not only the structural engineers, but also the architects, who are closely related with the design of high-rises, must be aware of the tubular system, to design contemporary tall buildings.

References

1. B.S. Taranath, *Steel, Concrete and Composite Design of Tall Buildings*, McGraw-Hill, Inc., New York, (1997).
2. L.S. Beedle, and D.B. Rice, (Ed.), *Structural Systems for Tall Buildings*, Council on Tall Buildings and Urban Habitat Committee 3, McGraw-Hill Inc., New York, (1995).
3. M. Fintel, "New Forms in Concrete", *Technique and Aesthetics in the Design of Tall Buildings*, Proceedings of the Fazlur R. Khan Session on Structural Expression in Buildings, (Ed.) Billington, D. P. And Goldsmith M., Institute for the Study of the High-Rise Habitat, Lehigh University, Bethlehem, PA, 39-56, (1986).
4. A. Ozgen, "The Historical Development of Multi-Storey Buildings and the Last Phase: Tubular Systems" (in Turkish), *Building*, No. 89, Building Industry Centre, Istanbul, 47-53, (1989).
5. CTBUH, *Structural Design of Tall Concrete and Masonry Buildings*, Volume CB, ASCE, Printed in USA, (1978).
6. F.R. Khan, F. Rankin,"Current Trends in Concrete High-Rise Buildings", *Tall Buildings*, Proceedings, Pergamon Press, Oxford, 571-590, (1967).
7. M.M. Ali, "Evolution of Concrete Skyscrapers: from Ingalls to Jin Mao". Electronic Journal of Structural Engineering, Vol. 1, No. 1, 2-14, available at (2005): www.civag.unimelb.edu.au/ejse/Archives/
8. F. R. Khan, Techniques and Aesthetics in the Design of Tall Buildings, Proceedings of Fazlur R: Khan Session on Structural Design Expression on Buildings, D.B. Billington and M. Gldsmith (Ed.s), Annual Fall Meeting, ASCE; Houston, (1986).
9. A. Ozgen, and A. Sev, *The Structural Systems for Multi-Storey High-Rise Buildings* (in Turkish), Birsen Publisher, Istanbul, (2000).
10. A. Sev, *The Analysis of Tall Buildings According to Architectural Design and Structural Systems in Turkey and at Abroad*, Doctoral Dissertation, Mimar Sinan University, Science and Technology Institute, Istanbul, (2001).
11. H. Iyengar, "Structural and Steel Systems", *Technique and Aesthetics in the Design of Tall Buildings*, Proceedings of the Fazlur R. Khan Session on Structural Expression in Buildings, (Ed.) Billington, D. P. And Goldsmith M., Institute for the Study of the High-Rise Habitat, Lehigh University, Bethlehem, PA, 57-69, (1986).
12. USA High-Rises. Available at (2005): http://www.geocities.com/PicketFence/5192/60statest.html
13. W. Schueller, *High-Rise Building Structures*, John Wiley and Sons, New York, (1977).
14. A. Coull, and B.S. Smith, *Tall Building Structures Analysis and Design*, John Wiley and Sons, New York, (1991).
15. T. İ. Beyazoglu, T. I., *Tubular Systems in Tall Buildings and Case Studies* (in Turkish), Master Thesis, Mimar Sinan University, Institute of Science and Technology, Istanbul, (1997).
16. D. Tekeli and S. Sisa, *Projects: Buildings 1954-1994*, Building Industry Center, Istanbul, (1994).
17. I. Balioglu, I., "The Structural System Project of Is Bank General Headquarters Complex" (in Turkish), *Design Construction*, No.160, Istanbul, 74-76, (1999).
18. P. Saliga, (Ed.), *The Sky's The Limit: A Century of Chicago Skyscrapers*, Rizzoli, New York, (1990).
19. D. Guise, D., *Design and Technology in Architecture*, Revised Edition, Van Nostrand Reinhold Company, New York, (1991).
20. I. Zacnik, M. Smith and D. B. Rice, *100 of the World's Tallest Buildings*, Council on Tall Buildings and Urban Habitat, Gingko Press, Hong Kong, (1988).

DIFFERENTIAL DEFORMATIONS OF COLUMNS AND CORES IN TALL BUILDINGS – ASSESSMENT, MONITORING AND CORRECTION MEASURES

SADUDEE BOONLUALOAH SAM FRAGOMENI YEW-CHAYE LOO
School of Engineering, Griffith University, Gold Coast, Queensland, Australia

DAKSH BAWEJA
Readymix Holdings, Australia

An extensive literature review has been carried out on the differential shortening measurement and behaviour of columns and cores in tall buildings. This covers various assessment methods for determining different types of shortening including: time dependent and seasonal reversible influences, instrumentation techniques required to acquire field data, and correction measures undertaken to overcome differential shortening in actual structures. Also presented is a comparison of predicted and measured axial shortening of a particular column in an 80-storey building currently under construction in Australia.

1. Differential Shortening and Associated Problems in Tall Buildings

Axial shortening of columns and cores is inevitable in every reinforced concrete building. Over the service life, all concrete members undergo deformations initiated by a number of natural phenomena including creep, shrinkage and elastic modulus. Some of the deformations are permanent whilst others occur temporarily or are due to seasonal changes.

Differential shortening refers to the difference in shortening rate of adjacent columns and walls in a building. Differential movements in columns and wall elements arise from differences in amounts of reinforcement, volume-to-surface ratio of the structural elements, environmental exposure conditions and relative loading history of the adjacent vertical structural members. Bast et al [1] reported the need to address and quantify differential shortening in taller high-rise buildings as differential shortening accumulates over many floors eventually affecting the flatness of floor slabs at higher levels (see Figure 1).

If differential shortening is not accounted for the effect is both structural and architectural [1]. Structurally, extra bending moments will be induced to columns/core walls in the affected frame member resulting in reduced safety factors and robustness of the structure. Architecturally, non-structural service elements such as finishes and partitions are affected as they could be easily damaged, consequently giving rise to certain serviceability and aesthetic issues [2].

It is therefore essential that a reasonable structural evaluation is made and suitable correction measures are used to tackle the anticipated differential shortening in high-rise buildings.

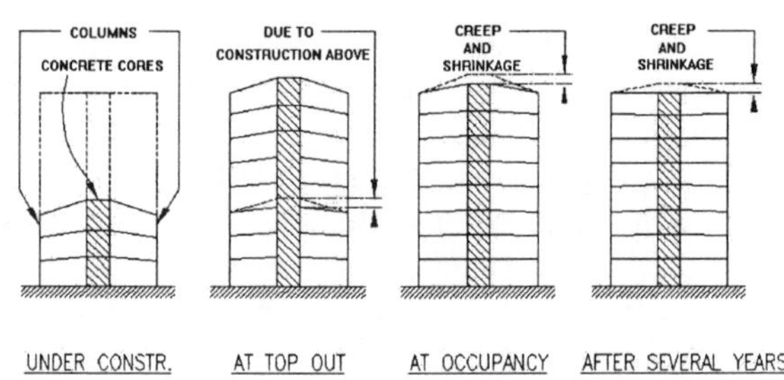

Figure 1. The effects of differential movements between vertical members on a high-rise building [1].

2. Assessment of Differential Shortening

2.1 *Reinforced concrete*

The critical elements that are most likely to initiate significant differential movements should be initially identified. They are normally adjacent members that vary in terms of loading, exposure conditions, geometric and material properties. The anticipated differential shortening of these elements should then be assessed to ensure adequate provisions are made to limit the phenomenon.

For analysis purposes, deformations that occur over the service life of concrete consist of six components of strain as follows; elastic strain (ε_{el}), creep strain (ε_{cr}), shrinkage strain (ε_{sh}), thermal strain (ε_{th}), wind strain (ε_{wi}) and reinforcement interaction (ε_{re}).

Elastic strain or instantaneous strain is the deformation that takes place immediately after the loading. Creep strain is the deformation developed over long-term due to the sustain loading. Shrinkage strain is the strain that occurs due to the loss of water in a concrete specimen and is independent from the loading [3]. Thermal and wind strain are the strain that occurs due to differences of exposed daily ambient temperature and wind gust respectively.

To simplify the calculation, these components are usually treated independently despite their interactions [3]. Axial shortening at time t (days) can thus be expressed as:

$$\varepsilon(t) = \varepsilon_{el} + \varepsilon_{cr} + \varepsilon_{sh} + \varepsilon_{th} + \varepsilon_{wi} + \varepsilon_{re} \quad (1)$$

Required concrete properties such as compressive strength, elastic modulus, creep and shrinkage can be determined most accurately by standard tests. However, since most of the material property data from laboratory tests may not be available at the design stage, major code recommendations, i.e. ACI Committee 209 [4], CEB-FIP Model Code 1990 [5], and Australian Standard for Concrete Structures AS 3600 [6] or accepted recommendations from researchers [7], [8], can be employed in absence of the test data.

2.2 *Time dependent effects*

The time dependent strains develop and accumulate over the life of building. They are permanent and depend mostly on the loading history and ageing process of concrete.

Shrinkage strain is known to be independent from other shortening components. It can be determined directly from any of the regional code models [4], [5], [6]. Elastic, creep and reinforcement restraint are know to be related and two particular approaches may be considered for the estimation of these three components.

The first approach is to determine elastic strain, creep strain and reinforcement effects separately using material models for plain concrete. Creep strain can be estimated directly, in a similar manner to shrinkage, from codes of practice. Reinforcement restraint effects can be determined by any of the methods given in Brook *et al* [9]. Elastic strain occurs at the application of a load, *P*, and is given by [10]:

$$\varepsilon_{el} = \frac{P}{A_c E_c + A_s E_s} \quad (2)$$

where A_c and A_s are sectional area of concrete and steel respectively, E_c and E_s are elastic modulus of concrete and steel respectively.

The second approach is to determine elastic strain, creep strain and reinforcement effects altogether using an analytical model for reinforced concrete members such as the method suggested by Fintel and Khan [11], the Age-Adjusted Effective Modulus Method or AEMM [12], and Koutsoukis and Beasley [13].

The use of an analytical model is considered theoretically more accurate than the first approach mentioned as the models take into account the effects of relaxation, ageing and amount of reinforcement altogether.

For example, the Age-Adjusted Effective Modulus Method [12] takes the following form:

$$\varepsilon^*(t,s) = \frac{\sigma_0}{E_c(s)}\left(1+\phi(t,s)\right) + \frac{\Delta\sigma(t)}{E_c(s)}\left(1+\chi(t,s)\phi(t,s)\right) \quad (3)$$

where $\varepsilon^*(t,s)$ is total strain at time t of a reinforced concrete member loaded initially at time s. σ_0 is initial stress applied to a concrete specimen. $\Delta\sigma(t)$ is the change in concrete stress at time *t* due to concrete relaxation. $\chi(t,s)$ is the ageing coefficient. Ø(t,s) is the creep coefficient.

2.3 *Temporary effects*

Unlike time dependent effects, thermal and wind effects are non-permanent as the intensity changes daily. In practice, thermal strain and wind strain can be determined independently. Both effects depend largely on the yearly cycle of weather conditions in the vicinity of the building.

Stain due to ambient temperature, ε_{th}, is related directly to the thermal coefficient of concrete (α) and is given by [10]:

$$\varepsilon_{th} = \alpha\Delta t \quad (4)$$

where Δt = ambient temperature in °C

To determine the design range of ambient temperature corresponding to the geographic location and size of columns, methods have been developed [14], [15] with local maps for lowest winter and highest summer mean daily temperature.

Wind effects can be determined by the analysis of the building when subjected to various scenarios of wind exposure, and on local weather history. The analysis method for wind effects involves either a simplified analysis from relevant codes of practice or by a more accurate finite element analysis of a 3-D model of the building structure and is beyond the scope of this paper.

3. Monitoring of Concrete Columns and Cores in Buildings

Structural monitoring is becoming more and more accepted as a suitable approach to predicting behaviour as building structures are usually subjected to conditions that greatly differ from the test conditions in the laboratory [16].

A number of tall buildings have previously been instrumented in columns and cores to obtain axial shortening readings during and after construction. In the USA, Lake Point Tower [17], Water Tower Place [18], [19], and a 38-storey building, [11] all in Chicago were instrumented for shortening measurements. In Australia, work has been conducted at the New South Wales Institute of Technology Tower Building [20], [21] and recently constructed World Tower, Sydney [22]. Elsewhere Concordia University's library in Canada [23], a 19-storey building in Singapore [16], and Baiyoke II Tower in Bangkok [24] have been monitored. A tall building currently being constructed in Australia [25] has been monitored by the authors [26], [27].

The monitoring processes and experimental details of the research completed in the past varied in many aspects including duration of the data collection, instrumented details, number of monitored parameters and method of instrumentation.

There are three major generally utilised methods depending on research budget and the required accuracy of the data collection to measure strains in buildings using:

1. Conventional survey techniques.
2. Mechanical strain gauges.
3. Sensors and data lodging systems

3.1 *Monitoring axial shortening using survey techniques*

Electronic distance measuring equipment can be used to measure relative length changes of a concrete member as compared to collimation marks installed on control floors [20].

This method is simple and economical, as it provides a direct means of measurement when overall shortening is of interest. However the accuracy is considered low and it is not practical for long-term measurements, particularly when frequent data collection is required. It is also not suitable in obtaining early-age strains, as the access required and survey methods can not be undertaken for sometime after the pouring [20].

In practice, the method has been used in buildings [20], [24] to verify the predicted movements. However, it has not been used as much in research projects except to double check measurements using other methods.

3.2 *Monitoring axial shortening using mechanical strain gauges*

The demountable mechanical strain gauge (Demec gauge) has been used by many researchers [11], [17], [18], [19], [22], [26], [27]. To obtain the length change, two reference points are attached to columns/cores and at regular intervals the length between the two points is measured and compared with the original length difference at installation.

This method is relatively and data from many locations can be collected using only one set of the Demec gauge. Also, unlike the survey method, it is possible to take readings as soon as formwork is removed.

The accuracy depends on the type of Demec gauge used and gauge length (an accuracy between one to ten microstrains can be normally registered).

The drawbacks of the method are that the accuracy of the measurement can be affected by climatic factors such as ambient temperature and humidity. Therefore temperature and humidity also needs to be monitored to aid with data adjustment. The method is also unable to measure initial strains in columns prior to the removal of the formworks. For long-term monitoring, the method can prove to be ineffective due to the loss of points [17],

[19], [26]. To overcome this, a number of measuring stations are placed on each column/wall.

3.3 *Monitoring axial shortening using sensors*

This method incorporates electronic strain gauges or sensors such as vibrating wires, acoustic gauges, strain transducers and extensometers that are embedded in the columns to measure strain over time. Additional sensors such as the Carlson stressmeter, temperature sensor and Anemometer are used for the measurement of other factors [21].

The instrumentation using sensors is undoubtedly state-of-the-art amongst the three methods. In this approach, data for many shortening components can be obtained simultaneously with extreme accuracy and unlike the other two approaches it is possible to obtain strain data during the initial period after casting of the member and prior to the removal of formworks [23]. Another advantage over the other two methods is that internal column strains are measured instead of surface strain which most resembles the actual behavior of members.

Although the performance of many sensor systems over long-term measurements has proved excellent in research ([21], [23], [16]), the implementation of such sophisticated systems demands a large investment and needs extensive of involvement of specialists or researchers thus making it unfavorable in-practice.

4. Axial Shortening Monitoring on a Tall Residential Building

The authors have been involved in the instrumentation of a tall residential building located on the Gold Coast in Australia. The DEMEC gauge techniques were used to monitor axial shortening in selected columns and core walls. An example of reasonably good agreement between measured shortening of a particular basement level column (TC06) of the building and predicted values, using the AS3600-2001 code models [6] and the AEMM [12], is given in Figure 2. The circular column has: a dimension of 2000mm diamx2850 mm height, 65 MPa design compressive strength concrete at 28 days and an amount of vertical reinforcement equal to 2.4%.

For analysis purposes, stress history of the column was derived from the concrete registry and a simplified construction sequence used [28]. This allows realistic estimation of shortening according to an actual loading condition.

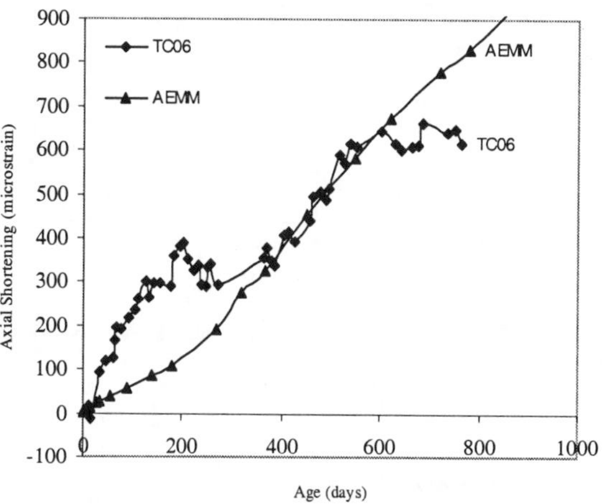

Figure 2. Measured and predicted shortening of a basement column.

It can be seen that initially shortening is not well predicted but later predictions seem to correlate reasonably well until an age of 600 days when overestimation begins. The shortening at 600 days when about 60 storeys were constructed was 620 microstrains or equal to 1.77mm for the 2.85 m high column.

5. Structural Correction Measures in Practice – Case Studies

Many techniques have been practiced to ease the differential shortening problem. All methods have one common aim, to maintain a safe slab evenness under the circumstances of differential shortening. However, the suitability of each particular method depends largely on the type of vertical movement to be resolved.

5.1 *Accommodating for severe environmental effects*

The eradication of differential movement due to severe exposure conditions is based on the following philosophy that; the intensity of differences in terms of exposed temperature at exterior and interior columns should be first minimized by providing substantial insulation for the exposed exterior column (for an example, see Figure 3), and then special joist details can be implemented to accommodate the discounted differential movement (for an example, see Figure 4).

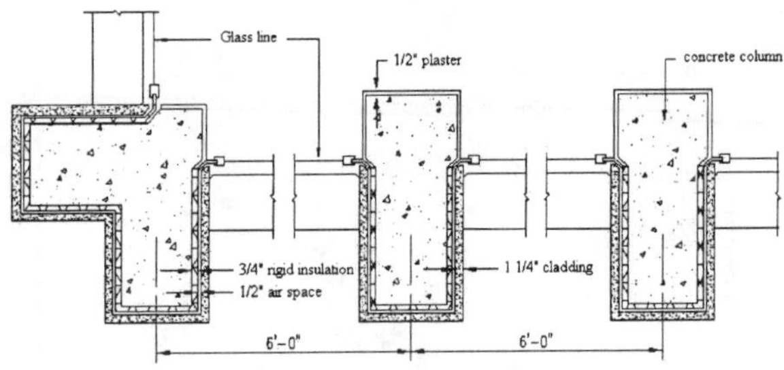

Figure 3. Detail of an insulated exterior wall system, National Life Building, Nashville, USA, [29].

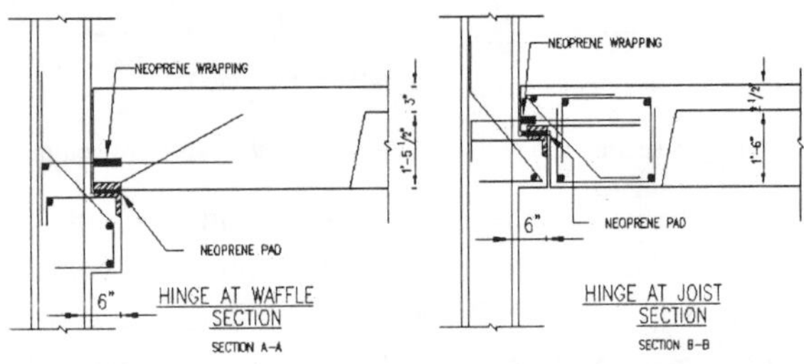

Figure 4. Hinge details of Brunswick Office Building, Chicago, USA, [29].

5.2 *Accommodating for creep and shrinkage effects*

Apart from thermal effects, the most important time dependent effects of creep and shrinkage also need to be accommodated to ensure an acceptable structural behaviour.

Unlike thermal effects that alter seasonally, time dependent effects are permanent and accumulate over long periods. Permanent adjustments to the structure are therefore necessary if significant differential shortening due to time dependent effects is expected in addition to the extra structural flexibility already provided for thermal movement.

The differential movement may be minimized by designing all vertical members with the same material and design parameters for a similar stresses. When this is not the case, it may be necessary to implement a methodology that will compensate for the expected differential movements of vertical members in order to ensure an acceptable evenness of slab during the service life of the building.

5.2.1 Slab precambering

In the case where a certain amount of post-construction differential shortening is predicted, the slab may be precambered to a given offset to compensate for the predicted amount of slab tilting (see Figure 5).

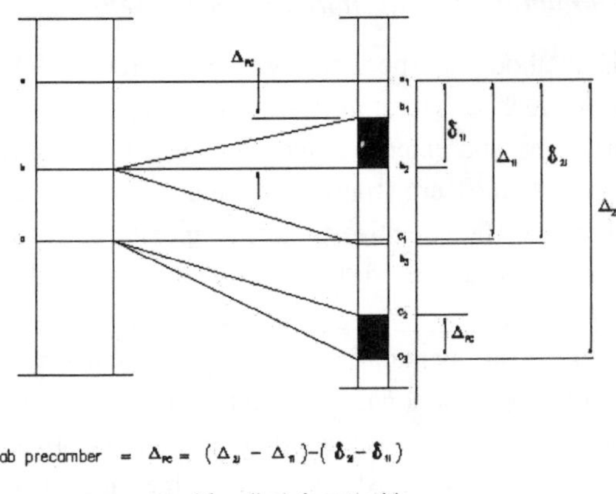

Slab precamber = $\Delta_{PC} = (\Delta_{3I} - \Delta_{II}) - (\delta_{3I} - \delta_{II})$

$a - a_1$ = zero deformation reference point

$b - b_2$ = survey controlled level columns at slap installation

$b - b_3$ = slab position with columns built exactly to length

$b - b_1$ = slab at installation, column adjusted to include precamber

$c - c_1$ = slab position is level at N days post-construction

$c - c_2$ = slab position beyond N days post-construction

$c - c_3$ = consequence of building to length with no precamber

Figure 5. Slab precambering model by Beasley [2].

The advantage of slab precambering is that the approach is very straightforward and economically implemented. However, the drawbacks are: the tilted slab exists before and after the time that the differential shortening attains its predicted amount, and the efficiency of this approach depends solely on the accuracy of the prediction which relies on the estimated material properties used as the input.

5.2.2 *Implementation of a passive structural system*

According to the literature [24], it is possible to implement a passive system which will respond to limit the rotation of slabs at the occurrence of differential shortening.

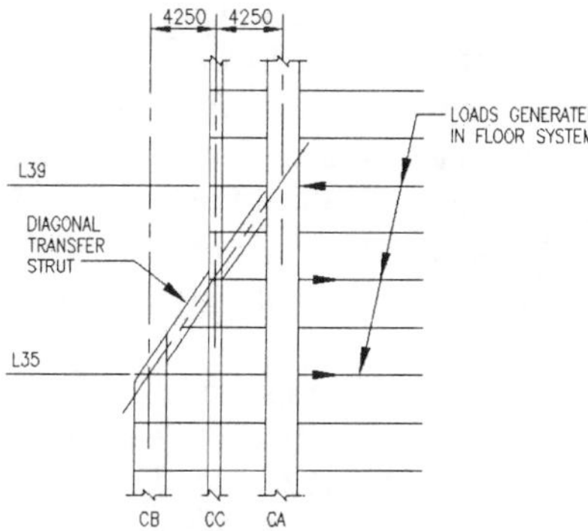

Figure 6. Detail of diagonal transfer strut at Baiyoke II Tower, Bangkok, Thailand [24].

For example in the Baiyoke II Tower (see Figure 6), the system comprises of diagonal composite steel box struts connecting three different columns together. When differential movements develop, the strut acts to transfer the induced loads between columns (up to 300 ton from CA to CB and CC) and the floor rotations are then controlled.

The case study at Baiyoke II is a good example of how the structure can be engineered to cope for the inevitable differential movement. However, it is not an easy task to incorporate such features into the design of other buildings since every building and their shortening behaviour is unique.

6. Conclusions

A literature review has been carried out on various matters related to shortening. It has been found that differential shortening in tall buildings can be predicted, monitored and eradicated effectively and economically to some extent when appropriate measures are implemented as presented in this paper.

Acknowledgments

A financial support provided under R&D Project by Readymix Holdings Ltd. is gratefully acknowledged.

References

1. W. D. Bast, T. R. McDonnell, L. Parker and S. P. Shanks, *"Measured Shortening and Its Effects in a Chicago High-rise Building"*, Proceedings of the Third Congress, Oct. 19-21, 2003, San Diego, CA, Technical Council on Forensic Engineering of the American Society of Civil Engineers, pp. 580-594, (2003).
2. A. J. Beasley, *"High Performance Concrete-Implications for Axial Shortening in Tall Buildings"*, Department of Civil and Mechanical Engineering, University of Tasmania-Hobart, Tasmania, (1997).
3. R. I. Gilbert, *"Time Effects in Concrete Structures"*, Elsevier Science Publishers B. V, 1988.
4. ACI Committee 209, *"Prediction of Creep, Shrinkage, and Temperature Effects in Concrete Structures (ACI 209R-92)"*, American Concrete Institute, Farmington Hills, Mich, 47 pp, (1992).
5. CEB-FIP Model Code 1990, Bulletin d'Information No 213/214, Comite Euro-International Du Beton, 436 pp, (1993).
6. Australian Standard for Concrete Structures AS 3600, Standard Association of Australia, Sydney, Australia, 176 pp, (2001).
7. Z. P. Bazant and S. Baweja, *"Creep and Shrinkage Prediction Model for Analysis and Design of Concrete structure-Model B3"*, Materials and Structures, V. 28, No. 180, pp. 357-365, 415-430, and 488-495, (1995).
8. R. I. Gilbert, *"Creep and Shrinkage Models for High Strength Concrete-Proposals for Inclusion in AS3600"*, Australian Journal of Structural Engineering, Vol.4, No.2, pp. 95-106, (2002).
9. J. J. Brooks, A.E. Gamble and M. M. Chouman, *"Assessment of Methods of Predicting Time-Dependent Deformations of Plain and Reinforced Concrete"*, Journal of the Institute of Structural Engineers, London, June, V.70, No.2, (1992).
10. H. G. Russel, *"High-Rise Concrete Buildings: Shrinkage, Creep, and Temperature Effects"*, ACI Special Publication SP-97, pp. 125-137, (1985).
11. M. Fintel, and F. R. Khan, *"Effects of Column Creep and Shrinkage in Tall Structures – Analysis for Differential Shortening of Columns and Field Observation of Structures"*, Designing for effects of Creep and Shrinkage Temperature in Concrete Structure, SP-27, American Concrete Institute, Detroit, pp. 95-119, (1971).
12. Z. P. Bazant, *"Prediction of Concrete Creep Effects Using Age-Adjusted Effective Modulus Method"*, ACI Journal, V.69, April, pp. 212-217, (1972).
13. M. Koutsoukis and A. J. Beasley, *"Second-Order Moment Dischinger Model for the Axial Shortening in Tall Concrete Buildings"*, EASEC-5 Proceedings, Griffith University Gold Coast Campus, Australia, pp. 525-530, (1995).
14. M. Fintel, and F. R. Khan, *"Effect of Column Exposure in Tall Structures – Temperature Variations and Their Effects"*, ACI Journal, Vol. 62, No. 12, December, pp. 1533-1556, (1965).

15. F. R. Khan and M. Fintel, *"Effects of Column Exposure in Tall Structures – Analysis for Length Changes of Exposed Columns"*, ACI Journal, V.63, No.8, pp. 843-864, (1966).
16. B. Glisic, D. Inaudi, K. C. Hoong and E. J. M. Lau, *"Monitoring of Building Columns During Construction"*, EASEC-9 Proceedings, Bundung Institute of Technology, Indonesia, pp. 593-606, (2003).
17. D .W. Pfeifer, D. D. Magura, H. G. Russell and W. G. Corley, *"Time Dependent Deformations in 70 Story Structure"*, Designing for effects of Creep and Shrinkage Temperature in Concrete Structure, SP-27, American Concrete Institute, Detroit, pp. 159-185, (1971).
18. H. G. Russel and W. Corley, *"Time-Dependent Behavior of Columns in Water Tower Place"* , Douglas Mchenry International Symposium on Concrete and Concrete Structures, SP-55, American Concrete Institute, Detroit, pp. 347-373, (1978)
19. H. G. Russel and S. C Larson, *"Thirteen Years of Deformations in Water Tower Place"*, ACI Structural Journal, V.86, No.2, pp. 182-191, (1989).
20. E. A. Brady, *"Monitoring the New South Wales Institute of Technology Tower Building"*, Symposium on Surveillance of Engineering Structures, University of Melbourne, 14-15 Nov, (1983).
21. S. L. Bakoss, A. Burfitt, L. Cridland and J. L. Heiman, *"Measured and Predicted Long-Term Deformations In a Tall Concrete Building"*, Deflection of Concrete Structures, SP-86, American Concrete Institute, Detroit, pp. 63-94, (1984).
22. S. Bursle, N. Gowripalan, D. Baweja and K. Kayvani, *"Refining Differential Shortening Predictions in Tall Buildings for Improved Serviceability"*. Concrete 2003, Concrete Institute of Australia, Sydney, (2003).
23. B. Miao, O. Chaallal, D. Perraton and P. Aitcin, *"On-site Early-Age Monitoring of High-Performance Concrete Columns"*, ACI Materials Journal, V.90, No.5, pp. 415-420, (1993).
24. P. Lukkunaprasit, E. Zyhajlo and C. Levy, *"Differential Column Shortening of Baiyoke II Tower"*, EASEC-5 Proceedings, Griffith University Gold Coast Campus, Australia, pp. 211-216, (1995).
25. B. Whaley, *"Q1 Tower-The Proposed World's Tallest Residential Building at Surfers Paradise, Australia"*, ACMSM-17 Proceedings, Vol.1, pp. 23-29, (2002).
26. S. Boonlualoah, S. Fragomeni, Y.C. Loo and D. Baweja, *"Monitoring Deformations of Vertical Concrete Members in Tall Building"*. Presented at the fourth International Conference on Concrete Under Severe Conditions: Environment & Loading 27th-30th June 2004 in Seoul, CONSEC04 Proceedings, Seoul National University, Korea, (2004).
27. S. Boonlualoah, S. Fragomeni, Y.C. Loo and D. Baweja, *"Aspects of Axial Shortening of High Strength Concrete Vertical Elements in a Tall Concrete Building"*, ACMSM-18 Proceedings, University of Western Australia, Perth, (2004).
28. M. Koutsoukis and A. J. Beasley, *"Idealising the Construction Cycle of Tall Concrete Buildings for Axial Shortening Analyses"*, Presented at the Concrete 95 Toward Better Concrete Structures Conference 4th-7th September 1995 in Brisbane, Concrete-95 Proceedings, Concrete Institute of Australia, Brisbane, Volume 1, Pages 447-453, (1995).
29. F. R. Khan and M. Fintel, *"Effects of Column Exposure in Tall Structures – Design Considerations and Field Observations of Buildings"*, ACI Journal, V.65, No.8, pp. 99-110, (1968).

CHINA CENTRAL TELEVISION NEW HEADQUARTERS, BEIJING, CHINA

CHRIS CARROLL XIAONIAN DUAN CRAIG GIBBONS RICHARD LAWSON, ALEXIS LEE ANDREW LUONG RORY MCGOWAN CHAS POPE

Ove Arup and Partners Hong Kong Limited, Level 5, Festival Walk, 80 Tat Chee Avenue, Kowloon Tong, Hong Kong

CCTV new headquarters building is a 234m tall building in the form of a three-dimensional continuous cranked loop formed by a 9-storey podium structure joining two 50-storey high leaning towers which are linked at the top via a 13-storey cantilevered "overhang" structure at 36 storeys above the ground. This innovative and iconic shape of the building is exploited to provide the primary structural support system, as well as achieving the architectural vision of high-rise occupancy with occupant interface maintained within short distances in a continual loop. The building's primary support, in high-seismic intensity Beijing, is achieved by its external skin of leaning columns, horizontal beams and triangulated bracings forming a network of diagrids in an extremely strong closed braced tube structure. This external diagrid structure is also expressed boldly in the building's façade. It reinforces the transparency between structure and architecture, a central philosophy to the building's design. The internal structure is supported by vertical columns and steel cores which diminish in size progressing up the building height, in tune with the shape of the leaning towers. The columns emerge and terminate up the height of the building, again influenced by the angled towers. Transfer trusses are located at various levels to collect these column loads into the cores and external structure.

1. Introduction

The China Central Television new headquarters building will be located in the Central Business District on the east side of the East Third Ring Road, north of Guang Hua Road and south of Chao Yang Road.

When completed, the building will form the new headquarters to China Central Television (CCTV), the principal state-run broadcaster in China. It will provide approximately 450000 m² of floor area, (equivalent to three standard high-rises) to sufficiently house the studios, facilities and offices for CCTV's projected operations expansion to over 200 broadcast channels by 2008.

The building was initially unveiled in the winning design in an international competition in 2002 attracting some of the biggest names in architecture. It is one of the synergistic conceptions between Rem Koolhaas' practice Office for Metropolitan Architecture (OMA) and Cecil Balmond of Arup.

Arup is the engineering force behind the building, providing consultancy for structure, building services, geotechnical design, fire, communications and security. The East China Architecture and Design Institute (ECADI) in China will act as the Local Design Institute (LDI) of record.

Figure1: CCTV New Headquarters Building

2. Structural System

The innovative and iconic shape of the building is capitalised upon to provide its main structural support and stability system. The form warranted a primarily structural steel building, for a "light-weight" solution and enhanced seismic performance. As such, all the structural support elements in the building are of structural steel, except some external columns are steel-reinforced concrete columns due to the magnitude of loads they are designed to carry. The floors are composite slabs on steel beams.

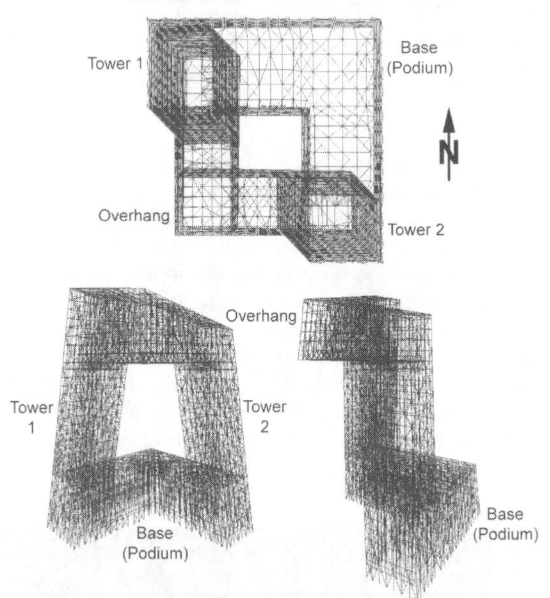

Figure 2: The Continuous Loop Structure of CCTV

2.1 *External primary structure*

Form the onset, it was decided to adopt an external skin of leaning columns, horizontal edge beams and triangulated bracing on a two-storey pattern to form an enclosed tube structure to support the building. Furthermore, the braced tube structure affords a multitude of alternative load paths. Such a robustness feature is highly desirable, especially in seismically

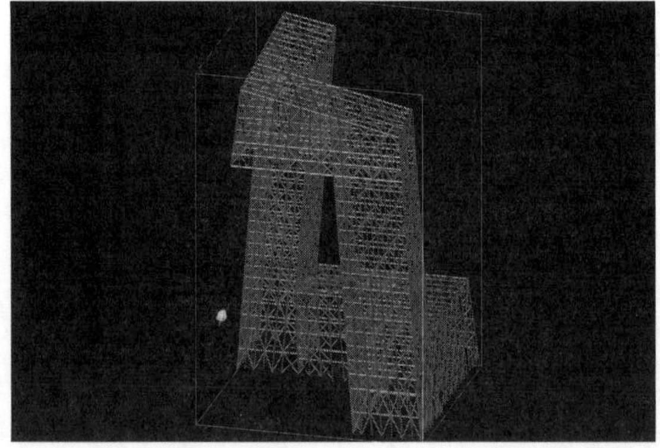

Figure 3: External Primary Structure (drawn using Xsteel)

sensitive Beijing. It also provides safety in the event of an extreme design incident, such as blast removal of a major column in the building. The external diagrid structure is also boldly expressed in the building's façade. It visually expresses the pattern of forces in the external tube, reinforcing the transparency between structure and architecture, a strong philosophy in the building's design. The unique diagrid pattern in the external structure was arrived at after extensive iteration and optimization, in close collaboration with the architect.

2.2 *Internal structure*

The internal structure is supported by vertical steel columns and cores. Sloping cores were initially considered in the design, in order to align with the sloping towers and to allow consistency of the floor plate layout. However they were ruled out for various reasons including constraints on the procurement of the lift systems. In addition, in supporting as much of the structure as possible on vertical structure, the overturning moment on the foundations is reduced. The sloping external tube walls and the vertical internal elements combine to create unique floor configurations for every floor of the building. That is, the floor span between cores or internal columns and the changes on every floor. Moving up the building, the floor spans increase on two adjacent sides of the building and decrease on the other two opposite sides. As a consequence, where floor spans decrease on the inward sloping sides of the building, some internal columns can be removed once the distances become structurally manageable in one span. For the same reason, once the increasing floor spans on the outward leaning sides of the building start to impact on the finished floor height due to increased beam depth, additional columns are introduced.

2.3 *Transfer trusses*

The looping form of the building, combined with the sloping external faces and the need for large open internal spaces for studios and facilities, lead to the introduction of transfer trusses at strategic locations in the building.

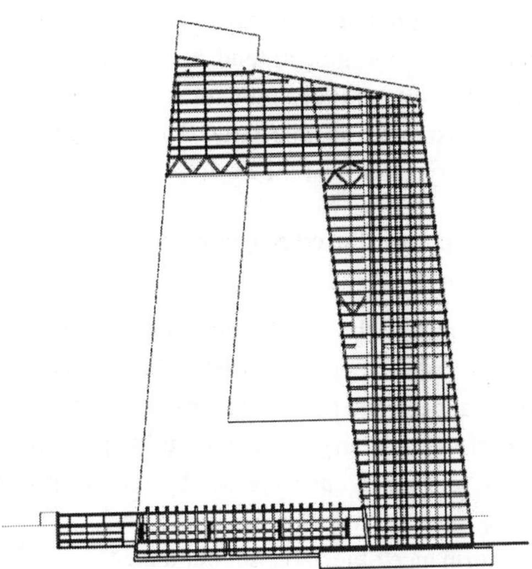

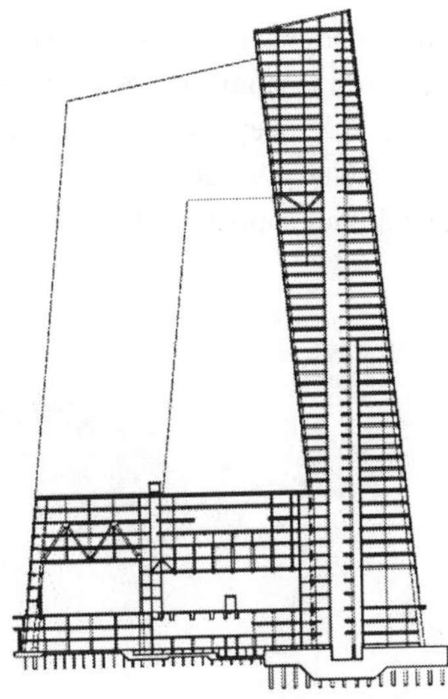

Figure 4: Sections Through the Building Showing Vertical Internal Structure and Transfer Trusses

2.3.1 Transfer trusses in the towers

Transfer trusses are introduced to collect the columns required at intermediate heights in the towers to cope with the increasing floor spans. The transfer trusses span between the internal core and the external tube structure. They are typically located in plant floors in the building so as to be hidden from view and minimize impact to the architecture and floor usage.

The large member sizes of the transfer trusses mean a potential for these trusses to act as outriggers as they link up the external tube with the internal steel cores. An outrigger effect would be undesirable because this would then complicate the primary seismic load path. (The seismic stability of the building is achieved through the diagrid framing of the external tube structure). To prevent this condition, the transfer trusses are connected to the internal cores and the external columns at singular pin-joint locations only. Detailed analyses were carried out to verify no outrigger effects result from the transfer truss geometry.

2.3.2 Transfer trusses at the underside of the overhang

At 36 storeys above the ground, the two leaning towers crank horizontally and cantilever 75m outwards in the air to join together forming the continuous loop defining the building shape. This 75m cantilever structure encompasses 13 storeys and is known as the "overhang". The overhang floor are supported by columns landing on transfer trusses. These trusses span the bottom two storeys of the overhang in two directions connecting back to the external tube structure. Thus the overall overhang structure is ultimately supported off the external tube structure.

2.3.3 Transfer trusses in the studios

Major transfer trusses are also located in the base (podium) part of the building. These trusses span over the major studios to support the columns and floors above them in the building.

2.4 Foundations

The bearing capacity of the subsoil around the main towers of CCTV is not sufficient to support the entire load from the superstructure whilst remaining within acceptable settlement limits. A piled raft foundation has been adopted under the main towers. The piles are 1200mm in diameter and up to 35m in length following optimisation. The total settlement of the building is estimated to be less than 100mm and differential settlement is kept to 1:500.

As the foundation loads in the towers are concentrated at the inside face, the piled raft is designed to be up to 7m thick and extends beyond the footprint of the towers in order to distribute forces more favourably into the ground. The foundation system is arranged such that the centre of the raft is close to the centre of load at the bottom of the tower, and no permanent tension is allowed in the piles. Limited tensions are permitted in some piles only in seismic events.

Away from the tower, for the 9-storey Base and the 3-storey basement under the rest of the site, a traditional raft foundation is used, with tension piles between column locations to resist uplift due to water pressure on the deep basement. Additional deep 1200mm diameter piles are required under secondary cores and columns supporting large transfer trusses.

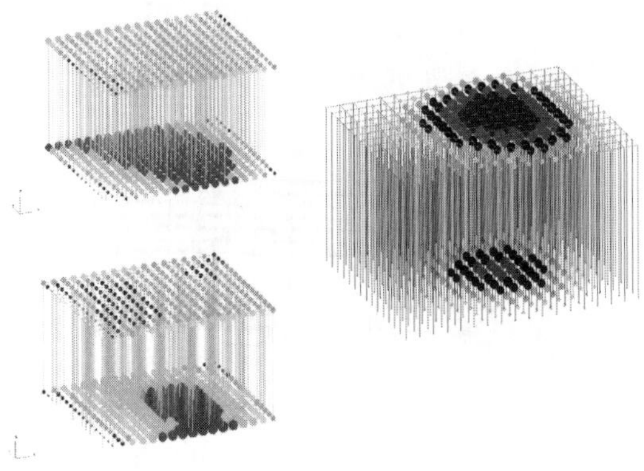

Figure 5: Tower Base Analyses Results

The design of the foundation requires that the loads are redistributed across the pilecap, and the soil properties of the site were non-linear. Therefore the analysis of the piled-raft solution became highly iterative.

3. Expert Panel Review and Approvals

The Chinese code for seismic design of buildings, GB50011 – 2001, prescribes a set of limits to the heights of buildings depending on their structural system and limits to the degree of plane and vertical irregularities in the building. The design of buildings exceeding these code limits must obtain approval from a project-specific seismic design expert panel review, as set out by the Ministry of Construction's Ordinance 111 - Regulations on Seismic Fortification for Buildings Exceeding Code Limits.

Although the height of the CCTV building of 234m is within the code's height limit of 260m for steel tubular structural systems (framed-tube, tube-in-tube, truss-tube etc.) in Beijing, the gravity-defying shape of the building means it is non-compliant in irregularity requirements. As a result, the seismic design of the CCTV building is required to pass through a project-specific expert panel review process for approval. For this purpose, the Seismic Administration Office of the Beijing Municipal Government appointed an expert panel consisting 12 eminent Chinese engineers and academics to closely examine the structural design with special focus on the seismic resistance, seismic structural damage control and life safety aspects of the design.

4. Performance Based Seismic Design Approach

As the design of the CCTV building is well outside the scope of the prescriptive Chinese codes of practice, a performance-based design approach was used. Performance targets for the building at different levels of seismic event were set by the Arup design team in consultation with the Expert Panel.

The performance objectives set out that:

When subjected to the design frequent earthquake (level 1) with an average return period of 50 years (63% probability of exceedance in 50 years), the building shall not sustain structural damage.

Under the design intermediate earthquake (level 2) with an average return period of 475 years (10% probability of exceedance in 50 years), the building may undergo repairable structural damage.

When subjected to the design rare earthquake (level 3) with an average return period of 2500 years (2% probability of exceedance in 50 years), the building is permitted to sustain severe structural damage but must not collapse.

5. Non-Linear Superstructure Design and Performance Verification

To demonstrate that the performance objectives are achieved in the building design, linear and non-linear seismic response simulation methods were carried out to verify the building's seismic performance under the three levels of design earthquake.

The process of establishing the inelastic deformation acceptance limits of various structural members (braces, SRC columns, steel beams and steel columns) is illustrated by the non-linear numerical simulation of the post-buckling behaviour of the concentric braces. These braces in the external tube are critical members both in the lateral as well as the gravity systems of the building. They are also the primary sources of ductility and seismic energy dissipation. Non-linear numerical simulation of the post-buckling behaviour of the braces was necessary in order to establish:

(i) The post-buckling axial force – axial deformation degradation relationship to be used as input in the global structure non-linear seismic response simulation; and

(ii) The inelastic deformation (axial shortening) acceptance limit. The post-buckling inelastic axial force – axial deformation degradation relationship curve illustrates strength degradation as the axial shortening increases under the cyclic "saw tooth" type axial displacement time history prescribed at the top end. The brace inelastic deformation acceptance limit was then established from the strength degradation backbone curve based on the required strength to support the gravity load.

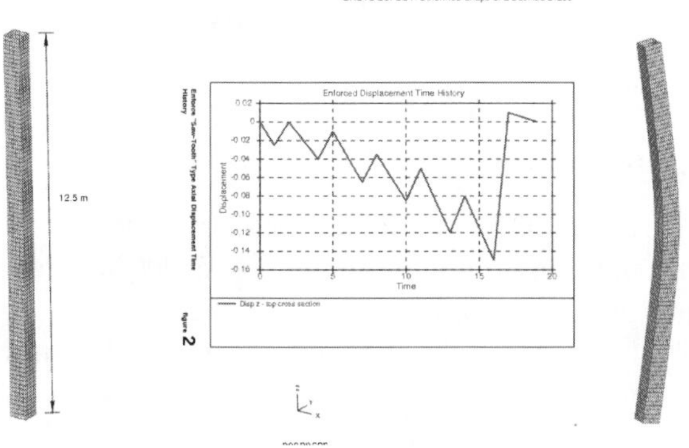

Figure 6: Non-Linear Modelling of Inelastic Deformation Acceptance Limits of Steel Braces

Having established the inelastic global structure and local member deformation acceptance limits, the next step was to carry out non-linear numerical seismic response simulation of the CCTV building subjected to the Level 2 and the Level 3 design earthquakes. Both the non-linear static pushover analysis method and the non-linear dynamic time history analysis method were employed to perform this task. Finally, the seismic deformation demands were compared against the structure's deformation capacities to verify seismic performance. This verification was carried out on a storey-by-storey and a member-by-member basis. For the CCTV building, all global and local seismic deformation demands are shown to fall within their respective acceptance limits, confidently demonstrating that the building achieves the quantitative performance objectives when subjected to both the design intermediate (Level 2) and the design rare (Level 3) earthquakes.

6. Connection Assessment

As explained the braces in the external tube structure are critical members in the structural system. The connection of these braces to the columns requires careful design and detailing consideration to ensure a "strong joint – weak member" capacity.

The connections must resist the maximum probable load delivered to them from the braces with minimum yielding and a relatively low degree of stress concentration. The force from the braces and edge-beams must be transferred into the column sections with minimal disruption to the stresses already present in the column. "Butterfly" plates were conceived to facilitate smooth load transfer.

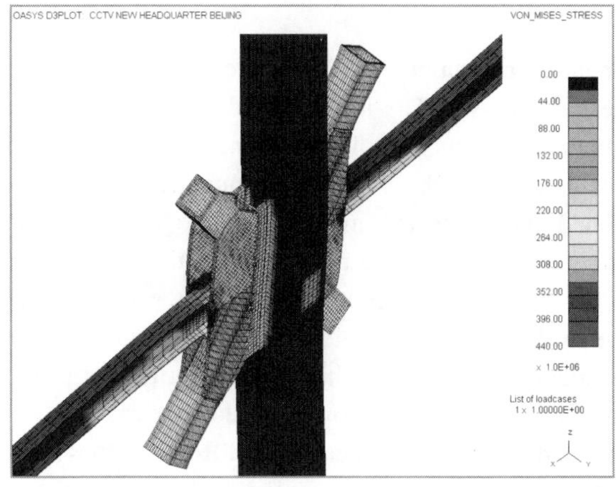

Figure 7: Brace Connection Detail Studies

It is important to prevent brittle fracture at the welds under cyclic seismic loading (a common cause of failure in connections observed after the 1994 Northridge earthquake in Los Angeles). "Butterfly" brace connections were modelled and analysed using a 3D finite element code MSC/NASTRAN to evaluate the stress magnitude and the degree of stress concentration in the joints subjected to the full range of forces that can be developed before the braces buckle or yield.

7. Construction Sequencing and Thermal Analyses

The chosen method of construction and sequencing of the works have an important impact on the dead load distribution and final locked-in stresses in the structural elements. Therefore, staged construction sequencing forms an integral part of the design and analyses of the structure.

To allow flexibility in the contractor's construction method and programme, upper and lower bound construction conditions are considered in the design analyses. The lower bound conditions assume minimal construction of the leaning towers before the construction of the overhang commences. The lower bound thus puts the most loads into the overhang structure as it acts as a prop between the two towers. The upper bound conditions assume the two towers are well constructed before the construction of the overhang commences. This assumption results in the largest stresses into the tower since they carry more of the load in bending as a cantilever. Between these two extremes, the contractor can choose his program and erection procedures to suit.

The construction joining of the two towers to form the closed loop overhang link was also assessed and designed to great detail. The successful connection of the link is especially sensitive to wind and solar path thermal effects and rigorous checks and measures are outlined in the project technical specifications to facilitate the proper implementation of this part of the works.

Despite its size, the continuous loop of the completed building structure, framed by the 160m x 160m x 234m volume of space, contains no movement or expansion joints. This is against code requirements but is necessitated given the shape and structural stability system of the building. To justify the omittance of movement joints, analyses were performed to check the design against the extreme temperature ranges in Beijing (the 30 year maximum temperature range is –27.4°C to +40.6°C). The analyses and design involved assessing the effects of locked-in thermal stress in the building elements during construction, taking consideration of the season and temperature at which the elements are erected, and seasonal temperature fluctuations post-construction. The thermal design also particularly focused on the sizing and construction linkage of the overhang link structure.

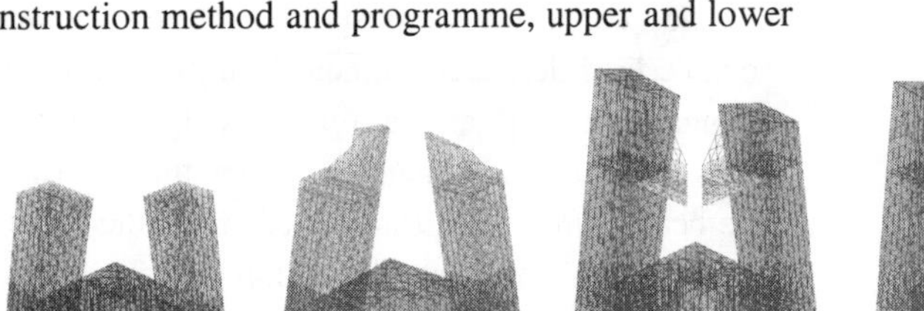
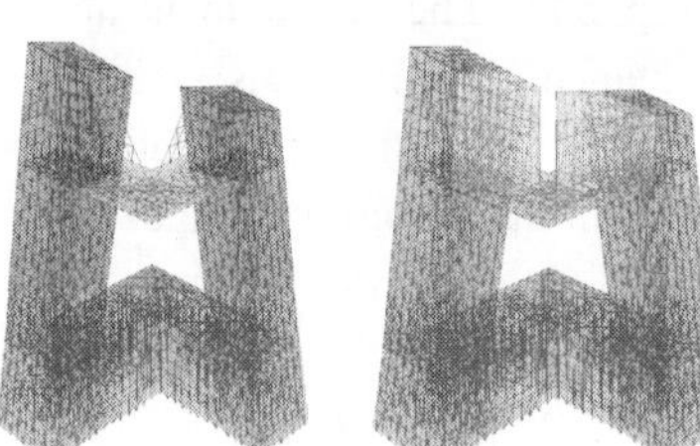
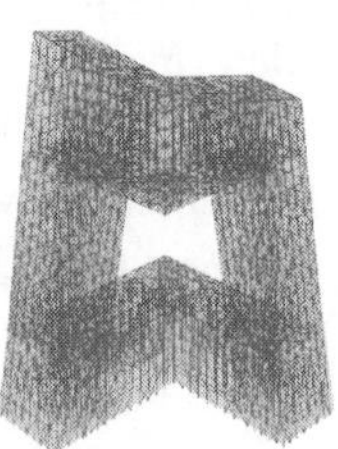

Figure 8: Construction Sequence Studies

8. Conclusion

The entire CCTV development covers a site area of 187000m² and will provide a total of 550000m² gross floor area. The project as a whole includes:

- The China Central Television headquarters building - presented in this paper
- The Television Cultural Centre (TVCC)
- A service & security building
- A landscaped media park with external features

The official ground breaking ceremony for the CCTV building was held on 21st September 2004, and construction work of the foundations is progressing.

This paper has attempted to introduce the technical complexities and innovative structural solutions in achieving such an astounding design. The rigorous nature of Arup's design work is of vital importance to the success of this building, with a core team of international engineers who travelled with the project to 4 cities and in the process delivered the Scheme Design in 4 months and Extended Preliminary Design in 6 months, and finally gaining approval from the Expert Panel Review.

It is not possible to cover the extensive details and all the interesting technical aspects of the building and its design within a single paper. Each of the technical topics touched upon here are themselves worth an entire dissertation. This paper has only commenced to tell the design story of the CCTV building.

THE STRUCTURAL DESIGN AND CONSTRUCTION OF "THREE PACIFIC PLACE" HONG KONG SAR, CHINA*

NG SHUNG LUNG, GORDON

Associate Director, Ove Arup and Partners Hong Kong Limited

TSANG SAU CHUNG, PAUL

Director, Ove Arup and Partners Hong Kong Limited

Three Pacific Place is recognized as a successful urban redevelopment project in Wanchai District of Hong Kong. The development has taken over a decade from inception to completion. The development comprises a 38-storey Grade A+ office tower on top of three levels of podium and three levels of basement. During construction, the site was divided into four parcels due to phased land acquisition. This imposed significant challenges to the construction and the engineering design.

1. The Development

Three Pacific Place (Figure 1) is recognized as a successful Urban Redevelopment Project in Wanchai District of Hong Kong. The development has taken over a decade from inception to completion.

Figure 1. Aerial view of Three Pacific Place

The development comprises a 38-storey Grade A+ office tower on top of three levels of podium and three levels of basement (Figure 2) in phase 1. Structural provision for possible future extension (Phase 2) was allowed for in Phase 1 construction.

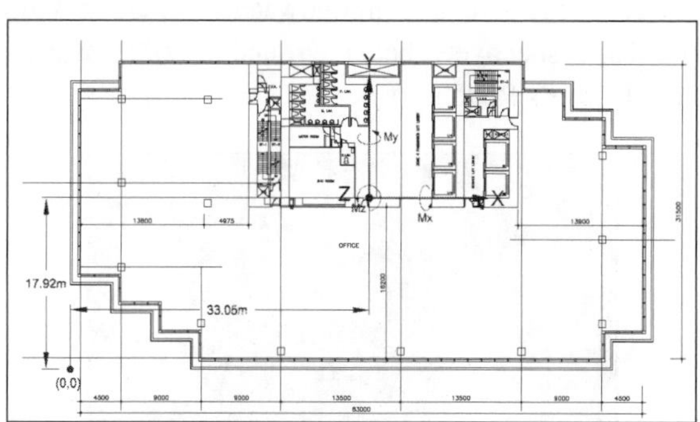

Figure 2. Plan view of Phases 1

A pedestrian subway connecting the development to the subway between MTRC Admiralty Station (Figure 3) and Pacific Place is due to be completed in 2006. This pedestrian subway will greatly improve the pedestrian comfort and safety of the existing unsheltered footpaths in the Justice Drive and Queen's Road East Area.

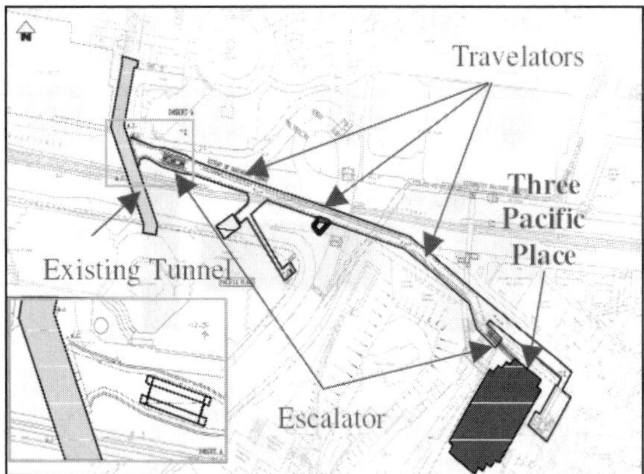

Figure 3. Layout of pedestrian subway

2. The Site

The site is bounded to the East by Wing Fung Street, to the South by Star Street, to the West by Monmouth Path and to the North by Queen's Road East. Wing Fung Street West is running across the site from East to West (Figure 4). The site dips from Star Street at +16.0mPD towards Queen's Road East at +6.5mPD. The rockhead level and groundwater level are at 3.5m and 1m respectively below ground level measuring from Queen's Road East.

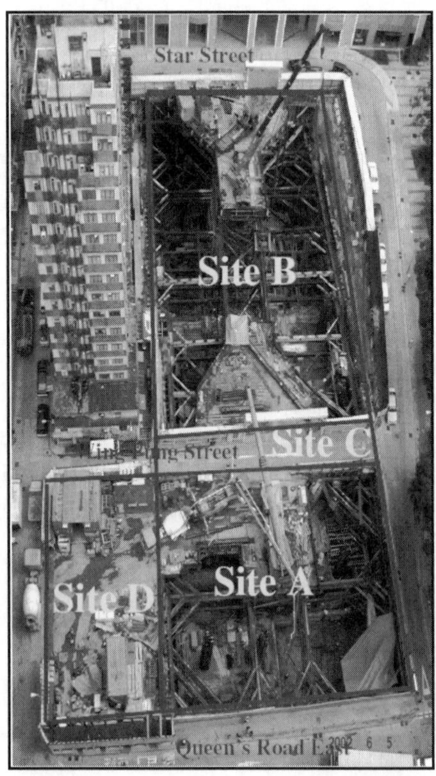

Figure 4. Plan view of site

3. The Site Formation

The site was physically divided into four parcels (Figure 4) during the superstructure construction stage due to the reasons of site possession, site formation works and foundation construction.

3.1 *Site A*

The site formation work of the northern part of Wing Fung Street West, namely Site A (Figure 4), took eleven months to complete.

3.2 *Site B*

Originally, two separated office towers were proposed to be built on the southern and northern side of Wing Fung Street West. The site formation works of the tower on the southern side of the Wing Fung Street West, namely Site B (Figure 4), was completed before the developer decided to build a single office tower covering the whole development site except site D.

3.3 *Site C*

Wing Fung Street West, namely Site C (Figure 4), was the last parcel of land acquired for the construction.
Due to the time lag of the site acquisition, the site formation works for Sites C and D were handled by the Main Contractor of the superstructure.
This severe site constraint rendered the completion of the foundation and core wall construction on Wing Fung Street West prior to the superstructure construction impossible. Consequently, retrofit construction technique for the concerned portion of the foundation and the core wall was adopted.

3.4 *Site D*

The area bounded by Queen's Road East, Wing Fung Street and Wing Fung Street West, namely Site D (Figure 4), was the third parcel of land acquired for the development.

3.5 *Excavation and lateral support*

The excavation sequences and shoring were designed for the scenarios that the excavation works were proceeded separately within individual parcels as well as the parcels will be joined together at different construction stages. The most changeling was to ensure the continuity of the struts forces between individual parcels as well as deigning for the de-propping sequences of the struts.

4. Wind Tunnel Test

The stability of the tower is provided by a highly eccentric core wall, which is located at the eastern side of the building (Figure 2).

Wind tunnel test (Figures 5 & 6) was carried out to verify the design wind forces on the tower, particularly the torsional component of the wind force as well as the acceleration of the tower due to torsional component.

Figure 5. Proximity model of wind tunnel test

Figure 6. Proximity model for Phase 1

4.1 *Wind induced structural response*

Two structural response tests have been carried out to determine the design wind load on the tower under Phase 1 and Phase 2 conditions.

The peak accelerations of the building under different wind climate and return periods also lie within the ISO criteria (Table 1).

Table 1. Peak acceleration of the building

Return Period (Years)	Peak Total Accelerations without Typhoons (milli-g)	Peak Total Accelerations with Typhoons (milli-g)	ISO Criteria (milli-g)
1	1.3	2.5	14.2
5	2.6	8.1	19.8
10	3.2	11.5	23.7

A comparison of the wind loads obtained from the wind tunnel tests and Hong Kong Wind Code is shown on Figures 7 to 9.

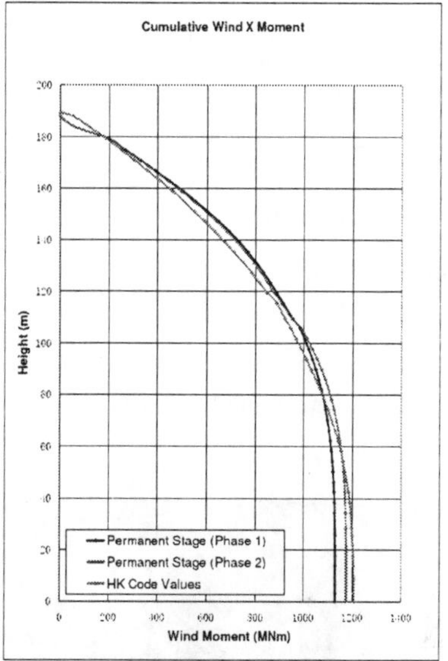

Figure 7. Wind load (Fx) from wind tunnel and wind code

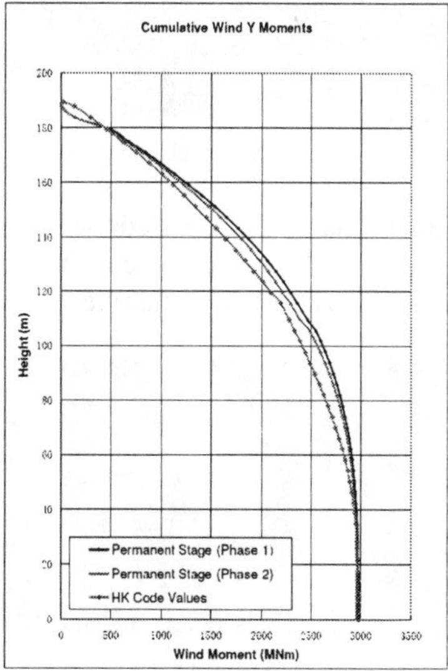

Figure 8. Wind load (Fy) from wind tunnel and wind code

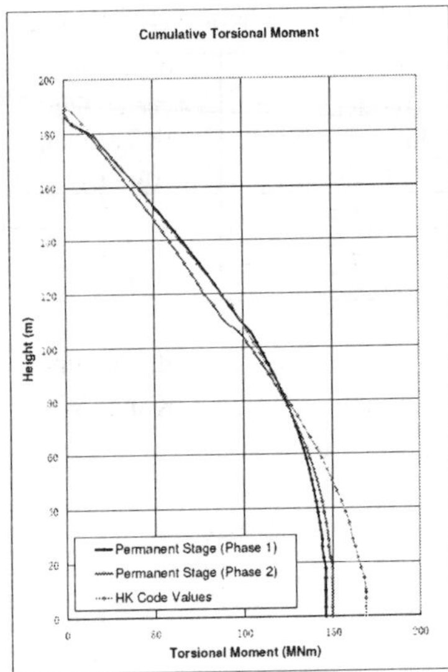

Figure 9. Wind load (Fz) from wind tunnel and wind code

4.2 *Cladding wind load study*

The design wind pressures on the curtain wall and glass wall have determined by wind tunnel tests as well (Figure 10).

Figure 10. Pressure probes for cladding pressure study

Pressure integration method was adopted to determine the design wind loads for the three-dimensional cable net glass wall system on the podium levels (Figure 11).

Figure 11. Cable net at the podium

5. Structural Stability

The overall stability of the tower is provided by a highly eccentric core wall. The dimension of the core wall is 34.9m by 15.7m on plan up to L22 and reduced to 29m by 15.4m from L23 to the roof (Figures 2 & 12). The thicknesses of the core wall also reduce gradually up the building. Grade 60D and Grade 45D concrete are adopted for the construction of the core wall at the lower and upper portions respectively.

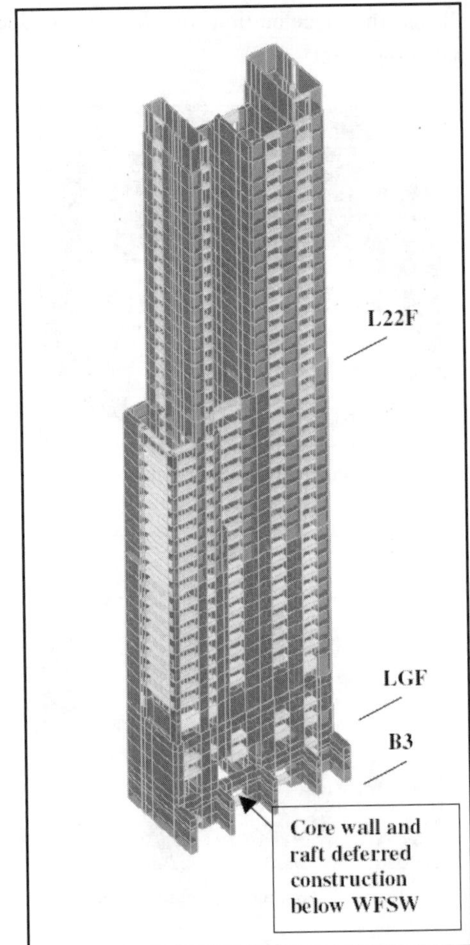

Figure 12. Core wall of the building

Three structural models have been created to verify the stability of the building under different stages, namely permanent stages – phase 1 & 2 and temporary stage.

5.1 *Permanent stage – phase 1*

The model includes completion of the 38-storey office tower, 3 levels of podium and 3 levels of basements.

As the core wall is highly eccentric, there is insufficient dead load on the eastern edge panel to counterbalance the tension force induced by wind moment. The stiffness of this wall panel was therefore downgraded by 20% to take into account of the net tension.

5.2 *Permanent stage – phase 2*

The model includes possible future extension. The wind load and gravity load due to future extension will be transmitted to the core wall constructed in Phase 1. The design of the core wall and foundation in Phase 1 have allowed for the additional loadings of future extension.

5.3 *Temporary stage*

During the construction stage, the stability of the tower with the omission of structure from the raft to Lower Ground Floor at Wing Fung Street West had been checked.

In order not to affect the construction of the rest of the tower structure above Upper Ground Floor. The core wall structure on Level 2 has been designed to temporarily span across the WFSW.

The following two design assumptions have been adopted to determine the design wind load for the tower stability check during construction:

1. 70% of the permanent design wind pressure is adopted;
2. The effective frontage area of the floor without installing any curtain wall is determined.

The findings indicate that the whole tower structure could be built without any additional strengthening works before the completion of the structure below the WFSW provided the curtain wall installation does not proceed above L27/F.

The storey drifts of the tower under various stages are shown in Figures 13 & 14. The limiting criteria for inter-storey drift and total drift are h/300 and H/450.

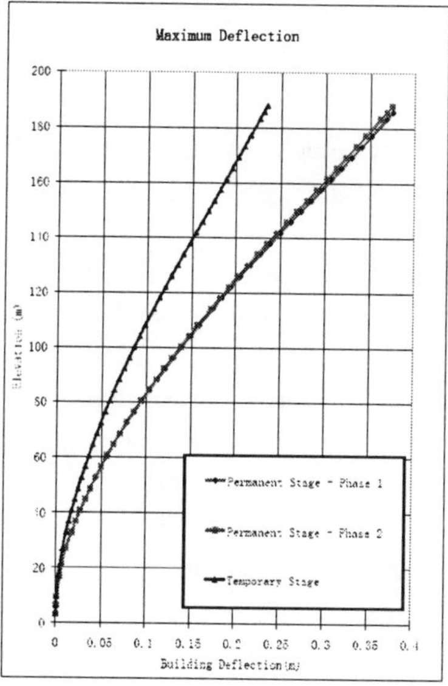

Figure 13. Inter-storey drift of the building

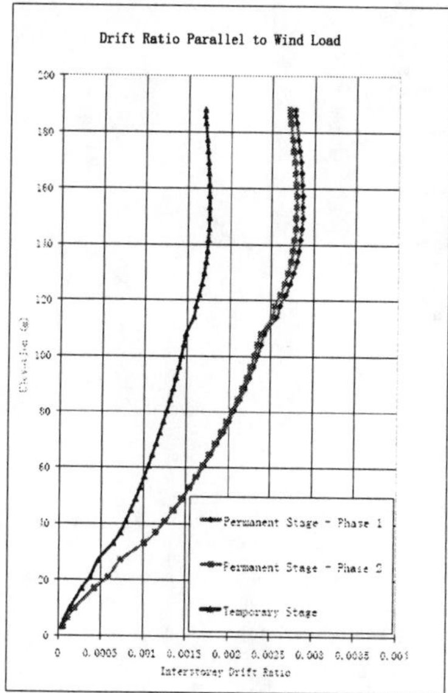

Figure 14. Storey drift of the building

6. Structural Optimization

After the stability model of the tower in permanent stage is finalized, structural optimization of the stability system of the tower is carried out by the Software, Optima.

The graphical interface of the software can easily indicate the work efficiency of the individual members.

The objective of the structural optimization process is to allocate the structural material from the inefficient locations to locations that are more efficient such that the overall construction material can be reduced.

This exercise demonstrates that the advantages of the reduction of the construction material are threefold. These includes:

- reduction in the construction cost;
- reduction in the building weight;
- reduction in the area occupied by the construction materials, i.e. increase usage floor area.

The parameters that could be adjusted during the optimization process include the following:

- Concrete grade
- Wall thickness
- Lintel Beam Depth

The efficiency of the structural elements after optimization is shown in Figure 15.

Figure 15. Work efficiency of core wall members
(Hotter colour represents elements with higher efficiency)

The following modifications have been incorporated into the stability model after the optimization process:

1. The core wall is divided into five zones along building height;
2. The thickness of the flange walls at the back of the core is increased;
3. The thickness of the web walls of the core is reduced, particularly those on higher levels;
4. C60 concrete are adopted for all walls and beams on levels below L23 and C45 concrete for those on levels above L23 respectively.

The benefits of the optimization process are summarized as follows:

1. The total volume of reinforced concrete reduced by 3502 m³.
2. The usable floor area of the tower has increased by 606 m².
3. Lighter structure also reduces the cost of foundation.
4. Reduction in the concrete volume marginally expedites the construction programme.

7. Gravity Load System

7.1 *Horizontal system*

Post-tensioning Slab

350mm thick post-tensioning flat slab with 500mm thick drop panels and 2400mm wide by 600mm deep band beams along the western edge is adopted for typical floors. The horizontal dimension of the floor plate measuring from the slab edge to the core wall is about 16m. The maximum cantilevered span of the flat slab is 4.5m (Figure 16).

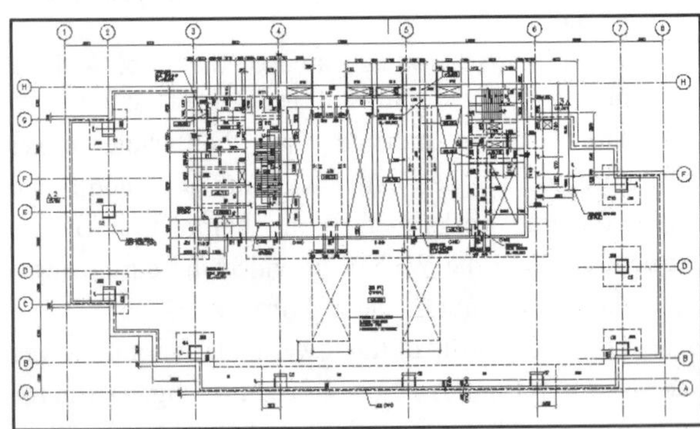

Figure 16. Typical floor at lower zone

The critical criteria of the post-tensioning slab are to maintain the long-term deflection of the floor slab within L/500 or 20mm whichever is smaller. It is assumed 25% of the live load as sustained load for the post-tensioning slab design.

The creep coefficient is determined based upon the Structural Use of Concrete published by the Buildings Department of Hong Kong. Different creep coefficients for different basic load cases are determined corresponding to the concrete age when the loads are applied. The design creep coefficients are summarized in Table 2 below.

Table 2. Creep coefficients for different loads

	DL	SDL	LL	PT
Concrete Age	3 days	28 days	90days	3 days
Creep Coefficient	2.42	1.51	1.13	2.42

Provision for future knock out floor panels has been allowed for in L4 to L20 (two locations) and L24 to L38 (one location) respectively for inter-storey staircase.

The additional bending moment induced onto the columns due to the pull-in effect of the post-tensioning slab has also been accounted for.

7.2 *Vertical system*

Ten numbers of concrete encased composite columns are used for the gravity support of the tower. Four standard universal column steel sections are used in each column below Level 23 (Figure 17). They are reduced to one standard universal column steel section above Level 23. The structural steel is Grade 50 and the concrete is Grade 60D. The maximum length of the steel section is limited to 11m. The column section sizes change seven times from bottom upwards. Some of them accommodate the changes of section sizes by eccentric alignment. Sophisticated transfer details of the steel sections are adopted.

8. Foundation

As intact rock is encountered on the final formation level, 3m thick raft on rock has been adopted the foundation system of the tower.

The finite element analysis software SAFE has adopted for the foundation analysis and design. Three SAFE models are established for the checking of the Permanent Stage – Phase 1 & 2 as well as the temporary stage.

The design criteria of the foundation are:

1. Allowable bearing capacity for Grade II/III rock is 5000kPa and 7500kPa under wind load;
2. The angular distortion of the raft is limited to 1:500.

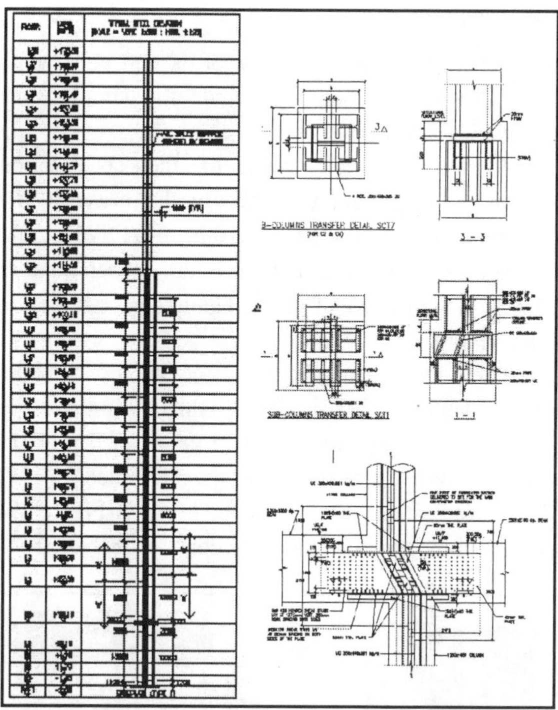

Figure 17. Composite column and transfer details

8.1 *Permanent stage – phase 1*

Owing to the eccentric core wall arrangement, net tension is encountered on the eastern side of the raft. The rock springs underneath the raft is eliminated in the analysis at those locations.

8.2 *Permanent stage – phase 2*

The additional load from the future extension is included in the raft design. The design force envelope of the raft for Phase 1 and Phase 2 were used for the reinforcement design of the raft.

The bearing stress of the raft in Phase 1 is shown in Figure 18.

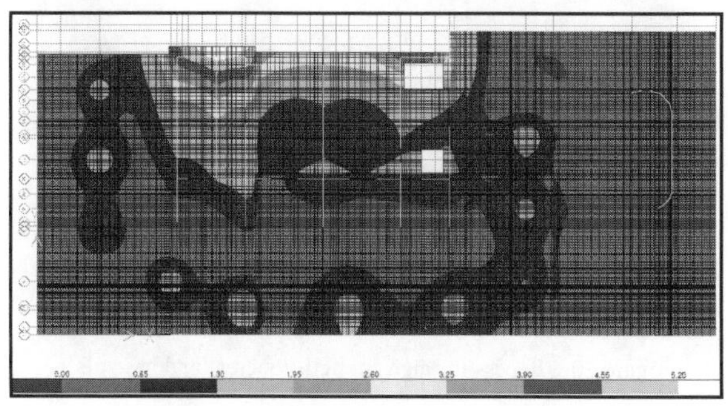

Figure 18. Bearing stress of the raft in Phase 1

8.3 *Temporary Stage*

The construction of the portion of the raft underneath the WFSW has been deferred due to the site possession schedule.

After a detailed study of the construction sequences and programme, the time lag arisen in the construction of the raft foundation in WFSW should be resolved by planning on the superstructure construction instead of strengthening the raft or the superstructure.

This has resulted in a constraint on the superstructure construction, which limits the installation of the curtain wall to level 27/F prior to the completion of the portion of the raft under WFSW. This constraint can eliminate overstressing (Figure 19) of the incomplete raft during the construction stage.

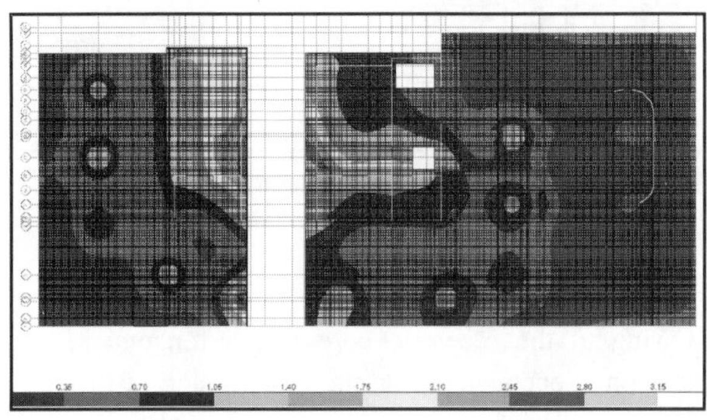

Figure 19. Bearing stress of raft in temporary stage

Grade 60D, high strength concrete, is adopted for the construction of the vertical element of the superstructure and Grade 45D, normal mix concrete, were adopted for the raft design.

Figure 20. Recess for high strength concrete and base plate

Recess on the raft (Figure 20) is proposed at the interfaces with the high strength concrete as well as bases of composite columns to eliminate the lengthy procedures of high strength concrete trial mixes and procurement period of the structural steelwork before the award of the advance foundation contract.

9. Construction

Owing to the situation that the site has been divided into four parcels during construction stage, the construction activities and sequences of this development are significantly different from other conventional development in Hong Kong.

The Demolition works, Foundation works, Site formation works, Excavation and Lateral Support Works and Superstructure works of the development (Photo 1) were proceeded concurrently to different extent. The packages of the contracts were also carefully designed to cater for this construction requirement.

Photo 1. Mini-piling at Site D and superstructures construction

9.1 *Core Wall*

Self-climbing form was adopted for the construction of the core wall. The cycle time of the core wall construction was 5 days.

Photo 2. Construction of tower

9.2 *Floor*

Two sets of small table forms adopted for the construction of the floor slabs. Two sets of Holland hoists installed for the lifting of the table forms. The cycle time of the floor construction is 5 days.

As the self-weight of the floor slab is exceeded the design imposed load, three levels of floor slabs were mobilized to resist the self-weight of wet concrete and construction load (Photo 2).

10. Conclusion

Although the construction of the development involved very complicated phasing requirements due to the phased site possession and allowance for future extension, it still can be handled efficiently and cost effectively by sophisticated engineering design, careful planning and construction programming.

The development commenced in May 2002 and completed on time and within budget in August 2004.

Acknowledgement

The authors would like to thank Swire Properties Limited who is the developer and project manager of the Three Pacific Place Development for their permission to submit this paper.

STRUCTURAL DESIGN CHALLENGES FOR TWIN TOWER OF "GATE TO THE EAST", SUZHOU CHINA*

JOEY TSUI CRAIG GIBBONS MICHAEL KWOK GORDON NG

Ove Arup and Partners Hong Kong Limited, Level 5, Festival Walk, 80 Tat Chee Avenue, Kowloon Tong, Hong Kong

The 280m high twin tower is the key component of the prominent development, "Gate To The East", Suzhou China. With the theme of being a "Gate to the East", two towers, which are of same height but have different number of storeys, will stand 60m apart at ground level and link together at top 8 floors. The underground light rail transit system will pass right through the "gate" formed by the twin tower. Each tower mainly comprises a reinforced concrete core and composite perimeter columns, linked together by belt trusses and outriggers system at three different mechanical and refuge floors. The continuous change in tower profile from bottom to top necessitates the extensive transfer structures in the middle and high zone of the tower for supporting the additional columns. Among those transfer structures, the long span linking structure located at level of approximately 240m above ground for supporting the top 8 floors between two towers provided the greatest engineering challenge of the project to structural engineers, primarily resulting from that the top linking structure would be subjected to a variety of structural responses arising from gravitational load, wind load, seismic load as well as differential shortening of the towers. Extensive studies have been carried out to evaluate the structural behaviour of the twin tower structure, prior to arriving at the final design of the twin tower structure. This paper will outline how the structural design resolved these engineering challenges, with the focus on those driven by the top linking structure, and meanwhile addressed the architectural, functional as well as economic considerations.

1. Introduction

The Suzhou twin tower sitting right in front of the west shore of the Jinji Lake, where is the most prominent location of the new Central Business District (CBD), is the key component of the mixed commercial development comprising a 278m tall twin towers for mixed commercial use, 6-storey retail podium and 5-storey basement mainly for retail and car park. Tower 1 has 61 storeys for Grade A office and 6-star hotel, and Tower 2 has 67 storeys for luxury serviced apartment.

The twin tower concept design was the result of the design competition with the theme of creating a 'Gateway' to Suzhou. This principal architectural concept is strongly expressed by rising the iconic linking tower facing to the west of Jinji Lake with the under-ground light rail line passing through the 'gate'. In addition to the sound structural concept, joint RMJM/Arup design team was chosen from an acclaimed field in a design competition to become the design team of the landmark project.

The profile of tower is unique and continuously changing from bottom to top. Along the height of the tower, the sections of towers through east-west direction will gradually shrink and hence look like the 'Rocket', while the sections through north-south direction will gradually increase in length and eventually connect together to form the 'gate' that face the Jinji Lake. The shapes of the floor plates of two towers at ground level are two squares with the dimensions of 45mx45m. The centres of two squares are 105m apart. The floor plates will become two rectangles in middle zone, and they eventually will join together to form a long thin floor plate at approximately 240m above ground. There are in total 8 storeys in the linking structure, with top occupied floor of dimensions of 156mx21.8m.

2. Engineering Design Challenges

To most of peoples, the architectural design of the twin tower was no doubt very unique and aesthetically pleasing. Meanwhile, owing to the continuously changing profile and the 8 storeys deep linking structures, the twin tower posed great engineering design challenges to the structural engineers. The principal engineering design challenges were attributed to the reasons as follows:

- Shrinking of floor plates in east-west direction in high zone of the towers causes the significant reduction in dimension, and hence stiffness, of cores in that direction.
- Floor plates in upper half of the towers expands toward centreline of linking structure to form the 'gate', resulting in significant increase in distance from core to perimeter up to about 40m on the floor right below the linking structure.

Courtesy of RMJM

Figure 1. Rendering of the Twin Tower of 'Gate to the East'.

- Expansion of floor plates in upper half of the towers and the linking structure contribute to considerable enlargement of frontal area in region of larger wind pressure, resulting that the twin tower would be subjected to much larger wind loads than two typical towers of same height.
- Such expansion of floor plates and the linking structure also leads to the centre of mass gradually deviating from the centreline of the core along the height of the tower as well as the mass in high zone of twin tower much larger than that of two typical towers of same height. Consequently, the twin tower has to resist much larger seismic loads, in particular twisting forces, than two typical towers of same height
- The linking portion of the tower with the distance of 90m between two cores has to support 8 floors of hotel and serviced apartment at the height about 240m above ground. On those linking floors of the towers, the width of floor plates would shrink most rapidly in the whole tower. Also, the linking portion would be narrower than the sections of towers at two ends. The top linking structure would certainly be subjected to a variety of structural responses due to both vertical and horizontal loads on the towers.

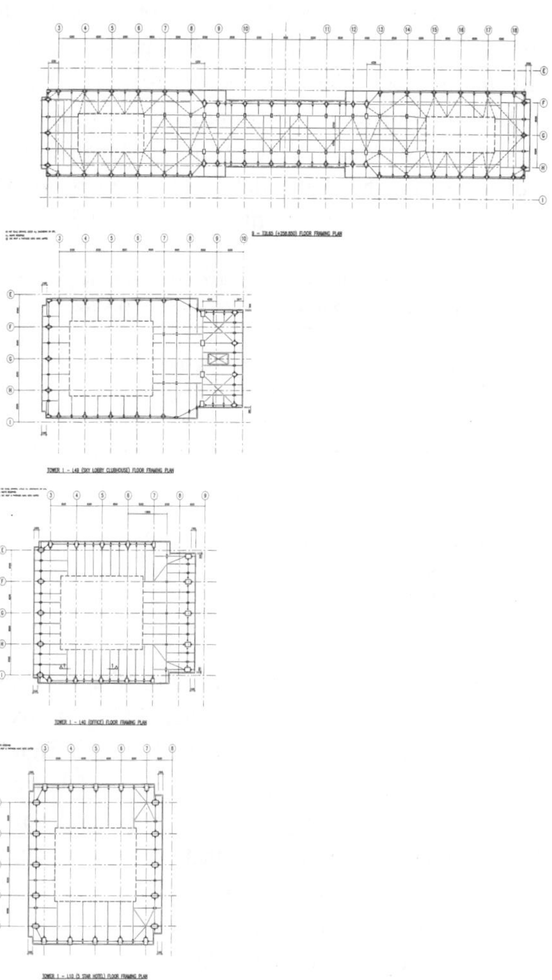

Figure 2. Sizes of floor plates in different zones of Tower 1 (Tower 2 is similar).

3. Design of Structural System

3.1 *Design Load and Design Criteria*

Design of the twin tower was performed considering both vertical loads and lateral loads, in which lateral loads had greater influence on the structural design. Both vertical loads and lateral loads are calculated based on the relevant Chinese codes.

For determination of wind loads, ground roughness of A (Lakeshore) and 100-year return period basic wind pressure of 0.55 kN/m^2 was used. Wind tunnel tests are underway and the results will be incorporated into the structural design at later stages.

The response spectrum was determined based on soil characteristic period of 0.45s (group 1, type III soil) and maximum horizontal seismic force coefficient of 0.072 that was suggested for this project in seismic hazard report. The site is in region of seismic fortification of level 6 in accordance with the Chinese code. Yet, the site-specific results as shown in table 1 reflected that the horizontal seismic forces would be close to that of seismic fortification of level 7.

Table 1. Comparison of horizontal seismic force coefficient between codified and site specified values

Seismic Fortification Level	Level 6	Site-specific Results for Twin Tower	Level 7
Horizontal Seismic Force Coefficient under Minor Earthquake (Frequently Occurred)	0.04	0.072	0.08

In addition to the height of tower that exceeds the code limits for composite building, the highly irregular profile and linking structure located at such high level are also identified as irregularity in accordance with relevant Chinese regulations. Approval authority imposed the design criteria that were more stringent than the code. The linking structure was therefore required to satisfy the requirement of remaining elastic under moderate earthquake, which is level 2 (moderate earthquake) of three seismic fortification levels.

3.2 *Lateral Stability System*

Apart from the structural considerations, key architectural design intents also had great influence on the development of lateral stability systems. The principal architectural design intents were to provide reasonably large unobstructed view through curtain wall, to allow planning flexibility for office and to be commensurate with the layout of apartments, hotel rooms as well as basement car parks. To achieve all the above architectural design requirements, reinforced concrete core with composite perimeter frame was selected as the lateral stability system. Being about 280m high, the twin towers would be subjected to considerably large lateral load. In addition, the owner and architect did not prefer the perimeter frame with closely spaced columns. Consequently, the steel outriggers and belt trusses were used to link up the cores and the perimeter frames to enhance the overall lateral stiffness of the towers.

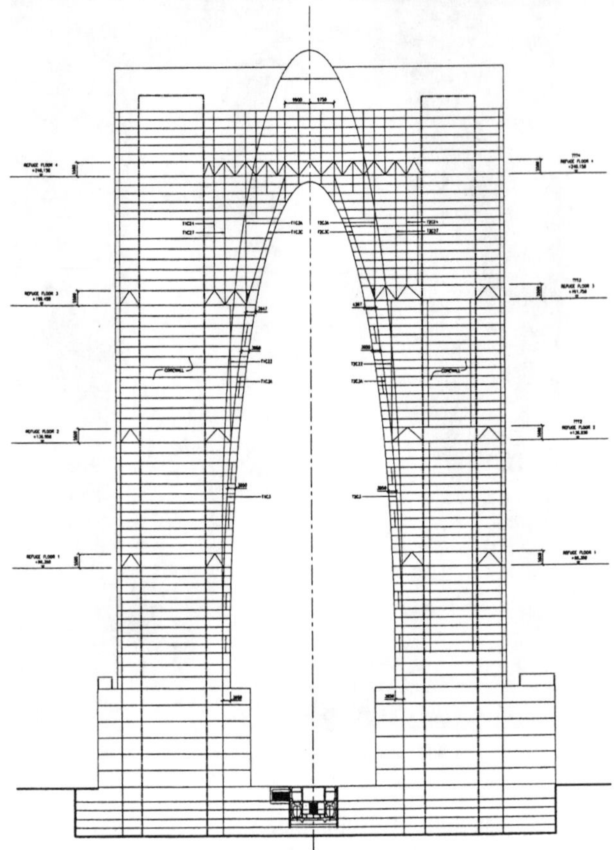

Figure 3. Section through the twin tower 1which illustrates the lateral stability system, transfer columns, linking truss between cores and its support system.

Central core, being the reinforced concrete structure, would resist not only the gravitational load, but also the overturning moment and most of the shear force arising from wind or earthquake. The concrete core provides the required stiffness, especially shear and torsional stiffness, and strength of the towers cost-effectively.

To be commensurate with the layout of towers and basements, structural layout with five number of perimeter columns on one side of the floor plates was adopted. For the high zone of towers with the long thin rectangular sections and supporting the top linking structure, four number of perimeter columns on two inner sides in the zone right below the linking structure were used. The perimeter columns were spaced generally at 8.5m centres and were generally of rectangular sections with the shorter side parallel to the curtain walls. Concrete encased steel composite sections were adopted as such sections were easy to meet the requirements of rectangular section and fire-proofing materials were not needed. Transfers of columns for supporting the enlarged floor plates and linking structure will be presented in later sections.

The outriggers and belt trusses of approximately 5.5m deep would be placed at lower three levels of refuge floors. Making use of the outriggers and belt trusses, the overall lateral stability could no longer rely on the moment frame action provided by perimeter frame, thus achieving the requirement of column spacing. Such structural system could effectively mobilise the axial stiffness of the perimeter composite columns for sake of increasing the overall bending stiffness of the tower, fulfilling the codified requirements of lateral deformations of the towers in an economical way. Also, bending moment acting on the core and the pile foundation underneath the core could also be reduced.

Pile foundation with base grouting was proposed for the twin tower, with the piles being 1000mm in diameter and 61m in length. The toe level of piles is 83m below ground level. One large pile cap of 4.5m thick supports both the core and perimeter structures of one tower. The adoption of base grouting can enhance the pile bearing capacity and reduce the settlement.

3.3 *Gravity Floor System*

Floor system above ground level will be steel/concrete composite beam and composite slab system for floors outside of core. The typical beam spans vary for this tapering building range from 8 to 13m. The composite beams would generally span from the core to the perimeter structure on the typical floors.

In the upper half of towers, the sections of the towers change dramatically from squares to long thin rectangles, resulting in significant increase in distance from core to perimeter up to about 40m. As a result, additional gravity columns became essential in the upper half of towers. In addition to the transfer trusses, some outrigger trusses located at the 2nd and 3rd refuge floors were also used as transfer trusses to support the columns. There are also some additional gravity columns that were provided to support the floors above the transfer trusses in the top linking structure.

Conventional reinforced concrete beam and slab system was proposed for the floors of towers at and below ground to make those floors compatible with the basement structure.

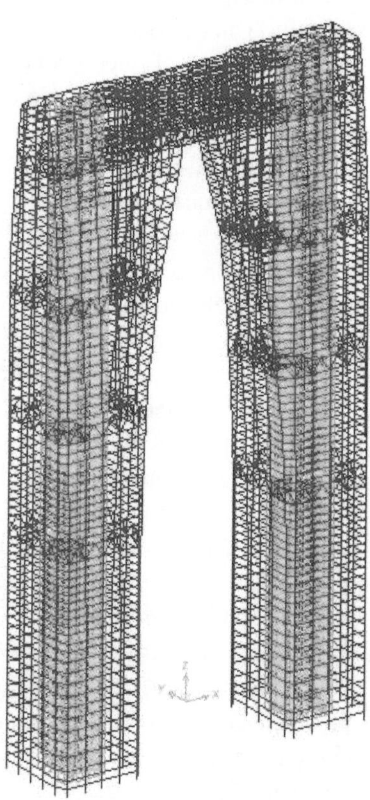

Figure 4. Structural model for lateral stability analysis.

3.4 *Top Linking Structure*

One of the principal engineering challenges is to overcome the problems due to the top linking structure arising from both vertical and lateral loads. As aforesaid, the linking structure has to span minimum 50m and support 8 floors of loading. The design of the linking structure was primarily based on principle of providing the linking structure with sufficient stiffness, but not making it over-stiff that may cause larger member forces. The considerations of stiffness and strength were given separately to the vertical and lateral direction of the linking structure.

For the linking portion is the prime location of the hotel and serviced apartment, diagonal members that will block the view through the curtain walls were not acceptable due to the architectural and functional requirements. Moreover, the columns must be slim to minimise the protrusions from the partitions. As a consequence, one storey deep steel trusses located at the 4th refuge floor, which are of 5.5m floor height and the lowest floor of the linking structure, became the basic structural form to support the linking portion of tower. There are totally 4 numbers of trusses adopted as the vertical support system of the linking structure, in which two of them are also part of the belt trusses to resist the lateral forces. Taking into account the considerations of build-ability, cost-effectiveness as well as structural adequacy, the vertical trusses with clear distance of 90m between two cores were divided into five continuous spans that are approximately 16m to 20m long. Along circumference of void underneath the linking structure, 5 number of perimeter columns in each tower were changed to 8 number of perimeter columns through the transfer belt trusses located at the 3rd refuge floor. On the east or west elevation, it looks as if one much larger perimeter column supports two smaller perimeter columns, with one of them being vertical and the other one that will lean toward its support point to the middle part of linking structure. As a result, the continuous expansion of floor plates beneath the linking structure was obviously a great disadvantage in the structural design, such disadvantage however provided a change of adding two more vertical supports to bring the span length of the linking structure down to a more manageable range.

Under the lateral loads due to wind and earthquake, two towers would deflect laterally and twist in different amount and even in different direction. Being the structure to link up two towers with enormous stiffness, the linking structure should be in possession of both stiffness and strength adequately accommodating the effects of shear, bending and torsion attributing to the effects of restraining two towers. The lateral system comprises the belt trusses surrounding the whole 4th refuge floor, composite slabs at those 8 linking floors, horizontal steel trusses located at underside of slabs at top 3 floors and bottom 2 floors of the linking structure. The belt trusses were used to supplement the required torsional stiffness to the linking portion of the twin tower. The composite slabs and the horizontal steel trusses would provide the sufficient shear, bending and axial stiffness and strength to accommodate the effects of restraining two towers and to transfer the lateral loads to the primary lateral stability systems of two towers.

4. Elastic Analysis

Three-dimensional computer models were set up to evaluate in detail the structural behavior of the twin tower. The models truly reflected the real level of all floors, in spite of the floors of two towers below the linking structure being not at the same levels. To avoid unrealistically over-estimating the stiffness of part of structure of the lateral stability system, rigid floor diaphragms were not tied to belt trusses, outrigger trusses, and all floors in the linking structures. In lieu of rigid diaphragms, flexible floor elements were adopted at all floors in the linking structures to achieve more accurate results.

The analysis was done in two stages. In the first stage, the structural behavior of each tower was evaluated separately since two towers have lots of differences, such as number of floors, usages, size and layout of cores. Three-dimensional computer models were set up accordingly to gain the full understanding of the structural responses of each tower under lateral loads, including lateral stiffness, lateral deformation shapes, inter-storey drift ratio, mass distribution along the height of towers and dynamic properties. Besides above reasons, one of the key purposes of performing the first stage study was to ensure that one of the towers would not be considerably stiffer than the other tower in resisting lateral loads, eventually resulting in the higher demand of both stiffness and strength of linking structure. Subsequent to the appropriate tuning, two towers possess similar stiffness and dynamic properties.

In the second stage, the whole twin tower was modeled. Based on the results of second stage analysis, the periods of first two translational modes were 6.14s in east-west direction and 5.85s in north-south direction respectively, while the period of first overall torsional mode was 4.79s. It should be noted that, in the first overall torsional mode, two towers behaved as they move in opposite direction in east-west direction. The maximum inter-storey drift ratio arising from wind and seismic effect were 1/592 and 1/917.

Apart from the aforesaid models for investigating the effects under gravitational, wind and seismic loads, separate analysis was carried out to study the structural behaviour induced by differential shortening and the analysis results were also taken in the member design considerations.

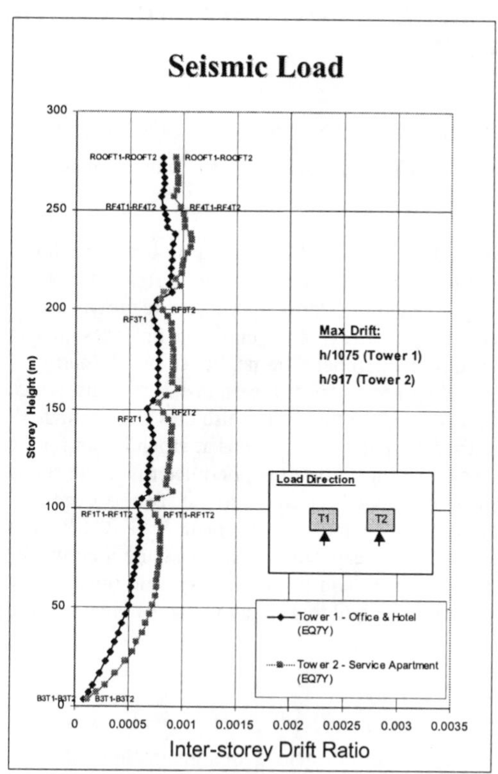

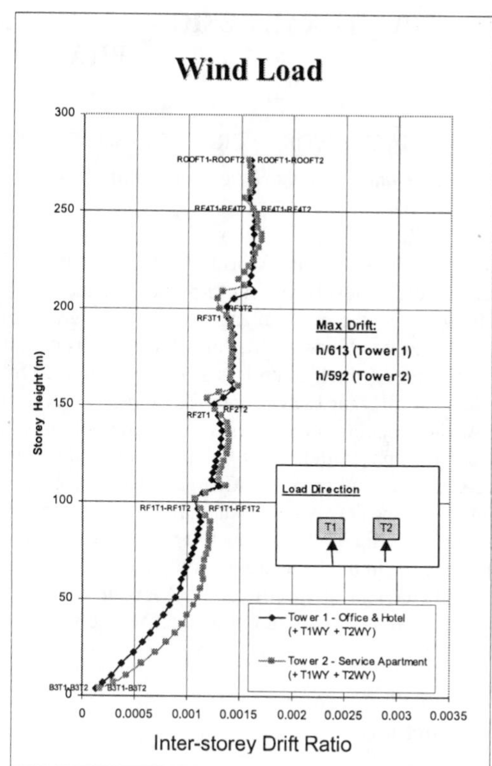

Figure 5. Inter-storey drift ratio of both Tower 1 and 2 under wind load and seismic load in Y-direction.

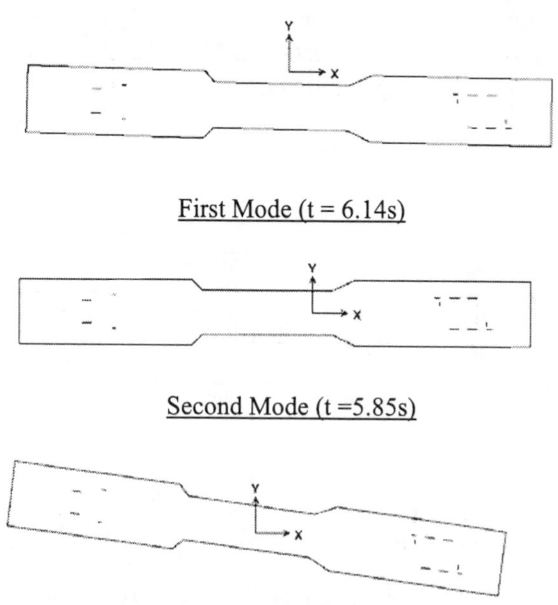

Figure 6. First 3 mode shapes of the twin tower

5. Conclusion

The detailed exploration of structural scheme has been performed for Suzhou Twin Tower, rendering the smooth development of the structural design for later stages. The structural design of Suzhou Twin Tower has also been an excellent demonstration of the balance of aesthetic, functional, economic, build-ability and structural considerations with the participation of the structural engineers at the outset of the architectural concept design development of the highly complicated project.

References

1. GB50011-2001 Code for Seismic Design of Buildings.
2. JGJ3-2002 Technical Specification for Concrete Structures of tall Buildings.
3. GB50009-2001 Specifications for Loading in Building Structure

THE STRUCTURAL DESIGN OF THE MEGA TOWER, CHINA WORLD TRADE CENTRE PHASE 3, BEIJING CHINA*

MICHAEL KWOK CRAIG GIBBONS JOEY TSUI PENG LIU YANG WANG GOMAN HO

Ove Arup and Partners Hong Kong Limited, Level 5, Festival Walk, 80 Tat Chee Avenue, Kowloon Tong, Hong Kong

The Mega Tower, China World Trade Centre Phase 3, Beijing China is 330m high and composed of a five-star hotel, grade-A office levels and multi-purpose spaces. The height of the building and the high seismic design intensity in Beijing poses a great challenge to the structural engineers, especially considering various stringent requirements by Chinese codes. Various structural types, utilising different materials, have been explored until the final design was accomplished, which comprises a composite braced frame core and a composite perimeter frame, linked by two outrigger systems at relevant E&M floors. The tapered elevation of the tower building necessitates three transfer belt trusses that allow the reduction in number of perimeter columns in middle and high zone of the tower. The use of 8 storeys high V-shaped columns with transfer belt truss admit of the grand entrances with wide spacing of columns at ground floor for the five-star hotel and grade-A office, and meanwhile provide a smooth structural transition to the perimeter moment frame above. Composite elements of various types are extensively used and positioned carefully to satisfy the combined requirements for stiffness, ductility, redundancy, and cost-effectiveness. The composite steel plate wall (C-SPW) is implemented in the structure, possibly the first time in China, to increase the shear capacity and stiffness and improve the ductile behaviour at specific zone. Accurate finite-element analysis and advanced non-linear elasto-plastic time history analysis have been carried out to evaluate the structural behaviour and ensure the building safety under different seismic levels.

1. Introduction

China World Trade Centre Phase 3 is located in the Central Business District (CBD) of Beijing. The site is surrounded by the East 3 Ring Road on the east, China Grand Hotel to the south and Kerry Centre to the north. China World Trade Centre Phase 3A includes:

1) A 330m tall mega tower
2) A ball room
3) A retail block

The whole phase 3 will be linked together by means of a basement. Basement level 1 will be used for retails. The structural design of the Mega Tower is introduced below.

With the elevation of helipad of 330m, the Mega Tower are totally 74 storeys above ground, in which 1/F to 4/F are for atrium and multipurpose, 5/F to 56/F are office levels, and above 56/F are for hotel. There are 3 underground levels, B3 is for parking and mechanical spaces, while B2 and B1 are reserved for commercial and mechanical spaces.

2. Structural Challenges and Design Criteria

Lateral stability under seismic load and wind load is the key issue for the design of such a tall building.

The seismic design code of China [1] classifies the seismic resistance requirements by earthquake fortification intensity, which is basically VI, VII, VIII and IX, corresponding to a design ground acceleration of 0.05g, 0.10g, 0.20g and 0.30g respectively with 10% probability of exceedance in 50 years. This project is located in Beijing where the design intensity is VIII, equivalent to zone 2B in UBC97. The seismic design philosophy of the Chinese code is to design to "three levels" of seismic fortification, as described in Table 1 specific for buildings in Beijing.

Table 1 Seismic Design Criteria

Seismic Fortification Level	1	2	3
Description	Minor (frequently occurred)	Moderate (seismic protection)	Severe (rarely occurred)
Design PGA	0.07g	0.20g	0.40g
Probability of Exceedance	50yr 63%	50yr 10%	50yr 2~3%
Fortification Criteria	No damage and remaining elastic	Damage allowed but repairable	No collapse

The structural design of the Mega Tower has to satisfy the fortification criteria for all the three seismic fortification levels. This is further complicated by the fact that the building of this nature does not fall within

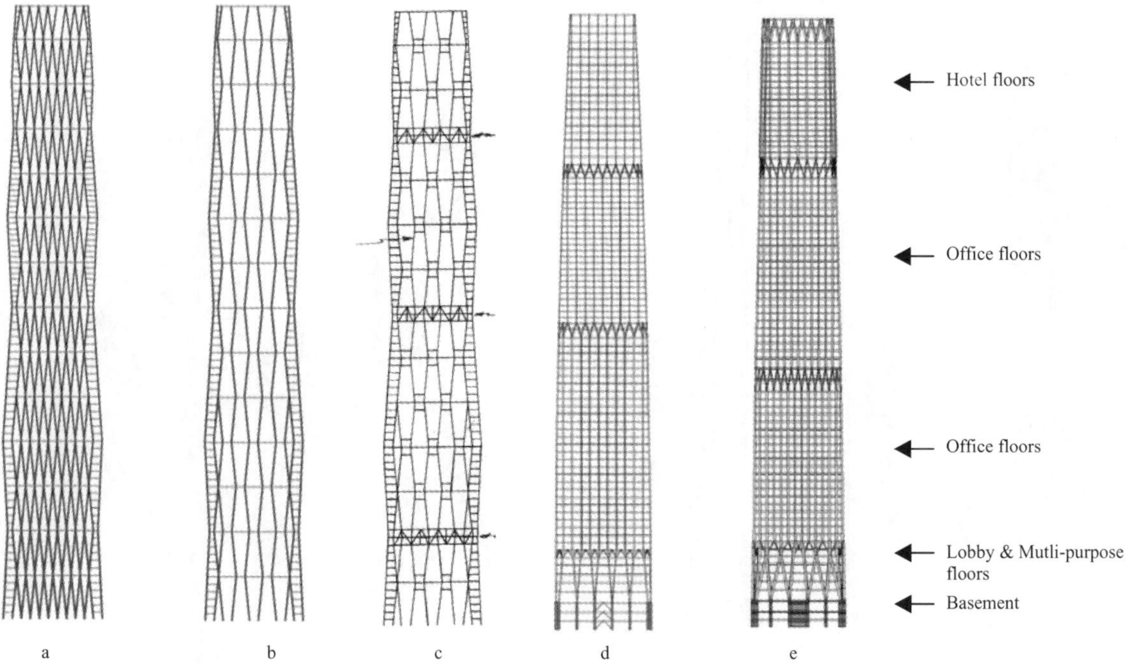

Figure 1: Schemes for the perimeter tube

the limits for which general design approaches are provided in Chinese design code. Hence both the applicable design criteria and the consequent structural preliminary design have to be reviewed by an expert panel organised by the National Seismic Design Review Committee for Buildings Exceeding Limits to get the consent for further work. This was accomplished for the structural preliminary design of Mega Tower, CWTC3 in September 2004.

Wind tunnel tests were carried out in Rowan Williams Davies & Irwin Inc, Ontario Canada. The 100 year return period wind load was sought for the structural design, which by Chinese code is equivalent to 10 minute mean wind speed at 10m height of 28.2m/s.

3. Structural System Design History

The Mega Tower, with a mixed usage for hotel and office, has undergone several major changes in terms of floor area and space division while the exterior look with an undulating profile has been basically kept unchanged. Various structural systems have been proposed and studied to be commensurate with the architectural layout, with the attempt to make the most structurally safe and cost-effective use of different materials. The schemes for the perimeter tube and the central core were effectively developed separately and then combined to a dual system. For the perimeter structure, several concepts with variations have been studied:

1. Steel diagrid frame (Fig 1a). This may have been the most structurally efficient option by combining the lateral and vertical load transfer system together. Different shapes of triangular unit were explored for best compromise between structural efficiency, build-ability and architectural design intent. The corner members were removed and replaced by flexural link beams to avoid fabrication problems of the large and shallow interconnection.
2. Steel hexigrid frame (Fig 1b, 1c). In order to improve buildability and views from the building, a diagonal hexigrid frame solution was derived from the diagrid option by removing pairs of diagrid frame members. This dramatically improved the open window space and also reduced the complex cross connections thus improving build-ability. The advantage of this system is that it was a hybrid system and achieved strength and stiffness from the axial stiffness of the inclined members and the flexural stiffness of the joints between the inclined members and the edge beam members.

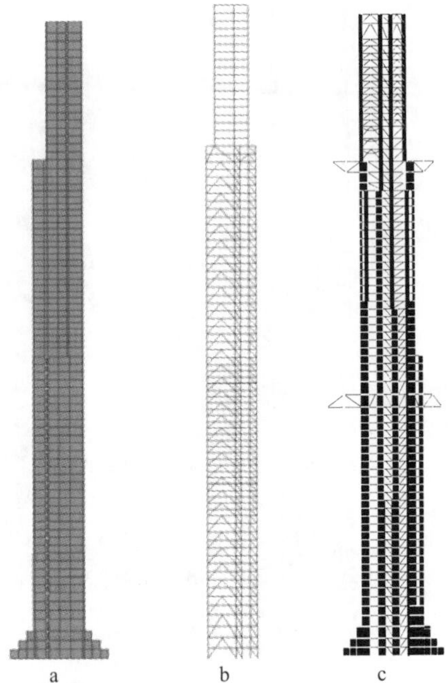

Figure 2: Schemes for the central core

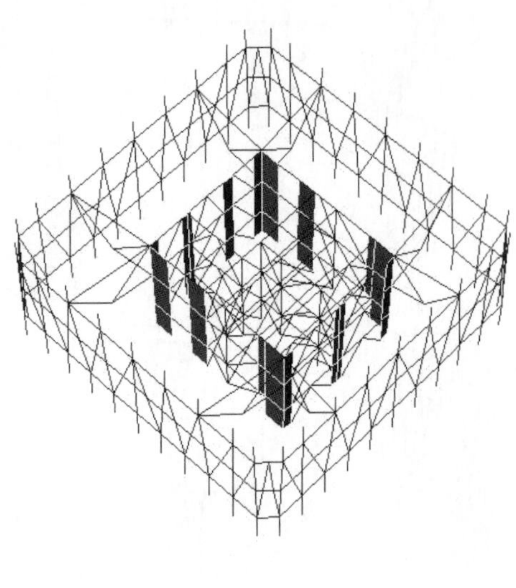

Figure 3: Isometric view of outriggers and the belt truss

3. Moment frame. Among all the schemes, moment frame provided the best exterior view for all profitable floors. The column grid was initially 5m (Fig 1d) for all levels but finally 5.6m spacing in hotel zone and 4.2m in office zones were adopted (Fig 1e). Belt trusses located in the mechanical levels of 6-8, 28-30 and 55-57 facilitated column shifts to cater for the reduction in the number of columns resulting from the gradual tapering of the building shape. Coupling with the outriggers, the belt trusses improved the shear lag effect of the perimeter tube and mobilised the perimeter columns more efficiently. At the bottom beneath the two-storey transfer belt truss V-shape braced perimeter columns achieved double column spacing for the entrance lobby level, providing a strengthened base for the perimeter frame instead of weakening it as many transfer structures.

The central core options, which were largely confined by the architectural layout and building service requirements, differentiated themselves primarily by the usage of the material. During the development history, three possible schemes, the concrete shear wall (Fig 2a), the pure steel braced frame (Fig 2b) and the composite braced frame (Fig 2c) have been studied in depth. While all structurally feasible they are, the key issues were how structural redundancy and safety margin could be adequately provided under earthquake events in a cost-effective way. The composite braced frame, which is composed of composite columns and steel beams and braces, was deemed to be the most appropriate. Both concentric bracing and eccentric bracing were used not only to accommodate openings and ducts, but also to supply more ductile mechanism within the stringent requirement on the lateral stiffness.

The final design is then the combination of the composite moment frame perimeter tube shown in Fig 1e and the composite braced frame core shown in Fig 2c. They are further linked and enhanced by two sets of single-storey outriggers between Level 28-29 and level 56-57, each set of 8 outrigger arms (Fig 3). The tips of outriggers arms connected to the junctions between the belt truss and the moment frame result that one outrigger arm would actually mobilise two adjacent columns. Studies have proved that such arrangement, in conjunction with the belt trusses, improves the shear lag effect of the perimeter tube significantly because the central few columns on four faces of the perimeter frame, which are originally inefficient in the framed tube action due to their locations far away from the corners of the frame and perimeter beams with insufficient stiffness, are hence efficiently mobilised by outriggers.

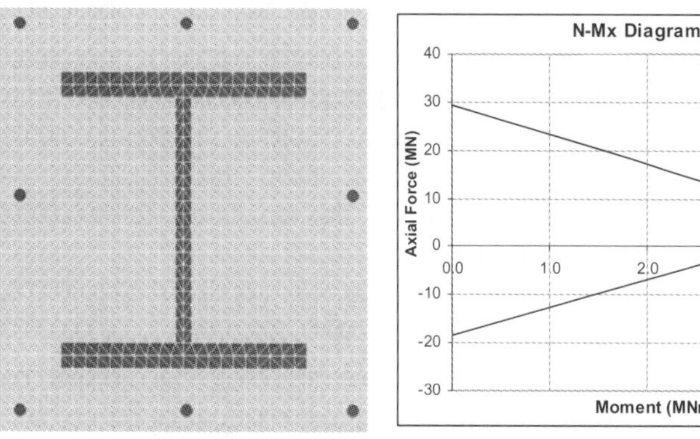

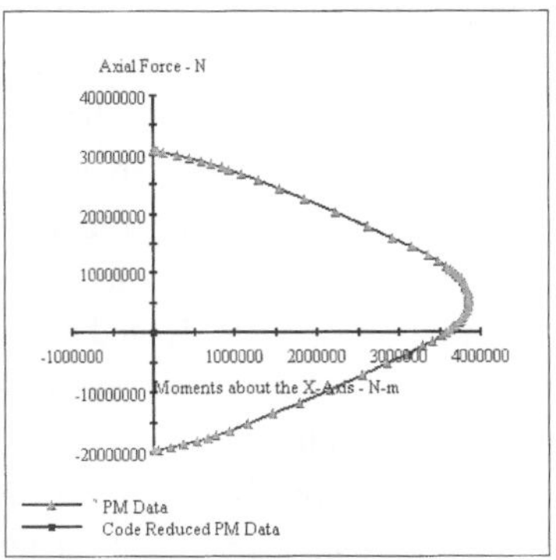

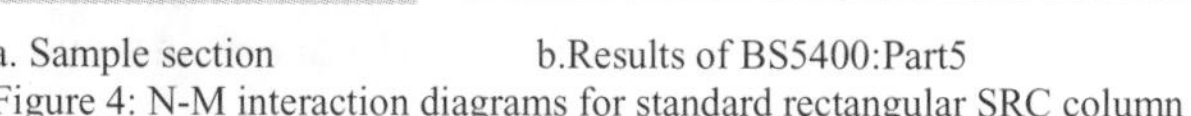

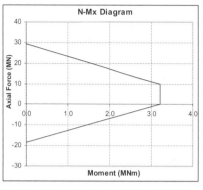

a. Sample section b.Results of BS5400:Part5 c. Results of finite element analysis

Figure 4: N-M interaction diagrams for standard rectangular SRC column

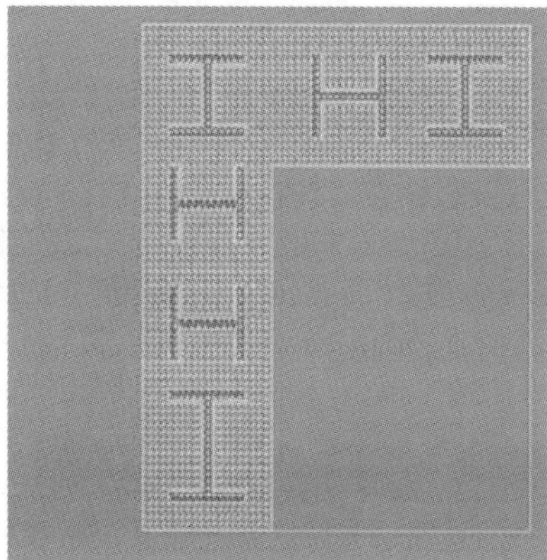

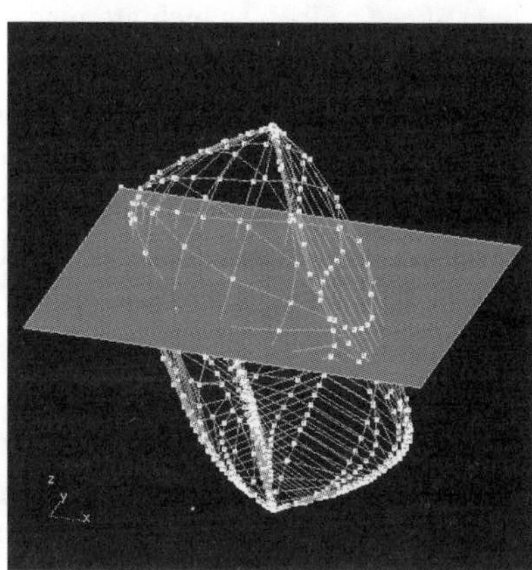

a. Finite element model b. Three-dimensional N-Mx-My interaction surface

Figure 5: Modelling and Capacity Curves for complicated L shape SRC columns

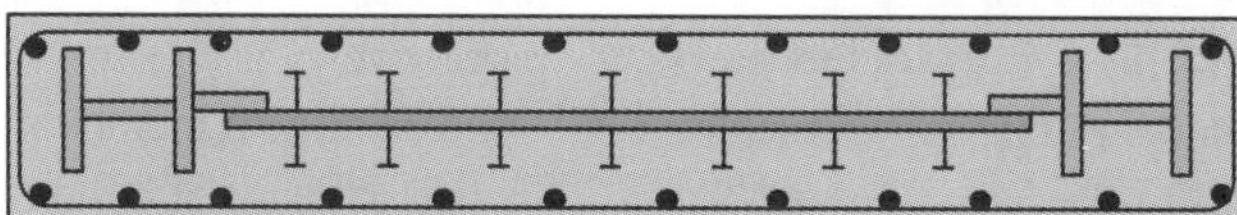

a. Typical configuration

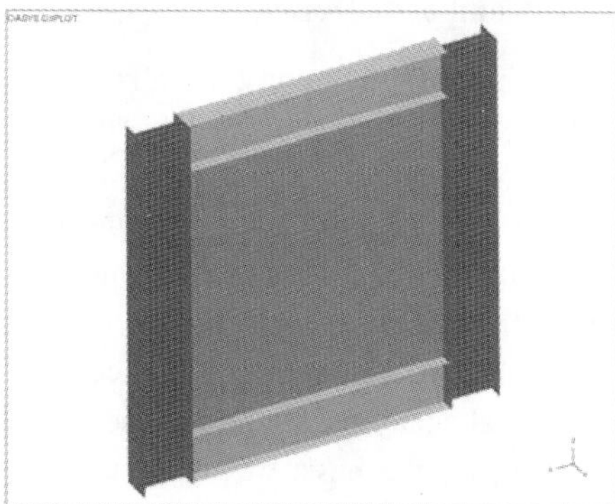

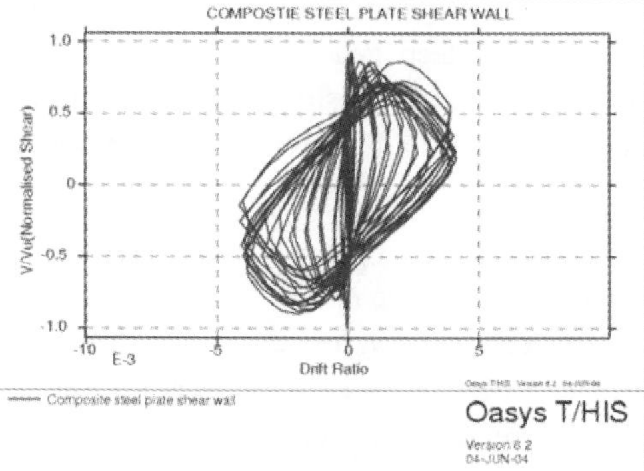

b. FE model for non-linear verification analysis

c. Hysteresis curves of C-SPW by numerical analysis

Figure 6: The composite steel plate wall (C-SPW)

Unlike the regular rectangular shape of the composite columns in perimeter frame, the composite columns in the core are of L shape or rectangle with multiple (nos 2 to 7) steel stanchions cast in (Fig. 5). Steel beams and braces are connected to these steel sections directly. The design complexity of such columns has been fully envisaged and verified with both international codes and finite element analysis, which will be presented later.

The internal core is further enhanced at the lower levels by the composite steel plate walls (C-SPW) which provide great lateral shear resistance under severe earthquake events and contribute stiffness to resisting wind load and frequently occurred earthquakes (seismic fortification level 1). The C-SPWs, which are placed in the internal partition areas of the core and between the composite columns, would actually integrate with those columns via steel connections, concrete and reinforcement (Fig 6a and 6b). In basement levels C-SPWs are introduced on the perimeter of the core and further project out at corners in each side to act as stiffeners for the pile cap.

The structural steel composite floor system is adopted. The typical floor beam span varies from 8m to 16m due to the tapered profile of the building. Human comfort due to the vibration of the floor system has been carefully checked especially for the hotel zone where the depth of the concrete slab is increased accordingly.

As the bearing capacity of the soil under the Mega Tower is not sufficient to support the superstructure, pile foundation was proposed for the Mega Tower. The piles are 1200mm in diameter and 55m in length and the toe level is 75m in depth from the ground level. In addition to the walls projecting out at core corners, the pile cap, 4.5m deep generally and 4m deep at some perimeter area locally, distributes the loading from the columns and walls in the basement to the pipes evenly.

4. Elastic Analysis

Three-dimensional computer models utilising two different analysis software according to the requirement of Chinese design code have been set up with all structural elements of the superstructure and the basement modeled explicitly. It is assumed in the model that the Mega Tower is fixed at the top of pile cap with the lateral restraint effect arising from basement floors outside the footprint of the tower considered properly. In the model, monolithic sections with equivalent sectional properties sensibly represent the composite elements. Rigid diaphragms may reduce the degree of freedom significantly, resulting in use of elastic slab elements at the area where this assumption would unnecessarily stiffen the structures, for example the outriggers and the braces.

Response spectrum analyses according to Chinese seismic design code (GB50011-2001) have been carried out, with the analysis results being verified by the average results of multiple elastic time history analyses using one artificial TH curves and two natural seismic acceleration records, as shown in Fig. 7.

The periods of the primary translational modes in two principal axes are 6.82s and 6.55s respectively while the period of the first torsional mode is 3.66s. The base shear forces in X and Y direction of the response spectrum analysis are 50.3MN and 51.0 MN respectively, which have been rectified to 67.6MN in compliance with minimum base shear requirement for specific element capacity checking. The seismic overturning moment in X and Y direction are 10297MNm and 10087MNm, becoming 13828MNm and 13360MNm respectively for the same reason. The base shear and overturning moment of the resultant wind load is 33MNm and 6666MNm, which is far less than the counterparts of the seismic effect.

The maximum inter-storey drift ratio resulting from the seismic and wind effect is 1/559 and 1/833 respectively, as shown in Fig. 8.

The critical members such as belt trusses and the V-shaped columns at bottom have been designed to remain elastic at seismic fortification level 2.

5. Composite Design

The composite columns, especially those located at the core corners, are subjected to the combined effect of axial force and bi-axial bending moment. As being not covered in Chinese code, the design methodology in BS5400 Part 5 has been employed with proper modifications to incorporate the material properties and the like in Chinese code. Accurate finite element sectional analyses assuming each mesh as a fiber to deform on the basis of plane section remaining plane has been carried out to verify the code-based results. It is proved that, in comparison with the FEA results, the BS5400 methodology would give relatively

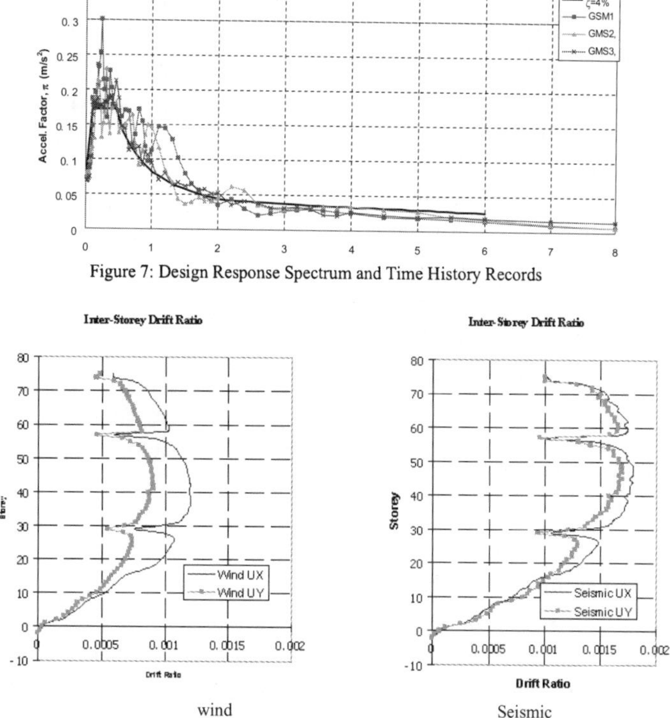

Figure 7: Design Response Spectrum and Time History Records

wind Seismic

Figure 8: Inter-storey Drift Ratio under Seismic and Wind Load

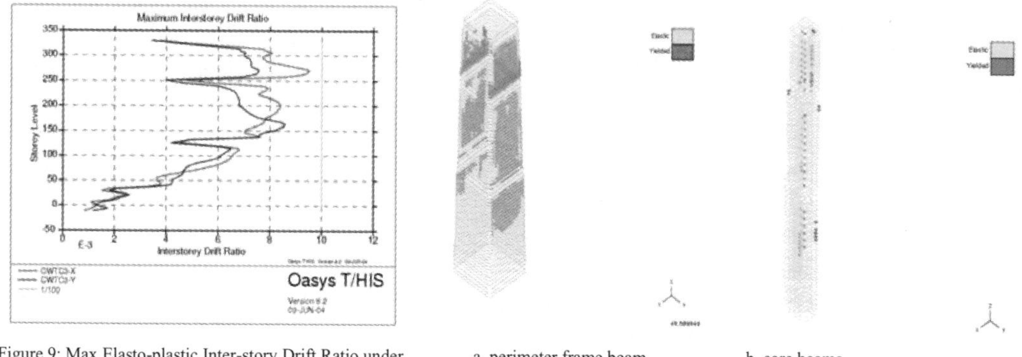

Figure 9: Max Elasto-plastic Inter-story Drift Ratio under Severe Earthquake (Artificial Time History Record)

a. perimeter frame beam b. core beams

Figure 10: Beam Plasticity under Severe Earthquake (Artificial Time History Record)

conservative results for the capacity of rectangular SRC columns.

However the L-shaped composite columns are beyond the coverage of the codified design as these columns are of unsymmetrical sections and the finite element analysis indicates that the N-Mx-My interaction curving surface is distorted spatially (Fig. 7). Therefore, the capacity curves produced by FEA are used in design, which will be further verified by specimen tests.

The composite steel plate wall (C-SPW) has been utilised in seismic zones of North America and Japan for a period and relevant provisions have been included in IBC/AISC[2,4] and Japanese codes. It is defined as[a] structural walls consisting of steel plates with reinforced concrete encasement on one or both side of the plate and structural steel or composite boundary members. With the restraint of the reinforced concrete, the shear capacity of C-SPW is able to be up to the shear yield capacity of the whole steel plate. This is significantly greater than the shear capacity of pure steel plate wall, where only the diagonal tension field is effective[3]. This composite mechanism of C-SPW provides sound ductility capacity (Fig. 6c) in addition to lateral stiffness. Fire protection is not necessary if the reinforced concrete is on both sides of steel plate.

As such elements are introduced in China possibly for the first time, the design criteria have been carefully considered. The member force adjustment factors for the composite walls follow those designated to RC walls originally and the shear capacity of steel plate is designed accordingly. On the other hand to comply with the design philosophy of "no damage in minor earthquake", the concrete encasement is reinforced to resist the seismic fortification level 1 shear force distributed by its stiffness proportion in the composite section.

6. Non-linear Elasto-plastic Time History Analysis

In an event of a rarely occurred earthquake, the structural element is allowed to deform plastically but P-Delta effect due to the gravity load acting on the building geometry with excessive lateral deformation would result in collapse of the entire building, which should be prevented. Therefore, an analysis with the structural model subjected to according ground motion excitation history, considering both material non-linearity and geometric non-linearity, has been carried out using LS-DYNA to evaluate the behaviour of the Mega Tower in such circumstance. The maximum elasto-plastic inter-storey drift and the plastic deformation of various structural elements are the criteria adopted for such evaluation. As suggested by the expert panel during the review, the limit for this inter-storey drift ratio has been set as 1/100, which should be used for RC frame-core system. For limits of plastic deformation of single element, FEMA356[5] is referred to as Chinese code does not provide such provisions.

The results of the 3-D elasto-plastic time history analysis using LS-DYNA reveal that plastic deformation occurs mostly at beams (Fig. 10) while the critical area such as transfer trusses and bottom V-shaped columns remains elastic. The plastic deformations of all elements are within their corresponding limits and the maximum inter-story drift ratio is 1/105, less than the preset limit.

7. Conclusion

As from making use of innovative structural systems with eccentric braces and composite steel plate walls to performing accurate finite element analysis and advanced non-linear elasto-plastic time history analysis for ensuring both structural behaviour and safety under different seismic levels, the structural design of the CWTC Phase 3 Mega Tower has not only been a comprehensive and detailed exploration of the innovative structural schemes, but also demonstrated that the probably best balance between the economy and the structural safety for a super high-rise building located in a highly seismic zone in China is capable of achieving by extensive use of concrete and steel as composite material for structures with sufficient ductility.

References

1. GB50011-2001 Code for Seismic Design of Buildings.
2. AISC (2002), Seismic Provisions for AISC (2002), Seismic Provisions for Structural Steel Buildings, American Institute of Steel Construction Inc., Chicago.
3. Astaneh-Asl, A. (2002), Seismic Behaviour and Design of Composite Steel Plate Shear Walls, www.aisc.org.
4. ICC, (2000), The International Building Code, IBC-2000, International Code Council, Falls Church, VA.
5. Federal Emergency Management Agency, FEMA 356 Prestandard and Commentary for the Seismic Rehabilitation of Buildings, 2000.

THE CONSTRUCTION TRACING AND SIMULATING ANALYSIS FOR HIGH-RISE LEANING STEEL BUILDING STRUCTURES

Y.L. GUO Q.L. DONG
Civil Engineering Department, Tsinghua University
Bejing, 100084, China

D.B. HU
Office of CCTV New Sites Construction
Bejing, 100084, China

The leaning tower of Pisa in Italy, although not intentionally designed, had been a well-known landscape all around the world. Several modern high-rise leaning buildings including Puerta de Europa (Gateway of Europe) in Madrid - the first modern leaning buildings, Barcelona Communication Tower and CCTV New Headquarters have been worked out by contemporary architects and engineers. During the construction of a high-rise leaning building structures, both the P-Δ effect and geometric nonlinearity must be considered because the continually added gravity loads cause additional bending moment to the structure. The construction of a leaning building structure is a challenging task involving in the determination of pre-set displacement amount as well as implementation of construction schemes. Key issues about construction tracing and simulating analysis of leaning structures are outlined and studied; some helpful conclusions which have been used in CCTV New Headquarters project are presented.

1. Introduction

1.1 *Leaning buildings in the world*

The 54-meter high leaning tower of Pisa in Italy was initially built as the bell tower of Cathedral from the year of 1173. But before it was completed, the tower began to tilt by reason of the asymmetric settlement of its foundation, and the tilting continued in the following years, which had made it one of the best-known landscapes all over the world. Leaning structures are more attractive than ordinary vertical-formed structures as they give a strong visual shock to the public. For this reason several other high-rise leaning structures have been worked out by contemporary architects and engineers, among which the Puerta de Europa (Gateway of Europe) twin towers in Madrid, Barcelona Telecommunication Tower and the CCTV New Headquarters in Beijing are the most famous ones.

The Puerta de Europa twin towers (Figure 1) in Madrid, Spain, was built during the year of 1989 to 1996, and was the first modern leaning building in the world. It is comprised of two 26-storey, 114m high towers leaning 15° off vertical toward each other across the street. Figure 2a shows the 88m high Barcelona Telecommunication Tower, Spain, another leaning steel structure constructed in the early 90's with the antenna height of 136m. Recently, the CCTV New Headquarters is under construction. The 230m high building with two towers learning 6° off vertical in both directions and a tremendous 14-story-high, 75-meter-long, 14,000-metric-ton-weight cantilever at the top joining the two leaning towers (Figure 2b), have really attracted the public attention all over the world because of its design and construction difficulties.

Figure 1. The Puerta de Europa twin towers in Madrid

(a) (b)

Figure 2. Barcelona Telecom-tower and CCTV New Headquarters

1.2 *Key issues in construction of leaning buildings*

Leaning building structures are against the law of gravity: the additional bending moment caused by consecutively added structural members and gravity loads will lead to considerable structural deformation. Therefore, P-delta effect should be considered in the construction analysis. Furthermore, the construction analysis and implementation of a leaning structure are challenging works. They involve in the determination of pre-set displacement amount in manufacture and erection stage, as well as internal forces in every erection step during the construction process.

The displacement amount of the pre-constructed structure due to the P-delta effect of the leaning structure during construction process is predominantly dependent on the construction procedure and consequence. It is of great importance to carry out an internal force and deformation tracing estimation of the full construction process so as to check if the completed structure configuration and internal force distribution could satisfy the design requirements and be locked in an acceptable level. Key issues to be solved in construction of a leaning building include: determination of construction scheme, establishment of the initial construction configuration, stage erection configuration and fabrication configuration of members. A 5-storey leaning line-wise FEM model and a simplified CCTV Headquarters FEM model were carried out to demonstrate the idea how these problems will be solved.

2. Determination of the Pre-set Amount of Leaning Structure

2.1 *Configurations in different stages and pre-set amounts of member fabrication*

Unlike traditional up-straight buildings, the construction and erection of leaning structures involves in different configuration analysis, corresponding with different construction stages.

(1) Design Configuration (DC) is defined as the structural configuration given on the construction drawings. It is structural configuration usually after taking account of the action of self-weight and super-imposed dead load (finishes, ceiling, services, cladding and partitions), but without live load of occupants. In other word, it is a completion state after the whole building construction finished.

(2) Initial Construction Configuration (ICC): For a leaning building structure, it is evident that the structure deforms continuously when members and superimposed gravity loads are added consecutively. In order to satisfy the design configuration when the construction is completed, it is necessary to give some pre-set amounts of displacement to the DC, such that deformation arising due to subsequent construction will result in the position of structure as well as members conforming to those stated on the construction drawings. By superimposing the pre-set amount of displacement to the DC, we got the Initial Construction Configuration (ICC).

(3) Stage Erection Configuration (SEC) is the coordinate positions of members in the space in certain erection stage. Theoretically speaking, the structural members deform under their gravity loads the very moment they are installed. In this way, the determination of installation coordinate position of the second set of structural members should consider the deformation of the first set of structural members at installation end state. Similarly, the later sets of structural members are checked in the same way.

(4) Fabrication Pre-set Amount of Members (FPM): dimensions of members in fabrication are different from that obtained according to design drawings. Due to the deformation caused by consecutively added loads during the erection process, the actual dimensions of structural elements, namely fabrication lengths of structural elements, should be made up. For example, the columns will differ slightly from the design dimensions. The difference rises with the increase of the building

construction height: the lower-storey column will sustain larger compression deformation than that of higher-storey columns. Therefore, the actual column compression shortenings should be compensated in the fabrication by an additional elongation increment due to actual compression shortenings. For a high-rise leaning structure, as a result of P-delta effect, the columns near the leaning-ward side will experience a larger compressive deformation than the columns in the other side, so the fabrication length should be longer than designed; while the columns against the leaning-ward side will perhaps have tension deformation if the tension caused by additional bending moment is greater than the compression arose by gravity. Therefore in the fabrication stage, pre-set amount for structural members should be determined to make up structural member length difference due to structural deformation during erection stages.

2.2 *Computation approach*

By employing commercial program ANSYS, a trial-and-error iteration program is carried out and the ICC, SEC and FPM of a leaning structure is found. The obtained results have been successfully used in the construction guide of CCTV new Headquarters.

A simple example of a 5-story leaning line-wise FEM model is selected to illustrate the procedure of construction analysis, see Figure 3a. This building is assumed to have a leaning angle of 45°off the ground level and with storey height of 4 meter each. Li (i=1~5) represents columns of i-th storey, while concentrated loads Pi (i=1~5) is applied on each node to stimulate gravity loads of i-th storey.

2.2.1 *Trial-and-error method to get the ICC*

A trial-and-error program to determine ICC of the leaning line-wise model is carried out and a sketch map is shown in Figure 3b. The displacement vector, $-[\delta_1]$, can be calculated when all the dead loads and self-weight were applied to the design configuration given as $[v_0]$, then an ICC curve, namely $[v_0]+[\delta_1]$, can be obtained by superimposing displacement $[\delta_1]$ to the design configuration. (C_1 in Fig. 3b).

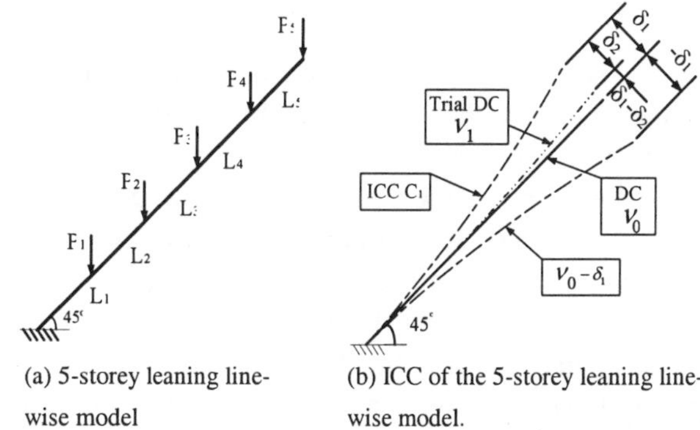

(a) 5-storey leaning line-wise model (b) ICC of the 5-storey leaning line-wise model.

Figure 3. A 5-story leaning line-wise FEM model and its ICC

If a structure is stiff enough and its geometric nonlinearity is negligible, the error $[\delta_1]-[\delta_2]$ will be in an acceptable level, then curve C_1 (Fig.3b) is just the ICC we are looking for. Otherwise repeat the process by replacing $[\delta_i]$ with $[\delta_{i+1}]$ (i=1, 2 …), until the error is less than the given criteria ε. Fig.4 is the flow chart of the whole computation process.

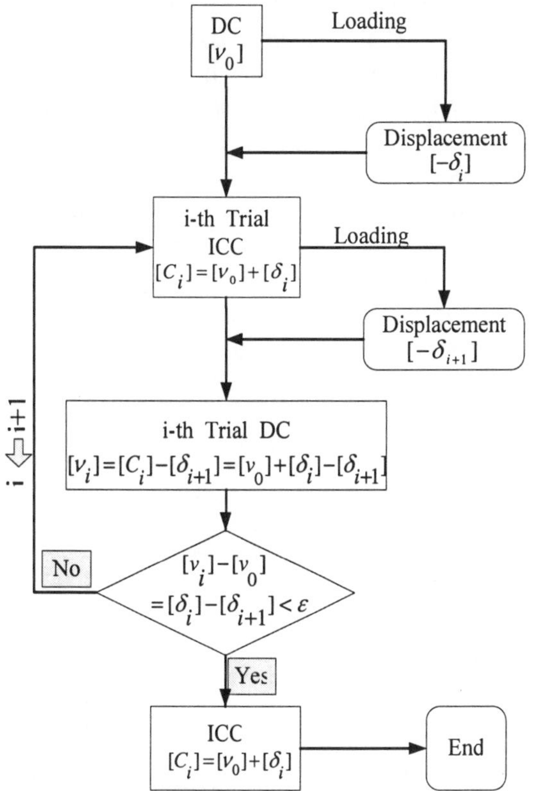

Figure 4. Flow chart of trial-and-error process for getting ICC

2.2.2 *Computation of the SEC*

Fig. 5 shows different configuration variation for a leaning building with each member erecting step-by-step. The ICC, namely the line 1'-2'-3'-4'-5', will approach the design configuration, namely the line 1-2-3-4-5 when all members were installed completely and after their gravity loads were applied. By taking off each member and gravity load in the reversed construction sequence, one can get the SECs and corresponding internal forces in every erection step. This analysis method is called 'taking off method', and it is frequently used in bridge construction analysis. While each member and gravity load is removed one by one, the leaning line-wise configuration will "rebound" from the design configuration to ICC step by step. It is obvious that when all members and gravity loads were eliminated, the configuration will coincide with the ICC.

Figure 5 also illustrates the analysis process by using taking off method and by employing the death and birth elements in ANSYS program. As stated above, the line 1-2-3-4-5 stands for the design configuration and the line 1`-2'-3'-4'-5' represents the ICC, while all others between the two lines denote the SECs of installation for every erection step. The footnote 'b' indicates such a state where the concentrated force was removed, while footnote 'a' gives a configuration where the corresponding column was taken off from the structure. For example, the line 'Step 5b' indicates the situation where the load P_5 was removed, and 'Step 5a' represents the case where the column of the 5th storey was taken off, which is exactly the SEC of the 5th step.

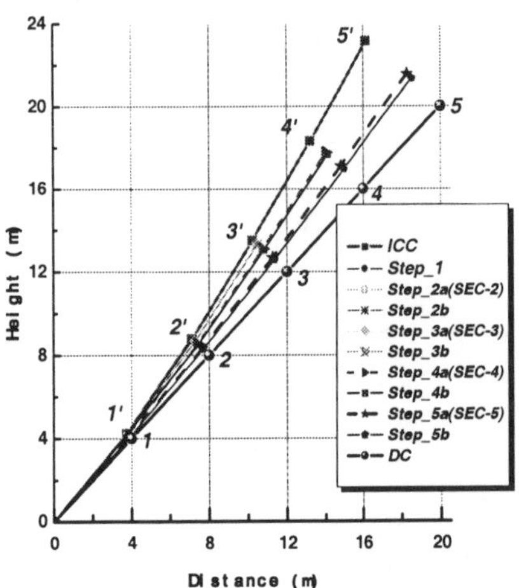

Figure 5. ICC and SEC of the 5-story leaning line-wise model.

2.2.3 *Computation of the FPM*

The FPM is the actual lengths of members to be fabricated. It is obtained as long as the SEC is obtained. As seen in Figure 5, 'Step 5a' is the SEC of the 5th step, while the length corresponding to the line 4'-5' is exactly the length according to which the member should be fabricated. Fabrication lengths of other columns can be obtained in the same way.

3. Analysis of the Simplified Model of CCTV New Headquarters

3.1 *Erection sequence of CCTV new headquarters*

As stated above, the structural scheme of CCTV new Headquarters make it one of the most complicated buildings ever constructed. During the formation of the whole structure from individual members, the internal forces and displacements of members and the pre-constructed parts will change continuously and dramatically, depending on different erection sequences and approaches. So it is of great importance to establish an appropriate construction scheme to ensure that the completed structure configuration and internal force distribution could satisfy the design requirements and be locked in an acceptable level.

Figure 6 illustrates the construction sequence of the cantilevers: the external tube frame of the cantilevers begins to stretch out from two leaning towers after they were erected to tops (Fig. 6a); then the 2-story-high transfer trusses was lifted to design location piece by piece (Fig. 6b); Fig. 6c shows the moment when the two parts of transfer trusses and external tube frame are about to join together. After that, the stories over the cantilevers are erected in a ordinary way (Fig. 6d).

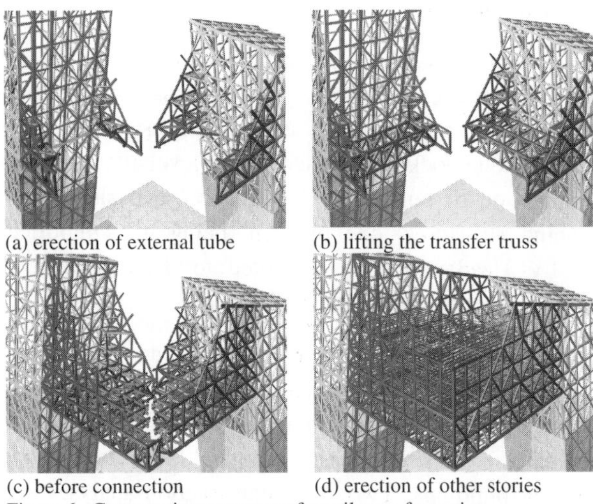

(a) erection of external tube (b) lifting the transfer truss

(c) before connection (d) erection of other stories

Figure 6. Construction sequence of cantilevers formation

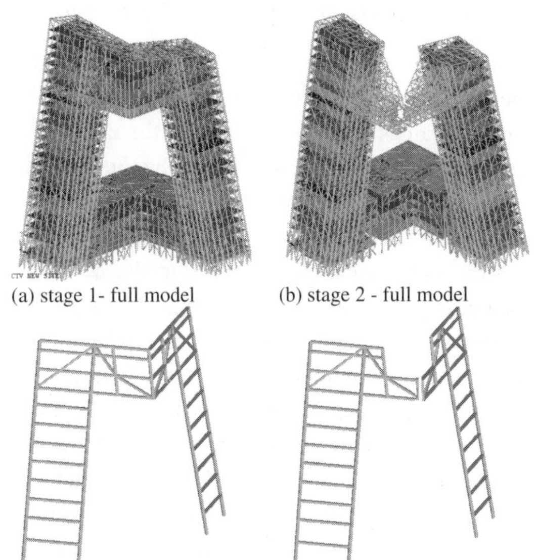

(a) stage 1- full model (b) stage 2 - full model

(c) stage 1- simplified model (d) stage 2 - simplified model

Figure 7. Full and simplified models used in two stages in determination of the ICC.

3.2 *Two-stage method to determine the initial construction configuration*

In terms of influencing extent on the behavior of the constructed structure, the cantilevers are the most critical aspects, as the structure will behave differently prior to and after formation of links. The presence of links will change the behavior of the pre-constructed towers: before the formation of the cantilevers, two towers deform in an independent way which is in the tendency of both leaning directions, but after the links are med, they behave as a whole structure and deform harmoniously. There exists an abrupt change in deformation as well as internal forces before and after the formation of the links. Unlike the 5-storey leaning line-wise model discussed above, it's unlikely to get directly an accurate ICC without considering the obvious difference of the two different construction stages.

A two-stage analysis process was put forward which taking account of the special feature of CCTV new Headquarters. Both a full model and a simplified analysis model corresponding with the two construction stages are shown in Fig.7: stage-1 refers to completion of the building construction, with full fit out, installation of facade enclosure, but prior to occupation (Fig. 7a, 7c); stage-2 indicates the moment when two parts of transfer trusses and external tube frame are about to connect (Fig. 7b, 7d).

3.3 *Implementation of two-stage process*

To find the ICC of CCTV new Headquarters by using the two-stage procedure, some points are outlined as follows: firstly, the ICC of stage-1 can be relatively easily obtained by using trial-and-error iteration method (see configuration A in Fig.8), which is the target configuration of stage-2; secondly, the ICC of stage-2 can be got by the same way (see configuration-B in Fig.8), which is also the ICC of the whole structure. In other words, in order to reach the configuration-A before connection of the links, coordinates of members must accord with the configuration-B. Only in this way, can the final design configuration be satisfied when the links were formed and loaded.

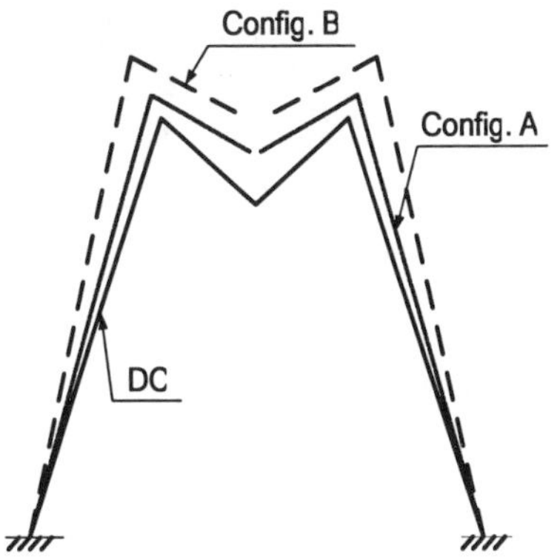

Figure 8. ICCs of two-stage process

3.4 *Analysis process of simplified CCTV model*

A simplified CCTV model was employed to demonstrate the whole process of construction analysis. Two plane frames was taken out from the external tube frame of the integrated model, where Fig.7c and 7d are stages-1 and stage-2 respectively.

Fig. 9 demonstrates an erection sequences simulation with dashed lines represent the SEC of the corresponding erection step. Step (a) and step (b) shows the erection of the individual towers, and cantilevers begin to stretch out simultaneously form two towers at step (c); step (d) illustrates the formation of the links, at this moment, the two separate towers joined together and begins to displace harmoniously. Step (e) and step (f) shows respectively the erection of top stories and completion of the whole building. The finished configuration got by this way will be exactly the design configuration.

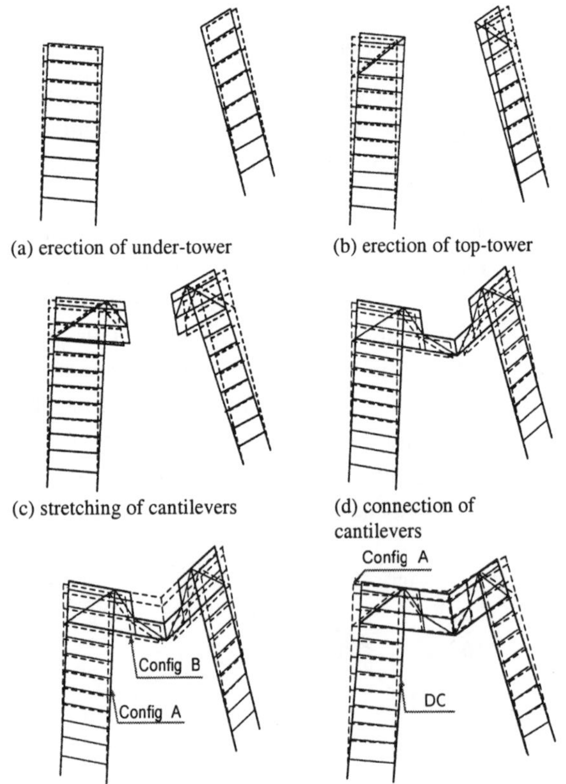

Figure 9. Construction sequence with construction erection configurations

4. Conclusions

As stated above, the construction of leaning buildings are challenging tasks. A full process construction analysis is necessary to ensure displacements and stresses to be locked in an acceptable level after erection completion of buildings. Construction analysis of two leaning structures, one of them being the simplified CCTV Headquarter, is presented to demonstrate the analysis procedure of construction stage.

Some useful technical concepts in construction analysis were given, for example, determination of ICC, SEC, FPM of a leaning building structures. They are valuable to guide the construction and fabrication of the leaning buildings. Two stage construction analyses are set up for determining the ICC, SEC, FPM of CCTV New Headquarters-a complex building comprised two leaning towers and super cantilevers.

References

1. Leslie E. Robertson, (Leslie E. Robertson Associates), Leaning high-rise buildings, Modern Steel Construction, v 31, n 5, May, 1991, p 25-28
2. Ove Arup & Partners Hong Kong Ltd, China Central Television New Headquarters, Basis of Design-Construction Stages, tender issue, (2004)
3. ANSYS, Inc. Theory Reference
4. ANSYS, Inc. ANSYS Guide to User Programmable Features.
5. ANSYS, Inc. ANSYS online help

STEEL STRUCTURE DESIGN OF ZHONGGUANCUN FINANCIAL CENTER TOWER

FAN ZHONG

China Architecture Design & Research Group, No.19, Che Gong Zhuang Street
Beijing, 100044, P. R. China

HU CHUNYANG

China Architecture Design & Research Group, No.19, Che Gong Zhuang Street
Beijing, 100044, P. R. China

Frame-bracing structure system is adopted in the high-rise steel structure of Zhongguancun Financial Center Tower. The objective of this paper is to give a brief introduction of the structural arrangement, lateral force-resisting system, local composite structure, kinky outer columns, floor horizontal trusses, excellent-property steel material Q345GJ for columns, and foundation design of the building.

1. Introduction

Located in the western section of Zhongguancun, Haidian, Beijing, Zhongguancun Financial Center is composed of the tower, bridge-way with roof and auxiliary building (see figure1). The building plan, south and north facade of the tower is in an arc shape. The tower has 35 stories above the ground and 4-story basement with the total floor area reaching 79318m^2. The full glass curtain wall has 35 levels above the ground with a total height of 150m, and the emergency shelter floor is located in Level 12 and 24 of the building.

The design reference period of the project is 50 years, and the seismic fortification intensity is 8 degree. The reference wind pressure is 0.50 kN/m^2 with a return period of 100 years, and the terrain roughness is class C.

The architectural scheme design of Zhongguancun Financial Center was completed through the partnership of KPF and CAG. The construction design of the tower steel structure has already completed in May 2003 by CAG. American MEINHARDT Company contributes to the work in the structure scheme and preliminary design period.

The erection of steel members began in June 2003, and the main steel structure was completed by the end of December 2003. The mounting of equipment and decoration of the building is now ongoing and the whole works are expected to be completed and put into use by October 2005.

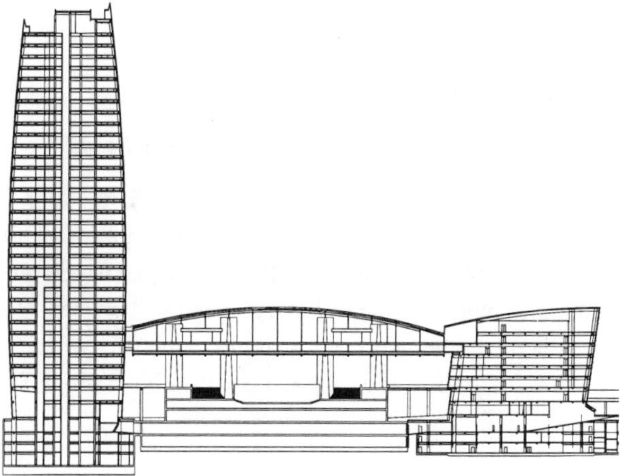

Figure 1. Building façade of Zhongguancun Financial Center Tower

2. Arrangement of Building Plan and Facade

Both the south façade and north façade of the tower are composed of vertical arc whose radius is 750m and whose center is 50m above the ground; the boundary of the south and north side of the building plan is made up of an arc whose radius is 93 ~ 99m. With the change of building height, the boundary of the east and west sides is in a form of broken line with the central barrel slightly in a rectangle form, resulting in the irregularity of the building façade. The geometrical configuration method of the building plan and façade are shown in Figure 2. Figure 3 is the real exteriors photo of the west façade of the tower.

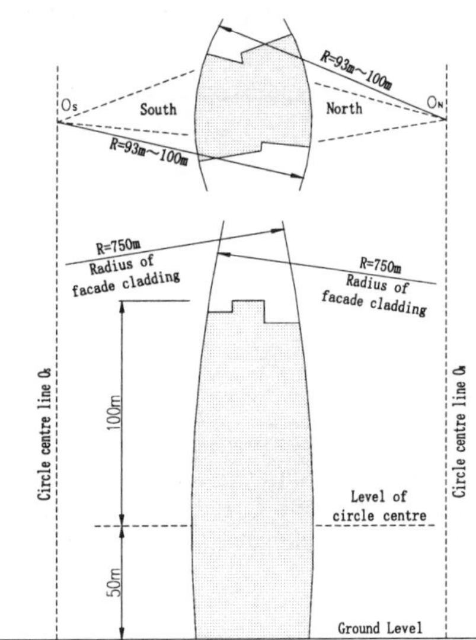

Figure 2. Geometric configuration of building plan and façade

Figure 3. Real exteriors photo of the west facade

3. Floor System Design

3.1 *Girder, primary beam, and secondary beam*

The structural arrangement of the standard floor is shown in Figure 4. The peripheral frame girder of the tower is mainly used hot-rolled H-shaped steel with 700mm height. The primary beams between the peripheral frame and core is rigidly connected with the outer columns, and hinged with the core. The 700mm-height welded H-shaped steel is adopted and the beam height is changed to 400mm close to the core so as to allow the pipeline to pass through. The material of the girder and primary beam is Q345C.

The secondary beam is 350 ~ 400mm high and mainly hot-rolled H-shaped steel which is Q345B. The interval between the secondary beams is about 2700mm for convenience of placing floor steel deck.

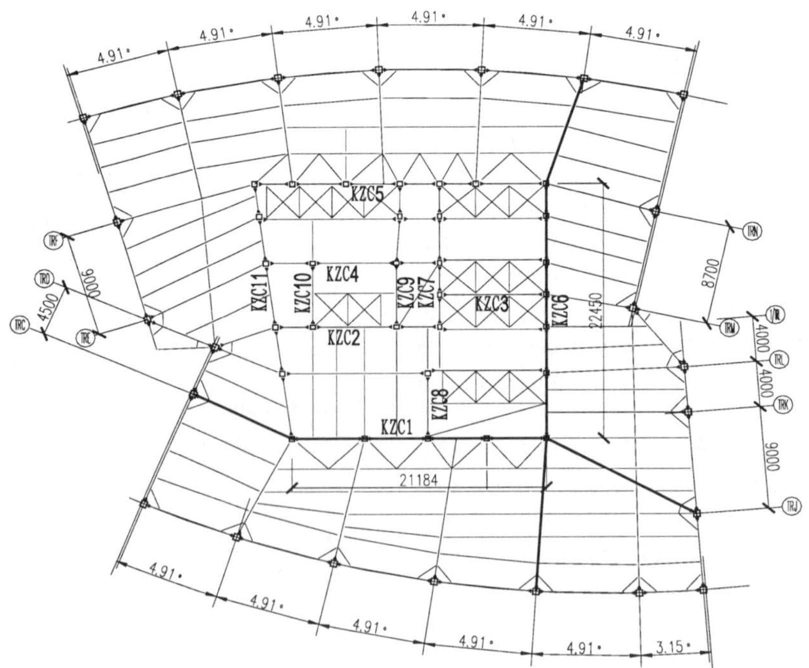

Figure 4. Structural layout of standard floor

3.2 *Arrangement of horizontal truss*

Cast-in-situ RC slab, with steel deck as construction form panel, is used for standard floors. The thickness of composite floor slab on the standard floor is 130mm, exception of the 12th and 24th floor and roof for 180mm for increasing the safety of the emergency shelters.

Sectionalized kinky line approach is adopted for the outer columns in the south and north peripheral frame in order to satisfy the curve façade effect of the building. The inclined angles of the outer columns change every four floors for standard floor, making it easier for fabrication and installation. Since the sectionalized kinky outer columns may result in horizontal unbalanced forces during the construction and use, and a part of the primary beam connecting the columns are unable to link directly with the bracing-frame columns of the core, therefore, the floor horizontal trusses are arranged on the north and south sides of the core according to the turning points of outer columns every four floors. In this way, the unbalanced forces and lateral loads from the outer columns transfers from the exterior of the building to the bracing-frame of the core directly and the diaphragm stiffness of the floor system is effectively enhanced.

4. Vertical Structural System

4.1 *Lateral force-resisting system*

Steel frame is adopted as the gravity load-supporting system in this project. The lateral force-resisting system is composed of the bracing-frame of the core and the peripheral frame.

The peripheral frame is composed of welded box-section columns, dimensions of which are 600× 600mm for the 20th floor and below, 600×500mm for floor 21 to 28 and 500×500mm for 28th level and above. The wall thickness of columns is 20 to 80mm. As for the steel columns of bracing-frames in the core, the sections are 500×500mm and 500×400mm and the wall is 20 to 85mm thick.

Q345GJ is used for the welded box-section columns. Comparing with ordinary Q345, it has evidently lower contents of phosphor and sulfur, thus enjoying excellent welding performance. Also, the yield strength is visibly increased, which is 325MPa for sheet of 50～100mm in thickness. The 34J impact energy at -40℃, 20% tensileratio, and cold bending property, etc. of the material all meet the requirements in Code for Seismic Design of Buildings (GB50011－2001).

The axes of outer columns in north and south façade are consisted of pieces of inclined straight lines and the others are vertical straight lines. The bending point on the kinky level of outer column is 300mm below the top flange of the main beam, shown in Figure 5.

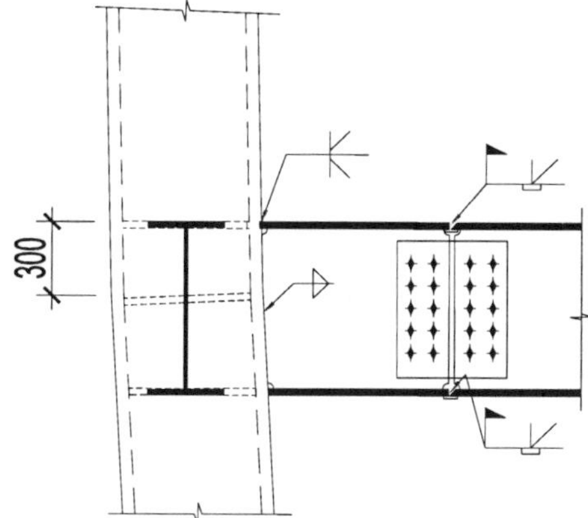

Figure 5. External column connection

Since the column section is strictly limited during the design, the spacing of bracing columns in the service core are properly increased for controlling maximum thickness of steel plate.

The arrangement of the bracing in the core is shown in Figure 4, and the west-east elevation of KZC1 and north-south elevation of KZC6 are shown in Figure 6. The plan layout of the bracing frames should be symmetrical as much as possible and access of door openings and equipment pipelines should be taken into account when the direction of the diagonals is being determined. The bracing diagonals are made of welded Q235C and Q345C H-section steel. The web height and flange width are 320mm and 370mm respectively.

In order to improve the lateral force-resisting performance of the structure, vertical crossing bracings are configured on the 12th and 24th levels (equipment floor) to coupled adjacent bracing-frames together for enhance the lateral stiffness of the structure and reducing the story deflection ratio.

4.2 *Part of steel-concrete composite structure*

The basement is cast-in-situ RC structure and the levels 1～3 above the ground is steel-concrete composite structure, Level 4 and above is in steel frame support system.

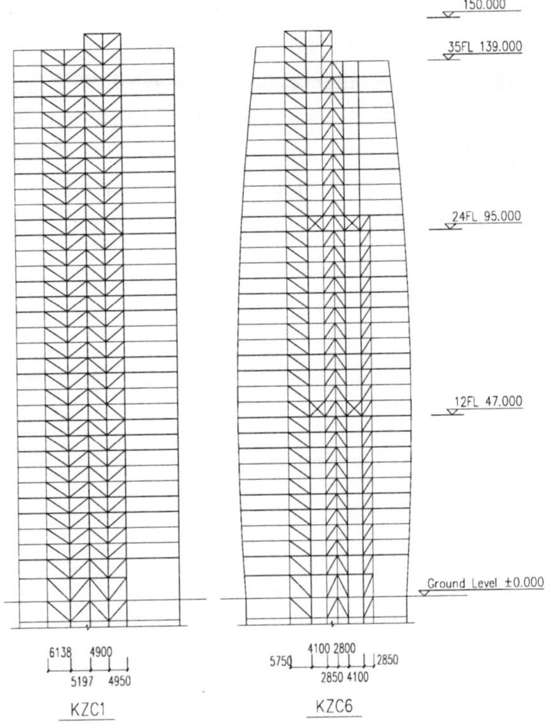

Figure 6. Facade of west-east KZC1 and north-south KZC6

The story height of the ground floor, 2nd floor and standard floor are 6.0 m, 5.0m and 4.0 m respectively. The ground floor and 2nd floor are merged into an 11.0m-height lobby. The dimension limitation of the steel column cross-section for architectural utilization results in large length/width ratio. And from the ground level to level 3 some diagonals of the bracing cannot be placed due to the door openings on the cores. Therefore, the service core and the peripheral frame in the ± 0.000 ~ 15.000m height range are altered to SRC elements to form a part of steel-concrete composite structure which may enhance the lateral stiffness of the substructure.

5. Analytical Results of Structure System

5.1 *Analytical model and major parameters*

The finite element analysis software SATWE for multi- and high-rise structure developed by China Academy of Building Research Institute is used for numerical analysis of the structure. The three-dimensional computing software ETABS and STAAD-PRO2002 are also adopted for checking and comparison.

The response spectrum method is used in seismic analysis. The first forty modes of vibration shall be taken into calculation. The CQC approach is adopted and reduction coefficient of period is 0.90. The occasional eccentricity, two-direction earthquake torsion and P-Delt effect shall be considered respectively in analysis. Damping ratio of the structure is 0.02 and 0.05 for frequent earthquake and severe earthquake respectively.

The frame-bracing structure is a dual lateral force-resisting system, which is composed of peripheral frame and bracing frame. In the steel frame-support system, the steel support bears the majority of horizontal earthquake action. To ensure the structure still has a certain lateral force-resisting capacity after the failure of the diagonals, the rigid frame shall bear no less than 25 % of the overall seismic shear force of each level during seismic design.

5.2 *Results of analysis*

The periods of free vibration mode for the analytical modal are listed in Table 1. As for the floors of composite structure at the elevation of 15.0m or below, the maximum inter-story displacement rotation is 1/981 under frequent earthquake; as for the steel frame-bracing above 15.0m, the maximum inter-story rotation is 1/366, shown in Table 2 and Figure 7.

The Seismic horizontal shear force to gravity load ratio in X-direction (east-west) is 3.20% and that in Y-direction is 2.84% respectively.

The total steel tonnage of the tower is 11205 tons and steel tonnage per unit area is about 164kg/m^2, including 41 % for columns, 47 % for girders and beams and 12% for bracing diagonals.

Table 1. Dynamic Modal Periods of the structure

No. of modes of vibration	1	2	3	4	5	6	7	8	9	10
Period (second)	4.2680	3.4373	2.9436	1.3323	1.1110	1.0586	0.7752	0.6871	0.5462	0.4476

Table 2. Inter-story drift displacement and inter-story displacement rotation under earthquake

Computing level (building level)	Height h (m)	X-direction				Y-direction			
		Max. (mm)	Mean (mm)	Max./mean	Interstorey drift ratio (max./h)	Max. (mm)	Mean (mm)	Max./mean	Interstorey drift ratio (max./h)
Small roof Floor	3.60	5.03	4.92	1.02	1/ 716.	5.66	5.24	1.08	1/ 636.
Equipment Floor	5.00	6.84	6.69	1.02	1/ 731.	6.98	6.89	1.01	1/ 716.
Floor 35	4.00	5.66	5.55	1.02	1/ 706.	5.99	5.70	1.05	1/ 668.
Floor 34	4.00	5.91	5.79	1.02	1/ 677.	6.40	6.00	1.07	1/ 625.
Floor 33	4.00	6.08	5.97	1.02	1/ 658.	6.73	6.26	1.08	1/ 594.
Floor 32	4.00	6.25	6.15	1.02	1/ 640.	7.09	6.55	1.08	1/ 564.
Floor 31	4.00	6.39	6.31	1.01	1/ 626.	7.42	6.81	1.09	1/ 539.
Floor 30	4.00	6.52	6.46	1.01	1/ 613.	7.72	7.05	1.09	1/ 518.
Floor 29	4.00	6.63	6.59	1.01	1/ 603.	7.98	7.27	1.10	1/ 501.
Floor 28	4.00	6.68	6.65	1.00	1/ 599.	8.13	7.38	1.10	1/ 492.
Floor 27	4.00	6.77	6.76	1.00	1/ 591.	8.22	7.47	1.10	1/ 487.
Floor 26	4.00	6.78	6.79	1.00	1/ 590.	7.89	7.25	1.09	1/ 507.
Floor 25	4.00	6.66	6.27	1.06	1/ 600.	6.59	6.35	1.04	1/ 607.
Floor 24	4.00	6.83	6.40	1.07	1/ 586.	8.75	8.08	1.08	1/ 457.
Floor 23	4.00	6.94	6.44	1.08	1/ 577.	9.62	8.80	1.09	1/ 416.
Floor 22	4.00	6.38	6.02	1.06	1/ 627.	9.67	8.86	1.09	1/ 414.
Floor 21	4.00	6.99	6.41	1.09	1/ 573.	10.47	9.46	1.11	1/ 382.
Floor 20	4.00	6.95	6.34	1.10	1/ 575.	10.69	9.62	1.11	1/ 374.
Floor 19	4.00	7.25	6.19	1.17	1/ 552.	10.92	9.66	1.13	1/ 366.
Floor 18	4.00	6.84	6.20	1.10	1/ 585.	10.80	9.59	1.13	1/ 370.
Floor 17	4.00	7.09	6.01	1.18	1/ 564.	10.71	9.35	1.14	1/ 374.
Floor 16	4.00	6.58	5.96	1.10	1/ 608.	10.18	8.94	1.14	1/ 393.
Floor 15	4.00	6.46	5.83	1.11	1/ 619.	9.81	8.57	1.15	1/ 408.
Floor 14	4.00	6.14	5.65	1.09	1/ 651.	8.61	7.47	1.15	1/ 465.
Floor 13	4.00	5.78	5.31	1.09	1/ 692.	5.27	4.39	1.20	1/ 759.
Floor 12	4.00	5.94	5.31	1.12	1/ 673.	8.56	7.36	1.16	1/ 467.
Floor 11	4.00	6.49	5.11	1.27	1/ 617.	9.49	7.98	1.19	1/ 421.
Floor 10	4.00	5.95	5.03	1.18	1/ 673.	9.66	8.15	1.19	1/ 414.
Floor 9	4.00	6.39	4.76	1.34	1/ 626.	9.56	7.94	1.20	1/ 418.
Floor 8	4.00	5.65	4.59	1.23	1/ 708.	9.30	7.75	1.20	1/ 430.
Floor 7	4.00	5.47	4.35	1.26	1/ 731.	8.97	7.39	1.21	1/ 446.
Floor 6	4.00	5.25	4.07	1.29	1/ 761.	8.57	6.90	1.24	1/ 467.
Floor 5	4.00	4.66	3.27	1.43	1/ 859.	7.14	5.60	1.27	1/ 561.
Floor 4	4.03	1.54	1.13	1.36	1/2599.	4.08	3.26	1.25	1/ 981.
Floor 3	4.97	1.42	1.10	1.28	1/3530.	4.55	3.47	1.31	1/1099.
Floor 2	6.00	1.26	0.95	1.33	1/4776.	3.86	2.79	1.39	1/1553.
Ground Floor	5.30	0.00	0.00	—	0.00	0.00	0.00	—	0.00
B 1	3.60	0.00	0.00	—	0.00	0.00	0.00	—	0.00
B 2	3.60	0.00	0.00	—	0.00	0.00	0.00	—	0.00
B 3	4.80	0.00	0.00	—	0.00	0.00	0.00	—	0.00

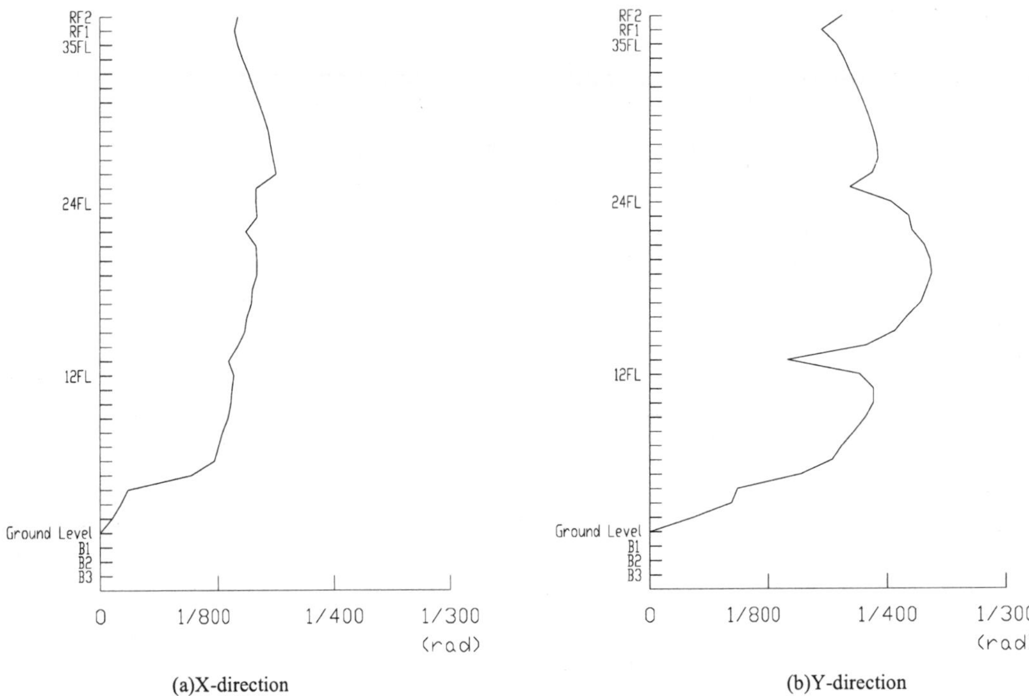

(a)X-direction (b)Y-direction

Figure 7. Inter-story rotation displacement subjected to the earthquake action

6. Foundation design

6.1 *Geology and foundation system choice*

This project is located in the northwest of Beijing plain area, which, generally flat, is at the upper part of the diluvial fan and the east of Yuquan Mountain. As indicated in the geotechnical investigation report, no active fault fracture with engineering significance occurs within the site and the 10km around. It is concluded that the soil on the proposed construction site will not be liquefiable under a seismic event of the intensity of eight (8) and under highest possible groundwater level. The thickness of overlaying layer at site is 48m, the construction site category b is Class II and at ground level the predominant period is 0.35 second.

In the construction design stage, the natural foundation is after repeated technical and economic comparison. The raft is 2800mm thick and the bearing stratum is the 5th one, powdered clay, f_{ka}=300kPa. After depth and width correction, the load bearing capacity of the stratum can reach 656 kN/m^2, which can satisfy the need of the high-rise steel structure.

The analysis shows that the average and maximum differential settlement of the building will decrease with the increase of the raft thickness. The estimation settlement values for the tower are average 56mm and maximum 65mm respectively; for the adjacent pure basement are average 33mm and maximum 47mm respectively. Such measures are taken in the design as to change the profile of the foundation and to fill soil around the edge so that the foundation center and the gravity center of the supper structure may keep coincident. The plan of the foundation appears as a quadrangle as shown in Figure 8.

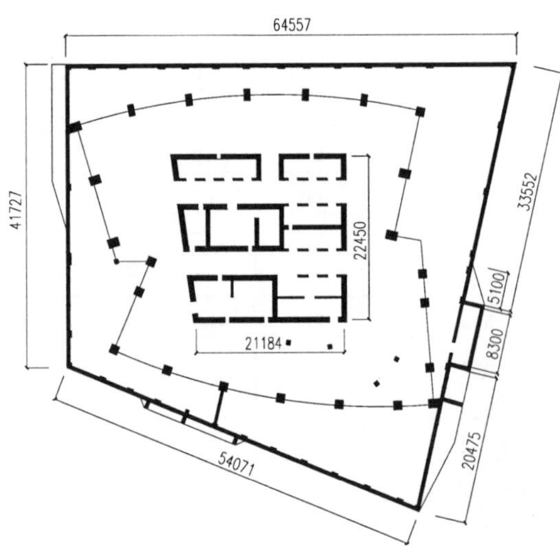

Figure 8. Foundation Plan

6.2 *Observation of foundation settlement*

The bottom elevation of the foundation is -20.1m. Average settlement and time relation curve of the building is shown in Figure 9. The observed settlement one month after the structure completion shows that the settlement of the peripheral frame and that around the core are similar, the average settlement of all points is 22mm which is 39% of the estimated final settlement; that the maximum settlement is 29mm occurring at the middle of the core and is 45%of the estimated final sedimentation; and that the minimum settlement is 8.5mm occurring at the northeast corner of the basement.

In the preliminary design phase, borehole cast-in-place concrete pile is selected with 900mm diameter, the pile end stratum is the 12th layer, gravel, the pipe length is 15 to 20m. Considerable economic benefit has been gained after changing into the natural foundation scheme. The total foundation cost has been cut down by RMB 2.87million against the estimated cost of RMB 4 million for the pile foundation. The construction period of the structure has also been shortened by 2 to 3 months.

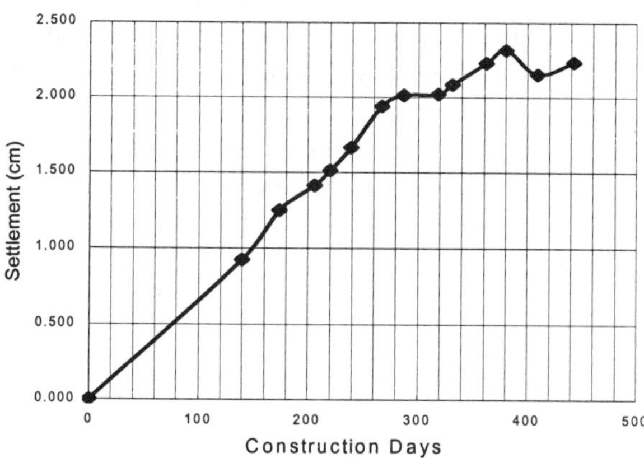

Figure 9. Average settlement and time Relation curve

7. Conclusion

The plan layout and facade arrangement of Zhongguancun Financial Center Tower is quit complicated for high-rise steel structure. Q345GJ steel with excellent properties was used for the first time in the high-rise building, the steel-concrete composite structure properly configured from ground level to 15.0m for reinforcing lateral stiffness of the structure. The every four level-kinky line outer columns were adopted for the satisfying the demand for architectural façade profile, manufacture and construction easily. Floor horizontal trusses arranged at outer column bending level for increasing the diaphragm stiffness of the floor slab. Changed pile foundation scheme into raft foundation scheme saved the cost and minimized construction period significantly.

References

1. PRC national standard, Code for Seismic Design of Buildings (GB50011-2001)
2. PRC metallurgical industrial standard, Steel Plates for High-rise Buildings (YB4104-2000)
3. PRC national standard, High Strength Low Alloy Structural Steels (GB/T1591-94)
4. PRC industry standard, Technical specification for Steel Structural of tall buildings (JGJ99-98)
5. Steel Structural Design of High-rise Buildings, by Chen Fusheng, Qiu Guohua and Fan Zhong, China Architecture & Building Press, July 2004

THE CHOICE OF STRUCTURAL FORMS AND USE OF STRUCTURAL MATERIALS OF THREE LANDMARK TALL BUILDINGS IN HONG KONG: CENTRAL PLAZA, TWO INTERNATIONAL FINANCE CENTRE, INTERNATIONAL COMMERCE CENTRE

JOHN W.K. LUK

Sun Hung Kai Properties Group

PETER AYRES CRAIG GIBBONS ALEXIS LEE ANDREW LUONG
GEORGE CHAN KELVIN LAM JACKIE YAU

Ove Arup & Partners Hong Kong Ltd

The Sun Hung Kai Properties Group is the developer/co-developer of the three captioned landmark tall buildings in Hong Kong, and Ove Arup & Partners Hong Kong Ltd. is the structural consultant. With a height of 368 metres and 78 storeys, the Central Plaza became the tallest reinforced concrete building in the world upon its completion in December 1992. The 88 storeys Two International Finance Centre was completed in 2004. It stands 420 metres and is now the 5th tallest in the world. The upcoming International Commerce Centre, which will be of 118 storeys and 484 metres in height, could be the tallest building in the world when it is completed (given that there are many tall buildings now under design and/or construction). In properties developments, financial, marketing and business goals are key considerations of the developers. This paper reviews how experienced and innovative structural and civil engineering, apart from providing sound solutions to technical matters, can collaborate to achieve the developer's such goals in tall building developments.

1. Introduction

The Central Plaza, the Two International Finance Centre and the International Commerce Centre are three landmark buildings in Hong Kong, setting many new records in tall buildings in both Hong Kong and the world. Although developed by the same developer and with the same Structural & Civil Engineering Consultant; due to different physical, economical, financial, market and technical factors and considerations at different time faced by the project teams, different structural design concepts and philosophy were adopted for these three mega tall buildings. The authors would like to share in the following their experience in the choice of structural forms and use of structural materials, for the purpose of providing optimum solutions to such factors and considerations.

Development	Central Plaza	2 IFC	Union Square – International Commerce Centre
Top Level of Building	374 mPD	420 mPD	490 mPD
No. of Storeys	78	88	118
GFA	173,000 m^2	180,000 m^2	270,000 m^2
Foundations	Large diameter caissons bearing on rock	Raft footing bearing on rock	Shaft-grouted barrettes
Basement	3 levels with perimeter diaphragm wall, top-down construction was adopted	5 levels with a 61.5m diameter cofferdam formed by 1.5m thick diaphragm panels	4 levels with a 76m diameter cofferdam formed by 1.5m thick diaphragm wall panels
Floor System	Reinforced Concrete	Steel Composite	Steel Composite
Stability System	Perimeter reinforced concrete frame plus central corewall	R.C. corewall + 8 composite perimeter columns with 3 steel outriggers	R.C. corewall + 8 R.C. perimeter columns with 4 outriggers (3 in steel and 1 in concrete)
Highest Concrete Grade	60 MPa	60 MPa	90 MPa (with volcanic aggregates)
Design Wind Speed	64 m/s	64 m/s	59.5 m/s

2. Central Plaza

Most high-rise building in Hong Kong has historically been based on the use of reinforced concrete, and building contractors there have always been among the fastest in the world at erecting such structures. Central Plaza is a prime example of the construction speeds which could be achieved with sensible planning from the commencement of the design. The most enduring memory for all of those who were involved in its completion would be how the development of extremely close working relationships between the developer, his designers and the contractor enabled the remarkably fast programme to be achieved. Despite a very short design lead time and the late decision to change to reinforced concrete, a four-day construction cycle for the typical floors was achieved, allowing the final completion of the building date matched that which would have been achieved by a structural steel scheme, but without the long lead time to order, fabricate and ship steelwork. Superstructure from ground floor to the base of the mast was completed in only 20 months.

The site for the Central Plaza was acquired at a land auction in Hong Kong on 18 January 1989 with a Hong Kong record price of US$430M for the 7000 square metre site. Daily interest charges of US120,000 provided a major incentive to complete building works as quickly as possible.

Arup were appointed for the structural and geotechnical engineering for the proposed office development shortly after the site had been purchased and Ng Chun man & Associates were appointed architect.

2.1 *Substructure*

The site is typical of a recently reclaimed area with sound bedrock lying between 25m and 40m below ground level. This is overlain by decomposed rock and marine deposits with the top 10 to 15 metres being of fill material. A permitted bearing pressure of 5.0MN/m² is allowed on sound rock. The maximum water table rises to about 2 metres below ground level.

It was initially clear that a multi-level basement of maximum floor area would be required. The design of a diaphragm wall, extending around the whole site perimeter, and constructed down to and grouted to rock, was completed in the first week after the site was acquired. This enabled construction to commence three months later.

An initial planning assessment had indicated that up to four levels of basement could be required and the design produced catered for this. By the time construction commenced it had been decided that only three levels would be necessary and the construction drawings were amended accordingly.

The diaphragm wall design allowed for the basement to be constructed by the top down method, this providing three fundamental advantages:-

1. The superstructure could be constructed at the same time as the basement, thereby removing time consuming basement construction from the critical path.
2. It used the permanent floor plates for stabilizing the wall, thereby reducing likely ground movements to below those expected for open cut execution.
3. It created a watertight box to enable traditional hand dug caissons to be constructed down to rockhead under controlled conditions.

The arrangement of the caissons was finalized once the superstructure was fixed. The maximum size of caisson required was 7.4m in diameter.

2.2 *Steel scheme*

The initial concept for the tower was a steel scheme having a height to the base of the mast of 340M, with a floor plate area of approximately 1800 square metres and slenderness ratio of more than 1:6. It was considered that an externally cross-braced framed tube fabricated from structural steel, in filled with high strength concrete, would provide the optimum structural solution. The floors were conventional with primary / secondary beams carrying metal decking with a 160mm thick reinforced concrete slab with a floor to floor height of 4m. The core was also of steelwork, designed to carry vertical load only. The core arrangement, developed by the Architect with the Developers input, made it ideally suited for construction in steelwork modules.

In August 1989, there was a sudden reduction in land values in Hong Kong, which necessitated a financial review of the development. It was decided to reduce the height of the tower base. This meant that a concrete structure became possible, with a higher strength of concrete than that previously used in Hong Kong to limit member sizes to acceptable proportions.

2.3 *Concrete scheme*

The construction of the diaphragm wall was under way by the time the concrete scheme was selected. Detailed design of the reinforced concrete scheme commenced immediately to avoid delaying the construction programme. The developers decided that their two in-house contracting firms, Sandfield and Tat Lee, should form a joint venture, Manloze Ltd, to act as main contractor.

The scheme that constructed has 78 storeys, and the height of the building is 308 metres above ground level, with the mast extending to a total height of 368 metres. The floor plate area has been increased to 2214m² in this case.

Above the tower base, 30.5 metres above ground level, stability is provided by the external facade frames acting as a tube. These comprise columns at 4.6 metres centres and floor edge beams 1.1 metres deep. The floor to floor height is 3.6 metres. The core has a similar arrangement to the steel scheme and, just above the tower base, it carries approximately 10% of the total wind shear.

The tower base structure edge transfer beam is 5.5m deep by 2.8m wide around the perimeter. This allows alternate columns to be dropped from the facade, thereby opening up the public area at ground level. The increased column spacing together with the elimination of spandrel beams in the towers base, results in the external frame no longer being able to carry the wind loads acting on the building. Over the height of the tower base, the core transfers all the wind shears to the foundations. A one metre thick slab at the underside of the transfer beam transfers the toal wind shear from the external frame to the inner core below.

The wind shear is taken out from the core at the lowest basement level, where it is transferred to the perimeter diaphragm walls. In order to reduce large shear reversals in the core walls in the basement, and at the top of the tower base level, the ground floor, basement levels 1 and 2 and the 5th and 6th floors, the floor slabs and beams are separated horizontally from the core walls.

To complete the dramatic impact of this building the tower top incorporates a mast, which was constructed of structural steel tubes with diameters up to 2 metres.

The performance of tall building structures in the strong typhoon wind climate is of particular importance. Not only must the structure be able to resist the loads generally and the cladding system and its fixings resist higher local loads, but the building must also perform dynamically in an acceptable manner such that predicted movements lie within acceptable standards of occupant comfort criteria. To ensure that all aspects of the building's performance in strong winds will be acceptable, a detailed wind tunnel study was carried out by Professor Alan Davenport at the Boundary Layer Wind Tunnel Laboratory at the University of Western Ontario.

When completed, this project became the tallest reinforced concrete building structure in the world and it would not be appropriate if use was not made in its design of modern technology. For a fast track project of this nature, it has also been important however, to ensure that the advances proposed were not such a significant departure of Hong Kong construction techniques and methods that the programme could be jeopardized.

2.4 *High strength concrete*

It was essential that a higher strength concrete be used for the vertical structure than normally permitted for private building works by the Hong Kong Government (28 day cube strength of 45N/mm2). It was decided to seek for approval for 60N/mm2 which was considered readily achievable using locally materials and was the first time applied to a building. Considerable research took place into materials and mix design and many trials were carried out including mock-ups of the large diameter columns to check on temperature effects. As a result of this, cooling was introduced into the major pours.

The use of higher strengths was considered, but it was decided to limit it to this value as it was considered by the development team that this material could be produced without difficulty from materials readily available in Hong Kong. Steps were in hand to further refine the mix design by using PFA as a cement replacement up to an allowed maximum of 25%, to reduce temperature and shrinkage effects.

For this project the structural engineer's responsibility has been to respond to the Developer's requirements, as expressed in the Architect's design, with a concept that is both economical in cost and quick to build. It has relied upon the latest methods of analysis and model testing and made effective use of normal construction techniques whilst using specially developed materials, such as grade 60 concrete.

3. International Finance Centre

Two International Finance Centre (2IFC) is the tallest building in Hong Kong and the 5th tallest in the world. Completed in 2004, the tower is 88 storeys tall, with the lowest basement floor at –32mPD rising to the top level of 420mPD.

The tower is just one element of a major new commercial development in central Hong Kong. The development as a whole provides office, retail, hotel and serviced apartment accommodation and accommodates Phase 2 of the MTRC Hong Kong Airport Railway Station in the 5 level basement below. It also houses major transportation facilities for the local bus companies together with several floors of car parking. The total constructed floor area of the Hong Kong Station Development amounts to 650,000 square metres.

Architects for the project are Cesar Pelli & Associates and Rocco Design Limited with Arup providing structural and geotechnical engineering consultancy services. Mechanical and electrical consultancy services were provided by J. Roger Preston Ltd.

3.1 *Tower primary structural scheme*

One of the main requirements of the brief was to provide open plan floors and to incorporate a large degree of flexibility such that the requirements of major tenants (such as financial institutions) could be accommodated. This, and the desire to maximise the panoramic views from the tower, necessitated that the building perimeter be as free of structural obstructions as possible. The requirement led to the structural solution with only eight main columns on the building perimeter (two per face).

The structural system utilizes a reinforced concrete core mobilized with outrigger trusses located at three zones in the building height. The outriggers connect to the eight main mega-columns to form an effect lateral stability system.

The reinforced concrete core was adopted following studies which concluded that there it was a significant cost advantage compared to steel and steel/composite core alternatives. Cost studies were also conducted on core designs for high strength concrete grades. These revealed that no great advantage was gained by using other than grade 60, as savings in lettable floor space did not significantly compensate for the additional cost and construction implications of using a higher grade concrete.

3.2 *Outriggers*

The outrigger trusses were located around the mechanical floors to maximize lettable floor space.

A punched concrete wall type outrigger was initially considered. However, due to the need for large openings through the core wall to accommodate the requirement for significant M&E access, it became apparent that a steel truss would be required to supplement the strength and stiffness of the perforated core. Thus the outriggers are in the form of complete structural steel trusses.

3.3 *Mega-columns*

The mega-columns comprise composite steel/concrete in which the steel elements are encased with reinforced concrete.

Extensive optimization studies were carried out to establish the most appropriate form of the columns in terms of cost and buildability and to optimise the size and steel/concrete ratios.

A performance-based approach was adopted in order to understand the massive mega-column's buckling behaviour and the restraint forces exerted on the connecting floor plates. The method involved a series of iterative analyses comprising two parts – modeling the buckling behaviour of the mega-columns; and modeling the floor diaphragm stiffness. The results enabled optimum sizing of the columns for strength and stiffness and detailing of the floor restraints.

3.4 *Construction-led initiatives and buildability design*

One of the key issues concerning steel outriggers in composite tall building construction, particularly with such large outriggers as required in the 2IFC, is their potential impact on the construction programme. The core of such buildings can be constructed with relative speed and efficiency using a climbform system. Typical cycle times achieved on the core were 3-4 days. Clearly, stopping the climbform at the outrigger levels to permit the steelwork contractor to install the outrigger is not conducive to optimising continuity of labour usage, or minimising the construction programme.

The solution came in the form of retro-outrigger installations. This involved building only part of the core wall thickness (300mm) at the outrigger zone such that the core construction can progress upwards and with sufficient support. The steel erection and

Installation of the outrigger could then be carried out in parallel with the advancing core construction above it. Once the steel installation was complete, the full core thickness at the outrigger zone (1000mm) was then retro-cast.

Another construction initiative involved detailing the mega-columns such that the structural steel components could be split into a number of subsections that could be lifted and connected with ease. In addition, it was determined at the outset the concrete encasement of the mega-columns would be formed using self climbing formwork. This initiative effectively removed the reliance on the cranes in lifting column formwork between floor levels, thereby maximising their efficiency for other construction lifting.

3.5 *Geology and foundation design*

With a total tower load of 5,200MN, foundations obviously had to be taken to rock head.

Surrounding the tower is a five level basement with a low-level commercial podium built above it. The southern most wall of the basement box also forms part of the adjacent Hong Kong Station Phase 1 structure. With trains under operation, it was essential that movements in this vicinity should be kept to a minimum. To achieve this, top down methods of construction were adopted for the general basement construction outside the footprint of the tower, enabling potential movements to be minimised.

On the other hand, in order to optimise the construction programme, it was considered appropriate for the tower to be built using a more conventional bottom up technique. A large diameter cofferdam, encompassing the plan form of the tower, was constructed using diaphragm walling techniques to enable excavation down to the pile cap level. This enabled the cap to be cast in open excavation and the whole tower constructed from the pile cap level within the cofferdam. The circular (compression ring) nature of the cofferdam eliminated the need for internal props to provide lateral support the excavation.

With a reasonably constant rock head level (as determined from site investigation) the foundation sub-contractor proposed to construct the 61.5 metre diameter cofferdam down to bedrock using a 1.5 metre thick diaphragm wall keyed into the rock. Using three reinforced concrete ring beams, and lowering the external water table by eight metres, excavation then took place to rock head level and the pile cap/raft was cast bearing directly on rock, thus omitting the need for bored piling. However, conditions on site proved slightly different from those anticipated, with a localised depression to the South East of the footprint of the rock head level down to a depth of approximately 55 metres, a depth too great to allow open excavation in this area. In order to overcome this a mixed foundation solution was finally adopted. Over much of the area the original raft solution was adopted. Mass concrete fill was then used locally, between rock head level and the underside of the cap, in the areas where the rock head sloped away.

Locally, at the location of deepest rock, barrettes were installed from ground level to transfer the pile cap/raft loads to the bearing stratum, use also being made of the cofferdam panels to transfer vertical load.

3.6 *Floor system*

The footprint at the base of the tower is 57m x 57m. Towards the top, a series of staggered step-backs reduce the plan dimensions to 39m x 39m at roof level. In all, it provides a gross floor area of over 180,000m2 of grade A office accommodation.

Design for the floor plates considered various materials including post-tensioned concrete and steel/composite options. Studies showed that the cost of a concrete floor was slightly less than a composite floor, and that the anticipated cycle times for the two systems were similar. However, the additional costs for the columns and foundations due to the increased dead loading showed that, overall, the composite solution was preferable.

The composite schemes comprises 125mm thick slabs acting compositely with permanent decking supported at up to 3m intervals on a variety of steel beam options. The weight of floor beam steel is 36 kg/m2 based on gross constructed floor area.

3.7 *Differential axial shortening*

For a super tall tower as 2IFC, a key design and construction consideration is the differential shortening between the core and the perimeter columns during construction and the long term effects post-completion. Differential shortening is particularly important in composite tall buildings in which the deformation characteristics of the steel intensive columns and the large concrete core can be markedly different.

Prior to construction a number of specific material tests were carried out to quantify the modulus and shrinkage characteristics of the proposed concrete mix. These tests enabled detailed axial shortening analyses of the tower elements.

The outrigger connections to the megacolumns incorporated a series of packing shims at the contact surfaces. This enabled the outriggers to be effective during construction in resisting any typhoon winds that may have occurred, whilst at the same time allowed the packing shims to be removed, or added, as required to adjust for the differential movements between the outrigger and columns. Using this approach, the build-up of very large internal forces in the mega-columns, outriggers and the core were prevented.

3.8 *Robustness and safety of the tower*

The 9/11 tragedy in New York occurred when the 2IFC tower was constructed up to the 33/F. In the days following 9/11 many questions were raised with regard to the robustness, integrity and egress provision within the 2IFC Tower.

Whilst it was recognised at an early stage that it would be impracticable to design a tower specifically to resist such extreme events, Arup undertook an extensive series of studies focused on the assessing the safety of the design. It should be noted that the studies where carried out in the 4 weeks after 9/11 – speed being essential so as to inform the ongoing construction of 2IFC.

The studies involved performing comparative assessments of international practice and code requirements with regard to structural integrity and escape, as well as detailed analytical dynamic simulations of a range of aircraft engine impact scenarios – the latter used software developed by crash simulations for the automotive industry.

The key structural findings from these studies were as follows:-

- The large composite steel/concrete megacolumn solution resulted in key elements which offered an extremely high resistance to extreme impact scenarios.
- The concrete core offers a hardened escape route to evacuees. The walls of the core are particularly good at maintaining integrity and redistributing load in the event of localised damage.
- The outrigger solution is capable of providing alternative load paths in the event that critical elements are removed. It was shown that the integrity of the building was maintained in the event that two of the megacolumns were removed at low level. The outrigger effectively acts as a 'gravity prop' to support the megacolumns above.
- Tall buildings in Hong Kong are designed to resist significant lateral loads (typhoon winds) compared to other tall building centers around the world. As a result, under non-typhoon conditions, the elements exhibit significant residual strength to enhance the resistance to impact events and to tolerate load redistribution when in a damaged state.
    - It is evident that the provision of exist staircases in the Hong Kong code is larger than that which exist in accordance with other international codes. The main reason is that the Hong Kong code requirement is largely based on a simultaneous evacuation of the building whilst other codes permit a phased evacuation strategy (thereby reducing the overall egress requirements).

In addition, the studies also showed that the exit staircase and escape provisions required in Hong Kong are much more stringent than those elsewhere.

The conclusion from these studies was that the structural form and planning of the 2IFC Tower provided a robust form with enhanced resistance to potentially damaging events. As a consequence of these studies, the construction of the 2IFC tower continued with no changes to the design.

The iconic 2IFC is newest and tallest addition to the Hong Kong tower-scape. Its design has combined innovative solutions with effective use of material, and tailored to efficient construction methods, to produce a robust landmark to the pride of the building's owners, end user and the general public.

4. International Commerce Centre

The International Commerce Centre is a key element of the Kowloon Station Development (Union Square), which is of 7 packages. Packages 5, 6 & 7 consists of:

Package 5	-	GFA 70,000 m^2 retail development located above the Kowloon Station.
Package 6	-	Two towers of 260m tall of total GFA approximately 150,000m^2, planned for suite hotel, residential flats and service apartments.
Package 7 (International Commerce Centre)	-	A 484m tall tower for office and hotel of total GFA around 270,000m^2.

The International Commerce Centre is located at the south-western corner of the Kowloon Station Development site. It consists of 118 floors of height 484m. The low zone of the building is for office and the high zone is planned for a six-star hotel. It also consists of a 4-level basement for plantrooms, carparking and loading and unloading service and a 4-level podium for plantrooms, carparking, loading and unloading and retails. The project started in early 2001 and is scheduled to finish in 2010.

There were a number of challenges in the design development of the project. Apart from technical challenges, control of construction cost is also a key concern in the design. Described below are the special features of the project to tackle various challenges.

4.1 *Foundations*

The footprint of the proposed tower is located above a major fault zone, where competent rockhead level is in general 80m below ground but down to 130m at the deepest area. Normal end-bearing piling system is not viable for such difficult geological conditions. Following a series of detail studies and comparison on various foundation types, shaft grouted friction barrettes were chosen as the foundation system.

Shaft grouting is a new technique in Hong Kong for barrettes construction to enhance the friction capacity of the piles. A designated volume of cement grout is injected into the soil-pile interface to improve the contact between the barrettes and the surrounding soil. The lateral pressure due to this injection grouting will compact the surrounding soil at the interface, which have softened or loosened during trench excavation. With this enhancement in the lateral pressure, the frictional capacity of the barrettes is significantly increased.

International Commerce Centre is the first private development in Hong Kong adopted this special shaft grouted friction barrettes system. There is very limited record for such a technique with in-situ tropically weathered rocks in Hong Kong. Therefore, 5 trial piles had been constructed to collect technical data for the actual performance of shaft grouting in different soil strata inside the site. Finally, one of the 5 trial piles has been loaded beyond the design capacities of the piles to verify the design against the actual performance.

The trial pile results form the basic design parameters of a single barrette. The group effect is then deduced by analytical finite elements model.

After the foundations were completed, 4 working barrettes were selected for load testing by the Buildings Department with the test load up to the maximum working load to verify their actual performance.

The trial piles and working piles were all tested to be satisfactory in the load carrying capacity as well as the settlement performance.

4.2 *Basement construction*

The 4-level basement under the tower footprint, which requires a total 26m deep excavation, is located only 3 metres away from the operating MTR tunnel. Very stringent control is therefore necessary on the movement of the tunnel during basement construction. In order to minimize the movement and to allow maximum flexibility for the construction of basement structure, circular cofferdam of 76m diameter, made up of 1.5m thick diaphragm wall panels, were used.

With the very high hoop compression stiffness provide by the circular cofferdam, this scheme enables a 'strut free' excavation and meanwhile achieves very small ground movement. The actual measured movement of the existing MTR tunnel is well within the allowable limit.

This 'strut free' excavation scheme also significantly reduced the time of excavation and basement construction and hence enabled the early completion of the pilecap and vertical elements.

In order to achieve the designated hoop stiffness of the cofferdam, diaphragm wall panels have been constructed with a stringent control of verticality so as to ensure adequate effective contact area of the 'hoop'. This stringent control in verticality also ensured that the toe of the deep diaphragm wall panels do not intrude into the 3m protection zone of the existing MTR tunnel.

Upon completion of the 4-levels basement under the tower footprint, the central core will provide a stiff lateral support. The remaining portion of the 10m deep basement outside the tower footprint is excavated using ordinary method with steel shoring propped to the central core.

4.3 *Structural system*

The typical floor plan size is approximately 60.5m by 60.5m, with a core of internal size around 30m by 30m, housing the lifts, staircases and services. Composite steel beams spanning between the core and perimeter edge beam with permanent steel deck for reinforced concrete slab are employed as the typical floor system. To maximize the flexibility of facade design and to allow the best view for the tenants, only 8 mega columns (2 numbers per face) are provided at the perimeter to support the building. The perimeter edge beams made of built-up steel sections, which span 27m clear between mega columns.

The stability system of the building comprises the core wall coupled with 8 mega columns through 4 outrigger trusses located at different levels. The mega columns were originally designed as composite columns with embedded steel sections to minimize the column area. However, due to the dramatic rise in steel price in the course of the project, the mega columns were eventually changed to reinforced concrete. To minimize the area of the columns core wall, high strength high stiffness grade 90 concrete is used.

Since there is sufficient structural depth allowed in the first outrigger zone (4 M&E floors and 1 refuge floor with total depth 21.5m), reinforced concrete is used for the first outrigger to minimize the demand on expensive structural steel. The concrete outrigger members are post-tensioned in order to control cracking of concrete under the high tension induced by lateral wind load. In order to minimize the lock-in stresses due to the differential settlement between the central core and the mega columns, temporary joints are introduced between the mega columns and the reinforced concrete outrigger panels. These temporary joints are to be filled with high performance grout at the late stage of superstructure construction when major portion of differential movement has occurred.

For the other outrigger zones, structural steel is adopted to accommodate the limited structural depth allowed, i.e. 2 M&E floors and 1 refuge floor with total depth of 14.1m.

4.4 *High grade concrete*

High strength high stiffness grade 90 concrete with volcanic aggregate is used for the vertical stability structure, that is, core wall, lintel beams and mega columns. The volcanic tuff rock is superior in terms of its elastic modulus value in comparison with the typical granite rock found in Hong Kong. With the use of volcanic aggregate, an elastic modulus of 39MPa is achieved. This has substantially increased the concrete stiffness as compares with ordinary concrete. Since stiffness is always the controlling structural properties for super-highrise building, this enhancement in concrete stiffness has resulted in a significant reduction in the construction cost as well as the size of vertical elements.

At the outrigger zones, the core walls are highly congested with structural steel and reinforcement. This causes difficulties to ensure good compaction of concrete. Self-compacting concrete is therefore developed to improve concrete compaction.

Self-compacting concrete (SCC) is defined as a highly flowable, yet stable concrete that can spread readily into place and fill the formwork without requiring any external consolidation energy and without undergoing any significant segregation. SCC is particularly important for tall building structures where high strength concrete is required to ensure good compaction during placing to achieve the strength and durability. Insufficient compaction dramatically lowers ultimate performance no matter how well it has been produced and how good the mix is designed. SCC is particularly useful for structural elements of high reinforcement density (such as composite column or core wall) since vibration of concrete in such areas will be difficult. By using SCC, concrete quality will be enhanced, construction time saved and environmental benefits of reduced concrete vibrator noise also result.

Various tests have been carried out to ensure the performance of the grade 90 concrete, which include trial mixes, trial columns with temperature monitoring and core tests, pumping trials, accelerated test for alkali-silica reaction, etc.

4.5 ***Design wind speed***

Wind load is a key parameter for design of high-rise building.

The old version of Hong Kong wind code has based on those wind data collected before 1970s. The 50-year return period hourly mean wind speed of 64m/s with gradient height 200m as specified in the code is therefore outdated and considered conservative.

Since 1970s, the database of wind data has continuously been enhanced. Based on the database published from the National Hurricane Centre (NHC), recent advances in Monte-Carlo modeling of typhoons have been made by Applied Research Associates (ARA). ARA predicts that the upper level 50-year return period hourly mean wind speed could be of 52-55 m/s instead of 64m/s whilst the gradient height should be at 500m above the mean sea level instead of 200m.

Upon detail discussion with the Buildings Department (BD) alongside with technical substantiations, a reduced gradient wind speed of 59.5m/s with gradient height at 500m was agreed with BD for the design of the International Commerce Centre.

The reduced gradient wind speed and the realistic gradient height were eventually incorporated in the new Hong Kong Wind Code released in 2004.

This reduction in wind speed has resulted in reduction in sizes of stability members, which ended up a very substantial cost saving of around HK$100M.

4.6 ***Optimization study***

In order to achieve a cost-effective structure for the building, detail optimization studies using sophisticated computer programme have been carried out. The studies have based on the unit costs of various stability elements and constraints imposed by the client, architect and contractor.

Two different objectives are defined to determinate the optimal sizes of the core wall, mega columns and outrigger trusses. They are:

a. To achieve the optimal cost for the stability structure
b. To achieve the optimal value of the building

The first study is solely to deduce the optimal combination of the stability elements so as to minimize the structural cost of those elements.

In the second study, in addition to the structural cost the estimated floor value relating to the net floor area has been introduced so as to deduce the most optimal combination of key stability elements.

Owing to the continuous change in material costs and floor value, the optimization results would be reviewed at different stages of the project.

With the tallest building in Hong Kong situated at a site of very difficult ground condition, meanwhile to satisfy various design requirements and keeping the structural cost within budget, the structural engineer has to pioneer on a number of key features governing the design of the building. This includes the shaft-grouted barrette foundations, grade 90 concrete using volcanic aggregate, use of reduced design wind speed, etc. All these breakthroughs have not just benefited the project itself, but also set the milestones for the building industry in Hong Kong.

5. Conclusion

This paper reviews the importance of experienced and innovative engineering in meeting the various financial, marketing, planning, architectural, aesthetic and other technical requirements in Mega Tall Building developments. Experienced and innovative engineering on one hand would provide sound technical solutions in its own respective discipline. On the other hand, it could provide more options which would be useful for the developers to do the necessary balancing so as to arrive at optimum solutions for their property developments – when facing all those multi-factors and rapidly changing conditions. It should also be noted that, apart from specialist expertise in various professional fields, teamwork is equally important if not more for successful property developments. These would include the collaborated inputs from all disciplines involved, including marketing, planning, architectural, environmental, civil, structural, building services engineering, construction engineering, and those from the ultimate users and facility management, taking all known factors into account.

In conclusion, we would need both experienced and innovated professionals, and well-seasoned business counterparts to work closely together as a team in order to ensure success to any sizeable property development project.

Wind Engineering

WIND TUNNEL STUDIES ON THE BEHAVIOR OF A HIGH-RISE BUILDING

ZHENG SHIXIONG[1]
School of Civil Engineering, Southwest Jiaotong University, Chengdu, 610031

XUE SHANGLING
Institute of Constructional Engineering, CISDI Engineering Co. Ltd., Chongqing, 400013

The behavior of Chongqing International Exhibition and Convention Mansion under wind is systematically studied. The Mansion is a steel structure tall building, its height is 303m, and the first frequency of which is 0.1234Hz. To obtain the wind load and wind-induced responses and assess the living comfort and safety of the building, wind-tunnel model test is done in the turbulent boundary layer, which simulates the natural wind on the site of the Mansion. The distribution of wind pressure on every surface of the Mansion is measured by the static model. the static wind load is calculated. The wind-induced responses, including the displacements, acceleration and angular velocity responses, are measured by the aero-elasticity model. The dynamic wind load is also acquired. The effect of surrounding buildings and structures was discussed. It is shown, that the building's structural stiffness and dynamic properties, which is vibration period and damping, strongly affect the wind-induced responses of the building, and the surrounding buildings influence the distribution of wind pressure and wind load seriously. no galloping will occur under the design velocity, the acceleration and angular velocity on the top of the building is smaller than the allowance and the living comfort of the building meets the need of China national code.

1. Preface

The design of high-rise building is often influenced by wind-induced response such as acceleration and lateral deflections. With the buildings getting taller, they become more flexible and lightly damped. As a result, these buildings are more sensitive towards wind-induced excitation. The wind loads may govern the design. The designers must pay attention to the wind-induced response of high-rise building.

Generally, the wind load is composed of static wind load and dynamic wind load. The static wind load can be acquired by multiplying the average wind pressure by the area of surfaces. The dynamic wind load relates to the character of wind, dynamic property of structure, etc. When the fluctuation wind attacks the tall building, two dynamic responses will occur. One is the background response, the other is resonance response. To the structures with larger stiffness, the resonance response is very small and background response is the major component. The flexibile structure, such as super high-rise building, the resonance response is the major component and background response is small.

As a case study, the behavior of Chongqing International Exhibition and Convention Mansion under wind, a high-rise building with a height of 303 meters, is investigated. The Mansion is regarded as the highest building of Southwest China. Its main structure is constructed by steel structures. There are seven-storied podiums in the Mansion and every bedding surface of main body structure above-seven-storied is a quadrate of 41 meters by 41 meters, and there is also an aircraft parking area at the top. It is located at the center area of Chongqing city, where high buildings are densely located. Most of the surrounding buildings have the height ranging from 100 to 150m. The surrounding buildings and land form have a notable influence on the wind field. So, the wind loads on the building and wind-induced responses of the building are difficult to calculate according to Chinese standard. To obtain the wind loads and the responses, wind tunnel test is needed. The distribution of wind pressure on every surfaces of the building was measured with the static model. And using the method of integrating, the static wind load shape coefficients are acquired. The first frequency of the Mansion is only 0.1234 Hz, so its dynamic effect under strong breeze allows of no ignorance, but the reliable way to ascertain the wind-induced response is the aero-elastic wind-tunnel model test. Under the simulating turbulent flow, the acceleration response and lateral deflections responses on the top of the building were measured in different windward and wind velocity.

[1] Professor and Ph.D. E-mail: zhengsx@swjtu.edu.cn

The effects of surrounding buildings and structures are also discussed. The safety and living comfortably of the building under wind was investigated.

Referring to the data provided by the meteorological bureau and Chinese codes [3][4][9] comprehensively, the basic wind pressure value is 450Pa..

2. The Simulation of Atmospheric Boundary Layer and the Determination of Model Scales

Considering the need of the physical dimension of the Mansion and its surroundings, the dimension of cross-section of wind tunnel and the simulation of the wind field comprehensively, the 1:250 geometric scaling factor of Chongqing International Exhibition and Convention Mansion's static wind pressure measured model and aero-elastic model is adopted. Considering their influences on the wind field, the buildings and structures around the Mansion are also simulated by the scaling factor of 1:250(only simulating the outline).

The wind-tunnel test is being done in the first working section of XNJD-1 industrial wind tunnel which belongs to Wind Engineering Research Center of Southwest Jiaotong University. It is a low-speed closed circuit wind tunnel. Its cross-section size of first working section is 3.6 meters by 3.0 meters (width by height) and the cross-section size of second working section is 2.4 meters by 2.0 meters (width by height). The experimental wind speed in the first working section can be up to 20.0m/s. At the bottom of the wind-tunnel, there is a revolving tray that can revolve 360^0 so that the wind angle can be changed.

On the basis of the similar law of wind-tunnel test, the coming wind field in the test should simulate the atmospheric boundary layer. The boundary layer must according to the characteristic of the site the Mansion located. The wind characteristics, which are the mean wind speed profile, turbulent intensity profile, velocity fluctuations spectrum and the integral scale profile, must be in accordance with the model scale and meet the need of the Chinese Code.

According to Chinese code, the catalogue of atmospheric boundary layer of the site where Chongqing International Exhibition and Convention Mansion locates is about C type. To simulate the turbulent boundary layer, the spires and the roughness are used. A mean velocity profile, of which the power law exponent α=0.22, is get. Fig.1, Fig.2 and Fig.3 show the characteristics of simulated atmospheric boundary layer and the comparing them with that in Chinese code. It shows the wind characteristics are satisfied.

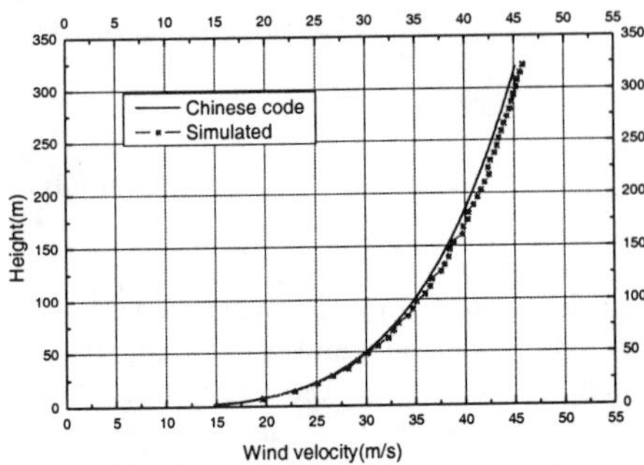

Fig.1 The mean wind speed profile

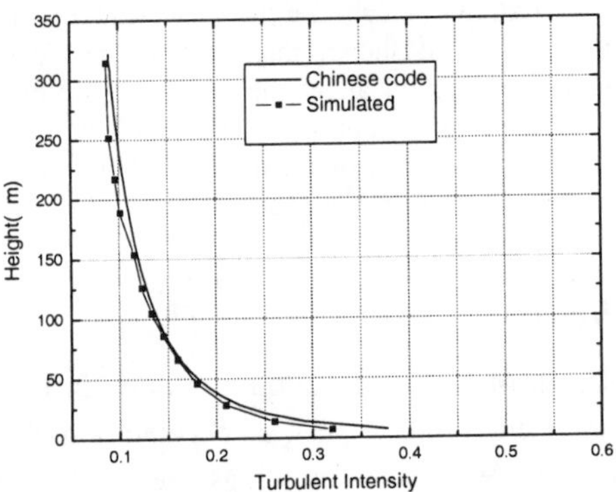

Fig.2 Turbulent intensity profile

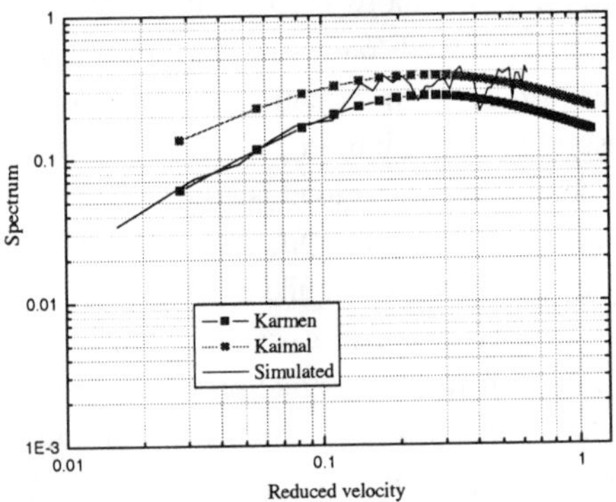

Fig.3 Velocity fluctuations spectrum

3. Measuring the Wind Pressure Coefficients using Static Model

The purpose of wind pressure measuring model test is to measure the pressure distribution in each surface of the building, and calculate the body shape coefficient on the basis of it. The model is processed by organic glass, and the measuring points are distributed on each surface of it. In the wind tunnel test, wind pressure value of each measuring point is measured at a given wind velocity.

The average pressure coefficient of each pressure hole can be calculated

The wind pressure coefficient is dimensionless factor. It can be used to calculate the wind pressure on the building's surface. The wind tunnel is carried out in 16 wind angles, and the conclusions are as follows: 1)The pressure on the windward side of the Mansion is mostly positive pressures, and the negative pressure only exists in some low parts of it because of the influences of surrounding buildings. However, the pressures on the other sides (side, leeward side and roof side) are all negative pressures. 2)The maximal positive pressure happens when the windward side is rightly opposite to the wind; the minimum pressure happens when the sideward wind is rightly opposite to the wind; The pressure on the roof of the Mansion, it is negative pressure under any wind angle, and it's maximal negative pressure is -2.325. 3)As to the rightly windward positive pressure of each side, the pressure along the cross direction and the edges is smaller than that in the middle. Besides, the maximal rightly windward positive wind pressure happens in the area which is about 15~30m away from the roof. However, because of the influences of the surrounding buildings, the positions of each side at which the maximal negative pressures happen vary. Besides, due to the influences of the surrounding buildings on the wind, some higher wind pressure may partially happen on each side of the Mansion. Therefore, more attention should be paid when designing the curtain wall. 4)Because the surrounding buildings are relatively near to the Mansion, they have relatively great influences on the wind pressure coefficient of each measuring point of the Mansion, especially those areas below the heights of the surrounding high buildings. Their influences include both the effects of amplification and reduction, and sometimes the opposite sign of the wind pressure would even appear.

4. Aero-elasticity Model Test

The structural dynamic character is the basic parameter for designing the aero-elasticity model and making wind-tunnel test. A structural three-dimensional finite element computing model is set up. The dynamic characters of the Mansion are calculated. In order to be convenient for description, the relations between the side faces, systems of axes and coming wind directions of the Mansion is defined as in the following: in which X-axial is from A side to B side. When wind comes in X-axial, the wind angle is $\beta=0^0$. Y-axial is from C side to D side. When wind comes in Y-axial, the wind angle is $\beta=90^0$. As to Z, it is vertical. The first six frequencies can be seen in the Table1.

Table 1 First six frequency and its modal shape

Serial No.	Frequency (Hz)	Modal shape
1	0.1234	Y-axial lateral deflection
2	0.1295	X-axial lateral deflection
3	0.1950	twist
4	0.4001	Y-axial lateral deflection
5	0.4202	X-axial lateral deflection
6	0.5511	twist

The purpose of the aero-elasticity model test is to study the wind-induced vibrating response of Chongqing International Exhibition and Convention Mansion, and to make an evaluation of its security and assessment of its comfort. The aero-elasticity model used in wind tunnel test should be designed in consistence with the following regulations, to keep the conformity of the following dimensionless faction between the prototype and model: Elastic parameter $E/\rho V^2$; mass parameter ρ/ρ_s ; damping parameter ζ ; Strouhal number $S_t = fL/V$. Where, ρ is air density; ρ_s is the structural material density; V is the wind velocity; L is the structural characteristic size; E E is elasticity model; ζ is structural model damping ratio.

In aero-elasticity model design, the conformity of the elastic parameter and mass parameter mainly means to make the vibration character of the model and the prototype meet the similar conditions. As for the aero-elasticity model of Chongqing International Exhibition and Convention Mansion, synthetically considering the influences of the elastic parameter and mass parameter, the vibration character of the model and the prototype shall be assured similar while designing the model so as to guarantee that the displacement, acceleration and such mechanical parameters of the model are similar with those of the prototype. Besides, according to the conformity principle of Strouhal number, if the geometric scaling factor of the model is 1:250, and its frequency scale is 25:1, then the wind velocity ratio of the wind tunnel test would be 1:10.

The structural wind-induced response is quite sensitive to damping parameter and the damping ratio in the real structure is unknown. According to Chinese code [4], the damping ratio may be 0.02 for the kind of the Mansion. So, the damping ratio of the aero-elasticity model shall be also designed and produced with 0.02 to meet the need of the conformity of the damping parameters, and needs to be checked by the modal test of the model.

It is shown by the results of the modal test, the damping ratios of each modal shape of the model are slightly smaller than 0.02. And according to the structural wind engineering theory, the vibration amplitude of the structure wind-induced response has an inverse relation with the damping ratio, and the results obtained are slightly larger. The damping ratio of the aero-elasticity model is proved to be rational.

In test, displacement sensor and acceleration sensor shall be installed on the top of the model in order to measure the displacement response and acceleration response of the model in the X and Y directions.

The wind tunnel test of aero-elasticity model would be made in the above-mentioned simulated atmospheric turbulence boundary layer, and the wind-induced response under the effect of 16 wind angles (0°~360°with intervals of 22.5°) would be examined. At each wind angle, the wind velocity in the test is from 1.0 to 9m/s, the interval of wind velocity is 0.5m/s,

With help of the similarity law, the test's results of model can be transformed into the prototype. The wind-induced response of the Mansion versus the wind velocity would be acquired. The results from test prove that the displacement response, acceleration response and angular velocity response increase as the wind velocity increases and any wind angles.

Fig. 4 shows the variation of displacement peak value of vibration response at the top of the Mansion versus wind angle. The wind velocity refers to the wind pressure with the 100 years' return period. From Fig. 4, we can see that the vibrating displacement response in the direction of cross wind of the Mansion is generally larger than that in the direction of along wind. According the above-mentioned definition of wind direction : X axis is identical with β=0°, Y axis is identical with β=90°. The displacement peak value of vibration response in the X direction at the top of the Mansion is the larger when the wind angles are β=90° and β=270°, and the wind comes across with the X –axial direction. The displacement peak value of vibration response in the Y direction is the larger when the wind angles are β=0°(β=360°) and β=180°, and the wind comes across with the Y-axial direction.

From Fig.4, we can also see that displacement along the wind is the bigger when the coming wind current exerts a head-on influence on one side of the structure, i.e. when the four kinds of specific wind are β=0°(β=360°), β=90°, β=180°, and β=270°. The corresponding displacements along wind are similar. Among them, it's the largest when β=270°. The displacement responses across wind are also the larger when wind comes with the four above-mentioned wind angles. Among them, it's the largest when β=180°.

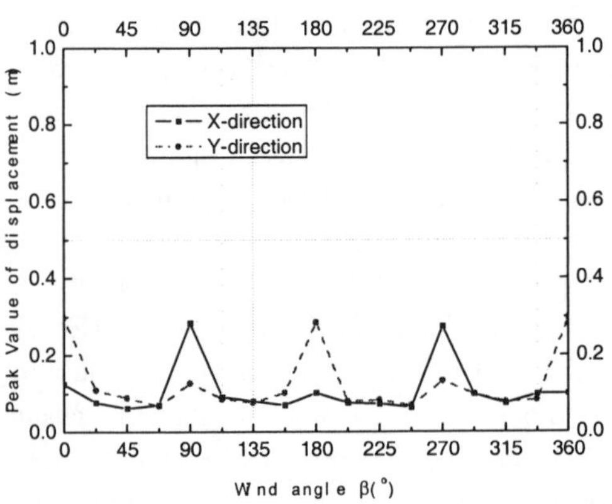

Fig.4 peak value of the displacement on top of the Mansion versus wind angle

Fig. 5 offers the variation of acceleration peak value of vibration response at the top of the Mansion with wind angle under the wind pressure of 10 years' return period.

From Fig.5, we can see that the regular pattern of the acceleration response being same as that of displacement response, the beam wind vibrating acceleration response of the Mansion is generally larger than that of following wind. And the acceleration peak value of vibration in the X direction on the top is bigger when the wind angles are β=90°and β=270°, and the acceleration peak value in the Y direction is bigger when the wind angles are β=0° (β=360°)and β=180°.

It is shows from Fig.5, when the wind comes with β=0°(β=360°), β=90°, β=180°, and β=270°, the acceleration of the following wind and acceleration of the beam wind are larger than that of other wind angle. Among them, it is the largest when β=90°.

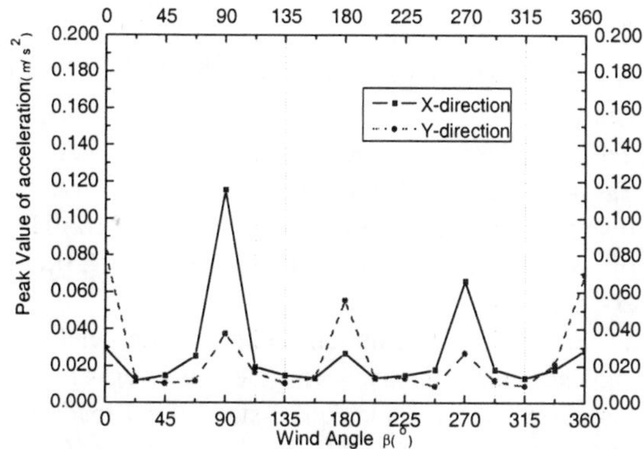

Fig.5 Peak value of the acceleration on top of the Mansion versus wind angle

Fig. 6 offers the variation of the angular velocity peak value with wind angle at the top of the Mansion under the 10 year's return period. From Fig.6, we can see that different from response of displacement and acceleration, the angular velocity is relatively larger when there is a certain angle of declination and it is the biggest when β=157.5°.

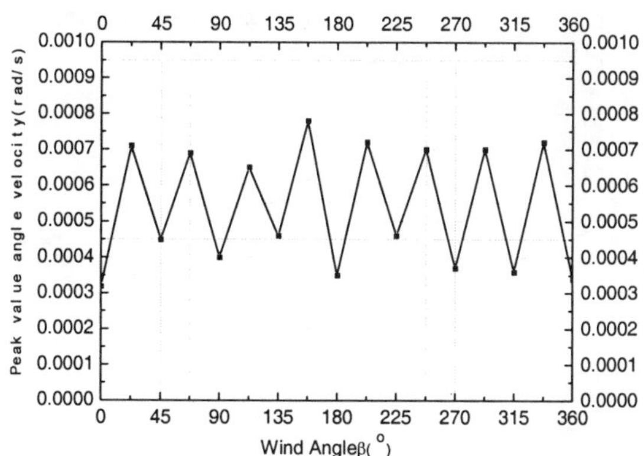

Fig.6 peak value of the angular velocity on top of the Mansion versus wind angle

The test also indicates that in each wind angle, the vortex vibration doesn't happen to the Mansion. Neither does the phenomenon of oscillating divergent galloping. Using the peak value of displacement response in Fig.4, the peak value of wind vibration load corresponding to the peak value of displacement can be obtained by the finite-element method. The total wind-induced load can be obtained by adding wind vibrating load to static wind load got from the preceding static test, and security evaluation can be made by loading the total wind-induced load on the structure. Or dividing the total following wind load by static wind load, we can get the wind vibrating coefficient. Limited by the space of the paper, the result of the wind vibrating coefficient will not be given.

When the wind orthogonally intersects the sides of the building, the oscillation of the beam wind is bigger than that of the following wind. Therefore, the dynamic load of the beam wind is bigger than that of the following wind, and the influence of the vibration load of the beam wind should be considered when checking calculation of design.

The result of the aero-elasticity model wind tunnel test indicates: 1) when the coming wind exerts a head-on influence on one side of the structure, the displacement and the acceleration vibrating response are bigger than that of the following wind. 2) Under wind pressure of 10 years' return period, among the 16 wind angles, the maximal vibrating displacement of the beam wind on the top of the structure is 0.297m when β=0°. the maximal vibrating displacement of the following wind is 0.133m when β=270°. 3) Under the wind pressure of 10 years' return period, the maximal vibrating acceleration response of the beam wind on the top of the structure is $0.116m/s^2$ when β=90°. The maximal vibrating acceleration response of the following wind is 0.038 m/s^2 when β=90°. Refer to Chinese code[9], that the acceleration response peak value of the following and beam wind of steel-structured building should be smaller than $0.2m/s^2$, it is satisfied. At same condition the maximal vibrating angular velocity response on the top of the structure is 0.00078rad/s when β=157.5°. Though there is no regulation on the limit value of angular velocity in Chinese code, its reversing vibrating angular velocity should be smaller than 0.001rad/s referring to the data both home and abroad[2][4], therefore, the reversing angular velocity at the top of the Mansion also satisfies the requirement. 5) At each wind angle, in the range of the design wind velocity, there is no sign of resonance vibration of vortex and oscillating divergent galloping.

5. Conclusion

The static wind tunnel model test and aero-elasticity model test of Chongqing International exhibition and Convention Mansion are done. The model scale is 1:250. The wind pressure coefficient of each side and wind-induced vibrating responses of the Mansion at 16 wind angles are obtained. Analyzing the results of the test, the following conclusions are achieved:

1) At each wind angle, in the range of the design wind velocity, there is no sign of resonance vibration of vortex and oscillating divergent galloping.
2) At each wind angle, and under the wind pressure of 10 years' return period, the maximal vibrating acceleration at the top of the Mansion is smaller than $0.2m/s^2$, the angular velocity torsional oscillation is smaller than 0.001rad/s, and the requirements for comfort are satisfied.
3) When the coming wind current exerts a head-on influence on one side of the structure, its displacement of beam wind, vibrating response of acceleration is bigger than that of the following wind. Therefore, the beam wind load effect of such kind of high building structure should not be ignored.
4) Because of the influences of the surrounding buildings on the wind, the phenomenon of high wind pressure would partially happen to each side of the Mansion. Therefore, we should pay great attention to this problem when designing the curtain wall. Besides, the influences of the surrounding buildings on the Mansion's wind pressure are relatively evident in the range of their own heights, and are relatively insignificant in the area of the Mansion's top.
5) The maximal negative pressure of each side of the Mansion is bigger than the maximal positive pressure.

References

1. Simiu,E., and Scanlan, R.H.," Wind effects on structures" , 2d Ed,. John Wiley and Sons, New York, N.Y.229-263(1986)
1. 2.Xiangtin,Zhang, " Anti-seismic and anti-wind design and calculating for high buildings" .Shanghai: Tongji Press,9-10(1997)(in Chinese)
2. Bureau of P.R. of China, " Load code for the design of building structures" ,GB50009-2001. Beijing: China architectural engineering press, 24-47(2002)(in Chinese)
3. Bencai,Huang, " Anti-wind analysis Theory and its apply" .Shanghai: Tongji Press,80-102 (2001)(in Chinese)
4. Yasushi Uematsu, Keisuke Watanabe, Akihiri Sasaki,et al. Wind-induced dynamic response and resultant load estimation of a circular flat roof[J].J. of Wind Eng. and Ind. Aerodyn., (83),251-261(1999).
5. W. Schneider, W. Zahlten. Load-bearing behavioue and structural analysis of slender ring-stiffened cylindrical shells under quasi-static wind load[J]. J. of Constructional Steel Research,,60,125-146(2004).
6. M.J.Palush, A.M.Loredo-Souza, J.Blessmann. Wind load on attached canopies and their effect on the pressure distribution over arch-roof industrial buildings [J]. J. of Wind Eng. and Ind. Aerodyn., (91),975-994(2003).
7. W. Zahlten, C.Borri. Time-domain simulation of the non-linear response of cooling tower shells subjected to stochastic wind loading [J]. J. of Eng. Struct., Vol.20,No.10,881-889(1998).
8. Bureau of P.R. of China, " Technical specification for steel structure of tall building" ,JGJ99-98. Beijing: China architectural engineering press,30-31(1998) (in Chinese)

WIND TUNNEL AND CFD STUDIES ON WIND FLOW AROUND A HIGH-RISE BUILDING WITH A REFUGE FLOOR

CHARLES C.K. CHENG[†, 1]

Department of Building and Construction, City University of Hong Kong, Tat Chee Avenue, Hong Kong, China

K.M. LAM

Department of Civil Engineering, The University of Hong Kong, Pokfulam Road, Hong Kong, China

RICHARD K.K. YUEN

Department of Building and Construction, City University of Hong Kong, Tat Chee Avenue, Hong Kong, China

S.M. LO

Department of Building and Construction, City University of Hong Kong, Tat Chee Avenue, Hong Kong, China

Under the Building Codes of Hong Kong Special Administration Region, the provision of refuge floors has been an indispensable element in their high-rise building design since 1996. A refuge floor is a designed temporary safe place to improve high-rise evacuation during fires. It is a fire-protected semi-permeable feature in buildings taller than 25 storeys. Direct cross natural ventilation by natural wind is an important design criterion of a refuge floor. It helps to prevent any smoke entering the refuge floor from logging and impairing the safe conditions of the floor. This paper reports a study of the wind flow patterns around a high-rise building with a refuge floor at wind incident angle perpendicular to the building. The study is based on CFD simulations which are qualified by wind tunnel data. The results revealed a smoke hazard potential if smoke is dispersed from the rear face of the building at a level closely below the refuge floor. Smoke may be logged inside the refuge floor area behind the internal blockage.

1. Introduction

Refuge floor requirements were introduced in the Building Code [1] of Hong Kong Special Administration Region (HK) since 1996. In buildings taller than 25 storeys, floors at every 25th level are to be totally separated from the other areas in the building by fire rated constructions. The separation region acts as an intermediate temporary safe place for evacuees to take a brief rest in their long descends during fires. At least 50% of this floor area is designated for refuge. At least two external walls must be opened and the semi-permeable refuge floor should allow direct cross natural ventilation of the space. This arrangement is a critical design feature to prevent smoke, if entered, from becoming logged inside the floor. Obviously, logging of smoke can fail the desired safe condition of the refuge floor.

Natural ventilation is a popular design tool in improving the thermal comfort condition of the internal built space [2]. While a substantial amount of studies have been done to identify the wind structure over the building envelope, there is scare detailed information on the behaviors of wind flow through an internal floor space. In early 1990's, computational fluid dynamics (CFD) study carried out by Satoh and Kuwahara [3] predicted that shearing to the buoyant gas flow by wall surface took up the gas to re-enter a floor at a higher level where an opening was opened on the path of buoyant flow. In 1992, wind tunnel study carried out by Lam [4] pointed out that wind flow onto a tall building was not significantly affected by having added a permeable floor at its mid-height level. After the refuge floor requirement was launched in Hong Kong, a CFD study was carried out by Yuen *et al* [5] and it was showed that if a constant low-velocity flow hits a three-storey building section where a refuge floor was built in its middle level, smoke dispersed from the lowest level might be brought by the velocity field to enter and

† Work fully supported by a grant CityU 1022/01E of the Research Grant Council of Hong Kong.

1 currently PhD candidate at School of Design, Faculty of Built Environment and Engineering, Queensland University of Technology, Brisbane, Australia

become logged under the ceiling soffit of the refuge floor. Furthermore, CFD studies carried out by Lu *et al* [6-8] showed that natural wind might escape the refuge floor via the rear face opening if only one internal blockage, i.e., services core, is built at its center. But wind will be logged between the blocks surrounding the center services core, if added, on the floor.

Although some important flow features were revealed by the CFD results, the fundamental physics of wind flow has not yet been studied and discussed. There is a lack of physical information of wind flow to assure the quality of CFD predictions. In this study, wind tunnel data of wind flow at an angle perpendicular to a high-rise building with a refuge floor at its mid-height level is presented. The data provide a basis of evaluation of the quality of CFD predictions carried out previously by some authors of this paper. Through an analysis of the CFD patterns of velocity and pressure coefficients, we attempt to provide an understanding of the effect of wind in ventilating a semi-permeable building feature and the desirable ventilation effect in a refuge floor design.

2. Setup and Procedures

2.1 *Wind tunnel test*

Wind tunnel experiment is a popular and accepted method in wind studies [9-11] and assessment [12]. The results provide qualitative physical data of wind flow and wind effects on buildings. This study was made in the boundary layer wind tunnel of the Department of Civil Engineering at the University of Hong Kong. It is a recirculating tunnel with a working section 3 m wide, 1.7 m tall and 11 m long. In the section, spires and roughness elements were used to provide at the model location a simulated atmospheric boundary layer of the open to suburban terrain type. In order to provide a large model size of the relative small space of a refuge floor in a high-rise building model, a large scale at 1:150 was used for the building model. The model was constructed with wooden panel. In full-scale, the building was a 40-storey high-rise square building of height 124.5 m (H) and width 20 m (B). The frontal area of the building model covered 3% of the cross sectional area of tunnel working section. A refuge floor was built on the 25th storey of the building. It has the basic feature of a rectangular block representing a main services core in the center covering 50% of the floor area. Two sides of its external walls were opened. However, to enhance the number and location of measurement points inside the gap-like refuge floor area, a distorted scale was used for the refuge floor. The height of the refuge floor provided in the model has a full-scale value of about 3.75 m (0.03 H; height of the building) which was taller than the minimum dimension of 2.5 m as required by the Code (Figure 1).

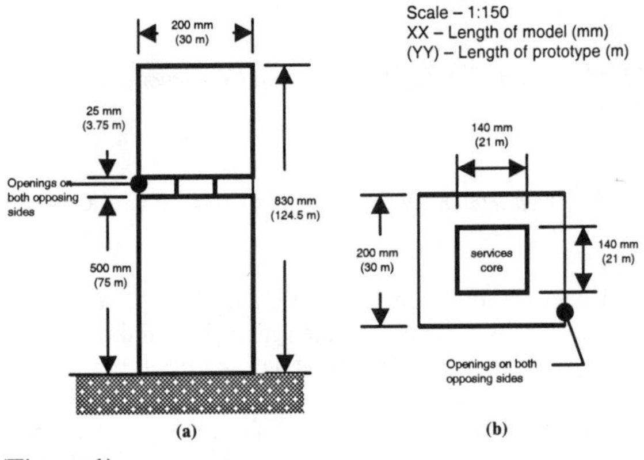

Figure 1. Model building

Wind velocities on the vertical mid-width plane of the building model, and on the half horizontal plane at 0.62 H level; i.e. 0.515 m above ground, were measured by miniature X-wire probes (Dantec 55P61, 55P63 and 55P64). The selected probes possessed miniature sensors which could measure instantaneous two-component wind velocities on the selected measurement planes. The hot wires were operated with a constant temperature anemometer. The output voltages from the anemometer were collected and converted inside a computer, which also performed probe calibration and analysis of wind tunnel data.

Tests were performed with the mean wind speed (U_h) and turbulence intensity (I_h) equal to 10.5 m/s and 0.088 at the roof height H_h of the building model 0.85 m above the wind tunnel floor. The measured mean wind speed and turbulence intensity profiles were found to follow the power law with an exponent of 0.19 for the wind speed and with an exponent of −0.47 for turbulence intensity (Figure 2).

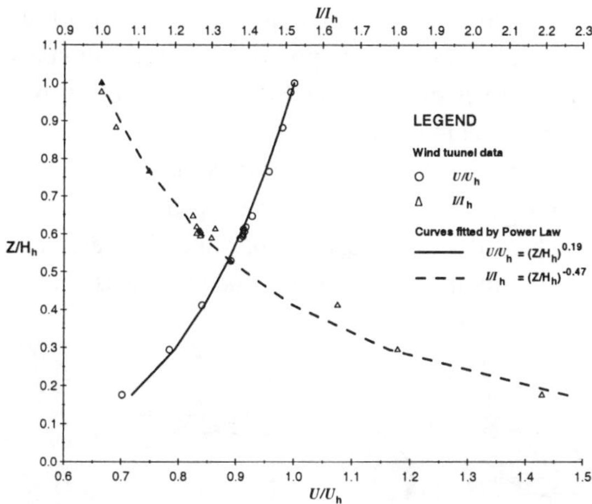

Figure 2. Measured wind profile

In each measurement, the reference wind speed was taken by a pitot-static tube mounted at some height above the building model. This enabled the wind velocity components U, V and W to be normalized by the mean wind speed at H_h (U_h) as:

$$\frac{U \ (or\ V\ or\ W)}{U_h} \tag{1}$$

2.2 *CFD simulation*

Apart from physical model tests, CFD has been increasing used as a tool for wind studies [13-17]. The commercial Code; Ansys CFX 5, was used in this study. To optimize the computing resource required in this study, the length of the flow domain was arranged to cover 1H in front and 6H behind the building model. The width and height of the flow domain were set according to the test section size of the wind tunnel. The domain was digitized into unstructured meshes comprising 3,294,714 tetrahedrons. Flow entered the domain with inlet boundary conditions according to the power law equations of U and I showed in Figure 2. The mean velocity components V and W were taken as zero at the inlet. At the outlet, influences to the flow due to blunt building surfaces vanished so that the relative static pressure was taken to be zero and the turbulence intensity returned to that at the inlet.

Turbulence of flow was modeled by two closure equations following the k-ε model [18]. The model used the gradient diffusion hypothesis to relate the Reynolds stresses to the mean velocity gradients and the turbulent viscosity. Turbulent viscosity was modeled as the product of a turbulence velocity and turbulent length scale. While the turbulence velocity scale was calculated from solving the transportation equation for turbulent kinetic energy, the turbulent length scale was estimated from solving the transportation equations for the turbulent kinetic energy (k) and its dissipation rate (ε).

The list of governing equations in dimensionless form is shown below [19]:

- Continuity equation:

$$\frac{\partial \rho}{\partial t} + \nabla \bullet (\rho U) = 0 \tag{2}$$

- Momentum transportation equation:

$$\frac{\partial \rho U}{\partial t} + \nabla \bullet (\rho U \otimes U) - \nabla \bullet (\mu_{eff} \nabla U) = \nabla p' + \nabla \bullet (\mu_{eff} \nabla U)^T + B$$

where $p' = p + \frac{2}{3} \rho k$, $\mu_{eff} = \mu + \mu_t$, and $\mu_t = C_\mu \rho \frac{k^2}{\varepsilon}$ (3)

- Transportation equation of k:

$$\frac{\partial (\rho k)}{\partial t} + \nabla \bullet (\rho U k) = \nabla \bullet \left[\left(\mu + \frac{\mu_t}{\sigma_k} \right) \nabla k \right] + P_k - \rho \varepsilon \tag{4}$$

- Transportation equation of ε:

$$\frac{\partial (\rho \varepsilon)}{\partial t} + \nabla \bullet (\rho U \varepsilon) = \nabla \bullet \left[\left(\mu + \frac{\mu_t}{\sigma_\varepsilon} \right) \nabla \varepsilon \right] + \frac{\varepsilon}{k} (C_{\varepsilon 1} P_k - C_{\varepsilon 2} \rho \varepsilon) \tag{5}$$

- Model constants:

$C_{\varepsilon 1} = 1.44$, $C_{\varepsilon 2} = 1.92$, $C_\mu = 0.09$, $\sigma_k = 1.0$, and $\sigma_\varepsilon = 1.3$

Predicted velocities on the sectional planes as in the wind tunnel tests were normalized by U_h produced in the simulations. Patterns of pressure coefficient (C_p) on these planes were calculated from the static pressure as:

$$C_p = \frac{P - P_h}{\frac{1}{2} \rho (U_h)^2} \tag{6}$$

where P_h is the pressure at height H_h at the inlet

3. Analysis and Implementation

Figures 3 and 4 show a comparison between CFD and hot-wire wind velocities on the two selected horizontal and vertical planes of flow around the building model. Hot-wire measurement is known to suffer from uncertainties in recirculating, re-attaching flows and in situations where mean wind velocity hits the direction of measuring probe with an angle of attack larger than 30°. However, results of some bench-mark studies of wind flow around building-like bluff bodies showed that hot-wire measurements were able to capture the main flow features [20, 21]. These features include separation of wind flow over the front roof corner and past the external side wall corners, and re-attachment of flow after the point of separation and in the building wake region. The CFD simulation results in this paper are numerically acceptable as the root mean square residuals of mass and velocities reported are below 0.000001.

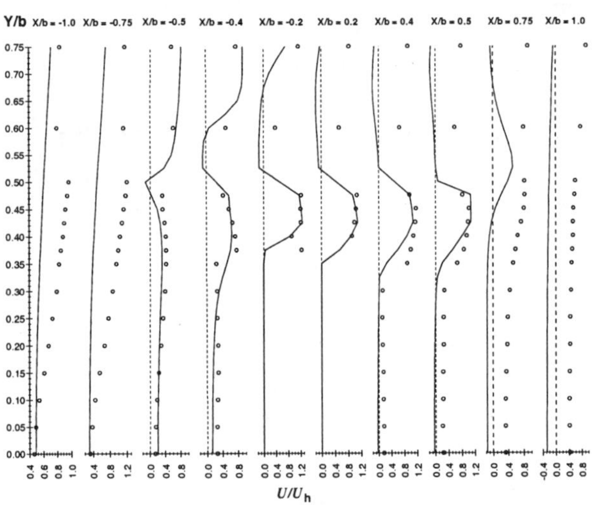

o wind tunnel data,; —— CFD prediction;········· "0" reference line

Figure 3. Pattern of U/U_h of vertical mid-width plane of the high-rise building

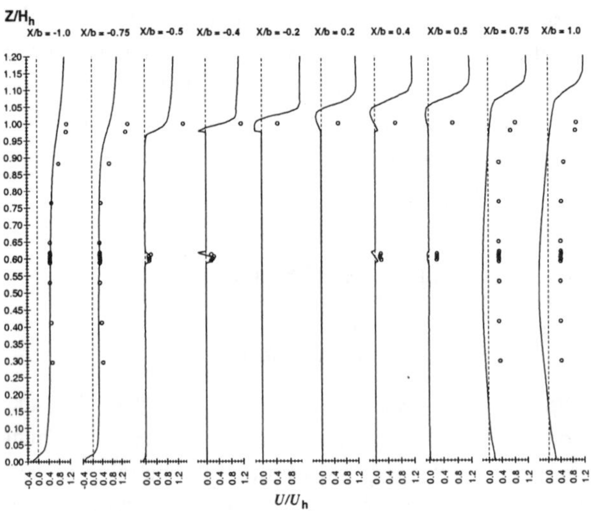

o wind tunnel data,; —— CFD prediction;········· "0" reference line

Figure 4. Pattern of U/U_h of horizontal half plane at 0.62 H_h level

In Figures 3 and 4, CFD under-predicts U/U_h when flow starts to separate from the front surface of the building. Apart from the regions of flow separation and wakes, wind velocities from CFD and wind tunnel show good agreement to each other.

Figures 5 and 6 show the wind vectors and streamlines from CFD on the two planes. The CFD results show the main features of wind flow around a high-rise building as in past studies results in the literature.

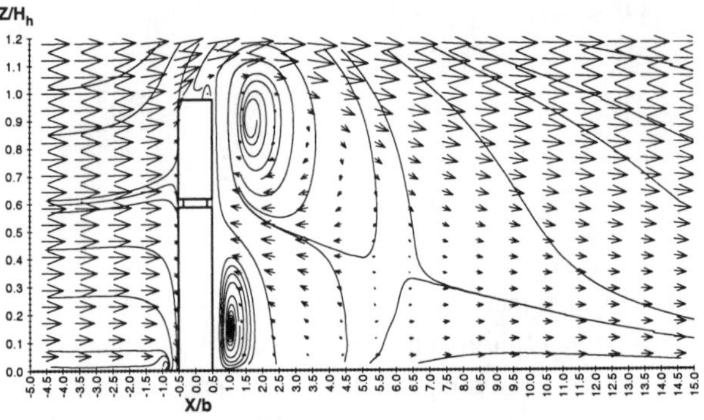

Figure 5. Pattern of mean velocity vector and streamline on the vertical mid-width plane of the building

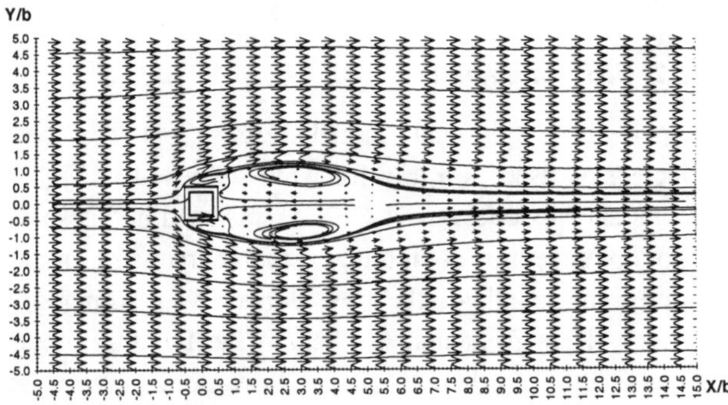

Figure 6. Pattern of mean velocity vector and streamline on the horizontal plane at 0.62 H level

3.1 *Wind flow on selected vertical plane*

Figure 5 shows that when wind blows normal to the front face of the building, stagnation of wind on the front wall is observed at Z/H_h = 0.73 (30th storey). Below this point, wind is brought down along the building face to low levels. On the building rear wall, another stagnation of wind is observed at Z/H_h = 0.52 (22nd storey). Above this point, wind is brought up to the high levels of the building. In Figure 6, a pair of counter-rotating vortices is observed in the building near-wake region on the vertical plane on the refuge floor level where wind re-attaches onto the building rear face. Figure 7 shows the detailed CFD flow patterns inside the refuge floor on the vertical plane. While wind flows down from the point of stagnation on the front wall, it enters the refuge floor and a counter-clockwise rotation of flow is observed under the ceiling soffit at mid-width section of the floor. On the other hand, while wind flows up along the rear wall, low-velocity wind re-enters the refuge floor into the near wake of the internal blockage and forms weak clockwise flow circulation.

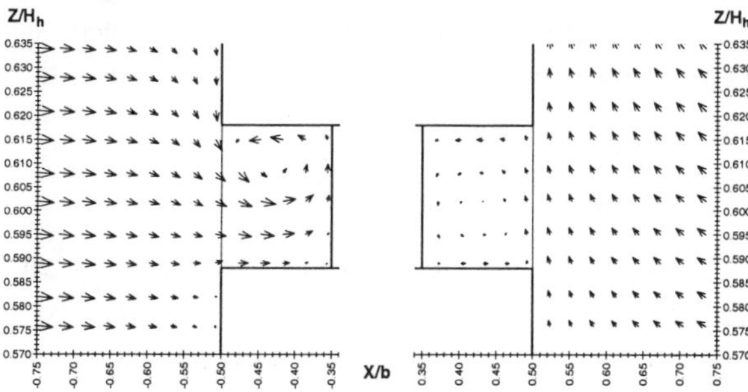

Figure 7. Expanded view of mean velocity vector pattern of vertical plane on the refuge floor

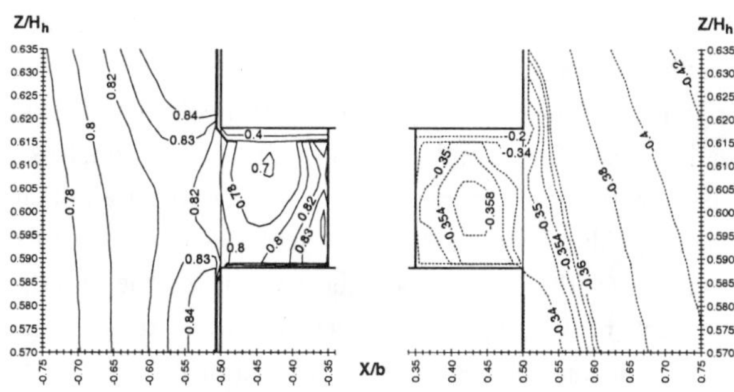

Figure 8. Expanded view of pressure coefficient pattern of vertical plane on the refuge floor

Wind pressure distributions are shown in Figure 8. A comparison with Figure 7 suggests that flow does follow the pressure distributions to enter and re-entering the refuge floor via its front and rear openings, respectively. Wind is induced by the high pressure region, $C_p = 0.82$, near the front opening of the refuge floor to flow into the floor space where the pressure is low at $C_p = 0.7$. Then as wind hits onto the front face of the internal blockage, wind pressure increases to $C_p = 0.83$. At the rear side of the building, a small pressure difference exists between $C_p = -0.35$ at the rear of the internal blockage and $C_p = -0.34$ at the rear opening of the refuge floor.

3.2 *Wind flow on selected horizontal plane*

In Figure 6, it is observed that separation of wind flow occurs at the building side corners. There exists a large wake behind the building which extends a distance of some building widths. Flow recirculation is observed inside this wall region from the mean flow field. It is in form of a counter-rotating vortex pairs. In between the pairs of vortices at mid-width of the building, wind is flowing backwards to the rear face of the building.

Figure 9 shows the computed flow through the refuge floor. A stagnation point is observed at the center of the front face of the internal blockage. From this point, wind entering the floor flows sideward and escapes around the sides of the internal blockage. The flow is then channeled into the two corridors formed by the internal and external walls of the refuge floor.

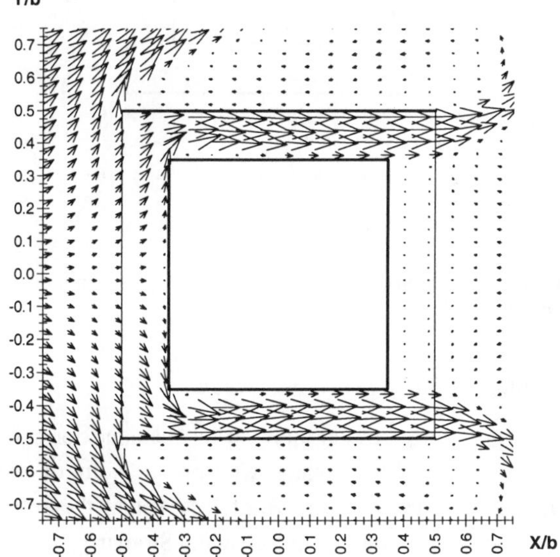

Figure 9. Expanded view of mean velocity vector pattern of horizontal plane at 0.62 H_h.

On leaving the corridors, the computed flow I observe in Figure 9 to curve outwards into the rear of the building. There is little flow into the region behind the internal blockage but the vectors show backward flow towards the rear face of the internal blockage. This flow then flowing sideward and join the escaping flow from the corridors.

Figure 10 shows the computed pressure distribution on the refuge floor. Similar to the findings observed in Figure 8, wind pressure at C_p = 0.8 in front of the front opening pushes wind to enter the refuge floor. The large pressure difference along the corridors is responsible for the fast channeled flow through the corridors. Behind the building, the small wind pressure difference between C_p = −0.4 in the near wake and C_p = −0.355 behind the internal blockage drag wind backwards into the refuge floor.

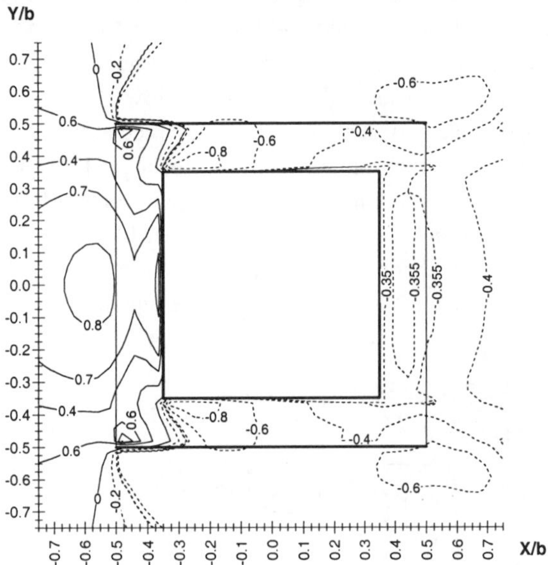

Figure 10. Expanded view of pressure coefficient pattern of horizontal plane at 0.62 H_h

3.3 *Implication to smoke movement*

From the above results under wind incident normal to the building, wind brought down from the point of wind stagnation at the building front face enters the refuge floor at the building mid-height. Similarly, at the building rear face, flow brought up from the stagnation point also takes wind into the floor space. Channel-like of flow created in the side corridors on the refuge floor sweeps air through the building. However, in the space behind the internal blockage, the flow re-entering into the refuge floor has very low speed.

If smoke is dispersed from a level just above the refuge floor level on the building front wall, it may be taken by the wind flow to enter the refuge floor space. According to the wind flow patterns found in this study, most wind flowing into the refuge floor is led to escape to the rear of the building through the side corridors. This will relieve smoke logging on the refuge floor. This postulation is similar to the prediction done by Lu *et al* in the early 2000. However, when the approaching wind has very low speed, wind flow may not have sufficiently large ventilation capacity to clear the smoke inside the space behind the internal blockage. This postulation gives a supplement to the prediction done by Yuen *et al* in the late 1990's.

More importantly, if smoke is dispersed from a level just below the refuge floor on the building rear face, wind flow may bring the smoke into the refuge floor inside the space behind the internal blockage. The almost stagnant flow condition there may keep the smoke logged for a long time. As a consequence, this area is a potential space of unsafe refuge.

4. Concluding Remarks

In this study, the behaviors of wind flow around a high-rise building with a refuge floor are investigated with wind tunnel and CFD experiments. CFD predictions are validated through a comparison with wind tunnel data and past data in the literature. Wind profiles in the wind tunnel are used in the CFD computation.

With wind blowing normal to the high-rise building, wind entering into the refuge floor from the front face opening is found to escape the building at its rear face opening mainly in region close to the two external side walls. The flow path is through the channel-like passages inside the corridors bridging the front and rear spaces of the floor. However, whether wind can totally escape the floor space is questioned by the almost stagnant flow condition observed in the region just behind the internal blockage. This region is a potential point of unsafe refuge if smoke is dispersed from the building rear face at a level just below the refuge floor. For wind incidences other than normal, different flow patterns will result and more studies are needed.

Acknowledgments

The first author wished to thank Professor Eddy Leonardi of School of Mechanical and Manufacturing Engineering, University of New South Wales, Sydney, Australia, Dr Nur Demirbilek and Mr Jack Williamson of School of Design, Queensland University of Technology, Brisbane, Australia. Their invaluable mentoring and advises are very important in his study.

References

1. Building Authority of Hong Kong, MOE: 1996, *Code of Practice of the provision of means of escape in case of fire*, 29-33 (1996).
2. Aynsley, R.M., Melbourne, W. and Vickery, B.J., *Architectural aerodynamics*, 180-236 (1977).
3. Satoh, K. and Kuwahara, K., *Proceedings of the 3rd International symposium on Fire Safety Science*, 355-364 (1991).
4. Lam, K.M., *J. Wind Engg. & Ind. Aerodynamics*, 41-44, 2313-2314 (1992).
5. Yuen, K.K., Lo, S.M. and Yeoh, G.H., *Proceedings of Interflame 99'*, 1273-1280 (1999).
6. Lu, W., Lo, S.M., Fang, Z. and Yuen, K.K., *Building and Environment*, 36, 219-230 (2001).
7. Lu, W., Lo, S.M., Yuen, K.K. and Fang, Z., *Intl. J. CFD*, 14, 327-337 (2001).
8. Lu, W., Lo, S.M., Fang, Z. and Yuen, K.K., *Intl. J. CFD*, 15, 169-176 (2001).
9. Souster, C.G. and Lee, B.E., *Dept. of Building Science, U. Sheffield, Research report No. BS 26* (1975).
10. Pitts, A.C. and Ward, I.C., *Dept. of Building Science, U. Sheffield, Research report No. BS 69* (1983).
11. Lam, K.M. and To, A.P., *J. Wind Engg. & Ind. Aerodynamics*, 54-55, 345-357 (1995).
12. Buildings Department of Hong Kong, PNAP 150 (1994).
13. Paterson, D.A. and Apelt, C.J., *Dept. of Civil Engg., U. Queensland, Research report No. CE 63* (1985).
14. Paterson, D.A. and Apelt, C.J., *Proc. Intl. Sym. Computational Fluid Dynamics*, 589-598 (1987).
15. Stathopoulos, T. and Baskaran, B.A., *Engg. Structure*, 18(11), 876-885 (1996).
16. Murakami, S., *J. Wind Engg. & Ind. Aerodynamics*, 74-76, 1-24 (1998).
17. Senthooran, S., Lee, D.D. and Parameswaran, S., *J. Wind Engg. & Ind. Aerodynamics*, 92, 1131-1145 (2004).
18. Launder, B.E. and Spalding, D.B., *Computer Methods In Applied Mechanics And Engg.*, 3, 269-289 (1974).
19. CFX-5 reference manual, 63-65 (2001).
20. Aroussi, A. and Ferris, S.A., *Proc. 2nd Intl. Conf. on Laser Anemometry – Advances and Applications*, 175-188 (1987)
21. Rodi, W, *J. Wind Engg. & Ind. Aerodynamics*, 69-71, 55–75 (1997).

WIND-INDUCED VIBRATION RESPONSE ANALYSIS OF LONG SPAN STEEL SPATIAL CORRIDOR

YIMIN ZHENG

Architectural Design & Research Institute of Tongji University, Siping Road 1239 Shanghai 200092, China

XIONGTAO CHENG

Department of Building Engineering, Tongji University, Siping Road 1239 Shanghai 200092, China

Aimed at the wind-induced vibration of the long span steel spatial corridor, based on FEM and data obtained from the wind tunnel test, an analysis of time domain is used to calculate the wind-induced vibration factor of along wind displacement of the corridor for Hangzhou citizen. Furthermore, the changing law of wind-induced vibration factor of the displacement of long span steel spatial corridor is also obtained. Characteristics of the dynamic wind-load factor are drawn.

1. Introduction

Along with the improved requirements for tall building structural types and usage functions, single building is sometimes designed to be connected to each other through different joining spatial corridor. One of the commonly used joining structure is the spatial corridor. Because the corridor is usually long-spanned and high in the air ,the corridor structure is similar to a bridge supported by the single building. The natural periods of this kind of corridor structures are commonly closely spaced and the first natural period is very close to the predominant period of the wind. Meanwhile, the corridor structures are lightly damped and relatively flexible, and are very sensitive to the wind force. Thus the wind-induced vibration of corridor structures subjected to the fluctuating wind should be considered in the design. The determination of wind vibration factor is very important for an effective and economical design of the corridor structures. Unfortunately, although the wind vibration factors of ordinary high-rise buildings and tower structures are stipulated in Chinese Load Code for the Design of Building Structure(GB50009-2001), the wind vibration factor of long span steel spatial corridor is not specified. This paper analyses the corridor in time domain by treating Hangzhou citizen center as engineering background, and get the wind vibration response of corridor in all wind direction.

The main body structure of Hangzhou citizen center is composed of arc-shaped towers, which distribute around the center wreath. The six towers are connected to each other through six steel spatial corridors at the 23rd and 25th story of each tower, which is at an altitude of about 90 meters. Similar structures are so far rarely met in the domestic and foreign engineering practices. This building is located near the qiantang river and wind load is very large, so the wind load is dominating in the structure design. It is necessary to conduct rigid body model based on wind tunnel test for the building. The surface wind pressure and shape coefficient of the building are obtained in the wind tunnel test. Wind vibration response analysis is then performed based on the wind tunnel test in time domain.

Figure 1 The wind tunnel test model of Hangzhou citizen center

2. Wind-induced Vibration Response Analysis in Time Domains

2.1 *Theory of structural dynamics*

This paper does dynamic analysis of structure using Hilber-Hughes-Taylor alpha method. According to motion-equation of structure, the recurrences of structure response vectors are derived from t to $t+\Delta t$. We start from $t=0$, the response vectors at every time in that can be obtained step by step.

There are three unknown vectors at $t+\Delta t$. They satisfy the motion-equation, and can be expressed as follow:

$$[M]\{\ddot{q}_{t+\Delta t}\}+(1+\alpha)[C]\{\dot{q}_{t+\Delta t}\}+[K]\{q_{t+\Delta t}\}=\{F_{t+\Delta t}\} \quad (1)$$

The velocity and displacement at $(t,t+\Delta t)$ can be expressed as follow:

$$\{\dot{q}_{t+\Delta t}\}=\{\dot{q}_t\}+\Delta t\left[(1-\gamma)\{\ddot{q}_t\}+\gamma\{\ddot{q}_{t+\Delta t}\}\right] \quad (2)$$

$$\{q_{t+\Delta t}\}=\{q_t\}+\Delta t\{\dot{q}_t\}+0.5\Delta t^2\left[(1-2\beta)\{\ddot{q}_t\}+2\beta\{\ddot{q}_{t+\Delta t}\}\right] \quad (3)$$

where $\alpha\in\left[-\frac{1}{3},0\right]$, $\gamma=(1-2\alpha)/2$ and $\beta=(1-\alpha)^2/4$. We can get $\{\ddot{q}_{t+\Delta t}\}$ and $\{\dot{q}_{t+\Delta t}\}$ using $\{q_{t+\Delta t}\}$、$\{\ddot{q}_t\}$、$\{\dot{q}\}$ 和 $\{q_t\}$ from equation (2)and (3).

$$\{\ddot{q}_{t+\Delta t}\}=\frac{1}{\beta\Delta t^2}(\{q_{t+\Delta t}\}-\{q_t\})-\frac{1}{\beta\Delta t}\{\dot{q}_t\}-\left(\frac{1}{2\beta}-1\right)\{\ddot{q}_t\} \quad (4)$$

$$\{\dot{q}_{t+\Delta t}\}=\frac{\gamma}{\beta\Delta t}(\{q_{t+\Delta t}\}-\{q_t\})+\left(1-\frac{\gamma}{\beta}\right)\{\dot{q}_t\}+\left(1-\frac{\gamma}{2\beta}\right)\Delta t\{\ddot{q}_t\} \quad (5)$$

Substitute equation (4) and (5) in equation (1) at $t+\Delta t$, we get equation (6)

$$[K^*]\{q_{t+\Delta t}\}=\{Q^*_{t+\Delta t}\} \quad (6)$$

Where

$$[K^*]=[K]+\frac{1}{\beta\Delta t^2}[M]+\frac{\gamma}{\beta\Delta t}[C] \quad (7)$$

$$\{Q^*_{t+\Delta t}\}=\{Q_{t+\Delta t}\}+[M]\left[\frac{1}{\beta\Delta t^2}\{q_t\}+\frac{1}{\beta\Delta t}\{\dot{q}_t\}+(\frac{1}{2\beta}-1)\{\ddot{q}_t\}\right] \quad (8)$$
$$+[C]\left[\frac{\gamma}{\beta\Delta t}\{q_t\}+\left(\frac{\gamma}{\beta}-1\right)\{\dot{q}_t\}+(\frac{\gamma}{2\beta}-1)\Delta t\{\ddot{q}_t\}\right]$$

So we get $\{q_{t+\Delta t}\}$ by solving the above equations.

2.2 *Displacement wind vibration factor calculation formula*

Wind vibration factors are usually largely different in different parts of structure, and it also apparently varies with the change of wind direction. But this difference is small using displacement wind vibration factor; we can use one wind vibration factor for the whole or local part of structure. Supposing along-wind static displacement response induced by the mean wind is $R_s(z)$ and along-wind dynamic displacement response induced by fluctuating wind is $R_d(z)$. The total response induced wind load is:

$$R_a(z)=R_s(z)+R_d(z) \quad (9)$$

Static displacement induced by mean wind is commonly amplified a coefficient which is greater than 1.Its result is equal to the total response by induced mean wind and fluctuating wind .we set the factor β as displacement wind vibration factor, which means the ratio between the total response induced wind load and static response induced by mean wind load:

$$\beta(z)=\frac{R_s(z)+R_d(z)}{R_s(z)}=1+\frac{R_d(z)}{R_s(z)} \quad (10)$$

Where mean wind load is static force, and the static displacement induced is easily obtained by static force analysis method. The dynamic response induced fluctuating wind load can be computed using time domain analysis method. This paper does transient analysis to structure and get structural response of time history based on FEM and the wind load time history data of every measuring point. At last, we compute displacement wind vibration factor. Some research indicates that the same internal force of structure can be obtained by using displacement wind vibration factor instead of load wind vibration factor of load code .So if displacement wind vibration factor can be computed, we still can do structure wind-resistant design according to the formula used in china load code (GB50009-2001)

$$w(z)=\beta(z)\mu_s(z)\mu_z(z)w_0 \quad (11)$$

2.3 *FEM modelings*

The long span steel spatial corridor of Hangzhou citizen center is mainly made up of two arc huge longitudinal trusses. The corridor is about equal to the height of two floors, seeing Figure2. To assure the lateral stability and resist the torsion effect of structure, we arrange four lateral trusses between two arc longitudinal trusses instead of architectural partition wall. By arranging intercross diagonal brace to form horizontal trusses, structure becomes "closed body system". Because of the requirement of architectural function, there is a large hole in the middle floor. It results in a weak constraint of the horizontal truss. To enhance the lateral stability of the corridor, the horizontal truss is arranged by the side of the middle floor hole. The corridor is pinned to the main structure. The parameters of the truss are listed as follow:

Table 1 Corridor of long span (52.114m)

	Outer belt truss	Inner ring truss
Span (m)	57.280	47.012
Height (m)	7.4	7.4
Ratio of span and height	7.74	6.35
The biggest section of top boom	□700×400×28	□700×400×24
The biggest section of bottom boom	□700×400×28	□700×400×24
The biggest section of web member	□400×400×28	□400×400×24

□ -square steel pipe

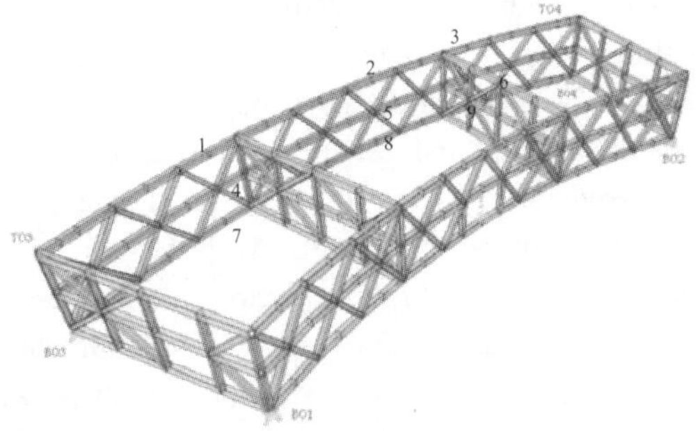

Figure 2 Longitudinal and lateral trusses and bearings

According to the arrangement of measuring point on rigid body models in wind tunnel tests, we create FEM model of the total corridor using SAP2000. All members are modeled using frame element, curtain wall and slab are modeled using shell element in finite element analysis. There are totally 1436 frame elements, 450 shell elements, and 447joints. The model is shown in figure3:

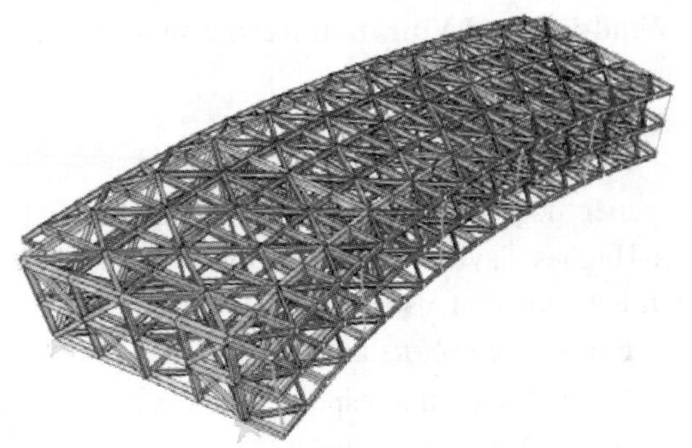

Figure 3 FEM model of long span steel spatial corridor

2.4 *Load information*

Wind pressure exists on the surface of both outer ring and inner ring of corridor. The time step is 0.209s, and the total number of time step is 6000. The static pressure coefficient on outer and inner ring of corridor surface can be computed as follow:

$$C_{pi}=\frac{p-p_{r\infty}}{p_{ro}-p_{r\infty}}\left(\frac{Z_r}{Z}\right)^{2(\alpha-\alpha_i)}=\frac{p}{q_r}\left(\frac{Z_r}{Z}\right)^{2(\alpha-\alpha_i)} \quad (12)$$

Where C_{pi} is static pressure coefficient of measuring point i; p is wind pressure acting on the measuring point i in wind tunnel test. p_{ro} and $p_{r\infty}$ are respectively the total pressure and static pressure; $q_r = p_{ro} - p_{r\infty}$ is dynamic pressure of reference point; Z is the height of measuring point above the ground. Z_r is the height of reference point corresponding to reference point of prototype above the ground.; α_i is mean wind speed profile power function exponent needed simulation i category wind terrain(i is B or C); α is actual mean wind profile power function exponent in terrain simulated in wind tunnel test. There is possibly little difference between α and α_i, these difference can effect the wind pressure coefficient, $\left(\frac{Z_r}{Z}\right)^{2(\alpha-\alpha_i)}$ can be used to correct these affect. Because the height of the reference point is different from the height of gradient wind, the following equation can be used to obtain the wind pressure time history acting on the real building. This equation is expressed as follow:

$$p_{ri}(t)=\left(Z_r/H_{G,i}\right)^{2\alpha_i}C_{pi}(t)w_{G,R} \qquad (13)$$

Where $H_{G,i}$ is the gradient wind height corresponding to $i(i=A,B,C,D)$ terrain; $w_{G,R}$ is real gradient wind pressure corresponding to R years recurrence period.

After the FEM model is created, the wind load time history can be spaced interpolated to obtain the wind pressure time history on the inner and outer ring surface of the corridor to satisfy the accuracy requirement of the finite element analysis.

By time domain analysis, we can get time history response U_i and response of standard variance σ of all joints:

$$\sigma^2=\frac{\sum_i^n\left(U_i-\bar{U}\right)^2}{n-1} \qquad (14)$$

Through equation (10):

$$\beta=1+\mu\frac{\sigma}{\bar{U}} \qquad (15)$$

where μ is peak value factor, $\mu=2.5$ here, $\bar{U}$ is displacement induced mean wind.

Except displacement wind vibration factor, we can also get some other statistics information, for example: dynamic displacement of structure, acceleration mean and root mean square, and some character about along-wind vibration response can also be obtained. The wind pressure time history of joint 5 is shown in Figure 4:

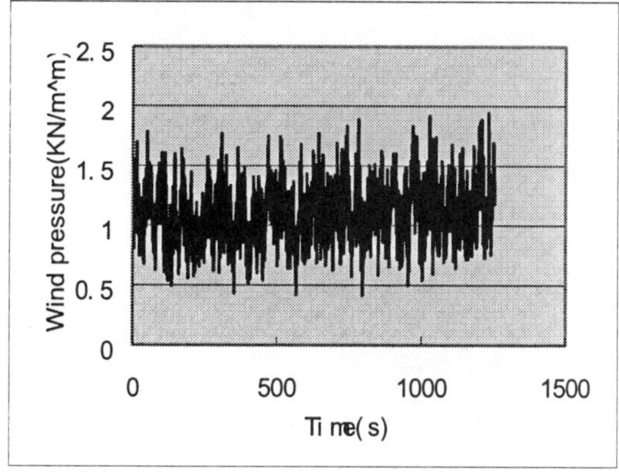

Figure 4 The wind pressure time history of joint 5

2.5 *Calculation result and wind vibration response analysis*

The wind directions of 0^0、30^0、60^0、90^0、120^0、150^0、180^0、210^0、240^0、270^0、300^0、330^0 are considered. Besides, two wind directions of juncture between B and C category wind field is also considered. So we totally computed wind vibration of the corridor in 14 wind directions. The wind direction 120^0 is normal to the corridor. For illustrating the character of structure along wind vibration response, we select 9 typical joints located in the middle and 1/4 span in analysis (seeing Figure1). The calculation results are listed are in Table 1. The wind vibration displacement time history of joint 5in the wind direction 120°C of is shown in Figure 5:

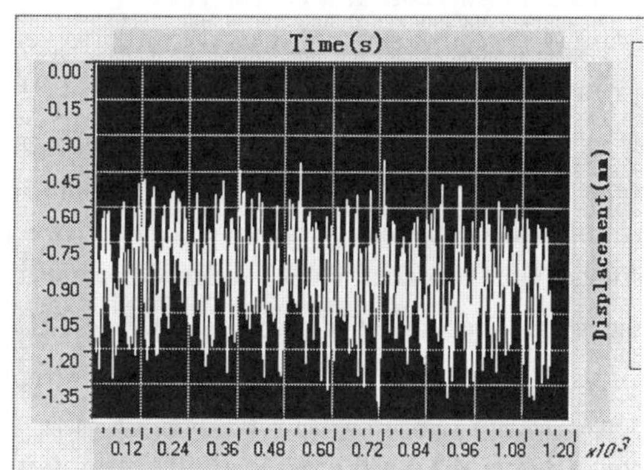

Figure 5 The wind vibration displacement time history of joint 5

Table 2 Displacement wind vibration factor of all joints in all wind direction.

	Joint1	Joint 2	Joint 3	Joint 4	Joint 5	Joint 6	Joint 7	Joint 8.	Joint 9
0°C	2.70	2.63	2.50	2.56	2.14	2.22	4.39	3.68	3.26
30°C	3.29	3.87	4.03	2.60	2.33	3.00	26.56	23.70	20.29
60°C	9.09	14.60	6.76	3.06	4.23	6.63	5.04	4.06	3.45
90°B	1.74	1.64	1.61	2.01	1.70	1.61	1.69	1.60	1.57
90°C	1.71	1.63	1.61	1.94	1.68	1.61	1.67	1.60	1.57
120°C	1.45	1.45	1.45	1.45	1.48	1.46	1.42	1.42	1.42
150°B	1.44	1.44	1.45	1.44	1.48	1.52	1.41	1.42	1.45
150°C	1.47	1.47	1.48	1.47	1.51	1.55	1.43	1.44	1.47
180°C	2.42	2.71	3.56	2.31	3.53	12.33	2.11	2.36	2.92
210°C	3.71	3.52	2.94	3.80	2.44	2.06	57.51	11.31	5.82
240°C	2.94	3.02	2.89	2.76	2.18	2.31	7.74	6.49	5.51
270°C	1.91	1.91	1.88	1.91	1.73	1.77	2.21	2.15	2.11
300°C	1.66	1.66	1.67	1.64	1.57	1.61	1.74	1.73	1.74
330°C	3.16	3.17	3.22	3.16	3.27	3.28	2.96	3.01	3.06

Table 3 Mean wind pressure of all joints in the wind direction of 0^0、30^0、60^0、210^0、240^0、330^0

	Joint1	Joint 2	Joint 3	Joint 4	Joint 5	Joint 6	Joint 7	Joint 8	Joint 9
0^0C	-0.19	-0.18	-0.19	-0.19	-0.18	-0.19	-0.16	-0.18	-0.19
30^0C	-0.20	-0.17	-0.17	-0.19	-0.17	-0.17	-0.20	-0.17	-0.16
60^0C	-0.31	-0.19	-0.11	-0.26	-0.19	-0.18	-0.31	-0.20	-0.11
210^0C	-0.17	-0.17	-0.22	-0.19	-0.19	-0.22	-0.14	-0.21	-0.25
240^0C	-0.20	-0.21	-0.22	-0.21	-0.21	-0.21	-0.19	-0.21	-0.22
330^0C	-0.28	-0.26	-0.25	-0.29	-0.28	-0.27	-0.27	-0.31	-0.25

From the above Table 2, one can see that the displacement wind vibration factor is approximately 1.45 in 120^0 wind direction. According to engineering experiences, the wind vibration factor is usually between 1 and 2, so the wind vibration factor is reasonable. From Table 3, it is apparent that mean wind pressure is very low ,some of then even near to zero in the wind direction of 0^0、30^0、60^0、210^0、240^0、330^0. It causes wind vibration response become lower. Because the proportion of fluctuating wind component in total component is larger ,we find that wind vibration coefficient suddenly become large in the wind direction of 0^0、30^0、60^0、210^0、240^0、330^0. The fluctuating component of structure response is apparent. Due to this reason, the joint response due to the mean wind is smaller than those of the other wind directions. The difference of the joint responses of different directions due to fluctuation wind is small. One can see from equation (10) that the value of β increases when the responses due to mean wind become small. So one can see that the displacement wind vibration factor increase suddenly in the wind direction of 0^0、30^0、60^0、210^0、240^0、330^0.

3. Conclusions

For large scale and complex structure wind vibration response, it is a good method to combine rigid body model test with time domain analysis. It is very precise and cheaper than aereoelastic model test. It should be noted that the quasi-steady assumption should be abandoned This paper start from the definition of displacement wind vibration factor, and do time domain analysis using FEM method. Its result is reasonable and effective. The wind loads used in dynamic analysis should be recorded time histories from rigid body based wind tunnel test.

For long span steel spatial corridor, wind vibration factor is too sensitive to the change of wind direction. When mean wind pressure is lower in those directions, wind vibration factor is bigger in contrast. So we should evaluate the worst direction, in which the multiplication of wind vibration factor and mean wind pressure is maximum.

According to computation result and observation of wind pressure distribution figure, it is apparent that displacement wind vibration factor is related to wind pressure distribution and fluctuating wind. (1) If the wind pressure distributions of different wind directions are similar with each other, the wind pressure amplitudes of different wind directions are also very close to each other. (2) If the wind pressure greatly differs from one wind direction to another, and the sign of the wind pressure varies with the fluctuation of wind pressure, the structural response will violently fluctuate and cause large changes in the value of displacement wind vibration factor.

References

1. Chinese Code for Loading on Building and other Structures, GB50009-2001,2001(in Chinese)
2. Hangzhou Citizen Center Wind Tunnel Test Research report, State Key Laboratory of Civil Engineering , Tongji University
3. He yan-li, Dong Shi-lin Wind response analysis method of spatial lattice structures in frequency domain with mode compensation[J], Spatial Structures 7(2) (2001) 3-9

INTEGRATION OF GPS WITH ACCELEROMETER FOR MEASURING TOTAL DISPLACEMENT RESPONSE OF TALL BUILDINGS

W.S. CHAN

Department of Civil & Structural Engineering, The Hong Kong Polytechnic University
Hung Hom, Kowloon, Hong Kong, China

Y.L. XU

Department of Civil & Structural Engineering, The Hong Kong Polytechnic University
Hung Hom, Kowloon, Hong Kong, China

X.L. DING

Department of Land Surveying & Geo-Informatics, The Hong Kong Polytechnic University
Hung Hom, Kowloon, Hong Kong, China

Accelerometer is a common sensor used to monitor displacement response of tall buildings under gust winds. Owing to the incapability of measuring static or quasi-static displacement response with accelerometers, Global Positioning System (GPS) has become an emerging sensor for measuring and monitoring both static and dynamic (termed total) displacement responses of tall buildings. However, the accuracy of total displacement measurement with GPS is affected by multipath signals. The current GPS is also incapable of accurately measuring high frequency displacement response. This study explores the possibility of integrating GPS signals with accelerometer signals to enhance the measurement accuracy of total displacement response of tall buildings. An interactive data processing procedure using both empirical mode decomposition (EMD) and adaptive filtering techniques is developed. In order to assess the effectiveness of the interactive processing method, a series of field measurement tests is performed on a site using a motion simulation table to generate various types of wind-induced dynamic displacement motions of a tall building in the pre-defined static position. In these tests, the GPS antenna and an accelerometer are installed on the same motion simulation table. The GPS and accelerometer data are processed with the interactive data processing method and are compared with the simulated motion. The test results show that the developed method can significantly enhance the measurement accuracy of total displacement response.

1. Introduction

Structural displacement is a key parameter to assess safety conditions of structures. Before the deployment of Global Positioning System (GPS), the structural responses were mainly monitored by accelerometers where displacements were obtained through either mechanical instrumentations or programmable software. The former method involves a signal conditioner that is used to accept and process signals from an accelerometer. The signal conditioner can output acceleration, velocity or displacement signals; where the velocity and displacement are respectively derived from single- and double-integration of measured acceleration response using built-in integration filter electronically. The latter method is a data processing scheme used to integrate acceleration data numerically. Among all programmable schemes provided, the scheme developed by the National Strong-Motion Program (NSMP) of U.S. Geological Survey is a common one used in earthquake engineering for monitoring structural responses during earthquakes.

Due to the high sampling rate, an accelerometer is able to extract dynamic motions of structures with natural frequency up to several tens hertz [1]. However, an accelerometer is insensitive to low frequency acceleration changes. Also, due to the unknown integration constants, high pass filter is used to cope with low frequency trends introduced during the integration process. Therefore, it is well recognized that an accelerometer is incapable of measuring static or quasi-static displacement response of tall buildings. GPS is advantageous in that it can provide three dimensional absolute position of a measurement point. It has become an emerging sensor for measuring and monitoring both static and dynamic displacement responses of tall buildings under gust winds. However, the dynamic displacement measurement accuracy of GPS in two orthogonal directions for civil engineering applications is still considered too low when the frequency of the signals is higher than 1 Hz [2]. As the performances of GPS and accelerometers are complimentary to each other, this paper explores the possibility of integrating GPS signals with accelerometer signals to enhance the measurement accuracy of total displacement response of tall buildings. The concept of integrating signals from GPS and accelerometers for structural deformation monitoring has been proposed by Roberts et al. [1]. The main characteristic of their integration approach is to apply accelerometer data set for bridging the gap between two GPS absolute positioning solutions and to use GPS data set for calibrating accelerometer drift during numerical integration process.

Focusing on the deformation monitoring of tall buildings, an interactive data processing procedure using both empirical mode decomposition (EMD) and adaptive filtering techniques is proposed in this paper. EMD developed by Huang et al. [3] is a method that can decompose any complicated data set into a small number of intrinsic mode functions (IMF) and a final residue. The method has been used to extract time-varying mean wind speed from typhoon-induced non-stationary wind speed for tall buildings [4]. An adaptive filter used as a signal decomposer is a method that can extract information of interest from the contaminated signal using the cross correlation between the reference and primary time series [5, 6]. Since GPS multipath effect repeats itself largely every sidereal day, this repetition characteristic was explored by Ge et al. [5] in an application of adaptive filter based on least mean square (LMS) algorithm for reducing the effect of multipath. LMS algorithm is simple but it is hardly applicable to non-stationary environment and it is slow to converge. The recursive least squares (RLS) algorithm is applied in this study.

In order to assess the effectiveness of this interactive processing method, a series of field measurement tests is performed on a site using a motion simulation table [2], GPS, and an accelerometer. A static test with the antenna stationary was first performed on the site which was susceptible to multipath effect to assess the background noise of GPS. The motion simulation table was then used to generate wind-induced dynamic displacement response of a tall building around a pre-defined static position, and the GPS and accelerometer measurements are carried out within the same time period as the static test but on the next sidereal day. After having obtained all kinds of measurement data, the interactive data processing scheme is then applied. The effectiveness of this integrated method is finally assessed through the comparison of the integrated results with the original motion generated by the motion simulation table.

2. Empirical Mode Decomposition

The starting point of EMD is to consider $x(t)$as the measured structural response time history to be analyzed. The upper and lower envelopes of $x(t)$ are constructed by connecting the local maxima and minima of $x(t)$, respectively, using a cubic spline function. The mean of the two envelopes is then computed and subtracted from the original time history. This procedure is termed a sifting process. The difference between the original time history and the mean value, c_1, is called the first IMF if it satisfies the following two conditions: (1) within the data range, the number of extrema and the number of zero-crossings are equal or differ by one only; and (2) the mean value of the envelope defined by the local maxima and the envelope defined by the local minima is zero. The difference between $x(t)$ and c_1 is then treated as a new time history and subjected to the same sifting process, giving the second IMF, $c_2(t)$. The EMD procedure continues until the residue becomes so small that it is less than a predetermined value of consequence, or the residue becomes a monotonic function. The original time history $x(t)$ is finally expressed as the sum of the IMFs plus the final residue.

$$x(t) = \sum_{j=1}^{N} c_j(t) + r(t)_N \tag{1}$$

where N = number of IMF components; and $r(t)_N$ = final residue.

The concept of the aforementioned decomposition is based on the direct extraction of the energy associated with various intrinsic time scales of the time history itself. Therefore, a number of mode mixing during the sifting process would be possible. A criterion, termed the intermittency check, was thus suggested by Huang et al. [7] to separate the waves of different periods into different modes based on the period length. This criterion is designed as the lower limit of the frequency that can be included in any given IMF component. It can be achieved by specifying a cutoff frequency ϖ_c for each IMF during its sifting process, in which the data having frequencies lower than ϖ_c will be removed from the resulting IMF.

3. Adaptive Filter

An adaptive filter used as a signal decomposer (Fig. 1) is operated on the information from two measurement inputs with the same length *N*: a primary measurement $p(i)$ that contains the desired signal of interest $s(i)$ contaminated by noise $n(i)$ and a reference measurement $r(i)$ that supplies a signal $s'(i)$ polluted with noise $n'(i)$. In order to isolate the information of interest $s(i)$ from polluted primary input, three conditions have to be satisfied: (1) the desired signal $s(i)$ and noise $n(i)$ in the primary input are uncorrelated with each other; (2) the noise $n'(i)$ in reference measurement is uncorrelated with the desired signal $s(i)$ but correlated in some way with the noise component $n(i)$ of the primary signal; and (3) the signal $s'(i)$ in reference measurement is uncorrelated with the desired signal $s(i)$ and noise $n(i)$ in the primary input [5].

In order to extract the information of interest $s(i)$ from the polluted primary input, it is shown from Fig. 1 that a transversal filter which operates on the reference measurement input is involved in the adaptive filter to produce an estimation of the noise $\hat{n}(i) = \mathbf{w}^T(i-1) * \mathbf{r}(i)$, also termed as a coherent component (i.e., a composition of the primary measurement correlated with a composition of the reference measurement) where $\mathbf{r}(i)$ is the input of reference vector with the length equal to M (filter length) at time i and $\mathbf{w}(i-1)$ is filter coefficient with the same size of reference vector. The estimation of the noise $\hat{n}$ is then subtracted from the primary measurement input to produce an estimate of the desired signal $\hat{s}(i) = p(i) - \hat{n}(i)$, also termed as an incoherent component (i.e., a composition of the primary signal which is uncorrelated with any composition of the reference signal). The incoherent component $\hat{s}(i)$ is then used to control the adjustments applied to the filter coefficient in the transversal filter. For RLS algorithm, the filter coefficient vector $\mathbf{w}(i)$ is updated by considering all the output up to current time instant i,

$$\varepsilon(i) = \sum_{j=1}^{i} \lambda^{i-j} \hat{s}^2(j) \tag{2}$$

where λ is the forgetting factor for affording the possibility of statistical variations of the observable data. A recursive equation for updating the least-squares estimation $\mathbf{w}(i)$ of the filter coefficient vector is

$$\mathbf{w}(i) = \mathbf{w}(i-1) + \mathbf{k}(i)\xi(i) \tag{3}$$

where $\xi(i) = d(i) - \mathbf{w}^{\mathbf{T}}(i-1)\mathbf{u}(i)$;

$$\mathbf{k}(i) = \frac{\lambda^{-1}\mathbf{\Phi}^{-1}(i-1)\mathbf{u}(i)}{1 + \lambda^{-1}\mathbf{u}^T(i)\mathbf{\Phi}^{-1}(i-1)\mathbf{u}(i)}; \text{ and}$$

$$\mathbf{\Phi}(i) = \sum_{j=1}^{i} \lambda^{i-j}\mathbf{u}(j)\mathbf{u}^T(j).$$

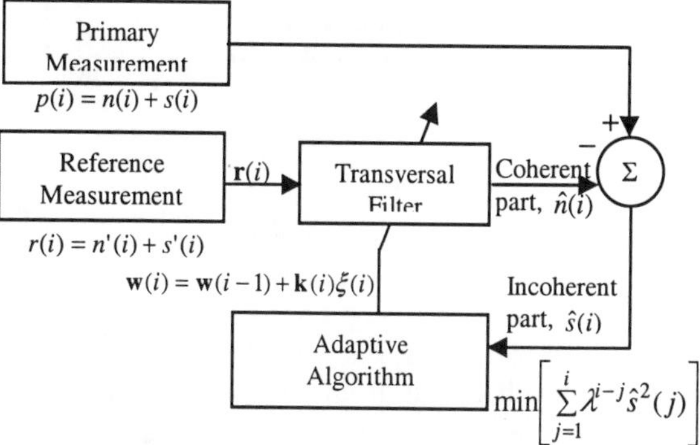

Figure 1. Adaptive filter.

4. Interactive Data Processing Approach

In the traditional wind speed model, the longitudinal wind speed $U(t)$ at a given height within the boundary layer is assumed to be composed of mean wind speed component $\overline{U}$ and a longitudinal fluctuating wind speed component $u(t)$. When a structure is placed in the atmosphere boundary layer, the flow of wind passing it will induce vibration and its displacement response can be regarded as sum of a mean static component $\overline{x}$ and a dynamic fluctuating component $x'(t)$ in alongwind direction.

$$x(t) = \overline{x} + x'(t) \tag{4}$$

The key issue in this paper is how to extract the aforementioned total displacement response from GPS and accelerometer measurements when used for structural deformation monitoring. Due to the inherent limitations of the two systems, accelerometer can be used for measuring the dynamic fluctuating component over a definite frequency ϖ_c whereas GPS can be applied to capture the completed static mean displacement $\overline{x}$ and the residue of dynamic fluctuating

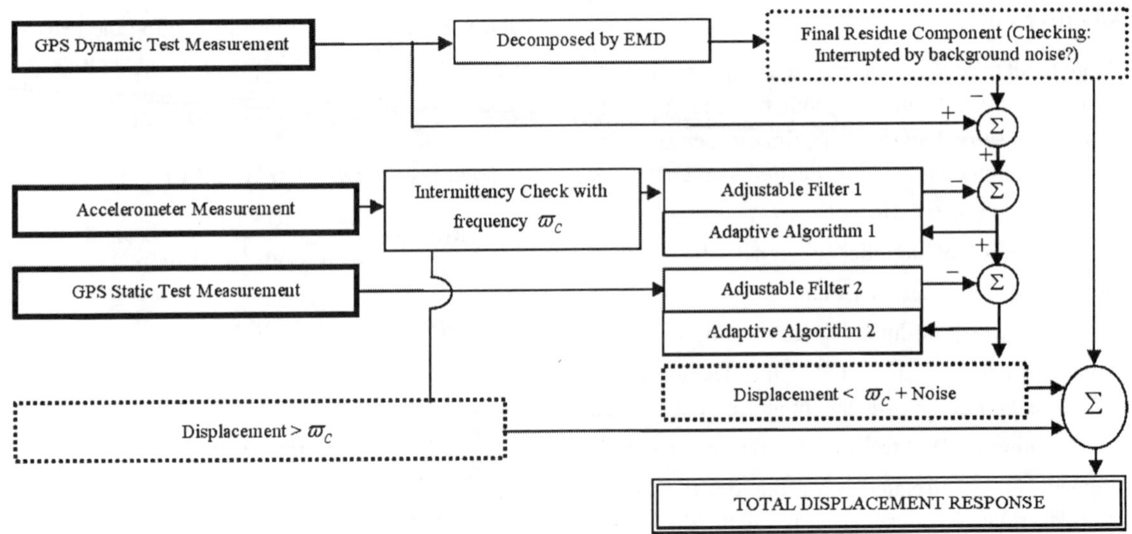

Figure 2. Algorithm of interactive data processing approach.

component. However, another main concern of how to mitigate multipath effects from the GPS measurements is addressed when the mean static displacement is involved. As multipath contains plentiful frequency components but distributes over a relatively low frequency range, the multipath effects are hardly mitigated in the process of adaptive filter when the multipath interference and mean static deflection are highly correlated with each other. To circumvent these difficulties, an interactive data processing algorithm using both EMD and adaptive filter is developed in this study.

Fig. 2 illustrates the interactive processing approach that combines signals from GPS and accelerometer together to obtain both static $\bar{x}$ and dynamic displacement responses $x'(t)$ of a large civil engineering structure. As indicated in Fig. 2, three types of data sets are involved: (1) dynamic; (2) static test measurements from GPS; and (3) displacement response obtained from accelerometer by double integrations of acceleration.

In the first stage, EMD is employed to a time history of the dynamic test measurement from GPS and decomposes it into a number of IMFs and final residue. As static deflection of a tall building is considered as a constant mean value, this static trend $\bar{x}$ must be the final residue $\bar{x}_d$ of the dynamic time history after EMD. In order to ensure this monotonic function $\bar{x}_d$ is not interrupted by the background noise (i.e. multipath), EMD is also employed to a time history of the static test measurement from GPS. If the final residue after decomposition of static measurement input is also a monotonic function $\bar{x}_s$, the mean static displacement $\bar{x}$ must be obtained by subtracting $\bar{x}_s$ from $\bar{x}_d$. The difference between the original time history of the dynamic test measurement from GPS and the mean static displacement $\bar{x}$ is considered as an untreated dynamic displacement. In the second stage, EMD with an intermittency frequency ϖ_c is implemented to extract high frequency signal measured from accelerometer.

After the processes of EMD, two stages of adaptive filter are required to obtain the low frequency dynamic displacement from GPS. In the first stage of filtering, the untreated dynamic displacement measured by GPS is used as a primary input while the accelerometer measurement after intermittency check is considered as a reference input. As the signal and noise components in both the primary and reference inputs listed in Table 1 fulfill all requirements mentioned in Sec. 3, the incoherent part from the first stage of adaptive filter is the dynamic displacement response with frequency not more than ϖ_c and noise (consisting of multipath and receiver noise) according to Fig. 1. This incoherent part is then used as the primary input in the second stage of filtering process in which the displacement with

Table 1. Primary and reference measurement inputs in two stages of adaptive filter.

Stage of Adaptive Filter		1	2
Primary Input	$p(i)$	untreated dynamic displacement time history	receiver noise+desired signal (low frequency)+multipath
	$s(i)$	receiver noise+desired signal (low frequency)+multipath	receiver noise+desired signal (low frequency)
	$n(i)$	desired signal (high frequency)	multipath
Reference Input	$r(i)$	Accelerometer measurement after intermittency check	GPS static test measurement
	$s'(i)$	high frequency noise	receiver noise
	$n'(i)$	desired signal (high frequency)	multipath in Day 1

stationary antenna measured from GPS within the same time period as the dynamic test but on the previous sidereal day is used as the reference input. Similar to the first stage of adaptive filter, the components in both measurement inputs (see Table 1) satisfy all conditions of adaptive decomposer. Therefore, the incoherent component in this stage of filtering would be the dynamic displacement measurement from GPS with the frequency less than ϖ_c and receiver noise. As the receiver noise is randomly distributed and small in amplitudes, this kind of error can be neglected [8]. The total displacement of a large civil engineering structure is thus a combination of final residue of the measurements from GPS obtained by EMD, high frequency dynamic displacement from accelerometer after intermittency check and low frequency dynamic displacement from GPS after adaptive filtering processes.

5. Data Analysis and Results

In order to assess the effectiveness of this interactive processing method in real application of tall buildings in urban environment, a series of field measurement tests is performed on the roof of Pao Yue Kong Library at The Hong Kong Polytechnic University using a motion simulation table, Leica GX1230 GPS receivers and a KYOWA ASQ-1BL accelerometer with a VAQ-500A signal conditioner at a sampling rate of 20 Hz. A static test with stationary antennae was first performed on the site to assess the background noise of GPS. 1-D white noise random wave, sinusoidal wave and the measured wind-induced dynamic motion of Di Wang Building during Typhoon York [9] was then simulated by the motion simulation table around a pre-defined static position at the same time as the static test but on the next sidereal day and recorded by GPS and accelerometer in which displacement was achieved electronically. In this paper, only the result of the measured wind-induced dynamic motion of Di Wang Tower is presented.

In this phase of testing, the interactive data processing method with the intermittency frequency of 0.5 Hz which is greater than the guaranteed frequency of the employed accelerometer (0.2 Hz) is applied to the measured wind-induced dynamic motion of Di Wang Tower during Typhoon York in 1999. Fig. 3 depicts the results of EMD obtained from the displacement time history recorded by GPS in the dynamic testing with pre-defined static position of 10 mm. The first curve (Curve 1) displayed in Fig. 3 is the original displacement time history measured by GPS. After applying EMD to this time history, a total of twelve IMF components (Curves 2 to 13) and one final residue (Curve 14) are obtained. The last curve (Curve 14) which is a linear line the amplitude of which varies from 7 mm to 8 mm must be part of the mean static displacement simulated by the table. In order to find the residue mean static displacement which is susceptible to the multipath effect, the time history of stationary antenna measured from GPS within the same time period as the dynamic test but on the previous sidereal day is decomposed by EMD. Again, a total of twelve IMF components and one final residue are obtained. As the last curve illustrated in Fig. 4a is a linear line varying from -3 mm to -2 mm, the mean static displacement generated by the table must be the difference between the two monotonic functions (Fig. 4b).This mean static displacement is then subtracted from the original dynamic displacement time history measured by GPS to obtain a new time history as the reference signal in the first stage of adaptive filter (Fig. 4c). The displacement measured by accelerometer with the frequency more than 0.5 Hz obtained from the intermittency check of EMD is used as the primary

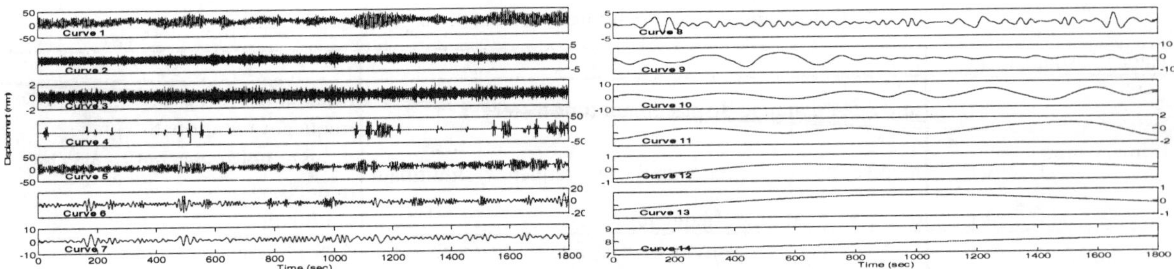

Figure 3. Empirical mode decomposition components for dynamic test measurement form GPS.

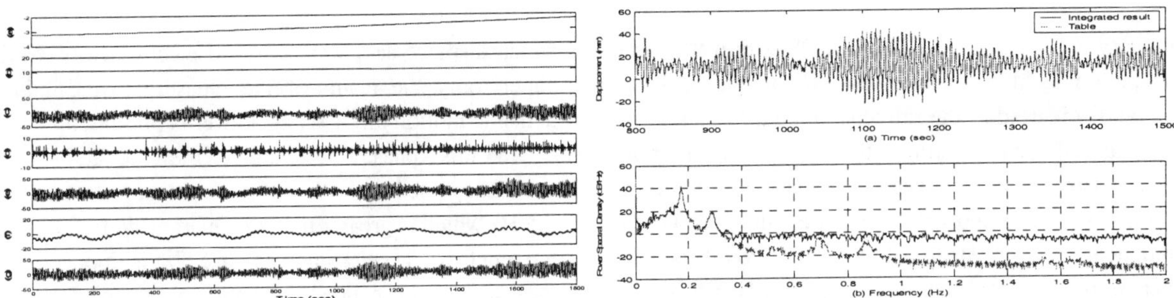

Figure 4. Results after interactive data processing approach

Figure 5. Comparison of (a) displacements and (b) PSD functions of measured motion of Di-Wang Tower.

signal in the first stage of filtering (Fig. 4d). This incoherent part of the adaptive filter as depicted in Fig. 4e is then treated as the primary signal in the second stage of adaptive filter. The objective of the second stage of filtering is to reduce the effect of multipath in the GPS measurement by using displacement time history measured by GPS with stationary antenna as reference. Again, the coherent and incoherent parts in the primary signal are separated and the results of them are illustrated in Figs. 4f and 4g, respectively. The total displacement is thus the combination of the final residue of the test measurements from GPS obtained by EMD (Fig. 4b), the displacement from accelerometer after intermittency check (Fig. 4d) and the incoherent part of displacement acquired by the second stage of adaptive filter (Fig. 4g). Fig. 5a depicts the motion extracted from 800 to 1500 seconds in the displacement time histories generated by the motion simulation table and derived by the proposed algorithm. It is noted that the time history of the derived result quite follows the motion generated by the table. The corresponding power spectrum density functions (PSD) shown in Fig. 5b also demonstrate that this interactive data processing approach can capture the major frequency components generated by the table.

Table 2 depicts the displacement results obtained from the interactive data processing approach as well as those recorded by the motion simulation table and measured by GPS and accelerometer. From the table, it can be seen that the mean displacement generated by the motion simulation table is not accurately captured by GPS due to susceptibility to multipath effects. However, after adopting interactive data processing, it is shown that all values are greatly improved; especially the mean displacement is improved by more than 46%. Although the accelerometer is incapable of capturing the mean static displacement response of the tall building, the fact that the standard deviation of dynamic displacement is not the same as the table indicates that the accelerometer cannot measure the dynamic displacement response accurately. After applying interactive data processing method, the standard deviation from accelerometer is improved by 36%.

6. Conclusions

A method of integrating GPS and accelerometer signals to measure the total displacement response of tall buildings has been presented in this paper in which an interactive data processing procedure using the empirical mode decomposition (EMD) and adaptive filtering techniques has been proposed. In the interactive

Table 2. Comparison of displacement measurements of Di Wang Tower.

Instrumentation	Mean (mm)	Std. (mm)	Max. (mm)	Min. (mm)
Table	10.75	11.69	45.25	-22.59
GPS	7.33	11.04	41.4	-24.5
Accelerometer	---	8.55	---	---
Interactive Result	10.71	11.62	46.83	-23.23

processing approach, the EMD is used to extract the static mean displacement from GPS and high frequency dynamic displacement from an accelerometer, while the adaptive filter is used to obtain low frequency dynamic displacement from GPS. 1-D measured wind-induced dynamic motion of Di-Wang Tower during Typhoon York was simulated by the motion simulation table in the pre-defined static position as one of the test cases to assess the effectiveness of this approach. From the results recorded by GPS and the accelerometer, it has been seen that GPS is susceptible to multipath during the test and the accelerometer was incapable of measuring dynamic displacement response of the simulated motion accurately. However, after applying the interactive data processing method, the final total displacement responses which were the combination of the mean static displacement and the low frequency dynamic displacement from GPS, and the high frequency dynamic displacement from the accelerometer demonstrated that all interested parameters are greatly improved through comparison with the original motion generated with the table. Therefore, it was believed that this approach did enhance the measurement accuracy of total displacement response of a tall building significantly.

Acknowledgments

The authors would like to express their sincere gratitude to The Hong Kong Polytechnic University for providing a postgraduate scholarship to the first writer for her PhD study and providing the second and third writers with a Research Grant (87HH) for the development of an integrated GPS-Accelerometer measurement system.

References

1. G. W. Roberts, X. L. Meng and A. H. Dodson, Integrating a Global Positioning System and accelerometers to monitor the deflection of bridges, *Journal of Surveying Engineering ASCE,* May 2004, (2004).
2. W. S. Chan, Y. L. Xu, X. L. Ding, Y. L. Xiong and W. J. Dai, Dynamic displacement measurement accuracy of GPS for monitoring large civil engineering structures, *Proc. of SPIE*, 5765 (2005).
3. N. E. Huang, Z. Shen, S. R. Long, M. C. Wu, H. H. Shih, Q. Zheng, N. C. Yen, C. C. Tung and H. H. Liu, The Empirical mode decomposition and the Hilbert spectrum for non-linear and non-stationary time series analysis, *Proc. R. Soc. Lond. A*, 454 (1998).
4. J. Chen and Y. L. Xu, On modelling of typhoon-induced non-stationary wind speed for tall buildings, *Struct. Design Tall Spec. Build.* 13 (2004).
5. L. Ge, H. Y. Chen, S. Han and C. Rizos, Adaptive filtering of continuous GPS results, *Journal of Geodesy*, 74 (2000).
6. G. W. Roberts, X. L. Meng and A. H. Dodson, Using adaptive filtering to detect multipath and cycle slips in GPS/accelerometer bridge deflection monitoring data, *FIG XXII International Congress*, (2002).
7. N. E. Huang, Z. Shen and S. R. Long, A new view of non-linear water waves: the Hilbert spectrum, *Annu. Rev. Fluid Mech.*, 31 (1999).
8. J. W. Lovse, W. F. Teskey, G. Lachapelle and M. E. Cannon, Dynamic deformation monitoring of tall structure using GPS technology, *Journal of Surveying Engineering* **121**, 1 (1995).
9. Y. L. Xu and S. Zhan, Field measurements of Di Wang Tower during typhoon York, *Journal of Wind Engineering and Industrial Aerodynamics,* 89 (2001).

PARAMETERS AFFECTING WIND-DRIVEN RAIN AND ITS EFFECT ON HIGH-RISE BUILDINGS

EDMUND C.C. CHOI

School of CEE, Nanyang Technological University, Singapore

Water leakage is a common problem during wind and rain for high-rise buildings. This problem arises due mainly to inadequate knowledge on the phenomenon of wind-driven rain acting on building faces. The design and testing criteria for water tightness are not well defined. This paper sets out a framework for the calculation of the effects of wind-driven rain on buildings using wind and rain information and based on computational fluid dynamics method. The dependency of the effects on the local climatic conditions is examined. Wind-driven rain characteristics are calculated for buildings situated in different geographic regions, e.g. Hong Kong, Singapore and Sydney, and are compared. The effects of building height and build geometry are investigated.

1. Introduction

One of the functions of the external envelope of a building is to protect its inside from wind and rain. Building envelopes should be adequately designed to resist water penetration due to rainstorms. Unfortunately, many buildings, including some relatively new ones, are subjected to water leakage problems. Water leakage is directly related to wind-driven rain (w.d.r.); and a good understanding of the wind-driven rain process will minimize the water leakage problems. Criteria for the watertightness testing of building envelope should be derived based on the wind and rain conditions on site.

2. Wind-driven Rain Calculation

Wind-driven rain (w.d.r.) has been studied for many decades. In the early years, wind-driven rain was studied with field measurements of wind and rain on an open site, e.g., Lacy [1971]. A Driving-rain Index, DRI, (product of average annual rainfall and annual mean wind speed) was proposed as a measure of the severity of the wind-driven rain. The importance of the co-occurrence effect of wind and rain was later studied by Choi [1994a].

More recently, the interaction between wind, rain and building was studied. Wind-driven rain falling onto the faces of a building was investigated. It was observed that w.d.r. around buildings behaved quite differently from the unobstructed w.d.r. on an open site. In the early 90's a systematic framework using numerical method was proposed by Choi[1994b] for the calculation of w.d.r. on building faces. Using this method, the characteristics of w.d.r. on building faces can be systematically analyzed. There are three components to this method; (a) using computational fluid dynamics method to obtain the wind flow pattern around a building, (b) solving the force equations of the raindrop moving in the wind-field to obtain its trajectory, and (c) based on the raindrop trajectories, obtain the w.d.r. intensity on the faces of a building as a function of the normal unobstructed rainfall intensity. This method and some result of the study are presented in various papers by Choi[2000]. This approach has been used in other studies of w.d.r. [Hangan, Blocken & Crameliet]. In the present paper, w.d.r. intensities on building faces are calculated by using this method. Studies are carried out on the w.d.r. intensities to investigate the controlling parameters.

3. Factors Affecting Wind-driven Rain

3.1 *Weather conditions*

As w.d.r. is the result of wind and rain, it is natural that the wind and rain conditions of a place are controlling parameters of w.d.r. Wind and rain conditions of Singapore, Hong Kong and Sydney are studied. The climatic conditions of these three cities are quite different. Singapore is situated close to the equator with mild monsoon winds and frequent thunderstorms. Hong Kong is located at about 22.5°N on the south east coast of Asia and it is subjected to the attack of typhoons. Sydney is situated at around 34°S at the eastern Australian coast and besides the large pressure system, it is subjected to severe thunderstorms and cyclonic depressions. The following table gives the mean annual rainfall, annual mean hourly wind speed and the Driving-rain index (DRI) for the different locations.

Table 1 Wind, rain and DRI values of different stations.

City	Station	Annual rainfall (mm)	Mean wind speed (m/s)	DRI	Annual WDR (Top corner)*
Singapore	CG	2144	2.09	4.48	319
	PL	2226	2.05	4.56	419
	TG	2421	1.88	4.54	366
Hong Kong	RO	2335	3.28	7.66	680
	KT	2048	3.97	8.13	1065
Sydney	Mascot	1138	3.40	3.87	467
	Syd Observatory	1315	2.48	4.10	465

* based on 15minutes duration and a patch area of 1/12 building face area

From the table it can be seen that both Singapore and Hong Kong receives slightly over 2m of rain in a year and is substantially more than those received in Sydney of just over 1m. Comparing the hourly mean speed, Hong Kong is the highest and comparable with that of Sydney. Singapore has the lowest mean wind speed. DRI, which reflects the wind-driven rain, or the combined wind and rain effect, for Sydney is around 4, Singapore is about 4.5 and Hong Kong is around 8. The last column of the table gives the annual total amount of w.d.r. on the top corner of a building calculated by numerical method as given in the previous section. Looking at the w.d.r. and the DRI a trend of higher w.d.r. for higher DRI can be observed. However there are discrepancies. Building in Sydney has higher w.d.r. than Singapore although DRI for Singapore is higher than Sydney. In Singapore, the DRI for PL and CG are more or less the same but the w.d.r. for PL is 30% higher than that of CG. Similarly, in Hong Kong, DRI for KT is not much greater than that of RO. However w.d.r. for KT is 57% higher than RO. On examining the data, it was observed that the higher w.d.r. are due to the simultaneous co-occurrence of high wind and high rain intensity.

3.2 *Directionality*

The values of wind speed, rain intensity and DRI in Table I are the aggregated value irrespective of wind direction. However, the faces of a building are facing specific directions. Since w.d.r. moves with the wind, the directionality effect is important. The following series of figures show the directional probability of the wind broken down into the 16 compass directions as shown in figures (a) and the directional DRI as shown in figures (b). The directional DRI is calculated as the product of the average wind speed of winds occurring at that sector and the total rainfall during the occurrence of such winds.

Looking at the figures 1, 2 and 3, there are several observations. First from Figure 1, it can be seen there is a very strong directional effect; there are a lot more wind from the easterly direction. Wind from the east contributed 29% of the wind from all directions. The DRI rose plot in Figure 1b shows similar pattern as in Figure 1a with 30% of the DRI from the east. This is a direct result of having more wind in the east direction. Thus a building face facing east in KT, Hong Kong will be much wetter than faces facing other directions.

The directional probability distribution of wind of Mascot in Sydney as shown in Figure 2a shows that winds are more frequent in the North-east, South and North-west directions and the highest probability from the South. However looking at the directional DRI of Mascot in Figure 2b, there is only one very strong peak in the South direction. Very small DRI values are observed in the NE and NW directions. On examining the data, it was found that there was relatively very little rain when winds were blowing from the NE or NW directions; whereas a lot of rain was found to accompany the southerly wind resulting the high south DRI value.

Figure 3a show the directional probability of wind in CG, Singapore. While a high peak is observed in the north direction showing winds are more frequent from the north, the directional DRI in the north direction is not the highest. The largest value for CG is from NNE. This shows that the most probable wind direction does not always give the highest directional DRI.

Another interesting point is noted when comparing Figures 3 and 4. Figure 4a and b are the directional wind probability and DRI for TG, Singapore. While CG (Changi airport) is on the east end of the Singapore

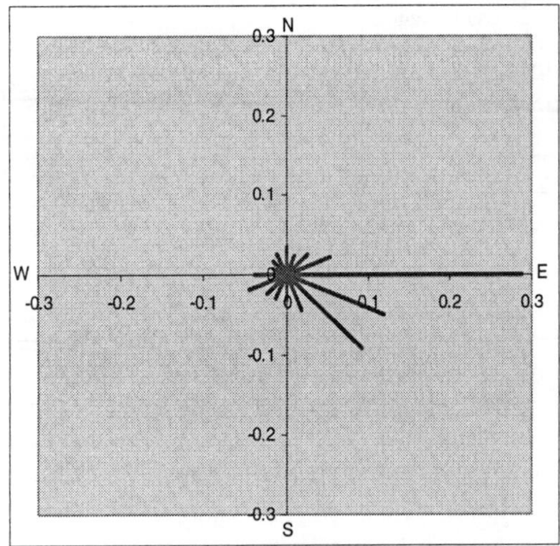

Figure 1a Directional probability of wind of KT

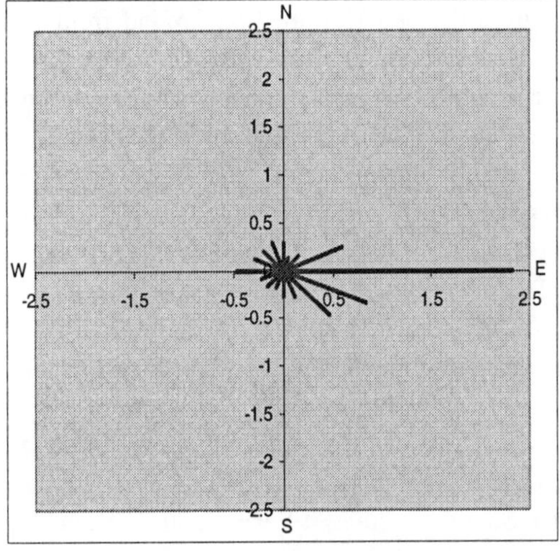

Figure 1b Directional DRI of KT

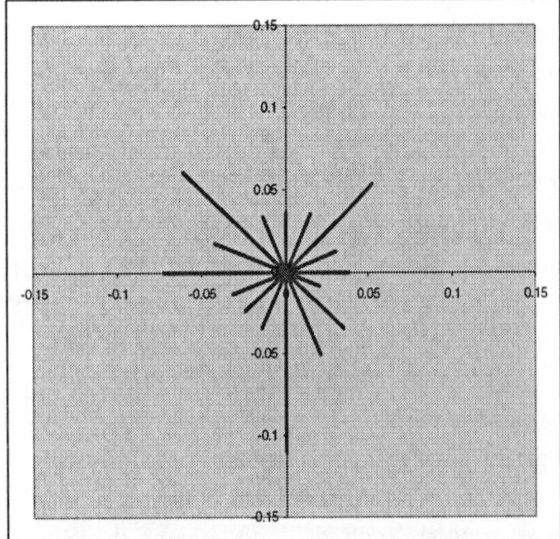

Figure 2a Directional probability of wind of Mascot

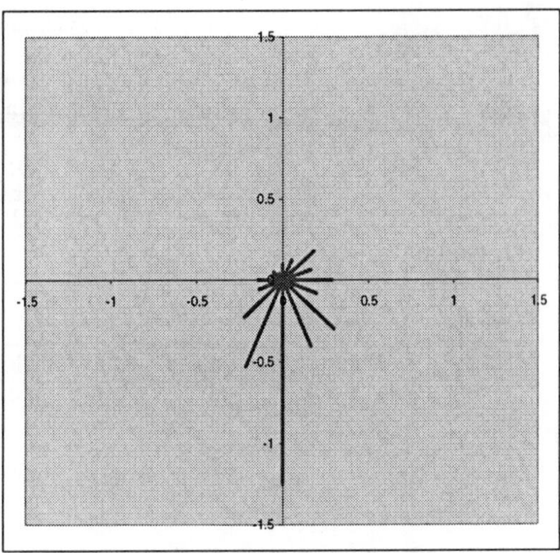

Figure 2b Directional DRI of Mascot

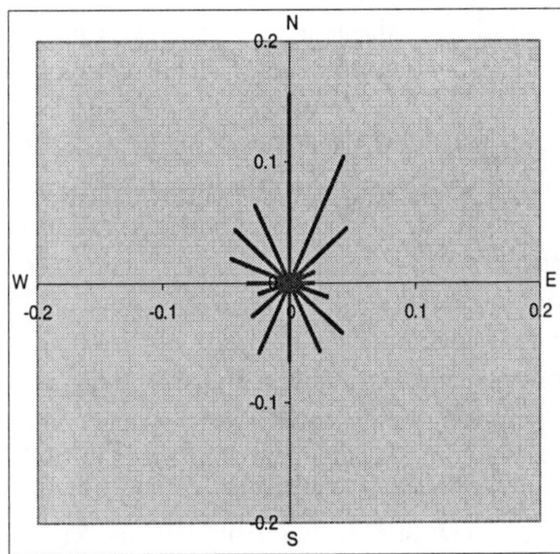

Figure 3a Directional probability of wind of CG

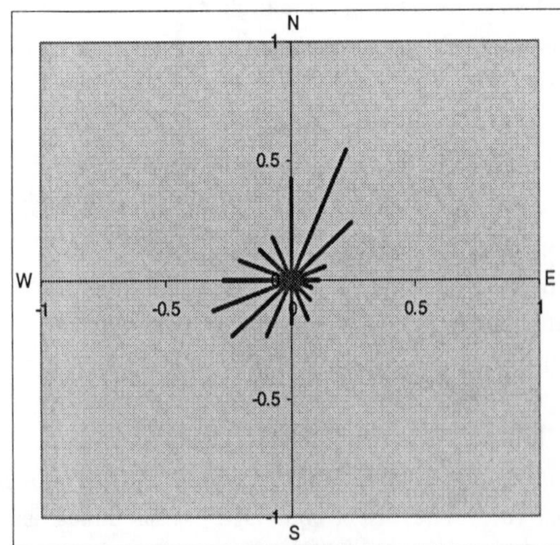

Figure 3b Directional DRI of CG

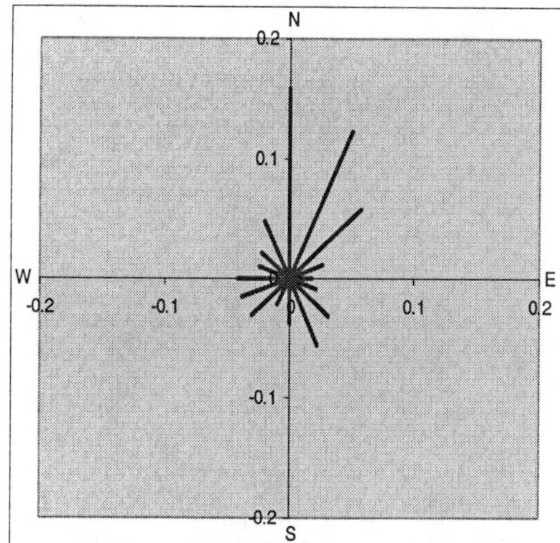

Figure 4a Directional probability of wind of TG

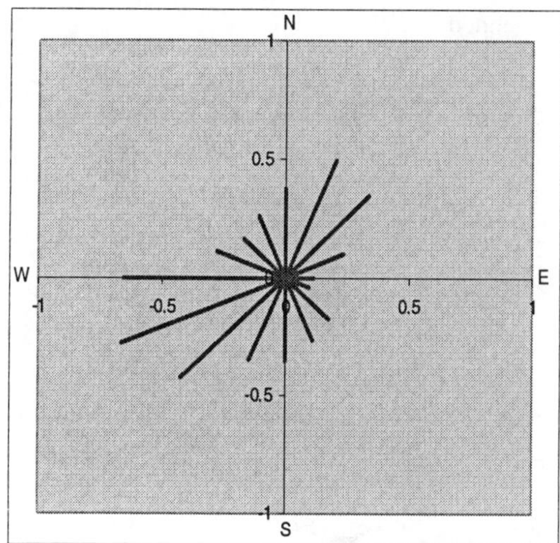

Figure 4b Directional DRI of TG

Island, TG (Tengah air field) is on the west end of the Island. Figure 4a is quite similar to Figure 3a showing that the directional wind patterns for the two sites are similar. However the directional DRI Figure 4b for TG is very different from Figure 3b for CG. Much higher values are observed in the South-West quadrant for TG than those of CG.

Furthermore, these higher values are not due to higher wind probabilities in these directions as evident from Figure 4a. Studying the data and the weather conditions reveals that many thunderstorms are observed to develop west of Singapore. They move in the general easterly direction; thus, hitting the west part of the Island first. As these storms dissipate as they move, heavy rains fall along their paths. Thus there are more rains accompanying the westerly winds at the western part of the Island than the eastern part.

3.3 *Extreme wind-driven rain*

The DRI and the annual w.d.r. intensity reflects the annual aggregated wind-rain condition; that is, they represent generally how wet is the building face. However, if the concern is the criteria for water penetration, then, one should look at the critical values occurring at any instance. In such cases, the extreme statistic analysis can be used. In doing so, obtain the highest w.d.r. value in a year and their distribution over the years is studied. Table 2 shows the annual highest w.d.r. (5-minute on the top corner of a building) value for the three stations CG, PL and TG in Singapore. The extreme distributions are shown in Figures 5a, b and c.

Table 2 Extreme w.d.r. values for Singapore stations

	Wdr(5-min on Top corner of building) (mm/hr)		
YR.	CG	PL	TG
1983	148	68	73
1984	70	82	176
1985	89	184	148
1986	99	98	103
1987	183	105	59
1988	90	200	147
1989	108	167	103
1990	62	445	55
1991	84	190	28
1992	117	107	137
10-year return	168	339	186

From the graphs, it can be seen that the data points for CG and TG are both in the range of around 50 to 200 and the plots are more or less similar giving a prediction of a 10-year return (reduced variate of 2.25) of 168 and 186 mm/hr respectively. However, the 10-year return prediction for PL as shown in the last row of table 2 is 339 mm/hr, much higher than those of CG and TG. Looking at the PL graph, it can be seen that all the data point (except one) are in the range of 50 to 200 similar to CG and TG. The one very high value is 445 mm/hr, occurred in year 1990; and due to this data point, the fitted distribution is having a much greater slope and thus giving a high 10-year return value.

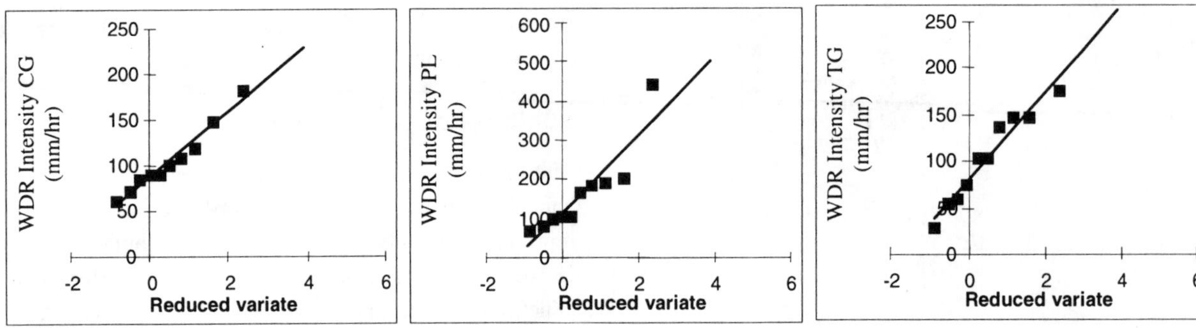

Figures 5a, b and c Extreme distributions of w.d.r. for CG, PL and TG

Examining the PL data reveals that there was one occasion in 1990 where high wind and very high rainfall intensity occurred simultaneously.

A study on the Hong Kong data reveals a similar phenomenon. Figures 6a and 6b show the extreme distributions for RO and KT. It can be seen that the data points of the lower branch (≤150mm/hr) of the KT plot are in the similar range as data in the RO plot. However there are three points in the KT plot which are much higher than the less. This are points occurred in 1989, 1991 and 1992. These high values pull up the fitted line and the prediction. Examining the data again shows that these were due to occasions of simultaneous occurrence of high wind and rain.

This study shows that the criteria for water penetration very much depend on the chance of co-occurrence of high wind and high rain. It only takes a few, or even just one, occasions of high wind and high rainfall intensity occurring simultaneously to give a high w.d.r. criterion. In the present study a return period of 10 years is used for the calculation of the criterion. For serviceability criteria, a return period of 10 or 20 years (reduced variates of 2.25 or 2.97) is generally recommended.

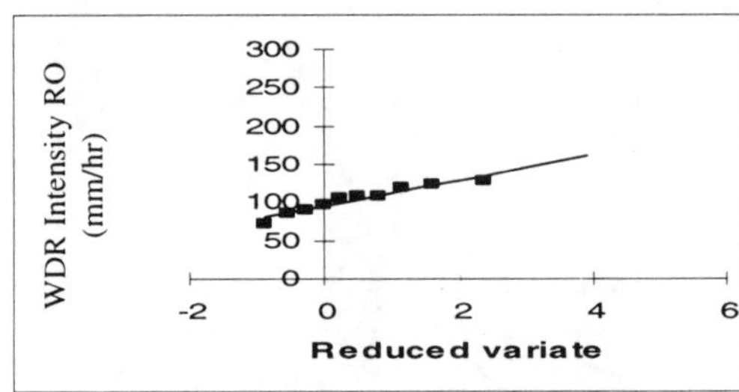

Figures 6a Extreme distributions of w.d.r. for RO

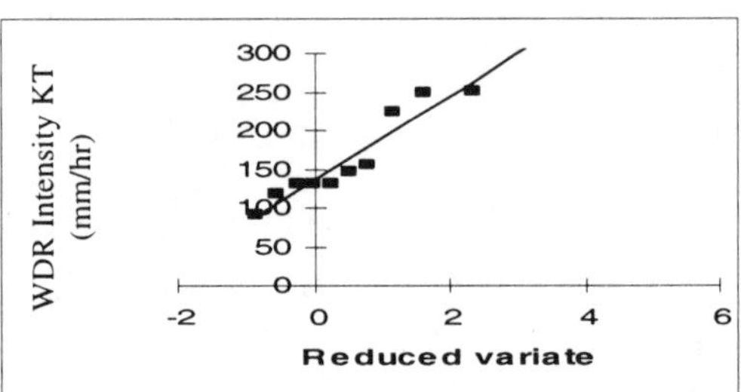

Figures 6b Extreme distributions of w.d.r. for KT

Table 3 Extreme w.d.r. values for Hong Kong stations

	Wdr(5-min on Top corner of building) (mm/hr)	
YR.	RO	KT
1988	128	91
1989	107	225
1990	90	156
1991	105	249
1992	72	
1993	118	147
1994	122	131
1995	85	119
1996	97	130
1997	107	132
10-year return	133	259

4. Building Geometry

In a wind stream, raindrops are carried along and move with the wind. However as the wind comes close to a building, the flow pattern is affected by the blockage effect of the building. Directly in front of the building, wind speed is slowed down; whereas stream of air with higher wind speeds will sweep around the building. Some raindrops are carried along by the faster wind stream off the building. Other raindrops traveling along by their own momentum will impinge onto the building. The wind-driven rain intensity hitting the building face

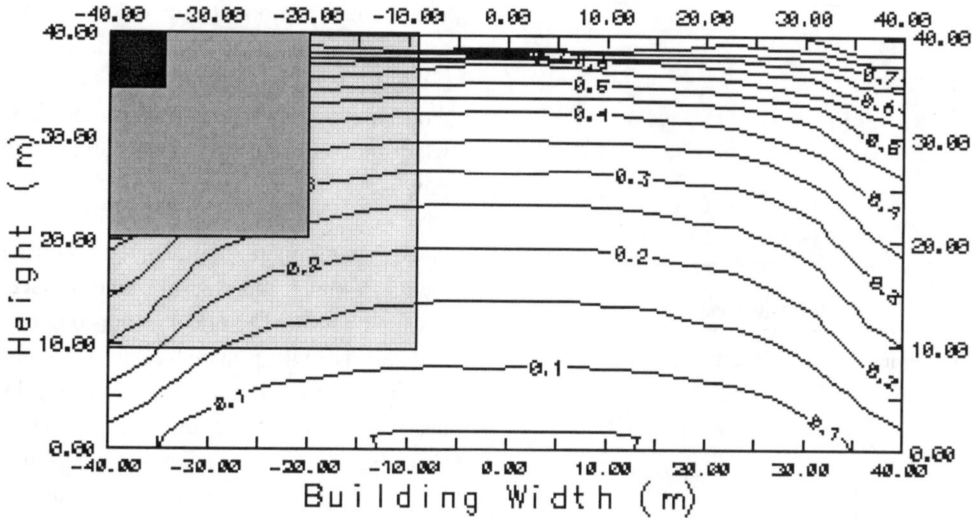

Figure 7 Contour of LIF values on the face of a wide building

will be different on different parts of the building and for different geometry of the building. Figure 7 is a contour plot of the w.d.r. distribution on the front face of a building with width two times the height. The value increases from about 0.1 at the lower central portion to about 1.0 at the top corners. The w.d.r. value is expressed by a Local Intensity Factor (LIF) which is defined as the w.d.r. intensity per unit normal rainfall intensity.

4.1 *Building height (roof height wind speed)*

As w.d.r. is the result of raindrops carried by the wind, the higher the wind speed the more raindrops will hit onto the building face. Therefore, the amount of w.d.r. impinging onto a building face would increase with wind speed. Studies are carried to obtain LIF values for different wind speeds. Figure 8 shows the variation of mean LIF (average LIF over the whole building face) with roof height wind speed. The fitted curve shows that the mean LIF is proportional to the 1.73 power of the wind speed at roof height. Using this figure, the mean LIF value can be obtained for a known roof height wind speed of the building.

4.2 *Building slenderness*

Values in Figure 8 are calculated for a building with height to width ratio, H:W, equals to 4:1. For a slender building with the same building height, the blockage effect to the wind is smaller. That is, the wind speed immediately in front of the building will not be slowed down as much and there will be more rain drops carried along to impinge on the building face. This results in a higher value of LIF. For a wider building, the reverse is true. That is, there will be larger blockage effect and resulting a smaller LIF value. The variation of the mean LIF (whole face) with the width to height ratio is plotted in Figure 9 where the mean LIF is given as a ratio of the mean LIF value normalized by that of a building with H:W=4:1. It can be seen that as the building gets narrower (for buildings of the same height), the mean LIF increases substantially. With the use of Figure 9, LIF for other building geometry can be obtained.

4.3 *Size of concerned area on building face*

In figure 7, the w.d.r. contours show sharp increase towards the edges of the building face, especially towards the top edge. The upper portion of the building face receives higher w.d.r. and the highest w.d.r. are at the top corners. If we take a patch of area at the top corner, it is evident from the figure that the average w.d.r. intensity over the patch increases as the size of the patch gets smaller, i.e. the average w.d.r. intensity varies with size of area under consideration. For example the w.d.r. on a joint between mullion and transom would be much higher than the w.d.r. on a strip of curtain wall; and that on a strip of curtain wall would be higher than the overall average for the whole building face. Study shows that the variation is a logarithmic relationship. Figure 10 is a plot of the Intensity Factor (the average LIF on a patch of area A divided by the mean LIF of the

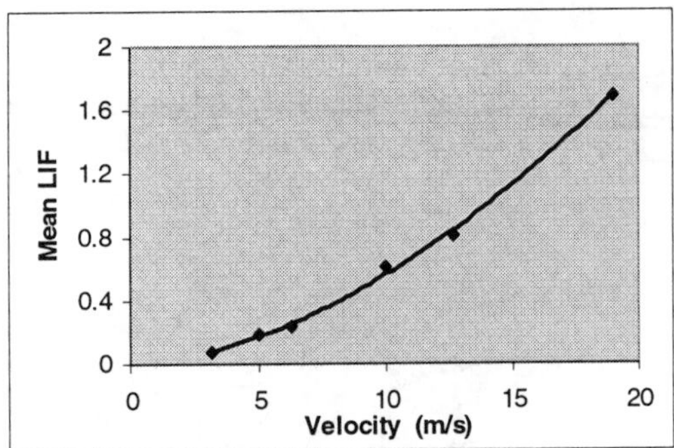

Fig 8 Mean LIF with roof height wind speed

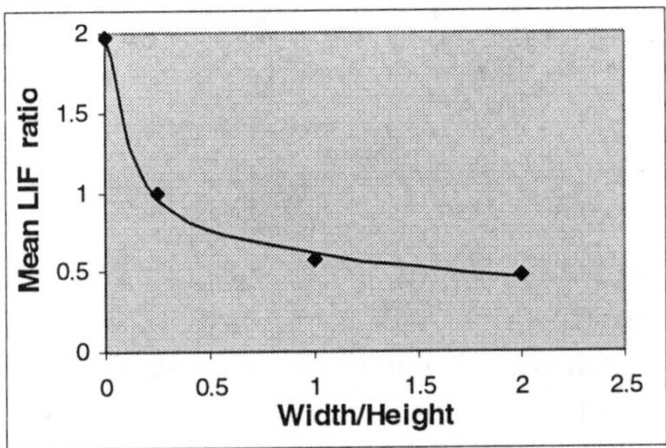

Fig 9 Mean LIF ratio with slenderness of building

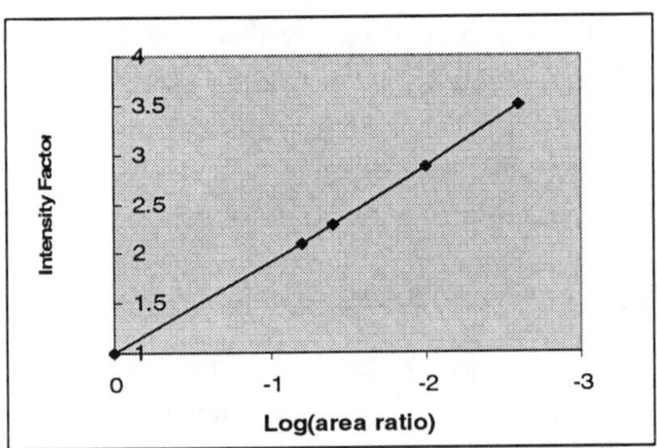

Figure 10 Variation of LIF intensity with area ratio

whole face) against the log of the area ratio (area A expressed as a ratio of the area of the whole face). From the graph, an area of size one one-hundredth of building face has an average LIF almost 3 times that of the mean LIF for the whole face.

5. Conclusion

It is shown that the amount of wind-driven rain acting on a building face is dependent on a number of parameters. The weather condition at the building site is important. The amount of normal rainfall, the wind speed and the driving rain index, DRI, gives a general idea on the severity of the wind-driven rain of a place. However, more importantly the co-occurrence characteristics of wind and rain will govern the amount of wind-driven rain. Two places having similar values of DRI could have very different amounts of w.d.r. Driving rain for a place having more high winds occurring during rain would be much more severe than a place having more dry high winds. The effect of the co-occurrence of wind and rain is more pronounced on the extreme w.d.r. on building faces. The study shows that directionality is also important. W.d.r. in a certain sector can be many times higher than w.d.r. from other sectors. W.d.r. on a building face depends further on the building height and the width to height ratio of the building. In general, the taller the building the larger is the w.d.r. value. For two buildings of the same height, the more slender building will have a higher w.d.r. value. which is contra to the general conception that a wider building will catch more rain. It is further established that the higher w.d.r. is located on the top strip of a building face and that the highest value is located at the top side corners. Furthermore the averaged w.d.r. intensity over a patch of building face increases as the area of the patch decreases. Therefore the criterion for a curtain wall joint should be much higher that that over a curtain wall panel.

The present paper shows that in order to estimate the amount of wind-driven rain, the size, location of the spot of interest on the building face as well as the orientation and geometry of the building have to be considered in conjunction with the wind, rain and their co-occurrence characteristics of the weather system at the building site.

References

1. Blocken B, Crameliet(2000) Driving rain on building envelope – I: numerical estimation and full-scale experimental verification, J. Thermal Env. & Building Sci. 24, 1 (2000) 61-85
2. Choi, E. C. C., (1994a), "Characteristics of the co-occurrence of wind and rain and the driving-rain index" Journal of Wind Engg. & Industrial Aerodynamics, 53 (1994), 49-62.
3. Choi, E. C. C., (1994b), "Determination of wind-driven rain intensity on building faces," Journal of Wind Engg. & Industrial Aerodynamics, 51 (1994), 55-69.

4. Choi E.C.C. (2000) "Variation of wind-driven rain intensity with building orientation", ASCE Journal of Architectural Engineering, USA, vol.6, No. 4, 122-128.
5. Hangan, H. (1999) 'Wind-driven rain studies. A C-FD-E approach', J. of Wind Engg. and Ind. Aerodynamics, Netherlands, 81(1999), pp323-331.
6. Lacy, R. E. (1971), "An index of exposure to driving rain" BRE Digest 127, Building Research Station, Garston, UK.

WIND-INDUCED DYNAMIC RESPONSE ANALYSIS FOR POWER TRANSMISSION TOWER BY PDEM [†]

LINLIN ZHANG

Department of Building Engineering, Tongji University, 1239 Siping Road Shanghai 200092, China

JIE LI

Department of Building Engineering, Tongji University, 1239 Siping Road Shanghai 200092, China

As taller- and longer-span power transmission towers (PTT) are recently being constructed in China, they tend to be more sensitive to wind action, and it is more important to accurately estimate their response to wind forces. For this purpose, a new method, called as probability density evolution method (PDEM), which had been proved to be of high accuracy and efficiency, is firstly described. Secondly, wind stochastic field is studied from a new viewpoint of stochastic Fourier spectrum (SFS). The basic random parameters of the wind stochastic field, that is, the roughness length z_0 and the mean wind velocity at 10m height U_{10}, as well as their probability density functions (PDF), are obtained. It provides opportunities to apply the PDEM in this problem here. The wind-induced dynamic response for a PTT was studied. The time process of mean and standard deviation of the top displacement response were obtained, and the instantaneous PDF of it was shown by means of two-dimensional and three-dimensional pictures. All of them verify that the PDEM is feasible and efficient in the dynamic response analysis of wind-excited PTT.

1. Introduction

Wind loading is a typical dynamic load taken into account in the design of power transmission tower (PTT). From field measurements it is obvious that the basic characteristics of the wind velocity can be regarded as random processes. In structural engineering it is reasonable to model these random processes as Gaussian ones [1].

However, due to the presence of the wind loading nonlinear function, it and the corresponding structural response aren't Gaussian. And the use of appropriate tools for nonlinear analysis is required even for linear-elastic structural behavior. This fact makes troubles in analyzing the dynamic response of PTT subject to the wind loading, as far as the second-order statistical quantities such as the mean, the covariance, etc., are concerned. It is worth noting that a class of probability density evolution method (PDEM), which had been verified to be applicable to stochastic response, has been developed [2-5]. In the PDEM, a one-dimensional probability density evolution equation (PDEE) is deduced for structural dynamic response analysis. This equation holds for any response or index of the structure. The solution will put out the instantaneous probability density function (PDF). The aim of this paper is to use the PDEM in dynamic response analysis of wind-excited PTT.

In the PDEM, the PDFs of random parameters, which reflect the uncertainty in the physical/geometrical parameters of the structures or the excitations, is supposed to be known beforehand. Unfortunately, those of the wind loading are still unknown now. Clearly, the traditional random vibration theory, which studies random process through their numerical characteristics, can't bear this burden. So in this paper, the wind stochastic field is studied from the viewpoint of stochastic Fourier spectrum (SFS), which was presented recently [6]. From this model, the roughness length z_0 and the mean wind velocity U_{10} at 10m height are proved to be the random parameters of the wind, characterized by lognormal distribution and extreme value-I distribution respectively.

Finally, as a numerical example for the application of the PDEM, a PTT subject to the wind-loading is investigated in detail without considering the effects of the wind-structure interaction. The mean and the standard deviation of the dynamic response are

† Work partially supported by grant 50321803 of the Natural Science Foundation of China for Innovative Research Groups.

evaluated by the PDEM. The PDF at certain time instants and the probability transition of structural responses are also depicted. It is founded that the PDEM is feasible and efficient in the dynamic response analysis of wind-excited PTT.

2. Stochastic Dynamic Response Analysis Based on the PDEM

2.1 *General evolutionary PDF equation of dynamic responses*

Without loss of generality, consider the equation of motion of a MDOF system subject to the wind loading as follows:

$$\mathbf{M}\ddot{\mathbf{X}}+\mathbf{C}\dot{\mathbf{X}}+\mathbf{K}\mathbf{X}=\mathbf{f}(\mathbf{Z},t) \tag{1}$$

where $\mathbf{X}$ is a $n\times 1$ displacement response vector, and $\mathbf{M}$, $\mathbf{C}$ and $\mathbf{K}$ are $n\times n$ mass, damping and stiffness matrices, respectively. The dot upon a variable denotes differentiation with respect to time, t. $\mathbf{f}$ is a $n\times 1$ stochastic dynamic excitation vector, and $\mathbf{Z}$ is a $n_Z\times 1$ random parameter vector which reflects the uncertainty in the excitation, with the known PDF $p_Z(z)$.

Usually, with a deterministic initial condition

$$\mathbf{X}(t_0)=\mathbf{x}_0,\ \dot{\mathbf{X}}(t_0)=\dot{\mathbf{x}}_0 \tag{2}$$

The structure responses $\mathbf{X}(t)$ is a random process dependent on and determined by $\mathbf{Z}$, i.e.

$$\mathbf{X}(t)=\mathbf{H}(\mathbf{Z},t) \tag{3}$$

where $\mathbf{H}$, existent and unique for a well-posed problem, is a deterministic operator.

Its component expression is

$$X_j(t)=H_j(\mathbf{Z},t),\ j=1,2,\cdots,n \tag{4}$$

Taking the formal solution as Eq. (4), the joint conditional PDF of $X_j(t)$ on the condition $\mathbf{Z}=\mathbf{z}$ is

$$p_{X_j|\mathbf{Z}}(x_j,t|\mathbf{z})=\delta(x_j-H_j(\mathbf{z},t)) \tag{5}$$

Thereby, the joint PDF $p_{X_j\mathbf{Z}}(x_j,\mathbf{z},t)$ yields

$$p_{X_j\mathbf{Z}}(x_j,\mathbf{z},t)=\delta(x_j-H_j(\mathbf{z},t))\,p_{\mathbf{Z}}(\mathbf{z}) \tag{6}$$

Differentiating Eq. (6) with regard to t on both sides, the PDEE will be derived [5]

$$\frac{\partial p_{X_j\mathbf{Z}}(x_j,\mathbf{z},t)}{\partial t}+\dot{H}_j(\mathbf{z},t)\frac{\partial p_{X_j\mathbf{Z}}(x_j,\mathbf{z},t)}{\partial x}=0 \tag{7}$$

where $\dot{H}_j(\mathbf{z},t)=\dot{X}_j(t)$ is the 'velocity' of the response for a prescribed $\mathbf{z}$. As long as it is numerically tractable, the PDF $p_{X_j\mathbf{Z}}(x_j,\mathbf{z},t)$ is solvable from Eq. (7) with the initial condition

$$p_{X_j\mathbf{Z}}(x_j,\mathbf{z},t)=\delta(x_j-x_{j,0})\,p_{\mathbf{Z}}(\mathbf{z}),\ \text{for } t=t_0 \tag{8}$$

where $x_{j,0}$ is the deterministic initial value of $X_j(t)$.

Finally, the PDF of $X_j(t)$ reads

$$p_{X_j}(x_j,t)=\int_{\Omega_{\mathbf{Z}}} p_{X_j\mathbf{Z}}(x,\mathbf{z},t)\,d\mathbf{z} \tag{9}$$

where $\Omega_{\mathbf{Z}}$ is the distribution domain of $\mathbf{Z}$.

It is worthwhile noting that the PDEE can be applied to not only the displacement response, but also any other structural response $X(t)$ such as the velocity, the internal force and the stress and the like.

2.2 *Numerical solving algorithm*

Because of the difficulties in computing the analytical solution of dynamic response, the numerical solving algorithm is recommended. The basic solving procedure is as follows:

1. Discretize $\mathbf{z}$ in the domain $\Omega_{\mathbf{Z}}$ and denote the lattice point as $\mathbf{z}_q$, $q=1,2,\cdots,N_s$, in which N_s is the total number of these discrete representative points.
2. For a given $\mathbf{z}_q$, solve Eq. (1) with a deterministic time integration method to obtain the velocity $\dot{X}(\mathbf{z}_q,t_m)$, in which $t_m=m\cdot\Delta t$ and Δt is the time step.
3. Solve the initial-boundary-value problem defined by Eqs. (7) and (8) with the finite difference method.
4. Carry out the numerical integration in Eq. (9) for the dynamic response analysis.

It is mentioned in the solving steps 1-4 that a routine deterministic analysis is applied in Step 2 to compute the coefficient of the PDEE, and a finite difference method is used in Step 3 to obtain the PDF. In this paper, Newmark-Beta time integration method [7] is adopted for the deterministic dynamic response analysis. Moreover, the TVD difference scheme [5] is embedded in Step 3.

3. Randomness of the Wind Field

As mentioned above, the PDFs of the random parameters must be known beforehand in the PDEE.

According to the traditional random vibration theory, the complete characterization of random process can be ensured by the knowledge of the power spectral density (PSD) function. Although a lot of empirical wind velocity models, such as Davenport spectrum, had been existed, they are just established on phenomenology level. In other words, they are dependant on the measured data. As a result, these PSD functions are unable to disclose the physical essence of the randomness of the wind stochastic field. To settle this quandary, physical random process was proposed in Ref. [6] for the first time. One of important concepts, that is, stochastic Fourier spectrum, is introduced herein.

3.1 *Stochastic Fourier spectrum*

Generally, two methods can be applied in describing the random process. One is time process, and another is spectrum. The former is defined in the time domain, and the latter is in the frequency domain. They connect with each other through Fourier transform. In other words, sample functions of the Fourier spectrum $F(\omega)$ can be computed from those of the time process by the Fourier transform, and on the contrary, the sample functions of the time path can also be calculated from those of the Fourier spectrum $F(\omega)$ by the Fourier inverse transform. According to the definition of the random process, the random Fourier Spectrum can be defined as a set of all possible sample functions of the Fourier spectrum obtained in a same experiment.

Usually, the Fourier spectrum $F(\omega)$ is often separated into two kinds of spectrum functions, that is, Fourier amplitude spectrum and Fourier phase spectrum. The former one can be used to compute the PSD function $S(\omega)$ numerically. The formula is

$$S(\omega)=\frac{1}{T}E\left[\left|F(\omega)\right|^2\right] \tag{10}$$

where $|F(\omega)|$ is the so-called Fourier amplitude spectrum, T is the duration of the sample functions, and ω is the angular frequency.

Eq. (10) suggests that the power spectrum $S(\omega)$, which substantially has certain physical meaning, represent the mean square value of the Fourier amplitude spectrum. So the PSD may be able to denote the Fourier amplitude spectrum. Based on this idea, the stochastic Fourier amplitude spectrum can be defined as

$$\left|F(\omega,\lambda,\xi;\cdots)\right|=\sqrt{T*S(\omega,\lambda,\xi;\cdots)} \tag{11}$$

where λ, ξ, $\cdots$ are basic measurable physical parameters, which are regarded as random variables in the stochastic Fourier amplitude spectrum according to practical physical background. Then, by means of statistics, the PDFs of these random variables can be acquired. Thereby, the function of the stochastic Fourier amplitude spectrum having certain physical meaning, which can reflect the probabilistic characteristics of the random process in a qualified sense, is obtained.

Herein, Davenport spectrum is used to define the stochastic Fourier amplitude spectrum for its extensive application in the wind engineering. Its two-sided PSD function is given as

$$S(\omega)=\frac{1}{2}\frac{u_*^2}{\omega}\frac{4f^2}{\left(1+f^2\right)^{4/3}} \tag{12}$$

where $f=1200\omega/(2\pi U_{10})$, U_{10} is the mean wind velocity at 10m height.

Commonly, derived from the logarithmic law of the mean velocity, the shear velocity u_* can be computed by the following equation.

$$u_*=\frac{k\bar{V}(z')}{\ln\left(z'/z_0\right)} \tag{13}$$

where $\bar{V}(z')$ is longitudinal mean wind velocity at z' height (usually taken equal to 10 m), $k\approx 0.4$ is the Von Karman constant, and z_0 is the roughness length.

It should be noticed that Eq. (13) isn't the definition equation for the shear velocity u_* in the wind engineering.

Substituting Eq. (13) to Eq. (12), the function of Davenport spectrum is simplified as

$$S(\omega)=\frac{11672.2\omega}{\left[\ln(10/z_0)\right]^2\times\left[1+(\frac{1200\omega}{2\pi U_{10}})^2\right]^{\frac{4}{3}}} \tag{14}$$

Substituting Eq. (14) to Eq. (11), the stochastic Fourier amplitude spectrum can be expressed as

$$\left|F(\omega)\right|=\sqrt{T\cdot\frac{11672.2\omega}{\left[\ln(10/z_0)\right]^2\times\left[1+(\frac{1200\omega}{2\pi U_{10}})^2\right]^{\frac{4}{3}}}} \tag{15}$$

3.2 *The PDFs of the random variables*

According to the definition in Eq. (11), the roughness length z_0 and the mean wind velocity at 10m height U_{10} should be regarded as the random variables. It is very reasonable because these two parameters are of randomness essentially. In the Ref. [6], the PDFs of them had been researched by means of statistics. The data used are collected from China. The details are omitted here for the limit of the paper, and the results are listed directly. The roughness length z_0 is characterized by lognormal distribution with the range of 0 ~ 4m, and the mean wind velocity at 10m height U_{10} is of extreme value-I distribution with the range of 20 ~ 50m/s. Their PDFs (shown in Figure. 1 and Figure. 2 respectively) are listed below.

$$f_{z_0}(z_0) = \begin{cases} \dfrac{0.40}{z_0}\exp\left(-\dfrac{(\ln z_0 + 3.51)^2}{2}\right), & z_0 \geq 0 \\ 0, & z_0 < 0 \end{cases} \tag{16}$$

$$f_{U_{10}}(U_{10}) = 0.26 * \exp\{-\exp[-0.26(U_{10} - 24.87)]\} * \exp[-0.26(U_{10} - 24.87)] \tag{17}$$

where the mean of the roughness length z_0 is 0.051, and its variance is 0.0042; those of the mean wind velocity at 10m height U_{10} are 27.05m/s and 23.43m^2/s^2.

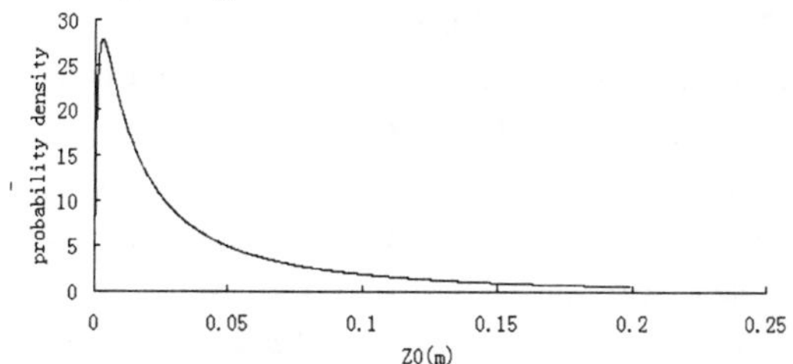

Figure 1. The PDF of the roughness length z_0

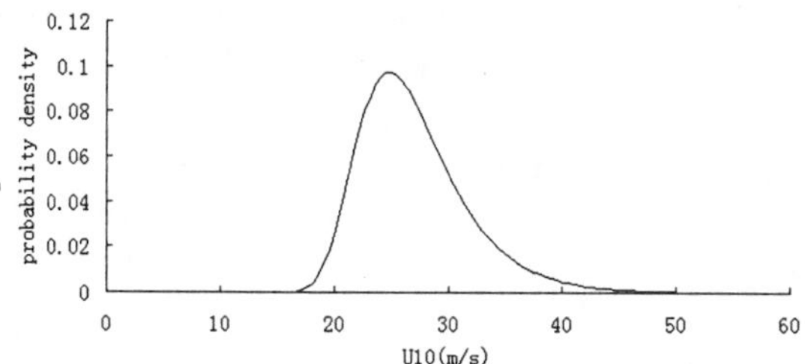

Figure 2. The PDF of the mean wind velocity at 10m height U_{10}

4. Example

In this paper, a PTT subject to the horizontal wind loads is investigated in detail without considering the effects of the wind-structure interaction. To reduce the compute cost, the finite element updating method based on Bayesian estimation and minimization of dynamic residuals [8] is used. As a result, the three-dimensional structure is simplified to a 2D lumped-mass model. Figure 3 shows both the 3D model and 2D model. The masses and heights of the ten lumped-mass points are shown in the Fig. 3 too. Two adjacent lumped mass points are jointed by a beam element, whose material and geometric parameters are shown in table 1 and table 2.

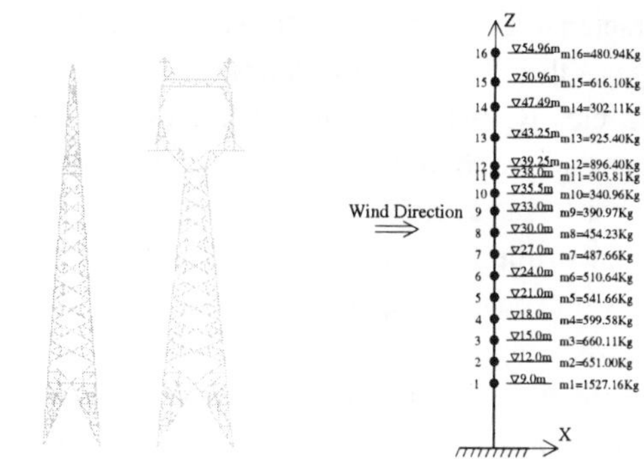

Figure 3. 3D finite element model and 2D lumped-mass model the transmission tower

Table1 Material parameters of the 2D lumped-mass model

Elastic Modulus(MPa)	Poisson Ratio
210000	0.3

Table 2 Geometry parameters of the 2D lumped-mass model

No. of column	Cross-sectional Area(m^2)	Moment of cross-sectional Area	
		I_{xx} (m^4)	I_{yy} (m^4)
1	0.8836	0.065062	0.0046295
2	0.8836	0.065062	0.14209
3	0.8836	0.065062	0.12718
4	0.8836	0.065062	0.090846
5	0.8836	0.065062	0.045616
6	0.8836	0.065062	0.0027532
7	0.8836	0.065062	0.00065062
8	0.8836	0.065062	0.00091667
9	0.8836	0.065062	0.0013784
10	0.8836	0.065062	0.00099922
11	0.8836	0.065062	0.00065062
12	0.8836	0.065062	0.032751
13	0.8836	0.065062	0.00065062
14	0.8836	0.065062	0.032892
15	0.8836	0.065062	0.058459
16	0.8836	0.065062	0.064310

In the analysis, as mentioned previously, the roughness length z_0 and the mean wind velocity at 10m height U_{10} are taken as the random parameters with the PDFs listed as Eq. (16) and Eq. (17). Rayleigh damping, i.e. $\mathbf{C} = a\mathbf{M} + b\mathbf{K}$, is assumed where a =0.30088, b =0.00109.

The mean and the standard deviation of the top displacement of the PTT, pictured in Fig. 4, are evaluated by the proposed method. The maximum coefficient of variation (COV) reaches 0.45.

Figure 5 shows the PDF at certain time instants, say, 30.0, 40.0 and 50.0s. From Fig. 5, the PDFs are similar to the widely used normal distribution or extreme value distribution. It maybe has relation to the Gaussian characteristics of the simulated wind field. However, the shape of them is varying with time. To show this change more clearly, the time varying process of the PDF is pictured in Fig. 6. It is seen that the PDFs vary with time irregularly and acutely. Additionally, Figure 7 shows the contour of the PDF with 20 levels, which seems like a river with some whirlpools. All these figures imply that the structural response process still can't be simply regarded as a stationary process, though the excitation is of Gaussian.

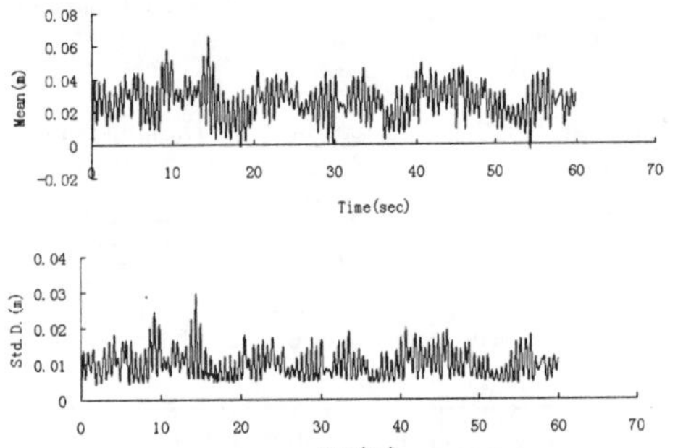

Figure 4. The mean and the standard deviation of the top displacement by PDEM

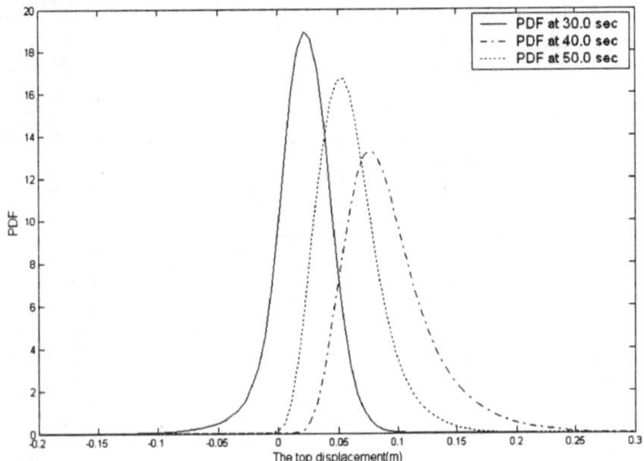

Figure 5. The PDF at certain time instants

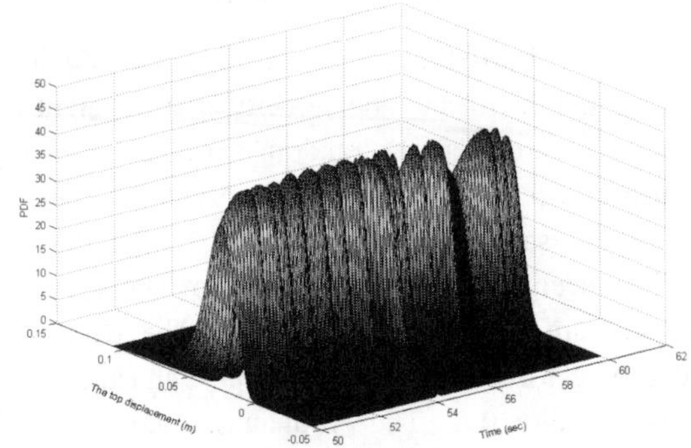

Figure 6. The PDF surface varying with time

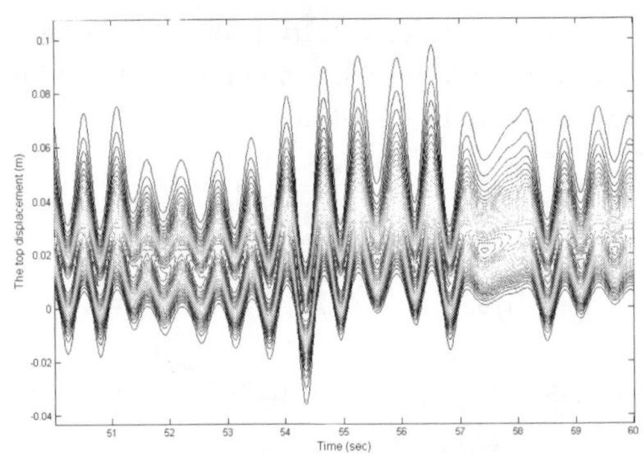

Figure 7. The counter of the PDF surface (20 levels)

5. Conclusion

The PDEM is adopted for dynamic response analysis of wind-excited PTT. The wind stochastic field is studied from the viewpoint of the stochastic Fourier spectrum. The roughness length z_0 and the mean wind velocity at 10m height U_{10} are proved to be the basic random parameters of the wind loading. Meantime, their probability density functions are obtained. At the end, a PTT subject to the random wind loading is investigated in detail without considering the effects of the wind-structure interaction. Some features of the responses of the PTT are observed and discussed. It is found that the PDEM can be successfully applied in the dynamic response analysis of wind-excited PTT with high efficiency.

Acknowledgments

The support of the Natural Science Funds for Innovative Research Groups of China (Grant No. 50321803) is gratefully acknowledged.

References

1. E. Simiu, R.H. Scanlan, Wind Effects on Structures: an Introduction to Wind Engineering, Wiley, New York (1978) .
2. Li J, Chen JB, Probability density evolution method for dynamic response analysis of stochastic structures. In: Zhu WQ, Cai GQ, Zhang RC, editors. Advances in stochastic structural dynamics. Proceedings of the Fifth International Conference on Stochastic Structural Dynamics-SSd03, May 26-28, Hangzhou, China, Boca Raton: CRC Press; p. 309-316 (2003).
3. Li J, Chen JB, The probability density evolution method for analysis of stochastic structural dynamic response. *Acta Mechnica Sinica*, **35(4)**, 437-442 (2003).
4. Li J, Chen JB, Dynamic response and reliability analysis of structures with uncertain parameters, *Int. J. Numer. Meth. Engng*, **62**, 289-315 (2005).
5. Chen JB, Li J, Dynamic response and reliability analysis of non-linear stochastic structures, *Prob. Engng Meth*, **20**, 33-44 (2005).
6. Li J, Zhang LL, A study on the relationship between turbulence power spectrum and stochastic Fourier amplitude spectrum (in Chinese). *Journal of Disaster Prevention and Mitigation Engineering,* **24(4)**, 363-369 (2004).
7. Clough RW, Penzien J., Dynamics of Structures; 2nd edn. McGraw-Hill: New York (1993).
8. Alvin, K.F., "Finite Element Model Update via Bayesian Estimation and Minimization of Dynamic Residuals". *AIAA Journal*, Vol. **35(5),** 879-886 (1997)

EFFECTS OF NONLINEAR DAMPING AND TIME CONSTANT ON WIND-INDUCED RESPONSES OF A 79-STOREY TALL BUILDING

J.R. WU P.F. LIU Q.S. LI

Department of Building and Construction, City University of Hong Kong, Tat Chee Avenue, Hong Kong

The full-scale tests on wind-induced response of Di Wang Tower during several typhoons were conducted during the past few years. The amplitude-dependent nonlinear damping ratios for this tall building were obtained from these full scale tests. Meanwhile a Monte Carlo simulation procedure was adopted in this study to generate the time series data of local fluctuating along-wind force acting on this building. The wind-induced response of this building was numerically analyzed in time domain based on the generated local fluctuating wind force and the established finite element model. The effect of nonlinear damping on the wind-induced response was evaluated under different load cases. Finally the effect of time constant on the wind-induced response was analyzed by comparing the time domain analysis results with those from conventional spectral analysis method. Some conclusions resulted from this study may be helpful to the wind resistant design of tall building structures.

1. Introduction

The natural frequencies and the damping ratios are very important parameters which affected the dynamic response of building structures under wind and earthquake excitation. The value of the natural frequencies can be normally obtained from traditional methods with high accuracy. However this situation doesn't apply to the determination of damping ratios. As it is very difficult to be determined prior to construction, the damping ratio is usually assumed to be constant value at design stage. Actually damping ratio is found to be a nonlinear parameter with amplitude-dependent property. Over the last three decades significant measurements for structural damping have been made throughout the world [1,2,3]. However, the amplitude dependent damping in exist literatures covered only relatively low amplitude area in structural response and mainly concerned for mid-rise buildings. For super high-rise buildings such as Di Wang Tower, there is a serious scarcity of damping data available especially from strong wind excitations. Fortunately much data on the amplitude-dependent damping ratios for this tall building were obtained from our field tests on wind-induced response of this building over the past few years.

Meanwhile the wind-induced response of a tall building in time domain was ever studied by Tsukagoshi [4] and Liang [5]. However the damping ratios are specified as constant values in their studies. The analysis of wind-induced response on tall buildings considering the amplitude-dependent damping effect has seldom been studied. Therefore, there is a need to evaluate this effect on the dynamic response of tall building excited from wind loads.

As previous description, the natural frequency and damping ratio are two parameters for determining the dynamic response of building structures. The product of these two parameters (related with time constant described by Jeary [6]), dictate how long the structure takes to response to an action. When time-constants become large then the response of a structure to an action may be delayed. The reduction factor of Root Mean Square (RMS) displacement response for structures with long time constants can reach to about 20% compared with the conventional spectral analysis result. The effect of time constant on the reduction of the magnitude in dynamic response of very large structures was studied by Jeary. However, the structural system was simplified as a single degree of freedom and a linear sway mode was considered in his study. This simplification may result in some deviation from its actual dynamic response for a tall building with many degrees of freedom.

This study presented a detailed investigation into the effects of nonlinear damping on the wind-induced dynamics response of a super tall building, named as Di Wang Tower in Shenzhen, China. First the amplitude-dependent nonlinear damping ratios and empirical nonlinear damping model for the first two vibration modes in both main-axis directions were obtained from the numerous field tests on this building. Then the time history of local fluctuating along-wind force was generated by the Monte Carlo simulation procedure termed as Weighted Amplitude Wave Superposition (WAWS) method [7,8]. Thirdly, the wind-induced along-wind response of this building was numerically analyzed in time domain. The effects of nonlinear damping model on the wind-induced response were studied. Finally the effect of time constant on the along-wind induced response was analyzed by comparing with calculated results from conventional spectral analysis method.

2. Structural System of Di Wang Tower

Di Wang Tower, a super tall building studied in this paper, is located in Shenzhen City, China. The main tower of this building includes 68 official storeys, plus 11 equipment and refuge floors as well as two top masts, as shown in Fig. 1a. Totally, the main structure of Di Wang Tower has 79 floors with 324m high above the ground level. The structural system utilises both steel and reinforced concrete (SRC), including SRC core walls and perimeter steel frames coupled by steel outrigger trusses at four levels. The length of the main building is 70m and 37m in longitudinal and transverse respectively. Therefore, the aspect ratio is 8.78 which has largely exceeded the criteria in current design specification. The detailed description of structural system of this tall building can be referenced from Li and Wu [9].

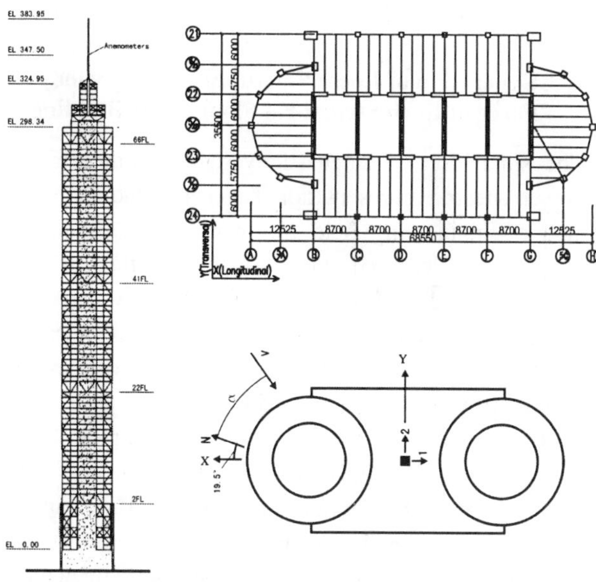

Fig. 1a Overview of Di Wang Tower

Fig. 1b Plan of standard floor and location of accelerometers

3. Field Measurement on Amplitude-dependent Nonlinear Damping of Shenzhen Di Wang Tower

The full scale test on wind-induced response of this building was conducted by Li et al. [10] during moderate Typhoon Sally in 1996. Xu et al. also [11] carried out field test on the wind-induced response during the strongest typhoon York attacked this building on 16 September 1999. A series of full scale tests on the wind-induced response of this tall building were made by Xiao et al. [12] during some moderate typhoons (Typhoon Compass, Imbudo and Dujuan in 2003) attacked this building. In their field tests, two accelerometers were installed at 68th floor of the main tower to measure the acceleration responses as shown in Fig. 1b. The series of full scale tests supply us a good opportunity to investigate the amplitude dependent damping ratios for this tall building. The amplitude-dependent damping ratios of the building were obtained by the random decrement technique with different amplitude levels. Based on the test results, a empirical amplitude-dependent nonlinear damping model was proposed in this study.

According to Jeary's proposed damping model for tall buildings [6], the damping values should include in two separate parts: those of low-amplitude damping and a rate of increase of damping with amplitude. A high amplitude plateau with constant damping ratio was also proposed by Jeary [6] when the response amplitude of the tall building reaches a specified higher range. As Di Wang Tower is classified as a steel-concrete composite structure, the upper limit of damping ratio with 3% is widely accepted in the current design specifications. Therefore a nonlinear damping model with two amplitude plateau characteristic was proposed in this study based on the series of full scale tests on this building, the proposed model is showed in Fig. 2 to 5 and described by the following formulas:

$$\xi_{NLY1} = \begin{cases} 3.1\times10^{-4}x + 0.0043 & 0 \le x \le 82.90mm \\ 3\% & 82.90mm < x \end{cases} \tag{1}$$

$$\xi_{NLY2} = \begin{cases} 0.55\times10^{-4}x + 0.00499 & 0 \le x \le 454.72mm \\ 3\% & 454.72mm < x \end{cases} \tag{2}$$

$$\xi_{NLX1} = \begin{cases} 8.81\times10^{-4}x + 0.004293 & 0 \le x \le 29.16mm \\ 3\% & 29.16mm < x \end{cases} \tag{3}$$

$$\xi_{NLX2} = \begin{cases} 8.57\times10^{-4}x + 0.0043 & 0 \le x \le 29.97mm \\ 3\% & 29.97mm < x \end{cases} \quad (4)$$

in which ξ_{NLX1} is amplitude-dependent nonlinear damping ratio for the first vibration mode in the X direction, the other three variables ξ_{NLX2}, ξ_{NLY1}, ξ_{NLY2} can be specified in the similar way. It was shown from Eqs. (1) to (4) that the damping ratios in the main X-axis direction are normally larger than those in Y-axis direction at the same amplitude level. Compared with other three modes, the damping ratio for the second vibration mode in the Y-direction increases much slowly with increasing amplitude.

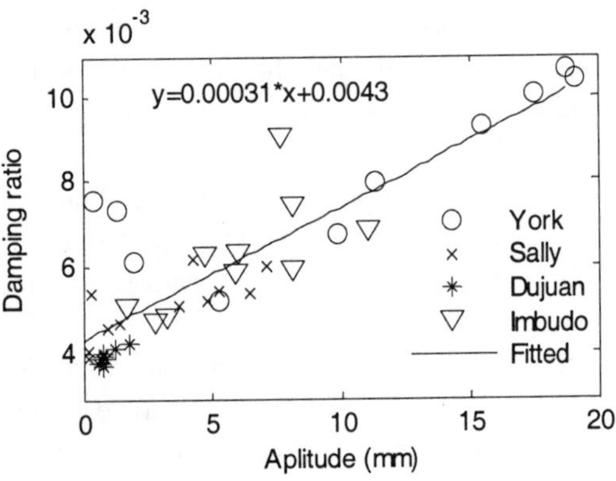

Fig. 2 The nonlinear damping model for the first mode in Y direction

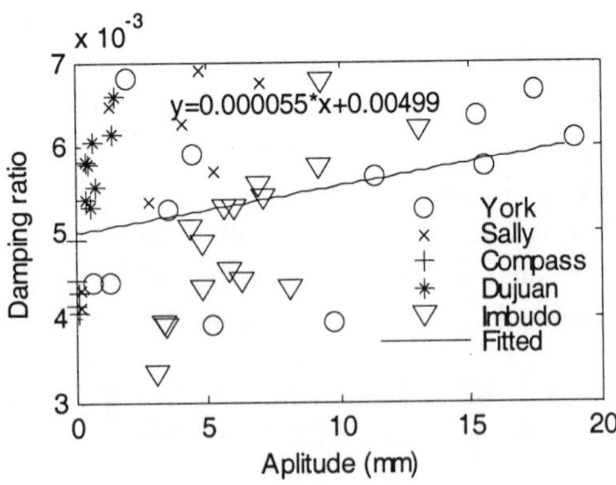

Fig.3 The nonlinear damping model for the second mode in Y direction

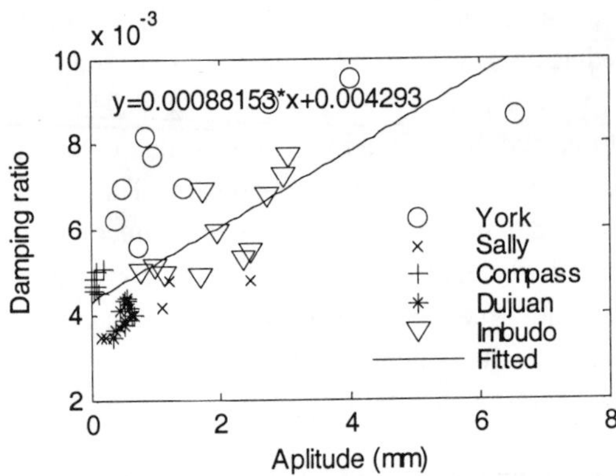

Fig. 4 The nonlinear damping model for the first mode in X direction

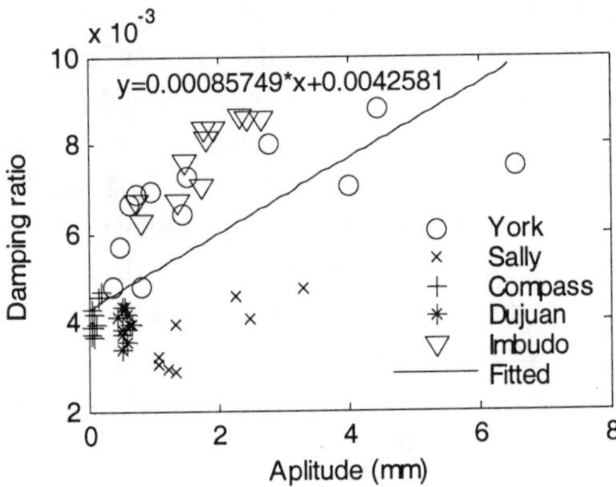

Fig. 5 The nonlinear damping model for the second mode in X direction

4. Generation of Local Fluctuating Along-wind Force at Each Structural Floor of the Building

The simulation of stochastic variables with arbitrary power spectra representation has been studied in the area of engineering field for many years. Shinozuka [13], Shinozuka and Jan [8] studied the efficient methods for digital simulation for the one-dimension multi-variable stochastic process by a series of cosine functions with weighted amplitude, almost evenly spaced frequencies, and random phase angles. This method, also termed as Weighted Amplitude Wave Superposition (WAWS) method [7], has an attractive feature that Fast Fourier Transform (FFT) can be adopted in the simulation procedure. Therefore it was used in this study to generate the local fluctuating along-wind acting on this building.

Based on the established 3D FEM model of Di Wang Tower [9], the lateral condensed stiffness and mass matrix in the main X- and Y-axis direction for this building are formed for wind-induced dynamic analysis. In each direction, the structural system of this tall building can be considered to be an equivalent cantilever structure with 79 lumped masses as shown in Fig. 1a. Normally the co-spectral density of the fluctuating along-wind drag force at each lumped mass can be expressed as the following:

$$S_{FDij}(z_i,z_j,f)=(\rho C_D)^2 A_{zi}A_{zj}V_{zi}V_{zj}\sqrt{S_V(z_i,f)S_V(z_j,f)}Coh(z_i,z_j,f) \tag{5}$$

where ρ is the air density, C_D is the drag coefficient. A_{zi}, V_{zi} are tributary area and mean wind speed at the specified height Z_i respectively. $S_V(z_i,f)$ is the auto-power spectral density of the along-wind fluctuating wind speed at the height Z_i, which is represented by Van Karman spectra with the following form [14]:

$$\frac{fS_V(z_i,f)}{\sigma_{Vzi}^2}=\frac{4f\times L_V/V_{zi}}{\left(1+70.8(4f\times L_V/V_{zi})^2\right)^{5/6}} \tag{6}$$

in which σ_{Vzi} and L_V are the RMS value of fluctuating wind speed and the turbulence integral length scale at specified height Z_i, respectively. The coherence function $Coh(z_i,z_j,f)$ of local fluctuating wind speed was represented by the following Eq. [4]:

$$Coh(z_i,z_j,f)=\exp(-8\times f\times\Delta z/V) \tag{7}$$

where $\Delta z=|z_i-z_j|$ and $V=(V_{zi}+V_{zj})/2$. Normally the following equation $S_{FDij}(z_i,z_j,f)=S_{FDji}(z_j,z_i,f)$ is always satisfied. Thus it implied that S_{FD} is a real symmetric matrix. The Choleskey decomposition to the real symmetric matrix S_{FD} can be therefore performed by the following form: $S_{FD}=HH^T$. The simulated fluctuating along-wind loads at the i-th lumped mass can be expressed as the following [8,13]:

$$F_D(z_i,t)=\sum_{j=1}^{i}\sum_{k=1}^{N}\left|H_{ij}(z_i,z_j,\omega_k)\right|\sqrt{2\Delta\omega_k}\left[\cos(\omega_k t+\psi_{jk})\right] \quad i=1,M \tag{8}$$

where $\Delta\omega_k$ is frequency interval, M is the number of lumped masses representing the total structural floors of the building structure, N is the number of frequency intervals, $\omega_k=(k-1)\Delta\omega_k$ and ψ_{jk} is independent random phase angle uniformly distributed between 0 and 2π. $H_{ij}(z_i,z_j,\omega_k)$ is the entry in Choleskey decomposed matrix H.

5. Time History Dynamics Analysis of Along-wind Response of Di Wang Tower

Based on the 3D FEM model of this building, the lateral condensed stiffness and mass matrix in both axis direction of this tower was established. The equation of motion of this tower in each main-axis direction can be expressed in the following equation:

$$[M]\{\ddot{y}\}+[C]\{\dot{y}\}+[K]\{y\}=\{F(t)\} \tag{9}$$

in which $[M]$, $[C]$ and $[K]$ are the 79×79 order matrix mass, damping and lateral condensed stiffness matrices of the structural system in each main-axis direction. The $\{\ddot{y}\}$ and $\{F(t)\}$ are vectors for the acceleration and local fluctuating wind force acting on the corresponding lumped mass respectively. The damping of the buildings is normally taken as Rayleigh damping model [15], in which the damping matrix are usually assumed to be proportional to the mass and stiffness matrices of the structural system, i.e

$$[C]=\alpha[M]+\beta[K] \tag{10}$$

where α,β are the proportional damping constants related to the dynamic properties of the structure, which can be determined from the damping ratio ξ_1 , ξ_2 and natural frequency ω_1,ω_2 for the first two vibration modes respectively [15]. For the time domain solution of Eq. 9, the step-by-step Newmark integration method can be used [16]. Only some minor modifications and iteration algorithms are needed for considering the amplitude-dependent damping effect.

6. Results of Time History Analysis for Along wind Response of Di Wang Tower

The time history analysis for along-wind response of this building in each main-axis direction was conducted in this study when different mean wind speeds atop of the building attacked the tower. The building was assumed as an isolated building located in Terrain type D as specified in the China load code. In order to evaluate the nonlinear damping effect on the wind-induced response, the empirical amplitude dependent nonlinear models proposed in this study were selected besides other three constant damping levels (1%, 2% and 3%). The along-wind response of this tower were conducted when the mean wind speed atop of the building was 15m/s, 27m/s, 32m/s and 48m/s respectively and wind is assumed to attack perpendicular to one side of the building. The time step was selected as 0.05 seconds and total time for step-by-step dynamics analysis is 10 minutes. In order to evaluate the time domain analysis results, the wind-induced response analysis in frequency domain for this building under the same load cases were also carried out for comparison purposes. The results of wind-induced response analysis were described in detail in the following sections.

6.1 *The along-wind response in X direction under different load cases*

The time history along-wind acceleration and displacement response atop the building were shown in Figs. 6 and 7 on condition that different constant damping ratios and proposed nonlinear damping model were selected and mean wind speed atop of the building is 15m/s. The wind-induced responses during the first 3-minutes were only shown in above figures for clarification. It was shown from these figures that the wind-induced response based on the proposed nonlinear damping model is less than all those responses resulted from constant damping ratios 1% and 3% respectively. The statistical results (RMS and maximum values) for the time history along-wind response of this tower under other load cases were listed in Table 1. It was shown in this table that the result obtained from frequency-domain analysis is normally a little greater than those from time domain. This showed that the accuracy of time-domain wind-induced analysis methods proposed in this study is generally reliable. The peak factor shown in Table 1 from time-domain analysis is normally between 2.7~3.3, which is a little less than the generally adopted value of 3.5 in wind engineering. From the results in Table 1, the effect of nonlinear damping on the along-wind response at different mean speed level was clearly demonstrated. It was shown that the actual damping ratios for the first mode in the X-direction of the tower were less than 1% when mean wind speed of 15m/s attacked the top of the tower. The actual damping ratios are almost 1%, 1%~2% and 2% respectively when other three mean speed levels (27m/s, 32m/s and 48m/s) attacked the building in this direction.

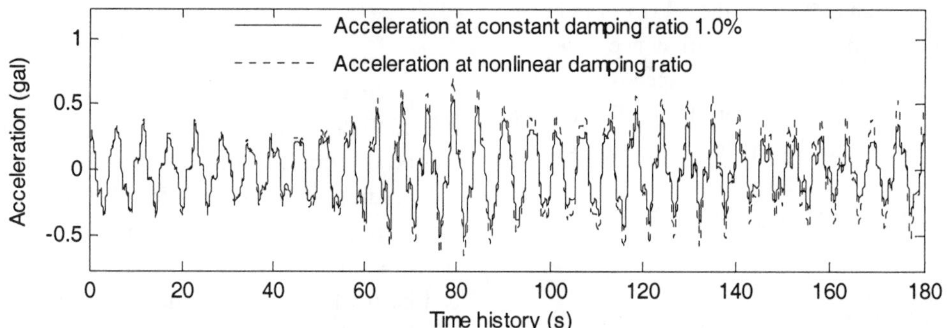

Fig. 6 Along-wind acceleration in X direction atop of the building (damping ratio: 1% and nonlinear damping model, mean wind speed 15m/s)

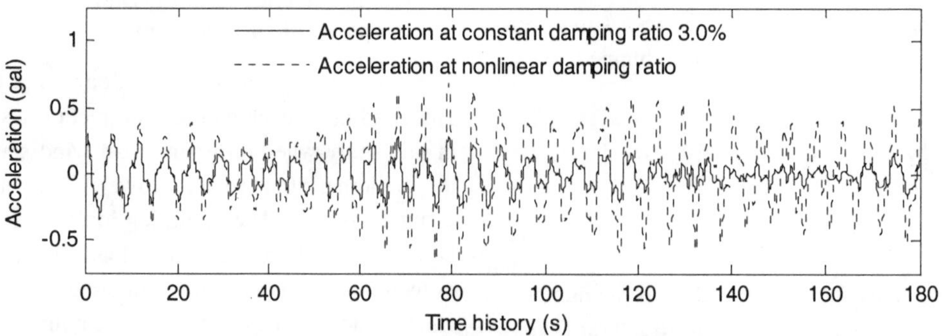

Fig. 7 Along-wind acceleration in X direction atop of the building (damping ratio: 3% and nonlinear damping model, mean wind speed 15m/s)

Table 1 The along-wind induced response of Di Wang Tower in X direction

Wind speed	Damping ratio	Analyzed results in time domain						results in frequency domain	
		Displacement (mm)			Acceleration (gal)				
		RMS value	Maximum value	Peak factor	RMS value	Maximum value	Peak factor	RMS displacement (mm)	RMS acceleration (gal)
15m/s	NL	2.32	6.62	2.85	0.24	0.70	2.91		
	1%	1.95	5.69	2.91	0.18	0.57	3.17	2.23	0.20
	2%	1.70	4.75	2.79	0.12	0.38	3.17	2.01	0.14
	3%	1.63	4.45	2.73	0.10	0.31	3.10	1.94	0.12
27m/s	NL	6.72	19.32	2.88	0.73	2.39	3.27		
	1%	6.43	18.49	2.87	0.69	2.26	3.27	7.39	0.75
	2%	5.34	15.01	2.81	0.49	1.61	3.29	6.14	0.59
	3%	4.98	13.25	2.67	0.41	1.29	3.15	5.43	0.44
32m/s	NL	11.04	29.50	2.67	1.22	3.31	2.71		
	1%	11.59	31.42	2.71	1.31	3.77	2.88	14.5	1.64
	2%	9.04	25.17	2.78	0.91	2.53	2.78	11.38	1.16
	3%	8.19	23.34	2.84	0.73	2.04	2.79	10.10	0.90
48m/s	NL	24.11	68.29	2.83	2.39	6.57	2.74		
	1%	27.20	71.65	2.63	2.93	7.82	2.67	32.08	3.26
	2%	23.72	68.91	2.90	2.31	6.71	2.90	26.73	2.58
	3%	22.52	59.31	2.63	1.98	5.68	2.87	24.28	2.13
Note: NL---proposed nonlinear damping model									

6.2 *The along-wind response in Y direction under different load cases*

The along-wind responses in Y-direction of this tower were also carried out in time domain by the same procedure as previous section. The time history acceleration and displacement response atop the building were shown in Figs. 8 and 9 when the mean wind speed atop of the building is 15m/s. Compared with Figs. 6 and 8, it seems that the difference in along-wind response between nonlinear damping model and constant damping ratio 1% is relatively less in this direction. This may show that the actual damping ratio for the first mode in this direction is approaching to about 1%, while the actual damping ratio in X –direction is relatively less than 1% when same mean wind speed attacked this building.

The statistical results for the along-wind response of this tower in Y-direction under other load cases were listed in Table 2. The peak factor shown in Table 2 is normally between 2.5~3.6, which is also a little less than the generally adopted value of 3.5. Compared with the calculated results in Table 1, it showed that along-wind response in Y-direction are much greater than those in X-direction when the same mean wind speed attacked this building. This showed that the lateral stiffness of the tower in Y-direction is relatively weak and more sensitive to wind excitation. It can be also seen from Table 2 that that the actual damping ratios for the first mode in this direction were almost 1% when mean wind speed of 27m/s attacked the tower. The actual damping ratios for other different mean speed levels (32m/s and 48m/s) are almost 1% and 2% respectively.

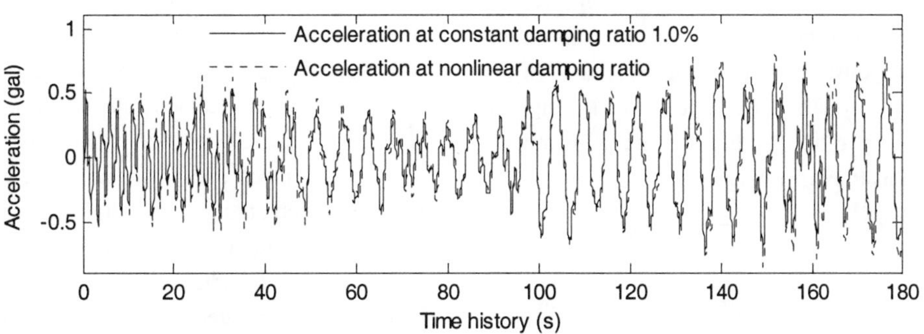

Fig. 8 Along-wind acceleration in Y direction (damping ratio: 1% and nonlinear damping model)

Table 2 The along-wind induced response of Di Wang Tower in Y direction

Wind speed	Damping ratio	Analyzed results in time domain						results in frequency domain	
		Displacement (mm)			Acceleration (gal)				
		RMS value	Maximum value	Peak factor	RMS value	Maximum value	Peak factor	RMS displacement (mm)	RMS acceleration (gal)
15m/s	NL	4.65	13.26	2.85	0.38	1.17	3.07		
	1%	4.28	13.20	3.08	0.31	0.97	3.12	4.85	0.39
	2%	3.97	13.00	3.27	0.24	0.75	3.13	4.08	0.28
	3%	3.79	12.29	3.24	0.20	0.66	3.30	3.89	0.22
27m/s	NL	25.17	67.58	2.68	2.49	6.83	2.74		
	1%	24.18	66.90	2.76	2.34	6.15	2.62	25.30	2.01
	2%	18.57	48.89	2.63	1.61	4.57	2.83	20.62	1.41
	3%	16.12	41.49	2.57	1.26	3.68	2.92	18.78	1.15
32m/s	NL	30.38	93.33	3.07	2.76	9.30	3.36		
	1%	31.00	95.59	3.08	2.76	9.24	3.34	38.90	3.19
	2%	26.74	85.17	3.18	2.06	7.32	3.55	31.10	2.25
	3%	24.61	77.69	3.16	1.70	6.18	3.63	28.06	1.83
48m/s	NL	67.41	197.30	2.93	6.64	22.05	3.32		
	1%	85.18	248.14	2.91	8.60	28.25	3.29	97.9	9.57
	2%	65.49	195.94	2.99	6.02	19.38	3.22	72.7	6.77
	3%	56.72	170.10	3.01	4.81	15.29	3.19	62.83	5.51
Note: NL---proposed nonlinear damping model									

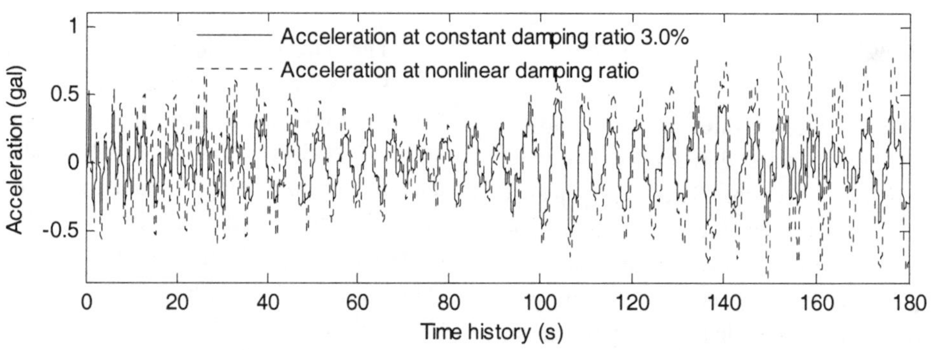

Fig. 9 Along-wind acceleration in Y direction atop of the building (damping ratio:3% and nonlinear damping model, mean wind speed 15m/s)

7. The Effect of Time Constant on the Wind Induced Response of Tall Building

As indicated in Tables 1 and 2, the statistical results of wind-induced response obtained from time-domain analysis are relatively less than those from frequency-domain analyzed results. The main reason may probably results from the effect of time constant on the dynamics response. As specified by Jeary [6], a structure with a resonance frequency of f_0 and an associated damping ratio ξ_0 has a time constant τ which is defined as the following form:

$$\tau = \frac{1}{2\pi f_0 \xi_0} \tag{11}$$

When the along-wind response of a tall building was evaluated in frequency domain, the ratio of the RMS deflection σ_x to the mean deflection $\bar{x}$ atop the building can be obtained by the following equation [16]:

$$\frac{\sigma_x}{\bar{x}} = 2I_u\sqrt{B + \frac{SE}{\xi_0}} \tag{12}$$

where I_u is the turbulence intensity atop the building, B is the background excitation factor, S is the size reduction factor to account for the correlation of pressure over a building, E is the gust wind energy factor which is defined by the Eq. $E = \pi/4 f_0 S(f_0)$, where $S(f_0)$ is the value of normalized fluctuating along-wind speed spectra at natural frequency f_0 and it can be calculated from the right terms in Eq. 6.

The acceleration response can be estimated by Eq. 13 in the way that the background displacement response contributed almost nothing to the RMS acceleration.

$$\frac{\sigma_x}{\bar{x}} = (2\pi f_0)^2 2I_u \sqrt{\frac{SE}{\xi_0}} \tag{13}$$

The relationships in Eqs. 12 and 13 assume that no effect caused by systematic response is presented [6]. In practice a systematic effect exist for a damped structural system. For a linear damped system responding to a specified load with single frequency sine wave, it will reach 99.3% of its final steady-state response only after 5 time constants (5τ). The structure may never reach the response estimated by Eqs. 12 and 13 because of the lags between the actual response and the "final" steady-state response. It was found that the reduction factor of RMS displacement of the structure with long time constant can reach to about 20% in Jeary's study. By taking this building as an example, the effect of time constant on wind-induced response will be studied and extended to multi-degree of freedom structural system.

Table 3 The along-wind induced response in Y-direction of the simulated system (mean wind speed is 32m/s atop of the building)

Damping ratio	Time constant (sec/rad)	results in frequency domain		results in time domain		Ratio of results in time domain to those in frequency doamin	
		RMS acceleration (gal)	RMS displacement (mm)	RMS acceleration (gal)	RMS displacement (mm)	acceleration	displacement
0.1%	930.1	10.29	108.5	7.59	79.21	0.739	0.730
0.2%	465.1	7.04	75.8	5.39	57.46	0.766	0.754
0.4%	232.5	5.05	56.43	4.13	44.03	0.819	0.781
0.8%	116.7	3.51	41.80	3.10	33.85	0.885	0.806
1%	92.9	3.19	38.90	2.76	31.00	0.869	0.810
1.5%	61.9	2.60	33.90	2.34	28.56	0.884	0.847
2%	46.5	2.25	31.10	2.06	26.24	0.917	0.862
2.5%	37.1	2.01	29.30	1.84	25.30	0.917	0.862
3%	31.1	1.83	28.06	1.70	24.61	0.934	0.877
3.5%	24.3	1.69	27.10	1.57	23.85	0.934	0.884
4%	23.3	1.58	26.30	1.48	23.32	0.943	0.893
4.5%	20.6	1.49	25.78	1.41	23.04	0.952	0.847
5%	18.7	1.41	25.30	1.35	22.79	0.962	0.901

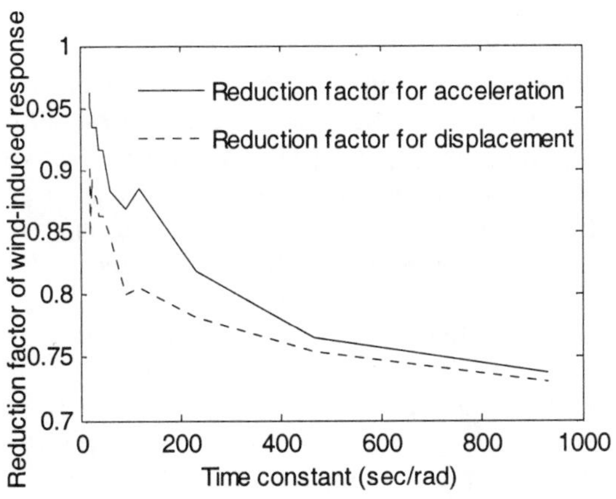

Fig.10 The effect of time constant on wind-induced response

Based on the generated local along-wind fluctuating force, the along-wind responses in Y direction of the building were tested in both time and frequency domain for a series of simulated structural system when the mean wind speed atop of the building is 32m/s. For the simulated series of system, it was assumed that the geometry and dynamics characteristic except damping ratio are the same. While damping ratios of the series of system were supposed to be at 13 different constant levels (0.01%~5% as indicated in Table 3). Therefore time constant for the series of system concerned were in range between 18.59~930 sec/rad. The ratios of RMS response obtained by time domain analysis to those by frequency domain method actually reflect the effect of time constant on the along-wind induced response for the investigated building. These ratios are also shown in Fig. 10. It was shown from this figure that there is a systematic decrease in the ratio of simulated time domain results to the "expected" frequency domain steady response as the time constants generally increase. The maximum of reduction factor can reach to 73% over the range of time constants investigated. As the tall building structures are constructed with more light weighted and stronger strength material, there is a natural trend that the natural frequency to become lower and the damping ratios to become smaller. Therefore the time constant of these kinds of tall buildings may be more greater than that of the exist buildings, the reduction of the wind-induced response by time domain analysis method among the certain time constants considered offers a potential benefit for security in wind resistant design of tall building structures.

8. Conclusion

The wind-induced response of Di Wang Tower was numerically carried out in time domain by generating the local fluctuating along-wind force and the proposed nonlinear damping model. The nonlinear damping effect on wind-induced response was evaluated under different load cases. The effect of time constant on the along-wind response was analyzed for this building. Some conclusions were obtained from this study.

(1) The peak factors of wind-induced response are normally between 2.5~3.6 from time-domain analysis. This value is less than the generally adopted value 3.5 in current design specification.

(2) The effect of nonlinear damping on the along-wind response at different speed levels was clearly demonstrated. From this study, it seems that the actual damping ratios were about 1% when mean wind speed atop of building is less than 27m/s. The actual damping ratios are generally between 1%~2% when the building was attacked by other mean speed levels investigated in this study.

(3) The along-wind responses in Y-direction are much greater than those in X-direction when the same mean wind speed attacked this building. This showed that the lateral stiffness in Y direction is relatively weak and more sensitive to wind excitation.

(4) The effect of time constant on wind-induced response was obvious when the value of time constant reached a certain range. There is a systematic decrease in the ratio of time domain results to the "expected" steady response as the time constants increase. The maximum of reduction factor can reach 73% over the range of time constants investigated.

Acknowledgment

The work described in this paper was fully supported by a grant from the Research Committee of City University of Hong Kong (Project No. 7001813). The financial support is gratefully acknowledged

References

1. A. P. Jeary, *Earthquake Engineering and Structural Dynamics*, **14**, pp.773-750, (1986).
2. Y. Tamura and S. Suganuma, *Journal of Wind Engineering and Industrial Aerodynamics*, **59**, pp.115-130, (1996).
3. Q. S. Li, K. Yang, C. K. Wong and A. P. Jeary, *Journal of Wind Engineering and Industrial Aerodynamics*, **91**, pp.1175-1198, (2003)
4. H. Tsukagoshi, Y. Tamura, A. Sasaki and H. Kanai, *Journal of Wind Engineering and Industrial Aerodynamics*, **46&47**, pp.497-506, (1993).
5. B. Liang, and Y. Tamura, *Computers & Structures*, **63**(3), pp. 601-606, (1997).
6. A. P. Jeary, *11th International Conference on Wind Engineering*,.(2003).
7. A. Iannuzzi and P. Spinelli, *Journal of Structural Engineering*, ASCE, **113**, pp.2382-2398, (1987).
8. M. Shinozuka and C. M. Jan, *Journal of Sound and Vibration*, **25**(1), pp. 11-128, (1972).
9. Q. S. Li and J. R. Wu, *Earthquake Engineering and Structural Dynamics,* **33**(14), pp.1311-1336, (2004).
10. Q. S. Li, Y. Q. Xiao, C. K. Wong and A. P. Jeary, *Engineering Structures*, **26**(2), pp. 233-244, (2004).
11. Y. L. Xu, S. W. Chen and R. C. Zhang, *The Structural Design of Tall and Special Buildings*. **12**, pp. 21-47, (2003).
12. Y. Q. Xiao, Q. S. Li and Z. N. Li, *The 3rd International Conference on Advances in Structural Engineering and Mechanics*, Seoul, Korea, (2004).
13. M. Shinozuka, pp. 93-133 in *Stochastic Methods in Structural Dynamics*, Martinus, (1987).
14. E. Simiu and R. H. Scanlan, *Wind effects on structures: fundamentals and applications to design*. John Wiley & Sons, Inc., (1996).
15. R. W. Clough and J. Penzien, *Dynamics of Structures*, McGraw-Hill, Inc. (1993).
16. J. W. Tedesco, W. G. Mcdougal and C. A. Ross, *Structural Dynamics: Theory and Applications*, Addison Wesley Longman, Inc., USA, (1999).
17. J. D. Holmes, W. H. Melbourne and G. R. Walker, *A Commentary on the Australian Standard for Wind Loads AS1170 Part 2, 1989*, (1990)

WIND LOAD ESTIMATION FOR A PRACTICAL TALL BUILDING CONSIDERING TOPOGRAPHICAL EFFECT

YUANQI LI, ZHENJUN LIU, JI XIA
Dept. of Building Engineering, Tongji University, 1239 Siping Rd., Shanghai 200092, China

YUKIO TAMURA
Wind Engineering Research Center, Tokyo Polytechnic University, Atsugi 243-0297, Japan

In this paper, according to the relative provisions of several important load codes/specifications in the world, estimation of the speed-up ratios of wind speed or wind pressure profile considering topographical effect has been discussed based on a practical important tall building being built on an island, and the effect of topography on the gust factor of wind speed has also been analyzed according to the related provisions of the new load code in Japan. Then, with an escarpment model and a ridge model, topographical effects on the estimated wind load for designing the main structural system and the maintenance components of this practical project were analyzed, respectively. Based on these investigations, the difference of the estimated values of the speed-up ratios due to topographical effect among these codes/specifications, including china load code, is presented, and the tendency in current revision of the corresponding provisions is discussed. The results can offer a convincible reference to wind load analysis in wind-resistant design for the practical tall building, as well as of other similar projects in the future.

1. Introduction

The building discussed in this paper is located at the west mountainside of an island, as shown in Figure 1. The elevation of the main building is +95.80m, and +59.40m for the outside annex buildings. The elevation of the base plane of the building is +40.50m. Figure 1 shows the topographical variations of two representative sections for the building in two direction, i.e., Section 1-1 and Section 2-2, as given by the continuous lines.

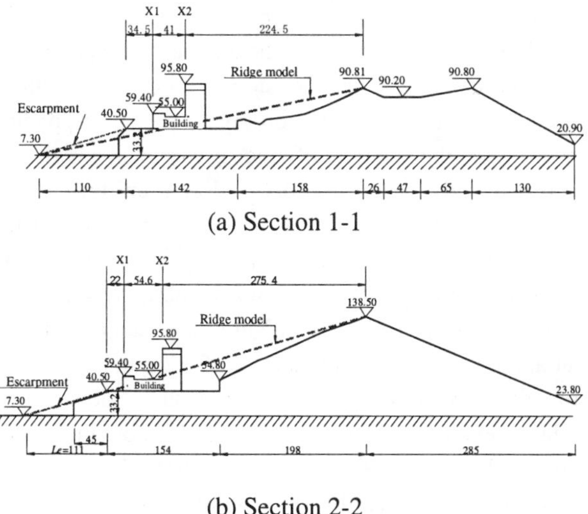

(a) Section 1-1

(b) Section 2-2

Figure 1 Two types of sections and the sketches of the escarpment model and the ridge model

As the administrative center of an important project, the importance of the building is evident. Considering the complicated topography in site, the possible amplification of the wind velocity or wind pressure due to the topographical effect should be estimated reasonably in wind-resistant design of the main structural frames, as well as the maintenance structural components of the building[1,2]. Since there is no information in details about the topography effect investigated by wind tunnel test, and the basic wind pressure itself in site is very strong, i.e., $w_0 = 1.3\,\text{kN/m}^2$, corresponding to a designed reference wind velocity of 45.6m/s. If a ridge model in China code is adopted, the speed-up ratio of wind pressure will induce very large wind load which may be very hard to be borne. Therefore, estimation on reasonable wind load value considering topographical effect is an important issue in the wind-resistant design of the building structure and the maintenance structural components.

In this paper, according to the relative provisions of several important load codes/specifications in the world, including ASCE (2000) [3], AIJ (1996&2004) [4,5], AS/NZS (2002) [6], EN (2002) [7], GB50009-2001 (2001) [8] and NBC (1995) [9], estimation of the speed-up ratios of wind speed or wind pressure profile considering topographical effect has been discussed based on the practical important tall building being built on an island, and the effect of topography on the gust factor of wind speed has also been analyzed according to the related provisions of the new load code in Japan. Then, with an escarpment model and a ridge model, topographical effects on the estimated wind load for designing the main structural system and the maintenance components of this practical project were analyzed, respectively. Based on these investigations, the difference of the estimated values of the speed-up ratios due to topographical effect among these codes/specifications, including china load code, is presented, and the tendency in current revision of the corresponding provisions is discussed. The results can offer a convincible reference to wind load analysis in wind-resistant design for the practical tall building, as well as of other similar projects in the future.

2. Basic Consideration

2.1 *Wind pressure variation ratio corresponding to height*

The wind pressure factor corresponding to height, μ_z, in mountainous area can be estimated according to the roughness category of the corresponding flat terrain considering topography condition in case. The effect of wind velocity or wind pressure profiles in upwind and downwind directions in an escarpment or a ridge area are commonly considered by speed-up ratios as[1~9]:

$$\eta_x = \frac{p'(x,z)}{p(z)} = (\eta_{Vx})^2 = (\frac{\overline{V}'(x,z)}{\overline{V}(z)})^2 \quad (1)$$

where, η_x and η_{Vx} are so-called speed-up ratios of wind pressure and velocity where the horizontal and vertical distances from the crest of an escarpment or a ridge are x and z, respectively; $p'(x,z)$ and $\overline{V}'(x,z)$ are the wind pressure and velocity where the horizontal and vertical distance from the crest of an escarpment or a ridge are x and z, respectively, $p(z)$ and $\overline{V}(z)$ are the approaching wind pressure and wind velocity at height z above the flat terrain.

Generally, when the slope in upwind direction of an escarpment or a ridge is larger than a certain degree, e.g., 1/10[3], and the height of a building is higher than a certain value (e.g. 4.5m for category A in NBC, 1995[9]), modification of the wind velocity/pressure should be considered when the building is built within a certain range in vertical and horizontal distances from the top edge of an escarpment or a ridge.

2.2 *Modification of the gust factor considering variation of turbulence intensity*

The relationship between the gust factor, G, and the turbulence intensity, I_u, can be expressed as follows[10]:

$$G = \frac{\hat{U}_{t\sec}}{\overline{U}_{T\sec}} = 1 + 0.5 I_u \ln(T/t) \quad (2)$$

where, $\hat{U}_{t\sec}$ is the mean maximal wind velocity within the statistic time interval for the gust velocity; $\overline{U}_{T\sec}$ is mean wind velocity within the statistic time interval for the mean wind velocity.

According to Eq.(2), in wind-resistant design of the maintenance structural component of a building, the gust factor, β'_{gz}, considering the variation of turbulence intensity can be expressed as:

$$\beta'_{gz} = [1 + E_{gI}(\sqrt{\beta_{gz}} - 1)]^2 \quad (3)$$

where β'_{gz} is the gust factor used in estimating the wind pressure for wind-resistant design of maintenance structural component.

3. Analysis Model

For this practical problem, because the whole building is built on a ridge, but partly it is also on an escarpment at 33.2m height, two types of analysis models are adopted respectively to each representative section, as shown in Figure 1 by the common dashed line and the bold dashed line, respectively. In the figure, the common dashed line is an escarpment model only considering the local topography effect, and the equivalent length can be determined by the upwind slope exceeding the upper limit, i.e., $tg\alpha = 0.3$ [8], according to the corresponding code, but the height of the escarpment model is approximately assumed equal to 33.2m uniformly; The bold dashed line is a 2-D ridge model considering the effect of the whole island, and the equivalent length corresponding to the escarpment part, $L_e = H / tg\alpha = H/0.3$, where, H is the height of the escarpment with an approximately consideration of 33.2m uniformly. As for the three-dimensional ridge model, there is no relative explanation in China load code[8], but the results are usually smaller than that using a 2-D ridge model. In this paper, the effect of the whole hill as a ridge on wind pressure/velocity for the building is analyzed by a 2-D ridge model conservatively instead of a 3-D model, and more disadvantageous result from the two models will be suggested for actual wind-resistant design.

At present, most load codes/specifications for building design in the world only give estimation formulas about the effects of actual topography on mean wind velocity profile. But for the variation of gust factor due to the topographical effect on the turbulence intensities, only in Japan it has relative provisions. Generally, the effects on the turbulence intensity by escarpments or ridges are not so important, even the increase of mean wind velocity will cause a little decrease of turbulence intensity. However, the related provisions in the newest load standard in Japan, which has just revised based on a great deal of wind tunnel investigations and other recent research achievements, show that the effect on the turbulence intensities by actual topography increases much in a range near to the ground. This will cause some disadvantageous effects on wind-resistant design of maintenance structural components. In other words, the value of the gust factor will also increase in the range of areas near to the ground. In this paper, newly revised Japan standard is referenced to analyze the topographical effect on turbulence intensity of wind field for wind-resistant design of maintenance structural components of this project.

4. Comparison of Relative Provisions in Different Load Codes/Specifications

The speed-up ratios for wind velocity or wind pressure profiles are specified in the load codes/specifications of different countries according to an escarpment model and a ridge model as shown in Figure 2.

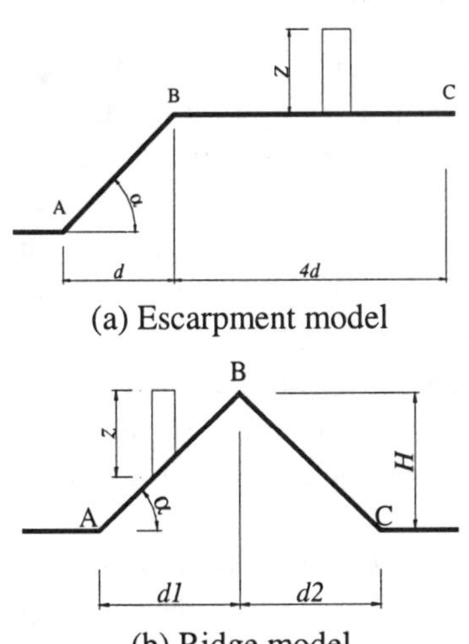

(a) Escarpment model

(b) Ridge model

Figure 2 Analysis models for topographical effects

4.1 *China load code* [8]

In China load code [8], terrain roughness is classified into four categories, i.e., A, B, C and D, respectively. The surface approaching sea, seaboard, lakeshore and desert area belongs to the A category.

According to China load code [8], speed-up ratios of wind speed/pressure profiles considering topographical effects are considered as follows:

The speed-up ratio of wind pressure at point B on the top of an escarpment or a ridge can be expressed as:

$$\eta_B = \left[1 + ktg\alpha\left(1 - \frac{z}{2.5H}\right)\right]^2 \tag{4}$$

where tg α is the upwind slope of the escarpment or the ridge , and it is equal to 0.3 when larger than 0.3; k is a constant, for escarpments, it equal to 3.2, and for ridges equal to 1.4; H is the height of the escarpment or the ridge at the crests; and z is the height of the location in question over the escarpment or the ridge, and equal to $2.5H$ when z>$2.5H$.

As for other location of the escarpment or the ridge, the speed-up ratios at points A and C are assumed equal to 1.0. Then between A and B, or B and C, they can be estimated by linear interpolation simply.

4.2 *Canada load code* [9]

In Canada load code, terrain roughness is classified into three categories: A, B and C, respectively. The surface approaching sea, seaboard, lakeshore and desert area belongs to the A category.

In Canada code, for the building on an escarpment or a ridge with a greatest upwind slope exceeding 1/10, topographical effect is necessary to be considered. A basal formula to estimate the speed-up ratios of wind pressure profile considering topographical effects is given as:

$$\mu_z^* = \mu_z(1 + k_1 k_2 k_3)^2 \tag{5}$$

where μ_z^* is the wind pressure factor on the escarpment or the ridge along height corresponding to difference roughness categories, μ_z is the wind pressure factor along height for the flat terrain on the bottom of the escarpment or the ridge, k_1 is a factor related to the slope of the escarpment or the ridge in windward side, H/L, as given in the corresponding tables in the code; $k_2 = 1 - |x|/(kL)$, and the coefficient k can be obtained in tables given in the code according to the type of topography; $k_3 = e^{-(\alpha z/L)}$, and α is a given constant. L is the horizontal distance from the crest to the half height location between the upwind foot and the top of the escarpment or the ridge, equal to $d/2$ in Figure 2. If H/L >0.5, then it equal to 0.5, and $L = 2H$.

4.3 *America load code* [3]

In America load code, terrain roughness is classified into four categories: A, B, C and D, respectively. The surface approaching sea, seaboard, lakeshore and desert area belongs to the D category.

In America load code, topographical effects is considered necessarily when $H/L_h \geq 0.2$, and $H > 15$ ft (4.5m) for the D category of the terrain roughness. A basal formula to estimate the speed-up ratios of wind pressure profile considering topographical effects is:

$$K_{zt} = (1 + K_1 K_2 K_3)^2 \tag{6}$$

where K_1 is a factor to account for shape of topographic feature and maximum speed-up effect, and given by table in the code; K_2 is a factor to consider reduction in speed-up with horizontal distance variation in upwind side or downwind side, as given by tables in the code according to the value of x/L_h; K_3 is a factor corresponding to reduction in speed-up with the variation of height, and given by a table in the code according to the value of z/L_h; L_h is defined as the same as in Canada load code[9].

4.4 *Australian/New Zealand load code*[6]

In Australian/New Zealand load code, terrain roughness is classified four categories: A, B, C and D, respectively. The surface approaching sea, seaboard, lakeshore and desert area belongs to the D category.

In Australia/New Zealand load code, within the range under topography effect (mostly prescribed by the horizontal distance from the crest and the height above the ground), the speed-up ratio of wind velocity is estimated by the following equations:

$$\begin{aligned} M_h &= 1.0 && H/(2L) < 0.05 \\ M_h &= 1 + \frac{H}{3.5(z + L_1)}\left(1 - \frac{|x|}{L_2}\right) && 0.05 \leq H/(2L) < 0.45 \\ M_h &= 1 + 0.71\left(1 - \frac{|x|}{L_2}\right) && H/(2L) > 0.45 \end{aligned} \tag{7}$$

where, x is the upwind or downwind horizontal distances of the structure to the crest of the hill, ridge or escarpment; L_1 is a length scale, in meters, to determine the vertical variation of M_h, to be taken as the greater of $0.36L_u$ or $0.4H$; L_2 is a length scale, in meters, to determine the horizontal variation of M_h, to be taken as $4L_1$ upwind for all types, and downwind for hills and ridges, or $10L_1$ downwind for escarpments.

The speed-up ratio of wind pressure considering topography effect can be approximately estimated by $(M_h)^2$.

4.5 *EU Code in revising* [7]

In Europe load code in revising[7], terrain roughness is classified into five categories, i.e., I, II, III, IV and V, respectively. The surface approaching sea, seaboard, lakeshore and desert area belongs to the I category.

The speed-up ratio factor of wind velocity considering topography effect is estimated by

$$C_0 = 1 + ks\phi \tag{8}$$

where, k is a constant determined by topography, when $\phi < 0.05$, k =0; when $0.05 < \phi < 0.3$, $k = \phi$ □ and when $\phi > 0.3$, k =0.3. s is a topographical location factor scaled to the length of the effective upwind slope length , L_e, which is given by a figure in the code, or calculated by a series of fitting equations; ϕ is the upwind slope, $\phi = H/L_u$; L_h is defined as the same as in America load code[3].

The speed-up ratio of wind pressure considering topographical effect can be approximately estimated by $(C_0)^2$.

4.6 *Old Japan code*[4]

In old Japan load standard, terrain roughness is classified into five categories: I, II, III, IV and V, respectively. The surface approaching sea, seaboard, lakeshore and desert area belongs to the I category.

This old version gives the corresponding equation just for estimating the effects of escarpments on wind velocity profile, and only some instructively explanations are given for buildings on ridges. Perhaps it is considered that the results for an escarpment model are usually more conservative than these from a ridge model.

The factor E_g to account for the topographical feature of an escarpment, i.e., the speed-up ratio factor of wind velocity profile considering topographical effect, could be estimated from the following equation:

$$E_g = E_a\left\{E_c\left(\frac{z}{D} - E_b\right) + 1\right\}\exp\left\{-E_c\left(\frac{z}{D} - E_b\right)\right\} + 1 \quad (9)$$

where, D is the height of escarpment, i.e. the height H in Figure 2 ; E_a, E_b and E_c are coefficients specified in the corresponding tables in the code.

Generally, the topographical effect factor, E_g, should greater than 1, and it is only used for the upwind slope greater than 5^0.

The speed-up ratio of wind pressure considering topographical effect can be estimated by $(E_g)^2$.

4.7 *New Japan code in use*[4]

At present, most load codes/specifications in the world do not have special provisions related to the gust wind factor induced by topographic effect on turbulence intensity. Only the newest Japan code give relative provisions based on a great deal of wind tunnel test and others newest investigations. It will affect the wind-resistant design for the maintenance components of a building on such sites, i.e. the gust wind factor will become greater near the ground. At the same time, the newest code replenished some corresponding formula to consider topographical effect of both escarpments and ridges on mean wind profiles.

The factor E_g to account for thetopographical feature of an escarpment, i.e., the speed-up ratio factor of wind velocity profile considering topographical effect, could be estimated from the following equation:

$$E_g = (C_1 - 1)\left\{C_2\left(\frac{Z}{H_s} - C_3\right) + 1\right\}\exp\left\{-C_2\left(\frac{Z}{H_s} - C_3\right)\right\} + 1 \quad (10)$$

where, H_s is the height of the escarpment, as H in Figure 2; C_1, C_2 and C_3 are the coefficients specified in the Eq.(9) as the same as $E_a + 1$, E_c and E_b, respectively. They can be estimated from the corresponding tables given in the code by the upwind slope and the ratio X_s / H_s. X_s is the horizontal coordinate of the location in question from the crest.

The speed-up ratio of wind pressure considering topographical effect can be approximately estimated by $(E_g)^2$.

The factor of turbulence intensities to account for the shape of topographic feature, E_{gI}, could be estimated by the following equation based on the amplifying coefficient of the standard deviation of the fluctuant wind velocity, E_I, obtained from topographic wind tunnel tests:

$$E_{gI} = \frac{E_I}{E_g} \quad (11)$$

$$E_I = (C_1 - 1)\left\{C_2\left(\frac{Z}{H_s} - C_3\right) + 1\right\}\exp\left\{-C_2\left(\frac{Z}{H_s} - C_3\right)\right\} + 1 \quad (12)$$

where, C_1, C_2, C_3 are coefficients given by some tables in the code according to the upwind slope at site and the ratio, X_s / H_s . X_s is the horizontal coordinate ,*x*, from the crest.

Generally, E_g and E_I should greater than 1.0, and it is only considered when the upwind slope is greater than 7.5^0.

5. Basic Analysis results

In this paper, based on two types of sections and two different analysis models, topographical effects have been analyzed according to the related provisions in different important load codes/specifications mentioned above. The corresponding results for the speed-up ratios of wind pressure profile and the amplifying factor of the gust wind are shown from Figure 3 to Figure 6, respectively.

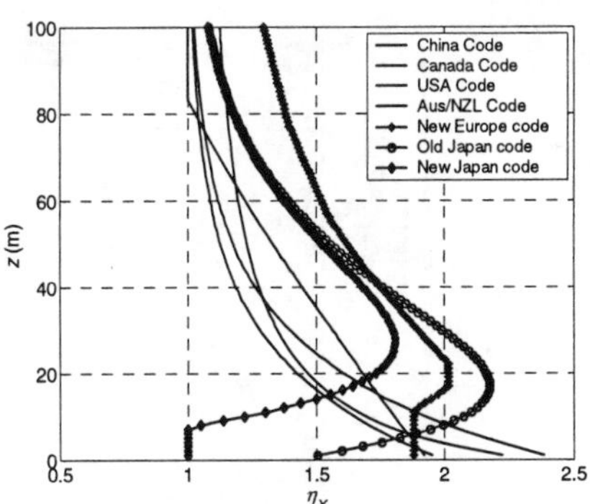

(a) Escarpment Model at X1

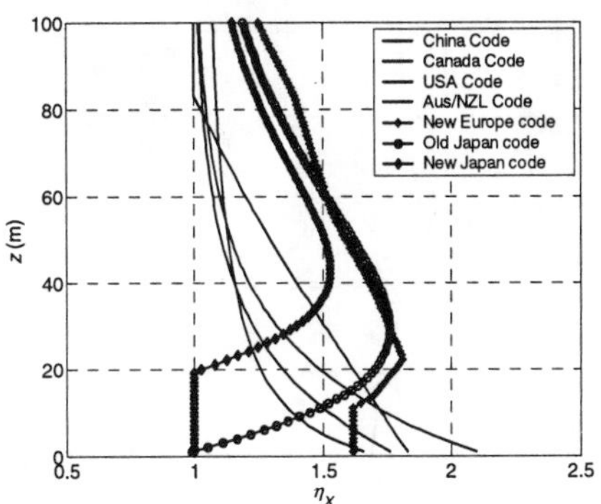

(b) Escarpment Model at X2

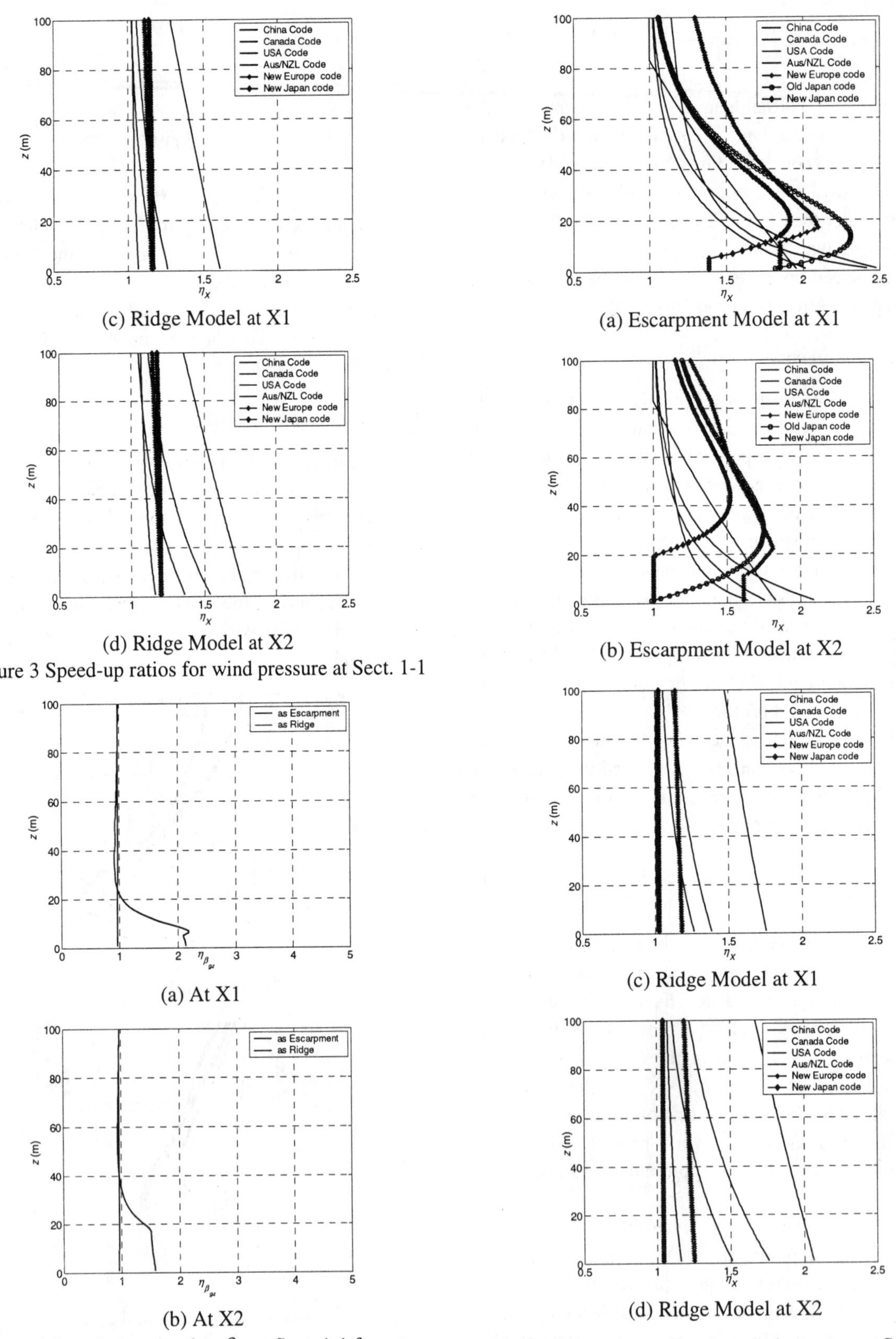

(c) Ridge Model at X1

(d) Ridge Model at X2

Figure 3 Speed-up ratios for wind pressure at Sect. 1-1

(a) At X1

(b) At X2

Figure 4 Speed-up ratios for β_{gz} at Sect. 1-1 from two models,□ β'_{gz}/β_{gz} □

(a) Escarpment Model at X1

(b) Escarpment Model at X2

(c) Ridge Model at X1

(d) Ridge Model at X2

Figure 5 Speed-up ratios for wind pressure at Sect. 2-2

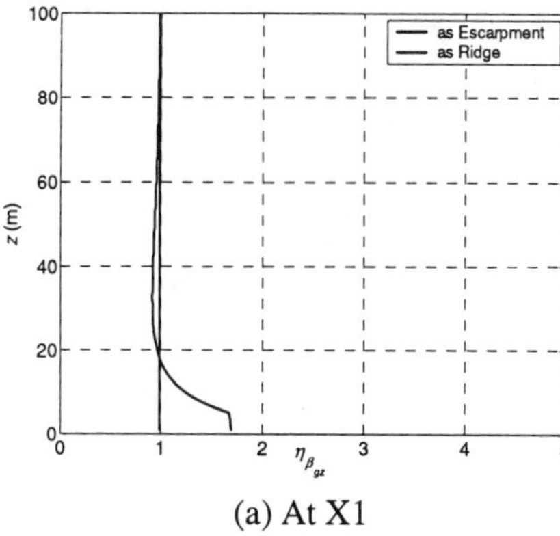

(a) At X1

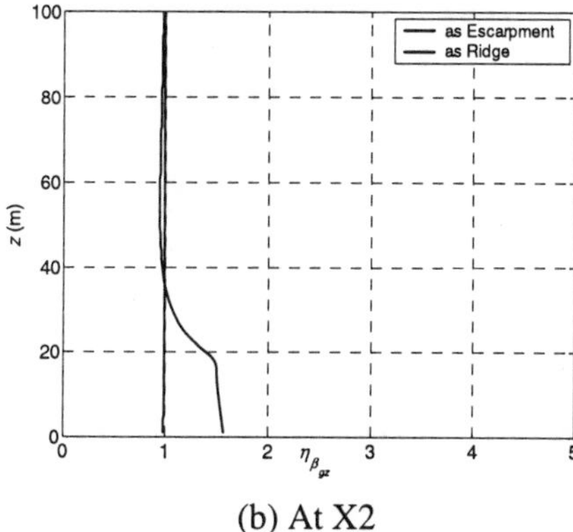

(b) At X2

Figure 6 Speed-up ratios for β_{gz} at Sect. 2-2 from two models, □ β'_{gz} / β_{gz} □

6. Conclusions

From above analyses, several conclusions can be drawn as follows:

1. It is shown that, for the building on an escarpment or a ridge, the horizontal distance of the building site from the crest and the height from the ground have dominant effects on the analysis results. Therefore, determination of the site at the initial stage of the project has important effects on the corresponding wind load in wind-resistant design for buildings built on those specific topographies.
2. For the case in this paper, results from the escarpment model are obviously greater than that using the ridge model. So it is more reasonable and credible to use an escarpment model in this project. With the escarpment model, the speed-up ratio of wind pressure profile based on the new European load code[7] is the greatest, if the results from the old Japanese load code[8] are not considered. Part of them may exceed 2.0. While with the ridge model, the speed-up ratio of wind pressure from China load code[3] is the biggest.
3. The results from the new Japan load code[9] are a little smaller than that from the old one, which shows a decreasing tendency of wind load under further exact investigations, especially that near to the ground.
4. For topographical effects on turbulence intensity of wind velocity, within a certain height in new Japan load code[9], turbulence intensity increases obviously, but it decreases a little out of the range due to the increase of wind velocity. In addition, this height depends on the horizontal distance of the analysis location from the crests of the escarpments or ridges.

Acknowledgments

This study was supported by the Scientific Research Foundation for the Returned Overseas Chinese Scholars, State Education Ministry, China.

References

1. J.D. Holmes, Wind loading of structures, Spon Press. (2001)
2. N.J. Cook, The designer's guide to wind loading on structures, Part 1, Butterworths. (1985)
3. GB50009-2001, Load code for the design of building structures, China Architectural & Building Press. (2002) (in Chinese)
4. National Commission of Building and Fire Code, National Building Code of Canada 1995 (NBC), National Research Council of Canada, 1995.
5. American Society of Civil Engineers, Minimum design loads for buildings and other structures, ANSI/ASCE 7-95, ASCE, New York. (2000)
6. Standards Australia, Structural design actions, Part 2: Wind Actions, AS/NZS 1170.2.2002, Standards Australia. (2002)
7. CEN/TC 250, EuroCode 1: Actions on structures, Part 1-4: General actions: Wind Actions, EN 2002-1-4. (2002)
8. Architectural Institute of Japan (AIJ), Recommendations for loads on buildings, published by Architectural Institute of Japan. (1997)
9. Architectural Institute of Japan (AIJ), Recommendations for loads on buildings, published by Architectural Institute of Japan. (2004) (in Japanese)
10. H., Ishizaki, Wind profiles, turbulence intensities and gust factors for design in typhoon-prone regions, Journal of Wind Engineering and Industrial Aerodynamics, 13:55-66. (1983).

Transfer Structure

SCHEME COMPARISON BETWEEN PRIMARY BEAM AND SECONDARY BEAM IN THE TRANSFER STRUCTURE OF TALL BUILDINGS

LIU YIWEI
Tsinghua-Yuan Architectural Design Ltd.
Shenzhen China

LIN RUTANG
Shenzhen Press Group
Shenzhen China

In this paper, the settlement of the vertical members supported by the secondary transfer beam is analyzed, and overturning resisting rigidity at the supporting point is studied for earthquake and the wind load. A comparison with those results obtained in primary beam case is made. It is reasonable to use secondary transfer beam to support the vertical member with the smaller lateral rigidity, if there is difficulty to use primary beam. Two secondary transfer beams can be used in crossing form rather than the single, to increase the settlement resisting rigidity and entirety.

1. Introduction

In the process of the engineering design, the structural engineer shall satisfy the requirement of the architects as far as possible. The structural engineers will face more grim challenges along with development of the variety of architectural design. As is known to all, the transfer design is a very important link in the whole process which the transfer is needed, and the use of the secondary transfer beam can make the overall structural arrangement more flexible in many occasions. The question we frequently meet is: what is the difference in mechanics characters between secondary transfer beam and primary transfer beam? In what occasion the secondary transfer beam can be used? And what are needed to pay attention in using of them?

2. Difference in Mechanics Character between Secondary Transfer Beam and Primary Transfer Beam

2.1 *Mechanics models adopted in the research*

In the different projects, the overall structural arrangement and the size of elements differ in thousands of ways actually. We have to take the simplest mechanics models of transfer beams supported by the columns, and adopt the most sketchy assumptions, to model the situation of the primary transfer beam and the secondary transfer beam separately. Even if they may differ from actual situation very much, as a relative comparison, they can express the general tendency of mechanics character and the regular in this kind of structure. Fig.1 shows the transfer of primary beam, Fig.2 shows the transfer of the single secondary beam, and Fig.3 shows the transfer of double secondary beams in crossing form. The planes of those models are symmetrical.

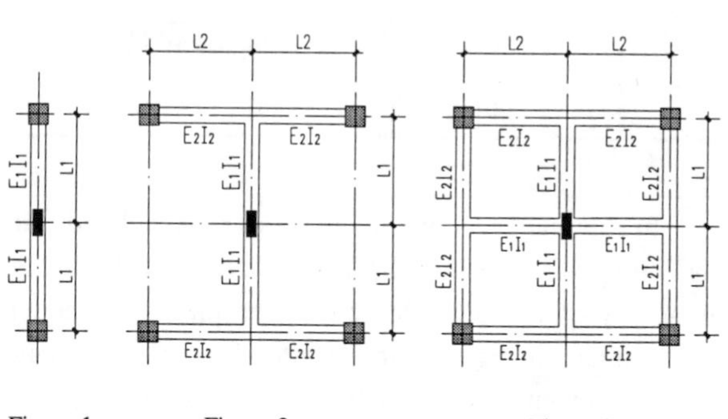

Figure 1 Figure 2 Figure 3

2.2 *Deduction of the formulas and the calculated samples*

According to three models mentioned above, neglecting the effect of the twist resisting rigidity of the beam, the formulas of the settlement flexibility and the overturning flexibility of the angle deformation at the bottom of the supported vertical members are deduced respectively. Two possible extreme situations are considered. The first one: the rigidity of the supporting column is very small; the other one: the rigidity of them is very large or the supported beams are continuous to the adjacent span. The two assumptions are adopted separately. One is neglecting the rigidity of supporting columns, another is considering the rigidity is infinitely large, to get the simple formulas. They can represent and contain the stage between both possible extreme states: upper limit and lower limit. The concrete formulas are showed in Table 1. At each formula, the first line shows the part caused by the bending of beam; and the second line shows which by the shearing.

Table 1. Approximate formulas to show deformation for symmetrical. models in three different supporting conditions

		According to Fig. 1	According to Fig. 2	According to Fig. 3
Vertical Settlement Flexibility $\Delta / P =$	upper limit	$\frac{L_1^3}{6EI_1} + \frac{L_1}{2GA_1}$	$\frac{L_1^3}{6EI_1} + \frac{L_2^3}{12EI_2} + \frac{L_1}{2GA_1} + \frac{L_2}{4GA_2}$	$\frac{L_1^3}{12EI_1} + \frac{L_2^3}{24EI_2} + \frac{L_1}{4GA_1} + \frac{L_2}{8GA_2}$
	lower limit	$\frac{L_1^3}{24EI_1} + \frac{L_1}{2GA_1}$	$\frac{L_1^3}{6EI_1} + \frac{L_2^3}{48EI_2} + \frac{L_1}{2GA_1} + \frac{L_2}{4GA_2}$	$\frac{L_1^3}{12EI_1} + \frac{L_2^3}{96EI_2} + \frac{L_1}{4GA_1} + \frac{L_2}{8GA_2}$
Overturning Rotation Flexibility $\Theta / M =$	upper limit	$\frac{L_1}{6EI_1} + \frac{1}{2GA_1L_1}$	$\frac{L_1^2}{6EI_1} + \frac{L_2^3}{12EI_2L_1^2} + \frac{1}{2GA_1L_1} + \frac{L_2}{4GA_2L_1^2}$	
	lower limit	$\frac{L_1}{24EI_1} + \frac{1}{2GA_1L_1}$	$\frac{L_1^2}{6EI_1} + \frac{L_2^3}{48EI_2L_1^2} + \frac{1}{2GA_1L_1} + \frac{L_2}{4GA_2L_1^2}$	

In this paper, simple examples are calculated with the computer software SATWE developed by China Academy of Building Research. The planes of models are similar to figures above, the beams have same section and same length, and the conjunctions between beams and columns are considered the hinged and the rigid separately, to get the different results expressing two extreme conditions. Taking some value of them as 1, the relative ratio of the vertical settlements at the middle points of the beams are showed in Table 2.

The value of the rigidity of the structure is the reciprocal of the flexibility. It is obviously that the rigidity resisting vertical settlements of the transfer of double secondary beams in crossing form is superior, but the rigidity resisting overturning deformation of the angle is improved little.

Table 2. The relative ratio of the vertical settlements at the middle points of the transfer beams

		According to Fig. 1	According to Fig. 2	According to Fig. 3
Vertical Settlement Flexibility $\Delta / P =$	upper limit	1.00	1.44	0.73
	lower limit	0.25	1.13	0.56

For the same models and taking the same assumptions mentioned above, neglecting the shearing influence, according to the formulas in Table 1, the relative ratio of the upper limit and the lower limit for the overturning deformation of the angle at the bottom of the supported vertical members are showed in Table 3.

Table 3. The relative ratio of the overturning deformation of the angle at the bottom of the supported vertical members

		According to Fig. 1	According to Fig. 2	According to Fig. 3
Overturning Rotation Flexibility $\Theta / M =$	upper limit	1.00	1.50	
	lower limit	0.25	1.13	

3. Plane Arrangement and Section Design

3.1 *Significance of Double Secondary Beam in Crossing Form in Construction and the Mechanics of the Plane*

As is known to the all, there is huge shear force in the structural transformer plane. The double beams in crossing form combined with the slab in the transformer plane, act as if the prone framed shear wall, the strength and the rigidity resisting shear force will be improved very much. Adopting of the double beams in crossing form will promote the entirety of the whole structure.

3.2 *Optimum of the Section Design*

We regard the formula " L_1^3/EI_1 " in the equation in Table 1 independent variable D_1, and the formula " L_2^3/EI_2 " the independent variable D_2, for the expression of the upper limit value of vertical settlement, in the case of Fig2, we can get:

$$\Delta = \frac{L_1^3}{6EI_1} + \frac{L_2^3}{12EI_2} = \frac{D_1}{6} + \frac{D_2}{12} \qquad (1)$$

obtained the partial derivative, there should be:

$$\frac{\partial \Delta}{\partial D_1}=\frac{1}{6}, \qquad \frac{\partial \Delta}{\partial D_2}=\frac{1}{12} \tag{2}$$

It is obviously that we can get twice the result with half the effort, by the increase in the rigidity of secondary transfer beam, to reduce the settlement.

4. Discussion

It is unreasonable to use secondary transfer beam to support vertical members with the large lateral rigidity, for the loss of rigidity is very large in this kind of structure.

ANALYSIS OF TRANSFER PLATE STRUCTURES USING 3D COUPLING ELEMENT

S. H. LO Y. K. CHEUNG
Department of Civil Engineering, the University of Hong Kong, Pokfulam, Hong Kong.

K.Y. SZE
Department of Mechanical Engineering, the University of Hong Kong, Pokfulam, Hong Kong.

A transfer plate is a very thick and large solid concrete slab, which is placed between shear wall and column supports, with a heavy self weight and large amount of reinforcement. Transfer plates give architect and structural engineer a great flexibility in modifying the upper commercial and residential tower blocks and the supporting system underneath the transfer plate in the design process. The tower blocks, the transfer plate and the supporting system, which are meshed separately, have to be put together to form a finite element model for structural analysis. However, the merging of the load bearing walls and columns will be more challenging if tower blocks of arbitrary shapes and orientations are allowed to load on the transfer plate. In order to cope with the most general situation of arbitrary number, position, size and orientation of load bearing members on a transfer plate, the idea of coupling element is proposed. By means of a penalty integral, the coupling element ensures that load bearing members are in full contact with the transfer plate. As the integral can be evaluated based on a Gaussian point scheme, the node for node matching over the interface of two solid elements is not required. This effectively allows solid hexahedral meshes to be connected without mesh modification over the contact surface. Examples of transfer plate structure will be given to demonstrate the effectiveness of the coupling elements compared with results where no coupling element has been used.

1. Introduction

Common examples of transfer structures are in the form of deep beams, thick plates and pile cap raft. A transfer plate can be regarded as a very thick and large solid slab, which is usually placed between shear wall and column supports, with a heavy self weight and a large amount of reinforcement. This structure is often critical and forms the key component of the whole structural supporting system. As the transfer plate extends almost to the site boundary and is of thickness up to several metres, it is a huge massive structure of reinforced concrete of volume in the order of several thousand cubic metres. Due to its structural importance and the construction cost involved, an accurate assessment of its structural behaviour will be very helpful in the design of the transfer plate and hence improve the overall structural performance of the entire building.

Due to the limitations in computational power and the finite element technology based on displacement approach, various simplifications and assumptions have to be made in the structural analysis of transfer plate. The existing methods of analysis, namely, the rivet group approach, grillage analysis, thick plate theory and strut-and-tie model, etc. seem to be inadequate to meet the increasing demand in construction technology to build massive transfer structures of irregular geometry with openings supported by solid walls and mega columns to take on intensive loading from very tall tower blocks. Almost without any underlying assumption, robust and efficient solid hybrid stress elements allow accurate solutions be obtained by a fairly coarse mesh in the most versatile manner for the analysis of massive structures of arbitrary shape. The most important aspect of using only hexahedral elements rather than a mixture of beam, plate and shell element is the complete removal of the difficulty in matching rotational and translational degrees of freedom, which is inevitable if various types of elements are used in the same structure. Another difficulty for the structural engineer is to judge what kind of structural members, beam, plate, shell, etc should be used for various parts of the structure in the idealization.

The objective of the paper is to devise a rigorous yet practical analysis procedure for the design of transfer plate structures based on the recently developed high-performance hybrid stress 3D solid elements. A floor plan section concept will be employed, in which the supporting wall/column system, the transfer plate and the load exerting walls and columns will be represented by three different section plans respectively. These systems will be integrated automatically through modern mesh generation techniques and interactive graphics interface to form the structural model for analysis. The resulting forces at the interface between elements will be collected and its effects will be summed over the plate thickness so as to construct bending moment and shear force contours on the plan of the transfer plate for engineering design. Local effects can also be checked on an element by element basis if necessary for a more accurate placement of reinforcement bars. Unlike other methods of analysis, no assumption has been made in the formulation of solid elements, hence more reliable solution can be obtained through an adaptive refinement procedure. Local mesh refinement can be carried out at strategic positions under very heavy loading to find out more detailed information about the force distribution across the thickness of the transfer plate. Torsional effects can also be estimated as the horizontal shear forces are available between elements. The idea of coupling element will be presented, by which two structures separately meshed into solid elements can be merged together without remeshing. This new technique is especially useful in connecting arbitrarily oriented residential tower blocks onto the top surface of the transfer plate.

2. Existing Methods of Analysis

2.1 Rivet group approach

In the rivet group approach [1], the three-dimensional continuum of the transfer plate or the pile cap is assumed to be a rigid body. Under this assumption, the transfer plate does not deform and remains as a plane such that the axial shortening of the supporting columns could only vary in a linear manner along a given direction. Once the global equilibrium of the applied load and the reactions are calculated, the internal load distribution in the plate can be determined. However, this assumption may not be valid when the thickness of the plate reduces or the plate has large openings. Designer may have to increase the thickness and hence the weight of the plate in order to reduce the local deformation of the plate and justify the assumption made.

2.2 Grillage analysis

By grillage analysis [2], the transfer structure is modeled by an assembly of beams in two orthogonal directions. The beams will be arranged such that they can support the load bearing walls and columns above the transfer structure and carry the load down to the supports. The advantage of this approach is that complicated geometry with large openings through the plate for lift shaft and light well can be modeled easily. As flexural deformation is allowed, rigid plate assumption need not be justified, and a thinner plate can be used. Based on this approach, it is quite easy to obtain fairly accurate piecewise linear variations of bending moments for design purpose. However, same accuracy for shear force distribution is difficult to achieve, as there is often a sudden jump in shear force across the beam junctions. Furthermore, the arrangement of the grillage beams would be very complicated when the plate is irregular. The situation would become more onerous when there are changes of walls/columns positions above and below the plate. Sometimes, the whole grillage system may have to be revised to cater for minor amendments in support locations. The arrangement of beams in a grillage analysis could be quite arbitrary, and has to be done manually. Change of beam sizes or the connection pattern could produce quite different results.

2.3 Analysis using plate/shell elements

With the rapid development in the plate and shell finite elements [3], the transfer plate can be modeled by plate elements specially developed for the analysis of thick plates. The inter-element boundaries of the plate have to be arranged such that the wall and column nodes are directly connected with the plate elements to allow for force transfer and displacement compatibility between elements. Using plate elements, higher accuracy of shear force, bending moment and torsional moment can be obtained. However, analysis using plate finite elements is essentially a two-dimensional simplification of a three-dimensional problem, the idealized model compresses the real structure with a thickness of several metres to zero thickness. As a result, the wall and column elements above and below the transfer plate have to be increased in their clear height in order to have proper connections at the mid-surface of the transfer plate. Similarly, columns and walls are idealized with zero lateral thickness, and as a result, unrealistic high concentrated loads would be imposed on the plate. Moreover, as the plate has zero thickness, the stress variation along the plate thickness due to concentrated loads cannot be correctly represented. Similar to the grillage method, minor changes in the positions of walls or columns above and below of the plate may lead to a re-meshing of the entire finite element model.

2.4 Strut-and-tie model

Strut-and-tie model is particularly suitable for the analysis of simple non-flexible shear dominant structural components such as pile caps and deep transfer beams. This approach assumes force equilibrium in the form of an internal truss inside the massive concrete structure. The internal compressive force is taken by the concrete strut whereas the tensile force is provided by the reinforcement bars. By a suitable choice of truss model, various geometrical discontinuities such as large opening and tapered section can be handled. The design criteria [4] and the design procedure [5] of the strut-and-tie model with special emphasis on Hong Kong construction practices have been reported. However, for general transfer structures with complicated loading cases, geometry and support conditions, it would be very difficult to set up the corresponding strut-and-tie models. As only force equilibrium is considered, structural deformations cannot be determined by the model.

2.5 Hybrid stress solid elements

Recent advances in the development of versatile and efficient hybrid stress 8-node, 12-node and 18-node hexahedral elements (HSH) offers a new possibility of structural modeling in a unified and consistent manner [6-9]. Benchmark test examples showed that HSH are not sensitive to element distortions, and they can be stretched into shapes of large aspect ratios without revealing any numerical difficulties. Apart from their high ability in geometrical adaptations, results on standard tests on structural behaviours are also very promising. Take for example the lowest order element of the HSH family, it is seen that the 8-node HSH element is flexible and responsive to the three major structural actions, direct tension, bending and twisting modes even in degenerated forms as shown in Figure 1 and Table 1. Typical building structures were also analysed by the HSH elements [10], it can be seen that very accurate results could be obtained by a relative coarse mesh such that essentially each structural member was modeled by one single HSH element as shown in Figures 3 and 4.

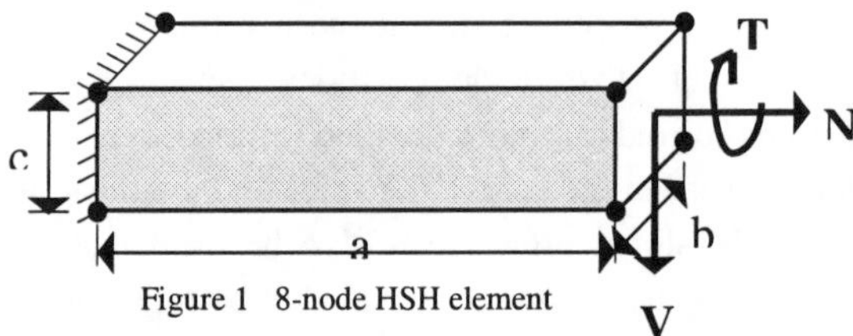

Figure 1 8-node HSH element

The ability to control the discretization error is crucial to the reliable application of the finite element method. This can be achieved through an automatic mesh generation procedure that adjusts the element size according to the result of previous analyses and thereby improved the solution quality in an optimal manner [11-12]. Building structures divided into classical structural members such as beams, plates and shells do not enjoy the benefit that a more accurate solution be obtained by a refined mesh. Solid hexahedral elements employing solely displacement degrees of freedom, however, lend themselves naturally to adaptation solution process, in such a way that critical stress points are captured and resolved in a fully automatic manner.

Shape	Mode	a	b	c	N	V	T	δ/δ_{ref}
Beam	Tension	10	1	1	1	0	0	100%
	Bending	10	1	1	0	1	0	90%
	Twisting	10	1	1	0	0	1	>95%
Plate/Shell	Tension	10	10	1	1	0	0	100%
	Bending	10	10	1	0	1	0	90%
	Twisting	10	10	1	0	0	1	>95%

Table 1. Structural behaviour of one 8-node HSH under tension, bending and twisting.
(δ_{ref} = Reference solution from the strength of material)

3. Coupling Element

Three distinct sections of the transfer plate structure are presented by floor plans which will be read by an interface program for mesh generation as individual structures. This arrangement gives the architect and the structural engineer the greatest flexibility in modifying the upper residential blocks, the transfer plate and the supporting system in the design process. This model separation into three parts also allows the engineer to make specific changes on the relevant floor plan view by means of interactive graphics drafting software. However, individual parts which have been meshed separately have to be integrated to form the finite element model for analysis. As for the mesh generation of the transfer plate, major supporting systems have all been taken care of and regular hexahedral elements have already been generated at the required positions, the matching of the supporting structures with the transfer plate is straight-forward and could be done in a natural manner. Nevertheless, the merging of the load bearing walls and columns will be more challenging if residential blocks of arbitrary shapes and orientation are allowed to load on the transfer plate. For some large scale residential and commercial development projects, commercial and residential towers consist of several blocks, each of which may point at a different direction. To cope with the most general situation of arbitrary number, position, size and orientation of load bearing members, the idea of coupling element is proposed. Figure 2 shows the super-position of a load bearing column/wall (represented by a rectangle) with the meshed transfer plate. In order to make sure that the load bearing member is in full contact with the transfer plate, it is required that

$$u_p - u_c = 0 \qquad (1)$$

where u_p and u_c are respectively the displacements of the transfer plate and the loading column at the interface. Equation (1) can be enforced by augmenting the functional of the problem domain with the following term,

$$\int_A \lambda (u_p - u_c)^2 \, dA = 0 \qquad (2)$$

in which λ is the penalty factor usually represented by a very large positive scalar, A is the area of the loading member. In terms of numerical integration, equation (2) can be written as

$$\sum_{i=1}^{m} \lambda w_i (u_{pi} - u_{ci})^2 = 0 \qquad (3)$$

in which m is the number of integration points (usually the Gaussian points at the face of the column element, m=9 in the example shown), and u_{pi} and u_{ci} are respectively the displacements of the transfer plate and the loading column at the ith integration point. There is a physical interpretation for equation (3) that each integration point represents a tie between the transfer plate and the loading column, which in reality could be provided by a steel reinforcement bar. Based on the coupling element concept, the merging process of the loading members onto the transfer plate can be done on an element by element basis. What is required is a relatively simple procedure to identify for each loading member all the affected elements on the transfer plate and the corresponding matching positions of the integration points.

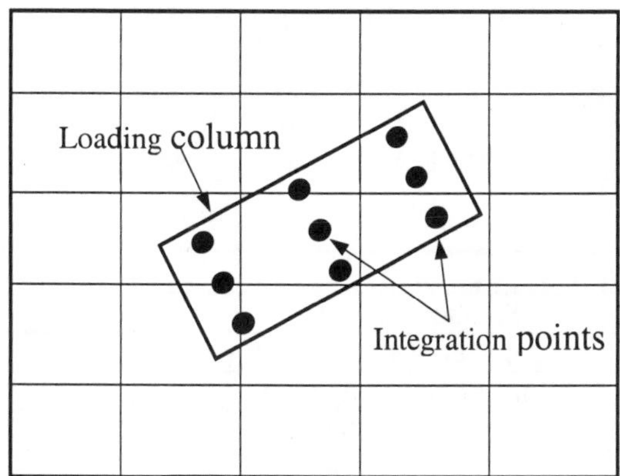

Figure 2. Coupling element

4. Examples

Two examples of a typical transfer plate structure consisting of three sections, the supporting frame, the transfer plate and the residential tower block are considered. In the first example, the residential block is in the same orientation of the supporting frame. As shown in Figure 3, two models A and B were constructed, whose characteristics are shown in Table 2. Model A consists of fewer elements and coupling elements are required to connect the upper block with the transfer plate as shown in Figure 3a. In the second model B, more elements are used so that element interface node to node compatibility can be achieved in a natural way without using coupling elements as shown in Figure 3b. For the structural analysis, both models were fixed at the bottom of the supporting frame and subjected to a uniformly distributed load on the longer side at the back of the upper block. From Table 2, it can be seen that model A and model B give very similar deflections of 134.79 and 137.56 respectively for the two models. As for member forces, columns at lower corners of the supporting frame and the upper block were checked. It is found that the differences are quite small as shown in Table 2.

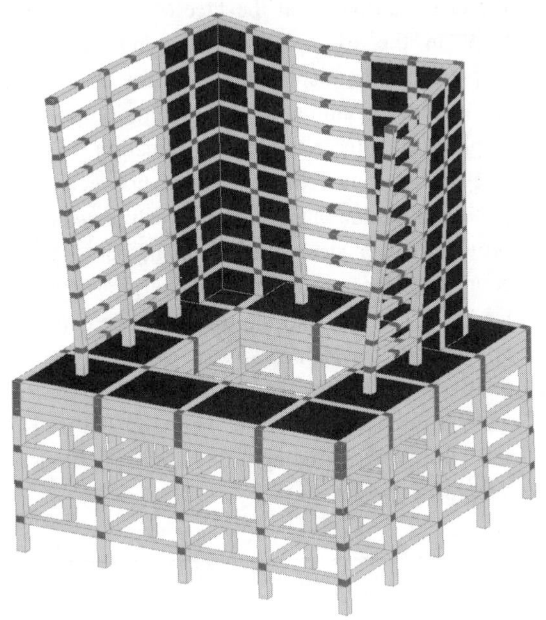

Figure 3a. Model A, deformed shape

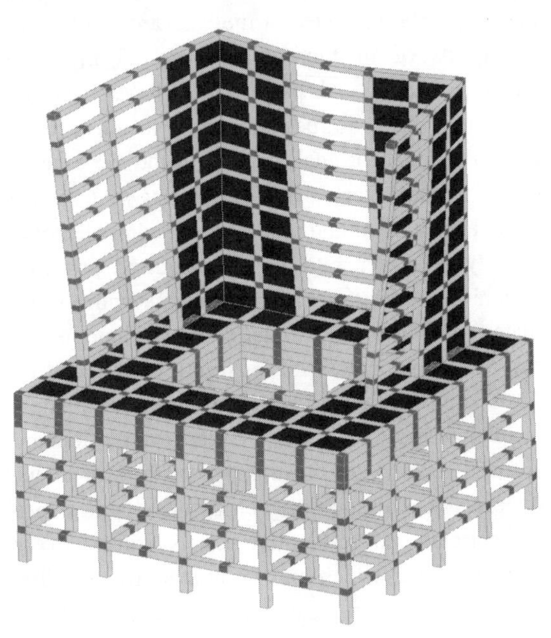

Figure 3b. Model B, deformed shape

	Model A		Model B	
Number of nodes NN	5124		8152	
Number of elements NE	1113		1884	
Number of coupling elements	21		0	
Maximum deflection	134.79		137.56	
CPU Time	10.8s		32.9s	
Location of column	Force	Moment	Force	Moment
Upper block (left)	142	664	165	671
Upper block (right)	199	929	214	938
Supporting frame (left)	684	1018	672	1016
Supporting frame (right)	627	1018	603	1017

Table 2. Characteristics, deflection and member forces of models A and B.

In the second example, the residential frame is rotated by 5 degrees from its original orientation. Three models C, D and E were built for structural analysis as shown in Figure 4. The first model C is a coarse model consisting of fewer elements and coupling elements were used for the connection between the upper block and the transfer plate. The second model D is a refined model for which connection is achieved also by means of coupling elements. The third model E is more or less the same as model D. However, some nodes of the transfer plate were shifted so that node to node compatibility was restored between the upper block and the transfer plate without using coupling elements as shown in Figure 4c. Under the same boundary and loading conditions of the previous example, all the three models give similar maximum deflection with a difference less than 2% at the same location. The discrepancy is probably due to the difference in stiffness of the supporting frame corresponding to a coarse mesh and a refined mesh. It is also noted that for compatible models such as models B and E, if we take away the node to node connection and use coupling elements instead at all element interfaces between the upper block and the transfer plate, the analysis results would be identical.

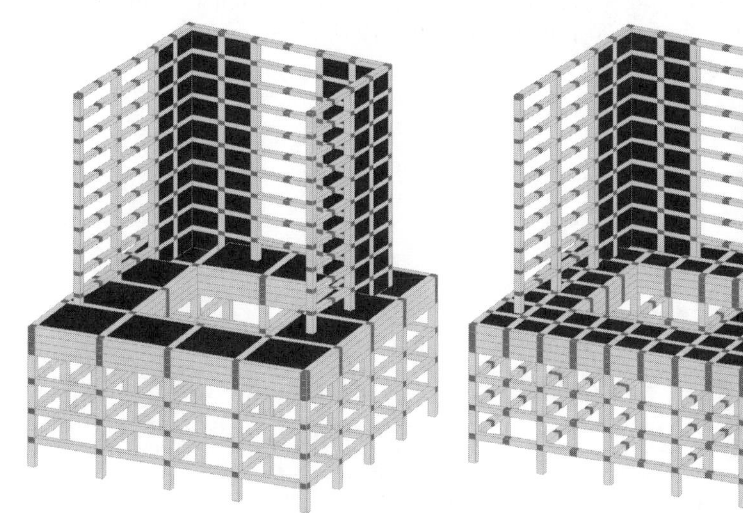

Figure 4a. Model C

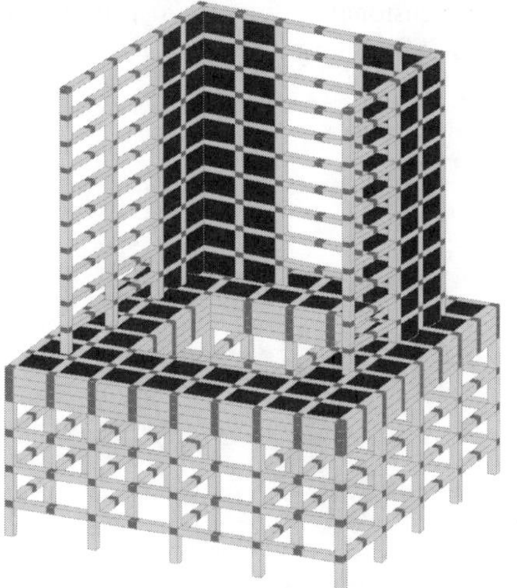

Figure 4b. Model D, a refined mesh

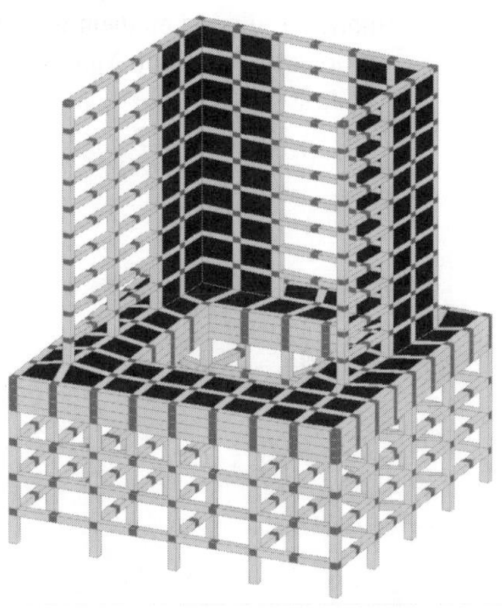

Figure 4c. Model E, nodes shifted

5. Conclusions and Discussions

In this paper, various methods for the analysis of transfer plate structures are discussed. In particular, it is shown that hybrid stress solid elements provide a robust and efficient method for the analysis of massive solid engineering structures. Due to the three-dimensional nature of solid elements, the connection between structures meshed into solid elements can be easily combined into one piece by means of coupling elements, largely alleviating the burden for the generation of compatible meshes over general irregular domains with sectional characteristics. The connection effect provided by the coupling element virtually makes no difference compared to natural conformable meshes with node to node compatibility between element interfaces. The flexible use of coupling elements allows sub-structures to be meshed and modified individually and put back together into one piece in arbitrary manner.

Acknowledgement

The financial support from the CERG research grant, HKU7117/04E "Analysis of transfer plate structures using high-performance solid 3D hybrid stress hexahedral elements" is greatly appreciated.

References

1. GEO (1997) Pile Design and Construction, GEO Publication No. 1/96, Geotechnical Engineering Office, Civil Engineering Department, Hong Kong.
2. Hambly EC (1976), Bridge Deck Behaviour, Chapman and Hall, London.
3. K.Y.Sze, "Three-dimensional continuum finite element models for plate/shell analysis", Prog. Struct. Engrg Mater. 2002, **4**:400-407
4. **Su RKL** and Chandler AM, "Design Criteria for Unified Strut and Tie Models", Prog. Struct. Engrg Mater., 2001, **3**(3), p288-298.
5. **Su RKL**, Wong PCW and Chandler AM, "Design of Non-Flexural Components Using Strut and Tie Models", Transactions of Hong Kong Institution of Engineers, 10(1), p31-37. (2003)

6. K.Y. Sze and A. Ghali, "An hexahedral element for plates, shells and beam by selective scaling", Inter. J. Numer. Methods Engrg., **36**, 1519-40, 1993
7. K.Y. Sze, **S.H. Lo**, "A 12-node hybrid stress brick element for beam/column analysis", Engineering Computations, Vol. 16, No. 7, 1999, 752-766
8. K.Y. Sze, **S.H. Lo** and L.Q. Yao, "Hybrid-stress solid elements for shell structures based upon a modified variational functional", Inter. J. Numer. Methods Engrg., **53**, 2617-2642, 2002
9. **S.H. Lo** and Chen Ling, "Improvement on the 10-node tetrahedral element for three-dimensional problems", Comput. Methods Appl. Mech. Engrg., **189** (2000) 961-974
10. **S.H. Lo**, K.Y. Sze and Y.K. Cheung, "Three-dimensional structural frame analysis using solid finite elements", The Fifth International Conference on Tall Buildings, Hong Kong, 1998
11. **S.H. Lo** and C.K. Lee, "On using meshes of mixed element types in adaptive finite element analysis", Finite Elements in Analysis and Design, **11** (1992) 307-336
12. C.K. Lee and **S.H. Lo**, "An automatic adaptive refinement finite element procedure for 2D elastostatic analysis", Inter. J. Numer. Methods Engrg., **35**, (1992) 1967-1989

THE THEORETICAL ANALYSIS AND ENGINEERING PRACTICE OF OVERLONG JOINTLESS TRANSFER STRUCTURES

ZHOU SHULING
Beijing CanBao Institute of Architectural Design, Beijing 100850

GU WEIJIAN
Beijing CanBao Institute of Architectural Design, Beijing 100850

Some factors influencing the crack control of overlong jointless transfer structures had been synthetically analyzed including the layout of a building, the structural system and the transfer forms of the high-rise building. The comparative research on the transfer form for the overlong jointless structures was carried out through a practical engineering project. With the aim of the finite element method for numerical calculation, the mechanical properties of the globe structure and the transfer beam was analyzed. Some fortification measures for overlong jointless transfer structure were put forward lastly. The result obtained from this research may serve as a reference for the similar project design.

1. Preface

Overlong jointless transfer structure is defined as a length of a building exceeding the maximum length of thermal expansion joint or seismic joint set as per relevant regulations rather than any form of fixed joint structure. This overlong jointless transfer structure can be used to satisfy the requirements for purposes and facade continuity and decoration styles of the whole building; to overcome problems on traceability, durability, fire resistance, water tightness, construction properties and maintainability, etc. and building defects, which are caused by setting transfer joints; to avoid torsional damages inconsistent with prescribed values caused by the misalignment of stiffness center and mass center of constitutional member in transfer joints; to make layout of pipelines in electro-mechanical constructions (wind, water and electricity) easier and wiring more flexible for the purpose of avoiding any influence from horizontal and vertical positions; to reduce limitations of the maximum distance of any expansion joint in reinforced concrete structure as prescribed in current regulations and provide analysis of constructional theories for reference and all kinds of effective design and technical measures for the revision of relevant clauses in these regulations.

Based on a breach of current regulations, this overlong jointless transfer structure is required to transform multi-layer vertical members above medium-high ground of a building for the purpose of achieving a large space in building construction, resulting in many bad influences on crack control of this overlong structure and transfer control of a building. Therefore, special methods and quality assurance measures are very important for the prevention of building structures from cracking to avoid any form of harmful cracks adverse to normal construction of a building and ensure the safety of its structure.

2. Project Overview

Located on No. A-23 Fuxing Road, Haidian District, Beijing City, the comprehensive building of China Institute Of Water Resources & Hydropower Research is composed of Units A, B, C and D mainly used for official business, research and education added to by fewer commercial services or guest rooms. Unit A is constructed in a structure composed of reinforced concrete frame-shear wall bottom with a transfer story and a beam slab rafted-shape frame. It has three underground stories, twelve overground stories and nine stories in its local part; the part overground is 75.6m high, 42.0m wide, and the standard height of its cornice is 44.7m (as shown in Fig.2-1and 2-2).

It has a total area of 52828m^2, including 5014m^2 area for people's air defence, which is located on the third floor underground. During the war it is used as a Grade 6 material warehouse, while in the peacetime it is a mechanical stereo parking garage. The safety level of this project is Grade 2, and this building has a design life of 50 years, Grade 1 fire resistance rating and its seismic fortification intensity is 8°. It is designed as the first earthquake echelon, 0.20g in basic earthquake accelerated speed, Class C in seismic fortification category for structures, of which the frame is designed according to Grade 2 in resistance against earthquake, and the bracket frame and shear wall is designed according to Grade 1 in resistance against earthquake. The basic weight-bearing layer is composed of conglomerate rock and clay rock, and the characteristic value of subgrade bearing capacity is 300kPa. The building field earth is classified as medium-hard field earth, and the site is Class II. Its basic foundation rating is Class B and its basic beam is designed according to Grade 3 in resistance against earthquake.

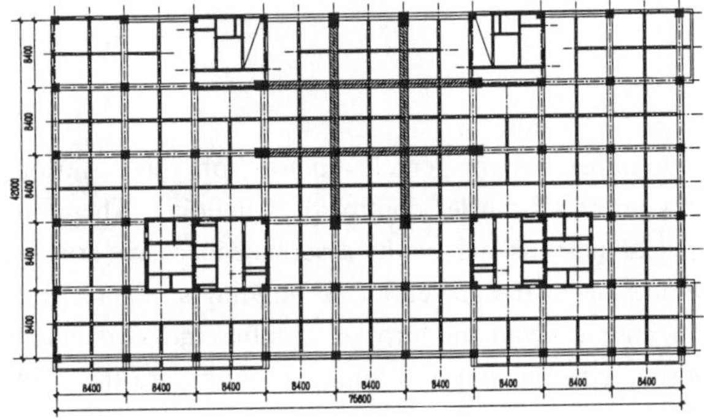

Fig. 1 Plan of Unit A Transfer story

3. Analysis for Overlong Jointless Transfer Structure Systems

3.1 *Choosing overlong jointless transfer structures*

The members of this transfer structure mainly include such six basic structures as beam, truss, vierendeel truss, box-like structure, inclined strut and thick board. A comparative analysis is carried out to all kinds of transfer forms as per details of this project.

Due to more advanced calculations and detailed constitution techniques of beam transfer structures as well as their many successful examples, the frame bracket has been proven to bear upper frame columns to avoid the secondary transfer. However, due to limited layer height, the transfer span of this beam transfer structure is limited to some extent.

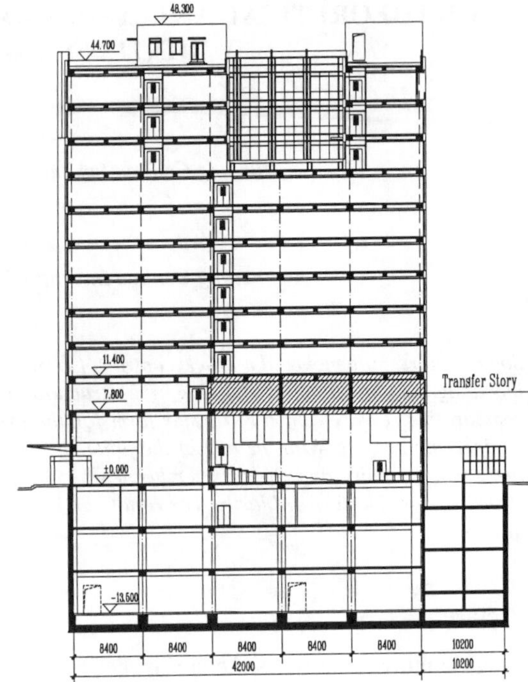

Fig.2—2 Profile Chart of Unit A

By combining the floors and transfer beams with limited thickness between stories, the box–like transfer structure can achieve a transfer structure with larger stiffness and bearing capacity. Appropriate positions in the box-like beam web board can be chosen to open a channel and a passage for personnel to make full use of all the space. The thickness of each upper or lower alar plate (floor) should be determined according to the thickness of each floor and load on it. Their thickness can vary due to different loads and tension faces.

The thick-board transfer structure is applied under out-of-order corresponding relations of upper or lower vertical member axes of low-rise area or transfer story. However, due to super rigid layers formed by thick boards to form sudden changes of vertical rigidities, weak layers, and plastic hinges in interconnected columns and walls, therefore it isn't suitable to adopt this thick-board structure.

It isn't suitable to adopt the transfer structure of truss or vierendeel truss by reason of complex structure for connecting members, constructional difficulties and inconvenience of using internal space of the transfer story.

Inclined strut isn't suitable for this project, so the feasibility of this structure is not taken into account.

Based on above-mentioned analysis, a generalized box-like beam transfer structure is chosen for this project to assort with the structure of the whole building and apply to special features of this overlong jointless transfer structure. The thickness of an upper alar plate of this box-like beam is 250mm, the thickness of its lower alar plate is 200mm, and the height of its well beam is 4500mm. Its integrated results are calculated in Tables 3.1, 3.2 and 3.3. These results show that earthquake effects (base shear and overturning moment) are a bit larger in bidirectional earthquake torsion coupling than in incidental eccentric influences, while it works the other way around for maximum layer-displacements.

3.2 *Design for frame bracket structure*

According to basic stipulations for frame bracket of JGJ3-2002 "Technical Specification for Concrete Structures of Tall Building", the well transfer beam is designed into bonded prestressed structure. Common reinforcing steel bars are constructed as per this prestress $\lambda = 0.55$. The calculated section dimension of a beam (bxh) is 1000mm×4500mm, its span L is 25200mm. The calculated results for all kinds of internal forces and bearing forces are shown in Tables 3.4 and 3.5.Be sure to provide steel wire strands 6x12ϕS15.24 with bonded prestressed forces. The standard strength of each strand is 1860Mpa, its designed strength is 1320Mpa, its stretching stress is 0.75fptk(1375Mpa), its pre-stress loss is 25 %; it should be deployed according to a quadratic parabola $y=ax^2$,and its contraflexure point is Position 0.15L. Calculated results for width of each beam joint are shown in Table 3.6. The width of each joint in bracket or mid-span is the prescribed value 0.2mm, and resistance against crack of each transfer beam should be up to the prescribed standard.

Table 3.1 Self-Sustained Oscillation Periods

Oscillation Type		1	2	3	4	5	6	7	8	9
Bidirectional torsion coupling	Period	1.1050	1.02133	0.9967	0.2873	0.2673	0.2577	0.1450	0.1336	0.1283
	Torsional factor	0.46	0.49	0.04	0.57	0.36	0.07	0.42	0.43	0.15
Incidental eccentric influence	Period	1.1049	1.0123	0.9959	0.2867	0.2646	0.2570	0.1450	0.1329	0.1281
	Torsional factor	0.46	0.49	0.04	0.56	0.37	0.06	0.42	0.44	0.15

Table 3.2 Base Shear, Shear-weight Ratio and Overturning Moment

Structure Effects		Base Shear (kN)	Shear-weight Ratio (%)	Base Overturning Moment (kN-m)
bidirectional torsion coupling	Direction x	36949	3.80	1468356
	Direction y	32463	3.34	1257727
incidental eccentric influence	Direction x	36932	3.80	1467918
	Direction y	32040	3.30	1240646

Notes: the total weight of a building W=978569kN

Table 3.3 Maximum Displacements

Structure Effects		Maximum Layer-Displacements		
		u (mm)	u/h	Position
bidirectional torsion coupling	Direction x	2.29	1/1570	Story 10
	Direction y	2.83	1/1272	
incidental eccentric influence	Direction x	2.45	1/1471	From Story 10
	Direction y	3.27	1/1102	Story 9

Table 3.4 Calculated Results for Internal Forces of Frame Beams

Item	Design value of bending moment（kN-m）			Nominal value of bending moment（kN-m）			Remarks
Internal forces	M_L	M_R	M_m	M_L	M_R	M_m	
	25000	25013	47115	9879	9833	24546	Ppermanent load
				1060	1060	4048	Variable load

Table 3.5 Calculated Results for Bearing Forces of Frame Beams

Item / Position	$M_{prestressed}$（kN-m）	M_{common}（kN-m）	Non-prestressed reinforcement	Reinforcement ratio（%）
Abutment	24120	21544	28ϕ28	0.39
Mid-span	42578	33630	42ϕ32	0.75

Table 3.6 Calculated Results for Width of Each Frame Beam Joint

Position	Abutment	Mid-span
Joint width ω_{max} （mm）	0.02	0.14

Under the normal loads, the deflection of upper beam (f_1) is 8.16mm; under the pre-stressed equivalent loads, the inverted camber of a beam (f_2) is 2.65mm, and its final deflection (f) can be calculated as per the following formula: $f = 8.16 - 2\times 2.65 = 2.86mm$ $f/L = 2.86/25200 = 1/1181 \prec 1/400$,which is up to the prescribed standard.

Because the transfer beam is stressed by complex forces and very important for this project, SATWE is used to calculate the whole structure, and then the transfer story is used to make finite element analysis for the structure by means of ANSYS program. Calculated results show that: the maximum displacement of a floor is produced on the transfer beam, x-direction displacement is 2.725mm, y-direction displacement is 2.831mm, Z-direction displacement is 15.146mm, mid-span bending moment of a beam is 28000kN-m,bending moment of an abutment is 13900kN-m, the top bending moment of the column is 12000kN-m, the bottom bending moment of the column is 1710kN-m. the bending moment and elastic displacement of the transfer beam are adjacent to results calculated by SATWE.

4. Main Fortification Measures for Overlong Jointless Transfer Structure

The control of overlong jointless transfer structure is comprehensive system engineering with respect to designs, building styles, structure and formation, construction procedures, building materials, climate environment, etc, which affect and restrict one another, and produce different control results of building cracks finally.

4.1 *Technical measures for design*

1. Setting one last-cost joints (ca. 0.8m wide), which should be poured again during the cold seasons after sealing the structure, and ensuring the maximum length of 40 meters of each structure member exposed to the air before the construction of outer heat-preserving materials, and its single length should be up to the prescribed standard during the construction.
2. Adopting shrinkage-compensating massive concrete, i.e. adding micro-expansion agent in a certain

proportion into common concrete, which can produce an appropriate extent of expansion during the hydration to form a certain prestress due to limitations of reinforced steel bars and adjacent positions. This pre-stress can compensate the tensile stress caused by the shrinkage of concrete in principle to avoid or reduce cracks of concrete members.

3. Add composite fiber in a certain proportion into the concrete to increase the tensile strength of a base, prevent original defects (micro-cracks) from expanding, obstruct the formation of new cracks, improve its distorting capability and enhance the brittleness and tenacity of concrete.
4. Adopting high-performance concrete to carry out the construction of transfer beams.
5. Adopting wall filling materials for heat preservation and insulation to ensure non-exposition of most of building members, reducing the influences of temperature changes on the main body to a minimum, and reducing cracks caused by temperature changes.
6. Laying bidirectional unbonded prestressed steel bars in the top and the first floor of this main body and bringing pre-stress to bear on them according to technical requirements for crack control to limit the distortion of floors and horizontal members caused by temperature changes.
7. Taking measures for heat preservation and insulation of structural rooves. Using reliable extruding polystyrene boards for heat preservation to reduce the temperature on the structural rooves for the purposes of heat preservation and insulation.
8. By considering influences of thermal stress, taking appropriate measures to reinforce horizontal distribution bars and reinforcement bars of horizontal members.
9. Avoiding centralized stress caused by sudden changes of generalized structure sections (e.g. section and linear stiffness of a member or structure stiffness, etc) as possible.

4.2 *Technical measures for construction*

1. Making a strict control of quality and technical indices of concrete raw materials. Cement fineness (its specific area is controlled in the range from 2500 to 3500cm2/g), and silt content of rough or fine aggregate should be reduced to a minimum (not more than 2%);
2. Paying attention to the time of removing scaffolds and tensile order of pre-stress during the construction of pre-stressed transfer beams. No stretching the pre-stress to the design value at a draught. Being sure to stretch the pre-stress in steps before finishing the construction of the roof of upper structure. This prestress must be orresponding to the load on the upper part of a beam in synchronism.
3. Low-temperature molding and care should be used for concrete pouring to reduce the temperature of final solidification of concrete to a minimum, hydration heat and shrinkages.
4. Vibrating is necessary for pumping fluid concrete. The most appropriate time for vibrating concrete is 5～15 seconds.
5. Cracks in walls, beams or boards are mainly caused by shrinkages, so layered heat sink pouring measures should be taken to hold concrete humidity or care while these members are poured. Do not take apart a mold ahead of time because it will lead to too early water loss of concrete.
6. Rabbet joints are suitable for construction joints of concrete. These rabbet joints must be roughened and cleaned.
7. After taking part molds of basement sidewall structures, being sure to backfill in time to reduce the expansion of forepart or metaphase cracks of walls.
8. After an integral overlong jointless transfer structure is formed in the upper part, all the temporary measures for heat preservation and insulation should be taken to the whole overlong structure to avoid the formation of structure cracks under least favorable loads.
9. Connections of main structure and filling wall or other non-structural members should be filtrated with a steel wire mesh and mortars to avoid any form of cracks in the main structure.

5. Conclusions

Overlong jointless transfer structure is a special structure with respect to all kinds of technical problems in designing overlong jointless transfer structures and difficulties of comprehensive transfer structures to the whole structure design. Choosing proper transfer structures is a key factor for ensuring none of horizontal or vertical harmful cracks in any form produced in this structure on the premise of not affecting purposes of a building, very important for avoiding any sudden change of structural stiffness and structural safety. This project is designed to test these theories and actual constructions at an early stage so that preparations will be made for more and better projects in future.

References

1. Gu Weijian, The Quality of Projects for Overlong Structure Crack Control Technologies, Project Quality, supplemented in 2002 (upper part).
2. Li Guosheng, Design for Transfer storys of Shear Wall Structure with Large Space Bottom, the Sixteenth National Joint Symposium on Tall Building Structure, Shanghai, 2000.

MEASURES TO ALLEVIATE SHRINKAGE PROBLEMS IN LARGE REINFORCED CONCRETE PODIUM STRUCTURES

C. H. LIU FRANCIS T.K. AU PETER K. K. LEE

Department of Civil Engineering, The University of Hong Kong
Pokfulam Road, Hong Kong, China

Shrinkage of a concrete structure as the concrete dries after hardening could lead to substantial shortening movement and, if the movement is restrained, serious cracking of the concrete structure. This is a common problem in Hong Kong. If a large podium is structurally connected to two or more core walls supporting tower blocks above the podium, the comparatively stiff core walls will prevent the shortening of concrete structures and shrinkage stresses will develop. In many cases, the shrinkage cracks formed have been causing aesthetic, water leakage and durability problems. To minimise the effect of shrinkage, it is a common practice to adopt stage construction by providing late-cast strips to be cast later. In this case, various structural members may have different ages, and hence different strengths and tendencies to shrink. The effects of stage construction on the long-term response of reinforced concrete podiums are investigated by the finite element method using the step-by-step time integration and based on the conventional creep coefficients, time-dependent concrete moduli and shrinkage strains. Investigation into various problems of a typical podium arrangement is also carried out. Factors affecting the effectiveness of late-cast strips including the optimum time lag between the first-cast area and late-cast strips, and the locations and number of late-cast strips are evaluated. The tensile stress criteria of concrete are used to predict possible cracking of concrete. Results show that stage construction is an effective measure to reduce the risk of shrinkage cracking. The maximum shrinkage stresses are reduced because a significant part of shrinkage movements have taken place relatively freely before various parts are stitched together. The above cases are analysed by using the shrinkage models suggested by the MC-90 Code, the ACI-209 Code, BS5400 and the Hong Kong Code, which are based on different material parameters. An approximate shrinkage model for concrete with shrinkage reducing agent is also proposed. To investigate the effectiveness of shrinkage reducing agent, the shrinkage behaviour of a typical podium is analysed. Finally, some design recommendations including the arrangement of late-cast strips and the possible methods to mitigate the adverse effects of shrinkage of concrete are made.

1. Introduction

Shrinkage of a concrete structure as the concrete dries after hardening is a natural phenomenon. It could lead to substantial shortening movement and, if the movement is restrained, serious cracking of the concrete structure. This is a common problem in Hong Kong. If a large podium is structurally connected to two or more core walls supporting tower blocks above the podium, the comparatively stiff core walls will prevent the shortening of concrete structures and shrinkage stresses will develop. In many cases, the shrinkage cracks formed have been causing aesthetic, water leakage and durability problems.

There are relatively few publications on the theoretical background of the effect of stage construction on the long-term behaviour of reinforced concrete podiums. To address the problem, the entire period of time of interest is divided into time steps of reasonable size for analysis. The investigation is to evaluate the stresses and strains of reinforced concrete members subjected to the effects of shrinkage and creep.

An effective design strategy is necessary to avoid the cracking of concrete in tension. Although this phenomenon is predictable, little advice has been given in the common codes of practice.

2. Rationale of Time Integration for Analysis of Shrinkage and Creep Effects

The time-dependent behaviour of concrete structures can be analysed using a suitable constitutive model of concrete that takes into account creep and shrinkage, and an appropriate algorithm for time integration. This problem has attracted interest of various investigators and is well covered by the literature [1, 2].

2.1 *Step-by-step analysis*

The calculation of creep caused by time-varying stress can be determined by the principle of superposition. According to this principle, the strain caused by stress history $\sigma_c(t)$ is obtained by summation of the responses of $d\sigma_c(\tau)$ applied at time τ. When the magnitude of applied stress changes with time, the total strain of concrete $\varepsilon_c(t)$ due to the applied stress and shrinkage is given by

$$\varepsilon_c(t) = \sigma_c(t_0)\left[\frac{1+\varphi(t,t_0)}{E_c(t_0)}\right] + \int_0^{\Delta\bar{\sigma}_c(t)} \frac{1+\varphi(t,\tau)}{E_c(\tau)} d\sigma_c(\tau) + \varepsilon_{cs}(t,t_0) \qquad (1)$$

where $\varepsilon_{cs}(t,t_0)$ is the shrinkage of concrete during the time period (t,t_0), $\varphi(t,t_0)$ is the coefficient for creep at t due to a stress applied at time t_0, $E_c(\tau)$ is the modulus of elasticity of concrete at the age of τ days and time t_0 is the start of analysis. For practical engineering problems, Eq. (1) has to be solved numerically. The entire period of time of interest is divided into time steps of reasonable size for analysis.

2.2 *Finite element formulation*

The basic idea and formulation of the two-dimensional triangular plane stress element are well established and can be found for instance in Cheung and Yeo [3]. In the present finite element analysis, each node is assumed to have two translational degrees of freedom. As the stresses and strains are dependent on time, the incremental stresses $\Delta\sigma$ within a typical time step Δt are given by

$$\{\Delta\sigma\} = [\bar{D}_c(t)](\{\Delta\varepsilon(t)\} - \{\Delta\varepsilon_\varphi(t)\} - \{\Delta\varepsilon_{cs}(t)\}) \qquad (2)$$

where $[\bar{D}_c(t)]$ is the elasticity matrix, $\Delta\varepsilon(t)$ is the incremental strain of concrete, $\Delta\varepsilon_\varphi(t)$ is the incremental creep strain and $\Delta\varepsilon_{cs}(t)$ is the incremental shrinkage strain at time t. Neglecting body forces and considering incremental quantities, the incremental element nodal force vector $\{\Delta q^e\}$ can be expressed in terms of the incremental displacement vector $\{\Delta\delta\}$ as

$$\{\Delta q^e\} = \int_{V^e} [B]^T \{\Delta\sigma\} dV \qquad (3)$$

$$\{\Delta q^e\} = [K]\{\Delta\delta\} - TA[B]^T[\bar{D}_c(t)](\{\Delta\varepsilon_\varphi(t)\} + \{\Delta\varepsilon_{cs}(t)\}) \qquad (4)$$

where A is the area of triangular element, $[B]$ is the strain matrix, T is the thickness of triangular element, and $[K]$ is the element stiffness matrix given by

$$[K] = TA[B]^T[\bar{D}_c(t)][B] \qquad (5)$$

2.3 *Modelling of steel reinforcement*

The slab of a podium structure may be provided with several sets of steel reinforcement. As far as the analysis of the effects of shrinkage and creep is concerned, the following assumptions on the steel reinforcement can be made:

- The steel reinforcement is assumed to be linearly elastic.
- Each set of steel reinforcement is assumed to be smeared and considered as an orthotropic continuum possessing stiffness in the direction of reinforcement only.
- The concrete and steel reinforcement in the slab are modelled separately. Their interaction within each element is ignored.

3. Computer Programme and Modelling of Restraints

The finite element analysis of reinforced concrete podium structures subjected to time-dependent loads due to shrinkage and creep effects is implemented by a computer programme specifically written. The shrinkage stresses in the floor slab depend very much on the arrangements of supporting structural members such as core walls and columns. As large reinforced concrete podium structures are often constructed in several stages, the programme also accounts for concrete of different ages.

3.1 *Column and wall restraints*

For column supports, it is assumed that the column is restrained at both ends from rotation. Under these assumptions, the lateral stiffness is $12EI/L^3$ where E is the elasticity of mature concrete, I is an appropriate second moment of area and L is the length of column. Alternatively the actual support stiffness may also be evaluated taking into account the possible end rotations. As the core walls are generally very stiff, it would act as movement restraint against the shrinkage deformation of the floor slab. The lateral stiffnesses of the core walls are modelled as follows:

- The node closest to the centre of the wall segment is given infinite (i.e. extremely large) in-plane stiffness.
- All nodes are given out-of-plane stiffnesses appropriate to the wall thickness and the wall height.

Figure 1 shows the movement restraints associated with a few typical wall segments.

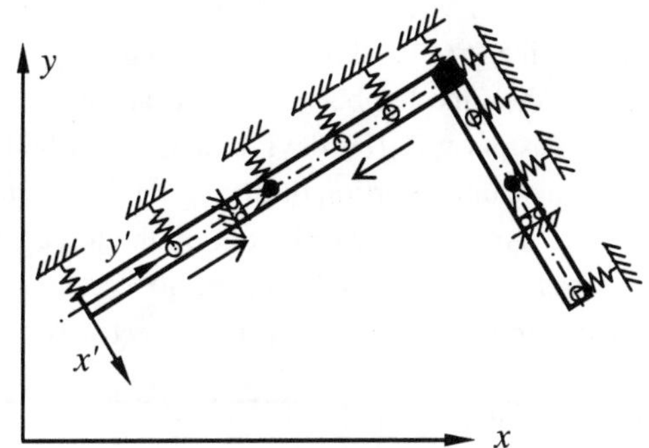

Figure 1. Support condition at wall nodes.

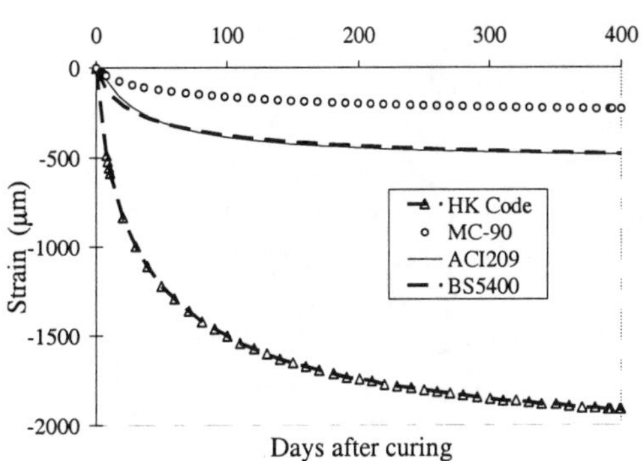

Figure 2. Shrinkage predicted by various models.

4. Reviews of Prediction Models

Realistic prediction of the shrinkage and creep of concrete is very important for analysis of the time-dependent deformation of concrete structures. Inappropriate estimation of these coefficients may lead to excessive deformation and cracking of the structure. Müller and Hilsdorf [4] compared coefficients by various models with a series of experimental results and found that the creep and shrinkage obtained from different methods vary widely.

The models suggested by MC-90 [5], ACI-209 [6] and BS5400 [7] for the prediction of shrinkage and creep have different emphases and capabilities as shown in Table 1. The discussions on BS5400 [7] also apply to the Hong Kong Code [8] as this model is largely based on that in BS5400 [7].

Figure 2 shows the predicted values by various models based on the concrete mix data as shown in Table 2. It demonstrates that estimates of shrinkage obtained from different methods vary widely. The Hong Kong Code [8] gives the highest values, but MC-90 [5] provides the smallest estimate. The additional multiplicative factor of 4 in the estimation of shrinkage strain according to the Hong Kong Code [8] has caused these estimates to be much larger than those based on other codes. However it should be pointed out that the estimates of creep and shrinkage are heavily influenced by the properties of local constituent materials in making concrete.

Table 1. Factors considered in various models for prediction of shrinkage and creep.

Factors	MC-90	ACI-209	BS5400
Relative humidity	Yes	Yes	Yes
Age of loading (maturity)	Yes	Yes	Yes
Size	Yes	Yes	Yes
Type of cement	Yes	Yes	No
Cement content	No	Yes	Yes
Water-cement ratio	No	No	Yes
Type of curing	No	Yes	No
Unit weight of concrete	No	Yes	No
Slump	No	Yes	No
Ratio of fine to total aggregate	No	Yes	No
Air content	No	Yes	No

Table 2. Concrete mix data (Grade 40).

W/B ratio	Paste volume (%)	Fine to total aggregate ratio	Water content (kg/m^3)	OPC content (kg/m^3)
0.44	35	0.38	202	459

5. A Typical Podium Structure

5.1 *Shrinkage-induced tensile stresses*

Table 3 shows the structural design information of a typical podium structure with two core walls adopted in the study. Apart from the calculation of shrinkage stresses, various options of the location of late-cast strips are examined. The reinforcement is provided in two orthogonal directions, namely along the *x*- and *y*-axes (Figure 3).

Figure 4 shows the areas identified to be prone to cracking as shaded based on the tensile strength criteria of concrete. As the core walls are very stiff, they effectively act as movement restraints against the shrinkage deformations. As a result, tensile stresses are induced near the core walls and tend to spread to the area between the core walls.

Gupta [9] observed that the crack direction might be taken to be perpendicular to the direction of the major principal stress. Figure 5 shows the orientations of potential cracks based on this idea. The cracks are roughly parallel to the *y*-axis because the major principal tensile stresses are generally along the *x*-axis.

Table 3. Structural design information of a typical podium structure.

Member	Grade	Size	Steel (%)
Beam	30	100000 mm^2	1.0
Slab	30	200 mm (thick)	0.4
Column	40	800mm×800mm	—
Wall	40	200 mm (thick)	1.0

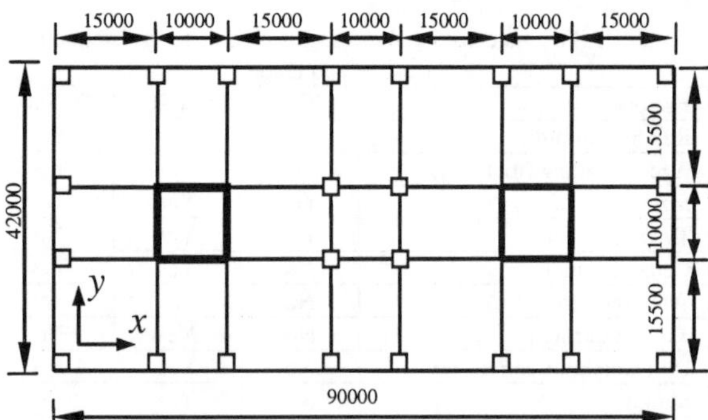

Figure 3. Structural plan of a typical podium (dimensions in mm).

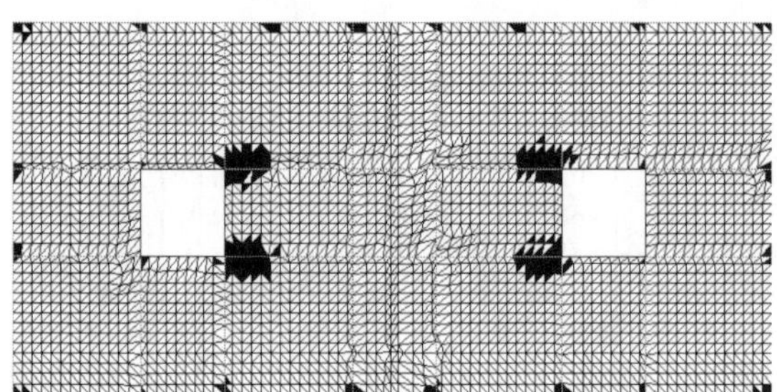

Figure 4. Areas identified to be prone to cracking after 365 days.

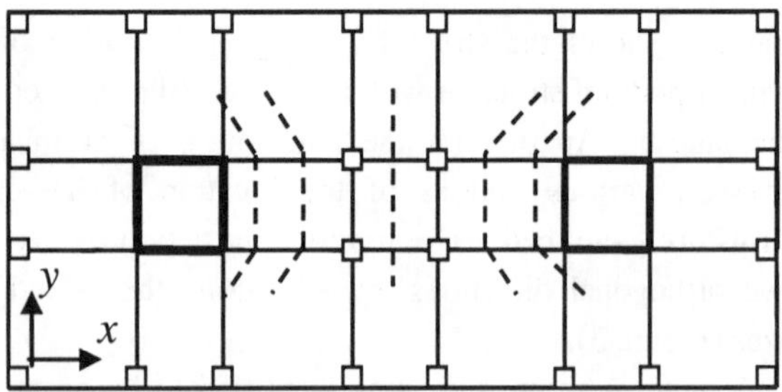

Figure 5. Potential crack orientations for the podium.

5.2 *Arrangement of late-cast strips*

In order to alleviate the high tensile stresses in the floor, three representative arrangements of late-cast strips are studied, as shown in Figures 6, 7 and 8. The late-cast strip is 1 metre in width so as to avoid the excessive deformation due to its subsequent shrinkage. It is assumed that the time lag between the first cast portion and the late-cast strip is 180 days which should be long enough for substantial portion of drying shrinkage strain to take place. Analysis is carried out on the three cases and the areas identified to be prone to cracking are shown shaded in Figures 9, 10 and 11 respectively.

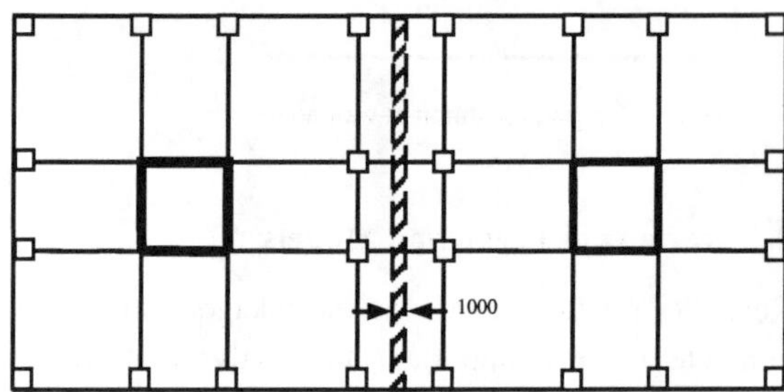

Figure 6. Arrangement of late-cast strips: Case 1 (dimensions in mm).

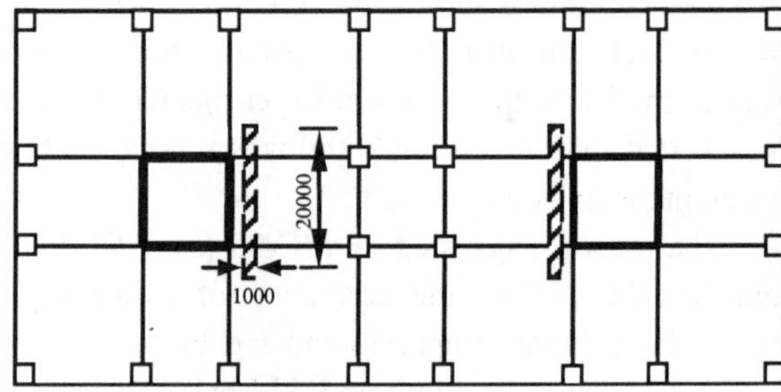

Figure 7. Arrangement of last-cast strips: Case 2 (dimensions in mm).

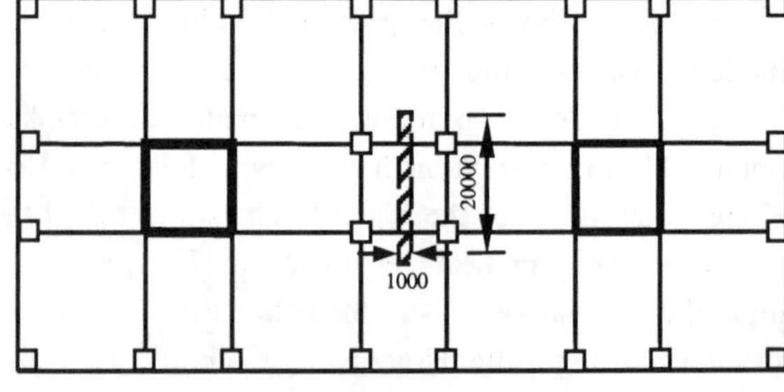

Figure 8. Arrangement of last-cast strips: Case 3 (dimensions in mm)

5.2.1 *Arrangement of late-cast strip: Case 1*

In Case 1 (Figure 6), most of the podium is cast at one go, leaving only a 1m gap between the two halves. In such an arrangement, before casting the stitch, the two halves are relatively free to shrink with each centre of core wall as the effective point of fixity as the columns are much more flexible compared with the core walls. Figure 9 shows that the provision of the late-cast strip alleviates the tensile stress of almost the whole of concrete podium except for some areas near the columns. It is because most of the drying shrinkage has taken place before casting the stitch. In this case, the late-cast strip can be considered as very efficient.

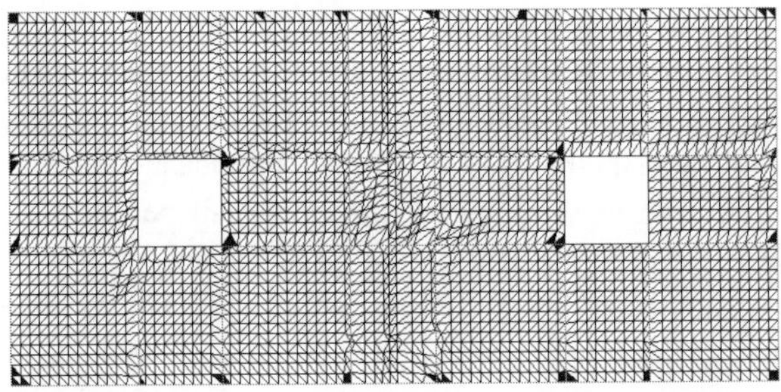

Figure 9. Areas prone to cracking at the end of 365 days (Case 1).

5.2.2 *Arrangement of late-cast strips: Case* 2

A full-length late-cast strip is very effective in alleviating shrinkage stresses but it is not always possible because of the reduction of stiffness in the interim period. The arrangement in Case 2 (Figure 7) is a compromise that consists of only two narrow strips each located next to a core wall. Figure 10 shows that the tensile stresses are slightly relieved but high tensile stresses still occur at the ends of the late-cast strips.

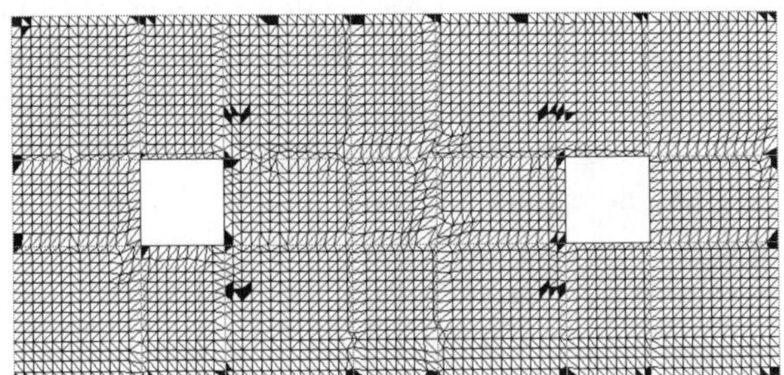

Figure 10. Areas prone to cracking at the end of 365 days (Case 2).

5.2.3. *Arrangement of late-cast strip: Case* 3

In Case 3 (Figure 8), only a late-cast strip is provided at the middle of the podium and the temporary reduction of stiffness is therefore minimal. Figure 11 shows that the relief of tensile stresses is rather small. High stresses still occur at the corners of core walls and the ends of the late-cast strip.

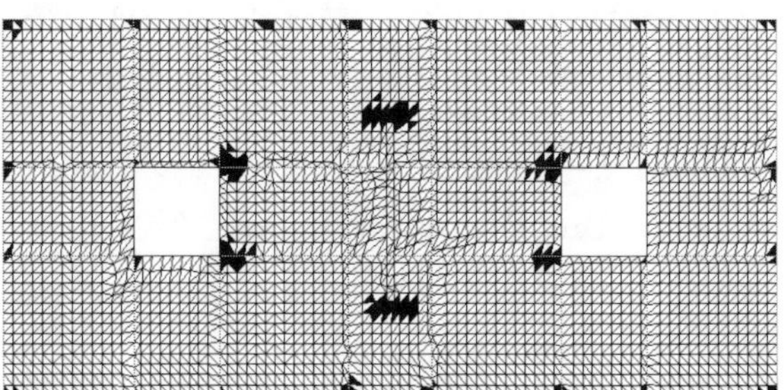

Figure 11. Areas prone to cracking at the end of 365 days (Case 3).

6. Effect of Stitching Time on Shrinkage Stresses

To find the optimum time lag between the first-cast area and the late-cast strips, the variation of shrinkage stress at 1000 days is studied. Case 1 of late-cast strip arrangement is selected and the stitching time studied includes 30, 60, 90, 120, 180 and 210 days. It is observed that if the stitch is cast too early, there are large areas prone to cracking. Figure 12 shows that the late-cast strip may be considered effective if it is cast after 180 days. Further increase of the time lag will reduce the stress but it will also hinder the subsequent construction works. In addition, it should be noted that the optimum stitching time will be different if the design parameters are changed. In addition, the optimum stitching time also depends on the design code adopted to model the shrinkage and creep behaviour. Although the late-cast strips can effectively reduce the shrinkage stresses, the resulting construction joints of course still require good workmanship to avoid subsequent water leakage problems.

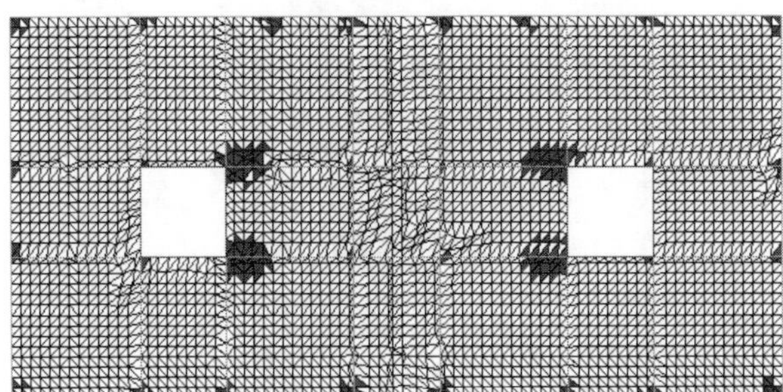

Figure 12. Areas prone to cracking at the end of 1000 days with stitching time of 180 days (Case 1).

7. Effect of Shrinkage Reducing Agents

Shrinkage reducing agents (SRA) are capable of lowering the surface tension of pore water in concrete, thereby reduces the evaporation of water. Shah *et al.*

[10] reported that the SRA significantly reduced the free shrinkage of concrete.

7.1 *Proposed shrinkage model*

For the standard 150mm cylinder tests, the shrinkage strain according to ACI-209 [6] is

$$\varepsilon(t) = \frac{\varepsilon_u(t - t_0)}{(t - t_0 + 55)} \tag{6}$$

where t is the time, t_0 is age when curing is stopped, ε_u is the ultimate shrinkage of concrete. If it is assumed that the ultimate shrinkage of concrete is 720×10^{-6} and the reduction of ultimate shrinkage strain is approximately 35%, an approximate shrinkage function can be obtained by curve fitting as

$$\varepsilon(t) = \frac{470(t - t_0)}{(t - t_0 + 30.64)} \times 10^{-6} \tag{7}$$

7.2 *Effects of SRA on the podium structure*

The results shown in Figure 13 indicate that even though SRA is applied to the construction of the podium structure, there are still large areas prone to cracking at the end of 365 days. The SRA does reduce the shrinkage stress to certain extent but the problem of shrinkage cracking cannot be completely solved by the addition of SRA alone. Anyway the addition of SRA to the concrete mix can be considered as a good measure for reducing early age cracking when the concrete has not developed its full tensile strength.

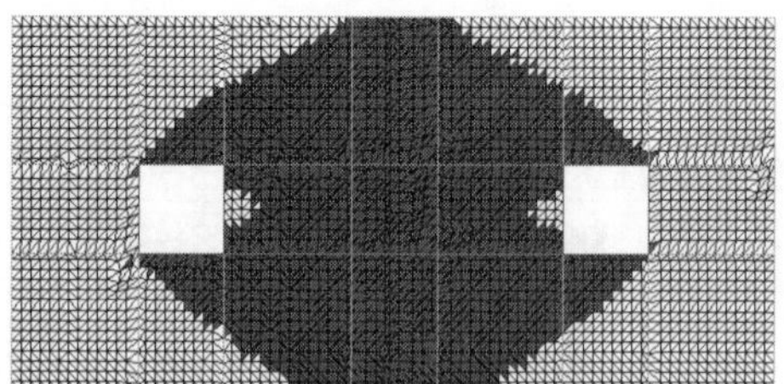

Figure 13. Podium with SRA: areas prone to cracking at the end of 365 days.

8. Reinforcement for Crack Control

When the tensile strength of concrete is reached, it may crack resulting in a sudden drop in the stiffness. In general, whether further cracking occurs after the formation of the first crack depends very much on the reinforcement ratio. If the steel percentage is too low, yielding of steel reinforcement occurs and the cracks will widen, and no further cracks can be formed. If the reinforcement is enough, the crack width is much smaller while the number of cracks increases.

For different exposure conditions, the limiting crack width varies from 0.25mm to 0.4mm. The Hong Kong Code [8] suggests limiting crack widths of 0.3mm under quasi-permanent loads and 0.2mm in the case of members with prestressing. The ACI Building Code [11] has adopted the principle that the reinforcement to resist shell membrane forces shall be provided so that the design strength in every direction shall be enough to resist the component of the principal member forces. Without carrying a rigorous shrinkage analysis taking into account the reduction of stiffness after cracking, this may be a sensible solution for areas prone to cracking.

9. Progressive Construction of Multi-storey Structures

In a multi-storey podium that is cast floor by floor, the stresses of the upper floors due to shrinkage will not be the same as those of the lower floors because of the difference in residual shrinkage strains. To discuss this problem, the term $\varepsilon_{sh}(t_3,t_2,t_1)$ is introduced as the shrinkage of the floor during the period (t_3,t_2) where t_1 is the time when the floor was cast and $t_3>t_2>t_1$. Suppose the first floor is constructed at time t_1 while the second floor is constructed at time t_2. In the period (t_3,t_2), the shrinkage strain of the first floor is $\varepsilon_{sh}(t_3,t_2,t_1)$ while that of the second floor is $\varepsilon_{sh}(t_3,t_2,t_2)$. Figure 14 shows that $\varepsilon_{sh}(t_3,t_2,t_2)$ is always higher than $\varepsilon_{sh}(t_3,t_2,t_1)$. The resulting differential shrinkage between the two floors after casting the second floor can be worked out as $\varepsilon_{sh}(t_3,t_2,t_2)\text{-}\varepsilon_{sh}(t_3,t_2,t_1) = \varepsilon_{sh}(t_3,t_2,t_2)\text{-}[\varepsilon_{sh}(t_3,t_1,t_1)\text{-}\varepsilon_{sh}(t_2,t_1,t_1)]$.

For example, if the second floor is cast 20 days after the first floor, one has t_1 = 0 day and t_2 = 20 days. It is found that the differential shrinkage of the second floor is effectively less than one-sixth of the free shrinkage strain as shown in Figure 15. Shrinkage analysis is again carried out using the differential shrinkage strain. It is observed that no area is identified as prone to cracking.

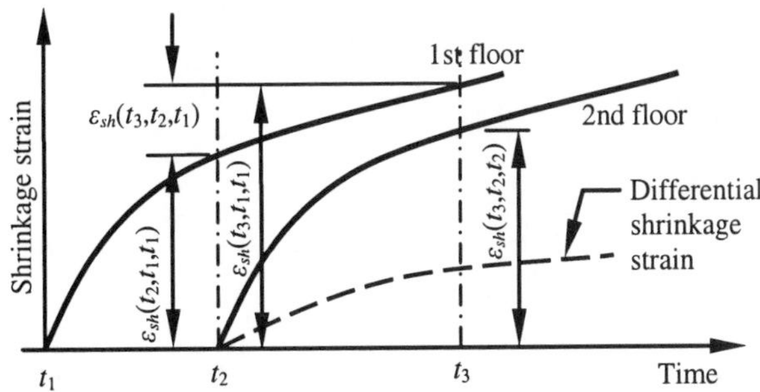

Figure 14. Shrinkage strains of different floors.

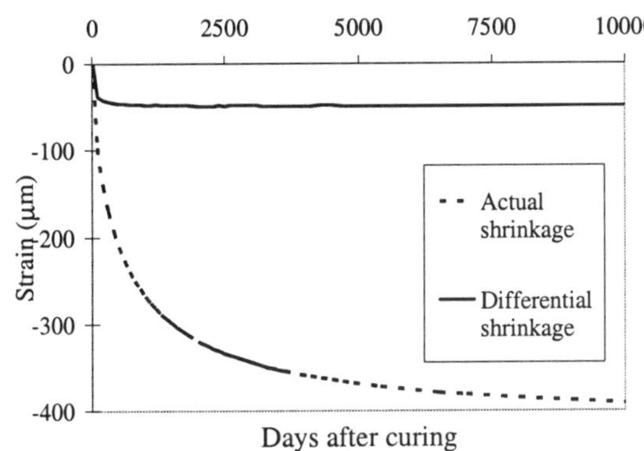

Figure 15. Differential and free shrinkage.

10. Conclusion

A step-by-step time marching analysis using the finite element method has been developed for the shrinkage analysis of podium structures. It is found that the use of full-length late-cast strips is the most effective in reducing shrinkage stresses. Although partial-length late-cast strips can reduce the shrinkage stresses to certain extent, cracking may still occur in the vicinity of the ends of such late-cast strips. It is suggested that, for large podium structures without movement joints, thorough investigation by means of finite element method or similar should be carried out before construction is commenced so that the designers can take the necessary precautionary measures to prevent excessive cracking. Early in the conceptual design stage, the vertical supports should be wisely arranged to avoid imposing unnecessary restraints to movements.

Results also show that SRA can alleviate but cannot completely prevent cracking. The choice of concrete mix with low shrinkage strain and high tensile strength is beneficial. The use of SRA together with the proper provision of late-cast strips can further minimise the shrinkage stresses.

In multi-storey podium structures which are cast floor by floor, the problem of shrinkage stresses mainly lies with the first floor while the problem is much less serious with the upper floors.

Acknowledgement

The authors are grateful to Mr. P. L. Ng for helping with the mesh generation for finite element analysis.

References

1. Z. P. Bažant, Mathematical models for creep and shrinkage of concrete, Z. P. Bažant and F. H. Witmann (eds.), *Creep and Shrinkage in Concrete Structure*, John Wiley and Sons, New York, pp. 163-256 (1982).
2. A. Ghali, R. Favre and M. Elbadry, *Concrete Structures: Stresses and Deformations*, 3rd Ed., Spon Press, London (2002).
3. Y. K. Cheung and M. F. Yeo, *A Practical Introduction to Finite Element Analysis*, Pitman, London (1979).
4. H. S. Müller and H. K. Hilsdorf, Comparison of prediction methods for creep coefficients of structural concrete with experimental data, F. H. Wittmann (ed.), *Fundamental Research on Creep and Shrinkage of Concrete*, pp. 269-278 (1982).
5. Comité Euro-International du Béton, *CEB-FIP Model Code for Concrete Structures* (MC-90), Paris-London-Berlin (1990).
6. ACI Committee 209, *Prediction of Creep, Shrinkage and Temperature Effects in Concrete Structures*, American Concrete Institute, Detroit (1992).
7. British Standards Institution, *BS5400: Part 4, Steel, Concrete and Composite Bridges, Code of Practice for Design of Concrete Bridges, London* (1990).
8. Building Authority, *Code of Practice for the Structural Use of Concrete, Hong Kong* (2004).
9. A. K. Gupta, Membrane reinforcement in concrete shell: A review, *Nuclear Engineering and Design*, **82**, pp. 63-75 (1984).
10. S. P. Shah, M. E. Karaguler and M. Sarigaphuti, Effects of shrinkage-reducing admixtures on restrained shrinkage cracking of concrete, *ACI Materials Journal*, **89**(3), pp. 289-295 (1992).
11. ACI Committee 318, *Building Code Requirements for Structural Concrete*, American Concrete Institute, Detroit (2002).

Computer Modelling & Analysis

THE RESEARCH AND DEVELOPMENT OF POST-PROCESSING SYSTEM FOR CAD OF HIGH-RISING STEEL BUILDING BASED ON 3D SOLID MODEL

ZHIGUO CHANG

College of Civil Engineering, Tongji University, 1239 Siping Road Shanghai 200092, China

HUIZHU YANG

College of Civil Engineering, Tongji University, 1239 Siping Road Shanghai 200092, .China

QILIN ZHANG

College of Civil Engineering, Tongji University, 1239 Siping Road Shanghai 200092, China

The research and development of post-processing system is the bottle neck of the CAD of steel structures. In this paper, the post-processing system for high-rising steel building based on 3D solid model is introduced. In this system, the complex spatial relationship between bars and detailed constitution of joints of steel structure is shown accurately in 3D solid model. Based on the 3D solid model, the drawing is created automatically in this system by using the geometrical and structural information of the solid model, namely the blue-print drawings, the designing drawings, and the manufacture drawings. Presently, this system has used in projects successfully.

1. Introduction

Nowadays , the domestic post-processing systems for high-rising steel building are all based on planar model, therefore, the complex spatial relationship between bars and detailed constitution of joints of structure cannot be shown accurately in these models. And, being the deficient information of these models, the drawings created by these systems cannot satisfy the requirements for the manufacture of the structure. Then, the manufacture designer must draw the manufacture drawings based on the designing drawings manually, which are inefficient and fallible processes.

In post-processing system based on 3D solid model, the bars are shown in 3D solids, and the components of joints, such as, gusset plates, bolts and welds, are shown and modified in AutoCAD modeling space. Furthermore, some structure information, which can be used to create the labels of the drawing, including the material type of the gusset plates, the grade of the blots and the type of the welds, etc. is affiliated to these solids.

According to different demands, three kinds of drawings with different levels can be created automatically in this system by using the geometrical and structural information of the solid model, namely the blue-print drawings, the designing drawings, and the manufacture drawings.

2. The Traits of the Post-processing System based on 3D Solid Model

2.1 *The discussed post-processing system is developed on the AutoCAD platform*

The post-processing system is developed in visual C++ programming language by using ObjectARX platform which is the most efficient DSK of AutoCAD. ObjectARX is the Object Oriented Programming interface, but the other DSKs, such as ADS and AutoLisp, are Process Oriented Programming interface. Therefore, the custom entity derived from the AutoCAD 3DSolid, by which the CAD system is customized and extended with the additive geometrical and structural data, can be accessed directly in the AutoCAD database.

2.2 *The structure details is shown in 3D solid*

In this system, some derived classes, which include some additive data, are derived from the basic class of ObjectARX named AcDb3dSolid. All of the 3D solids in this system are actually the objects of the derived classes. The modeling of solids is actually a process of creating and modifying of the additive data and the drawing process is the use of the data. The derived bar classes' structure is shown in chart 1.

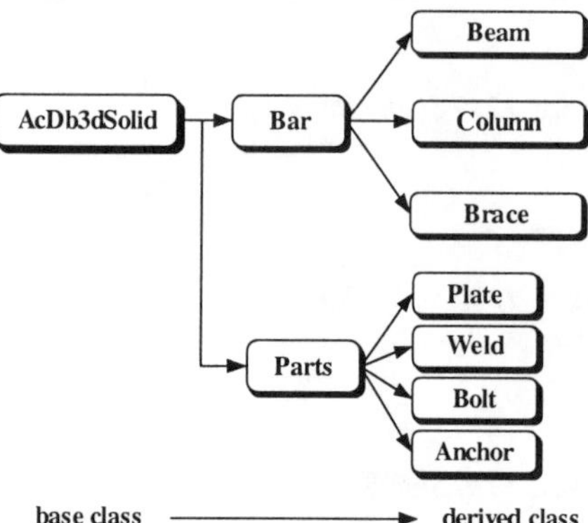

Chart 1. The derived classes' structure

The custom entity is the actual geometrical representation of the structure solid, and can be shown and modified in AutoCAD. Further more, the modification in the designing process, for instance, the designing or redesigning of the joint, can be shown immediately.

2.3 *The drawings are created automatically based on the 3D solid model*

Because the sufficient structural information is included in the solid model, the large numbers of drawings, which can meet the manufacturer requirements, can be created directly by using the geometrical and structural data in the model. The planar drawing is created by projected and hided, and then the labels are added using the structural data. The blue-print drawings and the designing drawings are actually created by using the partial geometrical and structural data in the model. Therefore, the contents of these three kinds of drawings are consistent.

3. The Main Processes of the Post-processing System

The main functions of the discussed post-processing system are to design joints and create drawings based on 3D solid model which is transformed from the calculation model. Therefore, the main processes of the system are model transforming, joints designing and drawings creating. The processes are shown in flow chart 1.

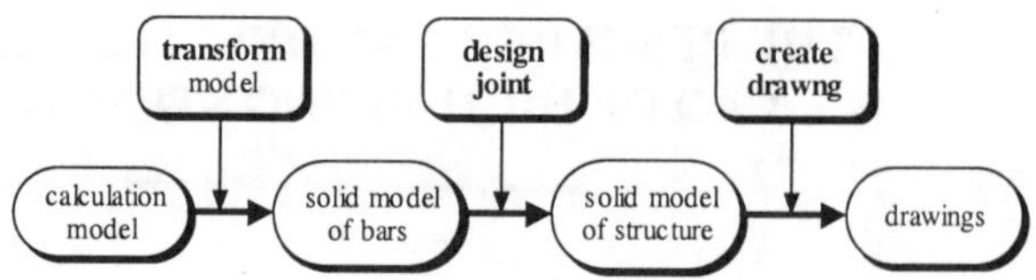

Flow chart 1. The main processes of the system

When the calculation model is transformed into the solid model, the section data of the bars is used to create the 3D solid in AutoCAD. A 3D solid model of a column intercrossing with four beams is shown in figure 1.

Figure 1. 3D solid model (without connections)

In the second step, the type of the joint is selected from joint lib and the relative bars are selected, and then the joint is calculated and designed automatically. Finally, the 3D solid model of the joint is created. The result is shown in figure 2. The components of the joint can also be modified and redesigned.

Figure 2. 3D solid model (with connections)

In the last step, the geometrical data, including lines, planes is collected, projected and hid, and the planar drawing is created; then the labels of the drawing are created using the structural information from the solid; finally, the drawing blocks are relocated to form drawing paper. The detailed process of drawing is shown in flow chart 2.

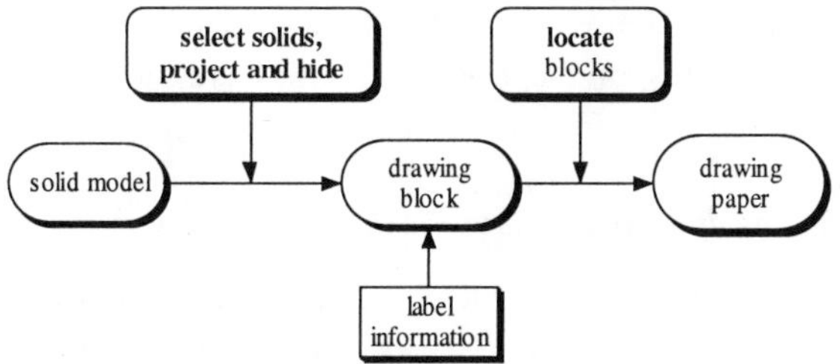

Flow chart 2. The detailed process of drawing

The 3D wire frame of the above solid model is shown in figure 3 and the drawing is shown in figure 4.

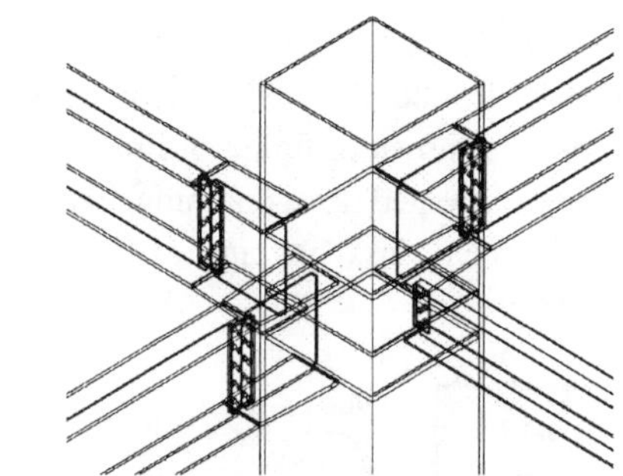

Figure 3. 3D solid model (3D wireframe)

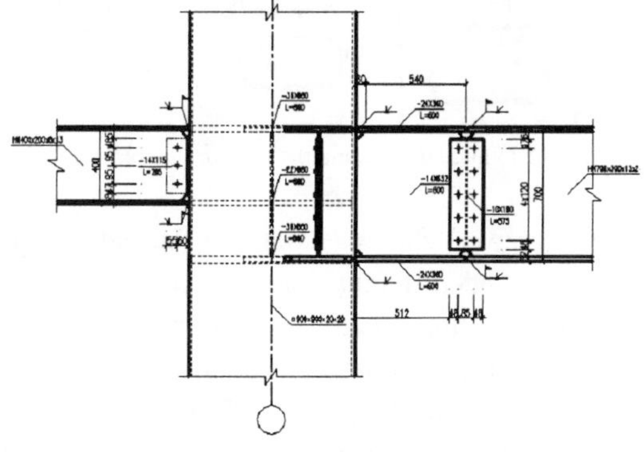

Figure 4. drawing of the joint(Front view)

4. The Creation of 3D Solid Model of Bars

There are two main contents in the constructing of the 3D solid model of bars.

In the first place, the result of the FEM (Finite Element Method) calculation model is read by the post-processing system to create the 3D solid model on the basis of the manufacture requirements. In the structural calculation model, bars are defined as finite element. Some geometrical information, such as, the section and location of the bar, and some physical information, such as, the material of the bar, are appended to the element. In the FEM calculation model, the eccentric location of the elements is not considered, and the elements are break at the intercross points, but the solid model is identical with the manufacture and assembling. Therefore, the model has to be remodeled. The changes include: (1) the spacial locations are changed to make the beams which are on the same floor have the same elevation. (2) The girders which are break by the secondary beam are connected to form the integrated bar. (3) The column elements on the floor are connected in term of the manufacture length of the column which is defined by the designer. It is the advantages of transforming calculation model to solid model that the solid model is identical with calculation model. But the manual modeling is onerous and fallible.

In the second place, the solid bars are created in three methods: (1) Extruding a region distance of height, the extrusion direction is along the normal of the region. (2) Extruding a region extruding region along the path curve. (3) Performing Boolean operation, such as, unite operation and subtract operation, between several simple solids.

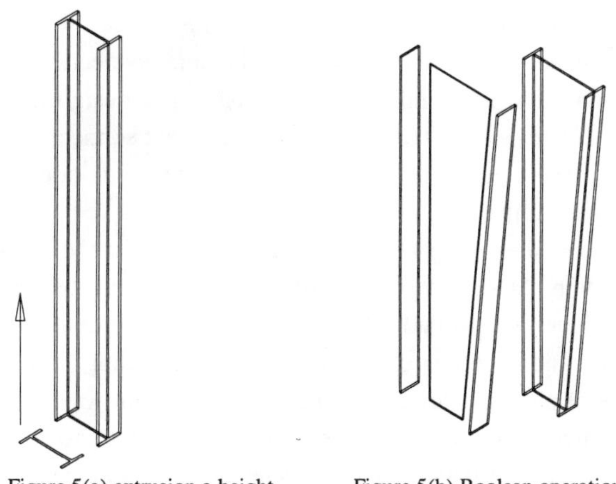

Figure 5(a) extrusion a height　　Figure 5(b) Boolean operation

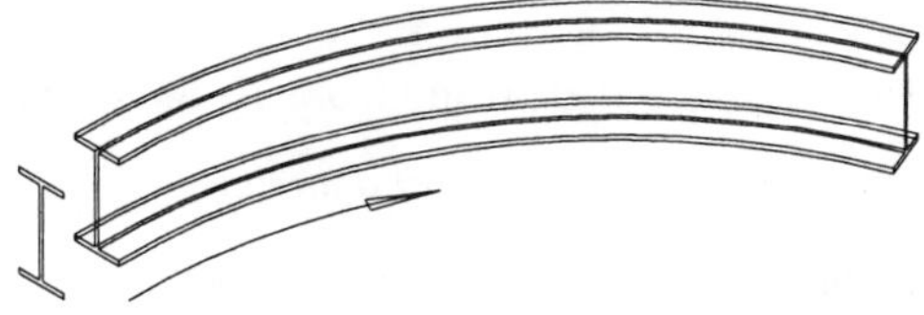

Figure 5(c) extrusion along path

Figure 5. creation methods of solid model of bar

5. Joint Designing

The steel structure is formed by assembling the section steel or prefab bars, and there is a joint at every intercross point of the bars. And the detailing design of joints is very important and complicated in steel structure designing by reason that: (1) there are various kinds of forms of joint, and the section and location of the bars connecting to the joint are diverse. (2) The conformation of the joint is complicated. There are many parts in the joint other than the connected bars, such as, plates, weld, and bolts.

Considering the complexity of the joint designing, in this system, a joint is divided into several connections, and a connection is also divided into up flange, web and down flange basic connection. By this means, the difficulty of joint designing is simplified sharply.

The basic connection is the minimum calculation element, and there are one or several connection parts in a basic connection. The size or location of the connection parts are defined by the basic connection. There are two kinds of connection parts: bolts connection and welds connection. For instance, the web blots connection of I-shaped section is a typical basic connection. It includes two connection parts, connection plates and web bolts, connected plates and welds.

When the location of the basic connections, including bolts connection and welds connection, is changed, the calculation formula for connection is changed correspondingly. If the system is designed to calculate directly the joint with the deferent connections, the program code will be redundant and not be expandability.

In the interactive graphical user interface of this system, firstly, the type of connect is selected by designer according to the constraint of the joint, and the size and location of the joint parts, including plates, welds and bolts are calculated and the relevant solids are created. Then, the joint can be redesigned, for instance, the size of a plate or the diameter of a blot is changed. Finally, the modifications can be checked automatically.

6. The Creation of Drawing from 3D Solid Model

6.1 *The interface to the 3D solid model*

Every entity in AutoCAD database, including custom entity, has only ObjectId (ObjectId is an object of AcDbObjectId which is a class in ObjectARX class lib). The ObjectId of an entity can be used for the only interface between the drawing creation module and the other modules. The drawing data can be prepared as following steps: first of all, an entity in AutoCAD database is opened using function acdbOpenObject to gain a pointer to the entity, and the structural information in the entity can be accessed by this pointer. In this way, the relative entities are selected. And then, the ObjectIds of the selected entities are stored in a linked list.

6.2 *The data structure of the drawing system*

The geometrical and structural data can be accessed by the ObjectIds of the selected entities in the above-mentioned method to create the data structure of the drawing system. The data structure of solid of the drawing module is analogous to that of the solid modeling module, which is shown in chart 1. Otherwise, class joint is another important class, which describes the joint constitute and the relationship between bars and joint, which is shown in figure 6. From the point of view of Object Oriented Programming, the combined relationship between class bar and class joint is also shown in figure 6.

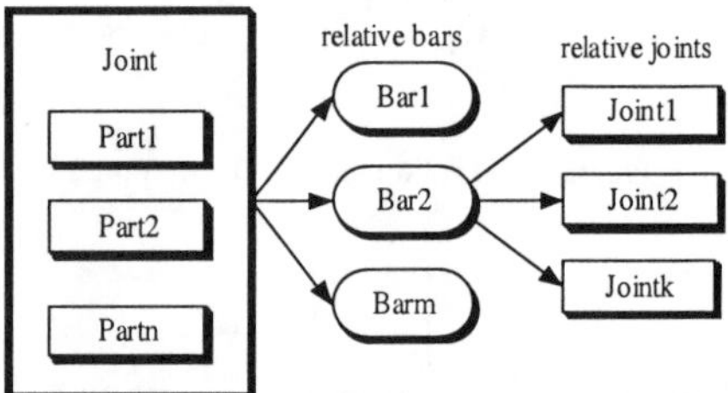

Figure 6. The joint constitute and the relationship between bars and joint

The above classes contain the all required information, and different drawings can be created by integrating this information in different way. As an illustration, the creation process of bar drawing is presented. As is shown in figure 7, the bar drawing is composed of the front view, the subsidiary view, and the section drawing of the bar, joint plates drawing, the location drawing which describes the planar location of the bar, as well as, material table. These drawing blocks can be classified into two classes, solid drawing block and planar drawing block. The former is created by projecting and hiding, but the latter is created directly using the partial geometrical data of solid (for example, the vertexes of joint plate). Besides, the deriving relationship between these blocks is also shown in figure 7.

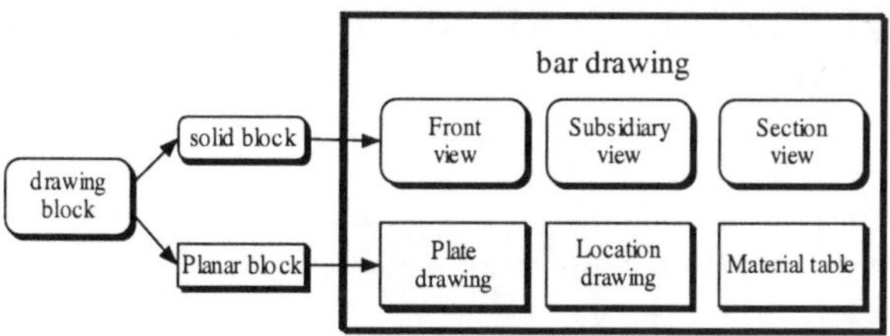

Figure 7. The joint constitute and the relationship between bars and joint

6.3 *The projecting and hiding of the drawing system*

The obstructing relationship between boundaries of the solids should be exhibited in the engineering drawings, the obstructed lines are shown in dashed, and the others are shown in the real line. Therefore, the projecting and hiding algorithm is applied to the boundaries of the solids as following three steps: first of all, every entity is analyzed to find hiding faces and hiding lines of it. And then, the hid solids are compared each other to find the lines hid by these solid. Lastly, the coincident lines are deleted and the break lines are joined.

Supposing that solid V_i is composed of several surfaces $\{S_{ij}^T\}$, which include forward direction surfaces $\{S_{ij}\}$ and backward direction surfaces $\{S_{ij}^{'}\}$. When the angle of the outside normal direction of a surface to the projection direction is within the range of $0 \le \alpha < \pi/2$, the surface is defined as forward surfaces, and when the angle is within the range of $\pi/2 < \alpha \le \pi$, the surface is defined as surfaces. The vertical planes are not considered in projecting and hiding work. By distinguishing the forward surface and backward surface, unnecessary operations are avoided and the efficiency of hiding is raised.

6.4 *The labeling automatically of the drawing system*

It is important to label the drawing block in the drawing system, therefore, a feasible method need to be used to describe the relation between it and drawing block. In respect that there are a lot of parts in steel structure, the following means is presented to avoid labeling to overlap each other.

1. Grouping parts. The parts are classified in term of the traits of the parts, for instance, the grade, diameter and length of the blots. The parts with the same traits are labeled at only point to avoid the overlapped labeling.
2. Labels are classified into two classes, bar label and joint label. The attributes of bar is labeled by a common module, which is called bar label. But the joint is labeled by deferent modules according to the type of the joint.
3. The number of blots is counted in term of the traits of blots, maximal

7. Conclusion

This paper presents the key techniques of the modeling and the drawing of steel structure. The details of the steel structure are expressed and the drawings are created automatically. The software system based on the 3D solid modeling can improve the work efficiency of the designer.

References

1. CHEN Ying-shi, "Solid Modeling Techniques for CAD Post-processing of Space Steel Structures"(in Chinese),Journal of Tongji University, Shanghai, China , vol28 , No3 , P297-300.
2. CHEN Shan, "The Development of AutoCAD applications by ObjectARX"(in Chinese), People Post Press, Beijing, China
3. SUN Jia-guang, "Computer Graphics(New Edition) Tsinghua University Post Press, Beijing, China.

OPTIMIZATION OF REINFORCED CONCRETE STRUCTURES WITH GENETIC ALGORITHM

K.I. WONG

Department of Civil Engineering, the University of Hong Kong, Pokfulam, Hong Kong

S.H. LO

Department of Civil Engineering, the University of Hong Kong, Pokfulam, Hong Kong

This paper investigates the potential of using genetic algorithms (GA) in optimization of reinforced concrete (RC) structures. The optimum cost design of three RC structures, a beam, a low-rise frame, and a high-rise single core wall building, using GAs with variable operators are presented. The results show that GAs help to efficiently locate a near optimal solution offering lower overall cost designs compared to traditional design methods.

1. Introduction

Engineering design often involves a number of decision variables to give optimal structures, which could be in terms of volume, weight, cost, or other performance. Traditionally, preliminary member sizing is carried out by inspection, trial and error, experience or rules of thumb for convenience. However, these methods may overlook the global picture on costs, and designs are on the conservative side. With the assistance of modern computers, optimum sizing is possible to achieve large savings in overall cost.

The optimization of RC structures conventionally requires complex formulations. The methods of optimality criteria and mathematical programming often necessitate the integration of optimization algorithm into the analysis and design process. It is difficult for engineers not familiar with optimization theories to put optimum design into practice. The emerging of genetic algorithm (GA) has enabled the complete separation of the optimization process and the analysis of structure. GA only operates on the results of analyses (such as displacements and forces) and adjusts the variables (such as dimensions) to be passed to the analysis in an iterative manner.

GA has found many applications in optimization of RC beams (Coello, Hernandez, Farrera 1997) and frames (Lee & Ahn 2003) (Rajeev & Krishnamoorthy 1998). However, most examples have not verified the optimality of the final design within the design space, nor have they investigated the performance of the GA to work as a true algorithm better than random searches or complete enumerations. Various literatures have doubted the performance of GA. The "no free lunch" theorem (NFL) (Wolpert & Macready,, 1995) pointed out that GA may not work as an algorithm for certain problems. The concept of "competent GA" (Goldberg, 2002) has also been brought forward meaning that GA is not elixir to all optimization problems.

This paper investigates the performance of GA with respect to the cost landscape of RC structures. Three illustrative examples are presented. In the fixed end-beam design, the performance of GA on a simplified two-variable fixed beam with varied precision is studied. The final design is verified as a near-to-optimal solution, more economical than traditionally designed beam with span-depth ratios commonly adopted. In the high dimensionality frame and single core wall examples, a modified version of GA, the Micro-GA (Krishnakumar 1989), has been interfaced with a finite element analysis code and has shown higher efficiency over traditional trial and error approach and ordinary GA.

2. Genetic Algorithm

The idea of GA originates from the natural evolution process. GA is an iterative method that works with the following steps:

1. The initial population of strings for the 1st generation is randomly generated.
2. The fitness of all individuals in the current population is evaluated.
3. Selection: According to fitness, two individuals are chosen for exchange of genetic information.

4. Crossover: The genetic information between two selected individuals is exchanged forming two new individuals for the next generation.
5. Mutation: The genetic information of a new individual is randomly altered, introducing new information absent in parents.
6. Step 4 to 6 is carried on until the next generation is filled with new individuals.
7. Check if stopping criteria is satisfied. Repeat the fitness evaluation, selection, crossover, mutation for the next generation.

The operations are not discussed in detail. The present paper adopts binary coding, tournament selection of 2 individuals, uniform crossover (p_x=0.8), jump mutation (p_j=0.01) and creep mutation (p_c=0.02).

3. Computer Implementation

In this paper, GA works with a modified finite element analysis code FACILE (Cheung, Lo, Leung 1996) using 8 nodes (for joints), 12 nodes (for beams, columns) and 18 nodes (for walls) hybrid stress hexahedral elements (Sze, Lo 1999). The reinforced concrete design follows the code of practice (BS8110,1990).

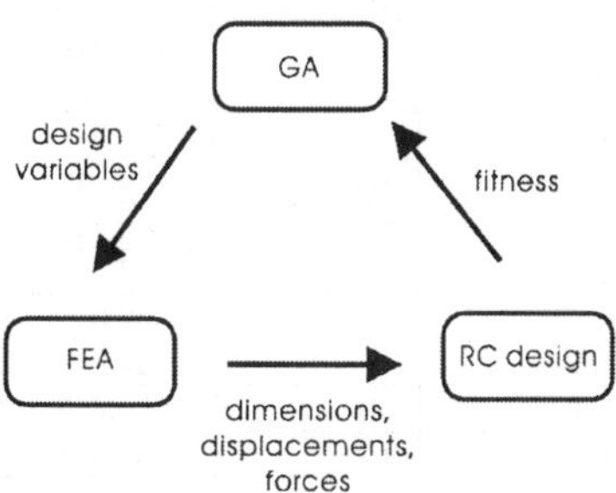

Figure 1. Diagram showing iterative relationship of the GA, finite element analysis and RC design

4. Cost Function and Fitness

The definition of cost function has a crucial influence on the optimum design. Most cost functions are expressed as a linear function of volume (Balling and Yao 1997) (Ceranic and Fryer 2000). In this paper, the cost is defined as only the material cost. When information is sufficient, the cost function can be revised to take into account of the formwork, labour and handling costs, while the optimization proceeds in the same way.

Cost function:

$$C_m = L[C_s(A_{ts} + A_{cs}) + C_s A_{sv}(B + H - 2D) + C_c BH]$$

where

C_m= total material cost of a member

C_s= cost of steel per unit volume (=\$28768/m^3)

C_c= cost of concrete per unit volume (=\$145/m^3)

A_{ts}= tension reinforcement area

A_{ts}= compression reinforcement area

A_{sv}= shear link area per unit length

B= breadth of member

H= height of member

D= cover to reinforcement

L= length of member

Traditional optimizations are minimization problems. However the GA is based on "maximizing the fitness". A conversion is required to map the cost (C_m) to fitness (f), and hence from a minimization to maximization problem.

$$f = \frac{k}{C_m}$$

A scaling constant k (10^3 or 10^6) is multiplied to correct the fitness to normal order of magnitude. The fitness then can be interpreted as the reciprocal of cost in thousands or millions.

5. Example #1 - fixed end beam

5.1 *Problem description*

A simple optimization problem was considered to verify the efficiency of GA. With simplifications to the design problem (figure 2), the variables are reduced to 2 only (breadth and height), enabling the complete fitness landscape to be calculated by enumeration of all variable combinations. The result is plotted as a contour map for convenient visualization and verification of optimization results (figure 3).

Problem parameters:	
No of variables:	2
Upper bound:	1000mm
Lower bound:	200mm
Precision:	5mm, 1mm, 0.1mm, 0.01mm
Span:	10m
Loading:	20kN/m imposed load + Self weight of beam (unfactored)
Deflection limit:	20mm
Penalty coeff:	0.1 (cost penalty= cost x 0.1 (δ-20)/20)

5.2 *Fitness landscape*

The fitness landscape of the 2-variable beam problem is plotted with complete enumeration method with precision of 5mm. Deflection penalty corresponds to the draw down of fitness for small depths. The 10 fittest set of variables and the corresponding fitness (precision=5mm) are shown in Table 1.

Table 1 Ten sets of variables giving maximum fitness found by complete enumeration (precision =5mm)

Height (mm)	830	845	835	850	840
Breadth (mm)	200	200	200	200	200
Fitness $(\$M)^{-1}$	1859.74	1856.84	1853.44	1850.62	1847.18
Height (mm)	855	840	830	860	845
Breadth (mm)	200	205	205	200	205
Fitness $(\$M)^{-1}$	1844.45	1841.56	1838.57	1838.32	1835.33

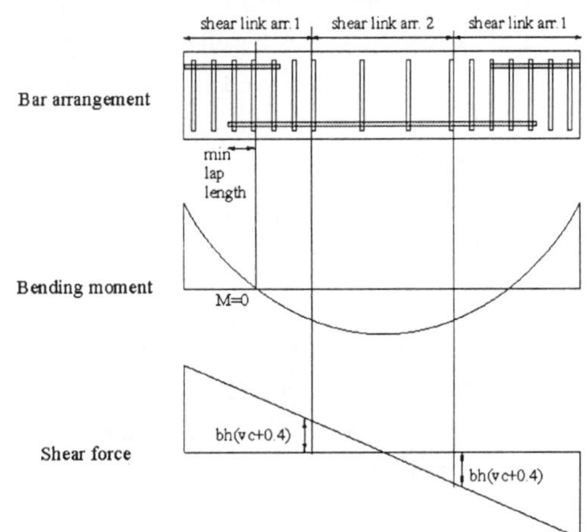

Figure 2 Design simplications for bar arrangements.

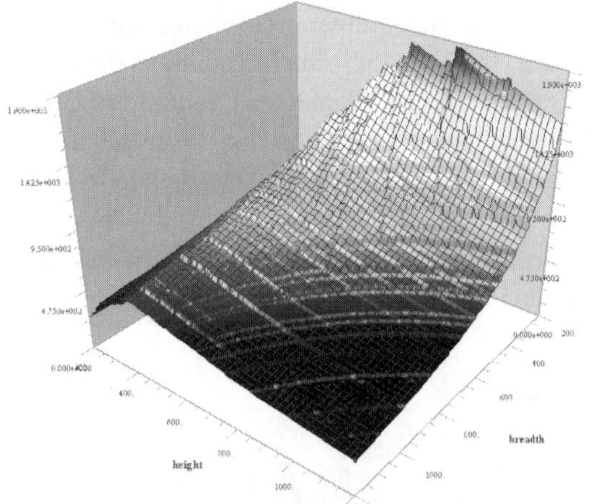

3 Fitness landscape of the fixed end beam

Figure 3 Fitness landscape of the fixed end beam

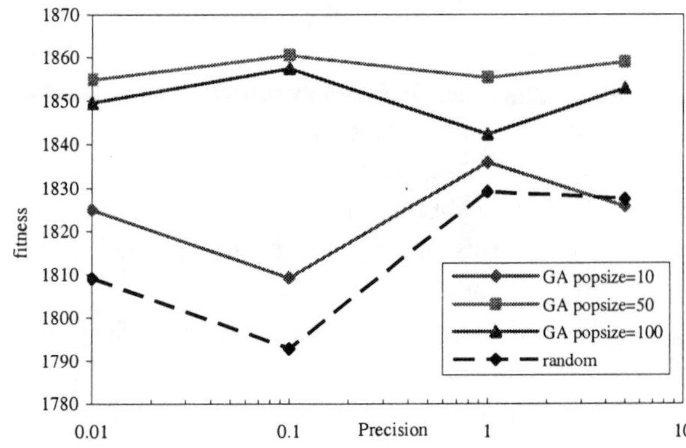

Figure 4 Maximum Fitness in 1000 Fitness Evaluations (average of 10 runs)

Table 2 Average of best design fitness after 1000 fitness evaluations (average of 10 runs).

	Precision (mm)	Population Size	Fitness $(\$M)^{-1}$	% deviation from max fitness
Random	5	-	1827.332	-1.74
GA	5	10	1825.674	-1.83
GA	5	50	1858.872	-0.05
GA	5	100	1852.903	-0.37
Random	1	-	1829.033	-1.65
GA	1	10	1835.932	-1.28
GA	1	50	1855.316	-0.24
GA	1	100	1842.363	-0.93
Random	0.1	-	1792.844	-3.60
GA	0.1	10	1809.215	-2.72
GA	0.1	50	1860.452	+0.04
GA	0.1	100	1857.400	-0.13
Random	0.01	-	1809.090	-2.72
GA	0.01	10	1824.958	-1.87
GA	0.01	50	1854.868	-0.26
GA	0.01	100	1849.543	-0.55

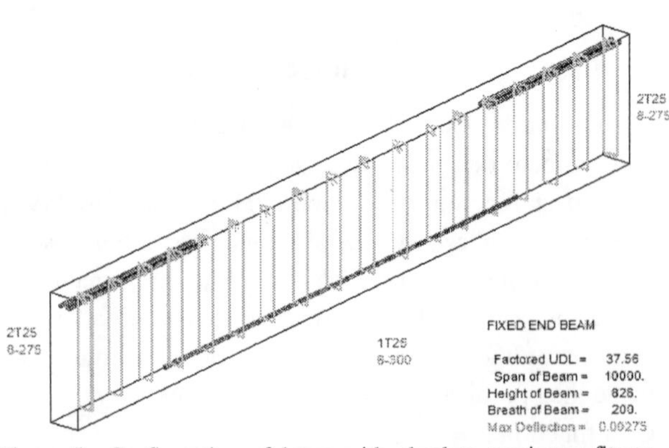

Figure 5 Configuration of beam with absolute maximum fitness (d=828, h=200mm, fitness=1862.277, corresponding cost=$536.98)

5.3 *Results*

The optimization is run with precision of dimension variables set as 5mm, 1mm, 0.1mm and 0.01mm. The last two precisions do not have practical significance but increase the complexity of problem and reduce the probability of finding an optimal solution completely by random. For each precision the number of fitness evaluations is limited to 1000. To gauge the performance of GA against random search, the offline performance indicator (De Jong, 1975) was adopted:

$$P_t = \sum_{i=1}^{t} \frac{f_{max}(i)}{t}$$

where t is the number of fitness evaluations. Pt is the offline performance at t. fmax(i) is the maximum fitness of individual up to the i-th fitness evaluations. The offline performance is the averaged maximum fitness up to the current fitness evaluation.

5.4 *Discussion*

Effect of problem complexity

The precision defines the problem complexity by increasing the number of combination of design variables. A conversion can show that:

Precisions (mm)	Number of possible combinations
5	25921
1	641601
0.1	6.4016 x 106
0.01	6.40016 x 109

GA did not outperform random searches in the problems of precision =5mm and 1mm in offline performance. For such small problems, random search (without replacements) or complete enumeration of the design space are perhaps more efficient methods to locate absolute maximum fitness than using GA given the computation cost is affordable. For precision=0.1 and 0.01, random search showed higher initial offline performance but later surpassed by GA with generation sizes 50 and 100.

Comparison with traditional design methods

A comparison of cost of the optimum beam with a traditionally designed one is made for illustrative purpose. Traditionally the sizing of beam for a large span (10m) may follow the suggested span depth ratio set by design guidelines. Adopting 15 as the span/depth ratio for continuous beam (ISE Manual, 2002), the depth of beam is 667mm (take height=710mm if 25mm cover + 20mm for half bar dia.), an estimation of breadth is calculated by maximizing the use of singly reinforced section:

$$0.156 = \frac{M}{f_{cu}bd^2}, \quad b = \frac{\frac{(1.6(20)+1.4(3))10^2}{12} x10^6}{0.156(50)(665)^2} \approx 90 \text{ mm, take b} = 200\text{mm}$$

The beam (figure 6) is analysed exactly defined above. The optimal beam has achieved a 5.3% saving in cost of the traditionally designed beam. This example only provides a microscopic optimization of material cost with the given design configurations. It should be noted that the form of cost function is always crucial for the definition of optimum. In this example, a larger beam size being more economical (due to less steel usage) is against our intuitive judgment, hence the determination of cost for RC members by inspection is often deceptive.

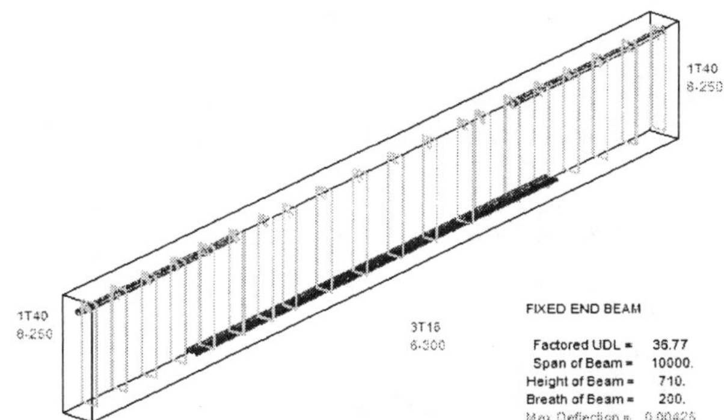

Figure 6 Configuration of a traditionally designed beam (d=710, h=200mm, fitness=1768.034, corresponding cost=$565.60)

6. Example #2 - three-storey frame

6.1 *Problem description*

The member sizes of a three storey frame building are optimized. This particular high dimensionality optimization is a challenging problem for any optimization algorithm as 24 variables are involved (figure 7). The performance of Micro-GA is compared with ordinary GA. Micro-GA makes use a small population size and the population will be restarted with random individuals when diversity is lost.

Problem parameters:
No of variables: 24
Upper bound: 1000mm
Lower bound: 200mm
Precision: 10mm
Loading: 10kN/m^2 IL (one way slab)

GA parameters:
Population size: 300
Max generations: 20
Micro-GA parameters:
Propulation size: 10
Max generations: 60

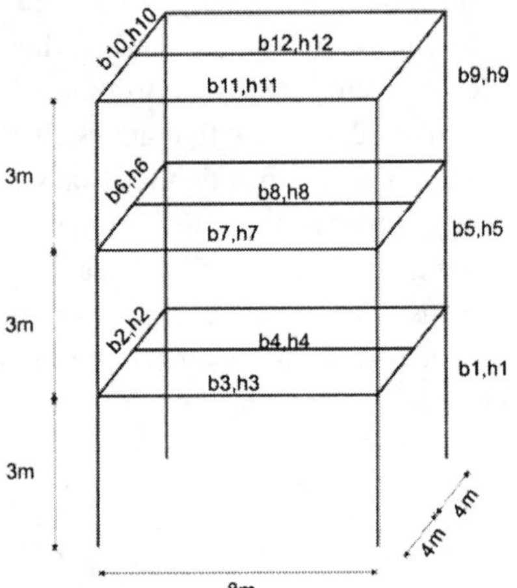

Figure 7 Three storey frame design variables

6.2 *Results*

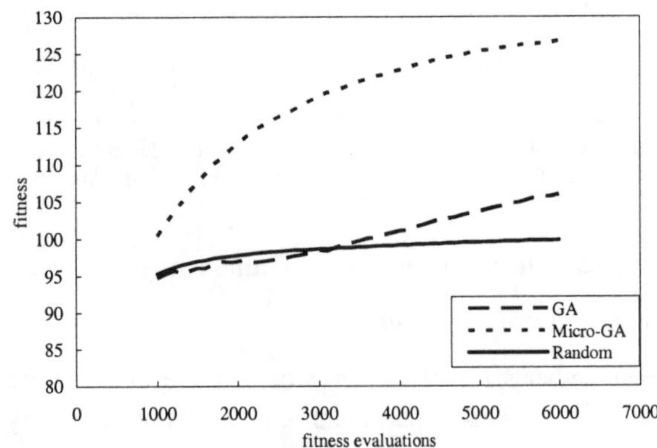

Figure 8 Offline performance for the RC frame optimization

Tables 3 Maximum fitness of RC frame optimization

	Max fitness attained (\$M)$^{-1}$	Max fitness appear after fitness evaluation
GA	120.1229	5744
Micro-GA	139.0567	4733
Random	101.1887	4265

Table 4 Results of optimized variables for different algorithms

GA

floor		3/F	2/F	1/F
column	b	0.21	0.21	0.21
	h	0.2	0.2	0.62
short side edge beam	b	0.2	0.2	0.37
	d	0.37	0.47	0.72
middle beam	b	0.2	0.2	0.28
	d	0.36	0.7	0.7
long side edge beam	b	0.2	0.2	0.2
	d	0.64	0.64	0.76

Micro GA

floor		3/F	2/F	1/F
column	b	0.21	0.21	0.21
	h	0.2	0.21	0.69
short side edge beam	b	0.2	0.2	0.2
	d	0.46	0.46	0.77
middle beam	b	0.2	0.2	0.2
	d	0.46	0.69	0.69
long side edge beam	b	0.2	0.2	0.2
	d	0.63	0.8	0.8

Random

floor		3/F	2/F	1/F
column	b	0.21	0.21	0.21
	h	0.32	0.32	0.48
short side edge beam	b	0.24	0.24	0.35
	d	0.7	0.84	0.84
middle beam	b	0.3	0.3	0.3
	d	0.46	0.46	0.59
long side edge beam	b	0.2	0.2	0.94
	d	0.64	0.65	0.65

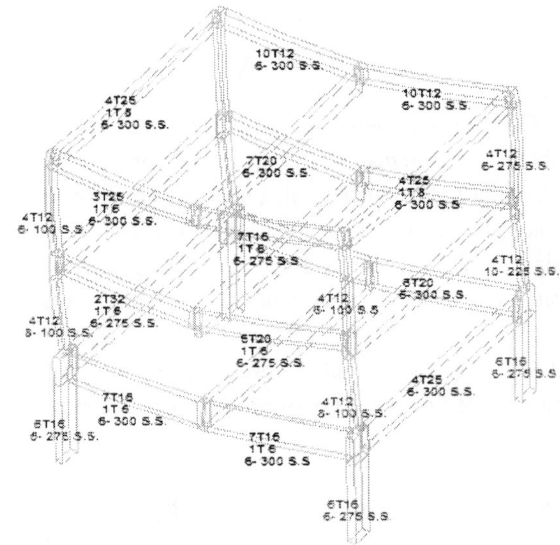

Figure 9 Optimal RC frame design by micro-GA showing part of the reinforcements (at mid span and near sides only, fitness = 139.0567, corresponding cost = \$7191.31)

7. Problem #3 – single core wall building

7.1 *Problem description*

In this example the lateral stiffness of a 20 storey building (figure 10) is optimized. Ignoring the contribution of stiffness from slabs and non-structural walls, the lateral stiffness is provided by the U-shaped single core wall and the concrete frame. The member sizes and wall thicknesses are variables to be decided (figure 11). The variables are allowed to vary every 5 stories. Each story has a height of 3m.

7.2 *Problem settings*

With considerations of the similar dimensionality and design space, the micro-GA parameters of the previous example are inherited to this tall building problem.

Problem Parameters

No of variables:	24
Upper bound:	800 mm
Lower bound:	300 mm
Precision:	10 mm
Loadings:	10kPa imposed load (factored) 10kPa wind load (factored)

MicroGA Parameters

Population size:	10
Maximum generation:	100
Horizontal deflection limit:	120mm (1/500 height)
Penalty coeff:	0.0001

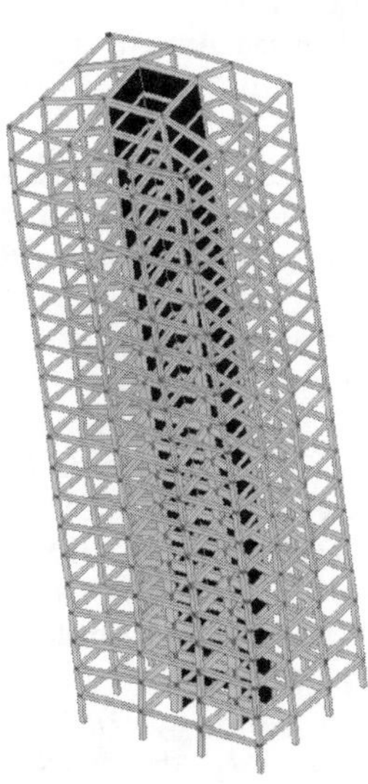

Figure 10 Final optimized single core wall building under deflection (max. hori. deflection = 105.38mm)

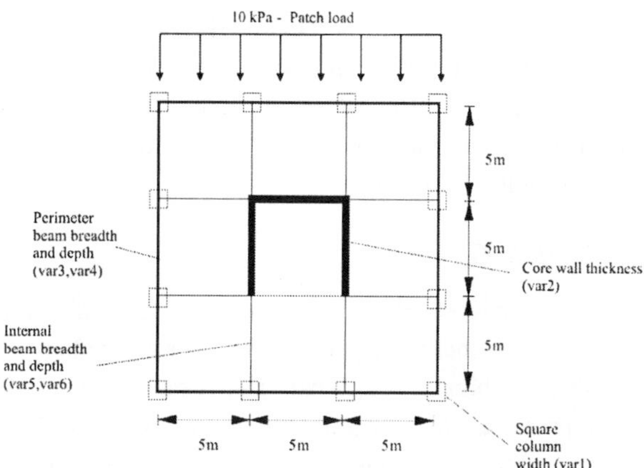

Figure 11 Typical plan of single core wall building showing the variables

7.3 *Results*

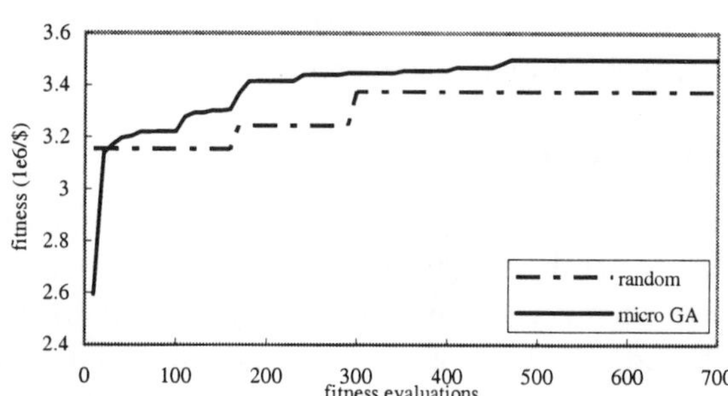

Figure 12 Maximum fitness for single core wall building optimization

Table 5 - Final optimized design variables (see figure 10 also)

	column width	wall thickness	internal beam breadth	internal beam depth	perimeter beam breadth	perimeter beam depth
16-20/F	0.300	0.300	0.300	0.460	0.300	0.300
11-15/F	0.620	0.300	0.300	0.490	0.320	0.300
6-10/F	0.630	0.360	0.320	0.620	0.360	0.360
1-5/F	0.630	0.360	0.320	0.680	0.360	0.390

7.4 *Discussions*

7.4.1 *GA vs. trial and error*

For this particular high dimensionality problem, the efficiency of GA over random search is more differentiable. The offline performance of GA has been increasing steadily due to constant hill-climbing in the fitness landscape. Similar to the beam problem the initial offline performance of random search is better than GA. However at later stages, random search has not been able to explored new variables states that gave fitness higher than certain value and showed convergence at lower fitness. The random search method was not efficient in obtaining optimal points given only small number of fitness evaluations for such a large number of variables.

7.4.2 *Micro-GA vs GA*

Micro-GA has outperformed GA in the example. For high dimensionality problems with large number of possible combinations, ordinary GA is more susceptible to trapping at local maximum points. Without high mutation rates to supply new information, the whole population may converge at regions far from the absolute optimal. Micro-GA, which replaces the population once convergence is detected, can avoid such local convergence maintaining diversity. The number of population size (10 in this example) should be large enough to explore the local optimal available from the building blocks of the current population.

8. Conclusions

A fixed end beam was optimized using GA. The fitness landscape of the design space was plotted by complete enumeration and the location of maximum fitness was identified. The beam is then subjected to optimization using GA with various precision definitions to adjust the complexity of problem. GA has not shown clear advantage over random search for the beam problem in offline performance especially for large precision such as 5mm and 1mm. However, in all runs, the maximum fitness attainable by GA was higher than random search, showing the power of GA in locating a near-optimal solution. The beam designed with GA has achieved economy compared to using traditional design methods.

GA and micro-GA has been compared in a 24-variable frame problem. Micro-GA was able to achieve higher offline performance and higher maximum fitness compared with normal GA and random search. Micro-GA has been proven more efficient in handling multi-dimensional and large number of possible variable combinations.

A 20 story single core building was optimized with micro-GA. Constraints of horizontal deflection limit was included and the excess deflection was penalized by a linear function. The resultant structure was a optimum cost design within deflection limits.

Acknowledgement

The financial support from the CERG research grant, HKU7117/04E "Analysis of transfer plate structures using high-performance solid 3D hybrid stress hexahedral elements" is greatly appreciated.

References

1. Balling, R.J.; Yao, X. (1997) *Optimization of reinforced concrete frames.* Journal of Structural Engineering, v 123, n 2, Feb, p 193-202
2. British Standards Institution (1990) *BS 8110: Structural Use of Concrete.* BSI, London
3. Camp, C.V., Pezeshk, S., Hansson, H. (2003) *Flexural design of reinforced concrete frames using a genetic algorithm.* Journal of Structural Engineering, v 129, n 1, January, p 105-115
4. Cheung, Y. K., Lo, S. H. and Leung, Y. T. (1995) *Finite Element Implementation* Blackwell Science.
5. Ceranic, B., Fryer, C. (2000) *Sensitivity analysis and optimum design curves for the minimum cost design of singly and doubly reinforced concrete beams.* Structural and Multidisciplinary Optimization, v 20, n 4, Dec, p 260-268
6. Coello, C.C., Hernandez, F. S., Farrera, F. A. (1997) *Optimal design of reinforced concrete beams using genetic algorithms.* Expert Systems with Applications, v 12, n 1, p 101-108
7. De Jong, K. A. (1975) *An analysis of the behavior of a class of genetic adaptive systems.* Doctoral Thesis, Department of Computer and Communication Sciences. University of Michigan, Ann Arbor
8. Goldberg, D. E. (2002). *The Design of Innovation.* Kluwer Academic Publishers
9. Institution of Structural Engineers (2002) *Manual for the design of reinforced concrete building structure.*
10. Krishnakumar K. (1989). *Micro genetic algorithms for stationary and non stationary function optimization.* SPIE Proceedings Intelligent control and adaptive systems, vol. 1196.
11. Lee, C., Ahn, J. (2003) *Flexural design of reinforced concrete frames by genetic algorithm.* Journal of Structural Engineering, v 129, n 6, June, p 762-774
12. Rajeev, S., Krishnamoorthy, C.S. (1998) *Genetic algorithm-based methodology for design optimization of reinforced concrete frames.* Computer-Aided Civil and Infrastructure Engineering, v 13, n 1, Jan, p 63-74
13. Sze, K.Y. and Lo, S.H. (1999) *A twelve-node hybrid stress brick element for beam/column analysis*, Engineering Computations, Vol. 16, pp. 752-766
14. Wolpert, D. and Macready, W. (1997). *No Free Lunch Theorems for optimization.* IEEE Transactions on Evolutionary Computation 1(1), 67–82.

OPTIMUM RC STRUCTURAL DESIGN SYSTEM BUILDS-C IN AN INTELLIGENT BUILDING DESIGN SYSTEM *i*-BUILDS[*,†]

CHANG-KOON CHOI

Chair Professor, Dept. of Civil & Environ. Eng., Korea Advanced Institute of Science and Technology, Daejeon 305-701, KOREA

HYO-GYOUNG KWAK

Professor, Dept. of Civil & Environ. Eng., Korea Advanced Institute of Science and Technology, Daejeon 305-701, KOREA

JI-EUN KIM

Graduate Student, Dept. of Civil & Environ. Eng., Korea Advanced Institute of Science and Technology, Daejeon 305-701, KOREA

This work is concerned with technical development of the optimum RC structural design. Based on the algorithm developed, BUILDS-C which is for the RC structural design program and one of subsystems in *i*-BUILDS (an intelligent BUILding Design System) has been developed. This optimum design algorithm, the structural behavior of beams and columns is implemented in the design sections of members satisfying code requirements are stored in the database and are manipulated by one major variable, i.e. the section identification number(N) linked with all other design variables. In design procedures, a continuous preliminary solution is obtained first, on the basis of regression curve equations, and then an effective direct search for the discrete optimum section is carried out. In advance, automatic design of walls, slabs and SRC columns for given design condition is followed.

1. Introduction

Intelligent building design system *i*-BUILDS bas been developed for effective and economic design of building structures. *i*-BUILDS is integrated system which includes all the series of procedures related to the structural analysis (BUILDS-A), foundation design (BUILDS-F), steel structural member design (BUILDS-S), RC structural member design (BUILDS-C) and quantity calculation of using materials according to the design results. Moreover, the adoption of STEP based algorithm in controlling the input data and output results makes it easy to communicate with other programs. In this paper, optimum RC structural design program BUILDS-C, one of the subsystems in *i*-BUILDS will be introduced. The basic idea adapted in developing the system is based on the works of Choi & Kwak [1, 2, 3].

Surely, all the RC members in building structures have dimensions of discrete sizes. The dimensions of concrete sections are usually increased by a certain size, e.g., 5 cm (2 in.) per a step, thus making the section dimensions discrete. The amount of reinforcement is also determined by the number of bars of standard sizes put into a RC section. Therefore, from a practical point of view, the design of RC member may be defined as the optimization problem with discrete variables rather than with continuous ones.

In this paper, a simplified, yet effective, algorithm for practical application of optimum design techniques to RC member design is proposed. After determining a continuous section from a regression formula which shows a relation between the section identification number and the member force, an effective direct search is followed to find an optimum member section from the predetermined section database arranged in an increasing order of the resisting capacity of section. Unlike beam and column optimization using database, the walls and slabs use direct design approach based on design criteria for the given loading condition. Moreover, SRC column design procedure has been also developed.

* This work is supported by the Ministry of Construction and Transportation.

2. Construction of Database

In a theoretical point of view, there can be an infinite number of RC sections designed to resist different levels of applied forces. In practice, however, one can easily observe that the dimensions of concrete sections usually have practical limitations; in many cases, the ratios of widths to depths do not exceed 2.5 in beams and 2 in columns, the dimensions of concrete sections are usually increased by 5 cm (2 in) per step, the most frequently used sizes of reinforcing bars in RC building structures are D19(#6), D22(#7), and D25(#8), and so on.

Complying with the aforementioned restrictions and, in addition, the current design code requirements such as the maximum and/or minimum steel ratio in a section, etc., constructing the database of RC sections that are frequently used in current practice is now possible.

A systematic construction of section dimensions for beams and columns by combining discrete widths and depths is suggested in this study. Figs. 1 and 2 represent all the possible beam and column sections that can be generated. The sections A in Fig. 1, for instance, represents the rectangular beam section of 35 cm × 65 cm (14 in. × 26 in.) and the section B in Fig. 2 represents the column section of 50 cm × 50 cm (20 in. × 20 in.).

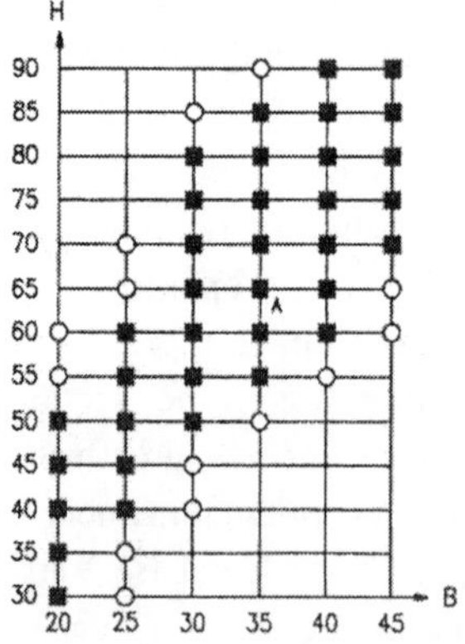

Figure 1. Construction of Section Dimensions for Beams

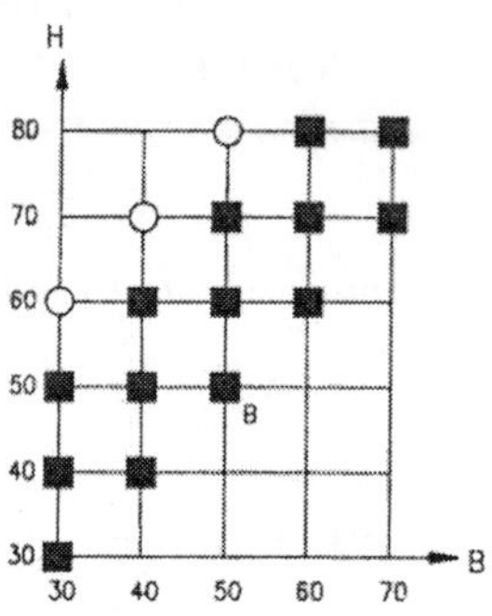

Figure 2. Construction of Section Dimensions for Columns

Once the overall dimensions of a section (i.e., width and depth) are decided, the number of reinforcing bars to be put into the section is determined with the restrictions of the design code in use. Providing different levels of reinforcement in a section ranging from the minimum to maximum steel ratio stated in the design code, it is possible to construct a number of different sections with a set of identical overall dimensions. The reinforcing bars used in beam sections are D19(#6) and D22(#7), and those used in column sections are D19(#6), D22(#7) and D25(#8).

Then, the resistant capacity of each section in terms of the maximum allowable moments or ultimate strength moments for the beam and the normalized volume of *P-M* Diagram for the column is calculated either by the working stress design method or by the ultimate strength design method, depending on the design code in use. A part of the database constructed in this study is given in Table 1~3 to show the general format of the contents of the data.

The cost per unit length of the member is calculated for each section by the following equations:

beam:

$$A_{\cos t} = C_s \times A_{st} + C_c \times B \times H + C_f \times (B + 2H) \tag{1}$$

column:

$$A_{\cos t} = C_s \times A_{st} + C_c \times B \times H + C_f \times (2B + 2H) \tag{2}$$

where C_s, C_c, and C_f is the unit cost of steel, concrete, and formwork including the related labor; and B and H is the section width and height, respectively. These sections are arranged in the database in order of increasing moment-resistant capacities for beam and increasing normalized volumes of the *P-M* diagram for columns and the expansion itself can be done very easily if so required.

The sections in the database are manipulated by one major variable (i.e., the section identification number), which is linked with all other design variables, such as the overall section dimensions, steel ratios, positions of reinforcement and resistant capacities, cost per unit length, etc. In the design procedures, only the stress constraints are considered, since the geometric constraints of the section were already considered during the construction of the database.

2.1 *Beam*

A part of database of the beam sections constructed through the procedure as shown previously is shown in Table 1.

Table 1. Beam Sections with D19(#6) Reinforcing Bar in Database

ID	B[a]	H[b]	A_{st}[c]	A_{sc}[d]	Moment[e]	Cost[f]
1	20	30	2	0	196,070	67.14
2	20	35	2	0	237,790	75.04
3	20	40	2	0	279,830	82.94
4	20	30	3	2	281,250	84.11
5	25	40	4	0	283,250	92.44
6	20	45	4	0	322,110	90.85
7	25	45	4	0	325,890	101.27
8	20	30	5	3	333,110	101.08
9	20	35	3	2	352,440	92.01
10	20	50	2	0	364,600	98.75

[a]Width in centimeters.
[b]Height in centimeters.
[c]Number of tension bar.
[d]Number of compression bar.
[e]Moment-resisting capacities in kgf/cm.
[f]Cost per unit length (Won/cm);U.S. $1.00 = 1,000Won.

2.2 *Column*

2.2.1 *Rectangular Tied-Bar Column*

A part of database of the rectangular tied-bar column sections constructed was shown in Table 2.

2.2.2 *Circular Spiral Column*

A part of database of the circular spiral column sections constructed was shown in Table 3.

Table 2. Rectangular Column Sections with D19(#6) Reinforcing Bar in Database

ID	B[a]	H[b]	BN[c]	HN[d]	CN[e]	Cost[f]
1	30	30	0	0	4	106.37
2	30	30	2	0	4	117.68
3	30	30	4	0	4	129.00
4	30	30	2	2	4	129.00
5	30	30	4	2	4	140.31
6	30	40	4	4	4	151.63
7	30	40	0	0	4	125.87
8	30	40	0	2	4	137.18
9	30	40	2	0	4	137.18
10	30	40	0	4	4	148.50

[a]Width in centimeters.
[b]Height in centimeters.
[c]Number of B-side bar.
[d]Number of H-side bar.
[e]Number of corner bar.
[f]Cost per unit length (Won/cm);U.S. $1.00 = 1,000Won.

Table 3. Circular Spiral Column Sections with D19(#6) Reinforcing Bar in Database

ID	B[a]	N[b]	Cost[c]
1	30	6	99.71
2	30	7	105.37
3	30	8	111.03
4	30	9	116.68
5	30	0	122.34
6	30	1	128.00
7	30	2	133.65
8	35	6	115.75
9	35	7	121.41
10	35	8	127.07

[a]Diameter in centimeters.
[b]Nunber of bar.
[c]Cost per unit length (Won/cm);U.S. $1.00 = 1,000Won.

3. Algorithm for Beam Design

3.1 *General*

To find out the optimum section from the database in this study, we must establish a relationship between the section identification numbers and the moment-resistant capacities of the sections. Such a relationship can be established by regression.

In Fig. 3, the moment-resistant capacities of the sections arranged in order of increasing capacities are plotted with the section identification numbers. These relationships obtained for the sections that use D19(#6) bars are as follows:

$$N = 0.6868 \times 10^{-4} \times M - 16.0 \qquad (3)$$

$M \leq 2,700,000\, kgf \cdot cm$

$$N = 0.0673 \times (M - 2,500,000)^{0.5} + 139.2 \qquad (4)$$

$M > 2,700,000\, kgf \cdot cm$

where N and M is the section identification number and the moment-resistant capacity of the section, respectively.

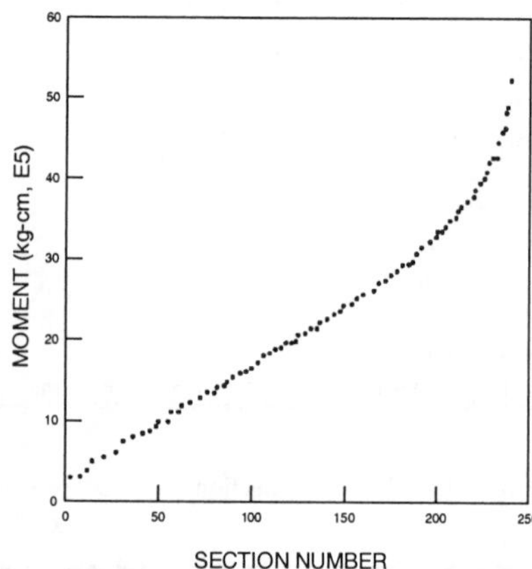

Figure 3. Maximum Allowable Moment versus Section Identification Number

3.2 *Continuous solution*

As the first step solution, the initial continuous solution (section) that satisfies the moment-capacity requirement only is obtained directly by using one of the relationships in Eqs. 3 and 4.

3.3 *Discrete solution*

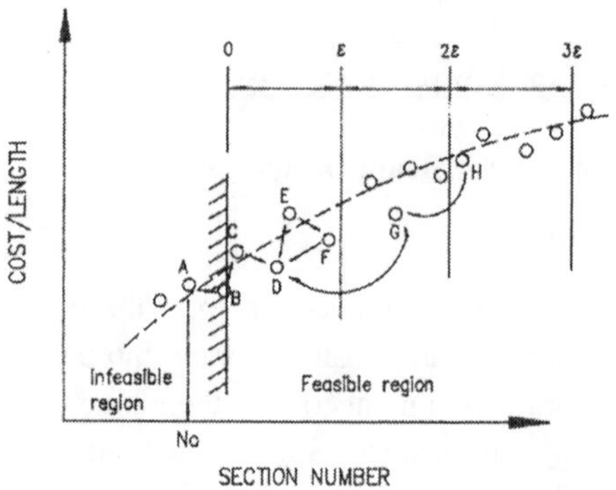

Figure 4. Search Process for Optimum Discrete Beam Section

Assuming that there exists the optimum section near the initially selected one or the first step solution, a direct search is conducted to find the discrete optimum section that satisfies all the design constrains, such as bending, shear, stability, and serviceability requirements.

Fig. 4 shows the procedure of searching for the discrete optimum size in the beam section table.

3.4 *Procedure of beam design*

A flowchart that shows the basic organization of the algorithm for the optimum beam section design is given in Fig. 5.

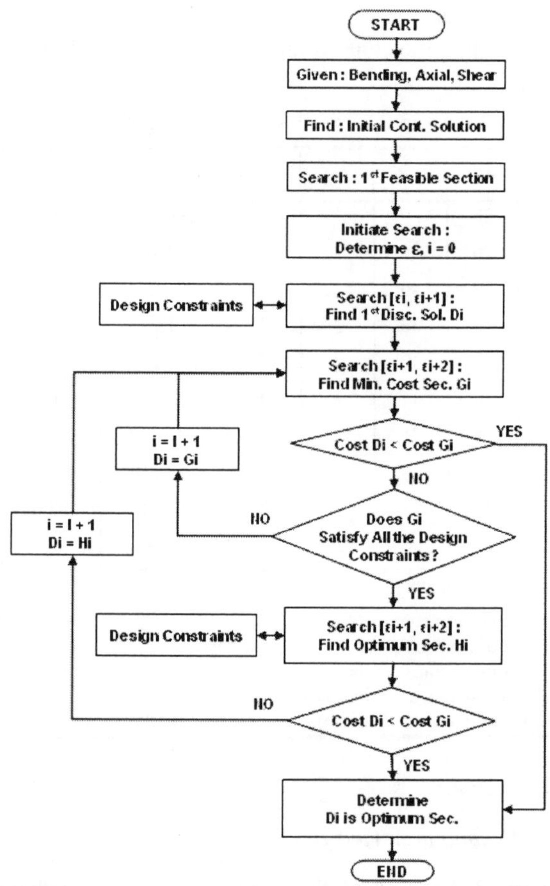

Figure 5. Flowchart for Searching for Optimum Beam Section

4. Algorithm for Column Design

4.1 *General*

A column is a structural member that supports the axial load and biaxial bending moments. The column section is designed to ensure that at service loads, the point A(*P*,

M_x, M_y) that represents the forces in a member exists inside the space *P-M* diagram of the column (Fig. 6).

In this paper, the initial continuous solution, which is used as the starting point for the elaborate search afterward for the discrete optimum section, is determined approximately by neglecting the biaxial bending effects for simplicity. Then, the biaxial effects can be considered in the later process for determining the final discrete solution by the Bresler's method.

In this study, the column section in the database are arranged in order of increasing normalized volume of *P-M* diagram. This means that the greater section identification number has the greater resistant capacity, making the search for the optimum section in the database more effective.

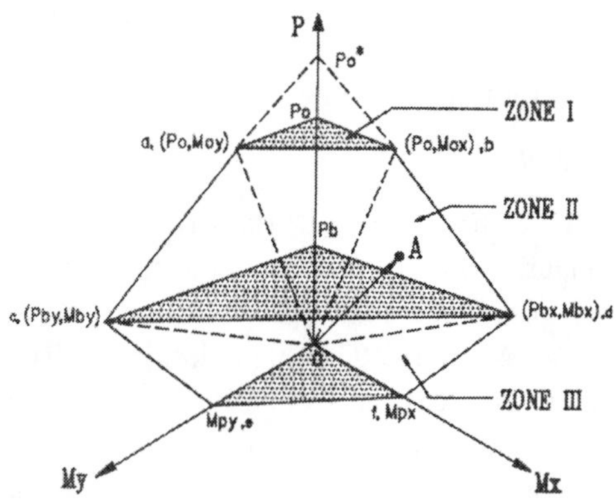

Figure 6 Space *P-M* Interaction Diagram

4.2 *Continuous solution*

Once the eccentricities are determined, the column section is designed generally in the same way as the beam section. The necessary information and the procedures for determining the continuous solution in each design zone are summarized in Table 4.

Table 4. Determining Continuous Column Section in Each Zone (Refer to Fig. 6)

Zone	Controlled by	Enveloped by surface	Inequality condition
I	*P* only	P_0ab	Eq. 5
II	*P*, M_x, and M_y	*abcd*	Eq. 6
III	M_x and M_y	*cdef*	Eq. 7

$$P \le P_0$$

(5)

$$P + \frac{P_0 - P_{bx}}{M_{bx} - M_{0x}} \times M_x + \frac{P_0 - P_{by}}{M_{by} - M_{0y}} \times M_y \le P_0^*$$

(6)

$$\frac{1}{M_{px}} \times M_x + \frac{1}{M_{py}} \times M_y \le 1$$

(7)

By substituting and rearranging on the section identification numbers, the following can be obtained, respectively:

$$N_1 = 0.1 \times (P - 160{,}000)^{0.6282} \quad (8)$$

$$N_2 = \left(\frac{P + 0.1443M_x + 0.1969M_y + 982{,}300}{95{,}450 + 0.0056M_x + 0.0083M_y} \right)^2 \quad (9)$$

$$N_3 = \left(\frac{0.655M_x + 0.907M_y - 100{,}000}{0.0582M_x + 0.0822M_y} \right)^2 \quad (10)$$

where N_1, N_2, and N_3 = the section identification numbers that are determined as the continuous solutions in zone I, II, and III, respectively.

4.3 *Discrete solution*

Assuming that there exists an optimum section near the continuous solution, a direct search is conducted to find the discrete optimum section in a way similar to that of the beam section discussed in the previous section.

Safety check for the adjacent design section in database to the initially selected continuous solution should be conducted. To consider axial forces and bi-axial bending moment, safety is calculated from the position of given load condition to the boundary of three dimensional P-M interaction diagram of the design section by using Bresler load contour method and calculation method by static analysis. If safety criteria is satisfied, final discontinuous solution can be determined through searching the adjacent regions whether more economic sections are present.

4.4 *Procedure of Column Design*

A basic algorithm for the optimum column section design is like as the procedures shown in Fig. 5.

5. Wall

The design of Wall under the action of axial force, in-plane bending and out-plane bending was conducted based on the code requirements for reinforced concrete structure established by the Ministry of Construction and Transportation [4]. Both vertical and horizontal reinforcing bars were arranged in equal spaces and 4-reinforcing bars added to both end of wall section. Assuming that axial load acts by point load, design is conducted for that in-plane and out-plane design strength of wall are sufficient for the given load condition.

Especially for out-plane bending, according to the range of the eccentricity calculated from axial force P and thickness direction moment M_y, design method is differently applied as the case of considering axial force only, practical design method or compressed member design method.

$e \leq 0.1h$

$$P_n = 0.8[0.85 f_{ck}(1{,}000h)] \quad (11)$$

$$M_n = P_n(0.1h) \quad (12)$$

$0.1h < e < \frac{h}{6}$ (practical design method)

$$P_n = 0.55 f_{ck}(1{,}000h)\left[1-\left(\frac{kl_c}{32h}\right)^2\right] \quad (13)$$

$$M_n = P_n\left(\frac{h}{6}\right) \quad (14)$$

$e > \frac{h}{6}$

The out-plane nominal strength for unit width (1m) of the wall of thickness h is determined from the equilibrium equations of the section internal forces when the axial force with eccentricity e acts.

6. Slab

The design of slab is conducted also based on the code requirements for reinforced concrete structure differently according to the type of slab, 1-way or 2-way, divided by the ratio of the length of short side to the length of long side.

For 1-way slab design is conducted according to the support condition of both ends and 4-support conditions are; both end continuous; both end simply supported; 1-end continuous and another end simply supported; and cantilever. For the design of 2-way slab, 9-supprot conditions were introduced according to the support condition of each 4-sides of slab, whether it is simply supported or fixed. 1-way slab is regarded as a beam of unit width and designed applying the practical design method, and for 2-way slab design, equivalent frame method that the frame constructed by slabs formed by center lines of slabs adjacent to column and columns that are the upper and downer part of the floor is analyzed and designed according to that results and direct design method that is used to approximately analyze the typical frame under vertical load can be applied. But, previous two methods are difficult to satisfy the constraints for applying and be programmed and that in BUILDS-C, slab design was conducted according to the coefficient method of ACI 318-63 [5]. This method apply the coefficient values for positive and negative moment in each case of 2-way slab to the results of analysis and determine design member forces. Moreover, in the procedure of determining the minimum depth of slab the stiffness of adjacent edge beams were considered.

7. SRC Column

7.1 *General*

When RC column can not support the given load condition, user can select SRC column design option. In this case, one select assumed initial H-steel section from the database used for SRC column design and based on this member SRC column section is designed according to the code requirements for steel structure by allowable stress design method established by Korean Society of Steel Construction [6]. Once initial design section is determined, safety factor check is conducted and the search for finding the optimum section in program is performed automatically to confirm whether more economic section is available or not changing the section of H-steel and concrete cover accordingly.

7.2 *Construction of H-Steel Section Database*

H-steel members in database are produced for structural use ordinarily in manufactory and a part of database of H-steel sections is shown in Table 5 to show all properties describing the each section.

Table 5. Database of H-Steel Sections

ID	H[a]	B[b]	t1[c]	t2[d]	r[e]	A[f]
1	150	75	5	7	8	17.85
2	100	100	6	8	10	21.9
3	198	99	4.5	7	11	23.18
4	148	100	6	9	11	26.84
5	200	100	5.5	8	11	27.16
6	125	125	6.5	9	10	30.31
7	248	124	5	8	12	32.68
8	250	125	6	9	12	37.66
9	194	150	6	9	13	39.01
10	150	150	7	10	11	40.14

[a]Height in mm.
[b]Width in mm.
[c]Thickness of web in mm.
[d]Thickness of flange in mm.
[e]Radius of Gyration in mm.
[f]Sectional Area in cm^2.

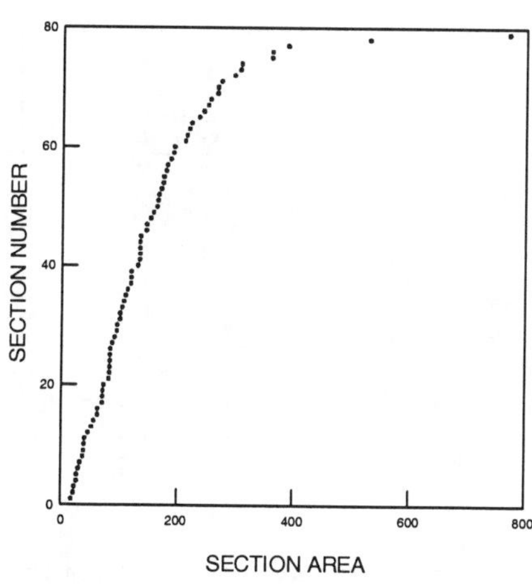

Figure 7. Section Number of H-Steel Section vs. Sectional Area

And moreover, from the fact that user initially select the section of minimum area capable of resisting the given axial load and start design, for the user's convenience of search, database is arranged in ascending powers of sectional area and numbered. Sectional area and section number shows an aspect of proportional monotonic increase shown in Fig. 7, it is not necessary to constitute of regression equation and one can search an appropriate initial H-steel section by bi-section method.

7.3 *Procedure of Assumption of Initial Section and Design*

Once H-steel section is selected, the minimum depth of concrete cover and typical reinforcing bar arrangement are adapted. From the initially designed section, capacity of composite section is calculated from the composite sectional properties, that are, composite sectional area, composite yield strength, secondary radius of composite section, composite elastic modulus and composite section coefficient.

After that, the allowable stress of composite column is calculated. From the calculated values of allowable compressive stress, allowable bending stress for both direction of axis and allowable shear stress, sectional strength of composite column is calculated and checked for given load condition by using Eq. 15.

$$\left(\frac{f_c}{F_c}\right)^2 + \frac{a_x f_{bx}}{F_{bx}} + \frac{a_y f_{by}}{F_{by}} \leq 1.0$$

(15)

where, f_c, f_{bx} and f_{by} = applied stresses and a_x and a_y = x and y direction moment magnification coefficients.

8. Example

First of all, an example of a RC office building is presented to show the result of optimum design of columns and beams using BUILDS-C.

A 10-stofy RC office building constructed with beams and columns of variable sections and slabs 12 cm (5in.) thick was tested in this example. The story height of the first floor is 5.1m (17 ft) and those of other floors up to the roof are 3.6 m (12 ft). The applied loads are the dead load plus the distributed load of 180 kgf/m^2 (37psf) on the floors. The geometry and numerical properties of the structure are given in Fig. 8 and the beam and column identification numbers in the vertical direction are given in Table 6. The reinforcements of column sections are designed with D22(#7) bars and those of beam sections with D19(#6) bars.

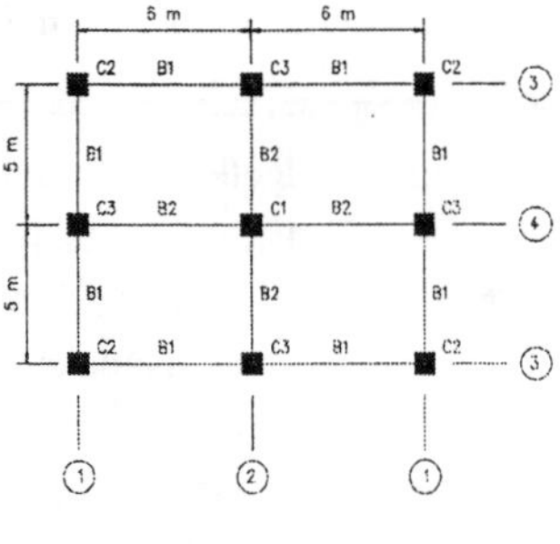

(a)

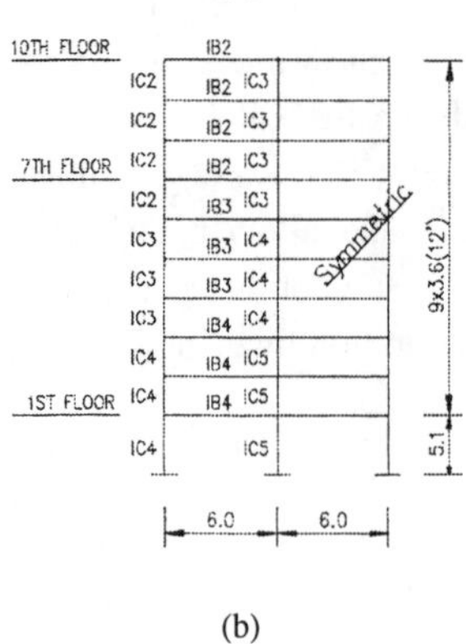

(b)

Figure 8. Configurations of 10-Story RC Building: (a) Plan View; and (b) Frame 4

Table 6. Column and Beam Line for 10-Story RC Building

Floor	Column Line			Beam Line	
	C1	C2	C3	B1	B2
7-10	IC3	IC1	IC2	IB1	IB2
4-6	IC4	IC2	IC3	IB2	IB3
1-3	IC5	IC3	IC4	IB3	IB4

The initial sections are assumed almost arbitrarily, and results from the final discontinuous design are presented in Tables 7 and 8. In Figs. 9 and 10, the variations of the section moment of inertia of beams and of the cross sectional areas of columns are shown as the iteration continues. As illustrated in these figures, the member sizes of beams and columns rapidly converge to the local optimum values after a few oscillations in the early iterations.

Table 7. Summary of Design Results: Column

ID	Initial Value 1		Initial Value 2		Member Forces		
	B[a]	H[b]	B	H	P[c]	M_x[d]	M_y[d]
IC1	40	40	70	70	33.11	210.61	178.27
IC2	45	45	70	70	55.80	337.50	1.07
IC3	50	50	70	70	97.79	290.39	1.01
IC4	55	55	70	70	185.65	0.51	2.64
IC5	60	60	70	70	270.84	0.29	1.34
Final Design							
ID			B	H	BN[e]	HN[f]	CN[g]
IC1			30	40	0D22	4D22	4D22
IC2			30	40	0D22	0D22	4D22
IC3			30	40	2D22	0D22	4D22
IC4			30	50	10D22	10D22	4D22
IC5			40	60	12D22	D1222	4D22

[a]Width in centimeters.
[b]Height in centimeters.
[c]Forces in metric tons.
[d]Moments in metric t/cm.
[e]Number of bars in width direction.
[f]Number of bars in height direction.
[g]Number of bars at corners.

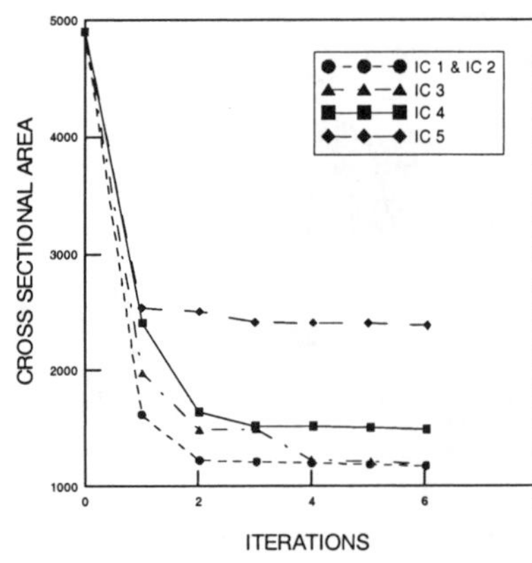

Figure 9. Variation of Cross-Sectional Area of Column with Respect to Number of iterations

Table 8. Summary of Design Results: Beam

ID	Initial Value 1		Initial Value 2		Member Forces	
	B[a]	H[b]	B	H	Me[c]	Mc[c]
IB1	30	60	40	80	518.11	323.88
IB2	30	65	40	80	894.52	667.85
IB3	30	70	40	80	851.35	596.72
IB4	35	75	40	80	1,015.14	607.77
Final Design						
ID			B	H	End	Center
IB1			20	50	3(0)[d]D19	2(0)D19
IB2			25	50	6(2)D19	4(0)D19
IB3			25	50	6(2)D19	4(0)D19
IB4			25	55	6(2)D19	4(0)D19

[a]Width in centimeters.
[b]Height in centimeters.
[c]Moments in metric t/cm.
[d]Number of parentheses is in compression region.

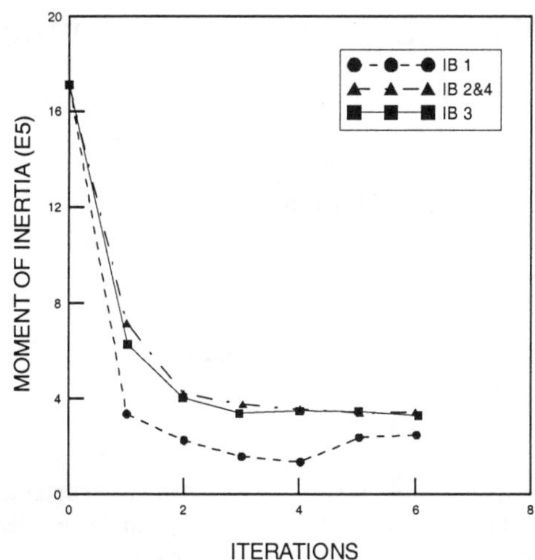

Figure 10. Variation of Moment of Inertia of Beam with Respect to Number of Iterations

Moreover, the design results of wall, slab and SRC column is shown in Table 9.

Table 9. Summary of Design Results: Wall, Slab, SRC Column

Member	Member Forces	Design Result
Wall	P_u = 600 ton M_{ux} = 500 ton-cm M_{uy} = 300 ton-cm	Thickness = 18 cm Vertical bar = D10@45 Horizontal bar = D10@31 End bar = 4D10@10
Slab	L_x = 600 cm L_y = 600 cm W_{dead} = 0.050 kg/cm^2 W_{live} = 0.030 kg/cm^2 W_{ult} = 0.121 kg/cm^2	Thickness = 18 cm Short span M- = D13@17 M+ = D13@29 Long span M- = D13@17 M+ = D13@29
SRC Column	P = 225 ton M_x = 15000 ton-cm M_y = 0 ton-cm	H-steel Section 400×400×20×30 Concrete Section B = 70 cm H = 70 cm Corner bar = 3×4D22

9. Conclusion

A simple, yet effective, algorithm for structural optimization is presented in this study. The algorithm uses a two-step procedure: determining the continuous solution by the relationships between variables of the predetermined sections in the database and determining the discontinuous solution by direct search within the database.

As shown in numerical examples, the optimum design process using the algorithm is conducted at the level of elements. Therefore, the design results satisfy stress constraints, section width or depth, and reinforcements quality, and the entire structure is optimized approximately through individual element optimization.

Acknowledgments

The work presented in this paper was sponsored by the Ministry of Construction and Transportation. Their support is gratefully acknowledged.

References

1. C. Choi and H. Kwak, *Optimum RC Member Design with Predetermined Discrete Sections, Journal of Structural Engineering, ASCE,* **Vol. 116**, No. 10, 2634 (1990).
2. H. Kwak, C. Choi and G. Chung, *Direct Search Approach to Optimum Spiral Column Design, Engineering Structures,* **Vol. 18**, No. 5, 371 (1996).
3. C. Choi and H. Kwak, *Integrated Building Design System (Part 2) Optimal Design Design of R/C Structures, Report of Structural Engineering and Mechanics Research, KIAST,* **SEMR 89-02** (1996).
4. *Code Requirements for Reinforced Concrete Structure, Ministry of Construction and Transportation* (1999).
5. *Building Code Requirements for Reinforced Concrete, ACI 318-63, American Concrete Institute, Detroit, Michigan* (1963).
6. *Code Requirements for Steel Structure by allowable Stress Design Method, Korean Society of Steel Construction* (2003).

MULTILEVEL OPTIMIZATION FOR STRUCTURAL DESIGN OF TALL BUILDINGS

GONG HAI †
Department of Civil Engineering, Tongji University, Shanghai SiPing Road 1239
Shanghai, 200092, China

TSE-YUNG P. CHANG GUO-QIANG LI
Department of Civil Engineering, Tongji University, Shanghai SiPing Road 1239
Shanghai, 200092, China

An application of multi-level optimization technique for structural design of tall buildings is presented. The strategy of optimization analysis is conducted by considering the structural behavior of a building at three levels, namely, element-by-element, sub-structural and global levels. In addition to the live and dead loads, both the wind and earthquake loading are considered in accordance with the Chinese design codes. From these, constraint requirements are formulated by including member's strengths, local and global buckling, maximum and inter-story drift limits. The building types treated include steel frames, concrete frames, shear walls, and hybrid structures (concrete and steel). In the paper, mathematical formulations of the optimization problem with constraint conditions are given. Two demonstration examples, i.e. a steel braced-frame and hybrid structure, are included to illustrate the utility of the methodology presented.

1. Introduction

In recent years□many tall buildings have been built in big cities such as Hong Kong, Beijing and Shanghai, and still many more are in the planning along with the economical expansion in China for years to come. Furthermore, these buildings are becoming taller and taller due to the scarcity of available lands. Correspondingly, large sizes of structural members have to be utilized in engineering design in order to satisfy the safety requirements. This can have significantly impact on the cost of a building. As such, a structural optimization technique may be employed to arrive at a cost-effective design solution.

In a traditional engineering design, engineers do utilize the optimization concept intuitively on a trial-and-error basis. That is, structural engineers assume members' dimensions of a building based on their experiences or intuitively, then going through the structural analysis and redesign of members' dimensions in several cycles before reaching a satisfactory solution. This "manual" process is obviously labor-intensive and time-consuming, one may not be able to do a thorough job. Instead, an automated structural optimization procedure would be much more cost-effective.

Research work of structural optimization has been in existence for many years and the technique has been widely used by designers in the aerospace and mechanical industries where the demand in using lighter weight materials are far more stringent. For building design, there can be four levels of structural optimization; namely, structural form or type, topology, shape and size optimization. Although some research has been reported in shape and topology optimization [1-3], much of the applications are in size optimization, e.g. in [4-11]. Even so, most of the works are focused on optimal lateral stiffness of monolithic structures without considering any strength limits. Obviously, the optimal solution so obtained may not be entirely useful, for example, in the case of reinforced concrete structures, in which the optimized dimensions could be too small to accommodate the reinforcements. For most tall buildings, their structural behavior is controlled by three major factors, namely, strength, local and global buckling (in the case of steel and composite structures), and stiffness. In formulating a structural optimization problem, these factors are treated as the constraint conditions. The constraint conditions are further divided into local levels and global levels. The local level includes the member's strength check, local buckling and inter-story drift limits of a building. The global level constraint mainly refers to the maximum drift at the top of a building. For difference in disposing method, the

† Work partially supported by grant 03-2-131 of the P. R. China Building Department Science Foundation.

first kind of constraint is usually called Local constraint, and the other two kinds of constraint are called Global constraint.

In this paper, both the local and global constraints are considered in our optimization analysis. For the analysis strategy, a displacement-based rule design method is adopted for the global constraints whereas an imitative full-stress method is employed to deal with the local constraints. In addition, the driving forces are the wind and earthquake loadings in accordance with the Chinese design codes are adopted..

2. Formulation of the Displacement Rule Design Method

The formulation of the optimization is represented as the following [12]:

$$C=\sum_{i=1}^{N}(C_iA_i(H_i,B_i))+\sum_{i=1}^{M}(C_{wi}t_i)$$

Minimize
which is subjected to:

$$d_j=(\delta_j-\delta_{j-1})\le d_j^U \quad (j=1,2,\dots,M)$$

Sub-structural lever constraint

$$d_{top}\le d_{top}^U \quad (j=1,2,\dots,M)$$

Global lever constraint

$$Strength(H_i,B_i)<Limit_i\,(i=1,2,\dots,N)$$

Element-by-element local constraint

$$Stable(H_i,B_i)<Limit_i\,(i=1,2,\dots,N)$$

Element-by-element local constraint

where H_i and B_i are vectors for beam design variables, and t_i are vectors for wall design variables; C_i,C_{wi} are cost coefficients respectively for beams and walls, where

$$\delta_j^x=\sum_{i=1}^{N}\left(\sum_{n=1}^{K_i}\int_0^{L_{in}}\left(\frac{(N^{in}*n_j^{in})_x}{E_iA_i}+\frac{(V_x^{in}v_{xj}^{in})_x}{G_iA_{ix}}+\frac{(V_y^{in}v_{yj}^{in})_x}{G_iA_{iy}}+\frac{(M_x^{in}m_{xj}^{in})_x}{E_iI_{ix}}+\frac{(M_y^{in}m_{yj}^{in})_x}{E_iI_{iy}}\right)dx\right)$$

$$=\sum_{i=1}^{N}\left(\frac{C_{0ij}^x}{E_iA_i}+\frac{C_{1ij}^x}{G_iA_{ix}}+\frac{C_{2ij}^x}{G_iA_{iy}}+\frac{C_{3ij}^x}{E_iI_{ix}}+\frac{C_{4ij}^x}{E_iI_{iy}}\right)$$

For the global constraints, we use the method of the displacement-based rule design method. It can be primarily represented as the following.

$$L(H_i,B_i,\lambda_j)=\sum_{i=1}^{N}(C_iA_i(H_i,B_i))$$
$$+\sum_{j=1}^{M}\lambda_j^x\left(\sum_{i=1}^{N}\left(\frac{C_{0ij}^x}{E_iA_i}+\frac{C_{1ij}^x}{G_iA_{ix}}+\frac{C_{2ij}^x}{G_iA_{iy}}+\frac{C_{3ij}^x}{E_iI_{ix}}+\frac{C_{4ij}^x}{E_iI_{iy}}\right)-d_j^U\right)$$
$$+\sum_{j=1}^{M}\lambda_j^y\left(\sum_{i=1}^{N}\left(\frac{C_{0ij}^y}{E_iA_i}+\frac{C_{1ij}^y}{G_iA_{ix}}+\frac{C_{2ij}^y}{G_iA_{iy}}+\frac{C_{3ij}^y}{E_iI_{ix}}+\frac{C_{4ij}^y}{E_iI_{iy}}\right)-d_j^U\right)$$

In the mathematical programming algorithm, a procedure is defined to find the direction vector for in each iteration cycle.

$$\frac{\partial L}{\partial H_i}=0\Rightarrow\frac{\partial(C_iA_i)}{\partial H_i}$$
$$+\sum_{j=1}^{M}\lambda_j^x\left(\frac{-C_{0ij}^x\frac{\partial(E_iA_i)}{\partial H_i}}{(E_iA_i)^2}+\frac{-C_{1ij}^x\frac{\partial(G_iA_{xi})}{\partial H_i}}{(G_iA_{xi})^2}+\frac{-C_{2ij}^x\frac{\partial(G_iA_{yi})}{\partial H_i}}{(G_iA_{yi})^2}+\frac{-C_{3ij}^x\frac{\partial(E_iI_{xi})}{\partial H_i}}{(E_iI_{xi})^2}+\frac{-C_{4ij}^x\frac{\partial(E_iI_{yi})}{\partial H_i}}{(E_iI_{yi})^2}\right)$$
$$+\sum_{j=1}^{M}\lambda_j^y\left(\frac{-C_{0ij}^y\frac{\partial(E_iA_i)}{\partial H_i}}{(E_iA_i)^2}+\frac{-C_{1ij}^y\frac{\partial(G_iA_{xi})}{\partial H_i}}{(G_iA_{xi})^2}+\frac{-C_{2ij}^y\frac{\partial(G_iA_{yi})}{\partial H_i}}{(G_iA_{yi})^2}+\frac{-C_{3ij}^y\frac{\partial(E_iI_{xi})}{\partial H_i}}{(E_iI_{xi})^2}+\frac{-C_{4ij}^y\frac{\partial(E_iI_{yi})}{\partial H_i}}{(E_iI_{yi})^2}\right)$$
$$=0$$

$$\frac{\partial L}{\partial H_i}=0\Rightarrow$$

$$\left\{\begin{array}{l}\sum_{j=1}^{M}\frac{\lambda_j^x}{\kappa_{Ai}^H}\left(\kappa_{0i}^H C_{0ij}^x+\kappa_{1i}^H C_{1ij}^x+\kappa_{2i}^H C_{2ij}^x+\kappa_{3i}^H C_{3ij}^x+\kappa_{4i}^H C_{4ij}^x\right)\\ +\sum_{j=1}^{M}\frac{\lambda_j^y}{\kappa_{Ai}^H}\left(\kappa_{0i}^H C_{0ij}^y+\kappa_{1i}^H C_{1ij}^y+\kappa_{2i}^H C_{2ij}^y+\kappa_{3i}^H C_{3ij}^y+\kappa_{4i}^H C_{4ij}^y\right)\end{array}\right\}$$

$$=1$$

$$\frac{\partial L}{\partial H_i}=0\Rightarrow\left\{\sum_{j=1}^{M}\frac{\lambda_j^x K_{ij}^{Hx}}{\kappa_{Ai}^H}+\sum_{j=1}^{M}\frac{\lambda_j^y K_{ij}^{Hy}}{\kappa_{Ai}^H}\right\}=1$$

So we can get the iteration equation as the following:

$$H_i^{v+1}=H_i^v\left\{1+\frac{1}{\eta}\left(\sum_{j=1}^{M}\frac{\lambda_j^x K_{ij}^{Hx}+\lambda_j^y K_{ij}^{Hy}}{\kappa_{Ai}^H}-1\right)_v\right\}$$

$$B_i^{v+1}=B_i^v\left\{1+\frac{1}{\eta}\left(\sum_{j=1}^{M}\frac{\lambda_j^x K_{ij}^{Bx}+\lambda_j^y K_{ij}^{By}}{\kappa_{Ai}^B}-1\right)_v\right\}$$

where v is the current iteration time, and η is the looseness coefficient.

For earthquake load in accordance with the Chinese design codes [13], it can be deducted as the following:

$$\delta_j^x=\sqrt{\sum_{p=1}^{R}\sum_{q=1}^{R}\rho_{pq}\delta_j^{px}\delta_j^{qx}}$$

$$=\sqrt{\begin{array}{l}\sum_{p=1}^{R}\sum_{q=1}^{R}\rho_{pq}\\ \left(\sum_{i=1}^{N}\left(\frac{C_{0ij}^{px}}{E_iA_i}+\frac{C_{1ij}^{px}}{G_iA_{ix}}+\frac{C_{2ij}^{px}}{G_iA_{iy}}+\frac{C_{3ij}^{px}}{E_iI_{ix}}+\frac{C_{4ij}^{px}}{E_iI_{iy}}\right)\right)\\ \left(\sum_{i=1}^{N}\left(\frac{C_{0ij}^{qx}}{E_iA_i}+\frac{C_{1ij}^{qx}}{G_iA_{ix}}+\frac{C_{2ij}^{qx}}{G_iA_{iy}}+\frac{C_{3ij}^{qx}}{E_iI_{ix}}+\frac{C_{4ij}^{qx}}{E_iI_{iy}}\right)\right)\end{array}}$$

where

$$\rho_{pq}=\frac{8\xi_p\xi_q(1+\lambda_T)\lambda_T^{1.5}}{(1-\lambda_T^2)^2+4\xi_p\xi_q(1+\lambda_T)^2\lambda_T}$$

$$\frac{\partial\left(\delta_j^x\right)}{\partial H_i}$$

$$=\frac{\sum_{p=1}^{R}\sum_{q=1}^{R}\rho_{pq}\delta_q^x\left(\kappa_{0i}^H C_{0ij}^{px}+\kappa_{1i}^H C_{1ij}^{px}+\kappa_{2i}^H C_{2ij}^{px}+\kappa_{3i}^H C_{3ij}^{px}+\kappa_{4i}^H C_{4ij}^{px}\right)}{\delta_j^x}$$

$$=\kappa_{0i}^H S_{0ij}^x+\kappa_{1i}^H S_{1ij}^x+\kappa_{2i}^H S_{2ij}^x+\kappa_{3i}^H S_{3ij}^x+\kappa_{4i}^H S_{4ij}^x=Z_{ij}^H$$

So we can obtain the iteration equation as the following:

$$H_i^{v+1}=H_i^v\left\{1+\frac{1}{\eta}\left(\frac{\sum_{j=1}^{M}\lambda_j Z_{ij}^H}{\kappa_{Ai}^H}-1\right)_v\right\}$$

3. Strategy in Combination with Imitative Full-Stress Design

At first, all the structural members of a building in question are logically divided into several groups and the cross section of each group of members (or elements) is treated as an independent optimization variable. Each cross section is identified by its size characteristics as H and B (global optimization variables), Tf and Tw, the mechanical area and inertial moment of the section. So the variable domain is defined as a section library that is more suitable to the practical design. Second, one will find the new iteration results through the global optimization as described in the above. At the end of an optimization cycle, a 'optimal' cross section closest to that listed in the section library (or cross section database) is identified from the principle of the closest mechanical characteristic rule.

The flow chart of the combined method is shown as the following:

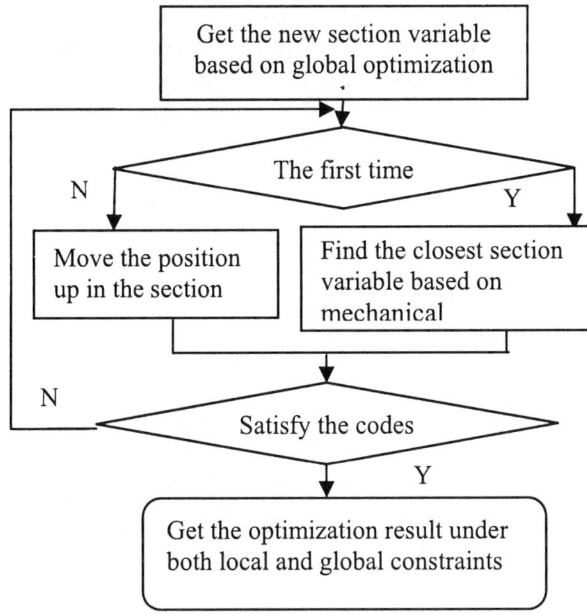

Figure 1. The flow chart of the combined method

4. Examples

The above described optimization formulation and computation strategy has been implemented into a computing program for design applications. Example 1 shows the optimum process of a steel braced-frame by the computing program, and example 2 shows the optimum process of a hybrid structure.

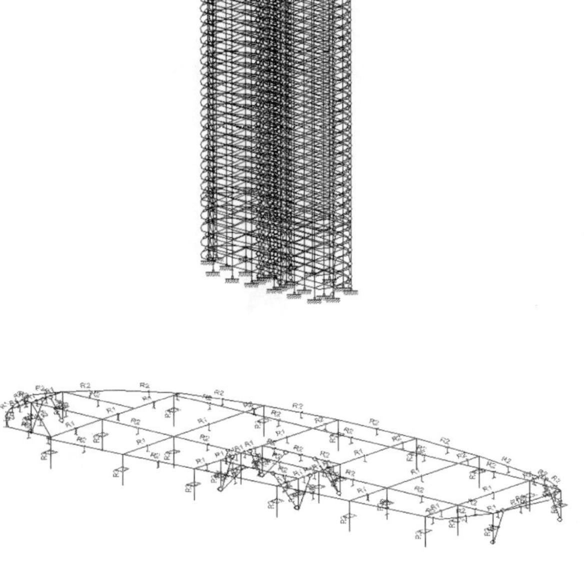

Figure 2. the model of example 1

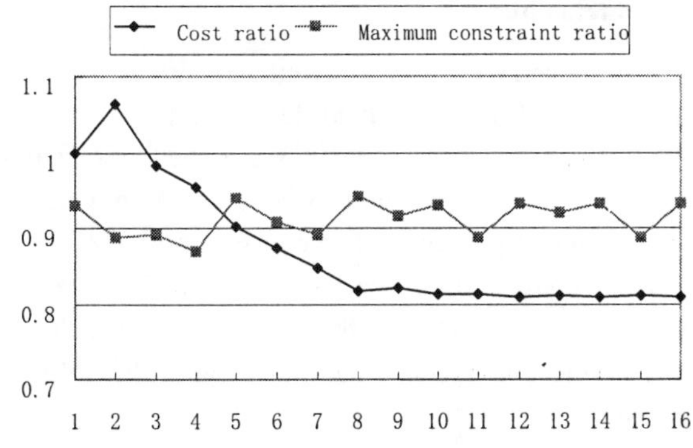

Figure 3. The optimization process of example 1

From the optimization process, we can find the optimization is converged at the 8th time, and the cost drop down about 19 percent, and the process is mainly controlled by the local constraints.

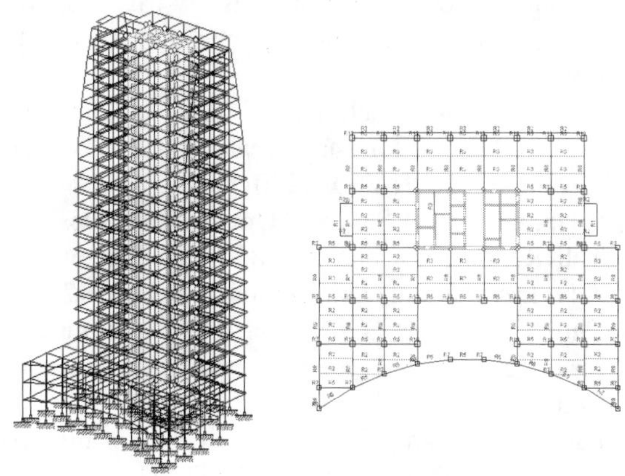

Figure 4. the model of example 2

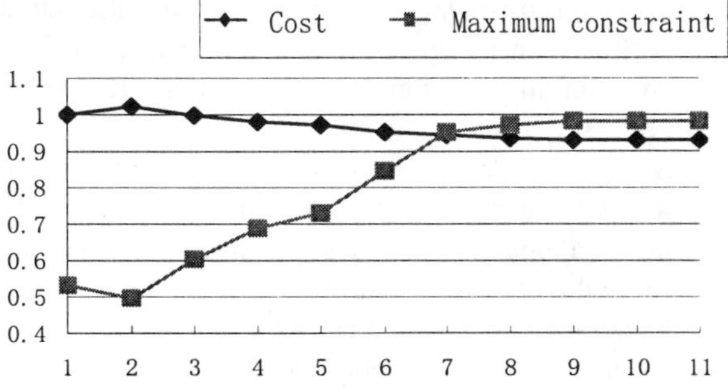

Figure 5. The optimization process of example 2

From the optimization process, we can find the solution has converged at the 9th cycle, and the cost drops down by about 9 percent, and the process is mainly controlled by the global constraints.

5. Conclusions

In the proposed method, both wind and earthquake loads are considered in the global lever, and a combined method that is also suitable to solve the optimization problem involving discrete variables. Moreover, the methodology is put forward to consider both the global and local constraints. From the two examples given in the paper, we found that there were no difficulty in terms of the convergence and stability of the solution algorithm.

References

1. Kirsch, U., Topping, B. H. V., "Minimum weight design of structural topologies", J. Struc. Eng., 118(7):1770-1785 (1992).
2. Salajegheh, E., Vanderplaats, G. N., "Optimum design of trusses with sizing and shape variables". Structural Optimization, 6: 79-85 (1993).
3. Cheng, G. D., "Some aspects of truss topology optimization". Structural Optimization, 10:173-179 (1995).
4. Berke, L, and Khot, N. S., "Use of Optimality Criteria Method for Large Scale Systems", NATO AgARD-Ls-70, NATO (1970).
5. Chan, C. M. and Sun, S., "Optimal Drift Design of Tall Reinforced Concrete Building Frameworks", Adv. In Structural Optimization, ASCE (1997)
6. Choi, C. K., Lee, D. G., Lee, H. W., Optimization for large scale steel structures with discrete sections. Comp. & Struc., 39, (5) (1991).
7. Park, H. S., Park, C. L., "Drift control of high-rise buildings with unit load method", The Structural Design of Tall Buildings, Vol. 6, 23-35 (1997)
8. Hwang, J. S., Hong, S. M., "Quantitative lateral drift control of high-rise building using the lateral stiffness-sensitivity analysis", Computing in Civil and Building Engineering, Balkema, Rotterdam (1995).
9. Seong Jin Park, Tai Hun Kwon. "Optimization method for steady conduction in special geometry using a boundary element method", International Journal for Numerical Methods in Engineering, Vol. 43, No. 6, November 1998
10. Balaur S. Dhilllon., "Optimum design of Composite Hybrid plate girders. Journal of structural engineering", Vol.117, No.7, July 1991.
11. Chang, T.Y.P., Liang, J ,and Chan, C.M., "An Integrated system of Computer Aided Design for Tall Building", Proceedings of the Seventh International Conference of Computing in Civil and Building Engineering , Seoul, Korea (1997)
12. Gang Li. "Multi-objective and multi-lever optimization for steel frames", Engineering Structures, Vol. 21 (1999).

NONLINEAR FINITE ELEMENT ANALYSIS OF REINFORCED CONCRETE FRAME UNDER DIFFERENT LATERAL LOADING PATTERNS

XIN JIA

Geotechnical Department, Tongji University, 1239 Siping Road
Shanghai 200092, China

XIAN LIU

Geotechnical Department, Tongji University, 1239 Siping Road
Shanghai 200092, China

YONG YUAN

Geotechnical Department, Tongji University, 1239 Siping Road
Shanghai 200092, China

The behaviors of reinforced concrete frame under different lateral loading patterns are investigated by using nonlinear finite element method in this paper. A constitutive model of reinforced concrete based on continuum damage mechanics is proposed. The Non-local Damage Model is employed as the constitutive equation specialized to continuum damage, and to prevent spurious strain and damage localization. The Smeared Crack Model is adopted to offer automatic generation of cracks and complete generality in crack orientation, without the need of redefining the finite element topology. Traditional bond-link element is modified to save the time in calculation by condensing out the degrees of freedom associated with the reinforcing steel. The reinforced concrete frame was modeled, calculated, presented using the program Virtual Digital Laboratory v1.0 (VDL v1.0) that is developed by the authors. From the numerical calculation results, the pushover method is used, and the results of analysis well accorded with the conclusion of correlative literature.

1. Introduction

The reinforced concrete frame is a common used structure in industrial and civil building engineering. The current engineering practice for seismic analysis of reinforced concrete frames is moving away from simplified linear-elastic method of analysis, and towards a more complex nonlinear inelastic technique. The time–history analysis is a powerful tool for the study of structural seismic response, but it increases the computational effort significantly when it is used to analyze the nonlinear behavior of RC frame. A pushover analysis is a simple method to estimate the seismic response of frame with applying lateral load on the nodes of frame structure, and it is becoming more popular in the field of performance-based design philosophy [1].

In order to give a finer nonlinear analysis, the finite element method was investigated, and a program was developed. A constitutive model has been proposed to simulate the cracking development of reinforced concrete structures, based on the continuum damage mechanics. A few strategies were adopted in the program as follows: Firstly the Non-local Damage Model is employed as the constitutive equation specialized to continuum damage, and to prevent spurious strain and damage localization. Secondly the probabilistic description of damage initiation (thresholds) is considered to reflect the effect of specimen size on damage initiation. Thirdly condensing out the degrees of freedom associated with the reinforcing steel modifies the traditional bond-slip element. This new element simulates the bond-slip behavior between concrete and steel very well. At last the smeared crack model is adopted to offer automatic generation of cracks and complete generality in crack orientation, without the need of redefining the finite element topology.

The organization of this paper is as follows. Section two describes the procedure of the pushover analysis, and the lateral loading pattern applied on the reinforced concrete frame. In this paper, three loading patterns are used. Section three gives a description of the constitutive model of concrete, steel and bond slip between concrete and steel. The strategies used in numerical simulation are introduced in section four, such as how to simulate crack, how to deal with bond slip behavior. Section five presents a numerical experiment to validate pushover method and damage theory. Finally, the summary is presented in section eight.

2. The Basic Principle and General Procedure of Pushover Analysis

2.1 *Basic principle*

FEMA 273 introduces the Nonlinear Static Procedure (NSP) for the modeling and analysis of seismically rehabilitated building frames [2]. The NSP, also known as a pushover analysis, is a relatively simple way to explore the design of a structure. It consists of pushing a mathematical model of a building frame over to a prescribed displacement in order to predict the sequence of damages in the inelastic range.

There is not a perfect theory about pushover analysis. The primary assumes are as follows [3]:

1. There are some relationship between the single degree freedom system and the multi-degree freedom system. The response of structure is controlled by single vibration mode.
2. The shape vector could control the vertical deformation of structure, and the shape vector does not change during the response progress.

It is investigated that the seismic response of structure whose performance depends on the first order vibration mode, is reasonable when using the pushover analysis.

2.2 *Procedure of push over analysis*

It is a key operation to choose a reasonable lateral loading pattern. The loading pattern should satisfy two requirements when it is imposed on the structure. It not only shows the inertia force distribution, but also gives the displacement of structure. There are several lateral loading patterns we can adopt [4]. In comparative analysis, three different lateral loading patterns were used:

Here, F_i, w_i, and h_i represents lateral load, weight, and height of floor *i* respectively. The total lateral load is V_b, and n is the number of floor.

- Load case JB: The lateral force applied at each node is proportional to the mass tributary to that node. When all nodes have the same mass, this load case results in a uniform distribution of the lateral loads. The lateral load can be calculated from the Equation as follows, which is numbered as Eq. (1):

$$F_i = \left(w_i \Big/ \sum_{m=1}^{n} w_m \right) V_b \tag{1}$$

- Load case DSJ: The lateral force applied at each node is proportional to the product of the height of the node times the mass tributary to that node. When all nodes have the same mass, this load case results in an inverse triangular distribution of the lateral loads. The lateral load can be calculated from the Equation as follows, which is numbered as Eq. (2):

$$F_i = \left(w_i h_i \Big/ \sum_{m=1}^{n} w_m h_m \right) \tag{2}$$

- Load case DJ: The lateral force is totally applied on the top node. The lateral load can be calculated from the Equation as follows, which is numbered as Eq. (3):

$$F_n = V_b \tag{3}$$

After choosing the lateral loading patterns, a pushover analysis can be done as the following steps [2]:

1. Build a finite element model, including the geometrical model, physical model, and the information of nodes and structure members.
2. Impose the lateral load on the frame structure.
3. Apply the load step by step, until the displacement of the frame meet the demand. Then the progress is over, and save the data.

3. Constitutive Model

3.1 *Concrete*

Kachanov firstly proposed the damage theory in 1958, when he studied the creep behavior of metal. The research on concrete damage began in 1979[5,6], and then many constitutive models were erected. A concrete constitutive model was proposed based on the continues damage theory, and the scalar damage parameter was used to show the damage status when the concrete is loaded monoaxially. The model can be expressed by the equation (4) and equation (5). In these equations, D is the parameter of damage; $\dot{D}$ is the increase of damage; $K(D)$ is the secant stiffness of concrete; and K_0 is the initial stiffness.

$$\sigma = K(D) : \varepsilon = K_0 (1 - D) : \varepsilon \tag{4}$$

$$-\left(\frac{1}{2} \frac{\partial K(D)}{\partial D} : \varepsilon : \varepsilon \right) \dot{D} \geq 0 \tag{5}$$

As the stiffness of material will decrease with the increase of damage, it yields $\frac{\partial K(D)}{\partial D} < 0$, then we obtain $\dot{D} \geq 0$.

Figure 1 gives the curve of constitutive law in uniaxial direction. From figure, we can get that the unload path is along the secant stiffness, and the concrete damage does not change [7].

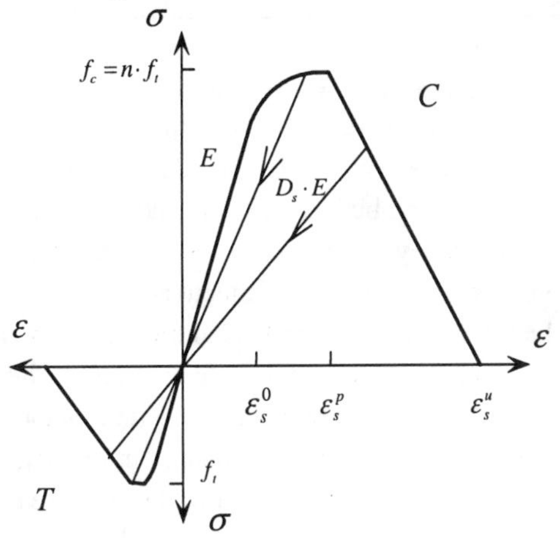

Figure 1 The stress-strain curve of constitutive law in uniaxial direction

3.2 *Steel*

In this paper, we assume the steel is linear elasticity and linear strengthening. The relationship of stress and strain can be obtained from figure 2.

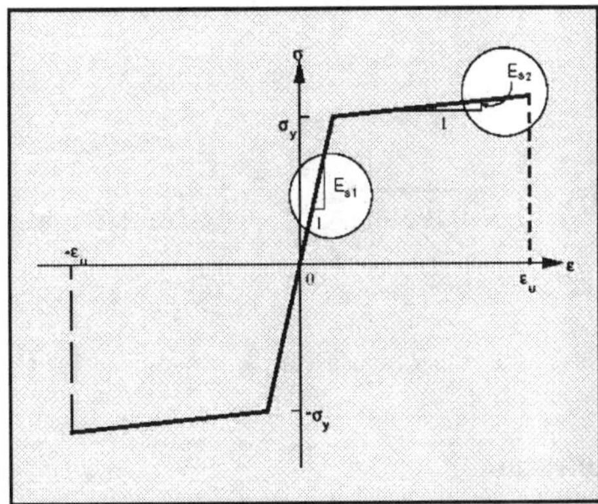

Figure 2 Steel bars stress-strain relationship

The yield stress is σ_y . The reasons to use this steel model are as follows:

- To simplify calculation.
- The behavior of reinforced concrete structural elements mainly depends on the steel yield, especially for the bending member.
- There is much plastic deformation when steel yields, if the perfect elasticity- plasticity constitutive model was adopted, it will lead to the divergence problem.

3.3 *Bond slip between concrete and steel*

Generally, the bond slip happens on the anchor surface and near the area of crack. It is caused by the deformation coordination of steel and concrete, and it is also a character of reinforced concrete. So we should consider the bond slip behavior when simulating the reinforced concrete. Figure 3 gives the bond stress-slip relationship for plane-stress problem [8,9].

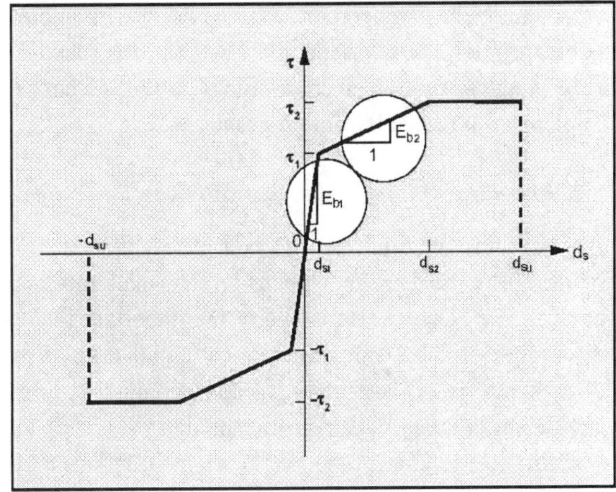

Figure 3 Bond stress-slip relationship for plane-stress problem

4. Nonlinear Finite Element Analysis

4.1 *Smear crack model*

The crack always causes the nonlinear response of concrete. It is a difficult problem to simulate crack in reinforced concrete. Researchers have proposed many crack models, but there are few efficient ones. As we investigate, there are three main kinds of crack models, such as discrete crack model, smear crack model, and fracture mechanics [10]. During the finite element analysis, how to choose a crack model is a problem, and it depends on the structure analysis destination goal. If you want to simulate the local cracks exactly, discrete crack model may be suitable. On the other side, if you don't care the structure crack status, but the load-slip behavior, smeared crack model will be qualified. In this paper, smear crack model is adopted. Figure 4 shows the difference between discrete crack model and smeared crack model [11].

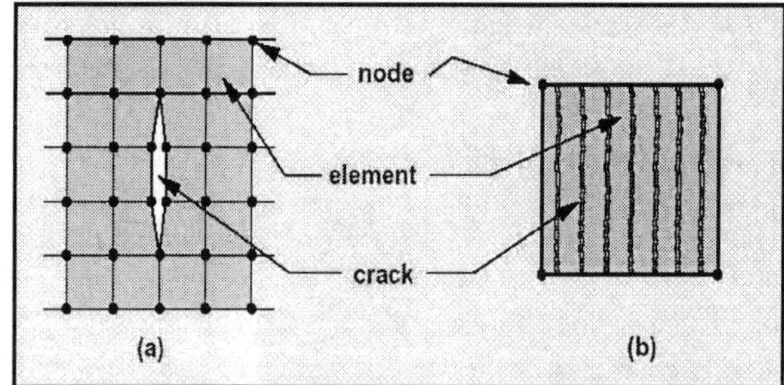

Figure 4 Representations of crack model: (a) discrete model; (b) smeared crack model

Otherwise, the smeared crack model is adopted to offer automatic generation of cracks and complete generality in crack orientation, without the need of redefining the finite element topology. It will be easy to implement in finite element program.

4.2 *Non-local Damage Model*

Although the material damage is a global variant, it is obvious that the damage happens on the small and local area. So the element size is one of the important factors that influence the result of numerical simulation. If the element size is too large, it will lead to computing error. But the grid division work during simulation is complex. It is difficult to control the size of element to suit local area. Then the non-local damage model is proposed to solve this problem in an easier way. The Non-local Damage Model is employed as the constitutive equation specialized to continuum damage, and to prevent spurious strain and damage localization [12,13]. Also the probabilistic description of damage initiation (thresholds) is taken into account to reflect the effect of the volume of the specimen on the damage initiation.

4.3 *Element*

The purpose of this paper is to present a finite element structural analysis methodology of the whole cracking propagation and coalescence process of reinforced concrete structure, so we select the 8 node isoparametric element to simulate concrete. Generally, 2×2 Gauss integral scheme is used. Here in order to consider the non-local model, the 3×3 Gauss integral scheme is selected, so that there are enough integral points. Figure 5 gives the diagrammatic sketch of 8-node isoparametric finite element.

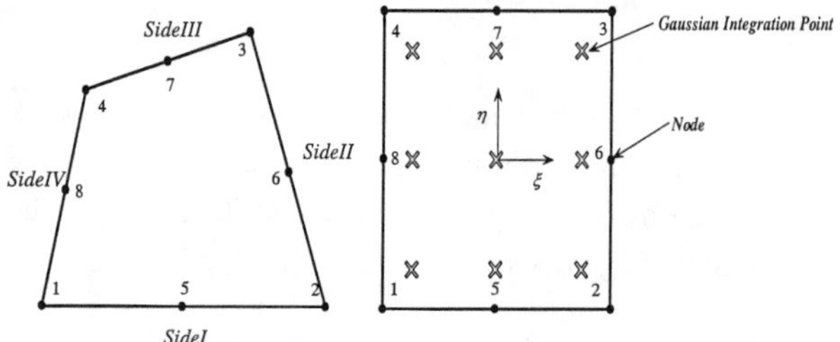

Figure 5 8-node isoparametric finite element

In the finite element method, the steel in concrete usually has three numerical models: distributed or smeared model, embedded model, and discrete model. In this paper, a new type of steel model was proposed. It is assumed that the steel is embedded in the concrete, but unlike the traditional embedded model, the bond slip between concrete and steel is concerned. The steel element has two nodes, and the bond link element also has two nodes [14]. After condensing out the degrees of freedom associated with the reinforcing steel, we call the new type element super steel element. We can get the information from figure 6. Points 1,3 are the nodes of concrete, and become the nodes of super element after condensing. Points 2,4 are the nodes of steel element.

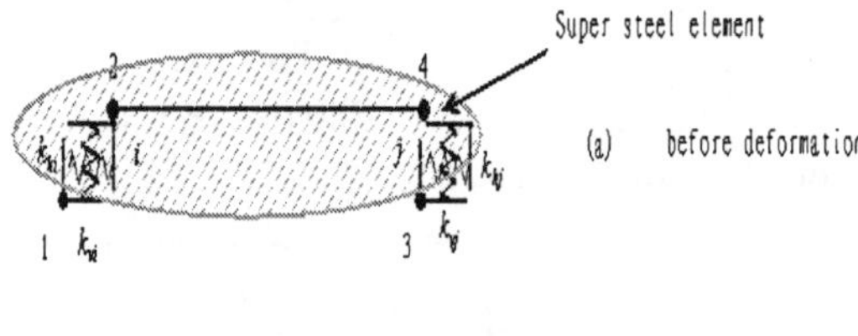

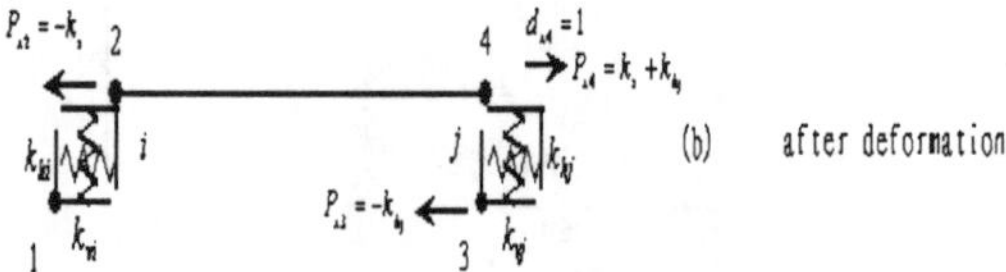

Fig.6 Steel rebar (super condensed) element with bond-slip

5. Application

5.1 *Program introduction*

Based on the theory stated above, an infinite element program was developed which is called virtual digital laboratory, version 1.0. It can be used to simulate the concrete frame structure test, and give a pushover analysis. The OpenGL standard is used. It is easy to build a graphic and a finite element structure model. The postprocessor is powerful, and the stress, strain, deformation, damage distribution can be obtained from the postprocessor. All the analysis bellow is based on the postprocessor.

5.2 *The numerical test model*

A reinforced concrete frame with two span and two layers is built in the VDL v1.0. Figure 7 gives the graphic model. The height of the first layer is 3m, and the height of the second layer is 2m. The first span is 6m, and the second span is 3m.

Figure 7. The model of a R.C. frame with two-span and two -layer

Table 1 gives the loading pattern and load value applied on the structure. The total lateral load is 100KN.

5.3 *Test results and analysis*

The relation between the bottom shear and top displacement is shown on figure 8, figure 9 and figure 10. From the figures, the failure displacements are close under different loading patterns. The curve has a yield period in figure 10. The frame has the largest displacement and gives a sign before failure. The frame performs a good ductility behavior, and its anti-seismic behavior is excellent. The frame under JB loading pattern has the smallest displacement when it began to failure, which shows a brittle behavior and the nonlinear character is not obvious. Its anti-seismic behavior is bad.

In addition, the failure load of frame under DSJ loading pattern is greatest, and that under DJ loading is smallest. But the stiffness of frame under JB loading pattern is largest which can be obtained from the slope of the curve.

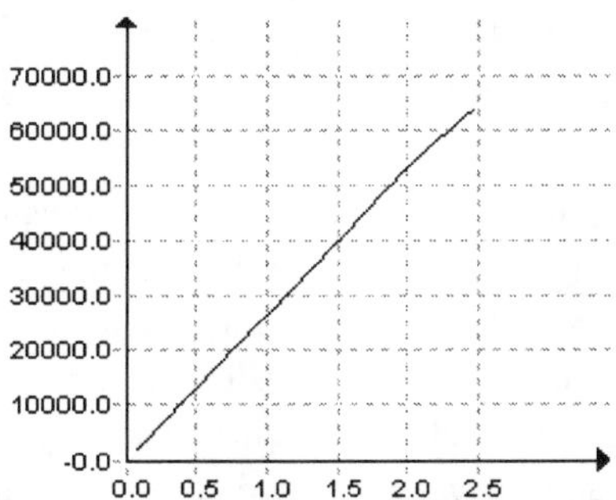

Figure 8 Bottom shear-top displacement under DJ

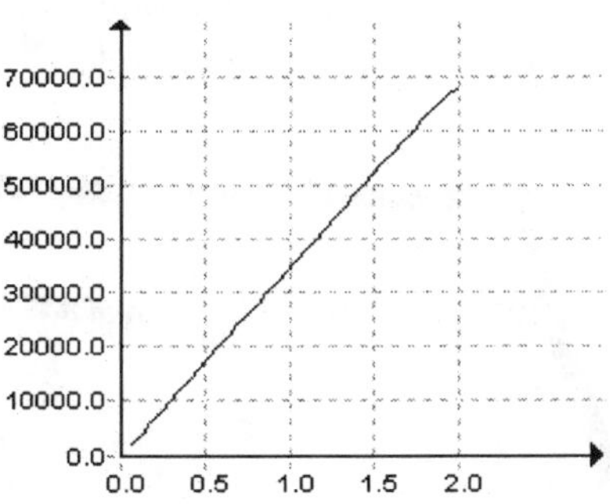

Figure 9 Bottom shear-top displacement under JB

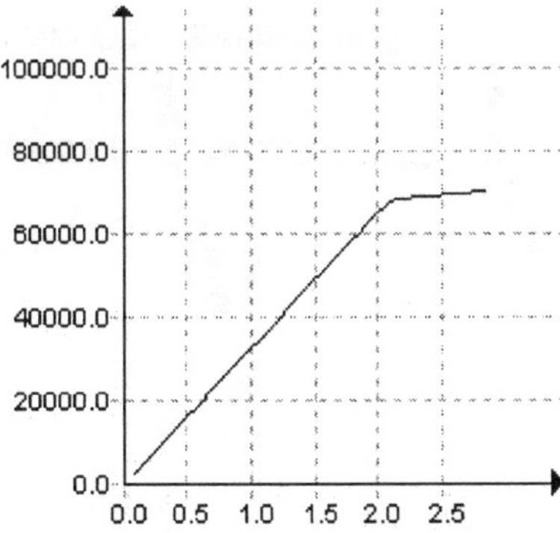

Figure.10 Bottom shear-top displacement under

Table 1. The load information

Loading pattern	*First layer load value (KN)*	*Second layer load value (KN)*	*Load step*
JB	0	100	50
DSJ	54.2	45.8	50
DJ	47.3	52.7	50

The damage distribution can be obtained from figure 11, figure 12 and figure 13. The red color shows the maximum value of damage, and the blue color shows the minimum value of damage. The deformation shapes of the frames under different loading patterns are uniform, and they are according to the first order vibration mode of the structure. Under the DSJ loading pattern, the frame has most plastic pins, and its members work sufficiently. The frame under JB loading pattern has the least damage, and it deforms little at the failure moment. Its anti-seismic behavior is bad.

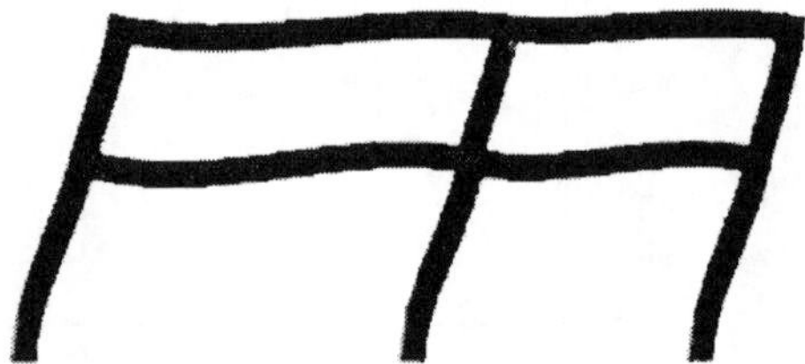

Figure 11 Damage distribution under DJ

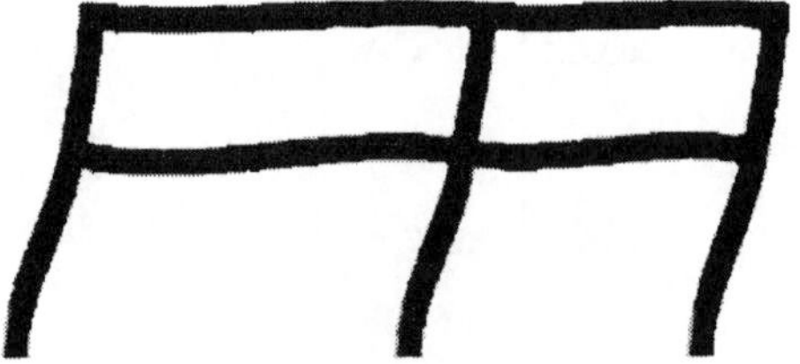

Figure 12 Damage distribution under JB

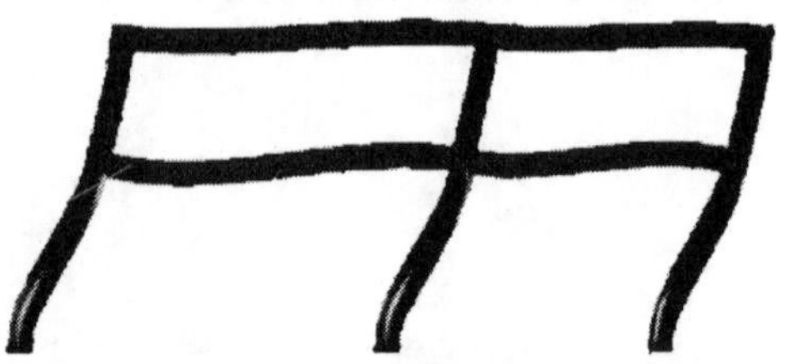

Figure 13 Damage distribution under DSJ

6. Conclusion

The applicability and accuracy of push over analysis in predicting the seismic response of RC frame are investigated. The continuum damage theory is proposed in the numerical simulation. The present investigation leads to the following conclusions:

- The pushover method is becoming popular more and more. It is a simple and efficient to evaluate the seismic behavior of reinforced concrete frame.
- The concrete constitutive model is based on the continuum damage theory, and it can give a fine numerical simulation for concrete.
- The smeared crack model is adopted to simulate the crack in concrete, without the need of redefining the finite element topology, which makes the program easy to implement.
- The non-local damage model is employed as the constitutive equation specialized to continuum damage, and to prevent spurious strain and damage localization.
- A super steel element was proposed, and it considers the bond-lip relationship between the concrete and steel.
- From the numerical test results, the reinforced concrete frame performs differently under different lateral loading patterns. During the practice analysis, it is proposed that two kinds of lateral loading patterns are used at least.
- The seismic behavior of the frame under DSJ loading pattern performs well.

References

1. Zhihao Lu, Hanbin Ge and Tsutomu Usami, Applicability of pushover analysis-based seismic performance evaluation procedure for steel arch bridges, Engineering structures. 26(2004). 1957-1977.
2. FEMA, NEHRP Guidelines for the Seismic Rehabilitation of Buildings [R], FEMA-273, Washington D. C., 1997

3. A.M. Mwafy and A.S. Elnashai, Static pushover versus dynamic collapse analysis of RC buildings, Engineering Structures, 23 (2001), 407–424
4. WANG Dasui , HE Junli and ZHANG Fengxin, The basic principle and a case study of the static elastoplastic analysis (pushover analysis), World earthquake engineering. March 2004, 45-53
5. Zhu Bolong and Dong Zhenxiang, Nonlinearity analysis for Reinforced Concrete Structure. Tongji university press, 1985.
6. Wu. H. C., and Nanakorn, C. K., A constitutive framework of plastically deformed damaged continuum and a formulation using the endochronic concept, Int. J. Solids Struct. Vol.36, 1999, pp5057-5087.
7. Chow, C. L., and Wang, J., A finite element analysis of continuum damage mechanics for ductile fracture, Int. J. of Fracture, 38, 1988, pp83-102
8. ASCE Task Committee on Finite Element Analysis of Reinforced Concrete Structures. State-of-the-Art Report on Finite Element Analysis of Reinforced Concrete, 1982, ASCE Special Publications.
9. Eligehausen, R., Popov, E.P. and Bertero, V.V,Local Bond Stress-Slip Relationships of Deformed Bars Under Generalized Excitations. Report No. UCB/EERC 83-23, Earthquake Engineering Research Center, 1983, University of California, Berkeley.
10. Bazant, Z.P. and Cedolin, L., Finite Element Modeling of Crack Band Propagation. Journal of Structural Engineering, ASCE, Vol. 109, No. 1, 1983,pp69-93.
11. Hayashi, S. and Kokusho, S., Bond Behavior in the Neighborhood of the Crack. Proceedings of the U.S.-Japan Joint Seminar on Finite Element Analysis of Reinforced Concrete, 1985,Tokyo, Japan, pp364-373.
12. Pijaudier-Cabot, G. and Bazant, Z.P., Nonlocal damage theory. J. Eng. Mech., Vol.113, 1987,pp1512–1533.
13. Pijaudier-Cabot, G., Non-local damage. M¨uhlhaus, H.-B. (ed.), Continuum models for materials with Microstructure, Wiley, Chichester, U.K., 1995, pp105–143.
14. Helmut Hartl Christoph Handel, 3D Finite Element Modeling of Reinforced Concrete Structures, ed H. Hartl (W-387) pp 1-10。

STUDY ON INTERACTIVE GRAPHICAL INPUT AND CALCULATION SYSTEM OF FRAME JOINT

QIANHUA XU
College of Civil Engineering, Tongji University, 1239 Siping Road Shanghai 200092, China

QILIN ZHANG
College of Civil Engineering, Tongji University, 1239 Siping Road Shanghai 200092, China

HUIZHU YANG
College of Civil Engineering, Tongji University, 1239 Siping Road Shanghai 200092, China

In this paper the development of a CAD system for detailing designing of frame is introduced and presented. Using this system, user can shape new joint type and calculate the new type. This paper studies the joint location and constraint solution. The result of joint detail design is shown with 3D solid.

1. Introduction

In steel structure, elements are made of steel plate and shaped steel, such as beam, column and brace. The whole structure is made of these elements which are connected by joint shop assembling or joint field assembling. Detailing design of frame structure is very important in civil engineering. Joint detailing design and calculation is a important step in steel structure design. The detailing design impact on degree of reliability of steel structure, overall stability, construction period, construction cost, and so on.

With the development of steel structure type, the joint type and construction technique get a rapid development. Civil engineers study and discuss characteristic of joint force, detailing of joint relation and influence of joint type on structure. Different design method of joint is applied in codes for steel structure of different country. Correspondingly , the software of joint detail design is different.

For most of CAD system, a data base of joint designing is supplied. The users design joint connection, whose type is offered by the data base. For any of joint connection type, the location and condition of stress of joint element, including joint plate, bolt and weld, is fixed. The user intervenes in or change designed joint hardly. As development and widely applied of steel structure in China, engineers ask for more flexible joint designed CAD system to be adapted to changeable connection type and increase worked efficiently.

In this paper the development of a CAD system for detailing designing of frame is introduced and presented. There are two parts in this CAD system: parametric entity formation of architectural component and connection element calculation bases on element's location and relationships.

2. Constraint in Joint CAD System

CAD (Computer aided design) is a new semester with computer used widely. CAD problem is constraint problem substantially. A steel joint is made of plate, weld and bolt according to some constraint. The constraint is classified to two type: geometric constraint and engineering constraint. Geometric constraint means geometric relationship of joint detailing, such as parallelism of two plates, perpendicular of two plates, and so on. Engineering constraint means strength and construction requirement of joint detailing design. These requirements give a expression about design course and design result. In joint detailing design, geometric constraint and engineering constraint affect each other. In a CAD system of joint design, these constraints are taked into account. Geometric constraint decides on detail location. Detail location decides on engineering constraint. Sometimes engineering constraint has effect on geometric constraint. For a data base of joint designing CAD system, joint detailing location is fixed. The system give the location and condition of stress of joint element, including joint plate, bolt and weld. For a interactive graphical input and calculation system, geometric constraint is given by user. The system must assay geometric constraint to get engineering constraint. Based on engineering constraint, a joint detailing designing is carried out.

2.1 *Geometric constraint*

Geometric constraint decides on detail(plate, weld, bolt) location. If joint detail location is recorded by location coordinate, some flaw must be thought about: the geometric constraint can not be recorded by coordinate; coordinate parameter can not be get tersely and exactly; when a detail(such as a plate) location is changed, other relative detail location can not be changed corresponding. Joint detail location is recorded by comparative location, as shown in Figure 1. The detail location is get by comparative location with the shape steel. The shape steel, as a element of the structure, has a location coordinate.

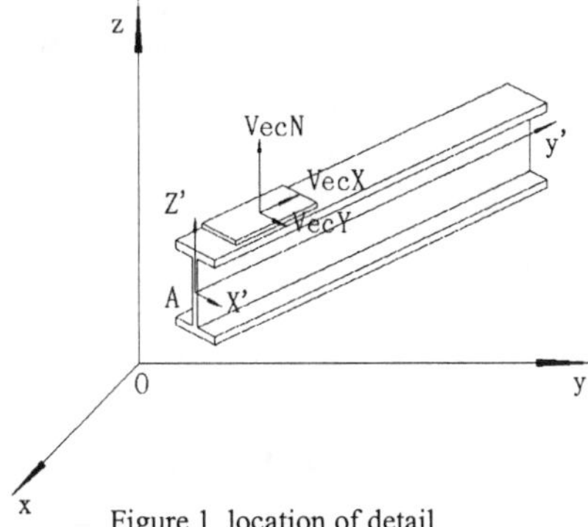

Figure 1. location of detail

All detail of a joint is connected each other. A plate is located by other plate or shaped steel. In finally, all joint detail location is connected with the shaped steel, which's location is fixed.

2.1.1 *Geometric constraint of plate*

As shown in figure 2,a plate is located with six amounts: three direction vector and three coordinate. The system need get these amounts to locate. In fact, when a civil engineering talk about a plate location, he means the comparative location with other plate, such as a plate which is parallel with up flange is connected by weld. Through observation and analysis to lots of joint type, system use three kind of geometric constraints to location plate: perpendicular, parallel, concatenation.

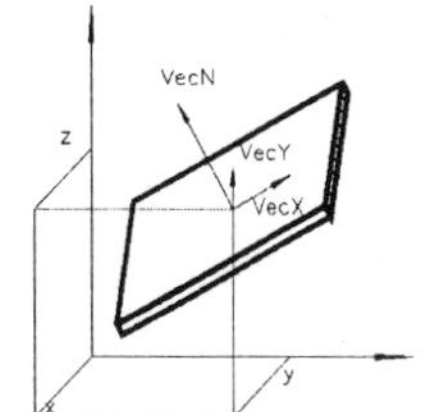

Figure 2. location of plate

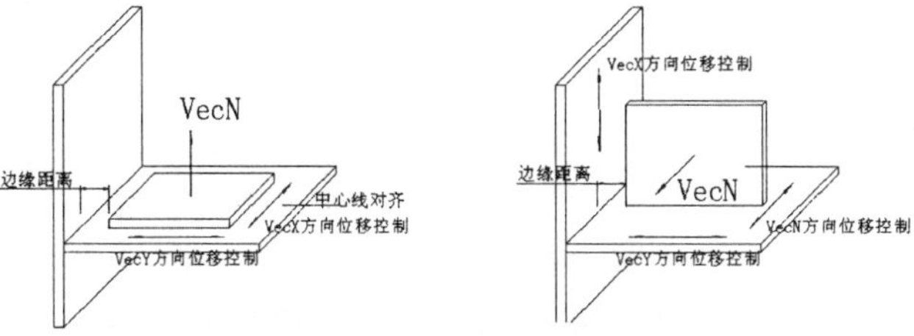

Figure 3. location of plate by other plate

Based on these geometric constraints, a mesh of relationship layer comes into being. The basic element of the mesh is shaped steel which's coordinate is known.

2.1.2 *Geometric constraint of bolt*

Bolt, as a connection detail, is using widely in frame structure. Bolt connect two plates or more plates which are parallel. The location of bolt could be recorded with the connected plate, as the connected plate is always existent. There are some especial geometric constraints for bolts , as Figure 4.

- Normal vectors of connected plates is parallel;
- Normal vector of bolts is parallel with connected plates;
- There is a surface which is common of connected plates;
- Bolts intersect with connected plates;

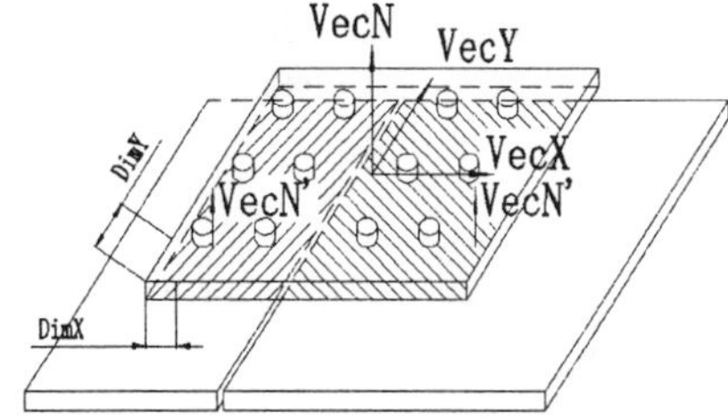

Figure 4. location of bolt by plate

Based on these rules, bolts could be located by a connected plate and other connected plates are found.

2.1.3 *Geometric constraint of weld*

Weld connection is cardinal connection type in model steel structure. Weld connection is applied random type structure. Its construction is simple.

Weld connection is classified to three type: flat weld, lap weld and top weld. Weld is located in flange of a connected plate. So the location of weld could be recorded with the connected plate.

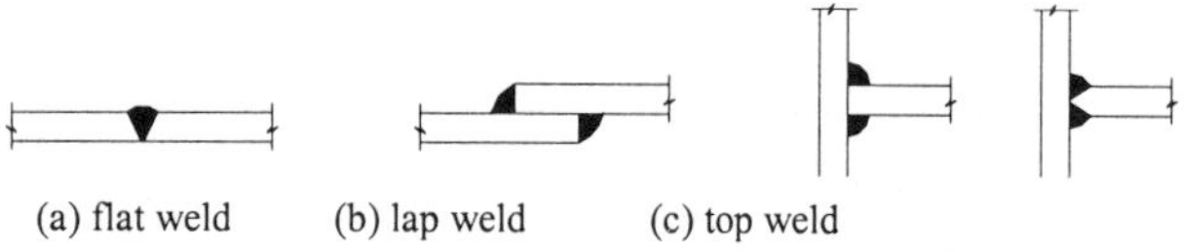

(a) flat weld (b) lap weld (c) top weld

Figure 5. The type of weld connection

There are some especial geometric constraints for weld, as Figure 6.

- There is a surface which is common of connected plates;
- There is a line which is common of connected plates and weld;
- Weld intersect with connected plates;

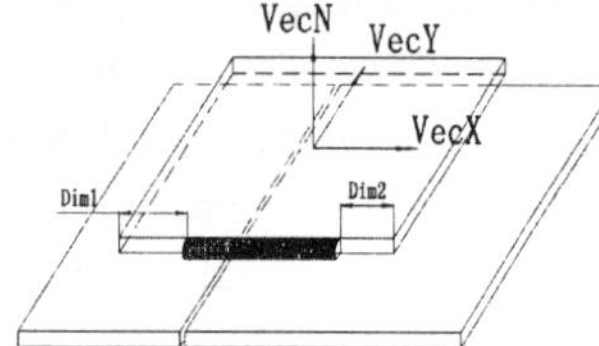

Figure 6. location of weld by plate

Based on these rules, weld could be located by a connected plate and other connected plates are found.

2.2 *Engineering constraint*

Plate connects other two parts with connection tool: weld and bolt. For example, the plate of joint connects web of beam with bolts and connects flange of column with weld, as figure 7.

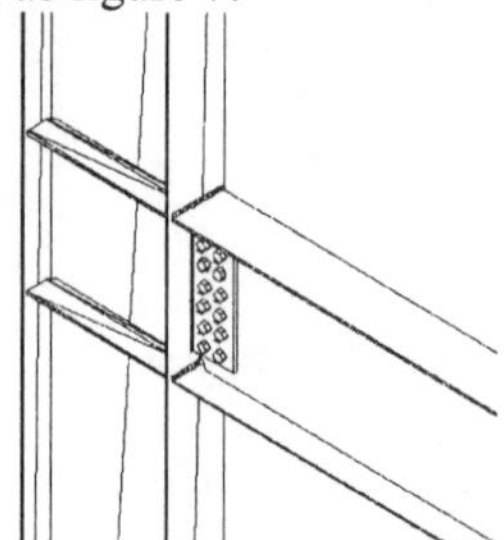

Figure 7. plate connection

In other word, a plate should include two or more connections, which connect different plate. Then a plate can work in a joint.

2.2.1 *Engineering constraint of plate*

As a important detail, plate is checked in strength and construction. In steel codes, plate calculation is rules. In a interactive graphical input and calculation system, plate colliding is checked.

2.2.2 *Engineering constraint of bolt*

Bolts which connect same plates are checked in strength as a whole body. Engineering constraint of bolt include two sides: construction and strength.

Construction Engineering constraints are some rules of codes for joint connection, such as pitch of bolts. Strength Engineering constraints make joint connection safe. When force of bolts and direction of bolts are known, bolts can be calculated.

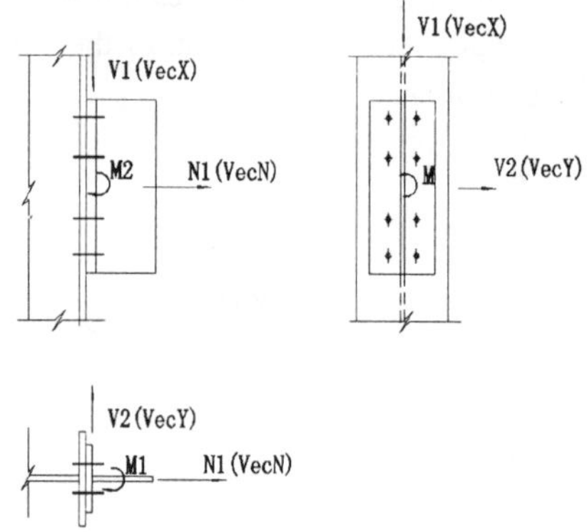

Figure 8. forces of bolts

$$V_{\max} = \sqrt{\left(\frac{V1}{n} + \frac{M \times y_{\max}}{\sum(x^2+y^2)}\right)^2 + \left(\frac{V2}{n} + \frac{M \times x_{\max}}{\sum(x^2+y^2)}\right)^2}$$

$$T_{\max} = \frac{N1}{n} + \frac{M1 \times y_{\max}}{\sum y_i^2} + \frac{M2 \times x_{\max}}{\sum x_i^2} \quad (1)$$

2.2.3 *Engineering constraint of weld*

A weld connected two plates is calculated with codes easily. The Engineering constraint of a single weld is simple : when construction and strength are satisfying, the height and length of weld are fixed. Sometimes, several welds as a whole carry loads. When locations of welds are known, system calculates with different rules by different type, such as surrounded weld and H shape weld.

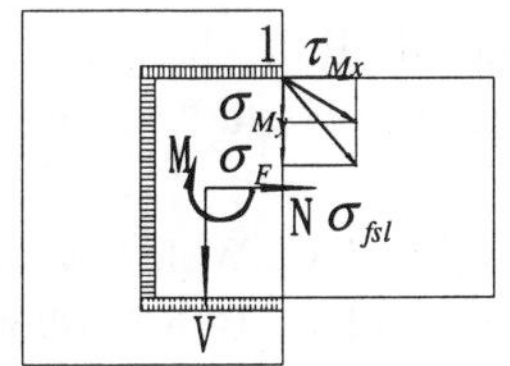

Figure 9. surrounded weld

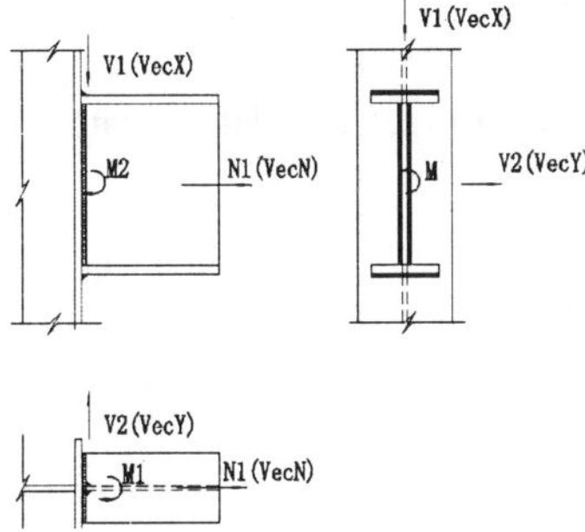

Figure 10. H shape weld

2.3 *Text and equations*

Please preserve the style of the headings, text font and line spacing in order to provide a uniform style for the proceedings volume.

Equations should be centered and numbered consecutively, as in Eq. (1). An alternative method is given in Eq. (2) for long sets of equations where only one referencing equation number is wanted.

3. Constraint Solution Procedure

In joint type database system, the location and stress direction of joint element is known and fixed. So program calculates and designs all bolts and welds easily. But all of them are unknown in a interactive graphical input system. A joint type which is defined by user needs be analyzed in this step.

Joint is connection of structure elements. Joint pass the forces of structure elements. When calculating a joint, we could consider that force from a connected element is passed to another connected element by joint. The plates of joint are the bridge of passing force, as the flow chart1.

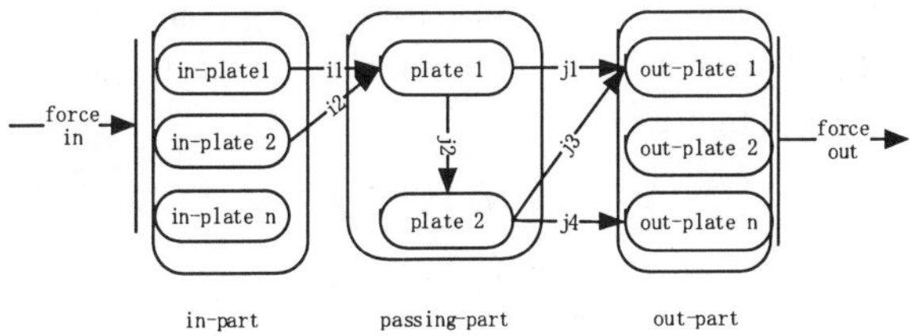

Flow chart 1. force passing

The joint detail as network passes element forces. When a new joint connection type is created, the joint detail geometric constraints are known. The system need analyzed geometric constraints and get engineering constraints for joint design. Based on analyzing of joint detail, this is realizable.

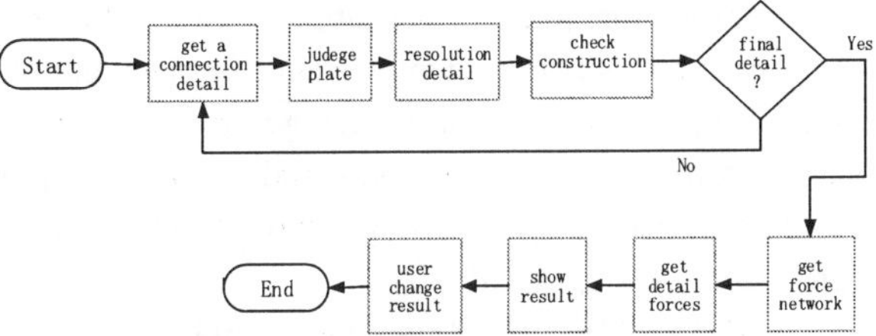

Flow chart 2 The joint assay and design

4. Joint CAD System Development

AutoCAD is a piece of software developed by Autodesk Co. It has been widely used and prevailed in civil engineering. At the same time, AutoCAD® software provides a flexible development platform for specialized design and drafting applications. Its open architecture enables developers to customize AutoCAD for unique purposes. Using ObjectARX (AutoCAD development platform) and VC++ program, the CAD system is customized and extended with direct access to AutoCAD database structures, the graphics system, and native command definition . The extend system creates a 3D solid view of steel roof in AutoCAD interfaces. Adopting solid model is the unique way to make users observe and perceive the resultant joints easily and directly.

4.1 *Data processing*

Beam, column and brace, which compose the frame structure, is showed with solid in different color for distinguish. At the same time, as interactive graphical input system, efficient method is applied for joint designing type which user defined. As a method of connection of member, elements of joint work on each other and connection each other. According to constraint equation and space characteristic , bolt and weld is defined to a member of joint plate or basic joint, as shown in figure 11.

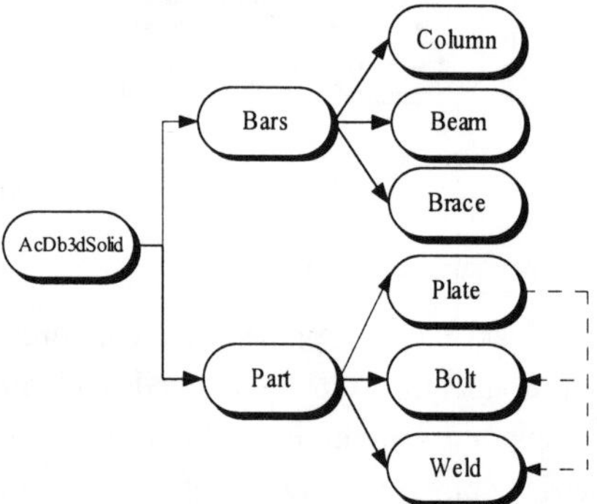

Figure 11. Basic and derived class relation

A class for joint is set up, in which the joint detailing message, connected element message and joint function as well as the other message are included. All the message in the joint class would be used in joint designing, data file writing and solid graphics constituting.

```
class TJ_POST_API Joint : public AcDbEntity
{
.......
AcArray<FreePart_Plate*> NodePlatePtr;
AcArray<FreePart_Bolt*> NodeBoltPtr;
AcArray<FreePart_Weld*> NodeWeldPtr;
........
}


class FreePart_Plate : public FreePart
{
public:
FreePart_Plate();
virtual ~FreePart_Plate();
int PlateNo;
bool UserOrModel;
double getPlateL();
double getPlateW();
double getPlateT();
void   setPlateL(double l);
void   setPlateW(double w);
void   setPlateT(double t);
void   CheckPart(FreePart_Bolt* pBolt);
void   CheckPart(FreePart_Weld* pWeld);

int    GetBoltCor(FreePart_Bolt * boltptr);
int    GetWeldCor(FreePart_Weld * weldptr);
int    SetBoltCor(FreePart_Bolt * boltptr);
int    SetWeldCor(FreePart_Weld * weldptr);

bool   IntersectOrNot(FreePart_Weld * weldptr);
bool   IntersectOrNot(FreePart_Bolt * boltptr);

AcArray<FreePart_Bolt*> BoltPtr;
AcArray<FreePart_Weld*> WeldPtr;
char* name;
int copyFrom(const FreePart_Plate* pPart);
protected:
double m_PlateL;
double m_PlateW;
double m_PlateT;
};
```

4.2 *Joint design result*

In the interactive graphical input and calculation system, user can create new joint type, as shown in figure 12.

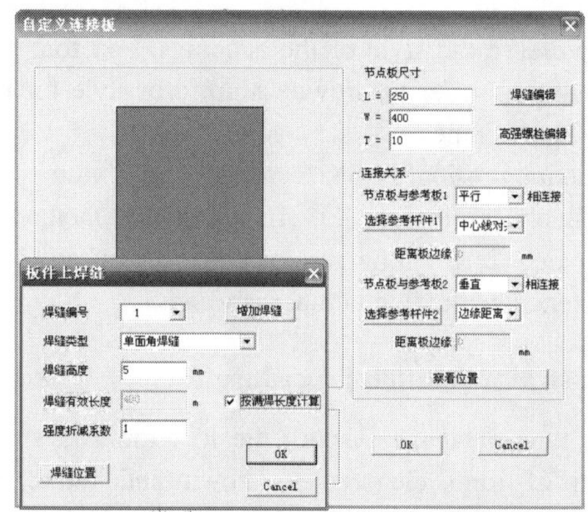

Figure 12(a) definite detail connection

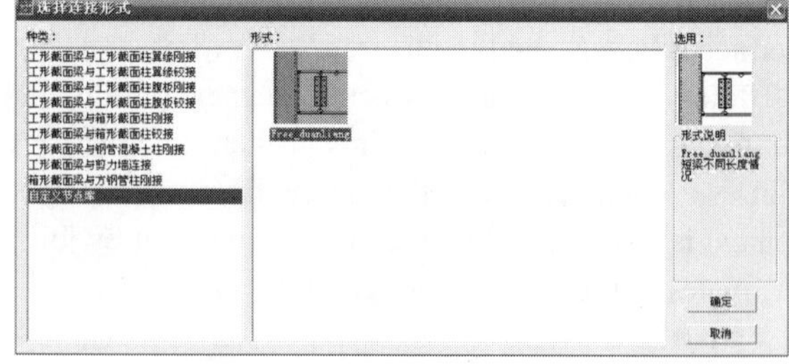

Figure 12(b) joint designing

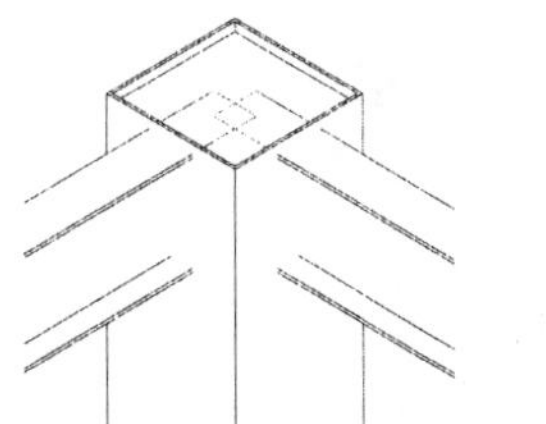

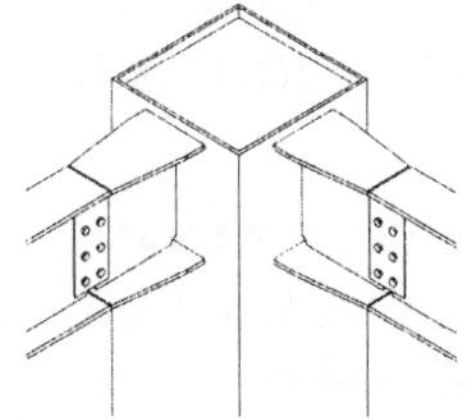

Figure 13. solid model of steel frame beam-column joint

5. Conclusion

Based on Chinese design codes for steel structures, a system for joint detail designing is constructed. It is a flexible CAD system. User can create new joint type. The solid graphics make civil engineers understand joint designing result easier.

References

1. Charles Mcauley: Programming AutoCAD 2000 Using ObjectARX, China Machine Press, 2000,152-165.
2. http://www.autodesk.com.cn/adsk/servlet/home?siteID=1170359&id=1810417.
3. Xing-Rong Li: Enchiridion of Joint Design for Steel Structure, China Architecture and Building Press, 2005:41-45.

IFC BASED DESIGN INTEGRATION BETWEEN ARCHITECTURAL AND STRUCTURAL MODELS*

XUE YUAN DENG TSE-YUNG P. CHANG+

Department of Civil Engineering, Shanghai Jiao Tong University, Xuhui
Shanghai, 200030, P R China

GUO JIAN WANG

Shanghai Xian Dai Architectural Design (Group) Co. Ltd.
Shanghai, 200041, P R China

AT present, engineers create a structural model for a tall building based on paper drawings of an architectural CAD model and it is usually difficult to engage in design collaboration between the architects and structural engineers. The future trend in design technology is to promote the use of object-oriented CAD model in which a building is represented by 3D objects such as beams, columns, walls, floor slabs, windows, doors, stairs, etc. These objects can contain not only the geometric information, but also physical behavior and design data associated with them. As a result, one may construct a unified common design platform so that information can be shared across different disciplines in Architecture Engineering and Construction (AEC) industry. With this objective in mind, we have initiated a development program for direct data integration between the architectural and structure disciplines as the first step in building design.

1. Introduction

Building information modeling (BIM) is the latest emergence of a 25-year-old idea that architects should create intelligent 3D models instead of relying paper drawings to communicate design ideas and guide construction [1]. Information Modeling is now a widely accepted approach to the development of specifications for information exchange and sharing in several industrial sectors, including AEC/FM. It has progressed from being a research topic to providing a basis in commercial software implementations for the exchange of object-based information.

Most of the research on information modeling in the building industry concentrates on building product modeling [2], which is intended to design a building product data model to represent a wide range of building-related information. A Building Product Model is defined [3] as a digital information structure of the objects making up a building, capturing the form, behavior and relations of the parts and assemblies within the building. It is potentially a richer representation than any set of drawings and the related data can be kept in different forms, including an ASCII file or file in a database. The data in the model can be created, manipulated, evaluated, reviewed and presented using any CAD tools. Traditional reports including drawings may be generated from the digital information representation. Building product data model is a conceptual schema for a building, which structures the information needed to describe a building. Moreover, the model describes a process relating to the design, construction and other associated activities.

The objective of this paper is to provide a means for data integration between the architectural CAD model and structural design model of a building without using the paper drawings.

2. Industry Foundation Classes (IFC)

In 1995, the International Alliance for Interoperability (IAI) was formed to provide interoperability between the software used by all building project participants. The intent is to provide a means of passing a complete, thorough and accurate building data model from the computer application used by one participant to another, with no loss of information. Industry Foundation Classes (IFC) consists of data elements that represent the parts of a building, or elements of the process, and contains the relevant information about those parts. IFC is used by computer applications to assemble a computer readable model of the facility that contains all the

*This work is supported by Shanghai Xian Dai Architectural Design (Group) Co. Ltd.

+Also, Professor Emeritus, Department of Civil Engineering, Hong Kong University of Science & Technology

information of the parts and their relationships to be shared among project participants. The project model constitutes an object-oriented database of the information shared among project participants and continues to grow as the project goes through design, construction and operation.

2.1 *Implementation*

The implementation of IFC [5] is based on the IFC schema provided by IAI with express language. The physical file of an IFC model is based on ISO 10303, part 21. The most famous CAD applications, such as Autodesk's Architectural Desktop, Graphisoft's ArchiCAD, and Bentley's Microstation TriForma, have provided the support of importing and exporting IFC model. The IFC model of a building structure consists of all building elements. These building elements are defined with non-abstract descendants of the generic object class.

An object in IFC is the abstract super-type, IFC-Object, and stands for all physically tangible items, such as wall, slab, beam or column, physically existing items, such as spaces or conceptual items, such as stories, girds or virtual boundaries. The property definition of these building elements is the generalization of all characteristics of building element objects. It reflects the specific information of an object type. The property definition is applied to the objects using the concept of relationships.

The concept of relationships in IFC is the relationship class, or called IFC-Relationship. The relationship class is the preferred way to handle relationships among objects. This allows relationship specific properties to be kept directly at the relationship object and enables the relationship semantics to be separated from the object attributes.

Any building element objects in IFC that derives from IFC-Product should have geometric representation that defines object placements, either using the absolute or relative axis locations, or referring to data relative to grid definitions. The geometric representation also defines standard profile (or cross section) definitions to be used in models of swept surface or swept solids. In another words, the geometric representations describe the location and shape of all building element objects.

In addition, material properties are defined and assigned to the corresponding objects with non-abstract descendants of the relationship class.

3. Methods of Integration Design

The main objective of this research is to create an integrated architectural design and structural design environment under which information for building design can be exchanged directly. Under this environment, engineer can construct a structural model for a building in the structural domain by directly capturing the necessary information from the corresponding architectural CAD model in the architectural domain without relying on paper drawings.

A set of procedures, together with computer tools, is developed to retrieve the information of an architectural CAD model and convert the related data to form a preliminary structural model. Using this preliminary model, the engineer will then layout the structural form (or skeleton) of a building, define preliminary dimensions of all structural members and then proceed to define all other structural analysis and design parameters for further structural analysis. After all the necessary data for a structural model are defined, a data file will be created automatically to perform structural analysis using a computer program such as ETABS, SAP2000 or others. Finally, the updated structural model can be imported back to the original architectural model for design collaborations.

In this project, the AutoDesk's Architectural Desktop (ADT), but not limited to ADT, is used as the computer aided design (CAD) platform for both the architectural and structural design of a building. Once the object-oriented architectural 3D model has been established using ADT, an IFC-Utility 2x is employed to export the corresponding IFC model as an IFC file.

Based on IFC schema, a structural model server is developed to retrieve the related data on structural element objects, such as wall, slab, beam, column, etc. For all of these structural element objects, their geometric information, like local placements and shape representations, material information and connection/ boundary information are acquired under the structural model server, subsequently exported into a Microsoft Access database.

The structural model stored in the Microsoft Access database is imported into a pre-processor to perform several preliminary tasks, such as model re-definition and editing, data check and visualization, defining loads and load analysis, defining finite element analysis data and design parameters, etc. At the end, an updated structural model is exported as an input data file to perform structural analysis, say using ETABS. Subsequently, the structural model with verified sectional properties and design information is converted as an updated IFC structural model in accordance with the IFC standards. Finally, the original architectural model is further revised by importing the updated IFC structural model. Fig. 1 shows the algorithm of IFC based design integration between architectural and structural models.

4. Case Study

We consider a 10-story reinforced concrete building with a typical architectural floor plan as shown in Fig. 2. All stories are assumed to have the same height of 2.8 m. The architectural CAD model in Fig. 2 shows several layers of entities, such as partitions, windows, doors, stairways, kitchen and toilet facilities, etc. Only the structural related entities are selected to create a structural model, i.e. the partitions, openings of windows and doors, locations of slabs, etc. All these members are picked element-by-element in the form of IFC data and these data are placed into the MS ACCESS format. The related data are then retrieved into a model server to further define the geometric data (such as members' cross sectional shapes and dimensions) needed for the structural analysis (shown in Fig. 3). In addition, one has to define all the finite element analysis and preliminary design parameters. Finally, the server automatically creates an ETABS data file (.e2k file) for subsequent structural analysis.

5. Conclusions

In this paper, we have presented a procedure for data integration so that the architectural CAD model of a building can be directly converted into a structural model without relying on the paper drawings. In order to achieve this purpose, all data must conform to the IFC standards so that data transfer can be done in a seamless manner. It is envisioned that the same approach can be readily extended for data integration between structural and foundation design, architectural and mechanical or electrical design for a building.

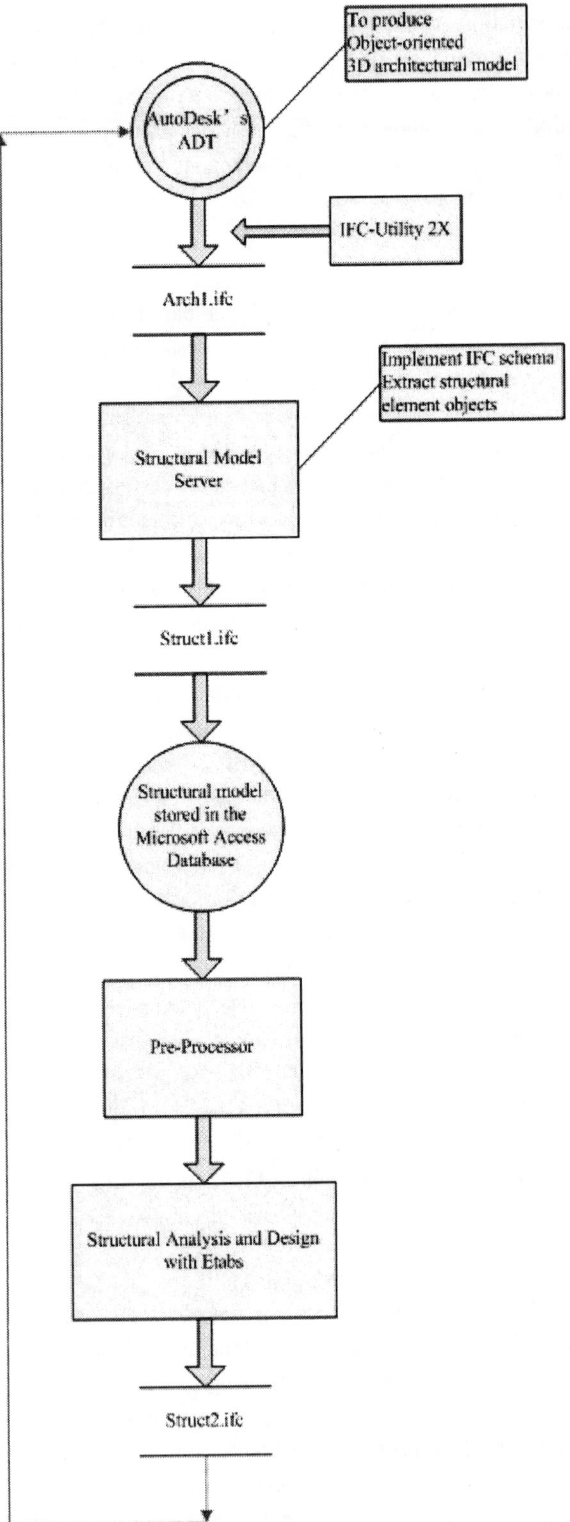

Fig. 1. Algorithm of IFC based design integration between architectural and structural models.

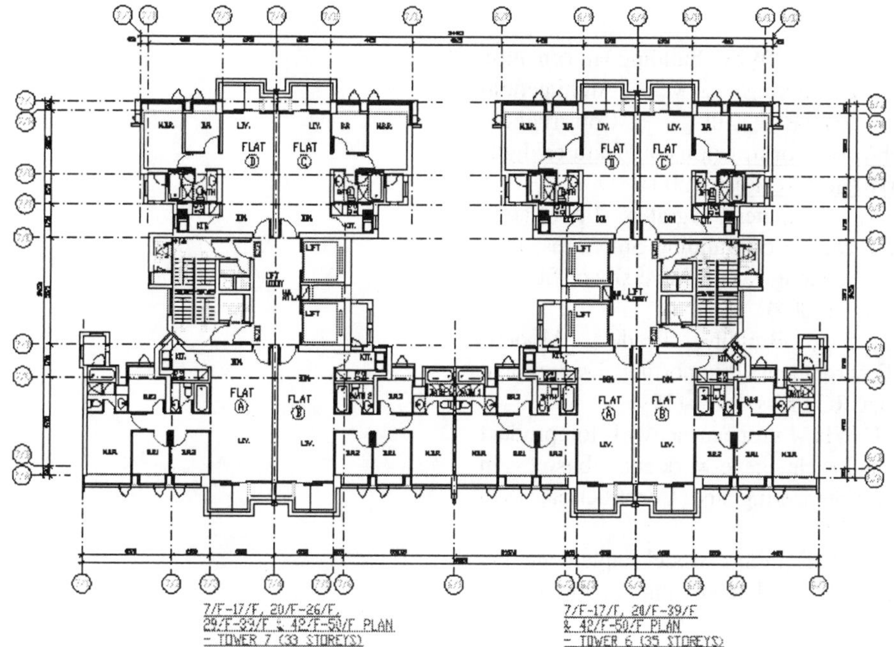

Fig. 2 Typical Architectural Floor Plan of A building

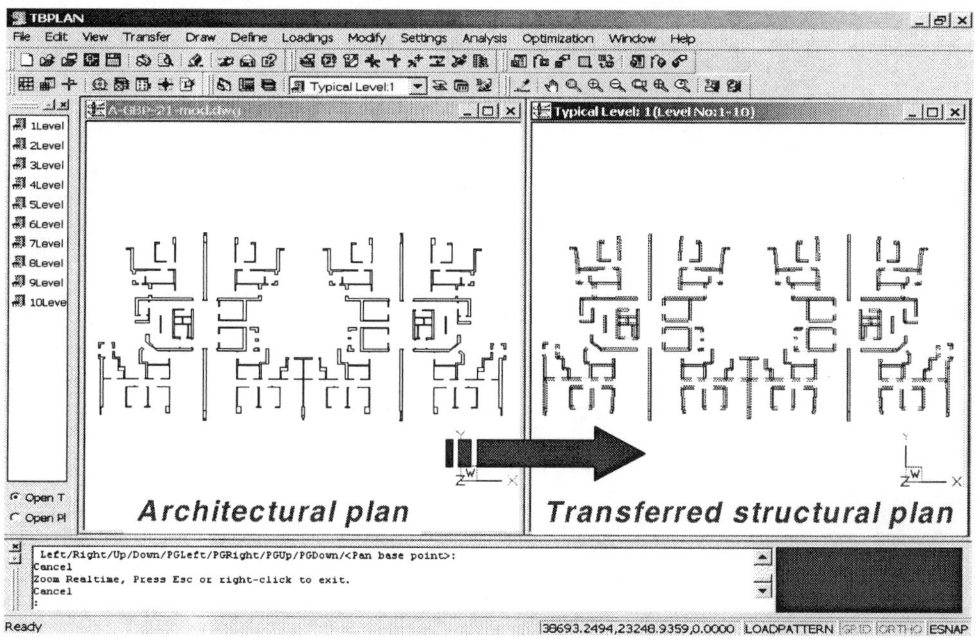

Fig. 3 Direct Data Conversion of the Architectural CAD Model for Formation of a Structural Model

References

1. K. Sanders, FAIA, "Why building information modeling isn't working ... yet", Architectural Record, McGraw_Hill Construction, http://qa.archrecord.construction.com/features/digital/archives/0409feature-1.asp, (2004).
2. J. H. Long, Computer-Integrated Information Modelling for Design of Building Structures, PhD Dissertation, The Hong Kong University of Science and Technology, (2004).
3. C. M. Eastman, Building Product Models: Computer Environments Supporting Design and Construction, CRC Press, LLC. (1999).
4. T. Liebich, J. Wix, European network for product and project data exchange, e-work and e-business in Architecture, Engineering and Construction, IST-2001-32035, (2002).
5. T. Liebich, IFC 2x Edition 2 Model Implementation Guide, Version 1.7, IAI modelling support group, (2004).

STUDY ON TALL BUILDING STRUCTURE ANALYSIS

LIU HUAFENG ZHAO NING
China Academy of Building Research, Beijing 100013, P.R.China

TBSA (Tall Building Structure Analysis) is a three-dimension analysis program for tall buildings. Version 6.0 is a new version based on new Building Codes of China. In version 6.0 there are many new features added to enhance the usability and quality. This paper illustrates some of the new features and discusses some points that should pay attention to when we analyze tall building structures with the program.

1. New Features

1.1 *Introduction*

TBSA (Tall Building Structure Analysis) is a three-dimension analysis program for tall buildings based on a member structure model. In the program, beams, columns and braces are defined as space bar members, with 6 freedoms on each end. Shear walls are defined as thin-walled column members with buckling, with 7 freedoms on each end. It is assumed that floors have infinite in-plane stiffness and no out-plane stiffness.

The program has no specific requirements on structural system. The structures may be unsymmetrical and non-rectangular in plan. TBSA can analyze not only general frame structures, frame-shear wall structures, tube structures, but also those more complex such as multi-tower structures, Staggered floor structures and connected buildings.

For nearly 20 years, the program has been used in thousands of buildings all over the country.

Version 6.0 is a new version based on new Building Codes of China. In version 6.0 there are many new features added to enhance the usability and quality.

1.2 *New graphical user interface*

The graphical user interface is used to model, analyze, and display the results for your structure.

Figure 1 shows the main window of the new graphical user interface. This window may be moved, resized, maximized, minimized, or closed using standard Windows operations.

The main title bar, at the top of the main window, gives the program name and the name of the data file. The menus on the Menu Bar contain almost all of the operations that can be performed. The buttons on the toolbars provide quick access to many commonly used operations. Status Bar shows the information about what the program is currently doing, the coordinates of the mouse cursor, the current floor and the current standard floor.

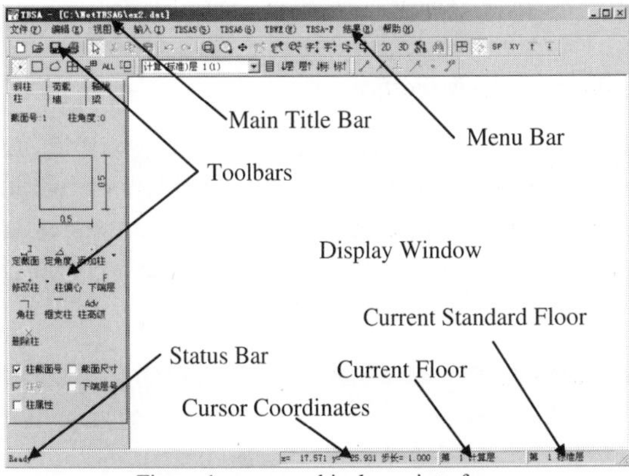

Figure 1. new graphical user interface

Display window shows the geometry, loading, or analysis results of the structure in 2D or 3D view. A 2D view shows a single floor. A 3D view shows the entire structure in an orthographic projection.

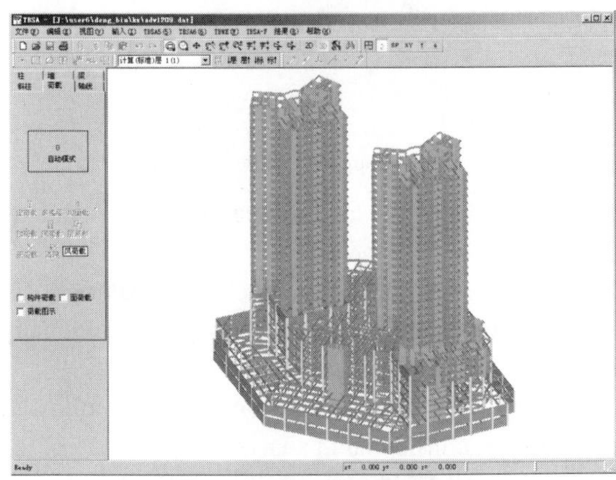

Figure 2 . OpenGL rendering

The OpenGL rendering feature depicts the structure in realistic 3D lighting and texture. The animate feature allows users to graphically view displacements, mode shapes in an animated mode.

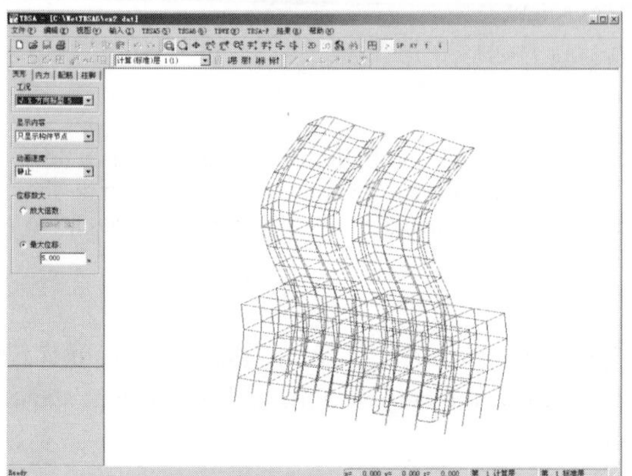
Figure 3 mode shapes in an animated mode

1.3 *New analysis features*

Version 6.0 can analyze steel structures, profile steel concrete structures and special-shaped column structures in addition to the reinforcement concrete structures and steel tube concrete structures that previous version can do.

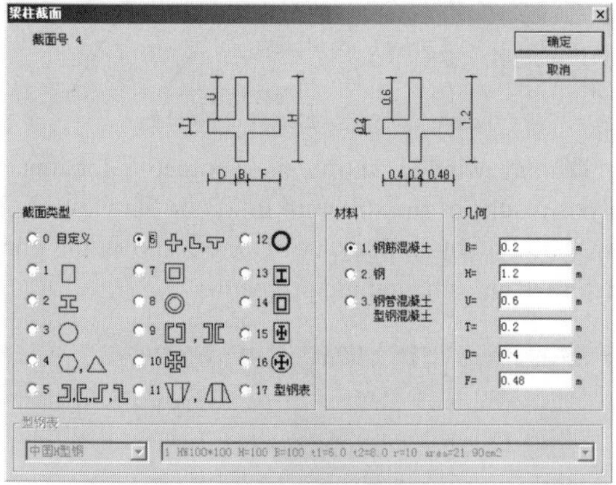
Figure 4 . sections

Dead load and live load can be input and calculated separately.

Properties of members such as material and earthquake-resistance class can be assigned individually. It makes the design work more accurate.

Version 6.0 offers the options that the following factors may be considered in analysis:

- o Basement with or without lateral restraint.
- o Bi-directional seismic load.
- o Random decentration effect.
- o Independent members and nodes.
- o P-Delta analysis.

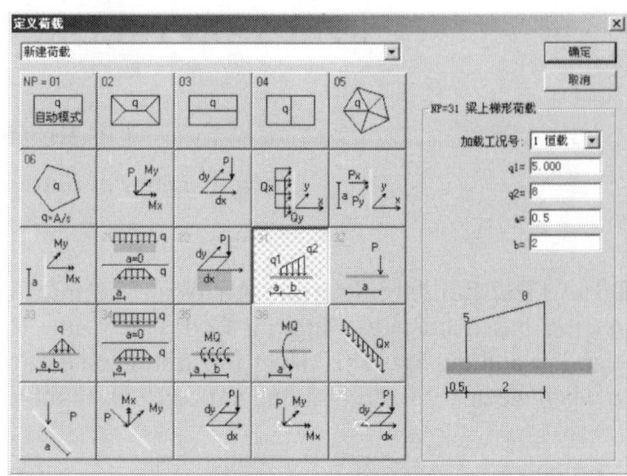
Figure 5. loads

2. Points that Should Pay Attention to

Analysis is an important part in the procedure of tall building structures design. The analysis results have great influence to the dimension and reinforcement of the members, as much as, to the project. So we think much of analysis. Now we discuss some points that should pay attention to when we analyze tall building structures with the program.

2.1 *Selecting analysis model*

The first step of any structure analysis is to select an appropriate analysis model that can objectively represent the stress of the structures.

In mechanics, the accuracy of a model needs two preconditions, first the displacement must be continuous, secondly the load transfer path must be correct. Most of the buildings with regular plane and elevation layout can meet the requirements. But for those with irregular elevation, they can hardly make the displacements continuous.

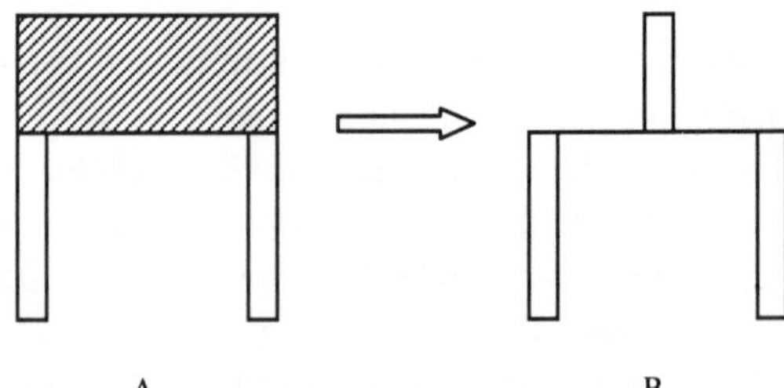

Figure 6 . frame-supported shear wall

For example, in a frame-supported shear wall structure, a single piece of shear wall supported by a beam as shown in figure 6A. Actually, the load of the shear wall distributes uniformly onto the beam then transfers to the columns. According to the simplified model shown in figure 6B, the load of the shear wall applies as a concentrated force to the frame-supporting beam then transfer to the columns below. This simplification ensures that the load transfer correctly. To make sure that the displacement of the model approaches that of the real structures, the stiffness of the frame-supporting beam should be considered larger than it is. When modeling, we can assign the beam's width as its actual width and assign its height as half of the story.

If it is a multiple shear wall on the frame-supporting beam, we can select the column that most close to the shear center of the shear wall as loading transfer point, connect each frame-supporting column by deep beams, to make sure that the load of the shear wall be supported by these columns together.

2.2 *Shear force adjusting of multi-tower frame-shear wall buildings under seismic load*

To adjust shear force of single-tower frame-shear wall buildings, we first find the maximum value of total shear force of all frame columns in each story, then multiply this value by 1.5, compare the result with 20% of basis shear force, get the less one. This adjustment is made for the horizontal seismic load.

To adjust shear force of multi-tower building, which consisting of several towers rising from a common base, we should take notice of that the adjustment is made in each tower separately and the $0.2Q_0$ refers to basis shear force of each tower.

2.3 *Basement with lateral restraint*

The program is for superstructures of tall buildings considering the base of building as fixed. It offers the options that the coactions of the superstructure and the basement may be considered.

The users can input the number of stories of the basement, wind load of these stories will be set to zero, the height of each story for wind load will be counted from the story above the basement.

The users can also input the number of stories of the basement with lateral restraint, the lateral displacement of these stories will be set to zero. The height of these stories will be removed from the height of the building. Load combinations are applied with the new height of building.

2.4 *Random decentraction and bi-directional seismic load*

According to the new code for seismic design of buildings, the torsional effect of horizontal bi-directional seismic loads should be considered. In the new technical specification for concrete structures of tall buildings, the random decentration effect should be considered while calculating single-directional seismic loads

If we analyze a building with random decentration and bi-directional seismic loads considered simultaneity, the reinforcement may be much more than reasonable. Therefore, if the bi-directional seismic loads are calculated, the random decentration effect may be ignored.

2.5 *Construction sequence loading*

In most analysis of tall buildings, it is assumed that the structure is not subjected to any load until it is completely built. In reality the dead load of the structure is continuously being applied as the structure is being built. The vertical members such as the columns and the shear walls continuously deform axially during the construction. Their force should be calculated in the case that dead loads are applied floor by floor as shown in figure 7.

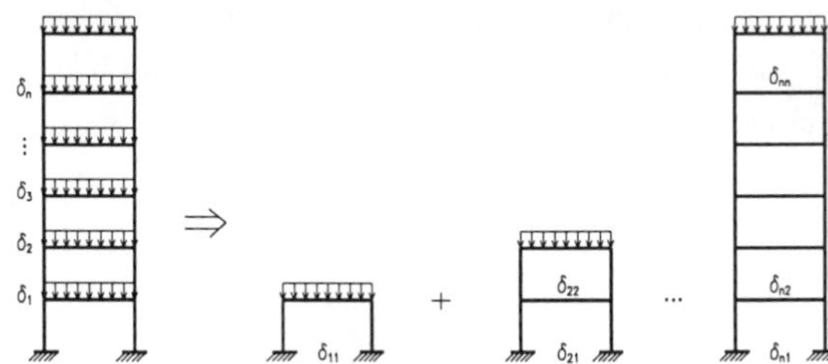

Figure 7 Construction sequence loading

However, after the building is built, the dead loads applied on it are long-term, moreover, the columns have larger values of axial compression ratio than the shear walls, their time deformations are obviously larger than that of the shear walls. After redistributed, the internal forces of the structure approach to the results of one-time loading. So the internal forces of the structure are intermediate between these two situations.

The program introduces an approximate construction sequence loading method as shown in figure 8. The stiffness matrix is completely formed at one time and the loads are applied floor by floor. In other words, when the loads are applied to a certain floor, the deformations of this floor and the floors below are effected by the stiffness of this floor and the floors above.

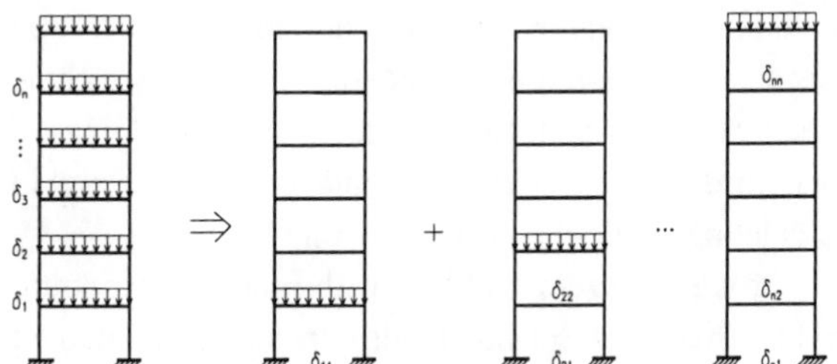

Figure 8 Approximate construction sequence loading

For those complex buildings such as multi-tower buildings, the users may specify different construction sequence loading in the program according to the different construction sequence in reality.

2.6 *Enlarging seismic loads of weak floors*

It is prescribed in the code for seismic design of buildings and the technical specification for concrete structures of tall buildings that the seismic shear force of weak floor should be multiplied by a factor of 1.15.

Here is the definition of "weak floor". According to definition of vertical irregular structure in the code for seismic design of buildings, a floor meets one of the following conditions will be defined as weak floor:

(1) The lateral stiffness of this floor is less than 70% of that of the adjacent upper floor or 80% of the average of that of the adjacent upper three floors.

(2) The shrinkage in horizontal dimension of this floor is large than 25% of the dimension of the adjacent lower floor, except that this floor is the top floor or it is in a multi-tower building.

(3) The internal forces of the vertical members resisting lateral loads transferred by horizontal transferring members.

(4) In a structure resisting lateral loads, the shear bearing capacity of this floor is less than that of the adjacent upper floor.

Besides, the seismic shear forces of the floors with staggered or connecting structures or with large cantilever members also should be enlarged appropriately.

In the program, the lateral stiffness of each floor and the stiffness ratio of adjacent floors are given. The designers select the weak floors according to the situations of their projects, the program enlarge the seismic shear forces of the selected floors.

References

1. The Ministry of Construction of China, Code for Seismic Design of Buildings.
2. The Ministry of Construction of China, Technical Specification for Concrete Structures of Tall Buildings.
3. China Academy of Building Research, TBSA Analysis Reference Manual.

APPLICATION OF INTEGRATED CAD FRAMEWORK IN TALL BUILDING DESIGN*

JIANGHUA LONG† J.S. KUANG T.Y.P CHANG

Department of Civil Engineering, Hong Kong University of Science & Technology
Clear Water Bay, Kowloon, Hong Kong

The design efficiency of tall buildings depends largely on the advancement of technologies, specifically Computer Aided Design (CAD) technologies. To days, CAD tools have become a routine practice for architectural and engineering designs and significantly improved quality and productivity in the building design. Nevertheless, a fragmentation characteristic of these CAD tools barriers the further improvement of productivity of the design in building industry. To address the fragmentation problem, an integrated CAD framework for the design of building structure is proposed in this paper, which is based on a process model and building product model. The proposed framework benefits the development of integrated CAD systems and potentially promotes design automation in building industry.

1. Introduction

With the constant concern on the improvement of efficiency and productivity in building industry, the professionals have paid special attentions to the studies of building design automation. Although the studies have been carried out for decades, the researchers haven't found any successful solution to implement the automation in the whole building design lifecycle. It is particularly true that cooperation between architectural and structural designs and collaboration of internal activities of structural design by the aid of CAD tools have not been sufficiently addressed in the studies.

As the current practice, cooperation between architectural and structural designs is mainly carried out by the methods of manual interpretation based on drawings and face-to-face communication between architects and engineers. The manual process is often time-consuming and ineffective, because repeated structural modelling causes the low efficiency in structural design. Nowadays, there are many CAD tools utilised in architectural and structural designs. These CAD tools may replace designers in individual tasks of building design, but can't share data and information with each other. For lack of an integrated software platform, these tools can't allow the automatic cooperation between architectural and structural designs in a computing environment.

In fact, structural design is a complicated process, which includes some complicated activities, such as modelling, analysis, optimisation, detailing and code compliance checking, etc. These activities are aided by numerous structural design CAD tools. Because the CAD tools are developed by different software vendors and have proprietary data models, they prohibit the smooth data sharing and information exchange with each other. Successful communication among the heterogeneous CAD tools is a big challenge to the design automation of building structure.

To address the above problems involved in structural design, this paper proposes an integrated CAD framework, which supports various CAD tools to share data and exchange information in a collaborative environment. The proposed process modelling, building product modelling and framework development are introduced in the following sections.

2. Computer-integrated CAD System

System integration in computer environment is the fundamental part of product design and manufacturing automation. Computer-Integrated Manufacturing (CIM) has been extensively studied in mechanical engineering for a few decades. CIM's successful implementation significantly improves the design and manufacturing productivity of mechanical products, and it is impacting the whole construction industry. The professionals in building industry have recognised the importance and feasibility of Computer-Integrated Construction (CIC). Some studies on CIC have been conducted [1,2]. The purpose of these studies is to integrate all construction processes including briefing, planning, design, construction and maintenance in an

* This work is supported by grant HIA03/04.EG01 of the Hong Kong University of Science & Technology.

† Corresponding author: drlongjh@ust.hk

effective manner. An integrated CAD system is considered as the critical links in implementing CIC and the integrated CAD system for the design of building structure is particularly the indispensable part of the whole CIC system.

Some CAD integration projects have been carried out by the professionals and organisations in building industry for the past decades. For example, the ATALAS project was launched to develop architectures, methodologies and tools for computer-integrated Large Scale Engineering (LSE) system [3]. The COMBINE project established a data infrastructure and tools to manage the information in a building design team for HVAC design [4]. The CIMSTEEL project defined product model-based standards for exchanging information relating to structural steelwork [5]. Other similar projects include ICON [6], GennCOM [7], etc. The feasibility and validity of the application of computer-integrated system in building industry have been evaluated by these projects. Although they seldom cover the domain of building structure design, the developed concept and methodologies on the system integration are valuable to implement the integrated CAD system for building structural design. The basic methodologies involved in the development of integrated CAD system are presented as below.

- **Process modelling.** This methodology is utilised to model the relevant activities in a design process. The deliverability of the processing modelling is a process model with the sequence of activities, which presents the main tasks carried out by an integrated CAD system. These tasks are referred in developing the main components of an integrated CAD system.
- **Building product modelling (BPM).** BPM intends to define an object-oriented representation of a building product. The defined object model is the basis of database design of the integrated CAD system. A building product model will facilitate data integration and enable data and information to share and exchange among fragmented CAD tools.
- **Component development.** Component-based software engineering approaches allow the development of large-scale software system. Various CAD tools interact with standardised components, which can communicate with other components in the integrated environment.

The above methodologies are necessary to design an integrated CAD framework.

3. Process Modelling

A process model is proposed to support the development of the integrated CAD framework. The main activities and information flow of the proposed model are identified as shown in Figure 1. Generally, a structural design process may be divided into structural modelling, analysis, detailing and documentation. These activities should be further refined to suit to the state-of-art of CAD tools utilised in structural design. For example, some advanced CAD tools have brought intelligent reasoning and structural optimisation functions into the structural design of buildings. In the proposed process model, intelligent modelling and structural optimisation are considered as individual activities for the newly introduced functions.

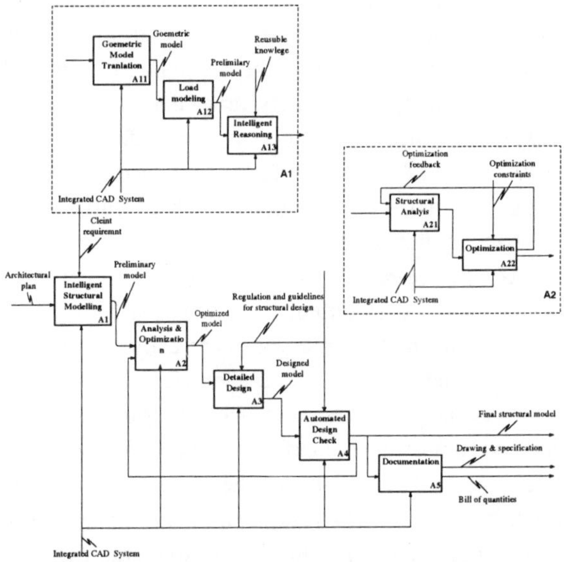

Figure 1. Proposed process model of structural design.

To demonstrate the process model, a usage scenario is presented here. After an architect finishes the architectural design of a building, an architectural plan in the computer-readable format is prepared. Then, the integrated CAD system receives the architectural plan and starts intelligent modelling. In intelligent modelling, a model translation tool directly interprets geometric information of the architectural plan to build a structural model. A load-modelling tool then assigns the model with load information. Finally, an intelligent reasoning program refines the model to get a feasible preliminary structural scheme according to the code constraint and design experience accumulated in past projects.

After getting the preliminary structural model, the integrated CAD system starts iterative sub-processes, namely, structural analysis and optimisation. The sub-processes are repeated until an optimised structural scheme is achieved. After optimised structural plan comes out, a detailed design tool carries out the model detailing. Finally, the detailed structural model is automatically checked by the code compliance-checking tool according to code requirements. The final documents may be prepared in the documentation activity if the structural model passes the checking.

4. Building Product Modelling

A building product model is an object-oriented representation of buildings. It is the basis of database design for the integrated CAD system. In this study, a building product model is proposed. The concept for the development of methodologies and processes of the proposed building product model has been elaborated in [8]. The schema of the proposed model is represented by a few UML class diagrams as shown in Figure 2.

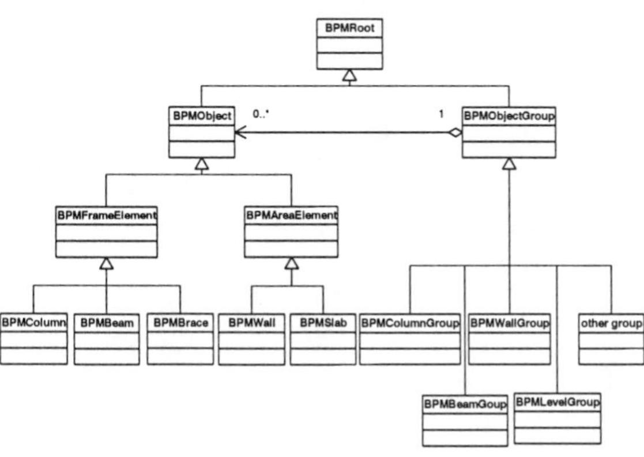

Figure 2. Inheritance relationships.

The diagrams in Figure 2 show the inheritance relationship between the classes of the model. It can be found that column, beam and brace class inherit the property of frame element class, while wall and slab inherit the property of shell element class. Both frame element and slab element classes are derived from the object class. For the class structure in the proposed model, apart from inheritance relationship, other relationships such as composition and association among all classes are also addressed, and these relationships are presented in the corresponding class diagrams.

5. Integrated CAD Framework

An integrated CAD framework for tall building design is proposed as shown in Figure 3. The framework is designed based on the distributed application technologies in software engineering and can be applied in the integrated CAD system. It mainly consists of three levels, namely, database, core, and presentation levels. The database level is controlled by a database management system, which is responsible for persistent data storage. The central database is designed according to the schema of the proposed building product model. The core level provides service components, which enable object-based data accessing, object management and application interfaces. The presentation level is mainly responsible for information representation and CAD component-side data transaction.

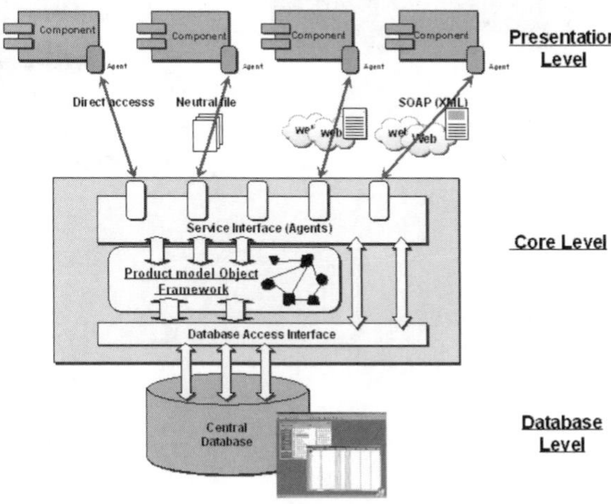

Figure 3. The integrated CAD framework.

After the framework is formed, a physical structure of the integrated CAD system can be established as shown in Figure 4. The structure consists of the following eight main components, which are corresponding to the activities in the proposed process model:

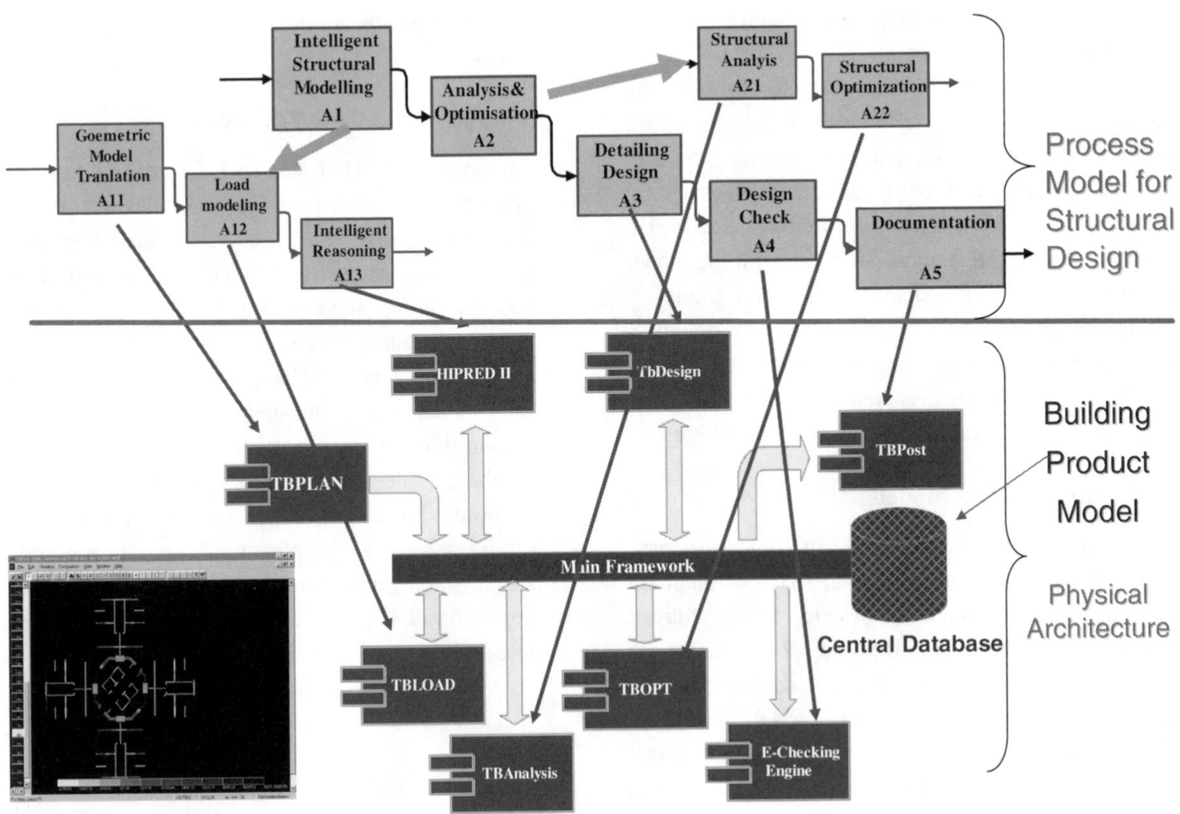

Figure 4. Physical structure of the integrated CAD system.

Component 1: TBPlan. This component is responsible for geometry modelling and structural model translation from architectural plan to structural plan. Automatic data capturing tools are also provided to interpret the geometric information of structural members in architectural plan.

Component 2: TBLoad. This component is designed for load modelling. It provides tools to assign different types of loading on a building structure. All load information is stored in the central database.

Component 3: HIPRED-II. This component is a case-based reasoning component for supporting engineers to make decisions through the past design experience.

Component 4: TBAnalysis. This component controls the structural analysis of buildings by commercial analysis software. It supports various finite element programs, such as ETABS, SAP2000, etc. The analysis results are stored in the central database.

Component 5: TBopt. It is designed as a provider of structural optimisation function. It can evaluate the original structural scheme and propose an optimised scheme by an advanced optimisation algorithm.

Component 6: TBDesign. It is designed mainly controls the detailed design. CAD tools involved are used to design structural members such as beams, columns and slabs in detail.

Component 7: E-Checking Engine. This engine is responsible for checking the code compliance of the structural plan according to codes of practice automatically.

Component 8: TBPost. This component is responsible for demonstrating and reporting. It provides interactive interfaces for users to review the results of analysis and design. Various documents such as cost reports, drawings and specifications can be generated by this component.

The above physical structure has already been implemented in an integrated CAD system for tall building design, named TBCAD. The validity of this integrated CAD system has been tested by a group of practical projects in Hong Kong.

6. Conclusion

This paper presents the development of an integrated CAD framework for the design of tall building structures. The framework supports a flexible, scalable software environment and enables various CAD tools to integrate each other and collaborate among them effectively. A process model and a building product model are introduced, which are referred to as the basis of component and data integration in the integrated CAD environment. The physical structure of an integrated CAD system for the design of building structure is also elaborated.

A group of practical tall building projects have been conducted to evaluate the feasibility of the proposed integrated CAD framework. It is shown that The proposed framework effectively benefits the development of integrated CAD systems and potentially promotes design automation in building industry. This study can be taken as a basis of the further development of an integrated CAD system for the whole building design lifecycle.

References

1. P. Brandon and M. Betts, *Integrated Construction Information*, E&FN Spon, London (1995).
2. T. Cornick, *Computer Integrated Building Design*, E&FN Spon, London (1996).
3. T. Cornick, *Computer Integrated Building Design*, E&FN Spon, London (1996).
4. G. Augenbroe, *COMBINE I Final Report*, EU DG XII JOULE (1993).
5. A. Watson and A. Crowley, *European Conference Product Data Technology Days*, Berkshire UK, 39 (1997).
6. G. Aouad, et al., *Integration of construction information*, University of Salford (1994).
7. T. Proese, *Integrated computer-aided project management through standard object-oriented methods*. PhD thesis, Dept. of Civil Engineering, Stanford University (1992).
8. J.H. Long, T.Y.P. Chang and J.S. Kuang, Proc. *9th East Asia-Pacific Conf. on Struct. Engg and Construction*, Indonesia (2003).

VIRTUAL IMPLEMENTATION OF RC FRAMES PUSH-OVER TEST BASED ON THE OBJECT-ORIENTED PROGRAMMING METHOD

YUSHI ZHOU

Civil Engineering College, Tongji University, 1239 Siping Road Shanghai 200092, China

XIAN LIU

Civil Engineering College, Tongji University, 1239 Siping Road Shanghai 200092, China

YONG YUAN

Civil Engineering College, Tongji University, 1239 Siping Road Shanghai 200092, China

Inelastic static pushover analysis is a powerful tool for the study of seismic evaluation of structures. It has been proved by overseas practical projects to be accurate and simplified. It can save both financial cost and computing time. But, there is few software packages programmed towards the analysis of pushover test. In this paper, a series of studies are carried out on the theory of reinforced concrete frame pushover test, and the whole process of pushover static test including concrete frame virtual model building, load applying, sensor installation and final numeric analysis is implemented with object-oriented method. Overall programming idea and implementation method of concrete frame pushover test are also presented in the paper.

1. Introduction

Nonlinear static analysis method is a kind of simplified method that can estimate seismic capability of structure. Under certain condition, the static pushover analysis method can estimate the seismic capacity of structures exactly and simply. The method is also described and recommended as a tool for design and assessment purposes by the guidelines of the National Earthquake Hazard Reduction Program 'NEHRP' (FEMA 273) for the seismic rehabilitation of existing buildings [1]. In Japan, this technique is also an analytical method to estimate the seismic capacity of structures. The Chinese new code for seismic design of buildings has introduced pushover analysis method too. Therefore, there is great academic and practical value to study the static pushover analytical method and apply it to practice.

At present, pushover analysis can be done in some commercial software. But these software can't show you how pushover test works. The user must change the physics model to the geometry model and form finite element model in the software. The user should be familiar with physics model and geometry model of structure and know how to simplify the physics. With the software "Virtual Digital Lab" (VDL), the users need not to know how to change the physics model to the geometry model, because we use Virtual Reality technology to simulate the whole process of pushover analysis test. If you want to do the pushover analysis test, you can only put the geometry data of the structure into the software and the program can do the residual things. VDL also can consider the damage of structure by using the Smeared Crack Model.

There are two main problems in implementing the R.C Frame pushover analysis test. One is how to implement the part of core analysis; the other is how to display the whole process of test and analyze the data. In order to make the code of the procedure reusable and better expansible, the article has selected object-oriented method to analyze it. And the program is implemented by using C++ on the basis of object-oriented programming method.

2. Object-oriented Programming Method

Object-oriented programming design emphasizes that the program should be divided into different levels and Classes. Based on this, we can define and describe the object in detail.

2.1 *The definition of object-oriented method*

The object refers to the combination of the data and code. In other words, one object or instance is a combination of data that can show the current state of the object with some functions. Object-oriented is a type of programming method which uses object, class, inheritance, capsulation, aggregation, message passing and polymorphism to construct a system. Class is the object combination, which have the same character. The class also gives open interface among the objects, through which objects interact with each other.

2.2 *The essence of object-oriented method*

In short, object-oriented program method bases on the objective things, which accords with the way of peoples' thinking. The object-oriented program method including two main parts: First, it defines the attributes of the object by using the data structure, and second, it designs the special code to implement the arithmetic.

3. RC Frame pushover test based on OOP analysis

3.1 *The principle, applied range and performance process of RC frame pushover test*

3.1.1 *The principle and applied range of pushover method [2]*

Pushover method is always used in forecasting the stress and deformation caused by seism. Its basic principle is: First, applying the lateral forces that simulate the seism horizontal inertia force in a mode of uniform load or inversed triangle load and so on to the structure analysis model, and the force is monotonic increased step by step, till it reaches the predefined state of the structure (reaching or exceeding the target displacement), and then evaluates the performance of the structure.

Pushover method is also used in seismic estimation and reinforcement of building, as well as seismic design and performance evaluation of new building structure. It can provide sufficient information for structure seismic design, such as evaluates the demand of inelastic deformation and impact on structure system behavior when the intension of some element degenerates, as well as the interformational deformation (with the considerations of intensity and height discontinuation), and verifies the loading way and so on, some of which can't be attained from elastic static or dynamic analysis [3][4].

3.1.2 *Operational procedure of pushover method*

Operational process of pushover method:

1. Prepares structure data. Includes setting up structure model, physics constant of structure and restoring force model and so on;
2. Calculates the structure internal force when applied with vertical load, and superimposes the structure internal force applied with some level of horizontal load, to judge whether the structure is crack or yield or not;
3. In the center of every structure layer, applies some kind of horizontal load distributing along the height. The size of the applied horizontal load is determined with following principles: the internal force produced by horizontal load superimposes the one calculated by step 2, which makes some or a set of components crack or yield;
4. After modifies the stiffness of the cracked or yielded components, applies next grade load, and makes another one or another a set of components crack or yield;
5. Continuously repeats step 3 and 4, till the displacement of the structure apex is large enough or the structure reaches the predefined damage limiting state.
6. Plotting the relational curve of base shearing force-top displacement, namely the pushover analysis curve.

3.2 *Implement of RC frame pushover test based on the object-oriented programming*

3.2.1 *Establishment of system class library*

Object-oriented programming technique has gradually been the main flow in the development of software technique. Object-oriented method is based on the object and class, and modular programming idea is applied in the whole software development process. We decompose the function of the target system based on the comprehensive analysis, and generalize some independent-function entity as well as its general attribute and operation, then encapsulate them into a class individually. Class that can be used repeatedly as a component has the characteristics of polymorphism, succession, high modularization and little redundancy and so on. Object-oriented method can reduce the complex degree of software, and realize the sharing code, avoid repeated code design. Object-oriented program is agile, easily maintainable, and easy to be extended and add the system function. Structure model of general class library of the system is as following figure 1.

- To romance the virtual model by calling graph engine;
- To turn the frame geometry information into finite element model and saved.

3. Finite element analysis module

Finite element analysis is the core of the reinforced frame pushover virtual test system. While finite-element

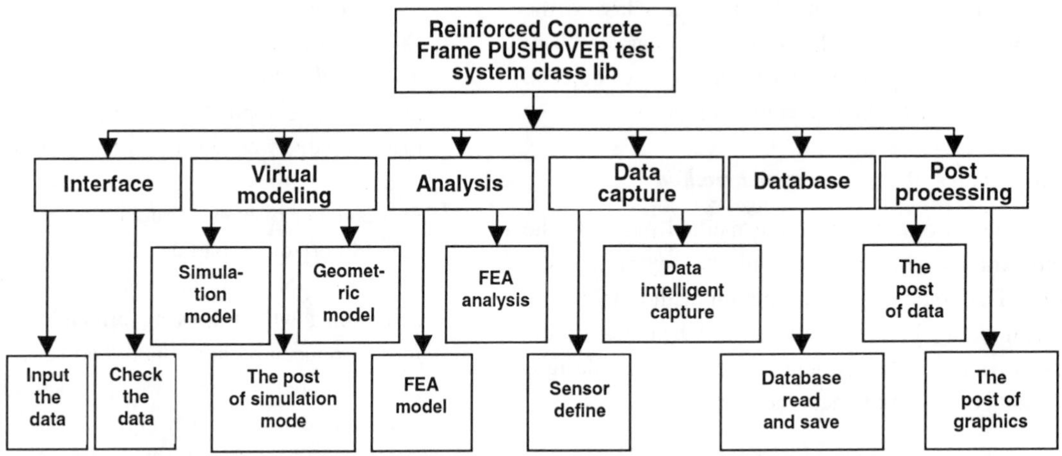

Fig.1 Generic class library of structure modeling

3.2.2 *Function module sort*

1. User inputting and checking interface

This module is an interaction window of the user and the program. User can set up the frame structure form, geometry size, material attribute of the virtual test, and confirm the collected special data information of different place on the structure. The module has special functions to check the input information. If the input information is not reasonable, the program will warn automatically.

2. Module of virtual model building

Virtual reality technique has become the focus of research recently. With the development of the capability of computer's hardware and software, it becomes possible to apply the technique to the old study of civil engineering. This module is namely applying the advanced technique to the reinforced concrete frame push over test, and set up a virtual test platform. Through the information inputted by the user, the system set up three-dimensional virtual model of the reinforced concrete frame. The module is made up of three functional parts:

- To form the three-dimensional virtual model according to the inputted information;

method develops to now, it has a powerful tool of engineering numeric analysis. Especially in the field of solid mechanic and structure analysis, finite-element method has acquired great progress. It has solved successfully many significant issues. A lot of general program and dedicated program are applied in practice. With the development of finite-element numeric method and application technology electronic computer, we can make relatively accurate elastic-plasticity analysis of reinforced structure, and present the whole developmental process of the structure internal force and the deformation, also it can describe fissure's form and spread, and the destroy process and its configuration of the structure. The system can make elastic-plasticity analysis of the frame under different loading mode, and evaluate damage level of the loaded frame and presenting the destroy process and its configuration of the structure through the introduced continuous damage model.

4. Data intelligent collecting module

In the process of the virtual test or after the finite element analysis, there will be a lot of data in the database. How to pick-up the useful data from the mass of data? It's a key problem, if the data is not well selected, it won't reflect the performance of the structure correctly and can't reach the purpose of the test. Data intelligent collecting module completes the work successfully through defining type and position of the virtual sensor, by which the point picked-up data can be conveniently found its position in the structure and it also can selectively pick-up data type according to the requirement. Correct orientation and right data type will make matter preparations for the next step's data correct analysis.

5. Graph data representation module

The two main function of the graph data representation module are:

- Presents the structure damage effect by dynamic rosy clouds graph in response to the load step number, and the structure damage progress can be seen in the cartoon;
- Presents the shear value and the displacement value of the special point by graph form.

4. The Key Technology of System Development Process

4.1 *Object-oriented finite element method*

The research of object-oriented finite element method started in the 1990s of last century. In 1990, Fenves put forward the advantages of the way of developing the project software based on OOFEM. In the same year, the research work of this field has been launched like a raging fire since B. W. R. Fordes firstly explained the concept of Object-oriented Finite Element completely in his article [5]. In reference 5, the author pointed out, the most basic object of the finite element method is some of element, node, material, shape function, gauss points, border condition and load, etc., and this thought was still adopted by most researchers so far. On this basis, a lot of foreign researchers have put forward their own object-oriented finite element model in succession. The focal point of their research concentrates on mainly:

1. The class of finite element model. Namely is used to describe the class of the finite element model, such as node, element, material, loads, the border, etc.
2. The class of finite element analysis. Namely is used to describe the class of the finite element calculation.

The key calculation part of RC Frame pushover test in this paper is completed on the basis of forefathers' research results, and it considers the particularity of RC Frame pushover test.

4.1.1 *Finite element model*

Similar to traditional finite element analysis, finite element modeling is generally divided into node, element, constraint, load and FEA domain in the object-oriented finite element method. The relationship of the classes of the finite element models in this text is as following figure2[6][7][8].

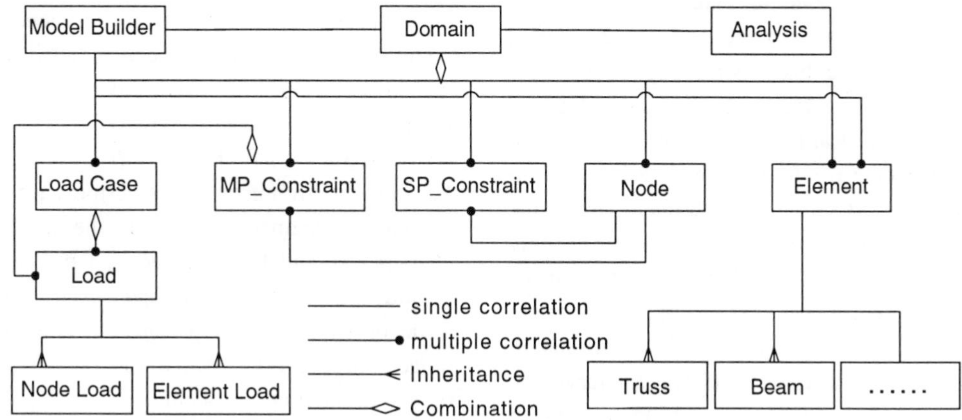

Fig.2. Class hierarchy of FEM model

4.1.2 *Finite element analysis*

To linear question, finite element analysis contains:

1. Analysis of regional discretization.
2. Formation of element stiffness matrix.
3. Formation of system global equation.
4. Introduction of boundary condition.

5. Solving the system equation to get the freedom degree of node.
6. Calculating the element stress according to the freedom degree.

The difference between non-linear question and linear question is that the system equation group is non-linear, and incremental method or iteration method is usually used to solve the equation. The relationship of the classes in the finite element analysis in this text is as following figure 3[6][9][10].

Virtual reality is the multidimensional information space formed by multimedia, and someone is participated in directly, so the establishment of the virtual environment is large-scale and complicated. Object oriented technology has good level structure, we can use the recursive analysis to these complicated system. According to the different contents in the multidimensional information space, cuts this big system of the virtual environment apart into several subsystems by certain rule. The subsystem can also continually be cut apart, until be a minimum independent object. The whole environmental system is divided into a lot of independent objects in this way. Consequently, grounds the base for the next step of describing the virtual environment.

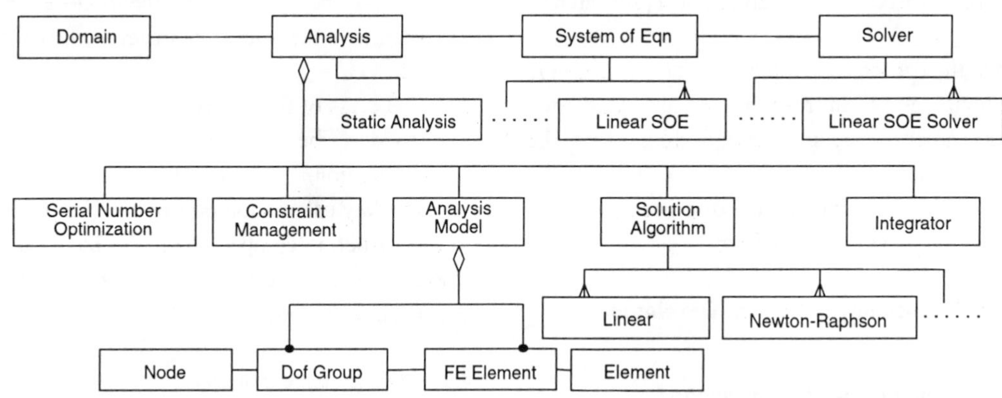

Fig.3 Class hierarchy of FEM analysis

4.2 *Virtual reality technology*

Computer simulation, which is independent of theory analysis and physics test, becomes a research method. Base on the development of the software and hardware, it already becomes possible to make use of "virtual reality technology" of the computer to build the virtual testing laboratory. As far back as the beginning of the 1980's, American VPL Research Company founder Jaron Lanier put forward the concept of "Virtual Reality". The meaning of "Reality" was realistic world or environment. "Virtual" illustrated that the environment was fictitious, structural, and existed only within the computer. Users can "enter" in this virtual environment. What is called "entering", means users are mutual with this environment by way of nature (including perceiving the environment and intervening the environment), and thus come the feeling faced unrealistic in the corresponding true environment. In other words, virtual reality is a kind of method that people make complex data visualization and interactive operation. Therefore, virtual reality can be defined as the virtual world that can be created and experienced, or the high-tech simulated system created by computer.

Because each object is influenced with the environment or the other objects, it has its own attributes and behaviors in the virtual environment. At the same time, different objects have the same or similar characteristics. All those forms rich and colorful fictitious environment. To describe each object in the virtual environment in detail, the attribute property character should be extracted and the behavioral code should be decided of each object. And then sorts out the same or similar behavior and attribute from different objects in order to realize the design philosophy of object-oriented programming.

In a word, through the object-oriented method, firstly the numerous and complicated virtual environmental data become logistic secondly the reusable degree of the data model is improved and the structure of the system is simplified. The advantage of the kind of procedure analytical method is embodied during the process of programming the RC Frame pushover virtual test system[11][12].

The overall system design model of RC Frame pushover test introduced in this paper is as following figure 4.

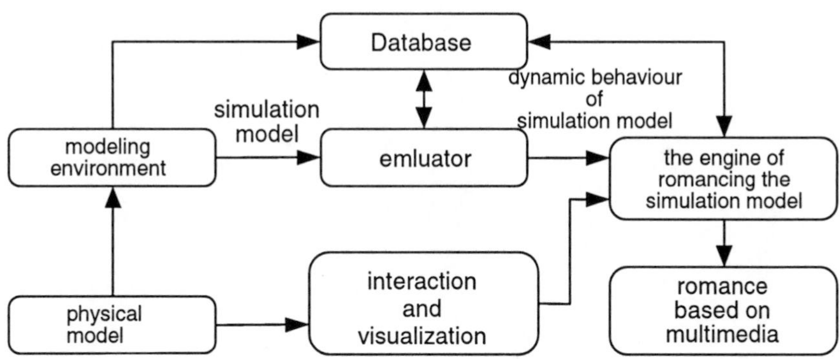

Fig.4 The system model of Virtual digital static test lab

5. Conclusion

In fact, virtual implement of the RC Frame pushover tests is the complicated system engineering, it includes not only virtual simulation of process of the test, but also the analysis of the finite element and data visualization. If software development is based on the structure design method, there will be lots of code should be modified when the function of program is changed. The maintenance and expansion of the procedure are also a question that should be faced all the time. Object-oriented method is a kind of powerful software development approach that has good feature such as capsulation, inheritance, polymorphic. Based on these features, the software can be reused and expanded. By using the object-oriented method in the software module, the design of software of structure is simplified; the ability of software meeting the demand is strengthened. Because of the encapsulation of the data, the object is relatively independent, and software gets more easily to maintain and expand too.

References

1. FEMA. NEHRP guidelines for the seismic rehabilitation of buildings. FEMA 273, Federal Emergency Management Agency (1996)
2. Yao_Qianfeng ,Chen_Ping, civil engineering structure test, China construction industry press (2001)
3. M.R. Maheri, R. Kousari, M. Razazan, Pushover tests on steel X-braced and knee-braced RC frames Engineering Structures 25,1697–1705 (2003)
4. A.M. Mwafy, A.S. Elnashai Static pushover versus dynamic collapse analysis of RC buildings Engineering Structures 23 407–424 (2001)
5. B. W. R.FORDES, FOSCHI, R. O., STEIMER, S. F. Object-oriented finite element analysis, Computers & Structure. , V. 34, n. 3, pp. 355-374, (1990)
6. Nguyen Hong Ha, Parallel Finite Element Method for Structural Analysis. PhD thesis, Tongji University (2004)
7. Carlos A. Felippa. Nonlinear finite element methods. Department of Aerospace Engineering Sciences and Center for Aerospace Structures University of Colorado (2001)
8. F.T. McKenna. Object-oriented finite element programming: frameworks for analysis, algorithms and parallel computing. PhD thesis, University of California at Berkeley (1997)
9. G.C. Archer. Object-oriented nonlinear dynamic finite element analysis. PhD thesis ,University of California at Berkeley (1996)
10. Lichao Yu,Ashok V.Kumar. An object-oriented modular framework for implementing the finite element method, Computer & Structures 79(2001)919-928
11. Yuan_XiangRong, Gao_YongLi, structural engineering and virtual reality technology, JOURNAL OF SHIJIAZHUANG RAILWAY INSTITUTE, 60~63 (2000)
12. Mason Woo Jackie Neider Tom Davis Dave Shreiner, The Official Guide to Learning OpenGL, Version 1.2

using the process of [illegible] the NC Frame [illegible] Virtual test [illegible]

The overall system design model of the [illegible] in this paper is as follows [illegible] figure 1.

Fig. 1 [illegible]

Conclusion

[illegible] structural analysis [illegible] based on the [illegible] the analysis of the [illegible] element and data [illegible] based on the structure [illegible] there will be lots of code [illegible] program [illegible] The maintenance and expansion [illegible] should be based on the [illegible] method [illegible] approach that [illegible] and [illegible] the [illegible] provide [illegible] design of software of structure [illegible] the [illegible] and software [illegible]

References

1. FEMA [illegible] for [illegible] of buildings, FEMA [illegible] Federal Emergency Management Agency (19[illegible])
2. [illegible] engineering [illegible] computation [illegible] press
3. M. [illegible] R. [illegible] M. [illegible] Frames [illegible] Engineering Structures 2[illegible]
4. A.M. [illegible] Structures [illegible] analysis of RC Buildings, [illegible] (2009)
5. T.W. [illegible] Computers & Structures [illegible]
6. [illegible] Structural [illegible] (2009)
7. [illegible] University of [illegible]
8. [illegible] object-oriented [illegible] programming [illegible] and parallel computing [illegible] University of California at Berkeley (1997)
9. [illegible] object-oriented [illegible] finite element [illegible] University of California at Berkeley (1998)
10. [illegible] object-oriented [illegible] for implementing the finite element [illegible] Computers [illegible] (2001) [illegible]
11. Yang [illegible] and [illegible] JOURNAL OF [illegible] INSTITUTE [illegible] (2003)
12. [illegible] Studio [illegible] Guide to Learning [illegible] Version)

Innovative Technology

OPTIMIZATION OF SCAFFOLD SUPPORTING SYSTEM FOR SLAB TRANSFER STOREY

SUQIN CAI
Shanghai COSCO-Liangwan Property Development Co. Ltd
Shanghai 200061 , China

ZHIMING CHEN
Shanghai Yanlord property Co. Ltd
Shanghai 200122 , China

YULIN HUANG
Department of Geotechnical Engineering, Tongji University, 1239 Siping Road
Shanghai 200092, China

Transfer storey itself is studied more than supporting system in constructing it. This paper bases on an actual project to study scaffold supporting system of thick reinforced concrete slab, which transfers loads of superstructure to the lower frame. Through measuring the stresses in the slabs and forces of supporting bars on the spot, the characteristic curve of regularity of developing of concrete strength and unloading of steel pipe supporting are acquired. Analyzing model is set up in the light of finite element method. During analyzing with FEM, contribution of concrete-filled tube to the stiffness of supporting system is included, while Young's modulus and strength of concrete evolving with various ages are considered. Contrasting the results with FEM with data from measuring, the same conclusions are arrived at. Based on this, Scaffold supporting system of top floor of basement is optimized and some constructive suggestions for the actual project are provided.

1. Preface

High-rise buildings often set up transfer storey to meet demands of structure function [1]. Transfer storey itself is studied more than supporting system in constructing it. Slab transfer storey has a great self-weight; a large number of scaffolds must be laid while constructing it in order to ensure quality and security of slab transfer storey. Those scaffolds are often extended till base plate of basement to shift load of transfer storey. Whether basement has already formed structure system can well function and partly bear load of transfer storey needs to be verified through testing on the spot and analysis in theory. With ages of concrete of transfer storey increasing, strength of it is improved and plays function of structure gradually; load burdened by transfer storey is worth studying. How to lay scaffolds is economized under ensuring structure safety.

Based on "old distribute reconstruction of two gulf one apartment" project, testing on the spot and research in theory were carried out to solve those problems.

2. Project Background

In the first period of "old distribute reconstruction of two gulf one apartment" project, five thirty-four-storey buildings are all frame-shearwall structure. The lower frame column is steel tube concrete column and transfer structure is slab transfer storey. The load of superstructure is firstly transferred to the lower frame column, then to base plate of basement. This position is a throat of the whole structure and has quite high demand in constructing.

While constructing slab transfer storey, load of it is greater than carrying capacity of lower structure platform and ordinary scaffold supporting system, so load of transfer storey should be transferred. Construction unit decides the following construction scheme through computing [2-5]: top plate of basement, middle slab and transfer storey slab have same scaffold supporting layout, which require scaffolds to be aligned from top and bottom and vertical. Scaffold space is 300 ×600mm, 300mm in structure length direction and 600 in vertical to length direction, which is shown in Figure 1.

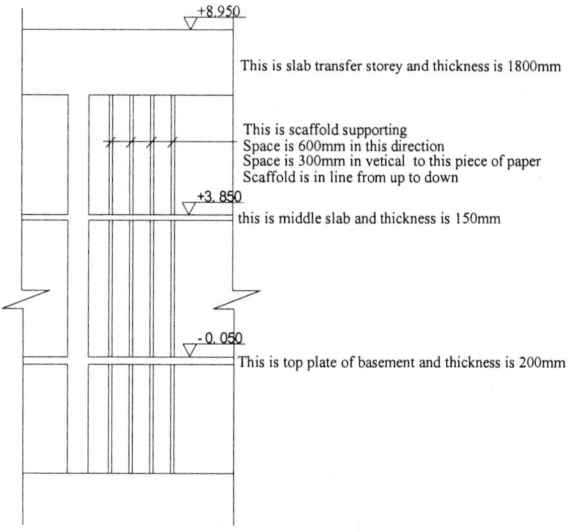

Figure 1. Scaffold supporting system

3. Optimization of Scaffold Supporting System

3.1 *Analysis organization*

1. Strain gauges are laid on scaffolds of top plate of basement, middle slab and transfer storey slab, testing stress and force of scaffold supporting system and acquiring the characteristic curve of regularity of developing of concrete strength and unloading of scaffold supporting.
2. Finite element models of top plate of basement, middle slab and transfer storey slab are established through FEM software—SAP, computing stress and force of scaffolds in various ages of concrete of transfer storey. The computing results are contrasted with testing results, if the two results are identical, optimization of scaffold supporting system is being done, if not, FEM model and parameters will be modified until computing results are identical with testing results, then optimization of scaffold supporting system continue to be run.
3. Scaffold supporting is laid again, the optimized scaffold supporting scheme is chosen through analysis of FEM software.

3.2 *Testing on the spot*

The main content of testing on the spot includes stress and force of scaffold supporting with various ages of concrete of transfer storey. The testing positions of top plate of basement are from 1 to 13, as shown in figure 2. The testing positions of middle slab and transfer slab are the same to ones of top plate of basement in plane face and serial numbers are from 14 to 26 and from 27 to 39.

Strain gauges are installed after moulding boards of transfer storey are set up. Before concrete of transfer storey is cast, initial strain of strain gauge is noted. After concrete is cast, the strain is noted as the beginning day's strain. When concrete reached final set, the strain is noted as the first day's strain. Then noting times of strain is the third day, the seventh day and the fourteenth day after concrete of transfer storey has been cast.

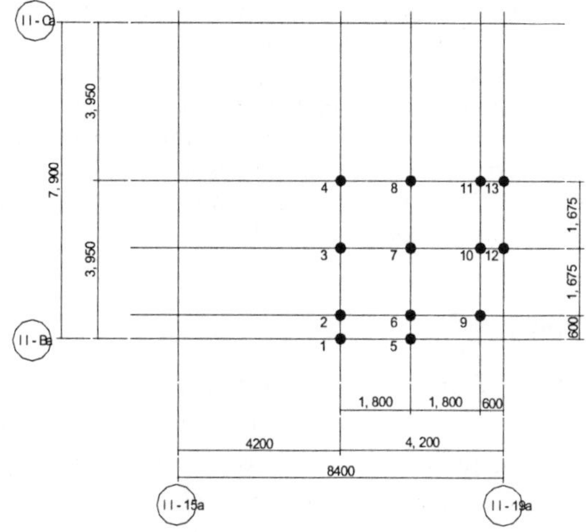

Figure 2 . Layout of scaffold supporting

3.3 *Analysis in theory*

3.3.1 *Element mesh*

One fourth slabs ranging from axeII - 15a to axeII - 19a and from axesII - Ba to axes II - Ca are finite element computing models of top plate of basement, middle slab and transfer storey slab. Based on layout of scaffold supporting, elements are meshed, space 300mm in structure length direction and space 600 in vertical to structure length direction. Thus, element of top plate of basement is 300×600×100mm, element of middle slab is 300×600×75mm and element of transfer storey is 300× 600 × 300mm. Figure 3 shows preprocessor model transfer storey slab after element is meshed in finite element software—SAP.

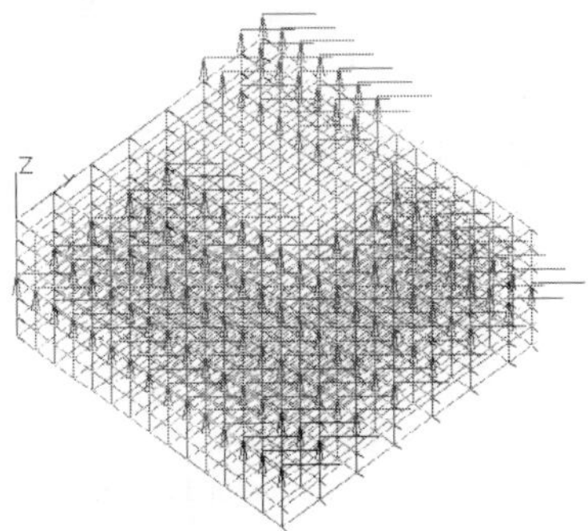

Figure 3. Preprocessor modal after meshing

3.3.2 *Boundary condition*

1. Scaffold bracket

Upright brackets provided by scaffold supporting function in positions of scaffolds, that is, positions of element nodes. Scaffolds are act as elastic spring and elastic spring stiffness values adopt the following formula's values.

$$f_s = k\frac{E_s A_s}{l} \quad (1)$$

Inside this formula:

k——reduction coefficient based scaffold fastener and flexible plate, value is 0.5;

Es——elastic modulus of scaffold;

As——section area of scaffold;

l—— length of scaffold.

2. Reinforced concrete column bracket
   a) Reinforced concrete column of top plate of basement and middle slab

When concrete of transfer storey is cast, ages of concrete of top plate of basement and middle slab both reach 28 days. Their strength adopts 28 days' strength. $E_C = 3.6\times10^{10}\text{N/m}^2$. Column section $A_C = k\times0.6\times0.6\text{m}^2$. k is expanding coefficient of column section, whose value is 1.5. Provided that slab has 6(2×3) nodes around it, the formula(1) is modified as the following formula.

$$f_c = k\frac{E_c A_c}{6\times l} \quad (2)$$

b) Steel-tube concrete column of transfer storey

With ages of concrete of transfer storey increasing, bonding strength of steel tube and concrete evolves. The following formula gets bracket stiffness.

$$f_{sc} = k\frac{\alpha E_s A_s + \beta E_c A_c}{6\times l} \quad (3)$$

inside this formula α、β is time-related coefficient and stand for contribution of concrete-filled tube to the stiffness of supporting system in various ages, their values is decided by the following formula.

$$\alpha = \frac{t}{5}\times 1.0 \quad t\in(0,5) \quad \text{others}\ ,\ \alpha = 0.8 \quad (4)$$

$$\beta = \frac{t}{14}\times 0.8 \quad t\in(0,14) \quad \text{others}\ ,\ \beta = 0.8 \quad (5)$$

3. Stiff bracket

Symmetrical simplification of concrete slab/plate is considered and stiff brackets are added around it.

3.3.3 *Load*

Load of basement and middle slab uses testing force of above scaffold and Load of transfer storey uses self-weight.

3.3.4 *Material parameter*

Using SAP to compute, in preprocessor interface size of modal, bracket and load information must be inputted, and then computing file is compiled. Element information need input Young's modulus of C40, top plate of basement and middle slab use 28 days' Young's modulus and Young's modulus of transfer storey is various with ages of concrete.

3.3.5 *Computing content*

1. Stress of scaffolds of top plate of basement is computed under testing force of above scaffolds. Computing time is 0, 3, 6, and 14 days. Middle slab is same to top plate of basement.
2. Stress of scaffolds of transfer storey is computed under self-weight for 1, 3, 6, 14 days after casting of concrete of transfer storey.

3.4 *Contrast analysis*

Figure 4~Figure 9 shows fitting curve of testing value and computing value of scaffolds of top plate of basement, middle slab and transfer storey. The fitting curve of testing value reflects the regularity of developing of concrete strength and unloading of scaffold supporting. Comparing testing value with computing value, both curves are almost identical, so concrete strength is accord with real condition. It is reasonable that the above finite modal and boundary conditions analyze scaffold supporting system of this project.

From testing strain curve and computing strain curve, we may see than concrete strength develops quickly and scaffolds unload quickly for six days after casting of concrete. After these six days, unloading of scaffolds becomes slowly with concrete developing. The result is expected.

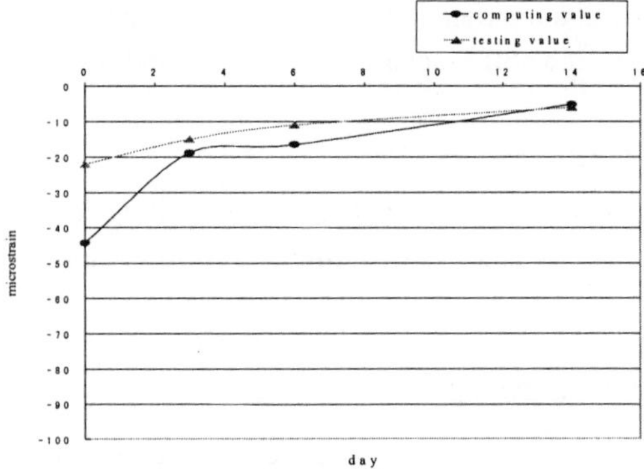

Figure 4. Fitting curve of testing value and computing value testing point 4 of scaffolds of top plate of basement

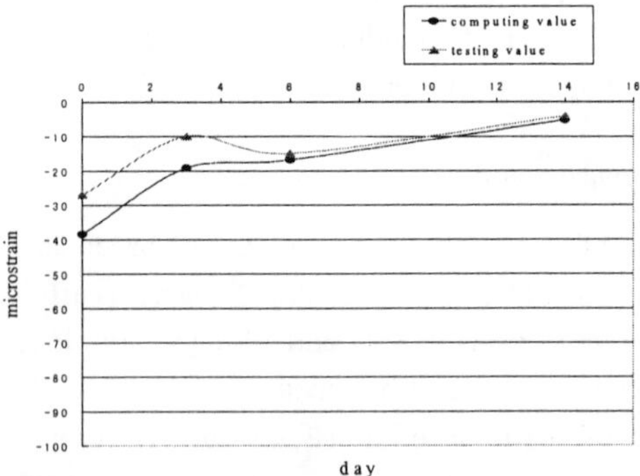

Figure 5. Fitting curve of testing value and computing value testing point 13 of scaffolds of top plate of basement

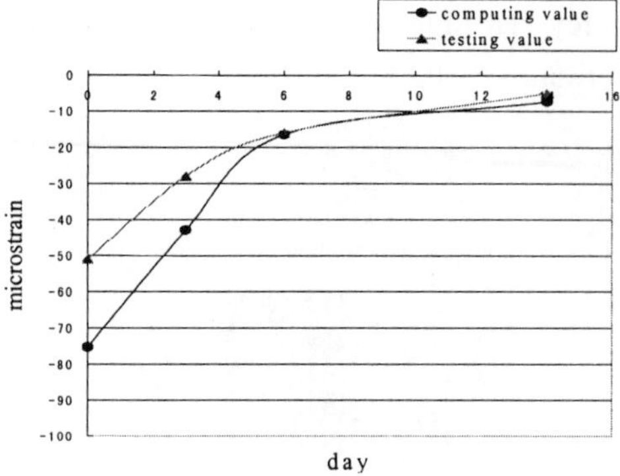

Figure 6. Fitting curve of testing value and computing value testing point 21 of scaffolds of middle slab

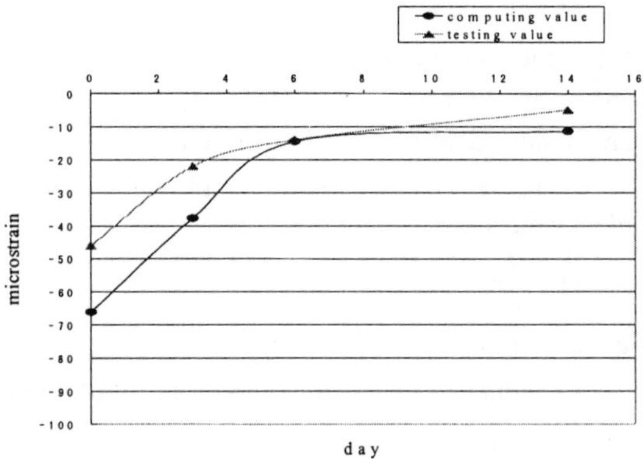

Figure 7. Fitting curve of testing value and computing value testing point 23 of scaffolds of middle slab

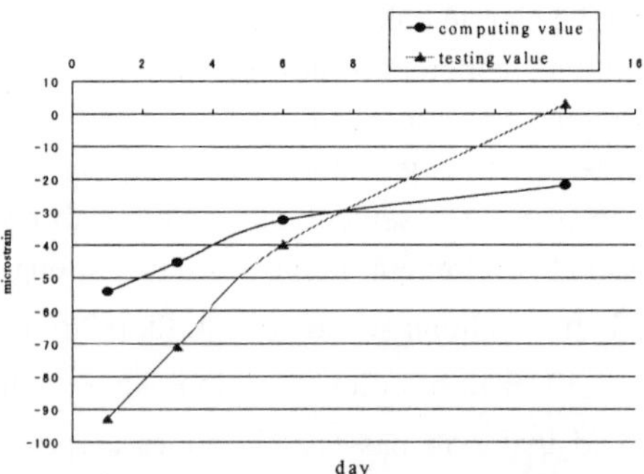

Figure 8. Fitting curve of testing value and computing value testing point 27 of scaffolds of transfer storey

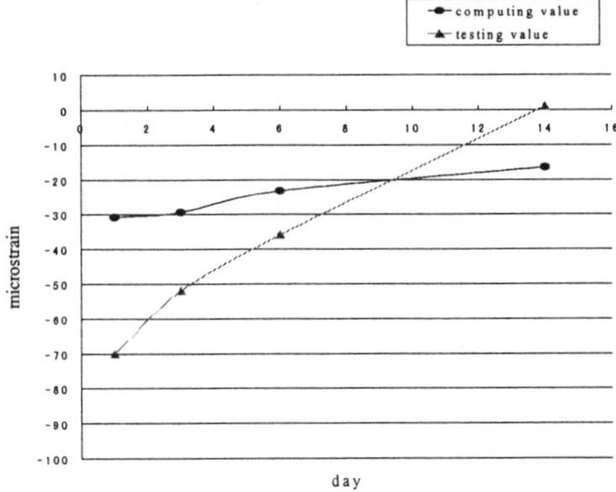

Figure 9. Fitting curve of testing value and computing value testing point 30 of scaffolds of transfer storey

3.5 *Optimization of scaffold supporting system*

When transfer storey is cast completely, load of scaffold supporting system is greatest; this time is the most dangerous period for scaffold supporting system. It is necessary that scaffold supporting layout of top plate of basement was analyzed at this time. Firstly, whether scaffold supporting system meets structure safe demand should be checked. Secondly, scaffold supporting layout should be optimized.

When top plate of basement is analyzed, its computing modal uses the above finite element modal. The top plate of basement is applied testing force of middle slab as soon as transfer storey is cast completely and the greatest moment and displacement of span centre of top plate of basement are computed under several scaffold supporting forms (including no supporting). Table 1 displays computing results.

Table 1. Computing results of various supporting forms

supporting form	greatest moment of span centre	Greatest displacement of span centre
no supporting	73.5kN·m	2.53×10^{-2}m
900×600	2.05kN·m	1.42×10^{-4}m
600×600	0.53kN·m	9.19×10^{-5}m

According table 2, reinforcing bars of top plate of basement is computed under several scaffold supporting forms. Reinforcing bars will be 2197mm2/m without supporting system, but actual reinforcing bars are 753mm2/m and this kind of supporting system doesn't meet reinforcing bar demand. Under 900×600 scaffold supporting, reinforcing bars will be 57.3mm2/m and meet reinforcing bar demand. So 900 × 600 scaffold supporting system is suggested to this project.

4. Conclusions

Scaffold supporting system of slab transfer storey is analyzed and scaffold supporting layout of top plate of basement is optimized. The valuable suggestions are provided for actual project. Thus, under ensuring structure safety, construction is more convenient and construction cost is more economized. The kind of scaffold optimization of transfer storey is worth using reference by other same project.

References

1. Tang Xingrong. Structure design and construction of transfer storey of high-rise building [M]. Beijing: China-building industry press (1987)
2. I. A. Macleod, D. R. Green. Frame idealization for shear walls supports systems, the structure Engineering No.2 Vol.51(1983)
3. F. R. khan, N. R. Analysis and Design of Framed Tube structures for Tall concrete Buildings, Publication SP-36, ACI (1973)
4. L. Gerny, R. Leon. Column-supported Shear walls, Concrete Framed structure. Stability and Strength (Edited by R. Narayanan), Elsevier Applied Science Publishers (1986)
5. S. Y. Xue, W. L. Du. Design and Analysis of Transitional Storey in Tall Buildings, From 《International Conference on Tall Buildings》, Vol.2 (1988)

DYNAMICAL PROPERTY ANALYSIS OF SELF-CLIMBING SCAFFOLD IN HIGH-RISE BUILDING

CAN CHEN

College of Civil Engineering, Tongji University, 1239 Siping Road, Shanghai 200092, China

XIAOSONG XIE

College of Civil Engineering, Tongji University, 1239 Siping Road Shanghai 200092, China

WEI XU

College of Civil Engineering, Tongji University, 1239 Siping Road Shanghai 200092, China

Self-climbing scaffold attached to structure is widely used in high-rise building construction. It is significant for wind resistance and construction safety to learn of the dynamical property of self-climbing scaffold attached to structure. In this paper, the finite element model of scaffold is founded and the dynamical property under several constructing conditions is calculated. The results indicate the dynamical loads under self-climbing state should be considered in the design of self-climbing scaffold in high-rise building. At last, some advice is brought forward: that we can choose only a scaffold as the calculate object and the exaltation state maybe the most disadvantage loads state.

1. Preface

Self-climbing scaffold attached to structure is widely used in high-rise building construction. The self-climbing scaffold is attached to the structure, depending upon own lifting equipment to rise and descent. The suspending scaffold can meet the need of the craftwork and safety defenses of the workers' operation outside of the building when the building is structure constructing, fix constructing and decorate constructing. It is developed in the late of 1980' and the beginning of 1990' from the suspending and hanging scaffold. It is a new type of scaffold, which is suit for the constructing of high-rise building, especially in the super high-rise building. Mastering the mechanics performance of self-climbing scaffold is significant to the safety in the construction of high-rise building.

In this paper, self-climbing scaffold is used in an emporium structure construction (235.4 m). The scaffold unit is composed of eight vertical frames and the horizontal truss. It is 17m high. The frames are welded with φ48×3.5 seamless steel pipes. The connection between the frames is by screw. The bearing point is set at the beam of cast concrete in site, overhanging beam is 2M24, braces are connected with structure by 2M30 through-wall screw and the braces all are used in φ30 round steel, when scaffolds are lifted and used.

Guiding wheel and guiding track in self-climbed scaffold can guide, prevent declining. At the same time, we can set four layers of hard-drawn knots to reinforce the scaffold. The knots are composed of scaffolds pipes and right-angle fasteners.

2. Dynamic Property Analysis

This paper calculated the dynamical property of scaffold under two load cases by the finite element method. Under load case 1, the scaffold is lifted to the end and the cantilever is 8.00 meter long. Under load case 2, the scaffold is used in construction. In order to analyze the difference of dynamic property, different numbers of scaffolds from 1 to 3 are calculated in every case. The finite element model of every case is shown in Figure.2.

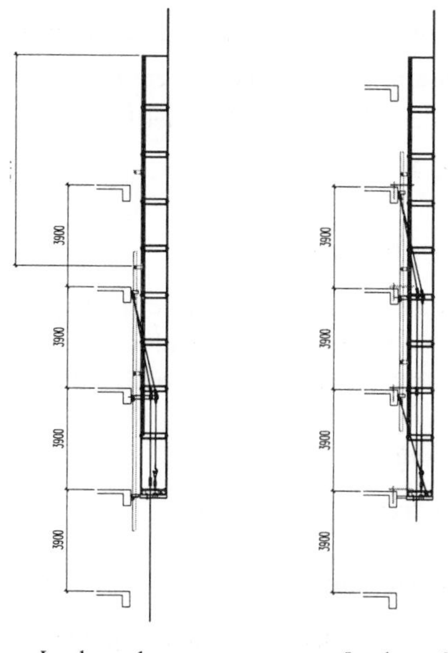

Load case 1 Load case 2

Figure.1 Calculation load case

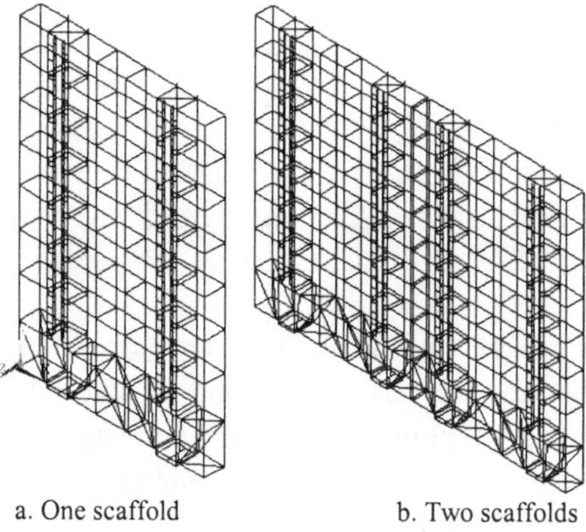

a. One scaffold b. Two scaffolds

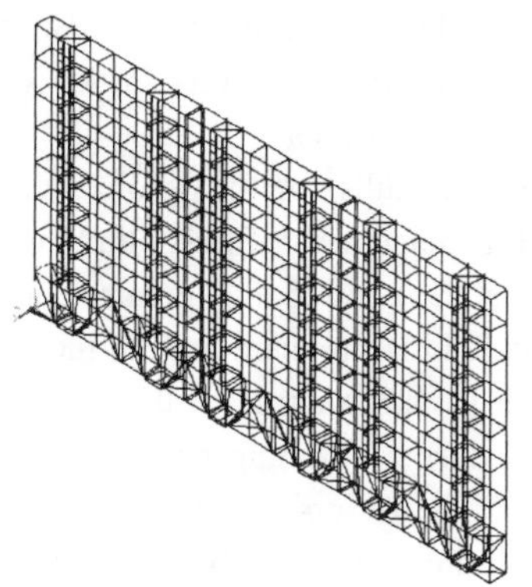

c. Three scaffolds

Figure2. The finite element model

The former 3 vibration frequencies of every case are shown in Table 1. The result indicates that the vibration frequency in the lifting stage is larger than it in the using stage. The scaffold in lifting stage has a longer cantilever and less attached bearing than in using stage, so the stiffness is comparatively less. Therefore, two cases must be considered in the mechanical calculation of self-climbing scaffold.

Table 1. Vibration frequency of scaffold under all kinds of load case

Load case		Vibration frequency (HZ)		
		ω 1	ω 2	ω 3
One scaffold	①	1.996	2.093	2.519
	②	3.116	3.259	4.136
Two scaffolds	①	1.960	1.990	2.480
	②	3.118	3.163	4.136
Three scaffolds	①	1.961	1.974	2.483
	②	3.118	3.142	3.886

The result is more accord with the reality when more scaffolds are considered in calculation, in theory. However, when many scaffolds are considered in calculation, too much computer resource is used. The result shows: more scaffold considered in calculation, the integer effect is lager, the stiffness is bigger, and the vibration frequency is less. The results indicate that the difference of vibration frequency between a scaffold and two scaffolds is 1.80% in lifting, 0.06% in construction application; that the difference of vibration frequency between a scaffold and three scaffolds is 1.75% in lifting, 0.045% in construction application. So, when we found a finite element model to analyze scaffold, we can choose one scaffold to calculate.

3. Integer Mechanical Analysis

In order to analyze the integer mechanical property, the finite element model is founded with one scaffold.

3.1 *Calculation of wind load*

The wind load standard value is calculated as formula (1):

$$w_k = k\beta_z \mu_s \mu_z w_0 \quad (1)$$

3.1.1 Basic wind pressure and the coefficient of wind pressure along altitude

According to regulations [1] and code [2], scaffold should be able to withstand class 6 wind in case 1. So the basic wind pressure is $w_0 = 0.11 kN/m^2$.

However, scaffold should be able to withstand class 11 wind in case 2. So the basic wind pressure is $w_0 = 0.55kN/m^2$ and the coefficient of wind pressure along altitude is μ_z=2.54.

3.1.2 Integer bodily form coefficient

According to the fourth literature reference, considering the most disadvantageous effect of security web and baffle to ward off wind, we assume that the counter-wind coefficient of scaffold is $\varphi = 0.55$. So $\mu_z = 1.3$, $\varphi = 0.715$.

3.1.3 wind-shake coefficient

According to literature 3, the wind-shake coefficient β_z of scaffold in lifting can be calculated as following formula:

$$\beta_z = 1 + \frac{\sqrt{\eta^2\xi_1\lambda_1 + 2\varphi\eta\xi_2\lambda_2 + \varphi^2\xi_3\lambda_3}}{\mu_{zn}} \quad (2)$$

Here, η is the ratio of integer bodily form coefficient of structure and the wind-shake coefficient of scaffold; φ is the counter-wind coefficient of scaffold; ξ_1、ξ_2、ξ_3 are the pulsatile dynamical coefficient when building and scaffold act together and λ_1、λ_2、λ_3 are the pulsatile dynamical coefficient considering pulsatile wind pressure and the space relativity. The base vibration frequency ϖ_1 can be calculated according a calculation result table in literature 3. In the end, the wind-shake coefficient under load case 1 is $\beta_z = 1.76$.

The wind-shake coefficient β_z under load case 2 is calculated as following formula:

$$\beta_z = 1 + \frac{\xi\nu\varphi_z}{\mu_z} \quad (3)$$

Here ξ is the pulsatile accretion coefficient, ν is the pulsatile accretion coefficient of high-rise building, φ_z is the vibration modal coefficient. In the end, the β_z is 1.95.

3.2 *Construction load*

Under load case 1, in the level projection area of busywork, the construction load is 0.5 kN/m^2, and under load case 2, the construction load is 3.0 kN/m^2

3.3 *Result*

The results are show in table 2.

4. Conclusions

Calculation results for dynamical properties show that: natural vibration frequency of Self-climbing scaffold during climbing is greater than it in construction. And restraints of scaffold in two cases are different. So in mechanical analysis on this kind of scaffold, two cases including climbing and application stages should be considered. And also the number of scaffold in calculation model has little effect on analysis results. Thus a scaffold calculation model can be used in the mechanical analysis of this kind of scaffold in engineering applications.

Analysis Results for the whole structure show that horizontal displacement of Self-climbing scaffold under construction and wind load is greater than the vertical of it. So wind load is the main load in the structural design of scaffold. In the example of this article, displacement of scaffold during climbing stage is bigger than that in application. The results approve that as for this kind of scaffold, climbing stage is likely the worst case during all cases.

Table 2. The results under whole loads

Load case	Standard value of wind load (kN/m^2)	Max level displacement (mm)	Max vertical displacement (mm)	Max stress (MPa)
①	0.351	102.46	5.41	200.30
②	1.363	48.85	4.06	210.16

References

1. Code for Engineering Construct in Shanghai. Safety Technology Regulations for Self-climbed Scaffold

Attached to Building in Building Construction (DG108-905-99) [S], Shanghai, 1999
2. Ministry of Construction P.R.CHINA, Load Code for Design of Building Structures (GB50009-2001). [S], Beijing, 2001
3. Yue Feng and Li Guoqiang. Wind load calculation of self-climbed scaffold attached building in high-rise building. Architecture Technology. 2004(8)
4. Zhang Yi and Yuan Yong. Analysis and calculation of self-climbed scaffold attached building in super high-rise building. Building Construction. 2000(6)

WORKING STATE ANALYSIS AND HYDRAULIC SYSTEM DESIGN FOR EXCAVATOR

RI YONG UK, ZHAO LIU
College of Mechanical Engineering, Tongji University, 1239, Siping Road Shanghai 200092, China

CAIPING HUANG
Qidong High Pressure Oil Pump co., 188, South Heping Road Qidong 226200, China

Excavator is a kind of multifunctional build-construction machine, which is widely used. Its working condition is bad and its working state is complicated. Hydraulic system is its kernel parts, which plays a very important role to the capability of an excavator. On the basis of analysis excavator's usual working state, we made a detailed analysis about some particular working states and hydraulic systems, after that, we make a comprehensive conclusion about the design requirements of an excavator. According to the analysis of the design requirements for the hydraulic system, this thesis integrates loading sensing and electric control and any other technologies, designs a new kind hydraulic system of excavator, promotes the function of energy-saving and manipulative characteristics more.

1. Status of the Development of Excavator Hydraulic System

The excavator's development is with a long history, it can retrospect to 1840. But there was nearly no development after then, just be used in the field of mine.

The main reason of the slow development of excavator is: its working equipment's motion is complicated, its working area is wide, and need to apply multi-freedom mechanism, the age-old machinery drive is not proper for it. Of that time, engineering construction is mainly used in the development of country, and cosmically road-building and field-refitting, they were all large-area's horizontal operation, so there was little application to excavator, which limited the development of excavator in some degree.

Since 1960s, the technology of hydraulic drive was applied on excavator, the excavator's hydraulic system has developed to a quite matured stage. In recent years, there are five main aspects of the development of hydraulic system:

1. Hydraulic system changed from open load sensing system (OLSS) to close load sensing system (CLSS).
2. The technology of energy-saved becomes the main aspect of study.
3. The development of high –pressure and high-reliability in system is more and more obvious.
4. The system's control characteristic has been up to a key position
5. The integer of hydraulic system and electron control is becoming popular.

2. The Working Condition of Hydraulic Excavator

The working procedure of a hydraulic excavator include the following actions (Figure 1): boom's up and down, arm's holding out and in, bucket's loading and unloading, swing platform's turning, the moving of the whole machine and some other assistant action, assistant actions need not use full-power driving, the other motions are main actions, which need full-power driving.

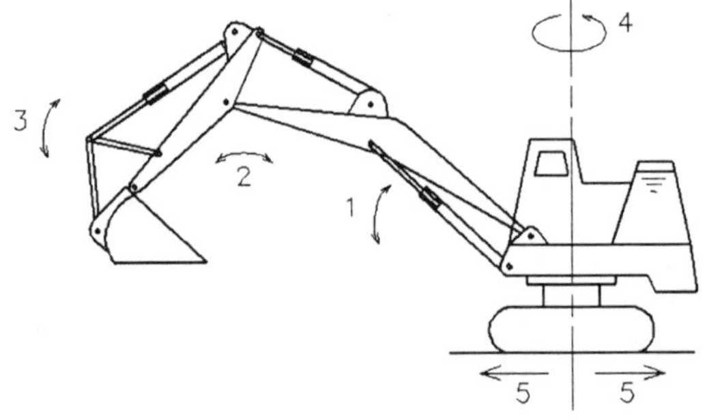

Figure 1. The working motion of a hydraulic excavator.

(1-boom's up and down; 2- arm's holding out and in; 3- bucket's loading and unloading; 4- swing platform's turning; 5- whole machine's moving)

Because the working objects and condition a hydraulic excavator are changeable, there are two special requirements about host machine:

1. When carry out all main actions, resistance and working speed are changed at any moment, so we have to make the pressure of hydraulic cylinder match hydraulic motor's pressure and flux.
2. In order to make good use of the power of the engine and shorten the time of circle act, it need two main motions (for example, excavator and boom moving, raise and spin) to be compounded at the same time.

Composition of a circle task and the compounding of motions of a hydraulic excavator include:

1. Excavation: usually apply hydraulic cylinder of bucket or arm carry out the excavation, or this two work together, so we can say, during this process, it is a compounding action between bucket and arm, if necessary, we can match the motion of moving boom.
2. Fully-loaded bucket raising and swing: after excavation, boom hydraulic cylinder lift the boom, fully-loaded bucket is raised, at the same time, swing hydraulic motor makes the swing platform turn to the position of unloading, at that time, it is the compounding action of boom and swing platform.
3. Unloading: turn to the position of unloading, swing platform is applied the brake, adjust the radius of unloading with the arm hydraulic cylinder, then the bucket hydraulic cylinder retracts, bucket is unloaded. In order to adjust the position of unloading, it need the assistant of boom hydraulic cylinder, at that time, it is a compounding of arm and bucket.
4. Backing out of empty bucket: after unloading, swing platform turns back inverted, boom hydraulic cylinder and arm hydraulic cylinder match each other, put the empty at a new site of excavation, at that time, it is a compounding of swing platform, boom and arm.

3. The Designing Requirement of the Excavator's Hydraulic System

The motions of hydraulic excavator are very complicated, and there are many mechanics, for example, there are travel mechanism, swing mechanism, boom, arm and bucket and so on, it is a kind of construction machine with multi-freedom. These mainly mechanisms are always started, braked and commutated, often meet impacting and vibration, so it is demanding to excavator's hydraulic system. According to the working characteristic of a hydraulic excavator, the design of the hydraulic system needs satisfaction of hereinafter requirement.

3.1 *The capability of power*

The requirement of power capability is under the premise of non-overloading of the engine, try to make good use of the power of engine, improve the efficiency of excavator's production. Especially, when the load change, it needs all right matching of hydraulic system and engine, try to advance output power of engine. When the outside loading is small, we hope improve the component's moving speed by increasing the output flux of oil. Double-pump hydraulic systems always apply the confluent flux to improve the efficiency of engine's power.

3.2 *The capability of power*

3.2.1 The demanding of speed-adjusting

Excavator is demanding to the control capability of speed-adjusting, how to fulfill the action of speed adjusting control according to the purpose of the driver. It becomes a very import part of the designing about hydraulic system that if it is reliable to operate the speed-adjusting of working component. Excavator has big change range during working process, all different working conditions require changeable power, so all component needs good capability of speed-adjusting.

3.2.2 Requirement of compounded maneuverability

It needs every executive unit work alone when excavator is carrying out some tasks. However, it requires that the executive units work together to implement some compounded tasks. Therefore, how to achieve the compounded action of the multi-executive units is the requirement of excavator hydraulic system maneuverability.

When multi-executive units work together, it requires that they won't interference each other. They can be assigned the flux of executive units to achieve ideal compounded action. Especially the compounded action of left and right of travel motor for running gear, the road holding is an important aspect in designing. If the speed of one side motor decreases because of hydraulic pump oil diffluence supply during the excavator runs, it will lead to excavator crooked running. It is easy to cause an accident.

On the other hand, when multi-executive units work together, every control valve all works under the condition of big-open , and often appear the condition of the total flux demand exceeding the maximal supplied flux of pump, in such case, executive units will reduce their motion speed, even stop, in order to supply the oil to low-pressure executive units. Therefore, when multi-executive units work together, how to supply flux to each component, is to need advisement in the design of excavator's hydraulic system.

3.2.3 Requirement of energy-saving

Because excavators have to keep long time working, large energy consumed, they need high-efficiency hydraulic system, we have to reduce the energy-consumed of every executive unit and pipeline, so we should fully consider every kinds of measure about energy-saving.

When we control every executive unit's speed-adjusting, the demanding flux of the system is larger than that of the oil pump's output, then consequentially caused some losing of flux. System hopes this losing flux as little as possible. When the excavator is under the condition of non-loading, how to reduce the output of the pump, reduce the pressure of the return oil when empty-loaded, it is also a key measure of reducing energy-consumed.

3.2.4 Requirement of security

Excavators always work under bad condition, the change of load and impacting vibration are big, the hydraulic system need well-protected measure about over load, in order to prevent the oil pump from being damage because the oil pump is over load and outer load impact every hydraulic executive unit.

3.2.5 Some other requirement of capability

Achieve all components and parts' standardization, groupware and generalization, and reduce the cost of manufacturing of excavators. Hydraulic excavators always work under bad condition, every functional units need very high reliability and durability; because excavators are widely used in city's construction, we should improve its work performance, reduce its vibration and noise, pay more attention to its capability of environment protection.

4. The Design of Hydraulic System of Excavators

Load-sensing and pressure compensated system is attached importance control way in our times, this system can satisfy power units and executive units match each other. The main characteristic is that it can control the oil pump according to the load and the demand of speed-adjusting, consequently, supplying oil with the demand, which make the losing of speed adjusting and flux-saved is under a small value, and the effect of energy-saving is obvious.

This design aims at applying general and multiple unit valve system, design a suit of load- sensing and pressure compensated system equipped with special control valve and simple electro- control system.

The condition of hydraulic excavators are complicated and multiform, if we want the system can supply the pressure and flux which the working equipments demand, we have to limit the highest pressure of the system, power and control the executive units timing, so we hope the performance of output as the show in Figure 2.By analyzing the figure, in order to gained such output characteristic, the hydraulic system excavators should have the following control function: supplying oil according to demand; when the demanded flux is reduced to zero, the pressure of the system is down to the lowest, and reduce the losing of flooding; the limit of the highest-pressure; prevent the system from impacting of pressure; when the system need flux is too big, pro rate limit every equipment's flux ; limit the engine's torsion.

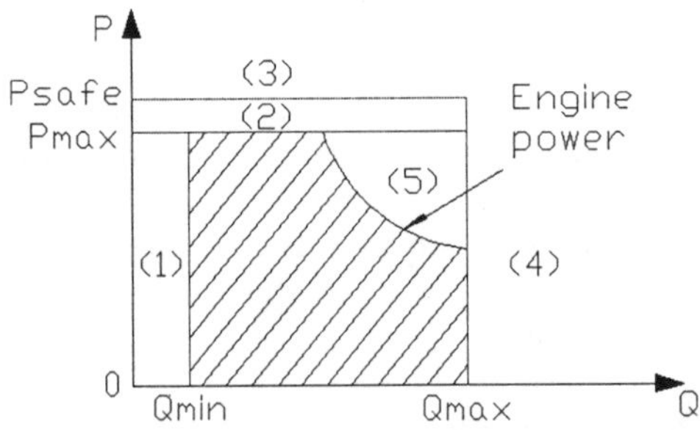

Figure 2. Output characteristic of a excavator's hydraulic system.

The designed load-sensing and pressure compensated system composed of five parts: combined multiple unit valve, control system's integrated valve, variable pump control system, the loop of swing and the loop of traveling. The whole system applies a variable pump with load- sensing control to supply pressure oil for every working equipments. In order to supply stable control pressure for every control system, make the control pressure avoid being affected by the load of main oil lines, and make the control more reliable, all forerunner oil line's pressure oil are served by one quantitative pump.

Combined multiple unit valve choose the multiple unit valve slice with quadratic pressure feedback LS control oil orifice.

As the Figure 3 shows, control system integrated valve combine the control of secondary pressure feedback LS, reducing the losing of system's flooding, limiting of the highest pressure and any other functions.

As the Figure 2 shows, when the working point of load demand enter part (1), because of the structure of the body of pump, the flux of the pump is impossible to down to zero, it always keep a minimal flux value. Then the rest flux will flooding through relief valve, which cause the losing of flooding energy. In order to reduce this losing, we should reduce the flooding pressure. in order to reduce the losing of flooding energy, consider setting a on-off valve in the system. When the system pressure Pp is bigger than the highest pressure of load, the on-off valve will be open, and unload the oil pump. Under normal condition, this valve will not be open. No other than the condition of very small flux which the system need, the on-off valve will be open, then the system pressure Pp is far smaller than that of opening the relief valve. In that case, the losing of flooding energy is reduced obviously.

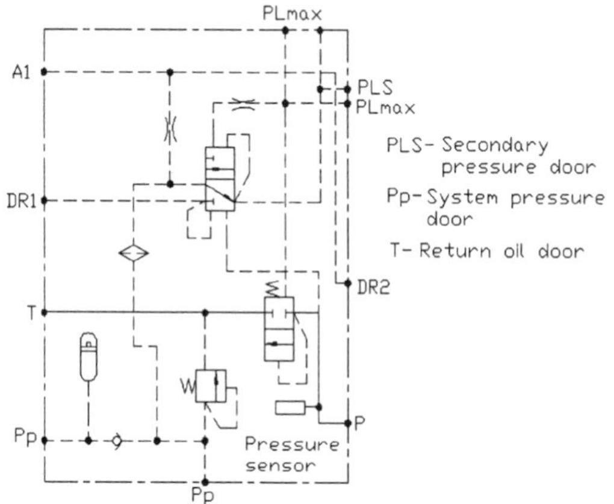

Figure 3. Principle of integrated valve of a control system

In order to prevent the hydraulic system components from being damage because of too high pressure, only we set a pressure sensor at the exit oil line to test the system's pressure, change it through A/D and put it into electronic controller, compare it with the system' s highest pressure, get a pressure difference P, through power zoom out, then put it into electromagnetism proportion decompress valve, fulfill the action of control adjust to the flux of output of the hydraulic pump.

When the pressure of the hydraulic system is improved instantaneous, though the highest pressure has been limited in design, instantaneous change of the system pressure can't response, for security, a relief valve should be set at the pump exit oil line.

In this design, the secondary pressure feedback LS control system is adopted. The normal mid position closed load-sensing pressure compensated hydraulic system is to give the maximum load pressure feedback to tilting mechanism of the pump, however, the secondary pressure feedback LS control system regards the pressure difference between pump pressure and the maximum load pressure as secondary pressure. It is detected by the reducing valve and feedback to tilting mechanism of the oil pump, then to increase or decrease the delivery capacity of the pump automatically. For the former, the tilting mechanism is controlled by the pressure difference between exit pressure of the pump and multiple unit valve. When the oil temperature is low, the pipeline pressure loss between pump and multiple unit valve increases, it makes decrease fore-and-aft pressure difference of the multiple unit valve. Therefore, the speed of execute component also decreases. But in the secondary feedback LS control system, the secondary pressure difference used in the pump is pressure difference between entry pressure of the multiple unit valve and exit pressure of maximal load, it is independent of pipeline pressure loss between multiple unit valve and oil pump. So it can achieve better speed adjusting properties and operation feeling. The system uses the precursor pump as the tilting mechanism to control the pressure oil, we will first induct the hydraulic oil to the side of pressure reducing valve, and then we induct the secondary pressure to precursor control oil circuit. Thus, the tilting mechanism of the pump can get a stable LS feedback pressure; it won't be affected by the fluctuation of the external load.

In order to make the hydraulic excavators' engine match the oil pump reasonable, can make good use of the power of the engine, but also avoid flameout of the engine, let the engine works well-balanced in the range of its capability, the torque of the engine need controlling, make the engine work at the place of biggest power as the Figure 2 shows.

This system composed of output flux control mechanic of pump, electromagnetism proportion decompress valve, electro-controller, rotate speed sensor and the degree of oil door open sensor. The sensor of oil door opening tests the oil door's open, and then puts the information into electronic controller, works out the rotate speed in this condition through Micro Processing Unit (MPU), and regards this rotate speed as control target. When the power the pump absorbed is too large, and make the rotate speed of engine is lower than ideal rotate speed, the controller will send out signal, make the power the pump absorbed reduce through the electromagnetism decompressing valve, reduce the load of the engine, make raise the rotate speed of the engine; when the power the pump absorbed is too small, make the rotate speed is higher than the ideal rotate speed, the electronic controller still send out signal, make the power the pump absorbed increase through electromagnetism decompressing valve, and increase the load of the engine, make the rotate speed of engine reduce, keep the engine working under the biggest power.

5. Conclusions

On the basis of analyzing the excavator's every working condition in detail, this paper has summarized the design requirement of the excavator's hydraulic system, power capability and operating capability.

In order to satisfy the requirement of design of hydraulic system, we design a set of load sensing and pressure compensated system of the excavator. This system adopts general multiple unit valve, equipped with special control valve and simple electronic control system, which achieves a good performance of timing and compounded operating performance.

References

1. Liu Zhao, Zhang Shanshan, Variable Pump's Control Modes and Their Application, Chinese Journal of Construction Machinery, Vol.2(3):304-307(2004).
2. Huang Zongyi, Ye Wei, Li Xinghua, A Survey of Hydraulic System in Hydraulic Excavators, Construction Mechanization, 12-16,9(2003).
3. M. Haga, W. Hiroshi and K. Fujishima, Digging control system for hydraulic excavator, Mechatronics, 665-676, 11(2001).
4. E. Budny, M. Chlosta and W. Gutkowski, Load-independent control of a hydraulic excavator, Automation in Construction, 245-254, 12(2003).

RESEARCH ON INTELLIGENT CONTROL OF BULLDOZER'S TRANSMISSION

CHAO ZHANG

College of Mechanical Engineering, Tongji University, 1239 Siping Road. Shanghai 200092, China.

ZHAO LIU, FANG DU

College of Mechanical Engineering, Tongji University, 1239 Siping Road. Shanghai 200092, China.

Information construction technology is a developing trend of construction machine. As a typical earthwork machine, bulldozer usually works with high loads and shift frequently in complex situations and bad surrounding. The existing machines are mostly equipped with manual shift transmissions, which are inefficient and need driver's hard work. The intelligent control of bulldozer's transmission has many advantages: ①It will release the drivers from hard work, make the drive easy and comfortable; ②The transmission will match with engines better during running, which will improve the dynamic performance and efficiency of the machines and reduce the discharge for saving energy and protecting environment; ③The intelligent control system will decrease the crash of the shift and prolong the life of the transmission; ④It will become the foundation of information construction. In this paper, some improvements are made on the basis of the traditional hydraulic control system and a fully new electronic control system for bulldozer's transmission is developed. On this base, the hardware of the control system is build up. According to the requirement of bulldozer's work situation, a suitable shift strategy is scheduled. After a series of simulation studies and tests, the system is verified and modified. The result shows that the control system of the transmission is effective.

1. Introduction

Information construction technology is a developing trend of construction technology, which reflects the technical level of current construction machine. As a typical earthwork machine, bulldozer usually works with high loads and shift frequently in complex situations and bad surrounding. Transmission is an important component of bulldozer and also is a key part of information construction technology. Traditional bulldozers usually use the manual shift transmission, which provide a low efficiency and need driver's hard work.

This paper rebuilds the traditional transmission's hydraulic control system, designs an electronic control system accordingly and presents a new shift strategy for the bulldozer to realize the automatic shift control.

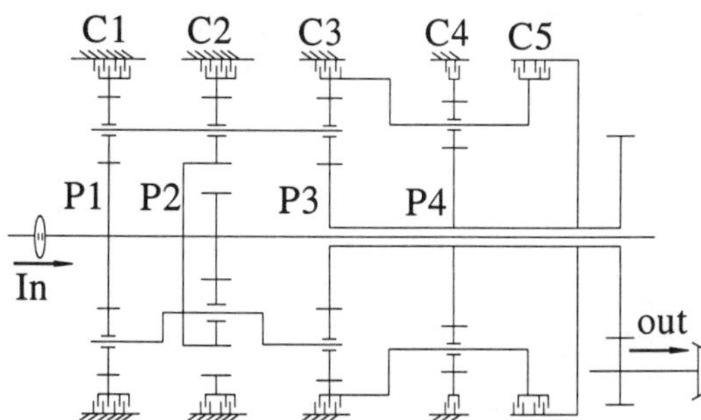

Figure 1. Structure of transmission

Table 1. Relation between shift and clutch engagement

Shift/Ratio \ Clutch	C1	C2	C3	C4	C5
F1/2.337	O				O
F2/1.500	O			O	
F3/0.676	O		O		
R1/0.984		O			O
R2/0.631		O		O	
R3/0.285		O	O		

Note: O represents engagement of clutch

2. Transmission System of Bulldozer

2.1 *Structure of transmission*

Figure 1 is the transmission structure diagram of bulldozer PD220Y. As shown in Figure 1, the transmission is mainly consisted of 4 sets of epicyclic gear (P1, P2, P3, P4) and 5 shift-clutches (C1, C2, C3, C4, C5). The second set (P2) have two planetary gears which make the rotation conversed. The planet carriers of front three sets (P1, P2, P3) are connected together and the third ring is connected to the forth planet carriers. Different combination of clutch engagement will cause different speed ratios and rotate directions. The relation between the actual shift and clutch engagement is listed in Table 1.

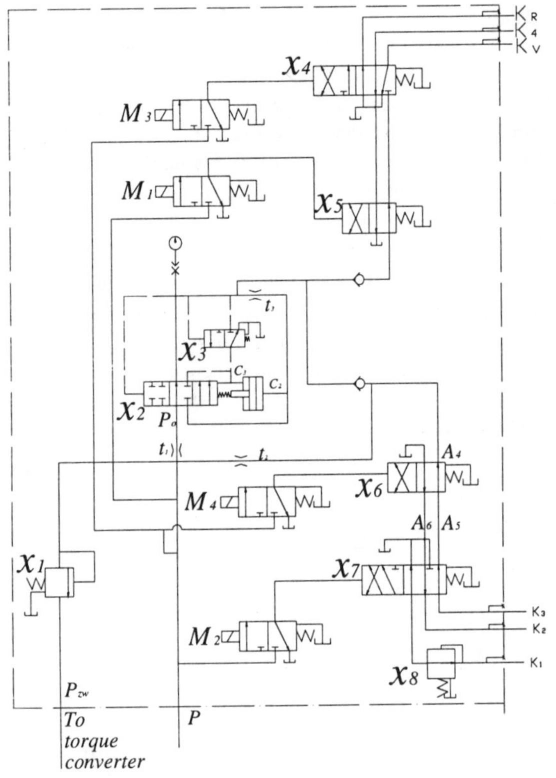

Figure 2. Hydraulic control system

2.2 *Hydraulic control system*

According to the control requirement of transmission, a hydraulic control system is developed. The hydraulic control system is composed of the following five major parts, as shown in Figure 2:

- Pilot magnet valve (M1, M2, M3, M4): The pilot magnet valve realizes the transformation from electrical signal to hydraulic action. It is controlled by electrical signal and offers pilot control pressure for the directional valve.
- Directional valve (X4, X5, X6, X7): The directional valve controls the oil feed and return of the clutches. It is divided into two groups and each group controls three shift-clutches.
- Main spillover valve (X1): The main spillover valve keeps the main pressure of the system at a fixed level.
- Pressure relief valve (X8): The pressure relief valve reduces the drive pressure of the clutch (C1) to meet the requirement of transmission.
- Buffering valve (X2, X3): The buffering valve absorbs the crash of shift clutch in the engagement process. It is a key component that influences the shift performance. The character of the buffering valve is shown in Figure 3.

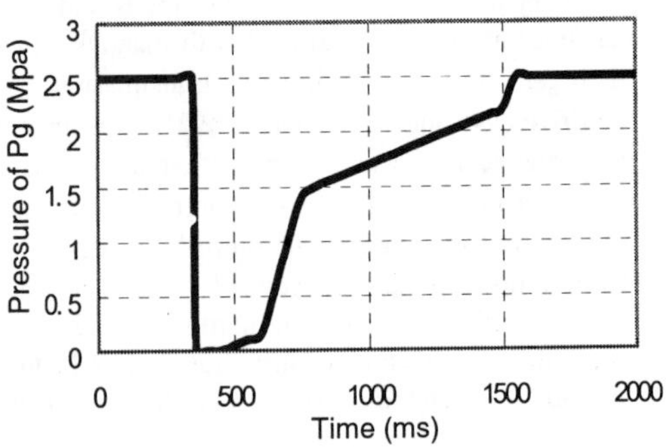

Figure 3. Performance of buffering valve

3. Electronic Control System

3.1 *Diagram of control system*

For those bulldozers in working state, the most important thing is to distinguish different running states and acquire driver's intention, and then make judgment and send command to control shift action. So the electronic control system is required to gather input information and send commands to control pilot magnet valve and system pressure. Figure 4 shows the components of control system.

3.2 *Function description*

The main function of the electronic control system can be described as follow:

1. Gather and transfer data: Gather input information from sensors and CAN bus, such as rotate speed of engine, speed of vehicle, gasoline throttle opening, position of control lever and so on. The current state of vehicle is the basis of data processing and shift control. The accuracy of the signal flow directly affects normal working. CAN bus optimizes arrangement of resources and coordinates the entire vehicle.

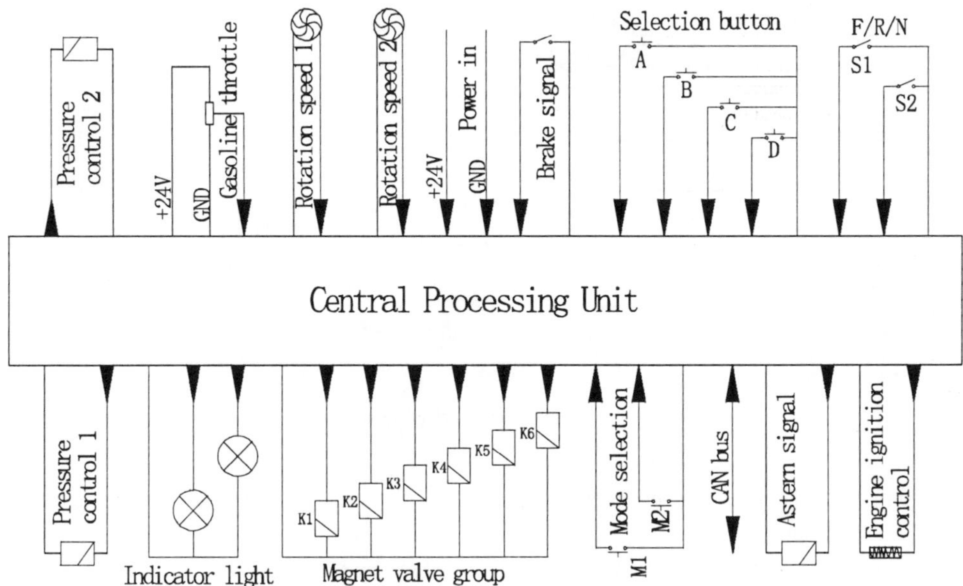

Figure 4. Diagram of the electronic control system

2. Data processing and shift control: After getting vehicle state, central processing unit (CPU) will calculate according to the shift strategy and give the result of current shift, pressure and time sequence of actions.
3. Shift performance control: Shift performance is affected by many factors, such as main system pressure, time sequence of clutch's engagement and the engagement pressure of clutch and so on. The main system pressure can be changed by regulating the duty ratio of proportioning magnet valve. The time sequence of clutch's engagement and the engagement pressure of clutch can be acquired by calculation and bench test, which is stored in database.
4. Failure tolerance and diagnosis: In order to enhance the reliability and security of the automatic transmission control system, there should be failure tolerance and diagnosis. When failure happens in some components, the system still keeps the basic running function and the bulldozer can drive back for maintenance. There are also some indicator lights to show the failure and then take emergency measures. During the maintenance process, it can be diagnosed by failure diagnosis device.

Each component has been calculated to fit the pressure and function requirements.

4. Shift Strategy

Shift decision is the core problem of automatic transmission. Bulldozer is a complex "man-machine-environment" integrative dynamics system. The running state is related with driver's intention, variable surrounding and the machine's parameter or technical condition. So the shift strategy should be decided by several key parameters induced from bulldozer's all side state. And then it can lead to intelligent control.

4.1 *Description of bulldozer's state*

There is much information to describe the bulldozer's state:

- Vehicle speed: It is related with the transmission system, driver's operation, working task and working condition. So it is a key parameter reflecting the vehicle's running state.
- Rotation speed of engine: It is the main parameter of engine, which reflects the torque-convert condition in some degree.
- Gasoline throttle opening: It is the symbol of driver's intention.
- Torque of engine: Output torque of crankshaft (or flywheel) of engine.

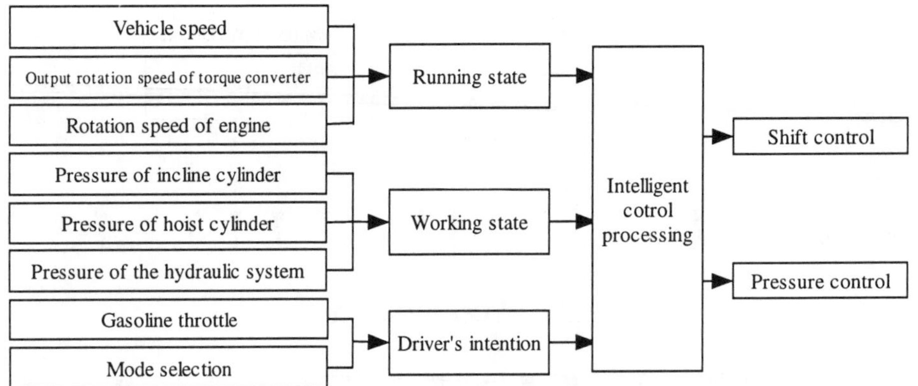

Figure 5. Flow diagram of shift strategy

- Output rotation speed of torque converter: Rotation speed of torque-converter turbine shaft.
- Output torque of torque converter: Output torque of torque-converter turbine shaft.
- Working condition: Main working condition of bulldozer, such as bulldozing, scraping, earthmoving and so on.
- Shovel state: The shovel's position and obliquity.
- Pressure of the hydraulic system: Including the pressure of working hydraulic system and turning hydraulic system.

4.2 *Shift decision*

There are several types of shift strategy, such as single-parameter equidifferent velocity shift type, divergent shift type, convergent shift type, compound shift type and so on. In the steady running situation, most of them can work well. But during the working process, the effect is not so good. The shift strategy should reflect the acceleration information and working condition.

When the bulldozer's state has been acquired, it is very important to choose the control parameters.

First, choose the running state parameter. As shown in previous statements, vehicle speed is a key parameter reflecting the vehicle's running state, and gasoline throttle opening is the symbol of driver's intention. So we choose these two parameters as the main running state parameters. For well reflecting the vehicle's running state, rotation speed of engine and output rotation speed of torque converter are also chosen to distinguish the engine's power dividing and running supervising.

Second, choose the working state parameter. Movement tendency of bulldozer's work devices can be judged by the hoist cylinder and incline cylinder's move direction, and the position of the shovel can be acquired from the displacement of these two cylinders. So we choose the displacement of the hoist cylinder and incline cylinder to reflect the shovel's condition. Further more, from the oil pressure of hydraulic system, we can get the load condition; from the oil temperature of the torque converter, we can get the torque converter state; from the oil pressure of turning cylinder, we can get the turning situation.

On the basis of the acquired parameters, CPU can judge the current state of the bulldozer according to the definite formulations and send out control command. In order to realize the intelligent control, optimum control and self-taught control are adopted in the control system. During the running process, bulldozer can adjust the shift time and shift point according to driver's style and working condition. The diagram of shift strategy is shown in Figure 5. All the function of shift strategy is carried out by the software of the control system.

5. Conclusions

In this paper, we present a bulldozer transmission's intelligent control system, which include analysis of transmission structure, hydraulic control system, electronic control system and shift strategy. After preliminary test, the system is identified and more detail performance can be gotten after further test.

References

1. Ge AnLin. Theory of automatic transmission and Design. Machine Press, 1993.5.
2. Kaoru Kondo and Hitoshi Goka, Adaptive shift scheduling strategy introducing neural network in automatic transmission, JSAE9538041.
3. Huang Zongyi, Li Xinghua. Introduction to construction with information technology, Construction Machinery, 2005.1.
4. A. Haj-Fraj and F. Pfeiffer, Optimal control of gear shift operation in automatic transmissions, Journal of Franklin Institute 338 (2001) 371-390.

INNOVATIVE DESIGN AND CONSTRUCTION TECHNOLOGIES FOR BUILDING INTERNAL PARTITIONS FOR SUPER HI-RISE BUILDINGS ON AN INTERNATIONAL BASIS

S. ABOMOSLIM

PhD Candidate, Department of Civil Engineering, University of British Columbia, Vancouver, V6T 1Z4, Canada, abomoslm@civil.ubc.ca

ALAN D. RUSSELL

Professor & Chairholder, Computer Integrated Design and Construction, Department of Civil Engineering, University of British Columbia, Vancouver, V6T 1Z4, Canada, adr@civil.ubc.ca

Construction of high rise buildings is no longer a local industry due to improved methods of transportation and information exchange which means the Global Village is becoming a reality. Moreover, the design and construction functions for very large hi-rise buildings tend to draw from the best practices and technology available on a worldwide basis because of the scale of the investment and the stakes involved. Witness the recent decisions to proceed with two of the tallest buildings in the world – one in Makkah, Saudi Arabia, and the other in Dubai, United Arab Emirates. In attempting to produce a quality product which meets capital cost and time expectations as well as life cycle expectations, amongst a number of challenges that need to be addressed, two very important ones are (i) how do firms gather and manage knowledge pertaining to new concepts in design and technological developments in terms of materials, physical systems, information systems, management systems, and construction methods in a way that makes it readily accessible to designers and constructors alike; and, (ii) what factors must be considered (e.g. logistics, codes, cultural norms, etc.) when assessing the suitability of these concepts and technologies for a given building and geographical context? This paper explores these questions by using a case study pertaining to the methods of design and construction of washrooms in the Middle East, an area of the world where the frontier of hi-rise buildings in terms of size and quality of systems is constantly being pushed. Current solutions adopted for building these elements constitute a significant part of the total cost of high rise building construction and have a direct impact on the speed of delivery. In this paper, we describe the rationale behind current practice, and how solutions used elsewhere can be adapted to meet local needs and sensitivities. We then generalize our discussion to describe elements of a framework currently being developed as part of a research program that allows design and construction technologies to be classified and represented in a way that facilitates their speedy assessment as to technical feasibility, environmental feasibility (including social norms), and impact on time, cost, and quality performance. Traditionally, high-rise building wet areas (e.g. washrooms and toilets) are made of blocks, plastering and in-situ finishes and enjoy social acceptance in the Middle East. Such acceptance has created a cultural barrier that inhibits the use of wet areas which are made of moisture resistant dry-wall partitions and in-situ finishing works which are easier and faster to install with minimal cleaning after work completion. The challenges of the traditional method include wet activities with low productivity rate, longer construction and curing times, large crew sizes, and logistical challenges in terms of materials handling and storage. The question becomes: what other technologies exist that can provide the sense of permanence and quality offered by existing partition systems, while addressing the objectives of cost, speed of delivery and quality. One potential alternative is the use of a concrete Pods system (a system introduced to the construction industry at a construction exhibition in Europe) which consists of a fully prefabricated bathroom including all internal finishing, electro-mechanical systems and accessories. The Pod's design in terms of area, finishing works and electro-mechanical systems can be changed as necessary to meet the project requirements. Pods can be manufactured using fibreglass, steel and normal concrete mix or lightweight concrete. Although fibreglass and steel Pods are easier and faster to install than concrete Pods, they have yet to gain social acceptance. As compared to wet areas built of block, plaster and in-situ finishes, concrete Pods offer the following advantages: moisture resistance, a fire rating of more than 2 hours, sound proofing up to 50 dB, a factory made product with minimum tolerance and high-quality finishes, a short site installation process which leads to faster project completion, comparable construction cost with the traditional method, efficient use of raw materials in production with no site waste, and easy logistical terms. A case study is used to illustrate the use of a framework that facilitates the representation of design and construction technologies, their interaction with other physical systems and construction processes, and their performance in terms of a number of metrics and constraint conditions.

1. Introduction

Super high-rise building construction is booming worldwide especially in the Arabian Gulf Area, an area of the world where the frontier of high-rise buildings in terms of size and quality of systems and design is constantly being pushed. Witness the recent decisions to proceed with two of the tallest buildings in the world, one in Makkah, Saudi Arabia, and the other in Dubai, United Arab Emirates. Moreover, the design and construction functions for super high-rise buildings tend to draw from the best practices and technology available on a worldwide basis because of the scale of the investment, the potential for economies of scale, and the reputation and image on an international basis that can accrue to all project participants, but especially to the client, for the successful execution of a unique, signature project.

To produce a quality product which meets capital cost, life cycle cost and time expectations as well as client satisfaction requirements, many decisions are required about the suitability of a large range of design and construction technologies available on a world-wide basis for the project and geographical context at hand. Therefore, development of an assessment framework for identifying and representing both the technological state-of-the-art and potential advances can aid in: (i) simplifying the choice process for decision makers in the early stages of the project; and, (ii) reducing the uncertainty associated with adapting the state-of-the-art from other venues in terms of cost, quality, time and local acceptance. We seek to develop and validate such a framework. To do so, we have conducted an extensive review of the literature on topics related to innovation and its diffusion, evaluation of construction methods, technology evaluation frameworks, and project success factors and associated criteria. In addition, we are compiling a number of case studies to help ensure completeness of the framework and to validate it.

In the academic arena, a number of researchers have addressed the issue of which factors determine the success of a construction innovation. Toole's work in classifying the related literature into two streams, namely, (i) construction innovation, and (ii) diffusion of innovation, is particularly helpful. For the first category, he identified five headings for grouping findings in the literature, these being: (1) the overall rate of diffusion of innovation in the industry and the characteristics of the industry that influence that rate; (2) the characteristics of organizations and projects most often associated with the adoption of innovations and actions organizations should take to facilitate innovation; (3) whether an innovation is more technology push or market pull; (4) the industrialization of building systems; and, (5) the seven national construction goals established in the 1990s in the USA [1]. With respect to the diffusion of innovation literature, Rogers identified five characteristics of an innovation which he asserts are the most significant ones for explaining the rate of diffusion, namely, relative advantage, compatibility, complexity, trialability, and observability [2]. Toole considered four technological trajectories for construction innovation: location of the work; means of production; materials used; and system design, all of which can lead to cost, time and performance advantages [1].

Other literature relevant to assessing technologies and innovation includes the following. In terms of evaluating construction methods, Rosenfeld analyzed three methods of construction by considering four aspects: (i) manufacturing vs. construction method on site; (ii) functionality and performance; (iii) process logistics; and, (iv) strengths and limitations of the method [3]. From the general project literature dealing with project success factors and success criteria, Wateridge identified six project success criteria: it meets user requirements; it achieves its purpose; it was delivered on time; it was delivered within budget; project users are happy; and it satisfies quality requirements [4]. Other success criteria were identified by Chan and Westerveld. They are: project delivered on time, within budget and with the quality required; and, in terms of the dimensions of project appreciation, positive appreciation is expressed by each of the client, project personnel, project users, contracting partners, and stakeholders [5, 6]. The foregoing success criteria have direct relevance to the task of technology and innovation assessment.

As a classification system for innovative construction technologies, Tatum developed a classification system based on the following four principals: material and equipment resources, construction-applied resources, construction processes, and project requirements and constraints. He also provided a tool for measuring technological change and analyzing specific operations for potential improvement [7]. Ioannou and Liu implemented the Advanced

Construction Technology System (ACTS) as a computerized data base system for classification, documentation, storage, and retrieval of information about emerging construction technologies. ACTS consists of a custom Microsoft (MS) Windows application, classification, and keyword files and a data base of emerging technologies [8].

Based on an extensive literature review, we observe that no comprehensive assessment tool exists for evaluating innovations and technologies for buildings in general or for our particular area of focus. Valuable contributions have been made in the literature, however, with respect to specific factors that should be considered when assessing construction technologies and design innovations. We have made use of such findings in order to develop an assessment framework for high-rise buildings on an international scale, and which treats groups of factors that deal with: (i) stakeholder acceptance; (ii) technical feasibility; (iii) logistics, and degree of prefabrication; and (v) the quantification of potential benefits. As discussed in the next section, we use these factors to form three filters to screen for promising and preferred technologies in a way that brings clarity to the evaluation process. In section 3, we present an example context for application of our framework – wet areas construction methods used in the Arabian Gulf Area for high profile high rise projects. Topics treated include the traditional way of building the wet areas (block, plastering and in-situ finishes), why the use of a dry wall system and in-situ finishes is not acceptable locally despite cost and time advantages, and use of the pods system as an innovative system. In section 4, we apply our assessment framework to evaluating the pros and cons of the traditional way of building versus use of the concrete pods system, which was ultimately chosen for the case study project examined.

2. Assessment Framework for Identifying Design and Construction Technologies of High-Rise Buildings on an international basis

Our assessment framework [9], involves dividing the evaluation process for current and proven innovative design and construction technologies into the application of three main filters. The first filter leads to a proceed to the next step or a no-go decision based on considerations of stakeholder acceptance and technical feasibility factors. Given success with the first filter, the second filter involves a focus on design, production, installation and logistical details and issues in order to clarify the advantages and disadvantages of the technology/method under consideration. Finally, the third filter deals with the choice of the most suitable option, based on quality, time, cost and risk factors. These filters are elaborated upon as follows.

2.1 *Filter (1): stakeholder acceptance and technical feasibility*

Filter (1) involves the application of a number of criteria under the headings of stakeholder acceptance and technical feasibility, which are judged on a pass-fail basis. Stakeholder acceptance criteria include: (i) cultural acceptance of the facility's features by future clients/users – for example, there is no cultural acceptance for the use of wood in permanent building construction in the Arabian Gulf Area; (ii) designer and developer acceptance; and (iii) compliance with local

Table 1: Definition of possible states of filter (1) factors

Factors	Type	Factors and States	
		Pass, (P)	Fail, (F)
1. Stakeholder acceptance:			
1.1 end-users	hard	acceptable	not acceptable
1.2 designers & developer	hard	acceptable	not acceptable
1.3 regulations and codes	soft	acceptable/can be negotiated	not acceptable/can't be negotiated
2. Technical feasibility, local:			
2.1 human resources	soft	available or can be imported	not available from any source
2.2 local materials	soft	available or can be imported	not available from any source
2.3 technology	soft	available or can be imported	not available from any source
2.4 infrastructure (roads, ...)	hard	exists	not available

authority and codes. A pass mark must be achieved on these criteria before examining technical feasibility criteria which embrace the ability to manufacture and/or construct the technology using local resources of material and manpower and/or the feasibility of importing related production equipment and technical expertise, and the availability of local infrastructure to facilitate the technology used. Some criteria may be hard, while others may be soft or act as constraints, the nature of which may be subject to negotiation – e.g. local regulations and codes. Failure on a hard criterion results in immediate dismissal of the technology/ innovation being examined. If a soft criterion cannot be made into a pass by negotiation if the initial state is fail, then again evaluation is terminated. Summarized in Table 1 are the criteria applied for the first filter step, along with as assignment of type of criterion – hard or soft (i.e. subject to negotiation) and applicable states for each criterion.

2.2 *Filter (2): engineering, production, construction, in-use factors and logistical issues*

Criteria dealing with engineering factors, scope of work and logistics under filter 2 have enjoyed the most attention in the literature. With respect to engineering factors, Toole introduced two factors and their sub details: (i) natural vs. engineered materials; and (ii) systems design based on fundamental engineering principles (mechanics, fluids, thermodynamics, building science, etc.) and principles dealing with the integration of physical systems [1]. As an example of the former, engineered materials can provide better performance in use and lower costs than conventional natural materials because they lack the variation found in natural materials. An example of the latter factor is an integrated structural and building envelope system

Table 2: Definition and possible states of filter 2 factors

Criteria	Factors	States, descriptors, values of the factors			
		High	Moderate	Low	Not Applicable
Engineering factors	Materials	context dependent	context dependent	context dependent	N/A
	Systems design	context dependent	context dependent	context dependent	N/A
	Compatibility	high	average	low	N/A
	Constructability	high	average	low	N/A
	Engineering functions	high	average	low	N/A
Production factors	Production location	on site & not in-situ	off site	in-situ	N/A
	Production means	automated production line	equipment intensive	labour intensive	N/A
Construction factors	Productivity	high	average	Low productivity	N/A
	Safety	low hazard potential	average hazard potential	High hazardous	N/A
	Material wastage	minimum	average	High wastage	N/A
	Prefabrication %	≥ 50 %	< 50 %	none	N/A
In-use factors	Sound proof	context dependent	context dependent	context dependent	N/A
	Fire proof	context dependent	context dependent	context dependent	N/A
	Water proof	context dependent	context dependent	context dependent	N/A
	Durability	high	average	low	N/A
	Maintainability	high	average	low	N/A
Logistic Issues	Material handling	Low – just in time to work face	average	High – double and triple handling	N/A
	Manpower supply	low	average	high	N/A
	Equipment	low	average	high	N/A

whose components serve dual functions and which can be manufactured off site. Constructability of the design offers many benefits for a project such as cost reduction, shortening of the schedule, improved quality and increased safety [10, 11]. Compatibility of the innovative system with other systems of a building is crucial [12]. The engineering functions of the innovative system are used as the basis of measures for meeting the structural and/or architectural requirements of the new technology [13].

With respect to scope of work and associated production factors, Toole introduced two factors of importance: (i) location of production - on-site vs. off-site; and (ii) production means – labour intensive vs. capital or equipment intensive [1]. In general, there is a belief that offsite production can lead to lower unit costs, faster production, a better quality product, improved safety, and much less interference with other ongoing-site activities. Consideration must also be given to construction factors, and in-use factors for permanent material, equipment or system being considered. Logistics is a major consideration for evaluating any innovative design and/or construction technology. Jang divided the logistics function into supply logistics and site logistics, and further categorized logistic processes under six headings: (i) material supply, storage, processing, and handling; (ii) manpower supply; (iii) schedule control; (iv) site infrastructure and equipment locations; (v) material flow management on a job site and (vi) management of information related to all physical and service flows [13]. Our primary focus is on material and manpower supply and equipment.

Summarized in Table 2 are the factors that need to be considered as part of applying the second filter. Several factor descriptors are not given in this table (indicated as context dependent), as they are application or context dependent. To date, we have not fully addressed how best to evaluate relative performance. What is shown in Table 2 is simply a 4-state evaluation system. Our main preoccupation at present is to ensure that the full dimensionality of a technology or innovation is considered in the evaluation process. More detailed attention will be directed later at various multi-criteria evaluation approaches.Filter (3): Assessing quality, time, cost and risk

Filter (3) deals with mainly quantifiable criteria as they relate to time or speed of production, cost (capital and life cycle), quality (possibly measured by degree of rework), and uncertainty or risk, measured in terms of the degree of variability that describes time, cost, quality and safety metrics. We are currently exploring how best to combine the quantitative metrics used at the filter (3) level with the evaluation schema of filter (2).

3. Alternative Methods for Construction of Wet Areas of High-rise Buildings

In general, wet areas of high-rise buildings are on the critical path of the project, since many of the associated activities have to be done in a critical sequence: internal partitions, electrical and mechanical works, and internal finishing. The configuration of most high-rise buildings consists of a few non-typical floors at the lower and uppermost levels and a very large number of typical floors in the middle. As a result, a very large number of typical toilets and washrooms have to be built, and it is thus important to seek economies of scale – hence the motivation for seeking innovative solutions, even those that require substantial investments in production facilities. The scale of super high rise facilities allows one to contemplate technologies that can't be considered for smaller buildings because of the different mix between fixed and variable costs. Here we examine the choice of prefabricated concrete toilets and washrooms called Pods, which are applicable to the high-rise building context in the Middle East. The design of Pods, in terms of area, finishing works and electro-mechanical systems can be modified, as necessary, to meet project requirements.

3.1 *Wet area traditional construction method: block, plastering and in-situ finishing*

In the Arabian Gulf Area, the traditional method of building wet areas consists of three main activities: partitions using block work and plastering, electro-mechanical works, and finishing works. Partitions are built either by using bricks or concrete blocks, mortar, bonding accessories, reinforcement, and lintels on the top of doors and internal window partitions. Cast in situ concrete tie beams may be required depending on the total area of the wall. When block partitions are built from scratch, a lengthy series of low productivity steps is involved, as follows: (i) preparing the mortar with bonding agents as required; (ii) placing reinforcement such as steel mesh, rebar and corrugated wall ties vertically and horizontally in the joints between the block layers; (iii) conducting random inspections to ensure quality for all of the building process steps such as mortar mixing, partition verticality, wall bonding accessories and the configuration of openings; and (iv) curing each block wall for at least three days before plaster can be applied. Accompanying the foregoing is a constraint on the maximum height of wall that can be built in one day - between 1.3 m to 1.5 m for initial curing to maintain the stability of the fresh wall. The average mason production rate of block work is between 15 m^2 and 20 m^2 per day.

The plastering activity starts after the block walls are cured and electrical and mechanical systems are roughed in. The plastering involves four main steps: affix plastering accessories (e.g. expanded metal lathing, corner bead and stop bead), lay out guide points, apply mortar slurry, and apply the finishing mortar application. As a curing period, there are two weeks between the overall plastering activity and painting. The production rate for plastering varies between 12 and 16 m^2 per day depending on the wall area. Block work and plastering are wet activities which require mortar mixes and curing periods. Both of these activities have low tradesmen productivity and production rates and must be executed in-situ. They are labour intensive, space intensive, and involve a significant logistics burden in terms of lifting requirements for the input materials. Considerable effort is required to clean up after the work is done, and there can be considerable wastage of materials.

Electrical and mechanical systems and their wares and fixtures are commonly executed in four steps: rough-in, 2nd fix, testing and commissioning and final fix. Finishing works include: water proofing, tiles for walls and floor, false ceiling, paint and toilet accessories. Figure 1 indicates an approximate estimate of the time required for building a wet area with grid pattern 2.0 X 2.0 m^2, perimeter length of 8.0 lm, height 3.20 m, and total partitions area of 26 m^2 with the assumption being that all required resources and materials are available on site. The foregoing provides considerable motivation for exploring alternative methods for building toilets and washrooms for projects with thousands of wet area units.

Figure 1: Construction Period for Wall with Area 26 m^2

Task Name	Duration	W-1	W2	W4	W6	W8
Wet Area traditional finishing:	36 days					
Block work activity:	4 days					
E/M rough-in	1 day					
Plastering activity:	12 days					
Floor water proofing	4 days					
Tiles for walls and floor	2 days					
E/M second fix	1 day					
False ceiling	1 day					
Final paint, testing E/M	7 days					
Door leaves	1 day					
Sanitary fixtures & ware	2 days					
Accessories	1 day					

3.2 *Wet area alternative construction method: moisture resistant dry wall and in-situ finishing*

One possible solution for building the wet areas is using moisture resistant dry wall and in-situ finishes. Moisture resistant dry wall partitions are built in three continuous steps - studding, sheeting and jointing with only the last step involving a wet activity that requires a curing cycle. Dry-wall partitions, as compared to block walls and plastered partitions can be built in a few days as opposed to weeks, enjoy high erection productivity, require neither lintel nor tie-beam, are cost effective, and involve relatively simple logistics. The in-situ finishing works follows the internal partitioning in the same sequence as is done for the block and plastering partitions. The main problem for using a dry wall system is that there is no public acceptance since it is perceived to be a non-rigid and non permanent system.

The question becomes: what other technologies exist that can provide the sense of permanence and quality offered by the existing traditional system, while addressing objectives of cost and speed of delivery? Optimally, one seeks a high quality, cost effective, technically feasible, fully prefabricated system with simple installation logistics that is suitable for super high-rise buildings, and at the same time enjoys public acceptance. This paper explores this question by examining the use of a proven innovative engineering and construction technology of the pods system.

3.3 *Decision for adopting pods system*

The King Abdul-Aziz Endowment project is a design, build, operate and transfer project located in Makkah, Saudi Arabia. It consists of a 17 floor podium topped by seven mixed use high-rise buildings which range in height from 213 m to 465 m. The total number of washrooms and toilets for the project is 7000 units. Such a scale potentially warrants the adoption of a new technology with up front factory costs. As a result, top management sought state-of-the-art, a proven design and construction technologies which would be culturally acceptable. One potential alternative identified was concrete Pods, a technology developed in Italy and used in Italy, Germany and the UK [14].

3.4 *Pods system as an innovative solution*

The Pods system is an innovative engineering and construction technology for wet areas: toilets and washrooms. The concept originated in Italy and they are now distributed throughout Europe. As shown in figures 2 and 3, Pods are fully prefabricated bathrooms which include all internal finishing, electro-mechanical components and accessories. There are many materials which can be used for manufacturing Pods: fibreglass,

Figure 2: Pods external view [16]

Figure 3: Pods internal finish [16]

steel and normal concrete mix or lightweight concrete. All the modules are structurally designed to have strengthened reinforcement and lifting eyes for site installation [15]. A concrete Pods system offers many advantages: moisture resistance, fire resistance of more than 2 hours, sound proofing up to 50 dB, a factory made product with tight tolerances and high finishing quality, a short site installation process which leads to construction time saving, comparable construction costs to the traditional method, efficient use of raw materials in production with no site waste, and easy logistical terms [16].

As a production method, concrete pods have four steps: (i) the casting stage includes preparing the steel mould, installing the reinforcement, bedding in the first fix of electrical/mechanical components, casting with steam curing and tagging each pod; (ii) the rough finishing stage consists of water proofing, the second fix of electrical/mechanical components, installing window and door frames, tiles for walls and floors, preliminary painting and installing the false ceiling; (iii) the final finishing stage involves installing windows and doors, final painting, electrical/mechanical components testing and final fix, installation of sanitary ware (lavatories, urinals, bathtubs and so on, sanitary fixtures (mixers, shower heads and so on) and toilet accessories; and (iv) the protection and storing stage consist of protecting doors, windows, electro-mechanical hook-ups and electrical boards and storing the pods until requested by the site [16]. The area and the number of bathroom units included in each Pod is mainly governed by weight which is required to be less than or equal to the on site crane capacity. In the King Abdul-Aziz Endowment project, every pod houses one toilet or washroom and every trailer transports four pods per trip as shown in figure 4 [17].

Figure 4: Pods lifting method [17]

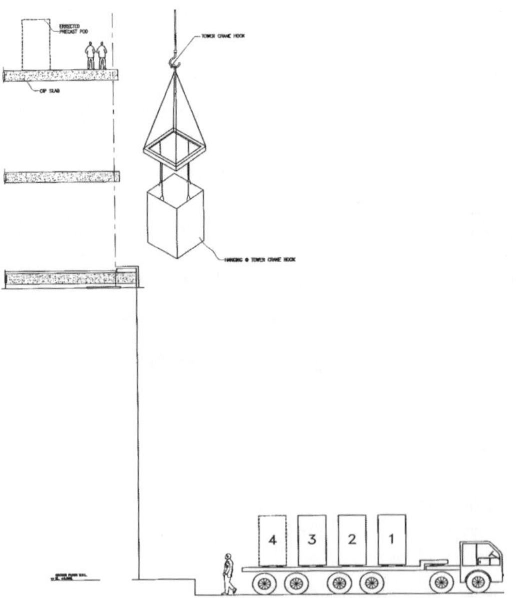

There are three methods for casting the concrete pods: (i) in a panel mould system, all panels for walls and ceiling are cast individually and assembled together by welding followed by casting the floor at the end. This method is really applicable when the number of typical pods is small since it has a low set up cost and productivity rate, and; (ii) in an adjustable moulds system, pods walls are cast in two L shaped operations welded together and followed by casting both roof and floor slabs. This method is suitable for producing a medium number of pods, since it has an average set up cost and is more efficient than the first one and (iii) in a fixed mould system, walls and ceiling are cast in one operation followed by the floor. This system is suitable for the mass production of pods of the same size since it has an expensive set up cost and offers construction time saving and high quality pods with accurate sizes [16].

As a site installation process [17], installation is carried out straight from a truck onto the slab by a tower or mobile crane using a special lifting beam as shown in figure 4. Pods can be installed according to the project design into the building either vertically from above before casting the slab or, horizontally with a roller platform after casting. There are two stages for erecting and adjusting the pods: in the first stage, the pods are hoisted to the right location and temporarily levelled by rubber shims to allow for structural creep and deflection of the slab to take place. In the second stage, final levelling is done in the finishing stage using a hydraulic jack as shown in figure 5. After final levelling, grouting under the pods as well as electro-mechanical connections are executed.

Figure 5: Pods final levelling [17]

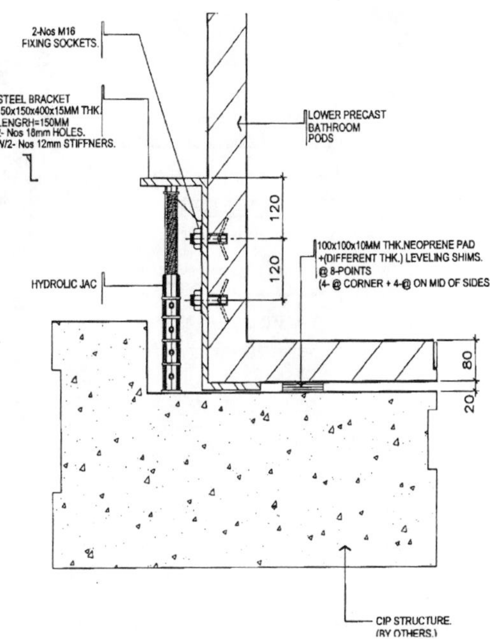

Table 4: Application of filter (2) for the alternative methods of building wet areas

Criteria	Factor	Traditional method	Concrete Pods
Engineering factors	Materials	low	high
	Systems design	low	high
	Compatibility	high	high
	Constructability	low	high
	Engineering function	high	high
Production factors	Location of production	low	high
	Production means	low	high
Construction factors	Productivity	low	high
	Safety	low	high
	Material wastage	low	high
	Prefabrication degree	low	high
In-use factors	Sound proof	moderate	high
	Fire proof	moderate	high
	Water proof	moderate	high
	Durability	moderate	high
	Maintainability	high	high
Logistic Issues	Material handling	Low	High
	Manpower supply	low	high
	Equipment	low	moderate

Figure 6: Summary of Assessment Process used for the alternative methods of building wet areas

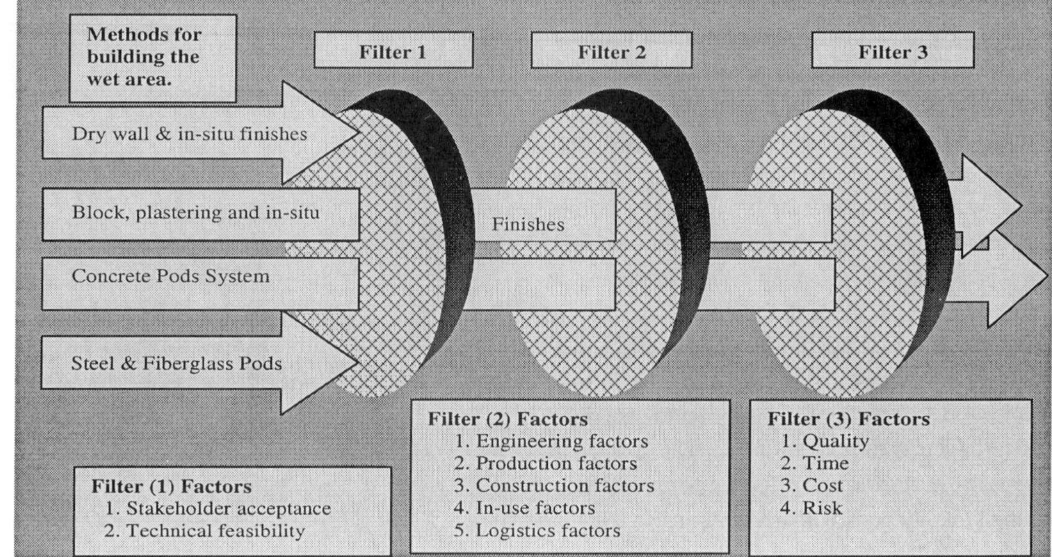

4. Applying the Assessment Framework

Application of the previously described framework is presented here for the four alternatives for wet areas construction identified. Applying the first filter as shown in Table 3, the dry wall and in-situ finishes in addition to fibreglass and steel Pods systems fail as a result of not meeting end-user acceptance, while the traditional method and concrete Pods system pass. Applying the second filter as shown in Table 4, both the traditional method and concrete Pods pass through, but the Pods system has more high state values than the traditional method based on our preliminary evaluation system. The main concern about the Pods system is the capacity of the crane which is required to lift pods of 5 to 7 ton of weight. Thus, based on these criteria, Pods is the dominant alternative, although all alternatives that successfully pass all criteria advance to application of the third filter.

Applying the third filter, in terms of quality, the pods system has a higher standard of quality over the traditional method, which is mainly attributable to its fully automated production in a factory under controlled environmental conditions. From a time perspective, the Pods system requires approximately one working day of site work versus 36 days for the traditional method for building toilets using block, plastering and in-situ finishing works. In terms of cost (not described herein), Pods are 5 to 15% less expensive than block and plaster. And finally, in terms of risk, the predictability of time, cost and quality is higher for the pods solution. Based on the assessment framework results, Pods is a preferred solution for building wet areas provided that sufficient scale of repetitions and availability of installation equipment.

Figure 6 summarizes the application of the evaluation framework to the case study. The preferred choice mirrors the actual choice. For the actual project, the process used for making the decision to use Pods involved two important issues: end-user acceptance had to be guaranteed and precast concrete elements were preferred to cast in place.

5. Conclusions

In this paper, a preliminary assessment framework for evaluating innovative design and construction technologies relevant to the design and construction of super high-rise buildings was introduced and applied to a case study. The work described in the paper constitutes is a step toward the development of a comprehensive evaluation framework and supporting knowledge management system in which technologies and their attributes can be stored for reuse. Significant work remains to be done to ensure a comprehensive identification of relevant evaluation criteria, how best to measure these criteria as a function of the building component, system or construction technology being examined, how to quantify this measurement, and how to combine different measurement scales. Validation of the framework will be pursued through a number of detailed case studies on actual projects.

Acknowledgements

The authors gratefully acknowledge the valuable input received from King Abdul-Aziz Endowment project Staff, Saudi Arabia: Architecture Muataz Al-Sawaf, COO, engineer Ahmad Fathi, finishing department general manager; engineer Haitham Al-Jayyousi, project director; Engineer Wageh Al-Baz, Premco general manager; engineer Khalid Waheed, Turner Arabia project deputy director, and architect Hani Hamdi.

References

1. Toole, T. M. (2001). "Technological Trajectories of Construction Innovation." Journal of Architectural Engineering, 7(4), 107-114.15.
2. Rogers, E. M. (1983). The diffusion of innovations, The Free Press, New York.
3. Rosenfeld, Y. (1994). "Innovative construction methods." Construction Management and Economics, 12, 521-541.
4. Wateridge, J. (1998). "How can IS/IT projects be measured for success?" International Journal of Project Management, 16(1), 59-63.
5. Chan, A. P. C., Scott, D., and Lam, E. W. M. (2002). "Framework of Success Criteria for Design/Build Projects." Journal of Management in Engineering, 18(3), 120-128
6. Westerveld, E. (2003). "The Project Excellence Model(R): linking success criteria and critical success factors." International Journal of Project Management, 21(6), 411-418.
7. Tatum, C. B. (1988). "Classification System for Construction Technology." Journal of Construction Engineering and Management, Vol. 114(No. 3), pp. 344-363.
8. Ioannou, P. G., and Liu, L. Y. (1993). "Advanced construction technology system - ACTS." Journal of Construction Engineering and Management, 119(2), 288-306.
9. AboMoslim, S. F., and Russell, A. D. (2005). "Evaluation of engineering and construction technologies of high-rise building on an international basis." 6th Construction Specialty Conference, CSCE, Toronto, Ontario, Canada.
10. Fischer, M., and Tatum, C. B. (1997). "Characteristics of Design-Relevant Constructability Knowledge." Journal of Construction Engineering and Management, 123(3), 253-260.
11. Rosenfeld, Y., Navon, R., and Cohen, L. (1998). "Methodology for Prioritizing Cost-Effective Construction Technologies." Journal of Construction Engineering and Management, 124(3), 176-184.
12. Lutz, J. D., Chang, L.-m., and Napier, T. R. (1990). "Evaluation of new building technology." Journal of Construction Engineering and Management, 116(2), 281-299.
13. Jang, H., Russell, J. S., and Yi, J. S. (2003). "A project manager's level of satisfaction in construction logistics." Canadian Journal of Civil Engineering, 30(6), 1133-1142.
14. King Abdul-Aziz Endowment project, Pods adaptation method, Makkah, Saudia Arabia
15. www.europeanensuites.co.uk/concrete.htm, Retrieved on July 6, 2005
16. Premco, Saudi precast company, Pods fabrication method statement, Jeddah, Saudia Arabia.
17. Premco, Saudi precast company, Pods installation method statement, Jeddah, Saudia Arabia.

RC Structure

CONCURRENT FLEXURAL STRENGTH AND DUCTILITY DESIGN OF NORMAL- AND HIGH-STRENGTH CONCRETE BEAMS

JOHNNY C.M. HO

Assistant Engineer, Ove Arup and Partners Hong Kong Ltd
Hong Kong, China

ALBERT K.H. KWAN

Professor, Department of Civil Engineering,
The University of Hong Kong, Hong Kong, China

The post-peak flexural behaviour of reinforced normal- and high-strength concrete beams has been investigated using complete moment-curvature analysis taking into account the effect of strain reversal in the tension steel. The results reveal that the major factors influencing the flexural ductility of reinforced concrete beams are the concrete grade, compression and tension steels ratios, and compression and tension steels yield strengths. Generally, flexural ductility increases with increase of compression steel and its yield strength, but decreases with increase of tension steel and its yield strength. However, the effect of concrete grade on flexural ductility is more complicated. To quantify how these parameters affect the flexural ductility, the authors have proposed a parameter λ as a measure of the degree of a beam section being under- or over-reinforced. Based on λ, an equation for direct evaluation of flexural ductility is established. Furthermore, the authors have studied the interrelation between the flexural strength and ductility that could be concurrently achieved and presented the interrelation in the form of charts for design applications.

1. Introduction

Conventional design of reinforced concrete beams requires the design of sufficient strength to sustain the ultimate design load as well as sufficient stiffness to limit the elastic deflection during service. The flexural strength can be obtained easily by elementary beam bending theory. However, there is no simple method to evaluate flexural ductility, apart from a nonlinear moment-curvature analysis that takes into account both the complete stress-strain curves (including post-peak region) of the constitutive materials and the stress-path dependence properties of the tension steel [1]. The evaluation of flexural ductility is important to protect the structures from accidental impact, sudden overloading and earthquake attack. In fact, for flexural members, the ductility allows a more rational redistribution between sagging and hogging moments. The moment redistribution requires the formation of anticipated plastic hinges in the beams, at which the section should possess adequate flexural ductility to allow for rotation to take place for redistribution of moment [2,3]. This is in fact mentioned by many design codes on concrete structures [4-6], which state that any implied redistribution of force and moment should be compatible with the ductility of the member concerned. Although the requirement of flexural ductility is implied in the design codes, little information is provided on the evaluation of flexural ductility of reinforced concrete sections, e.g. beams and columns.

Recently, the application of high-strength concrete (HSC) has increased drastically worldwide. Production of HSC of compressive strength of about 100-150 MPa can be easily achieved with the use of superplasticizer and fine minerals such as pulverized fuel ash, silica fume and ground granulated blastfurnace slag [7]. The advantages of adopting HSC in structural design are obvious, which include its high compressive, shear and tensile strengths as well as elastic modulus. The higher compressive strength increases the allowable span-depth ratio in the design, while the higher elastic modulus is able to reduce the elastic deflection for the serviceability limit state. Nonetheless, there is also concern with the use of HSC in regards to its low ductility. In the past, the flexural ductility of NSC beams is designed by a deemed-to-satisfy provision, which either limits the tension to balanced steel ratio or the depth to neutral axis of the beam section [4-6, 8].

Moreover, the use of high-strength steel (i.e. yield strength > 460 MPa) is also allowed in many of the concrete codes [5-6]. Therefore, the same application of the deemed-to- satisfy method to HSC beams and/or with the use of high-strength steel in providing the same flexural ductility capacity is questioned.

In this paper, the authors have conducted a parametric study on the effects of various structural parameters: concrete grade, tension and compression steels ratio, tension and compression steels yield strength, on the flexural ductility of NSC and HSC beams. The parametric study was carried out based on the proposed method of nonlinear moment- curvature analysis taking into account the strain reversal in the tension steel. From the study, it is evident that the beam's flexural ductility increases with the increase of compression steel and its yield strength, but decreases with the increase in tension steel and its yield strength. The variation of the beam's flexural ductility with concrete grade is more complicated. In order to evaluate the effects of these parameters on the ductility, the authors have proposed a parameter λ, which indicates the degree of the beam section being under- or over-reinforced. Further studies on λ showed also that the flexural ductility decreases with the increase of λ until it reaches unity, after then the flexural ductility remains constant on the ground that the tension steel does not yield. Based on the result, an equation relating the flexural ductility to λ and other parameters has been established for rapid evaluation of flexural ductility of NSC and HSC beams.

Apart from using the proposed equation, the authors have also studied the interrelation between the flexural strength and ductility of NSC and HSC beams. The interrelation has been presented in the form of charts for prescribed concrete strength, tension and compression steels yield strength. For a given set of flexural strength and ductility requirements, the designers could determine the respective maximum allowable value of λ from the charts. The ratios and yield strengths of the tension and/or compression steel(s) could be subsequently determined from the maximum value of λ and the flexural strength requirement.

2. Nonlinear Moment-curvature Analysis and Parametric Study

The authors have previously established a new method of nonlinear analysis of complete moment- curvature curves of NSC and HSC beams that incorporate the stress-path dependence properties of the tension steel. To compare the flexural behaviours of NSC and HSC beams, a complete stress-strain curve model that is applicable to a broad range of concrete strength from 20 – 130 MPa proposed by Attard and Setunge [9] was adopted in the analysis. The stress-strain curve for steel reinforcement is elasto-plastic. However, in order to account for the reduction of tension steel strain in the post-peak range, the conventional elasto-plastic stress-strain curve has to be modified with the addition of unloading path, which follows the initial elastic slope until reaching zero stress at residue strain ε_p. Details of these stress-strain curves could be found in another paper [1].

The analysis is based on the assumptions of ordinary beam bending theory, i.e. (1) plane sections remain plane before and after bending; (2) tensile stress of concrete is negligible; and (3) no bond-slip between the rebars and the concrete. These assumptions are widely accepted and nearly exact [2]. In this study, rectangular beam section with constant dimensions of breadth=300mm, overall depth=600mm and effective depth=550mm were adopted. The moment-curvature relationship of the assumed beam section is analysed by applying prescribed curvature to the beam section incrementally in very small steps starting from zero curvature. The strain and stress in concrete and steel are subsequently evaluated based on linear strain distribution and the respective stress-strain curve. The neutral axis depth of the section is then calculated iteratively such that the section equilibrium is attained. Based on the calculated neutral axis depth, the resisting moment of the beam section is computed using moment equilibrium method, which produces a point in the moment-curvature curve. The process is repeated until the moment has declined to 50% of the maximum moment in the post-peak region.

To observe the difference between the conventional analysis method (i.e. without stress-path dependence allowed for) and the proposed method (i.e. with stress-path dependence allowed for), the moment-curvature curve for a particular beam section evaluated using both methods were presented in Fig. 1. It is observed that the negligence of the stress-path dependence properties of the tension steel would lead to a significant deviation in the post-peak moment-curvature curves and hence flexural ductility of the beam section.

3. Ductility of Singly- and Doubly-reinforced Beams with Identical Steel Yield Strength

A parametric study investigating the flexural ductility of NSC and HSC beams using the proposed nonlinear moment-curvature analysis on previously stated beam section was carried out. The variations of parameters include: f_{co} from 30 to 100 MPa; ρ_c $(=A_{sc}/bd)$ from 0 to 2%; ρ_t $(=A_{st}/bd)$ from 0.4 to 2.0 balanced steel ratios to cover both under- and over-reinforced sections. Herein, the balanced steel ratio for a singly-reinforced concrete beam ρ_{bo} is defined to be the tension steel ratio of the beam section at which strain reversal just occurs.

The flexural ductility is measured in terms of a curvature ductility factor defined as $\mu = \phi_u/\phi_y$, where ϕ_u is the ultimate curvature and ϕ_y is the yield curvature. The ultimate curvature is taken as the curvature when the resisting moment has dropped to 80% of the peak moment in the post-peak range. The yield curvature is taken as the curvature at the hypothetical yield point of an equivalent elasto-plastic system whose initial elastic stiffness and yield point are set to the secant stiffness at 75% of the peak moment and the peak moment respectively [2].

Typical moment-curvature curves for under- and over-reinforced beam sections without compression steel is shown in Fig. 2. The moment-curvature curves for doubly RC beam is similar but of higher peak strength and larger ductility. From the figure, it is evident that the under-reinforced section is more ductile than that of over-reinforced section. The variation of flexural ductility with concrete strength is not so straightforward. For a fixed tension and/or compression steel ratio, the flexural ductility increases with the concrete strength albeit that HSC *per se* is more brittle. However, for the same degree of section being under reinforced, i.e. $(\rho_t - \rho_c)/\rho_{bo}$, μ decreases with the increase of f_{co}. The variation of μ with $(\rho_t - \rho_c)/\rho_{bo}$ is shown in Fig. 3.

4. Ductility of RC beams with Different Steels Yield Strengths

To study the effects of tension and compression steels yield strength on the flexural ductility of NSC and HSC beams, the parametric study is repeated to include cases when f_{yc} and f_{yt} = 250 and 600 MPa. Similar to the previous case for beam sections with equal steels yield strength, the beam's flexural ductility is actually determined by the degree of section being under- or over-reinforced, which is denoted hereafter as λ for simplicity. In previous section, it has been established that $\lambda = \rho_t/\rho_{bo}$ and $(\rho_t - \rho_c)/\rho_{bo}$ for singly- and doubly-reinforced section with identical steel yield strength respectively. In cases where unequal steel yield strength is provided in a beam section, $\lambda = (\rho_t - (f_{yc} / f_{yt})\rho_c)/\rho_{bo}$. The variation of μ with λ for different combinations of f_{yt} and f_{yc} for the analysed beam section are presented in Fig. 4. From Fig. 4, it is observed that μ decreases with the increase in λ until $\lambda = 1$, which is similar to those in Fig. 3. For $\lambda > 1$, μ remains constant. Together with Fig. 3, it is evident that the relationship between μ and λ is not unique, but is dependent on concrete grade and steels yield strength. Basically, at a given value of λ, μ decreases as f_{co} or f_{yt} increases but increases as f_{yc} increases.

5. Proposed flexural ductility Equation

From previous discussion, it is clear that the major factors affecting the flexural ductility of NSC and HSC beams are: (1) f_{co}; (2) ρ_t and ρ_t; and (3) f_{yc} and f_{yt}. These factors have been grouped into a dimensionless parameter defined as λ indicating the degree of section being under- or over-reinforced. It has also been observed that the variation of μ and λ is consistent in all cases including doubly-reinforced section and sections with different steels yield strength, in which μ decreases with increasing λ until reaching unity (see Figs. 3 and 4). Nevertheless, as shown in the same figures, the relationship between μ and λ is not unique but dependent on other factors such as concrete and steels yield strength. Consequently, the following mathematical equation has been set up to relate μ and the studied parameters:

$$\mu = k\,\alpha(\lambda)\,\beta(f_{co})\,\gamma(f_{yt})\,\delta(f_{yc}) \qquad (1)$$

where k is a real constant, $\alpha(\lambda)$ $\beta(f_{co})$ $\gamma(f_{yt})$ and $\delta(f_{yc})$ are functions to be determined based on the results obtained from the parametric study. By performing a nonlinear regression analysis of all the numerical data obtained, with $0 < \rho_t < 2\rho_{bo}$, $0 < \rho_c < 2\%$, $30 \le f_{co} \le 100$MPa, and $f_{yc} = f_{yt}$ = 250, 460 and 600 MPa, the following expressions for k, $\alpha(\lambda)$ and $\gamma(f_{yt})$ is obtained:

$$k = 10.7 \quad (2)$$

$$\alpha(\lambda) = \lambda^{-1.25} \quad (3)$$

$$\gamma(f_{yt}) = (f_{yt} / 460)^{-0.25} \quad (4)$$

It is found that in most cases within the above range of parameters that $\delta(f_{yc})$ is close to 1.0. For simplicity, the value of $\delta(f_{yc})$ is therefore always taken as 1.0. Conversely, it is not possible to obtain a simple expression for $\beta(f_{co})$ in Eq. (1) since the variation of μ and f_{co} is relatively complicated. Alternatively, $\beta(f_{co})$ is modified to include the effects of other parameters, and expressed in Eq. (5):

$$\beta(f_{co}) = (f_{co})^{-0.45} (1 + 95.2 (f_{co})^{-1.1}((f_{yc}\, \rho_c)/(f_{yt}\, \rho_t))^3) \quad (5)$$

Substituting Eqs. (2) to (5) into Eq. (1), we have the following final equation for the estimation of μ:

$$\mu = 10.7 (\lambda)^{-1.25} (f_{co})^{-0.45} (1 + 95.2 (f_{co})^{-1.1}((f_{yc}\, \rho_c)/(f_{yt}\, \rho_t))^3) (f_{yt} / 460)^{-0.25} \quad (6)$$

Within the ranges of structural parameters studied, the values of μ obtained by the above equation are accurate to within ±10% error when compared to those obtained from rigorous moment-curvature analysis.

6. Interrelation between Flexural Strength and Ductility

From Eq. (7), it could be observed that the flexural ductility could be increased by reducing ρ_t and f_{yt}, increasing ρ_c and f_{yc} and reducing λ. However, these measures would also at the same time influence the beam's flexural strength. Normally speaking, an increase in flexural ductility is always beneficial to the structures; however, any reduction in the resisting moment is undesirable. Conversely, an increase in flexural strength should also be handled with caution since it would increase the shear demand and the flexural strength of the neighbouring columns, the latter of which would violate the principle of "strong columns-weak beam". As a result, a design that allows simultaneous consideration of both flexural strength and ductility is essential to avoid the above problems. In connection with this, the interrelation between flexural strength and ductility of beam sections should be studied.

A series of charts plotting strength (in dimensionless form of M/bd^2) against μ is shown in Fig. 5 for various concrete grades and steels yield strength. From the figure, it is observed that the use of HSC would increase the maximum limit of flexural strength and ductility that could be simultaneously achieved, i.e. shifting the curve upward and to the right. It is also seen that the use of lower f_{yt}, higher f_{yc}, and the increase of ρ_c would increase the limit of flexural strength and ductility that could be achieved simultaneously. By looking at the intermediate curves representing constant λ, it is evident that the flexural strength and ductility limit would increase when λ, f_{co} and f_{yt} decrease, but when f_{yc} increases. The charts shown in Fig. 5 could be employed as design chart for designing beam to meet a pair of flexural strength and ductility requirement. Example illustrating their use is given next.

7. Design Example

The flexural strength and ductility requirements of a beam section having f_{co} = 60MPa are: M_p/bd^2 = 12.0 MPa and μ = 5.0. As a first attempt, Fig. 5(c) is used. Plotting the point (12.0, 5.0) on the graph, it is found that the required flexural strength and ductility can be simultaneously achieved for f_{yt} = 250 MPa, ρ_c = 0%, λ = 0.45 and ρ_t = 5.40% (ρ_{bo} = 12.01% [10]). To reduce the tension steel ratio, tensile steel strength may be increased. From Fig. 5(c), f_{yt} = 600 MPa, ρ_c = 0%, λ = 0.45 and ρ_t = 1.67% (ρ_{bo} = 3.71% [10]). Alternative method would be to add ρ_c. From Fig. 5(f), f_{yt} = 460 MPa, f_{yc} = 250 MPa, ρ_c = 1.0%, λ = 0.40 and ρ_t = 2.70% (ρ_{bo} = 5.39% [10]).

8. Conclusions

A parametric study based on the author's proposed method of nonlinear moment-curvature analysis, which takes into account the stress-path dependence of tension steel, on the flexural ductility was carried out for NSC and HSC beams with f_{co} from 30 to 100MPa, ρ_t from 0 to 2 ρ_{bo}, ρ_c from 0 to 2% and f_{yc}, f_{yt} of 250, 460 and 600MPa. From the study, it is evident that the major factors influencing the flexural ductility of NSC and HSC beams are, amongst other things, the degree of section being under- or over-reinforced defined as λ. λ has been derived for singly- and doubly-reinforced sections including the variation of steels yield strength. Finally, a theoretical equation that predicts the flexural ductility of NSC and HSC beams were derived based on

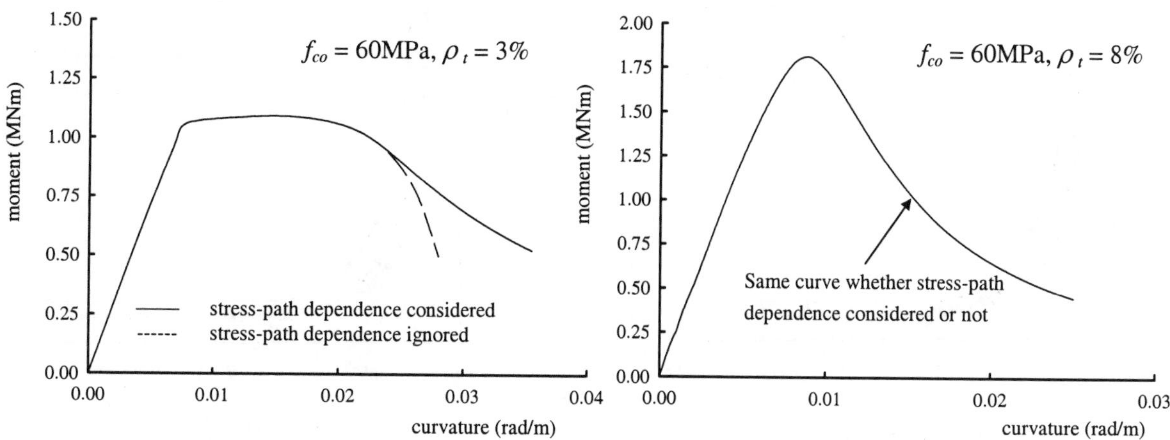

Fig. 1 Effects of strain reversal on moment-curvature curve

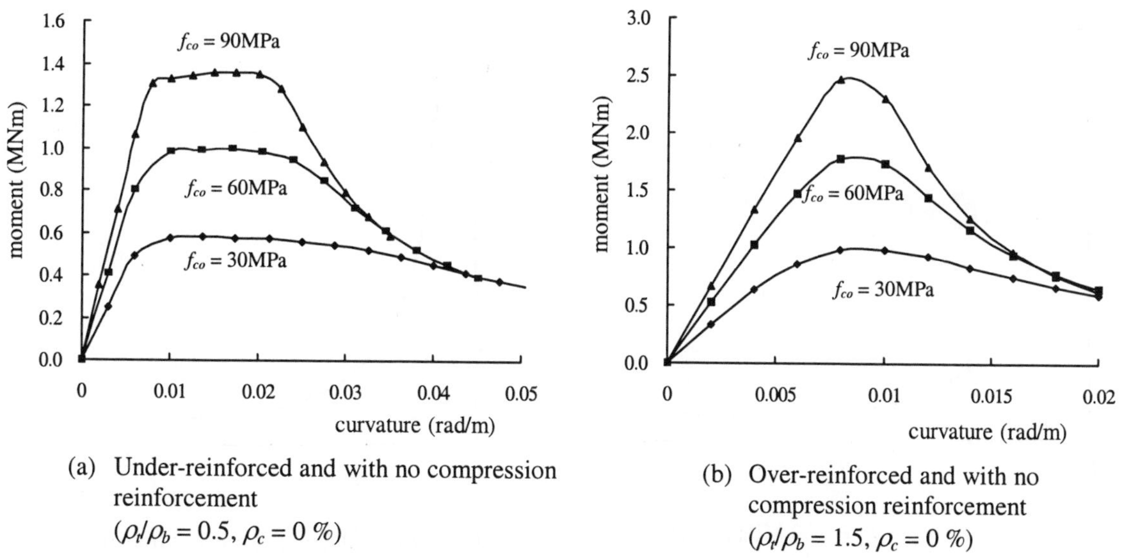

(a) Under-reinforced and with no compression reinforcement ($\rho_t/\rho_b = 0.5$, $\rho_c = 0$ %)

(b) Over-reinforced and with no compression reinforcement ($\rho_t/\rho_b = 1.5$, $\rho_c = 0$ %)

Fig. 2 Moment-curvature curves of some beam sections analysed

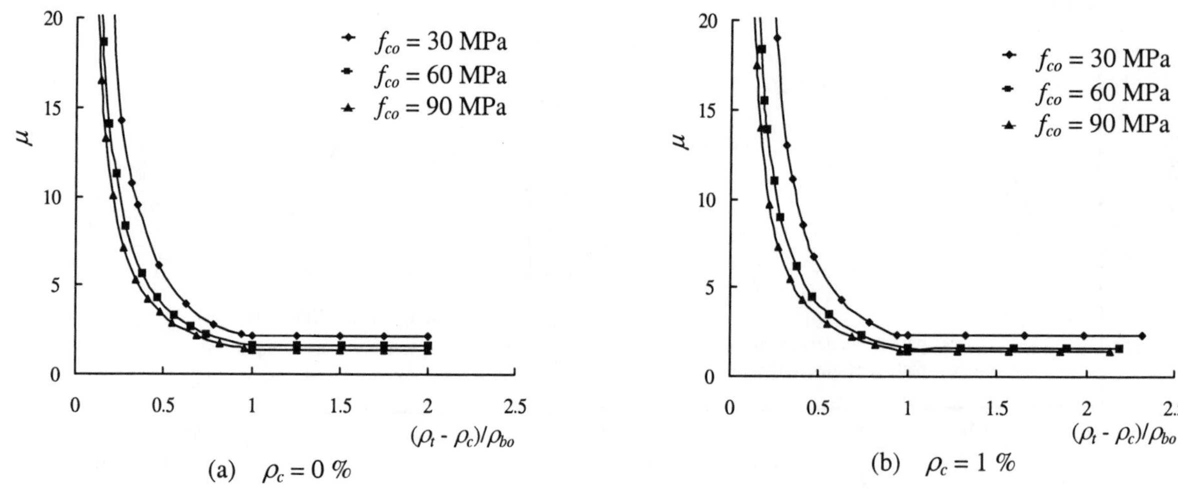

(a) $\rho_c = 0$ %

(b) $\rho_c = 1$ %

Fig. 3 Ductility factor plotted against the ratio $(\rho_t - \rho_c)/\rho_{bo}$

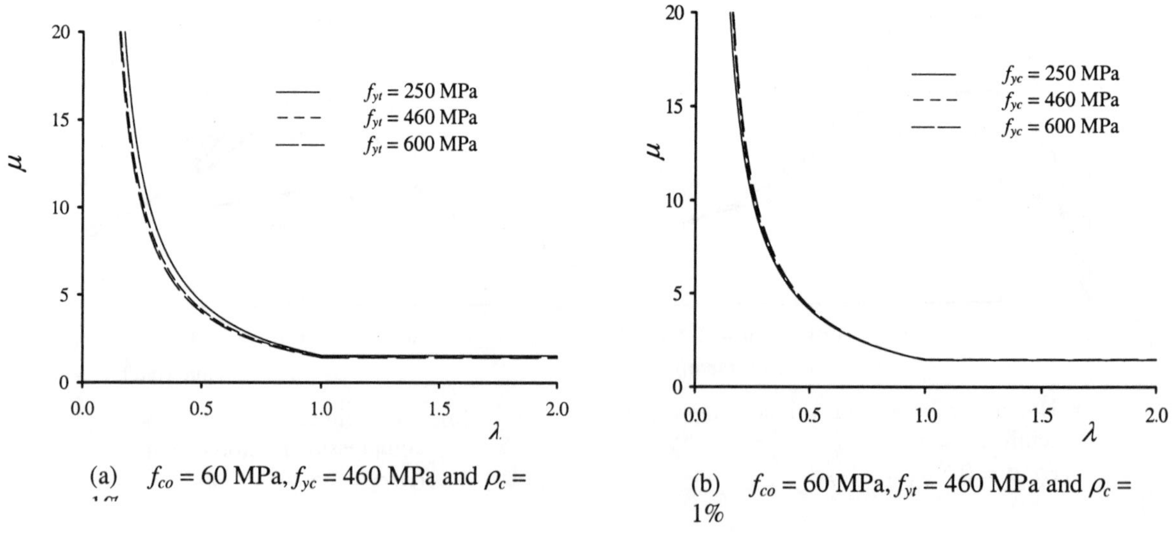

(a) f_{co} = 60 MPa, f_{yc} = 460 MPa and ρ_c = 1%

(b) f_{co} = 60 MPa, f_{yt} = 460 MPa and ρ_c = 1%

Fig. 4 μ versus λ for beam sections with f_{co} = 60 MPa, ρ_c = 1%, f_{yt} and f_{yc} = 250, 460 or 600 MPa

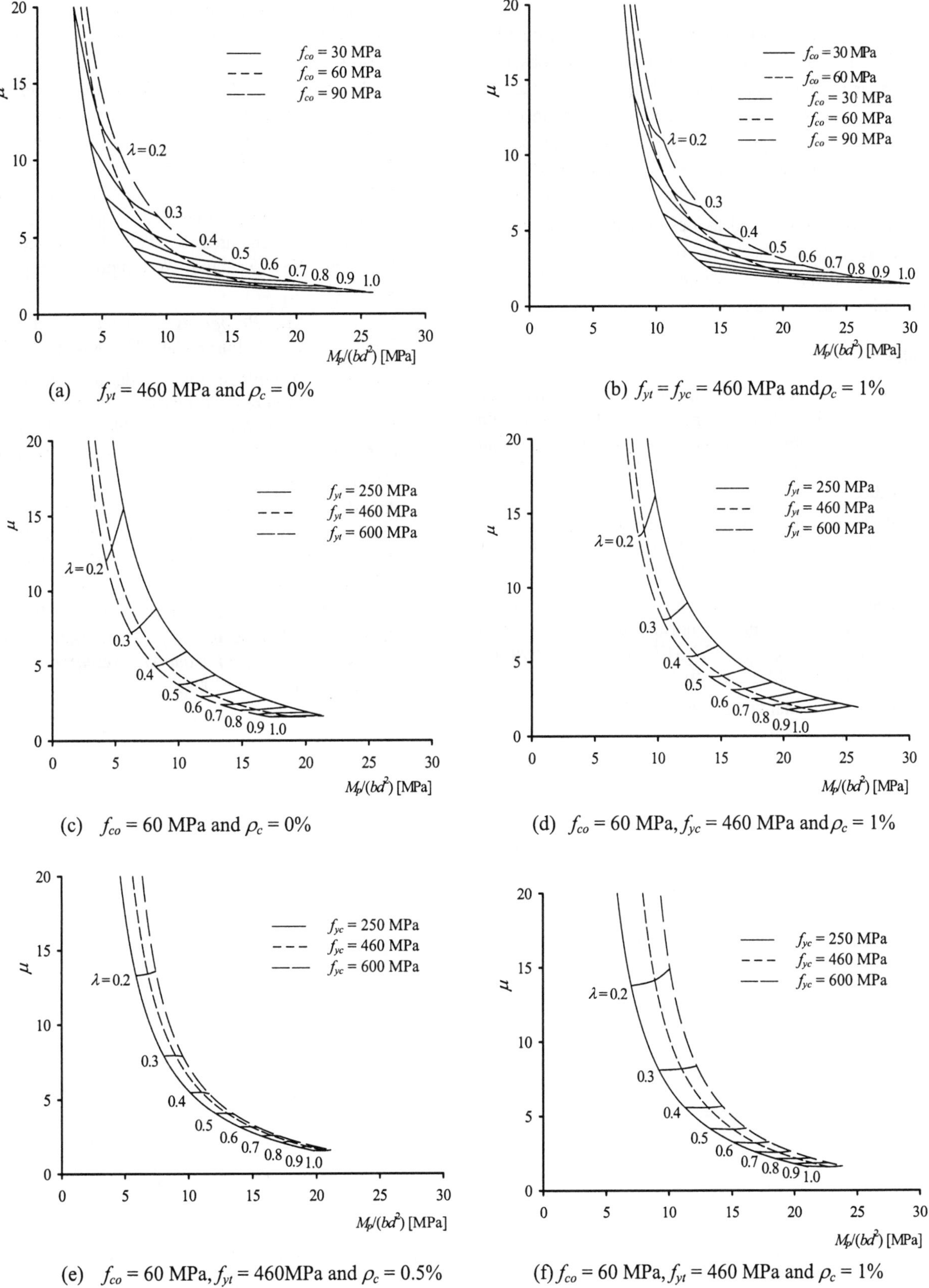

(a) f_{yt} = 460 MPa and ρ_c = 0%

(b) $f_{yt} = f_{yc}$ = 460 MPa and ρ_c = 1%

(c) f_{co} = 60 MPa and ρ_c = 0%

(d) f_{co} = 60 MPa, f_{yc} = 460 MPa and ρ_c = 1%

(e) f_{co} = 60 MPa, f_{yt} = 460MPa and ρ_c = 0.5%

(f) f_{co} = 60 MPa, f_{yt} = 460 MPa and ρ_c = 1%

Fig. 5 μ versus $M/(bd^2)$ for beam sections with various concrete grades and steels yield strengths

the parametric study.

By plotting charts of flexural strength and ductility that could be achieved simultaneously, it was observed that the maximum limit of flexural strength and ductility that could be achieved simultaneously is enhanced by using HSC and higher f_{yc}. Conversely, the use of higher f_{yt} would decrease the strength-ductility performance of NSC and HSC beams. The charts could also be employed as design chart for designing beam section to meet a pair of specified flexural strength and ductility. A numerical example has also been illustrated for this purpose.

References

1. Ho J.C.M., Kwan A.K.H. and Pam H.J., "Post-peak flexural behaviour of normal- and high-strength concrete beams", *The Structural Design of Tall and Special Buildings*, Vol.12, No.2, Jun, pp.109-125 (2003).
2. Park R. and Paulay T., *Reinforced Concrete Structures*, John Wiley & Sons, New York, U.S.A., 769pp (1975).
3. Pecce M. and Fabbrocino G., "Plastic rotation capacity of beams in normal and high-strength performance concrete", *ACI Structural Journal*, Vol.96, No.2, Mar-Apr, pp.290-296 (1999).
4. British Standards Institution, *BS 8110: Part 1, Structural Use of Concrete: Code of Practice for Design and Construction*, London, U.K., 122pp (1997).
5. ACI 318-95, *Building Code Requirements for Structural Concrete, and Commentary (ACI 318-R-95)*, American Concrete Institute, Farmington Hills, USA (1995).
6. NZS 3101, *The Design of Concrete Structures*, Concrete Structures Standard, Part 1: Code of Practice, supplemented by Part 2: Commentary, Standards New Zealand, Wellington, 255pp. and 264pp (1995).
7. Neville A.M., *Properties of Concrete*, Pearson Prentice Hall, Edinburgh Gate, UK, 844pp (2002).
8. European Committee for Standardisation, *EC 2: Design of Concrete Structures - Part 1: General Rules and Rules for Buildings*, European Committee for Standardisation, Brussels, Belgium, 207pp (1992).
9. Attard M.M. and Setunge S., "The stress strain relationship of confined and unconfined concrete", *ACI Materials Journal*, Vol.93, No.5, Sep.-Oct. 1996, pp.432-442 (1996).
10. Ho J.C.M., Kwan A.K.H. and Pam H.J., "Minimum flexural ductility design of high-strength concrete beams", *Magazine of Concrete Research*, Vol.56, No.1, Feb, pp.13-22 (2004).

COMPLETE NONLINEAR BEHAVIOUR OF NORMAL- AND HIGH-STRENGTH CONCRETE BEAMS UNDER CYCLIC LOADING

BILL Z.Z. BAI

Department of Civil Engineering, The University of Hong Kong,
Pokfulam Road, Hong Kong, China

FRANCIS T.K. AU

Department of Civil Engineering, The University of Hong Kong,
Pokfulam Road, Hong Kong, China

Even in a structure that is loaded monotonically, some parts of it may experience curvature reversal, not to mention the much more complex loading history imposed by earthquakes. Therefore to get a reliable estimate of the safety of a structure under earthquake loading, it is worthwhile to carry out full-range analysis under cyclic loading. In this paper, the complete nonlinear behaviour of reinforced concrete beams made of normal- and high-strength concrete under cyclic loading is studied using a numerical method that employs the actual stress-strain curves of the constitutive materials and takes into account the stress-path dependence of the concrete and steel reinforcement. The loading history considered includes loading, and repetitive cycles of unloading and reloading. In particular, loading past the peak strength into the post-peak stage is investigated. It is found that the complete moment-curvature relationship, which is path-dependent, is similar to the material stress-strain relationship with stress-path dependence. Degradation of section stiffness is observed as the cyclic loading extends to large curvatures, especially at the post-peak stage. Upon reloading, it may not follow the original unloading path, but the difference between the two paths is not significant for the cases studied. Different aspects of behaviour between under- and over-reinforced sections have been observed which include, for example, the characteristics of moment-curvature envelope curve and the variation of neutral axis depth.

1. Introduction

In the design of structures, both strength and ductility are important for structural safety. Very often the strength aspect is given much attention while it is simply assumed that the design code used will provide a certain minimum level of ductility to avoid brittle failure. With the increasing use of high-strength concrete [1] that is inherently more brittle than normal-strength concrete, engineers have started to pay more attention to the ductility and complete flexural behaviour of reinforced concrete (RC) beams made with such materials. To accurately assess the safety of a structure, a full-range analysis to obtain the behaviour of both the pre-peak and post-peak stages is necessary.

Carreira and Chu [2] presented a general non-linear method to compute the moment-curvature relationship of RC members. A numerical method for the full-range moment-curvature analysis of RC beams under monotonic increase of curvature, which takes into account the non-linear stress-strain relationship and stress-path dependence of the constitutive materials, has been developed and applied to rectangular beams [3] and flanged beams [4]. Consider an RC beam on which the deformations at certain sections are imposed monotonically. Normally a plastic hinge will sooner or later form at a loaded section, where the curvature and deflection increase monotonically as well. The bending moment there increases accordingly until it reaches the peak strength, after which it drops gradually until the structure completely collapses. However the equilibrium of the beam requires that the other sections not directly loaded will experience unloading and consequently curvature reversal without reaching their respective peak moments. Therefore even in a structure that is loaded monotonically, some parts of it may experience curvature reversal, not to mention the much more complex load history imposed by earthquakes.

A few investigators have studied the behaviour of reinforced concrete beams under complex loading typical of seismic motions. Kent and Park [5] investigated experimentally and theoretically the inelastic behaviour of reinforced concrete members under cyclic loading, which were made of normal-strength concrete and mild steel reinforcement. In particular, they studied the Bauschinger effect for cyclically stressed mild steel reinforcement and the influence of rectangular steel hooping on the stress-strain behaviour of concrete. Brown and Jirsa [6] carried out experiments on RC cantilever beams and investigated the effect of load history on the strength, ductility, and mode of failure. They concluded that the behaviour of the specimens under load reversal was influenced primarily by shear.

A new method of full-range moment-curvature analysis for RC beams under non-reversed cyclic loading is employed in this paper. It employs the actual stress-strain curves of the constitutive materials and takes into account the stress-path dependence of the concrete and steel reinforcement. The method is applied to the full-range analysis of RC beam sections to study their complete nonlinear response. Special attention is paid to the complete moment-curvature relationship. Other related issues are also discussed.

2. Stress-strain Relationship of Concrete with Stress-path Dependence

In the present study, the stress-strain curve of concrete developed by Attard and Setunge [7], which has been shown to be applicable to a broad range of *in situ* concrete strength from 20 to 130MPa, is employed. In this model, the parameters used to establish the stress-strain curve are the initial Young's modulus E_c, the peak stress f_{co} and the corresponding strain ε_{co}, and the stress f_{ci} and strain ε_{ci} at the inflection point on the descending branch of the curve. The stress of concrete σ_c is related to the strain ε_c by

$$\sigma_c / f_{co} = \frac{A(\varepsilon_c/\varepsilon_{co}) + B(\varepsilon_c/\varepsilon_{co})^2}{1 + (A-2)(\varepsilon_c/\varepsilon_{co}) + (B+1)(\varepsilon_c/\varepsilon_{co})^2} \quad (1)$$

where A and B are coefficients dependent on the concrete grade. Two sets of coefficients A and B are required, with one set for the ascending branch and another set for the descending branch of the curve. For the ascending branch where $\varepsilon_c \leq \varepsilon_{co}$,

$$A = \frac{E_c \varepsilon_{co}}{f_{co}}\ ;\ B = \frac{(A-1)^2}{0.55} - 1 \quad (2)$$

For the descending branch where $\varepsilon_c > \varepsilon_{co}$,

$$A = \frac{f_{ci}(\varepsilon_{ci} - \varepsilon_{co})^2}{\varepsilon_{co}\varepsilon_{ci}(f_{co} - f_{ci})}\ ;\ B = 0 \quad (3)$$

The parameters E_c, ε_{co}, f_{ci} and ε_{ci} are related to the peak stress f_{co} by

$$E_c = 4370(f_{co})^{0.52} \quad (4a)$$

$$\varepsilon_{co} = 4.11(f_{co})^{0.75}/E_c \quad (4b)$$

$$\frac{f_{ci}}{f_{co}} = 1.41 - 0.17\ln(f_{co}) \quad (4c)$$

$$\frac{\varepsilon_{ci}}{\varepsilon_{co}} = 2.50 - 0.30\ln(f_{co}) \quad (4d)$$

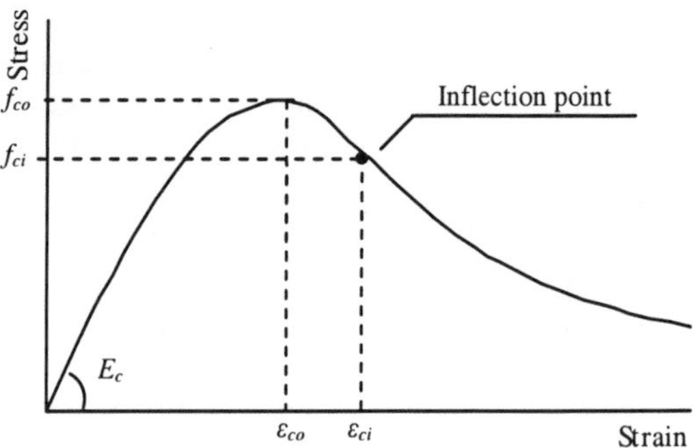

(a) Parameters to define stress-strain curve

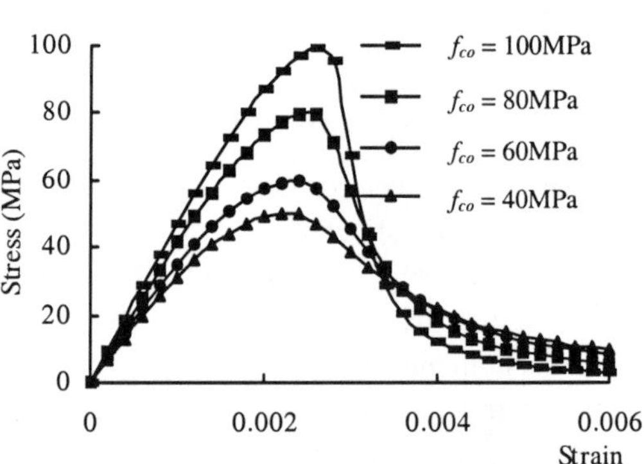

(b) Examples of stress-strain curves

Fig. 1 Stress-strain curves of concrete according to the Attard-Setunge Model

Fig. 1(a) shows the parameters used to define a stress-strain curve using this model, while some typical stress-strain curves for *in situ* concrete strength from 40 to 100MPa are shown in Fig. 1(b).

To cope with the unloading and reloading in the material, the stress-path dependence of the stress-strain relation is taken into account. A typical path 0-1-2-2'-3-4 is shown in Fig. 2(a). After following the loading path along the envelope to point 1 (ε_{un}, σ_{un}), unloading starts through an initial unloading region till point 2 (ε_{pc}, 0), which is then followed by a zero stiffness region 2-2'. The initial unloading region is a straight line starting from point 1 (ε_{un}, σ_{un}) and having a slope E_{un}, which is taken as a function of the unloading strain ε_{un} and the strain at peak compressive stress ε_{co}. The slope E_{un} can be determined by [8]

$$\frac{E_{un}}{E_c} = 1 \text{ for } \frac{\varepsilon_{un}}{\varepsilon_{co}} \le 1 \tag{5a}$$

$$\frac{E_{un}}{E_c} = 1 - 0.3\left(\frac{\varepsilon_{un}}{\varepsilon_{co}} - 1\right) \ge 0.25 \text{ for } \frac{\varepsilon_{un}}{\varepsilon_{co}} > 1 \tag{5b}$$

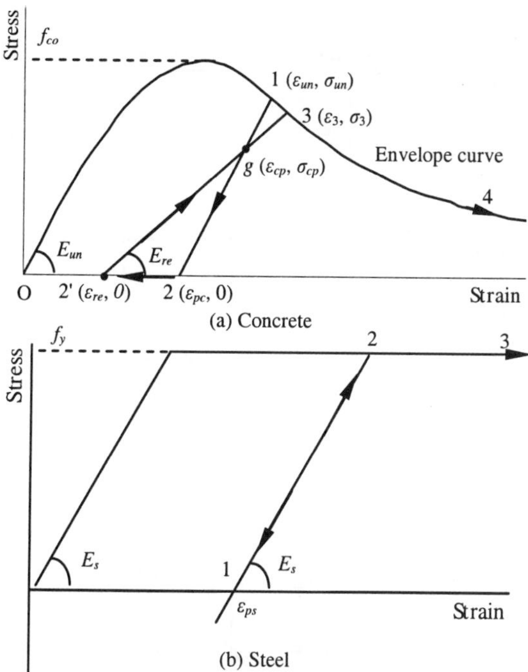

Fig. 2 Stress-strain curves of material constituents with stress path dependence

If reloading takes place in the initial unloading region, it retraces the same unloading path until it reaches the envelope. While reloading starting from a point on the zero stress region follows a different path. A typical path 2'-g-3 is shown in the figure. The paths 2'-3 and 1-2 intersect at the common point g (ε_{cp}, σ_{cp}) of the reloading curve in compression. The common point is assumed to be the one on the initial unloading curve with $\sigma_{cp} = 0.7\sigma_{un}$ [8] and therefore the stiffness E_{re} on reloading can be worked out accordingly. During unloading and reloading, the stress σ_c is related to the strain ε_c by

$$\sigma_c = E_{un}(\varepsilon_c - \varepsilon_{pc}) \text{ for } \varepsilon_{pc} \le \varepsilon_c \le \varepsilon_{un} \tag{6a}$$

$$\sigma_c = 0 \text{ for } \varepsilon_{re} < \varepsilon_c < \varepsilon_{pc} \tag{6b}$$

$$\sigma_c = E_{re}(\varepsilon_c - \varepsilon_{re}) \text{ for } \varepsilon_{re} \le \varepsilon_c \le \varepsilon_4 \tag{6c}$$

3. Stress-strain Relationship of Steel with Stress-path Dependence

A linearly elastic-perfectly plastic stress-strain model as shown in Fig. 2(b) is used for the steel reinforcement. When the strain ε_s is increasing, the stress σ_s in the steel is given by

$$\text{at elastic stage: } \sigma_s = E_s \varepsilon_s \text{ for } \varepsilon_s \le f_y/E_s \tag{7a}$$

$$\text{after yielding: } \sigma_s = f_y \text{ for } \varepsilon_s > f_y/E_s \tag{7b}$$

where E_s is the Young's modulus and f_y is the yield stress.

It is assumed that unloading from a typical point 2 on the yielding plateau in Fig. 2(b) follows the straight line 2-1 following the initial elastic slope. Reloading from point 1 then follows the path 1-2-3. The stress σ_s and the strain ε_s along path 1-2 are related by

$$\sigma_s = E_s(\varepsilon_s - \varepsilon_{ps}) \tag{8}$$

where ε_{ps} is the residual strain of path 1-2.

4. Non-linear Analysis Method

The three basic assumptions made in the analysis are (a) plane sections before bending remain plane after bending; (b) the tensile stress in the concrete may be neglected; and (c) there is no bond-slip between the reinforcement bars and the concrete. They imply that the longitudinal strains developed at various points of a section are proportional to the distances from the neutral axis. They are commonly accepted and are nearly exact except in deep beams or in the vicinity of cracks. Fig. 3 shows a beam section having a breadth b and total depth h, with the tension reinforcement area A_{st} provided at a depth d and the compression reinforcement area A_{sc} provided at a depth d_1 from the top. For convenience in analysis, the sign convention adopted is such that the strain and stress quantities are normally positive as follows: (a) compressive strain and stress in concrete are positive; (b) compressive strain and stress in compression reinforcement are positive; and (c) tensile strain and stress in the tension reinforcement are positive. When the curvature of the beam section is increased to ϕ as shown in the strain distribution diagram in Fig. 3, the strain ε developed is given by

$$\varepsilon = \phi\, x \tag{9}$$

where x is the distance above the neutral axis. Therefore, the compressive strain ε_{ce} at the extreme concrete compression fibre, the compressive strain ε_{sc} in the compression reinforcement and the tensile strain ε_{st} in the tension reinforcement can be written respectively as

$$\varepsilon_{ce} = \phi\, d_n \tag{10a}$$

$$\varepsilon_{sc} = \phi\,(d_n - d_1) \tag{10b}$$

$$\varepsilon_{st} = \phi\,(d - d_n) \tag{10c}$$

in which d_n is the neutral axis depth. The corresponding stresses σ_c, σ_{sc} and σ_{st} developed in the concrete, compression reinforcement and tension reinforcement can then be evaluated from the respective stress-strain curves of the materials taking into account stress path dependence. To cater for the subsequent unloading and reloading, the latest residual strains ε_{pc} of the concrete section, ε_{ps} of the tension reinforcement and the compression reinforcement should be stored and updated. The residual strains ε_{pc} of the concrete section at regular intervals along the depth are recorded so that the values at intermediate positions can be obtained by linear interpolation.

The stresses developed in the beam section must satisfy the conditions of axial equilibrium and moment equilibrium. The applied axial load P can be obtained from axial equilibrium as

$$P = \int_0^{d_n} \sigma_c\, b\, dx + \sum A_{sc}\,\sigma_{sc} - \sum A_{st}\,\sigma_{st} = 0 \tag{11}$$

where compressive force is taken as positive. Similarly, the resisting moment M can be obtained from moment equilibrium as

$$M = \int_0^{d_n} \sigma_c\, b\, x\, dx + \sum A_{sc}\,\sigma_{sc}(d_n - d_1) + \sum A_{st}\,\sigma_{st}(d - d_n) \tag{12}$$

where sagging moment is taken as positive. In the evaluation of integrals in Eqs. (11) and (12), Romberg integration [9], which can significantly improve the accuracy of the simple trapezoidal rule when the integrand is known at equispaced intervals, has been adopted. The axial equilibrium condition as shown in Eq. (11) can be used to determine the neutral axis depth d_n. Normally, given a specified curvature ϕ and a trial value of neutral axis depth d_n, the equilibrium condition is not immediately satisfied and there is an unbalanced axial force P. Since the relation between the unbalanced axial force P and the neutral axis depth d_n is nonlinear, an iterative scheme is required to determine the value of d_n which will give a zero value of P. The scheme adopted here is the modified linear interpolation method [9]. The complete moment-curvature relationship including loading, unloading and reloading is then obtained through a prescribed variation of curvature by a suitable step size.

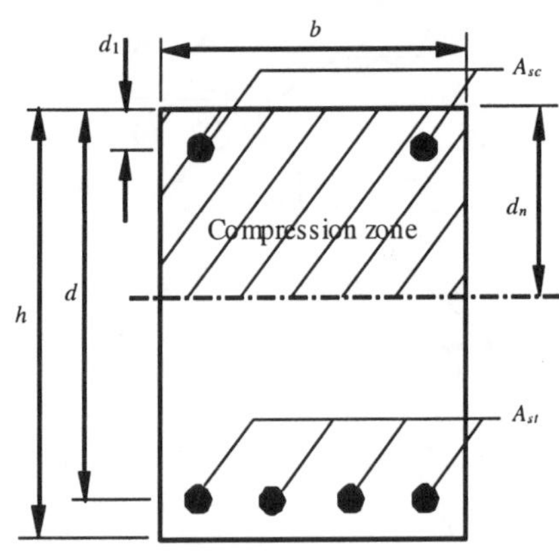

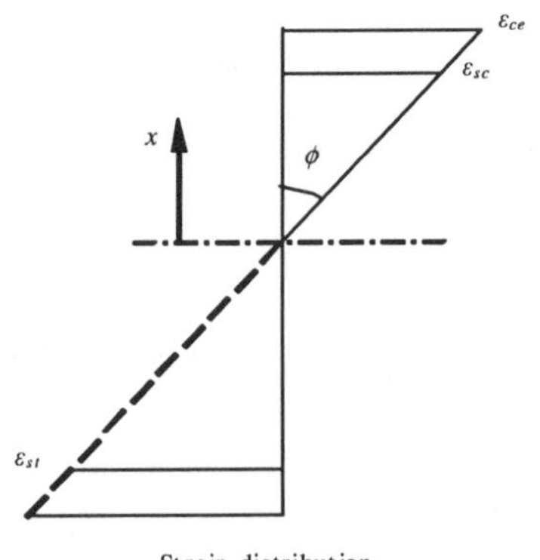

Fig. 3 A beam section subjected to bending moment

5. Results of Analysis

5.1 *Sections analyzed*

The beam sections analyzed are rectangle in shape as shown in Fig. 3. A typical beam section has a breadth b = 300mm and total depth h = 600mm, with the tension reinforcement provided at a depth d = 550mm and the compression reinforcement provided at a depth d_1 = 50mm from the top. The steel reinforcement has a yield strength f_y = 460MPa and Young's modulus E_s = 200GPa. Altogether twelve cases have been investigated as shown in Table 1.

The *in situ* concrete compressive strength f_{co} ranges from 30 to 90MPa to cover both normal- and high-strength concrete. The beam sections may be singly or doubly reinforced with the compression steel ratio ρ_c ($\rho_c = A_{sc}/bd$) up to 1%. The tension steel ratio ρ_t ($\rho_t = A_{st}/bd$) varies from $0.5\rho_b$ to $1.5\rho_b$ in terms of the balanced steel ratio ρ_b to cover both under- and over-reinforced cases.

Table 1. Cases investigated

f_{co} (MPa)	ρ_c (%)	ρ_b (%)	ρ_t (%)
30	0	3.19	1.60
	0	3.19	4.79
	1	4.19	2.10
	1	4.19	6.29
60	0	5.39	2.70
	0	5.39	8.09
	1	6.39	3.20
	1	6.39	9.59
90	0	7.30	3.65
	0	7.30	10.95
	1	8.30	4.15
	1	8.30	12.45

5.2 *Complete moment-curvature relationship of beam sections under cyclic loading*

The complete moment-curvature curves under monotonic loading for the sections in Table 1 are plotted in Fig. 4. They are dependent mostly on the reinforcement content. Curves for under-reinforced sections are characterized by fairly long yield plateaus while those for over-reinforced sections have sharp peaks indicating their brittleness. Under-reinforced sections are more ductile and a section tends to be ductile when normal strength concrete is used.

To investigate the typical response to non-reversed cyclic loading, the sections in Table 1 are analyzed. In particular, the imposed curvature increases from zero up to an unloading curvature (i.e. the unloading point) and then decreases until the moment becomes zero when only the residual curvature remains (i.e. the residual point). The section is then reloaded until the reloading path rejoins the envelope curve for monotonic loading at the returning point. Thereafter cyclic loading is repeated so that each new cycle reaches a higher unloading curvature. Fig. 5 shows the moment-curvature curves obtained for *in situ* concrete strength 90MPa.

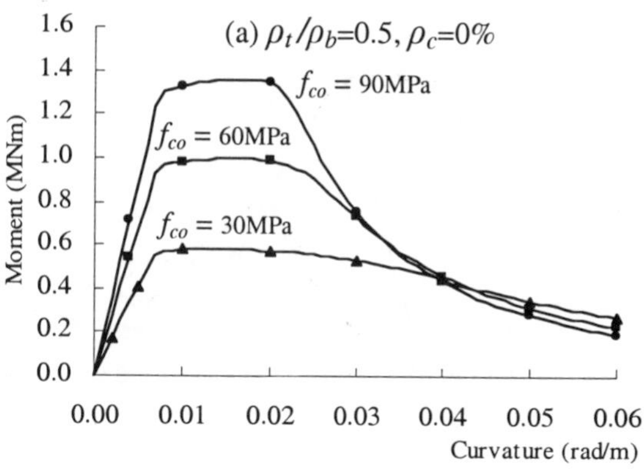

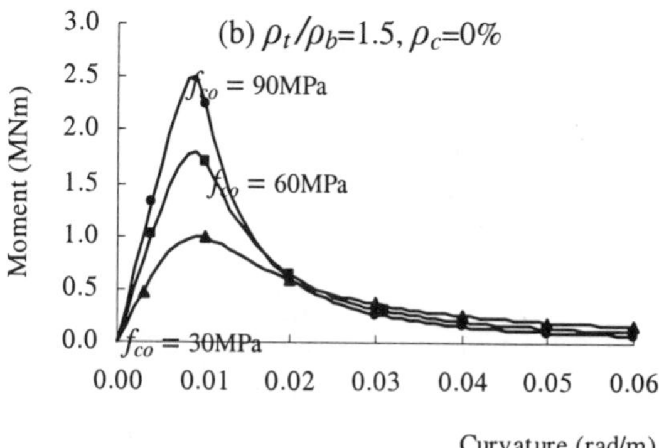

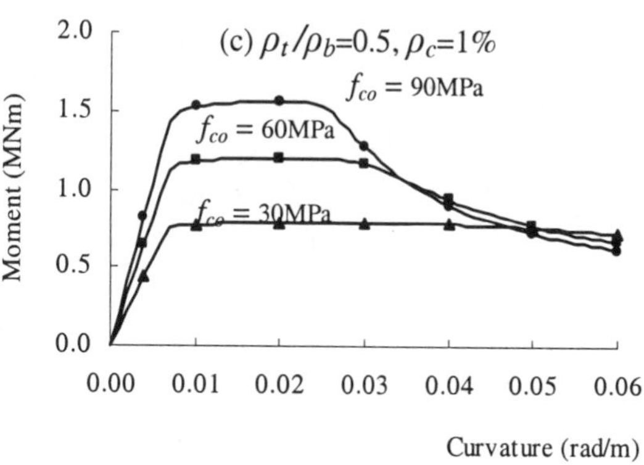

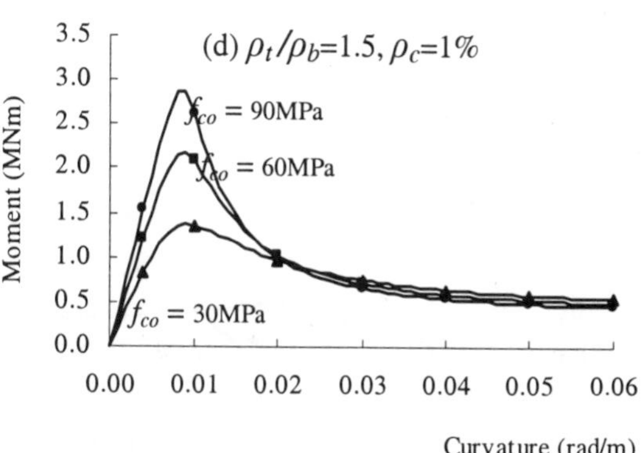

Fig. 4 Complete moment-curvature curves under monotonic loading

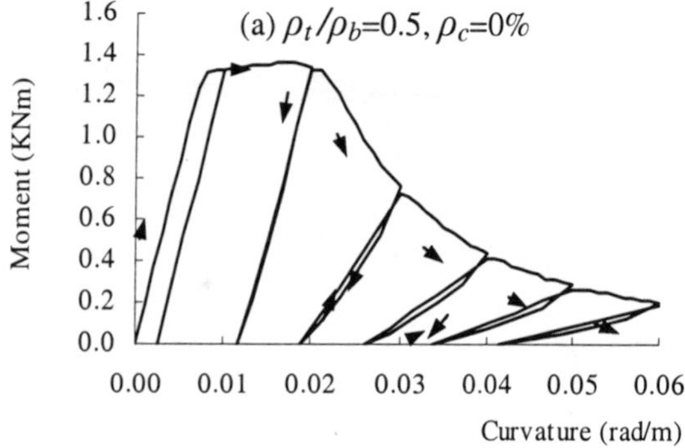

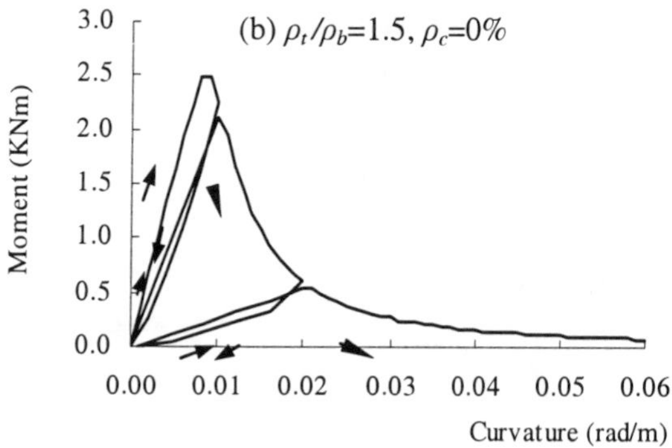

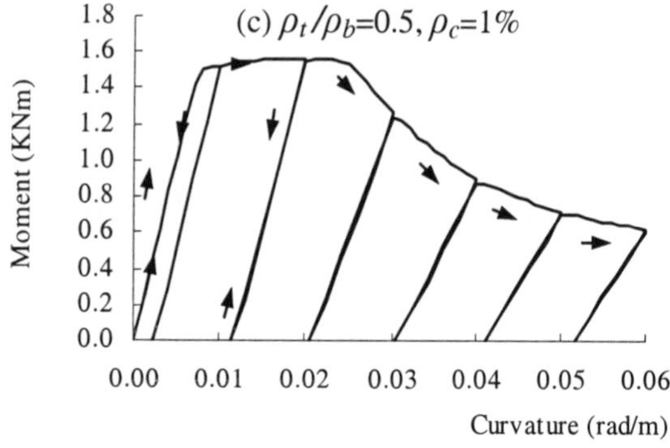

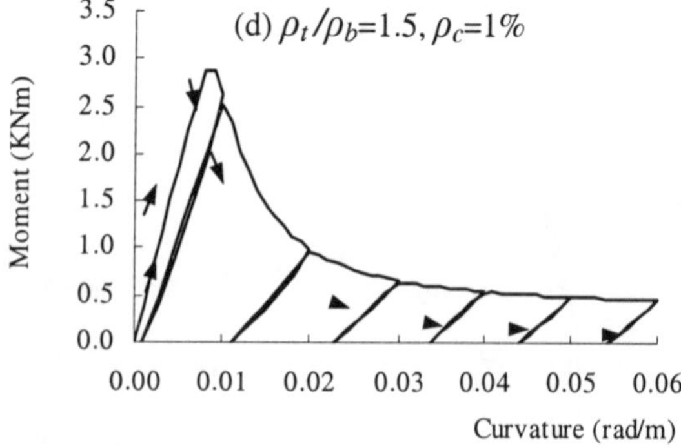

Fig. 5 Moment-curvature relations under cyclic loading of some sections with *in situ* concrete strength 90MPa

In general, the reloading path is quite close to the unloading path and the difference increases with the imposed curvature for singly reinforced sections. This can be attributed to the large difference in unloading and reloading paths of concrete in compression. The presence of compression reinforcement tends to reduce this difference. Upon unloading and reloading, the path always returns to the envelope curve for monotonic loading. Unloading at small curvatures before reaching the peak moment is normally along a path broadly parallel to the initial tangent. However, unloading at large curvatures after reaching the peak moment is along paths at reduced inclinations, suggesting deterioration in stiffness. This is particularly noticeable in singly reinforced sections that are over-reinforced.

There are residual curvatures in most cases when the applied bending moment is removed. Notable exceptions are over-reinforced sections without compression reinforcement, which result in very small residual curvatures, as shown in Fig. 5(b). It is noted that the residual curvatures are predominantly caused by plastic strains in the steel reinforcement. For over-reinforced sections without compression reinforcement, the steel reinforcement remains elastic throughout and that explains why the residual curvatures are very small. Similar phenomena are also observed in the sections with *in situ* concrete compressive strength f_{co} of 30 and 60MPa.

5.3 *Variation of neutral axis depth*

The variations of the neutral axis depth d_n for sections with *in situ* concrete compressive strength f_{co} of 60MPa are plotted in Fig. 6, where the solid lines denote the monotonic loading curves and the dash lines stand for the paths for unloading and reloading.

The variations for under- and over-reinforced sections are markedly different but the presence or not of compression reinforcement is immaterial. For both under- and over-reinforced sections, the neutral axis depths remain nearly constant initially. As the curvature increases more, for the under-reinforced sections, the neutral axis depth decreases and then increases as the bending moment enters the post-peak stage. However, for the over-reinforced sections, the neutral axis depth keeps on increasing with curvature.

The difference between under- and over-reinforced sections is again observed in the variation of neutral axis depth upon unloading and reloading. When under-reinforced sections are unloaded, the neutral axis depth keeps decreasing with the curvature. On the contrary, when over-reinforced sections are unloaded, the neutral axis depth increases as the curvature decreases. Similar phenomena are also observed in the sections with *in situ* concrete compressive strength f_{co} of 30 and 90MPa.

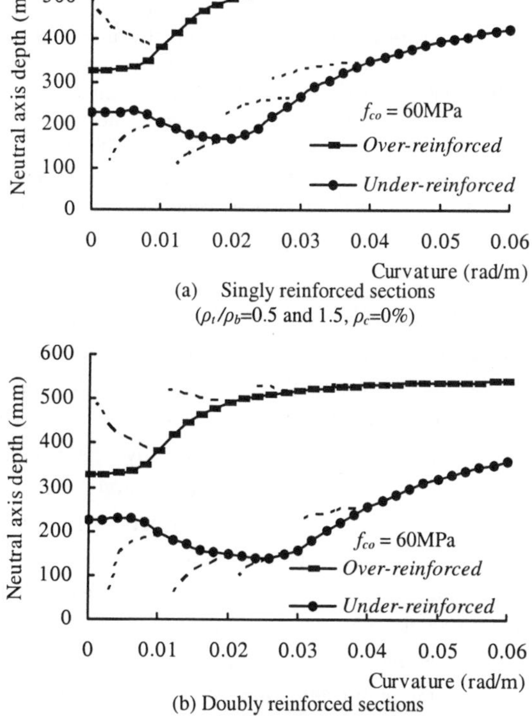

(a) Singly reinforced sections (ρ_t/ρ_b=0.5 and 1.5, ρ_c=0%)

(b) Doubly reinforced sections (ρ_t/ρ_b=0.5 and 1.5, ρ_c=1%)

Fig. 6 Variation of neutral axis depth of sections with *in situ* concrete strength 60 MPa

6. Conclusions

The moment-curvature relationship of reinforced concrete beams made of normal- and high-strength concrete under monotonic and cyclic loading is studied using a numerical method that employs the actual stress-strain curves of the constitutive materials and takes into account the stress-path dependence of the concrete and steel reinforcement. The results show that the complete moment-curvature relationship is path-dependent, which is similar to the material stress-strain relationship with stress-path dependence. It has also been observed that the monotonic curve and the variation of neutral axis depth are quite different for under- and over-reinforced sections.

References

1. ACI Committee 363, *State-of-the-art report on high strength concrete*, ACI 363-R92, American Concrete Institute, Detroit, U.S.A. (1992).
2. D. J. Carreira and K. H. Chu, The moment-curvature relationship of reinforced concrete members, *ACI Journal* **83**(2), 191-198 (1986).
3. H. J. Pam, A. K. H. Kwan and J. C. M. Ho, Post-peak behavior and flexural ductility of doubly reinforced normal- and high-strength concrete beams, *Structural Engineering and Mechanics* **12**(5), 459-474 (2001).
4. A. K. H. Kwan and F. T. K. Au, Flexural strength-ductility performance of flanged beam sections cast of high-strength concrete, *The Structural Design of Tall and Special Buildings* **13**(1), 29-43 (2004).
5. D. C. Kent and R. Park, Inelastic behavior of reinforced concrete members with cyclic loading, *Bulletin of New Zealand Society of Earthquake Engineering* **4**(1), 108-125 (1971).
6. R. H. Brown and J. O. Jirsa, Reinforced concrete beams under load reversals, *ACI Journal* **68**(5), 380-390 (1971).
7. M. M. Attard and S. Setunge, The stress-strain relationship of confined and unconfined concrete, *ACI Materials Journal* **93**(5), 432-442 (1996).
8. M. Elmorsi, M. R. Kianoush and W. K. Tso, Nonlinear analysis of cyclically loaded reinforced concrete structures, *ACI Structural Journal* **95**, 725-739 (1998).
9. C. F. Gerald and P. O. Wheatley, *Applied Numerical Analysis*, 6th Ed., Addison-Wesley, USA, 698pp (1999).

NUMERICAL STUDY OF PLATE STRENGTHENED DEEP RC COUPLING BEAMS WITH DYNAMIC SET ANCHORS *

R.K.L. SU

Department of Civil Engineering, University of Hong Kong, Pokfulam Road, Hong Kong, China

Y. ZHU

Department of Civil Engineering, University of Hong Kong, Pokfulam Road, Hong Kong, China

Coupling beam is an important member that has great influences on the behavior of coupled shear-wall structures under lateral loads. In conjunction with the society and economic development, more and more old reinforced concrete structures need to be strengthened, retrofitted, refurbished or rehabilitated to satisfy the modern requirements. This research aims to study the use of steel plates to strengthen deep RC coupling beams with dynamic set anchors. The numerical model developed for the nonlinear finite element analysis (NLFEA) is validated by the previous experimental results. Parametric study is then conducted to investigate the effects of varying thicknesses of steel plates and bolt group arrangements on the load-rotation behaviors of the strengthened beams. A set of design recommendations has been developed for strengthening deep RC coupling beams by attachment with steel plates using dynamic set anchors.

1. Introduction

Reinforced concrete (RC) is one of the most commonly used construction materials in modern society as it is easy to form and relatively inexpensive. However when a building is torn down simply because of it is no longer meets the modern requirements, the disposal of the waste material (concrete) obviously places a great strain on the environment. RC material is not good in recycling, to reuse the buildings and to protect the environment, substantial strengthening, retrofitting, refurbishment or rehabilitation of different structural components may be required. Strengthening of RC floor beams, slabs and columns by fibre-reinforced polymer [1] and steel plate [2] has been studied comprehensively. Using steel plate to strengthen medium long RC coupling beams has recently been conducted [3-4]. The improved inelastic responses of the strengthened beams supported the use of bolted connections for the case of high seismic loading and displacement demands. However, deep RC coupling beams, which are structurally different from the medium long coupling beams [5-6] have not been studied so far. In this paper, nonlinear finite element analysis (NLFEA) has been used to study deep RC coupling beams strengthened with steel plates using dynamic set bolts that were specially designed by Hilti Corporation to transfer dynamic loads. The dynamic set anchor is advantageous to reduce bolt-slip between the clearance hole and the bolt shank as the gap is filled up by injection adhesive [8]. In this study, the numerical model developed will be validated by the previous experiments [5]. Then parametric study will be conducted to investigate the effects of varying thicknesses of the steel plates and bolt group arrangements on the load-rotation behavior of the strengthened beams. A set of design recommendations will be developed for strengthening deep RC coupling beams by attachment with steel plates using dynamic set anchors.

2. Installation Procedure and Load-slip Behavior of Dynamic Set Anchors

Dynamic set anchor is particularly suitable for fixing steel members to existing RC structures. The anchor consists of a spherical washer, which permits injection adhesive to be dispensed into the clearance hole, a spherical washer, a nut and a lock nut as shown in Figure 1. To minimize the bolt slip, special installation procedures for dynamic-set anchors should be adopted. Firstly, the specified locations of the anchor bolts have to be identified on the surfaces of the RC walls or beam. A diamond coring machine is then used to bore holes

* This work is supported by Research Grants Council of Hong Kong SAR (Project Nos. HKU7129/03 & 7110/05E).

through the concrete. It is noted that coring is preferable rather than drilling of holes using rotary hammer drill machine, as the latter would lead to undesirable concrete damage around the drilled hole as well as at the far face of the shear wall. The diameter of the bolt holes should be slightly larger than the shank diameter of the anchor bolt by 4 to 5mm to ensure an easy installation of the anchor bolt and injection of adhesive mortar. The holes are cleaned up to remove all the dust and standing water. After that adhesive mortar is injected into the hole using foil pack holder until half of the depth of the hole is filled. The anchor bolt is then slowly inserted into the hole to ensure that the bolt is correctly aligned and the annular gap is filled up by adhesive mortar. After curing of the adhesive mortar, the steel plate is fixed onto the wall by tightening the dynamic set. Finally, adhesive mortar is injected into the clearance hole from the injection washer (see Figure 1) to fill up the gap between the clearance hole and the bolt at the steel plate.

	X	Y
A_1	0.40	175
B_1	4.20	285

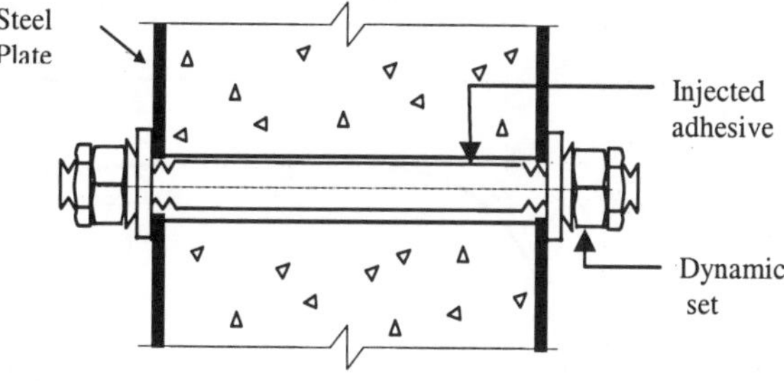

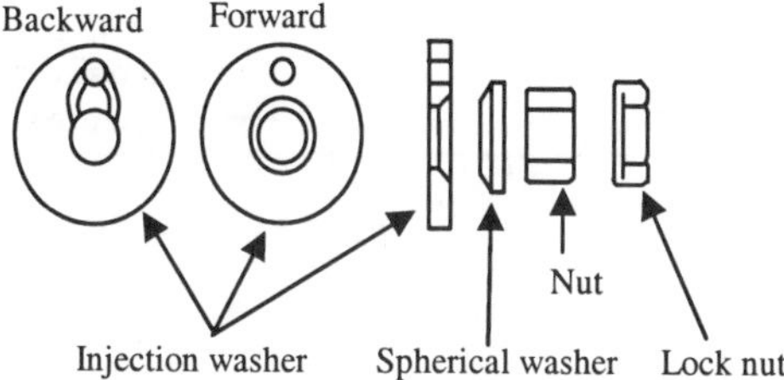

Figure 1. Dynamic set details.

Double shear bolt tests for dynamic set anchor have been conducted. It is found that all the bolts eventually underwent shear failures and the bolt heads were torn off from the bolt shanks. The measured load-slip relationships are presented in Figure 2. Very high initial stiffness was found as the gap between the bolt and the clearance holes had been completely filled up by the high strength adhesive. The experimental result demonstrates that the dynamic set connection is effective to control the bolt slip.

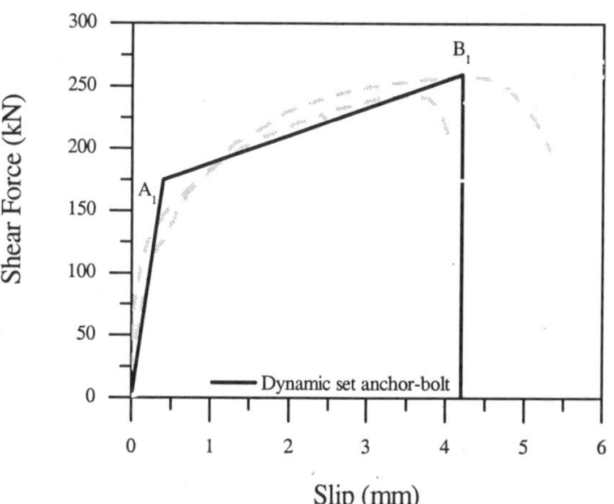

Figure 2. Idealized load-slip relationship of double shear of a M20 dynamic set anchor bolt.

3. Finite Element Analysis of Deep RC Coupling Beams

Extensive experimental studies on deep RC coupling beams had been conducted in The University of Hong Kong and the results of deep RC coupling beams subjected to monotonic or cyclic loading had been presented [5-6]. In the present study, Unit MCB1 [5] of the coupling beam specimens was chosen as a control RC coupling beam to compare the behaviors with that of the steel plate strengthened deep RC coupling beams with dynamic set anchor bolts.

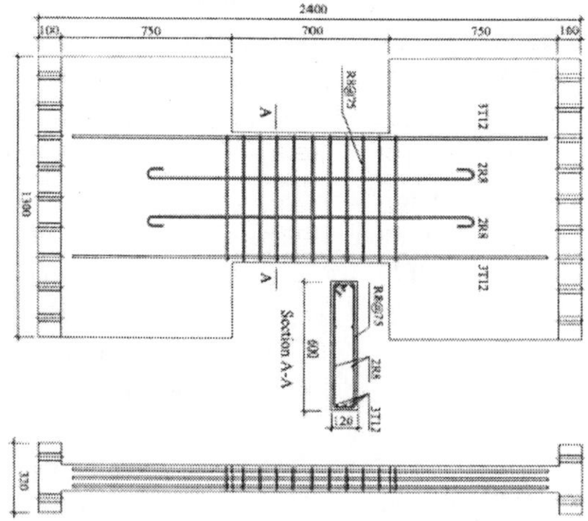

Figure 3. Dimensions and details of Unit MCB1.

The dimensions and reinforcement details of Unit MCB1 are shown in Figure 3, in which the span depth ratio is 1.17, the thickness and clear span of the beam are 120mm and 700 mm respectively. The RC coupling beam with shear wall panels was modeled by ATENA [7], which is a nonlinear finite element program. The numerical results are then compared with the experimental results to validate the accuracy of the numerical model. After that, this RC beam model together with the models of the steel plate and dynamic set anchors is used for the subsequent parametric study of steel plate strengthened deep RC coupling beams.

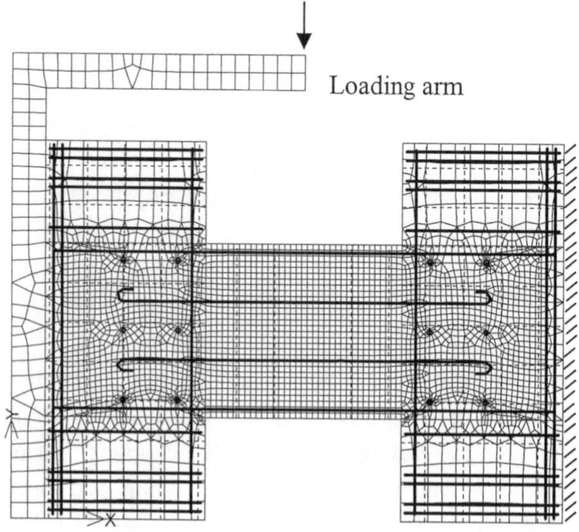

Figure 4. Mesh configuration of Unit MCB1 for NLFEA.

Figure 4 shows the two-dimensional finite element mesh of Unit MCB1 in reference [5]. Both smeared and discrete reinforcement models have been used in the analysis. The concrete parts were modeled by using four and three node iso-parametric plane stress elements. The main reinforcements and shear reinforcements were modeled by discrete two-node bar elements with consideration of bond-slip effect. Smeared reinforcement finite elements were used to model the distributed bars in the wall panels. The concrete model adopted in the NLFEA was according to the models available in ATENA [7] and the test results of the local concrete properties [9]. The factors have been considered in the nonlinear concrete material model including: nonlinear behavior in compression including hardening and softening; fracture of concrete in tension based on nonlinear fracture mechanics; biaxial strength failure criterion; reduction of compressive strength after cracking; and reduction of the shear stiffness after cracking (variable shear retention). A bi-linear stress-strain relationship with yielding was assumed for the reinforcement bars. The effect of bond-slip between the reinforcement and surrounding concrete was considered and the relationship shown in CEB-FIB [10] model code 1990 was adopted in this analysis. Monotonically increase loading was adopted in the present numerical simulation. Revised Newton-Raphon solution strategy was employed to calculate the nonlinear solution.

3.1 *Load-displacement relationships*

Based on the modeling methodologies as described in the last sections, numerical simulation by NLFEA for the RC coupling beam was carried out. The results were compared with those available from tests to validate the numerical model. The predicted and measured shear force-chord rotation relationships are shown in Figure 5. The stiffness and the peak loads of them are in very good agreement. The peak loads from numerical and experimental studies are 336 and 350kN, respectively. The error between the numerical and experimental results is less than 5%. The results show that the NLFEA is accurate and reliable for analysis of RC structures.

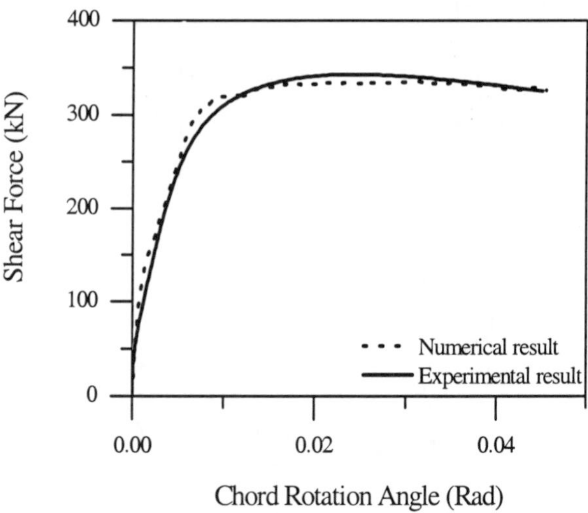

Figure 5. Predicted and measured load-rotation curves of Unit MCB1

3.2 *Crack patterns of concrete*

The experimental results [5] of Unit MCB1 reported that flexural cracks were formed at the tension area of the beam-wall joints and extended parallel to the beam-wall interfaces; combined flexural and shear cracks were formed around the beam-wall joints and diagonal shear cracks were appeared in the span of the deep coupling beam. Figure 6 shows the comparison of the calculated and observed crack patterns in Unit MCB1. The crack patterns are qualitatively agreed between the experimental and numerical results.

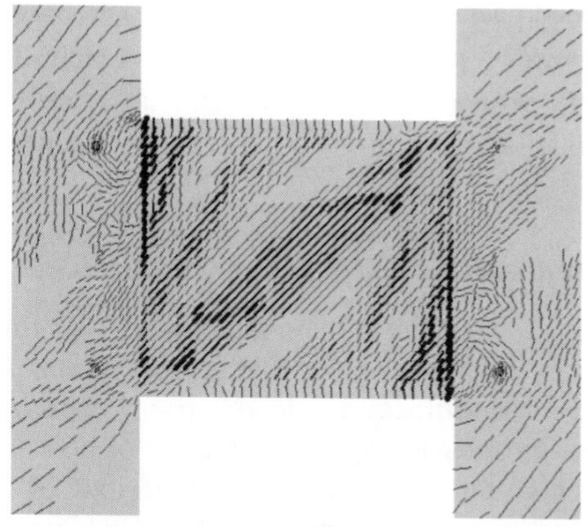

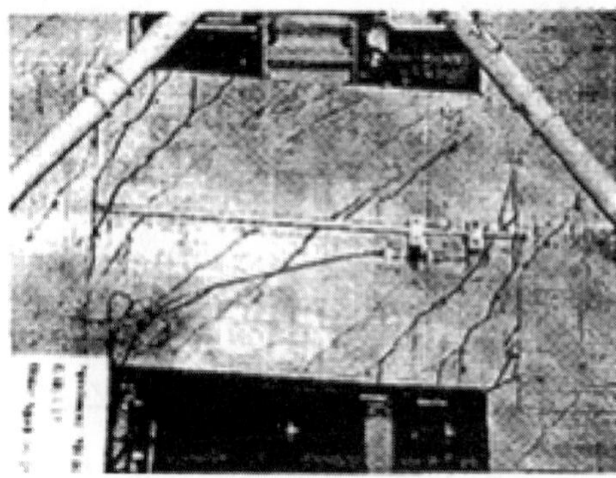

Figure 6. Predicted and observed crack patterns of Unit MCB1 at peak load.

4. Finite Element Analysis of Plate Strengthened Deep RC Coupling Beams

In this study, numerical instead of experimental method is used to study deep RC coupling beam (Unit MCB1) strengthened by steel plates with dynamic set anchor-bolts to avoid the cumbersome and costly laboratorial work. Similar approach has been used in Reference [8] to study the medium long RC coupling beams. The study had revealed that the beams strengthened by steel plates with cast-in or dynamic set anchor-bolts could sustain higher loading and larger deformations than the non-strengthened beams.

In the present study, the plate strengthened RC coupling beam and shear wall panels were modeled by NLFEA. The material parameters used for modeling the RC components are same as those used in the previous model as described in the last section. Accurate simulation of the load-slip effect of dynamic set anchor-bolts is essential [8] to obtain reliable structural responses of the strengthened beams. A tri-linear bolt-slip model of a M20 dynamic set anchor (as shown in Figure 3) obtained from double shear bolt test is used to simulate the load-slip behavior. Figure 7 shows a special mesh configuration developed for modeling the dynamic set anchor-bolt connection. In this figure, the inner circular finite element mesh represents the shank of the steel anchor-bolt, the outer ring in the finite element mesh having bi-linear softening material is used to model the load-slip behavior of the dynamic set anchor-bolt connections. Elasto-plastic plane elements were used to model the steel plate. The number of bolts used in strengthening deep RC coupling beam was designed to transfer all the bending and shear forces from the steel plate to the wall anchor by using classical bolt group theory.

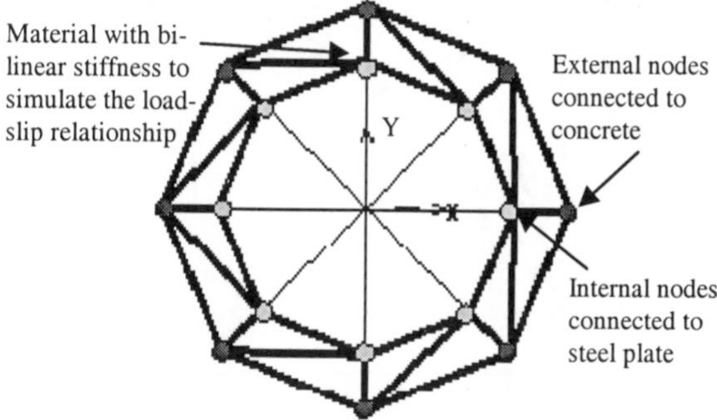

Figure 7. Finite element modeling of a dynamic set anchor bolt.

4.1 *Effect of varying thicknesses of the steel plates*

In this parametric study, different thicknesses of the steel plates were used to strengthen the deep RC coupling beam. The bolt group arrangement of Unit D6A, as shown in Figure 10, was adopted throughout this case study.

Figure 8 shows the calculated load-rotation curves of Unit MCB1 strengthened with different thicknesses of steel plates. The measured load-rotation curve for the control specimen Unit MCB1 [5] is also displayed for comparison. The models of Units 6MCB1, 3MCB1 and 2MCB1 are strengthened with steel plate of thickness 6mm, 3mm, and 2mm, respectively, on each side face of the beams. The peak loads of Units 6MCB1, 3MCB1 and 2MCB1 from the numerical analyses are 592kN, 576kN and 504kN, respectively. From Figure 8, it can be seen that the initial stiffness, peak load and yield displacement of the strengthened deep RC coupling beam are significantly increased. However, when 6mm steel plates are used, the initial stiffness of the strengthened deep RC coupling is same as that strengthened with 3mm steel plates and the peak loads of them are almost the same. There are two possible reasons for explaining these phenomena. Firstly, the concrete around the dynamic set shank could only provide limited bearing force due to local damage of concrete. As a result, using steel plate with excessive thickness will not further increase the beam strength but just waste of material. Secondly, yielding of steel around the clearance holes in the steel plates would limit the load carrying capacity of the beam. However, in this case, the saturation of peak loading of the strengthened beam is mainly due to the former reason.

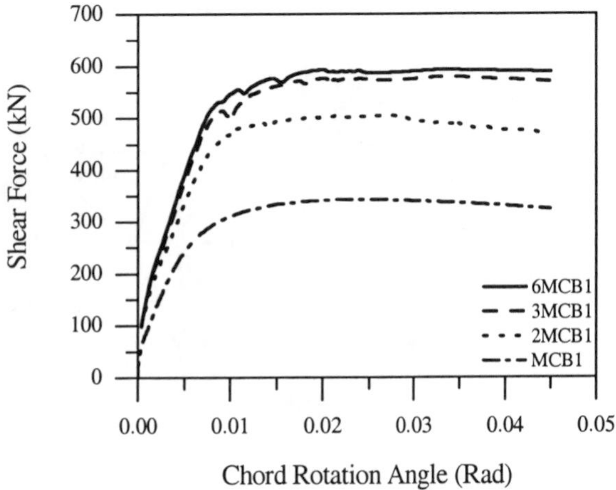

Figure 8. Load-rotation curves of strengthened beams with different thicknesses of steel plates.

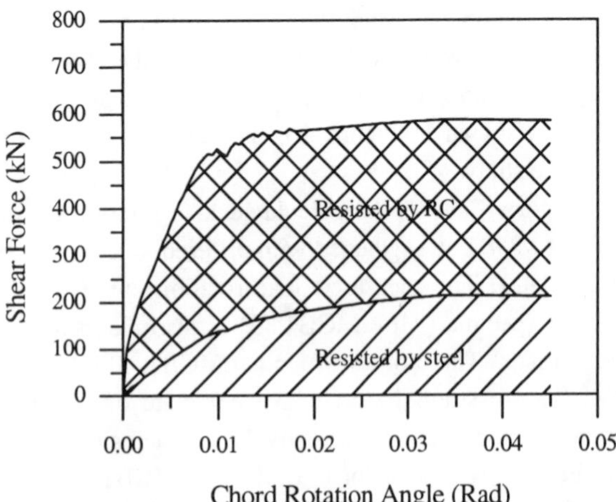

Figure 9. Comparison of shear force taken by the steel plates and the total applied shear force of Unit 3MCB1.

Figure 9 shows the variation of the internal shear force at the mid-span of the steel plates of Unit 3MCB1. The internal shear force obtained from the numerical analysis was calculated. The peak load of the strengthened beam is 576kN and the peak internal shear force of the steel plates is 201kN. The steel plates fixed by dynamic set anchor-bolts could take up almost 35% of the total shear load when the beam reached its ultimate loading.

4.2 *Effect of varying bolt group arrangements*

Bolt group arrangement is often the primary concern for designing external steel plate strengthened beams as it might affect the stiffness and load carrying capacity of the beams. In order to evaluate the effect of bolt group arrangements on the structural behavior of the strengthened coupling beams, Unit 3MCB1 was re-analyzed using four different bolt group arrangements, which consisted of four to six dynamic set bolts and are namely, Units D6A, D6B, D4A and D4B as shown in Figure 10. The load-rotation curves of the strengthened beams are presented in Figure 11.

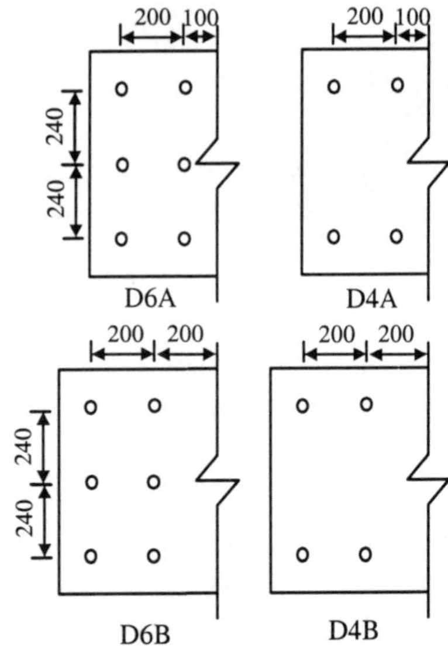

Figure 10. The bolt group arrangements

Units D6A and D6B were both composed of six bolts and the distances from the beam-wall interface were 100mm and 200mm, respectively. Numerical analysis revealed that the stiffness and the peak loading of the two bolt groups were very similar (see Figure 11) and were not affected by minor shifting of the bolt group.
To obtain a more efficient bolt group arrangement with minimum number of bolts, all the interior bolts for Units D6A and D6B were taken away and new bolt group

arrangements D4A and D4B were formed. NLFEA found that the peak loadings of Unit D4A and D4B were slightly reduced when compared with its original forms (Units D6A and D6B). The maximum shear forces at Units R6A, R6B and R4A, R4B were however very similar with differences less than 5%.

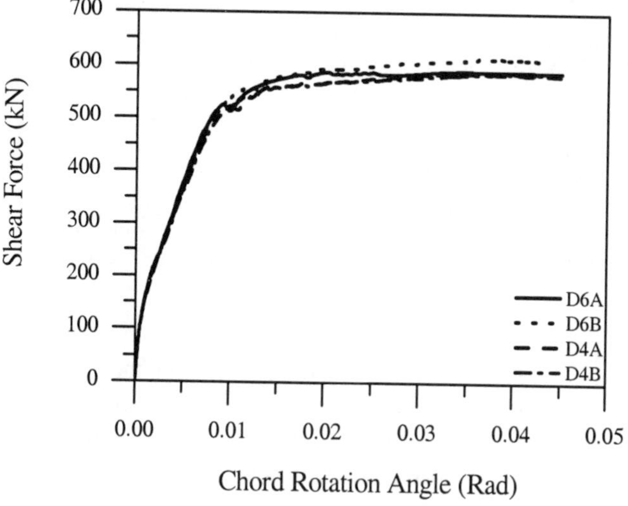

Figure 11. Load-rotation curves of different bolt group arrangements.

According to the NLFEA, design recommendations of using M20 dynamic set anchor bolts for fixing the steel plates onto deep RC coupling beams are listed as follows:

(1) The spacing of the anchor bolts should be a multiple of the wall reinforcement spacing so as to avoid clashing with the wall reinforcement.

(2) Actual spacing of the wall reinforcement (typically ranged from 100mm to 300mm) adjacent to the anchorage should be determined on site by cover meter.

(3) All the anchor bolts should be located at 100mm or more away from the beam-wall joint to avoid the location where the plastic hinge at joint would be formed during loading.

(4) Minor shifting of the bolt group location would not significantly affect the loading capacity of the strengthened beams.

(5) To avoid concrete premature damage and yielding of steel plates around the anchors, more number of bolts and longer anchorage length of the steel plates with wider spacing between the anchor bolts may have to be used in order to reduce the bolt forces.

5. Conclusions

In this study, NLFEA has been conducted to simulate the plate strengthened deep RC coupling beams using dynamic set anchor-bolts. The load-slip relationships of the dynamic set bolts have been measured from the laboratory test and simulated in the numerical analyses. The numerical study has shown that the strength and stiffness of the steel plate strengthened deep RC coupling beams could be significantly improved. The effects of varying plate thicknesses and the bolt group arrangements have also been investigated. It is found that minor shifting of the bolt group location would not significantly affect the loading capacity of the strengthened beams.

Acknowledgments

The generous technical supports from Hilti Corporation are gratefully acknowledged. Special thank is expressed to Dr. H.J.Pam who permitted this project to use the licensed finite element package ATENA for the numerical analysis.

References

1. J.G. Teng, J.F. Chen, S.T. Smith, L.Lam, *Chichester, West Sussex, Wiley,* (2002).
2. D. Oehlers, R. Seracino, *Oxford, Elsevier* (2004).
3. R.K.L. Su, Zhu Y, *Pro. of the Inter. Conf. on Struct. and Foundation Failures, 2-4 August, Singapore, Edited by Wang CM and Murugappan K, IES/IStructE* (2004) pp. 365-374.
4. R.K.L. Su, Y. Zhu, *J. of Eng. Struct.* (Accepted in May 2005)
5. A.K.H. Kwan, Z.Z. Zhao, *Proc ICE Struct Build,* **152**, 67-78, (2002).
6. A.K.H. Kwan, Z.Z. Zhao, *Proc ICE Struct Build,* **152**, 283-293 (2002).
7. V. Cervenka, J Cervenka, *Cervenka Consulting, Prague, Czech Republic*, (2002).
8. Y. Zhu and R.K.L. Su, *Submitted to FE in Analys. and Design*, June (2005).
9. L.S.B. Cheng, *Final Year Project, Department of Civil Engineering, HKU,* (2000-2001).
10. Comité Euro-International du Béton, *London, Thomas Telford* (1993).

A METHOD FOR DETERMINING OPTIMUM LOCATIONS AND NUMBER OF STIFFENING COUPLING BEAMS IN COUPLED SHEAR WALL STRUCTURE

LIU QINGSHAN LIANG XINGWEN DENG MINGKE

School of Civil Engineering, Xi'an University of Architecture & Technology, Shanxi Xi'an, 710055, China

It is shown in experimental study that the coupled shear walls with stiffening coupling beams is superior to the common coupled shear walls in aseismic performance. But the available method of determining the optimum locations and number of stiffening coupling beams is not presented currently. In this paper, several hypotheses based on uniformly coupled shear walls are made. The restraining effect of weak coupling beams is incorporated to walls whose rigidity is expressed with equivalent rigidity. Thus, the coupled shear walls is simplified to a coupled shear walls which only includes stiffening coupling beams. The restraining moment applied to walls is acquired through the equation of deformation compatibility at the location of stiffening coupling beam. The top drift of structure is got with unit-load method. The optimum locations are obtained by maximizing the drift reduction caused by the stiffening coupling beams, while the needed number n of stiffening coupled beams is worked out according to the top drift of structure meeting the serviceability limit state.

1. Introduction

It's well known that horizontal load such as wind load and horizontal seismic load the major loads in high-rise structure. However, the demand of stiffness of structure is different for resisting wind load and earthquake load. Wind loading can't be reduced by plastic deformation of structure, so the structure needs bigger lateral stiffness under wind load, while, such situation is disadvantageous for earthquake resistance. In order to solve the problem, the depth of the coupling beams in coupled shear walls is designed very low or they are directly made of strengthened slabs. This type of structure is characterized by good ductility than frame structure with strong columns and weak beams. The reduction of stiffness for the structure is caused by the weak coupling beams which decrease the restraining effect to walls, which tend to become two cantilever walls, as brings about another probability. It is very likely that the lateral stiffness of the structure can't meet serviceability limit state, the crack of the coupling beams may be widely open under serviceability load, and sometimes, there are even plastic hinges in them. Therefore, several stiffening coupling beams are setup at certain locations such as plant levels to strengthen restraining effect to walls and to improve the lateral stiffness of the coupled shear walls. This kind of structure is termed by coupled shear walls with stiffening coupling beams.

Test results in literature [1] shown that the kind of structure is superior to the structure with common R.C. coupling beams, and comparing with them, the top drift decreases by 50% after cracking of the structure and the ultimate strength increases by 58%. There is large deformation capacity, ductility ratio and many position of energy dissipation. Hence, it is suitable for the coupled shear walls whose integrity is inferior. The structure is endowed with superior seismic performance when the stiffening coupling beams have preferable capacity of deformation and energy dissipation.

But the available method of determining the optimum locations and number of stiffening coupling beam is not presented currently. In this paper, under several hypotheses based on uniformly coupled shear walls, the calculation method of the optimum locations and number of stiffening coupling beams is given by the author.

2. Simplified Calculating Model and Analysis of Structure

2.1 *Simplified calculating model structure*

To simplify calculation, the method of analysis is based on the following assumptions:

1. The structure is linearly elastic.
2. The stiffening coupling beams are rigidly attached to the walls and the walls are rigidly attached to the foundation.
3. The sectional properties of walls, stiffening coupling beams and weak coupling beams are uniform throughout their height.

4. The axial deformation of coupling beams is neglected.

The coupled shear walls with the stiffening coupling beams is shown in Figure 1(a), it can be regarded as superposition of the coupled shear walls only with the weak coupling beams[Figure 1(b)] and that only with the stiffening coupling beams[Figure 1(c)]. The analysis method of the former is presented in the literatures [2,3]. The contribution of the weak coupling beams to the lateral stiffness of the structure can be considered with equivalent rigidity of the walls. Thus, the Figure 1(a) is equivalent to the Figure 1(d). In the Figure 1(d), the stiffness of the walls is expressed with equivalent rigidity EI_{eq} which considers the restraining effect of weak coupling beams. The simplified calculating model of structure, the Figure 1(d), is only used for determining the optimum locations and number of stiffening coupling beams.

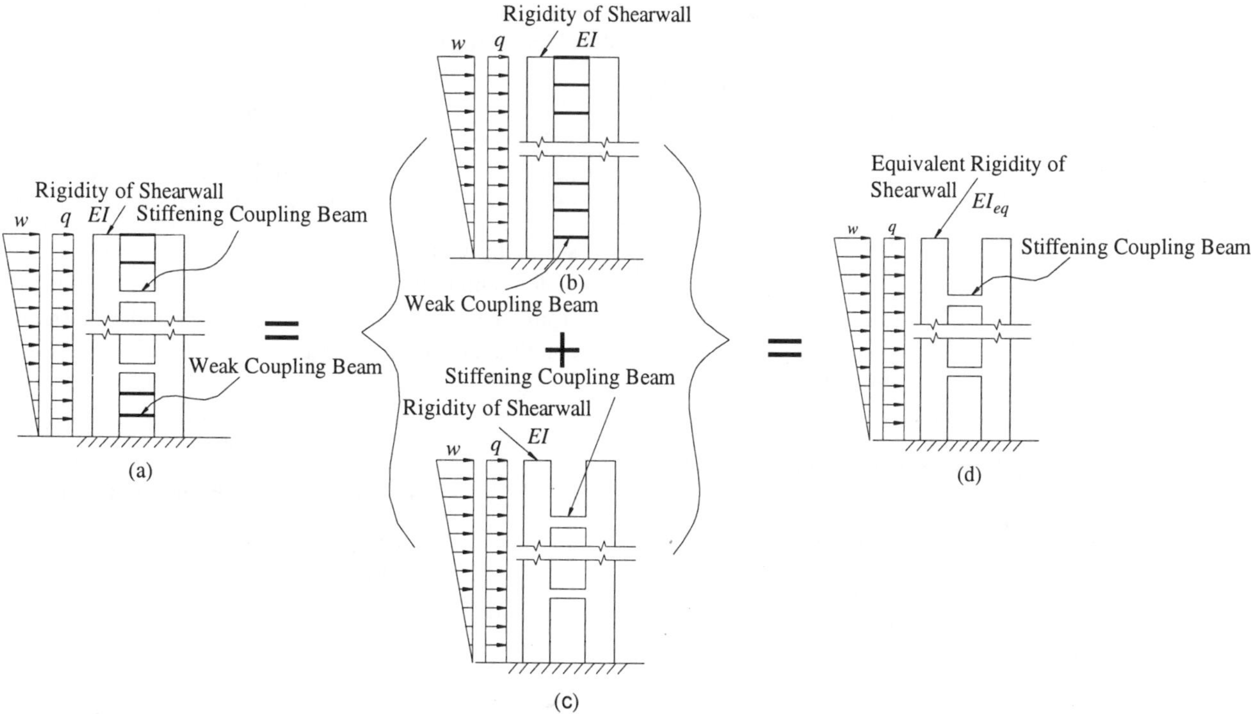

Figure 1 Simplified Calculating Model of Structure

2.2 *Analysis of structure*

A coupled shear walls structure with two stiffening coupling beams will be used to demonstrate the method of analysis because it includes all the necessary steps in their simplest form. The analysis of structures with more or fewer than two stiffening coupling beams can then easily be performed on the basis of the method for the two stiffening coupling beams case. The model for analysis is shown in Figure 2(a), subjected to a uniformly distributed horizontal load. A triangular load distribution with its apex at the base has also been made.

The bending moment diagram for the walls [Figure 2e] consists of the external load moment diagram [Figure 2b] reduced by the stiffening coupling beam restraining moments that, for each stiffening coupling beam, are introduced at the stiffening coupling location and extended uniformly down to the base [Figure 2c and 2d].

In Figure 2(b), 2(e), the bending moment $M(x)$ is $qx^2/2$ under uniformly distributed horizontal load, and $w(x^2-x^3/3H)/2$ under triangular load in which q is the intensity of the uniformly distributed horizontal load, and w is the maximum value of triangular load.

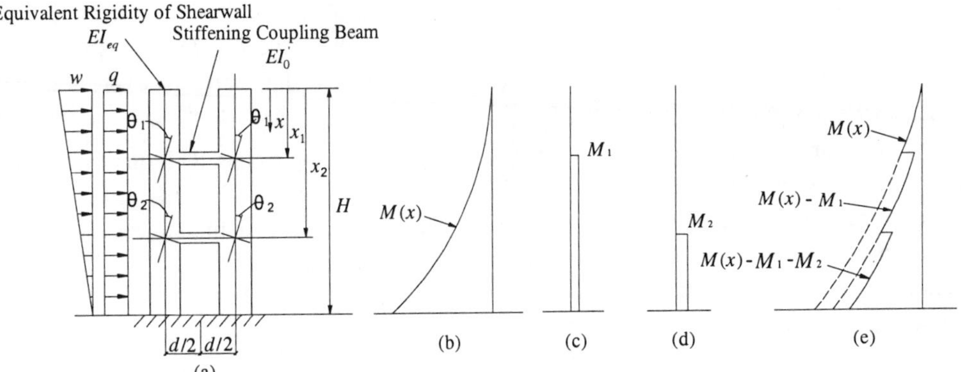

Figure 2 (a) Two Stiffening Coupling Beams (b) External Moment Diagram (c) M1 Digram (d) M2 Digram (e) Walls Resultant Digram

Under uniformly distributed horizontal load, from unit-load method, the wall rotations at locations 1 and 2 are, respectively

$$\theta_1 = \frac{1}{EI_{eq}} \int_{x_1}^{x_2} (\frac{qx^2}{2} - M_1)dx + \frac{1}{EI_{eq}} \int_{x_2}^{H} (\frac{qx^2}{2} - M_1 - M_2)dx \tag{1}$$

$$\theta_2 = \frac{1}{EI_{eq}} \int_{x_2}^{H} (\frac{qx^2}{2} - M_1 - M_2)dx \tag{2}$$

in which EI_{eq} and H are the equivalent flexural rigidity and total height of wall, q is the intensity of the uniformly distributed horizontal load, x_1 and x_2 are the respective height of the stiffening coupling beams 1 and 2 from the top of the wall, M_1 and M_2 are their respective restraining moment on the wall.

Expressions for the rotations of the stiffening coupling beams at the points where they are connected to the walls are, respectively,

$$\theta_1 = \frac{M_1 d}{12EI_0} \tag{3}$$

$$\theta_2 = \frac{M_2 d}{12EI_0} \tag{4}$$

in which d is the horizontal distance from the centroid of walls, EI_0 is the effective flexural rigidity of the stiffening coupling beam [Figure 3b], allowing for the wide-column effect of the wall, which can be obtained from the actual flexural rigidity of the stiffening coupling beam EI_0' [Figure 3a] as

$$EI_0 = \frac{1}{(1-2c)^3(1+\gamma)} EI_0' \tag{5}$$

in which γ is the coefficient of the shear deformation influence, $\gamma = \frac{12\mu EI_0'}{GAl_n^2}$, c is the ratio of the rigid zone length to d, l_n is the net span of the coupling beam.

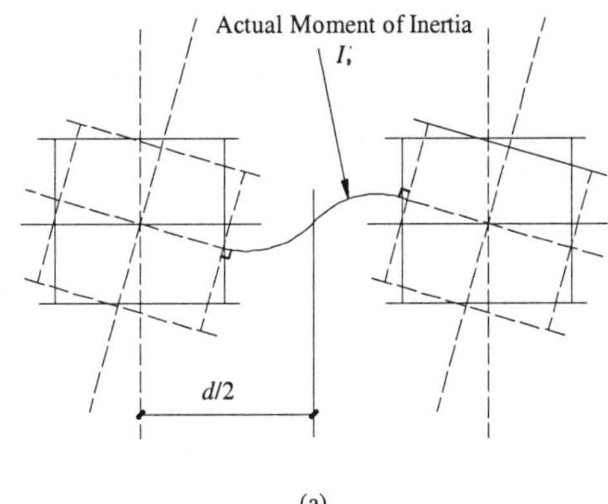

(a)

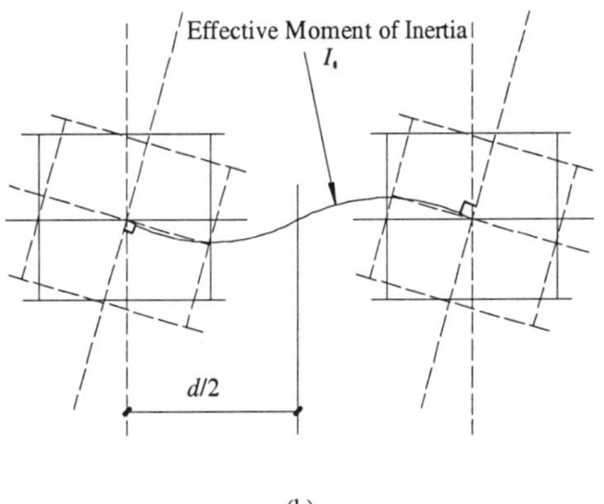

(b)

Figure3 (a) Stiffening Coupling Beam Attached to Edge of Walls
(b) Equivalent Stiffening Coupling Beam Attached to Centroid Walls

Equating the rotations of the wall and the stiffening coupling beam at location 1, Eq.(1) and Eq.(3) ,respectively

$$\frac{1}{EI_{eq}}\int_{x_1}^{x_2}(\frac{qx^2}{2}-M_1)dx+\frac{1}{EI_{eq}}\int_{x_2}^{H}(\frac{qx^2}{2}-M_1-M_2)dx=\frac{M_1 d}{12EI_0} \tag{6}$$

and similarly for the rotations at location 2, equating Eq.(2) and Eq.(4),

$$\frac{1}{EI_{eq}}\int_{x_2}^{H}(\frac{qx^2}{2}-M_1-M_2)dx=\frac{M_2 d}{12EI_0} \tag{7}$$

Rewriting Eq.(6) and Eq.(7) gives

$$M_1\left[(H-x_1)+\frac{1}{12}\beta H\right]+M_2(H-x_2)=\frac{1}{6}q(H^3-x_1^3) \tag{8a}$$

and

$$M_1(H-x_2)+M_2\left[(H-x_2)+\frac{1}{12}\beta H\right]=\frac{1}{6}q(H^3-x_2^3) \tag{9a}$$

Similarly, under triangular load, the equations of deformation compatibility are

$$\begin{aligned}&M_1\left[(H-x_1)+\frac{1}{12}\beta H\right]+M_2(H-x_2)\\&=\frac{w}{24}\left[4\left(H^3-x_1^3\right)-\frac{1}{H}\left(H^4-x_1^4\right)\right]\end{aligned} \tag{8b}$$

$$\begin{aligned}&M_1(H-x_2)+M_2\left[(H-x_2)+\frac{1}{12}\beta H\right]\\&=\frac{w}{24}\left[4\left(H^3-x_2^3\right)-\frac{1}{H}\left(H^4-x_2^4\right)\right]\end{aligned} \tag{9b}$$

in which variable β is

$$\beta=\frac{(EI_{eq})/H}{(EI_0)/d}=\frac{I_{eq}d}{I_0 H}$$

Introducing the following functions

$$f_1(x)=\left[(H-x)+\frac{1}{12}\beta H\right]$$

$$f_2=H-x_2$$

$$f(x)=\begin{Bmatrix}\frac{1}{6}q\left(H^3-x^3\right)\\ \frac{1}{24}w\left[4\left(H^3-x^3\right)-\frac{1}{H}\left(H^4-x^4\right)\right]\end{Bmatrix}$$

The first function of the function $f(x)$ is suitable for uniformly distributed horizontal load and the second for triangular load. With two load cases, the equations of deformation compatibility are simplified as follows:

$$f_1(x_1)M_1+f_2M_2=f(x_1) \tag{8}$$

$$f_2M_1+f_1(x_2)M_2=f(x_2) \tag{9}$$

The simultaneous solution of Eq.(8) and Eq.(9) gives the restraining moments applied to the walls by the stiffening coupling beams at location 1 and 2 as follows:

$$M_1=\frac{f_1(x_2)f(x_1)-f_2f(x_2)}{f_1(x_1)f_1(x_2)-f_2^2} \tag{10}$$

$$M_2=\frac{f_1(x_1)f(x_2)-f_2f(x_1)}{f_1(x_1)f_1(x_2)-f_2^2} \tag{11}$$

The top drift of the structure may be determined from the resulting bending moment diagram for the walls using unit-load method and can be expressed as

$$u=u_0-\frac{1}{2EI_{eq}}\left[M_1\left(H^2-x_1^2\right)+M_2(H^2-x_2^2)\right] \tag{12}$$

in which the first term u_0 on the right-hand represents the top drift of the walls subjected to the full horizontal load, under uniformly distributed horizontal load, $u_0 = qH^4/(8EI_{eq})$, and under triangular load, $u_0 = 11wH^4/(120EI_{eq})$, while the two parts of the second term represent the reductions in the top drift due to the stiffening coupling beams restraining moments M_1 and M_2. When the coupled shear walls have n stiffening coupling beams, the top drift of the structure is

$$u = u_0 - \frac{1}{2EI_{eq}}\sum_{i=1}^{n} M_i\left(H^2 - x_i^2\right) \qquad (13)$$

3. Optimum Locations of Stiffening Coupling Beams

The preceding analysis is useful in not only providing estimates of the walls moment and the top drift, but also allowing an assessment of the optimum locations of the stiffening coupling beams to minimize the horizontal top drift. This is achieved by maximizing the drift reduction [i.e., the second term on the right-hand side of Eq.(12)].

Illustrating the procedure by continuing to consider the two stiffening coupling beams structure, the second term of its top drift equation [Eq.(12)] is maximized by differentiating with respect first to x_1, then to x_2, thus

$$\left(H^2 - x_1^2\right)\frac{dM_1}{dx_1} + \left(H^2 - x_2^2\right)\frac{dM_2}{dx_1} - 2x_1M_1 = 0 \qquad (14)$$

and

$$\left(H^2 - x_1^2\right)\frac{dM_1}{dx_2} + \left(H^2 - x_2^2\right)\frac{dM_2}{dx_2} - 2x_2M_2 = 0 \qquad (15)$$

in which dM_1/dx_1, dM_2/dx_1, dM_1/dx_2, dM_2/dx_2 are the derivatives of M_1 and M_2 from Eq.(10) and Eq.(11), respectively, with respect to x_1 and x_2. Substituting for M_1 and M_2 and their derivatives, Eq.(14) and Eq.(15) can be solved simultaneously for the values of x_1 and x_2 that define the optimum locations of the stiffening coupling beams. The solution of Eq.(14) and Eq.(15) is, obviously, complex; therefore, a numeral method of solution by computer is necessary.

In Eq.(14) and Eq.(15), there are variables H, β, q. To calculate x_1 and x_2, variable q can be eliminated, so the optimum locations x_1 and x_2 are only related to the variables H, β and the type of the load, independent of the variable q. Given the type of the load, the total height H and variable β, Eq.(14) and Eq.(15) can be solved simultaneously for the values of x_1 and x_2 with a computer program composed by the first author. For two load cases, the results are listed in Table 1. In the table, $H_1 = H - x_2$ and $H_2 = H - x_1$, i.e., H_1 and H_2 express the optimum locations x_1 and x_2, which are measured from the base to the stiffening coupling beams 1 and 2 locations. It is shown in Table 1 that H_i/H varies with variable β, when $\beta=0$, i.e., the stiffening coupling beams are flexurally rigid. For uniformly distributed horizontal load, H_1/H should be at approximately one-third heights , while, H_2/H should be at approximately two-thirds heights, H_i/H increases slightly with the increase of β. The optimum locations H_i/H of the stiffening coupling beams for the minimum top drift of structure subjected to a triangularly distributed load, with its apex at the base, are slightly higher than for structure subjected to a uniformly distributed load. Concerning the coupled shear walls structure with more or fewer than two stiffening coupling beams, in an analogous manner, the optimum locations of the stiffening coupling beams can also be given. Comprehensive analysis of various circumstances, the relation between x_i and n is

$$x_i = \frac{i}{n+1}H \qquad (i = 1 \sim n) \qquad (16)$$

in which n is the number of the stiffening coupling beams.

Table 1. The Optimum Locations under uniformly Distributed Horizontal and Triangular Loading

β	uniform load		triangular load	
	H_1/H	H_2/H	H_1/H	H_2/H

0.0	0.3148	0.6878	0.3374	0.7084
0.2	0.3332	0.6954	0.3556	0.7156
0.4	0.3492	0.7021	0.3713	0.7219
0.6	0.3632	0.7082	0.3851	0.7276
0.8	0.3758	0.7139	0.3975	0.7328
1.0	0.3872	0.7191	0.4087	0.7376
1.2	0.3977	0.7241	0.4191	0.7422
1.4	0.4074	0.7287	0.4287	0.7464
1.6	0.4165	0.7331	0.4377	0.7504
1.8	0.4250	0.7373	0.4461	0.7542
2.0	0.4330	0.7413	0.4540	0.7578
2.2	0.4406	0.7450	0.4614	0.7612
2.4	0.4478	0.7487	0.4685	0.7645
2.6	0.4547	0.7521	0.4753	0.7676
2.8	0.4612	0.7554	0.4818	0.7706
3.0	0.4675	0.7586	0.4879	0.7735

4. Number of Stiffening Coupling Beam

As we all know, one purpose of setting the stiffening coupling beams is to improve the lateral stiffness of the shear walls, which should make the structure meet the serviceability limit state. The lateral top drift may be limited within a special value as the serviceability limit state of the shear walls structure, i.e.,

$$u \le [u/H]H \qquad (17)$$

in which $[u/H]$ is allowable value for the top drift of structure, taken as 1/1100 ~ 1/1200[4].

Substituting Eq.(13) into Eq.(17) and simplifying results in

$$u_0 \le \frac{1}{1-\alpha}[u/H]H \qquad (18)$$

in which $\alpha = \delta/u_0$,$\delta = \sum_{i=1}^{n} M_i(H_i^2 - x_i^2)\Big/\left(2EI_{eq}\right)$,δ is the reduction of top drift by the stiffening coupling beams

Generally speaking, the number of the stiffening coupling beams needed is usually fewer, so in designing, the number n is first given, then the variable α can be determined by consulting table according to the type of the load, the number n and variable β. Substituting variable α into Eq.(18), if Eq.(18) is tenable, the number n is suffice, otherwise, another variable n is assumed and Eq.(18) is checked again until tenable. As the range of n is limited, so the number n can be determined by two times trial calculation in most cases. For two load cases, the relation between variables β and α is listed in Table 2.

Table 2. The Relation Between β and α

β	α	
	uniform load	triangular load
0.2	0.9373	0.6891
0.4	0.9201	0.6770
0.6	0.9040	0.6655
0.8	0.8888	0.6547
1.0	0.8744	0.6445
1.2	0.8606	0.6347
1.4	0.8475	0.6254
1.6	0.8350	0.6164
1.8	0.8230	0.6078
2.0	0.8115	0.5995
2.2	0.8004	0.5916
2.4	0.7897	0.5839
2.6	0.7794	0.5765
2.8	0.7694	0.5693
3.0	0.7598	0.5624

5. Conclusions

1. The optimum locations of the stiffening coupling beams are only related to the variable β, the total height H and the type of the load, independent of the variable q . When $\beta=0$, under uniformly distributed horizontal load, the optimum locations should be at approximately one-third and two-thirds heights, in a three stiffening coupling beams structure, they should be at approximately the one-quarter, one-half and three-quarters heights, and so on. Generally, for the optimum locations of an n stiffening coupling beams structure, the stiffening coupling beam should be placed at the 1/(n+1),2/(n+1) up to n/(n+1) height locations. when $\beta \ne 0$, The optimum locations increases slightly with the increase of variable β, they can be determined by consulting the Table 1.

While the optimum locations of the stiffening coupling beams for the minimum drift of structures subjected to a triangularly distributed load, with its apex at the base, are slightly higher than for structures subjected to a uniformly distributed load.

2. In designing the coupled shear walls with the stiffening coupling beams, the optimum locations and the number are coupled each other. Therefore, the number n is needed to be given first, the variable α can be determined by consulting table 2 according to type of load, number n and variable β.

References

1. Han Xiaolei, Liang Qizhi. *Experimental study on coupled shear walls with rigid coupling beam [J]*. Journal of Civil Engineering (in Chinese), 1994.27(1):21-28.
2. Bao Shihua, Fang Ehua . *Design of Tall Building Structures.* Tsinghua University Press (in Chinese), 1990.
3. Liang Xingwen, Shi Qingxuan, Tong Yuesheng. *Design of Reinforced Concrete Structures.* Science and Technology Documentation Press (in Chinese) , 1999.
4. Profession Standard of People's Republic of China. *Design and Construction Specification for Reinforced Concrete Structures of Tall Building (JGJ3-91).* China Architecture & Building Press, 1991.
5. B.S. Smith. *Tall Building Structures Analysis and Design.* John Wiley & Sons, Inc

A BREAKTHROUGH IN PRECASTING OF PUBLIC HOUSING BLOCKS IN HONG KONG*

S.C. LAM
Chief Structural Engineer
Hong Kong Housing Department

K.C. CHUNG
Senior Structural Engineer
Hong Kong Housing Department

S.W. SHAM
Structural Engineer
Hong Kong Housing Department

The Hong Kong Housing Authority (HA) pioneered the precast concrete construction in high-rise residential buildings in the early '90s by successful introduction of planar precast elements, i.e. precast façades, semi-precast floor slabs and precast staircases, which altogether amount to some 18% of the total concrete volume of a building. Seeing the immense potential of precasting in building construction to enhance quality and sustainability, the HA has been dedicated to bringing greater benefits to the public through more precasting. To increase the percentage of precasting, to precast structural walls (given that a standard housing block is a wall-slab structure) and to leap from planar to volumetric precast construction has thus become the prime and foremost task. The HA has recently made a breakthrough in this area and developed an Enhanced Precast and Prefabrication System (EPPS) for the construction of standard high-rise housing blocks. The EPPS contains a lot of innovative initiatives in prefabrication and precast construction that include the key precast structural walls and volumetric precast construction as well. With the EPPS, the precast concrete volume will substantially increase to 70% of the total concrete volume of a building. Evolution of the EPPS has satisfactorily progressed stage by stage – from design development, a site mock-up to the present stage that the precast buildings are under construction.

1. Introduction

Hong Kong's building industry has been continually striving to keep up with the latest developments around the world. It is always ready to explore new technologies and try out innovative construction methods.

In recent years, the HA has been actively pursuing the mechanized precast concrete system in its public housing developments with an aim to satisfy customers' need, upgrade built quality, improve construction safety, enhance environmental protection and increase cost effectiveness.

Back in the early 90's, the HA pioneered the precasting technique in high-rise residential buildings by introduction of precast façades, semi-precast floor slabs and precast staircases. Over the past decade or so, improvements have been made to both the design and construction of these precast elements. Now, these precast elements amount to some 18% of the total concrete volume of a building and the façades construction cost has substantially dropped to about half of that when they were first used. The resultant benefits in quality, safety, environmental protection and cost-effectiveness are quite noticeable.

However, the HA is not complacent with this level of precasting and is dedicated to bringing greater benefits to the public through more precasting. Drawing on overseas experience, consultants' expertise and in-house knowledge, the HA has recently developed an Enhanced Precast and Prefabrication System (EPPS) for an existing standard block type that comprises innovative precast structural walls and volumetric precast construction. This is a breakthrough achievement as the precast concrete volume will substantially increase to 70% of the total concrete volume.

The HA chose the Kwai Chung Flatted Factory Redevelopment Project as a pilot to try out the EPPS. Evolution of the EPPS has satisfactorily progressed stage by stage – from design development, construction of a mock-up on site to the current stage that the precast buildings are under construction.

2. The Project

The 2-hectare Site is surrounded by two residential estates - Kwai Chung and Kwai Hing Estates - with some industrial developments to the north. It is for a housing development that comprises 2 no. 41-storey New Harmony 1 blocks (NH1) and 1 no. 36-storey New Harmony Annex 5 block, providing 1,983 flats for a population of approximately 6,346 persons and a total domestic GFA of 90,518 m².

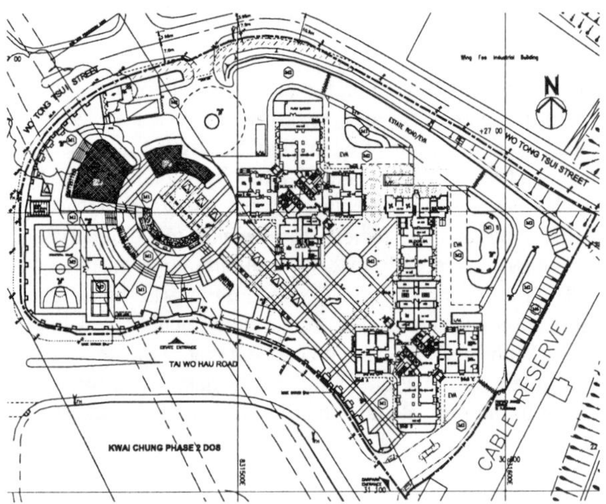

Figure 1 Site Location Plans

There are very severe site constraints - the Mass Transit Railway (MTR) protection zone (two MTR tunnels running underground at the west of the Site) and a cable reserve zone to the east. Ample open space thus provided renders the Site particularly suitable for this pilot project where on-site precasting of the crucial precast structural walls is mandatory to ensure close site supervision and quality control.

3. The Enhanced Precast and Prefabrication System (EPPS)

The EPPS is developed for the construction of the two standard 41-storey NH1 blocks with prime consideration not to change the standard structural layout for better cost effectiveness (minimizing associated changes to the architectural and building services works upon general application of the EPPS to other projects). It should be noted that the EPPS is not only restricted to the standard NH1 blocks; the various precast initiatives can also be adapted individually or in combination to other non-standard structural layouts.

In addition to the existing precast elements such as precast façades, semi-precast floor slabs and precast staircases, the EPPS contains the following new initiatives:-

Innovative precast structural walls

- Type A precast wall ('Welded' Wall)
- Type B precast wall (Hollow Wall)
- Type C precast wall (Semi-precast Wall)

Innovative volumetric precast components

- Precast stair-core
- Precast lift core
- Precast bathroom
- Precast bathroom-cum-kitchen

3.1 *The precast structural walls*

The precast structural walls are crucial to structural safety of the standard block which is a wall-slab structure. Unlike precast floor slabs, façades and staircases which carry the load of one single storey and therefore concern local structural safety only, the precast structural walls are to carry all the gravity loads and wind loads from the floors above and therefore impinge on global structural safety. Based on structural consideration, waterproofing requirements, buildability

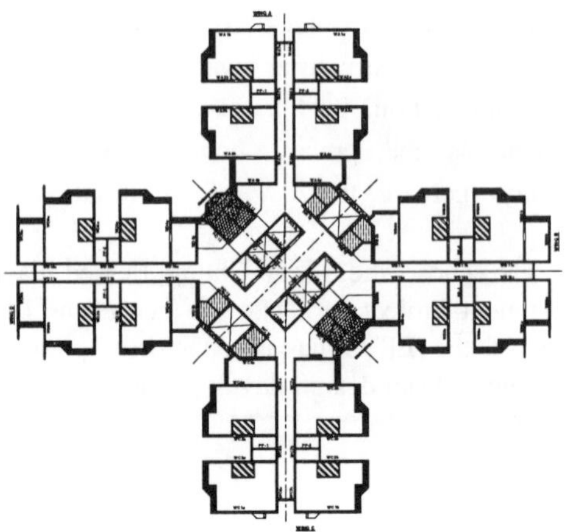

SCHEDULE OF PRECAST COMPONENTS

LEGEND	DESCRIPTION
	PRECAST WALL TYPE A
	PRECAST WALL TYPE B
	PRECAST WALL TYPE C
	PRECAST LIFT SHAFT
	PRECAST STAIR-CORE
	VOLUMETRIC PRECAST BATHROOM (INTEGRATED WITH PRECAST WALL)
	VOLUMETRIC PRECAST BATHROOM-CUM-KITCHEN (INTEGRATED WITH PRECAST WALL)
	PRECAST FACADE
	PRECAST STAIRCASE
	SEMI-PRECAST SLAB
	PRECAST BEAM / PARAPET

Figure 2 Key Plan of Precast Components

and cost factors, three types of precast shear walls are developed, namely Type A, Type B and Type C.

Type A Precast Wall is mostly used for external walls and specially designed with multiple waterproofing features. Joining upper and lower precast walls of this type is by welding of steel couplers that are cast in each end of the walls, i.e. structural steel sections at the two ends of the counterparts are welded together for load transmission. This connection neither requires special materials such as proprietary couplers nor special skills. Qualified welders are commonly available in the market to carry out high-quality jointing work. Quality-wise it is comparable to proprietary couplers; cost-wise, it is more favourable. Besides, it has several more advantages: on-site quality control tests can be performed, and its price and supply position can be better controlled in contrast to proprietary couplers that have only limited suppliers.

Type B Precast Wall is mainly used for inner walls such as corridor walls, where waterproofing requirement is less stringent and thus a cheaper precast option can be used. There are two or three vertical cavities in each

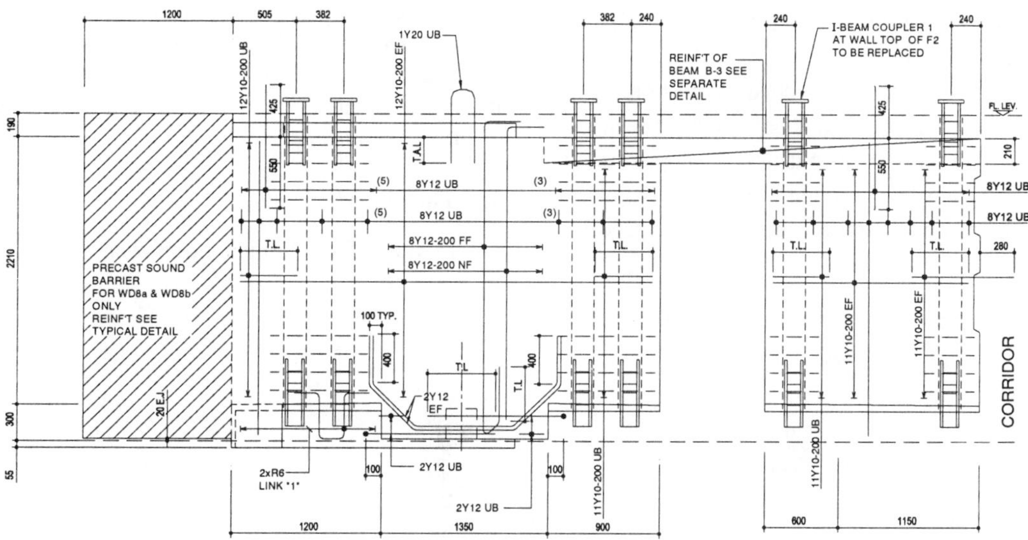

Figure 3 Type A Precast Wall

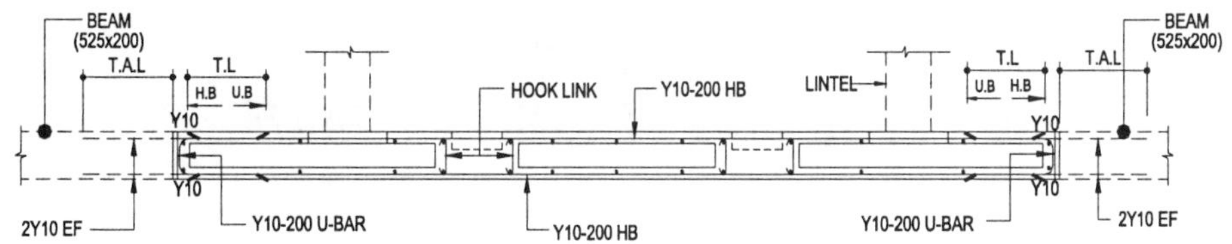

R.C. DETAIL OF PRECAST PORTION

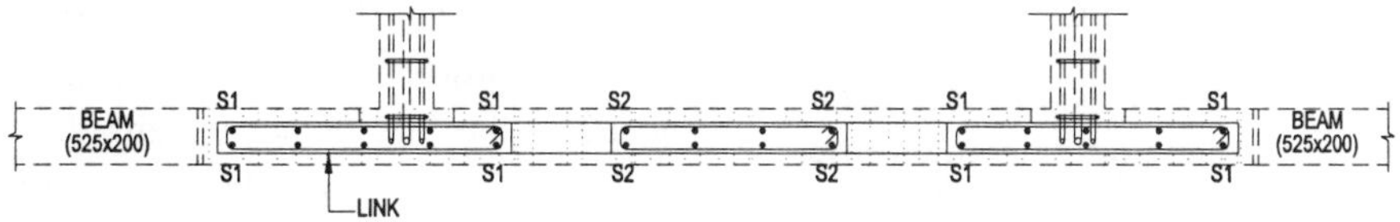

R.C. DETAIL OF CAST IN-SITU PORTION

Figure 4 Type B Precast Wall

wall. On-site installation begins with the insertion of reinforcement cages into the cavities when the wall is hoisted to position at working floor. The wall is then properly set, followed by filling the cavities with in-situ concrete. The overall wall thickness is only 200 mm, which fits into the current design of the standard block.

Type C Precast Wall is specifically designed for the gable end walls, furthest away from the tower crane that is usually located near the center of the block. Considering the normal lifting capacity of those tower cranes commonly available in the market, i.e. no need to upgrade the crane, the weight of the precast components must be properly controlled. Therefore, this wall type adopts a semi-precast design with an external precast wall panel and an internal in-situ concrete portion. Such semi-precast construction can also avoid undesirable vertical joints if smaller wall panels are adopted as an alternative to control the weight since these joints at external walls certainly require expensive waterproofing measures.

The innovative precast bathroom and bathroom-cum-kitchen have three special features. Firstly, they integrate intelligently with the adjoining shear walls, i.e. there will not be any double walls. Secondly, most of the architectural works are done in the precast factory, i.e. better quality control and less wet trades at working floor. Thirdly, a unique fibre reinforced polymer floor pan is used to replace the traditional waterproof membrane at the base of the bathroom and bathroom-cum-kitchen. It allows us to dispense with double floor slabs. As opposed to other proprietary products, more usable space can be provided with less construction costs because no double walls or double floor slabs are used. The floor pan is a seamless machine-moulded pan encased into walls, which will prevent water leakage and associated concrete spalling problems. This will mean much less maintenance and in turn minimize nuisance or disturbance to residents.

4. The Site Mock-up

In view of the limited experience of local building

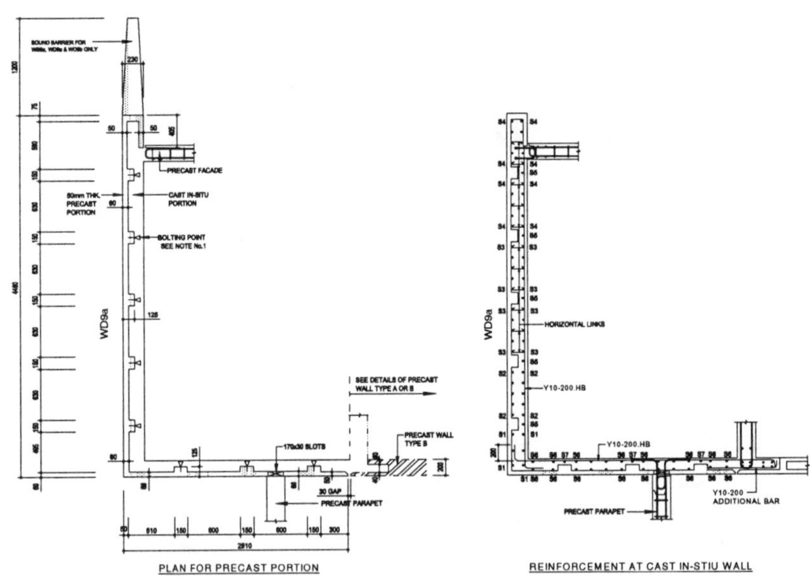

Figure 5 Type C Precast Wall

3.2 *The volumetric precast construction*

The breakthrough achieved in the precast structural walls has enabled it to explore and develop volumetric precast construction. Of the four types of volumetric precast components as mentioned in Section 3 above, the last two – precast bathroom and bathroom-cum-kitchen – are elaborated below.

contractors in the EPPS, the HA tried out the new system by constructing a two-storey one-wing mock-up structure at the Site before starting the main building project. By simulating the process of building the standard block with the EPPS, the HA could check out all the details involved in the production, installation and testing. During the course of the mock-up construction, the HA worked closely with the contractor in identifying problems, proposing improvements and developing feasible and practical solutions. The final, improved design was then let out for construction.

5. Structural Design Philosophy of the EPPS

5.1 *Structural framing*

The NH1 block is a reinforced concrete building that comprises basically slabs and walls and has to take up high wind loads. It is geometrically symmetrical with aspect ratio (height to width) less than 4.

Since design of the precast structural wall is gravity load sensitive, a more accurate load run-down is required. Current tributary area method commonly used in high-rise reinforced concrete building design may not be suitable for the purpose. Instead, finite element analysis was adopted; floor slabs and walls of the entire building were mesh-modeled. Further, for a more accurate assessment, construction stage analysis was also performed. The loading distributions in partially completed structures, built up floor by floor, due to gravity loads (mainly the self-weight) was carefully studied. Significant differences in member forces were found in columns and walls at low level and lintels at top few floors when compared to results obtained from conventional analysis of the whole building based on an elastic model.

5.2 *Precast structural walls*

The precast design was carried out to British Standard BS8110 with reference made also to precast standards and practices adopted in Singapore, New Zealand and USA.

As a pre-requisite to the precast concrete design, conscious effort was given to ensure the overall stability and safety at all stages of the building construction, particularly the structural continuity when precast elements were connected. Attention was paid also to the design of local effects and detailing of connections between precast units and between precast and in-situ concrete.

In addition, in consideration of the special design and locations of the joints, the behaviour of the precast walls was compared with that of cast-in-situ ones. Finite element analysis was used to find out whether precast walls and cast-in-situ walls could attain comparable stiffness and to check for presence of local weak spots that might cause local cracking of concrete and decrease the 'monolithic' behavior of the precast wall. With increasing loads, concrete stress was allowed to increase until a threshold value. Beyond this level, either local failure or cracking would occur. Although cracking in concrete is a very complex mechanism, it can be generally visualized to start at highly stress concentrated points. Mesh elements at these points were deleted from the analysis model. Similarly, supports were released in the model where tension reactions occurred which, in practice, no continuity of reinforcement through the joints would be provided and the grouting material assumed not to take any tension. Original continuum of the finite element model was disturbed since discrete voids were formed when some mesh elements were deleted. The 'transformed' model was subject to the same loads again in further analyses. New set of results (stresses and reactions) was obtained and the trimming exercise was carried out again. The iteration process repeated until, after several iterations, the model converged to an equilibrium state.

Results were then compared with those obtained from the 'control' cast-in-situ models for deviation in stiffness of precast walls. Any significant deviation was considered in the analysis of the superstructure.

5.3 *Volumetric precast bathroom*

Among the many volumetric precast components, volumetric precast bathroom is used as a representative to illustrate the structural design philosophy behind.

The precast bathroom is integrated with a precast structural wall, i.e. one of its walls is a structural wall. The provision of a joint at top of the non-structural walls on the other three sides, to be sealed up only at a later stage, defines a distinctive load path that the self-weight and imposed load are taken up by the integrated precast structural wall. It should be noted that premature sealing the joint may cause cracking to the non-structural walls since part of the imposed load will be transferred through the walls and accumulated at low floors.

6. Construction of the NH1 Block

The construction of the blocks started in March 2005 and reached the transfer floor in July 2005. Concurrently, production of the precast elements is actively underway in line with the progress for the construction of typical floors.

For the purposes of close monitoring and better quality control, in this pilot project, the HA has made it mandatory to have on-site manufacture of the precast structural walls. The set up of the on-site fabrication yard takes into consideration the logistics for production, the storage and erection of the precast elements with an aim to minimize double-handling in 'internal' transportation of the precast elements, which is costly.

It is anticipated that the typical 6-day construction cycle can be achieved when it reaches fifth floor later this year. Considering that there is less construction activities at working floor as compared to conventional cast-in-situ construction method, it is also evaluated that the construction cycle could be shortened to four or five days when the contractor becomes familiarized with the EPPS. The faster construction rate is one of the major benefits of the system.

7. Benefits of the EPPS

The EPPS has a number of major benefits.

7.1 *Effective quality control*

Better quality is ensured. As precast components are mass produced by moulding in factories, quality control is easier and construction errors are lesser.

Under the EPPS, not only can aesthetic finishes be applied to the precast components, but also aluminum windows and building services devices can be installed in advance. These will greatly reduce potential damage to the materials arising from on-site installation while improving durability and cutting down future maintenance costs. The use of the seamless polymer floor pans also effectively solve water leakage problems in bathrooms and kitchens, in turn reducing repair and maintenance costs and also minimizing nuisance or disturbance to the residents.

7.2 *Enhanced site safety*

The EPPS is much safer as it can reduce a large amount of work at height in the construction site such as reinforcement fixing, formwork erection and concrete pouring.

7.3 *Greater efficiency & shorter construction time*

As the precast components are factory-made, site work will only involve simple assembly of the components and a small amount of concrete pouring work. Site progress is thus sped up. It is estimated that the current construction cycle of typical floor can be shortened to four or five days. It is estimated that a normal 28-month building project (normally adopted for HA housing projects comprising 41-storey blocks) can be shortened by at least five months when the industry is familiar with the EPPS. Besides, the works progress is also less susceptible to inclement weather, which helps to ensure timely completion of a construction project.

7.4 *Better environmental protection*

The mechanized production of precast components in factories is much more environmentally friendly when compared to the conventional insitu construction. It will reduce both material wastage and construction waste. Also noteworthy is that the EPPS can substantially cut down the use of temporary external working platforms.

At the same time, as less concrete pouring work is carried out on-site, noise, air pollution and other associated nuisance to nearby residents can be minimized.

7.5 *Increased cost effectiveness*

All these benefits, when added together, can bring about a substantial saving in the overall construction and maintenance costs when the EPPS is widely used in public housing developments.

8. Conclusion

As the EPPS has so many valuable potentials, the HA should strive for its wider application. This will certainly contribute to the quality and sustainability of public housing developments and help to meet the ever-increasing expectation of the public.

STEEL STRUCTURE

EXPERIMENTAL INVESTIGATION AND SIMULATION ANALYSIS OF NEW STEEL SHEAR WALLS WITH SLITS

WEI DE-MIN

Civil Engineering Department, South China University of Technology, Guangzhou, Guangdong, 510640, China

WEN PEI-GANG

Guangzhou Kecheng Architecture Design Company Limited, Guangzhou, Guangdong, 510640, China

LI LIAN-BIN

Civil Engineering Department, South China University of Technology, Guangzhou, Guangdong, 510640, China

The experimental investigation and simulation computation for steel shear walls with slits under the low cyclic horizontal loading are presented. The experimental data show that the stability is the most important influence factor concerned with the load-bearing capacity of the steel shear walls without stiffening ribs. The simulation computation for the test shows that the thin steel shear walls with slits and stiffening ribs can increase the ductility of the structure and consume large amount of the earthquake energy through the plastic deformation produced.

1. Introduction

The steel shear walls as the lateral force resisting systems have been earlier used for the tall building structures. But the early steel shear walls with no slits are made up of the heavy steel plates or the stiffening ribs to guarantee the plastic yield takes place before the out-of-plane buckling. This type of steel shear walls is primarily used in Japan [1] and china, such as the construction of the 44-storey Jinjiang Hotel in Shanghai City [2,3]. The steel shear walls can provide the comparatively larger horizontal resistance and the dissipation rate of seismic energy. However, the cost is very expensive to reduce the out-of-plane deformation and obtain full-scale hysteresis loops for the large interstorey drift. Hence, due to the economic reason, the steel shear walls without stiffening ribs are generally used in the engineering practice, and under this circumstance the steel shear walls are allowed to buckle out of plane, even though only small interstorey drift takes place. By this way, the structures can be in the plastic state and the earthquake energy can be consumed by the plastic deformation of the structures. Caccese[4] and Driver[5] found that there is great pinching response in the hysteresis response. Only through proper design, the strength dropping of the steel shear walls without stiffening ribs can be avoided, and large ductile deformation can take place with the development of the tensile field. Since the great cost for manufacturing the stiffening ribs can be reduced in the steel shear walls without stiffening ribs, this type of steel shear walls are commonly used in North America. Recently, low yield point thin steel panel shear walls are developed in the Research Center of Seismic Engineering of Buffalo University in US. Combined with the openings and steel dog bone connections, this type of steel shear walls presents better seismic resistance capability.

In recent years, a new type of steel shear walls with slits began to be used in Japan. This type of steel shear walls can effectively reduce the interstorey drifts of the tall building structures, and also have some other advantages, such as the convenience in the construction, the flexibility in the layout and the broad usage area etc [7]. In this paper, the finite element method is used to analyze this new type of steel shear walls and the corresponding simplified calculation method for the seismic response [8,9] is put forward. Through the low cycle loading test and the corresponding simulation computation, the characteristics of this type of steel shear walls with slits, i.e. the lateral elastic stiffness, the bearing-load capacity, the hysteresis loops and the failure feature etc are obtained. The results of experimental investigation and theoretical analysis show that the new steel shear walls can be used for the tall buildings at the seismic regions.

2. Test Research

The dimension of the full-scale specimens in the experiment is shown in Table 1. They are made of Q235B steel; the thickness of the steel panel is t=12mm. The specimen I is a frame structure with its span and height as 3.6m and 2.9m respectively. Based on specimen I, specimen II employs the steel shear wall with slits, and the steel panel is connected with the steel beams by the friction-type bolts with high strength. Specimen III and IV are the steel shear walls without slits after removing the frames. Figure 1 indicates specimen II and IV.

Table 1. Dimension of specimens

specimen	n	w(mm)	B(mm)	h(mm)	h_u(mm)	h_b(mm)	h_m(mm)
II	10	1635	150	750	300	300	500
III	10	1635	150	750	300	300	500
IV	13	1740	120	600	380	380	640

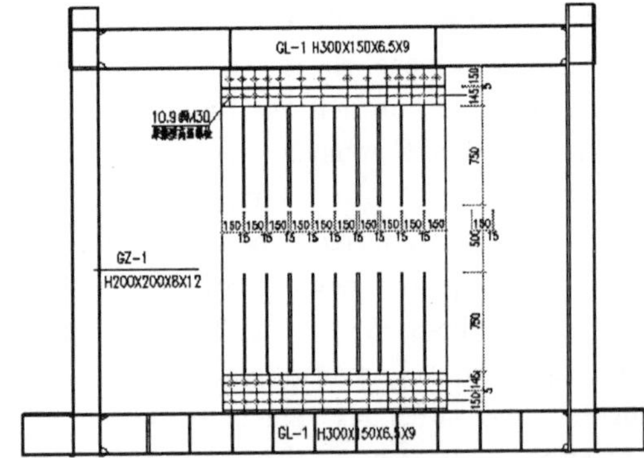

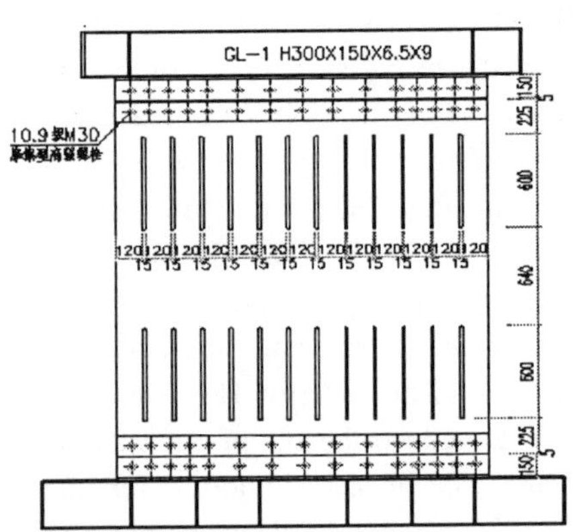

Figure 1. Geometrical dimension of Specimens: (a) specimen II; (b) specimen IV

In the test, the two scaffolds fixed at the two sides of the specimens were used as protective out-of-plane supports; 500kN tension-compression jacks placed at the two ends of the beams on top of the specimens were used to exert horizontal load; displacement sensors located at the middle of the beams were used to mark the horizontal displacements; load-displacement curves were recorded by X-Y function recorders; 4 mechanic-electric centigrade meters located in the steel panel were used to record the corresponding out-of-plane displacements; CM-2B stress survey systems were used to record the strains of the corresponding points of the steel panels. In the course of the tests, the vertical concentrated loads exerted on the specimen I and specimen II preserved as constant, i.e. 250kN. The primary measurement contents include the displacements on the top of the specimens, i.e. Δ, P-Δ hysteretic curves, the out-of-plane displacements δ of the steel shear walls and the strains of the frames and steel panels at the key positions. The loading schemes of the specimen II and III are very similar with that of the specimen I, in which forces and displacements are all used to control the course of loading. As for the specimen IV, step cyclic loading scheme is used in the elastic stage, and when nearly reaching the yield point, unloading and then loading again until the specimen failed.

The test results of initial rigidities, yield loads, failure loads and yield displacements of the specimens are all shown in Table 2; Figure 2 shows the results of load-displacement tests of the specimen IV. In the terms of the test results, we can find the specimen II , III and IV all fail when twice of the yield displacements take place, and the corresponding failure characteristics are out-of-plane buckling. Before yielding, comparatively large out-of-plane displacement, which can nearly reach

Table 2. Part of the test results

Specimen	Initial Rigidty (kN/mm)	Yield Loads (kN)	Yield Displacement (mm)	Failure Loads (kN)
Specimen I	13.45	172.4	21.15	268.9
Specimen II	43.94	467.57	15.38	642.2
Specimen III	14.65	146.46	23.07	169
Specimen IV	16.70	264.77	17.95	281.67

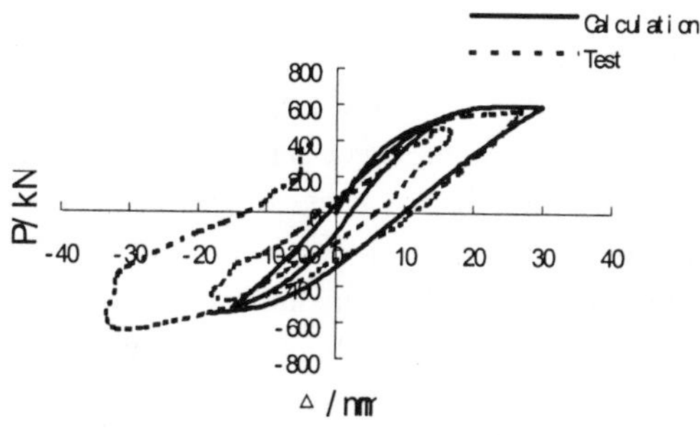

Figure 2. Load-displacement curve of the specimen IV

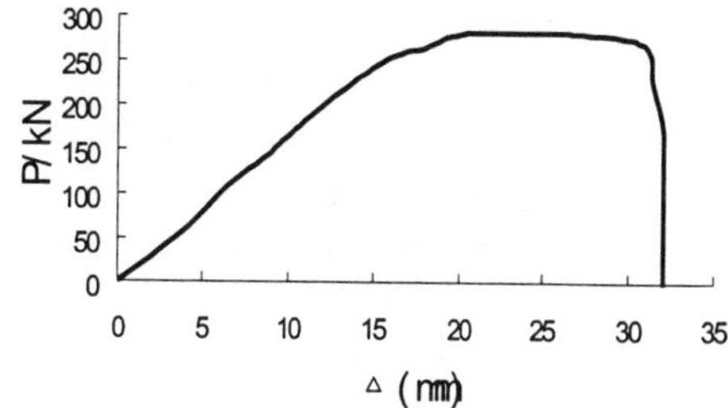

Figure 3. Comparison of computed and tested hysteretic curve of the specimen II

5mm, emerge at the right bottom corner of the panel. However, after unloading, the out-of-plane displacements nearly return to be zero. When the specimens fail, unresumable out-of-plane plastic displacement will take place with its magnitude reaching 12cm. The test results show that under the test conditions of this paper shown the out-of-plane displacements can not be avoided. Hence, the key problem of the simulation analysis for the tests is just how to assure the steel shear walls with slits do not buckle before the yield point of the material is reached, or buckle only when comparatively large yielding displacements take place.

3. Simulation Analysis

In order to further investigate the seismic properties and the failure mechanism of the steel shear walls with slits, simulation analysis of the tests will be done using the software ANSYS, in which the module will be chosen as the one based on the specimen II and adding stiffening ribs to resist the out-of-plane buckling.

In the simulation analysis, the material nonlinearity and geometric nonlinearity are all considered. The shell 181 element in ANSYS, which has the double nonlinearity characteristics, is used to simulate steel shear walls with slits; beam 188 element is used to simulate the outer frame structures; the constraints below the columns and steel shear walls with slits are all assumed as fixed supports. Horizontal loads are exerted at the end of the frame beams on top of the columns; the steel shear walls with slits are connected with the frame beams by friction-type high-strength bolts without relative displacement; the real thickness of the connection part is 36mm. The yield stress of the material is σ_y=285N/mm; the elastic modulus is E=206000N/mm^2; strengthen modulus is 1/100 of elastic modulus.

Table 3. Carrying capacity of the steel shear walls with stiffening steel panels

Loads (kN)	Height of stiffening ribs *hs*(mm)					Thickness of stiffening ribs *t*(mm)				Width of stiffening ribs *Ws*(mm)			
	0	1250	1625	2000	2600	6	12	18	24	100	150	200	250
Buckling loads	255.15	474.5	674.57	827.01	942.11	791.45	942.11	1067.2	1151.5	942.4	1067.2	1139.6	1194.1
Ultimate Carrying capicity	599.59	764.57	835.23	867.74	898.92	764.57	835.23	867.74	898.92	824.8	867.74	898.8	931.1

The elastic eignvalue buckling analysis is firstly done, and the obtained 1st order buckling load is 255.15kN. If the first bucking mode of the specimens is taken into nonlinear analysis as the initial flaw of the specimens and its magnitude was taken as 1/10 of the thickness of the steel shear walls, the obtained buckling load is 592.17kN. This obtained buckling load is close to the ultimate carrying capacity 642.2kN obtained from the tests. From this viewpoint, it is reasonable to use the calculated ultimate carrying capacity as its theoretical value. The results of tests and calculation are shown in Figure 3.

In order to increase the buckling capacity of the structures, stiffening ribs are added at the two sides of the steel shear walls with slits in the simulation analysis. The influence of dimension of the stiffening ribs on the buckling load and the ultimate carrying capacity of the steel shear walls is shown in Table 3.

The thickness of the stiffening ribs is t_s=12mm; its width is W_s=150mm. When its height is 1250, 1625, 2000 and 2600mm respectively, the calculation results of the horizontal load-displacement curves are shown in Figure 5. Figure 6 shows the calculation results of the horizontal load-displacement curves with the stiffening ribs of the height h_s=2600mm, width W_s=150mm and thickness t_s=6, 12, 18 and 24 respectively. The horizontal load-displacement curves with the stiffening ribs of the height h_s=2600mm, the thickness t_s=18mm and the width W_s=100, 150, 200 and 250 respectively are shown in Figure 7.

From the calculation results shown above, the following conclusion can be made that after employing stiffening ribs, the carrying capacity of the steel shear walls are obviously increased; the most important factor in the increase is the heights of the stiffening ribs; the second important factor is the thickness of the stiffening ribs; the smallest influence factor is the width of the stiffening ribs. When the height of the stiffening ribs is equal to that of the steel shear walls, the effect is the most obvious; the thickness of the steel ribs should be chosen in the range of 1~1.5 times of the thickness of the steel shear walls with slits because when its value is beyond 1.5 times of the thickness of the steel shear walls, the increase of the carrying capacity is very limited. When the width of the stiffening ribs reaches 1~1.2 times of that of the column-shape part of the steel shear wall with slits, the increase of carrying capacity reach limit and further increase of the width of the stiffening rib will have no obvious effect. Hence, it is optimal that the dimension of stiffening ribs of the specimen II is taken as the thickness t_s=18mm, the width W_s=150 and the height h_s=2600mm.

In this paper, nonlinear analysis is done for 5 kinds of steel shear walls with the same layout dimension, frame dimension and material parameters. These 5 kinds of steel shear walls are corresponding to the circumstance of without slits and stiffening ribs, without slits and with stiffening ribs, with slits and without stiffening ribs, with slits and stiffening ribs, as well as the circumstance of closely ribbed without slits. The obtained horizontal load-displacement curves are shown in Figure 8. From Figure 8, one can see that the initial rigidities of steel shear walls without slits, with or without stiffening ribs are all very large and the corresponding carrying capacities are very large too.

Figure 4. Calculation model of the stiffened specimen

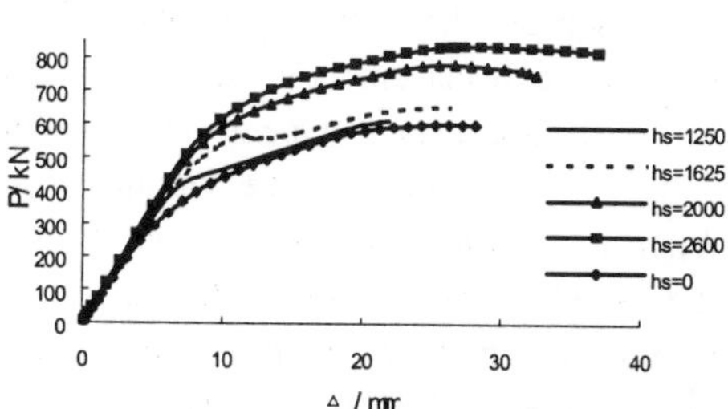

Figure 5. Load-displacement curves for different heights of the stiffening ribs

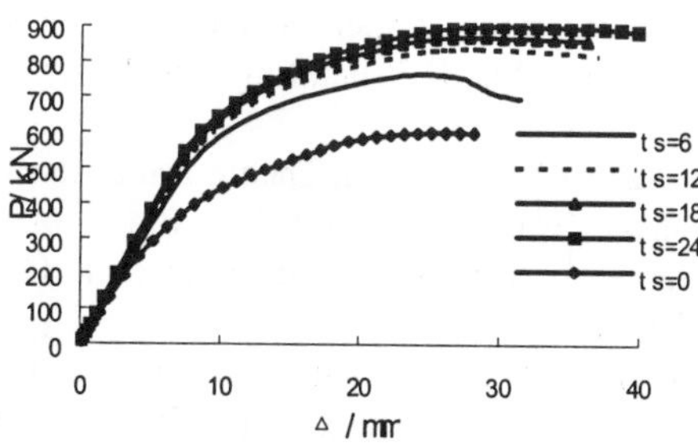

Figure 6. Load-displacement curves when changing the thickness of the stiffening ribs

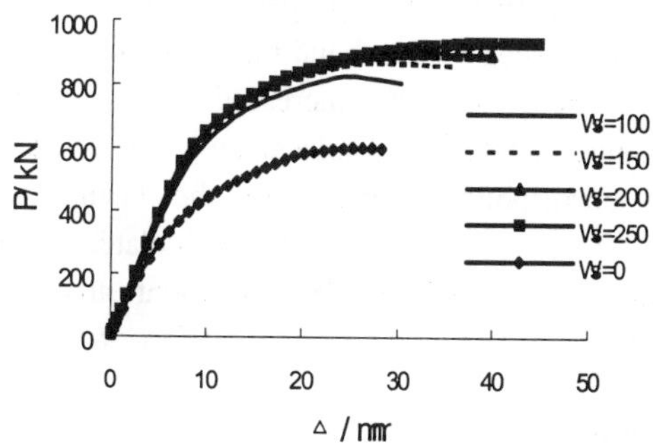

Figure 7. Load-displacement curves when changing the width of the stiffening ribs

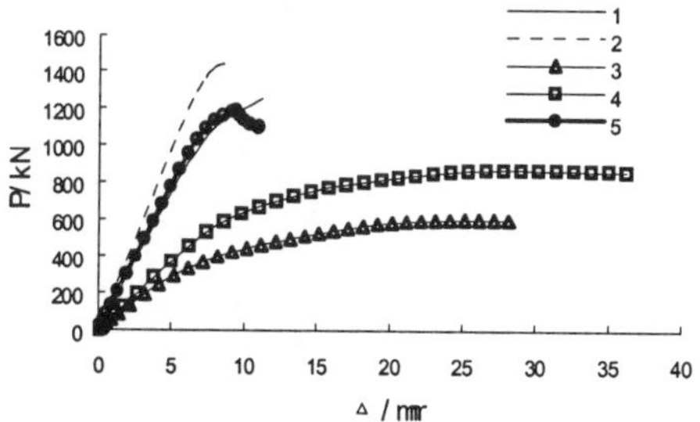

Figure 8. Comparison of load-displacement curves of different steel shear walls
(1. without slits and stiffening ribs
2. without slits and with stiffening ribs
3. with slits and without stiffening ribs
4. with slits and stiffening ribs
5. closely ribbed without slits)

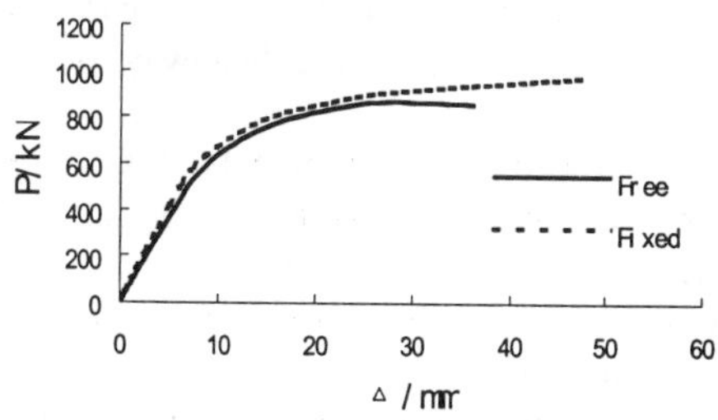

Figure 9. Load-displacement curves of different vertical constraints

However, the load-displacement curves are close to a straight line. When buckling capacity is reached, the corresponding ultimate displacement is only 10mm. The ductility is obviously not enough. Opening slits on the steel shear walls can help to increase the ductility of the structures. When employing stiffening ribs at the two sides of the steel shear walls with slits, the displacements will nearly reach 40mm when failing, which greatly improves the ductility and seismic energy dissipation property of the steel shear walls. And the carrying capacity is also much better than that of the steel shear walls with slits and without stiffening ribs.

When investigating the specimen II, the following conclusion can be made that the change of the upper boundary conditions can greatly influence the carrying capacity and ductility of the steel shear walls. Hence, when vertical displacements of the beams are constrained, the mechanical property of the specimen becomes pure shear from bending-shear so that the carrying capacity will be increased and the yield displacements and failure displacements will all larger than the corresponding values of the beams with vertically free supported. The calculated load-displacement curves with different constraints are shown in Figure 9.

4. Conclusions

Through the test and the numerical simulation analysis, the following conclusions can be obtained:

1. Stiffening ribs must be used at the two sides of the steel shear walls with slits, whose widths should be taken as 1-1.2 times of those of the column-shape part of steel shear wall with slit, thickness should be taken as 1-1.5 times of those of the steel shear walls, heights should be taken as the headway of the frames.
2. Constraining the vertical displacements of the upper boundary of steel shear walls with slits can improve the seismic property of the structures.
3. Sufficient spaces must be concerned for welded residual deformation when assembling the frames, otherwise the assembly of the steel shear walls will produce large residual stress, which can greatly reduce the buckling capacity of the steel shear walls with slits.
4. There are some open problems, such as optimization of the slit parameters, the equivalent strength and rigidity of the steel shear walls with slits and the model of simplified analysis and so on that need to be further investigated.

References

1. T. M. Roberts, Seismic resistance of steel plate shear walls, Engineering Structures, **17**, 344 (1995).
2. B. N. Gong, Y. Q. Qian, The application situation of tall steel structure in china, Building Science, **4**, 69 (1987).
3. S. Z. Yan, Construction of the main structure of Jinjang Hotel of Shanghai, Architecture Technology, **12**, 5 (1987).
4. Caccese, V., et al., Experimental study of thin steel plate shear walls under cyclic load, J. Struct. Engrg., ASCE, **119**, 573 (1993).
5. Driver R. G., et al., Cyclic test of four-story steel plate shear wall, J. Struct. Eng. ASCE, **124**, 112 (1998).
6. Darren Vian, et al., Cyclic performance of low yield strength steel panel shear walls, Proc. of 16th KKCNN Symposium on Civil Engineering, Korea, 379 (2003).
7. Toko Hitaka, et al., Experimental study on steel shear wall with slits, J. Struct. Eng. ASCE, **129**, 586 (2003).
8. Wen pei-gang, Wei De-min, Theoretical analyses of steel shear walls with cracks, Proc. of the computational mechanics conference of China, 1047 (2003).
9. Wei De-min, Wen Pei-gang, Seismic response analysis of frame structures with slit steel shear walls, Earthquake Engineering and Engineering Vibration, **24**, 63 (2004).

EFFECTS OF TRANSVERSE WELDS ON ALUMINUM TUBULAR COLUMNS*

JI-HUA ZHU

Department of Civil Engineering, The University of Hong Kong, Pokfulam Road, Hong Kong

BEN YOUNG

Department of Civil Engineering, The University of Hong Kong, Pokfulam Road, Hong Kong

A test program was conducted to study the effects of transverse welds on the strength of aluminum columns. A total of 42 fixed-ended column tests were conducted, which included 23 columns with both ends welded to aluminum end plates using Tungsten Inert Gas method, and 19 columns without welding of end plates. The column lengths were either 300 or 1000 mm. Local and overall initial geometric imperfections of the columns were measured. The material of the column specimens were 6063-T5 and 6061-T6 heat-treated aluminum alloys. Non-welded and welded material properties were obtained from various coupon tests. The observed failure modes of the columns include local bucking, overall bucking, interaction of local and overall bucking, and material yielding in the heat-affected zone. The test strengths are compared with the design strengths predicted by the American, Australian/New Zealand and European specifications for aluminum structures. It is shown that the design strengths predicted by the three specifications are conservative for the non-welded columns. The welded column design strengths predicted by the three specifications are generally more conservative than the non-welded column predictions.

1. Introduction

Aluminum tubular members are used in space structures, curtain walls and other structural applications. These structural members could easily be joined by welding. Heat-treated aluminum alloys by extrusions suffer loss of strength in a localize region when welding is involved, known as heat-affected zone (HAZ) softening. The heat-affected zone (HAZ) shall be taken to extend 1 inch (25.4 mm) to each side of the centre of a weld [1]. With the 6000 series alloys, the heat of welding can locally reduce the parent metal strength by nearly half [2]. The current American Aluminum Design Manual [1], Australian/New Zealand Standard [3] and European Code [4] for aluminum structures provide design rules for members containing transverse or longitudinal welds. However, there are not many test data available on the strength of aluminum alloy columns with transverse welds. Therefore, it is important to obtain test data of aluminum alloy columns containing transverse welds.

The purpose of this paper is firstly to present a test program on fixed-ended aluminum alloy columns with and without transverse welds; secondly, to examine the material properties of the welded and non-welded materials, and lastly, to compare the test strengths with the design strengths predicted using the American Aluminum Design Manual (AA) [1], Australian/New Zealand Standard (AS/NZS) [3] and European (EC9) [4] specifications for aluminum structures.

2. Experimental Investigation

2.1 *Test specimens*

The tests were performed on square (SHS), rectangular (RHS) and circular (CHS) hollow sections aluminum columns. The test specimens were fabricated by extrusion using the materials of 6063-T5 and 6061-T6 aluminum alloys. The test specimens were supplied by the manufacture in un-cut lengths of 4000 mm. The section sizes of the specimens were chosen to ensure that a range of practical cross-sections were covered in the test program. Each specimen was cut to a specified length of 300 or 1000 mm. The test program includes 23 fix-ended columns with both ends welded to aluminum end plates and 19 fix-ended columns without welding of end plates. In this paper, the term "welded column" refers to the specimen with transverse welds at the ends of the column to aluminum end plates. The term "non-welded column" refers to the specimen without transverse welds at the ends of the column, but aluminum end plates were used in the test. Therefore, the testing condition of the welded column and non-welded column is identical, except no welding in the non-welded column. The ends of the non-welded column were milled flat by an electronic milling machine to an accuracy of 0.01 mm to ensure full contact between specimen and the end plates.

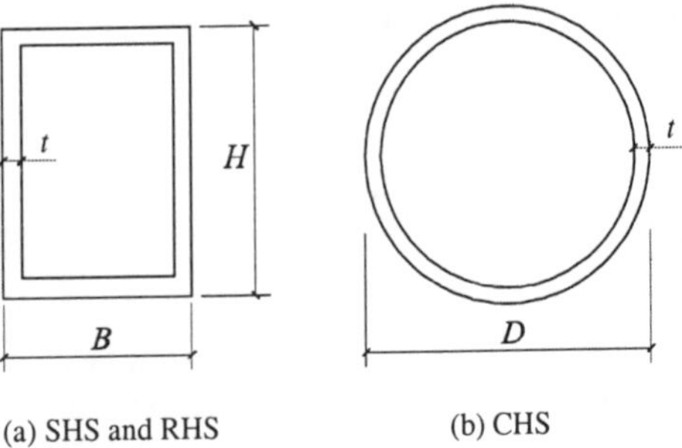

Fig. 1. Definition of symbols

The specimens were separated into nine series of different cross-section geometry and aluminum alloy as shown in Tables 1 and 2 using the symbols illustrated in Fig. 1. The specimens were tested between fixed ends. The test specimens were labeled such that the series, welds and specimen length could be identified from the label, as shown in Tables 6 and 7. For example, the label "N-R1-W-L300-R" defines the following specimen:

Table 1. SHS and RHS column test series

Test series	Material	Dimension $H \times B \times t$ (mm)
N-S1	6063-T5	44×44×1.1
N-R1	6063-T5	100×44×1.2
N-R2	6063-T5	100×44×3.0
H-R1	6061-T6	100×44×1.2
H-R2	6061-T6	100×44×3.0

- The first letter indicates the material of the specimen, where "N" refers to the normal strength aluminum alloy of 6063-T5 and "H" refers to the high strength aluminum alloy of 6061-T6;
- The second part of the label indicates the cross-section shape of the specimen, where "R1" refers to rectangular hollow section with nominal cross-section dimension of 100×44×1.2. The cross-section shape and dimension for other sections are shown in Tables 1 and 2;
- If a specimen was tested with transverse welds to aluminum end plates, then the letter "W" indicates the welded column specimen; if a specimen was tested without welding of end plates, then "NW" indicates the non-welded column specimen;
- The following part of the label "L300" indicates the length of the specimen, where the letter "L" refers to the column length and the following digits are the nominal length of the specimen in millimetres (300 mm); and
- If a test was repeated, then the letter "R" indicates the repeated test.

Table 2. CHS column test series

Test series	Material	Dimension $D \times t$ (mm)
N-C1	6063-T5	50×1.6
N-C2	6063-T5	50×3.0
H-C1	6061-T6	50×1.6
H-C2	6061-T6	50×3.0

2.2 *Non-welded material properties*

2.2.1 *General*

Non-welded material refers to the material of the specimen without being heat-affected by welding. Longitudinal tensile coupon tests were performed to determine the non-welded material properties. The coupon specimens were cut from the flat faces of the SHS and RHS, and the curved faces of CHS. Longitudinal compression coupon tests were also performed on test pieces cut from RHS. All coupons were taken from the specimens belonging to the same batch of specimens as the column tests.

2.2.2 *Tensile coupons*

The flat tensile coupon dimensions conformed to the Australian Standard AS 1391 [5] and the ASTM Standard [6] for the tensile testing of metals using 12.5 mm wide coupon and a gauge length of 50 mm. The coupons were also tested according to the AS 1391 Standard in a 250 kN capacity MTS displacement controlled testing machine using friction grips. The dimensions of tensile coupons cut from the curved faces of CHS were 6 mm wide coupon and a gauge length of 25 mm. Holes having a diameter of 8.5 mm were drilled at 20 mm from the ends of the curved coupons, and the coupons were tested from pin to pin. This avoids the bending stresses that could be introduced into the unsymmetrical shaped coupons during the test upon application of tensile stress.

A calibrated extensometer of 50 mm gauge length was used to measure the longitudinal strain of the flat tensile coupons. For curved tensile coupon tests, a 25 mm gauge length extensometer was placed on the side of the coupons, since the faces of the coupon were curved. In addition, two linear strain gauges were attached to each coupon on the midpoint of each face. The strain gauges readings were used to determinate the initial Young's modulus.

The Young's modulus of the material was determined using an accurate unloading method presented by Mennink [7] as illustrated in Fig. 2. This is because the stress-strain curves of aluminum alloys have a rounded curved knee and almost have not any linear portion even when the stresses are quite low [2]. In the tensile coupon tests, the force initially applied using load control until a stress of approximately 0.8 times the 0.2% proof stress ($\sigma_{0.2}$) of the specimen was obtained. The stress on the specimen was then unloaded until approximately $0.2\sigma_{0.2}$ of the specimen. The stress on the specimen was reloaded to approximately $0.9\sigma_{0.2}$. The specimen was loaded using displacement control with a loading rate of 0.03 mm/min until 0.8% of the tensile strain, then the loading rate was increased to 0.8 mm/min, and the specimen was test to fracture. The Young's modulus of the material was determined using the slope of the unloading portion of the stress-strain curve, as shown in Fig. 2.

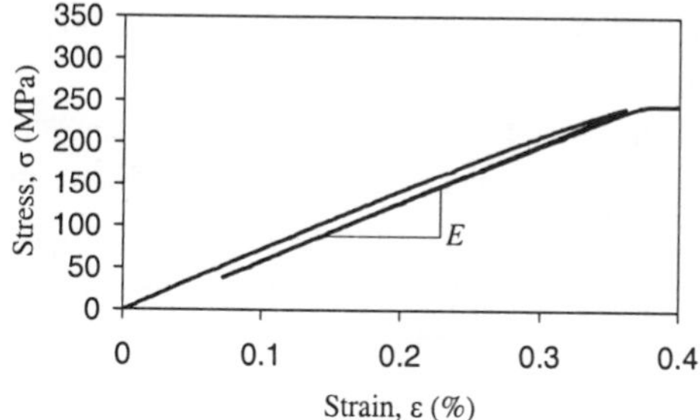

Fig. 2. Method of determining Young's Modulus

The readings of load and strain were recorded using the data acquisition system. All data were recorded at 0.5 second intervals during the tests. The static load was obtained by pausing the applied strain for 1.5 min near the 0.2% proof stress and the ultimate tensile strength. This allowed the stress relaxation associated with the plastic straining to take place. The measured material properties are the static 0.01% ($\sigma_{0.01}$), 0.1% ($\sigma_{0.1}$) and 0.2% ($\sigma_{0.2}$) proof stresses, the static tensile strength (σ_u) as well as the initial Young's modulus (E_0), and the elongation after fracture (ε_f) based on the gauge lengths of 50 and 25 mm for flat and curved coupons, respectively, as shown in Table 3.

Table 3. Measured welded and non-welded material properties of tensile coupons

Specimen	E_0 (GPa)	$\sigma_{0.01}$ (MPa)	$\sigma_{0.1}$ (MPa)	$\sigma_{0.2}$ (MPa)	σ_u (MPa)	ε_f (%)
N-S1-W	71.1	42.6	62.5	64.8	111.4	10.0
N-S1-NW	70.4	156.5	187.4	188.5	209.5	9.9
N-R1-W	65.1	59.9	81.6	83.7	121.9	8.1
N-R1-NW	69.1	142.1	194.2	195.9	218.7	9.9
N-R2-W	67.5	55.5	67.2	70.4	137.7	17.8
N-R2-NW	67.5	152.8	184.4	189.1	213.2	12.3
N-C1-W	73.0	49.6	68.7	71.3	120.9	9.9
N-C1-NW	66.7	162.9	193.9	194.6	214.4	10.0
N-C2-W	71.6	53.9	73.4	75.3	109.7	6.9
N-C2-NW	67.1	129.5	180.3	185.9	207.7	10.4
H-R1-W	67.7	90.4	105.0	109.3	159.8	8.9
H-R1-NW	70.3	229.3	254.6	260.0	276.3	7.6
H-R2-W	69.6	85.8	93.7	99.1	156.0	11.6
H-R2-NW	68.9	252.2	273.4	275.4	283.1	10.1
H-C1-W	72.6	76.5	87.4	92.5	148.3	10.7
H-C1-NW	67.1	258.1	285.7	286.7	310.1	10.7
H-C2-W	71.7	78.6	90.3	94.3	161.2	10.9
H-C2-NW	70.2	255.2	276.2	278.9	284.3	11.7

2.2.3 Compression coupons

The longitudinal compression coupon tests were performed on coupons cut from the slender face of the RHS. The nominal dimensions of the coupons were 72 mm long and 16 mm wide. The ends of the compression coupons were milled flat in order to provide uniform compressive stress to the coupon specimens. Two coupons for plate thickness of 1.2 mm were glued together before testing. A bracing jig was designed to prevent minor axis buckling of the test pieces, and this method have been used by Rockey and Jenkins [8], Rasmussen and Hancock [9], and Gardner [10]. The height of the bracing jig was 70 mm, and the coupon specimens protrude 1 mm either side of the central part of the bracing jig, without applying load to the jig. A thin layer of lubricating paste applied to the contact surfaces of the coupon specimen and the jig to reduce friction. Before the tests, the jig was relatively tight to prevent buckling of the coupons, but also loose enough to allow some movement and the unrestrained expansion due to Poisson's effect.

The compression coupon tests were performed using MTS displacement controlled testing machine. The compressive load was applied with a loading rate of 0.024 mm/min. Two strain gauges were attached to the edges of the coupons at mid-height. The same data acquisition system as the tensile coupon tests was used. In the test, an initial load was applied to the coupon up to a maximum of approximately 10% of the 0.2% proof stress ($\sigma_{0.2}$) of the material. The position of the coupon was adjusted until the strains in the two edges were almost identical. This ensured the load applied approximately concentric to the coupon specimen. Tests were terminated when 1% of the compressive strain has reached. The measured static 0.2% proof stress ($\sigma_{0.2}$) and initial Young's modulus (E_0) of the compression coupons are shown in Table 4.

Table 4. Measured non-welded material properties of compression coupons for RHS

Specimen	E_0 (GPa)	$\sigma_{0.2}$ (MPa)
N-R1-NW	73.0	205.1
N-R2-NW	69.1	199.9
H-R1-NW	70.1	299.6
H-R2-NW	72.5	280.9

2.3 *Welded material properties*

2.3.1 *General*

Heat-treated aluminum alloys suffer loss of strength in a localize region when welding is involved, known as heat-affected zone (HAZ) softening. The heat-affected zone (HAZ) shall be taken to extend 1 inch (25.4 mm) to each side of the centre of a weld [1]. In this paper, the term "welded material" refers to the material in the HAZ that being heat-affected by welding. Two types of welded longitudinal tensile coupons with gauge length of either 25 or 250 mm were tested to determine the welded material properties.

2.3.2 *Tensile coupons of 25 mm gauge length*

The welded tensile coupon specimens of 25 mm gauge length were cut from the centre of the faces of the SHS, RHS and CHS between two longitudinal welds, as shown in Fig. 3. The distance between the centerlines of the two longitudinal welds was approximately 32 mm. This ensured the coupons were fully heat-affected based on previous research [11] that the heat-affected zone is limited to approximately 3 inches to either side of the weld centerline, and is most pronounced within 1 inch of the weld. The dimensions of the flat tensile coupons cut from the SHS and RHS conformed to the Australian Standard AS 1391 [5] and the ASTM Standard [6] for the tensile testing of metals using 6 mm wide coupon and a gauge length of 25 mm. The dimensions of the curved tensile coupons cut from the CHS were identical to the flat coupons.

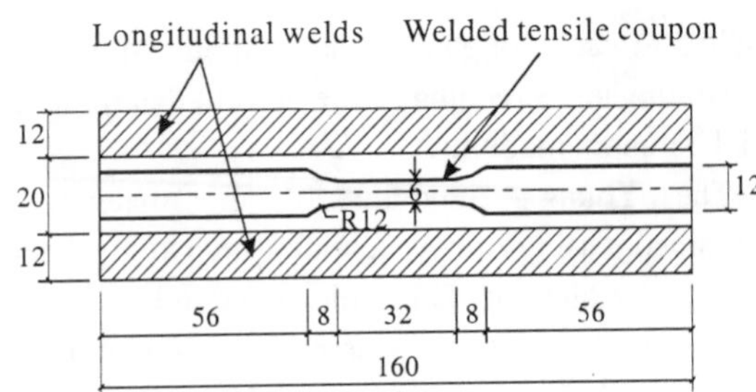

Fig. 3. Welded tensile coupon of 25 mm gauge length

2.3.3 *Tensile coupons of 250 mm gauge length*

The welded tensile coupon specimens of 250 mm gauge length were cut from the centre of the slender faces of the RHS with a butt weld across the mid-length of the coupon specimens. The coupon dimensions conformed to the American Structural Welding Code – Aluminum [12] for the tension tests using 38 mm wide coupon and a gauge length of 250 mm.

All welded tensile coupons were welded by Tungsten Inert Gas (TIG) method. The filler metal was 4043 alloy according to the American Aluminum Design Manual [1]. When the welding temperature on the surface of the coupons was approximately 200 °C, the voltage of 20V and the current of 120-200A depending on the different plate thickness were used. The welding was performed with a constant speed of 120 mm/min. This welding procedure is identical to the welded column specimens for the transverse welds at the ends of the columns.

Table 5. Measured welded material properties of tensile coupons of 250 mm gauge length for RHS

Specimen	$\sigma_{0.2}$ (MPa)
N-R1-W*	93.0
N-R2-W*	97.5
H-R1-W*	122.5
H-R2-W*	122.7

Note: * 250 mm gauge length coupon across a butt weld.

The welded tensile coupon tests were performed using the same testing method and data acquisition system as the non-welded tensile coupon tests. Test results of the welded tensile coupons of 25 mm gauge length are shown in Table 3. The measured static 0.2% proof stress ($\sigma_{0.2}$) of the welded tensile coupons of 250 mm gauge length is shown in Table 5.

2.4 *Column tests*

The tests were performed on SHS, RHS and CHS aluminum columns. The test program includes both the *non-welded* and *welded* columns. The nominal length of the column specimens was 300 and 1000 mm. The test rig and the test setup of typical column tests are shown in Fig. 4. A servo-controlled hydraulic testing machine was used to apply compressive axial force to the specimen. A moveable upper end support allowed tests to be conducted at various specimen lengths. A rigid bearing plate was bolted to the upper end support, which was restrained against the minor and major axis rotations as well as twist rotations and warping. Hence, the rigid bearing was considered to be a fixed-ended bearing. A special bearing was used at the lower end support.

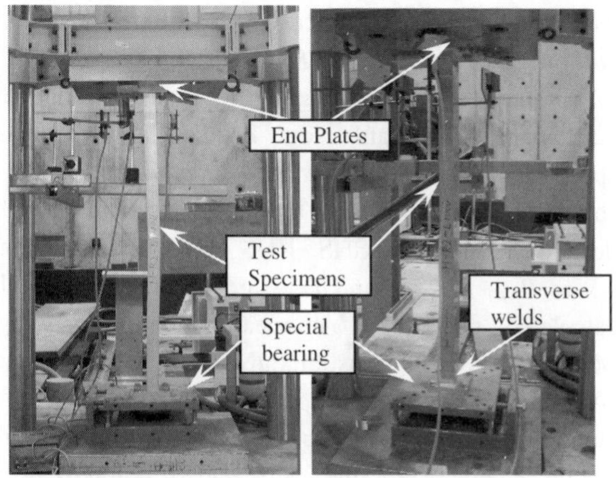

(a) Non-welded column test H-R2-NW-L1000 (b) Welded column test H-R2-W-L1000

Fig. 4. Typical column test

In the non-welded column tests, two aluminum end plates were bolted on the rigid bearing and the special bearing. The ends of the non-welded column were milled flat by an electronic milling machine to an accuracy of 0.01 mm to ensure full contact between specimen and the end plates. After each specimen was positioned on the bottom end plate, the ram of the actuator was moved slowly toward the specimen until the top and bottom end plates were in full contact with the ends of the specimen having a small initial load of approximately 1 kN. This procedure would eliminate any possible gaps at the ends of the specimen, since the special bearing was free to rotate in any direction. The special bearing was then restrained from twist and rotations by using horizontal and vertical bolts, respectively. Hence, the special bearing became a *fixed-ended* bearing. The fixed-ended bearing was considered to restrain both minor and major axis rotations as well as twist rotations and warping. The non-welded column specimens were tested between fixed ends.

The welded column tests were performed using the same test rig as the non-welded column tests. However, the two aluminum end plates were transversely welded to the ends of the specimen using the same welding procedure as the welded tensile coupons. Initially, the top end plate of the specimen was bolted to the rigid bearing plate. The load was then applied at the upper end through the rigid bearing plate. The ram of the actuator was moved slowly until the special bearing was in full contact with the bottom end plate of the specimen having a small initial load of approximately 1 kN. Once again, this procedure would eliminate any possible gaps between the special bearing and the bottom end plate of the column. The bottom end plate of the specimen was then bolted to the special bearing, and the bearing was restrained from twist and rotations by using horizontal and vertical bolts, respectively. The welded column specimens were tested between fixed ends. The testing condition of the welded and non-welded columns was identical.

Strain gauges were attached in axial direction at mid-length of the 300 mm length columns. Three laser displacement transducers were used to measure the axial shortening of the specimens. Displacement control was used to drive the hydraulic actuator at a constant speed of 0.1 mm/min. The use of displacement control allowed the tests to be continued into the post-ultimate range. A data acquisition system was used to record the applied load and the readings of the laser displacement transducers as well as the strain gauge readings at regular intervals during the tests.

2.5 *Measured overall geometric imperfections*

Initial overall geometric imperfections were measured on all 1000 mm length specimens prior to testing. The overall geometric imperfection measurements comprised the flexural imperfections about both the major x and minor y axes of the SHS and CHS specimens, and only the flexural imperfections about the y-axis of the RHS specimens. After the specimen was properly positioned in the test rig, axial force of approximately 1 kN was applied to hold the specimen in place. Theodolites were then used to obtain readings at mid-length and near both ends of the specimens. The geometric imperfections were measured near the plate junction of the SHS and RHS specimens. The maximum measured overall geometric imperfections at mid-length were 1/5251, 1/1608 and 1/1432 of the specimen length for SHS, RHS and CHS, respectively.

2.6 *Measured local geometric imperfections*

Initial local geometric imperfections were measured on five SHS and RHS specimens. The specimens were cut from those specimens belonged to the same batch of specimens as the column tests. Hence, the measured local geometric imperfections are considered nearly the same order of imperfections as the column specimens. Local geometric imperfections of each face were measured at mid-length and 30 mm away from both ends of the specimens having 1000 mm in length. A Mitutoyo Co-ordinate Measuring Machine with an accuracy of 0.001 mm was used to measure the initial local geometric imperfections. A typical measured local imperfection profiles for the SHS of Series N-S1 is shown in Fig. 5. The vertical axis is plotted the measured local geometric imperfections and the horizontal axis is plotted against the normalized plate width or depth (*d*) of the cross-section of the specimen. Readings were taken at 2 mm intervals across the width or depth of the plates. The negative values of local imperfection measurements indicated the concave profiles and the positive values indicated the convex profiles. The maximum measured local geometric imperfections were 0.183 mm for SHS of Series N-S1, and 0.377, 0.070, 0.207, 0.469 mm for RHS of Series N-R1, N-R2, H-R1 and H-R2, respectively.

2.7 *Test results*

The experimental ultimate loads (P_{Exp}) obtained from the tests of columns with and without transverse welds are shown in Tables 6 and 7. Six tests were repeated and the test results for the repeated tests are close to their first test values, with differences ranging from 1.7% to 6.7%. The observed failure modes include local bucking, overall bucking, interaction of local and overall bucking, and material yielding in the HAZ. It should be mentioned that failure of all welded columns were observed approximately 20 mm near the ends of the columns, which was failed in the HAZ.

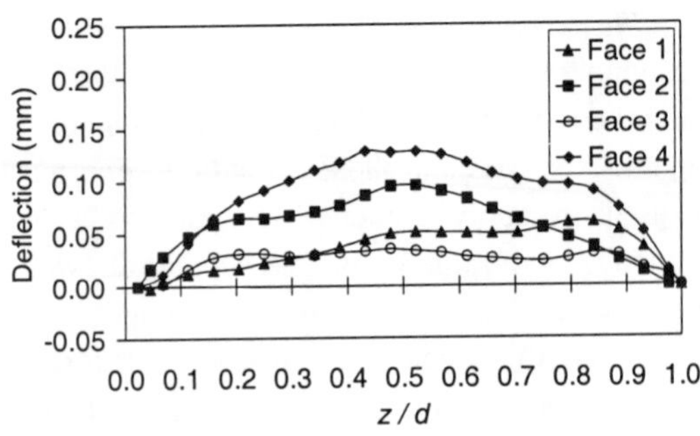

Fig. 5. Measured local geometric imperfection profiles for SHS of Series N-S1 at mid-length

3. Comparison of Test Strengths with Design Strengths

The test strengths of fixed-ended columns with and without transverse welds (P_{Exp}) are compared with the *unfactored* design strengths predicted using the American, Australian/ New Zealand and European specifications for aluminum structures. Tables 6 and 7 show the comparison of the test strengths with the design strengths for SHS, RHS and CHS specimens, where P_{AA}, $P_{AS/NZS}$, and P_{EC9} are the design strengths calculated using the American Aluminum Design Manual (AA) [1], Australian/New Zealand Standard (AS/NZS) [3] and European Code (EC9) [4] for aluminum structures, respectively. The AA and AS/NZS design strengths of the non-welded columns were calculated using the material properties obtained from tensile coupon tests of the non-welded material, whereas the design strengths of the welded columns were calculated using the material properties obtained from tensile coupon tests of the welded material using 25 and 250 mm gauge lengths. It should be noted that the AA Specification and AS/NZS Standard specified the welded material properties of 250 mm gauge length are used in calculating the design strengths of the welded columns, and these design strengths P^*_{AA} and $P^*_{AS/NZS}$ are shown in Table 7. The EC9 Code specified the design strengths of both non-welded and welded columns should be calculated using the non-welded material properties. In this paper, the EC9 design strengths of the non-welded and welded columns were calculated using the material properties obtained from tensile coupon tests of the non-welded material. The 0.2% proof stress was used as the corresponding yield stress, and the effective length was taken as one-half of the column length in calculating the design strengths. This is because the fixed-ended bearings are restrained against the major and minor axes rotations as well as twist rotations and warping. The design strengths were calculated using the measured cross-section dimensions for each specimen and the measured material properties as summarized in Tables 3 and 5.

In the AA and AS/NZS specifications, the design rules for column with transverse welds (welded column) are identical to those design rules for column without transverse welds (non-welded column), except that the welded column strength is limited by the yielding of the welded material. Furthermore, the design rules for welded column of CHS consider the column as fully heat-affected and using the buckling constants of the non heat-treated materials regardless of temper before welding in the AA and AS/NZS specifications. In the EC9 Code, the design rules for the welded column depends on the HAZ softening factor (ρ_{haz}) that allow for the weakening effects of welding. The values of ρ_{haz} are 0.6 and 0.5 for 6063-T5 and 6061-T6 aluminium alloys as specified in the EC9 Code, respectively. Generally, it is shown that the welded column design strengths for 6061-T6 aluminium alloy predicted by the EC9 Code are more conservative than the welded column design strengths for 6063-T5 aluminium alloy, as shown in Table 7. This could be due to a conservative value of ρ_{haz} is used for 6061-T6 aluminium alloy.

Table 6. Comparison of test strengths with design strengths for non-welded columns

Specimen	Test	Comparison		
	P_{Exp} (kN)	P_{Exp}/P_{AA}	$P_{Exp}/P_{AS/NZS}$	P_{Exp}/P_{EC9}
N-S1-NW-L300	34.1	1.08	1.08	1.13
N-S1-NW-L1000	33.7	1.06	1.06	1.20
N-R1-NW-L300	42.3	1.08	1.08	1.02
N-R1-NW-L1000	41.7	1.03	1.03	1.07
N-R2-NW-L300	147.9	1.03	1.10	1.08
N-R2-NW-L1000	145.8	1.01	1.08	1.14
H-R1-NW-L300	53.3	1.08	1.08	1.06
H-R1-NW-L1000	51.6	1.06	1.06	1.10
H-R1-NW-L1000-R	49.5	1.01	1.01	1.05
H-R2-NW-L300	209.2	1.06	1.14	1.14
H-R2-NW-L1000	202.4	1.04	1.12	1.21
N-C1-NW-L300	48.5	1.04	1.16	1.07
N-C1-NW-L1000	45.9	1.08	1.12	1.11
N-C2-NW-L300	102.4	1.20	1.37	1.24
N-C2-NW-L1000	86.1	1.10	1.16	1.14
H-C1-NW-L300	75.9	1.11	1.25	1.13
H-C1-NW-L1000	71.7	1.16	1.17	1.20
H-C2-NW-L300	129.6	1.04	1.16	1.06
H-C2-NW-L1000	119.6	1.06	1.07	1.10
Mean	---	1.07	1.12	1.12
COV	---	0.044	0.072	0.053

The experimental-to-design strength ratios P_{Exp}/P_{AA}, $P_{Exp}/P_{AS/NZS}$, P_{Exp}/P_{EC9}, P_{Exp}/P^*_{AA} and $P_{Exp}/P^*_{AS/NZS}$ are shown in Tables 6 – 7 for comparison purpose. For the non-welded columns, it is shown that the design strengths predicted by the AA, AS/NZS and EC9 specifications are conservative for all columns of SHS, RHS and CHS, as shown in Table 6. The mean value of P_{Exp}/P_{AA}, $P_{Exp}/P_{AS/NZS}$ and P_{Exp}/P_{EC9} ratios are 1.07, 1.12 and 1.12 with the corresponding coefficients of variation (COV) of 0.044, 0.072 and 0.053, respectively.

Table 7. Comparison of test strengths with design strengths for welded columns

Specimen	Test	Comparison				
	P_{Exp} (kN)	$\frac{P_{Exp}}{P_{AA}}$	$\frac{P_{Exp}}{P_{AS/NZS}}$	$\frac{P_{Exp}}{P_{EC9}}$	$\frac{P_{Exp}}{P^*_{AA}}$	$\frac{P_{Exp}}{P^*_{AS/NZS}}$
NS1-W-L300	18.8	1.45	1.62	1.15	---	---
NS1-W-L300R	19.7	1.53	1.71	1.23	---	---
NS1-W-L1000	19.2	1.51	1.65	1.24	---	---
NR1-W-L300	26.4	0.88	0.95	1.16	0.86	0.86
NR1-W-L1000	27.7	0.94	1.01	1.30	0.91	0.91
NR2-W-L300	101.0	1.77	1.98	1.29	1.28	1.43
NR2-W-L300R	97.4	1.71	1.91	1.24	1.23	1.38
NR2-W-L1000	89.7	1.61	1.61	1.23	1.17	1.27
H-R1-W-L300	37.5	1.02	1.03	1.65	0.98	0.98
HR1-W-L300R	40.0	1.10	1.10	1.76	1.05	1.05
HR1-W-L1000	37.9	1.04	1.04	1.76	1.00	1.00
HR2-W-L300	118.0	1.48	1.66	1.38	1.20	1.34
HR2-W-L300R	120.0	1.51	1.69	1.40	1.22	1.37
HR2-W-L1000	139.3	1.80	1.96	1.76	1.41	1.58
NC1-W-L300	35.9	2.41	2.15	1.33	---	---
NC1-W-L1000	30.3	2.02	1.86	1.22	---	---
NC2-W-L300	69.4	2.30	2.06	1.40	---	---
NC2-W-L300R	65.1	2.15	1.92	1.31	---	---
NC2-W-L1000	65.1	2.16	1.99	1.44	---	---
HC1-W-L300	47.2	2.45	2.18	1.43	---	---
HC1-W-L1000	39.0	2.03	1.88	1.35	---	---
HC2-W-L300	88.0	2.34	2.09	1.44	---	---
HC2-W-L1000	84.5	2.24	2.10	1.56	---	---
Mean	---	1.63	1.78	1.39	1.12	1.20
COV	---	0.246	0.246	0.132	0.246	0.194

Note: * Calculated using the welded material properties obtained from tensile coupon tests of 250 mm gauge length.

For the welded columns, it is shown that the design strengths predicted by the AA, AS/NZS and EC9 specifications are generally more conservative than the non-welded column predictions for SHS, RHS and CHS, except for specimen N-R1-W-L300 predicted by the AA and AS/NZS specifications and specimens N-R1-W-L1000 predicted by the AA Specification, as shown in Table 7. The mean value of P_{Exp}/P_{AA}, $P_{Exp}/P_{AS/NZS}$ and P_{Exp}/P_{EC9} ratios are 1.63, 1.78 and 1.39 with the corresponding COV of 0.246, 0.246 and 0.132, respectively. It is also shown that the design strengths calculated using the welded material properties of 250 mm gauge length are generally conservative for the AA and AS/NZS specifications, except for specimens N-R1-W-L300, N-R1-W-L1000 and H-R1-W-L300. The mean value of P_{Exp}/P^*_{AA} and $P_{Exp}/P^*_{AS/NZS}$ ratios are 1.12 and 1.20 with the corresponding COV of 0.246 and 0.194, respectively.

The effects of transverse welds at the ends of aluminium alloy columns are investigated in this study. A comparison of the welded and non-welded column test strengths is shown in Table 8, where the ratio of the welded to non-welded column test strengths is presented. It is shown that the welded columns reached 54 – 76% of the test strength of the non-welded columns for SHS, RHS and CHS.

Table 8. Comparison of welded and non-welded column test strengths

Test series	Weld column test strength / Non-welded column test strength	
	L = 300 mm	L = 1000 mm
N-S1	0.56	0.57
N-R1	0.62	0.66
N-R2	0.67	0.62
H-R1	0.73	0.75
H-R2	0.57	0.69
N-C1	0.74	0.66
N-C2	0.66	0.76
H-C1	0.62	0.54
H-C2	0.68	0.71

4. Conclusions

An experimental investigation on the effects of transverse welds of aluminium alloy columns has been described in this paper. Tests were conducted on fixed-ended columns with and without transverse welds at the ends of the columns for square, rectangular and circular hollow sections. The specimens comprised two different materials of 6063-T5 and 6061-T6 heat-treated aluminium alloys fabricated by extrusions. The initial overall and local geometric imperfections for the column specimens were measured. In addition, tensile coupon tests of the non-welded and welded materials were also conducted, and the material properties have been presented. The column test strengths were compared with the unfactored design strengths predicted by the American, Australian/New Zealand and European specifications for aluminium structures. It is shown that the design strengths predicted by the three specifications are conservative for the non-welded columns. The welded column design strengths predicted by the three specifications are generally more conservative than the non-welded column predictions. It is also shown that the welded columns reached 54 – 76% of the test strength of the non-welded columns.

Acknowledgments

The authors gratefully acknowledge the Asia Aluminum Manufacturing Company for supplying the test specimens.

References

1. AA. Aluminum Design Manual, The Aluminum Association, Washington, D.C. (2000).
2. F. M. Mazzolani. Aluminium alloy structures, 2nd Edition, E & FN Spon, London (1995).
3. AS/NZS. Aluminum structures Part 1: Limit state design, Australian/New Zealand Standard, AS/NZS 1664.1:1997, Standards Australia, Sydney, Australia (1997).
4. EC9. Eurocode 9: Design of aluminum structures – Part 1-1: General rules – General rules and rules for buildings, European Committee for Standardization, DD ENV 1999-1-1:2000, Final draft Oct 2000, CEN (2000).
5. AS. Methods for tensile testing of metals, Australian Standard, AS 1391 – 1991, Standards Association of Australia, Sydney, Australia (1991).
6. ASTM. Standard test methods for tension testing of metallic materials, American Society for Testing and Materials, E 8M-97, West Conshohocken, PA (1997).
7. J. Mennink. Cross-sectional stability of aluminium extrusions. PhD thesis, Technical University of Eindhoven, Eindhoven, Netherlands (2002).
8. K. C. Rockey and F. Jenkins. The behaviour of webplates of plate girders subjected to pure bending, *The Structural Engineer*, 35, 176-189 (1957).
9. K. J. R. Rasmussen and G. J. Hancock. Design of cold-formed stainless steel tubular members. I: columns, *Journal of Structural Engineering*, 119(8), 2349-2367 (1993).
10. L. Gardner. A new approach to structural stainless steel design, PhD thesis, Department of Civil and Environmental Engineering, Imperial College of Science, Technology and Medicine, London (2002).
11. M. L. Sharp. Behavior and design of aluminum structures, McGraw-Hill, New York (1993).
12. AWS. Structural Welding Code – Aluminum, American Welding Society, ANSI/AWS D1.2-97, Miami, Florida (1997).

EXPERIMENTAL INVESTIGATION OF COLD-FORMED HIGH STRENGTH STAINLESS STEEL COMPRESSION MEMBERS*

BEN YOUNG

Department of Civil Engineering, The University of Hong Kong, Pokfulam Road, Hong Kong

WING-MAN LUI

Department of Civil Engineering, Hong Kong University of Science and Technology, Clear Water Bay, Kowloon, Hong Kong

The paper describes a test program on cold-formed high strength stainless steel compression members. The duplex stainless steel having the yield stress and tensile strength up to 750 and 850 MPa, respectively was investigated. The material properties of the test specimens were obtained from tensile coupon and stub column tests. The test specimens were cold-rolled into square and rectangular hollow sections. The specimens were compressed between fixed ends at different column lengths. The initial overall geometric imperfections of the column specimens were measured. The strength and behaviour of cold-formed high strength stainless steel columns were investigated. The test strengths were compared with the design strengths predicted using the American, Australian/New Zealand and European specifications for cold-formed stainless steel structures. Generally, it is shown that the design strengths predicted by the three specifications using the material properties obtained from tensile coupon and stub column tests are conservative for the cold-formed high strength stainless steel columns. In addition, reliability analysis was performed to evaluate the current design rules.

1. Introduction

Stainless steel sections have been increasingly used in architectural and structural applications. This is due to their aesthetic appearance, superior corrosion resistance, ease of maintenance and ease of construction. Since its inception during the early part of the twentieth century, designers, engineers and architects alike, have used stainless steel in both practical and imaginative ways, with further use certain to arise as we enter a global transition towards sustainable development and reduction in environmental impacts [1].

Tests of cold-formed stainless steel columns have been conducted by Rasmussen and Hancock [2], Talja and Salmi [3], Macdonald et al. [4], Young and Hartono [5], Gardner [1], Young and Liu [6] and other researchers. These researchers proposed design rules and design recommendations for stainless steel columns. The current design rules in the American Society of Civil Engineers Specification (ASCE) [7], Australian/New Zealand Standard (AS/NZS) [8] and European Code (EC3) [9] for cold-formed stainless steel structures as well as the design rules proposed by the aforementioned researchers were mainly based on the investigation of cold-formed austenitic stainless steel type 304. However, little test data are available on cold-formed high strength stainless steel columns, such as duplex material. The current design rules may not be applicable to high strength material. Therefore, there is a need to investigate the appropriateness of the current design rules in the specifications for high strength material.

A test program was performed to examine the strength and behaviour of cold-formed high strength stainless steel columns in this study. The test specimens were cold-rolled from flat strips of duplex stainless steel. The square and rectangular hollow section columns were compressed between fixed ends. The column test strengths were compared with the design strengths obtained using the American [7], Australian/New Zealand [8] and European [9] specifications for cold-formed stainless steel structures. Reliability analysis was also performed to evaluate the current design rules.

2. Test Specimens

2.1 *General*

High strength stainless steel square hollow section (SHS) and rectangular hollow section (RHS) columns were investigated. The high strength material used in this study was duplex stainless steel. The test specimens were cold-rolled from flat strips. The test program consists of four test series that include two SHS (Series SHS1 and SHS2) and two RHS (Series RHS1 and RHS2). The nominal section sizes of the SHS are 40×40×2 and 50×50×1.5 mm, and the nominal section sizes of the RHS are 140×80×3 and 160×80×3 mm. The overall depth (D) to thickness (t) ratios are 20.0, 33.3, 46.7 and 53.3 for 40×40×2, 50×50×1.5, 140×80×3 and 160×80×3 sections, respectively. Tables 1 – 4 show the measured cross-section dimensions and column length (L) of each test specimen using the nomenclature defined in Fig. 1. The cross-section dimensions shown in Tables 1 – 4 are the averages of measured values at both ends for each test specimen. Each specimen was cut to a specified length ranging from 300 mm to 3000 mm, and both ends were welded to steel end plates to ensure full contact between specimen and end bearings. The longest specimen lengths produced l_e/r_y ratios of 99, 77, 46 and 44 for Series SHS1, SHS2, RHS1 and RHS2, respectively, where l_e is the effective length of fixed-ended column ($l_e = L/2$) and r_y is the radius of gyration about the minor y-axis.

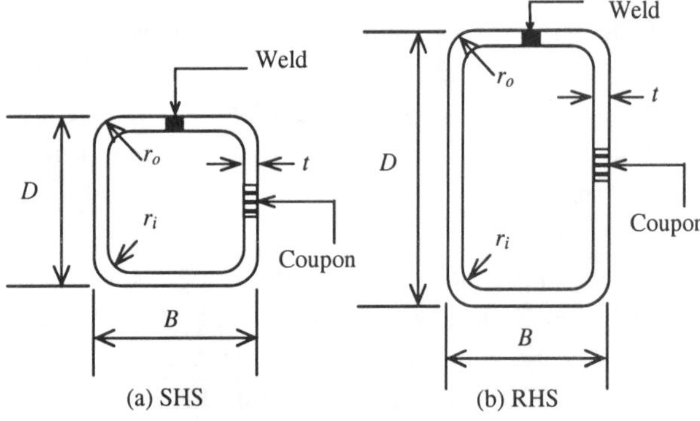

Figure 1. Definition of symbols and location of tensile coupon in cross-section.

2.2 *Specimen labeling*

The column specimens were separated into four test series of different cross-section dimensions. All specimens were tested between fixed ends at various column lengths. The test specimens were labeled such that the test series and specimen length could be identified from the label. For example, the label SHS1L300$^{\#}$ defines the specimen belongs to test Series SHS1, where the letters "SHS" refer to square hollow section. The third letter "L" indicates the length of the specimen followed by the last three or four digits that are the nominal length of the specimen in mm (300 mm). If a test was repeated, the last symbol "$^{\#}$" in superscript indicates the repeated test. The four test series SHS1, SHS2, RHS1 and RHS2 involved columns with section sizes 40×40×2, 50×50×1.5, 140×80×3 and 160×80×3, respectively.

3. Material Properties

3.1 *Tensile coupon tests*

Tensile coupon tests were conducted to determine the material properties of the test specimens. Longitudinal tensile coupons of each test series of specimens were tested. The coupons were extracted from the untested specimens belonging to the same batch of specimens as the column tests. The coupons were taken from the centre of the face at 90° angle from the weld, as shown in Fig. 1. The coupon dimensions conformed to the Australian Standard AS 1391 [10] for the tensile testing of metals using 12.5 mm wide coupon and a gauge length of 50 mm.

The coupons were tested in a 250 kN capacity MTS displacement controlled testing machine using friction grips. Displacement control was used to drive the machine at a speed of 0.04 mm/min within the elastic range of the stress-strain curve, then the speed changed to 0.8 mm/min for the range from proportional limit to a stress level beyond the 0.2% proof stress, and finally the speed changed to 1.35 mm/min until the fracture of the coupon specimen. The static load was obtained by pausing the applied straining for 1.5 minutes near the 0.2% proof stress and the ultimate tensile strength. This allowed the stress relaxation associated with plastic straining to take place. A calibrated extensometer of 50 mm gauge length was used to measure the longitudinal strain. Two linear strain gauges were attached to each coupon at the center of each face. The strain gauges readings were used to determinate the initial Young's modulus. A data acquisition system was used to record the load and the readings of strain at regular intervals during the tests.

The measured stress-strain curves were used to determine the parameter n using the Ramberg-Osgood expression (Ramberg-Osgood) [11]. The parameter n is used to describe the shape of the curve, which was obtained from the measured 0.01% ($\sigma_{0.01}$) and 0.2% ($\sigma_{0.2}$) proof stresses using $n = \ln(0.01/0.2) / \ln(\sigma_{0.01}/\sigma_{0.2})$.

Table 1. Measured test specimen dimensions for Series SHS1.

Specimen	Depth	Width	Thickness	Outer Radius	Inner Radius	Length	Area
	D	B	t	r_o	r_i	L	A
	(mm)	(mm)	(mm)	(mm)	(mm)	(mm)	(mm^2)
SHS1L300	40.1	39.9	1.945	3.8	1.8	300	288
SHS1L300#	40.1	40.0	1.947	3.8	1.8	300	289
SHS1L650	40.1	39.5	1.933	3.8	1.8	650	285
SHS1L1000	40.1	40.0	1.937	3.8	1.8	1000	288
SHS1L1500	40.1	40.0	1.924	3.8	1.8	1501	286
SHS1L2000	40.1	40.0	1.916	3.8	1.8	2000	285
SHS1L2500	40.2	40.0	1.918	3.8	1.8	2500	286
SHS1L3000	40.2	40.0	1.940	3.8	1.8	3000	288
Mean	40.1	39.9	1.932	3.8	1.8	---	287
COV	0.001	0.005	0.006	0.000	0.000	---	0.005

Note: # Second test; 1 in = 25.4 mm; COV = Coefficient of variation.

Table 2. Measured test specimen dimensions for Series SHS2.

Specimen	Depth	Width	Thickness	Outer Radius	Inner Radius	Length	Area
	D	B	t	r_o	r_i	L	A
	(mm)	(mm)	(mm)	(mm)	(mm)	(mm)	(mm^2)
SHS2L300	50.1	50.3	1.584	2.8	1.5	300	295
SHS2L300#	50.0	50.3	1.548	2.8	1.5	300	289
SHS2L650	50.3	50.2	1.596	2.8	1.5	650	298
SHS2L1000	50.2	50.2	1.535	2.8	1.5	1000	287
SHS2L1500	50.0	50.2	1.533	2.8	1.5	1499	287
SHS2L2000	50.1	50.2	1.533	2.8	1.5	2000	287
SHS2L2500	50.2	50.2	1.543	2.8	1.5	2500	289
SHS2L3000	50.1	50.2	1.539	2.8	1.5	3000	288
Mean	50.1	50.2	1.551	2.8	1.5	---	290
COV	0.002	0.001	0.016	0.000	0.000	---	0.015

Note: # Second test; 1 in = 25.4 mm; COV = Coefficient of variation.

Table 3. Measured test specimen dimensions for Series RHS1.

Specimen	Depth	Width	Thickness	Outer Radius	Inner Radius	Length	Area
	D	B	t	r_o	r_i	L	A
	(mm)	(mm)	(mm)	(mm)	(mm)	(mm)	(mm^2)
RHS1L600	140.0	78.8	3.075	10.0	7.0	600	1258
RHS1L1400	139.9	79.9	3.070	10.0	7.0	1400	1262
RHS1L2200	139.9	80.0	3.066	10.0	7.0	2200	1262
RHS1L3000	140.1	79.9	3.011	10.0	7.0	3000	1244
Mean	140.0	79.7	3.056	10.0	7.0	---	1257
COV	0.001	0.007	0.010	0.000	0.000	---	0.007

Note: 1 in = 25.4 mm; COV = Coefficient of variation.

Table 4. Measured test specimen dimensions for Series RHS2.

Specimen	Depth	Width	Thickness	Outer Radius	Inner Radius	Length	Area
	D	B	t	r_o	r_i	L	A
	(mm)	(mm)	(mm)	(mm)	(mm)	(mm)	(mm^2)
RHS2L600	160.1	80.8	2.869	9.0	6.3	600	1305
RHS2L1400	160.1	80.8	2.868	9.0	6.3	1400	1304
RHS2L2200	160.1	80.8	2.864	9.0	6.3	2200	1303
RHS2L3000	160.4	80.9	2.878	9.0	6.3	3000	1311
Mean	160.2	80.8	2.870	9.0	6.3	---	1306
COV	0.001	0.001	0.002	0.000	0.000	---	0.003

Note: 1 in = 25.4 mm; COV = Coefficient of variation.

The measured static 0.2% proof stress ($\sigma_{0.2}$) was 707, 622, 486 and 536 MPa, static tensile strength (σ_u) was 827, 770, 736 and 766 MPa, initial Young's modulus (E_o) was 216, 200, 212 and 208 GPa, and parameter n was 4, 5, 6 and 5 for Series SHS1, SHS2, RHS1 and RHS2, respectively.

3.2 *Stub column test*

Stub column tests were also conducted to determine the material properties of the test specimens for the complete cross-section in the cold-worked state. The shortest specimen lengths complied with the Structural Stability Research Council guidelines [12] for stub column lengths. A total of six stub columns were tested which consisted of specimens SHS1L300, SHS1L300$^{\#}$, SHS2L300, SHS2L300$^{\#}$, RHS1L600 and RHS2L600. The measured cross-section dimensions and specimen length of the stub columns are shown in Tables 1 – 4. A typical stub column test of specimen RHS2L600 is shown in Fig. 2. Four longitudinal strain gauges were attached at mid-length of the stub columns. The strain gauges were located at the corners of the sections. Three laser displacement transducers were used to measure the axial shortening of the specimens.

The stub columns were compressed between fixed ends using a 2500 kN capacity servo-controlled hydraulic testing machine. The fixed-ended bearings were restrained against the minor and major axis rotations as well as twist rotations and warping. Displacement control was used to drive the hydraulic actuator at a constant speed of 0.5 mm/min for all stub column specimens. The static load was recorded by pausing the applied straining for 1.5 minutes near the ultimate load of all the specimens, except for specimens SHS2L300 and SHS2L300$^{\#}$. The static ultimate loads for specimens SHS1L300, SHS1L300$^{\#}$, RHS1L600 and RHS2L600 were approximately 4 – 5% lower than their ultimate loads without pausing the applied straining. The measured 0.2% proof stress ($\sigma_{0.2}$) was 757, 608, 441 and 390 MPa for Series SHS1, SHS2, RHS1 and RHS2, respectively. The measured initial Young's modulus (E_o) was 226, 200, 214 and 213 GPa for Series SHS1, SHS2, RHS1 and RHS2, respectively.

Figure 2. Local buckling of specimen RHS2L600.

Figure 3. Interaction of Local and overall buckling of specimen RHS2L3000.

4. Geometric Imperfection Measurements

Initial overall geometric imperfections of the column specimens were measured prior to testing. Minor axis flexural imperfections were recorded for all specimens, except for the stub columns of Series SHS1 and SHS2. Theodolites were used to obtain readings at mid-length and near both ends of the specimens. The average overall geometric imperfections at mid-length were 1/2670, 1/1630, 1/4668 and 1/5614 of the specimen length for Series SHS1, SHS2, RHS1 and RHS2, respectively. The maximum initial local geometric imperfections of the specimens were 0.113, 0.164, 0.343 and 0.460 mm for Series SHS1, SHS2, RHS1 and RHS2, respectively. The local geometric imperfections measurements are detailed in Young and Lui [13].

5. Column Tests

The duplex stainless steel SHS and RHS columns were compressed between fixed ends. The tests were performed over a range of column lengths from 300 to 3000 mm. Therefore, column curve for each series of tests were plotted. A 2500 kN capacity DARTEC servo-controlled hydraulic testing machine was used to apply compressive axial force to the column specimens. Two steel end plates were welded to the ends of the specimen. A moveable upper end support allowed the tests to be conducted at various column lengths. A rigid flat bearing plate was connected to the upper end support, and the top end plate of the specimen was bolted to the rigid flat bearing plate, which was restrained against the minor and major axis rotations as well as twist rotations and warping. Hence, the top end of the column was fixed in position. A special bearing was used at the lower end support. Initially, the special bearing was free to rotate in any direction. The ram of the actuator was moved slowly toward the specimen until the special bearing was in full contact with the bottom end plate of the specimen having a small initial load of approximately 2 kN. This procedure eliminated any possible gaps between the special bearing and the bottom end plate of the specimen. The bottom end plate of the specimen was bolted to the special bearing, and the bearing was then restrained from rotations and twisting by using vertical and horizontal bolts, respectively. The vertical and horizontal bolts of the special bearing were used to lock the bearing in position after full contact was achieved. Hence, the special bearing became a fixed-ended bearing. The fixed-ended bearing was considered to restrain both minor and major axis rotations as well as twist rotations and warping.

Three laser displacement transducers were used to measure the axial shortening of the specimen. Displacement control was used to drive the hydraulic actuator at a constant speed of 0.5 mm/min for all specimens. The use of displacement control allowed the tests to be continued into the post-ultimate range. The static load was recorded by pausing the applied straining for 1.5 minutes near the ultimate load of all the specimens, except for the specimens of Series SHS2. The static ultimate load was approximately 2 – 5% lower than the ultimate load without pausing the applied straining for the specimens. A data acquisition system was used to record the applied load and the readings of the laser displacement transducers at regular intervals during the tests.

The experimental ultimate loads (P_{Exp}) and failure modes of the stub and long columns are shown in Tables 5, 6, 7 and 8 for Series SHS1, SHS2, RHS1 and RHS2, respectively. The stub column tests were repeated for Series SHS1 and SHS2. The test results for the repeated tests are very close to the first test values, with a difference of 3.0% and 1.1% for Series SHS1 and SHS2, respectively. The failure modes observed at ultimate load of the specimens involved yielding of material (Y), local buckling (L), flexural buckling (F) and interaction of local and overall flexural buckling (L + F), as shown in Tables 5 – 8. Flexural buckling was observed for all specimens of Series SHS1, except for the shortest specimens that failed by yielding of material. Pure local buckling was observed for specimens with effective lengths less than or equal to 700 mm ($l_e \leq 700$ mm) for Series SHS2, RHS1 and RHS2. Interaction of local and overall flexural buckling was observed for specimens with effective lengths greater than or equal to 1100 mm ($l_e \geq 1100$ mm) for Series RHS1 and RHS2 as well as for specimen SHS2L1500. Flexural buckling was observed for specimens SHS2L2000, SHS2L2500 and SHS2L3000. Fig. 2 shows the local buckling of specimen RHS2L600. The interaction of local and overall flexural buckling of specimen RHS2L3000 at ultimate load is shown in Fig. 3.

6. Design Rules

The design strengths of the cold-formed stainless steel concentrically loaded compression members were calculated using the American Specification (ASCE) [7], Australian/New Zealand Standard (AS/NZS) [8] and European Code (EC3) [9] for the design of cold-formed stainless steel structural members.

According to the design rules in the ASCE Specification, tangent modulus (E_t) was determined using Equation (B-2) in Appendix B of the ASCE Specification. The Ramberg-Osgood parameter n, initial Young's modulus (E_o) and 0.2% proof stress ($\sigma_{0.2}$) are required to determine the tangent modulus, thus an iterative design procedure is involved. The design rules of the AS/NZS Standard allow the use of Euler column strength that is identical to the ASCE Specification or the Perry curve, and the latter has been used for the purpose of comparison. In calculating the AS/NZS design strengths, the values of the parameters α, β λ_o and λ_1 are required, which depend on the type of stainless steel. These parameters were determined from the equations given by Rasmussen and Rondal [14], and the values of the parameters are shown in Table 9. The material properties obtained from both the tensile coupon tests and stub column tests were used to determine these parameters. The design rules of the EC3 Code, values of imperfection factor and limiting slenderness were taken as 0.49 and 0.4, respectively, which were obtained from Table 5.2 of the Code. The three specifications require the determination of effective cross-section area (A_e) of the column. In the ASCE, AS/NZS and EC3 specifications, the effective areas were generally found to be equal to the gross areas of cross-section for Series SHS1 and SHS2, except for the specimens with effective lengths less than or equal to 750 mm ($l_e \leq 750$ mm) for Series SHS2, whereas the effective areas were less than the gross areas for Series RHS1 and RHS2. This is in agreement with the local buckling observed in the tests.

The fixed-ended columns were designed as concentrically loaded compression members and the effective length (l_e) was assumed equal to one-half of the column length (L) for fixed-ended columns ($l_e = L/2$). The design strengths were calculated using the average measured cross-section dimensions and the measured material properties. The 0.2% proof stress ($\sigma_{0.2}$) was used as the corresponding yield stress.

7. Reliability Analysis

The reliability of the column design rules was evaluated using reliability analysis. Reliability analysis is detailed in the Commentary of the ASCE Specification [7]. A target reliability index (β_o) of 3.0 for stainless steel structural members is recommended as a lower limit in the ASCE Specification. The design rules are considered to be reliable if the reliability index is greater than 3.0. The resistance factors (ϕ_o) of 0.85, 0.9 and 1/1.1 for concentrically loaded compression members as recommended by the ASCE, AS/NZS and EC3 specifications, respectively, were used in the reliability analysis. The load combinations of 1.2DL + 1.6LL, 1.25DL + 1.5LL and 1.35DL + 1.5LL as specified in the ASCE, AS/NZS and EC3 specifications, respectively, were used in the analysis, where DL is the dead load and LL is the live load. The statistical parameters were obtained from Clause 6 of the ASCE Specification for structural members, where $M_m = 1.10$, $F_m = 1.00$, $V_M = 0.10$ and $V_F = 0.05$ which are the mean values and coefficients of variation of material and fabrication factors. The statistical parameters P_m and V_P are the mean value and coefficient of variation of tested-to-predicted load ratios respectively, as shown in Tables 5 – 8. The values of the reliability index (β_o) of the design rules were determined using the respective resistance factors and load combinations are shown in Tables 5 – 8. For the purpose of direct comparison, a constant resistant factor (ϕ_1) of 0.85 and a load combination of 1.2DL + 1.6LL as specified in the ASCE Specification were used to calculate the reliability index (β_1) for the AS/NZS and EC3 specifications, and the values of the reliability index are also shown in Tables 5 – 8.

8. Comparison of Test Strengths with Design Strengths

The column test strengths (P_{Exp}) are compared with the *unfactored* design strengths predicted using the American (ASCE) [7], Australian/New Zealand (AS/NZS) [8] and European (EC3) [9] specifications for cold-formed stainless steel structures. The design strengths were calculated using the material properties obtained from both the tensile coupon tests and stub column tests. Tables 5 – 8 show the comparison of the column test strengths with the design strengths, where P_{ASCE}, $P_{AS/NZS}$ and P_{EC3} are the design strengths calculated using the material properties obtained from the tensile coupon tests for American, Australian/New Zealand and European specifications, respectively. The P^*_{ASCE}, $P^*_{AS/NZS}$ and P^*_{EC3} are the design strengths calculated using the material properties obtained from the stub column tests.

The design strengths predicted by the ASCE, AS/NZS and EC3 specifications using the material properties obtained from the tensile coupon and stub column tests are generally conservative, except for a few specimens with long column lengths in Series SHS2 and RHS2 calculated using the material properties obtained from the tensile coupon tests, and some of the specimens with intermediate and long column lengths in Series SHS1. The mean values of P_{Exp} / P_{ASCE}, $P_{Exp} / P_{AS/NZS}$ and P_{Exp} / P_{EC3} ratios are 1.05, 1.08 and 1.11 with the coefficients of variation (COV) of 0.106, 0.077 and 0.054, and the corresponding reliability indices (β_o) of 2.69, 2.52 and 2.75 for Series SHS1, respectively, as shown in Table 5. The mean values of P_{Exp} / P^*_{ASCE},

Table 5. Comparison of test strengths with design strengths for Series SHS1.

Specimen	Test			Comparison					
	P_{Exp} (kN)	Failure Mode	l_e (mm)	$\frac{P_{Exp}}{P_{ASCE}}$	$\frac{P_{Exp}}{P_{AS/NZS}}$	$\frac{P_{Exp}}{P_{EC3}}$	$\frac{P_{Exp}}{P^*_{ASCE}}$	$\frac{P_{Exp}}{P^*_{AS/NZS}}$	$\frac{P_{Exp}}{P^*_{EC3}}$
SHS1L300	245.3	Y	150	1.21	1.21	1.21	1.13	1.13	1.13
SHS1L300#	238.0	Y	150	1.17	1.17	1.17	1.10	1.10	1.10
SHS1L650	222.8	F	325	1.10	1.10	1.10	1.03	1.03	1.03
SHS1L1000	197.8	F	500	1.03	1.03	1.12	0.93	0.91	1.04
SHS1L1500	136.0	F	750	0.98	1.00	1.02	0.93	0.91	0.95
SHS1L2000	106.3	F	1000	1.04	1.09	1.14	1.00	1.02	1.06
SHS1L2500	71.3	F	1250	0.93	1.01	1.08	0.90	0.96	1.00
SHS1L3000	52.1	F	1500	0.90	0.99	1.07	0.87	0.95	0.99
Mean, P_m	---	---	---	1.05	1.08	1.11	0.98	1.00	1.04
COV, V_P	---	---	---	0.106	0.077	0.054	0.096	0.083	0.057
Reliability index, β_0	---	---	---	2.69	2.52	2.75	2.51	2.22	2.46
Resistance factor, ϕ_0	---	---	---	0.85	0.90	0.91	0.85	0.90	0.91
Reliability index, β_1	---	---	---	2.69	2.95	3.18	2.51	2.64	2.88
Resistance factor, ϕ_1	---	---	---	0.85	0.85	0.85	0.85	0.85	0.85

Note: # Second test; 1 in = 25.4 mm; 1 kip = 4.45 kN; COV = Coefficient of variation.

Y = Material yielding; F = Flexural buckling.

* Calculated using material properties obtained from stub column tests.

Table 6. Comparison of test strengths with design strengths for Series SHS2.

Specimen	Test			Comparison					
	P_{Exp} (kN)	Failure Mode	l_e (mm)	$\frac{P_{Exp}}{P_{ASCE}}$	$\frac{P_{Exp}}{P_{AS/NZS}}$	$\frac{P_{Exp}}{P_{EC3}}$	$\frac{P_{Exp}}{P^*_{ASCE}}$	$\frac{P_{Exp}}{P^*_{AS/NZS}}$	$\frac{P_{Exp}}{P^*_{EC3}}$
SHS2L300	175.7	L	150	1.09	1.09	1.09	1.11	1.11	1.11
SHS2L300#	177.6	L	150	1.11	1.11	1.11	1.12	1.12	1.12
SHS2L650	181.0	L	325	1.13	1.13	1.13	1.14	1.14	1.14
SHS2L1000	175.1	L	500	1.09	1.09	1.10	1.11	1.11	1.12
SHS2L1500	156.8	L + F	750	1.11	1.14	1.11	1.11	1.11	1.12
SHS2L2000	124.7	F	1000	1.01	1.05	1.06	1.05	1.06	1.07
SHS2L2500	95.1	F	1250	0.92	1.00	1.06	1.00	1.04	1.06
SHS2L3000	72.4	F	1500	0.86	0.96	1.05	0.95	1.01	1.05
Mean, P_m	---	---	---	1.04	1.07	1.09	1.07	1.09	1.10
COV, V_P	---	---	---	0.097	0.059	0.026	0.064	0.043	0.030
Reliability index, β_0	---	---	---	2.72	2.58	2.74	3.00	2.69	2.78
Resistance factor, ϕ_0	---	---	---	0.85	0.90	0.91	0.85	0.90	0.91
Reliability index, β_1	---	---	---	2.72	3.01	3.18	3.00	3.13	3.22
Resistance factor, ϕ_1	---	---	---	0.85	0.85	0.85	0.85	0.85	0.85

Note: # Second test; 1 in = 25.4 mm; 1 kip = 4.45 kN; COV = Coefficient of variation.

L = Local buckling; F = Flexural buckling.

* Calculated using material properties obtained from stub column tests.

$P_{Exp} / P^*_{AS/NZS}$ and P_{Exp} / P^*_{EC3} ratios are 0.98, 1.00 and 1.04 with the COV of 0.096, 0.083 and 0.057, and the corresponding reliability indices (β_o) of 2.51, 2.22 and 2.46 for Series SHS1, respectively. The mean values of the test strength to design strength ratios, COV and reliability indices for Series SHS2, RHS1 and RHS2 are shown in Tables 6, 7 and 8, respectively.

Generally, the reliability indices (β_o) are less than the target value of 3.0 for all test series, except for Series SHS2, RHS1 and RHS2 that calculated using the material properties obtained from the stub column tests for ASCE Specification. The reliability indices (β_1) based on the same resistant factor and load combination are generally less than the target value for all test series,

Table 7. Comparison of test strengths with design strengths for Series RHS1.

Specimen	Test			Comparison					
	P_{Exp} (kN)	Failure Mode	l_e (mm)	$\frac{P_{Exp}}{P_{ASCE}}$	$\frac{P_{Exp}}{P_{AS/NZS}}$	$\frac{P_{Exp}}{P_{EC3}}$	$\frac{P_{Exp}}{P^*_{ASCE}}$	$\frac{P_{Exp}}{P^*_{AS/NZS}}$	$\frac{P_{Exp}}{P^*_{EC3}}$
RHS1L600	558.2	L	300	1.04	1.04	1.04	1.13	1.13	1.13
RHS1L1400	553.1	L	700	1.03	1.03	1.03	1.12	1.12	1.12
RHS1L2200	525.1	L + F	1100	1.06	1.08	1.02	1.14	1.15	1.09
RHS1L3000	513.5	L + F	1500	1.17	1.22	1.12	1.26	1.30	1.19
Mean, P_m	---	---	---	1.08	1.09	1.05	1.16	1.18	1.13
COV, V_P	---	---	---	0.060	0.080	0.043	0.055	0.072	0.037
Reliability index, β_0	---	---	---	2.89	2.39	2.50	3.22	2.73	2.83
Resistance factor, ϕ_0	---	---	---	0.85	0.90	0.91	0.85	0.90	0.91
Reliability index, β_1	---	---	---	2.89	2.78	2.92	3.22	3.13	3.26
Resistance factor, ϕ_1	---	---	---	0.85	0.85	0.85	0.85	0.85	0.85

Note: 1 in = 25.4 mm; 1 kip = 4.45 kN; COV = Coefficient of variation.

L = Local buckling; F = Flexural buckling.

* Calculated using material properties obtained from stub column tests.

Table 8. Comparison of test strengths with design strengths for Series RHS2.

Specimen	Test			Comparison					
	P_{Exp} (kN)	Failure Mode	l_e (mm)	$\frac{P_{Exp}}{P_{ASCE}}$	$\frac{P_{Exp}}{P_{AS/NZS}}$	$\frac{P_{Exp}}{P_{EC3}}$	$\frac{P_{Exp}}{P^*_{ASCE}}$	$\frac{P_{Exp}}{P^*_{AS/NZS}}$	$\frac{P_{Exp}}{P^*_{EC3}}$
RHS2L600	537.3	L	300	1.00	1.00	1.00	1.28	1.28	1.28
RHS2L1400	537.2	L	700	1.00	1.00	1.00	1.28	1.28	1.28
RHS2L2200	515.3	L + F	1100	1.02	1.01	0.99	1.32	1.38	1.23
RHS2L3000	439.4	L + F	1500	0.99	1.02	0.93	1.21	1.26	1.13
Mean, P_m	---	---	---	1.00	1.01	0.98	1.27	1.30	1.23
COV, V_P	---	---	---	0.013	0.011	0.033	0.036	0.042	0.056
Reliability index, β_0	---	---	---	2.85	2.41	2.24	3.74	3.35	3.03
Resistance factor, ϕ_0	---	---	---	0.85	0.90	0.91	0.85	0.90	0.91
Reliability index, β_1	---	---	---	2.85	2.87	2.68	3.74	3.78	3.44
Resistance factor, ϕ_1	---	---	---	0.85	0.85	0.85	0.85	0.85	0.85

Note: 1 in = 25.4 mm; 1 kip = 4.45 kN; COV = Coefficient of variation.

L = Local buckling; F = Flexural buckling.

* Calculated using material properties obtained from stub column tests.

Table 9. Values of parameters α, β, λ_0 and λ_1 for design strength calculation.

Test Series	Tensile Coupon Tests				Stub Column Tests			
	α	β	λ_0	λ_1	α	β	λ_0	λ_1
SHS1	1.262	0.191	0.701	0.404	1.363	0.269	0.705	0.347
SHS2	1.066	0.117	0.686	0.488	1.304	0.205	0.695	0.383
RHS1	1.023	0.116	0.649	0.435	1.083	0.124	0.640	0.423
RHS2	1.067	0.117	0.666	0.467	0.721	0.132	0.597	0.291

except for Series SHS2, RHS1 and RHS2 that calculated using the material properties obtained from the stub column tests for ASCE, AS/NZS and EC3 specifications. Therefore, it is shown that the design rules in the three specifications using the material properties obtained from stub column tests are generally reliable.

9. Conclusions

A test program on cold-formed high strength stainless steel columns has been presented. The duplex stainless steel was investigated in this study. The test specimens had the yield stress and tensile strength up to 750 and 850 MPa, respectively. Two series of square hollow section and two series of rectangular hollow section columns were compressed between fixed ends. The fixed-ended columns were tested at different column lengths, and column curves were obtained for each test series. The test strengths were compared with the design strengths predicted using the American, Australian/New Zealand and European specifications for cold-formed stainless steel structures. It is shown that the design strengths predicted by the three specifications using the material properties obtained from tensile coupon and stub column tests are generally conservative for the cold-formed high strength stainless steel square and rectangular hollow section columns. Reliability analysis has been performed to evaluate the reliability of the current design rules in the specifications. It is shown that the design rules in the American, Australian/New Zealand and European specifications using the material properties obtained from stub column tests are generally capable of producing reliable limit state design for the resistance factor of 0.85.

Acknowledgments

The authors are grateful to STALA Tube Finland for supplying the test specimens. The authors are thankful to Mr. Chung Lee and Mr. Pak-Kin Cheung for their assistance in this project.

References

1. L. Gardner. A new approach to structural stainless steel design, *PhD thesis,* Department of Civil and Environmental Engineering, Imperial College of Science, Technology and Medicine, London (2002).
2. K. J. R. Rasmussen and G. J. Hancock. Design of cold-formed stainless steel tubular members. I: Columns, *Journal of Structural Engineering,* ASCE, 119(8), 2349-2367 (1993).
3. A. Talja and P. Salmi. Design of stainless steel columns, VTT *Research Notes* 1619, Technical Research Centre of Finland, Espoo (1995).
4. M. Macdonald, J. Rhodes and G. T. Taylor. Mechanical properties of stainless steel lipped channels, *Proceedings of the Fifteenth International Specialty Conference on Cold-formed steel structures,* St. Louis, University of Missouri-Rolla, Mo., USA, 673-686 (2000).
5. B. Young and W. Hartono. Compression tests of stainless steel tubular members, *Journal of Structural Engineering,* ASCE, 128(6), 754-761 (2002).
6. B. Young and Y. Liu. Experimental investigation of cold-formed stainless steel columns, *Journal of Structural Engineering,* ASCE, 129(2), 169-176 (2003).
7. ASCE. Specification for the design of cold-formed stainless steel structural members, *American Society of Civil Engineers,* SEI/ASCE-8-02, Reston, Virginia (2002).
8. AS/NZS. Cold-formed stainless steel structures, Australian/New Zealand Standard, AS/NZS 4673:2001, Standards Australia, Sydney, Australia (2001).
9. EC3. Eurocode 3: Design of steel structures – Part 1.4: General rules – Supplementary rules for stainless steels, European Committee for Standardization, ENV 1993-1-4, CEN, Brussels (1996).
10. AS. Methods for tensile testing of metals, Australian Standard, AS 1391 – 1991, Standards Association of Australia, Sydney, Australia (1991).
11. W. Ramberg and W. R. Osgood. Description of stress strain curves by three parameters, *Technical Note No.* 902, National Advisory Committee for Aeronautics, Washington, D.C. (1943).
12. T. V. Galambos (ed.). Guide to stability design criteria for metal structures, 5th Edition, John Wiley & Sons Inc., New York (1998).
13. B. Young and W. M. Lui. Behaviour of cold-formed high strength stainless steel sections, *Journal of Structural Engineering,* ASCE (In press).
14. K. J. R. Rasmussen and J. Ronald. Explicit approach to design of stainless steel columns, *Journal of Structural Engineering,* ASCE, 123(7), 857-863 (1993).

STATIC BEHAVIOR OF BUCKLING-RESTRAINED STEEL PLATE SHEAR WALLS

Y.L. GUO Q.L. DONG
Department of Civil Engineering, Tsinghua University
Beijing, 100084, China

As a promising lateral force resisting system, Steel Plate Shear Walls (SPSW) have been widely used in new buildings as well as in retrofit of existing buildings since the early 1970s . It has demonstrated many advantages such as large elastic stiffness, stable hysteretic behavior and good energy dissipation capacity. But in case of severe earthquakes, steel plate shear walls may loss part of their stiffness and energy dissipation ability after buckling of the infill steel plate. A new kind of steel plate shear wall system, namely Buckling-Restrained Steel Plate shear Walls (BR-SPW), is put forward in this paper. This system is composed of the infill steel panel and the reinforced concrete slabs (RC slabs) on both sides. The RC slabs can restrain the buckling of the steel panel in out-of-plane so that the steel panel can reach full sectional yield stress without buckling. Meanwhile, the connection of the steel panel and reinforced concrete slabs can be deliberately designed to ensure their freely relative slide on the contact layer in case of strong ground motions. In this way, the concrete slabs can be protected from being crashed and the steel panel is prevented from buckling. Numerical studies presented reveal a good static behavior of BR-SPWs and some useful conclusions are presented.

1. Introduction

1.1 *Background*

For the past few decades, steel plate shear walls (SPSW), as a new kind of lateral force resisting system in high-rise buildings, have been widely used in newly built structures as well as in retrofit of existing building structures, because it has the advantage of large elastic stiffness, stable hysteretic behavior, good energy dissipation capacity and higher strength/weight ratio. The main disadvantage of SPSW is that the infill steel panel is prone to buckling along the diagonal compression field, and will cause the deterioration of elastic stiffness, reduction of shear strength and loss of energy dissipation capacity.

One of the most commonly used way to prevent the overall buckling of SPSW is to weld or fabricate stiffeners to it on one side or both sides. Different types of stiffeners, for example, cross stiffeners and diagonal stiffeners have been worked out by researchers and engineers, but at extra cost of materials and fabrication expenses. Figure 1 illustrates two kinds of stiffener layouts, namely cross and diagonal stiffener systems[1] (G. D. Chen 2002). Numerical and experimental studies demonstrate that the SPSWs with stiffeners have a better strength ability and more stable hysteretic behavior.

Composite Steel Plate shear Walls (C-SPW), as an improvement of a pure SPSW, consist of infill steel planes with prefabricated reinforced concrete encasement on one side or both sides of the steel panel. Some basic provisions about C-SPWs can be found in the AISC Seismic[2], 2002 edition. The main disadvantage of C-SPW is that both RC slabs and steel panel are active and provide stiffness and strength from the beginning of lateral displacement. As a result, RC slabs are prone to cracking and crushing due to tension and compression forces under relatively small lateral displacement.

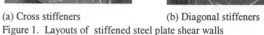

(a) Cross stiffeners

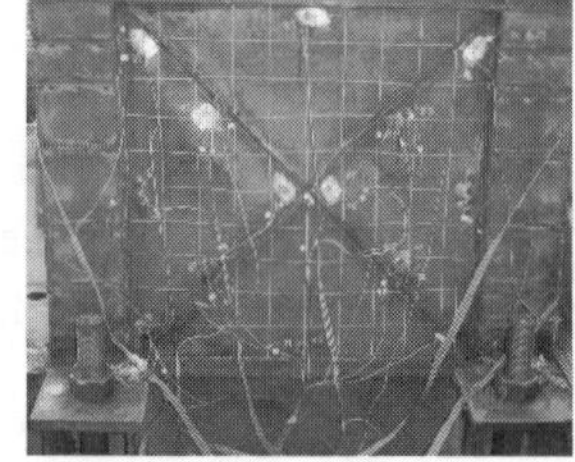

(b) Diagonal stiffeners

Figure 1. Layouts of stiffened steel plate shear walls

Recently, an innovative composite shear wall system, namely the innovative C-SPW was put forward by researchers [3] (Astaneh-Asl 2002). The so-called "innovative" C-SPW differs from the "traditional" system only in the way that reinforced concrete slabs do not contact with the boundary beams and columns directly. In a innovative C-SPW system, where the gap between the RC slabs and boundary beam and column elements ensures that the RC plate serves only as out-of-plane constraint rather than lateral force resisting elements in case of small lateral displacement. Under large lateral displacement, the gap is approached closely at the corners and both RC slabs and steel panel become active to resist lateral force together.

The comparisons of cyclic tests of both traditional and innovative C-SPWs[4] (Q. Zhao et al. 2004) shows that due to the presence of gap between RC slabs and boundary elements, the innovative system experiences less RC plate damage than the traditional one. But in case of strong earthquakes, RC slabs are also damaged severely. To overcome this disadvantage of RC slab's severely damage and to prevent steel panel from buckling significantly, a Buckling–Restrained Steel Plate shear Wall (BR-SPW) system is invented by the authors and preliminary studies are presented in this paper. Here it is confirmed that both its static ultimate strength capacity and its hysteretic behavior exhibit better than the others.

1.2 *Buckling-restrained steel plate shear walls*

A new kind of steel shear wall system, namely the Buckling-Restrained Steel Plate Shear Wall (BR-SPW) is brought forward by the author in this paper. Figure 2 demonstrates the brief sketch of this system. In a BR-SPW system, RC slabs are attached to the steel panel in both sides with high strength bolts. The gaps around the RC slabs, like the gap in the C-SPWs, are still kept and are determined by the inter-storey drift limits under severe earthquakes. Under service loads (wind force, etc) or during a small earthquake, the RC slabs provide out-of-plane restrains for the steel panel so as to prevent its overall buckling before full yielding. In case of large lateral displacement, the RC slabs and steel panel are deliberately designed to slide relatively so that RC slabs could be prevented from severe damage and the steel panel could not buckle and keep on dissipating energy in strong earthquakes. Numerical studies presented in the study reveal that the BR-SPW possesses a good static behavior under monotonic in-plane lateral shear force.

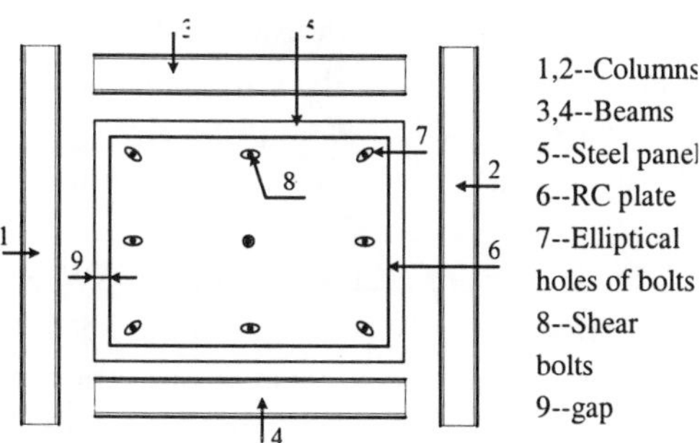

Figure 2. Schematic illustration of BR-SPW

2. Finite Element Analysis

2.1 *Analytical model*

Figure 3(a) shows a typical BR-SPW segment to be analyzed. The I-shaped columns are arranged 3m apart center-to-center with the size of 100X200X10mm. The gap of 60 mm around the RC slabs is selected, corresponding to an inter-storey drift limit of 1/50. Nine bolts are simulated to attach the RC slabs to the steel panel. The infill steel panel is welded to the boundary columns and beams. The node displacements on the top edge of steel panel are coupled to simulate a rigid beam; this is a rational treatment taking into consideration of the fact that the resultant of horizontal forces on the beam caused by wall panels of the adjacent stories is very small. The intersection corner points of beams and columns are assumed to be hinged to exclude the influence of frame members.

Two major geometric parameters, λ and t_c are selected in the study of BR-SPW behavior, where λ denotes the slenderness ratio of steel panel and t_c is the thickness of RC slab. Other parameters used are modulus of elasticity, E=206,000 MPa, Poisson's ratio, υ=0.3, yield stress fy =235 MPa and yield strain ε_y = 0.00114. The trilinear elastic-plastic strain-hardening relationship as shown in Fig. 3(b) is used for the the structural steel plate, and the strain-hardening starts at a strain value of $\varepsilon_{st} = 12\varepsilon_y$ with strain-hardening modulus Est =6,180MPa.

The two systems, namely SPSWs and BR-SPW, with the same geometric parameters and boundary conditions are analyzed respectively to outline their differences.

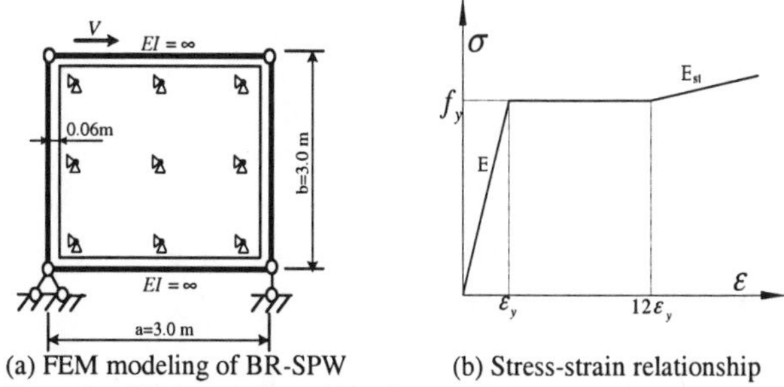

(a) FEM modeling of BR-SPW (b) Stress-strain relationship

Figure 3. FEM modeling of BR-SPW and Stress-strain relationship of steel panel

2.2 *Elastic buckling of BR-SPW under in-plane shear force*

Elastic buckling analysis with variable slenderness ratios from λ =100 to λ =600 are conducted for both BR-

SPWs and SPSWs. Shear buckling stresses obtained by using finite element analysis (ANSYS) are plotted in Figure 4 and classical shear buckling stresses for rectangular plates with all four edges being simply supported are also presented[5] in the figure evaluated after the equation (1).

$$\tau_{cr} = k\frac{\pi^2 E}{12(1-\upsilon^2)}\left(\frac{1}{\lambda}\right)^2 \qquad (1)$$

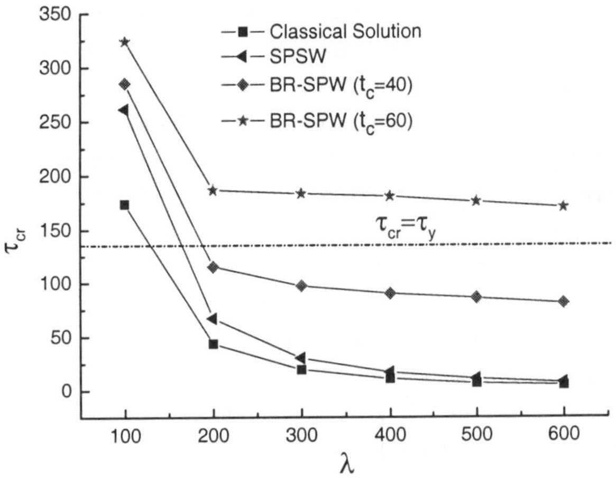

Figure 4. Critical shear stresses for BR-SPW and SPSW

It is evident from figure 4 that the critical stress of BR-SPW is higher than that of SPSW and it increases with the thickness of RC slabs. But as each side RC slab reaches a thickness of 60 mm, critical stresses of BR-SPWs with various slenderness ratios are always greater than shear yielding stress, $\tau_y = f_y/\sqrt{3}$. This means that the BR-SPWs can reach their shear yield stress prior to overall buckling. AISC Seismic gives a requirement of minimum thickness of the RC slabs as 102 mm when RC slab is provided on both sides and 203 mm when RC slab is provided on one side of the steel panel. These requirements are defined for C-SPWs, but they seems also reasonable considering not only for the fabrication details and minimum protection layer of concrete, but also for protecting steel panel from buckling.

2.3 *Ultimate strength analysis of BR-SPW under monotonic in-plane shear force*

Thin steel panels under the action of in-plane shear force will buckle at very small loads, but if properly reinforced by RC slabs in one side or both sides, they can reach steel yield stress prior to overall and local buckling. RC slabs with certain stiffness can provide sufficient out-of-plane restraints and improve significantly the performance of the steel panels both in stiffness and load-carrying capacity. Analytical model of BR-SPW are developed to study the ultimate strength behavior of steel panels up to failure. The BR-SPW is defined to reach its ultimate state usually due to overall buckling as a whole when RC slab restraint is not enough, or due to in-plane full yielding of steel plate when RC plate restraint is strong enough. Therefore, the analysis should include the effects of geometric and material nonlinearities. In this paper, the nonlinear finite element program ANSYS is employed in the analysis. And it is very important that contact element of ANSYS are used to simulate the relatively slide between the contact layer of steel panel and RC slabs. This means that the steel panel and RC slabs may slide relatively in the interface of contact layer and share the common displacements in out-of-plane direction.

As the initial geometric imperfection caused by welding and other production processes are always exist in steel panels, it should be considered in ultimate strength analysis both for SPSWs and BR-SPWs. In the case of square plates, as presented in this paper, the initial geometric imperfection is approximately in the shape of sinusoidal half-wavelength in both directions. The average maximum out-of-plane initial imperfection amplitude w_0 is taken as[6] (Smith et al):

$$w_0 = 0.1b\lambda\frac{f_y}{E} \qquad (2)$$

Where parameter b denotes the width of the steel panel.

For a moderate slenderness ratio, λ =300, the amplitude of initial geometric imperfection can be as large as 3 percent of the width of the steel panel. It is a very high value compared to the 1/1000 of plate width usually considered.

Previous numerical analysis (G. D. Chen) shows that initial geometric imperfection exerts little influence (usually less than 3 percent) on the ultimate strength of steel panels under the action of in-plane shear force. Numerical studies in this paper also confirm this conclusion.

The ultimate strength analysis of both SPSW and BR-SPW are carried out respectively and the obtained results are compared between them. Geometric and material nonlinearities as well as initial geometric imperfection are involved in the analysis. Figure 5 and 6 demonstrate the load-displacement relationships obtained. Geometric parameter studies include slenderness ratios ranging λ =600, 300, 200, 100, and RC slab thickness with a range of t_c=40, 60, 80, 100mm.

It is evident that, compared to SPSW, BR-SPW system can improve the structural behavior of steel panels efficiently. Both the elastic stiffness and ultimate strength are all increased evidently. Furthermore, steel panels with higher slenderness ratios tend to experience more stiffness and strength increase than those with lower slenderness ratios.

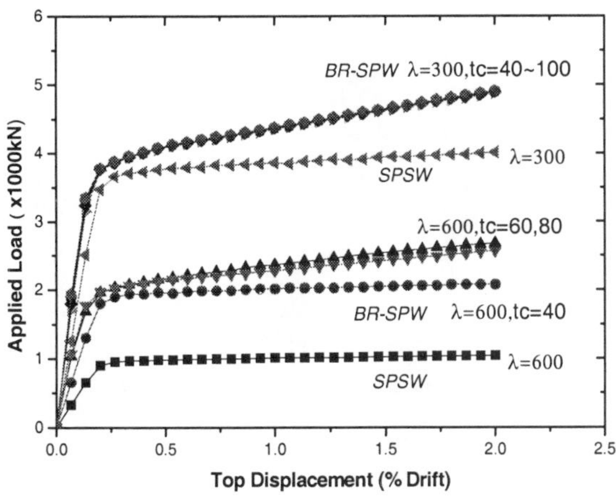

Figure 5. Load-displacement relationship of SPSW and BR-SPW with λ =300 and λ =600

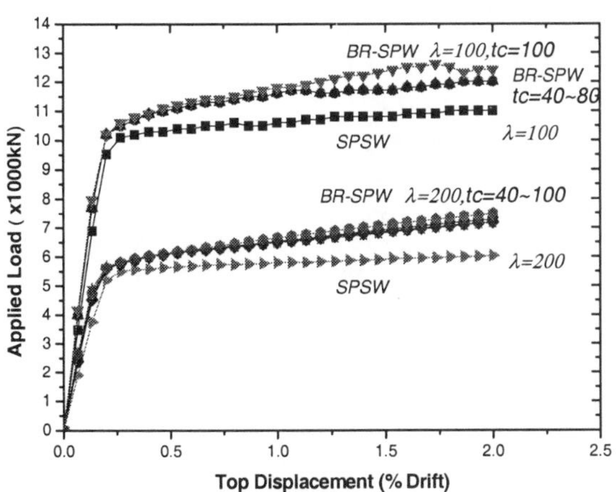

Figure 6. Load-displacement relationship of SPSW and BR-SPW with λ =100 and λ =200

In Figure 5, for steel panel of BR-SPW with slenderness ratio, for example, λ =600, the stiffness and ultimate strength increase nearly 100 percent when RC slabs of 40 mm in thickness (tc=40) are adopted. For RC slab thickness above tc=40, the strength increase slows down with RC slabs' thickness becoming larger. The maximum RC slabs' thickness, tc =100 mm, was found , beyond which the increase of RC slabs' thickness will no longer be rational. This also coincides with the minimum thickness requirements of RC slabs in C-SPWs provided by the AISC Seismic.

On the other hand, relatively compact steel panels are not as sensitive as their slender counterparts to the thickness variation of RC slabs. This is verified by the examples in Figure 6, with slenderness ratio, λ =100 and λ =200.

The failure modes of SPSWs and BR-SPWs are essentially different one another. Figure 7 illustrates their distinctively failure modes in ultimate strength state: it is obvious that the distinctive diagonal tension field was formed in SPSW system (Fig. 7a), while in the BR-SPW system, no tension field was detected and the full section of the steel panel tend to reach their yielding strength without any buckling (Fig. 7b).

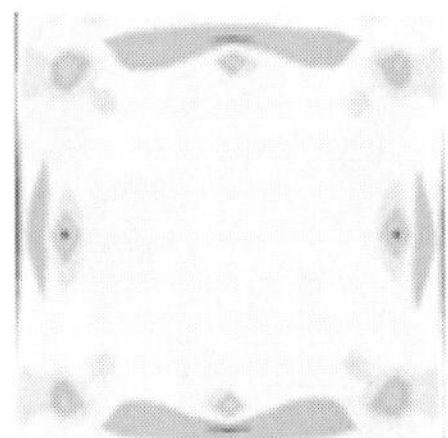

(a) failure mode of SPSW (b) failure mode of BR-SPW

Figure 7. Failure mode of SPSW and BR-SPW

3. Conclusions

A new type of steel plate shear wall, BR-SPW is proposed and preliminary study on its static strength behavior is presented.

Elastic buckling analysis shows that critical stress increase more rapidly with the increase of RC slab thickness for slender steel panels than thicker ones. Minimum thickness requirement of RC slabs for BR-SPW is found in the design.

Numerical analysis of ultimate strength of both BR-SPW and SPSW under monotonic in-plane shear force was carried out. The obtained numerical results reveals that slender steel panels are more suitable to be used in BR-SPW system as its stiffness and strength can be increased efficiently because of the restrains of RC slabs to the steel panel.

As a matter of fact, the most prominent advantage of BR-SPW lies in its hysteretic behavior and energy dissipation capacity. Further researches on BR-SPW both by numerical analysis and tests are under way to bring forth its hysteretic and dynamic behavior.

References

1. G. D. Chen, The investigation to structural behavior of steel plate shear walls, PhD dissertation, Tsinghua Univ. (2002).
2. AISC, Seismic provisions for structural steel buildings, American Institute of Steel Construction, Inc, Chicargo, IL. (2002).
3. A. Astaneh-Asl, Seismic behavior and design of composite steel shear walls, *Steel TIPS Report,* Structural Steel Educational Council, Moraga, Calif. (2002).
4. Q. Zhao and A. Astaneh-Asl, Cyclic behavior of traditional and innovative composite shear walls, *J. Struct. Engrg.*, ASCE, Vol. 130, 271(2004).
5. S. P. Timoshenko and J. M. Gere, Theory of Elastic Stability, 2nd Edn. McGraw-Hill, New York. (1961).
6. C. Smith, P. Davidson, J. Chapman et al, Strength and stiffness of ships' plating under in-plane compression and tension. Transactions RINA (1987).

OPTIMIZED LOCATION OF ENERGY-DISSIPATION BRACES IN TALL BUILDINGS WITH IRREGULAR FLOORS

SI CHAO

Architectural Design & Research Institute of Tongji University, 1239 Siping Road , Shanghai 200092 , China

FULEI YANG YUBING HU

Architectural Design & Research Institute of Tong ji University, 1239 Siping Road , Shanghai 200092 , China

It is a common problem in earthquake resistant design that lateral-torsion vibration is coupled in tall buildings. Especially, when the building appears obviously irregular, the seismic response would be magnified because of its highly coupled. It is an advanced study in seismic design that irregular tall buildings worked with energy-dissipation devices. This paper discusses the steel frame-brace system with energy-dissipation braces (velocity-depended viscoelastic dampers), and analyze its non-scaled structure damper ratio and the seismic response. Based on the optimal energy-dissipation principle and a recently finished shaking table test of a 21-story steel frame-brace structure, it discusses the optimized location of braces with added viscoelastic dampers. It finds that the structure with the optimized location of energy-dissipation braces changes the distributing of structure damping factor, gains higher damping level, and has acceptable seismic performances, while improves its economy. This paper may provide reference for the application and development of this technology.

1. Work Principle and Restoring Force of the Energy Dissipation Brace

1.1 *Work principle of the energy dissipation brace*

The energy dissipation brace is composed of the damper (only the velocity-depended viscoelastic dampers in this paper) and the braces joined to the damper. The energy dissipation brace works since the structure begins shaking with tiny disturb, and will reach the maximal damper force in a small displacement. While the story drift exceeds the bound (1/300 for example), the equivalent stiffness (secant stiffness) of the energy dissipation brace will no longer increase but decrease with the increase of the relative displacement so that it keep on dissipating energy and don't contribute more to the structure stiffness simultaneously. Namely in the common circs, the contribution of the energy dissipation brace to the structure stiffness is limited. With the to-and-fro movement of the stopcock in the damper, the viscoelastic resistance of the viscoelastic fluid in the damper works. And using the nonlinear hysteretic characters of the damper resistance and relative displacement to dissipate the seismic energy, the structure damper ratio and the structure mode damper energy increased; at the same time, the energy absorbed by the structure and the seismic performance of the structure decreased.

1.2 *Restoring force of the energy dissipation brace*

The mechanical characters of the damper in this paper is shown as the equation (1):

$$F_d = C_v Sign(V)|V|^{\alpha} \qquad (1)$$

Where the C_v is the dampness constant designed based on the needs($kN/(mm/s)^{\alpha}$); the V is the relative velocity between the stopcock of the thrust member and the crust of the damper(mm/s);and the α is the velocity index, constant designed by the need which varies from 0.1 to 1.0.

If the value of C_v ,A or α changes, the shape of the force –displacement curve of the damper will changed as the figure 1(where the A is the maximal distance of the damper in the unit of mm). Figure 1 shows that:

1. With the increase of the velocity index α , the curve change from rectangle to ellipse and the damper resistance augment immediately. While F_d (the damper resistance) is close to A (the maximal

distance of the damper), the area of the ellipse is smaller than that of the rectangle, namely the energy dissipation is less than the rectangle case.

2. With the increase of the damper constant C_v ,the shape of the curve remains the same ,but the damper resistance increase. It should notice that the resistance shouldn't be too high, otherwise the stiffness is close to that of the brace and the energy dissipation advantage won't be apparent. If the resistance is too low, the needed damper will be more and difficult to dispose and diseconomy.

So the damper resistance of the damper used in building is more than 500kN and the α is between 0.12 to 0.16.

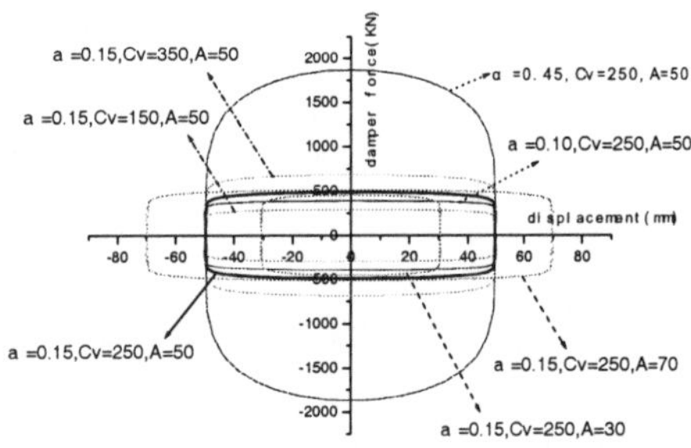

Figure 1 the theoretic curve of the viscoelastic damper

2. Optimal Locations of the Energy Dissipation Brace

The damper braces are usually fixed between the stories, and the displacement between the stories arouses the deformation of the damper and makes it work. Knowing from the figure 1 that the energy dissipated by the damper is directly related with the distance of the damper (or the relative displacement between the stories), the dampers should be fixed at the location where the story drift is maximal in order to make the dampers work efficiently. So the location of the dampers should follow these steps:

1. Analyze the original structure, and fix the braces at the story where the maximal story drift occurs.
2. Analyze the structure with the braces, and fix the braces at the story where occurs the maximal relative displacement.
3. Repeat 2 until the seismic performance match the design needs. This is the optimal energy-dissipation principle.

3. An Example

In this chapter we will illuminate the theory with a complex tall steel building. With the absence of part floor, the building's plane shape is like the" L", and rotates clockwise every three stories. It rotates six times to reach the 21-story. The story is 4m high except some is 6m, the plane is 48.6m×48.6m, the total height is about 100m. The physical parameter of the viscoelastic damper used in this project is that the velocity index $\alpha = 0.15$,the damper constant $C_v = 250kN/(mm/s)^{\alpha}$ .The Shanghai Wave SHW1 is used to time-history analysis of the structure. Figure 2 shows the plane of the structure, and the overstriking dashed means the location where the braces could be fixed.

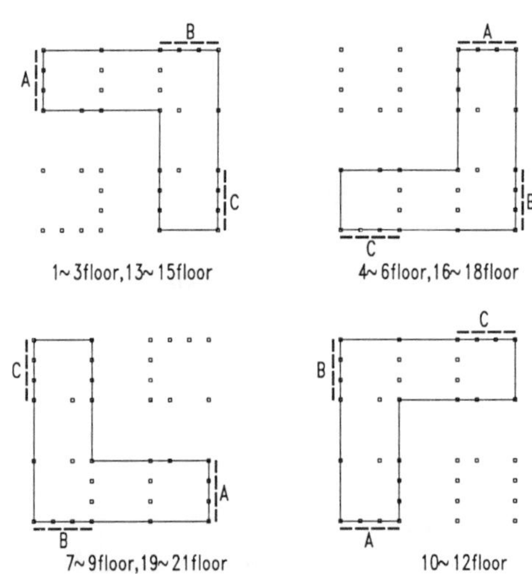

Figure 2 location where the braces can be fixed

Firstly choose the location of the brace. In order to make the lateral stiffness of the energy-dissipation braces at two principal axis is close to avoid the torsion of the structure, the braces should be disposed at both the two directions because of the plane shape is a square and the structure's stiffness is close at the two principal axis. We chose the A, B to fix the brace(equal to B,C).

We use the optimal method to dispose the braces in the upright direction. Based on the analysis result of ETABS program, the maximal story drift occur at the 3, 6, 8, 12, 14 and 19 stories. So the braces were disposed in the stories related to these six stories. To compare the optimal effect, we chose two comparable cases. Case 1 is to set braces at all the feasible places, and case 2 is to set braces at the non-optimal places. The time history analysis results of the two cases are shown in figure 3 and figure 4. figure 3 shows the time history of the top story displacements of the original structure and the structure with the optimal disposed braces. It tells that the reducibility of the seismic performance is apparent, that is to say the energy dissipation brace have well energy dissipation performance. Figure 4 shows the maximal displacement between stories in different cases. It tells that the displacement of the structure with suffusion of braces is the minimal, and the seismic performance of the structure with optimal disposed braces is smaller than that with the non-optimal disposed braces. As to this project, the structure with suffusion of braces needs 84 dampers, and the optimal and non-optimal case both need 56 dampers. Comparing the feasibility and the economy, the optimal disposed method has great advantage.

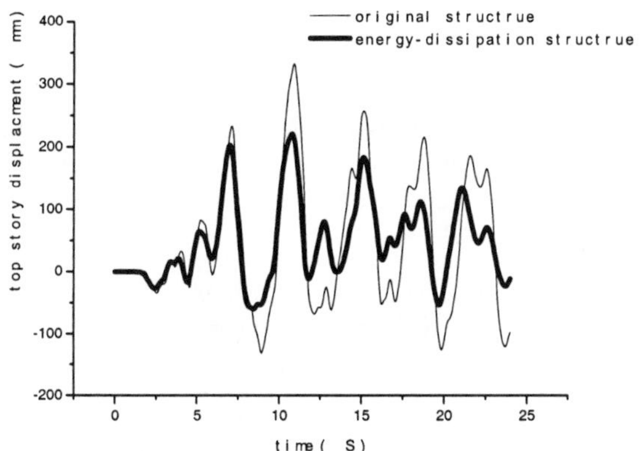

Figure 3 time history performance of displacement of the top story

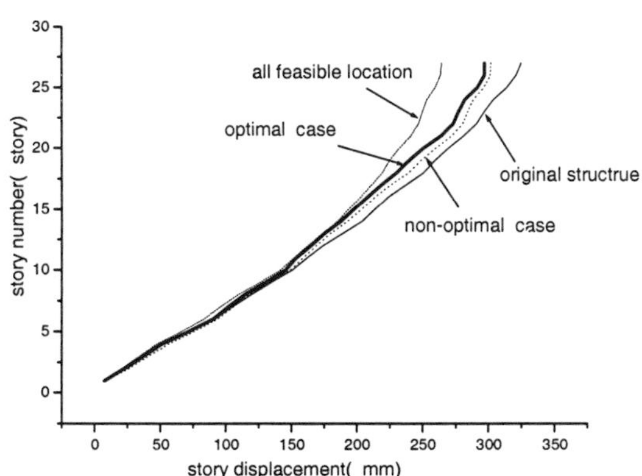

Figure 4 maximal relative displacement in different cases

4. Shaking Table Test

The shaking table test of this project is held in the State Key Laboratory for Disaster Reduction in Civil Engineering, Tongji University in 2004, (figure 5)

Figure 5 whole model in the shaking table test

As part of the experiment, we tested the model's performance with and without the dampers in the seven degree frequent earthquake, the results are shown in the figure 6 and figure 7.

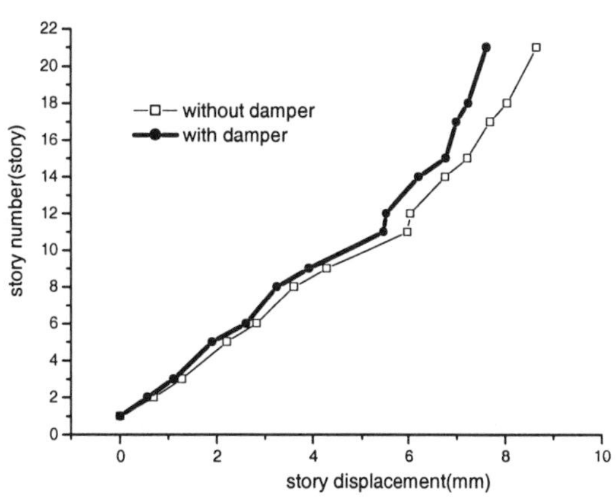

Figure 6 displacement in X axis

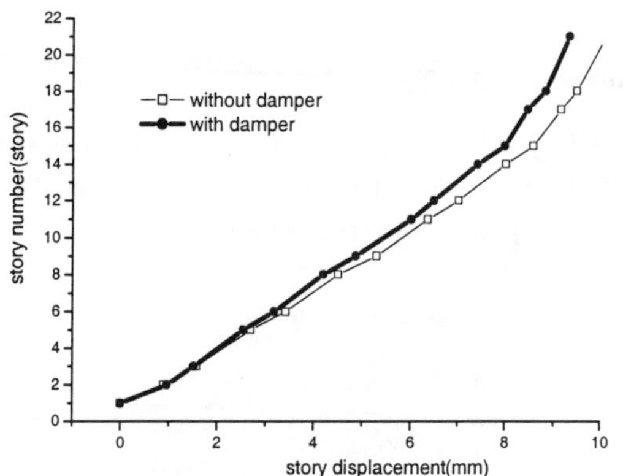

Figure 7 displacement in Y axis

It is rather apparent that the seismic performance of the structure with the optimal disposed braces is lower than that without the energy dissipation braces. This is accordant with the analysis results of ETABS. Limited by the experimental condition, we didn't hold the compare experiment between the structure with optimal and non-optimal disposed braces. But based on the previously accordance, we have enough confidence to believe the variety of seismic performance between the optimal and non-optimal disposed braces in the chapter 3 is correct.

5. Conclusion

This paper manifests the work principle of the viscoelastic dampers and the method based on the optimal energy-dissipation principle to optimal disposed the energy dissipation braces. Through the analysis and shaking table test of a practical tall building, it tells that the energy dissipation brace can well control the seismic performance of the structure. In practical projects, the optimization of the braces can make the energy dissipation device more economical and rational.

References

1. Zhang Rihui,soong TT. seismic design of vicoelastic dampers for structure applications. Journal of structural engineering, ASCE,1992,19(2):389-417.
2. Weng Dagen, Lu Zhuhui, etc., experimental study on the mechanics characters of the viscoelastic dampers. World Earthquake Engineering , 2002 , 18 (4) .
3. Chang K C , Lai M L , Soong T T, etc, Seismic behaviors and design guidelines for steel frame structures with added viscoelastic dampers. Technical Report NCEER-93-0009.SUNY at Buffalo .
4. Tsai C S, Lee H H. Applications of Viscoelastic Dampers to high-rise buildings, Journal of Structural Engineering ASCE,1993,119(4)
5. Fulei Yang , paper of master's degree 《Aseismic Analysis of Complex Tall Buildings with Irregular Floors》 , 2005.3.

Management Issues

APPLYING WIRELESS TRANSMISSION TECHNIQUE TO IMPROVE CONSTRUCTION MANAGEMENT IN TALL BUILDINGS

LEE-KUO LIN CHING-LUNG CHENG SHANG-XIAN LI PEI-CHIN LIN

Civil Engineering Department, National Taipei University of Technology, Taipei, China

In general, public tall building is a kind of hardware facilities for each citizen's use. The constricting quality of each public tall building construction program during the working process will have highly relative influence with construction management. Construction job-sites have too many workers and lots of stuff that become complexity. Such situation causing construction management has full with challenge. Meanwhile, once the performance of construction management doesn't work well, it not only makes worse quality but also threatens citizen's life. So how to transfer the high technology of wireless transmission into tall buildings construction job-sites to improving and promoting construction management will be an useful research. This paper will introduce relative techniques of wireless transmission, discuss how to apply the wireless transmission technique as a management tool for each tall building inspection engineer. Also will demonstrate and test the performance with practice application.

1. Preface

Construction projects use the professional technology in civil engineering and relevant facilities to change land renovate the environment and prevent calamity from occurring (Lin, 2004). This specialty not only is in close relation with the people's life but also is one of the key index in influencing the nation's public construction and economic development. In each country, most construction projects are built for every citizen's use. The quality of such construction projects will directly influence all users once these projects were finished. Meanwhile, if the performance of construction management doesn't work well, it not only makes worse quality but also threatens each citizen's life. Construction projects contain aspects such as economy, society, culture and education, medical treatment, and etc.. Because most tall buildings are built for public use, therefore the quality of tall building's constructions is an important index for the living quality of people. By following the feedback and analysis of the advanced countries, government should make discussions about how to apply public tall buildings constructions in a suitable way for the operation, management, safeguard and mechanism upgrade to the construction industry. So how to transfer the high technology of wireless transmission into construction management for each tall building construction project to improving and promoting construction quality will be an useful research.

The development and performance of construction industry are strongly relative with citizens' living quality in each country. During the whole life cycle with each construction project of tall building, including conceptual phase, designing phase, biding phase, construction phase and maintenance phase, every civil engineer always concentrates on those procedures before construction phase (Lin, 2005).

Nowadays, people always wants to get the real-time and reliability information immediately. Also for the inspecting information of construction management should be post into the internet to make information connection more fast and convenient. So it will be an appropriate solution to develop a real-time wireless transmission environment at construction job-sites (figure 1). Based such purpose, this research will investigate the feasibility of applying the wireless transmission of PDA or GPRS as a communication tool to inspection engineers. Then illustrate the performance of this application.

Figure 1. Practice picture's wireless transmission at construction job-sites

2. Current Situation and Management of Construction Industry

In the past 40 years, the global economy made big influence to the social economic development, resource allocation and quality of the life. From 1951 to 1969, due to the shortage of funding and the insufficient labor, public construction projects in Taiwan are focusing to improve the agriculture and light industry production (Lin and Kuo, 2001). Then from 1971 to 1979, government is development 12 national construction projects, and from 1981 to 1989, also continuously develops 14 national construction projects in sequence (Chang, 2004; Lin, 2005). Besides the basic construction, government begins to focus on improving the quality of life. But due to the international skyrocketing price and the bubble economy, caused the price of land and worker rapidly go up. This will also cause a lot of project progress to lag behind.

At command process, every inspection engineer at construction job-sites always walks between inspection offices and working activity place during all day long (Lin, 2002). For example, if an inspection engineer wants to check the design draw picture, then he/she needs to go back to the office. According to an investigation research, each inspector spends about 1.38 hours every day for such invalid walking, which will cost about 322 NT dollars (10.3US dollars) by changing to inspector's salary (Huang, 2003). To solve such disadvantage, developing an appropriate wireless transmission system at construction job-sites will be an effective solution. If an inspector at working activity pot wants to get or transfer design draws, pictures, document, and etc., he/she can directly do it by using wireless hardware to pass the signal to the computer at inspection office, then connecting to the PC server with wireless transmission. Therefore, the efficiency of construction management can be increased by applying this wireless technology into construction job-sites. Such philosophy of this algorithm is shown in figure 2.

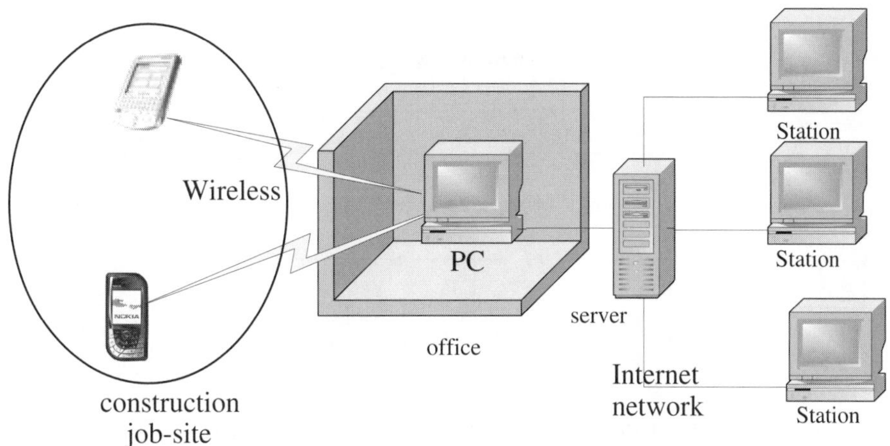

Figure 2. Structure of wireless transmission

3. Wireless Transmission Technology

3.1 *Development of wireless transmission and 802.11 communication protocol*

In the beginning, the launch application of wireless technology was first used for military implement in World War II (Hung and etc., 2004). A radio communication system developed by the Motorola was used by the American army to transmit the wireless telecommunication signals. They developed a wireless communication technology and used a certain degree of the coding technique to allow the soldiers to communicate through radio when they are in operation (Jolly, 1972). This shows the essentiality of the wireless technology to the future.

Up to today, the wireless transmission technology is a new milestone for the development of the data communication. It provides a convenient transmission without the limit of the particular range. This will impel the use of network and accelerate the speed of exchanging data with each other (IEEE, 1999). The transfer point to promote the ability of the wireless transmission is to increases of the data transmission rate, ensure of the quality effects on the exchanging data, ensure of the key knowledge and choose the suitable method to appraise the effects.

For global criteria of wireless communication, IEEE 802.11 is a communication protocol made to suit the wireless environment to formulate the Media Access Control Layer (MAC Layer) and the Physical Layer (PHY). In order to transparence the wireless network, the Wireless Local Area Network (WLAN) hopes to reach the infrastructure mode in the Logical Link Control Layer (LLC Layer). The WLAN must deal with the mobile station and maintain the reliable data transmission ability with in the MAC Layer. The function needed in the MAC Layer is different form the traditional wired backbone network and it is more real time and convenient (MacLaurin, 2002). This is the trend of the international automation (figure 3).

IEEE 802.11 is taking air as medium in the wireless regional network standard rule. For example, there is a metal desk in the square room (figure 4), scientists can catch and measure the distribution signal then identifies its difference with power in the static environment. Standard definition transmission technology of IEEE802.11 is including: (1) frequency hopping spread spectrum; (2) direct sequence spread spectrum; (3) infrared (Calhoun,1992) . When wireless local network with internet local network, setup the internet service farm (figure 5), every transfer stations can use the e-mail by the wireless transmission framework (IEEE, 1999).

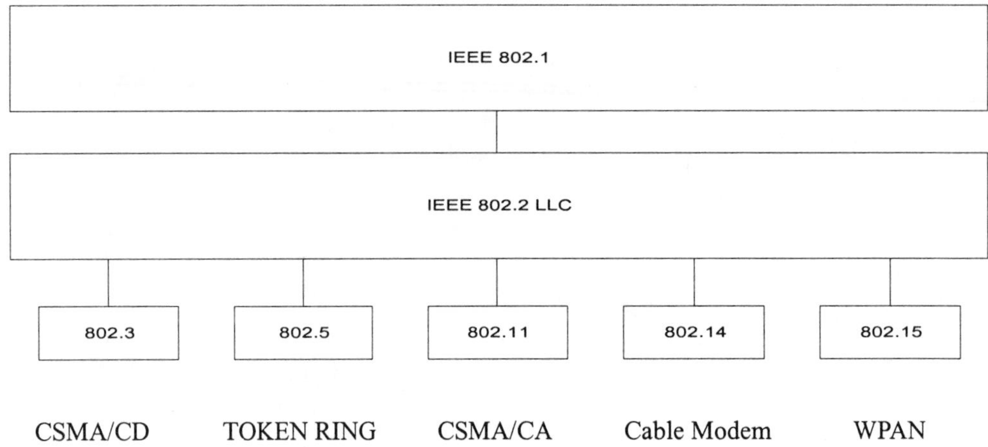

Figure 3. IEEE802 series

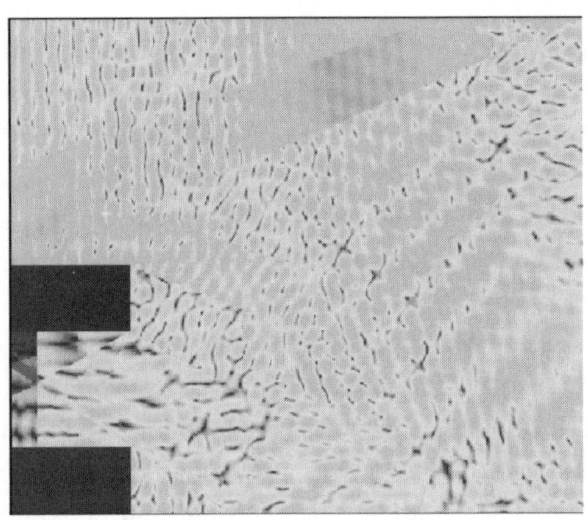

Figure 4. The wireless of 802.11 signal transmission strength degree diagram

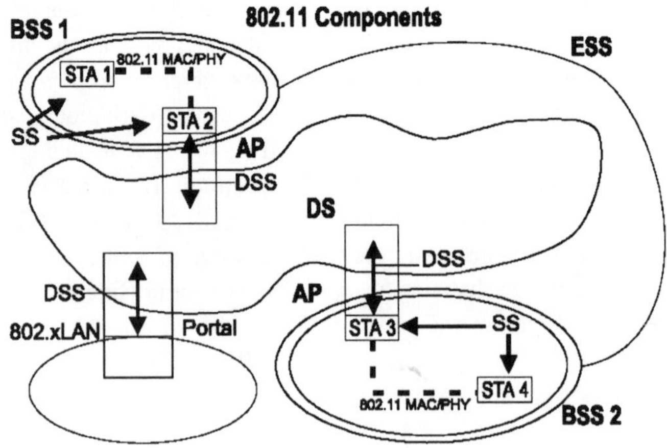

Figure 5. The wireless transmission framework

3.2 *Possible use of wireless transmission at construction job-sites*

Based on this study, there are three possible hardwares can be used by inspectors, including mini-notebook, personal digital assistant (PDA) and GPRS (General Packet Radio Service) mobile phone. The relative information of GPRS and PDA will be brief introduced in the following:

The most popular wireless communication system is the GSM mobile phone communication system and the internet is the most popular wired communication. Therefore, with GPRS mobile phone will be able to transmit sounds, images, words and other multimedia information. GPRS combines the above two independent communication network. The GPRS is a packet-linked technology that enables high-speed wireless internet and other data communications. Using a packet data service, subscribers are always connected and always on line so services will be easy and quick to access (Calhoun, 2003).

PDA is the abbreviation of personal digital assistant. It is a kind of electronic product that users can take to everywhere because of its' portable size. In general, an ordinary PDA function includes phone-book, note, schedule and etc (Lin, 2002). PDA can be input data by user and transferred information from computer that is called Hot Synchronize. Meanwhile, it also can be transferred between PDA by infrared data port.

If any user does not satisfy with the original function of PDA provided, he/she can download appropriate software from specific internet with his/her selection. On the other hand, it also can compensate by added module, such as MP3, GPS and etc.. The first PDA in the world was released in 1996 by Palm-Computing Cooperation Company, called "Pilot 1000" (Bidpalm, 2001). After hardware investigation, this research chooses Palm PDA of iPAQ H4150 as the studying hardware. The consideration reason includes energy saving, speed, size, multimedia and weight (table 1). In this research, a PDA software called "HanDBase" was selected due to it was a convenience database software for PDA hardware. The advantage of this software is it supports Windows CE from HanDBase 3.0 (Huang, 2003). Based on this advantage, program designer can build database on Windows platform first, then synchronize to PDA when the program was finished.

Table 1. General information of PDA with wireless

Category	iPAQ H2210	iPAQ H4150	O2 XDA□
Central processing	Intel® Xscale™ 400MHz	Intel® Xscale™ 400MHz	Intel Intel® XScale PXA263 400MHz
Memory	32MB ROM/ 64 MB RAM	64MB ROM/ 32 MB RAM	128MB ROM/ 64 MB RAM
Monitor	3.5" TFT 240×320	3.5" TFT 240×320	3.5" TFT 240×320
Input	Pen	Pen	Pen
Floppy	None	None	None
Sound	Built-in	Built-in	Built-in
Speaker	Built-in	Built-in	Built-in
Keyboard	None	None	None
Output	RS 232, Infrared	RS 232, Wireless	RS 232, Wireless
Battery	Lithium Battery	Lithium Battery	Lithium Battery
Battery Durability	12 hours	12 hours	12 hours
Weight	144.2 g	132g	190g
Operation System	Microsoft® Windows® for Pocket PC 2003 Mobile Premium	Microsoft® Windows® for Pocket PC 2003 Mobile Premium	Microsoft® Windows® for Pocket PC 2003 Mobile Premium
Characteristic	Fast speed, energy saving	Operate easy, energy saving, multimedia, cheaper	Multimedia, operate easy, energy saving, expensive
Price(USD)	330	400	838

4. Performance Evaluation of Practice Application

4.1 *Image application for inspection engineers*

In general, once an inspection engineer is working on his/her career of construction management at any specific construction job-site, he/she usually needs to monitor crew workers constructing process, and takes image photographs for document record. But at some special situations, for example, if the inspection engineer wants to check design draws while he/she needs to go back to the inspection office to check the figure, such disadvantage could be solved by supplying wireless instruments to inspection engineers (figure 6). With the same algorithm, inspection engineers can take practice pictures then transmits it to the PC server by using wireless transmission technique (figure 8). Therefore, the efficiency and effectiveness of construction management can be increased by developing an appropriate wireless transmission system at the construction job-site.

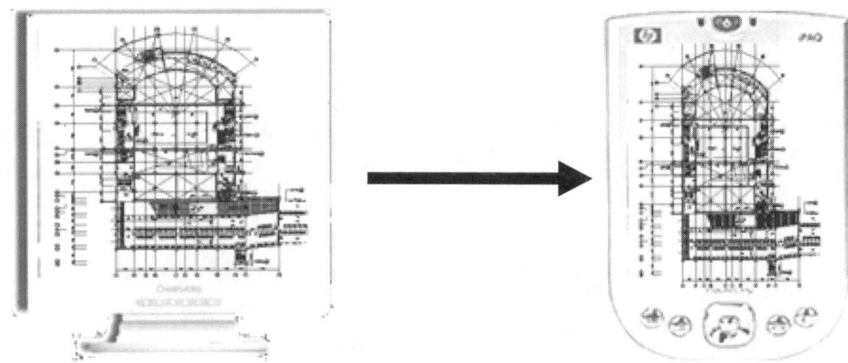

Figure 6. Design draw transmission between PC and PDA

Figure 7. Practice picture transmission between PDA and PC

4.2 *Introduction of practice project*

This research is going to illustrate the application of wireless transmission technology to tall building's construction job-sites. To evaluate the performance of this study, this research chooses a practice construction project to test the application result. The specific practice case is the Science and Technology Building construction project. Such project is still constructing in the National Taipei University of Technology (NTUT) right now. Its general construction information is shown as follow and figure 8:

1. Building area: 8412m^2
2. Total construction floor: 44,440m^2
3. Maximum length of construction area: 175.45m

4. Total width of construction area: 137.5 m
5. Basement: 4 floors
6. No. of floor: 16 floors.

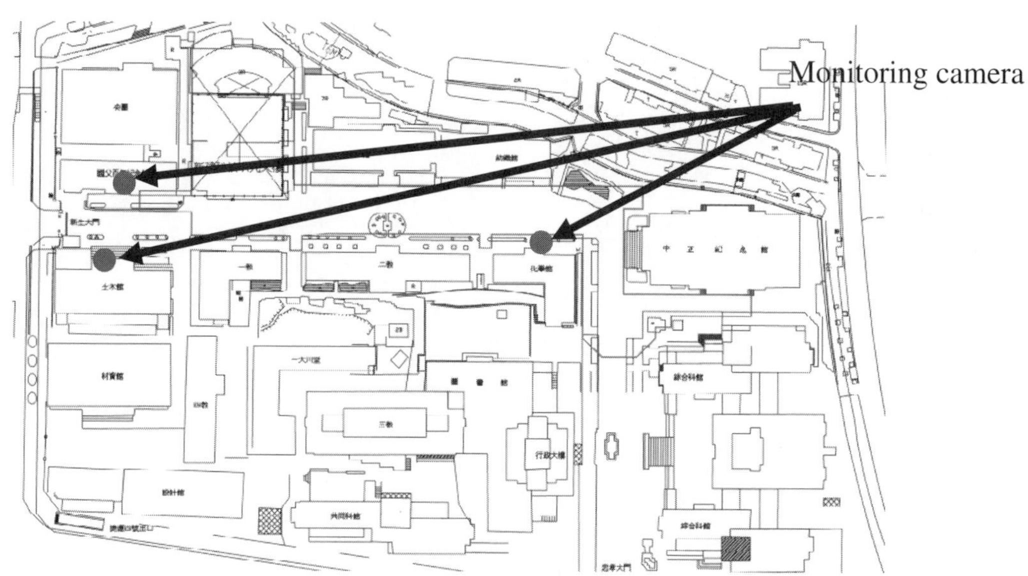

Figure 8. The construction area of Science and Technology Building at NTUT

To compare the performance of applying wireless transmission technique into this tall building's construction job-site, this research also sets up a monitoring system at this practice project, then measures the transmission situation by using wireless PDA and GPRS (figure 9 and figure 10).

Figure 9. Video picture of wire monitoring system

Figure10. Practice test of wireless transmission at the construction job-site

4.3 *Practice test of wireless transmission*

The career of each construction engineer is important and complicate. In order to manage the construction work effectively. There are many managerial skills and methods developed to help the engineer to control the construction work in recent years. This study illustrates the performance measurement of wireless communication result. During practice test fotm November 1, 2004 to August 31, 2005, this research investigated several conditions as shown in the following:

(1) The distance of wireless transmission will affect the transmitting time. If the distance is farther, the transmitting time is longer (figure 11).

(2) Once there is any barrier between wireless sending point and receiving point, such as RC buildings or big trees, it will cause transfer wireless signal breakdown.

(3) The memory of the transmitting image file will influence the transferring time, the bigger the slower (figure 11).

(4) Although it is needed to identify in detail, from the practice measurement can recognize that weather and environmental conditions are influence factors of the wireless transmission speed.

4. Conclusions

From this research can illustrate that, application of PDA or GPRS as a management tool at any tall building's construction job-sites can help all inspection engineers make a better construction report. Based on the development of PDA and GPRS with wireless system, all kinds of information regarding the construction activities such as standard specifications, figures, tables, equations, pictures and so on, can be developed and supplied to construction inspectors. It is a valuable study to promote the benefit of the construction management for using a real-time recording in construction industry.

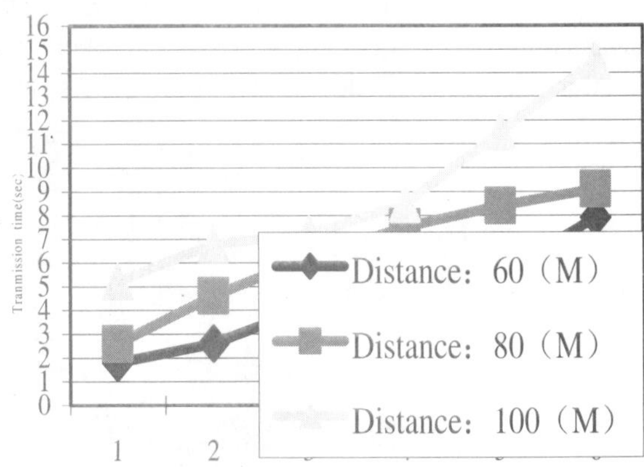

Figure 11. Wireless transmission speed with different distances

Up to today, if the inspection activity of construction job-sites can be combined with the wireless transmission technology, it will be more efficiency and effectiveness for the construction management. The beneficial result will fully depend on how to develop the structure of wireless transmission system during the construction process. In information-developed generation, digitized information is used to be limited by the traditional internet frame. It is inconvenient for the user to get information from linking to the interior network in a particular range. Therefore, this research developed a wireless network environment to have the internet access when ever it is possible and to pass over the information for inspectors right at the construction job-site by using the wireless network to the server on the internet.

This research investigated the feasibility of applying PDA or mobile phone as a wireless transmission hardware, which if can be integrated the video monitoring system then the performance of construction quality and safety management will be improved the inspection quality in construction industry (figure 12 and figure 13).

Based on the practice measurement, the distance between wireless sending point and receiving point will be an influence factor of transmission time. The regression equations of transmission time y (sec) and memory size x (kb) are:

(1)60m distance: $y=1.14x + 0.46$ (figure 14(a)).

(2)80m distance: $y = 1.29x+1.88$ (figure 14(b)).

(3)100m distance: $y = 1.77x+2.76$ (figure 14(c)).

Figure12. Un-safety behavior and operation behavior can be caught and shown on computer screen

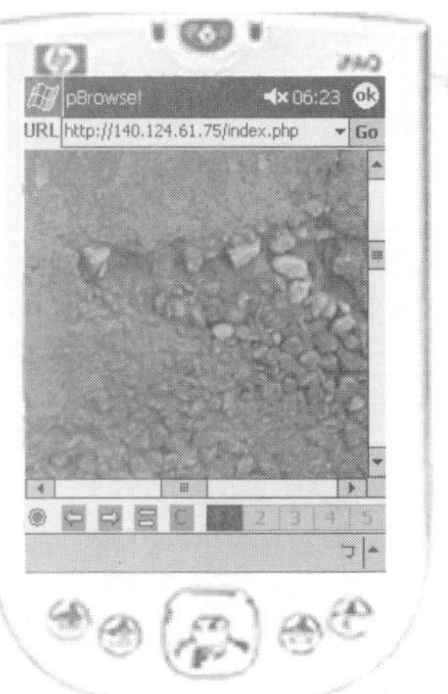

Figure 13. A concrete honey-comb picture can be caught and shown on the PDA screen

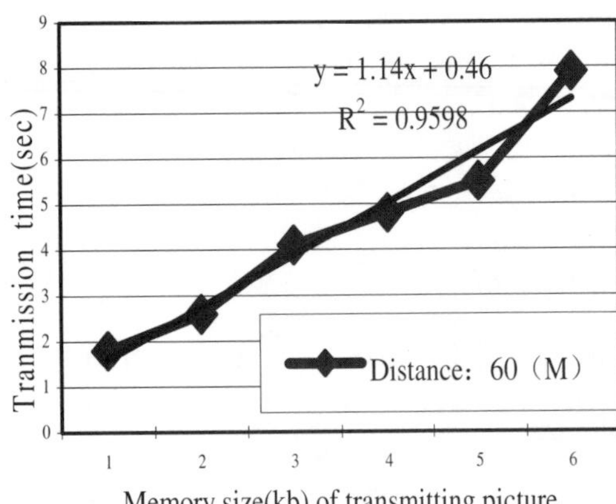

(a) 60m distance

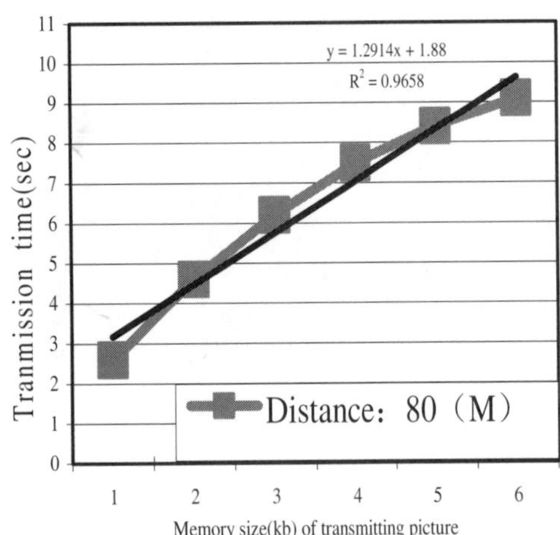

(b) 80m distance

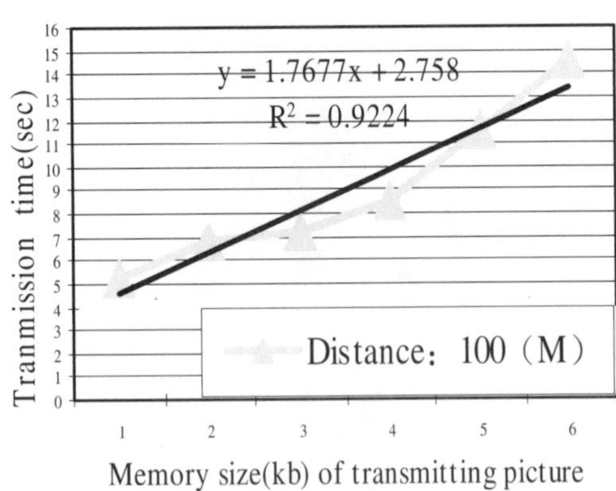

(c) 100m distance

Figure 14. Regression equations of transmission time and memory size of transmitting picture

References

1. Bid-palm web site (2001), http://www.bid-palm.com/doc/bidpalm_doc/article.htm.
2. Calhoun (1992), G., Wireless Access and the Local Telephone Network, Norwood, MA: Artech House.
3. Chang Chih-chiang (2004), A Study on the Life Cycle Management and Engineering Techniques of the Infrastructure --- such as Public Buildings, Report of Summer Study Program, Paragraph, National Science Council, Taipei, R.O.C..
4. George M. Calhoun (2003),"Third generation wireless systems", Artech House Publishers, London.
5. Huang Chun-long. and Lin Lee-kuo (2004), "The applying of construction automation management", the project report of Construction and Planning Agency Ministry, Construction and Planning Agency Ministry of the Interior, R.O.C, Taipei.
6. Huang Jin-ting (2003),"A Research of IEEE 802.11 Communication Protocol Improving Real-time Construction Management", Master Thesis, Institute of Civil and Disaster Prevention Engineering, National Taipei University of Technology, Taiwan , R.O.C..
7. http://www.csie.ncu.edu.tw/~cs882008/document/project/802_11.html.
8. IEEE (1999), *IEEE Std 802.11b-1999 (Supplement to ANSI/IEEE Std 802.11, 1999 Edition).*
9. Jolly, W.P (1972), "Marconi", New York: Stein & Day.
10. MacLaurin, W.R., and R.J. Harman,(2002) CTIA's Wireless Industry Indices: Semi-Annual Data Survey Results, 1985~2001.
11. Lin Lee-kuo and Kuo An-hsiang (2001), The Construction Knowledge Management in Taiwan, The Conference of Construction Knowledge Management, Taipei, R.O.C..
12. Lin Lee-kuo (2002), "The development of personal digital assistant with iInspection engineer in construction industry", the Ninth International Conference on Computing in Civil and Building Engineering, April 3-5, 2002, Taipei, Taiwan.
13. Lin Lee-kuo and etc. (2004), The Management of Life Cycle Cost in Construction Industry, Journal of Building and Construction Technology, Taipei, R.O.C..
14. Lin Lee-kuo (2005), The Management and Assessment of Life Cycle Cost in Construction Industry, HKIVM 7th International Conference, June 2005, Hong Kong.
15. Yang Shih-Hung (2004), Preliminary Research on the Energy Consumption Evaluation Method of Detached House Production Processes, The Conference of Japan Architect, Japan.

DISTINGUISHING THE DECREPIT FROM THE OLD: IS BUILDING AGE A GOOD PROXY FOR BUILDING PERFORMANCE?

WONG, S.K CHEUNG, A.K.C. YAU, Y. CHAU, K.W. HO, D.C.W.

Department of Real Estate and Construction, The University of Hong Kong, Pokfulam Road, Hong Kong, China

People often take building age as a proxy for building performance, and thus equate old buildings with decrepit buildings. In Hong Kong, a number of public policies also use building age as a criterion to screen off eligible applications for subsidies. The belief in the conveyance of information on building performance by building age rests on the traditional wisdom that building fabrics and building services will wear and tear and ultimately become obsolete over time. However, if building maintenance and management is in place, it is uncertain how much information building age can still convey. There has seldom been any empirical test in this regard, due to the lack of an objective method to quantify building performance. This paper aims to fill this gap by examining the relationship between building age and its performance in respect of health and safety based on results from a standardized assessment of the health and safety conditions of 134 residential buildings in Hong Kong. Our preliminary results confirmed that there was a significant negative relationship between building age and its health and safety conditions. However, the 20-year cutoff adopted in most public policies was not supported by the empirical results. In addition, we analytically show that the use of building age as a decision tool for resource allocation may lead to inefficiency, and accordingly propose three solutions.

1. Introduction

Buildings, like people, naturally age with time. All buildings inevitably start to deteriorate as soon as they are completed. Therefore, it is a common belief that when buildings get older, their conditions worsen because of wear and tear and functional obsolescence. Moreover, many old buildings agglomerate in old areas that are occupied by low income groups and characterized by poor environmental quality, which further reinforce the perception that these buildings are decrepit. This applies to Hong Kong, where old buildings in some areas (e.g. Sham Shui Po and Wanchai) are dilapidated. According to the Housing, Planning and Lands Bureau, Hong Kong now has over 42,000 blocks of private multi-storey buildings, out of which about 20,000 buildings are situated in the urban area (i.e., Hong Kong Island, Kowloon, Tsuen Wan, and Kwai Tsing).[1,*] Since these areas were developed in the earlier stages of Hong Kong's development, close to half of their buildings were built over 30 years ago, and this figure is anticipated to increase by 50% in ten years' time. As such, people often take building age as a proxy for the condition of a building, and thus equate old buildings with decrepit buildings. The older the building, the poorer its health and safety performance is assumed to be.

However, if proper building maintenance and management were already in place, old buildings would not necessarily pose a problem.[2] This is analogous to people, in that the older members who keep fit may be healthier than the younger members who do not. Given the above circumstances, how much information the age of a building can convey about its condition is non-trivial. Yet, there has seldom been any empirical test in this regard, due to the lack of an objective method to quantify building performance. To fill in this gap, this paper aims to examine the relationship between a building's condition and its age, and see empirically whether the use of building age as a selection criterion of public policy is justified. For the purpose of this paper, we define building condition as the performance of a building in respect of health and safety.

2. Conventional Wisdom about Building Age and Condition

2.1 *Their perceived association*

The association of old building and dilapidation has long been well documented. Loo makes a general statement that aged buildings have more maintenance problems and an increasing number of breakdowns in building services.[3] Leung and Yiu suggested that ageing buildings have posed threats to public safety in Hong Kong.[4] This was exemplified in fatal accidents of fallen external renderings and spalling concrete pieces in aged buildings. As showed in Table 1, which enlists the

[*] Data from "Database of Private Building in Hong Kong", compiled by the Home Affairs Department, http://202.67.130.69/eng/buildingsearch_submit.php.

reported incidents of falling building fabrics since the beginning of 2005, most of the incidents occurred in buildings that were close to or more than 30 years old. Other than the safety issue, old buildings are more prone to pest and hygienic problems. Previous studies found a statistical significant association between the degree of building disrepair and the cockroach allergen level.[5] This is because a dilapidated building is a suitable breeding ground for pests and vermin. Large cracks that developed over time serve as their habitats, and dripping pipes provide water for their survival.

Table 1. Incidents of falling building fabrics reported since 1/1/2005

Date	Building Name	Building Age	Incident
24/2/2005	Hankow Apartment in Tsim Sha Tsui	45 years	Falling concrete pieces hitting a policeman.
26/2/2005	Sin Hua Bank Building in Shamshuipo	29 years	Falling mosaic tiles hitting a passer-by.
27/2/2005	Windsor Mansion in Tsim Sha Tsui	40 years	Falling concrete pieces.
1/3/2005	32 Wood Road in Wanchai	48 years	Falling concrete pieces
11/3/2005	Champion Building in Yaumatei	27 years	Falling concrete pieces hitting a passer-by.
13/3/2005	Lung Wah Building in Tsim Sha Tsui	40 years	Falling concrete pieces
13/3/2005	Kam Tao Building in North Point	31 years	Falling concrete pieces damaging a car.
7/4/2005	Hankow Apartment in Tsim Sha Tsui	45 years	Falling concrete pieces.

Source: Various local newspapers.

The use of age as a proxy for building condition is not only a social belief, but is also widely applied on the policy level. In Hong Kong, a number of public policies (as shown in Table 2) use, *inter alia*, building age as a criterion to screen off eligible applications for grants or subsidies. A common objective of these schemes is to help low income flat owners carry out repairs or improvement works to multi-owned private apartment buildings to maintain health and safety standards. It is interesting that all these policies have adopted 20 years as the screening cutoff.

Table 2. Loan/subsidy schemes in Hong Kong

Scheme	Administrated by	Building Age Requirement
Home Renovation Loan Scheme	Hong Kong Housing Society	At least 20 years old
Building Maintenance Incentive Scheme	Hong Kong Housing Society	At least 20 years old
Building Rehabilitation Trial Scheme	Urban Renewal Authority	Around 20 years old and up
Building Rehabilitation Loan Scheme	Urban Renewal Authority	Around 20 years old and up

2.2 *Reasons for the association*

2.2.1 *Difficulties in getting finance*

It is theoretically true that improvement works can be made to old buildings to avoid or delay dilapidation. It is often said that with proper management and adequate investment, these costly failures can be kept to a minimum. One way to finance improvement works is by means of owners' equity. Yet, experience has revealed that sinking fund is absent for many old buildings. In the case of major improvement works, owners are therefore required to contribute a large sum of money, which they are often unwilling or unable to do. The other way is to raise funds through mortgaging. This is, however, not always feasible for owners of old buildings. Not all commercial banks will accept old buildings as collateral, due to their low liquidity. Even if a mortgage is offered, the terms (i.e., the interest rate, the amount and the repayment period of mortgage) are harsher than those for newer buildings. The shortage of funding for maintenance worsens an older building's maintenance problem. Moreover, old buildings are more difficult to insure, or can only be insured at a very high premium. This means that the owners of old buildings are at greater risk for accidents, especially those that involve third parties. A real-life illustration of this point is the Albert House case, in which the owners of the 21-year-old building in Aberdeen were held responsible for compensating $300 million to the dependants of the victims of an accidental canopy collapse in 1994.

2.2.2 *Maintenance quality*

Despite the fact that repairs can be made to old buildings to curb the problems, Leung and Yiu questioned the general effectiveness of repairs to buildings of poor durability.[4] They argued that inadequate standards of repair, inappropriate methods of repair, and insufficient considerations of long-term durability were contributing factors to poor-quality repair works. This was well exemplified by the collapse of a canopy on Marble Road in 1997. Before the accident, four reinforced concrete repair works had been carried out to the building, one of which was carried out by a government contractor. However, the underlying cause of the recurrent failure of the building's fabrics was the misplacement of steel bars in the concrete. As only superficial repairs were carried out, the structural problem was aggravated, and finally resulted in an accident. From this case, it was observed that many poor-quality repairs have been applied and these repairs have concealed, rather than reversed, a building's deterioration.[6]

2.3 *Lack of empirical support*

Although the claims above are widely conceived, they lack support from empirical studies. Ho *et al.*, based on a face-to-face questionnaire survey of 61 people, found that when compared to structural and system defects, building services, and building management, building age is less important for determining the value of a building.[7] This poses a question of whether building age can be taken as a proxy in the determination of a building's condition. Furthermore, a survey conducted on building maintenance and management in Hong Kong revealed that among those respondents supporting mandatory building inspection, about 39% of them suggested that buildings over 20 years old should be mandated for building inspection; and about 30% of them suggested a stricter requirement to mandate the inspection of buildings over 10 years old.[8,†] In other words, there have been divergent views on building cutoff age for building conditions.

Practicably speaking, building age is a convenient and exogenous variable for limiting the number of eligible applicants in public policies. But, the choice of cutoff age seems rather arbitrary. It has been presumed that properties of less than 20 years old pose no serious health and safety problems. As mentioned above, this presumption has never been tested empirically. Should those schemes aim to help low-income group people, the rateable value of a property alone would serve their purposes. It is, however, unclear why building age has to be a criterion.

3. Empirical Tests

3.1 *Methodology and data description*

Our study evaluated the degree of association between building condition and age using regression analysis. Before we arrived at the results, it was necessary for us to have the data for building conditions and age for analysis. We define the age of a building as the duration between the day on which an occupation permit was issued for it and the day of analysis, measured in years. This information can be obtained from Home Affairs Department. On the other hand, the health and safety conditions of a building are not readily available. A site visit has to be carried out to reveal its performance. In lieu of classifying a building's condition in an ordinal manner, such as "good," "fair", and "poor", we constructed a Building Conditions Index (BCI) to measure the existing conditions of residential buildings in a continuous cardinal scale. This provides us with finer measurement on building performance and enhances the accuracy of the regression analysis.

In the BCI assessment framework, a number of building factors was assessed, including hygienic conditions, structural conditions, building services conditions, and fire safety performance. The BCI is essentially an aggregate figure of ratings and weightings of all building factors related to the health and safety of occupants and passersby. Mathematically,

$$\text{BCI} = g\,(\,w_1, w_2, \ldots, w_n\,;\, F_1, F_2, \ldots, F_n\,) \quad (1)$$

where w_i (i=1, 2, ..., n) denotes the relative importance of the ith building factor in affecting the health and safety conditions of a building; F_i denotes the rating of the ith building factor collected using the assessment framework; n is the total number of building factors; and g is a function that combines all w_i's and F_i's. Without prior knowledge of the specific functional form for the BCI, we adopted the simplest form (i.e., the equally-weighted arithmetic mean) with all w_i's summed to unity. The rate achievable for each factor for a particular building ranges from 0 (the worst) to 1 (the best), and thus, the BCI for that building also ranges from 0 to 1.

Comprehensive condition surveys were conducted on 134 multi-storey residential buildings sampled in the

† In total, 1,435 Hong Kong people living in self-owned properties were interviewed in telephone opinion survey. About 88% of respondents agreed that buildings at or above a certain age should be mandated to undertake a building inspection.

Yau Tsim Mong mega-district (Yau Ma Tei, Tsim Sha Tsui, and Mongkok). The ages of these buildings ranged from three years to 50 years. Based on the findings from the surveys, the respective BCIs of these buildings were compiled. Table 3 presents the descriptive statistics of the data.

Table 3. Descriptive statistics of the data

	Mean	Std. dev.	Min	Max
Building Conditions Index (BCI)	0.49	0.11	0.23	0.72
Age (years)	30.72	11.84	3	50

The value of the BCI reflected the actual health and safety conditions of a building. With this information, we tested whether a relationship exists between building conditions and age. If such a relationship is established, is the use of the 20-year-old cutoff by the government agencies justified? In these regards, two hypotheses were set up:

- *Hypothesis 1*: There exists a negative relationship between building conditions and age.
- *Hypothesis 2*: Buildings aged at least 20 years deteriorate faster than newer buildings.

For testing the hypotheses, two regression models were formulated:

$$BCI = \alpha_0 + \alpha_1 AGE + \varepsilon \quad (2)$$

$$BCI = \beta_0 + \beta_1 AGE + \beta_2 D20 + \beta_3 AGE \times D20 + \varepsilon \quad (3)$$

Table 4 describes the meanings of the symbols used in Eqs. (2) and (3). In Eq. (2), the coefficient α_1 measures the effect of building age on building conditions. As for Eq. (3), our focus is put on the coefficients, β_2 and β_3, which measure the change in the effect of building age on building conditions, if any, after a building reaches the age of 20.

Table 4. Definitions of the variables

Variable	Definition
BCI	The Building Conditions Index of a building
AGE	The age (in months) of a building
D20	A dummy that equals 1 when a building is 20 years old or above, and zero if otherwise
α and β	Coefficients to be estimated
ε	The error term

3.2 *Results and discussion*

The results of the models in Eqs. (2) and (3) from the Ordinary Least Squares estimation are shown in Table 5. Both equations found that *AGE* is -0.006, confirming a significant negative relationship between building age and its health and safety conditions. Therefore, *Hypothesis 1* was confirmed. On the other hand, in Eq. (3), the coefficients of variable *D20* and the interaction term *AGE* x *D20* were found to be insignificant. This implies that there is no evidence for a change in the relationship between building conditions and building age after a building reaches the age of 20 (i.e. *Hypothesis 2* was not supported). The adjusted *R*-squared value for both equations was only close to 50%. That meant over half of the variations in building conditions were not explained by building age and further research is needed to explore other possible determinants of building conditions.

Table 5. Estimation results of the models in Eqs. (2) and (3)

		Eq. (2)		
Variable	**Coefficient**	**Std. Error**	**t-Statistic**	
Constant	0.6880	0.0192	35.9060	*
AGE	-0.0063	0.0006	-10.8333	*
R-squared		0.4706		
Adjusted R-squared		0.4666		
F-statistic		117.3605		
		Eq. (3)		
Variable	**Coefficient**	**Std. Error**	**t-Statistic**	
Constant	0.6765	0.0445	15.1941	*
AGE	-0.0064	0.0035	-1.8205	#
D20	0.0763	0.0576	1.3247	
AGE x D20	-0.0016	0.0037	-0.4277	
R-squared		0.4866		
Adjusted R-squared		0.4748		
F-statistic		41.0719		

Note: * significant at the 1% level; # significant at the 10% level

The results also yield important policy implications. We can obtain insight into the efficiency of a loan/subsidy policy using building age as a proxy for building conditions by referring to the scatter plot of building conditions against age in Figure 1. Without loss of generality, we assumed that 0.5 is the passing score for the BCI, and thus, buildings with a BCI of 0.5 or above were deemed satisfactory with respect to their health and safety conditions. This means that an ideal public policy should only allow those buildings with a

BCI of below 0.5 to apply for subsidies. This ideal "horizontal policy" will sever the plot in Figure 1 horizontally into two sections: the upper section contains buildings of satisfactory condition, while the poorer buildings fell into the lower section. However, due to information asymmetry, implementing a horizontal policy is very costly. A simpler alternative (called "vertical policy") is to use proxies for BCI, such as age. If all the observations are exact fits (i.e., they all lie on the regression line), then the use of a vertical policy, which draws a vertical line to cut through the intersection point of the horizontal line and regression line (i.e. point *X*), will yield equally efficient outcomes. The better ones scatter to the left, while the poorer ones scatter to the right. Therefore, the same outcome as if the horizontal 0.5 BCI cutoff is used can be achieved.

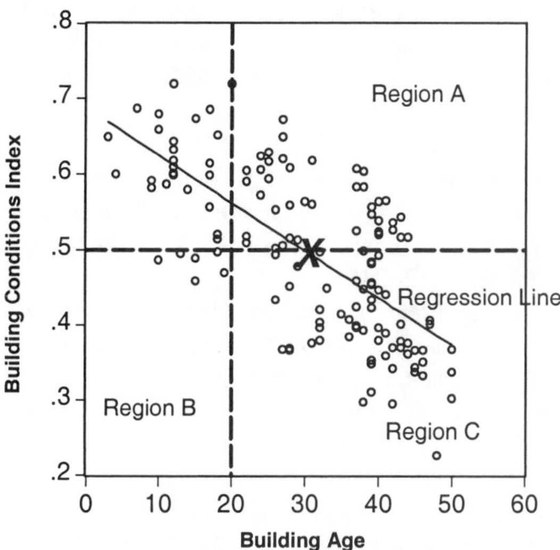

Figure 1. Scatter plot of the BCI against building age

However, as shown in Figure 1, the conditions of buildings of a given age did not fall exactly on the regression line. Also, the location of the intersection point may not be known. These rendered the use of any vertical line to partition the pool of buildings inefficient. Let's go back to the real-life situation. Without considering the actual health and safety conditions of buildings, the government and quasi-government organizations provided loans or subsidies for building maintenance or repair to owners of buildings as long as their buildings were at least 20 years old (referring back to Table 2). Buildings within Regions A and C were therefore covered by the loan/subsidy schemes, while those in Region B were ineligible. As a matter of fact, the conditions of the buildings in Region A were satisfactory, while those in Region B were not. This was a case of public resources being directed not to the neediest buildings (Region B), but to those in the best shape (Region A). Resource allocation was therefore inefficient.

4. Conclusions and Suggestions

Despite the widespread use of building age as a proxy for building condition, there has long been a lack of empirical study of their relationship. Based on a standardized assessment scheme, 134 residential buildings in Hong Kong were inspected and their conditions were measured by the BCI. The preliminary results from the regression analyses confirmed that there was a significant negative relationship between building age and its health and safety conditions. On the whole, however, the explanatory power of age was not very high. In addition, the rate of change of building conditions with respect to building age did not budge when the buildings turned 20. This indicated a lack of empirical support for the use of the 20-year cutoff.

From our findings, we analytically elucidated that the mere use of building age as a criterion for screening off buildings eligible for certain loan/subsidy schemes inevitably led to inefficient resource allocation. In view of this problem, we hereby propose three approaches, which are not mutually exclusive from each other, to deal with the problem. First and foremost, building age has the power to explain building conditions, but only at a medium degree. Clearly, building age matters, but it is not enough to reflect a building's condition. In fact, there may be some other factors that can supplement building age in conveying information on a building's condition. The Housing, Planning and Lands Bureau reported that about 83% of respondents agreed to mandate the inspection of buildings without owners' corporations by property management companies.[8] This provides us with insight into the management structure of a building, which may be perceived as an important determinant of the condition of the building. Therefore, further research is needed to verify whether factors other than building age (e.g. building management structure and the scale of development) can explain the remaining variations in building conditions.

The second recourse in the resource allocation problem is the market mechanism. The crux of this approach is that the subsidies should be in the form of vouchers, with restricted use in building maintenance, but tradable in a free market.[‡] As illustrated in Figure 1, inefficiency arises because poor buildings in Region B were not subsidized, whereas good buildings in Region A were subsidized. What if the subsidy vouchers are allowed to trade? Suppose the marginal benefit of carrying out maintenance to buildings in Region B is $50,000 per household, whereas the marginal benefit for buildings in Region A is $10,000.[§] The lower marginal benefit for Region A is due to premature maintenance. In such a case, building owners in Region B should be willing to purchase vouchers from those in Region A at a price between $10,000 and $50,000 before transaction costs are taken into account. Eventually, with this trading mechanism, buildings in Region B can be upgraded, whilst building owners in Region A can have more cash for their own disposal. Both parties will be better off through such trading. Of course, the use of this market mechanism incurs transaction costs (e.g. the difficulty of reaching collective decisions by co-owners to sell or buy vouchers). Therefore, this solution may be more applicable to buildings with owners' corporations.

Finally, as a long-prone strategy, the government or organizations administrating those loan/subsidy schemes should consider directly measuring building conditions instead of using building age for screening purposes. With the use of actual building conditions as screening criteria, buildings in Region B, rather than those in Region A, will be automatically covered by public policies, thereby resulting in a more efficient allocation of funds. Information on a building's condition can be obtained through periodic building inspections, be they mandatory or voluntary, using low-cost assessment schemes such as the Building Health and Hygiene Index and Building Safety and Conditions Index.[11,12]

Acknowledgments

We gratefully acknowledge the financial support provided by the Research Group on Sustainable Cities of The University of Hong Kong.

References

1. Housing, Planning and Lands Bureau, *Addressing Long-standing Problems*, http://www.hplb.gov.hk/eng/policy/csbstm.htm (2004).
2. K.J.K. Chan, Maintenance of Old Buildings, *The Hong Kong Surveyor*, **11**(2), 4-7 (2000).
3. F. Loo, *A Guide to Property Management in Hong Kong*, Hong Kong University Press, HK (1991).
4. A.Y.T. Leung and C.Y. Yiu, A Review of Building Conditions in Hong Kong, in *Building Dilapidation and Rejuvenation in Hong Kong* (eds. A.Y.T. Leung), CityU Press and HKIS, HK (2004).
5. V.A. Rauh, G.L. Chew and R.S. Garfinkel, Deteriorated Housing Contributes to High Cockroach Allergen Levels in Inner-City Households. *Environmental Health Perspectives*, **110**(Suppl. 2), 323-327 (2002).
6. Buildings Department, *Consultancy Study on Steel Corrosion and Material Deterioration of Buildings Complete between 1959 and 1980*, Taywood-Arup-Vigers JV, HK (1999).
7. D.C.W. Ho, C.M. Tam and C.Y. Yiu, Criteria and Weighting of a Value Age Index for Residential Use, in *Building Dilapidation and Rejuvenation in Hong Kong* (eds. A.Y.T. Leung), CityU Press and HKIS, HK (2004).
8. Housing, Planning and Lands Bureau, *Report on the Public Consultation on Building Management and Maintenance*, Hong Kong (2005).
9. R. Coase, The Problem of Social Cost. *Journal of Law and Economics*, **3**, 1-44 (1960).
10. K.W. Chau, A.Y.T. Leung, C.Y. Yiu and S.K. Wong, Estimating the Value Enhancement Effects of Refurbishment. *Facilities*, **21**(1/2), 13-19 (2003).
11. D.C.W. Ho, H.F. Leung, S.K. Wong, A.K.C. Cheung, S.S.Y. Lau, W.S. Wong, D.P.Y. Lung and K.W. Chau, Assessing the Health and Hygiene Performance of Apartment Buildings, *Facilities*, **23**(3/4), 58-69 (2004).
12. D.C.W. Ho and Y. Yau, Building Safety & Condition Index: Benchmarking Tool for Maintenance Managers, *Proceedings of the CIB W70 Facilities Management and Maintenance Symposium 2004*, 7-8th Dec 2004, HK (2004).

‡ This idea is borrowed from the Invariance Theorem of Ronald Coase.[9]

§ One method to estimate maintenance benefit is the use of hedonic pricing models.[10]

FIRE DISASTER AND PREVENTION MANAGEMENT OF TALL BUILDINGS

LEE-KUO LIN CHUAN-HSU WANG CHUN-TING CHAN HUI-CHING LIN

Civil Engineering Department, National Taipei University of Technology, Taipei, China

In Taiwan, the population is unceasing growth and the useable area is continuously decreasing. These situations are causing construction projects of tall building getting more and more. In each year, there are lots of fire disasters happened in tall buildings. Such tragedy always caused serious properties damage and many citizens' life lost. The characteristics of tall buildings are main factors to cause that serious damage, including too high to rescue, too complex in use, too many flammable materials, and etc. For example, in 2001, the fire disaster of the building of Eastern Science Park in Taiwan, burn for 42 hours and properties lost about US$ 300 million dollars. Therefore, it is really important issue in disaster prevention research. This paper will illustrate this specific practice fire disaster case of tall building to derive the prevention key factors of fire disaster in tall buildings, also will explain the collection of questionnaires from 200 firefighters, and then analyze the appropriate prevention policy of fire disaster of tall buildings.

1. Preface

There are lots of potential hazardous areas in Taiwan due to high-density land-use and construction building. Up to today, every city area in Taiwan is still continuously expending. Natural disasters such as earthquake, typhoon, flood etc. are happened from time to time within all year long. In addition, based on the high density of ecological environment, the careless or deliberate hazard such as arson fire, gas explosion and etc. always randomly occurs. Such situations will cause serious life loss and citizen's properties damage once any kind of disaster happened. Therefore the hazard mitigation and prevention research is getting popular and important in Taiwan.

On May 12, 2001, a fire took place at the tall building of Eastern Science Park in Si-Jhih City where is in the northern Taiwan. This fire burn more than 42 hours to make this disaster created the longest burning building time record in Taiwan (Lin, 2000) .It caused damage of more than 200 high tech companies and the estimated damage cost more than billions of NT. dollars. It also brought out the problems of the existing fire prevention system. In recent years, people pay more concern to the fire safety issues for tall buildings. As the economic growth and the rising of the land price therefore in order to make the best use to the limited land, construction designs tend to be in high-rise and diversification style. For example, there are the 51 storey high Shin Kong Life Tower, the 85 storey high Tuntex Sky Tower and Taipei 101 with island wide. The height, widely spread areas, complex structures, special construction materials, full of wire pipes, people and complicated purposes often make fire rescuing difficult.

Fire fighting and rescuing is timeliness. Once the fire occurred, people only have limited time to escape before it will do much damage to peoples body (Chang, 1996) . This is the main reason why tall buildings need to have its own fire safety and rescuing strategies.

2. Case Study

2.1 *Eastern Science Park fire disaster*

2.1.1 *Fire duration*

The Eastern Science Park located at Si-Jhihi city in Taipei county. It is a steel structure building and has 26 floors above ground and 5 floors underground. Each floor is about 36000 square feet and it is divided into A, B, C, D, four separated buildings (Lin and Lai, 1999) . On May 12, 2001, a fire suddenly took place. Fire started from the Good Fortune Buddha shop in the third floor of building A. Because there were too many flammable materials and it is an airtight glass curtain tall building, the high temperature heavy smoke was stored inside. When the fire fighters arrived, the whole building was full with smoke and suddenly was happened a flashover which caused the tempered glass of windows to explode when fire fighters were making damage to open the ironic rolling door (Yeh,2001) .

Because the shafts connect through over every floor of the building, the flames scurried out fiercely from the sixteenth floor around 7 p.m. and through the channel and out side the windows, the flames spread upward. The emergency electricity generator was unable to generate which caused the emergency elevator and the pump unable to operate. These increased the difficulty to rescuing of the high-rise tall buildings. Around 8:45 p.m. on May 12, the flames spread upward fast to the twentieth floor and due to the strong wind bowled the fire became out of control. Around 2:30 a.m. on May 13, the fire spreaded from the twentieth floor to the twenty-fourth floor. At 3:40 a.m. the fire extent to the twenty-fifth and twenty-sixth floor and at 4:10 a.m. the fire began to spread toward the twenty-fifth and twenty-sixth floor of B and C building. the fire was in control at 2:50 p.m. May 13 and was all put off at 10 p.m (figure 1, 2 and 3.

Figure 1 Pictures of Eastern Science Park after the fire hazard

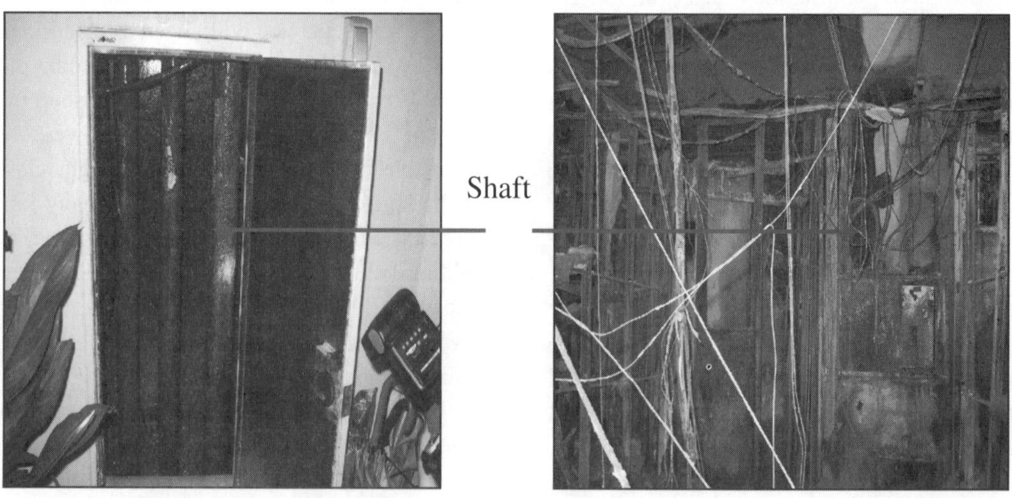

Figure 2 The fire spreading route of the Eastern Science Park

Figure 3 Damage situation of the Eastern Science Park

2.1.2 *Fire expend spreading route (figure 4 & 5)*

(1) Fire spreading in the third floor

1. Started from the Good Fortune Buddha store (A1) and spreader through the partition wall to A2 and B building.
2. There is a 40cm gap at the partition wall between A1 and A2 which cause the fire to spread horizontally.

(2) Fire spreading from third floor to the forth and fifth Floors

1. The fire control pipe shaft in A1(Shaft-1), the drain pipe shaft in A2 (Shaft-2) and the electric wire shaft (Shaft-3) are the main cause of the fire spread vertically.
2. There was a complete flashover occurred in the third floor which caused the glass curtain of the north side of A1 and A2 destroyed. Because of the double effects of the internal high temperature pressure and the outside air flow pressure, high speed wind caused by typhoon at that day, the flames directly heat up the glass curtain above causing the inflammables near the glass curtains of the forth floor to ignite. It is the same reason for the fire to spread from forth floor to fifth floor.

(3) The fire spreading from six to fifteenth floor

1. There is smoke in shaft 1, 2 and 3 of A1 and Shaft of A2. Because there were flammable stuffs near the overhauling door of the shaft, only some of the areas in the seventh, ninth and fifteenth floors were burned out.
2. There is no external wall to postpone the spreading of the fire and the A1, A2 of the third to fifth floor was all burned down. This proves the importance of the external wall which can prevent fire spreading.

(4) The fire spreading from sixteenth to twenty-sixth floor

1. Because of the higher temperature and pressure, the fire smoke from third to fifth floor spreads through the shaft and floats up than subsides around the sixteenth floor.
2. Because there are plastic pipes in shaft 2 and 3 in between the third to the fifteenth floor, the high temperature inside the shaft will cause the pipes to coke and decompose or even to burn.
3. Based on the outdoor wind theory, the building sustain the biggest wind pressure of the external wall at two-third of the building height. This building has twenty six floors therefore the biggest wind pressure is located near the sixteenth floor.

(5) The fire spreading from A to B building in the twenty-fifth and twenty-six floor

There were too many flammable materials and had no fire walls which caused the fire to spread from A to B building.

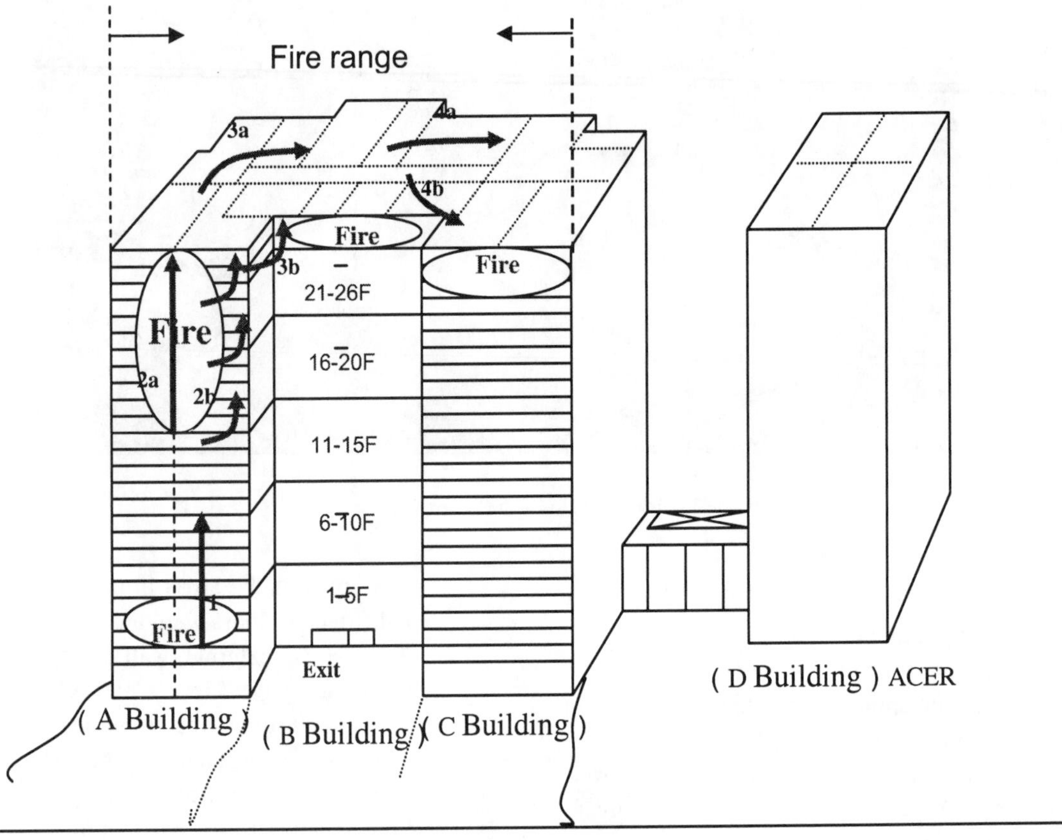

Figure 4 The fire route of a building in Eastern Science Park

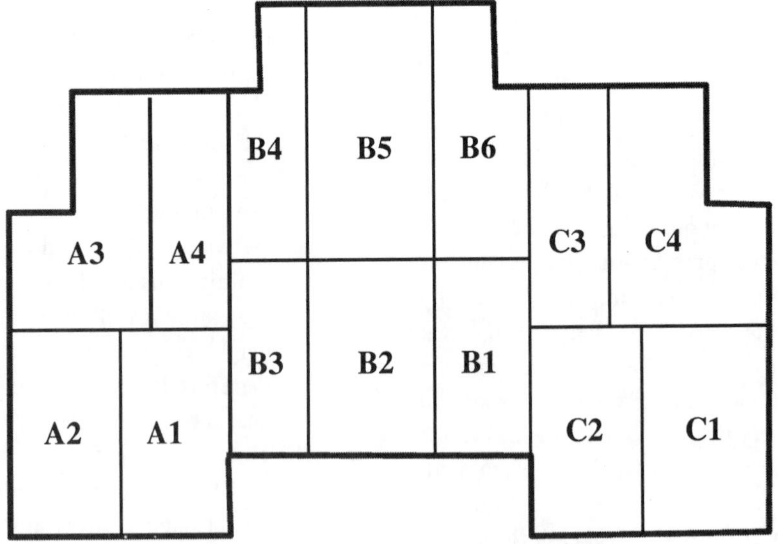

Figure 5 Plan picture of the Eastern Science Park

2.1.3 *Problems analysis*

(1) This fire was rescued by eighteen fire fighting unit in Taipei county and several other fire fighting unit from Keelung, Taoyuan and Taipei city (Yeh, 2001) . It had revealed the difficulty of tall buildings' fire rescuing.

(2) This fire disaster cause damage to six incorporation companies and many other private companies. It even made some of the companies stop operating. Also lost billions of dollars and affected the goodwill of social economy.

(3) It is often affected by the chimney effect, thermal expansion effect and the attraction of the interior air current when fire occur in tall buildings (Bruton, 1990 ; Birthed,1994) . Therefore above the sixteenth floor in the Eastern Science Park was seriously damaged. The main rout for fire spreading upward was through out the building shaft and the opening of the external wall. This shows the importance of the vertical fire isolation design (Miller, 1993) .

(4) The use of flammable materials was the major cause for fire spreading.

(5) When rescuers arrived the third floor was already full with smoke and when fire fighters destroyed the iron rolling door, the flashover occurred so the fire spread out fast. This meant the fire had been burning for a while before it was discovered. Also meant the automatic fire alarm system did not work.

(6) It is an airtight SRC structure high-rise glass curtain building. The internal shaft passed through every floor and the air condition ventilating shaft passed through the building horizontally. The imperfect shaft isolation design and the malfunction of the smoke preventing gate caused this fire to spread upward.

(7) Since the fire occurred on the weekend, the buildings' control center did not have any spare keys and could not contact the store owners to open the door. Fire fighters had to break into every door to put out fires which increases the difficulty of the fire rescuing.

(8) It is a SRC structure glass curtain compound building and has 26 floors above ground and 5 floors underground. Each floor is about 36,000 square feet and it is divided into A, B, C, D, four separated buildings. The heat and smoke made the ventilating shaft to form the chimney effect and also with the help of the wind outside of the tall building made the rescuing even more difficult.

(9) The fire safety facilities of this building did not work and the emergency elevator was broken. So the rescue team needed to climb up form the emergency stair for rescue which made it difficult to rescue.

(10) The securities did not place importance on fire safety management. When the fire fighters arrived the securities did not give them the location map of the Eastern Science Park, therefore the fire fighters did not have a clear image of the building. That is why they missed the best timing of rescuing.

3. Safety Factors of Tall Building Fire Disaster

The deputy chief of the Taipei fire bureau also used Analytic Hierarchy Process (AHP) method to establish a hierarchy structure according to the safety factors when fire hazard occurred in high-rise compound buildings. He made proper questionnaire for the specialists in fire prevention design, safety inspection and fire rescuing to investigate the most important influence factors are 「vertical region design」, 「horizontal region design」, 「automatic fire equipments」, 「management organization」, 「Hazard prevention Plan」 (Hsu, 2001). From this study can find the important factors which experts from different field thought. From enhancing on "fire separation", "fire management" and 「fire equipment」, can efficiently promote the safety of human lives for the high-rise compound building. On the other hand, the capabilities of fire resistance and fire rescue have no help to the safety of human lives when fire occurred in high-rise compound building. This paper precedes another site investigation for ten tall buildings in Taipei. The factors which scored low was 「interior decoration material」, 「manual fire equipment」,「rescue facility」,「horizontal region design」, 「rescue and escape equipment」 and 「hazard prevention and command sense」.

4. Fire Rescue and Management of Tall Buildings

Most tall buildings are fire resistant building. Once a fire disaster occurred on a specific tall building, the flammable behavior will totally different because this behavior will be affected by the chimney effect, thermal dynamic and wind blow mechanic. In general, the air situation of high elevation will make the fire spread out fast to cause it is harder for the rescue. The success of fire rescuing depends on the rescuing unit. This is the fire rescuing procedure for high-rise tall buildings (figure 6).

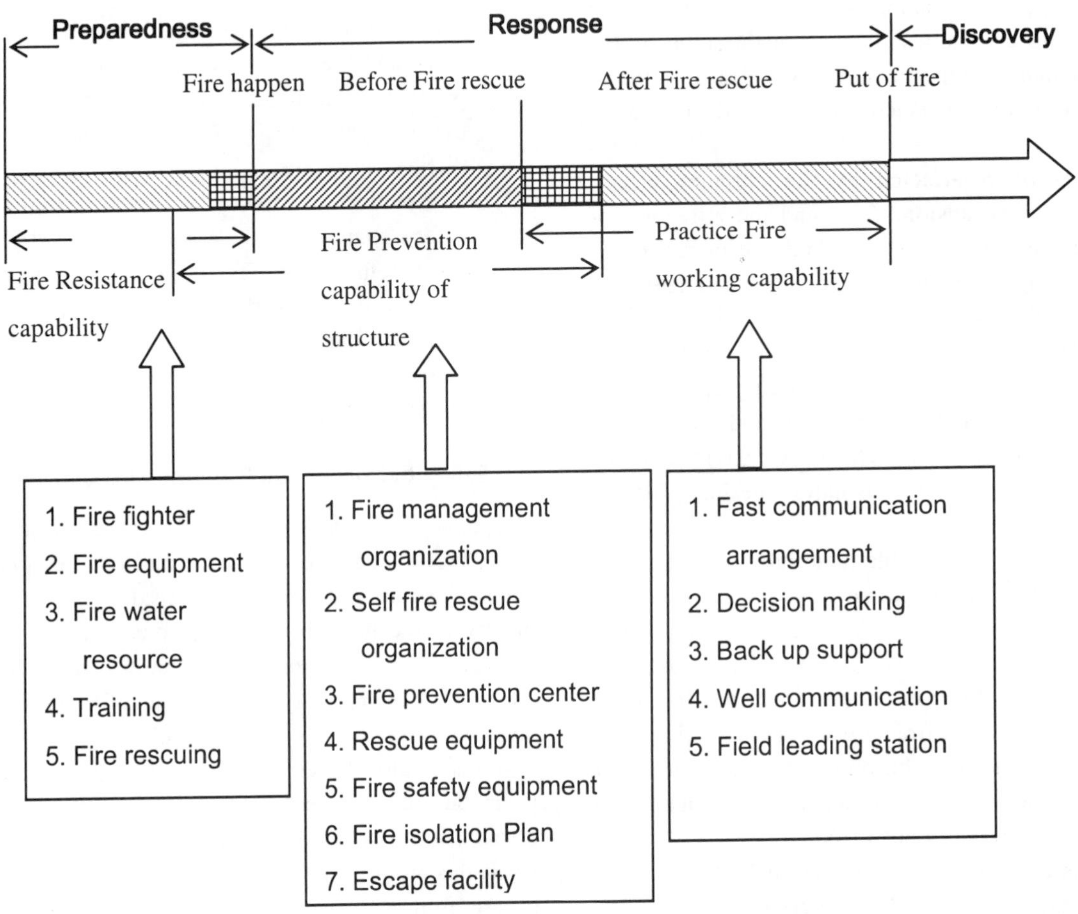

Figure 6 Relationship between fire duration and rescue of tall buildings

5. Conclusions

In order to maintain the public safety, protect human life and reduce the harm of fire disaster. Many developed countries had their detailed regulations of the construction fire safety. The regulations only demand engineers to conform with its lowest limits. However the legal design is not absolutely safe and the invest for the fire safety is endless. From this study, finding a reasonable range to invest and focusing on improving the major point can make appropriate fire disaster management of tall buildings.

(1) Identifying from the fire disaster of the Eastern Science Park. There are several high-tech companies, Buddha stores, markets, restaurants and etc. Such situation is hard to control the potential danger of fire hazard for multiple purpose of tall buildings. Hence, engineers need to re-assess the danger of the multiple purpose building in order to have a safer environment.

(2) The insufficient fire preventing ability of the building. For example, the Eastern Science Park did not have a complete fire separation system which is one of the general problems of most tall

buildings, so professional engineers needed to have a total safety check to increase the fire prevention ability for constructions.

(3) On the account of maintaining the ability of fire rescuing. This paper suggests that build the disaster preventing and central controlling room one floor above or under the ground level and make a sign to locate the fire alarm control room at the entrance of the building. Also suggests making an emergency evacuation plan and intensify the fire preventing abilities.

(4) Engineers should combine both construction and fire facility floor plan and transform them into the information of computer data base. This will make fire rescuing more efficient.

(5) Each tall building needs to establish a joint fire prevention and rescue plan. Meanwhile, in this building everyone is responsible of preventing fire.

(6) Both of fire safety management and prevention design for each tall building construction were very important.

References

1. Lin, Lee-kuo, *Disaster Prevention and Risk Management*, National Taipei University of Technology, 2000.
2. *Disaster Prevention and Rescue Plan*, 1994.
3. *Architecture Technique Regulation*, Jams Book Company, 1996.
4. Chang, H. I., *Collection of the Law and Regulation of Fire Prevention*, Mouchan Book Company, 1996.
5. Bruton, M. and Nicholson, D., *Local Planning in Practice*, Tonley Thomes, 1990.
6. Birthed, W. and Stein, R., *Planning and Environment Law*, Longman, 1994.
7. Miller, C. and Fricher, C., *Planning and Hazard*, Pergamon, 1993.
8. Lin, Lee-kuo. and Lai, Y. Y., Management System of Building Disaster Prevention, Final of National Science Council, 1999.
9. Hsu, li-Jen. A Study of Fire Prevention & Safety Strategy for the Multi-Purpose of Superhigh-rise Buildings ,Master Thesis, National Taipei University of Technology, Taipei R.O.C.,2001.
10. Yeh, Chun-Hsing. Proposing a Fire Rescue Management Strategy for High-rise Superstructure, Master Thesis, National Taipei University of Technology, Taipei R.O.C.,2001.

A STUDY ON THE DEBONDING OF EXTERNAL WALL TILE FOR HIGH-RISE BUILDINGS IN HONG KONG

S.M. LO

Associate Professor, Department of Building and Construction, City University of Hong Kong

C.W. HO

Associate Professor, Department of Real Estate and Construction, University of Hong Kong

EDWARD C.Y. YIU

Associate Professor, Department of Construction and Real Estate, Hong Kong Polytechnic University

D. Q. YANG

Associate Professor, Department of Naval, Architecture and Ocean Engineering, Shanghai Jiaotong University

ADA S.K. MAN

Research Student, Department of Building and Construction, City University of Hong Kong

Tile finishes have been extensively used in the decoration of the external walls of various buildings in Hong Kong. Debonding of such external finishes not only affects the appearance of buildings but also endangers the pedestrians. The aim of this paper is to discuss the failure mechanism and reasons of failure of external wall finishes that is a combination of concrete surfaces, rendering and tile finishes. Standard sizing wall panel is obtained and strain gauges are attached on each interface to measure the mechanical strain (i.e. the ratio: elongation / original length). While applying onto the wall panel, strain and stress distribution can be measured.

1. Introduction

While most ceramic tiling systems perform excellent in appearance and cost effective in appearance and cost effective in quantities, any failure compromises the reputations and growth of the industry. This indirectly has an adverse impact upon all manufacturers, merchants and installers, on the other hand, has triggered again public concern on the endangerment of the lives of public.

A typical wall tile system (Fig. 1) usually comprises of layers of finishing tiles, a substrate (render and spatterdash), a necessary fixative to adhere the tiles onto the substrate (adhesive), and a grouting material to seal up the pointing gap between tiles [1].

Tile finishes is used to protect concrete surface from directly explosion against climatic and environmental conditions. Moreover, it helps in increasing the durability of the external wall structure and providing sufficient aesthetic expression [2] & [3].

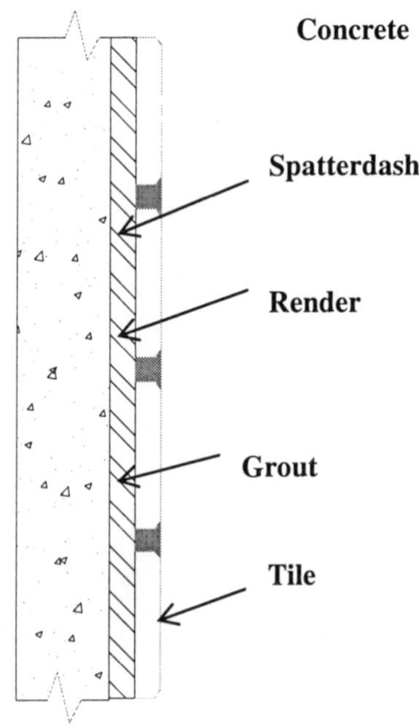

Fig. 1 A typical wall tile system

The result of delamination or bond faiure is typically pieces or sections of tile or other components of the wall which fall off and pose a serious risk to safety. A numbers of experiment have been carried [4] which found that if the strength is weak, more differential moverments occurs, leading to a disintegration of composite systems at the interface between layers. The causes of tile failure are several as summarize as shrinkage of adhesive (or mortar); differential movements between interfaces, e.g. tile, adhesive, substrates, due to thermal and moisture attacks; structural movements, shrinkage and creep, vibrations and settlement problems; and improper surface preparation, designs, selection of materials and sequence of works.

However, widespread adequate specification of tiling systems is a complex matter that has been partly addressed by the development of the existing product and installation standards. It is also being addressed by the computer-based expert systems [5] as previously advocated [6].

Computer modeling of tiling systems offers a cost-effective means of determining the stains and stresses that may develop when the system is subjected to specific loading conditions [7]. However, there is still the fundamental underlying requirement for comprehensive engineering data to determine appropriate compliance limits and to permit the development of engineering design codes that can support the project decision making process. While there is obvious need for such information, it is expensive to obtain, and there is no implicit requirement for any individual party to provide it.

The project aims to investigate differential movements of each interface of tile system and to look into the reasons for external wall finishes failure. Finite element models (FEM) are developed to verify the possible failure modes of external wall finishes under lateral loads.

2. Experimental Set Up and Measurement Procedure

The formation of the tiled wall system and measuring technique were reviewed as fundamentals. Besides, a series of preparation works were carried out before the destructive experiment starts. A reinforced concrete panel of dimensions 3000mm x 1000 mm x 100 mm is constructed and located in the steel frame for each time testing. The concrete is moist cured for a minimum period of 4 days and the formwork is strike after a minimum period of 7 days after concreting. Spatterdash of 3 mm thick with cement mix ratio of 1:1. After a 5 days curing period, a 20mm thick cement/sand rendering with mix ratio of 1:3 is applied subsequently. After a further curing time of 5 days, 95mm x 45mm x 7mm glazed ceramic tiles are fixed with the use of 3mm thick cement slurry. Finally, the panel is left for air curing by 5 days, under constant room temperature, before carrying out the test. The specification of tile finishes is tabulated in Table 1.

Table 1: Specifications of tile system for sample concrete panel

Background	Concrete
Formwork	Timber
Spatterdash	3mm (1:1)
Rendering	20mm (1:3)
Cement slurry	3mm
Type of tile	95mm x 45mm x 7mm glazed ceramic tile

Testing frame of structural steel with sections of 152 x 152 x 23 kg/m UC is constructed, and installed in the Heavy Structure Testing Laboratory of City University of Hong Kong. It is equipped with stainless steel bolts for locating the sample wall panels into position. A line load is applied horizontally at approximately mid-height of the concrete panel by an actuator installed with a steel rod at the front. The lateral deflection of the concrete panel during the test is measured by means of two transducers installed at the back of the panel and at the same horizontal level as the applied load (Fig. 2).

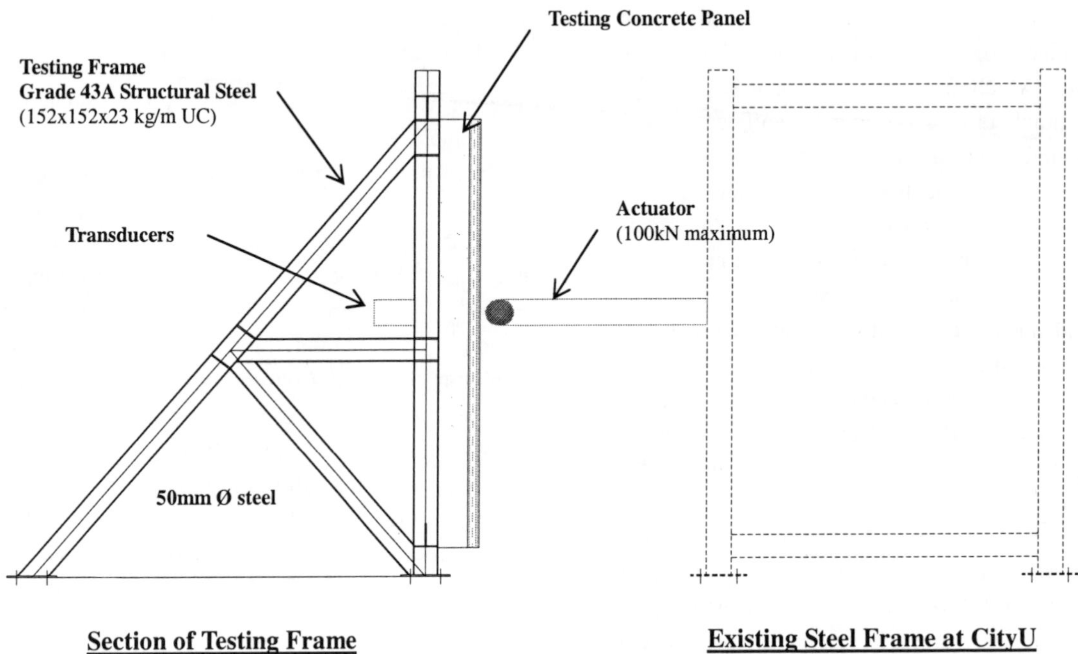

Fig. 2 Set-up of the testing concrete panel system

Strain gauges are attached on the surfaces of concrete, rendering and tile layers to investigate mechanical strain (i.e. the ratio: elongation/ original length). On each surface, three strain gauges are installed approximately at the edge, quarter and middle position of the panel respectively. The exact positions of the strain gauges installed on the concrete panel are shown (Fig. 3a). The stain gauges are numbered in ascending order from left and right and then bottom to top for further identification purposes (Fig. 3b). The data obtained from the actuator, transducers and strain gauges are recorded by a data logger for interpretation and analysis after the experiment.

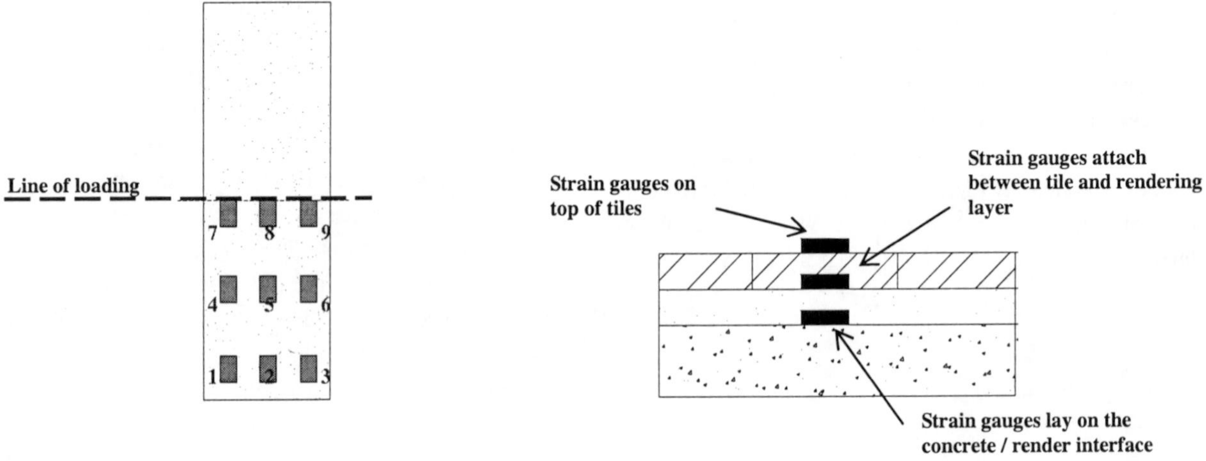

Fig. 3a. Positions of strain gauges on concrete panel

Fig. 3b. Strain Gauges installed on each material layer

2.1 Strain gauges

Strain Gauges are widely used for physical force measurement to determine the degrees and behavior of forces such as stress or load. Strain gauges are easy to use and offer a high degree of accuracy and stability.

When external force applied to an elastic material generates stress as shown which subsequently generates strain in the deformation of the material. The length L extends to L + ΔL, or if it compresses, it conversely reduces to L − ΔL. The amount indicated by the following equation is called normal or longitudinal strain [8].

$$\varepsilon = \frac{\Delta L}{L} \qquad (1)$$

The strain gauge is constructed by bonding a fine electric resistance wire to an electrical insulation base using an appropriate bonding material, and attaching gauge leads. The strain generated in the specimen is relayed through the base to the fine wire, where expansion or contraction occurs. As a result, the fine wire experiences a variation in resistance. This variation is exactly proportional to the strain.

$$\varepsilon = \frac{\Delta L}{L} = \frac{\Delta R/R}{K} \qquad (2)$$

Normally, this resistance changes is very small and requires a data logger to convert it to voltage output.

After all of the above items, concrete wall panel tiled with finishes, testing frame, transducers and strain gauges are well prepared, the test is then carried out according to the following procedures as in figure 4:

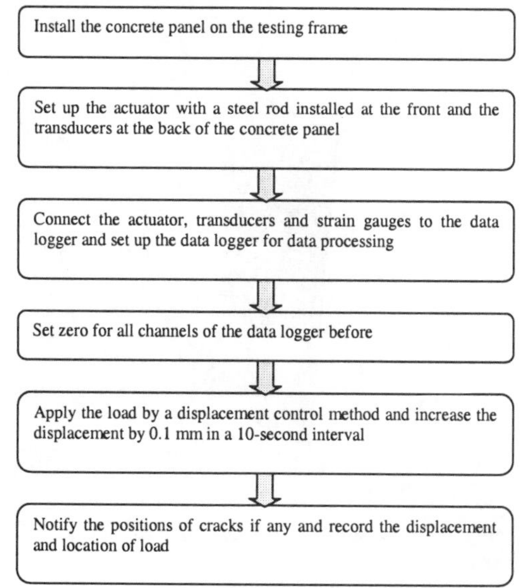

Fig. 4. Flowchart shows the experimental setting up procedures

3. Experimental Results

The testing load was applied in the form of a line load by means of a 50mm ϕ steel rod at approximately middle of the panel by a displacement-control method until the lateral deflection reached about 5mm.

It is found that hairline cracks were obtained along the concrete/ rendering interface [9]. As it can be verified according to the strain gauges no. 4 to 6 and no. 7 to 9 installed on concrete surface are at a long distance from the points restraint, the local restraining effect of the bolts at the panel edges did not effect the deformation of strain gauges no. 4 to 9 (Fig. 5 and Fig. 6).

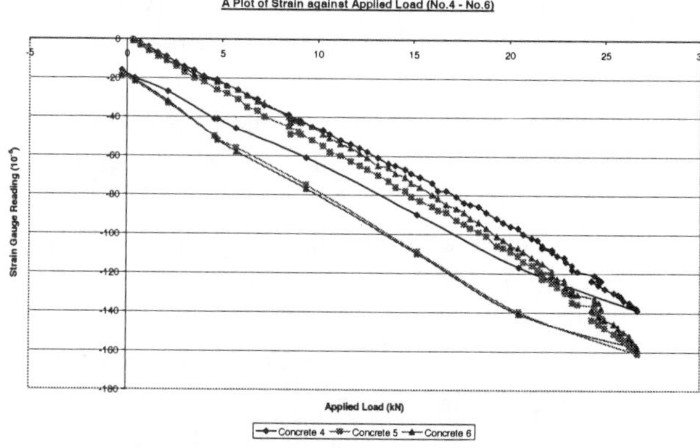

Fig. 5 Strain gauges plotting no.4 to 6

The test results showed that the concrete surface at the positions of strain gauges no. 4 to 9 was under compression. In addition, the strain was the largest at the middle of the panel.

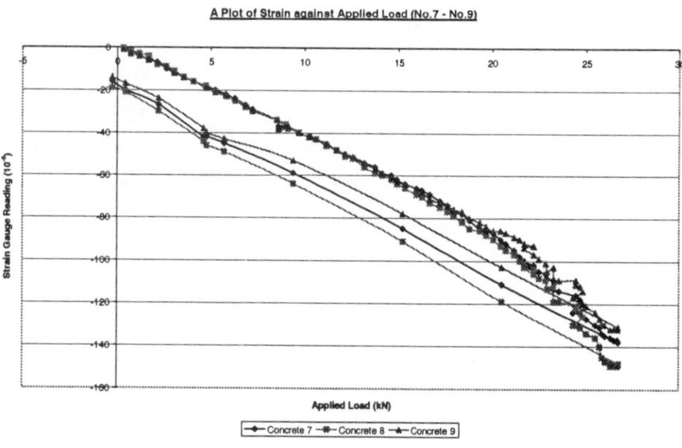

Fig. 6 Strain gauges plotting no. 7 to 9

On the cross section approach, failure occurred at the bottom and side edges of the panel. A crack was noted at the concrete/ rendering interface. The crack originated from the bottom edge and extends up to 330 mm towards the middle. Hammer tapping showed noticeable sound difference between the failure zone and the upper part of panel.

4. Discussion

In summary, the failure is unsymmetrical and occurs only on the bottom part of the panel. The reasons for this scenario could be due to the uneven distribution of spatterdash and uneven roughness of concrete surface.

In addition, airline cracks were obtained along the concrete/ rendering interface. There are several reasons for these phenomena. They are:

- differences in drying shrinkage characteristics between the top coat and the undercoat of the render
- poor preparation of the concrete background, e.g. lack of mechanical key and contaminated concrete surface
- weak concrete background without sufficient strength to hold the combined weight of rendering and tiles
- presence of impurities such as clay, silt & organic substances in the rendering
- inadequate protection and curing of the rendering

4.1 Finite element model analysis for tiling system

Computer modeling of tiling systems offers a cost-effective means of determining the strains and stresses that may develop when the system is subjected to specific loading conditions. It is also a quick way to get the stress distribution in wall finishes by computer methods under varied mix ration of materials, workmanship and size parameters. The FEM model consisted of steel frame, concrete panel, spatterdash layer, rendering layer, adhesive layer and ceramic tile. The strains obtained from the test are shown in table 2. The elastic modular, Poisson's ratio and density of steel are 200GPa, 0.28 and 7850kg/m^3, respectively.

Table 2: Strain values at measurement position

STRAIN GAUGE NO.	STRAIN VALUE ($\times 10^{-6}$)
7	-64
8	-66
9	-64
7A	-81
8A	-61
9A	-87
7A	-92
7B	-92

The physical properties of composed materials are calculated inversely by FEM simulation and listed in the following table 2.

Table 3: Physical Properties of Materials

NAME OF MATERIAL	ELASTIC MODULAR (GPA)	POISSON'S RATIO
Concrete	35	0.23
Spatterdash	0.8	0.2
Rendering	18	0.23
Adhesive/slurry	10	0.2
Ceramic tile	70	0.25
Grouting material	8	0.2

From the experimental results, the failure occurs in the concrete/ rendering interface, hence the spatterdash layer when the external force exceeds 24.5 kN. By simulating this condition by numerical method and get the corresponding critical stress strength of wall external finishes. The computational results are as follow:

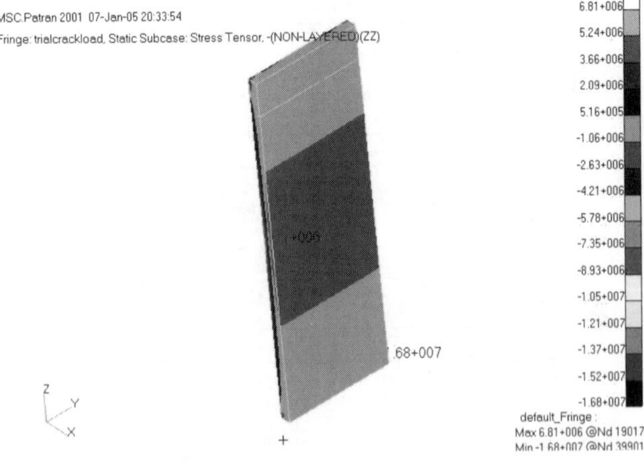

Fig. 7 Normal stress (in Z direction) in concrete panel

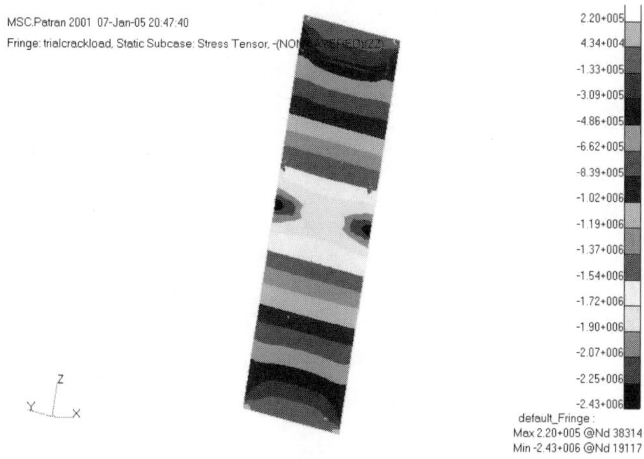

Fig. 8 Normal stress (in Z direction) on spatterdash layer

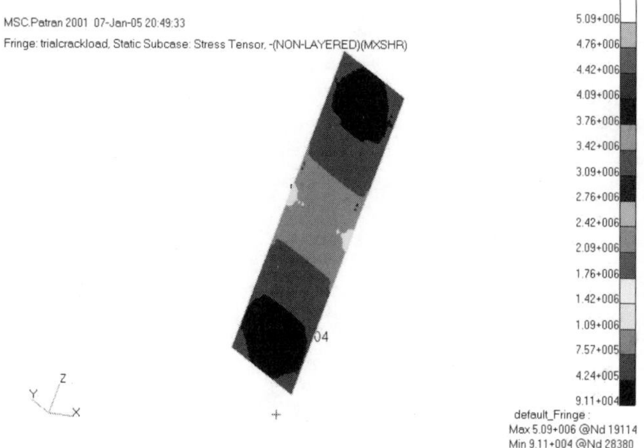

Fig. 9 Shear stress on spatterdash layer

5. Concluding Remarks

It is concluded that this experiment provides an alternative means to prove the failure of wall finishes apart from visual inspection. By the ways, feedback on problems is also a vital requirement to study about the peculiarities and specifications of wall tiles system. A sharing of experiences through research and case study analysis must eventually lead to high standards of performance.

The findings of these test confirms that falling-tile is properly verified in the spatterdash layer due to sources of reasons as discussed above. In order to obtain more detailed results of the analysis, a series of layered composite concrete cubic specimens will be designed and fabricated by applying material specifications with various substrates and bonding agents, shall be done in the future. Furthermore, the specimens are subjected to place under the influence of rain, wind and thermal influence after curing, then test destructively by pull-off and shear-off test.

Besides, finite element method is developed to conduct a detailed analysis of the mechanisms of stress transfer between the inner surface layer and the tiles. Finite element models (FEM) are developed in advance to verify the failure patterns of external wall finishes under lateral loads and establish the methods of inspection during the test and interpretation of the results after the test.

Acknowledgements

The authors acknowledge the full support of the Hong Kong Housing Authority.

Notation

The following symbols are used in this paper:

ε	=	Strain
L	=	original length
ΔL	=	increment due to force P
R	=	gauge resistance
ΔR	=	resistance change due to strain
K	=	gauge factor

References

1. Guan, W. L., Member, ASCE, Alum, J., Liu, Z. J. & Yang, T., (1997). Performance of external tiled-wall systems under tropical weathering. Journal of Performance of Constructed Facilities, 24- 34.
2. Wilkins, B., (1991). Guidelines for the selection of external protective wall finishes for smooth concrete surfaces in Hong Kong. Energy and Buildings, 15-16, 957- 962.
3. Ho, D. C. W., Lo, S. M., Yiu, C. Y. & Yau, L. M., (2004). A survey of materials used in external wall finishes in Hong Kong. Hong Kong Surveyor, 15(2), 7- 11.
4. Chew, M. Y. L., (1999). Factors affecting ceramic tile adhesion for external cladding. Construction and Building Materials, 13, 293-296.
5. Bowman, R. & Cass, C., (1995). Ceramic Tiles Today, Nov., 38-39.
6. Bowman, R. & Leslie, H. G., (1990). Proc. 1st World Congress on Ceramic Tile Quality, 83-84.
7. Goldberg, R. P, (1998). Direct Adhered Ceramic Tile, Stone and Thin Brick Facades – Technical Design Manual. U.S.A.: Laticrete International, Inc.

8. TML. Strain Gauges. Tokyo: Tokyo Sokki Kenkyujo Co., Ltd.
9. Development Performance Specifications and Testing and Acceptance Criteria of External Wall Finishes in High-rise Residential Buildings – Report on Trial Concrete Panel Test. Hong Kong Housing Authority Research.

SPRINKLER SYSTEM DESIGN FOR AUDITORIUM IN THE THEATRE

ZHENGRONG LIU
Architectural Design & Research Institute of TongJi University,1239 Siping Road Shanghai 200092 , China

SEJIN FAN
Architectural Design & Research Institute of TongJi University,1239 Siping Road Shanghai 200092, China

The fireproofing compartmentation measure for all the openings of a stage in the theater is one of the most important part in the fire design of the stage, and the metallic fire curtain that we usually adopt will make the collocation of the craftwork of the stage and the design of the architecture more complicated. So if we can use the drencher system replacing the fire curtain is to resolve this problem. Otherwise, the sprinkler system for the auditorium is another difficult problem in the fire design of the stage. In this article several system have been presented, we will recommend you the one in reason and more economical.

1. Preface

Stage is the hard core of a theatre. There are a great deal of curtain, scenery and flammable properties on top of the stage, at the same time, there are also a lot of electriferous equipment, such as lighting, stage machine, etc. All these factors make the stage more easy to be fired. Therefore, the sprinkler system design of stage is very important.

In the "Code of design for sprinkler systems" GB50084-2001,it prescribes the fire criticality of the grid is severe criticalityII .If there is no fireproofing compartmentation measure between main stage, bay area and back stage ,when one of them is fired, the fire will spread rapidly, even endanger the auditorium where is full of people. So, we should think much of the fireproofing compartmentation measure in our fire design.

The auditorium is usually more than 8 meters in height. It is impossible for the ordinary sprinkler to control or put out a fire. But the auditorium is a place which is full of people, and not easy for the people inside to evacuate, it is also a place that the code prescribes to have a sprinkler system. Therefore, we have to choice a practical sprinkler system for such a specifical place.

2. Fireproofing Compartmentation Measure

2.1 *Metallic fire curtain*

Around the main stage, there are four openings: proscenium opening, openings between the main stage and the two bay area, opening between the main stage and the back stage. According to the rule 8.1.1 in the "Design code for theater"J67-2001,fire curtains should be seted for the four openings around the stage. But in our practical design, these four metallic fire curtains will influence the layout of the stage. There are three aspects:

1. The metallic fire curtains (between the main stage and the two bay area, the main stage and the back stage) will cut off the connection between the fly gallery and the wall, hence the fly gallery should be hanged from the superstructure. But it will make the structure complicated, also reduce the stability of the fly gallery.
2. Narrow the limited fly gallery.Usually the space on the first fly gallery is not enough. If it is narrowed by the installation of the metallic fire curtain, the stage machine control room can not be put on the first fly gallery.
3. The metallic fire curtain need guideway and some machine bearing its weight so that it can works well. All these machine should be fixed upon the wall. They occupy the space of the stage.

On the other side, though the metallic fire curtain has better capability of fireproofing compartmentation, it has its own limitation:

1. When fire, the metallic curtain itself is heated unequally, it will deformed. People can not put it down to prevent the fire from other side. Actually, it will not work when needed.
2. Be difficult to maintain. For example, in the People's stage and the Changjiang theater, the metallic curtain all have damaged.
3. The metallic fire curtain costs a lot.

2.2 *Drencher system*

Instead of metallic fire curtain, drencher system can work well. The advantages of drencher system are obviously:

1. It has little influence on the layout of the stage. For the cause of the designing shape of a theater, the area for the bay area and the back stage is limited. We have no more space for the metallic fire curtain, then the drencher system is the only measure you can take.
2. Easy to maintain. There are no machines in the drencher system, only pipes and sprinklers. So you do not have to maintain all the time.
3. It is more economical.

2.3 *Design parameter for the drencher system*

Place	Name of system	Design parameter	Working time
Proscenium opening	Drencher for cooling protection	1L/S.M	1hour
Back stage opening	Water curtain for fire compartment	2L/S.M	1hour
Two bay area openings			

2.4 *Conclusion*

As we see, instead of metallic fire curtain ,drencher system play an effective role in the fireproofing compartmentation for the stage. Considering the proscenium opening is just face to the audience,and the safety of the people is most important,we suggest keeping the metallic fire curtain while with the drencher system for cooling effect to protect the proscenium opening, and the other three openings around the main stage can just be protected by the drencher system (water curtain for fire compartment).

3. Sprinkler System for Auditorium

3.1 *Choice of sprinkler system*

Usually the auditorium in a large theater is more than 8 meters in height. The ordinary sprinkler is not suitable for such a space. But we can choose other measures. They are:

1. Early suppression fast response sprinkler.
2. Deluge system.
3. Automatic-scanning elevated fire monitor extinguishing device.

3.2 *Early suppression fast response sprinkler system*

There is a fast response element to heat in the ESFR sprinkler. When in a large space more than 8 meters, it is very important to have such a sensitive element. It makes the sprinkler spurt out water as soon as fire. So it can control the fire in the earliest time. Otherwise the ESFR sprinkler has a lowest working pressure at 0.34Mpa,in fact such a pressure makes the sprinkler have enough water to extinguish fire. A large number of fire examples show that only 1~2 sprinkler of ESFR work, the fire can be in control. Thus we can see the ESFR sprinkler system will extinguish the fire efficiently in the earliest time of a fire, so that it can reduce the casualty and the loss of property. For the auditorium in a theater, it must be fine decorated and there is a lot of person there, ESFR sprinkler system is a better choice.

On the other hand, this system is comparatively simple. You just need to change the ordinary sprinkler into the ESFR sprinkler, and keep the system pipes. So you don't have to add any alarm valve and feed mains. As a result we can save the space of the pump room and the riser.

Design parameter of the ESFR sprinkler system:

1. flow coefficient of the sprinkler:K=200
2. lowest working pressure: 0.34Mpa
3. the number of working sprinkler in the active area:12

Then we can figure out the flow of every working sprinkler is 6.5L/S and the total flow of the working system is 78L/S.

3.3 *Deluge system*

Deluge system can also be used in extinguishing fire in a large space. It is the system that there is no water in the system pipes at ordinary times, but when fire occurred, all the sprinklers of the deluge system will spout out water at the same time. You can easily imagine that such a large area of water spraying would make the audience feel panic. Obviously, it is not benefit for people to run out. Though the fire is extinguished, but more life and property should be sacrificed.

Another character of the deluge system is there is no water in the system pipes at ordinary time, only after the alarm valve works, the sprinkler could spout out water. So there is a short time between the fire affirmance and the working of the sprinkler. Comparatively deluge system is slower in spouting out water than ESFR sprinkler system.

The fire criticality of the auditorium is middle criticality I ,so the spraying intensity of the deluge system is 6L/min.m^2,and the minimum working area is 160m^2.For a large theater, the seating capacity of the auditorium is more than 1500,accordingly the area of the auditorium is nearly 1000 m^2.Therefore such an area should be divided into 7 working area, videlicet we have to design 7 deluge systems in the auditorium.Centainly,7 systems must have 7 alarm valves in the pump room. It takes a lot of area in the pump room, moreover the entire system will become complicated. Considering the worst situation, the fire takes place at the cross of the four working area, in order to control the fire efficiently, all the sprinklers belong to the working area will have to work at the same time. The maximum flow of every working area is Q=1.2*6*160=1152L/min=20L/S.And as a result, the entire flow of the whole working deluge system is Q=4*20=80 L/S.

3.4 *Automatic-scanning elevated fire monitor extinguish device*

Though the automatic-scanning elevated fire monitor extinguish device can control fire in the large space, it is more suitable to the place such as gymnasium, exhibition hall etc. The auditorium is comparatively smaller.

On the other hand, the flow and the pressure for this elevated fire monitor extinguish device are very large, so it will hurt the audience. For all the reasons above, we can see that this system is not proper for the room with a large number of people such as auditorium. In another way, this system will cost a lot.

4. Conclusion

In conclusion, when we choose the fire compartmentation measure for the opening of the bay area and the back stage, it is better to take the drencher system than the metallic curtain. Because it has little influence on the layout of the stage and the structure, besides this system is easy to be maintained, and also it costs less. So it is not only satisfied the code but also a kind of measure which can compart the fired room from others efficiently.

The auditorium has a large space and a large number of people, the ESFR sprinkler system is a reasonable choice for the control of a fire. This system is comparatively simple to the deluge system and also comparatively economical to the automatic-scanning elevated fire monitor extinguish device. It can save the area of the pump room, moreover it reacts very quickly, this characteristic makes the system control the fire effectively at the early time of a fire, therefore it reduced the losing of the property and the number of casualties to the minimalist.

References

1. Code of Design for Sprinkler Systems GB50084-2001
2. Design Code for Theater JGJ 57-2000

A FRAMEWORK FOR MONITORING REWORK IN BUILDING PROJECTS*

EKAMBARAM PALANEESWARAN
Department of Civil Engineering, The University of Hong Kong, Pokfulam Road Hong Kong, China

M. MOHAN KUMARASWAMY
Department of Civil Engineering, The University of Hong Kong, Pokfulam Road Hong Kong, China

S. THOMAS NG
Department of Civil Engineering, The University of Hong Kong, Pokfulam Road Hong Kong, China

PETER LOVE†
School of Management Information Systems, Edith Cowan University Perth, Australia

The construction industry is project-based, and design and construction transactions within projects are generally separated. Effectively balancing the diverse interests of multiple stakeholders (e.g. clients, consultants, constructors) along with efficiently managing multiple risks across various project interfaces is a challenging task. Rework often arises from the unnecessary efforts of redoing/ rectifying incorrectly implemented processes or activities. In poorly managed construction projects, rework can significantly impact the time, cost and quality performance. Thus, from a business sustainability perspective of constructors, the control of rework is a great concern. A pilot study on the effective management of rework in construction projects was recently launched to address this concern. The research methods of this study exercise include (a) focused interviewing and questionnaire surveying of various experts and experienced persons in the Hong Kong construction industry and (b) conducting some case studies. The main focus areas of this study are (i) to identify the common root causes of significant rework items and (ii) to develop suitable structured frameworks for effective rework control and management. This paper presents (a) an interim summary of relevant findings from this study that relate to rework management in building construction projects; and (b) a proposed framework for monitoring rework in tall building works.

1. Introduction

1.1 *Overview of significance and study initiation*

Construction projects generally include several discrete tasks involving multiple stakeholders with diverse interests and differentiated activity streams, e.g. the design and construction activities are mostly segregated. Various inherent complexities in construction projects e.g. the difficulties in managing many interfaces are at the root of common problems such as changes/ variations, conflicts/ disputes. Hence, several industry-wide reform initiatives and resulting reports e.g. [1] and [2] sought to address these problems, so as to enhance performance, productivity and relationships and reduce costs, time and wastages.

Rework is a non-value adding activity or process that is related to redoing or rectifying incorrectly implemented items of work [3]. Earlier studies such as [4] and [5] found that the costs of rework in poorly managed scenarios could be very high (e.g. 25% of the contract value [6] and 10% of the total project costs [3]). Likewise, some other studies e.g. [7] and [8] identified rework as a significant contributor to time wastage and time/ schedule overruns, which will eventually impact on costs (e.g. indirect costs such as overheads), resources and quality as well [9].

In Hong Kong, a significant part of construction industry transactions are related to the construction of tall buildings. Thorough knowledge on rework causes and impacts are essential for developing effective

* This work is supported by HKU Seed Funding for Basic Research (Ref: 10205236).
† Work partially supported by the Australian Research Council

rework reduction measures and other management strategies. Appropriate monitoring mechanisms for rework should be considered extremely significant in tall building works so as to enable early detection and minimization of potential rework. The impacts of undetected as well as late detected rework items (e.g. errors, omissions, defects) could be particularly serious and severe in tall buildings i.e. safety related issues

Realizing the significance of rework reduction, a pilot study was undertaken. Despite the fact that numerous studies have attempted to quantify quality costs and analyzing time/ cost overruns, only very a few focused research initiatives have specifically targeted to the identifying the root causes of rework, analyzing the resultant impacts and developing relevant management frameworks as well as necessary control measures, e.g. [10], [11] and [12]. The main focus of the ongoing research exercise includes (i) exploring the causes and effects of major rework items in Hong Kong based construction projects; and (ii) developing a structured framework for effective rework reduction.

1.2 *Synopsis of the research approach*

The research that is being undertaken in Hong Kong forms part of a wider international study that is seeking to develop a staged life-cycle framework which can be used to reduce the incidence and costs of rework in construction and other infrastructure projects. A triangulation approach is adopted in this research that mainly includes (i) knowledge networking with domain experts and experienced practitioners in the construction industry through exploratory correspondences and semi-structured interviews, (ii) targeted questionnaire surveying, (iii) knowledge mining from relevant literature and forensic case-studies on recently completed projects. A taxonomy modeling for mapping rework (e.g. causes and impacts) and developing a structured framework for rework monitoring are aimed.

The questionnaire used in this study was mainly derived from a recent Australia-based study [9] and pilot tested in Hong Kong before distributing to its potential respondents (e.g. contractors, consultants) in the local construction industry. The questionnaire included a set of questions to be answered against a 5-point Likert type scale, for 87 rework related variables and a cluster of questions for 18 other information sets such as time and cost related datasets, project parameters and respondent details. The targeted respondents were asked to relate their responses to a particular recently completed project that they were involved with and they were encouraged to submit multiple responses if they have been involved in more projects recently. In this pilot research, 109 questionnaire survey responses were collected (of which 72 belong to rework in building projects) and 16 key interviews were completed so far. All the respondents are well-experienced persons from the Hong Kong construction industry (e.g. majority of them are contractors who have more than 10 years experience).

1.3 *Outline of the main contents in this paper*

This paper mainly presents the following:

- An interim consolidated summary of study findings extracted from the questionnaire survey responses on 33 rework related variables. The survey findings are further strengthened by some interviewee perceptions.
- A synopsis of reference works on monitoring rework in construction projects that is distilled from various rework related research initiatives. This section also includes an outline of the conceptualization and development of the proposed framework for monitoring rework in construction projects.

2. Essential extracts from the survey findings

This section presents some key extracts consolidated from the set of 72 questionnaire responses relevant to the appraisal of rework in building projects. Tables 1 and 2 summarize the rankings of various rework factors under site-related and sub-contractor related headings. The rankings are derived using the following formula as described in [13].

$$RI = \frac{\sum_{x=1}^{N} w_x}{AN}, \quad (0 \le RI \le 1). \tag{1}$$

In this, *RI = index for ranking, w = weighting assigned to each rework factor (ranging from 1 to 5), x = numeric index for counting, A = highest weighting used (i.e. 5), N = total number of respondents.*

Table 1. A ranking of site management-related factors causing rework.

Site management related factors	Index	Rank
Poor planning and coordination of resources	0.592	1
Ineffective use of quality management practices	0.533	2
Setting-out errors	0.525	3
Failure to provide protection to constructed works	0.511	4
Staff turnover or reallocation to other projects	0.483	5
Ineffective use of information technologies	0.431	6

Among the six site management related factors included in the questionnaire, 'poor planning and coordination of resources' and 'ineffective use of quality management practices' are top ranked. It may be interesting to note that the ineffective use of information technologies secured least consideration as a rework factor in the site management consideration, whereas it may be deemed necessary for overall organization.

Table 2. A ranking of subcontractor-related factors causing rework.

Subcontractor related factors	Index	Rank
Inadequate managerial and supervisory skills	0.647	1
Carelessness-based damages by subcontractors to works of other trades	0.606	2
Multi-layered subcontracting	0.583	3
Low labour skill level in subcontracted works	0.569	4
Ineffective use of quality management practices by subcontractors	0.556	5
Use of poor quality materials by subcontractors	0.447	6

Among the subcontractor-related factors, 'inadequate managerial and supervisory skills' and 'carelessness based damages by subcontractors' are the top-ranked items. It is envisaged that additional training arrangements and the recently introduced voluntary subcontractor registration scheme should be improving these subcontractor-related concerns. Also, it is interesting to know that the rework due to 'use of poor quality materials by subcontractors' is not a major concern for rework in Hong Kong based building projects.

Similarly, a set of summaries on rework impacts in building projects relating to various aspects relating to project performance is consolidated in Table 3.

Table 3. A summary of impacts of rework on project performance.

Impact on project performance	Index	Rank
Time overrun	0.692	1
Contractual claims	0.636	2
Cost overrun	0.619	3
Client dissatisfaction	0.617	4
Contractor's dissatisfaction	0.594	5
Design team's dissatisfaction	0.528	6

The consolidated summary on impacts of rework on project performance in building projects indicated that time overrun, contractual claims and cost overrun are more significant top ranked items. Table 4 and 5 provide more detailed break-up summaries relating to cost impacts from design-related and construction-related rework sources respectively.

Table 4. A summary of cost impacts of design-related rework sources.

Potential impacts on costs	Index	Rank
Changes initiated by end-users or regulatory bodies	0.686	1
Changes made at the request of the client	0.683	2
Revisions, modifications and improvements of the design initiated by contractor or subcontractor	0.503	3
Changes made at the request of the contractor during construction	0.483	4
Omissions from the contract documentation	0.456	5
Errors made in the contract documentation	0.444	6

Table 5. A summary of cost impacts of construction-related rework sources.

Potential impacts on costs	Index	Rank
Changes initiated by the client or an occupier after some commencement of a work	0.706	1
Changes initiated by the client or an occupier after completion of a work	0.681	2
Changes in construction methods due to site conditions	0.561	3
Changes to the method of construction to improve constructability	0.525	4
Changes made during the manufacture of a product	0.494	5
Damages caused by a subcontractor	0.492	6
Errors due to inappropriate construction methods	0.483	7
Changes initiated by a contractor to improve quality	0.453	8
Omissions of some activity or task	0.417	9

In these design-related rework sources for cost impacts, the top-ranked source is 'change(s) initiated by end-users or regulatory bodies'; and the lowest ranked source is 'errors made in the contract documentation'.

Among these construction related cost impact sources, the changes initiated by the client or occupier 'after commencement of work' and those 'after completion of a work' are top ranked. The 'omissions of some activity or task' is ranked as the least impacting source.

3. Monitoring rework in construction projects

3.1 *An overview of referable frameworks*

This subsection presents a basic review of an array of previously proposed rework monitoring models in many countries. They are grouped under two headings such as (i) rework tracking models and (ii) quality cost tracking frameworks.

3.1.1 *Rework tracking models*

Based on an array of international studies (such as in Australia and UK) on rework reduction, useful frameworks such as a procurement model for reducing rework [9] and a rework-benchmarking framework [13] have been developed. The following operational definition for rework was used for this research: "*Rework is the unnecessary effort of redoing a process or activity that was incorrectly implemented the first time*" [10].

The Construction Industry Institute (CII) in USA established a basic tool for early warning field rework and cost growth which mainly includes a simple Field Rework Index (FRI) scoring model and a FRI-based Rework Danger Chart. The CII definition of field rework is as follows: "*activities in the field that have to be done more than once in the field or activities which remove work previously installed as part of the project*" [4].

The Construction Owners Association of Alberta (COAA) in Canada commissioned a pilot study with the University of Alberta and developed a Field Rework Data Collection System (FRDCS) [14] and a Field Rework Reduction Tool (FRRT) [15]. In these rework management tools, the COAA definition for field rework is as follows: "*total direct cost of redoing work in the field regardless of initiating cause or source*" (Source: http://rework.coaa.ab.ca/default.asp). Accordingly, the field rework mainly refers to "*the activities in the field that have to be done more than once in the field, or activities which remove work previously installed as part of the project regardless of source, where no change order has been issued and no change of scope has been identified* (by the owner)" [16]. According to the COAA perspectives, field rework do not include items such as change orders, design changes or errors that do not affect field construction activities, off-site fabricator errors (i.e. that are corrected off site), and on-site fabrication errors that do not affect direct field activities (i.e. that are corrected without disrupting the flow of construction activities).

3.1.2 *Quality cost tracking frameworks*

In general, quality costs include costs for prevention, appraisal, and rectifying failures. Thus, additional quality costs are incurred for undertaking rework in construction projects. A Singapore based research initiative developed a construction quality costs quantifying system (CQCQS) which mainly includes a documentation matrix with components such as cost code, work concerned, cause(s), problem area(s), time expended, cost incurred, site record reference [17].

Similarly, a prototype Project Management Quality Cost System (PROMQACS) was developed to track quality costs, since rework is often considered as a form of quality failure [5].

3.2 *Potential derivatives for the local industry*

This section outlines some potential concerns identified in this research e.g. in case of directly adopting or adapting any (or relevant parts) of the above-mentioned frameworks in the local construction industry:

1. Terminology-oriented issues – The study identified that there is no common standard for terms and definitions in the construction industry. For example, several overlapping or conflicting definitions are available for defining the rework items.
2. Practice-oriented issues – The interviews and correspondence with various practitioners in the Hong Kong construction industry revealed that there is no standard practice for recording and reviewing rework occurrences. For example, in some cases, the rework was considered as a quality issue and under some other perspectives rework was treated as a project management issue. Moreover, in several cases, the causes and impacts of rework were not given due attention.

3. Subcontracting related issues – In Hong Kong, multi-layered subcontracting is a common practice. Thus, the direct costs and associated impacts from rework instances were mostly transferred to the lower layers of the supply chain. In most cases, the impacts of rework related additional indirect costs on main contractors (as well as subcontractors) were not realized.
4. Other concerns include, lack of awareness, cultural inertia (e.g. resistance to change, reluctance to introduce innovations/ new systems), evolution/ prototype status of earlier-mentioned research products, lack of motivation, lack of (additional) resources, etc.

Therefore, benchmarking good practices, useful learning from various research findings and knowledge-mining from available rework related frameworks are essential for developing a structured framework suitable for the local construction industry practices [19].

3.3 *Proposed framework for monitoring rework*

The conceptualized outline of the presently proposed framework for monitoring rework is portrayed in Figure 1. The main parts of the framework include (i) rework items, (ii) category mapping using a taxonomy-based classification, (iii) responsibility tracking, (iv) impact assessment, (v) compensability checking, (vi) recording lessons and transactions

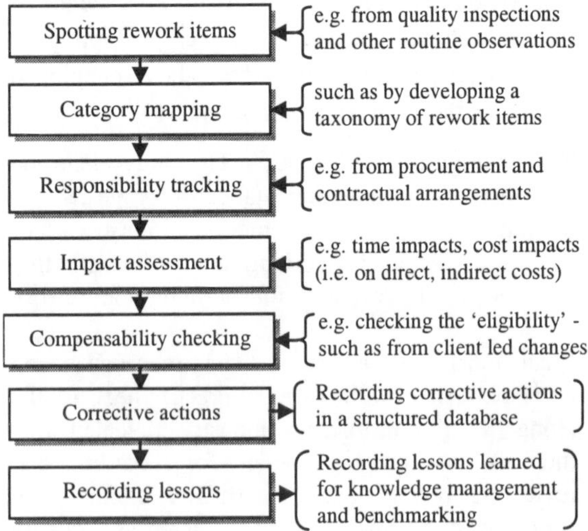

Figure 1. Proposed framework for rework monitoring in building construction projects.

Normally, the rework occurrences from non-compliance items (e.g. defects, errors, and omissions) and changes/ variations can be identified from quality inspections and other routine observations. A centralized project database with web-based interface is recommended for this purpose. Rework items could be classified in different ways, e.g. according to the source such as client-related, design-related, construction-related, site-management-related, and subcontractor-related sources and under each source, various factors (such as errors, omissions and changes) could be causing a rework occurrence.

A common taxonomy-based classification is proposed for structured and standardized classification of rework items and their root causes. A wide variety of procurement/ project-delivery approaches (e.g. design-bid-build, design-build) and various contractual arrangements (e.g. contract conditions, subcontracting, incentives/disincentives) are being used in construction projects. Appropriate tracking/ tracing of the roles and risks is essential for (a) fixing the responsibilities for rework items as well as their impacts, and (b) avoiding/ minimizing rework-related conflicts and disputes.

The rework could affect the time (e.g. schedule progress) and costs (such as direct and indirect costs). It should be noted that whatever the results of responsibility tracing for allocating risks and impacts, there could be some residual non-transferable impacts on every stakeholder in the construction supply-chain (e.g. part of indirect costs). Hence, introducing appropriate impact assessment arrangements, to be combined with structured time and cost management systems is suggested. Some rework items are compensable e.g. changes initiated by client or end-users after completion of a work, damages caused by a subcontractor. Developing suitable decision support systems, such as for 'eligibility' checking and 'quantification' of the compensations will be useful.

Also, it is recommended to register all the corrective actions and related transactions for (i) various controls and management measures in the current project, (ii) post-project analysis and (iii) pre-project planning in subsequent projects. Similarly, recording the lessons learned in structured databases is also recommended, which will be useful for knowledge management and benchmarking purposes.

4. Discussions and Conclusions

The observations from the ongoing rework research mainly indicated that there had been no structured approaches for monitoring rework occurrences and their impacts in the Hong Kong construction projects. Rework related studies in other countries such as Australia, Canada, UK and USA indicated that tracking and controlling rework will be beneficial for contractors through (a) cropping down the unnecessary efforts of redoing/ rectifying in those incorrectly implemented processes/ activities, and (b) reinforcing the essentials for business sustainability such as performance, productivity and profit margins.

The rework study presented in this paper has yielded a structured rework monitoring framework. Further efforts are necessary for system development and validation. Also, conducting some case studies will be useful in verifying the real impacts. Besides, this pilot research has opened up further research directions in related aspects such as rework in design and consultancy interfaces and related stress-based impacts on productivity and performance.

Acknowledgments

The ongoing research study is supported by a URC research grant (under Seed Funding for Basic Research') from the University of Hong Kong (Ref: 10205236). The authors are also grateful for the valuable knowledge-based contributions from many Hong Kong construction industry practitioners who shared their valuable experiences with the research team.

References

1. Egan, *Rethinking construction*, Department of Environment, Transport and Regions, London, UK (1998).
2. Construction Industry Review Committee, *Construct for Excellence*, CIRC, The Government of Hong Kong Special Administrative Region, Hong Kong (2001).
3. P. E. D. Love, *Journal of Construction Engineering and Management* **128(1)**, 18-29 (2002)
4. Construction Industry Institute, *The field rework index: Early warning for field rework and cost growth*, Research Summary **153-1**, CII, Austin, Texas, USA (2001).
5. P. E. D. Love and Z. Irani, *Information & Management* **40**, 649-661 (2003).
6. P. Barber, D. Sheath, C. Tomkins and A. Graves, *Int. J. Qual. Rel. Manage.,* **17(4/5)**, 479-492 (2004).
7. M. M. Kumaraswamy and D. W. M. Chan, *Construction Management and Economics* **16**, 17-29 (1998).
8. Construction Industry Institute, *An investigation of field rework in industrial construction*, Research Summary **153-11**, CII, Austin, Texas, USA (2001).
9. P. E. D. Love, Z. Irani and D. J. Edwards, *IEEE Transactions of Engineering Management* **51(4)**, 426-440 (2004).
10. P. E. D. Love, J. Smith and H. Li, *International Journal of Quality & Reliability Management* **16(7)**, 638-658 (1999).
11. P. E. D. Love, *Determinants of rework in Australian Construction Projects*, Unpublished PhD Thesis, Monash University, Australia (2001).
12. P. E. D. Love and D. J. Edwards, *Engineering, Construction and Architectural Management* **11(4)**, 259-274 (2004).
13. P. E. D. Love and J. Smith, *Journal of Management in Engineering* **19(4)**, 147-159 (2003).
14. Construction Owners Association of Alberta, *Field Rework Data Collection System*, Available at: http://www.coaa.ab.ca/ .
15. Construction Owners Association of Alberta, *Project Rework Reduction Tool*, Available at: http://www.coaa.ab.ca/ .
16. M. Dissanayake, A. R. Fayek, H. Wolf and A. V. Tol, Developing a standard approach for construction field rework assessment, *Proceedings of the 2003 Construction Research Congress*, Hawaii, ASCE, CDROM, 8 pages (2003).
17. S. P. Low and H. K. C. Yeo, *International Journal of Quality &Reliability Management* **15(3)**, 329-349 (1998).
18. P. E. D. Love and D. J. Edwards, *Civil Engineering and Environmental Systems* **21(3)**, 207-228 (2004).
19. E. Palaneeswaran, M. M. Kumaraswamy, S. T. Ng and P. E. D. Love, Management of rework in Hong Kong construction projects, *Proceedings of the 2005 QUT Research Week Conference*, CIB W89 and TG53, RICS COBRA, AUBEA, Queensland University of Technology, Brisbane, Australia, 6 pages (2005).
20. Construction Industry Research and Information Association, *Managing Project Change – A best practice guide*, CIRIA, London (2001).
21. P. E. Josephson, B. Larsson and H. Li, *Journal of Management in Engineering*, **18(2)**, 76-83 (2002).

COST COMPARISON STUDY OF PLASTERBOARD DRYWALL PARTITIONS VS. TRADITIONAL MASONRY/CONCRETE PARTITIONS IN HIGH-RISE RESIDENTIAL BUILDINGS, SHANGHAI

SHIMING CHEN†

School of Civil Engineering, Tongji University, 1239 Siping Road
Shanghai 200092, China

A cost comparison study of plasterboard drywall partitions vs. traditional masonry/concrete partitions in High-rise Residential Buildings in Shanghai is summarized. The study was purposed at demonstrating the cost advantages by using plasterboard drywall partitions over the traditional masonry/concrete partition system in high-rise residential buildings in Shanghai region. With the drywall plasterboard partitions substituted for the masonry/concrete partition system, it is found that the main structural construction cost can be reduced by 10% for the typical building studied. Besides, there is a massive reduction of laboring, and the construction period for the building is expected be reduced by 40 days approximately. An extra usable floor area of 360 m^2 is also expected for the building redesigned. The study reveals the positive cost advantages of high-rise residential building constructions with plasterboard partitions over those with masonry/concrete partitions. Although the cost comparison was based on a real-life high-rise residential building in Shanghai, the results can be applied to any regions where the material and labor rates are similar to those in Shanghai.

1. Introduction

Beginning from early nineties of the last century, there has been a numerous new high-rise buildings erected in Shanghai, and this trend has been continuing in Shanghai, as the golden value of its land and its national and international economic importance in China. However, even in the latest new buildings in construction, the wet masonry partitions are still found to be the first priority for the internal wall, particularly in the high-rise apartment buildings. The domestic designers and developers have an erroneous belief that the traditional partitions (for example: masonry /clay /concrete) are the best economical. In fact, apart from its excessively using land sources, these traditional partitions also have extra gravity load on the foundation of buildings therefore, about 10-20% of the structural resistance will be in account for the extra weight.

The study was based on a real life high-rise apartment building project, which was under way in Shanghai in 1997. The building targeted was a 28-storey concrete shear-wall structure and has a gross floor area of approx. 630 $m^{2.}$ The building consisted of one semi-basement level, a ground level, typical levels 2 to 24, penthouse levels 25 to 27 and the lift motor room on the top. This investigation was limited to the application of plasterboard drywall partitions for internal walls/ partitions on typical floor levels comprising 8 apartments per floor.

2. General Description of the Project

The study was commissioned by an international building sector with the purpose to further demonstrate the cost advantages by using plasterboard drywall partitions over the traditional masonry/concrete partition systems in high-rise building construction in Shanghai region.

Figure 1 shows the plane and elevation of the selected building. The main features of the building are summarized in Table 1

Table 1 The main features of the high-rise apartment building

No. of items	content and quantity
building position	Shanghai
construction period	2 years
gross area	15300 m^2 (approx.)
number of apartments	192
storey	24 typical floor stories; a semi-basement for parking; penthouse and lift-motor level
overall height	78.2 m
storey height	2.8 m
structural system	structural wall system
design seismic intensity	scale 7 (near field)

† corresponding author: e-mail: chensm@mail.tongji.edu.cn

The project specification called for traditional masonry/concrete partitions is the type of wet masonry partition of 240 mm and 120 mm or internal concrete walls of 200 mm in thickness with 20 mm hardplaster both sides. However, in the structural system, the internal concrete walls also function for the load-resistance purpose. Therefore, possible replacement of the internal concrete walls with the non load-bearing light partitions is based on the structural analysis and seismic assessment of the building in the redesign stage.

The study has been concentrated on the typical floor level 1-24, though the foundation parts including pilings under ground level also contribute to the cost saving of the construction. The penthouse units and stories above the 25th storey are not in consideration for cost comparison. The light internal partition systems being considered for comparison purposes are:

drywall partitions consisting of 75 mm metal frames with 1 layer of 12 mm firestop plsasterboard each side to achieve 0.75 HR fire rating (99 mm overall);

drywall partitions consisting of 75 mm metal frame with 1 layer of 12 mm firestop and 1 layer of 12 mm wet area firestop plasterboard to achieve 0.75 HR fire rating for bathroom areas (99 mm overall);

drywall partitions consisting of 75mm metal frames with two layers of 16mm firestop plsasterboard each side to achieve 2 HR fire rating.

Throughout this comparison study of using light weight drywall plasterboard partition system against traditional masonry/concrete partition system, the following terms will be used:

Original design: Design of the real-life building, with internal masonry and concrete wall for the partitions;

Redesign: Design totally based on the real-life building, without altering the building functions, but with light weight drywall plasterboards for the internal partition system.

Four specific areas are addressed as follows:

1. Concrete, formwork and reinforcement material and labor savings for internal concrete walls reduced, and reinforcement and thickness reduction in the external concrete walls;
2. Concrete, formwork and reinforcement materials and labor savings based on reducing thickness of floor slab from 120 mm to 90/100 mm;
3. Cost savings from reduced depth of the bottom foundation slab and reducing quantity of the piles;
4. Reducing construction time for the whole construction period, which will affect the final investment return..

3. Methodology

Comparison of physical properties of plasterboard drywall and traditional wet masonry/concrete partitions indicates that lower weight and reduced thickness are major advantages of the plasterboard option in the high-rise building construction. Table 2 illustrates the

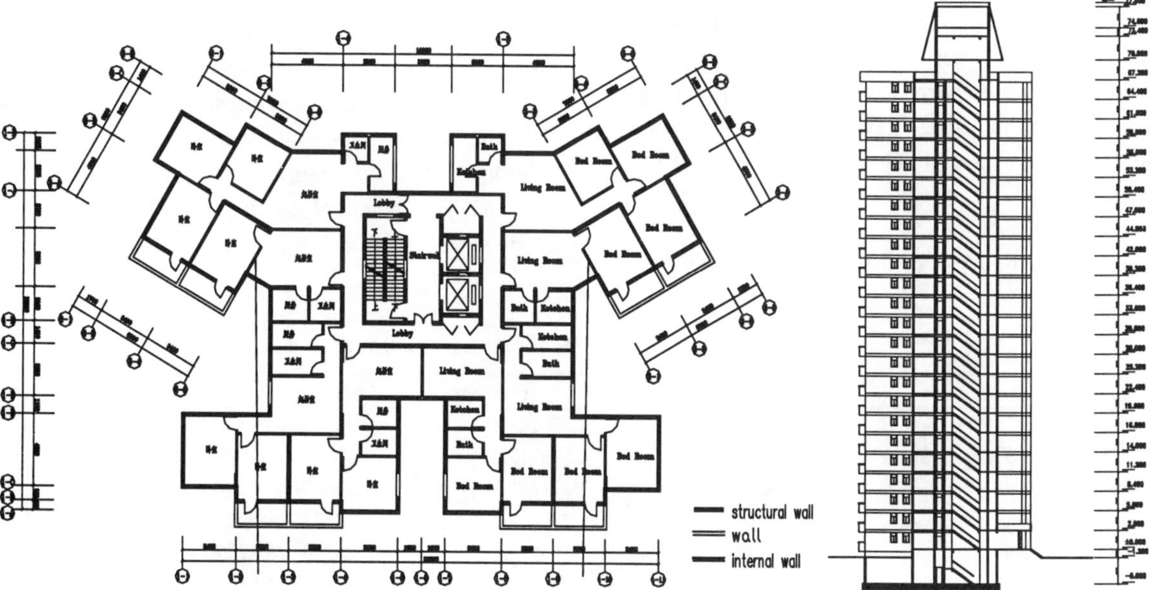

Figure 1 Floor plane and elevation of the building

Table 2 Physical properties of the alternative types of internal partitions

type of partition	Weight (kg/m²)	Thickness (mm)
solid masonry and hardplaster	530	280 (20 mm hardplaster double sided)
solid masonry and hardplaster	302	160(20 mm hardplaster double sided)
multi-holes masonry and hardplaster	422	280(20 mm hardplaster double sided)
multi-holes masonry and hardplaster	240	160(20 mm hardplaster double sided)
internal concrete wall	520	240 (20 mm rendering double sided)
plasterboard drywall	32	99
plasterboard drywall (with wet firestop for bathroom area)	42	99

alternative types of internal partitions and the corresponding physical properties for the selected building.

The comparison has been carried out based on the structural analysis of the building with masonry internal partitions and a part of the internal concrete wall substituted by light-weight drywall plasterboards. The structural analysis including a seismic assessment is in accordance with the relevant Chinese Design Codes and specifications, which are GBJ9-87, the loading code for buildings, GBJ10-89, the specification for design of concrete building structures, JGJ3-91, the design and construction specifications for high-rise building, and DBJ08-9-89, the seismic design code for buildings.

Structural member sizes and reinforcement requirement determined by structural analysis have been compared with the original design and the corresponding cost savings calculated. Replacement of internal concrete walls and masonry partitions by drywall plasterboard partitions based on a sound structural and seismic analysis offers a sharp reduction in the construction cost. Savings in pilings and foundation costs owing to total weight reduction of the building structure have also been considered in the cost assessment.

Reduction in thickness of the internal partitions has been accounted for provision of an extra effective usable floor area for the same gross floor area. This increases the ratio of effective usable floor area to gross floor area, being a bonus in for replacement of the traditional internal partitions with the thinner light weight plasterboard

Potential saving attributed from quicker construction time and reduction of 6296.5 on-site man day for plasterboard drywall partitions has also been assessed. Possible banking and loan plan will be compacted, and therefore a further reduction in the cost of investment strategy is expected.

Labor rates and material costs used through out this comparison study are those for Chinese market price issued on February, 1997. All costs are expressed in Chinese RMBYuan.

4 Results

The cost comparison of plasterboard drywall partitions against the masonry/concrete and rendering partitions for the selected high-rise apartment building is based on a redesign of the building, with the lighter drywall plasterboard for the internal masonry and concrete partitions. The potential savings for the construction cost are given in Table 3.

Table 3 Summary for structural cost comparison (in RMB yuan)

cost items	original design	redesign	Savings/ %
structural floor system	2452653.8	2076148.2	376505.6 /(15.35%)
external concrete wall	2930548.8	2626713.9	303834.9 /(10.39%)
internal walls/partitions	3252143.0	3071261.6	180881.4 /(5.56%)
slab beams	197276.9	178903.6	18373.3/(9.3 %)
foundation	2859597.1	2595914.9	263682.2/(9.22%)
Total	11692219.6	10548942.2	1143277.4/(9.78%)

Table 4 Reduction in construction laboring (man-day)

construction items	original design	redesign	labor savings
structural floor system	15168.8 man-day	13553.2 man-day	1615.6 man-day
external wall	20354.5 man-day	20231.4 man-day	123.1 man-day
internal wall/partition	23499.5 man-day	20113.4 man-day	3386.2 man-day
slab beam	1447.3 man-day	1312.5 man-day	134.8 man-day
foundation	11365.1 man-day	10328.2 man-day	1036.9 man-day
total	71835.2 man-day	65538.7man-day	6296.5 man-day

Table 5 Comparison in the floor areas for the typical floor

item	extra usable floor area	typical floor area
One typical floor level	15 m^2	630 m^2
Total in the building	360 m^2 (approx.)	15300 m^2 (approx.)

In Table 3, costs for the structural floor systems, the external walls, the internal walls/partitions, the slab beams, and the foundations of the original building and the redesigned building are given. As these five items are main parts affected when the lighter drywall plasterboard partitions are adopted, we found that about 10% of savings would be achieved for the total cost of these five items in the sample building. However, the total construction cost for the building will comprise electric facilities, hydraulic pipes, gas tubes, elevator, doors and windows, water-proof roofing and etc. Figure 2 illustrates the total construction cost for the redesigned building (before contractor overhead fees, profits and etc).

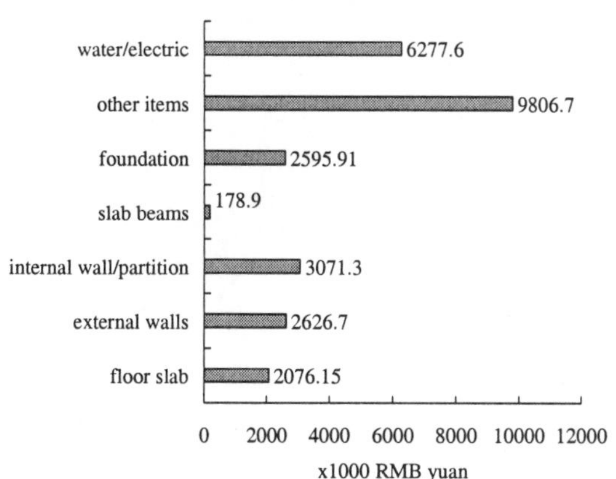

Fig. 2 Total construction cost for the building (before contractor overhead, profits, tax etc.)

Table 4 shows the reduction in the construction labor for the redesigned building. The direct savings due to the reduction of the laboring is already included in the cost per item, and reducing laboring in man-day, however will basically make the construction period for the building erection shorter. Although reduction of the construction time is very complex, which depends on the construction management, individual process organizations, or even the weather, a reasonable time reduction of 40 days is expected, and this will have the positive benefits on the investment return.

The extra useable floor areas was assessed by comparing the typical floor area of the redesigned building with that of the building originally designed. Because the drywall plasterboard partitions are thinner, and no hardplasters double sided, more space room can be achieved in the redesigned building with the drywall plasterboard partitions. Table 5 illustrates the figures that have been achieved in the redesigned building. 2.38% more usable floor areas are obtained in the building designed with thinner plasterboard partitions than the building with the masonry/concrete partitions.

A comparison of steel and concrete quantities used in the structural construction of the buildings is briefed in Table 6. The quantities of steel and concrete used for structural construction are dropped by 22.98% and 13.92% respectively, when the drywall partition system is adopted in the sampled building.

Table 6 A comparison of steel/concrete quantities used in the structural constructions

items	original design	redesign	savings
steel (ton)	1025.04	789.53	235.51
steel(kg/m2)	67.00	51.60	22.98%
concrete(ton)	5065.58	4360.51	705.07
concrete(kg/m2)	331.08	285.00	13.92%

5. Conclusions

This study has demonstrated that use of plasterboard drywall partitions in lieu of traditional wet masonry /concrete and hardplaster partition walls in high-rise residential buildings in Shanghai offers substantial reduction in the construction cost. Though the building selected is erected in Shanghai and 28 stories high, the assessment results will cover a wider variety.

To ensure the building redesigned comparative to the building originally designed, the internal structural walls for service wells remain. However, potential savings can be achieved, if an internal frame system integrated with shaftline plasterboard partitions is used for the internal wells.

With respect to land saving, and high efficiency in the investment return, benefits will be achieved in the savings of interest payment, and in the early income, when construction time is compacted. Besides, the extra floor area by using thinner partitions is also a bonus for both developer and property owners. The results of this study show the erroneous belief that the wet masonry/concrete partitions are the most economic for high-rise buildings be in the wrong direction.

In conclusion, we use two charts (Figure 3 and Figure 4) to summarize the overall savings owing to replacement of the drywall plasterboard partition system for the internal masonry/concrete with hardplaster partition system in the sample residential building.

For the typical high-rise residential building, substitution of the drywall plasterboard partition system for the internal partition system of traditional masonry/concrete materials, would have a potential overall saving of 6.8 million RMB yuan, which means that a saving of 20.84% of the total construction cost, or 7.88% cut in the total investment could be achieved.

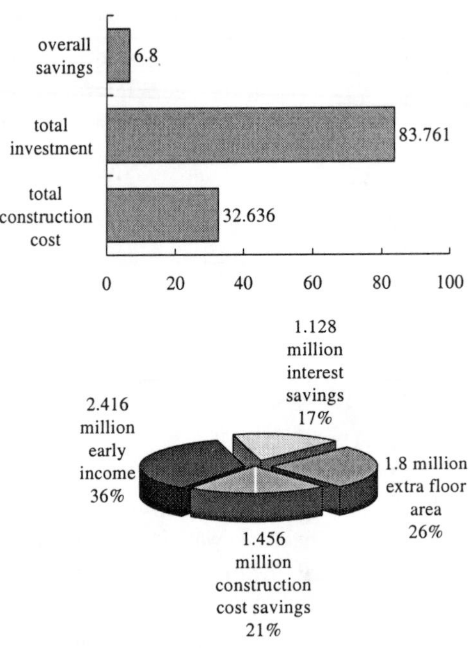

Figure 4 Items in overall savings for the redesigned building

STUDY ON THE ECONOMICAL EFFICIENCY AND PERFORMANCE COMPARISON BETWEEN REINFORCED CONCRETE HOLLOW GIRDERLESS FLOOR AND COMMON FLOORS

XUESHU WU

College of Civil Engineering, Tongji University, 1239 Siping Road, Shanghai 200092 , China

XUEYU XIONG

College of Civil Engineering, Tongji University, 1239 Siping Road, Shanghai 200092 , China

DINGYE HUANG

College of Civil Engineering, Tongji University, 1239 Siping Road, Shanghai 200092 , China

This paper introduces the calculation method of reinforced concrete hollow girderless floor. With this method, the quantity of the reinforced bar and concrete needed in reinforced concrete hollow girderless floor is computed. And unit material quantity of reinforced concrete hollow girderless floor is compared with that of waffle floor, flat slab and ribbed floor, unbonded prestressed concrete floor with the same span and loads. On this basis, unit square costs of floors are calculated according to market price of materials and economical efficiency comparison is conducted for floors. Then the performance of the hollow floor is presented, compared with that of other floors. Through comparison and analysis, it is concluded that service performance and economical efficiency of hollow girderless floor are better.

1. Introduction

At the present time, tall buildings are built one after another and the height of buildings increases, which demands the reduction of the dead load of floor so as to reduce the dead load of the whole building. On the other hand, the tall buildings tend to expect large space and free division. Through imbedding into floor light fillers such as thinwalled tube and hollow box which take place of middle concrete of the floor which contributes less to endogen force, reinforced concrete hollow girderless floor reduces dead load, increases span and meets the demand of tall buildings.

The engineer of the former Federal Republic of Germany, Leoplod Muller developed girderless floor system and first put forward reinforced concrete hollow girderless floor, named "B-Z system", tracing back to beehive-type reinforced concrete hollow girderless floor in German [2]. In 1964, professor GFranz made experiment on the floor and raised new computation method for this kind of floor [3]. In 1967, American engineer, EdgarH.Hendler, put forward Cellular Flat Plate, which is a kind of hollow girderless floor using plastic foam as filler, of which hollow rate was high. But he only studied construction methods and special structure measures of this structure instead of making experimental investigation for design theory further. Moreover, plastic foam made the construction not very convenient and gave rise to pollution. So this kind of floor didn't spread extensively [1].

Since the middle of 90s, Chinese engineer has improved filler to avoid pollution and reduced cost so that reinforced concrete hollow girderless floor develops in China, of which one using GBF tubes as filler is most widely used. This paper presents computation methods and then computes an example. Based on the example, economical efficiency and performance of the hollow floor using GBF tubes as filler are compared with that of other common floors.

2. Computation Method of Reinforced Concrete Hollow Girderless Floor

Arranging GBF tubes in single direction makes different direction's rigidities of slab different. But test has proved that unidirectional tubes don't change two-way bending character of the slab. Common computational methods of the floor are imitating plate method, virtual intersection girder, direct design method and the equivalent frame method.

This paper uses direct design method to calculate reinforcement quantity of the floor. Firstly, strip between two central lines of lattice slab is considered as computational strip and static moment are calculated with the formula:

$$M_0 = q l_2 l_n^2 / 8 \quad (1)$$

where q is even load, l_2 width of the computation strip, l_n clear span length of the computation strip [5]. According to coefficients in the paper [5], the static moment is divided into the sagging moment and the negative moment, which are then distributed to the middle strip and the column strip. One section of the slab perpendicular to tubes is equivalent to I-shaped cross section on the principle of the equivalent area and inertia moment (Figure 1.a). Another section parallel to tubes simply takes the section of the least rigidity as equivalent section (Figure 1.b). Then quantity of reinforcing steel needed is calculated for the equivalent section according to strip moment.

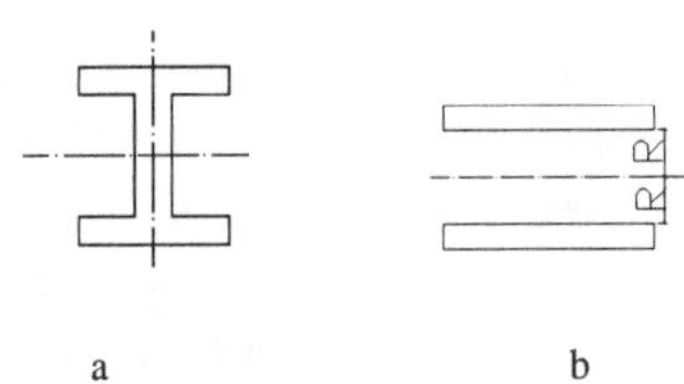

Figure 1. Equivalent section.

3. Economic Comparison

In order to compare economical efficiency of the hollow floor with that of common floors, the computational work should be conducted for flat slab, ribbed slab, waffle slab, unbonded prestressed concrete girdless floor and reinforced concrete hollow girderless floor on the the condition of the same span and loads. The paper uses the example in the paper [4]. For this example, the structure has three spans in both x and Y directions and length of every span is 9m(Figure2). Loads are defined as follows: apart from dead load of beam and slab, additional load containing surface and ceiling of structure and partition wall is 2.5 KN/m^2, live load 2.5 KN/m^2; size of the column section of is $700mm \times 700mm$; Strength grade of concrete used in unbonded prestressed concrete girdless floor is C35, in other floors C20; the grade of bearing reinforcing steel in slab is I, in beam II .

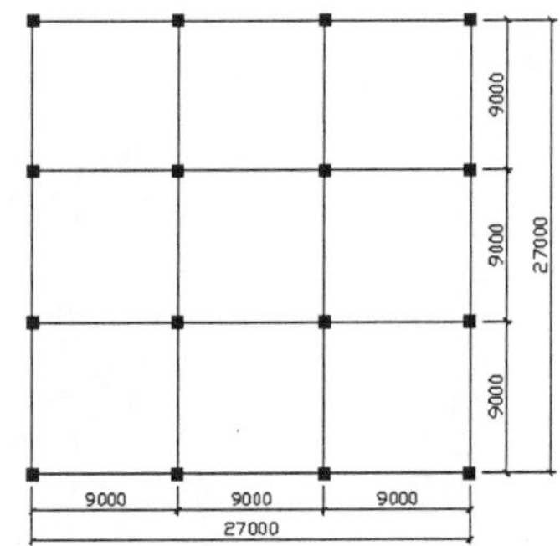

Figure 2. Column gird (mm) .

GBF tubes are arranged as Figure 3. The floor is under vertical load and meets the condition of direct design method, so the floor is computed with direct design method. The reinforcement quantity calculated is shown in Table 1 and Table 2.

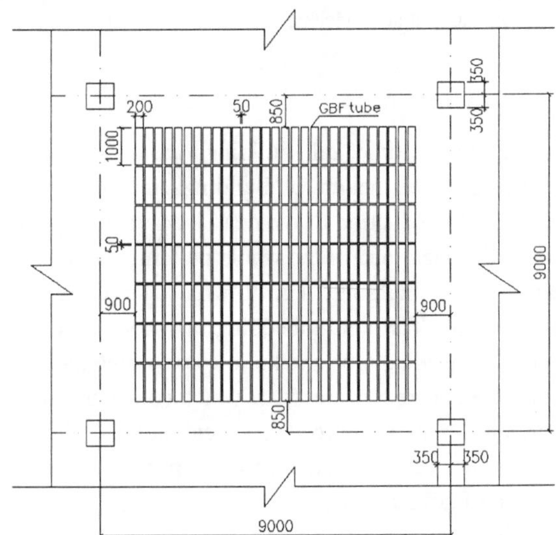

a. Arrangement of GBF tubes

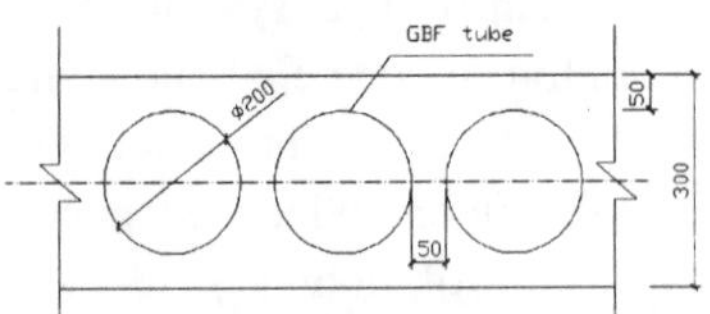

b. Section of slab perpendicular to tubes.

Figure 3. Reinforced concrete hollow girderless floor (mm).

Concrete quantity and steel quantity of other floor systems were calculated in the paper [4], and the data is compared with the calculation result of the hollow floor.

The economic comparison is made for the intermediate plan. The total costs of different floor systems are calculated according to unit price. The unit cost of concrete of c20 is 281 $yuan/m^3$, C35 294 $yuan/m^3$; reinforcing bars is 3400yuan/t; unbonded prestressing tendon is 7700 yuan/t; GBF tube is 22 yuan/m. The calculation result is shown in Table 3. Through the comparison, it is concluded that economical efficiency of the hollow floor is the best on the condition of 9m span. Furthermore, the dead load of the floor reduces so that the cost of the bearing column and foundation decreases; the bottom surface of slab is so flat that the cost of ceiling fitting and ceiling replacement decreases; this kind of floor avoids drawing tendons on site; formwork installation and formwork removal of the hollow girderless floor is easier than waffle floor and ribbed floor and the cost of formwork reduces.

Table 1. Summary of reinforcing bars perpendicular to tubes.

Strip / reinforcement	Negative moment of inner support		Sagging moment		Negative moment of external support
	Column strip	Middle strip	Column strip	Middle strip	Column strip
$A_s^{'}(mm^2)$	6647.7	2163.4	3892.5	2577.0	3232.4
Reinforcing bars	ϕ 14@100	ϕ 12@180	ϕ 12/20@150	ϕ 12@180	ϕ 12@150
Actual amount $A_s(mm^2)$	6925.5	2827.5	3794.8	2827.5	3393.0

Table 2. Summary of reinrcing bars parellel to tubes.

strip / reinforcement	Negative moment of inner support		Sagging moment		Negative moment of external support
	Column strip	Middle strip	Column strip	Middle strip	Column strip
$A_s^{'}(mm^2)$	6380.2	2078.3	3738.2	2475.5	3104.8
Reinforcing bars	ϕ 14@100	ϕ 12@180	ϕ 12/20@150	ϕ 12@180	ϕ 12@150
Actual amount $A_s(mm^2)$	6925.5	2827.5	3794.8	2827.5	3393.0

Table 3. Comparison of unit square cost for floors of 9m span.

Floor system	concrete		reinforcement		Prstressing tendon		GBF tube		Total cost (yuan)
	amount (m^3)	Cost (yuan)	amount (kg)	cost (yuan)	amount (kg)	cost (yuan)	amount (m)	cost (yuan)	
Waffle floor	0.198	55.6	15.10	51.3					106.9
Flat slab	0.260	73.1	14.50	49.3					122.4
Ribbed floor	0.234	65.8	19.52	66.4					132.2
Unbonded prestressed floor	0.220	64.7	4.74	16.1	7.52	57.90			138.7
the reinforced concrete hollow girderless floor,	0.148	41.59	6.52	22.2			1.67	36.75	100.54

4. Performance Comparison

Architectural performance of reinforced concrete hollow girderless floor is as follows:

1. Thickness of the hollow floor is smaller than beam-and-slab floor, so this kind of floor can reduce the height of the storey. The reduction of the height of the storey can cut down the cost of electric elevator and fittings of inner wall and external wall.
2. The hollow floor has the good performance of thermal insulation and sound insulation. The closed cavity in the floor impede the transmission of noise and heat, so performance of thermal insulation and sound insulation of the floor is improved evidently, which can reduce the cost of air-conditions and noise.
3. The bottom surface of the slab is flat and its visual effects are better.

The structure character of reinforced concrete hollow girderless floor is as follows:

1. The dead load of the structure reduces so the seismic action on the structure decreases. In the tall building, the total mass of the floors occupies the high proportion of the mass of the whole building, so reducing the mass of the floor can reduce the mass of the building efficiently. Seismic action on the building is proportional to the mass of the structure, so reinforced concrete hollow girderless floor can improve earthquake-resistance ability of the building.
2. The concrete is cast on site so the floor has good integrity and large rigidity. Under the horizontal load, the tall building depends on roof and floor to form the integrity to resist lateral force, so the floor's integrity must be good enough and rigidity must be large enough to make sure it is feasible to distribute lateral force. So the reinforced concrete hollow girderless floor can meet the demand of the tall building resisting lateral force.

Since performances of other floors are familiar to engineer, the paper doesn't detail them.

On the base of part 3, performance of the reinforced concrete hollow girderless floor and that of other floors is compared (shown in Table 4). It is concluded that the performance of the hollow floor is better.

Table 4. Comparison of building performance.

Floor system / performance	Waffle slab	Ribbed slab	Common flat slab	The hollow floor	Unbonded prestressed flat slab
Clear height of structure	0	Increase by5cm	Increase by24cm	Increasing by20cm	Increasing by 28cm
Dead load (KN/m^2)	0	0	0	About 0.95 lighter	About 0.95 lighter
Sound insulation	ordinary	ordinary	ordinary	Improved by10~22db	General
Fire resistance	ordinary	ordinary	ordinary	Fire resistance time3h	bad (prestress relaxation)
Utilization ratio of architecture area	bad	bad	good	good	Good
Ceiling	bad	`bad	bad	good	Good

5. Conclusions

Through analysis of economical efficiency and performance of the reinforced concrete hollow girderless floor, it is concluded that the hollow floor has good service performance and economical efficiency and is suitable to be used in tall buildings. However, the paper only studies the economical efficiency preliminarily. In future we should further study how to choose suitable length and diameter of the tube and how to reduce the cost of the tube so that the economical efficiency will be improved further.

Reference:

1. Edgar H.Hendler, Cellular Flat Plate Construction. ACI Journal, February, 1968.
2. Fertigteil-Vertrieb Gmbh and Mannheim, B-Z Reinforced Concrete Celluar Plate for One-way and Two-Way Stress Directions for High Loads and Large Spans. Engineering Design Brochure, 1965.
3. G.Franz, Test Report Extract on a Model of the Cellular Flat Plate. Mar.9, 1965.
4. Ji Tongqiang, Technique-economic comparison of common floor system [master of engineering dissertation]. South China University of Technology, China, 2001.
5. Technical specification for reinforced concrete girderless floor with tube core (draft). Mar, 2004.

ENABLING AUTOMATION IN TALL BUILDING WORKS WITH WIRELESS TECHNOLOGIES AND TAGGING DEVICES

EKAMBARAM PALANEESWARAN

Department of Civil Engineering, The University of Hong Kong, Pokfulam Road Hong Kong, China

S. THOMAS NG

Department of Civil Engineering, The University of Hong Kong, Pokfulam Road Hong Kong, China

MOHAN M. KUMARASWAMY

Department of Civil Engineering, The University of Hong Kong, Pokfulam Road Hong Kong, China

DAVID W. CHEUNG

Department of Computer Science, The University of Hong Kong, Pokfulam Road Hong Kong, China

The construction industry is often criticized for its sluggishness, if not inertia, in deriving early benefits from various innovative developments in enabling technologies. Recently, various applications of wireless technologies (e.g. Bluetooth) as well as radio frequency identification (RFID) and barcode tags are being increasingly used e.g. in automated inventory control, procurement tracking and supply chain management. This paper is based on a pilot study that aimed to map such advanced technologies enabling automation in construction industry. This research exercise is mainly based on an extensive literature review from various sources. A summary of basic information regarding an array of such enabling technologies is presented in this paper. Furthermore, a proposed framework for automation in tall building works, e.g. facilities maintenance and asset management using RFID applications is presented in this paper.

1. Introduction

Normally, the automation initiatives in the construction industry are aimed to enhance productivity, profitability and safety aspects through various potential targets such as (i) supporting the acceleration of works, (ii) avoiding the conflicts and errors, (iii) reducing the rework and wastages, and (iv) avoiding/ reducing hazards and risks. Rapid advancements in various technologies enable the suitability of those technology applications with enhanced viability for automation in the construction industry. Some such automation enabling technologies include information and communication technologies, automatic identification and data capture technologies, smart materials and artificial intelligence technologies, micro and nanotechnologies, and robotics. For example, recent advances in information and communication technologies such as internet and web based systems, mobile communication, wireless communications and portable thin-client devices (e.g. tablet computers, pocket PCs) have provided a good foundation for automation in the construction industry [1], [2], [3].

Similarly, several automatic identification and data capture systems are being advantageously used for automation in various industries (e.g. manufacturing, transportation). Popular automatic identification technologies include barcodes, radio frequency identification (RFID), optical character recognition (OCR), smart cards, biometrics (e.g. speech/ voice recognition, finger print identification, eye retina identification) [4], [5], [6].

In general, the inherent risks, complexities, problems and necessities associated with tall building works are also tall (i.e. large). Several researchers studied the impacts of height (in high rise constructions) on aspects such as costs and productivity e.g. [7], [8], [9], [10], [11], [12], [13]. A pilot research was recently initiated to explore the advanced construction technology topic such as RFID applications in the construction industry with a primary focus on tall building works. The pilot research mainly includes (i) a desk study based knowledge mining exercise (e.g. from published papers, reports and case studies) and (ii) knowledge networking with domain experts (e.g. exploratory correspondences, brainstorming discussions and semi-structured interviews). This paper mainly presents a basic summary from the pilot research in the following main sections:

- An overview of automatic identification and data transfer technologies and their sample applications.
- Automation in tall building works with RFID applications.

2. Overview of Automatic Identification and Data Transfer Technologies

This section provides an overview of some useful technologies for advancing automation in the construction industry through (i) automatic identification and data acquisition, (ii) advanced communication channels for data transfer, (iii) standardization measures, and (iv) privacy and security considerations.

2.1 *Automatic identification and data capture*

The popularly used technologies for automatic identification and data capture (AIDC) are barcodes, smartcards and RFID [6]. Earlier, various application areas for barcodes in the construction industry were studied e.g. [4], [14], [15], [16], [17]. More recently, many ADIC related research initiatives focused on various RFID applications, e.g. [18], [19], [20], [21]. This subsection provides a basic summary of RFID technology and its ADIC applications in several areas.

In RFID applications, radio frequencies are used for capturing and transmitting data. The principal components of RFID based systems include (i) tags or transponders which are attached to the items that are to be identified or tracked, (ii) readers which are used for identifying, reading and writing the tags as well as for communication with data collection applications, (iii) data collection applications that receives data from readers and enters into appropriate databases, (iv) suitable back end software applications for mining, analyzing and interpreting the captured data. The RFID tags can be either active (i.e. with dedicated internal power supply) or passive (i.e. without any dedicated power supply, but obtain operating power from the electrical field of the readers). In general, active tags can hold more information than passive tags. Also, the tags can be either 'read only' or 'read/ write' type.

Depending on the specific needs of different RFID applications requirements, the tags are available in various forms/ shapes e.g. some tags are available in sticky label type thin films and some animal tracking tags are even in injectable forms. Also, various operating frequencies are being considered for RFID devices, e.g. low frequency (30 to 300 KHz), high frequency (3 to 30 MHz), and very high frequency (300 MHz to 3 GHz). The following are some sample applications of RFID technologies: (i) low frequency range such as animal tracking, anti-theft systems, (ii) high frequency range such as access control and asset tracking in a built facility e.g. library, baggage control in airports, retail product tracking in departmental stores (iii) ultra high frequency range applications such as vehicle tracking, and pallet/ container tracking, and (iv) microwave range applications such as vehicle access control [20].

The main advantages of RFID applications include non-contact based and non-optical based identification and data capturing capacities (e.g. without requiring any line of sight or unobstructed view), relatively secure and tamperproof, ability to withstand adverse environments (e.g. dirty, damp, extreme temperatures such as -40°C to +200°C) and less proneness to damage, ability to store and retrieve data, faster and real-time data identification and capture (e.g. in milli-seconds). Comparing such attributes with common construction processes, RFID is recognized as a suitable technology for construction industry applications and some such applications identified in this study include wireless concrete maturity monitor, steel component tracking, pipe tracking (e.g. in pipe collars, pipe supports), quality control and quality assurance (e.g. in tracking ready mixed concrete trucks, concrete cubes) [5], [6], [18], [19], [21], and [22].

2.2 *Advanced data communication platforms*

The viability of RFID technologies are further enhanced by appropriate integration with advanced data transfer mechanisms through suitable communication platforms. This section provides a basic summary of some such advanced communication platforms:

- Web-based communications – e.g. broadband connectivity, Internet Protocols (IP).

- Wireless communications – e.g. IrDA (using infrared based communications), IEEE 802.11b (using radio waves for high speed data transfer such as 11 Mbps)
- Mobile communications – e.g. General Pocket Radio Systems (GPRS), High Speed Circuit Switched Data (HSCSD), Global System for Mobile communications (GSM), 3rd Generation mobile systems (3G).

The applications potential of RFID systems are further enhanced through other conditions such as the affordability of personal computers (PCs) with high speed processors and large memory capacities, mobility-based handheld thin client devices such as Pocket PC, Personal Digital Assistant (PDA), and tablet computers, and developments in other technologies e.g. Global Positioning System (GPS), Geographical Information System (GIS), embedded or onboard sensors.

2.3 *Outline of related concerns and other issues*

Lack of standardization is a main limitation for RFID applications, e.g. various standards were followed by different manufacturers/ suppliers of tags and readers. The emergence of various standardization initiatives (e.g. Auto-ID Centre, EAN International, Uniform Code Council, EPCglobal, International Standards Organization) around the globe yielded useful standardization products such as Universal Product Code (UPC), European Article Numbering (EAN), Japanese Article Numbering (JAN), Electronic Product Code (EPC). However, appropriate data models should be developed separately for specific applications e.g. facilities management [23].

The RFID systems could be deemed as cost-prohibitive. The choice of the tags and readers depends on the specific requirements of applications. For example, the active tags, high frequency ranges, read/write facilities are normally more expensive. Some economy could be realized if the tags could be reused or any mass scale applications could be mobilized. The proneness of interference from steel objects could be overcome through right selection of tags and careful external mounting of the tags (e.g. in ceramic or glass casings). Similarly, potential interference from other radio signals could be eliminated by employing appropriate encoding (e.g. Manchester encoding), collision management and anti-collision protocols [20].

The security and privacy aspects of automatic identification and data capturing with RFID systems is another area major area of concern. For example, the life span of data stored in the tags could be a critical issue for privacy and security aspects and some focused attention is necessary [24], [25].

3. Automation with RFID applications in Tall Building Works

In general, the RFID based automation arrangements adopt relevant RFID applications as front-end data retrieval mechanisms that are integrated with appropriate backend database driven applications, decision support systems and robotics. Also, suitable sensor technologies, communication systems and security systems are also advantageously combined for such RFID based automation applications. This section outlines an overview of RFID based automation constructs for the construction industry.

3.1 *Automation in procurement and project delivery stages in construction projects*

RFID applications are useful in the procurement and project delivery stages such as materials management, equipment tracking, security, safety, engineering and design. A basic overview of some such usage in the construction industry and proposals for potential applications are summarized in this section.

3.1.1 *Procurement and management of materials*

Suitable RFID tags could be attached to most of the packed or unpacked materials used in the construction industry. Such tagging will potentially facilitate automation through storing and identifying and acquiring of appropriate datasets in common business case applications such as: (i) materials procurement related transactions e.g. ordering, billing and keeping track of shipments; (ii) warehouse/ storage related transactions e.g. storing, stock level checks and inventory control; and (iii) other construction related transactions e.g. materials identification (e.g. checking for the right materials, detecting counterfeits) and assembling (e.g. checking components)

3.1.2 *Tracking and management of equipments*

In general, active/ semi-active tags are embedded in the equipment for enabling automation through AIDC. Such tags will be useful for various purposes e.g. tracking the availability and location, tracing the supplier/ manufacturer information and warranty tracking, condition assessment, scheduling operation and maintenance, operating instructions, asset management and security. In some cases, the readers are also attached with the equipment to read the user/ operator tags so as to support the access control and safety measures. Also, such readers or tags will be useful for exact positioning of equipment (i.e. placing at precise locations) in the site locations during construction, maintenance/ refurbishment, demolition/ deconstruction tasks.

3.1.3 *Other potential applications in design, construction engineering and management*

By suitably combining the RFID tagging with appropriate instrumentation systems (e.g. sensors, detectors and gauges), some other useful automation provisions could be enabled in tall building works. For example, embedding RFID tags that hold design and layout information and combining with suitable instrumentation devices (such as smoke detectors and thermal sensors) could be useful for automating emergency evacuations in tall buildings (during construction or occupancy). Some other such conceptualized RFID applications in the design, construction engineering and management of tall building works are listed below:

- Active and passive condition assessment of structures e.g. for vibration, wind loads.
- Dynamic condition assessment of materials and fittings/ fabrics/ furnishings.
- Assembling or disassembling structural components and other decorative elements
- Quality control, quality assurance, inspections and tracking compliance records
- Human resource management and labor/ workforce management
- End-user applications such as safety, access control, security.

3.1.4 *Using prefabricated/ off-site manufactured products*

In addition to the earlier-mentioned procurement, logistics and material management applications (in 3.1.1) other RFID applications could be useful in identifying and assembling tasks (e.g. for components/ locations), which would minimize/ eliminate errors and rework. Also, the RFID technology could be applied for checking the assembly/ installation status.

If suitable, such RFID tags could be even reused economically. If so, the passive/ active/ semi-active tags should be suitably housed in removable casings attached to the pre-cast or prefabricated materials.

3.1.5 *Site safety*

Site safety problems and associated issues could be addressed by adopting appropriate innovative technological solutions [26]. Specific advantages of RFID technology such as non-contact and non-optical based data identification and transfer capacities could be harnessed for improving site safety in certain unsafe/ hazardous scenarios in tall building works. Some potential instances are: (i) RFID integrated robotics applications in hazardous working conditions, (ii) RFID tagging based integration of health and safety information in equipment and unsafe locations (e.g. for access control to certain risk groups of employees, and automatic warning systems to workers approaching unsafe locations in the site).

3.1.6 *Project Security*

Recently, with increasing terrorism and other security related vulnerability concerns, project security is emerging as one of the critical focus areas. The main aim of project security considerations during the project delivery stages is to assess and improve the physical security, personnel security and information security aspects [27]. RFID technologies will be useful in enhancing physical, personnel and information security aspects. For example, (i) RFID applications will be useful in detecting and deterring pilferage/ theft as well as destructions/ damages; (ii) RFID tagging could be used in personnel security practices such as screening, hiring/ firing, informing security requirements and controlling access at specific 'secure' locations, (iii) RFID tags are mostly secure against malicious attacks such as hacking.

For improving security in tall buildings, scanners for reading RFID tags could be installed in doorways/ entrances, corridors, elevators, escalators and staircases. Also, the RFID technology could be integrated with other surveillance systems (e.g. security cameras).

3.2 *Automation in facilities and asset management*

In general, buildings/ facilities management is mainly related to strategies for maintaining the quality and services of constructed facilities, whereas asset management covers the full life cycle management of assets in order to maximize their advantage.

Facility management functions include (i) maintenance and operation management related functions such as monitoring (e.g. functional performance, operational efficiency), maintenance/ alteration/ repair (e.g. preventive maintenance), and space management that include space use management (e.g. post-occupancy evaluations), (ii) property management related functions (e.g. building assessment, rent/ lease management), and (iii) other services related functions (e.g. hazardous materials, recycling, emergency planning) [23]. Thus, facilities management in tall buildings include (i) building related items such as structural elements (e.g. roofs, walls, columns beams, stairs), non-structural elements (e.g. windows, doors, ceilings, floors), exteriors and interiors (e.g. painting), utilities (e.g. gas, water), electrical and mechanical building services (e.g. lighting, fire protection), security systems (e.g. alarms, locks), and (ii) non-building related items such as external access, road, external works (e.g. slope protection).

Asset management could broadly cover management strategies of various aspects such as capital assets, fixed assets and tangible physical assets [28]. The physical assets could range from complex underground facilities and transportation networks to simple buildings. The operational definitions for facilities and asset management considered in this section cover only infrastructure management of tangible physical assets relating to tall buildings.

A pilot research in Canada revealed that the costs related to operation, maintenance, repair and renewal/ refurbishment of 'built environment' represents a major and rapidly growing cost to Canada and the USA [29]. Asset management and facilities management transactions are complex and the costs and impacts of poor management could be very high, e.g. health and safety risks, poor quality and image, physical deterioration of built assets and reduced property value as well as reduced revenues from those properties. Especially, in some prominent urban locations and rapidly developing economic zones, the asset value (in whole life perspective) will be much more than the original investment value (for both land and building). Thus, in several cases, the upfront costs of design and construction of built facilities may be small and less significant when compared to the asset maintenance costs throughout the life of the built facility (e.g. tall buildings in central business districts of urban cities).

The service life of different building components and engineering works could be different and moreover their service quality and residual life would vary according to the maintenance strategies adopted. For example, the utility fixtures, furniture fittings, external finishing, damp-proofing and water-proofing treatments will have lower life-spans than the main structural elements. The differential maintenance requirements pose further risks and challenges to the facilities and asset management strategies. Additional complexities include the end-user considerations such as security and serviceability requirements. Inefficient maintenance strategies, poor monitoring arrangements and lack of adequate control systems will severely impact on the whole-life value of the assets. Therefore, advanced technical solutions (e.g. decision support systems, real-time monitoring mechanisms) will facilitate the technical and financial decision making such as (i) 'what' and 'when' to maintain and (ii) 'whether' to repair or renew/ refurbish or replace.

Design revisions and changes/ variations are common in construction projects. The disconnected linkages between the design, construction and maintenance interfaces normally introduce additional difficulties for obtaining right information sets [30]. For example, the asset/ facility managers may find difficulties in obtaining the correct "as-built" information of building components and structural elements.

RFID technology will be useful for identification, tracking and management of various tangible physical asset items through automation of the data collection and accurate entry of collected data in appropriate database systems for further processing. For example, suitable functions in asset and facilities management include:

- Identification of location and other particulars of assets
- Condition assessment and determination of operations required to be carried out

- Reporting maintenance related particulars (e.g. maintenance history, warranties).

Thus, RFID applications will be useful in the following aspects of asset and facilities management in tall building works: (i) effective tracking of assets, (ii) less damages, errors and stoppages (e.g. from real-time condition assessments and checking the availability status), (iii) optimized maintenance and procurement transactions, (iv) better security, (v) enhanced safety, (vi) other benefits such as reduced disturbance to occupants, better communications and information exchange in the supply chain management transactions. For example, asset tracking functions such as tracking the movements of men, materials and machinery/ equipment for various purposes including anti-theft, correct usage confirmation, detecting unauthorized handling (i.e. only authorized persons could be allowed to handle certain items).

Essential information related to assets (e.g. model, owner particulars, maintenance history) could be directly stored in the RFID tags, which could be attached to the assets. In addition, suitable data capture (i.e. tag readers/ scanners) and communication infrastructure (e.g. antenna, wireless gateway) should be arranged. Also, suitable integration with other backend applications should be considered, e.g. with relational databases, data encryption architecture, decision support systems for maintenance management, and other enterprise management applications such as resource management, supply chain management, claims management, work scheduling, materials management and storage. Furthermore, remote linking with web-based integration and mobile communication platforms will be useful for further harnessing the potential benefits from such RFID applications.

3.3 *A framework for automation with RFID applications*

Figure 1 portrays the proposed framework for introducing RFID applications for automation in tall building works. According to the proposed framework, the business case for RFID should be developed initially. The business case development includes following transactions: (i) needs analysis and suitability analysis (e.g. identifying applications and processes for automation), (ii) risk assessment (e.g. technological risks, legal risks, operational risks), (iii) assessment of costs and benefits, (iv) impact assessment, (v) technology evaluation (e.g. selecting type of tags and readers) and planning integration with other technologies.

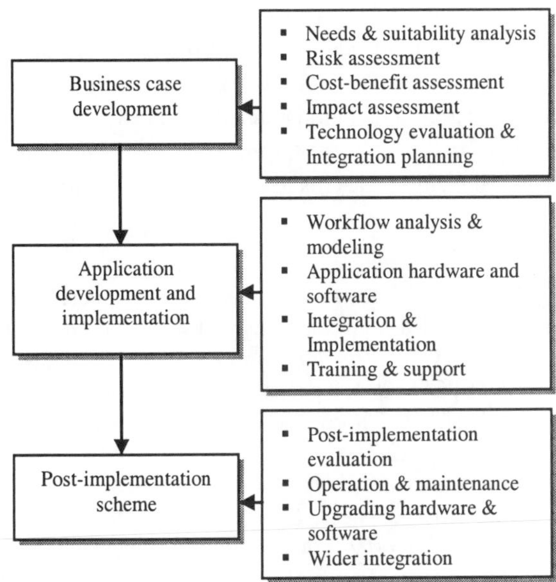

Figure 1. Proposed framework for automation with RFID applications

The application development and implementation functions include (i) workflow analysis and modeling the requirements (e.g. determining the product placement strategies by considering the interferences and shadows for RFID systems), (ii) selecting/ assembling the application hardware and developing the application software (e.g. front-end and backend applications), (iii) integration arrangements (e.g. with other wired and wireless systems) and implementation strategies, (iv) learning packages (e.g. online manuals), training arrangements and planning user support services. The recommended post-implementation scheme includes (i) post-implementation evaluation, (ii) operation and maintenance of services, (iii) assessment of needs based upgrading of hardware and software, and (iii) planning for additional/ wider integrations that may be subsequently considered (if necessary later).

4. Conclusions

Nowadays, numerous automatic identification systems are available for enabling automation of several processes in the construction industry. This pilot study reviewed an array of useful constructs for automation in tall building projects. Due to various advantages, the applications of RFID technologies are attracting greater attention in the construction industry. Although such advanced technologies may not completely replace the human interfaces, the applications should enhance the whole life value in more sustainable ways.

An array of initial concepts for various RFID applications in tall building works has been outlined in this paper. Further research is recommended to explore the application integration potentials and examine costs, benefits and associated problems of adopting such advanced technologies in the construction industry.

References

1. G. Elvin, Tablet and wearable computers for integrated design and construction, *Proceedings of the 2003 Construction Research Congress*, Hawaii, ASCE, CDROM, 9 pages.
2. M. Ward, T. Thorpe, A. Price and C. Wren, *ITcon* **9**, 297-311 (2004).
3. J. Sommerville and N. Craig, The digital revolution with pen and paper, *Proceedings of the 2005 CIB W92/TG23/W107 International Symposium on Procurement Systems – The Impact of Cultural Differences and Systems on Construction Performance*, Las Vegas, USA, 315-323 (2005).
4. Construction Industry Institute, *Bar code applications in construction*, Source Document **33**, CII, Austin, Texas, USA (1988).
5. E. J. Jaselskis, M. R. Anderson, C. T. Jahren, Y. Rodriguez, S. Njos, *Journal of Construction Engineering and Management*, **121(2)**, 189-196 (1995).
6. Construction Industry Institute, *Radio frequency identification tagging*, Research Summary **151-1**, CII, Austin, Texas, USA (2001).
7. A. Ashworth, *Cost Studies of Buildings* 2nd Edition, Addison Wesleyl Longman, Harlow (1994).
8. W. Tan, *Construction Management and Economics* **17**, 129-132 (1999).
9. D. H. Picker and B. D. Ilozor, *Construction Management and Economics* **21**, 107-111 (2003).
10. W. Tan, *Journal of Construction Engineering and Management* **129(4)**, 421-430 (2003).
11. P. F. Kaming, P. O. Olomolaiye, G. D. Holt, and F. C. Harris, *Construction Management and Economics*, **15(1)**, 83-94 (1997).
12. D. G. Proverbs, G. D. Holt, and P. O. Olomolaiye, *Construction Management and Econmics*, **17(1)**, 45-52 (1999).
13. D. G. Proverbs, G. D. Holt, and P. E. D. Love, *International Journal of Physical Distribution & Logistics Management*, **29(9)**, 659-675 (1999).
14. L. C. Bell, B. G. McCullouch, *Journal of Construction Engineering and Management*, **114(2)**, 263-278 (1988).
15. Construction Industry Institute, *Bar code standardization in industrial construction*, Source Document **47**, CII, Austin, Texas, USA (1989).
16. L. Blacky, *Journal of Construction Engineering and Management*, **116(3)**, 468-479 (1990).
17. L. E. Bernold, *Journal of Computing in Civil Engineering* **4 (4)**, 381-395 (1990).
18. E. J. Jaselskis and T. El-Misalami, RFID's Role in a fully integrated automated project process, 2003 Construction Research Congress, Hawaii, ASCE, CDROM, 9 pages (2003).
19. E. J. Jaselskis and T. El-Misalami, *Journal of Construction Engineering and Management* **129 (6)**, 680-688 (2003).
20. S. Shepard, *Radio Frequency Identification*, McGraw-Hill, New York (2005).
21. J. Yagi, E. Arai and T. Arai, *Automation in Construction* **14**, 477-490 (2005).
22. M. Clark, *Hong Kong Engineer*, **July 2004**, 18-21 (2004).
23. K. Yu, T. Froese, and F. Grobler, *Automation in Construction*, **9** 145-167 (2000).
24. D. Henrici and P. Muller, Tackling security and privacy issues in radio frequency identification devices, *Proceedings of the 2nd International Conference on Pervasive Computing - PERVASIVE 2004*, LNCS 3001, Springer-Verlag, Berlin, 219-237 (2004)
25. J. Saito, J. C. Ryou and K. Sakurai, Enhancing privacy of universal re-encryption scheme for RFID tags, *Proceedings of the International Conference on Embedded and Ubiquitous Computing* – EUC 2004, Aizu-Wakamatsu City, Japan, Springer-Verlag, Berlin, 879-890 (2004).
26. S. Rowlinson, Safety and Technology, *Construction Safety Management Systems*, Spon Press (Taylor & Francis Group), London, **Chapter 23**, 355-362 (2004).
27. S. R. Thomas, J. R. Sylvie, C. L. Macken *Best practice for project security*, Construction Industry Institute, USA **BMM 2004-10** (2004).
28. Scottish Government, *Asset management under best value*, Local Government in Scotland Act 2003 – Advisory Note, http://www.scotland.gov.uk (2003).
29. D. J. Vanier, Why industry needs asset management tools, *Proceedings of the APWA International Public Works Congress – NRCC Seminar Series on "Innovations in Urban Infrastructure"*, 11-25 (1999).
30. L. Mervi and H. Veli, *ITcon* **7**, 213-244 (2002).

CRITERION TO EVALUATE THE HIGH OFFICE BUILDING

WANG JIE

Department of Architecture, Zhejiang University, Zhijinggang Campus, Hangzhou, Zhejiang, China

ZHU HU FEI

Department of Architecture, Zhejiang University, Zhijinggang Campus, Hangzhou, Zhejiang, China

Nikken Design has built three high-rise office buildings in Pudong of Shanghai, that is the bank of China Mansion in Pudong, Shanghai Information Mansion and the City Bank Mansion in Shanghai. By studying these three buildings, this paper tries to examine the change of the tendency of the value for high office building in Shanghai these years, and then give the criterion to evaluate the high office building.

1. Change of the Social Circumstance Surrounding Office Buildings

1.1 *Establishment of economical logos*

After twenty years' of exploration and practice, the design of high office buildings in China has made great progress, with a number of high-rise office buildings emerging. The design of high office building is the counter-economy product. Endowing the building with additional value that exceeds pure economic efficiency and engineering technology out of many restrictions is the task of the designer.

In recent years, the owners of office buildings in Shanghai have began to pay attention to the consumption cost in the construction of buildings. Their cognition of economic buildings not only limits to the minimization of construction cost, but also includes the consumption cost of energy sources and resources in the process of using. Economic office buildings are not the cheapest buildings, but good-looking buildings with low cost in construction, management and manpower.

1.2 *Consider the value of office building from the building itself*

For a long time, the location has been regarded as the omnipotent key to a successful high office building, which has led to the emergence and virulent competition of many similar office building with identical products and target clients at the same area. The recent flooding of CBD is a reflection of this problem. While in developed countries in Europe, the exterior figures and the locations of top grade office buildings are not fixed. There are high-rise office buildings and multilayer office buildings, office buildings in the center of the city and garden-kind office buildings in the suburb.

With increasing of high office buildings year after year, good location with luxurious inside and outside fitment is no longer the predominance that the present high office building can flaunt. Only those high office buildings with integrated economic efficiency can remain invincible.

This situation urges the advent of the age of considering the value of office building from the building itself. There are several ways to raise the value of office buildings other than "location". It is more significant to consider the value of office building from the building itself, attaching more importance to the individuation, characteristic of office buildings and corresponding technical methods. From the viewpoint of asset value, the time has passed that the office building at the center of the city would involuntary be expected to be capital source. Shanghai has begun to enter the age that the office buildings are evaluated not only by location but also by the effect with using.

1.3 *Multi-appraising main bodies*

At this time, there are all kinds of main bodies who will calculate the value of the building on the side of their own position, including the owner of the office building, the leaser, the service administer, the insurance company, the visitors and so on. During the time of corporation-regroup and capital- appraisement, activities such as deals of the office building and stock converts always happen frequently. On such occasion, investors and appraising departments will calculate the value of the building with an impersonal view. So the design of office building now will be asked to response the various value and the clear design concept which can keep high estimate.

2. Increase the Value of Office buildings from Orientation and Service

2.1 *Create the developing value of office buildings by building orientation*

Promising office buildings are regarded to be located near railway stations, business center easy of access to airdromes and place close to finance department and shopping malls, etc. In the future, the predominance of new centers of big cities and large development areas with full city establishments and areas with plural high speed communication nets will increase. On the other hand, with the development of information communication method, office buildings near the outskirts will also increase.

The base of building planning is the orientation, such as the space economization, the agility of rentable units, the geniality of the room, automatization of the building, the continuance of zoology, the agility of building entrance, the management of parking lots, the security system and service range, etc.

For example, the three buildings in Pudong which designed by Nikken Design are all use transparent glass walls. At the foot of the buildings, there are full-open entrance halls which makes the residents and visitors deeply impressed by the buildings' entrance. They take the halls as the space to increase the overall levels. The City Bank Mansion in Shanghai has strong market competition power. It is consciously increasing the existence in high – building dense area of Pudong.

2.2 *Increase the value of office buildings by orienting the demands and optimizing the client interests*

The value per square meter lies on the use strategy of office buildings, so it is of great significance to make sure the real demands of clients. During this course, the investor must know the effect elements to establish relation with clients and also acquire valuable information so that they can meet the demand of clients and solve the practical problems while drafting out the lease contract.

The use planning of different office buildings brings the problem of how to economize cost. Some adopt the frequently-used solutions of building planning, and some use the cost-saving methods pertinent to some buildings. At the beginning of planning, ask potential lessee to participate in planning under the guide of target audience in the market. For example, at the early stage of the planning of City Bank Mansion in Shanghai, people of City Bank was asked to participate in the designing so that the designing of the building were timely adjusted according to the specific requirements of the user.

2.3 *Service*

The land agents may change their roles – from providers of builders and office buildings to problem solvers and providers of enterprise-pattern service. This change of roles calls for a new type of thinking mode and an appropriate method to take new challenges, from orientation by investor's benefit -oriented propaganda to service-guided building management.

All the service provided need to consider the competitive equipment from the cost and the interests of lessee, to consider the productiveness and economic dependence of the service and to raise money for the benefit of all lessees. For example, consider their integral purchasing power or consider the shared expensive basic facilities.

3. Ensure and Increase the Value of Office Buildings with Global Criterion

3.1 *Agility of space and facilities*

The agility of office building is to easily meet the demand of the owners and also meet the demand of building system aging. The building system is made up of many subsystems with different lifecycle. The shorter the life-span of the subsystems is, the more necessary is the change of technology. If possible, the subsystems should be independent of each other. Or even if they are connected, the joints should be easily connected.

Considering the structure integral, the moving efficiency and the agility of the building layout, the office building need integral big room, more depth and height floor. The burdens of floors are mainly from 300kg/sq.m to 500kg/sq.m. And the development of intensive information communication in the future, the electric power capacity and major facilities supporting office rooms such as the information communication facilities and power supply information wiring method should also be considered. According to the various demands of users, it is necessary to offer more room for perpendicular pipeline and office facilities layer.

Planning design must consider possible reconstruct and decoration, for example going close to future official handle style and so on, more and more working spots must be possibly changed from special function areas, or change reversly.

For example, Shanghai Information Mansion's vertical streamline and centralized equipment pipelines are set at the two head to conform with the narrow long site. Two-head core add super construct in Nikken Design style ensured security, supplied large space(43.2m×31.6m)and open none- pillar hall as high as 33.5metre.

3.2 Insurance of security

In the wave of internationalization and information, comparatively high security rank is needed, key-management is changing from inserting card to un-contact style. Safe and connected illuminating system,air-condition,and elevator control is essential, serious action as safe examination at the entrance commonly seen in the Europe and America is gradually adopted. As trust of working and data management is higher, electrical machine used at sudden time and insurance of install space is necessary.

Control the movement of the building in middle and small earthquake, use quakeproof structure to insure performance of the building in big earthquake. Then, leasehold corporations can perform as calamity monitor, can insure electrical source and water supply, can have high security and trust.

3.3 Considering comfortable environment

As the personalize of the working place, previous air-condition and illuminating system are changing from absolutely departed to floor-setting air-condition and intelligentized illumination. From the aspect of saving energy sources and ensuring the comfort of places around windows, it is praised that not only high-airproofed surface is insured but also open-close windows is used. Windows should use high hot-proof and double layered glass, individually controlled wall air-condition to alleviate hot burthen so comfortable environment is insured, also air-conditions in small units can be controlled by VAV(air- controllable setting).

3.4 Universal design

Henceforth, as the coming of aged society and acceleration of disabled engagement, required elevators usable to wheelchair, not only wheelchair lavatory should be set near the entrance, but also lavatories and doors for the disabled should be designed on the office floors.

4. Epilogue

Shanghai, a flourishing city where high buildings stand in great numbers and merchants gather, has become hotspot area to invest in top-grade office buildings. At the same time, the high office buildings in Shanghai, after the high-speed developing period, are stepping toward sense and maturity. People have been continuously thinking about the design logos and evaluation criterion of high office buildings. By summarize the three buildings designed by Nikken Design, this paper examine how to evaluate the value for high office building in Shanghai these years.

References

1. Tong Hui, Exploration on idea and Principle of Architecture Design from the View of Economics, <New Architure>, 2002, 2.
2. Yoshinobu Miyamoto,Design Conception of an Architect:A Record of the bank of China Mansion in Pudong,<Time Architecture>, 2002, 4.
3. Thomas Arnold, <A Design Manual Office Building>, the publisher of DaLian Science, 2005, 5.

Composite Structure

DESIGN AND RESEARCH FOR A TALL BUILDING OF CONCRETE FILLED SQUARE STEEL TUBE

JIEMIN DING

Architectural Design & Research Institute of Tongji University, 1239 Siping Road Shanghai 200092, China

SI CHAO

Architectural Design & Research Institute of Tongji University, 1239 Siping Road Shanghai 200092, China

JIANBIN CHEN

Architectural Design & Research Institute of Tongji University, 1239 Siping Road Shanghai 200092, China

The concrete filled square steel tube structures (CFST) become more and more popular nowadays. In this paper the design and research for a tall building of this kind was introduced. On the basis of calculating and adjusting of mass and stiffness for many times, braced-frame system was adopted to reduce the torsion effect brought by architectural irregularities of plan and elevation. The modal analysis, response spectrum analysis and time history analysis was carried out by several software. The period, displacement and story shearing force etc. were obtained and compared with each other. Relative design methods and research conclusions could be referenced to similar projects.

1. Introduction

The tall building of concrete filled square steel tube structures become more and more popular because of the its regular shape, simple joint and well seismic performance. This paper discussed the design, calculation and analysis of such a project.

The building height is 100m, including one story basement and 27 stories above ground. It was braced-frame system made of concrete filled square steel tube, steel beam and steel tube brace.

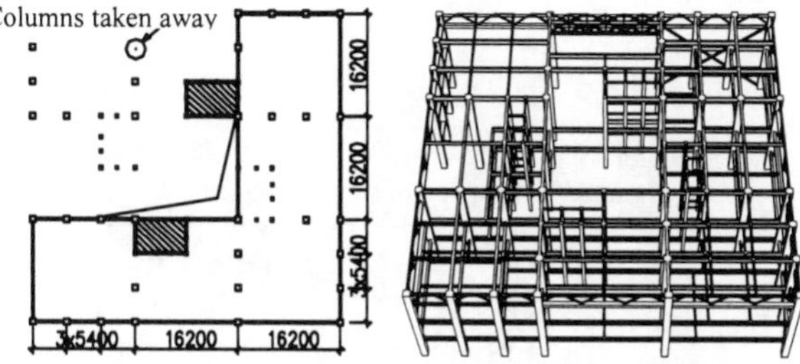

Figure1. Typical plan and structural model

The building was "L" shape and the floor rotated 90 degree every three stories. Figure.1 illustrated the typical plan and structural model. A 2m high device story was set during every rotation. Web members were set in the steel truss device story in order to strengthen connection between components. The standard story height was 4m and device story height was 6m.

The primary members were shown in table 1. Some columns were taken away. According to the code [1], this building was an irregular structure.

Table1. Primary members (mm).

Frame column	900×900×20×20 500×500×14×14 400×400×14×14	Primary girder	HN 700×300×13×24 HN 500×200×10×16 HN 400×200× 8×13
Braced-frame	Φ 299×16	Minor girder	HN 350×175×7×11 HN 250×125×6×9
Ring truss web members	Φ 180×12 Φ 140×10	Floor slab	YX50-245-735

2. Control of Torsion Effect

In the initial design stage, inner elevator and stairwell were made into reinforced concrete shearing force wall core, which form the frame-wall system with the peripheral columns.

But through calculation, the torsion period became the 1st structure period, which did not meet the requirement of code [1].

After adjusting wall thickness, location, number and some column's sections, it still couldn't satisfy the requirement that the ratio of 1st torsion period T_t of structure to the 1st self-vibration period T_1 of structure is not larger than 0.9 for A class tall building and not lager than 0.85 for B class tall building.

At the same time the pure frame system were ever adopted, but it did not satisfy the story drift (1/300) requirement of code [2].

Summarizing the above calculation results, the frame-braced system was adopted. The bracing was shown in figure 1, located orderly at the building periphery and rising spirally. The outside bracing can obtain bigger arm of force and effectively transfer shearing force through axis force.

3. Finite Element Calculation and Analysis

3.1 *Modal analysis*

First, the structure mode analysis was completed. Two typical modes were shown in figure2. The results from different software were given in table 2.

The vibration shape of the building was unitary and regular. All three software's results satisfied the period ratio requirement of the code [1].

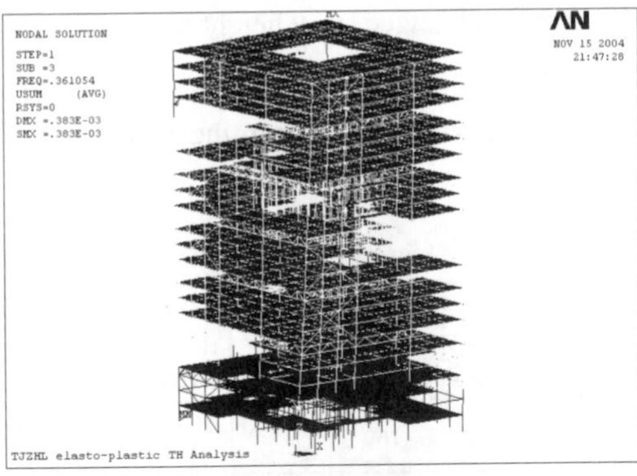

(a) Mode 3: 1st order torsion

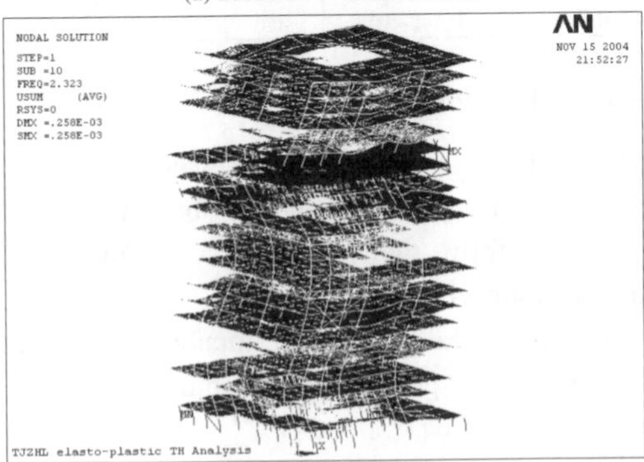

(b) Mode 10: 4th order translation

Figure2. Typical modes of the building

Table 2. Comparison of free-vibration period (s).

Mode	ANSYS	ETABS	SATWE	Vibration shape
1	3.3947	3.2678	3.4198	1st translation (x)
2	3.3392	3.2411	3.3923	1st translation (y)
3	2.7599	2.6304	2.6998	1st torsion
4	1.1526	1.1092	1.1464	2nd translation (x)
5	1.1089	1.0741	1.1136	2nd translation (y)
6	0.8836	0.8462	0.8672	2nd torsion
7	0.6143	0.6033	0.6181	3rd translation (x)
8	0.6059	0.5972	0.6090	3rd translation (y)
9	0.5074	0.4919	0.4949	3rd torsion
10	0.4305	0.4469	0.4286	4th translation (x)
11	0.416	0.4237	0.4194	4th translation (y)
12	0.3658	0.3909	0.3484	4th torsion

3.2 *Response spectrum analysis*

The response spectrum analysis results of X and Y direction were shown in figure.3.

It can be seen there were six laddery transitions, which correspond to the six steel truss device stories.

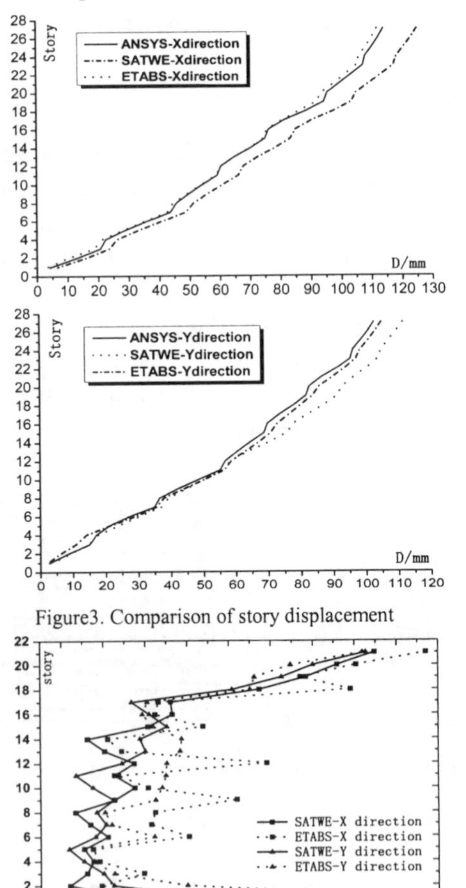

Figure3. Comparison of story displacement

Figure4. Story drift under frequently occurred earthquake action (horizontal axis: 1/story drift)

The maximum top floor displacement along X direction by ANSYS is 113.94mm and 102.53mm along Y direction. Compared with each other, results come from these software were correct.

The newly issued specification [3] for CFST structure prescribed that the limitation value of elastic story drift under frequently occurred earthquake action is not more than 1/300 and not more than 1/400 under wind loads. This project's story drift satisfied these requirements. Figure.5 gave the shearing force of the building story.

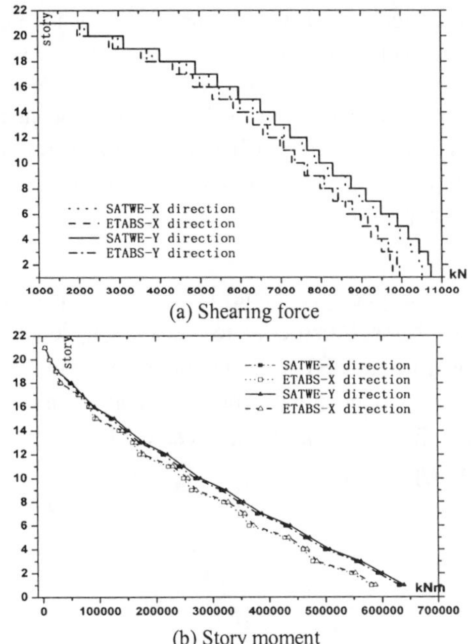

Figure5. Story shearing force and moment under frequently occurred earthquake action

3.3 *Time-history analysis*

According to the request of earthquake intensity of seven degree and IV field, El Centro earthquake wave, Pasadena wave and SHW2 wave (artificial wave) were adopted. Newmark -β method was use for numerical integral.

Figure.6 showed the maximum displacement curve of the building under basic earthquake action of seven degree (110gal) of El Centro earthquake wave. These results could be compared with those of response spectrum analysis. In this case, the maximum displacement appeared at southwest corner on the top floor.

Conclusions under other intensity, earthquake wave action cases were the same that the corner columns, steel truss web and chord members should be strengthened. These results were verified by the shaking table test of this project. And sections of upper stories' columns could be reduced.

The later design improved connection of steel truss to increase out of plane bending rigidity of chord members. The later results of elasto-plastic time history analysis still shown that lower parts of corner columns and some beams should be strengthened.

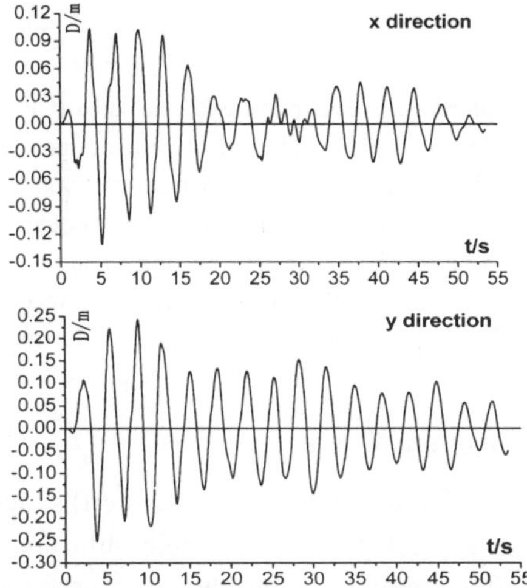

Figure6. Displacement time history curve under basic earthquake action of seven degree (110gal) of El Centro earthquake wave

4. Conclusions

This paper presented a design and analysis procedure of a tall building of concrete filled square steel tube.

On the basis of calculating and adjusting of mass and stiffness for many times, braced-frame system was adopted to reduce the torsion effect brought by architectural irregularities of plan and elevation. The period, displacement and story shearing force etc. were provided. And the seismic performance of the building was discussed.

Relative design methods and research conclusions could be referenced to similar projects.

References

1. Technical specification for concrete structures of tall building (JGJ3-2002 J186-2002).
2. Code for seismic design of buildings (Gb50011-201).
3. Technical specification for structures with concrete-filled rectangular steel tube members (CECS159: 2004).

EXPERIMENTAL STUDY ON CONCRETE-FILLED STEEL TUBE STRUCTURE SUBJECTED TO COMPRESSION AND BENDING IN HIGH RISE BUILDINGS

SI CHAO

Architectural Design and Research Institute of Tongji University, 1239 Siping Road Shanghai 200092, China

MINFANG WU

Department of Building Engineering of Tongji University, 1239 Siping Road Shanghai 200092, China

The application of concrete-filled steel tube (CFST for short) structure in high rise buildings becomes more and more extensive, and in the whole structure the design of joints takes a great part. In the article, several types of common joints are briefly discussed. Static experiment of compression and bending on joints with outer ring connections generally used in CFST structures has been undertaken. Through the experiment, the load-deformation curves and the stress distribution are both raised. Finite element method is also adopted to verify the feasibility of experiment, and the analysis indicates that the results of both methods agree well. According to the characteristic of compression bending test on joint and on the base of the experimental results, their value may emerge in the design of projects.

1. Introduction

In concrete-filled steel tube (CFST for short) structure, joints take a great part in the whole structure. The joints of CFST mean the connection of concrete-filled steel tubular columns and the beams. There are several kinds of common beams, and they are steel beams, general reinforced concrete beams and prestressed concrete beams. According to transferring loads, the commonly used CFST joints are outer ring connection, ring-beam connection, rooting connection and steel-transfixion connection. Table 1 shows the different types of joints.

Table 1 four types of CFST joints

Name		Graphics of connection
outer ring connection		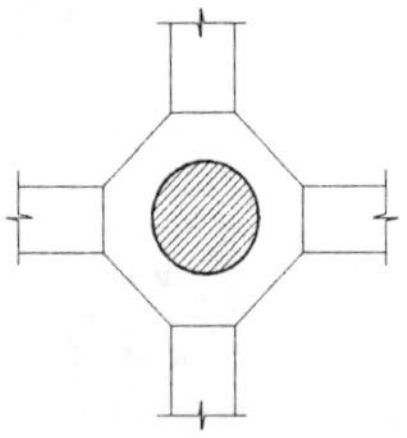
ring-beam connection	steel ring-beam connection	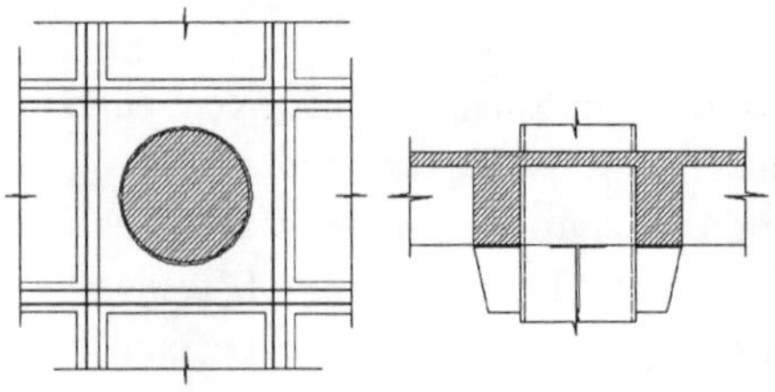

	firm ring-beam connection	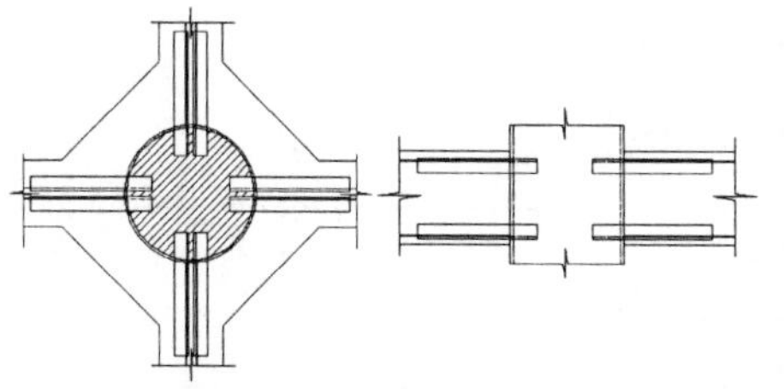
	RC ring-beam connection	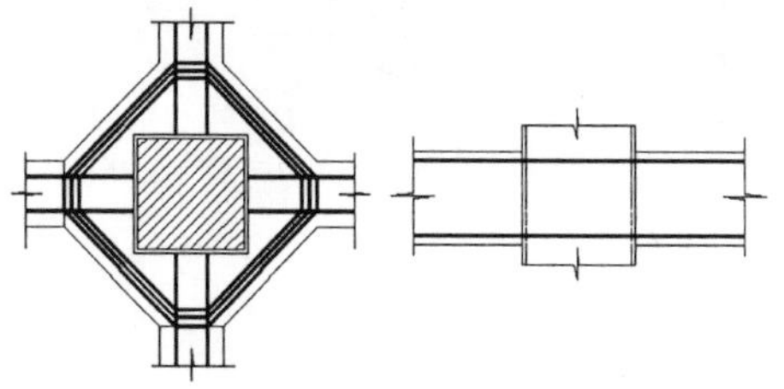
rooting connection		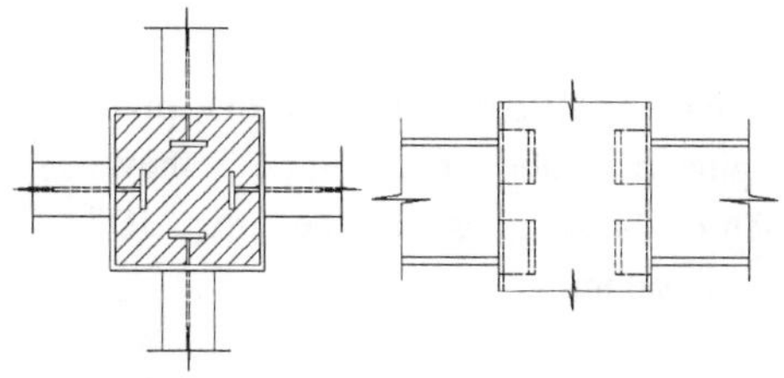
steel-transfixion connection		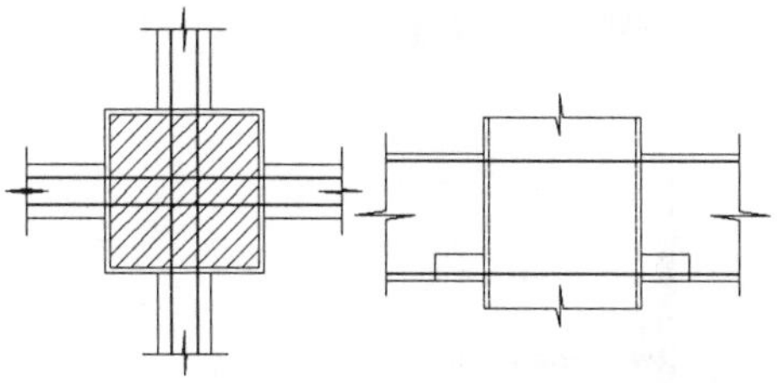

2. Experiment Design

In order to carry out the application of CFST structure in projects, static experiment of compression and bending on joints with outer ring connections in CFST structure has been undertaken in the article. And the bearing capacity of this kind of joint can be measured initially and approximately.

2.1 *Testing Sample Design*

In the testing samples, Q235 is used for steel, and C40 for the filling concrete. Two samples are tested in the experiment, and they are marked by JD1 and JD2. The difference of JD1 and JD2 is that inner cross board is added to JD2, and JD1 without. Besides, their dimensions and loads are absolutely the same. JD1 and JD2 are denoted by figure 1 and figure 2.

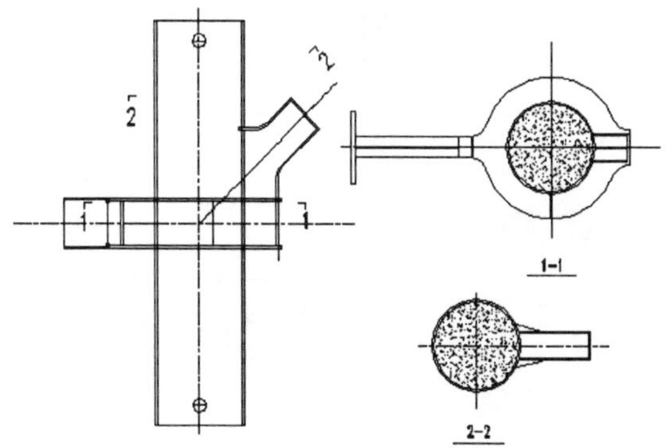

Figure 1 Sketch map of JD1

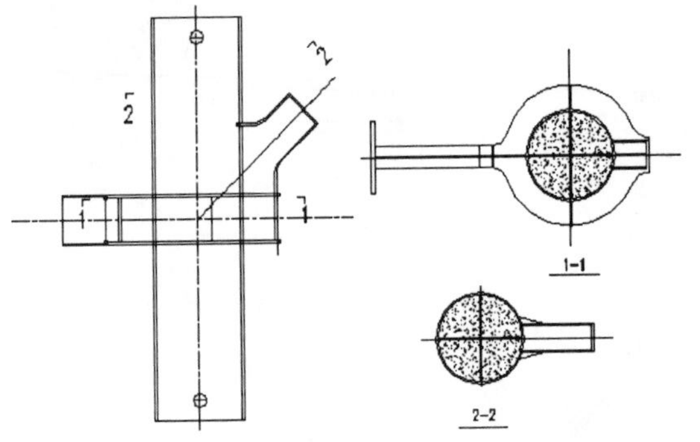

Figure 2 Sketch map of JD2

2.2 *Testing method and testing ontents*

2.2.1 *Loading device*

The experiment has been completed in the pivot national civil engineering laboratory. Figure 3 shows the loading sketch of the samples, and the letter "JD" stands for the testing sample.

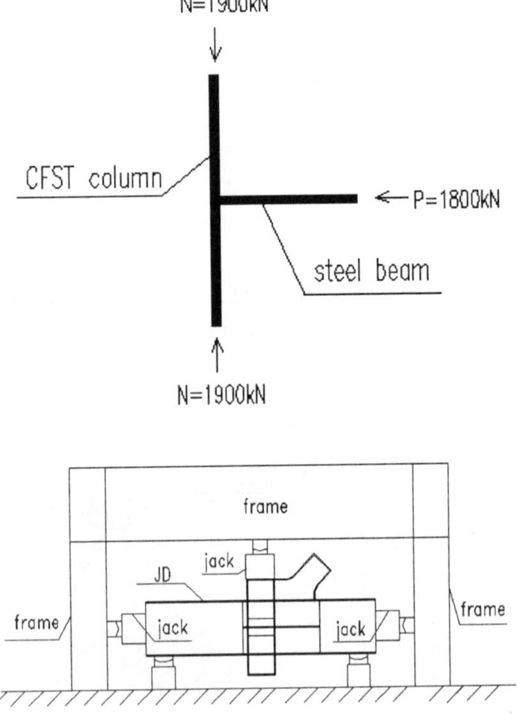

Figure 3 Loading sketch and loading device

2.2.2 *Testing contents*

The main purpose of the experiment is to measure the landscape orientation displacement Δ for short) of the joints in the landscape orientation force, and also to measure the axial strain (ε for short) of the joints in the axial force.

2.2.3 *Loading system*

Loads are divided to several steps. Firstly, axial force is inflicted to the CFST columns, and keep it at the maximum. Secondly, the landscape orientation force is added to the steel beam. Loading of both directions follows figure3.

2.3 *Main experimental results and analysis*

2.3.1 *Experimental phenomenon*

In the loading process, no crack is found in JD1 and JD2, and the possible reason is that although loads have reached the maximum supplied by the laboratory, the loads are still far from what can destroy the samples.

2.3.2 *Experimental results*

According to the measured data of JD1 and JD2, the axial force-strain curves and the landscape orientation force-displacement curves are worked out, and figure 4 and figure 5 give the results.

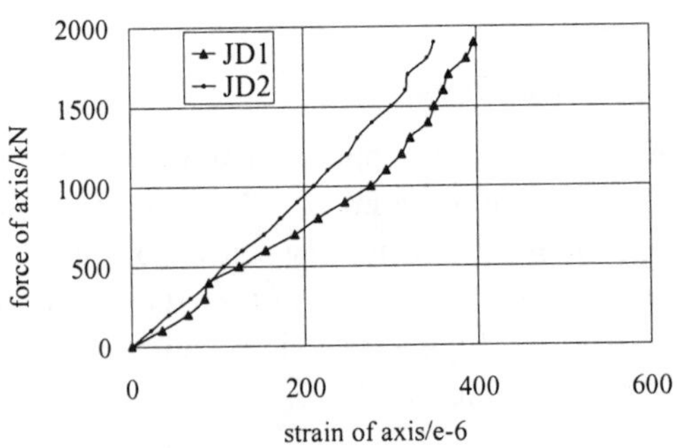

Figure 4 Axial N - ε curves

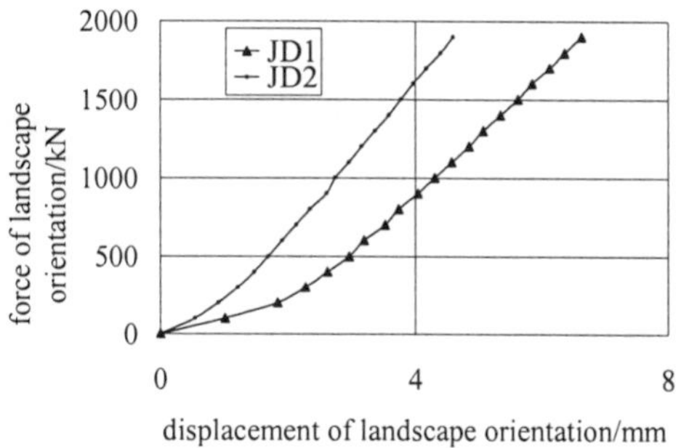

Figure 5 Landscape orientation P - Δ curves

From the N - ε and P - Δ curves, we know that JD1 and JD2 are both at their elastic process, and the displacements of JD2 is less than that of JD1, so CFST structure with inner cross board can enhance its stiffness.

3. Finite element method analysis

3.1 Modeling process of finite element method

Software ANSYS is used for finite element method (FEM for short). In setting up the model of the CFST joints, SOLID65 element is used for concrete material, and SOLID45 for steel material. The inputting material property data are the elastic modulus of concrete and steel E_c and E_s, intensities f_c and f_s, and ratio μ_s and μ_c, and their stress-strain curves. Through analysis of the models, N - ε and P - Δ curves are also figured out, and the results will be compared with the experimental results.

3.2 Analysis of FEM

3.2.1 Modeling elements

Figure 6 and figure 7 show element models of JD1 and JD2.

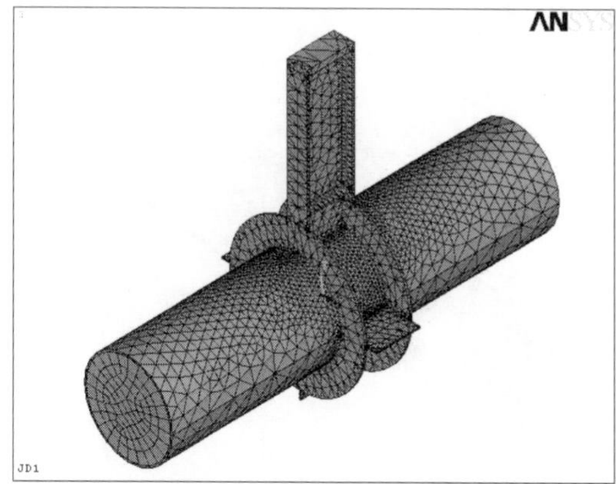

whole element

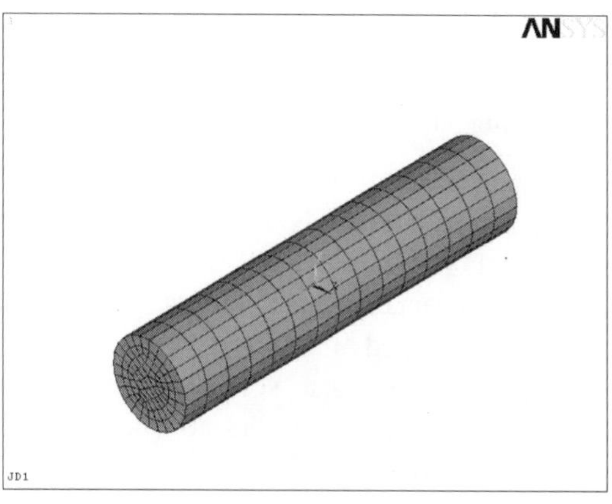

concrete element

Figure 6 element models of JD1

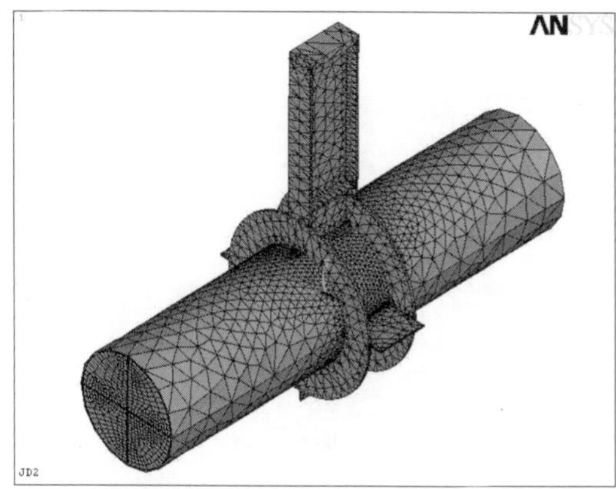

whole element

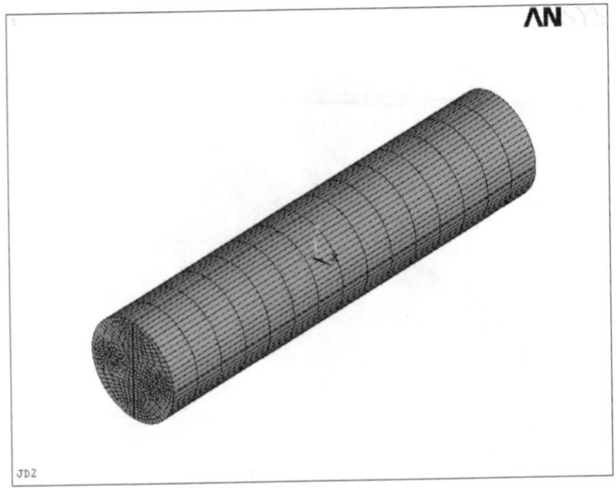

concrete element

Figure 7 element models of JD2

3.2.2 Loading method

Constraint loads of FEM is close to that of experimental method, and both ends of CFST columns are reamed. Force loads of FEM is almost the same as that of experimental method, that is to say, firstly, axial force is inflicted to the CFST columns, and keep it at the maximum, and secondly, the landscape orientation force is added to the steel beam. Loading of the models also follows figure3.

3.2.3 FEM results

According to the FEM results of JD1 and JD2, axial stress-strain curves and landscape orientation force-displacement curves are also worked out, as figure 8 and figure 9 show.

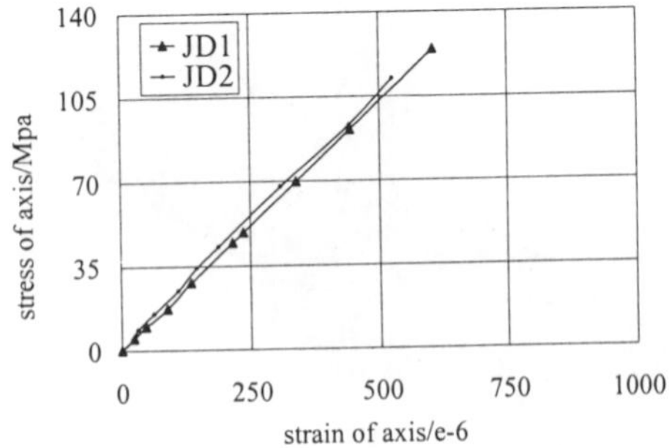

Figure 8 Axial σ - ε curves

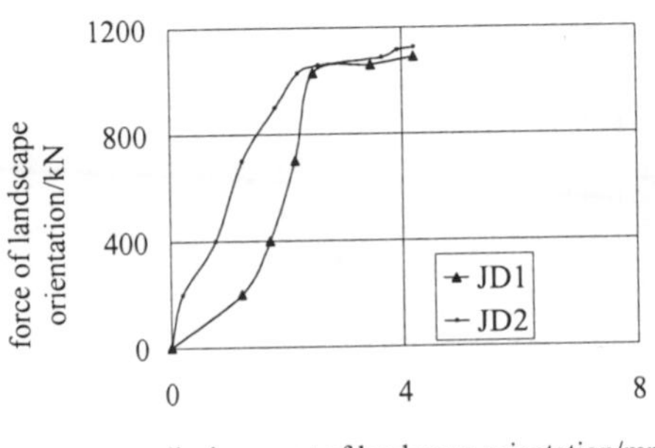

Figure 9 Landscape orientation P - Δ curves

3.2.4. FEM results analysis

From the σ - ε and P - Δ curves, we know that JD1 and JD2 are also at their elastic process, and the stiffness of JD2 is stronger than JD1, which agrees well with the experimental results.

4. Conclusion

1. The testing samples are selected from projects. The samples are still at their elastic process in the experimental loads which is bigger than the design loads, so the bearing capacity of the CFST structure is rather strong, which explains why the CFST structure is increasingly used in the high-rise buildings.
2. The results of experimental method and the finite element method match well, which demonstrates that both methods are appropriate to analyze CFST structure.
3. Both methods tell us that inner cross board can help CFST columns subjected to compression and bending enhance the stiffness, which means that bearing capacity of CFST structure can be improved just from the structure.

References

1. Tao Zhong, Han Lin-hai. Design calculations of load-bearing capacities for concrete filled square steel tubular members. Journal of Harbin University of Architecture. Vol.33-3, 2000, 6.
2. Zhong Shantong. Application and development in china of concrete filled tubular structure. Architecture Technology. Vol.32-2.
3. Zhong Shantong, Concrete-filled steel tube Structure, Tsing Hua University Press,2002
4. Jin Gang. Study on structural behavior of the new

type of connections between concrete-filled rectangular tubular column and steel beam. Dissertation submitted to Tongji University in conformity with the requirements for the degree of master of philosophy, Tongji University, 2005,3.

EXPERIMENTAL INVESTIGATION OF CONCRETE-FILLED HIGH STRENGTH STAINLESS STEEL TUBE COLUMNS FOR TALL BUILDING CONSTRUCTION*

BEN YOUNG

Department of Civil Engineering, The University of Hong Kong, Pokfulam Road, Hong Kong

EHAB ELLOBODY

Department of Structural Engineering, Faculty of Engineering, Tanta University, Tanta, Egypt

Concrete-filled steel tubes have been commonly used in the construction of tall buildings. Stainless steel tubes can be used to provide high corrosion resistance, ease of maintenance and construction as well as aesthetic appearance. This paper presents an experimental investigation of concrete-filled high strength stainless steel tube columns. A series of tests was performed to investigate the effects of the shape of stainless steel tube, plate thickness and concrete strength on the behaviour and strength of concrete-filled cold-formed high strength stainless steel tube columns. The high strength stainless steel tubes were cold-rolled into square and rectangular hollow sections. The concrete-filled high strength stainless steel tube specimens were subjected to uniform axial compression. The column strengths, load-axial strain relationships and failure modes of the columns were presented. The test strengths were compared with the design strengths calculated using the American specifications and Australian/New Zealand standards that consider the effect of local buckling using effective width concept in the calculation of the stainless steel tube column strengths. Based on the test results, design recommendations were proposed for concrete-filled high strength stainless steel tube columns.

1. Introduction

Concrete-filled steel tube columns have been commonly used in the construction of tall buildings. It provides high strength, high ductility, high stiffness and full usage of construction materials. In addition to these advantages, the steel tubes surrounding the concrete columns eliminate permanent formwork which reduces construction time. Furthermore, the steel tubes assist in carrying axial load as well as provide confinement to the concrete. In recent years, stainless steel tube members have become popular due to its high corrosion resistance, ease of construction and maintenance as well as aesthetic appearance. However, investigation of concrete-filled stainless steel tube columns are rarely found in the literature, especially using high strength stainless steel tubes.

Tests of concrete-filled carbon steel tube columns were conducted by Schneider [1], Uy [2, 3 and 4], Huang et al. [5], Han and Yao [6], Mursi and Uy [7], Liu et al. [8], Uy [9], Sakino et al. [10], Giakoumelis and Lam [11] and many other researchers. These tests were carried out on concrete-filled carbon mild steel and high strength steel tube columns using circular, square and rectangular hollow sections. Up to date, no test data are found in the literature on concrete-filled stainless steel and high strength stainless steel tube columns. The behaviour of stainless steel sections is different from that of carbon steel sections. Stainless steel sections have a rounded stress-strain curve with no yield plateau and low proportional limit stress compared to carbon steel sections. Recent experimental investigation of stainless steel columns without concrete infilled were conducted by Young and Hartono [12], Young and Liu [13], Liu and Young [14], and Gardner and Nethercot [15, 16], while experimental investigation of high strength stainless steel columns were conducted by Young and Lui [17, 18].

The local buckling of the steel tube has a considerable effect on the strength and behaviour of concrete-filled steel tube columns. Once the local buckling occurs, the steel tube would not be able to provide confinement to the concrete. The column capacity would be governed by local buckling failure mode. Uy [2] conducted a set of experiments for the local and post-local buckling behaviour of concrete-filled steel box columns, where the steel box sections are considered to be slender sections. A method for determining the required slenderness limits for inelastic local buckling in concrete-filled thin-walled box columns was presented. The results of the investigation were compared with

Australian and British standards that consider local buckling of the concrete-filled box columns. It was concluded that the Australian Standard method for determining the post-local buckling behaviour based on the effective width concept for a column under pure compression is suggested for use in an ultimate strength analysis. However, there are no clear guidelines by codes of practice for designing concrete-filled stainless steel tube columns for both slender and non-slender sections. Available design rules for concrete-filled stainless steel tube columns are limited to the general design rules specified in the American specifications [19, 20] and Australian standards [21, 22] for cold-formed stainless steel and concrete structures.

An experimental investigation was performed in this study to investigate the behaviour and strength of concrete-filled high strength stainless steel tube columns. A series of tests was conducted on square and rectangular hollow sections using different concrete strengths. The columns were subjected to uniform axial compression. The dimensions of the stainless steel tube cross-sections were chosen so that it includes both compact and relatively slender sections. The test strengths were compared with the design strengths calculated using the general design rules specified in the American specifications [19, 20] and Australian/New Zealand standards [21, 22] for stainless steel and concrete structures. The material properties of the stainless steel tubes were obtained using tensile coupon tests and stub column tests.

2. Experimental Investigation

2.1 *Test specimens*

Concrete-filled high strength stainless steel tubes using square hollow sections (SHS) and rectangular hollow sections (RHS) were tested. The tubes were cold-rolled from flat strips of duplex and high strength austenitic stainless steel material. The test program consisted of five test series that included two series of concrete-filled SHS tubes (SHS1 and SHS2) and three series of concrete-filled RHS tubes (RHS1, RHS2 and RHS3). The nominal section sizes ($D \times B \times t$) of series SHS1, SHS2, RHS1, RHS2 and RHS3 are 150×150×6, 150×150×3, 200×110×4, 160×80×3 and 140×80×3 mm, respectively, where D is the overall depth, B is the overall width and t is the plate thickness in mm, as shown in Fig. 1. The measured inner corner radius (r_i) are 5.3, 4.6, 9.1, 6.3 and 7.0 mm for series SHS1, SHS2, RHS1, RHS2 and RHS3, respectively. The measured average overall depth-to-thickness (D/t) ratios are 25.8, 54.1, 48.9, 55.4 and 45.3 for the concrete-filled tube columns of series SHS1, SHS2, RHS1, RHS2 and RHS3, respectively. The lengths (L) are 450, 450, 600, 480 and 420 mm for the concrete-filled tube columns of series SHS1, SHS2, RHS1, RHS2 and RHS3, respectively. The lengths were chosen so that the length-to-depth ratio (L/D) is generally remained at a constant of 3 to prevent overall column buckling. The specimens were tested using nominal concrete cylinder strengths of 40, 60 and 80 MPa. The measured dimensions of the concrete-filled high strength stainless steel tube column test specimens are shown in Table 1.

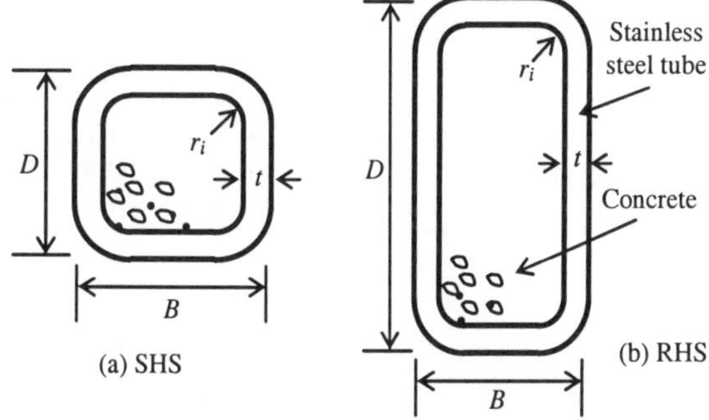

Figure 1. Definition of symbols for concrete-filled square and rectangular hollow section specimens.

The concrete-filled high strength stainless steel tube column test specimens are labeled such that the shape of stainless steel tube and concrete strength could be identified from the label. For example, the label "SHS1C40" defines the specimen has square hollow section that belonged to test series SHS1, and the letter "C" indicates the concrete strength followed by the value of concrete strength in MPa (40 MPa). Five tests were conducted on high strength stainless steel tube columns without concrete infilled that are denoted by "C0" in each series.

2.2 *Material properties of stainless steel tubes*

The material properties of the high strength stainless steel tube specimens were determined by tensile coupon tests as well as stub column tests. The tensile coupon test specimens were taken from the center of the plate at 90° from the weld in the longitudinal direction of the flat portion of the untested specimens. The coupon dimensions conformed to the Australian Standard AS

Table 1. Measured test specimen dimensions.

Specimen	Depth	Width	Thickness		Inner Radius	Length		Stainless steel Area	Concrete Area
	D	B	t	D/t	r_i	L	L/D	A_s	A_c
	(mm)	(mm)	(mm)		(mm)	(mm)		(mm^2)	(mm^2)
SHS1C0	150.6	150.2	5.855	25.7	5.3	601	4.0	3303	---
SHS1C40	150.5	150.5	5.834	25.8	5.3	450	3.0	3292	19250
SHS1C60	150.6	150.6	5.829	25.8	5.3	450	3.0	3293	19281
SHS1C80	150.5	150.5	5.839	25.8	5.3	450	3.0	3296	19247
SHS2C0	150.5	150.5	2.796	53.8	4.6	600	4.0	1623	---
SHS2C40	150.5	150.5	2.782	54.1	4.6	450	3.0	1615	20988
SHS2C60	150.5	150.5	2.780	54.1	4.6	450	3.0	1614	20989
SHS2C80	150.6	150.6	2.780	54.2	4.6	450	3.0	1615	21018
RHS1C0	196.2	108.5	4.010	48.9	9.1	600	3.0	2303	---
RHS1C40	197.1	109.5	4.060	48.5	9.1	600	3.0	2346	19088
RHS1C80	197.6	109.6	4.000	49.4	9.1	600	3.0	2317	19192
RHS2C0	160.1	80.8	2.869	55.8	6.3	600	3.7	1311	---
RHS2C40	160.5	81.3	2.920	55.0	6.3	480	3.0	1339	11637
RHS2C60	160.7	80.7	2.900	55.4	6.3	480	3.0	1328	11568
RHS2C80	160.6	80.4	2.900	55.4	6.3	480	3.0	1326	11514
RHS3C0	140.0	78.8	3.075	45.5	7.0	600	4.3	1263	---
RHS3C40	140.2	80.1	3.100	45.2	7.0	420	3.0	1282	9861
RHS3C60	140.2	80.1	3.100	45.2	7.0	420	3.0	1282	9861
RHS3C80	140.3	80.0	3.100	45.3	7.0	420	3.0	1282	9855

1391 [23] for the tensile testing of metals using 12.5 mm wide coupons of gauge length 50 mm. The initial Young's modulus (E_o) was measured as 194, 189, 200, 208 and 212 GPa as well as the measured static 0.2% proof stress ($\sigma_{0.2}$) was 497, 448, 503, 536 and 486 MPa for series SHS1, SHS2, RHS1, RHS2 and RHS3, respectively. The measured elongation after fracture (ε_f) based on a gauge length of 50 mm was 52, 52, 36, 40 and 47% for series SHS1, SHS2, RHS1, RHS2 and RHS3, respectively. The Ramberg-Osgood parameter (n) that describes the shape of the stress-strain curve (Ramberg and Osgood [24]) was 3, 4, 4, 5 and 6 for series SHS1, SHS2, RHS1, RHS2 and RHS3, respectively. The tensile coupon tests of the flat and corner portions are detailed in Young and Lui [17]. Table 2 shows the measured material properties of the flat portion of series SHS1, SHS2, RHS1, RHS2 and RHS3.

Five stub column tests of the cold-formed high strength stainless steel SHS and RHS tubes have been conducted to determine the material properties of the complete cross-section. The stub columns tests are detailed in Young and Lui [17]. The initial Young's modulus (E_o) was measured as 195, 198, 222, 213 and 214 GPa as well as the measured static 0.2% proof stress ($\sigma_{0.2}$) was 506, 250, 394, 390 and 441 MPa for series SHS1, SHS2, RHS1, RHS2 and RHS3, respectively. The Ramberg-Osgood parameter (n) was 3, 7, 4, 9 and 6 for series SHS1, SHS2, RHS1, RHS2 and RHS3, respectively. Table 3 shows the measured material properties obtained from the stub column tests for series SHS1, SHS2, RHS1, RHS2 and RHS3.

Table 2. Measured material properties from tensile coupon tests.

Test Series	Section $D \times B \times t$ (mm)	$\sigma_{0.2}$ (MPa)	σ_u (MPa)	E_o (GPa)	ε_f (%)	n
SHS1	150×150×6	497	761	194	52	3
SHS2	150×150×3	448	699	189	52	4
RHS1	200×110×4	503	961	200	36	4
RHS2	160×80×3	536	766	208	40	5
RHS3	140×80×3	486	736	212	47	6

Table 3. Measured material properties from stub column tests.

Test Series	Section $D \times B \times t$ (mm)	$\sigma_{0.2}$ (MPa)	σ_u (MPa)	E_o (GPa)	n
SHS1	150×150×6	506	570	195	3
SHS2	150×150×3	250	254	198	7
RHS1	200×110×4	394	418	222	4
RHS2	160×80×3	390	411	213	9
RHS3	140×80×3	441	444	214	6

2.3 *Material properties of concrete*

The material properties of concrete were determined from standard cylinder tests. The concrete cylinder dimensions and test procedures conformed to the American Specification [20] for concrete testing. The concrete were produced using commercially available materials with normal mixing and curing techniques. Twenty-five concrete cylinder tests were conducted. The mean compressive cylinder strengths of concrete at the time of the concrete-filled stainless steel tube column tests were determined as 46.6, 61.9 and 83.5 MPa with the corresponding coefficients of variation (COV) of 0.045, 0.111 and 0.026 for nominal concrete cylinder strength of C40, C60 and C80, respectively. Table 4 summarizes the measured concrete cylinder strengths and the number of tests.

Table 4. Measured concrete cylinder strengths.

Nominal concrete strength (MPa)	Mean value of measured concrete strength (MPa)	Coefficient of variation COV	Number of concrete cylinder tests
C40	46.6	0.045	10
C60	61.9	0.111	9
C80	83.5	0.026	6

Figure 2. Test setup of concrete-filled stainless steel tube specimens.

2.4 *Instrumentation*

Three transducers (LVDTs) were used to measure the axial shortening of the columns, as shown in Fig. 2. The axial shortening was obtained from the average readings of the LVDTs for each column specimen. Seven strain gauges on one side and two strain gauges on the opposite side of the columns were used to monitor the axial strain and plate deformation of each specimen. All strain gauges were placed on the outside surface of the high strength stainless steel tubes. The top and bottom gauges were located at a distance of 35 mm from both ends of the columns. The middle strain gauges were positioned at mid-height of each column equally spaced in line on one face of the columns. A date acquisition system was used to record the LVDTs, strain gauges and applied load readings at regular intervals during the tests.

2.5 *Column testing procedure*

The concrete-filled high strength stainless steel tube column tests are shown in Fig. 2. A 4600 kN capacity of 815 Rock Mechanics Test System servo-controlled hydraulic testing machine was used to apply compressive axial force to the column specimens. Prior to testing, both ends of the columns were strengthened by steel brackets so that the failure would not occur at the ends and the column strength would not be influenced by end effects. The ends of the columns were cast in plaster with large thickness of rigid steel plates being used. The load was applied to the columns in axial uniform compression over the concrete and steel tube. The values of the longitudinal strain gauges near the four corners of the columns were checked carefully while loading is applied to the columns within the elastic limit. This is to ensure a concentrically load applied to the columns. Displacement control was used to drive the hydraulic actuator at a constant speed of 0.5 mm/min. This allowed the tests to be continued in the post-ultimate range.

3. Test Results

The test strengths and load-axial strain relationships were measured for each column specimen. The test strengths (P_{Test}) of the concrete-filled high strength stainless steel tube columns of the SHS and RHS are shown in Tables 5 and 6, respectively. The load-axial strain relationships of the concrete-filled tube columns for Series SHS1 are shown in Fig. 3. The average strains near the four corners of the columns were plotted. It can be seen that the ductility of the columns generally decreases with the increases of concrete strength.

Generally local buckling failure mode of the high strength stainless steel tubes was observed for specimens with relatively slender sections for series SHS2, RHS1, RHS2 and RHS3. Concrete crushing failure mode

Table 5. Test strengths and design strengths of concrete-filled square hollow section columns.

Specimen	P_{Test} (kN)	$P_{ACI/ASCE-1}$ (kN)	$P_{ACI/ASCE-2}$ (kN)	$P_{ACI/ASCE-3}$ (kN)	$\frac{P_{Test}}{P_{ACI/ASCE-1}}$	$\frac{P_{Test}}{P_{ACI/ASCE-2}}$	$\frac{P_{Test}}{P_{ACI/ASCE-3}}$
SHS1C0	1927.4	1641.4	1671.1	1927.4	1.17	1.15	1.00
SHS1C40	2768.1	2399.4	2429.1	2689.9	1.15	1.14	1.03
SHS1C60	2972.0	2651.2	2680.9	2941.9	1.12	1.11	1.01
SHS1C80	3019.9	3004.4	3034.0	3293.5	1.01	1.00	0.92
SHS2C0	408.6	497.4	343.2	408.6	0.82	1.19	1.00
SHS2C40	1381.5	1324.5	1171.7	1239.9	1.04	1.18	1.11
SHS2C60	1620.0	1596.9	1444.3	1512.9	1.01	1.12	1.07
SHS2C80	1851.3	1984.4	1831.8	1900.4	0.93	1.01	0.97
Mean	---	---	---	---	1.03	1.11	1.01
COV	---	---	---	---	0.114	0.064	0.058

Table 6. Test strengths and design strengths of concrete-filled rectangular hollow section columns.

Specimen	P_{Test} (kN)	$P_{ACI/ASCE-1}$ (kN)	$P_{ACI/ASCE-2}$ (kN)	$P_{ACI/ASCE-3}$ (kN)	$\frac{P_{Test}}{P_{ACI/ASCE-1}}$	$\frac{P_{Test}}{P_{ACI/ASCE-2}}$	$\frac{P_{Test}}{P_{ACI/ASCE-3}}$
RHS1C0	957.0	965.2	808.1	957.0	0.99	1.18	1.00
RHS1C40	1627.2	1742.6	1581.9	1713.1	0.93	1.03	0.95
RHS1C80	2180.0	2329.5	2172.1	2319.2	0.94	1.00	0.94
RHS2C0	537.3	542.2	423.0	537.3	0.99	1.27	1.00
RHS2C40	881.5	1018.3	895.7	998.2	0.87	0.98	0.88
RHS2C60	1014.5	1159.0	1038.0	1146.0	0.88	0.98	0.89
RHS2C80	1280.1	1366.6	1245.9	1354.5	0.94	1.03	0.95
RHS3C0	558.2	538.6	498.5	558.2	1.04	1.12	1.00
RHS3C40	1048.7	939.2	898.3	948.8	1.12	1.17	1.11
RHS3C60	1096.9	1067.4	1026.5	1077.0	1.03	1.07	1.02
RHS3C80	1258.8	1247.8	1206.9	1257.7	1.01	1.04	1.00
Mean	---	---	---	---	0.98	1.08	0.98
COV	---	---	---	---	0.076	0.087	0.066

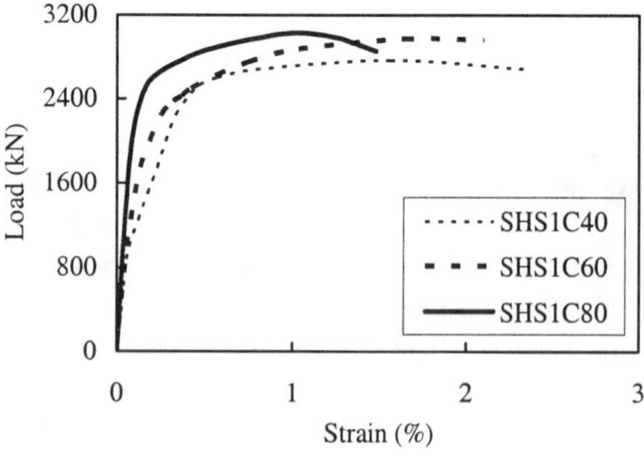

Figure 3. Load versus average strain curves for concrete-filled specimens of Series SHS1.

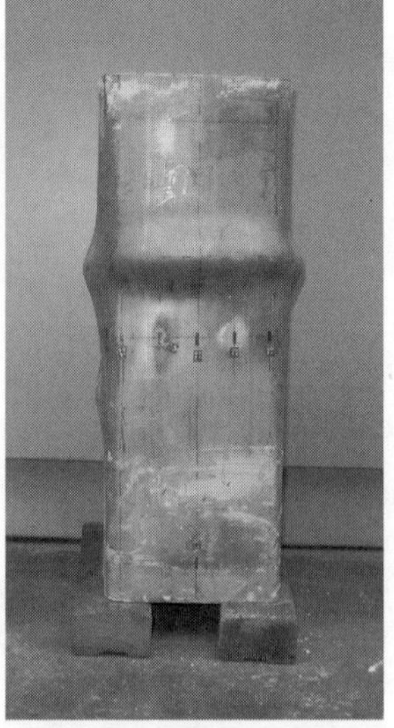

Figure 4. Failure mode of column specimen SHS1C60.

together with local buckling of the high strength stainless steel tubes for specimens with compact section for Series SHS1 was also observed. The failure mode of column specimen SHS1C60 having square hollow section of nominal dimensions 150×150×6 mm and nominal concrete cylinder strength of 60 MPa is shown in Fig. 4. For this specimen, concrete crushing as well as local buckling of high strength stainless steel tube were observed.

4. Design Rules

4.1 *General*

The concrete-filled cold-formed high strength stainless steel tube column test strengths (P_{Test}) are compared with unfactored design strengths predicted using the general guidelines specified in the American specifications [19, 20] and Australian/New Zealand standards [21, 22] for cold-formed stainless steel and concrete structures. These specifications consider the effect of local buckling of stainless steel tubes using effective width concept in the calculation of the design strengths. The American specifications and Australian/New Zealand standards for cold-formed stainless steel and concrete structures are generally using the same formula to calculate the design strengths. The design strengths ($P_{ACI/ASCE}$) for the concrete-filled stainless steel tube columns were obtained by determining the strength of the stainless steel tube (A_eF_n) using the specifications [19, 21] of cold-formed stainless steel structures as well as the strength of the concrete infilled ($0.85A_c f_c$) using the specifications [20, 22] of the concrete structures, as shown in Equation (1).

$$P_{ACI/ASCE} = A_e F_n + 0.85 A_c f_c \tag{1}$$

where, A_e is the effective cross-section area of the stainless steel tube that is equal to the full cross-section area (A_s) in the case of compact cross-sections and less than A_s in the case of slender cross-sections due to the effect of local buckling, F_n is the flexural buckling stress determined according to Sections 3.4.1 and 3.4.2 of the American Specification [19] and Australian/New Zealand Standard [21], respectively, A_c is the area of concrete and f_c is the measured concrete cylinder strength. In the calculation of the strength of the stainless steel tubes, it was found that the values of the design stresses (F_n) are equal to the yield stresses (F_y) for all columns. This is due to the short column lengths. In this study, the yield stress (F_y) is taken as the measured static 0.2% proof stress ($\sigma_{0.2}$).

The American and Australian/New Zealand specifications use the same Winter effective width equations in calculating the effective area (A_e) of stainless steel tube cross-sections. In the calculation of buckling stress (F_n), the design rules specified in the American Specification are based on the Euler column strength that requires the calculation of tangent modulus (E_t) using an iterative design procedure. The design rules specified in the Australian Standard adopts the Euler column strength or alternatively the Perry curve that needs only the initial Young's modulus (E_o) and a number of parameters to calculate the design stress. In this study, the Euler column strength method is used in the calculation of the design strengths for the Australian/New Zealand Standard. Hence, the design strengths calculated using Equation (1) is identical for both the American and Australian/New Zealand specifications. The columns were designed as concentrically loaded compression members. The term $0.85A_c f_c$ in Equation (1) represents the contribution of the concrete infilled in the calculation of the column design strengths.

Three design approaches were investigated in the calculation of the column design strengths using Equation (1). The calculated unfactored design strengths are denoted by $P_{ACI/ASCE-1}$, $P_{ACI/ASCE-2}$ and $P_{ACI/ASCE-3}$. The three design approaches are shown in the following Sections.

4.2 *Design approach 1 ($P_{ACI/ASCE-1}$)*

The design strengths ($P_{ACI/ASCE-1}$) are calculated using the material properties obtained from the tensile coupon tests for high strength stainless steel tubes in the calculation of the term A_eF_n in Equation (1). The measured material properties obtained from the tensile coupon tests are tabulated in Table 2. The calculation of the strength of the concrete infilled for the term $0.85A_cf_c$ in Equation (1) is carried out using the measured concrete cylinder strengths tabulated in Table 4.

4.3 *Design approach 2 ($P_{ACI/ASCE-2}$)*

In this approach, the calculation of the term A_eF_n in Equation (1) is carried out using the material properties obtained from the stub column tests for high strength stainless steel tubes. The measured material properties from the stub column tests are tabulated in Table 3. The design strengths calculated from Equation (1) in this case are denoted by $P_{ACI/ASCE-2}$. The strength of the concrete infilled is calculated in the same way as the design approach 1.

4.4 *Design approach 3 ($P_{ACI/ASCE-3}$)*

In this approach, the term A_eF_n in Equation (1) is replaced by the test strengths of the high strength stainless steel tubes without concrete infilled for each series of test. Hence, the strengths of the high strength stainless steel tubes are 1927.4, 408.6, 957.0, 537.3, and 558.2 kN for series SHS1, SHS2, RHS1, RHS2 and RHS3, respectively. Once again, the strength of the concrete infilled is calculated in the same way as the design approach 1. The design strengths calculated from Equation (1) in this case are denoted by $P_{ACI/ASCE-3}$.

5. Comparison of Test Strengths with Design Strengths

The comparison of test strengths (P_{Test}) with design strengths ($P_{ACI/ASCE-1}$, $P_{ACI/ASCE-2}$ and $P_{ACI/ASCE-3}$) is shown in Tables 5 and 6 for the concrete-filled high strength stainless steel tube columns of SHS and RHS specimens, respectively. The mean values of $P_{Test}/P_{ACI/ASCE-1}$, $P_{Test}/P_{ACI/ASCE-2}$ and $P_{Test}/P_{ACI/ASCE-3}$ ratios are 1.03, 1.11 and 1.01 with the corresponding coefficients of variation (COV) of 0.114, 0.064 and 0.058 for the SHS columns, as shown in Table 5. The mean values of $P_{Test}/P_{ACI/ASCE-1}$, $P_{Test}/P_{ACI/ASCE-2}$ and $P_{Test}/P_{ACI/ASCE-3}$ ratios are 0.98, 1.08 and 0.98 with the corresponding COV of 0.076, 0.087 and 0.066 for the RHS columns, as shown in Table 6.

The comparison of the test strengths with the design strengths for concrete-filled high strength stainless steel compact section tube columns of Series SHS1 has shown that $P_{ACI/ASCE-1}$ and $P_{ACI/ASCE-2}$ are conservative for the columns having nominal concrete cylinder strengths of 40 and 60 MPa, and accurately predicted the strength of the column having nominal concrete cylinder strength of 80 MPa. It is shown that $P_{ACI/ASCE-3}$ accurately predicted the test strengths of the columns having nominal concrete cylinder strengths of 40 and 60 MPa, but overestimated the strength of the column having nominal concrete cylinder strength of 80 MPa.

The comparison of the test strengths with the design strengths for concrete-filled high strength stainless steel relatively slender sections tube columns of series SHS2, RHS1, RHS2 and RHS3 has shown that $P_{ACI/ASCE-1}$ and $P_{ACI/ASCE-3}$ are unconservative for most of the columns having nominal concrete cylinder strengths of 40, 60 and 80 MPa, except for the columns SHS2C40, SHS2C60 and the columns in Series RHS3. On the other hand, the design strengths $P_{ACI/ASCE-2}$ that uses the material properties obtained from the stub column tests in the calculation of the strengths of the stainless steel tubes are conservative for most of the columns for the three concrete strengths, except for the columns RHS2C40 and RHS2C60 that are slightly unconservative. Hence, it is recommended to use the material properties obtained from stub column test in the calculation of the design strengths for concrete-filled high strength stainless steel tube columns.

6. Conclusions

An experimental investigation of concrete-filled cold-formed high strength stainless steel tube columns has been presented in this paper. Tests on concrete-filled high strength stainless steel square and rectangular hollow section columns were concentrically loaded. The overall depth-to-plate thickness ratio of the tube sections varied from 25.7 of compact sections to 55.8 of relatively slender sections. Different concrete cylinder strengths varied from 40 to 80 MPa were investigated. The column strengths, load-axial strain relationships and failure modes of the columns have been reported.

The test strengths were compared with the design strengths predicted using the American and Australian/New Zealand specifications for cold-formed stainless steel and concrete structures. The material properties of the high strength stainless steel tube specimens obtained from tensile coupon tests and stub column tests were used to calculate the design strengths. The test strengths of the high strength stainless steel tubes without concrete infilled were also used to calculate the design strengths. It is shown that the design strengths calculated using the material properties of stainless steel obtained from the tensile coupon tests as well as the test strengths of the high strengths stainless steel tubes without concrete infilled are generally unconservative for the tested concrete-filled stainless steel tube columns. It is demonstrated that the design strengths calculated using the material properties of stainless steel obtained from the stub column tests are generally conservative for both compact and slender sections with different concrete strengths. Therefore, it is recommended that the design rules in the American and Australian/New Zealand specifications for cold-formed stainless steel and concrete structures can be used for the design of concrete-filled cold-formed high strength stainless steel tube columns provided that the design strengths are calculated using the material properties of stainless steel obtained from stub column test.

Acknowledgments

The authors are grateful to Mr Kin-Chung Tsang and King-Hung Poon for their assistance in this project. The authors are also grateful to STALA Tube Finland for supplying the test specimens.

References

1. S. P. Schneider. Axially loaded concrete-filled steel tubes, *Journal of Structural Engineering,* ASCE, 124(10), 1125-1138 (1998).
2. B. Uy. Local and post-local buckling of concrete filled steel welded box columns, *Journal of Constructional Steel Research,* 74(1-2), 47-72 (1998).
3. B. Uy. Static long-term effects in short concrete-filled steel box columns under sustained loading, *ACI Structural Journal,* 98(1), 96-104 (2001).
4. B. Uy. Strength of short concrete filled high strength steel box columns, *Journal of Constructional Steel Research,* 57(2), 113-134 (2001).
5. C. S. Huang, Y. K. Yeh, H. T Hu, K. C. Tsai, Y. T. Weng, S. H. Wang and M. H. Wu. Axial load behavior of stiffened concrete-filled steel columns, *Journal of Structural Engineering,* ASCE, 128(9), 1222-1230 (2002).
6. L. H. Han and G. H. Yao. Influence of concrete compaction on the strength of concrete-filled steel RHS columns, *Journal of Constructional Steel Research,* 59(6), 751-767 (2003).
7. M. Mursi and B. Uy. Strength of concrete filled steel box columns incorporating interaction buckling, *Journal of Structural Engineering,* ASCE, 129(5), 626-639 (2003).
8. D. Liu, W. M. Gho and J. Yuan. Ultimate capacity of high-strength rectangular concrete-filled steel hollow section stub columns, *Journal of Constructional Steel Research,* 59(12), 1499-1515 (2003).
9. B. Uy. High-strength steel-concrete composite columns for buildings, *Structures & Buildings,* 156, 3-14 (2003).
10. K. Sakino, H. Nakahara, S. Morino and I. Nishiyama. Behavior of centrally loaded concrete-filled steel-tube short columns, *Journal of Structural Engineering,* ASCE, 130(2), 180-188 (2004).
11. G. Giakoumelis and D. Lam. Axial capacity of circular concrete-filled tube columns, *Journal of Constructional Steel Research,* 60(7), 1049-1068 (2004).
12. B. Young and W. Hartono. Compression tests of stainless steel tubular members, *Journal of Structural Engineering,* ASCE, 128(6), 754-761 (2002).
13. B. Young and Y. Liu. Experimental investigation of cold-formed stainless steel columns, *Journal of Structural Engineering,* ASCE, 129(2), 169-176 (2003).
14. Y. Liu and B. Young. Buckling of stainless steel square hollow section compression members, *Journal of Constructional Steel Research,* 59(2), 165-177 (2003).
15. L. Gardner and D. A. Nethercot. Experiments on stainless steel hollow sections - Part 1: Material and cross-sectional behaviour, *Journal of Constructional Steel Research,* 60, 1291-1318 (2004).
16. L. Gardner and D. A. Nethercot. Experiments on stainless steel hollow sections-Part 2: Member behaviour of columns and beams, *Journal of Constructional Steel Research,* 60, 1319-1332 (2004).
17. B. Young and W. M. Lui. Behaviour of cold-formed high strength stainless steel sections, *Journal of Structural Engineering,* ASCE (In press).
18. B. Young and W. M. Lui. Experimental Investigation of cold-formed high strength stainless steel compression members, Proceedings of the 6th International Conference on Tall Buildings, Hong Kong, China (2005).
19. ASCE. Specification for the design of cold-formed stainless steel structural members, American Society of Civil Engineers, SEI/ASCE-8-02, Reston, Virginia (2002).
20. ACI. Building code requirements for structural concrete and commentary, ACI 318-95, American Concrete Institute, Detroit, USA (1995).
21. AS/NZS. Cold-formed stainless steel structures. Australian/New Zealand Standard, AS/NZS 4673:2001, Standards Australia, Sydney, Australia (2001).
22. Australian Standards AS3600. Concrete structures, AS3600-1994, Standards Australia, Sydney, Australia (1994).

23. AS. Methods for tensile testing of metals, Australian Standard, AS 1391 – 1991, Standards Association of Australia, Sydney, Australia (1991).
24. W. Ramberg and W. R. Osgood. Description of stress strain curves by three parameters, Tech. Note No 902, National Advisory committee for Aeronautics, Washington, D.C. (1943).

STRUCTURAL ANALYSIS OF THE APARTMENT TOWER OF FORTUNE CENTER PHASE SITED IN BEIJING

TONG-YI ZHANG ZI-QIANG XIAO LIANG LI RUI ZHANG BIN LUO

IPPR Engineering International, Beijing 100089 P.R.China

The tall apartment building of Fortune Center phase□sited in Beijing is 193.4m height which structure system is mixed frame-corewall structure. In this paper, some analysis methods used to solve the problems presented because of the composite structure beyond the code-specification are introduced. The design criteria and analysis results in this paper can be referenced in the design of similar structures.

1. Introduction

The engineering of Fortune Center phase□which includes the hotel with 100m structure height, the tall apartment with structure height being 193.4m and the shop as the podium is sited in Beijing as Fig.1.

The tall apartment building with 55 stories is the tall building beyond the code-specification which structure limited height is 150m for the mixed frame-corewall structure system according to the codes of China. The total breadth of the apartment building is 52.249m.

Figure 1. Fortune center phase□

In this paper, some analysis methods used to solve the problems presented according to the demand of the beyond the code-specification structure are introduced. The design criteria and analysis results in this paper can be used in the design of similar structures.

2. Structure System and Elements

The structure system of the apartment is mixed frame-corewall structure in which the tube is concrete structure and the frame is composed of SRC column and steel beam.

The structure above the ground level is as Fig.2, and the standard story under the floor 30 is as Fig.3. The section of the concrete wall and the frame elements is shown in Table 1.

Table 1. Section of the elements(mm)

Story	Concrete strength	main wall thickness	Column	Beam
1-3	C55	1000,500	1000×800, 800×800, 600×1350	700×1250, H200×450
4	C55	900,500	1000×800, 800×800, 600×1350	700×900, H200×450
5-7	C55	800,500	800×800, 600×1350	H200X550 H200X400
8-12	C50	800,500	800×800, 600×1350	H200×650 H200×550 H200×400
13-20	C50	800,700, 500	800×800, 600×1350	H200×550 H200×400
21-26	C50	700,500	700×800, 600×1350	H200×550 H200×400
27-39	C45	600,500	600×800, 600×1350	H200×550 H200×500 H200×400
40-49	C40	500	600×800, 500×1350	H200×500 H200×400
50-55	C40	400,500	600×800, 500×1350	H200×500 H200×400

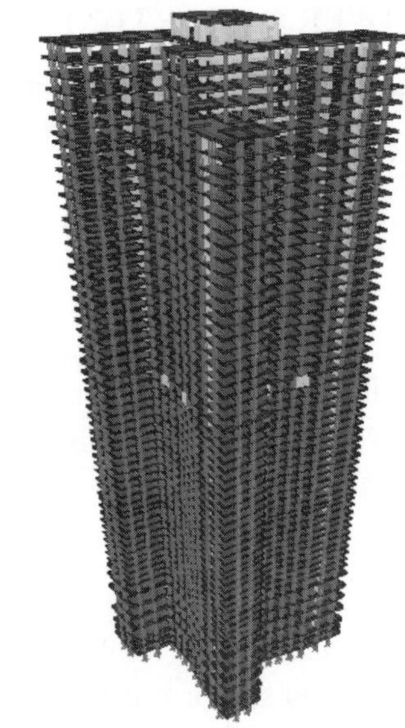

Figure 2. 3D structure analysis model

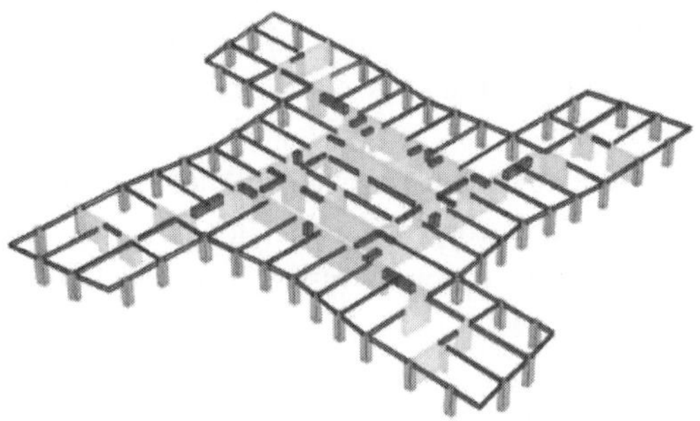

Figure 3. 3D model of the standard story

The story with transfer members or belt members was designed in utility stories which are in the 30th and 49th stories.

The steel transfer belt truss was designed for the 49th story because of the steel columns being above the truss and the axis force of the steel columns being small. Two transfer structure forms for the 30th story can be used as shown in Fig.4. One is transfer truss and the other is transfer brocket. The whole structure analysis was carried for the different transfer structures. The deflection and the displacement of the two kinds structure are shown in Fig.5.

The transfer brocket was used in the design considering the following reason. One is the stiffness of the whole structure is similar, on the other hand, the inter force of the elements which include shear walls, columns, beams and transfer elements of the transfer story and the adjacent stories is very complex for the transfer truss structure while that is very simple for the transfer brocket structure.

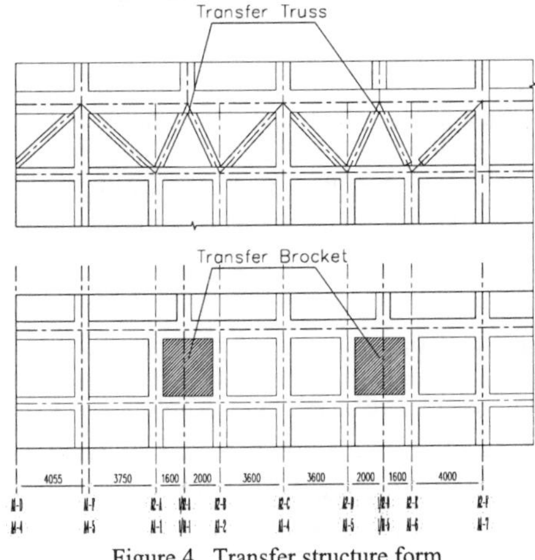

Figure 4. Transfer structure form

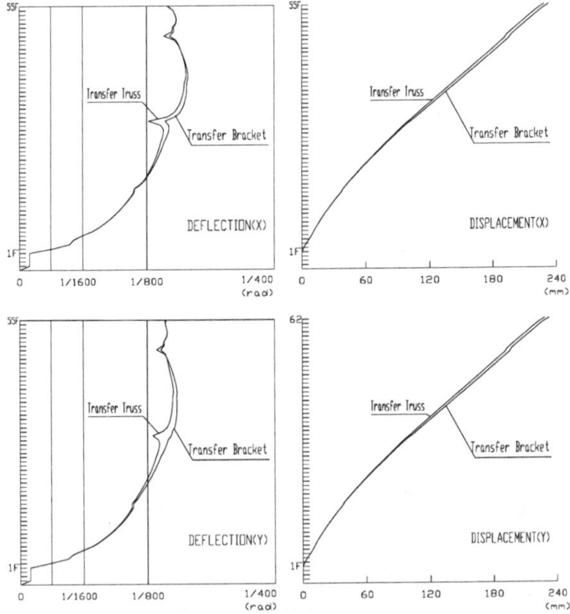

Figure 5. Deflection and displacement of the structure

3. Foundation Settlement

The pile-raft foundation in which the driven cast-in-place pile diameter is 800mm and the thickness of the raft is 2800mm. The post-grouting method is used around the piles.

The settlement analysis was carried by the cooperation method considering the upper structure, the pile-supported raft foundation and the ground.

The calculation formula is defined as:

$$S = \psi\psi_e \sum_{j=1}^{m}\sum_{i=1}^{n} \frac{\sigma_{j,i}\Delta H_{j,i}}{E_{Sj,i}} \qquad (1)$$

Where the parameter meaning is defined in *Technical Code for Building Pile Foundation* and the parameter value is as the following.

The settlement calculation depth is about 37m from layer 9 which is cobble to layer 17 which is sand and cobble. The average E_S value is about 43.0MPa. The Ψ value is 0.3 and the Ψe is 0.7. The average compression stress under the concrete tube is about 1914 kPa.

The maximum settlement of the foundation is 45mm under the concrete tube and the minimum settlement is 30mm under the outer retention wall. The average settlement is about 40mm and the maximum slope of the raft is 0.001 which is smaller than the limitation which is 0.002.

4. Structure Design Rules

The capacity of the horizontal elements such as the frame beams and the coupling wall-beams are controlled by the load combination which includes the dead load, live load, and the frequently occurred earthquake action, and the capacity of the vertical elements such as the shear wall and the frame columns are controlled by the load combination which includes the dead load, live load, and the seismic action under the basic intensity.

The maximum deflection under the frequently occurred earthquake is no more than 1/638 according to the codes, and the maximum elastic-plastic deflection under the seldomly occurred earthquake is no more than 1/100.

5. Structure Analysis under the Frequently Occurred Earthquake

The whole structure static analysis under the wind loads, the dead loads and the live loads taking into account the construction sequence is accomplished. The free vibration characteristic and the seismic actions obtained by the response spectrum method and the step-by-step method with two software ETABS and SATWE.

5.1 *Structure free vibration characteristic*

The free vibration characteristic is shown in Table 2. The dates in Table 2 can be satisfied the demand of the codes.

Table 2. Structure periods of free vibration

	SATWE(S)	ETABS(S)
T1	4.24□pendulum motion□	4.22□pendulum motion□
T2	4.11□pendulum motion□	4.18□pendulum motion□
T3	3.23□torsion□	3.07□torsion□

5.2 *Base shear and the displacement under the earthquake*

The maximum deflection and the base shear accused by earthquake is shown in Table 3. The deflection in Table 3 can be satisfied the demand of the codes.

Table 3. Deflection and base shear

		SATWE	ETABS
Deflection	X	1/666	1/676
	Y	1/690	1/694
Base shear (kN)	X	43194	40983
	Y	42885	42132

5.3 *Elastic analysis by the step-by-step method*

The structure elastic analysis by the step-by-step method was carried. There are four earthquake acceleration waves which include two artifical waves and two recording waves being used in the analysis. The response spectrums of four waves are shown in Fig.6. which prove that the earthquake action of the structure under artifical waves will be bigger than that under the spectrum of the code.

Figure 6. Response spectrum of the acceleration waves

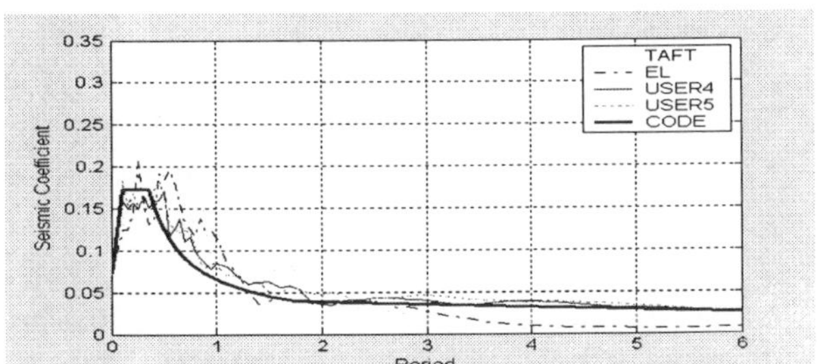

From the curves in Fig.7 and Fig. 8, it can be deduced that the calculation results by the step-by-step method can be satisfied the demand of the codes. On the other hand, The action and the inter force of the elements above floor 40 is larger than that by CQC method. In the

capacity checking, those inter force by the step-by-step method will be used.

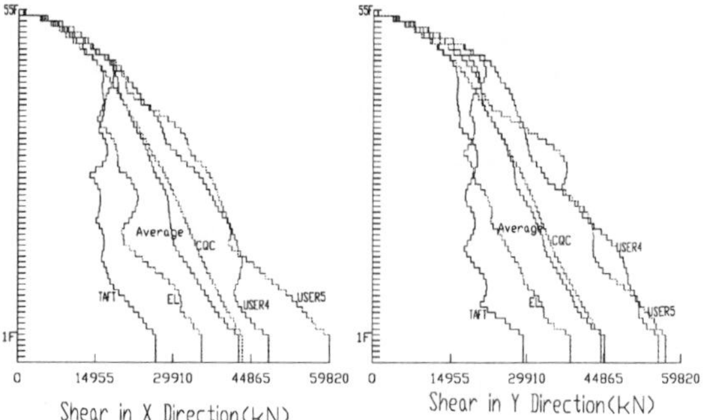

Figure 7. Earthquake shear

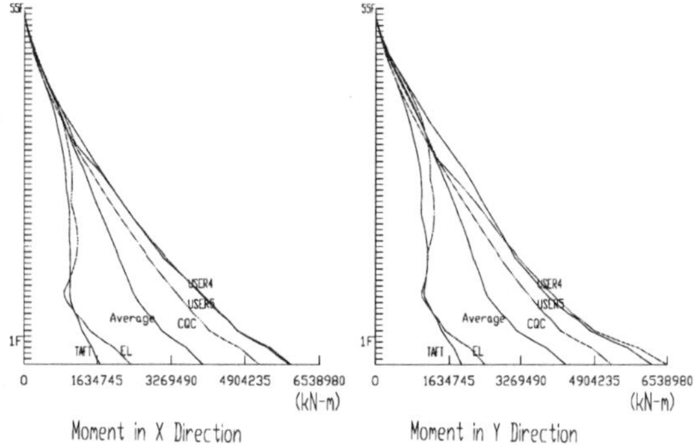

Figure 8. Earthquake moment

5.4 *Analysis of the corner structure under the earthquake*

Because there are four small parts extruded out the main plan of the building, The especial analysis for the extruded part which is shown in Fig. 9 is putted into practice.

The 3D calculation model is shown in Fig.10. The first period of the independent extruded part is 11.43S

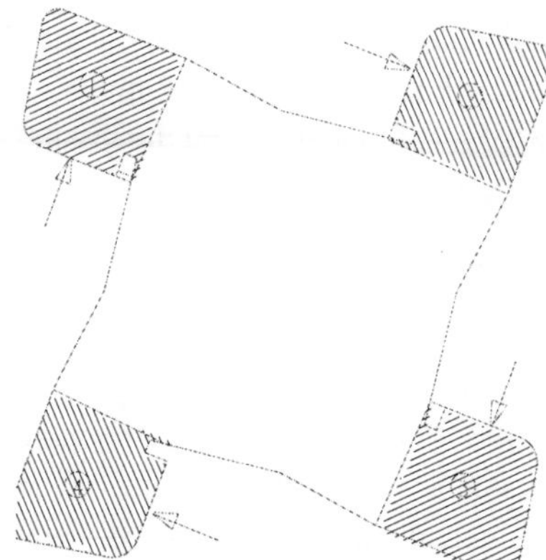

Figure 9. Extruded parts plane

and the second period in the arrowhead direction is 9.23S. So, the reduced coefficient can be calculated by

$$4.11/9.23=0.45$$

The inter force of the columns below the floor 4 by this method is bigger than that obtained from the analysis of the whole structure.

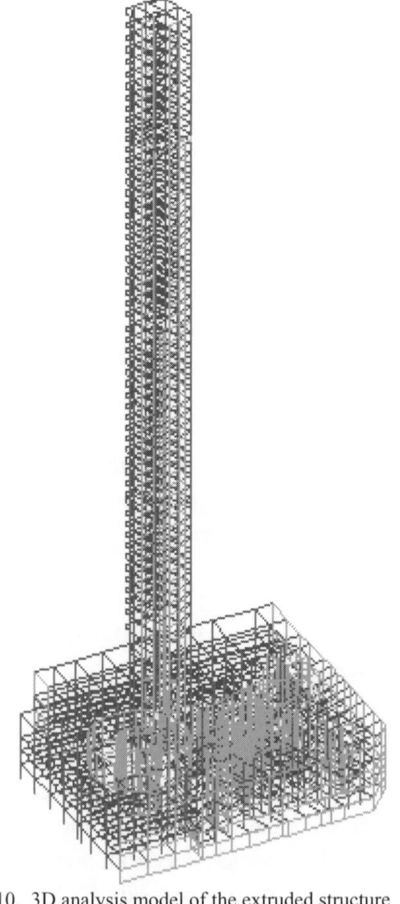

Figure 10. 3D analysis model of the extruded structure

6. Structure Analysis under the Basic Intensity

6.1 *Vertical Elements Inter-force Calculation Method*

According to the design rules, the whole structure analysis under the basic intensity earthquake was carried. In the analysis, the parameters were changed as the following.

The rigidity of section of the frame beams is not magnified thinking about the plastic zone having been occurred in the concrete plane above the ends of the steel beams. On the other hand, the magnification coefficient is 1.7 used in 5.1 for the beams linked to the concrete plane on both sides.

The stiffness of the coupling wall-beam was considered of the reduced coefficient 0.8 relatived to that in the analysis under the frequently occurred earthquake.

The damping coefficient is 0.05 instead of 0.04 which was used in the analysis under the frequently occurred earthquake considering shear wall's cracking and the plastic zone having been occurred in the concrete plane above the ends of the frame steel beams and plastic zone having been occurred in the composited coupling wall-beams.

The results including the periods, the deflection and the base shear being shown in Table 4.

periods(S)		Deflection		Base shear(kN)	
T1	4.35	X	1/225	X	115931
T2	4.25	Y	1/233	Y	118852
T3	3.37				

Table 4. Calculation results(SATWE)

In the design, the steel in the vertical elements will be not yielded under the standard load combination including the dead load ,the live load and the basic intensity seismic action.

6.2 *Problem from the different earthquake fortification level*

According to the frequently occurred earthquake demand of China codes, the element inter force will be adjusted by the magnified coefficient which is obtained from code taking into account the structure system, the earthquake intensity, the structure height and so on. On the other hand, the axis force will not be changed in the design while the moment and the shear will be adjusted. So, the shear wall maybe change from compression member with small eccentricity to compression member with large eccentricity.

If the earthquake fortification level presented in 6.1 being used in the design, all of the inter force will be magnified by the coefficient 2.86 for this apartment structure. The shear wall changed from compression member with small eccentricity to tension member with eccentricity. So, The amount of the steel bar will be largely magnified.

The conclusion can be obtained from the analysis that for the high-rise frame core-wall structure with the big value of H/B as the apartment building , where the H is the height of building and the B is the breadth of the tube, the design according to the level presented in 6.1 is very difficult especially for the building in the strong seismic zone.

Of course, the rule of the China code is not enough security especially to the high-rise building because of the axis force of the vertical element is smaller than that caused by the base intensity earthquake.

6.3 *Shear supported by the out frame*

According to *Code for seismic design of building* in China, the story shear supported by the frame is no less than the mini value of the 25 percent of base shear and the 1.8 times of the calculation story shear. On the other hand, the shear supported by the frame must be not less than 25 percent of base shear according to the advice of the examination committee.

The simple calculation was done for estimating the shear supported by the out frame when the core wall had been cracked. On the assumption that the concrete wall from story 1 to story 8 will be cracked under the base intensity earthquake and the rigidity of concrete wall section will reduced. The reduced coefficient was defined as

0.50 for the walls from 1F to 4F

0.60 for the walls form 5F to 8F

The shear supported by the out frame under the base intensity earthquake is shown in Table 5.

Table 5. Shear supported by frame

Floor		Calculation shear by the 6.3 method	Calculation shear by the 5.2/3method	Ratio(%)
X	1F	13862	43194	32
	2F	16305		38
	3F	15764		36
	4F	12311		29
	9F	8814		20
	14AF	9807		23
	25F	10862		25
	31F	14370		33
	33F	5744	28754	20
	35F	7116		25
	40F	6960		24
	45F	6903		24
	50F	4833	18896	26
	53F	2158	10684	20
Y	1F	10622	42185	25
	2F	12833		30
	3F	17388		41
	4F	11128		26
	9F	9371		22
	14AF	10637		25
	25F	10550		25
	31F	14269		34
	33F	5495	28728	19
	35F	6792		24
	40F	6671		23
	45F	6646		23
	50F	4459	19415	23
	53F	3128	10986	28

The date shown in the Table 7 proved that the out frame must have enough capacity for the basic intensity earthquake. So, the 25 percent of seismic shear of base shear was used to check the frame capacity from story 1 to story 31 except for special stories such as transfer stories and bottom story and so on. Similarly, the 25 percent of story 33's seismic shear was used to check the frame capacity from story 33 to story 49A, the 25 percent of story 50's seismic shear was used to check the frame capacity from story 50 to story 52 and the 25 percent of story 53's seismic shear was used to check the frame capacity from story 53 to story 55. It is for balancing the security and the cost.

7. Structure Analysis under the Seldomly Occurred Earthquake

The elastic-plastic analysis under the seldomly occurred earthquake by the step-by-step method will be carried for two kinds of structures. One is the structure designed according to the frequently occurred earthquake and the other structure is designed according to the rules presented in 4.

The plastic deflection of the first kind structure under the seldomly occurred earthquake can not be satisfied the demand of 1/100. The analysis for the second kind structure is being carried and the result will be presented in the future.

8. Conclusion

The different fortification level for the vertical elements and the horizontal elements is appropriate for the high-rise building structure.

The earthquake fortification level of this apartment structure is the steel in the vertical elements which include all of the shear walls and the columns are not yielded under the basic intensity earthquake. This level is carried difficultly for supper tall building in the strong intensity zone especially for the reinforced concrete structure and the mixed structure because of the great tension being in the vertical elements.

The earthquake action of the upper stories and the inter force of those elements should be obtained by the step-by-step method.

The outer frame which is the second sustainment should have enough capacity to support at least 25 percent of the whole shear.

Acknowledgments

The first version preliminary design of the apartment structure is finished by ARUP and the foundation settlement analysis is finished by CABR. The advice of the examination committee was adopted for the design. It's appreciated of their works for this engineering.

References

1. JGJ 3-2002 *Technical specification for concrete structure of tall building, China architecture & building press*, (2002).
2. GB 50011-2001 *Code fir seismic design of building, China architecture & building press*, (2001).
3. JGJ 138-2001 *Technical specification for steel reinforced concrete composite structure, China architecture & building press*, (2002).

4. Xu Peifu et al, *Experiments Study on Seismic Performance of High-rise Hybrid Structures, Building Structure,* Vol.35 No.5, (2005).
5. Xu Peifu, Dai Guoying, *Performance-based Seismic Design of Tall Building Structure Beyond the code-Specification, China Civil Engineering Journal,* Vol.38 No.1, (2005).
6. Fu Xueyi, *Update Philosophy of Case-in-situ RC Structure Aseismic Design, Building Structure,* Vol.35 No.5, (2005).
7. Huang Jifeng, et.al, *On the Uniform Shearing Stress Regulation of the Frame-Shear Wall Structure Considering the Re-distribution of the Elastic-Plastic Internal Force, Building Science,* Vol.21 No.3, (2005).

COMPOSITE (HYBRID) SYSTEMS FOR TALL BUILDINGS

AYSIN SEV

Department of Architecture, Mimar Sinan Fine Arts University, Address
Findikli, Istanbul, 34427, Turkey

Different structural materials possess different characteristics. In order to maximize the respective benefits of different materials, they are often used in combination. The result of this combination is the composite or hybrid systems, which has come to have the specific meaning of steel and concrete components connected together. In this manner, they behave as a single structural unit. In current practice, composite construction is one of the most efficient construction types for multi-story buildings. It can be applied to both lateral and gravity load resistance system. Since buildings of the future will involve a variety of forms, and will be flexible for multiple use spaces, composite construction will enable the designers for their desire to design contemporary tall buildings. This paper is intended to give general information of composite structures employed in multi-story buildings by giving descriptions of a number of buildings related to architectural applications of steel and composite building systems.

1. Introduction

The steel and concrete building systems evolved independently of each other and up until the 1960s, when engineers were trained to think of tall buildings either in steel or concrete. Dr Fazlur Khan of Skidmore, Owings and Merrill, broke this barrier in 1969 by blending steel and concrete into a composite system for use in a relatively short, 20-story building, in which the exterior columns and spandrels were encased in concrete to provide the required lateral resistance. The system was basically a steel frame stabilized by reinforced concrete. However, today the advent of high-strength concrete as ushered in the era of super columns and mega frames where the economy, stiffness and damping characteristics of large concrete elements are combined with the lightness and constructability of steel frames [1].

The term "composite construction" has assumed several meanings in recent years as creative combinations of steel and concrete, both for gravity systems and for lateral load resisting systems. The basic idea of using composite systems is to combine the benefits of steel and concrete in a structure that one complements the other in terms of stiffness, member size, speed of construction, fire resistance, and other factors [2,3]. In many instances, these systems are more economical than either an all concrete or an all steel system [1, 4, 5, 6].

All high-rises can be considered as composite buildings, because a functional building cannot be built by using only steel or only concrete. For example, concrete is invariably used for floor slabs in an all steel building. Similarly, in a critical sense, use of mild steel reinforcement transforms a concrete building into a composite building.

2. Reasons for Combining Concrete and Steel

Each of the materials has its own characteristic properties, its advantages and disadvantages and special conditions of utilization. The material characteristics may be expressed by mutual comparison. Table 1 contains the principal characteristics of two materials from several points of view. The table contains the advantage of both materials for tall building structural systems. However the set of aspects and criteria does not contain the interactive couplings and arrangements occurring in each case, since all the aspects should be utilized for each case, and some characteristics when considered separately will have a different importance when analyzed in wider circumstances.

3. Composite Building Systems

Composite building systems that can resist the lateral loads caused by wind and seismic forces may be conveniently classified into the following categories:

Table 1. Comparison of principal characteristics of structures made of steel and reinforced concrete [7].

Criterion		Steel	Reinforced Concrete
Statics	Stiffness/mass ratio	X	
	Compression stiffness/costs ratio		X
	Tension stiffness/costs ratio	X	
	Rigidity		X
	Rigidity/mass ratio	X	
	Durable deformations	X	
Dynamics	Damping		X
	Durability against repeated loads	X	
Building physics	Fire resistance		X
	Thermal and acoustic insulation compared with mass		X
Structural	Connection of the load-bearing function with dividing function		X
	Access to installation	X	
	Production precision		X
	Simplicity of details		X
	Flexibility, adaptability	X	
	Maintenance, durability		X
Technology	Terms for preparation of production		X
	Time of building in situ	X	
	Prefabrication, transport, connections	X	
	Dependency on the weather	X	
	Liquidation, reversibility of the material	X	

(X= preferable)

3.1 *Steel frame and reinforced concrete core or shear wall system*

Core walls enclosing building services such as elevators, mechanical and electric rooms, and stairs have been used extensively to resist lateral loads in tall concrete buildings. Polygonal or simple forms such as C and I shapes round elevators interconnected with coupling or link beams are used extensively (Figure 1) [1, 4].

In the core-only systems, the total lateral load is resisted by the shear walls and hence the remainder of the building can be conveniently framed in structural steel. Whether or not concrete or steel comes first in the construction process. In one version, concrete core is cast first by using jump or slip forms, followed by erection of the steel surround. Although the structural steel may not proceed as quickly as in a conventional steel building, the overall construction time is likely to be reduced because elevators mechanical and electrical services can be installed rapidly in the core while construction outside the core proceeds. In another version, steel erection columns are used within the shear walls to serve as erection columns and steel erection proceeds as in a conventional steel building. After the steel erection has reached a reasonable level, concreting of core starts using conventional forming techniques [1].

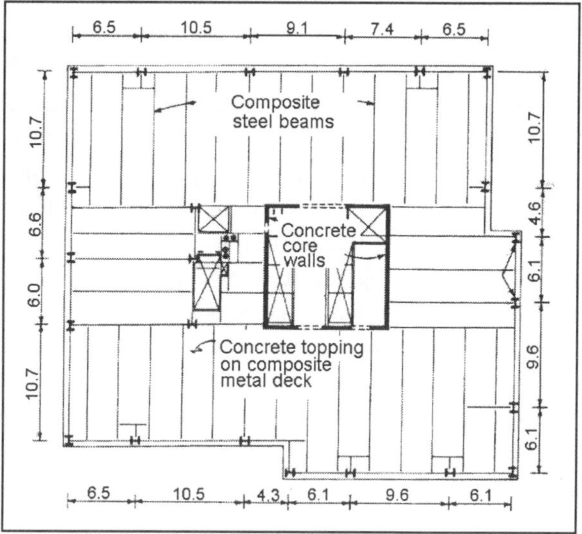

Figure 1. Typical floor plan of RC core and steel frame.

The cores can be separated as towers, as in the case of Roche Dinkeloo's design, Knights of Columbus Building in New Haven (Figure 2). These massive, windowless cores, in the form of corner towers, are connected to each other by ponderous rust-colored, exposed steel girders [8].

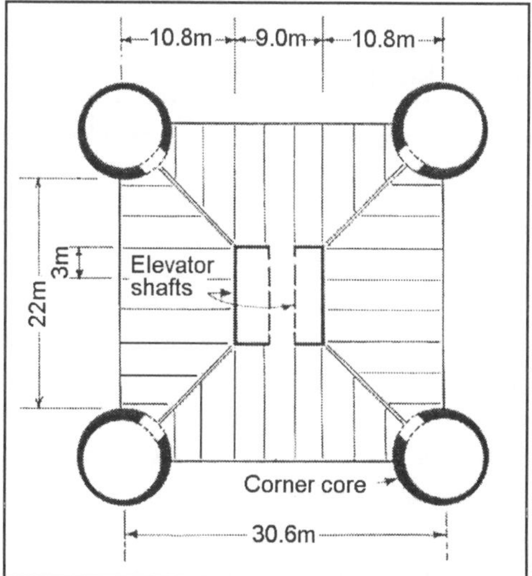

Figure 2. Knights of Columbus Building, New Haven, typical floor plan.

3.2 *Composite frame and reinforced concrete shear wall interacting systems*

This system has applications in buildings that do not have cores sufficiently large to resist the total lateral loads; interaction of shear walls with other moment frames located in the interior or at the exterior is called upon to supplement the lateral stiffness of the shear walls. When frames are located on the interior, usually the columns and girders are of steel because the cost of form work for enclosing interior columns and girders for composite construction far outstrips the advantages gained by additional stiffness of the frame. Interior columns typically have beams framing in four directions, making placement of mild steel reinforcement and formwork around them extremely cumbersome. On the other and, it is relatively easy to form around exterior columns, and if desired, even the exterior spandrels can be made out of concrete without creating undue complexity. If steel erection precedes concrete construction, it is usually more cost effective to use steel beams as interconnecting link beams between the shear walls. A schematic plan of a building using this system is shown in Figure 3 [1].

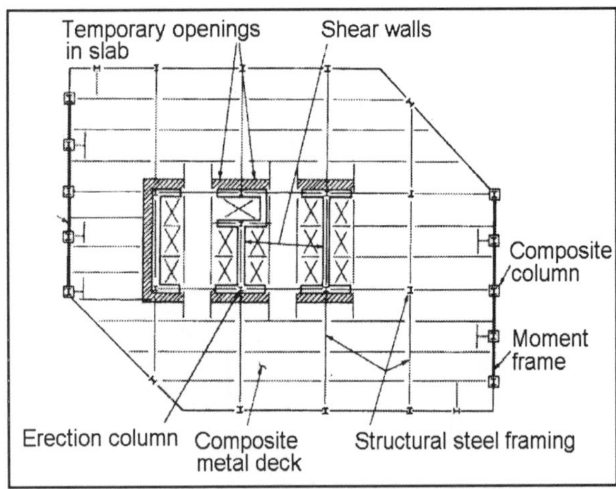

Figure 3. Typical floor plan of a composite building using shear wall and composite frame

Grosvenor Place in Sydney which is completed in 1988 utilizes this scheme as the lateral load resisting system (Figure 4). It consists of 46-storey office levels, with floor plans 14.6 m deep column free construction. This flexibility is achieved by a core in an elliptical plan form, which is made of reinforced concrete, a composite external frame, which consists of composite columns and spandrels, and a composite floor system. The core is made up of the two exterior walls, which are 1 m thick at the base, connected by the shorter interior cross walls and provides the primary lateral load resistance [9, 10].

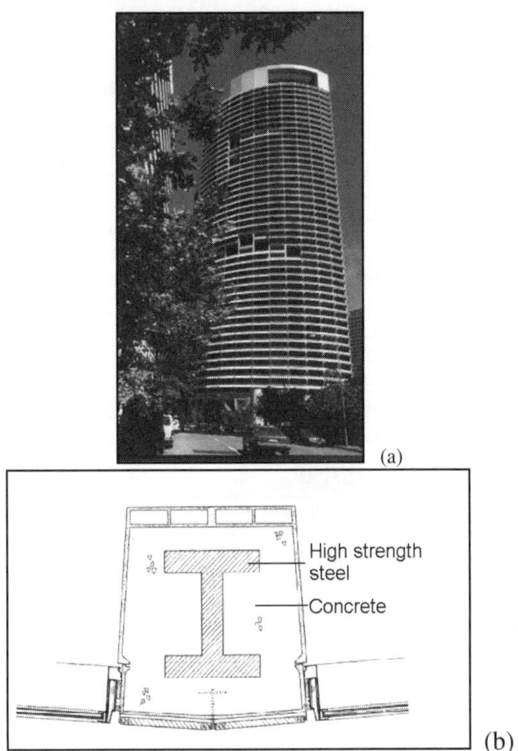

Figure 4. Grosvenor Place, Sydney: (a) General view of the tower; (b) Composite exterior column.

3.3 *Composite Tube Systems*

As in concrete and steel systems, closely spaced perimeter columns and deep spandrels form the backbone of the composite tube systems. Two versions are currently possible. One system uses composite columns and concrete spandrels and the other uses structural steel spandrels in place of concrete spandrels. A small steel section can be used as a steel spandrel in the former scheme to stabilize the steel columns. However, in the design of concrete spandrel, its strength and stiffness contribution is generally neglected because of its relatively small size. Schematic and plan sections for the two versions of tubular systems are shown in Fig. 5 and 6. The key to the success of this type of construction lies in the rigidity of closely spaced columns, which together with deep spandrels, results in an exterior façade that behaves more like a bearing wall with punched windows than a moment frame [1, 11, 12].

AT&T Center in Chicago can be considered as one of the most remarkable composite tall buildings. The 61-story office tower is supported by a composite structure, which consists of an exterior composite perimeter framed tube. This tube resists all of the lateral loads, and has setbacks at various levels to accommodate the articulated façade. The floor framing system is a combination of built up floor trusses and rolled beams. The structural system also involves a central core of reinforced concrete shear walls with a perimeter steel moment connected frame. Two-story outrigger trusses interconnect the concrete core and perimeter structural steel frame to provide a composite structure [10, 13]. The other remarkable tall buildings around the world include the America Tower in Houston, First Interstate World Center in Los Angeles, Society Tower in Cleveland, and Scotia Plaza and Canada Trust Tower both in Toronto. All these examples are constructed in the past decade.

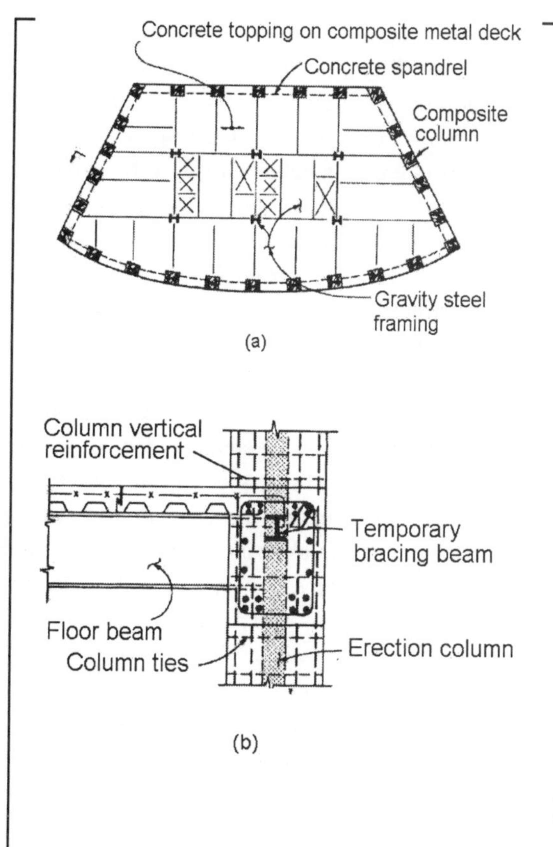

Figure 5. Composite tube with concrete spandrels: (a) Typical floor plan; (b) Typical exterior cross section.

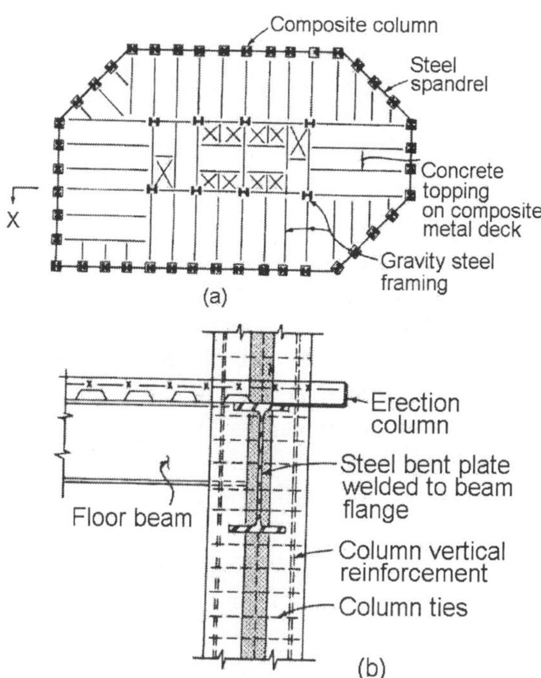

Figure 6. Composite tube with steel spandrels: (a) Typical floor plan; (b) Typical exterior cross section.

3.4 *Vertically mixed systems*

Mixed-use high-rise buildings provide for two or more types of occupancies in a single building by vertically stacking different amenities. For example, lower levels of a building may house parking; middle levels, office; and top levels, residential units, such as apartments and hotel rooms. Different functions need different arrangement of vertical and horizontal structures. For example, beamless flat ceilings are preferred in residential occupancies because of minimum finish required underneath the slab. Also, large spans are required for optimum lease space for office buildings. Due to these varieties in the function of the levels, different types of construction must be mixed within a structure. In this case, additional columns can be introduced without unduly affecting the architectural layout of residential units. The decrease in span combined with the requirement of flat-plate construction gives the engineer an opportunity to use concrete in the upper levels for apartments and hotels.

In certain types of buildings, use of concrete for the lower levels and structural steel for the upper levels may provide an optimum solution. The bracing for the concrete part of the building is provided by the shear walls, while a braced steel core provides the lateral stability for the upper levels. A suggested technique of transferring steel columns onto a concrete wall consists of embedding steel columns for one or two levels below the transfer level. Shear studs shop-welded to the embedded steel column provide for the transfer of axial loads from the steel column to the concrete walls [1, 4].

900 Michigan Avenue in Chicago is the most extreme example of this combination (Figure 7). It comprises high-rise flats, with a hotel rising from the 25th story to the 66th floor, and this part of the structure is concrete with in situ frame employing flat slab floors. Steel framing is utilized when non-cellular spaces are required for the office and retail zone. The reasons are linked to the changing grid that occurs between offices and the hotel/ residential construction. The transfer structure occurs within the reinforced concrete walls and floor/beam system [2, 4, 14]. The eight story base of the building is intended to lessen the impact of the building's enormous bulk and to recall the height and scale of Michigan Avenue buildings in early 1920.

3.5 *Mega frames with super columns*

The most efficient method of resisting lateral loads in tall buildings is to provide super columns, placed as far apart as possible within the foot print of the building. The columns are connected with a shear-resisting system such as welded steel girders, vierendeel frames, or diagonals.

The construction of super columns can take many forms. One system uses large-diameter thin-walled pipes or tubes filled with high-strength concrete generally in the range of 41 to 138 MPa compressive strength. Generally, neither longitudinal nor transverse reinforcement is used thereby simplifying construction. Another method is to form the composite column using conventional forming techniques, the only difference is that steel columns embedded in the super column are used for the erection of steel framing and also for additional axial and shear strength [1, 10, 12].

The 57-story Nations Bank Plaza in Atlanta is one of the remarkable examples of the structures utilizing mega columns. It has a square plan with the corners serrated to create the desired architectural appearance and to provide for more corner offices (Figure 8). A five-level basement provided below the tower is reinforced concrete construction. The gravity load is primarily supported by 12 composite super columns. Four of these are located at the corner of the core, and eight at the perimeter. The core columns are braced on all four sides with diagonal bracing. Steel girders are moment connected between the composite columns, to transfer part of the overturning moment to the exterior columns.

3.6 *Reinforced concrete construction to podium with steel frame above*

Reinforced concrete is frequently used for the construction of basement and sub-basement with possibly ground and first floor included to provide a platform or podium for a steel frame above [1, 12]. Early examples of this composite construction include the pioneer structures by Le Corbusier and his partner Pierre Jeanneret [4]. Amongst these is the Pavillon Suisse Paris (RC columns and deck with steel structure at first floor and above) (Figure 9). This subvision of the structure is often reflected in the contractual arrangements whereby the foundation and groundwork's specialist is different from the main contractor responsible for the frame and shell of the building. Another advantage in this divided responsibility is the limitation of claims for delay spreading from the more uncertain aspect of building below ground to the critical and expensive finalizing stages.

A rather different example of this form of composite structure was adopted for the Nat West Tower (Figure 10), where the deep basement and podium work in reinforced concrete was let as a separate contract to the superstructure. The upper 52 stores are carried by three large staggered reinforced concrete cantilever brackets of cellular section. The office floors of the 183 m high structure are steel framed with the outer ends of the universal beams spaced at 3 m centers with spans varying from 7 to 9 m. The inner ends of the beams are supported on steel brackets attached to the reinforced concrete core walls. The floor construction is composite with 55 mm trapezoidal steel decking overlaid by 120 mm thickness in situ concrete slab with 22 mm shear studs. Wind loads are absorbed by the curved core walls through the rigid action of the floor slabs on all three sides [4].

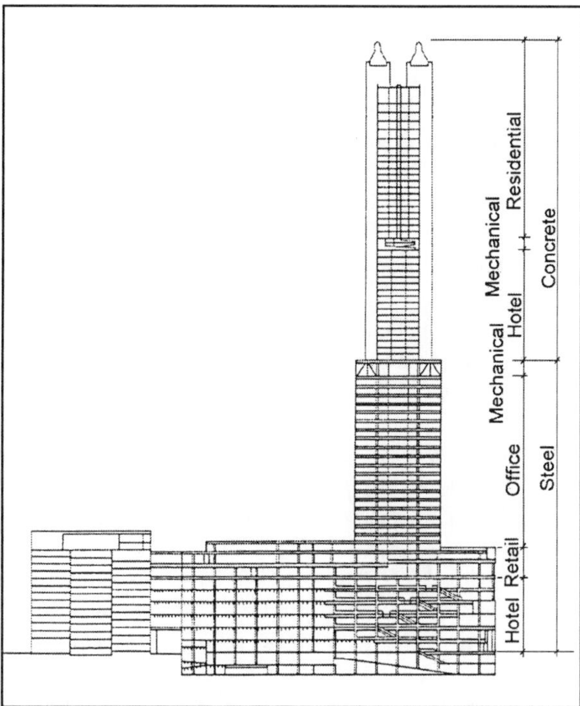

Figure 7. 900 Michigan Avenue, Chicago.

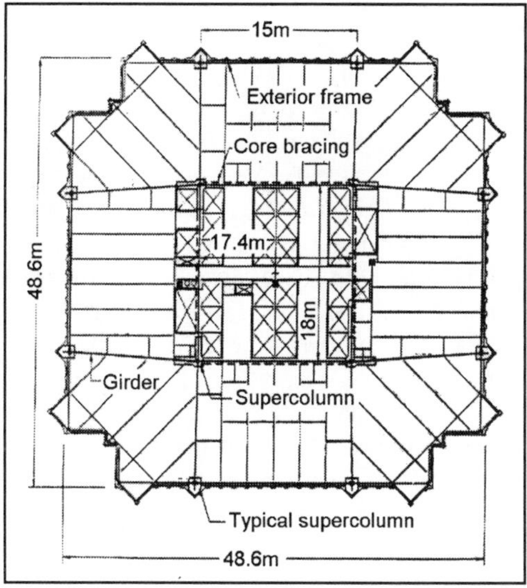

Figure 8. Nations Bank Plaza, Atlanta: typical floor framing.

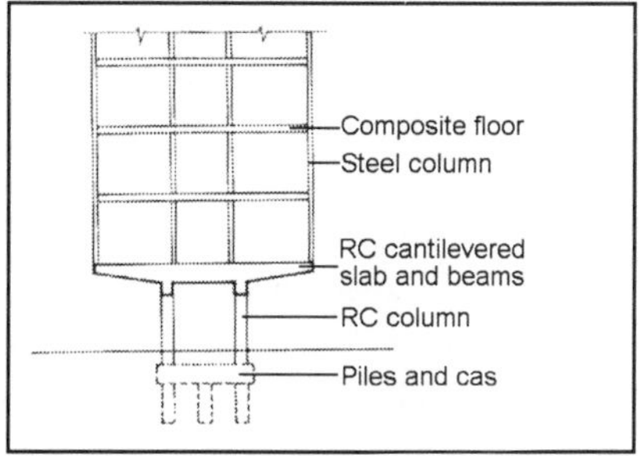

Figure 9. Pavillon Suisse Paris; section showing steel framing above RC podium.

4. Conclusions

The trends of modern architecture sometimes force the designers away from conventional design methods in a search for a structure that will accommodate aesthetic and functional demands while meeting the structural requirements. The result may be a structure, which on one face of the building is of a different type than the other face, or a structure with a number of quite different elements forming its lateral load-resisting frame.

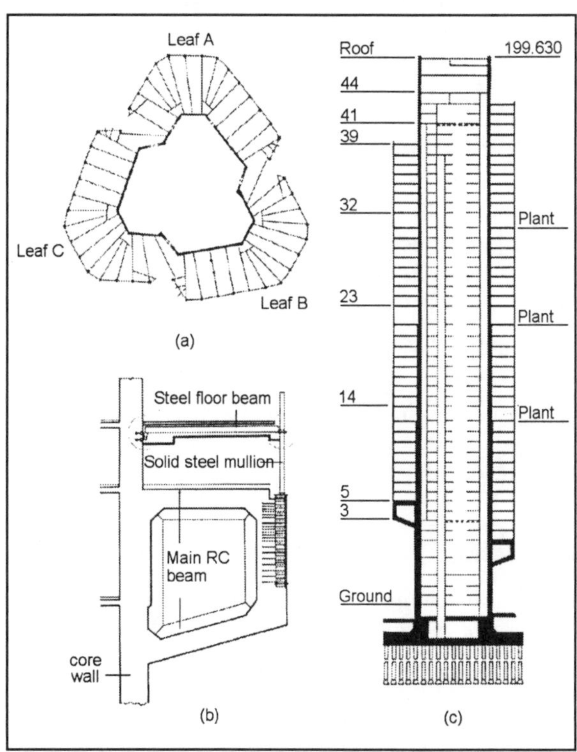

Figure10. *Nat West Tower, London: (a) Steelwork arrangement of typical floor plan; (b) Typical section through main RC beam; (c) Section through tower block.*

With the advent of high-strength concrete has become the era of the "super column" , where the stiffness and damping capabilities of large concrete elements are combined with the lightness and constructability of steel frames. High-strength concrete exhibits significantly lower creep and shrinkage and is therefore more readily accommodated in a composite system. The relative cheapness of high-strength concrete together with the fact that large members do not require large cranes means that the columns can be economically designed for stiffness rather than for strength.

Future high-rise structures will involve a variety of forms and shapes, and will demand considerable flexibility for creating exciting habitable and multiple use spaces. In the design process the problem lies on the selection of the structural system. While various systems can be selected on the basis of structural efficiency, the real process involves a variety of disciplines including architects and engineers as well as clients, service engineers etc. The most efficient structural system may not be effective for architectural solution or vice versa. Under these terms each concept should be checked with a variety of constructional systems using methods, which take account of the availability of materials, building economics and above all constructability.

In the future, although it seems to be more expensive than concrete or steel construction, the number of composite constructions in tall buildings will increase due to the many advantages which results in reduced construction time. For instance, the steel deck for the composite construction provides a working platform during construction, and eliminates the need to either prop or strip form work and the attendant delays resulting from these operations. The expense of composite construction has to be discussed by all the design team and owners, since it provides many advantages.

References

1. B. S. Taranath. *Steel, Concrete and Composite Design of Tall Buildings*, McGraw-Hill, Inc., New York, (1997).
2. M. M. Ali, P.J. Armstrong, P. J., *Architecture of Tall Buildings*, Council on Tall Buildings and Urban Habitat Committee 30, McGraw-Hill, Inc., New York, (1995).
3. H. Iyengar, H. *Composite of Mixed Steel – Concrete Construction for Buildings*, American Society of Civil Engineers, New York, N.Y. (1977).
4. R. Plank, P. Gregson, P. "Composite Floors and Structures", *Architecture and Construction in Steel*, A. Blanc, M. McEvoy, P. Roger, (Ed.), Chapman & Hall, London, 235-252, (1993).
5. J. G. Nut. "The Composite High-Rise Building – An Interaction of Planning, Structure, Speed and Economy", *Second Century of The Skyscraper*, Council on Tall Buildings and Urban Habitat, Lynn S. Beedle (Ed.), Van Nostrand Reinhold Company, New York, 339-351, (1988).
6. A. Sev. "Recent Trends in High-Rise Buildings", *Fourth International Congress on Advances in Civil Engineering*, Vol. 2, (Ed.) Toklu, C. Y. And Erbatur, F., Gazimagusa, 1-3 November, 625-636, (2000).
7. J. Kozak, J. *Steel-Concrete Structures for Multistory Buildings*, Elsevier, Amsterdam, (1991).
8. D. Guise. *Design and Technology in Architecture*, Van Nostrand Reinhold, New York, (1997).
9. Grosvenor Place. Available at (2000): http://www.arch.su.edu.au/kcdc/caut/html/GP/
10. A. Sev. *The Analysis of Tall Buildings According to Architectural Design and Structural Systems in Turkey and at Abroad*, Doctoral Dissertation, Mimar Sinan University, Science and Technology Institute, Istanbul, (2001).
11. W. Schueller. *High-Rise Building Structures*, John Wiley and Sons, New York, (1977).
12. A. Ozgen, A. Sev, A. *Structural Systems of Multi-Story High-Rise Buildings*. (in Turkish), Published by Birsen Yayınevi, İstanbul, (2000).
13. L.S. Beedle, D. B. Rice. *Structural Systems for Tall Buildings*, Council on Tall Buildings and Urban Habitat Committee 3, McGraw-Hill, Inc., New York, (1995).
14. I. Zacnik, M. Smith, D. Rice. *100 of the World's Tallest Buildings*, Council on Tall Buildings and Urban Habitat, The Images Publishing, Hong Kong, (1997).

Other Structural Aspects

CONSTRUCTION TECHNOLOGY FOR SUPER HIGH-RISE BUILDINGS

YAO YI-PING ZHANG SONG-LIN HUANG HONG-ZHENG

Jiangsu Zhongxing Construction Ltd., Chongqing Company
Chongqing 400010

This paper describes emphatically several construction technologies used in the third phase construction of Chongqing Jufucheng Building, which include the steel structured support technology for form-work suspension of building transition floor, the large-diameter HRB400 reinforced bar connection technology, the application technology of high-strength and high-performance fiber concrete, the single-side form-work support technology in "converse operation" process of concrete wall in basement as well as the technology of "resistance" and "release" for the construction of large-area building floor.

1. Outline of the Project

The third-phase construction project of Chongqing Jufucheng (now called Chongqing International Trade Center) is one of the main buildings located in Chongqing downtown area — Jiefangbei CBD and it was ranked as one of ten major projects in Chongqing in 2003. The project was designed as a 43-storey super high-rise building with tower building A and B and is a multipurpose building which merges supermarket, entertainment place, five-star hotel and office building into a whole.

The total building area of the project is 161,000 square meters. Its underground four storeys are 17.1 meters below the ground, and the a building area is 32,000 square meters. The annex on the ground is a nine-storey one, with a building area of 55,000 square meters. The two tower buildings are 34-storey ones with the height of 160 meters and total building area of 74,000 square meters. This project is a super high-rise building and was designed for resisting the earthquake intensity of 6 degrees. The main part of the project was designed into a frame-tube structure, the part below the minus No.1 floor can resist earthquake of 3 magnitude and the part above the minus No.1 floor can resist the earthquake of 2 magnitude.

The underground No.4 storey is 17.1 meters below the ground, Chongqing's original civil air defence main tunnel (with elevation of –25.2 meters, width of 12 meters and clear height of 7~8.5 meters) passes through this project from the south-west corner to the north-east corner of the project (at an angle of 52.5° with X-axis) and is located under the core tube area and the annex of tower building B (hotel).

2. Main Construction Technologies for Building Transition Floor

The roof of the civil air defence main tunnel is designed as a transition floor, the part at the annex is designed as transition crossbeam with the height of 2800~3500mm and the part at the core tube is designed as box-type structural transition floor with the height of 4310mm , the upper and lower boards of the transition crossbeam are 200mm thick, the roof elevation of the transition floor is –12.3m of the minus No.3 floor, and its bottom elevation is about 9m to the ground of the civil air defence works. Since a 4m-wide passageway must be reserved during construction, the construction of form-wok is extremely difficult.

2.1 *Steel-structured support technology for formwork suspension of building transition floor*

2.1.1 *Determination of formwork scheme*

The included angle of the civil air defence main tunnel with the transition floor is 52.5°, so the actual span of the transition crossbeam is 15 meters. The overhead span of the transition crossbeam above the 4-m high bridge crane lane is 5.05 meter. Since the transition crossbeam has greater dead weight, shape steel and steel pipe are finally selected as the support system of the formwork through optimum selection of different schemes and the cost comparison, and the transition crossbeam adopts the construction method of congruent beam.

2.1.2 *Main technical points for construction of support system of steel formwork*

1. Selection of shape steel: the existing I20b is selected as the steel beam of the support form-work over the passage according to the first pouring height of the congruent beam and the actual conditions at the site, the pouring height of the congruent beam should be fully consistent with the idea of the design personnel and be appropriately adjusted according to the actual bearing capacity of I-steel I20b, and the first pouring height is 2 meters.
2. Selection of I-steel I20b bearing seat: it uses Φ 48 ×3 steel pipe bent as the bearing body, and the number and space of steel pipe of the steel pipe bent are calculated by the load of the bearing seat, however, the key to this problem is how the force can be exerted uniformly on all pipes at the bearing seat. If the elastic deformation of shape steel makes the local force exerted on the steel pipe bent in excess of its bearing capacity, the bent may be broken down one by one. In the construction, bi-directional U-steel 14 is used under I20b I-steel bearing seat for the change of force distribution in order to make the force evenly acted on the steel pipes at the seat.
3. Edge restriction of steel pipe bents on two sides of the passageway. The erection height of the steel pipe bent of the transition crossbeam is 9 meters, it belongs to the construction with highly-supported form-work, so the stability of bent body is very important. In normal construction, the space of horizontal steel pipes of the bent body with highly-supported form-work should be determined by calculation , however, the bent bodies on two sides of the bridge crane passage lose restriction in the direction vertical to the passage due to the existence of the passage.
4. The steel form-work system used for this structural transition floor uses the local materials, it is easy for construction, economic and applicable, I20b maximum bend being 10mm, better results are obtained in construction, all of these cut down the investment for the owner.

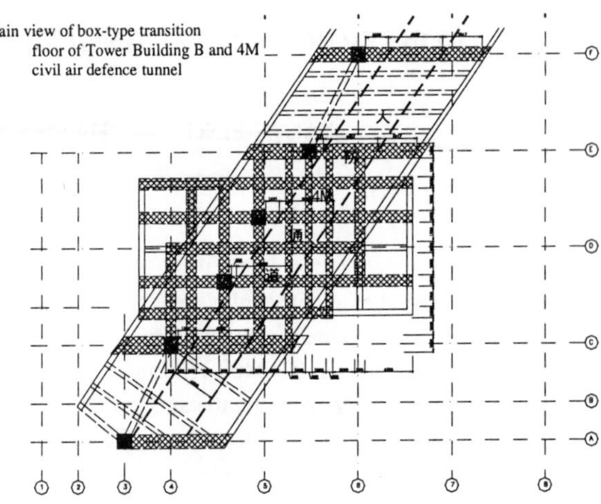

Civil air defence tunnel
Plain view of box-type transition floor of Tower Building B and 4M civil air defence tunnel

2.2 *Application and connection technology of large-diameter HRB400 reinforced bar*

1. The design of this project is set up defence by Grade 2 anti-seismic requirement, and HRB400 reinforced bar should conform to relevant requirements for anti-seismic design. When HRB400 reinforced bar is used for Grade 1 and 2 anti-seismic design, the ratio of the actually measured value of the tensile strength of HRB400 reinforced bar to that of its yield strength should not be less than 1.25, and the ratio of the actually measured value of the yield strength to the standard value should not be greater than 1.30 in order to guarantee the distortion ability after the appearance of the plastic articulation at a certain part in the structure and also to guarantee the implementation of the structural system with strong column and weak beam, strong shearing and weak bending, and strong node and anchor according to the requirements in GB50204-2002.
2. The reinforced bars under longitudinal force in the structure transition floor at the location of the civil air defence works use φ 22, φ 25, φ 28 and φ 32 large-diameter HRB400 bars. The connection of the reinforced bars is a key construction technology in concrete structure, so mechanical connection technology with stripped bars and rolled direct thread is used.

3. The connection technology with stripped bars and rolled direct thread is a new connection technology of reinforced bars, its technological principle is to make the diameter of the reinforced bar reaching the same size before thread rolling after stripping treatment of both transverse and longitudinal bars at the end of bar to be connected, and then to roll screw thread, this thus guarantee that the rolled thread has a consistent accuracy, has no spurious tread, has good mechanical properties and is safe and reliable in connection.
4. The connection technology with stripped bars and rolled direct thread will be used in the building transition floor of this project, it is easy to operate, connection operation can be done in all directions and all weather, bars can be easily and fast joined, the joint performance completely reaches Grade 1 requirement in standard JCJ107-2003 and the joint has good anti-fatigue property.

2.3 *Application technology of high-strength and high-performance fiber concrete*

1. The structure of the transition floor of this project uses C60 high-strength and high-performance fiber concrete and belongs to large bulk concrete construction. To control well the shrinking deformation of commercial concrete in construction is a key technology to reduce the cracks of structures in the transition floor.
2. The optimization of mixture ratio of concrete: To reduce dry and temperature shrinkage of the concrete by adding high-efficient water-reducing agent, Grade I or II coal ash and fine slag powder into the mixture to substitute partial cement.
3. Strengthening the secondary tamping of concrete: The secondary tamping of the concrete will be completed before the final set of concrete in about 2 hours (it should be determined according to retarded set time of concrete) after the first tamping. It is necessary to strengthen temperature control and do well the curing work of concrete. In construction, two layers of plastic film and straw bags will be used to control the temperature differential inside and outside the concrete (it should be controlled within 25℃).
4. The measures being taken in design of this project include: To improve the tensile strength of concrete by mixing with 0.85kg/m^3 of fiber and adding 20x20x3 wire net in the surface layer of the concrete structure; to counteract the early shrinking stress of the concrete by adding UEA-1 micro-expansion agent (the adding amount is 12% that of cement) into concrete.
5. The overall construction effect of the structural transition floor of the project is very ideal, and no cracks are found in the structure of the transition floor. However, since the transition floor and minus No.2 floor in the basement are at the same elevation, the impact of the integral shrinking of the transition floor upon the structure of the minus No.2 floor is not taken into consideration, which causes the slabs and frame around the structure of the transition floor to be ripped, so it is necessary to set up a pouring zone between two structural systems so as to release the shrinking stress of the structure of the whole transition floor through this pouring zone, or changing 100mm-thick floor-slab into 200mm-thick ones may have a better result.

3. Main Construction Technologies for Basement and Annex

3.1 *Single-side formwork support Technology in "converse operation" process of concrete wall in basement*

1. The foundation pit of this project is excavated by vertically blasting, since the outside concrete wall the basement is close to the rock wall, the construction of the outside concrete wall can be conducted only by plane support formwork. The floor height of basement is 4.5m and 5.0m and a large side pressure is exerted by concrete on the formwork, so the formwork needs a high support coefficient.
2. Multi-layer board in 18 thick and square wood in 50 x 100 are used as the form-work of the outside concrete wall of the basement of this project andΦ 48×3 steel pipes as the support system of the single-side form-work. In the early construction of outside concrete wall, we follow the construction sequence of the wall body first and then the slabs and frame. The pressure of wall body on the formwork is transferred to the pre-buried anchor rods on the floor by steel pipe support. Since the force transferred to the steel pipe support by the formwork is very large, and the upper part of the support body is a free end and only its lower part

is under restriction, it is easy to cause the support body to be deformed and displaced and the formwork to be floated.

3. The "converse operation" construction of the outside concrete wall of the basement is to first construct the upper and lower floor structures of the outside concrete wall and then the outside concrete wall of the basement. Though only the construction sequence changes, the result is entirely different, and it has following advantages:
4. The upper and lower floor structures of the steel pipe bent body are used as the support points of bearing force and the force on the support point is reduced by 50%. If the original steel pipe bent body is regarded as an overhanging force-bearing system, the present steel pipe bent body is an equi-span simply-supported force-bearing system, so the force-bearing system is much more rational.
5. The steel pipe bent body is simultaneously restricted by both the upper and lower floor structures, so the rigidity of the bent body increases by a big margin, deformation and displacement reduce with it, the floating phenomenon of form-work disappears and the shaping quality of wall concrete is thus improved.
6. The "converse operation" construction of the outside concrete wall changed the position of construction joint between the wall body and upper floor structure, the position of the construction joint is changed from wall body onto the floor structure, this better solves water leakage from the construction joint of the concrete wall.

3.2 *Technology of "resistance" and "release" for the construction of large-area building floor*

1. The building area of a single floor of the basement and the skirt building is about 7100 square meters, so how to prevent the appearance of shrinkage joint on the floor structure is a difficult technical problems puzzling technical personnel over a long period of time.
2. Improving the performance of pump commercial concrete and reinforcing the crack resistance of concrete. Less medium-sized and coarse aggregate, small particle size and high sand proportion in the commercial concrete as well as the extra-fine sand in Chongqing area and large collapse degree are the main reasons leading to the shrinkage and deformation of concrete itself. Therefore, in the construction of this project, we adopt appropriate "resistance" measures for improving the tensile strength of concrete.
   a) Optimizing the mixture ratio of concrete: Using mixed sand, i.e.70% of extra-fine sand and 30% of washed sand so as to reduce the content of aggregate in concrete; Mixing in efficient retarded water-reducing agent to substitute partial cement so as to increase the flow-ability of the concrete; mixing in Grade I and II coal ash in the concrete and also mixing in active material such as fine slag powder in the concrete above C50 so as to increase the

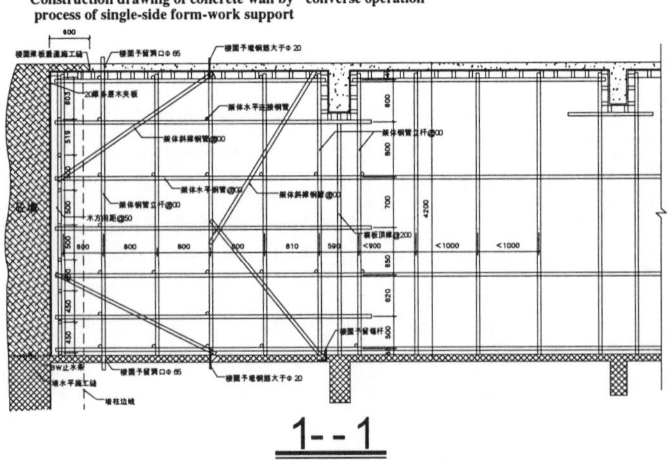

砼墙 Concrete wall 楼面梁板垂直施工缝 Vertical construction joint of floor beam 20 厚多层木夹板 20 thick multiple wood clamp plate 架体钢管立杆@600 frame body steel pipe pole setting @600 木方间距@250 wood space @250 BW 止水条 BW stripping 墙水平施工缝 horizontal wall construction joint 墙柱边线 side line of wall column 楼面予留洞口@65 floor reserved opening @65 架体斜撑钢管@600 frame body slant support steel pipe @600 架体水平钢管@600 frame body horizontal steel pipe @600 楼面予留洞口@65 floor reserved opening @65 架体水平连接钢管 frame body horizontal joining steel pipe 架体钢管立杆@600 frame body slant support steel pipe @600 楼面予留钢筋大于@20 floor pre-buried reinforcement>@20 模板顶撑@1200 form board roof support @1200 楼面予留锚杆 floor pre-set anchor bar

Figure 1. Construction drawing of concrete wall by "converse operation" process of single-side formwork support

workability and pump-ability of the concrete, substitute partial cement and reduce hydrated heat; mixing in UEA-H micro-expansion agent so as to compensate the early shrinkage of the concrete.

b) Anti-split steel bar and reinforced bar net should be provided in the plain concrete area and at the weak part so as to increase tensile strength of the concrete, this is a structural measure in design, but it is often neglected or is not completely considered in design.

3. As it is mentioned above that it is unavoidable to achieve the shrinkage of the concrete and the shrinkage is inversely proportional to the strength of concrete, i.e. it is large in the early stage and small in late stage. The shrinking deformation of the concrete (including creepage) in the late stage is unable to be controlled in construction, however we can "release" the early shrinking stress of the concrete to reduce its early shrinking deformation.

   a) In the construction of large-area basement and the floor structures of the skirt building, it is necessary to rationally divide the construction section according to the pouring zone and construction joint in order for flow process. The area of each construction section should be controlled within 1000 square meters so as to release the early shrinking stress of the concrete through the time differential in construction, this is an effective method for preventing the generation of cracks in large-area floor structures.

   b) In the pouring process of the concrete, it s necessary to strengthen the secondary tamping and ramming of the concrete. Water content of the concrete itself will lose and volumetric shrinking will occur due to the escape of the absorbed water (about 20%) of the early gelatin of the concrete , The inside of the structure shrunk slower than its surface, so the external and internal restriction causes the generation of cracks on the concrete surface, in this case, it is necessary to strengthen the secondary ramming so as to prevent the generation of cracks.

   c) It is necessary to do well the curing work of the concrete so as to make the internal and external stress of structures to be evenly released and prevent the generation of cracks due to the inside and outside temperature and humidity differential. It is also necessary to control well the charging operation time and prevent cracks to be artificially generated.

4. By taking above measures in this project, the cracking problem in large-area floor structures is better solved and good results are obtained in control of the vertical cracks on the concrete wall body of the core tube. This is well received by the owner, supervision and design units in the examination and acceptance of the main part of the project.

4. Construction Technology for Standard Floors

4.1 Application technology of attached externally-climbing scaffold

The technology of the attached externally-climbing scaffold has been found wide application in the construction of high and supper high buildings due to its safe and fast construction. The attached externally-climbing scaffold is designed into a multi-piece scaffold body according the external form of the building and takes the edge beam of the building as the attached hoisting point, and the scaffold body is hoisted synchronously by a electric hoist. When we apply the attached externally-climbing scaffold in the construction of this project, we focus the technical management on following aspects:

1. The rigidity of the scaffold is the base of this system, the integral rigidity of the bottom seat of the scaffold and the shaped-steel guide rail skeleton should be under control. In construction, careful calculation and recheck should be done according to the load and force on the scaffold while climbing.
2. The attached hoisting point is the technically control point of this system, it is necessary to first select the rational position of the hoisting point on the edge beam and then check the overloading force of the edge beam. It is necessary to analyze the internal force of the hoisting point and the force exerted on the edge beam under unfavorable operation conditions, and calculate the required minimum strength of the reinforced bar and concrete of the edge beam. Appropriate measures should be taken in construction so as to guarantee the safety of structures and scaffold body system.

3. The climbing of the attached externally-climbing scaffold should be emphatically under control in construction, while climbing, the attached points of the scaffold body gradually reduce and the safety factor of this system also lowers, so operation should be conducted according to the procedures. It is very important to keep synchronously hoisting of the system during climbing.

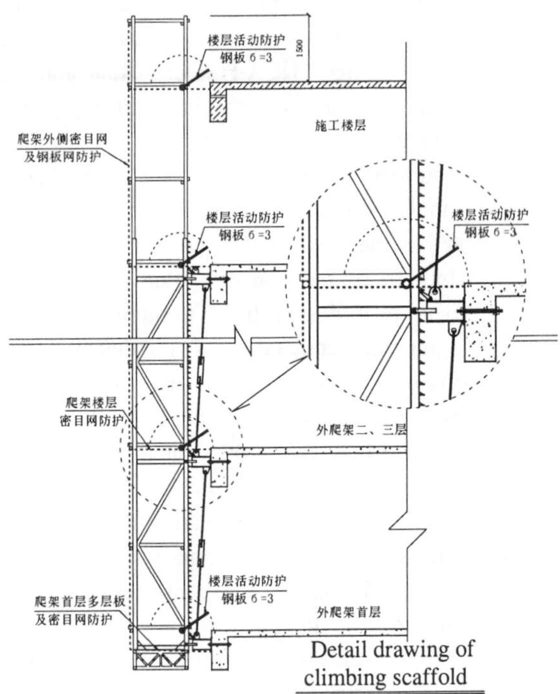

Detail drawing of climbing scaffold

楼层活动防护钢板： mobile protective steel plate of floor
楼层活动防护钢板： mobile protective steel plate of floor
施工楼层: floor under construction
爬架首层多层板及密目网防护: Multi-layer board and dense-mesh net protection for the first layer of the climbing scaffold
爬架楼层密目网防护: Dense-mesh net protection for the floor of the climbing scaffold
爬架外侧密目网及钢板防护: Dense-mesh net and steel plate protection for the outside of the climbing scaffold
外爬架二三层: Layer II & III of external scaffold
外爬架首层: First layer of external scaffold
爬升脚手架大样图: Detail drawing of climbing scaffold

5. Conclusions

The main body structure of the third-phase construction project of Chongqing Jufucheng (now called Chongqing International Trade Center) was completed in August 2004, 50 days ahead of the contract schedule. This project was appraised as Chongqing "Three Gorges Cup" good-quality project and was one of Chongqing's ten excellent projects in 2004.

COLLAPSE OF A PILED INDUSTRIAL BUILDING DUE TO SURCHARGE LOADS*

BI TANG ZHU MIN YANG

Department of Geotechnical Engineering, Tongji University, 1239 Siping Road, Shanghai 200092, China

A collapse of a piled industrial building in Shanghai, China is reported herein. Based on a field test of pile foundation adjacent to surcharge loads in the same area with soft clay soils, causes of the collapse were investigated. The field test indicated that (1) pile foundation would experience considerable deflection due to lateral soil movement generated by surcharge loads nearby; (2) differential settlement between the front and rear piles would be initiated by the negative friction due to soil settlement. The negative friction could also give rise to reduction of net pile capacity. As studied elsewhere, secondary stresses generated in the superstructure by the lateral displacement and differential settlement of pile foundation, and the loss of the pile capacity would be the main causes for the collapse. Thus, the response of pile foundation subjected to soil movement should be involved in design of piled industrial buildings.

1. Introduction

Except for lateral and axial loads transferred from superstructures, pile foundations may often be exposed to lateral loads from soil movement usually generated by adjacent surcharge loads, slope sliding or pile installation etc. Such lateral loads may induce bending moments and deflections in the piles, which in the worst instance may lead to structural distress or failure of the piles or superstructures (Stewart, 1992). In practical design, however, effects of such loads are often ignored due to some reasons. As a result, the foundation may be exposed to potential failure. In this paper, a typical failure of one factory building in a steel company in Shanghai, China due to adjacent surcharge loads is presented. Based on a field test in the area, thereafter, causes of the collapse are discussed.

2. Collapse of a Piled Industrial Building

At dawn of 10 Jan. 1998, a piled industrial building in Shanghai, China collapsed (see Figure 1(a)). The roof from the columns ① to ⑤ fell down and the collapse center was located at the bracket beam SA3-1 between the columns ② and ③ as enclosed by a rectangle in Figure 1(b). The frame system between the columns ① and ⑤ experienced severe damage: some steel members twisted, some torn, some tensed to failure and some compressed to yield (see Figure 1 (c) and (d)). The columns connecting the pile caps with the frame system were inclined to the collapse center and some bolts or structure nodes were broken as shown in Figure 1(d). As this building constructed for storing steel members and transferring scoria in 1980, played a pivotal role in the steel making, the whole manufacturing system was slowed down and heavy economical loss was thus made.

In order to restore the production immediately, the building was repaired with an expedient measure, in which the displaced piles were jacked back to the original stations and the connections between the columns and pile caps were reinforced. Because of the emergent clean-up of the debris and immediate repair, evidence for the collapse was not well protected and hence other buildings in the company with similar load conditions, i.e. long-term and heavy surcharge loads on the inner ground, were investigated hereafter.

Shown in Figure 2 are typical lateral displacements and settlements of pile caps in a building with 18-year service. It could be seen that the pile foundations had undergone remarkable lateral movement and settlement. As the loads via the superstructure were relatively small, the excessive displacements would be indebted to the long-term surcharge load inside the factory building. To be further clarified, a field test on the response of piles subjected to adjacent surcharge loads in the area is to be reported later on.

* This work is supported by Shanghai Key Fundamental Research Project (02DJ14062) and Doctoral Base Funding of the Ministry of Education (No. 02DJ14062) .

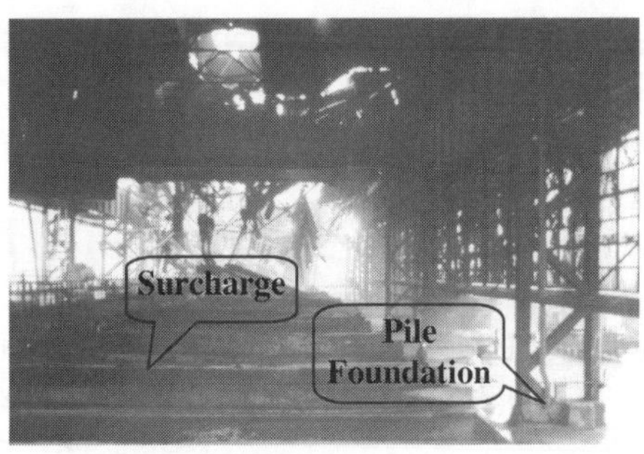

(a) Picture of an industrial building collapse

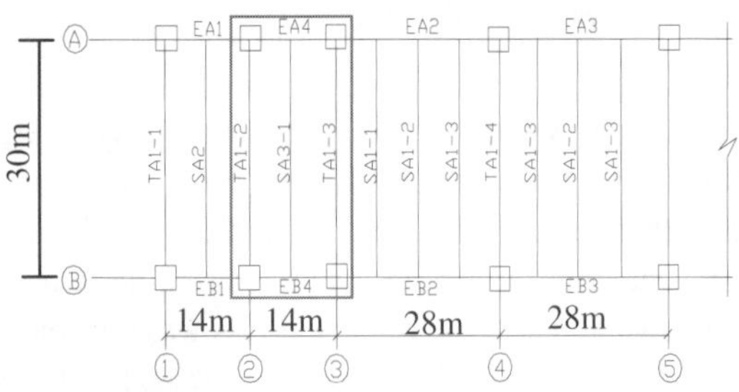

(b) Collapse location and failure center, TA- Crossbeam; EA, EB- longitudinal bracket; SA- Roof

(c) Collapse of the frame

(d) Failure of the frame node

Figure 1. Pictures of an Industrial Building Collapse

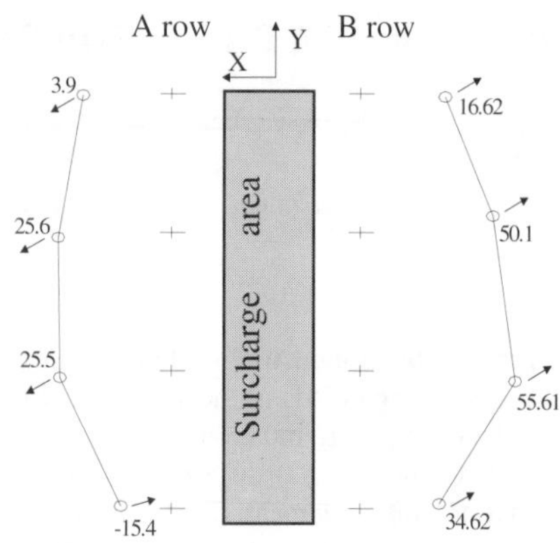

(a) Lateral movements at the pile heads

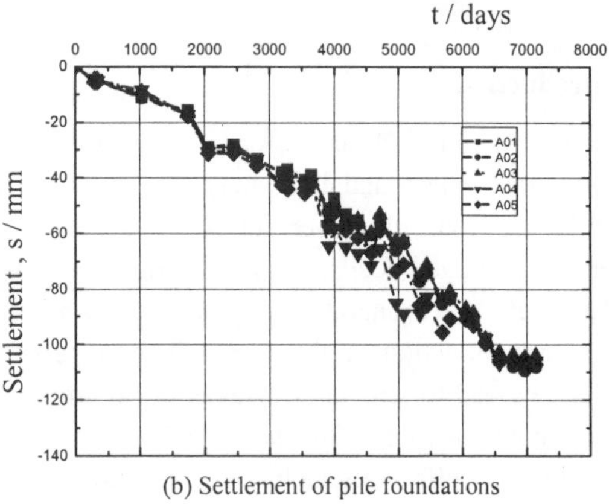

(b) Settlement of pile foundations

Figure 2. Lateral Movements and Settlements of some Pile Foundations in a Building (not to scale)

3. Soil Conditions

The site of the steel company about $1500 \times 6500\text{m}^2$ covered a very flat delta plain and the soil profile was fairly uniform in the whole area, which primarily consisted of (see Figure 3(a)) three categories of soils viz: upper clayey soil, lower sand soil and the transition layers between the clayey soil and the sand. The upper clayey soil comprised a stratum of soft silty clay underlying a 2-3 m thick deposit of stiff silty clay. Below this layer, a 10~12 m thick deposit of soft clay was encountered. In its lower part, the presence of organic matter in the silty clay matrix was quite frequent. The upper clay strata stretched up to 40m below the ground level. The soil profile was shown in Figure 3(a) along with the cone tip resistance, q_c of a typical static cone penetration test in the upper clayey soils. For comparison, the soil profile and the cone tip resistance, q_c in the building after 18-year service are also presented in Figure 3 (b). It can be seen that, (1) the upper stiff silty clay was replaced by backfill; (2)

decrease of the elevations at the bottom of the soft silty clay and soft clay occurred, and (3) the cone tip resistance, q_c increased remarkably. These demonstrated the upper soils experienced consolidated settlement and strength increased under surcharge loading.

Considerable field and laboratory tests were carried out since the building construction including: (1) UU (Unconsolidated, undrained) and CIU (Isotropically consolidated, undrained) triaxial compression tests; (2) Quick direct shear tests (DST); (3) Cone penetration tests (CPT); (4) Standard Penetration Tests (SPT); (5) Consolidation tests; and (6) Creep tests. Corresponding index and mechanical properties (including the average values, number of samples, minimum and maximum values) are tabulated in Table 1. Also provided in the table are values of blow count of SPT tests, N63.5.

It may be see that the upper clayey soils are much weaker than the lower sandy soils. The soft silty clay is of average compressibility, Cc of 0.26 and undrained shear strength, c_u of about 24 kPa (half of the unconfined strength), and the soft clay with Cc of 0.45 and c_u of about 25 kPa. Both of them are highly compressible and rather weak in strength. They are thus susceptible to excessive movements even instability under surcharge loading.

4. Field Study

4.1. *Test program*

Actually, since the first-stage buildings built in 1978 ~1980 were commissioned, the effect of surcharge loads on soil deformation and adjacent piles had attracted sufficient attention, such that a field test was carried out from November 1980 to January 1982 in the area (Yan et al., 1981). The configuration of the field test is schematically shown in Figure 4.For convenience, the orientation is assumed, which may be different from the real orientation of the site.

In the experiment, 22.0m wide by 30.0m long embankment loads were exerted on the ground in the east of the site. Along the length edges, two 3 m high slabs were placed in each side for retaining the embankment, which consisted of gravels for the first three load levels (60 kPa, 90 kPa and 120 kPa) and steel ingots for the fourth load level (150 kPa) increment. To the west of the embankment about 50 cm away, a foundation was installed with four open-ended steel pipe piles with 609.6mm in diameter and 11mm in wall thickness. The piles were arranged as 2×2, 4.0m apart in both directions and connected with a 5.4m × 5.4m × 2.45m (height) pile cap. Each of them had a total length of 60m. The front/lead row was about 1.2 m away from the edge of the embankment. From the symmetrical line (A line) of the embankment to the west, settlement and pore pressure at different depths were monitored in seven sections viz.: sections B, C, D, E, F and G. In the sections of B, C, D, E and the four piles, inclinometers were also instrumented. In the test program, stepwise surcharge loads were exerted on as given in Table 2.

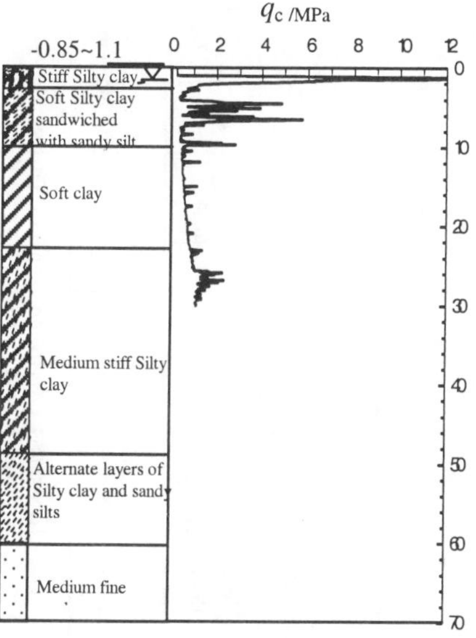

(a) Before surcharge loading

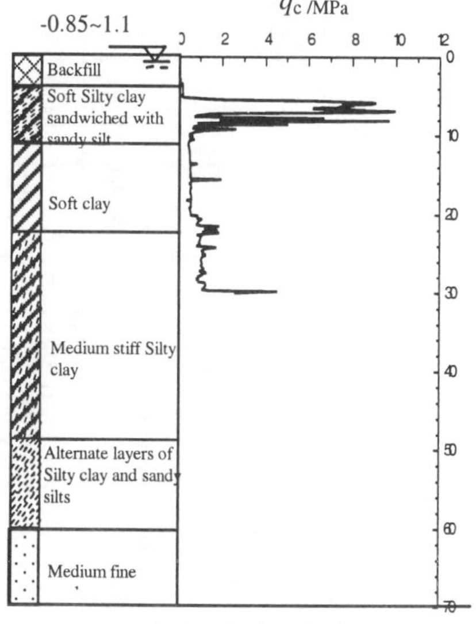

(b) After surcharge loading

Figure 3 Soil Profile, Ground Water Level and Typical Electrical Cone Penetration Test Profile

Table 1 Index properties and mechanical proprieties of soil strata

Soil type	Height (m)	Water content	Unit weight	Void ratio	Liquid limit	Plasticity index	SPT	Compressibility coefficient	Compressibility modulus	Compressibility	Permeability		CU Triaxial Test		Quick Direct Shear Test		Unconfined strength		Sensitivity
		W (%)	γ (kN/m^3)	e	W_L (%)	I_P (%)	$N_{63.5}$	a_{v1-2} (MPa^{-1})	Es (MPa	Cc	K_v (m/s)	K_h (m/s)	φ_{cu} (°)	c_{cu} (kPa)	φ (°)	c (kPa)	Undisturbed (kPa)	Disturbed (kPa)	St
Silty clay	2~3	34.6(126) 24~50#	18.6(125) 17.0~20.0	0.99(125) 0.70~1.40	36.7(125) 31~47	12.9 (4) 12.3~14.0	4.1(149) 0.5~9.5	0.40 (115) 0.10~0.90	5.75 (114) 2.0~14.0	0.22 (112) 0.06~0.44	8.49×10^{-6}(2) $0.33\sim1.37\times10^{-5}$	4.07×10^{-6} (2) $3.57\sim4.57\times10^{-6}$	18(42) 10~34	29(42) 4~61)	16.5 51(4/28)	13(51) 2~32	65(102) 10~150	20(90) 2~51	3.6(94) 1.0~9.0
Soft silty clay	6~8	41.0 (119) 29~57	17.8(111) 16.2~18.8	1.13(123) 0.70~1.70	36.7(121) 30~46	15.6 (7) 13.1~19.9	3.3(432) 0~11.5	0.59 (113) 0.10~1.70	4.13 (84) 1.0~9.0	0.258(79) 0.13~0.51	1.07×10^{-6} (2) $0.34\sim1.81\times10^{-6}$	2.91×10^{-6} (2) $2.19\sim3.62\times10^{-6}$	21(34) 5.7~43	20(34) 10~75	19.2(49) 6~29	5(49) 1~34	48(17) 7~130	19(52) 7~70	3.8(51) 1.0~10.0
Sand-wiched Sandy silt		32.2(35) 25~41	18.9(33) 17.8~20.0	0.90(35) 0.70~1.20	33.6(32) 25~39	-----	8.2(106) 1.5~16.5	0.17 (24) 0.07~0.50	13.46(27) 6.0~26.0	0.090(25) 0.03~0.216	4.05×10^{-5} (3) $0.66\sim10.2\times10^{-5}$	7.02×10^{-5} (3) $1.81\sim17.9\times10^{-5}$	27(5)	75(5)	24(13) 17~31	14(13) 3~25	57(10) 10~323	11(12) 2~22	3.3(14) 1.0~7.0
Soft clay	10~12	48.3(287) 38~60	17.2(289) 1.62~18.0	1.36(290) 1.10~1.70	45.3(289) 35~55	20.9 (17) 17.1~26.7	1.4(906) 0~6	0.91(252) 0.40~1.60	2.59(753) 1.4~4.6	0.446(254) 0.180~0.74	3.61×10^{-7} (4) $1.25\sim9.72\times10^{-7}$	7.68×10^{-7} (4) $1.4\sim23.0\times10^{-7}$	12(112) 7~22	27(112) 5~50	8(125) 3~19	12(125) 1~16	50(233) 10~197	9(199) 4~18	5.8(206) 1.0~15.0
Silty clay	30	33.9(720) 26~42	18.2(723) 17.0~19.8	1.02(708) 0.62~1.24	36.3(735) 21~46	14.6 (27) 12.3~17.1	8.8(1856) 0~22	0.38(642) 0.11~0.76	5.49(536) 2.5~13.5	0.199(623) 0.080~0.48	3.67×10^{-6} (3) $2.03\sim4.79\times10^{-6}$	1.34×10^{-5} (3) $0.80\sim2.39\times10^{-5}$	25(276) 6~40	42(276) 5~100	20(319) 4~34	11(319) 2~50	75(537) 10~160	26(512) 10~90	3.5(480) 1.1~11.5
Sandy silt	Altern-ate layer about 10m	25.3(40) 18~34	18.5(29) 17.1~20.3	0.89(24) 0.68~1.04	27.1(41) 18~36	-----	30.3(214) 5~60	0.20(17) 0.10~0.30	9.56(17) 6.0~14.0	0.112(15) 0.060~0.18	1.89×10^{-4} (5) $1.03\sim5.01\times10^{-4}$	1.83×10^{-4} (5) $0.28\sim2.87\times10^{-4}$	31(6) 18~33	45(6) 40~86	23.5(13) 4~30	12(13) 1~49	37(15) 10~70	17(18) 6~36	3.6(10) 1.0~9.7
Silty clay		32.9(65) 26~42	18.3(56) 17.4~19.4	0.94(57) 0.66~1.14	34.8(67) 24~42	13.2 (4) 12.2~14.8	15.4(217) 6~40	0.36(46) 0.14~0.62	5.19(42) 3.0~8.0	0.193(49) 0.10~0.32	2.09×10^{-6} (2) $2.07\sim2.11\times10^{-6}$	2.91×10^{-6} (2) $2.39\sim3.43\times10^{-6}$	25(11) 16~35	35(11) 22~86	20(26) 8~28	10(26) 2~54	63(33) 10~140	20(23) 4~36	3.7(30) 1.3~8.0
Fine sand	20	24.9(159) 16~34	18.9(58) 17.2~21.2	0.80(60) 0.46~1.06	-----	-----	43.8(1911) 10~90	0.11(17) 0.07~0.16	15.88(17) 6.0~24.8	0.106(11) 0.10~0.212	4.44×10^{-5} (3) $3.03\sim7.08\times10^{-5}$	7.02×10^{-5} (3) $4.37\sim9.69\times10^{-5}$	-----	-----	28(14)	7(14)	-----	-----	-----

#: 34.6(126) —— Average value (number of samples);
24~50 ——Minimum value ~ Maximum value.

Table 2. Loading Program and Measure Period

Load level	Max load (kPa)	Shape of loading	Loading date	Measure period	Days
I	60	$21.6\times29.6m^2$ Gravel $22\times30m^2$	1980.11.23-2.2	1980.12.2-12.24	32
II	90	$16.8\times24.8m^2$ Gravel $22\times30m^2$	80.12.25-2.26	80.12.26-81.4.1	98
III	120	$12\times22m^2$ Gravel $22\times30m^2$	81.4.2-4.6	81.4.6-8.5	126
IV	150	$12\times22m^2$ Ingot Gravel $22\times30m^2$	81.8.6-8.10	81.8.10-82.1.13	161
V	120	$12\times22m^2$ Gravel $22\times30m^2$	82.1.14-1.16	82.1.17-2.9	23
VI	0	-------------	82.2.10~2.15	82.2.27~4.5	---

4.2. Settlement of Soils and Pile Foundations

In Figure 5, relations of settlement (s_v) from items AS0, AS1, AS2, AS3 and AS4 with time are presented. Also included is the load scheme with time. Figure 6 shows the ground settlement profile, obtained by the settlement gauges on the ground surface AS0, BS0, CS0, DS0, ES0, FS0 and GS0, under different load levels. From Figure 5, it seems that the soil settlement mainly occurred in the upper clayey soil. Below 22.5m under the ground level, the settlement increased very little as the time and load increased. In general, the settlement in the soft silty clay and soft clay was greater than two-thirds of the overall settlement. Meanwhile, it may be seen from Figure 6 that at about 22m ($\approx 1.0\ B$) away from the edge of the surcharge, even if the surcharge increased from 60kPa to 150kPa, the ground settlement was negligible. It was only within about 9m ($\approx 0.5B$) away from the edge of the surcharge that the ground settlement was relatively remarkable. Therefore, the primary influence range of the embankment loads was thus about 1.0B in depth and 0.5B in the distance away from the edge of the embankment. It was hence anticipated that the piles within this range would be subjected to soil settlement. The measured settlements at the four corners of the pile cap and corresponding incline ratio are given in Table 3. Since the pile foundation was close to the surcharge, the relative settlement between the piles and soils around the piles was comparatively large. As a consequence, the negative skin friction would be initiated, which would give rise to the decrease of the allowable bearing capacity of pile foundation. In effect, as the upper soft soils were of remarkable creep behavior, negative skin friction could also be developed from secondary compression alone (Fellenius, 1998). It would be expected that the pile foundation would experience long-term settlement, which had been validated by the latest investigation on a building (Figure 2(b)). In addition, because of the shading effect of the front row piles, the piles in front row and rear row underwent different settlements, that is, incline ratio existed between the front and rear row piles (Table 3). Although the mean incline ratio was not so large in this test, the differential settlement might result in secondary stress in the superstructure as studied elsewhere.

5. Lateral Movements of Soils and Piles

Soil lateral movements (s_h) are shown in Figure 7. It demonstrated that lateral soil movement decreased with the distance away from the embankment but increased with time. Near the edge of the embankment, the inclinometer BI experienced big lateral movement when the load was beyond 90 kPa, which was above $3C_u$ (C_u =24~25 kPa for the upper soft clays). The maximum lateral movement at the load of 150kPa was 81.2mm at a depth of 6.3m below the ground level. Under such excessive soil movement, the piles in the soil deformation field underwent considerable deflection (y_h) as well (See Figure 8). At the load of 150kPa, the maximum deflections of the front row piles PI1 and PI3, reached 40.8mm and 36.6mm respectively, at about – 7.5m below ground level after 147 days. Correspondingly, the lateral displacements at the top of the front row piles PI1 and PI3 were 18 mm and 19mm. Nevertheless, the lateral movement was so dependent on time (see Figure 7 and 8) that the pile deflection would increase to excess the acceptable limits.

To date, unfortunately, the acceptable lateral movement of a pile foundation has not been well established because it is highly dependent on the superstructure stiffness, pile foundation configurations and soil conditions. For instance, U.S. Department of Transportation (1985) observed that horizontal abutment movements in excess of 25-50mm are more damaging than differential vertical settlements up to

100mm in causing most types of structural distress. Similarly, piled industrial buildings may also be sensitive to excessive lateral displacement of their pile foundations. For any reliable design of piled industrial building, therefore, the lateral displacement prediction of pile foundation should be evaluated as well as the settlement of the pile foundation.

From the analyses above, the field test demonstrated that:

1. Soil lateral movement could induce fairly large deflection and moments in the piles;
2. Soil settlement could impose huge downdrag forces on adjacent piles, which would reduce the bearing capacity of the piles. The difference in downdrag forces between the front and the rear piles could aggregate differential settlement;
3. The remarkable creep behavior of the soft clays implied that soil movement and pile settlement would increase under long-term repeated loading.

Even so, no measurements had been taken to reinforce the ground or the piled buildings in the steel company mentioned previously. Meanwhile, as the production scale increased, the surcharge loads were unexpected higher than the allowable capacity. Therefore, the collapse described previously and structural distress in other buildings were unavoidable.

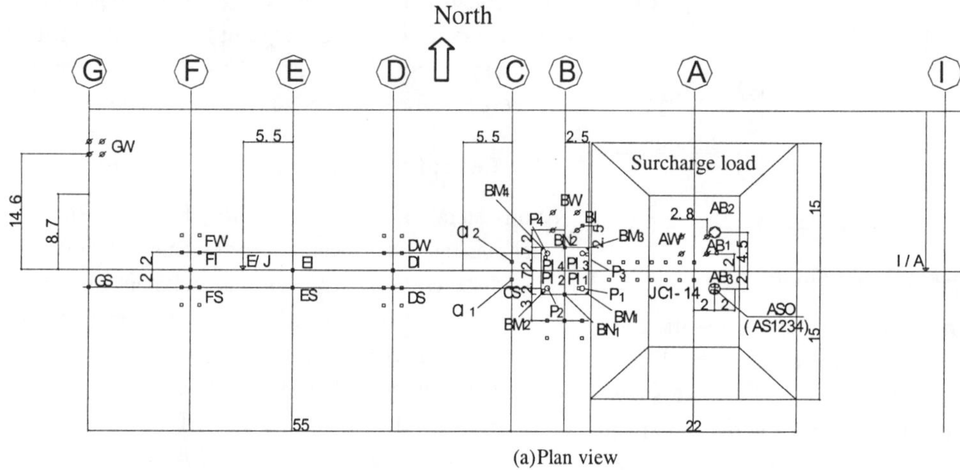

(a)Plan view

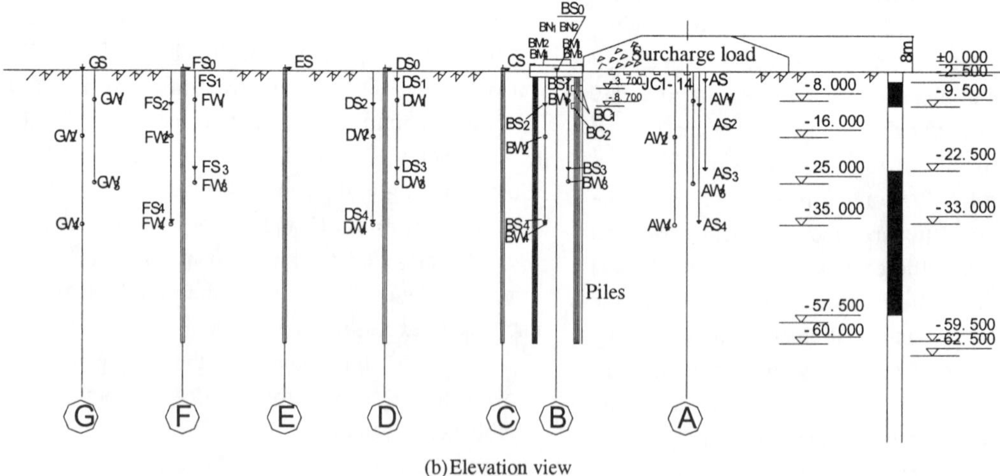

(b)Elevation view

AS0, AS1, AS2, AS3, AS4, BS0, BS1, … , FS4, GS0 — Settlement dial gauge; AW1, AW2, AW3, AW4, BW1, … , GW4 — Water pressure cell; BI, CI1, CI2, DI, EI, FI, PI1, PI2, PI3, PI4 — Inclinometer; BM_1、BM_2、BM_3、BM_4— Settlement gauge for the four corners of the foundation; BN1, BN2 — Settlement dial gauge for the southern and northern sides of the foundation, respectively; AB1, AB2, AB3 — Pressure cell under surcharge loads.

Figure 4. Configuration of the Field Test (Yan, 1981)

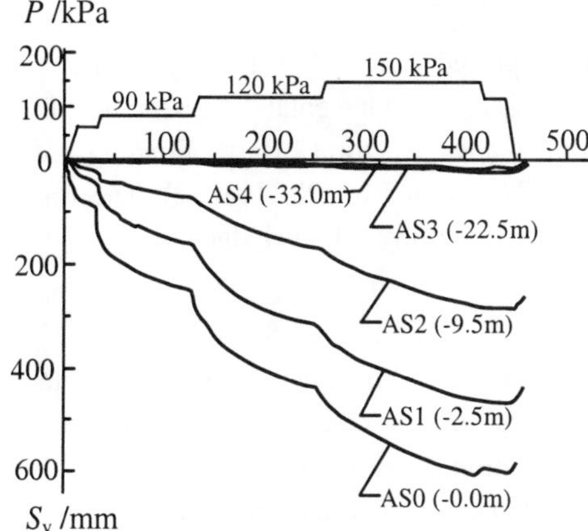

Figure5 P~t Curve and S_v~t Curve

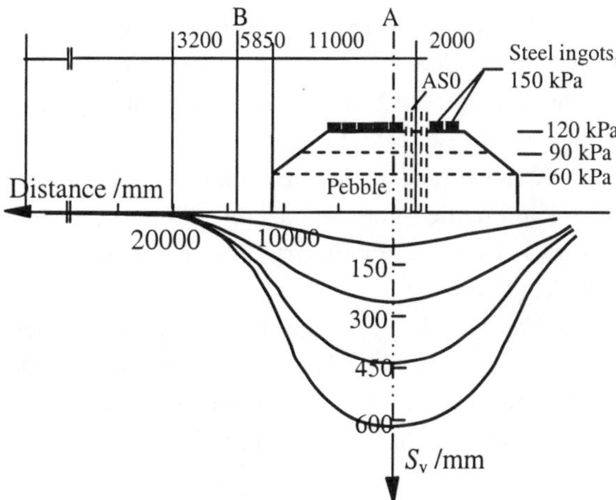

Figure 6 Settlement Profile along the Direction of Width

Table 3. Settlements and Incline of Pile Foundation

Point	Loads (kPa)			
	60	90	120	150
BM1	6.2	8.3	10.9	15.6
BM2	3.1	4.1	5.2	9.8
BM3	6.1	7.9	10.4	15.7
BM4	3.3	4.0	4.8	9.8
Mean settlement	4.7	6.1	7.8	12.7
Mean incline ratio tgφ (‰)	0.59	0.81	1.13	1.17

6. Conclusion

A collapse of a piled industrial building in Shanghai, China is reported in this paper. Based on a field test in the same area with soft clay soils, the causes of the collapse are investigated. The field test of pile foundation adjacent to surcharge loads indicated that (1) the pile foundation may be subjected to negative skin friction load induced by the soil settlement and the downdrag load would lead to the reduction of pile capacity and the differential settlement between the front and rear piles; (2) the pile foundation may produce considerable deflection due to soil lateral movement caused by the surcharge loads. The excessive secondary stress in the superstructure induced by the differential settlement and lateral displacement of pile foundation, and the reduction of the pile capacity may be the main reason for the collapse (studied elsewhere). Therefore, the response of pile foundation subjected to soil movement should be examined in the design of piled industrial buildings.

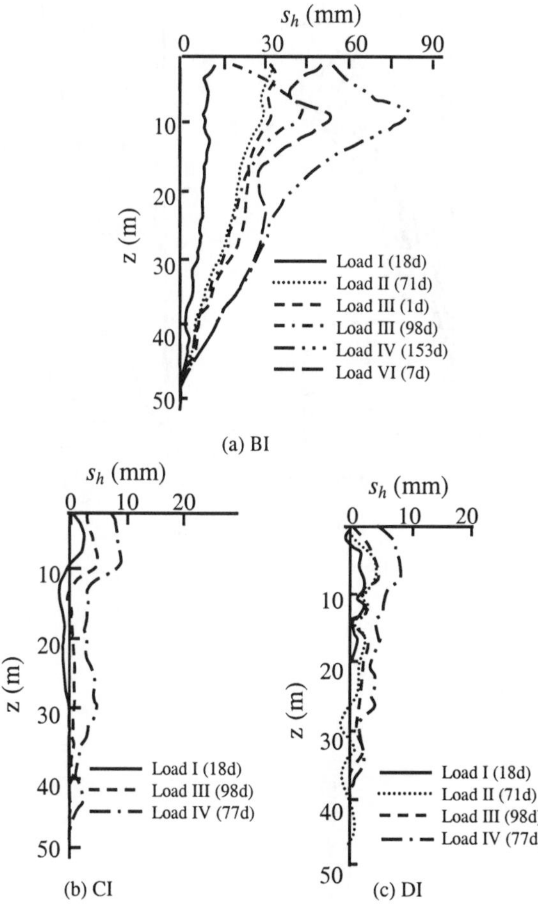

Figure 7. Lateral Soil Movement

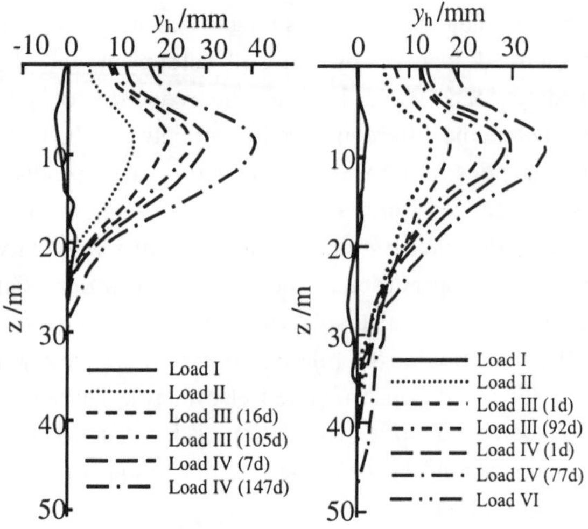

(a) Front row pile (PI1 and PI3)

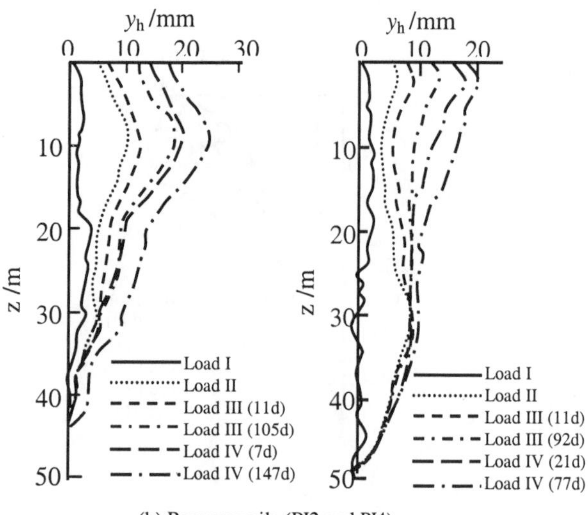

(b) Rear row pile (PI2 and PI4)

Figure 8. Pile Deflection

Acknowledgements

The work reported herein is currently sponsored by Shanghai Key Fundamental Research Project (02DJ14062) and Doctoral Base Funding of the Ministry of Education (No. 02DJ14062). The financial assistances are gratefully acknowledged.

References

1. M. Bozuzuk. "Bridge foundations move," Transportation Research Record 678, Transportation Research Board, Wash., 17 (1978).
2. B. Fellenius. "Recent advances in the design of piles for axial loads, dragloads, downdrag, and settlement," (1998) http://www.unisoftltd.com/techpapers /dragload_downdrag_211.pdf.
3. D. Stewart. "Lateral loading of piled bridge abutments due to surcharge construction," PhD Thesis, Dept. of Civil and Environmental Eng., Uni. of Western Aus.(1992).
4. F. Kulhawy and P. Mayne. Manual on estimating soil properties for foundation design. Research Report-EL-6800 for Electric Power Research Inst., USA (1990).
5. R. Yan et al. Full-scale test on passive piles. Proceedings of 4th National conference on Soil Mechanics and Foundation Engineering, China Architecture Publisher (1981) (in Chinese).

RESEARCH ON THE APPLICATION OF PRESTRESSED TECHNOLOGY TO FLOOR OF SUPER TALL BUILDING

XUEYU XIONG

College of Civil Engineering, Tongji University, 1239 Siping Road Shanghai 200092, China

YUE CAI

Profession Technology Education Institute, Tongji University, 1239 Siping Road Shanghai 200092, China

JISHU WANG

College of Civil Engineering, Tongji University, 1239 Siping Road Shanghai 200092 China

Prestressed technology is more and more applied to floor of super tall building because of its one's own characteristic. This paper takes a super tall building in shanghai as an example to discuss the application of prestressed technology to floor system of super tall building. Comparison among three prestressed floor designs is shown. Earthquake resistant performance in the entire structure and secondary interior force under restraint of tube structure are analyzed.

1. Introduction

With the development of modern structural material and design theory of the structure and computer method, a large number of tall building are built up in succession. And prestressed technology is more and more applied to these building, especially used for floor. In super tall building or the buildings with wide room, if the span of floor is enough long and the floor height is limited, the thickness and weight of the floor without prestressed technology are surprising.

One of the main reasons prestressed technology is adopted in tall buildings is that it can reduce structure height. The reducing of the structure height, on one hand, can increase headroom; on the other hand, can increase the quantity of storey in the situation that the total height does not change. It is always special economical meaning that we adopt prestressed technology when the total height of building is limited to certain height as a result of limit of fire control or urban planning. Therefore, the fabrication cost of one-sided concrete has risen after prestressed technology is adopted, but the comprehensive economic benefits are very remarkable. In addition, due to compress building storey height, the expenses of air conditioner and maintaining is remarkably decreased. Compared with reinforced concrete floor, the weight of prestressed concrete floor is lessened because prestressed technology can reduce the thickness of floor and the size of beam section. On account of it, the application of prestressed technology can reduce the section of support (wall and column), reduce earthquake load, improve the using area, and cut down fabrication cost of foundation and footing. Besides, adopting prestressed technology can improved the using function of structure. Under the weight of itself and permanent load, the structure deflection is little, and the crack is inconspicuous. Just because of these outstanding advantages, prestressed technology is widely applied to the floor of tall buildings. Familiar manner of prestressed concrete floor applied to tall buildings is as follows: prestressed concrete plat slab without beam , prestressed shallow beam-frame , prestressed shallow beam-plat slab ,prestressed intersecting beam ,combined slab etc. But there is much dispute about the application of prestressed technology to floor system of super tall buildings, including the prestressed technology influence on aseismic performance in the entire structure and the problem of secondary interior force under restraint of tube structure etc.

The paper takes a building as an example to discuss the application of prestressed technology to floor system of super tall buildings.

2. Project General Conditions

The Building is located at the intersection of west Nanjing road and Changde road. The comprehensive building for trade, office and hotel covers 21,308 square meters, with a total floor space of 202,686.6 square meters. The height of office building is 188.9 meters, with 43 aboveground storeys and 3 underground storeys with floor space of 54920.6 square meters. The height of hotel is 99 meters, with 24 aboveground storeys and 3 underground storeys with floor space of 54920.6 square meters. This paper only concerns the structure design of the office building. The design for structure plane arrangement of office building is as follows:

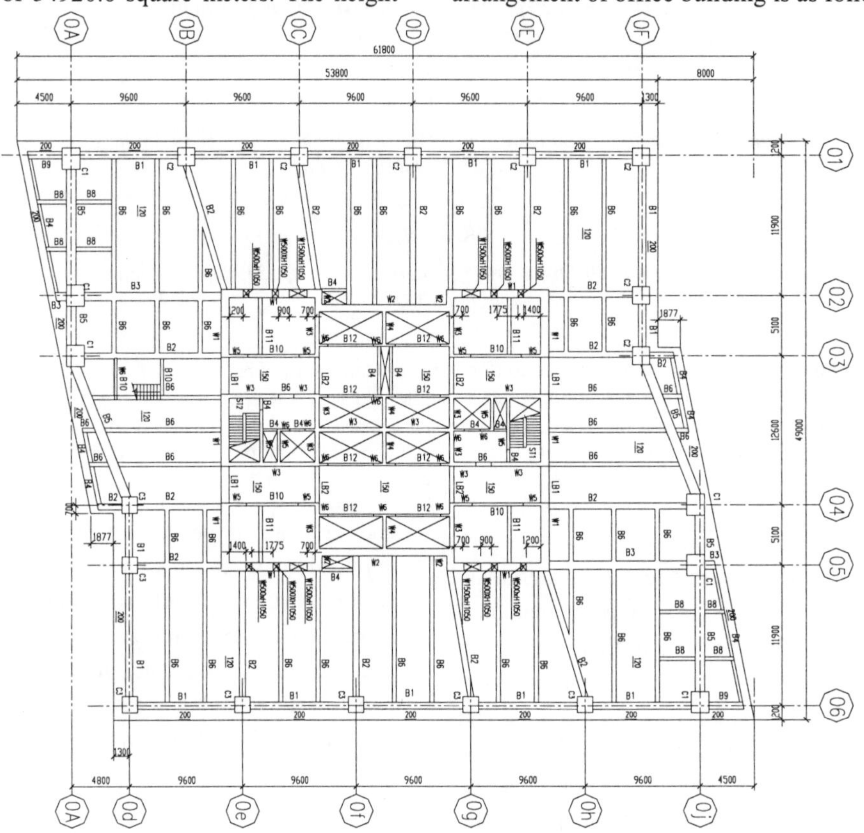

Figure 1. The design for structure plane arrangement of office building

3. The Comparison among Prestressed Structure Designs

3.1 *The first prestressed structure design*

The height of former beam, apart from boundary beam and inner tube, is reduced to 400 millimeters as result of adopting prestressed technology. And the span of slab is shortened. The thickness of slab is 120 meters. Compared with the former scheme, there is a decrease of 23 cubic meters per storey in the concrete consumption storey is and an increase of about 7.5 tons per storey in the steel bar consumption. According to past project experience, the ordinary steel bar consumption can be reduced by a large degree due to adopt this scheme.

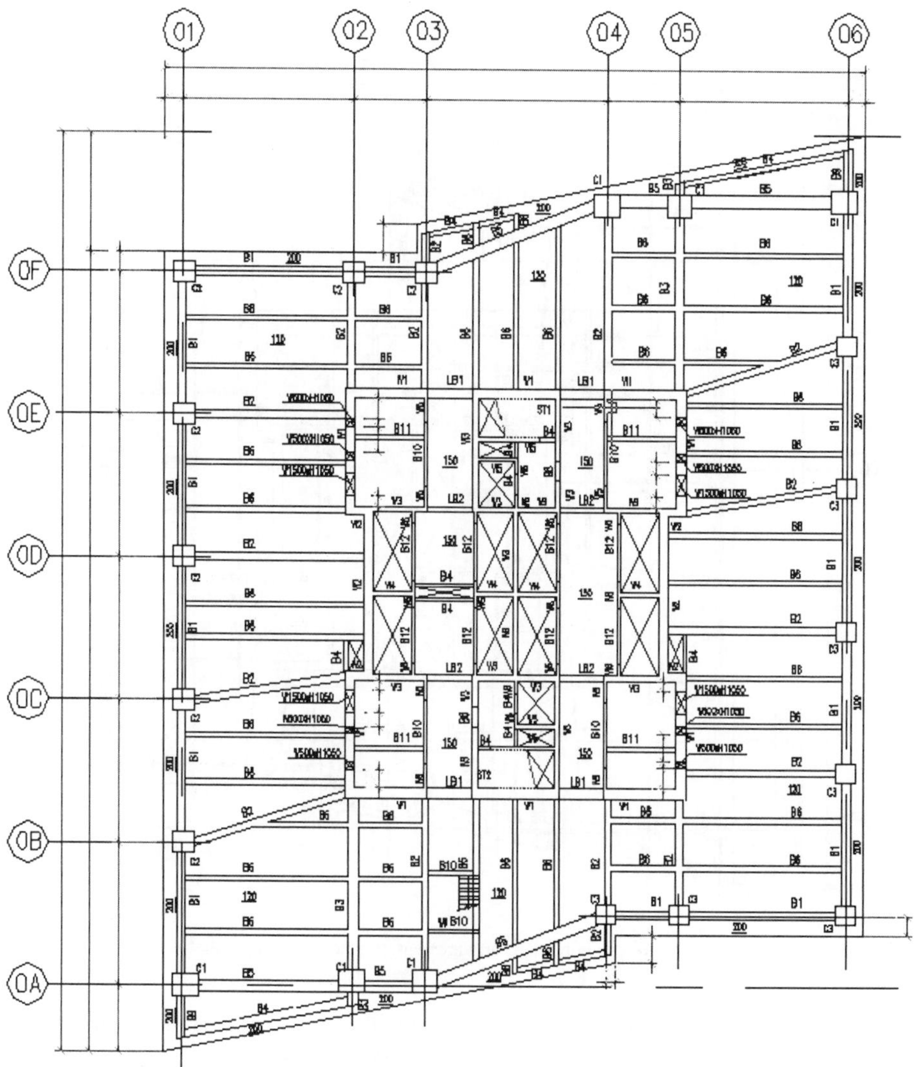

Figure 2. The design for structure plane arrangement of office building of the first scheme

3.2 *The second prestressed structure design*

The thickness of slab can be 160 as a result of adopting the second scheme to adopt wide shallow beam instead of sub-beam based on the former scheme. There is an increase in concrete consumption of slab, but the concrete consumption of beam can be reduced. Therefore, the total concrete consumption is slightly reduced than the ordinary concrete scheme.

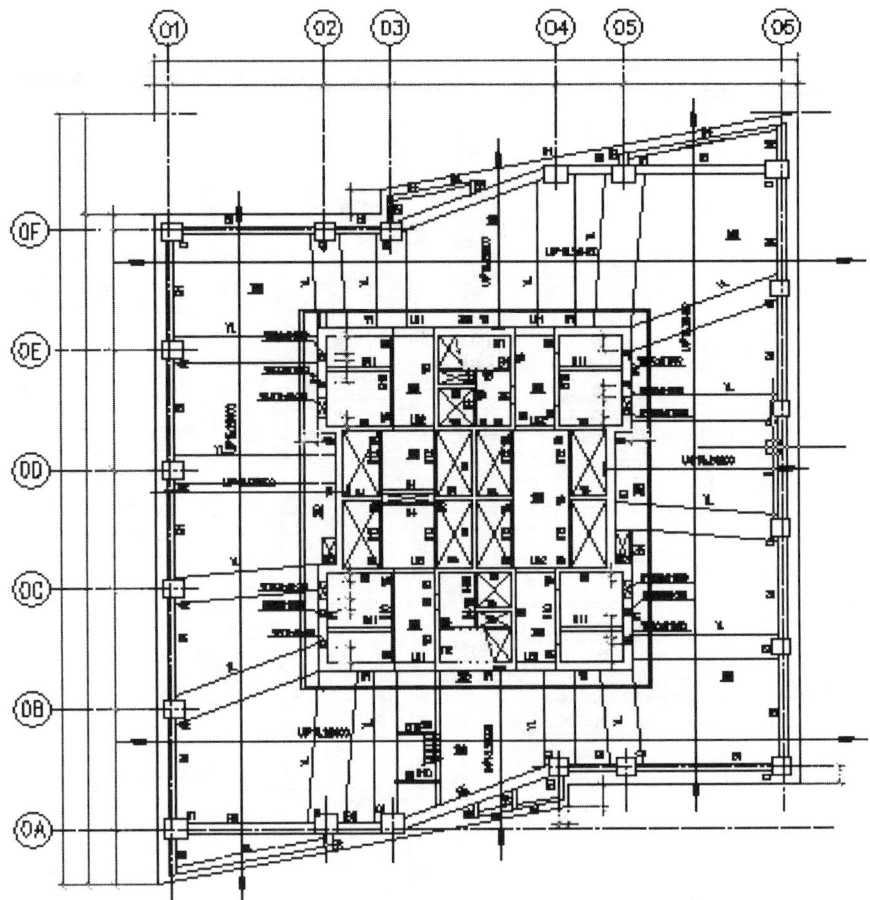

Figure 3. The design for structure plane arrangement of office building of the second scheme

3.3 *The third prestressed structure design*

The thickness of plat slab can be 190 millimeters, with the armpit thickness 280 millimeters, as result of adopt the third scheme to arrange concealed beam on the four corners of plat slab with wide shallow beam. Compared with the ordinary concrete scheme, there is a decrease of 15 cubic meters per storey in the concrete consumption storey and an increase of about 19 tons per storey in the prestressed steel bar consumption.

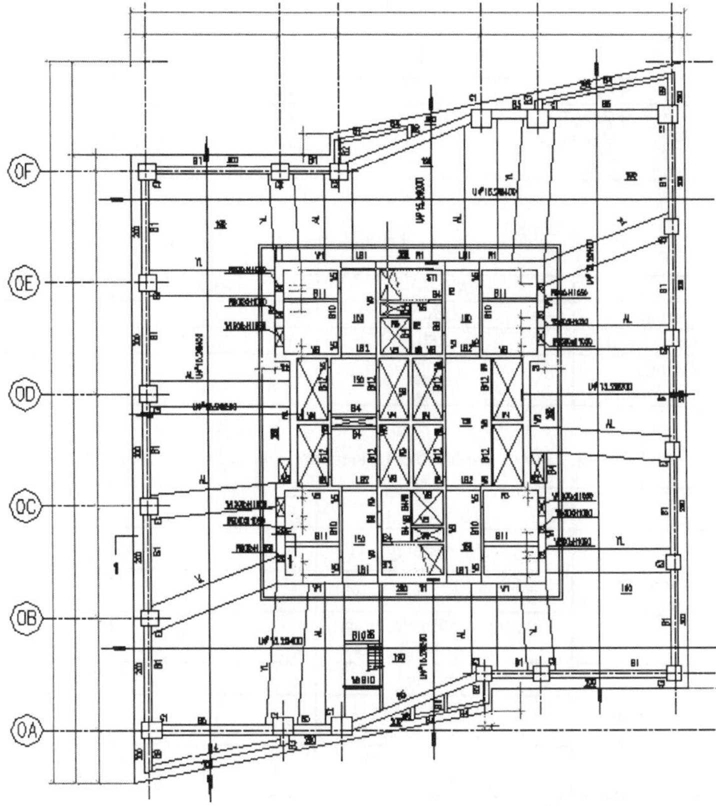

Figure 4. The design for structure plane arrangement of office building of the third scheme

3.4 *Comparison among structure designs*

The first scheme adopts un-boned prestressed shallow beam. Compared with ordinary concrete scheme, the storey height can be lower by 400 millimeters, the concrete consumption can be decreased, and the span of slab can be shortened, but the steel bar consumption is increased.

The second scheme adopts the un-boned prestressed shallow beam-slab. The slab thickness is 160 millimeters. Compared with ordinary concrete scheme, the concrete consumption can be decreased.

The third scheme adopts un-boned prestressed plat slab, with arranging wide shallow beam. Variable cross-section is adopted in order to lower the slab thickness and make full use of construct arrangement. Compared with ordinary concrete scheme, the storey height can be reduced by 610 millimeters, the steel bar consumption is increased, and the concrete consumption is almost kept. But as result of adopting this floor system, the storey height can be reduced. And the construction procedure can be convenient; therefore, the construction period can be remarkably reduced. By arranging wide shallow beam on four corners, the storey height is reduced.

The total height is reduced in the situation that the level number does change. Therefore the expenses of air conditioner, maintaining and hydroelectricity are remarkably decreased. If the plot ratio is allowed, the floor space can be increased. The comprehensive economic benefits are remarkably.

4. The Analysis of the Earth Proof Performance of Prestressed Structure Design

In analysis of earthquake response to the third scheme, the damping ration of reinforced concrete is 0.05, and the seismic fortification intensity is 7 degree, the max influence factor is 0.08, the field type is Ⅳ.The gravity load combination contains permanent load and variable load (50%). CQC of model combination is adopted. The results drawn from SATWE and PMSAP are as follow: table 1、table 2、 table 3.

Table 1. period(s)

	period(s)		SATWE Torsion coefficient (%)
	SATWE	PMSAP	
T1	4.8807	4.7636	0
T2	3.4675	3.4610	4
T3	2.9605	2.6587	93

Table 2. torsional period、translational period

	SATWE	PMSAP	code
torsional period	2.9605	2.6587	--
first translational period	4.8807	4.7636	
torsional period / translational period	0.61	0.56	<0.85

Table 3. storey displacement under wind and earthquake function

			SATWE	PMSAP
Max interstory drift	X Direction earthquake	position(storey)	35	41
		displacement/storey height	1/1160	1/1280
		First storey's displacement /storey height	1/3268	1/4347
	Y Direction earthquake	position(storey)	29	35
		Displacement /storey height	1/957	1/1016
		First storey's displacement /storey height	1/2632	1/2557
	X Direction wind	position(storey)	35	38
		Displacement /storey height	1/3457	1/3623
	Y Direction wind	position(storey)	28	28
		displacement/storey height	1/1357	1/1560

Known from the result of calculation the ratio of torsional period to translational period meet the demand of code, and even coupling influence is very small. The mentioned above interstory drift also meet the demand of code. Therefore prestressed technology applied to (super) tall buildings has little effect on the earth proof performance in the entire structure.

5. Secondary Interior Force under Restraint of Tube Structure

The frame-tube structural system is used in the building. The stiffness of column and wall are relatively large. Meanwhile, axial force will be weakened while prestressed reinforcement is drawn. Therefore, prestressed component produce hyperstatic axial force, namely the weakening degree of prestressing, which depends on the vertical bending stiffness. To prestressed beam, hyperstatic axial force make beam drawn, which is so large that it influence prestressing. Accordingly, the bearing performance, including crack resistance of beam, is influenced. On strength, crack, and deflection, influence of hyperstatic axial force is completely considered in the current norm. The special attention is paid to the influence which hyperstatic axial force has on entire structure while prestressed technology is applied to tall buildings.

In order to analyse this unfavorable factor, the space integral analytical method is adopted. The result drawn from this method is considerable rational. This method is different from the simple method, whose precision is higher than two-dimension analytical one. The hyperstatic axial force diagram of standard storey is as follow:

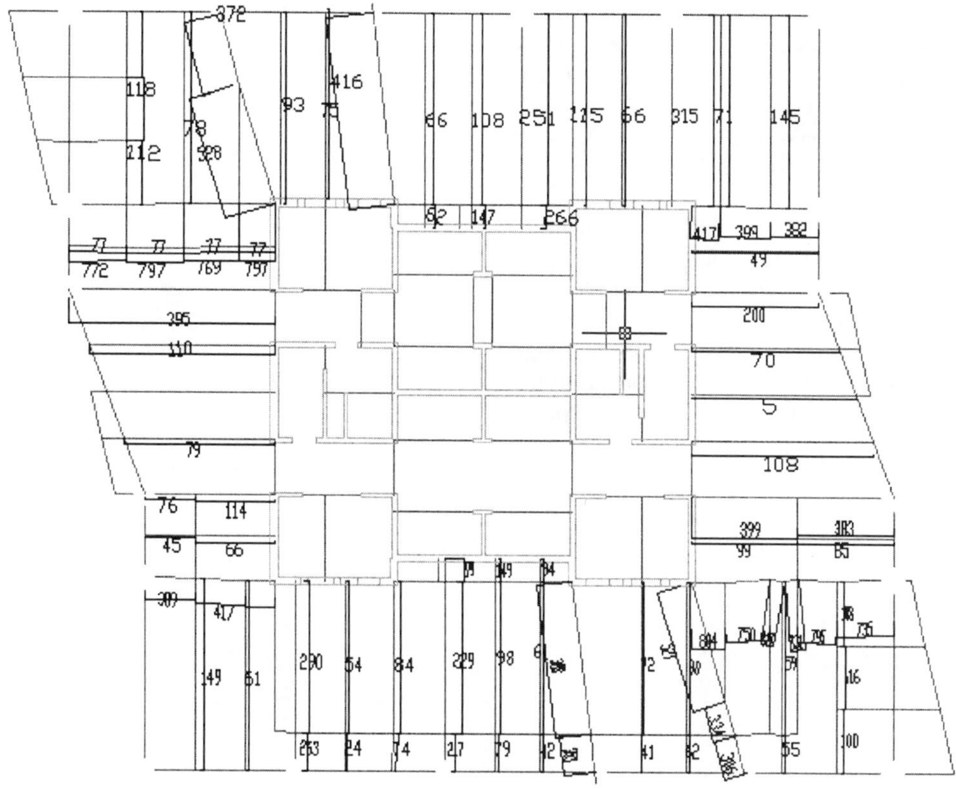

Figure5. hyperstatic axial force of standard storey

Shown from the above figure 5, the max hyperstatic axial force is 804 KN. The total tension of beam is 3754KN. The hyperstatic axial force has accounted for 21% of effective prestressing force. So the unfavourable factor that hyperstatic axial force cause is considered in the design

6. Conclusions

In the past 20 years, the development of tall building of our country has attracted people's attention. A set of (super) tall buildings are built up in succession. The dead load of structure is reduced and storey height is lowered as result of adopting prestressed technology. The comprehensive economic benefits are remarkable. Recently application of prestressed technology to tall buildings has been still in dispute. But known from the analytical result of this paper, it is feasible to apply prestressed technology to floor system of (super) tall buildings.

References

1. Lu Xilin, Tall Buildings Structure (second edition). Wuhan: Wuhan University of Technology Press 2003.
2. Jiao Zhengang and Tao Xuekang, Research of aseismic Performance and Response Spectrum of Prestressed Concrete Structures. Engineering Earthquake, 2000(30).
3. Alfred A. Yee, Social and Environmental Benefits of Precast Concrete Technology. Building Structure, 2004 34(1).
4. Xu Jinsheng and Xue Llihong, Modern Prestressed Concrete Floor. Beijing: China Building Industry Publishing Compony, 1998.
5. Chen Yufeng, Application of Prestressed Concrete Shallow Beam-slab System to Tall Building. Fujian Architecture, 2000(2).
6. Wu Jing, Design of Unbonded Prestressed Concrete Slab Floor System in Multi-storey and High-rise Buildings. the Paper for Ph.D of Southeast University.

7. Qian Zhong, Liu Jinhui, Fen Liangci and Yuan mu, Application of Unbonded Prestressed Wide Shallow Beam to Tall Building. OVM Communication, 2001(1).
8. He Sixun, Construction of Pretensioned Prestressed Precast Beam and Slab in Tall Building. Building construction, 1998(4).
9. Du Gongchen, Application and Development of Presetressed Concrete For Buildings in China in the Past 40 Years. China Civil Engineering Journal, 1997,30(1).
10. Xiong xueyu and Huang dingye. Guide of design and construction of prestressed structure. Beijing: China Building Industry Publishing Company, 2003.

EFFECT OF CONCRETE STRENGTH ON SHEAR RESISTANCE OF STUD CONNECTORS AND MOMENT RESISTANCE OF COMPOSITE BEAMS

HU XIAMIN YANG WEI ZHANG QI

College of Civil Engineering, Nanjing University of Technology, NO.200 Zhongshan north Road

Nanjing, 210009, P.R. China

In recent decades, composite steel-concrete beams have been more and more used in tall buildings. The mechanical connectors are commonly used to transfer longitudinal shear forces across the interface of the steel beam and concrete slab in composite beam design. The empirical equation for the shear strength of stud connectors, which was presented by Ollgaard et al in 1971, has been incorporated in codes of various countries. In this paper, based on experimental data of 95 push-out specimens and 11 composite beams with partial shear connection from home and abroad, the effects of concrete strength on the shear resistance of stud connectors and moment resistance of composite beams are analyzed and investigated. It was found that the shear strength calculated according to Ollgaard's empirical equation is lower than that of test results when the concrete strength is lower, while higher when the concrete strength is higher. And the same conclusions are got in composite beams with partial shear connection. A modified formula for the shear resistance of stud connectors is presented by means of regression analysis of the test data. From comparing calculated values with the results of push-out specimens and composite beams with partial shear connection, it is observed that the error of equation proposed by Ollgaard can be eliminated when used the modified formula proposed this paper. And the calculated value is in good agreement with the experimental results.

1. Introduction

In recent decades, composite steel-concrete beams have been more and more used in tall buildings. And in composite beams shear connectors are commonly used to transfer longitudinal shear between the steel beam and concrete slab and to prevent uplift of the slab from the steel beam. The headed stud is the most widely used type of connector in composite construction. The mechanical properties of shear connectors in composite beams are usually described by the results of push-out tests, which are shown in Fig.1.

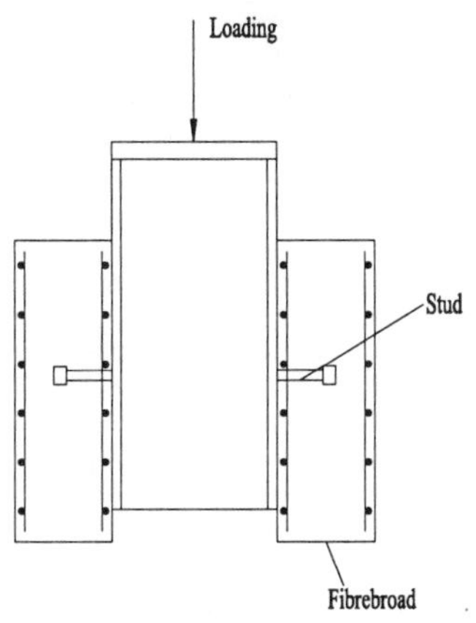

Fig.1 Push-out test specimen

Since the 1950s, a lot of push-out tests have been done and analyzed systemically in many countries, such as USA, Western European, Japan and Canada. And in 1971, based on experimental study and academic analysis, Ollgaard, Slutter and Fisher presented an empirical regression formula[1], which is shown in Eq.(1), when the height-to-diameter ratio for stud *h*/*d* is greater than 4.

$$P_t = 1.83 A_s f_c'^{0.3} E_c^{0.44} \quad (1)$$

Where P_t is the predicted value for shear resistance of the stud connector; A_s is the cross-sectional area of the shank of stud; f_c' is the cylinder compressive strength of the concrete; E_c is the elastic modulus of the concrete.

For the convenience of design, Ollgaard et al. suggested that Eq.(2)should replace Eq.(1). And Eq.(2) is given as following:

$$P = 0.5 A_s \sqrt{f_c' E_c} \le f_u A_s \quad (2)$$

Where f_u is the ultimate tensile strength of the stud.

The formula proposed by Ollgaard et al. has been wildly incorporated in codes of various countries because of its simple form, uniform dimension and can be applied to both normal weight concrete and light-weight aggregate concrete.

Since the mechanical properties of push-out tests are complicated and not amenable to simple calculations, the present formula for shear resistance of the stud connectors doesn't consider some factors comprehensively and correctly. As a result, the present formula is conservative in some case, while it is unsafe in other case[2].

In this paper, based on 95 push-out experimental data in Ref.1 and Ref.3-11 (exclusive of push-out specimens with light-weight aggregate concrete), the effects of concrete strength ($f_c' = 16.6 \sim 91.2\,\text{N/mm}^2$) on the shear resistance of stud connectors are analyzed and investigated. And then the disadvantages of Eq.(2) are pointed out and a modified suggestion are put forward. At the same time, on the basis of discussion about shear resistance of stud connectors, the effects of the calculated error of stud connectors shear resistance on moment resistance of composite beams should be analyzed and discussed through checking 11 composite beams with partial shear connection from home and abroad.

2. Effects of Concrete Strength on Shear Resistance of Stud Connectors

2.1 *Failure modes of stud connectors*

According to experimental observation, the failure modes of stud connectors can be classified two types, as described in Ref. 12.

1) Failure of concrete around the stud connectors. If the concrete in the slab is relatively weak to the stud, the concrete near the root of the stud is cracked or crushed due to local compression in failure. The ultimate capacity increases with the increase of the diameter of the stud and the concrete strength. And the stud has a better ductility. However, if the length of the stud is not enough, the ultimate capacity will be reduced because the studs may be pulled from concrete slab. The experimental results shows the effects of stud length on the shear resistance of the stud may be neglected when the length-diameter ratio of the stud *h*/*d* is greater than 4.

2) Failure of studs due to combination of tensile and shear. If the stud is relatively weak to the concrete, the studs will reach the ultimate strength due to combination of tensile and shear in failure. The shear resistance increases with the increase of the diameter of the stud and the stud ultimate tensile strength, and it is not depend on the concrete strength. In addition, the welding seam of the stud and the friction between steel and concrete slab will resist some shear. However, if the welding quality is not good, the failure may occur at the welding seam.

2.2 *Evaluation of experimental data with push-out specimens*

As mentioned above, it is very difficult to consider all the factors that affect the shear resistance of the stud in theory because of its complex behavior. In practice, the shear resistance of the stud is determined from the statistical analysis method. It is observed from Eq.(1) and Eq.(2) proposed by Ollgaard et al. that the shear resistance of the stud is mainly depend on the concrete compressive strength, the concrete elastic modulus, the stud shank cross-sectional area and the ultimate tensile strength of the stud. In order to check Eq.(2), 95 push-out experimental data (exclusive of push-out specimens with light-weight aggregate concrete) in Ref.1 and Ref.3-11 , in which 76 push-out specimens were used to determine the design strength of the shear studs connectors in Eurocode 4[13], are analyzed in this paper. The relation for concrete failure between the concrete compressive strength f_c' and the ratio $x = P_e/P_t$ is given in Fig.2, where P_e is the experimental determined shear strength and P_t is the shear resistance calculated according to Eq.(2). And the

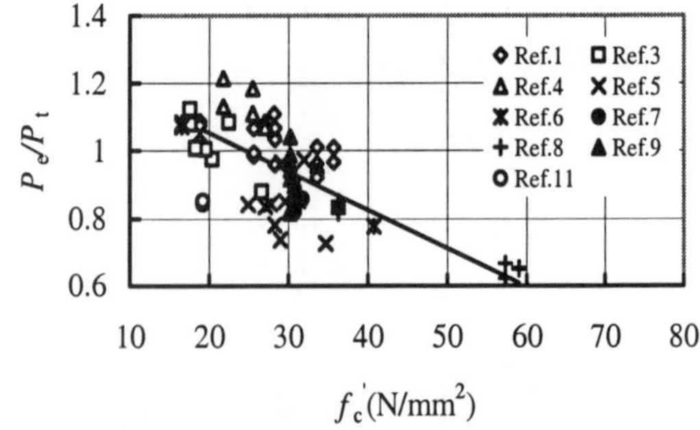

Fig.2 P_e/P_t vs. f_c' for concrete failure according to Ollgaard' s equation

relation for stud failure between them is shown in Fig.3.

When concrete failure, the mean value of the ratio $x = P_e/P_t$ in Fig.2 is 0.932, with a standard deviation σ of 0.140 and a coefficient of variation C_v is 0.150. It is found from Fig.2 that Eq.(2) underestimated the stud shear resistance when the concrete strength is lower, and overestimated it while the concrete strength is higher. And Fig.2 also shows the calculated values according to Eq.(2) are bigger than the experimental results when f_c' is greater than 24N/mm^2, which is equivalent to f_{cu}>30N/mm^2. For the upper limit of Eq.(2), the concrete strength doesn't affect the stud shear resistance basically, as observed in Fig.3. The mean value of the ratio $x = P_e/P_t$ for 20 push-out specimens is 1.099, with σ= 0.059 and C_v=0.054. It can be observed from tests that the ultimate tensile strength of the stud is approximately equivalent to the mean value of the stud failure due to combination of tensile and shear[2,13].

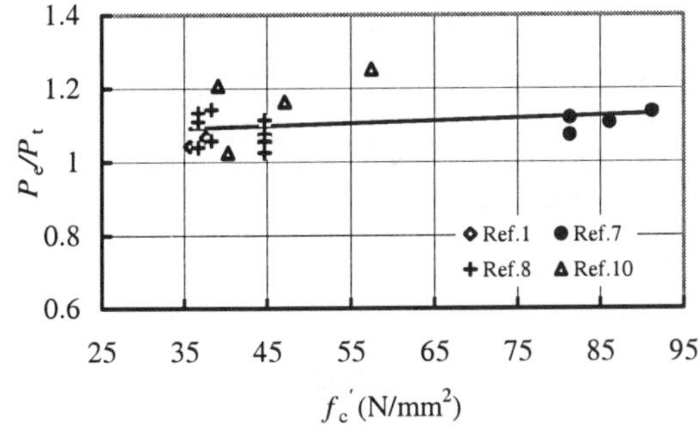

Fig.3 Pe/Pt vs. for steel failure according to Ollgaard's equation

It is observed from above analysis that the effects of concrete strength on the shear resistance of stud connectors will be better reflected if Eq.(2) will be revised as following:

$$P_t = 0.5\alpha_c A_s\sqrt{f_c'E_c} \le f_u A_s \quad (3)$$

Where α_c is a modified coefficient considering the effect of concrete strength on the shear resistance of stud connectors, and

$$\alpha_c = 1.28\,(1-0.9f_c'/100) \quad (4)$$

Fig.4 gives the relation between the ratio of the calculated value P_e/P_t and concrete strength f_c', where P_t is calculated according to Eq.(3). And the mean value of P_e / P_t for concrete failure is 1.005, with σ=0.092 and C_v=0.092.

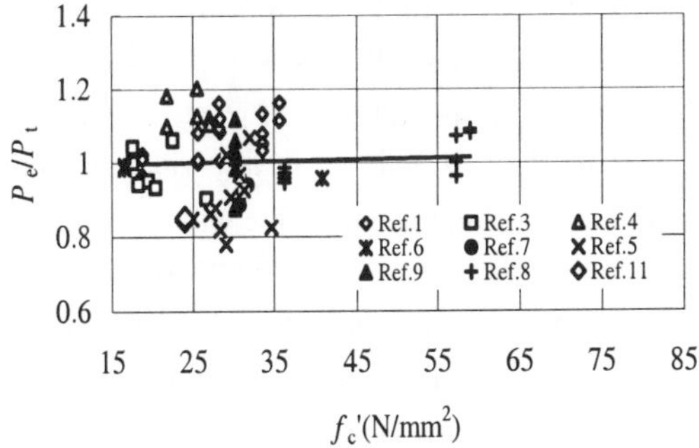

Fig.4 P_e/P_t vs. f_c' for concrete failure according to proposed equation

3. Effects of Computed Error of Shear Resistance of Stud Connectors on Moment Resistance of Composite Beams

In order to check the effects of computed error of shear resistance of stud connectors on the moment resistance of composite beams, 11 composite beams with partial shear connection from Ref.11 and Ref.14-15 are checked according to the plastic theory[16]. When neutral axis lies in the upper flange of steel beam (Fig.5), the plastic moment resistance of composite cross-section is given as following:

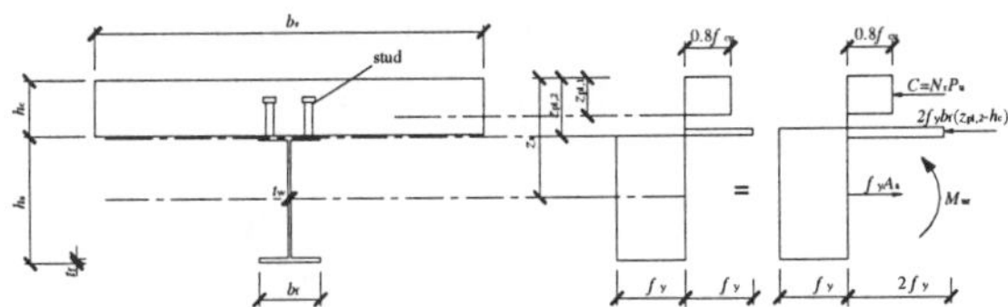

Fig.5 Plastic stress distribution of composite cross-section when neutral axis lies in the upper flange of steel beam

$$M_{ur} = f_y A_a(z_a - h_c) + N_r P_u(h_c - z_{Pl,1}/2) - f_y b_f\left(z_{Pl,2} - h_c\right)^2 \quad (5)$$

$$z_{Pl,1} = N_r P_u / 0.8 f_{cu} b_e \quad (6)$$

$$z_{Pl,2} = h_c + (f_y A_s - N_r P_u)/2 f_y b_f \quad (7)$$

Where f_y is the yield strength of steel beam; A_a is the cross-sectional area of steel beam; h_c is the thickness of concrete slab; z_a is the distance from the centroid axis of steel beam to the top of concrete slab; $z_{Pl,1}$ is the compressive depth of concrete slab; $z_{Pl,2}$ is the distance from the neutral axis of steel beams to the top of concrete slab; b_f and b_e is the width of upper flange of steel beams and the effective width of concrete slab respectively; f_{cu} is the cube compressive strength of concrete slab; t_f is the height of upper flange of steel beams; N_r is the number of stud connectors; P_u is the shear resistance of stud connectors.

When neutral axis lies in the web of steel beam (Fig.6), the plastic moment resistance of composite cross-section is calculated according to the following equation:

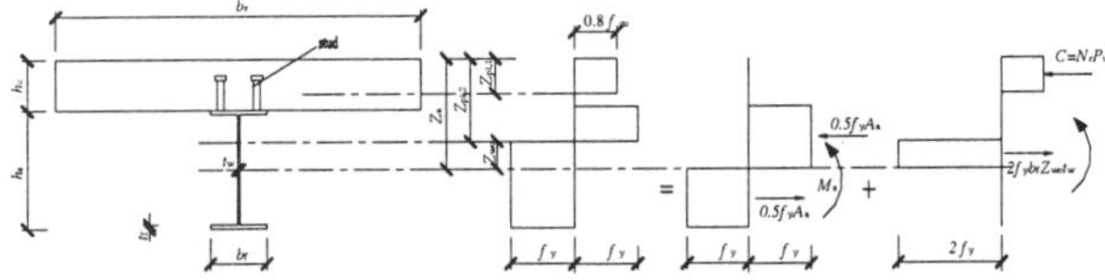

Fig.6 plastic stress distribution of composite cross-section when neutral axis lies in the web of steel beam

$$M_{ur}=M_s+N_rP_u(z_a-z_{Pl,1}/2)-(N_rP_u)^2/(4f_yt_w) \quad (8)$$

Where M_s is the plastic moment resistance of steel beams alone and the other symbols have a same mean as above.

The results calculated according to Eq.(5) or Eq.(8) are given in Fig.7 and Fig.8. And P_u in Fig.7 is calculated according to Eq.2, while P_u in Fig.8 comes from Eq.3. It is observed from the figure that the calculated moment resistance of composite beams increases with the increase of concrete strength when the shear resistance of stud connectors P_u is calculated according to Eq.(2) and when concrete strength is higher the calculated value may be greater

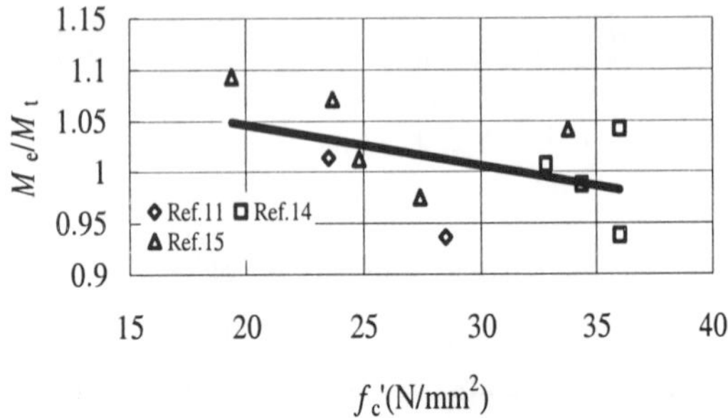

Fig.7 M_c/M_t vs. f_c' when P_u is calculated according to Ollgaard's equation

than the experimental value, which is the same as that of push-out specimens. While Eq.(3) which is modified from Eq.(2) is applied, the calculated moment resistance of composite beams remains unchangeable with the increase of concrete strength, but it is slightly smaller than the experimental results. The reason is that the mechanical properties of stud connectors between push-out specimens and composite beams are different. In most cases, the shear resistance of stud connectors in push-out specimens is the lower limit of that in composite beams approximately. Therefore, it may underestimate the moment resistance of composite beams when adopting the shear resistance of push-out

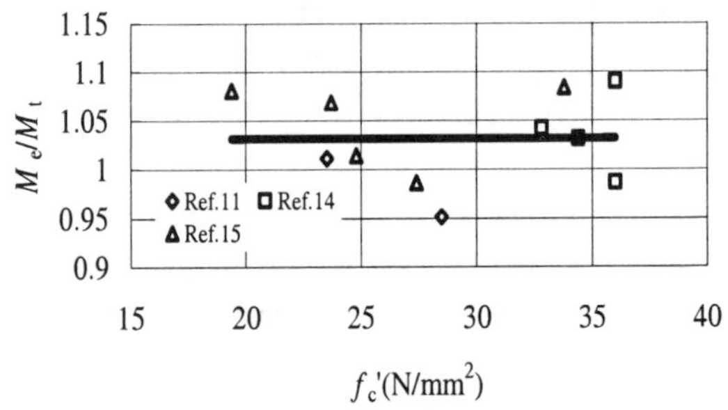

Fig8 M_c/M_t vs. f_c' when P_u is calculated according to proposed equation

specimens to calculate the longitudinal shear of composite beams.

4. Conclusions

1) In this paper, the effects of concrete strength on the shear resistance of stud connectors are discussed. Based on 95 push-out specimens, it is found that the values calculated according to the simplified equation proposed by Ollgaard et al. are higher than the experimental results when the cube compressive strength of concrete slab f_{cu} is greater than 30N/mm^2, while lower when f_{cu} <30N/mm^2. However the shear resistance of stud connectors according to Eq.3 proposed in this paper is in good with the test results

2) It is observed from the analysis of composite beams with partial shear connection, the calculated moment resistance increases with the increase of the concrete strength when the shear resistance of stud connectors is calculated according to the simplified equation proposed by Ollgaard, and when concrete strength is higher the calculated value may be

greater than the experimental value, which is the same as that of push-out specimens. While Eq.(3) suggested in this paper is applied to modify the calculated equation of stud connectors, the error of moment resistance of composite beams with partial shear connection caused by strength calculation of stud connectors can be eliminated.

References

1. Ollgaard H G, Slutter R G, Fisher J D. Shear strength of stud connectors in lightweight and normal-weight concrete[J]. Engineering Journal of American Institute of Steel Construction,1971, 8(2): 55-64
1. 2 Bode H, Hu Xiamin. Untersuchungen zur Schubtragfähigkeit von Kopfbolzendübeln, Bericht zum DFG-Vorschungsvorhaben[R]. Universität Kaiserslautern, August 1993
2. Sun Shaoyun. Studies on shear resistance and behaviour of stud connectors in steel-concrete composite beams (in Chinese)[D]. Zengzhou: Zhengzhou Institute of Technology, 1987
3. Oehlers D J. Results on 101 Push-Specimens and Composite Beams[R]. University of Warwick, Department of Civil Eng., Research Report CE 8, January 1981
4. Mainstone R, Menzies J B. Shear Connectors in Steel-Concrete Composite Beams for Bridges. Part 1: static and fatigue tests on push out specimens[J]. Concrete, 1967, 1(9):351-358
5. Menzies J B. CP117 and Shear Connectors in Steel-Concrete Composite Beams[J].The Structural Eng., 1971, 49(3):137-153
6. Li A, Cederwall K. Push Test on Stud Connectors in Normal and High Strength Concrete[R]. Chalmers Tekniska Högskola, Institutionen för Konstruktionstechnik Betongbyggnad, Report 91:6, November 1991
7. Roik K, Hanswille G. Beitrag zur Tragfähigkeit von Kopfbolzendübeln[J]. Der Stahlbau, 1983,52(10): 301-308
8. Yamamoto M, Nakamura S. The Study on Shear Connectors. The Public Works Research Institute. Construction Ministry Japan, Vol. 5, Research Paper 9, March 1962
9. Hiragi H, Miyoshi E, Hurita A, Ugai M, Akao S. Static Strength of Stud Shear Connectors in SRC Structures[J]. Transactions of the Japan Concrete Institute, 1981, 3: 453-460
10. Zhang Qi. An experiment study on longitudinal shear of steel-concrete composite beams(in chiese)[D]. Nanjing: Nanjing university of technology, 2005
11. Hu Xiamin. Design Method of Eurocode 4 for Steel- Concrete Composite Beams (in Chinese)[J]. Industrial Construction, 1996, 26(2):50-55
12. Roik K, Hanswille G. Eurocode 4 – Bemessungswerte der Tragfähigkeit für Kopf-bolzendübel[A], Festschrift Stefan Polony[C]. Köln: Verlagsgesellschaft Rudolph Müller GmbH, 1990
13. Wang Hongquan. Experimental studies of longitudinal shear and ultimate moment resistance in composite steel-concrete beams(in Chinese)[D]. Beijing：Tsinghua University，1996
14. J.C.Chapman, S.Balakrishnan. Experiments on composite beams[J]. The structural engineer, 1964, 42(11):369-383
15. Hu Xiamin. Design method of Eurocode 4 for steel-concrete composite beams (7) - shear connection design in beams (in Chinese)[J]. Industrial construction, 1996, 26(3):46-5

RESEARCH ON CALCULATING METHODS OF STOREY DRIFT FOR REINFORCED CONCRETE SHEAR WALL STRUCTURES

DENG MINGKE LIANG XINGWEN LIU QINGSHAN

School of Civil Engineering, Xi'an University of Architecture and Technology, Shanxi Xi'an 710055, China

According to different structure systems, the lateral displacement modes of tall buildings are classified. The difference of storey drift of frame and shear wall, which are two different structure systems, is analyzed in this paper. Then, the potential problem that the storey drift angles, which are calculated by the storey lateral displacement difference, is regarded as the deformation control parameter of shear wall structure is presented. Based on the characteristic of shear wall structure, the floor displacement is divided into destructive floor displacement and nondestructive floor displacement. Similarly, the storey drift is divided into nominal storey drift Δu_i and destructive storey drift $\Delta\tilde{u}_i$. Nondestructive floor displacement is produced by rigidity rotation of the floor, and the proportion of innocuous floor displacement is accumulated gradually from the base to roof for shear wall structure. In this paper, applying the inverted triangular distribution of lateral force to the cantilever wall with identical section, the displaced shape is regarded as the approximate lateral displacement curve. The relationships among nominal storey drift Δu_i, destructive storey drift $\Delta\tilde{u}_i$, interstorey rotational drift angle $\hat{\theta}_i$ and bending curvature of section $\phi(\xi)$ is derived, and practical formulas to calculate the destructive storey drift is presented. Finally, by the sample analyzing, the relationship of nominal storey drift, destructive storey drift, inter-storey rotational drift angle and bending curvature of section is explained. Moreover, the practical formulas to calculate the destructive storey drift are confirmed, and the validity and feasibility of the method is verified.

1. Introduction

Reinforced concrete shear walls have always been recognized for their large lateral stiffness and strength, which provides for good control over horizontal displacements and storey drifts. Thus, reinforced concrete structural walls are commonly used in tall buildings for resisting lateral forces imposed by wind or earthquakes.

With the increase of the height of structures, controlling over the deformation of tall buildings is especially important for structural design. Currently, the roof displacement and storey drift are limited in many seismic design codes. In the Chinese Code for Seismic Design of Buildings [1], the allowable values of elastic and inelastic drift ratio are specified according to different levels of earthquake loads.

The interstorey deformation of tall buildings includes shear deformation and flexural deformation in storey members under horizontal force action. Usually, there are two reasons for controlling over elastic interstorey deformation of building structures [2]: first, to prevent cracking and obvious damage of nonstructural members; then, to ensure the elastic design principle is satisfied under minor earthquake loading. In order to realize the two objectives, different deformation analysis methods should be adopted for different structural systems. For shear deformation dominating frame systems, the lateral deformation induced by storey shear should be limited. For flexural deformation dominating shear wall structural systems, the flexural deformation of the structural members in the storey should be limited. And for frame-wall structural systems, the flexural deformation in the storey below the point of contraflexure of the drift curve and the shear deformation above it should be limited.

In the Chinese Technical Specification for Concrete Structures of Tall Buildings, the storey drift ratio, which is calculated by the ratio of floor displacement difference to storey height, is regarded as the deformation controlling index of building structures. According to different structural systems, the corresponding storey drift ratio limit is provided. However, the method of calculating the storey drift is controversial. For shear wall structure, with flexural deformation dominating lateral displacement modes, the rotational displacement, which is produced by the rigid rotation of the floor, is included in the storey drift. But the rigid rotation of the floor can not produce destructive deformation. Thus, the storey drift, which is calculated by the method mentioned above, cannot account for the force-induced destructive deformation of the structural members. Therefore, it is necessary to distinguish the destructive storey drift with the nominal storey drift. In this paper, several different deformation analysis methods are employed to express the storey deformation of shear wall structure.

2. Floor Displacement and Deformation of Structure

The floor displacement under the action of horizontal earthquake is divided into three parts shown as follows [3]: (1) The total shear lateral displacement of the floor which is induced by storey shear and is called translational displacement. (2) The total flexural lateral displacement of the structure which is induced by overturning moment and is called rotational displacement. (3) The rotational displacement of the base. The third displacement, which is produced by the total rigid rotation of the structure due to the deformation of soil, is irrelevant to the deformation and destructive storey drift of structures. Consequently, only the first two displacements are discussed in this paper.

The total shear lateral displacement of the floor, which causes destroy of the structure, is called destructive floor displacement. While the displacement, which is produced by the rigid rotation of the floor induced by the total flexural deformation of the structure, is called nondestructive floor displacement for it can not result in destructive deformation of the structural members in the storey.

In frame systems, the nondestructive floor displacement can be ignored when the aspect ratio of the structure is limited to the allowable value. In shear wall structure with flexural deformation dominating lateral displacement mode, the proportion of shear lateral displacement is minor. However, the rotational displacement is the main lateral displacement component and the proportion of the nondestructive floor displacement among the floor displacement increases rapidly over the height of the structure. Moreover, with the increasing of the height and aspect ratio of the structure, the accumulated nondestructive displacement is aggravated.

Consequently, for the frame structure, the code-specified interstorey drift ratio limit is valid because the rotational displacement is negligible. Considering a shear wall structure, in which the shear lateral displacement is negligible, ignoring the dominating rotational displacement will result in large error and affect the accuracy of calculating the storey drift ratio. Therefore, the results obtained from above will not in agreement with the fact [4].

By analyzing the relationship between structural deformation and member's deformation, we find that the total shear lateral displacement of frame is produced by bending deformation of the structural members in the storey. Thus, controlling over the total shear displacement is equal to limiting the bending deformation of structural members. By contrast, the total flexural lateral displacement of shear wall is produced by the accumulating of the bending deformation of structural members, and the deformation of the structure is proportional to the bending curvature of members and the rotational angle of the floor. Consequently, in seismic design, controlling over the bending curvatures of shear wall sections or rotational angles of the floors is a more effective method to reflect the destructive deformation.

3. Nominal Storey Drift and Destructive Storey Drift

Under the horizontal earthquake action, the total storey drift of building structure is composed of the destructive storey drift, which is produced by the deformation of structural members, and rigid storey drift obtained by rotation of the floor. From Figure 1, the total storey drift can be expressed as

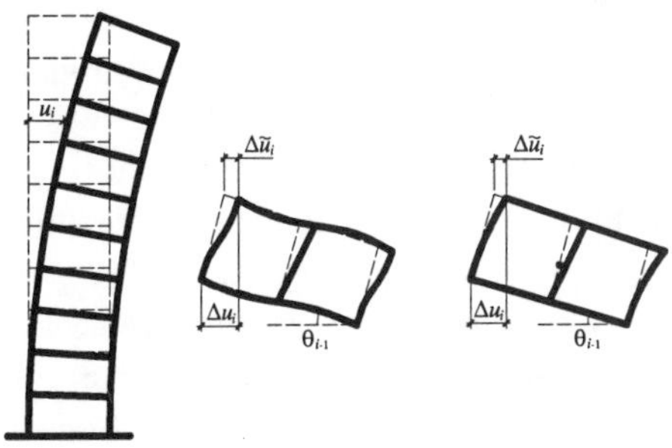

Figure 1. Storey drift of tall building

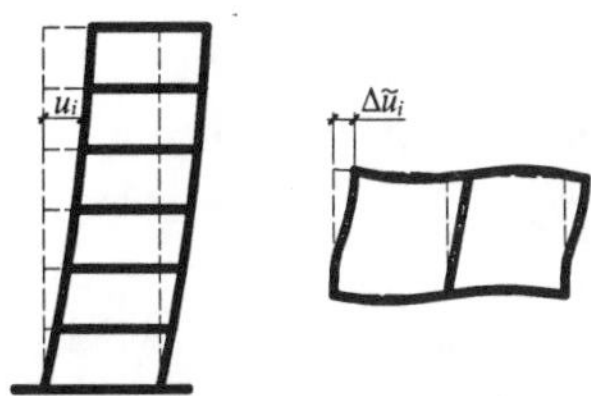

Figure 2. Storey drift of frame

$$\Delta u_i = \Delta \tilde{u}_i + \theta_{i-1} h_i \quad (1)$$

where Δu_i denotes the difference of floor displacement for adjacent floors which is called nominal storey drift, $\Delta \tilde{u}_i$ denotes the destructive storey drift produced by the flexural and shear deformation of members in the i-th storey, h_i denotes the storey height, θ_{i-1} denotes the rotational angle of the $(i-1)$-th storey produced by the total bending deformation, and $\theta_{i-1}h_i$, which is called the nondestructive storey drift, is produced by rotation of the $(i-1)$-th floor.

Assuming that $\overline{\theta}_i = \Delta u_i / h_i$ and $\tilde{\theta}_i = \Delta \tilde{u}_i / h_i$, from equation (1), we find

$$\overline{\theta}_i = \tilde{\theta}_i + \theta_{i-1} \quad (2)$$

where $\overline{\theta}_i$ is the nominal storey drift ratio, $\tilde{\theta}_i$ is the destructive storey drift ratio produced by the flexural and shear deformation of members in the i-th storey.

In frame structure, if the nondestructive storey drift produced by total bending deformation of the structure is neglected, the destructive storey drift is equal to the nominal storey drift. That is,

$$\Delta u_i = \Delta \tilde{u}_i, \quad \overline{\theta}_i = \tilde{\theta}_i$$

which is in accordance with the specified storey drift ratio in the Chinese Technical Specification for Concrete Structures of Tall Buildings.

Generally, this method is reasonable for frame with a few storeys and small aspect ratio.

4. Storey Drift and Storey Drift Ratio of Shear Wall Structure

As is stated above, for shear wall structure with flexural deformation dominating drift curve, the total storey drift of structure under the horizontal earthquake action is produced by rotational displacement of the structure if the shear deformation is neglected. The rotational displacement includes the rigidity displacement induced by floor rigidity rotation and destructive rotational displacement produced by flexural deformation of structural members in the storey.

As the rotational angle of the floor accounts for the accumulative effect of all the structural members in each storey, the proportion of rigidity displacement among the total storey drift increases rapidly in the upper storeys. Hence, the influence of the rigidity displacement cannot be neglected and the destructive storey drift, which is obtained by deducting the rigidity rotational from the nominal storey drift, can really represent the actual deformation characteristic of shear wall structure. This viewpoint has been accepted by many researchers in earthquake engineering.

A well-known example is the case of 68-storeyed Diwang building [5], in which the destructive storey drift ratio in the 58-th storey under the lateral force imposed by wind accounts only one percent of the nominal storey drift ratio. The nominal storey drift ratio calculated by the ratio of floor displacement difference versus storey height has exceeded the code-specified allowable value of storey drift ratio. However, the bearing capacity and lateral stiffness of the structure are satisfied, and both the cracking and deflection are adequately controlled because the destructive storey drift angle was satisfied. In the literature [3], the author mentioned that the storey drift should be strictly controlled only in the lower storeys for flexural deformation dominating structures. In the literature [6], the difference between nominal storey drift and force-induced storey drift was pointed out, and it is suggested that the allowable value of storey drift ratio should be enlarged for shear wall structure. Consequently, it is necessary to provide a practical method to calculate the destructive storey drift in current seismic design code.

According to the deformation characteristic of shear wall structure, the nominal storey drift on the upper storeys is much greater than that on the lower storeys. Hence, the code-specified allowable value of storey drift ratio only controls over the nominal deformation on the upper storeys of the structure.

If the storey drift ratio is regarded as the deformation controlling index, the stiffness on the upper storey of the structure should be increased, which is not accord with the actual deformation characteristic. Extensive testing and researches about the storey drift of shear wall structures, which are confined to low-rise shear walls, cannot reflect the characteristic of high-rise buildings. Recent shaking table tests have indicated that the failure is concentrated on the first or second storey for regular high-rise structures. Failure has hardly taken place on the upper storeys whose nominal storey drifts exceed the limit. In the literature [4], by analysis of the 30-storeyed frame-brace structure, it has shown that the distribution of storey drift ratio is opposite to the magnitude of the force imposed on structural members.

Usually, the limit of destructive storey drift has influence on the lower storeys, and the plastic hinges of shear wall form at the bottom. The reason is that the actual destructive storey deformation reduces gradually from the base to top. That is, the nominal storey drift increases, but destructive storey drift reduces gradually on the upper storeys for shear wall structure, which is different from the general knowledge [7-9]. According to the distribution of destructive storey drift over the height of the structure, it is reasonable to reduce the stiffness of all structural elements step by step. Further theoretical analysis is presented to illustrate the storey deformation in this paper.

4.1 *The relationship among lateral displacement, rotational angle and curvature for shear wall structure*

According to the characteristic of shear wall structure, applying the inverted triangular distribution of lateral force to the cantilever wall with identical section, the displaced shape is regarded as the approximate lateral displacement curve. As is shown in Figure 3, the moment on arbitrary section is expressed as

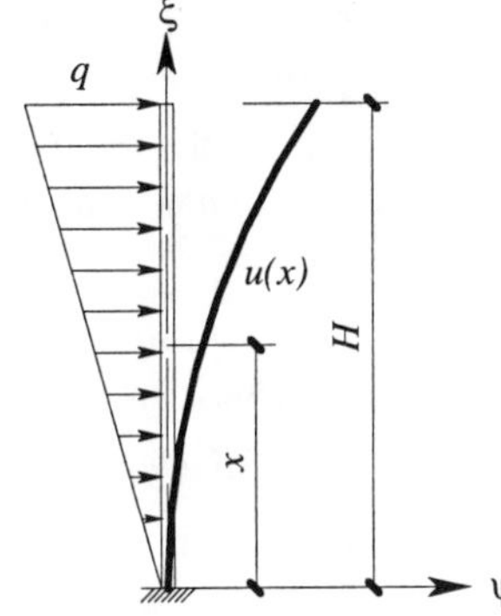

Figure 3. The approximate lateral displacement curve

$$M(x)=\frac{q}{6}(\frac{x^3}{H}-3Hx+2H^2)$$

giving

$$EIy''=M(x)$$

imposing boundary conditions, the curvature, rotational angle and displacement on arbitrary section are obtained as follows:

$$\phi(x)\approx y''=\frac{q}{6EI}(\frac{x^3}{H}-3Hx+2H^2) \tag{3a}$$

$$\theta(x)=y'=\frac{q}{24EI}(\frac{x^4}{H}-6Hx^2+8H^2x) \tag{3b}$$

$$u(x)=y=\frac{q}{120EI}(\frac{x^5}{H}-10Hx^3+20H^2x^2) \tag{3c}$$

giving $\xi=x/H$, equation (3) can be rewritten in the form

$$\phi(\xi)=\frac{qH^2}{3EI}\cdot\frac{1}{2}(\xi^3-3\xi+2)=\phi_b\cdot\Phi(\xi) \tag{4}$$

$$\theta(\xi)=\frac{qH^3}{8EI}\cdot\frac{1}{3}(\xi^4-6\xi^2+8\xi)=\theta_t\cdot\Psi(\xi) \tag{5}$$

$$u(\xi)=\frac{11qH^4}{120EI}\cdot\frac{1}{11}(\xi^5-10\xi^3+20\xi^2)=u_t\cdot U(\xi) \tag{6}$$

where $\Phi(\xi)$ denotes the curvature function, ϕ_b is the curvature at the bottom section of shear wall, $\Psi(\xi)$ denotes the rotational angle function, θ_t is the rotational angle on the top, $U(\xi)$ denotes the displacement shape function, u_t is the roof displacement of structure. Hence, u_t, θ_t, ϕ_b can be obtained in the form

$$u_t=\frac{11qH^4}{120EI}$$

$$\theta_t=\frac{qH^3}{8EI}$$

$$\phi_b=\frac{qH^2}{3EI}$$

From equations (4),(5)and(6), the relationship between u_t, θ_t and ϕ_b can be expressed as

$$u_t=\frac{11}{40}H^2\cdot\phi_b \tag{7a}$$

$$u_t=\frac{11}{15}H\cdot\theta_t \tag{7b}$$

$$\theta_t=\frac{3}{8}H\cdot\phi_b \tag{7c}$$

Similarly, from equations (4),(5),(6)and(7), we obtain

$$u(\xi)=\frac{11H}{15}\cdot\frac{U(\xi)}{\Psi(\xi)}\cdot\theta(\xi) \tag{8}$$

$$u(\xi)=\frac{11H^2}{40}\cdot\frac{U(\xi)}{\Phi(\xi)}\cdot\phi(\xi) \tag{9a}$$

$$\theta(\xi)=\frac{3H}{8}\cdot\frac{\Psi(\xi)}{\Phi(\xi)}\cdot\phi(\xi) \tag{9b}$$

4.2 *The rotational angle of floor and the interstorey rotational angle*

According to the height of each floor H_i, the corresponding height ratio ξ can be calculated, and then substituting it into equation (5), the slope angle θ_i of the wall at the *i*-th floor is obtained. Introducing the sign $\hat{\theta}_i$, which is called the interstorey rotational angle at the *i*-th storey, denotes the rotational angle of the *i*-th floor relative to the ($i-1$)-th floor. The rotational angle produced by deformation of the wall is well represented by $\hat{\theta}_i$. As shown in Figure 4, we obtain

$$\hat{\theta}_1 = \theta_1 \qquad (i=1)$$
$$\hat{\theta}_i = \theta_i - \theta_{i-1} \qquad (i \geq 2) \qquad (10)$$

Based on the relationship between floor displacement and rotational angle in equation (8), the floor displacement can be controlled by the maximum allowable interstorey rotational angle, which can be expressed as

$$\hat{\theta}_i \leq [\hat{\theta}]$$

Letting $$\Delta\hat{u}_i = \hat{\theta}_i h_i \qquad (11)$$

where $\Delta\hat{u}_i$, produced by the bending deformation of the storey walls, denotes the supposed storey drift induced by storey rotational angle of the wall and reflects the real magnitude of destructive storey deformation. Although $\Delta\hat{u}_i$ is not equal to storey drift, it is accord with the displacement profile of shear wall structure and much more effective than the nominal storey drift calculated by floor displacement difference.

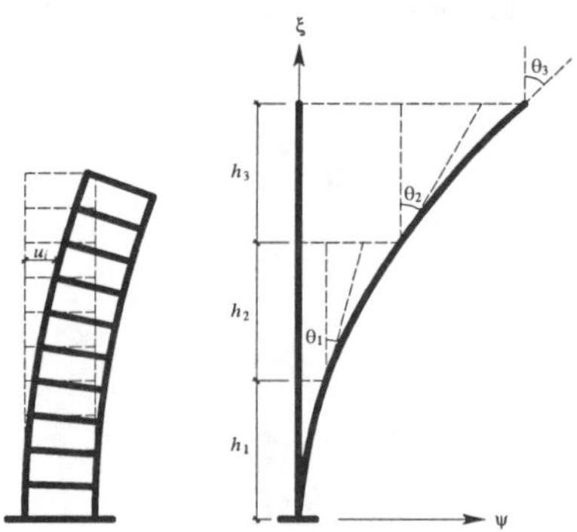

Figure 4. The rotational angle of floor

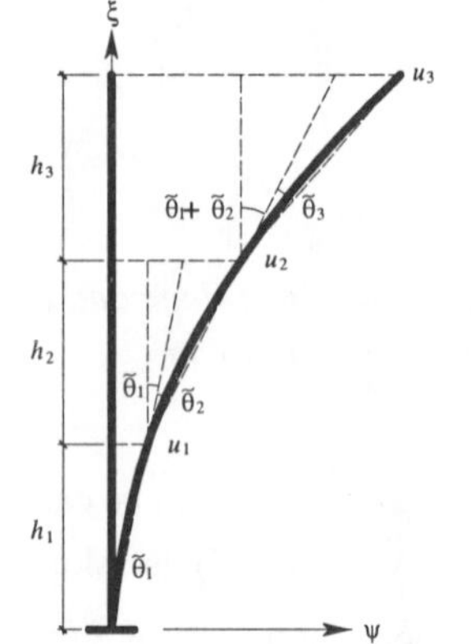

Figure 5. The nominal storey drift ratio and interstorey rotational angle

4.3 *Destructive storey deformation*

Based on the analysis above, the destructive storey deformation of shear wall structure, which represents the bending deformation of storey members, can be controlled over by the following two methods:

(1) Controlling over the destructive storey drift

The destructive storey drift, which is obtained by deducting the rigidity displacement from the nominal storey drift, can be given as

$$\Delta\tilde{u}_1 = \Delta u_1 = u_1 \qquad (i=1)$$
$$\Delta\tilde{u}_i = \Delta u_i - \theta_{i-1} h_i \qquad (i \geq 2) \qquad (12)$$

(2) Controlling over the supposed storey drift

Based on equation (11), the supposed storey drift can be obtained by the interstorey rotational angle at the *i*-th storey. Then, the magnitude of destructive storey drift is calculated.

4.4 *The relationship between nominal storey drift and destructive storey drift*

Actually, it is very difficult to calculate the destructive storey drift by equation (12) because of the inconvenience of obtaining the rotation angle of the floor θ_i. Thus, an approximate analysis method is presented as follow.

For shear wall structure, neglect the shear deformation of storey members, considering

$$\tilde{\theta}_i = \Delta\tilde{u}_i / h_i$$

$\tilde{\theta}_i + \theta_{i-1}$ is equal to the rotational angle of the *i*-th floor produced by total flexural deformation of structure, that is

$$\tilde{\theta}_i + \theta_{i-1} \approx \theta_i$$

from equation (2), we obtain

$$\theta_i \approx \bar{\theta}_i$$

Consequently, on the condition of small deformation, the nominal storey drift ratio can be used to substitute the rotational angle of the floor, namely,

$$\theta_i = \bar{\theta}_i$$

considering

$$\bar{\theta}_i = \Delta u_i / h_i$$

from the relationship shown in Figure 5, we obtain relationships as follows:

(1) The relationship of storey drift ratio from equation (2)

$$\bar{\theta}_1 = \tilde{\theta}_1$$
$$\bar{\theta}_2 = \tilde{\theta}_2 + \theta_1 = \tilde{\theta}_2 + \bar{\theta}_1 = \tilde{\theta}_2 + \tilde{\theta}_1$$

$$\overline{\theta}_3 = \tilde{\theta}_3 + \theta_2 = \tilde{\theta}_3 + \overline{\theta}_2 = \tilde{\theta}_3 + \tilde{\theta}_2 + \tilde{\theta}_1$$

$$\cdots$$

$$\overline{\theta}_i = \tilde{\theta}_i + (\tilde{\theta}_1 + \tilde{\theta}_2 + \cdots \tilde{\theta}_{i-1}) \quad (i \geq 2) \tag{13}$$

$$\tilde{\theta}_i = \overline{\theta}_i - (\tilde{\theta}_1 + \tilde{\theta}_2 + \cdots \tilde{\theta}_{i-1}) \quad (i \geq 2)$$

(2) The relationship of storey drift

$$\Delta u_1 = \Delta \tilde{u}_1 = u_1$$

$$\Delta u_i = \Delta \tilde{u}_i + (\tilde{\theta}_1 + \tilde{\theta}_2 + \cdots \tilde{\theta}_{i-1}) h_i \quad (i \geq 2) \tag{14}$$

in which

$$\Delta u_i = \overline{\theta}_i h_i$$

and

$$\Delta \tilde{u}_i = \tilde{\theta}_i h_i \tag{15}$$

where $\Delta \tilde{u}_i$ denotes the destructive storey drift, produced by flexural deformation of structural members in the i-th storey, where the rigidity displacement is deducted.

From equation (13) and (14), we obtain

$$\Delta u_i = \Delta \tilde{u}_i + (\Delta \tilde{u}_1 + \Delta \tilde{u}_2 + \cdots \Delta \tilde{u}_{i-1}) = \Delta \tilde{u}_i + \Delta u_{i-1}$$

$$\Delta \tilde{u}_i = \Delta u_i - \Delta u_{i-1} \tag{16}$$

Equation (13) indicates that the nominal storey drift ratio at the i-th storey includes the entire destructive storey drift ratio below the storey. Equation (16) provides the relationship between nominal storey drift and destructive storey drift and also presents a simple method for calculating the destructive storey drift.

If $\theta_i = \overline{\theta}_i$, equation (16) is equivalent to equation (12). To verify the reliability and accuracy of equation (16), it is necessary to analyze the range of error. From Figure 5, we obtain

$$\overline{\theta}_1 < \theta_1 < \overline{\theta}_1 + \tilde{\theta}_2$$

$$\overline{\theta}_2 < \theta_2 < \overline{\theta}_2 + \tilde{\theta}_3$$

$$\overline{\theta}_i < \theta_i < \overline{\theta}_{i+1} = \overline{\theta}_i + \tilde{\theta}_{i+1} \tag{17}$$

Consequently, calculating θ_i by the approximate method mentioned above will not result in accumulating of the error. The range of error depends on the magnitude of the destructive storey deformation at the $(i+1)$-th storey. In regular shear wall structure, with the increasing of storey number, the destructive storey drift reduces gradually, which diminishes the difference between θ_i and $\overline{\theta}_i$.

5. Deformation Controlling of Shear Wall Structure

The deformation of structures must be controlled to prevent excessive deflections affecting the appearance and efficiency of the structure. Thus, several different methods controlling over the deformation of shear wall structures are presented as follows.

(1) Controlling over the roof displacement

(2) Controlling over the nominal storey drift

(3) Controlling over the destructive storey drift

(4) Controlling over the interstorey rotational angle

(5) Controlling over the bending curvature of sections

The roof displacement of structure, which reflects the total deformation of structural members under lateral force, represents the average storey drift ratio. Hence, controlling the roof displacement is much reasonable than controlling the maximum nominal storey drift specified in the code. If the stiffness of structure is variable over the height of the structure, the roof displacement cannot represent the deformation of individual storey.

Controlling over the nominal storey drift is the method in current design specification (e.g., the Chinese Technical Specification for Concrete Structures of Tall Buildings). Based on the analysis above, the rigidity displacement is included in the nominal storey drift, and thus, it exaggerates the deformation on the upper storeys of shear wall structure. Consequently, this method is not applicable to shear wall.

Controlling over the destructive storey drift, which is obtained by deducting the rigidity displacement from the nominal storey drift, as shown in equation (12), well represents the storey deformation of structural members and accords with the actual displacement profile of shear wall structure. Furthermore, the deformation of individual storey can be effectively controlled by the destructive storey drift. Then, we can realize the objective of preventing cracking and obvious damage of nonstructural members and ensuring the elastic design principle to be satisfied under minor earthquake loading. In addition, the maximum destructive storey drift of shear wall structure occurs at the lower storeys. Hence, the allowable value of destructive storey drift controls the storey deformation at the lower storey. With the increasing of the height of the structure, the roof displacement will control the deformation of the structure because the destructive storey drift at upper storey is minor. Consequently, the allowable value of destructive storey drift should be reduced on the condition mentioned above.

The value of destructive storey drift can be obtained from equation (12) or (16). Equation (12) provides an accurate solution, and equation (16) offers an approximate solution. Take notice that the storey drift ratio, which is used to control over the deformation of structure, is calculated by the relative storey drift. That is, $\bar{\theta}_i = \Delta u_i / h_i$ and $\tilde{\theta}_i = \Delta \tilde{u}_i / h_i$. Thus, the obtained storey drift ratio can not accurately represent the flexural deformation of structural members at the storey.

Actually, storey deformation and storey drift are two different concepts. Controlling over the storey deformation is the objective for designing a structure, and controlling over storey drift is an effective way to achieve the objective. That is, controlling over destructive storey drift is an effective method to control the destructive storey deformation. In the method above, the real flexural deformation of shear wall can not be reflected for the reason that: for shear deformation dominating structure, the storey drift reflects the flexural deformation of beams and columns and represents the actual deformation; for flexural deformation dominating shear wall structure, the accuracy of destructive storey drift is affected by the rigid rotation of storey. Therefore, rotational angle or curvature is better than displacement for reflecting the deformation of shear wall structure.

Hence, another method for controlling the deformation is to control the interstorey rotational angle $\hat{\theta}_i$ or the supposed storey drift $\Delta \hat{u}_i$. Only if the rotation of storey is determined, can $\hat{\theta}_i$ and $\Delta \hat{u}_i$ be obtained from equation (10) and (11). From the relationship shown in equation (8), the corresponding floor displacement can be obtained. It is obvious that the relationship between rotation angle and displacement is determined by the equation of drift curve. Hence, controlling over the rotational angle is equivalent to controlling over the floor displacement. But it is very difficult to calculate the rotation angel in engineering structures. Thus, the scope of application is limited.

The deformation of shear wall is associated with the bending curvature of storey structural members. Thus, the destructive storey deformation can be measured by the bending curvature. From equation (9), the relationship between curvature and displacement can be determined by the equation of drift curve. Furthermore, the curvature reflects the rotational property of section, and it also establishes the relationship between deformation capacity of the member and curvature of the section. For the reason that the yield and failure of shear wall are associated with the curvature of section, the deformation ability, stress state and failure mode can be well controlled by curvatures.

6. Sample Analysis

A 10-storey reinforced concrete shear wall structure design example is considered, with storey heights of 3.0 meters in all cases. Assuming the lateral stiffness of structure keeping constant in the direction of vertical, applying the inverted triangular distribution of lateral force to the cantilever wall, and adopting the lateral displacement mode as shown in figure 3, the storey deformation can be analyzed.

(1) Controlling over the roof displacement

Giving the roof displacement ratio is 1/1000, namely, $u_t / H =$ 1/1000, then the maximum roof displacement is $u_{max} = u_t$ =30mm

(2) Controlling over the storey drift

Supposing the destructive storey drift at the first floor reaching 1/1000, we can obtain $u_1 = \tilde{\theta}_1 h_1$ =3mm, and the corresponding ξ is equal to 0.1. From equation (6), we have

$$u_t = \frac{11u(\xi)}{(\xi^5 - 10\xi^3 + 20\xi^2)} = 173.675\text{mm}$$

which exceeded the allowable value of roof displacement. Giving u_t=30mm and substituting u_t and the corresponding ξ at every floor into equation (6), each displacement u_i can be calculated. Hence, the nominal storey drift can be obtained by

$$\Delta u_i = u_i - u_{i-1}$$

Then, from equation (16), the approximate destructive storey drift $\Delta \tilde{u}_i$ can be obtained. The result is shown in table 1.

Table 1. The nominal storey drift and approximate destructive storey drift of shear wall structure

storey mumber	storey height	ξ	u_i (mm)	Δu_i	$\Delta \tilde{u}_i$
10	30	1	30	4.090	0.0183
9	27	0.9	25.910	4.071	0.0633
8	24	0.8	21.839	4.008	0.1345
7	21	0.7	17.831	3.874	0.2285
6	18	0.6	13.958	3.645	0.3423
5	15	0.5	10.313	3.303	0.4724
4	12	0.4	7.010	2.830	0.6155
3	9	0.3	4.179	2.215	0.7685
2	6	0.2	1.965	1.446	0.9281
1	3	0.1	0.518	0.518	0.518

(3) controlling over the interstorey rotational angle

Giving u_t=30mm, from equation (7b), we obtained $\theta_t = \frac{15}{11H} u_t$ =0.001364. Substituting u_t and the corresponding ξ at each floor into equation (5), the rotational angle θ_i at each floor is obtained. Then, from equation (10), the interstorey rotational angle $\hat{\theta}_i$ is calculated and the corresponding $\Delta\hat{u}_i$ can be obtained from equation (11). In addition, the accurate destructive storey drift can be calculated from equation (12). The result is shown in table 2. The destructive storey drift, nominal storey drift and interstorey rotational angle are compared in Figure 6 and Figure 7, respectively.

Table 2. The interstorey rotational angle of shear wall structure

storey mumber	storey height	ξ	θ_i (m^{-1})	$\hat{\theta}_i$	$\Delta\hat{u}_i$ (mm)	Δu_i	$\theta_{i-1}h_i$	$\Delta\tilde{u}_i$
10	30	1	0.001364	0.0000018	0.005	4.090	4.087	0.0029
9	27	0.9	0.001362	0.000012	0.036	4.071	4.051	0.0208
8	24	0.8	0.00135	0.000032	0.095	4.008	3.956	0.0523
7	21	0.7	0.001319	0.000059	0.178	3.874	3.778	0.0958
6	18	0.6	0.001259	0.000094	0.282	3.645	3.495	0.1500
5	15	0.5	0.001165	0.000135	0.405	3.303	3.090	0.2125
4	12	0.4	0.00103	0.000181	0.542	2.830	2.548	0.282
3	9	0.3	0.000849	0.000230	0.691	2.215	1.857	0.358
2	6	0.2	0.000619	0.000283	0.848	1.446	1.009	0.437
1	3	0.1	0.000336	0.000336	1.009	0.518	0	0.518

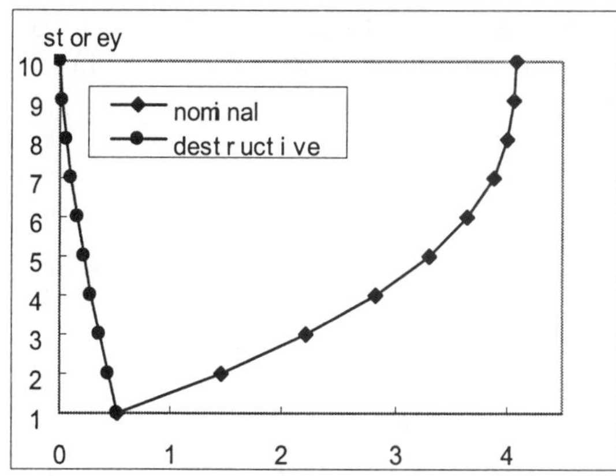

Figure 6. The comparison of destructive storey drift and nominal storey drift

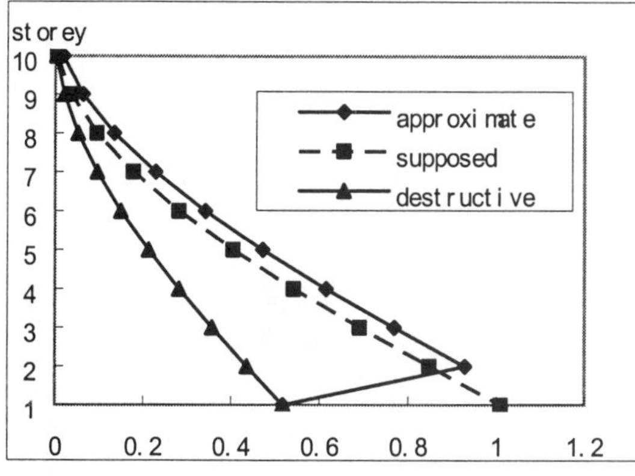

Figure 7. The comparison of destructive storey drift and supposed storey drift

(4) Controlling over the bending curvature of sections

From the lateral displacement mode shown in Figure 3, the maximum bending curvature is appeared at the bottom of shear wall structure. Only if the curvature of shear wall sections at the first storey is controlled, can the storey deformation be controlled. When u_t=30mm, the rotational angle θ_1 at the first floor is about 1/3000, then, from equation (7a), we obtained

$$\phi_b = \frac{40}{11H^2}\theta_t = 0.121\times10^{-3}\,\mathrm{m}^{-1}$$

(5) Summary

For regular shear wall structure, the storey shear decreases gradually from the base to the top. Thus, the destructive storey deformation reaches the maximum at the first storey. As shown in Figure 6, the distribution of nominal storey drift is opposite to that of destructive storey drift. Consequently, controlling the nominal storey drift of shear wall structure is unreasonable.

7. Conclusion

By analyzing the deformation and storey drift of shear wall structure, several conclusions can be obtained as follows:

(1) It is unreasonable to controlling the nominal storey drift of shear wall structure.

(2) The destructive storey drift is accord with the

characteristic of shear wall structure, and controlling over the destructive storey drift is reasonable.

(3) It is very difficult to obtain the accurate drift curve of structure and the rotational angle of the floor under horizontal loads. Thus, from equation (12), the destructive storey drift cannot conveniently obtained, which will obstruct calculating the interstorey rotational angle from equation (10). But the floor displacement is easy to be obtained. Then, from equation (16), it is convenient for calculating the destructive storey drift.

(4) The method in this paper is only applicable to regular shear wall structure. Considering the variation of the lateral stiffness of the structure, further research is needed to derive the expressions of drift curve and storey drift.

References

1. The Chinese Code for Seismic Design of Buildings GB50011-2001. Beijing: China Architecture and Building Press, 2002
2. Xu PF, Huang XK. Understanding and application of the Chinese Technical Specification for Concrete Structures of Tall Building. China Architecture and Building Press, 2003
3. Liu DH. Seismic Design for Tall Buildings. Beijing: China Architecture and Building Press, 1993
4. Yang ZY, He RQ. Calculation for Elastic deformation of High-rise Buildings. Journal of Harbin University of C.E. & Architecture, 2001, vol34(5): 14-18
5. Wei L, Gong AJ, Sun HZ, Zhu JX. Several Problem about Structural Design for Diwang Tower. Building Structures, 2000, Vol 30(6): 31-36
6. Wei L. Research on Displacement Control for Tall Buildings. Building Structures, 2000, Vol 30(6): 27-30
7. Lu XL, Wang YY, Guo ZX. Elastoplastic Deformation Checking of Building Structures under Different Seismic Levels. Building Science, 2002, Vol 18(1): 11-15
8. Guo ZX. Several Problems about Seismic Deformation Check of High-rise RC Building under Frequent Earthquake. Journal of Huaqiao University (Natural Science), 1998, Vol 19(3): 284-289
9. Xing DK, Hu SL, Jiang LX, Yan QZ, Chen ZL, Li B. Limit Value of Interstorey Drift of Reiforced Concrete Super Highrise Buildings. Journal of Building Structures, 2000, Vol 21(3): 10-15

Urban Planning & Development

BETTER UNDERGROUND UTILITY SERVICES DESIGN FOR SUSTAINABLE CITIES*

K.C. LAM

Dept of Building Services Engineering
The Hong Kong Polytechnic University

The development of many of the underground services has a direct impact on the planning of sustainable buildings development and environmental concerns. In promoting a more sustainable approach to the planning of our city (e.g. the West Kowloon complex and the large Kai Tak development), there is a strong desire to adopt a better practice in the design of the location of the principal utility services. This short paper addresses the major problems we face regarding the underground infrastructure: social and economic disruption caused by performing engineering works. In the second part, the author discusses how we engineers can solve the problems with the use of unusual services subways and the employment of some trenchless construction for more greener underground services. It is hoped that the discussions made will provide a forum for all construction stakeholders to explore the suggestions for the needs of a greener city.

1. Introduction

Hong Kong's population will continue to grow. Economic activity will also increase. If the growth of population and economic activity is matched by the existing patterns of resource use and consumption, pollution levels that are already high will rise. The increase in pollution etc. will affect the life of Hong Kong's people, weaken the vigour of its economy and cripple the ability of the city to function as a metropolitan city.

In parallel with a concern for public health and safety, construction of building should be extended to address public well-being in the long term. The design, construction, operation and maintenance of Hong Kong's buildings are at the heart of this equation.

Designing better building and its building services; more efficient construction techniques; intelligent and energy efficient electrical and mechanical services; improved maintenance and operation of these dynamic systems all can bring economic, social and environmental benefits to our city. Hence, we need to have innovative and yet practical designs for installing and maintaining all utility services for a sustainable future.

This short paper attempts to highlight:

- That the present designs of buildings for our cities are far from perfect in terms of complete integration of buildings, building services and planning of city;
- That improvements are feasible if we put more efforts into more integration of building design with our modern city; and
- That teamwork is needed in promoting the development of a modern city and its buildings.

In this paper, the author will focus on the integration of underground services as it is apparent that this topic is another Cinderella of the construction industry for too long.

2. Sustainability and Sustainable Design

Sustainability requires that the exploitation and allocation of resources, the direction of investment and technology development, and meeting of increasing quality of expectations for economic, social, amenities and the physical environment, are made consistent with future as well as present needs.

Within the building industry, the phrase "sustainable design" is coined to describe a process that looks at minimizing a building's environmental impact from cradle to grave. The evolution of sustainable buildings that are eco-efficient, energy saving, and designed for a better quality of life would contribute to a more liveable city. In spite of an increased awareness of environmental issues, sustainable building and city design is still rare. It is the author's viewpoint that the fundamental to a sustainable approach is the notion of integrated construction and engineering, looking at town planning, building design, construction and operation and maintenance as an integrated whole as opposed to series of independent construction and mechanical/electrical systems.

3. Underground Infrastructure

It is a fact that perfect buildings cannot be designed in isolation, teamwork is essential from all sides of the building industry and government agencies. All buildings and their building services must be fully integrated and designed together with the planning of all utility services such as water supplies, power supply, telecommunication, information technology systems, sewers, refuse disposal system, gas supply. In some cases, the provision of large district heating and cooling systems for environmental comfort must also be considered. It is evident that all the aforesaid mechanical and electrical systems will have a significant impact on the quality of a green city since all these engineering services require resources and energy, and produce waste and pollutants which all can directly or indirectly affect the quality of a modern city.

All in all, the planning of a city and all utility services has a direct bearing on the design of its buildings and their building services. It is, therefore, necessary to have a coherent design of a modern city and all services. Hence, the theme of this paper is a discussion of underground infrastructure and utility services.

4. Traditional Arrangement of Underground Services

In our cities an extensive network of pipes, ducts and cables can be found in the underground. This network provides all buildings with facilities such as water, gas, electricity, telecommunications, cable TV and sewerage. Although we do not always realize it, this network has become vital to the normal day to day life in our city.

For different good reasons these pipes, ducts and cables are installed in the underground. The underground offers a certain protection to weather conditions and to mechanical damaging. Moreover the underground makes it possible to construct large invisible networks without forming a physical obstacle at street level.

The demand for improved quantity and quality of community services such as utilities and communications has placed an ever-increasing burden on planning, engineering, installing and maintaining these services. Further, the increased value and density of improvements above these underground services increase the complexity and cost of maintaining what may be an aging plant or expanding system capacity and the capability to meet new customer requirements. As the network of these systems grows, space to install them becomes increasingly tight or congested and the difficulty of working on them becomes increasingly complex.

In order to keep such networks up to date and function properly, engineering works are regularly necessary for reasons such as:

- Marking new connections, expansions to the system
- Making adjustments
- Repairing defects
- Replacement or renovation

Disadvantages of underground installation then appear, the pavement has to be broken up and trenches have to be dug. When traffic intensities are low, traffic can be easily diverted and the necessity of undertaking these works is commonly accepted. When traffic intensities are high, the impacts of disturbance have become more significant, and the willingness to accept them is rapidly reducing.

It is very clear that any improvements as roadways, streets, buildings, and so on make it almost impossible or prohibitively expensive to gain access to the facilities by open-cutting or to secure permission to disrupt the natural flow of workplace for such excavations.

Social and economic disruption arising from performing underground engineering works have become a big problem.

4.1 *The problems*

The digging of trenches for installing cables, pipes and other services below ground is a civil engineering method of great antiquity.

Open-cut methods are most suited for trenching in space free of obstructions and unhampered by traffic or existing utilities. Despite modern improvements in mechanical plant, trenching inevitably suffers three fundamental disadvantages. In an urban setting, open-cut method can be slow, inefficient and cumbersome to operate (because of the large volumes of soil that have to be removed and then replaced), and the large equipment can be disruptive to the community and nearby business. Furthermore, there will be problems like:

- Traffic disruption (e.g. delays and disruptions to pedestrians and vehicles as the process sterilizes the ground surface for any use while the trench is open)

- o Visual intrusion
- o Environmental impacts such as noise/vibrations, air pollution, dust, dirt and mess etc.
- o Increased accident level
- o Safety
- o Damages to other underground utilities
- o Damage to pavement and its effect on accessibility
- o Disturbance of the surrounding ground

Three major economic costs are direct costs, indirect costs and environmental and social costs.

1. Direct costs are paid by the utility or government and include payments to contractors and suppliers; planning design and supervision; diversion of existing utilities; and permanent reinstatement.
2. Indirect costs which are sometimes paid by the utility or client and include compensation for damage to land and property; compensation for damage to utility plant; compensation to businesses for loss of profit; reduction in pavement life; and increased road maintenance.
3. Environmental and social costs are more difficult to evaluate financially and may not be directly charged to the utility or owner of the installed facilities.

These costs can be further subdivided into several factors. Materials, capital equipment, site investigation and survey, labor, overhead, and restoration are the major elements of the direct and indirect costs. Restoration will be affected by pavement cutting (both asphalt and concrete), excavations for entry and exit pits, resurfacing, fill, and tamping requirements. The digging rate will be affected by the type of soil (e.g. rock-free), length of trench, depth, number of obstacles, the need to support the opening, and the physical size.

In general, the intensity and nature of social and economic disruption depends on the site and the extent of the work to be done.

To continue to apply this traditional technique to the installation and rehabilitation of services is inconceivable. The complexity of existing services precludes the use of economical mechanical equipment in most cases. Existing pipes and cables must be accurately located and carefully supported as work proceeds and the crucial backfilling operations must be undertaken with equal care if later settlement damage is to be avoided. Altogether a labour-intensive job (>60% of the costs) which is time-consuming and therefore excessively disruptive and excessively expensive.

Already disruption arises on a significant scale due to the decay of existing services. The direct costs of repair are increased by up to a factor of perhaps four times by the social and other consequential costs of the work, particularly when emergency collapse repairs are involved. Traffic diversions may be required and in any case traffic congestion will always force frustrated drivers into diversionary side streets where the increased loading can compound the problem through further services failures.

Add together the direct costs of repair, the cost of additional policing, the loss of taxable profit for shopkeepers, commerce and industry, increased accident rate with medical and hospital services for the injured, the loss of production through injury, consequential damage to other services, inefficient employment of capital in vehicle delays – and numerous other factors which together place an incalculable burden on the economy. That other straw of an overhead which we can ill afford to carry.

4.2 *Management of engineering works*

The problems of social and economic disruption aggravates because of the fact that different utilities all have their own maintenance planning and construction schedules. Typically engineering works are performed independently by each utility.

City authorities who aim at preventing or decreasing problems related to engineering works on the underground infrastructure have roughly three options:

4.2.1 Limiting the social and economic disruption by integrating the underground services initiated by the different utilities

It is not so much as question of finding technical solutions; it is more of an organizational problem. Roads are usually dug up for the purpose of carrying out work on a single utility. A lack of coordination implies that a road will have to be dug up various times to carry out work for different utilities. Work carried out individually is more expensive, often unnecessary and causes a great deal of inconvenience. A coordinated approach should avoid many problems already discussed. Nonetheless, a high level of coordination is necessary.

The objective is to let the different utilities conduct their maintenance/installation works simultaneously or successively. In this way the number of times the public road has to broken up can be reduced. This approach does not reduce the actual disruption caused by individual engineering works but the duration and the number of times disruption occurs.

4.2.2 Reducing or preventing excavation by performing engineering works with trenchless technology (i.e. No-Dig technique)

As most problems arise from the trenches which form a physical obstacle, a solution must be sought in reducing or preventing excavation.

The growth of the use of trenchless technology is spurred by a number of major factors:

- o Increased urbanization and urban operation simplification plus less inconvenience to the public and continuous service operations above the bore
- o A aging utility infrastructure
- o Increasing demand for a variety of services
- o Concern for the physical and social environment
- o Labour cost reduction
- o Reduced surface damage

Trenchless technology has existed in the construction for many years. Changing market demands have called for trenchless methods which have the capability to undertake complete line installation or rehabilitation. The most dramatic technical changes have been based on making use of the existing pipe run. Replacing by on-line method or structural rehabilitation to provide a virtually new line in terms of service life has inspired a whole array of new technology.

Advantages and disadvantages of trenchless methods

The three major advantages are:

- o Less surface disruption
- o Better quality of completed installation
- o Speed of installation

The main advantage is generally agreed to be the reduced surface disruption, with speed of installation second and quality third.

The drawbacks suggested by the users are:

- o Risk of damage to existing buried services
- o High construction cost
- o Inability to achieve specified tolerances on line and level
- o Risk of tunneling/moling machine becoming stuck in the tunnel
- o Difficulty of making lateral connections

The risk of damage to existing services is agreed to be the major disadvantage (better and more accessible records of existing services must be provided as we do not know accurately the positions of existing underground services), whereas there are different views on the inhibiting effect of high cost. The question of accuracy is not generally regarded as a major problem but contractors consider the risk of the tunneling machine becoming stuck to be a serious drawback, while users express that the difficulty in making lateral connections is more significant.

There are many factors affecting use of trenchless methods. They are soil/rock type, hole size, hole or pipe protection, accuracy (line and grade), depth of burial, distance required, proximity of other services/utilities, surface access, surface improvements and restoration costs. The readers are advised to consult the specialist contractors for advice before making a decision.

Trenchless methods have clear advantages in reducing inconvenience to the general public; in purely economic terns trenchless construction can in many cases, and particularly below certain depths, compete favourably with traditional methods. However, trenchless methods of pipe and cable installation are no panacea.

Trenchless construction for underground services has been proved to be technically viable, and is already in use in Hong Kong.

4.2.3 Improving the accessibility of underground services by placing the different pipes and cables in a common facility or subways/tunnels

As stated in the introduction the positioning of pipes, cables and other services in the underground has many advantages in term of protection and functioning. The only major disadvantage is that to reach the pipes etc. the protective layer of ground has to be removed with considerable consequences for the daily life in the city (even with a services trench design). If we want to improve the accessibility a solution must be sought in creating a subsurface space which enables us to reach the engineering network without having to dig. The different pipes and cables may therefore be integrated in common facilities which can vary from a simple common duct to an accessible utility tunnel/subway.

From time to time it has been advocated that in heavily-trafficked streets in the centres of towns, subways should be constructed to accommodate utilities' mains so as to facilitate access to the mains for the purpose of maintenance and repair, and to avoid the necessity for repeated disturbances of the road surface and the heavy cost and great inconvenience occasioned thereby.

To achieve these objects subways would be required to comply with certain standards as regards general layout, design, construction and use, and a high capital cost of construction would be inevitable. It might, for instance, be necessary for a subway to be constructed on each side of the street (with or without cross connexions) and located so as to be as close as practicable to the adjacent buildings. In the design attention would have to be paid to the provision of adequate working space for the introduction of pipes into the chambers, for the laying of mains and their repair from time to time, and to such additional accommodation as the various utilities might require for future installations. Such matters as ventilation, drainage, lighting, flooding or fire gas explosion would also require special attention.

The principal advantages which a well designed subway system would have over the normal method of laying mains include:

- o Easier inspection of plant with a consequent longer life and a reduction in the number of breakdowns.
- o Greater facilities for some repairs.
- o Minimizing the number of road openings and elimination of the effect of such openings upon road surfaces and foundations.

The principal disadvantages include:

- o Damage or breakdown of one service more likely to affect other services, such as leakage of gas, fusing of electric cables, flooding, etc.
- o Greater difficulties for some repairs, such as replacement of large gas and water mains.
- o Difficulties in the provision of some service connexions to mains which would be deeper than if laid in the ground.
- o Legal and statutory restrictions on enforcing the use of subways.
- o A substantial reduction in the load-carrying capacity of electricity cables when they are laid in subways on account of the higher ambient temperature.
- o Interference, electricity cables can cause interference to telecommunication.

Most of the drawbacks can be overcome by good engineering and at a cost. Summarizing the assembly of pipes and cables in a common utility imposes a number of additional requirements which demand a change in traditional design.

The need for the services tunnel construction for underground service in congested areas is becoming increasingly apparent in Hong Kong. Besides the application in building complexes (e.g. large hospital, Science Park, large commercial development like Harbour City and the airport project), this solution can be chosen for a number of interesting environmental-friendly/sustainable projects like the West Kowloon cultural complex, the East Kowloon Tai Tak development, etc.

Dedicated utility tunnel is still not 100% perfect. The engineering works on sewers are the most far-reaching. Large sewers in tunnel are not quite flexible as the design of tunnel line or level and depth may conflict with the requirements of sewer (e.g. depth and uniform gradients between manholes). Further, the separation of foul-and surface-water drainage in some degree also makes the use of service subway for sewer difficult. Furthermore, sewers are generally larger than other services and therefore would increase the cost of subways disproportionately. It is of the greatest importance that the planning engineer should bear this in mind so as to ensure the satisfactory design of utility tunnel be made. Hence, the best system would be the use of two tunnels running in parallel on both side of the street. The size of each tunnel is large enough to cope with the expected population and further increase in capacity if one also wants to accommodate the foul and rain water sewers in this tunnel. It is by no means a simple design but a more workable solution. Nonetheless, if cost is of great concern, it would reduce the cost by locating the sewers outside the tunnels and leaving ample space for all cables, gas pipes, water pipes, HVAC installation as well as the pneumatic refuse chute.

5. Conclusions

The adoption of the foregoing recommendations will facilitate the work of the various underground services and result in an orderly disposition of pipes and cables and minimize the confusion and delay in vehicular traffic which often occur when repairs to mains are carried out or new mains are laid.

Provision of adequate underground services spines of tunnel/street throughout the ideal city for the efficient reticulation of all essential utility services is technically feasible. This will eventually eliminate the common nuisance of continuing digging of roads throughout the city. These spines will also improve the supply of each of the essential services to the buildings as the M&E

systems can now be seen and maintained for better service. Furthermore, any new development of high-tech information technology can be provided easily as this can be added in the tunnel without too much difficulties.

This short paper concludes that for a better city, all buildings and its building services must be fully planned and integrated with town planning. The design of underground services should not be taken lightly. It will provide better living standard and greener buildings. As such, our city will be a better one.

Sustainability has risen to prominence rapidly throughout the world over the last five years and will become a defining factor for construction in the 21st century. New paradigm of sustainability and sustainable city will emerge gradually to maintain its place among the world's top international cities. Hong Kong should be directed towards sustainable development.

Practical but not imaginative solutions like services tunnels and No-Dig techniques are available for the creation of better buildings and cities. To achieve the objectives of a modern and greener city, all developers, government officials, planners and engineers must work together and take this challenge – building one of the world's dynamic and sustainable great cities beyond the 21st century with better underground infrastructure and utility services.

A sustainable services tunnel will be designed to have a minimum impact on the environment. It will contribute to good engineering as well as a good quality of life for all people. The realization of utility tunnels will only be feasible in large infrastructure projects (e.g. a new town on the surface as well as underground). The application of tunnels can only form a solution in the long term and then if we are willing to make investments for the future.

To close this short discussion, I offer the follow:

"It comes to the conclusion that our environmental conditions have not got better and are likely to get worse unless we can improve our management of the integration and planning of the underground infrastructures." It is clear that what is happening now can affect the betterment of our next and future generations.

Reference Diagrams

Attached to this paper is a typical diagram of services tunnel and services trench which can be adopted for large construction projects.

Bibliography

1. Trenchless Construction for Utilities, (1985), Proceedings of the First International Conference, Institution of Public Health Engineers, U.K.
2. Trenchless construction for underground services, (1987), Technical Note 127, CIRIA, U.K.
3. Steven R. Kramer, William J. McDonald, James C. Thomson, (1992), An Introduction to Trenchless Technology, Van Nostrand Reinhold, New York.
4. Data obtained from Taiwan (Services tunnels for Taipei).
5. Data obtained from construction industry in Hong Kong between 1980 to present.

EARTH ENVIRONMENTAL PROBLEMS IN CITY CONSTRUCTION

MA BAOMIN LIN LEI
Architectural Design & Research Institute of Tsinghua University, Beijing, China

With the development of science and technology, peoples' living style is getting farther away from what used to be in nature, and city development is a good illustration of this change. As an artificial system, city has a way of development of its own, which is quite different to the natural system. This inconsonance brings about much disarrangement, such as pollutions, heat island, overburden of resource and waste, etc. From the view of evolvement and balance of ecosystem, this paper discusses the relationship between city construction and environment protection. Pluralistic industry based on multi-dependence is suggested to construct a sub-circulation in city, and a sustainable development pattern could thus be achieved. By partly converting scraps into building materials, cities could reduce its negative influences on environment. In the circulation, particular attention is given to tall buildings, for which there is a rising market demand. Some suggestions on tall building construction were proposed.

1. Preface

The earth environment is the basis upon which human beings rely their living and is a result of natural evolution of the earth over 4.6 billion years. It brings all the forms of life to the world, in particular human beings. Once the environment suffers any change or damage, however slight it may be, human beings could immediately face a deadly danger. The damage or change is usually driven by forces from forces from outer space or from inner earth factors, and equally possible and undeniable, forces from human beings themselves. City construction can be a good illustration of this human disturbance on earth environment.

The ability of disturbance of mankind grows together with the science and technology, a unique and by all means powerful gift enjoyed by the human race, a gift with upside and downside. On the one hand, it is constructive; on the other hand, it can be very highly destructive. With the help of science and high-tech, human beings could ruin what has being formed by natural forces during millions of years in a matter of hours.

Earth environment is a system of self-organization, spontaneously formed in the history of earth and with a built-in capability of evolution and sustainable development, a capability man-made systems usually lack. In other words, man-made systems are counteracting the effects of natural forces. To understand this, we may summarize man's unique way of survival in an equation like this: Labor (hand) + Creation (brain) + Natural Resources = The Man-made Environment of Survival. Farms, factories and cities can all be regarded as the man-made environmental systems. Whereas man is changing the natural world out of self –need, nature has its counter-effects on man-made systems. This explains why a farmland is apt to return to desolation only in one year's time after people stop cultivating it, indicating the nature's desire for freedom from foreign influence. Likewise, it is the case with the cities, a typical man-made system resulted form the human counteraction against the natural earth environment. In absence of human management, a city can be reduced to a wasteland due to the action of natural forces upon it.

The history of urban development witnesses that cities have been making people more and more distant from nature, and the target of the city planners is to make an artificial world of living exactly detached from its natural counterpart. We may break down this history into three periods. In its early years of development, cities did a good job in providing people with a safe shelter from wind and rain, helping people fighting such natural forces as the heat or coldness of weather, hunger and beasts. Then comes the middle period, in which cities functioned more as a means to develop economy. Later and finally, we have the third period, in which cities function more to provide people with a luxurious way of life characteristic of excessive consumption. Due to a negligence of the destructive effects of such a way of development upon the environment, cities more and more constitute a threat to the survival of mankind.

City construction grows fast all over the world at an amazing speed. In China, the scale of city construction equals to constructing a super city like Beijing or Shanghai per year. There are about 3000 square km land transferring to city in USA, and 2430 square km land in Egypt, as which it is the same in other countries. Cities enlarge while countries, farmlands, land of wild lives shrink, and natural environments are more and more replaced by artificial ones. Cites are a crux of the earth environment problems related to human influences.

2. Environmental Problems and Ways to Solve

The earth environment, including atmosphere, hydrosphere and biosphere, is a self-organized system evolved during 4.6 billion years without resource deficiency, and all this is because it is an extremely diversified and complicated self-balance network, capable of self-adjustment and sustainable development.

The mechanism of working for big cities, however, is a man-made system, going in manner off the course of nature. Man's will plays a decisive role in the origin and development of cities, and all those natural systems that against the will of mankind are readily destroyed. Although man indeed learns to observe some natural and thus scientific laws in technical details of construction projects in cities, little concern has been shown for the importance of earth environment in the general course of city development. City construction has been doing direct or indirect severe damages to the atmosphere, hydrosphere and biosphere, etc., with such problems as the drain on material resources, and waste disposal and other problems of sustainable development.

2.1 *Influence of city construction on atmosphere*

For atmosphere, the three most serious problems caused by city construction include increasing CO_2 concentration, ozone gap and air pollution, most of which directly or indirectly result from the burning of large amounts of mineral fuels used in city construction and operation.

Seen from figure 1, in the last 15 thousand years, CO_2 concentration increased and reached maximum 10 thousand years ago, and keeps constant till now. It is said that CO_2 concentration has reach a balance. Seen from figure 2, the record indicates the durative increase when CO_2 concentration reached maximum. The cause of the increase belonged to the economic development around the world. The CO_2 concentration increases at 10^{-6} rate per year. Figure 2 illustrates that CO_2 concentration caused by peoples' activities increases at a high speed and breaks the ecological balance. Increasing CO_2 concentration could cause greenhouse effect, heat island of cities, which has asphyxiated many thousands of people.

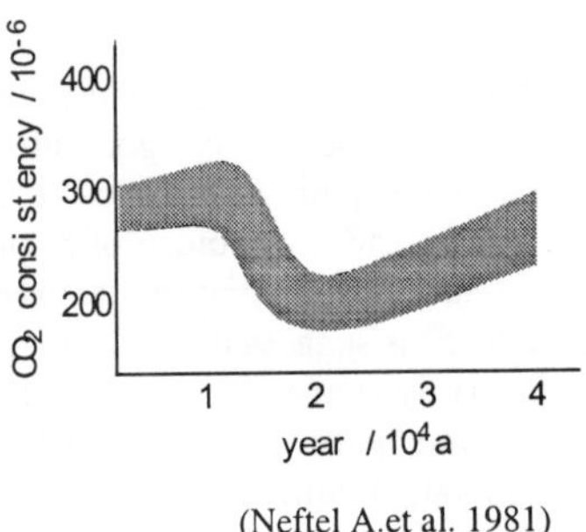

(Neftel A.et al. 1981)

Figure 1. The transformation of CO_2 concentration in atmosphere during 40,000 years.

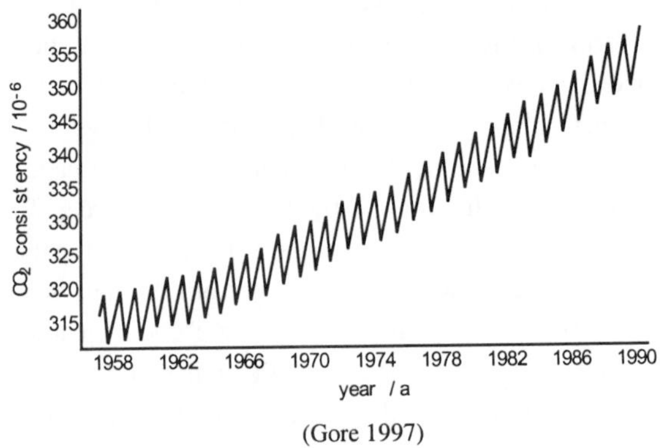

(Gore 1997)

Figure 2. The record of CO_2 concentration in Hawaii after 1957

The other effects such as depletion of ozone layer, air pollution are all caused by city life and city construction. Without ozone layer, green plants will be threatened, and air pollution causes diseases of breath system and cardiovascular system.

Most of the materials used in construction consume resources and send out great amount of heat and waste. For example, 0.235 t-c of CO_2 is produced while 1 ton of silicate concrete is made and 0.436 t-c of CO_2 is produced while 1 ton of steel is made. Air condition becomes worse in winter, because much more coal and oil are burned for heat.

An effective way to clean the atmosphere is to use cleaning energy, such as solar energy and wind energy, instead of mineral fuel, at the same time, to maintain enough plant in the city. It is forecasted that the mineral fuel is to fall into disuse in 2050. The research institute of solar energy, Beijing, China has built the Sangpu solar energy building of China. This four-floor building makes use of solar energy for illumination, heating and air-condition. As we know, there are large numbers of buildings in city. There is no technological difficulty in using the solar energy to heat water. If we do so, lots of mineral fuel could be saved, and it could also reduce a lot of air pollution. To design the new buildings that use solar energy, or to rebuild the old buildings in order to fit to the solar energy usage could improve the atmosphere. All this could rank as one of most important changes in the history of human activities.

2.2 *Influence of city construction on hydrosphere*

Many large cities, even including those near Changjiang River, are short of water. Ground water level drops because of overuse of ground water. However, when the flood season comes, disaster comes. It is the aftermath of that city function is incompatible with nature.

There is a perfect natural self-organized system in hydrosphere. Figure 3 illustrates self-cleaned system in river. This system is large, effective and its mechanism has not been well known yet. But it's certain that when large scopes of land were turn into concrete or asphalt, ways of transportation of water was interrupted. So the delicate mechanism was disturbed and overburdened. These unsolved wastes in cities are all carried into rivers, which could not clean it but be polluted. Pollution cuts off water supply and the level of city ground water drops rapidly.

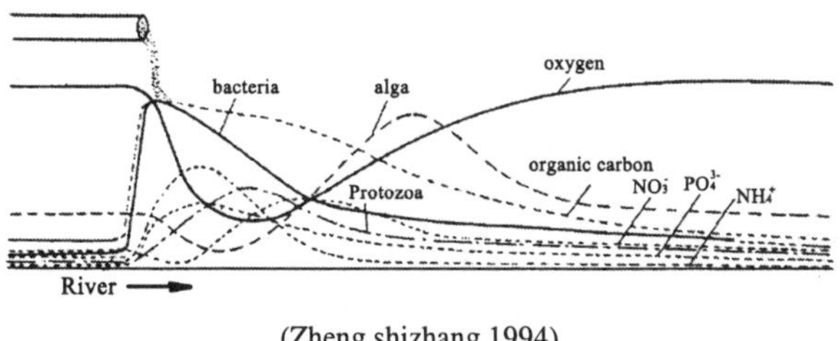

(Zheng shizhang 1994)

Figure 3. The self-cleaned system in river

An effective measure might be planning more greens in city. And special concrete elements are used in places to ensure the rain could renew underground water. We can also design mid-water circulation system to fit part of requirement. As tall buildings take up more and more proportion, more land could be liberated from hard ground. Mid-water system could be more effective, because waste water is concentrated and easy to collect and use.

2.3 *Influence of city construction on the biosphere*

The earth is colorful with a diversity of lives. Biosphere evolves for 3 billion years. The relationship among various lives forms the complex ecological network. No life could live without biosphere, nor could human beings. However, the human beings have destroyed the biosphere severely since industrialization. The Figure 4 shows the extinction rate of lives in the past 300 years.

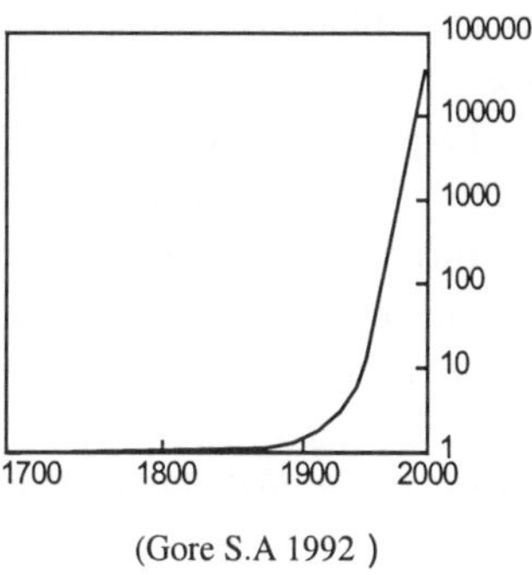

(Gore S.A 1992)

Figure 4. The extinction rate of lives

The biosphere consists of various ecologic balance systems. Every one has the ability to clean itself, and maintains the balance of economy by such self-regulation. Plants change the non-living element in soil and atmosphere to living element, animals eat plant, and the microorganisms discompose the remains of animal and plant. It is a perfect cycle without waste and hangover. Moreover, this system can regulate itself. It could rebalance after disturbed.

However, the mass construction of city is a big disturber. It destroyed the economic balance completely because ecosystems lose the ability of rebalance. As is different from the working mechanism of a normal eco-system with a four-level make-up, there is only the consumer but not the producer and the disintegrator. We need the other man-made system, such as refuse dump, to solve the problem. Obviously, it is not only that working mechanism of cities is incomplete, but that this incompleteness affects at the same time the balance of working of the suburban and rural areas.

2.4 *Influence of city construction on geology*

City construction is the course of digging, exploding, moving and accumulating. The process that matters are moved greatly is a huge damage to geology. The research indicates that only leveling up ground could make the soil eroded rate increase 40000 times.

Influence is more than that. Construction materials come from mountains and fields where damage is seldom considered. In order to construct a building for 50 years usage, people easily change the situation of soil existing there for several million years.

Flat surface of land is needed in city, but natural land has many curves. The basement of building should be stiff and strong enough, but soil near surface is mostly soft and suitable for planting. When buildings become taller and taller, deeper and stiffer basement is required. One thing we could do is to make better location plan. Large construction project should not be located on those areas that are important to the surroundings. Environment influence should be given priority in consideration rather than economic profits in the feasibility research, especially in sensitive areas.

2.5 *Influence of city construction on resource and waste*

In the process of city construction, large amount of resources are used and meanwhile lots of waste is produced. Three fourths of the city's waste is produced and 86 percent of energy is consumed in construction. Energy and waste problem becomes remarkable. Building more construction dumps is not a good idea. Three fourth of dumps in USA have been shut down because of being full.

Solution could be made by imitate the nature system. In nature, there is no 'waste', because waste to a species could be useful to other species. There are many species in city too, such as tall building, large span structure, flats and etc. each have different demand for material and other facts, for example, base. Circulation from waste to resource could be set up among these. Tall buildings have higher strength concrete and steel bars, while flats have lower requirements. Once a tall building is abandoned, it could be a material resource to those flats. And flat's waste could be used by road or ground. It is possible that requirements to nature system could be greatly reduced.

3. Summary

9000 years ago, the first city appeared. Till 1900, population proportion in cities reached 13.6%, and it turned out to be 41.6% in 1985 and over 50% till now. City is developing, but it must conquer the appearing difficulties. The environment problem is considered to be one of the main problems in today's world, and comes practically as a byproduct of the inappropriate living styles of people, featuring a desire for luxury and excessive consumption. As people change their attitude towards way of living and thus change their living styles fundamentally, a better solution for city construction, featuring a stress on a harmonious relation between city and earth environment, between man and nature, is highly expected.

References

1. Guan Guangyue, Earth At the Chaos Edge (2000).
2. Wang Changqin, etc, Green Architectural Manual. (1999).
3. Wang Xinwen, Globe Science Generality (1995)

PILOT SURVEY ON PERCEPTIONS AND USE OF OPEN SPACE IN TWO SLECTED MIXED USE RESIDENTIAL DEVELOPMENTS IN HONG KONG

S. B. A. COOREY

Department of Architecture, University of Hong Kong, Pokfulam HKSAR, China.

S. S. Y. LAU

Centre for Architecture and Urban Design for China and Hong Kong, Department of Architecture, University of Hong Kong, Pokfulam, HKSAR, China.

The compact urban form implies intensification, high density, mixed land uses, efficient public transport and dimensions that encourage walking and it is argued to have both negative and positive implications on urban living space and thereby quality of life and sustainability. Perceptions on lack of open space, urban greenery, poor environmental quality, crowding, lack of privacy, noise are some negative implications while urban variety, vitality, robustness, accessibility, permeability, social interaction, safety, efficiency and reduced travel time to services are some positive implications. The problems presented in the compact city debate can be sumarised as; claims about sustainability of the compact city; the feasibility, or the social acceptability of the compact city and the requirement of tools or measures that ensure successful implementation of the compact city [1]. Claims on social acceptability of a compact city is related to issues such as congestion, lack of privacy, poor environmental quality, constrained living spaces, high rise living, lack of open space and urban greenery. It is argued that individual proximity to a concentration of people increases social interaction [2] but some argue that crowding leads to weaker ties [3]. But the response towards such implications and arguments are based on human perceptions, preferences, adaptability and tolerance which differ across each cultural and urban context. This paper focuses on implications and arguments on compactness in relation to open space within two selected mixed use residential developments in Hong Kong. Critical implications of compactness are analysed in a pilot survey conducted among a total of 120 respondents in 3 open spaces located within the selected developments. Questionnaire interview is conducted to gather user's response to the following; perceptions on lack of open space, importance of open space, social interaction in open space, aesthetic and environmental quality of open space, crowding, time spent in open space, travel time to open space and the use of alternatives or substitutes to open space. Conclusions are drawn from survey findings in order to facilitate hypothesis for future research where regression analysis will be used for further investigating the implications of a compact mixed use urban context on the perceptions and use of open space with reference to compact residential mixed use developments in Hong Kong.

1. Introduction

1.1. *Background*

The compact city implies intensification, high density, high rise, mixed land use and efficient public transport as oppose to low density, low rise, mono use urban sprawl. Due to constrained private living spaces people in compact cities such as Hong Kong spend most of their time in public living spaces which constitutes both indoor and outdoor public spaces. A Hong Kong resident's private life goes beyond his home into public gathering places like shopping malls, restaurants, karaoke bars, parks, open spaces known as borrowed spaces [5]. Such urban living spaces both indoors and outdoors are vital land uses in high density compact cities.

Open space is an important land use which contributes to better quality of life and environmental and social sustainability of a city. But as argued by many researchers urban living spaces, which include urban open spaces, are being challenged by the high density compact development pattern. Such challenges are both negative and positive and needs to be discussed in detail to asses the implications of compactness on urban open space. But it must be noted that the user's perceptions, adaptability and acceptance of such challenges play an important role when arguing the claims and counter claims on the implications of compactness on urban living space.

1.2. *Research Aim*

This pilot study is limited to the survey of users' perceptions and use of public open spaces within two selected compact mixed use developments in Hong Kong. The paper aims to analyse the users' perceptions on open space provisions, importance of open space, social interaction, aesthetic and environmental conditions, crowding, time spent in open space, travel time and the use of alternatives to open space. Selection of two developments with variation in population size, number of households, mix use helps to analyse and discuss the variation in respondents' perceptions and use of open spaces. Such discussion will form the hypothesis for future research on the implications and influence of urban compactness on perceptions and use of open space in compact mixed use residential developments in Hong Kong.

1.3. *Hypothesis*

A compact urban context is claimed to have both positive and negative implications on urban living space. The positive claims are urban variety, vitality, robustness, high accessibility and permeability, social interaction, safety, efficiency and reduced travel time. Negative claims are perceptions on lack of urban open space and greenery, poor environmental quality, crowding, lack of privacy and noise. [7, 8, 9, 10, 11, 12, 13, 14, 15, and 16]. This paper is limited to speculating the following issues within two selected compact mixed use residential developments.

1.3.1. *Testing issues*

1. It is perceived that compact city form suffer from lack of greenery, open spaces, parks and privacy [7]. But the user's satisfaction with open space provisions in mixed use developments will be influenced by the availability of substitutes and alternatives to open space created by other recreational land uses within close proximity.
2. Compact form is claimed to induce better social interaction - User's satisfaction with social interaction in open space within the development.
3. High density and mix land uses create crowded public spaces, but the negative or positive response to such an attribute is based on attitudes and perceptions of user groups within a mixed use development.
4. A compact city is claimed to have better accessibility to facilities and services and therefore reduced travel time to open space must prevail among majority of users in a compact zone.
5. Frequency of use of open spaces and thereby the demand or needs for open space is influenced by urban compactness and mixed land uses.

1.4. *Methodology*

A pilot survey was conducted among a total of 120 respondents in Telford Gardens and Mei Foo Sun Chuen developments. Telford Gardens and Meifoo sun Chuen mixed use high rise residential developments are selected as representing typical high rise mixed use residential developments in Hong Kong. Respondents were randomly interviewed by a questionnaire survey in and around 3 public open spaces within the developments namely open space 1 (OS1) and open space 2 (OS2) in Telford, and open space 3 (OS3) in Meifoo Sun Chuen. Survey focused on a mix of user groups such as residents, commercial users, office users, Government, Institutional and Community (GIC) users and visiting population to the development. The open spaces selected represent public open space within high rise mixed use residential developments in Hong Kong. Questionnaire focused on the respondent's demographic and socioeconomic background, their perceptions and satisfaction with open space based on the research aim and testing issues and the users demand for open spaces. Intensity of use (number of visits per month) is used as an indicator for evaluating the respondents demand for open space.

2. Open space - benefits

Open space has its many contributions to a city. It is used to link neighbourhoods and buffer incompatible uses, when left natural it helps control floods, purify run off, recharge groundwater, support life and afford scenic views valued by residents, and if bound and amenitised it provides gathering places for social interaction, recreation, and civic function. Three key benefits of local open spaces are the manner in which it facilitates recreation; its psychological benefits and its role as a habitat for flora and fauna [17]. Urban nature provides environmental and ecological services as well as important social, psychological and economic benefits which contribute to city space and the quality of life and is a key indicator for sustainability [18].

2.1. *Open space-social and psychological benefits*

Empirical evidence shows that presence of natural areas, open spaces and greenery contribute to quality of life in many ways. The availability of open spaces, parks, greenery encourages outdoor recreation. Many sociologists, psychologists, social workers, outdoor recreation specialists, and others have emphasized the psychological and emotional need for outdoor recreation. – need for relief from tension and emotional strains which modern urban living place upon the individual. The recreation in general and to outdoor recreation in particular they ascribe great value, of almost a therapeutic kind. Some stress that in outdoor recreation the individual can test his physical fitness and his ability to cope with nature. Still others emphasize the opportunity which outdoor recreation gives for self fulfillment and individual choices. Recreation is also considered by some to have significant value in combating or preventing juvenile delinquency.

Much research has been conducted on stress reduction and mental health benefits of open space and greenery. Park experience reduces stress and serves as a psychological benefit [19]. Open space enhances contemplativeness, rejuvenates city dweller, and provides sense of peacefulness and tranquility [20]. Frequent visitors to local parks are found to be in better health than those who did not and the natural environments, vegetation and water is found to induce relaxed and less stressful states in observers compared with urban scenes with no vegetation.

2.2. *Open space-environmental and economic benefits*

The importance of open space or green space within urban areas has been recognized for over 100 years. In the 19th century, physicians subscribed to pathogenic theory relating the incidence of disease to bad air quality [21]. In the Victorian age open areas in cities were considered literally as the lungs of the city and were formally designed as public open space. At the end of the 20th century urban air remained as an unhealthy mix of sulphar dioxide, nitrogen oxide, ozone and carbon monoxide, carrying dust which is laced with particles of lead, arsenic, asbestos and cadmium. The provision of natural features, particularly trees act as important filters which significantly reduced such pollutants [22]. The environmental benefits are air and water purification, wind and noise filtering and microclimate stabilization. Therefore it could be said that the current policies aiming at intensifying urban land use may be posing a threat to these urban lungs.

Also environmental contributions lead to economic benefits. Air purification by trees lead to reduced cost of pollution reduction and prevention measures and aesthetic and recreation values of urban space also increases property values and therefore tax revenues [23].

3. Opens pace in Hong Kong

Hong Kong offers itself as a good specimen to study compact urban form because of its urban system with high density, high floor area ratio(plot ratio), mixed land use, short distance between different uses and efficient public transport [10, 14, 15 and 24]. HKSAR is a relatively small and high density city, in which available land for building is scarce with minimal natural resources. The total land area of Hong Kong is 1,108 square kilometers, with 21.8% of built up area which is concentrated in the triangular tip of Kowloon and the coastal strip of northern Hong Kong Island. The built up area consists of 2.2% residential, 1.9% government Institutional & community facilities, 0.3% of commercial, business and offices, and 1.8% of open space (Hong Kong Planning Department, 2004).

The total population is 6,803,100 with the median age rising from 30 in 1988 to 36 in 2003. The median monthly household income is HK$15,000(US$1920) (Census and Statistics Department, 2004). The population density is 6, 630 persons/sq.km with urban areas holding a staggering population density of over 55,000 persons/sq.km where certain districts rank among the most densely populated places in the world. The plot ratio is up to 15 for commercial uses and up to 10 for residential uses have led to buildings of up to 80 storey. The density of public housing reaches at least 2,500 residents/ha which is twice the density of the most crowded residential areas in mainland china [25]. Tight living space in Hong Kong results in overuse and high demand for public open space. But due to competing demands of high population and scarcity of land, the provision of good quality urban open space is a critical issue.

Designers in Hong Kong have taken advantage of the high population density to generate prolific mixed use designs with efficient infrastructure and higher order connectivity to urban services. The intensification of mixed use is found in a majority of city centers. Hong Kong has several mixed-use developments linked by efficient public transport which are mostly built around mass transit nodes. Each development acts as a primary node for mixed – use development with uses ranging from five to eight types. Each primary node is dependent on the neighbouring nodes for missing land uses. Such interdependency leads to a secondary mixed use zone that consists of several primary nodes and the combination of two or more secondary zones form a larger area of a tertiary mixed use zone. The entire Hong Kong SAR acts as a network of such primary, secondary and tertiary mixed use zones and such zones are linked by a relatively cheap and speedy transport network [4].

Open space in Hong Kong exists at regional, district and local or neighbourhood scale with active and passive uses. The most popular form of leisure among the Hong Kong population is passive recreation in planted areas and sitting out areas within neighbourhoods (local open spaces) and households earning less than HK$5,000 (US$ 640) per month spend more than 36h/w on leisure, more than any other income group [6]. The most deprived open space in terms of quality and quantity are in the older, more densely developed urban areas of Hong Kong and exhibits serious shortcomings in open space standards.

Limitations of land for open space and high population density and increasing demand for high building density have critical implications on the quantity and quality of open space in Hong Kong

4. Pilot survey

4.1. *Background of case studies*

Telford Gardens and Meifoo Sun Chuen were selected respectively from Kowloon Bay and Lai Chi Kok urban areas, in Kowloon side of Hong Kong. Kowloon Bay and Lai Chi Kok falls within Kwun Tong and Sham Shui Po districts having a district population of 5,727,000 and 3,556,000 respectively. The open space provisions within the Kowloon Bay and Lai Chi Kok outline zoning areas are 15% and 22.13% respectively (Outline zoning plans S/K13/21 and S/K16/13). The two developments Telford Gardens and Meifoo Sun Chuen were selected on the basis of having a similar site area and variation in population size, number of dwelling units, net population density, net residential density, gross floor areas(GFA) of mixed land uses, see Table 1.

Two open spaces in Telford Gardens and one open space in Meifoo Sun Chuen were selected for the survey. Easy access to transport terminals and MTR, linkage to residential, commercial, office and GIC land uses and access to the public are characteristics common to all three open spaces, with variation in hard and soft landscaping. Open space 1 (OS1) and open space 2 (OS2) in Telford Gardens is located adjacent to each other on podium level, while open space 3 (OS3) in Meifoo Sun Chuen is on ground level. All three open spaces are accessible to the public and all user groups within the development and are also accessible to people from outside the residential development. The three open spaces represent open space seen within compact mixed use residential developments in Hong Kong, see Figures 1, 2 and 3. The study focuses on public open spaces that are common to all user groups within the development and provides access to visitors. Such open spaces are commonly located on podium or ground level within the development. The open spaces commonly located on podium and restricted to only the residents of the development is not considered for the study, as such spaces does not incorporate a users of different land uses and therefore the perceptions of different user groups cannot be evaluated in such semi public open spaces.

Figure 1. Open space 1 (OS1) in Telford Development

Figure 2. Open space 2 (OS2) in Telford Development

Figure 3. Open space 3 (OS3) in Meifoo Sun Chuen Development.

4.2. *Survey methodology*

Respondents were randomly interviewed on three consecutive weekdays from 10.30 a.m. to 6.30 p.m. 40 respondents were interviewed from each open space, with a total of 120 questionnaires from all open spaces.

The questionnaire gathered the following; demographic and socioeconomic data: user group, age, sex, income, perceptions: user's perception on open space provisions (was measured on a 2 point scale indicating as 1-adequate and 0-inadequate), importance of open space (was measured on a 5 point scale), social interaction, aesthetic and environmental conditions (was measured on a 3 point scale), crowding (was measured on a 5 point scale), time spent in open space (was measured on a 4 point scale ranging from less than 15 mints, 15-30mints, and 30-1 hour to more than 1 hour), travel time (was measured on a 5 point scale ranging from less than 5 mints, 5-10mints, 10-15mints, 15-20mints and more than 20mints), use of alternatives to open space (was measured on a 2 point scale indicating as 1-Yes and 0-No) and open space use: frequency of use (number of visits to the open space per month-considering week day visits only). Frequency of use was measured on a 3 point scale ranging from "daily number of visits" (more than 20 times per month), "weekly number of visits" (4-20 times per month) and "monthly number of visits" (less than 4 times per month). Background data collected of each development was population size, number of dwelling units, net residential and population density, FAR of mix land uses, primary: secondary and public: private land uses, see Table 1.

JMP software is used for basic descriptive statistical analysis and summarizing data. Respondents' perceptions and use of open space is quantitatively and qualitatively evaluated in order to draw explanatory conclusions and hypothesis for further research.

4.3. *Discussion of case studies*

Table 1. Background information of Telford and Meifoo Sun Chuen Developments.

	TELFORD	MEIFOO
Population Size	19,968	43,509
No. of Dwelling Units (DU)	4962	13,109
Net Pop Density (persons/ha)	1,248 per/ha	2,559 per/ha
Net Res Density (DU/ha)	310.12 DU/ha	771.11 DU/ha
FAR Res Land Uses	173.05	539.95
FAR Com Land Uses	51.66	44.67
FAR Off Land Uses	32.58	1.85
FAR GIC Land Uses	0.56	2.34
FAR Primary and Secondary Land Uses	205.64 52.22	539.04 44.67
FAR Public and Private Land Uses	52.22 205.64	44.67 539.04

Socioeconomic background of respondents: majority of respondents are female, age 15-40, with income level below 8,000 HK$(1038US$) per month, see Figure 4.

The respondents consist of a mix of user groups such as residents living in the development, people visiting the shopping mall and other commercial uses, office workers, GIC users such as school going, users of other academic, sports and recreation facilities who are using the open space during their breaks etc. Therefore as seen in Figure 5, the total respondents interviewed represent all users of the mix land uses within the development.

Survey also shows that OS3 in Meifoo Sun Chuen is used by a majority of residents while OS1 and OS2 have mix user groups of commercial/visitors, Office and GIC users, see Figure 6.

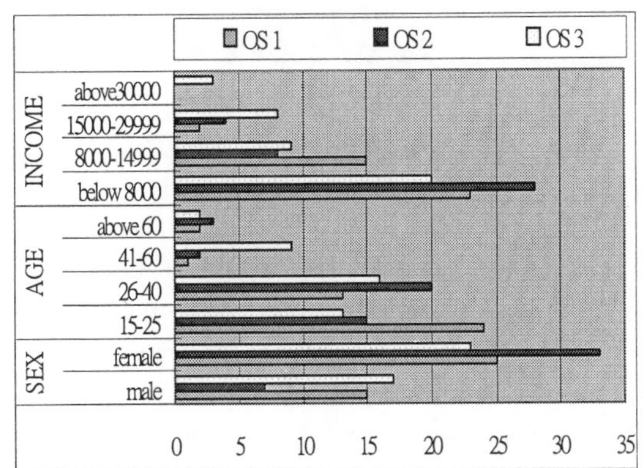

Figure 4. Distribution of sex, age and income groups in OS1, OS2 and OS3

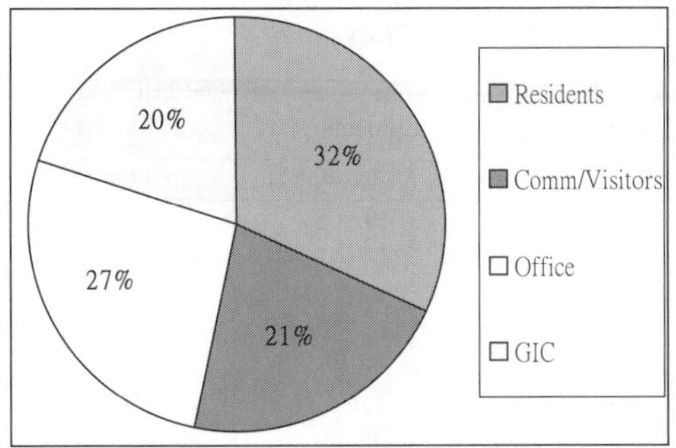

Figure 5. Distribution of user groups across all respondents.

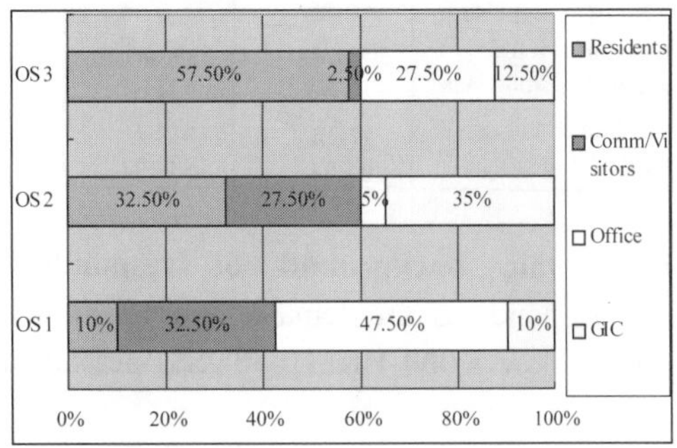

Figure 6. Distribution of user groups across OS1, OS2 and OS3

Open space provisions: OS1 and OS2 in Telford shows respectively 87.5% and 75% respondents satisfied with open space provisions while OS3 in Meifoo Sun Chuen shows 60% respondents satisfied with open space provisions, see Figure 7.

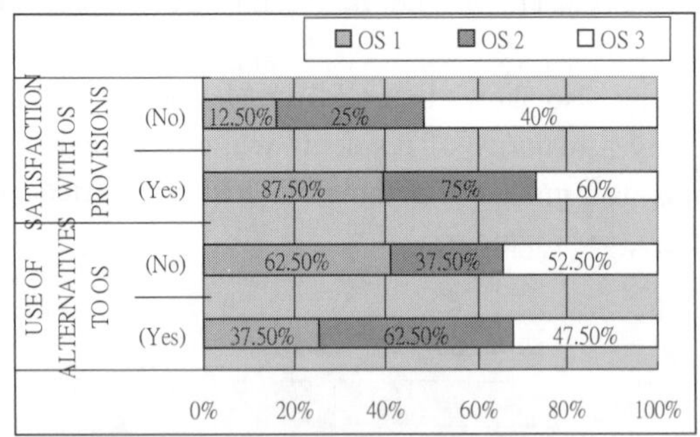

Figure 7. Comparison of respondents' satisfaction with open space provisions and the use of alternatives to open space in OS1, OS2 and OS3

Use of alternatives: Survey shows OS1 majority do not use alternatives but OS2 majority uses alternatives. But in both cases respondents are satisfied with the provisions of open space, see Figure 7. It could be speculated that alternatives to open space can be used even in cases where users are satisfied with open spaces. Other land uses such as shopping malls, indoor recreation, and cinema are used during leisure time instead of using open space and such habits are not merely generated due to lack of open space but are preferred choices of leisure by the respondents.

Telford has high FAR of commercial and GIC land uses or secondary land use (52.22) to support the population size (19,968) and primary land uses (205.64). But in comparison Meifoo Sun Chuen has low FAR of commercial and GIC land uses or secondary land uses (44.67) to support a comparatively larger population size (43,509) and FAR of primary land uses (539.04), see Table 1 and Figure 8. It could be said that high availability of secondary land use in Telford gives users a variety of choices for leisure and recreation other than using open space. But Meifoo Sun Chuen lacks in the availability of secondary land use to support a large FAR of primary land uses resulting in less alternative choices for recreation.

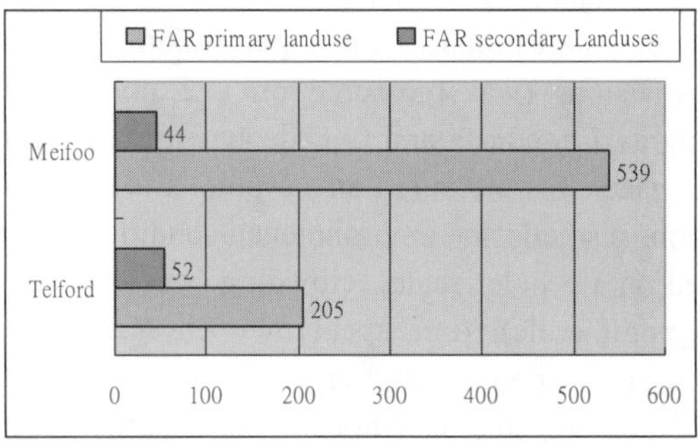

Figure 8. Comparison of FAR of primary land uses and supporting secondary land uses in Telford and Meifoo Sun Chuen

Frequency of use: OS1, OS2 in Telford shows low frequency of use respectively 14.4 (14visits/month) and 15.4 (15visits/month) mean values with a majority being less frequent users while Meifoo Sun Chuen OS3 respondents show a high frequency of use with a mean value of 20.45 (20visits/month) with a majority being high frequent users, see Figure 9. This shows that Telford respondents have low demand for open space while Meifoo Sun Chuen has higher demand.

Survey shows that in mixed use zones the high availability of other recreational land uses create a range of alternative recreational choices to using open space which will reduce the demand and need for open space and counteract lack of open space provisions.

But it must be noted that the variation in demand for opens space is not merely due to reduced availability of choices but will also be influenced by the distribution of user groups within the development.

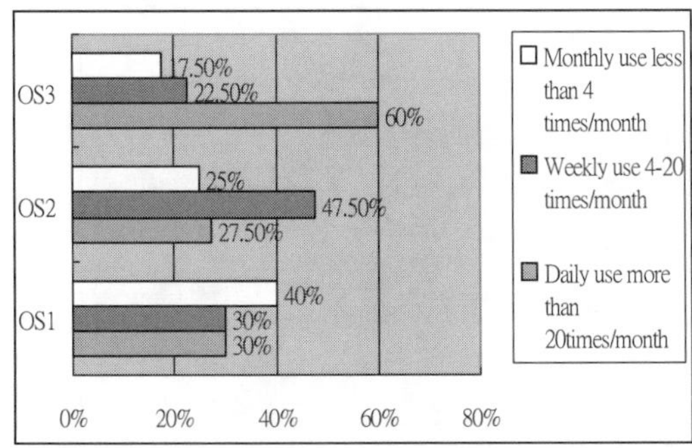

Figure 9. Comparison of frequency of use of open space in OS1, OS2 and OS3

Demand for open space among different user groups: Survey also shows that the high frequent users are residents with a mean frequency of 27visits/month while commercial/visitors, office and GIC users show lower frequencies respectively 11 visits/month, 13 visits/month and 12 visits/month, see Figure 10. This shows that in compact mixed use zones the frequency of use of open space and thereby the demand and need for open space varies according to user groups.

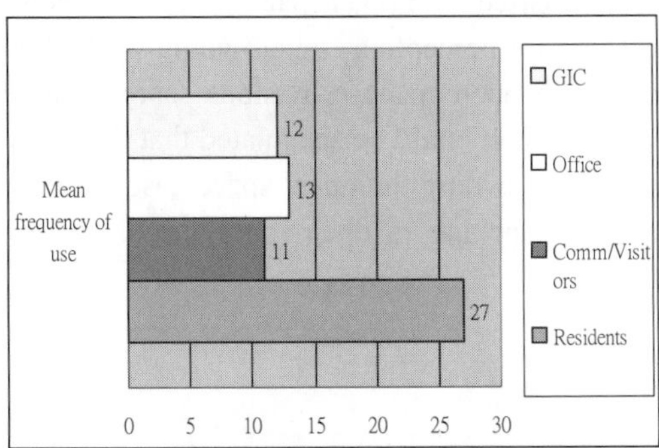

Figure 10. Variation in the frequency of use (no. of visits/month) and thereby demand for open space across different user groups.

The provisions of open space within such a mix use zone must take into consideration the variation in demand by different user groups. Developments with high resident population will have higher demand for open spaces while ones with more commercial, GIC and office land uses will have a reduced demand for open space, therefore the provisions, supply of open space must be coherent with the distribution of user groups within the population, see Figure 10.

It can be suggested that the proportion of primary to secondary land uses will influence the ratio of resident to commercial, office and GIC users, Telford has a larger proportion of mixed user groups, while Meifoo Sun Chuen has a majority of resident users, which coordinates with the FAR of primary and secondary land use in each case, see Figures 6 and 9.

Importance of open space: Survey shows respectively residents, commercial/visitors, office and GIC users score 4, 3.9, 3.7 and 3.5 equally high importance values for open space. This shows that all user groups share common high importance value for open space although it is noted that only residents are high frequent users while all other user groups are low frequent users. This shows that although users have high importance for open space their demand for open space is not governed by such perceptions.

Time spent in open space: When comparing the time spent in open space residents scores highest mean of 3.0 which is in the range of 30mints-1hr, while commercial/visitors and GIC users score mean of 2.4 and 2.3 respectively ranging from 15-30mints and office users score 1.9 ranging less than 15mints. It could be speculated that the availability of time to use open space differs among the user groups within a mix use context, see Figure 11.

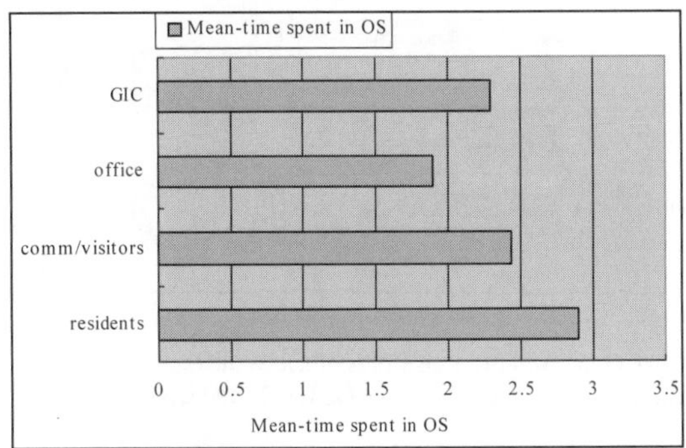

Figure 11. Variation in time spent in open space across different user groups in all three open spaces.

Social interaction in open space: The social values of OS1, OS2 and OS3 shows mean scores of 2.2, 2.3 and 2.3 respectively showing similar perceptions on social interaction in all three open spaces, see Figure 12. The respondents scoring high social value in OS1, OS2 and OS3 are respectively 35%, 45% and 32.5% while 50%, 47.5% and 67.5% of respondents scores average satisfaction. This shows that a larger proportion of respondents are some what hesitant about the social value of the open spaces. It could be suggested that mix used contexts offer different choices for leisure, social interaction and recreation in close proximity, which d-values the demand for open space as social, or recreational spaces.

Aesthetic value of open space: 22.5%, 35% and 57.5% respondents respectively in OS1, OS2 and OS3 scores high aesthetic values and 67.5%, 62.5% and 40% respectively score average aesthetic values. The aesthetic mean value scored in OS1, OS2 and OS3 are respectively 2.12, 2.32 and 2.55, showing that Meifoo Sun Chuen respondents perceive a higher aesthetic value in comparison to Telford, see Figure 12.

Environmental value: Only 32%, 35% and 45% of respondents in OS1, OS2 and OS3 respectively scores high environmental perceptions. The mean scores for environmental conditions in OS1, OS2 and OS3 are respectively 2.17, 2.25 and 2.42, see Figure 12. This shows that both aesthetic and environmental conditions of the open spaces in Telford are considered unsatisfactory while Meifoo Sun Chuen scores higher in comparison to Telford.

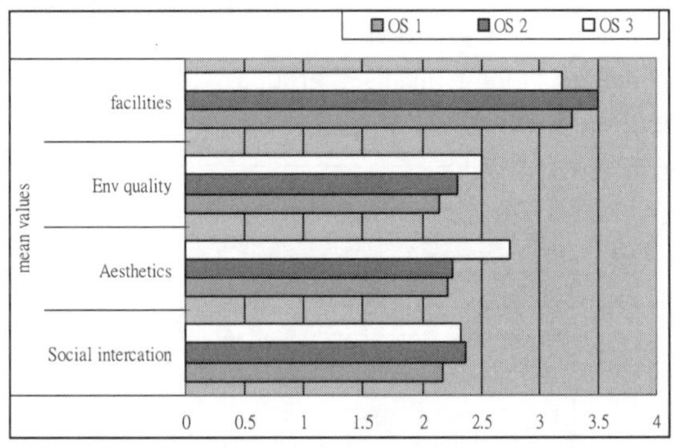

Figure 12. Comparison of mean scores of social interaction, aesthetics, environmental quality and facilities in OS1, OS2 and OS3.

Survey shows that respondents perceiving high mean values for social interaction perceive high mean values for aesthetic, environmental conditions and facilities. Social interaction is positively influenced by user's perceptions on aesthetic quality, environmental conditions, facilities and crowding in the open space, see Figure 13. According to [26] Social interaction takes place when necessary and optional activity is given good physical conditions, such as good environmental quality, aesthetics and facilities. A compact mixed use context has the potential for generating high density of necessary and optional activity that can generate high social activity if good physical conditions exist.

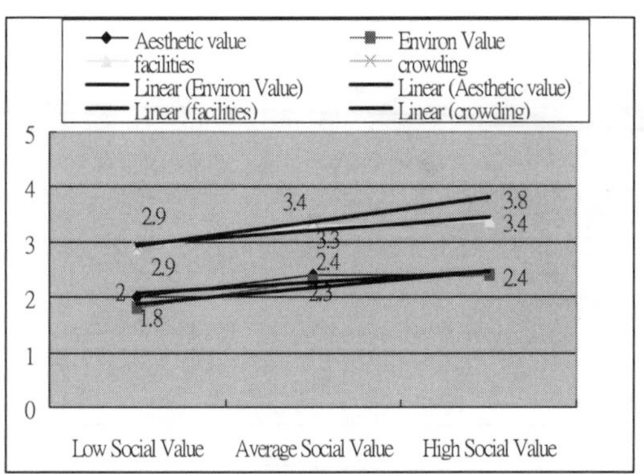

Figure 13. Relationships between respondents' perceptions on social interaction and the mean scores on aesthetics, environmental quality, facilities and crowding levels in all three open spaces.

Crowding: It must be noted that those having high social value for open space also perceive those spaces as high in crowding levels. It could be suggested that there is appositive relationship between perceptions on social interaction and crowding. In this case crowding can be hypothesized as beneficial for inducing better social interaction, see Figure 13. Survey also shows that respondents satisfied with open space provisions show a 3.4 mean score for perceptions on crowding while those dissatisfied with open space provisions score a higher mean score of 3.7. It could be speculated that those who perceive high crowding in open space also perceive those spaces as lacking in open space provisions, see Figure 14.

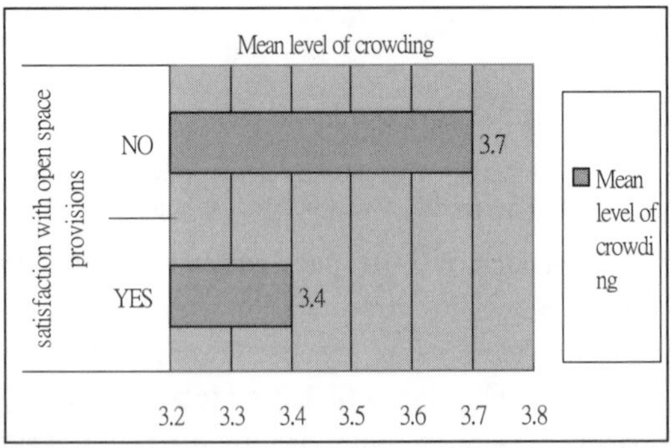

Figure 14. Relationship between respondents' satisfaction with open space provisions and their mean scores on crowding in open space.

Travel time: When comparing the travel time to open space in all three open spaces 48% travel less than 5 mints, 24% travel 5-10mints, 13% travel 10-15mints, 5% travel 15-20mints and 10% travel more than 20mints to get to open space. This shows that in general a majority of 85% users spend less than 15mints to travel to open space. Although the respondents consists of residents (33%), commercial / visiting population (21%), office (27%) and GIC users (19%) reduced travel time is beneficial for a wide range of user groups, see Figure 15.

Survey shows that reduced travel time to open space is a benefit in compact mixed use zones.

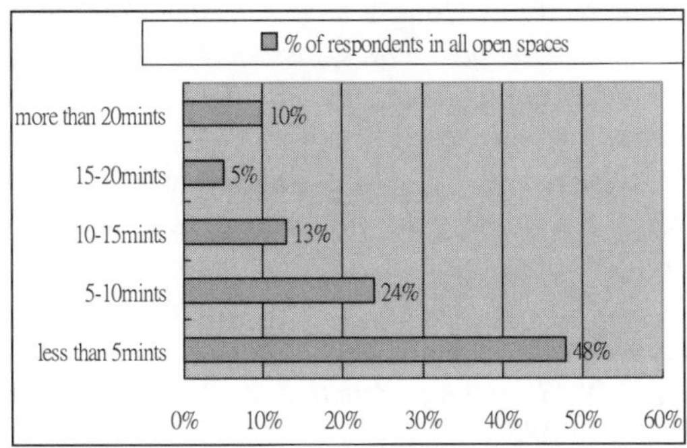

Figure 15. Distribution of respondents' travel times to open space across all three open spaces.

5. Conclusion

Pilot survey established the following hypothesis for future research on the implications of compactness on perception and use of open space is mixed use residential developments in Hong Kong.

Although crowding is claimed as a negative attribute in compact cities survey shows that respondents having high perceptions on crowding also perceives high social interaction. It could be hypothesized that high density has a positive influence on social interaction in open space.

Respondents perceiving high crowding also perceive a lack of open space. It could be hypothesized that high density also has a positive influence on perceived lack of open space.

In mixed use developments high availability of mix land uses or secondary land uses to support the primary land use gives users a range of alternative choices for recreation within a close proximity. Therefore the demand for open space for recreation is counteracted by substitute indoor and outdoor land uses. In such instance the need for open space may relate more to its environmental contributions rather than its social contributions. The social value of open space needs to be re examined in a mixed use urban context.

In a mixed use context the reduced demand for open space, which is created by high availability of alternatives will also counteract users' perceptions on lack of open space within a development.

Majority of interviewed respondents spend less time to access open space. But since survey was conducted in and around the open spaces it could be argued that majority visiting the open space were those who had easier access to it and those who would spend more time to visit the open space were not part of this sample population. This fact is a limitation in this pilot survey. But among those who would have less travel time to visit open space it must be noted that still a large majority of different user groups spends less than 15 mints to access the open space, which is beneficial in terms of high accessibility and reduced travel time.

Provision for open space is generally calculated by the population size taking into consideration area per person living within a given location. But in a time where increase in population and scarcity of land is high the efficient provision of open space is critical. In such instance the need, or demand for open space within a specific user population must be evaluated. Survey shows that within a mix user population the demand for open space and time spent in open space differs among user categories, highlighting the variation in demand for open space among different population groups. In a mixed use context such specific requirements of the mixed population must be considered in order to provide and plan open space that will meet the correct requirement of the present generation, saving for the demand that will be generated in the future.

Such estimates can introduce an urban design tool that could take into account the open space demands and provisions needed for a mixed user population. Such a tool will be essential for sustainable management of open space where the efficient provisions of open space is a critical issue in compact cities such as Hong Kong where populations is rapidly increasing and land for development is scarce.

This survey was conducted as a pilot study to investigate some selected issues as mentioned above. The survey findings are limited to establishing concluding hypothesis for future research. It must be noted that this survey was limited to respondents of a specific socioeconomic group. As shown above the respondents were majority female, within the age of 15-40 years with an income of less than 8,000HK$ per month and does not represent the population at large and it was limited to the study of only two residential mixed use developments in Hong Kong.

A well distributed sample population of demographic and socio-economic backgrounds and a substantial cross section of mixed use residential developments will be selected as the cases for future study. Hypothesis will be tested using co-relation analysis while regression will be used to analyse the relationship between urban compactness and mix use on the perceptions and use of open space within compact mixed use residential developments in Hong Kong.

References

1. K. Williams, E. Burton and M. Jenks (eds) *Achieving sustainable urban form.* E & FN Sponn. (2000).
2. C. S. Fischer, *To dwell among friends: Personal networks in town and city.* Chicago: University of Chicago (1982).
3. A. Churchman, Disentangling the concept of density, *Journal of Planning Literature.* **13**, 389-411 (1999).
4. S. S. Y Lau, R. Ghiridaran and S. Ganesan, Policies for implementing multiple intensive land use in Hong Kong, *Journal of Housing and the Built Environment.* **18**, 365-378 (2003).
5. R. Hughes, *Hong Kong Borrowed Place – Borrowed Time*, London: Andre Deutsch Ltd (1968).
6. L. Davies, *Study of leisure habits and recreation preferences and review of chapter 4 of the Hong Kong planning standards and guidelines, Final report*, Planning Department Hong Kong (1998).
7. M. R. Manavi, The compact city in practice: the new millennium and the new urban paradigm. *Achieving sustainable urban form.* E & FN Sponn (2000)
8. E. Burton, The compact city: just or just compact? A preliminary analysis, *Urban studies*, Edinburgh: **37**, 1969 (2000).
9. E. Burton, Measuring urban compactness in UK towns and citie, *Environment and Planning B: Planning Design*, **29**, 219-50 (2002).
10. M. Jenks, Sustainable urban form in developing countries? *Compact cities-sustainable urban forms for developing countries.* Spon press. London (2000).
11. J. Jacobs, *The death and life of great American cities.* Random house. New York (1961).
12. A. Coupland, *Reclaiming the city.* E & FN Spon, London (1997).
13. A. Rowley, Planning mixed use development; issues and practice. *Research report founded by Royal Institute of Chartered Surveyors.* UK (1998).
14. R. Burgess, The compact city debate: A global perspective. *Compact cities-sustainable urban forms for developing countries.* Spon press. London (2000).
15. Q. M. Zaman, S. S. Y. Lau and S. O. Mei, The compact city of Hong Kong: A sustainable module for Asia. *Compact cities-sustainable urban forms for developing countries.* Spon press. London (2000).
16. M. Roberts and T. L. Jones, Mixed uses and urban design. *Reclaiming the city-Mixed use development.* E & FN Spon, London (1997).
17. C. Nicol & R. Blake, Classification and Use of Open Space in the Context of Increasing Urban Capacity. *Planning Practice and Research.* **15**, 193-210 (2000).
18. A. Chiesura, The role of urban parks for the sustainable city. *Landscape and urban planning.* **68**, 129-38 (2004).
19. R. S. Ulrich, Natural versus urban sciences: some psycho-physiological effects. *Environmental Behaviour.* **13**, 523-556 (1981).
20. R. Kaplan, The analysis of perception via preference: a strategy for studying how the environment is experienced. *Landscape and urban planning.* **12**, 161-76 (1983).
21. S. E. Walker and B. S. Duffield, Urban parks and open spaces: an overview, *Landscape Research.* **8(2)**, 2-12 (1983).
22. A. Flores, S. T. A. Pickett, W. C. Zipperer, R. V. Pouyat, and R. Pirani. Adopting a modern ecological view of the metropolitan landscape: the case of a green space system for the New York City region. *Landscape and Urban Planning.* **39**, 295-308 (1998).
23. J. Luttik, The value of trees, water and open spaces as reflected by house prices in Netherlands. *Landscape and urban planning.* **48 (3-4)**, 161-67 (2000).
24. S. Ganesan, S. S. Y Lau, Urban challenges in Hong Kong: Future directions for design, *Urban Design International*, 3 – 12 (2000).
25. C. Q. L Xue et al, The quest for better public space: A critical review of Hong Kong. *Public places in Asia pacific cities.* Kluwer Academic publishers, Netherlands, 171-89 (2001).
26. J. Gehl, *Life between buildings: Using public space.* Copenhagen, Arkitektens Forlag (1996).

THE EFFECTS OF COMPACT NEIGHBORHOOD DESIGN ON PEDESTRIAN BEHAVIOR IN CONTEMPORARY CHINESE CITIES – CASES STUDIES FROM URBAN GUANGZHOU

OU YINGQING JIA BEISI S. S.Y. LAU
The University of Hong Kong

Suburban sprawl has been criticized by many to produce many costly and unfortunate impacts – environmental, economic, and social. Planners in communities are increasing turning to neighborhood environment as potential solutions to reduce automobile dependence. New Urbanism suggests that the right design would encourage walking, thereby foster a greater sense of community. Calthorpe (1993) argues that "Pedestrian are the catalyst which makes the essential qualities of communities meaningful". New Urbanism proposes neighborhood design elements consists of: compact, walkable neighborhoods; a diverse mix of activities and housing options; an interconnected street network. These claims about urban form and pedestrian choices have initiated a lot of empirical studies to investigate the difference of residents' pedestrian behavior in traditional vs. conventional suburban development. Most of these studies have supported these claims by founding less auto use and more walking trips in traditional neighborhoods (for instance, McNally and Ryan, 1993; Ewing et al. 1994; etc). These studies have found some sort of association between mixed-use attributes (compact, mixed-uses, and pedestrian oriented environment) and more pedestrian choices. The main objective of this paper is to clarify what aspects of neighborhood environment influence what aspects of pedestrian behavior through systematic evaluation. First, a model for urban form characteristics about pedestrian trips is proposed. Second, the results of a study of two neighborhoods in Guangzhou are presented. The characteristics of neighborhood form are analyzed by a variety of techniques and described, and the factors influencing choice about pedestrian trips are explored through an analysis of the results of a neighborhood travel survey. These analyses support the proposed model and suggest that certain different aspects of urban form can play an important role in different aspects of pedestrian behavior.

1. Literatuer Review

Many studies have explored the relationship between urban form and travel patterns in existing developments. Newman and Kenworthy (1989) compared and found significant negative correlations between density and energy use in international sample of cities. Friedman et al. (1993) studied travel patterns in neighborhoods in the San Francisco Bay Area and found that residents in older, denser neighborhoods travel less in terms of total miles and numbers of trip than residents of newer, lower density neighborhoods.

Some researchers found that use of local facilities especially shopping areas helped residents to walk around their neighborhood, and increased interact with other residents. Handy (1996a) use case studies of four neighborhoods in the San Francisco Bay Area, two "traditional" and two "modern" to explore the link between urban form characteristics and travel behavior. She found higher accessibility, in terms of short distance as well as qualitative factors that may lead to higher perceived levels of accessibility, is associated with a greater number of walking trips to destinations. Shriver (1996) also attributes higher walk rates in traditional neighborhoods to the more direct routes and the greater number of destination choice in these neighborhoods.

Studies also links the characteristics of the street environment influence the choice to walk. Lynch (1981) and Gel (1986), through their acute observation, suggest that sidewalk widths, the size of trees along the street, building setbacks, variations in building materials, orientation of buildings to the street, the design of buildings, and the amount and nature of human activity all influence perceptions of the pedestrian environment and may thus influence the choice to walk. Appleyard (1981) shows that the more traffic flow in the street result in a less inviting street for pedestrians.

These studies seem to support the growing belief that higher densities, mixed-use and pedestrian oriented environment can be effective in promoting pedestrian behavior. However, these studies ignore the differential effect of urban form on different types of travel. What seem to be lacking are studies on the systematic evaluation method and residents' requirement and degree for community.

Two types of pedestrian trips can be differentiated. The first type, walks to shopping, the primary motivation is to reach the destination to engage shopping there. The second type, strolling trips are about simply walking. They may be motivated by a desire to get some exercise, to get out of the house, etc. The primary activity is walking itself instead of whatever might be done at the destination (Handy, 1996b).

2. Case Studies

A set of case studies was used to test this model and explore the relationship between urban form and pedestrian choice. Two neighborhoods in Guangzhou have been selected for the cases studies. For the purpose of this study, firstly, the neighborhood population density should differ significantly. Another consideration is the era of development and median property value as controlled variables. The final consideration in selecting case study neighborhood was the socio-demographic characteristics of the residents in these neighborhoods.

A questionnaire was prepared for the residents. The method of performing the questionnaire research was direct interviewing between the researcher and respondents. The main contents of the questionnaire are composed with questions of residents' socio-demographic characteristics, residents' evaluation about the neighborhood environment, and their pedestrian behavior. Then the evaluation results of the residents have been put in order, and an analysis done using the plan condition, the design characteristic, field survey, and the evaluation result in each housing area of study object.

2.1 *Characterization of urban form*

Urban form was characterized by a variety of techniques, including analysis of hard-copy maps, aerial photographs, and data collected through site visits. The focus of this effort was on differentiating the neighborhoods and quantifying observable differences. In addition, factors suggested by the urban design literature as influencing the quality of the sense of community were studied. The question to be tested in the survey that followed, as will be described, was whether these factors or some sunset of them could explain differences in psychological and behavioral aspects of sense of community.

The general characteristics of Lijiang Garden and Biguiyuan were showed in Table1. Floor Area Ratio of Lijiang Garden is higher than Biguiyuan. Both neighborhoods have well designed open spaces. There are not only plenty of trees and grass but also some facilities such as seats in the open space. The main garden in Riverside Garden is placed in the central of the neighborhood. However, there are more and bigger open spaces in Biguiyuan.

Table1. General Characteristics of Lijiang Garden and Biguiyuan

	Lijiang Garden	Biguiyuan
Area	810,000m^2	666,000m^2
Households	10000	4000
Residents	30000	13000
FAR	2.4	<1
Green rate	34%	85%

Neighborhood commercial areas were evaluated in terms of their pedestrian-oriented qualities and their relationship to residential areas. In Lijiang Garden, a majority of houses are within walking distance of commercial activity. In contrast, the farthest house to commercial area takes 45 minutes walk in Biguiyuan. Generally, commercial development in Lijiang Garden tends to be pedestrian oriented: continuous line of buildings, wide sidewalk with front porches. Biguiyuan tend to be automobile oriented: on-street parking, large parking site in the front of commercial areas. Lijiang Garden has greater number of locally oriented commercial establishments.

Neighborhood street environment were analyzed. Street maps for these areas portray very different street layouts (Figure 1): Lijiang Garden has rectilinear grids, Biguiyuan has curvilinear layouts. In all neighborhoods, T-intersections rather than four-way interactions predominate. The level of transit services is low in both of neighborhoods, reflecting their location at the fringe of the city where transit services tends to be sparse. The bus stop in Lijiang Garden is located inside the neighborhood, thus the all households in the neighborhood are within walking (a quarter of a mile or about 0.4 km) distance of a bus stop.

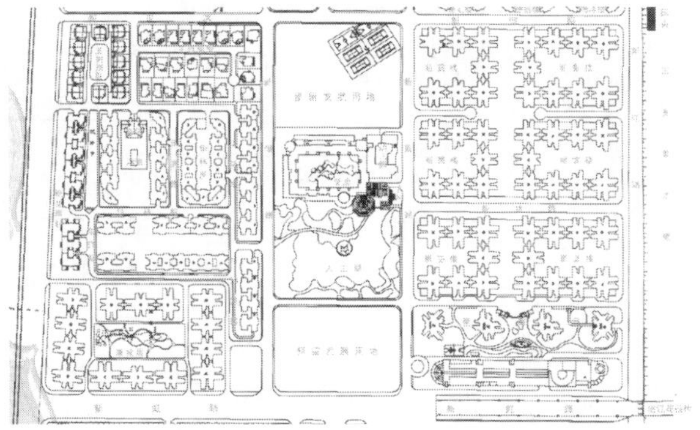

Figure 1 (a) Neighborhood site plan of Biguiyuan

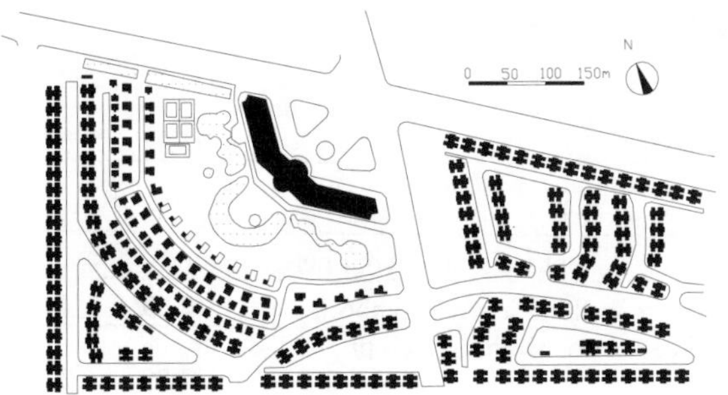

Figure 1 (b) Neighborhood site plan of Biguiyuan

Two streets in each neighborhood that were typical of other residential streets in the neighborhood were selected. Streets in Lijiang Garden are significantly narrower than the street in Biguiyuan. Lijiang Garden also has the higher level of shade provided by tree canopy. Sidewalk infrastructure is both more complete and in better condition in the Lijiang Garden.

The shop house facing the street in Lijiang Garden provides front porches and stores. Lijiang Garden has the greatest variation in the design of houses along any one street. These commercial areas have more complete and higher quality sidewalk infrastructure and front porches (Figure 2).

Figure 2 Main streets in Lijiang Garden and Biguiyuan

In sum, compared with Biguiyuan, Lijiang Garden has higher densities. It also has more local stores, with many kinds of goods and better accessibility. The street environment seems to be more pedestrian friendly in Lijiang Garden. However, the open spaces are bigger and in better condition in Biguiyuan.

2.2 *Perceptions about neighborhood environment*

These data were collected through an on-street survey. The survey included sections on dimensions of sense of community, and socio-demographic characteristics, as well as questions on perceptions of a variety of urban form characteristics.

The characteristics of the respondents were presented in Table 2. One interesting result is that the respondents are more likely to be female than the general population. The median age for this survey is 25-64 because the survey was conducted to adult members of the household. A majority household income is between 6000-9999. Overall, the income, household type and age are similar between these two neighborhoods.

Table 2. Neighborhood survey respondent characteristics

	Lijiang	Bigui		Lijiang	Bigui
10--24	4%	13%	<2000	7%	1%
25--44	49%	47%	2000-5999	26%	21%
45--64	32%	31%	6000-9999	39%	53%
65--	13%	7%	10000-14999	12%	10%
Female	56%	43%	15000-19999	7%	10%
Male	60%	40%	>20000	7%	2%

To test residents' perceptions about the walking environment in the neighborhood, respondents were asked to indicate whether they agreed or disagreed (on a five-point scale) with a series of statements about the walking environment (Table 3). The subjective physical variables selected for these analyses are respondents' level of satisfaction with local parks, satisfaction with local retail, and perception of walking environment. The results of these questions give some indication of the factors that may encourage or discourage walking in the neighborhood. Few large differences can be seen across neighborhoods.

From Table 3, respondents in Lijiang Garden feel more satisfied about the open space, walk environment and commercial areas. Respondents in Lijiang Garden are more likely agree that they feel comfortable walking in hot whether. They also are more likely to agree that the neighborhood has interesting houses to look at than respondents in Biguiyuan. It is interesting that respondents in Lijiang Garden are also more likely to agree that there is too much traffic in the neighborhood. Residents of Lijiang Garden were more likely to agree than Biguiyuan the following statement: Stores are within walking distance, local stores are of high quality, I feel comfortable walking around local shopping area. These results can help to explain the higher frequency of walks to local store and support the proposed model in which urban form factors – those related to commercial areas – play a significant role in encouraging walks to destinations.

Table 3 Perceptions about neighborhood environment

Statements about neighborhood garden	Lijiang	Bigui
The closest garden is within a reasonable walking distance	4.12	4.04
I feel safe playing in my garden	4.17	4.21
There are plenty of entertainment facilities in my garden	3.64	3.46
There are enough seating areas in my garden	4.12	4.18
The environment is attractive in my garden	4.12	3.89
There are plenty of trees or waters in my garden	3.98	3.81
The environment is maintained well in my garden	3.96	3.83
Statement about your neighborhood shopping area		
The closest store is within a reasonable walking distance	3.94	3.90
Local store meet many of my needs	3.54	3.27
I feel safe walking from my house to local stores	3.97	3.83
I feel comfortable walking aloud local shopping area	3.85	3.8
Statements about walking in your neighborhood		
I feel safe walking in my neighborhood	4.09	4.01
The trees in my neighborhood provide enough shade	3.80	3.52
There are enough walking facilities such as light, sidewalks	3.98	3.61
The houses in my neighborhood are interesting	4	3.8
I often see neighbors in my neighborhood	3.41	3.36
There is too much car traffic in my neighborhood	3.44	3.35

2.2 *Analysis of Pedestrian Trips*

2.2.1 *Walks to local store*

Residents were also asked about the frequency with which they walk to a local store or commercial area. Respondents in Biguiyuan walk to store (2.44 times per week) less than Lijiang Garden (2.48 times per week). This difference can be explained by differences in urban form: various commercial areas are within walking distance of most households in Lijiang Garden.

Correlations between answers to perceptions of neighborhood environment and the frequency of walks to the store were higher than those for strolling trips (see Table 4). Neighbors in the store had the highest correlation, followed by variety of goods, store proximity, and trees in the park.

Table 4 Regression models of walk to local store

Coefficients(a)

Model		Unstandardized Coefficients		Standardized Coefficients	t	Sig.
		B	Std. Error	Beta		
1	(Constant)	1.804	1.271		1.420	.158
	Age	.269	.198	.198	1.361	.176
	Family type	-.026	.136	-.021	-.189	.851
	Children	-.142	.146	-.154	-.976	.331
	Income	.217	.081	.230	2.688	.008
	Degree	-.485	.170	-.243	-2.848	.005
	Attit for Inter	.608	.210	.304	2.892	.005
	Attit for SOC	-.051	.198	-.027	-.259	.796
	Access to open space	-.041	.138	-.026	-.298	.766
	Access to store	-.245	.113	-.230	-2.166	.032
	Access to work place	.089	.080	.093	1.107	.271
	Density	.024	.104	.022	.236	.814
	Open space	-.210	.188	-.147	-1.118	.266
	Convenient of OS	.080	.254	.056	.314	.754
	Safety of OS	.068	.348	.042	.197	.844
	Facilities of OS	-.268	.234	-.203	-1.144	.255
	Sport Facilities of OS	.095	.308	.071	.308	.759
	Seats of OS	.165	.244	.108	.674	.502
	Comfort of OS	-.013	.256	-.009	-.051	.959
	Green of OS	.455	.224	.290	2.032	.045
	Management of OS	-.285	.213	-.203	-1.338	.184
	Local store	-.055	.125	-.040	-.438	.662
	Convenient of LS	-.011	.155	-.007	-.069	.945
	Variety of goods	-.335	.144	-.258	-2.334	.021
	Safety of LS	.099	.154	.075	.641	.523
	Comfor. for Pedestrian	-.083	.177	-.059	-.470	.639
	Neighbors	.402	.149	.296	2.701	.008
	Street environment	.055	.208	.039	.266	.791
	Safety of SE	-.213	.245	-.152	-.867	.388
	Tree shade of SE	.213	.245	.140	.868	.387
	Facilities of SE	-.257	.229	-.177	-1.121	.264
	Interesting of SE	.037	.229	.025	.163	.871
	Neighbors of SE	.056	.171	.044	.328	.743
	Traffic of SE	-.038	.146	-.030	-.258	.797

a Dependent Variable: VAR00006

Respondents indicated that they walk to local store because local store are convenient to get to and there are variety of goods (see Table 5). Those who do not use local store suggest that the goods are of poor quality and expensive. Theses indicate that factors other than urban form play a role in the residents' decision to local store.

Table 5 Reasons for walking to local store or not

	Lijianghuayuan	Biguiyuan
Reasons for walking to local store – percent of respondents who walk		
Convenient	72%	51%
Variety of goods	10%	11%
Reasons for not walking to local store – percent of respondents who don't walk		
Poor quality of goods	12%	22%
Expensive	5%	15%

2.2.2 *Strolling trips*

Residents were asked about the frequency with which they stroll around their neighborhoods. The frequencies are somewhat lower in the Lijiang Garden (2.75 times per week) than Biguiyuan (2.81 times per week). Although the differences in these results are not statistically significant, they suggest either that residents in Biguiyuan maybe somewhat more motivated to stroll or that the urban form of the neighborhood may be somewhat more encouraging of walking, or both.

Subjective physical variables played a relatively important role in predicting individual walking behaviors (see Table 6). The results of the hierarchical regression models for strolling walk trips shows that facilities in the open space had the highest influence on strolling frequency. Street environment also had significant and negative correlations with strolling walk, followed by variety of goods in the store, trees in the open space. This result shows that open space in the neighborhood has significant contributors to the decision to stroll.

Table 6 Regression models of strolling walk

Coefficients(a)

Model		Unstandardized Coefficients		Standardized Coefficients	t	Sig.
		B	Std. Error	Beta		
1	(Constant)	2.102	1.123		1.871	.064
	Age	.150	.175	.122	.858	.393
	Family type	-.126	.121	-.112	-1.043	.299
	Children	-.039	.129	-.047	-.305	.761
	Income	.204	.071	.240	2.863	.005
	Degree	-.253	.150	-.141	-1.686	.095
	Attit for Inter	.475	.186	.263	2.558	.012
	Attit for SOC	-.132	.175	-.077	-.752	.454
	Access to open space	-.024	.122	-.017	-.200	.842
	Access to store	-.174	.100	-.181	-1.742	.084
	Access to work place	-.027	.071	-.031	-.382	.703
	Density	.107	.092	.105	1.166	.246
	Open space	.141	.166	.109	.848	.398
	Convenient of OS	-.422	.225	-.329	-1.879	.063
	Safety of OS	.501	.308	.336	1.628	.106
	Facilities of OS	.681	.207	.570	3.291	.001
	Sport Facilities of OS	-.998	.272	-.832	-3.665	.000
	Seats of OS	-.090	.216	-.065	-.416	.678
	Comfort of OS	-.006	.226	-.005	-.027	.978
	Green of OS	.461	.198	.325	2.328	.022
	Management of OS	-.159	.188	-.125	-.844	.400
	Local store	-.178	.110	-.146	-1.613	.109
	Convenient of LS	.236	.137	.178	1.725	.087
	Variety of goods	-.306	.127	-.260	-2.405	.018
	Safety of LS	-.036	.136	-.030	-.263	.793
	Comfor. for Pedestrian	.136	.156	.106	.868	.387
	Neighbors	.276	.132	.225	2.099	.038
	Street environment	.532	.184	.414	2.900	.004
	Safety of SE	-.425	.217	-.337	-1.959	.053
	Tree shade of SE	-.040	.217	-.029	-.183	.855
	Facilities of SE	-.025	.202	-.019	-.126	.900
	Interesting of SE	-.170	.202	-.128	-.842	.402
	Neighbors of SE	.011	.151	.010	.074	.941
	Traffic of SE	-.091	.129	-.079	-.706	.482

a Dependent Variable: VAR00007

The reasons why respondents walk, as indicated in an open-ended question, covered a wide range. Residents of Lijiang Garden were mostly likely to indicate "Relaxation" and "Pleasure environment" as reasons for walking. The most important motivation appears to be the desire for exercise and concerns about health in Biguiyuan. One of the reasons might be that there are more elderly people in Biguiyuan.

Almost most of the respondents who do not walk in the neighborhood indicated they do not have enough time. Few respondents indicated they do not enjoy walking. These results suggest that most people would like to walk, all else equal, and that personal limitations more than characteristics of the neighborhoods are likely to limit their walking.

Table 7 Reasons for strolling walk or not

	Lijianghuayuan	Biguiyuan
Reasons for walking – percent of respondents who walk		
Relaxation	49%	8%
Pleasure environment	23%	21%
Exercise/health	10%	47%
Play with child	6%	8%
Talk with others	6%	13%
Reasons for not walking – percent of respondents who don't walk		
Not enough time	90%	90%

From above analysis, sharing of public space is very important for residents because of the way Chinese people use these spaces, as shaped by the characteristics of Chinese neighborhoods. First, the high density in Chinese cities limits the private living space in residents' apartments. Chinese residents need to rely on parks and other urban public space for conducting social activities which could have taken place in one's backyard in the US. Second, people's frequent visits to public space rely on walking and other public transit, so they tend to use the public space nearby.

3. Conclusions

Density seems to have low correlation with pedestrian choice. However, neighborhood with high density tends to enhance the accessibility to local facilities, increase the neighbors within the neighborhood. With these characteristics, density might increase the choice to walk indirectly.

Urban form plays an important role in encouraging or discouraging walking, giving the motivation to walk and absence of limitations. The results suggest that urban form plays a great role in the choice to walk. Specifically, the distance from home to the local store plays an important role in encouraging the decision of shopping walking. And facilities and trees in the park and the street environment would promote more strolling walks.

Some other factors also influence the decision to walk. Quality of the goods in commercial areas is an important factor influencing the walk to shopping. Personal motivations such as for relaxation or exercise seem to encourage people to stroll in the neighborhood. Having no time is an important limitation in the decision to stroll. What this means for planners is that there is potential for enhancing the pedestrian behavior through New Urbanism, but that these neighborhoods will be more successful if they are able to meets the residents' needs.

Reference

1. Appleyard, *D. (1981). Livable Streets. Berkeley: University of California Press.*
2. Calthorp, P (1993). *The Next American Metropolis: Ecology, Community and the American Dream*, Princetown Architectural Press, NY.
3. Ewing, R., Haliyur, and Page, G. W. (1993). Getting around a traditional city, a suburban PUD, and everything in-between. *Transportation Research Record* 1466: 53-62.
4. Friedman, B., Gordon, S. P. and Peers, J. B. (1993). Effects of neotraditional neighborhood design on travel characteristics. *Transportation Research Record* 1466: 63-70.
5. Gehl, J. (1986). *Life Between Buildings*. New York: Van Nostrand Reinhold. Translated by Jo Koch.
6. Handy, S. L. (1996a). Understanding the link between urban form and non-work travel behavior. *Journal of Planning Education and Research*, 15, 183-198.
7. Handy, S. L. (1996b). Urban form and pedestrian choices: study of Austin Neighborhoods. Transportation Research Record 1552, 135-144.
8. Lynch, K. (1981). *Good City Form*. Cambrige, Massachusetts: MIT Press.
9. McNally, M. G., and Ryan, S. (1994). Comparative assessment of travel characteristics for neotraditional designs. *Transportation Research Record* 1400: 67-77.
10. Newman, P. and Kenworthy, J. R. (1989). *Cities and Automobile Dependence: A sourcebook*. Brookfield, Vermont: Gower Technical.

11. Shriver, K. (1996). Influence of environment design in pedestrian travel behavior in four Austin neighborhoods. *Transportation Research Record*, 1578, 64-75.

THE RELATIONSHIP BETWEEN WATER AND LANDSCAPE

LIU JIN
Architectural Design & Research Institute of Tongji University
1239 Siping Road, Shanghai 200092, China

The collection and irrigation of rainwater, including waterproof and infiltration, is the key point of the large area vegetation of building. As the combination of waterscape with biology can create a harmonious create sustainable ecosystem environment.

1. Meaning of Frame of Ecotypic Landscape

1.1 *Natural order of city obstructed by concrete*

During the process of urbanization the natural soil is covered with buildings, roads and plazas. The nature vegetation disappears much faster than the speed of regeneration. Nature environment is badly destroyed. The nature circle is cut and the ability of nature regulation decline

In order to change the condition, we ought to make the surface of 3D space including city architecture, construction and square covered with nature vegetation and recreate manual greenbelt and manual lake during the design of the urban landscape. If the green field and waterscape we created are tender and flimsy, the clean water, energy and medicament put in to protect these landscape increase the second burden of the environment. In order to maintain the eco-environment and life chain, we ought to use less water and energy sources to decrease the depress of wastewater and rainwater polluting the environment and city facilities .We also should hold the notion of living harmoniously. Then, the manual created environment will be changed into a zoetic ecosystem environment.

2. Create Ecological Landscape

2.1 *The merits and ways of the vegetation of buildings*

The vegetatieren of building is good for the building itself. It can save water, filter water, clean, adjust the temperature, radiation and humidity, reduce the noise and so on. The vegetation building have some merits to buildings itself .It can be sun proof in the summer and improve insulation of the roof; preserve heat in the winter and the super stratum of planting layer lengthen the life of the waterproof rubber; decrease the rain flux. Besides, the roof garden can be a recreation area.

2.2 *The structure of the vegetation*

The planting layer is made on the preservation stratum and the waterproof layer. The practicing of the preservation stratum is as the ordinary ways .The waterproof layer include waterproof sand slurry, paint, curly material and so on .The down most of the vegetation is the drain layer, which on it successively is filter fabric, potsherd and the planting soil. There are two material of the drain layer. One is named plastic drain and stockpile board or NYLEX Cord raw Drain , which is concavo-convex. Moreover, it is communicate within. The water will be discharged when it is too much in the soil, or it will be used when the water is lacking. The other is named plastic NYLEX Board Vantage, which main function is to support and to drain. The water, which is filtered by the soil, can be drain away. Potsherd will finish saving water. The formation of Vegetation protect layer of roof is described in figure 1. In order to form a zoetic eco-system after planting .The selection of the plant is very important. Plants like those which can adapt the city environment, being easy to grow, can resist cold and drought, need the less conservation, the lamina is thick, having a good ability to saving water, can live with rainwater are appropriate .the arranging of its kind should be logical. Usually it is more than nine kinds. Rainwater can be absorbed to 50% after the roof being planted because there are many holes under the planting layer, which can be used in the plants in the roof in a fine day. So it has no use for any special management.

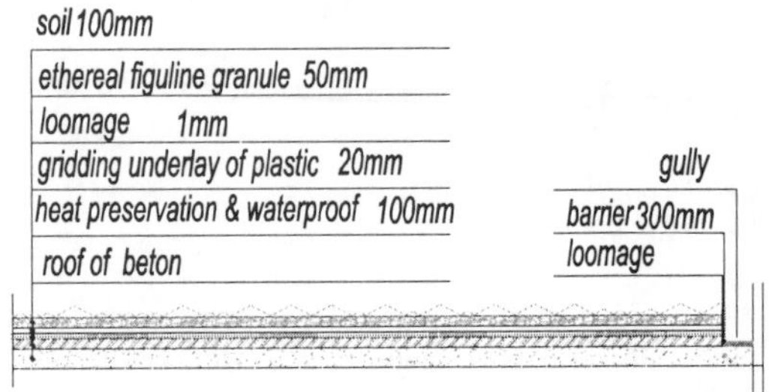

Figure 1. Section of the Roof top Vegetation

2.3 *Empty roof and the biologic disposal of rainwater*

Rainwater from empty roof and hard road should be treated before it is used. It can be used for drainage after disinfection and elementary filtering. It also can be used for sanitary water or gettable water after sterilization. Rainwater will be collected into reservoirs to finish the sedimentation and elementary filtering. Filtering can adopt filters with stainless steel net. The second decontamination can be finished in the ground "Depurate Microorganism Environment". It is a water environment with a kind of aquatic plants such as bulrush which being planted as fencing and accretes Microorganism to be cultivated on it. From the experience of the Potsdam Square in Berlin, every 100 square meters can deal with water of 3 stereo meters per hour. The water level of the sight pool controlled by the controller is connected with the " Depurate Microorganism Environment ". The water will be compensated automatically when the water level decline. Because alga will grow in the water we could put some carps in it to balance the system. A good ecology system will make the city sight more attractive. We will not only have good scenery, but also feel harmonious with the nature in the city. It is described in figure 2

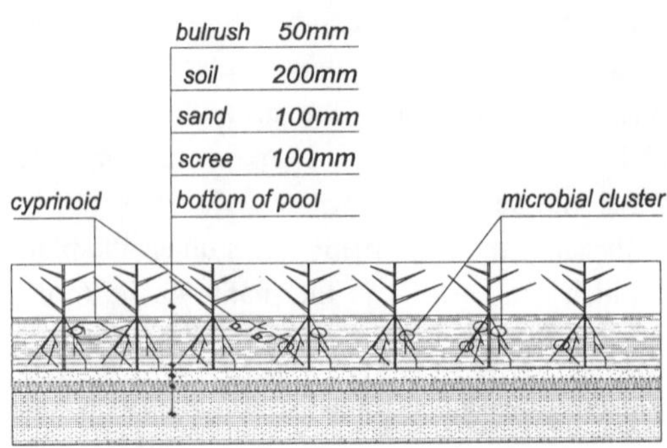

Figure 2. Section of Depurate Microorganism Environment

2.4 *The collection, using and soakage of rainwater in partial zone*

Using of cubage in district as villas is always low capability and its area is big ,so the cost of the storm sewer net is high. If rainwater can be well collected and used, not only will the cost be reduced, it will also store up water and compensate the underground water .There is a small kind of construct , which can filter, reserving water and penetrating. It is described in figure 3.

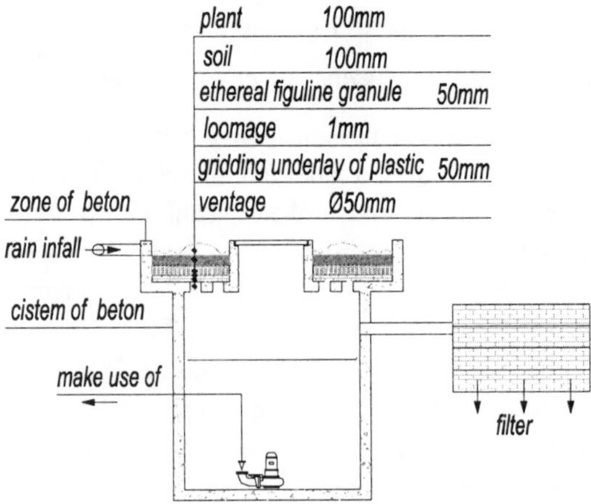

Figure 3. Section of cistern & Filter with Vegetable

The construct can collect rainwater from independent building and area around. Rainwater flow through the filter layer which match the root of flora into the reservoirs. The water in it can be used. The part which is over reservoirs cubage will flow through the pipe of overfall to the penetrating object. It will eventually penetrate into the ground. Above the penetrating object can be covered with soil to make ground or virescence. We should plant at least 9 kinds of vegetation in the ring, including some flowerer, to form a particular flower ring.

2.5 *The territorial collection, using and filter of rainwater*

The collection and transportation of rainwater in the region of public road can adopt ecology drain. The bottom of the ecology drain should be made treat with impervious stuff to keep the steady surface. If water is very much, it will penetrate underground. Ecology drain is devious as nature water and shrub in two sides grows naturally, which becomes a sight of an uptown. The section is figure 4th. Wastewater after cleaned could be supply of sight water if only it flow through the ecology drain. If possible, we can build a permeable pool, a sight in the down fold or a center in an uptown, which the water from ecology drain can flow into. There are two advantages. Firstly, the microclimate can be improved. Secondly, rainwater will penetrate to underground so that ground water can be compensated. Bulrush is a better plant to be planted in the permeable pool. The down most are carpolite, sand is on it and soil is on the sand .In the Science and Technology park in Germany, rainwater from the 9000m^2 of valuable office and laboratory and the 1800m^2 of office and work space is collected in the lake through a bigger ecology drain to adjust weather and to beautify the environment. Set the ecology drain in the grassland, then through raising water to the start of the ecology drain to flow in a circle. It will have an irrigation effect. The section of Filter is figure 5th.

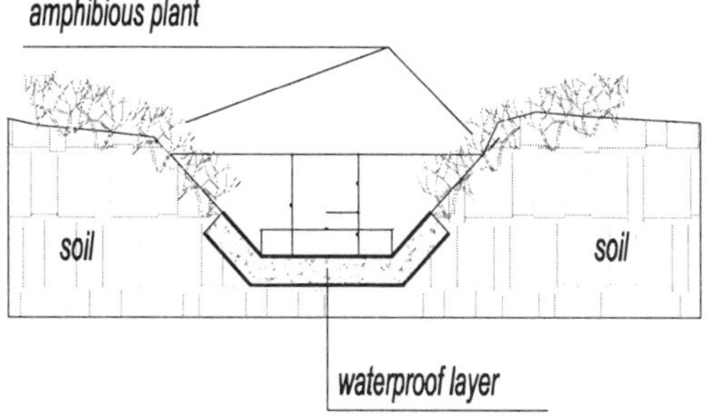

Figure 4. Section of Rainwater collection system

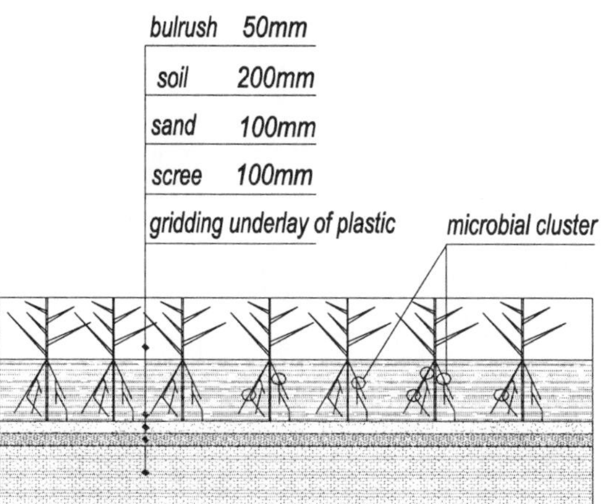

Figure 5. Section of Pool of Filter

2.6 *Increase O_2 of waterscape*

Fish can live only when O_2 in water is more than 4mg/l. When the concentration of oxygen in water is subter-2mg/l, there is color in water, green or black. When the concentration of oxygen in water is less than 0mg/l, there is unpleasant smell from the water. When the temperature is higher than 20°, alga will develop, in summer alga abnormal grow fleetly.

So it is necessary artificial aeration into the water. Going with via waterfall and fountain fill up some oxygen into water. Besides we can use as in wastewater treatment ventilation system, through fan and aerator fill up air into water, then there is a higher level of oxygen concentration, so that water quality can be maintenance.

3. It's Our Responsibility for Create City-Ecology

This is our partial labor, that the vegetarieren of building, collection and irrigation of rainwater and the oxygen transfer into waterscape .If the people can use the concept of Ecology and the engineering principle of City Ecology in the urban design ,and can develop appropriate measures by considering the applied principle, it is can be expectant ,to recover the good quantity of city environments, to create the City-Ecology.

References

1. Rong An, *World Architecture* . 12/2002 59(2002).
2. Lupp J, Philoppi H, Sonnenschein & Baick, Architects, *World Architecture* .12/2002 40(2002).
3. Lu Weiya, *Water & Wastewater Engineering* N0.10 Vol.30 2004 75 (2004)

URBAN LIVING SPACE

FANTASY AND SPATIALITY
HIGH-RISE LIVING BUILDINGS IN CONTEMPORARY CHINESE CITIES

HUA LI

Graduate School, Architectural Association, 36 Bedford Square
London WC1B 3ES, United Kingdom

This paper is a socio-cultural analysis of high-rise living buildings in contemporary Chinese cities. The considerable construction of high-rises has dramatically transferred Chinese cityscape, and has provoked a critical question of how to understand the Western influence in Chinese architecture and urban pattern. From the viewpoint of urban culture and spatial arrangement, this paper examines the way in which Chinese people inhabit in the modern city. This investigation is based on a case study of a high-rise residential development in Shenzhen Central District. By looking at the image making and spatial organisation in city construction, residential development and household space, this study reveals the re-interpretation and transformation of the Western importations in terms of Chinese fantasy of the 'modern', and an ambivalent relationship between the individuals and the city which is accommodated and constructed by a series of mediating spaces.

1. Introduction

In contrast to traditional horizontal living patterns, vertical cities and high-rise residential buildings are becoming normal in China nowadays. The increasing height of buildings and juxtaposition of various Western architectural styles, materials and elements, has disrupted the harmony and changed the continuity of traditional cities in China to a degree that many are now virtually unrecognisable. The situation has provoked the attention of Chinese architects and scholars and Zheng Shiling, a professor of Tongji University, has argued:

Figure 1. Cityscape of Luohu District, Shenzhen. Source: H. Li, 2002.

Under the impact of global culture, a number of Chinese cities gradually lose their identity and become more and more alike. Under the influence of international culture, especially American culture, urban space [of China] more and more resembles Manhattan of New York. Cities stretch higher and higher without considering the potential factors of social and historical context, infrastructure, population and internal explosion.*

In Zheng Shiling's statement, he raises three relevant issues. Firstly, the current construction of urban space in China is a reproduction of Western cities, such as Manhattan, which are regarded as a prototype for the modern city. Secondly, mimicry of the 'others' eliminates the character of Chinese cities and urban space. The last point is that, as an imported architectural typology, the construction of high-rise is promoted as a symbol of the modern city in practice and causes the urban problem of high density. In consequence, the three issues lead to a question of how to absorb these influences within Chinese urban space and architecture. This paper is not an answer this question, but an attempt to understand the way in which high-rise construction works in the current urban conditions of China, in terms of urban culture and spatial arrangement. In other words, what is the fascination within Chinese people for the kind of modern life represented by high-rise? How do Chinese people inhabit in the city through organising space? This paper will look at high-rise apartment building and development in contemporary Chinese cities as way of approaching these questions.

The prevalence of high-rise residential buildings in contemporary Chinese cities results from two main

* Zheng, Shiling, (2003) The Chinese Urban Space and Architecture Under the Influence of Globalisation, *Architectural Journal* (Beijing), 02/2003. pp. 10.

causes. Firstly, Chinese cities have faced the pressure of land shortage and increasing population. High-rise residential building is a way of meeting the requirements of saving land and of accommodating high density population. [†] Secondly, the housing supply policy of China has shifted from welfare to commercialised marketing since the 1980s. The housing industry gradually has grown to be one of the main economic developments within Chinese cities. Particularly since the 1990s, driven by increasing land rent and pursuit of profit in real estate markets, high-rise apartment building and development has proliferated. Around city centres, high-rise developments build 'luxurious' or high quality living space which embodies a sort of 'modern' life style. Thus, at the social and economic level, high-rise residential development is a means of accommodating high population density in Chinese cities and of economic production. At the cultural level, it reflects the Chinese fantasy of 'modern' life. Relating to the questions of this paper, my investigation is based on a case study of a high-rise residential development in Shenzhen Central District (SZCD) -- Huangpu Garden.

Figure 2. Huangpu Garden, Shenzhen. Source: H. Li, 2005

Located in the northwest of new Shenzhen Central District (SZCD), Huangpu Garden is the biggest residential development in this area, and contains more than 4000 households. Its construction started in 1998 alongside that of the new city centre project and was completed by four phases till 2003. In the property market of Shenzhen, it was a popular development and was listed as one of the top ten properties in 2002 and 2003. And it has been identified as 'luxurious houses'.[‡] The research on Huangpu Garden in this paper is an empirical study conducted by fieldwork and unstructured interviews.

2. Fantasy of the 'Modern' and Chinese urban culture

In the history of modern Chinese architecture and city construction, Chinese high-rises were built up mainly during two periods, the 1920s and 1930s, and the late 1970s onwards. The emergence and re-emergence of high-rise in Chinese urban landscape corresponds with the process of social modernisation in China. The earlier construction was concentrated on Shanghai. The later trend penetrates the whole country. Although the first construction cannot compete with the second in quantity, and had different objectives, between them they show a certain continuum in terms of socio-cultural context. According to Stephan Feuchtwang, 'there have been three modernisations in China'.[§] From 1850 to 1953, the first modernisation combined imperialist impact and republican modernisation. City construction and urban culture based on modern commodification was brought up. The second is characterised as collective foundation and production based upon state-owned enterprise from 1953 to 1978. From then onwards, the third modernisation and fast urbanisation takes the form of governmental administration and of market economy. Consumption penetrates into daily life, which mixes private ownership and collective organisation. Consumerist urban culture of the first modernisation comes back and flourishes.

'High buildings, electric light and telephone' were regarded as symbols of modern life in Chinese mass culture. All the products were imported from the West and accompanied the start of Chinese modernisation and formation of urban consumption culture – material fantasy. As Leo Ou-fan Lee pointed out, when the material forms of Western modernity were introduced into China in the early years of the twentieth century, they caused the natives 'a typical pattern of shock, wonder, admiration and imitation.' This 'typical pattern'

[†] Lü, Junhua, Rowe, Peter G., and Zhang, Jie, (2001) Modern Urban Hosing in China 1842-2000, Munich, London, New York: Prestel Verlag. pp. 180-181.

[‡] In 2000, the average price of residential housing was 5,698 yuan/sq. m. (Shenzhen Statistics Bureau, 2003, pp.158). The average price of the Phase Three is about 9,000 yuan/sq. m.

[§] Feuchtwang, Stephan, (2002) Tales of Territoriality: the Urbanisation of Meifa Village, China, *études rurales*, juillet-décembre 2002, 163-164. pp. 252-253.

indeed reflects a process in which the Chinese accept the foreign objects, and unfolds the Chinese fantasy of the 'modern'. The initial feeling of shock takes the form of astonishment at the novelty of the aesthetic experience. The skyscraper as an architectural typology is one such example. Without rooting in the development of traditional Chinese architecture and society, skyscrapers intervened into the Chinese landscape as a magic representation of imperialist wealth and power. While translated into Chinese society, it also signifies social status – 'the high and the low, the poor and the rich'. In terms of socio-cultural meaning, skyscrapers associate with the Chinese modern ideology of social progressiveness which is embedded in national consciousness and urban daily life.[**] Thus, skyscrapers as a cultural object convey the Chinese fantasy of the 'modern' – novelty, wealthy and progressive.

If skyscrapers or high-rise building represents the cultural re-interpretation in China during the first trend of modernisation, its cultural meaning continues and is even amplified in the third phase. In the mid and late 1970s, skyscrapers with modernist appearance appeared in China along with loosely diplomatic connections with the outside world, such as the Baiyun Hotel in Guangzhou in 1975 and the Great Wall Hotel in Beijing in 1983. The horizontal strip windows and the glass curtain wall were 'new' architectural characters which contrasted with the Chinese urban landscape at that time.[††] As most of the high-rise hotels and office buildings were linked to international business in the 1970s and 1980s, skyscrapers became an intermediate representation of the West to the Chinese, and of Chinese national modernisation to the outside world. Although this implies a paradoxical relationship between the West and China, it soon, and undoubtedly, becomes the material embodiment of success and achievement of modernisation, hence the progress of the Chinese nation and cities both now and in the future – rich and powerful. The previous symbol of imperialist power has been replaced by the 'new' national expression. Along with proliferation of commercialisation, the national fantasy is not only embedded within society, but in turn is transferred into an individual's imagination and identification of modern life – wealthy and outstanding. In popular culture of China, the material richness and significance are often referred to the fantasy of 'luxury'.

At this point, the Chinese fantasy of the 'modern' or 'Western modern' has little to do with western conception of modernity. In fact, it deeply reflects the anxiety of Chinese society for progress and security in the future. Regarding high-rise development and apartment building in contemporary Chinese cities, what I am suggesting here is not a literal representation of the 'Western modern' or Chinese modernity, but a vehicle conveying the fantasy of lifestyle within living patterns. For the purpose of this paper, I look in particular at the relationship between the individual and modern city through image making and spatial organisation.

3. A Case Study: Huangpu Garden in Shenzhen

3.1 *Urban setting: Shenzhen central district*

Figure 3. Model of Shenzhen Central District. Source: H. Li, 2002.

Shenzhen Central District (SZCD) is the urban location of Huangpu Garden. Located in the geographic centre of Shenzhen Special Economic Zone, SZCD occupies 607 hectares land. Its total construction area will be 7,500,000 square metres, accommodating 310,000 working people and housing 77,000 inhabitants.[‡‡] For urban planners, SZCD is an ambitious urban project which construction is expected to trigger the trajectory of the whole city towards an internationalised metropolis and a regional economic

[**] Lee, Leo Ou-fan, (1999) Shanghai Morden: the Flowering of a New Urban Culture in China, 1930-1945. Mass. Cambridge; London: Harvard University Press, pp.5-81.

[††] Zou, Denong, (2001) History of Chinese Modern Architecture, Tianjin: Tianjin Science and Technology Press. pp.313-374.

[‡‡] Chen, Yixin, (2000) The Course of SZCD Planning and Urban Design, *The World Architecture* Review (Shenzhen), 2001 Supplementary Issue, pp.6-7.

centre in the city-chain of the Pearl River Delta from Hong Kong to the province capital Guangzhou.§§

The layout of SZCD is basically a symmetrical grid network. Along the central axis a 250 metres wide green belt stretches from north to south in terms of constructing an ecological and garden city. To the north of the district is the administrative centre characterised by low-rise public buildings, such as Shenzhen Citizen Centre, Shenzhen Cultural Centre, etc. To the south is located the Central Business Centre (CBD). Around them there are commercial residential areas. Apart from the public buildings, all office buildings, hotels and living buildings are committed to commercial use, and are high-rise or super high-rise. The ordered plan, green land and high-rises propose a renewed image and environment of Shenzhen -- modern and international.

If the high-rises in the CBD aim at constructing an image of international commerce and at increasing the use value of the land, residential buildings are an instrument which does not only bring people into the area, but also encourages following investment in the CBD.*** Due to high land prices, planning density and market profit, high-rises are the inevitable choice. Huangpu Garden is invested by the Hutchison Whampoa Property Group, an international company based in Hong Kong. Owned by one of the richest Chinese in the world, Ka-shing Li, the participation of Hutchison in SZCD is supposed to enhance market confidence. In the property market, and in order to obtain more profit, Huangpu Garden provides a certain 'high quality' environment and service to meet consumers' fantasy of 'modern' life which is normally identified as 'luxury'. So, at both urban planning and administration level and commercial culture level, Huangpu Garden, as well as other residential areas in the district, is a part of urban production.

§§ Wang, Peng, (2003) Introduction: A Summary of Urban Planning and Architectural Design for the Shenzhen Central District, 1996 – 2002, in: Shenzhen Urban Planning and Land Resource Bureau (Ed.) *The International Urban Design Consultation at Core Area of Shenzhen Central District*, pp. 13 – 14, Beijing: China Architecture & Building Press.

*** Interview with Xin Gu by H. Li, 2002. Ms. Gu is an urban planner in the Shenzhen Urban Planning and Urban Design Institute and worked for SZCD project from the Master Plan to the Statutory General Plan. Interview with Xiaohua Fei by H. Li, 2005. Mr. Fei is the Principle architect of the X-urban Consultants (Shenzhen) Co. Ltd., and was the vice Chief Architect in the Architectural Department of the Shenzhen Urban Planning and Design Institute from 1991-1998.

3.2 *The residential area -- an sub-urban condition*

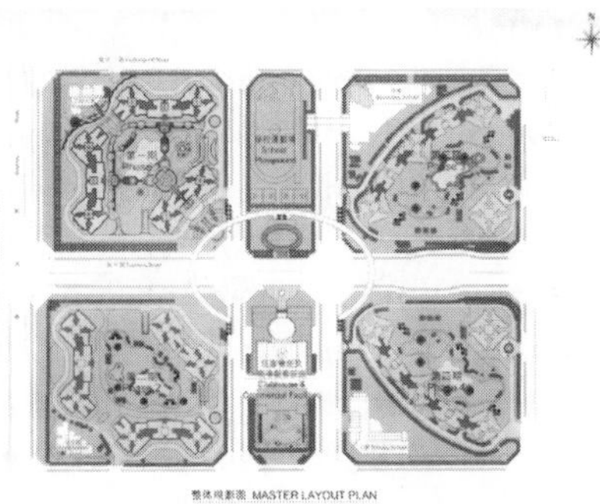

Figure 4. Master layout plan of Huangpu Garden, Shenzhen. Source: catalogue of the property.

Figure 5. The green fence of Huangpu Garden, Shenzhen. Source: H. Li, 2003.

The Huangpu Garden represents a typical micro-residential area – enclosure and self-contain. Physically, Huangpu Garden is divided into four residential blocks and two communal services blocks by the city roads. Each block is surrounded by green fencing and the entrance gates are guarded. The separated compounds are linked by a round pedestrian bridge for the residents, which is built over the roads, and which accesses each block. In terms of security and convenience, the strangers and intervention of the city are kept away, and an internal integration is managed.††† To the city, the boundary markers -- green fences and pedestrian bridge do not only protect the internalised living area, but also become the images of Huangpu Garden communicated to the outside.

††† Shenzhen Urban Planning and Land Resource Bureau, (2003a) A Collection of Residential Designs in Shenzhen Central District, Beijing: China Architecture & Building Press. pp. 41 – 48.

Themed gardens are one of the main characters of Huangpu Garden and the image of 'luxurious' living. Within the high-rise residential development, there has been a tension between high density and the quality of life. In the Chinese property market, communal ground is a main site for improving environmental quality of residential areas.‡‡‡ And it is also a space to express the uniqueness of each residential development. In Huangpu Garden, the themes of the communal gardens compose different images which refer to different cities of the world. According to the 'original' styles which are represented, the four-phase compounds of Huangpu Garden are individually named Parc Rome, Parc Barcelona, Parc Vienna and Parc Paris. However, they are neither reproductions of the 'originals', nor do they follow their stylistic and ideological principles.

Figure 6. The bird view of the communal garden of Parc Paris, Shenzhen. Source: H. Li, 2005.

Phase Four Parc Paris is a French style garden. Its symmetrical layout is a geometrical pattern within which every scenic spot is named, such as: entrance - Versailles plaza, swimming pool - Baroque colonnade, etc. Instead of identifying a French garden by spatial sequence and coherence between the elements, the composition corresponds to its own spatial circulation and functions, and makes the functional facilities and space figurative. The reference to the 'other' is not to form an exotica 'there', but to create a novel image distancing it from 'here'. By adopting these images and composing the material fragments, the differentiation of the life within the development from the outside city is intensified. In addition, the communal area does not only create an imaginary uniqueness and 'luxurious' lifestyle, but also sets up an imaginary collective identity for the individuals' identification.

As a living area, Huangpu Garden provides substantial facilities for the residents, such as kindergarten, secondary school, etc.§§§ The Clubhouse is the communal centre amongst the residential blocks. The architectural feature of the Clubhouse is figurative and is a highly 'European style' (*oufeng*) which employs a range of materials and elements, such as Greek Caryatid, Roman Corinthian capital and glass curtain wall.**** Although classical 'Western' elements are adopted, there is no implication of harmony, order or nostalgia for antiquity. It is important to build up a conspicuous figure which articulates wealth and uniqueness.

Figure 7. The appearance of the Clubhouse of Huangpu Garden, Shenzhen. Source: H. Li, 2003.

Within the Clubhouse are convenient amenities for daily life, such as supermarket, restaurant, library, gymnasium etc. Most of the provision is for 'additional' activities in spare time, such as musical classes for children, table tennis, beauty centre etc. The marriage of the hybrid architectural style and the internal function does not only have a strong visual identity among the

‡‡‡ Lü, Junhua, Rowe, Peter G., and Zhang, Jie, (2001) Modern Urban Hosing in China 1842-2000, Munich, London, New York: Prestel Verlag. pp.230-235.

§§§ According to the Regulation of the Urban Residential Area Planning (GB 50180-93), Huangpu Garden is a micro-residential district and should provide public facilities and programmes -- kindergarten, primary school, middle school, cultural station, restaurant and daily life facilities, such as grocery, rice and cooking oil shop, etc.

**** European Style has no strict definition and is not an academic category. The term emerges as a fashion in the real estate market of China since the early 1990s. It neither makes reference to particular style in European history nor refers to any particular time or country. There is no coherence in the formal composition and transformation and employment of elements. It generally takes architectural styles in Europe before modernism, such as classical, eclectic and Art-deco, as indifferential sources. From the mass cultural point of view, European Style symbolises wealth and luxury.

high-rises, but also projects a standard of 'luxurious' life. By integrating all the services into one building, it promotes an idea of healthy and successful life. To some extent, the 'additional' leisure activities compensate for working life in the city and enhance the collective identity and the fascination of 'modern' life.

From the three aspects, physical arrangement, image making and spatial organisation, Huangpu Garden constructs a two-fold relationship between the development and the city. On the one hand, Huangpu Garden draws itself out of the city and organises an internalised and sufficient domain. On the other, Huangpu Garden as a residential area is a social organisation for living. At this point, Huangpu Garden establishes a sub-urban condition mediating between the city and domestic life, and hence also the individuals.

3.3 *The household space*

Figure 8. The guarded entrance of a living building in Huangpu Garden, Shenzhen. Source: H. Li, 2005.

Due to security, individual households in Huangpu Garden do not directly connect to the city, but are distanced from the threat of the city through a series of spatial sequences -- from guarded entrance gate, communal garden, guarded entrance hall of the buildings, lift lobby, to the household. In this section, I will look at two household spaces in particular – living room and household garden.

3.3.1 *The living room*

The living room is a vital space in a Chinese household nowadays. The household layout in Huangpu Garden is arranged around the living room. In most cases the living room is combined with the dining room, and contains circulation space. As a family space, the living room acts as a lobby or hall which can not be avoided in moving around the household, and onto which all the functional spaces open. This spatial and functional combination is not only to save space in an economic sense and forms a sense of 'spaciousness' which the householders take as a sign of 'luxurious' life style, but also promotes coherence in family life. By integrating all the family activities into one space, such as watching television, having meals, cooking, playing games etc., a feeling of the family as a unit in the city is enhanced, though the individual's privacy is respected.

Figure 9. Typical floor plan of the Block 1 in Phase Four compound. Source: catalogue of the property.

Figure 10. Mr. Zhang's living room at 23rd floor, Huangpu Garden, Shenzhen. Source: H. Li, 2005.

In recent years, large window in the living room has become a popular feature in the real estate market. The architectural element associated with plated glazing started with some 'luxury' living buildings in Shenzhen, such as Baishida Garden, Donghai Garden, etc., and has filtered through to normal living buildings. In the case of Huangpu Garden, the size of the window is far more for day lighting rather than for ventilation. For instance, most of the window cannot open. At the same time, particularly for the households on the upper floors, large window becomes a frame of the outside city. Thus, beyond the physical confinement at ground level, the living room with the framed cityscape becomes a place connecting the household space with the city, whether favourable or undesirable.

Figure 11. Security bars and doors, Huangpu Garden, Shenzhen. Source: H. Li, 2005

At this point, living room in Huangpu Garden has twofold meaning and function in terms of the relationship between the individuals and the city. On the one hand, it forms a family space which to some extent reinforces differentiation between the outside and inside, and makes internal life coherent. On the other hand, it becomes a place which brings the city into the domestic domain at an intellectual level.

3.3.2 *The household garden*

Figure 12. Ms. Zhang's roof garden, Huangpu Garden, Shenzhen. Source: H. Li, 2005.

In the real estate market, household garden is a sort of 'luxurious' element to the householders. Huangpu Garden basically has two types of private garden, roof garden and sky garden. Ms. Zhang has a roof garden which is arranged in detail and is carefully maintained, such as a 'Roman style' fountain with underneath lighting, cobble paved path and grapevine etc. She said, 'the garden means a lot of things to us. Because of it, we can have more diverse family activities and our capacity of hosting friends is improved. Importantly, we have a natural space far away from the noisy city.'[††††] Mr. Zhang has a sky garden in his flat. Preferring the Chinese style and natural feeling, his garden is decorated with bamboo, wooden floor and plants. In the garden he has his tea and chats with his friends. To him, the garden is an extension of interior space and another expression of his personal taste, which could not be realised in the city and in the interior household space.[‡‡‡‡]

Figure 13. Mr. Zhang's sky garden, Huangpu Garden, Shenzhen. Source: H. Li, 2005

In both Ms. Zhang and Mr. Zhang's cases, the private garden is a mediating space between individuals, nature and the city. While it forms a more personal and complete internal domain against the impersonal city, it opens to the city with more personal consciousness.

4. Conclusion

From urban fabric and residential development to interior space, Huangpu Garden plays different roles between the city and the individuals – an instrument of urban production, protection of domestic domain, and negotiator of the internal space and the outside world. This complex relationship and function is organised through a series of mediating spaces and architectural elements: green fences, communal gardens, living rooms and household gardens. The intermediate spaces project individuals' fantasy of the 'modern' life to the image making and material arrangement of the surfaces and provide the convenience and security to the residents. Although facing the threat of the city, the sequence of the mediating spaces does not appear acutely confrontational between the individuals and the city, but suggests a more compromised and

[††††] Interview with S. Zhang by H. Li, Shenzhen, 2005.
[‡‡‡‡] Interview with H. Zhang by H. Li, Shenzhen, 2005.

sophisticated way to accommodate their desires and anxieties from city life. Through buying the property and services, the individual is somehow integrated into collective identification. In Huangpu Garden, although a number of Western architectural elements and materials are adopted, they do not have much relation to the original conception in the West. They are re-interpreted and serve for a Chinese fantasy of the 'modern' – production and economic success at the city level and personal identification of social status at the social level. This kind of mediating spaces form a particularly Chinese urban pattern. As fragmenting the continuity of urban landscape, they inhabit the individuals' ambivalence to the modern city.

SZCD, Huangpu Garden and the households I have presented above are not exceptional in China nowadays. The construction and inhabitation experience could be generalised into a representation of urban situation in present day China. With regard to the architectural question of localisation, they offer a potential view to rethink the Western or international impact, and Chinese inhabitation in Chinese cities. From my investigation, although Western technology and elements are largely employed in contemporary design, they are actually transformed into Chinese interpretations and instruments which are folded into Chinese urban culture and daily life. Nonetheless, I have not intention to hold back the problems in Chinese architecture and urban space. Taking Zheng Shiling's argument further, I would finish by posing these questions: How do we understand local identity of architecture and space in terms of life pattern in a city? To what extent is Chinese tradition preserved or changed in terms of spatial and social organisation? Do we have a more complex understanding of the Western effect and of localisation during a time of globalisation?

References

1. Chen, Yixin, The Course of SZCD Planning and Urban Design, *The World Architecture* Review (Shenzhen), 2001 Supplementary Issue, pp.5-9 (2000).
2. Le Corbusier, translated by Frederick Etchells, The City of To-Morrow and Its Planning, New York: Dover Publications (1987).
3. Feuchtwang, Stephan, Tales of Territoriality: the Urbanisation of Meifa Village, China, *études rurales*, juillet-décembre 2002, 163-164, pp. 249-266 (2002).
4. Lee, Leo Ou-fan, Shanghai Morden: the Flowering of a New Urban Culture in China, 1930-1945. Mass. Cambridge; London: Harvard University Press (1999).
5. Lü, Junhua, Rowe, Peter G., and Zhang, Jie, Modern Urban Hosing in China 1842-2000, Munich, London, New York: Prestel Verlag (2001).
6. Shenzhen Statistics Bureau, Shenzhen Statistics Year Book, Beijing: China Statistics Press (2003).
7. Shenzhen Urban Planning and Land Administration Bureau, Introduction to Shenzhen Comprehensive Plan 1996 – 2010 (2000).
8. Shenzhen Urban Planning and Land Resource Bureau, A Collection of Residential Designs in Shenzhen Central District, Beijing: China Architecture & Building Press (2003a).
9. Shenzhen Urban Planning and Land Resource Bureau, The International Urban Design Consultation at Core Area of Shenzhen Central District, Beijing: China Architecture & Building Press (2003b).
10. Wang, Peng, Introduction: A Summary of Urban Planning and Architectural Design for the Shenzhen Central District, 1996 - 2002, in: Shenzhen Urban Planning and Land Resource Bureau (Ed.) *The International Urban Design Consultation at Core Area of Shenzhen Central District*, Beijing: China Architecture & Building Press, pp. 13 – 27 (2003).
11. Wu, Jiang, A Hundred Years History of Shanghai Architecture, Shanghai: Tongji University Press (1997).
12. Zheng, Shiling, The Chinese Urban Space and Architecture Under the Influence of Globalisation, *Architectural Journal* (Beijing), 02/2003, pp. 7-10 (2003).
13. Zhong, Botao, An Approach to the Urban Closed Residential Area, *Architectural Journal* (Beijing), 09/2003, pp.14-16 (2003).
14. Zou, Denong, History of Chinese Modern Architecture, Tianjin: Tianjin Science and Technology Press (2001).

PROPERTY PRICE, FLOOR LEVEL, AND BUILDING DENSITY

WONG, S.K.* CHAU, K.W. YAU, Y. CHEUNG, A.K.C.
Department of Real Estate and Construction, The University of Hong Kong

High-rise high-density building developments are one of the most spectacular features in Hong Kong's urban areas. Buildings are built as tall as possible, subject to statutory and contractual controls, in order to maximize development profits. Developing tall apartment buildings has at least two implications in the occupancy stage: 1) the availability of more shared facilities and 2) overcrowding. This paper aims to study the economic impacts of building density and the disposition of building services on property prices. It also sheds light on the important question of how floor levels are priced by property buyers in high-rise buildings. Our sample came from the transaction records of a unique estate-type development in Hong Kong. It comprises buildings which, except for their heights (ranging from 11 to 26 floors), are highly homogeneous. This means that other factors affecting property prices, such as location and views, can be effectively held constant. Based on hedonic pricing analysis, we found that taller (and thus denser) buildings were sold at a discount compared to shorter buildings, ceteris paribus. These results suggested that while developers build taller buildings for better returns, they should be reminded that it also comes with a decrease in sales for units on lower levels.

1. Introduction

High-rise high-density building developments are one of the most spectacular features in Hong Kong. With a total land area of 1,100 km^2 and a present population of 6.7 million, our overall population density is about 6,100 persons per km^2. However, when considering urban areas alone, the gross population density is four times higher – about 26,000 persons per km^2. In some old urban areas like Mongkok, the population density at the street block level even reaches 400,000 persons per sq.km.[1] It is beyond doubt that Hong Kong is the most densely populated city in the world. In this regard, a limited supply of developable land signifies the optimization of developmental density. There is often little incentive for developers to sacrifice valuable development potential for lower density developments, given the very high property prices in Hong Kong. In general, low-rise developments can only be found in places where 1) statutory or lease restrictions do not allow high-rise developments; 2) the location is very remote with low property prices; or 3) redevelopment is too costly due to fragmented ownership in existing buildings. Otherwise, buildings are built as tall as possible with a view to maximum development profits.

Given a site coverage, developing tall apartment buildings has at least two implications to its residents. One is that a higher density living environment results in more shared facilities and thus more social interactions among residents. The provision of these shared facilities (e.g. shopping centres and transportation hubs) is only justifiable when there is a sufficient population base. The other implication is that residents living in high density developments may feel overcrowded because of stress and the loss of privacy. It is not unusual to see conflicts among residents over the use of common areas in condominiums. How this high-rise mode of living affects residents' satisfaction, and thus property prices, is not well understood. This paper therefore aims to study the economic impacts of building density on property prices. More specifically, we will test: 1) whether low-rise housing commands a premium over high-rise housing and 2) whether the pricing of floor levels in high-rise buildings is different from that in low-rise buildings.

2. Literature Review

This section explores the implications of density on the quality of the living environment. At the macro-level, high density urban development is beneficial to society. This facilitates overall city planning for avoiding urban sprawl, thereby conserving the natural environment and optimizing the use of public transportation.[2-4] However, once the micro-level view is reached, population density becomes a negative environmental externality.[5] Although this relationship is not straightforward, high density is often believed to be associated with social ills like crime, mortality, and mental illnesses.[1] Living in densely populated areas attracts the following comments:

* *Contact e-mail:* skwongb@hku.hk.

> The close living together and working together of individuals who have no sentimental and emotional ties foster a spirit of *competition*, *aggrandizement*, and *mutual exploitation*... The necessary frequent movement of great numbers of individuals in a congested habitat causes *friction* and *irritation*. *Nervous tensions* which derive from such personal frustrations are increased by the rapid tempo and the complicated technology under which life in dense areas must be lived (italics ours, p.15-16).[6]

This density-pathology argument was further elaborated on by a number of theoretical studies. Living in places with a high population density would likely to have a variety of negative effects.[7] These include physical, psychological, and social illnesses;[8] stress and stress-related illnesses;[9] and crime and intergroup violence.[10]

Although there have been many empirical studies on the density-pathology argument, their findings were mixed and no general conclusion has yet been drawn. Some studies found a weak positive or even no association between population density and social well-being after social-economic status had been controlled.[11] Similar results were found in studies in Hong Kong.[12,13] Nonetheless, there were reports showing contrary results. It was found that population density was positively correlated with mortality rate and psychiatric disorders after controlling for social structural variables.[14] This was supported by a recent study in Japan.[15] As a result, the evidence so far was inconclusive, casting doubt on the validity of the density-pathology argument.

Stokols tried to reconcile the inconsistent findings by distinguishing the concept of "density" and "crowding".[16] He interpreted density as a physical condition, while crowding was a psychological reaction to the restricted space by those individuals exposed to it, implying that there is no absolute relationship between density and crowding. Empirical works in Hong Kong supported the above proposition. It was also found that occupants' health and well-being were strongly correlated with an intolerance of high density (i.e., a feeling of crowding).[13] Forced social interaction between non-relatives as a result of flat-sharing tended to create stress and tensions.[12] Therefore, the health of occupants depends not solely on the physical density of the living environment, but how occupants cope with the physical limitations that will also play a role in moderating their health. Nonetheless, a high population density is viewed as a necessary, though not sufficient, condition for the experience of crowding.[17]

Moreover, a high living density also facilitates the transmission of communicable diseases, which was revived after the Severe Acute Respiratory Syndrome (SARS) outbreak in Hong Kong in 2003. A direct relationship between urban design and the spread of SARS has yet to be established; it prompted discussions about the role of a congested living environment in transmitting diseases.[18,19] In Hong Kong, it was revealed that the infection rate of tuberculosis was positively associated with density at the household level.[20]

Given its high population density, Hong Kong should be a good place to study the social and economic impact of density and crowding. In this paper, we will focus on how property prices vary with buildings of different densities. We postulate that given a similar floor plan, living in a higher-density building (e.g. a high-rise building) is less preferable to living in a lower-density building (e.g. low-rise building). Apart from the possible link between higher living densities and occupants' health, a high-density building also necessitates a greater sharing of communal facilities, like staircases, corridors, lifts, and lobbies. This creates a more crowded communal space, reduces one's sense of privacy and tranquility, but enhances the value of the building. (For example, a longer waiting time is needed for lift services in high-rise buildings.) All these suggest lower property prices for the residential units in high-density buildings when compared to those in low-density buildings, all other things being equal.

3. Methodology

Before carrying out any analysis, a workable definition for the density of a building is needed. In this paper, we suppose that there are two groups of building that are exactly the same except for their height (i.e., one group is high-rise and the other is low-rise). Since they have the same number of flats and the same floor area on each storey, a taller building would have a higher population density than the shorter one. This means we can simply measure building density by building height. In the following sections, we will use the terms building density and building height interchangeably.

To evaluate how building height (and thus density) is priced, hedonic pricing analysis is applied. According to Rosen's seminal work, hedonic pricing models can extract the implicit price of property attributes from property transaction prices.[21] As such, they can be used to estimate the implicit price of building height. But in the real world, buildings with variations in height usually have very different designs and location characteristics, rendering the separation of the implicit price of height from other design and location attributes difficult.

To solve this problem, we identified a housing estate in Kowloon (Telford Gardens) that has buildings of drastically different heights (ranging from 11 to 26 storeys). Telford consists of 41 apartment building blocks with the same typical floor plan and building services systems. While the majority of the blocks are 11 or 12-storeys tall (low-rise), there are 10 blocks that are 25 or 26 storeys (high-rise). This unique housing development offers us a very good sample to study the relationship between building height and property prices.

Our hypothesis is that, for two otherwise identical flats, the one in a shorter building will be priced higher than the one in a taller building. To test the hypothesis, the transaction data of flats on or below the 10/F was collected. We excluded units on or above the 12/F, as their transactions were only available for the taller buildings only. Flats on the 11/F were also truncated because they are located just beneath the roof and are more prone to water leakage and noises and vibrations generated by lifts. The quality of the top floor is therefore likely to be different from that of intermediate floors. After sample truncation, a total of 671 transactions remained, covering the period from January 2000 to October 2004. Table 1 presents the descriptive statistics of the data.

To estimate the implicit price of building density, a linear hedonic pricing model was set as follows:

$$P_{it} = \beta_1 LR_i + \gamma_1 FLR_i + \gamma_2 FLR_i^{\,2} + \alpha_0 + \alpha_1 AGE_i + \alpha_2 SIZE_i + \alpha_3 ST\_V_i + \alpha_4 GDN\_V_i + \alpha_5 DPT\_V_i + \alpha_6 MTR\_D_i + \sum_{k=1}^{T} \chi_k TIME_{ik} + \varepsilon_{it} \quad (1a)$$

The variables used in Eq. (1a) are defined in Table 2. Our focus is on β_1, which measures the relative price of a low-rise building to a high-rise building (or simply a low-rise premium). Its sign is expected to be positive because, as aforementioned, living in a dense environment may lead to crowding and a higher risk of health problems (mentally and physically). Other variables, including building age, flat size, floor levels, views, distance to the MTR station, and time factors, were controlled. The base period for the time dummies is January 2000. The base view type for the view dummies is a building view. A square term for floor levels was added to cater for any non-linear effect.[a] In fact, a more flexible specification for the price-floor relationship was to use floor dummies, *FLR_D* (see Table 2 for its definition). Building upon Eq. (1a), we replaced the continuous floor variable with floor dummies, as shown below:

$$P_{it} = \beta_1 LR_i + \sum_{j=2}^{10} \gamma_j FLR\_D_{ij} + \alpha_0 + \alpha_1 AGE_i + \alpha_2 SIZE_i + \alpha_3 ST\_V_i + \alpha_4 GDN\_V_i + \alpha_5 DPT\_V_i + \alpha_6 MTR\_D_i + \sum_{k=1}^{T} \chi_k TIME_{ik} + \varepsilon_{it} \quad (1b)$$

Table 1. Descriptive statistics of the property transaction data

	Mean	Std. dev.	Min	Max
Telford Gardens (No. of Transactions = 671)				
Property price (HK$mil)	1.48	0.26	0.60	2.28
Age (months)	255.48	17.96	216.00	292.00
Flat size (ft^2)	601.85	29.03	392	681
Floor level	5.37	2.85	1.00	10.00
MTR distance (m)	168.10	56.36	64.35	248.70
Street view (%)	0.29	-	-	-
Podium Garden view (%)	0.34	-	-	-
MTR depot view (%)	0.09	-	-	-
Low-rise building (%)	0.82	-	-	-

We assumed in Eqs. (1a) and (1b) that the premium for low-rise buildings, if any, was constant across all flats within the buildings. But if the premium exists, will it change with a change in floor level? For instance, will the premium for a low-rise flat on the 10/F be different from that on the 1/F? To test this assumption, we added an interaction term of the low-rise dummy (*LR*) and floor levels (*FLR* and *FLR_D*) to Eqs. (1a) and (1b). The resulting equations were:

$$P_{it} = \beta_1 LR_i + \beta_2 LR_i \times FLR_i + \beta_3 LR_i \times FLR_i^2 + \gamma_1 FLR_i + \gamma_2 FLR_i^2 + \alpha_0 + \alpha_1 AGE_i + \alpha_2 SIZE_i + \alpha_3 ST\_V_i + \alpha_4 GDN\_V_i + \alpha_5 DOT\_V_i + \alpha_6 MTR\_D_i + \sum_{k=1}^{T} \chi_k TIME_{ik} + \varepsilon_{it} \quad (2a)$$

[a] *We tried to add the square terms of AGE and SIZE to the equation, but they were insignificant.*

Table 2. Definition of the variables

Variable	Definition
P_{it}	The transaction price (in HK$ million) of property *i* at time *t*
LR_i	A building height dummy that equals 1 when property *i* is low-rise, and zero when it is high-rise
FLR_i	The floor level of property *i*
FLR_D_{ij}	A floor dummy that equals 1 when property *i* is on the jth floor, and zero if otherwise (base floor = 1/F)
AGE_i	The building age of property *i* in months
$SIZE_i$	The gross floor area of property *i* in square feet
ST_V_i	A view dummy that equals 1 when property *i* possesses a view facing the street, and zero if otherwise
GDN_V_i	A view dummy that equals 1 when property *i* possesses a view facing the podium garden, and zero if otherwise
DPT_V_i	A view dummy that equals 1 when property *i* possesses a view facing the MTR depot, and zero if otherwise
MTR_D_i	The distance between property *i* and the MTR station in metres
$TIME_{ik}$	A monthly time dummy that equals 1 when property *i* was transacted at time *k*, and zero if otherwise
α, β, γ and χ	Coefficients to be estimated
ε_{it}	The error term

$$P_{it} = \beta_1 LR_i + \sum_{j=2}^{10} \beta_j LR_i \times FLR_{ij} + \sum_{j=2}^{10} \gamma_j FLR_{ij} + \alpha_0 + \alpha_1 AGE_i + \alpha_2 SIZE_i + \alpha_3 ST\_V_i + \alpha_4 GDN\_V_i + \alpha_5 DPT\_V_i + \alpha_6 MTR\_D_i + \sum_{k=1}^{T} \chi_k TIME_{ik} + \varepsilon_{it} \quad (2b)$$

4. Results

Based on the hedonic pricing analysis, the results of the two models without interaction terms (i.e. Eqs. (1a) and (1b)) will be discussed first. As shown in Table 3, the R-squared values of these models were around 80%, with the value in Eq. (1b) slightly higher than Eq. (1a). This indicated that our models successfully explained a great deal of variability in the data. It should be noted that apart from the floor specification, the two models were similarly structured. For both models, most of the coefficients of physical attributes (except *MTR_D*) were highly significant with expected signs. Moreover, the magnitudes of the coefficients estimated were very close to each other. For our main hypothesis, we found that units in taller (and thus denser) buildings were sold at a discount compared to those in shorter buildings, *ceteris paribus*. The estimated price differentials, captured by the coefficient of *LR*, were HK$28,073 in Eq. (1a) and HK$19,469 in Eq. (1b), representing about 1-2% of the mean transaction price in the sample. They were significant at the 5% and 10% levels, respectively. This confirmed that there is a positive and significant premium for shorter buildings over taller buildings. In addition, Eqs. (2a) and (2b) were estimated to test whether the premium would change with a change in floor level. In fact, some people may argue that the premium is larger for lower floor flats for the following reasons:

- Given the same number of lifts, a taller building will necessitate a longer lift waiting time and a larger occupancy rate because the lifts have to serve more people. So, people living on the lower floors of a high-rise building would be more disadvantaged in enjoying lift services than those living in a low-rise building.
- Lower floor flats in a shorter building are less prone to the problem of back pressure in sanitary fitments.

The regression results of Eqs. (2a) and (2b) are shown in Table 3. As the results suggest, all interaction terms of LR with floor levels were insignificant. There is no evidence to support the argument that a low-rise premium is floor dependent. Since more correlated terms were added, the significance level of the coefficients of LR dropped beyond 10%.

5. Conclusion

In a city where super-high-rise buildings as human habitats are common, we have evaluated homebuyers' preferences as they pertain to high density living in Hong Kong. Based on the hedonic pricing analysis, we found that although high-rise high-density living is the norm, Hong Kong people still prefer living in shorter, hence less dense, buildings, with floor level remaining constant. Our results indicated that the implicit premium of living in less dense buildings is invariant under the function form of the floor variable, and not floor dependent. The implication is that while developers build higher buildings for better returns, they should be reminded that this also comes with a decrease in sales for units on the lower levels. Although the current data set did not allow us to determine an optimum building height, we believe that revenue will not strictly increase with building height, but rather comes down.

Acknowledgments

We gratefully acknowledge the financial support provided by the Research Group on Sustainable Cities of The University of Hong Kong.

Table 3. Estimation results of Eqs. (1a), (1b), (2a), and (2b)

	Eq (1a)		Eq (1b)		Eq (2a)		Eq (2b)	
Variable	Coefficient		Coefficient		Coefficient		Coefficient	
LR	0.028073	**	0.019469	*	0.005513		0.045620	
FLR	0.093792	***	-	-	0.079172	***	-	-
FLR^2	-0.006214	***	-	-	-0.004554	***	-	-
FLR_D_2	-	-	0.221088	***	-	-	0.279765	***
FLR_D_3	-	-	0.276484	***	-	-	0.333099	***
FLR_D_4	-	-	0.269176	***	-	-	0.306244	***
FLR_D_5	-	-	0.279909	***	-	-	0.257462	***
FLR_D_6	-	-	0.295252	***	-	-	0.319566	***
FLR_D_7	-	-	0.325746	***	-	-	0.327751	***
FLR_D_8	-	-	0.315682	***	-	-	0.320130	***
FLR_D_9	-	-	0.325992	***	-	-	0.347470	***
FLR_D_{10}	-	-	0.341126	***	-	-	0.409844	***
*LR*FLR*	-	-	-	-	0.018865		-	-
$LR{*}FLR^2$	-	-	-	-	-0.002122		-	-
$LR{*}FLR\_D_2$	-	-	-	-	-	-	-0.069473	
$LR{*}FLR\_D_3$	-	-	-	-	-	-	-0.064560	
$LR{*}FLR\_D_4$	-	-	-	-	-	-	-0.045489	
$LR{*}FLR\_D_5$	-	-	-	-	-	-	0.029682	
$LR{*}FLR\_D_6$	-	-	-	-	-	-	-0.029519	
$LR{*}FLR\_D_7$	-	-	-	-	-	-	0.001041	
$LR{*}FLR\_D_8$	-	-	-	-	-	-	-0.007631	
$LR{*}FLR\_D_9$	-	-	-	-	-	-	-0.028094	
$LR{*}FLR\_D_{10}$	-	-	-	-	-	-	-0.087369	
CONSTANT	0.735229	***	0.734827	***	0.764767	***	0.776275	***
AGE	-0.003161	***	-0.002629	***	-0.003167	***	-0.002837	***
SIZE	0.002603	***	0.002461	***	0.002588	***	0.002433	***
ST_V	-0.035492	***	-0.041425	***	-0.035511	***	-0.041232	***
GDN_V	0.070449	***	0.072914	***	0.070331	***	0.073595	***
DPT_V	-0.059584	***	-0.061142	***	-0.058634	***	-0.058060	***
MTR_D	6.35E-05		-4.03E-05		5.98E-05		-2.88E-05	
R-squared	0.817323		0.842351		0.818150		0.844757	
Adjusted R-squared	0.797362		0.823073		0.797609		0.823108	

Notes: Time dummy results are not reported here, but available upon request.

White's standard errors were used to adjust for heteroskedasticity

*significant at the 10% level; **significant at the 5% level; ***significant at the 1% level

References

1. A. G. O. Yeh, The Planning and Management of a Better High Density Environment, in A.G.O. Yeh and Ng, M.K. (Eds.) *Planning for a Better Urban Living Environment in Asia*, Ashgate, Hampshire, 117-143 (2000).
2. C. S. Chau, High Density Development: Hong Kong as an Example, in R.Y.W. Kwok and K.S. Pun (Eds.), *Planning in Asia: Present and Future*, Centre of Urban Studies and Urban Planning, Hong Kong, 1-14 (1981).
3. P. Newman and J. Kenworthy, *Cities and Automobile Dependence: An International Sourcebook*, Gower, Aldershot (1989).
4. L. H. Wang and A. G. O. Yeh, *Keep A City Moving: Urban Transport Management in Hong Kong*, Asian Productivity Organization, Tokyo (1993).
5. J. A. Mirrlees, The optimum town, *Swedish Journal of Economics*, **74**, 114-135 (1972).
6. L. Wirth, Urbanism as a Way of Life, *American Journal of Sociology*, **44**, 1-24 (1938).
7. H. M. Choldin, Urban Density and Pathology, *Annual Review of Sociology*, **4**, 91-113 (1978).
8. S. Zlutnick and I. Altman, Crowding and human behaviour, in J.F. Wohlwill and D.H. Carson (Eds.) *Behavioural Science and the Problems of Our Environment*, American Psychologists Association, 300 (1972).
9. J. B. Calhoun, Social welfare as a variable in population dynamics, *Cold Spring Harbor symposia on quantitative biology*, **22**, 339-356 (1957).
10. S. F. Hartley, *Population: quantity vs. quality: a sociological examination of the causes and consequences of the population explosion*, Prentice-Hall, Englewood Cliffs, N.J. (1972).
11. R. C. Schmitt, L. Y. Zane and S. Nishi, Density, health, and social disorganization revisited, *Journal of American Institute of Planners*, **44**, 209-211 (1978).
12. R. E. Mitchell, *Levels of Emotional Strain in Southeast Asian Cities*, Orient Cultural Service, Taipei (1972).
13. S. E. Millar, *The biosocial survey in Hong Kong*, Centre for Resource and Environmental Studies, Australian National University, Canberra (1979).
14. O. R. Galle, W. R. Gove and J. M. McPherson, Population Density and Pathology: What Are the Relations for Man? *Science*, **176**, 23-30 (1972).
15. A. Tanaka, T. Takano, K. Nakamura and S. Etkeuchi, Health Levels Influenced by Urban Residential Conditions in a Megacity – Tokyo. *Urban Studies*, **33**, 879-894 (1996).
16. D. Stokols, A social-psychological model of human crowding phenomena, *Journal of American Institute of Planners*, **38**, 72-83 (1972).
17. D. Stokols, A Typology of Crowding Experiences, in A. Baum and Y. M. Epstein (Eds.) *Human response to crowding*, L. Erlbaum Associates, Hillsdale, N.J. (1976).
18. Department of Health, *Main Findings of an Investigation into the Outbreak of SARS at Amoy Gardens*, http://www.info.gov.hk/info/ap/ (2003).
19. WHO, *Final Report - Amoy Gardens - WHO Environmental Investigation*, http://www.iapmo.org (2003).
20. C. C. Leung, W. W. Yew, C. M. Tam, C. K. Chan, K. C. Chang, W. S. Law, M. Y. Wong and K. F. Au Socio-economic factors and tuberculosis: a district-based ecological analysis in Hong Kong. *The International Journal of Tuberculosis and Lung Disease*, **8**, 958-964 (2004).
21. S. Rosen, Hedonic Prices and Implicit Markets: Product Differentiation in Pure Competition, *The Journal of Political Economy*, **82**, 34-55 (1974).

AIR QUALITY AND PROPERTY PRICES IN HIGH DENSITY URBAN AREAS

ANDY T. CHAN
Department of Mechanical Engineering, The University of Hong Kong

K.W. CHAU
Department of Real Estate and Construction, The University of Hong Kong

S.K. WONG
Department of Real Estate and Construction, The University of Hong Kong

K. LAM
Department of Mechanical Engineering, The University of Hong Kong

This study examines the relationship between air quality and property prices at a microscopic level to eliminate district level noise. A densely populated area dominated by high-rise developments was chosen for a case study. The street level air quality of the chosen area was estimated by computational fluid dynamics software using a three-dimensional Reynolds stress turbulence model. Property transaction prices in the same areas were then collected to construct a hedonic price model, which includes simulated air quality as an explanatory variable. Our findings indicated that air quality was reflected in property prices. This suggested that the property market was more efficient than any studies have implied so far, since there was no publicly available information on street level air quality. Further research and field measurements may be conducted to validate the findings.

1. Introduction

Good air quality is a highly valued good, especially in a compact city with high density development. In a world with zero information cost, variations in air quality in different areas should be reflected in property prices. However, in the absence of publicly available information on air quality, it is uncertain whether variations in air quality will be factored into property prices even in a very liquid and efficient market. Also, empirical evidence adduced from the literature was mixed. Some researchers found a significant negative relationship between air pollution and property values,[1-6] while others produced insignificant results.[7-9]

There are two main reasons for this inconclusiveness. One is the use of different air pollution measures such as sulphation levels, suspended particulate levels, and visibility by previous studies. It is obvious that some measures are brought to the public's attention through public announcements (public information), while other measures, notably those compiled by researchers, cannot be observed by the public (private information). Failure to take this into account may result in divergent conclusions. The second reason is that most studies had a macroscopic focus, and were conducted at the district or metropolitan level. This likely caused estimation problems because the wide variation in the many attributes of properties in the same district may introduce too much noise to their models, and thus invalidate their results.

Previous studies showed that Hong Kong's property market is very efficient, due to its high liquidity and transparency. This makes Hong Kong a good laboratory for testing the effect of air quality on property prices. Empirical findings in Hong Kong have suggested that the district level Air Pollution Index (API), which is compiled and published daily by the HKSAR Government,[*] has an impact on relative property prices in different districts. However, a recent survey conducted by Greenpeace revealed that Hong Kong people did not believe that the API is accurate.[†] Moreover, it is uncertain whether the variations in property prices are due to the availability of the district level API to the general public or a genuine response of market participants to variations in air quality. If the latter is true, then a much stronger form of market

[*] http://www.epd-asg.gov.hk.
[†] http://www.greenpeace.org.

efficiency than what is known today is implied. It is beyond doubt that further tests are needed.

This study examines the relationship between air quality and property prices at a microscopic level to eliminate district level noise. A densely populated area dominated by high-rise development, Mongkok, was chosen for a case study. According to Demographia,[‡] the population density of Mongkok was 40,404 persons per km^2 in 2001. Street level air quality of the chosen area was estimated by computational fluid dynamics (CFD) software using a three-dimensional Reynolds stress turbulence model. Property transaction prices in the same areas were then collected to construct a hedonic pricing model, which included simulated air quality as an explanatory variable. This model has been widely applied to study the implicit price for air quality.[10-13,§] If air quality was found to be significant, the property market would be more efficient than any studies have implied so far, since there is no publicly available information on street level air quality. If the effect of air quality is not significant, there is potential for assisting market players (buyers, tenants, developers, and the government) to make more informed decisions by making more air quality information available to the general public through research and field measurements.

2. Simulation of Air Pollution Levels

In this part, a simulation model is created in order to analyse the relationship between air quality and property prices in high density urban areas. A computational method is used as a means to estimate the street level air quality of the chosen area. This method is a popular and reliable way to predict air quality. Moreover, it saves the time needed to set up a wind tunnel experiment.

2.1 *The simulation model*

The Reynolds Stress Model (RSM) was chosen as the model used in this simulation. The model can solve differential transport equations individually for each Reynolds stress component. No matter how popular the application of k-ε model is, this turbulence closure model does not always ensure an accurate prediction of the air pollutant dispersion in the urban domain. Thus, RSM is chosen in this simulation task.

[‡] http://www.demographia.com.

[§] See Chau, *et al.* for a review of the hedonic price modelling of environmental attributes.[14]

A three dimensional numerical model with the commercial code CFX 5 was applied to accomplish this task. This commercial software has the advantage of being able to solve all the hydrodynamic equations as a single system with an advanced solver. It boasts a high processing speed for solving all necessary transport equations compared to other software. CFX-5 needs fewer iteration processes to arrive at a converged solution than other commercial codes. It can solve a set of equations that are the unsteady Navier-Stokes equations in their conservation form.

The governing equations of mass, momentum, and energy conservation in CFX-5 include the *Continuity Equation* (Eq. (1)), the *Momentum Equation* (Eq. (2)), and the *Energy Equation* (Eq. (3)):

$$\frac{\partial \rho}{\partial t} + \nabla(\rho U) = 0 \tag{1}$$

$$\frac{\partial \rho U}{\partial t} + \nabla(\rho U \otimes U) = \nabla(-\rho\delta + \mu(\nabla U + (\nabla U)^T)) + S_M \tag{2}$$

$$\frac{\partial \rho h_{tot}}{\partial t} - \frac{\partial p}{\partial t} + \nabla(\rho U h_{tot}) = \nabla(\lambda \nabla T) + S_E \tag{3}$$

2.2 *Boundary conditions*

In this simulation task, the simulation model is considered a rectangular domain with the dimension LxWxH: 925m x 590m x 270m (see Figure 1). It was reasonably believed that traffic emissions are the major source of air pollutants in Mongkok. The locations of the air pollution source were set at the two main roads in the chosen area, namely Yin Chong Street and Kwong Wa Street. Line source pollutants were assigned to simulate traffic emissions in the domain.

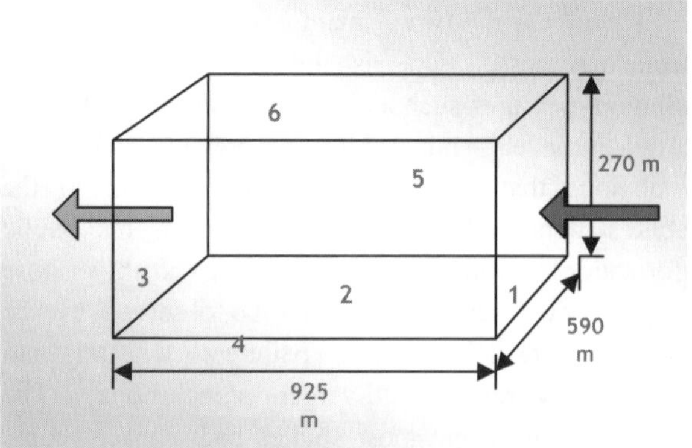

Figure 1. Urban model domain dimensions

The boundary conditions for the system are listed in Table 1. The system was set to the non-buoyant buoyancy mode. The reference pressure was $1.01\text{x}10^5$Pa. The domain temperature was 288K in isothermal heat transfer mode. The template fluid was set as air at a standard temperature and pressure. All these boundary conditions were set with regard to the general environment of Hong Kong. The maximum finite mesh element with size varied from 3m to 6m (see Figure 2). The Reynolds number for this simulation was kept at the order magnitude of 10^7. Based on these boundary conditions, the RSM model was used to generate the air pollution level for each flat of the developments in the chosen area (Figure 3).

Table 1. Boundary conditions

Surface	Nature	Input Values
1	Wind Inlet	3.0 m/s
2	Ground	Roughness height = 0.01m
3	Wind Outlet	Relative Pressure = 0.0 Pa
4	Atmosphere	Free slip
5	Atmosphere	Free slip
6	Atmosphere	Free slip

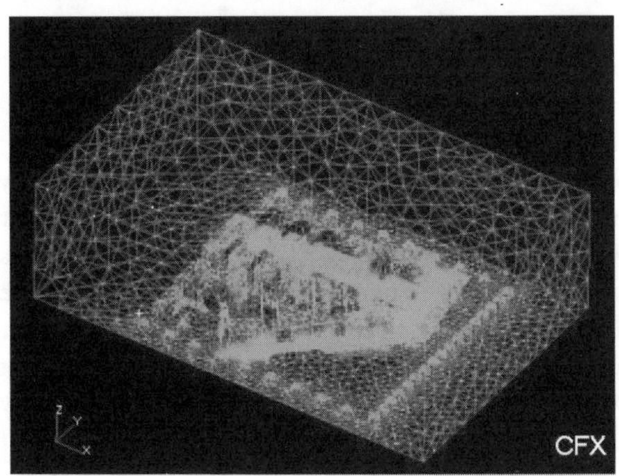

Figure 2. Mesh plots for the complete computational domain

Figure 3. Pollutant concentration slice plane – Mongkok

3. Hedonic Price Analysis and Results

3.1 *Hedonic pricing model*

Based on the flat-specific air quality data simulated in the previous section, a hedonic pricing model was constructed to examine the relationship between air quality and property prices. This model assumed that property prices (*P*) are a function of property attributes, one of which is flat-specific air pollution levels (*AP*). Other relevant attributes of apartment buildings include building age (*AGE*), the floor level of a flat (*FLR*), flat size (*SIZE*), and development scale (*EST*).** Since property transactions do not occur at the same time, time effects have to be controlled with a residential property price index (*TIME*), which is compiled by the Rating and Valuation Department of the HKSAR Government. The buildings in our sample are in very close proximity to each other, sharing highly similar neighbourhood characteristics (e.g. views and public transport). This meant that they need not be incorporated into the model. More importantly, investigating air pollution levels in such a microscopic level helps eliminate district level noise.

To allow functional form flexibility, two hedonic pricing models were estimated:

$$\ln P = \beta_1 AP + \alpha_0 + \alpha_1 AGE + \alpha_2 FLR + \alpha_3 SIZE + \alpha_4 EST + \alpha_5 TIME + \varepsilon \quad (4)$$

** For simplicity, we defined development scale as a dichotomous measure indicating if a development has more than one tower. *EST* was set to equal 1 if a development has at least two towers (so called an "estate"), and zero if otherwise.

$$\ln P = \beta_1 AP + \beta_2 AP^2 + \alpha_0 + \alpha_1 AGE + \alpha_2 AGE^2 + \alpha_3 FLR + \alpha_4 FLR^2 + \alpha_5 SIZE + \alpha_6 SIZE^2 + \alpha_7 EST + \alpha_8 TIME + \alpha_9 TIME^2 + \varepsilon \quad (5)$$

In both equations, the dependent variable is the log selling price of a property.[††] On the right hand side of the equations, β and α are the unknown coefficients to be estimated, and ε is the error term. Eq. (4) is a log-linear model, while Eq. (5) is a log-quadratic model. The latter is more flexible because quadratic terms were added to cater for any non-linear price effects. This allowed us to check the robustness of the results across different functional forms. Our central interest lay in the marginal effect of AP on $\ln P$ (i.e., β_1 for Eq. (4) and $\beta_1+2\beta_2 AP$ for Eq. (5)), while other variables were for control. Both equations are linear in coefficients and can be estimated by the Ordinary Least Squares (OLS) method.

Table 2. Descriptive statistics

	Mean	Std dev	Min	Max
Property price (HK$m)	1.58	0.72	0.25	5.41
Air pollution level	0.15	0.11	0.00	0.72
Building age (months)	179.32	92.28	1.00	391.00
Floor level	13.43	7.42	1.00	39.00
Flat size (sq. ft.)	409.55	134.40	226.00	890.00
Estate development	0.32	-	-	-
Price index (1999=100)	100.40	30.44	53.32	172.90

Our sample of property transactions was collected from 24 residential buildings in Mongkok, totaling 1,700 valid units of data during the period April 1991 to August 2004. Table 2 presents the descriptive statistics of the data.

3.2 *Results*

The OLS estimates of Eqs. (4) and (5) are shown in Table 3. The explanatory power of the models was fairly high, with adjusted R-squared values of 84% and 87%, respectively. Most coefficients were also statistically significant at the 1% level (those with ***). The White's standard errors were used to adjust for heteroskedasticity. Since the dependent variable was in log scale, the coefficients should be interpreted as a percentage change in property prices, given a unit change in an independent variable. A detailed discussion of the results is given below.

Table 3. OLS results

	Eq. (4) (log-linear)	Eq. (5) (log-quadratic)
Variable	**Coefficient**	**Coefficient**
Constant	-1.010 ***	-1.929 ***
AP	-0.128 ***	0.174 *
AP^2		-0.608 ***
AGE	-0.002 ***	-0.003 ***
AGE^2		2.38×10^{-6} ***
FLR	0.008 ***	0.005 **
FLR^2		1×10^{-4}
SIZE	0.002 ***	0.005 ***
$SIZE^2$		-2.82×10^{-6} ***
EST	0.031 ***	0.042 ***
TIME	0.007 ***	0.013 ***
$TIME^2$		-2.73×10^{-5} ***
R-squared	0.843	0.867
Adj. R-squared	0.843	0.866

***, **, and * indicate significance levels of 1%, 5%, and 10%, respectively.

Let's start with Eq. (4) first. The negative and highly significant coefficient of AP shows that air pollution levels bear a negative relationship to property prices. Specifically, an increase of one unit in the air pollution level lowers property prices by 12.8%. This profound result suggested that homebuyers are concerned about street level air quality, and their concerns, as reflected in property prices, were consistent with our simulated air pollution levels even though they did not have any public information on the air quality of the flats they purchased. This also meant that the property market was more efficient than any studies have implied so far. Such a result is not unreasonable, as people are likely to care more about air quality in densely populated areas such as Hong Kong. Moreover, the Hong Kong property market has been very liquid, thereby facilitating the transmission of information through property prices.

[††] Taking log gives marginal effects a convenient interpretation of percentage changes.

The coefficients of the control variables also showed plausible signs. The negative coefficient of *AGE* meant depreciation. The positive coefficients of *FLR* and *SIZE* confirmed the common sense that property prices rose with floor level and flat size. The positive coefficient of *EST* supported the presence of an estate premium due to, perhaps, better management and facilities. The positive *TIME* coefficient indicated that individual property prices co-moved with the general property price level.

Eq. (5), in fact, produced similar conclusions. Since the quadratic terms (except for FLR^2) were significant, the additional insight from Eq. (5) was that the price effects were not linear. For instance, buildings depreciated at a diminishing rate, and flat size rose at an increasing rate. Perhaps more important is the significance of the AP^2 term. This finding suggested that property prices dropped faster in the face of higher air pollution levels (see Figure 4).

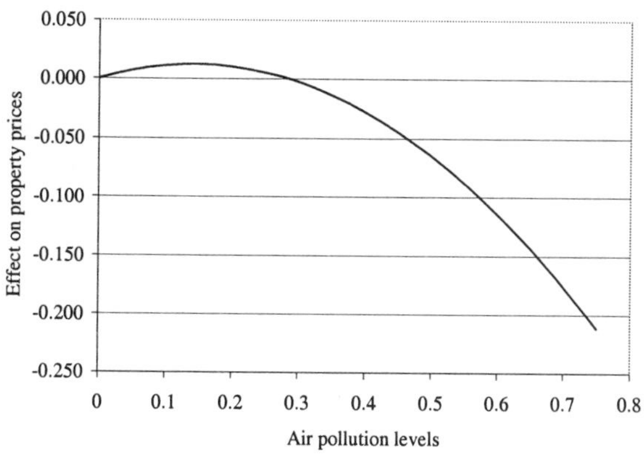

Figure 4. The effect of air pollution levels on property prices

4. Conclusion

This study was not merely about the confirmation of many previous studies that air quality is reflected in property prices. Rather, the main contribution of this study was more general in nature – the property market is more efficient than any studies have implied so far, since property prices reflected street level air quality information that was not publicly available. Apart from this, our study used air quality and property price data at a flat level so as to eliminate the district level noise inherent in previous studies. This microscopic level of analysis would not have been possible without the interdisciplinary synthesis between engineering and real estate economics. We expect to conduct further research, such as field measurements of air quality, to further validate the findings.

Acknowledgments

We gratefully acknowledge the financial support provided by the Research Group on Sustainable Cities of The University of Hong Kong. The authors are very grateful to Miss Astor Chung for providing us with the building plans and Mr. Patrick Wong for his research assistance in compiling the transaction data.

References

1. R.G. Ridker and J.A. Henning, "The determinants of residential property values with special reference to air pollution," *Review of Economics and Statistics*, Vol.49 No.2, pp.246-257 (1967).
2. R.J.Jr. Anderson and T.D. Crocker, "Air pollution and residential property value," *Urban Studies*, Vol.8, pp.171-180 (1971).
3. T.A. Deyak and V.K. Smith, "Residential property values and air pollution: some new evidence," *Quarterly Review of Economics and Business*, Vol.14 No.4, pp.93-100 (1974).
4. D.B. Diamond, "The relationship between amenities and urban land prices," *Land Economics*, Vol.56, pp.21-32 (1980).
5. J.C. Murdoch and M.A. Thayer, "Hedonic price estimation of variable urb an air quality," *Journal of Environmental Economics and Management*, Vol.15 No.2, pp.143-146 (1988).
6. P. Graves, J.C. Murdoch and M.A. Thayer, "The robustness of hedonic price estimation: urban air quality," *Land Economics*, Vol.64 No.3, pp.220-233 (1988).
7. K.F. Wieand, "Air pollution and property values: a study of the St. Louis area," *Journal of Regional Science*, Vol.13 No.1, pp.91-95 (1973).
8. V.K. Smith and T.A. Deyak, "Measuring the impact of air pollution on property values," *Journal of Regional Science*, Vol.15 No.3, pp.277-288 (1975).
9. M.M. Li and H.J. Brown, "Micro-neighborhood externalities and hedonic housing prices," *Land Economics*, Vol.56 No.2, pp.125-140 (1980).
10. D.Jr. Harrison and D.L. Rubinfeld, "Hedonic housing prices and the demand for clean air," *Journal of Environmental Economics and Management*, Vol.5, pp.81-102 (1978).
11. J.P. Nelson, "Residential choice, hedonic prices, and the demand for urban air quality," *Journal of Urban Economics*, Vol.5 No.3, pp.357-369 (1978).

12. S. Chattopadhyay, "Estimating the demand for air quality: new evidence based on the Chicago housing market," *Land Economics*, Vol.75 No.1, pp.22-38 (1999).
13. J.E. Zabel and K.A. Kiel, "Estimating the demand for air quality in four U.S. Cities," *Land Economics*, Vol.76 No.2, pp.174-194 (2000).
14. K.W. Chau, C.Y. Yiu, S.K. Wong, and L.W.C. Lai, "Hedonic price modelling of environmental attributes: a review of the literature and a Hong Kong case study," in *Understanding and Implementing Sustainable Development* (eds. L.W.C. Lai and F.T. Lorne), Nova Science, 87-110 (2003).

A STUDY OF MICROCLIMATE CONDITIONS IN OUTDOOR SPACES OF HIGH-RISE RESIDENTIAL DEVELOPMENTS

R. GIRIDHARAN J. TAI
Department of Architecture, University of Hong Kong, Pokfulam HKSAR, China.

S.S.Y. LAU
Centre for Architecture and Urban Design for China and Hong Kong, Department of Architecture, University of Hong Kong, Pokfulam, HKSAR, China.

The urban environment modifies microclimate in numerous ways. In general, urban climates are warmer and less windy than that in rural areas. A review of literature shows that the modification to urban climates is highly variable, and depends on particulars such as topography, regional wind speeds, urban morphology, time of year, and other physical factors. Measurement work around buildings, scale models, and simulation work have been used to try and identify the processes that govern the microclimatic changes. Alteration to temperature, as exemplified by Urban Heat Island Effect (UHI), is a particularly important aspect of microclimatic changes, as this has a direct impact on domestic fuel use: in a sub-tropical climatic region like Hong Kong, warmer temperatures increase air conditioning use substantially, particular during the hot humid summer months. The Urban Heat Island also affects the human comfort levels which can influence human behaviour. Literature reveals a considerable body of research work describing the conditions in cities and how they are different from their more rural surroundings. Most studies are directed at understanding urban climatology, rather than on the impact UHI on urban areas and the lives of urbanites. In order to mitigate the possible adverse effects of UHI, the impact of UHI needs to be investigated. This paper is a study into the extent of UHI in three high-rise high-density residential developments in Hong Kong. The differences in UHI between these developments are then explained in terms of urban design factors.

1. Introduction

Many studies in the last decades have demonstrated the climate change within the cities in comparison with rural areas. The resulting urban heat island (UHI) is a consequence of a number of anthropogenic and physical factors. The physical factors include higher heat absorption, heat production, pollution and impervious surfaces, reduced ventilation and vegetation of the urbanized areas. Recently researches have shown that the geometry of urban form also has a significant contribution to the forming of UHI. These factors include density, height to floor ratio and skyview factors of urban buildings.

The 'urban canopy layer' describes the spaces in the streets and between buildings. In an urban setting most people spend their days indoors, any outdoor activities are likely to be taken place within this urban canopy layer. In this study, the microclimate measurements are taken within this urban canopy layer. Different microclimates cause different physical and psychological sensation on man, and these sensations inevitably have influence on human behaviour. Therefore, there exists a highly complex relationship among tall buildings, climate, and human behaviour.

Some architects have made significant advances in designing tall residential buildings that are responsive to these microclimatic conditions, thus increasing the comfort of people living *inside*. However, there appears to be very little research into the microclimate of public spaces at podium or ground level of these high-rise, high-density residential developments. This paper describes the microclimate of such spaces in terms of their heat island intensity.

With a population of 6.7 million, and a total land area of just 1,100 sq. km, Hong Kong is one of the most densely populated cities in the world. Due to the scarcity of available land for building in Hong Kong, most of the population lives in high-rise high-density residential developments in urban areas, where the population density averages at around 26,000 persons per sq. km. The urban fabric is dominated by tall residential blocks (Figure 1a & 1b). Unlike the situation in Northern American or European cities, where most high density residential blocks cater for the less privileged, tall residential buildings in Hong Kong range from cheap public housing units provided by the government to privately development luxury apartments. In such high-density living conditions, the provision of open space at neighborhood level is of vital importance to enable residents to enjoy better quality of life with the social, psychological and environmental benefits that they bring. However, when designing these public spaces, architects and planners need to consider how the overall build forms and materials used would affect the microclimate conditions of these spaces. Only when the conditions allow for human comfort would these spaces serve their purposes properly.

Neighborhood level open space in these residential developments are mostly provided at ground level, or podium level, the latter being the dominant trend in recent years. The benefits of podium level open spaces include better privacy, security and ease of management, as only residents of the particular developments have access to these spaces. In the past few years, with the relaxation of government regulations in order to encourage the design of more environmentally friendly buildings, many tall residential buildings have introduced sky gardens to provide recreational spaces at higher levels.

Fig. 1a & 1b High-rise, high- density urban fabric of Hong Kong

2. Hypothesis

One significant factor that has adverse effect on human comfort is heat stress. The main source of heat stress in urban areas is described by Urban Heat Island intensity. Lowering of UHI will increase the human adaptability to environment and reduce the dependence on high degree of air conditioned space and encourage the use of outdoor public spaces.

This study investigates the impact of design-related variables on outdoor micro level daytime heat island effect in residential developments in Hong Kong in order to understand the design implications involved in making comfortable outdoor public spaces for residents. The paper hypothesizes that a significant part of the differences in outdoor temperatures within and between residential developments can be explained by the impact of design-related variables on the overall residential environment.

3. Pilot Study

In this paper, only ground level and podium level open spaces are surveyed for their microclimate conditions. The studied residential estates have high canyon geometry ratio in the order of 2 to 3. According to Oke (1987), 70–80% of daytime radiant energy surplus within canyon is dissipated to air through turbulent transfer. The balance 30 to 20% is stored and released in the night. The sky view factor is very low in most of the residential developments in Hong Kong. This is largely due to high-rise and high-density environment. In principle, as described by Oke, Givoni and Santamouris, low sky view factor should lower the nocturnal cooling of the environment but increases the daytime shadow effect. But there has been no empirical research done to indicate the real impact of the low sky view factor on the outdoor environments in residential developments in Hong Kong

The initial investigations are carried out on three large residential developments in Hong Kong, i.e. Belchers, Wah Fu I and Wah Fu II. All three developments are located on the south western coast of Hong Kong Island. The Belchers is a new generation private development with public spaces for residents provided at podium level. While Wah Fu I and Wah Fu II belong to the first generation high-rise high-density developments of the 1960s provided by the government to cope with the population boom. The public spaces are provided at ground level and access is not restricted to residents of the developments. A detailed analysis of the Belchers is presented, as this scheme is more illustrative of the impact of urban design on outdoor thermal conditions.

3.1 *Selection of variables*

Changes in the quality of façade material, non-residential to total floor area ratio in the development, average non-residential to total floor area ratio of the building fabric, plot ratio and location quotient are negligible between the developments and the variables are constant within each development. Therefore, implications of these variables are discussed without incorporating them into the model. Other changes in design variables will lead to changes in thermal, moisture and aerodynamic characteristic (Oke 1987, Givoni, Golany). The study will analyze the impact of urban design variables on summer time urban heat island intensity in the three selected high-rise high density residential developments of coastal Hong Kong.

3.1.1 *Dependant variables*

The dependent variable (UHI) for multiple regression analysis is the temperature difference between a station point in the estate and Wong Chuk Hang (south of Hong Kong Island) observatory (HKO) at the particular time.

3.1.2 *Independent variables*

Considering Hong Kong's geography, topology and built forms, the following independent variables are included in the multiple regression models: *Glass to surface area, total height to floor area ratio, surface albedo, local green area, width to height ratio, proximity to heat sink, sky view factor, surrounding built area ratio, altitude of the site, wind velocity and solar radiation.*

The independent variables chosen are likely to explain the variations in the dependent variable, the UHI.

3.2 *Measurement protocol*

In each of the three residential developments chosen for case study, two-fixed stations were placed. The two points were located so as to represent fairly the differences in layout of the development, thermal properties and vegetation cover. Each fixed point had a HOBO micro logger placed inside a mini weather station (Figure 2a & 2b).

Figure 2a Mini weather station at Belcher's

Figure 2b Mini weather station

The micro loggers were at a height of approximately 1.5m above the ground. In addition, mobile measurements were recorded on six to twelve points using an anemometer. Anemometer measurements were also taken at the fixed points. The mobile measurements were executed at 2m above ground.

The micro loggers measured air temperature, relative humidity and absolute humidity. The anemometer measured air temperature and wind velocity. Due to restrictions by the management of the residential developments, the data collection period was limited to 13:00 hrs to 20:00 hrs. Once every hour measurements were conducted at all points. Generally all the points in a development were covered within thirty minutes. Data between 15:00 and 18:00 were used for analysis of daytime UHI as the highest temperatures were recorded during these hours. The nocturnal UHI research work done by Kolev et al indicate that the air temperature data between 18:30 hrs and 21:00 hrs is fairly reliable for studying the impact of urban design variables on nocturnal UHI. The current study uses data collected between 18:00 hrs to 20:00 hrs of three days for analysis of nocturnal UHI due to field measurement restrictions mentioned above. Temperature measurements over three days were taken for analysis. Oke, too had used temperature measurements over three days to analyze the influence canyon geometry on temperature differences.

The data for most of the independent variables were collected from the developer's briefs, electronic maps and land use documents from the Planning Department of Hong Kong SAR. The sky view factor (SVF) is calculated based on a graphical method suggested by Watson and Johnson. The calculated sky view factor value is further reduced by 0.08 to 0.1 depending on the location and allowing for obstructions *such as mountains and other buildings* from the immediate surroundings. The albedo levels for the surfaces are based on the research work of Oke, Santamouris, Fathy, Sailor and Taha. The local green area ratio is calculated for the area lying within approximately 30 m radius from the point of measurement (1000 sq. m). For each measurement point, the mean value of glass to surface area ratio (GS), total height to floor area ratio (HA) and sky view factor (SVF) of all four sides of the station points were considered in the analysis.

4. Observations

The residential developments, Wah Fu I, Wah Fu II and Belchers were studied during July to September of 2002. All three estates are situated in the South of Hong Kong Island and are subjected to quite similar geographical conditions such as topography, proximity to sea etc. (Table 1) Wah Fu I and Wah Fu II are public housing estates completed in 1960. Belcher is a private development completed in 2001. Table 2 presents a summary of the field observations collected.

The atmosphere during the study period was fairly *stable* in the night especially with lower wind velocity and clear skies. In general the summer climatic conditions during 2002 especially temperature, cloud cover and humidity, were not normal as per Hong Kong Observatory report. The August and September mean daily global solar radiation levels were less by 10% and 21% respectively.

Table 1 Summery of field observations

Description	Belchers	Wah Fu-1	Wah Fu-2
Location		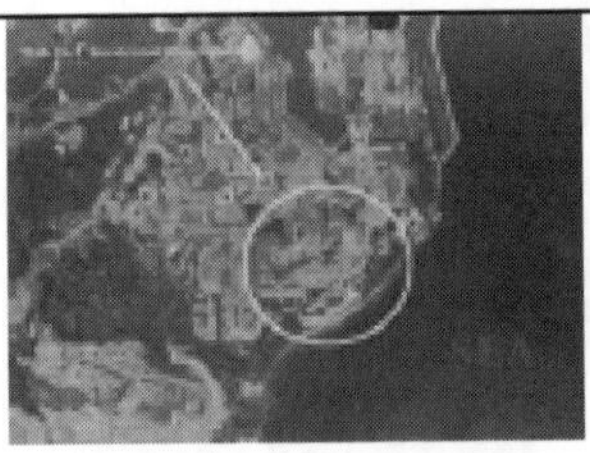	
Layout Plan	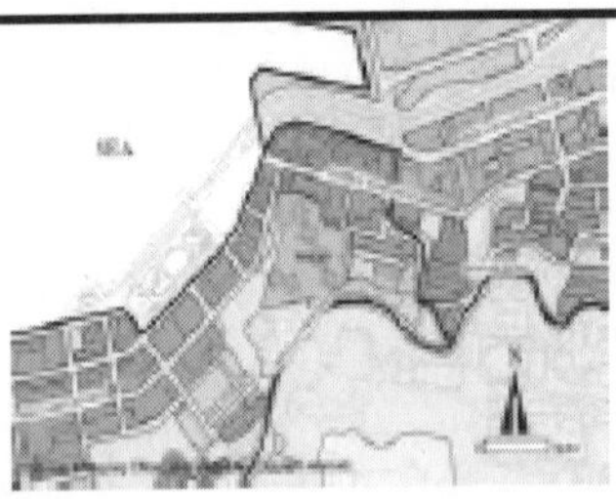	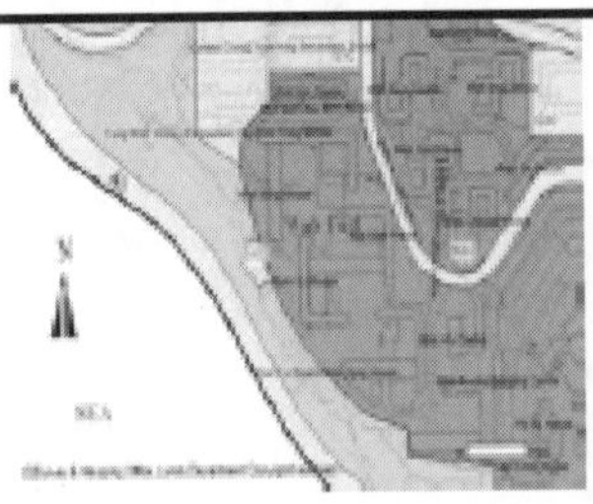	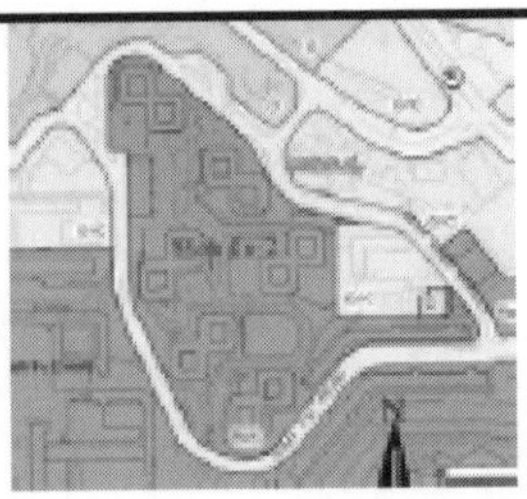
Temperature range	28.3 deg C to 30.5 deg C	28.3 deg C and 30.8 deg C	29.1 deg C and 30.8 deg C
Humidity range	69% to 83%	56% to 76%	89% to 68%
Wind velocity range	0.25 m/s to 4.25 m/s	0.0 to 1.05 m/s	0.02 m/s to 2.0 m/s

5. Results and Discussions

5.1 *Differences in UHI*

Daytime mean UHI at Belchers, Wah Fu-1 and Wah Fu2 to be of the order of 0.23 °C, 1.18 °C and 1.13 °C respectively. The highest difference within the developments was found in Belchers. On any given day, Wah Fu II is warmer during daytime and it has a mean UHI of 1.319 °C during the study period.

The difference between maximum and minimum nocturnal UHI within Belchers, Wah Fu I and Wah Fu II are 1.38 °C, 0.20 °C and 0.32 °C respectively. But the mean UHI in Belchers, Wah Fu I and Wah Fu II are 0.09 °C, 0.5 °C and 0.24 °C respectively. At Belchers, except for the Station ST1 UHI (1.33 °C), which is closer to Pokfulam Road, the UHI at other stations are in the region of -0.01 °C to 0.25 °C. Generally, most of Belches areas are cooler than the observatory in the night. (Table 2) This condition is analyzed below. This pattern of nocturnal UHI within Belchers makes the environment pleasant to live in.

Table 2. Mean UHI at different measurement stations

sdf	UHI Between 18 hrs to 20 hrs										
	ST1	ST2	ST3	ST4	ST5	ST6	ST7	ST8	ST9	ST10	ST11	ST12
Belchers	1.33	-0.05	0.05	-0.05	-0.13	-0.15	0.01	-0.01	0.01	-0.15	0.08	0.25
Wah Fu I	0.45	0.46	0.53	0.48	0.65	0.53	-	-	-	-	-	-
Wah Fu II	0.38	0.35	0.4	0.16	0.06	0.2	0.2	0.2				

Note: 'ST' means a measurement station

The nocturnal mean UHI is less than the daytime mean UHI by 150%, 136% and 370 % respectively. Therefore the heat stress is high in daytime compared to night. Nighttime UHI in residential developments in inland areas like Wan Chai and Kowloon could be of higher order due to higher residential densities. In the case study areas, even with low nocturnal UHI, the night is more uncomfortable compared to daytime due to higher relative humidity (in the order of 30% to 40% more compared to the daytime). Since occupancy level and usage of air conditioning is high in the residential developments during night times, mitigating nocturnal UHI is of great importance. In all three estates the mean UHI shows good correlation to population density but the sample is too small to establish a correlation coefficient.

In general, during both day and night of the summer months, Wah Fu I and Wah Fu II outdoor environments were observed to be uncomfortable despite being close to the sea which normally functions as a heat sink.

5.2 *Multiple regression analysis*

5.2.1 *Description of analysis*

A regression model involving eight independent variables was developed for Belchers since it has substantial variation in design variables between station locations compared to other two developments.

Thereafter, the regression analysis was extended to a combine model incorporating ten independent variables.

5.2.2 *Stage 2 analysis*

In Belchers, increase in glass area (glass to surface area ratio) will reduce the nocturnal UHI since glass will help to transmit outdoor heat into relatively cooler thermal mass due to good cross ventilation (wind speed of 1.6 to 2 m/s) and usage of air-conditioning. Similarly being closer to a heat sink (sea) tends to reduce the UHI in Belchers but an increase in green area will increase the nocturnal UHI due to blocking of long waves and transpiration. But the effect of increase in green area on nocturnal UHI is really insignificant. Belchers is closer to a heat sink (sea) and this generally increased the daytime UHI and reduced nocturnal UHI.

However, those parts of the Belchers closer to the sea were obstructed from receiving sea breeze at all times, and this contributed to a higher UHI during day and night. The opposite locations within Belchers were free to receive mountain winds (land breeze), and therefore recorded a lower nocturnal UHI.

In ten variables combine model, an increase in albedo level, height to floor area ratio, glass to surface area ratio and proximity to heat sink by 1% will lead to decrease in nocturnal UHI by 1.3%, 8.7%, 2.2%, and 4.3% respectively. Increase in local green area by 1% will increase nocturnal UHI by 0.15%. The influence of the sky view factor is not correctly revealed by the model due to limited period of measurement. The influences of proximity to heat sink and local green area are marginal. Therefore, design variables such as surface albedo, height to floor area ratio and glass to surface area ratio have significant influence on nocturnal UHI level of Belchers, Wah Fu I and Wah Fu II. By manipulating the above mentioned three design variables, the nocturnal UHI in all three developments could be mitigated significantly.

Theoretically, sky view factor and altitude have significant influence on nocturnal UHI, but the ten variables model was unable to capture the influence due to small set of data. For all practical reasons these variables should be considered in the formulation of design guidelines. Further, the climatological parameters change directions of influence regularly in an environment, i.e. increase in sky view factor increases daytime heat island intensity but decreases nocturnal heat island intensity. Therefore vector characteristics of urban design variables are important in formulating design guidelines.

The late evening solar radiation shows an abnormal influence (negative) on UHI. It is noted that the solar radiation measured at Hong Kong Observatory is used in the regression equation. Ideally the models should use solar radiation measured on the sites. This could explain largely the differences between the measured and predicted UHI. Therefore, the regression model developed in this study should be refined further.

6. Conclusions

Ideally urban design guidelines developed from a study of this nature should be applicable throughout the Territory of Hong Kong, and during all seasons and all times of the day and night. The geographical location and seasonal variations have an impact on UHI measurements (Oke 1990-91, Chandler 1976). It is practically impossible to study the entire geography of Hong Kong throughout the entire calendar year. As a result, there is a need to control the seasonal and geographical variables as much as possible in different phase of a long term study in order to understand the influence of variables related to physical design and then develop design guidelines. This study reports the result from such phase. The control over geographical variations is done by focusing on coastal area residential estates. The control over seasonal variations is achieved by focusing on summer. Further, by looking at day and night separately, research shows which variables influence both day and night.

Although the nocturnal UHI intensity is low relative to daytime UHI, thermal distress is high in night especially in Wah Fu I and Wah Fu II. Proximity to heat sink has insignificant influence on nocturnal UHI compared to daytime UHI. However, changes in surface albedo, height to floor area ratio, sky view factor and altitude have better possibility in mitigating nocturnal UHI than daytime UHI. Nocturnal UHI is largely due to daytime heat storage. Therefore designers should focus on design variables that mitigate daytime UHI and do necessary modifications to enhance nighttime ventilation.

Substantial variation in the ground surface albedo of Belchers has enhanced the ventilation level and created a pleasant outdoor environment. Therefore designers should maximize the variation in albedo level within the environment. This study did not cover the benefits of changes in the vertical surface characteristics of buildings. It is clearly logical that designers should seek to maximize the albedo level of these surfaces as well. The influence of some of the variables is not correctly represented in the above study, especially sky view factor. This problem could be avoided if the measurements are done for a longer period of, say, 18 to 22 hours.

Acknowledgments

The authors would like to express their thanks for the sponsorship from the HKU Research Group on Sustainable Cities towards the registration fee for the Symposium on sustainable cities at the tall building conference.

This research work is sponsored by the Hong Kong Government Research Grants Council, Project number HKU7309/03H.

References

1. T. R. Oke, T. R. Boundary layer climates. Methuen, USA. 1987.
2. B. Givoni, Climate considerations in building and urban design, John Wiley & sons, USA. 1998.
3. T. R. Oke, Canyon geometry and the urban heat island: Comparison of scale model and field observations. Journal of Climatology 1, 237-254. 1981.
4. A. D. Asimakapoulos, Climate and climate change, in: M. Santamouris, (Eds), Energy and climate in the urban built environment, chapter 2. James & James, UK. 2001.
5. R. Giridharan, S. Ganesan, S. S. Y. Lau, 2004. Daytime urban heat island effect in high-rise high-density developments in Hong Kong. Energy and Building 36 (6) (2004) 525-534.
6. T. R. Oke, R. Taesler, L. E. Olsson, The tropical urban climate experiment (TRUCE). Energy and Building 15/16 (1990-1991) 67-73.
7. T. J. Chandler, Urban climatology and its relevance to urban design. World Meteorological Organisation, Technical notes No.149, Geneva. 1976.
8. H. Sawid, M. E. Hoffman, Climatic impacts of urban design features for high and mid latitude cities, Energy and Building 14 (1990) 325-336.
9. M. Santamouris (Ed), Energy and climate in the urban built environment, James &James, UK 2001 (Chapters 1,3,4,5,6,7,10,11).
10. S. Ganesan, S. S. Y. Lau, Urban challenges in Hong Kong: Future directions for design, Urban Design International (2000) 3-12.

11. T. R. Oke, The urban energy balance, Progress in physical geography 12 (1988) 471-508.
12. T. J. Chandler, The climate of London, Hutchinson, London, 1965.
13. B. Stone, M. O. Rodgers (Jr), 2001. Urban form and thermal efficiency, Journal of American Planning Association 67 (2) (2001).
14. G. S. Golany, Urban design morphology and thermal performance. Atmospheric Environment 30 (1996) 455-465.
15. R. Emmanuel, Summertime heat island effects of urban design parameters. PhD thesis, The University of Michigan. 1997.
16. Z. Bottyon, J. Unger, The role of land-use parameters in the development of urban heat island in Szeged, Hungary. 4th Symposium on urban environment. 2001.
17. http://ams.confex.com/ams/AFMAPULE/4urban/program.htm.
18. U. Wienert, W. Kuttler, Statistical analysis of the dependence of urban heat island intensity on latitude. 4th Symposium on urban environment. (2001).
19. http://ams.confex.com/ams/AFMAPULE/4urban/program.htm
20. R. L. Knowles, Energy and Form: An Ecological Approach to Urban Growth. The MIT Press, USA, 1977.
21. D. Pearlmutter, A. Bitan, P. Berliner, Microclimatic analysis of "compact" urban canyons in an arid zone. Atmospheric Environment 33, (1999) 4143-4150.
22. I. Kolve, P. Savov, B. Kaprielov, O. Parvanov, V. Simeonov, Lidar observation of the nocturnal boundary layer formation over Sofia, Bulgaria. Atmospheric Environment 34 (2000) 3223-3235.
23. I. D. Watson, G. T. Johnson, Graphical estimation of sky view-factors in urban environment. Journal of Climatology 7 (1987) 193-197.
24. H. Fathy, in:W. Shearer and A. A. Sultan (Eds), Vernacular Architecture: Principles and examples with reference to hot arid climates. The University of Chicago. USA. 1986.
25. D. J. Sailor, Simulated urban climate response to modifications in surface albedo and vegetation cover. Journal of Applied Meteorology 34 (1995) 1694-1704.
26. H. Taha, Urban climate and heat islands: albedo, evapotranspiration, and anthropogenic heat. Energy and Building 25 (1997) 99-103.
27. Hong Kong Observatory (HBO): http://www.hko.gov.hk/wxinfo/

AN INITIAL ENVIRONMENTAL-EPIDEMIOLOGICAL ANALYSIS OF THE SPREAD OF SARS AT A HIGH-RISE HOUSING ESTATE IN HONG KONG

LAWRENCE W.C LAI K.W CHAU DANIEL HO GEORGE LAM VERONICA LIN

Department of Real Estate & Construction
University of Hong Kong

Previous studies on the SARS outbreak in Hong Kong in 2003 that sought to determine the causes of SARS transmission assumed that the living environments of SARS patients were critical. This article tests the null hypothesis that the SARS incidents in a local housing estate were random, and hence unrelated to the built environment. This test was executed by: (a) using the chi-square method to evaluate eight empirical hypotheses about the relationship among the built environment variables of floor level, block location, and flat location; and (b) employing the Gini Coefficient and Lorenz Curve to measure SARS information aggregated by residential unit for analysis of inequalities and concentration of dates of SARS incidence. The results show the disparity of SARS incidence distribution is significantly large, i.e. the probability that the SARS incidents in the sample area occurred randomly was insignificantly low. Using the same data set that has been used to establish Yu's theory (2003), the results were not inconsistent with the theory that SARS was possibly transmitted by contaminated domestic moisture carried by air currents, so the limitations of the analysis will be discussed.

1. Background: a Planning Enigma

Severe Acute Respiratory Syndrome (SARS) was a new and lethal epidemic disease that first hit Mainland China and Hong Kong in Spring 2003 (SARS Expert Committee 2003), and spread to other countries and regions, notably Canada, Singapore, and Taiwan, during the same period. The global and local research community, led by medical researchers under huge public pressure and great expectations, sought to identify the causes of the disease in the shortest possible time and, if possible, find a cure.

From 15 March 2003 to 23 June 2003, Hong Kong recorded 1,755 SARS cases, in which 299 were fatal (with a mortality rate of 17%). All patients identified as SARS patients had to be hospitalised by law beginning 27 March 2003 (orders were gazetted with immediate effect, with SARS added to the list of infectious diseases specified in the *First Schedule of the Quarantine and Prevention of the Disease Ordinance*, Chapter 141 of the Laws of Hong Kong) to avoid the further spread of the disease in the community. The deaths of all SARS patients happened in hospitals.

Most researchers who undertook a causal analysis of SARS transmission presumed that the living environment of the patients was a major cause of the rapid spread of the disease in the community. This was because in a private housing estate in East Kowloon that totalled 19 building blocks, 329 SARS cases were recorded. The number of deaths in this private estate was 42 (with a mortality rate of 12.7% of patients who lived in the estate). Such a high concentration of SARS incidents in just one housing estate, almost 20% (18.7%) of the city's total, suggested to researchers that the built environment had to be a major cause of the rapid spread of the disease within the housing estate, if not also beyond it.

To the urban planner, the general geographic distribution of SARS and its specific concentration in this particular housing estate is surprising, as the disease did not claim as many victims among residents of older buildings in the high-density (Lai 1993) old urban core of the city or low income public housing estates. The estate in question happened to consist of relatively modern private buildings that were governed by modern layouts and believed to be properly managed. Moreover, most SARS cases occurred among residents of medium to high income private housing. The housing estate in question is about 20 years old and much younger than the first of its kind, in West Kowloon, which is almost 40 years old. Furthermore, the mortality rate of SARS patients living in the private housing estate was lower than the territorial average.

Hitherto, other than the building health analysis of Ho, *et.al.* (2004), most works on SARS were produced by medical researchers. In this context, there were many competing hypotheses for the media of transmission of SARS in a built environment. These included face-to-face contacts in the public areas of building blocks, rodents, domestic insects, internal sewage systems, and bathroom moisture spread by powerful ventilation fans.

2. The Yu's Theory

The leading logit regression-computational fluid-dynamic and multizonal modeling article by Yu, *et.al.* (2003; Yu's theory) provides powerful support for the role of moisture in the spread of SARS in the housing estate in question. Yu, *et. al.* (2003) did not formally specific the regression model or tabulate the statistical estimates in their exercise to estimate "the probability of infection for each apartment unit (not each resident) as the dependent variable and applied the logistic-regression model to explore the association between location (i.e., apartment unit and floor in each building) and the probability of infection." We shall reconstruct the model used by Yu *et.al.*, not specified in their original article, for the benefit of the reader.

The logit model solves the problem by a linear transformation with the following procedures: -

$$\ln[P/(1-P)] = a + BX + e \qquad (1)$$

$$P-(1-P) = \exp(a + BX + e) \qquad (2)$$

where

ln is the natural logarithm,

exp = 2.71828...,

P = Probability that event Y occurs [p(Y=1)],

P/(1-P) is known as the "odds ratio", and

ln[P/(1-P)] is the log of odds ratio, or 'logit'

The result of the transformation is a logistic model (1) and (2). It is an S-shaped distribution function and is similar to the standard-normal distribution.

Yu *et. al.* (2003) applied the logistic regression model to estimate the probability of infection of apartment units related to the spread of airborne transmission associated to airflow dynamics. From the analysis in their paper, we can reconstruct the model they used should be: -

$$\ln[P/(1-P)] = f(a + BiXi + ei) \qquad (3)$$

where

P =Probability of infection of an apartment unit, assuming the number of residents be 0, 1, 2, 3, or 4 with assigned probability of 0, 0.25, 0.50, 0.75 and 1.00,

Xi = Association between apartment unit.

X1 = Building blocks, A through S

X2 = Building floors, 3 through 36 (Floors 1, 2 and 3 are non-residential):

Lower 4th to 13th floors,

Middle 14th to 23rd floors, and

Upper 24th to 36th floors.

X3 =Apartment unit numbers, representing the direction each faced, upper case letters denote 'front facing' and lower case letters denote 'site facing windows', where

A, a = southeast

B, b = southwest

C, c = northeast

D, d = northwest

Bi = Coefficient of Xi

ei = Error term, where ei ~ N(0,1)

(Normal distribution unit μ = 0 and ? = 1)

In the analysis, risk was calculated as the number of cases divided by the assumed number of resident. The odds ratios for SARS infection at different floor levels are compared with units facing direction (Aa, Bb... etc.). The confidence interval (CI) through out the analysis is 95%.

If Yu's theory is correct, then the design of the building blocks in terms of the widths of the recesses to which bathroom aerial exhausts are drained and the disposition of the building envelopes of the housing estate would be critical environmental factors for the spread of the disease. This would mean that if, in the future, a resident in the housing estate would contract SARS or a similar type of disease, the same outbreak would repeat itself in the housing estate in question unless there was a major alteration to the bathroom ventilation system. While there has been no scientific analysis to eliminate the relevance of other factors and possible causes[1], including evidence of airborne and animal transmissions, which have been and are still being studied, Yu's theory is presently the "received

[1] See Yu, et al (2003) and Ng (2003).

view".

There is a need to reconstruct the scientific footing for Yu's theory, for it would be too presumptuous to "find reasons" for the spread of the disease simply by believing that the built environment or building design *per se* was the culprit. To satisfy this need, a researcher should first seek to refute the assertion that the built environment is a factor in the spread of SARS. When such refutation is not tenable, then the search for reasons of building design is valid. To achieve the purpose of justifying Yu's theory, this article[2] uses the chi-square method to evaluate eight empirical hypotheses about the relationship among the built environment variables of floor level, block location, and flat location using the same data set that was obtained from the Hong Kong Hospital Authority. (Hospital Authority) and used by Yu, *et.al.* (2003) to establish the claim that SARS is possibly transmitted by domestic moisture spread by air currents.

3. Data and Methodology

The raw data for analysis was based on that appearing on the table "Dates of onset – confirmed cases in Block (the block no, the subject housing estate)(as of 4:30 PM, 3.4.03)". The blocks refer to Blocks A, B, C, D, E, F, G, J, N, Q, and S, and the data was for the period 21 March 2003 to 1 April 2003. SARS cases happening after 1 April 2004 were excluded from both this analysis and Yu's analysis because of the isolation order. In the data set, there are 11 blocks of flats or living units in the housing estate that recorded SARS cases. The analysis by Yu, *et.al.* (2003) used data for Blocks B, C, D, and E, and excluded the blocks that had fewer than 10 cases each. In the same data set, there were 187 SARS cases altogether and Block E had most of the cases (99 or 53%) during the period.

The first goal of this epidemiological study was "to determine associations between patterns of disease occurrence and patterns of exposure to risk".

The first step we have to take note of and handle is the contrast between SARS *incidence* and SARS *prevalence*. Using the data supplied by the Hospital Authority, which only provides information on location (by block, floor, and flat) and dates of incidence, our analysis will examine whether the incidences were randomly distributed. In other words, we will determine if, within the housing estate in question, t categories mentioned are *independent* of each other.

We shall evaluate this null hypothesis through two groups of empirical hypotheses. The first group will be tested by the Chi-square model and the other by the Lorenz curve model.

The model in the first group in this study is Chi-square goodness-of-fit statistics. The model compares all observations (the number of SARS cases) to a constant value (the estimated number of SARS cases by the sample mean). The Chi-square test for goodness-of-fit compares the observed frequencies of a multi-cell classification with those expected from some hypotheses. They may be frequencies associated with classifications of discrete variables (e.g. marital status), or with continuous var ıbles which have been scaled into class intervals (e.g income) (Sprowls, 1964). The statistical method is widely understood and applied in a variety of situations which call for such a test. For example, Wolfe *et. al.* (2001) used Chi-square tests with two categories of each variable to examine the relationship between past maltreatment, current adjustment, and dating violence in a community sample of adolescents. Zeanah *et. al.* (2001) compared an intervention group (IG) with a comparison group (CG) to examine outcomes for infants and toddlers in foster care. The four outcome categories for foster children were reunification with birth parents, termination of parental rights, surrender, and placement with a relative.

A lot of studies, particularly in medical research also used Chi-square test to evaluate the data of patients with diseases, such as human cancer cells, prostate cancer, bipolar disorder, headgear compliance and digestive diseases, etc (Agar, *et. al.* 2005; Bradley, *et. al.* 2005; Ottaiano, *et. al.* 2005; Vieta, *et. al.* 2005). The Chi-square test is useful when the data considered are frequencies, specifically, when the interest is in how many people (the frequency count) fall into a particular category, relative to a different category. The Chi-square test for independence is the appropriate statistical test for determining the association between

---

[2] This article is a condensed version of a full paper of the same title by the authors. For details of this paper, please contact Lawrence Lai wclai@hkusua.hku.hk.

two nominal scale variables, where each participant can be included in only one cell of the contingency table (Gliner, *et. al.* 2002).

The problem to be tackled by the Lorenz Curve Model is whether the confirmed SARS cases in the housing estate were evenly distributed during a specific period of time. Here, the period includes the first day (March 21, 2003) when SARS cases were discovered in the area and all the subsequent days until new cases were no longer being discovered (April 1, 2003).

4. Tests and Findings

The chi-square method described above was used to evaluate eight empirical hypotheses on the relationship among the built environment variables of floor level, block location, and flat location. The estimates for the eight tests for these hypotheses are summarised in Table 1. In other words, all eight hypotheses on the randomness of the SARS incidents in the data set were rejected, and hence, we have statistical grounds for the claim that they assume a pattern.

Table 1. Summary estimates for the eight tests for the Chi-square method

	Chi-square (χ^2)	Degree of Freedom	p-value
Test one	68.2500	24	0.0000
Test two	269.0160	32	3E-39
Test three	449.3636	28	2E-77
Test four	61.6768	24	0.0000
Test five	41.2222	24	0.0030
Test six	53.8182	21	1E-04
Test seven	162.3636	24	1E-24
Test eight	32.4054	16	0.0035

As for the Lorenz Curve analysis, the findings are in Table 2 and Figure 1. The analysis shows a high degree of inequality (G = 0.9834) in the distribution of SARS cases. This means that the temporal occurrence of SARS in the housing estate had a low degree of randomisation, and thus our null hypothesis is also refuted. There thus is statistical evidence that the SARS incidents in the data set assume a pattern.

5. Discussion and Conclusion

Our findings support a theory that assumes that the built environment in the housing estate in question has a bearing on the internal spread of the epidemic, thus lending support to Yu's theory. However, our results are constrained by the limitations of the data set and must therefore be regarded as preliminary.

References

1. U. Agar, C. Doruk, A. A Bicakci, N. Bukusoglu, The role of psycho-social factors in headgear compliance. *European Journal of Orthodontics*, **27(3)**, 263-267 (Jun 2005).
2. C. J. Bradley, D. Neumark, Z. Luo, and H. Bednarek, Employment outcomes of men treated for prostate cancer. *Journal of National Cancer Institute*, **97(13)**, 958-965 (Jul 2005).
3. D. C. W. Ho, S. K Wong., A. K. C. Cheung,, S. S. Y. Lau, W. S. Wong, D. P. Y. Lung, and K. W. Chau, Assessing the health and hygiene performance of apartment buildings. *Facilities*, **22(3/4)**, 58-69 (2004).
4. L. W. C. Lai, Hong Kong's density policy towards public housing in Hong Kong. *Third World Planning Review*, **15(1)**, 63-85 (1993).
5. A. Ottaiano, A. Palma, M. Napolitano, and C. Pisano, Inhibitory effects of anti-CXCR4 antibodies on human colon cancer cells. *Cancer Immunology, Immunotherapy*, **54(8)**, 781-791 (Aug 2005).
6. SARS Expert Committee, SARS in Hong Kong: from experience to action. *Report of the SARS Expert Committee*, Hong Kong Government, 1-300 (October 2003).
7. C. R. Sprowls, Sample sizes in Chi-square tests for measuring advertising effectiveness. *Journal of Marketing Research*, **1**, 60-64 (Feb 1964).
8. I. T. S. Yu, Y. Li, T. W. Wong, W. Tam, A. T. Chan, J. H. W. Lee, *et al.* Evidence of airborne transmission of the severe acute respiratory syndrome virus. *The New England Journal of Medicine*. **350(17)**, 1731-173 (2004).
9. E. Vieta, J. Mullen, M. Brecher, B. Paulsson and M. Jones, Quetiapine monotherapy for mania associated with bipolar disorder: combined analysis of two international, double-blind, randomised, placebo-controlled studies. *Current Medical Research and Opinion*, **21(6)**, 923-932. (Jun 2005).
10. D. A. Wolfe, K. Scott, C. Wekerle, and A. Pittman, Child maltreatment: risk of adjustment problems and dating violence in adolescence. *Journal of American Academy of Child and Adolescent Psychiatry*, **40**, 282-289 (2001).
11. C. H. Zeanah, J. A. Larrieu, S. S. Heller, *et. al.*

Evaluation of a preventive intervention for maltreated infants and toddlers in foster care. *Journal of American Academy of Child and Adolescent Psychiatry*, **40**, 214-221 (2001).

Table 2. Lorenz curve and Gini Coefficient estimates of SARS cases

1	2	3	4	5	6	7	8	9	10
	Order	**Cases**					**A**	**B**	**A x B**
Date:03	**X**	**Y**	**Cum X**	**Cum Y**	**Rel Cum X**	**Rel Cum Y**	**X(i+1)-X(i)**	**Y(i+1)+Y(i)**	
20-Mar	0	0	0	0	**0.00**	**0.00**	0	0	0
21-Mar	1	3	1	3	**0.01**	**0.02**	0.0110	0.0163	0.0002
22-Mar	2	7	3	10	**0.03**	**0.05**	0.0220	0.0007	0.0000
23-Mar	3	9	6	19	**0.07**	**0.10**	0.0330	0.0016	0.0001
24-Mar	4	53	10	72	**0.11**	**0.39**	0.0440	0.0049	0.0002
25-Mar	5	37	15	109	**0.16**	**0.59**	0.0549	0.0098	0.0005
26-Mar	6	34	21	143	**0.23**	**0.78**	0.0659	0.0137	0.0009
27-Mar	7	15	28	158	**0.31**	**0.86**	0.0769	0.0164	0.0013
28-Mar	8	13	36	171	**0.40**	**0.93**	0.0879	0.0179	0.0016
29-Mar	9	6	45	177	**0.49**	**0.96**	0.0989	0.0189	0.0019
30-Mar	10	3	55	180	**0.60**	**0.98**	0.1099	0.0194	0.0021
31-Mar	11	3	66	183	**0.73**	**0.99**	0.1209	0.0197	0.0024
1-Apr	12	1	78	184	**0.86**	**1.00**	0.1319	0.0199	0.0026
2-Apr	13	0	91	184	**1.00**	**1.00**	0.1429	0.0200	0.0029
Total	**91**	**184**							**0.0166**

G = 1 - (Col 8)(Col 9)
= 1 - Col. 10
= 1 - 0.0166
= 0.9834

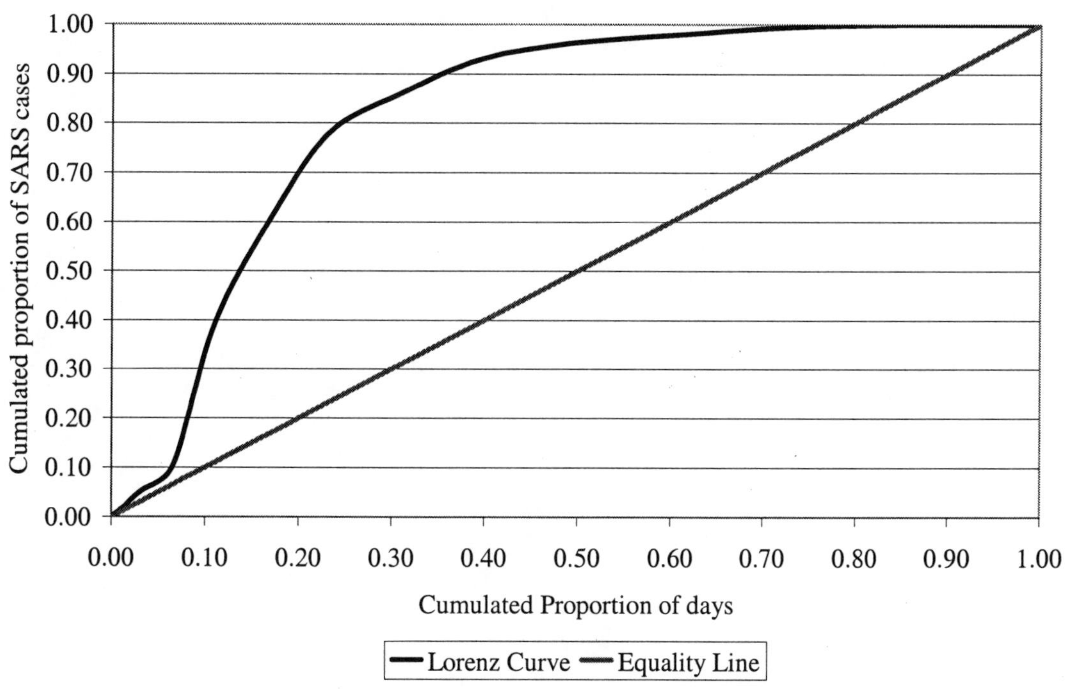

Figure 1: The Lorenz Curve and the Deviation of the Lorenz Curve from the Equity Line

Energy Efficiency, New & Renewable Energy

A SYSTEMIC SIMULATION MODEL AND EXPERIMENTS ON VERTICAL GROUND HEAT EXCHANGERS IN GCHP SYSTEMS

PING CUI

Department of Building Services Engineering
The Hong Kong Polytechnic University, Hong Kong

HONGYING YANG

Department of Building Services Engineering
The Hong Kong Polytechnic University, Hong Kong

ZHAOHONG FANG

The Ground Source Heat Pump Research Center
Shandong Institute of Architecture and Engineering, Jinan 250014, P.R. China

Ground coupled heat pumps (GCHPs) are a promising new technology that has been used in recent years in Chinese market. The most important component of the GCHP systems compared to conventional air conditioning systems is the ground heat exchanger (GHE). This paper mainly presents the entire simulation model of heat transfer of GHEs in GCHPs. The explicit analytical solution of the finite line source in semi-finite medium is derived for convenient calculation of the thermal resistance outside the borehole for long time steps. The resistance inside the borehole is also better expressed accounting for the thermal interference between the two legs of the U-tube. The techniques of spatial superimposition for multiple boreholes and sequential temporal superimposition for arbitrary heating/cooling loads of the systems are also presented in this paper. The validation of the models by experiments in the actual project of GCHP system in China is carried out. The results of these tests have shown that the measured data on site agree quite well with predicted data according to the model. The simulation model is suitable for engineering applications.

1. Introduction

More and more primary energy resources are consumed in heating, ventilating, and air-conditioning (HVAC) systems with very low exergetic efficiency. It is necessary to develop a new kind of HVAC technique which can employ the renewable energy. Ground coupled heat pump (GCHP) systems are perhaps the most widely used "green" HVAC systems, with an estimated 1.1 million GCHPs installed worldwide [1,2]. These systems, which are environmentally friendly, can make significant contributions to reductions in electrical energy usage and have very low maintenance requirements.

In GCHP systems, also referred to as closed loop ground source heat pump (GSHP) systems, heat is extracted from or rejected to the ground via a closed loop through which pure water or an antifreeze solution circulates. The ground heat exchangers (GHEs) used in closed loop system typically consists of high-density polyethylene pipes installed in vertical boreholes or horizontal trenches. In horizontal GCHP systems, GHEs typically consist of a series of parallel pipe arrangements laid out in dug trenches about below 1-2meter deep. One of the most disadvantages is that the horizontal systems are more affected by weather and air temperature fluctuations because of their proximity to the earth's surface. The other disadvantage is installing the horizontal systems needs much more ground area than vertical systems. In vertical GCHP systems, GHE configurations typically consist of one to tens of boreholes each containing a U-tube pipe through which the heat exchange fluid is circulated. Typical U-tubes have a diameter in the range of 19mm to 38mm and each borehole is normally 40m to 200m deep with a diameter ranging from 0.1m and 0.2m. The borehole annulus is generally backfilled with a material that prevents contamination of ground water. Vertical GCHP systems are the most widely used in HVAC systems currently compared to other GSHPs. It can be installed at any location where drilling or earth trenching is feasible.

This paper mainly systemically presents the mathematical models of vertical GHEs in the heat transfer process with the surrounding soil. These analyses provide better understanding of the thermal processes in GHEs, and can serve as useful tools in design and simulation of their performance as well. In order to verify the simulation models, the relative tests has been carried out on an actual GCHP system.

2. Simulation Models of Heat Transfer in GHEs

A main challenge for GCHP applications is the design and performance simulation of the GHEs. As is known, the construction costs of GHEs are critical for the economical competitiveness of GCHP systems in the heating and air-conditioning market. It is important to work out sophisticated and validated tools by which the thermal behavior of any GCHP system can be assessed and then, optimized in technical and economical aspects. The schematic diagram of a single borehole in vertical GHEs is illustrated in Figure 1. Heat transfer between a GHE and its surrounding soil/rock is rather complicated and difficult to model for the purpose of sizing the exchanger or energy analysis of the system. Besides the structural and geometrical configuration of the GHE a lot of factors influence its performance, such as the ground temperature distribution, soil moisture content and its thermal properties, groundwater movement and possible freezing and thawing in soil.

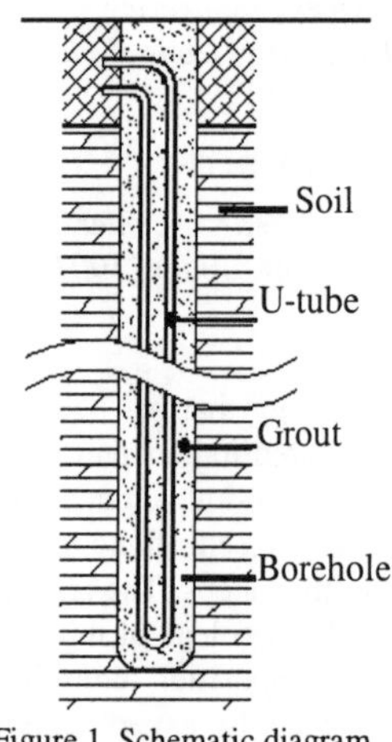

Figure 1. Schematic diagram of a grouted borehole

Models of varying complexity have been presented for practical applications in design and performance prediction of GHEs. Numerical simulations have been used [3] to obtain more exact solutions for research and parametric analysis of GHEs, but are of limited practical value for use by designers of GCHP systems. A general design manual for GCHP has been compiled by ASHRAE [4]. In recent years the focus has been shifted to commercial and institutional applications [5, 6]. A group of Swedish researchers have made detailed analysis of the heat transfer problem in boreholes mainly for heat storage [7, 8]. A model suitable for predicting hourly or minutely response of the GHE has also been presented to facilitate detailed energy analysis of GCHP systems [9].

The main objective of GHE thermal analysis is to determine the temperature of heat carrier fluid based on certain operating conditions. A design goal is then to control the temperature rise of the ground and the circulating fluid within acceptable limits over the life of the system. A fundamental task for application of the GCHP technology is to grasp the heat conduction process of a single borehole in the GHE. Heat transfer in a field with multiple boreholes may be analyzed on this basis with the superposition principle. Involving a time span of several years, the heat transfer process in the ground around the vertical boreholes is rather complicated, and should be treated, on the whole, as a transient one. The heat transfer process may usually be analyzed in two separated regions. One is the solid soil/rock outside the borehole, where the heat conduction has to be treated as a transient process. With the knowledge of the temperature response in the ground the temperature on the borehole wall can then be determined for any instant on specified operational conditions. Another sector often segregated for analysis is the region inside the borehole, including the grout, the U-tube pipes and the circulating fluid inside the pipes. The following analysis is to solve the inlet and outlet temperatures of the circulating fluid according to the borehole wall temperature, the thermal resistance inside the borehole and the heat rate of the GHE.

2.1 *Heat conduction outside borehole*

In analysis of GHE heat transfer some complicating factors, such as ground stratification, ground temperature variation with depth [7] and groundwater movement [10,11], are proved usually to be of minor importance and may be analyzed separately. As a basic problem, it is assumed here that the ground is homogeneous in its thermal properties and initial temperature.

Most practical models use the Kelvin line-source theory, in which the ground is regarded as an infinite medium and the borehole as an infinite line source. This model is characterized by its simplicity, and is proved to be appropriate for response to step heat rejection/extraction of a few hours to a few months [4, 5]. A shorter time step response is distorted by the finite radial dimension of the heat source, and should be treated by more accurate models [9] accounting for the configuration and composition of the borehole.

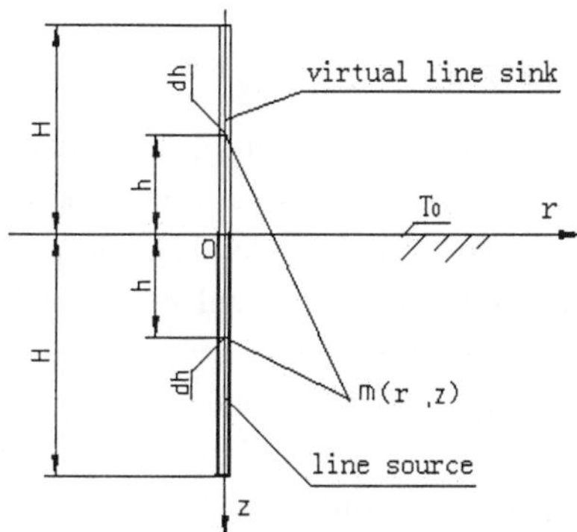

Figure 2. The geometry of a finite line source system

Our research group has derived an explicit analytical solution of the finite line source model taking the finite length of the borehole and the ground surface as a boundary into account [12]. The merit of the analytical solution is that the temperature response of a borehole field can be calculated with arbitrary combination of parameters within a small fraction of a second on a microcomputer. Therefore the algorithm can be integrated directly into simulation programs.

A diagram of the theoretical model for a single borehole in the GHE is illustrated in Figure 2. As accepted in most models for GHE thermal analysis, assumptions are taken in following discussions:

1. The ground is regarded as a homogeneous semi-infinite medium, and its thermophysical properties do not change with temperature;
2. The medium has a uniform initial temperature, T_0.
3. The boundary of the medium, i.e. the ground surface, keeps a constant temperature same as its initial one throughout the period concerned;
4. The radial dimension of the borehole is neglected so that it may be approximated as a line-source stretching from the boundary to a certain depth, H;
5. The heating rate per unit length of the source, q_l, is constant since a starting instant, $\tau = 0$.

Set a virtual line-sink with the same length H but a negative heating rate $-q_l$ on symmetry to the boundary as shown in Figure 2. The isothermal boundary condition is complied due to the symmetry of the line-source and the virtual line-sink.

The solution of the temperature excess at any point m on any time τ can be obtained by integrating contributions of all the increments of the line-source and sink [12]. That is:

$$T(r,z,\tau)-T_0=\frac{q_l}{4k\pi}\int_0^H\left\{\frac{\operatorname{erfc}\left(\frac{\sqrt{r^2+(z-h)^2}}{2\sqrt{a\tau}}\right)}{\sqrt{r^2+(z-h)^2}}-\frac{\operatorname{erfc}\left(\frac{\sqrt{r^2+(z+h)^2}}{2\sqrt{a\tau}}\right)}{\sqrt{r^2+(z+h)^2}}\right\}dh \tag{1}$$

It is obviously noticed that the temperature on borehole wall, where $r = r_b$, varies with time and depth. The temperature at the middle of the depth ($z = 0.5H$) is usually chosen as its representative temperature. An alternative is the integral mean temperature along the borehole depth, which may be determined by numerical integration of Eq. (1). The difference between them were analyzed, and found to be insignificant [12]. For the convenience of application the former is accepted as the representative temperature in our design and analysis program. Then, the temperature response to a step heating from a single borehole can be determined by the radial distance and time. Following the definition in Eskilson's approach [7], we have the expression of g-function for a single borehole

$$g\left(\frac{a\tau}{H^2},\frac{r}{H}\right)=\frac{2\pi k\left[T(r,0.5H,\tau)-T_0\right]}{q_l}$$
$$=\frac{1}{2}\int_0^1\left\{\frac{\operatorname{erfc}\left(\frac{\sqrt{(r/H)^2+\left[0.5-(h/H)\right]^2}}{2\sqrt{a\tau/H^2}}\right)}{\sqrt{\left(\frac{r}{H}\right)^2+\left[0.5-\left(\frac{h}{H}\right)\right]^2}}-\frac{\operatorname{erfc}\left(\frac{\sqrt{(r/H)^2+\left[0.5+(h/H)\right]^2}}{2\sqrt{a\tau/H^2}}\right)}{\sqrt{\left(\frac{r}{H}\right)^2+\left[0.5+\left(\frac{h}{H}\right)\right]^2}}\right\}d\left(\frac{h}{H}\right) \tag{2}$$

Once the temperature distribution from a single borehole is determined, it can be superimposed in space to obtain the temperature distribution of the whole borehole field. For a GHE of m boreholes the representative temperature rise on the wall of a borehole concerned can then be calculated accordingly.

$$T_b - T_0 = \frac{q_l}{2\pi k}\left[g\left(\frac{a\tau}{H^2}, \frac{r_b}{H}\right) + \sum_{i=1}^{m-1} g\left(\frac{a\tau}{H^2}, \frac{r_i}{H}\right)\right] \quad (3)$$

Here r_i is the distance between the i-th borehole and the borehole concerned.

2.2 *Heat transfer inside borehole*

Compared to the infinite ground outside the borehole, both the dimensional scale and thermal capacitance of the borehole are much smaller. Moreover, the temperature changes inside the borehole are usually slow and minor. Thus, it is a common practice that the heat transfer in this region is approximated as a steady-state process. Such simplification has been proved appropriate and convenient for most engineering practices except for analyses dealing with dynamic responses within a few hours [9].

Simplified one-dimensional models [4, 13] and the two-dimensional model [14] presented quantitative expressions of the thermal resistance in the borehole have been recommended for engineering design. However, because these models did not take into account the heat transmission by the axial flow of the circulating fluid, and, then, discerned no distinction between the entering and exiting pipes.

On the basis of the 2-D model mentioned above a new approach has been presented to take account of the fluid temperature variation along the borehole depth. Being minor in the order, the conductive heat flow in the grout and ground in the axial direction, however, is still neglected so as to keep the model concise and analytically manageable. This model is referred as quasi-three-dimensional.

The energy equilibrium equations can be written for up-flow and down-flow of the circulating fluid [15, 16]. That is:

$$\left.\begin{aligned} -Mc\frac{dT_{f1}}{dz} &= \frac{(T_{f1}-T_b)}{R_1^\Delta} + \frac{(T_{f1}-T_{f2})}{R_{12}^\Delta} \\ Mc\frac{dT_{f2}}{dz} &= \frac{(T_{f2}-T_b)}{R_2^\Delta} + \frac{(T_{f2}-T_{f1})}{R_{12}^\Delta} \end{aligned}\right\} \quad (0 \le z \le H) \quad (4)$$

Where, R_1^Δ and R_2^Δ are the relative thermal resistances between the circulating fluids and the borehole wall, and R_{12} the resistance between the pipes, and their detailed expressions can be referred to the reference [14-16]. Two conditions are necessary to complete the solution, they are

$$\left.\begin{aligned} z=0,\ T_{f1} &= T_f' \\ z=H,\ T_{f1} &= T_{f2} \end{aligned}\right\} \quad (5)$$

The general solution of this problem is derived by Laplace transformation, which is a little complicated in form, and can be found elsewhere [17]. Then, the inlet and outlet temperatures of the fluid can be calculated according to Eq. (4) and Eq. (5).

For the purpose of practical application an alternative parameter $\varepsilon = (T_f' - T_f'')/(T_f' - T_b)$ is derived from the temperature profiles, which we refer to as efficiency of the borehole [16].

From the derived temperature profile the more accurate heat conduction resistance between the fluid inside the U-tube and the borehole wall can be worked out.

$$R_b = \frac{H}{Mc}\left(\frac{1}{\varepsilon} - \frac{1}{2}\right) \quad (6)$$

The more accurate quasi-3-D model is highly recommended for design and thermal analysis of ground heat exchangers.

2.3 *Temperature response to a pulse train of heating load*

The heat extracted from or rejected to GHEs varies with time because the GCHP load usually varies with time. The variable heat flow can be approximated by a series of continuous rectangular of heating or cooling pulses as shown in Figure 3.

For continuous rectangular heating pulses, as shown in Figure 3, the borehole wall temperature rise at time τ

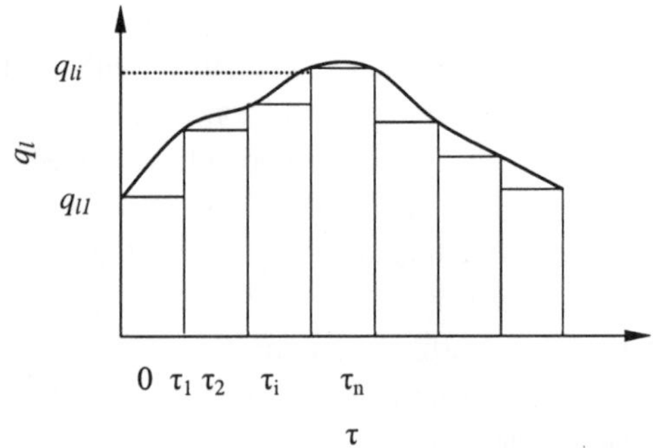

Figure 3. Continuous heating approximated by a rectangular pulse train

can be obtained by superposition as [7,18]

$$T_b = T_0 + \sum_{i=1}^{n} \left(\begin{array}{c} \frac{1}{4\pi k}\left(q_{l_i} - q_{l_{i-1}}\right) \cdot \\ \left(g\left(\frac{a\tau}{H^2}, \frac{r_b}{H}\right) + \sum_{i=1}^{m-1} g\left(\frac{a\tau}{H^2}, \frac{r_i}{H}\right) \right) \end{array} \right)$$

$$(q_{l_0} = 0) \tag{7}$$

These new algorisms discussed above have been incorporated into a computer program (named as GeoStar), developed by the authors for engineering design and thermal analysis of vertical GHEs. The program has been employed in design of a pilot GCHP project in China [19], which has been operating successfully since its completion in May, 2001. The following test is also carried out on the pilot CCHP project.

3. Validation of the Model with Experiments

The experimental GCHP project locates in Jinan, Shandong Province, China. It was originally planned for the lecture hall and later expanded to cover a reading room and some library offices. The ground heat exchanger system consists of 25 vertical U-tubes 60 meters deep for a total of 1500 meters of vertical boreholes. The U-tubes are constructed of nominal 1-inch (SDR 11) high-density polyethylene pipe connected to headers up to 4 inches.

Monitoring and measuring devices have been installed in the project in order to validate the model, as shown in Figure 4. All the testing instruments are calibrated in laboratory. A data acquisition system is developed to record these data including the water flows, outlet and inlet temperatures of the two fluids and the power consumptions in the GCHP system. During the test, all the data were recorded at the intervals with three minutes.

Based on the measured data, the loads on the GHE can calculated according to the principle of the heat balance. Hence, the entering water temperature (EWT) to the heat pump from the GHE and the outlet temperature (OWT) from the heat pump to the GHE can be predicted by the program (GeoStar) which was based on the mathematical model discussed above.

The test began at 8:40 and ended at nearly 20:00 on June 21, 2003. From the tested data, it can be seen that the water temperatures of the GHE rose rapidly, because the additional cooling loads were added to this system that was not considered in the original GCHP system design. But it doesn't influence the test results.

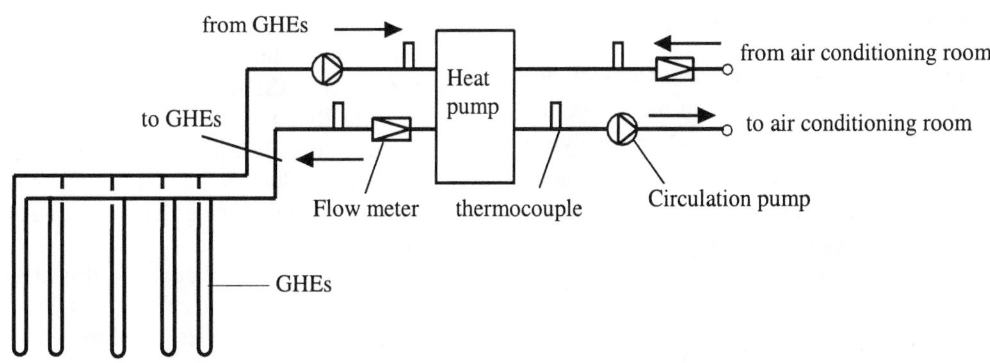

Figure 4. Diagram of the in-situ measurement apparatus in the GCHP system

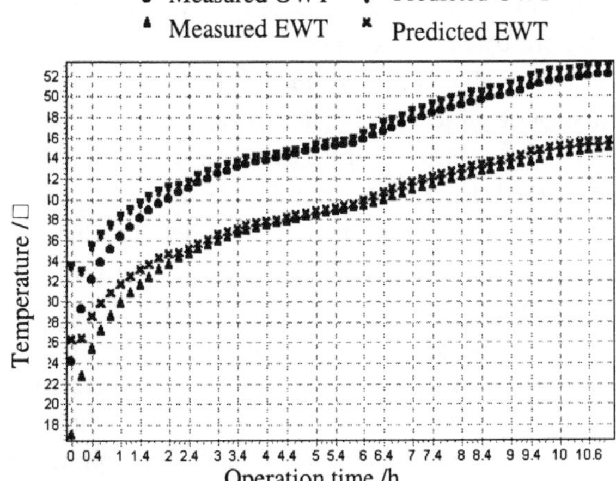

Figure 5. Comparison of predicted and measured EWT and OWT

Figure 5 presents an hourly comparison of the predicted and measured temperatures including EWT and OWT with time. In general, the mathematical model matches the tested data quite well for the whole operating period. From Figure 5, it is noted that the difference between the predicted and measured data is relatively large in the first two hours due to the assumptions in the model, such as neglecting the thermal capacities of the material in boreholes. But the predicted data are more close to the measured with the increase of operating time when the influence of the thermal capacities of material on the heat transfer gets smaller. The maximum relative error between the predicted and measured data is no more than 2% after a short period of operation time (i.e. the first two hours in this test), which is enough accurate for engineering applications. It is also detected from Figure 5 that the predicted temperatures were always higher than the measured, this is also because the assumption of line source for the borehole.

4. Conclusions

This paper mainly presents the entire model of heat transfer of GHEs which can be used in the design and simulation GCHP systems. The explicit analytical solution of the finite line source in semi-finite medium is derived for convenient calculation of the thermal resistance outside the borehole for long time steps. The resistance inside the borehole is also better expressed accounting for the thermal interference between the two legs of the U-tube. The techniques of spatial superimposition for multiple boreholes and sequential temporal superimposition for arbitrary heating/cooling loads of the systems as proposed by Eskilson [7] are also presented in this paper.

The validation of the models by experiments in the first pilot project of GSHP system in China is carried out. The results of these tests have shown that the monitored data on site agreed quite well with predicted data according to the improved model. The maximum relative error between the predicted and measured data is no more than 2% after the first two hours. The predicted temperatures were always higher than the measured due to the assumption of line source for the borehole. Hence, it is concluded that the entire simulation model is suitable for engineering applications.

References

1. Lund. J, et al. Geothermal (Ground-Source) Heat Pumps-A World Overview. Geo-Heat Center Bulletin, September 2004.
2. Spitler J.D., Ground-Source Heat Pump System Research-past, present, and future. HVAC&R Research, 2005, V11, n2, 165-167.
3. Mei VC, Baxter VD. Performance of a ground-coupled heat pump with multiple dissimilar U-tube coils in series. *ASHRAE Transactions*, 1986; 92 Part 2; 22-25.
4. Bose JE, Parker JD and McQuiston FC. Design/data manual for closed-loop ground coupled heat pump systems. Oklahoma State University for ASHRAE, 1985.
5. Kavanaugh SP. Ground source heat pumps, Design of geothermal systems for commercial and institutional buildings. American Society of Heating, Refrigerating and Air-Conditioning Engineers (ASHRAE), 1997.
6. Caneta Research Inc. Commercial/institutional ground source heat pump engineering manual. American Society of Heating, Refrigerating and Air-Conditioning Engineers (ASHRAE), 1995.
7. Eskilson P. Thermal analysis of heat extraction boreholes. Doctoral thesis, Department of Mathematical Physics, University of Lund, Sweden. 1987.
8. Hellstrom G.. Ground heat storage, Thermal analysis of duct storage systems, Doctoral thesis, Department of Mathematical Physics, University of Lund, Sweden. 1991.
9. Yavuzturk C, Spitler JD, Rees SJ. A transient two-dimensional finite volume model for the simulation of vertical U-tube ground heat exchangers. *ASHRAE Transactions* 1999; 105(2); 465-474
10. Chiasson AD, Rees SJ and Spitler JD, A preliminary assessment of the effects of groundwater flow on closed-loop ground-source heat pump systems, *ASHRAE Transactions*, 106(1) (2000) 380-393.
11. Diao NR, Li QY, Fang ZH, An analytical solution of the temperature response in geothermal heat exchangers with groundwater advection, *Journal of Shandong Institute of Architecture and Engineering*, 2003, 18(3) :1–5.
12. Zeng HY, Diao NR, Fang ZH, A finite line-source model for boreholes in geothermal heat exchangers, *Heat Transfer Asian Research* 2002, 31(7): 558-567.

13. Gu Y, O'Neal DL. Development of an equivalent diameter expression for vertical U-Tubes used in ground-coupled heat pumps, *ASHRAE Transactions* 1998; 104; 347-355.
14. Diao NR, Cui P, Fang ZH. The thermal resistance in a borehole of geothermal heat exchanger, Proc. 12th International Heat Transfer Conference, 2002 August 18-23; France, 339-343.
15. Zeng HY, Fang ZH. A fluid temperature model for vertical U-tube geothermal heat exchangers, *Journal of Shandong Institute of Architecture and Engineering* 2002; 17(1); 7-11.
16. Zeng H Y, Diao N R and Fang Z H, Efficiency of vertical geothermal heat exchangers in ground source heat pump systems, Journal of Thermal Science, 12(1): 77-81, 2003
17. Carslaw HS, Jeager JC. Conduction of Heat in Solids, second ed., Oxford Press, Oxford, 1959.
18. Fang ZH, Diao NR, Cui P, Discontinuous operation of geothermal heat exchangers, *Tsinghua Science and Technology* 2002; 7(2); 194-197.
19. Yu MZ, Diao NR, Su DC, Fang ZH. A pilot project of the closed-loop ground- source heat pump system in China, Proceeding of IEA 7th Heat Pump Conference, 2002 May 19-22; Beijing, 356-364.

FUTURE LIVING – SAFE, SMART AND EFFCIENT USE OF ENERGY

T. F. CHOW
Marketing & Customer Services Business Group
CLP Power Hong Kong Limited

Energy efficiency and indoor air quality have been the focuses for green and sustainable building design project. Whilst energy-efficient air conditioning, water heating and lighting systems are subjects of interest in the commercial and industrial sectors, households also have a growing quest for a safe, healthy and quality living. The preference for a combustion free indoor environment is therefore emerging. This paper discusses the latest technologies in electrical cooking, water heating, clothes drying and indoor air quality (IAQ), and the benefits they bring to consumers.

1. Introduction

The two local power companies, CLP Power Hong Kong Limited and Hongkong Electric Co., Ltd., in conjunction with the Government, launched a three-year Demand Side Management Program (DSM) in 2000 to encourage the use of energy-efficient lighting and HVAC systems. The public's response was overwhelming.

Subsequent to the outbreak of SARS, there are escalating concerns about health and hygiene, and such concerns have extended to building operations. IAQ related audits are now becoming popular. In 2003, the Environmental Protection Department launched a certification scheme to promote good IAQ in offices and public places.

In the residential sector, in association with the promotion of the DSM, compact fluorescent lamps and electrical appliances with energy labels are available at reasonable prices. Consumers can make purchase decisions based on simple cost and benefit calculations. With an increasing demand for a combustion free indoor environment for better safety, health and living quality reasons, more varieties of electrical cookers, water heaters and fresh air units are coming to the market. It is worthy to note that in residential flats, it is very often that for cooking and water heating only a single energy source is provided by the property developers. Should the energy source be gas, the electricity supply capacity may not be sufficient for the residents to switch to electrical appliances.

This paper discusses the latest technologies in cooking, water heating, clothes drying and IAQ, and how these contribute to the Safe, Smart and Efficient Use of Energy.

2. Cooking

Hong Kong citizens are used to flame cooking, which is commonly considered to be the only cooking option for Chinese cuisines. Flame cooking relies on combustion to heat up the cookware to a high temperature. In this process, less than 40% of energy is transferred to the food. The rest is dissipated to the surrounding, resulting in a hot and uncomfortable cooking environment.

Pollutants generated by combustion are potential hazard to health. An article by Department of Community and Family Medicine of The Chinese University of Hong Kong revealed the association of household gas cooking and respiratory illnesses in preschool children [1]. Other epidemiological studies in U.S. also reported the associations between increased respiratory symptoms and the presence of gas cooking [2]. To improve indoor air quality and hence health, an emission-free cooking technology with sufficiently high power is desirable for cooking Chinese cuisines. Electric induction cooker satisfies such wants.

Induction cooker produces a high frequency, alternating current magnetic field through a ceramic-glass cooking surface to reach the cookware. This changing magnetic field induces eddy currents at the ferrous base of the cookware. The energy transferred to the cookware is instantaneous and can be controlled easily. Heat loss to the surrounding is minimal, keeping the kitchen cool. As no combustion is required in the cooking process and with overheat protection, induction cooking reduces fire hazard and combustion pollutants substantially.

In recent years, induction cookers rated above 2kW are brought to the market. These cookers have superior performance and are suitable for all cooking methods including steaming, roasting, boiling, stewing, frying and stir-frying. There is also a variety of induction friendly cookware, with materials ranging from iron, stainless steel, and even clay with plated ferrous material at the bottom.

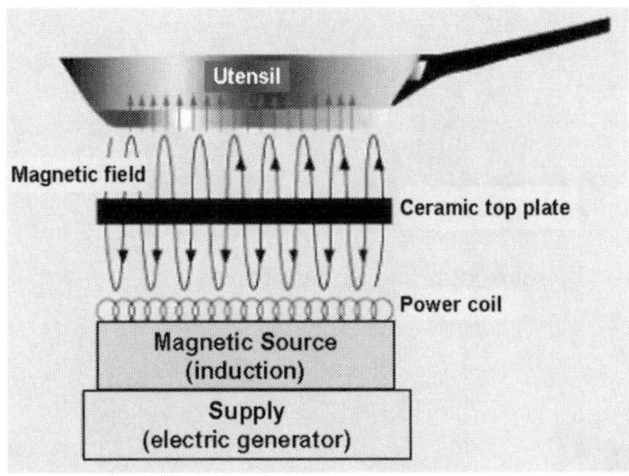

Figure 1. Principle of Induction Cooking

Total : 2.8 kW (shift power)
Right : 2.8 kW
Left : 2.8 kW

Figure 2. Examples of Advanced Induction Cookers

Residential flat sizes in Hong Kong are mostly in the small to medium ranges. Open kitchen concept is a popular interior design to enlarge the perceived living space through integrating the kitchen and dining space together. However, this kind of design does not favor the use of gas cooker, as it would lack fire containment, expose residents to harmful pollutants and create higher air conditioning demand. Instead, no-flame induction cooker, with its excellent safety, health and environmental benefits, is the ideal choice for a flexible and open kitchen design. As a matter of fact, an increasing number of induction cookers have been installed in new apartments. It is believed that induction cooker is entering a high growth era in Hong Kong. As a matter of fact, an increasing number of induction cookers have been installed in new apartments. It is believed that induction cooker is entering a high growth era in Hong Kong.

3. Water Heating

Residential water heaters are broadly classified into storage-type and instant-type. While the storage-type water heater is cheaper in price and requires only single-phase power supply, instantaneous water heater, due to its convenience and less installation space requirement, is the preferred choice in modern homes. Driven by the quest for energy conservation, new water heating technologies are emerging. In the following section, we will discuss an advanced energy efficient instantaneous water heater and two innovative storage water heaters, viz., the Air-Link storage water heater and the integrated heat pump water heater.

3.1 *Instantaneous water heater*

Residential instantaneous water heaters can be powered by either electricity or gas. Three-phase power supply is needed for instantaneous electric water heater because of the high power required. To compare the performance of electric and gas instantaneous water heaters, 21 models of electric and 15 models of gas water heaters were tested by an accredited laboratory in the period from December 2001 to June 2002. The test results indicated that on average, all electric water heaters were found to be 24% more efficient than the gas water heaters. When translated into running cost, the consumers pay 5 to 9% less.

There are other advantages of instantaneous electric water heater. It is compact in size and does not require flue aperture or windows in the bathroom. It can be located anywhere closest to the point of use to give quick and instant hot water supply. Such flexibility allows more freedom for the interior designer to optimize the use of space, in comparison with gas heaters. The electronic controlled electric models also provide quick and accurate stabilization of preset hot water temperature. As it requires no combustion, no thermal combustion pollutants would be emitted to the local environment.

Figure 3. Examples of Apartments with Integrated Kitchens, Dining Rooms and Living Rooms

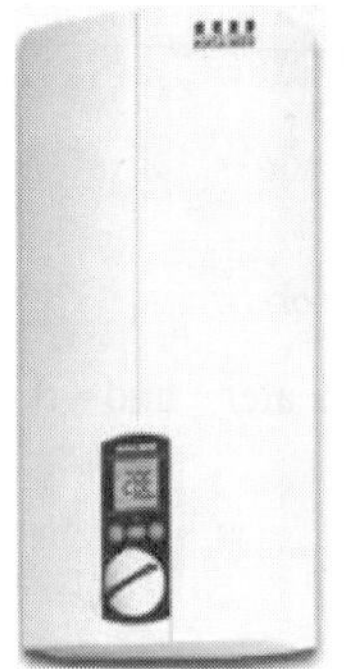

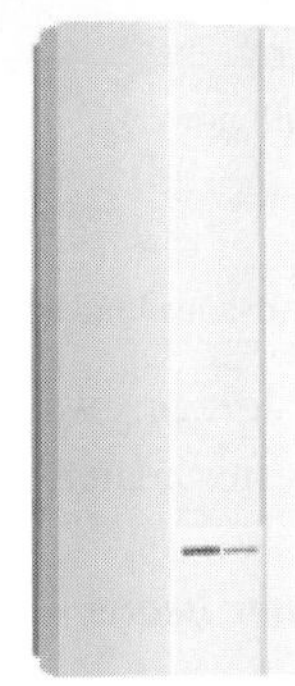

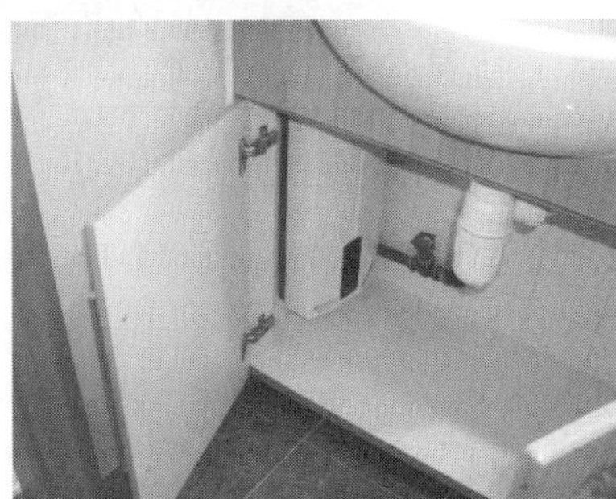

Installations inside cabinets and under sinks

Figure 4. Examples of Instantaneous Electric Water Heaters

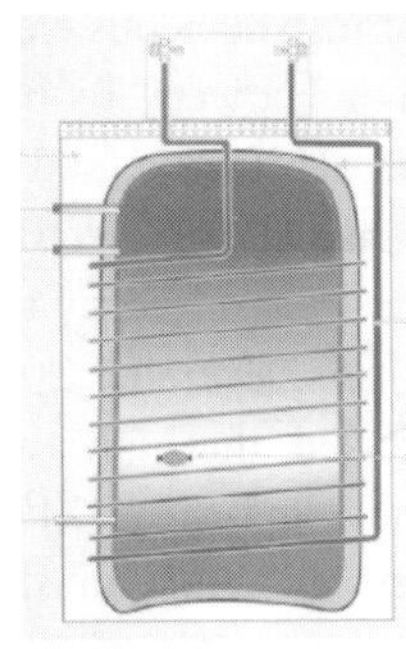

Figure 5. Air-Link Water Storage Tank

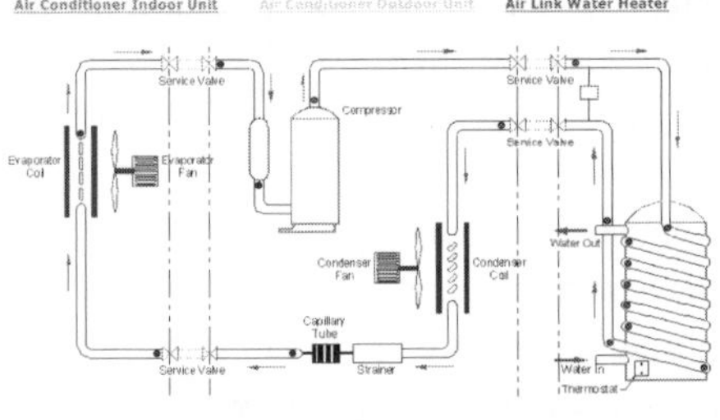

Figure 6. Refrigerant and Water Circuit of Air-Link Storage Type Water Heater

3.2 *Air-link storage water tank*

The air-link water storage tank (as shown in Figure 5) utilises the waste heat from split type air conditioners to raise the temperature of water up to 60°C without consuming additional electricity. The obvious benefits are that the heating energy is free while air conditioning is on, and there is much less hot air discharged into the atmosphere.

The water storage tank is linked to a split-type air conditioning system, which can be explained by using two loops as shown in Figure 6.

Firstly, heat is absorbed from an indoor environment by the evaporator to heat up the refrigerant. Then, the refrigerant evaporates into the compressor where it is compressed and superheated. Next, the superheated compressed refrigerant passes through the copper tube coiled around the water storage tank. The coil itself acts as the condenser and rejects heat energy to the water in the tank by conduction. The refrigerant gas then passes through a condenser where it is liquefied. Upon passing through the expansion valve, the refrigerant is super-cooled and returned to the evaporating unit to complete the cycle.

Secondly, there is a water flow circuit. The circuit stretches from the inlet of the cold water from the water source to the outlet of the hot water for the bathroom or kitchen.

The Air-Link water storage tank can be used to supply hot water for shower, washing, cleaning and other purposes in residential flats as well as commercial premises such as hair and beauty saloons, food courts, and restaurants. The storage capacities of the water tank range from 60 to 340 litre, and can be suited to different cooling capacities.

Despite being a relatively new application in the household market, Air-Link Storage Water Tank has a good market potential due to its higher energy efficiency and environmentally friendly benefits.

3.3 *Integrated heat pump water heater*

Figure 7 shows another innovative type water storage tank – Integrated Heat Pump Water Heater (IHPWH), which consists of a compact sized heat pump and a storage water tank. The heat pump uses ambient air as its energy source and simultaneously generates hot water and cool air for spot cooling in non-air conditioned areas such as the kitchen.

The upper section of the water heater is a removable compact-sized heat pump (600(diameter) x 450mm(height)). Its heating capacity ranges from 2 kW to 4 kW. The compressor of the heat pump compresses the refrigerant vapour and directs the hot pressurised vapour refrigerant to the copper tube coiled around the water storage tank.

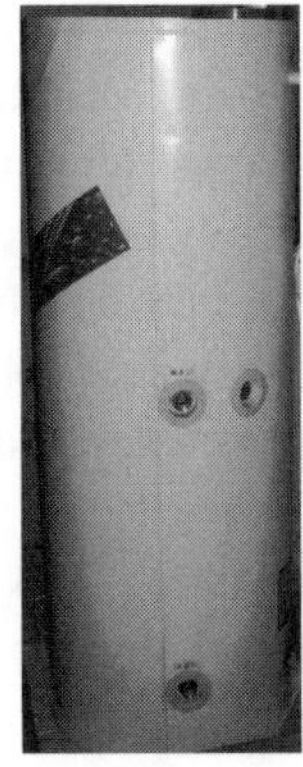

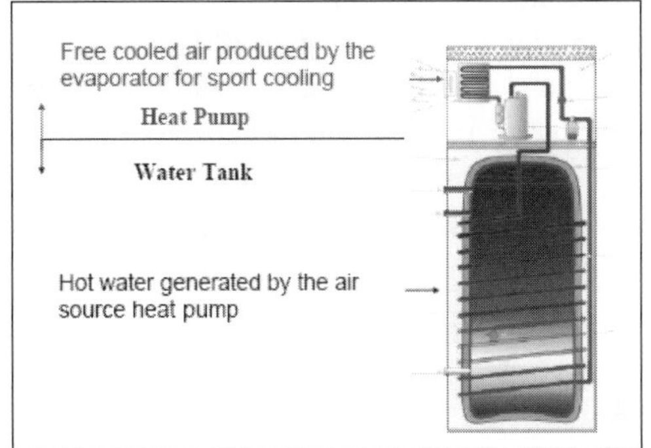

Figure 7. Integrated Heat Pump Water Heater

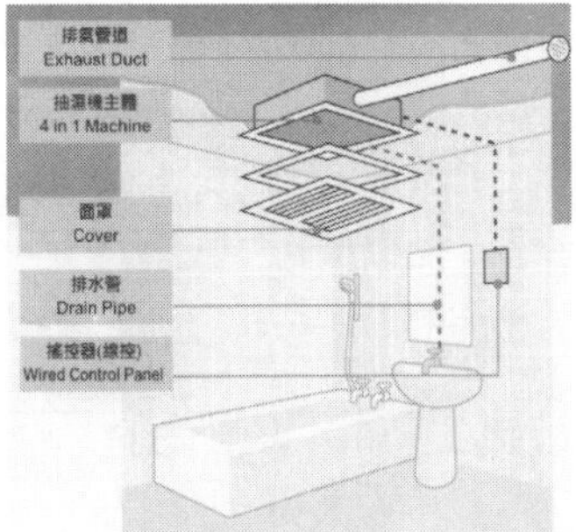

Figure 8. Examples of Multifunctional Dehumidifiers

The coil acts as a condenser and rejects heat energy to the water in the storage tank by way of conduction. The cooled refrigerant is then passed through the liquid receiver and expanded through the throttle valve into an evaporator where heat from the surrounding is absorbed into the vapour. The refrigerant then flows back to the compressor to complete the heat-exchange cycle.

The lower section of the heater is a hot water storage tank with water storage capacity ranged from 120 to 340 litres.

A performance test was conducted by an accredited laboratory for a 2 kW heat capacity of heat pump water heater with a 120 litre water storage capacity. The test showed that in comparison with a typical water storage tank using an electrical water heater, the IHPWH saved 70% of the energy consumption by heating up 120 litres of water from 20°C to 60°C. The average temperature of cooled air generated during the operation of the heat pump was ~18°C and can be used for spot cooling.

The product has good market potential for small to medium sized commercial premises such as hair saloons, saunas, or large residential apartments.

4. Clothes Drying

Due to aesthetics consideration, new residential buildings rarely provide practical facilities for natural clothes drying. Residents generally rely on clothes dryers to remove moisture by heat. This clothes drying process is highly energy consuming and will occasionally cause shrinkage or even damage to clothes fabrics.

Many households in Hong Kong do use dehumidifiers in a close room to dry out clothes at night time. However, traditional dehumidifiers occupy the scarce space inside the house. Another handicap is that the moisture removal capacity of home use dehumidifiers is generally lower than 15 L/day, which may not be sufficient for effective clothes drying, not to mention that the condensate has to be removed manually from time to time.

Some ceiling-mounted or wall-mounted dehumidifiers have therefore been introduced to the market. Drain pipe is provided for condensate removal. The dehumidification capacity of these dehumidifiers is higher than that of conventional dehumidifier. When installed in a bathroom, they also help maintain a dry environment. Some multifunctional dehumidifiers are equipped with heating elements for space heating in bathrooms during cold weather. This feature is highly welcomed by households with children or the elderly as they may easily catch a cold when bathing in winter.

Table 1: Estimated Annual Energy Consumption Comparison of an all electric home and Traditional Home for a Household of 3 Persons

	Energy (kWh)	
	All Electric Home	Traditional Home
Cooking	2,037	4,634 (G)
Water Heating	2,007	2,490 (G)
Clothes Drying	524	812 (G) / 57 (E)
Total	4,568	7,936 (G) / 57 (E) /7,993 (T)

Remark : E – Electricity, G – Gas, T – Total

To compare the technical performance and fuel economy of an electric dehumidifier and gas clothes dryers, tests were carried out by an accredited laboratory in 2003 and 2004. It was found that using multifunctional dehumidifier for clothes drying could save energy and cost by 40% and 32% respectively, albeit a longer drying time. Users who opt for clothes drying by dehumidifier are likely to do the laundry at night when the usage of bathroom is low. The added advantages are that shrinkage of clothes will not happen as the clothes are dried "naturally", and that the relative humidity of the bathroom is maintained at a comfortable level.

5. Electrical Installation for Residential Flats

According to CLP Power Supply Rules 203.2 and 203.3, electricity demand less than 60A is normally supplied using single-phase. Installations requiring more than 60A may require a three-phase supply [3].

The standard of living in Hong Kong is rising. A wider range of high power electric appliances that increase the level of comfort and living quality are brought into the market, such as the aforesaid induction cookers, instantaneous electric water heaters and multifunctional dehumidifiers. These trendy items, together with other basic appliances like air conditioners and washing machines, add to the total electrical loading. A single-phase supply to these households may impose restriction on the residents on how many appliances they would like to install.

A residential building generally lasts over 30 years. A good design of the fixed electrical network in a flat should enable the residents to use a wide range of modern electric appliances and ensure adequate capacity for future needs, avoiding costly and cumbersome modification / reinforcement works afterwards.

We therefore recommend developers, architects and E&M consultants to adopt three-phase electricity supply for flats in new residential developments.

6. Benefits to End Users

The above sections demonstrate the superior performances of induction cooker, instantaneous electric water heater, air-link water heater, integrated heat pump water heater and multifunctional dehumidifier in safety, convenience and energy efficiency aspects, as well as a higher living standard. Sophisticated control and protection features give the users a peace-of-mind. Most importantly, the reduction of local gaseous emissions, such as particulate, carbon monoxide contributes to an improved community environment.

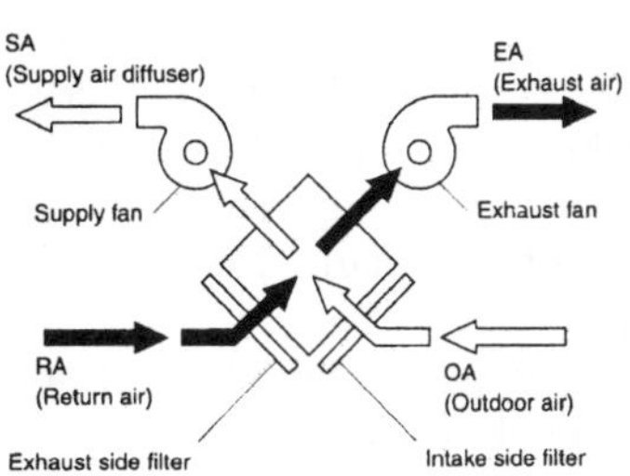

Figure 9. Energy Recovery Mechanism of Fresh Air Unit

Figure 10. Fresh Air Unit Installed in School

Cost saving is also a key driver. In this regard, a calculation was conducted to compare the annual energy consumption of an all electrical home and Traditional Home for a household of 3 persons. The calculation result is shown in Table 1.

For a typical household of 3 persons, an electric home could reduce the total energy consumption in cooking, water heating and clothes drying by 43%. Based on the latest energy tariff, the resulting fuel cost saving can be up to 36%, equivalent to about $2,500 per annum.

7. Indoor Air Quality (IAQ)

Awareness of IAQ has heightened significantly after the SARS incident. The Environmental Protection Department has also published guidance notes to promote good IAQ in offices and public places.

One of the most effective ways to improve IAQ is to provide adequate fresh air to dilute the concentration of pollutants inside an occupied area. To do so, additional energy is needed to condition the hot and humid outdoor air. Energy saving can be achieved by using heat recovery ventilators, in which a fraction of the energy is recovered from the exhaust cool air to the supply hot air (as shown in Figure 9).

There are two commonly used techniques for heat recovery ventilators - sensible heat exchanger and energy wheel. The sensible heat exchanger is simple and reliable, but it cannot recover latent energy. Energy wheel, which can save latent energy, has maintenance issues due to the rotating moving parts.

An alternative is the Fresh Air Unit (FAU), which uses hydroscopic material as the heat transfer medium. In addition to heat transfer, such material has selective hydroscopicity and moisture permeability that permits the passage of water vapour only. This can help save both the sensible and latent load. The FAU is widely applied in primary and secondary schools (as shown in Figure 10). In comparison with using a ventilation fan to supply the same amount of outdoor fresh air to the classroom, FAU can save about 40% of energy consumption.

Household residents usually close windows and switch on the air conditioning unit during summer night. In particular, for those flats in the rural area, residents normally close all the windows to prevent contracting diseases such as Dengue Fever, Japanese Encethalitis. In such situation, FAUs with energy saving features would be an efficient way to provide adequate fresh air to dilute the concentration of pollutants such as carbon dioxide generated by the residents. With advanced technology, FAU has been developed into a very compact size that can be installed for residential flats as shown in Figure 11. The supply air flow rate ranges from 65–105 m^3/hr.

Figure 11. Fresh Air Unit for Household

8. Conclusion

This paper discusses the energy efficiency, safety, health, economy, and environmental benefits of a number of home-used products, including induction cookers, electric storage water heaters, instantaneous water heaters, multifunctional dehumidifiers and fresh air units. They are the essential components for a safer, healthier and cleaner home environment with great cost saving. We therefore recommend developers, architects and building services consultants to make these provisions in new residential development projects to give consumers a better choice and create a better living environment for Hong Kong

References

1. T W Wong, T S Yu, H J Liu, and A H S Wong. Household gas cooking: a risk factor for respiratory illnesses in preschool children. Archives of Disease in Childhood, Jul 2004; 89: 631 - 636.
2. Roy Fortmann, Peter Kariher, and Russ Clayton. Indoor Air Quality: Residential Cooking Exposures. Final report prepared for State of California Air Resources Board. November 2001. Introduction, pp.1.
3. CLP Power Hong Kong Limited, March 2001, Supply Rules, pp. 6.

IMPLICATIONS OF SOLAR CHIMNEY SIZE FOR VENTILATION

MATTHIAS HAASE†
Department of Architecture, Hong Kong University, Pokfulam Road
Hong Kong, HKSAR, China

ALEX AMATO
Department of Architecture, Hong Kong University, Pokfulam Road
Hong Kong, HKSAR, China

The Hong Kong climate is sub-tropical with hot and humid weather from May to September and temperate climate for the remaining 7 months period. A mechanical ventilation and air conditioning (MVAC) system is usually operated to get rid of the high peak cooling loads. The support of the MVAC system by means of natural ventilation are in the summer period limited due to the very high humidity. There is nevertheless a possibility for supporting the mechanical ventilation system by using the buoyancy effect in a 'solar chimney' to extract exhaust air. A 'solar chimney' is a chimney integrated in the façade with Photovoltaic (PV)-arrays on the outside for electricity production. High-rise buildings of 35 stories and above are common in all building sectors in Hong Kong, providing adequate pressure differences, ensuring the efficacy of solar chimneys. In this report different dimensions of a solar chimney and its implications for extracting air from an office building were investigated. It could be demonstrated that the geometry of solar chimneys with a fixed airflow are more efficient when designed with maximum width and depth. From the examined sizes it could be shown that a width of 1.5m and a depth of 1.5m are the most appropriate geometries for a solar chimney with a fixed airflow rate of 48l/s.

1. Introduction

1.1 *Hong Kong climate*

The Hong Kong climate is sub-tropical with hot and humid weather from April to September and temperate climate for the remaining 6 months (winter) period. Detailed analysis of the weather conditions and their main characteristics have been reported [1-4]. The results indicate a hot and humid climate which makes cooling evitable and limits the use of natural ventilation [5,6].

A mechanical ventilation and air conditioning (MVAC) system is usually operated to get rid of the cooling loads and consists of several components. Fig. 1 shows the typical energy consumption distribution of a commercial building in Hong Kong. It also gives an overview of the different components of the MVAC system and its part of energy consumption within the building.

1.2 *Hong Kong buildings*

The ventilation in the air conditioning system uses 10.9% of the energy consumption in office buildings in Hong Kong [7]. This indicates the huge potential of energy savings by enhanced natural ventilation.

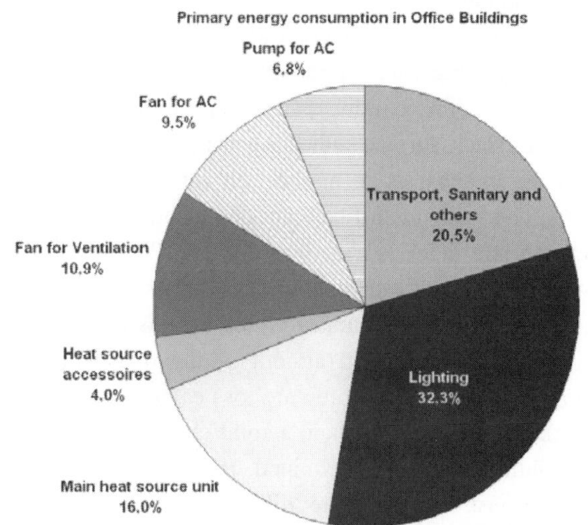

Figure 1. Energy consumption in commercial buildings in Hong Kong [7]

A Photovoltaic (PV) solar stack has been attached to a high-rise building of 35 floors which is a common height in all building sectors in Hong Kong. The

† contact tel.: ++(852) 2241 5839, email address: mathaase@hkusua.hku.hk

electricity production of an installed PV solar stack has been reported [8].

Previous studies focused on evaluating the yield of the PV solar stack. But even with the additional wind behind the modules no significant rise in electricity outcome could be shown [9]. Another possibility to add value to the solar stack is by making use of the buoyancy effect in the stack. The idea of a solar chimney is not new. Bansal and others developed the idea of solar chimneys since 1992 [10-12]. But combining a PV stack with the buoyancy effect of a solar chimney has not been investigated before.

Subsequently a model with a PV solar stack has been attached to a high-rise building of 35 floors which is a common height in all building sectors in Hong Kong. The buoyancy effect was analyzed under steady weather conditions using Computational Fluid Dynamics (CFD) analysis [13].

It indicated a possibility for supporting the mechanical ventilation system by using the stack effect to draw air out of the building. This possibility has also been examined by Wong et al. for a double-skin façade system [14]. Priyadarsini et al. modeled a single-floor building with a wind supported chimney and tested it in the wind tunnel [15].

Recent work focused on an analysis of the impact of the dynamic weather pattern consisting of wind over the top of the chimney and solar radiation on the vertical stack surface indicating that the solar chimney has advantages as an application for office buildings [16]. The study identified 0.6m depth and 1.5m width as appropriate stack dimensions for the PV faced solar stack. These dimensions were also used in further studies for a solar chimney which acts as a ventilation chimney drawing used air out of the office building [17]. By applying the pressure loss calculation for duct work the pressure losses in a solar chimney attached to a building could be determined. It reports that for 50% of the year the pressure differential created by the combined natural forces at the top of the stack is 8Pa during the 8am-8pm day. This indicated sufficient pressure to ventilate the attached rooms with minimum air change rates. This revered to a room with length, width and height of

$$l_R = 8m; w_R = 3m; h_R = 2.4m$$

following the area and volume and thus the airflow rate Q to

$$A_R = l_R \times w_R = 24m^2 \quad (0.1)$$

and

$$V_R = A_R \times h_R = 57.6m^3 \quad (0.2)$$

$$\dot{Q}_R = q_P \times p = 48l/s \;\square\; 0.048m^3/s \quad (0.3)$$

with

$$q_p = 8 \text{ l/s/p} \quad (1.4)$$

$$p = 6 \text{ persons} \quad (1.5)$$

(data taken from ASHRAE [18])

1.3 *Objective of this study*

This paper investigates different dimensions of a solar chimney and its implications for extracting air from an office building. A comparison of the pressure gains due to solar radiation with the pressure losses gives the possibility to further evaluate the potential for enhanced ventilation in buildings in a hot and humid climate. Further integration of the solar chimney to the HVAC system depends on the capacity of the solar chimney to work as an extract duct. This could allow reductions of the ventilation system size and also reduce the energy consumption. The implications of the geometry to the pressure distribution in the solar chimney can be established. This will help to further develop a hybrid ventilation system.

2. Methodology

The study uses the pressure differences calculated by Close et al. as the natural forces which drive the enhanced natural ventilation and the pressure losses due to airflow through duct work [16,17].

Figure 2 shows the pressure losses for different airflow rates over the height of the building. It shows that higher airflow rates tend to increase the pressure losses due to the higher air velocities.

The results are highly dynamic. While the pressure gains depend upon the dynamic processes of wind and sun the calculation of the pressure losses in the chimney depend upon the velocity.

By applying the results of the pressure loss calculations to the calculation of pressure gains due to solar radiation the potential for enhanced natural ventilation can be calculated. This work focuses on different geometries in order to determine the size of solar chimneys that are strong enough to “drive” the ventilation.

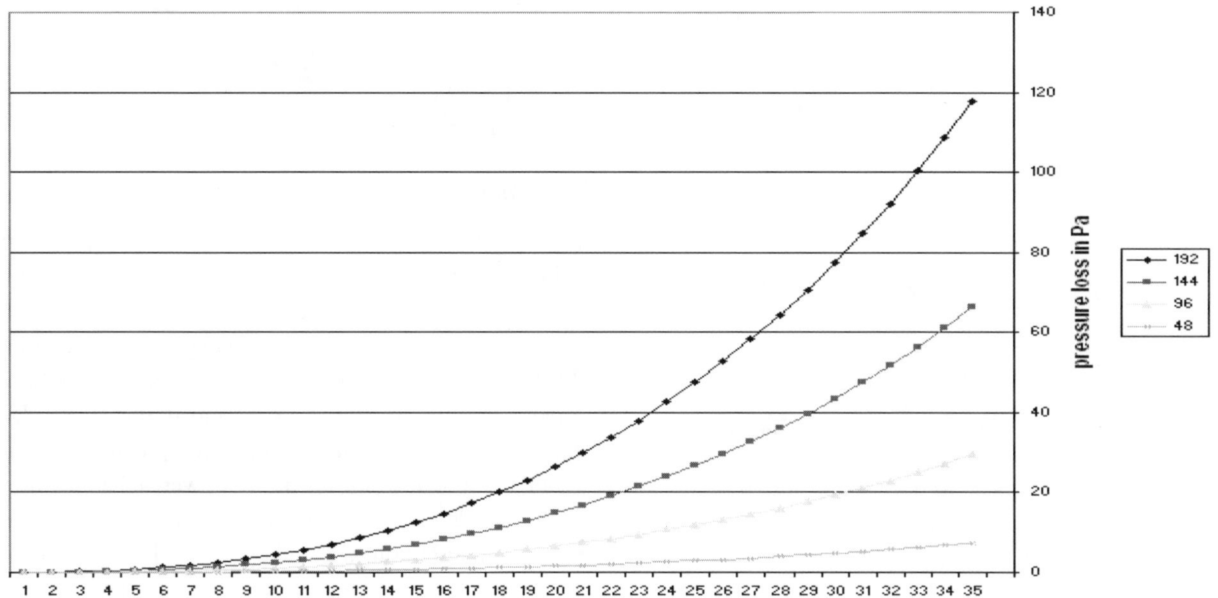

Figure 2. Pressure losses for different airflow rates over height

2.1 *Calculation of pressure difference gain from natural forces*

The procedure for calculating the pressure gain has been described in Close et al. in detail. The formulas for pressure gains were found to be:

$$\Delta p_{gain} = \Delta p_{sun} + \Delta p_{wind} \quad (2.1)$$

with

$$\Delta p_{wind} = \frac{1}{2}\rho v^2 \quad (2.2)$$

$$\Delta p_{sun} = gh\Delta\theta_{avg}\rho / T_{amb} \quad (2.3)$$

with

$$\Delta\theta_{avg} = \frac{\Delta\theta}{2} \quad (2.4)$$

$$\Delta\theta = \frac{P_1}{c_p Q\rho} \quad (2.5)$$

$$P_1 = P_0(1-\eta)(Ar_b + t) \quad (2.6)$$

$$P_0 = Gwh \quad (2.7)$$

$$\Rightarrow \Delta p_{sun} = \frac{gh^2wG(1-\eta)(Ar_b + t)}{2c_p QT_{amb}} \quad (2.8)$$

with v = wind velocity [m/s]
ρ = density of air = 1.2kg/m³
h = floor height = 3.4m
T_{amb} = outside temperature = 300K
g = gravity = 9.81 kg/
cp = specific heat capacity of air = 1006 J/kg
n = efficiency of solar cells = 0.2
rb = reflection coefficient = 0.3
A = absorption factor = 0.11
t = transmission of solar cell = 0.8
G = solar radiation [W/m²]

2.1.1 *Pressure losses and airflow*

As can be seen from the above formula the pressure gain from solar radiation depends on the airflow Q. The larger the airflow the smaller will be the amount of pressure gain from solar radiation.
If the solar chimney is meant to be used as an exhaust chimney for a room in each floor the amount of airflow rate Q will increase with each floor and therefore with building height.

2.2 *Calculation of pressure difference losses*

The pressure losses in ductwork can be calculated with

$$\Delta p_{total,i} = \Delta p_{loss,i} + \Delta p_{fitting} \quad (2.9)$$

with

$\Delta p_{fitting}$: losses in the chimney inlets, taken from [19]

$$\Delta p_{loss,i} = f_c \times \frac{l}{d_{hydr}} \times \Delta p_{v,i} \quad (2.10)$$

with:

$$\Delta p_{v,i} = \frac{1}{2} \times \rho \times v_i^2 \quad (2.11)$$

and

$$\Delta p_{fitting} = C \times \Delta p_v \quad (2.12)$$

with
C = fitting loss factor, taken from [20]

The losses depend on the velocities which again depend on the amount of air flowing through the duct. With an active hydraulic diameter of the stack of

$$d_{hydr} = \frac{2 \times a \times b}{(a+b)} \quad (2.13)$$

with
a = solar chimney width
b = solar chimney depth
the actual velocities and subsequently the pressure losses can be calculated.

With the introduction of a overall loss factor K it is possible to rewrite the losses

$$K_i = \frac{\Delta p_{loss,i}}{Q_i^2} \quad (2.14)$$

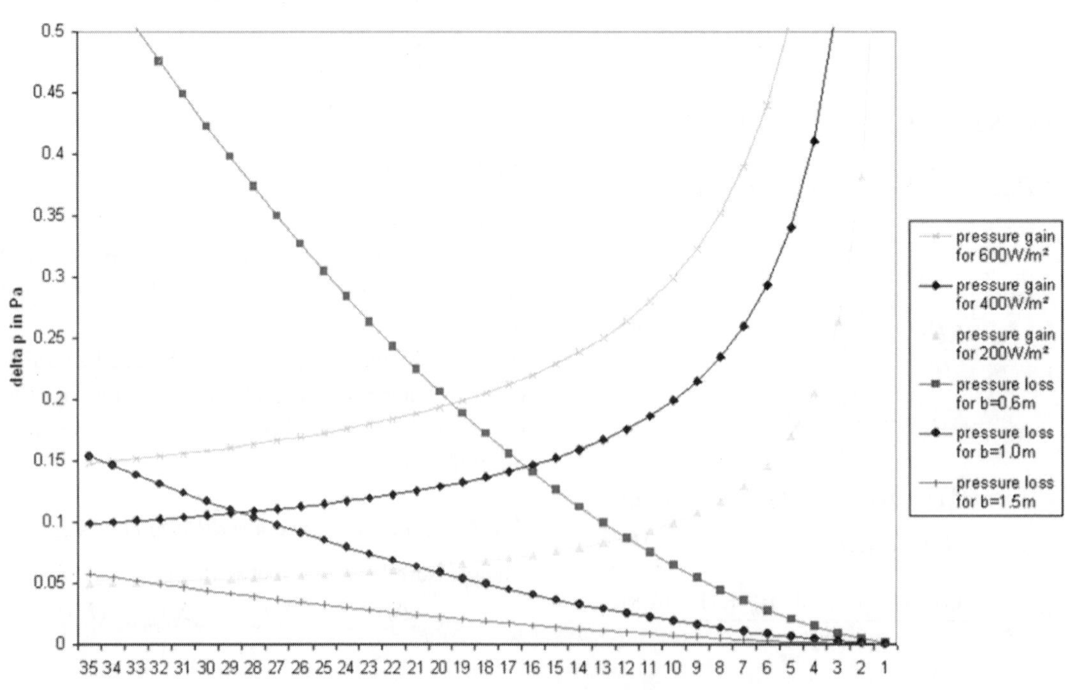

Figure 3a. Pressure loss and gain for chimney size a=1.5m

with
Q_i: airflow rate
K_i: overall loss factor in each floor

2.3 *Calculation of natural forces potential*

In order to be able to access the potential for solar chimneys it is important to understand the influence of a hybrid mode for the chimney.

Previous models of solar chimneys tried to describe the velocities in the chimney as a result of natural forces. In this work it was assumed that the solar chimney is part of the ventilation system. A fixed ventilation rate of Q=48l/s per floor were taken in order to examine the supporting influence of the sun on the pressure distribution in the chimney [17]. This leads to increasing airflow rates over the height of the building. A series of calculations were done in order to determine the friction losses in the chimney depending on the geometry of the chimney.

Then the influence of the geometry on the airflow rate and subsequently on the pressure gain were determined. Table 1 gives an overview of the different geometries that were examined.

Table 1. Examined different geometry of solar chimney and solar radiation

Solar rad. G	width a	depth b
W/m²	m	m
200	0.6	0.6
400	1.0	1.0
600	1.5	1.5

3. Results

There are two main results. One is showing the pressure distribution over the height of the building. The other is concentrating on the cumulated pressure on top of the chimney.

It can be seen from Fig. 3a to 3c that a solar chimney with a width of a=1.5m has higher pressure gains than chimneys with a width of a=1.0m and a=0.6m.

At the same time it provides the smallest amount of pressure losses. It should be noted that the pressure gains are much higher in the lower floors between floor 1 and 10. After that the pressure gains become less due to the increase in airflow rate. At the same time the pressure losses increase with an increase of airflow rate.

A chimney width of a=1.5m is providing more area for solar radiation than chimneys with smaller widths. The depth of the chimney does not increase the pressure gains but decreases the pressure losses due to an decrease of velocity.

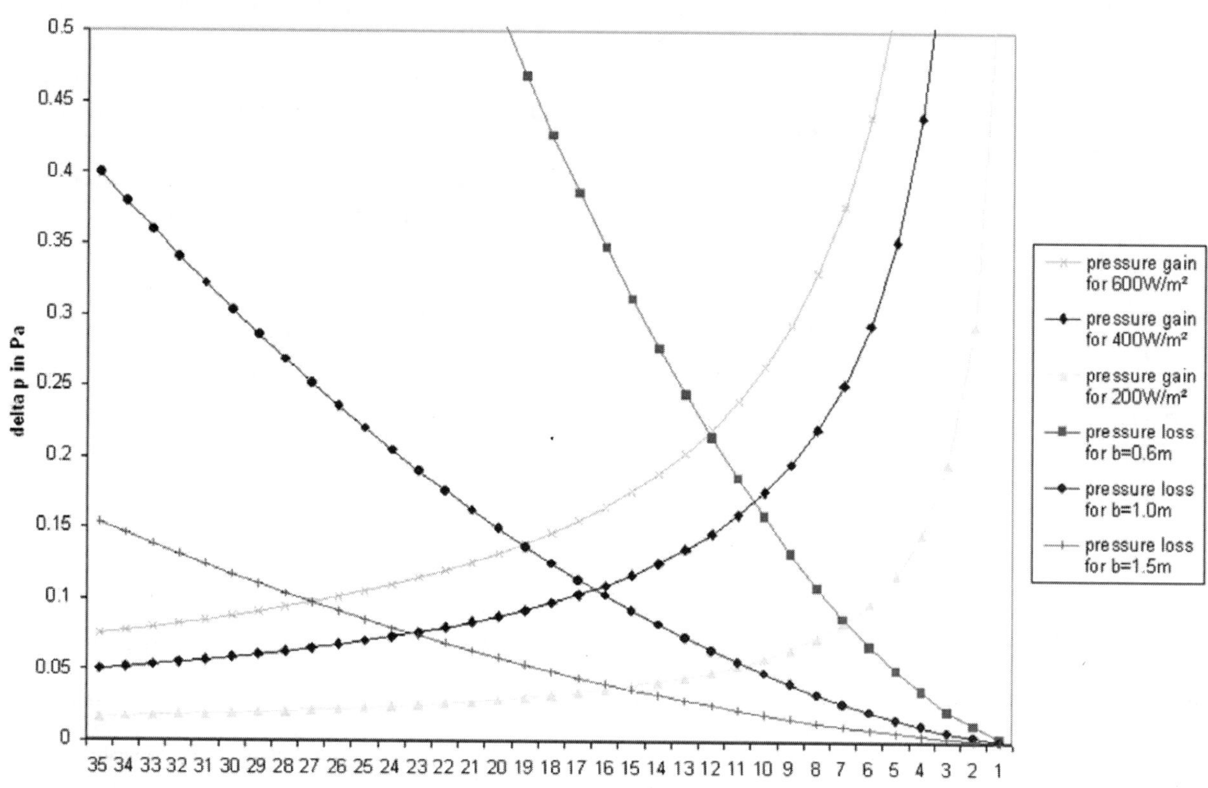

Figure 3b. Pressure loss and gain chimney size a=1.0m

3.1 *Pressure distribution*

The results of the study are given in Fig. 3a-c. It can be seen that the geometry of the solar chimney influences the pressure gain and losses. The solar chimney has to have a certain size in order to be able to contribute to the MVAC system.

3.2 *Cumulated pressure at chimney top*

The cumulated pressure at the top of the chimney was determined and the results are shown in Tab. 2.

It can be seen that the cumulated pressure is highest for a chimney with a width of a=1.5m and a depth of b=1.5m. A chimney with a depth of b=0.6m can provide more cumulated pressure gain than pressure loss for a depth of 1.5m and a solar radiation of 400W/m².

A chimney with a depth of b=1.0m and a width of a=1.0m can supply more pressure than its losses with a solar radiation of G=400W/m².

A chimney depth of b=1.5m and a width of a=1.0m can give a positive pressure at the top of the chimney at 200W/m².

Table 2. Cumulated pressure on top of chimney

cumulated pressure at chimney top in Pa			
b=0.6m	G in W/m²		
a	200	400	600
0.6m	-59.66	-58.20	-56.74
1.0m	-15.94	-11.07	-7.42
1.5m	-3.41	0.54	4.49
b=1.0m			
a			
0.6m	-16.91	-15.45	-13.99
1.0m	-2.87	2.00	5.65
1.5m	1.88	5.83	9.78
b=1.5m			
a			
0.6m	-5.90	-4.44	-2.98
1.0m	0.37	5.23	8.88
1.5m	3.16	7.11	11.05

4. Conclusions and discussions

When determining the chimney geometry for a given airflow rate Q the pressure losses and gains can be calculated and it can be shown that a wider and deeper chimney is beneficial. While a wider chimney provides a higher pressure gain due to an increase in exposed area the chimney with a greater depth reduces the pressure losses and thus provides higher pressure gains at the chimney top.

A small chimney tends to be easier to integrate into the architectural design but for an airflow rate of Q=48l/s it is not advisable to apply.

The assumed 48l/s are a minimum ventilation rate and become much higher when cooling load is to be removed. The highly dynamic wind pattern on top of the chimney were not taken into consideration in this work. But they tend to increase the pressure difference and thus are beneficial for the exhaust airflow in the chimney.

Future work will focus on the wind forces and the dynamic behavior of the natural forces.

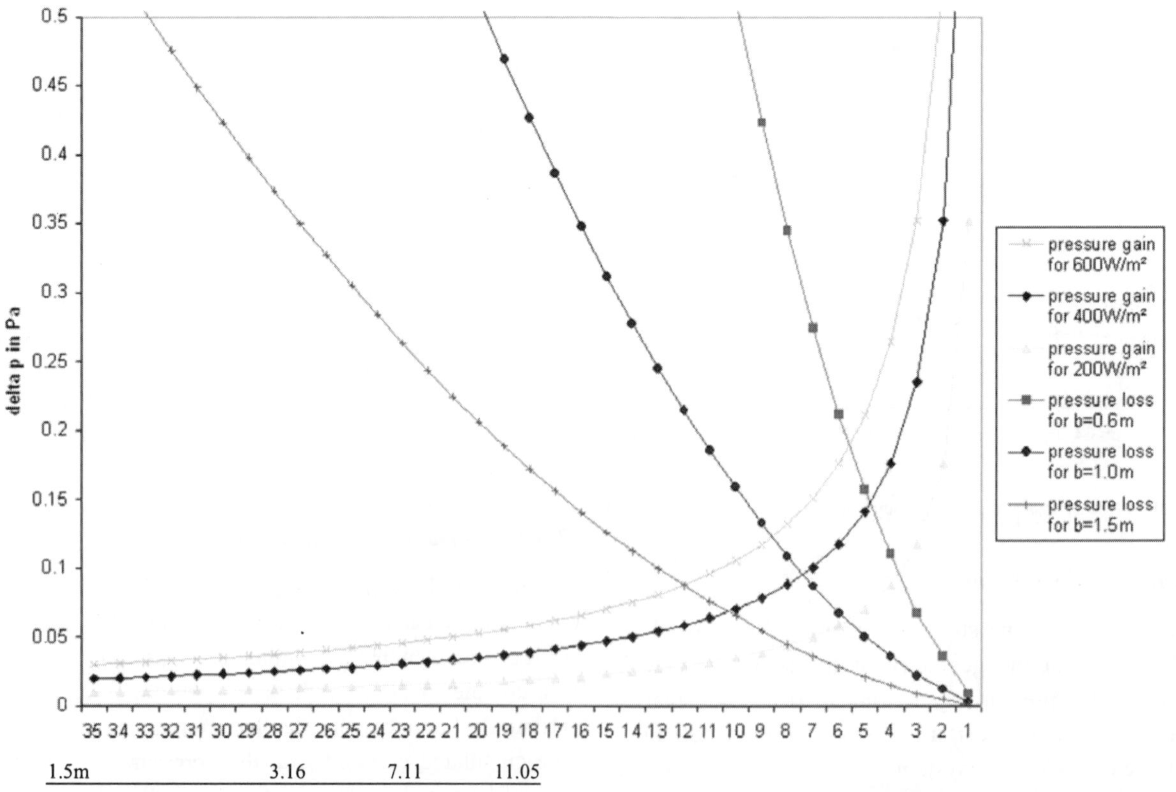

Figure 3c. Pressure loss and gain for chimney size a=0.6m

References

1. [1]J.C. Lam, S.C.M. Hui, Outdoor design conditions for HVAC system design and energy estimation for buildings in Hong Kong, Energy and Buildings 22 (1995) 25-43.
2. [2]J.C. Lam, D.H.W. Li, Study of solar radiation data for Hong Kong, Energy Conversion and Management 37 (1996) 343-351.
3. [3]J.C. Lam, C.L. Tsang, D.H.W. Li, Long term ambient temperature analysis and energy use implications in Hong Kong, Energy Conversion and Management 45 (2004) 315-327.
4. [4]D.H.W. Li, J.C. Lam, A study of atmospheric turbidity for Hong Kong, Renewable Energy 25 (2002) 1-13.
5. [5]J.C. Lam, Building envelope loads and commercial sector electricity use in Hong Kong, Energy 20 (1995) 189-194.
6. [6]J.C. Lam, D.H.W. Li, S.O. Cheung, An analysis of electricity end-use in air-conditioned office buildings in Hong Kong, Building and Environment 38 (2003) 493-498.
7. [7]J.C. Lam, Energy analysis of commercial buildings in subtropical climates, Building and Environment 35 (2000) 19-26.
8. [8]K.H. Lam, J. Close, E.W.C. Lo, Construction of two 25m tall vertical arrays with 2nd generation amorphous silicon photovoltaic technologies, in: 17th European Photovoltaic Solar Energy Conference and Exhibition (Munich, Germany, 2001).
9. [9]T.T. Chow, J.W. Hand, P.A. Strachan, Building-integrated photovoltaic and thermal applications in a subtropical hotel building, Applied Thermal Engineering 23 (2003) 2035-2049.
10. [10]N.K. Bansal, R. Mathur, M.S. Bhandari, Solar Chimney for Enhanced Stack Ventilation, Building and Environment 28 (1993) 373-377.
11. [11]G.S. Barozzi, M.S.E. Imbabi, E. Nobile, A.C.M. Sousa, Physical and Numerical Modeling of a Solar Chimney-Based Ventilation System for Buildings, Building and Environment 27 (1992) 433-445.
12. [12]G. Gan, S.B. Riffat, A numerical study of solar chimney for natural ventilation of buildings with heat recovery, Applied Thermal Engineering 18 (1998) 1171-1187.
13. [13]J. Close, Y.H. Wong, W.C. Lee, A.T.Y. Chan, CFD analysis of the potential updraft from a vertical PV array to provide a secondary energy source in the high-rise context, in: PV in Europe: PV Technology to Energy Solutions Conference (Rome, Italy, 2002).
14. [14]N.H. Wong, S. Heryanto, The study of active stack effect to enhance natural ventilation using wind tunnel and computational fluid dynamics (CFD) simulations, Energy and Buildings 36 (2004) 668-678.
15. [15]R. Priyadarsini, K.W. Cheong, N.H. Wong, Enhancement of natural ventilation in high-rise residential buildings using stack system, Energy and Buildings 36 (2004) 61-71.
16. [16]J. Close, M. Haase, H. Pang, "Added value" for a PV-stack, in: World Renewable Energy Congress (WREC-VII) (Denver, USA, 2004).
17. [17]M. Haase, J. Close, H. Pang, Natural ventilation with solar chimneys, in: 2nd CIB Student Chapters International Symposium, Sustainability and Innovation in Construction and Real Estate (Tsinghua University, Beijing, China, 2004).
18. [18]R.H.R.H. Howell, 1935-, Principles of heating, ventilating and air conditioning : a textbook with design data based on the 1997 ASHRAE handbook-fundamentals (American Society of Heating, Refrigerating and Air-Conditioning Engineers,, Atlanta, Ga. :, 1998) 1 v. (various pagings) :.
19. [19]R.R.R.R. Johnson, 1948-, Fundamentals of HVAC systems (American Society of Heating, Refrigerating and Air-Conditioning Engineers Inc.,, Atlanta, GA :, c1999.) 1 v. (various pagings) :.
20. [20]Z.D. Chen, P. Bandopadhayay, J. Halldorsson, C. Byrjalsen, P. Heiselberg, Y. Li, An experimental investigation of a solar chimney model with uniform wall heat flux, Building and Environment 38 (2003) 893-906.

BUILDING DEVICES CONTROL SYSTEM BASED ON ETHERNET AND CAN AND ITS APPLICATION IN LIGHTING SCENE PROJECT

AN XU

School of Electronic and Information Engineering, Tongji University, 1239 Siping road Shanghai 200092, China

ZHENFU ZHANG

School of Electronic and Information Engineering, Tongji University, 1239 Siping road Shanghai 200092, China

JUN LIU

School of Electronic and Information Engineering, Tongji University, 1239 Siping road Shanghai 200092, China

RUTONG LI

School of Electronic and Information Engineering, Tongji University, 1239 Siping road Shanghai 200092, China

CAIYUN DING

School of Electronic and Information Engineering, Tongji University, 1239 Siping road Shanghai 200092, China

The feature of building devices control system can be described as multi-point, exact demands of synchronization and real-time control, dynamic connections among the control points and so on. Aim at it, this paper presents a complex control system based on Ethernet and CAN (Control Area Network). Hardware and software of the field-controllers, area-controllers and control center are designed in the paper. In CAN Field-bus, field-controllers are in charge of receiving instructions from the host computer, showing real-time control effect, obtaining fault information of lighting devices and sending it to the host computer. Area-controllers realize the conversion between the data transmission protocol of Ethernet and that of CAN Field-bus. Meanwhile, it makes the data transmission between host computer and field-controllers transparent. The host computer will calculate the matrixes in control instructions, send them to the field-controllers, change the control strategy according to the fault information from field-controllers and send the alarm signal out as well. The whole system is designed as a modularization configuration, flexible, easy to be integrated and extended. Alarm function is introduced into the control system, in which the alarm message is transmitted by GSM wireless network. It can also receive remote wireless control instructions. The project has been applied into a 29-story building's lighting scene project successfully.

1. Introduction

With the accelerating process of urbanization and construction, increasing request of intelligent building control, a controllable, reliable, and stable building control system is developed to realize the intellectual, effective and modern building management. However, most of the building control systems used now are made by abroad companies without intellectual property, which are high cost, little finished product, difficult to integrate and redevelop.

This paper presents a complex control system based on Ethernet and CAN, which makes use of advanced computer technology and field bus technology with independent intellectual property. Meanwhile, our products are low in cost, stabile in system, flexible in structure, easy to be integrated and secondary exploiture, convenient to be applied in various building device control system.

This system has been applied into a building's lighting scene project successfully.

2. The general design of system

According to the feature and requirement of building devices control, the complex control system based on Ethernet and CAN in this paper, takes the Ethernet as the main network of the control system, which ensures the system has high data transfer bandwidth, and chooses CAN network on the field, which makes the wire arrangement easier, cuts down the cost, and also gets reliability and high transfer speed between the two network, area-controller is installed to complete the conversion between the data transmission protocol of Ethernet and that of CAN control network. In the topside the industry controlling computer is adopted as the host computer. The system structure shows as follow: (see Figure 1)

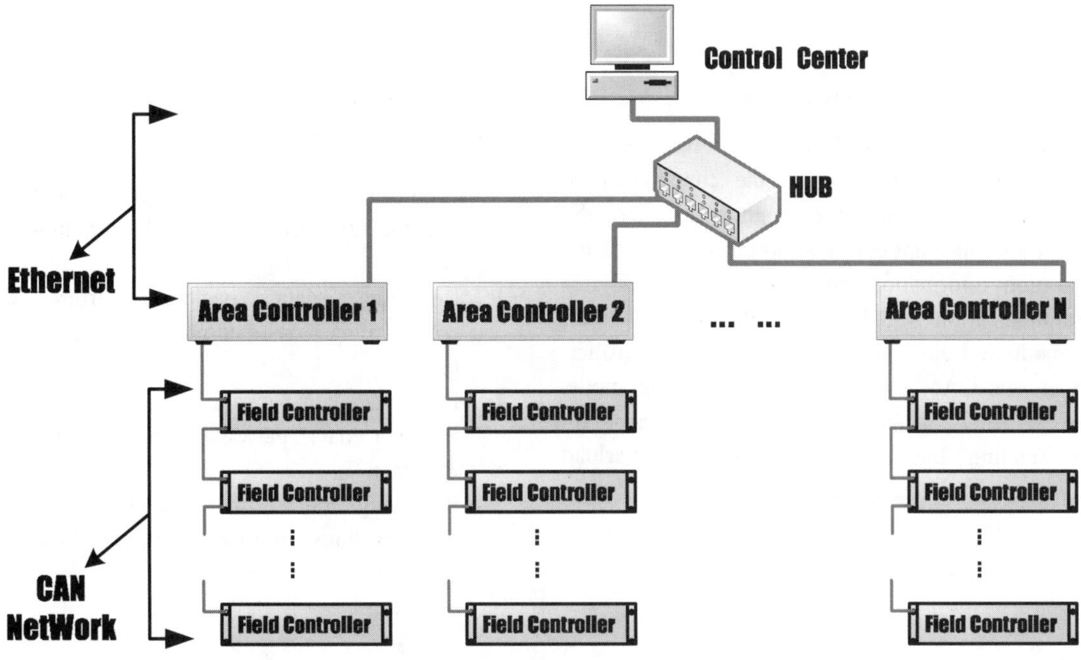

Figure 1. The structure of the building devices control system

2.1 *Field-controller*

2.1.1 *Field-controller hardware design*

As the ultimate part to execute the host computer's control order, field controller has such functions as follow:

1. CAN interface module, which completes the data conversion with CAN network
2. Control module, which controls the switch or adjustment according to the command from the host computer
3. Detection module, which detects the situation of the device, and reports the fault message
4. Serial port communication module, which burns the program into the flash memory via the serial port
5. EEPROM, saving the configuration parameter

The structure of field-controller show as follow. (see Figure 2)

2.1.2 *Fault detection hardware design*

In the building device control system, fault detection in the system is a reliable guarantee of system maintenance and running. The keystone in the fault detection design is reliability and accuracy. The following figure is the signal flow chart of field-controller detection part. (See Figure 3)

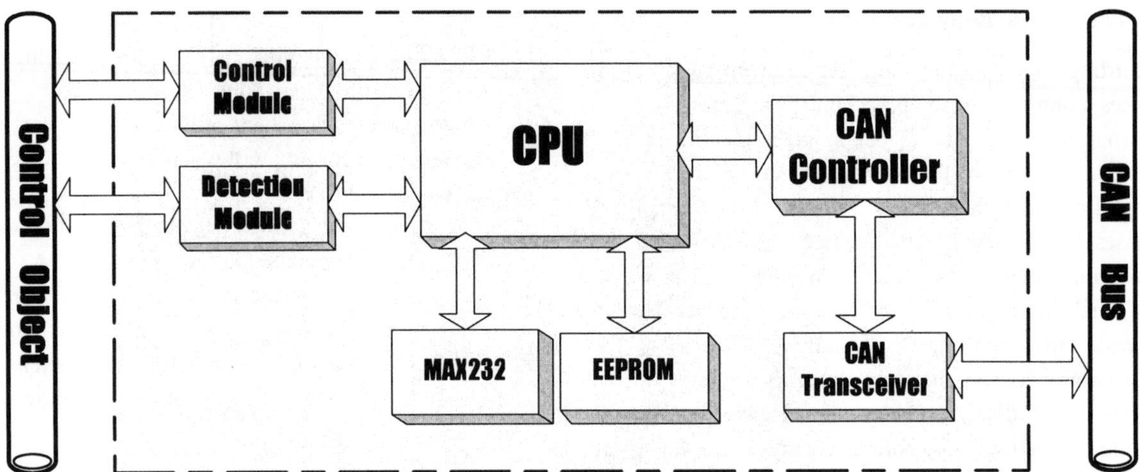

Figure 2. The structure of field-controller

2.1.3 *Field-controller Software design*

As a node of CAN bus, field controller receives data from CAN, sends data, disposes the communication fault and overload. Firstly, the program initializes the serial port and interrupt configuration. Then it reads the configuration information from EEPROM, configures the communication speed, check filter, shield filter, output mode and clock divided of the CAN controller. Finally it sets CAN controller in the working mode, waiting for receiving data. In this mode, the program detects sending interrupt, fault interrupt, overload interrupt and fault information in turn.

2.2 *Area-controller*

2.2.1 *Area-controller hardware design*

Area controller mainly takes charge protocol transition between the Ethernet and CAN. Its functions listed as follow:

1. Receiving data packages from Ethernet, transforming CAN data packages and sending them to CAN;
2. Receiving data packages from CAN, transforming Ethernet data packages and sending them to Ethernet;
3. Receiving control commands and parameter configurations from the main control computer and

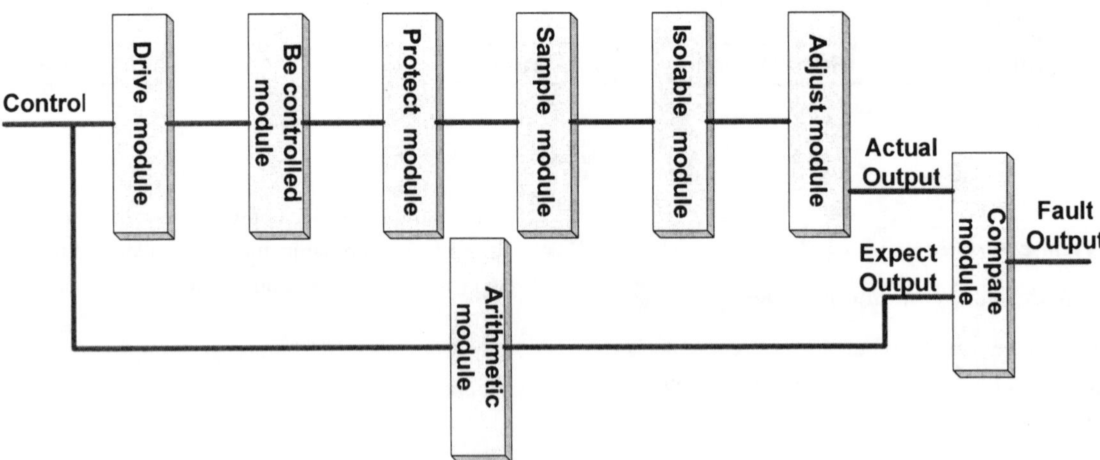

Figure 3. Signal chart of fault detection

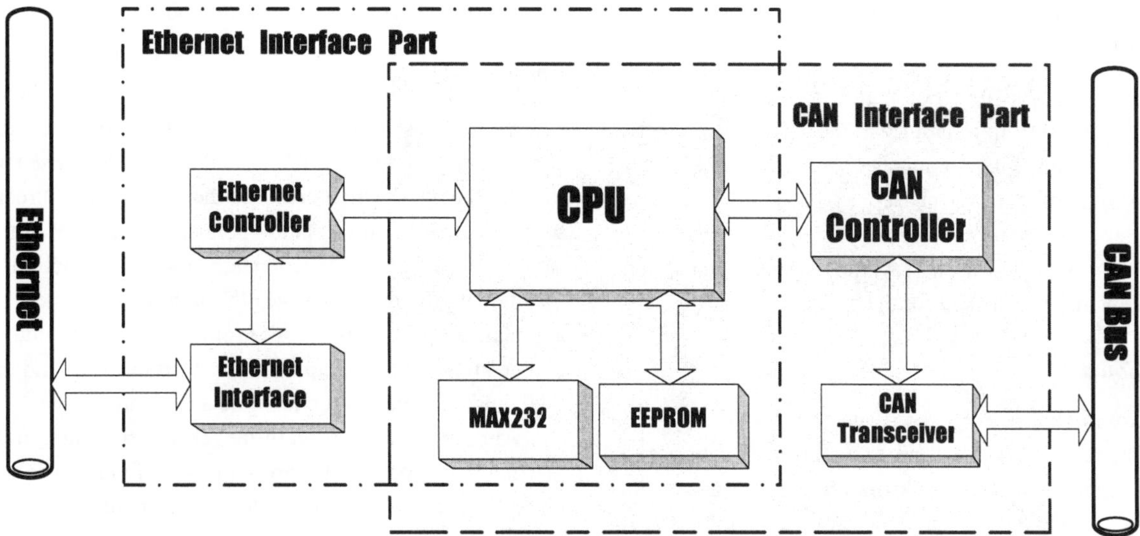

Figure 4. The structure of area-controller

disposing corresponding handling. Structure diagram of system sees Figure 4:

In the structure diagram, area controller mainly consists of two parts: Ethernet interface and CAN interface. Ethernet interface part contains Ethernet control chip and Ethernet interface chip; CAN interface part contains CAN control chip and CAN transceiver chip. In addition, it includes CPU, serial port and EEPROM. CPU chosen is the high performance 51 compatible 8 bytes processor; serial port uses MAX232 chip for debugging and programming.

2.2.2 *Field-controller software design*

The tasks of the field controller can be divided into two aspects: firstly it receives the data packages from the Ethernet, converts them into CAN packages, then sends them to the CAN network; secondly the CAN network receives the data packages, converts them into the Ethernet data packages, then sends them to the Ethernet network.

The whole program design of the field-controller is: when the data in the application layer need to be sent to the node in the CAN network, integrated data, which need to be sent to the CAN nodes later, are parsed by the Ethernet interface protocol module from the output layer message. Then the data will be stored in the data buffer until the main program is informed. The CAN interface module is transferred by the main program in order to send CAN protocol data to the CAN bus. The other way round when the data need to be sent from the CAN nodes to the controlling host computer, the CAN protocol integrated data packages are stored in the data buffer by the CAN protocol module until the main program is informed. The main program transfers the Ethernet interface protocol module; the Ethernet interface protocol module encapsulates the integrated CAN protocol data packages as the data of the application layer, then sends them to the Ethernet.

2.3 *Program of the controlling host computer*

2.3.1 *The functions of controlling host computer*

Among the whole building facility controlling system, the controlling host computer stands in the kernel position. All the controlling commands are sent out by it. All the scene status messages and fault messages are collected in the controlling host computer, and then disposed by it.

The main functions of the controlling system are:

1. Send out the on-off dictates, control the action of devices;
2. Send out the configuration parameters; configure the parameters of the configuration area-controllers and the field-controllers;
3. Send out the inquiry commands; inquire the status of the range controllers and the scene controllers.
4. Receive the fault messages from the field-controllers; then send short messages to the mobile telephone of administrator, so that it is convenient for the system maintenance.

2.3.2 *The hardware configuration of controlling host computer*

According to the functions and at the same time satisfying the system reliability in the bad environments, IPC-610 controlling computer of the TAIWAN YANHUA Company is chosen as its controlling host computer hardware platform. The Intel P4 1.8G processor is also used for its strong processing capability and convenient for the system upgrade.

2.3.3 *Software*

Software modularization can reduce the complexity-y of the system, make the system easy for exploitation and modification, and meanwhile increase the productivity of the software, reduce the exploitation cost and time. It has the virtues of being convenient for writing the interfaces, consistent with the interface of the Windows, easy for operation and so on.

This design can divides the system into several modules: Flash interface module, status display module, commands sending module, TCP communication module, UDP (User Datagram Protocol) communication module, and short message sending module.

2.3.4 *Data transmission*

Due to the vast amount data transferring between the client port and the server port, the TCP is chosen to transfer the data, which provides the reliable data stream service facing to conversation connection. The IP of the each data package is independent. Its basic function is transferring the data packages through the network. The application programs of the client/server provide a simple and convenient approach to visit the TCP and the UDP and achieve bidirectional data exchange.

Aiming at higher real-time character, when field matrix control commands are sent to the field controller, UDP (User Datagram Protocol) communication is chosen to improve the transmission efficiency and enhance the real-time character greatly. It makes less effect for the system even that the data packets have been lost, for the data are sent periodically and continually by the frame unit.

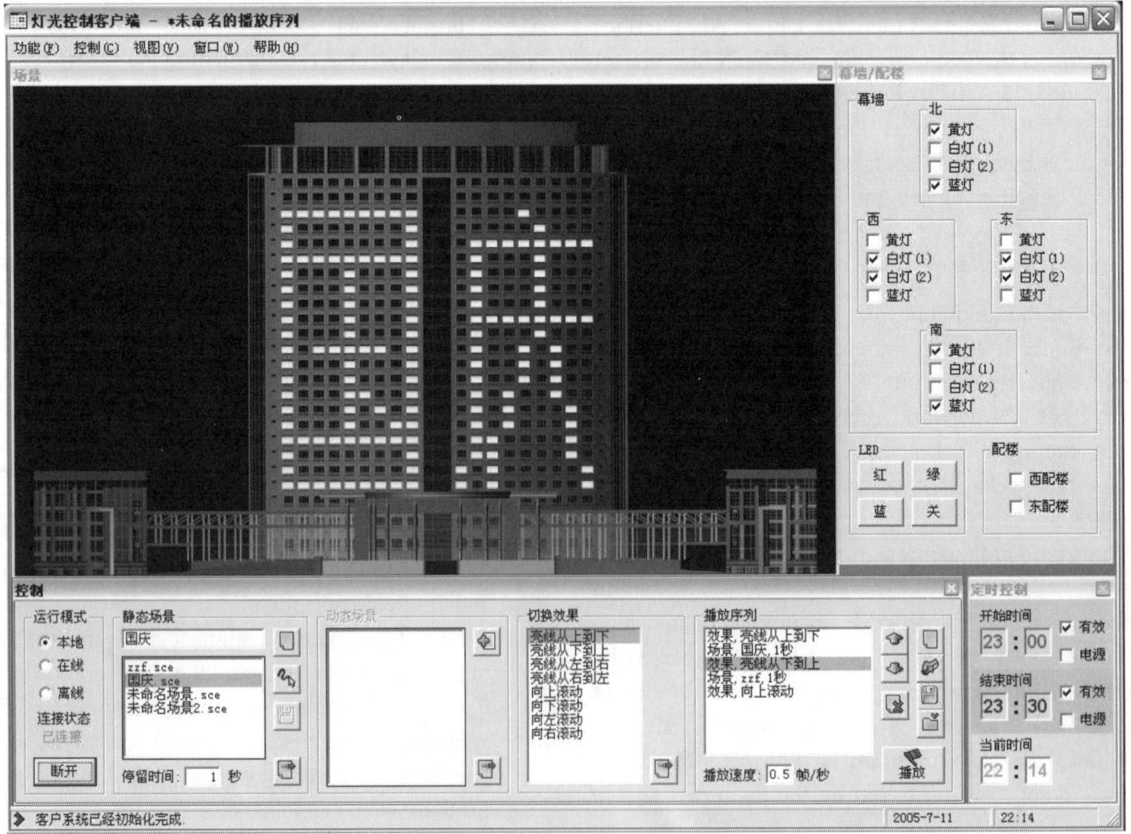

Figure 5. Scene display interface

2.3.5 *The short message program for fault-alarming*

A wireless transceiver module is used as our short message sending hardware. It connects to the control computer by the serial interface. The fault detection information of field-controller will be sent to the mobile telephone of administrator in the form of short message. The software of sending short message has been made as a component. As long as adding this component in the main program, it's easy to set the short message center number, the serial interface and the baud rate by invoking the property of this component. After that the short message can be sent easily through a transparent process. It reduces the process of short message sending.

2.3.6 *Interface design*

The interface part mainly includes FLASH scene, system situation display and control command sending. The FLASH scene has been divided into two parts: display scene and edit scene. The display scene mainly takes charge of displaying current frame synchronously when the system runs normally. The edit scene takes charge of editing new scene or modifies and saves the original scenes.

The control function of the system mainly includes: editing scene, playing, timing playing, and timing closing. The control window is consists of running mode, static scene, dynamic scene, switch effect and playing sequence.

3. System Performance Analyze

The network deferment, network load and its reliability is vital to the whole building devices controlling system. The system packs all the controlling nodes and sends the controlling messages uniformly in order to reduce the average deferment and improve the system performance. The messages transferring in the CAN network also adopts the pack method, which effectively reduces the transferring time and the transformer handling time.

As far as the network load, since the host computer sends data through the Ethernet, expect the fault messages the controllers of the different layers do not send any other messages. So that it can effectively reduce the possibility of the conflictions in the network.

The system reliability is considered from the system level and the parts of an apparatus level. At the scene level, the high-powered CAN bus is used to satisfy the request of the reliability in the bad situation. Meanwhile the system adopts the redundancy method, increases the terminal matching devices, which reduces the fault's influences. The reliability is guaranteed through three methods: reducing the disturbing sources, cutting off the disturbing spreading paths and increasing the anti-jamming abilities of the parts. At the same time the software adopts the dictates redundancy, software traps, software watchdog and other anti-jamming methods.

Synthesizing the hardware and software anti-jamming technology effectively improves the anti-jamming ability of the parts and reinforces the system reliability. Through the above methods, all the controllers of this system meet the needs of the scene and achieve success in the practices.

4. Applications

The building facility controlling system of this design has the following characteristics:

1. Upper bandwidth, easy to realize the seamless connection among the information network.
2. Aims at the weak points of the Ethernet controlling, adopts the multi-nodes data pack to send data, effectively reduces the transferring amount of the data packages on the Ethernet, greatly decreases the possibility of the conflicts, improves the using efficiency and guarantee the system real-time performance.
3. The communication has high reliability, high rate and far distances.
4. The transferring medium is the shield twisted-pair, with strong anti-jamming ability, greatly reduces the cost and satisfies the client's request.
5. Adopts the modularization structures, easy for the system integration and enlarge.

The building devices control system designed in this paper has been applied into a 29-story building's lighting scene project successfully.

5. Conclusions

The building devices control system based on Ethernet and CAN adopts modularization method. So, it is flexible, easy to be integrated and extended. Alarm function is introduced into the control system, in which the alarm message is transmitted by GSM wireless network. It can also receive remote wireless control instructions. In the application, the whole system behaves well with a good reliability, little delay in the main network, less network load and so on.

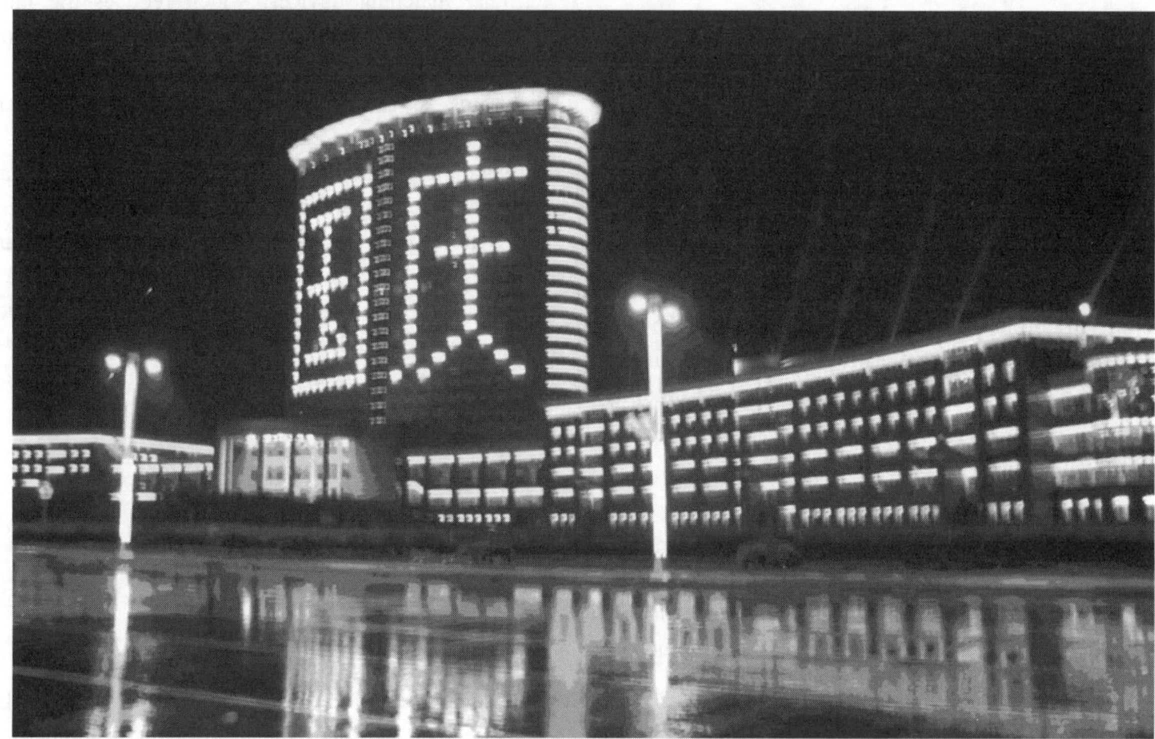

Figure 6. The night piece of a government building

References

1. WindRiver Systems, Inc. Tornado User's Guide (Windows Version) [Z]. U.S.A (1999)
2. ANSI/IEEE STD 802.3. Information technology-local and metropolitan area network.
3. D.T. Miklovic. Real Time Control Networks [J/OL]. http://www.isa.org/Bookstore/(1998)

DESIGN CHARACTERISTICS OF THE ELECTRIC SYSTEM FOR PROVINCIAL MOBILE TELECOMMUNICATIONS HUB BUILDING

HONGYU NA

Architectural Design & Research Institute of Tongji University, 1239 Siping Road Shanghai 200092, China

This paper describes a distinctive case design of the electric system design for the provincial mobile telecommunications hub building, which located at Ji Nan City of Shan Dong Province, has a two-storey basement and twenty-one storeys reaching a height of 86 meter and covering a total building area of 46000 square meters. For such a high building with its complete architectural functions it is necessary to have a rational electric design for various power systems as well as advanced electric facilities. This paper, while omitting the power design for the routine systems, gives an elaborate description of the electrical design for the building.

1. Design of Power Supply and Distribution

Power supply and distribution system for the provincial mobile telecommunications hub building must be completely safe and reliable. The system is the most important part of the whole electric system for the building.

You can delete our sample text and replace it with the text of your own contribution to the proceedings. However we recommend that you keep an initial version of this file for reference.

1.1 *Level of power load*

In the light of "the Fire Protection Design Code for Civil High Buildings ", the mobile telecommunications hub building in Shan Dong Province with its special requirements in terms of fire danger, fire evacuating and fire extinguishing is a tower of first grade. Also according to "the Electric Design Code for Civil Buildings", its power load is graded as:

1.1.1 *Power load level*

Power supply of equipment rooms, power supply of fire protection, and power supply for lighting of plant rooms, among them the power supply of plant rooms and the power supply of fire protection the most important.

1.1.2 *Power load level 2*

Domestic water pumps, guest elevators and various low voltage facilities.

1.1.3 *Power load level 3*

All the rest power loads.

1.1.4 *Power supply*

Facilities of Power Load Level 1 must have at least three different kinds of power supply. In the completed electric design for power supply of the equipment rooms there are two circuits of municipal power supply of 10kV + diesel generator + centralized UPS (4 power sources); for fire protection there are two circuits of municipal power supply of 10kV + diesel generator (3 power sources).

1.2 *Locations of transformer stations and diesel generators*

In the light of "the Electric Design Code for Civil Buildings" and "the Code for Transformer stations of below 10kV" and some other relevant codes, transformer stations must be located in the first level or the second level of the basement, for the convenience of wiring and maintenance. As the first level of the basement is occupied by the client for use of reception and business display, the place chosen for transformer stations is located at the first level of the basement in the vicinity of the exterior power supply, with enclosure walls. Adjacent to the place is the diesel generators room. Inside the transformation station are provided spaces for high-voltage distribution, transformers, low-voltage distribution, emergency distribution and duty room. The emergency distribution boxes are arranged by themselves in a separated room away from ordinary distribution boxes. In addition, the locations for distribution boxes, transformers, diesel generators, the design for cable channels under the distribution boxes, the design height and fire protection of the transformer station, the opening direction of doors and the outlet of the transformer room, the hole for hoisting generators and so on are all designed according to the relevant codes.

1.3 *Selection of distribution systems*

Possible power failure of mobile communication equipment will bring about huge economic loss and possible failure of the mobile communication equipment will depress credit to mobile communication corporations causing serious social problems. So the power supply of this sort should, with exception of force majeure, never fail even in case of conflagration. Power failure for fire emergence lighting should not last more than 5s. Taking all these into account, the author of this paper conceived several schemes in the preliminary design and then decided, after careful analysis and selection, on the following two schemes for further comparison:

1.3.1 *Scheme 1*

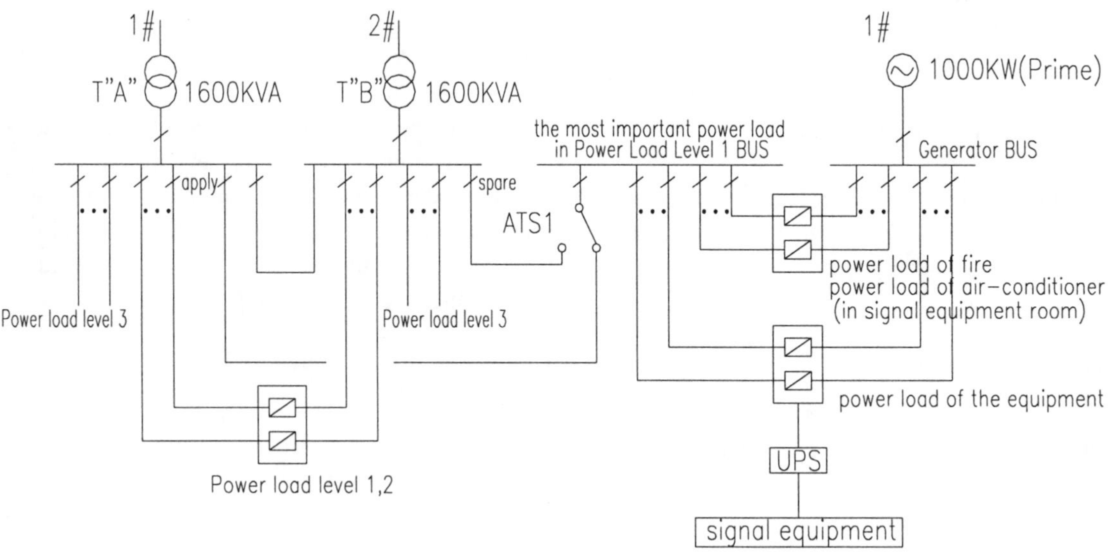

Figure 1.

Restrained by the capacity limitation of diesel generators (as prescribed by the client), for the lighting of equipment rooms of Power Load Level 1 and of offices the design adopts two circuits of municipal power supply + lamps with storage battery, allowing diesel generators for support of the most important power supply in Power Load Level 1.

The advantage of the design: in case of failure of Municipal Power Supply No. 1, ATS1 automatically switches to municipal power supply No. 2 to provide continuous power supply; when Municipal Power Supply No. 2 fails, the switch at the terminal automatically changes over to diesel generator section. Within 15s the diesel generator starts to supply electric power. For both Power Load Level 1 and Level 2, two circuits of municipal power supply with automatic switch at the terminal are provided. As for Power Load Level 3, one circuit of Municipal Power Supply is conceived.

1.3.2. *Scheme 2:*

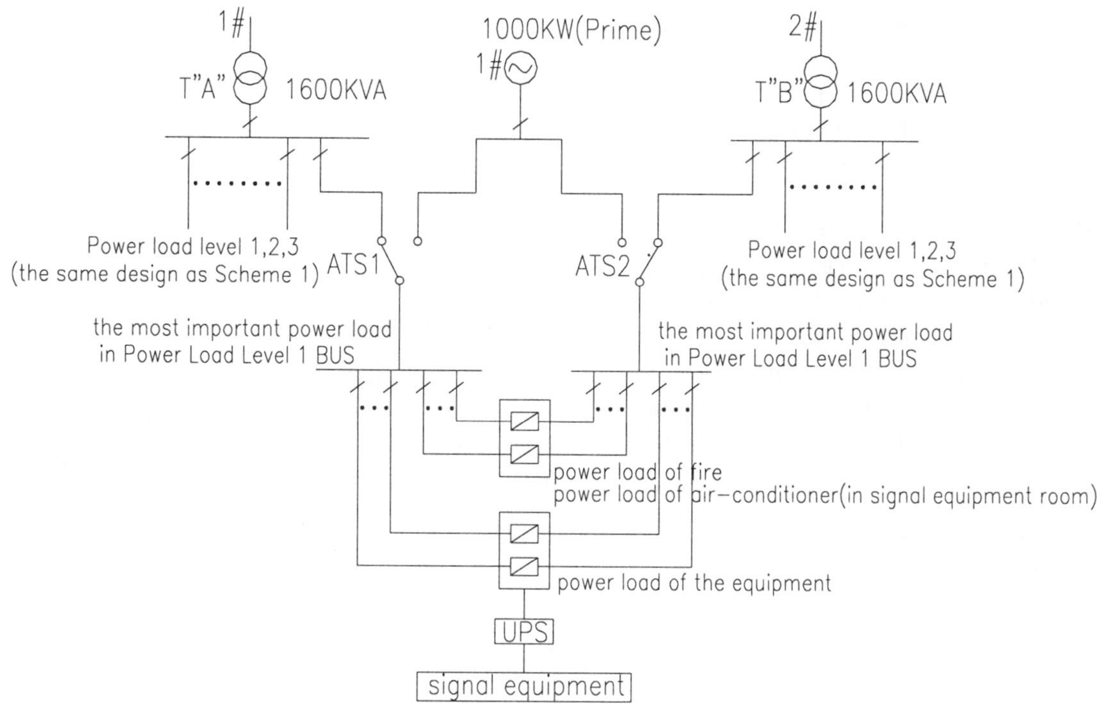

Figure 2.

The advantage of the design: for the most important power load in Power Load Level 1, when Municipal Power Supply No. 1 fails, ATS1 switches automatically to the diesel generator section to put the generator into a standby state while Municipal Power Supply No. 2 starts to supply power at the terminal. When Municipal Power Supply No. 2 fails, the diesel generator is automatically put into operation by ATS2 and the diesel generator starts to supply electric power for the terminal. The same power supply is designed for the other power load levels.

1.3.3 *Scheme comparison*

Comparing the two schematic designs, the major difference between the two designs lies in the fact that in Scheme 1, when both municipal power supplies fail and before diesel generator starts into function there is a interval of 15s; in Scheme 2, uninterrupted power supply is assured with only a break of only 2s for ATS to switch. Power supply for equipment rooms is also assured by UPS with a switching break of less than 4ms and even less than 1ms by high-speed UPS. Considering that no power failure is permitted even during the time when the generator is being put into operation, the author finally decided to adopt Scheme 2. Although ATS has to be added into the scheme, for the sake of absolute safety it is worthwhile.

With design ideology conceived and scheme decided upon, the work remained is for improvement and perfection. After power load computation, for the whole building are provided schematically four transformers of 1600kVA, two water-cool automatic diesel generators of 1000kW, and one measuring screen. Considering insufficient supply of power in China, an intelligence system to control the transformer station is conceived in order to provide an integrative monitoring platform for high and low-voltage facilities and transformers. The system not only assures an automatic control of power supply but also connects itself to the BA system of the building.

2. Design of Lighting

The substructure is an integral work consisting

equipment rooms and underground parking lots while the up-structure comprises a mechanical tower and an office tower. The whole building, except for the parts of the second decoration, adopts power-saving lamps. Electronic current regulator is basically incorporated for all fluorescent lighting fixtures in the office towers. With use of lighting grills, the designed luminance is 400Lx. Lighting fixtures for the underground parking and equipment rooms are designed according to the relevant codes. Double power supply circuit is incorporated, in addition to lamps with storage battery, for all the emergency lighting and the lighting of the mechanical tower. Details of this part of electric design is given below:

2.1 *Types of control for lamps with storage battery (precondition: three-wire system)*

2.1.1 *Continuous lighting type*

The lamp is on all the time and when power supply fails it switches to emergency lighting automatically.

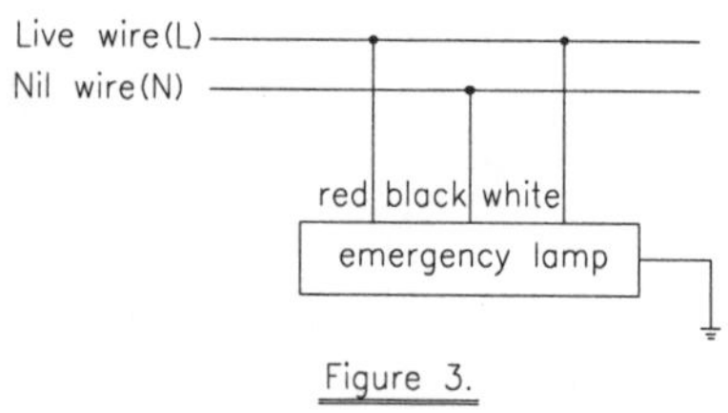

Figure 3.

2.1.2. *Non-continuous lighting type*

The lamp is off at ordinary times. During fire-fighting, the lamp is forced into function. When power supply fails the lamp changes over to emergency lighting.

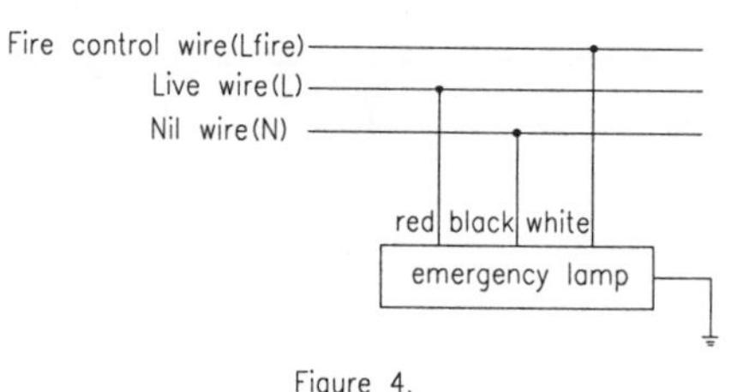

Figure 4.

2.1.3 *Controllable lamps type*

The lamp is controlled by a switch K. During fire-fighting, the lamp is forced into function. When power supply fails the lamp changes over to emergency lighting.

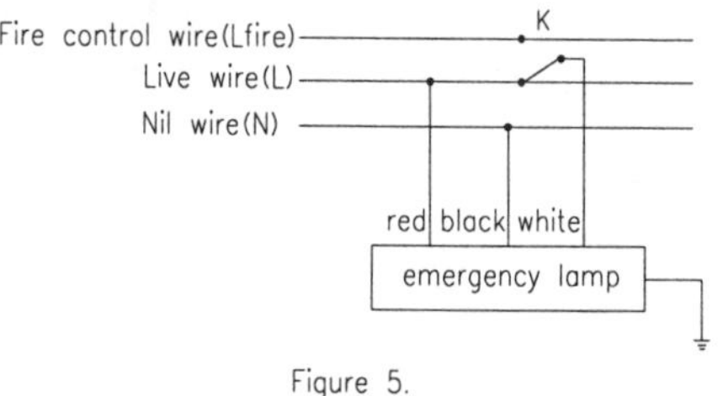

Figure 5.

2.2 *Forced-on lamps in fire emergency*

The circuit for the above-mentioned fire-fighting lamps makes it possible to turn on all emergency lamps in exit passageways without any external control. The system is actualized by the following design.

2.2.1 *An random example of distribution system for emergency lamp:*

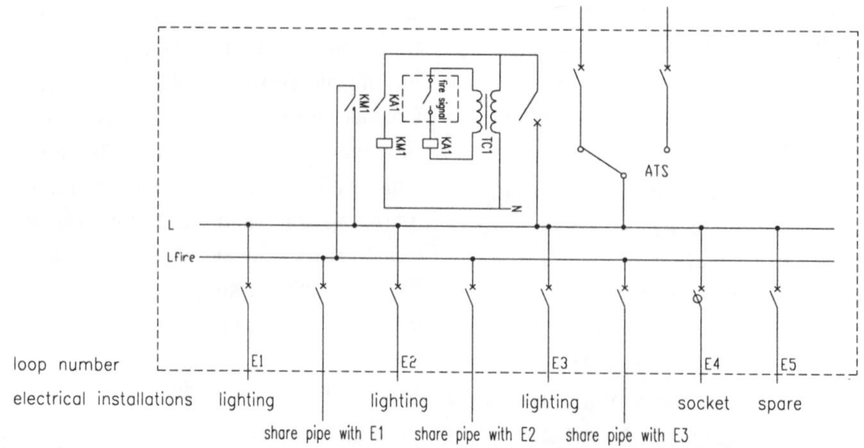

Figure 6. distribution system for emergency lamp

As indicated in the circuit diagram, when fire-fighting signal comes up, Lfire changes from non-electrified line over to electrified line.

2.2.2 *Types of control for lamps on site*

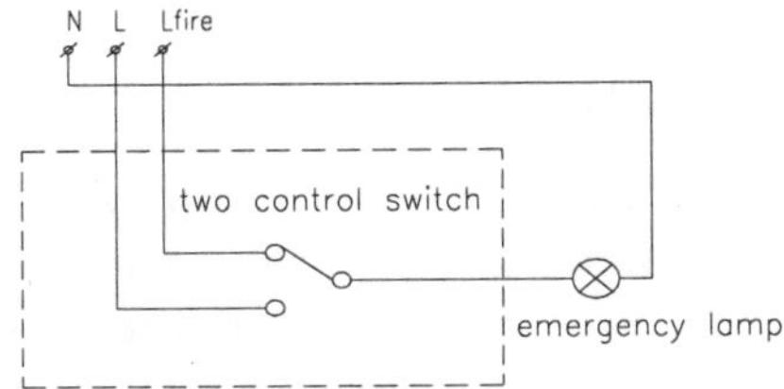

Figure 7. corridor emergency lamp circuit diagram

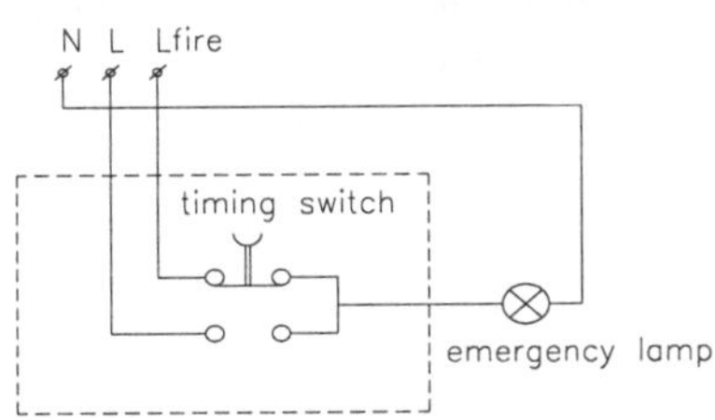

Figure 8. staircase emergency lamp circuit diagram

2.3 *Types of light fixtures*

Adhering to the above principle and taking the actual conditions of the building into consideration, the author reduced the lamps with storage battery to four categories and implemented them in the actual design.

2.3.1 *Indicator lamp for evacuation*

Indicating escape direction and exit, the lamp is off at ordinary times but forced on during fire-fighting. When municipal power supply fails, battery power comes into operation. The lamp, with an energy-saving power of 8W, is only for use of fire-fighting. Its circuit diagram is as blows:

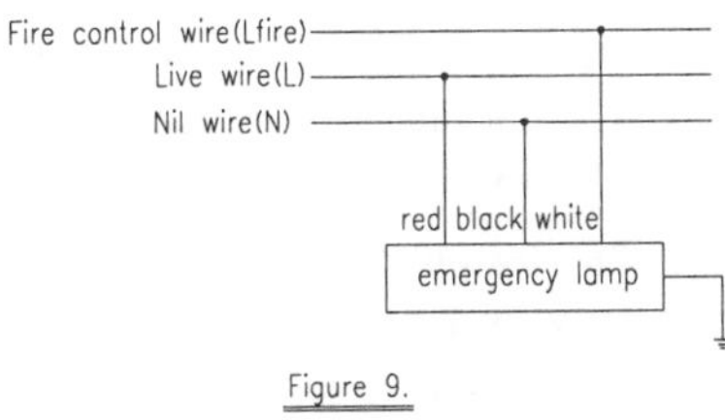

Figure 9.

2.3.2 *Stand-by lamp for evacuation I*

It refers to the lamp for staircase lighting. The indicator lamp alone is not adequate for escape in a fire emergency. When there is heavy smoke, fleeing people may be unable to see the indicator lamp. This possibility fire tragedy must be taken into account in design of electric system. The author therefore conceived for staircases lamps reinforced with storage battery. The lamp is for normal lighting of staircases at ordinary times and forced into action in fire emergency. When municipal power supply fails, it is powered by the storage battery power. To save energy, an alternative of using municipal power supply and timing switch is thus conceived. The following is the circuit diagram for the lamp.

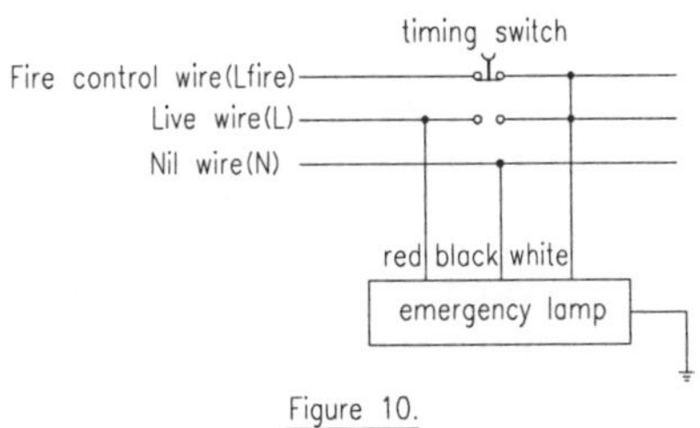

Figure 10.

2.3.3 *Stand-by lamp for escape II*

It refers to the lamp for passageways and lobbies of fire lift. It functions like the stand-by lamp for escape I, but as it is for the use of lighting passageways, it is unnecessary to provide storage battery for all of the kind. For the purpose of adequate luminance in fire emergency, the author adopted alternative placement of the lamp. The circuit diagram for the lamp is as follows.

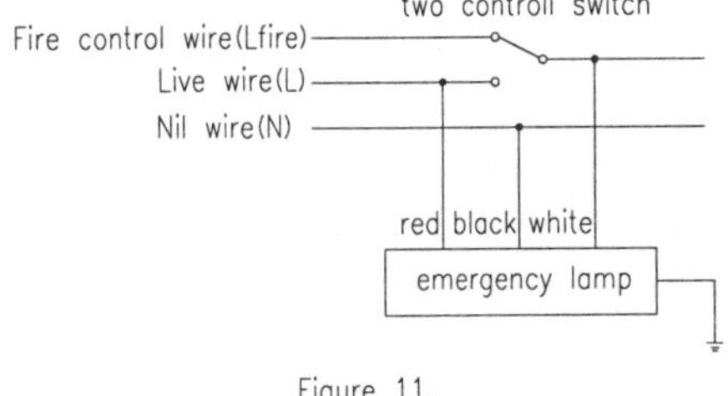

Figure 11.

2.3.4 *Stand-by work light*

Stand-by work light includes lamps and lighting fixtures for Communications Plant Room, Fire Control Room, Generator Room, UPS Room, Transformer Station, Fire Pump Room, Fire-fan Room and so on. The light is a reserve lighting service to ensure continuous work (or continuous work for time being). It can be switched on and off at ordinary times. When municipal power supply fails, storage battery comes into action. The circuit diagram for the lamp is as follows.

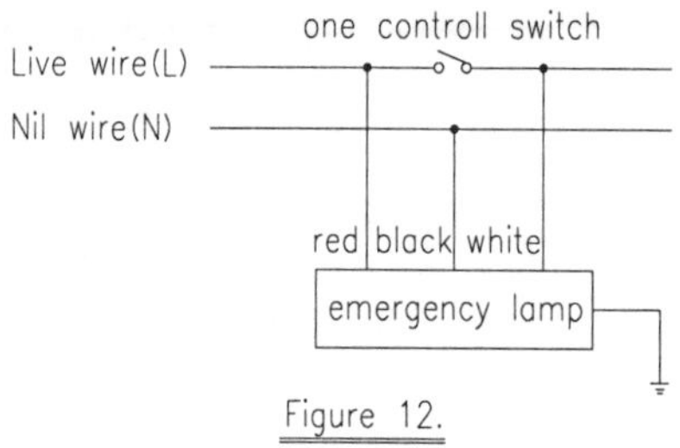

Figure 12.

2.4 *Notes*

2.4.1 *Emergency duration*

The article 9.2.6 of "the Fire Protection Design Code for Civil High Buildings " (Version 2001) states, "emergency lighting and exist indicator should adopt storage battery as their power supply, with a supply duration of not less than 20min". The relevant article of 24.7.5 in "the Electric Design Code for Civil Buildings" prescribes, "indication lighting for fire evacuation should be no less than 20min and for stand-by work light no less than 1h". Designing according to the prescription will surely pass muster with fire check-up. But the lower limit as prescribed in the codes may not meet the requirements of fire evacuation in China where far more time is needed to fight against a fire. The author therefore conceived extra working power supply of storage battery: evacuation lighting duration of 90min and stand-by work lighting duration of 2h.

2.4.2 *Central power supply system or dispersed power supply system*

In China there are available several emergency lighting systems using central power supply from storage batter, called EPS or FEPS. The author had the pleasure of attending the symposium presided by Mr. Chai Jun and exchanging opinions with him. According to the experience of the author, the central power supply system comes into existence because lighting fixtures in the dispersed power supply system demand large amount to maintenance to avoid possible failure of lamps when municipal power supply fails. But the central power supply system has a fatal weakness of breakdown at a large area. Since there is more one lamp in an emergency lighting loop, if in a fire emergency one lamp breaks down, trips or cause short-circuit, the whole loop has to stop functioning, creating partial blackout which must be absolutely avoided in fire evacuation. After overall analysis, the author adopted the design of dispersed power supply system. To attack the problem of maintenance, lamps of maintenance-free and long-life can be incorporated and they can be well replaced at intervals of four years. It is a suggestion of the Author for suppliers in China to prolong life-span of lamps and solve the problem.

3. Design of Earthing

Earthing is always an important issue in the system of power supply and distribution for a building. It concerns safety and reliability of the system and therefore most essential. The mobile telecommunications hub building as different from normal office building has special and strict requirements for earthing system. The project has a building area of approximately 8000 square meters, consisting of 8 storeys of electronic communications equipment and facilities and thus making earthing a particularly hard job.

3.1. *Earthing system*

Incorporated in the building is an independent transformer and distribution room which normally adopts the "TN-S" system. The system is characteristics of absence of any electric connection between the neutral Wire N and ground protection Wire PE except for the neutral place in the transformer where both wires share an earthing. The neutral Wire N is charged while PE is not charged. The earth connection is absolutely safe as a reliable basic electric potential and therefore chosen by the author as the best earthing system.

3.2. *Earthing measures*

Prescriptions in various codes are incorporated: "when a communications hub building is built together with office buildings or high civil buildings, the earthing installation for the communications should be separated from the lightning protection installation of the building. If the separation is impossible because of terrain limitation, the earthing installation for communications can be connected with the lightning protection installation of the building and the earthing installation of alternating power supply system. But the electrical resistance of the earthing system should not be more than 1Ω", "Unless otherwise stipulated, electrical resistance of an electronic earthing, connected into the ground at one point, should not be bigger than 4Ω. The earthing should be connected with the lightning protection earthing and electrical resistance of the combined earthing should not be bigger than 1Ω. When separated from the lightning protection earthing, the gap distance should be no less than 20m." and "When the combined earthing is adopted for electric installation, the electrical resistance of the earthing should refer to the minimum requirement of the kind. When combined with lightning protection earthing, the electrical resistance should equal to or less than 1Ω." The hub building in question is faced by city streets in two sides and by other buildings in the other sides. The base site is hardly bigger than the foundation pit of the building. It is impossible to have an isolated earthing 20m away.

After consulting with the client and the head of the local Postal and Telecommunications Design Institute, it was decided to adopt the combined earthing: galvanized flat steel of 40X4(mm) joining all ground plates which is connected to the steel bar at the base of the foundation. The combined earthing serves as the basic electric potential earthing connecting by wires to all other ground protection. Master equipotential and secondly equipotential are also linked to form an integral earthing system. The electronic resistance of the earthing system meets the requirement of ≤1Ω. Various earthing installations of the hub building are explained as belows.

3.2.1 *Earthing of lightning protection*

The purpose of the earthing of lightning protection is to guide harmful thunder current fast into the ground. Inside the hub building there are millions of electronic facilities with their complicated wirings meandering in top slabs, floor slab, walls and suspended ceilings. All the electronic facilities and their wirings, low in grade of anti-voltage, have high anti-interference requirements. They are especially vulnerable to thunder hit whether it is direct, indirect or reflected. A highly rigorous and completed earthing system must be installed for lightning protection. In the light of relevant lightning protection codes and after careful calculation (The hub building is designed according to the lightning protection Standard II) the author designed a lightning arrester combining lighting rods and lightning protection belts, with the lightning protection belts forming a grid of ≤10X10(m). The grid is electronically connected to metal roofing components and then, through the main bar inside the structural columns, to the combined earthing. All metal components of exterior walls are also connected to the ground installation, thus forming a cage of lightning protection system with a multilevel shield. Hence effectively protecting all electronic facilities inside the building from thunder hit as well as fending external electromagnetic interference.

3.2.2 *Earthing of alternating current work*

The connection at one point of an electronic power system through metal wires either directly or via special device (such as impedance, resistance) to the ground is called earthing of alternating current work. The earthing of alternating current work of the building refers to the ground connection of the neutral point or the neutral Wire N in the transformer. The four points of the four transformers are respectively connected to the combined earthing. Ground connection of the neutral point can prevent excursion of nil voltage so as to ensure equipoise of three-phase voltage. It is very useful for low-voltage system and can be conveniently applied to single-phase power source.

3.2.3 *Earthing o f safety protection*

The earthing of safety protection refers to the connection of non-charged metals of electronic facilities to the ground protection installation, i.e., metal casings of electronic facilities connected to the PE wire. There are a large number of electronic facilities in the hub building. Without earthing protection the casing of these facilities may carry electronic charge. When the neutral point of an electronic power system is directly connected to the ground, if a person touches casings of electronic facilities with poor insulation, short circuit current of the ground connection may flow through human body and then via the ground back to the neutral point, causing electroshock. With the earthing of safety protection, short circuit current of the ground connection will have two passages: the earth connection and human body. As the electronic resistance of the earth connection is small, the pressure drop of the short circuit current is weak, and the voltage of facilities casings is low, when a man standing on the ground touches a casing, he is exposed to a very low voltage and hence little danger. Therefore all electronic facilities of either high or low-voltage must have the earthing of safety protection. Only in this way can the functioning of electronic facilities in the building assured and the personal safety protected.

3.2.4 *Earthing of direct current work*

The earthing of direct current work refers to the logical earth connection at equipotential surface in digital circuit and signal earth connection at basic electric potential in simulated circuit. In the hub building there are a large number of telecommunications facilities, computers and other building autoimmunization facilities. They are work very fast, imputing and outputting data, transforming and amplifying signals all the time, using micro electric potential or micro electric current. So in addition to a reliable power source, a reliable basic electric potential must be also provided to assure electronic precision and stability. In the electronic design the author used the ground as the basic potential for earthing of the direct current work. The telecommunications equipment is connected via independent wires to the ground. The terminal strip for direct current work is also provided in the telecommunications equipment room.

3.2.5 *Earthing of shielding and earthing of electrostatic protection*

The earthing of shielding refers to the ground connection of casing and shield wires or metal conduits in order to fend external electromagnetic interference; the earthing of electrostatic protection refers to the earthing of prevention of static electricity that arises in a dry environment of the equipment rooms. In a building for telecommunications, design consideration of electric power and electromagnetism is a very important. To avoid dysfunction or even damage of facilities, wiring system of any fixtures should be able to prevent electronic interference from both the internal and the external. This kind of interference may come from wire coupling or complex coupling, but typically from extra high voltage, high-duty electromagnetic radiation field, natural thunder hit, and electrostatic discharge. Shielding and proper earth connection is the best solution to the problem of electromagnetic interference. Connection of casings to PE wire, connection of both ends of shielding conduits to PE wire, and multi-connections of interior shielding to Wire PE also serve as an effective means to fend and avoid electromagnetic interference. Protection of electrostatic interference is also important. In a clean and dry interior environment, any walking and movement of facilities will create a great amount of static electricity. Proper multiple connections of casings and facilities (including floors) to Wire PE become most essential.

Note should be made that any ground wires must by no means be connected to Wire N.

3.3 *Coupling of master equipotential*

In the transformer station is provided master equipotential end board (MEB). One terminal of the board is connected to the combined earthing, and the other terminal is linked, via different interconnecting terminals, respectively to Wire PE, metal components of the building, all metal pipelines in and out of the building (including power supply wires, data ducts, water pipes, air-con conduits, heating tubes, and so on). Secondly end board (SEB) is also provided at Refrigeration Plant Room, Water Pump Room, Telecommunications Room, Fire-control Room, TV Room, Fixed Telephone Programme Control Room, Computer Network Room and so on. In addition, in places like Bath Room is provided local end board (LEB). SEB is connected, via MEB lead, either to terminals of MEB or Wire PE in high and low-voltage shafts in the vicinity. LEB is connected, via LEB lead, to reinforcing mesh and Wire PE inside rooms. This arrangement ensures that there is no electric potential difference between any two points so as to ensure safety of personnel. The arrangement will be made successful only through concerted efforts of all builders.

In a word, the earthing system of the telecommunication hub building should have a combination of single earth connection and master equipotential.

4. Conclusion

The above is what the author has learned from the design of the electronic system for the telecommunications hub building in Shan Dong Province. It is a piece of knowledge for sharing. As Vercors says, “There is no work of art without a public.”

Sustainable Development Cases

THE NEW POLICE HEADQUARTERS – ARSENAL HOUSE AT ARSENAL STREET, WANCHAI: SMALL FOOTPRINT, BIG EXPECTATIONS

ANTHONY WILSON ALICE YEUNG ANDREW FUNG
Architectural Services Department
41/F, Queensway Government Offices, 66 Queensway, Hong Kong

1. Introduction

This paper briefly expands on experiences in the New Police Headquarters project sustainable design as a case study covering elements most of the sub themes. Common sustainable design challenges for most high rise buildings generally fall into the following categories.

- Urban planning and development
- Optimising site potential
- Energy efficiency, renewable energy
- Environmental friendly materials
- Construction and waste management
- Flexibility of internal spaces and its environment
- Sustainable culture and social responsibility
- Partnership in the project
- Post occupancy matters

2. Urban Planning and Development

To find a location for 110,000m2 Gross Floor area of office space and a variety of other spaces to suit different complex functions is not easy. Space is always tight in Hong Kong and best use has to be made of any available sites. Site search planning could not identify a suitable location for the new complex, therefore using the normal tradition in Hong Kong and other major city centres, the demolition of an older smaller building on an existing site, made way for the new.

Urban setting

The analysis of the compact city centre site, immediately identified the many typical challenges that lay ahead. How to set the new development into the urban setting? What to do with the traffic problems during construction and upon completion? How to provide utility connections to service the project with the surrounding existing constraints? These were some of the major issues and challenges common to many high rise buildings.

3. Optimising Site Potential

The tight site is surrounded on one side by a heavily trafficked road, one side faces a park and the other two abut against existing fully operational buildings. There was no scope for expansion outside the small site footprint which had an area of 7 400 square meters. The estimated number of floors expected to accommodate the client requirements was at least 42.

Two conflicting functional issues, easy public access and adequate security required to be factored in. Based on the engineering conditions and most suitable arrival points, the access and exit points were fixed. The maximum site development was calculated and conceptual design options explored and finalised.

4. Urban Setting and Functional Response

The selected sustainable design solution which met functional and urban design massing wants, was to minimise the building impact as far as possible and excavate several basement floors to accommodate delivery service areas, vehicle parking, refuse collection, building services plant and the force armoury. The open air parade ground was set facing the park with the main tower located behind. The local police station is independently expressed and faces directly onto the street and the district for easy public access. Included are two pedestrian connections to the existing buildings on either side, access for the public from the main street and a special staff entrance with a feature boundary wall on the side facing the park. High rise buildings have to respect their surroundings and if possible create open spaces within their own footprint.

The Police Headquarters benefits from being highly prominent with views from many directions including the open vista next to the landscaped Harcourt Garden. The architect took the opportunity to depart from the norms of hemmed-in architecture of the neighbourhood, to present both a varied building envelope and more inviting entrance points.

The detail design developed the earlier sustainable concept by providing a building envelope that was modern, easy to maintain and clean, and provided excellent thermal insulation to reduce future operational costs. The building envelope is clad in a combination of metal cladding, double-glazed curtain wall and natural stone, with the Overall Thermal Transfer Value (OTTV) being 19.10w/m2.

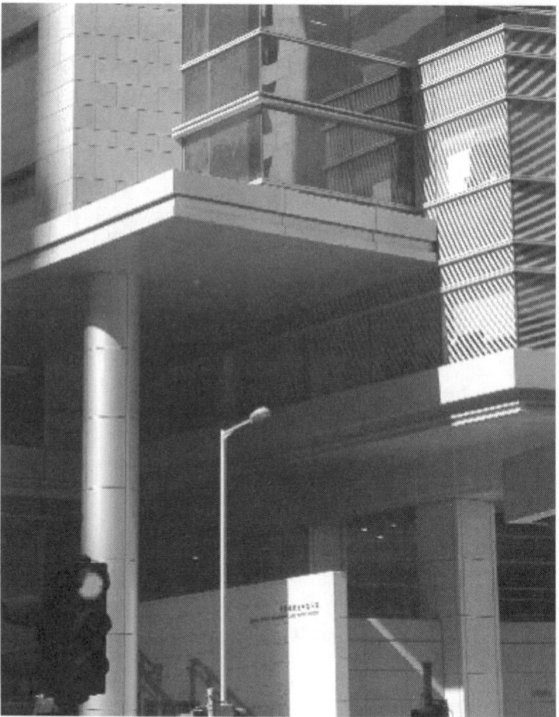

Façade treatment – play with stone and glass cladding

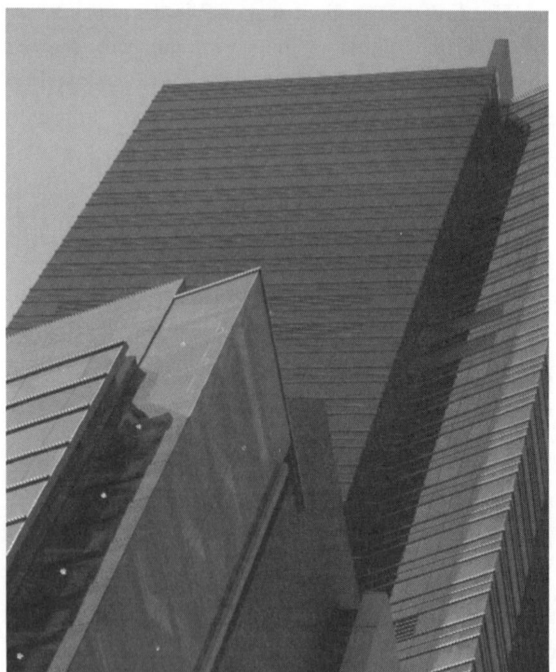

Façade treatment – play with different plans and materials

To break down the scale of the building, which features large economical floor plates, different elevational approaches are used. The south-facing facade is clad with strip windows reminiscent of the compound's adjacent West wing tower, while the elevation towards the harbour is clad in reflective glass with fins for a refined appearance. The Harcourt Garden-facing façade features differing treatments combined to lift and express the height of the tower above the columns at the base. Stone is introduced at the lower levels, presenting a variety of materials and articulating the building's form.

In Hong Kong we have to maximise our site utilisation. Rather than the most economic square or rectangle plan around a central core, slight modulation and different treatments to the facades have enhanced the visual interest yet ensured the maximum use of daylight reducing the need for artificial lighting. The sculptural affect is seen and appreciated from the harbour, the surrounding roads, the park, high rise residences, the major hotels and the hillsides behind.

Roof terraces are provided for the police station, at the senior management floor and on the very top roof. Other roof space is required for building services plant and communications equipment, which were considered and grouped in a more orderly and sculptural manner. The care and attention to detail of roof design is important, as although it is high rise building, it is still looked down upon from the hillsides behind.

Roof garden of Central Police Station

5. Environmental Friendly Materials

Sustainable design commences from day one of a project, therefore, care was taken in salvaging materials from the building that was demolished. Granite stone was reused on the new external garden wall, bronze doors were used as a stunning feature at the new auditorium and old hardwood strips were reused for feature decorations. Other more basic rejected materials were recycled or used for public fill.

The selection of major new materials was also based on a balance between time frame for construction, cost and reuse or suitability for recycling in future. Reinforced concrete, insulated glass curtain wall, metal cladding and selected granite stone are the main feature material.

Materials chosen for the internal fitting-out are generally environmentally friendly including PVC-free carpet, computer floor made from recycled steel, low VOC paints and timber from sustainable cources. They are hard-wearing to minimise the demand for maintenance. Contemporary, modern, transparency and spaciousness are the aesthetic themes for the interior design.

The relocable partitions and computer floor system throughout the complex allows for more flexible layouts, with the building services being equally adaptable for change of use. Open plan offices add to the spaciousness and foster communication among staff ensuring flexibility of use in future.

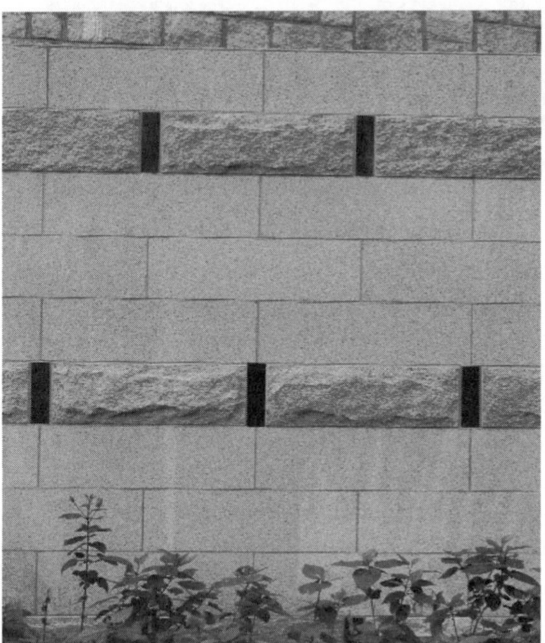

Reused granite stone demolished from the existing old fence wall

Bronze door reused for Auditorium entrance door

Old hardwood strips reused for feature decorations

6. Sustainable Culture and Social Responsibility

The Architectural Services Department with ISO 14000 Certification and the Hong Kong Police Force had environmental targets and objectives for the project. Both are fully aware of the impact of construction projects and the ongoing maintenance and operating costs. The input and effort to create a more sustainable design contributing to the urban skyline and supporting public safety were considerable.

One of the greatest challenges in the design was to integrate the functional requirements with the very high security requirements. The brief from the Police Force called for special areas such as the force armoury and firing range to suit their operational needs. Different security requirements were required for these special areas. These requirements were "factored in" during the planning stages of the new complex, after consultation with experts with regard to the established regulations, guidelines and legal requirements. To give an indication of the size, user complexity and security requirements for this complex, over thirty lifts servicing, under over 20 no departments, are provided as well as 24 escalators. Good vertical transport is a must for successful high rise buildings.

7. Construction Challenges and Waste Management

Construction commenced with the demolition contract in May 1999, which was immediately followed by the foundation and superstructure contract, in November 2000. The project was completed on schedule in April 2004.

The foundation work involved many challenges including the removal of much rock material and 148,000 cubic metres of soil for the four level basements. This amount of excavated material was the equivalent of 14 football pitches. A diaphragm wall providing water prevention was built around the whole basement area. During excavation, several old cannons were discovered, removed and restored for relocation at the parade ground area.

Superstructure work followed in full swing and the contractor worked on a concrete floor completion cycle of 4 days per floor. 92, 000 cubic metres of concrete was poured. The roof level was reached in June 2003.

Another major challenge was the excavation of the existing adjacent road. It could only be closed on weekdays after midnight until 6 am, and on Sundays to minimise disruption to the public. The utilities were so packed under the road that several sequences of levelling had to take place to accommodate them all with most work being carried out at night. The shoring work for the trench was extensive due to depth of the trenches and the amount of excavation. Problems of the surrounding existing utilities are common to all high rise buildings.

The call for sustainable practices within the construction sector has demanded an approach that is less disruptive and less polluting as well as making economic sense. The use of cost-effective prefabricated materials and modular designs for the finishes and for the building services installation, are not only beneficial to the construction programme but also significantly reduced the material wastage during the course of construction work on the tight site. For example, pre-insulated pipes are a modular thermal insulated pipe product, which is factory prefabricated. This type of material has been used for the chilled water piping installations of the air-conditioning system. Construction time can be reduced compared to conventional piping insulation methods, which rely heavily on workmanship and require workspace.

Basement excavation

Normally high rise buildings have refuse chutes to Refuse Collection Points (RCP) but for PHQ we with the client designed a recycling collection system of paper and other waste to be transported by lift to the RCP. This encourages recycling, uses existing lifts and removes the additional space requirements for the chutes.

The Contractor used an Environmental Management Plan to set out procedures for activities and requirements during the construction of the project. Measures were taken to control water runoff by providing temporary ditches, drainage pipe and sump pits to direct water to nearby drainage. Prior to the discharge, sedimentation for treatment of effluent was provided to comply with discharge standards.

Superstructure construction of the tower block

Superstructure construction of the low block

A wheel washing bay was located at site exit and the sand and silt was removed on a weekly basis.

In handling the underground water, de-watering wells and three sedimentation tanks were used to preliminarily screen the underground water which was then connected to the site's walter treatment facility.

The site was provided with sufficient number of portable chemical toilets for handling sewage from the very large construction workforce of around 1500 workers at the peak.

8. Use of Seawater

Three remote seawater pumps are located at the harbour front at Fenwick Pier, designed to supply seawater for the water-cooled chiller plant in the complex. To avoid disruption to the busy 6 lane Harcourt Road traffic, pipe jacking construction was used consisting of a 70m long, 2.4m diameter sleeve pipe. It was located 7 m below ground and contains the seawater ducts, cooling mains and cables. An extensive site investigation was carried out to identify the best possible pipe location and to ensure it was free of conflicting embedded utilities and structures. Exemplary challenges for the employment of this construction method included, accurate control of line and level of the pipe route, underground water tables being linked to tides and uncertain ground conditions consisting of old foundations, sea wall and any other materials obstructing the pipe alignment. This approach had the particular advantage of avoiding a full open trench cut which would have disrupted the surroundings and used only a small works area where the pipe jacking launch pit was located. Minimum disruption to traffic flows and the public are required when building high rise projects.

9. Energy Efficient Building Services Design

With the low OTTV created by the building envelope, the building service were given the opportunity to further improve efficiency and the indoor environment. Given the irregular operating hour of the client and that the demand on air conditioning and mechanical ventilation (ACMV) systems would be high, the building services system design had to consider energy effectiveness. Energy efficient equipment reduces operational cost and will in future consumes less energy.

The following were incorporated into the energy efficient building services design for the project:

1. Sea water-cooled heat rejection method for the air-conditioning plant;
2. Heat reclaim technologies, such as desiccant total energy wheel and heat recovery chiller;
3. Variable speed drives for water pumping equipment and air fan installations;
4. Presence sensors to regulate the air-conditioning supply automatically;
5. Flexible zoning of lighting control and air-conditioning supply for offices;
6. Energy efficient light sources, including T5 fluorescent lamp, LED and induction lamp;
7. Presence sensor controlled lighting system;
8. Harmonic filter installation;
9. Service-on-demand escalators; and
10. ACVVVF lift system drive

The air-conditioning system installed has been built in with various control measures to achieve the indoor air quality up to "Good" class standard of the "IAQ Certification Scheme for Offices and Public Places".

Reliability is vitally important for the Police 24 hours operation, with special attention given to the quality and reliability power supply. Conditioning of harmonic distortion generated by non-liner electronic equipment load becomes a prime design consideration for assurance of high quality of power supply for the building. With the estimation of the harmonic pollutant content, hybrid harmonic filter to achieve a balance between cost and performance was designed. Resilience of power supply is vulnerable without back-up facilities, therefore three back up generators are installed for emergency use.

10. Completion and Post Occupancy Operational Maintenance

In view of the tight construction programme, the different functions of the building and the complexity of the building services provided, an efficient and effective testing and commissioning process was essential. The onset of testing and commissioning works proceeded in a vertical zonal manner during the early stages of construction work. Most of the vertical backbone services risers were tested in sections in pace with the building structure progress. For fire services installations, arrangements were made with statutory bodies for early inspection of high ceiling voids and concealed spaces before covering with suspended ceilings.

Near to completion of the building, a series of detailed discussions took place with over 20 different user departments to work out a suitable decanting and phased moving in plan. This was finalised one year in advance of the actual move-in dates. Experience was drawn from the Singapore experience to minimise the disruption to those who were the first to move in by scheduling a few selected lifts specially for the moving of goods. By all accounts this went very well and the logistics kept under control. This is also an important area for attention for high rise projects with multi users.

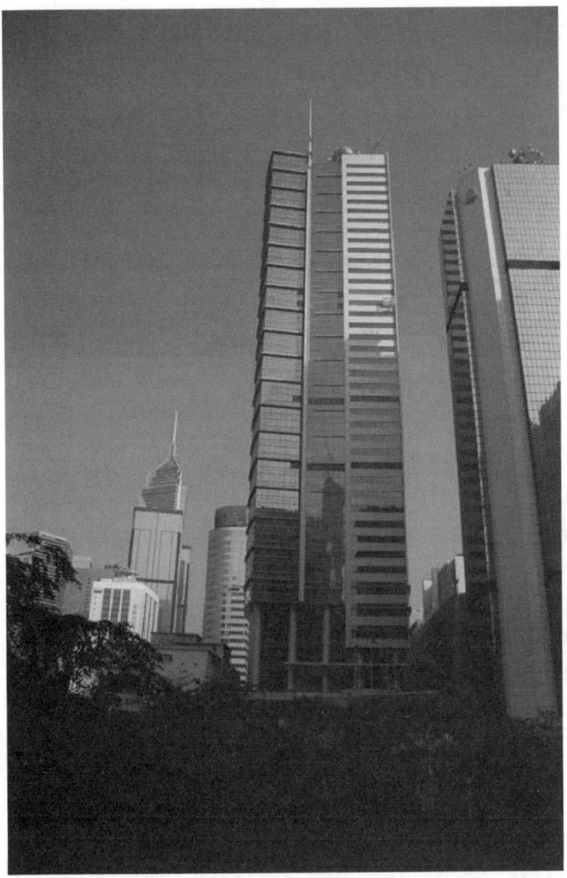

Elevation from Harcourt Garden

11. Sustainable Assessment

The project was assessed by HK BEAM in 2004. The building achieved the highest HK-BEAM rating of Excellent for best practices in building environmental performance. It is a perfect illustration of Architectural Services Department drive towards building environmental excellence with high standards of "green" and innovation design, on site practices, management and maintenance provision's integrated into the planning, conceptualisation and construction of the project.

HK-BEAM Society
香港環保建築協會

This is to certify that the
評估證書

ARSENAL HOUSE
1 Arsenal Street, Wanchai
警政大樓
灣仔軍器廠街1號

has been assessed in accordance with the Hong Kong Building Environmental Assessment Method HK-BEAM for New Office Designs (Version 1/96 R), achieving the Rating of

EXCELLENT

評估工作按香港建築環境評估法新辦公大樓設計(1/96R版)進行,
已獲評定爲以下級別
優秀

this day the 7th September 2004
於2004年9月7日簽發

Mr. Kevin Edmunds, 顧啓榮先生
Secretary, HK-BEAM Society
香港環保建築協會秘書長

Business Environment Council

Certificate Reference (證書編號): C1209a

HK-Beam certificate rating of excellent

12. Partnership in the Project

The enormous challenges on this project were overcome by the joint planning and efforts of an excellent project team, a helpful and knowledgeable client, and contractors who adopted a partnering, problem solving attitude to obtaining a quality building as the end result. EMSD, APB members, Contractor and Client together ensured a smooth handover. Social events including the ground breaking, topping out and official opening ceremonies allowed all to share and enjoy the success of this major sustainable design high rise contribution to the skyline of Hong Kong.

Opening ceremony of Arsenal House

SUSTAINABILITY OF A TALL BUILDING IN HONG KONG – TWO INTERNATIONAL FINANCE CENTER

EVIA WONG

Senior Programme Director
HKU School of Professional and Continuing Educating
The University of Hong Kong

BONNIE LAM

MPhil Student
Department of Real Estate and Construction
The University of Hong Kong

Two International Finance Centre (2IFC) is the tallest building in Hong Kong and the 6th tallest building in the world today. With its outstanding height of about 420 meter, which is almost the same level as The Peak in Hong Kong, 2IFC might have been the only tall building of this height to be set up right next to the waterfront in the world today. This paper has looked at the sustainability of the software and hardware of 2IFC more from the users and management point of view.

1. Introduction

"Tall people make big money – each inch of height make about $789 a year in pay over shorter co-workers", an interesting discovery found by the University of Florida in year 2003.[1]

If tall people make more money, tall buildings might also able to help the users to work smarter or happier or there should not be so many taller and taller buildings being constructed in these years. The trend of developing tall buildings is noticeable, especially in the Asia Pacific region. Among the top 10 tallest buildings in the world today, eight are located in the Asian countries. China, Hong Kong and Taipei occupied six out of the top ten: Taipei 101 in Taipei (no. 1 tallest), Jin Mao Building in Shanghai (5th tallest), Two International Finance Centre in Hong Kong (6th tallest), CITIC Plaza in Guangzhou (7th tallest), Shun Hing Square in Shenzhen (8th tallest), and Central Plaza in Hong Kong (10th tallest)[2]. From the construction point of view, constructing very tall buildings might not be economy to both the landlord and the building mangers. Also, in the world of increasing concern on the long-term value and environmental friendliness issues today, tall buildings should indeed be accountable in their long run. This paper is going to look at the sustainability of Two IFC from the users and management point of view.

2. The Development

With the building height of about 420 meters which is almost the same height as The Peak, Two International Finance Centre (2IFC) is the tallest building in Hong Kong and the 6th tallest building in the world today. Same as most of the other cities in the world, this tall building is located in the center of the CBD. It lies above the Airport Express Hong Kong Station, which is not only a transportation focal point but also within a well-established commercial area – Central District.

2IFC is the second phase of the IFC development. The whole of this development has a total site area of 430,000 sq.ft., being constructed in three phases:

Phase I - A 3-storey shopping mall (South IFC Mall) and a 38-storey office tower office space (One IFC). It is opened to the market at the end of 1998 after the opening of Chek Lap Kok Airport.

Phase II - Another 3-storey shopping mall (North IFC Mall) and an 88-storey office tower (Two IFC). It has been completed in year 2003.

Phase III- A 5-star hotel (Four Season Hotel) and a serviced apartment block which is expected to come to service by the end of year 2005.

The IFC development also provides a 4-storey basement parking of 1,800 spaces and a total floor area of 4.68 million sq.ft.. To minimize the management labour cost and to quicken the payment procedure, both

[1] http://www.shortsupport.org/News/0404.html
[2] http://www.infoplease.com/ipa/A0001338.html

One and Two IFC carparks are adopting the credit card and Octopus (the most popular travel and purchase stored value smartcard system in Hong Kong) auto-payment system. Drivers can simply "touch and go" while they are driving through the entrance and exit gates. Thus, no counter cashier is required in these carparks. More labours can therefore be distributed to carpark patrol and administration.

2IFC, the tallest tower of the IFC development, has 88 office floors. It provides totally 1.9 million sq.ft. GFA at a construction cost HK$2.6 billion (about US$333.33 million). This is a joint development of five big property developers and financiers in Hong Kong (Sun Hung Kai Properties Limited, Henderson Land Development Company Limited, The Hong Kong and China Gas Company Limited, Bank of China, and MTR Corporation). Because of its prime location and special designs, 2IFC has also become the most expensive office tower in Hong Kong today (up to about HK$70 dollars per sq.ft. or US$96.60/sq.m.).[3] To maintain this high value in the market, 2IFC has indeed owns some uniqueness besides its tall building height:

3. Lifts

2IFC has display monitors and directory at its reception counter, and also inside lift cabs. These monitors provide news headlines, financial information and flight schedules to the tenants and pedestrians of the building. 2IFC has 63 high-speed passenger lifts in 7 zones. Each passenger lift has a loading of 1,600kg in which 21 people can be carried per service. To increase the security control, Octopus card access control is being used in the passenger lifts. Also, there are 2 service lifts serving all levels. (One 1,350kg at 2.m/s and another on 3,000kg at 4.m/s). Besides elevators, there are escalators for transit from the ground floor main entrance to the third floor main lobby, and there are 2 carpark lifts for transit between third floor main lift lobby and basement carpark. It also has Hong Kong's first double-decker lifts design. Under such circumstances, the transferred lift lobbies are located at 32/F, 33/F, 53/F and 54/F. In 2IFC, the lifts run at a speed of 6m/second for the low zone and 8m/second in the high zone. The management staff also got to remind the visitors to go to the right floor for lift transference whenever necessary.

[3] DTZ Research. Hong Kong Update. 1st quarter 2005.

4. Electrical

2IFC also got a well-designed electricity supply system. It has a 60kVA/floor generator back up and a 500 sq.ft. UPS room at mechanical floor. Besides the dual sources feed high voltage power supply by the Supply Authority, there is a dual power supply riser for each floor. Like other intelligent buildings, a standby generator provides emergency power supply to the tenants.

To enable tenants can work under a more comfortable environment, an electronic ballast for office lighting for eye comfort and eliminating harmonic distortion to vision display unit (VDU) was installed. There is also the Chartered Institute of Building Services Engineers (CIBSE) Cat 1 and Cat 2 imported type ultra low brightness luminaries suitable for VDU's application.

During emergency, 2IFC's standby generator can provide half of the usual lift services and air-conditioning services for about ten hours. In the office tower, there are four numbers of fire services generators at a 6.2 MVA installed capacity and 13 numbers of non fire services generators at a 18.9 MVA installed capacity. Also, a standby oil tank, which claims to be the largest oil tank in Hong Kong, is being constructed in the building for emergency fuel storage during urgent situation.

To ensure there could be sufficient oil supplied to the building during emergency, 2IFC management has to contractually agree with the oil company for the provision of oil within two hours in the event of electrical disasters.

5. Air-Conditioning

Air-conditioning is another important facility in this building. 2IFC is using seawater pumps for air-conditioning cooling. It has a total number of 6 pump cells in which three duty plus one standby pumps are built for the office Tower. 2IFC has two air-handling units per floor. It has a 35 cooling tons/floor Essential Chilled Water Supply. Also, there is the VAV Ceiling Ducted System with perimeter zone heating in winter. A fully distributed intelligence Direct Digital Control for HVAC system is also installed. As a contemporary intelligent building, each of its VAV box can be switched on and off independently and remotely. There is also an Overall Thermal Transfer Value (OTTV) around 24 W/M^2. It is better than many of the local code requirement of not exceeding 35 W/M^2. As well, a supplementary cooling system is available.

In the whole office tower, a total number of ten 980 tons (LV) chillers (Normal) that contents 9,800 refrigeration tones and another eight 350 tons (LV) chillers (essential cooling)that contents 2,800 refrigeration tonsare being constructed.

To maintain a good indoor air quality (IAQ), Primary Air Units (PAU) are installed on the mechanical floors on 6/F, 33/F, 55/F and 56/F where the PAU unit would pre-treat the incoming air and allow fresh air supply to the AHUs, which can dilute the concentration of CO_2 level and keep 800 ppm fresh air in office environment.

To allow the tenants to have more flexibility in adjusting the air-conditioning temperature, each 23,000 sq. ft. office floor can set into 72 individual air-conditioning control zones at maximum. Each of the individual air-conditioning zones can be remote controlled by the management staff in the BMS room. Although the building's standard hours of air-conditioning supply are according to the majority Hong Kong office hour (8am - 7pm on Mon – Fri; 8am - 1pm on Sat), air-conditioning hours can be extended by 2 hours per day at no additional charge to the tenants subject to agreement with the landlord. For some of the major tenants of 2IFC are international financial institutes, long and flexible air-conditioning supply is especially important for such users. Management of such tall buildings should therefore learn to be more flexible in dealing with their clients.

6. Human Comfort

Comfort environment concept has been brought into 2IFC since its construction.

First, it has to maintain good air quality for each typical floor and provide a fresh air rate of 9 liter per person. It adopts the CO_2 Control for Primary Air Quality Moderation.

Second, there is a telephone switching for the after hour A/C services via the building management system (BMS).

Third, IFC has also considered the elimination of Sick Building Syndrome. There is also a high efficiency air filter and provision of access door in ductwork for duct cleaning.

Also, the building adopted a ceiling return system, which meant no return air duct but the supply air duct is being set. This setting can reduce some ceiling void and make the ceilings higher. As a result, the towers have a high ceiling of about 2.7 metres for the general office floors and 3.3 metres for the trading floors are provided.

Fourth, the building has the low solar heat gain and low noise transmission through the double-glazed curtain wall. In other words, bigger windows and more natural lighting are available in the building.

Fifth, the office lighting is set with electronic ballast allowing flicker-free VDU operation hence eliminating eyestrains.

Also, it also has a raised-floor of 190mm on the office floors and 340mm on the trading floor. These raise area provide space for the tenants' cable and wire networking so as to allow users to have a more effective working area. The whole design is to provide a physical and psychological comfy environment to the tenants of the building.

7. Structural Design

Two IFC, with its 88-floor height, provided horizontal fire separation with four refuge floors and four emergency exit staircases. 2IFC's central concrete core combines with eight huge external columns and eight secondary columns to carry the structure's weight. The structure of the building is strengthened by four large, three-storey, steel outriggers that are reinforced to withstand windy conditions.

8. Building Management System

IFC has a central system for security management, fire services, HVAC Control, and Building Services Supervision. It is also one of the few buildings in Hong Kong that got a full year 2000 compliance for passing through the Y2K computer bug challenge. (One IFC is built in year 1998).

There are about 1000 CCTV cameras, which are accessed via turnstiles, installed in and around this development. The building has also adopted the Octopus Card access control system.

As many landlords have aware that facilities manager is the one that run the building for most of the building life, property management input should be invited into the building during its development stage. In IFC, the building management team is also able to add opinions to meet the technical requirements in some technical tenders, and also offer opinions on the kind of materials to be used for floors, carpeting and wall coverings.

9. Octopus Card System

Octopus Card is the smart card system that was introduced by MTR Corporation in 1997. Within these years, the Octopus has become the most popular "touch and go" traveling and purchase smartcard in Hong Kong. In June 2005, there are already 12,000,000 cards being issued by Hong Kong Octopus. In the other words, it is believed every Hong Kong people should at least own one Octopus Card. (As per the figure of Census and Statistics Department of Hong Kong, the population of Hong Kong in year 2004 is 6.883 million).

2IFC has adopted this most popular smartcard system as its building's access and lifts operation control. With this contactless smartcard system, all the entrances and exits of the tenant company's staff, visitors, contractors, workmen of the building are recorded. This face recognition system is also adopted to provide a better control at the entrance/exit and lifts operating control during the off hours. Thus, building managers today are required to update their knowledge on new products from time to time without intermission.

9. Security

After the September 11 incident, people begin to aware of the security management of tall buildings. On the anti-territories attack provision, 2IFC has designed to limit car sizes to reach some structural column so as to prevent bomb attack to the building.

Also, an X-ray checking machine is available at the mailing and postal packages of the in-house post office. In addition, several trained dogs are used for guarding and detecting dangerous goods in both the building lobby and carparks. The need to find suitable dog trainers is another special requirement for the building management of 2IFC.

10. Cleaning

Besides the daily office cleaning, it also provides a tailor-made and flexible cleaning schedule that fits individual tenants and business requirements.

To ensure the required services can be handled in due course, a Real Time PDA/PC housekeeping report/work with bar code system is being used for daily management.

The cleaning team undertakes year-round cleaning of the building's glass curtain wall to enhance the spectacular views of Hong Kong and Victoria Harbour.

With its huge size window frontage, the management of 2IFC has difficulties in finding a suitable automatic window cleaning system. To complete one round of window cleaning in Two IFC, it takes a team of window cleaners operating three gondolas from 7:00am – 12:00 mid-night everyday for window cleaning. As the building is slightly narrowing up on the upper floors, the flexible extension arms gondolas are also adopted for this cleaning purpose.

In addition to the routine cleaning of public areas, the management adds value by providing individual daily office cleaning, with tailor-made and flexible schedules that suit the tenants' preferences or business needs.

11. Concierge

Other than the services provided by most of the commercial buildings, the concierge provides additional services to their tenants – such as restaurant reservations, travel arrangements to floral bouquets ordering.

Rather than traditionally processing all enquiries through a remote management office, the concierge services are directly accessible at the Ground Floor and First Floor entrance lobbies.

A Duty Manager is also available 24 hours a day, seven days a week to answer enquiries and provide assistance at a more senior level in 2IFC.

12. Telecommunication

Besides the well-designed fixed telecommunication system, 2IFC also has a dual dedicated riser duct for tenant's telecommunication cables, an in-house Voice and Data Backbone System, and an obstruction free raised floor for the telecommunication systems of the tenants.

As well, the building has installed a mobile phone reception amplifier at the individual floor lobbies. Tenants of 2IFC can thus use the mobile phone freely even when they are inside the lift cabs.

13. Green Construction Techniques

2IFC has put in green construction techniques since the designing stage. The building has worked to maximize its use of mechanical construction and system formwork, the use of prefabrication, the use of recycled material, and the use of steel construction so as to minimize formwork consumption of the building.

From its first day of construction, the owner of the building has carried out an interactive approach. Numerous discussions were held with potential tenants to identify their specific requirements.

To ensure the clients requests can be solved properly, an in–house contractor on board is also available from day one.

14. Genuine Energy Savings and Sustainability

Besides the functional and cosmetics items mentioned above, the building has indeed got a real concert of energy savings and sustainability.

The direct seawater–cooling chiller plant in lieu of air–cooled has indeed saved 30% of the energy consumption. The extensive use of variable speed drives saved up to 10%. The fluorescent light fittings with electronic ballast and energy saving lamps saved 15% of the electricity. Also, the fully integrated control for air–conditioning saved another 10%.

Also, 2IFC got a simple double-glazed low-e façade, an energy efficient building envelope of OTTV 23w/m, a CO_2 control for primary air quality moderation, a primary air supply run around coil that provides sufficiently dry air supply to the building.

15. Prizes

With these different kinds of value-added hardware and software, 2IFC received prizes and awards from the industry.

In year 2004, Two IFC received the Commendation Award from the Hong Kong Institution of Engineers. It has also achieved the rating of excellence from The Hong Kong Building Environmental Assessment Method (HK-BEAM). Also, 2IFC was chosen to be the Intelligent Building of the year 2004 by the Asian Institute of Intelligent Buildings (AIIB). It received a historical high score of 95 out of 100 (distinction score) from the AIIB assessment.

16. Conclusion

As a building of the 21st century, 2IFC has put users' requirements at a very high rating since its stage of design. Both the physical and psychological concerns of users are being put into consideration. The participation of the management is definite crucial to upkeep the sustainability of the building.

According to the information of this building's design team, IFC is designed built to last up to 120 years of life.

References

1. Construction & Contract News. Issue no. 2, 2004.
2. DTZ Research. Hong Kong Update. 1st quarter 2005.
3. International Finance Centre sales brochure (1998).
4. Vertical Challenges. *Building Services Professional*. April 2004.
5. http://www.shortsupport.org/News/0404.html
6. www.shortsupport.org/News/0404.html

SUGGESTING A NEW LIFESTYLE IN ASIA ROPPONGI HILLS IN THE VERY HEART OF TOKYO, A COMPACT NEIGHBORHOOD WHERE WORKPLACES AND RESIDENCES ARE WITH CLOSE PROXIMITY---A MODEL FOR THE NEXT DECADE

MASAHIRO WATAHIKI[1] HITOMI NAMBA[2]

Mori Building Co., Ltd., 6-10-1 Roppongi, Minato-ku,Tokyo, 106-6155 Japan

During the past 25 years, cities throughout Asia have expanded more rapidly than ever before. Tokyo is no exception. The urbanization tide that began in the latter half of the 1960s made Tokyo one of the world's largest cities; nevertheless, the era of industrialization is now over, and we have shifted to a post-industrial society based on knowledge and information. Now, with the change of times, urban planning in Tokyo also needs to evolve out of its priorities of the industrial era. It is true that, compared to the West, Asia has lagged behind in terms of investment into cities. By employing new technologies and concepts, however, Asian cities undoubtedly have the ability to become leading cities of the 21st century. The future direction calls for multi-functional, compact cities where workplaces, residences and leisure are within close proximity. In order to realize this, the present urban structure, which is horizontally dense and vertically sparse now needs to be remodeled in order to build multi-functional cities which can achieve high density and favorable environment simultaneously. By taking Roppongi Hills (opened: spring 2003) as an example, let us look at a new lifestyle for the 21st Century that makes this compact city concept a reality. If a newly attractive urban lifestyle for the 21st century is established in Asia, the world will look to Asia for a new urban model.

1. The Era of the "City State"

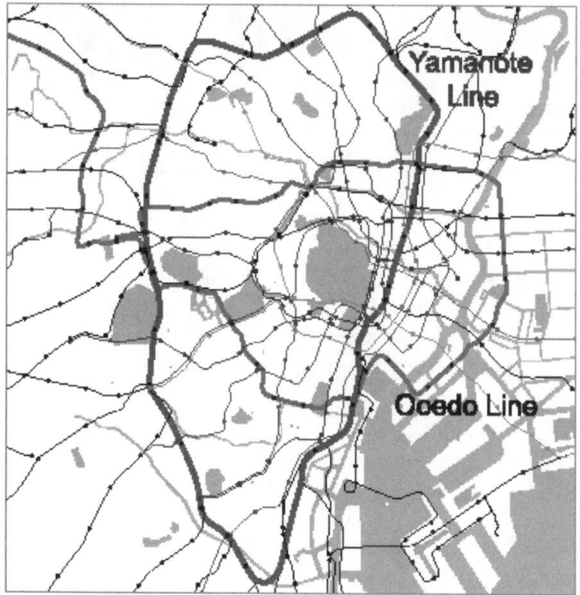

Figure 1. Railroad system network in central Tokyo

1.1 *An era of mega-competition between global cities*

Cities throughout the world are competing intensely with one another and the fiercest global competition is not in the agricultural or industrial sectors, but is, in fact, between global cities.

In the past, cities have played the role of representing their nations. Today, however, human resources, information and capital travel easily across borders and have the liberty of choosing their own bases. Thus, cities have become the entrances or the first contact points. The key to vitality and development is whether a city has the ability to attract human resources, information and capital — this is true not only for cities themselves but also for their countries.

1.2 *History of city formation in Tokyo*

During the era of Japan's industrialization, the excess population from agricultural districts was absorbed by secondary industries, with the result that a large number of people flowed into cities throughout the 20th century. In the 1960s, when Japan was at its peak of economic growth, 400,000 to 500,000 people annually moved into Tokyo; a number equal to the population of an entire

[1] Senior General Manager, Advisory Staff to C.E.O.
[2] Assistant Manager , Advisory Staff to C.E.O.

small city. Urban policy at that time, therefore, was to refine the use and curb density. Due to noise, vibration and contamination, factories back then had to be located away from zones dedicated to other functions. Commercial districts also were separated from other city functions.

In addition, as a result of trying to place a limitation on density, efficient use of land in city centers (through high-rise construction) was restrained. Consequently, residences, workplaces and leisure districts were separated by considerable distances, leaving cities with external inefficiencies.

In Tokyo, a belt line railway, the Yamanote Line, encircles the central wards, with subway networks serving the area inside the belt. From stations on the Yamanote Line, trains reach out to suburban areas.

Basically, the land inside the Yamanote Line is considered to be mainly for business use and Yamanote Line stations are surrounded by commercial facilities. Fanning out from these stations are suburban residential areas.

Now that we have entered the 21st century, however, a reconsideration of this urban structure is required in Tokyo. Urban planning is necessary to ensure quality urban life that combines residential comfort, good working and leisure environments. Tokyo must become a more attractive city in which to live, and has to offer ample space and satisfying time. Achieving such goals would bring about, not only high quality urban living, but also continuous development, as human resources, information and capital would be attracted to Tokyo from all over the world.

2. Features and Problems of Tokyo

2.1 *Tokyo city structure — horizontally dense and vertically sparse*

The structure of Tokyo is horizontally dense and vertically sparse compared to other major global metropolises. Because efficient land use has not made any progress in central Tokyo, the overflow residents and industries both spread far into the suburbs, resulting in one of the rare cities in the world sprawling endlessly in a low-rise horizontal way, with extremely long movement distances for people.

2.1.1 Unbalanced daytime and nighttime population ratio

Tokyo's four central wards and Manhattan, New York City, have approximately the same areas, 6,000 ha (60 sq km). And the daytime populations of the two districts are also roughly the same — both approximately 3 million. But the nighttime populations of the two districts, however, are 0.55 million and 1.54 million, with, population densities of 92 people/ha and 250 people/ha, respectively.

The daytime and nighttime population ratio in Manhattan is 2 to 1; whereas in Tokyo's four central wards, it is an extremely unbalanced 6 to 1. Because of this, a massive commuting movement occurs every morning and evening, resulting in long and infernal commuting hours that shorten leisure time for suburban residents.

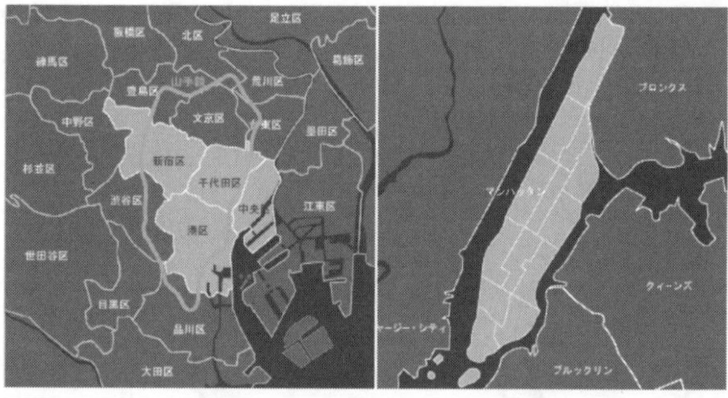

Tokyo central four wards	New York, Manhattan
Area 6,000 hectares(60 km^2)	Area 6,000 hectares(60 km^2)
Day-time population 3 million (500 per hectares)	Day-time population 3 million (500 per hectares)
Night-time population 500,000 (83 per hectares)	Night-time population 1500,000 (250 per hectares)
D/N population ratio 6:1	D/N population ratio 2:1

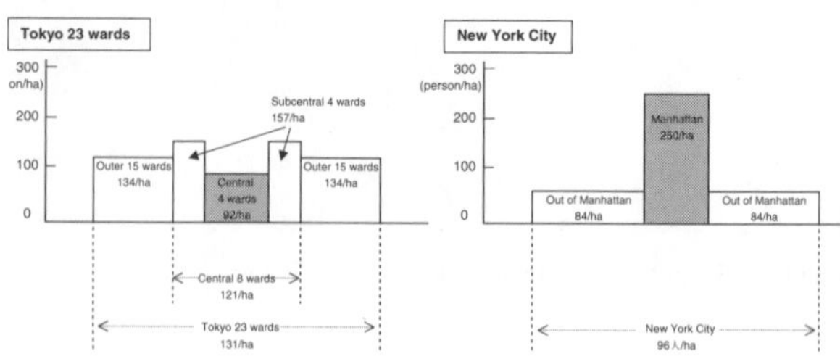

Figure 2. The daytime and nighttime population of Tokyo and New York

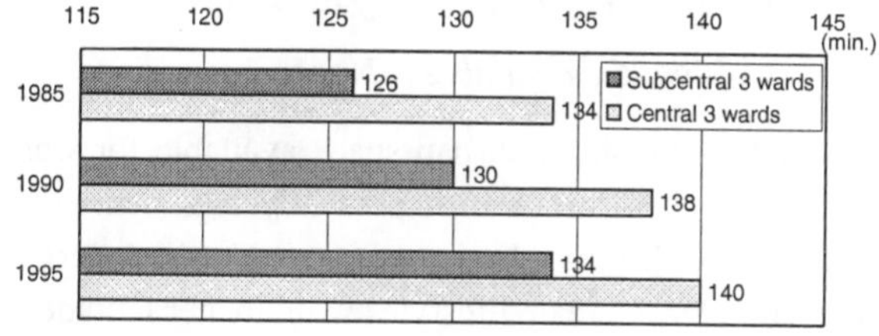

Central three wards: Chiyoda, Chuo, Minato
Sub-central three wards: Shinjuku, Shibuya, Toshima

Figure 3. Average commuting time of person who commutes to Tokyo central three wards (round trip and train use)

2.1.2 Depopulation of central Tokyo

Seen Tokyo and New York from the wider viewpoint, the population density of Manhattan is approximately three times that of the four central wards of Tokyo. In New York the population is denser in the heart of the city; whereas in Tokyo, it is sparser.

2.1.3 Poor urban living space with insufficient residential space

Spaces available for residences, offices and cultural facilities in Tokyo are generally smaller than in the cities of advanced Western nations. As to its residential space, it is not only far inferior to that of cities in Western countries but also those in East Asian countries.

Table 1. Area of house and park per capita by city

	Floor area per housing unit (m^2/unit)	Park Space (m^2/person)
Shanghai	--	11.1
Hong Kong	--	10.2
Tokyo	55	5.3
Sigapore	85	7.5
New York	80	29.1
London	--	25.3
Paris	90	10.2

2.1.4 Inefficient land use

The lower nighttime population, smaller residential spaces and less greenery in central Tokyo all result from inefficient land use.

The average effective FAR (Floor-Area ratio) of Tokyo's 23 wards is 136.1%, and only 337.3% even in the four central wards.

Table 2. The average effective Floor-area ratio of Tokyo's 23 wards

(ha,%)

	Building land area A	Gross Floor Area B	Existing FAR C=B/A×100	Legal FAR D	Rate of filled vacancy E=C/D×100
Chiyoda	352	1,986	563.9	538	104.9
Chuo	390	1,874	480.2	569	84.3
Minato	915	2,766	302.3	407	74.3
Shinjuku	972	2,243	230.8	386	59.7
central wards	2,629	8,869	337.3	475	71.0
Bunkyo	590	1,105	187.3	338	55.5
Taito	453	1,239	273.6	485	56.4
Shibuya	807	1,753	217.3	327	66.4
Toshima	791	1,411	178.4	352	50.7
central wards	5,269	14,376	272.8	425	64.2
23wards	31,379	42,692	136.1	254	53.5

Source:Tokyo no tochi(Land in Tokyo)2002

2.1.5 Small blocks surrounded by narrow streets

Intensive land use, of course, requires large building sites; however, in central Tokyo, small blocks, often approximately 8m wide, are surrounded by narrow streets, even in re-zoned districts.

The size of a block in Tokyo is approximately 0.1 ha; extremely small compared to the average 1.5 ha blocks of Manhattan, Paris and London. Moreover, Tokyo blocks are often further subdivided into smaller lots, which clearly hampers efficient land use.

To promote the kind of city-making for the 21st century, it is, therefore, absolutely vital to reorganize Tokyo's narrow streets, build wider thoroughfares, and create as many large sites as possible to allow larger constructions.

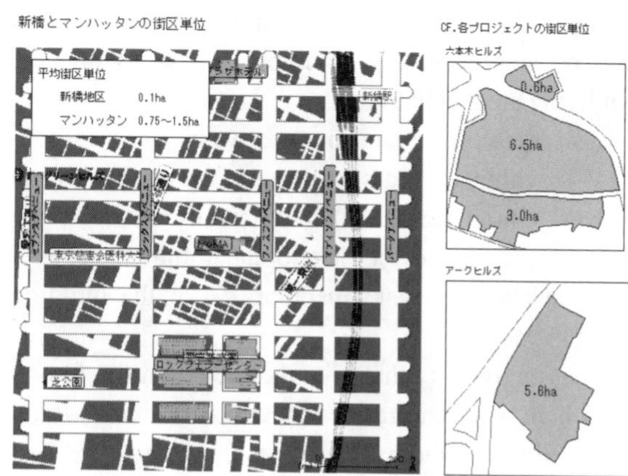

Figure 4. Comparison of city blocks

3. The Future Way of Modern Cities (Building cities where elements of daily life can be found within walking distance)

Along with the change of social and economic structures from an industrial society to a knowledge- and information-based society, urban structural reform becomes indispensable. Since the Industrial Revolution, and particularly during the 20th Century, an industry-based society developed, in which work, residential and entertainment areas were separately zoned. As our society transforms into a post-industrial, knowledge- and information-based society, however, a new urban structure, better suited to this era becomes absolutely necessary.

In Europe and North America we are already seeing the construction of attractive cities that meet the needs of the times, and also in Asia, many metropolises are moving ahead with rapid reforms; however, Tokyo and other Japanese cities have not freed themselves from outmoded industrial-age urban planning policies. The myth of single-family houses, combined with an urban policy that has separated work and residential zones, have resulted in a massive horizontal urban sprawl with population and industry both overflowing into the suburbs. As a result, Japan cannot now afford to delay reforming its urban policies from those based on the priorities of the industrialization era into those that fit a highly advanced information-oriented society. To achieve this goal, subdivided small lots of land must be consolidated into large-scale blocks. Furthermore, living, working and leisure areas should be concentrated in high-rise complexes; thereby improving safety and the urban living environment.

It is possible to create an attractive city — a re-born Tokyo — within 25 years by drawing up and following an urban 'Grand Design' that efficiently channels the energy of private sector development initiatives, and which plans the redevelopment of 3% of the building stock each year.

An attractive city that suits the society of the future, is one where work and residential areas are found within close proximity; where quality time, living space and urban environment are plentifully available to urban dwellers. If, however, we are to create an attractive, re-born Tokyo, the following five elements need to be considered.

3.1 *Objectives to be achieved by the re-born city*

3.1.1 Doubling the available urban space

We need to double the urban space available for work, residence and leisure, which presently lags behind the standards of leading European and North American cities, in order to create lifestyles with greater latitude.

3.1.2 Creating leisure time

We need to create mixed-use urban complexes that provide work, residence and entertainment facilities within walking distance of each other, and where urban residents are free from unbearable commuting; thereby creating increased leisure time.

Concomitantly, these urban complexes will draw full-time residents back to the center of Tokyo; as is the case in Manhattan.

3.1.3 Wider range of choices for citizens

To meet the personal desires of residents, we need to create a cultural city capable of doubling the choice of urban amenities for those who live, work and visit.

3.1.4 Improving safety

Tokyo still has many areas that are densely packed with wooden housing and other buildings constructed to old anti-earthquake standards, despite the fact that vibration-control engineering has made great strides since the Kobe earthquake. In fact, the new standards now make it safer to remain inside buildings built with modern technologies than take refuge outside in the event of a major earthquake.

3.1.5 Serving the global environment

Increasing open space and covering rooftops with greenery helps prevent the heat island phenomenon; constructing buildings to last more than 100 years preserves resources; and building compact cities, where work and residences are in close proximity, conserves energy that would otherwise be spent on transportation.

3.2 *Changing the Basic Paradigm of Urban Policy*

The social environment has improved dramatically since the current Urban Planning Law and Building Standards Act were promulgated. The main transportation networks, such as subways and JR, are much better, and approximately 80% of the districts inside the Yamanote Line are now located within 500 m of a subway station. The daytime population in the center of the city, however, has stagnated and currently shows little growth.

Approximately 70% of households live in apartments.

Taking these changes into account and in order to gain the competitiveness as an international city, the paradigm of urban policy must change as follows.

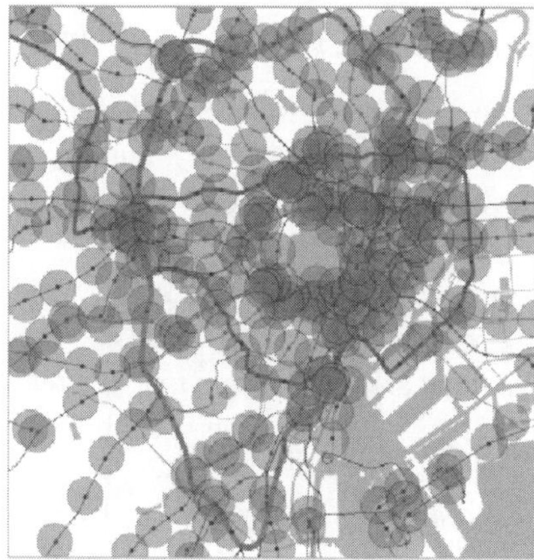

Figure 5. Circles within 500 meters from railway or subway station

3.2.1 Accelerating the consolidation of land lots

Restructuring districts through the consolidation of small, subdivided land lots and the abolishment of narrow roads can create more spacious areas, wider roads, increased public spaces and parks, and provide a better urban environment.

The previous urban policy placed no restrictions on the division of land, while the consolidation of land, when deemed as a special case, required multilayered complex procedures. As it is indispensable for the provision of an international-level urban framework, we must change our urban policy to one that promotes the consolidation of land, by acknowledging the precedence of public interest over the excessive protection of private ownership

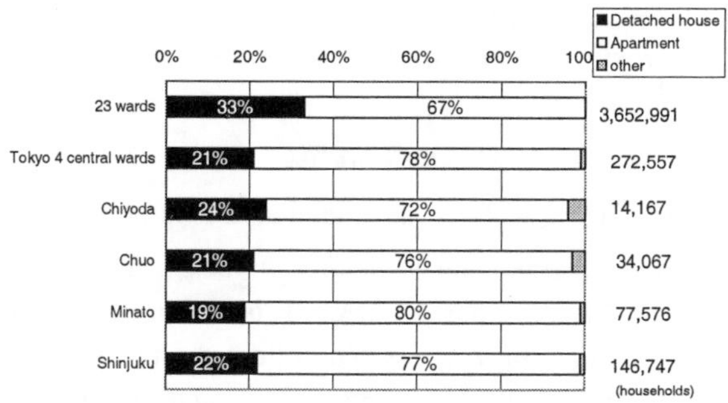

Figure 6. Type of existing house in Tokyo (by unit)

3.2.2 Promoting efficient use of land

As public facilities were previously improved only to meet the demands of the growing city, the efficient use of land in the center of city was restricted under the former urban policy. We now need to promote the efficient use of land in the center of city, taking into account the underlying changes such as improvements in public transportation, the decreasing population in the near future, the desires of the populace to return to urban housing and increased space per capita in the center of city. We also need to attract more people back to the center of city, perhaps setting the population ratio of Manhattan as a goal, and encourage an expansion of urban space per capita.

The construction of high-rise buildings is the most efficient way to create open spaces in the center of the city, if we are to try to achieve such apparently contradictory targets as a good environment and a high population density.

3.2.3 Shifting from simple use to mixed use

The previous urban policy, which encouraged the separation of working and living places, was good for an industrializing society, but is not suitable for the current highly advanced information society. In fact, an environment with continuous space and time, within which the boundaries of work, residence and leisure become blurred, is more suited to our modern society. In the future, we can expect less children, more elderly people, and more working women, so urban functions, such as work, living, leisure, recreation, schools, hospitals and childcare facilities, all need to be within walking distance of each other.

3.3 *Creating a highly-dense but high quality urban environment through the utilization of high-rise buildings*

3.3.1 Tokyo's four central wards

(i) Rebuilding with the same total area as at present

As mentioned in table 2, the FAR in the privately owned land of Tokyo's central four wards is 337%. If, however, all the buildings were consolidated into 50-story constructions, the percentage of land needed for buildings would amount to a mere 6.7% of the total land currently built, enabling the remaining 93.3% to be turned into green space.

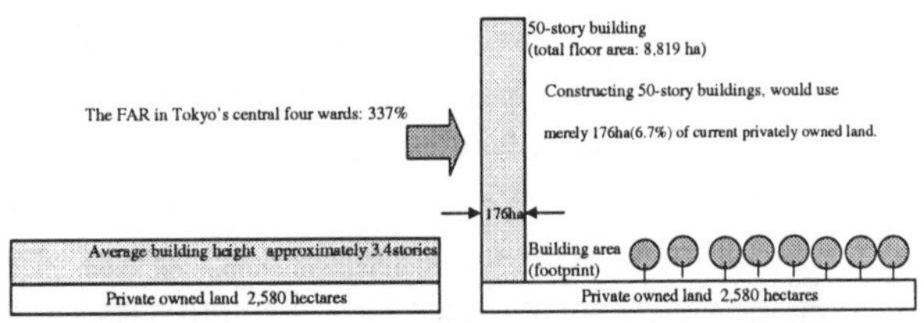

Figure 7. The case of rebuilding with the same total area as at present

(ii)Rebuilding with triple the nighttime population and double the living space per capita

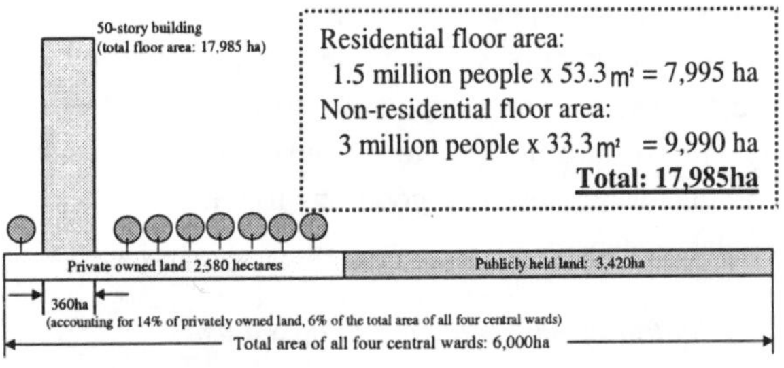

Figure 8. The case of rebuilding with triple the nighttime population and double the living space per capita

In order to achieve a more spacious urban environment and improve the ratio of daytime to nighttime population to 2:1 in Tokyo's four central wards, we need to assume a trebling of the nighttime population to 1.5 million, with 40 m² residential area per person and 25 m² non-residential space (for work, leisure, recreation and schooling) per person.

The above numerical target requires a total area of 53.3 m²/capita and a non-residential area of 33.3 m² /capita, including common use spaces such as elevators, hallways, etc. A total floor area of 17,985 ha (currently 8,819 ha), comprised of a residential floor area of 7,995 ha (currently 2,743 ha) and a non-residential floor area of 9,990 ha (currently 6,076 ha), is needed.

By concentrating the total floor area of 17,985 ha into 50-story buildings, the constructed area stands at 14% of the privately owned residential land, or 6% of the total area of the four central wards, and the remaining area could be used for public open spaces and green areas.

We can see the creation of a three-dimensional city with treble the nighttime population (1.5 million) and significantly increased living space in Tokyo's four central wards.

Therefore, not only would living space be significantly increased, but also more parks and green areas could be created, even with a trebling of the nighttime population.

3.3.2 Distance between neighboring buildings

We can see the distance between neighboring buildings, assuming that 50-story buildings, with the same floor plates as Mori Tower, were constructed.

17,985 ha (total floor area) ÷ 5,000 m² (Mori Tower) ÷ 50F = 720 buildings
6,000 ha (total area of Tokyo's four central wards) ÷ 720 buildings = 8.3 ha/per building

The area of 8.3 ha is a 290m-sided square, and the distance between neighboring buildings would be 220m, as shown in the figure below.

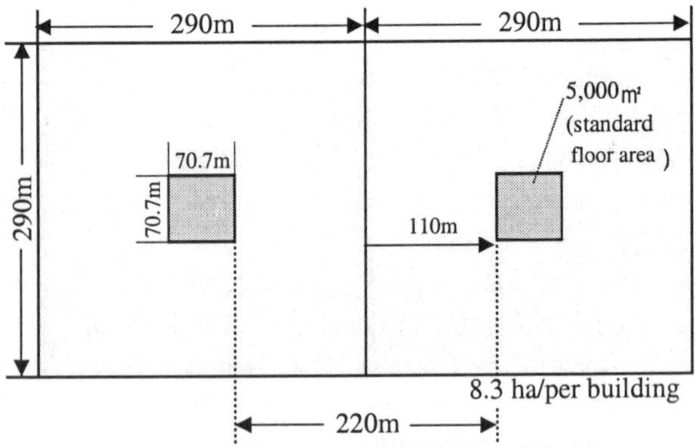

Figure 9 . The case of rebuilding to 50-story buildings which have the same building area as RH Mori tower

Now, we see the distance between neighboring buildings, assuming that 50-story buildings, with the same floor plates as the 43-story Roppongi Hills Residential Tower, were constructed.

17,985 ha (total floor area) ÷ 1,000 m² (RH Residence)
÷ 50F = 3,600 buildings
6,000 ha (total area of Tokyo's four central wards)
÷ 3,600 buildings = 1.7 ha/building

The area of 1.7 ha is a 130m-sided square, and the distance between neighboring buildings is 98m.

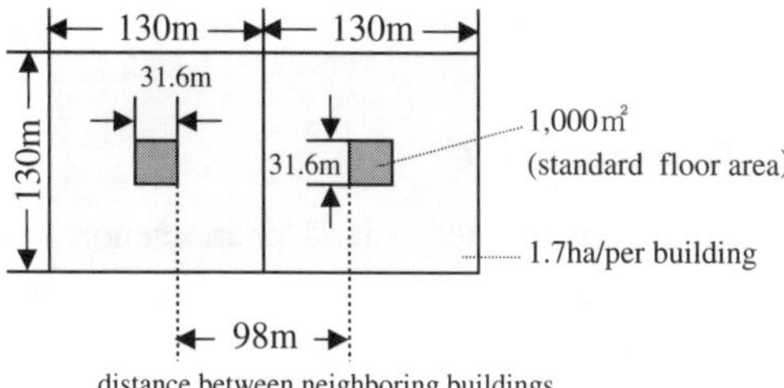

Figure 10 . The case of rebuilding to 50-story buildings which have the same building area as RH residential tower

As can be clearly seen, city creation based on proper planning provides ample distances between neighboring buildings. According to our calculations, distances of between 100 and 200 meters can be created between neighboring buildings, despite the differences in residential floor areas.

4. Roppongi Hills as a solution

4.1 *Before and after Roppongi Hills development*

Roppongi Hills, which opened on April 25, 2003 after 17 years of planning, negotiations and construction, instantly became one of the most popular destinations in Tokyo. We propose Roppongi Hills as a model solution to the urban problems Tokyo currently faces.

(i) Before the development

Before Roppongi Hills was built, the land on which it stands consisted of the former TV Asahi Roppongi Center and small commercial buildings and detached houses along Roppongi Street and TV Asahi Street. The area was on a gently undulating hill and was comprised of subdivided lots of land with a mesh of 4-meter or so wide roads. There were approximately 500 landlords with rights to the 500 lots of subdivided land.

Figure 11. Aerial view of the site before development

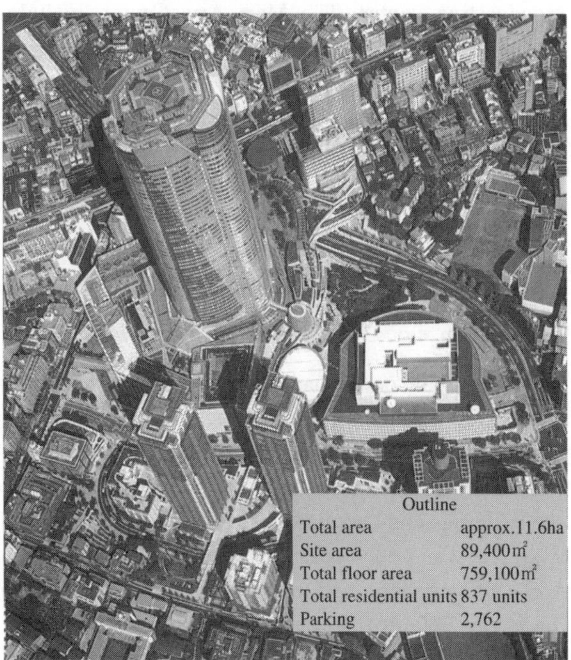

Figure 12. Aerial view of Roppongi Hills development

History of Roppongi Hills development

1986	Tokyo Metropolitan Government designated the Roppongi 6-chome area a"Guided Redevelopment Area"
1988	Minto-ku conducts studies in order to establish a basic urban area redevelopment project promotion plan
1990	"Community Building Council" established
1995	Final city plan (Redevelopment district plan, Type 1 Urban Redevelopment Project, et.al)
1998	"Roppongi 6-chome District Urban Redevelopment Association" establishment approval
2000	Rights Conversion plan approval by Tokyo metropolitan government Construction begins
2003	Completion

(ii) After completion of Roppongi Hills

Roppongi Hills, which was built in collaboration with approximately 400 rightholders on a huge, consolidated lot of land with an area of approximately 12 ha, consists of 11 buildings, including a 54-story office tower, two 42-story residential towers, a 21-story hotel and a broadcasting center. The Mori Arts Center, including Mori Art Museum, an observation deck, a membership dining club and Academy Hills, occupies the top five floors of the office tower. Roppongi Hills is a huge, multi-use complex, including offices, residences, a hotel, a broadcasting center, a museum, an observation deck, a cinema complex, restaurants and retail shops.

4.2 *Greenery effect by the redevelopment*

4.2.1 Green space increased by the redevelopment

Previous urbanization had destroyed ponds and green spaces and had created a horizontally overcrowded area. The Roppongi Hills project, however, uses the land much more efficiently through the construction of high-rise buildings and has succeeded in providing huge areas of greenery and open spaces at the foot of the buildings. Areas of greenery and open spaces, such as Mohri Garden (4,300 ㎡), Roppongi Hills Arena (4,900 ㎡) and 66 Plaza (2,000 ㎡), were either recreated or newly produced throughout the redevelopment. A total of 68,000 trees were planted, including 51 zelkova (keyaki) trees along the newly built road, Keyakizaka-dori, and 63 cherry (sakura) trees along Sakura-dori to the south of the residential towers. Furthermore, the rooftops of the Keyakizaka complex housing the multiplex cinema, the residential towers and the broadcasting center were covered with greenery in order to prevent the heat island phenomenon. The redevelopment project increased the area's green spaces by approximately 9,000 square meters; one-and-a-half times that of before the project.

Figure 13.Mohri garden, roof garden of Keyakizaka complex building

4.2.2 Greenery Effect

The progress of the heat island phenomenon in the city in recent years has had a major effect on society in Japan. Tokyo's average temperature went up about three degrees in the past 100 years in Tokyo. Promotion of greening and the conservation of the green are paid to attention as one means to improve such a situation.

The Ministry of Land, Infrastructure and Transportation in Japan investigates greenery effect in various ways to make the best use of for a green policy in the city. Figure 14, taken by thermal imagery camera, shows the land surface temperature of Roppongi Hills and its surroundings. It shows that Roppongi Hills' temperature is apparently lower than the surrounding areas, suggesting the effectiveness of greenery.

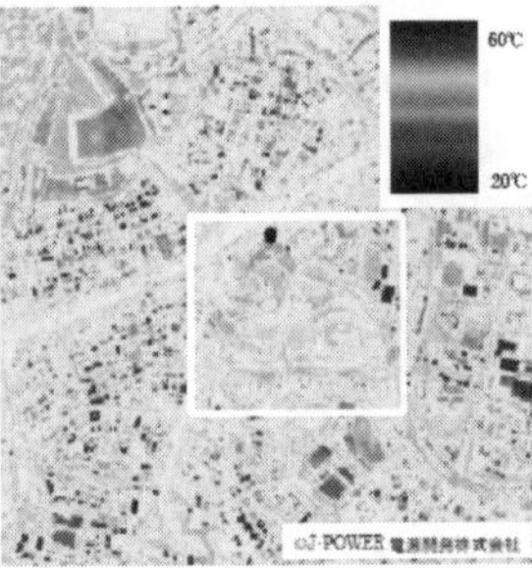

Figure 14. Aerial shot of RH; thermal imagery and actual photograph

4.3 *Disaster prevention conscious city planning*

4.3.1 Transportation infrastructure improved by the development

The Roppongi redevelopment project included the construction of new roads and the improvement of existing thoroughfares. Before the completion of Roppongi Hills, Ring Road No. 3 did not cross Roppongi Street, as it was blocked by an underground tunnel. The Roppongi Hills project, however, provided a new intersection linking the two roads (one of which is 15m higher than the other), by constructing a sideway inclined at 8%. This contributed to easing traffic congestion in the areas surrounding. However, the construction could have cut off the land, so a continuous flow of people was secured by siting artificial ground on the second-floor level of the building. This made the newly provided intersection a multi-level crossing, with a second underground level for the tunnel on Ring Road No. 3, the first underground level for the inclined sideway, the first level for the intersection between the sideway and Roppongi Street and the driveway connected to Roppongi Hills, and the second level with artificial ground with open space used by pedestrians.

Because the roads had been too narrow and steeply sloped for emergency vehicle, Keyakizaka, a 16m-wide, 400m-long street was created to improve access both to TV Asahi-Streer and Ring Road No. 3. The 24-meter wide road and walkway area, created by backing the buildings' wall lines away from the road, was united with the shops on the artificial ground to produce a pleasant and comfortable environment.

Figure 15. Improvement of transportation infrastructure by the development

Tokyo central four wards	New York, Manhattan
Area 6,000 hectares(60 km^2)	Area 6,000 hectares(60 km^2)
Day-time population 3 million (500 per hectares)	Day-time population 3 million (500 per hectares)
Night-time population 500,000 (83 per hectares)	Night-time population 1500,000 (250 per hectares)
D/N population ratio 6:1	D/N population ratio 2:1

4.3.2 Consideration to energy conservation and saving resource

The energy of necessary electricity and heating, etc is produced with the gas in Roppongi Hills. And a regional air conditioning is done by using the waste energy. As a result, the reduction in the primary energy of about 20% becomes possible by using a business electric power compared with the case of the individual air-conditioning, and the amount of the exhaust of CO_2 and NO_X has been greatly reduced.

Figure 16. Regional air conditioning system (steam boiler) and a large-scale gas co-generator (gas turbine)

4.3.3 Earthquake-resistant systems

Each building of the Roppongi Hills development is equipped with earthquake-resistant systems, such as seismic isolation and vibration control; thereby protecting buildings against damage from all earthquakes, including those of the level of the Great Hanshin earthquake.

Moreover a new earthquake vibration-control technology was introduced into the Keyakizaka complex by structurally separating its greenery-covered rooftop from the building itself.

Figure 17. Mechanism of Green Mass Damper

4.3.4 Installation of Well for disaster and fire fighting water supply

The wells for the disaster are dug up from the stop of the supply of waterworks to restoration when the earthquake occurs to maintain the vital function at least.

Moreover, the fire fighting water supply of 100 cubic meters is set up by eight places in total to contribute for in the region and the surrounding area in a time of disaster. Moreover, Pond Mouri that existed before development is maintained, and the route where the fire fighting pump car can approach the pond is secured.

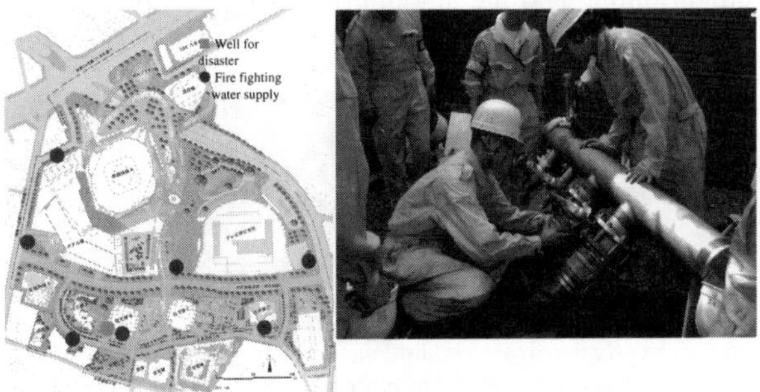

Figure 18. Disaster preparedness equipment in Roppongi Hills

4.4 *Job creation*

Roppongi Hills is a mixed-use complex comprising offices, residences, a hotel, a broadcasting center, a museum, an observation deck, a multiplex cinema, restaurants and shops. Offices and residences occupy 60% of the total floor area with the remaining being used for leisure, recreation, educational, cultural and informational functions. While the employees of the offices and residences were relocated from other areas, approximately 6,700 new jobs were created in other facilities.

Because most visitors to Roppongi Hills come for purposes other than to visit offices or residences, the leisure, recreation, educational, cultural and informational facilities contribute considerably to the attractiveness of the area.

These facilities provide visitors with enjoyable times, encourage them to spend and create new employment; thereby working to revitalize the city.

usage			total floor space		employed person* (resident)	increase
work	office		277,400㎡	(37%)	13,000	-
housing	residence (approx.850units)		124,600㎡	(16%)	2,000	-
leisure	retail	26,800㎡	87,800㎡	(12%)	2,900	2,900
	restaurant	58,100㎡			2,000	2,000
	support facilities (DPE etc)	2,900㎡			30	30
culture	observatory, art museum, cinema complex, conference, salon etc.	34,800㎡	35,300㎡	(5%)	1,000	1,000
	temple	500㎡			-	-
recreation	hotel (approx.400rooms)		53,200㎡	(7%)	700	700
healthcare	medical facility		1,600㎡	(0.2%)	20	20
press	TV station		63,900㎡	(8%)	2,000	-
public	communal facilities, DHC etc.	16,300㎡	114,300㎡	(15%)	-	-
	parking	98,000㎡			-	-
total floor space			758,100㎡	(100%)	employed 21,650 resident 2,000	6,650

*) Number of employed person and residents are estimated figure.

table 3. Number of residents and employed person by usage

5. Problems of large scale redevelopment projects

Some Asian cities, including Shanghai and Hong Kong, are energetically involved in urban renovation in order to compete with other international centers. For the urban regeneration of Tokyo, however, urban policy has to be reconstituted in a way that best utilizes the know-how and abilities of the private sector and which adequately uses the capital, human resources and technologies accumulated over the second half of the 20th century.

5.1 *Shorter implementation timetable*

Any business, regardless of its nature, is developed based on forecasts of the scope of work necessary, financial planning and a timetable. However, forecasts for redevelopment projects are hard to complete, which considerably hampers the process. The 17-year period spent on the Roppongi Hills project, from its planning to completion, is, in fact, deemed to be rather short in Japan. Time, however, is the main risk factor in a major redevelopment project, and needs to be significantly reduced.

5.2 *A 'grand design' for Tokyo*

If we are to achieve an efficient restructuring of Tokyo, the private sector must be encouraged to pour its energy into drawing up a grand design aimed at the construction of an attractive city, which should make the forecast of districts and FAR possible. Municipal governments create objectives and simple ground designs based on specific numerical targets which are set up according to the nature, location and role of each district.

- Balance of daytime and nighttime population
- Urban space per person
  (work, housing, leisure, child care, commerce, recreation, schools, medical facilities)
- Average commuting time per employee
- Green space ratio
- Anti-earthquake systems

6. Conclusion

In many Asian cities, urban restructuring was neglected for the sake of economic growth. Those cities often looked at Western cities as models. As we now face the transition from the industrialization of the 20th century to the new age of intelligence and information, however, no new city model has yet been found.

As we enter the 21st century, many Asian cities have succeeded, or are currently succeeding, in the accumulation of human resources, capital and technologies through economic development. Now, however, the time has come for Asia to create and transmit new ideas of urban regeneration, which has unfortunately been on the back burner, and provide the world with city models for the 21st century.

AIR QUALITY AND HIGH-RISE HIGH-DENSITY RESIDENTIAL ENVIRONMENT IN HONG KONG

PRIYANTHA S. EDUSSURIYA ARLEN M. YE
Department of Architecture, The University of Hong Kong

ANDY T.Y. CHAN
Department of Mechanical Engineering, The University of Hong Kong

Air quality is a major concern for quality of life in high-rise high-density urban environments like Hong Kong. The physical fabric of this environment is suspected to have contributed to poor dispersion of air pollution caused primarily by emissions from automobiles. This paper reports an ongoing research that investigates the relationship between street level air pollutant concentration and the morphology of hyper dense residential environments in Hong Kong. The research postulates that urban morphological attributes, such as locality factor, development intensity and built-form configuration affect the concentration or dispersion of air pollutant within an urban locality. A method has been developed to model the relationship between morphological attributes, microclimatic conditions and air pollution phenomena in real situations. Preliminary findings from empirical studies provide evidence supporting theoretical assumptions and validating the method. Conclusions and new hypotheses arising from the study shall be used to guide extensive empirical survey and further development of theory and research methodology.

1. Introduction

1.1 *Air quality and the hyper dense residential environment in Hong Kong*

According to the Environment Protection Department of Hong Kong, high-density motor vehicle usage constitutes the biggest source of air pollution in Hong Kong (HKEPD 2001) (Figure 1), with the high daily volume of automobile emissions on roads contributing to high air pollution levels in the urban areas (Figure 2). Conventional wisdom would suggest that high air pollution levels in Hong Kong could always be found in urban districts located close to main polluting sources, such as highways and industrial centers. However, studies show that street-level air pollutant concentration in downtown residential areas is almost as high as in the industrial areas. This phenomenon indicates that sources of pollutant air may not be the sole factor contributing to poor air quality in certain urban settings, and there must be other factors that also cause high level of air pollution in urban areas distant from major polluting sources (Figure 3).

Researchers have for sometime been suspecting that air pollution in urban settings is not only a function of the rate of emission but also that of pollutant dispersion; and the movement of air pollutants generated by vehicle traffic could also be influenced by the characteristics of a built-up area (De Haan et al., 2000; Mayer, 1999; Theurer, 1999; Weber, 2001). While the physical form of a thinly developed urban center may not pose significant problems for pollutant air dispersion, the urban fabric of Hong Kong composed largely by high-rise high-density building masses could prove to be a crucial factor affecting the concentration or dispersion of air pollutants.

Figure 1: Sources of Air pollution in Hong Kong

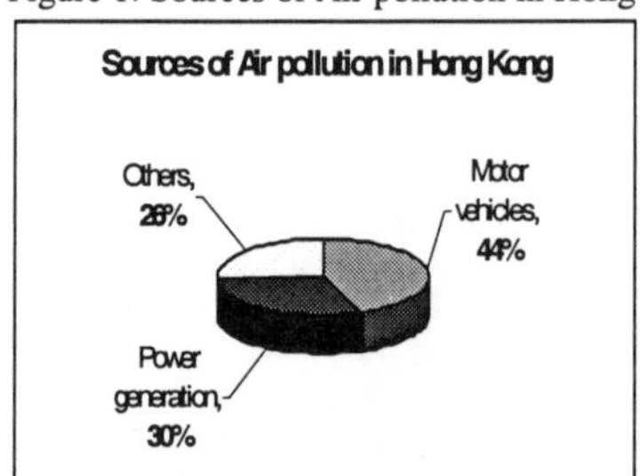

(Source: SCMP 22nd April 2003 & HKEPD,)

Figure 2: Diurnal variation of respirable suspended particulates (RSP) and traffic flow in HK.

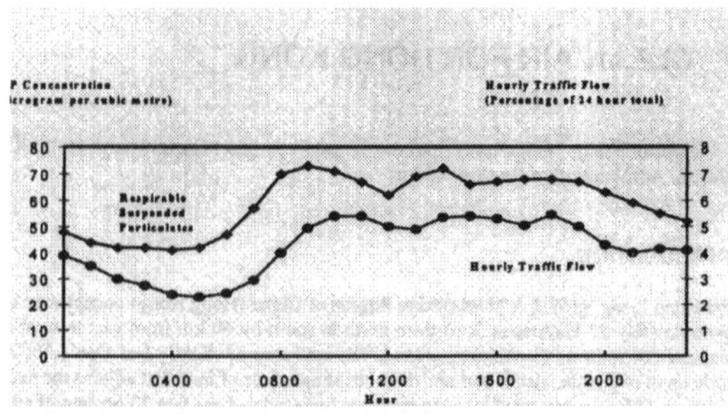

(Source: Extract from Hong Kong EPD 2001)

The hyper dense residential environment of Hong Kong is quite unique in terms of population and development densities, building height and building volume, and intensity in land-use mix. More than two third of Hong Kong's 6.8 million population, a considerable part of its transport network and most daily (economic) activities are agglomerated within an urban area that measures less than 20% of the total land coverage (1,068 sq. km.) of the region. Constrained by land scarcity, most residential developments in Hong Kong take the form of high-rise apartment houses, which are often tightly grouped together, leaving only narrow gaps between the tall structures for lighting and ventilation. In most urban districts, where residential developments normally take up 60% to 70 % of the built spaces, different uses or activities are zoned vertically rather than horizontally, with commercial and service activities routinely occupying the lower levels of residential tower blocks. The unique structure of Hong Kong's residential environment thus on the one hand has greatly increased peoples' exposure to polluted air by concentrating considerable volume of vehicle traffic as well as population within a limited urban area; and on the other seems to have a direct or indirect impact on air pollutant concentration by everywhere creating closely-knit building walls that prevent airflow.

Figure 3: Daily RSP concentrations in residential and industrial areas in Hong Kong.

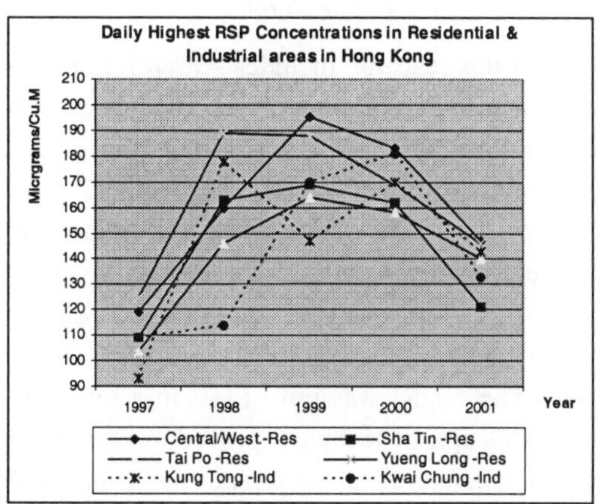

(Source: Extract from Hong Kong EPD reports 1997 to 2001)

1.2 *Physics of street level air pollution concentration*

Street level air pollutant concentration at any given urban location (micro level) is the function of two factors: rate of (vehicular) emission and rate of dispersion and dilution. While the rate of emission normally depends on the sources of air pollutant, the rate of dispersion and dilution is governed primarily by a number of factors. In general, researches on urban atmospheric environment suggest a correlation between air pollution dispersion and the climatic conditions including temperature, wind velocity & humidity, etc. (Cogliani, 2001; Givoni, 1998; Hawkes et al., 1998; Marvroids and Griffiths, 2001).

However, more recent researches indicate that both natural (topological features, vegetation, etc.) and man-made (built forms) factors also have a direct or indirect impact on urban microclimate, which in turn affects the rate of dispersion and dilution of air pollutant (Givoni, 1998; Hawkes et al., 1998; Santamouris M, 2000). It has been found, for example, that building geometry changes urban "aerodynamic roughness" (friction of an urban area), "surface albedo/emissivity" (sun light reflectivity due to surface materials) and "sky view factor" (degree of sun-light penetration between built masses) etc. that can affect wind speed, temperature, humidity and so on in micro urban environments. Therefore, both urban morphological factors and micro-meteorology are sequentially related aspects that affect air pollution concentration or dispersion (Figure 4).

Figure 4: Sequence of effects to demonstrate the urban air quality relationship with urban morphology

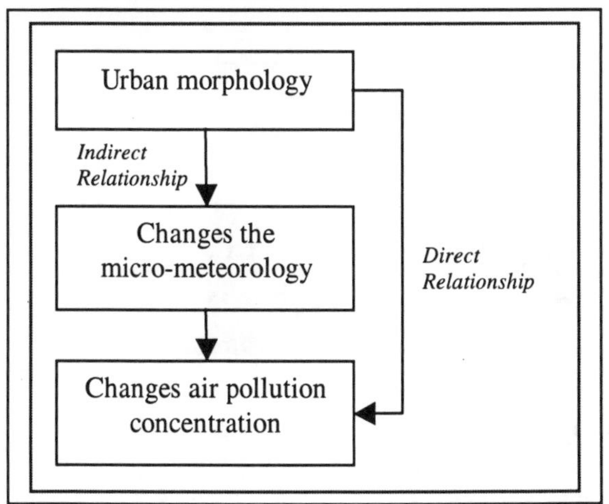

Modeling the impact of morphological factors on microclimates is crucial to understanding the complex relationship between air quality and hyper dense residential environment in Hong Kong. The first step towards such a model requires that the urban environment at issue be considered as an "air quality box" (De Nevers, 2000). According to this model, an air quality box can be demarcated horizontally by compact land use surface obstacles (buildings) and vertically by tall buildings (walls up to roof level). With reference to this box, air pollution phenomenon below the roof level of a high-rise high-density environment can be distinguished from air pollution phenomenon above the roof level (background air pollution or the urban smog). This tri-dimensional medium is also suggested by Adolphe (2001) as a "porous medium with rigid skeleton". The concept distinguishes the climate above the roof level as 'background conditions' and that below the roof level as 'local conditions' (Adoplhe, 2001; cited in Eernes, 1994). The layer between the ground and roof levels is referred to as the "building canopy layer" (or urban canopy layer) and its air pollution phenomenon is defined as "street level air pollution" (Figure 5). Since the unique urban profiles of Hong Kong have created pockets of micro environments that have their own microclimatic and air pollution conditions, modeling these micro environments as *local air quality boxes* may enable research to link street level air pollution concentration to the morphological attributes of an urban locality, via the intermediate layer of microclimatic conditions.

Figure 5: Urban building canopy layer & street level air pollution phenomenon

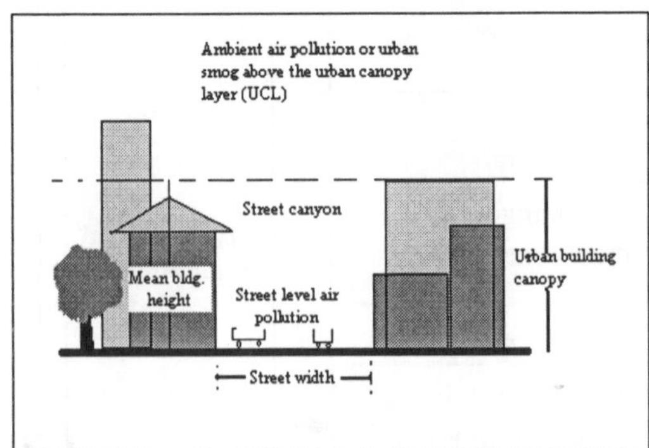

1.3 *The present research*

Previous researches have shown that air pollutant dispersion in urban areas depends largely on building geometry and microclimatic conditions within an urban building canopy layer (Chan et al. 2002; Cionco, R.M & Ellefsen, R, 1997; Hawkes et al., 1998; Hong, 1999; Marvroids and Griffiths, 2001; Ratti et al. 2000). These studies indicate that air pollutant may get trapped and thus increase air pollutant concentration amidst urban structures (tall buildings). However, while attempting to link the movement of polluted air to the physical urban form at the macro level, these studies in general do not concern themselves in identifying physical factors at the micro level, which may prove crucial to studying of street-level air pollution in an urban environment like Hong Kong. In what follows, an on-going research that investigates the relationship between urban morphology and air quality in the hyper dense residential environment of Hong Kong is reported. The study starts with the search of a suitable approach to modeling the complex relationship between street level air pollution concentration, micro-meteorology, and the morphological dimension of an urban locality, and proceeds to validate such a model through theoretical experiments and limited empirical observations. The method is then applied to a sample of residential areas in Hong Kong to search for patterns of correlation between morphological factors and street level air pollutant concentration, within the urban building canopy layer of these study areas.

2. Method of Study

2.1 *Development of theoretical model*

Early work of this research includes search for an initial hypothesis that relates air quality to the hyper dense residential environment in Hong Kong and relevant morphological attributes for empirical studies. These lead to a crude model postulating the relationship between street level air pollution, general and specific morphological factors such as geographical location, development density and built forms, and climatic conditions. The model was put in test in a pilot study that involved design of technical procedures for on-site sampling of pollutant air concentration, and theoretical experiments simulating and validating postulations on relationships between some key morphological variables and airflow / air pollutant movement (Edussuriya et al. 2004).

Findings and insights that emerged from the work informed the development of a tighter theoretical model that postulates interactions between street level air pollution concentration and the most relevant causal factors (Figure 6). Specifically the model identifies effective urban morphological attributes under the urban building canopy layer of near field regimes (a near field regime for Hong Kong is defined as a 200m x 200m urban grid; see discussions below) and a set of key attributes for micro-meteorology to be related to street level air pollutant concentration (Table 1). Out of many possible morphological attributes, the assimilation of the effective ones however is assumed to be also dependent on the actual context of the study areas and should be continuously tested in empirical analysis.

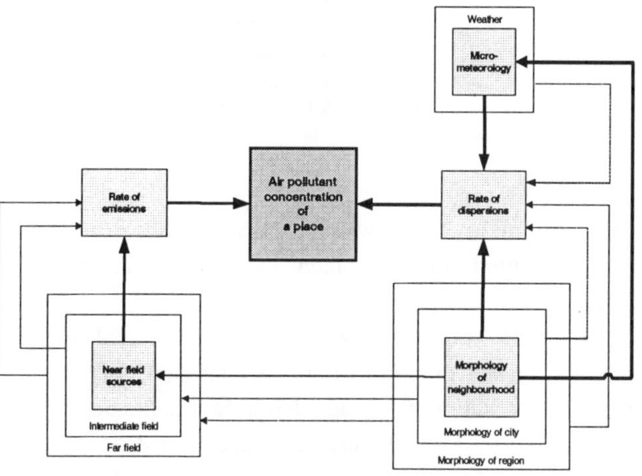

Figure 6: Generic relationship model depicting street level air pollutant concentration of a place and the most relevant causal factors for Hong Kong

2.2 Design of empirical studies

2.2.1 *Defining the near-field regime of study areas*

Assimilation of the spatial dimension of an urban fabric is fundamental to the monitoring of street level air pollution concentration in empirical studies. It is therefore not uncommon for previous researches to set the boundaries of a study area at certain scale range, typically in a 1km x 1km urban grid or a diameter of areas covering a similar range (Theurer, 1999, Weber et al. 2001). Most of these ranges however are based on rules of thumb and are difficult to be used for measurement of street level air pollution in Hong Kong's circumstance. In contrast, the spatial field regime used by Santamouris (2000) suggests a more realistic range for the Hong Kong environment. For according to this and other researches, air pollutant concentration of an urban location depends upon sources of emissions in intermediate field regime (in a range of few hundreds of meters) and near field regime (within tens of meters). And air pollutant emissions from far field regime (distance beyond 1 km) tend to have negligible impact on street level air pollution within a particular urban location due to their inability to reach that location, caused by surface obstacles such as a compact built form. With references to these facts the near field regime of an urban location in this study is set as a 200m x 200m grid, with all concerned morphological variables obtained within that range only.

2.2.2 *Representation of morphological indicators*

The morphological indicators about a study area can be obtained from secondary sources such as maps and on-site measurements, based on the morphological factors as identified in Table 1. Since it is postulated that street level air pollutant concentration is most likely affected by urban fabric at micro level, only morphological indicators at the neighbourhood level are considered for this study. These indicators are based on planning and design parameters commonly used in Hong Kong like, location-related factors, development and population densities, and built form configuration factors, etc.

Although the near field regime is limited to a smaller urban fabric, there can be quite distinct variations in buildings or building blocks. These variations in urban fabric can have direct and indirect impacts on the microclimates within an urban location, and hence also affect street level air pollutant concentration in that area. In order to indicate the morphological attributes of a study area as a whole the mean effects of a heterogeneous urban fabric will be used to represent variations in that fabric (Adolphe, 2001).

The building and street geometries of an urban location are key parameters for calculation of its morphological indicators. In this study, computations of these indicators thus will be based on the mean effects of parameters such as mean building height, width, and etc. To get the most representative values for a study area, the mean values of relevant building parameters are weighted either by plan area method or frontal area method depending on the situations (Weber et al. 2001, Grimmond & Oke, 1998). The key morphological indicators of the mean effects of an urban fabric used in subsequent case studies are presented in Table 2 and explained later.

Table 1: Key variables that represent and measure of micro-meteorology and urban morphology for street level air pollution concentration

	Macro level	Meso level	Micro level
Urban micro-meteorology	***Weather impact indicators:*** • Wind speed distribution & direction • Temperature distribution • Humidity distribution • Solar radiation and Precipitation, etc.		***Microclimate impact indicators*** • Wind speed & direction • Temperature • Humidity Solar radiation and etc.
Urban morphology	***Regional impact indicators:*** • Topography /land topology • Altitude • Distance to sea and etc	***City or Peripheral impact indicators:*** • Urban layout: built, street, roof form • Proximity to sources & its intensity • Proximity to air pollutant sinks • City size	***Neighborhood impact indicators:*** ***Locality factors -*** • Population density quotient • Urban land use: % of residential use • Traffic load: vehicle fleet/ area, etc. ***Development factors -*** • Land utilization • Land development intensity, etc. ***Built form configuration factors -*** • Street & building geometry • Street & building orientation, etc.

2.2.3 Monitoring vehicle emissions, air pollutant concentration, and microclimate indicators

Data about air pollutant concentration in a study area will be collected mainly from vehicle emissions measured for that area, represented by vehicle traffic volumes and the quantity of some major air pollutants. The vehicle traffic volume of a study area is obtained from the database maintained by the Transport Department of Hong Kong (Annual Average Daily Traffic: AADT) (HKTD, 2002), expressed as Average Vehicle Density.

The average vehicle density for a study area of 200m x 200m grid is calculated according to the following sampling technique. Every half an hour, the numbers of vehicles passing a number of sampling points are recorded for 10 minutes and taken as the local traffic volumes (vehicle km/hour for 50m distance). All these traffic volumes are taken by vehicle type and weighted by coded value assigned for each vehicle type, based on respective air pollutant emissions by vehicle type. Four hours average vehicle densities (cars/hr) taken for a study area will be used as traffic volume indicator for that area (see Table 2).

Both particulate (suspended particulate matters-$PM_{2.5}$, NOx, O_3 and CO) and gaseous air pollution data will be monitored for this study, as these would provide good indication of vehicular-traffic-related street level air pollution (Theurer, 1999, Weber et al. 2001). Site-specific microclimatic conditions like temperature, relative humidity (RH), wind speed and direction will be also measured and be taken into consideration for analysis[1].

Portable equipment will be used for real time monitoring and duration of monitoring is restricted to repeated measurements conducted in four hours per day, in three days. The measurements are carried out from October to March/April as atmospheric condition is reasonably stable during this cooler season in Hong Kong. All sampling results are calculated for four hour average concentration/readings based on statistics methods.

3. Case studies

Data on street level air pollution from three representative residential areas are first analyzed in relation to the morphological indicators of their urban fabrics. The three cases share some common morphological characteristics such as terrain condition, altitude, area size, proximity to air pollution sources and pollution sinks etc. but differ markedly in building geometries and plan layouts and some other crucial aspects. The similar nature of their micro level morphologies helps to minimize any location effects that might have a distorting influence on climatic

[1] Wind speed & direction measured on site are not used in this study as they do not give accurate readings due to local turbulence effects generated by vehicles. Instead readings obtained from the nearest weather station (Hong Kong Observatory survey results) are used for analysis.

conditions and level of air pollution concentration at the local scale.

Table 2 is a summary of major morphological attributes of the three urban fabrics represented in mean effects for comparison, key meteorological indicators and average air pollution data registered over a period of four hour per day repeated three days measurements. The readings of $PM_{2.5}$ and other key air pollutants show that Tai Koo Shing's air pollutant level of concentration is high, as compared to other two cases. The average vehicle density varies among the three areas significantly, which is not in proportion to observed vehicle density and $PM_{2.5}$ measured in the three areas.

3.1 Micro-climatic conditions and air pollutant dispersion

Preliminary site observation shows that the behaviours of gaseous pollutant (CO, NOx) and particulate pollutants ($PM_{2.5}$) are different. This seems to indicate that different types of pollutants interacted with micro-meteorological and/or morphological conditions differently. That in turn suggests the need to consider both particulates and gaseous pollutants for a comprehensive study of street level air pollution phenomenon in real situations. On the other hand, no Ozone (O3) concentrations are found in any of the three study areas, as their dense built structures could have effectively prevented the penetration of ambient air into the urban building canopy layer.

As mentioned above, the location effects of the three study areas have been minimized due to the common or similar conditions of their macro level morphological characteristics. But at the local level, some common morphological attributes seem to have caused similar microclimatic conditions in all three areas. Specifically, land utilizations (ground coverage and effect of open spaces) in all three cases display almost identical values. For although according to plan layouts there appear to have more open spaces in Hung Hom and Tai Koo Shing as compared to Jordan, most ground areas in the three cases are covered by 2-to-4 storey podium structures with tower blocks at top, which substantially increases the actual ground coverage by buildings. In the meantime, the effects of paved open spaces including local parks appear to have no positive impact on microclimate, due to impervious surface materials. Therefore, in addition to the role played by building mass and road net-works, these paved open spaces seem to be also contributing to the total thermal mass of an urban fabric. In other words, the largely identical *development factor effect* - i.e. high albedo/emissivity of built form surface among the three areas - in effect had a similar impact on increase of temperature within these urban fabrics. That in turn could have caused poor air pollutant dispersion.

3.2 Urban morphology, micro-meteorology, and street level air pollutant concentration

3.2.1 Development factors

Urban building density and porosity are two key land development intensity indicators that enable us to understand air pollution concentration levels in these three cases. A higher urban building density and relatively low porosity level found in Tai Koo Shing (case 3), seem to offer a plausible explanation for its higher air pollution level, despite the lowest vehicle density and relatively better climatic conditions registered in that area. Likewise, the high air pollution level found in Hung Hom (case 2) can be also related to its higher building density and extremely low porosity. But a lower building density and a high porosity as found in Jordan (case 1) seems to suggest a different tendency that contradicts the above relationship. Since the higher porosity of an urban fabric - in the case of Jordan it seems to be the effect of a regular grid street network that acts as a continuous porous medium - tends to allow air pollutant to escape from a densely built area. The high level of air pollutant concentration in Jordan thus should be explained by some other factor than porosity[2].

---

[2] The porosity of an urban fabric is not necessarily the function of the ratio of open space in a development. This is clearly shown in this study. For although Hung Hom and Tai Koo Shing have similar portion of open spaces as compared to Jordan, they cannot act as an equally effective evacuation medium. The open spaces (pores spaces) in both areas are either blocked by buildings or are shaped in court yard-like semi-enclosed or enclosed open spaces amidst tall buildings. Therefore, they tend to have lower drag-off effect on air pollutants dispersion than continuous open spaces as created by the street network of Jordan.

Table 2: Empirical survey analysis of three urban residential environments in Hong Kong

	Variables/Descriptions	Case 1 – Jordan	Case 2 – Hung Hom	Case 3 – Tai Koo Shing
1	**Study area – 200m x 200m grid**			
	Photo Images of study area			
2	**Air Pollutant Concentration** (hourly average, micro g/m^3)			
	$PM_{2.5}$	55	51	53
	NOx	317	369	224
	CO	2600	2890	1200
	Av. Vehicle Density (cars/hr)	857 – (within grid 6,310)	614 – (within grid 4,134)	252 – (within grid 3,163)
3	**Micrometeorology**			
	Temp (°C)	25.8	20.2	18.1
	H %	51.3	48.8	32.3
	Wind V (m/s)	6.2 m/s @ 90 degree	6.1 m/s @ 60 degree	8.3 m/s @ 80 degree
4	**Location factor effect**			
	a) Built- Street form quotient	compact - rectangular grid pattern	clusters - larger square grid pattern	disperse - irregular pattern
	b) Population density quotient (/km^2)	40,932 people – 14,550 households	38,059 people – 12,031hoesholds	33,147 people – 10,547 households
5	**Development factor effect**			
	Land utilization			
	a) Ground coverage	0.52	0.57	0.54
	b) Effect of open spaces	0.98	0.93	0.99
	Land development intensity			
	a) Rugosity/ur. bldg density	12.4	16.5	27.3
	b) Contiguity (compactness)	0.47	0.23	0.77
	c) Frontal area density	1.01	0.85	1.44
	d) Porosity (Po)	51.7	3.32	19.3
6	**Built form configuration factor effect**			
	Street & building geometry			
	a) Canyon ratio	2.0	1.77	3.5
	b) Urban roughness	28 m or 14.1m	43.3 m or 23.1 m	115 m or 54 m
	c) Roughness density	0.46	0.48	1.32
	d) Mean bldg. height	28m	53 m	75 m
	Street & building orientation			
	a) Sinuosity	0.45	0.48	0.51

The frontal area density and compactness are two other development intensity-related indicators that, in theory, could also cause blocking effects on airflow and hence reduce the rate of air pollutant dispersion in a densely built area. But a comparison of these among the three cases against their air pollution levels suggests a mixed pattern that demands further exploration.

3.2.2 *Built form configuration factors*

In addition to the development factors, a comparison of the built form configuration factors among the cases also reveals relationships that at once suggest some interesting tendency and raise new questions concerning the complex interactions between morphology, microclimatic conditions and air pollutant concentration. The mean building height of an urban fabric would create a high building canopy this could have impacted on higher mixing height of airflow in Tai Koo Shing and Hung Hom as compared to Jordan. As (the vertical plane created by) building height together with narrower street widths could create more vortices and thus decrease airflow within an urban street canyon. This seems to be particular the case in Tai Koo Shing, where the largest mean building height is found among all three areas, despite the low vehicle density registered for the area. As a contrast, the low mean building height in Jordan could be one of the configuration factors that, despite a very high vehicle density and a compact built geometry, seems to have exerted less impact on air pollutant concentration within that area. Meanwhile the relatively high mean building density in Hung Hom may offer another explanation for the poor dispersion of air pollutants found in that area even though it is less exposed to vehicle traffic as compared to Jordan.

The perhaps most illuminating fact arising from the comparison is about the role played by the building height factor in affecting morphological indicators like street canyon ratio, urban roughness and roughness density etc. For higher street canyon ratio and low urban roughness (expressed in high value) would create multiple vortices within a street canyon that reduce airflow and hence exacerbate the concentration of air pollutant like the particulates matters. This is already confirmed by theoretical experiments and earlier empirical studies (Edussuriya et al., 2004). The very high urban roughness reading found in Tai Koo Shing (115m) and the reverse order of this indicator and vehicle density among all the three cases, strongly suggest that there may exist some causal relationship between morphological factors and air pollution concentration in dense urban environments dominated by high-rise buildings, mediated by climatic conditions.

4. Concluding remarks

The above analyses show that the development factor and built form configuration factor are relevant morphological aspects that permit us to look into the complex relationship between urban form and air quality in the hyper dense residential environment of Hong Kong. Out of the few indicators of the development dimension, while land utilisation offers little explanation for variations in microclimatic conditions and air pollution concentration, most land intensity indicators suggest some describable relations to both aspects. Similar and even a clearer pattern is found in the relationship between built form configuration indicators such as street canyon ratio and urban roughness, and climatic conditions and air quality among the three cases. On the other hand, preliminary results suggest that location factors have no clear relation to air pollutant concentration in the study areas. The impact of these factors however cannot be readily dismissed as street layout form in a larger context may have direct or indirect influence on climate, exposure to polluting sources, and level of air pollutant concentration of a local area.

In conclusion, despite limited scope the preliminary results of the study enabled validation of our theoretical framework that postulates a relationship between urban morphology and air quality, and an effective test of a research method that relates urban morphological attributes to level of air pollutant concentration, via the mediating force of micrometeorological conditions. The comparative analysis of the small sample of cases has enabled the narrowing down of research focus on key morphological attributes and other variables whose relation to street level air pollution can be examined through more extensive empirical survey and be precisely determined with more statistical rigor.

Acknowledgments

This research is financially supported by the Department of Architecture and Department of Mechanical engineering, The University of Hong Kong for technical equipment and operation cost.

References

1. Adolphe, L, 2001, A simplified model of urban morphology; application to an analysis of the environmental performance of cities, Environmental and Planning B; Planning and Design, Vol. 28, pp. 183-200

2. Chan Andy T Y, Au William T. W, So Ellen S.P, 2002. Strategic guidelines for street canyon geometry to achieve sustainable street air quality – part II: multiple canopies and canyons, Atmospheric Environment, Vol 37, pp. 2761-2772
3. Cionco, R.M & Ellefsen, R, 1997. High resolution urban morphology data for urban wind flow modelling, Atmospheric Environment, Vol 32, pp. 7 – 17
4. Coliani, Euro: 2001, Air pollution forecast in cities by an air pollution index highly correlated with meteorological variables, Atmospheric Environment, Vol 35, pp. 2871 – 2877
5. De Haan P, Rotach M.W & Werreli M: 2000, Modification of operational dispersion model for urban applications, *Journal of Applied Meteorology*, vol 40, pp 864 – 879
6. De Nevers. N., 2000, Air pollution control engineering, 2Ed, Imprint Boston: McGraw-Hill, Boston
7. Edussuriya P.S., Ye, A.M. and Chan, A T.Y., 2004. Urban Building Density and Air Quality: the case of Mong Kok, Hong Kong, In Proceedings: First International Conference on Tropical Architecture, National University of Singapore, February 2004, Singapore.
8. Grimmond & Oke, 1998, Aerodynamic properties of urban areas derived from analysis of surface form; Journal of applied meteorology; Vol 38, pp. 1262-1292
9. Givoni, B, 1998, Climate Considerations in Building and Urban Design. USA
10. Hawkes D, McDonald J & Steemers K, 2002. The selective environment – An approach to environmentally responsive architecture, Chap 1, 3 & 8, UK
11. Hong Kong Environmental Protection Department (EPD); Air Quality in Hong Kong 1997, 1998, 1999, 2000, 2001, Annual Reports, Hong Kong.
12. Hong Kong Environmental Protection Department (HKEPD), 2001. Clean Air for Hong Kong: Fung Christopher, A Report, Hong Kong
13. Hong Kong Transport Department. 2002. Traffic and Transport Survey - Annual Traffic Census 1997, Hong Kong
14. Hong, H: 1999, A Two-Dimensional Air Quality Model in an Urban Street Canyon: evaluation and sensitivity analysis in Atmospheric Environment. Vol 34, 2000. University of Tokyo, Tokyo, pp. 37-42
15. Marvroidis I, Griffiths R.F, 2001. Local characteristics of atmospheric dispersion within building arrays, Atmospheric Environment, Vol 35, pp. 2941 – 2954
16. Newton P, 2000. Urban form and environmental performance, Ed, Williams K, Burton E, Jenks Mike, Achieving sustainable urban form, USA
17. Ratti C, Canton F, Di Sabatino s & Britter R, 2000. Morphological parameters for urban dispersion models, In Architecture, city, environment: Proceedings of PLEA International Conference July 2000, Cambridge, UK, ed. Koen Steemers and Simos Yannas. London: James & James
18. Santamouris M, 2000. Environmental planning in urban areas, UK
19. South China Morning Post (SCMP), 2003, 19th September 2002a, 20th September 2002b and 22nd April 2003, Hong Kong
20. Theurer, W., 1999; Typical building arrangements for urban air pollutant modelling, Atmospheric Environment 33, 4057-4069
21. Weber, C, Hirsch, J, Schnell, L, Perron, G, Kleinpeter, J and Ranchin T; Urban fabric and measures of variability; Neighborhood effects on proximity pollution, 2001; 12th world clean air and environment congress, Greening the new millennium, August 2001, Seoul, Korea.

TWO IFC

JOSEPH LEUNG* ALBERT TO†
J. Roger Preston Limited – Consulting M&E Engineers

The skyline of central district area has been re-drawn by the completion of an 88-storey skyscraper called Two IFC which stands at 420m off the ground and the tallest building in Hong Kong. Two IFC is a grade A office building with a total gross floor area (GFA) of more than 180,000m^2 and an estimated working population of 15,000 person. The tower is divided into 7 lift zones with twin sky lobbies and 4 mechanical floors to service each individual zones of the building in the most economical and energy saving manner. Two IFC tower is also the icon building of the integrated development which comprises of the Airport rail link extension on the northern concourse, a mega retail mall of 70,000 m2 linking to the 1000 rooms Four Seasons Hotel complex, totalling over 4,500,000 sq. ft. development area. The sustainable M&E services design features in Two IFC includes (i) achieving building thermal comfort and energy efficiency through detailed OTTV study and external façade treatment; (ii) pre-fabrication and pre-commissioned technology employed for AHU, PAU and others ready for "plug-and-play" installation; (iii) energy comfort and controls through advance BMS DDC control running on TCP/IT fast speed ethernet; (iv) hydraulic treatment in M&E system design in super highrise building; (v) continuous energy monitoring / auditing for energy saving opportunities.

1. Introduction – Sustainable Design

At the outset of the project, the aim towards making Two IFC a sustainable, energy efficient and environmental building was well endorsed and blessed by the top management of the developer. In brief, sustainable design for tall building can actually be divided into three main categories of design considerations

- Sustainable Design for Energy Efficiency and Environment.
- Sustainable Design for Adaptation to Future.
- Sustainable Design for Building Operation.

1.1 *Sustainable design for energy efficiency and environment*

The building has over 2,000,000 sq ft of usable floor area and in light of the building symmetry and core design, best opportunity existed to maximize the opportunity for pre-fabrication and pre- Engineering design approach. This would not only minimize site wastage, less construction noise, faster construction, but also optimize maintenance and plant space at the outset of preliminary design.

For Two IFC, the pre-fab design (among many others) includes over 160 units of AHU / PAU, miles long of high velocity spiral air ducts, over 8,000 units of VAV boxes. All AHUs, PAUs, VAV boxes are equipped with built-in DDC controllers (BMS contract was tendered and awarded separately) which are pre-installed and pre-commissioned (to design performance) before they leave the factory floors for shipment to Hong Kong. Pre-shipment performance and acoustic acceptance checks (Factory Assembly Test) were witnessed by project team to ensure compliance on quality standards and design specifications.

Figure 1. Two IFC Development Model

---

* Director, BSC(ENG) MBA CENG FIEE FHKIE FIEAUST FCIBSE RPE
† Technical Director, BENG CENG MIEE MCIBSE MHKIE RPE

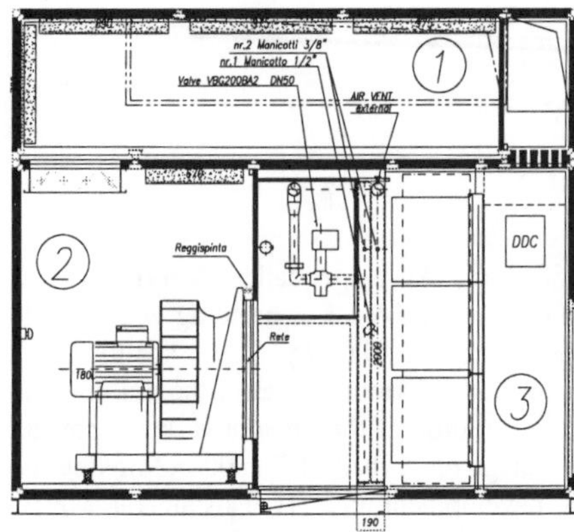

- All actuators, DDC, pipe works are pre-fab/integrated at factory, plug & plat at site
- Use direct drive plug fan to eliminate maintenance of belt drive
- Use fan chamber acoustic treatment to eliminate supply air silencer
- Built-in control equipment to improve system reliability
- 3 main components per AHU using knock-down delivery

Figure 2. AHU Construction – Fully Prefab and Pre-commissioned at Factory

This would further ensure the production quality, minimizing human error and facilitate fast pre-commissioning using a "Plug-and-play" approach (see Fig. 2). This was proved to be very cost and time effective and have saved a lot of manpower in the initial trouble-shooting of the commissioning trail. Other sustainable design includes,

- Centralised seawater cooling plant design for Two IFC, IFC Mall and the Four Seasons Hotel blocks totaling over 20,000 refrigerant tonnage are running on seawater as the heat rejection media. Despite high quality plant equipment being specified and used, the energy out turn has been very satisfactory in both the running cost and operating efficiency.
- Solar heat shading efficiency for the fenestration of building façade – through numerous studies and highly involved design resolutions with the architect, a OTTV value of 23W/m^3 were achieved with great delight.
- This demonstrates that even with a tall building of this height (almost no beneficial shading contribution by surrounding buildings) and with all four façade directly exposed to solar radiance, sustainable energy design has been achieved by careful selection of double-glazed façade with low-e coating and beautifully incorporated with the external architectural fins and horizontal shading device. This has resulted in a tremendous saving in air-conditioning costs along with its life span (see Fig.3).

Figure 3. Façade Design with Shading Features

An extensive use of VFD were applied to Two IFC MEP design including all pumps, fans, AHU and PAU. Each AHU is equipped with CO_2 sensor which could modulate and control the drawing of fresh air from the respective PAU to satisfy the IAQ and occupation density. This will ensure the appropriate amount of cooling energy is smartly brought to the space that are needed while not allowing it to be unsympathetically wasted (see Fig. 4).

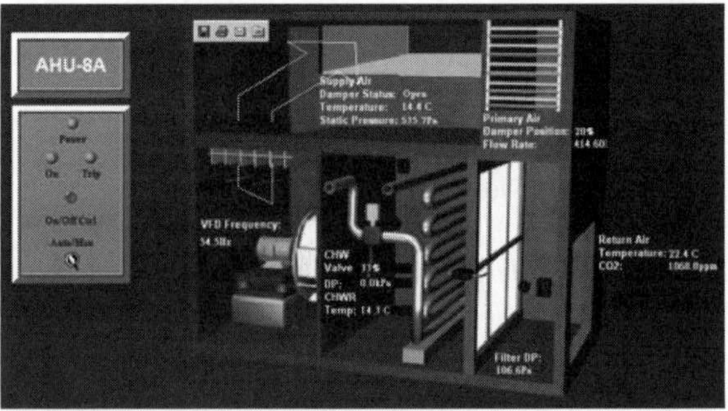

Figure 4. AHU – 3D BMS Dynamic Readout

Every PAU is also pre-fab with a run-around coil compartment pre-assembled as a separate energy wheel which can be easily put together at site during construction. The run-around coil will give an energy saving of around 15-20% loading for use in fresh air pre-treatment before passing the treated air downstream to all AHUs.

1.2 *Sustainable design for adaptation to future*

The Two IFC development is a new comprehensive financial landmark city of Central, Hong Kong. The HVAC controls for such a massive air-conditioning system are put under the hand of a true TCP / IP based BMS system which now handle over 120,000 digital / analogue / software points every day. Despite facing some engineering reluctance on the initial design proposal (using an IP based Ethernet network instead of the conventional industrial communication network protocol) due to cost and yet-to-prove technology (six years back on drawing board), the adoption of a true intranet based BMS network proved not only to be necessary, but also essential to the daily operation of the Two IFC.

- Fast Response - All BMS control and monitoring points at field level, system level and head-end level are actually linked by a fast Ethernet with over 100Mbps transmission capability. Both the response to command locally by the DDC controllers and the ultimate response / reports feedback to operators level were done within splits of a second. Normal individual network communication protocol (Arcnet, token ring, etc.) simply cannot suffice such a level of operating performance.
- A fully computerized standalone Power Analyser system (PA) was installed to monitor all the electrical characteristics of all main electrical circuits to both the landlord equipment as well as the tenant distribution. They can spy on with high level intelligence of all harmonize contents and the associated power quality at any one time, reporting through the BMS system to the central Building control office. With this, no energy culprits can be escaped from the facility management radar screen and positive measures can be implemented in the most educated manner to eliminate energy wastage (see Fig. 5).

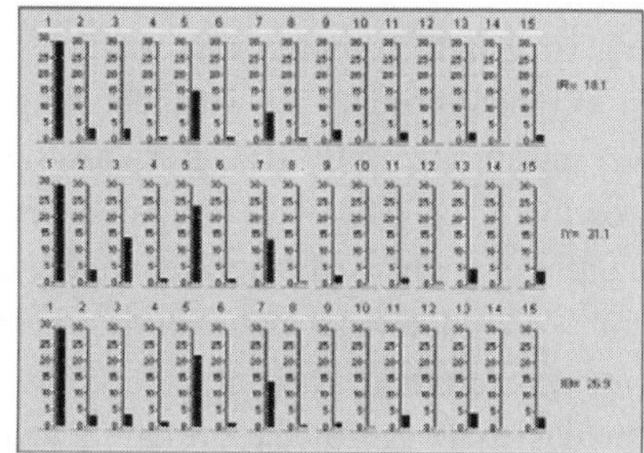

Figure 5. Power Analyzer – up to 15th Harmonic Content Capture

- Reserved Space and Riser - Lots of services riser, reserved duct space and tenants plantroom space are reserved in both the mechanical floors and within the existing M&E services core to cater for the eventuality.

1.3 *Sustainable design for building operation*

1.3.1 *Interaction between tenants/customers and Facility Management (FM)*

A multi-million dollar video wall was also installed in the central building control room which interact with all other building services systems such as lift and escalator, security, fire services etc. for best real time response and capture of abnormal building incidents, or tenants complaints.

One Good thing is that the tenant cannot understate the building services system performance to the building management any more as almost all operating conditions can be immediately displayed on the video wall (see Fig. 6). Furthermore, the FM can identify the emergence of problems much earlier than the tenants lodging the complaints.

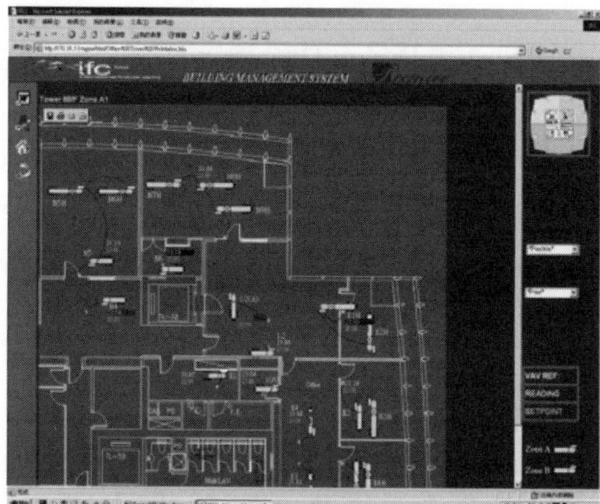

Figure 6. VAV – Typical BMS Dynamic Readout

1.3.2 *Hydraulic zoning*

Many tall building M&E design took on different approach and Two IFC has adopted the view that the hydraulic zoning for both the chilled water and other pumping shall not exceed an operating pressure of 16 bar over 80% of the system design. Key reasons are,

- Most conventional equipment can satisfy such system hydraulics and are commercially available for any tenant fast track fit-out installations.
- The building management team will be able to conduct regular maintenance/repairing work in an economical manner rather than using high pressure rated devices, equipment and regulators.

- Most of the power driven device/ equipment such as motors and water pumps are within conventional range to manage the hydraulic delivery with the highest efficiency motor commercially available (see Fig. 7).

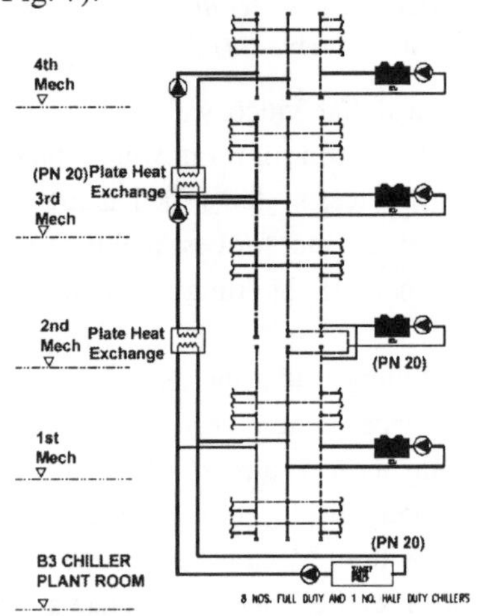

Figure 7. Simplified Chilled Water Distribution Hydraulic Zone

2. MEP System Design Overview

The following sections describe in more details of the system employed to achieve the goal of sustainability design for tall buildings.

2.1 *HVAC system and seawater cooling plant*

The Two IFC office tower HVAC system consists of seawater cooled chiller plant serving VAV air conditioning system on individual floors. Air-cooled chillers installed on various mechanical floors with essential power backup provide a reliable source of chilled water supply for mission critical tenant facilities.

The chiller plant seawater cooling system delivers seawater from the pump house located at the harbour water front via dual plus standby seawater supply mains to seawater cooled chiller plant located in basement providing over 20,000 TR cooling capacity for the entire development (office tower, retail podium and hotel) (see Fig. 8). The seawater cooled chiller plant for the Two IFC Tower has 10 nos. 1,000 TR direct seawater cooled chillers with titanium tube condenser. Chilled water distribution is separated vertically into 3 circuits coupled by plate type heat exchangers on the 2nd and the 4th mechanical floors. This limits most of the chilled water distribution operating pressure on office floors to within 16 bar for cost effectiveness and system reliability.

Each floor is served by 2 air-handling units supplying conditioned air to VAV units. Perimeter VAV units have electric heater for winter heating. Fresh air is provided by primary air handling units on mechanical floors.

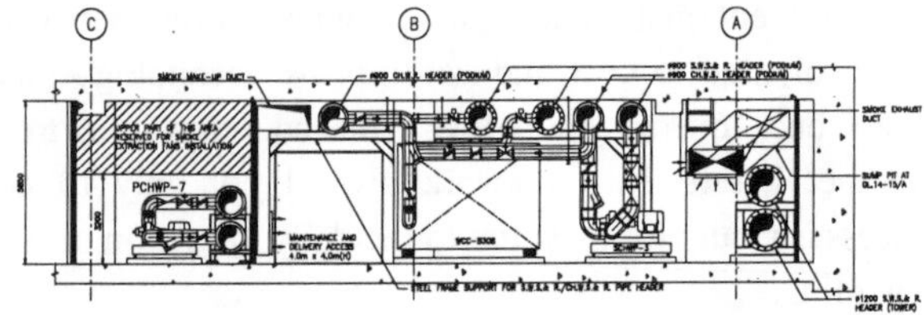

Figure 8. B3 Main Chiller Plant Room – Chiller & Pump Arrangement

2.2 *Reliability for building operation*

With the envisaged tenant requirement for 24 hour uninterrupted operation, system reliability is a major design focus and the following features are incorporated:

- Chiller plant seawater cooling system has dual supply mains with standby and split seawater pump / chiller headers.
- Primary chilled water riser mains has standby pipe.
- Office floor has two stacks of AHU each with individual chilled water rising mains.
- The 2 nos. of AHU on individual floor are connected by ring mains supply air duct with section damper.
- Air cooled chillers with essential power backup (2 x 300 TR on each of the mechanical floors) and independent supply mains provides essential chilled water supply for 24 x 7 critical tenant facilities.

2.3 *Indoor air quality and energy conservation*

All primary air handling units are installed with run around coil to reheat primary air with incoming outdoor air to deliver unsaturated (70%RH) fresh air to suppress bacteria growth in the otherwise saturated primary air stream. Quantity of primary air supplied to individual air handling unit is modulated to maintain desirable carbon dioxide content in return air stream.

A particular challenge of the project is to maintain adequate separation between air intake and discharge on mechanical floors due to the limitation of external façade louver area for aesthetic reasons. This is further complicated by air intake / discharge requirement for air cooled chillers / generators on mechanical floors which together occupy more than 75% of available louver area. To achieve desirable air intake / discharge separation, partitioned cavity voids were erected behind external louver to form an array of 3 dimensional air paths to direct intake / discharge ducts to appropriate locations of external louver. Due to the complexity, mechanical floor "design model " (see Fig. 9) was being built to confirm ventilation plenum arrangement prior to construction.

Figure 9. Test Design Model

2.4 *Energy efficiency*

The installation has extensively utilized variable speed drives (VFD) including seawater / chilled water pumps and air handling units to optimize operating efficiency throughout the range of load profile. To further enhance energy efficiency, equipment operating speed is being controlled to maintain opening position of modulating control valve / damper within desired range instead of commonly adopted control strategy of maintaining constant pressure at supply mains and / or index circuit. This allows supply pressure control to be based on actual system requirement and adoption of lower design pressure drop for modulating control device both of which contribute to reduction of system supply pressure.

Plate type heat exchangers have been selected for variable flow operation by reducing the dependence of heat transfer on turbulent flow and hence reducing pressure drop under partial load condition. The effect has been noticeable since resistance of plate type heat exchanger being significantly higher than that of air handling unit dominate system supply pressure requirement. Similarly, chillers operate on variable chilled water flow to reduce the dependence on using bypass which increase pumping energy and reduce the capability of system control to maintain chiller operating under optimum condition.

3. Building Management System (BMS)

Due to its sheer size of various M&E engineering systems, it is a challenge to the building management team to control, monitor and maintain a satisfactory operating environment for such a complex building. In total, over 77,000 analogue points and 36,000 binary points are installed and connected to other M&E systems. Such arrangement represents a challenge to design team to achieve a reliable, inter-operative and efficient building management system (BMS).

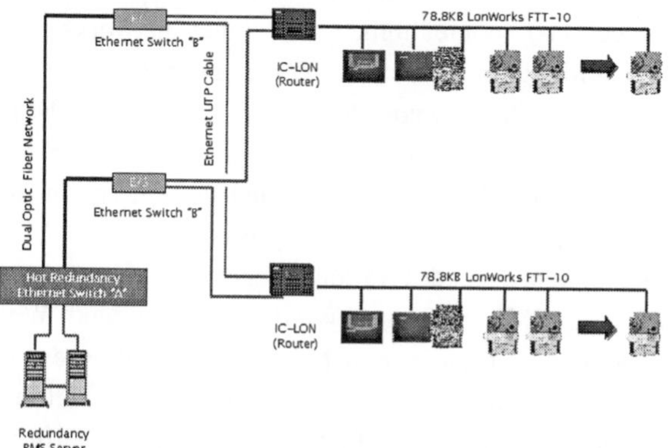

Figure 10. BMS Schematic – TCP/IP Ethernet Network Hot Redundancy, Dual Path

The BMS system is designed for the future based on a powerful communication system which is a vital part of the BMS system in view of the substantial amount of information being managed. The communication network is composed of 1,000 BASE-T single mode optic fibre backbone connecting Ethernet switches to headend server. Ethernet switches in turn are connected to supervisory DDC controllers with Cat 5e UTP cables. Communication system is fully addressable TCP/IP for efficient data traffic management.

3.1 *Reliability*

The communication network has dual optic fiber backbone individually connected to hot standby ethernet switches together with hot standby server. In addition, there is an individual network for major HVAC equipment (seawater supply system, chilled water system, primary air handling units), which operates in parallel with the BMS system as hot standby in case of BMS system failure. Central supervisory DDC controllers and communication network equipment are provided with distributed UPS to ensure all critical equipment will still be functioned in case of power failure.

System architect is based on distributed intelligence with independent controller for individual equipment (such air AHU, VAV, heat exchanger). This limits the impact of communication network failure to disabling of interaction between different items of equipment but automatic operation of individual equipment remains fully intact.

3.2 *Connectivity*

The BMS design emphasizes on the ability to communicate with user (both internally and for external connections), compatibility with other networks, and adaptability to future modification and expansion.

Extra efforts were devoted to developing BMS interactive graphics reflecting as far as possible the actual location / image of individual item of equipment to facilitate instant recognition by user. In addition, the BMS system provides various means of remote access including SMS message, WAP, telephone and Web Browser. All these aim at ready and instant access to BMS system and to draw necessary attention of building management team at the first instance to fault alarm or other situations. To further enhance the ability to enable user to focus on particular M&E system, the BMS control has a large size video wall displaying selected BMS graphics and other M&E engineering system including fire services, vertical transportation, security system and power management system.

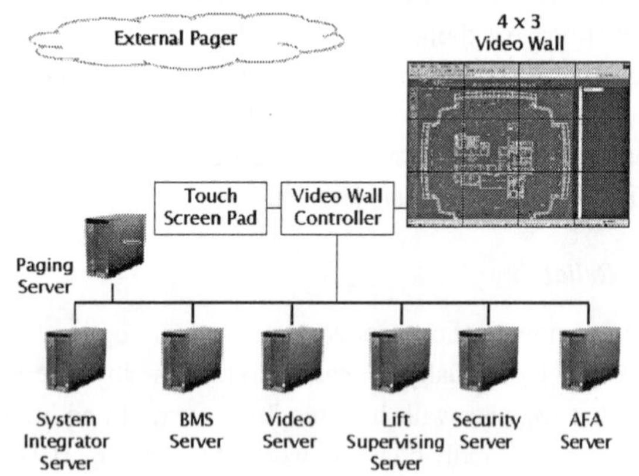

Figure 11. Open System Platform – System Integration

The field level network operate on LonWorks while headend controller operates on open protocol platform which supports a variety of industrial standards including BACnet, Modbus, OPC, DDE, etc. This enables the BMS to serve as a nerve centre integrating a variety of M&E systems via high level open platform interface allowing abundant and convenient data exchange in between (see Fig. 11).

4. Electrical Services Design

The electrical power distribution plays an important role in providing a reliable and efficient energy source for the proper function of all the mechanical and electrical engineering systems.

4.1 *Servicing strategy*

Reliability of power supply system is of paramount importance to ensure the normal operation of the building with over 15,000 design working population. A number of measures have been adopted as follows:

- HV dual supply : Currently, the Power Company is feeding their power supply from their separate sub-stations to ensure dual HV infeed are maintained at all time.
- Dual HV riser : Power Company's 11kV cables are running in two independent risers rising to eliminate the possibility of total blackout.
- Dual tenant riser : Each tenant floor will be served by dual riser LV busduct system feeding from different transformers sources and floor LV rooms.
- Reduced loading capacity : Each transformer shall be loaded to only up to ~ 60% (approx.) of rated capacity of the transformer. This would allow additional loading transfer from other tenant riser in case of emergency situation.

4.2 *Emergency power supply*

Two categories of emergency generators are provided, namely the FSI generator and non-FSI generator (including Tenant's backup supply).

All FSI generators are connected to a centralized fuel supply system which consists of 2 x 20,000 litre bulk fuel tank at G/F, upfeed pumps in fuel tank room; intermediate transfer tanks are provided on mechanical floors with transfer pumps to transfer fuel to the higher zones and so on.

As Two IFC is targeted to serve financial institutions who will demand a highly reliable power supply system. In this respect, non-FSI generators are provided within the building to cater for:

- All essential air-cooled chillers and essential chilled water circuits for base building and tenants mission critical equipment.
- At least one of the lifts in each zone, including the shuttle lifts for sky lobbies, are backed up by generators for tenants emergency operation.
- Base building critical equipment are also backed up by generators. This includes the security system, CCTV system, carpark control systems, emergency broadcast, telecom and the building automation system.

5. Power Quality Monitoring

The tremendous increase in the use of information technology has dramatically increased the importance to monitor the quality of power supply system throughout the building. Traditionally, power quality could be measured by a standalone type harmonic analyzer which can only measure data for a short duration of time on a needed basis. Such approach will be expensive and ineffective. Two IFC has adopted a state-of-the-art fully computerized power analyzer system which can monitor the power quality on real time basis continuously for almost all major loads and circuits.

All power analyzers are connected to a dedicated fast Ethernet network which in turn connects to building management system via DDE link for effective cross-transfer of useful energy trail audit data enabling building operation monitored and controlled at its peak efficiency. The following parameters are measured at real time:-

- Current, voltage, frequency, power factor
- KVA, kVAr, maximum demand, kWh
- Harmonic content up to 15th order

6. Vertical Transportation System

Two IFC has divided into 7 passenger lift zones with 2 sky lobbies serving by double deck shuttle lifts to cater for a maximum population of 15,000 people . In total, there are 64 off high-speed lifts serving different floors.

6.1 *Lift zoning*

Zone 1 and Zone 2 shall be accessed directly from 1/F main lobby. A bank of double deck lifts shall be provided to serve the first sky lobby which will then serve the floor for local lift zone 3 and zone 4. Another group of double deck lifts shall be provided to serve the second sky lobby, which will then serve the local, lift zone 5, zone 6 and zone 7. As a single tenant will occupy zone 7, 2 off passenger lifts with rated capacity of 1,600kg at 2.5 m/s are provided to serve solely for these floors to handle internal traffic.

To provide a high level of user comfort, car shrouds to improve the overall aerodynamic performance are installed at top and bottom of lift car in order to achieve a ride comfort of +/-10mg (where 'm' is 10^{-3} and 'g' is acceleration due to gravity which is 9.8 m/s^2) for car speed from 2.5 up to 8m/s.

6.2 *Building sway*

The resultant tall and flexible structure has made Two IFC susceptible to wind effect that in turn creates a very complex oscillating / torsional movement of this building. This movement (or called building sway) causes not only perceptible within human thresholds but also impact on the vertical transportation system from safety point of view. As a consequence of this building sway, the damage to lift equipment, such as the oscillation of roping systems is a function of the following factors: -

- Height of lift shaft
- Tension on roping systems
- The natural frequencies of this building

Whilst the height of lift shaft is a given architectural form and tension on roping system is a 'controllable' factor, it is therefore of paramount importance to understanding the natural frequencies of this building in order to assess its implications, possible damages to lift equipment and its lift response proposal (see Fig. 12).

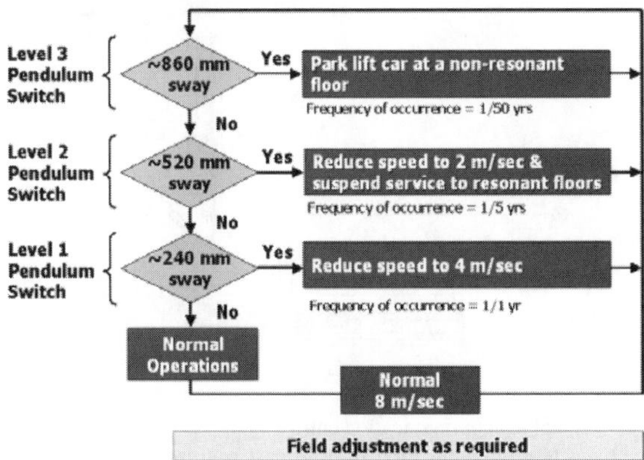

Figure 12. Sway Control Algorithm for high speed lifts

7. TELECOMMUNICATION

With the advance of technology, both in terms of hardware and software, business are now more and more dependent on the information technology to maintain a sustainable growth. The telecommunication infrastructure therefore plays an important role in maintaining a seamless data transfer even in the case of system and material failure.

7.1 *TBE room, sub TBE rooms and external incomers*

Disposition of TBE rooms and connectivity to outside world is fundamental in maintaining the proper operation of telecommunication system. Two IFC was provided with 4 nos. of TBE room with the following design objectives in mind:-

- Dual external incomers at north and south-east of the development.
- Each TBE rooms are interlinked by optical fibre backbone, each serves as a back up centre.
- All TBE rooms are properly protected from fire as well as water flooding.
- In order to provide adequate telecommunication facilities to this grade A office building, sub TBE rooms were provided at each mechanical floors with a minimum of 1 voice circuit per m^2 plus allowance for fax and data (see Fig. 13).

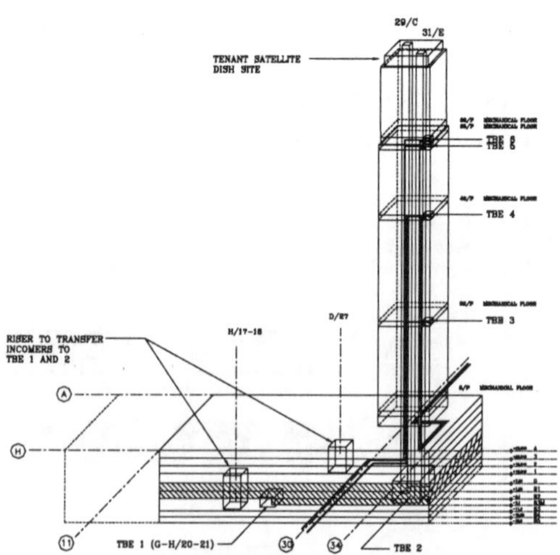

- Dual telecommunication entries
- 4 telecommunication & broadcast equipment (TBE) rooms directly connected to external services
- Twin telecommunications risers
- Provision for copper & fibre optic cabling to each floor (suitable for broadband / VOD services)
- Minimum provision of 1 voice circuit per 10 m^2 + allowance for fax & data services at all sub TBE rooms

Figure 13. TBE Room Locations Telecommunications Pathways & External Incomers

8. Security System and System Integrator

Security is one of the prime concern of those tenants who work within this building and after the 9-11 tragedy. This will not only affect their office locally in Two IFC but also their global corporation policy that will also be one of the major factors in deciding whether to move into this premises or not.

8.1 *Security system*

A state-of-state of security system is provided to facilitate the safe and efficient operation of this building. The system design must be user friendly so that the FM team could respond to any possible incident within a very short duration of time. It should also be resilient enough to be able to respond to any possible system failure so that the building could be monitored on a continuous basis. With this in mind, the following are some of the key features of the security system (see Fig. 14):

- The system is running at TCP/IP protocol so that any future change in security provision could be easily modified. On top of this, the system could be monitored off-site so that senior manager could assess the system at anywhere and also be able to set up a remote command center in case of most adverse condition within this building.
- All images from CCTV cameras are digitally stored in a tailor made storage server, which is capable of storing up to 48TB of data.
- Fully redundancy and hot standby configuration, including dual fiber backbone and redundant servers, to avoid single point of failure.

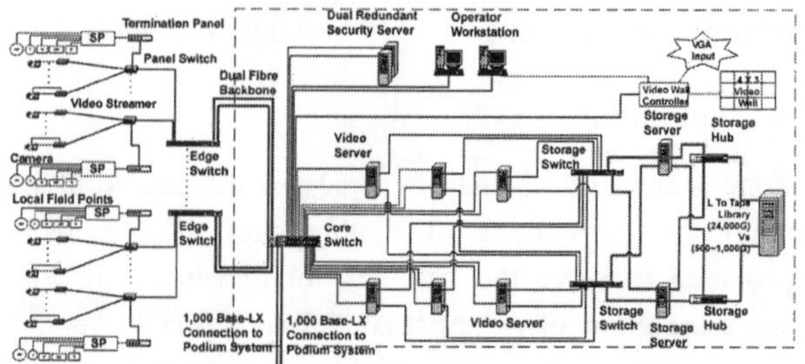

- Full Digital Storage for over 800 CCTV (to be continued)

- Fibre back bone as high speed data path (TCP/IP)
- Dual redundancy and diversified routes
- Easy data retrieval

Figure 14. Intelligent Security and CCTV System

8.2 *System integrator*

Imagine there are more than 20 different M&E engineering systems connecting to various field points around the whole building need to be monitored at all time. The situation will become even worse as there are gigabit of CCTV camera signals continuously pumping in to the central control room. Such huge data makes the building operation virtually impossible unless an unorthodox engineering solution is provided. At the onset of design stage, the management team has provided a wish list how they would operate this building. Based on the interactive discussion, it came to a conclusion that conventional BMS room arrangement will not work for this building. It shall be a tailor made system integrator so designed that all different engineering systems, including BMS, fire services, vertical transportation system, security system, power management system, etc, could be integrated onto an open protocol platform so that critical information could be controlled through a single user interface (see Fig. 15 and 16).

The photo below shows the general arrangement of the BMS room in which it consists of security work console reserved for all security systems whilst a separate system integrator work console is reserved for the system integrator as well as other M&E engineering systems. Such an advance control and monitoring system is unprecedented for commercial building in Hong Kong.

Figure 15. BMS Room Layout

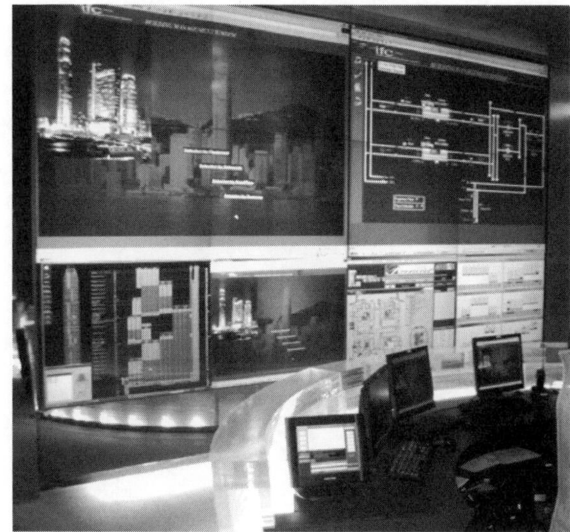

Figure 16. BMS Room – Interactive Video wall with system integration

Conclusions

Two IFC development is an icon tall building as well as a financial building in Hong Kong. It is built to echo the need of a city within a city concept. Many sustainable features were attempted, tested and proved. It is our belief that the use of high rise tall buildings, despite their cost and complex construction, will continue to perform and outbid the scarcity of land in Hong Kong. Hong Kong will continue to be a dynamic and vibrant world city, Two IFC is a landmark for people to remember as one of the permanent images of Hong Kong's success, from the past and into the future.

References

1. HKBEAM Assessment Criteria for Commercial Building
2. Two IFC Quality Building Award Submission (2004)
3. J. Leung, A. To, & PK Yip, Electrical Services Design in the New 88-Storey - Two International Finance Centre, HKIE, Electrical Division, The 21st Annual Symposium, Paper No.3, 3.1-3.5 (2003).
4. A. To & PK Yip, The Challenge of Vertical Transportation System for Two IFC, IEE, First IEE Int'l Conference on Building Electrical Technology (BETNET), 152-157 (2003).
5. J. Leung, A. To & V. Leung, Building Service Development in Hong Kong - Private Sector, CIBSE HK Branch 25th Anniversary Booklet, 76-92 (2004).

Sustainability Assessment

The New Intelligent Building Index (IBI) for Buildings around the World – A Quantitative Approach in Building Assessment and Audit Experience with the Hong Kong Tallest Building, Two International Finance Centre (420m and 88-storey High)

LEONARD K.H. CHOW

Asian Institute of Intelligent Buildings (AIIB)

The world first most comprehensive quantitative assessment method to audit intelligent buildings was published and released on 18 January 2005. It is known as the Intelligent Building Index (IBI) Manual version 3.0. The technical tool consists of 10 important constituting index modules (M) in any intelligent buildings namely Green, Space, Comforts, Working Efficiency, Culture, High-tech Image, Safety & Structure, Management Practice & Security, Cost Effectiveness, and Health & Sanitation. Under the index modules, there are 378 constituting elements (x). The manual employs the concept of "Relative Importance" of each element (x) contributing to the corresponding module (M). The relative importance is reflected by the weighting (w) of the elements using the Cobb-Douglas function in calculation. Similarly, the manual can be applied to audit various types of buildings but with different weight per module. Again, the concept of relative importance or weight (Y) is used for each modules contributing to the overall IBI index (I). In November 2004, the IBI manual v3.0 was used for the intelligent building audit for the 88-storey office building "Two International Finance Centre", the tallest building in Hong Kong. The overall score is 95% which is the highest score so far audited by the IBI manual. The building is very strong in the areas of Working Efficiency, Cost Effectiveness, Space, Culture, and particularly outstanding in Management Practice & Security. It is believed that the IBI Manual is the most up-to-date, objective and comprehensive assessment tool for auditing intelligent buildings of the world.

1. INTRODUCTION

The Intelligent Building Index (IBI) manual is developed by the Asian Institute of Intelligent Buildings (AIIB). AIIB is the first non-profit making organization in Asia committed to promoting public understanding and industry adoption of the principles and technologies of the intelligent buildings (IBs). The Institute was officially established in Hong Kong in 2000 and it has the following missions:

- ✧ To develop Asia's definition and standards for IBs.
- ✧ To act as an independent certification authority for IBs through the use of The Intelligent Buildings Index (The IBI).
- ✧ To educate and promote to the community benefits of IBs.
- ✧ To work with international counterparts to bring Asia up to date on developments related to IBs.

AIIB adopts an official definition of IB, "*An Intelligent Building is designed and constructed based on an appropriate selection of Quality Environment Modules to meet the User's Requirements by mapping with the appropriate building facilities to achieve a Long-Term Building Value*". Based on this official definition, AIIB is able to develop the world's first quantitative and comprehensive assessment method for an IB, called the IBI. All buildings can be assigned the IBI which is a score within the range from 1 to 100. "100" means that it is a perfect IB while "1" means that it is certainly a non-IB. The former IBI version 2 was published in 2001.

In early 2005, the new version 3.0 is released in Asia with the adoption of some revisions to enable greater practicality and better presentation. It is the intention that the document will be used not only in Asia, but also in the US, Europe and all parts of the world.

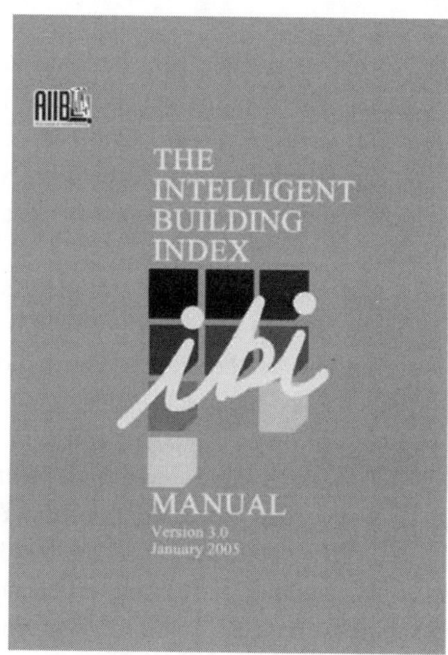

IBI Manual V3.0

The new IBI model consists of 10 **Modules (M)**, namely: Green, Space, Comfort, Working Efficiency, Culture, High-tech Image, Safety and Structure, Management Practice & Security, Cost Effectiveness, and Health & Sanitation. Within each module, there are various **Elements (x)**. The overall numbers of elements are 378. These elements are assessment criteria questions in 10 different building disciplines, namely, Architectural, Structural, Building Services (Air-conditioning, Heating & Ventilation, Electrical & Building Management System, Fire, Lift & Escalators, Plumbing & Drainage), Acoustics, Environmental, and General.

1.1 *History and background of IBI audit*

The Asian Institute of Intelligent Buildings (AIIB) published her Intelligent Building Index (IBI) manual version 2.0 in October 2001. This IBI manual has been used as a tool to audit building intelligence since 2001. This manual had been employed in various Asian cities such as Singapore, Taipei, etc. in intelligent building auditing.

In particular, the IBI manual had been selected as a study module in universities courses, such as Master of Science and some Diploma courses in Hong Kong. It has also been employed as a study reference guide in some tertiary education institutions in Singapore and Hong Kong.

Since 2002, AIIB has been awarding the "Building of the Year" to the most intelligent building in Hong Kong every year. The building which has the highest score in the IBI audit in a particular year will be awarded. So far, according to the auditing record, twenty two (22) buildings were audited in Asia using the IBI manual.

In 2002, AIIB gave an award to the Kadoorie Biological Sciences Building at the University of Hong Kong as the Building of the Year. The building is strong in the modules of high-tech image, comfort and space. The audit was based on IBI version 2.0.

In 2003, the Hong Kong Science and Technology Park Phase I was awarded the Building of the Year by AIIB using the IBI version 2.0. The building is outstanding in the modules of working efficiency, cost effectiveness, high-tech image, comfort and green.

From March to May 2003, Hong Kong was seriously suffered from the attack of Severe Acute Respiratory Syndrome (SARS) virus. AIIB anticipated the need to introduce a new index module no. 10 "Health & Sanitation" in addition to the IBI version 2.0. The six-star residential estate, The Leighton Hill, was awarded the Distinction Award on Module 10 in 2003.

In 2004, AIIB awarded the Building of the Year to the 88-storey high Two International Finance Centre using the IBI version 3.0. The building is of excellent performance in modules of working efficiency, cost effectiveness, space, culture, management practice & security.

2. Intelligent Building Index (IBI)

With reference to the previous sections, IBI consists of 10 Modules and 378 Elements. Please refer to the Figure 1, the IBI Model.

2.1 *Weighting concept in elements and modules*

The 378 elements will be assessed for different intelligent buildings. Each element is an "assessment criteria question" and it is given a weight for its contribution to the module. The score of each element x is an integer from 1 (lowest score) to 100 (highest score). The weight is assigned from an integer range of 1 to 9.

From the IBI manual, in some situations, the same element question occurs in 1 module and repeated in another module(s). This is normal. For example, Element no. 1.24 "Thermal comfort: Temperature and Relative Humidity" under the M1 "Green" module is of weight 5. The same element occurring in element no. 3.18 under the M3 "Comfort" module weighs 9, and that of element no. 4.16 under the M4 "Working Efficiency" weighs 8. From this, it is reflected their relative importance of an element contributing to different modules. It understandable that the "Temperature and Relative Humidity" score is more important in the "Comfort" module of as it weighs 9, while it is relatively less important in the "Green" module since it weights only 5 there.

One may ask a question, "With the similar weighting concept, will different weightings be also applied to the contributions to different types of buildings?" The answer is definite. It is readily realized that each module will also have a different relative importance, or weightings, contributing to the overall IBI score for different types of buildings.

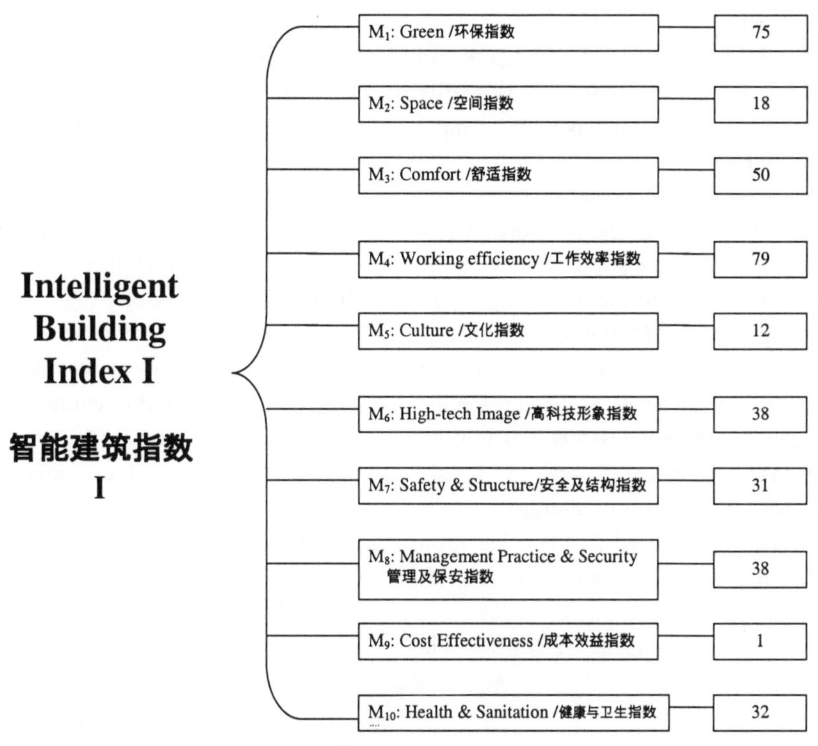

Index/指数: 1   Module/模块: 10   Element/元素: 378

Figure 1: The IBI Model

For example in a situation, if a hospital has all 10 module scores exactly the same as that for an office building, should the 2 buildings obtain the same scores for the overall intelligent building index (I)? From our intelligent judgment, the 2 building should have 2 different scores of I. It is readily understood that the module of "Working Efficiency" is more important in "Office Building", while the module of "Health & Sanitation" is more important in "Hospital". Therefore, the concept of weightings for each module corresponding to each "Types of Building" is readily established. Table A shows the weighting of modules for different types of buildings.

It is readily recognized that the concept of weightings, or relative importance, is required in the 2 stages of calculations:

- ✧ Calculation of the 10 Module scores (M) from Element score (x) using Element Weights (w); and
- ✧ Calculation of IBI score (I) from Module score (M) using the Module Weights (Y).

Table A: The module weightings, Y, for different types of buildings

Type of Building 大厦类型	M_1	M_2	M_3	M_4	M_5	M_6	M_7	M_8	M_9	M_{10}
Commercial (office) Buildings 商用办公楼	7	8.5	7.5	9	6	8.5	6.5	7	6	8
Hospitals 医院	7	5	7	6	2	4	8	7.5	1	9
Residential Buildings 住宅楼	6.5	2	9	4	7	2	7	8	3	9
Hotels 酒店	4	6	9	5	4	7	6.5	8	3	8
Educational Institutions 教育机构大厦	7	8.5	6.5	9	8	5	6.5	6	4	8

2.2 *Calculation method*

With the acceptance of the concept of weightings within the 10 Modules **M** and the Overall Index **I**, it is readily realized that the calculation is not a simple multiplication and summation.

The use of the "Cobb-Douglas" utility function is adopted by AIIB four years ago in the IBI version 2.0. The methodology employs the calculation of raising the power of the score with respect to the weightings. To calculate the Module Score **M**, the Cobb-Douglas function is to be used with respect to the Element Weightings **w** for each Element Score **x**. To enable the calculation of **I**, again Cobb-Douglas function is used with respect to the Module Score **M** and the Module Weightings **Y** designed for the type of buildings.

The following formulae are used for the calculation of 10 Module Score **M**, and subsequently the Overall IBI **I**.

M_1 to M_{10} are calculated using the general formula M.

$$M_n = x_1^{\frac{w_1}{w_1+\ldots+w_n}} \ldots x_n^{\frac{w_n}{w_1+\ldots+w_n}}$$

...... Formula ***M***

where w_1 is the weight corresponding to x_1,
w_n is the weight corresponding to x_n.
n is the number of elements under module M_n.

Note: If any element that is not applicable, the score x is reset to 1 and the corresponding weight w is reset to 0.

Having all the M_1 to M_{10} calculated, the calculation of I employs the formula below.

$$I = M_1^{\frac{Y_1}{Y_1+\ldots+Y_{10}}} \ldots M_{10}^{\frac{Y_{10}}{Y_1+\ldots+Y_{10}}}$$

...... Formula ***I***

where Y_1 is the weight corresponding to M_1,
Y_{10} is the weight corresponding to M_{10}.

Note: The weights Y_1 to Y_{10} depend on the type of building and can be found in the Table A above.

2.3 *The modules and elements*

The Module 1 "Green" consists of 75 Elements. This module includes the environmental friendly constituents in a building. The constituting elements can categorized in groups of Environmental, Lift & Escalators, Lavatories and Provisions of Appliances, Thermal Comfort, Electrical Services, Heating services, Ventilation and Air-conditioning, Lighting, Drainage, Waste Disposals, etc.

In Module 2, the "Space" Module includes groups of elements in Architectural Design, Carpark and Transportation, Flexibility in Internal Arrangements, Building Provisions, etc.

For the Module 3, "Comfort" Module consists of various groups of elements like Architectural Design, Lift, Lavatory and Provisions of Appliances, Thermal Comfort, Ventilation and Air-conditioning, Lighting, Access, Acoustics, Colour, Entertainment Facilities, etc.

The "Working Efficiency" module 4 includes various element groups in Architectural Design, Lift & Escalators, Lavatories and Provisions of Appliances, Thermal Comfort, Lighting, High Technology, Sign and Directory, Carpark and Transportation, Building Facilities, etc.

The Module 5 "Culture" is considered to be the "soft" side of an intelligent building. It is believed that the module consists of the most valuable elements and these elements have never been identified in any intelligent building assessment tools. The elements include Feng Shui, Entertainment Facilities, Privacy Requirements, Colours, Interior Decoration, Food and Beverage, External Landscape and View, Indoor Plants, Religious Facilitation, Culturally Based Interior design, Promotion Activity, etc.

The Module 6 "High-tech Image" has been interpreted for many years to be the only provision and/or criteria of an intelligent building. In the IBI manual, it only constitutes a module since there are 9 other modules to go with it. However, it weights high in an office building. The element groups are Broadband Internet, Electrical Services, Office Automation, Electronic Facilities, Artificial Intelligent Control, Webpage, Hotlines, Telephone Provisions, Fibre-optic Network, Building Services Automation, Mobile Phone Coverage, Advanced Carpark Facilities, etc.

The "Safety and Structure" module 7 consists of various important parameter elements in an intelligent building. The elements include Earthquake Monitoring, Building Structural Condition Survey, Structural Monitoring, Tile Debonding Monitoring, Terrorist Attack Precaution Plan, Fire Protection,

Electrical Safety, Elevator Reliability, Public Announcement, Total Egress, Escape Plan, Essential Electrical Power, Preventive Maintenance of Building Systems, Indoor Air Quality, Safety Management System, Settlement Monitoring, Risk Management, Security and Crowd Control, etc.

The Module 8 of "Management Practice and Security" indicates some very important parameters for successful implementation criteria of an intelligent building. A truly intelligent building must require a very good building management to operate and maintain. The essential element groups are Building Management Practice, Building Services Operation and Maintenance, High Technology, Security and Monitoring, Controls, Management Planning, Water Supply and Drainage, Environmental Protection, etc.

The Module 9 "Cost Effectiveness" consists only 1 formula to calculate the life cycle costing of a building.

The last Module 10 "Health and Sanitation" is one independent module that can be used separately to examine a building as a measurement tool for healthy buildings. It is particularly popular for residential building assessment. The element groups includes Potable and Flushing Water System, Drainage, Toilet, Carpark and Lift Ventilation, Odour, Cleanliness, Refuse Handling, Pest and Mosquitoes Controls, Filtration in Swimming Pools, Jacuzzi and Sauna, etc.

2.4 The calculation tool

The complete calculation is best to be carried out by computer. By making use of computer software of spreadsheet, e.g. Excel, the whole calculation can be done systematically. Furthermore, the sensitivity of each element contributing to each module can be easily identified.

3. Two International Finance Centre (2 IFC)

An intelligent a building assessment was carried out in November 2004 using the new IBI manual V3.0 for the most significant building in Hong Kong. This building is the 88-storey Two International Finance Centre at Central District. The building has many outstanding design features and the major building provisions are highlighted as follows.

Photo 1: Two International Finance Centre

Development:

- Two IFC has 2,000,000 sq. ft. area development.
- Total development including One IFC 785,000 sq. ft., IFC Mall 800,000 sq. ft., Four Seasons Hotel 800,000 sq. ft., Public Open Space 140,000 sq. ft., Total Car Parks 1,800.
- Integrated with Underground Lines: Airport Express and Tung Chung lines.

Structure:

- Building height of 420 metres.
- Central concrete core with 8 external mega columns.
- Foundation of 61.5 m diameter, 38 m deep cofferdam.
- Floor loading up to 10kPa.
- Design wind load of mean wind speed of 230 km/h, highest typhoon in the world.

Office Space:

- Extra high ceiling height of 3.3 m for trading floors with 340mm raised floor space.
- 2.7 m typical office floor with 190mm raised floor space.
- 15 m core wall to window depth.

Fire:

- Horizontal fire separation at 4 refuge floors of 7th, 32nd, 52nd and 64th floor with 20 escape staircases.
- 2 fully fire-protected "fireman lifts" serving all floors.

Lift:

- 42 high speed passenger lifts up to 6 m/s.
- 8 double deck shuttle lifts at 8 m/s and 6 double deck shuttle lifts at 6 m/s.
- 3 service lifts of 4,000 kg up to 8 m/s and 2 car park lifts.

- 2 executive express lifts.

Air Conditioning:

- Variable Air Volume (VAV) system with VAV box Direct Digital Control (DDC) on heating and cooling.
- Two air handling units per floor with interconnecting ductwork.
- Overall thermal transfer value (OTTV) at 24 W/sq.m. better than 35 W/sq.m. required.
- 10 numbers of 980 RT seawater cooled chillers.
- 8 numbers 320 RT air-cooled chillers, backed up by emergency generators, located on 9 mechanical floors for essential cooling.
- 24-hour chilled water supply at 17 W/sq.m. for office floor and 51 W/sq.m. for trading floor.

Electrical & Telecommunication:

- Dual feed power company incoming power supply with 48,300 kVA capacity.
- 17 numbers of standby generators of capacity from 800 to 1,800 kVA for fire fighting and tenant use.
- Fibre-optic cables for tenants.
- More than 30,000 pairs of Cat. 5 copper and 1,000 core fibre-optics.

3.1 *Audit result*

The Two IFC was audited and has the overall Intelligent Building Index (IBI) at a score 95.06%. The breakdown of module scores is given in Table B.

Module	Score
M1 – Green	93.28
M2 – Space	96.39
M3 – Comfort	91.42
M4 – Working Efficiency	97.00
M5 – Culture	95.31
M6 – High-tech Image	94.78
M7 – Safety & Structure	94.76
M8 – Management Practice & Security	96.66
M9 – Life Cycle Costing	96.00
M10 – Health & Sanitation	94.84
Overall IBI Score	95.06

Table B: Module Score of Two IFC

Apart from the outstanding building design features described above, there are other special provisions from the building worth further illustrations. They are given extra bonus points in the audit process. These features are:

- Smart Card Access System
- Face Recognition System for Access Control
- X-ray checking machine and post office
- Trained dog for guarding and dangerous goods detection
- Anti-territories attack provision e.g. limiting car size to reach some structural column to prevent bomb attack
- Car Park patrol security
- Real Time PDA/PC housekeeping report/work with bar code system
- Concierge Building Management Service
- ATM banking service in building lobby

3.2 *Bonus features*

1.2.1 *Smart Card Access System*

The most popular smart card in Hong Kong is called "Octopus". In Hong Kong, every citizen has at least one Octopus store-valued card. This is a coin-less purse which is widely used in traveling, buying newspapers and breakfast on the journey into the office, and many more applications. The customized Octopus Card is used to identify the tenants and their employees of Two IFC when accessing the building and its facilities through a fully computerized access control system during normal and after office hours. The tenants and their employees are assigned a unique Personal Identification Number (PIN) on the Octopus card. The first assess control is at the concierge. The PIN is identified at the Octopus machine installed at the concierge desk for authorizing access to the lift lobby. The second access control is at the lift lobby. Especially during after office hours, the card is used by the tenants to operate lift to their own authorized office floors. For building visitors, registration is required at the concierge desk. A pass with bar code is issued to a visitor with designated assess in the building. It records the visitor's name and identity with access in and out time in the computer.

Photo 2: Octopus Card Access

1.2.2 *Face recognition system for security*

A face recognition system is employed to check if there is any un-welcome person entering into the building. Contract-based workers are required to line up for recognition daily before they start to perform installation and repair works at the designated authorized floors. There is a camera installed at the line up location where all office decoration workers, delivery people, etc. will need a registration in order to enter the building via service lifts. In an un-noticeable way, their faces are automatically captured and be compared by the computer database if the are any black listed trouble makers.

The face recognition system is an intelligent video surveillance system that automatically detects and matches individual facial features for identification and authentication. The system extracts an image of the face from a video data stream and compares it to the pre-registered system database in real-time. The model-based coding of faces is operated through an original learning algorithm, which is a type of neural network model. It combines extraction of a region that resembles eye and facial features. Matching is possible when part of face is hidden due to a high-speed and high-precision face detection method focusing on uncovered parts. It compares the similarity of partial facial features with an adaptive regional blend matching system. The accuracy of result is up to 80% when approximately 200 pix is analyzed. This technology will definitely help in the terrorist and security control and should be widely employed in intelligent buildings.

Photos 3: Face Recognition System

1.2.3 *Computerized real time update housekeeping system*

A computerized real time reporting housekeeping system is developed by the management company. The system consists of a number of hand held PDAs to report of any tasks completion and working schedule. This wireless system enables an 'instant' reporting and communications among staffs and the central management system which allows them to operate in a highly efficient and secure manner. Barcodes set in the PDA which recognizes to offer control access and use of the facilities. This facilitates an intelligent housekeeping system.

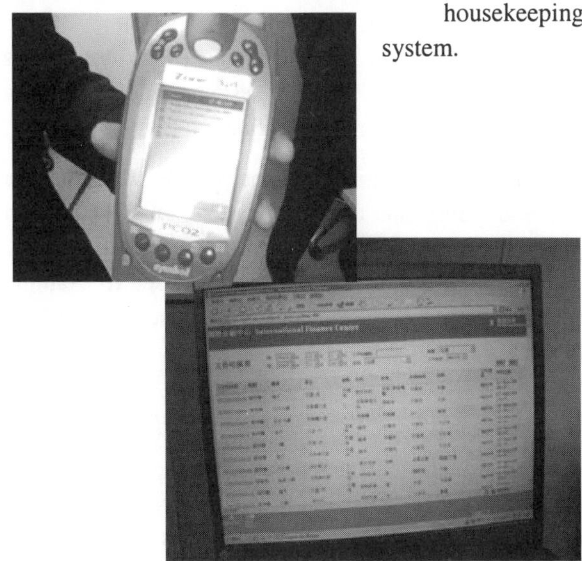

Photo 4: Computerized Housekeeping System

1.2.4 *Highly efficient security system*

An integrated security programme is developed to safeguard tenants with the provision of anti-territories attack through proactive prevention methods and crisis response management.

Monthly car park tenants are designated at an area with barrier free optical turnstiles that can be installed at short notice. Suspected vehicles can be immediately blocked by restricted headroom. The visitor car park is located at designated car park perimeter zones. These parking zones are of low headroom and are well designed that any visitor car carrying explosives will not be able to damage any structural columns in case of any explosion event. In addition, bomb sniffer dogs with trained handlers patrol the loading bay and car park for detection of explosives during times of heightened security risk. To prevent thievery of cars at car park, there are patrol cars carrying out duty on a 24-hour basis.

Ten senior staffs form a crisis response team to deal with incidents ranging from fires and criminal acts through to catastrophic scenarios. They are equipped with pagers and supported by the building's internal public address system. This ensures the team can react at short notice in the event of a crisis or emergency. At each office floor, there are floor wardens who are nominated by tenant to liaise with the team during any emergency. In an event of an evacuation of the building, the team executes the evacuation plans through the internal paging system, guiding tenants efficiently to the assembly locations.

In addition, routine foot and vehicle patrols of the entire building are executed by both uniformed and plain clothes security personnel 24hours a day. X-ray equipment and bomb sniffer dogs are used to examine suspicious packages or mail at the post office of the building.

1.2.5 *Concierge building management service*

Concierge management service is provided for tenants for hotel type of services. Services include air tickets reservations, flight confirmations, restaurant reservations, transportation arrangement, ferry ticket reservations, helicopter reservations, private jet arrangement, hotel reservations, visa assistance, mail service, courier service, sightseeing arrangement, gift and floral service, express printing, interpretation and translation service, mobile phone rental, language tutorial, residential properties search, short-term staff employment service, etc. The concierge service of Two IFC is extremely outstanding and receiving a lot of praises.

Photo 5: Night Scene of Two IFC

4. Conclusion

With the rapid construction together with the increasing demand for intelligent buildings in Asia and the world, the need for an independent and objective intelligent building assessment is immediate. The Intelligent Building Index (IBI) Manual Version 3.0 developed by the Asian Institute of Intelligent Buildings (AIIB) can be widely used as a practical tool for IB assessment. For details of the assessment method and calculation, it is recommended for readers to refer to the manual directly.

The experience of the assessment result on the Two International Finance Centre (Two IFC), the tallest building in Hong Kong, with the bonus features presented in this paper are solely for sharing of technical knowledge. There are many more further information on technical and management practices for the building. Please refer to the AIIB web site www.aiib.net for details.

5. Acknowlegement

The authors would like to thank Mr. Eric Chan and Mr. Christopher Poon and their colleagues at Premier Management Services for offering technical and management information for Two IFC.

6. Author

Dr Leonard Chow is the chairman of AIIB Executive

Committee. He was graduated at the Mechanical Engineering Department with First Class Honours at the Imperial College of Science and Technology, University of London, U.K. He practices as a Mechanical and Building Services Engineer for over 20 years. Subsequently, he obtained his Master MSc and PhD degree in Engineering at the University of Hong Kong and University of Western Australia. Dr Chow is currently a chartered engineer in UK, a registered professional engineer in Hong Kong and a chartered professional engineer in Australia. He establishes his own company ISPL Consulting Ltd in the mid 1995 and celebrates the 10th year anniversary in 2005. His company is the consultants in Mechanical, Structure, Safety and Environmental (Energy & IAQ) aspects with about 25 employees. Dr. Chow is actively involved in professional institution contributions. Apart from AIIB, he works for HKIE, ASHRAE, HKBCxC, HKAEE, etc.

References

1. Chow, K. H. L. et al., January 2005, The Intelligent Building Index Manual Version 3.0.
2. So, A. T. P. et al, October 2001, The Intelligent Building Index Manual Version 2.0.
3. So, A. T. P., Wong A. C. W., Wong, K. C., 1999, A New Definition of Intelligent Buildings for Asia, Extracted and revised from Facilities, Vol. 17, No. 12/13.
4. Wong, K. C., The IBI Arithmetic.

TALL BUILDINGS: MEASURES OF GREEN AND SUSTAINABLE!

JOHN BURNETT
Department of Building Services Engineering, Hong Kong Polytechnic University, Kowloon, Hong Kong SAR

Hong Kong people live, work and spend leisure time in tall buildings. Nowadays 'green and sustainable' buildings are commonplace in the vocabulary and literature, although clear concepts tend to be absent. So-called building environmental assessment methods (BEAMs) provide some measure of the greenness of a building, but it remains to be clarified what measure of greenness is provided by an eco-label such as for example BREEAM-Excellent or HK-BEAM-Gold. Some BEAMs now includes aspects of building performance that go beyond environmental aspects to include assessment of social and economic issues, raising the question as to how a labelling system can provide a measure of sustainability. Based on an examination of several tools, this paper examines the extent to which a given BEAM assessment grade provides a measure of the 'greenness' and 'sustainability' of a building.

1. Background

Most Hong Kong people live in tall buildings, work in tall buildings, and may enjoy leisure time in tall buildings, and spending on average some 86% of their time indoors [1]. Hong Kong is a services dominated economy. Offices are the facilities in which much of Hong Kong's business is conducted, so office buildings are important resources in sustaining economic development. The quality or lack of quality of office buildings will have an impact on the productivity of individual businesses, and the economy overall. As with office buildings, the quality of residential buildings also impacts on the quality of life of Hong Kong residents. The social dimension of tall buildings extends beyond the impact on the users of buildings to the impact on neighbours and neighbourhoods. Buildings place a momentous burden on the natural environment, and buildings are a major reason why cities are environmentally unsustainable.

The descriptions 'green buildings' and 'sustainable buildings', often used interchangeably, are now well-embedded in the popular and technical literature but are usually ill-defined. Tall buildings cannot be environmentally sustainable so being 'green' implies a building that is *more* environmentally friendly than the norm, but this begs the question, by how much? If they cannot be sustainable, it is appropriate to think of such buildings that are *more* sustainable than the norm.

The emergence of so-called building environmental assessment methods (BEAMs) in the early 1990's provides some measure of the 'greenness' of buildings not available previously. Two of the most often quoted methods are BREEAM [2] from the UK, and LEED [3] from the US, although as Table 1 shows the HK-BEAM scheme [4] has had comparable success in terms of impact and market penetration. Within the popular, technical and academic press buildings that are labelled (certified and graded) under a BEAM are often regarded as being green, or sustainable, or green and sustainable! However, it remains to be seen exactly what measure of green is provided by an eco-label such as 'BREEAM-Excellent' or 'LEED-Gold'. This discussion seeks to provide an insight as to the 'greenness' of such an eco-label, as well as the extent to which 'sustainability' is embraced by such tools, with the emphasis on high-rise office buildings.

2. Green Buildings

Many construction and real estate professionals perceive green buildings as buildings having green features, such as sky gardens, green materials, photovoltaics, etc, but which may have relatively insignificant impact on the environment. Through ignorance, much 'greenwash' abounds in marketing and promotional literature (20 kW of PV's on a building with three 1500 kVA transformers does not a green building make!). Hong Kong professionals tend to refer to international standards. One standard defining green buildings is ASTM International's [5]:

> sustainable building, n—see **green building**
>
> **green building,** *n*—a building that provides the specified building performance requirements while minimizing disturbance to and improving the functioning of local, regional, and global ecosystems both during and after its construction and specified service life.
>
> DISCUSSION—A green building optimizes efficiencies in resource management and operational performance; and, minimizes risks to human health and the environment.

Table 1 BEAMs – coverage, weightings and awards

BEAM	Grouping of issues	Points/Credits	Award Levels	Assessments
Initial launch 1991. BREEAM 98 for offices (Ref: 2005 checklist for new buildings)	Management Health/Well-being Energy Transport Water Materials Land Use Ecology Pollution	160 Pts (16%) 150 (15%) 136 (14%) 98 (10%) 48 (5%) 88 (9%) 30 (3%) 126 (13%) 144 (15%) Total 980 points	Pass = 235+ (24%) Good = 385+ (39%) Very Good = 530+ (54%) Excellent = 675+ (69%) (Each issue is weighted 4, 8, 10, etc. so equivalent number of credits is around 90)	Prior version BREEAM for offices – 25% of new buildings. BREEAM 98 for offices – estimated 260
Initial launch 1996. Ref: HK-BEAM 4/04 New Buildings (data assumes office building)	Site Aspects Materials Aspects Energy Use Water Use Indoor Environmental Quality (Sustainable issues) Innovation	25 credits (21%) 20 credits (17%) 29 credits (25%) 12 credits (10%) 32 credits (27%) Total 118 credits (7 credits) 5 bonus credits	Bronze = 40% Silver = 55% Gold = 65% Platinum = 75%	All versions from 1/96 (1996) to 4/04 (2004) total 100 new and existing residential and commercial buildings
Initial launch 2000 Ref: LEED NC 2.1 (November 2003)	Sustainable Sites Water Efficiency Energy/Atmosphere Material/Resource IEQ Innovation	1 Pre + 14 (20%) 5 (7%) 3 Pre + 17 (25%) 1 Pre + 13 (19%) 2 Pre + 15 (22%) 5 (7%) Total 69 points + 7 pre-requisites	Certified 26+ (38%) Silver 33+ (48%) Gold 39+ (57%) Platinum 52+ (75%) (Points are simi9lar to credits in HK-BEAM)	Including version 2.0 around 180 certified projects, 1800 registered projects

This concept leads to the idea of life cycle efficiency in which the 'output' is the indoor environmental quality (IEQ) and the 'input' is the resources consumed and the environmental loadings to air, land and waters, thus:

$$\text{Life cycle efficiency} = \frac{\text{Indoor Environmental Quality (IEQ)}}{\text{Resource Consumption + Environmental Loadings}}$$

IEQ embraces the indoor 'climate', i.e. thermal comfort conditions, indoor air quality (IAQ), lighting quality, and noise, and some aspects of hygiene. The external impacts focus on consumption of non-renewable resources, global and local air pollution, waste recycling, etc over the building life. Typically, IEQ would form 25-30% of the issues assessed.

The aforementioned BEAMs are voluntary schemes, and unless endorsed by governments or provided with incentives rely on take-up by the market for their impact. Therefore, a quandary for BEAM developers is the performance standards that can be set whilst maintaining take-up. BEAMs have been criticised in terms of coverage, lack of scientific rigour [6], undemanding performance standards [7], or for not being sufficiently relevant [8], yet at the same time are praised for raising standards [9]. Notwithstanding, at least BEAMs provide some legitimacy to claims that a building is 'green' [10], or greener than otherwise!

3. Measures of Greenness

A BEAM assessment awards points or credits for compliance with the performance criteria for each of the issues assessed. The number or percentage of points/credits obtained determines the over grade (e.g. Excellent or Platinum) as given in Table 1. Essentially, the higher the grade achieved the 'greener' the building. Even so the depth of green depends on a number of factors, including:

i. how comprehensive is the range and number of environmental issues covered by the BEAM;
ii. the extent to which assessments are performance-based or feature specific;
iii. standards of performance required for compliance relative to prevailing standards or benchmarks;
iv. the extent to which assessment is based on actual measurements, simulation, check-lists or other means;
v. the weighting (points/credits) assigned to each issue.

3.1 *Coverage*

BEAM assessments generally cover the most significant environmental issues - global impacts (global warming, ozone depletion, deforestation, etc), regional impacts (NOx, SOx, particulate emissions, river pollution, etc), local impacts (waste, water pollution), neighbourhood impacts (overshadowing, noise, etc), and indoor environmental performance, but grouping of the various issues differs amongst the tools.

Most of the assessments tend to focus on what the client and the project team can achieve in the given circumstances. In addition, a number of performance issues that may be outside the client's influence are included, some for the purpose of enhancing credibility and others that have relevance to building users. The matter or relative or absolute performance remains central to the debate about the relevance and future role of BEAMs [9].

Clearly, an assessment that is based only on design and specifications (e.g. BREEAM) may not see the performance realised due to poor construction and/or lack of commissioning. Construction performance can be assessed, as can the outcome of commissioning (e.g. LEED), but additional tests on the completed building (e.g. HK-BEAM), especially in respect of IEQ can additional evidence of the quality of the final product.

3.2 *Performance criteria and assessment*

Objectivity of assessments is important for the credibility of a BEAM certificate. Generally, issues are only included if there is a good measure of objectivity, else there is room for argument between assessor and client, and for variability in assessment outcomes. In the absence of quantifiable criteria check-lists are used. Where features such as provisions of meters are prescribed it would be relevant to specify numbers required and specifications.

Although not high on the agenda for investors, owners or even tenants, energy used by buildings is significant and obviously a key concern in respect of global warming. Energy assessments usually include an estimation of energy efficiency (e.g. for air-conditioning, lighting, equipment), performance of particular systems and/or equipment not covered by the estimation (e.g. lifts), and inclusion of features (e.g. meters) or actions (e.g. commissioning, energy management manual) that can allow improved energy management during operation.

Computer simulation is used to estimate how efficient the assessed building design is in comparison with a benchmark building design [11]. Since the energy use depends, not only on the design of the building and systems, but also on weather data and usage, the modelling will assume a typical weather year, and use an appropriate occupancy density and schedule for both designs. The assessment seeks to quantify the extent to which the enhancements included in the design improve on a defined benchmark (a design meeting minimum local code requirements), i.e. assessments of design features of the buildings and systems that are under the control of the designer.

Performance criteria may include dimensions that are outside the influence of the designers. For example, where energy is provided by a mix of fuels assessment of energy performance may be based on quantified emissions, e.g. $kgm^{-2}y^{-1}$ CO_2, NOx, etc. The assessment outcome in this case depends also on the efficiency of the utility supply, which is outside the designers control.

Given that office buildings exist to provide facilities to house people and various activities, assessment not only covers design, but also provisions management, operation and maintenance, but excludes the impacts of the user themselves (e.g. office waste), although provisions to better manage user impacts (e.g. facilities for waste sorting and recycling) are covered.

Whilst structural safety is not an issue given the inherent design safety margins used for tall buildings, but embodied energy, modular design, buildability, durability and maintainability now figure prominently in BEAMs, given that they can enhance whole-life performance and flexibility in building use.

3.3 *Weighting of points/credits*

Weighting, i.e. the allocation of points or credits to the various issues assessed is a controversial matter in the development of BEAMs. In the absence of sufficient data to assess importance it is usual practice to survey stakeholders on the relative weights to be given to the various issues. In future LCA will no doubt provide a measure of the relative importance of the various external environmental impacts, but this will not be applicable to IEQ aspects.

Given the importance of indoor environments the relative weight given to external environmental impacts (resource consumption and environmental loadings) when compared to IEQ is significant. Table 1 shows the percentages for the latest versions of BREAM, HK-BEAM and LEED, assuming application to a new air-conditioned office building.

3.4 *Overall standard*

The overall standard of a BEAM grade, i.e. the difficulty in complying with the individual assessments will only be relative to the benchmarks that are used, e.g. compliance with the local environmental regulations. For example, assessments under LEED are almost entirely performance-based, yet there is criticism that the criteria are often too prescriptive. Where local benchmarks already achieve relatively high standards, such as in California, the eco-label (LEED-Silver) is very easy to obtain [7,18].

Based on the latest on-line documentation for BREEAM it may be thought that there is a good deal of subjectivity in the assessments (although it is understood that BREEAM assessors would use various assessment check-lists). In HK-BEAM 4/04 performance-based criteria are used wherever possible and all assessments seek to be reasonably objective.

Improving market acceptance may influnce overall standard. For example, the 2002 version of the BREEAM checklist [2] "marks an advance is the marketing and promotional activities of BREEAM. For BREEAM clients being seen to be green is a key motivator. For exemplar buildings some means of giving extra recognition is often sought by design teams". The outcome is that BREEAM now gives more points for the process as opposed to the outcome. Perhaps it is a good thing to focus more on process given that planning and design [12] is key to the ultimate success of a building, providing that commissioning of the completed works is also of high standard [13].

4. BEAMs in practice

Cost, particularly initial cost is the main barrier to take-up of BEAM assessments, but concerns that the cost premium for building green is excessive are unwarranted. When one considers these costs ratios: cost of operating a business in an office building 100-200, cost of the space 10, construction cost 5, energy cost 2, design cost 0.5, commissioning 0.2, it is obvious that the more time, effort and money spent on the planning and design phases, and on commissioning, will be rewarded by faster construction, and more efficient and effective buildings [7,14,15].

Even for owner occupied buildings it is rare for clients to consider life cycle cost or to undertake a benefit-cost analysis. Furthermore, it is noted that the cost of complying with certain credits are likely to change when other related measures are put in place [16], and costs will certainly be higher if measures are introduced late in the design process. Where building regulations change or incentives are introduced by government, the viability of certain measures can also change. For example, relaxation in the counting of Gross Floor Area allowed under the HKSAR Government's Joint-Practice Notes [17] enables certain green features to become cost-effective.

In practice it seems that clients will seek a particular grade and will ask their project team the approach and costs to achieving the target [18,19]. Developers of prestige buildings would most likely target 'Platinum' or the equivalent, not wanting to advertise anything less. Public sector clients tend to seek a lower grade (e.g. Silver or equivalent) since public buildings tend to be more functional or on lower budgets, and policy rather than marketing is often the main consideration. A project team would then examine each point or credit on offer to determine feasibility and cost of achievement, targeting the issues that come at low-cost [20]. Then, points or credits thought to be within reach would be considered to see if the accumulation of points or credits is sufficient to achieve the target label (e.g. 65% of available credits to achieve 'Gold'). If the targeted rating or grade is thought to be unattainable or perceived to be too expensive, the client would likely abort the formal assessment.

Focussing effort on more detailed planning and better integration of designs will ultimately lead to better quality, and 'greener' buildings [12,13]. However, given the tendency for clients to 'seek' points or credits if the scoring system in a BEAM is overly complex, i.e. using many sub-criteria and multiple weightings it will be difficult for the project team to assess the likely outcome, to cost alternatives, or to judge the value of any additional investment to improve the grade. The client is faced with an outcome that is unclear and an increased investment risk.

Similar grades from different BEAMs are not necessary the same 'shade' of green, and even two similar buildings with the same grade from a particular BEAM can differ in their relative 'greenness'.

5. Sustainable Building Issues

Office buildings that are healthy and comfortable are essential to sustainability given the impact of productivity [21,22] and the quality of life generally. The main physical qualities included are air quality, thermal comfort, lighting and noise, but the interplay amongst these and the broader aspects of the working environment are very complex. Evidence shows that failure to provide conditions that satisfy the majority of employees will impact on productivity due to discomfort [23]. Failures that result in unhealthy buildings will result in increased absenteeism, lower productivity, and maybe building related illnesses [22].

Within the concept of life cycle efficiency IEQ can be broadened to include aspects of building performance that go beyond comfort and well-being to include the quality of services, such as lift service (important in tall buildings), IT provisions (important is offices), amenities for building users, etc, and not forgetting safety and security. In the context of tall buildings, safety is 1st priority, and for businesses security is usually 2nd. However, even with the application of comprehensive building safety codes, residual problems remain, sometimes attributable to lack of design integration between safety systems and other building services systems [24], and sometimes with aspects of green design [25].

Means of access to buildings for disabled persons is also covered by legislation, but usually this does not go far enough and neglects the social dimension, especially in respect of public buildings. The amenities provided to building users also have an impact on office work experience and job satisfaction [26].

When a tall building is erected, particularly in a redevelopment and mixed-use area, the negative impacts on neighbours and neighbouring facilities can be regarded by the locals as antisocial, so development should seek to minimise negative impacts and where possible compensate by including attributes that benefit the neighbourhood, i.e. the concept of being a 'good neighbour' building.

Taking these broader performance issues into account the life cycle efficiency concept for green buildings can be expanded, thus:

Life cycle efficiency =

$$\frac{\text{Indoor Environmental Quality + Services + Amenities}}{\text{Resource Consumption + Environmental Loadings}}$$

IEQ now includes safety, security and hygiene issues, 'services' covers issues such efficiency of vertical transportation, and 'amenities' include those provided to both building users, neighbours, and even the wider community. Such 'sustainability' issues are now incorporated in HK-BEAM 4/04 [4], and the HKSAR Government's CEPAS scheme [27], which is somewhat similar to the Japanese CASBEE method [28]. Notwithstanding, the extent to which these additional economic and social issues can be included in an assessment scheme still depends on the assessment of each issue being reasonably objective.

6. Developments in Hong Kong

A cornerstone of the latest HK-BEAM scheme is to incorporate most if not all good practices in the design and operation of buildings, including those for existing buildings (Figure 1). These good practices include initiatives by various HKSAR Government departments (covering IAQ, energy, water quality, waste recycling, etc), which are promoted through separate labelling and certification schemes, such as for building energy efficiency [29], indoor air quality [30], and the guidance given in through various practice notes for professional persons. Albeit slowly, professional groups in Hong Kong are also providing 'good practice' guidance materials. A very good example is HKCA's extensive guide on construction practice [31].

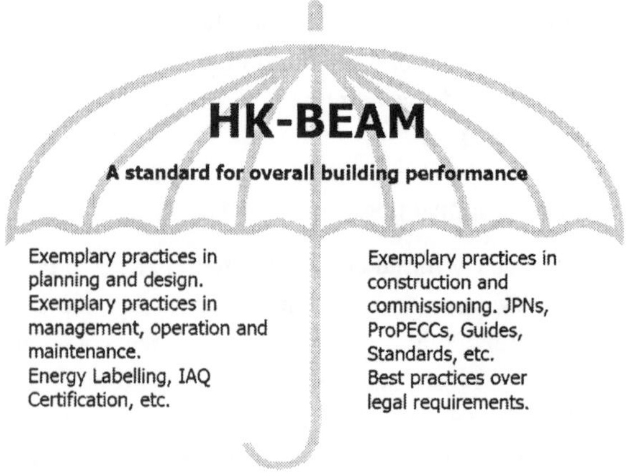

HK-BEAM
A standard for overall building performance
Exemplary practices in planning and design.
Exemplary practices in management, operation and maintenance.
Energy Labelling, IAQ Certification, etc.
Exemplary practices in construction and commissioning. JPNs, ProPECCs, Guides, Standards, etc.
Best practices over legal requirements.

Figure 1 Coverage of HK-BEAM 4/04

The rationale for the HKSAR Government to develop a second building environmental assessment method - CEPAS [27] is unclear, but comparisons between HK-BEAM and CEPAS are inevitable. At the time of writing, the final draft of CEPAS was available for review. In terms of issues covered, there is some 85% overlap, and those that are additional to CEPAS tend to be somewhat minor aspects of performance. In HK-BEAM 4/04 about 10% of the assessments may be regarded as being somewhat subjective than objective. Examination of CEPAS suggests that at least 40% of the assessment criteria to be rather subjective.

Not all the weighting factors required to derive the final score in CEPAS were finalised, so it is difficult to say whether a CEPAS certificate will be based on more objective, performance-based criteria or otherwise. Like CASEBEE the scoring system in CEPAS to derive the final grade is relatively complicated, so prediction of the assessment outcome becomes more uncertain.

7. Summary and Conclusions

Much is written about green and sustainable buildings, but often this amounts to 'greenwash'. Building environmental assessment methods provide a more reasonable measure of the greenness of a building, although the greeness very much depends on the grade achieved which in turn depends on the coverage of environmental issues, the performance standards, as well as the objectivity of the assessment. Criteria and standards change over time so an eco-label is fixed in time.

IEQ aspects of performance are very important for office buildings, given the potential to enhance or detract from worker productivity. Building performance issues such as safety, security, quality of services, and amenities can be included in a BEAM to further embrace social and economic dimensions. Currently, some BEAMs are moving towards becoming sustainable building assessment methods, although for environmental sustainability it is considered that the number of performance criteria would be limited to key impacts such as land use, energy use, etc [6,9].

The tenure of this paper is that a sustainable office building is one that positively support and certainly do not detract from the business activities of users, whilst minimising resource inputs and environmental loadings.

A good assessment grade from a BEAM that is comprehensive in covering design, construction, commissioning and handover, and sets high (but attainable standards) should considerably enhance initial building performance, and permit good management, operation and maintenance practices to maintain the performance over the whole life of a building. In the final analysis it is a matter for the market (client) to decide, to obtain a green label, or a label that identifies a building as being truly green!

Given that sustainability, particularly the environmental dimension, is becoming increasingly critical demanding cooperation amongst nations and communities for the common good, it is interesting to observe that the development of BEAMs seems to be going in the opposite direction. Besides Hong Kong, there are now competing schemes in the US, Canada and Australia.

Acknowledgement

This paper arises from the authors work on the development of HK-BEAM and research on sustainable buildings. Funding support from the HK-BEAM Society, Construction Industry Institute Hong Kong, and the Hong Kong Polytechnic University is acknowledged.

References

1. C.K. Chau, E.Y. Tu, D.W.T. Chan, J. Burnett. Estimating the total exposure to air pollutants for different population age groups in Hong Kong. Environment International 27 (2002) 617–630.
2. Building Research Establishment. BREEAM 98 for Offices. http://www.breeam.org/offices.html
3. US Green Building Council. LEEDTM NC 2.1. http://www.usgbc.org/.
4. HK-BEAM Society. HK-BEAM 4/04 'New buildings'. http://www.hk-beam.org.hk/
5. ASTM International, Designation E 2114 – 01, Standard Terminology for Sustainability Relative to the Performance of Buildings, 2001
6. Cooper I. Which focus for building assessment methods - environmental performance or sustainability? Building Research & Information, Vol. 27, No. 4/5, 1999, pp 321–331.
7. Northbridge Environmental Management Consultants, Analyzing the Cost of Obtaining LEED Certification. April 16, 2003, pp14.
8. Close J, Lo S. Report on HK-BEAM (Residential) Version 3/99 for HK Sustainable Development Forum. August 1999.
9. Cole R J. Building Environmental Assessment Methods: A Measure of Success. International

Electronic Journal of Construction. The Future of Sustainable Construction, 14 May, 2003, p 1-8.
10. Schendler A. Greenwashing Construction: LEED Certification Helps Avoid 'Building Makeovers'. October 2003, pp 1-2.
11. Lee W L, Yik F W H, Burnett J. Simplifying energy performance assessment in the Hong Kong Building Environmental Assessment Method. Building Serv. Eng. Res. Technol. Vol. 22,No. 2, 2001, pp. 113–132.
12. Office of Government Commerce, UK. Achieving Excellence in Construction, Procurement Guide 05 - The integrated project team. 2003.
13. US Department of Energy. The Business Case for Sustainable Design in Federal Facilities. Resource Document. October 2003.
14. Bartlett E, Howard N. Informing the decision makers on the cost and value of green building. Building Research & Information. Vol. 28, No. 5/6, 2000, pp 315-324.
15. Altwies, J E, McIntosh, I B D. Quantifying the Cost Benefits of Commissioning. National Conference on Building Commissioning. May 9-11, 2001. pp 1-13
16. Lee W L, Yik F W H, Jones P. A strategy for prioritising interactive measures for enhancing energy efficiency of air-conditioned buildings. Energy, Vol. 28, 2003, pp 877–893.
17. Buildings Department. Joint Practice Note No. 1, Green and Innovative Buildings, HKSAR Government. Joint Practice Note No. 2, Second Package of Incentives to Promote Green and Innovative Buildings.
18. Portland Energy Office. Green City Buildings: Applying the LEED™ Rating System. June, 2000. Portland, Oregon
19. Kats, G. 2003. The Costs and Financial Benefits of Green Buildings. A Report to California's Sustainable Building Task Force. October 2003.
20. Lee W L, Chau C K, Yik F W H, Burnett J, Tse M S . On the study of the credit-weighting scale in a building environmental assessment scheme. Building and Environment, Vol. 37, 2002, pp 1385-1396.
21. Heerwagen J. Green buildings, organizational success and occupant productivity. Building Research & Information, Vol. 28, No. 5/6. 2000, pp 353-367.
22. Sensharma N P, Woods J E, Goodwin A K. Relationships between the indoor environment and productivity: A literature Review. ASHRAE Transactions; Vol. 104, 1998, pp 686-700.
23. Building Owners and Managers Association (BOMA) International and Urban Land Institute. What Office Tenants Want. BOMA/ULI Office Tenant Survey Report. 1999.
24. Marchant E W. Fire safety systems - interaction and integration. Facilities. Vol. 16, No. 4. 1998. pp 229-235.
25. Chow C. L, Chow W K. Assessing Fire Safety Provisions For Satisfying Green Or Sustainable Building Design Criteria: Preliminary Suggestions. International Journal on Architectural Science. pp 141-146.
26. Myerson J. MT, Stanhope, ICM Research. "Workplace Heaven? The MT Workplace Satisfaction Survey". June 2003.
27. CEPAS – Comprehensive Environmental Performance Assessment Scheme. Final Draft Report. March 2005. Buildings Department, HKSAR Government.
28. Institute for Building environment and Energy Conservation. Comprehensive Assessment System for Building Environmental Efficiency. CASBEE Manual, October 2003.
29. Electrical & Mechanical Services Department, HKSAR Government. The Hong Kong Energy Efficiency Registration Scheme for Buildings. 2000.
30. Indoor Air Quality Management Group, HKSAR Government. A Guide on Indoor Air Quality Certification Scheme for Offices and Public Places. 2003.
31. Hong Kong Construction Association (HKCA). Best Practice Guide for Environmental Protection on Construction Sites. November 2002.

HONG KONG BUILDING ENVIRONMENTAL ASSESSMENT METHOD

PETER CHAN
HK-BEAM Society

CHARLES CHU
HK-BEAM Society

Hong Kong Building Environmental Assessment Method (HK-BEAM) was introduced to Hong Kong in 1996 and is an environmental performance based assessment scheme purpose designed for high rise buildings in Hong Kong. This scheme is completely voluntary and is owned by the HK-BEAM Society, a non-profit making organisation consisting of industry professionals. The latest HK-BEAM standards (Version 4/04 and Version 5/04) covers all "New" and "Existing" building types, including office, residential, mall, hotel, school, hospital, institutional and mixed complexes – centrally air-conditioned, naturally ventilated or mixed mode. The HK-BEAM assessment embraces a range of good practices into a pool of criteria using a life cycle approach. The comprehensive assessment framework encompasses exemplary environmental practices in planning, design, construction, commissioning, operation, maintenance, and management. However, more emphasis has now been placed on the wider attributes of a sustainable building, particularly safety, hygiene, security, amenity and overall quality. This approach encourages buildings that fulfil their intended functions whilst minimizing the resulting impacts on the environment. Finally, the scheme provides a benchmark for sustainable buildings and third party recognition to building performance excellence. This paper outlines the latest framework of HK-BEAM, including rationale on the assessment approach and criteria, and a synopsis of its achievements and benefits is also provided. Reference and correlation will be drawn to case studies of buildings which have been submitted for certification.

1. Introduction

Generally speaking, sustainable city is closely related to economic development, social stability as well as environmental considerations. Strategic balance of these factors is crucial to maintain quality of living both for us and our future generations. Physically, the dominant components of a modern city are buildings and associated occupants. Substantial benefits will therefore be rewarded if due considerations have been taken in the planning, design, consideration, commissioning, management, operation, maintenance and management of buildings, with respect to environmental sustainability. In view of these, Hong Kong Building Environmental Assessment Method (HK-BEAM) was introduced to Hong Kong in 1996 and is an environmental performance based assessment scheme purpose designed for high rise buildings.

HK-BEAM is the private sector initiative to promote environmental sustainability in different stages of buildings. One of the prime objectives of HK-BEAM is to address sustainability in buildings and its definition of a sustainable building is one which adversely effects neither the health of its users nor that of the larger environment. Therefore, HK-BEAM focuses on indoor environmental quality and amenities, as well as local, regional and global environmental impacts. HK-BEAM is a comprehensive, fair and transparent assessment scheme, with the following attributes :

- embraces many areas of sustainability, particularly social and environmental;
- recognises best practices;
- provides for a comprehensive method of quantifying overall performance;
- demonstrates performance qualities to end users; and
- provides economic benefits to stakeholders.

HK-BEAM sets best practice criteria for environmental sustainability in buildings, against which performance is verified through a voluntary and independent assessment. 'Credits' are awarded where these criteria are satisfied, and guidance provided where performance can be improved. The assessment outcome is communicated through the voluntary HK-BEAM green building label as a rating of "Bronze", "Silver", "Gold" and "Platinum".

2. The HK-BEAM Framework

HK-BEAM seeks to measure, improve and label the performance of buildings over their whole cycle through adoption of a set of best practice criteria. The first HK-BEAM standards were established in 1996, covering newly built office buildings and existing office premises. In 1999, a standard was introduced to appraise new residential developments. The latest advancement from HK-BEAM has lead to the establishment of two standards which covers all local buildings according to their life cycle stage : -

- *HK-BEAM Version 4/04 for New Buildings* (for planning, design, construction and commissioning, with design and specification provisions for deconstruction);

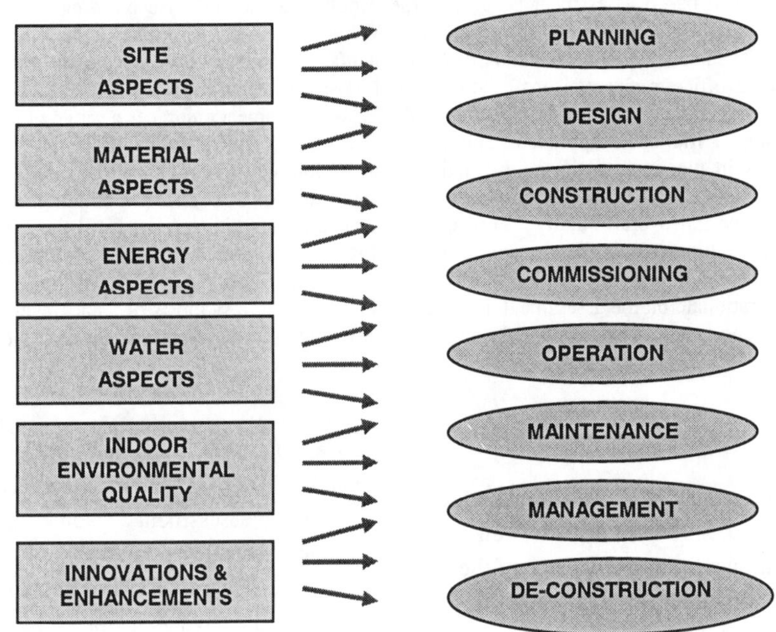

Figure 1 The HK-BEAM Whole-Life Assessment Framework

- *HK-BEAM Version 5/04 for Existing Buildings* (for management, operation and maintenance, with some overlap on commissioning and inherent building design).

All buildings under single ownership can be assessed by HK-BEAM, including but not limited to offices, retail, catering and service establishments, libraries, educational establishments, industrial, hotels and residential. Within each building type, HK-BEAM caters for central air conditioning, mechanical ventilation, and natural ventilation with supplementary cooling, or any combination of these.

Over 100 best practice criteria, described in more detail in the following sections, are applied to the assessment of each building at each life stage, as illustrated in *Figure 1* above.

The intention is that HK-BEAM for new buildings dovetails with that for existing buildings, such that a building rated under HK-BEAM 'new' and suitably operated and maintained would achieve a similar rating under HK-BEAM 'existing' some years later.

Both versions of HK-BEAM aim to reduce impacts of buildings using the best available techniques and within reasonable cost. It is not expected that a building will meet all of the requirements. By meeting some of the criteria, however, the building will have less impact than one in which the requirements have not been met.

3. The HK-BEAM Coverage of Issues

Despite its name, HK-BEAM extends beyond the environment. Since its inception and over time HK-BEAM has covered an increasing number of building related socio-economic issues such as, for example, land utilisation, site amenity and accessibility, pedestrian comfort, environmental health and hygiene, aesthetics, occupant health and well-being, etc. As such, HK-BEAM continues to evolve as an "environmental sustainability" assessment method and, as parameters become better understood and defined, ultimately towards a "building sustainability" assessment method.

The best practice criteria put forward by HK-BEAM are grouped within a general framework similar to other schemes in use worldwide, with due consideration of their relative importance :

- *Site Aspects* – land use and location, site layout optimisation, transportation, accessibility, ecology, amenity, site and neighbourhood interfaces, site emissions and management, etc
- *Materials Aspects* – optimisation in design and operation, innovative construction methods, building flexibility and durability, the avoidance of environment-damaging materials, waste minimisation, etc
- *Energy Aspects* – passive/low-energy design, microclimate, plant/equipment efficiency, renewable energy, annual energy consumption reduction, etc
- *Water Aspects* – potable water quality, water economy and recycling, and effluent management, etc
- *Indoor Environmental Quality* – safety, security, hygiene, amenities, thermal comfort, ventilation effectiveness, indoor air quality (internal and external pollutants), natural/artificial lighting, acoustics and vibration, etc
- *Innovative Techniques* – innovative techniques and enhancements beyond those stipulated in the HK-BEAM criteria indicated above

Credits allocations within the assessment framework have been formulated by taking into account of international and local consensus as given by analysis of weightings used in similar overseas assessment methods, as well as surveys and informed opinions of stakeholders whom have contributed to the development of HK-BEAM.

For more detail information on each of the aforementioned aspects and criteria, full versions of the two latest HK-BEAM standards can be downloaded from the HK-BEAM Society website (www.hk-beam.org.hk).

4. Establishing the HK-BEAM Performance Criteria

Notwithstanding its environmental labelling of buildings, the success of HK-BEAM should no doubt be measured in the extent to which it actually improves their performance. Several key factors are taken into account when defining criteria with a view to encouraging and facilitating such improvement:

(1) *Validity of issues, practicality of assessment* – only those issues for which there is good evidence of the problems they cause, and for which effective performance criteria can be defined, are embraced. Certain aspects attributable to buildings and their use have yet to be included, either because their impacts are not well defined, or because performance criteria have not been established (these may be included in future updates when information becomes available to enable their objective assessment).

(2) *Transparency and clarity* – all criteria and assessment methods need to be transparent, allowing an understanding of the benchmarks (baselines), data, assumptions and issues taken into account in the assessments and consequent rating. This is particularly the case when taking a holistic view of building performance, with an emphasis on life-cycle impacts, with priorities established and weighted for global, local and indoor aspects.

(3) *Setting the baselines* – legal requirements and industry norms constitute the baseline for (zero) credit in HK-BEAM (in certain cases, legal compliance is a prerequisite). First level credits are achieved for good practice in line with international and local guidelines, and further credits achieved commensurate with building size, complexity and age. Criteria are updated periodically as new information becomes available and legal requirements evolve (issues becoming subject to legislation no longer count for the award of credits, and these are deleted or amended in subsequent revisions of the scheme).

(4) *Adopting recognised standards* – where available, HK-BEAM embraces codes of practice, technical circulars, and practice notes issued by authorities and institutions in Hong Kong, with a view to building upon existing approaches and educating practitioners to became familiar with these local professional guidelines. Examples of these include

EPD's Indoor Air Quality Certification Scheme and EMSD's Energy Efficiency Design Codes which have been adopted as assessment criteria. Where such local standards are not available, international standards are used (with relevant tailoring to the Hong Kong context). As a last resort, where such standards are not in existence new criteria are developed by the HK-BEAM Society and its researchers, as is the case, for example, for simulating and benchmarking the energy usage performance of residential buildings.

(5) *Prescriptive versus performance based* – HK-BEAM is as far as possible performance based. This is because prescriptive assessments can preclude buildings that don't have the features prescribed from obtaining a good assessment result (regardless of actual performance), and tend to encourage feature-based design, construction and operating practices. Prescriptive measures put forward by HK-BEAM are only done so with requirements for their effective implementation to thereby improve performance. Furthermore, for issues on energy consumption performance, HK-BEAM has communicated with EMSD to set up a mutual recognition where the results from HK-BEAM can be applied for the EMSD's Performance-Based Building Energy Code.

(6) *Self-assessment methods* – where appropriate, assessments to demonstrate compliance with the prescribed criteria in HK-BEAM are undertaken by suitably qualified persons acting on behalf of the Client, submitting evidence in the form of documents, data and reports. This has been found to be effective in terms of self-learning, enabling the practices and processes recommended by HK-BEAM to be adopted more effectively by clients, designers, contractors and managers. In other cases, issues assessed are based on evidence collected by the assessor in the interest of efficiency.

(7) *Flexible assessment methods* – as mentioned above, HK-BEAM does not seek to be overly prescriptive in setting criteria and compliance methods, instead encouraging building owners and operators to consider alternative approaches that meet the same objectives. Client representatives are invited to submit a 'method statement' if a credit is sought using an alternative approach, stating the proposed alternative criteria and assessment method.

(8) *Raising the performance bar* – as a voluntary scheme without financial or other incentives, the extent to which performance can be enhanced is determined by market acceptance of the assessment criteria, the cost of undertaking assessments, the relative weighting of the credits counting towards the overall assessment, and the perceived benefits to clients. The criteria included in HK-BEAM are considered to be realistic and attainable in practice.

(9) *Flexibility for change* – since collective knowledge on the environmental sustainability of buildings continues to increase, the HK-BEAM framework is designed to respond, resulting in a dynamic system subject to incremental change and update. With wider implementation it is also expected that the scheme will be subject to further scrutiny by an increasing number of stakeholders.

All of the above considerations allow HK-BEAM to evolve over time as new practices and techniques are introduced by industry, enabling standards for environmentally sustainable buildings to increase on an incremental basis as illustrate in Figure 2.

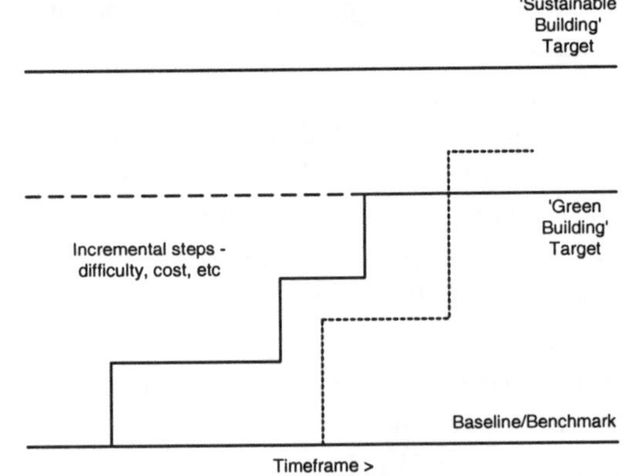

Figure 2 Incremental Improvements in Building Performance

5. The HK-BEAM Certification Process

HK-BEAM is a voluntary scheme and the most important aspect of the assessment process is that it includes the provision of information and guidance from assessors (originating from ongoing research and previous assessments) for which clients and designers and building managers can use in their own projects, and future assignments. This two-way process is often cited as the most beneficial element of the assessment process by those that have been involved.

Premises are assessed against the criteria and 'credits' awarded where the criteria are satisfied. The assessment basically includes two stages, the "provisional assessment" and "final assessment" which is geared for buildings to improve their performance and upgrade their HK-BEAM rating.

Clients are provided with a questionnaire and supporting checklists that details the information required. In the *Provisional Assessment,* an initial scoping process is conducted to identify appropriate criteria for the premises under assessment, since HK-BEAM embraces over 100 criteria designed to cover all types of local buildings. For criteria which are not relevant due to particular circumstances or building type of the assessed premises, these are allocated as *Not Applicable*, and are excluded from the total number of credits available.

Based upon intended design features and construction/management practices, the performance of the premises is identified provisionally. The client and project team are also able to achieve credits through commitment at this stage if they are confident of fulfilling credit requirements during the course of the assessment. These credits will be designated as "*Provisional Credits*" which require submission of additional materials from the project team, site inspections and building surveys for further confirmation prior to the final assessment. In addition, recommendations for improvement are also presented at this stage through which the performance of the premises could be enhanced.

Prior to the Final Assessment, the client and designers have the opportunity to undertake the recommended improvements of their choice to upgrade the building's HK-BEAM rating. For a new building development, the assessors perform inspections during construction and upon completion of the property to verify allocation of credits. In the case of existing premises, the assessors appraise documentations and other relevant information submitted to demonstrate recommended improvement measures have been taken on board, and subsequently designate additional credits achieved. The final outcome of the assessment is presented on the HK-BEAM Final Assessment certificate as a rating of *Bronze, Silver, Gold and Platinum*, according to the number of credits achieved.

HK-BEAM intends to encourage clients to pursue outstanding environmental features, innovative technique or requirement above the norms by providing *Bonus Credits* for certain criteria throughout the assessment framework. Bonus credits *are included as the total number of credits gained but excluded in the total number of credits available i.e. failure to comply will not be penalized.*

When assessing a building complex with different parts e.g. a premises combining hotels, retails and office accommodations, an *Area Weighting Method* is introduced for the *Energy Use* and *IEQ* categories only, as the differences in these categories is much significant than others. Under the area weighting method, the credits achieved are weighted by the areas (in percentage of the overall normally occupied building areas) of a particular part of the building complex. Summation of credits of individual parts of the building complex results in the overall weighted credits.

The HK-BEAM assessment process is summarised in *Figure 3* below :

Client: Approaches HK-BEAM assessors with their selected building for evaluation, receives easy-to-use information checklists with which to collect data from the designers / builders / managers.

↓

Assessor: Appraises the project against the HK-BEAM criteria. The *Provisional Assessment Report* identifies those credits that have been achieved and exactly performance could be improved.

↓

Client: Pursues additional credits through refinements in building design, specification and management. Details submitted for reassessment, advice provided on an ongoing basis.

↓

Assessor: Evaluates the proposed improvements, undertakes construction site / building visits for verification. The Final Assessment Report and HK-BEAM certificate issued on building completion.

↓

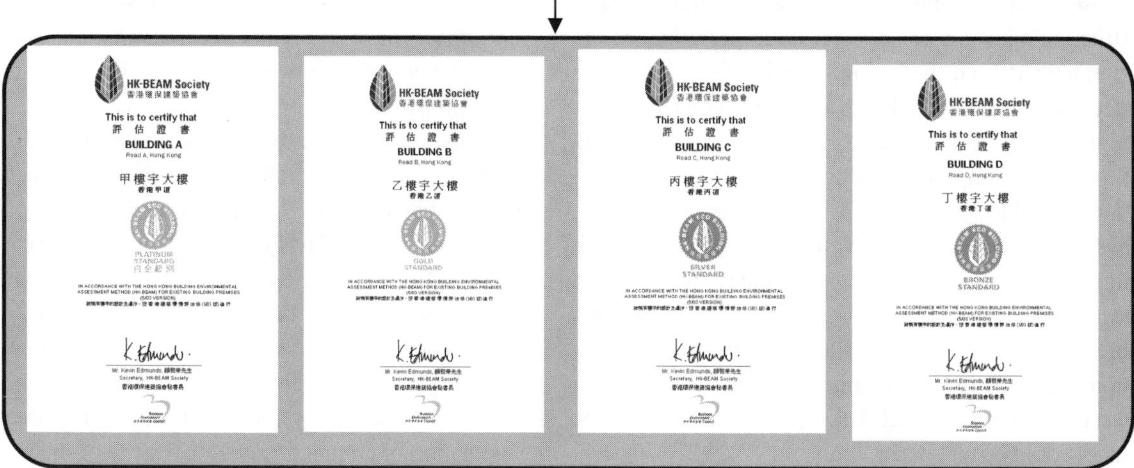

Figure 3 The HK-BEAM Certification Process

6. HK-BEAM Achievements to Date

Initiated in 1996, HK-BEAM has become Hong Kong's most widely recognized industry benchmark to measure, improve and label the environmental sustainability of buildings. Over 100 private and public sector premises covering 6.2 million square meters (including residential, office, commercial, educational, health and mixed use) and 52,000 residential units have so far been submitted for voluntary certification to HK-BEAM. This level of adoption makes HK-BEAM one of the most widely adopted voluntary green building labelling schemes in the world, particularly in terms of number of buildings, percentage of total stock assessed, and buildings per capita.

The benefits of HK-BEAM to its users have been many and varied. In addition to gaining third party recognition for their achievements, most clients cite improvements in building performance, economic gains, increased knowledge and integration amongst their design, construction and management teams. Examples of financial benefits from HK-BEAM include:

- annual cost savings of \$0.7 million for a new office premises, through enhanced envelope specification to obtain a particular HK-BEAM rating, achieving a payback of less than 1.5 years on the additional capital investment (HK\$1 million);
- reward of additional saleable space, through Government's GFA exemption scheme, for sky and podium gardens, prefabricated construction, solar

shading and other features (all encouraged by HK-BEAM) at a residential development; and

- annual electricity cost savings of HK$1 million, with a pay back period of less than 3 years, for the owner of a large existing office building through a series of energy improvement measures identified and recommended by HK-BEAM.

In addition to the environmental, social and financial gains recognised by the assessed buildings described throughout this paper, HK-BEAM has also done much to stimulate partnership between stakeholders, and introduce innovative technologies and techniques which may not otherwise be adopted by clients.

7. HK-BEAM Society

The HK-BEAM Society is the non-profit organisation which owns and operates, on a self-financing basis, HK-BEAM. The society oversees the on-going development and implementation of HK-BEAM standards for building assessment, performance improvement, certification and labelling.

HK-BEAM Society is a membership base organization. Individual members, more than 160 in number, originate from all industry disciplines – designers, contractors, suppliers, consultants, clients, users and managers – in the private and public sectors. Corporate members include architectural and engineering practices, developers, product and service providers, construction companies, and various trade and professional associations. Society members work to continually enhance and refine HK-BEAM to meet the expectations of all interested parties (in particular building users), and to encourage and assist the local building industry to move towards sustainable development.

The development of HK-BEAM has so far been sponsored by The Real Estate Developers Association and managed by the Business Environmental Council, with research undertaken principally by the Hong Kong Polytechnic University's Department of Building Services Engineering and University of Hong Kong's Department of Architecture. Research has made ongoing reference to overseas initiatives such as UK's Building Research Establishment Environmental Assessment Method, and involved widespread consultation (through peer review, various working groups, and knowledge sharing seminars, etc) with building professionals from all disciplines. The final outcome therefore represents general consensus, although unanimous agreement from all stakeholders on all issues has not necessarily been reached.

Day-to-day implementation of HK-BEAM is undertaken by an elected Executive Committee on behalf of the HK-BEAM Society. Assessments are conducted by specialists at the Business Environment Council, itself a not-for-profit and membership-based business association. Further information on the Society and implementation of HK-BEAM is available at www.hk-beam.org.hk including downloadable HK-BEAM standards, case studies, newsletter, supporting guidelines, and membership application forms. Enquiries can also be made to the HK-BEAM Society secretariat at the Business Environment Council:

HK-BEAM Society Enquiries:
Mr. Kevin Edmunds
HK-BEAM Society Secretary
email: kpe@bec.org.hk
Telephone: +852 2784 3910

HK-BEAM Assessment Enquiries:
Mr. Peter Chan
Lead HK-BEAM Assessor
email: peter@bec.org.hk
Telephone: +852 2784 3932

8. Conclusion

In order for Hong Kong to evolve as a sustainable city, one of the key issues is the embracement of environmental sustainability of its buildings.

As a community we continue to grapple with the concept of Sustainable Development - *meeting the needs of the present without compromising the means of the future* - and what it means for Hong Kong. The design, construction and use of our buildings consume enormous quantities of resources; creates untold environmental pollution and generates much of our waste. Building design and management shape our urban environment and control our overall quality of life. In short, as a large sector of the economy, the construction industry can make a significant contribution to a more sustainable environment in Hong Kong.

HK-BEAM has provided a unique stepping-stone towards embracing such sustainability issues, through partnership of stakeholders in the building industry. The inherent comprehensive nature of the life cycle approach adopted in HK-BEAM has lead to the adoption of many building designs and management measures, which are environmentally beneficial, and often financially rewarding, which might otherwise have been overlooked. HK-BEAM's comprehensive assessment framework encompasses exemplary environmental practices in planning, design, construction, commissioning, operation, maintenance, and management of properties. This approach encourages buildings which fulfil their intended functions, with strong emphasis on occupant health and comfort, whilst minimizing the resulting impacts on the environment. Finally, the scheme provides a benchmark for sustainable buildings and third party recognition to building performance excellence.

THE CULTURAL SHIFT OF THE CONSTRUCTION INDUSTRY OF HONG KONG UNDER THE INFLUENCE OF SUSTAINABLE DEVELOPMENT

C.S. POON Robin C.P. YIP

Department of Civil and Structural Engineering
The Hong Kong Polytechnic University, Hong Kong, China

Sustainable development is a continuous career and a forward-looking business for the future development of human society. Developing a sustainable society requires the recognition and support of all members of the community. It is a top down movement from policy making, system development to task implementation. Recognition of the importance of sustainability starts with the development of a sustainable culture in provision of goods and services. A systematic measurement of the trend of cultural changes would provide valuable information for decision makers, system developers and stakeholders of every industry to formulate their strategy in meeting with the requirement for a sustainable society. This paper provides a method, the T-model, to survey and measure the cultural movement among stakeholders in response of the requirement of sustainable development. The T-model is a mechanism that converts abstract elements of cultural factors into numerical scores and synthesizes them by using a space diagram. The method is verified applicable in a pilot test proceeded in the construction industry of Hong Kong. Data obtained from a questionnaire survey of the pilot test were analyzed by the T-model and provided the result of cultural shift of the construction industry of Hong Kong.

1. The Need to Understand Cultural Shift of the Construction Industry

Sustainable development concept established in the Brundtland Report (1987) attempts to support continuous development of human society for the present and future generations. The concept emphasizes balance of growth in social, economic and environmental aspects, which was regarded as key attributes in developing a sustainable society.

The construction industry is one of the leading industries that supports the economic and social stability of Hong Kong. Rowlinson & Walker (2003) stressed that *"The construction industry has always played a major role in Hong Kong's economy and if its contribution to gross domestic product is analyzed in detail it can be shown that its peal property and construction contributed almost 25% of Hong Kong's GDP. In 2000 the gross value of construction work undertaken in Hong Kong was over HK$120 billion"*. The construction industry is therefore a major economic support of Hong Kong. The number of labour employed in real estate and construction industry as recorded in the June 2004 Census Department website (Census 2004) is 144,200, 6.4% of the total employment of labour (participants involved in design, administration, management and civil servant are excluded). It is an important source of employment that strengthens the social stability of Hong Kong. The construction industry contributes significantly to the environmental issues like pollution and waste. According to the recent report issued by the Environmental Protection Department (EPD, 2005), the construction waste accounts for 6,595 tonnes per day which represents 38% of the total waste in 2004. Therefore, the construction industry of Hong Kong fundamentally encompasses economic, social and environment attributes and is one of the most important influential industries affecting the successful implementation of sustainable development.

Implementation of sustainable performance of the construction industry starts with the development of a sense of sustainable culture of all construction participants. Therefore understanding the cultural movement of the construction industry would facilitate policy making of the government and system development of stakeholders approaching sustainable performance.

2. Development of Sustainable Construction Culture

The culture development of the construction industry in Hong Kong is a growth of consciousness in basic assumptions and perceptions that formed the common views among the participants. The evolution of culture of an industry was gradually transformed from basic attitude to sub-sequential behaviour in ideological judgment and performance.

The construction industry of Hong Kong composed many different disciplinary stakeholders. These stakeholders are presumably consisted of the government; the developers; architects, structural engineers, electrical and mechanical engineers, quantity surveyors (collectively the consultants); main contractors, subcontractors, suppliers (the contractors); site agents, foremen and workers (individual non professional participants).

Implementing Sustainability is a career that cannot be achievable instantaneously. The extent of sustainable implementation commences with the consciousness of stakeholders with respect to their ***awareness*** and ***concern*** of the products they produce and the services they provide that affect the society as a whole. The needs of improvement towards more sustainable achievement are the driving forces that conglomerate the attitude and cognition to ***motivation*** and ***implementation***. Stakeholders of different disciplines discharge different influential powers that promote the performance of sustainability.

Stakeholders, as categorized above, can be derived mainly into five composite groups:

- □ government
- □ developer
- □ consultant
- □ contractor
- □ individual non professional participant.

Each group has different extent of influential power that steers implementation towards sustainable development.

Cognitive and behavioral patterns are the initiative elements stimulating the driving force towards sustainability. The degree of cognition of each group is proportional to their degree of awareness and concern. Likewise, motivation and implementation are reflective attributes of attitude and behavior. ***Awareness, concern, motivation*** and ***implementation*** are therefore the basic ingredients that constituted the culture of the construction industry. Sequential migration of these cultural ingredients is the mind-set of all stakeholders from cognitive attitude (thinking) into reflective action (behavior) within their respective cultural environment of each group. It is the driving force towards sustainable development that initiated from the culture of the industry.

Implementation of sustainability is the result of accumulative movements in mind-set and in action (Figure 1). Each stakeholder group in recognition (mind-set) of the necessity of sustainable development will proactively perform (action) their respective duties accordingly. It is the cultural tendency of change in "thinking and acting" (cultural shift) towards sustainable development.

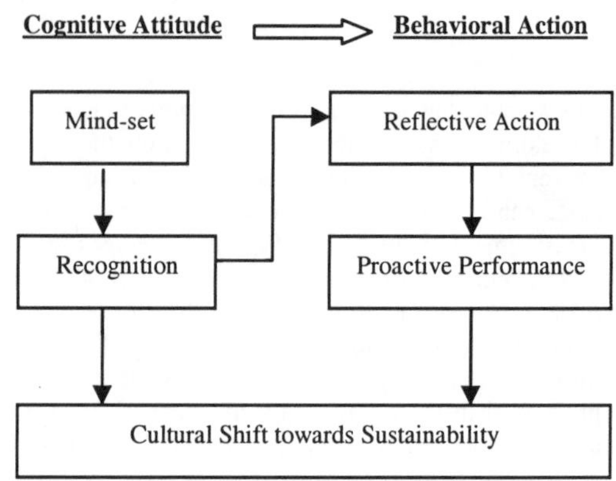

Figure 1: Cultural Changes in Thinking and Action

3. Measurement of Cultural Shift

Awareness, concern, motivation and implementation are abstract elements. They composed most, if not all, basic ingredients of the culture of the construction industry. The movements of these abstract elements from ideological mind-set to behavioral action represent the cultural shift among stakeholders. Understanding the extent of cultural shift of the construction industry would provide a picture of how sustainability is evolving. It is an important information to evaluate the pace of sustainability of Hong Kong towards a sustainable society within the 21st century.

To evaluate the extent of cultural shift is equivalent to measure the trend of movement of awareness, concern, motivation and implementation among the stakeholders. A method of measurement is therefore established along with the changing process of these abstract elements. The measurement method is called the **T-model**. The measurement starts with the identification of a suitable year as the start point – the base year of the measurement. Data of the attitude and behaviour of stakeholders are obtained from questionnaire survey and all the abstract elements are then converted into measurable values. To evaluate the cultural shift for a particular year, i.e. the research year is achieved by assigning scores to every question of a questionnaire survey. Data obtained from questionnaire survey of the research year will be calculated according to the scores assigned to each question. Comparing the scores obtained from data of the questionnaire survey of the research year with the scores of the base year would tell numerically the cultural shift between the base year and the research year. A series of investigation take place at suitable interval would reveal a tendency of cultural movement of sustainability. The result of tendency of cultural movement is very useful for decision makers to evaluate the current situation and envisage their policy formulation.

Since cultural shift is a continuous movement that reflects the trend of sustainable development, continuous measurement should therefore be processed regularly.

To formulate the method of measurement, the process is illustrated by the mechanism of the T-model as follow:-

Since awareness (**x**), concern (**y**), motivation (**z**) and implementation (**w**) are abstract elements showing the attitudinal direction of ideological mind-set and behavioral action.

The tendency of cultural change of any stakeholder group "i" is represented by $\mathbf{S_i}$

The ingredients of $\mathbf{S_i}$ of any stakeholder group is represented by the collection of "a set of abstract elements", $\mathbf{[x, y, z, w,]_i}$

Stakeholders of different disciplinary groups discharge different influential power that affects the process of culture movement. It is represented by an influential factor ($\mathbf{a_i}$) of different stakeholder groups that affects the implementation of sustainable development in the construction industry. Therefore, the tendency of cultural change is

$$[S_i] = a_i\,[\,x,\ y,\ z,\ w,\,]_i \ldots\ldots\ldots\ldots\ldots\text{(formula 1)}$$

Where **S = Tendency of cultural change**
i = Disciplinary Group
a = Influential Power
x = Awareness
y = Concern
z = Motivation
w = Implementation

Summation the tendency of cultural change of the prescribed five stakeholder groups would tell the total tendency of cultural shift of the industry towards sustainability. The total tendency of cultural shift is a synergy of **all** stakeholders' awareness, concern, motivating effects, and the power of implementation and is expressed by Formula 2.

$$T = \sum_{i=1}^{5} S_i$$

Where T = Total Tendency of cultural change towards Sustainability

$$T = \sum_{i=1}^{5} S_i = \sum_{i=1}^{5} a_i\,[\,x,\ y,\ z,\ w,\,]_i \ldots\ldots\ldots\text{(formula 2)}$$

It is important to note that **Formula 2, in combination of the generalized questionnaire survey forms the proposed mechanism - the "T-model"**

The T-model has been verified as a useful tool for the investigation of the Tendency of Cultural Change towards Sustainability in a numeric valuation by using an Orthogonal Space Diagram which synthesizes values of cultural elements of the T-model.

4. Assessment of Cultural Shift by Means of Orthogonal Space Diagram

Since a series of survey can be carried out periodically in different research years with respect to the base year, data of cultural elements (x, y, z, w) from different research years could be converted into comparable scores with respect to the base year.

The scores of each cultural element x, y, z, w, are plotted according to its research year with respect to the base year on the assigned planes of the orthogonal space diagram (see Figure 2).

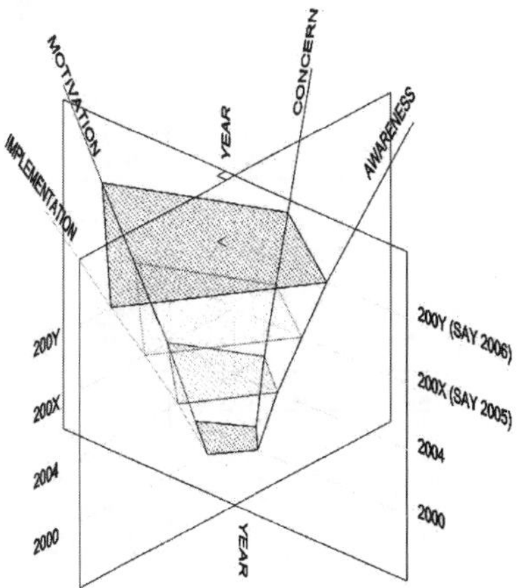

Figure 2: Synthesizing Cultural Elements with Orthogonal Space Diagram

Each point of cultural element is linked up with respect to their research year forming a space polygon (quadrangle) on the same year plane of the space diagram. The area of each quadrangle can be calculated numerically, because of every quadrangle is set on the same plane formed by 4 right-angled triangles with their bases and heights made up by the corresponding scores of the same cultural element. Compare the area of each quadrangle formed by each research year with the area of the base year would numerically provide the cultural shift of that year with respect to the base year.

5. Validation of the Effectiveness of the T-model by Pilot Test

A pilot test has been carried out in 2004 in Hong Kong by sending questionnaires via email and post to 40 targeted stakeholders who practiced construction works in Hong Kong representing different groups of stakeholders in various disciplines. 21 responses were received and the return rate is 52 %. One of a returned questionnaire is discarded because it was damaged and incomplete. Data of other returned questionnaires (20 number) were scored according to the marking scheme and analyzed by using the T-model and the orthogonal space diagram synthesis method.

5.1 *Data analysis: data sorting and score calculation*

A marking scheme is methodically established in a carefully designed generalize questionnaire which was tailor-made for the study. Data of the returned questionnaire survey are entirely sorted and tabulated according to their nature and function of each question which is shown in Table 1. The name of each respondent is anonymously replaced by alphabets and the sorting result is summarized in the summary portion shown at the bottom of the table.

5.2 *Calculation of cultural shift*

Data obtained from the pilot test were converted to scores according the established Marking Scheme for each question and calculated by using formula 2 of the T-model.

$$T = \sum_{i=1}^{n} [S_i] = \sum_{i=1}^{n} a_i [\, x, y, z, w,]_i$$

Table 1: Sorting and Scoring of Cultural Shift Elements between 2000 - 2004

Government

Name	Profession	Influential factor G	D	C	C	I	Awareness	Shift 00 - 04		Concern	Shift 00-04		Motivation	Shift 00-04		Implementation	Shift 00 - 04	
A	HKIA	0.55	0.25	0.15	0.03	0.02	4 + 3 + 7 + 1 + 3 = 18	6	14	21	6	10	25	6	12	24	6	15
B	HKIA	0.20	0.20	0.20	0.10	0.30	3 + 3 + 6 + 1 + 3 = 16	6	11	13	7	10	26	6	11	28	6	10
C	HKIE	0.40	0.30	0.20	0.10	Nil	2 + 1 + 4 + 1 + 2 = 10	6	12	7	10	11	31	6	11	27	3	12
		1.15	0.75	0.55	0.23	0.32	44	18	37	41	23	31	82	18	34	79	15	37

Developer

Name	Profession	Influential factor G	D	C	C	I	Awareness	Shift 00 - 04		Concern	Shift 00-04		Motivation	Shift 00-04		Implementation	Shift 00 - 04	
D	HKIE	0.40	0.30	0.15	0.10	0.05	3 + 3 + 7 + 1 + 5 = 19	6	11	12	6	10	18	6	10	20	6	9
E	HKIA	0.20	0.20	0.20	0.20	0.20	2 + 3 + 5 + 1 + 4 = 15	6	13	12	6	10	24	6	11	21	6	10
F	HKIS, HKICM	0.30	0.30	0.20	0.15	0.05	2 + 3 + 7 + 1 + 5 = 18	10	13	14	6	7	26	6	9	19	7	10
G	HKIA	0.60	0.20	0.10	0.10	Nil	1 + 3 + 6 + 1 + 4 = 15	5	14	14	9	11	28	5	8	33	3	9
		1.50	1.00	0.65	0.55	0.30	67	27	51	52	27	38	96	23	38	93	22	38

Consultants

Name	Profession	Influential factor G	D	C	C	I	Awareness	Shift 00 - 04		Concern	Shift 00-04		Motivation	Shift 00-04		Implementation	Shift 00 - 04	
H	HKIE	0.40	0.30	0.20	0.08	0.02	4 + 3 + 7 + 1 + 5 = 20	4	9	16	6	9	23	3	8	18	3	9
I	HKIA	0.35	0.35	0.20	0.05	0.05	4 + 3 + 7 + 1 + 5 = 20	10	11	9	9	9	24	6	7	19	7	9
		0.75	0.65	0.40	0.13	0.07	40	14	20	25	15	18	47	9	15	37	10	18

Contractor

Name	Profession	Influential factor G	D	C	C	I	Awareness	Shift 00 - 04		Concern	Shift 00-04		Motivation	Shift 00-04		Implementation	Shift 00 - 04	
J	HKICM	0.50	0.30	0.14	0.04	0.02	3 + 3 + 5 + 1 + 4 = 16	3	7	15	6	8	17	3	6	14	3	6
K	HKICM	0.50	0.30	0.10	0.08	0.02	3 + 3 + 3 + 1 + 3 = 13	3	6	10	6	9	15	6	9	11	3	7
L	HKICM	0.15	0.20	0.40	0.15	0.10	2 + 3 + 4 + 1 + 4 = 14	7	12	9	6	9	16	6	9	18	6	9
		1.15	0.80	0.64	0.27	0.14	43	13	25	34	18	26	48	15	24	43	12	22

Non-Professional Participants

Name	Profession	Influential factor G	D	C	C	I	Awareness	Shift 00 - 04		Concern	Shift 00-04		Motivation	Shift 00-04		Implementation	Shift 00 - 04	
M	Foreman	0.25	0.20	0.30	0.20	0.05	1 + 1 + 3 + 1 + 1 = 7	5	7	7	7	8	13	4	5	13	3	6
N	QS	0.15	0.25	0.30	0.20	0.10	1 + 2 + 3 + 1 + 1 = 8	5	9	9	5	8	15	6	6	11	3	6
O	AHKIE, AHKICM	0.60	0.20	0.10	0.10	Nil	4 + 1 + 2 + 1 + 5 = 13	4	7	-8	8	8	16	3	4	8	3	6
P	AHKIS	0.50	0.15	0.20	0.10	0.05	4 + 3 + 5 + 1 + 1 = 14	7	9	0	7	8	17	6	7	13	8	9
Q	Property Mgt	0.50	0.20	0.10	0.10	0.10	1 + 3 + 7 + 1 + 2 = 14	9	12	4	9	10	19	9	9	18	8	11
R	AHKIS	0.60	0.20	0.10	0.10	Nil	1 + 3 + 4 + 1 + 3 = 12	6	9	4	6	10	21	3	7	16	3	9
S	AHKICM	0.70	0.15	0.10	0.08	0.02	1 + 1 + 2 + 1 + 2 = 7	3	9	3	8	10	25	3	5	25	3	10
T	Sub-con	0.70	0.15	0.10	0.05	Nil	1 + 1 + 2 + 1 + 2 = 7	3	7	3	6	9	14	3	4	13	5	9
		4.00	1.50	1.30	0.93	0.32	82	42	69	22	56	71	140	37	47	117	36	66

Summary

		Influential factor Gov	Dev	Cons	Cont	INP	Awareness		Shift 00 - 04		Concern	Shift 00-04		Motivation	Shift 00-04		Implementation	Shift 00 - 04	
	Total:	8.55	4.70	3.54	2.11	1.15	276	Gov.	18	37	174	23	31	413	18	34	369	15	37
	20	All divided by 20						Dev.	27	51		27	38		23	38		22	38
		0.428	0.235	0.177	0.106	0.058		Cons.	14	20		15	18		9	15		10	18
								Cont.	13	25		18	26		15	24		12	22
								INP	42	69		56	71		37	47		36	66

The calculated cultural shifts were summarized and tabulated in Table 2 and Table 3 below:

Table 2: Calculation of Cultural Shift between 2000 and 2004

Stakeholder Groups	Cultural Elements		Shift		Cultural Score		Influential Factor a_i	Cultural Score of each Stakeholder Group (a_i) x (Cultural Value)	
			2000	2004	2000	2004		2000	2004
Government	Awareness (x)	44	18	37	21.41	44	0.428	9.16	18.83
	Concern (y)	41	23	31	30.42	41		13.02	15.55
	Motivation (z)	82	18	34	43.41	82		18.58	35.10
	Implementation (w)	79	15	37	32.03	79		13.71	33.81
Developer	Awareness (x)	67	27	51	35.47	67	0.235	8.34	15.75
	Concern (y)	52	27	38	36.95	52		8.68	12.22
	Motivation (z)	96	23	38	58.11	96		13.66	22.56
	Implementation (w)	93	22	38	53.84	93		12.65	21.86
Consultant	Awareness (x)	40	14	20	28.00	40	0.177	4.96	7.06
	Concern (y)	25	15	18	20.83	25		3.69	4.43
	Motivation (z)	47	9	15	28.20	47		4.99	8.32
	Implementation (w)	37	10	18	20.56	37		3.64	6.55
Contractor	Awareness (x)	43	13	25	22.36	43	0.106	2.37	4.56
	Concern (y)	34	18	26	23.54	34		2.50	3.60
	Motivation (z)	48	15	24	30.00	48		3.18	5.09
	Implementation (w)	43	12	22	23.46	43		2.49	4.56
Non-Professional Participant	Awareness (x)	82	42	69	49.91	82	0.058	2.90	4.76
	Concern (y)	22	56	71	17.35	22		1.01	1.28
	Motivation (z)	140	37	47	110.21	140		6.39	8.12
	Implementation (w)	117	36	66	63.82	117		3.70	6.79

Table 3: Movement of Cultural Elements between 2000 and 2004

Cultural Elements	Σ of Cultural Score of year 2000	Σ of Cultural Score of year 2004
Awareness	9.16+8.34+4.96+2.37+2.90 = 27.73	19.83+15.75+7.08+4.56+4.76 = 50.98
Concern	13.02+8.68+3.69+2.50+1.01 = 28.90	17.55+12.22+4.43+3.60+1.28 = 39.08
Motivation	18.58+13.66+4.99+3.18+6.39 = 46.80	35.10+22.56+8.32+5.09+8.12 = 79.19
Implementation	13.71+12.65+3.64+2.49+3.70 = 36.19	33.81+21.86+6.55+4.56+6.79 = 73.57

5.3 *Use Orthogonal Space Diagram to Synthesize Cultural Shift*

The values of cultural elements in Table 3 were used to plot the orthogonal space diagram to identify the shift of each element between years 2000 to 2004. The space diagram is illustrated in Figure 3. Quadrangles are formed by the calculated scores of each cultural element. Area bounded by the quadrangle of year 2000 = 2425.58 Area bounded by the quadrangle of year 2004 = 7353.91 Square root these areas provide a linear measurement of the cultural shift from 49.25 in 2000 to 85.76 in 2004.

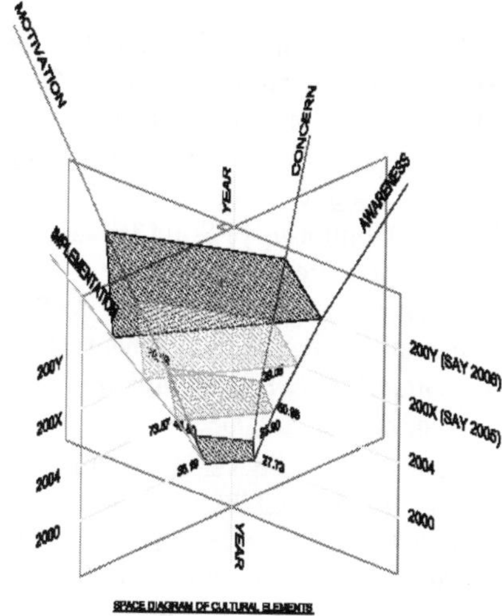

Figure 3: Data Analysis of Pilot Study

6. Conclusion

The study has successfully demonstrated that the T-model is a useful tool to measure cultural shift of the construction industry, and it can also be applied to any other industries of the same purpose.

It is clear that the findings of the study illustrate a positive movement of cultural shift in the construction industry of Hong Kong from 2000 to 2004 with a culture shift value changes from 49.25 to 85.76, i.e. approximately 75% of improvement in 2004 with respect to 2000.

The applicability of the T-model to work out the cultural movement of sustainability is verified. The cultural shift of the construction industry could be obtained in a continuously survey in consecutive years to identify the trend of cultural movement. The result is very useful for decision makers to evaluate the current situation and envisage their policy formulation to establish Hong Kong a sustainable city.

References

1. Abeysekera, V. (2002). Understanding 'Culture' in An International Construction Context. *CIB Report, Perspectives on Culture in Construction (2002) Publication* 275, 39-51.
2. Avery & Baker (1990), *Psychology at Work*. New York: Prentice Hall.
3. CIRC Report (2001), *Construct for Excellence*, Report of the Construction Industry Review Committee
4. Dresner S (2002), *The Principles of Sustainability*, Earthscan Publication Ltd. p. 22, ISBN 1 85383

842
5. EPD (2002), Environmental Protection Department, HKSAR, *Environmental Assessment & Planning*, Section 2, last revision, 26 March 2002, www.epd.gov.hk/epd.english.environmentinhk/eia_planning/sea/baseline_2htm#
6. EPD (2005), Monitoring of Solid Waste in Hong Kong- Waste Statistics for 2004.
7. EPD-Environmental Baseline Report (2003), Environmental Protection Department, HKSAR, *Environmental Assessment & Planning*, Section 2, last revision, 26 March 2003,
8. www.epd.gov.hk/epd.english.environmentinhk/eia_planning/sea/baseline_2htm#
9. Environmental Resources Management (2001), *Sustainable Development for the 21st Century, Executive Summary*, August 2000.
10. EPD-Environmental Baseline Report (2003), Environmental Protection Department, HKSAR, *Environmental Assessment & Planning*, Section 2, last revision, 26 March 2003,
11. Final Report (2004), the Executive Summary of the Final Report for SUSDEV 21, HKSAR, 2004,
12. www.info.gov.hk/planning/p_study/comp_s/susdev/ex_summary/final_eng/chl.htm.
13. Freeman, R. E. & Evan, D. L. (1983), *Stakeholder and Stakeholder: A New Perspective on Corporate Governance*, California Management Review, 25(3), 93-94
14. Hendriks (2001) *Sustainable Construction*, Aeneas Technical Publishers, 2001, p.11.
15. Jones, K. et.al. (2003). Improving the Sustainability of Existing Building. *Second International Conference on Construction in the 21st Century.*
16. Policy Address (1999), Address by the Chief Executive, the Honourable Tung Chee Hwa – *Quality People, Quality Home*, HKSAR
17. Poon C. S. et al (2001), *On Site sorting of Construction and Demolition Waste in Hong Kong*, Resources Conservation & Recycling, 32, 2001 157-172
18. Poon C. S. et al (2002), Use of Recycled *Aggregates in Molded Concrete Bricks and Blocks,* Construction and Building Materials, v 16, n5, July 2002, p281-289
19. Poon, C. S. (2003), *Waste Management: Editorial,* Waste Management, V23, n4, 2003
20. Rowlinson & Walker (2003), - *The Construction Industry in Hong Kong*, Longman Asia, 2003
21. Samuelson P.A. and Nordhaus W. D. (2001), *Economics*, 7th edition, 2001, McGrow Hill, p.32.
22. Shen et.al. (2003). Paradigm Shift of Project Management Organization in Implementing Environmental Management. *Hong Kong Surveyor.*
23. Sustainable Development in Hong Kong for the 21st Century.
24. Tang (2001), Construct for Excellence, *Report of the Construction Industry Review Committee*, January 2001, p.138
25. Teh C. S. (2003) *Sustainable Development in the Construction Industry in Hong Kong Special Administrative: An Exploratory Study*, MSc. Dissertation (2003), Department of Architecture and Civil Engineering, University of Bath, U.K. p.80-81
26. The Canadian Chamber of Commerce & City University of Hong Kong. (2003). Hong Kong Sustainable Development Index (HKSDI) 2003.
27. WCED (1987), The World Commission on Environment and Development, *Our Common Future*, Oxford and New York: Oxford University Press, 1987
28. WFEO (2002), *Engineers and Sustainable Development*, the World Federation of Engineering Organization Committee on Technology, August 2002
29. World Fact Book (2004), www.cia.gov/cia/publications/factbool.hk.htm
30. Yip, C.P.R. & Wong, E.O.W. (2004). Promoting Sustainable Construction Waste Management in Hong Kong. *Construction Management and Economics.*

Websites:

a. Census (2004), www.info.gov.hk/censtatd/eng/press/labour2/lb2_latest_index.html

b. http://www.sustainablemeasures.com/Sustainability/index.html

ECOLOGICAL ENERGY CONSUMPTION SERIALIZATION OF HIGHRISE HOUSING IN SHANGHAI

DEXUAN SONG
College of Architecture and Urban Planning, Tongji University, 1239 Siping Road Shanghai 200092, China

FEI GUO
College of Architecture and Urban Planning, Tongji University, 1239 Siping Road Shanghai 200092, China

The article pays close attention to the regional climate features and the particularity of the community in Shanghai and cities around. In order to realize ecologicalization in the mass construction of high-rise housing, the multi-objective programming strategy is put forward which enables us to meet the demands of different time, season and residence in different high-rise houses. The optimization strategy grades, rearranges and serializes energy consumption to raise its efficiency. In this serialization, the use of new energy resources ensure the enhancement of comfortableness of high-rise housing, creation of healthy, harmonious and comfortable community and while reducing the conventional energy consumption simultaneously.

1. High-rising Housing in Shanghai

The UN forecast data shows that population of china will grows into 1.5 billion by 2050, up to 1/6 of world's population. Another research shows that population of Shanghai then may be more than 25 million. As the rapid population increase with land resource shortage, high-rise houses and high density of land using is an inevitable choice.

Although the municipality is trying to lower the floor area ratio, but it is till very high. The residential and commercial floor area ratio is respectively 2.5 and 4.0. According to the statistics from Shanghai Bureau of Housing and Land, up to the end of 2001, there are 4,226 high-rise buildings with the area of 74.1 million m^2, far more than that of in Hong Kong, rank first in the world. In the 90 years ever since the first high-rise in 1913, three climax of building high-rise has emerged, in which the first two there were 1,095 high-rise built, while the third, 3,181, much more than summation of the first two (as figure1shows).

In the foreseeable future, most housing in shanghai is going to be high-rise. How to meet the demands of different residents and families by different flats and to create inhabitation and surroundings suitable for living in is the new proposition for architects.

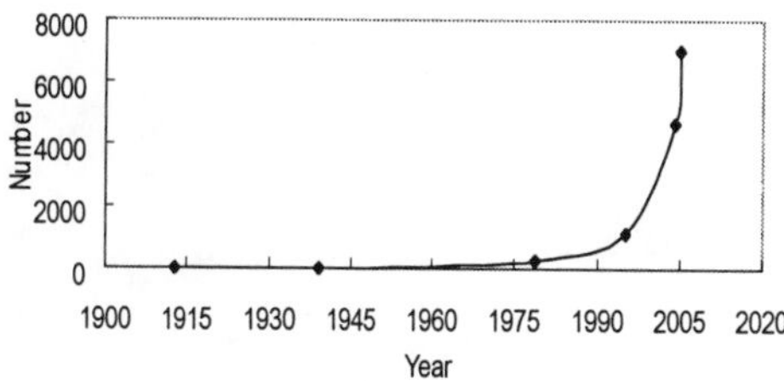

Figure 1. The Development of High-rise in Shanghai from 1913 to 2005.

Shanghai locates in Yangtze River drainage basin in the Middle of China, where it is sweltering in summer and cold in winter. In the hottest month daily mean temperature is 25-30°C. There are 10-20 days in which the maximum temperature is higher than 35°C. In the most cold month daily mean temperature is 0-10°C. There are 50-60 days in which the minimum temperature is lower than 5°C. However the high temperature in summer is tend to be more and more severe. In 2003, there are 40 days in which the maximum temperature is higher than 35°C; from July 19 to August 6, high temperature continues for 19 days with the maximum temperature of 39.6°C (as figure2 shows).

Some researches indicate that in the midsummer the high-rise and high density land using of shanghai causes severe urban heat island effect especially in downtown area, and that the residential land contribute most to it.

In winter difference between the interior and exterior temperature is only from 2.4 ~ 3.1°C.

In the extremely adverse environment the high-rise housing in Shanghai has to consume large amount of energy to ensure the comfort, which in reverse intensify the deterioration of the urban environment.

The energy saving of architecture is carrying out and the municipality proposes the aim of reducing the energy consumption by 50%, but the main practice is still some certain technologies, lacking of unification and conformity. For example in the energy saving review in Shanghai a project measures up when each single item reaches the standard.

How to pick out suitable ones for shanghai high-rise from many technologies and products and exert best effect by organizing them well? How to optimize the energy consumption by low-tech and exercisable technologies?

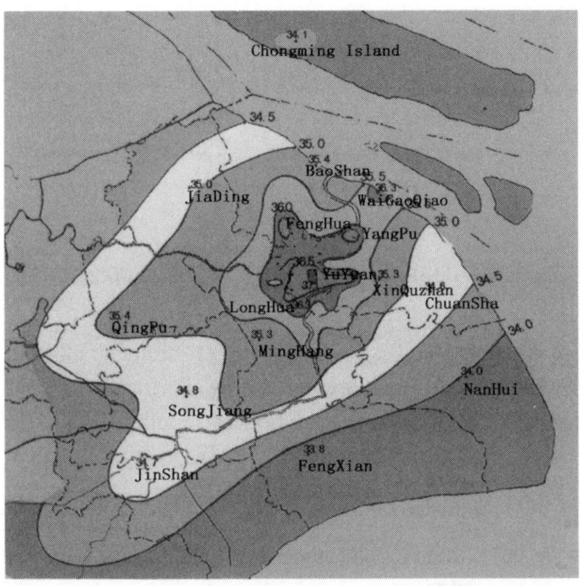

Figure 2.The maximum temperature synthesis of Shanghai summer high temperature observation period

The energy-saving work of high-rise housing is not satisfying at present. At the same time large energy consumption can not ensure the comfort of the users. All of the height, flat patterns, the higher and lower floors, the residents and the distinct four seasons consist of the base of variation. After clear understanding of it has been achieved, energy consumption in the high-rise housing can be serialized. The essence of the serialization is to optimize energy consumption according to the different objectives and variations, which can ultimately ensure the comfort. The study is going to be based on the analysis of the data from surveys, with an eye to rational organization and optimization of the methods in existence and design procedures and method, providing practical guidelines for architects. Consequently, the serialization of energy consumption will save energy greatly and brings an ecological, healthy and harmonious life style for our city.

2. The Dynamic Multi-objective Programming of Energy Saving Strategy

The dynamic outer environment caused by great diversification between seasons in Shanghai brings different comfort demands of the residents; so does the great diversification of residents and their activity status. To satisfy most of the different demands, the multi-objective programming strategy has to be introduced (figure3).

The multi-objective programming strategy allows architects to bring forward one optimal chief objective according to the many sub-objectives when making his decision. From the many requirements and the limitations, the strategy will evaluate the preference of each to come into a chief one that enables the partial differences of sub-objective the least.

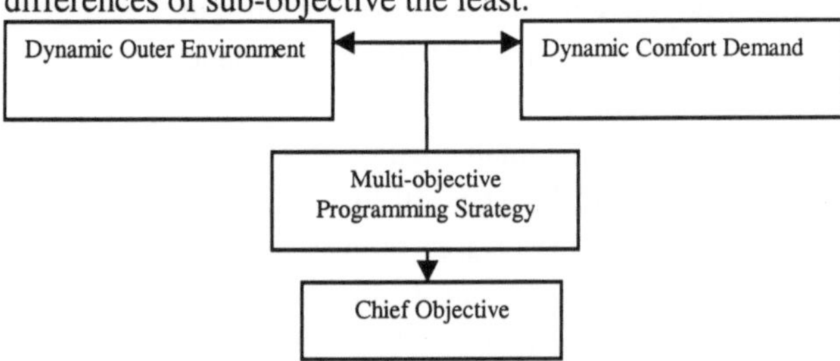

Figure 3. Multi-objective Programming Strategy

To achieve comfortable interior environment of the different families, the different condition of which needs to be specified. Families vary in their members, income and cultural background, activity status and life, which determine their different preference of comfortable environment. They are the basis of the difference in time and space of the energy consumption. Once the multi-objective programming strategy is set up, the whole is in a most satisfied state according to the dissimilar or adverse requests. That the whole is satisfied doesn't mean all the part may be satisfied; some parts and some interests are sacrificed. In this case the notion of satisfaction degree can solve the problem.

$$\min f = \sum_{i=1}^{k} p_i \bullet \sum_{j=1}^{m} p_i(\omega_{ij}^{-} d_j^{-} + \omega_{ij}^{+} d_j^{+}) \quad (1)$$

$$\sum_{j=i}^{n} a_{ij} x_j \leq (or \geq, or =) b_i$$

$$\sum_{j=1}^{n} c_{ij} x_i + d_i^{-} - d_i^{-} = e_i \qquad i=1,2,3,\ldots,l$$

$$x_i \geq 0 \qquad i=1,2,3,\ldots,n$$

$$d_i^{-}, d_i^{+} \geq 0 \qquad i=1,2,3,\ldots,m$$

Equation 1 Multi-objective Programming Model Equation

Suppose there are n decision-making variables, m objectives, 2m partial differences, l resources constraints in the equation.

x_i——decision-making variable

c_{ij}——coefficient of x_i in the No. i Objective

e_i——expectation of the No. i Objective

a_{ij}——coefficient of x_i in the No. i constraint

b_i——restriction of No. i constraint(resource constraints)

d_i^{-}——partial differences of the No. i Objective

d_j^{+}——partial differences of the No. i Objective

p_i——PRI of Objective

ω_{ij}^{-}——weight coefficient of d_i^{-} in NO. p_i Objective

ω_{ij}^{+}——weight coefficient of d_j^{+} in NO. p_i Objective

For the same family users, suppose the different constraints of exterior environment are l_1, l_2, l_3…, decision-making variables (x_1, x_2, x_3, x_5, x_6…) are their different conditions and activities; and the objective variables allow a deviation range.

Establish the PRI and weight coefficient of the each objective. If importance of objectives is six to one, some are rather sacrificed for other.

Different period and condition the users' objective is m, while constraint of the objectives variable range is d (which can also be regarded as an objective). Classified according to their importance, every objective multiplies their weight coefficient.

Lastly deal with the objectives and the constraints all together according to the objective, decision-making variables and restrictions (as figure4 shows).

How to deal with the dissimilarities of different families may be deduced by analogy.

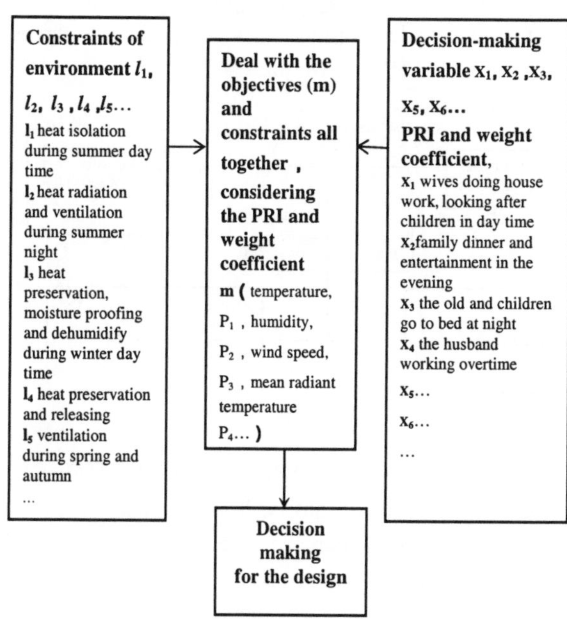

Figure 4. Multi-objective Programming of Energy Saving Strategy

3. The Optimizing of Energy Consumption

In the adverse weather condition as hot in summer and cold in winter in Shanghai, comfort is based on the large amount of fixture energy consumption. How to utilize the energy sources and satisfy the users' demands most efficiently and effectively? In the traditional method, the answer of an architect is to adjust elements of his design without further discussion of regularities in energy movement. Saving energy, reducing the non-renewable energy resources consumption and ensuring the comfort at the same time is our goal. However when we take energy as our research object, there are some new method to achieve the goal (as figure5 shows).

The exterior environment exchanges energy with high-rise housing and influences the interior thermal environment. It also happens between high-rise housing and its fixture system when it accommodates the interior thermal environment. During the exchanges in a house there happens a series of energy exchange processes including storage, transformation, transition, until its diminishing and disappearing. Through the rationale organization and conformity of the process, we can optimize positively the energy consumption and ensure the comfort of the user at the same time.

Utilizing the regularities in the exchanges in the high-rise housing, energy consumption optimization strategy can respond to chief objective best, sets up the new way of thinking as grading, storing and transforming, developing new clean resources, rearranging the PRI as a whole.

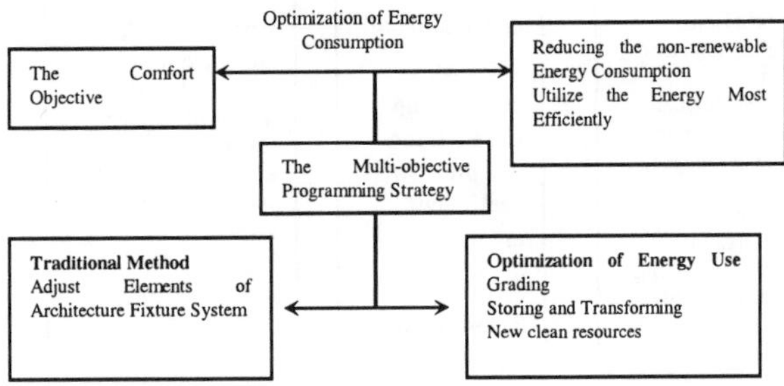

Figure5. Optimization of Energy Consumption Strategy

For an ecological high-rise housing, the application of the optimization can be extended with rational arrange of space and time, period and season (as figure6 shows).

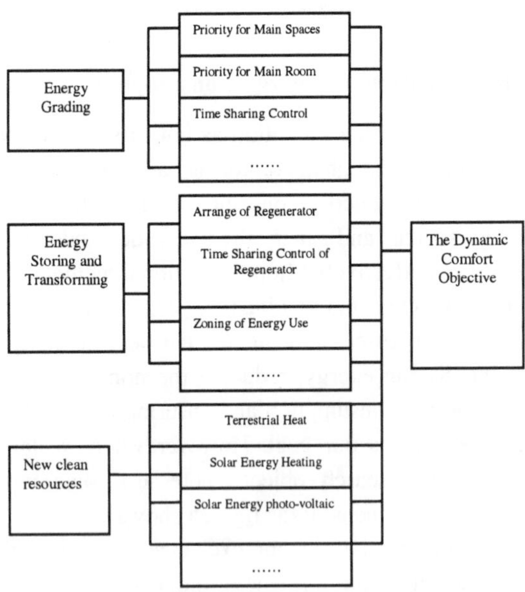

Figure 6. The Application of the Optimization of Energy Consumption Strategy

4. System and Modeling

4.1 *The system thinking*

The study of exchange of energy between high-rise housing and the fixture system or environment, which puts multi-objective programming strategy and optimization of energy consumption strategy into practice, was penetrated with system theory. In the study the complexity of many objectives and their relationships is no more disorderly and unsystematic but mutual affiliated, dependent, constrained and interact in an organism. Through the integration rules applied here including all the parts with their structures and performance and their relationships are to be studied all together to find out the intercommunity and regularity.

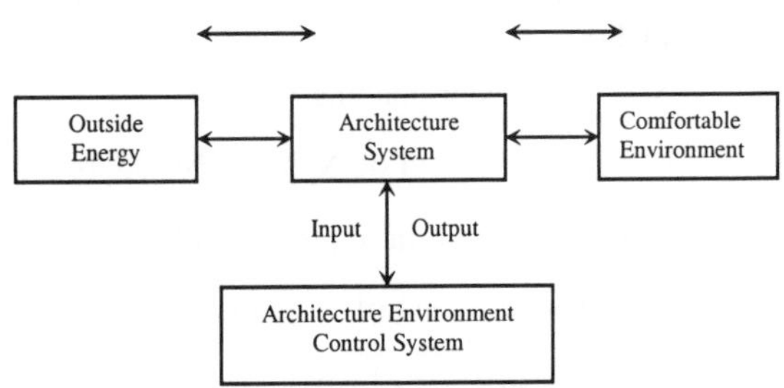

Figure 7. Relationship between Energy and Systems

4.2 *Modeling*

Modeling is the process to establish a model which interconnects information of various relevant components across the system. Abstracted with mathematical, physical and other methods, this model could precisely mimic the original system's behavior and characteristics. Not only the harvest of these information offered quantitative evidence for our research, also, it gives us a panorama image to tackle complex scientific decision. Thus, we need things mathematical to get quantitative representation of the relationships in our information, and to reflect its development. The quantitative process could be a complement to a mere qualitative analysis in that it provided a shortcut to further recognize our research object or the energy consuming process (as figure8 shows).

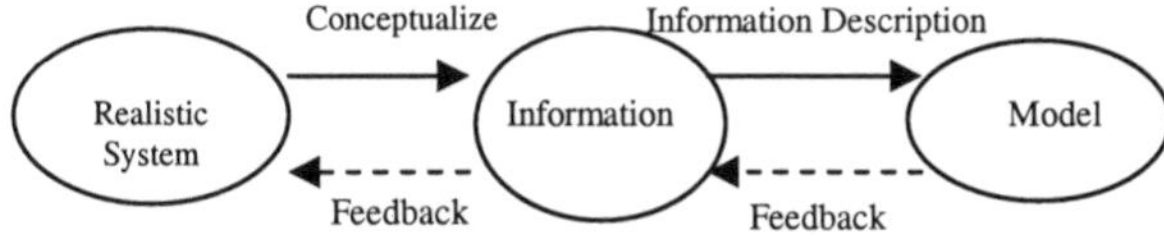

Figure 8 .Modeling Process

The energy is constantly interacting dynamically with external environment and the houses. And the definition of a "comfortable environment" surely differs among the users. The myriads factors involved, dynamic states per se, could prove to be too much for a simple qualitative analysis of a human brain. And facing these monstrous amounts of data, it is computer modeling to the rescue. Only when a model was substantiated using enough precise data, could we matched, individually, the numerous factors determining the flow of energy in the building with the states we discovered, therefore, announcing a case closed.

5. Future Researches' Interface

5.1 *Research on supporting multi-objective programming strategy decision*

The research requires investigation of the elementary information of Shanghai high-rise housing users. The earlier stage of the research is sorting the basic information of Shanghai high-rise housing, which includes distribution, height, floor number, flat- pattern, apartment-pattern, proportion and so on.

This work advances the starting point of objective programming, contributes to summing up of the relationship between the style of high-rise housing and its users.

The research also requires surveying on the users' activity status and their different request for comfortable environment according to various season, time and room. The data from the research show how the request for comfortable environment varies in time and space are determined by users' family numbers, occupation, income and habitude. It is the base of the farther research of optimization of energy consumption and improving the comfort.

5.2 *Modeling the energy consumption of high-rise housing*

Data from the research on supporting multi-objective programming strategy decision, which illuminates the decision and aim we made, is the base of optimization of energy consumption.

The next work should regard energy as an object, discuss how to improve the way of architects' work to satisfying the users' different request and optimizing the energy utilization at same time. The conclusion should include designing guidelines which architecture designers are familiar to, used to and operable.

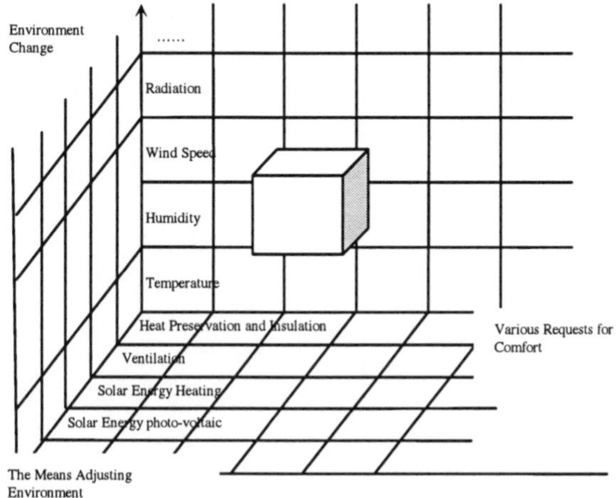

Figure9. High-rising housing's Energy Serialization

6. Figure of High-rising Housing's Energy

According to Holl systems engineering researching method, set the elements (temperature, humidity, radiation, air pressure) affecting energy using in three-dimensional coordinate. The means adjusting environment system are heat preservation, heat insulation, aeration, solar power utilization, adumbral architecture and so on. The third dimension is users' request for comfortable. Base on the chosen location, a three-dimensional space is created, we call it "black box" temporarily .Strategy of multi-objective programming and energy optimization guarantee borderline of the "black box". Then a system in which energy payout, save and release appropriately built up. Abundance technology and information are conserved in the system, which will be an effective means of the realization of high-rise housing's energy serialization.

Acknowledgments

This work is supported by National Natural Science Foundation of China, Integration of Solar Energy Using and Architecture Design for Ecological High-rise Housing in Shanghai (Project No. 50478101).

References

1. H. Su and J. Wang, *Computer Application.* **V23**, 411 (2003).
2. W.J. Liu, *Shanxi. Architecture*, V30 (1995).

Sustainable Design

A STUDY OF ARCHITECTURAL DESIGN FACTORS FOR TALL OFFICE BUILDINGS WITH REGIONAL CLIMATES BASED ON SUSTAINABILITY

JONG-SOO CHO, Ph.D

Assistant Professor, Graduate School of Architecture, Konkuk University, 1 Hwayang-dong, Gwangjin-gu, Seoul, Korea

Throughout history, buildings have been interrelated with certain indigenous characteristics such as regional climate, culture and religions. In particular, the control of regional climate has been primarily a concern for compatibility with nature. In our modern age, technologies to control climate have been successfully developed in architecture but the consumption of large quantities of natural resources can also produce environmental problems. This study is based on the proposition that this negative trend can be minimized with architectural design that is motivated to coexist with a regional climate. This study develops these design strategies for tall office buildings by analyzing various combinations of building design configurations based on regional climates. The objective is to determine the optimum architecture of tall office buildings during the initial design process that will reduce energy consumption for regional climatic conditions. The eQUEST energy simulating program based on DOE-2.2 was used for this comparative analysis study of the energy use in tall office buildings based on architectural design variables and different regional climates. The results are statistically analyzed and presented in functional architectural design decision-making tables and charts. As a result of the comparison of architectural design consideration for tall office buildings in relation to regional climates, buildings physically need less energy consumption when the architecture is concerned with the regional climate and it produces a more reasonable design methodology. In reality, imbalanced planning which is architectural design's lack of regional characteristics requires additional natural resources to maintain desired comfortable indoor conditions. Therefore, the application of integrated architectural design with regional nature should be the first architectural design stage and this research produces the rational. This architectural design language approach must be a starting point to sustaining long-term planning.

1. Introduction

1.1 *Prelude*

The tall building and the twenties century can be synonymous since the tall building style is one of the landmarks in our age. This is a dramatic phenomenon and a powerful expression of architecture in the modern civilization. The height of building is evident on certain structures and forms from ancient times in the various civilizations. Higher towers have been built in almost all cultures from time immemorial. Conceptually, the significant evolution of tall buildings from ancient towers is a "change of function" from some religious symbols to a commercial concept that has aesthetically become acceptable with the changing of modern society and culture driven by a technological development.

Generally, tall buildings greatly impact the scale and context of the urban environment due to their proportional mass and height, and tall buildings are a natural response to the scarcity and high cost of land and concentrated population growth. Physically, tall buildings have a concentration of built space placed on a small site area. Functionally, they enable usable floor spaces to be stacked high. In addition, growth in population and economic resources along with limited building sites and infrastructure in major business districts have motivated the rise of the construction of high-rise building to maximize the use of urban space in the big cities around the world.

The architecture of tall building style has been developed with mutual contributions of architectural aesthetic form and advanced technologies. In particular, the technologies to control climate in architecture has been primarily a concern for compatibility with nature but it can also produce environmental problem by use of large quantities of natural resources. That is, the evolution of the mechanical system for heating, cooling and air handling systems provides for demanded comfort in a tall office building which causes higher energy demand. Thus, this situation has allowed architectural design to be free from adapting to a regional nature system.

Consequently, the great paradox is that modernized nations have succeeded in uncoupling humankind from nature though modern technology and the exploitation and depletion of fossil fuels. Yet civilizations still strongly depend on nature; not only for energy and materials but also for the vital life process. Therefore, if human life is to be sustained in the future, it must depend on knowledge and intelligent action to preserve and enhance the natural condition by means of a harmonious rather than disruptive technology. Architects can be forced to develop design strategies for optimized design considerations based not only on low energy and low cost, but also on minimizing the environmental crisis for sustaining long-term human life. That is, architectural design should be concerned not only with interrelationship of space, but also with external factors such as climate.

1.2 *Definition of tall buildings*

The concept of tall buildings can be defined as a relative concept of our time and contextual conditions. The tall building cannot be defined by its height or the number of stories of the building since the outward appearance of tallness is a relative matter, but it does depend on the effect and quality of the tallness of building. The multi-story building is not generally defined by its overall-height or number of stories, but only by the necessity of additional operation and technical measures due to the actual height of the building.(Beedle, 1971)

The significant criterion is that the design is influenced by some aspects of "tallness". That is, the height of a building differently impacts on architectural, structural, and mechanical designs, as well as construction and building use when compared with "common" buildings of a certain region and period.

1.3 *Office buildings in architectural design*

Office buildings should generally be designed by considering the inter-relationship between a user's functional need and regional characters such as microclimatic system since illumination and thermal comfort must coexists during most of the day throughout the year. It will have physically and mentally effect on human comfort. The 1950's, the architectural design strategy was to physically use as much natural lighting and ventilation as possible. Although the engineering for air-conditioning systems to control temperature and humidity were developed during 1920's and the fluorescent bulb became commercially available in the late 1930's, the limited abilities of these initial mechanical systems kept the office building floors very narrow. However, since the 1950's the high land cost in major cities along with increased equipment reliability and low energy costs have produced the concept of large open spaces with office landscaping and sealed buildings.

Developed equipment reliability has to take care of the direct influence of climate on the internal space and provide all of the required comfort conditions at the expense of non-renewable energy sources. That is, before the energy crisis of the early 1970's, the annual energy consumption for an office building was a minor concern in relation to other design considerations, because of the technological development in equipment capacity and relatively inexpensive energy supply, it was not considered necessary for building design to respond to local climate characteristics and solar effect.

This method of designing office buildings that were dependant on mechanical system technology resulted in unnecessary large energy consumption to achieve human comfort. Consequently, these situations led to architectural design becoming dependant on mechanical systems.

1.4 *General approach to the problem*

The population growth and non ecological technology development have generally brought the rapidly required energy consumption, and the simultaneous destruction of CO_2 diminishing forest regions of the earth. From IPCC (Intergovermental Panel of Climate Change), 1992, the global mean temperature will surpass the 1990 average by approx. 1.8-4.2K by the end of the 21st century which is based on the assumption that the atmospheric CO_2 content, 360ppm in 1992 will rise to double the pre-industrial rate, i.e. to 560ppm.(Daniels, 1997) For the situation, one of the essential variables was the emission of carbon dioxide as a parameter. In fact, the combustion of fossil fuels since the industrial revolution has led to an excessive build-up in CO_2 and these CO_2 levels will continue to increase with the demand of energy consumption. As result of these air-polluting elements, there may be climate changes which will show various kinds of the greenhouse effect and have a negative influence on the ecosystem. (Daniels, 1998) From IPCC, 1995, global warming would impact temperature change, rise in sea levels, extreme weather, human health, forests, coasts, snow and glacier formation.

Therefore, the destroying the balance nature clearly means having concerns for not only human life but also all other living organisms. In other words, the crisis of a natural system is not one of human comfort, but a question of survival.

1.5 *General impact factors in architecture*

General impact factors relating architectural design with sustainable issues can be categorized by "policies and social responsibility with globalization", "lack of commitment" and "large population growth" in the Study.

The energy policies and social responsibility are generally related to environmental protection and renewal. As a result of the oil crisis in 1973, many countries have adopted some kind of policies to actively improve the production and application of renewable energy systems based on an economical purpose. This effort to actively promote the energy efficient systems may first come from governments that initiate taxation incentives, introduce energy regulations and codes of practice and promote renewable energy. Secondary, it may also come from communities and industries that are motivated to take advantage of energy polices. And, it may be generated by the individual who becomes aware of his of her social responsibility in environmental protection and renewal. (Lim, 1999) Globalization can also generally provide for energy and environmental protection with a positive political point of view. In fact, the world is being changed with presenting ethnic, religious, or culture in almost all societies. Globalization is being spread and absorbed in different religious and culture by the growing interdependence of the world's nation economies, and the integration of the financial and telecommunications markets. "Globalization is also driven by the universal drive to respect human rights and to preserve the environment". (Serageldin, 1997) Therefore, the polices or rules from governments may be a desirable way to control and promote energy consumption patterns. Economic viability and social acceptance are also equally significant opportunities to achieve a low energy/ sustaining world in the future.

Despite a common agreement on the benefits of energy efficient/ecological design regarding the protection of the environment, reduced energy bills and increased comfort, architects do not appear to incorporate its features to a notable extent in their works. "One major reason for this disregard may be architect's lack of commitment towards energy efficient/ecological architecture".(Wittmann, 1998) Generally, architects and other designers did not consider the climate as a defining inevitable design factor in the modern age after the industrial era. This contrasted with the pre-industrial architecture with a unique society and culture without high technology and an architectural design that responded to the regional climate conditions. Conceptually, this traditional design was based on a local culture which expressed the local constraints such as location, topography, climate, culture, heritage, building materials, and available technology. After the rapid evolution of technology, architects tended to depend on available technology to solve the local constraints problem. This attitude generally accepted an architecture that did not consider the local natural system (weaken local characteristics), but designed the building to maintain human comfort by technology which, in turn, created an environmental problem because of the higher energy consumption to support the mechanical systems in a building.

Large population growth may accelerate the depletion of energy resources because of the demanded increased energy consumption and then it will have a negative impact on the global environmental systems and human ecosystem. In developing countries, if the population of these countries double as expected, their required commercial energy would increase 7.5 times from 1980 to 2020, and the increment in energy consumption would be 1.3 times the world's total energy usage in 1980.(Shea, 1999)

1.6 *Necessity of designing with climate*

The climatic condition of each building site depends on the influence of the regional climate and the microclimate modified by the surroundings of the building. Regional climate is defined by geographic location - latitude, longitude, and altitude. Local deviation on the regional climate is caused by site topography, vegetation, water bodies, land slope, neighboring man-made structure, etc. This study focuses on the impact of the regional climate because it makes possible the comparison between diverse regional environments and their effects on architectural principles and building elements.

Throughout history, local builders have used ingenuity in working within the exigencies and constraints of nature and a regional climate. This has been true ever since humans first constructed shelters to create a comfortable space and developed their traditional buildings to preserve heat in a cold outdoor climate or as protection from in a hot outdoor climate. Even protection from precipitation was done by designing with a maxim adaptability to nature. Then, with the rapid growth of science from the Industrial Revolution, modern man began to believe that these shelter problems could be solved by technology and that it was no longer necessary for human life to be in harmony with nature. These trends are in conflict with the local natural system and have produced negative unbalanced environmental problems.

For the future of architecture, it is desirable to use an integrated design to achieve an optimal combination of regional characteristics and a fundamental understanding of nature and technology. This research is based on the proposition that design information from a regional climate is significant for energy-conscious architectural building design to minimize the negative environmental impact, more efficiently distribute the daylight to the interior space and to absorb the winter solar gain or keep out summer sun. Therefore, the building designer must optimize the early solutions by understanding the relationships of individual energy characteristics.

1.7 Research framework

This study generally will develop fundamental concepts and applications by incorporating different regional climate conditions. Through a hypothetical construction of a tall office building model positioned within regional climates, architectural design considerations will be examined. Based on the examined design data, a set of quantitative analysis is pursued with two design considerations – building sizes and lease span depth[1] as major internal design factor – for the development of tall office building design methodology.

The results demonstrate general integrated design information for the initial design determinations of tall office buildings in different regional climate conditions, not as a forecast or exact value but rather as a quantitative comparison with available alternatives. These quantitative analyses contribute to the first stage of the design process with a climate condition to measure whether the alternatives are becoming more or less energy efficient within regional climates. Building energy use is influenced considerably by early design decisions such as building scale and internal design variations within a specific climate which can rarely be reconsidered close to or at the end of the design process. Efforts toward optimization of their energy impact should be undertaken as early as possible.

To perform such analysis, computer-aided energy modeling helps the building designer by providing information on the relationship between potential design solutions and their performances.

2. Design Ethic with Nature

Designing with adaptability to nature strongly depend on providing human comfort from the regional climate and the microclimate modified by the surroundings of the buildings in architecture. Based on the concept, providing human comfort can be found from the traditional town structure, building material, building form, building façade, orientation, room depth, and etc. since there are strongly interrelated with local climate condition.

With the development of an industrial society, the new forms of industrial production and capital organization were demanded. The new society also affected architectural building design and constriction based on new technologies to control regional climate.

Architecturally, a comfortable human life in a building space can be achieved by the use of local environmental characteristics, by applying artificial technology, or by combining both of them, as shown in Figure 1.

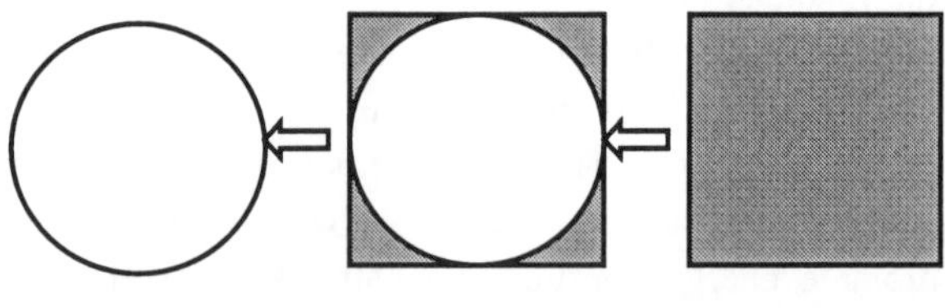

Figure 1. Society and energy use pattern

Low-energy societies have traditionally needed low energy consumption to maintain comfort in a space by their use of local environmental system. High-energy societies have demanded a higher energy supplement of fossil fuel to provide comfort in a space. That is, these high-energy societies after industrial revolution with their developed technologies to control climates, have brought convenience to human comfort by depending on high-energy requirement from nature. This has resulted in unexpected impacts on global nature.

---

[1]Lease span depth means the length from exterior wall to interior structural wall such as the core in the study.

Thus, the necessity of physical designing with nature for sustaining long-term human life on the earth can start with a conceptual basis of ecology[2] in architecture. Ecological design conceptually means a partnership with nature for the long-term survival of all species on the earth, and it must clearly address the design with nature in architecture.

3. Approach of a Concept and Hypothetical model

The energy design of an architectural environment to achieve an optimum human comfort can conceptually be divided into a relationship between an architecture designed with nature and one with engineering control. The end result generally provides an occupant's comfort in a space that is sensitive to the global environmental and the issue of artificial energy generation and consumption. Conceptually, there should be an increasing portion of architectural designing with nature(passive design) and a decreasing portion of engineering(active design) to maintain the desired human comfort in a tall office building. The relationship between both factors is ideally an inverse proportion. Therefore, a performance of maximizing passive control for human comfort in a space can achieve not only minimizing active control, but also minimizing the environmental consumption as a positive result for sustaining long-term human life. Therefore, to approach the target, the hypothetical model of the tall office building for the experimental study has been constructed based on architectural and engineering assumptions in the study.

3.1 *Architectural considerations*

To establish a pragmatic study model of a tall office building, this study made use of the quantitative design analysis of the existing tall office buildings. It is to analyze commonly used design pattern and method for constructing adaptable model. From the quantitative design analysis, commonly used design factors piloted to set the criterion of the constructing the hypothetic tall office buildings which have the same total area, the same volume, the same typical floor area, the same rentable space ratio, the same core ratio, and the same interior conditions in the space. This objective was to have the same comparison in the study for architectural design and planning considerations. From these considerations and analysis, the architectural factors are constructed as shown in Table 1 based on Figure 2 in the study.

Table 1. Construction of architectural factors

	Characteristics	Conditions
General Description of Modeled Tall Office Building	Building function	Office with open plan
	Building Gross area	2,166,000 sqft
	Building Typical floor	36,100 sqft
	Number of floors	60
	Percent of occupied area	70% of total building area with open office plan
	Percent of core and others	30% of total building area with utility core
	Modeled building forms	Square
Zone 0 (Outdoor)	Cool climate	Chicago, Illinois
	Temperate climate	New York, New York
	Hot-Arid climate	Phoenix, Arizona
	Hot-Humid climate	Miami, Florida
Zone 1 (Façade)	External walls	Aluminum exterior finish + 1in. polyurethane, and for roof, 2in. polyurethane + medium exterior color
	Windows	Double insulating 1/4in, 1/2in with curtainwall type Frame type: Aluminum without brick
	Modeled window areas	100%
Zone 2 (Indoor)	Lease span depth	Based on 45 ft, variation span depth: 30 ft, 35 ft, 40 ft, 50 ft, 55 ft, 60 ft, 65 ft, 70 ft
	Perimeter depth	20 ft
	Floor-to-Floor height	13 ft
	Ceiling height	9ft
	Floor structural material	Metal deck + concrete slab
	Typical floor finish	Carpet (no pad)
	Ceiling material	Suspended ceiling system with recessed fluorescent luminaries/ Lay-In Acoustic title (no ceiling insulation)
Zone 3 (Core)	Core type	Central utility core system
	Core material	Concrete composite

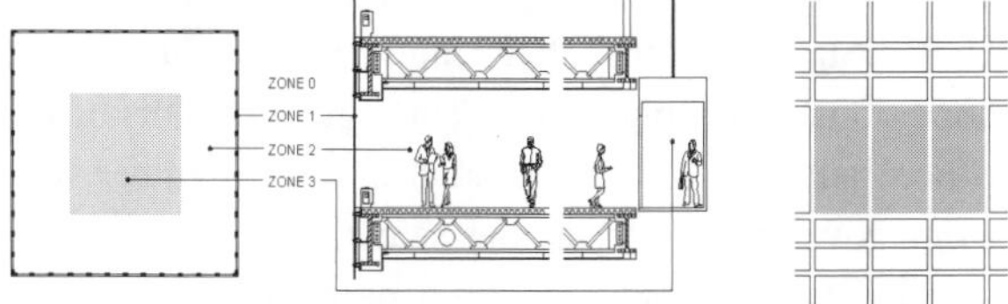

Figure 2. Organization of Tall Office Buildings in the Study

3.2 *Engineering considerations*

The engineering decisions for the model construction are based on general programmatic requirements and design criteria or common engineering practices which are assumed and applied to the hypothetical models with the same conditions as the systems in the study.

[2]The meaning of the word, ecology is derived from the Greek "oikos" which means household, and "logos" which means study. Thus, the study of the environmental house includes all the organisms in it and all of the functional processes that make the house habitable. (Odum, 1983)

External thermal and solar loads are assumed as transmission through exterior walls and windows. Internal thermal loads are assumed as heat gain from occupants, lighting fixtures, and commonly used office equipment. Mechanical and electrical systems are hypothesized to provide the services required to constantly maintain temperatures and required lighting levels. A variable air-volume system is modeled for the HVAC system. And engineering assumptions for mechanical and electrical systems are constructed in relation to typical office building requirement as shown in Table 2.

Table 2. Assumption of engineering factors

Factors	Assumptions
HVAC system	Standard variable air volume system with electric zone reheat, return air path by duct Electric resistance for heating equipment Chilled water coils for cooling, open tower Chiller type: electric centrifugal hermetic
Occupancy schedule	9:00am to 6:00pm, Monday through Friday
Space requirement	150 sqft per person 52.5 sqft for restroom
Indoor thermal	72 °F for heating set point (occupied) 75 °F for cooling set point (occupied) 50% relative humidity for cooling
Indoor lighting	50fc for ambient perimeter zone lighting level 3.0w/sqft for artificial light design 1.5 W/sqft for equipment density 0.4 W/sqft for task light Lighting controller type: Continuous, Maximum glare: 20
Indoor ventilation	20 cfm per person for outdoor air requirement in typical space 15 cfm per person for ventilation in corridor 7.35 cfm per person for restrooms 45 cfm per person for mechanical room

3.3 *Others*

This study selects region to perform the research as the United States. Because that country has various climate conditions according to different region, has lots of tall office buildings in cities and has exact weather data from government. From the selection, climate zones selected based on four different climatic conditions in the United States; cool, temperate, hot-arid, and hot-humid.

Considered regions selected by four major cities in the United States according to climate zones; Chicago - Illinois, New York - New York, Phoenix - Arizona, and Miami – Florida as shown in Table 1.

For comparative analysis in the study, results of total annual energy consumption of modeled buildings used for the quantitative analysis. And output data for energy use in modeled buildings indicated as kWh.

The study is based on using photo-sensors in the occupied zone to evaluate daylighting from outside through the windows into a space. The height of installed photo-sensors are above 2.5 feet from floor, continuous lighting type controller, maximum glare is 20 and the design light level is set as 50 fc.

4. Analyzing Strategy of the Model

4.1 *Tool for the achievement of experimental study*

The objective of this research is to develop the design methodology for tall office buildings with regional climate adaptability. To measure regional climate adaptability, the criterion of the comparative analysis for evaluation of considered architectural design factors in each of the regional climate is total annual building energy consumption.

For the analysis, the simulation of energy consumption is a useful design tool for designers who want to evaluate the impact of building design on energy performance. The complexity of energy analysis and the high level of interaction between many variables require a computer to link design and analysis. With computer aids, energy simulation has become increasingly powerful, quickly providing information on design solutions and their performance characteristics for decision-making.

The process makes an experimental study by means of computer simulating as the tool for experiments to achieve comparative analysis between model to be considered and regional climates in the study. This computational parametric analysis can assure accuracy to achieve its objectives. This study uses the eQUEST program based on DOE 2.2 and it will be utilized to obtain expected data between design considerations and energy performance to measure annual energy consumptions in relation to regional climate conditions.

4.2 *Analysis of sequence and structure*

As a result of the previous discussion, this study has developed an integrated method of building design and planning, incorporating heating, cooling and lighting that use a unique combination and sequence of computer simulations and regional data related to different climate conditions. The study then produces energy consumption according to building sizes and interior design(lease span) considerations. Therefore, this method is characterized by the following:

- Analysis problems
- Selecting design considerations for office buildings
- Model definition by empirical research database
- Computer simulation (analytic engine)
- Analysis of collected data
- Development of integrated design information

5. Analysis of Design Character

5.1 *Design analysis of building size with climate*

The purpose of this experiment is to analyze the energy use pattern and differences according to different tall office building sizes with regional climates.

To achieve the information of energy use patterns and differences in tall office buildings according to climates, three tall office buildings having different sizes are modeled as shown in Table 3. The selected buildings are constructed to have the same architectural and mechanical conditions to make the equal comparative analysis that is analyzed by kWh per square foot/year. Generally, as the number of floors increase, the core also requires more space for elevators, shafts, stairs, and services so that the ratio of rentable space to core space in all three cases is constructed as the same since it is significant for the exact measurement and evaluation in the study. Based on the concept of model construction, the foot print size increases as the number of floors increase as shown in Figure 3.

Table 3. Constructing different size office buildings

Building sizes	Mega	Large	Small
Foot print(ft)	220 X 220	190 X 190	155 X 155
Storeys	90	60	30
Typical floor(sqft)	48,400	36,100	24,000
Total area (sqft)	4,356,000	2,166,000	720,750
Rentable ratio (%)	70%	70%	70%
Core ratio (%)	30%	30%	30%
Measured unit	kWh/sqft,yr	kWh/sqft,yr	kWh/sqft,yr

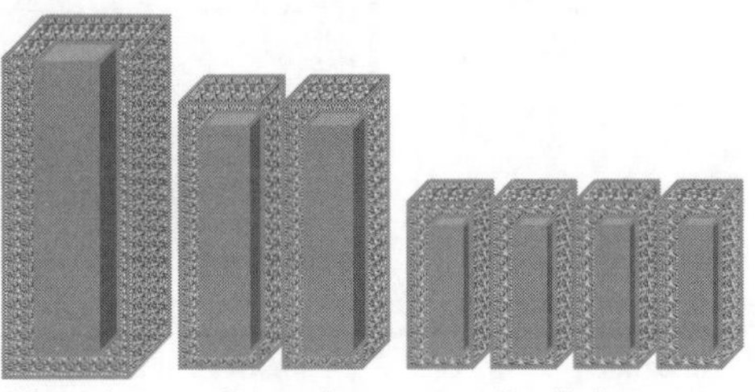

Figure 3. Assumption of Different Size Buildings

From the result, larger size office buildings generally consume less energy than smaller size buildings for all climate conditions and the heating performance especially shows greater different energy use related with building size as shown in Table 6. Otherwise the individual energy use patterns and differences for heating, cooling, and lighting in three modeled tall office buildings are different according to regional characters as shown in Figure 4. To approach optimum spaces in the tall office building design with different climate condition, the understanding of individual energy performance is important.

From the study, integrated energy use is that larger size buildings makes smaller surface area per square foot. Architecturally, this situation reflects the relationship between surface/volume ratio and building design as shown in Table 4. That is, smaller size buildings have large surface/volume ratios than mega size buildings by as much as 47.4%, and they than large size buildings by as much as 27.3% which produces different energy use in building. And it generally makes same energy use patterns according to building orientations as shown in Table 5.

Table 4. Surface area and volume by building size

Building size	Mega size	Large size	Small size
Building surface area(ft^2)	1,078,000	628,900	265,825
Building volume(ft^3)	56,628,000	28,158,000	9,369,750
Surface/volume ratio	0.019	0.022	0.028

In point of architectural design, energy use pattern indicates the condition of space environment which is different according to climate conditions. Therefore, for the optimum spaces in the tall office building's design, the understanding of individual energy performance's character according to different climates is important since it is can support to confirm of the proper architectural design characteristics for regional climate adaptability based on building energy use

Table 5. Energy use pattern between mega and large size office buildings with climate

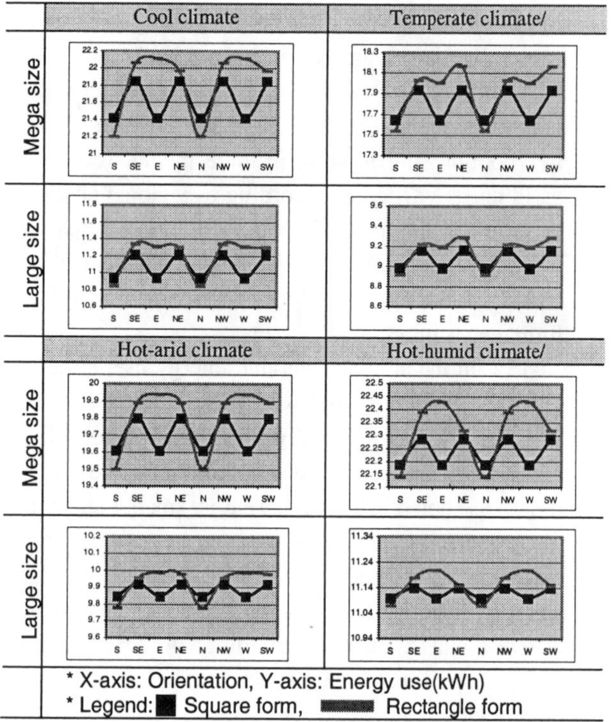

Table 6. Energy use differences between mega and large size office buildings with climate

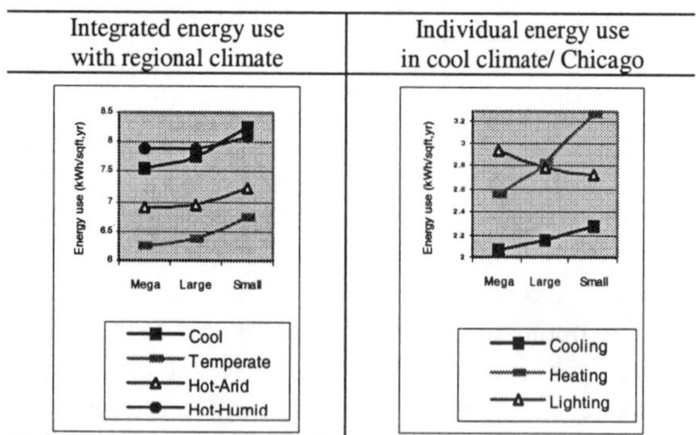

in climate conditions. From the analysis of the data, the hierarchy of architectural design factors based on building size in cool climate area is heating, lighting, and cooling. In case of temperate climate area, the order is lighting, cooling, and heating. In case of hot-arid climate area and hot-humid climate areas, the order is cooling, lighting, and heating. Therefore, space design must be considered by integrated design factors with climate. From the result, energy use between cool area and hot-humid area is equivalent because of thermal requirement. Therefore, in the design process, building size can be controlled to control the proper spaces with climate. In case of temperate area and hot-arid area, major energy use factor is lighting. Therefore, in design process, window size can be considered to use natural daylighting.

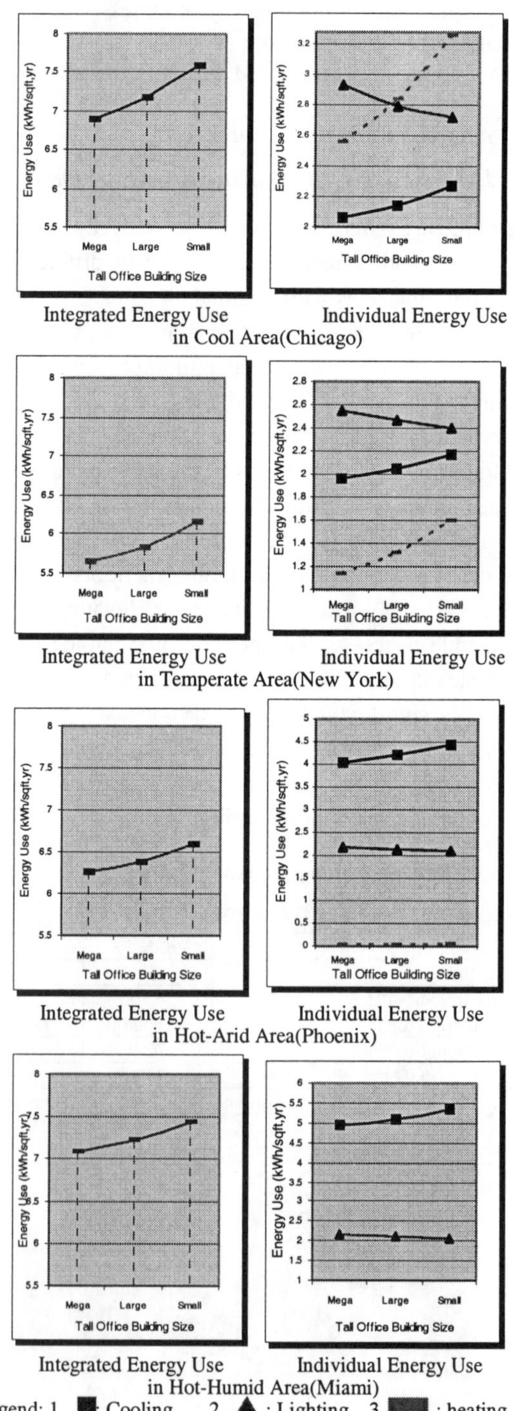

Legend: 1. ■: Cooling, 2. ▲: Lighting 3. ▬: heating

Figure 4. Energy Use Pattern by Building Sizes

5.2 *Design analysis of lease span and building height*

Lease span is a significant design factor to composite tall building space and form. There are no international standards that determine lease span depths for office buildings, since the determination of optimum lease span depth in tall office buildings have the complicated inter-relationship between external factors -such as local codes, local social and cultural pattern, client and user needs, economical support, local environmental systems and internal factors- such as site size, lobby circulation, modulation for office landscaping, mechanical system efficiency, structural system efficiency, and current technology. For instance, according to regional code or custom, some countries may have a requirement that all offices must have an outside exposure to light and air.

In such cases, office function arrangements may affect the dimension of the lease span. Adherence to this requirement may also affect the aspect ratio of very tall buildings. On the other hand, Rubanenko (1973) points out that flexible floor layout allows quick alteration of the office structure and rearrangements, it is basic considered concept of this experiment.

Based on the assumption that the required building area is the same, lease span determination automatically determines building height and aspect ratio. Furthermore, it effects to land use, that is, higher building height can be proportionally achieved lower land use.

In the architectural space, lease span can be composed by perimeter area and interior area based on environmental system. And lease span variation also effects the ratio between perimeter area and interior area from an occupied area as shown in Table 7.

The objective of this experiment is to analyze a tall office building design influenced by the relationship between lease span and building height, under the condition that modeled buildings have the same total building area in relation to a climate adaptability measured by the annual energy consumption.

From the results of area analysis, a building which has a longer lease span produces a lower building type and a building which has a shorter lease span makes a higher building type under the condition that they have the same building area. For the energy analysis, the

Table 7. Constructing model

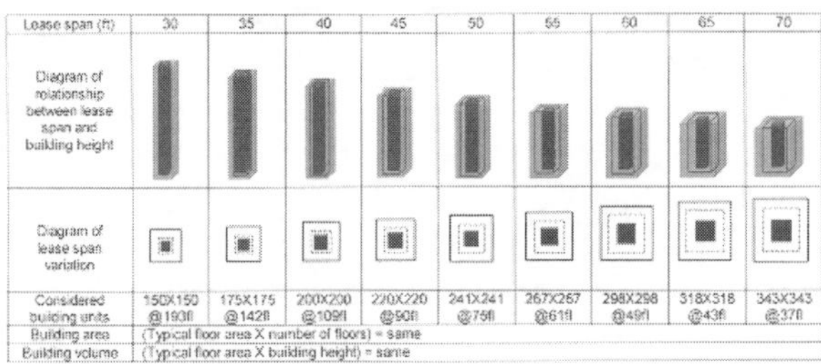

Lease span (ft)	30	35	40	45	50	55	60	65	70
Diagram of relationship between lease span and building height									
Diagram of lease span variation									
Considered building units	150X150 @193ft	175X175 @142ft	200X200 @109ft	220X220 @90ft	241X241 @75ft	267X267 @61ft	298X298 @49ft	318X318 @43ft	343X343 @37ft
Building area	(Typical floor area X number of floors) = same								
Building volume	(Typical floor area X building height) = same								

total building energy consumption gradually decreased from a building type which has a shorter lease span to a building type which has a longer lease span; that is, a building type which has a shorter lease span (30 to 40 feet) consumed more energy than a building type which has a longer lease span (55 to 70 feet). The shorter lease span building consumes more than 12% to 20% energy in cooler areas and uses more than 10% energy in warmer areas than the longer lease span building. In addition, energy use differences according to building types varies slightly between a building which has a 55 feet lease span and a building which has 70 feet lease span as shown in Table 9.

Based on the assumption that the required building area is the same, lease span determines building height and affects the aspect ratio and surface area in a tall building. Furthermore, leases span variation also affects the ratio between perimeter area and interior area from an occupied area as shown in Table 8. Shorter lease span determination in a design increases the perimeter area and decreases the interior area which simultaneously produces a larger surface area. Longer lease span determination decreases the perimeter area and increases the interior area which results in a smaller surface area. Therefore, from the study, two significant design considerations based on the evaluation of building types in relation to energy use were observed. It is firstly that variation of parameter area and interior area in a space have an effect on energy use in a building. And it is secondary that the surface area affected by the relationship between lease span and building height directly causes the energy use pattern.

Consequently, the lease span determination is a tall office building directly affects the variation of the relationship between perimeter and interior space, total building height, and building surface area. These factors affect the energy requirement as much as 10.6% to 20.7% in relation to region. The lease span

Table 8. Analysis of building parameters

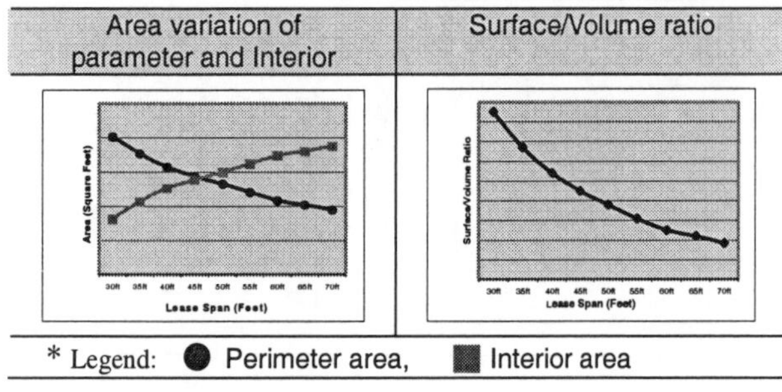

Table 9. Energy use according to regional climate

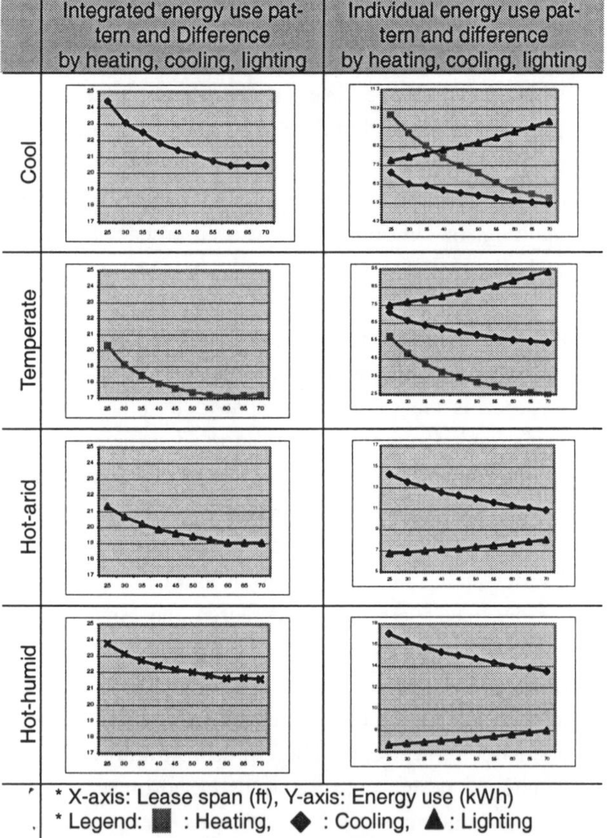

determination is related with architectural expression for aesthetics in building aspect ratio, and it creates an occupied ratio between perimeter and interior space. For determination of the lease span in a tall office building at the first design stage, the design should be compromised based on the relationship of the building height, the occupied space ratio, and energy use. From the analysis, a building which has less than 40 feet lease span can be profitable to create a large perimeter space and to reduce the light load. A building which has more 55 feet lease span can be advantageous for reduction of thermal loads, but the lighting load proportionally increases and much of the portion of the interior space is affected. The 1:1 ratio between perimeter and interior space can be created between a building which has 45 feet lease span and a building which has 50 feet lease span.

6. Conclusion

As a result of the previous comparison of design consideration for tall office buildings in relation to regional climates, buildings physically need less energy consumption when the architecture is concerned with the regional climate and it produces a more reasonable design strategy. That is, an architectural design and planning creates an internal working environment that is strongly related with regional microclimate. In reality, imbalanced planning (a design's lack of regional characteristics) requires additional environmental resources to control indoor working conditions.

Therefore, the application of integrated regional design information should the first architectural design stage and this research produces the rational. This design methodology focused on regional climate adaptability which simultaneously reduces the destructive environment in nature for sustainability. Consequently, this architectural design language approach must be a starting point to long-term planning.

References

1. Cho, J. S. (2002) "Design Methodology for Tall Office Buildings: Design Measurement and Integration with Regional Character", *Illinois Institute of Technology*, Chicago, USA
2. Council on Tall buildings and Urban Habitat, (1995) "Architecture of Tall Building"', *McGraw-Hill Inc.*, N.Y., N.Y., p. 85
3. Daniels, K. (1997) "The Technology of Ecological Building: Basic Principles and Measures Examples and Ideas" *Birkhauser/Princeton Arch Press*, Germany
4. Daniels, K. (1998) "Low-tech Light-tech High-tech" *Birkhauser/Princeton Arch Press.*, Germany, 1998, p.17
5. Olgyay, V. (1963) "Design with Climate: Bioclimatic Approach to Architectural Regionalism, Princeton University, Princeton, New Jerwey, USA

ARE RE-ENTRANTS GOOD OR BAD? AN EMPIRICAL TEST IN HONG KONG

WONG, S.K. YAU, Y. CHEUNG, A.K.C. CHAU, K.W. HO, D.C.W.

Department of Real Estate and Construction, The University of Hong Kong, Pokfulam Road, Hong Kong, China

Natural lighting and ventilation have long been a primary consideration in building design, particularly for those high-rise and densely packed apartment blocks where mechanical ventilation is normally secondary. In Hong Kong, there are prescriptive legal requirements governing the provision of natural lighting and ventilation in private buildings. This, coupled with developers' profit-maximizing incentives, often gives rise to re-entrant designs commonly found in apartment buildings in Hong Kong. Despite the popularity of re-entrants, little research has been done to evaluate the preference of homebuyers for them. Moreover, the revelation of the chimney effect of re-entrants by the mass outbreak of Severe Acute Respiratory Syndrome (SARS) in Amoy Gardens in 2003 might have influenced homebuyers' preferences. This paper aims to study the economic impact of the disposition of re-entrants on property prices. Based on the hedonic pricing analysis, we found that the preference for re-entrants is floor-dependent. Before SARS, homebuyers were not fond of re-entrants on low floor levels, but they were willing to pay more for re-entrants as the floor level increased. Yet, the outbreak of SARS did not significantly change their preferences for re-entrants.

1. Introduction

In a densely populated city, high-rise buildings compete with each other on natural sunlight and ventilation. Although the indoor environment can be modified by electrical and mechanical building services, natural lighting and ventilation is still important for apartment buildings. In Hong Kong, the governing rules with regard to the provision of lighting and ventilation in private buildings are Building (Planning) Regulations, or B(P)Rs. B(P)Rs stipulate that every habitable room (e.g. living room and bedroom) and kitchen shall be provided with natural lighting and ventilation through *prescribed windows*.[a] Apart from the requirements of glazed and easily opened window areas, each prescribed window shall face into either (1) a street that is at least 4.5m wide or (2) an uncovered and unobstructed space fulfilling the rectangular horizontal plane (RHP) and inclined angle requirements. For lavatories, B(P)Rs require that a window shall be provided and opened into *open air*.[b] As defined in the B(P)Rs, open air means a space which: 1) is uncovered and obstructed vertically; 2) is at least 1.5m in any horizontal dimension; and 3) in case it is enclosed on all four sides, has a horizontal cross-sectional area of at least 1 m^2 for every 6m of the mean height of the walls enclosing the space.

Since development density is extremely high in Hong Kong, it is rather difficult to have all prescribed windows facing a street that is at least 4.5m wide. Most developments rely on the RHP requirement for their habitable rooms and kitchens. Moreover, except for a few recent exempted cases, open air has to be provided for lavatories. This means that developers have to find a design solution that maximizes their development profit and satisfies both site constraints and statutory requirements. Also, it is widely conceived that the values of good scenic views are higher in habitable rooms than in kitchens and lavatories. These factors induce developers to increase the perimeters of their buildings to provide windows for kitchens and lavatories in less prominent positions.

Given the floor plate and total gross floor areas, one of the most efficient ways to provide a greater perimeter is to have recesses on the outer perimeter within the building enclosure. Communal areas, such as lift lobbies and staircases, are placed at the core, whereas the building's wings radiate out from the core. To fulfill the RHP requirements, the length of the base of the plane shall be at least 2.3m. Thus, it is advantageous to arrange lavatories and kitchens of adjacent flats, as in Figure 1. Two neighbouring units are arranged in such a way so that a deep and narrow recessed void is formed between them. This area is termed "re-entrant," which is a vertical open-air channel with three sides enclosed by the external wall and windows of adjoining units, while the remaining side is opened to outdoor air (see Figure 2). Both kitchens share the same RHP, and the open air for the lavatories' natural lighting and ventilation overlaps with the RHP. This explains why most re-entrants are around 2.3m wide. With such a

[a] See Regulations 30 and 31 of the B(P)Rs.

[b] See Regulations 2 and 36 of the B(P)Rs.

configuration, the re-entrant provides an open air outside a window to satisfy the requirements of natural lighting and ventilation for kitchens and lavatories. On the other hand, the enjoyment of scenic views in habitable rooms is not affected. This area is also widely used to accommodate the condensing units of split-type air-conditioners and other building services ductworks.

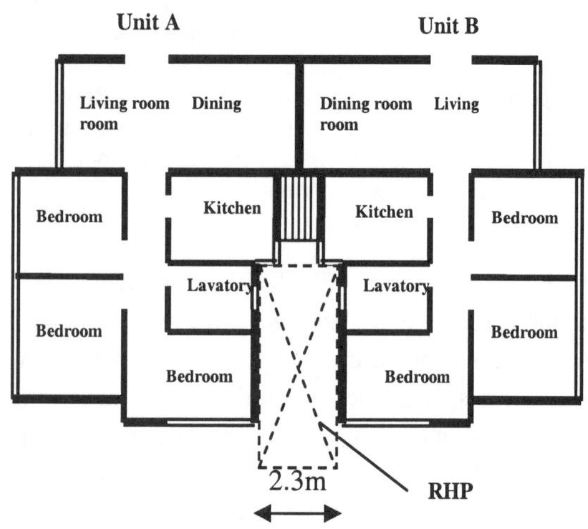

Figure 1. Typical arrangement for adjacent flats in a "cruciform" plan

In summary, the emergence of re-entrant design appears to be driven by developers' profit maximization incentives as well as building legislation in Hong Kong. However, from a homebuyer's perspective, are re-entrants good or bad? Did their preference for re-entrants change after SARS? To answer these questions, this paper will evaluate homebuyers' preferences by studying the relationship between property prices and re-entrant designs

2. A Review on the Environmental Effects of Re-entrants

As a result of the popularity of the re-entrant or light well designs, a lot of research has been done on their environmental impact, such as daylighting, natural ventilation, and perceived air quality. On the issue of daylighting, Kristl and Krainer studied how to enhance the illumination from light wells by applying reflective paints on its walls.[1] Ng identified the current problems of window design in Hong Kong practice.[2] He found that four common designs, even though in full compliance with prescriptive building regulations, did not provide adequate daylight to indoor spaces. In particular, two of them were re-entrants: 1) windows placed inside deep re-entrants, and 2) windows placed in the RHP, which is formed by two re-entrants closed to each other (see Figure 1).

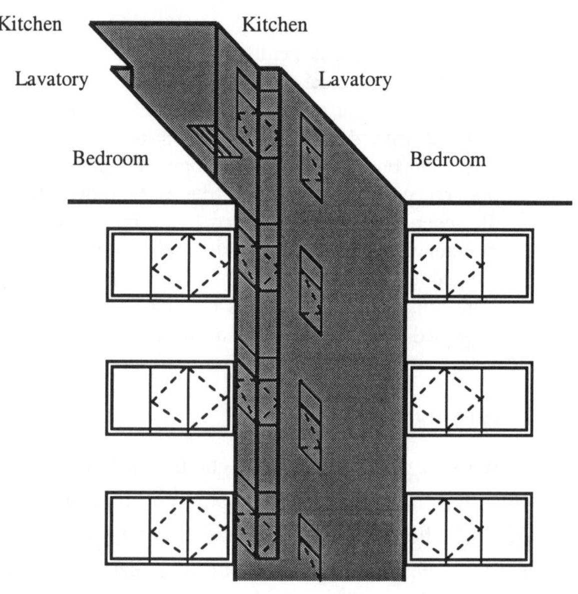

Figure 2. A re-entrant (the shaded part) in a "cruciform" plan

As far as ventilation is concerned, two teams of local researchers, Chow, *et al.* and Bojic, *et al.* used a computational fluid dynamics simulation to investigate the air flow characteristics, air temperatures and ventilation rates in re-entrants.[3-6] Both showed that heat energy dissipated by condensing units of split-type air conditioners induces a thermal buoyant airflow. In addition, a T-shaped re-entrant is better than the more common I-shaped re-entrant in terms of the energy performance of air-conditioners.[3] Bojic, *et al.* examined further how the depth of a re-entrant affects the environmental conditions inside it.[6] Their results suggested that the depth of a re-entrant did not affect the mass-flow rate and temperature inside. Kotani, *et al.* worked out a simple calculation model to predict the vertical temperature distribution and ventilation rates in light wells.[7] They conducted a wind tunnel test to validate the calculation model.

Whilst most of the above research placed a heavy focus on the physical characteristics inside re-entrant areas, there is a lack of research on how occupants evaluated the living environment in buildings with re-entrants. Kotani, *et al.* came out with a recently published piece of work on this topic.[8] They carried out a questionnaire survey from 1992 to 1994 and asked occupants to evaluate their living environments in buildings with light wells. Based on a sample of less than 400 replies in four buildings, they claimed that with the exception of air quality, the environment was satisfactory. However, the survey was conducted without a valid control group, so it was unable to conclude whether these respondents were more satisfied with flats with light wells than those without.

3. Methodology

Our study evaluated homebuyers' preferences on the re-entrant design by making use of property transaction data for residential units with and without re-entrants. The advantage of using property transactions is that a large part of one's wealth is put into housing, the purchase of which should be based on rational and careful considerations rather than impulse. Our approach is to examine what homebuyers have actually done and purchased rather than their opinions.

In this study, we used a hedonic pricing model to examine whether flats designed with re-entrants are sold at different prices from those without re-entrants. Theorized by Rosen, hedonic pricing models are a well established tool for studying the implicit price of heterogeneous goods.[9] They have been widely used in estimating the impact of environmental attributes on property prices[10] but seldom have they been applied to study building design.[c] The main reason is that building design is a qualitative building-specific attribute, making it very difficult to separate it from other important attributes, notably location.

Our solution to this problem was to draw a sample of property transactions from a popular single residential development with variations in re-entrant design. One obvious advantage of a single development is that its buildings: 1) are located in close proximity to each other, 2) are managed by the same property management company, and 3) share the same communal facilities. This controls the effects of many qualitative building-specific attributes. Also, the selected development must have variations in re-entrant designs across buildings, or preferably across flats in each building so that we could extricate the implicit price of re-entrants and the change in price differentials, if any, after the SARS outbreak. Finally, the development had to be popular so as to ensure sufficient data was available for analysis.

In Hong Kong, City Garden, a housing estate on Hong Kong Island, satisfied the above sample selection criteria. City Garden consists of 14 residential towers surrounding a local shopping centre. Except for Block 11, which is T-shaped, all towers have a "cruciform" with re-entrants. In particular, Blocks 4-10 have a floor plan of six flats per storey – four of which have narrow and deep re-entrants, whilst the other two flats do not have any re-entrant (see Figure 3). This gave us a sample of flats with variations in re-entrant designs in the same building block so that the effect of re-entrants can be estimated with building location being held constant. In addition, due to the disposition of City Garden, flats with re-entrants faced either the shopping centre, the harbour, or other buildings. As such, no particular view is correlated with the re-entrant design and the effect of re-entrants can also be separated from the effect of views.

Figure 3. Typical floor plan for City Garden (Blocks 4 to 10), taken from the Bo Fung Property Agency (http://www.bo-fung.com.hk/)

We excluded Blocks 1-3 and 11-14 from the sample because they have different floor layouts. For the remaining seven blocks (i.e., Blocks 4-10), property transaction records from January 1998 to November 2004 were retrieved for analysis. The properties transacted during the SARS period (i.e., from March 2003, when the first case of SARS was confirmed, to June 2003, when the WHO removed Hong Kong from

---

[c] Chau, *et al.* was one of the few studies that used hedonic pricing models to examine building designs. They evaluated whether a balcony, with its environmental filtering functions, added value to a property.[11]

the SARS list) were discarded because transactions were less frequent during that period, so the market would not have consistent or sufficient information to price the properties. Accordingly, a total of 357 transactions were used. Table 1 presents the descriptive statistics of the data.

Table 1. Descriptive statistics of property transaction data

	Mean	Std. dev.	Min	Max
City Garden (Nos. of Transaction = 357)				
Property price (HK$mil)	4.36	1.30	1.65	7.90
Age (months)	194.56	31.83	140.00	262.00
Floor level	14.69	7.79	3.00	28.00
Flat size (ft^2)	1,033.22	210.77	641.00	1,368.00
Distance to MTR station (m)	471.95	35.18	418.72	534.01
Seaview (%)	0.49	-	-	-
Shopping centre view (%)	0.32	-	-	-
Building view (%)	0.18	-	-	-
Flats with re-entrants (%)	0.64	-	-	-

A linear hedonic pricing model was formulated:

$$P_{it} = \beta_1 RET_i + \beta_2 RET_i \times SARS_{it} + \beta_3 RET_i \times FLR_i + \beta_4 RET_i \times FLR_i \times SARS_{it} + \alpha_0 + \alpha_1 AGE_{it} + \alpha_2 SIZE_i + \alpha_3 SEA_i \times FLR_i + \alpha_4 SHOP_i \times FLR_i + \alpha_5 BLDG \times FLR_i + \alpha_6 FLR_i \times SARS_{it} + \alpha_7 MTR_i + \sum_{j=1}^{T} \chi_j TIME_{ij} + \varepsilon_{it} \quad (1)$$

Table 2 describes the meaning of the symbols used in Eq. (1). The focus of this study was on β (i.e., the terms associated with *RET*), while the remaining variables were used to control other property attributes and macro time factors. The coefficients β_1 and β_3 measured the pre-SARS implicit price of re-entrants and its change with respect to floor levels, respectively. The coefficients of the interaction terms, β_2 and β_4, measured the changes in implicit prices, if any, after SARS. Since most property attributes were controlled, the differences in value, if any, between a flat with a re-entrant and one without should be attributable to re-entrant designs. In particular, after the renowned SARS investigation reports, the chimney effect of re-entrants was pinpointed as a contributory factor to the spread of SARS.[12] Interacting *RET* with *SARS* enabled us to test whether homebuyers would avoid flats with re-entrants, especially those on a higher floor level (i.e., to see whether β_2 and β_4 were negative).

Table 2. Definitions of the variables

Variable	Definition
P_{it}	The log transaction price (in HK$mil) of property *i* at time *t*
RET_i	A re-entrant dummy that equals 1 when property *i* has a re-entrant, and zero when it does not has any re-entrant
$SARS_{it}$	An event dummy that equals 1 if property *i* was transacted after SARS, and zero if it was transacted before SARS
AGE_{it}	The log age (in months) of property *i* at time *t*
$SIZE_i$	The log gross floor area (in square feet) of property *i*
FLR_i	The floor level of property *i*
SEA_i	A view dummy that equals 1 when property *i* possesses a panoramic Victoria Harbour view, and zero if otherwise
$SHOP_i$	A view dummy that equals 1 when property *i* possesses a view facing the shopping centre, and zero if otherwise
$BLDG_i$	A view dummy that equals 1 when property *i* possesses a view facing other buildings, and zero if otherwise
MTR_i	The log distance (in metres) between property *i* and the MTR station
$TIME_{ij}$	A monthly time dummy that equals 1 if property *i* was transacted at time *t*, and zero if otherwise (the base period is January 1999)
α, β and χ	Coefficients to be estimated
ε_{it}	The error term

4. Results

After Ordinary Least Squares estimation, the results of the model in Eq. (1) are shown in Table 3. The R-squared value is 83% and significant at the 1% level.

The pre-SARS findings will be discussed first. We found that the coefficients of *RET* and *RET*FLR* were -0.191 and 0.065, respectively. Both of them were statistically significant at the 10% level. This meant that homebuyers were not fond of re-entrants on low floor levels, but they were willing to pay more for re-entrants as the floor level increased. For example, for two similar flats on the lowest floor (3/F), the one with a re-entrant was 19% cheaper. This was probably because the re-entrant flat was subject to poorer daylight and ventilation,[2] and was usually less hygienic due to falling garbage from the floors above. But these problems were less severe for upper-floor flats with re-entrants. To facilitate an understanding of this phenomenon, Figure 4 plots the implicit floor premium against floor levels.[d] It confirms that the implicit floor price for re-entrant flats increases at a slightly faster rate than that for flats without any re-entrant.

Table 3. Estimation results

Variable	Coefficient	Std. Error	t-Statistic	Prob.	
RET	-0.1908	0.0957	-1.9942	0.0471	**
RET x SARS	0.1634	0.1523	1.0729	0.2843	
RET x FLR	0.0654	0.0365	1.7911	0.0744	*
RET x FLR x SARS	-0.0700	0.0609	-1.1499	0.2512	
Constant	-5.0461	0.8085	-6.2415	0.0000	#
AGE	0.0960	0.1457	0.6585	0.5108	
SIZE	0.8303	0.0487	17.0410	0.0000	#
SEA x FLR	0.0542	0.0318	1.7032	0.0897	#
SHOP x FLR	0.0335	0.0319	1.0509	0.2943	
BLDG x FLR	0.0321	0.0332	0.9658	0.3350	
FLR x SARS	0.1048	0.0509	2.0562	0.0407	**
MTR	0.1119	0.1744	0.6418	0.5215	
R-squared	0.8282	F-statistic		14.467	#
Adj. R-squared	0.7710	Prob (F-statistic)		0.0000	

Note: Time dummy results were not reported here.

*, **, #: significant at the 10%, 5%, and 1% levels, respectively

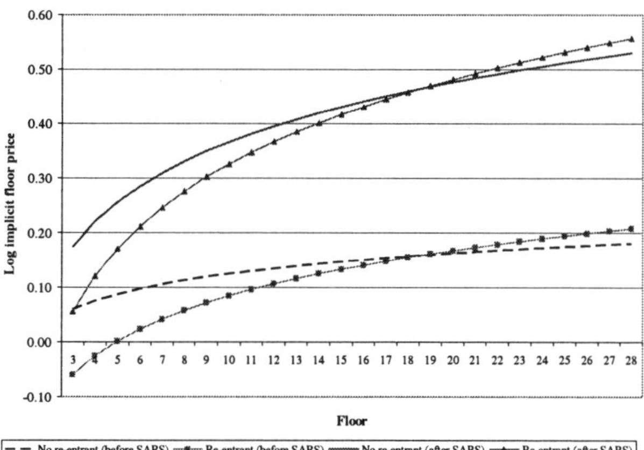

Figure 4. Floor premium for flats with and without a re-entrant (before and after SARS)

Next, the post-SARS results were given by the coefficients of *RET**SARS* and *RET**FLOOR**SARS*, which were 0.163 and -0.070, respectively. Yet, both of them were not statistically significant, and thus there was no evidence to support the notion that homebuyers had changed their preferences for re-entrants after the SARS outbreak. This could mean that: 1) homebuyers did not believe that re-entrants were more conducive to SARS transmission, or 2) the stack effect of re-entrants was already known to the public before SARS, as exemplified by the research work of Chow, *et al.* and Bojic, *et al.*[3-6] The only significant effect identified was an overall increase in the implicit floor premium by 10.5% after SARS, irrespective of the presence or absence of re-entrants (see Figure 4).[e]

The coefficients of other control variables generally showed plausible signs. The coefficient of *SIZE* was significant at the 1% level with an expected positive sign. For views, we found a significant floor premium for flats facing Victoria Harbour (i.e., homebuyers preferred high-level flats with a harbour view). The floor premiums for the shopping centre view and the building view were positive, yet insignificant. The coefficient of *AGE* turned out to be insignificant, probably because the two phases of City Garden were completed at around the same time and their depreciation effect has been largely absorbed by the time dummies. Also, the effect of distance to the MTR station was insignificant because Blocks 4 to 10 are in close proximity to each other. The travelling time to and from the MTR station differed little.

5. Conclusion

In this paper, we explored the popularity of the re-entrant design with reference to developers' profit maximization incentives and the prescriptive building legislation in Hong Kong. We also evaluated, from a homebuyer's perspective, whether re-entrants are good or bad. Based on the hedonic pricing analysis, we found that the preference for re-entrants is floor-dependent. Before SARS, homebuyers did not prefer re-entrants on low floor levels, but they were willing to pay more for re-entrants on higher floor levels. Yet, the outbreak of SARS did not significantly change their preferences for

[d] Assuming a sea view flat, the log implicit floor price for "No re-entrant (before SARS)" was computed by 0.054*ln(*FLR*), whereas that for "Re-entrant (before SARS)" was -0.191+(0.065+0.054)*ln(*FLR*).

[e] Similarly, the log implicit floor price for "No re-entrant (after SARS)" was computed by (0.054+0.105)*ln(*FLR*), whereas that for "Re-entrant (after SARS)" was -0.191+(0.065+0.054+0.105)*ln(*FLR*). Insignificant coefficients were not included in the calculation.

re-entrants. Further research is needed to understand this phenomenon.

Perhaps a more important message from this study is that the market has capitalized building design into property prices. Given these findings, developers and designers will know better which types of building design are more welcomed by homebuyers. Then they can further improve their designs to meet the changing needs of the market.

Acknowledgments

We gratefully acknowledge the financial support provided by the Research Group on Sustainable Cities of The University of Hong Kong.

References

1. Z. Kristl, and A. Krainer, Light Wells in Residential Building as a Complementary Daylight Source. *Solar Energy*, **65**(3), 197-206 (1999).
2. E. Ng, Studies on Daylight Design and Regulation of High-Density Residential Housing in Hong Kong. *Lighting Research and Technology*, **35**(2), 127-139 (2003).
3. T.T. Chow, Z. Lin and Q.W. Wang, Effect of Building Re-Entrant Shape on Performance of Air-Cooled Condensing Units. *Energy and Buildings*, **32**, 143-152 (2000).
4. T.T. Chow, Z. Lin and Q.W. Wang, Effect of Condensing Unit Layout at Building Re-Entrant on Spilt-Type Air-Conditioner Performance. *Energy and Buildings*, **34**, 237-244 (2002).
5. M. Bojic, M. Lee and F. Yik, Flow and Temperatures outside a High-Rise Residential Building due to Heat Rejection by its Air-Conditioners. *Energy and Buildings*, **33**, 737-751 (2001).
6. M. Bojic, M. Lee and F. Yik, Influence of a Depth of a Recessed Space to Flow due to Air-Conditioner Heat Rejection. *Energy and Buildings*, **34**, 33-43 (2002).
7. H. Kotani, R. Satoh and T. Yamanaka, "Natural Ventilation of Light Well in High-Rise Apartment Building. *Energy and Buildings*, **35**, 427-434 (2003).
8. H. Kotani, M. Narasaki, R. Sati and T. Yamanaka, Environmental Assessment of Light Well in High-Rise Apartment Building. *Building and Environment*, **38**, 283-289 (2003).
9. S. Rosen, Hedonic Prices and Implicit Markets: Product Differentiation in Pure Competition. *Journal of Political Economy*, **82**, 34-55 (1974).
10. K.W. Chau, C.Y. Yiu, S.K. Wong and L.W.C. Lai, Hedonic Price Modelling of Environmental Attributes: a Review of the Literature and a Hong Kong Case Study, in *Understanding and Implementing Sustainable Development* (eds. LWC Lai and FT Lorne), Nova Science, 87-110 (2003).
11. K.W. Chau, S.K. Wong and C.Y. Yiu, The Value of the Provision of a Balcony in Apartments in Hong Kong. *Property Management*, **22**(3), 250-264 (2004).
12. World Health Organization (WHO) *WHO Environmental Health Team Reports on Amoy Gardens*, http://www.info.gov.hk/info/ap/who-amoye.pdf (2003).

TOWARDS BIOCLIMATIC HIGH-RISE BUILDINGS: IS A BIOCLIMATIC DESIGN APPROACH APPROPRIATE FOR IMPROVING ENVIRONMENTAL PERFORMANCE OF HIGH-RISE BUILDINGS?

JOYCE LAW

s327104@student.uq.edu.au, MPhil(Arch), The Centre for Sustainable Design, Department of Architecture, The University of Queensland, St. Lucia, QLD 4072, Australia

This paper seeks to address the potential of the bioclimatic approach for improved energy efficient design and better comfort condition for the occupants in a hot humid climate in typical high-rise office buildings. An awareness of the need to conserve and minimise energy needs for heating and air-conditioning of buildings has arisen as a result of the energy crises. High-rise buildings which originated from North American are now found in almost every major city around the world, especially in the developing countries of Asia. The bioclimatic approach is concerned with problems of energy usage, air quality, amenity provided by the building to its users, and productivity, which can be affected by the environmental conditions in the building. The synthesis of all these listed important aspects is directly linked to the comfort criterion for the occupants.

1. Introduction

"If the bioclimatic skyscraper's architecture is to be justified as a new genre of building type, it must transcend being a clever reorganization of external building forms and superficial changes to facades. Essentially, the skyscraper's design must create a new and significant form of internal life for its inhabitants that has not existed before in other genres of the same type[1]".

The building-specific term, 'bioclimatic', relates to "an approach to design which is inspired by nature and which applies a sustained logic to every aspect of the project, focused on optimising and using the environment[2]". The potential role of bioclimatic high-rise buildings is their focus on thermal comfort and well-being of the individual.

The methods used to examine the bioclimatic approach involve a literature review related to the original notions of the bioclimatic approach to architecture. The paper reports on the performance assessment of two selected bioclimatic high-rise buildings of the Malaysian architect, Ken Yeang, against a base case conventional high-rise building also located in Malaysia. The assessment of the sample in this research, consisting of one conventional and two bioclimatic high-rise buildings, was conducted using two physical parameters: a checklist of bioclimatic strategies, constructed for this study based on the relevant literature, and the measurements of the LTV[3] study. In addition, a qualitative User Response Survey was used to gain the users' perceptions of the comfort levels and satisfaction with the indoor environment of their buildings. These assessments, conducted in Malaysia in 2001, have indicated that the performance of the bioclimatic designed high-rise buildings is promising.

Overall, the three bioclimatic high-rise buildings designed by Ken Yeang were found to achieve better results when compared to the base case traditional high-rise building. Yeang has adapted significantly more bioclimatic principles in the two buildings than were adapted in the traditional buildings. Also, not only are the measurements for daylighting, thermal and ventilation improved, the users' satisfaction with these buildings was found to be more positive than with the conventional high-rise building assessed. The research outcomes have implications for reconsidering current high-rise building designs. As a result, the paper also recommends architects to implement the bioclimatic approach in their next high-rise design by using the bioclimatic strategy mapping and checklist.

2. The Potentials of the Bioclimatic Approach to High-rise Design

In general, how can architects apply the bioclimatic approach more frequently to new high-rise buildings and

[1] Yeang, 1996:73.

[2] Jones, 1998:9.

[3] Lighting, Thermal and Ventilation

what strategies should be used? The high-rise buildings[4] reconsidered with bioclimatic design could secure the creation of climate responsive buildings since the basics of bioclimatic design are about climate and people[5]. Thus, the significance or argument for adapting bioclimatic strategies to high-rise design is to produce "a tall building with passive low energy benefits, achieved through design responses to the climate of the place and through optimising the use of the locality's ambient energies, to enhance the quality of life and comfort for its occupant[6]". In addition, bioclimatic strategies applied to high-rise buildings can produce bioclimatic architecture, which will minimise many of the problems associated with the high-rise buildings. This will become apparent from the case study results.

2.1 *Ken Yeang's 'Bioclimatic Skyscrapers'*

The Olgyay brothers are regarded as largely responsible for formalising bioclimatic or passive solar design as a discipline within the field of architecture during the 1950s. Today, Malaysian architect, Ken Yeang, has also adapted similar bioclimatic strategies but has applied them to the high-rise building type. He is pioneered for the 'bioclimatic skyscrapers'. The bioclimatic approach uses passive, low-energy techniques to produce buildings which are environmentally interactive and efficient. Nevertheless the most important criterion is to achieve increased occupant comfort.

Yeang suggests that the bioclimatic approach to high-rise design could be offering a better alternative to the current model and the bioclimatic design features could induce improved health and comfort conditions for its occupants. Clearly, sustainability is an outcome of bioclimatic design. The aim of the bioclimatic high-rise is simple, that is, to reduce the energy usage both in the making and the continuous operation of the high-rise buildings and at the same time generating a pleasant internal life for its inhabitants. This paper anticipates summarising the respond to three questions:

1. How bioclimatic are the selected high-rise buildings?
2. How well do these high-rise buildings perform, in terms of their LTV results?
3. What do the users of these high-rise buildings think?

[4] This research is mainly focussed on the commercial high-rise type since this type is the majority found in many major city centres today.
[5] Olgyay, 1963.
[6] Yeang, 1994: inside cover.

The method carried out for this research is through an empirical study of two high-rise buildings using a control building as base case for comparison.

2.2 *Building case study 1 – Menara IMC, 1994, Architect – Daya Bina Akitek*

Figure 1. Menara IMC, A conventional high-rise building in Kuala Lumpur, Malaysia.

Menara IMC, chosen as the base case control building for this research, is a 21st Century office tower in Malaysia that was conceived and built in preparation for the challenges of the new century. The Project Architect is Daya Bina Akitek, Kuala Lumpur, and the design consultant is Tsao & McKown Architects PC, New York. The building is twenty-nine-storey high and is designed especially for accommodating more flexibility for tenants' needs, such as large open work areas or more structured private offices. Thus, this building could represent the typical Malaysia office tower, which is to a certain extent, a commitment to the environment and for efficiency.

2.3 *Building case study 2 – IBM Plaza, 1983-1987, Architect - Ken Yeang*

Figure 2. IBM Plaza, Ken Yeang's 'Bioclimatic Skyscraper'.

The principal idea that Yeang has explored in the IBM Plaza in Malaysia was that of vertical landscaping. This building has illustrated one of his experimental approaches to the design of high-rise. The IBM Plaza was referred by Yeang at the stage of development as "an archetype suited to the tropical climate[7]". The building has incorporated the idea of escalating planter-boxes along the facade of the building.

The planters ascend one face diagonally, traverse across the floor in the mid-level and then ascend the other facade to the uppermost floor. The twenty-four-storey tower, with its distinctive silhouette, has quickly become a prominent landmark on the Kuala Lumpur skyline.

2.4 Building case study 3 – Menara UMNO, completed in 1998, Architect - Ken Yeang

Figure 3. Menara UMNO, Ken Yeang's 'Bioclimatic Skyscraper'.

Menara UMNO is a twenty-one-storey tower, the headquarters of Malaysia's ruling party (UMNO), built on the island of Penang, in Malaysia. It is a significant building for analysis in that it has won numerous awards, both locally and internationally. Furthermore, the UMNO tower is a more recent project completed in 1998 and has been claimed to adapt the most number of bioclimatic principles in practice. It is the first high-rise to have a 'wind wing wall', which is a fin that projects from the facade and channels the prevailing wind inside through an adjustable opening.

3. Research Methodology

The bioclimatic strategies mapping

For analysing the case study buildings, a combined strategies mapping of bioclimatic principles constructed by the author was utilised. The bioclimatic strategies mapping is developed from the author's undergraduate thesis (Law, 2001) which went through a detail survey of Ken Yeang's 'bioclimatic skyscrapers'. It is simply a quick way for indicating the number of bioclimatic strategies adapted for each high-rise building. It also aids the investigation of the effectiveness and defectiveness of such principles and perhaps indicating how bioclimatic is each of the high-rise building.

[7] Powell, 1999:24.

4. Measurements

The LTV studies will be cross checking against the performance of the buildings. The results may possibly correspond to the post occupancy evaluation.

POE (Post Occupancy Evaluation)

The questionnaire is based on the internationally recognised tool for measuring comfort and sick building syndrome[8]. Performance profiles were generated9 to allow clear visual comparisons between settings. There are twenty four questions presented to the building users and each question relates to a specific issue within a one to five scale between extreme perceptions. Twenty two of the questions are combined by simple summation into seven criteria. Two other questions relate to the users' satisfaction with the workplace and also whether the workplace enables or impedes the respondent's productivity.

5. Case Study Results

The bioclimatic strategies mapping

From Table 1, it can be seen that out of the possible total twenty-two bioclimatic strategies listed, Menara IMC, representing a conventional high-rise building, has only adapted six. On the other hand, both IBM Plaza and Menara UMNO, the two high-rise buildings by Yeang, have achieved eleven ticks or higher, equating to over 50 per cent. Menara UMNO in particular has adapted fifteen bioclimatic strategies, of which most are from the 'bioclimatic' category. In general, all three buildings have most ticks belonging to the 'bioclimatic' category relative to the 'form and envelope' category. This could be indicating that these bioclimatic strategies are easier to be utilised in high-rise design. Nevertheless, in this initial analysis, it is assume that all of the twenty-two bioclimatic strategies have the same weighting or importance. Refer to Law, 2001 for the full explanation and evaluation of the bioclimatic strategies mapping, the indicators for how to measure each of the bioclimatic strategy and the results.

[8] Vischer, J. 1989.

[9] These graphs have been processed according to David Rowe of Sydney University.

	Strategies mapping of Bioclimatic Strategies for Three Case Study Buildings in Malaysia			
	Strategies Adapted			**Bioclimatic Strategies:**
	Base case study 1 **Menara IMC**	Case study 2 **IBM PLAZA**	Case study 3 **Menara UMNO**	
B	√	√	√	**Plan / Use patterns / Ventilation**
I		√	√	**Recessed sun-spaces**
O				
C			√	**Balconies & terraces**
L		√		**Vertical landscaping**
I			√	**Site / Building Solar sky-courts**
M		√	√	**View out from lobby**
A				(lift lobby, stairways and toilets) end / side core
T		√	√	**Awareness of place**
I			√	**Environmentally-interactive wall**
C				
			√	**Transitional spaces**
	√	√	√	**Site Adjustment**
	√			**Curtain wall at N & S facades**
	√	√	√	**Open-to-sky ground**
F		√	√	**Shading devices**
O			√	**Wind-scoops**
R				**Wind-ducts**
M		√	√	**Insulative wall**
&		√	√	**Structural mass**
E				
N				**Solar-collector wall**
V				**Water-spray wall**
E				**Service core positions:**
L		√		- **end core** (double core)
O	√		√	- **side core** (single sided core)
P	√			- **central core**
E	Total: **6/22**	Total: **11/22**	Total: **15/22**	

Table 1. The combined bioclimatic strategies mapping for three case study high-rise buildings.

Measurements

5.2.1 Menara IMC

Room temperature: 28°C
Humidity: 61%
Window sill: average 15560 lux

Passive zone[10]: range from ~500 to 2500 lux
Daylight Factor: 0.77% to 3.85%
Active zone[11]: range from ~35 to 500 lux
Daylight Factor: 0.05% to 0.77%
External plaza: 32°C, 4 m/s external wind velocity.
External daylighting: approx. 125000 lux.

[10] Passive zone shall mean the zone in plan which can be affected by the natural environment. Its depth in plan is twice the floor-to-ceiling height.

[11] Active zone shall mean the zone in plan which is not in the passive zone.

5.2.2 IBM Plaza

Room temperature: 26°C
Humidity: 74%
Wind velocity: ~0.3-0.5 m/s
(Note: measurements taken with overcast sky condition)
Passive zone: range from ~100 to 500 lux
Daylight Factor: 0.33% to 1.67%
Active zone: range from ~10 to 100 lux
Daylight Factor: 0.03% to 0.33%
External plaza: 30°C, ~7.25 m/s external wind velocity.
External daylighting: approx. 65000 lux.

5.2.3 Menara UMNO

Room temp: 30°C
Humidity: 58%
Wind velocity: 0.1-2.3 m/s (large range)
Window sill: average 12550 lux
Passive zone: range from ~900 to 10000 lux.
Daylight Factor: 1.3% to 15.4%
Active zone: range from ~70 to 900 lux
Daylight Factor: 0.1% to 1.3%
Tenanted floor (with A/C): 24oC; Humidity: 42%
External balcony: 0.3-3.1 m/s
External plaza: 32°C
External wind velocity: 4-5m/s
External daylighting: approx. 60000-70000 lux

POE (Post Occupancy Evaluation)
On the radar graphs, the concentric rings represent the question score numbers and the points plotted on these are the average scores for each question. Large scores and subsequent larger profile shape represent better performance in the Key Performance Indicator profile. However, for the Sick Building Syndrome profile, lower scores and subsequent smaller shape shall represent better performance. The seven criteria are thermal comfort, air quality, activity noise, light quality, spatial comfort, environmental control, and prevalence Sick Building Syndrome (SBS) symptoms. Because of the limited data sample collected for Malaysia office high-rise buildings, the average scores are based on the results of twenty seven previous workplace surveys[12] in Australia by David Rowe of Sydney University[13]. It is expected that inaccuracy might be apparent because of cultural differences, different expectations and perceptions from the building users. Although many assumptions existed, the following POE results are only one qualitative creation considered for the building. This should however, indicate a sense of the users' satisfaction with the building.

5.3.1 The POE result for Menara IMC

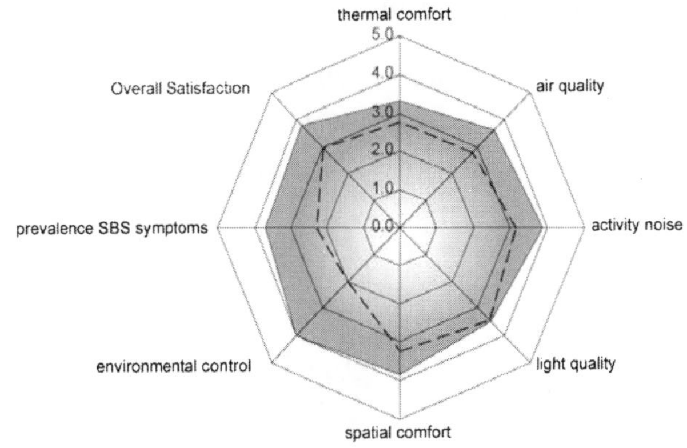

PREVALENCE OF SICK BUILDING SYMPTOMS - MENARA IMC

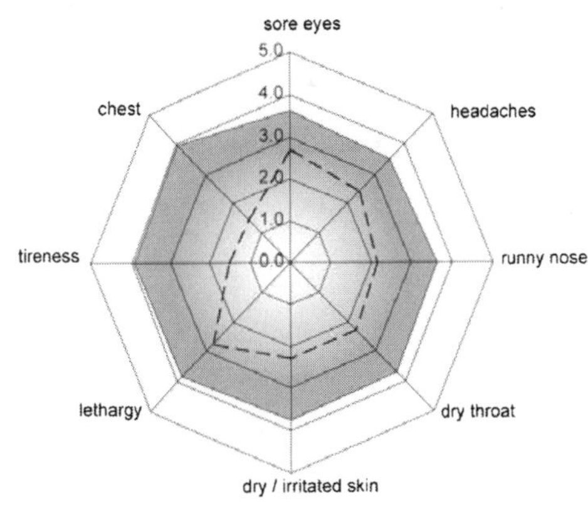

----- average all settings

Figure 4. Menara IMC

The POE result for Menara IMC shows that the environmental control creation has been rated the highest by its occupants, scoring approximately two above the average setting. Generally, all the scores are above the average settings. Although the score for the light quality criterion is the lowest, it is still slightly higher than the average setting.

The high prevalence SBS symptoms found in Menara IMC may possibly be caused by the lack of

[12] These twenty seven previously surveyed buildings are representing typical building stock in Australia.
[13] Rowe, 2000.

connection to the natural environment. Being sealed in a totally active environment, poor SBS symptoms ratings have been reported by its occupants. Tiredness and chest problems have achieved a scoring of roughly 2.5 worse than the average setting. One solution may involve the introduction of fresh air supply from outside the building. The overall satisfaction score for Menara IMC is approximately 3.8 which is 0.8 above the average score for a standard office building in Australia.

5.3.2 *The POE result for IBM Plaza*

KEY PERFORMANCE INDICTAORS - IBM PLAZA

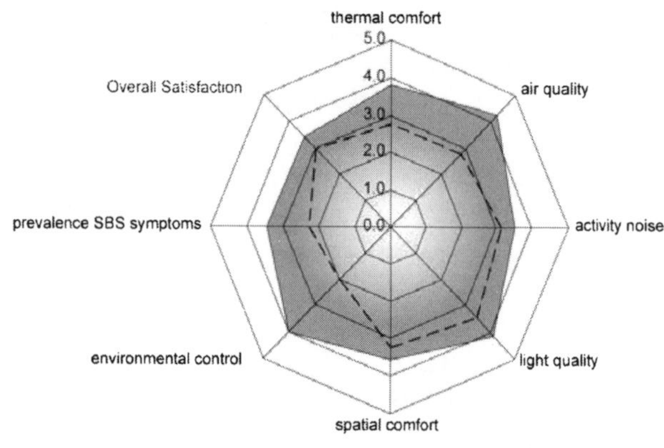

PREVALENCE OF SICK BUILDING SYMPTOMS - IBM PLAZA

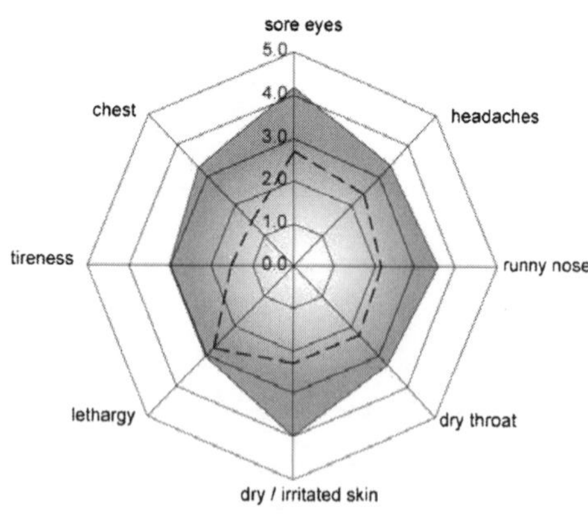

----- average all settings

Figure 5. IBM Plaza

The POE result for IBM Plaza shows that the highest level of satisfaction from the occupants of the building is in terms of its air and light quality. Similar to Menara IMC, the environmental control of this building is relatively high, scoring roughly two above the average setting. The overall satisfaction score for IBM Plaza is approximately 3.5 which is 0.5 above the average score for a standard office building in Australia. The survey result for IBM Plaza suggests that this building is relatively healthier than Menara IMC. The improvements are obvious especially for the prevalence of SBS symptoms. The respondents are generally having fewer problems with tiredness, chest and lethargy problems.

5.3.3 *The POE result for Menara UMNO*

KEY PERFORMANCE INDICTAORS - MENARA UMNO

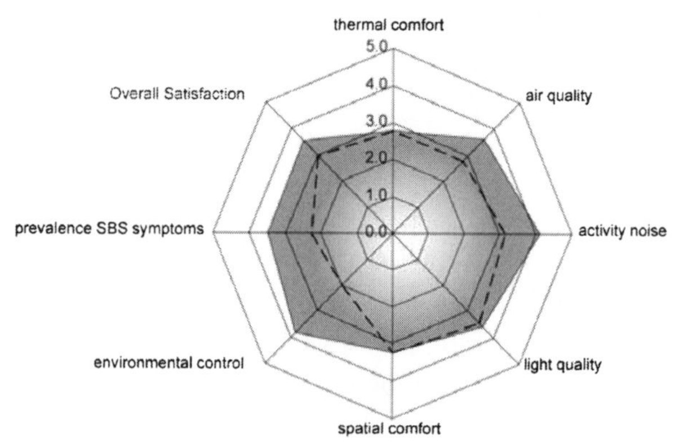

PREVALENCE OF SICK BUILDING SYMPTOMS - MENARA UMNO

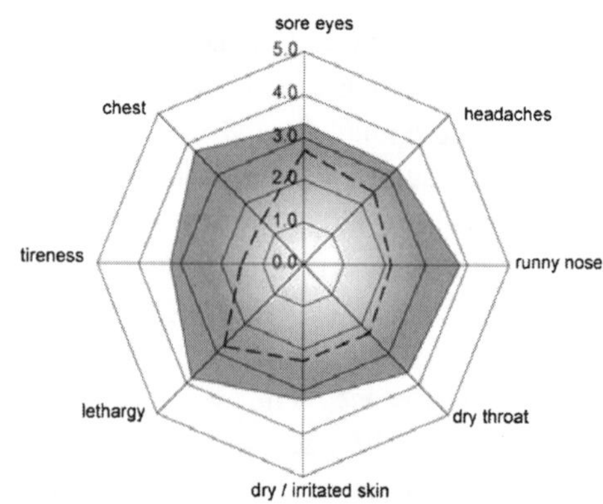

----- average all settings

Figure 6. Menara UMNO

The POE result for Menara UMNO shows that the highest level of satisfaction from the occupants of the building is for its control on activity noise. Similar to Menara IMC and IBM Plaza, the environmental control of this building is also scoring reasonably higher than the average setting. The overall satisfaction score for Menara UMNO is approximately 3.6 which is 0.6 above the average score for a standard office building in Australia. The survey result for Menara UMNO concludes that it is performing well especially in terms of its low SBS symptoms for the users. However, the occupants rating for thermal and spatial comfort are exceptionally low followed by the light quality. Yet, the respondents are having fewer problems overall.

5.3.4 Combined results

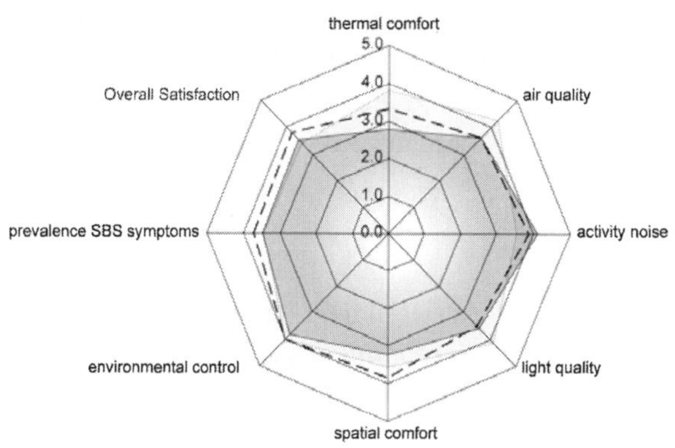

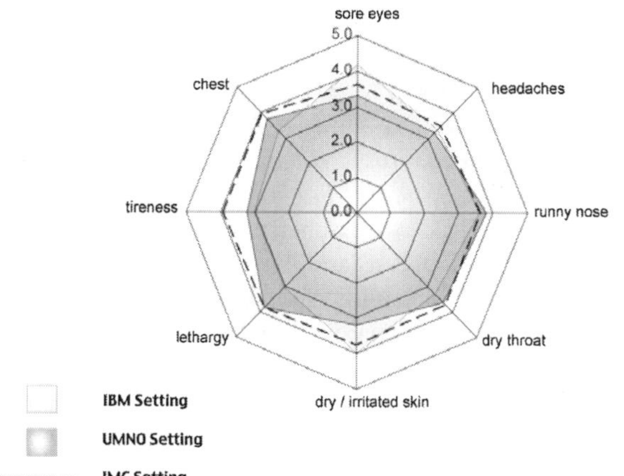

Figure 7. Combined Results

There are improved thermal comfort, air quality and light quality as experienced by the users in the IBM Plaza. For the prevalence of SBS symptoms, it is apparent that IBM Plaza is scoring better in terms of achieving lower SBS symptom scores in almost every aspect, except for the sore eyes and dry/irritated skin scores, when compared to the base case building.

For Menara UMNO, it is scoring better for activity noise only. However, for the SBS scores, it is again achieving better scores in all aspects (that is lower SBS scores), except for the runny nose symptom, as compared to Menara IMC.

It is noted that although both IBM Plaza and Menara UMNO has achieved better results in terms of the SBS scores, the overall satisfaction score for each of these buildings are lower than that of Menära IMC. This might be due to the fact that Menara UMNO is located in Penang, which has comparatively more extreme weather conditions as compared to Kuala Lumpur, where the rest of the case study buildings are located. Thus, the users of Menara UMNO will expect better comfort conditions. To proof that bioclimatic buildings are preforming better than conventional buildings, more samples and more extensive data will be needed. It should also be noted that the users' overall satisfaction for both the bioclimatic buildings is lower than that of the base building by only approximately 0.3. On the other hand, some of the SBS scores for the bioclimatic buildings are approximately on average 0.6 better than the base building.

ISSUE	MENARA IMC	IBM PLAZA	MENARA UMNO	AVG
Thermal Comfort	3.3	3.8	2.8	3.3
Air Quality	3.7	4.2	3.7	3.9
Activity Noise	3.9	3.6	4.1	3.9
Light Quality	3.5	4.1	3.6	3.7
Spatial Comfort	3.9	3.6	3.3	3.6
Environmental Control	4.0	4.0	3.8	3.9
Overall Satisfaction	**3.7**	**3.5**	**3.5**	**3.6**
SBS Symptoms	3.8	3.5	3.5	3.6

Table 2. The summarised POE results created from the graphs

The comparative POE results do still suggest, to a certain extent, that the bioclimatic strategies can play a positive role in contributing to a healthier environment for its users and possibly meaning an increased in productivity from the users. This might also mean that some of the bioclimatic strategies need to better integrated into future designs or more bioclimatic strategies are needed in place.

Summary

To sum up, the bioclimatic concept and attitude is evident from the case study results. From the bioclimatic strategies mapping exercise, with reference to Yeang's 'bioclimatic skyscrapers', it can be seen that Menara UMNO has the most numbers of ticks indicating that it has adapted the most numbers of bioclimatic strategies. IBM Plaza has scored second whilst Menara IMC has the least. It might also be worthwhile to note that the IBM Plaza was completed in 1987 while Menara UMNO is a more recently built high-rise by Yeang in 1998. Yeang has been experimenting the bioclimatic strategies in his 'bioclimatic skyscrapers' and Menara UMNO could be one of his high-rise buildings which has adapted the most bioclimatic strategies until that time. The lighting level and wind velocity for each case study reflect its floor depth ratio and hence, the bioclimatic principles adapted for each design. For Menara UMNO, it is clear that since its micro-climate is different to both IBM Plaza and Menara IMC and that it has employed the wind wing-wall device, it has performed reasonably better. From the measurement results, again Menara UMNO has achieved higher daylighting level and better ventilation which penetrate further into the inside of the building. Also, the POE response shows that it has done especially well in having lower SBS symptoms on the whole. From this initial research and analysis, it is reasonable to claim that bioclimatic high-rise buildings are capable of performing better than the conventional ones. Nevertheless, it is the creation of improved internal environments for the users of the buildings which is most important.

5. Discussion

The paper provides an overview of the bioclimatic approach for high-rise buildings in Malaysia. It is anticipated that a larger sample of survey will give a more precise representation of the performance and the sick building syndrome symptoms in high-rise buildings. However, it is intended that generating a bioclimatic strategies mapping checklist and perhaps a pre-design scorecard[14] of bioclimatic strategies for high-rise buildings can be a design primer and a tool for architects to use in their projects. In conclusion, bioclimatic approach is not a new concept. It takes advantage of climate and environmental conditions for achieving better thermal comfort inside. It is essentially a set of strategies relating to climate and people. Furthermore, the architecture that evolved out of the bioclimatic approach is not necessary to be prettier or uglier, cheaper or dearer, and most importantly there remain the freedom to design.

Acknowledgments

Ken Yeang; All the Malaysian occupants who have filled in the User Response Survey or interviewed; Dr. Richard Hyde, Dr. Steve Szokolay; and The continuous support from the Centre for Sustainable Design and the Department of Architecture, The University of Queensland.

References

1. Council on Tall Buildings, 1978-1981, *Planning and Design of Tall Buildings*, a Monograph in 5 volumes, ASCE, New York.
2. Huxtable, Ada Louise, *The Tall Building Artistically Reconsidered: The Search for a Skyscraper Style*, Pantheon Books, New York, 1984.
3. Jones, David Lloyd, *Architecture and the Environment: Bioclimatic Building Design,* Laurence King Publishing, London, 1998.
4. Law, Joyce Hor Yan, *The Bioclimatic Approach to High-rise Building Design: An Evaluation of Ken Yeang's Bioclimatic Principles and Responses in Practice to Energy Saving and Human Well-being"*, a Thesis for Bachelor of Architecture at the University of Queensland, December, 2001.
5. Noll & Scupelli, *Bioclimatic Skyscrapers*, <http://www.biocl/parole_bioclimatic_skyscraper.htm>, 1994.
6. Olgyay, Victor, *Design With Climate: Bioclimatic Approach to Architectural Regionalism*, Princeton University Press, New Jersey, 1963.
7. Schon, Donald A., *The Reflective Practitioner: How Professionals Think in Action*, Ashgate Publishing Limited, England, 1995.
8. Vale, Brenda and Robert, *Green Architecture: Design for a Sustainable Future,* Thames & Hudson Ltd., London, 1991.

[14] The checklist and scorecard of bioclimatic strategies has been developed, from the Bachelor of Architecture thesis (Law, 2001), through the extensive survey of four 'bioclimatic skyscrapers' by Ken Yeang.

9. Wittmann, Sabine, *Architects' Perceptions Regarding Barriers to Sustainable Architecture: A Survey Among Australian Architects,* a Thesis for Doctor of Philosophy in Architecture at the University of New South Wales, December, 1997.
10. Yeang, Ken, *Hamzah & Yeang: Bioclimatic Skyscrapers*, Ellipsis London Limited, London, 1994.
11. Yeang, Ken, *The Skyscraper bioclimatically Considered: A Design Primer*, Academy Editions, Great Britain, 1996.

DEVELOPMENT IN ENHANCING THE STANDARDS OF LIFT INSTALLATION FOR TALL BUILDING*

C.K. MOK EDDIE W.K. WU S.Y. MUI

Architectural Services Department, HKSAR Government
Queensway Government Offices, 66 Queensway, Hong Kong

The development of lift technology has contributed significantly to the history of tall building. Lifts are used in almost all buildings of more than a few stories high throughout the world as an effective and quick means of vertical transportation. To meet the demand of building's occupants on vertical transportation particularly in tall buildings, good vertical traffic analysis and planning as well as proper selection of vertical transportation equipment are very crucial. Apart from these, there are other important standards like lift ride comfort, fire safety in lift as well as monitoring of lift services that require to be addressed to. This paper discusses the recent development in the aspects of lift ride comfort, insulated lift doors for improving fire safety, and common protocol for communication with the central control and monitoring system..

1. General Introduction

The development of lift technology has contributed significantly to the history of tall building. Lifts are used in almost all buildings of more than a few stories high throughout the world as an effective and quick means of vertical transportation. The ability of lift systems in providing good and effective performance in meeting the anticipated demand of building's occupants on vertical transportation is essential particularly in tall buildings. Good vertical traffic analysis and planning as well as proper selection of vertical transportation equipment are therefore very crucial in the design of vertical transportation system for tall buildings.

In vertical traffic design, there are two key factors affecting the design and choice of lift system and these are: -

- the quantity of service factor which is the number of people that will use the lift system over a defined period of time (i.e. the handling capacity of lift system)
- the quality of service factor which is the capability of lift system in dealing with the passengers (i.e. the passenger waiting time)

Apart from the traffic analysis and planning of lift installations mentioned above, there are other important standards like lift ride comfort, fire safety in lift as well as monitoring of lift services that require to be addressed to.

The subject of lift ride quality or lift ride comfort has increasingly received attention from lift users in recent years. Lift ride quality in terms of lateral and vertical vibrations, acceleration, deceleration and jerk has become important criteria for judging a lift performance. Noise produced during lift operation is also another salient factor for consideration.

There have been concerns that the fire will spread to other floors by way of lift shaft particularly in tall buildings. The Code of Practice for Fire Resisting Construction issued by the Buildings Department of the Hong Kong Special Administrative Region (HKSAR) stipulates that where lift landings are not isolated by fire resisting enclosure or lifts that are completely surrounded by stairwells, the lift doors shall be constructed to satisfy the fire resisting criteria both in terms of integrity and insulation.

Good monitoring of lift services enables effective building management. It can also enhance the responsiveness, safety, availability and reliability of lift services. Most of the lift companies have their own micro-processor based or computerised lift monitoring systems available in their lift installations but the lift monitoring systems of different lift companies cannot communicate with each other or with the central control and monitoring system (CCMS) of the building directly due to the lack of common protocol universally adopted by the lift industry.

The recent development in the aspects of lift ride comfort, insulated lift doors for improving fire safety, and common protocol for communication with the central control and monitoring system in Hong Kong will be further elaborated below.

2. Lift Ride Quality

2.1 *Background*

Along with the development of modern cities, more and more skyscrapers are emerging from the horizon. People rely heavily on lift services as an indispensable means of vertical transportation in tall buildings. They tend to raise increasing concerns on lift ride quality or lift ride comfort of which their concerns are proportionate to the height of buildings. Lift ride quality would ultimately become a basic requirement as a part of specifications for modern lift systems and a symbol of quality lift service.

At present, performance of a lift, particularly in tall building, is usually evaluated based on the time for it to undertake the function of vertical transportation. The required performance includes the time of closing the lift doors, starting the lift car, traveling between departure floor and destination floor, stopping the lift car and opening the lift doors. To accomplish this series of lift motions, some noise and vibrations would inevitably be generated. All these parameters are then perceived as the lift ride quality by the passengers.

Physical movement of lift car is the major contributory factor affecting the quality of a lift ride. Lateral and vertical accelerations of a lift car are the two main motive elements and the perceptible levels of which will influence the ride quality perceived by lift passengers.

2.2 *Current practice in evaluating lift ride quality*

At present, the ride quality of a lift system is often evaluated by means of the personal feedbacks from the lift users that are based on their perceptions and subjective experiences. This is predominantly a qualitative approach to address the demand of the lift users on lift riding comfort performance. Hence, the quality of lift ride is "measured" without any systematic and/or scientific foundation.

In view of the drawbacks of the evaluation in qualitative approach, the first hurdle to be overcome in the assessment of lift ride quality is to filtrate perceptive sensational judgment of individual users. It is therefore required to quantify the lift performance assessment by defining exactly what lift ride quality is, how it is measured systematically, how it is interpreted scientifically and how it is used consistently.

2.3 *Quantitative evaluation of lift ride quality*

Vibration, acceleration, jerk, and noise level of a traveling car are the quantities generally adopted as the indicators of lift ride quality that need to be quantitatively assessed in an explainable perspective. To eliminate individual's subjective interpretation of lift ride quality, detailed definitional terminology and methodological system are therefore required to be devised and hence adopted universally. These two fundamentals are pillared by instrumental characteristics and measurement methodology of lift ride quality. The instrumentation used would be utilized for the measurement of lift performance and the interpretation of ride quality. Evaluation of physical motions and sounds inside a lift car relies on the direct measurement of vibration and noise levels which should be as close as possible to and consistent with the perception of a passenger inside the lift car.

Although numerous national and international standards on lift technology and safety have been developed and adopted worldwide, methodologies and standards for determining lift ride quality are still under development. Without a commonly adopted methodology and standard on lift ride quality, there will be no objective tools to gauge the ride quality achieved even though the demand for quality performance on lift ride is ever increasing. However there are no international standards established at present so far which clearly specify the acceptance criteria applicable to lift ride quality. One of the reasons why the development of lift ride standard lags behind its technical and safety counterparts may probably be due to the fact that there is a want of mutual agreement on this issue amongst the diversified stakeholders in the industry.

In response to the expectation of the users for improved quality of lift service, many lift manufacturers and associations have developed their own in-house methodologies and standards on lift ride quality for their products. However, no general consensus has yet come up with amongst the stakeholders until an attempt had been made in Australia that a standard on measurement of lift ride quality was developed as the first step addressing to this issue. It was the draft Australian Standard for Measurement of Lift Ride Quality (DR 99446) issued on 15 August 1999. Its objective was to encourage industry-wide uniformity in the definitions, measurement, processing and expression of the vibration and noise data which contribute to lift ride quality.

2.4 *Measurement methodology and lift ride quality standard*

As a Government department responsible for the delivery of public buildings of the HKSAR, the Architectural Services Department (ArchSD) has taken a leading role to cohere the stakeholders of Hong Kong lift industry in developing a general consensus on measurement methodology and performance standard for acceptance of lift ride quality. This is necessary so as to enhance the quality of lift service to meet the demand of high standard lift system from the public. This requires a set of measurement methodology and quality standards of lift ride be developed and applied locally. We aim to set up an objective, systematic, reliable and consistent evaluation and acceptance scheme. The quality standard of lift ride at public buildings should be strict but compatible with the technological capability available in the trade.

The ArchSD has adopted the draft Australian Standard for the formulation of the instrumentation framework in developing the lift ride quality measurement methodology. With respect to the acceptance standard, the approved lift contractors has been consulted to formulate a set of prototype performance standard for acceptance of lift installations which was unanimously assented by the eligible lift companies.

2.5 *Present development*

Having developed the measurement methodology and prototype performance standard on lift ride quality, the ArchSD had carried out a quantitative research on the lift ride quality of existing lift installations at public buildings of the HKSAR the results of which were then used for analysis and further review.

To achieve a broad coverage and wide representation of the research, 12 emblematic government premises including office buildings, institutional buildings, quarters and market complex were chosen for conducting the measurements of lift ride quality and a total number of 55 lift installations were measured. The selected lift installations were supplied and installed by different lift manufacturing giants which embraced two Japanese companies, one American company, two European companies and one Korean company.

From the analysis of site measurement results, the prototype performance standard for lift ride quality devised and used in this research has been proved to be pragmatic and reasonable and compatible with the present commercially available lift technologies. It is therefore considered that as an endeavor to achieve quality performance of lift ride, the lift ride quality standard could be used by public or private developers, consultants, designers or authorities for inclusion into their specifications for the procurement and acceptance of new lift installations.

2.6 *Future development*

To lead the lift industry on promoting quality performance of lift service delivered to the public, the prototype performance standard has been adopted as the acceptance criteria for all new passenger lifts in government buildings. This acceptance standard will be formally included into the next version of General Specification for Lift, Escalator and Passenger Conveyor in Government Buildings of the Hong Kong Special Administrative Region issued by the ArchSD.

It is of our view that the standard of lift ride quality is *dynamic* and will continuously be developing in the light of technological advancement in lift engineering. The lift ride quality standard will continuously be reviewed in line with the design of modern lift systems at tall buildings to cope with the sustainable development of the city.

3. Insulated Lift Doors

3.1 *Background*

The lift doors have to be designed and constructed so as to provide assurance that in the event of a fire on one floor, the spread of fire will not delayed or deferred from that floor to other floors by way of the lift shaft. This is particular important in tall building where the lift shaft can act as a built-in "chimney" to transmit heat and smoke quickly from the floor on fire to the topmost floors. Moreover, a lift moving within its shaft may act similarly to a piston within a cylinder and this "piston" effect will push the hot smoke and gases of a fire to floors not initially involved. This will expose occupants in the other floors particularly those in topmost floors to a very hazardous situation, as they may not be aware of the occurrence of fire.

The Code of Practice for Fire Resisting Construction 1996 issued by the Buildings Department has stipulated that the lift doors should have fire resistance period (FRP) of not less than one hour with regard to integrity. Moreover, if the lift landings are not isolated by fire resisting enclosures or lifts that are completely surrounded by stairwells, the lift doors shall be constructed to satisfy both the criteria for integrity and insulation in accordance with BS 476: Parts 20 to 24.

The insulation requirement of lift door stipulated in the above-mentioned Code of Practice has certain implications on the design of lift lobby and/or the provision of fire shutters, etc. The ArchSD had enquired the lift manufacturers in mid 2001 as to whether the lift industry could be able to supply and install lift doors that could satisfy both the insulation and integrity criteria. We were advised that not many of the lift manufacturers could provide lift doors that were tested to satisfy both criteria. In the meantime, the Buildings Department had also taken into account of the circumstances and agreed to waive the insulation requirement subject to the provision of enhanced fire safety measures. In this connection, three interim arrangements were considered to be accepted by the Buildings Department:-

- the floor area adjacent the landing doors are sprinklered, and
- the additional sprinkler heads shall be provided in front of the lift landing doors in form of a water curtain; and
- the homing device of lifts shall be upgraded such as it is activated by smoke detectors in addition to sprinklers.

Despite the above, the Architectural Service Department had adopted a practical approach in ensuring the requirements of the above-mentioned Code of Practice were complied with either by provision of protected lift lobbies or lift doors that complied with both the integrity and insulation criteria.

3.2 *Experience on adopting insulated lift doors*

The Police Headquarters Phase III at Arsenal Street of Wanchai was the first tall building government project in Hong Kong that had adopted the lift landing doors that met both integrity and insulation requirements. In this project, hundreds of insulated lift doors have been installed for the 36 numbers of lifts in the whole complex.

Although the lifts had been tested and commissioned satisfactorily in respect of safety and performance requirements before handover to the users for operation, many faults associated with lift door operating mechanism occurred and most of them were relating to fine-tuning and re-adjustment of lift components in the door operating mechanism.

Having carried out detailed analysis of the lift door faults occurred, it was found that the major reason was attributable to the heavy insulated lift doors installed. As the lift traffic in tall buildings was much greater than that in low-rise buildings, the frequency of opening and closing of heavy insulated lift doors were very high. Moreover, due to the high speed of lift for tall building, the relief of air pressure created by lift cars inside a lift shaft became more difficult resulting in great pulsating pressure being exerted on the lift door. All these contributed to the need to fine-tune and re-adjust the original settings of components of lift door operating mechanism. Having identified the problems and implemented appropriate measures, the lift insulated doors have been working properly.

3.3 *Present Development*

Since the first adoption of insulated lift door in the Police Headquarters Phase III at Arsenal Street of Wanchai, more and more lift manufacturers are now able to supply and install lift landing doors that meet with both integrity and insulation requirements. Moreover, lift companies have now been more acquainted with the installation of insulated lift doors. They are aware of the problems that may be caused by the heavy insulated lift doors and have the knowledge in carrying out proper testing and commissioning of the insulated lift doors.

3.4 *Future development*

Fire safety in lifts is very essential particularly in tall buildings where lift system is an indispensable means of vertical transportation for occupants and the "fire load" is usually high. The provision of lift doors that satisfy the fire resisting criteria both in terms of integrity and insulation is a very effective means in delaying/deferring the spread of fire from one floor to other floors by way of the lift shaft. Moreover, unlike active fire protection means such as sprinklers that may be subject to malfunction, passive fire protection means like fire resisting lift door and fire resisting construction are more reliable in terms of fire safety. As more and more lift manufacturers are able to supply and install insulated lift doors, it is the right time to consider the inclusion of insulated lift doors as a standard provision for lift installations in tall buildings to further enhance fire safety in lifts.

4. Common Protocol for Communication with CCMS

4.1 *Background*

In Hong Kong, lift installations are installed and maintained by Registered lift companies that are agents of different lifts and escalators manufacturers. In most of the projects, the lift companies only provide lift supervisory control panels to the building owners for monitoring of some basic lift operation status which normally include the lift-in-service status, position and travel of lift, fault indication etc. But in fact most of the lift companies have their own micro-processor based or computerised lift monitoring systems available in their lift installations which can be utilised for monitoring the lift equipment and providing operational and maintenance data for immediate or future analysis. These lift monitoring systems enable the lift installations be effectively managed. They can also enhance the responsiveness, safety, availability and reliability of lift services. The general features of basic lift monitoring system include: -

- Indication of lift-in-service status
- Trapped passenger alarms
- Inoperable lift alarms
- Performance malfunction (alarms)
- Automatic collection of lift performance data
- Two-way voice communication with trapped passengers
- Data analysis, etc.

4.2 *Need for common protocol*

However, due to the fact that different lift companies have developed or adopted their own proprietary communication protocol in their lift monitoring systems, lift monitoring systems of different lift companies cannot communicate with each other or with the building management system (BMS)/central control and monitoring system (CCMS) of the building directly due to the lack of common protocol universally adopted by the lift industry. Such deficiency is further amplified by the need of inter-communicability and integration of the lift monitoring systems with the BMS/CCMS for design of modern buildings towards the concept of intelligent high-rise building.

Lift manufacturers are normally not willing to release their protocols to others. It is also difficult to unify all lift manufacturers to use a common set of protocols. If a common protocol system for BMS/CCMS for access lift control and monitoring information can be developed and made available for use by the lift industry, the lift performance as well as the operation and maintenance data in the various lift monitoring systems of different lift companies can be easily accessed by BMS/CCMS.

4.3 *Benefit for integration with BMS/CCMS*

The availability of a common protocol system for the CCMS to access lift monitoring and control information for use by the lift industry can integrate the present lift control and monitoring functions available in individual lift monitoring systems of different manufacturers into the BMS/CCMS. Integration with BMS/CCMS has the following benefits:-

(i) Common user interface
(ii) Cost savings
(iii) Space savings
(iv) Multiple access points
(v) Use of common software packages

4.4 *Present development*

Having realised the benefits brought about by integrating the lift monitoring systems with the building BMS/CCMS, ArchSD has taken a leading role for the lift industry. A Task Group has thus been set up comprising ArchSD, allied professional bodies (Hong Kong Institution of Engineers (Building Services Division), Building Services Operation and Maintenance Executives Society and The International Association of Elevator Engineers (HK-China Branch)) and the Lift and Escalator contractor associations (The Lift and Escalator Contractors Association and The Registered Elevator & Escalator Contractor Association Ltd.) to review the existing situation and to develop a common protocol system which is acceptable for use by the lift industry. The agreed approach will be to adopt "protocol converter" which translate the required pre-defined information obtained from the lift monitoring systems and communicate with the BMS/CCMS so that the information can be read via the BMS/CCMS. The common protocol system to be developed shall be:-

(i) implemented by a standalone 'common protocol' converter
(ii) secure, reliable and of proven standard
(iii) of open architecture
(iv) integrated with lift monitoring systems
(v) flexible to cater for the existing and future lift monitoring systems
(vi) easily maintained and administered
(vii) able to receive control and monitoring information from lift installation and to convert it to the required format for display in CCMS

4.5 *Future development*

ArchSD has planned to develop a common protocol system for BMS/CCMS to access lift control system information. The 'common protocol' concept will be worked out. The 'common protocol' converter together with the required interfacing between the 'common protocol' converter and the lift installation as well as between the 'common protocol' converter and the BMS/CCMS system will be designed. The performance of the 'common protocol' converter will be evaluated through actual installation on some trial projects before acceptance.

5. Conclusion

With the increasing demand from lift users for quality of lift service, the ArchSD has been working closely with the lift industry with a view to enhancing the standards of lift installations in respect of good lift ride quality, adequate fire safety, as well as effective monitoring of lift service. Through the collaborated efforts of the ArchSD and the lift industry, we can provide better and better lift services that will continue to contribute to the success of the design of tall buildings.

References

1. 1.Abraham, Earl (1984) Performance Criteria: Car Ride Quality. Elevator World, I-19-23.
2. 2.Asian Institute of Intelligent Buildings (2001) The Draft Manual of the Intelligent Building Index (IBI).
3. 3.British Standard (2003) Lifts (elevators) Measurement of Lift Ride Quality, BS ISO 18738:2003
4. 4.Building Services Branch, Architectural Services Department, General Specification for Lift, Escalator and Passenger Conveyor Installation in Government Buildings of the Hong Kong Special Administrative Region, 2002 Edition.
5. 5.K.K. Li, Andy M.T. Suen, Eddie W.K. Wu (2004) A General Survey on Lift Ride Quality at Public Buildings of the Hong Kong Special Administrative Region.
6. 6.Lorsbach, Gregory P. Measurement of Lift Ride Quality. Physical Measurement Technologies, Inc.
7. 7.National Elevator Industry, Inc. (1994) Vertical Transportation Standards – Standards for Elevators, Escalators and Dumbwaiters, 7th Edition.
8. 8.NFPA Life Safety Code Handbook.
9. 9.Standards Australia (1999) Measurement of Lift Ride Quality, DR99446 - Draft Australian Standard for Comment.
10. 10.The Buildings Department (1996) Codes of Practice for Fire Resisting Construction.
11. 11.The Chartered Institution of Building Services Engineers (2000) CIBSE Guide D Transportation Systems in Buildings.

A DISCUSSION ON DESIGN & TECHNOLOGY OF ENERGY-SAVING ENVIRONMENT-FRIENDLY HIGH-RISE BUILDINGS

HONGLIANG YAN

Department of Architecture of Tong ji University, 1239 Siping Road Shanghai 200092, China

MIAOFANG CHEN

Department of Architecture of Tong ji University, 1239 Siping Road Shanghai 200092, China

High - energy - consumption (mainly for air - conditioning and heating) now is one of the general problem in high - rise buildings in cities of China. With the increasing number of high - rise buildings year after year ,a new architectural design and research subject is that how to use advanced and applicable technology to high - rise buildings turning high - energy consumption to low - energy consumption ,promoting energy efficiency ,and solving the problems such as light and thin exterior cladding ,poor insulation property ,small thermal capacity and storage coefficient and poor thermal stability . This article tries to put forward an energy - saving technical system of improved thermal environment and low energy consumption with research of architectural design building construction and thermal and equipment system.

1. Preface

1.1 *Present situation of domestic high-rise buildings*

Since the reform and opening, China's architecture has been experiencing rapid development, and a number of high-rise hotels, high office buildings as well as high residential buildings have become one of the symbols of city modernization. Especially during the recent years, with the reform of urban housing system, the improvement of urban residents' living standard and that of housing conditions, the energy-consumption for architecture has considerably increased. Relevant data shows that architectural energy consumption is 33.3% of the total energy consumption. Energy consumed for architecture includes that used in heating, air-conditioner, illumination, cooking, household appliances and elevator, etc., among which, building heating and air-conditioner consume the most energy, approximately 60%～70% of the architectural energy consumption. These high-rise buildings are generally poor in thermal insulation performance and the heating and air-conditioner systems are large in capacity and low in operating efficiency, therefore, high-energy consumption of domestic high-rise buildings is quite a general phenomenon. At present, due to lack of energy in domestic market, how to solve the problem of high-energy consumption for high-rise buildings has become a more and more important factor.

1.2 *Energy saving significance of high-rise building*

Based on the characteristics of high-rise buildings in domestic cities, this article makes deep study on aspects of exterior cladding structure concerned in high-rise buildings' energy-saving technical system such as thermal performance of exterior cladding, new construction material, new structure and new technology and building equipment, etc., applies feasible architectural energy-saving technology, with which not only the architectural energy consumption is reduced but the Urban heat island effect is decreased, and also air pollution and electromagnetic pollution relevant to architecture is reduced, thus protecting the environment and setting a good example for the economic and social sustainable development.

2. Building up Design Concept of Architectural Energy Saving

2.1 *Energy-saving technical design of high-rise buildings*

According to the characteristics of high-rise buildings, architectural energy-saving design should be strictly in compliance with local conditions, try to make use of physical factors, natural resources as well as new construction material, new structure and new technology, and comprehensively consider solving the energy saving problem of high-rise buildings with a view to the climatic and environmental characteristics of the local area.. First of all, appropriately select building orientation and structure considering characteristics of high-rise buildings, well design natural ventilation of buildings, design exterior elevations, door and window forms and window-control area according to different orientation of buildings, and calculate the range to set sunshade and shutter, trying to reducing radiation entering indoor, thus decreasing the capacity of heating and air-conditioner and reaching the aim of architectural energy saving.

First of all, appropriately select building orientation and structure considering characteristics of high-rise buildings, well design natural ventilation of buildings, design exterior elevations, door and window forms and window-control area according to different orientation of buildings, and calculate the range to set sunshade and shutter, trying to reducing radiation entering indoor, thus decreasing the capacity of heating and air-conditioner and reaching the aim of architectural energy saving.

In addition, as to the exterior building envelope of high-rise buildings, select new high-efficient and energy-saving exterior wall material and energy-saving door and window, and carry out reasonable detailed structural design, organically combining thermal insulation technology of envelope with ventilation, lighting and sun shading technology, which is one of the most important contents of the energy-saving design for domestic high-rise buildings and also the leading work for architects to participate in architectural energy-saving design. With the issuance, in succession, of legislations on domestic architectural energy-saving design, architectural energy-saving design work has been gradually attached more importance by extensive architects, engineers and developers.

2.2 *Analysis on energy-saving design of high-rise buildings*

Heightening architectural energy-saving awareness and designing energy-saving buildings will increase construction cost (the increased cost is approximately not more than 10% of the initial construction cost). Thanks to the application of architectural energy-saving technology, architectural energy consumption is reduced, so heating and air-conditioning load, installed capacity of equipment and daily energy consumption are accordingly reduced.

Furthermore, the design life of domestic high-rise buildings is specified as 50 to 70 years, therefore, when we analyze from life cycle cost of buildings, we find that though energy-saving design will increase initial-stage (one-time construction) cost during the prophase, the consumption and maintenance (running) cost within the long-time period of consumption should be attached even more importance to. As to some long-life buildings featured by high quality, low consumption and low maintenance, they essentially contain considerably low cost advantage.

3. Heat Transfer Coefficient and Load Characteristics of High-rise Building's Envelope

3.1 *Heat transfer coefficient of building envelope*

The heat transfer coefficient of building envelope "K" is an important index to measure thermal performance of buildings, and the size of heat transfer coefficient directly influences architectural energy consumption. We know from building thermal engineering that the value of "K" relates to heat transfer coefficient of envelope material "λ", material thickness "δ" and coefficient of thermal transform of inner and external surface of building envelope "α_n" and "α_w", while the coefficient of thermal transform of external surface of envelope depends on convection heat output and radiation heat output.

When outdoor air flows, the air velocity will gradually increase from floor with the elevation of buildings, which will cause increase of coefficient of thermal transform of the external surface of high-rise buildings. Because the high parts of high-rise buildings continuously radiate heat to the outside sky, radiation heat loss of the external surface of high-rise buildings increases, especially, most high-rise buildings adopt glass curtain walls or comparatively large windows (the surface radiated heat is even more, and the external surface temperature is higher than that of the wall body because thermal resistance of glass is lower than that of wall body), which leads to increase of heat transfer coefficient of high building's envelope. It goes against architectural energy saving.

What's more, airspeed increase outside high-rise buildings will lead to the increase of wind pressure applied on the windward side of the buildings. In case doors and windows of the buildings are of poor sealing property, it will not only increase air infiltration capacity but increase heating load of the high parts. In addition, outdoor cooling air enters the buildings through doors, windows or gaps of the lower parts of the buildings and is discharged outside through the buildings with the action of thermal pressure.

3.2 *Thermal capacity and thermal inertia of high-rise buildings*

Most exterior walls of domestic high-rise buildings adopt light and thin glass curtain wall, which makes the buildings "glass cases". Their thermal capacity is much lower than that of exterior walls of traditional buildings, and the coefficient of thermal capacity "S" and the thermal inertia "D" are also decreased. The indoor heating and cooling load will change rapidly with the outdoor temperature and solar radiation.

To sum up, due to the characteristics of high energy-consumption for high-rise buildings, their energy-saving technical design is even simpler with even higher requirements.

4. Improving Indoor Thermal Environment of High-rise Buildings and Promoting Energy Efficiency of Building Equipment

4.1 *Indoor thermal environment design of high-rise buildings*

Quality of indoor thermal environment is associated with local outdoor climatic environment. Severe cold and cold areas mainly depend on winter heating while areas hot in summer and cold in winter mainly depend on air-conditioners in summer and also winter.

Indoor thermal environment design mainly lies on selection of indoor temperature and humidity. Excessively low temperature in summer and comparatively high temperature in winter not only consumes excessive energy but also does harm to human body's comfort and health. Generally, suitable indoor temperature in winter for public buildings and residential buildings is 26℃. If the indoor temperature is 1℃ higher, then the energy saving effect of the whole year will be considerably perfect, in addition, discharge amount of CO2 will accordingly decrease and so does environmental pollution. Therefore, with the premise of meeting human body's comfortableness and according to different buildings in different areas, properly reducing design standard of indoor temperature and humidity, reasonably selecting design parameters of indoor temperature and humidity and avoiding blind low temperature in summer and excessively hot in winter will get favorable energy-saving effect. If a set of environment control system of indoor temperature and humidity is built up, i.e. adjust and control indoor temperature and humidity in accordance with change of outdoor temperature and solar radiation, then the energy-saving effect will be even more visible.

Another index influencing indoor thermal environment, comfortableness and health is air quality. Currently, most of the air-conditioning system design in domestic hotels and office buildings apply primary air fan-coil system. Air volume value selection of central ventilation system is generally determined according to detection of CO_2 and change of persons in the room. And taking energy-saving requirements into consideration, minimum fresh air requirement is preferably applied for winter and summer and full fresh air should be adopted for transition seasons, and meanwhile, air supply temperature should be appropriately controlled.

4.2 *Equipment selection and system running*

In order to match requirements of indoor thermal environment design, equipment selection should be carried out on condition that the indoor design parameters should be satisfied, and energy-saving, environment protection and highly efficient equipment should be selected in so far as possible. For example, new and highly efficient radiators, new auto-control heat pump units, variable refrigeration volume (VRV) air-conditioning units and variable air volume air-conditioning units, ice-on-coil tank air conditioning technology, water source and terrestrial heat source heat pump units, low temperature and heat water floor panel radiation heating system devices are selected to replace traditional radiator heating system and constant volume air conditioning units, etc. improving energy efficiency ratio of heat supply, ventilation and air-conditioning system devices.

Energy saving property and safety should also be taken into consideration so far as design of heating, air-conditioning system and hot water supply systems are concerned. Domestic high-rise residential building generally adopted dispersed air-conditioning and water supply system making per household as the unit, this system is quite low in operating efficiency, what's more, hanging of outdoor air-conditioners on the exterior walls affects safety as well as the appearance of the buildings. Therefore, high-rise residential buildings should, as much as possible, adopt central systems and set metering devices by household, so as to improve operating efficiency of the systems and reach the aim of energy saving.

Highly efficient central heating system, various energies central cooling system, gas air-conditioning units and triple-generation system (heating, power and cooling) should be selected for the design considering characteristics of high-rise buildings. In order to facilitate energy-saving operation of systems, various highly efficient technical instruments of temperature control and heat or refrigeration volume should be adopted, and intelligent building equipment management systems should be created so as to further improve the indoor thermal environment and reduce the architectural energy consumption.

5. Use of Regenerative Energy – Exploration of Integration Design of High-rise Buildings and Solar Energy

5.1 *Research on exploration of the integration design of high-rise buildings and solar energy*

*E*nergy-saving design of high-rise buildings is a comprehensive technical design. It not only takes into consideration architectural design itself, building equipment selection and operation but more importantly, it carries out technical and economic comparison and comprehensively consider climatic environments, energy policies, energy status and service efficiency, etc. in different areas.

It is well known that solar energy is a renewable and inexhaustible natural energy source with no pollution. Our country shares abundant solar energy and areas with yearly sunshine duration over 2,200 hours approximately cover more than 2/3 of the state area.

Using solar energy as the source has wide prospects and potentials so far as architectural energy saving concerned. Therefore, domestic economic conditions permitted, design of urban high-rise buildings should consider making maximal use of solar energy technology and perform integrated design of solar energy water heaters and buildings, applying thermal and photovoltaic technology of solar energy to sun shading, light and shadow and comfortable environment creation, using photovoltaic power generation and highly efficient power storage system of solar energy according to different areas, daily irradiation times and sun light intensities, and even making use of solar heating and unitized solar refrigeration technology, making it the heat source and cool source of central heating and refrigeration systems of high-rise buildings.

6. Conclusion

Energy-saving design of high-rise buildings covers a wide range of aspects with high technical requirements. An architect should not only take into consideration function of use and style of a building, but more importantly, he should have architectural energy-saving concept, and he should actively participate in the full cooperation with ecological engineers, structural engineers and plant engineers from the initial stage of building planning and conceptual design, make use of new construction material, new structure and new technology and comprehensively and consider energy-saving technology problem of high-rise buildings, contributing to China's sustainable development.

Acknowledgments

This article is supported by National Natural Science Foundation of China. (Project No. 50478101)

References

1. F. X. Tu and Q.Y. Wan, *Chinese Building Newspaper.* (2005).
2. Z. F. Liu, *Sustainable Development of Residential Industry Forum.* (2005).
3. C. H. Song, *Residential Industry Promoting Center of Ministry of Construction.* (2005).
4. B. Edwards, *Sustainable Architecture.* (2003).
5. H. L. Yan, *Special Construction of Architecture.* (2002).
6. M. F. Chen, *Building Equipment.* (2002).

Post-occupancy Operation, Maintenance & Management

CORPORATE ENVIRONMENTAL STRATEGIES – A STAKEHOLDER MANAGEMENT PERSPECTIVE

THOMAS K.T. WONG EVIA WONG
School of Professional and Continuing Education, The University of Hong Kong

Organisations can develop and implement a range of strategies, from reactive to proactive, to address environmental issues. This paper examines how the firms perceive the expectation of different stakeholders and integrate them into their company strategies to enhance the competitive advantages. Stakeholders do not only include shareholders, customers, employees but also government and communities. Effective stakeholder management can constitute intangible, socially complex resources that may allow the firm to outperform competitors in terms of long-term value creation.

1. Introduction

The definition of sustainable development is various and controversial. According to the World Commission on Environmental and Development (WCED, 1987), sustainable development '*is development that meets the needs of the present without compromising the ability of future generations to meet their own needs.*' The WCED asserts that sustainable development requires the simultaneous adoption of environmental integrity, economic prosperity, and social equity. Environmental integrity is to ensure that human activities do not erode the earth's land, air, and water resources. While, social equity is to ensure that all members of society have equal access to resources and opportunities and finally, economic prosperity is to promote a reasonable quality of life through the productive capacity of organisations and individual in society (Bansal, 2005). These three dimensions of sustainable development are internally consistent rather than at odds with each other.

Over time, corporate commitment to sustainable development has changed considerably. Shareholder, customers and policy makers are increasingly demanding in improving environmental performance of business firms. External pressures such as legislation and public concern, as well as market opportunities arising from environmental concerns, have compelled firms to integrate environmental issue into their strategic planning process (Banerjee, 2001).

It has often been argued that environmental regulation is instrumental to the introduction of better environmental management practices within firms, and that more stringent regulations are needed to further improve such practices (Banerjee, 2001; Newton and Harte, 1997; Porter and van der Linde, 1995). However, when crafting specific environmental strategies, firms undoubtedly attach importance to other stakeholders than government regulators (Neu, Warsame, and Pedwell, 1998). If the greening of corporate strategies can be interpreted as an attempt to meet these stakeholders expectation, then identifying salient stakeholder becomes a critical step in corporate strategy formulation. This assumption has led to various researches on analysis of the linkages between environmental strategy and stakeholder management during the past decade.

2. Stakeholder Management

Social responsibility of a firm, according to the traditional view, means maximizing profits for stockholders (Friedman, 1962). However, this perspective is challenged by the stakeholder theory (Freeman, 1984).

In his landmark work, Freeman (1984) defines stakeholders as '*any group or individual who can affect or is affected by the achievement of the organisation's objectives.*' Thus, stakeholders of an organization include stockholders, employees, customers, the local community, government agencies, public interest group, trade associations, and competitors. Under the stakeholder theory, a business firm, in making strategic planning, must consider the impact of its action on all stakeholder groups (Banerjee, 2002).

Following Freeman's work, there were numerous authors that have expanded on the stakeholder management concept. Basically, the stakeholder literature has two broad branches, namely, a strategic (Goodpastor, 1991) and a moral branch (Frooman, 1999).

The strategic stakeholder literature emphasizes the active management of stakeholder interests, whereas the moral stakeholder literature is interested primarily in balancing stakeholder interests. The former literature classifies stakeholders as primary or secondary, based on the type of relationships they entertain with the firm. The primary stakeholders refer to employees, suppliers, customers, and public agencies engaged in formal relationships with the organisation. The secondary stakeholder groups include actors such as the media and special interest groups, not engaged in formal transactions with the firm (Clarkson, 1995).

Mitchell, Age and Wood (1997) argue that the degree to which managers give priority to competing stakeholder claims is a function of managers' perceptions of three key stakeholder attributes; namely, power, legitimacy, and urgency. In their model, stakeholder salience, as perceived by managers, is positively related to the cumulative impact of these three stakeholder attributes. Since the degree to which managers give priority to competing stakeholders claims is dynamic, stakeholder salience can vary over time and depends on the issue considered. Managerial perceptions are critical in this model, because they ultimately determine stakeholder salience. These perceptions can be influenced by the managers (own) values (Agle, Mitchell, and Sonnenfeld, 1999; Sharma, 2000). Banerjee (2001) supports that managerial perceptions of legislation, competitive advantage, public concern and top management commitment have close association with environmental strategies in most of the business firms.

The moral stakeholder management approach thus suggests that corporations should not narrowly focus their strategic management decisions on creating shareholder value. They should broaden their objectives to address the expectations and interests of a wide variety of salience stakeholders (Garrod, 1997; McGee, 1998). Such objectives may include customer satisfaction, regulatory compliance, good corporate citizenship, and social and environmental responsibility among others.

At the micro level, a firm with a reactive environmental strategy may face an overall loss of competitive advantage if proactive environmental management becomes a common practice among its rivals (Garrod, 1997).

Hillman & Keim, 2004 have tested the relationship between shareholder value, stakeholder management and social issue participation. They conclude that stakeholder management leads to improve shareholder value, while social issue participation is negatively associated with shareholder value. However, they further remark that they are not making the normative assertion that firms should not engage in such social activities. Indeed, many firms have multidimensional performance goals that may include social issue activism. The conflict between these goals and shareholder wealth creation should be recognized. The use of a firm's resources always has an opportunity cost. Implementing a social issue participation strategy appears to come at the cost of foregoing opportunities to increase shareholder value.

One of the limitations of stakeholder approach is that it only focuses on 'what should be done' by organisations to take into account the needs of their stakeholders (Banerjee, 1999). However, different groups can have different or conflicting interests. To balance these competing interests may be a difficult task. Stakeholder theory does not attempt to provide too much detail on how to resolve these pragmatic difficulties.

3. Classification of Environmental Management Strategies

There are large numbers of literatures attempting to classify environmental management strategies from different perspectives.

According to the works of Hunt and Auster (1990), Azzone and Bertele (1994) and Roome (1992), firms can be identified according to their environmental management practices as:

i) reactive,
ii) defensive,
iii) accommodative, and
iv) proactive.

These strategies reflect an increasingly important focus on societal issues, both in terms of strategy formulation and implementation (Clarkson, 1995).

3.1 Resource-based theory

Hart (1995) developed a more grounded typology of environmental strategies based on the resource-based theory of the firm. This theoretical approach is receiving increasing attention as it attempts to link corporate environmentalism and competitive behaviour.

The theory suggests that corporate strategy will only lead to sustainable competitive advantage if it is supported by firm-level competencies. Such competencies reflect unique combinations of resources that are rare, non-substitutable, difficult to imitate, and valuable to customers. These resources combinations may build upon a wide variety of basic components, including physical assets, employee skills, and organizational processes.

In this context, Hart (1995) distinguishes four types of resource-based environmental approaches that can be summarised as follows:

Types	Approach	Resources Commitment
1) End-of –pipe	Reactive to environmental issues.	Only limited resources committed to solving environmental problem. Product and manufacturing process improvements are made to conform to legal requirements.
2) Pollution prevention or total quality management (TQM)	Continually adapt their products and production processes in order to reduce pollution levels below legal requirements. May be viewed as a cost leadership approach.	Resources allow firms to achieve regulatory compliance at a lower cost and to reduce liabilities.
3) Product Stewardship	Product differentiation. Products and manufacturing processes are designed so as to minimize the negative environmental burden during the products' entire life cycle.	Some forms of life cycle analysis (LCA) are implemented.
4) Sustainable Development	Minimize the environmental burden of firm growth through the development of clean technologies. A long-term version shared amongst all relevant stakeholders.	Proactive Investments in the following: i) Conventional green competencies; ii) Employee skills; iii) Management systems and procedures; and iv) Strategic planning process.

Different types of resource-based environmental approaches (Hart, 1995)

Hart (1995) argues that simultaneous investments in several linked resource domains are required to move from one environmental strategy stage to the next.

Buysse and Verbeke (2003) have conducted an empirical analysis based on a sample of 197 firms to see the significance of linkage between environmental strategy and stakeholder management. From their studies, they observe the following:

> *'First, it is shown that several simultaneous improvements in various resource domains are required for firms to shift to an empirically significant, higher level of proactiveness.*
>
> *Second, more proactive environmental strategies are associated with a deeper and broader coverage of stakeholders.*
>
> *Third, environmental leadership is not associated with a deeper and broader coverage of environmental regulations, thereby suggesting a role for voluntary cooperation between firms and government.'*

Banerjee (2003) in his study indicates that the firms in high-environmental impact industries like chemical and industrial manufacturing have a more proactive environmental management practice. He believes that stricter legislation and public concern for the environment have significantly influenced those firms' environmental strategies and environmental orientation.

3.2 Institutional theory

Institutional theory emphasizes the role of social and cultural pressures on organisations that influence organisational practices and structures (Scott, 1992). DiMaggio and Powell (1983) argue that managerial decisions are strongly influenced by three institutional mechanisms, namely, coercive, mimetic and normative isomorphism which in turn create and diffuse a common set of values, norms and rules to produce similar practices and structures across organisations that share a common organisational field. They define an organisational field as '... *constitute a recognized area of institutional life: key suppliers, resource and product consumers, regulatory agencies and other organisations that produce similar services or products.'*

Jennings and Zandbergen (1995) apply institutional theory to explain firm's adoption of environmental management practices. They argue that because coercive forces, primarily in the form of regulations and regulatory enforcement, have been the main impetus of environmental management practices, firms throughout each industry should have implemented similar practices. Milstein (2002) considers that as distinct levels of coercive pressures are exerted upon different industries, this may lead to different environmental strategies.

Firms can adopt various types of environmental management practices in response to institutional pressures (Sharma, 2000). These can be based on:

1. Environmental strategies of conformance that focus on complying with regulations and adopting standard industry practices; or
2. Voluntary environmental strategies that seek to reduce the environmental impacts of operations beyond regulatory requirements. For instances, firms adopting voluntary approaches can implement EMS elements by creating an environmental policy, developing a formal training programme or instigating routine environmental auditing. In addition, management can choose to have the comprehensiveness of their EMS validated by a third party by pursuing ISO 14001 certification. Companies can implement Total Quality Environmental Management (TQEM) to expand the conventional notion of quality to include environmental quality (GEMI, 1992). Companies can also seek to improve relations with regulators and signal a proactive environmental stance by participating in government or industry sponsored voluntary programmes.

4. Levels and Types of Strategy

Environmental issues can be integrated at different levels of strategy depending on the characteristics of the firm and the industry, regulatory forces and public concern.

In outlining the strategy process, Schendel and Hofer (1979) propose four hierarchical levels of strategy, namely, enterprise strategy at the top, followed by corporate, business and functional strategies.

These levels are hierarchical in nature that the definition and scope of the higher strategic level constrains the lower one.

STRATEGY LEVEL	STRATEGY TYPE	OUTCOME
Enterprise Strategy	Sustainable development	Ecologically sustainable organizations
Corporate Strategy	Leading edge innovative pre-emptive	Green business portfolio environmental Protection Businesses
Business Strategy	Accommodate excellence compliance plus ecologically sustainable strategies: least cost, niche & differentiation	Sustainable competitive advantage
Functional Strategy	Defensive compliance	Green Marketing emissions control

Levels and Types of Strategy (Banerjee, 2001)

4.1 Enterprise strategy

At this highest level, the focus is on the role of a firm that plays in society. This relates to a firm's governance, function and fundamental mission. Other than those organisations like Greenpeace where environmental protection is their basic mission, few business firms show evidence that they have integrate environmental concerns at this level.

Generally, the mission of business firms in providing value for shareholders and customers is more likely to be integrated at this level.

4.2 Corporate strategy

Corporate strategy is the next highest level. This level involves identifying the kind of business that the firm should enter to meet its enterprise strategic goals. Thus, if a firm has a fundamental mission in environmental protection at the enterprise strategy level, most likely he would choose to enter a business that could meet his primary goal, for instance, environmental management consultancy and services company.

4.3 Business strategy

Business strategy involves allocating organisational resources in an optimal way to achieve competitive advantages. This level of strategy also focuses on integrating different business functions such as production, marketing, etc into the business.

Competitive strategies are developed at this level of strategy in order to enhance a competitive advantage in the market. There are different strategies that can serve this purpose including cost leadership, product differentiation and niche marketing.

4.4 Functional strategy

Functional strategy is the lowest level and it involves planning operating procedures in the different functions, such as purchasing, marketing or research and development, etc. Simply speaking, functional strategy is to find ways to comply with business strategy to gain competitive advantage.

New product development, location of new manufacturing plants, increased research & development investments, technology development (especially in pollution prevention and waste management), and changes in product and process design are some strategic actions to be developed by a firm.

4.5 Implication of a hierarchical structure of strategy

From the hierarchical structure, it implies that the enterprise strategy level is the most important level of strategy that drives the lower levels.

Banerjee (1999) develops a model (Figure 1) to indicate the possible responses by a firm at different levels of strategy influenced from external environment including economic, legal, political, biophysical, technological and socio-cultural impacts.

5. Conclusion

Various studies have concluded that there is a close linkage between environmental strategy and stakeholder management. Effective environmental management requires the identification of important stakeholders.

Stakeholder perspective of corporate environmentalism involves recognition of stakeholders' environmental concerns, which are translated into strategic actions designed to improve a firm's environmental performance, as well as its relationship with key external stakeholders. The importance attached to specific sets of stakeholders which appears to be associated with a particular environmental strategy, as suggested by most of the studies, is ultimately determined by managerial values.

Given the complexity of environmental issues facing different industries, it is important to understand how pressures from external constituencies such as regulators and environmental organizations are internalized by decision-makers.

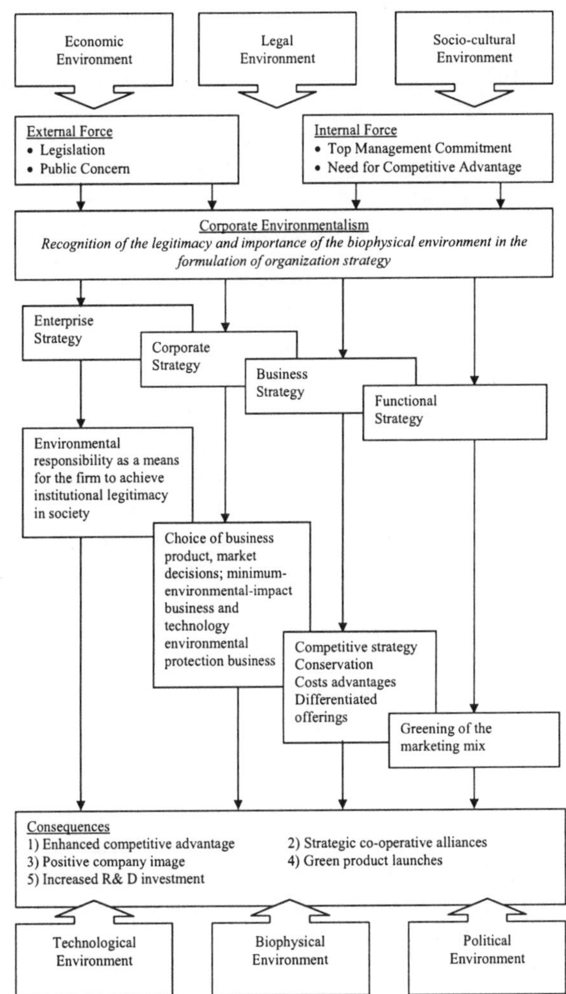

Figure 1 : Conceptual Framework of Corporate Environmentalism (Banerjee, 1999)

Research has shown that governments often enact environmental legislation as a result of pressure from the public and environmental organization. Legislation, in turn, has the potential to influence corporate environmental strategies.

There are various typologies designed to classify firms according to their environmental practices. It ranges from reactive, defensive, accommodative to proactive. Several simultaneously improvements in various domains are required for firms to shift to a higher level of proactiveness. More proactive environmental strategies are associated with a deeper and broader coverage of stakeholders.

References

Agle, B.R. Mitchell, R.K. Sonnenfield, J.A. 1999. Who matters to CEOs? An investigation of stakeholder attributes and salience, corporate performance, and CEO values. *Academy of Management Journal* 42(5): 507-525.

Azzone G, Bertele U. 1994. Exploiting green strategies for competitive advantage. *Long Range Planning* 27(6): 69-81.

Banerjee, S.B. 1999. Corporate environmentalism and the greening of strategic marketing, in Greener Marketing: *A Global Perspective to Greening Marketing Practice*, eds M. Charter & M.J. Polonsky, Greenleag Publushing, Sheffield, UK.. *Journal of Management Studies* 38(4): 39-55.

Banerjee, S.B. 2001. Managerial perceptions of corporate environmentalism: Interpretations from industry and strategic implications for organizations. *Journal of Management Studies* 38(4): 39-55.

Banerjee, S.B. 2002. Corporate environmentalism: The construct and measurement. *Journal of Business Research* 55(3): 171-191.

Banerjee, S.B. Lyer, E.S. Kashyap, R. 2003. Corporate environmentalism: Antecedents and influence of industry type. *Journal of Marketing* 67: 106-122.

Banerjee, S.B. 2001. Corporate environmental strategies and actions. *Management Decisions* 39(1): 36-44.

Bansal, P. 2005. Evolving Sustainably: A longitudinal study of corporate Sustainable development. *Strategic Management Journal* 26: 197-218.

Buysse, K. Verbeke, A. 2003. Proactive environmental strategies: A stakeholder management perspective. *Strategic Management Journal* 24: 453-470.

Clarkson, M.B.E. 1995. A stakeholder framework for analyzing and evaluating corporate social performance. *Academy of Management Review* 20(1): 92-117.

Delmas, M. Toffel, M.W. 2004. Stakeholders and environmental management practice: an institutional framework. *Business Strategy and the Environment* 13: 209-222.

Dinaggio, P.J. Powell, W.W. 1983. The iron cage revisited: institutional isomorphism and collective rationality in organizational fields. *American Sociological Review* 48: 147-160.

Freeman, E.R. 1984. *Strategic Management: A Stakeholder Approach*. Pitman/Ballinger: Boston, MA.

Frooman, J. 1999. Stakeholder influence strategies. *Academy of Management Review* 24(2): 191-205.

Garrod, B. 1997. Business strategies, globalization and environmental. In *Globalization and Environment*. OECD: Paris; 269-314.

GEMI. 1992. Total quality environmental management: the primer. Washington DC: *Global Environmental Initiative*.

Goodpastor, K. 1991, Business ethics and stakeholder analysis. *Business Ethics Quarterly* 1: 53-71.

Hart, S.L. 1995. Natural-resource-based view of the firm. *Academy of Management Review* 20(4); 986-1014.

Hillman, A.J. Keim, G.D. 2001. Shareholder value, stakeholder management, and social issues: what's the bottom line? *Strategic Management Journal* 22: 125-139.

Hunt, C.B. Auster, E.R. 1990. Proactive environmental management: avoiding the toxic trap. *Sloan Management Review*: 7-18.

Jennings, P.D. Zandbergen, P.A. 1995. Ecoloically sustainable organizations; an institutional approach. *Academy of Management Review* 20: 1015-1052.

McGee, J. 1998. Commentary on 'Corporate strategies and environmental regulation: an organizing framework'. *Strategic Management Journal* 19(4):377-389.

Mitchell, R. Agle, B. Wood, D. 1997. Toward a theory of stakeholder identification and salience: defining the principle of who and what really counts. *Academy of Management Review* 22(4): 853-886.

Neu, D. Warsame, H. Pedwell, K. 1998. Managing public impressions: environmental disclosures in annual reports. *Organizations and Society* 23(33): 265-282.

Newton, T. Harte, G. 1997. Green business, technicist kitsch. *Journal of Management Studies* 34(1): 75-98.

Porter, M.E. and VAN DER LINDE, C. 1995. Green and competitive: ending the stalemate. *Harvard Business Review* 73(5): 120-134.

Roome, N. 1992. Developing environmental management systems. *Business Strategy and the Environmental* 1:11-24.

Schendel, D.E. Hofer, C.W. 1979. *Strategic Management: A New View of Business Policy and Planning*. Boston: Little, Brown and Co.

Scott, W.R. 1992. *Organisations: Rational, National, and Open Systems.* Prentice-Hall: Englewood Cliffs, NJ.

Sharma, S. 2000. Managerial interpretations and organizational context as predictors of corporate choice of environmental strategy. *Academy of Management Journal* 43: 681-697.

WCED. 1987. *Our Common Future*. Oxford University Press: Oxford.